ISBN 978-1-5276-2897-7
PIBN 10876652

English
Français
Deutsche
Italiano
Español
Português

www.forgottenbooks.com

Mythology Photography **Fiction**
Fishing Christianity **Art** Cooking
Essays Buddhism Freemasonry
Medicine **Biology** Music **Ancient
Egypt** Evolution Carpentry Physics
Dance Geology **Mathematics** Fitness
Shakespeare **Folklore** Yoga Marketing
Confidence Immortality Biographies
Poetry **Psychology** Witchcraft
Electronics Chemistry History **Law**
Accounting **Philosophy** Anthropology
Alchemy Drama Quantum Mechanics
Atheism Sexual Health **Ancient History**
Entrepreneurship Languages Sport
Paleontology Needlework Islam
Metaphysics Investment Archaeology
Parenting Statistics Criminology
Motivational

39TH CONGRESS, } HOUSE OF REPRESENTATIVES. { EX. DOC.
2d Session. } { No. 109.

ANNUAL REPORT

OF THE

COMMISSIONER OF PATENTS

FOR

THE YEAR 1866.

VOLUME II.

WASHINGTON:
GOVERNMENT PRINTING OFFICE.
1867.

Rec. May 24, 1900.

DESCRIPTIONS AND CLAIMS OF PATENTS

ISSUED IN THE YEAR 1866.

ILLUSTRATED WITH ENGRAVINGS.

VOLUME II.

No. 54,407.—LEWIS REESE, Rolling Prairie, Ind.—*Windmill.*—May 1, 1866.—When the feed trough is full, a hook stops the revolution of the wind wheel.

Claim.—The combination and arrangement of the lever M, cords *b c*, weighted vessel Q, and vessel R, operating substantially as and for the purposes specified.

No. 54,408.—HELEN M. REMINGTON, Springfield, Mass.—*Paper Shirt.*—May 1, 1866.—A paper shirt; threads and size inserted between the layers of tissue paper; a collar loop is attached.

Claim.—First, as a new article of manufacture, a paper shirt.

Second, forming the same materials, composed of two thicknesses of paper, prepared by the insertion of the compounds described, or their equivalents, substantially in the manner and for the purpose described.

Third, in combination with the said shirt, threads *s s*, &c., applied substantially as set forth.

Fourth, the hook *b*, constructed and combined with the shirt, substantially in the manner and for the purpose set forth.

No. 54,409.—GEORGE E. REYNOLDS, Philadelphia, Penn.—*Furnace.*—May 1, 1866.—In each fore plate is a recess into which is slipped a removable cast-iron block, to occupy the position where the heat is most intense.

Claim.—A detachable block H, adapted to the fore plate of a furnace, substantially as and for the purpose herein set forth.

No. 54,410.—UEL REYNOLDS, New York, N. Y.—*Carriage.*—May 1, 1866.—The head block has a central pivoted projection which sets in the socket on the axle, the relative position being maintained by the brace which connects the head block to the coupling.

Claim.—The pivot *f* and socket *k*, applied substantially as specified, between the axle and head block, in combination with the brace *m* and pivot *o*, substantially as and for the purposes specified.

No. 54,411.—CELIUS E. RICHARDS, North Attleboro', Mass.—*Elastic Chain.*—May 1, 1866.—The loop at each end of a given link is connected to two sides and the more remote end of the box; the sections, being placed at an angle of 90°, enclose the resilient block, which is compressed by tension on the chain, conferring elasticity upon it.

Claim.—An improved elastic link, (or chain, composed of a series of such links,) the same having its parts A B C constructed, arranged, and applied together, and so connected as to operate, when applied to a cable, substantially in manner as set forth.

No. 54,412.—VAN BUREN RYERSON, New York, N. Y.—*Extracting Precious Metals from Ores.*—May 1, 1866.—The superheated steam is introduced into the charge of ore through a perforated coil, and the vapors of mercury condensed by the lid and finally by a stream of water.

Claim.—The process of decomposing the sulphurets found in admixture in the ores of gold and silver, by subjecting said ores in the pulverized or granulated state to the action of superheated steam, so as to convert the sulphurets into sulphates and sulphites, substantially as and for the purpose described.

Also, in combination with the said process for decomposing said ores, the process, substantially as herein described, of amalgamating the particles of the precious metals with mercury.

No. 54,413.—JOHN ROBERTSON, New York, N. Y.—*Atmospheric Hammer.*—May 1, 1866.—The height of the hammer above the anvil is graduated by the adjustment of its piston rod, and its stroke by the adjustment of the wrist on the crank shaft.

Claim.—First, in combination with the cylinder hammer head and piston, applied and operating as hereinbefore specified, the provision for raising and lowering the piston rod, and shortening or increasing its effective length while the hammer is in operation, substantially as herein set forth.

Second, in combination with the hammering apparatus, constructed as described, a variable crank, substantially as shown.

No. 54,414.—WILLIAM H. SANGSTER, Buffalo, N. Y.—*Apparatus for Refining Petroleum.*—May 1, 1866.—The purifying solution extends above the foot of the partition, and the crude oil is exposed to it as it passes from the induction to the eduction chamber.

Claim.—First, the partition A, or its equivalent, when constructed as and for the purpose herein substantially described and set forth.

Second, in combination therewith, the plate B, or the equivalent thereof, as and for the purposes described.

No. 54,415.—HORACE B. SCOFIELD, Now York, N. Y.—*Grate Bar.*—May 1, 1866.—Explained by the claim.

Claim.—A grate bar for furnaces, formed with a straight upper surface, and a rib on its under side, corrugated in the manner and for the purposes specified.

No. 54,416.—T. E. SEXTON, Wi'mington, Del.—*Frame for Printing Photographic Pictures.*—May 1, 1866.—The strip is adjusted against the edge of the plate by set screws, to insure its position and render it returnable to its exact place after examination.

Claim.—A strip F applied to and rendered adjustable in a photographic frame, substantially as and for the purpose herein set forth.

No. 54,417.—WM. A. SHARPE, Syracuse, N. Y.—*Filter for Well Tubes.*—May 1, 1866.—The strainer on the well tube consists of a series of rings with interposed lugs to preserve their distance.

Claim.—First, the rings A A, having the parts *e e* attached, for the purpose described.

Second, the combination of the rings A A and frame-work B B C D, substantially as described.

No. 54,418.—A. N. SHATTUCK, San Francisco, Cal.—*Wave Power Water Raising Apparatus.*—May 1, 1866.—The buoy is guided by vertical side supports, and its motion forces water through its central pipe into a stationary pipe leading to a reservoir; each pipe is furnished with check valves.

Claim.—The buoy pump, made to act by the agitation of the water, substantially as above described, consisting of a floating vessel A, its tube B, combined with a fixed tube C, both tubes being provided with check valves, and the floating vessel A being guided in any suitable frame, as above set forth.

No. 54,419.—HENRY F. SHAW, West Roxbury, Mass.—*Hoisting Apparatus.*—May 1, 1866.—The revolving frame has wheels of differential diameters on its central shaft, and chain connections from them to the loose pulleys on the other shafts.

Claim.—First, the general combination and arrangement of the differential pulleys and chains, substantially as described and for the purpose set forth.

Second, the latch I, or its mechanical equivalent, working as described and for the purpose set forth.

Third, the holding pin or lock H, or its mechanical equivalent, in combination with the driving pulleys R and S, substantially as described and for the purpose set forth.

No. 54,420.—JOHN W. SHEAFFER, Sterling, Ill., assignor to ALONZO SHERMAN and CHARLES DILLER, Jordan, Ill.—*Pump.*—May 1, 1866.—The foot valve may be lifted by the engagement of its bail with a hook depending from the plunger.

Claim.—The valve boxes L and J, provided with the hook N and bail O, when constructed and operating substantially as and for the purposes set forth.

No. 54,421.—A. H. SHEFFER, West Donegal township, Penn.—*Combined Harrow and Cultivator.*—May 1, 1866—The harrow has curved, forwardly projecting teeth, and adjustable handles and sides.

Claim.—The specific combination of the adjustable handles F on the side pieces A, also made adjustable at the apex by bolts H I and central jaw piece only, together with the arrangement of the curved, flat, and narrow shares or spikes S, inverted and operated in the manner shown and for the purpose specified.

No. 54,422.—P. B. SHELDON, Prattsburgh, N. Y.—*Fruit Basket.*—May 1, 1866; ante-dated April 21, 1866.—The lower ends of the splints are clamped in the recessed bottom by an interior hoop.

Claim.—A fruit basket, made substantially as herein described.

No. 54,423.—DAVID SHIVE, Philadelphia, Penn.—*Steam Engine Governor.*—May 1, 1866.—The axes of the mandrels, on which the ball arms vibrate, are tangential to the spindle at a point distant about the semi-diameter of the ball from the axes of revolution of the vertical spindle.

Claim.—Suspending the balls A A by rigid arms h h, connected to the spindle B by means of joints, consisting of the cylinders f f and the mandrels g g, or their equivalents, arranged in relation to the said spindle, arms, and balls, substantially as described and represented.

No. 54,424.—WARREN A. SIMONDS, Boston, Mass.—*Fluid Regulator.*—May 1, 1866.—The flexible envelope of the chamber is counter-weighted; when expanded by its fluid contents it rises and gives motion to gearing which closes its induction opening.

Claim—First, the combination of the cone f and rod g, as and for the purpose described.

Second, the arrangement of the bevel gears o and p with the shaft i and valve stem, substantially as described and for the purposes stated.

Third, the arrangement of the guide finger v with the slotted valve stem, to prevent rotation of the valve or stem.

No. 54,425.—ALBERT S. SKIFF, Trenton Falls, N. Y.—*Land Roller.*—May 1, 1866.—The rollers are so journalled in their respective hinged frames that their tracks overlap.

Claim.—First, the construction of a land roller in sections, one section in advance of the other, and the frame in sections, connected by pivot joints, and so arranged as that the bearings of the inner ends of the rollers are supported by the opposite frame, thereby allowing the ends of the rollers to lap, as and for the purposes described.

Second, the use of the pivoted journal box, in combination with the frame and roller journal, as and for the purposes set forth.

No. 54,426.—C. D SMITH, Chicago, Ill.—*Paint for Metallic Roofs.*—May 1, 1866.—Composed of asphaltum, shellac, coal tar, petroleum, benzine, and India-rubber.

Claim.—As a new article of manufacture, a composition made of petroleum oil, coal tar, asphaltum, gum shellac, India-rubber, and benzine, prepared in the proportions and manner as above described, and for the purposes specified.

No. 54,427.—Mrs O. SMITH, Chicago, Ill.—*Cooking Range*—May 1, 1866.—The two front covers are fire chambers, flanking a central front oven. The rear of the stove is occupied by two tiers of ovens.

Claim.—The combination of the fire boxes U T with the ovens R Q S 13 13, when the latter are arranged in relation to the former and to each other, as shown and described.

No 54,428.—JAMES F. SPENCE, Williamsburgh, N. Y.—*Revolving Condenser.*—May 1, 1866.—The exhaust steam from the cylinder is conducted to a revolving condenser, which acts as a fly wheel, and by its condensation relieves the back pressure on the piston.

Claim.—A revolving condenser, constructed and applied in combination with a steam cylinder, substantially in the manner herein described, for the purpose specified.

No 54,429.—A. B SPROUT, Hughesville, Penn.—*Tempering Steel.*—May 1, 1866.—The vessel is steam-heated, and has a registering thermometer and an alkalimeter.

Claim.—The use of a saponaceous or alkaline liquor, covered with a coat of oil, and heated to about the boiling point, and regulated and graduated in its strength to suit the different kind and quality of steel, article, or thing to be tempered, substantially as herein described and set forth.

No. 54,430 —ARIEL B. SPROUT, Hughesville, Penn.—*Horse Rake Teeth.*—May 1, 1866.—A spring tooth, tapering from a point near its middle towards each end.

Claim.—Constructing of steel or iron a curved tooth for horse rakes, said tooth being a spring within itself, and tapering from the line B to the line C, from the line B to the line A, substantially as herein described and for the purpose set forth.

No. 54,431.—ARIEL B. SPROUT, Hughesville, Penn.—*Horse Hay Fork.*—May 1, 1866.—The lower ends of the fork arms are pivoted like shears, and their entering edges when opened cut the hay in the mow or stack.

Claim—Constructing and arranging the bars A and B in such manner that they may be used either for elevating hay or as hay shears, substantially as herein described.

No. 54,432.—J. M. STANTON and S. F. STANTON, Manchester, N. H.—*Saw Mill Head Block.*—May 1, 1866.—An arrangement of shafts and their couplings with pinions, working and travelling in raaks, which are operated by a lever sliding in a notched guide so as to actuate the knees and slides which move the log to the saw.

Claim.—First, the operating of the uprights or knee pieces C through the medium of pinions D gearing into rack *b* at the under side of the knee pieces and into the rack *c* of the blocks B, substantially as and for the purpose specified.

Second, the arrangement of the sectional shaft E, clutches F, lever G, rods *d*, constructed and operating in the manner and for the purpose herein described.

Third, the combination of the lever Q, notched guide R, pawl P, pinion D, rack *b*, and rack *c*, all arranged in the manner and for the purpose herein specified.

No. 54,433.—NICHOLAS STARR, Jr., Homer, N. Y.—*Horse Power.*—May 1. 1866.—The wheel arms are in pairs and the rim consists of sections with cusps to engage the chain which communicates motion to the cone pulley. The weighted lever and shifting pulley maintain the tension.

Claim—First, the arrangement in the large reel of placing the spokes in pairs and connecting them by cross-pieces which shall extend beyond the spokes, as and for the purposes described.

Second, the weighted lever *l*, in combination with the pulley *m*.

Third, the adjustable cap piece *f*, in combination with the pulley *m* and lever *b*, substantially as described.

Fourth, the combination of the wheel G, cone R and reels *p p*, when the same are arranged and operated substantially as above described.

No 54,434.—J. STEVENS and W. B. FAY, Chicopee Falls, Mass.—*Curtain Fixture.*—May 1, 1866.—The weight of the curtain causes the angular edge of its spool rim to bind in the groove of the pawl and support the curtain at the required point.

Claim.—As a new manufacture the spool *a*, having a bevelled periphery *b*, in combination with the grooved lever pawl *c*, substantially as herein described and set forth.

No. 54,435.—JOSHUA B. STEWART, South Paris, Maine.—*Hay Fork*—May 1, 1866.—The shank of the tines acts as a plunger in the socketed end of the handle, which contains a spring; the collapse of the latter enables the finger on the shank to indicate on a scale the weight of the hay.

Claim.—A hay fork having a spring balance or weighing attachment applied to it, in the manner substantially as and for the purpose herein set forth.

No. 54,436.—J. C. STODDARD, Worcester, Mass., assignor to J. B. KNOX.—*Car-window Deflector.*—May 1, 1866.—The deflector is attached to the casing of the open window, and projects outward and rearward, causing an outward current of air from the car.

Claim.—First, a portable air and dust deflector, constructed and operating as set forth.

Second, the combination with the body of the deflector of a book at the top and spring at the bottom for retaining the same in place, substantially as set forth.

Third, the combination of the piece F, and elastic spring G, with the metal back E, flanges *c c* and loop I, substantially as shown and described.

No. 54,437.—CONRAD STOLL, Mokena, Ill.—*Cheese Box*—May 1, 1866.—The cheese is supported on a circular revolving platform, the sides are of glass and wire gauze, and a portion is hinged to expose the interior.

Claim.—First, the combination of the circular revolving support M with a cheese box, substantially as her-in described and for the purpose set forth.

Second, a cheese box, constructed and arranged substantially as herein described and for the purpose set forth.

No. 54,438.—WILLIAM MONT STORM, New York, N. Y.—*Submarine Explorer.*—May 1, 1866.—This diving bell has an elevated floor and a manhole in its apex; glazed openings for observation and light; a double casing, divided by a horizontal partition, the upper space forming a reservoir of compressed air, and the lower space a chamber for water ballast.

Claim.—First, the hinged or "toggle" bolts with their clutching jaws and binding nuts, all combined and operating substantially as described for fastening in place the scuttle O and trap Q

Second, the application of the three pressure gauges J K L, prepared and applied in the manner and for the purposes described.

Third, the lookouts F, constructed substantially as described, and combined with a watertight lens *f*, as described.

Fourth, making the ballast ring of the explorer compound, to wit, of a permanently fixed portion E, combined with a series of removable portions or sections E', in the manner and for the purpose set forth.

Fifth, the air-purifying lining R R' to the working chamber of my explorer constructed and operating substantially as described.

Sixth, the combination of the annular sprinkler h', cock g' and pipe i', operating together as described for the purpose described.

Seventh, the application in conjunction with the lookouts F, or their equivalent, of the reflectors V, in the manner and for the purpose described.

Eighth, in combination with my purifier the water space at its bottom and the cock S, constructed and operating in the manner and for the purpose described.

No. 54,439.—ALBERT STUCKENRATH, New York, N. Y., assignor to WILLIAM C. BARNEY, New York, N. Y.—*Adjustable Cut-off and Horse Power Indicator.*—May 1, 1866.—By turning the rod the cut-off valves are simultaneously moved toward or from each other to adjust their position relative to the ports. An exterior index on the valve rod indicates the point of cut-off of the valve, which is moved by contact with spring dogs on the main valve, and in an opposite direction.

Claim.—First, the right and left-handed screws g g', rod h, index j, and dial K, in combination with the cut-off valves E E', and main valve C, constructed and operating substantially as and for the purpose described.

Second, the dogs l, levers m, and arms O, in combination with the cut-off valves E E', and main valve C, constructed and operating substantially as and for the purpose set forth.

No. 54,440.—JÁMES SWEENEY, St. Louis, Mo.—*Machine for Bending Tubes.*—May 1, 1866.—The tube is placed in the mortise of the frame, and bent by the traction of the chain and the screw in the pivoted nut. The chain is flexed by passage around a sheave when a rectangular bend is required.

Claim.—First, the combination of the beam A, having a mortise a in it, with the screw f, when constructed as and for the purpose set forth.

Second, the sheaves d d, in combination with the chain g, or its equivalent, and the screw f.

No. 54,441.—SYLVANUS J. TALBOTT, Milford, N. H.—*Planing Machine.*—May 1, 1866.—The shafts of the cutter head are hung in a sliding frame. attached by a connecting rod to a swing frame, which allows the cutter head to be raised and lowered to suit the varying widths of boards to be tongued and grooved.

Claim.—The sliding frame P', swinging frame F', and connecting bar R', constructed as described, in combination with each other and with a board-matching machine, substantially as and for the purpose set forth.

No. 54,442.—ROBERT H. THURSTON, Providence, R. I.—*Magnesium Lamp.*—May 1, 1866.—The magnesium is burned directly as it issues from the feed rollers, which are divested of the accumulating ashes by a stationary scraper.

Claim.—First, the use of the feed roller B, as a surface on which to burn a strip or wire of magnesium, substantially as described.

Second, the combination of the stationary scraper with k the roller B, substantially as described and for the purposes specified.

No. 54,443.—LYMAN L. TINGLEY, Pawtucket, R. I.—*Spring Bed Bottom.*—May 1, 1866.—The hinged frame may be folded. The lower support bars of the springs are adjustable vertically to vary the resiliency. The tops of the springs are kept in place by a cord which is rove through eye bolts on the bed rails and staples on the disks above the springs.

Claim.—Connecting the tops of the several springs of a spring bed bottom by a cord going loosely through eye bolts in the inside faces of the frame A, substantially as described.

Also, adjusting the springs of a bed bottom so as to prevent unevenness when there is a disparity in the weight of its occupants, substantially as described.

Also, the heart plates K, in combination with the inside slotted bars J on each end of the frame, substantially as and for the purpose described.

No. 54,444.—WILLIAM H. TOWERS, New York, N. Y.—*Flour Barrel.*—May 1, 1866; antedated April 21, 1866.—A sliding cover and sieve at the opening in the barrel head.

Claim.—A flour barrel provided with a sieve agitator and sliding cover, arranged and operating substantially in the manner and for the purpose above set forth.

No. 54,445.—CLEMENS UNVERZAGT, Richmond, Ind.—*Shuttles and Bobbins for Looms.*—May 1, 1866; antedated April 30, 1866.—The projection on the side works against the side of the shuttle box, and assists in guiding; the bent spring is attached to the shuttle, and engages in a recess of the bobbin head; the spindle is fixed; the conical portion of the head lifts the spring and fastens the bobbin, and a flat portion permits its disengagement and removal.

Claim.—First, a shuttle with a projection or rib upon one side, substantially as and for the purposes set forth.

Second, the lever spring c, in combination with the bevelled or conical head of the bobbin B, when attached to a shuttle, in the manner and for the purpose described.

Third, the bobbin B, provided with a conical head c, substantially as and for the purpose described.

Fourth, the combination of the shuttle A, projection D, spring c, and bobbin B, all substantially as and for the purposes set forth and described.

No. 54,446.—GREY UTLEY, Petersburg, Va.--*Straw Cutting Knife.*—May 1, 1866.—The face of the knife has vertical and angular grooves, which form teeth on the edges of the plate.

Claim.—The construction of the blade of the knife, having the vertical and angular grooves on its face to form the diagonal-shaped teeth, as herein described and for the purposes set forth.

No. 54,447.—JASPER VAN WORMER and MICHAEL McGARVEY, Albany, N. Y.—*Base Burning Stove.*—May 1, 1866.—The central magazine is adjustable vertically. The diving flue surrounds the fire pot and ash chamber.

Claim.—First, an adjustable feeder, whereby the supply of coal may be increased or diminished, by raising or lowering the mouth of the feeder from or toward the grate of a stove, substantially as described and for the purposes set forth.

Second, the combination of an open flue extending entirely around the fire pot, and the outer shell of the stove with a magazine or feeder, as described and for the purposes set forth.

No. 54,448.—JOSHUA A. VARNEY, Alton, N. H.—*Horse Rake.*—May 1, 1866.—The teeth have rollers at their lower ends. They are tripped by the motion of the band lever, which, by the engagement of a pawl and ratchet, revolves a shaft on whose ends are pointed bars, which catch in the ground and lift the whole series of rake teeth to discharge the load.

Claim.—First, the shaft I, provided with the bars K K and shafts J J, in combination with the levers F F, arranged and applied substantially as shown and described, for raising the rake teeth so that they may discharge their load.

Second, the rollers d, in the lower ends of the rake teeth E, substantially as and for the purpose specified.

Third, the combination of the bars D with the teeth E, attached and fitted on the rod c, the shafts I J J, bars K K, and levers F F, all arranged on a mounted frame, to operate in the manner substantially as and for the purpose set forth.

No. 54,449.—GERRET VEDDER, Battle Creek, Mich.—*Straw Elevator for Threshing Machines.*—May 1, 1866.—The sides of the straw carrier have independent vertical adjustment by cords to drums on separate shafts, which are endwise to each other and actuated by a common gear wheel.

Claim.—First, the means substantially as herein described and shown, for levelling the straw carrier or stacker of threshing machines.

Second, the combination of the spur wheels f f', and ratchet wheels g g', with the adjustable spur wheel K, and the winding drums d d', separated and independently arranged and operating in conjunction with an adjustable straw stacker, substantially as described.

No. 54,450.—ENOCH WAITE, Franklin City, Mass.—*Machine Card.*—May 1, 1866.—The backing for the teeth is composed of a compound of ground leather and fibre enclosed between webs of fabric.

Claim.—The improved manufacture of wool cards as made with its body or teeth, supporting part composed of layers of paper and cloth arranged and cemented together, substantially as specified.

No. 54,451.—ELIAS M. WALKER, Gallatin, Mo.—*Corn Planter.*—May 1, 1866.—Two furrowers open the soil and are followed by a seed-box whose valves are opened and shut by treadles and springs under the control of the driver.

Claim.—First, the peculiar manner in which corn planters are constructed, as described in these drawings and specifications.

Second, the manner in which the plough stock, corn box, and slide are combined, as described in the drawings and specifications.

Third, the manner in which the treadles are applied, combining the device of dropping by the foot or by the action of the wheels.

No. 54,452.—WASHINGTON WALLICK, Philadelphia, Penn.—*Horseshoe Machine.*—May 1, 1866.—The bent lever is vibrated by a crank on the main shaft, and carries a cutter and a bending die, whose lower end is curved to correspond with the shape of the shoe; the severed portion is pressed over a former and the heels pressed by lateral jaws.

Claim.—First, the combination of the lever G G' H, and the bending and cutting die m, constructed and operating as described.

Second, the jaws P and P', and the former O, for compressing the sides of the shoe, constructed and operating as described.

Third, the creaser and presser R, in combination with the die m, the jaws P and P', and the former O.

Fourth, the mechanism for discharging the shoe, consisting of the lever G G' H, the pieces S S' and t, and the reacting spring, arranged and operating as described.

No. 54,453.—W. M. WATSON, Tonica, Ill.—*Gang Plough.*—May 1, 1866.—The plough beams are respectively connected by rings to the draught bar and laterally by adjustable and flexible braces to each other, so as to permit a certain independence of motion; the outrigger frame has the driver's seat, and a supporting wheel on the "land."

Claim.—The combination in a gang plough of the hinged braces and bolts *e e e e e e*, or their equivalents, and side seat and support *n n*¹ K, all arranged substantially as and for the purpose set forth.

No. 54,454.—CHARLES HUNTER WEBB, San Francisco, Cal.—*Rock Drill.*—May 1, 1866.—The cutters are expanded laterally by the pressure of a cone traversing in angular slots of 45° and bringing the whole length of their cutting faces to bear upon the sides of the hole.

Claim.—The arrangement at an angle of 45° of each end of the back of the cutters or dies, together with the arms or guides thereof, at the same angle of 45° of each edge of the wedge, by which the cutters or dies are driven at its points of contact with the cutters or dies, and also at the same angle of 45° of the various slots in which the cutters or dies are made to traverse in their propulsion by the blow toward the rock and their recoil therefrom, each separately and the whole collectively for the purpose described, namely the effective delivery of the blow with the least amount of friction.

No. 54,455.—SAMUEL WESSON, Worcester, Mass.—*Machine for Finishing Pen Handle Tube Ends.*—May 1, 1866.—The tube is sleeved on a mandrel and advanced through a collar at whose end an obliquely set cutter dresses the end of the tube.

Claim.—The combination and arrangement of the external and internal holders A B, and the cutter F, the latter and the internal holder being movable and provided with mechanism for operating them, substantially as described.

No. 54,456.—CHARLES O. WEST and JOHN CAREY, Martinsville, Ohio.—*Centrifugal Machine for Draining and Cleansing Sugar.*—May 1, 1866.—The sugar is fed from the valve-bottomed hopper to the disk, which delivers it inside the revolving, reticulated frustum, where steam can be used as an auxiliary.

Claim.—First, making the screen of a centrifugal sugar mill in the form of a frustum of a hollow cone, for the purpose described.

Second, the distributing head in combination with the screen of a centrifugal sugar mill.

Third, the adjustable hopper in combination with the screen of a centrifugal sugar mill.

Fourth, the induction pipe K, in combination with a centrifugal sugar mill for cleansing the sugar by steam.

No. 54,457.—JOHN WEST, Bethlehem, Penn.—*Cross-head.*—May 1, 1866.—The angular blocks are adjusted to the slides on which they reciprocate by means of the temper screws and jam nuts, in the frame of the cross-head to which the piston and rod are connected.

Claim.—The within described cross-head consisting of the portion *a*, to which the piston rod is secured, the side pieces *b* and *b'*, transverse pieces *c* and *c'*, the pin *d*, the sliding blocks B and B', and screw studs D D, the whole being arranged as and for the purpose herein set forth.

No. 54,458.—ABEL C. WHITTIER, Lawrence, Mass.—*Washing Machine.*—May 1, 1866.—To be placed in a wash-tub and secured to its sides. It has two grooved surfaces, the upper oscillating upon the lower and depressed by springs.

Claim.—The combination of the slotted arm *e*, spur H, slot in lever I, levers M, rods N, and spring O. for the purpose herein set forth and described.

No. 54,459.—ELI YORK, Windsor, Ill.—*Fence.*—May 1, 1866.—The sticks are arranged vertically between horizontal stringers, which are pinned to the sticks and rest upon the sills.

Claim.—The manner herein described of constructing fences, whereby a strong and durable fence may be put up without any essential preparation of the timber composing the fence, substantially as specified.

No. 54,460.—WILLIAM BAKER and GAYLORD MARTIN, Schenectady, N. Y., assignors to EMPIRE STATE BRICK MACHINE COMPANY, same place.—*Brick Press.*—May 1, 1866.—The press-box has a yielding gate and bar which present a sharp upper edge, and direct the clay to the moulds.

Claim—First, providing a press box which is attached to a pug mill with a yielding gate *i*, which will operate to relieve the press from obstructions, substantially as described.

Second, constructing the bottom of the press box of grate bars *g* of a lozenge shape, or of such shape that they will direct the clay, or other substance under pressure in said bore, towards the ends of the mould boxes, substantially as described.

No. 54,461.—EDWIN BATTLEY, Mont Clair, N. J., assignor to himself and JAMES CRANE, same place.—*Manufacture of Varnish.*—May 1, 1866; antedated April 16, 1866.—Creosote or carbolic acid is used as a solvent for the gums.

Claim.—A varnish compound in which creosote or carbolic acid is used as the solvent of the gum, such as rosin, substantially as set forth.

Also, the composition of a rosin dissolved in carbolic acid with lampblack, substantially as and for the purpose specified.

No. 54,462.—DAVID E. BREINIG, New York, N. Y. assignor to himself and A. C. CRON-DAL, same place.—*Water-proofing Cork and other materials.*—May 1, 1866.—Cork, canvas, leather, &c., are water-proofed by dipping in a composition prepared by treating with an alkali, and afterwards in a solution of a metallic salt, such as nitrate of lead.

Claim.—The use of metalline gum, such as herein described, for treating cork, leather, or other fabrics, in the manner and for the purpose, substantially, as set forth.

Also, forming the metalline gum on the fabric by first dipping it in the alkaline solution, and afterwards in the solution of the proper metallic salt, as described.

No. 54,463.—JOHN A. H. DUNNE, Boston, Mass., assignor to JAMES E. ROGERS, Chelsea, Mass.—*Picture-frame Clamp*—May 1, 1866.—The quadrantal portions are clamped on the slotted cross-pieces by sliding-blocks, which are actuated by an encircling cord and winding roller. The angle of the cross-pieces is adjusted to suit the frame whose sections are to be glued together.

Claim.—The combination of the adjustable cross or bars B C, the sliders, the rope, and the windlass, the whole being arranged and applied together and to a table, substantially as and so as to operate as and for the purposes specified.

No. 54,464.—LEMUEL P. FAUGHT, Foxboro', Mass., assignor to himself and WM. T. COOK, Boston, Mass.—*Brush.*—May 1, 1866.—The hollow frustum occupies an axial position and expands the surrounding bristles against the casing.

Claim.—The hollow metallic cone or thimble D, substantially as and for the purpose set forth.

No. 54,465.—LEVI FERGUSON, Lowell, Mass., assignor to himself and D. M. WESTON, Boston, Mass.—*Bobbin for Spinning, &c.*—May 1, 1866.—Elasticity is conferred upon the bobbin by the slits at the smallest part of the bore, where it embraces the spindle.

Claim.—The arrangement of the slit or slits entirely in the shank or body, and above the head of the bobbin, substantially as described.

No. 54,466.—B. G. FITZHUGH, Baltimore, Md., assignor to himself, J. M. GRIFFITH, and JAMES BREWSTER, same place.—*Harvester.*—May 1, 1866.—Gravitating bent arms engage the sprockets of the ground wheels when the driver takes his foot from the trigger which holds them from engagement.

Claim.—So combining locking arms with the wheels of a harvesting machine and with appliances substantially such as herein described, so that while the driver or conductor is in his seat, stand, or position the locking arms shall be held out of action, and when he is thrown from or leaves his seat, stand, or position said locking arms shall be immediately thrown into action, as and for the purpose herein described.

No. 54,467.—SAMUEL C. GOODSELL and DENNIS FRISBIE, New Haven, Conn., assignors to themselves, H. B. BIGELOW, and DAVID P. CALHOUN, same place—*Hoisting Apparatus.*—May 1, 1866.—The two pawls attached to their respective eccentrics act alternately upon the ratchet wheel, one retreating to catch a tooth while the other is advancing. The pawls are lifted by the revolving trips, the reversal of whose revolution changes the action of the hoisting drum.

Claim.—The combination of the trip T with the two pawls I and L, when constructed and arranged to operate so that the two said pawls act to hoist or lower, substantially as and for the purpose specified.

No. 54,468.—EDWIN E. MARSH, Providence, R. I., assignor to AMERICAN EYELET COMPANY, same place—*Eyelet Machine.*—May 1, 1866.—The lateral expansion and roughened surface of the punch cause the eyelet to adhere to and be retracted from the die.

Claim.—First, the use of a plunger J, having a roughened surface as described, in combination with the die in which the eyelet is formed, for the purpose specified.

Second, the use of a plunger capable of lateral contraction and expansion in combination with the die within which the eyelet is formed, substantially as described, for the purposes specified.

No. 54,469.—FRANK MILLWARD, Cincinnati, Ohio, assignor to himself and THOMAS H. FOULDS, same place.—*Combined Steam and Water Motor.*—May 1, 1866.—The force of the induction stream of water is increased by a jet of steam.

Claim.—First, a reaction water-wheel or turbine whose receiving end is provided with a nozzle for the discharge into said wheel of steam, substantially as and for the purpose set forth.

Second, in the described combination with the reaction wheel or turbine A b b', the injector D c, substantially as set forth and for the purpose specified.

Third, the tank B and supply-pipe C c in the described combination with the injector D c and wheel A b b', as set forth and for the purpose specified.

No. 54,470.—CHAS. TRUESDALE, Cincinnati, Ohio, assignor to himself and WM. RESOR & Co., same place.—*Cupola Furnace.*—May 1, 1866.—The area of the blast openings of the tuyeres decreases toward the upper portion of the series, and their mouths are protected by pillars of fire-brick.

Claim.—First, the provision in a cupola or melting furnace of one or more vertical series of tuyeres, with graduated or decreasing ventages towards the upper portion of the series, substantially as set forth.

Second, the arrangement of one or more vertical series of tuyeres which project beyond the common or general lining wall, and are protected by vertical pliers, substantially as set forth.

No. 54,471.—C. C. WEBBER, Springfield, Mass., assignor to himself and WARREN M. SMITH, same place.—*Wrench.*—May 1, 1866.—In addition to the movable jaw which slides on the shank, the latter is drawn into the socketed handle by means of a screw which has a coarse thread in the nut and a finer one in the end of the shank.

Claim.—The combination of the jaw B, bar A, screw-rod E E', handle c, and nut b, and head f, substantially as specified.

No. 54,472.—NATHANIEL G. WHITMORE, Mansfield, Mass., assignor to himself and EDWARD L. DAY, same place.—*Blacking Brush.*—May 1, 1866.—A drawer in the brush contains blacking.

Claim.—A blacking brush A provided with a sliding drawer a, substantially as and for the purposes set forth.

No. 54,473.—MOSES H. WILEY, Bucksport, Maine, assignor to himself, C. J. COBB, and J. P. AMES, same place.—*Mop.*—May 1, 1866.—The ends of the mop handle and the pivoted lever are respectively provided with rollers, through which the rag is drawn by the gripper to squeeze out the water.

Claim.—The arrangement and application of the mop and the two levers and one or more squeeze-rollers, the whole being applied together so as to operate substantially as specified.

Also, the arrangement of the mop and the two levers as described.

No. 54,474.—J. McGRIGOR CROFT, London, Great Britain.—*Rudder.*—May 1, 1866.—Oblique lateral flanges on the rudder blade.

Claim.—The application or form of diagonal, curved, or oblique blades to rudders, substantially in the manner and for the purposes above described.

No. 54,475.—CHARLES FALCK, Middlesex Co., England.—*Refrigerator.*—May 1, 1866.—The water is enveloped by ice in a chamber which forms part of the safe.

Claim.—First, the combination as well as the arrangement of the cooling chamber a, the chamber c for holding ice, the waste-water receptacle e, and the pipe or conduit f, connecting the chambers c and e, as set forth.

Second, the combination as well as the arrangement of the vessel d (for holding water or a liquid) and its eduction pipe d' with the ice chamber C and the cooling chamber a, as explained.

Third, the combination as well as the arrangement of the vessel d (for holding water or a liquid) and its eduction pipe d' with the ice chamber c and the cooling chamber a and the waste-water chamber e, the whole being substantially as set forth and represented.

No. 54,476.—G. ALBERT REINIGER, Stuttgart, Wurtemburg.—*Cigar Filler Machine.*—May 1, 1866.—The leaf tobacco is fed compactly and continuously to the knife, which cuts it into bunches ready for the wrappers. The present improvements on the patent of October 3, 1861, are in placing the moving portions above the conveyer-belt, to avoid the grit. The movable wings are actuated by arms on a vertical rock shaft, and cam on the main shaft. A gate presses on the tobacco while being cut.

Claim.—First, placing the journal boxes of the driving shaft above the endless aprons E F, instead of below as heretofore, substantially as and for the purpose specified.

Second, the vertical shaft a', with arms e' f' k', in combination with the cam b' and the driving shaft and with the rock shaft e', and the jaws of the receiving apparatus, constructed and operating substantially as and for the purpose set forth.

Third, the lantern p in combination with the eccentric h, on the driving shaft, and with the receiving apparatus K and endless aprons E F, constructed and operating substantially as and for the purpose described.

Fourth, the gear wheels s t v and cogs w, in the side of the rack H, in combination with the lantern p and the receiving apparatus K, constructed and operating substantially as and for the purpose set forth.

Fifth, the gate r in combination with the endless aprons E F, knife g, and receiving apparatus K, constructed and operating substantially as and for the purpose specified.

No. 54,477.—FRANCOIS STOKER, Lyons, France.—*Fuel Composition.*—May 1, 1866.—Composed of carbonaceous material, nitrate of potash, and gummy matter, formed into blocks, \

especially for use in sad-irons, foot warmers, &c., where heat without flame or smoke is desirable.

Claim.—The exclusive use, first, of the combustible substance or fuel, to whatever purpose it may be applied.

Second, of the foot warmer, smoothing iron, stir-up apparatus, and soldering iron, the whole substantially as hereinbefore described, and illustrated on the accompanying drawings.

No. 54,478.—NATHANIEL PULLMAN, New Oregon, Iowa.—*Switch for Replacing Cars upon the Track.*—May 1, 1866.—The jointed section forms a bridge from the ground to the track.

Claim.—First, the shoes E, attached to the cross-bar F, in combination with the bevelled rails C, and the movable rails B, all arranged to operate as shown and described.

Second, in combination with the rails B, the chair H, and the supplementary rails D, arranged to operate as and for the purpose set forth.

No. 54,479.—C. ALBERT, Harrisville, Ohio, and J. J. SHANK, West Salem, Ohio.—*Sheep Holder for Shearing.*—May 8, 1866.—The sheep is strapped by its neck and feet to the shaft and arms respectively, and the presentation is changed by rotation of the shaft.

Claim.—The shaft C C', arms d, loops d', strap m, in combination with the ratchet plate F, pawls b, and table, arranged in the manner and for the purpose set forth.

No. 54,480.—LOUIS ALEXANDER, New York, N. Y.—*Water Closet Seat.*—May 8, 1866.—A series of seats over one pan, each seat having its separate lock.

Claim.—The combination seat as constructed of several flaps a b c d e, &c., with projecting flanges, or ribs, in combination with suitable fastenings g, constructed and operated substantially as and for the purpose set forth.

No. 54,481.—JOHN F. ALLEN, New York, N. Y.—*Propeller.*—May 8, 1866.—A vertical, vibrating, oar-shaped propeller behind the stern post, with an adjustable range of motion to steer as well as propel.

Claim.—First, the arrangement of a propeller blade at the end of a vibrating shaft, which gives motion to said propeller blade, when said propeller does move at a regular and fixed pitch, during the whole length of its motion, and is capable of being operated by the action of the water against its surface, at the beginning of each stroke, in such a manner that thereby its forward edge shall always be brought forward to its line of motion, independent of the direction in which this motion may be communicated, substantially as described.

Second, the collar F, with a projection n, fastened to the propeller, in combination with the worm wheel J, provided with a projection or stop m, and operating together in the manner and for the purpose described.

Third, the arrangement of the worm wheel J, with a suitable stop or projection m, operated by suitable gearing, for the purpose of regulating and altering the pitch of the propeller blade, substantially as set forth.

Fourth, the arrangement described for changing the pitch of the propeller blade while in motion, so as to make said propeller act at the same time as a rudder for the vessel, substantially as set forth and specified.

No. 54,482.—ASA H. ALLISON, Charlottesville, Ind.—*Scrubber and Tender.*—May 8, 1866.—The mop head has a valved water tank above the scrubber.

Claim.—First, the tender C, with the trough D, and stop valve E, substantially as and for the purposes set forth.

Second, the combination of the tender and scrubber, constructed and arranged substantially as and for the purposes set forth.

No. 54,483.—R. H. ANDREWS, Elizabethtown, Penn.—*Pump.*—May 8, 1866.—A wrist on the rotary wheel is connected to a slide and the latter to the plunger.

Claim.—The combination and arrangement of the wheel d, slide e, connecting rod g, and cylinder h, for the purpose herein set forth.

No. 54,484.—FRANCIS ARMSTRONG, Pittsburg Penn.—*Apparatus for Protecting Pumps in Deep Wells.*—May 8, 1866.—A flange on the shell of the piston and a cup on the piston rod catch and retain falling bodies.

Claim.—The cups a and b, when arranged and operated as, or substantially as and for the purpose set forth.

No. 54,485.—HAINES AUSTIN, East Liberty, Ohio.—*Coupling for Carriage Felloes.*—May 8, 1866.—The box encases the ends of the felloes and the sectional block, which is expanded by the conical screw.

Claim.—The combination and arrangement of the tube or box A with the conical male screw C, and blocks D D, (with semicircular female screws,) working in said box for the purpose set forth, substantially as described.

No 54,486.—J. W. BALDWIN, Sidney, N. J.—*Grinding Mill.*—May 8, 1866.—The vertical position of the runner is adjusted by a choice of spindles or a screw extension spindle passing through the "balance-rynd."

Claim.—The screw E, or equivalent detachable points d, combined with the spindle F and bail D, and arranged with the runner A, substantially in the manner as and for the purpose herein set forth.

No. 54,487.—A. G. BARNARD, Seville, Ohio.—*Fence Gate.*—May 8, 1866.—The notched ends of the gate rails project between fence-posts, and the gate is sustained by the contact of its inclined slat with a post.

Claim.—The strip D and posts E E' in combination with the brace C and gate A, cut out as at b, substantially as and for the purpose set forth.

No. 54,488.—CARL BARTHOLMAE, New York, N. Y.—*Composition for Gold Size.*—May 8, 1866.—Composed of water, soda, rosin, shellac, glue, and boiled oil.

Claim.—First, the use of glue in a composition for gold sizing, substantially as and for the purpose described.

Second, a composition made of the ingredients above specified, and mixed together, substantially as and for the purpose set forth.

No. 54,489.—EDWIN BELL, St. Paul, Minn.—*Removing Sand Bars from Rivers.*—May 8, 1866.—Barges are attached and moored to form a sluice-way, the current through which rotates the wheels, which are armed with shovels.

Claim.—First, the removal of sand bars and similar obstructions, in the manner and by the means substantially as set forth.

Second, the wheels C having the shovels or prongs attached to the paddles, when arranged to operate in combination with the boats or barges A, as shown and described.

Third, a series of boats or barges provided with the lee-boards B, and arranged to form a channel for the passage of the water for the purpose of controlling or directing the current, as and for the purpose set forth.

Fourth, in combination with boats arranged as set forth, the wheel C and the cables c, arranged and operating as described, for the purpose of controlling the movements of the boats, as herein set forth.

No. 54,490.—HENRY BERRY, HUGO HOCHHOLZER, and FRANK DENVER, Virginia City, Nevada.—*Cages for Hoisting Purposes in Mines.*—May 8, 1866.—The cage rises in guides; the lifting rope is attached to hinged arms, which expand by springs if the rope break, and dig into guides to arrest the fall.

Claim.—First, the adjustable arms A A, India-rubber springs K K, or their equivalents, together with the head-piece B, pillow E, and the bevelled shoulders L L, in combination with the large arms F F and guides G G, substantially as described and for the purposes set forth.

Second, the cap C C, hung on hinges, in combination with the braces c' c', and head-piece B, substantially as described and for the purposes set forth.

No. 54,491.—J. F. BIRCHARD, Milwaukee, Wis.—*Apparatus for Forcing Air in Carburetting Machines.*—May 8, 1866.—The wheel is revolved by a spring in an open reservoir, and has curved wings, which carry the air beneath the surface and discharge at the trunnion.

Claim.—An air feeder open and exposed, and constructed and operated substantially as herein described.

No. 54,492.—CHARLES BIRDSALL, Jersey City, N. J.—*Compound Sirup.*—May 8, 1866.—Compound of gum tragacanth, gum-arabic, molasses, and honey, boiled with the addition of cream of tartar and isinglass, and skimmed.

Claim.—The compound sirup, prepared with the materials and as set forth.

No. 54,493.—JAMES BLAYNEY, Lowel, Iowa.—*Vertical Hand Spinning Machine.*—May 8, 1866.—The inclined endless feed apron gives facility for splicing. The devices of the second claim refer to means for notifying the operator that a certain twist has been attained.

Claim.—The application to the vertical spinner of all the apparatus delineated and shown in Fig. 2 of the accompanying drawings except that part marked z, and called in this specification the clamp.

Also, the application to the above-named spinner of all the gearing shown and delineated in Fig. 3 of the accompanying drawings, except those parts marked h and z, and called in this specification the main wheel and the clamp.

No. 54,494.—THOMAS H. BONHAM, Elizabethtown, Ohio.—*Portable Fountain.*—May 8, 1863.—A water reservoir in an ice chamber, with a non-conducting casing and a pump for the condensation of the air above the water surface to produce a cool jet.

Claim.—Herein as new, and of my invention, the arrangement of non-conducting refrigerator A B C, reservoir F, feeder G g, pan H, and nozzle I i, the whole forming a portable fountain, substantially as described.

No. 54,495.—HORATIO B. BRACE and WILLIAM T SWART, Canandaigua, N. Y.—*Burning Fluid for Illumination.*—May 8, 1866.—Composed of benzine, carbonate of potash, sulphate of ammonia, muriate of soda, gum benzoin, gum camphor and spirits of nitre.

Claim.—A burning fluid for illuminating purposes, composed substantially of the ingredients above described, and in about the proportions herein contemplated.

No. 54,496.—THOMAS H. BRADLEY, St. Louis, Mo.—*Ladies' Skirt Lifter.*—May 8, 1866.—Explained by the claim.

Claim.—A machine to be worn as a belt, for raising and lowering the skirts of ladies' dresses with cord and tape, by means of a wheel having a case, axle and crank, with pulleys and guides attached.

No 54,497.—HENRY A. BREED, Lynn, Mass.—*Quartz Pulverizer.*—May 8, 1866.—Improvement on patent No. 45,213. The rotating disk has openings and the interior of the casing a revolving clearer in each case to prevent clogging of the material.

Claim.—The construction of rotating disk *f*, with the openings *l*, in the manner and for the purpose substantially as described.

Also, the teeth or clearers *m*, upon the periphery of the wheel *g*, operating substantially as set forth.

No. 54,498.—HENRY BUCK, Harrisburg, Penn.—*Broom Head.*—May 8, 1866.—The butts of the brush are encased in a cap; a perforated encircling strap sustains the broom lower down.

Claim.—First, the arrangement of the bars C C, perforated as described, and the ends of said bars formed substantially as shown, when used with a sliding adjustable clamp, for the ends of said bars, as and for the purpose specified.

Second, the arrangement of the keeper D, with the socket A, the strap B, and the bars C C, the several parts being constructed and used as and for the purpose specified.

No. 54,499.—N. M. BUFFINTON, Fall River, Mass.—*Preparing Stone for Power Printing.*—May 8, 1866; antedated April 30, 1866.—The design intended to be left salient for surface printing is covered with an acid-resisting composition; the other portions are eaten away by acid to depress them out of contact with the inking roller.

Claim.—Preparing stone substantially as herein described, so that the drawing, lettering or design upon the same may be used for letter-press printing, and that stereotypes and electrotypes may be produced from the same.

No. 54,500.—E. P. H. CAPRON, Springfield, Ohio.—*Brick Machine.*—May 8, 1866.—The bricks are formed in cells of the horizontally revolving mould wheel; a roller presses any hard substance into the body of the brick out of the way of the scrapers in the curved finishing box.

Claim.—First, constructing the wings O, of the pressing wheel, as shown and described, for the purpose of drawing the clay inward, at the same time that it presses it downward into the cells.

Second, the roller *a*, located in the mouth of the finishing box T, substantially as shown and described.

Third, in combination with the roller *a*, the scraper *m*, arranged and operating as set forth.

Fourth, the finishing box T, provided with the scrapers *c* and *d*, constructed and arranged to operate substantially as set forth.

Fifth, the table H, having the cells N formed therein and provided with the projecting shoulders *r*, to support the follower P, as shown and described.

Sixth, providing the followers P with the set screws *j*, arranged to operate in connection with the shoulders *r* of the cells N, as and for the purpose set forth.

Seventh, the track I, extending only a part of the way around, for raising the brick out of the mould in combination with the revolving table, in which the followers are supported entirely by the table, while passing under the pug mill, or hopper, for the purpose of reducing the friction and power required to operate the same.

Eighth, constructing the followers P with the hinged lid *s* and lever *p*, arranged to operate as shown and described.

Ninth, in combination with the follower P, arranged as described, the roller *n*, for depressing the follower and opening the cell, as it passes under the hopper to receive the clay.

No. 54,501.—A. CARTER and R. SPAKE, Salem, Iowa.—*Hand Loom.*—May 8, 1866.—Explained by the claim.

Claim.—First, in hand-power looms operating the treadles through a system of slides R, which are moved by pins or cogs on the treadle shaft J, substantially as described.

Second, the combination of the rod G, which extends from the lay, the pawl H, the shaft J, armed with pins or cogs L, and the slides R, substantially as shown and described.

No. 54,502.—WILLIAM CAVEN, Cincinnati, Ohio.—*Teakettle*—May 8, 1866.—The rear end of the bail is secured to the kettle by a plug, which is also the pintle of the lid. The wings of the pintle pass through slits in the kettle top and are rotated to catch beneath it

Claim.—In the described combination with the sockets C *c c'*, and flanged and perforated lid D E *e*, the winged rear lug F G *g g'* H, adapted to be secured in place by the ordinary bail, in the manner set forth.

No. 54.503.—ORRIL R. CHAPLIN, Salem, Mass.—*Combined Square and Bevel.*—May 8, 1866.—The tongue slips within a frame having a stationary arm at 90°, and a hinged arm with a protracting arc.

Claim.—The combination of a movable tongue B, and beam C affixed thereto, with the movable or hinged beam D, substantially as described.

No. 54.504.—CHARLES S. CISNA, Burlington, Iowa.—*Buckle.*—May 8, 1866.—The tongue is pivoted to one cross-bar, and the buckle is attached by its ears and a screw bolt to the strap.

Claim.—There movable bolt *m*, parallel with and in rear of the fixed cross-piece *r*, to which the tongue B is pivoted, all being combined and arranged substantially as shown and described.

No. 54.505.—WILLIAM C. CLEVELAND, Cambridge, Mass.—*Scrubbing Brush.*—May 8, 1866.—Flexible rubbers on the end of the brush retain the water around the bristles.

Claim.—Combining with a brush block one or more flexible rubbers, applied and operating in connection with the bristles, substantially as described.

No. 54.506.—J. D. COCHRAN, Milford, N. H.—*Seeding Machine.*—May 8, 1866.—The vibrating hopper is actuated by a cam on the main shaft, oscillating the brush shaft by its own motion. The seed-conveying spout and coverer are adjustable.

Claim.—The vibrating hopper D, placed on the pivoted plate E, and operated through the medium of the cam F, and spring *c*, substantially as and for the purpose set forth.

Also, the oscillating cut-off brush *t*, in combination with the vibrating hopper D, all arranged to operate substantially as and for the purpose specified.

Also, the adjustable seed-conveying spout H, substantially as and for the purpose set forth.

Also, the coverer J, attached to the rear of the share I, and having the spring L applied to it, substantially as shown and described.

No. 54.507.—J. D. COCHRAN, Milford, N. H.—*Corn Planter.*—May 8, 1866.—The two handled planter has a reciprocating slide for seed and fertilizers, and opening jaws to make a cavity for the seed.

Claim.—The strips A A', with the slide F attached and provided with the plates B B, pivoted together as shown, in combination with the hopper D, provided with the partition E, the slide and partition being provided with holes, and all arranged to operate substantially as set forth.

Also, the manure or supplemental hopper H, applied to the strip A, in combination with the seed hopper D and slide F, substantially as and for the purpose specified.

Also, the plates J J, in combination with the plates B B, arranged to operate in connection therewith, substantially as and for the purpose set forth.

No. 54.508.—P. F. CROUCH, New York, N. Y.—*Shears.*—May 8, 1866.—The upper blade and handle move upon a slotted fulcrum, which gives a "draw" cut. The edges are kept in contact by a spring.

Claim.—First, the blade *e*, with a slot or mortise for the fulcrum *g*, in combination with the handle *c*, shear *a*, and handle *b*, as set forth, whereby the blade *e* receives an endwise motion to act with a drawing cut, as specified.

Second, the broad, thin blade *a*, blade *e*, and spring *i*, in combination with the handles *b* and *c*, joint and slotted fulcrum *g*, for the purposes and as set forth.

No. 54.509.—H. S. CUSHING, Blossburg, Penn.—*Scale Beam.*—May 8, 1866.—The scale beam is enveloped by a rotating cylinder, divided longitudinally and transversely, to indicate the price of fractions of weights.

Claim.—The combination of the scale beam A, the rotary cylinder B, marked and divided longitudinally and transversely, as shown, and the catch H, and ratchet plate G, for holding the cylinder stationary, substantially as shown and described.

No. 54.510.—JOHN W. DIXON and GEORGE HARDING, Philadelphia, Penn.—*Apparatus for the Manufacture of Paper Pulp.*—May 8, 1866.—The digester is heated by a furnace; has a perforated diaphragm, and a central, valved, pulp-discharge tube. An equilibrium pipe connects the upper and lower parts of the boiler, which receives from a rotary pump a current of water heated by exhaust steam.

Claim.—First, the pulp digester, in combination with a lower perforated diaphragm, having a central pulp passage, a circulating tube or tubes, connecting the upper and lower parts of the digester.

Second, the pump digester, in combination with a lower perforated diaphragm, a central pump passage, a circulating tube and pump, to produce a circulation through the mass from bottom to top, or *vice versa*.

Third, the two valves D and F, in the pump escape tube, for withdrawing a portion of the contents without interfering with the operation within the digester.

Fourth, the charging pipe S, with its double valve, for introducing fresh material without disturbing the operation.

Fifth, the combination of pulp digester, circulating tube and pump, when the fire is applied directly beneath the digester.

Sixth, the continual forcing in of fresh liquid below and upwards through the mass to be pulped, in combination with the introduction of the fresh charges of the material into the digester above, whereby the cleaner liquid would always be in contact with the pulp most cooked.

No. 54,511.—JOSEPH DIXON, New York, N. Y.—*Planing Machine.*—May 8, 1866.—The cutter head is rigidly attached to its shaft. The latter has a vertical adjustment, and an index indicates its vertical position.

Claim.—The combination of the cutter head A, shaft B, adjustable bed E, indicator *g f*, and adjusting screw I, substantially in the manner and for the purpose set forth.

No. 54,512.—M. B. DODGE, New York, N. Y.—*Quartz Crusher.*—May 8, 1866.—Wedge, between the inclined adjacent faces of the bearing boxes limit the approach of the roller faces.

Claim.—The combination of the wedges D with the boxes B of crushing rollers, when arranged to limit the approach of said rollers, as herein described.

No. 54,513.—J. E. P. DOYLE, New York, N. Y.—*Railway Switch.*—May 8, 1866.—The car strikes an arm of the revolving post, and by connecting rods, &c., actuates the switch, sets the signals, and rearranges them after having passed, by corresponding devices, further along.

Claim.—First, the combination of the revolving radial arms I J K L, the revolving cylinders M N, and the chains R S T U V W, or their or either of their equivalents, with each other and with the switch B, substantially as described and for the purpose set forth.

Second, the combination of the revolving signals G' with the revolving cylinders M N, or either of them, substantially as described and for the purpose set forth.

Third, the combination of lever C' and bar B' with each other, with car A', and with the radial arms I J K L, substantially as described and for the purpose set forth.

No. 54,514.—CHARLES A. DURGIN, New York, N. Y.—*Table Forks.*—May 8, 1866.—The two inner tines are depressed below the outer to give it greater capacity.

Claim.—A new and improved article of manufacture, a fork constructed and described as set forth.

No. 54,515.—ELI H. ELDRIDGE and THOMAS LEACH, Taunton, Mass.—*Machine for Embossing Napkin Rings.*—May 8, 1866.—The ring is placed on a mandrel and rotated in contact with a die roller.

Claim.—The above described chasing or embossing machine, constructed and arranged substantially as and for the purposes set forth.

No. 54,516.—JAMES FALLOWS, Philadelphia, Penn.—*Sheet Metal Spoon.*—May 8, 1866.— The handle is made of a piece of sheet metal, cut, swaged, folded, and finished in connection with the bowl.

Claim.—A sheet metal spoon, having a hollow handle, constructed of a single piece of sheet metal, substantially as herein described and set forth for the purpose specified.

No. 54,517.—JAMES FALLOWS, Philadelphia, Penn.—*Pepper Box.*—May 8, 1866.—The cover vibrates horizontally on a pivot and is fastened to the handle.

Claim.—A pepper box for table use, having its perforated cover constructed and applied so as to operate in combination therewith, substantially as and for the purpose described.

No. 54,518.—JOSEPH C. FIELD, Buffalo, N. Y.—*Tile Mould.*—May 8, 1866.—The cement pipes are moulded in a flask, and the core drawn out from above, forming the socket on the end without a loose collar or ring.

Claim.—First, the removable core cylinder *c* in a mould for making tiles or pipes, as set forth and for the purpose specified.

Second, the fixed shoulder on the bottom end of the core cylinder, in combination with the bell-shaped bottom of the outer flask or mould, for the purpose of forming a socket in one end of the tile or pipe, as and for the purposes described.

Third, the use of the circular shoulders or projections on the surface of the stationary base plate, in combination with the cover, with its cross-pieces on the top of the core, for the purpose of centring and preserving the requisite distance between the core and the outer flask or mould, in order to obtain a uniform thickness in the walls of the tile or pipe, arranged substantially as described and for the purposes set forth.

Fourth, the use of the cross-bars f and e, and the lever r r, for the purpose of starting the cylinder core, substantially as described and for the purposes set forth.

Fifth, the adjustable hopper o o, substantially as described and for the purposes set forth.

Sixth, moulding vertically, with an adjustable core c and the other means herein described.

No. 54,519.—WALTER FITZGERALD, Boston, Mass.—*Gearing for Lathes, &c.*—May 8, 1866 —The shipper lever places the pulley, in connection with the band shaft, by an intermediate and larger wheel, or by smaller and immediate wheel, so as to obtain a reversal of direction and change of speed.

Claim.—The overhead driving and reversing gearing, constructed and operating substantially as described.

No. 54,520.—WILLIAM H. FLINN, Nashua, N. H.—*Machine for Cutting Wire.*—May 8, 1866.—The perforated cutter plate and die are enclosed in a circular chamber and the former operated by a lever.

Claim.—The application of the case A and its perforated cutter or cap plate D to the perforated disk B, substantially in the manner as described, the case under such circumstances operating to support the plate B circumferentially.

Also, the combination and arrangement of the case A, the cutter plate B, the cap cutter plate D, the arm d, and the cam lever C.

Also, the arrangement of the gauge E with the cam lever, the perforated case A, and the perforated disk B, provided with an arm d on the same, and a spring f, as specified.

No. 54,521.—ADOLPH O. FORSBERG, Cincinnati, Ohio.—*Hernial Truss.*—May 8, 1866.—Explained by the claim.

Claim.—The hinged pad H, having a positive and non-elastic adjustment substantially as described.

No. 54,522.—J. L. FOUNTAIN, New Milford, Ill.—*Attaching Cultivator Blades.*—May 8, 1866.—The blade is attached to the standard by a block, dovetail piece, and screw-bolt.

Claim.—The offset B, slot D, and screw-bolt E, in combination with the standard C and blade A, arranged in the manner and for the purposes substantially as set forth.

No. 54,523.—C. L. FRINK, Rockville, Conn.—*Rubber Packing.*—May 8, 1866.—Metallic filings are added to the ordinary rubber composition and the compound vulcanized.

Claim.—A rubber compound made of the ingredients herein specified for the purposes set forth.

No. 54,524.—C. R. FRINK, Norwich, N. Y.—*Hay Spreader.*—May 8, 1866.—The metallic brackets have two arms arranged at right angles to each other to afford bearings for the wheels and fork shafts; the latter are moved in their bearings by levers. The spreader forks are also vertically adjustable in boxes placed upon the cranks of the driving shafts.

Claim.—First, the metallic brackets E attached to the bar A, and arranged as shown to serve as bearings for the axles F of the wheels G, and also as bearings for the shafts J L of the wheels I, and pinions K, substantially as and for the purpose set forth.

Second, the arrangement of the sliding bearings a, brackets E, crank shafts L, with levers N O, combined and operating in the manner and for the purpose herein specified.

Third, the fitting of the forks R in boxes Q, placed in the cranks M of the shafts L L, and arranged in such a manner as to admit of the adjustment of the forks nearer to or further from the surface of the ground as may be desired.

No. 54,525.—CLINTON T. FROST, Medfield, Mass.—*Straw Cutter.*—May 8, 1866.—The hay is cut between two knives, the lower one of which has a forward motion as the upper descends to make a "draw" cut.

Claim.—The combination of the spring d, the knives C D, the bar F, and the lever G, substantially in the manner and to operate as before described.

No. 54,526.—WILLIAM H. GANDEY, Lambertville, N. J.—*Coal Stove.*—May 8, 1866.—Air from the ash pit circulates upward through the grate and also around the fire pot, thence through passages in the frustum to the chamber, where it reverberates and dives through passages around the rim of the fire pot, reaching the space between the casings and the flue.

Claim.—The fire pot F and chamber G, in combination with the draught passages H, draught spaces d, external cylinder D, and base A, all arranged substantially in the manner and for the purpose herein set forth.

No. 54,527.—EDWARD GILLIAM, Allegheny City, Penn.—*Fruit Gatherer.*—May 8, 1866.—The wire basket oscillates in a bail at the end of a handle, when the spring is retracted.

Claim.—The combination of the basket A with the yoke B, spring C, and handle D, constructed, arranged, and operating substantially as herein described, and for the purpose se forth.

C P 2—VOL II

No. 54,528.—EDWARD GILLIAM, Allegheny City, Penn.—*Apparatus for Exterminating Insects.*—May 8, 1866.—The furnace chimney is connected with a bellows, worked by a lever, to eject through the nozzle the volatile results of combustion upon the insects and their grubs.

Claim.—The arrangement of the furnace B, bellows D, pipes c and g, and levers e and f, constructed, arranged, and operating substantially as herein described, and for the purpose set forth.

No. 54,529.—WILLIAM GREENLEAF,Terre Haute, Ind.—*Securing Boxes in Hubs.*—May 8, 1866.—The box consists of a threaded tube, engaging the caps at each end of the hub.

Claim.—The box B, provided with a screw thread a at each end, in combination with the thimbles E, provided with internal screws c to receive the screw threads a of the box, and connected to the bands C D, substantially as and for the purpose herein set forth.

No. 54,530.—GEORGE HADFIELD, Cincinnati, Ohio.—*Medical Apparatus for Treating Diseases by Vacuum.*—May 8, 1866.—The frame, with its air-tight curtain, is placed around the patient, and the air exhausted from its interior.

Claim.—First, in a portable apparatus for treating diseases by exhaustion of the atmosphere, the arrangement of a receiver with a vase F, the adjustable conical frames G, and sides H, and cloak I, all constructed substantially as and for the purposes set forth.

Second, the arrangement of an air pump A, with the pipes C and D, the pipe E, and receiver B, composed of several separable parts F G H, and cloak I, all constructed and combined, substantially as and for the purpose set forth.

No. 54,531.—A. F. HAMMOND, Houston, Ohio.—*Burglar Alarm.*—May 8, 1866.—The device is to be attached to a mantel piece or window-sill; a string connected to the door is the means of springing the hammer.

Claim.—The combination of the wire I and the lever trigger G H with the detent, hammer, and detonating match-lighting arrangement, substantially as described.

No. 54,532.—WASHINGTON H. HARBAUGH, Piqua, Ohio.—*Heating Stove.*—May 8, 1866.—The air chamber is placed centrally in the upper part of the combustion chamber, has an exterior door in the side, and hot-air exit above. Diaphragms with dampers deflect the current in the combustion chamber.

Claim.—First, the arrangement of parts, the oven H, door h, ventage K, and cap L, substantially as described, for a convertible cooking and warming and ventilating stove.

Second, the chambers H I, ventiducts J K K', doors h i, and caps L h', or their equivalents, for a convertible baking and ventilating apparatus, as set forth.

Third, the provision of the supplementary shell E and its described accessories, as and for the purpose stated.

No. 54,533.—PETER W. HARDWICK, Wayne county, Ind.—*Invalid Bedstead.*—May 8, 1866.—The frame is counterbalanced to assist in raising the invalid to a sitting posture.

Claim.—The independent frame B, with its counter-balancing attachment.

No. 54,534.—A. T. HARRISON, Clinton, Ill.—*Cultivator.*—May 8, 1866.—Convertible into a cotton seed planter or cultivator. The agitator revolves in the hopper and the seed escapes to the spout and feed wheel, which latter drops it into the furrow behind the winged share.

Claim.—First, the revolving agitator, with curved and pointed arms, in combination with the double inclined cylindrical hopper bottom, substantially as and for the purposes set forth.

Second, the hinged casing of the seeding wheel, in combination with said wheel f, as and for the purposes described.

Third, the combination of the wings w with the gopher h, as and for the purposes specified.

No. 54,535.—EDWARD HARRISON, New Haven, Conn.—*Grinding Mill.*—May 8, 1866.—The recesses between the grinding surfaces are composed of softer material, and by wearing keep the cutting edges salient and sharp.

Claim.—Providing the faces or grinding surfaces of millstones, metal plates, &c., with grooves or channels, corresponding in position to the furrows required for said surfaces, and having the grooves or channels provided with a filling composed of a material softer than the stone or plate.

No. 54,536.—W. H. HART, Jr., Philadelphia, Penn.—*Neck Tie.*—May 8, 1866.—The forks straddle the button attachment and the wings slip under the lappel of the collar.

Claim.—A cravat holder provided with wings a a, arms b b, and slot m, constructed and operating substantially as described.

No. 54,537.—CHARLES T. HARVEY, Tarrytown, N. Y.—*Propelling Cars on Railways.*—May 8, 1866; antedated April 6, 1866.—The wire rope is placed inside of concave guiding plates with interior projections. The chain has bulbs at regular intervals to protect from frictional wear. The joint ferrules are protected by adjustable rings.

Claim.—First, the use, in connection with a railway, of one or more series of continuous guide plates independent of (but parallel to) the rails, for containing and guiding a propelling rope or chain; said plates having one or more openings between or in them permitting car connections, and also egress of injurious accumulations, substantially as and for the purposes described.

Second, the combination and use of interior "friction slides" or projections with the aforementioned guide plates, substantially as and for the purposes described.

Third, the combination and use of the "ferrules," (so-called,) whether pointed or adjustable, or permanently attached to or with the wire or other rope or chain, substantially as described.

Fourth, the combination and use of the adjustable "friction rings," (so-called,) with the ferrules, and ropes or chains aforementioned, substantially as described.

No. 54,538.—E. A. HARVEY, Wilmington, Del.—*Neutralizing Acid on Sheet Iron.*—May 8, 1866.—Explained by the claim.

Claim.—The process substantially as herein described for removing or neutralizing the acid used in cleansing sheet iron or other articles by means of an alkaline solution applied under pressure in a close vessel to which rotary or other motion is applied to agitate the articles under treatment.

No. 54,539.—GEORGE HASTINGS, Waltham, Mass.—*Watch Dies.*—May 8, 1866.—The plates of which the cutting face of the die is composed are so grouped as to produce a piece of the required shape for a watch hand.

Claim.—The watch-hand die as made of the three pairs of plates A A B B C C, formed, arranged, and combined substantially as specified.

No. 54,540.—CHARLES J. HAUCK, Williamsburg, N. Y.—*Oiler.*—May 8, 1866.—The oiler has a false bottom depressed by a spring, but yielding to the pressure of the finger.

Claim.—An oiler provided with a false bottom *d*, protected and supported by the annular lip *c*, in combination with the spring *e*, applied substantially as and for the purposes described.

No. 54,541.—J. B. HAWLEY, New Haven, Conn.—*Knife and Fork Rack.*—May 8, 1866.—A knife and fork rack, a butter dish, a salt cellar and call bell combined.

Claim.—The combination of the rack F with individual "salt" and "butter" C D, substantially as and for the purpose specified.

No. 54,542.—LABAN HEATH, Boston, Mass.—*Optical Instrument.*—May 8, 1866; antedated May 1, 1866.—To convert the microscope into a telescope, the part containing the two double convex lenses is detached and the eye-glasses changed.

Claim.—As a new article of manufacture, the convertible optical instrument herein described, constructed, arranged, and operating substantially as set forth.

No. 54,543.—AUGUST HERMANN, New Haven, Conn.—*Mangle or Rolling Press.*—May 8, 1866.—The segment oscillates above the arc-shaped bed, the pressure being given by treadle and the motion by hand.

Claim.—The mangle or rolling press consisting of a segment, circular bed and cover, with a treadle frame and radial bars operated by a circular reciprocating motion, substantially in the manner and for the purposes as herein described.

No. 54,544.—G. W. HEWITT and J. B. HALEY, Cincinnati, Ohio.—*Low Water Detector.*—May 8, 1866.—The tube has a stopcock, fusible disk, and steam whistle.

Claim.—A low water detector, constructed as described, as a new article of manufacture.

No. 54,545.—ABNER HITCHCOCK, Wayne county, Mich.—*Bed Bottom.*—May 8, 1866.—The inner frame is supported by its transverse wires and horizontal springs upon the coiled springs which are supported on the longitudinal slats.

Claim.—The combination and arrangement of the frames A and B, the springs D and D', the beams C, and the cross-wires E, substantially as and for the purpose set forth.

No. 54,546.—CHARLES HOLTZ, Chicago, Ill.—*Chair and Walker for Children.*—May 8, 1866.—Explained by the claim.

Claim.—First, in combination with a child's chair provided with the circular frame E F, constructed as described, the employment of an annular revolving table G, substantially as and for the purposes specified.

Second, the combination and arrangement of the circular arm supports E F, the circular frame A, and removable seat C, substantially as and for the purposes set forth.

Third, the combination and arrangement of the frame A, the removable seat C, the arm supports E, the removable segment F, and revolving annular table G, substantially in the manner and for the purposes shown and described.

No. 54,547.—JAMES A. HOPKINS, Oxford, N. Y.—*Saw Set.*—May 8, 1866.—The sliding slotted stirrup acts in a direct line on the teeth of the saw, so that the lower face bears flatly on the teeth.

Claim.—A sliding stirrup D, substantially as herein described, when combined with the set screw c and stop b of a saw set, substantially in the manner and for the purpose herein set forth.

No. 54,548.—FRANK H. HOUGHTON, Brooklyn, N. Y.—*Apparatus for Measuring Liquids.*—May 8, 1866.—The liquid flows into the measuring globe, and by the rotation of the faucet is shut off from the induction pipe and drawn off by the eduction faucet.

Claim.—The globe B with'measure marks thereon, the valve and plug E, and cock C, with double ports, constructed, combined, and arranged substantially as and for the purposes specified.

No. 54,549.—DAVID HOWARTH, Portland, Me.—*Button.*—May 8, 1866.—The shank of the button has a penetrating coil like a corkscrew, which is flattened after insertion.

Claim.—As an improvement upon B. P. Coston's invention, patented August 27, 1850, the spiral wire connected with a button, when said piece of wire, after insertion into the cloth, is struck or pressed down flat on the inner side of the cloth, all substantially as and for the purpose set forth.

No. 54,550.—JAMES V. HULSE, Acra, N. Y.—*Medical Compound.*—May 8, 1856.—This prophylactic is composed of cayenne pepper, gum camphor, and sulphur.

Claim.—The within described medicated compound, made as set forth.

No. 54,551.—WILLIAM A. INGALLS, Chicago, Ill., assignor to WM. W. KIMBALL, same place.—*Piano Stool.*—May 8, 1866.—The removable legs, by the devices described, are secured to the column on which the seat rests.

Claim.—First, the employment of the keys or braces E in a piano stool, arranged and operating substantially as and for the purposes set forth.

Second, the combination of an extended standard A with an upper joint C and a lower joint D, when applied to a stool having removable legs, substantially as and for the purposes set forth and shown.

Third, the arrangement and combination of the upper lug C, provided with the projection c, and the lower lug D, provided with the T-shaped or double hooked projection d, with the corresponding openings and recesses in the column A, and the key or brace E, substantially as set forth and specified.

No. 54,552.—JOHN INSULL and THOMAS INSULL, New Haven, Conn.—*Cupola Furnace.*—May 8, 1866.—A horizontally divided chamber around the cupola receives air from two blast pipes, and discharges it by tuyeres into the furnace.

Claim.—The hollow hoop or belt C C, with its partitions b b, in combination with the blast pipes D and E, and their tuyeres, when the whole is constructed, arranged, and fitted to operate, substantially as herein described.

No. 54,553.—JOHN H. IRWIN, Chicago, Ill.—*Illuminating Apparatus.*—May 8, 1866.—Air is forced by the bellows through pipes and in contact with the hydro-carbon which rises through the wicking.

Claim.—The combination of the air pipes B, the oil reservoir C, and an apparatus for forcing air through said pipes, when constructed and arranged substantially as and for the purpose specified.

No. 54,554.—NATHANIEL JENKINS, Boston, Mass.—*Elastic Packing.*—May 8, 1866.—Composed of rubber, gum shellac, Paris white, French chalk, litharge, lampblack, and sulphur.

Claim.—The composition above set forth, and its process of manufacture, substantially as and for the purposes described, disclaiming the vulcanizing of rubber by means of heat, and the method of mixing the ingredients otherwise than as applicable to this composition.

No. 54,555.—EZRA JOHNSON, Hermon, Maine.—*Connecting the Joints of Stove Pipes and Water Conductors.*—May 8, 1866.—A link is attached by a riveted loop-piece to one section, and is connected to a corresponding loop on the next section by a wire key.

Claim.—The application to stove pipes and water conductors of the clasp and bolt as herein described, to secure and make fast the connection, and to secure also the principle of this form of direct attachment and connection by rivet, or its equivalent.

No. 54,556.—JOSEC JOHNSON, New York, N. Y.—*Fastening for Buttons.*—May 8, 1866.—A wire key formed like a figure 8 passes through the eye of the button to retain it in the fabric.

Claim.—First, the button fastening a, formed as shown in figures 1, 2, and 3, so as to compel the eye *b* of the button to be held in the centre of the fastening by the contact of the ends of said fastening therewith, as herein set forth.

Second, in connection with the above said ends, substantially as in the figure 1, so that both ends must be opened in order to liberate a button confined thereby.

No. 54,557.—A. T. JONES, Stamford, Conn.—*Water Elevator for Wells.*—May 8, 1866.— A cam made fast to the windlass shaft operates upon the ratchet or friction block, so as to raise the bucket, or detach and lower it under frictional pressure upon the drum.

Claim.—First, rotating the drum of a water elevator by means of a cam on the crank shaft acting on a radial moving friction block placed within a box on one end of the drum, constructed and operating substantially as described.

Second, the finger J of the cam I, in combination with the horn X of the friction block K, for the purpose of arresting the descent of the bucket, substantially as described.

Third, the socket N, within the box H of the drum, in combination with a spring M, for the purpose of bringing the cam I into action against the block, substantially as described.

No. 54,558.—F. H. JONES, Attica, N. Y.—*Water Drawer.*—May 8, 1866.—The lever holds the drum clutched, or when vibrated unclutches and forms a brake by frictional contact.

Claim.—First, the brake F, provided with lug *e*, for the double purpose of a brake, and for holding or securing the drum B to the shaft, or loosening it at will, for the purposes and substantially as described.

Second, the ratchet and pawl, when constructed and operated substantially as herein set forth.

No. 54,559.—MARTIN C. KILGORE, Washington, Iowa.—*Broom Head.*—May 8, 1866.— One toothed jaw has an attached socket; the other jaw hooks into the said socket to form a hinge thereon, and is fastened by clamping screws through the lower bar.

Claim.—A broom head constructed in two sections, connected by opening D and snake head E, and having screws and threads G, and projections H, constructed, combined, and arranged substantially as and for the purposes set forth.

No. 54,560.—EDGAR L. KINSLEY, Cambridge, Mass.—*Steam Hammer.*—May 8, 1866.— The piston is connected to the rear of the hammer helve by means of a lever and stirrup, so as to act directly upon it in each direction. The valve arm is connected by a curved slot and a rod to the lever, so as to adjust its range of motion when balls of varying sizes are under the tilt.

Claim.—First, raising the hammer by means of a lever operating upon the rear end of the hammer helve, and connected with the piston of the steam cylinder, arranged and operating substantially in the manner and for the purpose described.

Second, the stirrup *m*, operating in combination with the helve *f*, lever *l*, and piston rod *p*, and operating in the manner and for the purpose herein specified.

Third, the arrangement of the cam groove *d*, arm *s*, as applied to the valve of a steam engine and rod *v*, and connecting said arm with the lifting lever or beam *l*, or any part moving in unison therewith, substantially as herein described, for the purpose specified.

No. 54,561.—HENRY F. KNAPP, New York, N. Y.—*Machine for Making Paper Collars.*— May 8, 1866.—The platen is moved vertically; its face and the bed plate contain co-operative parts. The platen carries a pawl which operates a ratchet wheel upon one of the two feed rollers.

Claim.—First, operating the feed rolls by means of a ratchet wheel on one of the rolls, and a pawl actuated by the platen which carries the reciprocating cutters, substantially as herein specified.

Second, the combination of the angular edged adjustable cutters I I with knives *j j*, attached to a removable bar J, and operating to shear off the waste and form the ends of collars, substantially as herein set forth.

Third, the projection *h*, formed upon the reciprocating cutter G, and operating to prevent the said knife from striking or riding upon the stationary knife H, substantially as herein set forth for the purpose specified.

Fourth, the arrangement of the feed rolls E F, cutters I I and *j j*, and cutters G and H, in relation with each other and the bed A, and platen B, substantially as and for the purpose herein specified.

No. 54,562.—JOHN KREIGBAUM, Milton, Ohio.—*Tuyere.*—May 8, 1866.—Built up of four pieces, described and shown.

Claim.—First, the piece A, with its pipes *e d*, constructed substantially as described and for the purpose specified.

Second, the combination of the piece C with the piece A, substantially as described and for the purposes set forth.

Third, the part B, with its pipe *f*, and slide *g*, in combination with the piece A, substantially as described and for the purposes specified.

Fourth, the thimble D, and arm to support the part C, substantially as described.

No. 54,563.—JOHN J. LAHAYE, Reading, Penn.—*Railroad Axle Box.*—May 8, 1866.—
The oil from the chamber beneath is fed by a wick to the journal; a flange prevents its return. The oil chamber is supplied through an opening which is closed by the hinged lid.

Claim.—First, the combination of an axle box, having an oil chamber at the under side
and an opening *t*, when the said opening is arranged in front of the axle, and is surmounted
by a flange *t'*, as and for the purpose described.

Second, the passage *n*, in communicating with the chamber *p*, and so situated that its upper
end can be closed by the lid H.

No. 54,564.—LOUIS F. LANNAY, Indianapolis, Ind., and WILLIAM F. PARKS, Baltimore,
Md.—*Machine for Combing Bristles.*—May 8, 1866.—The bristles are clamped in a frame
which is reciprocated in a horizontal plane, so as to advance to and recede from the vertically
reciprocating combs.

Claim.—First, the reciprocating table F, for carrying the bristles or hog's hair to be
operated, in combination with the eccentric Q, and the nut and screw which regulates the
extent of the reciprocations produced by the eccentric, substantially as described.

Second, the devices for holding the bristles, arranged and combined as shown, to wit, the
plates L, the loose plate M, both having pins I, as shown, and the eccentric roller N for
clamping the bristles between said plates, substantially as described.

Third, so making and arranging the plate L, as that it is capable of being moved transversely by lever L², substantially as described.

Fourth, the reciprocating combs U V, moved from the same shaft in combination with the
reciprocating table F, substantially as described.

No. 54,565.—HENRY BEDELL LEACH, Boston, Mass.—*Rotary Pump.*—May 8, 1866.—The
rotary piston receives the water in its forward face and delivers it upward between the shaft
and the sleeve. An eccentric drum rotates with the piston and forms a bearing for the inner
edge of the abutment valve.

Claim.—First, one or more inclined spiral scoops E, revolving in a shell or casing A,
substantially in the manner and for the purpose set forth.

Second, the self-compensating balanced abutment F, the curvature of one of its surfaces
corresponding to and coming in contact with the outer surface of the scoop E, and drum B,
while the curvature of its opposite surface coincides and comes in contact with the surface of
the recess *y*, provided for it in the casing A, for the purpose set forth.

Third, the abutment F, or its equivalent, when used in combination with one or more inclined spiral scoops E, for producing behind the scoop or scoops a vacuum or partial vacuum.

Fourth, gradually decreasing the size of the spiral water way or ways, substantially as
and for the purpose set forth.

Fifth, revolving the drum B by the revolution of the screw cylinder C and scoop E, for
the purpose set forth.

Sixth, the segments G H, as arranged between the scoop E and drum B.

Seventh, a rotary pump consisting essentially of one or more spiral inclined scoops E, revolving drum B, segments G H, abutment F, and shell or casing A, substantially as described.

No 54,566.—WILLIAM P. LYON, Portchester, N. Y.—*Oyster Cracker.*—May 8, 1866.—
The toothed jaw is vibrated by a pivoted lever and a spring.

Claim.—The combination of the casing A, lever handle B, arm G, terminating in a jaw J,
fixed jaw K, of casing A, when connected together, and arranged as and for the purposes
described.

No. 54,567.—JOHN LYNN and JACKSON R. CROWELL, Seneca Falls, N. Y.—*Bed Bottom.*—
May 8, 1866.—The slats are suspended by hooks on their edges and by elastic loops from the
hooks attached to the bedstead rails.

Claim.—The combination of the hook C, double hook D, clasp E, elastic band F, and
shield G, when made substantially as specified, and applied as herein set forth.

No. 54,568.—LEWIS MACKEY, Dunmore, Penn.—*Apparatus for Treating Green Hides.*—
May 8, 1866.—The hides are suspended from the rounds of the cylinder, and by its revolution are alternately exposed to the air and to the liquor.

Claim.—The wheel D, in combination with the tan vat, said wheel being constructed substantially as described—that is to say, so constructed that the hides hang from the outside of
the wheel, and as the wheel revolves are alternately immersed in the tanning liquor and
exposed to the air, substantially as described and for the purpose set forth.

No. 54,569.—WILLIAM MALLERD, Bridgeport, Conn.—*Regulating Attachment for Gas
Burners.*—May 8, 1866.—The cap which regulates the sectional area of the passage way has
a screw stem and a spring click, which engages with notches to maintain it in position.

Claim.—First, the click A, or its equivalent, in combination with the regulating cap D, and
with the notches *i*, or their equivalents, constructed and operating substantially as and for
the purpose described.

Second, the cup-shaped depression e, in the bottom end of the screw f, in combination with the diaphragm b, gas chamber B, and burner C, constructed and operating substantially as and for the purpose set forth.

Third, the secondary gas chamber g, formed between the cap D and chamber B, in combination with any burner or burners, substantially as and for the purpose described.

No. 54,570.—CHARLES J. MALLON, Schoharie, N. Y.—*Sad-iron.*—May 8, 1866.—The under part of the divided handle acts as a button to secure it to the standards of the iron. A shield is suspended from the handle to protect the hand from heat.

Claim.—First, attaching the handle C, carrying a suspended shield to the standard B, by means of the button E, pivoted to the under side thereof, in the manner and for the purpose herein specified.

Second, the combination and arrangement of the handle C, standards B, arms c, button E, shield D, and body A, constructed and operating in the manner and for the purpose herein specified.

No. 54,571.—JEREMIAH A. MARDEN, Newburyport, Mass.—*Leather Splitting Machine.*—May 8, 1866.—The sectional rolls have internal teeth fitting loosely in the longitudinal grooves of the central shaft so as to have a certain independence of vertical motion.

Claim.—The shaft J, in combination with the sectional rolls K, the shaft driving the rolls by means of teeth and one or more grooves, substantially as described.

Also, the guide Q, substantially as and for the purpose set forth.

No. 54,572.—EDWARD MCARDLE, Cambridge, Mass.—*Tool for Holding Glass Bottles.*—May 8, 1866.—The angular jaws advance and recede by the rotation of the right and left handed screws which traverse their shanks and reciprocate them in the slotted holder.

Claim.—The combination of the jaws A, slotted plate B, screw D, and stock C, all constructed and arranged substantially as described.

No. 54,573.—SAMUEL MCCAMBRIDGE, Philadelphia, Penn.—*Car Brake.*—May 8, 1866.—The chain which vibrates the brake levers of one car is attached to a vibrating arm which transmits the motion by the continuous chain to the next car in the series.

Claim.—The combination of a continuous chain with the brake levers of a train of cars by means of the intermediate levers G, and the sheaves f and g, the said parts being arranged and operating substantially upon the principle and in the manner hereinbefore described and for the purpose specified.

No. 54,574.—FRANCIS MCCOY, West Philadelphia, Penn.—*Attaching Thills to Vehicles.*—May 8, 1866.—The yoke supports the thimble on the end of the thill, and is formed by the forward extension of the lower bar of the axle clip.

Claim—The yoke E, with one end curved upward, forming a concave bed E', for the head A of the shaft between the bars B of the band or clip F, constructed and applied in the manner and for the purpose specified.

No. 54,575.—WILLIAM C. MCGILL, Cincinnati, Ohio—*Flour Sifter.*—May 8, 1866.—The flour is driven through the reticulated diaphragm by the revolving spring scrapers.

Claim.—In the described combination with the box or case A, and sieve B. the scraper G g G' g', springs F, elastic arm E, shaft C, and crank c, arranged and operating substantially as and for the purposes herein set forth.

No. 54,576.—STEPHEN MCNAMARA, Chicago, Ill.—*Washing Machine.*—May 8, 1866.—The lower set of conical rollers is revolved by the central shaft, and the upper set by the sleeve shaft; the motion is effected by the bevel wheel engaging the pinions on the respective shafts and turning them in opposite directions. The contiguous faces of the conical rollers are parallel.

Claim.—First, the employment of two disks of conical rollers, arranged and operating substantially as herein shown and described and for the purposes specified.

Second, the combination of the two disks of rollers P Q, the shaft D, and the bevel wheels H I J, or their equivalents, arranged and operating substantially as described and shown.

Third, in combination with the above, the arrangement of the vertically adjustable frame c, operating as and for the purposes described.

Fourth, in combination with the above, the employment of the tub A, and stationary frame B, arranged as and for the purposes specified.

No. 54,577.—T. L. MELONE, Granville, Ohio—*Sewing Machine.*—May 8, 1866.—The presser bar and needle lever have a common fulcrum; the former is locked in its working position by a spring bolt beneath the table. The feed wheel does not revolve when feeding, but grips the cloth between itself and the presser foot.

Claim.—First, the cam shaft E, placed vertically beneath the bed of the machine with its cams G and C' and its band wheel H, and its crank for operating the connecting rod F, which drives the reciprocating shuttle, all substantially as described.

Second, locking the presser bar to its proper position for work, by means of the spring bolt B², constructed and applied substantially as described.

Third, feeding the material in sewing machines by means of a feed wheel working on a slide beneath the cloth bed and having reciprocating motion in horizontal directions, the same being locked so as not to rotate when it is moved forward to give the feed, but being allowed to rotate so as not to carry the cloth back when it is returned to its former position, substantially as described.

No. 54,578.—BENJAMIN F. MILLER, New York, N. Y.—*Washing Machine.*—May 8, 1866.—The box is rotated on journals attached to diagonally opposite angles; pins project inward from its ends, and strips from its sides.

Claim.—First, the fingers *g*, in combination with the box *a*, mounted as set forth for the purposes and substantially as specified.

Second, in combination with the box *a*, mounted as specified, the flanges *h*, for the purposes and as set forth.

No. 54,579.—ISAAC MINER and D. J. MINER, Freehold, N. J —*Water Wheel.*—May 8, 1866.—The wheels are arranged in succession in communicating chambers, and the induction and eduction passages are so arranged and valved that the water may act upon either or all of them.

Claim.—The arrangement of the wheels A A' A", placed within the curbs B B' B", communication passages *a a'*, supply passages *b b' b"*, gates *c c c*, discharge openings *d d' d"*, gates *e e e*, and penstock C, relatively to each other, and operating in the manner and for the purpose herein specified.

No. 54,580.—MARY JANE MONTGOMERY, New York, N. Y.—*Apparatus for Punching Corrugated Metals.*—May 8, 1866.—The tool has openings corresponding to the shape of the corrugated metal to be punched and is fed along to the transversely penetrating punch, which is guided by the openings through the tool.

Claim.—The apparatus for punching bolt or rivet holes in corrugated or curved metals, constructed and operating substantially as described.

No. 54,581.—JOHN J. MORRIS, Cincinnati, Ohio.—*Rotary Pump.*—May 8, 1866.—The contacting side of the cylinder with the rotating hub is adjusted by a set screw; the pistons have a common central pin and expanding screw.

Claim.—First, the tapering tongue A', forming part of the cylinder in the described combination with a set screw F, or its equivalent, substantially as described.

Second, the connecting pin or pins I, or mechanical equivalent, and coiled spring or springs J, as described and for the purpose specified.

No. 54,582.—WASHINGTON MOSHER, Pittstown, N. Y.—*Horse Rake.*—May 8, 1866.—To maintain the rake in working position the ends of vertical bars pivoted to the handles rest upon short bars projecting from the rake head. The pivoted bars are held in position by springs.

Claim.—The bars E E, pivoted to the front ends of the handles D D. in combination with the springs F' and the bars G G on the rake head, substantially as and for the purpose set forth.

No. 54,583.—HORACE MOTT, Bellevue, Mich.—*Churn.*—May 8, 1866.—The churn is on rockers, and has a central open frame of vertical and horizontal slats to act as a dasher.

Claim.—The rocking churn-box, the sliding dasher gates, and the handles, arranged and used as and for the purpose specified.

No. 54,584.—E. P. NEWBANKS, Lincoln, Ill.—*Shingling Gauge.*—May 8, 1866.—The butts of the shingles are placed against the bar, which is adjusted by the gauging strap after each course of shingles is nailed.

Claim.—The shingle gauge herein described, being composed of the frame A, adjustable bar B, and straps E E, constructed and used as and for the purpose herein set forth.

No. 54,585.—JOHN K. O'NEIL, Kingston, N. Y.—*Attachment for Sleeves and Leggings.*—May 8, 1866.—Inside the sleeve or legging is a lining attached thereto and embracing the limb by an elastic band at its edge.

Claim.—The attachment for sleeves and legs of garments provided with the elastic cord or braid *c* and the elastic or shirred strips *b b*, substantially as and for the purpose specified.

No 54,586.—ROBERT OXLAND, San Francisco, Cal.—*Process in Refining Sulphur.*—May 8, 1866.—A cold, saturated solution of nitrate of soda 5 pounds, muriate of soda 7 pounds, is combined with sulphuric acid 14 pounds, and water 3 pounds; the compound is stirred into 2,000 pounds of melted sulphur at 235° Fah. The result is melted in a steam-jacketed vessel and cast in moulds.

Claim.—The use of chemical agents in the manner hereinbefore described, either in substitution of the ordinary process of sublimation for refining sulphur or as supplementary thereto, for the completion of the refining operation.

No. 54,587.—MOSES W. PAGE, Franklin, N. H.—*Preparation of Leather.*—May 8, 1866.—The leather is brought to a uniform thickness by dressing off the grain or hair side instead of the flesh side.

Claim.—Buffing the grain side instead of shaving the flesh side of the butt or thick portion of the leather, substantially as and for the purpose set forth.

No. 54,588.—MOSES W. PAGE, Franklin, N. H.—*Process of Tanning.*—May 8, 1866.—The hides are limed, unhaired, and bated as usual, then treated with a tanning solution into which sweet fern enters; then in a saline liquor of alkaline salts and alum; then in a liquor of salts, acids, and a salt of tin. The result is to condense the fibres.

Claim.—The within described process of tanning leather, substantially as set forth.

No. 54,589.—S. W. PALMER and J. F. PALMER, Auburn, N. Y.—*Gearing.*—May 8, 1866.—The teeth of the different disks mesh into each other at such intervals as to have several in contact at all times.

Claim.—An improved gearing consisting of two or more disks or circular plates having teeth or cogs on their faces, so constructed and arranged that there may be two or more sets of teeth or cogs on each shaft to operate in transmitting motion from one shaft to another when said shafts are at a varying distance apart, substantially as herein described and for the purpose set forth.

No. 54,590.—C. C. PARSONS, Boston, Mass.—*Constructing Cast-iron Apparatus for Superheating Steam.*—May 8, 1866.—The sand core of the convoluted superheater is supported by a frame and transverse wires which are allowed to remain after the sand is withdrawn.

Claim.—A steam superheater having a construction substantially as described.

No. 54,591.—JOHN E. PHILLIPS, Philadelphia, Penn.—*Preparing Wisps for Brooms.*—May 8, 1866; antedated January 11, 1866.—The broom corn is bunched and pressed, then treated with glue and sand and dried ready for its position in the broom head.

Claim.—The herein described new article of manufacture consisting of a wisp for filling broom heads or clamps, constructed in the manner and retained in form by the means herein described or any equivalent means, substantially as set forth.

No. 54,592—J. F. C PICKHARDT, New York, N. Y.—*Sofa Bedstead.*—May 8. 1866.—The mattress is hinged and folds up, so that a portion forms the back of the sofa, while the foot portion by the longitudinal contraction of the bedstead becomes the sofa seat.

Claim.—First, the attaching of the cushion D by hinges *b* to a cross-bar C of the side pieces A A, in combination with the back board C and the cushion E, attached by hinges *e* to D, substantially as and for the purpose specified.

Second, the foot board J, attached by hinges *i* to a cross-bar *d* of the posts *c c*, in such a manner as to admit of the foot board being folded downward and inward, to be covered by the seat cushion, when the device is used as a sofa, substantially as described.

Third, the clothes receptacle I connected by hinges *f* to the under side of the seat cushion H, and arranged to work or rest upon the front pair of slides G, substantially as and for the purpose set forth.

Fourth, the pillow K, in combination with the cushions D E and cap piece L, arranged substantially as and for the purpose specified.

No. 54,593—MARTIN L. POWELL, New Castle, Ind.—*Universal Index*—May 8, 1866; antedated May 1, 1866.—The column of distances or fares is flanked by index fingers, pivoted on vertical rods, and capable of independent rotation, to be pointed to the required item in the column.

Claim.—The construction, arrangement, and combination of the movable and independent spring indexes E, operating between the partitions K on each side of the plate A against the wings J, as herein described and for the purposes set forth.

No. 54,594.—DAVID H. PRIEST, Watertown, Mass.—*Fruit Basket.*—May 8, 1866.—The bottom of the basket is in two sections, sustained by a book, with a tripper attached. Its purpose is to drop the bottom, after the basket is lowered to discharge the fruit, without descending to the ground.

Claim.—A fruit gatherer or basket, constructed substantially as and for the purposes above set forth.

No. 54,595.—A. D. PUFFER, Somerville, Mass.—*Mosquito Guard for Bedsteads.*—May 8, 1866.—The netting is stretched over the head board and the bows.

Claim.—The arrangement of the netting over the bars, or stretched bows, and so as to be distended by them around the head, a loose fringe or border of netting pending from the lower bar, and the frame being attached to the headboard of the bedstead, substantially as shown and described.

No. 54,596.—C. PURDY, Bedford, Ohio.—*Gate.*—May 8, 1866.—The swinging portion is pivoted by a central rod to the sliding portion of the gate, so as to open fully or make a small opening as required.

Claim.—First, centrally hanging the swing-gate B to the laterally sliling gate A, when constructed so as to operate conjointly as and for the purpose set forth.

Second, the sectional gate A, in combination with the swinging section B, catch L, rod I, and slats *g*, in the manner substantially as and for the purpose set forth.

No. 54,597.—THOMAS H. QUICK, New York, N. Y.—*Machine for Stirring and Dissolving Sugar in Sugar Refineries.*—May 8, 1866.—The attached engine vibrates the blade to stir the sugar in the semi-cylindrical steam jacketed vessel.

Claim.—The combination and arrangement of the vibrating agitating blade D with the heater A and engine E, substantially as herein set forth, for the purpose specified.

No. 54,598.—THEODORE F. RANDOLPH, Cincinnati, Ohio.—*Sewer.*—May 8, 1866.—The catch basin has several receptacles, which combine to fill the space, and are separately removable.

Claim.—The arrangement in the catch basin of sewers of movable receptacles F F', formed and adapted to operate substantially as set forth.

No. 54,599.—CHARLES H. RAYMOND, Woodstock, Vt.—*Machinery for Folding Tin.*—May 8, 1866.—In the machine for folding sheet metal is arranged an inclined plane against which the adjustable cap strikes to regulate its position with reference to the stationary "former."

Claim.—Not the combination of the folding bar F, and its adjustable cap A, the stationary and movable jaws A B, or the same and the adjustable gauge C, but the combination of the same and the cam *m*, and flange *l*, or equivalent mechanism for effecting the elevation of the cap up to the level of the top of the upper jaw, under circumstances and for the purpose substantially as specified.

No. 54,600.—HENRY REYNOLDS, New Haven, Conn.—*Breech-loading Fire-arm.*—May 8, 1866.—A cover is hinged to the barrel, and occupies the upper rear slot through which the cartridge is inserted. The hammer strikes through a hole in the cover, and locks it in position.

Claim.—The arrangement of the cover H, projecting piece I, lug *f*, hammer C, in combination with the barrel A, with its slot E, and operating in the manner and for the purpose herein specified.

No. 54,601.—H. A. ROBINSON, Cleveland, Ohio.—*Hat Block*—May 8, 1866.—The hollow-shaping block is capable of rotation on the core, whose supporting table may be inclined from the horizontal when required.

Claim.—Making the forming block C of shell-like construction, fitting upon a core C', and operating conjointly with said core, as and for the purpose set forth.

No. 54,602.—THOMAS ROBJOHN, New York, N. Y.—*Sewing Machine Braid Guide.*—May 8, 1866; antedated April 27, 1866.—Two pieces of wavy flat braid are laid together, their undulations coinciding, and are fed through a straight guide by wheels, whose projections correspond to the wavy edges, which are to be stitched together.

Claim.—The combination of the toothed and recessed guide wheels *f f*, in combination with the guide tube *d*, or its equivalent, substantially as and for the purpose herein specified.

No. 54,603.—WILLIAM ROBJOHN, New York, N. Y.—*Organ.*—May 8, 1866.—This is designed to give the organist the command of the stops without removing his hand from the keys; the finger pieces project above the bank of keys and actuate the slides and levers connected to the pneumatic devices.

Claim.—First, the composition board E, constructed of a series of longitudinal slides *g g' g''*, &c., and a series of transverse slides *n n' n''*, &c., when the same is applied in combination with buttons or finger pieces *q q¹ q²*, &c., and with pneumatic levers B B' B'', &c., substantially in the manner and for the purpose described.

Second, the use of an ordinary slide valve D, of that class commonly known as D valves, in combination with the two parts *a a'* of a pneumatic lever, and with rods *c*, connecting said parts, substantially as and for the purpose set forth.

Third, the spring pawl *w* and serrated standard *n*, in combination with the treadle A², pedal F, and with the slides *g g' g''*, &c., which serve to impart motion to the draw stops of an organ, substantially as and for the purpose described.

Fourth, the roller G, with wipers t t^1 t^2, &c., spring s, and strap r, in combination with the pedal F, levers x x' x'', &c., and slides g g' g'', &c., substantially as and for the purpose set forth.

Fifth, the oscillating spring-dog t^4, and button or finger piece i^2, in combination with the mechanism which serves to couple and uncouple the several keyboards A A', &c., constructed and operating substantially as and for the purpose described.

No. 54,604.—CHARLES ROSEFIELD, Washington, D. C.—*Tooth Brush.*—May 8, 1866.— The brushes face each other to cleanse the teeth inside and outside of the dental arch.

Claim.—First, the two brushes A and B, connected and used substantially as herein set forth, for the purpose of cleansing the teeth upon both sides at one operation, as specified.

Second, in combination therewith, the handle D, pivoted to the brush, substantially as and for the purpose specified.

No. 54,605.—AMBLER J. ROGERS, Kittanning, Penn.—*Machine for Rolling Spikes.*— May 8, 1866.—The upper roller has a cam which forms one side of the spike, while the other is formed by the shape of the bed which reciprocates beneath; the bed rests on two supporting rolls and a driving gear which engages in the rack beneath the bed.

Claim.—The combination of the roll C with the table D, said roll and table being constructed, arranged, and operating in the manner substantially as herein described and for the purpose set forth.

No. 54,606.—JOHN ROTHERHAM and JOSEPH HOLDEN, Middletown, N. Y.—*File Cutting Machine.*—May 8, 1866.—The blanks are secured side by side on the upper surface of the bed, which is automatically fed after each stroke by the feed screw; a separate chisel and hammer act upon each blank; the chisels are supported by springs on arms with roller feet, which bear upon the blank; the chisels are thrown back after each cut to raise a burr.

Claim.—First, giving to the chisel F a backward motion, in imitation of the hand cut, by means substantially such as herein described, for the purpose set forth.

Second, the swivelled roller n, applied in combination with the arms E, chisels F, and anvil B, all constructed and arranged substantially as and for the purpose set forth.

No. 54,607.—JAMES SANGSTER and ORAN W. SEELY, Buffalo, N. Y.—*Paper Neck Tie.*— May 8, 1866.—The bow is crimped upon the tuck and secured by an elastic band.

Claim.—First, the bow made of the parts B and E, and folded, substantially as described.

Second, in combination therewith, the elastic band A, as shown and described.

No. 54,608.—ANTON SAUR and ALBERT B. COLTON, Franklin, Ind.—*Sawing Machine.*— May 8, 1866.—The lever which straightens the toggle to raise the saw also brings the brake into action; the saw guide is attached to the end of the counter bar.

Claim.—First, the guide I, formed as described, when attached permanently to the end of the counter bar E, substantially in the manner and for the purpose set forth.

Second, raising and lowering the counter bar E, and saw K, by means of a toggle joint M, and hand lever L, the whole being arranged to operate substantially as set forth.

Third, the combination of the brake O, rod N, and lever L, with the toggle joint M, for the purpose of bringing the brake O into action on the fly wheel by the same movement that raises the saw, substantially in the manner set forth.

No. 54,609.—ELIJAH C. SEARS, Crystal Lake, Ill.—*Flour and Meal Sifter.*—May 8, 1866.— The upper drawer has a sieve bottom; by its reciprocation the flour is discharged into the drawer beneath.

Claim.—The arrangement of the drawer C, and the notched piece D, lever H, drawer B, and frame A, constructed and operating in the manner and for the purpose herein specified.

No. 54,610.—D. W. SEELEY and F. JACOBI, Albany, N. Y.—*Brick Machine.*—May 8, 1866.—Explained by the claim.

Claim.—First, a horizontal, or nearly horizontal, feeding and grinding shaft, with knives or blades of a screw form, so arranged, relatively, that they force the clay or other material from both ends of the machine into the press box, notwithstanding the clay is only introduced into the grinding box at one end, substantially as described.

Second, the combination of a shaft, with right and left screw blades or knives, and a brick moulding press, substantially as set forth.

Third, the means, substantially as herein described, for retaining the filled moulds under pressure, and, while under pressure, forcing them out from under the press box, for the purpose set forth.

Fourth, retaining the pressure on the filled moulds during their removal from under the press box, by means substantially as set forth.

Fifth, the means, substantially as herein described, for keeping the moulds up to the bottom of the press box and allowing the moulds to descend, clear of an obstruction, as they pass under the front part of the press box, as set forth.

Sixth, the slotted and perforated stem of the press follower, in combination with the link and walking beam, substantially as and for the purposes described.

No. 54,611.—JOHN F. SEIBERLING, Doylestown, Ohio.—*Harvester.*—May 8, 1866.—The end of the clutch lever plays between the deep flanges of the clutch collar as the platform is raised or lowered, and is retained by a spring and pin; the device permits a limited adjustment of the platform without interference with the devices for throwing the cutters in and out of action.

Claim.—The arrangement and combination with the broad flanged clutch J, of the shift lever G, pin p, and spring E, or their equivalents, substantially in the manner and for the purposes set forth.

No. 54,612.—ANDREW J. SELLON, Boston, Mass.—*Knife Cleaner.*—May 8, 1866.—The knife blade is reciprocated between the elastic roller and the tablet, whose groove contains the polishing powder.

Claim.—The knife cleaner, composed of the elastic roller, the covered tablet, and the groove thereof, such tablet being provided with a handle, as specified.

Also, the combination and arrangement of the abutment with the tablet and the elastic roller, arranged as specified.

No. 54,613.—J.'H. SEYMOUR, Hagerstown, Md.—*Spittoon for Railroad Cars.*—May 8, 1866.—The covered spittoon has a discharge valve operated by a rod and counterweight lever.

Claim.—The combination of the bowl A, cover C, valve D, lever f, and rod E, arranged in connection with the floor of the car, substantially as herein described.

No. 54,614.—JAMES SHEPARD and B. B. LEWIS, Bristol, Conn.—*Fruit Box.*—May 8, 1866.—The blank is cut "bias" from the veneer, the edges brought into contact and secured by wire.

Claim.—Cutting the blank for a box obliquely with the grain of thin wood or a veneer, substantially as and for the purpose herein specified.

No. 54,615.—D. M. SHAPLEY, St. Louis, Mo.—*Pressure Head for Siphons and Force Pumps.*—May 8, 1866.—The annular valve which encircles the exit opening of the feed pipe is movable longitudinally by a set screw to adjust the area of the opening between them. The check valve prevents the reflux of the liquid.

Claim.—First, the combination and arrangement of the feed pipe D and the annular valve E, as herein set forth.

Second, the valve E, in connection with the valve rods F and c, or their equivalents, and with the chamber A, when used as hereinbefore set forth.

Third, the check valve K, in connection with the valve E, as hereinbefore set forth.

No. 54,616.—SAMUEL SHERWOOD, New York, N. Y.—*Blacking for Harness.*—May 8, 1866.—Composed of beeswax, ivory black, prussian blue, shellac, spirits of turpentine, alcohol, and oil of lavender.

Claim.—The water-proof blacking composed of the ingredients herein specified in about the proportions herein set forth.

No. 54,617.—OLIVER P. SHINKLE, Covington, Ky.—*Glass Blower's Mould.*—May 8, 1866.—Designed for moulding jars with channelled tops. It has a downward opening bottom, two horizontally opening side parts, and two semi-circles which combine to form the crease.

Claim.—First, a mould for forming jars, &c., with channelled tops, having a downwardly opening bottom, two horizontally opening side parts, and two crease parts which open parallel with, but subsequently to said side parts, substantially as and for the purpose set forth.

Second, the arrangement of parts substantially as described, for the automatic opening first of the bottom and side parts, and subsequently of the crease parts, and for their simultaneous closure by the foot of the blowman or otherwise.

No. 54,618.—JOSEPH SLUSSER, Cincinnati, Ohio.—*Shutter Hinge.*—May 8, 1866.—The projection on the upper leaf is received in the recesses of the collar of the lower leaf to hold the movable leaf in position.

Claim.—The self-locking shutter hinge provided with the V-indented half collar B C D, and locking edge C, on its lower member, and one V-tooth E on its upper member, constructed and arranged as set forth.

No. 54,619.—BENJAMIN M. SMITH, Brooklyn, N. Y.—*Teeth of Machines for Picking Cotton.*—May 8, 1866.—Explained by the claim.

Claim.—The construction of the teeth of machinery for picking, ginning or opening cotton, or other fibrous material, with grooves or notches, substantially as herein specified.

No. 54,620.—HENRY G. SMITH and E. MINER POMEROY, West Meriden, Conn.—*Fastening Handles to Plated Ware.*—May 8, 1866.—The threaded socket in the cup forms a nut for the screw by which the handle is attached.

Claim.—Attaching handles and other projecting parts to cups, pitchers, and other analogons vessels made of plated or solid metal, by means of male and female screws, when the screws are made and fitted in the manner and for the purposes substantially as herein described and set forth.

No. 54,621.—CHARLES F. SPAULDING, St. Johnsbury, Vt.—*Machine for Spinning Metal.*—May 8, 1866.—The plate is secured by a fastening ring upon the die, into whose concavity it is spun by the wheels on the mandrel; this has a position at right angles to the shaft by which the forming wheels are adjusted toward and from the die.

Claim.—The combination of the rotary die wheel A, and the plate fastening ring E, and devices for holding it to the wheel with the slide mandrel G, its spinning wheel or wheels and mechanism for supporting and moving such mandrel longitudinally or toward and away from the die wheel, as specified; and in combination therewith, a mechanism for effecting the vertical adjustment of the mandrel as specified.

Also, the holding ring E, as made with the fastening arms extending from it, as and for the purpose set forth.

Also, the combination of the clamp ring F, with the holding ring E, the die wheel A, and the mandrel G, and its spinning wheel.

No. 54,622.—M. W. STAPLES, Saratoga Springs, N. Y.—*Washing Machine.*—May 8, 1866.—The clothes are carried upon the arms of the vertical revolving shaft in a cylindrical tub, which is rotated in a contrary direction; tub and shaft have flanges to agitate the water.

Claim.—First, the upright cylindrical revolving tub C, constructed as described, in combination with the stationary tub A, substantially as and for the purpose set forth.

Second, the combination with the shaft D. and bar E, with each other, and with the tubs C and A, substantially as described and for the purpose set forth.

Third, the combination of the gear wheels L and M with each other, with the shaft H and with the revolving tub C, substantially as described and for the purpose set forth.

Fourth, the combination of the gear wheels K and G with each other, with the crank shaft H and vertical shaft D, substantially as described and for the purpose set forth.

Fifth, the combination of the flanges N, arms O, uprights P, and flanges R with each other, with the shaft D, and with the tub C, substantially as described and for the purpose set forth.

No. 54,623.—TIMOTHY TERREL, Spring Hills, Ohio.—*Shovel Plough.*—May 8, 1866.—The rear downward prolongation of the beam supports the brace of the standard.

Claim.—The curved rear part of the beam A, in combination with the brace G* and standard F, the latter being of double bevel form to receive the share, and all arranged substantially as and for the purpose herein set forth.

No. 54,624.—DAVID C. THRESHER, Freetown, Mass.—*Gun, Stock, and Barrel Attachment.*—May 8, 1866.—The gun barrel has a dovetail tenon at its rear end to engage with a recess in the breech-plate, and is there locked by a sliding key-piece actuated by a screw.

Claim.—The combination and arrangement of the clamp and the lifting screw thereof with the dovetailed tenon or tenons, and the chamber applied to the stock, and the barrel or barrels, substantially as described.

No. 54,625.—EDWARD TOWN, Washington, D. C.—*Cotton Gin.*—May 8, 1866.--By the knobbed roller and the travelling belt the natural connection of the fibre and seed is severed before entering the separator proper, where the cotton is removed at one grating by the teeth, while the seed escapes at the other.

Claim.—First, the arrangement of the teeth in the separator, as and for the purposes described and set forth in this specification, viz: So arranging them that no two or more teeth can seize upon the same fibre at the same time, and thereby break it by drawing it across the bars of the grating; and yet so arranging the teeth that all the interstices between the bars of the grating will be penetrated alternately, so as to draw the cotton equally from all parts of the mass, as specified.

Second, the studded or knobbed roller F, or a set of such rollers operating in combination with the endless apron E, as described, and for the purposes set forth in this specification.

Third, in combination with the revolving cylinder G and cylindrical receiver H H, with its two gratings *a* and *b*, operating together as and for the purposes described and set forth in this specification.

No. 54,626.—ANDREW J. VANDEGRIFT, Cincinnati, Ohio—*Spirit Level.*—May 8, 1866; antedated April 25, 1866.—The liquid is contained in a continuous groove whose transparent face and graduated cover permit its indication to be read at two separate points.

Claim.—First, forming liquid chamber *t*, by turning or otherwise forming a uniform groove in metal or other suitable material, and providing the same with a transparent face, the same being fitted and hermetically sealed, forming a uniform liquid chamber, substantially in the manner and for the purpose set forth.

Second, stock A, liquid chamber *t*, circuitous liquid chamber *g g*, dial plate D, and segmentary dial plates E E, all contained, in the manner and for the purpose set forth.

No. 54,627.,—FRANK VAN ETTEN, Jackson, Mich.—*Hay Loader.*—May 8, 1866.—The rake lever is pivoted to the frame of the wagon, and gathers the hay as it advances; the upward vibration of the lever raises the hay, which is dumped by unlocking the jointed end of the handle.

Claim.—The bent lever E, having the fork G, or rake I, pivoted to the outer end of its lower arm *e*, in combination with the lever F, swivel standard D, the cords and catch, and the bedpiece B, connected to the arm A, all being arranged and applied to a wagon or cart, to operate in the manner substantially as and for the purpose set forth.

No. 54,628.—HARTWICH VON UNWERTH, New York, N. Y.—*Manufacture of Artificial Stones for Grinding and Polishing Metals.*—May 8, 1866.—Composed of litharge, 1 : red lead, 1; oxide of iron, ¾; alkaline silicate, in powder, 1½; calcined lampblack, 1¼; emery, 5, (of a grade adapted to the work,) mixed, dried and compounded with 6 parts soluble glass.

Claim.—The manufacture of artificial stone from the ingredients named, and by the processes and for the purpose and in the manner substantially as herein described and set forth.

No. 54,629.—CHRISTIAN WAHL, Chicago, Ill.—*Vacuum Pan.*—May 8, 1866.—The horizontal, cylindrical, steam-jacketed pan receives a continuous stream of liquid, and the vapor is removed by branch pipes to the exhaust pump.

Claim.—First, evaporating liquids by passing them in a continuous current through an airtight receiver or vacuum pan, placed in a horizontal position or nearly so, to which pan the requisite degree of heat is imparted by steam or any other suitable medium, substantially as and operating in the manner described.

Second, connecting the vacuum pan with the condenser for the vapors arising from the evaporation of the liquid within the pan, by and through a series of pipes, or their equivalent, communicating with the said condenser, substantially in the manner described, and for the purpose specified.

No. 54,630.—ALBIN WARTH, Stapleton, N. Y.—*Anti-reversing Attachment to Treadles.*—May 8, 1866.—The double cam on the shaft and the springs combine to throw the crank in a proper position for starting.

Claim.—The arrangement of the lever cam E, spring *b*, and the friction roller *b'*, and spring *c*, in combination with the crank shaft A, pin *a*, and treadle D, and operating in the manner and for the purpose herein specified.

No. 54,631.—HENRY WEHLE, Hoboken, N. J.—*Clothes Wringer.*—May 8, 1866.—Explained by the claim.

Claim.—First, the hollow metallic shafted rollers K L, substantially as herein described and for the purpose specified.

Second, the mechanism for fastening the clothes wringer to the wash-tub as shown in Fig. 2, consisting of the links *l m n p*, the longitudinal washer *q*, the smooth nut *r*, and pin *s*, substantially as described, and for the purpose specified.

Third, the mechanism for fastening the clothes wringer to the wash-tub, as shown in Fig. 6, consisting of the links *l m n p*, the washer *q*, in combination with the nut *z v w*, substantially as described and for the purpose specified.

No. 54,632.—HIRAM W. WHITE, Olney, Ill.—*Churn.*—May 8, 1866.—The dasher consists of disks of decreasing diameter towards the top of the series and with intervals for the escape of the cream against the corrugated sides of the churn.

Claim.—The dasher, with pyramidal or conical gradatory exterior and interior chamber, substantially as described and presented.

Second, the churn with its interior surface, corrugated or grooved, substantially as and for the purpose described.

No. 54,633.—MORRIS WILKINSON, Burlington, Mich.—*Corn Planter.*—May 8, 1866.—The seed is delivered in the axle to secure in connection with the casing of the tube and the curved recess in the stone standard, and the pivoted coverer in the rear is kept to its work by a spring on the frame.

Claim.—First, forming a chamber for the portions of the axle C, containing the seed cells *e*, or for any equivalent seed cylinder, by means of the plough stock G and seed tube J, when arranged relatively with each other, and with the axle and hopper, and constructed and connected substantially in the manner and for the purpose herein described.

Second, the employment of pivoted coverers S, in combination with the rods R and adjustable spring M, substantially as and for the purposes herein specified.

No. 54,634.—NIMROD E. WILSON, Central Station, West Va.—*Churn.*—May 8, 1866.—The reciprocating motion given to the dasher by the spring, cog-wheels, and spiral screws, causes the churn to oscillate on its bearings.

Claim.—First, actuating a churn by means of one or more spiral springs communicating motion through wheels M and I and endless screws K and H, substantially in the manner set forth.

Second, giving an oscillating motion to the barrel of a churn by means of the crank F and dasher E, the churn being suspended on pivots, substantially in the manner and for the purpose set forth.

No. 54,635.—JACOB WISTER, Greencastle, Penn.—*Grinding Bones for Manure, &c.*—May 8, 1866.—Plaster fed with the bones to the mill absorbs the oily matter and prevents "gumming" the mill.

Claim.—Mixing hard plaster, plaster of paris or gypsum, or its equivalent, with bones, and grinding such mixture for the purpose of facilitating the process of grinding and preventing the mill from gumming, substantially in the manner and for the purposes set forth.

No. 54,636.—DAVID S. WOOD, Delavan, Wis.—*Pump.*—May 8, 1866.—An annular packing on the upper edge of the lower stock and surrounding the plunger

Claim.—The annular packing P applied around the piston D, upon the upper end of the interior cylinder, substantially as and for the purpose specified.

No. 54,637.—BENJAMIN F. YOUNG, Toulon, Ill.—*Cultivator.*—May 8, 1866.—The inner frame is pivoted forward to the carriage frame and carries the pivoted plough standards, which are adjustable as to height and presentation.

Claim.—First, the inside frame B and the mode of connecting it with the frame A, constructed and operated substantially as described.

Second, the iron lever F hung upon a pivot by which the front shovels are swung to the right or left.

Third, the adjustable wooden wedges I, together with the crooked iron necks D, by which the front shovels are adjusted so as to throw the dirt to or from the corn.

Fourth, the side or gauge P, by which is adjusted the depth the shovels enter the ground

No. 54,638.—CALVIN YOUNG, Auburn, N. Y.—*Machine for Making Carriage Axles*—May 8, 1866.—The heated axle is upset between two pairs of gripping dies, the stationary pair of dies being inclined from a right line to give the "set" to the spindle.

Claim.—Forming a solid swell or collar on axles and giving the bed or arm thereof a "set" by one and the same operation by means of a stationary and a movable clamping die that seize the bar or blank at points remote from its ends, and leaves uncontrolled that portion of the bar or blank from which the enlargement is to be made until the dies are about to meet, when said enlargement is jammed up into the proper form by said dies, substantially in the manner herein described and represented.

No. 54,639.—CALVIN YOUNG, Auburn, N. Y.—*Machine for Making Carriage Axles.*—May 8, 1866.—One pair of gripping dies is stationary and the other movable ; the movement of the latter toward the former upsets the hot iron, and the adjacent faces of the dies give form to the collar produced.

Claim.—Forming a solid collar or shoulder on an axle by griping the bar or blank of which the finished axle is made, at two points remote from its ends by griping dies so as to leave a portion of said bar between the two sets of dies uncontrolled and free to expand laterally by end pressure applied to one of the griping dies until near the end of the movement of said dies, when the expanded or swelled metal may be driven into the dies to give it better form and shape, by means substantially as herein described and represented.

No. 54,640.—JOHN ADT, Wolcottville, Conn., assignor to ELISHA TURNER, same place.—*Corkscrew.*—May 8, 1866.—To fix for use, the head slips into the tube and the neck follows the slot.

Claim.—The slotted metallic case *a*, in combination with the T-head, *c*, of the instrument *c*, as and for the purpose specified.

No. 54,641.—D. B. BAKER, Rollersville, Ohio, assignor to himself and P. S. MILLER, same place.—*Gate*—May 8, 1866.—The braces are hinged, one on each side of the post on which the gate rolls, and support it in position.

Claim.—The braces E E, so arranged that they will act as supports for sustaining the weight of the gate after it has passed its balancing point, substantially as specified.

No. 54,642.—JAMES CAMPBELL, Harrison, Ohio, assignor to himself and WILLIAM CAMPBELL, same place.—*Corn Planter.*—May 8, 1866.—A rotating pocketed disk in the adjustable hopper is rotated at varying speeds by a worm on the shaft rotated by a pinion engaging with one or another of the gear wheels on the main shaft.

Claim.—First, a grain box whose front portion is supported and pivoted upon a curved fender plate F, and whose rear portion is supported in notched posts G *g* G' *g'*, substantially as described.

Second, the grain box E *o o'*, having the pivoted front and perforated rear portion in the described combination with the adjustable hook P Q *q q'*, for the purposes set forth.

No. 54,643.—JOSEPH FIELDHOUSE, Dighton, Mass., assignor to DIGHTON FURNACE COMPANY, Taunton, Mass.—*Dies for Forming Gas Pipe.*—May 8, 1866.—The die is in two parts, larger at the entrance, where a blade keeps the edges of the sheet metal straight and in line; the metal is at a welding heat, and the edges are joined at the narrower portion after passing the guide.

Claim.—The combination and arrangement of the edge guide C with the pipe forming and welding die, or its equivalent, substantially as set forth.

No. 54,644.—DENNIS FRISBIE and SAMUEL C. GOODSELL, New Haven, Conn., assignors to themselves and DAVID P. CALHOUN, same place.—*Baling Press.*—May 8, 1866.—The power is applied to alternate sides of the platen by the alternate engagement of the pawls on the main shaft, giving the platen a rocking motion.

Claim.—First, the combination of the two pawls S and S', having their eccentrics set opposite to each other and upon the same shaft, with the platen of a press, when arranged to operate in the manner described so as to actuate the platen alternately at each end, substantially as and for the purpose specified.

Second, the combination of the hook *w* with the operating parts of a baling press, substantially as and for the purposes specified.

No. 54,645.—J. A HANSON, Amsterdam, N. Y., assignor to himself and H S. McELWAIN, Amsterdam, N. Y., and H. K. KENT, Pittsfield, Mass.—*Water Wheel.*—May 8, 1866 — The water is admitted to the wheel by a scroll chute whose entrance is regulated by a gate moved by rack and pinion. The water strikes the vertical curved faces of the buckets and escapes between their downward and outward curved extensions.

Claim.—First, the buckets with the part *b*, as described, lip *c*, rim *d*, and supplementary portion *e*, in combination with the hollow hub *a* of the wheel B, arranged and operating in the manner and for the purpose herein described.

Second, the arrangement of the gate C, sunken rack *g*, pinion D, and box F, operating in the manner and for the purpose herein described.

No. 54,646.—JOHN W. HOARD, Bristol, R. I., assignor to S. W. YOUNG and R. A. DENNISON.—*Eyelet Stock.*—May 8, 1866.—The depressions are formed consecutively on a strip, without detaching.

Claim.—The stock prepared as herein described, that is to say, the strip or band in which a series of cups or depressions is formed, approaching the form and size required for the finished eyelets, substantially as and for the purposes herein set forth.

No. 54,647.—WILLIAM LA DEW, Norway, N. Y., assignor to himself and HORACE U. SOPER, Batavia, N.Y.—*Arranging Vats for Tanning.*—May 8, 1866.—The hides are suspended by books or bars so as to hang within the vat; a rotating shaft with arms and bars agitates the liquor and the hides.

Claim.—First, constructing a tan vat with slats or books so that the leather may be hung in the tanning liquor, in combination with a movable wheel or other movable machinery, located in the bottom of the vat, to be operated either by hand or power, for the purpose of equalizing the strength of the tanning liquor, and imparting a motion to the leather hanging therein, constructed and arranged as described and for the purposes set forth.

Second, placing in the bottom of the vat a movable wheel, with buckets or other movable machinery, so arranged as to be operated by hand or power, for the purpose of plunging up and agitating the tanning liquor, and also imparting a motion to the leather hanging therein, thus saving labor both in handling and laying away, constructed and arranged as described and for the purposes set forth.

No. 54,648.—JACOB K. MINICH, Mount Joy, Penn., assignor to himself and JACOB R. HOFFER, same place.—*Harrow and Cultivator Combined.*—May 8, 1866; antedated September 25, 1865.—Three shovels precede the cultivator share and its diverging wings, and three shovels follow in the rear.

Claim.—The construction and arrangement of the shovels 1, 2, 3, and 4, 5, 6, in combination with the cultivator G, and its slotted side wings H, all arranged and operating in the manner and for the purpose specified.

No 54,649.—AUGUSTUS T. MORRIS, Bloomfield, N. J., assignor to himself and JAMES CUMMINGS, New York, N. Y.—*Scoop for Excavating.*—May 8, 1866.—The sections of the semi-cylindrical scoop are presented open to the mud and by vibration on their hinges enclose a load.

Claim.—First, the construction of the frames *a'*, in combination with the cross plate *e'* and block 20, for the shaft *b*, substantially as and for the purpose specified.

Second, the combination of the cross bar or shaft *e*, with the frames *a'*, said shaft *e* being guided at its ends in said frames *a'*, as specified.

Third, the plates 24, attached to the frames *a'*, as set forth, and receiving the joints or hinges by which the scoops *g g* are attached, substantially as set forth.

Fourth, the combination of two separate pairs of hinges, with the quadrant buckets as specified, so that the buckets shall open wider than the diameter of the buckets when closed, and increase the efficiency of the said bucket in excavating, as specified.

No. 54,650.—OBED PECK, Southington, Conn., assignor to WILLIAM A. IVES, New Haven, Conn.—*Bit Brace.*—May 8, 1866.—The flange of the sleeve bears on the shoulder of the tool and moves longitudinally on the stock, the pin on the stock traversing a longitudinal and then an oblique slot in the sleeve.

Claim.—The combination of the sleeve C with the socket A, when the said sleeve is constructed with the vertical and inclined slot *a*, and the socket provided with a pin *e*, and arranged to operate in the manner and for the purpose substantially as specified.

No. 54,651.—HIRAM ROBBINS, Cincinnati, Ohio, assignor to himself and THOMAS H. FOULDS, same place.—*Roll for Wringers.*—May 8, 1866.—Explained by the claim.

Claim.—First, a wringer roll composed of an external cylinder of rubber, and the longitudinally divided shaft, composed of the central rod A *a a'*, and roughened or indented segments B B', substantially as and for the purpose set forth.

Second, in combination with the elements of the clause next preceding, the guard or sheath C, adapted to enable the insertion of the segments B B', in the manner explained.

No. 54,652.—ABEL SIMONDS, Fitchburg, Mass., assignor to himself, B. SNOW, Jr., ALVIN A. SIMONDS, and GEORGE F. SIMONDS, same place.—*Mowing Machine Guard.*—May 8, 1866.—A plate is inserted in the slot of the heated guard to keep the water used in hardening from the portion which it is not desired to harden.

Claim.—The above specified mode of hardening the knife supporting surface of a mowing machine guard, and protecting the lip of such guard from being hardened at the same time.

No. 54,653.—GEORGE L. WITSIL, Philadelphia, Pa., assignor to CHARLES A. WATERMAN.—*Ice Pick, Nut Cracker, &c.*—Explained by the claim.

Claim.—Combining and arranging in one tool or instrument an ice ick *d*, gas burner forceps, and pipe tongs, and nut cracker *e*, tack and nail extractor *f*, nailpand tack hammer *g*, meat thresher *h*, and plate lifter *k*, substantially as and for the purpose herein specified and described.

No. 54,654.—WILLIAM ZELLNER, New York, N. Y., assignor to himself, M. M. ROUNDS, New Haven, Conn., and J. E. JEROLD, Jersey City, N. J.—*Steam Generator.*—May 8, 1866.—The boiler is extended forward to overhang the grate.

Claim.—A steam generator having a portion of the boiler extended into the fire box, for the purpose of increasing the heating surface thereof and distributing the heat therein, substantially as and for the purpose described.

No. 54,655.—PAUL T. WARE, Toronto, Canada West.—*Compressed Air Bath.*—May 8, 1866.—The chamber has non-conducting walls, and the air therein is warmed by a steam coil. The temperature is adjustable by the patient and a limit of pressure afforded by a safety valve. Facilities are afforded for a shower bath.

Claim.—First, in combination with the compressed air chamber A, and exterior jacket B, the heating coil *a*, or its equivalent, for radiating warmth equally through the metallic sides or walls of said bath chamber, substantially as set forth.

Second, the combination and arrangement of the steam chest E, and compressed air coil *d d*, and valve *d'*, within the bath chamber A, for warming and regulating the temperature of the condensed air before it enters the bath, substantially as shown and described.

Third, in combination with a closed chamber or vessel for condensed air baths, and the pipes for the induction and eduction of air, water, &c., into and from the same, the employment of valves *r r'*, and *d'*, arranged within the bath chamber, so as to be controlled and operated exclusively by the patient confined therein, substantially as and for the purposes shown and described.

Fourth, in combination with a compressed air bath chamber, the safety valve *l'*, so arranged as to be inaccessible to the occupant of the bath, substantially as and for the purpose described.

Fifth, in combination with the air chamber A, and water reservoirs L and M, the pipes *u u*, or their equivalent, for the purpose of maintaining within said reservoirs a pressure corresponding with that within the air bath, as and for the purposes set forth.

No. 54,656.—JAMES H. CLAPHAM, New York, N. Y.—*Boring Oil Wells.*—May 8, 1866.—The drill rope is automatically and intermittingly turned in one direction and the feed screw rotated; the drill rope passes through the trunnion of the windlass drum to remove the twist incident to the rotation of the drill.

Claim.—First, the arrangement of the band wheel *d*, crank pin I, connecting rod *g*, and slide A, in the manner specified and for the purposes set forth.

Second, the friction clutch *p*, and rope barrel *n*, in combination with the shaft *e*, connect-lug rod g, and slide h, for operating the boring tool as specified, so that the said boring tool can be drawn up by the direct application of power to the tool rope, in the manner set forth.

Third, the fork or slotted lever r, actuating the temper screw progressively by motion from the pitman, as set forth.

Fourth, the fork or slotted lever for rotating the rope clamp progressively by motion from the pitman, as set forth.

Fifth, connecting the temper screw directly to the reciprocating crosshead, as specified, thereby dispensing with the slings heretofore employed for suspending the said temper screw.

Sixth, the combination of the rope m, clamp l, with its rotating device r' and s', temper screw k, with its rotating device r and s, and connecting yoke 14, and the windlass n, arranged, constructed and operated as described, and for the purpose set forth.

No. 54,657.—EDWARD L. ALLEN, Fair Haven, Vt.—*Vessel for Petroleum, &c.*—May 8, 1866.—Explained by the claim.

Claim.—First, the employment of the wooden box or casing A, in combination with the metal receptacle B, and surrounding the same in the manner and for the purposes hereinbefore described and set forth.

Second, the employment in a reservoir for the reception of kerosene and other volatile oils of the bottom E, so constructed as to incline downward in a conical or other shape or form, to the point where the discharge pipe F is inserted, substantially in the manner and for the purposes herein described and set forth.

No. 54,658—JOHN RANDOLPH ABBE, Providence, R. I.—*Bed Bottom.*—May 15, 1866; antedated May 7, 1866.—The metallic springs are founded on the end rails and support rods whose knobs fit the grooved slats.

Claim.—The springs B, applied to the head and foot rails a a of the bedstead in combination with the springs B, rods C, provided with knobs d and the slats D, having concave surfaces to receive the knobs, all being arranged substantially as and for the purpose set forth.

No. 54,659.—AUGUSTUS ADAMS, Sandwich, Ill.—*Corn Sheller.*—May 15, 1866.—The ears of corn are prevented from choking in the mouth of the sheller by a revolving shaft with beaters.

Claim.—In combination with a series of feeding throats in the hopper of a corn-shelling machine, the use or employment of a rotating bar or shaft M, provided with one or more suitable projections or angles, or a series of projections or angles corresponding with said series of feeding throats, arranged and operating substantially as herein specified and shown, and for the purposes set forth.

No. 54,660.—I. J. W. ADAMS, Galestown, Md.—*Water Wheel.*—May 15, 1866.—The wheel has 8-shaped buckets, receives the water from a scroll chute, is closed on one side and discharges the water at the other.

Claim.—The arrangement of the penstock E, scroll C, frame d, wheel B, with its sides a a', and buckets c, as described, shaft A, crank G, and box H, operating in the manner and for the purpose herein specified.

No. 54,661.—A. L. ANDERSON, Ware, Mass.—*Harness Motion for Looms.*—May 15, 1866.—Intended particularly for weaving straw, palm leaf, &c. The device secures the simultaneous motion of the harness and the lathe.

Claim—First, the combination for moving loom harnesses, consisting of the treadle E, arm G, pieces a a', wheel H, levers d d, and cords or similar connections e e, when arranged and operating substantially in the manner herein set forth.

Second, connecting the arm G both to the lathe D and to the operating devices as shown, so that it may actuate both simultaneously, substantially as herein described.

No. 54,662.—E. H. ASHCROFT, Lynn, Mass.—*Railway Car.*—May 15, 1866.—The car has a water jacket and a set of distributing perforated pipes ; a plug, fusible at a low degree of heat, is interposed. A pipe extends from the bottom to the top of the car, making a coil around the stove.

Claim.—A safety car constructed with a water space, one or more showering pipes, and a fusible plug apparatus, arranged substantially in manner and so as to operate with respect to the car chamber, as specified.

Also, in combination with a car so made, a heating apparatus and a circulation coil, or the equivalent or equivalents thereof, applied to the water heating space, substantially as and for the purpose, and to operate as explained.

No. 54,663.—EDWARD H. ASHCROFT, Lynn, Mass.—*Apparatus for Moulding and Drying Peat.*—May 15, 1866.—The hopper and spout feed the peat into the moulds, which traverse on an endless belt in the kiln ; the contents of the moulds are compacted by a pressure roller and discharged at the rear end.

Claim.—The herein described mode of preparing peat for fuel, namely, filling the moulds, compacting the same, drying, and discharging at one operation, in the manner substantially as set forth.

No. 54,664.—WILLIAM S. ATCHLEY, Williamsburg, Ohio.—*Skiving Machine.*—May 15, 1866.—The knife is in a reciprocating carriage, which traverses over the bed whereon the leather is laid. A pressure roller precedes the knife, and another roller is placed in the bed-plate.

Claim.—First, the combination of the adjustable bed-plate A² with the knife blade B, when the two are arranged together so as to operate substantially in the manner described and for the purposes specified.

Second, in combination with the above, the pressure roller S of the knife-carrying frame K, arranged just in front of the knife blade, as and for the purpose specified.

Third, hanging the roller Y in the bed-plate A, as and for the purpose described.

No. 54,665.—JAMES S. ATTERBURY and THOMAS B. ATTERBURY, Pittsburg, Penn.—*Mode of Applying Labels to Bottles.*—May 15, 1866.—The label is attached in the sunken panel on the side of the bottle.

Claim.—The bottle, having a label applied to a recess formed in it, and secured in such recess and covered by means of a soluble glass cement, as a new and improved article of manufacture.

No. 54,666.—WILLIAM S. AUCHINCLOSS, New York, N. Y.—*Axle Box Cover.*—May 15, 1866.—The cover is clamped against the inside of the box by a bolt, and an outer bridge piece over the opening.

Claim.—The arrangement of the square or polygonal-sided bolt b, attached to the cover B, and capable of manipulation by means of the bridge C, substantially in the manner described and represented.

No. 54,667.—DANIEL W. BAKER, Harwick, Mass.—*Meat Chopper.*—May 15, 1866—The devices described elevate and depress the knife above the meat block, which is intermittingly rotated beneath.

Claim.—The combination of the bent arm L with the post D, or its equivalent, the knife and its slide rod, and the mechanism for imparting to the knife'rod its reciprocating movements, such mechanism being the lever G, the connecting rods I K, and the gears on their shafts, the whole being substantially as specified.

Also, the arrangement of the lever R, its pawl P, and spring S, with the post D and lever G applied thereto, and to the knife shaft, as described.

Also, the combination and arrangement of the guard O with the bent arm L and the knife and its rod, as applied to such arm, and to operate in the tub, as specified.

Also, the combination of the spring latch M, or its equivalent, with the post D, the bent arm and the knife combined together, and with machinery for operating the knife, as specified.

No. 54,668.—W. K. BALDWIN, Chicago, Ill.—*Machine for Opening Tin Cans.*—May 15, 1866.—The can is clamped by a yoke and set-screw against a standard; a knife on a lever is forced through the metallic lid of the can, which is rotated beneath.

Claim.—First, the arrangement and combination of the standard B, clamp P, and guides V, when constructed and operated substantially as described.

Second, the combination of the parts H G and E, in combination with the bar K and lever O, substantially as described and for the purpose set forth.

No. 54,669.—ORLANDO BARR and FRANKLIN F. COX, Beloit, Wis.—*Cultivator.*—May 15, 1866.—The ploughs are adjustable laterally and vertically.—The sliding head blocks are moved on the transverse rod by means of a rack, sectional pinion, and foot lever; vertical slide rods are attached to the beams.

Claim.—The rod B' B', as shown in Fig. 1, the head blocks B B, sliding upon the said rod B' B', connected with the gear b b, and foot lever and section pinion D, the sliding rods A A attached to the beams A' A', and the device and arrangement of the draught rods g g, and chains e e, when constructed substantially as and for the purpose herein set forth and described.

No. 54,670.—W. B. BARTRAM, Norwalk, Conn.—*Sewing Machine for Stitching Cord to the Edge of Fabrics.*—May 15, 1866.—The zigzag stitch is produced by a combined reciprocating, lateral and intermittent, forward movement of the material, which passes between two surfaces subjected to the like motions.

Claim.—First, in combination with automatic mechanism capable of producing both a forward movement and a lateral reciprocation of the material to be stitched, a cloth-holding device, which partakes of the said lateral reciprocation, and between the upper and lower parts of which the said material and cord pass in their forward movement, for the purpose described.

Second, in combination with the laterally reciprocating table A, the clamp or holding device B, substantially as and for the purpose set forth.

No. 54,671.—W. B. BARTRAM, Norwalk, Conn.—*Sewing Machine for Stitching Eyelet Holes.*—May 15, 1866.—The cylindrical guide fits the eyelet to be worked, and about it, as an axis, the cloth is revolved by the feed movement; the radial slit in the guide is projected into the table and allows the needle to pass both within and outside the edge of the eyelet hole when the table is moved laterally.

Claim —The combination of the cylindrical guide A, and a presser-foot D, with a rectilinear forward feed movement, and a horizontally reciprocating mechanism, to effect a lateral feed of the cloth, for the purpose set forth.

No. 54,672.—JOSEPH BARTA, La Crosse, Wis.—*Raking and Binding Attachment to Harvesters.*—May 14, 1866.—The speed of the raking and binding mechanism is changed for adapting them to work in light or heavy grain; the rake is adjusted to adapt it to the length of the straw upon which it operates, and for the presentation of the gavel to the binding devices to be bound at its mid-length. The arrangement of the band carrier and sheaf-compressing devices, the mechanism for operating the rake, and the construction and arrangement of the devices for tying the band, will not admit of a brief description.

Claim.—First, the wheel E, provided with pins i, arranged in concentric circles, and the ratchet F, connected with said wheel, and both placed loosely on shaft D, in combination with the notched wheel G on shaft D, the pawl H, provided with the lever H', and the plate I, all arranged substantially as shown and described, for the purpose of rotating the shaft D, and consequently the raking and binding device, intermittently, at longer or shorter intervals, as may be desired.

Second, the endless chain S, arranged to operate first in one direction and then in the other, in combination with the plate P, having the rake head Q attached, the levers o q, plate R, and the staple x, all arranged substantially as shown, to give the reciprocating movement to the rake, and also the rising and falling movement, substantially as described.

Third, the adjustable frame O, on which the plate P works, for the purpose of adjusting the rake in such a relative position with the cradle as to cause the sheaves to be bound centrally, whatever the height or length of the grain may be, substantially as set forth.

Fourth, the pawls n' n', applied to the plate P, and arranged so as to be operated therein, in and out of gear, with rack O', through the medium of the bent lever O', at the time of raising and lowering of the rake, substantially as described.

Fifth, the cradle C', and bar E', arranged and combined as shown, so that the bar, as it is raised by the rotation of shaft D, will raise the cradle with it, the cradle holding the gavel in position, and the bar E drawing the cord r' around it, substantially as set forth.

Sixth, the twister J, provided with jaws w' w', one of which is furnished with a hollow square a'', in combination with the hook b'', and the clamps t' v', all arranged to operate substantially as shown, for the purpose of tieing the knot.

Seventh, the pivoted plate K', operated from the shaft D, and provided with slots, segment rack, &c., for the purpose of operating the hook b'', and twister J, as set forth.

Eighth, the apron L, operated by the lever or arm p'', cam q'' on shaft D, in combination with the cradle C' and bar E', all arranged to operate in the manner substantially as and for the purpose specified.

Ninth, the shears m'', placed on the bar E', and operated from the shaft D, to cut the cord, substantially as herein set forth.

No. 54,673.—MARSHALL M. BISHOP, Le Roy, N. Y.—*Saw Mill.*—May 15, 1866.—The arbor of the saw is adjusted vertically by the vibration of its frame, one leg of which is pivoted, and the other is provided with a rack, operated by a pinion and sustained by a pawl.

Claim.—Adjusting the saw by means of the pivoted frame D, racks d d, pinions E, and atchet device S, arranged to operate as herein described.

No. 54,674.—MARSHALL BLISS, Grinnell, Iowa.—*Mosquito Canopy.*—May 15, 1866.—The canopy is arranged for vertical adjustment against the wall of the room and over the head of the bedstead.

Claim.—The combination of the frame A, so connected to the wall of a room that it can be elevated or lowered, as desired, with the cord b, or its equivalent, and the netting or canopy B, substantially as specified.

No. 54,675.—JAMES BOOTH, St. Louis, Mo.—*Cotton Bale Tie.*—May 15, 1866.—The cast-iron piece riveted to one end of the hoop has three hooks, which catch within recesses on the other end of the hoop.

Claim.—An improved cotton-bale tie A, constructed with three hooks a¹ a² a³, and combined with the ends B and C of the hoop, substantially in the manner described and for the purpose set forth.

No. 54,676.—EDWARD H. BOSWELL, Philadelphia, Penn.—*Paper Tin.*—May 15, 1866.—The straight-edge, whose usual function is to assist in tearing off checks, bills, blanks, &c., has the additions described to increase its usefulness.

Claim.—The arrangement of the calendar holders A, elastic band D, and pen-holding block E, in combination with a paper tin, of which one side is graduated and the other raised, by projections, to act as a ruling guide, in the manner and for the purpose herein specified.

No. 54,677.—EDWARD H. BOSWELL, Philadelphia, Penn.—*Stair Rod and Fastening.*—May 15, 1866.—The stair rod has slots, which slip over the heads of the set screws. The latter, by partial rotation, retain the rods by the contact of the shouldered heads.

Claim.—The combination of the slotted rod B and the shouldered fastening D, which in one position will admit the passage over it of the rod, retaining the latter by a partial revolution, substantially as described.

No. 54,678.—THOMAS S. BOWMAN, St. Louis, Mo.—*Key.*—May 15, 1866.—The spring-follower in the barrel of the key prevents the accumulation of dust, &c., therein.

Claim.—The application to a key, having a hole drilled in its shank or arbor. of a spring and stopper, substantially in the manner as and for the purpose herein set forth.

No. 54,679.—FREDERICK BUCKNAM, Portland, Maine.—*Scoop, Sifter, Grater and Holder.*—May 15, 1866.—The nutmeg and its grater slip within the hollow handle of the scoop, which has a rotating sieve.

Claim.—The combination with a scoop, of the described form and kind, of the roller R, sieve B, handle H, having the grater and holder G, made as specified, all constructed, arranged, and operated as set forth.

No. 54,680.—JOHN BURKE, Courtland, Ill.—*Breech-loading Fire-arm.*—May 15, 1866.—The stock is hinged, and the barrel divided just in advance of the chamber, the forward portion sliding in the end piece of the stock. The chamber has a forward prolongation, which fits a socket in the rear of the forward portion.

Claim.—First, the combination of a telescopic joint in the barrel, with a hinged joint in the stock, when the barrel slides in the forward portion of the stock, substantially as and for the purpose set forth.

Second, the combination of the sliding barrel with the dovetailed bar and front portion of the hinge, substantially as and for the purpose set forth.

Third, the combination of the barrel with the front and rear portions of the hinge, substantially as and for the purposes set forth.

Fourth, the combination of the projection r, on the front part of the hinge with the slot h. with the rear part of the hinge, to hold the barrel at an angle when loading, substantially as described.

Fifth, the combination of the catch lever s, thumb bolt n, and bar f, substantially as and for the purpose set forth.

No. 54,681.—VICTOR H. BUSCHMAN, Baltimore, Md.—*Machine for Cutting Cloth.*—May 15, 1866.—The knife consists of an endless metallic strip, with serrated edges, and passes over pulleys above and below the cloth table.

Claim.—The endless belt knife, constructed and applied and operating substantially as described and for the purpose set forth.

No. 54,682.—A. S. CAMERON, New York, N. Y.—*Lock-up Safety Valve.*—May 15, 1866.—The single disk valve and its weighting arrangements are enclosed in a lock-up case, which permits egress to the steam, and the raising of the valve by the engineer, but not the means of adding to the load of the valves.

Claim.—First, the closed case A, constructed substantially as herein described, in combination with a single disk safety valve a and lever, substantially as and for the purposes set forth.

Second, the notched standard c, rising from the bottom of the case A, and arranged relatively to the internal screw k, applied substantially as and for the purpose described.

Third, the counter lever C, in combination with the slotted bar n, provided with a shoulder p, and with the main lever d of the safety valve, constructed and operating substantially as and for the purpose set forth.

Fourth, further, a cap applied to the valve port or discharge port of a lock-up safety valve apparatus, for the purpose and in the manner explained.

No. 54,683.- MILES H. CARD, Fulton, Ill., and JOHN W. STEWART, Lyons, Iowa.—*Car Coupling.*—May 15, 1866.—A link on each draw-head, as it enters the recess in its opposed draw-head, strikes a projection and throws a catch into the link, where it is secured by a falling pin.

Claim.—First, the combination and arrangement of the guards C, links B, and cams D with the draw-head A, substantially as and for the purposes specified.

Second, in combination with the links B, and cam D, provided with the shoulders m, the employment of the pin or stop c, arranged and operated substantially as specified and shown.

Third, providing the cam with the projection n, operating as set forth.

No. 54,684.—W. D. CHAPMAN, Theresa, N. Y.—*Fishing Tackle.*—May 15, 1866.—A spring hook is secured to the snood, and affords the means of attaching a bait or other hook.

Claim.—The combination of the fixed wire hook E, spring hook F, and snood G, when arranged together and so as to operate substantially in the manner described and for the purpose specified.

No. 54,685.—W. F. CLASS, Cleveland, Ohio., assignor to NATHAN PAGE, same place.—
Pump Sprinkler.—May 15, 1866.—The perforated base stands in a tub and the water is ejected
through the hollow plunger, handle and connected hose.

Claim.—First, the pipe F, nozzle C. and plunger B, in combination with the cylinder A,
valves H L, and perforated base A', when arranged in the manner and for the purpose set
forth.

Second, the lips i, wings n, and plunger B, in combination with the valves H L m, pipe
F, and nozzle C. when arranged in the manner and for the purpose described.

No. 54,686.—DANIEL CLAVIDGE, Indianapolis, Ind.—*Apparatus for Expelling Air from
Cans.*—May 15, 1866.—The cans are supported on a shelf, and pads are pressed against
them by an eccentric to eject the air.

Claim.—An apparatus for expelling the air from cans by pressure upon their sides, arranged
substantially as set forth.

No. 54,687.—JOHN COCHRANE, Wall Township, N. J.—*Machine for Forging Screws.*—
May 15, 1866.—Cannot be briefly described.

Claim.—First, the method of forming screw threads upon metal bolts by means of forging
dies of proper shape, in combination with a guide screw and rotating wheel, all constructed
and arranged substantially as described.

Second, the combination of the tongs and guide screw with the forging dies and templates,
constructed and operating substantially as described.

Third, the combination of the clevis with the sliding rods or templates, so as to sustain and
move with the bolt or blank while it is under the action of the screw-forging dies, constructed
and operating substantially as described

Fourth, controlling the space between dies, operating as described, as the operation progresses by means of sliding templates, or their equivalent, so as to forge screws of uniform
diameter, or taper, or gimlet-pointed, substantially as described.

Fifth, the combination of the sliding templates with the guide screw by means of the slide
rods and yoke, so as to be operated by it with an equal and simultaneous retractive movement, constructed and arranged substantially as described.

Sixth, the combination with screw-forging machinery of the open bearing or clevis in front
of the dies and the open screwed bearing in the pedestal for the guide screw, so that a blank
or bolt, and the detachable apparatus by which it is held and rotated, can be safely and expe-
ditiously placed in, or removed from, the machine while in motion, and without interruption
to the speed, substantially as described.

Seventh, the straight-threaded forging dies, in combination with the jumper c and anvil
block J, constructed and operating in the manner and for the purpose substantially as
described.

Eighth, the combination in a screw-forging machine of a screw-forging die that has a positive and definite reciprocating movement, and a screw-forging die that has a positive movement to or from the other, as required in the production of screws of varying diameter, when
such movement of the latter die is caused and controlled by means of sliding templates, or
their equivalent, substantially as described.

Ninth, the method of forming or making screws upon metal bolts by operating upon and
completing them from the neck or shank toward the point, by the means and in the manner
substantially as described.

No. 54,688.—GEORGE COFFIN, Boston, Mass.—*Device for Operating Ship's Windlasses.*—
May 15, 1866.—The toggle-jointed levers are worked by brakes or a winch on the top-gallant
forecastle; the veering of the cable is regulated by a lever on the chain pipe.

Claim.—First, the combination and arrangement of the windlass and its chain drums,
operated by the toggle-jointed levers, substantially as described.

Second, regulating and checking the veerings of the cable by means of a lever working in
the top of the chain pipes, as in the manner as described.

Third, the arrangement and construction, substantially as described, of the chain drums
and brakes, by which they can be made to turn with, or be disengaged from, the windlass.

No. 54,689.—WILLIAM COGSWELL, Ottawa, Ill.—*Ratchet Attachment for Harvesters, &c.*—
May 15, 1866.—The spring pawls are thrown into or out of gear with the ratchet on the shaft,
so as to admit of action in either direction, or a suspension of movement during reverse motion of the machine to which the device is applied.

Claim.—First, the combination with the pinion A and shaft B of the adjustable pawls
C C', and ratchet B², substantially as described, to operate in the manner and for the purposes set forth.

Second, the combination with the pawls C C', and ratchet B² of the screw bolts E E', and
nuts E E³, as and for the purpose described.

No. 54,690.—A. C. COOKE, Ohio, Ill.—*Neck Yoke.*—May 15, 1866.—Anti-friction spools
for the breast straps in the rings of the neck yoke.

Claim.—First, the combination with a neck yoke of friction spools C, substantially as and for the purpose specified.

Second, the manner of connecting the friction spool C to the bar A, as shown and described

No. 54,691.—LEWIS L. COON, Nunica, Mich.—*Shoemakers' Jack.*—May 15, 1866.—The jack with its heel and toe-rests revolves horizontally upon the arm of the shaft; the latter revolves vertically, and is maintained in position by the sliding ratchet and detents. The combination of adjustments gives every presentation required.

Claim.—First, the combination of the working shaft C, and the standards H H' H'' revolving at right angles to it, substantially in the manner and for the purpose set forth.

Second, so arranging the standard H' by means of a bevelled base and eccentrically placed screw, that by turning the nut K the last will be held firmly in place, substantially in the manner set forth.

Third, the combination of the spring D, sliding cone E, and detents F F, for the purpose of fixing the position of the shaft C, substantially as set forth.

No. 54,692.—THOMAS COREY, Marlboro', Mass.—*Machine for Cutting Pegs out of Boots and Shoes.*—May 15, 1866; antedated November 15, 1865 —The cutters at the ends of the shaft have a rotary motion, imparted by gearing and a hand crank.

Claim.—First, the arrangement in a machine for removing pegs from the inside of boots and shoes of cutters, in combination with suitable mechanism for imparting rotary movement thereto, substantially as shown and described.

Second, in combination with cutters and machinery for imparting rotary movement thereto, the frame provided with vertical and oblique arms or horns, substantially as and for the purpose set forth.

Third, so constructing the arms or horns of the frame as to enclose and surround the cutters and mechanism in rotating the same, leaving only the cutting edges of the former to project, as shown and described.

Fourth, the arrangement of the rotary cutters in relation to their respective driving machinery and horns or arms, so that they shall revolve on vertical axes, substantially as set forth.

No. 54,693.—THOMAS J. CORNELL, Decatur, Ill.—*Gang Plough.*—May 15, 1866.—The described devices are for raising the fore and bind ends of the beams for regulating their action and position, and for attaching the beams and tongue to the carriage.

Claim.—First, the combination of the lever M, and bent arm O for raising the beams from beneath at a point in the rear of their forward point of attachment, so as to be vibrated upon the said lifting device by a weight applied to the forward end, substantially as described.

Second, the combination of the link H attached to the forward end of the beams and to the carriage, and operating substantially as described.

Third, the combination of the beam and the curved arm, or its equivalent, which in its backward motion operates to keep the beam down under the circumstances described.

Fourth, the mode of attaching the tongue to the carriage. consisting of the laterally sliding socket and the guides, arranged and operating as described.

Fifth, the combination of the links H and standards G, whose holes admit of the vertical adjustment of the links therein.

Sixth, the arrangement of the foot lever R, the links H and beams A, operating as described.

No. 54,694.—RICHARD COX, Philadelphia, Penn.—*Pipe Tongs.*—May 15, 1866.—The outer jaw is pivoted to a nut, which is adjustable on the shank of the inner swivelled jaw, and their distance is thus adapted to different sizes of pipes.

Claim.—The jaw C, serrated and shaped as described, swivelling upon the screwed handle A, in the manner and for the purpose specified.

No. 54,695.—CHARLES CROSMAN, Tompkinsville, N. Y.—*Folding Hair Brush.*—May 15, 1866.—The hinged handle folds over upon the back.

Claim.—A hair brush, the handle of which may be folded over upon the brush head, the said handle being hinged or pivoted, and having suitable catches for holding the two parts in an extended position, substantially as described.

No. 54,696.—JOHN A. DAUM, Canton, Ohio.—*Fruit Gatherer.*—May 15, 1866.—A canvas frame on wheels for catching falling fruit, a conductor for leading the same to a basket, and adjustable supporter for steadying the frame.

Claim.—First, the adjustable bar D, with set screw F secured to the frame A, when used as and for the purposes herein set forth.

Second, the frame A, receiver E, pipe H, wheels B B, secured by levers c c, and bar D, arranged and used substantially as and for the purposes herein set forth.

No. 54,697.—THOMAS DAVEY, Houghton, Mich.—*Ore Separator or Jigging Machine.*—May 15, 1866.—The ends of the pump stock are connected to two boxes, covered with jig sieves, the water being forced through the latter alternately to agitate their contents; the ore is discharged at an opening; the sand falls over the tail-board, to be rewashed in the lower sieve.

Claim.—First, the combination of a double-acting plunger pump, with two sieves of a jigging machine, so that each motion of the plunger shall force water through one or the other of the said sieves, substantially in the manner and for the purpose set forth.

Second, arranging the jigging sieves of an ore washer and separator, one above the other, substantially as and for the purpose described.

Third, in combination with the sieves of a jigging machine, the receiving hopper B, provided with the perforated bottom, as and for the purpose set forth.

Fourth, in combination with the sieve of a jigging machine, the ore discharging chamber L, provided with a gate or gates, as described.

Fifth, in combination with the sieves and pump of a jigging machine, the partition S, as and for the purposes set forth.

Sixth, the adjustable pressure board T, in combination with the pump and sieve of a jigging machine.

Seventh, in combination with the sieves D and J of a jigging machine, the passage H, and shoot board I, substantially as and for the purpose set forth.

No. 54,698.—GEORGE DAVIDSON, Philadelphia, Penn.—*Sextant.*—May 15, 1866.—A spirit level, with a reflector and adjustable cross-wire for determining the horizontal plane, is attached to the sextant, together with a segment of a double convex lens for transmitting the images of the cross line and bubble of the spirit level in condition for distinct vision.

Claim.—The combination of the spirit level, cross-wire reflector and double convex lens, or their equivalents, with the sextant, quadrant, octant, reflecting circle, or their equivalents, substantially as above described.

No. 54,699.—WILLIAM C. DAVOL, Fall River, Mass.—*Spinning Mule.*—May 15, 1866.— The cam shaft is turned and operated to effect the required changes by the arrangement of geared wheels and clutch box and escape plate. A fixed latch moves the clutch box in and out of gear at intervals in combination with another escape plate and lever with a helper spring.

Claim.—First, the escape wheel or plate p' on the clutch P, constructed and operated substantially as described for the purpose of engaging and disengaging the clutch P at stated intervals, in combination with the fixed stud, the escape plate W, and helper spring 2, as above set forth.

Second, the fixed latch or stud t, in combination with the escapements of the escape wheel p', substantially as described, for the purpose of moving and holding the escape plate clutch box P out of gear at intervals.

Third, the combination of the escape wheel p' and the clutch box P with the stop finger 14 and stop plate 15 for turning the cam shaft E at intervals, substantially as described.

Fourth, the combination of the escape wheel p' and the clutch box P with the escape plate W, escape lever y and helper spring 2, substantially as described.

Fifth, in the combination with the escape plate W, the helper spring 2, the escape lever y, the stop finger 14, the stop plate 15, and the catch or clutch box P, with its escapement p', arranged substantially as above shown.

No. 54,700.—GEORGE W. DEVIN, Ottumwa, Iowa.—*Harness Snap.*—May 15, 1866.—Explained by the claim.

Claim.—The within described snap as an article of manufacture, said snap made of wire with a loop at one end and hooks at the other, the two parts of the wire being twisted together or around each other between the loop and the hooks for the purpose of making a firmer loop, and for binding the hooks together more securely, and thus throwing the tension upon both parts of the wire in whatever line the strain may be directed, substantially as herein specified.

No. 54,701.—ELLIS DOTY, Janesville, Wis.—*Plug for Washing Machines.*—May 15, 1866.—The point of the plug closes the water discharge opening in the bottom of the suds box, and it is maintained in position by the engagement of the shoulder on its upper end.

Claim.—The inside plug valve for closing the discharge opening of the tub or chest, actuated by a spring or lever, and maintained in open position by the engagement of a shoulder or pin with the ledge or other portion of the chest, substantially as described.

No. 54,702.—RUFUS DUTTON, New York, N. Y.—*Manufacture of Guard Fingers for Harvesters.*—May 15, 1866.—Explained by the claim.

Claim.—First, forming a guard finger of one uniform metal, as steel, cutting, sawing or milling a slot therein, and tempering such slot on its under surface or edges, while the body of the finger remains untempered or not hardened, substantially as and for the purposes set forth.

Second, forging a solid guard finger, cutting the slot therein and tempering such slot on its under surface or edges, substantially as and for the purposes set forth.

Third, forging the guard finger from a single piece of metal without a weld, and cutting a slot for the knife to work through, substantially as and for the purposes set forth.

No. 54.703.—GEORGE W. DUVALL, Norfolk, Va.—*Boiler Thimble.*—May 15, 1866.—The two sections of the ferrule are spread apart by the keys and caused to bear against the inner surface of the tube, pressing it out against the tube sheet.

Claim.—A cylinder of iron in two sections with scroll-shaped head and groove, as described, in combination with the wedges A, constructed and operated substantially as described.

No. 54.704.—WILLIAM EBERHARD, Sharon Centre, Ohio.—*Spinning Machine.*—May 15, 1866.—Does not admit of brief description. The spring in the spool holds it to the spindle with more or less force according to the position of the spool, but does not prevent rotation.

Claim.—First, the arrangement of the fusee and the devices connected with it, and the crank shaft and the means connected therewith, for giving the backward and forward motions to the carriage as herein recited.

Second, the arrangement of the lever *b'*, and the means connected therewith, for producing the change of gear and making and breaking the connection of the feeding apparatus with the carriage, as herein described.

Third, combining the spring *m'* with the spool and spindle, as and for the purposes herein set forth.

No. 54,705.—GEORGE W. FAIR, Dayton, Ohio.—*Smoke Consuming Heater.*—May 15, 1866.—A fire clay arch rests upon the base, and is sprung over the fire, an upper opening forming an exit. An outer sheet-iron casing encloses an air chamber, whose contents mix with the volatile results from the fire.

Claim.—First, the fire-clay arch A, with its long aperture B at top, as arranged and combined with its fire-place C and inside flue G, as herein described.

Second, the outside flue H, constructed of sheet iron or metal around the arch, as arranged and combined with the dome J, as herein described, and for the purposes set forth.

No. 54,706.—DANIEL FASIG, Rowsburg, Ohio.—*Lifting Jack.*—May 15, 1866.—The lever is pivoted on the end of the rack bar, which is supported by a pawl and on the leg, whose end engages the recesses of the standard.

Claim.—The combination of the standard A with the sliding rack bar C, lever D, and adjustable arm J, when the latter acts as a fulcrum and detent, its teeth *e e* engaging in corresponding recesses *h* in said frame A, all co-operating and constructed substantially as described, and for the purpose set forth.

No. 54,707.—JOSEPH FIRMENICH, Buffalo, N. Y.—*Beer Faucet.*—May 15, 1866.—The barrel of the spigot is made of caoutchouc and its cap of wood.

Claim.—The plug of a faucet made wholly of rubber, partially elastic, when arranged and combined with the wooden wrench, operating substantially as and for the purposes herein set forth.

No. 54,708.—ADALBERT FISCHER, New York, N. Y.—*Liquid Cooler.*—May 15, 1866.—The annular chamber has movable heads, secured by a bolt, with packing at their points of contact.

Claim.—The movable heads E, with annular recesses *a*, in combination with the centre bolt F, annular cylinder A, and vessel B, all constructed and operating substantially as and for the purpose described.

No. 54,709.—CHARLES P. FRAZER. Allowaystown, N J., and C. C. HINCHMAN, Clarksboro', N. J.—*Saw Filing Machine.*—May 15, 1866.—The saw is clamped against the front edge of the table, and the sliding bar carrying the file is reciprocated in the slotted uprights of the reversible swivel; this is adjusted to the required angle, and secured by a thumb screw.

Claim.—The combination of the reversible swivel plate G, file carrier D, groove M, tongue O, and clamp B, the whole being constructed and arranged in relation to each other, substantially as described and for the purposes specified.

No. 54,710.—J. FREDENBURGH and G. A. DAVIDSON, Greene, N. Y.—*Receiving and Delivering Mail Bags upon Railway Cars.*—May 15, 1866.—One scoop is pivoted to the car and the other to the platform; they are actuated by vibration imparted by the passing car, and respectively deliver the mail bags on to the platform or into the car.

Claim.—First, the scoops or hoppers B and C, constructed and arranged in the manner described, the same being secured to the mail car, and also one placed upon the front edge of each mail station platform, to be operated by passing the car, herein set forth.

Second, the method of tossing or throwing the mail bags from the car to the station platform, and from the platform to the car at the same time, by the means employed, substantially as herein described, for the purpose specified.

No. 54,711.—ROSCOE R. FRONOCK, Boston, Mass.—*Garden Digger.*—May 15, 1866.—The fixed and vibrating tines are together driven into the soil, and the hinged portion vibrated by a rack on the shaft, which engages with a quadrant rack on the fork.

Claim.—A digging implement, constructed with two blades *a* and *b*, arranged to operate together, substantially as described.

No. 54,712.—WILLIAM J. FRYER, Jr., New York, N. Y.—*Telegraph Post.*—May 15, 1866; antedated May 1, 1866.—Explained by the claim.

Claim.—The flanged tubular telegraph post, made of corrugated sheet iron B, resting upon a base A. the radiating arms D, and the cap C, all constructed substantially in the manner and for the purpose herein specified.

No. 54,713.—H. P. GENGEMBRE, Pittsburg, Penn.—*Fire Screen.*—May 15, 1866.—An extensible holder, to be attached by thumb-screw to the mantelpiece, and supporting a suspended screen.

Claim.—First, the combination of the flexible screen V, mounted in the case T, and supported by the piece P, arms M and K, and the clamp E, for the purpose specified.

Second, the combination of the oscillating disk G, washers I I, and rivet H.

Third, the arms K and M, articulated together, and also with the clamp E, for the purpose specified.

Fourth, the combination of the piece P with its pulley p p, balls S, and cords R R.

Fifth, the combination of the case T with its roller X and pins Y Y, as described and for the purpose specified.

No. 54,714.—CHARLES D. GIBSON, New York, N. Y.—*Mechanism for Adjustment of Head Lights*—May 15, 1866.—The head-light is adjustable by the engineer in a horizontal plane to direct the rays in the required direction.

Claim.—First, a movable locomotive head light, when its adjustment to any given angle is placed under the control of the engineer or driver, substantially in the manner and for the purpose herein described.

Second, the central plate B and support C, when combined with lever D, arm *f*, elbow joint E, shaft F, or their equivalents, substantially in the manner and for the purpose herein described.

No. 54,715.—JULIEN C. GIRARDIN, Philadelphia, Penn.—*Tension Pulley for Sewing Machines.*- May 15, 1866.—Explained by the claim. The object is to make the parts readily detachable for cleaning.

Claim.—The use of screws in place of rivets for fixing together the plates composing the tension pulley of a sewing machine, they having either ordinary heads or the same with a segment cut away, for the purposes hereinafter described.

No. 54,716.—VIRGIL D. GREEN, Watertown, Wis.—*Safety Valve for Steam Generators.*—May 15, 1866.—The pressure of steam is greater above than beneath the main valve until the closure of the lower small valve and the opening of the escape valve beneath the lever; the pressure is then released above and the steam pressure below the main valve opens it and makes a large area for escape.

Claim—The combination or the cup-shaped or hollow valve D, steam passage I, and weighted valve N, and boiler valve K, connected to a common stem L when arranged so as to operate together, substantially as herein described and for the purpose specified.

No. 54,717.—POWELL GRISCOM and J. H. MILLER, Baltimore, Md.—*Lime Kiln.*—May 15, 1866.—The draw pits and furnaces are on opposite sides of the kilns, and the heat of the furnaces passes into the kilns by lateral openings.

Claim.—The arrangement of the draw pits and furnaces on opposite sides as described; and secondly, the arrangement of the furnaces in respect of the kilns, so that the furnaces, with the exception of those at the ends, shall communicate with the kiln at each side.

No. 54,718.—E. HAMBUJER, Detroit, Mich.—*Wardrobe, Bureau, Desk, Washstand, and Bed Combined.*—May 15, 1866.—Explained by the claim.

Claim.—First, a combined wardrobe, bureau, writing desk, washstand, and bed, when arranged substantially as set forth.

Second, in the folding bed I I', the middle legs K, formed with a notch as described and hinged to one section l, in combination with a projection upon the end of the frame I', substantially as and for the purpose set forth.

No. 54,719.—THOMAS HANSBROW, Sacramento, Cal.—*Pump.*—May 15, 1866.—The top of the induction tube is enlarged to receive two clack valves; the chamber is placed below the level of the valves so as to prevent the pump cylinder losing its "priming," even though the valves leak.

Claim.—The arrangement of the air vessel G, nozzle *d*, side pipe C, valve chest F, valves *b b' c c'*, passage D, and cylinder A, operating in the manner and for the purpose herein described.

No. 54,720.—WILLIAM B. HARRIS, Springfield, Mass.—*Paper Collar.*—May 15, 1866.—The engagement of the lip at the end of the collar in the elongated button-hole maintains the consecutiveness of the lower edge at the junction.

Claim.—The notches or slits in the ends of the collar and the lengthened button-holes or slits, arranged in such a manner that when the notches are placed in the lengthened button-holes or slits, they will prevent the upper edge of the collar from turning in toward the neck, and by slipping along the button-holes, allow the collar to adjust itself exactly to the neck, as herein described.

No. 54,721.—DANIEL HEILIG, Chicago, Ill.—*Cooking Stove.*—May 15, 1866.—Water chambers line the inside of the wooden walls and the sheet-iron oven is removable from the fire chamber.

Claim.—The arrangement and combination of the oven D with the fire box P, when constructed as described and used for the purpose set forth.

No. 54,722.—FREDERICK HENKE, Scranton, Penn.—*Steam-gauge Cock.*—May 15, 1866.—The valve stem has an enlargement which acts as a piston in a chamber of the faucet. The valve is held to its seat by pressure of steam and a coiled spring.

Claim.—A gauge cock, framed by combining the piston I with the lever J, valve D, and spring G, substantially as described and for the purpose set forth.

No. 54,723.—S. R. HIGGINS, Parma, Mich.—*Hay Loader.*—May 15, 1866.—The gathering rake is attached by arms to a turn-table on a wheeled frame and operated by hoisting tackle. The hoisting rope passes through a slot, and is locked, when required, by a slide connected to the arbor of the caster wheel.

Claim.—First, the rake I, connected by joints to arms J J, which are secured by joints to a turn-table D on a mounted frame, and arranged with a hoisting tackle and upright on the turn-table to operate in the manner substantially as and for the purpose set forth.

Second, the sliding plate K, provided with an oblong slot g, through which the hoisting rope H passes, and connected with the lever B of the arbor of the caster pulley b, to serve as a clutch for the hoisting rope, substantially as described.

No. 54,724.—JOSEPH J. HILL, Xenia, Ohio.—*Potato Harvester.*—May 15, 1866.—The roller is made in two parts, and precedes the vertically-adjustable share and shovel, over which latter the toothed cylinder revolves.

Claim.—First, the double or compound roller A A', constructed and operating as described, in connection with the fifth wheel, for the purpose set forth.

Second, the pivoted and adjustable plough and cutting coulter, constructed and operating substantially as described and for the purposes set forth.

Third, operating the digging shovel H through the leverage frame J J K.

Fourth, constructing the toothed cylinder I of sections or disks, with recesses h' to hold the teeth, as described, for the purposes set forth.

No. 54,725.—CHARLES HIRES, Salem, N. J.—*Grate.*—May 15, 1866.—The upper edge of the grate rim has an annular projection, which fits a corresponding recess in the lower surface of the fire pot.

Claim.—The raised sloped annular portion d' on the rim of the grate D, and the corresponding annular groove c^1 c^2 in the bottom edge of the cylinder C, the said parts being arranged together so as to operate substantially as and for the purpose described.

No. 54,726.—JOHN A. HITCHINGS, Denver City, Colorado Territory.—*Machinery for Separating Metals from Ores.*—May 15, 1866.—A continuous series of operations in a descending grade, consisting of stamping mill; crushing rollers; roasting furnace, with a cold-water supply tank; an amalgamator, with mercurial fumes; an arastra, and an amalgamator.

Claim.—First, the arrangement of the mortars, rounded stamps, and slotted connecting openings, substantially as and for the purpose set forth.

Second, the arrangement, as a sequence to the subject-matter of the first claim, of the rollers N in the trough for the further cominution of the ore received from the stamps.

Third, a basin-shaped revolving roasting plate, Fig. 4, provided with scrapers, as described, and with a vessel containing salts of soda, alum, or potash, which are intermingled with the ground ore.

Fourth, the arrangement, with the revolving roaster, of the cold-water tank which receives the heated ore therefrom, as described.

Fifth, the arrangement of the roasting plate, cold-water bath, amalgamator, and arastra, as described.

Sixth, the quicksilver-coated copper amalgamator, acting as a final means of arresting non-mercurialized metals, arranged and operated as described.

Seventh, the condenser, arranged as described, consisting of the flue Z passing through the water chamber, the discharge pipe h, and the chamber b, the shower bath c, and the exit flue f.

No. 54,727.—M. D. HOTCHKISS, Sheboygan Falls, Wis.—*Clothes Dryer.*—May 15, 1866.—The arms are pivoted on a vertical spindle, and are supported, when expanded, each on its own step of the arc.

Claim.—In combination with the swinging arms of a rack for holding clothes while drying, the supporting arc D, substantially as set forth.

Also, in combination with the swinging arms of a rack for holding clothes while drying, the spring E, substantially as and for the purpose set forth.

No. 54,728.—CHARLES HOWARD, New York, N. Y.—*Bayonet Attachment.*—May 15, 1866.—
An open spring ring moves in a groove around the tubular shank of the bayonet; a slot is cut through at one point of the groove, and a corresponding stud is formed on the inner surface of the ring to engage with a hole in the barrel.

Claim.—First, the groove around the centre of the hilt or tubular part of the bayonet, with a slot cut through for the purpose of supporting and keeping in place an open spring-ring fastener.

Second, the open spring ring, a segment of which is made thick, so as to drop into a notch of the barrel to fasten the bayonet, for the purpose and substantially as above set forth.

No. 54,729.—ELBRIDGE HOWE, Malboro', Mass.—*Tethering Halter Apparatus.*—May 15, 1866; antedated April 13, 1866.—To prevent the coiling of the rope around the legs of an animal; the tether is strung with thimbles which resist a short flexure.

Claim.—The tethering apparatus, substantially as described, as composed of the halter and the series of cylinders as applied together, as and for the purpose specified.

No. 54,730.—DEWITT C. HOWELL, Goshen, N. Y.—*Steam Generator.*—May 15, 1866.—
The fuel is introduced by a piston beneath the furnace; the gaseous products are conducted through the boiler, mingled with the steam, and passed to the cylinder. Equilibrium of pressure in the steam space and fire box is attained by a valved connection.

Claim.—First, the conductor G G′ g g′, constructed and connected with the fire box and arranged with reference to the fire box and surrounding portion of the boiler, substantially as herein described for the purpose set forth.

Second, the fuel reservoir D, piston D′, and slide valve M, applied in combination with the grate C, or fire bed, substantially as herein described.

Third, the pipe P and valve v′, applied substantially as and for the purpose herein set forth.

No. 54,731.—LAFAYETTE HUNTOON, Milford, Mass.—*Steam Engine.*—May 15, 1866.—
Two cranks on the same shaft and of different radii are respectively attached to pistons of varying diameter working in cylinders where the steam is used directly in one and afterwards expansively in the other.

Claim.—As my invention the improved steam engine, constructed substantially in manner and so to operate as hereinbefore described, that is to say, as composed of the main and expansion cylinders A B, of different capacities or lengths and diameters, as stated, their separate pistons C D, piston rods E F, connecting rods G H, and cranks I K, steam chests V W, valves T U, and steam passages and valve-operating mechanism, arranged together and applied to a driving shaft, substantially as explained.

No. 54,732.—DANIEL HUSSEY, Nashua, N. H.—*Drawing Roller.*—May 15, 1866.—The two shafts are geared together; the upper roller is tubular and revolves with its shaft. The sleeve between the rollers enters sockets in their ends.

Claim.—First, the combination of the gears c d, or their equivalents, with the tubular top roller A, the weight shaft a, and the bottom roller B, or the shaft b, thereof, the whole being substantially as and for the purpose set forth.

Second, the combination of the weight-bearing sleeve C with the shaft a and the tubular top rollers A A, thereof.

Third, the arrangement of the sleeve C′ with respect to each of the top rollers, viz., so that one shall extend into the other a short distance and be encompassed by it, substantially as and for the purpose as specified.

No. 54,733.—WILLIAM ALLEN INGALLS, Chicago, Ill.—*Screw for Stools, &c.*—May 15, 1866.—The pattern screw is twisted out of the mould instead of making a two-part mould, and leaving a fin on the screw along the line of section of the mould.

Claim.—The above-described screw as an article of manufacture.

No. 54,734.—THOMAS JAMES, Baltimore, Md., assignor to PRUDENCIO DE MURGUIONDO, same place.—*Brick Mould.*—May 15, 1866.—The brick mould, follower, or piston are either or each made of or faced with glass or its equivalent.

Claim.—A mould for forming bricks or other articles of clay, made wholly or in part of vitreous material, or faced or lined with vitreous material for the purposes set forth.

No. 54,735.—JOHN P. JAMISON, New York, N. Y.—*Crimping Machine.*—May 15, 1866.—
The gate works in guides and is actuated by rack and pinion; it carries four jaws at the ends of as many springs and pushes the leather upon the "former" beneath.

Claim.—First, the four springs D and E, made of metal or other substances, adjusted, arranged, and secured to the cross-piece C, in the manner hereinabove substantially set forth and described.

Second, the four brass jaws or stretchers marked F, in combination with said four springs D and E secured to cross-bar C, as regulating and providing for the self-adjustment thereof, as herein shown and described.

No. 54,736.—JOHN M. JOHNSON, New York, N. Y.—*Button.*—May 15, 1866.—The button has a tubular shank and an enclosing spring. The barbed stud pierces the fabric, enters the button, and is secured by a partial rotation.

Claim.—The barbed piercing stud C, in combination with the ridged tubular shank of the button A and the spring D, and operating in the manner and for the purpose herein specified.

No. 54,737.—WARREN JONES, Berlin, Wis.—*Bed Bottom.*—May 15, 1866.—The camber of the slats is maintained by central coils and girder braces beneath. The bars under the ends of the slats rest on coiled springs.

Claim.—The combination of the braces E with the springs C C'¹ C'², slat B, bars D, and frame, the whole being arranged to operate substantially in the manner and for the purpose set forth.

No. 54,738.—CHARLES KANE, Pittsburg, Penn.—*Die for Heading Bolts.*—May 15, 1866.—The head of the staving die fits the space enclosed by the pressing dies, the space being a parallelogram in cross section.

Claim.—The heading or staving die M, when its cross section is a parallelogram in combination and made to correspond with and fits a rectangular space left between the pressing dies, the dimension of which is greater from ledge to ledge as above described.

No. 54,739.—J. B. KIBLER, Girard, Penn.—*Sand Pump.*—May 16, 1866.—The stem of the upper valve is prolonged through the seat of the lower valve and below the cylinder; striking the bottom it raises the upper valve and gives freedom to the lower valve to open to allow detritus to pass. When raised the lower valve seat is dropped to discharge the contents.

Claim.—First, the movable valve seat J in the bottom of a sand pump, substantially as described.

Second, the stopper or plug valve C' in the top of the pump in combination with a spindle extending downwards below the ends of the supports of the pump, substantially as described.

No. 54,740.—JACOB KINSER, Pittsburg, Penn.—*Handle for Coffee, Spice, and other Small Mills*—May 15, 1866.—The arm is cast on to the rivet after the knob is attached.

Claim.—The cast-iron handle for coffee, spice, and other small mills, substantially as shown and described as an article of manufacture.

No. 54,741.—JOHANN CHRISTIAN KNOEPKE, Philadelphia, Penn.—*Wash Basin.*—May 15, 1866.—The wash basin has a soap dish inside and a ring foot.

Claim.—As a new article of manufacture a tin wash bowl made of a combination of the bowl B, curved strip C, and ring E, in the manner and for the purpose described.

No. 54,742.—LUCIUS J. KNOWLES, Warren, Mass.—*Loom for Weaving Tape, Ribbon, &c.*—May 15, 1866.—The devices operate the lay rack, which gives motion to the various shuttles used in looms for weaving narrow goods; the object is to secure its easy action and free it from sudden stoppages.

Claim.—A combination composed not only of the straps e e, and their guide wheels as applied to the lay and its rack, substantially as described, but of the crank n, the shaft o, and the bevel gears t and u, or their mechanical equivalent or equivalents, operated by the cranked shaft of the lay.

Also, the straps e e, and the guide wheels f f g g, arranged and combined together, and with the lay and the rack, substantially in manner or so as to be operated by a crank or its equivalent, as specified.

Also, the arrangement of the shaft o, and its crank n, and gear t, with the crank shaft C, and its pinion n, and the straps e e applied to the lay and its rack, substantially as specified.

No. 54,743.—T. T. S. LAIDLEY, United States Army, and C. A. EMERY, Springfield, Mass.—*Breech-loading Fire-arm.*—May 15, 1866.—The independent, vibrating locking brace which sustains the breech piece in firing position is unlocked by a cam and thrown back by a pawl actuated by the hammer.

Claim.—First, locking the movable breech piece by means of a piece independent and separate from the hammer or tumbler, but moving on the same axis, and the same operated or brought into place as soon as the breech piece is closed, irrespective of the extent to which it may have been opened.

Second, the arrangement of a cam on the same axis of the breech piece, for the purpose of throwing back the locking piece, so that by means of a stop on the cam the breech piece may be opened by the simple motion of the said cam.

Third, the arrangement of a pawl attached to the locking brace and operated by the hammer or tumbler for the purpose of throwing back the said locking piece. The whole arranged and operated substantially as and for the purpose specified.

No. 54,744.—JAMES LEE, Milwaukie, Wis.—*Breech-loading Fire-arm.*—May 15, 1866.—The lever has studs which operate in grooves of the vertically sliding breech block, and a hook on its forward end to extract the cartridge shell.

Claim.—First, the breech block C, provided with the grooves c and f, substantially as and for the purpose set forth.

Second, providing the lever D with the studs m and l, to operate in connection with the breech block C, as shown and described.

Third, the lever D, having its front end provided with the lip or hook o, and arranged to operate within the chamber of the gun, substantially as and for the purpose set forth.

No. 54,745.—RODMAN LOVETT, Canton, Ohio.—*Cow and Sheep Rack*—May 15, 1866.—The parts described are adapted and fitted to be taken apart and put together without nails.

Claim.—First, the section A A', hay racks B B, revolving troughs D d, and guide board C, arranged substantially as and for the purposes set forth.

Second, the adjustable hay racks B B when arranged and used for the purposes set forth.

Third, the adjustable trough F, in combination with the sections A A', when used as and for the purposes specified.

No. 54,746.—OSBORNE MACDANIEL, New York, N. Y.—*Evaporating Solutions of Salt, Sugar, &c.*—May 15, 1866.—The cloth is attached to folding frames which are submerged when depressed and collapsed, and which expose it to the air when raised and extended.

Claim.—First, in connection with evaporating vats the vertical sliding and folding frames on which are fastened cloths, mats, or boards, so arranged as to fold over and upon each other at the bottom of the vat when the frames are lowered, and to hang in the air when the frames are raised, substantially as and for the purpose herein described.

Second, the elevated distributing troughs o o, and the return troughs v v, in combination with the tanks D and E, and the evaporating apparatus suspended in the vats B B, for the purpose and substantially as herein described.

No. 54,747.—A. D. MANLEY, Washington, Mich.—*Dumping Wagon.*—May 15, 1866.—The box is divided into pivoted sections which are capable of vibrating to dump their loads simultaneously or as required.

Claim.—The sliding bar r, with its pins w, operating in combination with the spring bars m, and the swinging boxes G H I, for joint or independent action in the manner and for the purpose herein specified.

No. 54,748.—W. C. MARSHALL, New York, N. Y.—*Mode of Extinguishing Fires.*—May 15, 1866.—Perforated pipes traverse the interior of the building, and distinct sets ramify from stand-pipes to which hose are connected by couplings on the exterior of the building.

Claim.—The arrangement of several series of perforated pipes a b c, (one or more for each floor of a building,) in combination with nipples a' b' c', situated on the outside of the building, and each communicating with a distinct series of pipes, substantially as and for the purpose set forth.

No. 54,749.—LAWRENCE MASON, Turin, N. Y.—*Tenoning Machine.*—May 15, 1866.—The knives on the reciprocating bars have saws on their edges to cut the shoulder, while the knives cut the tenon; the bars are adjustable for tenons of different thicknesses.

Claim.—The arrangement of the vertically moving frame B* B*, adjustable bars D* D*, and springs E* E*, saws F* F*, cutters G* G*, gauged by means of the tappets i* i*, and arms N*, operating substantially as described and represented.

No. 54,750.—LAWRENCE MASON, Turin, N. Y.—*Machine for Boring Wagon Hubs.*—May 15, 1866.—The hub is chucked in the angular jaws, which are simultaneously approached by the treadle lever and connecting rods. The angle of direction of the cutter mandrel is adjusted to give the required taper to the bore.

Claim.—First, the arrangement in the hub-boring machine of the adjustable cutter mandrel B, with its plate S, and set screw T, and the clamping and centering plates D D', with their triangular openings F F', constructed and operated as described.

Second, the combination and arrangement of the frame M, arbor K, slotted dog N, arm P, collar Q, disk R, shafts B, and adjustable plates D D', substantially as and for the purpose described.

No. 54,751.—WILLIAM McCORMICK, Muscatine, Iowa.—*Cultivator.*—May 15, 1866.—Explained by the claim.

Claim.—First, the cast iron standards Z, provided at their upper ends with lips i, to fit over the upper and lower edges of the plough beams I, in order to avoid the use of the braces to retain the standards in position, substantially as set forth.

Second, the arrangement of the plough beams I, as shown, to wit: the rear ends being provided with rods J, to fit into staples or guides K, at the lower parts of uprights G, at the rear of the frame A, and their front ends connected by chains R to uprights G, which are attached by universal joints g to the front part of the frame A, the uprights G passing down between rollers M, and the front bent parts a^* of one of the clevises L of each beam, substantially as and for the purpose specified.

Third, the attaching of the doubletree Q to the front ends of the beams I, when said beams are connected to the bar I*, through the medium of the uprights O, in the manner substantially as set forth.

Fourth, the arrangement of the treadles T T, with the beams I I, and levers W, substantially as and for the purpose specified.

No. 54,752.—A. R. McNair, New York, N. Y.—*Brick Machine.*—May 15, 1866.—The reciprocating plunger, actuated by the eccentric, feeds the moulds consecutively from the bottom of the pile to the press.

Claim.—First, feeding the moulds to a brick machine in a pile from the front side, substantially as specified.

Second, the eccentric wheels F and L L, in combination with shaft D, levers and hinges O, presser H, sliding boards E and E, as described, constructed and arranged, substantially as and for the purpose set forth.

No. 54,753.—Joseph Montgomery, Harrisburg, Penn.—*Axle Box.*—May 15, 1866.—The lid vibrates on a hinge and is secured above by a disk on the axis of the handle and engaging a recess on the upper edge of the box. The floor of the oil chamber is inclined upward and outward and has a loose perforated frame to hold the packing.

Claim—First, the journal box C, arranged and constructed substantially as described and for the purpose set forth.

Second, the lid A, hinged below and secured by handle B and jamming cam b, constructed and operating substantially as described and for the purpose set forth.

Third, the loose or hinged rack frame C', inside of the box, constructed and operating substantially as described and for the purpose set forth.

No. 54,754.—E. D. Montrose, Nashua, Iowa.—*Portable Fence.*—May 15, 1866.—The chain is composed of book links and each loop embraces a vertical picket.

Claim.—A portable fence constructed of two or more parallel chains A A, composed of links a, bent in the form and around the pickets B, in the manner substantially as herein set forth.

No. 54,755.—F. B. Morse, Milwaukee, Wis.—*Carriage Seat.*—May 15, 1866.—The rim-back or buggy top is made detachable by the disengagement of the post from the slots on the withdrawal of the lever fastenings.

Claim.—The lever fastenings G, when used in combination with the seat rail posts F, for the purpose specified and arranged, and operating substantially as described.

No. 54,756.—John Mott, Danville, Cal.—*Plough.*—May 15, 1866.—A right and a left hand plough are attached to a horizontal axis; either may be brought into working position, and the idler is inverted beneath the beam.

Claim.—First, as a new invention, the use of a double plough revolving upon a horizontal axis L, the two ploughs being placed one over the other in an inverted position, substantially as described and for the purpose set forth.

Second, the clasps M and N, for hinging the main rod L to the standards C and I, and the adjustment with washers (or their equivalents) of the clasps N, for turning the plough more or less to land, substantially as described.

Third, the set-screws s and s', placed in the upper end of the standard of the ploughs for steadying them and keeping off the land-side from the standard J', substantially as described.

No. 54,757.—Isaac Newton, New York, N. Y.—*Superheating Apparatus.*—May 15, 1866.—The steam from one division of the chamber passes through the annular space between the pipes, and then by the central pipe to the other division, being heated in its passage by the surrounding fire in the chamber. The rear ends of the tubes are secured together and the forward ends to their respective tube sheets.

Claim—First, the arrangement of a superheating apparatus composed of a system of tubes and a circulating chamber within an enlarged casing or box N O P Q, in relation to the outlet of the boiler and the base of the chimney, substantially as and for the purposes specified.

Second, bracing the circulating chambers A B C D, by securing one or more of the inner tubes to their corresponding outer ones at their ends H F, and their reverse ends respectively to the side C D, and dividing plate I K, of the circulating chamber, substantially as described.

Third, the arrangement of the superheating tubes within the casing N O P Q, in relation to the circulating chamber and to a door A' B', in the said casing, substantially as herein specified to provide for the removal and replacement of any of the said tubes.

No. 54,758.—O. W. NOBLE, Darlingson, Wis.—*Machine for Making Eave Troughs.*—May 15, 1866.—The rounded edge of the trough is secured by a slide to the groove of the roller; the latter is then rotated above the forming board, which is raised by a treadle and the plate assumes the shape of the forming roller.

Claim.—First, the slide E, in combination with the roller B, arranged and operating substantially as specified.

Second, the forming board F, in combination with the treadle G and roller B, substantially as and for the purpose specified.

Third, the combination and arrangement of the roller provided with a permanent rib D, and slide E, with the forming board F, and treadle G, substantially as specified.

No. 54,759.—EDMUND OLDHAM, Brooklyn, N. Y.—*Pentographic Machine.*—May 15, 1866.—The transparent tracer is an eye-glass with a dot on its surface and the motion is communicated to a diamond pointer which rests upon the plate. By the usual pentographic arrangement it is capable of adjustment for changing the scale; it has also an adjustment for parallel or "macbine" ruling.

Claim.—First, in pentographic machines the use of an eye-glass with a speck or dot on its face for the purpose of tracing outlines of plane or other surfaces, sustantially as above set forth.

Second, the means above described for giving independent movements to the diamond pointer, to wit: the bar E, the fusee pinion F, and its shaft G, and the toothed wheel H², on which the carriage H rests, substantially as above set forth.

Third, the ruling device R S T, constructed and operated substantially as set forth.

No. 54,760.—L. H. OLMSTED, Stamford, Conn.—*Chuck.*—May 15, 1866.—The nut has a conical opening in the end which operates against the inclined backs of the jaws, to clamp them upon the drill; when relieved they are expanded by springs.

Claim.—Making a chuck, substantially as shown and described.

No. 54,761.—S. B. PANGBORN and G. H. GRIFFIN, Boston, Mass.—*Egg Beater.*—May 15, 1866.—A vertical shaft, with perforated, serrated-edged flanges, rotates in the cylindrical vessel as the latter is swung in the hand, causing the weighted crank-arm to revolve.

Claim.—The improved egg-beater, as composed of the vessel or pot A, the shaft e, and its wings or beaters g g g, and the arm or projection e', and the weight f, fixed to such arm, the whole being arranged in manner and as to operate substantially as and for the purposes as hereinbefore specified.

Also, the combination of the centrifugal guard h with the vessel A and its cover, and with the shaft, its bearers and the arm and weight applied to such vessel, they being substantially as specified.

No. 54,762.—V. P. PARKHURST, Templeton, Mass.—*Chair.*—May 15, 1866.—The chair back is pivoted between the posts, and accommodates itself to the inclination of the back of the sitter, resuming its normal verticality when pressure is removed.

Claim.—The combination of the plates E, pivots a, ferrules F, caps G, and pins g, arranged relatively with the back D, and posts C, of the chair A, and operating in the manner and for the purpose herein specified.

No. 54,763.—EDWIN PARMELE and R. N. PATTERSON, Davenport, Iowa.—*Cultivator.*—May 15, 1866.—The slotted shovel standards are moved laterally by treadles, and vertically by levers in connection with a pivoted roller.

Claim.—First, applying shovel standards to a carriage in such manner that they are allowed to rise or fall bodily or independently of each other, and also of being moved laterally together by the feet of the attendant, substantially as described.

Second, The slotted shovel standards D D, in combination with the treadles r r, and the driver's seat, substantially as described.

Third, connecting the standards D D to the roller e, in such manner that their lower ends can be swung forward and upward in combination with contrivances for allowing said shovel standards to be adjusted laterally, and also moved up and down in a direction with their length, substantially as described.

Fourth, so constructing the two treadles r r, that each one serves as a lever, and also as a means by which both standards can be moved simultaneously either to the right or to the left, substantially as described.

No. 54,764.—JAMES N. PEASE, Panama, N. Y., assignor to METROPOLITAN WASHING MACHINE COMPANY.—*Clothes Wringer.*—May 15, 1866.—Explained by the claim.

Claim.—The improved method of gearing wringer rolls by doubling or multiplying the toothed wheels in the manner hereinbefore described, that is to say, by the employment upon each of the shafts geared together of two or more sets of teeth of like number, configuration or construction, each set being situate in different planes, but in such a fixed relation in respect to the other set or sets, that the spaces between the teeth of either set shall be equally divided by the teeth of the other.

No. 54,765.—DANIEL PETERS Eaton, Ohio —*Broom Head.*—May 15, 1866.—The jaw plates on each side are clamped against the cap by a bolt with a threaded tube and nut.

Claim—The tube E, which forms a guide and socket for the bolt D, in its described combination with and relation to the said bolt, the wrapper B, the handle A, and bars F F.

No. 54,766.—OLIVER S. PETTIT, Brooklyn, N. Y.—*Water Closet.*—May 15, 1866.—A valve in the discharge pipe is opened by pressure on the seat and closed by a counterbalance weight when the sitter rises; the simultaneous operation of a second valve admits water.

Claim.—The combination of the valve H, rod I, and lever J, with the discharge pipe G, and cover B, of a water closet, substantially as described and for the purpose set forth.

No. 54,767.—BARTON PICKERING, Dayton, Óhio.—*Ladder.*—May 15, 1866.—One portion of the double ladder has a horizontal platform whose catches engage the rounds on the other portion to secure the position when spread, and serving to lock the two when extended.

Claim.—The platform C, provided with catches *n n*, the orifices *o*, and bent round *f'*, in combination with the ladder A B, having projections on the part Δ, substantially as described and for the purpose set forth.

No. 54,768.—ALEXANDER POLLOCK, New York, N. Y.—*Safety Valve.*—May 15, 1866.— The valve arrangements are closed for security against tampering with the weight; but the valve may be raised and steam blown off at discretion. The devices are explained in the claims.

Claim.—First, the valve chest C, and its two valve seats *e f*, the double puppet valve E F, and the outlet D, arranged in relation with each other and with the steam box A, communicating with the steam space of the boiler, substantially as and for the purpose herein specified.

Second, the arrangement of the safety valve, the weight H, or its equivalent, and the lever J, through which the weight or spring acts upon the valve, all within a steam box A, which is always in communication with the boiler, substantially as herein described.

Third, the tripping lever M, outside of the steam box A, and the rod L, applied in combination with loaded rod I, connected with the valve lever J, within the steam box A, substantially as herein described, whereby the lifting or tripping of the enclosed valve for blowing off steam is provided for, without permitting the load on the said valve to be increased, or the said valve to be secured in a closed condition.

Fourth, the arrangement of the valve chest C, attached to steam box A, by the removable plate D', whereby by the removal of the pin *n*, the valves and valve chest lever J may all be removed from the steam box with the said plate D.

Fifth, the bar Q, staple or staples R, pin or pins *i*, and seal or seals *j*, applied and arranged in relation with each other and with the steam box A, and door P, substantially as and for the purpose herein set forth.

Sixth, the double puppet valve provided between its heads with a chamber *g*, which is open to the steam space in which the valve chest is situated, substantially as and for the purpose herein specified.

No. 54,769.—G. H. POOL, New York, N. Y.—*Churn Dasher.*—May 15, 1866.—The vertically reciprocating shaft has a series of dashers and a disk which drives air before it to act upon the cream. The air is introduced through the double cover.

Claim.—First, the dasher B, constructed substantially as described and for the purpose set forth.

Second, the combination with the dasher of the disk or fan D, substantially as described and for the purpose set forth.

Third, the combination of the double cover E F, constructed as described with the dasher and with the churn, substantially as and for the purpose set forth.

Fourth, the combination of the crank wheel I, arm H, and guides N, with each other, with the dasher B, cover G E F, and frame J. substantially as described and for the purpose set forth.

No. 54,770.—THOMAS B. PURVES, Greenbush, N. Y.—*Railroad Switch.*—May 15, 1866.— The clamp pieces are fastened to the rails, and pivoted bridge pieces form inclined planes for the wheels to climb.

Claim.—The construction of a portable switch, for the purpose of placing cars upon a railroad track by the combination of the bar *a b c*, secured as described to the rail, with a movable limb *c f g'*, formed as shown and hinged or pivoted to it over the centre of the rail, substantially as described, and for the purposes set forth in this specification.

No. 54,771.—THOMAS H. QUICK, New York, N. Y.—*Purifying Bone-black.*—May 15, 1866.—To remove the dust from re-burnt animal charcoal, it is fed into a blast flue, where the lighter particles are driven off.

Claim.—First, the flue G for receiving the descending currents of animal coal or bone-black, constructed substantially as and for the purpose above described, with the diaphragm B and shelf C in combination with the feed pipe A.

Second, the combination of the flue G with a Hancock steam blower, arranged substantially as and for the purpose above described.

No. 54,772.—JACOB REEDY, Toledo, Iowa.—*Portable Fence.*—May 15, 1866.—The dovetailed ends of the boards are received in receding mortises and clamped therein by wedges.

Claim.—Constructing the posts A with a double beveled mortise, and the boards B with corresponding notches in their lower sides, the fence being secured by such notches and by keys driven above the boards, substantially in the manner set forth.

No. 54,773.—HELEN M. REMINGTON, Springfield, Mass.—*Ladies' Paper Undersleeves.*—May 15, 1866.—The fibres of the two thicknesses of tissue paper are laid transversely; the paper has an exterior sizing of glue and an interposed layer of lard and white wax, dissolved in alcohol.

Claim.—First, as a new article of manufacture, a paper undersleeve, substantially as described.

Second, the material composed of two thicknesses of paper, with an oily substance, to render it water-proof and a suitable preparation to prevent this oil from striking through to the outside, substantially as described.

No. 54,774.—ALBERT L. RICE, Worcester, Mass.—*Attaching Circular Saws to Arbors.*—May 15, 1866.—As the annular bush slips upon the mandrel, its conical face penetrates the central orifice in the saw, and maintains the concentricity thereof; an elastic packing intervenes between the bush and the end collar.

Claim.—First, the improved mode herein described, of attaching and adjusting circular saws to their arbors by the combination with the stationary and movable recessed collars as herein described, of a conical bush, sliding on the saw arbor, and fitting in the recess formed in the movable collar under such an arrangement that the saw may be simultaneously adjusted, and secured on its arbor, as shown and set forth.

Second, in combination with the stationary and movable collar and conical bush fitting in a recess in the movable collar as described, the employment of a rubber or other elastic packing or washer, as and for the purposes herein set forth.

No. 54,775.—WILLIAM F. RIPPON, Providence, R. I.—*Axle Box.*—May 15, 1866.—A reservoir in the box above the axle has a movable cover and filtering tubes, which discharge the oil through the bottom and upon the journal in its bearing.

Claim.—The independent oil reservoir A, provided with a filter c, as described, in combination with the journal bearing D, substantially as described.

No. 54,776.—ALEXANDER M. ROBINSON, Boston, Mass., assignor to himself and HOWARD TILDEN.—*Egg Beater.*—May 15, 1866.—Serrated plates project inward from the sides, and the funnel-shaped ends, respectively salient and depressed, are closed by a cork and cover.

Claim.—The funnel-shaped ends in combination with the ribs soldered to the sides of the case.

No. 54,777.—GEORGE W. RODEBAUGH, Toledo, Ohio.—*Head Block for Saw-mills.*—May 15, 1866.—The main compound driving wheel gears into pinions on distinct shafts, which operate the knees against which the log is dogged; one or both ends can be set. The upper dog is attached to a shaft driving vertically in the knee standard. The lower dog catches the log beneath and is actuated by a lever.

Claim.—First, setting the log to the saw at either end or at both simultaneously by means of the independent shafts E E, when operated by compound cog wheel I and adjustable pinions K K, substantially in the manner and for the purposes herein specified.

Second, the construction of the dog, consisting of the bit a attached to a shank a^4, which fits and slides within the vertical groove of the knee, substantially as described and represented.

Third, operating the pivoted dog a' by means of the lever b and pins e e, as and for the purposes set forth.

No. 54,778.—ARTEMAS ROGERS, Painesville, Ohio.—*Extension Fruit Ladder.*—May 15, 1866.—One portion of the ladder slides upon the other, and the extension upper section is maintained at its elevation by a spring-pawl which engages the rounds.

Claim.—The lever I, brace D, in combination with the ladder A and B, loops C, arranged and operated in the manner and for the purpose set forth.

No. 54,779.—JONAS ROUSE, Dowagiac, Mich.—*Bed Bottom.*—May 15, 1866.—The upper slats rest upon the secondary camber slats, whose ends are supported on the bed rails.

Claim.—The combination of the slats C and secondary slat D, arranged to operate substantially as specified.

. No. 54,780.—STEPHEN P. RUGGLES, Boston, Mass.—*Steam Generator.*—May 15, 1866.—The steam generating chamber occupies a position at the lower part of the vessel, and receives water therefrom in thin films, which are converted into steam and removed by a pipe.

Claim.—Combining with an open boiler or reservoir of water a steam generator within said boiler, and at or near the bottom of the column of water therein, substantially as and for the purpose herein described.

No. 54,781.—WILLIAM H. SARGENT, Weymouth, Mass.—*Washing Machine.*—May 15, 1866.—The stationary rubber is flanked by anti-friction rollers which have their bearings in an open handled frame.

Claim.—The combination and arrangement of the stationary rubber *f*, the friction rollers *e e*, and their carrying frame A, the carrying frame A being constructed with its two opposite sides *b b* inclined toward the handle, and made with openings through its ends to enable the water and sapenaceous matter to flow freely through them into and out of the case while the machine may be in use.

No. 54,782.—LEONARD SAWYER, South Amesbury, Mass.—*Attaching Props to Carriage Bows.*—May 15, 1866.—The arrangement described is to prevent the accidental unscrewing of the prop from the plate.

Claim.—The plate B, provided with the square tubular projection C, and the prop or arm D, provided with a screw to fit into an internal screw thread in the projection C, in combination with the collar F, having a square interior and fitted on a square portion on the prop or arm and on the projection C, substantially as and for the purpose specified.

No. 54,783.—STEPHEN SCOTTON, Richmond, Ind.—*Device for Shooting Gravel at Cows upon Railroad Tracks.*—May 15, 1866.—The tube is charged with gravel, which is projected at the animals by a blast of steam.

Claim.—The construction and operation of the valve C in combination with the hopper A and tube B, or their equivalents, for the purposes described.

No. 54,784.—JACOB SEIBEL, Manlius, Ill.—*Harvester.*—May 15, 1866; antedated May 1, 1866.—The cutting apparatus and platform frame are supported and drawn forward by a jointed drag-bar attached, at its forward end, to the front cross-timber of the frame, and are braced in their proper relation to the frame by the shaft which communicates motion to both the cutters and the endless belt rake; the said shaft is jointed to allow the platform to conform to the surface of the ground. Any undue strain upon this shaft, resulting from this freedom of motion of the platform and cutting apparatus, is obviated by means of an arched plate centering in this shaft, attached to the platform and suspended upon a bar rigidly attached to the main frame.

Claim.—First, the arrangement of the jointed shaft H with the two shafts J N, with their attachments, operating substantially as and for the purposes herein specified and shown.

Second, suspending the platform upon the main frame by means of the bar R and the bent support W, attached to the platform, all being arranged and operating as and for the purposes specified and shown.

No. 54,785.—PERMELIA E. SHEFFIELD, Pontiac, Mich.—*Hoop Skirt.*—May 15, 1866.—Explained by the claim.

Claim.—Hoop skirts of steel wire, having an apron A extending down the front of the skirt covering the wires, substantially as and for the purpose set forth.

No. 54,786.—JOHN Y. SMITH, Alexandria, Va., assignor to himself and HERMAN HAUPT, Philadelphia, Penn.—*Triple-ribbed Metallic Bar.*—May 15, 1866.—These bars, applicable to drills, reamers, &c., are rolled by means of two grooved rollers.

Claim.—The manufacture of the metallic triple-ribbed bar in the manner and by the means substantially as herein described.

No. 54,787.—MARSHALL SMITH, St. Louis, Mo.—*Mail Pouch or Box.*—May 15, 1866.—The letters are confined in position in the mail box by a compressing follower, which is held to its place by racks.

Claim.—First, the combination of the box C D with the toothed ways *e f* and sliding follower or compressor A, or corrugated ways *h* and followers or compressor B, or their equivalent devices, when employed as and for the purpose set forth.

Second, attaching the reversible label M to the box C, substantially as described.

Third, interposing a rubber or any other suitable packing *z* between the different parts of wooden boxes, so as to make them water-tight, for the conveyance of mails.

Fourth, a box fitted with internal partitions and followers, arranged to separate parcels of mail matter, while being conveyed in the same box, the whole used as a package for enclosing mails for and in transportation, substantially as set forth.

No. 54,788.—EDWARD SPENCER, Philadelphia, Penn.—*Flour Sifter.*—May 15, 1866; antedated May 3, 1866.—The balls are journalled on a bar suspended from the axis of the rotary sieve, and mash the lumps of flour therein.

Claim.—The employment of the balls E in combination with the rotary cylindrical sieve A B, suspended in a suitable case C D, the whole being constructed and arranged together, so as to operate substantially as and for the purpose described.

No. 54,789.—ARA D. SPRAGUE and ASBURY DOCKUM, Caledonia, Minn —*Harvester Rake.*—May 15, 1866.—The rake is moved by a belt in a track corresponding to the shape of the platform. In connection with the slotted rake head is employed a horizontal shelf deflector on the outer grain-dividing board, arranged to pass between the teeth. A deflecting arm is attached to the rake head above the teeth. The interference of the rake with the grain is prevented till it reaches its initial point. The arrangement of the rake-clearing teeth relative to the slotted rake head, and the application to the latter of the adjustable rake teeth, is identified in the claim.

Claim.—First, the combination and arrangement of the elevated rake head G F and arm H with the shelf J, constructed and operating substantially as and for the purposes herein delineated and shown.

Second, in combination with a slotted rake head F G, the employment of the stationary fingers L, arranged and operating as and for the purposes set forth and shown.

Third, the adjustable teeth *c*, when constructed, arranged, and operating as and for the purposes shown and described.

No. 54,790.—JOHN STEINER and JOHN P. MILLER, Harrisburg, Ohio.—*Combined Harrow and Roller.*—May 15, 1866.—The drag-frame is hinged to the wheeled roller frame; the barrow is pivoted above the drag, and its teeth project through the latter so as to strip off accumulations thereon when the harrow is raised.

Claim.—First, the drag bar A, constructed and arranged with reference to the barrow B, substantially as described and for the purposes described.

Second, the combination of the harrow B, drag-bar A, the wheels D D E E, and the roller C, in the manner substantially as described and for the purposes specified.

No. 54,791.—A. P. STEPHENSON, Equality, Ill.—*Sorghum Stripper.*—May 15, 1866.—The blade tops the stalk, which then enters the neck between the knives; these strip the leaves as they descend, and at the butt the blade cuts the stalk near the ground. •

Claim.—A tool constructed substantially as described, for topping, stripping, and cutting sorghum and other canes.

No. 54,792 —CHRISTOPHER C. STILLMAN, Westerly, R. I.—*Water Wheel.*—May 15, 1866.—The power of the wheel is graduated by the operation of the circular series of gates which admit the water to the buckets. •

Claim —First, furnishing every one of the pitches of a turbine wheel with a separate gate, operated substantially as and for the purpose herein specified.

Second, in combination with a series of gates applied one to each of the pitches of a turbine water-wheel, a cam so applied as to open and close the said gates in pairs, successively, all around the wheel, one on each side of the centre thereof, substantially as herein described.

Third, the combination of the annular grooved cam E, the circular slotted guide frame D, and the stems of the gates, substantially as and for the purpose herein described.

Fourth, the springs *t*, applied in combination with the gates and their stems, substantially as and for the purpose herein described.

No. 54,793.—A. E. STRODEL, New York, N. Y.—*Weather Strip for Doors.*—May 15, 1866.—The weather strip enters a slot in the sill and fastens the door.

Claim.—The plate D, attached to the door and operating as a weather strip by projecting into and receding from the groove *c*, with the movement of the bolt I, in the act of latching or unlatching the door, in combination with suitable means for locking the plate within the groove, substantially in the manner and for the purpose herein set forth.

No. 54,794.—JOHN THOMPSON, Williamsburg, N. Y.—*Portable Fence.*—May 15, 1866.—Explained by the claim.

Claim.—The swivel pickets B and rails C D E, in combination with the slotted eyes G and hooks H, so arranged that the fence will be conformed to any irregularities of the surface, and the posts always retained perpendicularly in position, substantially in the manner herein represented and described.

No. 54,795.—THOMAS WALLACE, Chicago, Ill.—*Hot Air Furnace.*—May 15, 1866; antedated May 1, 1866.—The zig-zag form of the furnace chamber is made by horizontal partitions, and the air pipe follows the continuous, sinuous course, exposed to the surrounding heat.

Claim.—The arrangement of the zig-zag pipe A, and horizontal plates or partitions B, in a furnace, in such a manner that the heat from the fire in the furnace shall be made to follow the pipe, substantially as and for the purpose set forth.

No. 54,796.—H. M. WARD, Stone Church, N. Y.—*Gate.*—May 15, 1866.—The gate slides half-way open and then swings horizontally, the block supporting the gate traversing automatically the downward inclines of the socket.

Claim.—The self-acting joint made up of the socket D and bearing E, connected by the threads d g, and the guide G, with two or more friction rollers h h, when used in combination with the gate A, in such a manner as to balance it when slid half back and then swing it around automatically, substantially as set forth.

No. 54,797.—WILLIAM WEAVER, Phœnixville, Penn.—*Cherry Stoner.*—May 15, 1866.—The disk has curved radial ribs on both sides, and revolves in the centre of the double hopper with inclined channels.

Claim.—First, the disk C, having ribs on both sides, in combination with a double or with two hoppers arranged at opposite sides of the disk, substantially as and for the purpose described.

Second, the combination with the disk C, of one or more inclined channels d, for the purpose specified.

Third, the combination of one or more hoppers with a revolving disk having blunt-edged projections or ribs c on one side, and sharpened ribs c' on the other, for the purpose set forth.

No. 54,798.—W. S. WELLS, New York, N. Y., and S. B. WELLS, Middleburg, N. Y.—*Boiler Feeder.*—May 15, 1866.—A water chamber is connected with the steam boiler; the float in the former, as the height of the water fluctuates, opens or closes a valve by which boiler steam is supplied to the feed pump.

Claim.—The combination with a steam boiler of an auxiliary steam and water chamber, provided with a float so constructed with the valve through which steam is supplied to the feed pump, as to cause the starting up and stopping of the said pump, when the several parts are arranged as herein described.

No. 54,799.—WILLIAM H. WHITE, West River, Md.—*Fan and Parasol.*—May 15, 1866.—Explained by the claim.

Claim.—First, as a new article of manufacture, a fan, sun-shade, or canopy, composed of a hooped frame and wrapper, when the latter is provided with draw cases and draw strings, and constructed and applied in the manner hereinbefore set forth.

Second, as a new article of manufacture, a combination fan and sun-shade, constructed and operating substantially in the manner set forth.

No. 54,800.—DANIEL WILDE, Washington, Iowa.—*Corn Plough.*—May 15, 1866.—The plough beams are adjusted relatively by an extensible shackle bar which connects them; the slotted standards are adjusted by the brace bars, and the seeder is attached to the front of the frame by a hinge joint, perforated strap, and pin.

Claim.—First, adjusting the shovel stakes and shovels, by means of the slots q and bars r, as and for the purposes herein recited.

Second, in combination with the shackle bar t, between the handles and connected thereto, the slotted or space plated front ends of the beams m, affixed to the front bar u, substantially as described, as and for the purposes herein set forth.

Third, connecting the seeder to the plough frame, by means of the hinged straps z, and bar z', and pin y', as described.

No. 54,801.—ARETUS A. WILDER and WILLIAM GOODING, Detroit, Mich.—*Propeller.*—May 15, 1866.—The bucket arms oscillate by the motion derived from the cranks on the rotary shaft and the controlling effect of the pivoted rods above, which compel the upper end of the bucket arms to describe arcs.

Claim.—The combination of said double flanged buckets with paddles, crank shaft, and radial arms, constructed substantially as shown and described, and for the purpose set forth.

No. 54,802.—MOSES H. WILEY, Bucksport, Maine.—*White-washing Apparatus.*—May 15, 1866.—The bibulous larger roller distributes the wash from the reservoir upon the ceiling; the smaller roller acts as a guide, and the brush precedes to remove dust.

Claim.—The said apparatus or combination of the reservoir A, the rollers b and d, and the brush e, arranged substantially in manner and so as to operate as and for the purpose or object hereinbefore specified.

No. 54,803.—JOSEPH J. WILKINS, Virdey, Ill.—*Buckle.*—May 15, 1866.—The frame is smooth on the side next the animal, and the tongue self-operating by means of a cam.

Claim.—First, the buckle constructed and operating substantially as shown and described.

Second, providing the buckle tongue with the cam n, for the purpose of rendering it self-operating, as set forth.

No. 54,804.—JANE MARIA WILKINSON, Urbana, Ill.—*Adjustable Sandal.*—May 15, 1866.—The toe is secured in a socket; a strap passes over the instep; the heel is secured by a serrated front plate and counter; the shank is adjustable as to length.

Claim.—First, the plate B, in combination with the heel A, of the sandal, substantially as and for the purposes set forth.

Second, an adjusting device by which a sandal may be adapted to different sizes of shoes, substantially as and for the purposes set forth.

Third, the sole D, with the opening E, in combination with the adjustable heel A and plate B, substantially as and for the purposes set forth.

No. 54,805.—SETH C. WINSLOW and SAMUEL WINSLOW, Worcester, Mass.—*Floor Clamp.*—May 15, 1866.—In addition to the claws on the socket of the screw clamp, pivoted dogs are attached to be driven into the joist for further security.

Claim.—The combination of the dogs G G with the floor clamps, substantially as and for the purpose herein specified.

No. 54,806.—A. WIRSCHING and WILLIAM ZOEHE, Williamsburg, N. Y.—*Billiard Register.*—May 15, 1866.—Two sets of keys are used, one for each player, and operate their respective index hands. When 100 is reached they return to the initial point and the game is registered upon the upper dial.

Claim.—First, the use, in a billiard indicator, of concave keys or buttons C C', substantially such as herein described, so that the points of the cues are not liable to slip thereon, when the players wish to force the same in.

Second, the keys C C', in combination with rockers m, levers h, and stop pins i, substantially as and for the purposes set forth.

Third, the keys C C', in combination with elbow levers $q\ r\ q'\ r'$, racks $w\ w'$, and index hands $a\ a'$, constructed and operating substantially as and for the purpose described.

Fourth, the serrated arms $i^2\ i^3$, in combination with stops i, keys C C', racks $w\ w'$, and index hands $a\ a'$, constructed and operating substantially as and for the purpose set forth.

Fifth, the racks $w\ w'$ and pinions $c^2\ c^3$, in combination with the radiating plates $f^2\ f^3$, arms $c^2\ c^3$, pawls $b^2\ b^3$, ratchet wheels $a^2\ a^3$, and index hands $a\ a'$, constructed and operating substantially as and for the purpose described.

Sixth, the levers $m^2\ m^3$, and inclined planes $o^2\ o^3$, in combination with the stops $k^2\ k^3$, lifters $p^2\ p^3$, and ratchet wheels $a^2\ a^3$, constructed and operating substantially as and for the purpose set forth.

Seventh, the additional hook catches $s^2\ s^3$, in combination with the ratchet wheels $a^2\ a^3$, and index hands $a\ a'$, constructed and operating substantially as and for the purpose described.

Eighth, the pawls $a^4\ a^5$, and ratchet wheels $w^2\ w^3$, in combination with the ratchet wheels $c^2\ a^3$, and with the index hands $a\ a'$ and $b\ b'$, as and for the purpose set forth.

Ninth, the inclined planes $e^4\ e^5\ f^4\ f^5$, in combination with the pawls $a^4\ a^5$, and index hands $a\ a'\ b\ b'$, constructed and operating substantially as and for the purpose described.

Tenth, the keys $c^*\ c'^*$, in combination with ratchet wheels $a^2\ a^3$, and index hands $a\ a'$, constructed and operating substantially as and for the purpose set forth.

No. 54,807.—OLIVER E. WOODS, Philadelphia, Penn.—*Slinging Arms.*—May 15, 1866.—The devices refer to the mode of shifting and slinging the gun, bayonet, cartridge box, and canteen so as to counterpoise each other and adjust the equipage carried on the knapsack.

Claim.—First, slinging the gun by means of the strap C and one or more hooks D D', substantially as and for the purpose described.

Second, a book O attached to the cartridge box P, to adapt the latter to be attached to and supported by any part of the equipage.

Third, the method of holding open the flap of the cartridge box.

Fourth, providing the bayonet with a hook R^2 to enable it to be slung to the equipage, as and for the purpose specified.

Fifth, slinging the cartridge box behind to the knapsack, in any manner, substantially as described.

Sixth, the blind buckles I I, or their equivalents, attached to the front of the knapsack to enable the blankets or other equipage to be retained by the coat straps G G in a position forward of the centre of the knapsack, in the manner and for the purpose set forth.

Seventh, the ring K^2 attached to the under side of the knapsack to support the accoutrements when shifted to the rear.

No. 54,808.—M. B. WRIGHT, West Meriden, Conn.—*Foot Stove.*—May 15, 1866.—The lamp sets in a central chamber with perforated sides and exterior casing supporting a hanging plate and upper grating.

Claim.—The arrangement of the movable perforated plate G, rim B, lamp C, movable hanging plate H, and movable grate J, in combination with the perforated vessel A, constructed and operating in the manner and for the purpose herein described.

No. 54,809.—LOUIS ZISTEL, Sandusky, Ohio.—*Machine for Coiling Wooden Hoops.*—May 15, 1866.—By the combination of devices the hoop is wound around the coiling drum and held in its position until properly secured to remain coiled.

Claim.—First, the combination of the belt L, pulley J, self-adjusting pulleys M and U, carrying pulley T, drum K, weighted lever N, finger lever V, arranged relatively to each other, and operating in the manner and for the purpose herein specified.

Second, the combination of the swinging cover W, drum K, pushing fork E′, shaft D′, arms B′ C′, arranged and operating in the manner and for the purpose herein specified.

No. 54,810.—RICHARD S. ADAMS, Loyd, N. Y., assignor to himself and GEORGE E. PARROTT, same place.—*Machine for Tightening and Securing Hoops of Compressed Bales.*—May 15, 1866.—The ends are drawn together and overlapped by two pairs of gripers, operated by racks and a pinion ; a punch descends, cuts and punches out a loop in the overlapped portion, which receives a key to secure the connection.

Claim.—First, an improved machine formed by combining a pair of gripes B, constructed as described, with each other, with the cog wheel C, ratchet wheel E, pawl F, and with the frame A, in which they are placed.

Second, the combination of the punch G, constructed and operating substantially as described, with the machine, for the purposes set forth.

No. 54,811.—MATI ALI, Philadelphia, Penn., assignor to WILLIAM MAYER, same place.—*Clamp for Holding Cigars for Tying.*—May 15, 1866.—The pliable metallic band is made to enclose and compress the pack ready for tying.

Claim.—The within described clamp, composed of the elastic band B and block A, constructed substantially as and for the purpose herein set forth.

No. 54,812.—J M. ARMOUR, Craftsbury, Vt., assignor to NATIONAL KNITTING MACHINE COMPANY.—*Knitting Machine.*—May 15, 1866.—An endless chain of needle frames is used, each frame carrying a sliding hook needle and a sliding stitch holder or clamp. These frames are connected to each other by hooks and eyes. As the needle with its new loop recedes within its frame its hook is freed of its old loop which has been seized and held by the pressure of the stitch holder against the end of the needle frame, and in a cross groove just above and free from the traverse of the hook, the grasp of the stitch holder being released for that purpose. The length of the chain may be varied to vary the size of the tubular goods, or flat goods may be knitted by reverse motions of the chain.

Claim.—First, forming a stitch or loop by drawing the needle and thread through an opening in the end of the frame d, which supports the needle and stitch holder, substantially as described.

Second, the stitch holder b, arranged to press the stitch against the face of the frame d and hold it there while the needle with the new stitch is drawn into it, substantially as and for the purpose set forth.

Third, the carrier frame d, constructed and operating as and for the purpose herein shown and described.

Fourth, withdrawing the needle a by means of the spring i, or its equivalent, for the purpose of imparting to the needle a quick backward movement, and thereby insuring the drawing of the new stitch through the previously formed stitch before the latter is released by the stitch holder b, as described.

Fifth, providing the frame d, within which the needles traverse, with hooks and eyes e e, for the purpose of uniting more or less of them together, as and for the purpose set forth.

Sixth, the combination and arrangement of the cam E and the cams o and n with the needle a and stitch holder b, for the purpose of giving to said needle and stitch holder the required movement, substantially as herein set forth.

Seventh, constructing the form or endless chain for holding the needles of lags or sections, so arranged or connected together as to permit of enlarging or contracting the form or chain by adding more sections, or removing some of them, at pleasure.

Eighth, the combination of a revolving endless chain or form, constructed as above described, with one or more stationary cams for operating the needles.

No. 54,813.—JOHN G. BAKER, Philadelphia, Penn., assignor to HENRY DISSTON, same place.—*Saw Grinding Machine.*—May 15, 1866.—Two grindstones revolve vertically in contact with the saw blade, which is fed horizontally between them by friction rollers. The saw is supported on its edge by a central strip and side pieces, which guide it at points not in contact with the stones and the feed rollers.

Claim.—First, the vertically adjustable frame D for holding the saw, in combination with the grindstones B and B′, the whole being arranged and operating substantially as set forth for the purpose specified.

Second, the frame D, its detachable packing pieces e e, and detachable strips f, the whole being arranged for supporting and grinding the saw blade, substantially as described for the purpose specified.

Third, the combination of the adjustable frame D with the adjustable cross-head I, its spindles J, and the friction pulleys m.

No. 54,814.—ALBERT N. BEACH, Winsted, Conn., assignor to himself and EDWARD HATCH, Charlestown, Mass.—*Heating Stove.*—May 15, 1866.—The two concentric cylinders have an interposed air space and metallic tip and bottom plates; in the former are the apertures for fuel, draught, and exit.

Claim.—Improved stove as constructed with the fuel supply opening E, furnished with a door or cover F, and the draught or air induct or inducts, provided with a valve or valves a, arranged in the top of the stove and over the fuel chamber in the manner and for the purpose set forth.

No. 54,815.—BENJAMIN BRAZELLE, Nashville, Ill., assignor to himself, H. P. WALKER, and J. CLARK BROWN, same place.—*Self Rocking Cradle*—May 15, 1866.--The devices permit adjustment for a long or short rocking motion which is derived from the clockwork.

Claim.—The rocker wheel K, the sliding verge N, the lever L, the detent or lock lever P, the stud O, and the hinged foot M, in combination with each other, with the clockwork, and with the frame of the cradle or crib, substantially as herein described and for the purpose set forth.

No. 54,816.—GEORGE N. GOODSPEED, Winchendon, Mass., assignor to T. S. PAGE, Toledo, Ohio.—*Feeding Mechanism for Sewing Machines.*—May 15, 1866.—The rock shaft carries two cams, one to lift and the other to move the feed dog. The dog is free at one end, and at the other is pivoted to an arm which rocks on a fulcrum. The friction nut and spring prevent the feeder rebounding after a movement.

Claim.—The arrangement and combination of the cams fg, the feeder F, its arms $b c$, the regulating screw d, and the vibratory arm h.

Also, the combination of the same and the friction apparatus, substantially as hereinbefore described.

No. 54,817.—DAVID HOWARTH, Portland, Me., assignor to himself, W. R. JOHNSON, and SAMUEL C. RUNDLETT, same place.--*Dinner Pail.*--May 15, 1866.---Explained by the claim.

Claim.---A dinner pail composed of several rings so arranged that they may be pressed out when the pail is to be used, the rings connected with each other when so pressed out, by forming books or shoulders on their edges to overlap each other, and the lower or bottom ring having a compartment to contain liquid, all constructed and arranged in the manner and for the purposes herein set forth.

No. 54,818.—EDWIN LOCKWOOD, Bordentown, N. J., assignor to himself and WILLIAM H. CARRYL, Philadelphia, Penn.—*Sash Operating Device for Ventilating Windows.*—May 15, 1866.—The rod being raised is rotated, and its crank lever vibrates the window on its hinges, pressure on the rod being withdrawn the pawl catches in the ratchet to maintain the position.

Claim.—The combination of the rod A, lever D, spring L, pawl M, and ratchet or corrugated plate B, in the manner and for the purpose substantially as shown and described.

No. 54,819.—MATTHEW MORIARTY, Bangor, Me, assignor to himself and WILLIAM A. ULMER, Ellsworth Falls, Me.—*Clothes Sprinkler.*—May 15, 1866.—The contraction of the elastic bulb ejects the water through the jet pipe and foraminous nozzle over the clothes.

Claim.—As my invention and as a new article of manufacture for the special purpose for which it is designed, the clothes sprinkler composed of the elastic bulb, the jet pipe, and the foraminous nose, arranged and combined substantially as specified.

No. 54,820.—ALBERT G. PAGE, Fitchburg, Mass., assignor to PAGE, WILSON & Co., same place.—*Pipe Tongs.*—May 15, 1866.—The inner jaw is adjusted longitudinally by rotation in the threaded socket in the end of the short arm of the tongs.

Claim.—The arrangement of the arm B, with respect to the arm B', of the lever B B', in the manner substantially as described, and the application of the jaw D to such arm B, by a screw b, projecting from the jaw and screwed through the said arm in the manner and so that it may be revolved with and by means of the jaw, substantially as specified.

Also when the jaw D of the pipe tongs is connected to its screw b, so as to operate with it the arrangement of one of the two teeth of the jaw in and the other aside from the axis of the screw produced, the whole being as explained.

No. 54,821.—JAMES PARKER, Woburn, Mass., assignor to himself, L. HOLDEN, S. B. HOLDEN and L. L. HOLDEN.—*Machine for Boarding and Graining Leather.*—May 15, 1866. —The rubber board and the bolster are vertically adjustable by springs, one in the table and the other in the reciprocating holder.

Claim.—First, the combination in a boarding machine of the top board G, with the arm F, and slide C D, substantially as described and for the purpose set forth.

Second, the combination of the top board G, with the springs K, and rubber board H, substantially as described and for the purpose set forth.

Third, the combination of the rubber board H, and the bolster board I, substantially as described and for the purpose set forth.

Fourth, the combination of the bolster board I, with the springs S, and trencher board N, substantially as described and for the purpose set forth.

Fifth, the combination of the trencher board N, with the springs S, screws M, and board or frame L, substantially as described and for the purpose set forth.

Sixth, the combination of the arm U, and lever T, with the trencher board N, and bolster board I, substantially as described and for the purpose set forth.

No. 54,822.—JOSEPH T. PARLOUR, Buffalo, N. Y., assignor to himself and JAMES DEAN, same place.—*Grain Elevator.*—May 15, 1866.—Intended for scraping grain towards the hatchway from distant parts of the hold. Endless travelling belts with scrapers pass over drums under the hatch and over drums rotating in clamps temporarily attached to the floor and ceiling of the hold.

Claim.—The adjustable hinged screw clamps P, when used for the purpose of holding the endless bucket chain, constructed and arranged as herein described.

No. 54,823.—JACOB P. REES and R. A. GRAHAM, Greensburg, Ind., assignors to themselves and C. C. BURNS, same place.—*Ditching Machine.*—May 15, 1866.—The trench is formed by the sharp-edged bevelled wheel, which is pressed into the soil by the weight of the wheeled carriage in which its vertical position is adjustable.

Claim.—The ditching wheel F, or its equivalent, arranged in a suitable frame so hung in and to the body of the machine as to be raised or lowered, substantially in the manner described and for the purpose specified.

No. 54,824.—EDWARD A. TURNER, New York, N. Y., assignor to himself and JOHN A. LEGGET, Branford, Conn.—*Steering Apparatus.*—May 15, 1866.—The described devices sustain the position of the drum until relieved by the treadle and obviate the necessity for continuous exertion of the steersman and prevent the back lash of the rudder.

Claim.—First, the pawls n r, and the frames K, arranged and operating with reference to each other and with the two ratchet wheels g h on the drum of the apparatus, substantially as herein set forth for the purpose specified.

Second, the levers s provided with treadles u, or their equivalents, operating the pawls n r in connection with the ratchet wheels g h, substantially as hereinbefore set forth for the purpose specified.

No. 54,825.—G. W. WAIT, Wayland, Mich., assignor to himself and A. J. SMITH, same place.—*Tool for Making Spiles.*—May 15, 1866.—To the shank of the auger that bores the spile is attached an obliquely set knife to taper the end thereof.

Claim.—The combination of the shaping knife B with a wood-boring tool A in such a manner as to bore and shape a wooden spile at one operation, substantially as specified.

No. 54,826.—SETH WILLIAMS, Foxboro', Mass., assignor to HENRY C. WILLIAMS and C. T. WILLIAMS, same place.—*Mould for Casting Curved Pipes.*—May 15, 1866.—The core of the pipe is supported in the mould by the extension of the stock through openings in one end and one side of the flask, half in each part; the wings of the stock are supported by brackets.

Claim.—The arrangement and combination of the wings or extensions of the core stock with the two openings, or the same, and the brackets applied to the two parts of the flask, substantially as specified.

No. 54,827.—JOSEPH WOODROUGH, Hamilton, Ohio, assignor to WOODROUGH & McPARLIN, Cincinnati, Ohio.—*Saw.*—May 15, 1866.—The removable teeth are sustained laterally by movable gibs countersunk in the saw plate.

Claim.—The countersunk or imbedded gibs E, employed in combination with the teeth B C, substantially as and for the purpose set forth.

No. 54,828.—FRANÇOIS CARRÉ, Paris, France.—*Seat and Back for Chairs.*—May 15, 1866.—The seat is made of C-shaped metallic strips, a central disk, and supporting end pieces. The elastic back is made of C-shaped strips and cross-ties.

Claim.—First, the radiating strips a of sheet steel, or other suitable material, in combination with the central disk b and frame C, constructed and operating substantially as and for the purpose described.

Second, the secondary supporting strips e, in combination with the radiating strips a and central disk b, constructed and operating substantially as and for the purpose set forth.

Third, the back of a chair, or other similar article. composed of C-shaped strips a' of sheet metal, or other suitable material, in combination with horizontal rods h' and frame h', constructed and operating substantially as and for the purpose described.

No. 54,829.—WILLIAM CROOKES, London, England.—*Separating Gold and Silver from Ores.*—May 15, 1866.—To the mercury for amalgamation are added metals, such as tin, zinc, cadmium, &c., either combined directly or reduced electro-chemically from compounds, so as to unite with the mercury in the act of separation; cyanide of potassium is added to the mercury to keep it bright.

Claim.—The employment of zinc, tin, and cadmium, and such other metals as hereinbefore mentioned, and also of such several processes for the extraction and separation of gold and silver from the ores and substances containing them, and for the treatment of mercury employed for such purposes as hereinbefore substantially set forth and described.

No. 54,830.—RICHARD EATON, London, England.—*Ventilating Stove for Railroad Cars.*—May 15, 1866.—The air passes by the central pipes around the front of the stove into the fire chamber, then up by the serpentine route around the induction pipe to the exit flue.

Claim.—First, the arrangement of the air ducts F in the front corner of the stove and air passage D in the top thereof, in combination with the air pipe C, in the manner and for the purpose herein specified.

Second, the arrangement of the chamber A, air ducts F, air passage D, air pipe C, spiral chamber G, and chimney B, constructed and operating in the manner and for the purpose herein specified.

No. 54,831.—F. A. LANGE, Glashütte, Saxony.—*Watch.*—May 15, 1866.—Explained by the claim.

Claim.—First, the removable key E inserted within the pendant of a watch, and retained therein substantially in the manner described, in combination with the winding mechanism of the watch, when constructed and operating in the manner and for the purposes as set forth.

Second, the concave contrate wheel F upon the arbor D, and gearing with the crown wheel Z, said contrate wheel having its teeth constructed in the peculiar manner described.

Third, providing the ratchet cylinder H upon the arbor D, and gearing with the contrate wheel F, with a projection M upon its lower surface, which enters a slot or recess upon the face of the pinion O, and operates in the manner and for the purposes set forth.

Fourth, the loose pinion O upon the shaft D, in connection with the minute wheel, when provided with a recess N upon its face, and constructed and operating in the manner and for the purpose as set forth.

Fifth, the combination of the removable key B, arbor D, concave contrate wheel F, crown wheel Z, ratchet cylinder H, with its spring K, and the pinion O, when constructed and operating substantially in the manner and for the purpose set forth.

No. 54,832.—EDMUND SHARPE, Paris, France.—*Artificial Fuel.*—May 16, 1866.—The peat and coal are washed, crushed together, and mixed with coal tar.

Claim.—The herein described artificial fuel, consisting of peat, anthracite, and bituminous coal, with a coal-tar cement, combined in the proportion and in the manner substantially as set forth.

No. 54,833.—WILLIAM LOCKE, JOHN WARRINGTON, W. E. CARRETT, W. E. MARSHALL, and J. TILFORD, Kippax, England.—*Mining Coal, Minerals, &c.*—May 15, 1866.—The hydraulic engine gives a steady thrust motion to the tools; the position of the machine during the thrust is maintained by a holdfast in contact with the roof, and released to enable the machine to move forward.

Claim.—First, the system or mode of actuating the cutting tool or tools of machines for working coal and other minerals and earthy matters by the direct pressure of a flow of water or other practically non-elastic fluid medium, by the means and in the manner substantially as herein shown and set forth, so as to produce a steady, even, slotting action of the tool or tools at any angle with the substance to be cut.

Second, the application and use to and in machines for working coal and other minerals and earthy matters of a holding-on head or feeler, which is pressed against the roof during the cutting action of the tool or tools, and released therefrom to allow the machine to move forward during the return or back stroke of the tool or tools; the movements of this head or feeler being obtained from the pressure of water or other practically non-compressible fluid medium, substantially as herein shown and described.

No. 54,834.—GEORGE T. JONES, Cincinnati, Ohio.—*Manufacturing Bank Notes.*—May 15, 1866.—Explained by the claim.

Claim.—The process, substantially as herein described, of manufacturing bank notes, bonds, or analogous securities by applying ink or coloring matter to unsized or partially sized paper, so that it will penetrate or come into actual contact with the fibre thereof, and afterward applying a coating of size to permeate the paper, cement its fibres, and prevent counterfeiting or alteration by the transfer or removal of the ink.

Also, as a new article of manufacture, a bank note, bond, or other evidence of value, with the printed portion or signatures protected by size, for the purpose of preventing counterfeiting or alteration.

No. 54,835.—GEORGE T. JONES, Cincinnati, Ohio.—*Preparing Bank Notes to Prevent Counterfeiting.*—May 15, 1866.—Explained by the claim.

Claim.—Covering the impression with a transparent film of paper, or other fibrous material, substantially as described.

No. 54,836.—M. E. ABBOTT, Summit Hill, Penn.—*Drilling Machine for Wells.*—May 22, 1866.—The walking beam is operated directly by the piston rod, the length of stroke being controlled by tappets which actuate the valves. The devices for rotating, lowering, and withdrawing the drill are explained in the claims.

Claim.—First, the arrangement of devices herein described for imparting to the drill rod a movement in a vertical plane, the same consisting in securing it within the frame or cross-bar *o*, moving on suitable vertical guide rods or bars *p p*, and connected with the engine beam *g* in any proper manner, substantially as set forth.

Second, revolving the drill through the vertical movement of the same, by means of the sliding plate *z* moving in and between parallel inclined guides *b′ b′*, and provided with a spring pawl *a′* engaging with the ratchet wheel *s* of the frame *v*, arranged and operating substantially in the manner described.

Third, the combination of the jaws *q′ q′* and bent arms *s′ s′*, arranged and operating with the rod *v*, in the manner and for the purpose herein specified.

Fourth, the combination of the jaws *t′ t′* and lever frame *v′*, arranged and operating with the rod *v*, in the manner and for the purpose herein specified.

Fifth, in the described combination with the drill rod *v* and the beam *g* for operating the same, the crank arms *d′ h′ m′*, rods *f n′*, and adjustable tappets *g g′*, arranged and employed substantially as set forth, to regulate the stroke of the drill.

No. 54,837.—JOHN ADT, Wolcottville, Conn.—*Riveting Machine.*—May 22, 1866—The vertical reciprocating hammer stock is revolved slowly as the hammer gives the blows to the rivet.

Claim.—The reciprocating and rotating tube I, operated substantially as shown, in combination with the internal hammer rod O, arranged with the springs P Q and the nut R, or its equivalent, substantially as and for the purpose set forth.

Also, the adjustable or rising and falling disk or bed T, in combination with the reciprocating tube I and the hammer rod O and spring P, all constructed and arranged substantially as and for the purpose specified.

No. 54,838.—W. M. ARNOLD, New York, N. Y.—*Method of Combining Wrought Iron with Cast Iron.*—May 22, 1866; antedated May 18, 1866.—The iron is cast around a wrought-iron skeleton of the same general form.

Claim.—First, the mode, hereinbefore described, for combining and uniting wrought with cast iron—that is to say, forming a skeleton of wire, which has been coated or tinned, and so arranged and placed at such intervals that the melted metal can penetrate all portions of such skeleton, and then filling those interstices with cast metal, substantially in the manner and for the purpose above set forth.

Second, in this connection, the tinning or coating of iron wire in such a way as to prevent the molten metal, which shall afterwards be combined with it, from producing any injury to such wire, substantially as above set forth.

Third, forming a skeleton of metallic wire, and then filling up the interstices with molten metal, for the purpose and substantially in the manner above set forth.

No. 54,839.—EDWARD H. ASHCROFT, Lynn, Mass.—*Apparatus for Preparing Peat.*—May 22, 1866.—The peat is fed from a hopper into a cylinder with revolving heads, perforated revolving shaft and hollow screw blade; from thence it passes to a perforated steam-heated pipe, whence it is discharged by a reciprocating piston.

Claim.—First, the hollow perforated shaft B, and the screw C, or their equivalents, in combination with the cylinder A, the whole being constructed and made to operate together, substantially as described and for the purpose set forth.

Second, the combination of the steam jacket or cylinder F, with the inner or forming cylinder E, constructed and operating substantially as described and for the purpose set forth.

No. 54,840.—C. H. BASSETT, Derby, Conn.—*Coupling for Carriage Thills.*—May 22, 1866.—Conical surfaces on the bolt and its socket bear against the sides of the thill iron and prevent rattling.

Claim.—The employment of centres or taper bearings, in combination with the tongue piece and clip, and a suitable means of holding the parts in adjustment, substantially as set forth.

No. 54,841.—J. R. BATEY, New York, N. Y.—*Hand Protector for Sad Irons.*—May 22, 1866.—A coiled flexible wrapper around the handle and a shield suspended beneath it to intercept radiated heat.

Claim.—A hand protector for flat-irons, composed of a wrapper A and guard B, substantially as described as a new article of manufacture.

No. 54,842.—FRANCIS M. BIRDSALL, Martinsville, Ohio.—*Furrowing Machine.*—May 22, 1863; antedated May 16, 1866.—The carriage has two share frames adjustable vertically by levers and laterally by slots and screws, so as to furrow out the ground with balks of the required width.

Claim.—The plough frames *e*, longitudinal strips *e'*, and levers *d*, in combination with ploughs *g*, frame A, and sliding straps *h*, constructed as above described and for the purposes set forth.

No. 54,843.—ELI S. BITNER, Lock Haven, Penn.—*Sash Fastening.*—May 22, 1866.—A spring catch on the window casing engages notches in the sash and is retracted by depressing the exposed lever whose arm vibrates the spring.

Claim.—The lever D, with flange C, and knob E, constructed and connected with plate B, as shown in combination with the spring C, having the spur A at its end, all arranged and operating as herein shown and described.

No. 54,844.—H. E. BODWELL, Jr., Paterson, N. J.—*Sewing Machine Guide for Stitching Hat Tips to the Side Linings.*—May 22, 1866.—The side lining and bottom lining or tip of a hat are sewn together without previously basting, the stitch presenting an ornamental appearance.

Claim.—The combination of the hemmer C, and gauge plate and guide D F, constructed and arranged substantially as above set forth, for the purpose of stitching and also of connecting a side lining to a hat tip at the same operation.

No. 54,845.—B. P. BOWER, Cleveland, Ohio.—*Hydrant.*—May 22, 1866.—The valve chamber, branch pipes, and upper connecting tube are collectively detachable from the main. The central screw rod operates the valve to permit the flow into the service pipe or cut it off and empty the service pipe by the waste way.

Claim.—First, the adjustable toothed plate B, the corresponding depression in the plate B', and the pipes C' and D, arranged in the manner and for the purpose set forth.

Second, the adjustable tooth plate B, with the corresponding depressions in the plate B', as arranged in relation to the screw stem *h* and cap F', the washers *m*, valve L, packing *q*, and grooves *e*, in the manner and for the purpose set forth.

Third, the cylinder F, with flaring end G, pipes E and pipes D, in combination with the packing *t*, and induction pipe H, arranged as and for the purpose set forth.

No. 54,846.—J. C. BRIDGMAN, London, Ohio.—*Churn.*—May 22, 1866.—The dasher is rotated by engagement of the threaded shaft with the nut in the lid; the handle is swivelled.

Claim.—The combination of the cover B and pin H with the handle D, having spiral threads D', the loose collar E, and dasher C, the parts being severally constructed and arranged for use substantially in the manner and for the purpose set forth.

No. 54,847.—C. F. BROWN, Warren, R. I.—*Anchor.*—May 22, 1866.—The body is cylindrical; the flukes project transversely.

Claim.—First, forming the sides of anchors of continuous bands, with the flukes arranged across the inner sides thereof, substantially as described.

Second, strengthening the body of an anchor whose sides are made continuous by means of an inner rib or membrane E, and by a stay rod C, one or both, substantially as described.

Third, cutting away the upper parts of the body of an anchor, whose sides are continuous, as at D D, and forming points F F at one side of the centre of gravity, so that the anchor will right itself, substantially as described.

No. 54,848.—LEWIS BRADLEY and MILTON BRADLEY, Springfield, Mass.—*Croquetterie.*—May 22, 1866; antedated April 17, 1866.—The metallic pliable arch is socketed in pegs driven into the ground, and the balls are painted with designating bands.

Claim.—First, the combination of arch *a*, and sockets *b c*, to form a croquet bridge or arch, substantially in the manner and for the purpose herein set forth.

Second, an iron croquet bridge or arch coated with zinc, tin or other similar metal, substantially in the manner and for the purposes herein fully set forth.

Third, the method herein fully described of painting croquet balls, in which the light balls are designated by black or dark stripes and the dark balls by white or light stripes.

No. 54,849.—T. STUART BROWN, Philadelphia, Penn.—*Trunk Lock.*—May 22, 1866.—The lock is attached to the hasp and has two bolts, thrown simultaneously to the right and left respectively, to engage in recesses of the face plates.

Claim.—The combination of the two bolts *b* and *b'*, arranged in respect to the case *a*, and operating substantially as and for the purpose herein set forth.

No. 54,850.—PETER BUDENBACH, New York, N. Y.—*Head Light for Engines.*—May 22, 1866.—The plates stand out and cover the holes in the sides of the tube; air is also admitted between the radiating oblique plates in the bottom. The object is to make the air enter obliquely and avoid disturbance of the flame.

Claim.—The combination of the plates B, covering and extending beyond the margin of the aperture C, with the radial and oblique director plates in the lower opening of the tube of A, the combined devices operating to deflect the entering currents of air to prevent the direct impingement of the air upon the flame and the consequent disturbance thereof.

No. 54,851.—M. BURNETT and H. COMSTOCK, Boston, Mass.—*Steam Valve.*—May 22, 1866.—The valve seat is an inserted annular piece of resilient material, with a rounded top surface corresponding to the circular recess in the valve above.

Claim.—As an improvement in stop valves, the arrangement of the seat *a* and valve *b,* when constructed substantially as described.

No. 54,852.—E. G. BURNHAM, Bridgeport, Conn.—*Ice Pick*—May 22, 1866.—The weighted plunger acts as a hammer within the cylinder and upon its pointed extension.

Claim.—The pick iron or steel point formed on the tube which is held in the hand in combination with the solid metal rod fitted within the tube, so as to work up and down therein, and act as a hammer upon the pick, all substantially in the manner and for the purpose described.

No. 54,853.—JESSE S. BUTTERFIELD and M. S. GREEN, Philadelphia, Penn.—*Rail for Railroads*—May 22, 1866.—The upper, inverted U-shaped rail has transverse, horizontal pins which engage in the slots in the upper central pin of the standard.

Claim—The rail composed of the upper portion A, with its longitudinal groove and pins *f*, and lower portion B, with its longitudinal rib *b*, adapted to the said groove, and its openings *e*, all substantially as set forth.

No. 54,854.—H. F. BYERLY, Clinton, Ill., assignor to ISRAEL CAMPBELL.—*Cultivator*—May 22, 1866.—By the devices described the vertical and lateral adjustments of the shares are obtained.

Claim.—First, the sliding bolts *k k*, slots *i i*, and pendants *b b*, combined and operated substantially as and for the purpose specified.

Second, the roller L, lever *f*, and chain M and *h*, combined and operated substantially as and for the purpose set forth.

No. 54,855.—CHARLES E. BYERS, Mahanoy Plain, Penn.—*Switch Stand for Railroads.*—May 22, 1866.—A spring-bolt to detain the switch lever in the required position relatively to the notched upper edge of the switch stand.

Claim.—The combination of the bolt G, lever K, piece I, and spring J L, applied to the lever D, arranged relatively to each other and with the top F of the switch, in the manner for and the purpose herein specified.

No. 54,856.—WILLIAM BYFIELD, New York, N. Y.—*Needle Wrapper.*—May 22, 1866.—The needles are ranked and placed in a pocket of the wrapper which exposes their ends; the flaps cover them when wrapped.

Claim.—The combination of a wrapper having two flaps with a holder, substantially as described, so that the opening of two flaps will permit the points and eyes of the needles to be seen while they are held securely to be removed one at a time.

No. 54,857.—GEORGE G. CARVER, Roxbury, Mass.—*Flour Dipper, Measurer and Sifter.*—May 16, 1866.—The dipper has a semi-cylindrical bottom, in which rotates a crank armed with flexible rubbers. The ridges on the sides are measures of capacity.

Claim.—The combination of the dipper with ridges or lines indicating quarts and pints, the horizontal beater with radial flexible rubbers and concave sieve, all as specified.

No. 54,858.—ELISHA CHILD, Jr., Springfield, Mass.—*Paper Neck-tie.*—May 22, 1866.—• Explained by the claim.

Claim.—The ornamental neck-tie composed of a paper scarf, stock, and band, and the metallic clasp carrying an elastic loop, combined and arranged substantially in the manner and for the purpose set forth.

No. 54,859.—JOHN B CHRISTIAN, Mount Carroll, Ill.—*Apparatus for Inducing the Flow of Oil from Wells.*—May 22, 1866.—The well is tightly closed and the escape of gas prevented by the packing around the single tube which descends to the level of the oil which flows into the holes at the bottom of the tube. The packing is expanded by a wedging ring, driven by the base of the lowering tube.

Claim.—The combination of the flanged pipe A, with the slotted, double hollow cylinder B, and the hollow cylinder Y, as arranged in relation to each other, substantially as and for the purpose specified.

Also, lowering the pipe A, or its equivalent, and securing it firmly to its place in the well by means of the tube D, with its circular base E, in combination with the rod H, substantially as set forth.

No. 54,860.—CHARLES B. CLARK and O. S. GARRETSON, Buffalo, N. Y.—*Mop Head.*—May 22, 1866.—The screw socket of the stirrup jaw moves upon the threaded shank of the stationary jaw.

Claim.—Connecting the nut K with the collar that carries the movable jaw E E, by means of the lugs *g g*, or their equivalent, in combination with the threaded shank of the jaw C, arranged and operating substantially as set forth.

Also, connecting the movable jaw E E with the collar D, by means of the bows *l l*, constructed and operating substantially in the manner and for the purpose specified.

No. 54,861.—AMOS CLIFT, Jr., Mystic River, Conn.—*Churn.*—May 22, 1866.—The hollow trunnions admit a current of air, and their flaring inner ends prevent the escape of cream.

Claim.—First, the combination of the hollow journals or axles constructed with flaring enlargements on their inner ends, with the revolving churn box, substantially in the manner and for the purpose described.

Second, making the inner funnels removable, substantially as and for the purpose described.

No. 54,862.—LYMAN CLINTON and EZRA S. MUNSON, North Haven, Conn.—*Horse Rake.*—May 22, 1866.—To a rod fixed in the axle (which is also the rake head) is hinged a sliding bar, which moves in a socket and is actuated by a foot or hand lever. The foot lever, holding the rake in position, is pivoted to an arm which is thrown forward when the rake is tilted.

Claim.—First, combination of the slide *c*, the rod *h* A, and axle A with the double acting lever I, substantially as and for the purpose specified.

Second, the combination of the slide *c*, the rod *h*, and axle A with the lever G, substantially as and for the purpose specified.

Third, the combination of the lever L with the axle A, when arranged upon a movable fulcrum and constructed so as to operate substantially as and for the purpose specified.

No. 54,863.—J. W. COCHRAN, New York, N. Y.—*Chair, Sofa, and Car Seat.*—May 22, 1866; antedated May 7, 1866.—Explained by the claim.

Claim.—Employing the perforated or whole sheet of hardened or enamelled India-rubber or gutta-percha, for chair, sofa, car, and carriage seats, substantially as above described and set forth.

No. 54,864.—SAMUEL COLAHAN, Cleveland, Ohio.—*Apparatus for Preparing Hay for Market.*—May 22, 1866.—The follower ascends in the hay chamber by power communicated by ropes and pulleys. The devices are explained in the claims.

Claim.—First, the arrangement of the follower *a*, and sheaves with their respective ropes or chains, in combination with the side doors L M, trap door R, loops V, and clamps P, substantially as and for the purpose set forth.

Second, the followers *m*, bolts *s*, and catches *p*, in combination with the follower, constructed and arranged in the manner and for the purpose substantially as set forth.

Third, the adjustable ends *e* connected with the frame, as specified, in combination with the followers *a m*, and side doors, arranged as and for the purpose substantially as set forth.

No. 54,865.—L. O. COLVIN, New York, N. Y.—*Cow Milker.*—May 22, 1866.—The several milkers are attached by branch pipes to a main pipe connected to a pump. The water in the diaphragms under each teat tube being withdrawn the milk flows, is collected and discharged at one outlet.

Claim.—First, the construction of a cow milker in such manner that it may be operated by water or other fluid acting upon its flexible diaphragms, substantially as herein specified.

Second, so constructing the water spaces on the backs of the several flexible diaphragms of a cow milker that all are in free communication and made to communicate with a single pipe, substantially as herein specified.

Third, the combination of two or more cow milkers with a pump L, or its equivalent, by means of a pipe or pipes filled with water or other fluid, substantially as set forth for the purpose specified.

No. 54,866.—JOHN P. CONKLING, Batchellerville, N. Y.—*Detecting Check Box.*—May 22, 1866.—The tickets are placed in a pile in a locked box, are depressed by a follower sustained by spring pawls and racks, are withdrawn from the bottom one at a time, but cannot be replaced without unlocking the box.

Claim.—First, the combination of ratchets and pawl *c d*, and gravitating follower G, substantially as and for the purposes described.

Second, the open-bottom ticket boxes, in combination with the follower G, or its equivalent, substantially as described.

Third, the stops *g g*, substantially as and for the purposes described.

Fourth, the use of locking bars *a'* or J, applied substantially as described.

Fifth, the removable transparent graduated face plate E, in combination with ticket holder, substantially as described.

No. 54,867.—ROBERT CORNELIUS, Philadelphia, Penn.—*Sand Pump.*—May 22, 1866.— The cylinder has a ball valve at its base, a movable spiral blade running through its length, and an annular filter near its top. A drill may be attached to its lower end.

Claim.—First, the employment of wire gauze or perforated metal in the sand pump, in the manner and for the purpose substantially as described.

Second, the employment of the turning plate D E F in the interior of the sand pump.

Third, the combination of the sand pump, having an interior turning plate, with the drill.

No. 54,868.—D. W. CROCKER, Deposit, N. Y.—*Railway Chair.*—May 22, 1866.—The longitudinally divided chair has jaws on its inner edges griped by corresponding flanges of a longitudinally driven wedge, secured by a spring catch.

Claim.—First, dividing the chair into two similar parts or sections in the direction of its length, in combination with the wedge-shaped plate or regulator G, having the general form and construction herein above described, which plate is driven in and between the bases of the said chair, suitably constructed or formed therefor, and the rail extending across from one section to the other and operating upon the same, substantially in the manner and for the purpose specified.

Second, the combination with the two-part chair A, and its regulator G, of the spring catch *t*, interlocking with suitable notches made in the base of the chair, and arranged substantially as and for the purpose specified.

No. 54,869.—M. C. CRONK, Auburn, N. Y.—*Chair.*—May 22, 1866.—The tilting chair seat is pivoted to lugs on the frame; its backward throw is eased by springs and its reclining position maintained by the engagement of a crank rod with detent catches.

Claim.—First, the combination of the rod *s*, and plate *n*, constructed and operating as and for the purpose herein described.

Second, the combination of the rod *s*, the plates *n*, the hinges *h*, and rubber springs, the whole constructed and operating in the manner herein set forth.

No. 54,870.—JAMES W. CROSSLEY, Bridgeport, Conn.—*Making Pile Fabrics.*—May 22, 1866.—The wires form a temporary weft, and are withdrawn after the edges of the warp loops are cemented to a backing.

Claim.—The process herein described of producing the pile fabric by the aid of wires, which are withdrawn after cementing the warp upon the backing.

No. 54,871.—CHARLES F. DAVIS, Auburn, N. Y.—*Harvester Rake.*—May 22, 1866.—By the devices described the rake head is partially rotated, and while still performing its functions as a reel arm, the rake teeth are thrown into a horizontal position so as not to remove the grain in their passage over the platform. The act is performed by the depression of a lever as the rake falls into position to sweep off the grain.

Claim.—The combination in a harvester rake of the lever L, springs M, or equivalent latches I, and catches J K, with each other and with the rake heads F, arms D E, and frame or guide B, substantially as described and for the purpose set forth.

No. 54,872.—H. G. DAYTON, Maysville, Ky —*Stove for Railroad Cars*—May 22, 1866.—A jacket around the fire pot and central chamber has at its upper portion perforated diaphragms to detain and circulate the ascending current of air.

Claim.—The arrangement of the flanges F, the combination chamber B, air jacket E, and pipe G, substantially as and for the purpose described.

No. 54,873.—JONATHAN DENNIS, Jr., Washington, D. C.—*Machine for Turning and Pointing Screw Blanks.*—May 22, 1866.—The mandrel is hollow and contains the tool which points the blank while the head is turned; the head-turning tools have curved or angular ends to pass the nick gradually and prevent chattering.

Claim.—Pointing screw blanks, rivets, and other articles, at the same time their heads are turned or shaved, by the devices shown and described, or their equivalents.

Also, arranging the pointing tool behind the griping jaws for turning the heads of screw blanks, rivets, and other articles, so as to point them the same time their heads are turned or shaved.

Also, the rod H made hollow or otherwise, and arranged in the arbor to hold the pointing tool, substantially as described, in combination with a pointing apparatus.

Also, arranging the shanks of the griping jaws, the levers, and other devices in the arbors of screw machines, in such a way and manner as to allow the rod which holds the pointing tool to traverse in the arbor between them, substantially as described.

Also, in combination with a pointing tool arranged behind the griping jaws, the cross L, pins L', lever N¹, springs N³, and cam N¹, or such other equivalent mechanism as will hold and traverse the pointing tool to do its work, substantially as described.

Also, in combination with the pointing tool the mechanism described, or its equivalent, for pushing the blanks out of the griping jaws after they are pointed.

Also, making the cutting edges of tools for turning the scored, nicked, or grooved heads of screws, screw blanks, or other articles, curved or rounded, as shown and described.

No. 54,874.—MARSHALL S. DRIGGS, Brooklyn, N. Y.—*Hatchway Door.*—May 22,1866.—
Explained by the claim.
Claim.—Constructing a hatch E, with trucks D, running upon horizontal ways B, suspended beneath the floor so that the hatchway may be opened and closed from below, substantially in the manner set forth.

No. 54,875.—M. EASTERBROOK, Geneva, N. Y.—*Reaping and Mowing Machine.*—May 22, 1866.—Two bevelled gear wheels of different diameters are attached to a feathered sleeve, which slides upon and is driven by the pinion shaft; either wheel is made to gear with a pinion on the crank shaft, according to the speed required of the cutters.
Claim.—First, the sliding gears B and C, connected together by the feathered sleeve S, and arranged to slide upon the shaft D, without moving the latter longitudinally, as and for the purposes set forth.
Second, the arrangement of the hand lever E, ratcheted arch H, and yoke y, in combination with the feathered connecting sleeve S, and gears B and C, as shown and described, and for the purposes specified.

No. 54,876.—M. EASTERBROOK, Jr., Geneva, N. Y.—*Harvester.*—May 22, 1866.—Two loose pinions are attached to the forked arms of an adjusting lever and are at discretion made the means of transmitting the motion of the spur wheel to the counter shaft, engaging with pinions on the latter of different sizes, to vary the rapidity of the stroke of the cutter.
Claim.—First, the combination of the hand lever and two loose pinions p and p', with the double pinion b, and the spur wheel B, arranged and operating substantially as and for the purposes set forth.
Second, the two loose pinions p and p', (whether they are adjusted with a hand lever or other suitable device,) in combination with the double pinion b, and the spur wheel, substantially in the manner and for the purposes shown and described.

No. 54,877.—THOMAS FALLOON, Lyons, N. Y.—*Bed Bottom.*—May 22, 1866.—The yielding slats, elastic straps, links and friction rollers are arranged in a bed bottom with an expanding frame.
Claim.—The combination of the elastic slats B B, and the expanding frame A as a whole, substantially as and for the purpose herein set forth.
Also, the combination of the caps f f, covering the ends of the slats, the links g g, pins h h, friction rollers i l, and elastic loops m m, operating substantially as herein set forth.
Also, the cross strap or straps D, provided with loops q, or equivalent, and connected with the elastic loops s when used in combination with the slats B B, substantially as described.

No. 54,878.—BENJAMIN S. FLETCHER, Cornish, N. H., assignor to himself and SYLVESTER DAVIS.—*Churn.*—May 22, 1866.—The revolving dashers are triangular in cross section, and revolve between the adjustable inclined breakers.
Claim.—The form and arrangement of the dashers E E and the breakers G G.

No. 54,879.—JAMES H. FLETCHER, Lowell, Mass.--*Bed Bottom.*—May 22, 1866.—
Explained in the claim.
Claim.—The said improved bed foundation or combination and arrangement of the frame A, with its mortises a a a a, and India-rubber springs c c c c, the cross-bars B B, with their series of notches d d d, and the series of slats C C C, with their supporting springs c c c, or the same and the holding bands D D, the whole being substantially in the manner and for the purpose specified.

No. 54,880.—DELOS C. FLINT, Bushnell, Ill.—*Machine for Stripping Cane.*—May 22, 1866.—The cane is embraced between a stationary and a descending spring jaw plate, which removes the leaves as the cane is drawn through the throat.
Claim.—First, in combination with the strippers E' H, the auxiliary hinged or jointed stirrups F F, attached to the movable stripper E', and operated by the action of the upwardly divergent grooves h h, in the fixed stripper H, on the ribs or points f f, all arranged and constructed substantially as shown, to operate as set forth.
Second, the combination of strippers E, vertically slotted parallel guides B B, spring D, and screw C, all arranged and operating in the manner herein described.
Third, the slotted vertically adjustable angular plate G, when used for the purpose herein described.

No. 54,881.—NELSON T. FOGG, Lewiston, Me.—*Pump.*—May 22, 1866.—The cylinder is included between the heads, which are connected by screw-rods; the plunger rod is attached to the bail of the piston, which has upper and lower plates, and an interposed packing ring.
Claim.—The described arrangement and combination of the rod y, the bail A, the screw B, the cross-bar C, valve d, screw K, and fastening rods m m', as and for the several purposes herein set forth.

No. 54,882.—GEORGE F. FOOTE, New York, N. Y.—*Tool Holder for Dentists.*—May 22, 1866.—The tool is disengaged from its handle by the engagement of a flattened portion with a notch in the tool rack; this retains it in proper presentation for future attachment to the handle.

Claim.—The combination of the slot P, in the tool rack, with the flattened surface N of the tool, and its upper point C, fitting a corresponding socket of a tool holder, for the purpose herein set forth.

No. 54,883.—GEORGE H. FOWLER, New Haven, Conn.—*Manufacture of India-rubber Shoes.*—May 22, 1866.—The shoe is made of three thicknesses—a napped woolenet, a cottonet, and an exterior surface of rubber; the two fabrics are united by rubber cement.

Claim.—The napped woolenet, (or its equivalent,) in combination with the cottonet (or its equivalent,) and the India-rubber, for making an elastic ventilating shoe, when they are prepared, fitted together, and finished, substantially as herein described.

No. 54,884.—JOSEPH FOWLER, Rahway, N. J.—*Ore Crusher.*—May 22, 1866.—The jaws are simultaneously advanced and receded by connection with cranks on the same shaft; the cheeks are adjustable relatively to the jaws.

Claim.—First, the crank shaft e, and connecting rods k k and l, in combination with the jaws g and k, when the parts are arranged and operating, as and for the purposes specified.

Second, in combination with the jaws g and k, the adjustable side pieces n n, and screws o o, for the purposes specified and as set forth.

No. 54,885.—A. A. FRADENBURG, Nevada City, Cal.—*Animal Trap.*—May 22, 1866.—When the spring door is disengaged it revolves, sweeps the animal past the grated portion, and resets itself.

Claim.—In traps for vermin, the combination of the grating D on the platform P with the revolving door E, the stops N and F, substantially as shown.

No. 54,886.—WILLIAM GALAWAY, Sheboygan Falls, Wis.—*Clothes Drier.*—May 22, 1866.—The upper ends of the braces project into the annular groove in the staff, to prevent vertical displacement; the heads are strengthened by annular plates.

Claim.—The annular channel C, around the point or shank of the staff, so that when the arms D I are wholly or partially distended, the head cannot be lifted from the staff, substantially as set forth.

Also, the revolving head F, resting upon the projecting metallic shoulder E, in combination with said shoulder and the notched braces D, for the purpose specified.

Also, the collar G, attached to the heads, for the purpose of preventing the same from being split, substantially as described.

No. 54,887.—ELIJAH H. GAMMON, Batavia, Ill.—*Apparatus to Prevent Horses from Kicking.*—May 22, 1866.—Straps attached to the surcingle are bifurcated and buckled to the legs, above and below the houghs.

Claim.—The device to prevent horses from kicking, &c., constructed substantially as herein described, as a new article of manufacture.

No. 54,888.—GEORGE P. GANSTER, New York, N. Y.—*Wrench.*—May 22, 1866.—A thumb wheel in the side of the handle has teeth which engage the rack on the shank of the extensible jaw to project or retract it.

Claim.—A wrench, constructed of the hollow frame or handle A, the movable rack or hammer bar B, and the voluto toothed wheel C, these several parts being combined and arranged in the manner substantially as set forth, and operating as described.

No. 54,889.—JOHN C. GARDNER, Portland, Oregon.—*Shackle for Prisoners.*—May 22, 1866.—The weighted shackle ring is supported in its position around the ankle by an inverted stirrup whose bow passes through and is riveted to the heel.

Claim.—The invention of an improved shackle and the supporter, and the mode of fastening them upon the ankle, as hereinbefore described.

No. 54,890.—DAVID GARRISON, Philadelphia, Penn.—*Machine for Jointing Oval Frames.*—May 22, 1866.—The saw carriers are adjustable in the slotted rectangular frame, and rotate in planes at an angle of 90° with each other; the segment or blank is clamped to a table which descends and brings the blank against the jointing saws.

Claim.—The combination of two saws, the cutting planes of which are at an angle to each other, a table having a movement parallel with the cutting planes of the two saws, and a set of adjustable stops or guides arranged upon the said table, in relation to each other and to the saws, substantially as and for the purpose herein specified.

No. 54,891.—JOHN GELSTON, Cincinnati, Ohio.—*Churn Dasher.*—May 22, 1866.—Air is admitted to the cream through the tubular handle and between the dashers; the lower dasher has perforations and the upper one deflecting surfaces on its lower side.

Claim.—First, the provision in the tubular handle of a churn dasher of the apertures F, for passing compressed air through the body of the cream, at right angles to said tubular handle, substantially as described, and for the purpose set forth.

Second, the arms E, when provided with deflecting faces *e*, in combination with the perforated dash D, all arranged to operate substantially as and for the purpose herein described.

No. 54,892.—FRANKLIN A. GLEASON, Brooklyn, N. Y.—*Washing Machine.*—May 22, 1866.—The corrugated interior inverted frustum has a reciprocating rotary motion in the corrugated casing of corresponding form.

Claim.—Constructing the tub and cone of corrugated sheet metal, in combination with the balance wheel, gear wheel, and pinion, substantially as herein described, and for the purposes specified.

No. 54,893.—W. W. GRIER and R. H. BOYD, Hulton, Penn.—*Machine for Making Auger Bits.*—May 22, 1866.—The revolving and longitudinally moving shaft has a recess in its end for holding and twisting the blank in connection with a series of dies arranged to clasp and hold the auger as fast as it is twisted, completing the process in one operation.

Claim.—First, a machine for twisting auger bits, drills, and similar articles, and consisting essentially of the shaft B, and a series of dies, arranged to operate substantially as and for the purpose set forth.

Second, the rotating and longitudinally moving shaft B, having a hole made longitudinally therein, of proper size and form to receive the blank and hold it while being twisted and drawn out, as shown and described.

Third, in combination with said shaft B, the tongs L, having their jaws constructed as shown and described.

Fourth, in combination with the shaft B, the screw C, with its thread constructed as described, for the purpose of imparting to the shaft an intermittent longitudinal movement, while it has a continuous rotary movement during the operation of twisting the blank for the purpose of both twisting the blank and setting the lips at one operation.

No. 54,894.—FRANK P. GRIFFITHS, Philadelphia, Penn.—*Filter.*—May 22, 1866.—The lower edge of the inner frame holds the edge of the felt diaphragm upon the inner flange of the cylinder and the frame is clamped above by the ring.

Claim.—The frame or stand clamped between the permanent flange *b* and the movable flanged ring F, for the purpose of firmly clamping the felt diaphragm to its seat, and allowing it to be readily removed, cleansed and replaced, substantially as described and represented.

No. 54,895.—P. GRISWOLD, Hudson, Mich.—*Grain Separator.*—May 22, 1866.—The grain, &c., passes over a vibrating screen which removes the tailings, and thence to a riddle which detains the coarser offal; it next falls on a lower adjustable sieve.

Claim.—First, the arrangement of the frame L, with its vibrating bottom and trough N, attached with the screen I, substantially as and for the purpose herein specified.

Second, the arrangement of the shaft H, rod P, and curved plate *b*, with the frame L, for the purpose of giving the bottom of said frame and the screen I a vibratory as well as a vertical motion, as and for the purpose specified.

Third, the adjustable cleat O, as constructed when arranged with the screen J, as and for the purpose specified.

Fourth, the arrangement of the bar R, button S, and screen K, the same being constructed and used as and for the purpose herein set forth.

No. 54,896.—GEORGE HAMMER, Philadelphia, Penn.—*Machine for Cutting Corks.*—May 22, 1866.—The spring pressure-foot descends and bends the cork so that the cylindrical cutter removes a piece which assumes the shape of a conical frustum when released.

- *Claim.*—The use of pressure pad E, when arranged and operating in combination with a cutter and cutting block, substantially in the manner and for the purpose set forth.

No. 54,897.—ASAPH L. HARWOOD, Malone, N. Y.—*Broom Clamp.*—May 22, 1866.—The ring slips on the handle and the bent loops encircle the broom and preserve its shape.

Claim.—The mode of strengthening and supporting corn brooms by means of the ring above described, attached to springs made of wire and arranged as above specified.

No 54,898.—GEORGE HASTINGS, Waltham, Mass.—*Dies for Swaging Hands for Watches.*—May 22, 1866.—The contacting faces of the dies perform the requisite cameo and intaglio functions according to the pattern prepared, and raising the elongated socket.

Claim.—The combination and arrangement of the tube projection forming chamber *f*, with the finger and ring forming parts *a b d*, made as described.

Also, the combination and arrangement of the eye former, *c*, or *c c'*, the tube projection forming chamber *f*, ring former *d*, and the finger dies *a b*, the whole being substantially as and for the purpose specified.

No. 54,899.—THOMAS HAZARD, Wilmington, Ohio.—*Fodder Cutter.*—May 22, 1866.—The guide which determines the length of the feed is adjustable relatively to the mouth-piece by a slot and set screw. During the feed the spring pressure board is lifted by the back of the knife which strikes the roller on the pressure frame.

Claim.—First, the provision of the guard or mouth-piece H, having the slotted arm I *i*, and screw bolt N, and being formed and adapted to operate in connection with the throat G, and knife K, in the manner explained.

Second, the arrangement of obliquely moving knife K, pressure board 12, arm 13, and roller 14, for the purpose described.

No. 54,900.—ISSACHAR A. HEALD, Carlisle, Mass.—*Machine for Making Cigars.*—May 22, 1866.—The devices described in the claim are for rolling and forming the fillers of cigars, smoothing and cutting the wrappers, and putting the wrappers on the fillers.

Claim.—First, mounting the roller D, in such a manner that it can be moved in the arc of a circle, without throwing it out of gear, substantially as and for the purpose set forth.

Second, forming a filler or body of a cigar by rolling it between two disks more or less inclined, substantially as shown and described.

Third, smoothing and cutting the wrappers for cigars, substantially as herein shown and described.

Fourth, cutting the end of the cigar off, by moving it past a knife fixed in position, substantially as shown in figure 1, and herein described.

Fifth, the socket or cup *t*, arranged to operate substantially as described, to hold and form the end of a cigar ready to receive the wrapper.

Sixth, providing the surface of the rollers with a series of cavities increasing in size and depth toward the centre and decreasing towards the end of the rolls, as shown and described.

Seventh, the revolving and sliding fingers *e'*, for seizing and holding the wrapper, substantially as set forth.

Eighth, holding the end of the wrapper upon the cup *t*, or other device more unyielding than the cigar itself, said device being arranged opposite to the end of the cigar and caused to revolve with it, for the purpose of drawing the wrapper sufficiently tight.

No. 54,901.—J. L. HEISEY, Elizabethtown, Penn.—*Corn and Seed Planter.*—May 22, 1866.—The seeding devices are operated by gearing from the ground wheel, and adjusted by a hand lever on the main shaft and intermediate devices.

Claim.—First, the radiating rods M, with their markers N, across the periphery of the wheels, supported on and made adjustable by the loose cog wheels, G' G², in combination with the gear I, shifting and turning lever E, and side-toothed pulley D, all on the seeding or roller shaft B, arranged and operating substantially in the manner and for the purpose specified.

Second, the combination of the seed valves, when placed in the bottom of the spouts P, and operated in unison with the markers N, simultaneously actuated by the seeding or roller shaft B, with its gearing I, and depressing loops F, arranged substantially in the manner and for the purpose set forth.

Third, the adjustable mould-board-like side plate X, adjustable on the supporting bars T of the shovels R, constructed and operated in the manner specified.

No. 54,902.—A. R. HENDERSON and JAMES FORD, Andover, N. Y.—*Bed Bottom.*—May 22, 1866.—The double slats with intervening springs are suspended by webbing from the bedstead rails.

Claim.—The combination of the side rails C D, hanging webbing G, double slats E F, arranged in the manner and for the purpose herein specified.

No. 54,903.—FOSTER HENSHAW, Washington, D. C.—*Window Sash Spring.*—May 22, 1866.—The steel spring has an inclined slot at the top in which the journal of the roller descends as the sash is raised, and rises as the weight of the sash revolves it and jams the surface of the roller against the window casing.

Claim.—The construction of the steel wire spring with its incline slot at top, in which the grooved axle of the roller operates and is combined as herein described and for the purposes set forth.

No. 54,904.—MICHAEL HEY, Philadelphia, Penn.—*Bung for Beer Casks.*—May 22, 1866.—The described device is to render removable the valve of the self-venting beer barrel bung, and prevent its becoming gummed fast.

Claim.—The sharp edged disk figure 3, when applied to operate upon a flat seat in a ventilating beer bung, as and for the purpose described.

No. 54,905.—BIRDSILL HOLLY, Lockport, N. Y.—*Air-compressing Pump.*—May 22, 1866.—Water is urged by the piston to force air into the reservoir, which collects any water which passes over and returns it to the pump; the air is admitted after each pulsation at a valve way.

Claim.—First, the combination and arrangement of the air and water reservoir L, with the water extension B', of the pumping cylinder B, operating substantially as described.

Second, the cylinder B, constructed substantially as described in combination with the reservoir L, and water extension B'.

No. 54,906.—WILLIAM CLEVELAND HICKS, New York, N. Y.—*Steam Engine.*—May 22, 1866.—To one shaft are connected four pistons which with their cylinders are so provided with induction ports, that each piston and its cylinder performs the functions of a valve and steam passage for the adjoining cylinders.

Claim—The arrangement of the cylinders, valvular pistons, and steam admission and exhaust passages, substantially as herein described, to operate in the manner and for the purposes set forth.

No. 54,907.—FREDERICK HOFFMAN and ROBERT WENDLER, Brooklyn, N. Y —*Disinfecting Composition.*—May 22, 1866.—Composed of charcoal, carbolic acid, burned lime, and hypochloric acid, in combination with the bases of alkaline earths; vegetable absorbents, such as sawdust, may be added.

Claim.—A disinfecting compound made of charcoal and hydrated oxide of phenyl, substantially as described.

Also, a disinfecting compound of charcoal, hydrated oxide of phenyl, and burned lime as described.

Also, a disinfecting compound made of charcoal, hydrated oxide of phenyl, burned lime, and hypochloric acid, and with or without the addition of an indifferent vegetable cellulose as specified.

No. 54,908.—JOSEPH HORNER and SILAS VAN DOREN, New Brunswick, N. J.—*Window Shade Fixture*—May 22, 1866.—The brackets which support the ends of the rollers have flanges which rest on the top of the window frame; T-shaped projections on the spindles enter corresponding slots in the ends of the rollers.

Claim.—First, the brackets B B, for supporting a roller D, constructed with flanges which are applied to the top of the frame, substantially as shown and described.

Second, attaching the roller to its journals, by forming in its ends a T-shaped groove into which is inserted a T-shaped arm or plate extending from each journal, substantially as shown and described.

No. 54,909.—DAVID K. HOXSIE and THOMAS L. REED, Providence, R. I.—*Manufacture of Flexible Tubing.*—May 22, 1866.—Explained by the claim.

Claim.—In combination with flexible gas tubing, as a finish for the same, a covering of webbing or braid of many single strands, plaited in minute stitch-like wales, substantially as described for the purpose specified.

No. 54,910.—EBENEZER HUNT, Danversport, Mass.—*Building Concrete and Mortar Walls.*—May 22, 1866.—The walls are formed hollow by casting the concrete around cores placed against the inner face of the wall in the mould; the said cores are afterwards withdrawn.

Claim.—First, the construction of concrete or mortar walls for dwelling and other houses by the use of cores made of sand, loam, or some other cheap material, so arranged within the moulds which form the walls as to leave a space for the concrete or mortar forming the exterior surface of the wall, and from this outer face of the wall projections of the same material extending perpendicularly from top to bottom and inwardly as far as the contemplated thickness of the wall, strengthened, if required, by horizontal projections of the same material at proper distance, substantially in the manner and for the purposes set forth.

Second, the building of posts, fences, roofs, floors, flag-stones, and other like structures of concrete or mortar, or their equivalents, by the use of cores, substantially in the manner and for the purposes set forth.

No. 54,911.—FRANK P. HUNT, Medina, N. Y.—*Stove Pipe Drum.*—May 22, 1866.—The heater has two inner perforated cylinders; the outer one movable to make its holes register with the inner one, or otherwise; a damper at the other end closes or opens simultaneously with the rotary motion.

Claim.—First, the interior perforated cylinders I and K, and the perforated disks J and L, constructed and arranged substantially as described, in combination with each other and with the exterior cylinder A, substantially as and for the purpose set forth.

Second, the combination of the damper O with the interior cylinder I and the exterior cylinder A, substantially as described and for the purpose set forth.

No. 54,912.—JACOB B. HYZER, Janesville, Wis.—*Heat Radiator.*—May 22, 1866.—The volatile results pass up the front flues and rearward through an opening to an exit; or, if that be closed by damper, then downward to a chamber above the fire, and thence revert upwardly to the flue opening.

Claim.—The combination and arrangement of the flues F, and hot air chamber T, (when constructed with the partition U and the damper S,) with the flues G, hot air chamber I, and return flues H, substantially as and for the purpose set forth.

No. 54,913.—MARSHALL INGERSOLL, Grafton, Ohio.—*Car Coupling.*—May 22, 1866.— The coupling hook is depressed by springs and raised for uncoupling, by a lever, lifting hook, and pawl.

Claim.—The arrangement of the coupling hooks C, connected to the springs *g* and *a*, in combination with the lifting hook J', lever *m*, pawl E, and bumpers, operating in the manner and for the purpose set forth.

No. 54,914.—ALBERT JACKSON, Clifton Springs, N. Y.—*Hay Rack for Wagons.*—May 22, 1866.—The bars which support the bows to make the side extensions of the hay rack are detachable, and capable of forming the vertical standards of the wood-rack; enabling the rack to be converted as to its purpose, and removable by sections to facilitate handling.

Claim.—First, in connection with the frame A, the detachable bars B, constructed of two parts *a a*, and fitted to the frame in the manner substantially as shown and described.

Second, the removable wheel fenders G attached to the cleats E E' H, on the bars B B B I, substantially as and for the purpose set forth.

Third, the removable boards F F, placed between the cleats E E, substantially as described.

Fourth, the combination of the frame A, bars B, fenders G, and ladder or guard L, all arranged substantially as and for the purpose specified.

No. 54,915.—BERNARD JACOBS, New York, N. Y.—*Attachment to Gas Fixtures.*—May 22, 1866.—The spring valve closes automatically when the burner is removed.

Claim.—The spring valve *b*, or its equivalent, in combination with the pipes A and B, or in the joint between two pipes bearing a similar relation to each other as said pipes, substantially as and for the purpose set forth.

No. 54,916.—J. D. JENKINS, Jackson, Ill.—*Scuffle Hoe.*—May 22, 1866.—The rake teeth depend from a rod above the blade of the scuffle hoe.

Claim.—First, the combination of a rake and hoe blade A, when arranged to operate substantially as described.

Second, making the blade A with a concave bottom surface, and connecting this blade to the forked head at a point which is beneath the rake teeth, substantially as described.

No. 54 917.—JOHN JOHNSON, Saco, Me.—*Collecting Light Oils from Oil Wells.*—May 22, 1866; antedated May 10, 1866.—Explained by the claim.

Claim.—The separating and gathering at the wells a new article of commerce, viz: the light condensable vapors which rise with petroleum, and which may be pumped off, under any suitable seal, and condensed, substantially as herein set forth.

Also, volatilizing and condensing the condensable products from petroleum at the wells, by forcing air or gases through petroleum as carriers of the light products, for the purpose herein set forth.

No. 54,918.—FREDERICK JUDSON, Jersey City, N. J.—*Calk for Horse Shoes.*—May 22, 1866. The taper shanks of the calks are received in corresponding sockets through the enlarged toe and heel portions of the shoe.

Claim.—The tapering calks *c c'*, constructed as described, in combination with the projections B B and tapering sockets *a a* of the shoe A, arranged and operating in the manner and for the purpose herein described.

No. 54,919.—ERNST KAST, Waterbury, Conn.—*Reading Desk.*—May 22, 1866.—The reading desk is supported on a telescopic bar, vertically adjustable on the standard.

Claim.—The reading desk A, provided with an adjusting shank B, in combination with the tubular arm C, which is adjustable on the vertical column D, and with the table E, constructed and operating substantially as and for the purpose described.

No. 54,920.—JOEL F. KEELER, Pittsburg, Penn.—*Machinery for Forming Flanges upon Boiler Heads*—May 22, 1866.—The plate is clamped between two rotating tables, and flanged by a conical former, which is raised by a lever against the edge of the rotating plate.

Claim.—The rotary tables *b d* and *e e'*, in combination with the cone *g*, or their equivalents, for the purposes set forth.

No. 54,921.—JOHN KIEFER, Hamilton, Ohio.—*Broom Head.*—May 22, 1866.—The brush is clamped between the roughened surfaces of the central plate and the inner and outer caps.

Claim.—The two series of leaves *a a a* and *a' a' a'*, in combination with the central continuous leaf *b*, each of which is provided with teeth by punching the sheet metal as described, and so arranged as to be capable of self-adjustment upon the stalks of broom corn constituting the wisps, when compressed thereon, in the manner and for the purpose herein set forth.

No. 54,922.—PATRICK KILLEN, Mount Healthy, Ohio.—*Alarm Whistle.*—May 22, 1866.—A bellows attached to a whistle to sound an alarm or call.

Claim.—The bellows B in the cylindrical chamber A, the spring C, rod D, and whistle or horn G, constructed and operated as above described and for the purpose set forth.

No. 54,923.—AUGUST KOLLENBERG, Owensboro', Ky.—*Machine for Making Tin Fruit Cans.*—May 22, 1866.—The tin is wrapped around an extensible cylindrical former and held in position by a lever till the lapping edges are soldered.

Claim.—The combination with the adjustable cylinder A, of the retaining lever C, arranged to operate substantially as described.

No. 54,924.—LEONARD LAKIN and NORMAN HALL, Brodhead, Wis.—*Heating Stove.*—May 22, 1866.—Between the inner fire chamber and the exterior semi-annular flues are air passages traversed by pipes, which connect the former. The volatile results rise, descend in the flue which occupies one side, traverse the hollow base, and rise in the other side to the exit pipe.

Claim.—The construction of a stove, with its fire chamber surrounded by a smoke chamber, and an open space between said fire and smoke chambers, and the conveyance of heat and smoke from the one chamber to the other by means of the flues in such manner as to secure three heating or radiating surfaces and a draught of air between said fire and smoke chambers, substantially as herein set forth.

No. 54,925.—JOHN W. LATCHER, Albany, N. Y.—*Car Wheel Lubricator.*—May 22, 1866; antedated May 7, 1866.—The hollow in the wheel forms an oil reservoir, open toward the bearing on which it revolves, and with the supply boles plugged.

Claim.—First, forming an oil reservoir B in the body of independent revolving car wheels, and the oil retained therein for gradual use, by means of plugs d in combination with stationary journals C, substantially as described.

Second, forming an oil reservoir in such a manner as shall compromise unequal shrinkage, as shown and described.

No. 54,926.—WILLIAM M. LEAVENS, Philadelphia, Penn.—*Sewing Machine.*—May 22, 1866.—Improvement on Sibley's patent, No. 42,117. The object is to throw the looper for making the Grover and Baker stitch, in or out of action, without removing the Sibley attachment from the Wheeler and Wilson machine when exchange of stitch is required.

Claim.—First, preventing the operation of the thread carrier F by throwing it back and securing and retaining it in that position, so that the stitch may be changed without further modification or substitution of the parts, substantially as described.

Second, securing and retaining the thread carrier F in a position to prevent it from operating by means of lever P, pin n, arm m, or their equivalents, substantially as described.

No. 54,927.—W. K. LEWIS, Boston, Mass.—*Pea and Bean Sheller.*—May 22, 1866.—The pods are fed by an endless apron longitudinally, and cut by the serrated wheels; the toothed rollers then carry off the hulls, allowing the seed to drop into the receptacle.

Claim.—The serrated wheels i, placed on a shaft D, over the front part of the apron C, when used in combination with the endless apron and rollers B B, and as an auxiliary feed for said rollers, for the purpose specified.

No. 54,928.—HENRY LOWENBERG, New York, N. Y.—*Buckle.*—May 22, 1866.—Explained by the claim.

Claim.—A buckle, having sectional cross-bars with points turned toward each other, and that will hold and be adjustable on both straps without being permanently attached to either, and constructed substantially as herein described and represented.

No. 54,929.—WILLIAM W. LYMAN, West Meriden, Conn.—*Device for Opening Fruit Cans.*—May 22, 1866.—A bridge rests upon the shoulder of the jar and affords a fulcrum for the lever, which tears up the cover with its swinging hook.

Claim.—The fulcrum bridge or yoke d and lifting lever e, substantially as and for the purpose as described.

No. 54,930.—DAVID W. MAURICE, Springfield, Ohio.—*Churn Dasher.*—May 22, 1866.—Slats, triangular in their section, and revolving edge foremost, are placed between the disks on the horizontal shaft.

Claim.—An improved churn dasher, constructed and arranged substantially as herein described and for the purpose set forth.

No. 54,931.—JAMES McKELVEY, Ogdensburg, N. Y.—*Refrigerator.*—May 22, 1866.—The water from the ice chamber is filtered in its passage to the reservoir below.

Claim.—The filter J, in combination with the ice chest I and refrigerating chamber H, arranged substantially in the manner and for the purpose set forth.

No. 54,932.—HARRISON B. MEECH, Fort Edward, N. Y.—*Manufacture of Paper Pulp.*—
May 22, 1866.—The vegetable matter is cut into pieces a few inches long, and treated in a
rotary boiler with a watery solution of lime, soda ash, and common salt.
Claim.—First, the method of treating hay, hemp, flax, and other grasses for the manufacture of paper, substantially as described.
Second, using liquors in a rotary boiler prepared in the manner and for the purpose herein set
forth.

No. 54,933.—JOSIAH V. MEIGS, Washington, D. C.—*Apparatus for Making Coffee.*—
May 22, 1866.—The ground coffee is enclosed in a tube, with a perforated diaphragm at
each end; this is immersed in a boiler, the metallic top of the tube forming a joint; when
the steam is generated the water is forced through the stratum of coffee and discharged at a
spout connected to the coffee tube.
Claim.—First, making coffee by enclosing the ground coffee in an immersed chamber
through which hot water is forced, substantially in the manner herein described.
Second, the combination of the kettle or boiler with the coffee chamber, when arranged
and operating substantially in the manner and for the purpose set forth.
Third, the combination of the coffee tube with a check, substantially as and for the purpose described.
Fourth, cleansing the coffee tube by condensing the steam in it, as described.
Fifth, the combination of the coffee pot, coffee chamber, and inclined planes I, J, substan-
tially as and for the purpose described.

No. 54,934.—J. V. MEIGS, Washington, D. C.—*Breech-loading Fire-arm.*—May 22, 1866.—
The breech block descends in a transverse vertical slot in the stock and is thus moved by a
link connecting with the trigger guard, which, together with the trigger, has a longitudinal
sliding movement along the shank of the stock. A spring detent, which holds the guard
plate in position, is released by the action of the trigger after the discharge of the piece.
Claim.—First, reciprocating the breech block vertically by reciprocating the guard, or its
equivalent, horizontally.
Second, the combination of the trigger with the dog and sliding guard, to fire the gun and
lock or release the breech block, substantially in the manner described.
Third, the combination of the reciprocating guard and vertically sliding breech block with
the bar carrying a rebate to correspond with the recess of the barrel for the purpose of removing the cartridge case, substantially as described.

No. 54,935.—RUFUS S. MERRILL, Boston, Mass.—*Lamp.*—May 22, 1866.—The metallic
wick tube is insulated from the cap of the reservoir by an interposed non-conductor; the
drip is vaporized and ascends again to the burner.
Claim.—First, the method herein described of insulating the wick tube from the cap of the
vessel by the interposition of a non-conductor of heat, in the manner and for the purpose set
forth.
Second, the combination with a cap provided with a socket to contain the insulating
cementing substance of the wick tube provided with an annular plate, to prevent its being
moved within the cementing matter.
Third, as an attachment to and in combination with the wick tube, annular chambers or
reservoirs, arranged in the manner and for operation as described.

No. 54,936.—JAMES MADISON MILLER, Greenwood township, Penn.—*Loading Attachment to Hay Wagons.*—May 22, 1866.—The pinion wheel is advanced by a lever to the drive
wheel, and when engaged thereby winds up the rope to which the hay fork is attached, and
the rotating crane swings it over the wagon bed.
Claim.—The wheel m m attached to the wagon wheel in combination with the wheel N,
lever D, bar E E, and the crane P O O, when the same are constructed in the aforesaid combination and for the purpose set for thin the manner described.

No. 54,937.—JOHN C. MILLER, Amsterdam, N. Y.—*Water Wheel.*—May 22, 1866.—The
water is delivered on to the interior of the wheel by a helical scroll; the wheel revolves horizontally, has a close bottom, curved buckets, and peripheral discharge.
Claim.—First, as an improvement in water-wheels, the arrangement of the spiral or helical
scroll F and curved buckets C of the wheel D, when constructed and operating in the manner and for the purpose herein specified.
Second, the combination of the helical or spiral scroll, circular bottom plate, and curved
buckets, when applied in connection with a water-wheel which receives the water at its centre
and discharges it from its periphery, in the manner substantially as herein described.

No. 54,938.—GEORGE RODNEY MOORE, Lyons, Iowa.—*Heating Stove.*—May 22, 1866.—
The space included between the stove door and the fire and the diaphragm plate is used as a
food-warmer or oven.
Claim.—The construction of a fire chamber with a part H shelving inwardly and termi-

nating at any desired distance above the grate I, for the purpose and in the manner herein set forth.

Also, the recess or chamber between the fire J, the plate H, and door D, as and for the purposes described.

No. 54,939.—C. NEUMANN, New York, N. Y.—*Machine for Attaching Clasps to Skirt Hoops.*—May 22, 1866—Explained by the claim.

Claim.—The cup-shaped receptacle with flanged edges to its conducting chute, constructed, arranged, and operating substantially as described and represented.

No. 54 940.—JOHN NICHOLSON, Allegheny City, Penn.—*Feed-water Heater*—May 22, 1866.—The feed-water trickles over the surface of the steam chamber and unites with the water of condensation from the latter to pass out at the lower orifice.

Claim.—The steam chamber A, furnished with an inlet B and outlets C e, for steam, and placed within a water chamber D which is connected to the steam boiler by pipe X and to the supply of water by pipe O, the whole being constructed, arranged, and operating substantially as herein described and for the purpose set forth.

No. 54,941.—JOHN K. O'NEIL, Kingston, N. Y.—*Spring Balance.*—May 22, 1866—Several springs are arranged in the instrument, and they are capable of separate or conjoined use in connection with several scales of varying range.

Claim.—The arrangement of any desired number of springs, so that by detaching or attaching they may be used either separately or conjointly for a weighing instrument, substantially in the manner herein described for the purposes set forth.

No. 54,942.—ALBANY PACKHAM, Prestonville, Ky—*Cotton Seed Planter.*—May 22, 1866—The seed is stirred in the hopper by the bar which wipes the inner periphery and the toothed spiral which revolves therein. The seed drops into a furrow and is covered by a share and roller.

Claim.—The inverted conical hopper F, in combination with the rotary spiral toothed strip or seed distributor G*, with bar H attached, all arranged on a mounted frame to operate in the manner substantially as and for the purpose herein set forth.

No. 54,943.—HALSEY PELTON, Erie, Penn.—*Lime Kiln*—May 22, 1866.—Explained by the claim.

Claim.—First, making the furnaces of lime kilns with upper openings or doors for admitting coal in combination with horizontal doors or openings for admitting wood, so that either kind of fuel can be used, substantially as described.

Second, the block G, made substantially as described, for filling the front part of a furnace when coal is to be used as fuel, and for conducting the coal to the grate, as above shown.

Third, making the body of a lime kiln substantially as above described, that is to say, with a metallic outer cylinder, a fire-brick or equivalent lining, and a non-conducting material between the two, with a projecting base for the arrangement of flues.

Fourth, making a horizontal flange on the inside of the metallic cylinder I, for the purpose of separating the upper part of the inner lining from the lower part, substantially as above described.

Fifth, the arrangement of the gate H and base of the kiln, in the manner and for the purpose herein specified.

No. 54,944.—E. G. PERRY, Knoxville, Ill.—*Medical Compound.*—May 22, 1866.—For the cure of hog cholera; composed of unslaked lime, sal soda, mandrake root, garget root, saltpetre, copperas, ginger, and sulphur.

Claim.—The within-described compound, made of the ingredients set forth and mixed together, substantially in the manner specified.

No. 54,945.—FRANCIS W. PERRY, and JOHN H. PIERCE, Woburn, Mass.—*Apparatus for Making Extracts from Bark, &c.*—May 22, 1866.—The ground bark is placed in a revolving cylinder: steam is admitted through the trunnions; the decoction is drawn off through a strainer and lower duct.

Claim.—First, a revolving and self-emptying leach for extracting or displacing tannic acid from bark, substantially as herein described.

Second, the combined arrangement, substantially as herein described, for revolving the leach and introducing steam thereto, or any analogous device to produce the same results, when used for extracting tannic acid from bark for tanners' use.

No 54,946.—JOHN G. PERRY, South Kingston, R. I.—*Harvester.*—May 22, 1866.—The "floating" finger bar is arranged in line with the main driving wheels; the cross-timbers of the main frame are connected at the end next to the cutters by a rim which constitutes the axle of the inner drive wheel, and through which the drag bar is passed.

Claim.—First, the centrally open rim or axle z, or its equivalent, of the main wheel C, in combination with the frame B, constructed substantially as herein described and for the purpose set forth.

Second, the brace rod J, passing through the open wheel C or axle z, in combination with the drag bar or shoe V, connected to the frame B, substantially as described and for the purpose set forth.

Third, the combined hand and foot lever s a', arranged for raising either or both ends of the cutting apparatus, substantially as described and for the purpose set forth.

Fourth, the arrangement of the bevel gears a m, crank l, or its equivalent, and pitman S, in combination with the open wheel C, or axle z and main frame B, substantially as herein described and for the purpose set forth.

No. 54,947.—WILLIAM M. PIATT, West Liberty, Ohio.—*Horse Rake.*—May 22, 1866.—The rake is held to its position or raised for discharging by a bar and hooks respectively; these depend from a hand lever pivoted to a post on the axle and are swung back out of engagement with the teeth by a cord.

Claim.—The arrangement of engaging and releasing mechanism J K L M N P, all constructed and operating as and for the purpose specified.

No. 54,948.—ASHAEL PIERPONT, New Haven, Conn.—*Boot-jack.*—May 22, 1866.—The folding boot-jack is hinged in the rear of the legs.

Claim.—The combination of the two parts A and D for a folding boot-jack, when hinged laterally and constructed, arranged, and fitted for use, substantially as herein described and set forth.

No. 54,949.—J. C. PLUMER, Boston, Mass.—*Vegetable Slicer.*—May 22, 1866.—The end of the vegetable is presented through a circular socket to the knives on the revolving disk.

Claim.—The disk P with the raised cutting edges E E', as described, in connection with the support S and receptacle R, all substantially as and for the purpose specified.

No. 54,950.—JULIUS R. POND, New Hartford, Conn.—*Preparing Vacuum Vessels for Condensing Milk, &c.*—May 22, 1866—The pores of the vessel are filled with a fatty substance to prevent the accumulation of coagulated albumen thereon.

Claim.—Treating metal with an oily or greasy substance, substantially such as set forth, or its equivalent, in the manner and for the purpose substantially as described.

No. 54,951.—WILLIAM E. PRALL, Knoxville, Tenn.—*Corn Harvester.*—May 22, 1866.—The cutter is annular, is attached to the upper side of a balance wheel supported by an anti-friction roller and cuts against a finger and guard. The cutter can be stopped by the engagement of a lip with a notch in the edge of the balance wheel.

Claim.—The combination of the cutter provided with a ratchet edge upon its under side, cutter bar R, provided with a lip S and friction wheel P, when arranged and operated substantially as and for the purpose set forth.

No. 54,952.—D. A. PRATT, New York, N. Y.—*Lock.*—May 22, 1866.—The described devices are guards against the key being turned by an "outsider," or being displaced from the outside for the introduction of a pick.

Claim.—First, the employment or use of the movable lip c on the key C, in connection with a notch in the stationary plate B, attached to the lock, or to the escutcheon on the door, arranged in such a manner that the one will, as the key is turned in order to lock the door, engage automatically with the other, substantially as set forth.

Second, the tumbler E, arranged within the plate B, so as to be operated by the turning of the key in the manner shown, or in any equivalent way, so as to serve as a guard to the key-hole to prevent the passing of a rod or instrument through the key-hole to act upon the lip c, and thus admit of the key being turned from the outer side of the door.

Third, the combination of the lip c upon lever d, applied to the key C, the arm k, the notch b in the plate B, and the rotary tumbler E, all arranged and applied to operate in the manner substantially as and for the purpose set forth.

No. 54,953.—D. A. PRATT, New York, N. Y.—*Sash Fastening.*—May 22, 1866.—Spring dogs in the window casing engage the notched edge of the sash, and are withdrawn by pressing two levers together.

Claim.—First, the sash fastening in a sash frame with its catches d placed on separate rods d d', arranged and constructed as described, so that the point of contact with the stile B, of the window, is at the centre of the frame of the fastener b.

Second, the combination of the catches d, secured laterally to the levers c c', end springs e, applied to the frame A, arranged to operate with the notches on the stile B, of the sash, in the manner and for the purpose herein specified.

No. 54,954.—LEVI H. PROCTOR, East Saugus, Mass.—*Shoe-jack.*—May 22, 1866.—The upper plate has a circular adjustment on the lower and carries a reversible jack; the heel rest is movable, and slides upon a rack toward and from the toe rest. A quadrant bar depending from the upper rest affords inclined adjustment.

Claim.—The combination and arrangement of the arms *d e*, the spring *o*, the grooves *i l*, the racks *k m*, and the sets of teeth *f g*, with the heel and toe rests F G, the plate C, and its plates E E, the whole being substantially as and to operate as described.

Also, the arrangement and application of the pin *p*, the bolts *r*, and spring *s*, and nut *t*, with the heel rest F.

Also, the application of the curved bar H to the plate B, by means of the socketed projection *w*, the hole *b'*, and the ears *y y*, arranged as described, whereby the two may be connected or disconnected as circumstances may require.

No. 54,955.—GEORGE QUICK and JOHN N. WALLIS, Fleming, N. Y.—*Car Coupling.*—May 22, 1866.—A swinging tongue vibrates before the link, gravitates into a slot therein, and is detained by contact below. Slotted side pieces raise the tongue to release the link.

Claim.—First, the swinging bolt F, when combined with the link A, slotted piece K, and shaft G, in the manner and for the purpose described.

Second, the arrangement, as described, of shaft G, with its arms J and I, and slotted piece K, to release the coupling of the cars without passing between them, in the manner and for the purpose described.

Third, the combination of the hook N upon the link A, with the lug M on shaft G, so that on the withdrawal of the link the coupling will be restored to its normal position as herein described.

No. 54,956.—THOMAS D. READ, F. R. READ, and CYRUS DORREL, Rising Sun, Ind.—*Hay and Cotton Press.*—May 22, 1866; antedated May 3, 1866.—The clamps in the beater, and the notches for their reception in the frame, prevent the upward motion of the beater when the bale is compressed.

Claim.—The combination of the levers *d e* and *i*, rollers *f* and *h*, inclined plate *g*, and notch *j*, in the guide of the beater, constructed as above described and for the purpose set forth.

No. 54,957.—GEORGE H. REED, Boston, Mass.—*Dyes and Colors.*—May 22, 1866.—The liquid dyes are prepared with mordant and the addition of alcohol, glycerine, wood spirit, or other preservative, against freezing, fermentation, or the deposition of the ova of intusorial animalculæ.

Claim.—First, as a new manufacture or composition of matter, and as a new article of commerce, liquid dyes produced from vegetable or mineral coloring matters, so mixed and prepared with the proper proportion, or quantity, of mordants, and glycerine, alcohol, pyroxylic spirit, or wood naphtha, or either of them, that the same will endure both heat and cold, and may be kept for any period without undergoing change, and will produce a complete and effectual dye upon either silk or woollen cloth, or fabric, at one application.

Second, the process of making liquid dyes in all colors by the mixture and preparation of coloring matters, and mordants, either of vegetable or mineral origin, or both, with the addition of glycerine, alcohol, pyroxylic spirit, or wood naphtha, or either of them, in such manner that the dyes will endure both heat and cold, and may be kept for any length of period without undergoing changes, and will produce a complete and effectual dye upon either silk or woollen cloth, or fabric, at one application, in the manner substantially as above set forth.

No. 54,958.—UEL REYNOLDS, New York, N. Y.—*Fifth Wheel for Vehicles.*—May 22, 1866.—Upon the rim of the wheel are anti-friction rollers, which bear up the bolster.

Claim.—A stop guard for preventing the wheels coming in contact with the body, formed by projections or stops upon the fifth wheel, substantially as set forth.

Also, the anti-friction rollers and jaws, in combination with the fifth-wheel circles or arcs, as set forth.

No. 54,959.—THOMAS RICHARDS, Lansingburgh, N. Y.—*Balanced Valve for Steam Engines.*—May 22, 1866; antedated May 7, 1866.—Peripheral openings in the conical valve and interior openings in the corresponding face of the valve seat preserve an equilibrium of steam pressure.

Claim.—The within described arrangement of the ports in valve seat and valve, so as to make the steam and exhaust ports balanced, under any pressure of steam, substantially as set forth.

No. 54,960.—JOHN W. RICHARDSON, DANIEL L. DAVIS, and WILLIAM C. HOWELL, Sligo, Ohio.—*Cane Mill.*—May 22, 1866.—The cane is fed between rollers to a revolving cutter, and passes thence to a grinding hopper and to crushing rollers, being subjected in transitu to a jet of steam.

Claim.—First, the provision, in a cane mill, of the series of feeding, chopping, crushing, and expressing apparatus, arranged and co-operating as set forth.

Second, the combination of the revolving sickle-formed knives E, throat piece F, and intermittent feed mechanism G *g g'* H I J K, substantially as set forth.

Third, in a cane mill, arranged and operating substantially as described, the nozzle Z', when used for injecting steam upon the cane while in the operation of being crushed, for the purpose set forth.

No. 54,961.—W. J. Ross, Worcester, Mass.—*Sash Fastener.*—May 22, 1866.—The tooth of the pivoted spring tumbler engages holes in the casing to sustain the sash, and is with-drawn by the trigger, allowing the sash to be raised by the thumb-piece.

Claim.—First, the combination with the case A of the lift-piece C, lever D with its locking projection *d*, and spring *f*, substantially as set forth.

Second, making or coring out the case A, as seen at *h*, in combination with the use of spring *f*, as and for the purposes stated.

Third, the combination with the projections *d* and *e*, on the end F of lever D, of the edge *g*, as and for the purposes stated.

No. 54,962.—PHILIP C. ROWE, Boston, Mass.—*Tool for Removing Scale in Boiler Tubes.*—May 22, 1866.—The cutters are adjusted in the cutter head by a nut; the head is adjusted by an inner and a sleeve screw in the clamp, and the latter is secured by a turn-buckle.

Claim.—In combination with a cutter head, the tubular screw *d* and its internal screw *i*, arranged to operate substantially as described.

Also, in combination with the piece *c*, supporting the nut for the screw *d*, the turn-buckle clamping device, substantially as set forth, and the construction and arrangement of the parts described by which the cutters of the head are adjusted, substantially as specified.

No. 54,963.—JACOB L. RUNK, JAMES H. BROWN, and ELIAS M. MORGAN, Nashville, Ill.—*Gang Plough.*—May 22, 1866.—The described devices are for raising the ploughs from the soil, maintaining them in position, elevated or otherwise, the plough beam being attached to a bolster hinged to the axle.

Claim.—First, the plough beam attached to a bolster, the latter being hinged or pivoted to the axle or carriage, substantially as and for the purpose described.

Second, at'aching the draught-rod to the beam in such a manner as to utilize the draught of the team in raising the ploughs from the ground, when the forward support of the beam is removed.

Third, the combination of the beam A and tongue G, or its equivalent, affixed to the carriage, with the toggle V U, the latter forming a brace, when straight, and a means, by deflection, of depressing the fore end of the beam.

Fourth, the combination of the beam A, tongue G or its equivalent, and the notched stay-hook X.

Fifth, the arrangement of the hand-lever S, connecting rod T, beam A, and carriage or tongue, operating as described.

Sixth, the arrangement of the hand-lever S, connecting rod T, toggle U V, beam A, and tongue or equivalent, substantially as described and represented.

Seventh, attaching the draught rod to a point so far removed from the fore end of the beam, and above and near the axis of vibration of the latter, as to maintain the draught point, when the ploughs are out of the ground, at or about the same height as when in working position.

Eighth, the combination of the beam, hinged to the bolster of the foot-board and pivoted hand lever, as described and represented.

Ninth, the seat post pivoted to the carriage and beam, substantially as and for the purpose described.

Tenth, the arrangement for the lateral adjustment of the tongue, for the purpose described, consisting in the pivoting of the tongue on a vertical bolt in the axle, and the side braces adjustable as to length, as described and represented.

No. 54,964.—SARAH E. SAUL, New York, N. Y.—*Boiler for Culinary Purposes.*—May 22, 1866.—A segment of the cover is hinged, and from the other portion a descending pipe carries the steam and odor into the fire chamber.

Claim.—The combination of the hinged cover *b* and tube C, arranged relatively with the boiler A and escape-pipe *a a d*, constructed and operating in the manner and for the purpose herein specified.

No. 54,965.—EMERY SAWYER, Millbridge, Maine —*Washing Machine.*—May 22, 1866.—The slatted and weighted pounder is attached to the lid, which is hinged to the box and vibrated by a revolving cam. The wringer is a slatted box with bridge and pressure screw.

Claim.—The hinged cover, with its washing ribs C, when operated by the cam wheel G, as and for the purpose described.

Also, the wringing box, working between the arms *f* and furnished with the rubber bottom and screw pressure, as and for the purpose described.

No. 54,966.—CHARLES C. E. SCHWARTZ, Philadelphia, Penn.—*Instrument for Cutting Finger Nails.*—May 22, 1866.—The nail is inserted in the circular opening, the plate rests on the finger, and the sliding knife is pressed against and traverses across the nail.

Claim.—First, the plate A, the circular shaped opening *a*, and the straight opening or guide *b*, substantially as specified and set forth.

Second, the combination of the plate A, the circular-shaped opening *a*, the straight opening *b*, the sliding knife B and set screw *d*, and the serrated edges or files *c*, as specified and described.

No. 54,967.—DAVID W. SEELEY, Albany, N. Y.—*Brick Press.*—May 22, 1866.—Opposite sides of the brick are supported by the perforated ends of the reciprocating piston and platen respectively while being discharged from the mould box on to the endless apron.

Claim.—First, a brick press which is so constructed that the brick will be supported on two opposite sides during the act of releasing the other sides of the brick, substantially as described.

Second, the combination of a reciprocating mould box D, a reciprocating piston E, and a platen G, the piston and platen having each an irregular motion, and all operating so as to press and release bricks, substantially as described.

Third, discharging the bricks from the pressing devices, substantially as described.

Fourth, constructing the ends of the rectilinear-moving piston and platen, substantially as described.

No. 54,968.—JOHN F. SHEARMAN, Brooklyn, N. Y.—*Slide Valve for Steam Engines*—May 22, 1866.—The exhaust pipe passes through the cover of the steam chest, and its enlarged end rests on the back of the valve, so as to embrace the exhaust port through it, in any part of its stroke.

Claim.—First, the cover *g*, forming a termination to the exhaust pipe *h*, within the steam chest, in combination with the exhaust port 3 passing through the slide valve, substantially as and for the purposes specified.

Second, the arrangement of the exhaust pipe *h* and cover *g* to the exhaust port of the valve, and the expansion joint or stuffing box on the valve chest, around the exhaust pipe, as and for the purposes specified.

No. 54,969.—JOHN SHONE, Blackwoodtown, N. J.—*Ventilating Grain.*—May 22, 1866.—The perforated tubes rise from the perforated bottom; some of them are fixed to the sides and others pass centrally through the dryer and out at the top, where they are provided with ventilating caps.

Claim.—First, combining with the main tubes C C the auxiliary vertical tubes G and horizontal tubes H, constructed, arranged, and operating substantially in the manner and for the purpose hereinbefore described.

Second, the combination of the cap pieces C' with the main tubes C, substantially in the manner and for the purpose specified.

Third, combining the perforated tubes I with the sides of the bin, substantially as and for the purpose set forth.

No. 54,970.—S. P. SLEPPY, Wilkesbarre, Penn.—*Artificial Leg.*—May 22, 1866.—The broad metallic lever is strapped to the abdomen, hinged to the thigh, and connected by a rod to the knee; the lower leg is hinged to the thigh and connected by a rod to the heel. The forward motion of the stump flexes the knee and depresses the heel.

Claim.—First, the combination of the lever *a*, connecting rods *d* and *f*, thigh *c*, leg *e*, and foot *g*, in the manner and for the purpose described.

Second, the combination of the straps *j j j* and *i* with the lever *a*, thigh *c*, leg *e*, foot *g*, connecting rods *d* and *f*, all arranged and operating substantially as described.

Third, the combination of lever *a*, straps *i*, and connecting rod *d*, substantially as described.

No. 54,971.—J. LAWRENCE SMITH, Louisville, Ky.—*Gas Burner.*—May 22, 1866.—Explained by the claim.

Claim—A gas burner, constructed as herein described, with an aluminum tip, to adapt it to resist corrosion and to be readily and safely cleansed.

No. 54,972.—PETER J. SMITH, Philadelphia, Penn.—*Mechanical Movement.*—May 22, 1866.—The shafts are simultaneously vibrated in opposite directions. The crank of one shaft is connected by links with that of the other, so as to impart the required motion thereto.

Claim.—The combination of the two shafts A and A', arms D and D', rods E and E', and the radius rod F, the whole being arranged and operating substantially as and for the purposes herein set forth.

No. 54,973.—D. M. SOMERS, Washington, D. C.—*Tumbler Washer and Cooler.*—May 22, 1866.—The depression of the inverted tumbler and its rest opens the valve and exposes the interior of the tumbler to a jet of water.

Claim.—The combination of the tumbler rest D, the spring K, and valve I, when constructed and operated substantially as herein described and for the purposes set forth.

No. 54,974.—HIRAM STEVENS, New Haven, Conn.—*Hen's Nest.*—May 22, 1866.—The weight of the hen depresses the box and draws a shutter over the entrance.

Claim.—The construction and arrangement of the box C within a case A, so as to operate substantially in the manner specified.

No. 54,975.—HIRAM STRAIT, Cincinnati, Ohio.—*Railway Truck.*—May 22, 1866; antedated May 1, 1866.—Long axles are dispensed with; the wheels are journalled in longitudinal bridge pieces arranged in pairs, and stayed by longitudinal and transverse ties; the ends of the frame come near the rail, and have adjustable track clearers.

Claim.—First, the arrangement or combination of car wheels, brackets, springs, and cross and longitudinal ties, either with or without the rubbers, substantially as herein specified.

Second, the grooved or flanged rubbers, either adjustable or fixed, attached to either the ends of the brackets or to brakes to remove all obstacles from before the car wheels, substantially as herein specified.

No. 54,976.—DANIEL SWEENEY and STEPHEN SWEENEY, Constableville, N. Y.—*Stovepipe Damper.*—May 22, 1866.—A diaphragm of gauze arrests the sparks. The damper has an annular disk and concave cap, with an intervening space, and may be rotated on its axial rod and knocked against a transverse rod to clear it of soot, &c.

Claim.—First, the damper composed of the annular concave disk *a* and cap *e*, in combination with the cross-bar E, constructed and operating substantially as and for the purpose described.

Second, also the perforated diaphragm D, in combination with the damper C and cross-bar E, constructed and operating substantially as and for the purpose set forth.

No. 54,977.—BENJAMIN C. TAYLOR, Dayton, Ohio.—*Horse Rake.*—May 22, 1866.—The rake frame is pivoted above the axle; is maintained in working position by a foot piece, and rotated to discharge the load by a hand lever. The pieces referred to constitute the hinge joint between the frame and the carriage; the springs bearing upon the teeth allow an independent upward motion, the tooth traversing between the stay rods.

Claim.—First, the supporting pieces *p* and *q*, constructed and arranged substantially as described and for the purpose specified.

Second, the adjustable spring bar *b*, when arranged with reference to the springs *s s*, and the rods *o o*, substantially as and for the purpose specified.

No. 54,978.—H. K. TAYLOR and D. M. GRAHAM, Cleveland, Ohio.—*Treating Oils.*—May 22, 1866.—The oil is agitated at 90° Fahrenheit, with the addition of sulphuric acid and muriate of soda; the sediment is removed, the acid neutralized, and the resulting sulphate washed out with water.

Claim.—First, the treatment of petroleum and other similar hydrocarbons by means or nascent hydrochloric acids, chlorine, fluorine, or other equivalent chemical reagents, so as to change the constitution of the oil and purify it, substantially as herein described.

Second, the use of sulphuric acid, nitric acid, or salts containing these, or either of them, when used in combination with other materials such as herein described, or their equivalents, for the purpose of treating petroleum or other hydrocarbons, as described.

No. 54,979.—WILLIAM T. TILLINGHAST, Dayton, Ohio.—*Composing Stick.*—May 22, 1866.—The devices described provide for the adjustment of the slide in the stick.

Claim.—The gauging slide B, screw shaft *g*, and clamping nut *h*, in combination with check blocks *e e' c''*, and the checking bar *b*, constructed and operating as above shown and for the purpose set forth.

No. 54,980.—PETER A. VOGT, Buffalo, N. Y.—*Refrigerator.*—May 22, 1866.—The air enters the refrigerator through the waste pipe and passes to the ice chamber to be cooled, the water trap preventing the escape of air at that point. The falling door, supported by segments, forms a shelf for the entering ice.

Claim.—The arrangement of the induction air passage through the drip pipe *l* and extension pipe *m*, in combination with the trap *k*, or its equivalent, whereby the air entering is carried to the top of the ice chamber and cooled in its passage, substantially as set forth.

No. 54,981.—RIVERA WARD, Alder Creek, N. Y.—*Nail Machine.*—May 22, 1866.—This nail-forging machine cannot be briefly described.

Claim.—First, the arrangement of a movable fuel box C, with adjustable bottom *e*, in combination with the furnace A, feed rollers *n o*, hammers H H*, and shears M M*, substantially as and for the purpose set forth.

Second, imparting to the nail rods a retrograde motion before cutting, and a forward motion after cutting, by means substantially such as herein described, or any other equivalent means, for the purposes specified.

Third, the oscillating guide G, in combination with the M-shaped anvil I, constructed and operating substantially as and for the purpose set forth.

Fourth, the hammers H H*, in combination with M-shaped anvil I, constructed and operating substantially as and for the purpose described.

Fifth, the oscillating jaw M*, in combination with the feed lever F, clamps *s* and *t*, disk E, and feed rollers *n o*, constructed and operating substantially as and for the purpose described.

Sixth, the dog *k'* and pin *i'*, in combination with the arms *j' l'*, rod *m'*, lever *n'*, bolt *q'*, and jaws M M*, constructed and operating substantially as and for the purpose set forth.

Seventh, the rock shaft *q'* and tappet *t'*, in combination with the jaws M M* and cam K, constructed and operated substantially as and for the purpose described.

Eighth, the tappet *b²*, in combination with the studs *a²* on the cam K, and with the arms *j' l'*, jaws M M*, and rock shaft *g'*, substantially as and for the purpose set forth.

Ninth, the ratchet wheel *c²* and stud *g²*, in combination with the catch *h'*, spring *w'*, arm *j'*, and with the rods connecting said arms with the jaws M M*, substantially as and for the purpose set forth.

Tenth, the disk *k'*, with its cavity *l²*, in combination with the dog *j²*, catch *h²* and hammer H, substantially as and for the purpose described.

Eleventh, the hand lever N, with spring *o²*, in combination with the catch S² and hammer H*, constructed and operating substantially as and for the purpose described.

Twelfth, the combination of the catch *s²*, dog *j²*, disk *k²*, and catch *h²*, with the hammers H H*, substantially as and for the purpose described.

Thirteenth, the spring catch *s²*, in combination with the annular grooves *w² w²ª* in the cam K, and with the hand lever N, constructed and operating substantially as and for the purpose set forth.

No. 54,982.—OWEN G. WARREN, New York, N. Y.—*Rock Drill.*—May 22, 1866.—Explained by the claim.

Claim.—Loading the shaft with the collar weight to render unnecessary a high lift and for rapid boring.

No. 54,983.—GEORGE I. WASHBURN, Worcester, Mass.—*Steam Engine.*—May 22, 1866.—Each working piston rod is connected to the rod of the piston valve, which controls the admission of steam to the other working cylinder.

Claim.—The arrangement of the piston B B' and valves D D', the latter being respectively actuated by one piston and the means of actuating the other piston, substantially as described.

No. 54,984.—P WEISSENBERGER, Pittsburg, Penn.—*Refining Hydro-Carbon Oils.*—May 22, 1866.—The petroleum is distilled so that the combined distillate shall be about 60° B. The distillate is then treated with sulphuric acid, agitated, settled, and the acid drawn off; washed to remove traces of acid. It is then bleached by the addition of ten to twenty per cent. of boiling water, without alkali.

Claim.—The above described process of purifying distilled petroleum, or other liquid hydrocarbons, without the aid of any alkalies, by means and with the use of water at 212° Fahrenheit, or approximate degree of temperature, in the manner specified.

No. 54,985.—SAMUEL WHITE, Philadelphia, Penn.—*Shade Hook.*—May 22, 1866.—The screw-shanked hook is adjustable in the head of the holder.

Claim.—The combination of the socket A, screw rod B, and nut *d*, the same constituting an improved device for holding and regulating the tension of the cords used for raising window shades.

No. 54,986.—WILLIAM H. WHITE, West River, Md.—*Hat and Cap.*—May 22, 1866.—Explained by the claim.

Claim.—First, making metallic or hard rubber sweat bands, with flanges, as and for the purposes set forth.

Second, the combination of a sweat band with a hat rim or cap visor, when made in one piece.

Third, the combination of a metallic or hard rubber sweat band with a wrapper provided with draw cases and draw strings, substantially as herein set forth.

No. 54,987.—A. L. WHITNEY, Brooklyn, N. Y.—*Potato Masher.*—May 22, 1866.—A radially flanged foot on the end of the handle.

Claim.—An implement consisting of a suitable handle, and provided with a series of arms D, constructed, arranged, and operating substantially in the manner described, and for the purpose specified.

No. 54,988.—LEVI WILKINSON, New Haven, Conn.—*Tuyere.*—May 22, 1866.—The opening in the large portion in contact with the fire is cylindrical, and that in the other portion conical.

Claim.—A cast-iron tuyere, when constructed, shaped, both inside and outside, and fitted for use, substantially as herein described and set forth.

No. 54,989.—SAMUEL P. WILLIAMS, Sheridan, N. Y.—*Fence.*—May 22, 1866.—The fence is raised on an embankment of earth, between ditches. The oblique ends of the rails are scarfed together between the bars of the arched post, and rest on ties. The upper ends of the posts are connected by a continuous tie wire.

Claim.—In combination with the iron staple-shaped posts, the scarfed and scored rails, arranged substantially as described.

Also, in combination with the posts and rails, the iron rod or wires extended between the tops of the posts, as described.

No. 54,990.—WILLIAM WIMMER, Elizabethport, N. J.—*Securing Bits in Braces.*—May 22, 1866.—The teeth of the pivoted dogs are pressed against the shank of the tool by the collar, which slips on the socket.

Claim.—The combination of the entire socket A, the sliding ring C, and the catches B B, each of said catches being formed with a flange *a* and a nose *b* to dispense with the customary pivot pins, and all arranged as herein described, so that the torsional strain shall be sustained by a solid socket, and the bit firmly secured within said socket by the noses *b b* engaging beneath its head.

No. 54,991.—JOHN WOOD, Brooklyn, N. Y.—*Holding Glass Bottles.*—May 22, 1866; antedated May 10, 1866.—An adjustable tool for grasping the bottle by its base while finishing the top.

Claim.—The adjustable holder for receiving and retaining articles of glass while being heated and formed, substantially as specified.

No. 54,992.—JOHN WOOTTON, Boonton, N. J.—*Flying Machine.*—May 22, 1866.—Explained by the claim.

Claim.—First, in combination with a parachute A, constructed substantially as described, the arrangement of one or more screw propellers, placed underneath said parachute, and operated by a suitable engine, and arranged on a movable frame N, operated in the manner and for the purpose substantially as described and set forth

Second, the parachute A, composed of a large ring or hoop B, with numerous rods or braces G G running from the central frame to the ring, thereby making it into a continuous arch, or rigid circle, said circle to be covered with oil-cloth, or its equivalent, on the underside of the rods or braces, and to be held in its proper and rigid position, at right angles with the frame of the machine, by rods or braces C and C', fastened to the ring B and to the top D, and lower part D' of the frame of the machine, substantially as described and set forth.

Third, in combination with the parachute A, and the propeller wheels M M, arranged as described, the coil of their copper pipes P, when arranged in the manner and for the purpose as specified.

Fourth, in combination with a flying machine, constructed as described, the arrangement for hoisting up and lowering the car from the machine, substantially in the manner and for the purpose set forth.

Fifth, in combination with the car of a flying machine, arranged and constructed as set forth, the wheels or rollers R, and springs Q, when arranged in the manner and for the purpose substantially as described and specified.

No. 54,993.—LARNARD M. WRIGHT, Fort Edward, N. Y.—*Apparatus for Washing Paper Pulp.*—May 22, 1866.—The beater is adjustable vertically and revolvable in either direction in the tub.

Claim.—First, the adjustable arms S, in combination with a shaft A A, cog wheels B and C D F K L and M, with their shafts and the clutches H and H', and the pulleys 1 2 3 4 and 5, and the rack R R and the guides *z z z z*, substantially as and for the purposes described.

Second, broadly the application of an adjustable revolving beater or washer to a tub or vat, containing stock to be beaten or washed, in the manner and for the purposes above described.

Third, broadly the reciprocating or back and forward or right and left motion, which may be given to the revolving arms S, by which they rotate both ways.

No. 54,994.—CHARLES E. YAGER, Hudson, N. Y.—*Flour Sieve.*—May 22, 1866.—The frame which rotates above the foraminous surface of the sieve is journalled in a frame attached by a clamp and set screw to the edge of the sieve.

Claim.—The attachment consisting of the removable radial frame C, with the clamp and set screw *a b c*, carrying the rotary scrapers D, shaft *e*, and crank *f*, arranged and operating with reference to each other, when applied to a circular flour sieve, substantially as herein set forth, for the purpose specified.

No. 54,995.—ELIJAH YOUNG, Fayetteville, Mo.—*Seed Drill.*—May 22, 1866.—The rear extension of the shares forms drills, and the furrowing device is raised by a lever and link.

Claim.—Extending the mould boards of the ploughs D out behind the other parts of the ploughs, as and for the purpose set forth.

Second, the rock shaft H, the lever H', and links g g', or their equivalents, in combination with the beam E' and tubes C', when employed for the purposes herein set forth.

No. 54,996.—JAMES ABORN, Providence, R. I., assignor to HENRY H. GILES, same place.—*Blind Fastener and Operator.*—May 22, 1866.—The notched sliding stem is attached by a toggle to the hinge, which it vibrates, and is locked by engaging a notch in a holder plate.

Claim.—The hinge provided with the arm e, in combination with the connecting bar S and the sliding bolt G, the whole being constructed and operating substantially as described and for the purpose specified.

Also, the combination of the apparatus employed as described for swinging the blind, with a series of notches in the sliding bolt G, and a button or key d, substantially as described.

No. 54,997.—GEORGE A. ALGER, Manchester, N. H., assignor to himself and WALDO WHITNEY, same place.—*Punching Machine.*—May 22, 1866.—The vertical operating stem has two cams and levers; the cam brings the punch down to the thickness of the material, and the upper one forces it through.

Claim.—The use of two revolving cams and collars, or double cam on the punch bar, one cam to set the punch to the thickness of the metal plate or other material, and the other cam to force the punch through the metal plate or other material to be punched.

No. 54,998.—ANNA M. BARDWELL, Amherst, Mass., assignor to C. D. CLAPP and H. J. BARDWELL, same place.—*Hoop Skirt.*—May 22, 1866.—The hoops overlap each other in front, passing freely through the tapes next to the one to which their ends are fastened. The size may then be regulated by adjusting the relative position of these tapes. Elastic thimbles maintain the required position against accidental displacing pressure.

Claim.—The combination of the hoops C and tapes B B', the ends of the hoops being arranged as described, in the manner and for the purpose herein specified.

Also, in combination therewith, the tubular fastening D, with a loop at one end, made substantially as shown.

No. 54,999.—ORSON BILLINGS, Lagrange, Ohio, assignor to himself, RUSSEL H. PENFIELD, and HOMER PENFIELD, Elyria, Ohio.—*Hand Corn Planter.*—May 22, 1866.—Explained by the claim.

Claim.—First, the angularly shaped rubber spring, having the two upper surfaces plain and the lower surface convex, and having the inner end cut at right angles to its axis, and the outer end cut at such an angle that, when adjusted and the handle of the machine is extended, that it just corresponds to said angle, thus obtaining an equal pressure on all parts, as described and set forth.

Second, the spring D, inclined apron E, and seeding slide F, in combination with the inserters, for the purpose of delivering said seed to said inserters in a uniformly divided state, in the manner described and set forth.

No. 55,000.—ALEXANDER BOYDEN, Foxboro', Mass., assignor to ALFRED FALES and ERASTUS P. CARPENTER, same place.—*Attaching Handles to Hoes, Rakes, &c.*—May 22, 1866.—The bent flange on the upper edge of the blade is clamped between a collar jaw and a screw shank jaw attached to the handle.

Claim.—The stationary jaw B, in combination with a movable jaw E, operated by a screw nut C, or equivalent device, for receiving and holding the implement, substantially as described.

Also, in combination with the above, a blade or other implement provided with a lip, projection, or recess, substantially as and for the purpose set forth.

No. 55,001,—MELLEN BRAY, Newton, Mass, assignor to himself and WILLIAM WOODBURY, Chelsea, Mass.—*Guide for Boom Jaws.*—May 22, 1866.—Metallic guide pieces project downward over the saddle to keep the corner of the jaw from wearing the mast.

Claim.—The guide pieces b, in combination with the boom C and saddle D, substantially as and for the purpose set forth.

No. 55,002.—JOHN BROOKS, North Bridgewater, Mass., assignor to himself and JOSEPH BROOKS, same place.—*Welt Trimmer.*—May 22, 1866.—The long knife is adjusted in the longitudinal groove of the stock and maintained by set screws in the required relation to the guard.

Claim.—The improved welt trimmer as made not only of the guard C, and grooved handle or stock A, but with the long knife B, to fit into the groove of the stock, and be confined by means substantially as described.

Also, the arrangement of the lip f, with the guard C, and the knife B, when arranged with and applied to the stock substantially as described.

Also, the combination and arrangement of the adjusting screw g, with the knife stock and guard arranged and applied together, substantially as specified.

No. 55,003.—CHARLES E. CONDON, Harrisburg, Penn, assignor to himself and G. B. HAMMER, same place.—*Sash Supporter.*—May 22, 1865.—The notches toward the upper end of the bar are presented upward to prevent raising the sash, and the others downward to sustain it, the pivoted catch assuming the required positions for each duty.

Claim.—The bar a, with the reversed notches arranged to operate in connection with the hook e, substantially as shown and described.

No. 55,004.—C. B. COTTRELL, Westerly, R. I., assignor to himself and NATHAN BABCOCK, same place.—*Apparatus for Tempering Steel.*—May 22, 1866.—The articles are contained in a cage and immersed in a bath which is provided with a thermometer.

Claim.—In the draw tempering of steel the employment of a vessel constructed substantially as described so as to admit of the tempering process, and prevent the ignition of the tallow, oil, or other oleaginous tempering medium, as herein set forth.

Also, the employment of an independent perforated basket for holding the steel, in combination with the oil vessel, constructed substantially as herein described.

No. 55,005.—JOHN J. CURRIER, Boston, Mass., assignor to himself and JAMES H. PLAISTED, same place.—*Machine for Folding Paper Collars.*—May 22, 1866.—An arched bed-plate and a correspondingly arched platen, the bed-plate bevelling inwards beneath. As the platen descends to hold the paper, a spindle-shaped presser descends therewith, bends the paper over the edge, and passes automatically under the arch, folding the collar on a curved line.

Claim.—First, a machine consisting of plates C and D, a roller G, a beam F, and guides I I, constructed and operating substantially as described and for the purpose specified.

Second, the two plates C and D, in combination with the roller G.

Third, the plate C, in combination with the roller G, substantially as and for the purpose specified.

Fourth, the mode of perfecting the fold by guiding the roller G under the plate C.

No. 55,006.—JOHN DANE, Jr., Newark, N. J., assignor to W. J. DUDLEY and W. U. DUDLEY, New York, N. Y.—*Machine for Shaping Wood.*—May 22, 1866; antedated May 16, 1866.—The clevis holds the plank to the pivoted stock and is rotated thereon to avoid contact with the adjustable rotary cutter.

Claim.—First, the vice or clamp I, arranged to rotate upon a pivot K, which attaches it to the table H, substantially as and for the purpose herein specified.

Second, the vertically movable frame G, carrying the revolving cutter, spindle, and tracer, and the horizontally sliding cross head F, carrying the said frame, arranged in relation to the universally movable table H, upon which the work is placed, substantially as herein described, to operate as set forth.

No. 55,007.—JAMES EMERY and JOSEPH E. GOTT, Bucksport, Me., assignors to JAMES EMERY, same place.—*Fly Trap.*—May 22, 1866.—Explained by the claim.

Claim.—The said improved fly or insect trap, as composed of the box with a foraminous top and partition, or its equivalent, the tapering mouth or entrance frustum, the escape opening with its valve or damper, and the receiver or glass bell, arranged together substantially as specified.

No. 55,008.—WILLIAM P. GANNETT, Boston, Mass., assignor to himself and ISAAC GANNETT, Chicago, Ill.—*Clothes Sprinkler.*—May 22, 1866.—The water is expelled through the perforated bottom by shaking; the reservoir is filled by immersing the instrument in water and depressing the lower valve, which makes an air escape through the stem.

Claim.—The clothes sprinkler herein described, the same consisting of a water reservoir, provided with a valve, the air passage through the stem of which is so made that the opening of the valve establishes a communication through the air passage of the stem with the interior of the reservoir, and the closing of the same shuts off or closes such communication, substantially as and for the purpose specified.

No. 55,009.—WALTER HYDE, New York, N. Y., assignor to DEBORAH A. BALLOU, same place.—*Rock Drilling Machine.*—May 22, 1866.—The hammer is vibrated by the eccentric gearing, so as to give a quick forward stroke and a slower return motion, when the recoil is received on a spring; the drill is held by a guide rod.

Claim.—First, the eccentric gear J K, in combination with the platform A, handle D, and hammer E, constructed and operating substantially as and for the purpose specified.

Second, the slot E on the bracket F, in combination with the eccentric gear J K and hammer E, constructed and operating substantially as and for the purpose set forth.

Third, the curved rod G and double cranks H, in combination with the eccentric gear J K and hammer E, constructed and operating substantially as and for the purpose described.

No. 55,010.—BARTON H. JENKS and ROBERT B. GOODYEAR, Philadelphia, Penn., assignors to BARTON H. JENKS, same place.—*Heddle Motion for Looms.*—May 22, 1866.—The diverging slots in the rising and falling frames, in connection with the inclination of the cross-bars, lift the rear heddles successively higher than the forward ones, to form a more perfect shed. The notched ratchet and the pawl enable the operator to move the lathe and carry the ratchet and pattern cylinder back or forward correspondingly, without affecting the pattern when the weaving is resumed. The loose lifting cam and its projection being allowed to slip upon the shaft, permits the latter to be turned backward or forward without disturbing the proper time or position of the action of said cam.

Claim.—First, giving the drawbars J' a reciprocating motion of graduated degrees, diminishing throughout the series by means of the cross-bars I' l" and the slotted and removable frames E' E", substantially as described.

Second, the combination of the supporting bar X with the draw bars J', shaped substantially as described.

Third, giving to the ratchet S, which operates the cylinder Q, a double-acting motion, by which the operator can vibrate the lathe without making a mistake or losing a twilt, substantially as described.

Fourth, the combination of the ratchet S, pawl R', pin R", and incline S", substantially as described.

Fifth, the lifting cross head W, in combination with the loose cam W', operated substantially as described.

Sixth, the combination of the loose cam W' with the crank N and clutch pieces W³ attached to it, or their equivalents, substantially as described.

No. 55,011.—BARTON H. JENKS and ROBERT B. GOODYEAR, Philadelphia, Penn., assignors to BARTON H. JENKS, same place.—*Loom.*—May 22, 1866.—The devices cited in the claim are for giving the treadle motion and shuttle box changes, and for stopping the loom and preventing injury to the goods in case the shuttles should not be brought in line with the race board, or the picker straps should break.

Claim.—First, regulating the movements of the shuttle boxes, and also the picking motions, by two or more sets of pins on the same chain, when these motions are imparted from the chain through the intermediary means, substantially as described

Second, the combination of the box chain P, lever O, the double arm M. the forked swivel J, and eccentric disk H, or their equivalents, for operating the pick, substantially as described.

Third, the combination of the knuckle lever U, attached to the breast beam v, with the stop rod X', to stop the loom and so prevent the yarn from being broken when the shuttles fail to leave the box, substantially as described.

Fourth, the combination of the sliding rod E, arms G, sliding tappets C' C", and caps B' and B", operating substantially as described.

No. 55,012.—NELSON KING, Bridgeport, Conn., assignor to O. F. WINCHESTER, New Haven, Conn.—*Magazine Fire-arm.*—May 22, 1866 —The cartridge is transferred to the barrel from the magazine tube beneath by a carrier block, operated by the trigger-guard lever. Cartridges are supplied to the magazine through a side opening in the frame, (protected by a cover,) and a corresponding channel in the carrier.

Claim.—Forming an opening through the frame relatively to the carrier block and magazine, and in combination therewith, so that the magazine may be charged through the carrier block from one side of the arm, substantially in the manner herein described.

No. 55,013.- JACOB KUNSMAN, Sr., Reading, Penn., assignor to himself and FREDERICK J. NAGLE, same place.—*Apparatus for Preserving Dead Bodies.*—May 22, 1866.—The body lies upon a perforated metallic plate, with adjustable head rest and attachable foot extension. The chamber below contains ice, and the ice chest above the neck and viscera has a discharge thereinto.

Claim.—First, the combination of the perforated supporting plate B with the box A, substantially as described and for the purpose set forth.

Second, the combination of the perforated adjustable head support E with the perforated plate B and the box A, substantially as described and for the purpose set forth:

Third, the combination of the neck box O, provided with a cover o' and exhaust pipes and stop-cocks o", with the perforated plate B and box A, substantially as described, and for the purpose set forth.

Fourth, the combination of the detachable part N with the box A, substantially as described and for the purpose set forth.

Fifth, the combination of the board or shelf I with the hinged legs H and with the box A, substantially as described and for the purpose set forth.

No. 55,014.—JOSEPH MALONE, Temperanceville, Penn., assignor to himself and GEORGE WETTENGILL.—*Furnace for Annealing Sheet Iron, &c.*—May 22, 1866.—The volatile results pass from the tire chamber to the u per flue, thence by diving side flues to the lower flue, and by apertures and cross flue to the chimney.

Claim.—Constructing a furnace for annealing sheet iron and other articles, having a fire chamber at one end, and an annealing chamber, surrounded at top and bottom, and both sides with flues for the passage of the flame, hot air, and products of combustion, constructed and arranged substantially as and for the purposes herein described.

No. 55,015.—PETER H. NILES, Boston, Mass., assignor to himself and AUGUSTUS RUSS, Cambridge, Mass.—*Seam for Sheet Metal Ware.*—May 22, 1866.—A recess is formed in one edge and a bead in the other, being locked together and forming a flush outside joint; solder is melted into the crack.

Claim.—The socket joint *e*, formed by the articulation of the socket *a* and the bead *d*, substantially as and for the uses and purposes above described.

No. 55,016.—PETER H. NILES, Boston, Mass., assignor to himself and AUGUSTUS RUSS, Cambridge, Mass.—*Attaching Sides of Sheet Metal to Each Other.*—May 22, 1866.—The edges of the sheet metal are locked by the engagement of semicircular lips punched up on the respective portions, one projecting inward and the other outward, so as to allow one to slide over and the other under its opposite.

Claim.—The lips *b* and *c*, interlocked so as to form the rivet joint *a*, substantially as and for the uses and purposes above described.

No. 55,017.—EPHRAIM PARKER, Marlow, N. H., assignor to ALFRED A PARKER, Orange, Mass.—*Stove Pipe Damper.*—May 22, 1866.—A series of concavo-convex annuli, with intervals, and decreasing in diameter to the top, which is closed by a concavo-convex disk.

Claim.—The above described arrangement of annuli A B and a concavo-convex disk C, namely, in a conic frustum, and with the convexity of the disk extending within the damper, in the manner and for the purpose as specified.

Also, the damper as made with the series of annuli and the disk arranged in a conic frustum, and with each ring or annulus concavo-convex, as set forth.

No. 55,018.—JULIO H. RAE, Syracuse, N. Y., assignor to himself and EDWIN L. BUTTRICK, Milwaukee, Wis.—*Peat Machine.*—May 22, 1866.—The armed shaft revolving in the cylindrical casing conveys the peat to the box, whose screw conveyer pushes the disintegrated peat to the second box for condensation and delivery.

Claim.—First, enlarging the corners or angles of the opening R, through which the peat is received into the conveyer chamber P, substantially as described, in combination with the conveyers I.

Second, contracting the diameter of the condensing and delivering tubes for a portion of their length, and then enlarging their diameter towards their discharge ends, substantially as described.

Third, making the outer portions of the tubes J square or straight on their under sides, substantially as described.

Fourth, making the upper sides of the tubes J, at their discharging ends, of an elevated form, so that the peat discharged therefrom will shed falling water, substantially as described.

Fifth, the combination of the conveyers I with the condensing and delivering tubes J, substantially as described.

Sixth, the arrangement and combination of the driving shaft O and its spur wheel M with the conveyers I and the pinions N N of the conveyer shafts, in combination with shaft B, substantially as described.

Seventh, the arrangement and combination of the driving shaft O with the vertical shaft B, and the wings and blades C, the cutter arms E, and the propeller F, and the perforated steam pipe G, substantially as described.

No. 55,019.—HENRY SCHREINER, Philadelphia, Penn., assignor to himself and WILLIAM H. BOWYER, Tamaqua, Penn.—*Horseshoe.*—May 22, 1866.—The circumferential clips embrace the edge of the hoof and the encircling bands overlap and are secured to the upward prolongation of the front clip.

Claim.—A nailless shoe for horses and mules, consisting of the sole A, clips *b c c'*, belts *d d'*, and nutted screw bolt *e*, the same being constructed, combined together, and applied as and for the purposes described.

No. 55,020.—ROBERT SIMPSON, Boston, Mass., assignor to JOSEPH PRATT and CHARLES C. WENTWORTH, same place. — *Stove Door.*—May 22, 1866.—The damper or slide is moved in guides on the door by a rack and pinion. The catch of the door has stops at its open and closed positions.

Claim.—First, the arrangement in combination with the door D of the guide bars *x*, damper V, rack R, pinion *w*, and its shaft H, passing through slot P of the damper, substantially as described.

Second, the arrangement in combination with door D of the turn buckle F, boss B, and stops *s*, substantially as described for the purpose stated.

No. 55,021.—LEWIS SMITH, Foxboro', Mass., assignor to himself and GEORGE T. RYDER, same place.—*Spring for Beds.*—May 22, 1866.—Looped wire springs, rooted in the slat and presenting a cruciform bearing to the mattress.

Claim.—The duplex spring composed of the ring C and the two springs A B, formed and arranged substantially as described and represented.

No. 55,022.—E. L. STAPLES, Nashville, Tenn., assignor to JAMES E. RUST, same place.—*Pump.*—May 22, 1866.—The packing ring to which the valve is attached is connected by bars to the hollow plunger, which is packed at the end of the cylinder and discharges by a hollow piston rod.

Claim.—First, connecting to the piston B of a pump a chamber B′, open at its inner end, and connected to the piston by bars or rods C, said chamber having a hollow piston rod screwed into its closed outer end, substantially as described.

Second, in combination, the pump cylinder A, the piston B, and the chamber B′, connected with it the hollow piston rod G and the side eduction pipe I, substantially as shown.

No. 55,023.—AUGUSTUS J. THOMPSON, Malden, Mass, assignor to himself and A. B. LINCOLN, Roxbury, Mass.—*Cover for Sewing Machines.*—May 22, 1866.—Explained by the claim.

Claim.—The sliding hinged cover in combination with a sewing-machine table, when constructed and operating substantially as and for the purpose set forth.

No. 55,024.—HENRY THOMPSON, Rockland, Me., assignor to himself, D. C. HASKELL, and C. H. HASKELL, same place.—*Rotary Harrow.*—May 22, 1866.—Tangential wings on the ends of the arms, by contact with the soil, cause the barrow to revolve.

Claim.—The revolving barrow head, made of the series of rotary wings B, their studs *e*, the harrow teeth *a*, and arms A, or their equivalent, the whole being arranged substantially in the manner and so as to operate as described.

No. 55,025.—JOHN W. WHEELER, Cleveland, Ohio, assignor to H. H. WHEELER and S. R. BOARDMAN.—*Water Drawer.*—May 22, 1866.—The elevator has a windlass pawl and friction brake device to lower the bucket. The valve in the bottom of the bucket is opened by contact of a trigger with the top of the well curb and the water discharged by a spout.

Claim.—A water elevator as described, having a circular curb A and spout B, bucket J*ₜ* with valve K and tripping lever M, windlass G, with brake U, lever S, and hook r, all combined, arranged, and operating in the manner substantially as set forth.

No. 55,026.—THOMAS WHITTAKER, Buffalo, N. Y., assignor to himself and WILLIAM P. SIMPSON, same place.—*Apparatus for Raising Sunken Vessels.*—May 22, 1866.—Buoyant air casks are forcibly submerged and attached in sufficient numbers to the wreck to raise it. The transverse tube through the descending barrel forms a guide for the rope.

Claim.—First, the tube *n*, in combination with the air cask E, or its equivalent, constructed and arranged substantially as and for the purposes set forth.

Second, in combination with the cask E and tube *n*, also the pawls *o o*, operating in the manner and for the purpose described.

Third, the combination and arrangement of the pendent bars D D, provided with the curved arms *b b*, guide frame B B, with windlass F and guide rope *l*, for lowering casks to submerged wrecks, constructed and operating substantially as set forth.

No. 55,027.—JAMES G. WILSON, New York, N. Y., assignor to UNION SEAMLESS KNITTING MACHINE COMPANY.—*Knitting Machine.*—May 22, 1866.—This is an improvement on the McNary machine, No. 28,290. The object is to knit simultaneously on a number of immediately adjacent needles and with a separate thread for each of such needles. The needles have two hooks on the same shank; the shank has a tubular spring for the purpose of holding it firmly in its socket in the needle ring; the pressers move vertically, only, instead of having a compound motion, and their upward motion is a quick one to carry them out of the way of the yarn conductors. The inner books of the needles are to prevent the new loops being taken off the needles by the action of the stitch hooks whilst the latter are in the act of taking off the old loops. The machine is designed to knit a number of stockings in a continuous piece.

Claim.—First, providing the needles of that class of knitting machines in which stitch books are used to throw off the stitches, without any longitudinal movement of the needles or closing of their beards, with additional books or projections *b b* at their back, so constructed, arranged, and operating in combination with the stitch hooks as to confine the last formed loops on the needles, while the previously formed loops are being taken off from the needles by the stitch hooks, substantially as herein described, whereby I am enabled to dispense with the operation of the pressers at the time of taking off the loops, and to take off the loops simultaneously from many immediately adjacent needles.

Second, giving the pressers *v v* a movement in a vertical line only, but of such character that they are removed from between the needles and out of the way of the yarn guides and stitch hooks at the time the latter operate to throw off the loops.

Third, making the slots in the arms V′ V′ of the stitch-hook bar V of curved form, substantially as herein described.

No. 55,028.—PETER DINZEY, Island of St. Bartholomew, West Indies.—*Rudder.*—May 22, 1866.—A pinion on the lower end of the shaft which passes through the rudder engages with a toothed segment attached to a flange on the stern post. The shaft is turned by ropes connecting with the ordinary steering apparatus.

Claim.—The rudder B, with its shaft C and pinion e, in combination with the segment D, the whole being arranged and operating substantially as and for the purpose specified.

No. 55,029.—JAMES HAYES, London, England.—*Sewing Machine.*—May 22, 1866.—The conical stud is blunt and fixed on the under side of the table, near the needle hole; the hooked instrument receives the loop from the looper, then rises and lays it over the stud till the needle enters and then retreats.

Claim.—The stud I, in combination with the hooked instrument H, for opening and holding the loop of the looper thread, in the manner herein represented and described.

No. 55,030.—ROBERT W. THOMSON, Edinburgh, Scotland.—*Apparatus for Obtaining Motive Power.*—May 22, 1866.—The cylinder of this rotary steam engine contains two diaphragm pistons, one of which is keyed upon an axial shaft, and the other upon a tubular shaft: each shaft is connected to the driving shaft, which is disposed eccentrically to the steam cylinder.

Claim.—The obtaining and applying of motive power by means of pistons or diaphragms revolving in a cylinder in a continuous manner, but connected to an external shaft having a different axis from that of the cylinder, so that their velocities vary, that of one or more increasing whilst that of the other or others decreases, and vice versa, the parts being constructed and arranged and working substantially in the manner herein described.

No. 55,031.—HENRY VOELTER, Heidenheim, Wurtemburg.—*Reducing Wood to Paper Pulp.*—May 22, 1866.—Improvement on his patent of August 10, 1858.—Blocks of wood are fed to the revolving stone; the fibres are sorted into grades by revolving sieves; provision is made for regrinding between stones.

Claim.—First, the use of millstones N N', constructed and operating substantially as described, for reducing particles of wood to paper pulp.

Second, the combination of the rake H, projection I, sieve J, and pipes K and K', with the box B and stone D', the whole being arranged and operating substantially as and for the purpose set forth.

Third, the spring pawl b and ratchet wheel a, constructed and operating in combination with the shaft F and pulley 9, substantially as and for the purpose specified.

Fourth, the elastic cushion or spring d, in combination with the cross-head 15 and rod 11, for the purpose set forth.

Fifth, the block e formed substantially as described, in combination with the stone D' and boxes G G', the whole being arranged and operating substantially as and for the purpose set forth.

Sixth, the boxes or cross-pieces G G', constructed with parallel sides, for the purpose specified.

Seventh, the tanks L L' and L², with their gauze cylinders R R', and paddle wheels W W', the whole being constructed and arranged for joint operation, substantially as and for the purpose set forth.

Eighth, the vibrating basket P, arranged and operating in the tank L, for the purpose described.

Ninth, the rotating comb K and the scraping plate m, in combination with the cylinder R, the whole being constructed and operating substantially as and for the purpose specified.

Tenth, the hopper V, with its sliding door n and rotating fluted cylinder V, arranged in respect to the tank L² substantially as and for the purpose described.

Eleventh, the tanks Y Y' and Y², with their gauze cylinders R² and R², paddle wheels W² W², sieve v, rollers y, and plates z, the whole being constructed and operating substantially as and for the purpose set forth.

Twelfth, the bands h h of leather or other suitable material, secured to the rollers y y, for the purpose specified.

No. 55,032.—OSCAR F. MORRILL, Chelsea, Mass.—*Petroleum Stove.*—May 22, 1866.—Explained by the claims.

Claim.—Providing the perforated air-heating tube with a shield having an extended base and a contracted top, the shield surrounding the tube, with their tops united, and operating substantially as described.

Also, the base shield surrounding the wick tube, and the opening through the reservoir top, where this shield enters and discharges into the space between the perforated air-heating tube and its shield, substantially as described.

Also, the expansion of the wick tube k at its top into a cup l of such size, form and location as to receive and steady the perforated air tube m, holding it properly in place with reference to the wick, and also receiving any fluid overflow therefrom.

Also, the combination and arrangement of the spurred wick-tube wheels r on one side of the wick, with counteracting smooth-surfaced yielding presser rolls q on the other side of the wick, substantially as described.

No. 55,033.—OSCAR F. MORRILL, Chelsea, Mass.—*Apparatus for Cooking by Coal Oil Burners.*—May 22, 1866.—An inverted frustum deflector surrounds the wick tubes; a horizontal shield is fitted between the reservoir and point of combustion, and acts as a wick raiser; gas escaping from the reservoir is conducted to the burner.

Claim.—The deflector c^2, placed within the cylinder b, to give an upward tendency to the lateral currents of heated air, substantially as set forth.

Also, placing an interceptor plate between the fuel reservoir and the flame chamber, as set forth.

Also, making the interceptor plate n, the wheel for regulating the height of the wick, substantially as set forth.

Also, the vapor passages e' for conducting vapor from the reservoir into the cone.

Also, in combination with such passages, the ring b', for preventing lateral escape of such vapors.

Also, supporting the plate n, or wick tube r, upon points or projections v, to insulate the reservoir from the head of the flame, substantially as described.

Also, the construction of the top of the tube K, or the plate r, with projections l, to centralize and keep in position the outer disseminator, when they are arranged in the manner described.

Also, the conical or tapering tube, for so supporting the inner disseminator as to permit it to be adjusted, substantially as described.

Also, placing the inner chimney s upon a tube, which is insulated from the oil reservoir, substantially as set forth.

No. 55,034.—CHARLES ALLARDICE, Cohoes, N. Y.—*Self-Lubricating Knitting Machine Burr.*—May 29, 1866; antedated March 29, 1866.—The revolution of the burr carries the oil up the spiral grooves on the spindle, returning by the sides of the cup to the reservoir; the inclination prevents its escape.

Claim.—The shaft B, having its closed cap b, in combination with the spindle D, and its spiral grooves a a, and the reservoir C, substantially as described and for the purposes set forth in the above specification.

No. 55,035.—T. H. ARNOLD, Troy, Penn.—*Horse Hay Fork.*—May 29, 1866.—The slotted and pivoted hooks form the entering portion and maintain a general longitudinal direction while the fork is thrust into the hay, and are vibrated outwardly to hold the hay by the pivoted rod, which is actuated by a lever and auxiliary rope.

Claim.—A horse hay fork frame, by combining the bars A and C, the lever D, and the prongs or hooks E and F, the parts being constructed and arranged substantially as described, and for the purpose set forth.

No. 55,036.—ANDREW J. BAILEY, Charlestown, Mass.—*Machine for Cutting Corks.*—May 29, 1866.—The knives are segments of a cylinder attached to a revolving disk; the elastic grasping centring feeder is operated automatically and delivers the blank cork to the revolving mandrel, where it is shaved by the knife and then released.

Claim.—First, the rotating knife D, having its face in the form of an arc of a cylinder, and its edge parallel with its planes of rotation, when arranged in segments of a circle and operating in combination with the rotating mandrel J, substantially as herein shown and described.

Second, the centring feeder, constructed with elastic or self-adjusting gripers, for the reception of blanks of varying size, applied and operating in relation to the rotating mandrels and opposite centre, substantially as herein specified.

Third, the compound carriage K M, having attached to it the mandrel which carries and produces the rotation of the cork in its rotation, and made adjustable relatively to the cutter wheel, to vary the size and form of the cork, substantially as herein specified.

No. 55,037.—J. M. BAIRD, Wheeling, W. Va.—*Water Cooler and Purifier.*—May 29, 1866.—The service pipe is conducted to and from a deeply buried cooling and sediment-collecting chamber.

Claim.—A water cooler and purifier, formed by combining the tank D, constructed and arranged as herein described, with the induction and discharging pipes c and e, substantially as and for the purpose set forth.

No. 55,038.—JAMES S. BALDWIN, New York, N. Y.—*Collecting and Separating Carbonic Acid from Mixtures of Gases.*—May 29, 1866.—Carbonic acid gas for charging soda water, &c., is extracted from the gases evolved in combustion of fuel from lime-kilns, &c. The gases enter a strong chamber, under a pressure of seventy-five pounds to the square inch, where a shower of water absorbs the carbonic acid gas and the others escape at a valve way; as the water collects it is passed to another chamber, where, the pressure being removed, the gas escapes.

Claim.—Separating, purifying, and transferring carbonic acid gas from waste gases, by means of water, or its equivalent liquid, substantially in the manner and for the purpose herein set forth.

Also, driving the pumps, or other machinery, by the waste gases evolved in this process, substantially as described.

No. 55,039.—JAMES S. BALDWIN, New York, N. Y.—*Preserving and Storing Carbonic Acid Gas.*—May 29, 1866.—Solid carbonic acid is contained in a close vessel surrounded by a non-conducting material and communicating with a vessel containing lime, to absorb any vapors evolved.

Claim.—The combination of vessel A with vessel C, or the equivalent thereof, arranged and prepared substantially in the manner and for the purpose herein set forth.

No. 55,040.—JAMES S. BALDWIN, New York, N. Y.—*Charging Water with Carbonic Acid.*—May 29, 1866.—The vessel containing solid carbonic acid may be enclosed in the fountain reservoir, or may be separate and discharge into it by a conducting pipe.

Claim.—Charging water, or other equivalent liquid, with carbonic acid by enclosing said liquid with solid carbonic acid, in the same vessel, or series of vessels, substantially in the manner and for the purpose herein set forth.

No. 55,041.—ACHILLES BALLARD, Dublin, Ind.—*Shoe.*—May 29, 1866.—The vamp overlaps the quarter, and, inclining backward, contracts upon the axle above the instep; the fit is completed by a string passing behind the quarter.

Claim.—A shoe, the vamp A and quarter B of which are shaped and united substantially as set forth.

No. 55,042.—HAZEN J. BATCHELDER, Boston, Mass.—*Machine for Bending Horseshoes.*—May 29, 1866.—The former is made to correspond with the inner shape of the horseshoe and is reciprocated by a rack and pinion, thrusting the iron between bending lugs and then between curved levers which rotate on their fulcra to bend in the heels

Claim.—The benders H H, made and arranged in combination with the spring *f*, in manner and so as to operate with the former A, substantially as described.

Also, the combination and arrangement of the jaws *i k* with the former A, combined and to operate with the spring bending levers H H, substantially in manner as described

Also, the combination as well as the arrangement of the dischargers *k k* with the former A and the bending levers H H.

No. 55,043.—ADAM BAUMANN, Philadelphia, Penn.—*Pruning Shears.*—May 29, 1866.—The coil spring on the pivoted rod of the movable jaw acts against a stud on the other jaw and opens the mouth.

Claim.—The rod D and its spiral spring *h*, combined with and adapted to the hooked portion and knife of pruning shears, substantially as and for the purpose herein set forth.

No. 55,044.—ALEXANDER BECKERS, New York, N. Y.—*Stereoscope.*—May 29, 1866.—The pictures are attached to a revolving belt and consecutively presented to view; the concave, movable reflector throws the light upon the transparent picture. The clasp is attached to the wire frame to hold the picture.

Claim.—First, a concave mirror or reflector, combined with the stereoscopic box and movable reflectors *b* and *c*, substantially as and for the purposes set forth.

Second, the sheet-metal clasp *i*, attached to the wire frame of the picture holder, as and for the purposes set forth.

No. 55,045.—JACOB BEHEL and JOHN M. BUELL, Rockford, Ill.—*Horseshoe.*—May 29, 1866.—The shoe is secured by projections on its upper face and by angular clamps, whose heads rest against the walls of the hoof and whose shanks are secured by nuts in the recesses on the under side of the shoe.

Claim.—First, the curved clamp F P with roughened faces or edges, when constructed substantially as and for the purpose set forth.

Second, the clamp F P, in combination with a shoe having recesses E, and openings A with projections B, and the nut *y*, for securing and tightening the shoe, the parts being constructed and arranged for use, substantially in the manner and for the purpose set forth.

No. 55,046.—ANTHONY A. BENNETT, Norwalk, Conn.—*Operating the Doffer Comb of Carding Engines.*—May 29, 1866.—The wrist pin is adjustable on its driver to vary the length of throw as it traverses noiselessly in the slot of the vibrating arm of the doffer shaft.

Claim.—The arrangement with the comb shaft B, and its oscillating slotted arm A, of the crank pulley G, constructed as described, and its adjustable pin D, in the manner and for the purpose set forth.

No. 55,047.—WILSON BOHANNAN, New York, N. Y.—*Lock.*—May 29, 1866.—In this device for securing drawers the keeper is arranged below the plate, which is inserted in the upper drawer and directly in the rear of the socket, into which it enters in combination with the bolt of the latch.

Claim.—First, the arrangement of the catch or hook of the keeper C below the plate *a*, and in rear of the base of the socket, into which it enters in combination with the arrangement of the belt *e* of the lock, substantially as and for the purpose described.

Second, the construction of the hooked keeper C, with a bevelled nose, in combination with the bevelled nose of the self-locking bolt *c*, substantially in the manner and for the purpose described.

No. 55,048.—WILSON BOHANNAN, New York, and FRANK G. JOHNSON, Brooklyn, N. Y.—*Lock.*—May 29, 1866.—A permutation is formed by the construction of the case and the several cylindrical and grooved tumblers. The tumblers and their springs are enclosed in a solid case attached to the rear lock case; the celled case has a hooked bolt to receive the hasp and a spring bears on its periphery to restrain motion when the hasp is withdrawn.

Claim.—First, enclosing the grooved tumblers *g g*, together with their springs *h h*, within cells which are formed in the movable case C, constructed substantially as described.

Second, the construction of the celled case C, with a hooked bolt *c*, a groove for receiving the guard plate D, and a notch for receiving the nose of the hasp B, substantially as described.

Third, the movable cap *g*, in combination with the nose *c*, constructed for receiving the tumblers *g g*, substantially as described.

Fourth, the combination of the guard plate D, and movable tumbler case C, all constructed and arranged substantially as described.

Fifth, the combination of the notched cylindric case C', with the hasp B², the said parts being constructed and operating substantially as described and for the purpose set forth.

Sixth, the means substantially as herein described and shown for throwing the hasp open when it is unlocked, in combination with the tumbler case C, constructed and arranged substantially as described.

Seventh, the combination of the friction spring S, with the movable tumbler case C, for preventing this case from moving too freely when the hasp is unlocked, substantially as described.

Eighth, the construction of the tumblers with depressions in them for receiving and keeping in place the springs *h h*, substantially as described.

No. 55,049.—J. E. BRIGGS, West Randolph, Vt.—*Washing Machine.*—May 29, 1866—The hinged, wedge-shaped pressing block has a corrugated under surface and is adapted to be reciprocated horizontally over the grooved bed plate placed in the wash-tub.

Claim.—The combination with the grooved board B, of the perforated irregular pressing block C, joint G, and handle D, when the parts are constructed and applied to a tub A, in the manner and for the purpose specified.

No. 55,050.—CHARLES B. BRISTOL, New Haven, Conn.—*Snap Hook.*—May 29, 1866.—The loop and book are made in one piece and the spring tongue is pivoted thereto; the joint pin is cast to the tongue and the latter has a recess for the spring to work in.

Claim.—As a new article of manufacture, the hook and loop part A and B, cast on a plane in combination with the tongue *e* and spring *g*, when the parts are constructed, attached, and fitted for use, substantially as herein described.

No. 55,051.—ALBERT BROOKER, Atalissa, Iowa.—*Portable Fence.*—May 29, 1866.—The support at the junction of the rails of each panel consists of inclined posts united into a frame by cross slats and a leaning brace, which, together with the tallest post, forms a crotch for the rider.

Claim.—A fence having posts A B and D, with brace C, and cross slats and rails as represented, all constructed, combined, and arranged substantially as herein specified.

No. 55,052.—ASA T. BROOKS, New Britain, Conn., assignor to RUSSEL & ERWIN MANUFACTURING COMPANY.—*Reversible Lock or Latch.*—May 29, 1866.—The middle portion of the hub is allowed to move across the path of the spindle when the latter is withdrawn, and thereby permit the retraction of the bolt for reversal, the detents limiting the motion.

Claim.—First, a latch hub made in two or more parts, acting conjointly with each other, substantially as and for the purpose described.

Second, the arrangement of the several parts in the manner substantially as described, by virtue of which the spring is made to act in the double capacity of actuating and holding the latch and slide respectively in place.

Third, the combination of the latch bolt *b*, spring *e*, with or without the plate *d*, slide *i*, and two or more part hubs, substantially as and for the purpose described.

No. 55,053.—SAMUEL BROWN and LEON LEVEL, San Francisco, Cal.—*Apparatus for Lowering Ships' Boats.*—May 29, 1866.—The two ropes are embraced between the respective pairs of sheaves, and the lever which closes them together acts equally and simultaneously on each, to allow an equal amount to be payed out in a given time, to place the boat with an even keel in the water.

Claim.—A breaking block, for running out even lengths of two or more ropes, having an extra set of sheaves *c*, which are capable of being moved to and from the real sheaves *a* by a breaking lever F, the different sets *a c* of sheaves being secured rigidly to their pins *b d*, substantially in the manner and for the purpose set forth.

No. 55,054.—V. W. BROWN, Camden, N J., and D. FRANKISH, Philadelphia, Penn.— *Lamp Chimney.*—May 29, 1866.—A portion of the bulb is flattened to act as a reflector.

Claim.—The chimney A, having a portion at X flattened or depressed, as and for the purpose described.

No. 55,055.—NIRUM CADWALLADER, Birchville, Cal.—*Blasting.*—May 29, 1866.—Compressed air is forced into the chamber containing the charge, to increase the explosive force.

Claim.—The use of compressed air in combination with all explosive substances, for the purpose of increasing the explosive force of said substance, by the aid of a greater amount of oxygen, and assisting, by the force of pressure, as well as by the expansion of the air from the heat generated by the explosion, said compressed air to be confined in the chamber with the explosive material, or, when in a chamber or recess contiguous to it, to be let free at the time the explosion takes place, substantially as described and for the purpose set forth.

No. 55,056.—EDWARD A. CALLAHAN, Brooklyn, N. Y.—*Machine for Punching Paper for Telegraphs.*—May 29, 1866.—Improvement on Humaston's patent of September 8, 1857.— Stops are arranged on the surface of the type wheel, or a prolongation of its shaft, which are brought into contact with a stop on the extremity of a lever connected with a key when the said key is depressed, thus arresting the motion of the shaft. Each key corresponds to a certain letter; its depression arrests the type-wheel shaft at the proper place to enable the corresponding letters of the type wheel to control the action of the punches in perforating the paper.

Claim.—First, a type wheel, revolved by friction, in combination with revolving stops and keys, and an apparatus, substantially as specified, for punching fillets of paper for composing telegraphic messages, as set forth.

Second, the stoppage bar 39, escapement 33, and ratchet wheel, in combination with the shafts B' and *g'''*, as and for the purposes specified.

No. 55,057.—E. A. CAMPBELL, Williams Bridge, N. Y.—*Sash Supporter.*—May 29, 1866.— The levers are pivoted in the window casing, and their serrated ends bear against the sash.

Claim.—The lever sash supporters *f* and *i*, formed with their ends wide, in the manner shown, so as to be fitted upon fulcrum screws in recesses in the casing beneath the moulding, as and for the purposes set forth.

No. 55,058.—JARVIS CASE, Springfield, Ohio.—*Farm Gate.*—May 29, 1866.—The gate is supported by hanging it upon the projections of the posts, and rotates thereon when balanced.

Claim.—First, so constructing and arranging the rests or supports *r r*, with shoulders *c c* in combination with the notches *n n* and post A, that the gate may be readily placed on said rests or supports in its proper position in relation to the post, and moved forward to close the gate or swung round to fully open it, or, if so desired, may be taken off the rests or supports and removed from the posts, substantially as described.

Second, so combining and arranging the rests or supports *r r* and adjustable piece D with the gate and posts that, when closed, the gate is firmly held in place, or may be partially opened and still retained in its proper position, resting a portion of its weight on each post, substantially as set forth.

No. 55,059.—WILLIAM J. CASE and RHUTSON CASE, Pittstown, N. J.—*Harvester Rake.*— May 29, 1866.—The driving shaft of the rake is rigidly mounted on the main frame, and driven by means of a system of levers arranged to act as a parallel ruler, whereby the parallelism of the rake and hinged platform are preserved, while both the main frame and platform are allowed to conform, independently of each other, to the surface of the ground over which they are drawn; the arrangement of means for operating and for regulating the throw of said rake, for locking it when out of action, and for locking the rake in a vertical position for transportation, are identified by the claims.

Claim.—First, the combination of a vibrating-sweep rake, with a crane post rigidly secured to and turning upon the main frame, by means of a system of levers forming a parallel-rule joint, substantially in the manner described, for the purpose of enabling the rake to conform to the undulations of a hinged platform without the use of a connection between the rake and platform, as set forth.

Second, the combination with the rake of the bevel gear F, cam *f*, tappet *o*, rock shaft O, and arm P, as described, for the purpose of raising and lowering the rake.

Third, the combination of the rake and crane post, by means of inclined arms, arranged and operating substantially as described, to enable the rake to strike close to the finger beam.

Fourth, the combination of the clutch lever e with the locking arm e', when arranged and operating as described, for the purpose of simultaneously uncoupling the rake and locking its driving wheel.

Fifth, the guide R to support the rake after discharging the gavel.

Sixth, the arrangement between the cranks g i of the adjustable pitman H, as described, for the purpose of varying the stroke of the rake, as set forth.

Seventh, the combination of the levers K M with the rake arms, by the socket joints P, as set forth.

Eighth, the combination of the rake head with a system of levers, substantially as described, for the purpose of locking the rake in a vertical position over the frame for transportation, as set forth.

Ninth, the combination, substantially in the manner described, of the hinged platform with the self-locking automatic rake, so located upon the main frame as to enable the finger beam to be folded up for transportation.

No. 55,060.—P. CHARLES CHIPRON, Highland, Ill.—*Harvester Rake.*—May 29, 1866.— The rake head has two sets of teeth, one as usual, and the other curved downward and in advance. The object is to secure certainty of action.

Claim.—The combination of the teeth e, curved-wire teeth f, and the rake head E, arranged and operating in the manner and for the purpose herein specified.

No. 55,061.—W. CLEMENT, Charleston, Ill.—*Fruit Basket.*—May 29, 1866.—The two end pieces are grooved to receive the side pieces and bottom, and the parts are tied together by a wire.

Claim.—A fruit box, constructed substantially as herein described.

No. 55,062.—AARON J. COOK, North Branford, Conn.—*Self-tilting Bucket for Wells.*— May 29, 1866.—As the bucket rises its ears are caught upon the disks, which carry the bucket between them and tilt it, discharging the water into the spout.

Claim.—The combination and arrangement of the shaft C, drums D and E, and the bucket G, suspended on cords or chains of an equal length, the whole being constructed and arranged substantially as herein described and set forth.

No. 55,063.—ISAAC COOK, St. Louis, Mo.—*Flour Packer.*—May 29, 1866.—The cylinder stands in the bag while filling through the detachable funnel and compressing the flour, and is afterwards withdrawn.

Claim—The use and combination of the packing tube C, the connecting funnel B, and the feed spout A, with such flour packers as use a packing device which in packing rises out of the tube C, and when otherwise arranged as set forth.

No. 55,064.—WILLIAM G. COOK, New York, N. Y.—*Fastening for Garments.*—May 29, 1866.—A conical roll of metal with a tapering slit to clamp the edges of a garment introduced into the slit.

Claim.—A fastening for garments and other articles, constructed and operating substantially as herein described.

No. 55,065.—CLAY CRAWFORD, East Cleveland, Ohio.—*Open Coal Grate.*—May 29, 1866—The coal magazine is behind the open fire, is supplied from the top and discharges at the base on to the inclined grate; the air is admitted at the rear; gases may pass from the magazine by a pipe which discharges them beneath the grate.

Claim—First, the magazine B, in combination with an ordinary open fireplace, substantially as and for the purpose set forth.

Second, in combination with the magazine, arranged in the open fireplace as described, the grate D, having the bars inclined to throw the coal to the front, as shown and described.

Third, the air chamber H, or its equivalent, arranged to operate in combination with the magazine B, as set forth.

Fourth, the tube E, arranged to operate in connection with the magazine B, substantially as set forth.

No. 55,066.—JOHN L. CRIST, Harrisburg, Penn.—*Screw Wrench*—May 29, 1866.—The shanks of the two jaws are toothed so as to engage when pressed together to maintain their relative longitudinal position when locked by the slotted ferrule with its detent catch.

Claim.—First, the combination of A A' and B B', operating conjointly in the manner described.

Second, the slotted rotating ferrule D, serving also as a lock for the sliding jaw A, and operating in the manner shown and described.

No. 55,067.—JOHN DAVIS, Allegheny City, Penn.—*Car Brake*—May 29, 1866.—The motions are communicated to the brakes from the draw-heads, and the arrangement of levers and connecting rods, &c., is explained in the claim.

Claim.—First, applying brakes to wheels of railroad cars so that said brakes become operative by the action and inaction of the locomotive and cars, always operating with relation to the wheels, so that they will adjust themselves, and bring the proper force to bear on the wheels when the locomotive becomes inoperative from any cause, or when any undue or improper action is imparted to the cars, said brakes being constructed and arranged substantially as herein described and set forth.

Second, so arranging the cranks *o* and *o'* on the spring shaft 11, that it will require less power to hold the brakes off the wheels than is required to draw them off, as herein described and set forth.

Third, the combination of the pawl *m*, and ratchet wheel *n*, with the shaft 11, cranks *o* and *o'*, and c<il spring combined, arranged and operating substantially as herein described and for the purpose set forth.

Fourth, the combination and arrangement of the coil spring shaft 1 1, cranks *o o'*, rods *j k j' k'*, levers 18 and 16, and shaft *z*, with the rods *i i'* and *h h'*, brake-bars *e*, and brakes *d*, the whole being constructed, combined, arranged and operating substantially as herein described and for the purpose set forth.

No. 55,068.—GEORGE M. DENISON, New London, Conn.—*Washing Machine.*—May 29, 1866.—Elastic knobs are attached to the face of the rubber which vibrates above the transverse India-rubber ribs of the concave bed.

Claim.—The elastic knobs F, in combination with elastic ribs G, on the surfaces of the rubber and bed, respectively, of a washing machine, substantially as herein set forth and for the purpose specified.

No. 55,069.—LIBERTY B. DENNETT, Portland, Maine.—*Plough.*—May 29, 1866.—The mould board is supported from the rear so as not to expose any object upon which the stubble can accumulate; the landside and the brace connect the front and rear of the mould board with the standard, which is far in the rear of the breast of the plough.

Claim.—The standard C, supported by the wing or brace F, extending from the standard to the rear of the mould board, the standard C being so placed as to offer no resistance to the stubble or grass as it falls over the mould board into the central cavity of the plough.

No. 55,070.—JULIUS C. DICKEY, New York, N. Y.—*Rock Drill.*—May 29, 1866.—The middle bit occupies a zone of the cross section; has an elevated centre and peripheral cutting edges; the side bits are socketed in the upper enlargement of the middle bit and are secured by a transverse key.

Claim.—The combination of the drill bits A B and C, substantially as set forth.

No. 55,071.—SILAS R. DIVINE and CHARLES A. SEELY, New York, N; Y.—*Apparatus for Distilling.*—May 29, 1866.—In the still is a steam pipe which lies in and follows the track of the descending trough wherein the liquid flows.

Claim.—The apparatus and its modification, substantially as herein described.

Also, the combination of the shell A with the pipe B, and the gutter C, substantially as described.

No. 55,072.—L. W. DOTY, Aurora, Ill.—*Rain Water Cut-off.*—May 29, 1866.—The swinging gate is vibrated to one side or the other, being detained by a spring catch in either position and turning the water to such one of the branch pipes as is desired.

Claim.—The gate E, and spring *d*, in combination with the partition D, arranged relatively with the induction pipe B, and eduction pipes C C'.

No. 55,073.—ALPHEUS C. DUNN, New York, N. Y.—*Composition for Lining Barrels.*—May 29, 1866.—Composed of gum copal, three parts; boiled linseed oil, eight parts; and one-sixteenth part of an aqueous alkaline solution of 3° Beaumé.

Claim.—A composition substantially as herein described of gum copal, boiled linseed oil, and aqueous solution of soda or potash, either with or without glue.

No. 55,074.—RUFUS DUTTON, New York, N. Y.—*Construction of Finger Bars of Harvesters.*—May 29, 1866.—The blanks for finger-bars are rolled in strips wide enough to be cut in the middle to make two bars, turning up the edges against which the shoulders of the fingers will bear when secured thereto.

Claim.—First, forming the finger-bars of harvesting machines from a metallic piece rolled or formed by suitable machinery, of sufficient width for two finger-bars, and having the edges turned orrolled up, as described, when such metallic piece or plate is made thinnest at its edges and increasing in thickness toward the centre, substantially as and for the purposes set forth.

Second, making in the centre of such metallic plate a recess or thinner portion, for the purpose of facilitating the dividing or cutting such plate, substantially as set forth.

Third, making the finger-bar of harvesting machines wedge-shape or increasing in thickness from the front toward the back side, and rolling or turning up its thinnest edge, substantially as and for the purposes set forth.

Fourth, the combination of the finger and finger-bar, substantially as described, so that or by which the finger is braced and supported by the bar at the back end of the under side of the slot of the finger, substantially as and for the purposes set forth.

No. 55,075.—MICHAEL B. DYOTT, Philadelphia, Penn.—*Lamp.*—May 29, 1866.—Drip, overflow, and trimmings pass to the drip cup below, the annular diaphragm of wire gauze allowing the passage upward of the surrounding current of air, but arresting falling particles of wick.

Claim.—The air-deflecting drip cup E, constructed and arranged so as to operate in combination with the bowl C D and the shell A B of a lamp, substantially as and for the purpose described.

Also, the combination of the screen F with the said drip cup E, and the opposite sides of shell A B, substantially as and for the purposes described.

No. 55,076.—WINSLOW P. EAYRS, Nashua, N. H.—*Beer Pitcher.*—May 29, 1866.—The beer is poured into the larger division, passes through perforated diaphragms and beneath the vertical division plate to the discharge chamber; by this means the froth is withheld.

Claim.—The arrangement of the divisional plate B, perforated plate C C′ C″ D D′ D″, as described, mouth b b, in combination with the pitcher A, constructed and operating in the manner and for the purpose herein specified.

No. 55,077.—S. F. EMERSON, Seville, Ohio.—*Roller for Wringer Machines.*—May 29, 1866.—The end of the roller is divided and the two portions are wedged into the slot of the wooden cylinder which surrounds the central shaft. The wedges are held by a central and end bands and the rubber rolled upon the core to the required thickness.

Claim.—First, the hollow slotted cylinder A, constructed as herein described in combination with the wedges D, and with the rubber C, substantially as and for the purpose set forth.

Second, the combination of the central band E with the cylinder A, wedges D, and rubber C, substantially as described and for the purpose set forth.

Third, the combination of the washers or ferrules F, constructed as described, with the cylinder A, substantially as and for the purpose set forth.

No. 55,078.—DAVID EMONNOT, New York, N. Y.—*Numbering Stamp.*—May 29, 1866.—Pressure applied to an object sets the working parts in motion moving the unit wheel; this motion is communicated to the train of wheels carrying the tens, hundreds, &c., to bring them into action consecutively, and afterwards move them in regular sequence.

Claim.—The type-carrying wheels No. 6, the ratchet wheels 7 and 9, the pawls 10, 11, 11′, and spring catch or trigger 13, in combination with the guides 17 and 19, and the spring-mounted shaft 15, operating together in the manner substantially as and for the purposes set forth.

No. 55,079.—WILLIAM ENNIS, New York, N. Y.—*Furnace.*—May 29, 1866.—The water in the tubular grate bars is converted into steam which issues in the chamber above, and is passed down through the fire to the lower exit pipe.

Claim.—First, in a furnace having a downward draught, the introduction of the steam above the grate in such a manner that it may pass downward through the burning fuel and through the grate, substantially as and for the purpose herein set forth.

Second, the tubular steam-generating grate and the elevated steam chamber in combination with a chimney below the grate, substantially as and for the purpose herein set forth.

No. 55,080.—CALEB C. FORSHEY, Washington county, Texas.—*Steering and Turning Apparatus for Vessels.*—May 29, 1866.—The paddle wheel is attached to a portable frame which is fitted to hang over the side of the vessel; the wheel rotates at right angles to the keel by motion derived from the hand crank and chain.

Claim.—First, the application of a rotary steering apparatus to a portable frame C, which is so constructed that it can be readily applied to or removed from the side of a vessel at pleasure, substantially as described.

Second, the construction of the supporting frame C, so that it shall be self-holding when applied to the side of a vessel, substantially as described.

No. 55,081.—WILLIAM FUZZARD, Chelsea, Mass.—*Steam Trap.*—May 29, 1866.—The rising of the buoyant stopper allows the water of condensation to escape.

Claim.—The buoyant stopper C, in combination with the exhaust chamber A, and a suitable seat B in said chamber, substantially as and for the purpose described.

No. 55,082.—J. H. A. GERICKE, New Orleans, La.—*Turbinate Force Pump.*—May 29, 1866.—The turbinate force pump has three systems of buckets in vertical series, and revolves beneath a horizontal plate to relieve it of the pressure of superincumbent water.

Claim.—First, a turbinate wheel enclosed as shown, with three systems of vanes or paddles D E F, substantially as described.

Second, the fixed case H over the turbinate wheel or engine, which drives the water upward, fitted with three different systems of curved vanes to direct the water toward the centre of the said case, substantially as described.

Third, separating the column of water in the case H, and above it, from the turbinate wheel above described by means of a fixed plate M, substantially as described.

No. 55,083.—CHARLES GIBBS, Pittsfield, Vt.—*Hay Loader.*—May 29, 1863.—The rake, which gathers and then elevates the hay, is attached to the side of the wagon, and when filled is vibrated by a rope which passes over sheaves and thence to a pulley on the hub of a rear wagon wheel, which is thrown into action by a pawl when the elevation is required

Claim.—First, the loader P, constructed substantially as described, in combination with the hay frame of a wagon or cart, for the purpose set forth.

Second, the combination and arrangement of the pulleys S, W, and U, the arms O, S, and T, and the rope R, with each other and with the hay frame, substantially as described and for the purpose set forth.

Third, the combination of the pulley U, ratchet wheel X, stop B', spring pin G', and pawl A', with each other and with the wheel of the wagon or cart, substantially as described and for the purpose set forth.

Fourth, the combination of the levers D' and F', the bar E, pin G', and spring I', with each other and with the pawl A', and hay frame, substantially as described and for the purpose set forth.

No. 55,084.—SEWELL GILLAM, New York, N. Y.—*Harvester Cutter.*—May 29, 1866.—The reciprocating cutter is divided midway of its length and driven by parallel pitmen, and cranks set at 90° on the same shaft.

Claim.—The arrangement of the pitman *a* and crank *d*, pitman *b* and crank *c*, in combination with the sickles A and B, operating in the manner and for the purpose herein specified.

No. 55,085.—JOHN C. GOVE, Cleveland, Ohio.—*Refrigerator.*—May 29, 1866.—Pivoted louvre slats are placed in the ceiling of the refrigerator under the ice chest, to close the connection temporarily when the door is opened.

Claim.—First, the frame D, with the slats G, slides E and F, with spring *d*, constructed, arranged, and used as and for the purposes herein set forth.

Second, the box A, ice box B C, with shutter frame D, constructed in the manner substantially as and for the purposes herein specified.

No. 55,086.—ROBERT GRACEY, Pittsburg, Penn.—*Socket Joint for Oil Tools.*—May 29, 1866.—The key prevents the disengagement of the parts of the joint, and the sleeve keeps the key from dropping out.

Claim.—The use of the swivel head *d*, key *g*, and key seats *e* and *h*, when used in combination with the male and female parts of the joint and the sleeve or band C, the whole being constructed, arranged, and operating substantially as herein described and for the purposes set forth.

No. 55,087.—WILLIAM GREEN, Cleveland, Ohio.—*Cement Roofing.*—May 29, 1866.—Felt paper is saturated with oil, and covered with a cement composed of comminuted iron ore mixed into a paste with oil. The paper is then rolled and cut into shingles.

Claim.—The herein described cement roofing made substantially as set forth, for the purpose specified, being a new article of manufacture.

No. 55,088.—ADAM HAGNY, Keokuk, Iowa.—*Buckle.*—May 29, 1866.—The tongues operate separately, and pass into the same strap from the respective sides.

Claim.—An improved trace buckle, formed by combining the two tongues G and I, constructed and arranged as herein described, with each other and with the frame of the buckle, substantially as and for the purpose set forth.

No. 55,089.—ALBERT HALL, New York, N. Y.—*Toy Spring Gun.*—May 29, 1866.—The cup to contain the projectile is thrust back by a ramrod until the trigger engages it; when released the India-rubber spring projects the missile holder; the coil of wire keeps the restraining cord in place.

Claim.—The coiled wire *b*, cord *i*, plate *r* and spring *m'*, arranged and operating in relation with the barrel B and case C', as herein shown and described.

No. 55,090.—LESTER HALL, Richfield, Ohio.—*Sheep Rack.*—May 29, 1866.—Grain is fed to the troughs by a spout, and the hay in the box behind is drawn by the animals through the openings.

Claim.—The arrangement of the grain trough C, tubes or pipes D, manger E, and trough L, with the adjustable rack F, door G, and frame, constructed as and for the purpose described.

No. 55,091.—MANLEY HALL, Richfield, Ohio.—*Portable Field Fence.*—May 29, 1866.—. The panels fit together at their ends and are locked by diagonal cross-braces and stakes.

Claim.—The arrangement of the panels or sections A B C, and posts D, so connected to said sections as to lock them together, in combination with the cross-pieces G and stakes *d*, in the manner and for the purpose set forth.

No. 55,092.—SAMUEL G. HALL and GEORGE W. BUGBEE, Norwich, Conn.—*Rotary Steam Engine.*—May 29, 1866; antedated May 21, 1866.—The automatically acting valves in the body of the piston balance the pressure on the slides. The piston wheel revolves against anti-friction rollers in the cylinder heads. Projections on the sides of the piston enter grooves in the side of the piston wheel and are secured by screws.

Claim.—First, the valves *e* in the body of the piston wheel B, in combination with the slides D, substantially as and for the purpose specified.

Second, the friction rollers *h*, and intervening blocks *i*, in combination with heads *f* of the piston wheel and with annular grooves *g* in the cylinder head, constructed and operating substantially as and for the purpose set forth.

Third, the annular grooves *g* in the heads *f* of the piston wheel, in combination with corresponding projections *l*, rising from the quadrants composing said piston wheel, substantially as and for the purpose described.

Fourth, the thin layer of Babbitt's metal, or other suitable material, fitting steam-tight into the end of the cylinder, in combination with one of the cylinder heads and with adjusting screws *m m'*, substantially as and for the purpose described.

Fifth, the pipes *p p'*, with branches *q q° q' q'°*, in combination with the openings, *t t' t° t'°*, in the valve G, and with the valve chamber F, and steam and exhaust pipes *r s*, substantially as and for the purpose set forth.

No. 55,093.—WILLIAM HAMMOND, Marshall, Mich.—*Wheel Cultivator and Gang Plough.*—May 29, 1866.—The cultivator frame is raised by the rotation of the levers which are pivoted to the frame and have the wheel axles on their ends.

Claim.—The employment, independently, or in connection with each other, of the bent axle, bearing levers L, in combination with the fulcrum pins F, and adjustable stop-gauge plates G, connected and arranged relatively to and with the frame and wheels of a cultivator, or gang plough, substantially as and for the purpose herein specified.

No. 55,094.—GEORGE HARDY, Lawrence, Mass.—*Car Ventilator.*—May 29, 1866.—A dampered opening in the car side admits the air, which is deflected by a pivoted shutter and directed into the car by the shield.

Claim.—The vibratory ventor E, constructed as described, and its combination and arrangement with a passage *c*, leading into a car or structure to be ventilated, and with the guard A, made and arranged with respect to the said ventor and passage, as hereinbefore explained.

No. 55,095.—JOHN HARRAR, Philadelphia, Penn.—*Shutter Fastening.*—May 29, 1866.—The extensible shanks interlock and are adjustable as to length by the engagement of their notches on the spindles of the handles.

Claim.—The combination of the coupling shanks B and C, and their respective parts, the pin *p*, throat *t*, and holes *n n*, with the shanks A A, and their respective parts, the slits *s s'*, rest holes *v v*, and the arrangement of the spindle and its parts D *e f g h i m* and K, to operate together, substantially in the manner and for the purpose herein set forth.

No. 55,096.—GEORGE W. HARRIS and ANDREAS FRANZ, New York, N. Y.—*Attaching Handles to Tools.*—May 29, 1866.—The countersunk notches on the outer edge of the eye form sockets for the heads of the keys, whose notched shanks engage the handle and are clamped by a sleeve.

Claim.—The combination of the countersunk shoulders B, the wedges or keys figs 2 2', and 2'', and the clamp or ring R, as and for the purpose herein set forth.

No. 55,097.—GEORGE HASTINGS, Waltham, Mass.—*Block for Polishing and Glossing Watch Hands.*—May 29, 1866.—The steel block has longitudinal and transverse recesses in which the hand is placed, exposing its surface to the polisher and with a recess on its lower side to hold the eye.

Claim.—The carrier, constructed substantially as described, and also the employment of such, or its equivalent, in manner and for the purpose as herein explained.

No. 55,098.—A. W. HEANY, Doylestown, Penn.—*Machine for Raking and Loading Hay.*—May 29, 1866.—The rake precedes the wagon, and the accumulating hay is raised by rakes traversing on an endless belt and discharged into the hay rack.

Claim.—The arrangement of the frame C, shafts R F O, heads H, teeth I, rearwardly inclined, rotating straps S, and teeth T, all located at the front of a wagon, and operating in the manner and for the purposes set forth.

Also, the heads H, rod J, and arms K K, connected by a rod L, in connection with the driver's seat M, all arranged to operate in the manner substantially as and for the purpose set forth.

No. 55,099.—LUMAN F. HEATH, Lansing, Mich.—*Sawing Machine.*—May 29, 1866.—
The operator stands upon a spring stage and throws his weight to the right and to the left alternately, oscillating the stage and reciprocating the saw.
Claim.—First, the spring *q*, constructed and operating in the manner and for the purpose herein set forth.
Second, the combination of the spring *q*, the platform *p*, the rods *t*, and cross bar M, for operating a saw in the manner herein described.

No. 55,100.—JOHN HEGARTY, Jersey City, N. J.—*Screw Wrench.*—May 29, 1866.—The fixed cross-piece has an attached stationary jaw and a sliding jaw operated by a screw.
Claim.—The arrangement of the cross-piece A, fixed jaw C, sliding jaw D, and screw E, in relation with each other, and with the shank B, substantially as herein specified.

No. 55,101.—MILAN HICKS, Richmond, Ill.—*Stove-pipe Damper.*—May 29, 1866.—The three perforated disks and two smaller concave end disks are united by a metallic strap; the combined device is capable of revolution to bring the disks coincident with the air current or across its path.
Claim.—The band I, concave plates A A', perforated plates B B' and C, constructed and arranged substantially in the manner and for the purposes herein set forth.

No. 55,102.—AARON HIGLEY, South Bend, Ind.—*Lifting Jack.*—May 29, 1866.—This is a toggle; one arm rests on the base, the other against the object; one arm is prolonged to form a lever by which the straightening of the toggle and the lift is performed.
Claim.—The link D' pivoted to the standard A', and lever B hinged to said standard by the straps E and plates *c*, and to the base A by the shaft *a* and staples *a'* in combination with the rack C and bands C', arranged as and for the purpose set forth.

No. 55,103.—JONAS HINKLEY, Norwalk, Ohio.—*Knitting Machine.*—May 29, 1866.—In this machine a reciprocating eye-pointed needle and no other is used; it is driven by a crank-rod, and is removable with its stock for convenience of threading, &c.; its stock also carries a tension device, a looping instrument, a cast-off, and a slide or detent to steady the feeding rack. This rack is straight and forms part of a comb plate, on the teeth of which the work is automatically set up, and which also carries a rod to keep the work properly suspended on the comb; the plate has also a series of holes coinciding in number with its teeth, and in which removable pins may be placed to reverse the knitting at any point desired for narrowing or widening. These changes may be made by hand at will by means of a hand lever. Provision is also made for winding the yarn upon a spool.
Claim.—First, an eye-pointed needle, (for carrying a thread or yarn,) a work-supporting comb, (on which the work or fabric is set up or supported,) and a looper or looping mechanism, (by means of which the stitch or loop is taken from the needle and conveyed to or deposited on the comb,) or their equivalents, combined to operate for the production of knit fabrics, substantially in the manner set forth and specified.
Second, an organism or combination consisting of an eye-pointed needle, a work-supporting comb, and a looping mechanism, substantially and as described, and a traversing mechanism by which traversing movements of the said comb may be produced, essentially in the manner and for the purposes set forth and specified.
Third, an organism or combination consisting of an eye-pointed needle, and a work-supporting comb, and a looping mechanism substantially such as described, and a tension mechanism, by means of which a proper friction or tension is exerted on the yarn or thread while in the act of being knit, as and for the purposes set forth.
Fourth, an organism or combination consisting of a work-supporting comb and traversing mechanism, substantially such as described, and a shipping mechanism, by which the movements of the said comb may be reversed in the manner and for the purposes shown.
Fifth, an organism or combination consisting of a work-supporting comb, and its traversing mechanism, substantially such as described, and a detent mechanism, which operates to detain or hold the comb while the loop or stitch is being formed, in the manner and for the purposes set forth.
Sixth, an organism or combination consisting of an eye-pointed needle, a work-supporting comb, and a looping mechanism, substantially such as described, and a cast-off mechanism, by which the loop is pushed from the tooth, in the manner and for the purposes set forth.
Seventh, an organism or combination consisting of an eye-pointed needle for carrying the yarn being used; a work-supporting comb, on which the work is set up or supported; a looping mechanism, by which a loop or loops are taken from the needle and conveyed to or deposited on the comb; a traversing mechanism, by which traversing movements of the comb are produced; a shipping mechanism, by which the traversing movements of the comb are reversed; a detent mechanism, by means of which the comb is held while the loop is formed; a cast-off mechanism, by which the loop is pushed from the tooth of a comb; and a tension mechanism, by which tension is applied to the yarn, the said instrumentalities, parts, or mechanisms being substantially as hereinbefore described, and combined and used for the production of knit fabrics, essentially in the manner set forth and specified.

Eighth, the combination of a work-supporting comb, substantially such as described, and a device by which the shipping mechanism may be automatically operated, essentially in the manner and for the purposes set forth.

Ninth, the combination of a work-supporting comb, substantially such as described, and an index mechanism or means, as explained, (or its equivalent,) for indicating at what point the movement of the comb will be reversed, substantially in the manner and for the purposes specified.

Tenth, the arrangement or application of the rod a^2, with the comb teeth, in such a manner as to traverse therewith, substantially in the manner and for the purposes specified.

Eleventh, a work-supporting comb, constructed with a rod a^2, and a rack and teeth, substantially as set forth and described

Twelfth, the combination of a work-supporting comb, substantially such as described, and a slotted or grooved end or piece of metal b', or its equivalent, so arranged or disposed in relation to the comb, that the teeth of the comb may pass through the groove or slot, whereby the loops, especially in the vicinity of the needle, will be more securely kept upon the said teeth, substantially as and for the purposes set forth.

Thirteenth, the bracket l^2, for supporting or holding the parts disposed in or upon the same, as constructed and arranged, substantially as described.

Fourteenth, a combination composed of an eye-pointed needle and a work-supporting comb, substantially such as described, and a guard mechanism, by which the discharged loop is prevented from being carried too far from the comb, by the advancing needle, as described.

Fifteenth, a combination consisting of a work-supporting comb, an eye-pointed needle, and a looper, substantially as described, and the projecting piece $6'$, or its equivalent, so arranged as to prevent the needle from being sprung upward by the action of the looper, substantially as described.

Sixteenth, a combination consisting of an eye-pointed needle, a work-supporting comb, a looper substantially such as described, and a spring finger s^2, for assisting in the conveyance of the loops from the needle to the tooth, as specified.

Seventeenth, the finger s^2, so constructed and arranged as not only to assist in conveying the loop as stated, but also to act as a part of the guard mechanism, substantially as and for the purpose set forth.

Eighteenth, a guard mechanism, one part of which is a spring so arranged that it will yield to admit of the use of a large or small needle, as circumstances may require, substantially as shown.

Nineteenth, the combination of the cast-off G, directly with the needle carrier, so as to be operated by it in the manner set forth and described.

Twentieth, the combination of the tension spring E, directly with the needle carrier, so as to operate with and be operated by it, substantially in the manner shown.

Twenty-first, the combination of the arm or detent H, directly with the needle carrier, so as to be operated by it, substantially in the manner set forth,

Twenty-second, the combination and arrangement of an eye-pointed needle A, a tension mechanism or spring E, a cast-off G, and a detent H, or their equivalents, so that any two or more of them will have but one carrier, substantially as set forth and described.

Twenty-third, the extension of the tension spring E back to the pitman or rod h, so that the same may also act to keep the said rod on the crank wrist, substantially in the manner shown and specified.

Twenty-fourth, the needle carrier and looping mechanism, so combined and arranged with respect to each other that the looping mechanism will be operated by the needle carrier, substantially in the manner set forth.

Twenty-fifth, the needle carrier as constructed with a cam-shape slot i^2, or its equivalent, for operating the looping mechanism. substantially as shown.

Twenty-sixth, the combination of the slot i^2, tri-armed lever f^2, stud h^2, looper rod c^{30}, cam groove k^4, cam groove p^2, projection r^2, and stud q^2, or their equivalents, for producing the compound movements of the looper, substantially as set forth and described.

Twenty-seventh, a looping mechanism, considered or taken as a whole, constructed and operating substantially as set forth and specified.

Twenty-eighth, the combination and arrangement of the socket t^{20}, or its equivalent, with the clutch shaft, or its crank, for connecting the spool or bobbin to the said shaft, preparatory to filling it, substantially as described.

Twenty-ninth, a narrowing and widening mechanism, consisting of the row of holes w', the pins z' z', the slide rod q', provided with projections p' p' and z' to operate with the pins z' z', and clutch lever o', or their equivalents, substantially as set forth and described.

Thirtieth, the application of the shipping pins z' z', directly to the index plates y' y', substantially in the manner and for the purposes specified.

Thirty-first, the means of holding or keeping the loop on the tooth until the needle has fairly entered the loop, such being effected by the looper itself, substantially in the manner set forth and described.

Thirty-second, a combination composed of an eye-pointed needle, a tension mechanism, a work-supporting comb, and a looping mechanism, substantially such as described, and a yarn guide g, arranged and used in the manner and for the purposes specified.

Thirty-third, the peculiar construction of the eye-pointed needle, that is the bent or crooked form described, so that the old loop may have room to pass back under the tooth, when said needle is arranged and used substantially in the manner set forth.

Thirty-fourth, the arch k', or its equivalent, arranged with respect to the comb, as shown, for supporting the plate i' and bracket t², substantially in the manner set forth.

Thirty-fifth, a knitting machine, composed of all the instrumentalities hereinbefore mentioned, or their equivalents, when constructed, combined, and used substantially in the manner and for the purposes specified.

No. 55,104.—WILLIAM W. HINMAN and WILLIAM W. NEVILL, Jacksonville, Ill.—*Car Coupling.*—May, 29, 1866.—The sliding draw-head acts as a bumper, having a spring in the rear, and engages the coupling-bar by its spring-hooks. The vertical position of the draw-head is regulated by lever and detents.

Claim.—The frame or draw-head C, which contains the spring-hooks H H, secured within the frame B, attached to the under side of the car, substantially as shown, to admit of the vertical and horizontal movement thereof, in connection with the vertically sliding frame L, arranged as shown, or in any equivalent way, for the purpose of adjusting the frame or draw-head C higher or lower, as may be desired.

No. 55,105.—J. H. HOGAN, St. Louis, Mo.—*Nail Extractor.*—May 29, 1866.—The slide at the end indents the wood so as to allow the nippers to grasp the nail by the head to extract it.

Claim.—The combination of the handle A, the slide B, and the nippers C, substantially as described and set forth.

No. 55,106.—E. G. HOLLAND, Union Springs, N. Y.—*Beverage.*—May 29, 1866.—Wine made from the expressed juice of sumach berries, diluted with water, sweetened and fermented.

Claim.—The aforesaid discovery of making wine, vinegar, and brandy from the properties of the sumach drupes or berries.

No. 55,107.—L. S. HOYT and E. G. HOYT, Croton Falls, N. Y.—*Screw Wrench and Hammer.*—May 29, 1866.—Explained by the claim.

Claim.—The implement herein described, having at one end of its shank A a hammer C, and at its other end a stationary jaw B, in combination with the adjustable spring jaw D, as a new article of manufacture.

No. 55,108.—A. INGALLS, Independence, Iowa.—*Carriage Plough.*—May 29, 1866.—The plough is suspended from the tongue and axle of a carriage; it is adjusted by a lever, bar, and cord, assisted by a treadle on the beam.

Claim.—The lever m, standard E, rod J, pulleys n n', in combination with the chain or cord I, guide plate D, and plough, arranged substantially as and for the purpose set forth.

No. 55,109.—JOEL C. JACKSON, Rochester, N. Y.—*Screw Wrench.*—May 29, 1866.—The toothed shank of the movable jaw traverses a groove in the stock and is operated by a nut on the end of the handle.

Claim.—The sectional screw rack extending from the bracket c' of the moving jaw c, and occupying a groove in the bar b, in combination with the nut g, surrounding said bar b, contiguous to the socket h, substantially as set forth.

No. 55,110.—BARTON H. JENKS.—Bridesburg, Penn.—*Coloring Wood, &c.*—May 29, 1866.—The air is exhausted from the pores of the wood and the coloring matter injected.

Claim.—The process, substantially as described, of injecting wood and other porous substances with coloring matter.

Also, the new manufacture of wood, prepared substantially as described.

No. 55,111.—BARTON H. JENKS, Bridesburg, Penn.—*Treating Wood for the Manufacture of Carding Engines.*—May 29, 1866.—The wood is saturated with paraffine, or equivalent, to prevent expansion and contraction by moisture and drying.

Claim.—First, the construction of the cylinders and rollers of carding engines of wood, which is impervious to the action of moisture, substantially as described.

Second, applying card clothing to a base or backing which is composed of a combination of wood and paraffine or other equivalent substances, substantially as described.

No. 55,112.—NIELS JOHNSON, Ripon, Wis.—*Whiffletree.*—May 29, 1866.—Two clevises from the double-tree afford attachment for the clip-loop of the single-tree.

Claim.—The clevises C C', arranged with the double-tree and whiffletrees, as and for the purpose herein set forth.

No. 55,113.—GEORGE R. KELSEY, West Haven, Conn.—*Buckle.*—May 29, 1866.—The looped ends of the lever embrace one bar of the frame and fall a little short of the opposite bar to clamp the strap against the latter.

Claim.—The combination of the bow and lever, Figs. 3 and 4, when the two parts are constructed, put together, and fitted for use, substantially as herein described.

No. 55,114.—WASHINGTON KENDRICK, New York. N. Y.—*Amalgamator.*—May 29, 1866; antedated May 14, 1866.—The agitator operates in the bottom of the tank, the contents of which are heated by the steam-box therein.

Claim.—First, the use for amalgamating and separating ores of a tubular heater E, provided with suitable pipes for injecting and discharging wet or superheated steam or hot air in combination with the tank A, constructed and operating substantially as and for the purpose set forth.

Second, the agitator F, in combination with the tubular drum E and tank A, constructed and operating substantially as and for the purpose described.

No. 55,115.—THOMAS KENNEDY, Branford, Conn.—*Adjusting Door Knobs to Spindles.*—May 29, 1866; antedated May 20, 1866.—The adjustable knob slips upon the square and corner-threaded spindle and is retained by the engagement of the rose with projections on the flat nut which traverses on the spindle.

Claim.—The combination of the nut D, and spindle E, with the rose G, and neck F, constructed and arranged with the projections *a a*, and notches *c c*, substantially in the manner and for the purpose specified.

No. 55,116.—P. B. KILLAM, Alina, Mich.—*Wool Press.*—May 29, 1866.—The hinged leaves of the table-top swing upward, and the folded fleece is then compressed in the other direction between a stationary and a movable guard, the latter being moved by rope and pulley devices.

Claim.—The arrangement of the shafts having wheels or pulleys, or their equivalents, connected together by a chain or belt, knuckle-jointed levers for raising and lowering the leaves of the table, in combination with any suitable device for holding the leaves in an upright position, substantially as herein described and for the purpose specified.

Also, the combination of the fixed and movable guards when arranged together, and so as to operate in the manner and for the purpose specified.

Also, the use of the catch buttons for holding one end of the binding cords for the twine as described.

No. 55,117.—JACOB KING, Jr., Fort Wayne, Ind.—*Bent Lever Scale.*—May 29, 1866.—The scale lever is counterweighted, and its vibration actuates a toothed segment, a pinion, and an index finger on a dial-plate.

Claim.—An improved lever-weighing scales formed by combining the lever F, scale pan G, cog-wheels K and L, and index J, with each other and with the frame of the scales, substantially as herein described and for the purpose set forth.

No. 55,118.—RICHARD KITSON, Lowell, Mass.—*Picker House for Opening and Cleaning Cotton, &c.*—May 29, 1866.—The hollow spaces between the walls connect with chambers beneath the floors; the dust pipes from the machines may connect with the spaces at any part and the resulting dust is led to a place of deposit.

Claim.—The hollow walls or flues B B, and openings C, in combination with the enclosed spaces between the beams, or their equivalents, the whole arranged substantially in the manner and for the purpose specified.

No. 55,119.—A. H. KNAPP, Coxsackie, N. Y.—*Attaching Sole to Horseshoes.*—May 29, 1866.—The leather pad is secured against the sole of the hoof by a spring block with lips, which enter recesses on the inner upper edge of the iron shoe.

Claim.—The combination of the spring frame B, lips *b*, and web C, applied relatively with the recessed shoe A, in the manner and for the purpose herein specified.

No. 55,120.—DAVID J. KNAPP, Fallsburg, N. Y.—*Churn Power.*—May 29, 1866.—The dasher shaft is reciprocated vertically by power derived from the spring through a train of gearing.

Claim.—The slide G, fitted in suitable guides or a slot in a standard or support H, with the pitman F passing through it, substantially as shown and described, in combination with the gearing and spring, all arranged to operate in the manner substantially as and for the purpose specified.

No. 55,121.—NORBERT LANDRY, San Francisco, Cal.—*Coin Holder.*—May 29, 1866.—The leaves have panelled recesses, upon or between whose glass surfaces the medals or coins are placed, and rotate upon an axis to display them in succession.

Claim.—The arrangement of coins and medals upon a vertical inclined or horizontal axis, between or upon transparent substances, to be revolved at pleasure, for the purpose of displaying said coins or medals, substantially as herein described and set forth.

No. 55,122.—FINLEY LATTA, Cincinnati, Ohio.—*Steam Engine.*—May 29. 1866.—Ports connect the ends of the cylinders directly for the passage of the exhaust steam from the small high-pressure cylinder to the larger low-pressure cylinder.

Claim.—The arrangement of the direct steam passages in combination with steam cylinders of unequal diameter, in a steam engine, adapted to operate, as and for the purposes set forth.

No. 55,123.—H. H. LOCKWOOD, Madison, Wis.—*Medical Compound.*—May 29, 1866.— Cure for ringbone, compounded of sulphuric acid, nitric acid, mercury, tincture of cantharides, and tincture of iodine.

Claim.—As a new article of manufacture, the medicine herein described.

No. 55,124.—LOUIS LŒFFLER, East Cambridge, Mass.—*Self-Acting Decanter Stopper.*— May 29, 1866.—The fingers keep the ball in place.

Claim.—The self-acting ball valve decanter stopper as made with the series of fingers applied to the valve seat tube, and so as to extend therefrom and about the ball, when in place, substantially in the manner as hereinbefore specified.

No. 55,125.—Cancelled.

No. 55,126.—S. B. LOUGHBOROUGH, Canandaigua, N. Y.—*Fastening Keys in Locks.*— May 29, 1866; antedated May 18, 1866.—The bit of the key is turned so as to allow the bar to enter the lower part of the keyhole; a slotted lug on the bar passes through the handle of the key and is cottered by a wedge.

Claim.—The employment or use of the safety bar *b*, constructed substantially as and for the purpose herein shown and described, in combination with the clamping wedge *w*, or its equivalent.

No. 55,127.—ADDISON LOW, Albany, N. Y.—*Safety Valve.*—May 29, 1866.—The steam lifts the measuring valve in the usual manner, and lifts a piston attached to the long arm of a lever ; to the other arm a large valve is attached which is depressed against the pressure of steam beneath it to afford a large area of escape.

Claim.—First, the combination of the measuring valve H with the piston F, for the purpose herein set forth.

Second, the combination of the piston F with the safety-valve D, and the measuring valve H, as and for the purpose described.

Third, the combination of the piston F, lever G, and safety-valve D, as and for the purpose named.

Fourth, the combination of the measuring valve H with the piston F, lever G, and safety valve D, as and for the purpose set forth.

No. 55,128.—JOSHUA R. LUPTON and N. E. LUPTON, Stafford, Ohio.—*Fruit Jar.*—May 29, 1866.—The disk has a recessed face and a coating of sheet rubber, secured by nails or cement.

Claim.—An improved fruit-can stopper, formed by combining an India-rubber or equivalent surface or lining with a wooden or metallic cap, constructed and applied substantially as herein described and for the purpose set forth.

No. 55,129.—D. F. LUSE, Spring Mills, Penn —*Farm Gate.*—May 29, 1866.—The vertical suspensory slats are pivoted above and below, and the horizontal bars of the gate shut into their intervals when folded.

Claim.—The shoulders or offsets *d d*, and *f f*, in the bar H, and post A, so as to allow the pivoted rails C D D to shut by each other in folding, substantially as and for the purpose set forth.

No. 55,130.—GEORGE LUTZ, Lancaster, Ohio.—*Water Indicator for Steam Generators.*— May 29, 1866.—The ball-cock lever actuates a rod and segment rack which turns the index figure on the dial.

Claim.—The rod I incased within a tube J, and hung at its lower ends to the float rod G, and at its upper end hung to the crank arm O of the shaft P, in combination with the sector rack T of shaft P, interlocked with the pinion U of the index spindle, when arranged together substantially in the manner described and for the purposes specified.

No. 55,131.—JOHN LYDY, Georgetown, Ohio.—*Sash Fastening.*—May 29, 1866.—The sash is held down by the clip of the hook which is pivoted to the casing, or is supported by the engagement of the point of the hook with the teeth of the ratchet on the sash.

Claim.—The combination and arrangement of the pawls D and D', the serrated plates C and C', the plate *i*, and the springs *a* and *a'*, substantially as herein described and shown, and for the purpose specified.

No. 55,132.—SAMUEL L. LYFORD, Portland, Maine.—*Gauge for Augers.*—May 29, 1866.— Explained by the claim.

Claim.—A gauge for bits and angers composed of two pieces A A, firmed to fit the faces of the twisted blade of the bit or auger, so as to turn with the spiral twist, and adjustably united by set screws, substantially as and for the purpose set forth.

No. 55,133.—THEODORE LYMAN, Sandusky, Ohio.—*Car Truck.*—May 29, 1866.—The devices described are intended to prevent oscillation.

Claim.—The combination of the braces O P U V, and block R, with the swing beam D of an ordinary railroad car truck, substantially as described and for the purpose set forth.

No 55,134.—BENJAMIN MACKERLEY, Paint, Ohio.—*Atmospheric Governor.*—May 29. 1866 —The brake lever is brought into contact with some moving wheel of the machine to be regulated, by the pressure of air in the cylinder upon the piston.

Claim.—An atmospheric governor composed of a cylinder A, with two pistons B C, and brake D, arranged substantially as and for the purpose described.

No 55,135.—HENRY B. MALBONE, Middletown, Conn.—*Mop Handle.*—May 29, 1866; antedated May 21, 1866.—The mop cloth is clamped between wire loops, whose bows rest against each other, and whose ends are socketed in the handle and secured by a sliding ferrule

Claim.—The cloth holder for mop handles herein described, when constructed and arranged substantially in the manner set forth.

No. 55,136.—REUBEN A. MCCAULEY, Baltimore, Md.—*Hydrant.*—May 29, 1866.—The depression of the goose-neck forces a piston into a chamber which empties itself through the discharge pipe; the end of the pipe depresses a valve and allows the water to flow continuously from the main. The elevation of the goose-neck closes the valve, and the rising of the piston empties the discharge pipe into the piston chamber.

Claim —The combination with the elbow seat *c*, of the chamber *d*, the strainer and supporting disk *j″*, spring *i*, holding cup *h*, elastic ball *g*, and pipe F, arranged and operating substantially as described and for the purpose set forth. •

No 55,137.—JOSHUA MERRILL, Boston, Mass.—*Manufacture of Casks, Barrels and Kegs.*—May 29, 1866.—The staves are tongued and grooved together and the joints glued.

Claim.—The improved cask, substantially as described, having its joints made with tongues and grooves, substantially in the way and for the purpose hereinbefore described and set forth.

Also, in combination with the joints of a tongued and grooved cask, a coating or stuffing of glue or similar gelatinous cement in the joints, substantially as hereinbefore described.

No. 55,138.—ALBERT H. MERSHON, Philadelphia, Penn.—*Regulating the Draught of Furnaces.*—May 29, 1866.—By a cord attached to a weight and pulley in the upper story, and connected at its lower end to the ash-pit door, the draught can be regulated from any floor of the house.

Claim.—The ash-pit door and draught door, the chain I I, pulleys K K, and weight M, with the door and index gauge, when combined and constructed substantially as and for the purposes herein described.

No. 55,139.—WILLIAM S. MESSINGER, Roxbury Mass.—*Apparatus for Parlor Croquet.*—May 29, 1866.—The cloth, perforated for the wickets and stakes, is laid on the table and secured by the imposition of the marginal frame.

Claim.—First, the wickets *f*, or stakes *i*, with their sockets D or E, in combination with the cloth B, perforated to receive them, substantially as set forth.

Second, the perforated cloth B, and frame C, in combination with the wickets *f*, and stakes *i*, and their removable sockets D and E, substantially as described.

No. 55,140.—HENRY W. MILLAR, Utica, N. Y.—*Cheese Hoop.*—May 29, 1866.—The metallic lining renders the cheese more readily removable.

Claim.—A cheese hoop made of wood, hooped with iron bands, the inner surface whereof is lined with a metallic lining, as hereinbefore described, the invention consisting in the application of the metallic lining.

No. 55,141.—JOHN MILLER, Russellville, Ky.—*Washing Machine.*—May 29, 1866.—The rotating rubber is composed of balls strung upon elastic wires and rotating freely thereon; the bed is formed in the same way and has a furnace beneath.

Claim.—First, the combination of the loose balled drum, with the loose balled frame, constructed, arranged and operating substantially as set forth.

Second, the combination of the furnace trough, drum and frame, when constructed, arranged and operating as described, for the purpose set forth.

No. 55,142.—JOHN MILLER, Russellville, Ky.—*Seeder and Cultivator Combined.*—May 29, 1866.—The shares of the cultivator throw their furrows inward, and form a ridge which the succeeding tool plants and smooths.

Claim.—The combination of a cultivator which furrows and forms the ridge, with a planter that plants, covers, and finishes the ridge with a smooth, flat surface, all in once passing over the ground, substantially as set forth.

No. 55,143.—JACOB G. MINER, Morrisania, N. Y.—*Street Lamp.*—May 29, 1866.—An inverted bell-shaped glass is placed over the burner which enters the opening below; the glass is supported by a flanged rim on the metallic frame to which the lid is hinged.

Claim.—First, a cylinder, or its equivalent, when sustained substantially as described. Second, the combination of cylinder frame and lid, for the purpose set forth.

No. 55,144.—HENRY T. MOODY, Newburyport, Mass.—*Auger.*—May 29, 1866.—For centring the tool while enlarging the end of the hole, a collar fitting the hole is temporarily attached to the end of the auger below the cutting lips.

Claim.—The removable centring plug D, in combination with the bit or auger A, for the purpose set forth.

No. 55,145.—CHARLES A. MOORE, Westbrook, Conn.—*Kettle Bottom.*—May 29, 1866.—The bottom of the kettle is of the shape of an inverted funnel, to increase fire surface.

Claim.—The above described improvement in kettle bottoms, formed substantially as specified, and for the objects set forth.

No. 55,146.—ISAAC MORLEY, Pittsburg, Penn.—*Brick Machine*—May 29, 1866.—A series of plungers reciprocates in the bottom of the hopper and press the clay to alternate sides into moulds, which are raised into position and lowered by the periodic movements of the peculiarly constructed gearing. Wires travelling with the raising and lowering devices cut off the clay in the moulds from that in the hopper.

Claim.—First, the reciprocating slide I, and plunger K, arranged and operating in connection with the hopper A, as and for the purpose set forth.

Second, the combination and arrangement of the wheels F and G, and the eccentrics E_r for the purpose of operating the slide I, and plunger K, as shown and described.

Third, the frames d, for the purpose of holding the moulds in position while being filled and having the wire s arranged as described, for cutting off the brick as set forth.

Fourth, the wheel B, constructed as shown, and arranged to operate in connection with wheels C, as described, for the purpose of elevating and depressing the frames d, at the required intervals.

Fifth, the combination and arrangement of slide I, plunger K, and frames d, to operate in connection, and at the proper intervals, substantially as herein shown and described.

No. 55,147.—GEORGE MOULTON, Jr., Bath, Maine.—*Ship's Pump.*—May 29, 1866.—The open top of the chamber and the removable valves with ring handles give access to the pump for the removal of obstructions.

Claim.—Arranging the valves in or over the pump, tube or pipe, in combination with an open top above the valves, and a piston working in an auxiliary cylinder, substantially as described, so that the pump may be sounded by removing the valves, without removing or disturbing the piston.

No. 55,148.—CHARLES MURTHA, Philadelphia, Penn.—*Brick Machine.*—May 29, 1866.—The clay from the mud-mill is lumped by revolving recessed wheels, and is thus delivered in balls of a definite size to the trays which traverse on an endless apron beneath, being sanded before being dumped into the chambers of the revolving mould wheel. A cap presses on the exterior surface and the follower compresses and discharges.

Claim.—First, the intermittingly rotating cylinders H H, provided with blades or knives I, in connection with the endless carrier K, said parts being arranged in relation with a mud-mill, substantially as and for the purpose set forth.

Second, the plate M, suspended on pivots above the endless carrier K, to operate in connection therewith, and the sanding device, substantially as and for the purpose specified.

Third, the intermittingly rotating block P, provided with a series of radial moulds having plungers R fitted within them, and arranged in relation with the endless carrier K, substantially as and for the purpose set forth.

Fourth, the cap S*, operated in the manner shown, or in an equivalent way, and arranged in relation with the block P, so as to impart an intermittingly rotating movement to the block, and at the same time serve to confine the clay in the moulds while the clay is under pressure therein, substantially as set forth.

Fifth, the heating of the moulds by steam, substantially in the manner and for the purpose as set forth.

No. 55,149.—JOSEPH NASON, New York, N. Y.—*Surface Condenser.*—May 29, 1866.—The joints between the corrugated plates are formed by tongues and grooves with interposed packing and extend around the connecting openings. The combination forms narrow, non-communicating chambers through which the refrigerating liquid and the fluid to be condensed pass in opposite directions by sinuous courses.

Claim.—The within described construction and arrangement of the plates 1 2 3, &c., and coinciding holes G H, &c, whereby one of the fluids is allowed to flow freely through the holes into and through alternate spaces m m, &c., between the plates, and is prevented from communicating with the intermediate spaces n n, &c., while the other fluid is allowed to flow through said intermediate spaces, substantially in the manner and for the purposes herein set forth.

No 55,150.—ELIZA H. NEWCOMB, New York, N. Y.—*Fruit Gatherer.*—May 29, 1866.—Explained by the claims.
Claim.—First, the employment in a fruit gatherer of converging blades b b, applied to the frame or ring A, and operated by the simple withdrawal of the fruit gatherer, substantially as described.
Second, the combination of the ring A, converging blades b b, curved slats m m, pins n n, and spring s, by means of which a shear action of the converging blades is obtained by the simple withdrawal of the fruit gatherer, substantially as described.
Third, the combination of the hinged frame A, either square or circular, with four joints or more, converging blades b b, and springs s, by means of which a shear action of the converging blades is obtained by the simple withdrawal of the fruit gatherer, substantially as described.

No. 55,151.—A. F. W. NEYNABER, Philadelphia, Penn.—*Apparatus for Preserving Beer.*—May 29, 1866.—The weighted head is attached to the lining and descends in the cask to raise the liquid in the discharge tube.
Claim.—The method of preserving beverages or fermented liquors, as lager beer, ale, porter, &c., in air-tight India-rubber vessels, and discharging these liquors therefrom in parts, by means of compressing the vessel by weight, in such a manner as to avoid the admission of any air, substantially as set forth in the foregoing specification.
Also, the construction of the beer keeper, by the application, combination, and arrangement of the India-rubber vessel A, inlet i, with plate m, screw n, and bar o, outlet j, with valve k, tap l, and presser y, frame B with ring a', presser C, with pulleys j' j', and stand D, substantially in the manner and for the purpose as herein described.

No. 55,152.—HARRISON OGBORN, Richmond, Ind.—*Grain Screen.*—May 29, 1866.—The suspended shoe is oscillated longitudinally by a crank, the rollers at the bottom working in a fixed zigzag groove on the frame, to give a lateral shake to the shoe.
Claim.—First, the crooked groove 8, in combination with the friction rollers or pulleys n n, and arms m m, when used for the purposes set forth.
Second, the perforated crank c, shaft b, rod n, and hopper A, in combination with the nest of screens D E and P, and crooked groove 8, when arranged and operated substantially as set forth, and for the purpose specified.

No. 55,153.—PATRICK O'RORK, Norwalk, Conn.—*Composition for Stiffening Felt Cloth.*—May 29, 1866.—Explained by the claim.
Claim.—The union of alum with a solution either of shellac and borax, shellac and salsoda, or shellac and ammonia, for the purpose of stiffening hats, hat bodies, hats for ladies' wear, and other articles, in which a greater degree of stiffness is required than the material used possesses of itself.

No. 55,154.—F. M. OSBORN, Dover Plains, N. Y.—*Carpet Stretcher and Hammer.*—May 29, 1866; antedated May 22, 1866.—As the lever is depressed the stretcher is extended and the tilt hammer dropped on to the head of the tack.
Claim.—The combination, in one machine, of a stretcher w and hammer o, arranged and operating with regard to each other substantially in the manner described, and for the purpose specified.
Also, the arrangement of the stretcher w and hammer o, with the intermediate parts or devices, connecting them with a common lever handle g, substantially as and for the purpose set forth.

No 55,155.—ALONZO PALMER and JOHN BEAN, Hudson, Mich.—*Subterranean Reservoir for Wells.*—May 29, 1866.—The arms below the pump stock disturb the earth around them, and the muddy water is then withdrawn by the pump.
Claim.—The arrangement of the cylinder A, shaft B, disks 8 and 8', together with the agitators d d, when used as and for the purpose herein specified.

No. 55,156.—ALONZO PALMER and ROBERT GILLILAND, Hudson, Mich.—*Stump Extractor.*—May 29, 1866.—Horse power is applied to the rope to depress the lever and raise the clevis with the stump attached; the weight of the lever is sustained by the cross-bar and supporters, which are raised to throw the weight on the wheels for locomotion.
Claim.—The, beams A A, supports B B, clevis F, stays D D, lever E, braces C C, pulley I, cord H, and tongue G, arranged in the manner substantially as and for the purposes herein set forth.

No. 55, 157.—WILLIAM PIRSSON, Newark, N. J.—*Ratchet Wrench.*—May 29, 1866.—The spring lever is pivoted to the handle and engages a groove in the tool socket to maintain it in place but permit rotation. The lever is vibrated to release the socket for the substitution of another.

Claim.—First, the lever-like catch D, applied as described, with a spring f, in a mortise in the stock, and in combination with a circumferential groove d in the exterior of the ratchet socket, the whole arranged substantially as herein described.

Second, the screw s, in combination with the catch D, substantially as and for the purpose herein specified.

No. 55,158.—LOUIS PRANG, Boston, Mass.—*Manufacture of Caustic Alkali.*—May 29, 1866.—The carbonate of soda or potash is dissolved in water and poured upon the quick-lime in such quantity as to reduce the latter to a dry powder. The chemical reaction renders the alkali caustic and it becomes portable in its mechanical combination with the carbonate of lime.

Claim.—The mode herein described of combining any alkali with lime in such a way as to obtain a powder to serve when dissolved as caustic lye.

No. 55,159.—SAMUEL RICHARDS, Philadelphia, Penn.—*Manufacturing Glass.*—May 29, 1866.—Explained by the claim.

Claim.—The manufacture of merchantable glass from impure cullet and common fluxes by first melting them and straining the impurities out of them and then taking the product and mixing it with a small portion of ordinary batch, substantially as described.

No. 55,160.—MARGARET RICHARDSON, Norristown, Penn.—*Medical Compound.*—May 29, 1866.—This medicine for the cure of cholera and cognate disorders consists of a combination of the following tinctures : opium, camphor, peppermint, ginger, rhubarb, kino, capsicum.

Claim.—A medicine for the cure of cholera and other bowel complaints, made of the ingredients herein described and in about the proportions specified.

No. 55,161.—CHARLES ROBERTS.—Lake Village, N. H.—*Washing Machine.*—May 29, 1866.—The interior revolving cylinder has a periphery of spring pressure bars, and a gear wheel on its axis revolves the surrounding concentric series of rollers.

Claim.—A washing machine composed essentially of a concave of concentrically arranged rolls I I, nearly touching each other, except a wider space at the top, for the introduction of the clothes, and having a cylinder c, with elastic or yielding pressure bars, revolving inside thereof, the whole to be placed within a tub or other vessel holding the water or suds, substantially as herein specified.

Also, giving the rolls I I a positive revolving motion nearly equal to and in the same direction as the pressure bar surfaces of the revolving cylinder by means of pinions J J on said rolls, and a driving cog wheel H on the cylinder, or its shaft gearing into said pinions, substantially as herein set forth.

Also, the arrangement of the stationary bars F F, and yielding pressure bars E E, in the heads of the cylinder C, substantially as herein described.

No. 55,162.—THOMAS B. ROCHE, Folsom, Cal.—*Swivel Shackle.*—May 29, 1866.—The eye has a thimble, and its shank passes through the clevis bolt so that it may rotate and have a lateral play.

Claim.—The swivel shackle, constructed substantially as herein described.

No. 55,163.—BENJAMIN C. ROGERS, Stockbridge, Vt.—*Training Hop Vines.*—May 29, 1866.—The stakes in alternate rows are of different sizes, and they are stayed by wires.

Claim.—The arrangement of stakes and lines substantially as herein described, and as represented by the accompanying drawing.

No. 55,164.—ELBERT J. ROOSEVELT, Pelham, N. Y.—*Machine for Cutting Succulent Roots for Feed.*—May 29, 1866.—A curved knife is attached to a projection on the oscillating arch and at each end the throat is cleared by a lip.

Claim.—First, the combination of the curved knife E, attached to the projection G of the arch F, with the hopper C, the parts being constructed and operated substantially as and for the purpose herein recited.

Second, the groove M for the entrance of the edge of the knife, and the lip N under the groove, for clearing the throat or space under the edge of the knife, as herein described.

No. 55,165.—A. J. ROSS, Rochester, N. Y.—*Burglar Alarm.*—May 29, 1866.—The spring hammer is vibrated and the bell struck by the engagement of the trigger with the corrugations of the plate on the opening door.

Claim.—The portable burglar alarm, made up of the combination of the alarm bell C and base plate D, engaged and disengaged by simply changing the position, as described, and the cam plate G, provided with a series of corrugations or cams k k, for giving a succession of alarms, as herein set forth.

No. 55,166.—CYRUS W. SALADEE, Newark, Ohio.—*Padlock.*—May 29, 1866; antedated May 24, 1866.—The pivot of the dog constitutes the key stud; the hole in the dog is countersunk to receive the key, and has one straight side for the bit of the key to operate against.

Claim.—Securing the wards or guards B, for the key, in the body of the lever A, or its equivalent, in the manner and for the purpose substantially as set forth.

No. 55,167.—CYRUS W. SALADEE, Newark, Ohio.—*Padlock.*—May 29, 1866; antedated May 24, 1866.—The doubled-hooked, rotating bolt engages both ends of the baep stirrup, and by disengagement permits one end of the latter to be withdrawn from the lock case.

Claim.—First, the lever S, or its equivalent, in combination with the lock bar A, in the manner and for the purpose substantially as shown and described.

Second, the spring B, or equivalent, in combination with the lever S and lock bar A, in the manner and for the purpose substantially as shown and described.

No. 55,168.—THOMAS H. B. SANDERS, Philadelphia, Penn.—*Heating and Ventilating Railroad Cars.*—May 29, 1866.—The fan is driven by an exterior wind-wheel, and is placed in the upper part of the air chamber of the stove; it drives the heated air through pipes, which pass around the car and have registers at intervals.

Claim.—The arrangement of air pipes P and registers R with hot or cold air chambers c' c' of stove S and fan B B, combined together as and for the purpose hereinabove described.

No. 55,169.—JOHN F. SEAMAN, Cortlandville, N. Y.—*Churn.*—May 29, 1866.—The pendulum is connected by levers to the dasher shaft.

Claim.—First, the arrangement of bars O and Y, in their connection with the lever Q and pendulum weight L, for the purpose described.

Second, churning by an attachment to the bottom of a pendulum or pendulum weight, as set forth, and for the purpose specified.

Third, the arrangement of the springs 2 and 3, as set forth, and for the purpose specified.

No. 55,170.—E. SHOPBELL, Ashland, Ohio.—*Carpet Holder.*—May 29, 1866.—To the edge of the carpet is attached a triangular strip of wood, which is secured to the mop board by a button.

Claim.—The combination of the cleats and buttons, when used in connection with the carpet and mop board, as and for the purpose set forth.

No. 55,171.—AMOS B. SIMONDS, Youngstown, Ohio.—*Screw Cutting Machine.*—May 29, 1866.—The four threading dies are arranged in guides upon two movable jaws, which are closed by a cone in the rear and opened by a spring.

Claim.—First, the combination and arrangement of the dies H, movable jaws F, and stops J, constructed and operated as described, with each other and with the head d' of the shaft D, substantially as and for the purpose set forth.

Second, the combination of the spring M with the arms f' of the movable jaws F, substantially as described, and for the purpose set forth.

Third, the combination of the sliding conical block L, gauge rod R, arms P and N, and connecting bar O with each other, with shaft D, and with the movable jaws F, substantially as described and for the purposes set forth.

No. 55,172.—ANDREW J. SMITH, Greenville, Ohio.—*Steam Generator.*—May 29, 1866.—The water is introduced through water pipes and rose-heads into the perforated inner chamber, the object being to disperse it and prevent its reaching directly the highly heated sides.

Claim.—An improved steam boiler, formed by combining the perforated interior boiler B and the water induction pipe D d' E with each other and with the outer boiler A, substantially as described and for the purposes set forth.

No. 55,173.—H. M. SMITH, Richmond, Va.—*Machine for Cutting Smoking Tobacco.*—May 29, 1866.—The tobacco is stripped into shreds by the revolving toothed cylinder and the comb; the snuff and dirt are removed by a screen.

Claim.—The fine steel or iron comb, made single or with two edges to reverse, in combination with the fine steel teeth in the cylinder, and in combination with the screen, operating for the purpose and in the manner as above described and set forth.

No. 55,174.—GEORGE STEVENSON, Zionsville, Ind., and JOHN J. CRIDER, Greenfield, Ind.—*Grain Cleaner.*—May 29, 1866.—The grain, in tumbling around the screen, is caught by the flaring mouth of the inner casing and subjected to the action of brushes.

Claim.—The combination of the outer drum screen and the inner casing K K, provided with a revolving brush, the whole constructed and operated substantially as described and represented.

No. 55,175.—AMOS D. STOCKING, Dowagiac, Mich.—*Cultivator.*—May 29, 1866.—The rear caster wheel supports the cultivator; the plough standards are supported by stay rods and by transverse pins in themselves and in the frame, and are raised by the partial rotation of the board to which they are attached by chains.

Claim.—First, the application of a swivel wheel *a* to the rear end of a frame of a cultivator, which is constructed substantially in the manner described.

Second, keeping the shovel standards E in proper position during their passage through the soil by means of the transverse rods *h'* and notches *i*, combined with the forward braces *k*, substantially as described.

Third, the pivoted board G, or its equivalent, arranged transversely across the cultivator frame and connected to the shovel standards E, substantially as described.

No. 55,176.—SILAS STUART, Sterling, Mass.—*Rotary Brush.*—May 29, 1866.—The cylindrical brush is made of annular sections whose perforations form holders for the bunches of bristles, and whose corrugations enable them to be so associated as to preserve the continuity of the brush surface.

Claim.—The perforated corrugated toothed cylinder A, constructed and operating in the manner and for the purpose herein described.

No. 55,177.—HOWARD TILDEN, Boston, Mass.—*Tobacco Cutter and Nut Cracker.*—May 29, 1866.—The dog's viscera consist of gearing which oscillates his lower jaw when his tail is depressed, and at the same time brings a knife out of his belly to cut the tobacco. The dog cuts up and bites when you play with his tail.

Claim.—A tobacco cutter and nut cracker, having the operating lever in the form of the tail of a dog, or other animal, and the jaws that hold the nut corresponding to the mouth of an animal, the operating mechanism, or its equivalent, being contained in the case H, as herein described.

Also, the combination of the traversing knife or cutter, operated as herein set forth, with the jaw U and the case H, constructed as herein set forth.

No. 55,178.—HAMILTON E. TOWLE, New York, N. Y.—*Apparatus for Collecting Floating Oil from Streams.*—May 29, 1866.—The oil is collected by boo· and lifted upon an inclined endless belt passing over expressing rollers from which it is discharged into a receptacle.

Claim.—First, the combination with the squeezing roller of an apparatus constructed as herein described, of a swinging apron or scraper actuated by means of a spring, or equivalent mechanism, and arranged and operating as and for the purposes herein set forth.

Second, in combination with the squeezing roller and swinging scraper as described, the employment of a second scraper, arranged and operating as herein shown and set forth.

Third, in combination with the squeezing roller, the lever arms, in which it has bearings, and the rods or bolts, by means of which the pressure of the roller may be regulated, the whole being constructed and arranged substantially as herein shown and described.

No. 55,179.—B. T. TRIMMER, Rochester, N. Y.—*Grain Cleaner and Separator.*—May 29, 1866.—Air is circulated through the machine by a fan which receives at its centre and disperses at its periphery. The separating, scouring, and dusting of the grain is performed in one operation.

Claim.—The special construction and arrangement of the machine, consisting essentially of the draught passages *d g h i*, of the trunk E, fan C, shoe H, beater chamber G, and beater M *f*, arranged and operating substantially as and for the purpose herein set forth.

Also, the combination and arrangement of the passages *d g h i* uniting in a common discharge, having the turns or deflections *k l* and the pockets *m n*, the whole forming a trunk that is connected with the top and bottom of the machine in such a manner as to form a circuit through, substantially as described.

Also, exhausting both the inside and outside of the beater cylinder by the single fan C, operating above, the whole arranged substantially as herein specified.

Also, distributing the grain falling into the passage *i*, from the trough *g*, by the wall *w*, and floor *z*, the whole arranged and combined as described.

Also, making the beater cylinder in sections 1 2 3, &c, and using the same in combination with the grooves *a' a'*, and space *b'*, of the heads I I', so that said sections may be applied or removed without removing the outer casing of the frame, substantially as specified.

Also, forming the disks M M of the beater with the star points, having opposite sides *g' h'*, for attaching the wings *f f f* in opposite angles, substantially as described.

No. 55,180.—FRIEDRICH VILLARD, Mount Eaton, Ohio.—*Air-cooling Ventilator.*—May 29, 1866.—The air introduced into the room or wine vault is passed in contact with surfaces of cloth suspended between an upper and a lower reservoir, and a pump for returning the water from the lower to the upper

Claim.—First, the water grate C D, constructed substantially as described, for distributing the water over a number of conductors.

Second, in combination with the distributing water grate, constructed and adapted to operate as set forth, the two tanks A G, and water elevator J, for passing the water repeatedly over the evaporating surfaces, as explained.

No. 55,181.—C. PH. WAGNER, New York, N. Y.—*Quartz Mill.*—May 29, 1866.—The reciprocating motion in opposite directions of the lower ends is produced by hanging them on the respective sides of the pivot of rotation of the oscillating levers, and their upper ends are caused to approach by the rear supporting links, which are vibrated by the reciprocation of the jaws.

Claim.—First, communicating an alternate reciprocating movement bodily to the lower ends of the crushing jaws I J, and at the same time a vibrating movement to the upper ends of said jaws, so as to produce a rubbing or grinding action and a crushing action combined, employed for the purpose by means substantially as described.

Second, pivoting the lower ends of the vibrating jaws I J, or their equivalents, to bearings which have a vertical vibration, and communicating a lateral vibration to the upper ends of said jaws, by means substantially as shown and described.

No. 55,182.—ALBIN WARTH, Stapleton, N. Y.—*Sewing Machine.*—May 29, 1866.—The machine makes either the lock-stitch or the single thread chain stitch. The shuttle is wedge-shaped, and is carried in a shoe. The feed is by the needle, assisted by an elastic pad, which presses the cloth on a smooth roller and moves forward with it. Other features refer to sewing in undulating curved lines, to tension of the thread, &c.

Claim.—First, the combination of the rakes v w, teeth or pins v'', slip weight a' and adjustable dog b', arranged and operating in the manner and for the purpose herein specified.

Second, the elastic pad l, in combination with the needle slide C, pendulum b and needle n, constructed and operating substantially as and for the purpose set forth.

Third, the perforated flat sole shuttle S, in combination with the shoe G, and needle n, arranged and operating in the manner and for the purpose herein described.

Fourth, the combination of the curved tail piece b² and shuttle S, for the purpose herein described.

Fifth, the false shuttle S', with tongue e² and book d², in combination with the shoe G and needle n, constructed and operating substantially as and for the purpose described.

Sixth, the cavity h² d²*, in combination with the book d² and needle n, as described, so that the loop can be brought in such a position as to allow the needle to pass through it on its subsequent descent.

Seventh, the cavity c²*, located as shown, in combination with the sole of the shuttle and the shuttle race, constructed and operating substantially as and for the purpose set forth.

Eighth, the tracer y² and adjustable cloth guide M, in combination with the design wheel z², and with a sewing mechanism, constructed and operating substantially as and for the purpose described.

No. 55,183.—C. D. WALTERS and JOHN WILSON, Harrisburg, Penn.—*Lamplighters' Torch.*—May 29, 1866.—To the handle are attached wrenches of different sizes and angles of presentation, together with a torch.

Claim.—As a convenient article of manufacture the lamplighter's torch, herein described, and represented.

No. 55,184.—LEVI N. WARREN, Milwaukee, Wis.—*Grain Drill.*—May 29, 1866.—The seed is discharged from the slotted dish in the bottom of the hopper, and is fed by the rollers beneath to the elastic spout which hangs within the furrowers: the latter are lifted from the ground, and the feeding devices thrown out of gear simultaneously by a pivoted bar in the rear.

Claim.—The inclined metallic aperture plate k, or its equivalent, the metallic sub-tube y, with flexible joint, the attachment of the ground tube q to the falling bar z, all in combination.

Also, the combination of the detachment rod and lever in elevating the ground tubes and detaching the gearing wheels at one operation, substantially as described and shown, and for the purpose set forth.

No. 55,185.—TIMOTHY U. WEBB, Springfield, Ill.—*Cultivator.*—May 29, 1866.—Explained by the claims.

Claim.—First, the construction of the frame, the timbers A A and cross bars or timbers B B B, so notched and halved together on an angle that the face of the cross bars B will place the straight standards C C C C in the proper inclination for supporting and bracing the ploughs or cultivators, substantially as herein described.

Second, making corresponding series of holes e e through the cross bars B B B for the purpose of bolting and bracing the cultivator standards C C C C, so that the cultivators may be changed to work either right or left and the spaces adjusted between them, as and for the purposes herein set forth.

No. 55,186.—JOSEPH WELLER, Washington C. H., Ohio.—*Tuyere.*—May 29, 1866.—The throat is surrounded by a water or air chamber, and the blast passes through an annular opening around the plunger; the latter is raised by a lever to clear the throat of obstacles.

Claim.—First, the construction of a tuyere of an annular chamber B, surrounding a central throat which communicates with the double conical chamber G, substantially as described.

Second, the construction of the plunger H with a head b', cylindrical portions b^1 b^2, and a reduced stem b^3, in combination with the chamber G, substantially as described.

Third, the conical ring J' applied within the throat of the tuyere, in combination with a contrivance for clearing the throat, substantially as described.

Fourth, the oscillating ring J, provided with the wings e e, in combination with the plunger H, substantially as described.

Fifth, providing for giving a vertical, and also an oscillating, movement to the solid or hollow plunger H, by means of a lever K, substantially as described.

No. 55,187.—GILBERT D. WHITMORE, Boston, Mass.—*Spring Bolt.*—May 29, 1866; antedated May 20, 1866.—The bolt has a bevelled end, is contained in a case, and sprung by a spiral spring; stnds upon the shank of the handle and the base plate of the case interlock to keep the bolt retracted. Lifting the handle withdraws the bolt by contact with an oblique slot in the case.

Claim.—The improved spring lock as made not only with the bolt-retaining mechanism or catches g h, but with the handle F and the inclined or cammed slot c, arranged together and with the bolt and the case and the spring of the bolt, substantially as described.

No. 55,188.—HENRY WIGHT, East Cambridge, Mass.—*Boot and Shoe.*—May 29, 1866.— The edge of the sole is grooved, and the vamp is turned into it and nailed.

Claim.—The within described construction of a boot or shoe, consisting of the peculiar formation of the edge of the sole and the connection therewith of the vamp as shown with its edge turned outward, the sole and vamp being united by suitable fastenings driven through the upper into the sole at the bottom of the groove; also the employment of substantially the staple-formed fastening described in combination with the construction above claimed.

No. 55,189.—DANIEL H. WILLIAMS, Antwerp, N. Y.—*Clinching Iron.*—May 29, 1866.— Rectangular and more acute sided grooves are cut longitudinally on the periphery of the metallic cylinder, and a groove is held against the end of the nail while the head is hammered.

Claim.—The above-described cylindrical clinching iron, having grooves A A and B B, substantially as set forth.

No. 55,190.—WILLIAM W. WILLS, Janesville, Wis.—*Pipe Wrench.*—May 29, 1866.—The inner, toothed jaw is on the end of the permanent shank; the outer, pivoted jaw is adjustable longitudinally by a screw which traverses a socket attached to the sleeve of the outer jaw.

Claim.—The combination and arrangement of the toothed jaw E, sleeve D, spring a, jaw B, screw rod F, all constructed and operating in the manner and for the purpose berein specified.

No. 55,191.—A. WOEBER and G. WOEBER, Davenport, Iowa.—*Thill Coupling.*—May 29, 1866.—The slot in the thill iron which engages the coupling bolt is enlarged rearward to contain a plug of resilient material to prevent rattling; the bolt is held by a key and leather tongue.

Claim.—First, the construction of a slot a in the thill iron C, in the rear, and obliquely to the bolt, and the application of India-rubber a' therein as arranged, whereby the eye of the thill iron is thus made to have a bearing on the clip without any addition or device to either one or the other or heretofore, substantially in the manner and for the purposes as herein set forth.

Second, the application of the key E and strap D, as arranged in combination with the bolt c, substantially in the manner and for the purpose as herein set forth.

No. 55,192.—A. WOEBER and G WOEBER, Davenport, Iowa.—*Thill Coupling.*—May 29, 1866—The coupling bolt is retained by a key and a leather tongue; the key slips in the grooved sid- of th- clip and through a slot in the bolt; a spring bears on the rear of the thill iron to prevent rattling on the bolt.

Claim.—First, the application of a double-slotted key E, as constructed and arranged in connection with the strap g, as constructed and applied in combination with the grooved slid- f of the clip B and bolt F, substantially in the manner and for the purpose as herein set forth.

Second, the combination of the key E, groove f and bolt F, with the adjustable slotted metallic spring D, substantially in the manner and for the purpose as herein set forth.

No. 55,193.—LEMUEL WELLMAN WRIGHT, Thorndike, Mass.—*Ordnance.*—May 29, 1866; antedated May 16, 1866.—A thin metal tube is surrounded with a series of tubes, each about twelve inches long, their joints overlapped by similar tubes, and so on, until the thickness is obtained; the tubes are then united by pouring in an alloy.

Claim.—An improved cannon or piece of ordnance, as made with the breech f, the central tube a, and one- or more series of tubes or concentric cylinders, arranged and combined substantially as specified, the whole being united or braced together in the manner substantially as set forth.

No. 55 194.—J. YOUNG, Williamsburg, N. Y.—*Steam Bath.*—May 29. 1866.—The patient is surrounded by a blanket, supported on a skeleton frame, and steam is introduced through a perforated pan with a deflecting plate to merge the separate jets.

Claim.—The arrangement within the cage B, which supports the enclosing sheets, of the perforated steam chamber A, with deflector plate c, substantially as described and represented.

No. 55,195.—JOSEPH ZENTMAYER, Philadelphia, Penn.—*Combination of Lenses for Photographic Purposes*—May 29. 1866.—Explained by the claim.

Claim.—First, a doublet made of uncorrected meniscus lenses of different spherical curvatures, arranged concentrically, or nearly so, substantially in the manner and for the purpose specified.

Second, the arrangement of a series of uncorrected meniscus lenses of different spherical exterior, any two of which series, when set concentrically, form a corrected, or nearly corrected, doublet, substantially in the manner as specified.

No 55,196.—WILLIAM BACHELLER, West Newbury, Mass., assignor to himself and J. BRADBURY, Charlestown, Mass.—*Supporter for Ladies' Skirts.*—May 29, 1866.—Explained by the claim

Claim.—A skirt supporter A made of sheet metal or other suitable material, and in two parts or sections, secured together by a hinge having each of its wings or leaves fastened to the supporter sections in such a manner as to allow a vertical play or motion thereto, substantially as herein described and for the purpose specified.

No. 55,197.—FRANCIS D. BALLOU, Abington, Mass., assignor to ALFRED B. ELY, Newton, Mass.—*Buckle Clasp.*—May 29, 1866.—The hinged leaf of the buckle shuts down upon the lapped portions of the straps, bends them over the ridge beneath, and is secured by a catch.

Claim.—The buckle clasp, substantially as and for the purpose described.

No 55,198.—GEORGE L. BAUM, Bethlehem, Penn, assignor to himself and HERMAN A. DOSTER, same place.—*Shaft Coupling*—May 29, 1866.—The studs of the coupling fin enter holes in the respective portions of the divided axle, and a sleeve maintains the parts in position

Claim.—An improved shaft coupling, formed by combining the shell E, key D, and set screws F with the ends of the shafts to be coupled, substantially as described and for the purpose set forth.

No. 55,199.—EDWARD COOK, Valparaiso, Ind., assignor to himself and BRADFORD JONES, same place.—*Machine for Upsetting Wagon Tires.*—May 29, 1866.—The tire is clamped in two places by a stationary and a movable clamp, respectively; the latter being moved towards the former upsets the iron in the intervening space.

Claim.—In combination with the movable section I and the immovable section C, united by the long screw H, the inclined clamps K D and the anvil and clamping bars M N, the whole arranged to operate substantially as herein described and represented.

No. 55,200.—PETER CROWL, Brownsville, Penn., assignor to himself and H. H. FINLEY. same place.—*Vice.*—May 29, 1866.—The movable jaw is closed by pressure on a treadle which is connected by a cord to the lever on the axis to which the dog is pivoted; the dog engages the rack on the stock of the movable jaw.

Claim.—In combination with the movable jaw A, having a rack upon the side of its stock, the ratchet E, dog D, lever M, and springs F and N, arranged and operating substantially in the manner and for the purpose set forth.

No. 55,201.—WILLIAM T. CUSHING, New York, N. Y., assignor to SANFORD HARROUN & Co., same place.—*Machinery for Printing Railroad Tickets.*—May 29, 1866.—The pressure rollers for the several slips are in independent stocks, to permit them to operate upon varying thicknesses of paper; the type beds, arranged for receiving several forms and inking rollers simultaneously operated for the distribution of several colors, if required.

Claim.—First, in machinery for printing railroad tickets and other articles, the feeding roller, substantially as described, in combination with a series of pressure rollers or separate adjustable stocks, substantially as described, whereby tickets of different thickness can be printed at the same time and by the same machine.

Second, the one platen for receiving two series of forms of types in combination with two adjustable beds, substantially as described, whereby the two series of impressions are obtained by the movement of one platen alone.

No. 55,202.—BRANCH W. DURKEE, Livonia, N. Y., assignor to himself and J. BURNS WEST, Lakeville, N. Y.—*Sawing Machine.*—May 29, 1866.—The saws are arranged at right angles relatively, in an adjustable frame, and operated by a single driving pulley.

Claim.—First, the arrangement in an adjustable frame of one or more vertical with one or more horizontal saws, substantially in the manner and for the purpose set forth.

. Second, the combination of vertical and horizontal saws in an adjustable frame with the driving drum, arranged and operating in the manner and for the purpose set forth.

Third, the combination and arrangement of the main carriage with the toggle-lever mechanism and weight, to render its reciprocating motions automatic, substantially as set forth.

No. 55,203.—THOMAS JAMES, Medford, Mass., assignor by mesne assignments to THOMAS JAMES, JOHN MCCRELLISH, and CHAS. H. WHITE, same place.—*Blacking for Leather Harness, &c.*—May 29, 1866.—Composed of logwood, cutch, extract of hemlock bark, boiled, filtered, and mixed with neatsfoot oil, japan, and alcohol.

Claim.—The within described oil blacking for leather, composed of the materials specified, mixed in the proportions substantially as set forth.

No. 55,204.—PHILIP W MACKENZIE, Jersey City, N. J., and CHARLES W. ISBELL, New York, N. Y., assignors to the SMITH AND SAYRE MANUFACTURING COMPANY, New York, N. Y.—*Cupola and other Furnaces.*—May 29, 1866.—The separate blast chambers are arranged in a vertical series around the cupola, receive air by valved branch pipes from the main, and communicate by distinct rows of tuyeres with the interior.

Claim.—First, the two or more separate chambers *i i*, arranged one above another around the lower portion of the furnace, and each communicating with a separate tuyere or series of tuyeres, for conducting the air into the same furnace at different levels, substantially as herein set forth, for the purpose specified.

Second, in combination with such air chambers, the valves *u u*, forming separate communications with the blast pipes for the purpose of admitting the air, and controlling or stopping the admission thereof, to the several chambers, substantially as herein described.

No. 55,205.—SAMUEL MARDEN, Newton, Mass., assignor to himself and DUSTIN LANCEY, same place.—*Apparatus for Preparing Peat.*—May 29, 1866.—The grinder rotates in a chamber with a semi-spherical perforated bottom, and the peat passes to a trough which terminates in front of the moulding cylinder. The peat is driven by a reciprocating piston into the cells of the revolving mould wheel.

Claim.—First, the combination and arrangement of the corrugated or grooved breaker *q*, in connection with the perforated bowl at the bottom of the hopper.

Second, the revolving chambered cylinder *z* combined with the compressing plunger *u* and the expelling plunger *l'*.

Third, the combination of the lever *d'*, the pitman J, the tripping lever *i'*, and the vertical slider or actuator *e'*, connected with the cylinder *z*, as above arranged and described.

Fourth, the combination and arrangement of the forked lever *g'*, as connected with the frame A.

No. 55,206.—F. A. MORLEY, New York, N. Y., assignor to himself and J. W. MOUNT, Medina, N. Y.—*Potato Digger.*—May 29, 1866.—The rigid shovel is succeeded by a cylindrical screw rotated by intermediate devices from the master wheel. The depth is gauged by the adjustable draught pole and by change of position of the crank axle.

Claim.—First, a rotating cylindrical screen F, entirely open through its centre and free from obstructions, in combination with the digger or shovel E, substantially in the manner and for the purpose set forth.

Second, the combination of the shovel E, frame A, and crank axle N, substantially in the manner and for the purpose described.

Third, the standards O, having rotative adjustment on the crank axle N, to regulate the depth of digging, substantially in the manner and for the purpose specified.

Fourth, the combination of the shovel E, frame A, and pivoted draught pole C, substantially in the manner and for the purpose set forth.

No. 55,207.—GEORGE MÜLLER, New York, N. Y., assignor to himself and HENRY JOSEPH BAUG, same place.—*Carpenter's Plane.*—May 29, 1866.—The sliding box is placed above the cap and bit, and clamped thereon by the thumb-screw and forked lever, whose bearings are in the cheek pieces.

Claim.—The sliding box E, with clamping screw I, and the forked prize piece F, in combination with the cap plate D and bit C, together with placing the same in the throat *a* or forward end of the plane stock, substantially as and for the purposes set forth and described.

No. 55,208.—WILLIAM W. S. ORBETON, Haverhill, Mass., assignor to himself, GEORGE W. CAMPBELL, Bradford, and GEORGE P. RUSSELL, Haverhill, Mass.—*Clothes Drier.*—May 29, 1866.—The line is stretched around a wheel and through a pulley attached to a stationary object; the plane of revolution of the wheel is adjustable.

Claim.—The combination of the wheel A, the adjuster D, the pulley S, the posts H and P, or their equivalents, and endless line *o* with the lever *h* and its rope *r*, the whole being arranged so as to operate as and for the purpose set forth.

Also, the adjuster D, constructed substantially in the manner as set forth, and provided with a clamping screw and nut and a journal *d*, respectively applied to the post H and wheel A, the whole being arranged in manner and so as to operate substantially as set forth.

No. 55,209.— EDWARD PAYE. New York, N. Y., assignor to himself and CORNELIUS H. DELAMETER, same place.—*Machine for Turning Boiler Flanges.*—May 29, 1866.—The sliding hammers are arranged around a vertical sleeve and operated vertically by a pair of tappets and a cam groove; the metal is clamped on the top of an annulus and the descending hammers turn down the flange, the diameter being determined by the opening into which it is driven.

Claim.—First, the use, in a machine for turning boiler flanges, of a series of hammers F', acted upon by a cam b, so that the flanges are turned by a series of blows, as set forth.

Second, making the hammers adjustable on their shanks by means substantially as herein described, or any other equivalent means, so that the same can be adjusted for flues of different diameters.

Third, the arrangement of a revolving cam and cam grooves, in combination with the hammers F', said hammers being constructed and operating substantially as and for the purpose described.

Fourth, the arrangement of a centre in the bottom ring l, substantially as and for the purpose set forth.

Fifth, the clamp composed of two rings u l, in combination with the hammers F', constructed and operating substantially as and for the purpose described.

Sixth, the cam levers w, in combination with the rings l and u, and with the swivel bar r', constructed and operating substantially as and for the purposes set forth.

No. 55,210.—JAMES F. SEVERANCE, East Bridgewater, Mass., assignor to himself, ELEAZER C. BENNETT, and OLIVER H. WADE, same place.—*Stove Pipe Thimble.*—May 29, 1866.—The sliding jaws in the face are capable of embracing pipes of varying sizes and the opening is closed by a shutter; a damper is placed in the rear end.

Claim.—The arrangement and application of two or any other suitable number of adjustable slides c c, made substantially as set forth, or their equivalents, with the pipe thimble, composed of the plate A and the tube B.

Also, the combination and arrangement of the cover F with the pipe thimble and its adjustable slides c c, applied to it in manner and so as to operate substantially as specified.

Also, the combination and arrangement of the damper with the pipe thimble and the adjustable slides thereof.

Also, the combination of the damper, the pipe thimble, the adjustable slides with the cover, the whole arranged substantially as specified.

No. 55,211.—JAMES SUTHERLAND, East Hampton, Mass., assignor to himself and JOSEPH M. MUMFORD, same place.—*Driving Spindles in Spinning Frames.*—May 29, 1866.—The spindles of one row are opposite the intervals of the other row; the pulleys are in the same horizontal plane; the drum is placed either above or below such plane; each belt is passed from one pulley to an opposite one without touching the drum, and then returning, passes once around the drum and then around the first mentioned pulley; the belt has no inclination relatively to the pulleys and does not abrade itself.

Claim.—The combination of the drum A, spindles a a b b, and bands C C, the whole arranged and operating in the manner and for the purpose herein described.

No. 55,212.—S. B. H. VANCE and E. M. SMITH, New York, N. Y., assignors to MITCHELL, VANCE & Co., same place.—*Drop-light Fixture.*—May 29, 1866.—Between the curved arms of a stationary chandelier is an extensible gas tube, a combination of toggles or "lazy tongs" arrangement, which is protected from the heat of the burner beneath by deflector plates, and the handle whereby it is operated kept cool by a non-conducting collar.

Claim.—First, the arrangement of an adjustable drop light D centrally between two or more curved pipes B of a stationary chandelier, operating substantially as shown and described.

Second, the shield r n, constructed and arranged in combination with the drop light, substantially as herein set forth, for the purpose specified.

Third, the non-conducting block v, in combination with the drop-light burner b and handle E, substantially as herein set forth, for the purpose specified.

No. 55,213.—WILLIAM WOODBURY, Chelsea, Mass., assignor to himself and MELLEN BRAY.—*Bending Fore and Aft Sails.*—May 29, 1866.—One "jack rope" is rove through the hoops and the other through the eyelets on the edge of the sail; thimbles embrace both ropes between the points where they are connected to the mast or sail.

Claim.—The two jack ropes c d, in combination with the sail D and mast hoops a, arranged and operating substantially as set forth.

Also, in combination with the above, the rings f, or their equivalents, operating substantially as and for the purpose described.

No. 55,214.—ELLEN V. CONWAY, Richmond, Va., executrix of the estate of JAMES H. CONWAY, deceased.—*Medical Compound.*—May 29, 1866.—A cough mixture composed of sirup of tolu, acetate of morphia, acetic acid, chloroform, and gum-arabic.

Claim.—The compound or cough mixture, as herein described.

No. 55,215.—Thomas Henry Ince, Westminster, England.—*Attaching Shoes to Horses*—May 29, 1866.—Explained by the claim.

Claim.—The mode, herein described, of attaching a horseshoe by screws, whose heads shall be sunk within the body of the metal, as shown and described, and which penetrate the hoof in a direction parallel, or nearly so, with the outer wall thereof, but without piercing the latter.

No. 55,216.—Frederick Ransome, Ipswich, England.—*Preserving Timber*—May 29, 1866.—The pores of the wood are filled with an alkaline silicate, which is decomposed by a soluble chloride.

Claim.—Treating timber, for the purpose of preserving the same, with a solution of silicate of soda or potash, and afterwards with a solution of chloride of calcium or other soluble salt of an alkaline earth, or chloride of aluminum or iron, substantially as described.

No. 55,217.—G. Albert Reiniger, Stuttgart, Wurtemburg.—*Cigar Machine.*—May 29, 1866.—Improvement on his patent of October 29, 1861. The apron is rendered adjustable by a sliding screw clamp, to lengthen or shorten it; the spring hooks receive the cigar when wrapped.

Claim.—First, the adjustable screw clamp *d*, in combination with the flexible apron *b* and table A, constructed and operating substantially as and for the purpose described.

Second, the spring hooks *j*, in combination with the table A, apron *b*, and roller *a*, constructed and operating substantially as and for the purpose specified.

No. 55,218.—John Robinson and James Gresham, Manchester, England.—*Feed Water Injector.*—May 29, 1866.—The internal nozzle is moved by rack and pinion to regulate the annular opening and, consequently, the feed.

Claim.—Injectors, where the internal nozzles *e'* and *f'* are actuated by a toothed rack and pinion, and are arranged and combined with the external nozzles *e* and *f* and the steam nozzle *c*, substantially as hereinbefore set forth.

No. 55,219.—Elijah A. Andrews, New Britain, Conn.—*Snap Hook.*—June 5, 1866.—The pivoted spring tongue shuts in flush with the back and closes the mouth of the hook.

Claim.—The combination of the latch *e*, within and flush with the back of the recess *d*, relative to the opening *g*, with the pad *i*, spring *x*, and loop *c*, substantially as and for the purpose described.

No. 55,220.—Harrison Armstrong, Sparta, Wis.—*Horseshoe.*—June 5, 1866.—The calks are prepared in blank by rolling out a bar of steel with one edge sharp and a ledge on the side of the bar; the bar is divided into sections and a spur is formed on each by driving a portion of the ledge downward.

Claim.—Forming a spur on the calk by driving a portion of the ledge downwards, substantially as and for the purpose described.

No. 55,221.—John Ashcroft, Brooklyn, N. Y.—*Low Water Detector.*—June 5, 1866.—The fusible disk on the end of the valve-stem keeps the valve from its seat until the said plug is fused.

Claim.—First, the combination of the pipe G, connected to the boiler below the proper water-line, the discharge passage H, and the fusible disk C with the screw valve D, arranged substantially as and for the purpose herein set forth.

Second, the bearing piece F, in combination with the fusible disk C and the connections G and H, arranged as specified, so that the communication between the boiler and the said fusible part cannot be closed until after the fusible part has yielded to the temperature, as herein specified.

No. 55,222.—Abel T. Atherton Lowell, Mass.—*Tuyere.*—June 5, 1866.—The taper of the iron sleeve around the blast pipe and the elastic packing permit expansion without rupturing the tank when fire is applied, after the water has been frozen.

Claim.—The combination of the cistern *k*, tuyere iron *f*, pipe *c*, and packing *e*, constructed and arranged substantially as set forth.

No. 55,223.—Juan S. L. Babbs, New Albany, Ind.—*Dovetailing Machine.*—June 5, 1866.—The carriage has conical revolving cutters, arranged to cut their way through the mortise, between the tenons on the back, and then out between the tenons again, cutting the wood away from the latter on three sides.

Claim.—In combination with the conical cutters and the carriage by which the stock is presented to the action of the cutters of a secondary carriage or stock-holder, arranged to operate substantially as set forth.

No. 55,224.—D. Ballou, Havana, N. Y.—*Tire-bending Machine.*—June 5, 1866.—The wheel is revolved by a lever which engages the ratchet teeth, and the tire is wound on the periphery, the lower roller pressing it against the wheel.

Claim.—The combination of the wheel D with its holder N, or its equivalent, the ratchet G H, lever F, and spring-pressure roller J K L, substantially as described and represented.

No. 55,225.—A. C. BARSTOW, Providence, R. I.—*Cooking Stove.*—June 5, 1866.—Explained by the claims.

Claim.—First, the construction of stoves of otherwise ordinary or suitable arrangement for admission of interchangeable collars or closing caps for the flue and boiler holes, substantially in the manner herein described, dispensing with the use of rivets or other means of permanent attachment.

Second, the combination with a stove of otherwise ordinary or suitable construction of a hot closet arranged underneath the body of the stove, substantially as shown and set forth.

Third, in a stove, provided with a hot closet underneath the body thereof, in lieu of the usual legs, forming a base around the stove to enclose the said closet, in the manner and for the purposes set forth.

Fourth, the employment, in combination with a stove of otherwise ordinary or suitable construction, of a movable fire-pot or pan, having a closed bottom and holes at the sides, substantially as and for the purposes set forth.

No. 55,226.—WILLIAM O. BARTLETT, New York, N. Y.—*Breaking and Training Bridle.*—June 5, 1866.—The bars of the bit are levers whose fulcrum is afforded by a strap which passes over the neck behind the headstall.

Claim.—An improvement in bridles, by attaching a cord or strap to the same, so constructed that by pulling on the reins a pressure is produced downward on the sensitive part of the animal's neck, and upward in his mouth, substantially as herein described.

No. 55,227.—WILLIAM BEDLE, Keyport, N. J.—*Cradle.*—June 5, 1866.—A weighted arm beneath the cradle is rotated by clockwork; as it passes by rotation the sides become depressed alternately, giving a rocking motion.

Claim.—The weighted arm G, the brake I, and clock mechanism D E F, or their equivalents, when in combination with the cradle A, substantially as and for the purposes described and set forth.

No. 55,228.—CHARLES T. BELBIN, Baltimore, Md.—*Dredge Roller for Oyster Boats.*—June 5, 1866.—The device is placed on the rail or gunwale to help the dredge on board by lifting the rake teeth clear of the side of the vessel.

Claim.—The arrangement on the roller *d* of the ledge or flange *c* and the handle *e*, as and for the purpose described.

No. 55,229.—WILLIAM BELLAIRS and OLIVER D. BARTO, Atkinson, Ill.—*Snap Hook.*—June 5, 1866.—The inner hook slides in a groove in the shank of the outer one, and is closed by a spring; when the inner hook is retracted the opening for the bit-ring, &c., is exposed.

Claim.—The combination of the outer hook and the inner hook with its arm H, projection D, and dovetail-slide F, spring E, and shank B, arranged and operating in the manner and for the purpose herein specified.

No. 55,230.—WILLIAM N. BERKELEY, Cedar Rapids, Iowa.—*Turning Bridges.*—June 5, 1866.—The bridge revolves horizontally upon a pedestal resting on anti-friction balls. The power is applied by the revolution of a sheave, the rope from which encircles a stationary annulus on the abutment.

Claim.—First, in combination with a revolving pedestal B, the employment of the stationary sheave D and track E with the friction roller H and chain, or its equivalent *a*, arranged and operating substantially as herein specified and shown.

Second, the combination of the revolving pedestal B, sheave D, track E, roller H, pulley I, and chain *a*, arranged and operating substantially as shown and described.

Third, the employment of the adjustable pulley I, when arranged with the roller H and chain *a*, as and for the purposes set forth.

No. 55,231.—WILLIAM N. BERKELEY, Cedar Rapids, Iowa.—*Turning Bridges.*—June 5, 1866.—The bridge rotates horizontally on a pedestal upon anti-friction balls; it is moved by rotation of a shaft connecting with a sheave which has frictional contact with a stationary track ring on the supporting abutment. The levers referred to adjust the frictional contact.

Claim.—First, in combination with the revolving pedestal B, the employment of the circular stationary bearing D, the shaft F G, and friction wheels H I, arranged and operating substantially as and for the purposes specified.

Second, in combination with the above, the arrangement of the levers M N and connecting bar O, as and for the purposes set forth.

No. 55,232.—HENRY BLACK, Lewisburg, Ohio.—*Machine for Scutching Flax.*—June 5, 1866.—The swords are attached obliquely to the arms of the reel and operate consecutively upon the flax, which projects over the edge of the adjustable scutching board.

Claim.—The use of a reel, with swords attached, which operate transversely upon the flax upon a movable scutching board.

No. 55,233.—AMOS S. BLAKE, Waterbury, Conn.—*Metallic Cartridge.*—June 5, 1866.—The cylindrical sheet-metal cartridge is slipped into a cap containing the fulminate. When exploded, a portion of the inner cylinder is expanded into an opening in the cylindrical portion of the cap to attach the two together to facilitate retraction.

Claim.—As an improvement in cartridge cases, the perforated cap B, with its oblong opening F, arranged to operate with the case A, so that by the expansion of the metal in explosion of the powder a portion of the case A will be forced into the opening in the cap B, and the two parts be held together substantially as described.

No. 55,234.—GUSTAVUS BODE, Milwaukee, Wis.—*Medical Compound.*—June 5, 1866.—This electco-magnetic agent for the cure of rheumatism, &c., consists of magnetic iron, copper, brass, amber, and horn, comminuted and worn in a belt around the body.

Claim.—A medical compound, of the ingredients herein specified, and in about the proportions named, intended as a remedy for rheumatism, neuralgia, and other nervous diseases.

No. 55,235.—JAMES BRAIDWOOD, Wilmington, Ill.—*Coffer Dam.*—June 5, 1866.—The parts are adjustable telescopically upon each other; the extension or contraction is performed by screw rods; the packing consists of a surrounding flap membrane compressed by the surrounding water against the coffer dam.

Claim.—First, the making them of sections, that may be fitted within each other, and operated and detached from each other, substantially as herein recited.

Second, the elastic valve surrounding the points of union of the sections, and made tight by the outward pressure, as described.

Third, the detachable mechanical means or devices for forcing down the sections by pressure upon the surfaces, as set forth.

No. 55,236.—HIRAM BROWN, Cape Elizabeth, Maine.—*Sail.*—June 5, 1866.—The supplementary reefing yard is attached by eyelets to the spilling lines, and the sails are fastened thereto by a band of sail cloth stitched to the sail.

Claim.—First, in conjunction with the supplementary reefing yard, the use of the band D, to attach the sail to the centre yard as described, for the purposes set forth.

Second, the combination of the spilling lines, arranged as specified, with the eyelets i i, with the supplementary reefing yard, as described.

No. 55,237.—JAMES A. CAMPBELL, St. Louis, Mo.—*Steam Generator.*—June 5, 1866.—The coiled pipe in the furnace is connected with chambers above, the latter having an intercommunication whose valve is operated by a float in the lower chamber; the pressure is equally maintained in each portion; the lower chamber has the steam eduction pipe, and the upper chamber receives the water supply.

Claim.—First, the combination of the reserve water chamber E, the pipe M N, the stopper J, the regulating nuts L L L L, and the float I.

Second, the combination of the furnace K, the coil of pipe A B C, the float I, the regulating nuts L L L L, the reserve water chamber E, and the pipes M and N.

Third, the combination of the coil of pipe F G H with the reserve water chamber E.

Fourth, in combination with the reserve water chamber E and steam chamber D, the chamber O, the pipes R S T U, and the cocks P Q V W.

No. 55,238.—AARON CASEBEER, Sipesville, Penn.—*Machine for Jointing Staves.*—June 5, 1866.—The device has adjustment by which the required shape is given to the stave while jointing, to secure a future fit with its fellows correspondingly treated.

Claim.—A stave holder, with the arc gauges marked A A, with the screws B B B, by means of which a bevelled joint for any circumference may be made.

No. 55,239.—GEORGE F. CHAMBERS and SQUIRE ROBINSON, Worcester, Mass.—*Yarn-delivery Apparatus for Braiding Machines.*—June 5, 1866.—The yarn is delivered evenly, the increase of tension operating the stop weight, relieving the pressure of the friction lever upon the bobbin head. The weight stops the machinery of the thread brake, as the brake is thereby brought against the bobbin head.

Claim.—First, the combination with standard C of a friction lever G, substantially as and for the purposes set forth.

Second, the combination with standard C and sliding weight P of a friction lever having a projection h, as and for the purposes set forth.

Third, the combination with friction lever G of slide or stop and tension weight D, and spring j, substantially as set forth.

C P 8—VOL. II

No. 55,240.—ROBERT A. CHESEBROUGH, New York, N. Y.—*Separating Hydro-carbon.*—June 5, 1866.—The hydro-carbon is separated from the bone-black, which has been used in filtering, by subjecting the animal charcoal to the action of steam under pressure in a tight vessel. The oil rises to the top and is decanted.

Claim.—The application of steam in a closed receptacle to bone-black or other filtering media, for the purpose of extracting the hydro-carbon oils remaining therein after filtration, whether the steam is made in the receptacle or conducted therein.

No. 55,241.—WILLIAM W. CLAY, Philadelphia, Penn.—*Knitting Machine.*—June 5, 1866.—This machine uses double-ended hook needles, which ride in grooves on one or both sides of a central opening. The selection of the needles to be used from time to time is made by the jacquard. Several thread carriers are employed, either of which may be brought into use as desired, and each may carry a different colored thread. The fabric may be knitted with cross stripes by changing the threads and carriers, or with longitudinal stripes by using two carriers and different colored threads, and either ribbed or plain.

Claim.—First, the combination of the needles C, carriers D D', a jacquard apparatus, and the within-described devices, or their equivalents, whereby the needles can be transferred from one carrier to another by the operation of the jacquard apparatus, for the purpose described.

Second, the guides L L', in combination with the within-described devices or their equivalents, and with a jacquard apparatus, for the purpose specified.

Third, bars K K' K'', with their thread-guides L L' and recesses x, in combination with the adjustable plate O, and spring-catch A, or its equivalent, the whole being arranged and operating as and for the purpose set forth.

No. 55,242.—THEODORE COCHEN, Williamsburg, N. Y.—*Bottle Filler.*—June 5, 1866.—The tube connecting the tank with a reservoir has a regulating valve; when the long leg of the discharging siphon enters the bottle, the lip of the latter raises the sleeve, which is depressed by a spring, when the bottle is removed and stops the eduction opening.

Claim.—First, placing a valve at the discharging end of the siphon, and operating the same by an automatic movement, said valve consisting of the aperture P, the packing ring u, the sheath R r and spring v, or their equivalents, substantially as set forth.

Second, the peculiar arrangement of the parts which produce a uniformity of supply and discharge, viz., the valve D, valve-stem F, nut I, and float G, when in combination with the reservoir and siphon, substantially as and for the purposes set forth and described.

No. 55,243.—GEORGE E. COOPER, Baltimore, Md.—*Combined Drill and Fertilizer.*—June 5, 1866.—The seed slide openings are adjustable, and are kept clear by the pins on the reciprocating shaft, while other pins agitate the contents of the hopper; the slide apertures are closed when the drill tubes are raised.

Claim.—First, in combination, the hopper having two compartments $D^1 D^2$, the reciprocating rods E E', provided with two or more pins $e\ e\ e^2\ e^3$ to each of the holes, the pins being arranged so as to project horizontally over the seed apertures, and with pins e^1 projecting upwardly, as and for the purpose set forth.

Second, in combination with the above, the cam F, lever G, studs H H. and pins h, the whole being arranged and employed substantially as and for the purpose herein set forth.

Third, the combination of the lever J, bar I, link K, pivoted crank L, and slides M M, for the purpose of closing and opening the discharge apertures of the hopper simultaneously with the raising and lowering of the drill tubes.

No. 55,244.—WALTER CORBETT and WILLIAM BURNS, St. Louis, Mo.—*Lock.*—June 5, 1866.—The devices described give a peculiar sinuosity to the track leading from the keyhole to the tumblers as a defence against picks. The edges of the slots in the tumblers are notched irregularly to catch the passing bolt stud, when imperfectly thrown by a hesitating and imperfect actuation. A "night latch" bolt-stop is provided to arrest all motion of the bolt when desired.

Claim.—First, the combination and arrangement of the bridge B and wards a a°, and key-hole bushes a', as above recited.

Second, the tumblers C, when constructed with the toothed notches c and the notches c', and also in connection with the post b^3.

Third, the combination of the stop G with the above-described lock, as and for the purpose set forth.

Fourth, the construction of the bolt D with two notches E and E', as and for the purpose set forth.

Fifth, the combination and arrangement of the various parts of this lock, viz., the bridge B, wards a' and a°, and key-hole bushes a', the tumblers C', the post b^3, and the stop or catch G.

No. 56,245.—GEORGE G. COTTRELL, Sharon, Conn.—*Animal Trap.*—June 5, 1866.—The jaws have a double holding ridge; the plate has spits to hold bait; the set lever has a knife edge or narrow bearing.

Claim.—First, the double ridges or holding jaws M' N' M² N², arranged to operate substantially as and for the purpose herein specified.

Second, the knife edges or narrow bearings D' E', arranged substantially in the manner and for the purpose herein set forth.

Third, the spits formed on the pan D, substantially in the manner and for the purpose herein set forth.

No. 55,246.—C. N. CULVER, Bowling Green, Ohio.—*Horse Hay Fork.*—June 5, 1866.—
The upper ends of the loaded tines are suspended from the bead block by links, a ring and a catch; when the latter is unshipped by a trigger rope, this support fails, and by means of the chains the suspensory point being transferred to the tines below the pivoted point, they open and discharge.

Claim.—The arrangement of the levers H and I, in combination with the hooks G, ring *a*, links *p*, and forks A A, operating in the manner and by the means substantially as described.

No. 55,247.—LYMAN DAGGETT, Boston, Mass.—*Inner Sole.*—June 5, 1866.—The perforated sole has supporting flanges, with gaps therein to form connecting passages.

Claim.—The flexible inside water-proof elastic sole, as made with the supporting projections extending from its lower surface, and with passages around and between such, and holes leading therefrom up through the sole, the whole being substantially as described.

No. 55,248.—REUBEN M. DALBEY, Springfield, Ohio.—*Fruit Jar.*—June 5, 1866.—
Notched spiral lugs on the neck of the bottle are engaged by the hook ends of the yoke, and a pivoted cam bears upon the top of the lid.

Claim.—First, the lugs or projections *a*, having notches on their under surfaces, as and for the purpose set forth.

Second, in combination with a jar having the serrated projections, the yoke B, and the cam-headed lever D, arranged to operate as shown and described.

No. 55,249.—CLARENCE DELAFIELD, Factoryville, N. Y.—*Manufacture of White Lead and Saltpetre.*—June 5, 1866.—Improvement on his patent of April 3, 1866. A carbonate of lead, similar to the ordinary white lead, is obtained by precipitation, when a hot solution of carbonate of potash is added to a hot solution of nitrate of lead.

Claim.—First, the application of a jet of steam to a solution of the nitrate of lead or to a solution of the carbonate of potash, or their equivalents for this purpose, or to the united or combined solutions of nitrate of lead and carbonate of potash, or its equivalent for this purpose, for the purpose of aiding in the production of the white lead of commerce, substantially as set forth.

Second, the production of saltpetre or nitrate of potash as the residue of white lead manufactured after the process substantially as described.

Third, raising the temperature of the solutions of the nitrate of lead and the carbonate of potash, after their union or combination, either by the use of hot steam or by the application of other heat to aid in the production of the white lead of commerce, substantially as described.

No. 55,250.—MICHAEL DEPUE, Mattoon, Ill.—*Wind Power.*—June 5, 1866.—The fan wheel is enclosed in a drum, with one open side, a permanently closed portion, and a portion at which the wind enters, provided with a canvas door to adjust the size of the opening. The latter portion of the drum is presented to the wind by the controlling vane.

Claim.—The combination and arrangement of the sliding canvas door K, with the drum C I B, substantially for the purpose set forth.

No. 55,251.—JOSHUA DERR, Oley, Penn.—*Meat Cutter.*—June 5, 1866.—The spiral flanged feeder revolves more slowly than the cutters, which are arranged spirally, and have interposed clearing tongues.

Claim.—The construction, arrangement, and combination of the cutters H and I, and the cleaners J, when operated as herein described and for the purposes set forth.

No. 55,252.—ELLEN DEXTER, Quincy, Ill.—*Abdominal Supporter.*—June 5, 1866.—Explained by the claim.

Claim.—The combination of the supporter, substantially as described, with the pad *h*, the former so constructed as to envelop the entire abdomen, and sustain it both vertically and horizontally, and operating in such manner as, while the pad exerts a special pressure immediately above the "os pubis" on the "pelvic viscera," the lateral displacement of the same is prevented by the pressure of the supporter, and the adjacent parts of the abdomen are restrained from enlargement consequent on the pressure of the pad.

No. 55,253.—JOHN W. DIXON, Philadelphia, Penn.—*Process for Pulping Wood.*—June 5, 1866.—Explained by the claim.

Claim.—First, the circulation of a highly heated solution of lime in water or of magnesia in water, or a mixture of lime water and magnesia under pressure, through a mass of woody fibrous material contained in a digester, as a process or preparatory process for making paper pulp.

Second, the above process, in combination with a circulation of highly heated fresh water under pressure through the immersed mass, either as a precedent or subsequent operation to the one first above claimed.

No. 55,254.—A. DOBROWSKY, Shasta, Cal.—*Tatting Shuttle.*—June 5, 1866.—The pin passes through the central block and projects at the end.

Claim.—As a new article of manufacture, a tatting shuttle, composed of two sides A A, centre-piece B, and pin C, extending beyond the sides A A, at one or both ends, as shown and described.

No. 55,255.—S. P. DODGE, Boston, Mass.—*Lamp for Vehicles.*—June 5, 1866.—The lamp is supported upon springs, and its corners have sockets for the posts of the stand. Coils of wicking around the perforated cylinder prevent the disturbance of the flame by the swashing of the oil.

Claim.—First, supporting the lamp upon springs c, in combination with the posts d and sockets e, substantially as described.

Second, the perforated cylinder h, extending down into the oil chamber and surrounded with fibrous material, as and for the purpose substantially as set forth.

No. 55,256.—EDWARD L. DORSEY, Union, Ind.—*Field and Garden Cultivator.*—June 5, 1866.—The devices cited refer to the attaching and adjusting the main shares, and in addition to these an axle in the rear is provided with shares which, when reversed, form barrow teeth.

Claim.—First, the rod E, arranged and used as and for the purpose set forth.

Second, the reversible and adjustable ploughs and rakes M and N, secured to the axle K, said axle being attached to the upright B by means of an adjustable plate H, arranged and used as and for the purpose set forth.

Third, the block R with the ploughs P, arranged in the manner substantially as herein specified.

Fourth, the guides w, arranged and used as and for the purposes herein set forth.

Fifth, the plough-beam A, with the guide-wheel F, upright B, and shovel M, in combination with the axle K, wheels L L, shovels m n, and block R, when used as and for the purposes specified.

No. 55,257.—SIDNEY M. DUMONT, FRANK DUPRAZ, and J. DICKASON, Vevay, Ind.—*Steering Apparatus.*—June 5, 1866.—The rope is kept taut by passing over the grooved, spindle-shaped intermediate pulleys.

Claim.—In the described combination with the double cone-shaped pilot-wheel hub A A', the compensating sheave pulleys C C', as described and for the purpose specified.

No. 55,258.—JOHN E. EARLE, New Haven, Conn.—*Attaching Hoop Skirt Wire.*—June 5, 1866.—The hoops in the clasp have a short bend to prevent longitudinal retraction.

Claim.—A shoulder formed upon one or both ends of the wire so that the two ends may be secured together, substantially as herein set forth.

No. 55,259.—ROBERT R. EARNEST, Springfield, Ohio.—*Gate.*—June 5, 1866.—The outer end of the gate is raised, and the notched upper end of the braces engaged in the top of the heel post, the slats being pivoted at each end in the respective posts.

Claim.—A gate having its horizontal bars pivoted at each end, as shown, in combination with the brace c, pivoted at its lower front end, and having it notched and arranged to operate in connection with the loops f and block o, as set forth.

No. 55,260.—THOMAS ECKERT, New York, N. Y.—*Ladies' Muff.*—June 5, 1866.—Metallic hoops are placed between the covering and the annular elastic sack, which is inflated at the valved opening.

Claim.—As a new article of manufacture, a muff constructed with an air-chamber a, hoop c, and inflating valve b, as and for the purposes specified.

No. 55,261.—JOHN H. ELWARD, Polo, Ill.—*Apparatus for Separating Metals from Ores.*—June 5, 1866.—The lower part of the cupola communicates with a steam boiler and the upper with a fan-draught tube ending in a water tank. The ore is scraped into a series of pans below the cupola, which are filled with a solution of sulphate of soda and nitric acid; from the last of these it passes by an elevator to a chamber in the upper part of the cupola, and descends thence to the amalgamator.

Claim.—First, the process herein described for separating the precious metals from ores, and amalgamating them as set forth.

Second, the combination of a cupola, a series of saturating vats, and an elevator, substantially as and for the purpose set forth.

Third, the combination of a cupola with a heating chamber having crushers and scrapers, arranged and operating substantially in the manner and for the purpose set forth.

Fourth, the combination of a furnace, lead bath, and reservoir, arranged and operating substantially as and for the purpose set forth.

Fifth, the stirrers L, combined with the guide trough K, substantially as and for the purpose set forth.

Sixth, passing the products of combustion escaping from the cupola and heaters through an exhaust pipe kept charged with moisture, into a reservoir of water, substantially in the manner and for the purpose set forth.

No. 55,262.—JAMES E. EMERSON, Trenton, N. J.—*Swage for Sharpening Saw Teeth.*—June 5, 1866.—The transverse pin which passes through the end of the stock has notches or dies corresponding to the width and shape of the teeth. The dies operate upon the under side and two edges of the tooth, and the projecting jaw of the stock operates upon the upper side.

Claim.—First, a swage for sharpening the teeth of saws provided with a guiding groove *h*, substantially in the manner described and operating for the purpose set forth.

Second, the manner substantially as herein described, of forming and sharpening the teeth of saws and securing uniformity in their width and thickness by means of a pin adjustable transversely through the punch, as described, having dies to be used successively in combination with a punch or swage, as set forth.

No. 55,263.—J. E. EMERSON, Trenton, N. J.—*Forging, Shearing, and Punching Device.*—June 5, 1866.—Explained by the claim.

Claim.—An anvil shears and punching machine, arranged, combined, and operating substantially in the manner and for the purpose set forth.

No. 55,264.—OWEN EVANS, Alliance, Ohio.—*Wood Bending Machine.*—June 5, 1866.—By the staple or hook in the "former" the stuff to be bent is held in contact therewith at one end, while the pressure-block forces the wood on the "former" to bend it into the shape where it is secured by clamps and wedges.

Claim.—First, the arrangement of the staple *h'*, form B, and plate *n*, in combination with the head E, block F, clasp I, and key *d'*, as and for the purpose substantially as set forth.

Second, the construction of the lever L, guide P, arm *p*, in combination with the catch J and head E, as and for the purpose substantially as described.

No. 55,265.—E. E. EVERITT, Philadelphia, Penn.—*Sofa and Crib.*—June 5, 1866.—The bed sections are hinged together, and when slipped forward and bent to form a chair the end railings respectively become a leg support and a brace for the back support.

Claim.—First, the combination of the box or frame A, the adjustable frames C C and cushions D D', the whole being constructed and arranged for joint operation substantially as and for the purpose set forth.

Second, the combination of the cushion of a sofa bedstead and a frame E hung to the cushions so as to be turned beneath or outwards at right angles to the latter, substantially as and for the purpose set forth.

Third, the combination of the above and a leaf F, or equivalent device, for supporting the frame E at any required angle to the cushion, for the purpose set forth.

No. 55,266.—MARKS FISHEL, New York, N. Y.—*Hoop Skirt.*—June 5, 1866.—A wire hoop is passed through a single vertical tape without cutting the threads, and clasps at each side of the tape keep it distended.

Claim.—First, embracing the hoops of skirts within holes in tapes, or analogous material of a single thickness, when the holes are produced without removing or cutting any of the fabric and fit around the hoops, substantially in the manner and for the purpose herein set forth.

Second, combining with tapes punctured by the hoops, as shown, the clasps D, embracing the hoops and performing the double function of preventing the moving of the hoops in the tapes and of preventing the twisting or doubling of the tapes, as herein specified.

No. 55,267.—MARKS FISHEL, New York, N. Y.—*Instrument for Puncturing Fabrics and Introducing Flat Skirt Wire Therein.*—June 5, 1866.—The end of the wire to be introduced between the threads of a woven fabric is socketed into a pointed and flat cap of greater width than the wire.

Claim.—The sharp-pointed needle C, having a form flattened and hollow at the large end, smoothly rounded throughout, adapted to be used in inserting the flat springs through close material, substantially in the manner and for the purpose herein set forth.

No. 55,268.—CHARLES H. FITCH, Auburn, N. Y.—*Stove Pipe Thimble.*—June 5, 1866.—The inner and outer cylinders are fitted to the annular upper and lower plates; the inner cylinder fitting in rabbets and the outer lapped over the plates.

Claim.—First, the top plate D and bottom or end-plate C, with the cylinders A and B, constructed and combined as and for the purposes set forth.

Second, closing the ends of the cylinder A under the bottom and over the top, in the manner and for the purpose above described.

No. 55,269.—A. D. FORBES, Rockford, Ill.—*Stock for Broom Head.*—June 5, 1866.—The toothed malleable iron clamp has a flange to embrace the flattened end of a wrought-iron screwed spindle

Claim.—The combination in a broom-head of the malleable iron clamp with the wrought-iron stem, when formed and united substantially in the manner and for the purpose described.

No. 55,270.—THADDEUS FOWLER, Seymour, Conn.—*Horseshoe Nail.*—June 5, 1866.—Explained by the claim.

Claim.—As a new article of manufacture, a horseshoe nail, in which the point and junction of the body with the head are more dense or stiffer than the body portion of the nail; the point is bevelled upon one side (without being spread widthwise) and the head has a projection in the central parts, for the purpose and as set forth.

No. 55,271.—EDWARD J. FRAZIER, Peoria, Ill.—*Car Coupling.*—June 5, 1866.—Projections on a suspended key keep the coupling-pin elevated until the key is pushed to the rear by the link, when the pin falls into engagement.

Claim.—The grooved pendulated key when suspended in a slotted opening as described, in combination with the coupling-pin as specified for the purposes set forth.

No. 55,272.—ANDREW FULLER, Milford, N. H.—*Planing Machine.*—June 5, 1866.—Explained by the claim.

Claim.—The method herein described of working lumber across the grain by the use of one or more rotary cutters or scrapers fixed or secured in the cutter-head G, so that their planes shall pass through the axis of motion of the head G in combination with a table or platform to support the lumber to be fed to the cutter, the whole being arranged and operating as shown in Fig. 2 of the accompanying drawings.

No. 55,273.—H. L. FULTON, Chicago, Ill.—*Apparatus for Amalgamating Gold or Silver with Lead.*—June 5, 1866.—The ore is passed from the roasting furnace to the hopper and thence by the screw in the hollow cylinder is carried to the bottom of the lead bath in the kettle; the lead circulates around the lower edge of the division in the kettle and enters the screw cylinder at the gap beneath.

Claim.—The arrangement in combination of the amalgamating vessel B, its partition D, the hollow cylinder A with its helical shaft C, its hopper E, its spout F, and its opening G, in the manner and for the purposes specified substantially as described.

No. 55,274.—BENJAMIN M. GARD, Champaign county, Ohio.—*Brick Machine.*—June 5, 1866.—Explained by the claim.

Claim.—Making the bottom of the clay tub of a conical form when used in combination with a tapering screw for forcing the clay into the mould underneath the tub, in the manner and for the purpose herein described.

Also, a sectional plate *d* held to its place by wooden pins that will break when a stone gets jammed in by the screw, and give way, thus preserving the machine from greater damage, as described.

Also, making the scraper shank adjustable, in combination with the pivoting of the scraper on said shank, as and for the purpose described.

Also, in combination with the pressure-plate *g* on the top of the table, the rise *k* in the track *o* underneath said plate, for the purpose of giving the clay additional pressure over what it receives from being packed in by the screw, substantially as described.

Also, in combination with the mould a locking and unlocking mechanism, that is automatically operated for locking and unlocking the lid of the mould at stated intervals, as set forth.

Also, in combination with the pin *j*, having a bolt-head on its top, the cam *k* on the under side of the lid, so that these rounded surfaces may meet without liability of the lid jamming the bolt, and thus preventing it from falling into its proper position as described.

No. 55,275.—JOHN C. GARDNER, Hingham, Mass.—*Smith's Tongs.*—June 5, 1866.—Explained by the claim.

Claim.—The combination with a pair of tongs of a segmental strip B and pawl D, substantially as and for the purpose specified.

No. 55,276.—W. M. GARRISON, New York, N. Y.—*Cracker Machine.*—June 5, 1866.—The sheet of dough is laid upon the hopper and fed by the intermittent motion of the rollers, which divide it into cylindrical strings; from these they are taken in globular form by pressers, and delivered to the cavities in the traversing platform, which carries them by an intermittent motion to the dotters, which give the final shape.

Claim.—First, the combination of the cutting bars F with the grooved rollers C D, substantially as herein set forth for the purpose specified.

Second, the combination of the conducting tubes I with the cutting bars F and the recessed endless platform H, substantially as herein set forth for the purpose specified.

Third, the sliding frame L carrying the dotters T, and arranged in relation with the recesses r′ of the endless platform H, substantially as herein set forth for the purpose specified.

No. 55,277.—STEPHEN F. GATES, Boston, Mass.—*Rock Drill.*—June 5, 1866.—The cutting edges diverge from a centre, but divide unequally the circumscribing circle.

Claim.—A rock drill, when made with cutting edges, arranged substantially as and for the purpose specified.

No. 55,278.—EDWARD GILLIAM, Alleghany, Penn.—*Horse Hay-fork.*—June 5, 1866.—A prong is pivoted near the lower end of the stock, and its upper end has grooves which form guides for the sliding bar and its detent link. The motion of the sliding bar projects or retracts the prong.

Claim.—The rod A, provided with the head piece B, furnished with slots e and o, when used in combination with the lever D, rod C, and barb or prong 3, the whole being constructed, combined, arranged, and operating substantially as herein described, and for the purpose set forth.

No. 55,279. —THEODORE GILSON and NICOLAS MARTIN, Port Washington, Wis.—*Plough.*—June 5, 1866.—The flanges project from the inner surface of the mould board, to form points of attachment for the land-side and share.

Claim.—The land-side D, in combination with the flanges G b and share C, as and for the purpose specified.

No. 55,280.—R. A. GOODYEAR, New Haven, Conn.—*Lamp.*—June 5, 1866.—The cover of the reservoir has a turned-up flange, to form an annular drip cup.

Claim.—The construction of the drip cup attached to the metal cup which covers the fluid vessel or reservoir, as and for the purpose shown and described.

No. 55,281.—BENJAMIN GRAHAM, Lyons, Iowa.—*Churn.*—June 5, 1866.—The milk falls upon a rapidly-revolving fan beater, which acts upon the globules of the milk and delivers it to a screen, through which it passes to the shield and the ordinary dasher arrangement in the box below.

Claim.—The beater B, the screen P, and the shield S, all for the purposes as above set forth.

No. 55,282.—DARWIN A. GREENE, New York, N. Y.—*Machine for Bundling Wood.*—June 5, 1866.—The lever, by intermediate devices, operates the gripping bow upon the split wood that lies in a slotted cradle, and holds it firmly for binding.

Claim.—First, the short links $E^1 E^2$, short arms $D^1 D^2$, shaft d, and slots a, arranged as represented, relatively to each other and to the lever D and yoke C′, or their equivalents, substantially as and for the purpose herein specified.

Second, the hinged arm J, shaft g, spring H, and handle G, arranged for operation in connection with the lever D and its connections, substantially in the manner and for the purpose herein specified.

No. 55,283.—JOHN GRIFFIN, Louisville, Ky.—*Device for Picking Cotton.*—June 5, 1866.—Improvement on his former patents of November 22, 1859, July 3, 1860, and January 22, 1861.—The cup is placed over a cotton boll, and the air-exhausting operation is initiated by depressing a trigger, which makes the requisite connections, and the cotton is drawn through the tube.

Claim.—First, the tube D, with the cylinders E F and plates a within cylinder E, in combination with the cylinders N O, piston c, and plunger e, with the tubes Q R S, the tubes Q R communicating with an exhaust chamber and tube S, provided with an orifice e*, to be opened and closed by the thumb of the operator, substantially as and for the purpose set forth.

Second, the piston K working within the cylinder L, and connected with the flap H, as shown, and the cylinder L communicating with the cylinder F, substantially as and for the purpose specified.

No. 55,284.—CHARLES T. GRILLEY, New Haven, Conn.—*Capping Wood Screws.*—June 5, 1866.—The hemispherical nicked screw cap and screw head are placed in a concave die, through the bottom of which is projected a flat punch, which is supported by a spring beneath, and pierces the nick and head: a tubular punch descends and folds the edge of the cap upon the shoulder of the screw head.

Claim.—First, in the manufacture of capped screws, placing the screw which is to be capped in an inverted position, or head downwards, together with its cap, in the closing die, constructed and arranged as described, so that the cap, while held in its proper relation to, may be closed on the screw without disfiguring or closing up the nick in the cap, as herein shown and set forth.

Second, in machinery for capping screws, as described, providing the closing die with a punch, whereby the ready adjustment of the cap to the screw may be effected, as herein shown and set forth.

Third, in machinery for capping screws, as herein set forth, the combination of the closing die and centre punch with the recessed plunger or drop, for joint operation, as described, so that the cap may be adjusted to and closed in the screw, while the nick in the cap is kept open, as shown and set forth.

Fourth, in combination with the closing die, provided with a centre punch, the yielding support, as herein shown and described and for the purposes set forth.

Fifth, in combination with the closing die and centre punch and yielding support, as described, the employment of a screw for adjusting and holding in place the centre punch, as herein shown and set forth.

No. 55,285.—WILLIAM GURLEY, roy, N. Y.—*Water-spreading Nozzle for Fire Engines.*—June 5, 1866.—The deflecting or spreading device has concentric flaring orifices, is pivoted in the cap of the nozzle, and is brought into action as required.

Claim.—First, the placing of a spreader for the nozzles of engines, pumps, &c.. within a plate, so that the spreader may be adjusted over the orifice of the nozzle, or moved off from it bodily or entire, by the movement of a single part D, substantially as shown and described.

Second, a spreader, composed of a series of frustums of cones placed one within the other, and arranged so as to form a single or entire part, substantially as shown and described.

No. 55,286.—SAMUEL N. HAIGHT, Bedford Station, N. Y.—*Self-inserting Faucet.*—June 5, 1866.—On the entering end of the faucet is a bridge piece, with cutting edges and a salient gimlet-pointed screw.

Claim.—The combination of the open space E with the sloping cutting edges b b and the channel or groove C of the arched and detachable cutter, substantially as and for the purpose described.

No. 55,287.—ALEXANDER W. HALL, New York, N. Y.—*Churn.*—June 5, 1866.—Explained by the claim.

Claim.—The combination of a swinging or vibrating dasher with the swinging body of a churn in such manner that, by the act of moving the body in one direction, the dasher is moved in the opposite direction, substantially as herein set forth for the purpose specified.

No. 55,288.—HORACE P. HAMMOND and WILLIAM A. HATHAWAY, North Kingston, R. I.—*Step-ladder.*—June 5, 1866.—The notched segment bar maintains the legs at the required extension.

Claim.—The link D, provided with a series of notches a a a, in combination with the stud pin b, or its equivalent, and the frame of a step-ladder, substantially as described for the purpose specified.

No. 55,289.—THOMAS HANSON, New York, N. Y.—*Transmitting Motive Power.*—June 5, 1866.—The distant pump oscillates a column or columns of water, to communicate motion to another pump on the spot where the work is to be made effective.

Claim.—Giving the required reciprocating motion to a pump or pumps for raising water by means of pistons working in separate cylinders, and having water or other non-elastic fluid interposed as a medium for communicating the motions, substantially as described.

No. 55,290.—GEORGE HAVELL, Newark, N. J.—*Carpet Bag Frame.*—June 5, 1866.—The two jaws are pivoted to the hinge rod, and shut beneath the cap piece of the frame, which is T-shaped in cross section.

Claim.—First, the combination of the "frame" of carpet, travelling, and other similar bags of three pieces, when combined and arranged substantially as set forth.

Second, the particular form of the central piece E, when constructed in the manner and for the purpose set forth.

No. 55,291.—EZRA T. HAZELTINE, Warren, Penn.—*Lamp Chimney Cleaner.*—June 5, 1866.—By means of the elastic pad on the end of the handle, a piece of soft paper is rubbed over the inner surface of the glass chimney.

Claim.—The herein-described instrument for cleaning lamp chimneys by adhesion of an intervening fibrous substance, consisting of a pad A of India-rubber or its equivalent, rod C, and handle B, constructed and operating substantially as set forth.

No. 55,292.—SAMUEL C. HENSZEY, Jr., West Chester, Penn.—*Bronchial Troche.*—June 5, 1866.—Composed of muriate of ammonia, chlorate of potash, cubebs, extract of licorice, and gum arabic.

Claim.—The use of muriate of ammonia as compounded with certain other parts or ingredients herein named, or the component parts as combined, compounded and made into a troche or lozenge, substantially in the manner and for the purpose as herein specified. -

No. 55,293.—GEORGE W. HILDRETH, Lockport, N. Y.—*School Desk and Seat.*—June 5, 1866 —The seat folds up against the back of the desk behind it, and when down is supported by its rear extension, which contacts with studs on the desk frame.

Claim.—First, the lever *a*, moving upon the pivot *c*, between the centre of the lever and the back end thereof, when said lever (or seat) is sustained in a sitting position, by the back end of the lever bearing against the stop *d*, on the frame, as herein specified.

Second, the tapering pivot *c*, with the corresponding socket in a bolt *f* and jamb-nut *g*, as and for the purpose described.

Third, the spring *i* and stops *d d*, for the purposes herein specified and shown.

No. 55,294.—JAMES G. HOLT, Chicago, Ill.—*Making Moulds for Castings.*—June 5, 1866.—The sand-mould press has a follower which serves as a guide for keeping the patterns in place in the flasks, and applies a steady pressure on the sand in its descent.

Claim.—First, the devices, arranged and combined substantially as herein described, for regulating the descent of the plunger B, with respect to the height of the flasks in which the moulds are made, for the purpose set forth.

Second, the combination of the projections *k⁴*, with the follower of the press, for forming the branch sprue holes *k'*, substantially as described.

Third, the combination of the press and its follower with the cope H' and flask H, all constructed as described in the manner shown, so that moulds for axle skeins and boxes may be pressed, substantially as set forth.

Fourth, the perforated follower B', in combination with the sectional flask H H' and patterns *h h*, for the purpose substantially as described.

No. 55,295.—JAMES G. HOLT, Chicago, Ill.—*Machine for Preparing Axle Skein Moulds.*—June 5, 1866.—Standards are arranged upon a platform supporting a sliding frame capable of vertical adjustment by a rack and pinion. Attached to this frame are rotating tools or formers, which are furnished with cutters to produce the undercutting in the mould, and arranged to be thrown out to perform their work, and drawn back out of the way before the tool is elevated out of the mould.

Claim.—First, the machine constructed and operating substantially as herein described for cutting, sleeking, and packing the walls of sand-moulds for casting axle-skeins, which moulds have been previously prepared by patterns, substantially as described.

Second, the means, substantially as herein described, whereby sand moulds, which have been prepared by patterns substantially as described, are subjected to the operation of rotary and vertically moving and laterally sliding tools, which are adapted for finishing sand moulds, substantially as set forth.

Third, providing the tool E, which finishes the moulds, with one or more movable cutters, constructed and operated substantially as described, for producing the undercutting in the moulds, substantially as described.

Fourth, the combination of moulding tools E, with a vertically sliding frame C. and also with devices which will admit of these tools being rotated about their axes for finishing moulds, substantially as and for the purpose described.

No. 55,296.—JAMES G. HOLT, Chicago, Ill.—*Mould for Hub-boxes.*—June 5, 1866.—The rotating tool packs the walls of the sand-moulds for hub-boxes, the moulds having been previously made by the ordinary conical pattern.

Claim.—First, making the interior form of a sand-mould for hub-boxes by means substantially as herein described, the said means being constructed and operating substantially as set forth.

Second, the combination of movable plates *j*, and sliding plate *f*, with a tapering tool E, for finishing sand-moulds for hub-boxes, substantially as described.

No. 55,297.—S. J. HOMAN, Walden, N. Y.—*Cider Mill.*—June 5, 1866.—The flat adjustable knife forms a throat in connection with the scrapers on the rotary drum.

Claim.—The adjustable knife E, operating in combination with the scrapers C of the cylinder B, in the manner and for the purpose herein specified.

No. 55 298 —WILLIAM W. HOSFORD, New Britain, Conn.—*Sash Fastening.*—June 5, 1866.—Vibrating the lever brings the cam into action, and when in its central position the motion of the sash is unimpeded.

Claim.—The combination of the cams *k e* with the plate *a*, spring *i*, and arm *o*, arranged and operating substantially as and for the purpose described.

No 55,299 —THOMAS S HUDSON, East Cambridge, Mass.—*Construction and Manufacture of Printing Type* —June 5, 1866.—The letters are stuck up from sheet metal and the surface ground off until the roundness is removed and definite square sides and angles are obtained.

Claim.—The said manufacture of printing type, made substantially as described, viz., by the combined processes of stamping the letter or figure from a plate or piece of metal, and subsequently reducing the same in the manner and for the purposes set forth.

No. 55,300.—THOMAS S. HUDSON, East Cambridge, Mass.—*Construction of Printing Wheels, &c.*—June 5, 1866.—The sections containing salient letters are hinged together and the chain lapped around a core.

Claim.—The new manufacture of the printing or type-wheel prism, cylinder or plate, consisting of the supporting parts B, and the printing band or chain A, made with the concavo-convex types, and applied to such part B, substantially in manner as described.

No. 55,301.—ROBERT HUGHES, Dangerfield, Texas.—*Rotary Engine.*—June 5, 1866 — The cylinder revolves around the central, stationary head, having radial pistons operated by the curved head, through which the induction and eduction is accomplished.

Claim.—First, the construction of the stationary centre head E, with curved point G, and its steam channels D and T, as herein described.

Also, the construction of the valves K arranged in the valve box L, as herein described.

Also, the arrangement and combination of the valves K, centre-head E, cam G, and channels D and T, to operate in a cylinder as a rotary engine, whereby to utilize the exhaust steam *x*, and operating other machinery, as herein set forth.

No. 55,302.—P. F. HULBERT, Chatham, N. Y.—*Clamp.*—June 5, 1866.—The parts of the flask are clamped together by the deflection of the bar when its pivotal point is moved by vibrating the lever.

Claim.—The combination of the angular bar A, and serrated cam lever arm *c*, admitting of operation to the right or left, for the purpose herein specified.

No. 55,303.—THOMAS HUNTINGTON, New Rochelle, N. Y.—*Detaching Boats from their Davits.*—June 5, 1866.—Hooks are pivoted to the thwarts and are disengaged simultaneously by the rotation of the shaft.

Claim.—The hooks B, suspended or fitted in the plates A, in combination with the arms *e*, attached to the shaft C, all being applied to the boat, and arranged to operate substantially in the manner and for the purpose set forth.

No. 55,304.—JOHN JANN, New Windsor, Md.—*Harvester.*—June 5, 1866.—This arrangement of lever, slides. springs and clutches operates to throw the cutters into and out of action; the slotted slide covers the hinge of the cutter to make the attachment rigid when required. The third claim refers to the drag bar and the elevating apparatus of the finger bar.

Claim.—First, the arrangement of the lever H, bars I I, slides F, clutches E, ratchets D, and springs K, as and for the purposes specified.

Second, the slotted bar O, in combination with the pivot bolt M, for making a stiff joint between the cutting apparatus and its connections.

Third, the combination of the double bar N, hook Q, slotted link R, segment S, pinion U, and lever T, all arranged and operating as described.

No. 55,305.—E. H. JANNEY and E. J. HAMILTON, Fairfax, Va.—*Portable Door Fastener.*— June 5, 1866.—The tooth of the hold-fast is imbedded in the casing and a cross bar in the projecting portion arrests the motion of the door.

Claim.—First, the short stationary pivots of the case A, in combination with the slotted swinging and sliding hooked hold-fast B, substantially as described.

Second, the combination of the oblique slots, short pivots, hooked hold-fast B, and stop *d*, in construction of a portable night-lock, substantially as described.

Third, the combination of the notches *s*, and shoulder *k*, in the construction of the portable night-lock, substantially as described.

No. 55,306.—JAMES HERVEY JENKINS, Yorkville, N. Y.—*Washing Machine.*—June 5, 1866.—Explained by the claims.

Claim.—First, the arrangement of the cylinder B, and spring plate C, rollers M M, and board O, as and for the purpose herein specified.

Second, the revolving hollow brush cylinder P, provided with brushes at its exterior and interior surfaces, and open at one end, substantially as and for the purpose specified.

Third, the arrangement of the pressure rollers M M, spring K, perforated board O, and case A, for the purpose specified.

Fourth, the arrangement of the pounder I, rod K, and crank *g*, suitably operated, in combination with the perforated box J, within the partition *h*, in the manner and for the purpose herein specified.

Fifth, the arrangement of the treadle frame F, slotted upright bar G. crank *s*, and shaft D, constructed and operating in the manner and for the purpose herein described.

No. 55.307.—LEMUEL P. JENKS and GEORGE ARTHUR GARDNER, New York, N. Y.—
Rock Drill.—June 5, 1866.—This percussion drill is fixed in place by the engagement of the vertical screw clamp with the ceiling and floor, or other objects, and by elevation on its supporting post.

The inclination of the slide adjusts it to the required angle of presentation; the drill is approached to its work by the screw shaft, is projected by a spring, retracted by a revolving cam, rotated and fed by pawls.

Claim.—First, suspending the machine bearing-slide N, upon the same centre with the disk M, in combination with the arrangement whereby the slide N can be fastened at any angle, substantially as described.

Second, the standard block H⁰, in combination with the gibs or friction pieces J J, when the same is used with rock drills, and constructed substantially as shown, for the purpose of giving universality of motion, horizontally, all substantially as described.

Third, regulating the feed of a rock drilling machine toward the rock, according to the extent of penetration of the drill into the same, by the arrangement constructed and used substantially as described.

Fourth, the arrangement of a screw shaft and nut, for the purpose of raising and lowering the machine when used with rock drills, and constructed substantially as described.

No. 55,308.—EDWARD KAY, Philadelphia, Penn.—*Stop Motion for Knitting Machine.*—
June 5, 1866.—Intended to knit a chain of stockings linked together top to top and toe to toe. The pattern pin-wheel determines the distance to be knitted with a given colored yarn, in knitting fabrics having cross stripes. When another color is to be used the stop motion is brought into action; one of the pins at that stage lifting the plate and thereby releasing the slide plate from the notch and permitting the shifting lever to disconnect the gearing.

Claim.—First, the machine as a whole, composed of the parts combined, arranged, and operating substantially as set forth.

Second, the pin-wheel M, the sliding plate O, and the lifting plate P, combined and arranged substantially as set forth.

Third, the sliding plate O, the lifting plate P, and the shifting lever R, combined and arranged as herein specified and described.

Fourth, the sliding plate O, the shifting lever R, and the spring S, combined and arranged substantially as set forth.

No. 55,309.—JAMES B. KELLY, Kendallville, Ind.—*Row Lock.*—June 5, 1866.—The circular plates on the oar and gunwale have interposed anti-friction rollers. A central stud rises from the lower plate and protrudes through a slot in the oar.

Claim.—First, the combination of the circular plates B and C having rollers interposed between them, with an oar D, substantially as described.

Second, constructing the base plate B with a central stud *a*, for receiving, centring, and keeping in place the movable plate C, to which the oar is affixed, substantially as described.

Third, constructing the anti-friction rollers G G, with flanges *e e* on them, in combination with the annular projections on the plates B and C, substantially as described.

No. 55,310.—WILLIAM KELLY, Saranac, Michigan.—*Washing Machine.*—June 5, 1866.—The concave tub and the corrugated roller are suspended from the frame and reciprocated by double cranks; the roller is vertically adjustable.

Claim.—First, the rod *g*, with ratchet *j*, in combination with the slotted plates *m*, and rubber E, the whole constructed as and for the purpose herein set forth.

Second, the shaft P, furnished with double cranks, in combination with rods L and T, in connection with plates K, and tub B, the whole constructed and operating in the manner and for the purpose herein specified.

No. 55,311.—GEORGE R. KELSEY, West Haven, Conn.—*Buckle.*—June 5, 1866.—The buckle frame is struck from sheet metal; ears on it are lapped to form the hinge of the tongue.

Claim.—The combination of the frame (Fig. 3) with the tongres, (Fig. 4,) when they are constructed, attached, and fitted for use, substantially as herein described and set forth.

No. 55,312.—MARTIN KENNEDY, Boston, Mass.—*Plough.*—June 5, 1866.—Explained by the claim.

Claim.—First, the handle C, having the wedge-shaped opening *p* at its lower extremity, in combination with the pin *o*, sole A, standard B' and bolt *e*, for attaching the handle to the plough, substantially as described.

Second, the hooked beam D, in combination with the standards B B', and bolts *a* and *h*, for attaching the beam to the plough, substantially as described.

Third, the hinged mould-boards M M, constructed with their surfaces turned in at the bottom as described, and hinged to front standard B, in combination with the hinged plates or pieces *m m* and piece *d*, for adjusting the mould boards to any angle, the whole being constructed and operated in the manner and for the purpose set forth.

No. 55,313.—A. KIRLIN, Rock Island. Ill.—*Churn Dasher.*—June 5, 1866.—The tubular shaft is connected to the upper dasher and the inner plunger to the lower dasher; the plunger and dasher have valves and together act as a pump to drive air into the cream.

Claim.—The combination of the double dasher-shaft in combination with the pump rod when arranged together, substantially as and for the purpose described.

No. 55,314.—WILLIAM LANGDON, Langdon's Station, Penn.—*Stump Extractor.*—June 5, 1866.—The hook and chain are attached to an eccentric on the lever which is pivoted to a base piece.

Claim.—The eccentric short arm of the lever of the first kind in connection with the chain and hook, substantially as described.

No. 55,315.—ROBERT LAYNG, New York, N. Y.—*Machine for Pressing Lead Pipe.*—June 5, 1866.—The core holder depends from a bridge which traverses in the slot of the descending ram which expels the lead from the cylinder.

Claim.—The bridge *g*, attached to the end of the lead cylinder and sustaining the core holder *f*, in combination with the ram *e*, formed with an opening or mortise through which said bridge passes, substantially as and for the purposes specified.

No. 55,316.—JOSEPHUS LIGET. Posey Township, Ind.—*Broom Head.*—June 5, 1866.—The set-screw passes through the cap and the mortise in the handle to fasten them together.

Claim.—The handle D, provided with the mortise E, and the set-screw C, in combination with the socket A, all arranged substantially as set forth and described.

No. 55,317.—C. L. LOCKWOOD, New York, N. Y.—*Collar and Neck Tie Supporter.*—June 5, 1866.—The slotted retainer is secured to the shirt collar or button, and the pin attached to the bow or scarf is caused to interlock therewith.

Claim.—A plate or retainer, adapted for confinement in front of a shirt collar, a bow, or tie and devices, whereby the bow may be secured to or detached from the retainer, the whole being constructed and adapted to each other, substantially as described.

No. 55,318.—HENRY MAAS.—Homestead, Iowa.—*Rotary Pump.*—June 5, 1866.—The case has two equal parts united by flanges and has induction and eduction openings. The pistons are rotated on radial axes by a central cam and also revolve in the water passage of the casing, ejecting the water therefrom in succession.

Claim.—The disk or ring C, having a series of plates J hung in and to the same, when arranged within a casing having an outlet and inlet opening thereto, in such a manner that it can be rotated, and the said casing is so constructed as to operate the said ring plates J, substantially in the manner and to accomplish the purpose described.

No. 55,319.—G. S. MANNING, Springfield. Ill.—*Carriage Spring.*—June 5, 1866.—Improvement on his patent of July 31, 1863. The S-shaped springs are connected to the arms of the carriage body and to the axle by clips; may be strengthened by auxiliary leaves and have a secondary connection with the body at a point of flexure of the spring.

Claim.—First, the S-shaped springs D, either with or without the strengthening leaf or leaves K, constructed and connected to the axle and to the supporting arms of the carriage body, substantially as described and for the purpose set forth.

Second, the connection J, constructed as described, in combination with the supporting arms H I, substantially as and for the purpose set forth.

No. 55,320.—DAVID MARSHALL, Genoa, N. Y.—*Churn.*—June 5, 1866.—The cam wheel shaft is driven by a hand crank and the cam communicates a vertical reciprocating motion to the churn dasher.

Claim.—First, the arrangement of the spindle F, cam-wheel I, and slider J, when made substantially as specified and used for the purpose set forth.

Second, the sleeve K, ring L, and elastic rings M and M, when made and applied as herein specified.

No. 55,321.—RENÉ MASSON, Tremont. N. Y.—*Tobacco Pipe.*—June 5, 1866.—The pipes from the bowl and mouth-piece enter the chamber on the respective sides of the separating diaphragm and connect with the bubble-bubble chamber; a more immediate connection is afforded by the groove in the stop-cock if desired.

Claim.—First, the pipe *e*, extending down near to the bottom of the globe B, when the same is applied in combination with the bowl and stem of a smoking pipe and with the chamber C, substantially as and for the purpose described.

Second, the plug or stop-cock *f*, in combination with the channels *c d*, chamber C, globe D, and with the bowl and stem of a smoking pipe, constructed and operating substantially as and for the purpose described.

No. 55,322.—HENRY MAXELL, Canton, Ohio.—*Gate.*—June 5, 1866.—The gate is rotated on its hinge by the longitudinal motion of two rails of the fence which engage an arm on the heel post of the gate.

Claim.—First, the gate A, with the arm K, pivoted to the block *g* and upright L, by means of the L-shaped bar E, when arranged and used substantially as and for the purpose herein set forth.

Second, the bars B and B', with the slotted bar C, when used to slide in the slots of the uprights of the fence for the purposes of opening and closing the gate A, substantially as specified.

No. 55,323.—THOMAS McAULEY, San Francisco, Cal.—*Water Wheel.*—June 5, 1866.—This vertically revolving turbine-formed wheel receives the upward impact of the water from a nozzle at a point on its outer periphery and discharges it inward on to an apron which conducts it away.

Claim.—In combination with the water wheel described, the discharge nozzle N, arranged outside of the wheel and the dash-board D for joint operation, substantially as described.

No. 55,324.—S. T. McDOUGALL, New York, N. Y.—*Apparatus for Carburetting Air.*—June 5, 1866.—The carburetter has a series of connected diaphragms with spiral partitions which direct the air into a sinuous, reverting course in contact with fibrous material covering the partitions. A float shuts the valve of the liquid induction opening when the chamber is sufficiently charged. The air valve is operated by a wire and lever.

Claim.—First, the combination of the float E, and the valve D, with the carburetter A, for the purpose of regulating the admission of oil thereto, substantially in the manner herein described.

Second, in combination with a carburetter, containing a series of pans arranged one above another, the valve N, when operated by means of the wire P, or its equivalent, substantially as and for the purpose specified.

No. 55,325.—J. McKNIGHT and WILLIAM S. DEISHER, Reading, Penn.—*Motive Power.*—June 5, 1866.—The handles are moved alternately and wind up the spring, whose reacting force is transmitted by gearing to the desired point.

Claim.—First, the hub C, ratchet-wheels E E, cams F F, wheels G G, collars D D, and shaft B, arranged and used in the manner and for the purpose herein specified.

Second, the arrangement of the cams F F, the chains *f f* and springs *g g*, within the frame A, substantially as and for the purpose herein specified.

Third, the rods *h* and *i*, the arms *j j*, the shaft R, and the bar *e*, arranged and used as and for the purpose herein specified.

No. 55,326.—JAMES W. McLEAN, Indianapolis, Ind.—*Gate Latch.*—June 5, 1866.—The inclined projection on the upper arm of the latch is elevated by the pin when the gate closes, and the contact of the pin with the inclined projection on the lower arm brings the former incline down over the pin.

Claim.—The latch A, with projections D and E, when constructed and operated substantially as and for the purposes set forth.

No. 55,327.—FRANCIS McMANUS, Ellenburg Centre, N. Y.—*Lumber Register.*—June 5, 1866.—The unwinding of the cord sets in motion the train of wheels and actuates the index fingers of the series of dials. The cord is rewound by a spring, and the fingers indicate the aggregate result of many applications of the cord to the lumber.

Claim.—The barrel *n*, with pawls *n' n²*, ratchet-wheel *l*, and suitable train of wheels connecting the axes *a¹ b² c² d² e²*, &c., in combination with the cord *m*, spring *o*, and stationary cap *p*, constructed and operating substantially as and for the purpose specified.

No. 55,328.—JOHN McWILLIAMS, Pittsburg, Penn.—*Gas Generator.*—June 5, 1866.—The oil passes from an upper chamber by a pipe to the retort, and the gas is passed upward through a purifying stratum of lime to the exit pipe.

Claim.—The arrangement of the stove A, gas generator C, furnished with a coniformed bottom *e*, rack *f*, pipes *b* and *g*, and vessel *m*, arranged, constructed, and operating substantially as herein described and for the purpose set forth.

No. 55,329.—M. H. MERRIAM and E. L. NORTON, Charlestown, Mass.—*Manufacture of Shoe Binding.*—June 5, 1866.—The skin is slit into wide strips of even width, the ends of these are bevelled and united by cement; the continuous wide strip is then slit into ribons for binding.

Claim.—The improvement in the manufacture of shoe-binding, substantially as described.

No. 55,330.—M. H. MERRIAM and E. L. NORTON, Charlestown, Mass.—*Slitting Machine.*—June 5, 1866.—A wide strip is again divided into narrower strips to be connected into continuity for shoe-binding ribbon. The described devices refer to the rotary cutters, the stationary and pressure guides by which the strip is maintained in position, the fingers extending between the pressure blocks and the draw-rolls.

Claim.—The combination with the rotary cutters of the stationary and the pressure guides, or guide surfaces, by which the strip is maintained in position as presented to the action of the cutters.

Also, in combination with such cutters, the fingers extending between the cutter-blocks and keeping the material in position during the action of the cutters.

Also, combining with such rotary cutters the draw-rolls, operating substantially as set forth.

No. 55,331.—M. H. MERRIAM and E. L. NORTON, Charlestown, Mass.—*Sorting Machine.*—June 5, 1866.—The strips of leather cut from the skin to form shoe binding are sorted into parcels according to thickness. Connected with the contacting surfaces, between which the strips pass, is mechanism indicating by a finger and dial the thickness of the passing strip.

Claim.—The combination with contact surfaces, between which the strip of leather is placed or passed, of an index mechanism operated by the movement of the movable contact surface, substantially as set forth.

Also, the employment of the feed-rollers in combination with the index mechanism, one of the rollers operating the index by its movement, substantially as set forth.

Also, overhanging the rollers which feed the material and operate the index mechanism.

Also, making the rollers or the roller-frame adjustable as to height, so as to set the index-pointer to the starting point of the dial.

No. 55,332.—M. H. MERRIAM and E. L. NORTON, Charlestown, Mass.—*Machine for Winding Shoe Binding, Tape, &c.*—June 5, 1866.—The spool is formed of two independent disks, each having its independent arbor; a removable axis or core for the goods to be wound is inserted through the centre of one disk, the end of the binding slipped into a slit in said axis, and the latter being then pushed into the centre of the other disk, the spool is ready for winding. The material rests on the periphery of a revolving wheel and receives motion by frictional contact; a given number of revolutions of the wheel shifts the clutch pulley and stops the machine.

Claim.—The employment of the independent cheeks or disks, for supporting the tape or ribbon as it is being wound, in combination with the removable arbor i, upon which the coil is formed, substantially as described.

Also, the removable and slit arbor forming the core or spindle upon which the coil is wound.

Also, the employment of a measuring wheel driven by contact of the rotating coil, when such coil is formed and supported between disks, substantially as described.

Also, so connecting the measuring wheel with the mechanism by which the coil is wound that the rotation of the coil shall be automatically stopped when the determined length is wound, substantially as set forth.

No. 55,333.—M. H. MERRIAM and E. L. NORTON, Charlestown, Mass.—*Apparatus for Drying Hides.*—June 5, 1866.—The chamber is heated by a lower coil of steam pipe and a current of air circulates through it. The hides are hung upon the arms of a reel which rotates on a vertical axis, and the slats are capable on reaching a certain point in succession of being run out and in to deliver or receive the hide.

Claim.—First, a series of sliding panels, arranged within a drying chamber or case and made movable with respect to an opening or door in such case, so that each panel as it is brought opposite to such opening may be wholly or partially withdrawn from the casing for the attachment and removal of the skins.

Also, in combination with the series of panels, and the movable rails which support the same, a stationary frame having ways upon which each panel is guided and supported as it is slid from the drying chamber.

Also, the radial arrangement of the movable series of panels in the drying chamber, substantially as set forth.

Also, the combination with the stationary case and movable system of sliding panels of heating pipes or their equivalents, substantially as set forth.

No. 55,334.—M. H. MERRIAM and E. L. NORTON, Charlestown, Mass.· *Gluing Press.* · June 5, 1866.—The contiguous, bevelled, cemented ends of the leather strips are clamped to unite them firmly between the bed piece and the face of the pivoted platen.

Claim.—In combination with the platen and its bed, a guide-piece upon each side of the platen, for bringing the strips to be united into line, substantially as set forth.

Also, the employment of an identifying mark upon each of a series of platens, as and for the purpose specified.

Also, hanging the platen so that it may swivel to accommodate the acting face of the platen to the surface of the strip beneath it, substantially as set forth.

No. 55,335.—M. H. MERRIAM and E. L. NORTON, Charlestown, Mass.—*Gluing and Cementing Machine.*—June 5, 1866.—The lower portion of the roller revolves in the tank of cement; a scraper removes the redundancy, and its upper surface is coated with a thin pellicle, suitable for application to the end of the leather strip to be joined to its fellow.

Claim.—The combination of the cement-containing vat, the rotating cement cylinder, and the scraper, operating together in the manner and for the purpose substantially as set forth.

No. 55,336.—M. H. MERRIAM and E. L. NORTON, Charlestown, Mass.—*Cutting Machine.*—June 5, 1866.—The skins are cut into strips to be afterwards joined by their ends to form a continuous shoe binding ribbon. The skin is fed between two conducting aprons, slit by the upper and lower disk cutters, and is led away in strips by tapes.

Claim.—First, the arrangement of the cutter-blocks to slide upon their shafts when they are kept in relative position by a spring or springs upon the end of one or both of the cutter-shafts, substantially as described.

Also, the combination of two series of tapes holding the skins in position and feeding them to the action of the cutters, with two series of disk-cutters, which divide the skins into strips, substantially as set forth.

Also, the combination with cutters which divide the skins of a series of tapes conducting the material to and carrying it in strips from the action of the cutters.

Also, the employment of the auxiliary tapes which, in connection with the main tapes, keep all the strips in position until the skin is cut entirely through.

Also, hanging the drums around which the upper tapes pass upon swinging frames, so that pressure of the upper tapes upon the skin is maintained, substantially as set forth.

No. 55,337.—W. D. MILLER, Enon, Ohio.—*Potato Digger.*—June 5, 1866.—The plough opens the ground; the rake succeeds and gathers the vines; the revolving digger throws out the potatoes, vertical and lateral adjustments accommodate the machine to varying depths, uneven surface, and to the hills when out of line.

Claim.—First, the combination of a laterally adjustable mould-plough with the gatherer of a potato-digging machine, substantially as described.

Second, the combination of the rake D with the plough and gathering devices of a potato-digging machine, for the purpose of gathering the vines, substantially as set forth.

Third, the combination of a mould-plough, a rake, and a revolving fork or gatherer, for the purpose of digging potatoes, and substantially as set forth.

Fourth, in combination with the beam of the plough of a potato-digging machine, the crosshead I, and frames J and K, for the purpose of permitting an adjustment of the plough either vertically or laterally, substantially in the manner set forth.

Fifth, the runners R, in combination with the drag-frame, which carries the revolving fork or gatherer of a potato-digging machine.

No. 55,338.—WILLIAM MOORE, Kokomo, Ind.—*Automatic Boiler Feeder.*—June 5, 1866.—The ball float in the boiler turns the nozzle of the supply pipe so as to discharge into the boiler-feeder or otherwise, and at a low stage operates the whistle valve.

Claim.—First, the swinging nozzle or its equivalent, constructed and arranged substantially as described.

Second, the combination of the swinging nozzle with the shaft B, arranged in the support R, with the pivot L, substantially as described and for the purpose set forth.

Third, the combination and arrangement of the nozzle *a*, shaft B, and whistle S, all constructed and arranged as and for the purposes set forth.

No. 55,339.—ARTHUR MOFFATT, Washington, D. C.—*Peat Machine.*—June 5, 1866.—The peat passes between the hollow, steam-heated, revolving drums, and is ground between their surfaces and the concave faces of the steam-heated standard. From thence it passes by the hoppers to the throat, whence it is pressed by a reciprocating plunger into the cells of the revolving mould wheels, and is discharged therefrom by followers.

Claim.—First, the cylinders B and B', in combination with the standard C, arranged for the purpose of feeding, crushing, and grinding, substantially as and for the purpose herein set forth.

Second, the application of heat to the different parts of the machine, substantially as and for the purpose herein described.

Third, the cylinders *s* and *s'* having chambers *m*, in combination with the plunger *o*, operating substantially as and for the purpose herein described.

Fourth, the piston *o* having an endward movement forward and backward, in combination with the feeding openings *i* and *i'*, operating substantially as and for the purpose herein specified.

Fifth, the ejecting-rod *o'* provided with a disk in each chamber, and operating as herein described, for the purpose set forth.

Sixth, the vibrating rod *p*, in combination with the cogged-wheel *r*, and operating substantially as described.

No. 55,340.—OSCAR F. MORRILL, Chelsea, Mass.—*Heat Generating Apparatus for Cooking Purposes.*—June 5, 1866.—Explained by the claims.

Claim.—The employment of the inclined spur-wheel for effecting the adjustment of the wick by imparting a rotative spiral movement, substantially as set forth.

Also, the incasement of this wheel in the chamber or box *n*, substantially as and for the purpose set forth.

Also, bending the teeth of the inclined spur-wheel, so that they act at right angles to the surface of the wick, substantially as described.

Also, in combination with the spur-wheel and spindle, the tube *o* extending up through the reservoir, substantially as described.

Also, giving to the feed-wheel such construction with reference to the wick which it rotates that its teeth shall not drag the threads from the wick-tube or become entangled therein, each tooth freeing itself from the wick before it reaches the surface of the tube through which the wheel projects, as described.

Also, the employment, in combination with a tubular wick and two foraminous cylinders, of a non-conductive or slow-conductive packing *p*, over the top of the fluid-containing chamber, or between the flame and said chamber, substantially as described.

Also, in combination with a tubular wick and two foraminous cylinders, the packing *q* between the air-tube *l* and the wick-tube.

Also, supporting the inner flame-tube *g* upon this packing or a tube *r*, extending therefrom.

Also, the passage *t* communicating from the upper part of the reservoir chamber with the wick chamber, substantially as described.

Also, so applying the outer wick-tube that there shall be a space *s* between the outer wick-tube and the outer flame-tube, which space is filled with the packing *p*, substantially as described.

Also, making the supporting tube *r* and the upper part of the outer wick-tube of perforated metal, for the purpose described.

No. 55,341.—CYRUS H. MORSE, North Kingston, R. I.—*Vial for Holding Solutions.*—June 5, 1866.—The vial, with its funnel-shaped bottom and lower opening, sets in a cup in which the sediment is collected.

Claim.—A vial, bottle, or similar vessel A, made with a funnel-shaped diaphragm B, as described, in combination with the chamber C, for the purpose specified.

No. 55,342.—RICHARD M. MOYLE, South Manchester, Conn.—*Liniment.*—June 5, 1866.—This liniment for rheumatism is composed of vinegar, alcohol, ammonia, camphor, and turpentine.

Claim.—Australian rheumatic white liniment, composed of ingredients in about the proportions substantially as described.

No. 55,343.—PETER MURRAY, Milwaukee, Wis.—*Cable Stopper.*—June 5, 1866.—The jaws on the rail which clamp the cable against the stationary pin are moved simultaneously by an eccentrically slotted wheel and a lever.

Claim—The combination of the movable jaws or jaw C C' with the stationary pin or jaw D, eccentrically slotted wheel G, and lever H, all constructed and arranged to operate substantially as and for the purpose herein described.

No. 55,344.—G W. NELL, Philadelphia, Penn.—*Curtain Fixture.*—June 5, 1866.—The tubular button is screwed upon the shank to adjust the vertical position of the cord roller.

Claim.—A shade cord-holder composed of a sliding rod B, with screw-shank D, spring *b*, and tubular button E, constructed and operating substantially as and for the purpose set forth.

No. 55,345.—PETER H. NILES, Boston, Mass.—*Broom Clasp.*—June 5, 1866.—The wire frame which clasps the broom is fastened to it at its respective upper ends by the permanent and opening eyes of the spring pin.

Claim.—The spring pin in combination with the protector, for the purpose and in the manner substantially described.

No. 55,346.—WILLIAM H. NOBLES, St. Paul, Minn.—*Harbor Dredging Boat.*—June 5, 1866.—Explained by the claims.

Claim.—First, the construction of the hull of the dredging steamer containing the water-tanks P, and having upon its bottom surface the plough L, when the bow or forward part of said dredge is constructed so as to draw less water than the stern and admitting across the centre thereof an open space B, for the introduction of ploughs or scrapers, as herein described and set forth.

Also, the ploughs or scrapers C, placed spirally on its shaft E, operating on an upright movable frame D, with the endless chains W, sand-buckets T, and pulleys X, arranged and combined as herein described and for the purposes set forth.

Also, the mode of filling the buckets T, by means of the ploughs or scrapers C, as herein described.

No. 55,347.—WILLIAM H. NOBLES, St. Paul, Minn.—*Steam Dredging Boat.*—June 5, 1866.—The machinery is mounted upon an enlarged area of deck, and the scoops operate to throw the sand, &c., on each side of the hull.

Claim.—The wheels E, operating in square apertures, and located either obliquely or straight across the boat, having their ploughs L set spirally and obliquely, so as to make a clear channel, and throw the dirt to each side of the boat, when said devices are combined with the upper works or frame of a vessel greater in extent than the hull thereof, as herein described.

Also, the combination of the spirally set ploughs L, platform C, and centre plough T, when arranged upon a boat, as herein described and for the purposes set forth.

No. 55,348.—O. B. NORTH, New Haven, Conn., assignor to O. B. NORTH & Co —*Harness Saddle Seats*—June 5, 1866.—The seat is cast in one piece with a core print inserted from the open side.

Claim.—A saddle seat, of the form described, cast in one piece, substantially in the manner as herein fully set forth.

No. 55,349.—JOSEPH P. NOYES, Newark, N. J.—*Comb.*—June 5, 1866.—Explained by the claim.

Claim.—As an improved article of manufacture, the folding comb herein described, consisting of the two blades A B, metallic backs C D, and hinge-joint E, made solid with the said metallic backs, all as specified.

No. 55,350.—JOHN NUSBAUM, Alliance, Ohio.—*Car Truck.*—June 5, 1866.—A broad shoe with laterally extending flanges is suspended from the truck and is intended to support the latter if an axle or wheel break.

Claim.—The shoe D, with the flanges e, and braces E F, in combination with the truck, in the manner and for the purpose set forth.

No. 55,351.—WILLIAM S. O'BRIEN, Brimfield, Ill.—*Harrow.*—June 5, 1866.—The side barrows are attached by transverse rods and eye bolts to the side rods of the central section. The position of the side rods is adjusted by the middle threaded portions, and nuts in the middle bars of the central harrow.

Claim.—The arrangement and attachment of the eye rods a a a to the screw nuts b, in the centre barrow, for the securing and adjustment of the outer barrows A' A', substantially in the manner and for the purpose as herein described.

No. 55,352.—A. W. OLDS, Green Oak, Mich.—*Fence.*—June 5, 1866.—Each panel consists of horizontal bars, vertical slats and oblique brace slats ; the panels have stakes and braces for support and are keyed together.

Claim.—The arrangement of the strips A B C, and slats b d, in combination with the brace f, and mode of connecting the panels together, substantially as and for the purpose set forth.

No. 55,353.—WILLIAM ONIONS, St. Louis, Mo.—*Cotton Bale Tie.*—June 5, 1866.—The head slips through the larger opening in the corresponding ends of the tie, and the expansion of the bale draws the smaller slots around the neck of the button.

Claim.—The head C, having the hook upon one side, substantially as and for the purpose described.

Also, forming the head C, plate E, and rivet a, of one and the same piece, as and for the purpose specified.

No. 55,354.—JASON C. OSGOOD, Troy, N. Y.—*Hose Coupling.*—June 5, 1866.—A portion of the threads on the screw and socket are cut away to enable the parts to be slipped past each other and lock by a partial revolution, when the undivided threads catch and complete the joint.

Claim.—The screw and nut cut with double threads and leaving two or more of said double threads at the base of the screw and nut whole or uncut, in combination with the division into sections of the upper threads of the screw and nut, substantially as and for the purposes as herein set forth.

No. 55,355.—SETH E. PARSONS, Albany, N. Y.—*Vaginal Syringe.*—June 5, 1866.—The rubber sack has a metallic tube and pear-shaped discharge bulb, with vent holes.

Claim.—The peculiar shape of the perforated bulb A, and the long stem on the tube B, in connection and combination with the rubber bulb C, in the manner and for the purpose as herein set forth.

No. 55,356.—JOHN S. PATRIC, Rochester, N. Y.—*Cattle Pump.*—June 5, 1866.—A jointed, tilting frame, with a counter-balancing weight, is depressed by the cattle, sinking the hollow pump stock and discharging the water.

Claim.—The combination and relative arrangement of the base A, or its equivalent, and the jointed tilting frame B D, its fulcrum E, and counter-balancing weight H, with the hollow piston or pipe p of the pump, as and for the purposes set forth.

No. 55,357.—JOSEPH PAUDLER, Jr., and FRIEDRIK BAUSCHTLIKER, Washington, D. C.—*Hydrant.*—June 5, 1866.—The wrench which turns the stop-cock also turns the cap and curved piston, and forces the water from the chamber; the return of the wrench admits the water from the supply pipe to follow the semicircular piston and avoid freezing.

Claim—The arrangement and combination of the grooved circular air and water chambers R and S, valve T, air tube D D, with the upright lever or wrench L for letting on or cutting off the supply of water, and the upright levers or wrenches G G, to tighten the packing K, as herein described.

No. 55,358.—J. C. PLUMER, Boston. Mass.—*Furniture Caster.*—June 5, 1866.—An annular rubber disk, retained on the spindle by a groove, rests on projections in the socket to keep the spindle from falling out.

Claim.—First, the use of a disk of rubber, leather, or other similar flexible material, for the purpose of retaining the caster spindle in its place, in the manner set forth.

Second, the combination of the circular support E and the shoulders F, as and for the bjects specified.
o

No. 55,359.—E. A. POND and M. S. RICHARDSON, Rutland, Vt.—*Generating and Supplying Illuminating Gas.*—June 5, 1866.—Air from the blowing apparatus is conducted by mains and branches to the carburetting apparatus, which is located at each house or place where the gas is required.

Claim.—The method herein described of supplying gas generators or vaporizers, of whatsoever construction, from an independent air pump or air-forcing apparatus located at any place convenient, but so that the air shall enter the said vaporizer, and become charged with the hydro-carbon vapor at or near the point where the gas is to be consumed.

No. 55,360.—S. T. W. POTTER, Scott, N. Y.—*Churn.*—June 5, 1866.—The reciprocating, alternate dashers have corrugated beaters attached to curved arms, and scrapers which wipe the inside of the box.

Claim.—First, the floats F secured to the dasher rods g, constructed and arranged substantially as described.

Second, the scrapers G, in combination with the floats F, substantially as shown and described.

No. 55,361.—ANDREW RANKIN, New York, N. Y.—*Chamber Vessel.*—June 5, 1866.—The deodorizing agent is contained in the hollow lid and discharged into the vessel by rotating the slide, whose radial plates cover the slots in the bottom of the cover.

Claim.—The combination of the arms G and spindle H, operating relatively with the slotted plate P of the vessel B, having a nozzle D for the reception of the deodorizing agent, all arranged and applied in the manner herein described for the purpose specified.

No. 55,362.—DANIEL K. REEDER, Elliottsburg, Penn.—*Corn Sheller.*—June 5, 1866.—The wheels have obliquely set teeth on both faces; the ears are fed from a grooved hopper floor, and held by concave helical sockets on the spring slats; the grain is cleaned by an air blast on an inclined floor and grating.

Claim.—First, the combination of the double-toothed wheels H, the concave holders i, mounted on slats e, provided with yielding springs f, and the grooved hopper D, all constructed and arranged substantially as and for the purposes described.

Second, the helical gains h in the concave holders i, as and for the purposes described.

Third, the teeth upon the wheels E set at an angle, as described, to the radii of said wheels, in combination with the concave holders i, as and for the purpose described.

Fourth, the grooved floor of the hopper D, in combination with the apertures D', substantially as and for the purpose described.

Fifth, the shaking inclined floor G and slotted screen G', in combination with the shelling machinery and revolving fan, substantially as and for the purposes described.

No. 55,363.—WILLIAM REID, West Arlington, Vt.—*Machine for Compressing the Cylinders of Casks upon their Heads to form a Tight Barrel.*—June 5, 1866.—The cylinder cut from the log is compressed on to the heads by a conical ring, composed of sliding segments, which hold it while being hooped.

Claim.—The conical ring composed of sliding segments, operated substantially as described, for compressing the shell in combination with the movable disk, or the equivalent thereof, for holding and controlling the shell to be compressed, substantially as described.

No. 55,364.—CROMWELL O. RICHEY, Aurora, Ind.—*Brace for Bits.*—June 5, 1866; antedated May 16, 1866.—A pin on the shank of the bit fits a recess in the socket, and being turned therein, is locked by a spring bar.

Claim.—The retaining bar D, spring E, ferrule k, and slot C, in combination with bit B and pin b, arranged as above described and for the purpose set forth.

No. 55,365.—EDWIN RITSON and WILLIAM W. BRIGG, Mattaville, N. Y.—*Hoe.*—June 5, 1866.—Explained by the claims.

Claim.—First, a hoe composed of a back A, shank B, and ferrule C, all cast in one piece, and with teeth D secured to the back A, in the manner substantially as set forth.

Second, having the teeth D made in the form of a scaline triangle, with oblique sides, and attached to the back A in reverse position from the centre outward, so that the oblique sides will face the centre of the hoe, as shown and described.

No. 55,366.—D. D. ROBINSON, Niles, Mich.—*Rivet or Bolt Cutter.*—June 5, 1866.—The equal movement of the jaws is secured by the link connection of the eccentric levers.

Claim.—First, an improved machine for cutting off bolts or rivets, formed by combining and arranging the eccentric levers A B, the jaws C D, and plates E F G J with each other, substantially as described and for the purpose set forth.

Second, the combination of the toothed and stop plates H I with the eccentric levers A and B, substantially as described and for the purpose set forth.

No. 55,367.—CHARLES SAFFRAY, New York, N. Y.—*Manufacture of Artificial Fuel.*—June 5, 1866.—Equal parts of coal dust and rosin are mixed and agglomerated by the melting of the latter; being pulverized, it is mixed with more coal dust and pressed into cakes in slightly heated moulds.

Claim.—First, the within described process of aggregating coal dust or waste coal by first producing a powder of pitch or rosin and coal dust, which is mixed cold with the coal to be aggregated, and, after having been subjected to a suitable pressure, is dried, substantially as and for the purposes set forth.

Second, the product obtained by treating coal dust or waste coal in the manner above specified.

No. 55,368.—I. A. SALMON, Boston, Mass.—*Dentist's Chair.*—June 5, 1866.—Explained by the claims.

Claim.—The arrangement of the gimbal ring d, the journals b b e e, the bearings c c f f, curved bars g l, and their bolts i m, with the chair body B and the base or foot frame A.

Also, the arrangement and combination of the movable foot-board supporting racks r r and their operative slide rod n with the chair body B, the foot-board supporting arms p p with their cross-bar t, and the crank shaft s and band w, for raising and lowering the foot board, as specified.

Also, the mechanism for adjusting the elevation of the chair seat, the same consisting of the wheel a', with its scroll flange b', rack z, and the carrier y, arranged and combined with the seat and the chair body, substantially as set forth.

Also, the improved head-rest hinged joint, as made of the recessed cylinder f', the pin k', the clamp screw m', and the bearing h', constructed and arranged together as specified.

Also, the combination of the cross-slot n', in the adjuster, with the clamp screw m' and the head rest E, applied together substantially as described.

No. 55,369.—HOWARD SARGENT, Boston, Mass.—*Truss.*—June 5, 1866.—Attached to the elastic girdle is a pad of sponge, with a large base and a smaller raised boss to fit the umbilicus.

Claim.—The pad A as described, made of sponge, for the purpose set forth.

No. 55,370.—D. F. SEXTON, Whiting, Vt.—*Sheep Rack.*—June 5, 1866.—The rack has tapered division strips, and is vertically adjustable in its frame.

Claim.—Arranging the troughs in a sheep rack so that the same can be slid up and down in suitable guides, substantially as and for the purpose specified.

Also, so arranging the division strips C that a greater space shall be provided at the top than at the bottom, substantially as and for the purpose specified.

No. 55,371.—T. P. SHAFFNER, Louisville, Ky.—*Preserving Vegetable Fibre.*—June 5, 1866; antedated April 10, 1866.—The material to be preserved receives a coating of plumbago in a bath, and then a metal is precipitated upon it by electric action.

Claim.—The process of imparting a pure metallic coating to fibrous substances by precipitation, by electrical action of a metal upon a metallic surface, previously given by saturation with a metallic solution and subsequent removal or evaporation of the water of suspension or evaporation.

Also, saturating the fibre, fabric, or wood, as described, with a liquid containing plumbago in suspension for the purpose of imparting a metallic coating to said fibrous substances, upon which a film of metal may be afterward precipitated by electrical action.

No. 55,372.—N. B. SHERWOOD, Millville, N. Y.—*Cotton Seed Planter.*—June 5, 1866.—The clusters of teeth are placed diagonally on the delivery belt, a stationary brush limiting the delivery of the seed; the side of the hopper is adjustable toward the delivery belt as the seed decreases in quantity.

Claim.—First, the vertically revolving feeder-belt B, constructed and operating substantially in the manner and for the purposes herein shown and described.

Second, the fixed separating brush E, arranged and operating substantially as and for the purposes set forth, in combination with the vertically revolving belt B.

The revolving discharge-brush F, arranged and operating in connection with the belt B, substantially as and for the purposes herein shown and described.

Fourth, the employment or use of the automatically adjustable back G of the grain or seed hopper, in combination with the delivering devices.

Fifth, constructing and arranging the parts so that the belt B, whether vertical or inclined, shall constitute one side or end of the hopper or seed box, as set forth.

Sixth. so arranging and operating the toothed belt B, or delivering devices in this class of seed planters, as to convey the seed upward out of the hopper or box, for the purpose set forth.

No. 55,373.—N. B. SHERWOOD, Millville, N. Y.—*Cotton Seed Planter.*—June 5, 1866.—The separating brushes on the roller near the top of the toothed delivery belt sweep between the teeth of the latter to render uniform the passage of seed.

Claim.—The employment of the revolving separator-brush B, arranged and operating in combination with the toothed delivery belt D, substantially as and for the purposes herein shown and described.

No. 55,374.—FRANCIS SHINN, Rockford, Ill.—*Horseshoe.*—June 5, 1866.—The hooked ends of the calks slip into slots in the shoe and are retained by nails driven alongside.

Claim.—Attaching the toe or calks of a horseshoe to the plate, substantially as described, so that they shall be held to the plate by a single nail, which at the same time serves to hold the shoe to the hoof, and so that the removal of the single nail which holds a toe or calk shall permit it to be removed or replaced at pleasure without removing the plate.

No. 55,375.—CHARLES T. SHOEMAKER, Philadelphia, Penn —*Saw.*—June 5, 1866.—The slot in the plate for receiving the removable tooth is wider at its base than at the periphery, and the split base of the tooth after insertion is spread by a rivet to preserve its position.

Claim.—First, the tooth B, with its split projection *b*, and grooved edges adapted to a recess in the saw, and secured thereto by taper pins *m m*, or their equivalents, all substantially as set forth.

Second, a piece D, formed, fitted to an opening in the saw, and adapted and secured to the tooth for the retention of the same, all substantially as set forth.

No. 55,376.—JOHN T. SHRYOCK, Zanesville, Ohio.—*Brick Drying Press.*—June 5, 1866.—The bricks are ranked in a vertical series of trays on a car which travels on rails into the kiln.

Claim.—The brick drying apparatus herein described, consisting of the kiln A, track or rail B B, car C, tray G, fireplace D, heating flue E, and stack F, constructed and employed substantially as and for the objects specified.

Also, the brick bearing cars, consisting of the open frame C, wheels C¹, and axles C², constructed and arranged in the manner and for the purposes set forth.

No. 55,377.—LEBBEUS SIMKINS, Brooklyn, N. Y.—*Working Ship Pumps.*—June 5, 1866.—By intermediate gearing, crank, &c., the capstan is made the means of working the ship's pump.

Claim.—The arrangement of a suitable connection, such as the shafts F F I L and bevelwheels H J K, or any other equivalent means, in combination with the pump A and capstan O, constructed and operating substantially as and for the purpose described.

No. 55,378.—GEORGE SIMPSON, Waterbury, Vt.—*Burglar Alarm.*—June 5, 1866.—A string attached to the door sets in motion clockwork that opens a lid, erects and lights a candle, and rings a bell.

Claim.—The sliding head E, hinged taper holder *m* on the guide rod *l*, spiral spring *r*, friction plates *q* and *n*, lever I, and catch S, in combination with clockwork to produce the alarm and strike a light, substantially as described for the purposes specified.

No. 55,379.—E. W. SKINNER, Madison, Wis.—*Sugar Cane Mill.*—June 5, 1866; antedated December 5, 1865.—A guide between the lower rollers gives direction to the cane. Waste oil from the bearings is prevented from dripping into the juice.

Claim.—First, arranging a guide or shed N, provided with shoulders O O, between the lower rollers, to prevent the cane from running off the ends of the rear roller, as and for the purposes set forth.

Second, arranging beneath the journal bearings of the rollers the cups *a*, provided with the outlets *a'*, substantially as and for the purposes described.

No. 55,380.—EPHRAIM SMITH, Clinton, Penn.—*Rake Attachment to Harvester.*—June 5, 1866.—The claim identifies the peculiar construction of the rods forming the gavelling device, into which the grain is compressed by the rake, and by the tilting of which the grain is discharged in compact gavels upon the ground; also the arrangement of devices for automatically tilting the said rods.

Claim.—The lifting rods *d d d*, when provided with the guard-fingers *f f f*, projecting upward from their outer ends near the shaft on which they turn, substantially as and for the purpose herein specified.

Also, in combination with the above, the arrangement of the disk or wheel G with its cam pin and gear teeth, and the disk or wheel H with its cam projection and gear teeth, for giving the required movements to the lifting rods, substantially as herein set forth.

No. 55,381.—H. B. SMITH, Eureka, Ill.—*Cultivator.*—June 5, 1866.—A certain independence of motion is permitted to the two sides. The vertical and lateral motions are performed by manipulating the inner ploughs, assisted by the weight of the driver on the tilting carriage frame.

Claim.—First, the construction of the frame A in the manner substantially as herein shown and described, to admit of said frame being expanded and contracted laterally to adjust the ploughs nearer together or further apart, as may be required, and admit of a direct application of the draught of each animal to the device, as set forth.

Second, in combination with the frame A thus constructed, the pivoted beams C C, arranged or applied substantially as and for the purpose set forth.

Third, the connecting of the plough beams G G to the beams C C, by means of the universal joints H*, constructed substantially as shown and described, to admit of the vertical, lateral, and rolling motion of the ploughs, as set forth.

Fourth, the combination of the adjustable frame A, pivoted beams C C, plough beams G G, all arranged to operate in the manner substantially as and for the purpose set forth.

No. 55,382.—JAMES M. SMITH, Seymour, Conn.—*Hollow Auger.*—June 5, 1866.—The lower end of the revolving sleeve has teeth which rotate the pinions on the screw-shafts; these actuate the radial slides and adjust the cutters and guides.

Claim.—The combination of the hollow shank A, cylinder I, adjusting screws H, and slides, all constructed and arranged substantially as described, so as to adjust the cutters and guides as and for the purpose specified.

No. 55,383.—WILLARD H. SMITH, New York, N. Y.—*Vapor Burner.*—June 5, 1866.—The wick tube is isolated from the burner by the interposition of a sleeve, the two tubes being attached to a shield.

Claim.—The sleeve E, the wick tube F, and the shield G, combined and arranged in the manner and for the purpose herein set forth.

No. 55,384.—SIMON SOULES, Dowagiac, Mich.—*Potato Digger.*—June 5, 1866.—The inclined shovel has sides which guide the potatoes, &c., on to the double-inclined grating, which is vertically agitated by the tappets on the revolving shaft beneath.

Claim.—First, the construction of the sides A A, of the form substantially as shown, with fenders *g g*, in combination with the forward inclined shovel D and the inclined grating G, substantially as described.

Second, in combination with the upright sides A and inclined shovel D and the inclined grating G, the rollers *c c* and their tappets *d*, substantially as described.

Third, the double-inclined grating G, hinged and operated substantially as described.

No. 55,385.—WILLIAM STEINWAY, New York, N. Y.—*Piano-Forte.*—June 5, 1866.—Explained by the claims.

Claim.—First, the use in piano-fortes of a metal case cast in one solid piece, consisting of the plate *a*, braces *b*, rafters or brace frame *c*, and connecting piece or flange running round on three sides of the case and supporting the regulating apparatus, leaving one side open for the insertion of the sounding board, with its bars and bridges, substantially as described.

Second, the method herein described of supporting the sounding board by means of screws, springs, wedges, wire-draws, or any other equivalent means, bearing on and bracing against the edges thereof, substantially as and for the purpose set forth.

Third, supporting a number of the lowest steel strings and highest covered bass strings a second time between the regular sounding-board bridge and the hitch-pins, either upon a prolongation of the regular sounding-board bridge or upon an independent bridge, for the purpose of equalizing the transition from the steel strings to the covered bass strings, thus preventing any break in the tone.

No. 55,386.—JOHN J. STEVENSON, Auburn, N. Y.—*Lubricating Journal.*—June 5, 1866; antedated May 22, 1866.—To one journal bearing is attached a tight closing box filled with saturated fibre and communicating with the journal.

Claim.—The application of the self-oiling cup or box to the bearings of mowing and reaping machines, when used as and for the purpose above specified.

No. 55,387.—EDWARD SULLIVAN, Pittsburg, Penn.—*Piston Head for Steam Engine.*—June 5, 1866.—The requisite pressure is maintained in the piston for the expansion of the packing ring. The areas of the ends of the valve vary in proportion to the desired difference of pressure between the steam on the outside and that on the inside of the piston.

Claim.—First, the use in piston-heads of steam engines, furnished with expanding packing rings, of one or more self-acting valves, for the admission of steam therein, each valve consisting of a single valve piston or plunger, so arranged that the live steam shall press on both ends of the plunger at the same time, one from the inside of the piston head and the other from the outside, the area of the two ends of the valve plunger differing in proportion to the relative desired pressure of the steam on the exterior of the piston head and on the packing rings within the piston head, substantially as and for the purpose hereinbefore described.

Second, the combination of the valve cylinder *g*, piston *p*, and stem *q*, forming a self-acting valve for the admission of steam into the interior of the piston heads, for regulating the pressure of steam on the packing rings, constructed and operating substantially as hereinbefore described.

Third, combining with the piston head of a steam engine, having expanding packing rings, a device, constructed and operating substantially as hereinbefore described, so as to retain the live steam which enters the piston head, and obviate the frequent change of steam, thus preserving a constant and uniform pressure on the packing rings, and preventing the shuffling motion of the rings consequent on the continual passage of steam in and out of the piston head.

No. 55,388.—E. C. SUMMERS, Huntingdon, Penn.—*Apparatus for Collecting Floating Oil from Streams.*—June 5, 1866.—A jointed boom-trough is diagonally placed to direct the surface water to a chute and a strainer; a vertically adjustable dividing plate directs a given depth of stratum to the reservoir.

Claim.—First, the combination of the boards or planks B with the bent sheets of iron C, when arranged to form a floating boom for collecting surface oil, substantially in the manner hereinbefore described.

Second, the combination of the trough D with the trough or boom A, for confining the oil and water as it passes through the former, to effect the more perfect action of the chute E, substantially as described.

Third, the combination of the chute E with the trough D and tank I, when arranged to operate substantially as described and for the purposes specified.

No. 55,389.—JOHN B. TARR, Chicago, Ill.—*Planing Machine.*—June 5, 1866.—The mouth-piece bears upon the stuff immediately before and in the rear of the cutters, and yields to the uneven thickness without affecting the cutters.

Claim.—First, the construction of the mouth-piece B B *e f*, substantially in the manner herein described and shown.

Second, the arrangement of the mouth-piece B B, slide *b*, spring *c*, in relation to the cutter A, substantially in the manner herein described and shown.

No. 55,390.—PETER TAYLOR, Pawtucket, R. I, and WILLIAM A. GOVE, Charlestown, Mass.—*Bed Bottom.*—June 5, 1866.—The bed bottom is suspended inside the bedstead by elastic straps, which pass over rollers on the rails and are attached by clamping links to the brackets beneath the bed bottom.

Claim.—First, the combination and arrangement of the arms C (projecting from the frame A) with the elastic or India-rubber straps and their supporting rollers, applied to the frame B, the whole being substantially as specified.

Second, the combination of the buckles, or equivalent take-up mechanism, with the arms and the elastic straps applied to the frames A and B, substantially in manner as specified.

Third, the combination of the straining plates F, the elastic straps or springs, their supporting rollers, and the arms C, the whole being arranged together and with the frames A and B, substantially in manner and so as to operate as specified.

No. 55,391.—NICHOLAS THOMAS, Chicago, Ill.—*Tool for Cutting Off Boiler Tubes.*—June 5, 1866.—The lower end of the stock is inserted into the tube; the tool projects radially from the revolving stock and cuts off the boiler tube from the inside.

Claim.—An improved tool for cutting off boiler tubes, constructed and arranged substantially as herein described and for the purpose set forth.

No. 55,392.—HORACE A. THOMPSON, Hartford, Conn, assignor to GEO. T. THOMPSON.—*Call Bell.*—June 5, 1866.—The clapper is suspended from the hinged button on top; is struck by the depression of the latter, which is kept elevated by the weight of the clapper when the striking force is withdrawn.

Claim.—The hinge pad *d*, in combination with the bell *a*, striking knob *e*, arranged and operating substantially as and for the purpose described.

No. 55.393.—ROBERT H. THURSTON, Providence, R. I.—*Magnesium Lamp.*—June 5, 1866; antedated January 4, 1866.—The magnesium is fed from below in wire or ribbon; its section of flame at the point of combustion is circular, and the feed motion is derived from a wheel driven by a falling stream of sand.

Claim.—First, feeding magnesium upwards to the point where combustion is to take place, from below the flame, by a regular motion, to be determined by the rate at which combustion proceeds, in the manner substantially as described.

Second, the combination of curved guides *b b'*, or guides arranged in a curve with the mechanism for feeding a ribbon or ribbons of magnesium, so that the section of such ribbon or ribbons, where combustion takes place, shall be circular.

Third, combining with such guides a clearer *e*, constructed and operated substantially as described.

Fourth, the combination of a bucket-wheel A having a continuous rotary motion as described, with a set or sets of feeding rollers *a a'*, arranged and operated substantially as specified.

No. 55.394.—HOWARD TILDEN, Boston, Mass.—*Broom Clasp.*—June 5, 1866.—This clasp is placed around the broom corn to strengthen the broom.

Claim.—The improved broom strap or clasp described, to wit: a broom strap or clasp made of two loops, connected by two springs, substantially as described.

No. 55.395.—OAKES TIRRILL, Burlington, Mass.—*Gas Apparatus.*—June 5, 1866.—The liquid hydro-carbon is brought to the proper temperature by a burner operating upon a side coil through which the liquid flows.

Claim.—First, in generating illuminating gas from hydro-carbon fluids, the method described of heating the fluid in a coil, or other suitable apparatus, which shall be outside of and separate from the generator, and yet at the same time so connected and combined therewith as to cause the fluid to circulate freely and continuously through the said coil or other apparatus and generator, as and for the purposes herein shown and set forth.

Second, in combination with apparatus for generating gas, as described, the employment of a spiral coil of pipe or tubing outside of the gas generator, and connected therewith by means of pipes or other suitable devices, together with a burner for heating the said coil, substantially as herein described and for the purposes set forth.

Third, in combination with the spiral coil of pipe or tubing, as set forth, the double jacket and cap fitting over said coil, and arranged substantially as and for the purposes herein shown and described.

No. 55.396.—WILLIAM S. TROWBRIDGE, Milwaukee, Wis.—*Transit Instrument.*—June 5, 1866.—Explained by the claims.

Claim.—First, the attaching of the extra telescope by extending the shaft of the transit instrument, for the purpose of describing angles from one to ninety degrees.

Second, the attachment of the graduated quadrant or quadrants on the transit instrument, in combination with the telescope or telescopes and the table or tables for ascertaining the horizontal and perpendicular, by taking the angle or surface measurement of slope or incline of ground passed over.

No. 55.397.—CYRUS TUCKER, La Crosse, Wis.—*Till Lock.*—June 5, 1866.—The guards are changeable, to alter the combination, and the bell is rung by an attempted opening with a wrong combination.

Claim.—First, the independent guards D, constructed as shown and described, and mounted vertically in the frame C, as set forth.

Second, in combination with the guards D, the bolt F, arranged to move vertically and engage in the notches of the guards, as set forth.

Third, in combination with the bolt F, provided with the arm *r*, the pivoted dog E, arranged to operate as set forth.

Fourth, the pawl *n*, arranged to operate in combination with the dog E and arm *r*, as herein described.

Fifth, the bell H, having its hammer *g* provided with the lever *f*, arranged to operate in combination with the arm *r*, as set forth.

No. 55.398.—P. H. VANDER WEYDE, New York, N. Y.—*Apparatus for Inhaling Gases.*—June 5, 1866.—The gas from the neck of the retort passes to a purifier and thence to a strong reservoir; from this extend two tubes to as many purifiers, and from these, two tubes to a mouth-piece, whose valves, opening in opposite directions, allow gas to pass from one vessel to the mouth by inspiration, and from the mouth to the other vessel by expiration.

Claim.—First, the construction of the apparatus described for the generation, preservation, but particularly for the administration of anæsthetic gaseous substances, so that the injurious products of the respiration are absorbed by the passage of the expired gas, by a separate channel, through an appropriate alkaline solution, and all waste of the expired gas is thus avoided.

Second, the attachment of a strong cylinder, able to withstand the pressure of at least fifty atmospheres, containing the laughing gas condensed to its liquid form, of which a portion, by the simple, partial opening of a stop-cock, expands, cools, and supplies the common breathing bags, or, what is better, the above described inhaling apparatus with any desired quantity of fresh, cool, and perfectly pure nitrous oxyd gas.

No. 55,399.—CHARLES A. WAKEFIELD, Pittsfield, Mass.—*Hand Corn Planter.*—June 5, 1866.—The scrapers clean the soil from all sides of the rising plunger. The slide is operated by a feather, and the short zig-zag thereon shakes the slide to equalize the feed.

Claim.—First, the inside scraper *a*, applied to the rear surface of the plunger, in combination with the scraper *b*, applied to the front surface thereof, substantially as herein set forth for the purpose specified.

Second, the side scrapers *c* and *d*, applied to the plunger, substantially as herein set forth for the purpose specified.

Third, the oblique tongue or feather *e*, working in the notch *f* of the slide *g*, to operate the said slide, substantially as herein set forth for the purpose specified.

Fourth, the zig-zag deflectors *n* in the feather *e* operating the slide *g*, substantially as herein set forth and for the purpose specified.

No. 55,400.—PHILIP N. WOLISTON, Springfield, Ohio.—*Brick Machine*—June 5, 1866.—The kneader in the clay box delivers the clay through the die-plate in determinate quantities; the arrangement of the moulds and devices for removing the moulded brick are cited in the claims.

Claim.—First, the mixer or kneader shaft C, having a scraper K at its lower end, in combination with the scraper N, when arranged together and so as to operate substantially in the manner described for the purpose specified.

Second, so arranging the scraper N that its movement can be adjusted at pleasure, substantially as and for the purpose described.

Third, the arrangement of the moulds V², in combination with the followers or plungers F², when so operated as to move up and down, substantially in the manner described for the purpose specified.

Fourth, forming apertures in the moulds, as and for the purpose specified.

Fifth, the rock shaft N², connected with the sliding arm Q², in combination with the lug V², or its equivalent, upon the rotating disk D², when arranged together and with the endless travelling apron U, so as to operate substantially in the manner and for the purpose described.

Sixth, the disk D², having cam-shaped flanges, in combination with the moulds V² and followers F², connected with the said flanges, and all arranged together so as to raise and lower the said moulds and followers, substantially as and for the purpose specified.

Seventh, forcing the clay from the kneading mill through openings corresponding in shape to that which it is desired to impart thereto, according as bricks, tiles, &c., are to be made from it, substantially as described; the clay, as it is forced through said openings, passing to the moulds to be operated upon by them, for the purpose specified.

No. 55,401.—JOSEPH B. WARREN, Danversport, Mass.—*Cradle, Stool, and Chair.*—June 5, 1866.—The body of the cradle is pivoted upon the rocker frame; a box opens in the frame to make a place for the feet, and one end of the cradle is adjustable to form a back support.

Claim.—The cradle pivoted *j*, with box *h* in the rocker-frame, and the adjustable portions *j d*, A and B B, as above described, for the purposes herein set forth.

No. 55,402.—CHARLES WEBER, West Meriden, Conn.—*Table Mat.*—June 5, 1866.—The mat is made of pasteboard prepared with oil and varnish, covered with cloth or leather, and furnished with a metallic rim.

Claim.—A table mat, made as herein described, and provided with a metal frame, substantially as and for the purpose specified.

No. 55,403.—LEVI H. WEST, Cambridge, Mass.—*Car Truck.*—June 5, 1866.—The weight of the platform is thrown upon the sub-levers, each of which rests by one end upon the truck, and at the other end is suspended from the platform. Springs at the supporting points relieve the jar.

Claim.—The combination and arrangement of the elastic fulcra or the springs *e e* with the said frames A B and the levers D D, or the same and the springs F F, the whole being substantially as and so as to operate as specified.

No. 55,404.—JONATHAN WHEELER, Athol, Mass.—*Knife Polisher and Knife and Scissors Sharpener.*—June 5, 1866—Two polishing disks and several serrate guiding disks are associated by a compresser spring on a revolving mandrel.

Claim—An improved combined knife polisher and knife and scissors sharpener, formed by combining the emery-wheels F and G and tempered steel-cutters H I and J with each other, and with the shaft D and spring N, the whole being constructed and arranged substantially as described and for the purpose set forth.

No. 55,405.—NORMAN W. WHEELER, Brooklyn, N. Y.—*Steam Generator.*—June 5, 1866.—The water space below the furnace is connected with that above by a central tube passing through the furnace, being connected to the floor of the ash-pit and the tube-sheet of the boiler. The circular series of man-holes around the lower portion of the boiler gives access to the flues and tube-sheet for cleansing.

Claim.—First, the frames *c c c* when arranged around and secured to the shell of a vertical tubular boiler, substantially as and for the purposes described.

Second, the circulating pipe *g*, in combination with the water bottom *h*, and the crown-sheet of the furnace D D, when such crown-sheet is also a flue-sheet for the tubes B B.

No. 55,406.—NORMAN W. WHEELER, Brooklyn, N. Y.—*Packing Slide Valves for Steam Engines.*—June 5, 1866.—Steam is introduced by ports beneath the segments of packing on the periphery of the piston valve; it enters at one portion of its stroke, and acts expansively during the other portions.

Claim.—The ports *i i*, in combination with packing rings or segments thereof expanded by their own elasticity or by special springs, substantially as and for the purposes described.

No. 55,407.—H. WHISLER, New Market, Ohio.—*Churn*—June 5, 1866.—The dashers are rotated in contrary directions by the master-wheel and their respective pinions.

Claim.—The combination of the wheels D and *m*, the ears *s* and the dashers E, the whole constructed and operating in the manner and for the purpose herein set forth.

No. 55,408.—PETER A. WISE, Stockbridge, N. Y.—*Horse Hay Fork.*—June 5, 1866; antedated December 5, 1865.—The shank is attached to the fork head and passed into a slot in the handle to which it is pivoted, and held in working position by a spring catch. A brace extends from the bail to the handle, and is pivoted to both.

Claim.—First, the metallic shank *a*, attached to the head of the fork and passing into the slotted end of the handle *c*, to which it is attached by the cross-bolt *b*, in combination with the latch *d*, as and for the purposes specified.

Second, the combination of the fork hinged to the handle, a suspending bail, and a brace extending from the handle to the bail, substantially as specified.

No. 55,409.—JOEL WISNER, Aurora, N. Y.—*Broom.*—June 5, 1866.—The butts of the corn brush are bent across an inside bar, which is drawn tight to the inside of the cap by two thumb-screws, which pass through the upper portion of the cap.

Claim.—The employment of the adjusting bar E, in combination with the head-block B, substantially in the manner and for the purpose described.

Also, suspending the holding plate E by the screws D, in combination with the fixed head-block, as shown and described.

No. 55,410.—H. T. WOODMAN, Dubuque, Iowa.—*Sash and Door Bolt.*—June 5, 1866.—The bolt is reciprocated in its slotted sleeve by a crank, turned by a key and connected by a rod with a projection on the bolt.

Claim.—The combination of the bolt A, slotted sleeve or casing B, connecting link or rod D and crank-arm E of shaft or spindle F, suitably formed for receiving a handle-knob, key, or other equivalent device, when arranged together so as to operate substantially in the manner described and for the purpose specified.

No. 55,411.—E. R. BIGELOW, of Boston, and CHARLES H. WATERS, Groton, Mass., assignors to CLINTON WIRE CLOTH COMPANY.—*Machinery for Making Wire Sieves.*—June 5, 1866.—The devices shown and described are for cutting the bottoms of sieves out of wire-cloth, for forming them, and for attaching them to their hoops.

Claim.—First, the cutting apparatus, constructed and operated substantially as described.

Second, the forming apparatus, constructed and operated substantially as described.

Third, the expanding fastener, constructed and operated substantially as described.

Fourth, the method of manufacturing wire-cloth sieves by the successive use of the cutting apparatus, cutting the sieve-bottoms diagonally across the web of the cloth, the forming apparatus for turning up the edges of the sieve-bottom, and the expanding fastener for distending the sieve-bottoms, substantially as described.

No. 55,412.—L. L. CRANE, Cleveland, Ohio, assignor to LEAVITT CRANE & Co., same place.—*Lathe*—June 5, 1866.—The live and dead centres are arranged upon ways not parallel to the ways on which the rest traverses, so that a regular taper may be turned.

Claim.—The ways *m m*, when arranged in the same plane, but in an angular or inclined position in relation to the ways *n n* and centres *g e*, the ways *n n* being in line or parallel to each other and the centres *g e*, all combined and operating conjointly in the manner and for the purpose set forth.

No. 55,413.—JOHN M. ENOS, St. Joseph, Mich., assignor to himself and FRANKLIN SMEAD, same place.—*Mop and Wringer.*—June 5, 1866.—A bent wire is passed through tho mop and secured to the handle; to use the wringer, detach the wire and turn it in the hand, twisting the mop rag.

Claim.—The combination and arrangement of the wringer C, when constructed substantially as shown with the mop M, mop-head A. and handle B, provided with a pin *a*, operating substantially as specified, and for the purposes set forth.

No. 55,414.—CALVIN A. FOSTER, Winchendon, Mass., assignor to A. F. SPAULDING and S. M. SCOTT, same place.—*Meat Chopper.*—June 5, 1866.—The knives have a draw-cut, imparted by bent levers and crank shaft; the knife frame, together with the scrapers to push the meat to the middle of the tub, are pivoted on a hinge in the rear of the crank shaft; the tub is rotated by a bent lever and a pawl beneath.

Claim.—The combination of the raising arm I with either or both the knives and the operative mechanism and supporting frame thereof.

Also, the combination as well as the arrangement of the post *g* and its turn-button *h*, or the equivalent thereof, with the frame A and the raising arm I, and one or more knives G G, and mechanism to operate such knife or knives, substantially as described.

Also, the operative mechanism of each knife G, the same consisting of the knife-carrier H, the vibratory arm *e* and the rotary crank *d*, arranged substantially as specified.

Also, the application of the tub-cover to the raising arm I, so as to be lifted off the tub thereby when such arm is in the act of being elevated, substantially as set forth.

Also, the application of the plough K to the raising arm I, so as to be lifted out of the tub thereby when such arm is in the act of being raised, as explained.

Also, the mechanism for imparting to the tub an intermittent rotative motion and enabling the tub to be separated from the cross of such mechanism, the same consisting of the stud *z*, the cross or their equivalents, the internal ratchet *r*, the pawl *s*, the bent lever *u*, the connector *v²*, and the arm *w*, projecting from the carrier, as set forth.

No. 55,415.—MARTIN FREE, Philadelphia, Penn., assignor to ALFRED LOUDERBACK, same place.—*Planing Machine for Cutting Slats for Blinds.*—June 5, 1866.—The planer has a reciprocating knife; the knife carriage on its return strikes a lever connected to a pawl that operates gear wheels to feed down the block of wood for the next cut.

Claim—The reciprocating carriage C D, carrying the cutters *c′ d′*, and operating automatically the devices which feed the block E to the same as described, the said feeding devices consisting of the lever G, pawl H, springs I I, stud wheel J, screw-shaft *f′ f² f³*, and gear-wheels F F F, constructed and arranged substantially as described.

No. 55,416.—JOHN FYE, Hamilton, Ohio, assignor to himself and JOHN F. SUTHERLAND, same place.—*Mail Bag.*—June 5, 1866.—The bag has a water-proof lining with a weight attached, and a round mouth with a lid closing on it like a box cover.

Claim.—The construction of the mouth of the bag as shown at F, in combination with the cover as shown in Fig. 2, and the weight at the bottom of the lining D, for the purposes set forth.

No. 55,417.—WILLIAM H. HAWKINS, Rochester, N. Y., assignor to himself and JAMES D. ORNE, same place.—*Spool-thread Regulator for Sewing Machines.*—June 5, 1866; antedated May 30, 1866.—The described device prevents the slipping of the thread from the spool and its becoming wound upon the spindle.

Claim.—The adjustable thread-regulator herein described, the same consisting of the slotted arm C D *b*, sleeve *a*, sliding arm E, and plate F, arranged to operate in the manner and for the purpose specified.

No. 55,418.—H. L. JONES and D. S. FARQUHARSON, Rochester, N. Y., assignors to themselves, ALBERT M. HASTINGS, and ALEXANDER McVEAN, same place.—*Treating Wood, Straw, &c., for the Manufacture of Paper Pulp.*—June 5, 1866.—The wood is placed in a revolving cylinder with hollow trunnions, through which an alkaline solution is passed under pressure.

Claim.—First, the subduing of straw, wood, or any fibrous material to be converted into paper pulp, by subjecting the same to the action of alkali liquor of any desirable temperature, applied under the hydrostatic pressure of the liquid itself applied by a force-pump or otherwise, instead of using steam pressure, preparatory to the bleaching of such material in the ordinary methods, substantially as above described.

Second, the combination with the cylinder A, of the pump D, and pipe B, substantially as and for the purposes above set forth.

Third, the safety-valve K, in combination with the pump D, below the piston or plunger and in direct communication with the pump-barrel, substantially as above described.

No. 55,419.—WERNER KROEGER, Milwaukee, Wis., assignor to himself and CONSTANTINE RIES, same place.—*Stove Pipe Drum.*—June 5, 1866.—The devices described and shown deflect the air into a reverting sinuous course.

Claim.—A heat-radiator composed of three concentric cylinders *a b b′ c*, in combination with two horizontal partitions *g g′* and dampers *h h′ i*, all constructed and operating substantially in the manner and for the purpose herein set forth.

No. 55,420.—M. T. LAMB, Valparaiso, Ind., assignor to himself and ISAAC W. LAMB, same place.—*Knitting Machine.* —June 5, 1866.—Improvement on I. W. Lamb's patent of October 10, 1865. The dial indicates the number of rows knitted, and thereby informs the operator when to make the changes; as from the foot to the heel, and from the heel to the leg of a stocking &c. The changer wire can be inserted and removed readily, and enables the operator to shift the cams by which the tubular knitting may be changed into knitting with one side of the goods left open or disunited.

Claim.—First, attaching the counter to the sliding frame in such a manner that the movement of the sliding frame will carry the teeth of the dial against a stationary pawl or ratchet, substantially as and for the purpose herein described.

Second, the changer G, consisting of a bent wire, or its equivalent, attached at its ends to the cam-shifters p, and so placed that the operator can operate it with one hand while he turns the machine with the other hand, substantially as described.

Third, constructing and applying the changer G to the cam shifters p of a knitting machine, in such a way that it can be removed at pleasure without alteration in the machine, substantially as shown.

No. 55,421.—WILLIAM R. LANDFEAR, Hartford, Conn., assignor to himself and DAVID WHITTEMORE, same place.—*Making Eyelets.*—June 5, 1866.—The eyelet is placed within a die in the lever and its projecting edge cut off by the knife under which it passes.

Claim.—The combination and arrangement of the lever, or its equivalent, the socket and knife, or the same and the supporting plate, the same being to operate substantially in manner and for the purpose as set forth.

No. 55,422.—JOHN LESSELS, Troy, N. Y., assignor to CLARK TOMPKINS, same place.— *Taking-up Mechanism for Circular Knitting Machines* —June 5, 1866.—Improvement on Brockway's patent of November 8, 1864. The screw on the sliding shaft tends to force it upwards, and also to keep a stop lever out of contact with a stop connected with the pawl. But whenever the fabric offers enough resistance to being taken up to overcome the force of the spring, the screw, still revolving within the teeth of the cog wheel, which now refuses to revolve, causes a downward movement of the shaft, which consequently operates the lever so that its outer end arrests the movement of the stud, and locks the lever in such a position that the cam cannot move it until the slack of the fabric again allows the shaft to rise, the stop to be released, and the lever and its pawl and ratchet connection again to be brought into action by force of the spring.

Claim.—The combination in the take-up mechanism of a knitting machine of the following instrumentalities, viz., the take-up roller, screw-shaft, (having an endwise movement,) vibrating pawl, and movable stop operated by the screw-shaft, all operating substantially as set forth.

No 55,423.—JOHN LIPPINCOTT, Pittsburg, Penn., assignor to himself and THOMAS BAKEWELL.—*Saw.*—June 5, 1866.—The tapering slot in the saw plate has grooves in the sides, and a wider opening at the bottom into which the shoulders of the split spring tooth catch when it is forced into place.

Claim.—First, the use of tapering saw teeth, or teeth points inserted into correspondingly shaped slots in the saw plate, having their sides flush with the side of the saw plate, and secured from lateral displacement therein, without riveting or upseting, by the groove in the edges of the slot in the saw plate and correspondingly shaped edges of the tooth, (or groove in the edges of the tooth, and bevelled edges of the slot,) substantially as hereinbefore described.

Second, the use of saw teeth, or teeth points bifurcated at the rear end, and with or without a head or projection at the extremity of the prongs, for insertion into the plates of circular or long saws, substantially as and for the purposes hereinbefore set forth.

Third, the tapering slot in the saw plate for the insertion of the removable teeth, with an enlargement or opening at the rear end, for the purposes hereinbefore set forth.

No. 55,424.—AARON LLOYD, Mattoon, Ill., assignor to FRANCIS HAMBLIN.—*Floor Clamps.*—June 5, 1866.—The clamp straddles the joist; is retained by the serrated cams, and be forward thrust of the lever is maintained by a pivoted brace, which engages the joist behind it.

Claim.—Hinging the prop G to the handle A, when used for the purpose of a floor clamp, constructed and operating in the manner as herein described and represented.

No. 55,425.—ELI J. MANVILLE, Waterbury, Conn., assignor to himself and SIDNEY L. CLARK, Torrington, Conn.—*Operating Dies for Forming Articles of Metal.*—June 5, 1866.— Pairs of converging dies are arranged in the end of a mandrel, and their inner faces are approached by the contact of the adjustable revolving tappets which press upon their bevelled outer ends; as each pair is pressed in, it forces the other out by the contact of the inclined faces.

Claim.—First, two or more dies fitted to move radially in a stock, in combination with an adjustable tappet or tappets revolving around the said stock and acting directly upon the outer ends or edges of said dies substantially as se forth

Second, pairs of radial dies, formed with their contiguous edges bevelled or at an inclination, as specified.

No. 55,426 —C. L. MOREHOUSE, Cleveland, Ohio assignor to himself and J. B. MERRIAM, same place.—*Process for Preparing Stuffing for Currying* —June 5, 1866.—Crude petroleum is steamed: pumped into a vat; treated with sulphuric acid while agitated and cooled by an air blast; settled; decanted: washed with hot water, and afterwards with an alkaline solution. and the paraffine separated by a cold blast.

Claim.—First, the above described mode or process of clarifying paraffine oil by the use of a blast of air in jets for agitating the oil while treating the same with a large proportion of sulphuric acid thus checking the excess of chemical heat, substantially as set forth.

Second. the use of hot water in washing the oil, substantially in the manner and for the purposes set forth

Third. the use of a blast of air from the ice chamber for crystallizing the paraffine, substantially as set forth.

No 55,427 —DAVID R PAISTE, Willistown, Penn., assignor to REESE, LAKE, MELICK & Co.—*Harvester.*—June 5, 1866. —The described devices are for connecting the hinged finger bar and platform with the main frame.

Claim.- First, the finger bar F, with its bar J, in combination with the brace K' and the grooved pulley a secured to the frame of the machine, the whole being arranged and operating substantially as and for the purpose described.

Second, the combination with the above of the adjustable lever F, substantially as and for the purpose herein set forth.

No. 55,428 —S. H. ROPER, Roxbury, Mass, assignor to ELMER TOWNSEND.—*Hot Air Engines* —June 5, 1866.— The auxiliary air passage communicates with that leading from the pump to the furnace, and has a safety valve which opens to relieve pressure and allows air to enter and cool the induction valve.

Claim.- Providing the auxiliary air passage j with the safety valve k, when arranged to operate substantially as described, and in combination with the inlet pipe e, and its throttle valve g.

No. 55,429 —JAMES SANGSTER and MILTON BOYD, Buffalo, N. Y., assignors to JOSEPH B LICHTENSTEIN, same place.—*Mop.*—June 5, 1866.—The endless mop cloth is held between the rollers and is wrung by rotating one roller in contact with the other, the latter being adjustable and depressed by a spring.

Claim.—First, the combination with said rollers F and G of the guide plates, or their equivalents, when constructed as and for the purposes herein substantially described.

Second, in combination therewith the thumb piece D, spring E, and bar H, as described.

No. 55,430 —THOMAS SHELDON, Enfield, Conn., assignor to himself, HENRY INMAN, Portland, and E. P. FURLONG, Westbrook, Maine.—*Lifting Jack.*—June 5, 1866.—The lever is suspended from a link and has a lifting socket which slips on the shank of the lifter, the retaining socket slipping on the standard.

Claim.—A lifting jack as herein described, and combining the various parts in the manner and for the purposes set forth.

No. 55,431.—WILLIAM H. SHURTLEFF, Providence, R. I., assignor to himself and HENRY A. CHURCH.—*Combined Hook and Button.*—June 5, 1866.—The button forms the head of the hook; the shank has flanges which pass through the object, and are clinched.

Claim.—The combination of a button with a shank secured at the periphery and bent under and down at a point diametrically opposite of the button to constitute the lacing stay, substantially as herein shown and set forth.

No. 55,432.—JOHN WATSON, Buffalo, N. Y., assignor to ORAN W. SEELY, same place.— *Brick Machine.*—June 5, 1866.—The slot of the rod rests upon a pin while the rod is reciprocated by the engagement of a stud with an irregular cam groove in the face of a wheel.

Claim.—The combination of the slotted rod D with the cam I, when used to give an irregular reciprocating motion to the moulds of a brick machine, for the purposes and substantially as herein described.

No. 55,433.—S. H. WHEELER, Dowagiac, Michigan, assignor to himself, RICHARD HEDDEN, JAMES THOMPSON and RONERT R. THOMPSON.—*Measuring Pump.*—June 5, 1866.— A ratchet with a train of wheels and a dial is attached to the pump stock, and is actuated by an adjustable cone on the valve stem, so as to register the pulsations of the piston in the cylinder of known capacity.

Claim.—First, the enlarged chamber C upon the upper end of the pump barrel, having a check valve *a* in it, as arranged in relation to the valve *c*, on the rod D, and caged piston *c'*, substantially as described

Second, providing the upper end of the piston rod D with a cone or circular wedge F for actuating the registering devices at every descent of the piston, substantially as described.

Third, the vibrating lever *i*, spring *j*. pawl *h*, and arm *k*. in combination with the ratchet wheel *g'*, disk *g²*, and a circular wedge F, substantially as described.

Fourth, in a pump having a registering device applied to it providing for regulating the length of the piston rod, by means substantially such as described.

Fifth, in a pump having the registering device applied to it, the use of a valve *e c'* applied to the piston rod, substantially as described.

No 55,434.—JAMES G WILSON, New York, N. Y., assignor by mesne assignment to UNION SEAMLESS KNITTING MACHINE COMPANY.—*Knitting Machine.*—June 5, 1866.— Designed for knitting stockings complete; a studded pattern wheel making but a single revolution to effect all the necessary changes. This wheel has no movement in a direction parallel with its axis. The needles are bearded and move longitudinally, while the whole system of needles, together with the needle-ring or cylinder, revolves in either direction and for any determined distance, or for any number of revolutions. Several yarns may be used simultaneously, one for each of several immediately adjacent needles.

Claim.—First, the employment for the purpose of controlling the direction and changes of direction of the rotary motion of the needle ring or needle bar of a knitting machine, of a studded wheel or drum, so constructed, applied and operating that it performs its duty without any movement in a direction parallel with its axis, as herein described.

Second, so constructing and operating the aforesaid wheel or drum, having no movement in a direction parallel with its axis, that it will make but one revolution during the operation of making a complete stocking or other knitted article, as herein specified.

Third, combining the aforesaid studded wheel or drum having no movement parallel with its axis with a slide H and pawls *h h'*, or their equivalents, which produce the rotary movement of the needle ring of a circular knitting machine by means of a three-armed lever K L L', constructed, applied and operating substantially as herein specified.

Fourth, giving the studded wheel or drum a compound rotary motion, viz., a slow motion to bring its studs to an operative position, and a quicker motion to produce the action of its studs upon the needle ring or needle bar, substantially as herein described.

Fifth, the employment in a circular or straight knitting machine of the specified kind herein described of a system of needles having longitudinal and lateral movements alternately, and operating substantially as herein specified.

Sixth, the combination with a system of needles having a longitudinal reciprocating motion, when used in a machine constructed specifically as herein described, of a system of yarn conductors so applied as to deliver yarn to immediately adjacent or consecutive needles, substantially as herein specified.

Seventh, the plate F, either fitting between two projections *c c'* on the needles, as represented in the drawing, or, what is equivalent, grooved to receive single projections on the needles, and operating substantially as and for the purpose herein specified.

Eighth, the vibrating presser applied and operating in combination with a laterally and longitudinally moving series of needles, substantially as herein specified.

Ninth, combining the yarn guides and presser substantially as herein specified, to be operated by the same mechanism.

Tenth, the combination of a system of needles and a system of yarn guides operating automatically in conjunction with a yielding compressor and bridge for carrying and removing the needles without the use of stitch books for taking off the stitches, in such manner as to knit simultaneously on two or more immediately adjacent or consecutive needles, substantially as herein described.

Eleventh, the combination of the studded wheel or drum M, having no movement parallel with its axis, and a series of needles and yarn conductors so applied as to effect the knitting simultaneously with separate yarns on immediately adjacent needles.

No. 55,435.—SILAS H. WILSON, Auburn, N. Y., assignor to WILLIAM H. BROWN, same place.—*Pitman Head for Harvesters, &c.*—June 5, 1866.—The wrist revolves in the divided box, which is supported on a screw at the end of the pitman, and a pivot from the opposite side of the oil cup; a determinate freedom of motion is allowed in a plane coincident with its points of suspension and also transverse thereto.

Claim.—So combining and uniting the crank wrist and pitman of a harvesting or other machine, through a box and oil cup, as to admit of a triplicate motion between the wrist and pitman, to prevent all binding or cramping, as also clatter between the parts, substantially as described.

No. 55,436.—GEORGE L. WITSIL, Philadelphia, Penn., assignor to himself and JOHN F. CABOT, Elizabeth, N. J.—*Lamp Chimney.*—June 5, 1866.—The chimney is corrugated by a spiral depression.

Claim.—A lamp chimney which has the winding or spiral groove formed in it, substantially as described.

No. 55,437.—HENRY C. WOODING, Wallingford, Conn., assignor to himself and L. W. TURNER, Galesville, Conn.—*Attaching Axes to their Handles.*—June 5, 1866 —The metal wedge which expands the halve in the tapered eye of the tool is maintained in its position by a longitudinal screw.

Claim.—The herein described wedge provided with a head and constructed so as to be secured, substantially in the manner specified.

No. 55,438.—HOWARD BUSBY FOX, Oxton, England.—*Bottle Stopper.*—June 5, 1866.— The cap is lined with an elastic material which will accommodate itself to the threads on the neck of the bottle.

Claim.—First, the within described soft-lined cap *b c d*, adapted to close the mouths of bottles or vessels having a screw-thread formed on the outer surfaces of said mouths by screwing thereon and to fit tightly not only upon the edge, but also along the surfaces of the screw-threads, substantially as and for the purpose herein specified.

Second, in connection with the above, the divided or serrated edge *e*, of the rigid portion of the cap *b*, to facilitate the bending inward thereof to confine the soft lining, substantially as herein specified.

No. 55,439.—JAMES GOODIER, Chester, ENGLAND, and J. F. KILSHAW, New Brighton, England.—*Paddle Wheel.*—June 5, 1866 —The paddles are feathered by the vibration of the radial arms to which they are attached ; cranks on the radial arms are connected to the cam on the hub.

Claim.—The cam or piece *b*, with groove *n*, elbow pieces *l*, links K, and gabs *j*, arranged and operated substantially as herein specified.

No. 55,440.—RICHARD HORNSBY, Grantham, England.—*Reaping Machines.*—June 5, 1866.—Diagonally arranged endless chains or bands provided with teeth for removing the grain are employed, in connection with an adjustable or tilting slotted platform, which is held up in an inclined position to receive the grain as it is cut, until a sufficient quantity has accumulated thereon to form a gavel, when it is dropped so as to allow the teeth on said bands to pass through the slots therein and remove the grain.

Claim.—The combination of diagonal chains or bands for removing the cut grain, with a movable platform provided with slots or openings, as described.

No. 55,441.—GEORGE LIONEL LECLANCHÉ, Paris, France.—*Galvanic Battery.*—June 5, 1866.—A plate of copper with a wire attached is covered in the jar with powdered carbonate of copper ; a superstratum of sand has a plate of zinc with the negative pole wire attached. The whole is saturated with a liquid such as chlorohydrate of ammonia.

Claim.—The use in electrical piles of insoluble or slightly soluble salts of copper or other equivalent material moistened with a liquid containing a salt in solution capable by its decomposition of rendering the said salts of copper or other equivalent material soluble, substantially as described.

No. 55,442.—ANTON LOHAGE, Unna, Westphalia, Prussia.—*Casting Steel.*—June 5, 1866.—The furnaces for superheating are so placed that by means of a crane the melted steel may be poured directly from the converting vessel into the intermediate furnace, where it is covered with slag, and tempered with cast iron of different degrees of carbonization if required.

Claim.—First, the process hereinbefore described, and called an intermediate process, whereby the melting process is continued till the mass is "overmelted" or "superheated," made uniform and ready for casting.

Second, the modes hereinbefore described of altering the temper of the molten steel in the intermediate process.

No. 55,443.—OLIVER SARONY, Scarborough, England.—*Photographic Rest.*—June 5, 1866.—The described devices afford adjustment for supporting the human figure in its various parts and positions.

Claim.—The combined arrangement of an upright sliding bar or bars *b*, (in a suitable stand,) the plate *g*, the part *h*, with means of receiving a curved slide *i*, the slide *n*, and the stems or bars *k* and *o*, substantially as herein described, and combined therewith the supports for the body and head, as described.

No. 55,444.—JOHN THORNTON and WILLIAM THORNTON, Pease Hill Rise, Nottingham, England.—*Knitting Machine.*—June 5, 1866.—The latch needle slides in and out in grooves for the purpose of taking and discharging its loops. Each needle is caused to make a semi-rotation on its axis by means of a screw thread on its stem. When its hook is turned upward it receives thread from one thread carrier, and when turned downward it is supplied from another carrier. A course is knitted with the needle in one of these positions, and another course while they are in the other position ; the fabric being so guided as to allow the needles to play on either side of it as they are knitting with one or the other of the threads.

Claim.—The combination of the looping instruments with the mechanism for operat'ng them and with the thread guides, in such manner that said instruments are caused to make a partial rotation upon their axes, and have the thread laid and loops formed on them in the two positions occupied by them and at opposite sides of the fabric, substantially as set forth.

No. 55,445.—VALENTINE WARD, San Francisco, Cal,—*Hitching Post.*—June 5, 1866.— The bar has a ring by which it is elevated so as to project above the hollow post to hold the bridle or hitching strap. A spring catch supports it.

Claim.—First, the post or shell A, sunk underneath a sidewalk or pavement, having an extension shaft or bar P. to be drawn from or inserted in said post or shell at will, substantially as described and for the purpose set forth.

Second, the spring E, or its equivalent, when arranged as above described, or by weights and pulleys, when a larger post is desired, substantially as described and for the purpose set forth.

No. 55,446.—DAVID S. ABBOTT, Ischua, N. Y.—*Flax-dressing Machine.*—June 12, 1866.— The curved form of the swords tends to keep the flax in the middle of the bench; the forward feed roller is removable for cleansing, and the flax is held by a clamp while being scutched.

Claim—First, making the blades of the beaters of flax-dressing machines with a curve that rises towards their ends, substantially as described.

Second, making one of the boxes for the journals of the apron roller S movable, so that that roller can be taken out and cleaned at pleasure, substantially as described.

Third, the clamp H, constructed substantially as described, applied to the feed table of a flax-dressing machine.

No. 55,447.—BOYD ALLEN and JOHN RIDDELL, Boston, Mass.—*Portable Gas Stand.*— June 12, 1866.—The sliding telescopic stem is vertically adjustable by rack and pinion, and has a stuffing box-joint.

Claim.—The sliding portable gas stand, constructed and operating substantially as described.

Also, the combination of the stuffing-box with the inner end of the sliding tube, as and for the purpose specified.

No. 55,448.—HORATIO ALLEN, New York, N. Y.—*Seat and Couch for Railroad Cars.*—June 12, 1866.—The separate seats are arranged in pairs diagonally on each side of the aisle; by raising certain parts and depressing others a couch is formed for one passenger, and the other is provided for by an upper diagonal couch suspended from the ceiling.

Claim.—First, the combination with the floor and sides of a railroad passenger car of couches of a rhomboid form, placed diagonally to the length of the car, as herein described, and constructed of two seat pieces A A', the corner pieces E E', and two central pieces F F', and supported by frames and legs, said couches being convertible into a pair of seats, by putting out of the way the two corner pieces E E', and securing in a vertical position the two centre pieces F F' by the cap piece G, said pair of seats having a relative position diagonal to the length of the car, all substantially in the manner and for the purpose herein described.

Second, the combination with the seats herein described, and sides of a railroad passenger car, of upper couches of rhomboidal form placed diagonally to the length of the car, as herein described, and constructed of a frame supported as herein described, all substantially in the manner and for the purpose herein described.

No. 55,449.—JOHN S. ANDERS, North Wales, Penn.—*Medicine.*—June 12, 1866.—This prophylactic for veterinary uses is composed of saltpetre, 2; antimony, 2; brimstone, 2; ginger, 4; cream of tartar, 1; fenugreek, 2; black brimstone, 2; alum, 4 parts.

Claim.—A cattle powder, made of the ingredients herein specified, and mixed substantially as set forth.

No. 55,450.—H. C. APPLEBY, Conneaut, Ohio.—*Lamp Chimney.*—June 12, 1866.—Explained by the claim.

Claim.—A spirally corrugated or fluted glass lamp chimney, constructed substantially as shown and described.

No. 55,451.—JAMES C. ARMS, Northampton, Mass.—*Paper Bosom and Collar.*—June 12, 1866.—The collar is attached to the bosom without intermediate band by means that determine its angular connection and conform it to the neck of the wearer.

Claim.—A new article of manufacture, consisting of a paper bosom and collar combined, when constructed as herein shown and described.

No. 55,452.—WILLIAM M. ARNOLD, New York, N. Y.—*Composition of Iron and other Metals.*—June 12, 1866.—The ladle which receives the melted cast iron from the cupola has 1 pound of carbonate of soda to 100 pounds of iron, and an alloy is added consisting of copper, 1 pound; tin, ¼ pound; zinc, 5 pounds; antimony, ½ pound.

Claim.—The composition produced by the mixture of the ingredients above described, when made substantially of the proportions and in the manner herein contemplated and set forth.

Also, the preparation of an alloy adapted and intended for use in the manufacture of my final composition, and which alloy is composed of copper, tin, zinc, and antimony, in the proportions above contemplated and set forth.

No. 55,453.—D. L. BABCOCK, St Charles, Minn.—*Bolster for Wagons.*—June 12, 1866.—The metallic caps on the ends of the bolsters form sockets for the standards.

Claim.—The metal cap B, constructed substantially as shown, and applied to the bolster of ordinary wagons and similar vehicles, as and for the purpose set forth.

No 55,454.—JACOB B. BAILEY, New York. N. Y.—*Curtain Fixture.*—June 12, 1866.—The grooved metallic cord ring slips on the roller, and is kept from turning by teeth; the roller rests at one end in a semi-cylindrical socket, and its other end is split and turns in a round socket.

Claim.—First, the cord ring *e*, formed with a groove around its periphery for an endless cord, an opening through it for the curtain roller to pass entirely through to its bearing, and with teeth to penetrate said roller, as and for the purposes set forth.

Second, the combination of the semi-circular shoe or bracket *h* with the cord ring *e* and roller *b*, for the purposes set forth.

Third, the friction spring for the curtain roller formed by slitting the end of the said roller and introducing it within a ring or bracket, as set forth.

No. 55,455.—HARRY T. BARKER, Napa, Cal.—*Fastening for Fruit Boxes.*—June 12, 1866.—The slotted end of the lid is held by a headed catch, and the other by a bolt and screw nut. The catch and bolt are immovably attached to the box and project above the lid.

Claim.—The arrangement shown and described, consisting of the headed catch at the slotted end of the lid and the threaded bolt at the other end, upon which the lid is screwed by the nut.

No. 55,456.—JOSEPH BRADT, Avon, N. Y., and JAMES HAYES Rochester, N. Y.—*Railroad Switch*—June 12, 1866; antedated March 6, 1866.—The switch rail is lifted by a tread-lever and dumped into such one of the cells in the ribbed chair as may be required, and is there laterally braced.

Claim.—The combination of the lifting tread-lever I with the ribbed chair or chairs H, having seats *d d d*, which correspond in number and position with the several diverging rails B C D, when arranged in connection with the ordinary switch-lever E and connecting-rod G, substantially as and for the purpose herein specified.

No. 55,457.—ALFRED BRIDGES, Newton, Mass —*Suspending Cars on Springs.*—June 12, 1866.—The strap below the pedestals has suspension bolts whose head plates rest upon the cylindrical springs; the latter upon flanges of the axle-box.

Claim.—First, the guide-pieces *e e*, combined and arranged relatively to the supporting bar E, box C, axle *b*, springs D¹ D², and suspension rods G¹ G², or their respective equivalents, substantially as and for the purpose herein set forth.

Second, the jaw or frame I, bar H, suspension rods G¹ G², supporting bar E, springs D¹ D², and the axle-box C, or their respective equivalents, combined substantially in the manner and for the purposes herein set forth.

Third, the method of locking the jaw or frame I, upon the removable bar H, by flanges *h'*, or their equivalents, so as to resist horizontal strains without throwing such strains on the bolts J J, or their equivalents, substantially as and for the purposes herein set forth.

Fourth, the arrangement of the shelves C¹ C², which support the springs D¹ D², so as to swing near, but not in contact with, the upright inner faces *h* of the enclosing frame, substantially as and for the purpose herein forth.

No. 55,458.—ALDEN BRIGHAM, Coldbrook, Mass.—*Dust Range or Receptacle.*—June 12, 1866.—This dust box is to receive the sweepings of a room; is inserted in the floor; has a register top, a sliding bottom, and is removable.

Claim.—A dust range and receptacle as a new article of manufacture, the same consisting of a box with top flanges, whereby it is held in the floor, and sliding bottom and register-cover, all the parts being constructed and arranged for use as set forth.

No 55,459.—JESSE BRINGHURST, Philadelphia, Penn.—*Artificial Leg.*—June 12, 1866; antedated June 1, 1866.—The ankle joint is formed by a metallic swivel-box resting in bearings and upon an elastic pad, and receiving a coupling bolt by which the leg and foot are united.

Claim.—First, the bolts C, as described, and for the purpose set forth.

Second, the swivel-box E, in combination with the bolt C and rubber packing H, as specified and for the purpose set forth.

No. 55,460.—B. V. M. BROUSE, Kokomo, Ind.—*Lock.*—June 12, 1866.—The bolt may operate as a latch or lock bolt. The detents are removed by the studs on the key, whose point removes the obstacle at the entrance. The conversion from a latch to a lock is by a trigger tumbler, which allows the bolt to come within the action of the detents.

Claim.—First, the combination of the pointed key K and the spring obstacle plate N, substantially as described.

Second, the combination of the toothed key K with the spring tumblers H H', spring bolt with notches *i i*, arranged and operated substantially as described and represented.

Third, the combination of the notched spring-bolt with notch *d*, and the spring tumbler F *f*, arranged and operating substantially as described and represented.

No. 55,461.—JOHN D. BROWNE, Cincinnati, Ohio.—*Centrifugal Machine.*—June 12, 1866.—The screen and spiral distributor are driven at different velocities.

Claim.—In the construction of a centrifugal sugar separating machine is the separate or variable motion of the screen and distributor, as herein substantially described.

Also, the oblique or spiral deflector as herein described and for the purpose set forth.

No. 55,462.—DANIEL W. BURBANK, New York, N. Y.—*Bed Bottom.*—June 12, 1866 — The end of the slat slips into a loop of a double helical spring, which is fastened by books to the bed rail: the spring is coiled around a double-headed bolt.

Claim.—First, the combination of a double helical spring C C with the double-headed mandrel O O, and the arm E E, when the coils of the spring are wound apart so that no part touches another part, substantially as specified.

Second, the enlarged loop D, in combination with the slat B and mandrel O, as specified.

Third, the double hook N N, in combination with the spring and rail, as specified.

Fourth, the corner lock L L K, in combination with the frame slats and springs, substantially as set forth.

No. 55,463.—JOHN BURGUM, Concord, N. H.—*Device for Keeping Meat under Brine.*—June 12, 1866 —Dogs are hinged to the disk and engage with the sides of the barrel to keep the cover down. They are attached by links to a sleeve on the central shaft, which raises them simultaneously.

Claim.—First, the platform or disk A, provided with dogs B, or their equivalent, constructed and arranged so as to operate substantially in the manner specified.

Second, the combination of the dogs B, platform A, collar C, and connecting link *b* and *b'*, or their equivalent, substantially as herein specified.

No. 55,464.—W. H. BUTLER and R. G. HATFIELD, New York, N. Y.—*Starting Cars.*—June 12, 1866.—The body has a limited movement upon the trucks, and compresses the springs when the brake is applied; the reaction of the spring when the brake is released assists in rotating the wheel.

Claim.—First, the combination and arrangement of the body of a vehicle, loose and portable, upon the truck which supports said body, and movable thereupon by means of rollers, roller-sheaves, wheels, pulleys, or their equivalents, substantially as and for the purpose herein described.

Second, the combination of the frame A with the guide *d*, springs F, flanges E, small wheels C, and body B, substantially as described and for the purpose set forth.

No. 55,465.—O. B. BUTTLES, Milwaukee, Wis.—*Dental Impression Cup.*—June 12, 1866.—An inflected flange runs around the upper edge of the cup.

Claim.—Combining with a dental impression cup an inner rim or flange *a*, for preventing the wax or other impressed material from moving or slipping in the cup as the latter is being removed from the mouth, substantially as described.

No. 55,466.—GARDNER CHILSON, Boston, Mass.—*Cooking Store.*—June 12, 1866.—The stove has rear-corner diving flues and bottom reverting flues, a registered connection between the oven and the exit flue, and a side opening in the fire chamber for the grate-shaking lever or for air.

Claim.—The arrangement of the separate heat-saving plate within each of the oven flues and case of the stove, so as not only to perform its function of saving heat, as described, but to gradually diminish the smoke passage through the flue, in manner as specified.

Also, the combination and arrangement of the arched chamber G with the oven, the vertical flues at the back of the oven, and with the three flues arranged beneath and against the oven, in manner as explained.

Also, the combination and arrangement of the bay A, and its side opening and its closing slide, with the stove body and with the grate journal, as described.

No. 55,467.—MORRLL CLARK, Castalia, Iowa.—*Reaping and Mowing Machine.*—June 12, 1866.—The forked finger bar is journalled to the drag bar, whose forward end is bifurcated, and has bearings on the revolving axle. The master wheel on the latter, by intermediate devices, drives the sickle.

Claim.—The arrangement of forked bar C, bar D, forked finger bar E, and journals *e e*, in.

combination with the axle A, bevel wheels F G, shaft H, wheels I J, shaft K, crank pulley L, pitman M, and sickle N, constructed and operating in the manner and for the purpose herein specified.

No. 55,468—JOHN E. COFFIN, Portland, Maine.—*Machine for Cutting the Backs of Books.*—June 12, 1866.—The various motions are consecutively effected by the six cams upon the main shaft.   The book frame rises to its position under the jaws; the regulator turns and adjusts the projection of the book through the jaws, and gives the desired curve to the book back; the jaws close and rise under the pressure bar, which oscillates in contact with the rounded back.

Claim.—First, the combination and arrangement of the cams F G H I J and K on the main shaft C, in the manner and for the purposes described.

Second, the combination of the shaft M', the truck o, the cog-wheels w and x, the screws z, the sliding boxes y, the posts L, the segment levers k, and the adjustable pressure bar O, as and for the purpose described.

Third, the combination of the cam F, lever N, link Q, lever R, regulator M h, and sliding mould f, as and for the purposes described.

Fourth, the combination of the cam G, lever g, shaft P, levers i, links J, segment levers I, and segment levers k, and the pressure bar O, as and for the purpose described.

Fifth, the combination of the cam I, lever r, frame p p, supports c c, and sliding frame s s, for the purpose of raising the jaws under the pressure bar and pushing the book through the jaws so it can be easily removed.

Sixth, the combination of the cam J, shaft S, levers t t, links u u, toggles z' z', links e' e', posts L L, as and for the purpose described.

Seventh, the combination of the cam K, lever o', shaft T, lever p', link e, frame p p, sliding frame s s, screws b b, and bar m', for the purpose of adjusting the position of the book and presenting it to the operations of the pressure bar.

Eighth, the combined use and arrangement of a solid adjustable pressure bar O with the jaws which rise under the bar when the bar is in operation.

Ninth, the use of the clasp holder d, the screws c' c', and the hinges or joints r' r', as and for the purpose specified.

No. 55,469.—HENRIETTA H. COLE, New York, N. Y.—*Fluting Machine.*—June 12, 1866.—The fluted rollers are rotated by hand crank and gearing; the upper one is raised clear of the lower one by spiral springs under its bearings, until the pivoted weighted lever depresses it for operation.

Claim.—The upper roller of a fluting machine in boxes or bearings, resting upon spiral or other suitable springs, and moving upon guides in combination with adjustable weighted levers for holding such boxes or bearings down, when arranged together and so as to operate substantially in the manner described and for the purpose specified.

No. 55,470.—BYRON D. COOK. Clarendon, Mich.—*Seeding Machine.*—June 12, 1866.— The contents of tho hopper are agitated by the toothed shaft and scraped toward the opening by the spiral wings; clearing rods vibrate in the openings and the falling fertilizer is distributed by a revolving winged shaft.

Claim.—First, the arrangement of the converging wings w, upon the shafts S¹ S², in such manner as to scrape seed or fertilizing material from each side and towards the centre of their respective discharge apertures O¹, O², substantially as herein set forth.

Second, the employment of the converging wings w in combination with the revolving agitator E, the vibrating cleating fingers f, and the distributor D, arranged and operated relatively with each other and with the rest of the machine, substantially in the manner and for the purpose herein described.

No. 55,471.—ELISHA H. COOK, Clarendon, Mich.—*Washing Machine.*—June 12, 1866.— The pendulous weighted beater is vibrated from the partition by the pressure of the revolving cam upon the adjustable roller, and returns by its own weight to pound the clothes between its corrugated surface and that of the perforated partition.

Claim.—First, the employment of the suspended beaters B, constructed with boxes V, to contain weights in combination with the perforated partition board P, substantially as and for the purpose herein described.

Second, the mode of actuating the suspended beaters aforesaid, by means of the crank shaft s, and cam C, in combination with an adjustable roller D, the same being operated in connection with the hangers and standards herein described, substantially as and for the uses specified.

No. 55,472.—JACOB CUSTER and CHARLES ROWLAND, Clinton, Ill.—*Plough Coulter.*— June 12, 1866.—The arc-shaped coulter has tenoned ends which fit indifferently into the share or sbeth, permitting its reversal when worn; a forked brace depends from the beam.

Claim.—The construction of a self-supporting coulter in the form of an arch resting on its abutments, the share and post, and which from its peculiar construction and application is reversible and equivalent to two single coulters, which form one arch, or arc of a circle.

Also, the construction of the rod in combination with the coulter, which rod passes through the beam and descends to and down at each side of the coulter in the form of a fork, substantially as shown and described.

No. 55,473.—J. G. DeCoursey, Philadelphia, Penn.—*Mosquito Net.*—June 12, 1866.— The net has an elastic marginal band and perforated corner pieces which afford a means of attachment by hooks or cords.

Claim.—First, a net A, in combination with the elastic cords or bands, one of which is applied to each edge of the net, as and for the purpose described.

Second, in combination with the above, the perforated corner strips or plates B, of metal or other equivalent material, for the purpose specified.

No. 55,474.—J. Peter Eisenhut, Monroe, Mich.—*Medical Compound.*—June 12, 1866.— Composed of olive oil, sulphur, angelica, elder and linden blossoms, turpentine and saffron.

Claim.—A medical compound made of the ingredients herein specified, and mixed together substantially as and for the purposes set forth.

No. 55,475.—J. W. Elliot, of Leicester, Mass.—*Sash Fastener.*—June 12, 1866.—A spiked plate socketed in the outer edge of the sash is retracted by a pivoted lever which acts as a cam.

Claim.—First, the combination of the bolt A, plate C, spring F, and lever H, the whole being constructed and arranged to operate as herein described, so as to constitute a sash lock and sustainer.

Second, the combination with the bolt C, and lever H, of the fulcrum *h*, and lever *c*, permitting the ready attachment and detachment of the lever, as and for the objects specified.

No. 55,476.—William Elliott, Stockport, N. Y.—*Farm Gate.*—June 12, 1866.—The gate is drawn up the inclined plane by means of a cord which hangs from the post within reach of the occupant of the carriage; is detained by a tumbler catch and released by pulling on another cord, the gate closing automatically.

Claim.—First, employing the inclined railway D, in combination with the cords, pulleys, and posts, arranged substantially as and for the purpose herein described.

Second, the combination of the knee-levers *d d'* with the drop catch *h*, arranged substantially in the manner and for the purpose set forth.

Third, the combination and arrangement of the posts, pulleys, and cords, as or substantially as herein described, when employed for opening and closing gates of any other construction.

No. 55,477.—Levi S. Enos, Almond, N. Y.—*Portable Door Fastening.*—June 12, 1866.— The case has a pivoted gimlet and screw which shut within it or occupy notches when projected for use; when screwed into the jamb it forms a button to fasten the door.

Claim.—The portable door fastening, constructed as herein described, as a new article of manufacture.

No. 55,478.—Josiah B. Fairchild, Covington, Ky.—*Animal Trap.*—June 12, 1866.— The doors have spiked edges and swing horizontally: they are held open by engagement with notches on the tripping platform, whose depression releases them to be closed by springs.

Claim.—The folding doors *a*, armed with prongs *d*, the springs *e*, and trip-floor *f*, in combination with the body A of the trap, all constructed and operating as above described and for the purpose set forth.

No. 55,479.—James M. Foss, Concord, N. H.—*Draught Pipe for Locomotives.*—June 12, 1866.—The openings in the rear of the draught-pipe have visors, which guide the products of combustion into the pipe, whence they are ejected by the exhaust steam.

Claim.—The draught-pipe as made with two or any other suitable number of holes *b b*, arranged in its rear, and having a visor *c* to each, and with the front of the pipe closed and its lower end open, substantially as described.

No. 55,480.—Thomas W. Fox, New London, Conn.—*Expelling Water from the Holds of Vessels.*—June 12, 1866.—A semi-cylindrical plate or semi-tube is passed through an opening in the bottom of the vessel, its open side toward the stern; its purpose is to lead out the bilge water when the vessel is in motion, and it is closed by a cap at other times.

Claim.—The combination of the semi-cylindrical vacuum producer, Fig. 4, with the tube A, when they are constructed, fitted together, attached to the vessel and used substantially as herein described and set forth.

No. 55,481.—Patrick Freeman, Benton County, Iowa.—*Gate.*—June 12, 1866.—A panel at the lower portion of the gate is revolved to make an opening for the passage of small stock, or locked shut by the two sliding bars.

Claim.—The reel or revolving part of the gate upon the shaft A, together with the locking bolts *r* and *p*, substantially as and for the purpose herein specified.

No. 55,482.—REUBEN B. FULLER, Norwich, Conn.—*Vase for Cultivating Strawberries.*—June 12, 1866.—This earthen pot has a flaring top, and its lower portion may sit in the earth or another vessel.

Claim.—The vase A, made substantially as herein described and for the purposes specified. Also, in combination with the above, the use of the vessel or pot E, or its equivalent, as and for the purpose described.

No. 55,483.—MICHAEL GATES, Pawpaw, Mich.—*Pruning Hook.*—June 12, 1866.—The sliding blade is projected against the hooked knife by a rack and pinion, actuated by one hand, while the stock is held by the other; the crook in the sliding bar gives a "draw" cut to the blade.

Claim.—Communicating a drawing cut movement to the chisel C by bending its bar as at e, and operating it between guides, in connection with the hook A and its attached stock, substantially in the manner and for the purpose herein described and set forth.

No. 55,484.—A. J. GIBSON, Cincinnati, Ohio.—*Churn Dasher.*—June 12, 1866.—The hollow shaft has a valve opening downward, and the dashers consist of conical, non-perforate frustums.

Claim.—First, the churn-dash A, constructed with the two truncated cones, placed base to base, one or more of which may be united, forming a more extended dash, (as shown in the drawings,) as above described and set forth.

Second, the churn-dash A, in combination with the churn-handle B, for the purpose above specified.

No. 55,485.—WILLIAM GILMAN, Ottawa, Ill.—*Corn Sheller.*—June 12, 1866.—Explained by the claims.

Claim.—First, the rail O, placed just in front of the open end of the cylinder concave I, for throwing or deflecting the corn escaping from the said concave to the screen or riddle J, arranged substantially as described.

Second, the front rail D, in combination with the rail O, substantially as and for the purpose specified.

Third, the combination of the deflector rail O, screen J, and fan or other suitable blower, when arranged together, and so as to operate substantially as described and for the purpose set forth.

No. 55,486.—R. GIPSON, Shelby, Ohio.—*Clothes Wringer.*—June 12, 1866.—The wheels on the ends of the rollers gear into pinions on a square shaft; the upper pinion slides on the shaft as its roller rises and falls; the upper end of the shaft has its bearings in the bridge tree of the upper roller.

Claim.—The arrangement of the shaft D, sleeve a, lugs C and C', in combination with the bridge-tree B, gearing D D' and H H', operating as and for the purpose set forth.

No. 55,487.—WILLIAM F. GOODWIN, Washington, D. C.—*Harvester Rake.*—June 12, 1866.—The devices described in the claim are for operating a vibrating sweep-rake.

Claim.—First, the standards G^1 and G^2, and cap G, arranged for the purpose, and to operate in the manner substantially as described.

Second, the cam M, flange g, yoke Y, bar I, guide-box D, link L, crank S^3, journals S^1 and S^2, rod S, bar F, with its projection F', and stud E, arranged to operate in the manner and for the purpose substantially as described.

Third, the pulleys P and P', chain N, shaft C, and crank C', arranged to operate in the manner and for the purpose substantially as described.

Fourth, the bars B^1 B^2 B^3 B^4 B^5 and B^6, constructed and arranged to form the jointed arm B, to operate in the manner and for the purpose substantially as described.

No. 55,488.—JOHN GREIVES, Brooklyn, N. Y.—*Rock Drill.*—June 12, 1866.—Explained by the claim.

Claim.—The drill constructed of a central polygonal rod, with cutting point and angular sectional cutters bolted to the sides of said rod, substantially as herein specified.

No. 55,489.—WARNER GROAT, Green Island, N. Y.—*Wooden Mat for Cars.*—June 12, 1866.—The slats and interposed washers are strung upon transverse bolts.

Claim.—A wooden mat for railroad cars and for other similar or suitable purposes, composed of wooden slats A, spaced or retained at a proper distance apart by washers B and bolts C, substantially as herein shown and described.

No. 55,490.—P. S. HAINES, Newburg, N. Y.—*Apparatus for Treating Wool for Picking, Carding, &c.*—June 12, 1866.—By the jets of steam the adhesive matter is dissolved, and the wool softened to render it more pliant and tractable; the water of condensation is removed by a pipe, which prevents the wool becoming charged with water.

Claim.—The perforated steam-pipe F, constructed and located as described, and continued beyond the line of its perforations, and connected to a water-pipe to run off the water of condensation continuously, substantially as set forth.

No. 55,491.—ALEXANDER W. HALL, New York, N. Y.—*Churn.*—July 12, 1866.—The churn dasher depends from a shaft which is oscillated by the vibrations of a pendulum attached thereto.

Claim.—Providing for the operation of the swinging dasher G, by means of a pendulum weight applied outside of the churn and in connection with the said dasher, substantially as herein set forth.

No. 55,492.—CHARLES HOPE, Springfield, Mass.—*Revolving Desk.*—July 12, 1866.—The polygonal desk revolves on an upright axis and has above it a series of pigeon holes, book racks, &c., also capable of being revolved.

Claim.—The revolving desk herein described, constructed with the standard B, supporting and forming a pivot for the part A, and having these and the other parts arranged substantially as shown and described.

No. 55,493.—BENNETT HOTCHKISS, New Haven, Conn.—*Construction of Screw Taps.*—June 12, 1866.—The thread of the tapered tap is concentric with the axis from cutting point to cutting point, instead of increasing in diameter.

Claim.—Forming reamers, taps, and dies, substantially in the manner and for the purpose herein set forth.

No. 55,494.—HORATIO WILLARD, Plainfield, Ind, assignor to himself and GEORGE F. ADAMS, Indianapolis, Ind.—*Transfer Switch.*—June 12, 1866.—The switch section is pivoted at one end and at the other is attached to a segment rack bar moved by a pinion, its inclined under surface resting on anti-friction rollers.

Claim.—Pivoting the movable or transfer section of track at one end and attaching the other end to a segmental tie or bar, formed as shown, and resting on friction rollers so arranged as to raise the section from its bed, for the purpose substantially as set forth.

No. 55,495.—STILLMAN HOUGHTON, Worcester, Mass.—*Hoop-skirt Wire.*—June 12, 1866.—Explained by the claim.

Claim.—As an improved article of manufacture skirt hoop-wire, first covered with a fibrous materiel and then a braided plated wire covering or casing applied thereto, substantially as shown and described.

No. 55,496.—EUGENE HUTCHINSON, Manchester, N. H.—*Invalid Bedstead.*—June 12, 1866; antedated June 1, 1866.—The construction and the various adjustments of the supporting and auxiliary movable frames and rests are described in the claims and illustration.

Claim.—First, the combination of the bed-supporting frame D, the back elevator G, the bedstead frame A, and mechanism, substantially as described, for operating or moving the said frame D, so as to elevate the frame G.

Also, the combination of the auxiliary frame E and its elevating mechanism, substantially as described, with the bedstead frame A and the back elevator G.

Also, the combination of the frame E and its elevating mechanism, the bedstead frame A, the back elevator G, the frame D, and mechanism for operating the latter, as described.

Also, the combination of the leg-rest or frame F, the frame E, the back elevator G, the elevating frame D, and the two bedstead frames A B, the whole being constructed and applied together in manner and so as to operate substantially as explained.

No. 55,497.—E. P. IRONS, Baltimore, Md.—*Saw Mill.*—June 12, 1866.—The saw is strained by elastic bands which are stretched over walking beams while the saw is controlled by vertical guides. The bands of the saw and of the tension rod pass across the walking beams and are fastened to the opposite ends. Other adjustments are cited in the claims.

Claim.—First, the described mode of connecting the elastic saw bands g g' to the walking beam by rods ff', attached at points e at the opposite ends of the beam from that upon which the said bands g g' are lapped, substantially as described.

Second, the manner substantially as herein described of attaching the flexible wire cords or bands to the ends of the saw by means of the stirrups q, in combination with the jaws t t, the adjustable gibs u u, and guides U, the parts arranged and operating substantially in the manner and for the purpose set forth.

Third, the combination of the connecting rods i, bell crank K K', graduating connecting rod k, lever M, feed hand m, rag wheel O, and shaft L, with pinions and racks arranged and operating substantially as and for the purpose set forth.

Fourth, the manner of graduating the length of the cut or feed by means of the graduating connecting rod K, moving up and down on the arm K' of the bell-crank lever, so as to be moved a long or short distance, for the purpose and substantially in the manner set forth.

Fifth, the combination of the adjustable steel arming plates z with a connecting rod k, arranged and operating as herein set forth, for the purpose of compensating for wear, and allowing lost motion to be readily taken up, as set forth.

No. 55,498.—RUSSELL JENNINGS, Deep River, Conn.—*Dies for Swaging the Ends of Auger Blanks.*—June 12, 1866; antedated December 19, 1865.—The end of the auger blank has thickened portions, which, by a subsequent upsetting, will form the lips and cutting edges without welding.

Cl·im.—The swaging of the ends of auger blanks so as to have thick masses or portions *c c* at the sides thereof, with a central thick portion for the pintle, by means of dies constructed substantially as described, for the purpose of enabling the heads of augers to be formed or swaged at one operation, and so avoid all welding and joining of parts, as set forth.

No. 55,499.—ALBERT JOHNSON, Putnam, Conn.—*Machine for Trimming Mitre Joints.*—June 12, 1866.—The jointer slides in contact with the sawed end of the stuff, which is supported in a rest adjustable to the required angle, the table itself being vertically adjustable in relation to the face of the jointer.

Claim.—The adjustable bed C, in combination with the adjustable bar or rest E and the plane H, substantially as and for the purpose set forth.

No. 55,500.—FRANK G. JOHNSON, Brooklyn, N. Y.—*Knob Lock.*—June 12, 1866.—The lock mechanism is enclosed within a hollow knob and is operated by a key whose pins enter the holes in the front of the handle and depress the tumblers, allowing the bolt to rotate when the key is turned.

Claim.—The movable tumbler case C, in combination with a knob A, having a stationary flanged guard B applied to it, substantially as described.

No. 55,501.—EDWARD KAYLOR, Pittsburg, Penn.—*Bolt-heading Machine.*—June 12, 1866.—One pair of hammers operates in a vertical, and the other in a horizontal plane, in connection with a sliding header. The latter pair act upon the bolt while the header is staving up the bead and form it on two sides; the other pair act upon it on the other two sides while the header is retreating.

Claim.—Forming the head on square head-bolts by means of a machine constructed and operating substantially as hereinbefore described, by staving up the iron with a heading tool between two side dies, which at the same time advance and compress the iron as and while it is being staved, and then compressing the other two sides of the head by another pair of dies which advance and compress the opposite sides of the head as the first pair recede, the operation being repeated until the head is properly formed.

No. 55,502.—AUGUSTUS G. and E. E KYLE, Newville, Penn.—*Preserving Eggs.*—June 12, 1866 —The eggs are packed in a vessel containing a mixture of bran, salt, and lime, and then hermetically sealed.

Claim.—The within-described compound as a packing for preserving 'eggs, substantially as set forth.

No. 55,503.—JESSE S. LAKE, Smith's Landing, N J., and EZRA B. LAKE, Bridgeport, N. J.—*Weighing Scale.*—June 12, 1866.—The weight and price are shown on indexes with varying scales of prices on the beam and rotating cylinder. A weight is suspended from the beam and is diverged from the perpendicular by the imposition of the article weighed. The graduated scale beam rises above the straight edge and by segment rack and pinion rotates the graduated cylinder.

Claim.—First, the combination with the beam or lever C, and beam or partition H, of a weighing scales, of the scales M N O, constructed and arranged substantially as described and for the purpose set forth.

Second, the combination of the cylinder J, and cog-wheel I, with the toothed end of the beam or lever C, substantially as described and for the purpose set forth

Third, the combination with the cylinder J, and beam K, of the weighing scales, of the tables P R S, constructed and arranged substantially as described and for the purpose set forth.

No. 55,504.—P. W. LAMB, Albany, N. Y.—*Apparatus for Moulding Castings.*—June 12, 1866.—The cope containing the impression is raised from the patterns by means of cams, which simultaneously act upon different portions, giving an even movement and preserving the mould intact.

Claim.—The cams J, and cam shafts I, in combination with the box A, frame B, and cope D, arranged substantially as herein set forth for the purpose specified.

No. 55,505.—HENRY LAST, West Lebanon, Ind.—*Gate Hinge.*—June 12, 1866.—The bifurcated piece on the gate has two bearings which rest against corresponding studs on the post. The gate vibrates upon one or the other as it is opened, and tilts in so doing.

Claim.—An improved double-jointed gate hinge, constructed and arranged substantially as herein described and for the purpose set forth.

No. 55,506.—OBADIAH B. LATHAM, Seneca Falls, N. Y.—*Rope-guard.*—June 12, 1866.—A hollow metallic knob is fastened by a transverse pin on the rope to prevent its abrasion.

Claim—The combination of a metallic knob with a rope of whatever material composed, such knob being concave in its inner surface, to allow for the expansion of the rope, fastened and operated in the manner described.

No. 55,507.—JOHN D. LEE, Trenton, N. J.—*Composition for Water Proofing.*—June 12, 1866.—Compound of rosin, 2 parts; beeswax, 1 part; mutton suet, 1 part; neat's-foot oil, 1 part; to be spread on muslin and placed between the insole and outsole, and between the upper and lining.

Claim.—The mixing of the ingredients and the applying them to boots and shoes to render them impervious to water, as herein described.

No. 55,508.—E. C. LEONARD, Binghamton, N. Y.—*Churn.*—June 12, 1866.—The oscillating suspended churn has at its upper end a segment rack which actuates a rack on the rock-shaft and vibrates the dasher in the reverse direction.

Claim.—The combining and arranging of the body and dasher, substantially as herein recited, so that the body and the dasher may be operated as described.

No. 55,509.—FRIEDERICH LIESCHE, East New York, N. Y.—*Apparatus for making Coffee.*—June 12, 1866.—The vessel is suspended on a counterpoised frame, when the water therein is ejected by the heat of the lamp into the second vessel, which contains the ground coffee; the former vessel rises, the spring cover rotates and closes the lamp; the condensation in the former draws back the infusion of coffee.

Claim.—First, the combination of the suspended vessel C, weighted frame D, curved tube G, strainer H, and stationary vessel F, the whole arranged with regard to a lamp or other burner, substantially as herein set forth for the purpose specified.

Second, the lever k, and spring or weight m n, operating in combination with the movable vessel A, to extinguish the lamp or burner when the said vessel is raised, in the manner substantially as herein set forth.

No. 55,510.—H. A. and D. E. LONGSDORF, Mechanicsburg, Penn.—*Paint.*—June 12, 1866.—Composed of lime, 50 parts; casein, 25; alum, 6, mixed with milk.

Claim.—The compound composed of the ingredients herein named, and mixed together in or about the proportions described for the purpose specified.

No. 55,511.—CHARLES LENTZ, Philadelphia, Penn.—*Speculum.*—June 12, 1866.—The instrument has four pivoted, expanding leaves actuated by closing the parts of the handle; the outer leaves are adjusted to the inner ones by pins passing through holes in the latter; springs beneath the flange heads of the pins preserving their position.

Claim.—The described improvement in speculums, consisting in the use of the springs I I, and notched pins G G, or their equivalents, when arranged relative to each other and to the leaves A A and F F, substantially as and for the purpose specified.

No. 55,512.—JOHN LETCHWORTH, Philadelphia, Penn.—*Fruit Jar.*—June 12, 1866.—The internal projections of the cover catch below the external flanges on the neck of the jar.

Claim.—The cap B, made of thin metal, and having internal projections made by external indentations, the whole being applied to the neck of a jar substantially as described.

No. 55,513.—JOHN LETZKUS, Pittsburg, Penn.—*Squeezer.*—June 12, 1866.—The gearing is placed under the rotating bed to protect it from the cinder; the driving shaft is out of the way of the workman, and the mouth is placed at any point.

Claim.—The combination of the top upset p, lower upset f, and drums e and n, with the main shaft s, geared directly or indirectly to the inner circumference of the lower upset, the whole being arranged and operating for the purpose of squeezing puddler's balls, substantially as hereinbefore described.

No. 55,514.—THALES LINDSLEY, Rock Island, Ill.—*Machine for Tunnelling Rock.*—June 12, 1866.—This machine is constructed to cut circular concentric channels in the rock, and thus form circular concentric rings in the heading; to detach these rings and the fragments; to drill holes in the crown of the excavation; to provide seats for brackets; to cut a trench in the floor, for drainage. The machine is driven by compressed air, and the drills subjected to water jets, to prevent heating and dust and to wash off the debris.

Claim.—The drill gauge, substantially as and for the purpose specified.

Also, the ram guide, in combination with said gauge, the ram, and the drill wheel, substantially as herein specified.

Also, constructing the drills and the drill shafts, and connecting the same, substantially as set forth.

Also, the combination of the compensating springs with the drill shafts, substantially as set forth.

Also, the drill-shaft guides and the notched collars between the compensating springs, as specified.

Also, it to water pipes and jets, in connection with the drill wheel and ram, substantially as set forth.

Also, the combination of parts forming the drill wheel, substantially as set forth.

Also, the grooved collar upon the long ram sleeve and the clutch attached to the rear face of said drill wheel and working into said collar, substantially as herein specified.

Also, the ram and the ram hammers, substantially as herein specified.

Also, the wedge index, in connection with said hammers, or their equivalents, together with the splitting apparatus, substantially as set forth.

Also, the cam wheel and its adjustable cams, for working the drills, substantially as set forth.

Also, the drain drill, and the collar upon the long ram sleeve, which serves as its guide, constructed and arranged substantially as described.

Also, the non-revolving of the ram sleeve aforesaid, and the non-revolving of the short ram sleeve of the rear frame of the machine, as specified.

Also, the combination of the valves receiving the compressed air to the ram cylinders, and the valves discharging it from them with a hand lever, so as to control the action of the ram, &c., by a touch of the engineer, substantially as set forth.

Also, the construction of the platforms upon the legs of the machine, substantially as described.

Also, the supporting of the machine upon friction wheels, bevelled upon their face, substantially as set forth.

Also, moving the drilling apparatus back and forward, by means of the ram cylinders and their connections, substantially as set forth.

Also, moving the ram back and forth at any velocity desired by the engineer, by means of the ram cylinders and their dependencies, substantially as specified.

Also, moving the machine, by means of said ram cylinders and the toggle levers, substantially as set forth.

Also, the toggle levers and their necessary appendages, substantially as set forth.

Also, the bracket drill, constructed and operating substantially as specified.

Also, the hauling out of the debris, by means of the drag-pulley and its appendages, substantially as specified.

Also, the hauling out of the debris, by means of the ram and the tackle and clamps appended, substantially as specified.

Also, the combination whereby the ram and the drill wheel are united and revolved, substantially as set forth.

Also, in combination with a machine, constructed substantially as herein set forth, the method of levelling the same transversely of the tunnel, and of adjusting it to the grade line of the excavation, as herein specified.

Also, the combinations by which the ram cylinders operate, without the oscillating cylinders or in conjunction with them, and *vice versa*; by which the bracket drill works independently of the drill wheel, or simultaneously with it; by which the drag-pulley hauls rock independently of, or contemporaneously with, the snag-pulley, and *vice versa*; by which the drill wheel revolves without the cam wheel or in conjunction with it; by which the ram cylinders, through the toggle levers, may move the machine forward and backward, whilst the oscillating cylinders, through the drag-pulley, are hauling out rock from the heading; by which the drills are kept home to their work and at the point of maximum action, and by which the bottoms of the concentric channels are kept relatively in the same plane, whatever the disparities in the hardness of the rock; by which the ram is permitted, at the will of the engineer, to move independently back and forth and without shock to the machine from the oscillations; by which the drills for the heading are kept cool, the dust from them laid, and their minute chips swept out of the concentric channels into the common drain; by which a drain is cut in the bottom of the tunnel parallel with, and directly under, the axial line of the same; by which the machine progresses forward and backward with or without the convenience of a railroad, and by which the tunnel, adit, &c., are supplied with an abundance of fresh air and water; by which, finally, the drill wheel, the cam wheel, the ram, the bracket drill, the drag-pulley, the snag-pulley, the ram cylinders, the oscillating cylinders, and other parts, may operate concurrently and otherwise; all of which are substantially as presented.

No. 55,515.—BARTHOLOMEW McGRATH, Gloucester, Massachusetts.—*Boom Connection for Masts.*—June 12, 1866.—A yoke is attached to the mast, and has a bow, on which the jaws of the boom are slipped.

Claim.—First, the mast and boom connection, as composed of the clasps F F, the curved rod C, its screws and nuts, and the arms D D, arranged substantially as specified.

Second, the combination of the wheel E with the clasps F F, the curved rod C, its screws and nuts, and the arms D D, the whole being arranged as explained.

No. 55,516.—WILLIAM LOUIS WINANS and THOMAS WINANS, London, Eng'land.—*Steam Engine.*—June 12, 1866.—The propeller shaft is arranged between the pis'ts 12, 1866.— bearings placed above the cylinder heads, the crank shafts working alongside.

Claim.—The arrangement of the propelling shafts *c* between the piston r t c, mounted in suitable bearings on the cylinders *a*, or frames attached to the cylinders, the cylinders being placed directly below the propelling shafts, whose cranks work down alongside of the cylinder or cylinders, for the purpose herein set forth.

No. 55,517.—DAVID McKANNA, Madison, Wis.—*Hand Corn Planter.*—June 12, 1866.—
The reciprocating bar has an inclined rod, which passes through and reciprocates the seed slide; this drops the seed into the duct in the stationary rod which carries the shoe.
Claim.—First, operating the seed slide D, by means of the inclined rod *b*, when placed at one end of the slide, as shown and described.

Second, the combination of the stationary bar B, having the groove H therein, the bar C, provided with the slot and the inclined rod *b* and slide D, all arranged and operating as set forth.

No. 55,518.—B. F. McKINLEY, Falmouth, Ky.—*Balanced Cut-off Valve.*—June 12, 1866.—
The upward pressure on the piston is equal to that downward upon the valve; the cut-off valve attached to the piston moves in a different direction to the main valve underneath.
Claim.—The adjustable cut-off valve E and balance piston F, in combination with the revolving valve C, constructed and operating substantially as and for the purpose described.

No. 55,519.—CHARLES MESSENGER, Chicago, Ill.—*Broom Head.*—June 12, 1866.—The toothed jaws which enclose the broom are retained by a ring which slips over them; the ring is fastened by springs.
Claim.—Making the head in two parts, as described, in combination with the ring K, springs N, teeth H, as and for the purpose set forth.

No. 55,520.—ISAAC M. MILLBANK, Greenfield Hill, Conn.—*Breech-loading Fire-arm.*—
June 12, 1866.—The hinged breech block, which swings upward and forward, is provided with a rotating bolt which passes vertically through it, and is bevelled off at its lower projecting end so as to enter a circular hole at the bottom or bed; when closed and rotated, the straight side of the bolt within the hole locks the breech block, and an arm on top of the bolt lies in the way of the hammer, so that the latter shall insure the rotation of the bolt before it can discharge the arm.
Claim.—First, a swinging breech, in combination with a transverse turning bolt, bevelled on one side to enter a recess, and acting to retain the breech by a partial turn of said bolt, substantially as set forth.

Second, the lever A for turning the bolt *e* of the swinging breech, in combination with the hammer *g*, the parts being fitted substantially as specified, so that the discharge of the hammer shall insure the proper turning of said bolt, as set forth.

Third, the latch *i*, in combination with the lever *h* and turning bolt *e*, as set forth, whereby the said latch is disconnected from the lever by the closing of the breech, as set forth.

Fourth, the spring K, in combination with the turning bolt *e* and swinging breech, to effect the locking or partial locking of the breech as soon as closed, as set forth.

Fifth, the claw or retractor *a*, formed with or attached to the supporting block 9, in combination with the swinging breech block, as and for the purposes specified.

No. 55,521.—HENRY MINTON, Brooklyn, N. Y.—*Vaccinator.*—June 12, 1866.—The perforator is operated by a spring, and when it is pressed against the skin the vaccine matter is forced into the wound by a plunger.
Claim.—The puncturing tube *b* and plunger *d*, in combination with the springs *h i*, trigger K, and barrel A, constructed and operating substantially as and for the purpose described.

No. 55,522.—J. J. MOORE, Little York, N. Y.—*Sleigh Brake.*—June 12, 1866.—Pulling back on the tongue vibrates a toggle and projects a locking tooth below the level of the runner.
Claim.—The tongue B, working in a slot in the bar C, being provided with the rods *b b* and *f f*, the same connected to jointed spurs *h h*, when arranged and used substantially as and for the purposes herein set forth.

No. 55,523.—LEOPOLD F. MORAWETZ and CHARLES VOLKMAR, Baltimore, Md.—*Automatic Heliotrope.*—June 12, 1866.—The optical axis of the instrument is brought parallel with the sun's rays, and kept coincident therewith by the adjustments described, a motor, and a pendulum, or other time keeper.
Claim.—First, the axis A, adapted to revolve on one or more fulcra, so that it can be adjusted to a position parallel to the axis of the earth, in conjunction with a solar instrument, substantially as specified.

Second, the driving wheel O, applied at a suitable point on the axis A, in combination with a weight, or its equivalent, and with or without an escapement, substantially as described.

Third, the construction of the main revolving axis of the heliotrope so that it supports a solar camera, for photographic and other purposes, and revolves said camera, as well as permits it to be revolved, in such manner that the main supporting axis and the camera may be brought into any required angle of inclination with relation to each other, substantially as described.

Fourth, automatically moving an optical or solar instrument in the plane of the daily course of the sun, and synchronic with the sun, so that the solar rays shall fall directly or continually with the same angle of incidence upon a certain point of said instrument, by means substantially as herein specified.

No. 55,524.—GEORGE L. MORRIS, Taunton, Mass.—*Nicking Screw Heads.*—June 12, 1866.—The two opposite edges of the head have flaring nicks and an intervening uncut portion.

Claim.—The improved two-nicked screw, having each of its nicks made so as to increase in width as it approaches the circumference of the head of the screw, the same being substantially as and for the purpose specified.

No. 55,525.—EMIL MÜLLER, New York, N. Y.—*Photographic Apparatus.*—June 12, 1866.—Improvement on patent No. 25,276. The article is held by angle plates or clamps on a table capable of vibration to give the required presentation and adjustment for distance.

Claim.—The apparatus for holding vases or other uneven objects for the purpose of photographing thereon, substantially as hereinbefore described.

No. 55,526.—JOHN MUMMA, Middletown, Ohio.—*Fanning Mill.*—June 12, 1866.—Explained by the claims.

Claim.—First, constructing the fans *e'* with curved ends for gathering the air and concentrating the blast in the manner described.

Second, the deflecting board *c'*, upon adjustable shaft *a'*, arranged so as to be set and retained in any desired position by means of the exterior handle *d'*, in the manner and for the purpose described.

Third, the combination of the toothed feeding roll Q with the hopper, arranged and operated as described.

Fourth, shaft R and fingers *c*, vibrated by the arm *e* and cam S, in the manner and for the purpose described.

Fifth, the shoe F, in combination with spring *g* and cams S, on shaft P, operating in the manner and for the purpose described.

Sixth, the combination of the double tapering tappet or cam L, with adjusting block 4, and spring *i*, in the manner and for the purpose described.

Seventh, the adjusting roll T, arms *m*, and staples *n*, in combination with the shoe suspending wires *h*, for regulating the motion of the shoe, in the manner described.

Eighth, the combination of the journal blocks V W with the bolt and thumb-screw *r*, with the friction rubber block interposed as described, for the purpose specified.

Ninth, the combination of the wires or rods across the inner end of the shoe, with the clamping rod and thumb screw *z*, for securing the sieves and screens in the manner substantially as described.

Tenth, the arrangement of the India-rubber blocks in the sides of the casing B, adjustable by set screws Z, to regulate the vibrations of the shoe and obviate wear and noise in running the mill, as specified.

Eleventh, the combination of the shoe F, rock shaft R, cams S, and spring *g*, with their connecting and regulating mechanism, operating substantially as set forth for the purpose specified.

No. 55,527.—CUTHBERT L. MUNNS, Philadelphia, Penn.—*Gas Inhaler.*—June 12, 1866.—The stop-cock has a valve which allows the gas to pass to the mouth, the other valve closing; these actions are transposed on exhalation to allow the breath to escape to the outside.

Claim.—The hollow tube C, in combination with the reversible plug, with its ball valve, and the ball valve O, the whole being constructed and operated substantially in the manner and for the purpose set forth.

No. 55,528.—JOHN K. O'NEIL, Kingston, N. Y.—*Horse Hay Fork.*—June 12, 1866.—The tines at points equidistant from their joint are connected by links to the centre of vibration of the cross-lever, and a rectangular arm from the latter is pivoted to the tines at their point of articulation. The vibration of the cross-lever opens and closes the jaws.

Claim.—Operating the tines or fingers of a grappling fork by means of a rod connecting said tines or fingers with a pivoted cross-lever, said rod being pivoted to both tines and lever at a distance from their centres of motion, as set forth.

Also, the cross-lever C, in combination with the arms B B, fingers or tines A A, and connecting rod D, as and for the purpose described.

No. 55,529.—P. PALLISSARD, Aroma, Ill.—*Baby Walker.*—June 12, 1866.—Explained by the claim.

Claim.—The arrangement of the hoop A, supported by the caster wheels F F, and connected with the ring E by the supports B B, having screw-threads cut upon their ends for the ready adjustment of the ring E to the height of the baby, the whole being constructed and operated substantially in the manner and for the purpose set forth.

No. 55,530.—JESSE K. PARK, Marlboro', N. Y.—*Fruit Basket.*—June 12, 1866.—The sides of the basket are formed by interlaced splints, vertical and horizontal, and the lower edge is nailed to a circular bottom board.

Claim.—The combination of the warp bands D with the interwoven weft or filling splints E, passing up and down, and bent over the upper edge of the basket to form the binding, substantially as herein set forth for the purpose specified.

No. 55,531.—JEFFERSON PEABODY, Dixmont Centre, Maine.—*Valve.*—June 12, 1866.—The valve has no stem and is confined within the box by the guide projections, which form a cage.

Claim.—The improved valve box as made with the case A, series of valve guides D D D, stops E E E, arranged with respect to the valve B, and the opening C of its seat, substantially in manner as described.

No. 55,532.—CHARLES F. PIKE, Providence, R. I.—*Refrigerator for Cooling Oil, &c.*—June 12, 1866.—The ice box above and the tank below are connected by vertical pipes and expose a cool surface to the fluid or liquid surrounding them.

Claim.—First, the construction of parallel open-mouth pipes or tubes fastened to the ice box or receptacle A, and water tank C, in combination with the ice box or receptacle A, and also in combination with the water tank or receptacle C, substantially as herein described and for the purposes hereinbefore set forth.

Second, the application of the pump or pumps, as herein described, for the purposes of keeping up an artificial circulation by raising the water from the water tank C, and throwing it into the ice box or receptacle A, and the using of the water so raised and thrown into the ice box or receptacle A, substantially in the manner and for the purpose hereinbefore stated.

No. 55,533.—J. N. PLOTTS, New York N. Y.—*Buckle.*—June 12, 1866.—The cross-bar slides upon the ends of the buckle frame and clasps the strap which is passed around it against the side of the frame.

Claim.—A buckle having the bar B, with its eyes *a a*, encircling the sides of the frame A, arranged to slide to and fro, in the manner and for the purpose herein described.

No. 55,534.—PETER POUCIN, Minneapolis, Minn.—*Medical Compound.*—June 12, 1866.—This febrifuge is composed of white ginger, 4 parts; aloes, 4; gum camphor, 3; beaver's gall, ½; sassafras, 3; gum myrrh, 3, mixed with wine or brandy.

Claim.—The medical compound composed of the ingredients united in the proportions and manner as above set forth.

No. 55, 535.—CHARLES F. REESE, Millersville, Penn.—*Lard Lamp.*—June 12, 1866.—The lamp has a tin-plate casing, and is adapted to burn lard.

Claim.—The combination of the lamp stand J K L N, with its vessel B, spout I, and arrangement of the cylinders A C and rings E G, in the manner and for the purpose set forth.

No. 55,536.—F. REYNOLDS and F. L. HILBRIGHT, Newark, N. J.—*Dies for Forming Metal Heads on Harness Nails.*—June 12, 1866.—The nail is held by two half-dies placed in a solid block, one secured by a wedge and the other movable by a lever.

Claim.—First, the half-sides B and C, constructed as described, in combination with each other and with the block A, substantially as and for the purpose set forth.

Second, the combination of the lever F with the block A, and with the half-die B, substantially as described and for the purpose set forth.

Third, the combination of the spring G with the block A, and with the lever F, substantially as described and for the purpose set forth.

Fourth, the combination of the wedge-stop E with the block A, and with the half-die C, substantially as described and for the purpose set forth.

No. 55,537.—NATHANIEL ROBBINS, Jr.—Rockport, Mass.—*Car Coupling.*—June 12, 1866.—The presentation of the link is controlled by a lateral extension bar, within reach of a person outside the track.

Claim.—In combination with the link and the draw-bar, a mechanism, substantially such as described, or its equivalent, for enabling a person, without going between the cars or taking a position where he will be liable to be crushed by and between them, to control and direct the link with respect to its entrance into the mouth of another draw-bar, as specified, such mechanism being the arm applied to the link and the rod or rods arranged on the draw-bar, the whole being as set forth.

No. 55,538.—AUGUSTUS ROEHRIG, Williamsburg, N. Y.—*Hanging Window Sash.*—June 12, 1866.—A wire extends vertically the length of the window frame and passes through staples arranged in zigzag order on the sash, so that the friction on the wire will hold the sash in position.

Claim.—The arrangement of a sinuous wire d, in the frame A, in combination with three or more staples a, b, c, fastened in a zigzag line in the sash B, substantially as and for the purpose described.

No. 55,539.—ROBERT E. ROGERS and JAMES BLACK, Philadelphia, Penn.—*Steam Generator.*—June 12, 1866.—The boiler is suspended within the fire chamber, is connected by tubes to the exterior water jacket and has bent tubes connecting its upper and lower portions and exposed to the fire heat.

Claim.—One or more water spaces or jackets surrounding or in combination with the boiler a, constructed substantially as herein recited, said water spaces or jackets having or not the tubes for circulation and exposure to heat, and being provided with fire tubes or not, substantially as described and for the purposes set forth.

No. 55,540.—P. G. ROSENBLATT.—Greenville, Tenn.—*Medical Composition.*—June 12, 1866.—Composed of socotrine, rhubarb, gentian, opium, ginger, and alcohol.

Claim.—The medical compound composed of the ingredients and mixed together in about the proportions herein described.

No. 55,541.—C. H. RUDD, Sandusky, Ohio.—*Telegraphic Repeater.*—June 12, 1866.—By the described devices fresh currents are brought into action at the necessary points to transmit the signal.

Claim.—The posts G L, with the spring M, post K, and extension of the lever E, beyond the line of the post F F, so arranged as to enable one to use extra force for holding the relay closed, when said extra power is obtained from the main current, which is unemployed just at the time when needed, substantially in the manner and for the purpose set forth.

No. 55,542.—H. V. SCATTERGOOD, Albany, N. Y.—*Cotton Gin.*—June 12, 1866.—Explained by the claim.

Claim.—The cylinder having a surface consisting of rounded needle-pointed teeth so curved and arranged as that the point of each tooth approaches nearer to its next preceding tooth than at any other point thereof, substantially in the manner and for the purpose described.

No. 55,543.—ALBERT J. SESSIONS, Bristol, Conn.—*Bag Frame.*—June 12, 1866.—A strip of sheet metal of the proper size is cut, folded and fastened so as to form the rib and side of the jaw, which constitutes one-half of the bag frame

Claim.—Slitting or cutting the strip, substantially as described, so as to form one-half of a bag frame in one piece of metal.

No 55,544.—JOSEPH SHOLL and JOHN COLLINS, Burlington, N. J.—*Alarm Funnel.*—June 12, 1866; antedated May 28. 1866.—The funnel being placed over the bung-hole of the barrel the rising liquid raises the float, which detaches the button from its stop and rings the alarm bell.

Claim.—The combination of the funnel A, with its tube b, the rod C, with its float D, and the bell H, as and for the purpose described.

No. 55,545.—FREDERICK RANSOME, Ipswich, England.—*Manufacture of Stone, Cement and Plaster.*—June 12, 1866.—Composed of a solution of silicate of soda or potash, quicklime and chalk, sand or clay, with or without a soluble salt of an alkaline earth, or of aluminum or iron.

Claim.—First, the manufacture of artificial stone, cement, or plaster, by mixing silicate of soda or potash with quicklime and chalk, or sand, or clay, or other similar substance, substantially as described.

Second, the manufacture of artificial stone, cement, or plaster, by mixing together in a paste chalk or sand, or other suitable mineral in a powdered state, soluble silicate and a soluble salt of an alkaline earth, or of aluminum or iron, substantially as described.

No. 55,546.—WILLARD H. SMITH, New York, N. Y.—*Lamp Burner.*—June 12, 1866.—The disks of cork are fastened together by a pointed metal plate, and are interposed as a nonconductor of heat between the metal of the burner and the cap of the reservoir.

Claim.—First, the metal plate, or its equivalent, with points between the two pieces of cork, entering both as represented, as and for the purpose set forth.

Second, the two parts A and B constructed and combined as and for the purpose set forth.

Third, the mode of fastening the wick-tube to the said metal plate, and of fitting the same snugly in the said cork or other non-conductor, so as to hold the same firmly in its place without touching the metal plates enclosing the non-conductor either above or below, as set forth.

Fourth, the combination of the parts A and B, filled with cork or other non-conductor, fastened together by means of the metal plate as shown in Fig. 4, as set forth.

Fifth, the combination of the sleeve G, and the tips E and F, as and for the purpose described.

No. 55,547.—ELIHU SPENCER and E. L. MEYER, Elizabeth, N. J.—*Water Meter.*—June 12, 1866.—The cylinders are measures of capacity, and the strokes of their pistons are registered upon the dial; the water induction and eduction pipes are connected to the chamber between the cylinders, and the valves of the latter admit water to them alternately.

Claim.—First, a water meter made and arranged substantially as above shown—that is to say, consisting of two cylinders having a common valve-chamber between them, within which valve-chamber are placed two slide-valves, operated substantially as described by means of a shaft rotated by proper connection, from the pistons of the cylinders, such shaft carrying a worm E, which rotates a gear wheel F, whose spindle carries a fixed index, all as above set forth.

Second, combining the said apparatus mentioned in the preceding clause with an enclosing box A, in the manner substantially as above shown.

No. 55,548.—A. W. SPRAGUE, Boston, Mass.—*Apparatus for Generating and Washing Gases for Inhalation.*—June 12, 1866.—When the heat becomes too intense the gas forces water from the vessel and raises the float, which allows the lever to drop and moderate the flame, and conversely.

Claim.—First, the combination of the float F and vessel E with the lever H, or its equivalent, as and for the purpose described.

Second, the perforated conical projections *o* formed in the tube B, or its equivalent device, substantially as and for the purpose specified.

No. 55,549.—MARTIN STACHELIN and HENRY YOUNG, Port Chester, N. Y.—*Lock.*—June 12, 1866; antedated June 8, 1866.—A bridge plate swings from a pin on the short bolt, and its slotted lower end engages the pin on the latch bolt. The key is turned on the long key-pin either above or beneath the plate, and is introduced through a particular orifice in the bridge plate to retract one bolt, and then being partially withdrawn passes over the bridge-plate and through another opening to reach the other bolt.

Claim.—The swinging frame E applied in combination with the shot-bolt B, latch C, and key D, substantially as and for the purpose set forth.

No. 55,550.—ISAAC H. STONE, St. Louis, Mo.—*Hose Protector.*—June 12, 1866.—Bridge rails rest upon the main rails and protect the hose which crosses the track.

Claim.—The combination of the rails A and A' with the hollow beam B, the dogs *e* and spurs *e'*, forming a secure protection for hose during the entire length of its passage over railway tracks, as set forth.

No. 55,551.—OLE O. STORLE, North Cape, Wis.—*Harvester Rake.*—June 12, 1866.—This arrangement of gearing for operating the rake will be understood from the claims and illustration.

Claim.—First, bevel wheels 5, and segmental wheels 6, and pinion 4, in combination, constructed and operated substantially as and for the purpose described.

Second, guide-wheel 18 in combination with guide-track 19 and rake 14, substantially as and for the purpose described.

No. 55,552.—THOMAS L. STURTEVANT, Boston, Mass.—*Nippled Cartridge for Breech-loading Fire-arms.*—June 12, 1866.—The tongue or projection in the rear of the nipple fills the opening in the barrel made to receive the nipple of the charger.

Claim.—The combination and arrangement of the tongue *d* with the nipple C, and the charging tube A, arranged substantially as specified.

No. 55,553.—MILES SWEET, Troy, N. Y.—*Currycomb.*—June 12, 1866.—The combs, which are strung upon the side rods proceeding from the handle, are held to their proper distance by interposed collars.

Claim.—First, a series of comb-bars A formed with perforations *b*, and fastened together by collars *d*, and rivet-bolts *c* extending through the said collars and comb-bars, substantially as herein described.

Second, a series of perforated comb-bars A, fastened together by rivet-rods *c* and collars *d*, and with the rivet-rods forming a shank *m* for a handle, substantially as herein described.

Third, a series of perforated comb-bars A fastened together by rivet-rods *c* and collars *d*, and having one or more of the collars extending beyond the ends of the comb-bars, substantially as herein described.

Fourth, a series of single perforated comb-bars A fastened together by rivet-rods *c* and collars *d*, and having a back-plate P secured to the said comb-bars, substantially as herein described.

No. 55,554.—JOSEPH C. THOMAS, Kennebunk, Maine.—*Planking Clamp.*—June 12, 1866.—The chain hooks catch against the ribs and afford a hold for the planking clamp to strain against.

Claim.—The combination of the screw *a*, the nut C, the rotary head *g*, the two chains

D D, and the hooks E E, the whole being constructed, arranged, and applied to the stock A, provided with the screw B, substantially as and so as to operate as and for the purpose hereinbefore specified.

No. 55,555.—CHARLES TIMMERMAN, Amsterdam, N. Y.—*Air-tight Burial Case.*—June 12, 1866.—The adjacent edges of the pieces are tongued and grooved together and have strengthening angle irons. The lid has a wooden panel and metallic frame tongued into a groove on the upper edge of the coffin.

Claim.—First, the use and arrangement of the tongue-pieces E and F and corner-plate D', for connecting and fastening the bottom and upright parts of the case together, substantially as described.

Second, constructing the cover or upper part of the burial case of iron and wood in combination, substantially as described.

No. 55,556.—FRANKLIN TRAXLER, Scottsburg, N. Y.—*Dog Churn.*—June 12, 1866.—The cam wheel within the annular tread wheel vibrates vertically the system of levers connected with the dasher rod; anti-friction rollers traverse on the upper and lower sides of the cam wheel.

Claim.—First, communicating a vibrating movement to the arm b by means of a cam wheel G applied to the shaft of the tread wheel, said wheel acting upon rollers c c', the bearings of which are allowed to have a movement independent of the arm to which the wheels are attached, substantially as described.

Second, the pivoted bearing plate d, having rollers c c' applied to it, in combination with a vibrating arm b, or its equivalent, and a cam wheel G, substantially as described.

Third, the construction of the main supporting frame of triangular supports A A, horizontal beams B B', and a stall E, substantially as described.

Fourth, the combination of an inclined tread wheel D, cam wheel G, rollers c c', vibrating arm b, pitman a, and vibrating lever F, arranged and operating substantially as described.

No. 55,557.—JOHN P. TUCKER, South Reading, Mass.—*Coal Elevator.*—June 12, 1866.—The coal bucket ascends until its director enters the slot in the arm which guides the bucket until it is brought up by the tripper and its contents tipped into the chute.

Claim.—The combination of the slotted arm E', the curved lever H, provided with the tripper c, the director J, the scoop I, and the rope K, with the gallows frame and its discharging chute, the whole being arranged and made to operate substantially as above set forth.

No. 55,558.—ANDREW TURNBULL, New Britain, Conn.—*House Bell.*—June 12, 1866.—The bell is screwed to a central post and the clapper is fastened to an oscillating block which is moved by a rod and retracted by a spring.

Claim.—The cast metal hub e, having the eyes g and rod h firmly secured therein, in combination with the spiral spring n, detents m, plate a, and bell d, substantially as and for the purpose described.

No. 55,559.—J. B. VAN DEUSEN, New York, N. Y.—*Steering Apparatus.*—June 12, 1866.—Springs are interposed between the internal flanges on the ring of the tiller and the projections on the rudder-head. to modify the effect upon the tiller of a blow upon the rudder.

Claim.—First, the interposition between the tiller and the rudder-head of a steering apparatus of a spring or springs, operating substantially as and for the purpose herein described.

Second, the combination of the inner and outer ring with their radial projections, and India-rubber or other springs, substantially as and for the purpose herein fully described.

No. 55,560.—FREDERICK VAN PATTEN and OREN A. ANTHONY, Ilion, N. Y.—*Dies for Welding Links into Chains.*—June 12, 1866.—The cavities in the opposed faces of the dies correspond to the impression of several consecutive links so as to hold the end of the chain steady while another link is welded between the dies.

Claim.—Forming the dies E and F with two or more cavities for the reception of two or more links, as and for the purpose specified.

No. 55,561.—WILLIAM WALKER, New Haven, Conn.—*Picture Holder.*—June 12, 1866.—The pictures are contained in frames of uniform shape arranged side by side in two series in a case; the turning of the latter on a pedestal or handle causes the frames to pass in succession from one row to another, bringing each picture in turn to the outside.

Claim.—First, a holder for pictures, cards, and for other analogous purposes, consisting of a series of frames, or their equivalents, placed one in front of another but in two rows or sections, when such frames are so connected through any suitable mechanism with the outer casing or box in which they are arranged, or with a pedestal or other portion of the same, that by either revolving such casing or box, or its pedestal or other portion thereof so connected, the said picture frames, in regular order and succession, can be brought to the end of each row or section in proper position for being viewed, substantially in the manner described.

Second, the combination of the reciprocating sliding carrier-plate H with the transverse sliding plate K, when both connected with and operated by a common revolving disk E of the box or casing A, or its equivalent, and arranged with regard to the double row of frames B so as to act upon the same, substantially in the manner and for the purpose specified.

No. 55,562.—THOMAS V. WAYMOTH, New York, N. Y.—*Machine for Gumming and Printing Envelopes.*—June 12, 1866.—The pile of blanks is placed upon the elastic bed of a hinged table and underneath the pickers; the uppermost blank is gummed and printed, separated from the pile, deposited on a carrier, and placed on an intermittingly revolving endless apron.

Claim.—First, the hinged table B, which swings back and forth on arms *a* to operate in combination with the gummer D, substantially as and for the purpose described.

Second, the movable separator G, in combination with the gummer D, substantially as and for the purpose set forth.

Third, the endless apron H, in combination with a suitable mechanism, imparting to it an intermittent motion, and with the reciprocating carrier F and gummer D, constructed and operating substantially as and for the purpose described.

Fourth, the finger *l'* and rollers *k'*, in combination with the apron H and carrier F, constructed and operating substantially as and for the purpose set forth.

No. 55,563.—R. L. WEBB, New Britain, Conn.—*Snap Hook.*—June 12, 1866.—The ends of the V-shaped outer spring engage projections on the hook and tongue and keep the latter closed.

Claim.—The employment of the spring *d* passing around the heel or joint formation, when both ends are secured to the outside edge of the hook-shank and the latch without a pivoted joint, substantially in the manner as and for the purpose described.

No. 55,564.—EDWARD WEBSTER, Hartford, Conn.—*Hot Air Furnace.*—June 12, 1866.— An annular diaphragm plate surrounds the air tubes near their lower tube-sheet, the intervening space being a flue where the volatile results revert downward to the lower exit. The described water vessel is in the upper chamber near the hot air exit.

Claim.—First, the employment of the plate *j* in combination with the arrangement of the exit flange or pipe *t*, substantially as and for the purpose described.

Second, the case *u*, constructed as described, in combination with the dish *v''*, tube *w*, and cover *v'''*, substantially as and for the purpose described.

No. 55,565.—EDWARD WEISSENBOHN, Hudson City, N. J.—*Apparatus for Moulding Peat.*—June 12, 1866.—The plunger attached to the steam piston in its descent drives the peat from the hopper into the mould box, where it is compressed, and whence it issues through the narrowed throats which form the divided discharge opening.

Claim.—First, the combination of a reciprocating ram F, having a hammer-like action, the compression-box E and the open-bottomed mould or moulds *c*, substantially as and for the purpose herein specified.

Second, the cylinder B and its piston, in combination with the ram F, compression-box E, and mould or moulds *c*, substantially as and for the purpose herein set forth.

Third, the construction of the partitions *b b* of the mould frame with sharp cutting upper edges, substantially as and for the purpose herein described.

Fourth, the open-bottomed mould or moulds *c*, constructed with a downward taper, substantially as and for the purpose herein specified.

No. 55,566.—AMOS WESTCOTT, Syracuse, N. Y.—*Churn.*—June 12, 1866.—The radial dashers on the shaft present bevels to the right and left on the respective sides of the shaft, and a blast of air is introduced to facilitate the operation.

Claim.—First, the use of dasher-paddles having their faces bevelled or cut away diagonally, substantially in the manner and for the purpose above described, when combined with the main shaft, as above described.

Second, the manner of connecting the main shaft with the gearing and body of the churn, in combination with said shaft and the dasher-paddles, constructed substantially as and for the purposes above described.

Third, the combination of the parts mentioned in the preceding claims, constructed as above described, with the fan-wheel or blower, substantially as above described.

No. 55,567.—JOHN N. WILKINS, Chicago, Ill.—*Caster for Sewing Machines.*—June 12, 1866.—The caster wheel is journalled in a plate attached by prongs and screws to the leg.

Claim.—The combination of the leg with the forked plate A and the wheel by means of the screw and lug, substantially as specified.

No. 55,568.—LEVI WING and DAVID MYERS, Chicago, Ill.—*Wardrobe and Bedstead.*—June 12, 1866.—Explained by the claim.

Claim.—The combination of the wardrobe A, provided with its regular and distinct compartment for wearing apparel, and the bedstead E, arranged so as to fold into a recess in the rear part of said wardrobe, substantially as herein specified and shown.

No. 55,569.—Tobias and John W. Wise, Adrian, Mich.—*Sheep Shears.*—June 12, 1866.—The central blade affords opposed cutting edges to each of the side blades, and diminishes the danger of cutting the skin of the sheep.

Claim.—First, the employment in a pair of sheep-shears of a central blade *c*, arranged in such relation to the two blades of an ordinary pair of sheep-shears as to effect the object herein specified.

Second, the combination of the guard D with the central blade C and stops *g g* on the blades A A, substantially as specified.

Third, attaching the shank of the central blade by a screw and nut, or an equivalent thereof, which will permit of the said shank and the central blade also being removed from the shears at pleasure.

No. 55,570.—J. H. Wonderly, Williamsport, Penn.—*Machine for Sand-papering Woodwork.*—June 12, 1866.—The table is vertically adjustable; the sand-paper disk is rotated at the end of a jointed tubular extension arm, which also conducts the air away, through the hollow arm, to the exhaust fan.

Claim.—First, the combination of the cap *d*, or equivalent, jointed connecting pipes, and exhaust fan, operating substantially as described.

Second, the adjustable table B, provided with the sliding frame *p*, having the inclined planes *s s*, and the regulating screw C, arranged and operated substantially as shown and for the purpose set forth.

No. 55,571.—Moses M. Young, Chelsea, Mass.—*Screw-cutter.*—June 12, 1866.—The die is readily detachable from the stock; a centralizing guide tube is rigidly attached to the latter.

Claim.—First, the improved arrangement of the guide C, or its application directly to the die B, and so as to project therefrom, as specified, in combination with the application of the die to the stock, so as to enable the two to be separated without the necessity of first detaching the guide or centralizer from the die.

Second, the combination and arrangement of chip-discharging passages *k k*, with the guide and the die applied together, as set forth.

No. 55,572.—George Philip Zindoraf, Philadelphia, Penn.—*Paint Mill.*—June 12, 1866.—The spindle is prolonged through the runner to carry a feeding screw, and its rounded shoulder supports the balance-rynd.

Claim.—First, rounding the shoulder *r* of the part *q'* of the mill spindle N, and forming a rounded recess or cup in the top of the rynd *q*, into which the rounded shoulder *r* is received, substantially as and for the purpose as herein specified and described.

Second, the extension of the mill spindle up through the balance-rynd of the lower running stone, so as to attach the feeding screw thereto, as specified.

No. 55,573.—Moritz Baumgarten, New Haven, Conn., assignor to himself, Jacob Heller, G. C. Clark, A. S. Keeler, and Morris Steinhert, same place.—*Valve Arrangement for Organs, &c.*—June 12, 1866.—A single rod passes through the several wind chests and through an arm on each valve, felt collars fitting against the holes in the partitions around the rods to cut off the passage of wind; the rod is operated from a single crank by its own key.

Claim.—The arrangement of the rod L, provided with packing collars *a*, in combination with the valves D E and F, more or less, substantially in the manner and for the purpose herein described.

No. 55,574.—Charles B. Bristol, New Haven, Conn., assignor to himself and Philippe Koch, same place.—*Spring Bed Bottom.*—June 12, 1866.—A hinged metallic frame sets within the bedstead frame, and has transverse coil springs united occasionally by wire hooks.

Claim.—The combination of the spiral or helical springs C C, &c., when placed horizontally with the connecting rods or hooks *b b*, &c., or their equivalents, and the frame A A and A' A', when the whole is constructed, arranged, and fitted for use substantially as herein described and set forth.

No. 55,575.—Felix Brown, New York, N. Y., assignor to A. & T. Brown & Co., same place.—*Oscillating Engine.*—June 12, 1866.—An eccentric on the main shaft operates the rock shaft, which is connected by two pairs of links to the induction-valve stem; these move more rapidly at the opening and closing points than at other times. Links connect the engine frame with the cranks of the exhaust valve, and oscillate by the motion of the cylinder, causing these valves to remain open during the greater portion of the stroke.

Claim.—First, the rock shaft *j* and arm *i*, or other equivalent mechanism, in combination with the links *h h'*, arms *g g'*, valves *b b'*, and oscillating cylinder A, constructed and operating substantially as and for the purpose described.

Second, the arms *o*, links *p*, and studs *q*, in combination with the exhaust valves *c c'* and with the oscillating cylinder A, constructed and operating substantially as and for the purpose set forth.

No. 55,576.—DANIEL BULL, Amboy, Ill., assignor to himself, C. D. VAUGHN, and F. A. GIBBS.—*Mitre Box.*—June 12, 1866.—Self-adjusting, slotted, cylindrical guides are pivoted on vertical axes in the sliding sashes of the frame; the sides of the latter are adjustable, but preserve their parallelism. An index marks the angular adjustment, and a catch-spring maintains it while operating.

Claim.—First, the parallel-moving saw-guide frame, arranged, constructed, and operating substantially as and for the purpose set forth.

Second, arranging the slotted self-adjusting cylinders or rollers in sashes *e e*, or their equivalents, in combination with the parallel-moving frame and the stationary frame, all constructed and operating substantially as described.

Third, the adjustable gauge-stops I in combination with the saw-guide frame and the lugs or pointers *p*, all constructed and arranged substantially as described.

Fourth, the combination of the index plate *m*, cross-tie H *n*, parallel-moving frame and notched segment C², and support C³, substantially as described.

Fifth, the combination of the segment notched-plate C², support C³, parallel-moving saw-guide frame, and combined thumb-catch and lever or link F, all constructed and arranged substantially as described.

Sixth, the stationary saw-guide frame, constructed with a self-adjusting roller, in combination with the parallel-moving saw-guide frame, also constructed with a self-adjusting roller and with a front board or piece D D, the said parts being applied together on a mitre box, which is constructed and furnished with the appurtenances described, substantially as set forth.

No. 55,577.—H. M. CLARK, Meriden, Conn., assignor to CHARLES BLANCHARD, same place.—*Blind Fastener.*—June 12, 1866.—The slotted arm is attached to the horizontally rotating wheel, and, by pressure on the pin, vibrates the blind on its hinges. The wheel is operated by a pinion and hand-shaft.

Claim.—The combination of the slotted bar E, pin A, the wheels C and D, and the knob n, or its equivalent, constructed and arranged to operate together substantially as herein described.

No. 55,578.—DURFEE W. COGGESHALL, Providence, R. I, assignor to GEORGE E. CHURCH, same place.—*Pipe Wrench.*—June 12, 1866.—The hinged clasp which embraces the pipe is pivoted at one end to the stock and at the other end has a projection engaged by the tooth of the stock, to form a fulcrum in grasping and rotating the pipe.

Claim.—The pipe wrench made substantially as described, viz., of the lever and hinged jaws or clasp, provided with the tooth *a* and bearing *f*, and combined and arranged in manner and so as to operate substantially as and for the purpose as hereinbefore described.

No. 55,579.—CHAUNCEY DOWD, New Haven, Conn., assignor to H. W. GEAR, New York, N. Y.—*Canvas Stretcher.*—June 12, 1866—The four sides of the frame are forced out by vibrating the ends of the buttons against the eccentric plates.

Claim.—The swivel buttons *c* and eccentric abutting plates *d* in combination with the strips *a b* of the frame A, constructed and operating substantially as and for the purpose described.

No. 55,580.—OTIS E. DROWN. Pawtucket, R. I., assignor to DARIUS GOFF and DARIUS L. GOFF, same place.—*Carrier of Braiding Machine*—June 12, 1866.—These devices are to be used in those machines in which wide flat braid is made, and wherein the threads are interbraided at a point aside from the centre of a single serpentine race-plate. The object is to reduce the height of the guide post, and so prevent the carrier from binding in the race-plate by reason of the lateral strain caused by the tension of the yarn, and also to prevent unnecessary abrasion of the yarn and afford greater convenience and despatch in resuming work after a strand breaks.

Claim.—Combining the weight constructed as described, to be lifted from the bottom by the surrounding bight of the yarn, with the pawl, constructed, as described, to slide on a separate guide, and permitting the weight partially to pass it before being lifted to let off more yarn, substantially as described and for the purpose specified.

Also, the combination of the hooks or guides *r* and *s*, as described, with the tension-weight constructed as described, with the groove or hook *t* on one side of the bottom thereof for reeving the yarn, substantially in the manner described and for the purpose set forth.

No. 55,581.—JOHN FOCER, Glassboro', N. J., assignor to T. H. and S. A. WHITNEY, same place.—*Fruit Jar.*—June 12, 1866.—The thin metal sleeve has spiral indentations which fit the threads upon the neck and cover.

Claim.—The thin metal ring D, having screw-threads adapted to similar threads on the cover B and neck A of a preserving jar, all substantially as and for the purpose herein set forth.

No. 55,582.—PORTER A. GLADWIN, Boston, Mass., assignor to himself and HORACE M. LEE, Dorchester, Mass.—*Key for Locks.*—June 12, 1866.—The plate slips into the key-hole and is hooked to the bow of the key, which is thereby prevented from being turned.

Claim.—The within-described safe-guard for locking keys, consisting of the shank B in combination with the hook C, substantially as described.

No. 55,583.—ISAAC GREGG and CHARLES GREEN, Philadelphia, Penn., assignors to ISAAC GREGG, same place.—*Brick Machine.*—June 12, 1866.—Improvement on I Gregg's patent, September 19, 1865. The lever arms which operate the sweeps are vibrated simultaneously, but in opposite directions, by the connections described, actuated by motion of the single lever arm on one of the rock shafts.

Claim.—Operating the alternating "sweeps or mould clearers" of the said brick machine by means of the device consisting of the lever arms D D' and H H', and the bars E E' and I', and the fixed joint G, in combination with the two rock shafts C C' and the lever arm I, the same being constructed, arranged, and applied to operate together, substantially as described.

No. 55,584.—AUGUSTUS S. HADAWAY, Plymouth, Mass., assignor to himself and W. S. HADAWAY, same place.—*Ice Creeper.*—June 12, 1866.—The spiked plate is fixed underneath the heel by a wire and a toggle-lever, the latter pivoted to a hook which catches beneath the upper leather on the back of the heel.

Claim.—The combination of the wire F, the toggle-lever L, and the plate D, all for the purpose set forth and substantially as specified.

No. 55,585.—FRIEDERICH H. LAUTERBACK, Boston, Mass., assignor to himself and NEWMAN S. WAX, same place.—*Preserve Jar.*—June 12, 1866.—The can cover has a packing underneath and is depressed by a yoke which engages an ear on one side of the can, and is drawn down by a loop on the other side.

Claim.—The combination as well as the arrangement of the ring D and its stud c and curved arm d, with the staple C and the hinged loop L' applied to the body of the can, as specified, the cover of the can being provided with an annulus b of India-rubber, or its equivalent, and the whole being substantially as hereinbefore explained.

No. 55,586.—JOHN McCLAY, Hartford, Conn., assignor to himself and J. W. BLISS, same place.—*Driving Piles.*—June 12, 1866.—The yoke on top of the pile clasps the bars of the frame and guides the pile when driven by the monkey.

Claim.—The yoke or clamp d, constructed as described, and operating as set forth.

No. 55,587.—EDWARD MINGAY, Boston, Mass., assignor to THE BOSTON AND MAINE FOUNDRY COMPANY, South Reading, Mass.—*Ash Sifting Device in Cooking Stores.*—June 12, 1866.—Explained by the claim.

Claim.—Improved construction of the sifting hod, as described, and the arrangement of it and its chamber so as to extend directly underneath the grate, in order to cause the coals or ashes, when discharged from the grate, to pass vertically into the hod without first falling on an inclined plane or such a spout leading into such hod as to require the said ashes to be raked from it into the hod.

No. 55,588.—W. G. OLIVER, Buffalo, N. Y., assignor to himself and C. K. REMINGTON. same place.—*Safety Stop for Gun Locks.*—June 12, 1866.—A flat spring catches against the rear of the hammer and locks it against the nipple; when the spring is pressed against the stock the hammer is free to be cocked.

Claim.—The combination with the hammer of the gun lock of the automatic spring guard, so arranged as, in its normal position, to engage with the hammer and act as a detent thereto, and relieved from the same for the purpose of cocking by pressure towards the stock, substantially as described.

No. 55,589.—THOMAS PIMER, New London, Conn., assignor to HENRY W. PUTNAM, Bennington, Vt.—*Bottle Stopper.*—June 12, 1866.—The metallic cross-bar has a semi-spherical piece of rubber beneath it and is sustained by a link and a hook connecting it with a collar on the neck of the bottle.

Claim.—The combination of the collar a, hook b, rigid cross-piece d, tapering mass of soft material c, and the c, arranged substantially as and for the purpose herein specified.

No. 55,590.—ROBERT F. SISSON, Wickford, R. I., assignor to CYRUS H. MOORE and ASA SISSON, North Kingston, R. I.—*Photographic Bath.*—June 12, 1866.—Below the funnel-shaped false bottom is a cup to collect sediment and prevent the solution becoming turbid.

Claim.—The use of the independent false bottom B B, in combination with a photographer's bath, substantially as and for the purposes described.

No. 55,591.—F. O. & WM. W. TUCKER, Meriden, Conn., assignors to themselves and N. C. STILES, same place.—*Toy Top.*—June 12, 1866.—The weighted ring gives steadiness of rotation; in the whirling jack two cords are attached to the cylinder within the barrel, winding in opposite directions, so that drawing upon one winds the other around the cylinder, the top being spun by the latter.

Claim.—First, the arrangement of the ballast ring D, substantially in the manner and for the purpose specified.

Second, the combination of the two cords L and P with the hooked tube F and barrel E, constructed and arranged to operate substantially in the manner and for the purpose specified.

No. 55,592.—WILLIAM WILLIS, Birmingham, England, assignor to VINCENT BROOKS, London, England.—*Photographic Process for Copying Drawings, &c.*—June 12, 1866.—The surface is prepared with a solution containing a soluble chromate mixed with an acid, which will combine and form a compound with the oxide of chromium formed by the action of light, and darkening the impression produced by the unaltered chromate by means of aniline or other organic substance capable of forming a dark and insoluble compound with the said unaltered chromate.

Claim.—The improvements in processes for copying or reproducing, by the agency of light, drawings, engravings, lithographs, and photographs, and written and printed documents, herein described—that is, preparing the sensitive surface to be acted upon by light by the use of a solution containing a chromate mixed with an acid which will combine with the oxide of chromium formed by the action of light, and with the organic base used for development, and developing the picture by means of aniline, pyrrol, and other organic bases, which, when applied either in the state of vapor or liquid, are oxidized by the chromic acid and form therewith a dark-colored compound.

No. 55,593.—PETER DINZEY, St. Bartholomew, West Indies.—*Anchor.*—June 12, 1866.—The fluke is pivoted in the forked end of the shank, and has stops to limit its vibration in either direction.

Claim.—First, the shank A with its branches *a a'*, in combination with the adjustable fluke B, the whole being constructed and arranged substantially as and for the purpose specified.

Second, the combination with the above of the cross-pieces *c c'*, for the purpose described.

No. 55,594.—E. D. AVERELL, New York, N. Y.—*Electro-magnetic Attachment to Ruling Machines.*—June 12, 1866.—The contact of the edge of the paper with a guide roller closes the circuit by which the armature of an electro-magnet and the pens attached are thrown down on the paper; contact with another roller breaks the circuit and the ruling pens rise.

Claim.—First, the circuit break, consisting of a metallic lever and finger, with a piece *x* of non-conducting material, in the hub of the lever, and a spring *l*, or its equivalent, the whole applied and operating substantially as herein specified, in combination with a galvanic battery, or other generator of an electric circuit, and a ruling machine.

Second, the lever R' and finger *i'*, applied in connection with the lever R and finger *i*, substantially as and for the purpose herein specified.

Third, the rollers Q Q', applied in connection with the levers R R' and fingers *i i'*, and adjustable relatively to the cloth B and rollers D D, substantially as and for the purpose herein described.

Fourth, the combination of the posts L L', adjustable about their axes, the arms M M', and the rods N N', carrying the foot-pieces P P', rollers Q Q', circuit-break R, and lever R', substantially as and for the purpose herein specified.

Fifth, the foot-pieces P P', carrying the rollers Q Q', circuit-break R, and lever R', adjustable relatively to the rods N N', or other equivalent supports, by screws *g g*, substantially as and for the purpose herein set forth.

No. 55,595.—J. P. ABBOTT, Cleveland, Ohio.—*Eave Trough.*—June 19, 1866.—The hanger of the trough is adjusted vertically by means of the notched shaft which passes through the openings in the strap and is secured by a key.

Claim.—The adjustable arm A, notched standard B, slotted lips C D, and key F, in combination with the cross-tree E, strap G, and gutter H, as and for the purpose set forth.

No. 55,596.—ETHAN ALLEN, Worcester, Mass.—*Heating and Soldering Gun Barrels.*—June 19, 1866.—The gun barrels are clamped together, the joining ribs in position, and are brought to a soldering heat by a hot blast from the nozzle of the fire-box to which they are attached.

Claim.—First, the mechanism, substantially as described, for heating and soldering gun-barrels by blowing the heating blast through them.

Second, the clamps J, composed of set-screws and springs, substantially as described, for holding gun barrels while being soldered.

No. 55,597.—HENRY E. ANTHONY, Providence, R. I.—*Bolt-heading Machine.*—June 12, 1866.—The swivel punch is suspended from the beam, and is surrounded by a socket die, suspended above a vertically reciprocating tubular die, which forces the blank against the punch within the socket; the support of the bolt blank is withdrawn to a certain extent to allow excess of metal in the head to be forced into the shank.

Claim.—First, the combination of a swivel-punch F, with the die or collar which confines the bolt at d point above the heading chamber, substantially as described.

Second, removing the thimble R and pin T, or any other equivalent device which may be used to support or confine the lower end of the bolt, after such bolt has been partially headed, in order to allow any surplus stock to be forced from the upper chamber down through the head of the bolt, substantially as described.

Third, the sliding die-holder G, operating substantially in the manner described.

Fourth, the combination of the screw E and swivel-punch F, operating substantially as described.

Fifth, the manner of securing the hand-tool by the cap *n*, substantially as described.

No. 55,598.—JOHN ASHCROFT, New York, N. Y.—*Covering Steam Boilers.*—June 19, 1866.—Explained by the claim.

Claim.—Covering a steam boiler pipe, or other heater, with felt or other non-conducting material, when the latter is supported on a framework removed from and surrounding the former, not being in direct contact, but having an air-space intervening between said felt and boiler pipe, or other heater, constructed and operated substantially in the manner described and for the purpose set forth.

No. 55,599.—LEONARD BAILEY, Boston, Mass.—*Spoke-shave.*—June 19, 1866.—The stock is constructed in two parts and the handles divided lengthwise; the lower portions of the latter are in one piece with the cutter-rest, and the other parts of the handles, with the adjustable shank piece.

Claim.—The combination of the screws *f f*, and nuts *g g*, or their equivalents, with the clamp-bar *e*, and with the stock divided lengthwise into two parts or portions, constructed substantially in manner and so as to operate as described.

Also the arrangement of the clamp-bar *e*, with the screws *f f*, and nuts *g g*, and with the stock divided lengthwise into two parts or portions, constructed substantially in manner and so as to operate with the said screws and nuts, substantially as set forth.

Also, the combination and arrangement of the shoulders *i i* with the screws *f f*, the clamp-bar *e*, the cutter *d*, and bed *b* or the stock, substantially as set forth.

No.55,600.—HAYDN M. BAKER, Rochester, N. Y.—*Manufacture of Carbonate of Soda, &c.*—June 19, 1866.—Carbonate of magnesia and chloride of sodium are mixed in equivalent proportions in a vessel, and the carbonate changed into a bicarbonate by the introduction of carbonic acid. The bicarbonate decomposes the chloride of sodium, forming chloride of magnesium, and leaving bicarbonate of soda, which is precipitated. The chloride of magnesium may be decomposed by heat and the evolved hydrochloric acid collected.

Claim.—The application of the combined processes as herein described for the formation of carbonate and bicarbonate of soda, muriatic acid, and caustic lime, using for the said purposes the aforesaid carbonate of lime, carbonate of magnesia, and chloride of sodium, in the manner herein set forth, or any other processes substantially the same and which will produce the same intended effects.

Also, the construction and application of the boiler and lime-retort in combination, as herein described and represented by the accompanying drawings, for the purposes duly set forth.

Also, the application of heat and pressure without limitation in the process of forming bicarbonate of magnesia, and subsequent double decomposition of chloride of sodium and bicarbonate of magnesia, forming chloride of magnesia and bicarbonate of soda.

No. 55,601.—CYRUS W. BALDWIN, Boston, Mass.—*Hot-air Engine.*—June 19, 1866.—The water is exposed in the air-duct leading from the furnace to the cylinder, and being vaporized and carried to the cylinder tends to preserve the working parts.

Claim.—A vessel or reservoir placed between the furnace and the working cylinder, into which jets of water are introduced at certain intervals and in regulated quantities, for the purpose specified.

Also, the combination of the vaporizing trough, or its equivalent, with the hot-air duct, essentially as above set forth.

Also, the peculiar construction of the trough *a*, as made pyramidal or pointed at one or both its ends, and with the partitions C C C, in manner and to operate as before set forth and explained.

No. 55,602.—HAZEN J. BATCHELDER, Boston, Mass.—*Machine for Making Horseshoes.*—June 19, 1866.—Each of the rolls is composed of two semi-annular portions and is fastened to its mandrel by lips, a ring and clamping devices; the lower roller shaft is vertically adjustable.

Claim.—The mode of constructing each of the rolls, viz., not only with each of its dies made in sections and with lips to each, as described, but with rings to encompass the lips, the said sections and rings being held in place by a shoulder *l*, and a channel *i*, and clamps N N, or their equivalents, the whole being substantially as and for the purpose specified.

Also, the separate creasing-die O, made as described, in combination with two of the rings H H, and a sectional die R, arranged between them, the whole being as set forth.

Also, the combination of the lateral die P, for obtaining variation of width of the blank with the two ring-dies H H, the dies E O and R, and the mechanism for moving the roll-shaft C vertically, for the purpose of effecting the variations of thickness of the blank, as described.

No. 55,603.—HORATIO K. BATES, North La Crosse, Wis.—*Grate Bar.*—June 19, 1866.—The middle portion of each bar is depressed below the level of the ends, which rest on bearers.

Claim.—The off-set or dropper grate-bar herein set forth and shown.

No. 55,604.—ALEXANDER BECKERS, New York, N. Y.—*Ornamenting Wood.*—June 19, 1866.—Thin layers of wood are steamed, cemented together, and then pressed in a mould formed of metal plates having counterpart cameo and intaglio ornamentation. After cooling under pressure it is removed and the salient portions of the upper stratum removed, exposing the lower and differently colored wood.

Claim.—Producing mosaics of different woods, &c., in the manner herein specified.

No. 55,605.—ALANSON DINGHAM, Sherry, N. H.—*Chair Seat*—June 19, 1866.—The ends of the flexible strips are passed through slits in the edges of the frame and clamped therein.

Claim.—Attaching a seat composed of a single thickness of strips of oak, ash, or other suitable wooden material to a frame by means of slits, or slits and grooves, substantially as described.

No. 55,606.—J. F. BIRCHARD, Milwaukie, Wis.—*Slide for Extension Tables.*—June 19, 1866.—The slide bar for the support of the end leaf forms a part of the extension slide and is drawn from thence through the frame bar.

Claim.—Adding thereto or combining therewith the additional slide-bar for supporting the leaves of the table, substantially as herein set forth.

No. 55,607.—W. W. BLAIR, Lebanon, Tenn.—*Cotton Cultivator.*—June 19, 1866.—The coulters and adjustable scrapers are in the advance, and are followed by harrow teeth and cutters which revolve upon a horizontal longitudinal axis to give tilth to a strip whereon to plant.

Claim.—First, the arrangement of the adjustable revolving harrow H with the revolving and adjustable chopping knives J J upon the shaft I, substantially as and for the purpose herein specified.

Second, the scrapers M M, pivoted near their inner ends, as represented and adjusted by means of the rods *d d*, levers *c c*, and rack-bars *g g*, substantially as and for the purpose herein fully set forth.

Third, the arrangement of the adjustable scraper-feet D D, with the scrapers M M, and the cutters or coulters L L, substantially as and for the purpose described.

No. 55,608.—ERASTUS BLAKESLEE, Plymouth. Conn.—*Corset and Skirt Supporter.*—June 19, 1866.—A hoop of skirt wire encircles and is sustained by loops on the outside of the corset, and the buttons which slip thereon afford points of attachment for the skirt.

Claim.—First, the stay C, when arranged upon the body of the corset, so as to secure and support the skirt, substantially as described.

Second, in combination with a corset, buttons constructed and arranged so as to be self-adjusting on the stay C, as and for the purpose specified.

Third, a corset supporter, constructed substantially as described, and attached to the corset in the manner herein set forth.

No. 55,609.—L. W. BROADWELL, New Orleans, La.—*Ordnance.*—June 19, 1866.—The re-enforce has a wide groove extending around its inner surface; this is slipped while hot over an enlargement on the breech so that the inner flanges of the sleeve bear upon the shoulders of the enlargement and bring a longitudinal tension upon the breech.

Claim.—The exterior re-enforce B, with a depression on its interior periphery corresponding to the enlargement on the gun, with shoulders *b b'*, substantially as and for the purpose described.

No. 55,610.—H. BUCKNALL, Darien, Wis.—*Washing Machine.*—June 19, 1866.—The concave bed is journalled loosely in the box; a cam wheel elevates and depresses the rubber.

Claim.—First, the combination of the rubbing concave B, hung loosely in the machine by journals or pivots, with the uprights G, cam H, and rubbing bar C, substantially as shown and described.

Second, the cam H, in combination with the uprights G G, and pin *q*, when constructed and operating substantially as and for the purpose specified.

Third, the thumb-screw *f*, for raising and lowering the rubbing-board C, in combination
with the rock-shaft F, and rod E, as shown and described.

Fourth, in a washing machine constructed substantially as described, the employment of
a joint *e*, at the connection of the rod E, and the cross-rod *d*, substantially as and for the
purpose specified.

No. 55,611.—JAMES BULLOCK, Rendsboro', Vt.—*Sap Bucket Hoop.*—June 19, 1866.—The
clamp by its hooks catches upon the trunk of the tree and affords a hook from which to suspend a bucket for collecting the drip of the spiles.

Claim.—The form of three hooks combined in one piece of wire in the circle.

No. 55,612.—JOHN A. BURCHARD, Beloit, Wis.—*Hoe and Corn Planter Combined.*—June
19, 1866.—A cylinder attached to the handle contains the seed, the feed cylinder is operated
by a rod extending beyond the end of the hoe.

Claim.—The rod D, hoe C, spring E, and roller B, in combination with the stirrer I, gauge-
plate K, and cylinder A, as and for the purpose set forth.

No. 55,613.—JOHN BURKE, Sycamore, Ill.—*Breech-loading Fire-arm.*—June 19, 1866.—
The barrel tilts vertically on a hinge to expose the rear of the bore for loading ; the hinge is
forward on the stock and its vibration actuates the retractor, which is retained by contact
with the breech block ; the locking bolt is recessed into the lock plate and breech.

Claim.—First, the combination of the tipping-barrel with the front part of the hinge and
stock, and of the rear part of the stock and hinge with the breech-piece or plug, when the
parts are arranged for joint operation, substantially as described.

Second, the jointed retractor, constructed substantially as described, to expel the cartridge
by a positive motion.

Third, moving the retractor in one direction by a cam, and closing it by grazing the breech,
as described, whereby the retractor is drawn in without using a spring, and friction prevented
on the flange of the cartridge.

Fourth, the combination of the tipping-barrel, the steady-pin I, the locking-bolt L, the
cone-plug K, and the shoulder K', substantially as described, to hold the barrel firmly while
firing.

Fifth, constructing the locking-bolt so that it shall slide in a groove partly in the lock and
partly in the breech, as described, to prevent strain on the lock, as set forth.

No. 55,614.—JONATHAN BURT and LEONARD F. DUNN, Oneida, N. Y.—*Cutting Green
Corn from the Cob.*—June 19, 1866.—The sliding frame is worked by a treadle and forces
the ear against spring cutting edges arranged circularly and expanded by the oblique rods.

Claim.—The sliding frame D, in connection with the springs *f*, cutters *e'*, scrapers *h*,
any or all them, and the tube C, or its equivalent, all arranged to operate in the manner
substantially as and for the purpose set forth.

Also, the oblique rods *b* of the frame D, connected with the plates or springs *d'*, for the purpose of expanding the cutters *e'* and springs *f'*, substantially as and for the purpose specified.

No. 55,615.—RUSSEL BURTON, South Adams, Mass.—*Steam Plug Valve.*—June 19,
1866.—The frustum plug has an oil reservoir and outlets to the contacting parts.

Claim.—First, the arrangement, in connection with the pipe A, of the conical enlargement
B, conical plug E, spring H, and screw-cover D, when the parts are constructed and combined as described and represented.

Second, the combination of the reservoir J, one or more, with the conical plug E, substantially as described and for the purpose set forth.

No. 55,616.—IRA D. BUSH, Detroit Mich.—*Lock.*—June 19, 1866.—The tumblers are
hung on one or both sides of the bolt so as to slide and, swing upon it. The knob-spindle
is secured to the casing by a collar and a yoke which carries the horns to actuate the bolt.

Claim.—First, the sliding and swinging tumblers G, hung upon the bolt of the lock, and
arranged and operating substantially in the manner and for the purpose specified.

Second, the collar K, of the knob, spindle, or shaft Q, in combination with the opening
S T in the lock-plate, the yoke U placed over such shaft and interlocking with the said collar,
substantially as and for the purposes described.

No. 55,617.—ENSIGN A. BUSHNELL, Dodge county, Wis.—*Machine for Sharpening Horseshoe Calks.*—June 19, 1866.—A cutting burr is attached to a shaft and revolved by a hand
crank, the stock maintaining it in position against the calk which is to be sharpened.

Claim.—The arrangement of all the parts as herein set forth, as and for the purpose described.

No. 55,618.—D. P. BUTLER, Boston, Mass.—*Lifting Bar.*—June 19, 1866.—In this lifting
bar for gymnastic purposes, the handles are so arranged that the inner surface or palm of

each hand is brought directly in line with the centre of the bar from which the weight is suspended.

Claim.—The construction of the lifting bar with the surfaces *c* in line with the centre of the bar *a*, substantially as set forth.

Also, the socket or socket-piece on the bar and the pivot on the ring, substantially as shown.

Also, the elastic cushion placed between the ring and bar.

Also, the construction of the bar with the sockets or socket-pieces on opposite surfaces, thereof, substantially as and for the purpose set forth.

Also, making the ring detachable from the weight-rod, substantially as set forth.

Also, combining with the ring the spring *l* by which the ring and bar are held together substantially as described.

No. 55,619.—JEREMIAH CAMPBELL, Lancaster, Penn.—*Machine for Pressing Cigars.*—June 19, 1866.—The press has side screws working horizontally and a vertical screw so placed as to be over the stack containing the cigars; the side press-board slides in the slotted sides of the horizontal boards. The cigars are arranged upon the boards with intervening slats.

Claim.—A press provided with a vertical screw I and horizontal on side screws H H, when the top and bottom cross-pieces C D are provided with grooves *d* on one end in combination with a vertical press board E, arranged, constructed and operating in the manner and for the purpose specified.

Also, the loose slates J provided with blocks K on one or both sides at the ends, and when of double or triple length, intermediate blocks, all half the thickness desired to form the chamber when employed in the manner and for the purpose set forth.

Also, the partition boards G L, with the extended arms and followers F, in combination with the vertical press board E, constructed and operating in the manner and for the purpose specified.

No. 55,620.—GEORGE J. CAPEWELL, West Cheshire, Conn.—*Spring Lancet.*—June 19, 1866.—The lips of this fleam rest against the side of the protuberance formed by the vein and hold it in position for the lancet.

Claim.—The slotted cap-piece K for the casing A having outward bent lips or flanges *h* upon each side of its slot or opening, substantially as and for the purpose described.

No. 55,621.—J. CARTON and WILLIAM RALPH, Utica, N. Y.—*Milk Pail.*—June 19, 1866.—Explained by the claim.

Claim.—As an article of manufacture a wooden milk pail with a tin lining, constructed as described.

No. 55,622.—S. CASE and A. W. PRATT, Pultneyville, N. Y.—*Atmospheric Churn Dasher.*—June 19, 1866.—The dasher has a concavity on its lower face and radial wings extending therefrom; the air is driven downward through the valved, tubular dasher shaft.

Claim.—The combination of the adjustable and removable valve D with the rod A and tube G, substantially as and for the purpose described, the rod being provided with the bulb B, having the seat *a* for the reception of different sized wings, as set forth.

No. 55,623.—RICHARD COLLIER, Springfield Ohio,—*Pruning Instrument.*—June 19, 1866.—Explained by the claim.

Claim.—A pruning tool the blade of which is sharpened upon the chisel-formed point and also upon both the edges, the same being attached to a handle, substantially as set forth.

No. 55,624.—DENNIS CONLON, Portland, Me.—*Portable Door Fastening.*—June 19, 1866.—The spurs of the plate enter the door casing and to its outer flange the button is pivoted.

Claim.—The combination of the plate B having the spurs *b* and shoulder *s*, with the pivot *n* and button A, constructed as described all as and for the purposes set forth.

No. 55,625.—MOSES G. CRANE, Boston, Mass.—*Egg Beater.*—June 19, 1866.—Segment gears on the axis come consecutively into action upon the pinion, and revolve it in reverse directions. The apparatus is temporarily attached to a plate fastened to the table.

Claim.—First, the combination of the segment gears and pinions with the spindle, beating-wires, and standard, when said segment gears are constructed and arranged to rotate horizontally, substantially as set forth.

Second, in combination with the standard *b*, spindle *a*, and wires *c*, arranged and operating as described, the plate *q*, with its finger *r*, arranged to hold the standard *b*, substantially as described.

No. 55,626.—JOSEPH H. DAVIS, Allegheny City, Penn.—*Pump for Deep Wells.*—June 19, 1866—The upper valve box is depressed into the chamber above the foot valve, in which place it is readily rotated to screw on to the lower valve box for the purpose of withdrawing it.

Claim.—The enlargement B of the valve chamber A, when constructed and operating for the removal of the foot valve D, substantially as herein described and for the purpose set forth.

No. 55,627.—SAMUEL F. DAY, Ballston, N. Y.—*Sounder Magnet.*—June 19, 1866.— The posts which support the armature lever and the contact screws are mounted upon a bridge plate supported at its ends upon an insulator board so as to give greater resonance to the blow of the armature.

Claim.—First, the combination of the posts *d d* with the raised plate or bridge C, in the manner and for the purpose set forth.

Second, the combination of the metallic frame A, sounding board or insulator B, and metallic bridge or arched plate C, for the purpose set forth.

No. 55,628.—LEWIS P. DECKER, Brooklyn, N. Y.—*Ferry Bridge Gate.*—June 19, 1866.— The gates are simultaneously opened by the revolution of their heel posts by means of the racks and pinions, which are actuated by windlass and weight.

Claim.—First, the combination of the gates C and D, constructed and arranged as herein described, with the shafts E and F, and with the bridge arches or other suitable supports, substantially as described and for the purpose set forth.

Second, the combination of the ratchet bar I with the gear wheels G and H, and with the shaft E and F, substantially as described and for the purpose set forth.

No. 55,629.—H. J. DEISSNER, Waukesha, Wis.—*Making Sirup from Corn.*—June 19, 1866.—Consists of a concentrated extract of malt and cornmeal sweetened.

Claim—The within described process of making sirup from corn, by following the various manipulations which are specified.

No. 55,630.—PAUL DENNIS, Schuylerville, N. Y.—*Shovel Plough.*—June 19, 1866.—The shovel has adjustable reversible wings of twisted form and double cutting edges.

Claim.—First, sharpening or providing the wings D D with double cutting edges, as and for the purpose described.

Second, the wings D D, constructed in such a manner as to be capable of being reversed in position, so as to throw the earth outward to a greater or less distance, and also to be capable of being expanded or contracted, as occasion may require, as and for the purpose set forth.

No. 55,631.—T. B. DOOLITTLE, Ansonia, Conn.—*Fruit Box.*—June 19, 1866.—Explained by the claim.

Claim.—A fruit box formed of two end pieces, in combination with a single piece bent around said blocks or end pieces, and overlapped, the whole constructed and arranged substantially as set forth.

No. 55,632.—JOHN D. EDMOND and WILLIAM W. WIRT, Washington, D. C.—*Table Fork.*—June 19, 1866.—Explained by the claim.

Claim.—Providing a table fork with a file edge or surface, substantially as specified, for the purpose of sharpening a knife.

No. 55,633.—J. S. ELKINS and J. T. GREEN, Marquette, Wis.—*Automatic Gate.*— June 19, 1866.—The pressure of the wheels on the track depresses the levers and by means of connecting rods vibrates the gates laterally on their pivots of suspension in the bridge piece above.

Claim.—First, the method of hanging the gate to the bars F F, by which they are kept in contact by positive force while closed, without the use of latches or catches of any kind, substantially as shown and described.

Second, the combination of the timbers H H, levers I I, and links J J, rods *l l*, bell-cranks K K, and links *m m*, all arranged and operating substantially as shown and described.

No. 55,634.—JOHN C. FELLOWS, South Adams, Mass.—*Washing Machine.*—June 19, 1866.—The oscillating rubber rests by its own weight upon the concave which is supported upon a second pivoted frame whose bars work upward between those of the concave bed above it.

Claim.—First, the combination of the vibrating frame E with the frame B, the slats of frame E rising between the slats of frame B, substantially as described.

Second, the vibrating roller whose frame has vertical motion, as described, with the fixed and vibrating frames B and E, substantially as described.

No. 55,635.—EDWARD A. FIELD, Sidney, Maine.—*Fishing Net.*—June 19, 1866.—This bag-shaped net has a lower sinker from which rises a bait holder with a float; the bait is exposed when the net lies collapsed, but is enclosed with the capture when it is raised.

Claim.—The improved manufacture of fishing net or apparatus made substantially as described, with the sinker, the ground-guard, and the mouth-hoop, the lip-hoops or the same and the bait-line and float, arranged and combined together and with netting, and so as to operate substantially as specified.

No. 55,636.—D. D. FOLEY, Washington, D. C.—*Postal Letter Box.*—June 19, 1866.—Explained by the claim.

Claim.—A postal letter box having an ante-chamber in which packages are deposited, when said ante-chamber has a sliding bottom, which is withdrawn when the entrance port is closed, and *vice versa*, said sliding bottom being connected to and operated by the valve which covers the entrance port by a lever or tilting bar and so adjusted that entrance to the receptacle below the ante-chamber will always be barred either by the valve over the entrance port or by the sliding bottom, substantially as described.

No. 55,637.—WILLIAM FREDERICK, Ashland, Penn.—*Boot and Shoe Stretcher.*—June 19, 1866.—The toe end is raised by a roller which is projected by a pinion and rack, and the instep is raised by a screw. The respective parts are actuated separately or unitedly.

Claim.—First, an improved boot and shoe stretcher formed in two parts, A and D, hinged together at their forward ends by a treble-jointed hinge E, substantially as described and for the purpose set forth.

Second, the combination of the rod J, pinion wheel I, rack B, and roller or rollers C, with each other and with the parts A and D, of the last, substantially as described and for the purpose set forth.

No. 55,638.—C. L. FRINK, Rockville, Conn.—*Gauge for Boilers.*—June 19, 1866.—A safety cock in line with the opening into the boiler allows a stream of water to pass and checks the downward passage through the glass of a steam jet from the upper boiler connection.

Claim.—The arrangement of the cocks C D, in combination with the socket B and tube A, and operating in the manner and for the purpose herein specified.

No. 55,639.—C. L. FRINK, Rockville, Conn.—*Safety Valve for Boilers.*—June 19, 1866.—The hinge support of the lever upon the valve stem avoids lateral strain thereon. The packing ring on the valve face is fastened by a clamping disk and screw.

Claim.—First, the hinged supporter D, in combination with the valve stem C and lever P of a safety valve, substantially as and for the purpose described.

Second, the central screw *f*, and clamping plate *e*, in combination with the packing piece *d*, and valve B, constructed and operating substantially as and for the purpose described.

No. 55,640.—C L. FRINK, Rockville, Conn.—*Gauge Cock.*—June 10, 1866.—By turning back the gate which is hinged to the handle, access for the removal of obstructions is had to the passage and the tubular elastic valve which closes by pressure on its conical seat.

Claim.—The flexible and elastic valve D, with a central passage *c*, in combination with the conical seat E, spindle C, adjustable gate G, and body A of a gauge cock, constructed and operating substantially as and for the purpose set forth.

No. 55,641.—SAMUEL GARDINER, Jr., New York, N. Y.—*Lighting Gas by Electricity.*—June 19, 1866.—The suspended tassel is connected to one of the poles of a battery, and by the application of its wire to the other pole attached to an insulator on the burner, an electric lighting connection is made.

Claim.—First, the combination with a gas-burner of an electrical conducting cord and tassel D D', connected with the poles of a battery, substantially as described.

Second, the combination with the above of the stiff wire G, igniting finger J, lever L, and spring M, substantially as described.

Third, in combination with the electrical lighting devices herein described, the non-conducting or insulating stud R, employed in the manner described.

No. 55,642.—SAMUEL GARDINER, Jr., New York, N. Y.—*Turning Gas-Cocks by Electro-Magnetism.*—June 19, 1866.—The ratchet wheel on the axis of the gas-cock is turned by the successive pulsations of the armature, which is alternately actuated by electric connection and a retracting spring.

Claim.—First, turning a gas-cock by means of a sliding-rod E, and click G, acting directly upon the toothed-wheel H, on the axis of said gas-cock, and employed in combination with an armature B, and magnet A A, substantially as described.

Second, the combination with the armature B, of the guides C C, and springs K, substantially as and for the objects specified.

Third, the stop or stud *e*, attached to the sliding-rod E, and employed to limit the motion of the wheel H, as set forth.

Fourth, the combination with the wheel H, rod E, and click G, of the retaining spring J, applied and operating in the manner and for the purpose substantially as described.

Fifth, in combination with the apparatus herein described, the helical spring D, adjustable collar M, and thumb-screw N, arranged as described and employed for the purposes specified.

No. 55,643.—O. S. GARRETSON, Buffalo, N. Y.—*Butt for Blinds.*—June 19, 1866.—The loose pintle has a cam and a notch which cause it to rise as the shutter opens and drop over a locking stop when open. The leaves have a conical seat and bearing to give a certain freedom of motion to the pintle.

Claim.—First, the loose pin C, provided with the cam *e*, and notch *f*, in combination with the wings A B, provided respectively with the flange *f*, and stop *g*, arranged and operating substantially in the manner and for the purpose set forth.

Second, in combination with the loose pin C, the conical seat and corresponding bearing of the parts A B, or their equivalents, arranged and operating as shown and described.

Third, the free axial pin or bolt C, operated by its own weight, in connection with suitable stops on the wings of the butt for forming a self-fastening hinge, substantially as set forth.

No. 55,644.—O. S. GARRETSON, Buffalo, N. Y.—*Shutter Hinge.*—June 19, 1866.—The cheeks of the leaves are made hollow to allow space between them when closed for the pivoted locking bar which is suspended from a notch in the upper edge of one leaf, and as the shutter is opened rides over and falls behind an inclined spur upon the other leaf.

Claim.—The locking-bar H, in combination with the parts B and D of a hinge, constructed and operating substantially as shown and described.

Also, forming the wings B D with concave faces in combination with the pendent bar H, when said bar moves on a plane with the wings, substantially as and for the purposes set forth.

No. 55,645.—DAVID GILSON, Nashua, N. H.—*Artificial Leg.*—June 19, 1866.—The stump is received in a padded socket which is supported by springs within the sheath.

Claim.—An adjustable socket or pad supported upon springs or their equivalent, for the purposes as herein set forth; I do not limit my claim to the particular form as herein shown, but extend it to any other substantially the same.

No. 55,646.—A. B. GLOVER, Derby, Conn.—*Bolt-heading Machine.*—June 19, 1866.— The bolt blank is placed between two vertical gripping dies; the slide carrying the headers has a lateral intermittent motion, first carrying a die over the bolt blank to stave it, and then the heading die which performs two operations upon the head, alternating with two pairs of converging hammers working horizontally against the four sides of the head.

Claim.—First, the two pairs of levers Y Y Y' Y', with the forming dies X X X' X', arranged so that one pair will operate at right angles to the other pair, in combination with the two heading dies O O*, all arranged to operate in the manner substantially as and for the purpose set forth.

Second, the attaching of the heading dies O O* to a transverse or laterally moving slide I, fitted to the vertically moving slide H, and operated through the medium of the arm J, rock-shaft K, arm L, and the cam M or their equivalents, for the purpose of bringing the dies O O* over the bolt-rod at the proper time, substantially as shown and described.

Third, the holding dies S S*, in combination with the lever V, operated by the notched wheel W, for the purpose of holding the bolt-rod during the heading operation, and releasing the same after said operation is performed, constructed and arranged substantially as described.

Fourth, the lever B*, in combination with the pawl *v*, in the driving shaft notches *w*, in the driving pulley A*, rod D*, spring *d*, and cam E*, all arranged substantially as shown to automatically stop the machine at the completion of the heading of the bolt, substantially as shown and described.

No. 55,647.—J. S. and H. F. GRAY, Chelsea, Mass.—*Hanging Mirrors.*—June 19, 1866.— The face of the bearing plate on each side of the mirror is out of the plane of its axis of rotation, and lies against a correspondingly inclined surface of the bed plate on the standard; the inclination of one bearing plate is the converse of that of the other, so that partial rotation of the mirror increases the bearing force and holds the mirror in position.

Claim.—The friction plates *d f*, having irregular contact surfaces, constructed and operating together as and for the purpose substantially as set forth.

No. 55,648.—ADAM HAGNY, Keokuk, Iowa.—*Snap Hook.*—June 19, 1866.—The spring passes through an opening in the shank, and is fastened by the projections which are hammered down upon its base.

Claim.—A snap hook A, having an eye *b* in its shank, through which passes the spring B, which is attached by projections *c c* on the side, and the projection *c* at the end, which holds the said spring by being flattened down upon it, constructed and arranged as described.

No. 55,649.—L. C. HAINS, Bedford, Ohio.—*Cheese Vat.*—June 19, 1866.—The pan is hinged to the vat and rests upon pins within it; the contents are warmed by a furnace beneath; the whey drawn off by a strainer; adjustable legs permit the inclination of the vat.

Claim.—First, the detachable hinges E, in combination with the pan C and vat B, when arranged as and for the purpose set forth.

Second, the brace-rib J, pins *n*, in combination with the pan C and vat, arranged as and for the purpose set forth.

No. 55,650.—ALEXANDER HAMILL, Baltimore, Md.—*Railway Chair.*—June 19, 1866.— Explained by the claim.

Claim.—The construction of the chair with its hinge L at the outside end, when arranged and fastened to the cross-tie B by a bolt F, with a flat bar-head G below, and a forked key H above, substantially as herein described and for the purposes set forth.

No. 55,651.—PRESCOTT V. HARRINGTON, Attleboro', Mass.—*Skirt Supporter.*—June 19, 1866.—A metallic ribbon is doubled to form a loop; the ends terminate in a catch and pin respectively; the latter is passed through the dress and engaged by the catch; the dress rests in the loop.

Claim.—A skirt-supporter consisting of a loop and pin tongue and catch combined, the article being substantially as specified.

No. 55,652.—DAVID GREENE HASKINS, Cambridge, Mass.—*Gas Stove for Heating.*—June 19, 1866.—Has an outer casing and inner radiating chamber supplied with air from the apartment or by pipe with air from outside. The gaseous results are carried off by a separate pipe.

Claim.—First, the combination of the air-heating chamber B with the chamber D, the concentric casing *a*, and interposed radiating material, substantially as and for the purpose specified.

Second, the combination of the tapering chamber G with the chamber D, substantially as and for the purpose specified.

Third, the combination of the air-heating chamber B, the escape pipe *o*, and the partitioned mantel or radiating chamber F, as and for the purpose set forth.

Fourth, the arrangement of the upper burners *g* with the chamber D, in combination with the chamber B and space H, as and for the purpose specified.

No. 55,653.—ANSON HATCH, New Haven, Conn.—*Fishing Reel.*—June 19, 1866.—The skeleton spool for holding the line is made of two pieces stamped out of sheet metal into form and enclosed in an encircling band.

Claim.—The skeleton spool in combination with the band A A, when the whole is constructed, arranged, and fitted for use, substantially as herein described.

No. 55,654.—ELEAZAR and DAVID HINCHLEY, Worcester, Mass.—*Mowing Machine.*—June 19, 1866.—The arrangement of devices for operating the cutter is described in the claim.

Claim.—The combination and arrangement of the cam-wheel F, rocker lever E, connecting rod D, the shaft G, and its bevelled pinion H, and annulus I, with the driving wheel K, and the knife C, the said rocker-lever being provided with friction rollers or projections for the cam-wheel to operate against, and the whole being substantially as specified, and for the purpose of operating the knife C, by the revolution of the wheel K.

No. 55,655.—AARON A. HINKLEY, Boston, Mass.—*Refrigerator.*—June 19, 1866.—The safe has an ice tray supported on rods and a coiled water discharge pipe; air is admitted through a registered opening in the cover and passes by an air duct therein to the sides of the chamber.

Claim.—The combination and arrangement of the chamber *d* and air passage *e* of the cover, with the passages *f f* and the ice box or pan C arranged in the case A, substantially as specified.

Also, the combination and arrangement of the cold water coiled pipe D and its discharging branches, with the case A and the ice pan, the chambered cover and its air passage and the air passages leading therefrom, the whole being substantially as described.

No. 55,656.—JOHN S. HOAR, West Acton, Mass.—*Bench Vise.*—June 19, 1866.—The bed plate of this rotary bench vise has a hole to receive the pivot of the jaw carrier, whose base has three curved slots with clamp screws. The shank of the movable jaw is grooved lengthwise to receive the screw and standard employed in giving motion.

Claim.—As a special improvement in bench vises of the kind described or those to turn horizontally on a bed plate, the combination of the long curved back slot *d* and its screw *f*, and two wide curved slots *e e* and their set screws in the stationary jaw carrier C and with the jaws C D, all constructed and arranged to operate together substantially as specified.

No. 55,657.—LYMAN J. HOLCOMB, Nunda, Ill.—*Potato Planter.*—June 19, 1866.—A device on the wheel rings a bell as a signal for the driver to drop a potato, which is conducted by a spout to the space behind the share which opens the furrow.

Claim.—First, the combination of the pole E, frame F, seat G, hopper H, tube I, beam M, and plough L, arranged and operating as and for the purposes specified.

Second, in combination with the above and the wheels A and axle B of a wagon, the circular plate R, pins *a*, bell S, lever T, and spring *d*, arranged and operating substantially as and for the purposes set forth.

No. 55,658.—GEORGE C. HOWARD, Philadelphia, Penn.—*Machine for Pressing and Moulding Pliable Materials.*—June 19, 1866.—The treadle depresses the vertically sliding frame, which is elevated by the springs when the foot is withdrawn. The intaglio attached to the yoke of the frame is brought down upon the cameo die on the table, each being adapted to its work and centred by the adjusting devices.

Claim.—First, the form of the housings A A, combining all the necessary bearings in one piece with the table B, constructed substantially as described.

Second, providing the treadle and housings with two or more fulcrum bosses, substantially as and for the purpose specified.

Third, the slots in the stand P at right angles to those immediately under it in the table B, substantially as and for the purpose described.

No. 55,659.—H. C. HUXT, Amboy, Ill.—*Window Sash Supporter.*—June 19, 1866.—The case has an angular slot and double inclines. The wedge which binds the sash is adjusted vertically by the knob and assisted in its motions by a roller.

Claim.—First, the case A with its double inclines b b'' and its corresponding slot e, all operating as and for the purpose shown.

Second, the anchor C with its double inclines $f f''$ and its knob or finger piece r, operating as and for the purpose shown.

Third, the friction roller E, operating as and for the purpose shown.

No. 55,660 —JAMES B. HUNTER, Ashley, Ill.—*Gang Plough.*—June 19, 1866.—The ploughs are adjusted laterally and vertically by means of crank screws, slides, and uprights, and are connected to the bolster above the axle by a rod extending from the rear of the plough stocks.

Claim —First, the bolster G, screws H H I, and plough beams J J when used in combination with the rods K L, and all arranged substantially as and for the purpose set forth.

Second, the attaching of the plough beams J J to the bolster G through the medium of the rods, K, placed at the under sides of the beams J and fitted loosely at their front ends on a rod L at the rear of the bolster, substantially as and for the purpose specified.

Third, the raising of the ploughs P out of the ground by means of a rod M, crank shaft N, and lever O, all arranged substantially in the manner as and for the purpose set forth.

Fourth, the adjusting of the shares or points of the ploughs in a greater or less inclination downward by means of the screw rods N connected to the upper parts of the standards Q, substantially as shown as described.

No. 55,661.—JOHN C. HURSELL Boston, Mass.—*Cutter for Dovetailing Machine.*—June 19, 1866.—The cutter head is slotted longitudinally and the blade is fastened therein, forming the two cutting edges.

Claim.—The construction of the cutters as herein described for polishing and condensing the surfaces of dovetail tenons or mortises.

Also, providing the upper outer corners of a cutter constructed as above claimed with outward cutting lips, as and for the purpose specified.

Also, constructing the conical cutter head a with a slot to receive a solid cutter with opposite cutting edges, the cutter being inserted in said slot and confined to the head, as described.

No. 55,662.—WELCOME JENKES, Manchester, N. H.—*Cleaner for Ring Traveller Spinning Machines*—June 19, 1866.—The cleaner is cast in sections of a length suitable for cleaning from eight to sixteen rings at once, and is fastened to the back of the rail and level with the top of it by screws passing through holes, so that the standards (one for each ring) appear above the rail. An iron pin is held horizontally on each of these standards in a plane with the top of the ring and set radially with it; each pin is adjustable by means of screws nearer to or further from the ring to adapt it to different sizes of travellers.

Claim.—An adjustable cleaner for ring travellers, made substantially as above described.

No. 55,663.—F. W. JENKINS, Brooklyn, N. Y.—*Railway Car for Preventing Accidents.*—June 19, 1866.—The roller is revolvable on a vertical axis dependent from a pedestal underneath the car platform, and is capable of vertical adjustments. Its office is to remove obstructions from the rail.

Claim.—A roller or rollers hung in a vertical plane in front of the wheels of a railway car, and so as to turn thereon, substantially as and for the purpose described.

Also, so hanging the roller C to railway cars that it can have a play in a vertical direction, substantially as described and for the purpose specified.

No. 55,664.—N. B. JEWETT and E. EVERSON, Haverhill, Mass.—*Machine for Facing Boot and Shoe Buttons.*—June 19, 1866.—The shaft runs the cylinder covered w th sand paper, and also the fan which draws in and removes the dust resulting from the facing operation.

Claim.—When combined and arranged as described and so as to operate in the manner and for the purpose specified, the shaft d, grinding cylinder b, (these being parallel with each other, and operated from wheel k, by one belt g,) the cases c and e, and the angular fan blades f.

No. 55,665.—WILLIAM JOHNSON, Milwaukee, Wiss.—*Caster for Furniture.*—June 19, 1866.—The slotted rose plate slips over the shank of the caster so as to rest upon one shoulder when in use, and suspend the caster by the other when the leg is lifted.

Claim.—First, the combination of the rose plate D, in one or more parts, the chamber h, and the collar G, for the purpose described.

Second, the shoulder i, in combination with the rose plate D, thimble l, collar G, and chamber h, substantially as shown and described and for the purpose set forth.

No. 55,666.—EDWARD E. JONES and G. L. KITSON, Philadelphia, Penn.—*Spinning Top.*—June 19, 1866.—The air enters at the axial opening, and is discharged by centrifugal action at the peripheral openings, vibrating a musical tongue in its passage.

Claim.—The holes A A, with the hole C in the centre of the top, in combination with a musical attachment, for the purpose herein described;—an æolian or musical top.

No. 55, 667.—FRANK JONES, Boston, Mass.—*Manufacture of Brick.*—June 19, 1866.—The bricks are saturated with oil, and covered on their facing portions with mastic.

Claim.—First, the process substantially as above described of applying to bricks a preparation of mastic or cement.

Second, the apparatus constructed and operating substantially as above described for applying mastic to bricks separately before being laid.

Third, as an improved article of manufacture, a mastic-covered brick prepared substantially as above described.

No. 55,668.—WILLIAM H. KARICOFE, Harrisonburg, Va.—*Corn Planter.*—June 19, 1866.—Explained by the claim.

Claim.—The combination of the several parts above described, in the construction of a machine that will furrow two rows and drop therein at regular intervals corn and ashes or similar fertilizer, and will cover the same, removing clods and small stones, that by means of the adjustable screws in the cups the amount of corn or of the fertilizer may be fixed by the operator; that by means of the cross-bars on the right wheel the land will have the appearance of being checked, and by means of the driver the machine may be thrown out of gear and removed to any part of the field without the shaft turning.

No. 55,669—CHARLES BRIGHT KEYS, Washington, D. C.—*Spark Arrester.*—June 19, 1866.—The hood has its mouth presented in the direction of progress, and the air draught carries the sparks through a tube to the tender, where they are extinguished in a water tank.

Claim.—First, the revolving cover *d d*, constructed substantially as described in paragraph 2, letters *d d* and *i i*.

Second, the combination of the revolving cover *d d*, with a shield or deflector *e e*, and an opened-mouthed trumpeter *g g*, substantially as described in paragraph 2, letters *g g*.

Third, the combination of the revolving cover *d d*, and the shield or deflector *e e*, and the open-mouthed trumpeter *g g*, the opening through *f f*, in connection with a pipe *b b*, and the water tank *e e*, and the arrangement *i i*, for turning the cover *d d*, substantially as described in paragraphs 3 and 4.

No. 55,670.—WASHINGTON H. KILBURN, Kennedy, N. Y.—*Car Brake.*—June 19, 1866.—The momentum of the train is the means of bringing the brakes against the wheels by a connection which may be brought into action between the brakes of the several cars, or detached, as required.

Claim.—First, bringing the brakes of a railroad train to bear against the wheels of the several cars, by the momentum and weight of the train itself, when the engine drawing such train, or any car of its series of cars, is arrested in its motion in any possible manner, substantially as herein described and for the purpose specified.

Second, the draw-head E, bar F, and lever G, when combined and arranged with regard to and connection with the brakes of a railroad car, substantially as described and so as to operate as and for the purpose specified.

No. 55,671.—JOHN KIMBALL, Boston, Mass.—*Boot and Shoe.*—June 19, 1866.—The interposed layer of cork is within the line of stitching.

Claim.—The improved manufacture or shoe as made with a layer of cork, so arranged between its inner and outer soles that the leather of the outer sole may come in contact with the leather of the upper where it laps over the inner sole (the same being so as to hide the edges of the cork,) and the soles, cork, and upper be united by sewing or nails, as specified.

Also, the improved manufacture of water-proof sole for boots or shoes, it being composed of leather and cork arranged with a border or a piece of leather circumscribing the cork, as and for the purpose set forth.

No. 55,672.—JACOB KING, Omaha, Neb.—*Horse Rake.*—June 19, 1866.—Swinging draught frames are pivoted to each end of the rake head, which has teeth on each side.

Claim.—The rake provided with sets of teeth on each side of the head and drawn by means of the reversible swinging frames K, constructed and operating substantially as described and represented.

No. 55,673.—DAVID KNOWLTON, Camden, Me.—*Winch Capstan.*—June 19, 1866.—Winch heads are arranged on the top of the capstan. The operative devices are explained in the claims.

Claim.—The combination as well as the arrangement of the capstan A, the capstan-head D, and two or any other suitable number of winches applied to such head and provided with mechanism for revolving them separately from the capstan.

Also, the arrangement of the head D, with the capstan A

Also, the combination as well as the arrangement of the internal ratchet P, and its pawl e, travelling pinion O, the driving pinion R, the rachet K, and pawl L, and also the combination of such additional power mechanism with each winch F, the head D, and the capstan A, or the same and the holding rachet G, and pawl H, of the winch.

Also, the combination as well as the arrangement of the head D, with the two winches and their operative mechanism as specified.

Also, the combination of the two holding rachets G G and their retaining pawls H H, with the two winches, their shaft and operative mechanism as applied to them, their shaft and the capstan head, substantially as specified.

No. 55,674.—ANDREW KLOMAN, Pittsburg, Penn.—*Upsetting Press.*—June 19, 1866.— The end of the shaft to be upset is placed in a swaging box with movable top and sides; the shaft rests on a table and is braced longitudinally by an adjustable buttress. A plunger having the proper form on its end moves in the swaging box, being actuated by a cam, an intervening wedge dropping between the cam and plunger to advance the latter.

Claim.—First, the top-piece C with its side projecting ledges d d, and inclined top, when used in an upsetting press, as described and for the purpose specified.

Second, the combination of the two wedges D D', screws S S, frame A, plate E, and piece C, arranged as specified and for the purpose already described.

Third, the combination of the two pieces G G' with the piece C, and bed B, as described and for the purpose specified.

Fourth, the mode of forcing down and holding stationary the piece C, and of preventing the pieces G G' from spreading apart at one and the same time by use and means of the screws S S; wedges D D', inclined top-piece C, plate E, and ledges d d'.

Fifth, the plunger I, having two points or projecting angles i i, and notches g' g', as described and for the purpose already mentioned.

Sixth, the combination of the fly-wheels K K, shaft k k, cam L, box b, yoke M, wedge W, and bolt J, for obtaining from a motive.power comparatively small an enormous pressure and long throw in a short space of time, variable in intensity to any desired degree of power or speed by the single motion of the wedge W.

Seventh, the combination of the bars q q, piece Q, pieces R R, plates T and S, and pieces r r and T' T', as described and for the purpose specified.

No. 55,675.—R. KNUDSEN and W. T LASSOE, Brooklyn, N. Y.—*Portable Boat.*—June 19, 1866.—The narrow floats are united by jointed bracing and the seat rests upon both boats.

Claim.—First, a portable boat composed of two water-tight floats connected substantially as herein described, whereby they may be drawn apart to give the necessary stability for use, and closed together to afford facility for transportation.

Second, the combination with a double boat of the movable frame B, constructed and applied substantially as herein set forth for the purpose specified.

Third, the seat C, having row-locks r arranged over the space between the two hulls of the boat when the said hulls are extended and held apart, substantially as herein set forth.

No. 55,676.—THEODORE T. S. LAIDLEY, Springfield, Mass.—*Priming Metalic Cartridges.*— June 19, 1866.—The anvil plate is placed edgwise in the base of the cartridge, and has a projection upon which the percussion cap is placed in contact with the end disk.

Claim.—The combination of the cartridge case with an anvil A, which is of such shape that it holds the percussion cap in a central position within the case against the head, and is held firmly in its place by resting against a shoulder formed in the case below the head, after the anvil has been inserted, in the manner and for the purpose above described.

No. 55,677.—WM. L. LANCE, Plymouth, Penn.—*Serving Table.*—June 19, 1866.—Above the annular table at which the guests sit is an annular series of shelves, arranged one above another, containing the viands, &c., and revolving to expose them to the guests.

Claim.—First, the moving table b, combined with one or more stationary tables a a, for dining and other uses, substantially as set forth.

Second, the moving tables b, in combination with one or more pantry or furnishing or receiving rooms P, in the manner described.

Third, the moving table b, and stationary tables a a, so arranged having an open space Q, forming a room on the inside or inner part of the tables a a b, as described.

Fourth, in combination with the tables, a passage way i, either under or over tables a a b, by stairs or otherwise, substantially as set forth.

Fifth, dividing the tables a a b into sections, substantially as set forth.

Sixth, in combination with the.tables, the application of a flanged wheel R, or its equivalent, to support, guide, and steady a moving table b, for dining and other uses, substantially as set forth.

Seventh, in combination with the moving table, the fixed shaft z to the legs H of the tables a a, supporting tables a a, upon which shaft z is wheel R to guide movable or moving table b, substantially as and for the purpose set forth.

Eighth, the application of a belt S, or its equivalent, to the driving of a movable or moving table b, for dining and other uses, substantially as set forth.

No. 55,678.--S. E. LAMPHEAR and H. H. BLAIR, Brunswick, Ohio.—*Bed Bottom.*—June 19, 1866.—An elastic loop passes over a staple on the bed rail and is fastened to the end of the slat; its ends are slipped endwise into the slotted slat, clamped by a pin and the slat clamped to prevent it from spreading.
Claim.—The elastic loop B, pin I, and collar D, as arranged and in combination with the slat A, staple F, and cleat G, for the purpose and in the manner set forth.

No. 55,679.—JOHN LANZA, New York, N. Y.—*Forming Metallic Characters on Paper, &c.*—June 19, 1866.—The letters are formed by a pen charged with an ink composed of gum-arabic, gum ammoniac, water and garlic juice; after drying the ink is made adhesive by breathing upon it and dusted with bronze powder.
Claim.—The above described writing fluid for the purpose of making metallic or other dust (or metallic or earthy mixtures) adhere to the writing, and thus give metallic letters, substantially in the manner and for the purposes set forth.
Also, the above-described writing as a new and useful improvement, when the above-described fluid is employed substantially as described.

No. 55,680.—RUFUS LAPHAM, New York, N. Y.—*Spring Bedstead.*—June 19, 1866.—Inserted in a central beam are elastic ribs projecting obliquely upward to receive the transverse slats in which the mattress is laid.
Claim.—The centre-piece *a*, or its equivalent, with the interstices *b b b*, the springs *c c c*, with their slots, the cross-pieces or springs *d d d*, with their pins *e e*, fitting in the slots of *e e e* in combination, operating substantially as described and for the purposes set forth.

No. 55,681.—CHARLES W. LE COUNT, Norwalk, Conn.—*Adjustable Mandrel.*—June 19, 1866.—The keys fit in inclined longitudinal seats in the mandrel, and are adjusted therein by a screw sleeve on the shank.
Claim.—The arrangement of the sliding keys F, socket B, and thimble C, in combination with the mandrel A, in the manner and for the purpose substantially as herein described.

No. 55,682.—CHAS. A. LEECH, Philadelphia, Penn.—*Photographic Apparatus.*—June 19, 1866.—This apparatus for taking photographic negatives by the wet process combines into one machine the several parts relative to the solar action on the plates, the nitrate bath, the developing bath, the washing bath, and fixing-material bath, and dispenses with the dark tent. The devices are cited in the claims.
Claim.—First, the combination and arrangement of the dark chamber F with the camera box by means of the sliding frame G, substantially in the manner above described and for the purposes set forth.
Second, the combination of the dark chamber F with the bath case A, and baths B C D E, by means of the sliding frame G, the said parts being constructed and arranged in relation to each other substantially as described; so that the said chamber may be brought successively in its vertical and horizontal positions with all the baths for the immersion of the plate into the same and its removal therefrom, without removing it from its fixed position in the dark chamber, as and for the purposes above specified.
Third, constructing the baths B C D E with the slots *b*, and hinged lids *d*, substantially as and for the purposes above described.
Fourth, the combination and arrangement of the opaque valve M with the dark chamber F, for shutting out the light from beneath the latter when it is in its elevated positions, substantially as described.
Fifth, the combination of the springs I I with the sliding frame G, for holding the latter in its vertical position with the baths by means of the recesses *i* in bath case A, substantially as described and for the purposes specified.
Sixth, the combination of the spring H, having a pin or projection *j*, with the sliding frame G, for holding the dark chamber F in its elevated position, substantially as described.

No. 55,683.—JEREMY E. LINDSLEY, Goshen, Ind.—*Sill and Weather Strip.*—June 19, 1866.—The strip covers the threshold, has downward projecting flanges, and the closing door comes in contact with the upward projection by which the strip is tilted, closing the aperture under the door and excluding the weather.
Claim.—The metallic strip D, having flanged edges *a a* and *i*, when applied to the sill of the door, as and for the purpose specified.

No. 55,684.—C. M. LUFKIN, Claremont, N. H.—*Plough.*—June 19, 1866.—The independent cutter attached to the beam is connected by a rod to the reversible mould-board and by the movement of the latter is swung on its axis to preserve its coincidence with the plane of the land-side when the mould-board is reversed.
Claim.—First, a cutter I, of any convenient form, operated and connected by an eccentric M, shaft N, tube H, and slide-rod K, to the mould-board F, in such a manner as to admit of the oscillation of the cutter by the adjustment of the mould-board, as herein set forth.

Second, a tube H, slide-rod K, latch L, spring e, and catches g g, operating and arranged substantially as and for the purpose herein set forth.

Third, the pivot f, on the cutter I, in connection with the eccentric M and socket J, all constructed, arranged, and operating substantially as and for the purpose specified.

No. 55,685.—T. W. MAHLER, Rome, N. Y.—*Water Wheel.*—June 19, 1866.—The buckets are hinged and connected by arms to a movable rim operated by a lever to adjust the area of the openings; the chute is deeper than the wheel, except at its contracted discharge opening.

Claim.—First, the buckets E, hinged on axes b in such a manner that they may be turned or adjusted, substantially as shown and described, and for the purpose specified.

Second, connecting the axes b, of the buckets E, by means of arms F, to a ring e, adjusted through the medium of the plate or lever G, shaft h, and a bit i on said shaft working in a hole g in a projection f of ring e, all arranged substantially as and for the purpose specified.

Third, the shaft l, provided at its lower end with a crank u, and pin o, in combination with the plate or lever G, substantially as and for the purpose set forth.

Fourth, the scroll A having greater depth than the wheel or with its top and bottom plates p p', respectively above and below the top and bottom rims c c' of the wheel, substantially as and for the purpose specified.

No. 55,686.—THOMAS T. MARKLAND, Jr., Philadelphia, Penn.—*Street Lamp.*—June 19, 1866.—The globe has an exterior flange which rests upon the perforated base and perforated roof with a cap showing the names of streets; upper and lower reflectors reflect the rays outward.

Claim.—First, the combination of the screen and reflector H with the burner B and reflector H', substantially in the manner and for the purposes set forth.

Second, the combination of the globe A with the base D and roof E, when said parts are constructed and arranged in relation to each other, substantially as described and for the purposes set forth.

Third, the combination of the reflector H' with the roof E, reflector H, and burner B, substantially as described and for the purposes set forth.

Fourth, constructing the globe A with the annular projection m, for turning the water from the lamp, substantially as specified.

Fifth, constructing the base D with perforations f. and the roof E with the slots or openings K, for causing a current of cold air to flow over the interior surface of the globe A, to counteract the heat from the burner B, substantially as described and for the purposes set forth.

Sixth, the combination of the cap L with the central tube K and the burner B, when constructed and arranged to operate in relation to the draught of said tube, substantially as described.

Seventh, the perforated names in the sides of the cap L in combination with the burner B, substantially as described and for the purpose specified.

No. 55,687.—WM. W. MARTIN, Alleghany City, Penn —*Bending Flanges upon Boiler Heads.*—June 19, 1866.—The plate upon which the flange is to be turned is clamped upon a rotary table, at the two opposite sides of which is a vertical roller having a tapered end which is gradually brought down upon the projecting edge of the plate, bending it at right angles to its face.

Claim.—The construction and arrangement of the revolving table B and rolls C, said table and rolls operating substantially as herein described and for the purpose set forth.

No. 55,688.—JOHN MCCLOSKEY, New York, N. Y.—*Button Hole Sewing Machine.*—June 19, 1866.—The devices are attachable to the Wheeler and Wilson machine. The needle forms part of an arm which, by the movement of the needle arm, is moved forward and back alternately by means of an eccentric and other parts secured to the arm and to the presser arm, thus causing the needle to descend first through the cloth and next through the button hole. The cloth is confined between a loose movable button-hole shaped piece of metal having pins on its lower surface and a movable bed of leather or other flexible material; the metal piece is held and guided by a hollow flange or ring projecting downward from the bottom of the presser foot, and about which ring it and the leather travel. The feed is of the usual kind, and by acting directly upon the leather moves with it both the cloth and the guide. The disk bobbin is not used, but the rotating hook is retained, and a lower thread carrier is introduced, which is operated by the same cam that operates the feed. When the upper thread only is used, this looper is dispensed with, and the hook patented to McClosky, June 20, 1865, is employed.

Claim.—The grooved cylinder H, constructed substantially as described, for moving the needle forward and backward alternately, attached to or moved with the needle-arm of a reciprocating needle.

Also, the grooved cylinder H, in combination with the yielding finger G, substantially as described.

Also, the eccentric on the lower end of the grooved cylinder for alternately moving the needle forward and backward, substantially as described.

Also, connecting the reciprocating needle O with the grooved cylinder by means of an arm J, applied substantially as above described.

Also, the hollow flange j on the under side of the presser foot, in combination with the guide Q. substantially as described.

Also, the combination of the movable bed S, with the loose guide Q, operating in conjunction, substantially as described.

Also, the combination of the supplementary book f with the needle O, and the devices which move it forward and backward, substantially as above described.

Also, the horizontal lower needle C, made and operated substantially as described, in combination with the revolving hook, and a reciprocating needle O, moved forward and backward alternately, substantially as described.

No. 55,689.—ALEX. McDONALD, Charlestown, Mass.—*Travelling Trunk.*—June 19, 1866.—The body of the trunk is made of thicknesses of wood glued together with the grain crossing and covered with leather; a replication of the lid straps around loops gives facility for straining it tightly.

Claim.—The combination of the leather covering e and the boards c and d, arranged with respect to each other, and the bands f and x, substantially as set forth.

Also, combining with the buckles upon the front side of the body and the straps, which fasten the lid thereto, the auxiliary straining loops, operating substantially as described.

No. 55,690.—S. S. MEILY, Lebanon, Penn.—*Shifting Buggy Top.*—June 19, 1866.—The rails of the buggy top are permanently secured to an upper seat which is removable from the main seat, to which it is attached by thumb-screws.

Claim.—The application of turn-buttons g g, which are constructed with screw-stems to the secondary seat B and main seat A, these two seats being constructed and fitted together substantially as described.

No. 55,691.—R. S. MORISON, Bangor, Me.—*Weighing Grain.*—June 19, 1866.—An oscillating chute delivers the grain alternately into either side of a divided receptacle. When one side or compartment of the latter has received a given amount it is depressed, the scale beam rises, and by a peculiar device the chute is turned to the other side or compartment of the receptacle.

Claim.—In combination with the scale-beam, the mechanism operated thereby to control the passage to the scale of material to be weighed, when constructed and arranged to operate in the manner shown and described.

Also, in combination with the scale-beam, a secondary lever or beam, as set forth, when arranged to suddenly release the scale-beam when the weight received by the scale equals the amount noted on its register, substantially as described.

Also, in combination with a scale-beam, as set forth, a slotted link to permit free motion of the beam to an extent sufficient to secure a momentum by which to actuate a controlling mechanism, as described.

Also, the arrangement of mechanism for changing the chute, substantially as described.

No. 55,692.—F. B. MORSE, Milwaukee, Wis.—*Rub-iron for Carriage.*—June 19, 1866.—The spring presses against the revolving rub-iron to keep it from rattling on its bolt, and the iron presents different portions of its length to the wheel according to the weight in the carriage.

Claim.—Spring D, in combination with shaft C, and revolving rub-iron A, substantially as and for the purpose described.

No. 55,693.—GERSHOM MOTT, Big Run, Ohio.—*Drilling Machine for Wells.*—June 19, 1866.—The vertical drill beam is raised and tripped by the contact of the wrist on the end of the crank with the incline on the beam.

Claim.—The tripping-beam E, having its toe E'' shaped as shown in combination with the crank D, substantially as and for the purpose set forth.

No. 55,694.—W. J. L. MOULTON, San Francisco, Cal.—*Protecting Piles.*—June 19, 1866.—The end of the pile is encased in a metallic sheath with an intervening layer of hydraulic cement or concrete.

Claim.—The mode of protecting piles by means of metallic covering and cement filling, as set forth and described.

No. 55,695.—CHARLES MURDOCK, Ellenville, N. Y.—*Stave Machine.*—June 19, 1866; antedated June 2, 1866.—The saw, nearly cylindrical in form, cuts the staves from the block in conformity to their ultimate shape in the barrel, and a planer in the disk of the saw joints their edges.

Claim.—First, the block-carrying frame Q, with its supplementary frame W, arranged together substantially in the manner described, and operating with regard to the saw, as and for the purpose specified.

Second, the arrangement of the swinging-arm b, with its spring pawl, ratchet-wheel Z, adjustable plate C, and fixed arm f, connected through a pinion and rack gear, or its equivalent, with the block-carrying frame Q, and operating together substantially in the manner described and for the purpose specified.

No. 55,696.—F. NEVERGOLD and G. STACKHOUSE, Pittsburg. Penn.—*Combined Drill and Blacksmith's Tongs.*—June 19, 1866.—A drilling apparatus is attached to one jaw, the plate to be drilled lies upon the other, the pressure upon the handles giving the required feed to the drill.

Claim.—First, the new and improved tool called a "drill tongs," constructed as described, or its equivalent.

Second, the combination of the tongs A B with the frame F, spindle G, wheels M M, shaft N, crank O, and drill K, as described and for the purpose specified.

Third, the disk D, on the tongs A, in combination with the disk E, the slots g' g', and the bolts H H, for holding the frame F in different positions in relation to the tongs A' B'.

Fourth, the combination of the piece R with the jaw B of the tongs A' B', constructed and applied one to the other as described and shown.

No. 55,697.—H. D. NILES and JAMES C. BROOKS, Bristolville, Ohio.—*Composition for Curing Rot in Sheep.*—June 19, 1866.—Composed of coal-tar, alcohol, benzole, venice turpentine, tincture of myrrh, oil of origanum, butter of antimony, and sulphuric acid.

Claim.—The aforesaid compound formed of the above named ingredients, in about the proportions and for the purpose herein set forth and described.

No. 55,698.—ONESIPPE PACALIN, New York, N. Y.—*Boot and Shoe.*—June 19, 1866.—The sole is riveted between the plates; the heel socket is secured to the lower plate; the heel plug is fastened in its socket by a bolt.

Claim.—The combination of the inner and outer plates A B, heel-socket E, plug G, bolt H, and fastenings D D, &c., constructed and arranged substantially as described and represented.

No. 55,699.—CLARK D. PAGE, Rochester, N. Y.—*Lime Kiln.*—June 19, 1866.—The cupola is divided at the bottom by a vertical partition; water is introduced by pipe and by exposure of surface in pans and in the ash-pit to aid the draught and the process of conversion. Air is introduced by side draught into the furnace above the grate bars.

Claim.—First, the employment of water in coal-burning lime-kilns for the purpose of first steaming the coal, to produce a more perfect and economical combustion, and the absorption of all sulphurous acid gas of the coal by said steam and using the gases of the decomposed water in producing a greater degree of heat, substantially as described.

Second, the combination of the pans c c, and water-pipes M, or equivalent, operating substantially as and for the purpose specified.

Third, the water receptacles f f, in combination with the ash-pits L L and grates b b so arranged that the steam that is produced by the fire will pass upward around and through the grates to keep them cool, substantially as described.

Fourth, the partition I, in combination with the particular form of the cupola at the base, the latter provided with the concaves a a, and having the chamber on each side of uniform thickness, substantially as described.

Fifth, the arrangement of the recesses N O and cold-air flues k k, in combination with the grates b b and the sides of the furnace, the same opening directly over the grates and so constructed as to furnish cold air and prevent clinkering, substantially as described.

No. 55,700.—C. C. PARSONS, Boston, Mass.—*Stop-cock.*—June 19, 1866.—Air-tight chambers in the barrel of the faucet connect with the opening through the plug when the latter is closed, and afford room for the contents of the plug to expand when freezing.

Claim.—A stop-cock constructed with one or more closed air-tight chambers d, operating in combination with the opening through the plug.

No. 55,701.—JOHN PERKINS, Providence, R. I.—*Cotton-seed Huller.*—June 19, 1866.—The concave shell of the huller has toothed sections having respectively teeth parallel with the axis and in a plane at right angles thereto.

Claim.—The combination and arrangement of the series of vertical ribs d', with the shell C, its series of horizontal ribs e, the cylinder A, and its case B, the whole being to operate together substantially as specified.

No. 55,702.—CHAS. L. PIERCE, Buffalo, N. Y.—*Shingle Machine.*—June 19, 1866.—The bolt carriage is reciprocated by means of a cross-head with a straight slot in which works the wrist pin of a crank which gives to the carriage an unequal progressive reciprocation; the diamond shaped slot in connection with the wrist pin of the crank working therein gives uniform feed to the bolt carriage. The dogging devices act automatically and operate by weight and spring respectively. The two double taper cams on one shaft alternately tilt the block to give taper to the shingle.

Claim.—First, reciprocating the block-carriage which feeds the block to the saw in an unequal progressive movement by means of the crank-pin D, working in the slot of cross-head D', substantially as described.

Second, imparting an equal progressive movement of the block-carriage by means of the diamond slot L, and crank D, and thereby giving an equal and uniform feed of the shingle block to the saw, substantially as set forth.

Third, operating the movable dog-bar G' of the dogging device by a weighted bell crank G², working in combination with the hinged lever G⁴, on the bed-frame, in the manner described.

Fourth, operating the movable dog-bar G' of the dogging device by the bell-crank J, and spring bar J¹, working in combination with the stop-piece J³, on the bed-frame, in the manner described.

Fifth, the arrangement and combination of the segment lever K' K², with the bell-crank G², carrying segment K, and movable dog-bar G', for the purpose of operating the dog-bar by hand power.

Sixth, the double taper cams I, in combination with the tilting block-tables, when arranged in relation to the block-carriage and operated thereby in the manner and for the purpose set forth.

No. 55,703.—CLARK POLLEY, Sinking Springs, Ohio.—*Straw-cutter.*—June 19, 1866.—By the described arrangement the pressure upon the knife is downward and also inward against the throat and the feed rake vibrated between each cut.

Claim.—First, the levers A' and V, in combination with each other, with the knife B, and sash-frame J K, and with the walking beam P, and driving-cam O, the whole being constructed and arranged substantially as described and for the purpose set forth.

Second, the combination of the spring X with the shaft T and walking-beam P, substantially as described and for the purpose set forth.

No. 55,704.—ALVIN POND, Southington, Conn.—*Tack Hammer.*—June 19, 1866.—The tack is held by the jaws until placed in the intended position; it is then released and driven by the hammer.

Claim.—The hammer herein described, consisting of the head A, constructed with a notched jaw C, and having a corresponding jaw D attached thereto, provided with their respective handles and constructed to operate substantially in the manner herein set forth.

No. 55,705.—T. W. PORTER, Bangor, Me.—*Sleigh.*—June 19, 1866.—The angle irons have sockets for the raves, benches, and knees, or knees and runners, as the case may be; or form knees to unite runners, benches, and raves.

Claim.—First, the metallic coupling or bar-end A, Fig. 1, substantially as and for the purposes specified.

Second, the metallic coupling or double T, marked B, Fig. 1, substantially as described and shown.

Third, forming metallic sleigh standards with the socket c', Figs. 1, 3, and 4, in manner substantially as described and for the purposes specified.

No. 55,706.—JOHN PRIESTLY, New York, N. Y., and THOMAS C. BRADBURY, Poughkeepsie, N. Y.—*Preparing Paper Pulp from Straw.*—June 19, 1866.—The dry straw is ground, between stones running at different velocities, boiled in alkaline solution at sixty pounds pressure, and pulped in a "Kingsland" or other engine.

Claim.—First, the process effected by a crushing machine used for the purpose of opening, splitting, and flattening the straw with a rotary steam boiler, as described.

Second, the process effected by a crushing machine used for the purpose aforesaid with the rotary steam boiler containing the paper stock, operated at about sixty pounds pressure, substantially as described.

Third, the process effected by a crushing machine used for the purpose aforesaid with the rotary steam boiler containing the paper stock, and with a pulping engine (Kingsland or other) for the purpose of disintegrating the fibres, substantially as described.

Fourth, the combination of the rotary boiler containing the paper stock operated at a pressure of about sixty pounds, for the period described, with a pulping engine, (Kingsland or other,) substantially as described.

Fifth, the combination of a crushing machine and boiler, containing the paper stock, operated at a pressure of about sixty pounds and with a corresponding temperature, for the period described, with a pulping engine (Kingsland or other) for the purpose of disintegrating the stock and producing a fibre suitable for the manufacture of paper without the addition of other stock, substantially as described.

No. 55,707.—A. PUTNAM, Owego, N. Y.—*Grain Drill.*—June 19, 1866.—The bar to which the teeth are attached is capable of two longitudinal adjustments : in one the seed is dropped behind the shares in the drills made thereby, and in the other the seed is dropped broadcast on the ground, and the shovels follow to plough it in.

Claim.—First, the changing of the machine from a grain drill to a broadcast sower, and from a broadcast sower to a grain drill, by moving or adjusting the tooth bars K to the feed box F, as above described, or its equivalent.

Second, forming the teeth G and H from two different shaped patterns, which is to incline one forward and the other backward alternately, in the manner already set forth and described.

No. 55,708.—JOHN J. RALYA, Alleghany, Penn.—*Stave Machine.*—June 19, 1866.—A pair of adjustable dressing knives are formed to dress the staves to shape and hung to an adjustable oscillating head stock : a ram, adjustable as to length of stroke, forces the stave to and between the dressing knives and allows the stave to be truly cut as regards its thickness notwithstanding the wind of its grain.

Claim.—First, setting the head stock or knife frame on journals, so that it may admit of a slight motion on its axis to accommodate its position to any twist or irregularity of shape of the stave which is forced between the knives in the operation of shaving.

Second, placing the knives in a head stock or frame susceptible of motion on its axis in such a way that the centre of motion shall be on a line between the inner face of the knives and between their back and edge.

Third, limiting and regulating the range of motion of the knives by means of set screws, substantially as hereinbefore described.

Fourth, the use of the movable head piece in the end of the ram, so constructed and arranged as to be susceptible of a limited motion on its axis for the purpose of allowing the stave to turn in its passage through the knives to accommodate any twist or irregularity of shape of the stave.

Fifth, the use of a spring in connection with the movable head of the ram, so as to permit of its yielding slightly in the operation of forcing the staves through the knives, substantially as described.

Sixth, crotching the end of the ram so as to hold the stave in place as it is being forced through the knives.

Seventh, the use, in combination with the cutters or knives and ram, of an upright pulley post to carry the rope and weight for withdrawing the ram after the stave is passed between the knives.

Eighth, so arranging the toothed rack of the ram as to be capable of adjustment towards or from the segmental gear wheel for the purpose of regulating the length of stroke of the ram toward the knives, substantially as and for the purpose hereinbefore described.

Ninth, giving to the knife blades a concave curve from their outer edge, so as to form a ledge or shoulder for the purpose of turning the shaving or chip outwards at such an angle as to break it off just above the edge of the knife, substantially as hereinbefore described.

No. 55,709.—CHARLES REESE, Baltimore, Md.—*Fruit Box.*—June 19, 1866.—Explained by the claim and cut.

Claim.—A box made of a single piece whose flaps are so bent up as to form sides, which are secured together by eyelets, substantially as described.

No. 55,710.—JACOB REESE, Pittsburg, Penn.—*Reducing Metallic Oxides.*—June 19, 1866.—The charge of ore and purifying agents is melted in the cupola and discharged into the reducer, where it collects in the belly. When the reducer is oscillated a stream of hydrogen gas or hydro-carbon vapor is introduced by a tuyere and passes through the molten iron, converting it into wrought iron, or into steel, or cast iron by continuing the process.

Claim.—First, deoxidizing metallic oxides while in a molten or liquid condition, by means of hydrogen gas or a vapor of carbon or of hydro-carbon, or a mixture of such vapor or gas, so that these oxides or ores may be reduced to a metallic condition without the use of additional fuel, substantially in the manner hereinbefore described.

Second, the use of liquid petroleum or other liquid hydro-carbon in the manufacture of iron or steel and other metals, substantially in the manner and for the purposes hereinbefore described.

Third, the use of hydrogen gas for the purpose of deoxidizing metallic oxides, substantially in the manner hereinbefore described.

Fourth, making liquid wrought or malleable iron from the ore by subjecting the ore while in a melted condition to the action of hydrogen gas or hydro-carburetted vapor, or a vapor of carbon, or a liquid hydro-carbon, substantially as hereinbefore described.

Fifth, making cast steel by deoxidizing iron ore while in a molten condition, in the manner hereinbefore described, and subjecting the pure, iron thus produced to a vapor of carbon or hydro-carbon, or adding thereto a liquid carburet or hydro-carbon, until the requisite amount of carbon is added, substantially as hereinbefore described.

Sixth, making cast iron by deoxidizing iron ore in a melted condition, in the manner described, and supplying the requisite amount of carbon in a gaseous or liquid form, substantially as hereinbefore set forth.

Seventh, refining iron and steel by means of a carbon in a gaseous or liquid form, to which, after the metal has been deoxidized thereby, a sufficient amount of air, water, or steam is added to support the combustion of the carbon thus added as fuel to the melted metal.

Eighth, making a belly in the lower side of the deoxidizing chamber or reducer so as to hold the charge of melted ore away from the tuyere holes in the bottom of the reducer until the reducer is raised to admit the deoxidizing vapor or liquid, substantially as hereinbefore described.

Ninth, the use of a valve on one of the trunnions of the reducer, constructed substantially as hereinbefore described, so as to shut off the deoxidizing vapor or liquid from entering the reducer when in position to receive its charge, and open the communication when the reducer is restored to its position for working.

Tenth, the use of gas meters, in combination with the air cylinder and vapor generator and reducer, for the purpose of measuring the amount of deoxidizing vapor or air admitted to the interior of the reducing chamber, substantially as hereinbefore described.

No. 55,711.—JOSHUA REGESTER, Baltimore, Md.—*Hydrant.*—June 19, 1866.—The cylinder is supported upon semi-annular flanges at the base of the longitudinally divided casing, the flanges embracing a neck on the base of the cylinder.

Claim.—Constructing the base of the cylinder B in such manner that this cylinder will be held in a permanent position within a divided case A A, between and upon base supports or collars *a a*, substantially as described.

Also, the combination of a divided case A A with a cylinder B, which is constructed with a contracted neck B', and a pipe D leading to this neck below the base *a a*, all substantially as described.

Also, the construction of the hydrant case of two sections A A, two half base-pieces *a a*, and perforated portion *g d*, the said perforated portion being below the base *a a*, substantially as described.

No. 55,712.—JOSHUA REGESTER, Baltimore, Md.—*Hydrant.*—June 19, 1866.—The flow of water is gradually cut off to prevent concussion of the valve on its seat and the rupture of the pipes ; the form of valve and the arrangement of parts is explained in the claims and cut.

Claim.—First, the construction of the nut D' with an external flange, and with an internal flaring passage, in combination with the tapering plug valve F, packing *f*, internal cylinder B, and plunger B', all substantially in the manner and for the purpose described.

Second, fitting the plug valve F in a recess formed in a conical seat E, which is perforated near its circumference, all substantially in the manner and for the purpose herein described.

Third, the combination of the conical seat E, plug valve F, packing *f f*, perforations in the seat E, and the nut D', all constructed and arranged substantially as described.

Fourth, securing the packing *f* upon the seat E by means of a flange *c* and plug valve F, substantially as described.

Fifth, applying the crank rod *i* to a tubular bearing *j*, having a flanged head *k* with stops *k' k'* on its outer end, in combination with a crank arm or handle *l*, which has a stop *k²* formed on it, all used in connection with the foregoing features of invention, substantially as and for the purpose herein described.

No. 55,713.—WILLIAM D. RINEHART, Pittsburg. Penn.—*Mould for Casting Pulleys.*—June 19, 1866.—The mould consists of a cope and drag, each provided with a water or steam jacket, through which a current passes to regulate the temperature of the same while casting.

Claim.—A flask made in two parts ("cope" and "drag,") each part being furnished with a chamber for heated air or steam, the whole being constructed, arranged, and operating substantially as herein described and for the purpose set forth.

No. 55,714.—JOHN ROBINSON, Calais, Vt.—*Butter Worker.*—June 19, 1866.—Explained by the claim and cut.

Claim.—The conical roller B *b* and handle C, fitted so that the handle may turn upon the roller, in combination with a sector shaped tray A A¹ and cross bar *d*, substantially in the manner and for the purpose herein set forth.

No. 55,715.—JULIUS A. ROTH, Philadelphia, Penn.—*Preparing Hides, Skins, Furs, &c., for use.*—June 19, 1866.—The hides are treated with a compound prepared as follows: A solution of sal soda is rendered caustic by lime and heated to 200° Fahrenheit; lard or other fats added, and the mixture cooled.

Claim—The softening of leather, hides, furs, and the hair or wool thereon, by treating the same in a saponified solution, made in the manner as set forth and for the purpose as specified.

No. 55,716.—WILLIAM RUDOLPH and A. BRAUN, San Francisco, Cal.—*Gun Lock.*—June 19, 1866.—The trigger is connected to the sear by a jointed arm; one spring with two arms operates the two triggers.

Claim.—An invention and improvement connecting the trigger to the sear by means of the hinged arm *g*, substantially as described.

Also, in combination with the trigger and hinged arm, the spring *i*, substantially as described.

No. 55,717 —LOOMIS W. RUSSELL, Galesburg, Mich.—*Buckle.*—June 19, 1866.—A coupling link unites the blocks, whose serrated faces are presented to the respective straps; these are clamped by tension, which causes the link to assume an angular position.

Claim.—The ring or clasp A, blocks *b* and *c*, provided with a groove, and corrugated on their inner surfaces, being made of wood or metal, substantially in the manner and for the purpose herein set forth.

No. 55,718.—GEORGE W. SANDERS, Springfield, Vt.—*Dough Kneader.*—June 19, 1866.— The roller is swivelled to a holder, which rotates horizontally and is retained in its socket by a pin.

Claim.—The combination of the roller B with its grooved end *a'*, the block C, pin *b*, staples *c c*, hook *d*, and board A, arranged and operating in the manner and for the purpose herein described.

No. 55,719.—ADOLF SAYER, Naubuc, Conn.—*Breech-loading Fire-arm.*—June 19, 1866.— The hook, which is seated in an eccentric groove in the horizontally swinging breech block, in opening, engages with a pin on the cartridge extractor to retract it; the pin becomes detached from the hook as the block continues to rotate, allowing the ejector plate to return.

Claim.—In combination with swinging breech block, the hook W, the groove *a* in which it is placed, and the groove *s*, substantially as and for the purpose described.

No. 55,720.—P. G. H. SCHAEFFER, West Meriden, Conn.—*Tea and Coffee Pot.*—June 19, 1866.—The tin plate body is embos-ed and bent, and the upper edge turned in to stiffen the body and form a flange to receive the rim-seat of the cover.

Claim.—The combination of the flange *k* with the embossed body, substantially as described.

No. 55,721.—PETER SCHOULLER, Boston, Mass.—*Billiard Table.*—June 19, 1866.—A rabbet in the face of the rail behind the caoutchouc cushion gives greater resiliency to the latter.

Claim.—The arrangement and combination of the channel *a* with the rail B, and the elastic or caoutchouc strip A of a billiard table, the same being substantially as specified.

No. 55,722. —THOMAS A. SEARLE, Providence, R. I.—*Nail Plate Feeder.* —June 19, 1866.—The cam groove in the sleeve acts in conjunction with the vibrating frame to turn the nail plate holder, turning the plate before it enters between the knives.

Claim.—The cam, substantially such as described, for giving the turning motion to the nail plate, in combination with the vibrating feeder frame for drawing back and lifting the nail plate, that it may be turned, and returning it to the required position on the bed knife, substantially as described.

No. 55,723.—WILLIAM SELLERS,' Philadelphia, Penn. —*Planing Machine.* —June 19, 1866.—The frame, with the adjustable tool, traverses back and forth upon ways, over the bed which supports the work. The motion is transmitted from shifting pulleys at each end of the bed to others on the traversing frame, in which are the pinions and necessary gearing for giving motion to the frame.

Claim.—First, the use in planing machines for metal of traversing uprights to support the cross-head, and upon which the cross-head may be elevated and depressed to suit the varying heights of the material to be operated upon, in combination with a fixed platform, provided with ways, slides, or their equivalents, arranged so as to avoid the necessity of raising the materials to be operated upon from the platform, so as to come within reach of the cutting tool, all constructed, arranged, and operating substantially as described.

Second, actuating the traversing uprights of the planing machine for metal herein described from a revolving shaft or shafts attached to and moving with the uprights, substantially as described.

Third, the use of an endless belt in combination with fast and loose pulleys, or their equivalents, when applied to the planing machine herein described, substantially in the manner and for the purposes specified.

Fourth, reversing the direction of the movement of the cutting tool by means substantially as described and for the object specified.

No. 55,724.—JACOB A. SHERMAN, New York, N. Y.—*Truss.*—June 19, 1866.—The pad is adjustable to give upward or downward pressure, and is attached to a spring piece; this is connected to the band, which has two spring portions conformable to the hips.

Claim.—First, the curved pressure-spring *a*, introduced between the bars *b* and *c*, and carrying the truss-pad, as specified.

Second, the clips *d e*, 3 and 4, to which the ends of the bars *b* or *c* are attached, in combination with the segmental slots and clamping screws for connecting the bars *b c* to the respective springs *a, f,* or *g*, and allowing for adjustment, as set forth.

Third, the inclined hinge 11 or 12, for uniting the pad to the truss-spring *a,* so as to allow the adjustment of the pad as specified.

Fourth, the lever *q* and screw 15, in combination with the lever *o* and diagonal hinges 11 and 12, as and for the purposes set forth.

No. 55,725.—LEVI SHULTZ, Upper Sandusky, Ohio.—*Fanning Mill.*—June 19, 1866.—The sides of the shoe form the lateral enclosure, and it is suspended in the open frame; springs receive its lateral impact, and the amount of blast is graduated by peripheral shutters on the fan case.

Claim.—First, constructing a fanning mill with a suspended riddle-shoe D, which in part extends above and below the fan-case B, and forms the closed sides of the mill, substantially as described.

Second, suspending the shoe D by a pivot *c,* and providing it with springs *g g,* for equalizing its movements, substantially as described.

Third, providing the blast opening through the fan-case with adjustable slides *a a',* for regulating the force and direction of the blast, substantially as described.

Fourth, the combination of a fan-case B and a shoe D, having side-boards applied to it, with an open supporting-frame A, substantially as described.

No. 55,726.—FELIX JOHN SIMEON, Brooklyn, N. Y.—*Feed-bag.*—June 19, 1866.—The foraminous bag has springs to hold the feed against the horse's mouth, and dispense with a stand.

Claim.—First, constructing both the sides and bottom of a feed-bag of perforated metal or wire-cloth, substantially as and for the purpose set forth.

Second, the combination, with an open or perforated feed-bag, of the springs F F and strap E, substantially as and for the purpose set forth.

No. 55,727.—HENRY C. SMALL, Portland, Me.—*Cattle Tie.*—June 19, 1866.—The bows are adapted to encircle the neck of the animal, and are fastened together by a hinge and spring hook.

Claim.—The combination of the bows *d* and *c,* the bow *d* having the ring *h* and hooks *f f* on its ends, the bow *c,* the two hooks *k k,* and spring *s,* all constructed, arranged, and operating as set forth.

No. 55,728.—SAMUEL SMITH, Philadelphia, Penn.—*Heating Stove.*—June 19, 1866.—The annular plate forms a connection between the sheet-iron cylinder and the base plate; the latter has openings to discharge ashes that escape from the cylinder.

Claim.—First, the ring B secured to the cylinder D, constructed and adapted for attachment to the base or cap plate of a stove, substantially as described.

Second, the combination with the above of the base plate A with its openings *z,* arranged as set forth, for the purpose specified.

No. 55,729.—WILLISON G. SMITH, Carlisle, Penn.—*Spittoon for Cars.*—June 19, 1866; antedated May 26, 1866.—The cover and the bottom are connected and open together; a spring maintains them in the required position.

Claim.—First, an improved self-cleaning spittoon, formed by combining the box or cup B, the cover H, the arm G, and sliding bottom D with each other, the parts being constructed and arranged substantially as herein described and for the purpose set forth.

Second, the combination of the spring F with the lower part of the spittoon and the sliding bottom D, substantially as described and for the purpose set forth.

No. 55,730.—THOS. W. SPEISSEGER, Charleston, S. C.—*Medical Composition.*—June 19, 1866.—This salve for cutaneous diseases is composed of lard, sulphur, tincture of myrrh, and aqua ammonia.

Claim.—A medical compound composed of the ingredients herein specified, and in about the proportions named.

No. 55,731.—EDWIN SPRAGUE, Alleghany City, Penn.—*Valve Gear for Steam Engines.*—June 19, 1866.—The cam and its adjustable point by means of the yoke and rod operate the hinged lifters on the rock shaft, the lifters being held in position by inclined planes; the arrangement opens the steam ports and closes them at the required point.

Claim.—First, the hinged lifters *s,* said lifters being operated by means of triggers *r* and inclines 1, or their equivalents, and one cam-rod, made in one or more parts; said lifters, triggers, inclines, and cam-rod being dependent for their action upon the cam 18 and adjustable point 17, substantially as herein described and set forth.

Second, the adjustable point 17, when used in combination with a cam and cam-yoke and a single cam-rod, for working a full stroke, and used for operating the cut-off gear of steam engines, as herein described and set forth.

Third, the inclines 1 on plate y, said plate and inclines being operated by lever o, through the medium of a governor, or otherwise, substantially as herein described and for the purpose set forth.

No. 55,732.—JOSEPH H. SPRINGER and WILLIAM M. BARTRAM, Philadelphia, Penn.—*Indicator for Steam Generators.*—June 19, 1866.—As the water sinks the cup descends, depresses the valve lever, allowing the steam to pass to the whistle.

Claim —First, a cup I open at the top and suspended within a steam generator, or in a tube communicating with the said generator, in combination with devices constructed and arranged substantially as herein described, whereby the said cup is caused to discharge a volume of steam when the water becomes low, as set forth.

Second, the combination of the tube A, its cup I, steam whistle J, spring valve G, and lever H, the whole being arranged substantially as and for the purpose described.

Third, the combination with the above of the glass tube F.

No. 55,733.—ISRAEL STEALY, Crestline, Ohio.—*Pessary.*—June 19, 1866.—Explained by the claim.

Claim.—The application of a rubber globe under the womb, to be filled with air after inserting the same, in order that it may press upward against the womb to prevent falling, weakness, and pain, thereby giving ease, comfort, and strength to the suffering patient.

No. 55,734.—ALDEN B. STILLINGS, Springfield, Mass.—*Knitting Machine.*—June 19, 1866.—Improvement on the Lamb knitting machine, No. 39,934. The slot in the crank-arm permits such an adjustment of its throw as that the sliding plates need not be made to traverse any further than the width of the goods being knitted actually requires, avoiding an excess of slack thread. The arrangement of the two right and left-hand screws and the four threaded slides thereon insures a simultaneous and equal adjustment of all the stops; while the chain and its wheels allow the operator at any time to revolve the shafts and set the stops as desired.

Claim.—First, constructing the crank in such a manner as to obtain a varying or adjustable thrust, substantially as described and for the purposes set forth.

Second, the use of the shafts E E', having screw-threads cut upon them, as described, in combination with the cam-stops, when arranged and operating substantially as set forth.

Third, imparting to the four cam-stops the same relative motion by means of the chain-carriers and endless chain, or by any equivalent means, substantially as described.

Fourth, the combination of a crank, constructed as set forth, with the mechanism described for adjusting the position of the cam-stops, substantially as set forth.

No. 55,735.—ALONZO STOW, Calais, Vt.—*Upsetting Tires.*—June 19, 1866.—The curved bar has a socket and vibrating clutch at each end and a depresser cam in the middle; the latter descends upon the periphery of the red-hot tire and, by straightening, shortens it.

Claim.—A tire-upsetting machine, with self-adjusting jaws or holders, and operated by an eccentric, or cam-lever, substantially as set forth.

No. 55,736.—WM. J. STOWELL, Baltimore, Md.—*Railroad Switch.*—June 19, 1866.—The movable guard has a swelled head and projecting tongue combined with a guide rail and inclined plane or bridge to pass the flange of the wheel over the rail if the switch be misplaced. The switch on one side alone is movable.

Claim.—First, the construction of the movable guard E with a projecting tongue f, in combination with the swelled head e of this guard and the two rail-sections A D, arranged to operate substantially as described.

Second, the combination of the rail-section D, which is constructed with an enlarged head e, and inclined plane c,' with the guard C and the web or bridge a, substantially as described.

Third, the movable switch rails D E, when constructed substantially as described and combined with stationary guide rails on the opposite side of the track, substantially in the manner and for the purposes specified.

No. 55,737.—CHRISTOPH SUSSEGGER, New York, N. Y.—*Beverage.*—June 19, 1866.—The solution is fermented, cleaned, decanted, and bottled.

Claim.—First, the "American sherbet," as a new article of manufacture.

Second, the manufacture of a beverage by a process substantially as hereinbefore set forth and described.

Third, the combination of water, sugar, tartaric acid, yeast, linden blossoms, and rose leaves, or their equivalents, for the production of a beverage or "sherbet," substantially as hereinbefore set forth and described.

No. 55,738.—ZURIEL SWOPE, Lancaster, Penn.—*Sawing Machine.*—June 19, 1866.—The operator stands upon a treadle, alternately throwing his weight to the respective sides of the rock-shaft, whose motion is communicated by toggles to the reciprocating saw-frame.

Claim.—Operating a reciprocating saw by means of a treadle and the devices connected thereto, when they are constructed and arranged to operate in the manner and for the purposes substantially as specified.

No. 55,739.—GEORGE C. TAFT, Worcester, Mass., assignor to THOMAS H. DODGE, same place.—*Screw Wrench.*—June 19, 1866.—The sleeve-screw which operates the movable jaw is secured to the latter by an internal flange which catches on a stirrup in retracting the jaw; the jaw is closed by the pressure of the sleeve against a shoulder.

Claim.—First, the combination with nut E, having a flange e, of the jaw D and its stirrup f, substantially as set forth.

Second, the combination with the front of nut E of the shoulder c on jaw D, as and for the purposes set forth.

Third, bevelling off the front of flange e, in combination with bevelling off the rear of stirrup f, as seen at 3 in the accompanying drawings.

No. 55,740.—WM. B. S. TAYLOR, New York, N. Y.—*Manufacture of Flexible Tubing.*—June 19, 1866.—A flexible tube, formed of animal membrane softened in glycerine and covered with glue.

Claim.—The combination in a flexible gas tube of a layer or layers of animal membrane, coated with glue or other suitable gelatinous cement, substantially as described, and for the purpose of resisting the penetrative action of the gas and its fluids.

No. 55,741.—JOHN B. TERRY, Auburndale, Mass.—*Apparatus for Carburetting Air.*—June 19, 1866.—The curved buckets of the two metre wheels are arranged opposite each other's spaces, so as to make a more uniform blast, gathering the air at their peripheries, and discharging it at apertures around the axis.

Claim.—The improved air-forcing apparatus, made substantially as described, viz., not only of the two wheels A B, having their buckets arranged as explained, but of the case E, provided with the chambers I K L, the connecting tube H, and the induction passage G, as set forth.

No. 55,742.—J. H. THOMAS and P. P. MAST, Springfield, Ohio.—*Grain Drill.*—June 19, 1866.—The drill is expanded or contracted laterally, without removing or detaching its parts; is especially adapted for use between rows of corn, nursery trees, &c. The operative devices are worked by a single ground-wheel.

Claim.—First, the hopper A attached to the central or stationary bar B, in combination with the adjustable bar E, substantially as shown and described.

Second, the adjustable tubes I and J, arranged to operate in combination with the stationary hopper and adjustable bars, substantially as and for the purpose set forth.

Third, the slide l, when arranged to operate in combination with the rod f and hopper A, as herein described.

Fourth, the agitator a, having a to-and-fro movement over the openings of the hopper A, as shown and described.

Fifth, the tube I, having its upper end enlarged and so connected to the hopper A as to cover the opening therein and receive the seed therefrom, at all adjustments of which the bars E are capable.

Sixth, pivoting the tube I to the hopper A, and the tube J to the bar E, and having the tubes I and J arranged to slide upon each other, substantially as shown and described.

No. 55,743.—WM. TIBBALS, South Coventry, Conn.—*Revolving Fire-arm.*—June 19, 1866.—A groove in the side of the chamber of the cylinder extends from the face to the base, and permits the insertion from the front end of a metallic cartridge, with a laterally projecting pin.

Claim.—Constructing the chamber of fire-arms with the groove or slot in its side for the reception of the pin cartridge, substantially as herein shown and described.

No. 55,744.—FRANCIS W. TILTON, New Bedford, Mass.—*Clothes Pole.*—June 19, 1866.—The clothes line passes between the inturned points of the hooks, and is engaged by one or the other of them when raised by the wind.

Claim.—A clothes pole having one of its ends provided with a double hook B, constructed substantially as shown and described.

No. 55,745.—CYRUS L. TOPLIFF, New York, N. Y.—*Lamp Wick.*—June 19, 1866.—The lamp wick is treated with alum and gum to preserve it from rapid combustion.

Claim.—The application to a lamp wick of a solution of gum and alum, for the purpose described.

No. 55,746.—JOHN TRAGESER, New York, N. Y.—*Apparatus for Cooling and Condensing Beer, Alcohol, &c.*—June 19, 1866.—A long sheet-metal plate is folded and then convolved, so as to form a helix with two distinct spaces; the edges of the plate are attached to the top and bottom disks which form heads, and one space in the convolution is for the refrigerating liquid, the other for the vapor or fluid to be condensed; induction and eduction passages admit and discharge the respective fluids.

Claim.—The mode herein described of constructing volute condensers or coolers by the introduction of a strip of metal between the edges of the sheet-metal divisions, and soldered together in the manner specified.

No. 55,747.—J. TRAGESER and J. G. SCHREIBER, New York, N. Y.—*Coil for Brewer's Boilers.*—June 19, 1866.—The supply steam pipe and the discharge water pipe enter at the same opening in the boiler, the latter within the former, and each is so coupled with the semicircular sections of coiled pipe as to form hinges upon which the said coils swing.

Claim.—The steam pipe *b* and condensation pipe *c*, provided with the couplings *e c* and *g g*, in combination with the coils *f f*, substantially as and for the purposes set forth.

No. 55,748.—B. VAN VRAUKEN, Schenectady, N. Y.—*Clay and Peat Press.*—June 19, 1866.—The pusher has an intermittent reciprocating motion on the table to advance the moulds to the throat of the pug-mill and past that point, and is operated by rack bars and segment wheels actuated by pitman, &c., from the main shaft.

Claim.—First, in a machine for moulding clay and peat, employing a vertical pug-mill or mixing box B and a press box D, the adjustable pusher F, applied to a table E, for moving the mould boxes from the rear to the front end of said table, substantially as described.

Second, the combination with an upright pug-mill B, press box D, and a table E, the rack bars G G, oscillating sectors H H, and a pitman rod J, connecting with the main shaft C, substantially as described.

No. 55,749.—WM. T. VOSE, Newton, Mass.—*Lathe-rest for Turning Balls.*—June 19, 1866.—The base rests on the ordinary lathe shears and the cap is secured above it so as to rotate thereon. To the cap is secured a projecting lug, to which the tool holder is journalled.

Claim.—The arrangement of the adjustable base A, revolving cap B, with the hinged stock H, set screw E, cutter F, and handle H, substantially in the manner and for the purpose set forth.

No. 55,750.—ARTHUR WADSWORTH, Newark, N. J.—*Watch.*—June 19, 1866.—The winding of the watch and the setting of the hands are performed by turning a part of the pendant of the watch case; the appropriate arm is pushed in to make the connection of either the winding or the setting gears with the worm on the pendant.

Claim.—First, the combination of the gear wheels O and Q, respectively hung to arms T and T', with the pendant spindle H, main-spring axis M, and arbor N of a watch movement, when arranged together so as to operate substantially as and for the purpose described.

Second, the combination of the spindle H and pusher sleeve E, arranged together substantially in the manner described and for the purpose specified.

No. 55,751.—GEORGE W. WALKER, Boston, Mass.—*Stove Damper.*—June 19, 1866.—The register has a gauze over the air-admitting openings.

Claim.—In combination with a foraminous plate or gauze, through which air is admitted at the front of the stove for the support of combustion, a provision for regulating or shutting off the admission of such air, substantially as set forth.

No. 55,752.—H. F. WHEELER, Boston, Mass.—*Breech-loading Fire-arm.*—June 19, 1866.—Combined with a single lock and hammer are two barrels of a different calibre, rotatable on a breech-pin to bring either into firing position.

Claim.—In combination with the breech block and single hammer, the two breech-loading barrels of varied calibre, when so hung upon the breech-pin as to permit either barrel, at will, to be brought into position, with respect to the hammer, for firing, and to be slid longitudinally on said pin for expulsion of the cartridge shell and insertion of a cartridge, substantially as set forth.

No. 55,753.—NATHANIEL T. WHITING, Lawrence, Mass.—*Button-hole for Paper Collars.*—June 19, 1866.—The button-hole has a transverse slit to admit the collet of the button more readily.

Claim.—The lateral slots *c*, or cuts *d d*, combined with the button-holes A, substantially as and for the purpose specified.

No. 55,754.—HORACE WICKHAM, JR., Chicago, Ill.—*Burglar Alarm.*—June 19, 1866.—The spring is wound up and the alarm placed so that the opening door or window trips the trigger and sets the clapper in motion.

Claim.—The operating of the clock-alarm by means of the arm H, support U, standard F, and screw I, when constructed substantially as set forth and operated as described.

No. 55,755.—JAMES L. WIGGIN, South New Market, N. H., assignor to JOHN W. HOARD, Bristol, R. I., and GEORGE B. WIGGIN, South New Market, N. H.—*Nail Plate Feeder.*—June 18, 1866.—The nail-plate feed-bar is arranged upon a vertically vibrating frame, and has an intermittent rotary and a back-and-forth movement imparted to it, by means of a cam disk working beneath the frame of the machine.

Claim.—First, combining in a movable frame with a nail-plate feed bar the wheels and other means of transmission whereby the rotary and vibratory up and down and back and forth motions are directly imparted to the feed-bar, the whole being constructed and arranged for operation substantially as herein shown and set forth.

Second, the sleeve for holding the feed-bar in the movable frame, as described, in combination with gear-wheels, one of which is mounted on said sleeve under the arrangement herein shown and described, whereby an intermittent rotary movement is imparted to the feed-bar.

Third, in combination with the sleeve for holding the feed-bar as described, the spring spline in combination with a slot or groove in said bar, substantially as set forth.

Fourth, pressing the feed-bar against the gauge plate of a cutting apparatus by spring power, mechanism, or the equivalent thereof, when applied through the intermediary of an arm mounted on a sleeve capable of a rotating and sliding movement on a rod parallel to the feed-bar, substantially as herein shown and described, so that the feed-bar may be instantly disengaged at the pleasure of the operator.

Fifth, effecting the movements of rotation, lifting up and drawing back of the feed-bar in the manner herein described, the various devices for this purpose used being actuated by a single disk provided with cams and pins, as herein shown and set forth.

Sixth, in combination with the movable frame which carries the feed-bar and the intermediate support, the disk provided with cams, whereby the movable frame is actuated to cause the feed bar to be lifted between each stroke of the cutting apparatus, as set forth.

Seventh, pivoting the dogs or jaws for grasping and drawing back the feed-bar as described, to a slide secured in the top of the vibratory frame, and constructed and arranged as set forth.

Eighth, in combination with the dogs or jaws pivoted to a slide in the top of the movable frame, the vibrating yoke hinged or pivoted to the stationary frame, and actuated by the lever as described, to cause the alternate opening and closing and drawing back of the dogs or jaws, as and for the purposes herein shown and set forth.

No. 55,756.—JOSEPH WORCESTER, Newport, Ky.—*Globe Cock.*—June 19, 1866.—The valve can be ground in its seat without rotating the screw stem by which it is vertically actuated.

Claim.—The valve C, adapted to rotate in the manner substantially as described on the stem G, and provided with a rigid stem D, operated from the outside, in the manner and for the purpose specified.

No. 55,757.—MAX ZABEL, Milwaukee, Wis.—*Fastening Ventilators in Place.*—June 19, 1866.—Curved springs are attached by one end to the periphery of the register, and hold it in place by contact with the casing in which it rotates.

Claim.—The springs e e e, securing the ventilator in place, when arranged and applied as shown and set forth.

No. 55,758.—WILLIAM ZIMMERMAN, Oskaloosa, Iowa.—*Harvester.*—June 19, 1866.—The claim describes the arrangement of devices which receive the grain from an endless apron and discharge it in gavels of regulated size upon the ground.

Claim.—The combination of the door B, screw-rod b, counterpoise C, spring catch E, cord d, bar f, rake D, and pivoted arm g, arranged relatively to each other and with the endless carrying apron-roller a, and operating in the manner and for the purpose herein specified.

No. 55,759.—JAMES W. BARBER, Cincinnati, Ohio, assignor to himself and ELISHA P. STOUT, same place.—*Tobacco Press*—June 19, 1866.—The box receives a charge of tobacco and a follower at each stroke of the reciprocating plunger; the wad is delivered at the open bottom of the trunk, being pressed by the frictional contact with the sides thereof. The trunk has clamp bands and either of two trunks may be brought in connection with the plunger.

Claim.—First, in the described combination with a reciprocating plunger, the bottomless pressing box or trunk, substantially as set forth.

Second, in combination with the elements of the clause next preceding the bands H, and set-screws I, or their equivalent.

Third, the arrangement of two or more shiftable pressing trunks adapted to be brought alternately or successively in connection with a single plunger.

No. 55,760.—PURMORT BRADFORD, New Haven, Conn., assignor to SARGENT & CO., same place.—*Sausage Stuffer.*—June 19, 1866—The axis of the sliding piston is eccentric to the case by contact with which it is operated; in its revolution it forces the meat from the induction at the hopper to the eduction at the spout over which the gut is slipped.

Claim.—The combination and arrangement of the outer case of the form described, constructed in two parts A and B, hinged together as specified, having a hopper H on the one part, and an outlet I on the other part, with the cylinder E and piston F, constructed and arranged to operate substantially in the manner and for the purpose herein set forth.

No. 55,761.—LEWIS WELLS BROADWELL, New Orleans, La., assignor to C. M. CLAY, Ky.—*Packing Projectiles for Rifled Ordnance.*—June 19, 1866.—This projectile is wrapped with several separated bands of cord, which are protected from too easy combustion by the application thereto of plumbago.

Claim.—The described method of wrapping the projectile by belts of cord, which occupy detached annular recesses around the ball.

Also, the application of the said fibrous covering of pulverized graphite or plumbago to serve as a partial protection to the fibre, as and for the purpose described.

No. 55,762.—LEWIS WELLS BROADWELL, New Orleans, La., assignor to C. M. CLAY, Ky.—*Breech-loading Ordnance*—June 19, 1866.—The bore extends through the cannon. The transverse breech-block slides in a mortise, and is wedged in position by a sectional screw. A conical gas-check fits within a chamber of the gun in front of the breech-block, and is tightened by the closing of the latter.

Claim.—First, the permanently located, self-acting, conical or curved gas ring, in combination with a wedge-shaped breech-block, which moves in a line at right angles to the axis of the gun to secure the gas ring in position.

Second, the combination of a conical or curved gas-ring, as described, with an adjustable bearing plate D, in the face of the wedge-shaped breech block.

Third, in combination with the adjustable bearing plate D, an intervening softer material or cushion inserted between the block and the bearing plate, for the purpose described.

No. 55,763.—FRANCIS C. COPPAGE, Terre Haute, Ind., assignor to himself and WILLIAM COPPAGE.—*Harvester.*—June 19, 1866—The described devices are for adjusting the length and height of the vibrating lever, which gives the reciprocating motion to the rake, so as to vary its stroke and maintain its relation to the platform; also, jointing the rod which, by its rotation, raises and lowers both ends of the platform simultaneously, so that it may, with the latter, conform to the surface of the ground.

Claim.—First, the jointed lever *e g*, adjustable vertically at the end by means of the slot, set-screw, and pivot *k'' k'*, and as to working length by the sleeve and set-screw *t t'*, substantially as described.

Second, the jointed connecting rod 8 9 10, in its relation to the winding devices at the inner and outer ends of the platform, substantially as and for the purpose described.

Third, the arrangement of the sliding frame 8, guides T T, leg R, caster-wheel Q shoe P, in combination with the jointed shaft 8, arranged and operated in the manner and for the purpose set forth.

No. 55,764.—ALANSON C. EASTABROOK, Florence, Mass., assignor to J. S. PARSONS, same place, and GEORGE A. SCOTT, Lansingburg, N. Y.—*Brush.*—June 19, 1866.—The bunches of bristles project through the perforated plate, and are imbedded in cement, which forms the finish of the back and handle.

Claim.—As a new article of manufacture, the brush constructed and arranged as herein described; that is to say, a brush in which the bristles, inserted through a perforated plate, are imbedded and held firmly in a cement of any suitable substance, as described, which cement shall at the same time, in combination with a strip of metal or other material, form the back and handle of the brush, as herein shown and set forth.

No. 55,765.—FRANCIS J. FISCHER, Hamilton, Ohio, assignor to himself and JOHN B. BERNING, Cincinnati, Ohio.—*Hay Knife.*—June 19, 1866.—Explained by the claim.

Claim.—The hay knife constructed as described, with a shank A, diverging blades B B', and a short middle blade C, the whole adapted to cutting hay from the stack, substantially as described.

No. 55,766.—F. W. FLAGG, Middletown, Conn, assignor to himself and E. B. MANNING, same place.—*Whirling-jack for Toys.*—June 19, 1866.—By pulling on the string, motion is communicated through the multiplying gearing to the spindle to which the top is temporarily attached, and when released the cord is rewound by the spring ready for another operation.

Claim.—The combination of the spring D with the barrel and spindle of a whirling jack, when constructed and arranged to operate in the manner herein described, so as to rewind the cord, as specified.

No. 55,767.—JOHN GARDNER, New Haven, Conn., assignor to CHARLES T. GRILLEY.—*Capping Wood Screws.*—June 19, 1866 —The cap is nicked after being closed upon the screw head. A longitudinal slot in the shank of the screw is engaged by a spring to determine the position of the screw under the punch, so that the nick in the cap shall correspond with that in the screw head.

Claim.—In the manufacture of capped screws for operation upon screws for this purpose specially provided with a notch in their shanks, the punching apparatus and spring catch, or equivalent mechanism, as described, so that the screw when capped may be adjusted in its proper relation to and its cap nicked by the punch, substantially as and for the purposes herein set forth.

No. 55,768.—THOMAS S. GILBERT, New Haven, Conn., assignor to himself and PERKINS, COOK, & Co.—*Hoop Skirt.*—June 19, 1866.—The associated wire and cord are enclosed in a woven covering, and when secured to the tapes the stitches pass through the cord.

Claim.—Enclosing a cord upon the wire substantially in the manner described, so that the hoop and tape may be secured as and for the purpose specified.

No. 55,769.—GEORGE W. GOLAY, Vevay, Ind., assignor to himself and ELI T. OGLE, same place.—*Broom Head.*—June 19, 1866.—The butts of the corn are pressed around the handle by a hinged collar whose ends are engaged by ratchet and pawl to maintain the closure attained by the tightening lever. Lower down on the broom is a band passing through a link which traverses an orifice in the stump of the handle.

Claim.—First, the hinged clamp B, composed of parts B' B'', loop D and pawl E, constructed and operating substantially as and for the purpose specified.

Second, in combination with the above, the sliding clamp C, composed of parts C C', and link *c*, as set forth and for the purpose specified.

No. 55,770.—WILLIAM H. HOLLAND, Boston, Mass., assignor to himself and WILLIAM GOODMAN, same place.—*Paddle Wheel.*—June 19, 1866.—From each of the main, parabolic floats extend auxiliary wing floats which act to propel the vessel astern when the motion is reversed.

Claim.—The arrangement and combination of the series of auxiliary floats D D, with the wheels A A, and the series of main floats C C C, formed and arranged substantially as described.

No. 55,771.—DAVID HOWARTH, Portland, Me., assignor to himself, W. N. GOURLEY, and S. C. RUNDLETT, same place.—*Lamp.*—June 19, 1866.—An annular transparent portion intervenes between the burner socket and the lower part of the metallic reservoir so as to exhibit the condition of the oil contents.

Claim.—Inserting into a lamp constructed of any opaque substance a transparent ring, of the form in the place and manner, and for the purpose substantially as set forth.

No. 55,772.—CHRISTOPHER R. JAMES, Jersey City, N. J., assignor to himself and N. W. CONDICT, Jr., same place.—*Operating Hammer and Stamp.*—June 19, 1866.—The pistons are depressed alternately by steam admitted by one valve, and are elevated by a constant pressure of air below, admitted from a reservoir.

Claim.—Operating a stamp or hammer by means of a piston working in a cylinder, the upper end of which is opened at proper intervals to a steam boiler and to the atmosphere alternately, and the lower end of which is in constant communication with a reservoir of compressed air, substantially as herein specified.

Also, in combination with two stamps or hammers so operated by pistons working in separate cylinders, a valve and passages so operated as to bring each cylinder alternately in communication with the boiler, and so produce the action of the pistons and their attached hammers, substantially as herein specified.

No. 55,773.—CHARLES KINKEL and MARTIN HUBBE, New York, N. Y., assignors to CHARLES WEHLE, Hoboken, N. J.—*Propellor for Steamships.*—June 19, 1866.—The water discharging nozzle pipes are arranged on each side of the vessel; each is connected with a pump worked by an engine and capable of being turned to eject the water in the desired direction.

Claim.—First, the combination of a number of nozzle pipes with the hull of the vessel, when each one of said pipes is connected with a pump and engine, substantially as described.

Second, the mechanism for connecting and turning the nozzle pipes, substantially as described and for the purpose set forth.

No. 55,774.—JAMES KIRK, Dover, Del., assignor to R. HOE & Co., New York, N. Y.—*Registering Apparatus for Printing Press.*—June 19, 1866.—The upper part of the feed-board is fixed and the lower part which adjoins the nipper cylinder is movable and has an arm underneath carrying a pin which works in a slot in the upper portion; the latter also has a pin which works in a slot in the lower portion, which is operated by the nipper cylinder and so arranged that the pins are withdrawn before being grasped by the nippers.

Claim.—A feed-board for printing presses composed of two parts connected by a joint, one part being fixed and the other part movable so as to work on the joint in connection with a fixed or stationary pin or point connected with the fixed part of the feed-board, and a pin or point connected with the movable part of the same, all arranged to operate in the manner substantially as and for the purpose set forth.

No. 55,775.—FREDERICK KLEE, Williamsburg, N. Y., assignor to LOUIS KLEE, same place.—*Instrument for Irritating the Skin.*—June 19, 1866.—The regulating screw adjusts the depth of penetration of the points which project from the diaphragm to introduce the liquid through the skin.

Claim.—First, the arrangement of the screw *f*, piston *d*, spring *a*, nut *g*, and handle *c*, applied relatively to the cylinder A, combined and operating in the manner and for the purpose herein specified.

Second, the diaphragm *h*, in combination with the cylinder A, and pricks *b*, constructed and operating substantially as and for the purpose set forth.

No. 55,776.—WILLIAM and DAVID McCAINE, Groton, Mass., assignors to themselves and DANIEL McCAINE, same place.—*Tree Protector.*—June 19, 1866.—The jointed frame has glass at its angles and a tent cover tightened by a draw-string by which it is suspended from the tree.

Claim.—As an improvement in the tree protector, the arrangement of its plates of glass so that each two which are next adjacent shall make an angle with each other in a vertical direction at their junction.

Also, the arrangement of two of such glass plates in a groove or rebate, and so as to meet together and make angles with each other in a vertical direction.

Also, the arrangement and combination of the passage *h*, and its holes *k k*, in the top-plates of glass *c c*, *d d*, arranged as specified.

No. 55,777.—JOHN C. McNULTY, San Francisco, Cal., assignor to himself and THOMAS LEE.—*Burning Gas.*—June 19, 1866.—Gas is passed up through and ignited above vitreous clinkers.

Claim.—The burning of jets of inflammable gas in combination with clinkers, substantially as described.

No. 55,778.—A. C. MESSENGER, Syracuse, N. Y., assignor to himself and A. T. SMITH, same place.—*Apparatus for Carburetting Air, Gas, &c.*—June 19, 1866.—The perforated double walls which divide the chamber enclose a woven fabric, which becomes saturated with the hydro-carbon, and through which the air is forced to pass.

Claim.—First, in an apparatus for carburetting gas for illuminating purposes the use of double perforated walls *d d*, having a suitable capillary substance confined between them, said walls being so arranged as to form a porous division through which the gas is forced, substantially as described.

Second, the forming of two apartments in a vessel A by means of an upright double wall partition, which is rendered sufficiently porous to allow of the absorption of the fluid in said vessel and the passage of gas through it, substantially as described.

Third, subdividing the chamber C by means of a partition, which is applied on one side of the induction passage *a*, for the purpose and in the manner substantially as described.

No. 55,779.—CHARLES NELSON, Newburgh, N. Y., assignor to himself, JAMES W. TAY-LOR, WILLIAM R. BROWN, and FREDERICK W. BANKS.—*Operating Beater or Power Press.*—June 19, 1866.—The arrangement provides for applying a steady downward movement to the follower subsequent to the beating operation without stopping or reversing the main driving shaft; also for stopping the descent of the follower at the desired moment; also for relieving the friction of the screw shafts by the interposition of anti-friction bearings at their lower ends; also for bringing the beater elevating drum into action.

Claim.—First, automatically throwing the clutch *d'* out of gear from the wheel *d* by the descent of the follower during the operation of pressing for the purpose of preventing said follower from descending too far, substantially as described.

Second, the combination of the follower *c*, rope *q*, drum *r*, screws *b² b²*, shaft *b²*, and a continuously revolving shaft *e*, all arranged and operating substantially as described.

Third, the combination of the loose wheels *a d* and clutches *a' d'* with the shaft *e*, and the two wheels *b c* on the reversible shaft *b²*, substantially as described.

Fourth, so constructing a baling press that a rapid vertical movement can be communicated to the follower, or the follower brought down upon the bale or compressed mass with a dead pressure from a main driving shaft, which has a continuous movement in one direction, substantially as described.

No. 55,780.—DANIEL G. NORRIS, St. Johnsbury, Vt., assignor to himself, JOHN H. PADDOCK, and RUFUS S. MERRILL.—*Harvester.*—June 19, 1866.—The arrangement of mechanism for driving the cutters and attaching the draught will be understood from the claims and cut.

Claim.—First, in mowing machines constructed and operating substantially as herein described, the combination with the recessed sliding bar, provided with friction rollers, as hereinbefore described, of the cam wheel on the pinion shaft, said sliding bar communicating its movement to the cutter bar through the medium of a connecting rod or pitman hinged to the sliding bar, as set forth, the whole being arranged and operating substantially as shown and described.

Second, in combination with the recessed sliding bar, as described, the pinion shaft revolving in bearings on the cutter frame, when arranged relatively to the driving pinions, as set forth, and having the cam wheel located within and the self-adjusting clutches without the cutter frame, upon the said shaft, as and for the purposes herein shown and described.

Third, in mowing machines, arranged and operating as described, attaching the whiffletree directly to the metal cutter frame and shafts to the wooden frame, as and for the purposes herein described.

No. 55,781.—DAVID M. OSBORNE, Auburn, N. Y., assignor to himself and WILLIAM A. KIRBY, same place.—*Harvester.*—June 19, 1866.—The conical pin at the end of the pitman passes through an eye in the sickle-bar head, and is held in place therein by a flat spring which is riveted at one end to the pitman and perforated at the other end to encircle the end of the pin and press it against the side of the head opposite the pitman.

Claim.—Connecting the pitman of a harvesting machine to the head of the cutter-bar or cutters, or connecting bars, to their supports by means of a single pin, and retaining said pin in place by means of a self-adjusting perforated spring, substantially as above described.

No. 55,782.—AGUR PIXLEY and JOHN ROBERTSON, Brooklyn, N. Y., assignors to ROBERTSON, DOW & CO., same place.—*Hydrant.*—June 19, 1866.—A ring valve on the neck, which is perforated for a waste way, is rotated so as to close that aperture during the season when freezing does not occur.

Claim.—The ring I, formed with a hole or opening n, and arranged upon the hollow valve stem D, with reference to the hole or opening r thereof, in combination with the valve stem D and valve box C, substantially as herein set forth for the purpose specified.

No. 55,783.—AUGUSTUS A. RANDALL, New Haven, Conn., assignor to SARGENT & CO., same place.—*Machinery for Finishing Coffin Nails and Screw Heads.*—June 19, 1866.—The nails are fed into the machine through a vertical channel, through the bottom of which a sliding pusher works, draws against the head of the nail, and forces the point into a revolving mandrel; when the pusher recedes a tool advances and turns the head to the proper shape, after which a milling tool and burnisher advance and finish the head.

Claim.—First, the combination of the spindles F and I, fingers N, and channel ℞, constructed and arranged to operate substantially in the manner and for the purpose specified.

Second, in combination with the above the cutter b, mill f, and a burnisher, substantially as and for the purpose specified.

No. 55,784.—FRANCIS ROACH, Boston, Mass., assignor to himself and JOSEPH ZANE, same place.—*Stop Cock.*—June 19, 1866.—The valve is pressed upward to its seat by the conjoint action of the spring and the liquid, and is depressed by the key, but does not revolve.

Claim.—The arrangement and combination of the socket e, the guide b, the valve A, the spring B, the valve seat i, the induction chamber E, the eduction chamber F, the pivot a, the key G, the neck H, the screws f g, and the induction and eduction passages or pipes l m, the whole being as specified.

No. 55,785.—M. M. ROUNDS, New Haven, Conn., assignor to himself and WILLIAM YELLNER, New York, N. Y., and J. E. JEROLD, Jersey City, N. J.—*Steam Generator.*—June 19, 1866.—A series of tubes proceed from the forward part of the crown sheet and pass to the rear parallel with the other tubing.

Claim.—The tubes d, arranged in the crown sheet c, so as to open into the fire-box, substantially as and for the purpose specified.

No. 55,786.—F. W. SCHROEDER, Philadelphia, Penn., assignor to himself and WILLIAM H. HOSKINS, same place.—*Sabot.*—June 19, 1866.—The sole and heel pieces are riveted to a metallic frame, which conforms to the shape of the sole; it is hooked to the toe, screwed to the heel, and fastened by transverse bars and clips to the tread and the shank respectively of the shoe sole.

Claim.—First, the sabot, composed of the sole piece A and heel piece B, connected together by a metal strip D, bent to conform, or nearly to conform, to the shape of the sole and heel of a boot or shoe, and having a shoulder a and turned-up end b, all substantially as described.

Second, the combination of the strip D and its hooked end e with the transverse strip E and its hooked ends and sole piece A.

Third, the transverse strip F, having hooked ends, and being arranged to slide on the strip D, as set forth, for the purpose specified.

No. 55,787.—CHARLES SPOFFORD and C. H. HERSEY, Boston, Mass, assignor to themselves, W. E. HAWES, and FRANCIS C. HERSEY, same place.—*Cotton Gin.*—June 19, 1866.—A bar triangular in cross-section is introduced between the cylinders. As the combs alternately move up and down they act upon the cotton which is drawn in by the continuously revolving rolls, and comb back the seeds by working against the upper and lower edges of the bar.

Claim.—The employment of the triangular-shaped bar K, in combination with the rolls G H and combs N, operating substantially as and for the purpose set forth.

Also, the triangular-shaped bar K, with its projecting ledge i, substantially as and for the purpose described.

No. 55,788.—R. C. SWANN, Brownsville, Ind., assignor to himself, JOHN L. RITER, and T. JEFFERSON WEST, same place.—*Smut Machine.*—June 19, 1866.—This scouring device for grain has a perforated bed, reciprocating beneath roughened rollers which are journalled in yielding bearings.

Claim.—The reciprocating perforated bed or screen A, in combination with the pressure rollers B, having a rough periphery and their journals fitted in springs C, all arranged to operate in the manner substantially as and for the purpose set forth.

No. 55,789—HENRY THOMPSON, Rockland, Maine, assignor to himself, D. C. and C. H. HASKELL, same place.—*Rigging Stopper.*—June 19, 1866.—The faces of the jaws have helical grooves to fit around the strands of a rope, and the movable jaw is clamped upon the other by the vibrating sector pinion on the lever journalled in coupling links.

Claim.—The rigging stopper made and for use substantially as hereinbefore described, that is to say, of the toothed sectional lever, the rack, and the movable supporting arm, and the two jawed levers, arranged and applied together in manner and so as to operate as explained.

No. 55,790.—JAMES H. TOBEY, Cranston, R. I., assignor to himself and ALFRED E. TENNEY.—*Swaging Machine.*—June 19, 1866.—One plunger surrounds the other and has a true reciprocating movement derived from an eccentric, while the other has a variable movement derived from a cam, each plunger performing separate functions.

Claim.—The duplex plunger, composed of two independent plungers D and E, having coincident axes operating relatively to each other, as described, in combination with the die C, constructed and arranged substantially as and for the purpose specified.

No. 55,791.—ISAAC F. VAN DUZER, Middletown, N. Y., assignor to self and R. M. SAYER.—*Milk Can.*—June 19, 1866.—Explained by the claims and cut.

Claim.—First, the manner herein set forth of attaching the breast to the body of the can, viz., by having the breast-piece confined between the body and hoop, as and for the purposes described.

Second, the manner of constructing the bottom part of the can of the two hoops, and body and bottom, as herein recited.

Third, the manner of securing the body-piece and the hoops together, by first riveting them and then dipping them in the melted solder, as set forth.

No. 55,792.—JULES AUBIN, Paris, France.—*Grinding Mill.*—June 19, 1866.—The metal boxes or compartments in the runner are covered with wire cloth to bolt the grain as it is ground, by carrying off the flour through the stone.

Claim.—Constructing millstones with metal boxes or compartments let into the stone and covered with metallic or other cloth, substantially in the manner and for the purposes hereinbefore described.

No. 55,793.—AUGUSTE BOISSONNEAU, Paris, France.—*Artificial Eye.*—June 19, 1866.—The caruncular portion of the ocular orbit has unguinal depressions on each side of the nasal extremity, so as to establish harmony between the circumference of the prosthetic shell and the organic sinuosities when used on either side.

Claim.—The shaping or forming artificial eyes in enamel, with a hollow c^4 in the lower internal section, so that the lower section is symmetrical to the upper one, for the purpose as hereinbefore set forth of using the said eyes on either side as substantially described.

No. 55,794.—JEAN BAPTISTE EVRARD and JEAN PIERRE BOYER, Paris, France.—*Machine for Making Hinges.*—June 19, 1866.—Each leaf of the butt hinge is made of a piece of sheet metal cut out and doubled up to make loops for the pintle. The machine acts automatically upon two strips from which the leaves are cut, doubled, united by their pintle, and their screw holes countersunk.

Claim.—First, the arrangement of the spring dogs b, reciprocating carriage c, in combination with the slotted rod e, with its cam slot d, the lever e^a, dovetailed strips h, operating in the manner and for the purpose herein specified.

Second, the punches l and knives r, in combination with the cams n, levers o, and with the guide grooves through which the blanks are fed, constructed and arranged substantially as and for the purpose set forth.

Third, the bending tools s, and grooves a', in combination with cams v and levers w, constructed, arranged, and operating substantially as and for the purpose described.

Fourth, the arrangement of the dog j, cam l, lever m', constructed and operating in the manner and for the purpose specified.

Fifth, the spring clamp $a^2 b^2$, in combination with the cam d^2 and lever c^2, substantially as and for the purpose set forth.

Sixth, the adjustable guide tube i, arranged in combination with the mechanism for feeding in and bending the plates and feeding the wire, substantially as and for the purpose specified.

Seventh, the countersinks $o' o'$, applied in combination with the mechanism for feeding in, bending, and cutting off the plates, substantially as and for the purpose set forth.

No. 55,795.—LOUIS H. G. EHRHARDT, Bayswater, England.—*Gunpowder.*—June 19, 1866.—Explained by the claims.

Claim.—The combination of mineral carbon with cutch, tannin, or gambier, substantially as herein described, to form a safety powder which may be rendered explosive at will by the addition of chlorate and nitrate of potash, in the proportions substantially as herein set forth.

Also, the combination of tannin, or its equivalent, with mineral carbon, chlorate of potash and nitrate of potash, substantially in the manner and for the purpose herein set forth.

No. 55,796.—SAMUEL H. HAYCOCK, Ottawa, Canada West.—*Projectile.*—June 19, 1866.—This elongated bullet of cylindro conoidal form has a conical rear extension.

Claim.—An elongated pointed projectile with a cylindrical portion to fit the lands of the bore, and a conical rear, the centre of gravity being in advance of the cylindrical portion.

No. 55,797.—S. G. ELLIOTT, San Francisco, Cal.—*Photometer.*—June 19, 1866.—The degree of opacity of an interposed medium which is required to intercept a ray of light is made the measure of the intensity of the latter. The inner tube in addition to its scale has a glass tube graduated by a photographic process, from transparency to opacity. The outer tube has an opening to admit light, which is reflected by a mirror along the axis, the scale indicating the point at which the light becomes visible through the interposed medium.

Claim.—The combination of the tube A, with the sliding, graduating tube B, reflector F and rod E, when said tubes, reflector, and rod are constructed in the manner and for the purpose as substantially described.

No. 55,798.—ISAAC LEWINE, New York, N. Y., assignor to himself and MARGARET M. PIPER.—*Button Attachment for Apparel.*—June 19, 1866.—A button shank is held in an eyelet; an elastic band passes through two eyelets on a stay piece and is looped over the button.

Claim.—The combination of the elastic loops F F and buttons D D both attached to the garment by eyelets in stay pieces, substantially as and for the purpose herein specified.

No. 55,799.—JOHN ADAMS, Kokomo, Ind.—*Cutting Boots.*—June 26, 1866.—Explained by the claim.

Claim.—As a new article of manufacture, a boot produced as follows, viz : by cutting the foot and leg portion out of a single piece of leather, the counter or heel piece being left out, and then constructing the uppers of the boot from said foot and leg portion and a separate counter-piece, by having a seam extend down the back to the top of the counter, then running the seam to the front edges of the counter, and from thence down to the shank of the boot, said horizontal and vertical side-seams serving to fasten the separate outside counter-piece of leather over the opening which was left in cutting the foot and leg of the boot, all substantially as and for the purpose described.

No. 55,800.—WILLIAM M. ARNALL, Sperryville, Va., assignor to himself and WILLIAM H. BROWNELL, New York, N. Y.—*Grain Separator and Cleaner.*—June 26, 1866.—The large cylinder has a hard periphery which is held in contact with the smaller rollers by springs; the latter have soft surfaces which catch the cockle, garlic, &c., and carry it away to be removed by brushes and separately discharged, while the dust is blown away by a fan.

Claim.—First, the arrangement of the cylinder E, the rollers F F F in the arc of a circle described by a radius from cylinder E, together with the adjustable brushes *a a a* and dividing board P, substantially as and for the purpose herein set forth.

Second, the throat G, with its gauge board *f*, rollers F, cylinder E, and fan D, arranged and used as and for the purpose specified.

No. 55,801.—ALFRED ARNOLD, Senafly, N. J.—*Device to Prevent Boiler Explosions.*—June 26, 1866.—The device is applied to that part of the generator most exposed to sudden increment of heat where the expansion of its disks moves the tube longitudinally and emits steam to the whistle.

Claim.—First, the mode of preventing steam boiler explosions, substantially as herein set forth.

Second, the construction and arrangement of the devices necessary to carry the mode into operation, substantially as described.

No. 55,802.—FREDERICK ASHLEY, New York, N. Y.—*Egg Beater.*—June 26, 1866.—The shaft is rotated by the engagement of its stud with the thread inside the reciprocating sleeve.

Claim.—The grooved or screw-threaded slide or sleeve E, in combination with the plain shaft A, having a fixed stud or pin *a*, and a beater B, arranged together and operating as and for the purpose specified.

No. 55,803.—SAMUEL W. AYRES, Monticello, Ind.—*Water Wheel.*—June 26, 1866.—The wheels are attached to the same vertical shaft, the upper receives the water from a surrounding spiral chute and discharges at the centre, the water passing thence to the lower wheel, which discharges at its periphery.

Claim.—In combination with the spiral scroll A, the double wheel B C, when so arranged that the water shall be discharged from the scroll through the upper section, towards its centre, and thence passing into the lower section be discharged from the centre through the periphery, substantially in the manner set forth.

No. 55,804.—H. A. BAILEY and A. R. BURDICK, Racine, Wis.—*Horse Rake.*—June 26, 1866.—The ends of the rake teeth pass through the metallic blocks and the rake head and are secured by nuts in front, the lips of the blocks resting on the rake head; the latter is rotated to discharge the load of hay by the engagement of clutches on the axle with the wheel hubs when the hand lever connected thereto is actuated.

Claim.—First, the metal heads J, provided with lips *g*, and grooves *i*, and retained on the rake head by the upper ends of the rake teeth K, passing through them and the rake head, substantially as and for the purpose herein set forth.

Second, the attaching of the rake head H to clutches D D, fitted loosely on the axle A, and operated by means of levers by the driver from his seat G, for the purpose of automatically raising the rake to discharge its load, substantially as shown and described.

No. 55,805.—WILLIAM W. BEACH, New York, N. Y.—*Roof of Buildings.*—June 26, 1866.—Explained by the claims.

Claim.—First, the use of mica or mineral isinglass in sheets or plates as a roofing material, transparent, translucent, opaque, and ornamental, substantially in the manner and for the purpose herein set forth.

Second, ornamenting upon or between the plates or thicknesses of the mica, substantially as set forth and described.

Third, making an elastic roofing by cementing the plates of mica with flexible cement, substantially as set forth herein.

Fourth, the combination of mica with wood, slate, or other equivalent substances, substantially in the manner and for the purposes herein set forth.

No. 55,806.—JACOB H. BEIDLER, Lincoln, Ill.—*Lamp.*—June 26, 1866.—The current of air ascending to the flame is heated by contact with the cylinder which contains steam generated in the boiler above the flame and conducted downward by pipes.

Claim.—First, the method herein described of creating an ascending current of air to feed the flame, by means of steam generated by the caloric emanating from the illuminating flame of the lamp.

Second, the combination of the boiler or heater D, the steam pipes D', D', the annular chamber C, and the feed tube B, with the oil cup A, perforated plate *g*, and deflector *f*, as and for the purpose described.

Third, the hot water tank C', in combination with the feed tube B, and the boiler or heater D, as and for the purposes described.

No. 55,807.—WILLIAM A. BEMIS, Spencer, Mass.—*Grater and Egg-Beater.*—June 26, 1866.—The divided cup has a diaphragm of wires which cut the egg shaken therein; outside the cup is a grater surface.

Claim.—First, the combination with the lower half B of the body of the beater, of the removable frame D, provided with the cutting wires *a*, substantially as and for the purposes set forth.

Second, in combination with the egg-beater, the grater moving in guides on the body of the egg-beater, the whole being constructed and arranged substantially as and for the purposes shown and described.

No. 55,808.—JACOB BENTZ, Brooklyn, N. Y.—*Machine for Framing Matches.*—June 26, 1866.—The loose box in the hopper has partitions to distribute the matches upon the hinged grooved bed, and a follower above to keep the splints down; the front part of the bed has guides to keep the matches in position after they are pushed into the frame; while the plunger is retreating, the frame is moved downward to bring another series of recesses opposite to the bed.

Claim.—First, in combination with the grooved bed and hopper, the loose hopper-box or frame, constructed and operating substantially as described.

Second, the construction of the grooved bed, in two parts; the front part being hinged to the rear fixed part for the purpose of being swung down below the plane of the fixed part when required to remove broken splints, slivers, or other obstructions, substantially as described.

Third, in combination with the grooved bed, the guides or slots in front of the grooved bed, substantially as described.

Fourth, in combination with the plunger frame and the front sliding frame which supports the splint frame, the rock-shaft and system of pawls and levers, whereby the down-feed motion of the sliding frame is communicated thereto by the back motion of the plunger frame.

No. 55,809.—JOHN W. BLODGETT, Plymouth, Ind.—*Stump Extractor.*—June 26, 1866.—The posts and sills are pivoted to the axle and are lowered into working position. The chain is wound upon an axle by wheel, pinion, and rotating sweep.

Claim.—First, the combination and arrangement of the posts C, and foot boards E and N, substantially as and for the purposes set forth.

Second, wheels M and P. frame L, windlass W, chain L, and lever K, in combination with wheels A, and foot-boards C, E, N, substantially as and for the purposes set forth.

No. 55,810.—JOHN M. BROSIUS, Liberty, Va.—*Utilizing Steam.*—June 26 1866.—The boiler of the locomotive is temporarily attached while at a station or depot, to a wood-saw or pump, so as to utilize the steam that would be blown off, thus dispensing with station engines.

Claim.—The arrangement; with a stationary depot, or water station engine, of the locomotive boiler, substantially in the manner and for the purpose described.

No. 55,811.—FREDERICK H. BROWN, Chicago, Ill.—*Cotton-Seed Planter.*—June 26, 1866 — To secure uniformity in the delivery, the seeds are torn from the mass in the hopper by the spikes of the endless, revolving band, and cleared from the teeth by the revolving wings on the lower shaft.

Claim.—First. in a cotton-seed planter, the combination of the shaft R, provided with arms V, and the conveying belt L, provided with teeth as described, arranged, and operating substantially in the manner and for the purposes herein specified.

Second, in combination with the said conveying belt L, the shaft U, provided with suitable wings or cleaners, arranged and operating substantially as specified and shown, and for the purposes set forth.

Third, the combination of the shaft R, provided with arrows or spikes, the toothed belt L, and the winged shaft U, arranged and operating as and for the purposes described.

No. 55,812 —DUNCAN BRUCE, Rossville, N. Y.—*Constructing Vacuum Vessels for Evaporating, &c.*—June 26, 1866.—A vessel of wood and cement, enclosed by frame work and masonry, as a substitute for the metallic vacuum pans used in the concentration of saccharine juices, vegetable extracts, &c.

Claim.—The method, substantially as herein described, of strengthening and rendering wooden vessels air-tight, for the purposes set forth.

No. 55,813.—JOHN T. BRUEN, New York, N. Y.—*Hammer for Bending Couplings.*—June 26, 1866.—Designed for coupling the ends of the wires of skirts, telegraphs, hat, or bonnet-frames, &c. The sudden impulse given to the hammer by the heel of the operator upon the treadle overcomes the force of the spiral spring which assists the hammer to rebound, and throws out the material acted upon. The arms drop by the side of the anvil when the blow is struck, and are held by a spring till the hammer rises, when a stub on the axle of the hammer actuates the arms, which remove the article from the dies.

Claim.—First, the anvil, the hammer, the connecting-rod, and the treadle, when constructed and arranged substantially as and for the purpose herein specified.

Second, the apparatus for throwing out the material after the successive operations of the hammer, applied, arranged, and operating substantially as herein specified.

Third, the annular spring or buffer *d''*, applied at the lower end of the connecting-rod, substantially as described, and serving the two purposes of regulating the stroke of the hammer, and of preventing destructive concussion and noise.

No. 55,814.—HENRY and SAMUEL W. BUDD, Philadelphia, Penn.—*Lock.*—June 26, 1866.—The motion of the sliding block is governed by the notched tumblers and detent projections in the barrel; when the block is depressed a wedge retracts the spring bolt.

Claim.—First, the bolt B and its spring *d*, in combination with the sliding block F and plate *b*, or its equivalent, the whole being constructed and operating substantially as and for the purpose described.

Second, a series of tumblers in combination with the block F, and with projections in the casing D, the whole being constructed and operating substantially as and for the purpose specified.

No. 55,815.—CHARLES H. BUSH, Fall River, Mass.—*Brush for Boiler Flues.*—June 26, 1866.—The semi-annular brushes are attached to springs by which they are expanded to the capacity of the tube.

Claim.—An expanding and contracting brush for clearing obstructions from the flues of steam boilers, constructed and operating as herein set forth and described.

No. 55,816.—WILLIAM H. BUTLER, Chicago Ill.—*Railroad Water Elevators.*—June 26, 1866.—Water is raised from a well by the condensation of steam admitted for that purpose into the chamber, and then ejected into the cistern of the tender by pressure of steam from the locomotive.

Claim.—A metal tank lined with wood, which has been previously saturated with oil or other resistant to rapid condensation of steam, or coated with such resistant on one or both sides, substantially as described and for the purpose mentioned.

Also, in combination with the follower and steam pipe, the rubber packing I, as described.

No. 55,817.—E. P. H. CAPRON, Springfield, Ohio.—*Step Ladder*—June 26, 1866.—The ladder is stayed at the required extension by a notched brace, which catches upon a round; the feet of the upper platform rest upon the rounds of the ladder.

Claim.—The ladders A and B, platform C, and the notched bar E, all combined and arranged to operate as shown and described.

Second, the slotted bar D provided with the series of holes and pins i, arranged to operate in combination with the ladders A and B and platform C, as herein set forth.

No. 55,818.—CHARLES F. CARPENTER, Louisville, Ky.—*Treating Ores.*—June 26, 1866.—Explained by the claim.

Claim.—The mode of using steam of any temperature for the purpose of facilitating the process for extracting gold and silver from ores, and consists in introducing said steam into a reverberatory furnace between the flame of said furnace and the ores containing gold or silver, which are spread upon the hearth of said furnace, as herein described, or any other substantially the same.

No. 55,819.—FRANKLIN CHALFANT, Lancaster, Penn.—*Water Indicator for Steam Generators.*—June 26, 1866.—The cylinder is attached by hollow couplings to the boiler, and rises and falls with the diminution or increase of the water therein, indicating its motions on a dial if desired.

Claim.—First, the vibrating column or cylinder F, in connection with a steam boiler A, when employed for the purpose specified.

Second, the soapstone disks K, when applied substantially in the manner and for the purpose set forth.

No. 55,820.—JOHN R. CHAMPLIN, Laconia, N. H.—*Ice Cream Freezer*—June 26, 1866.—The revolving arms scrape the cream from the sides of the can, which rotates in a contrary direction. The rotating gear is hinged to the pail in which the freezer sets, and couples to the beater shaft and cam, when brought into working position.

Claim.—First, the side-scrapers G, when made in the form and adjusted in the manner described, in combination with the horizontal arms E of the beater, substantially as and for the purpose set forth

Second, the frame R and support X, when constructed substantially as described, in combination with each other, and with the shafts T and L, as and for the purpose set forth.

Third, the use of a combined joint and swivel for coupling the gearing to the beater and cream-holder of an ice-cream freezer, substantially as described.

Fourth, the coupling device herein described, in which the coupling is accomplished by dropping the clutch upon the shaft and cream-holder, substantially as described and for the purpose set forth.

No. 55,821.—ROBERT COCHRAN, Morrison, Ill.—*Pump.*—June 26, 1866.—The water raised in each cylinder by the single acting plunger therein is discharged by the side pipes, where a single disk valve closes the apertures alternately.

Claim.—The combination and arrangement of the valve I, pipes O and P, and cylinders A and B, substantially as herein described and set forth.

No. 55,822.—ROBERT CORNELIUS, Philadelphia, Penn.—*Pump for Deep Wells.*—June 26, 1866.—The oil, &c., entering at the side apertures passes through the wire gauze to the pipe, while the debris, grit, &c., fall into the receptacle below.

Claim.—The combination of an outer case with apertures, an interior wire gauze or perforated screen, and a receptacle below for the debris, substantially as described.

No. 55,823.—ELLIOTT H. CRANE, Burr Oak, Mich.—*Door Bell and Burglar Alarm.*—June 26, 1866.—One hammer is operated by the bell pull outside the door, and the other is operated by a trigger set across the path of the door, when it is desired to act as an alarm.

Claim.—First, so arranging two hammers to a bell that, while the bell answers the purpose of an ordinary door-bell, its parts can be adjusted to act as a burglar alarm when the door is opened, substantially as described.

Second, the combination of the two hammers L and O, bell D, notched base plate C, swinging pawl or arm T, and knob handle, with a plate or tripper H upon its spindle, when the several parts are arranged together, and so as to operate substantially in the manner and for the purposes described.

No. 55,824.—THOMAS C. CRAVEN, Albany, N. Y.—*Horse Hay Fork.*—June 26, 1866.—The lifting prong works loosely in and is projected from the hollow end of the stock by a rod which is moved by the vibration of the toggle above; the latter is locked in either position and released by a tripping cam.

Claim.—First, applying a hook or barb A to the hollow point of a harpoon hay fork, so that this hook shall work loosely within said point, and be projected therefrom by the depression of a rod C, substantially as described.

Second, connecting the upper portion of the curved rod C to the upper portion of its hollow staff A by means of toggle joints which are adapted to serve as a locking device, substantially as described.

Third, the combination of the tripping lever or eccentric *f*. or its equivalent, with locking toggle levers and rod C, for the purpose of unlocking said rods.

Fourth, curving the upper portion of the rod C, for the purpose substantially as described.

No. 55,825.—F. J. CRISSEY, Leesburg, Va.—*Churn.*—June 26, 1866.—Explained by the claim and cut.

Claim.—The arrangement and combination of the centre dasher with fly-wheel at top and propelling cord on the shaft, so that the shaft is operated swiftly with a forward and reverse action, as herein described and for the purposes set forth.

No. 55,826.—JOHN CUSTER, Sandusky, Ohio.—*Cultivator.*—June 26, 1866.—This clod-crushing roller has a series of cutters on its periphery whose intervals are occupied by hinged clearing plates which prevent the accumulation of soil.

Claim —First, the hinged cleaners H, in combination with the levers I, arranged to operate with the cutting rollers D, in the manner and for the purpose herein specified.

Second, the combination of the cutting rollers D, hinged cleaners H, frame A, bar K, and chain J, arranged and operating as described.

No. 55,827.—SAMUEL CUSTER, Salem, Va.—*Marine Compass.*—June 26, 1866.—The battery of magnetic needles, placed below the indicator magnet, is capable of deflection on its axis to indicate the presence and strength of local disturbing influences. The cruciform magnet has two needles, the indicator occupying a wire bisecting their angles.

Claim.—First, the combination of the lower battery magnet with the upper indicator magnet, substantially as and for the purpose described.

Second, the construction of the cruciform indicator magnet with a pointer placed midway between the north and south poles of its two needles.

Third, the arrangement of the indicator magnet upon and eccentric with the main compass-card or its frame, for securing a longer radius to the pointer.

No. 55,828.—JAMES DEMPSTER, Naugatuck, Conn.—*Roller Feed for Carding and Picking Machines.*—June 26, 1866.—The upper and lower rollers of each pair revolve with the same velocity, but those of the successive pairs have an increasing velocity so as to draw and straighten the fibre.

Claim.—As an improvement in roller feed for cards and pickers the combination and arrangement of the rolls B C D E F G, and the gear wheels I L M N S R, and the gear wheels J K O P T U, with each other and with the frame A of the machine, substantially as described and for the purpose set forth.

No. 55,829.—DANIEL DENNETT, Buxton, Me.—*Hay Rack for Wagons.*—June 26, 1866.—The notches in the sides of the hay bed enable the wheels to turn on a shorter lock.

Claim.—The stretchers *b b b b* cut in two at *d d*, in the manner and for the purposes specified.

No. 55,830.—J. DENSMORE, Meadville, Penn., and G. W. N. YOST, Corry, Penn.—*Car for Transporting Petroleum.*—June 26, 1866.—The tanks are permanently attached to the car platform for the transportation of petroleum in bulk.

Claim.—The combination of the two-tank car with the one-tank car, being the three tanks B B B attached to and combined with the car platform A A A, by means of the frames C C C C C C C and C C C C, and the bolts D D D D D, D D D and D D D D, when constructed in the combination hereinbefore described, and for the application to the purposes hereinbefore written, or when done by any other mechanical construction substantially the same and which will produce the same results.

No. 55,831.—J. DENSMORE, Meadville, Penn., and G. W. N. YOST, Corry, Penn.—*Car for Transporting Petroleum.*—June 26, 1866.—A wooden tank is secured on the middle of the car platform.

Claim.—The one tank B, attached to and combined with the platform of a car A by means of the frame of bars C C C C and the bolts D D D D, over and upon the middle of the platform and car, when constructed and combined as and for the purposes hereinbefore described and set forth, or when attached and combined by any other mechanical construction substantially the same and which will produce the same results.

No. 55,832.—JAMES DENSMORE, Meadville, Penn., and AMOS DENSMORE and G. W. N. YOST, Corry, Penn.—*Car for Transporting Petroleum.*—June 26, 1866.—Explained by the claims and cut.

Claim.—First, the one tank B. square or oblong square, of wood planks bolted together and attached to an ordinary railway car A by means of the cleats E and the bolts D D D D, when the same are constructed and combined as hereinbefore described and for the purposes set forth.

Second, the two tanks B B, square or oblong square, directly over the trucks or abutting together in the middle of the car, of wood planks bolted together and attached to the car A by means of the cleats E E and the bolts D D D D, when the same are constructed and combined as hereinbefore described and for the purposes set forth.

Third, the three tanks B B B, square or oblong square, of wood planks bolted together and attached to the car A by means of the cleats E E E E and the bolts D D D D, when the same are constructed and combined as hereinbefore described and for the purposes set forth.

Fourth, the square or oblique square tank or tanks of wood planks attached to an ordinary car, when constructed and combined by any other mechanical contrivance substantially the same, and which will produce the same results.

No. 55,833.—THOMAS DICKENSON, Newark, N. J.—*Machine for Planing Mouldings.*—June 26, 1866.—Attached to the frame of the machine is a box table in which a cutter head revolves; a moulding is planed upon the top of the table by the cutter beneath it, and another moulding upon the bottom of the box by the same cutter above it.

Claim.—The arrangement described and represented for dressing two mouldings simultaneously by one cutter head, consisting of the adjustable box table W X, constructed and operated substantially as described.

No. 55,834.—JOHN W. DIXON, Philadelphia, Penn.—*Apparatus for Making Paper Pulp, for Bleaching and for other Purposes.*—June 26, 1866.—The horizontal digester has perforated partitions near its ends, a heating coil over a furnace, and a pump to cause circulation.

Claim.—First, the combination of the digester A, the manhole B, the diaphragm C, and the discharge pipe and aperture N.

Second, the combination of the digester A, the heating coil and the pump D, as described.

No. 55,835.—JOHN W. DIXON, Philadelphia, Penn.—*Apparatus for Making Paper Pulp from Wood, Straw, and other Materials.*—June 26, 1866.—The vertical digester has a lower discharge and a circulation of water induced by a pump and heated by passage through a furnace.

Claim.—The combination of the circulating pump, the pulp digester and heating coil, or its equivalent, for heating the liquid while being made to circulate by the pump.

No. 55,836.—JOHN W. DIXON, Philadelphia, Penn.—*Process of Making Pulp from Wood, Straw, and other Materials.*—June 26, 1866.—The material to be pulped is digested in an aluminate of soda or potash.

Claim.—The pulping of wood, straw, and other vegetable substances with a solution of highly heated aluminate of soda under pressure, substantially as described.

Also, the pulping of wood, straw, and other vegetable substances by circulating a highly heated solution of aluminate of soda through the mass to be pulped, substantially as described.

No. 55,837.—JOHN A. DODGE, Auburn, N. Y.—*Harvester.*—June 26, 1866.—The described devices relate to the cam which controls the movement of the rotating rake and reel arms, and the frame which supports said cam, located at the inner front corner of the hinged grain platform. Also, to the arrangement, in connection with the means employed for driving said rake and reel, of the clutch throwing the same into and out of action; and in the arrangement of a socket and arm for supporting the adjustable inner end of the platform, all as identified by the claims.

Claim.—First, the combination of the cam B', bed piece B, and frame A, resting upon the platform, the said several parts being respectively constructed and arranged for use, substantially in the manner and for the purpose set forth.

Second, the plate E and ratchet E', upon the outside of the driving wheel and pulley D, upon the projecting extension of the axle, in combination with the belt M, pulley C, and bevel gearing for actuating the rake with the forward motion of the machine, substantially as set forth.

Third, the device for supporting the arm H by means of the bracket L', attached to the socket I, secured to the frame, substantially as set forth.

Fourth, the arrangement as herein set forth for supporting adjustably the inner end of the cutter-bar and platform by means of the arm G, attached to the platform and connecting chain and projecting arm H, secured to the main frame by the bracket I' and socket I.

No. 55,838.—ALONZO DRUMMOND, Newark, N. J.—*Pocket Book Protector.*—June 26, 1866.—The metallic frame which holds the pocket has a flexible inside and metallic outside, the clasp closes on the latter and is retained by a sliding hasp, whose motion throws out a sharp spur to arrest the hand of the thief.

Claim.—First, the combination of the spring B or springs B B with the lever D and piece C, carrying c' c', in the manner and for the purposes set forth.

Second, the plate I, with its buttons or studs K, in combination with the springs B B, when constructed in the manner and for the purpose described.

Third, the lever D, the clasp or hinge E, slot e, and knob d'', constructed and operating substantially as set forth.

Fourth, the spur F, dog G, slide f'', lever D, and knob g'', in combination for operating the spur, substantially as set forth.

No. 55,839.—A. M. Duburn, Chicago, Ill.—*Lantern.*—June 26, 1866.—The bottom is divided and one portion opens on a hinge; the milled head of the wick-raising shaft is reached by a recess in the base.

Claim—First, the part H', secured by the hinge K, when constructed as described and used for the purpose set forth.

Second, the recess m m' m, for the purpose set forth.

No. 55,840.—Abraham Dyson and William N. Macqueen, St. Louis, Mo—*Corn Harvester.*—June 26, 1866.—The rotary cutters are constructed with separate blades fitted between two plates, which are held in contact by a screw and nut, and are arranged in movable frames; their driving gearing being placed between two plates, and arranged with rack and pinion to adjust their relative position. A swinging rack is hinged to each side of the frame for receiving the gavel; the upper bars of the racks are held together at will by clamps attached to a lever pivoted to the frame.

Claim.—First, the rotary cutters C', arranged at the front end of the machine, with the plates U, gearing, and the racks and pinion, substantially as shown and described, to admit of the lateral adjustment of the cutters to suit the width of the spaces between the rows of stalks or plants, as set forth.

Second, the particular manner of constructing the rotary cutters C', to wit, by having blades h fitted between the two parts i' i' of a hub D, which parts are secured in contact by a screw nut j, one of the parts being notched to receive the inner ends of the blades h, substantially as shown and described.

Third, the arrangement of the racks G G, fenders A", frame A, and cutters C', relatively with each other, and operating in the manner and for the purpose herein specified.

Fourth, the lever I, provided with the clamps J, and arranged relatively with the upper edges of the racks G, as shown and described, for the purpose of preventing the casual tilting of the racks, as described.

No. 55,841.—Josiah S. Elliott, East Boston, Mass.—*Brick Machine.*—June 26, 1866.—Improvement on his patent of February 14, 1860. The devices are for preventing choking of the hopper and carrier; for operating the carrier and mould, and a sweep for discharging the brick from the stationary bottom of the mould.

Claim.—First, operating of the carrier N by means of the levers P, connected with the sleeves K, to which the arm O of the carrier is attached, and actuated by the cam Q, substantially as shown and described.

Second, the operating of the mould G by means of the bars H H, arms c c, and toothed segment J, on shaft I, toothed segment K, and cam L, all arranged substantially as and for the purpose specified.

Third, the rods r, arranged and operated substantially as shown and described, for insuring the proper filling of the carrier, as set forth.

Fourth, the sweep U, arranged and applied to operate substantially in the manner as and for the purpose set forth.

No. 55,842.—Henry Essex, West Haverstraw, N. Y., and Job Johnson, Brooklyn, N. Y.—*Dress Elevator.*—June 26, 1866.—The dress is caught between the loops, one of which is pushed into the other; the ring is slipped down to hold the parts engaged.

Claim.—The fastening for garments formed by the spring a, eye 3, and pusher 2, combined with the sliding ring or slide b, as and for the purposes set forth.

No. 55,843.—S. S. Evans, El Paso, Ill.—*Broom Head.*—June 26, 1866.—The corn is wired around the handle above the spud, and again below it.

Claim.—An improved broom, formed by combining the ring A and handle B, constructed as described, with each other, and with the clasp D and corn C, substantially as described and for the purpose set forth.

No. 55,844.—Philip L. Fox and George P. Herthel, Jr.. St. Louis, Mo.—*Hydraulic Spindle and Turning Apparatus for Draw Bridges.*—June 26, 1866—Friction is reduced by supporting the pivot of the counterbalanced platform on the piston of a hydraulic engine. The platform is protected against unequal strain by anti-friction wheels which bear against the inner surface of a ring concentric with the turning pivot.

Claim.—First, the application of fluid pressure to the pivot of turning bodies, when used to raise the same and diminish the friction resistance to the turning motion, substantially as described and shown.

Second, the counter-balance, when operating in connection with such fluid pressure in raising the structure to be turned, substantially as shown and described.

Third, the combination and arrangement of the pivot C and cylinder D, in conjunction with the counter-balancing apparatus E and the pump F, all acting substantially as and for the purposes shown and described.

Fourth, the ring surface M, the friction rollers N, the gearing N', when used in combination substantially as and for the purposes set forth.

No. 55,845.—JOHN M. FRANCIS, Waldo, Ohio.—*Making Tin Fruit Cans.*—June 26, 1866.—The wooden cylindrical core around which the sheet metal is formed consists of several sections with the central angles removed and their place occupied by a wedge with a corresponding number of inclined faces and which forms a handle.

Claim.—The section A and wedge B, connected by strips *a* and *b*, arranged in the manner and used substantially as and for the purposes herein set forth.

No. 55,846.—GEORGE FYFE, Ottowa, Ill.—*Table Leaf Support.*—June 26, 1866.—Hinged beneath the table leaf is an arm which is borne up by a spring and engaged with a notch in the frame to support the leaf.

Claim.—In combination with the leaf of a table, the hinged arm F and the frame C, the peculiar arrangement of the spring G, operating upon said arm F, as and for the purposes specified.

No. 55,847.—ANTON GALLETH, New York, N. Y.—*Feeding Wheel for Sewing Machines.*—June 26, 1866 —The lever being reciprocated at one movement it clamps the block against the wheel and turns it, and on its return motion allows the block to slip in its seat in the inner periphery of the wheel.

Claim.—The arrangement of the wheel A, lever B, with its side G, as described, block E, spring F, and friction spring J, combined and operating in the manner and for the purpose herein specified.

No. 55,848.—ROLAND R. GASKILL, Mendota, Ill.—*Roller and Cornstalk Cutter.*—June 26, 1866.—Pins on the end of the roller trip the pivoted frame and give a chopping motion to the knife.

Claim.—First, the combination of the roller A and knife H, the two being so connected by intermediate mechanism as that the knife shall fall with the forward movement of the roller, substantially as and for the purpose set forth.

Second, the combination of the knife H, rod G and G', and levers E and arm F, substantially as and for the purpose set forth.

Third, the roller A, with pins B, in combination with the levers D and E, and knife H, substantially as and for the purpose set forth.

No. 55,849.—WILLIAM GIBSON, Fort Wayne, Ind.—*Baby Jumper.*—June 26, 1866.—The ends of the extensible post are in contact with the floor and ceiling respectively. The C-spring passes through a slot in the post, and a dependent cord supports the child.

Claim.—First, the extensible post A A', in combination with step and cap pieces B C, substantially as described.

Second, the combination of a spring D, or its equivalent, with an extensible post A A', substantially as described.

Third, the combination of a strap or cord G, a spring D, and a post which will rotate where it is erected between two fixed objects, substantially as described.

No. 55,850.—WILLIAM GILBERT, Bardstown, Ky.—*Hand Corn Planter.*—June 26, 1866.—The handles are collapsed by the grasp of the hand driving the slide into the seed box and dropping the corn; releasing the grasp retracts the seed slide and drops the corn above the openers ready for the next operation.

Claim.—The hand seed planter, constructed, arranged, and operating substantially in the manner described for the purpose set forth.

No. 55,851.—ISAAC C. GLEASON, Middletown, Conn.—*Fruit Basket.*—June 26, 1866.—Explained by the claim and cut.

Claim.—A polygonal fruit basket having tapering sides formed of a single piece cut partially through to form the angles of the sides, its ends being secured together by eyelets, and having a circular bottom held in place by its edge fitting into slits formed in the said sides near their lower edge, substantially as described and for the purpose set forth.

No. 55,852.—CHARLES GOODYEAR, Jr., New York, N. Y.—*Lacing for Boots and Shoes.*—June 26, 1866.—One end of the wire forms a loop for the lacing cord, and the other an eyelet which is riveted to the material.

Claim.—First, the method herein described of applying lacing eyelets to boots and shoes by inserting and fastening the same in between two sheets or thicknesses of leather, in such manner that the lacing eye alone shall protrude at and at right angles to the edge of said leather, substantially as herein shown and set forth.

Second, the lacing eyelet constructed substantially as herein shown and described, that is to say, forming the same of a single piece of wire, the ends of which are bent into eyes standing at right angles to each other, substantially as herein set forth.

Third, in combination with eyelets constructed substantially as herein described, the employment of rivets traversing one of the eyes of the lacer and either or both thicknesses of leather between which the said lacer is confined.

Fourth, the employment in combination with lacers and rivets, as heretofore described, of a folded leather welt, binding, or equivalent, the folded edge of which having at right angles to it incisions through which the lacing eyes protrude, substantially as herein shown and described.

No. 55,853.—CHARLES GOODYEAR, Jr., New York, N. Y.—*Lacing for Boots and Shoes.*— June 26, 1866.—The rings are strung upon a wire sustained in the fold of the material, which is notched for the insertion of the rings?

Claim.—First, the method herein described of applying lacing eyelets to boots or shoes by securing and holding a series of eyelet-rings in their proper positions, as set forth, by means of a single holding device which serves as the common support of said rings, as and for the purposes herein shown and described.

Second, as lacing eyelets, rings in combination with one or more fastening devices as set forth, holding the said rings to the edge or to the face of the welt or upper, at right angles thereto, and transversely to the length of the shoe, as and for the purposes herein described.

Third, in combination with rings for forming the lacing eyes as described and for the purpose of holding the same in their proper positions as set forth, the wire imbedded between the folds of the welt or two thicknesses of leather, so that the eyelets may be strung or held on a support common to all, as and for the purposes herein shown and described.

Fourth, in combination with the eyelet-rings, as described, the wire corrugated or bent at the points of support of the rings, substantially in the manner and for the purposes herein set forth.

No. 55,854.—J. F. J. GUNNING and ISAAC T. MEYER, New York, N. Y.—*Buckle.*—June 26, 1866.—The strap is looped to one side and passes around one bar and between the two bars of the other side, by which it is clamped.

Claim.—A buckle made of one piece of metal, wire, or other suitable material, bent so as to form a shank a, and two spring jaws b c, substantially as and for the purposes set forth.

No. 55,855 —C. H. HALL, New York, N. Y.—*Distilling Petroleum and other Liquids.*— June 26, 1866.—In this apparatus for continuous distillation the crude oil is run in a shallow stream into the retort, the vapor passes to the condenser by a trap which returns the liquid to the retort. The steam generated by the heat of the vapor draws the vapor from the still and frees the retort of the residuum, which is treated with water and steam and the lighter portion returned to the still.

Claim.—First, in the continuous distillation of petroleum or other liquids, the use of a retort B, in combination with furnace D and arch C, substantially as shown and described and for the purpose set forth.

Second, the series of scrapers h h h, or their equivalent, connected to the rod m, in combination with retort B, constructed and operating substantially as shown and described, or in any other manner whereby a scraper is used for the purpose specified.

Third, the device herein described for generating steam, consisting of the water jacket G, and water supply pipe Y, and steam tube I, in combination with condensing tube F, substantially as shown and described, or any other means whereby steam is generated by passing the vapors of oil or other liquids, being distilled, through a vessel containing water, or vice versa.

Fourth, the method herein described of separating the condensible from the non-condensible gases, or any other method whereby the condensible gases are made to collect in the lower part of a receiver, while the non-condensible gases are made to pass off by the suction of a current of steam, substantially as herein set forth and for the purpose specified.

Fifth, the water jacket G', connected with supply pipe H and water tank R, in combination with the tubular condenser F', operating as described, or in any other manner to accomplish the purpose specified.

Sixth, the receiver L, in combination with tubular condenser F', steam pipe K, and discharge pipe O, operating substantially as and for the purpose shown and described.

Seventh, the annular chamber E, composed by an inner and outer vessel, in combination with the condenser F, constructed and operating substantially as and for the purpose specified.

Eighth, the method herein described of freeing the retort B from residuum, or any other equivalent means, whereby a retort or still is freed of its residuum by the force of a jet of steam, operating substantially as shown and described.

Ninth, the within-described process of cleaning the residuum by treating it with steam and water, substantially in the manner described and for the purpose set forth.

No. 55,856 —MARTIN HAXELINE, Clear Creek, Ind.—*Broom Head.*—June 26, 1866.—The bent wooden cap extends down the edges of the broom and has metallic side plates; a toothed bar passes through the broom lower down and is secured to the extension prongs of the cap.

Claim.—First, the wooden frame A, and the metallic plates B, constructed and combined substantially as described and for the purpose set forth.

Second, the combination of the screw C and staple D with each other and with the wooden frame A, substantially as described and for the purpose set forth.

No. 55,857.—WILLIAM H. HART, Medfield, Mass.—*Butter Worker.*—June 26, 1866.—The various devices are arranged upon one stand.

Claim.—The combination consisting of the table A, inclined table e, sliding-receiver g, churn A, and salt receptacle a', when combined and arranged substantially in the manner and to operate as hereinbefore described. And, in connection with such a combination of parts, the heater surrounding the churn, essentially as set forth.

No. 55,858.—GILBERT HAWKES, Lynn, Mass.—*Manufacture of Boots and Shoes.*—June 26, 1866.—Explained by the claims and cut.

Claim.—First, an inner sole A, formed of a textile fabric, to be used either with or without a stiffening preparation, as set forth and described.

Second, a strip B, of a textile or any other suitable material, when used as and for the purpose specified.

Third, the combination of an inner sole A, formed of a textile material, with a strip B of any suitable material, as and for the purpose specified.

Fourth, the application to the lasting of boots and shoes sewed by machinery of an inner sole of textile material, substantially as described.

No. 55,859.—JOHN G. HITCHCOCK, New York, N. Y.—*Neck-tie Fastening.*—June 26, 1866.—A spring snap and a hooked socket are attached to the respective ends of the neck-band and are engaged behind the bow.

Claim.—First, the spring snap A, having the cross-top a² in the socket a, and the locking end b² on the catch b, arranged as represented and having an opening d in the front of the socket a, adapted to receive pressure as indicated, substantially in the manner and for the purpose herein set forth.

Second, a neck-tie, one end of which is held by a clasp behind the bow, and the other is held adjustably by a hook and loops, all substantially as shown and described.

No. 55,860.—AUSTIN D. HOFFMAN, Detroit, Mich.—*Machine for Cutting Fly-nets.*—June 26, 1866.—The leather is fed upon an intermittently moving cutting board underneath a pair of knives arranged in a guillotine sash; a spring block steadies the leather and prevents its being drawn up when the knives rise.

Claim.—First, the combination of two or more guillotine knives with an intermittingly moving cutting board for carrying the leather, which moves in a direction transverse to the length of the knives.

Second, the combination with the said guillotine knives of the spring block T, arranged substantially as described, for detaching the leather from the knives.

Third, in combination with a pair of guillotine knives a vertically moving strip between the blades to disengage the leather therefrom.

No. 55,861.—ELIJAH HOLMES, Lynn, Mass.—*Machine for Fine-cutting Tobacco.*—June 26, 1866.—The tobacco rests on the bed plate and is sliced by the reciprocating knife, the feed being regulated by the gauge plate on the lever.

Claim.—The knife D, the bed plate G, and the gauge plate E, combined and arranged in the manner and for the purposes specified, substantially as above set forth.

No. 55,862.—BENJAMIN F. HORN, Boston, Mass.—*Lamp Cleaner.*—June 26, 1866.—To the handle are attached covered bow springs and a loop spring whose contraction tends to expand the bows into the bulb of the chimney.

Claim.—The improved lamp chimney cleaner as made with the loop spring D, or its equivalent, and the hook E, arranged and combined with the series of covered bow springs and the handle, arranged and applied together substantially as specified.

No. 55,863.—JAMES A. and HENRY A. HOUSE, Bridgeport, Conn.—*Button-hole Sewing Machine.*—June 26, 1866.—The cloth is clamped by a presser upon a vibrating and travelling plate which supports the presser. A scroll cam revolvable in either direction by a shifting pawl causes the plate to travel a greater or less distance forward and back, according to the length of button-hole wanted; the position of a pin in the scroll cam at starting determines this distance. Motion is imparted to all the devices through a single jog-bar which is operated by a reciprocating movement of any sewing machine to which the apparatus may be attached. The devices may be attached to the cloth plate of a Wheeler and Wilson machine, a rotating switch cam being first attached to the spooling spindle which projects from the main shaft. This cam operates the jog-bar.

Claim.—First, working a button hole automatically and on both sides without turning the cloth by the devices, substantially as described.

Second, the scroll cam ratchet wheel to give a rectilinear reciprocatory motion to and govern the length of the button hole, substantially as described.

Third, the combination of the scroll cam ratchet wheel, shifting pawl, and jog bar, substantially as and for the purpose set forth.

Fourth, the combination of the scroll cam ratchet wheel, the sweep bent lever, and jog bar, to space the stitching, substantially as set forth.

Fifth, the combination of the reciprocating vibrating clamping plate, the adjustable switch wedge, and the jog bar, for the purpose of varying the length of stitching for a button hole, substantially as described.

Sixth, the combination of the jog bar, clamping plate, movable switch and switch block, to sew both sides of button holes automatically, substantially as described.

Seventh, the combination of an adjustable stitching plate with the fixed bed plate, substantially as and for the purpose set forth.

Eighth, the combination of the clamp to hold the cloth with an adjustable screw pivot to render the bearing surface of the clamp parallel with cloth of varying thicknesses, substantially as set forth.

Ninth, the combination of the vibrating and reciprocating plate with the adjustable clamp, when so arranged that both shall move together to hold the cloth smoothly between them, as set forth.

No. 55,864.—JAMES A. and H. A. HOUSE, Bridgeport, Conn.—*Button-hole Sewing Machine.*—June 26, 1866.—The apparatus is enclosed within a shell or case and is designed to be attached with either side upward to either the upper or lower side of the bed plate of a sewing machine; all the movements are communicated from any proper horizontal reciprocating motion of such machine. The cloth holder or clamp slides back and forth in a slot in the toothed disk, and also has at periods an intermittent semi-rotation.

Claim.—First, working button holes automatically above or below the bed plate of sewing machines, substantially as and for the purpose set forth.

Second, sewing button-holes automatically, giving the goods a progressive, a semi-rotary, and a vibrating movement from a centre, to stitch the sides and ends of the button hole with stitches of uniform length.

Third, the combination of a jog bar, a rack bar, and a ratchet wheel with the cloth holder, substantially in the manner and for the purpose set forth.

Fourth, the combination of a jog bar, a pawl lever, and toothed wheel with the cloth holder, substantially as and for the purpose set forth.

Fifth, the combination of a jog bar, a crank lever, and supporting beam L, with the cloth holder, substantially in the manner and for the purpose set forth.

Sixth, the combination of an adjustable slide with the feeding devices and cloth holder, substantially in the manner and for the purpose set forth.

Seventh, the combination of a jog bar, a pawl sweep, and a ratchet wheel with the cloth-holder, substantially as and for the purpose set forth.

Eighth, the jog bar, when combined with the cloth holder and pawl sweep, substantially as described to give the cloth holder all its required motions, to vary the spacing between the stitches, and to adapt the sewing to button holes of different lengths, substantially in the manner described.

No. 55,865.—JAMES A. and H. A. HOUSE, Bridgeport, Conn.—*Button-hole Sewing Machine.*—June 26, 1866.—Cannot be briefly described.

Claim.—First, operating the button hole mechanism from the spooling pin, substantially as and for the purpose set forth.

Second, the bed plate E, having a reciprocating motion and being laterally and longitudinally adjustable, substantially as and for the purpose set forth.

Third, the revolvable, reciprocating feeding plate, combined with the vibrating bed plate, substantially as and for the purpose described.

Fourth, the adjustable stitching plate, combined with its eyelet guide, substantially as and for the purpose described.

Fifth, the combination of the centre plate and gibs, substantially as and for the purpose set forth.

Sixth, the dog lever, combined with the centre plate, the block cramp, and the revolvable plate, substantially in the manner and for the purpose described.

Seventh, the combination of the clamping lever with the revolvable plate, substantially as described, to admit of the turning of the cloth, so that the button hole can be worked at a right angle or less than a right angle to the selvage, or parallel therewith.

Eighth, the combination of the grooves in the rotating plate with those in the clamping lever, substantially as and for the purpose set forth.

Ninth, the feeding screw to intermittently reciprocate the feeding plate, or hold it at rest, substantially as and for the purpose described.

Tenth, the combination of the feeding screw with the feeding plate by spring half-nuts, substantially as and for the purpose set forth.

Eleventh, the combination of the feeding screw with the bed plate and adjustable on the switch cam plate, substantially as and for the purpose set forth.

Twelfth, the combination of the switch, the jog bar, and the slotted strap, substantially as and for the purpose set forth.

Thirteenth, the combination of the jog bar with the adjustable slide bar, substantially as and for the purpose set forth.

Fourteenth, the combination of the bed plate and shield cap, substantially as and for the purpose set forth.

Fifteenth, the combination of the bed plate and table with the adjustable lever and stitching plate, substantially as and for the purpose described.

Sixteenth, the combination of the bed plate, jog bar, lever, and stitching plate, substantially as and for the purpose described.

Seventeenth, the combination of the friction spring and adjustable stops of the throw-lever g, with the jog bar and its stop pin, substantially in the manner and for the purpose described.

Eighteenth, the combination of the plate and its adjustable screws and pins with the switch cam and jog bar, substantially as and for the purpose set forth.

No. 55,866.—JAMES A. and H. A. HOUSE, Bridgeport, Conn.—*Sewing Machine Clamp.*—June 26, 1866.—Intended for stitching button-holes and dispensing with pins. The spring presser has a sharp lower edge which may tightly fit the inner edge of the rubber.

Claim.—First, surrounding the opening in the table of sewing machines through which the needle passes, with a lining of India-rubber, substantially in the manner and for the purpose set forth.

Second, the combination of the yielding presser with the rubber-lined opening, substantially as and for the purpose set forth.

No. 55,867.—M. V. B. HOWE, Ashburnham, Mass.—*Knife and Scissors Sharpener.*—June 26, 1866.—Knives are sharpened on the bevelled edge, and scissors against the ends in the angular notches of the handles.

Claim.—The combination of holder A, having angular notches E, and concave sharpener plate C, as and for the purposes described.

No. 55,868.—N. W. HUBBARD, Fulton City, Ill.—*Truss*—June 26, 1866.—The pad is fastened to an elbow of the almost rigid steel hoop which passes half around the body, is covered with leather and has a terminating strap which hooks to the stud on the other end.

Claim.—A truss in which a pad of the form and arrangement substantially as set forth is connected in the manner set forth with a main spring of the form and rigid character substantially as herein described.

No. 55,869.—SAMUEL HUFFMAN, Carthage, Ill.—*Puncturing Machine for Making Patterns.*—June 26, 1866.—As the lower end of the instrument is passed over the paper supported upon cloth, the needle is alternately projected and withdrawn by the action of clock work which is stopped and started as required.

Claim.—A machine, when so constructed with a puncturing point in combination with a spring, and when the machine entire is movable when in use, as is a pen or pencil tracing point, substantially as herein described and for the purpose herein set forth.

No. 55,870.—HIRAM HUGHES, Savona, N. Y.—*Car Coupling.*—June 26, 1866.—The spring latch is lifted by the contact of the entering link with the inclined face of the book; the bar beneath is pivoted out of centre and tends to keep the hook in engagement.

Claim.—The combination of the latch B, with its catch c, and pendants d d', of the form shown, and the pivoted bar D, operating in combination with a suitable link E, in the manner and for the purpose herein specified.

No. 55,871.—JOSEPH and ABRAHAM HURSH, Philadelphia, Penn.—*Fertilizer.*—June 26, 1866.—Explained by the claim.

Claim.—The application of ochre as a fertilizer, in either a raw or burnt state, substantially as described.

No. 55,872.—JOHN HUTCHISON, Three Rivers, Mich.—*Grinding Mills.*—June 26, 1866.—The central, vertical shaft has radial stirrers and revolves in the hopper, the slip coupling permitting the adjustment of the runner without affecting the feed.

Claim.—The slip coupling in combination with the central shaft having stirrers and the hopper, arranged and operating as described.

No. 55,873.—ANDREW IRION, Femme Osage, Mo.—*Washing Machine.*—June 26, 1866; antedated May 14, 1866.—A series of rubbers work within a tub from opposite sides, are arranged in pairs and operated by cranks.

Claim.—The combination of the wheels B C C', and crank shaft D, with the rubbers E', and box A, as and for the purpose set forth.

No. 55,874.—THEODORE E. KING, Painesville, Ohio.—*Fence.*—June 26, 1866.—The rails and pickets of the fence are supported and strengthened by the straps and plates described. The form of the hinge adapts it to be fastened to the gate by screws, and that of the catch to be secured to the post and engage the latch.

Claim —First, the plate A, constructed as shown in figures 1 and 2, and herein particularly specified and used in series by overlapping or interlocking each other consecutively, for the purpose set forth.

Second, the plate as shown in figures 3 and 4 and constructed as specified, the same consisting of the addition to the foregoing plate of a vertical or upright part and arched braces, L L, and used in series in like manner, as and for the purpose specified.

Third, the manner of plating or strapping the couplings of the sections and covering their abutting ends, together with the mode of adjusting and bracing the panel J, by means of the combined parts herein described, viz., plates figures 5 6 7 and 8, projecting plates L' M' and M, nuts O and T, and bolts R and S, arranged and operating as herein set forth.

Fourth, the adjustable hinge constructed with the depressions *a a a* and slots U U, substantially as shown in figures 10 and 11, in combination with the tooth *b*, and screw holes *c c*, of the lower plating of the gate, as herein described and for the purpose specified.

Fifth, the adjustable catch, as shown in figures 14 and 15, provided with the screw bolt W, and projecting stud X, in combination with the opening Y of the post, and constructed and arranged as and for the purpose set forth.

No. 55,875.—LOUIS KRATZER, Baltimore, Md.—*Water Wheel.*—June 26, 1866.—The water passes to the wheels consecutively, they being successively of smaller diameter and discharging at or about the level of their axis.

Claim.—The arrangement of the wheels A B and C, so as to make use of the whole body of water in its descent to the level below, and its action in passing off, as herein set forth and described.

No. 55,876.—JOEL LEE, Galesburg, Ill.—*Farm Gate.*—June 26, 1866.—The valve is journalled to a hanger in the post and supports the longitudinally sliding gate.

Claim.—First, the friction rollers A A, when used as set forth and described.

Second, the arrangement of rollers A A with the roller B and the roller support C, all arranged as set forth and described.

No. 55,877.—EDWARD J. LEYBURN, Lexington, Va.—*Harvester Rake.*—June 26, 1866.—The rake sweeps transversely to the line of draught, and deposits the gavel behind the draught frame and across the line of draught.

Claim.—First, combining a rake with a revolving reel in such manner that the rake shall receive a rectilinear movement across the platform at every revolution of the reel, and without stopping the movement of the latter, substantially as described.

Second, supporting and guiding a rake staff D, by means of a revolving reel, in such manner that the rake shall revolve with the reel after the termination of each raking stroke, and then move across the platform in a line which is at right angles, or nearly so, to the line of draught, substantially as described.

Third, supporting and guiding the rake staff by means of a slotted cross-piece *b*² applied to the reel, and arranged so that the rake D' shall be held at right angles to the finger beam in its passage across the platform, substantially as described.

Fourth, a rake which is applied to a revolving reel and moves in a right or nearly right line across the platform, in combination with means which finally discharge or deliver the gavels of grain with their length at right angles, or nearly so, to the path of the machine, and also out of the return track of the horses, substantially as described.

Fifth, a rake which moves in a right or nearly right line across the platform, so pivoted to the reel that it is capable of swivelling and vibrating, substantially as herein described.

Sixth, in combination with a revolving reel and vibrating rake, pivoting the rake to its staff, substantially as described.

Seventh, providing for moving the rake to the outer end of the reel previously to the action of the guide-arm *b'* upon the standing grain, substantially as described.

Eighth, the rake pivoted to the reel so as to revolve with it, constructed and controlled substantially in the manner and for the purpose described.

No. 55,878.—O. J. LIVERMORE, Worcester, Mass., assignor to himself and CLARK, SAWYER & CO., same place.—*Can Opener.*—June 26, 1866.—A curved projection beneath the handle has a sharp lip to engage and lift the cover.

Claim.—A can-opener, constructed substantially as above described.

No. 55,879.—E. A. LOCKE and L. HEUSER, Boston, Mass.—*Tags or Labels for Express Companies, &c.*—June 26, 1866.—The tag, of combined metal and paper, is secured to the loop of a strap which is adapted to be fastened to a package and secured by an eyelet, left projecting for that purpose.

Claim.—First, the combination of the two parts of the tag *a* and *b*, when united by the folding of the end of *a* over and with *b*, and perforated through the re-enforcement of the folds for reception of the band.

Second, in combination with a tag and band, a projecting unclinched eyelet, secured to the band by the confinement of its head or flange between the adjacent surfaces of the parts eyeleted together to secure the tag to the band, the eyelet being thus held in position for application and securement of the opposite end of the band, substantially as set forth.

No. 55,880.—DAVID MANSFIELD, Oshkosh, Wis.—*Burning Fluid.*—June 26, 1866.—.Composed of alcohol, naphtha, white oak bark, alkanet root, slippery elm, camphor, saltpetre, and salt.

Claim.—A burning fluid, for illuminating purposes, which is composed of the several ingredients, mixed together, in about the proportions herein mentioned.

No. 55,881.—JAMES McDONALD, Detroit, Mich.—*Carding Machine.*—June 26, 1866.—The claims indicate the devices. This is an improvement on what is styled the "Wool custom carding, or double machine." The object of the narrow fillets, whose card teeth are shorter than, and point in the direction opposite to, those of the ordinary roll doffer, is to catch any rolls of material that may chance to run over the "roller" and draw them into the machine, and so prevent their going between the roller and roll doffer and flattening the cards. The thin metallic shell is placed underneath the roller, encircling its lower half, its edge being in close proximity to the roll doffer.

Claim.—The use of a thin shell in the place of the thick and wooden one heretofore used, which thin shell is set in combination with the roller for the purpose of taking the wool from the roll doffer at once without comb crank or intermediate roller, said shell being set closer up to the roller than heretofore, and made adjustable nearer to or far from the roller, at will.

Also, as a component part of said improvement, above described, the insertion of narrow bits of wool cards (the teeth of which should be a trifle shorter than, and set contrariwise to, the ordinary card teeth of the roll-doffer) in the spaces of the roll doffer heretofore left smooth and unfurnished with cards or teeth, and the change above prescribed in the size of the pulleys, by means of which pulleys the band of the machine turns the roller and roll doffer.

No. 55,882.—J. J. MEEKER and LEVI CORTRIGHT, Columbia county, Penn.—*Gate for Forebays.*—June 26, 1866.—The sections of the gate are hinged together so as to render it flexible.

Claim.—The flexible gate B, constructed and operating substantially as and for the purposes set forth.

No. 55,883.—S. H. MITCHELL, El Paso, Ill.—*Fence.*—June 26, 1866.—The ends of the boards in the consecutive panels are strapped, scarfed together, and clamped between a slat and a post which is supported by a stake and brace.

Claim —The stakes B D, connected by bolts C, in combination with the upright posts A of the panels, with which posts the upper ends of the stakes B are connected by bolts E, substantially as and for the purpose specified.

Also, the covering of the notched ends of the slats F of the panels with sheet metal or strap iron *c*, substantially as and for the purpose set forth.

Also, the securing of the panels so as to prevent a vertical movement of the same by means of a screw *b*, or its equivalent, applied as shown, when used in combination with the notched ends of the slats F of the panels, substantially as described.

No. 55,884.—DAVID MORRIS, Bartlett, Ohio.—*Well-boring Apparatus.*—June 26, 1866.—The cutting faces are arranged at angles of 120° relatively on the tubular end of the drill rod, whose valve opens when the drill ascends. The devices for operating are set forth in the claims and cut.

Claim.—First, a drill, having two or more wings W, with their cutting edges constructed and arranged as herein described and represented.

Second, the tube H′, provided with a valve *h*, and attached to and moving with the drill rod *h*, substantially as and for the purpose explained.

Third, the combination of the circular frame A, cogged rim C, rotating and revolving shaft D′, pinion D′, crank wheel F, connecting-rod G, rope I, and drill H, the whole being constructed and arranged as herein described and represented.

Fourth, the combination of the cogged rim C and gearing 8 8 with the bevelled gear R, windlass Q, shifting-rod S′, rope N, and adjustable block L, the whole being arranged to operate in the manner and for the purpose set forth.

No. 55,885.—JOHN MUSSER, Frizellburgh, Md.—*Flour Packer.*—June 26, 1866.—The stamps are raised alternately by the tripping arms on the revolving shaft.

Claim.—The movable levers E, operating on the vertical stampers G, with the upright revolving cylinder J, when arranged, combined, and operated as herein described and for the purposes set forth.

No. 55,886.—GEORGE NEBEKER, Wilmington, Del.—*Apparatus for Agitating and Heating Substances.*—June 26, 1866.—The central pipe injects steam and carries with it a body of air, which is introduced at the other arm and passes into the vessel where the agitation is desired.

Claim.—A tube A, in combination with an internal pipe B and with a tube C communicating with the said pipe B and with a steam boiler, substantially as and for the purpose described.

No. 55,887.—FREDERICK NISHWITZ, Williamsburg, N. Y.—*Sugar Mould.*—June 26, 1866.—The body of the mould is made of papier-maché, and the tip of metal or other hard material.

Claim.—First, combining a sugar mould tip with a body of papier-maché, or other equivalent material, the two parts being constructed and arranged substantially as described.

Second, clamping the body of the mould to the tip, and holding the two parts in place by means of a wrapping of wire, substantially as described.

Third, combining with the dovetailed flange on the tip the shoulder on the body of the mould, when the latter acts as a stop to prevent the wrapping from slipping off the cone, substantially as described.

No. 55,888.—FREDERICK NISHWITZ, Williamsburg, N. Y.—*Sugar Mould Tip.*—June 26, 1866.—The body of the mould is made of papier-maché, and its apex is clamped between the inner perforated tip and the outer screw cap.

Claim.—First, a separate and separable tip for sugar moulds constructed and operating substantially as described.

Second, clamping the body of the mould between an inner metallic tip and an outer metallic cap, substantially as described.

No. 55,889.—MICHAEL NOLL, New York, N. Y.—*Eye Water.*—June 26, 1866.—Explained by the claim.

Claim.—The herein-described eye water, composed of sulphate of zinc and the white of an egg dissolved in distilled water, substantially as set forth.

No. 55,890.—E. D. NORTON, Bradford, Penn.—*Mill Sieves.*—June 26, 1866.—The corrugated overlapping strips form steps over which the grain, &c., is passed exposed to the jets of air which issue between the strips.

Claim.—An improved screen or upper sieve for a fan mill, formed of strips B of corrugated sheet metal, attached at their ends to the side bars of the frame A of the sieve, and so arranged that the forward edge of each strip B shall overlap the rear edge of the adjacent strip and that the concavities b' of each strip shall be in the same vertical plane with the convexities b' of the adjacent strips, substantially as herein described and for the purposes set forth.

No. 55,891.—RUFUS NORWOOD, Baltimore, Md.—*Roofing.*—June 26, 1866.—Explained by the claims and cut.

Claim.—First, a roof which is in part composed of several overlapping strips of felt, or its equivalent, applied in counter diagonal positions, substantially as described.

Second, the combination of the reveal strips c c and strips d with the felting, as described, so that gutters or channels at the junction of the roof with the fire wall can be formed, preventing thereby the cement covering from being detached from the roof and wall, substantially as described.

No. 55,892.—AARON S. OGDEN, Newark, N. J.—*Carbon Battery Connection.*—June 26, 1866.—To prevent the oxidation of the wire at its point of connection with the platina they are united at the bottom of a perforation in the carbon and surrounded with lead, hard rubber, or other unoxidable material.

Claim.—Surrounding the junction of the wire A with the platina, and a portion of the said wire A with the covering B, or its equivalent; also the combination of the carbon with the said covering B and platina D, substantially as described and for the purposes set forth.

No. 55,893.—JAMES OLD, Pittsburg, Penn.—*Hydrant.*—June 26, 1866.—The operative parts are attached to a removable slide-piece; the connection with the service pipe is made by a nozzle on the lower end, which enters a conical socket and depresses the valve, which closes the exit when the operative parts are withdrawn.

Claim.—First, connecting the operative parts of the hydrant to the supply or service pipe at a point below the working valve or cock by means of a joint, in such manner that the hydrant may be removed and separated from the supply or service-pipe without cutting any pipe or unscrewing any joint, substantially as hereinbefore described.

Second, so constructing the hydrant, substantially as hereinbefore described, as that all the operative parts above and including the working cock or valve are connected together as one piece, and may be attached to a removable slide, either placed in or forming part of the hydrant box, substantially as and for the purposes hereinbefore set forth.

Third, the check valve placed below the working valve of a fire-plug or hydrant, so constructed and arranged as that the valve will shut whenever the hydrant or fire-plug is removed from the supply or service pipe, and that it shall be opened to allow of the passage of water from the supply pipe either by the opening of the working valve whenever the fireplug is used, or be kept open by the plug or hydrant when it is placed in connection with the supply or service pipe, substantially as hereinbefore described.

No. 55,894.—JOSEPH OLD, Reading, Penn.—*Lifters for Stove-covers and Culinary Vessels.*—June 26, 1866.—Explained by the claim and cut.

Claim.—The described universal lifter, constructed with hooks C D, in combination with the fixed jaws A B, at one end, the other being prolonged into a handle E, in the manner and for the purpose specified, as a new article of manufacture in design and figure.

No. 55,895.—EDWARD E. and CHARLES T. PACKER, Philadelphia, Penn., assignor to WILLIAM E. LOCKWOOD.—*Drill Couplings.*—June 26, 1866.—A recess in one section receives the end of a spring secured to the other section; the spring is depressed while the screw joint is being tightened, and when released by engagement with the recess prevents revolution.

Claim.—A section having an external screw thread and a recess, in which is a spring spline or feather, in combination with a section having a socket with an internal screwthread and a slot adapted for the reception of the end of the said spline or feather, the whole being constructed substantially as and for the purpose specified.

No. 55,896.—ENOS PAGE, Streetsborough, Ohio.—*Sheep Rack and Holder.*—June 26, 1866.—The sides of the trough are adjustable as to relative distance and are fastened in position by cleats and buttons.

Claim.—The adjustable sides B B, lips b' b'', and grooves a a', in combination with the cleat D and buttons D', arranged as and for the purpose set forth.

No. 55 897.—OWEN B. PARKER, Woodville, Mass.—*Book Holder.*—June 26, 1866.—The two parallel supporting bars are connected by elastic cords and the upper one sustained by a brace. The leaves are held by spring fingers.

Claim.—The combination as well as the arrangement of the two bars A B, their lips a a b b, and leaf-holders or arms D D, with the elastic bands C C.

Also, the combination and arrangement of the brace hole and looped strap or ribbon with the two bars A B, and their lips a a b b, the elastic connections C C, and arms D D.

Also, the combination and arrangement of the brace or prop and the looped strap with the two bars A B, their lips a a b b, and elastic connections C C, and arms D D.

Also, the combination of the fingers g g with the arms D D, the bars A B, their lips a a b b, and elastic connections C C, the whole being as and for the purpose set forth.

No. 55,898.—ANNA E. PARROTT, Norfolk, Va.—*Medicine to Cure Cholera, &c.*—June 26, 1866.—Compound of red oak bark, cloves, allspice, cinnamon, white ginger, mace, sugar, rye whiskey, laudanum, spirits of camphor, Hoffman's anodyne, and oil of peppermint.

Claim.—The medicine prepared substantially as herein described.

No. 55,899.—BENJAMIN P. PENDEXTER, Mechanics' Falls, Me.—*Corn Sheller and Bean Thresher.*—The corn or beans is fed by the cups or books in the revolving drum to the rubbing device, which consists of a toothed cylinder and spring concave; it passes thence to the fanning mill and elevator.

Claim.—Combining the feeding drum C, having horizontal buckets m and hooks n on the outside, with the toothed beater D, the elastic concave toothed bed E, made in sections, the elastic bed plates F, supported and operated by the spiral springs b, the fan G, the sieve H, and the elevator I, the whole several parts being constructed, connected, and arranged substantially as and for the purposes herein described.

No. 55,900.—BARTON PICKERING, Milton, Ohio.—*Water Closet.*—June 26, 1866.—When the sitter rises the spring seat opens automatically and closes the valve; pressure on the seat vibrates the valve upward against the back of the vessel.

Claim.—First, the arrangement of the vessel A, seat B, and valve C, the whole constructed substantially as described and for the purpose specified.

Second, the arrangement of the spring f and bar p, to hold the seat B in a vertical position, and the valve C in a nearly horizontal position, substantially as described and for the purposes specified.

No. 55,901.—JAMES L. PLIMPTON, New York, N. Y.—*Skate.*—June 26, 1866.—Improvement on his patent of January 6, 1863, in respect to the construction of the hangers which fasten to the plates on the sole, the runners, wheels, and the rubber spring.

Claim.—First, the construction of the plates B and hangers C, arranged and applied to the stock or foot-stand, to operate in the manner substantially as and for the purpose set forth.

Second, the key F, provided with the slots *j l* and the button G, and arranged in connection with the pin K, substantially as and for the purpose set forth.

Third, the springs *p* applied to the screws *s*, which secure the hangers C to the plates B, for the purpose of preventing vertical or upward and downward play of the hangers, and controlling the turning, tilting, or canting of the stock or foot-stand, as set forth.

Fourth. the clamps H, composed of two parts *g r*, for holding the reversible runners I.

Fifth, the reversible runners, arranged substantially as shown, for the purpose specified.

No. 55,902.— ISAAC T. PRICE, Leesville, Ohio.—*Horse Hay Fork.*—June 26, 1866.—The fork is suspended by a hook which engages a pin on the fork head; a trigger forces the hook off the pin and allows the fork to rotate, discharging the load.

Claim.—First, the combination and arrangement of link C, hook H, catch G, and trigger Q, substantially as and for the purpose set forth.

Second, the construction and arrangement of the handle B, head A, and prongs *p*.

No. 55,903.—JAMES RADLEY, New York, N. Y.—*Locomotive Lamp.*—June 26, 1866.— The small tubes which convey the oil to the burner are surrounded with a non-conductor to screen them from the heat reflected from the flame.

Claim.—The combination of the lamp with the oil tank by means of one or more small oil tubes incased with a non-conductor of caloric, arranged and constructed in the manner substantially as described.

No. 55,904.—JAMES RADLEY, New York, N. Y.--*Engine Head Light.*—June 26, 1866.— Explained by the claims and cut.

Claim.—First, the foundation piece, with its recess and covering plate, in combination with the enclosing tube and the wick so arranged and constructed that the parts may be conveniently brazed together instead of being soldered, thereby rendering the lamp more safe and permanent. substantially as herein shown and described.

Second, the wick pinion and its spindle, in combination with the removable stuffing-box, arranged and constructed substantially as described.

No. 55,905.—AMOS RANK, Salem, Ohio.--*Mowing Machines.*—June 26, 1866.—The construction and arrangement of the stirrup bearings of the longitudinal gearing shafts are set forth in the claims and cut.

Claim.—First, the construction of the hanger E with a depressed stirrup E', which is open on all sides but one above the shaft *e* and below the shaft *b*, substantially in the manner and for the purpose described.

Second, the combination of the hanger E constructed with the open stirrup E', and the rear hanger F, all constructed and arranged substantially as described.

Third, constructing the front hanger E E' so that it can be attached to the top of the longitudinal beams A A, and at the same time will serve as a support for the two shafts *e b*, in the manner substantially as shown and described.

Fourth, the hanger E and open stirrup E', cast in one piece, substantially as shown and described.

No. 55,906.—F. RAYMOND and A. MILLER, Cleveland, Ohio.—*Fence Gate.*—June 26, 1866.—The weighted cord runs over a swivelled pulley, and is attached to an arm on the gate to close the latter or to hold it open when vibrated to a point where the weight is counterbalanced.

Claim.—The cord *h*, adjustable case *e'*, and sheave *e*, in combination with the gate A, arm F, when arranged and operating substantially as and for the purpose set forth.

No. 55,907.—HENRY READ, Ypsilanti, Mich.—*Window Shade.*—June 26, 1866.—The two cords proceed from a single tassel, passing upward and diverging: reverting downward they are passed below the roller and upward to two staples at the back.

Claim.—The combination of the staples or rings C C² C³, or their equivalents, cords E E, secured to a tassel D, either weighted or not, when arranged and connected together and to a window curtain or shade, so as to operate thereon, substantially as and for the purpose described.

No 55,908.—W. E. RICH, New Providence, Iowa.—*Corn Planter.*—June 26, 1866.—After the field has been furrowed out one way the machine is passed over it crossways; the seed dropper and covering share are attached to hinged bars supported on rollers which drop into the former furrow; the fall operates the seed dropper and covering share, and the field is thereby planted in check rows.

Claim.—First, the hinged bars D D, provided with the rollers *i* and boxes G, in connection with the wheels H, having seed cells *b* made in them and rods *c* attached with weights *f* at their upper ends, all arranged to operate in the manner substantially as and for the purpose specified.

Second, the boes F attached to the rear ends of the bars D, in combination with the seed-dropping mechanism attached to said bars D; for the purpose set forth.

Third, the rollers I I in combination with the seed-dropping mechanism and covering hoes, substantially as and for the purpose specified.

Fourth, the attaching of the rollers i to arms g pivoted to the bars D and retained in position by the bars i* and pins j, arranged substantially as shown and described for the purpose of adjusting the bars D, as set forth.

No. 55,909.—H. STONE RICHARDSON, Manlius, N. Y., assignor to himself and E. P. RUSSELL, same place.—*Automatic Machine for Lighting and Extinguishing Gas.*—June 26, 1866.—The time keeper is combined with the cog-wheel on the spindle of the gas cock to operate the latter at a determinate time, and by friction of a match on the cylinder with a roughened surface, to light the gas.

Claim —First, the setting wheel C, in combination with the hands d and e, shafts g and h, and cams D D', substantially as and for the purposes set forth.

Second, the match cylinder G, constructed and operating substantially as and for the purposes set forth.

Third, the dogs K K' and K², in combination with the cylinder G and plate J, substantially as and for the purposes set forth.

Fourth, the cams D D', cam levers N N', connecting levers O O', dog levers L L', and springs P P', constructed and arranged substantially as and for the purposes set forth.

Fifth, the setting wheel C', provided with cog wheel j, when so combined with cock cog-wheel i as to alternately turn the gas on or off, substantially as herein set forth and described.

Sixth, the draught pipe H, operating substantially as described and in combination with the match tubes k.

No. 55,910 —E. P. RUSSELL, Manlius, N. Y.—*Reaping Machine.*—June 26, 1866.—This machine has a dropping platform attached to the arms of the reel and operating in connection with a side dropper and an extended track clearer; the reel is tripped by a pedal rod and actuates the side dropper.

Claim.—First, the "side dumper" O, or its equivalent, substantially as and for the purposes set forth.

Second, the dumper reel H, in combination with the side dumper O, substantially as and for the purposes set forth.

Third, the combination of the dumper reel H, side dumper O, and track clearer U, substantially as and for the purposes set forth.

Fourth, the device used for tripping the dumper reel, substantially in the manner described.

Fifth, the device used for tripping the side dumper, substantially in the manner described.

No. 55,911.—W. T. SALIE, Bowdoinham, Maine.—*Medical Compound.*—June 26, 1866.—This compound for the cure of diphtheria, sore throat, &c., consists of rum, decoction of white oak bark, and alum.

Claim.—A medical compound composed of the several ingredients herein named, and either with or without a flavoring material or substance, when mixed together in about the proportions described, as and for the purposes specified.

No. 55,912.—GEORGE W. SARGENT, Fairhaven, Mass., and BENJAMIN P. RIDER, Chelsea, Mass.—*Machine for Making Horseshoe Nails.*—June 26, 1866.—The pivoted hammers are operated by sliding bars whose lugs take into cavities in the shanks of the hammers, and whose yokes are operated by eccentrics on the main shaft, which revolves in spring journal boxes.

Claim.—Operating the hammers C C C' C' by means of the sliding connecting pieces D and cam F, substantially as and for the purpose specified.

Also, the adjustable boxes G, in combination with the cam shaft E, connecting pieces D, and the hammers, as and for the purpose specified.

No. 55,913.—CHARLES H. SAWYER, Hollis, Maine.—*Railway Switch.*—June 26, 1866.—The arrangement of levers and springs is operated automatically by attachments projecting downward from the car platform.

Claim.—First, the combination of the jointed levers h with the sliding bar g, and curved arms a with the cross-bar b, all constructed, arranged, and operating as herein set forth and described.

Second, the arrangement of the springs y to lift the levers i, as described.

Third, the arrangement of the arms z z for the purpose of locking and unlocking the switch by hand, as described.

No. 55,914.—J. H. SCHENCK, Chicago, Ill.—*Automatic Gate.*—June 26, 1866.—Pressure on the platform vibrates the pendulous gates and opens the roadway, which is closed as the weight upon the platform is withdrawn.

Claim.—First, the gate A, pivoted at the top of the post so as to swing like a pendulum when the way is to be opened, substantially as described.

Second, in combination with the vibrating gate A, the reciprocating platform, for the purpose set forth.

Third, in combination with the vibrating gate A, the latch, substantially as and for the purpose set forth.

Fourth, the reciprocating platform, supported upon the lazy bars, and connected to the gate by the rod, substantially as described.

Fifth, in combination with the reciprocating platform the flanges, and substantially as described and for the purpose set forth.

No. 55,915.—JOHN SCHNEIDER, Williamsburg, N. Y.—*Apparatus for Producing Extract of Hops, &c.*—June 26, 1866.—The cover closes tightly by the dipping of its flange into the annular trough. The extract is passed through a coil immersed in a cool medium so as to be cooled before reaching the external air, to preserve the aroma.

Claim.—The siphon tube E, in combination with the tightly closed vessel A, which is provided with a horizontal partition or spurger D', false bottom C, and discharge coil F, substantially as and for the purpose described.

No. 55,916.—J. SCOVILLE, Buffalo, N. Y.—*Manufacture of Car Wheels.*—June 26, 1866.—About five per cent. of spiegeleisen is added to the cast iron in the cupola or ladle.

Claim.—The combination of spiegeleisen and iron in the manufacture of car wheels, in the manner herein described.

No. 55,917.—FREDERICK J. SEYMOUR, Walcottville, Conn.—*Picture Nail.*—June 26, 1866.—The portion of the nail head to which the thread of the shank is attached has an opening stamped up and a thread cut in it.

Claim.—An ornamental picture nail-head, made with a screw for the nail, in the manner specified.

No. 55,918.—FRIEDRICH SHALLER, Hudson, N. Y.—*Drilling Machine.*—June 26, 1866.—The drill spindle is rotated by band and pulley, adjusted vertically by an arm above, attached to a nut, which traverses on a fixed vertical screw on the frame.

Claim.—First, the fixed feed screw E, the hand wheel F, containing the nut z, and the arms i and b, in combination with the drill spindle and frame B of the machine, all arranged substantially as herein specified.

Second, the combination of the frame $f f$, the drill spindle C, the pulley D, and rod d, substantially as and for the purpose herein set forth.

No. 55,919.—P. P. SIMMONS, Davenport, Iowa.—*Machine for Cutting Window Shade Slats.*—June 26, 1866.—The circular revolving cutters are arranged upon nearly vertical shafts and are vertically adjustable to cut slats of differing thickness.

Claim.—First, the arrangement of the nearly horizontal rotary cutters C C, table A, and the guides a a, substantially in the manner and for the purpose described.

Second, the arrangement of the socket bearings D, cutter arbor b, cutter C, adjustable devices c and d, and table A, substantially in the manner and for the purpose described.

No. 55,920.—JAMES JOSEPH SLEVIN, New York, N. Y.—*Garbage Box.*—June 26, 1867.—Explained by the claims and cut.

Claim.—First, the garbage box, with its top constructed so as to form a part of the sidewalk of the street, and with a door or lid in said top, so that the said top with its door covers the garbage box, forms part of the sidewalk for pedestrians, and permits the contents to be readily thrown out, thus enabling the garbage box to be sunk below the level of the sidewalk, as described.

Second, the construction of the garbage box with two horizontal doors, to form parts of the sidewalk, the smaller within the larger, as described.

Third, the garbage box, with one of its sides constructed so as to form a part of the curb of the sidewalk, so that the box may be set as close as possible to the carriage way, for convenience of throwing its contents into a cart.

Fourth, the combination of the garbage box with a drain grating extending up to or near its top, so that the liquid matter may drain from the surface of the garbage, wherever that may be, as described.

Fifth, the combination in a garbage box of the following parts, viz: the body of the box, the top constructed with a door to form part of the sidewalk of the street, and one side constructed to form part of the curb of the street, as described.

No. 55,921.—JOSEPH D. SMITH, Peoria, Ill.—*Gang Plough.*—June 26, 1866.—The ploughs have an independent motion in the frame, which is forward of the axletree; the seat of the driver balances the frame.

Claim.—First, in a sulky gang plough, arranging the gang of ploughs between the forward and the rear points of support of the carriage frame to which said ploughs are attached, when they are free to rise or fall independently of each other, substantially as described.

Second, pivoting the forward ends of the plough beams D D to the carriage frame by means of rods b b, which are adjustable at their outer ends in slots c c, for the purpose of levelling the ploughs, substantially as described.

Third, the combination of the vibrating plough beams D, which are pivoted to a carriage frame at a point which is in front of the transporting wheels, with a device which will admit of the forward part of said frame being elevated or depressed at pleasure, substantially as described.

Fourth, in combination with ploughs which are arranged in advance of the axletree B and pivoted, so as to be capable of rising or falling, the gravitating catches d d, or their equivalents, substantially as described.

Fifth, the arrangement of levers E E, vibrating ploughs D D, and catches d d, substantially as described.

Sixth, providing for adjusting the forward end of the carriage frame for regulating the pitch of the ploughs by means of draught pole G and lever H, substantially as described.

Seventh, connecting the rear end of the pole G to the axle B by means of a pin e and a laterally adjustable plate f, substantially as described.

Eighth, the adjustable draught rod K arranged and applied to the plough carriage so as to operate substantially as described.

Ninth, the combination of a pole G, which is adjustable at its rear end, with a draught rod K, which is adjustable at its front end, substantially as described.

No. 55,922.—WM. G. SMITH and DANIEL HOOPER, New York, N. Y.—Topsail Reefing Rig.—June 26, 1866.—The loops of the stops on the reefing yard project through eyes in the sail and a binding rope is rove through them to secure the sail to the yard; the latter is checked by a line which connects it to the topmast and is raised by the lifts.

Claim.—First, in connection with a single topsail, a flowing middle yard or reefing boom D, checking line g g, bifurcated lifts M m m, and preventer lifts K, all arranged and operating as herein shown and explained.

Second, the combination of the binding rope f f and loops formed by stops e e, for securing the belly of a topsail to a boom D, or middle yard, substantially in the manner and for the purpose specified.

No. 55,923.—LEONARD A. SPRAGUE, New York, N. Y.—Lacing for Boots and Shoes.—June 26, 1866.—The lacing eyes are formed by loops of the wire which project through notches in the folded edge of the material.

Claim.—The lacing eyelet, in which the lacing eye is formed in one piece with the fastening device, substantially as herein shown and described.

Second, the combination of two or more lacing eyes with intermediate links, when the same are formed of one piece of wire, substantially as herein shown and described.

Third, the method of fastening lacing eyelets, substantially as described, by confining the shanks or links thereof in the folds or between two thicknesses of leather or other material to which the lacing is attached.

No. 55,924.—DAVID I. STAGG, New York, N. Y.—Book Holder.—June 26, 1866.—The base of the inclined portion has jointed arms with pivoted fingers which rest against the leaves and keep the book open.

Claim.—The arms C C, secured in the cleat or strip B by screws b b, and provided with pivoted fingers D at their outer ends, in connection with the inclined frame or base, all arranged substantially as and for the purpose specified.

No. 55,925.—DAVID I. STAGG, New York, N. Y.—Stand, Desk, and Book Holder.—June 26, 1866.—The base has two portions, hinged at opposite sides; the upper one forms a cover to the stand and when slanted forms a desk supported by a button, and the other portion is raised to a more nearly erect position to form a book stand with fingered arms to hold the book open.

Claim.—The stand, writing desk, and book holder, consisting of the bed piece D, hinged frame E, hinged board F, provided with the fingered arms G G, and hinged bar I, combined and operating in the manner and for the purpose herein specified.

No. 55,926.—JOHN K. STAMAN, Mifflin, Ohio.—Portable Fence.—June 26, 1866.—A letter A brace sustains each post, which has two rows of mortises to receive the ends of the rails of the adjacent panels.

Claim.—The braces d d' and counter-brace d'', in combination with the posts B and cross-rails e e', when constructed and arranged in the manner and for the purpose set forth.

No. 55,927.—MONROE STANNARD, New Britain, Conn., assignor to PRATT, WHITNEY & Co.—Needle for Sewing Machines.—June 26, 1866.—The needle has a cylindrical shank in a socket in the end of the needle bar; the device is intended to maintain a given proximity of the needle to the shuttle, notwithstanding variations in the size of the former.

Claim.—The employment of the needle as described, in combination with a shuttle, or its mechanical equivalent, substantially as and for the purpose described.

Also, a new article of manufacture, an eye-pointed sewing machine needle *a*, eccentric with reference to its shank *c*, substantially as and for the purpose described.

No. 55,928.—HENRY H. STAPLES, Woburn, Mass.—*Washing Machine.*—June 26, 1866.—The rubbing bed is formed of rollers in the bottom of the tub; the rubber is attached to a reciprocating slat, which has a handle at one end, and passes under an anti-friction roller at the opposite side of the tub.

Claim.—The arrangement and combination of the bar or lever D, its handle or handles *d d*, and the standards and roller *e*, with the rubber, the tub, and the series of rollers disposed in such tub, the whole being substantially as specified.

No. 55,929.—DANIEL STICK, Hughesville, Penn.—*Means for Fish to Pass Over Dams.*—June 26, 1866.—A chute with a sinuous track for diminishing the velocity and assisting the passage of the fish to the level above the dam.

Claim.—An improved device for facilitating the passage of fish over dams, formed by combining a series of inclined planes E I J K L with each other, with the partition walls G O P R, with the side walls F N, and with the end walls T and S, the whole being constructed and arranged in connection with the end of the dam and with the bank of the stream, substantially as described and for the purpose set forth.

No. 55,930.—HENRY STIMMEL, Canton, Ohio.—*Horse Rake.*—June 26, 1866.—A slotted arc is attached to the axle, and forms a guide for a set-screw on the lever to determine the distance of the teeth from the ground.

Claim.—The combination of the peculiarly shaped iron G, having a slot K therein, with the screw H, in connection with the lever F, substantially in the manner and for the purpose specified.

No. 55,931.—HERMAN STRATER, Jr., Roxbury, Mass.—*Stop-cock.*—June 26, 1866.—A cavity in the plug above the water way contains resilient material, to afford room for expansion of the liquid when freezing.

Claim.—In a stop-cock providing its plug with elastic yielding filling, substantially as and for the purpose described.

No. 55,932.—GEORGE C. TAFT, Worcester, Mass.—*Wrench.*—June 26, 1866.—The ferrule which sustains the thrust of the adjusting screw is itself sustained by a nut upon the tang of the stationary jaw inside of the ferrule.

Claim.—The arrangement of the nut I, in combination with the handle G, tang A, and ferrule H, with its shoulder *b* and step J, constructed and operating in the manner and for the purpose herein described.

No. 55,933.—D. S. THOMPSON, West Haven, Conn.—*Buckle.*—June 26, 1866.—The tongue frame is hinged to the main frame; a strap carried above the central bar, and thence down and inward between the two, is clamped by tension.

Claim.—The combination of the frame A with the tongue B and bar C, when the tongue and bar are hinged to the rear bar of the frame, the whole constructed and arranged to operate substantially in the manner herein set forth.

No. 55,934.—W. H. TILLOW and S. SHUMWAY, Le Roy, N. Y.—*Churn.*—June 26, 1866.—The curved cogs on the face of the driving wheel engage with the teeth of the pinion, and, being set in different directions, give a frequent reversal of the direction of revolution.

Claim.—The gear or cog wheel D, constructed with curved cogs or flanges on the side of its rim, substantially as herein described, in combination with the cog wheel C of the dasher handle B, for the purpose of frequently reversing the motion of the dasher A.

No. 55,935.—E. W. TWING, Springfield, Mass.—*Carpet-stretcher.*—June 26, 1866.—Two metallic sleeves are fastened together by a swivel joint, and the arms of the stretcher are adjustable therein by set screws.

Claim.—First, the coupling consisting of the slides D E, pivot *b*, socket *c*, and set screw *a*, or its equivalent, in combination with the shafts A B of a carpet-stretcher, substantially as specified.

Second, the rounded head C, in combination with the levers A B and slides D E, arranged and operating in the manner and for the purpose herein specified.

No. 55,936.—JAMES WASSON, Fairwater, Wis.—*Broom Head.*—June 26, 1866.—The metallic sides have roughened inner plates to hold the corn, and are fastened at their lower ends by barbed rods.

Claim.—First, an improved broom head, formed by combining with each other and with the socket A the two parts or halves B and C, constructed substantially as herein described and for the purpose set forth, said halves being held together at the lower part of their side edges by the barbed spike hooks F.

Second, the combination of the two perforated or roughened plates D and E, constructed as described, with the parts or halves B and C, substantially as and for the purpose set forth.

Third, the combination of the barbed spikes F with the halves or parts B and C, substantially as described and for the purpose set forth.

No. 55,937.—W. H. WATSON, Yonkers, N. Y.— *Tobacco Press.*—June 26, 1866.—The sides and ends of the inner box are placed loosely in the outer one, the box filled, and the cover clamped down. When the tobacco, &c., is to be removed, the clamp is taken off and the slightly tapering inside box pushed out of its strong case.

Claim.—First, the inner or secondary box, constructed and operating substantially as described for the purposes set forth.

Second, in combination with the same, the outer box, top plate, and clamps, when the same shall be constructed, combined, and operated substantially as shown.

No. 55,938.—JAMES E. WEAVER, Temperanceville, Penn.—*Apparatus for Oiling Journals.*—June 26, 1866.—The oil is led to the journal by a siphon from the reservoir, to which air is admitted by force pipe or by removing the stopper.

Claim.—The combination of the vessel *c* and *pi* es *d* and *e* with a "plumber block," or bearing of shafting, as herein described and for thæpurpose set forth.

No. 55,939.—HARVEY G. WHITTAKER, Brattleboro', Vt.—*Stove-pipe Drum.*—June 26, 1866.—The flue occupies the middle chamber, and has an air duct inside and on the exterior.

Claim.—The combination of the internal drum B, intermediate drum A and external drum C, with the pipes D E F G, when the parts are constructed and arranged in the manner and for the purpose herein described.

No. 55,940.—MARTIN V. B. WHITE, Troy, N. Y.—*Sash Supporter.*—June 26, 1866.—The roller occupies the angular recess in the casing, and supports the sash by frictional contact; the sash is freed when the roller is raised by the thumb-piece.

Claim.—The employment of the arm-piece or rod *c*, having the thumb-piece I thereon, and the opposite end thereof so constructed as to receive the roller D, and operate the same within the triangular recess E, each being arranged and combined in the manner substantially as herein described and set forth.

No. 55,941.—A. H. WHITNEY, Portland, Me.—*Measuring Funnel.*—June 26, 1866.—Explained by the claim and cut.

Claim.—The combination in a funnel of the graduated scale, the stop-cock, the flange and air vent all as and for the purposes specified.

No. 55,942.—J. J. WILKINS, Virden, Ill.—*Harness Hook.*—June 26, 1866.—The slot admits the passage of the transverse strap, to which it is attached by rivets through the perforated ears.

Claim.—The book A, provided with the transverse opening or hole *a*, and the horizontally projecting ears *a*, constructed and operating as shown and described.

No. 55,943.—JOHN A. WILLIAMSON, Lafayette, Ind.—*Railroad Switch.*—June 26, 1866.—The bridge straddles the rail, the long incline affords a track for the wheel, and the shorter directs it on to the rail.

Claim.—The combination of the bridge A, tapered guide B and tapered bar C, when the latter is securely pivoted to the guide B, and all constructed and arranged in the manner and for the purpose herein specified.

No. 55,944.—SAMUEL R. WILMOT, Bridgeport, Conn.—*Dies for Cupping and Raising Articles of Metal.*—June 26, 1866.—The blank is clamped by the holder in the annular rabbet of the die, while the stamp descends to form it into shape.

Claim.—The attachment of the blank holder B directly to the bed-plate or fixed die A, substantially as herein set forth for the purpose specified.

No. 55,945.—BENJAMIN F. YOUNG, Charlestown, Mass.—*Boring Machines.*—June 26, 1866.—Improvement on B. Merritt's patent, May 24, 1864. The face of the cone fits in inclined sides of the groove, and they are relatively adjusted to prevent looseness.

Claim.—A grooved cam constructed with inclined sides, in combination with a cone-traveller.

No. 55,496.—JOSHUA H. ZINN, Kingston, Tenn.—*Weighing Car.*—June 26, 1866—The bottom of the car rests upon levers connected to a weighing scale beam; to relieve the weigh beams of the constant pressure, a shaft with disks is arranged beneath to take the weight therefrom as required.

Claim.—First, the arrangement of the levers or frames B and C, within the bottom portion of the frame of a car and connected together, and with any suitable weighing lever or beam in combination with the inside box or casing resting at each end upon such f ames B and C, substantially as described and for the purpose specified.

Second, the shaft P, having cam-shaped disks R, so arranged as to lift the inside box from the weighing levers or frames, for the purpose set forth.

* No. 55,947.—ROBERT ANDERSON, Brooklyn, N. Y., assignor to S. O. RYDER, of New York, N. Y.—*Machine for Hulling Coffee*—June 26, 1866.—The coffee is divested of its husk by a hulling cone, with oblique teeth revolving in a toothed concave, and passes to a polishing chamber, against whose rough inside surface the grains are driven by the revolving wings.

Claim—The combination in a machine for hulling and polishing coffee of an obliquely ribbed wheel E, rotating within a slightly conical toothed concave A, constructed as described, and a series of wings G, rotating within a horizontal tapering or conical case A, the whole arranged to operate substantially as herein set forth for the purpose specified.

No. 55,948.—J. D. M. ARMBRUST, Apolloborough, Penn., assignor to himself and JACOB FREETLY, same place.—*Broom Head.*—June 26, 1866.—The metallic cap and the successive bands are lapped around the broom, and retained by bolts, which pass through the handle.

Claim.—The metallic band composed of thin metal and furnished with slots, in combination with the strengthening pieces and bolts and nuts, as and for the purpose described; also, the combination of the movable cap piece and the handle slotted at its lower end, for the passage of the bolts constructed as described, and for the purposes set forth.

No. 55,949.—FRANKLIN H. BROWN, Chicago, Ill., assignor to himself and JAMES F. GRIFFIN, same place.—*Apparatus for Carburetting Air.*—June 27, 1866.—No air passes from the blower to the carburetter when the motion of the former stops; but an automatic arrangement admits air by a valve to the carburetting chamber, so as to maintain the operation temporarily after the blower has ceased to operate.

Claim --First, the combination of a series of revolving buckets F, and a hollow stationary shaft G, provided with the openings *a* and *b*, arranged and operating substantially as and for the purposes specified.

. Second, the combination of a series of revolving buckets F and a stationary hollow shaft G, when provided with the partition *m*, and the openings *a* and *b* and *c* and *d*, arranged and operating as and for the purposes shown and described.

Third, in combination with a carburetter A, arranged above the burners, the employment of a device for compressing and forcing air into the same, substantially as and for the purposes herein specified and shown.

Fourth, the employment of an automatically closing valve V, in combination with a close carburetter, arranged above the burners, and a device for compressing and forcing air into the carburetter, and a weight or its equivalent, for operating the same, substantially in the manner and for the purposes specified.

No. 55,950.—FRANKLIN H. BROWN, Chicago, Ill., assignor to himself and JAMES F. GRIFFIN, same place.— *Apparatus for Carburetting Air.*—June 26, 1866.—The upper series' of pans are filled with heavy oil and the lower with lighter oil; each pan has a siphon for withdrawing the oil as required, and the air inlet has a cap and deflector.

Claim.—First, the combination in one device of two or more siphons, arranged and operating substantially as and for the purposes specified and shown.

Second, the arrangement of the pans of a carburetter in two or more independent sets, in combination with devices for drawing off the contents of said sets separately, and of the pans in each set simultaneously, substantially as specified and shown.

Third, the arrangement of a deflector V with the inlet tube N, substantially as and for the purposes described.

Fourth, providing the inlet tube N with a cap P, arranged as shown and for the purposes set forth.

Fifth, the combination and arrangement of the deflector V and cap P with the inlet pipe N, substantially as specified and shown,

No. 55,951.—LEWIS FRANCIS, New York, N. Y., assignor to himself and CYRUS H. LOUTREL, same place.— *Composition of Matter.*—June 26, 1866.—This elastic material, for lining of vessels, inking rollers, moulds for plaster casts, &c., is composed of glue, glycerine, soluble silicates, water and saccharine matter, in different proportions, according to purpose.

Claim.—Combining glue, glycerine, and silicates, with or without saccharine matter, to form a new and useful composition of matter for various purposes.

No. 55,952.—JOHN GREEN, Norwalk, Ohio, assignor to himself and JAMES W. BARKER, same place.—*Grain Separator.*—June 26, 1866.—Two separate shoes, with their respective fans, are driven simultaneously, the grain passing continuously from the upper to the lower. The inclination to each side of the upper screen gives a lateral discharge to coarse offal through the side openings in the shoe box.

Claim.—First, the two fans C C', in combination with the two shoes F F', operated by means of the gearing, connecting rods, and levers from the driving shaft of the machine, substantially as and for the purpose herein set forth.

Second, the double inclined ends of the upper screens *i*, of the shoes F F', in combination with the side openings *m* in the shoe boxes, substantially as and for the purpose set forth.

No. 55,953.—GEORGE H. HAMMER, Newville, Penn., assignor to himself, D. J. BROUGHER, and WM. A. MIDDLETON, Harrisburg, Penn.—*Fastening for Fruit Cans.*—June 26, 1866.—The ends of the bail catch in ears on the can and the cover is secured by the pressure of the bail upon it.

Claim.—Securing the cover *a* by means of the bail B attached to ears *e* or lugs *c*, substantially as shown and described.

No. 55,954.—JAMES C. HYDE, West Haven, Conn., assignor to himself and D. S. THOMPSON, same place.—*Butt Hinge.*—June 26, 1866.—The leaves are united without a pintle. The tongues of one leaf are passed through slots in the other and bent.

Claim.—A hinge, the two parts of which are formed and united, substantially in the manner herein set forth.

No. 55,955.—RALPH S. JENNINGS, New York, N. Y., assignor to himself and NORMAN G. KELLOGG, New York, N. Y.—*Metallic Seal Envelope.*—June 26, 1866.—A projecting tube on the part attached to the envelope is surrounded by the perforated disk, attached to the flap, the tube is spread over the disk, making an eyelet connection.

Claim.—First, the metallic seal applied to envelopes in their manufacture, substantially as and for the purpose described.

Second, the new article of manufacture herein described and shown, to wit, an envelope for letters furnished with a metallic seal, which envelope is ready for being sealed when on sale, as set forth.

No. 55,956.—JOHN L. KENDALL, New York, N. Y., assignor to himself, R. H. TRESTED, and LEWIS HURST, same place.—*Clothes Sprinkler.*—June 26, 1866.—The conical vessel has a handle and perforated cap.

Claim.—A clothes sprinkler, composed of a conical vessel A, handle B, neck C, and nipple D, as a new article of manufacture.

No. 55,957.—C. D. KUBACH, Philadelphia, Penn., assignor to himself and W. W. CLAY, same place.—*Fire Alarm*—June 26, 1866.—A piston is forced up against a coiled spring and secured by placing tallow in the casing. The fire melting the tallow, the piston is forced down by the spring, and in passing down moves a pin which opens the case and drops the connecting wires operating the alarm mechanism.

Claim.—The casing E. with its piston G, pin *u*, and spring *t*, in combination with the slotted plate F and the within-described alarm mechanism, or its equivalent, the whole being constructed and operating substantially as and for the purpose described.

No. 55,958.—CALEB S. NELSON, West Troy, N. Y., assignor to himself and HENRY I. SEYMOUR, Troy, N. Y.—*Setting Stationary Steam Boilers.*—June 26, 1866.—The boiler is suspended in the brick furnace in such a way as not to interfere with the integrity of the latter by its contraction and expansion, and is surrounded by flue space to prevent the radiation of heat.

Claim.—First, the annular chamber or surrounding chamber *a*, formed between the stationary boiler B and steam-dome C and the inner part or surface of the brick or mason work A, in the manner and for the purposes substantially as herein described and set forth.

Second, the suspending of the stationary steam boiler B within the brick arch or mason work A, surrounding the same, so as to form an air chamber *a* immediately upon and over the upper surface of such steam boiler so as to prevent the escape of heat, in the manner and for the purposes substantially as herein described and set forth.

No. 55,959.—GEORGE W. PATERSON, Newburyport, Mass., assignor to himself and H. M. PAYNE, same place.—*Picker-arm for Looms.*—June 26, 1866.—The described devices are intended to lessen the friction on the cam which imparts motion to the picker and prevent the lubricating oil of the anti-friction roller on the picker-arm from being thrown upon the loom or fabric.

Claim.—First, the picker-arm, constructed and arranged for operation substantially as herein shown and described.

Second, in a picker-arm the combination of a friction roller with a shell enclosing the same on top and sides, as herein described.

Third, in a picker-arm, and in combination with an enclosing shell, as herein described, the friction roller when of an ovoidal or equivalent shape and having its bearings in the shell, as herein shown and set forth.

No. 55,960.—THOMAS W. POMEROY, East Hampton, Mass., assignor to himself and WM. J. LAYMAN, same place.—*Churn and Ice-cream Freezer.*—June 26, 1866.—The cream in the inner vessel is stirred by a perforated, vertical, revolving dasher, and the surrounding jacket may contain a tempering or freezing mixture..

Claim.—The combination of the cylinder C with the case A and rotary dasher E, when the cylinder C is secured within the case A by means of horizontal arms *b* projecting from C and fitted between guides or lips *a* attached to the inner surface of A, and either with or without the feet *c*, to rest on the bottom of A, substantially as and for the purpose specified.

No. 55,961.—LOUIS C. RODIER, Springfield, Mass., assignor to himself and HENRY HAMMOND.—*Mop Head.*—June 26, 1866; antedated December 26, 1865.—The mop cloth is clamped between toothed jaws, whose shanks are hinged and fastened by a ring.

Claim.—The jaws B and C, having fingers or teeth, for the purpose herein specified, when combined with the handle and ring *d* in such manner as to form a floor mop, the whole being arranged substantially as herein described.

No. 55,962.—JOSEPH R. B. SCHWARZE, Washington, D. C, assignor to himself and DANIEL PFEIL, same place.—*Liquid Cooler.*—June 26, 1866.—The coil leads from the faucet through a vessel charged with ice, and passes out at the bottom.

Claim.—The chamber C, provided with a pipe B, and arranged for attachment to the faucet by the coupling E, as described and represented.

No. 55,963.—F. M. STRONG and THOMAS ROSS, Brandon, Vt., assignors to JOHN HOWE, Jr., same place.—*Platform Scale.*—June 26, 1866.—The light ends of the main levers are connected by compound links with the arm on one end of the extensible rod which connects the levers at the platform with the levers in the weigh box. The rod is made extensible to suit variation in distance between the platform and weigh box. The poise is operated from the outside; stop pieces on a graduated scale indicate the determinate position of the poise shifters.

Claim.—The combination of the levers C C and the links *k k* with their upper and lower bearings *m i* and the secondary lever D.

Also, the shaft F, in combination with the adjustable arms E E', as and for the purpose described.

Also, the combination of the poise S, lever P *z*, graduated plate *u*, and movable stop pieces *w*, as and for the purpose described.

No. 55,964.—JAMES TEMPLE, Selinsgrove, Penn., assignor to himself, JOHN SCHOCK, and H. E. MILLER.—*Car Brake.*—June 26, 1866.—The brakes are operated upon the wheel simultaneously by their rock shafts, which are connected by rods and operated by a single lever.

Claim.—The combination of the rock shafts D, arms *f f'*, and rods *a*, with the arms *s* and rubbers E, operated by the levers *h*, and arranged as described for the purpose set forth.

No. 55,965.—FREDERICK VOTTELER, Cincinnati, Ohio, assignor to JACOB W. HOLINSHADE, same place.—*Nut-tapping Machine.*—June 26, 1866.—The central tap revolves in one direction, and its gearing drives the surrounding taps in the other direction, the lever bringing several nuts simultaneously into adjustment for being tapped.

Claim.—First, the combination in a tapping machine of taps for right and left hand nuts, arranged and operating substantially as described.

Second, the combination of two or more taps with a mechanism for adjusting simultaneously a nut to each, arranged and operating substantially as described.

No. 55,966.—D. T. WARREN, New York, N. Y., assignor to himself and AMEDEE SPADONE, same place.—*Pencil Case.*—June 26, 1866.—By the described arrangement either the lead or pencil holder can be projected or retracted by rotating a certain portion of the case.

Claim.—The arrangement of the lead holder *a*, shaft *b*, tube *c*, straight slot *d*, stud or pin *k*, loose sleeve *m*, with its enlargement *o*, spiral slot *l*, tube *r*, pen-holder *t*. tube *v*, straight slot *w*, pin *z*, spiral slot *y* and loose sleeve *z*, as and for the purpose set forth.

No. 55,967.—DARIUS WELLINGTON, Boston, Mass., assignor to CORNELIUS WELLINGTON, same place.—*Water Closet.*—June 26, 1866.—The incline by which the pan is operated extends from the pan itself. The pan is sealed more effectually by the flexible chamber which expels water into the pan; the mode of connecting the hopper to the stand admits of its removal and replacement.

Claim—First, the arrangement of the shaft *d*, slotted plate *f*, arm *i*, and rod *k*, for supporting the pan by the connection between it and the operating rod, substantially as described.

Second, combining with the supply pipe and pan the flexible air ball or chamber, operating to seal the pan, substantially as set forth.

Third, in combination with the hopper *e* the divided nozzle and plate *o*, operating as and for the purpose described.

Fourth. the combination of the ring (serving as an elastic packing) with the flange *p* and stand *b*, substantially as described.

Fifth, connecting the supply pipe to the hopper by means of the flanged coupling *t*, the nut or screw in the supply pipe, and the interposed packing *u*, arranged and operating substantially as shown and described.

No. 55,968 — H. L. WILLIAMS, Seneca Falls, N. Y., assignor to NATIONAL KNITTING MACHINE COMPANY.—*Knitting Machine.*—June 26, 1866.—The latch needles are carried by and slide in a frame; a number of these frames are linked or pivoted together in an endless chain; a cog wheel, whose teeth mesh with teeth on the frames, gives motion to the chain, which is supported by a guide; proper cam grooves on a plate give endwise motions to the needles. The size of the tubular goods to be knitted may be varied by leaving out from or putting into the endless chain more or less of the frames as desired.

Claim.—First, the combination of the cogs *k k* with the lags constituting the "form" and with the gear-wheel by which they are driven, substantially as specified.

Second, the combination of the curved guide G with the form B and gear wheel C, substantially as described.

No. 55,969 —WILLIAM KIRRAGE, London, England —*Manufacture of Artificial Stone.*—June 26, 1866.—Silica, gravel, &c., is mixed with cement and sulphate of iron, or equivalent; water is added to corrode and bind the mass, which is moulded into blocks.

Claim.—First, the manufacture of artificial stone, bricks, and tiles, or other hard material, by mixing and using the sulphate of iron or other similar metallic sulphates, or other similar mixtures of acids with metallic bases, and the several materials hereinbefore described, in combination with cements and silica and limes and silica, contained in the materials such as are mentioned, for the purposes hereinbefore described.

Second, the application of the coloring matters before named in combination with the silica and other materials.

Third, in the manufacture of bricks and tiles from plastic materials, the application of the materials specified for the purpose of effecting a more perfect combustion, burning, and hardening, thus producing a superior form of article.

No. 55,970.—PAUL JACOVENCO, Bucharest, Wallachia.—*Petroleum Tank.*—June 26, 1866.—The inner covered cylinder is surrounded by a water space, with which it communicates by openi__ below ; pipes for the introduction and discharge of oil, the discharge of gas, and the introduction of water, connect with the central space.

Claim—The herein described apparatus for preserving or storing petroleum and other oils, the same consisting of two concentric cylinders, the space between them being divided by a transverse partition, and communicating with the interior cylinder by means of apertures, as described, in combination with pipes arranged for operation, as and for the purposes set forth.

* No. 55,971.—WILLIAM REINLEIN, Barcelona, Spain.—*Hot Air Engine.*—June 26, 1866.—Explained by the claims and cut.

Claim.—First, the arrangement of the support of an air engine, so that the axis of the motor cylinder shall be vertical, in the manner and for the purposes set forth.

Second, the combination and arrangement of a tubular boiler in the place of the dome which serves as a furnace in the Ericsson machine, as and for the purposes described.

No. 55,972.—E. HAMBUJER, Detroit, Mich.—*Corset*—June 26, 1866.—The corset does not open in the back, but expands the portion which is folded up inside when the corset is laced. Skirt supporting hooks are attached to the waist.

Claim.—First, a corset made substantially as herein described. with a continuous or connected back, forming a lap or gore when the lacings are drawn to tighten the garment around the person.

Second, in combination therewith, the skirt supporting hooks H H, arranged substantially as described.

No. 55,973.—CHARLES E. ABBOTT, Boston, Mass.—*Extinguisher for Lamps.*—July 3, 1866.—A sleeve is operated by rack and pinion, and slips upon the wick tube; when elevated the hinged lid shuts down upon the top of the wick and extinguishes the flame.

Claim.—The tube *c*, with its lid or cover *d*, and rack *g*, or equivalent, operated by the toothed wheel *h*, upon the rod C, substantially as set forth.

No. 55,974.—CHARLES ARNOLD, Chicago, Ill.—*Paper Cloth and Bosoms.*—July 3, 1866.—The plaited fabric is fastened by its edges to a paper backing.

Claim.—A bosom with cloth front and paper back, when secured together substantially as described and used for the purpose set forth.

No. 55,975.—JAMES ASHTON, Fall River. Mass.—*Oil Can*—July 3. 1866.—The bent wire has a valve at one end which governs the air inlet, one at the other which governs the oil discharge, and one at the bend; these are operated simultaneously when the air inlet is closed, and oil exudes from the nozzle.

Claim.—The combination of an air inlet I I I. supply tube T T, dripping chamber X X, spiral spring S, shut-off rod A B. valve V, and rod C D, all as applied to an oiler, or oil can, in the manner described and for the purposes set forth.

No. 55,976.—JEFFERSON AUGHE, Dayton. Ohio.—*Drop Hammer.*—July 3. 1866.—The arms on the drop weight project laterally and fit against vertical guide ways; the upper edge of the drop die is mortised to the drop weight, and fastened in position by a central, vertical bolt.

Claim.—First, the mode of connecting a die to the ponderous hammer E of a swaging machine substantially as herein described and represented.

Second, the arrangement of the dies F F', the hammer E, and bolt G, substantially as described, and for the purposes specified.

No. 55,977.—A. S BABBITT, Keesville. N. Y.—*Water Elevator.*—July 3, 1866—By the action of the springs and cams the forward motion of the handle produces friction between the ratchet wheel and the disk wheel, to raise the bucket and hold it; by depressing the handle the friction is removed and the empty bucket allowed to descend.

Claim.—The arrangement of the wheels D and B the wheel B being provided with a hub upon which cams are formed, with the lever A which is also provided with cams to correspond with those upon the wheel when used with the springs E E', and collar F, as and for the purposes specified

No. 55 978.—HENRY O BAKER. New York N. Y.—*Curtain Fixture*—July 3, 1866.—The cord is tightened by pushing down the holder in the case; a prong on the holder engages the holes in the side of the casing

Claim.—The lever button, in combination with the perforated side of the frame, substantially as and for the purposes herein set forth.

No. 55,979.—HORACE BAKER, Cortland, N. Y.—*Hay-raker and Loader.*—July 3, 1866.—The hay is gathered by dividers, a toothed roller, and a vibrating horizontal rake, and raised between two endless elevators which traverse in opposite directions; the devices are all actuated from the ground wheels.

Claim.—First, the rake teeth P², in combination with two positively actuated endless aprons W W', revolving in opposite directions, and so placed in relation to each other that the cut grass raised over the teeth shall be seized between the aprons and elevated, substantially in the manner set forth.

Second, the standards T T, when so constructed that by their elasticity they shall maintain the aprons W W' face to face, and permit variations in the quality of grass carried between them, substantially as set forth.

Third, in combination with the endless aprons W and W', the guides S² and T', when attached to the standards T T, and so constructed as to carry the cut grass over and beyond the endless aprons, substantially as set forth.

Fourth, the driving wheel B and cup-shaped wheel C, constructed as described, in combination with the spur wheels G H and K, substantially as and for the purpose set forth

Fifth, the shoe X, in combination with the wheel Y and roller X', arranged substantially as and for the purpose set forth.

Sixth, the combination of the teeth P², shaft N, and rollers P³, substantially as and for the purpose set forth.

No. 55,980.—J. C. BARRETT, Stamford, Conn.—*Water Elevator.*—July 3, 1866.—The lazy tongs or compound lever is used to depress and elevate the bucket, which is upset by the contact of its lip with a pendant.

Claim.—First, the lazy tongs B, in combination with the lever D, provided with the rocker E, all being arranged substantially as and for the purpose set forth.

Second, the pendent rod i, provided with the cross-head j, and attached to the lazy tongs in connection with the lip k on the bucket, substantially as and for the purpose specified.

No. 55 981.—H. DE BAUM, Paterson, N. J —*Sash Fastening.*—July 3, 1866.—A casing containing a pivoted spring-catch is attached to or in the edge of the sash and sustains the latter by engagement with a notched bar in the casing.

Claim.—The combination of the case E, latch D, and spring e, operating with the rack C, when constructed and arranged in the manner and for the purpose herein specified.

No. 55,982.—ALFRED BAYLEY, Newark. N. J.—*Wash Stand.*—July 3, 1866.—The cut shows the arrangement of the looking-glass in the lid, the plug-bottom basin, slop bucket and water bucket.

Claim.—As a new article of manufacture a portable metallic wash stand, constructed and its parts arranged substantially as described.

No. 55,983.—GEORGE BECK, Rochester, N. Y.—*Beverage*—July 3, 1866.—Compound of wine or spirit, 1 part; water, 3 parts; sugar to sweeten and induce a new fermentation; cask tightly to retain the carbonic acid and draw the ebullient liquor.
Claim—The production of a beverage by the mixture of wine or other similar fermented liquor with water, in the proportions set forth or thereabout, and the addition of sugar to produce a secondary fermentation, to be conducted and continued in a close vessel, so as to retain the gases produced thereby, for excluding the air and expelling the beverage in an effervescing state, substantially as and for the purpose herein specified.

No. 55,984.—SOLOMON BECKETT, Olive Branch, Ohio—*Plough*—July 3, 1866.—The front edge of the mould-board is protected by an overlapping "shin" plate attached to the cutter.
Claim.—First, the sheathing plate B fitted and secured to mould-board A of a plough, for the purpose above described and set forth.
Second, the sheathing plate B, in combination with cutter C.

No. 55,985.—B. M. BECKWITH, Plattsburg, N. Y.—*Carriage Spring.*—July 3, 1866.—The ends of the arched leaf springs are supported on boxes which have their bearings upon the axles.
Claim.—The boxes F, applied to the hind axle and bolster of a buggy to admit of the attachment of the springs and permit their extremities to vibrate about the points of support, substantially as and for the objects specified.

No. 55,986.—ORMUS D. BEEBE, Beaver Dam, Wis.—*Washing Machine.*—July 3, 1866.—Explained by the claims and cut.
Claim.—First, securing the slats c c, within the tub upon the metallic bottom, by having them extend under or mitring with the sides A A, substantially as and for the purposes herein set forth.
Second, the combination of the metallic bottom with the slats c c lying thereon and mitred with the sides, as herein set forth.

No. 55,987.—GEORGE W. BEERS, Bridgeport, Conn.—*Carriage Hinge.*—July 3, 1866.—The dovetail plate on the end of the pivoted leaf is received into the socket piece on the door, which can be lifted from its bearings at pleasure.
Claim.—The plates E and G and the catch F, constructed as described, in combination with the swinging arm of an ordinary concealed carriage-door hinge, substantially as described and for the purpose set forth.

No. 55,988.—EDWIN BENNETT, Philadelphia, Penn.—*Implement for Grooving the Mouths of Glass Bottles.*—July 3, 1866.—Two grooved wheels are attached to a glass blower's spring clamp and are rotated inside of the lip to make a depression therein as the clamp is revolved.
Claim.—The application to a glass blower's bottle clamp of the toothed wheels C C, the same being arranged and operated substantially as and for the purpose described.

No. 55,989.—WILLIAM BERGMANN, Philadelphia, Penn.—*Car Coupling.*—July 3, 1866.—The loose link of one car is secured by the pivoted dog in the draw-head of the next car; the dog is retained by a pivoted lever and freed by the vibration of the latter.
Claim.—The casings A and A', with their dogs H H', levers F F', and links M M', in combination with the within described devices or their equivalents, for operating the levers and securing the dogs, the whole being constructed and operating substantially as and for the purpose specified.

No. 55,990.—ALOIS BERNY, Williamsburg, N. Y.—*Cradle and Chair.*—July 3, 1866.—To make the conversion from a chair to a cradle, the seat is turned on its pivot 90°; the hinged seat laid over, the additional hinged arms brought into position as sides for the extension bed, the elevated portion of the back removed and placed between the ends of the hinged arms.
Claim.—A chair provided with a hinged additional seat, additional back, and hinged additional arm pieces, said seat being made to swivel on its frame D, and the arm pieces being secured to each other and to the seat A by catches c and screw e, or other equivalent fastening, substantially as and for the purpose described.

No. 55,991.—JOHN T. BEVER, Bethel, Ill.—*Clothes Tongs.*—July 3, 1866.—The hemispheres at the end of the tongs afford a means of handling and grasping the clothes.
Claim.—As a new article of manufacture the within described clothes tongs, viz: a pair of wooden tongs furnished at their ends with united wooden hemispheres as described.

No. 55,992.—WILLIAM R. BISHOP, Sherwood, Wis.—*Gib for Cross-heads.*—July 3, 1866.—The gib is provided with a working face of leather.
Claim.—A gib composed of a metal body A, and strip of leather a, substantially as and for the purpose described.

No. 55,993.—A. M. BLACK, Auburn, Ill.—*Cultivator.*—July 3, 1866.—The frame which includes the share beams is pivoted to the front cross-bar; this is rigidly attached to the tongue whose rear extension supports the seat; the shares are laterally and vertically adjustable and may be supported clear of the ground by a pivoted prop.

Claim.—First, the pole a having a cross-bar A rigidly secured thereto, and a framework composed of the plough beams B B, with cross-pieces b b' pivoted to the ends of said cross-bar A, substantially in the manner and for the purpose specified.

Second, the cross-bar A, plough-beams and frame B B b b', the adjustable prop C, and lever H, all arranged and operating in the manner and for the purpose set forth.

Third, the hook and eye-bolts, which project from the inside of the plough beams B B, and fasten the standards m m, to beams B B, the connecting rod g, eye and rod fastening in cross-piece b.

No. 55,994.—RICHARD C. BOCKING, Indianapolis, Ind.—*Gas Purifier.*—July 3, 1866 — The wheel in the reservoir has reticulated vanes, and is rotated by the induction stream of gas, which is purified by intimate contact with the liquid.

Claim.—First, the combination of the liquid reservoir, tube B, and the wheel provided with vanes having reticulated surfaces, and revolving on a horizontal axis under the impulse of the entering gas, substantially as described.

Second, the mode of hanging the said wheel in orifices in the sides of the chamber, capped on the outside, as described.

Third, the combination with the series of reticulated vanes of the bands, supported on the vanes by the pins, or equivalent means.

No. 55,995.—JOHN T. BONNELL, Columbia, Cal.—*Quartz Mill.*—July 3, 1866.—The vibration of the pendulum actuates the ratchets, pawls, and the drum; the motion of the latter is transferred to the tripping shafts, which raise the stamps.

Claim.—The drum C, provided with ratchets D D and a pendant rod A, with a weight or bob O attached, in combination with the spur wheels E E, placed loosely on the shaft of drum C, and provided with pawls F to engage with the ratchets D D, the above parts being used in connection with the pinions I I on the shafts J J, which actuate the weight or pounder rods L, substantially as and for the purpose specified.

No. 55,996.—JACOB BOYERS, Orrville, Ohio.—*Churn.*—July 3, 1866.—The shaft enters the churn obliquely, so as not to interfere with the lid or be below the level of the contents; it is connected with the horizontal axis of the dasher by a universal joint.

Claim.—First, the oblique shaft C, so arranged in its connection with the dasher shaft D, by means of the curved plate b and staples b', as to give an oblique or nearly rectilinear rotary motion when operating substantially in the manner and for the purpose herein set forth.

Second, the retention of the rod of the oblique shaft C in its axis by the flexible rod E and pivot c, as arranged and applied, substantially in the manner and for the purpose as herein set forth.

Third, the movable slide plate F, strips e e, and bridge plate d, in its connection with the dasher shaft D, as arranged substantially in the manner and for the purpose as herein set forth.

Fourth, the flexible rod f, pin f', and catch g, for retaining the slide plate F when in operation, substantially in the manner as arranged and for the purpose as herein set forth.

Fifth, the dish a' and air tube a, as arranged in combination with the lid B, substantially in the manner and for the purpose as herein set forth.

No. 55,997.—NATHANIEL A. BOYNTON, New York, N. Y.—*Coal Stove.*—July 3, 1866.—The arrangement is explained by the claim and cut.

Claim.—The combination of the flanged fire chamber B, cylinder C, cylinder D, with perforated bottom K, forming an annular space E, air flues F, semicircular flue G, and exit flue H, all arranged and operating in the manner and for the purpose herein described.

No. 55,998.—JOSEPH BRADT, Laporte, Ind.—*Beehive.*—July 3, 1866 —Explained by the claim and cut.

Claim —The arrangement of the doors B and E, flap d, incline plates g g, comb frames F, and hinged cap A, substantially as and for the purposes set forth.

No. 55,999.—T. E. C. BRIMLY, Louisville, Ky.—*Plough.*—July 3, 1866.—The mould board, share, and point consist of one piece of steel, having a hook on its under side to fit over the point of the landside, and holes by which it is fastened to lugs on the standard.

Claim.—A plough provided with a point E, welded, rolled, or otherwise formed or permanently secured to a steel mould board, with a hook or shoulder b at its under side to fit over the front end of the landside of the plough, substantially as herein shown and described.

No. 56,000.—HIRAM BROWN, Cape Elizabeth, Me.—*Hay Rake.*—July 3, 1866.—Without lifting the hand rake the load is discharged, the head being rotated by the oscillation of the frame which is connected thereto.

Claim.—The combination and arrangement of the frame A, connecting rod B, arm C, and hinges or bands D, attached to the tongue and head of a rake, as herein set forth, and operating as and for the purposes described.

No. 56,001.—WILLIAM BURNHAM, Union City, Mich.—*Washing Machine.*—July 3, 1866.— A wheel, bands, and pulleys rotate the tub on its vertical axis; fluted conical rollers are attached to a stationary centre post above the tub.

Claim.—First, the employment of a revolving tub *e* in combination with the fluted conical rollers R R, which rollers operate against the clothes by spring pressure, and are arranged relatively with and connected to the frame of the machine, substantially in the manner and for the purpose herein specified.

Second, the combination of the band M, pulley P P^1 and P^2, with the machine frame and tub shaft D, substantially as and for the purpose set forth.

No. 56.002.—EZRA BUSS, Springfield, Ohio.—*Apparatus for Raising Bread.*—July 3, 1866.—The chamber has shelves for the dough, and the entering air is heated by a hot-water vessel in the bottom.

Claim.—A box or chamber A for receiving the dough to be raised, when provided with a close vessel filled with heated water for communicating a very gradual, continuous, dry heat to the atmosphere within the chamber, and arranged and operating substantially as herein s ecified.

pAlso, locating the hot-water vessel H in a drawer G, closed except at the top, and provided with apertures near the bottom, substantially as and for the purpose herein set forth.

Also, providing the chamber with an opening C, controlled by a valve *e*, for the purpose of admitting fresh air therein, and of regulating the temperature inside, as herein described.

Also, the opening D in the top, controlled by a valve *d*, for the purpose specified.

Also, the use of a thermometer A in connection with the apparatus, arranged substantially as specified.

Also, the use of the glass *b* in the door or side, for the purpose set forth.

No. 56,003.—THOMAS BYRNE, New York, N. Y.—*Leeway Indicator for Vessels.*—July 3, 1866.—A graduated plate is attached to an oscillating support; the index finger is swerved by a suspended plummet when the leeway of the vessel gives it a list to starboard or port.

Claim.—An appara us for indicating the leeway of a vessel, consisting of a graduated plate attached to an oscillating support, whether such support presents a plane, conical or globular surface, and of an index finger, operated by a cord and a plummet, or other equivalent device in the water, and vibrating on a fixed pivot with a double support, or on double hinges, substantially as and for the purpose set forth.

No. 56.004.—GEORGE J. CAPEWELL, West Cheshire, Conn.—*Oiler.*—July 3, 1866.—A swivelled tube is attached by an elbow to the central discharge tube, and reaches the oil collected at the lower portion of the can when tilted.

Claim.—An oiler constructed with the tube C and the tube D in combination with the swivel joint, substantially as and for the purpose herein described.

No. 56,005.—L. R. CAVENDER, Eureka, Ill.—*Table Leaf Support.*—July 3, 1866.—When the leaf is raised a spring rotates the hinged support; this is drawn back by a cord when the leaf is to be lowered, and the weight of the closed leaf holds it in place.

Claim.—An improved table-leaf support formed by combining the pivoted arm F, the spring H, and cord I, constructed and arranged, substantially as herein described, with each other and with the frame A of the table, for the purpose set forth.

No. 56,006.—J. A. CHAMBERS, Ogdensburg, N. Y.—*Fire Escape*—July 3, 1866.—The ladder consists of hinged sections, projected by side racks and pinions, worked by hand-winch, locked by catches, and when in the box rest upon a frame sliding on rollers.

Claim.—First, the combination of the rollers C, sectional ladder E, hinged as described, and sliding frame D, substantially as and for the purpose set forth.

Second, the supports B, with the flanges P, in combination with the sectional ladder E, and the gearing for elevating the same, the parts being severally constructed and arranged for use, substantially as and for the purposes set forth.

No. 56,007.—JAMES M. CLARK, Jersey City, N. J.—*Pen and Pencil Holder.*—July 3, 1866. —The slotted tube fills and sustains the external ornamented case and guides the pen socket, which is operated by the screw tube within it; this is also supported by an inner slotted tube, within which is an extension handle to hold a pencil.

Claim.—The slotted tube *e*, with its ring *i*, in combination with the screw *c*, attached to the tube *b* and acting upon the pen socket *o*, introduced between the tubes *c* and *e*, in the manner and for the purpose set forth.

No. 56,008.—SAMUEL CLARKE, New York, N. Y.—*Bed Bug Protector for Bedsteads.*—July 3, 1866.—The foot-piece which is provided for the leg of each bedstead has a pendent plate whose edge is impassable to insects.

Claim.—First, the combination of the cup B and ring C, with its pendent flange D, substantially in the manner represented and described.

Second, the pendent flange D in combination with the cup or guard B and support A, and with the legs of a bedstead, substantially as and for the purpose specified.

No. 56,009.—ROBERT COKARROE, Camden, Ohio.—*Cane Stripper and Knife Combined.*—July 3, 1866.—Pivoted to the handle of the knife is a spring stripper which opens to receive the cane, and when closed is the means of removing the blades by scraping them from the stalk.

Claim.—The cane-knife and stripper, consisting of a handled blade provided with a spring gripping jaw, which is supported by the guard E and combined with the blade by means of the forked ends *w*, substantially as described and represented.

No. 56,010.—JOSEPH H. CONNELLY, Wheeling, West Va.—*Combined Lamp Chimney and Shade.*—July 3, 1866.—Explained by the claim and cut.

Claim.—The combination with the glass chimney A and metal top B of the shade or reflector C, when provided with supporting points *c c c* and otherwise constructed and applied as herein specified, to leave an air-space between it and the chimney, lessen the conduction of heat, permit its ready removal, and avoid confining the glass, substantially as set forth.

No. 56,011.—JOSEPH H. CONNELLY, Wheeling, West Va.—*Lamp-shade Supporter.*—July 3, 1866.—The shade has a metallic upper rim and is suspended by a collar and book from the top of the chimney. Interior springs on the rim rest against the glass and preserve the concentricity of the shade.

Claim.—A lamp shade or screen B, having a metallic rim *a* and a wire or other metal band *c* arranged substantially as set forth, in combination with a supporter consisting of the collar D and suspenders and hook *c d e*, constructed and operating substantially as described, for the purpose of adapting the shade B to use with a common bulged chimney or a straight wide chimney, as set forth.

No. 56,012.—CHARLES W. CONOLLY, Rochester, N. Y.—*Draught Attachment.*—July 3, 1866.—An adjustable bow is attached to the hames and to the neck yoke; pressure upon the latter draws the load by the end of the tongue.

Claim.—First, the adjustable bows B, when attached to the front of the hames and used without tugs, substantially as and for the purposes set forth.

Second, in combination with the bows B, the suspension straps *s*, and the steadying straps S.

Third, the swivelled clip C, constructed and arranged as described and for the purposes set forth.

No. 56,013.—WILLIAM W. COOPER, Washington, D. C.—*Goniometer.*—July 3, 1866.—The hollow spherical zone has an internal graduated spiral in which travels the outer end of a spirit-level case, hinging at a point coincident with the centre of the sphere; the latter is rotated on the bar or on the tube of a telescope till the bubble indicates a level, when the pointer shows the angle.

Claim.—An instrument constructed substantially as herein described, and which, when made of pocket size or any other size, serves for measuring vertical angles with great precision, by means of a spirit-level and graduation made along a spiral or zig-zag, grooved to receive the free ends of the case which carries the spirit-level bulb.

No. 56,014.—WILLIAM W. COVELL, Jr., New York, N. Y.—*Safety Watch Pocket.*—July 3, 1866.—Explained by the claim and cut.

Claim.—A watch protector formed of a case having a spring connected with the hinges by which it is held closed, and against the force of which it must be opened, and having loops upon the rear side by which it is to be attached to the interior side of the pocket, all substantially as described.

No. 56,015.—DANIEL M. CUMMINGS, Enfield, N. H.—*Heating Stove.*—Explained by the claims and cut.

Claim.—The combination and arrangement of air-flues or passages *c a d o n r*, together or singly, with the fire-box or chamber of a stove constructed for heating purposes, substantially in the manner and for the purpose herein set forth.

Also, the combination and arrangement of the direct air-flues *n* with the radiating flues *t f* of a heating stove A, substantially in the manner and for the purpose herein set forth.

No. 56,016.—THEODORE D. DAY, New York, N. Y.—*Hoop Skirt.*—July 3, 1866.—The bustle hoops are incomplete in front, the space between the ends narrowing downward. The ends are severally connected by a loose joint to a piece of spring wire which is also connected to the top hoop of the skirt proper.

Claim.—First, a spring forming the edges of the opening at the upper and front part of a skeleton skirt, for the purposes and as set forth.

Second, the edge-springs e, in combination with the bustle-springs b b, the ends of said bustle-springs being connected to the edge-springs by hinges or otherwise, as specified.

Third, attaching the ends of the edge-springs e to the upper body hoop or hoops a of the skeleton skirt, for the purposes and as set forth.

No. 56,017.—JULES DELERY, Parish of St. Bernard, La.—*Low Water Detector.*—July 3, 1866.—The gauge is opened automatically at given intervals by the connection with some running portion of the engine with the gauge cock, indicating to the engineer the state of the water in the boiler.

Claim.—The general arrangement of all the different parts composing the machine, and the combination of the device with the steam-boilers, in the manner and for the purpose described.

No. 56,018.—ANDREW DIETZ, New York, N. Y.—*Dust Brush and Broom.*—July 3, 1866; antedated June 22, 1866.—The leather strips have plumulose edges, and are attached to the handle by tapering pieces of rattan which form the quills.

Claim.—The use and application of leather in the construction of brushes and brooms, substantially as and for the purposes set forth.

No. 56,019.—CHARLES DIXON and S. H. CLOSE, Port Byron, N. Y.—*Gates.*—July 3, 1866.—The levers are arranged one on each side of the gate, are reached from the carriage, and operate gearing which opens or closes the gate.

Claim.—First, the spur wheel G, shaft E, and bevel gear-wheel H and K, constructed and arranged as described, in combination with each other, with the gate post D, and with the rear upright bar a' of the gate A, substantially as described and for the purpose set forth.

Second, the combination of the levers S and T, and band or chain R, with each other and with the spur wheel G, substantially as described and for the purpose set forth.

Third, the combination of the pin M, band L, and rod O, with each other and with the shaft E and bolt P, substantially as described and for the purpose set forth.

Fourth, the combination of the pin J with the shaft E and gear wheel H, substantially as described and for the purpose set forth.

No. 56,020.—GEORGE L. DULANEY, Mechanicsburg, Penn.—*Sewing Machine.*—July 3, 1866.—This machine is designed to make the single thread chain-stitch, the lock-stitch, and also a chain-stitch intertwined with a second thread. A shuttle is used and also a hooked looper to carry the needle thread around the shuttle. A removable guard bears off the loop from the path of the needle when the lock-stitch is to be made. The two feed-wheels form the table or support for the cloth; each wheel has an independent axis and one may be turned away from the other to get access to the shuttle and other devices. A succinct statement of details cannot be given.

Claim.—First, the construction of a looping hook formed of one individual piece of metal of the shape shown at f^1 g^2 g^3, and so connecting said hook in such a manner directly on to the extremity of a rotating axle as to dispense with all secondary joints, rods, elbows, or crank arms; and so also that while its pivoted end e^2 moves within a slot J^2, up and down with a reciprocating motion, the hooked or barbed end g^2 g^3 moves in and describes parts of an ellipse E E, and also diagonal lines R R, as shown in figure 11, the action of said looping hook being received direct from the end of a primary axle, the motions of both being around one common centre, as shown and described.

Second, the manner or mode of constructing, combining, and arranging the cloth feeding-wheel r^2 r^4 r^2 r^6, so that the hub or centre q^2 thereof revolves over and around a stationary sleeve p^2, within which sleeve rotates the driving axle a^2 a^4, figure 1, and through which construction the action of a primary driving wheel is communicated direct to said feeding wheel, and by which combination and arrangement the looping hooks f^2 g^2 g^3 receives direct motion, and whereby great simplicity, compactness, and durability of construction are attained, substantially as and for the purposes set forth and described.

Third, the manner or mode of hanging or attaching the cloth feeding wheel r^4 r^2 on to a jointed, hinged, or pivoted standard bracket, bearing, or an axle w w, so that said feeding wheel may be cast off or turned outwardly to one side out of position, or from beneath the material, in the manner as shown by the dotted lines in figure 1, as and for the purposes set forth and described.

Fourth, the construction and arrangement of the laterally working detached clutch devices d^3 d^3 e^3 e^3, with the yoke wires f^3 f^4 g^4 g^4, and the vibrating lever z^2 z^2 y^2 y^2, as set forth and for the purpose specified.

Fifth, the peculiar construction and arrangement of the thread tension wheel m m i^2 J^2, and the detachable lifting and intermittingly acting lever clutch n n k^3, figures 1 and 2, or its equivalent, for the purpose substantially as set forth and described.

Sixth, the peculiar construction of the stationary spool-case holder k k L^2 L^2, figures 2, 3 and 16, formed of two concave jaw-like pieces provided with (or without) the slot n^2, together with the latch or spool-case guard n^2 n^2, figures 1 2 3 and 16, or its equivalent, as set forth and described and for the purpose specified.

Seventh, the detachable shield n^5, or its equivalent, in combination with the spool-case holder, for the purpose as set forth and described.

Eighth, the construction of the enclosed spool-case, (or its equivalent,) figures 7 8 and 9, together with the encompassing wire-fillet, formed as shown in figures 7 and 10, for the purpose substantially as set forth and described.

Ninth, constructing a sewing machine needle with short abrupt grooves, one edge of one (or both) of the grooves being indented or cut away immediately at the side of the eye, as shown at i^5, figure 12, and by which form of construction any kink, knot, or inequality of thread is prevented from chafing, catching, or clogging between the needle and edge of the needle-puncture at the position where one thread overlays the other close to the eye of the needle, as herein described, as and for the purpose specified.

No. 56,021.—JAMES B. EADS, St. Louis, Mo.—*Operating Ordnance.*—July 3, 1866.—To maintain the centre of the muzzle at the same point, and secure the minimum port-hole, the trunnions of the gun are raised in arcs in the carriage cheeks, whose centre is the centre of the port-hole; the trunnions rest in a frame, supported between the cheeks upon an elevating screw.

Claim.—First, the arrangement and combination of the gun-trunnions a with the boxes a', levers $a^1_2{}^a$, the support c, screw-shaft d, nut e, and connections acting to permanently support the gun at d', and to raise and lower the same substantially as set forth.

Second, the arrangement and combination of the trunnions a, boxes a', slots b, levers $a^1_2{}^a$, pin $a^1_2{}'$, and guides f, as shown in Figs. 1, 2 and 3, or their general equivalent combination of the trunnions a, boxes a', and circular slots b, of Figs. 4 and 5, when used to produce a change in the axial direction of the gun, substantially as set forth.

Third, the combination and arrangement of the different parts mentioned in the foregoing first and second claims to produce a pendulous gun-motion about a horizontal axis lying in or near the face of the gun, as set forth.

No. 56,022.—THOMAS E. ELLETT, Monmouth, Ill.—*Cultivator.*—July 3, 1866.—The beam is attached to an axle mounted on two wheels, and drawn by a tongue; the position of the wheel is adjustable, so as to maintain the position of the plough on the surface of the side hill on which it traverses.

Claim.—First, the combination of wheel A, bar P, and axles G G, with the ends at right angles, for the purpose and substantially as described.

Second, the attachment of tongue with side movement, substantially as described.

Third, the bars H H for supporting axle-boxes F and clevis O, substantially as and for the purpose described.

Fourth, the rod 2, for the purpose described.

Fifth, the combination of the axle G, bars H H, wheel A and tongue, in a manner to produce side movement, for the purpose and substantially as described.

No. 56,023.—WILLIAM H. ELLIOT, New York, N. Y.—*Hay Loader.*—July 3, 1866.—The arm is pivoted to the revolving base, so as to adjust the point of suspension to the required height. The clasp attachment of the handle prevents the latter from turning when lifting.

Claim.—First, joining the arms c d e to revolving base b, by means of pivot bearings p, so that the arm may have a vertical movement, as described.

Second, the spring-clasp h, in combination with the fork handle g and cord m', substantially as and for the purpose set forth.

No. 56,024.—M. S. EVERY, Bridgewater, Mich.—*Sheep Feeding Rack.*—July 3, 1866.—The troughs are pivoted on end bearings, so as to be emptied by upsetting, and are protected by roof boards, which are hinged on the eaves, and may be vibrated in front of the rack to shut off access of the animals.

Claim.—The roof c c, hinged at the outer edges as described, so as to fall down and shut off access to the troughs, whilst the latter are being filled with the feed, substantially as specified.

Also, the manner of pivoting the troughs so that they can be emptied, substantially as specified.

No. 56,025.—B. FIGER, Cleveland, Ohio.—*Scrubbing Brush.*—July 3, 1866.—Explained by the claim and cut.

Claim.—The removable back or brush C, in combination with the head A and screw bolts, in the manner and for the purpose set forth.

No. 56,026.—ANTHONY L. FLEURY, Pittsburg, Penn.—*Manufacture of Steel.*—July 3, 1866.—Spongy wrought iron is immersed in a bath of cast iron, isolated from the atmosphere by soda, fluor-spar, &c., to which is added a little cyanide of potassium or chloride of ammonium. It is then converted into steel by hammering, rolling, &c.

Claim.—The processes herein described for the manufacture of bars, rails, or ingots, or steel-like iron, by treating wrought iron with melted cast iron, and subjecting the product to the varied treatment of squeezing, compressing, rolling, or hammering, substantially as set forth.

No 56,027.—ANTHONY L. FLEURY, Pittsburg, Penn.—*Desulphurizing Gold and Silver Ores.*—July 3, 1866.—The sulphurets or tailings are mixed with coal dust, and baked into a metalliferous coke, which may be heated again and treated with steam, after which the ore is to be amalgamated.

Claim.—First, to treat sulphurets containing gold or silver in the way and for the purpose specified.

Second, the compound obtained, denominated as metalliferous coke, when prepared as above specified.

No. 56,028.—H. L. FOLSOM, Upper Gilmanton, N. H.—*Tether.*—July 3, 1866.—The tether rope is fastened to the end of a pole, which is pivoted to a swivelled collar on the stake, so that the rope is kept taut, and freedom of range is permitted in every direction within the given radius.

Claim.—The construction and arrangement of the swivel ring G, and its swivel collar *a* and sleeve *b*, substantially as and for the purpose herein specified.

No 56,029.—ALBERT M. FORCE, Norwich, Conn.—*Steam Trap.*—July 3, 1866.—The perforated piston is moved longitudinally in its cylinder by the expansion or contraction of the pipe to which it is attached, closing or opening the exit aperture by which the accumulated water of condensation is discharged.

Claim.—First, the cylinder having escape-port G, piston B attached to a hollow stem or tube C, connected with an expanding tube fixed at one end, when combined and arranged together substantially in the manner described, so as to operate as and for the purpose specified.

Second, so hanging the cylinder to the bed-piece that it can be moved laterally thereon to adjust its escape-port to the piston, substantially as described.

Third, so constructing or forming the piston B that the steam within its chamber or cylinder will act upon both sides, and thus balance the same substantially as specified.

No. 56,030.—NELSON B. FORREST, Auburn, N. Y.—*Car Brake.*—July 3, 1866.—When the lever is dropped it falls upon the cam attached to the revolving shaft, and thus motion is given to a system of levers which bring the brakes in contact with the wheels.

Claim.—First, the cam C, placed on the axle of a railway car, for the purpose of actuating automatically a brake, substantially in the manner set forth.

Second, the device for automatically stopping and releasing the hook M' by means of a pulley H and levers I and L, constructed and arranged substantially as set forth.

No. 56,031.—GILBERT S. FOSTER, Sullivan, Me.—*Washing Machine.*—July 3, 1866.—The conical rollers are attached to a frame which is pivoted to a spring bearing on the edge of the washboard, and vibrates in arcs above it.

Claim.—Improved washing machine, as composed of the washboard, the sectional rubber, the centre bolt, and the spring, constructed, arranged, and combined substantially in the manner and so as to operate as specified.

No. 56,032.—CARLOS FRENCH, Seymour, Conn.—*Car Spring.*—July 3, 1866.—Explained by the claim and cut.

Claim.—A car or other spring made of a steel plate previously folded into two, three, four, or more folds, and then bent into form around a mandrel or over a former, substantially as and for the purpose described.

No. 56,033.—WILLIAM FROST, Newark, N. J.—*Tool for Cutting Twist Drills.*—July 3, 1866.—The cutting face of the milling tool has such a shape that it will be straight at that part which forms the cutting edge of the drill.

Claim.—A tool for cutting twist drills in which that surface or portion which cuts that side of the groove terminating in the straight cutting lip shall have a curvature, and the tool be otherwise constructed substantially as described and indicated in Fig. 3 of the drawings.

No. 56,034.—JOHN S. GETCHELL, Machias, Me.—*Ship's Windlass.*—July 3, 1866.—By the arrangement described the windlass may be worked with increased power or increased speed by the engagement of the respective sets of cams, nearer to or further from the centre of vibration of the lever.

Claim.—The combination of the two sets of single or double pawls L and M and bent lever stop N with each other, with the lever H, and with the ratchet wheels G of the windlass B, substantially as herein described and for the purposes set forth.

No. 55,035.—R. P. GILLETT, Sparta, Wis.—*Scrubbing Brush.*—July 3, 1866.—The layers of brushes are clamped between the alternate blocks and tightened by nuts on the central screw.

Claim.—The combination of the frame A, alternate layers of bristles *b b*, or their equivalents, and cross-block *c*, screw-bolts B, having nuts D and centre cross-bar C, when arranged together so as to operate in the manner and for the purpose described.

No. 56,036.—THEOPHILUS GILLMOR, St. Louis, Mo.—*Churn.*—July 3, 1866.—The cylinder is suspended on trunnions and oscillated by the reciprocating dasher shaft, which is connected to the revolving crank.

Claim.—First, the combination of the crank, connecting rod, and cylinder, substantially as described and for the purpose set forth.

Second, the combination and use of a strap joint on the connecting rod E with the crank D², when used to connect the churn cylinder C with the motive power, as set forth.

Third, the combination of the cylinder C with the head C′ and its scoop C², also with the inlet funnel c³, and the outlet faucet c⁴, substantially as described and set forth.

No. 56,037.—C. L. GODDARD, New York, N. Y.—*Wool-burring Machine*—July 3, 1866; antedated June 19, 1866.—The construction secures lightness, cheapness, avoids the springing action due to centrifugal force when running at high velocities, and prevents yielding to the forces applied or to the warping incident to atmospheric changes.

Claim.—The manner of constructing inner or skeleton cylinders for burring machines of thin bars or strips of metal inserted in slots in the heads on the shaft, substantially as and for the purpose described.

No. 56,038.—C. L. GODDARD, New York, N. Y.—*Cylinder for Wool-burring and Carding Machines.*—July 3, 1866; antedated June 19, 1866.—By the construction of the cylinder it is braced against the forces tending to spring or warp it.

Claim.—The manner of constructing cylinders for burring and carding machines of a series of wooden lags secured to two heads on the shaft, and with radial bars inserted in grooves or gains cut into the periphery, substantially as and for the purpose described.

No. 56,039.—JOHN H. GRAVES, Rochester, N. Y.—*Gate.*—July 3, 1866.—The gate has an extension guide running upon rollers, and having notches which rest upon the rollers in the open and closed positions. A weighted cord or spring assists in closing.

Claim.—The employment in connection with a sliding gate of the extension guide c, running on rollers d f, and provided with sockets or depressions i k l, for fitting over the said rollers, and closed by the reaction of a weight or spring, substantially as described.

Also, the combination with the above, forming the lower edge of the guide c and the periphery of the roller d, with the tongue and groove g h, substantially as and for the purpose specified.

Also, the combination of the catch q and handles r r with the projection o of the extension guide, when so arranged as to hold the gate from being raised and thrown back when closed, substantially as specified.

Also, the special arrangement of the wire w with the sliding weight v, cord t, and pulleys u f, for producing the reaction of the gate, as set forth.

No. 56,040.—WILLIAM GREEN, Cleveland, Ohio—*Roofing Cement.*—July 3, 1866.—Composed of pulverized iron ore, separate or combined with pulverized slate, coal, coke, &c., and mixed with petroleum or other oil to the proper consistence.

Claim.—A cement or paint for roofing, or other purposes for which it may be adapted, composed of pulverized iron ore in combination with pulverized slate, stone, coal, coke, or either of them, the said substances being mixed with mineral or other oils, as herein specified, applied, and used, as set forth.

No. 56,041.—WILLIAM A. GREENE, Troy, N. Y.—*Range and Furnace combined*—July 3, 1866.—A range and furnace so combined that either or both can be used. A shield is placed in the doorway of the furnace to render it inoperative; direct draught can be had, or a circuitous route around the ovens.

Claim.—First, in combination with the oven flues P′ P′, and combustion chamber A, the direct communication flue N, arranged in manner substantially and for the purposes as herein set forth.

Second, the removable partition plate or wall D, or its equivalent device, arranged in manner substantially, and for the purpose as herein specified.

Third, in combination with the removable partition plate D, or its equivalent device, and the flue C, the arrangement of the boiler-hole plate J, and the slicing doors L L, for the purpose of forming additional fire-chamber space, in manner substantially and for the purposes as herein set forth.

Fourth, in combination with the oven flues P′ P′, and direct flue N, the arrangement of the dampers b and c, to operate with reference to each other, in manner and for the purpose as herein shown.

No. 56,042.—ANSEL HAINES, Pekin, Ill., and JOHN KIRKMAN, Kickapoo, Ill.—*Construction of Metal Wheels.*—July 3, 1866.—The felloes of angle iron are surrounded by a tire, and attached by rivets to the angle-iron spokes, around whose inner ends the metal hub is cast. Thimbles at the ends of the box form bearings for the axle spindle.

Claim.—First, the combination of a wrought-metal felloe ring or rings, formed of angle iron or angle steel of the shape substantially as shown, with wrought-metal spokes formed of angle iron or angle steel, of the shape substantially as shown, the said felloe ring or rings being encircled by a tire, and the said spokes being fastened by their outer ends to the flanch of the felloe ring or rings, and their inner ends cast into the metal hub, all substantially in the manner and for the purpose herein described.

Second, the manner substantially as herein described of constructing and applying removable thimbles a a to the cast-metal hub of the wheel, for the purpose set forth.

No. 56,043.—DAVID HAMMOND, Canton, Ohio.—*Bridge.*—July 3, 1866.—The arch is formed of double T-irons, bolted to intervening clamp-irons, and sustaining by rods the longitudinal girder of the roadway. A plate above the arch excludes rain, and braces laterally.

Claim.—First, the peculiar combinations of the double T-irons b b, and clamping pieces D, or P, with bolts M M, and hole N, substantially in the manner and for the purpose herein set forth.

Second, the peculiar combination of the covering piece H, the double T-irons b b, the securing pieces J J, with bolt s, and nut K thereon, substantially in the manner and for the purpose herein set forth.

No. 56,044.—JOSEPH B. HARRIS, Germantown, Ky.—*Composition of Matter for Rendering Paint Fire-proof.*—July 3, 1866.—The material is powdered and freed from palpable grit before mixing with the paint.

Claim.—The combination of calcined schist, shale, or mineral coal, prepared as above described, with any kind of oil paint, to render the same uninflammable or fire-proof, substantially as above set forth.

No. 56,045.—R. FRANCIS HATFIELD, New York, N. Y.—*Steam Engine, Valve, &c.*—July 3, 1866.—The rotary valve supplies and exhausts two parallel double-acting steam cylinders, so as to alternate their supply and exhaust, and is also arranged to operate four double-acting cylinders.

Claim.—First, a rotary valve arranged to supply alternately the ports of two or four double-acting cylinders as required when the shaft cranks are placed at right angles to one another, substantially as shown and described.

Second, the rotary valve I, Fig. 10, constructed with two or more wings as shown at Figs. 10, 14, and 21, in connection with the arrangements of the ports 1, 2, 3, and 4, Figs. 9, 13, and 19, substantially as described, and for the purpose specified.

Third, the crank-pin brasses t and q, Fig 24, in combination with the screw bolts p p, and the screw channel in the piece q, substantially as described, and for the purpose set forth.

Fourth, arranging the upper part r r, Fig. 24, of the crank cross-head to serve as an oil cup, in combination with the supply wicks u u u, substantially as shown.

Fifth.—The combination of the above several improvements, as shown by the drawings.

No. 56,046.—HIRAM A. HAWKINS, Virden, Ill.—*Weather Strip.*—July 3, 1866.—The hinged edge of the closing door strikes the end of the lever arm, and raises the strip from the threshold into contact with the door.

Claim.—The strip a, in combination with the sliding lever c, rollers e and o, and spring m, and the incline n, when said parts are constructed and arranged to operate substantially as and for the purpose set forth.

No. 56,047.—ALEXANDER HERDLEIN, Egan Cañon, Nevada.—*Quartz Stamp Mill.*—July 3, 1866.—The stamps are raised consecutively by the tripping levers which are vibrated by toggle connections with cranks on a revolving shaft.

Claim.—First, the double-armed levers F, in combination with the stampers E, and cranks or other equivalent devices on the driving shaft C, constructed and operating substantially as and for the purpose described.

Second, the hinged tappets i, in combination with the adjustable heads G, levers F, and stampers E, constructed and operating substantially as and for the purpose set forth.

No. 56,048.—WILLIAM F. HEYWOOD, Cumberland. R. I.—*Machine for Folding Cloth.*—July 3, 1866.—The swinging folding frame delivers the edge of each fold to a set of spring nippers, which hold it until about to seize another fold. The motion of the frame opens the nippers. Springs bear the cloth-supporting table constantly upward to insure a smooth fold, yet allow it to yield as the number of folds increases. The machine is stopped when the cloth is all folded, by its ceasing to support the fingers, which then drop and actuate the shipping device. Provision is also made for registering and indicating the number of folds.

Claim.—First, the spring nipping jaws E E, in combination with the pendulum frame C³, and operating together in the manner substantially as described.

Second, an elastic table M, in combination with the pendulum frame C³, substantially as described.

Third, combining with the place of cloth to be folded the belt-shipping apparatus herein cribed, operating as and for the purposes set forth.

No. 56,049.—WILLIAM HINDHAUGH, New York, N. Y.—*Safety Pocket.*—July 3, 1866.—The spring frame which forms the mouth of the pocket is engaged by a sliding bolt and thus fastened.

Claim.—The bolt D and spring frame B in combination with the mouth of a pocket, constructed and operating substantially as and for the purpose described.

No 56,050.—EDWIN HOYT, Stamford, Conn.—*Churn.*—July 3, 1866.—Explained by the claim and cut.

Claim.—A churn in which all the parts described and represented are arranged in the manner set forth.

No. 56,051.—JOHN S. HULL, Cincinnati, Ohio.—*Soldering Lamp and Blow-pipe.*—July 3, 1866.—Air is compressed in the reservoir by the external pump, and expels the oil through a valved passage to the generator, whence the gas issues to the burner, a part of it escaping at an orifice to beat the retort.

Claim.—The arrangement of the condensing pump on the outside of the reservoir, substantially as and for the purpose herein specified.

Also, the employment of a cut-off valve c between the pump and the reservoir, for the purpose set forth.

Also, the employment of a regulating and cut-off valve E between the reservoir and the gas generator, as described.

Also, a regulating valve I between the gas generator G and the jet orifice L, as herein set forth.

Also, the employment of a gas generating burner, supplied by gas produced by the gas generator itself, together with the main supply, substantially as and for the purpose herein specified.

No. 56,052.—JOHN S. HULL, Cincinnati, Ohio.—*Gas Heater for Cooking, &c.*—July 3, 1866.—Explained by the claims and cut.

Claim.—The employment of atmospheric pressure to force the liquid to the burner, in combination with a self-generating gas-burner for cooking and heating purposes, so as to produce a blowing jet of flame at any distance from the reservoir, substantially as herein specified.

Also, the gas-generating burner retort M, constructed and operating substantially as described in combination with the condensing pump and reservoir for forcing the oil to the retort, for the purpose specified.

Also, the air-valve H having a cork cushion, covered with buckskin, or its equivalent, when applied to the condensing pump of the reservoir of a cooking apparatus supplied by the force of atmospheric pressure, substantially as described.

Also, surrounding the pump with the case or tube C, arranged in combination with the condensing pump and reservoir of a cooking apparatus supplied by the force of atmospheric pressure, substantially as set forth.

Also, the packing of asbestos around the stem of the jet regulator, for the purpose herein set forth.

No. 56,053.—D. HUMPHREYS, Oskaloosa, Iowa.—*Hand Corn Planter.*—July 3, 1866.—The motion of the plunger operates the compound slide, which is slotted in its vertical sides and perforated below in correspondence to the feed opening in the floor.

Claim.—The arrangement in a hand corn planter of the compound slide H I, having the described slotted wings J j, and K k, openings b' and i i', and rod h' in combination with the stud f on the plunger F on the openings b b' in the floor B, as set forth.

No. 56,054.—J. L. HUSBAND, Philadelphia, Penn.—*Fibrous Packing for Steam Engines.*—July 3, 1866; antedated June 19, 1866.—Strands of fibre with a mixture of elm bark are loosely braided together and covered with a net-work of twine.

Claim.—The manufacture of packing for steam engines, as hereinbefore described and more specifically set forth.

No. 56,055.—ANTHONY ISKE, Lancaster, Penn.—*Cigar Press.*—July 3, 1866.—The forms between which the cigars are pressed have flexible backs and the ribs are placed in alternate position, each occupying the space between two ribs on the opposite backing. When pressed the forms are clamped and removed.

Claim.—The arrangement of the combined cases A B, constructed by alternate slats o z, and spaces meshing one into the other, substantially in the manner and for the purpose specified.

Also, the use of clamping irons I, or their equivalent, for securing the pressure when removed from the press until the cigars have dried, substantially in the manner specified.

No. 56,056.—ALBERT JACKSON, Clifton Springs, N. Y.—*Milk Rack.*—July 3, 1866.—On the radial arms of the vertical post are rings in which the milk pans are set. The plate is an obstacle to the ascent of mice.

Claim.—First, the combination of the ring guards F with the supporting arms D, and wire E, substantially as described and for the purpose set forth.

Second, the combination of the tin plate C, or its equivalent, with the standard B, substantially as described and for the purpose set forth.

No. 56,057.—ABIEZER JAMESON, Trenton, N. J.—*Vice.*—July 3, 1866.—Concave-faced washers and convex collars on the sockets and screw, respectively, of a vice accommodate the bearing surfaces to the change in the direction of the strain at different positions of the movable jaw.

Claim.—First, the screw E, with its convex collar or washer *j*, adapted to the concave sliding washer K, all substantially as and for the purpose described.

Second, the combination of the above with the washers G and H intervening between the collar *f* of the socket F, and the fixed jaw A of the vice.

No. 56,058.—RUSSEL JENNINGS, Deep River, Conn.—*Tool for Finishing Augers.*—July 3, 1866.—Explained by the claim and cut.

Claim.—A rotary wheel or burr having a bevelled surface *a* at one side, and a semicircular edge or periphery corrugated to form a series of cutters *b*, which have a radial or nearly radial position and extend from the inner edge of *a*, to the outer edge of the same and entirely around the semicircular periphery of the wheel, in combination with the concave surface *c* at the opposite side of the wheel, substantially as and for the purpose herein set forth.

No. 56,059.—FRANK G. JOHNSON, Brooklyn, N. Y.—*Drill Chuck.*—July 3, 1866.—The two jaws are secured to springs, and the latter to the shank ; the end of the drill is socketed centrally in the hollow shank and the shaft of the drill clamped by the taper jaws, which are closed by set-screws.

Claim.—First, the double spring jaws A B, made with the tapering hole R P between them, in the manner and for the purpose substantially as described.

Second, the combination of the said jaws as described, with the shank C and its tapering bottomed socket L, by means of which different sized drills are at once both centred and held at the back end and about midway of their length by simply screwing up the nuts F F, essentially in the manner and for the purpose set forth.

No. 56,060.—ROBERT V. JONES, Canton, O.—*Railroad Switch.*—July 3, 1866.—The motion of the switch lever is against the force of a spring whose reaction returns it and the switch to their normal position when the holding force is withdrawn.

Claim.—First, the lever B, bar E, arms *a a*, crooked uprights D D, treadle M, and switch-rod C, arranged and used substantially as herein specified.

Second, the pitman H and spring K, working in the case F, the same being locked by the spring *f*, with plate *d* and key *g*. said pitman being connected to the lever B. the whole being constructed, arranged, and operating in the manner and for the purpose set forth.

No. 56,061.—WHITCOMB JUDSON, Galesburg, Ill.—*Device for Preventing Hogs from Rooting.*—July 3, 1866.—The plate is hooked behind the cartilage of the nose and presents a sharp overhanging point to the snout.

Claim.—The construction and application of the device on the top of the snout of a hog to prevent rooting, substantially in the manner as described.

No. 56,062.—EDWARD KAYLOR, Pittsburg, Penn.—*Machine for Making Nuts.*—July 3, 1866.—The bar is punctured by a taper punch, a piece severed and forced over a round punch and squeezed between two side dies which enclose it on all sides and give it its conformation. It is then swaged by a drop hammer which make a washer seat on the top, and the dies opening, it is discharged.

Claim —First, the combination of the side dies F F', so constructed as to enclose the nut on all sides, but not at the ends ; the square punch *d*, stationary eye punch *k*, and swage *f* fitting the cavity formed by the side dies, constructed and operating substantially as described, for the purpose of making nuts by cutting off a square blank, forcing it over a perforating punch, moulding it into shape, and finishing it with the blow of a hammer.

Second, finishing the nut after it has been perforated and moulded laterally into shape, and while still confined in the side dies by means of the swage and hammer, substantially as hereinbefore described.

No. 56,063.—J. M. KEEP and S. R. DUMMER, New York, N. Y.—*Sprinkler and Dredger.*—July 3, 1866.—The barbed valve stem forces the contained material through a hole in the bottom.

Claim.—First, the valve-stem D, when provided with teeth, barbs, flanges, or their equivalent, as and for the purposes herein described.

Second, a combination of the valve stem D when barbed or flanged, or provided with their equivalents, cup H, spring G, cover B, operating substantially as and for the purposes herein described.

No. 56,064.—JOHN S. KINYAN, Webster City, Iowa.—*Sofa Bedstead.*—July 3, 1866.—
A portion of the hinged bottom is vibrated upward and the hinged portions of the ends are closed in against it.

Claim.—First, the hinged bottom A A', in combination with the hinged head and footboards B B' C C', all arranged to operate substantially as herein specified.

Second, in combination with the above the catches E E, and recesses F F, substantially as described.

No. 56,065.—PETER A. LAFRANCE, Elmira, N. Y.—*Horseshoe.*—July 3, 1866.—A receiving plate is attached to the boof and a wearing plate is attached thereto and renewed as required.

Claim.—The slotted shoe B, when constructed with recesses in the upper face and elongated openings D, in combination with the receiving plate A, when constructed with a flange C, on its inner edge.

No. 56,066.—Cancelled.

No. 56,067.—JOHN H. LIGHTNER, Shirleysburg, Penn.—*Broom Head.*—July 3, 1866.—
The handle is pinned into the cap; the latter has a hinged door for the insertion of the corn, and toothed bars to penetrate and hold it securely.

Claim.—First, the toothed bars I constructed as described, in combination with the side-plate B, and door or flap H, of the broom head, substantially as described and for the purpose set forth.

Second, the combination of the band K, constructed and arranged as described, with the side-plate B, and door or flap K, substantially as described and for the purpose set forth.

Third, hinging the door or flap H to the solid part of the side-plate A, at the upper edge of the notch formed in said side-plate, substantially as described and for the purpose set forth.

Fourth, securing the handle E to the cap A B C, by the wire G, passing through the notch e', formed in the side of the said handle E, substantially as described and for the purpose set forth.

No. 56,068.—GEORGE H. LINCOLN, Providence, R. I.—*Toilet Soap.*—July 3, 1866.—
Designs in colors are formed in the body of the soap to form a trade-mark as lasting as the soap.

Claim.—The improvement in soap described, consisting of a bar or cake of soap made up of soaps of different colors, and arranged so as to exhibit an ineffaceable ornamental design or trade-mark in one or more contrasting colors, the article being substantially as specified.

No. 56,069.—JOHN B. LOGAN, Thorntown, Ind.—*Lock for Mail Bags and Carpet Sacks.*—
July 3, 1866.—The catches on one jaw are engaged simultaneously by the pins on the sliding bolts, which are operated by a bar; the bar is locked by the engagement of its plate and two plates on the frame, by the hasp of a padlock.

Claim.—The combination of the slide-bolts B C D, and stops e e, slotted sliding bar E of the jaw A, operating with the catches f g h of the jaw A', constructed and arranged in the manner and for the purpose herein specified.

Also, the combination of the plate G, and sliding bar E, plates F F, and jaws A A', the plates having suitable holes i to receive the shackle of a padlock, and preventing the withdrawal of the bolts unless the padlock is removed, arranged in the manner herein represented and described.

No. 56,070.—JOSHUA LOWE, New York, N. Y.—*Steam Gauge.*—July 3, 1866.—The stem on the apex of the inverted cup moves a pinion and index finger; the cup as it rises becomes loaded with additional weights of increasing size.

Claim.—First, the inverted cup C, working in the annular chamber D, in combination with a series of weights of gradually increasing size, and with a suitable index, all constructed and operating substantially as and for the purpose described.

Second, arranging the walls of the annular chamber D as and for the purpose set forth.

No. 56,071.—SAMUEL MACFERRAN, Philadelphia, Penn.—*Door Key Fastener.*—July 3, 1866.—A bent wire is suspended by a staple, its limbs passed on each side of the knob spindle and its ends through the bow of the key to prevent the revolution of the latter.

Claim.—First, constructing the fastener E with the hooks e e, substantially in the manner hereinbefore described and for the purpose specified.

Second, the combination of the fastener E with the slide F by means of the eye a of the former, substantially as described and for the purpose specified.

No. 56,072.—LUCINDA MARMADUKE, Shelbyville, Mo.—*Medical Compound.*—July 3, 1866.—For the cure of pulmonary and cognate disorders; composed of ground ivy, cohosh, hoarhound, polypody, valerian, Iceland moss, licorice root, spikenard, balm of Gilead, elecampane, Indian turnip, liverwort, comfrey, striped elder, water ash, blood root; boil, sweeten, add tincture of lobelia and spirits to preserve it from fermentation.

Claim.—The compounding of said herbs or vegetables as above described.

No. 56,073.—CHARLES S. MARTIN, Milwaukee, Wis.—*Wagon Spring.*—July 3, 1866.—Explained by the claim and cut

Claim.—Constructing the springs of vehicles of solid blocks of India-rubber in the form of the frustum of a cone or pyramid, or having only an opening sufficient for the passage of a bolt through them, substantially as set forth.

No. 56,074.—C. S. MARTIN, Milwaukee, Wis.—*Wagon Spring.*—July 3, 1866.—Springs are placed between the axle and bolster, and also under a bar resting on the bolster and supporting the bed.

Claim.—First, springs M, enclosed within cylinder K, substantially as and for the purpose described.

Second, springs O, in combination with bolster C, bar D, bolts I, and cups P, substantially as described.

No. 56,075.—JOHN L. MASON, Jefferson, Ky.—*Medicine for Hog Cholera.*—July 3, 1866.—Compounded of arsenic. 6; sulphate of copper, 60; pulverized poke root, 6 parts.

Claim.—A combination of the components above named in the proportion above set forth, (or of any other components when combined resulting in a compound with similar properties,) which components so combined or compounded is a preventive and cure of the disease popularly known as hog cholera.

No. 56,076.—IVES W. McGAFFY, Chicago, Ill.—*Seed Planter.*—July 3, 1866.—Explained by the claims and cut.

Claim.—First, the seed-distributing cylinder, having holes or cups in its periphery, with screws fitted into the cups for adjusting their capacity, and set-screw at the sides to hold them in place when adjusted in connection with the shifting slide N, constructed and operated in the manner and for the purpose substantially as specified.

Second, the adjustable cog or tooth-gears *g* and *h*, for regulating and controlling the seeding device while the machine is moving over the field, constructed and operated substantially as and for the purpose specified.

Third, in combination with the adjustable toothed or cog-gears *g* and *h*, and seeding device, a revolving pointer or marker P, for marking or indicating the position of the hills in automatic check-row planting.

Fourth, constructing a brush for seed-planting machines with cap, bolt, and nut for holding the bristles, in the manner specified or its equivalent.

Fifth, the detachable spur or sod-cutter *r*, fitted to the heel of the runner, in the manner and for the purpose specified.

Sixth, the oscillating plate W, provided with the sockets for the lever *y*, and having the arm W', provided with the segmental rack, arranged to operate in combination with the pinion X, on shaft K, substantially as and for the purpose set forth.

No. 56,077.—W. W. McGREGOR, Dedham. Mass.—*Harness Motion of Looms.*—July 3, 1866.—The harness is operated with great positiveness. Provision is made for varying the degree of elevation of the heddle frames, and consequently the opening of the "shed," and also for adjusting the "shed" to prevent the warps being chafed by the shuttle. The claim enumerates the devices.

Claim.—The combination or mechanism, substantially as described, for operating the harness carriers, the same consisting of two sets of levers *a b*, the connecting rods *g*, the vibratory arms *f*, and the lifter D, and its rotary cranked wheel S, with its crank pin *r*.

Also, the combination of the rest-frame E, or its equivalent, with the mechanism or combination, substantially as described, for operating the harness carriers.

Also, the combination of the stop-bar *y* and the series of studs or catches *z* with the rest-frame E, the harness carriers, and their operative mechanism, substantially as described.

Also, the combination of the slot *u*. and adjustable crank pin *r*, and the adjustable rest-frame E, with the harness carriers and the mechanism for operating them, substantially as described.

Also, the application of the slotted curved arm *l* to the remainder of the lifter D, so as to be adjustable with respect to the same, substantially as and for the purpose specified.

No. 56,078.—CHARLES R. MILKS, Waterford, N. Y.—*Composition for Coating Ships.*—July, 3, 1866.—To protect hulls of vessels from the adhesion of barnacles and the attacks of the teredo; composed of coal tar, benzine, linseed oil, asphaltum, shellac, soapstone, plaster of Paris, India-rubber, and potash.

Claim.—A composition of matter compounded from the hereinbefore named ingredients or their chemical equivalents, substantially in the manner and for the purposes set forth.

No. 56,079.—T. W. MIRICK, Boston, Mass.—*Bottle Stopper.*—July 3, 1866.—The wires pass through the rubber stopper, and are secured by hinge and catch wires proceeding from a band below the lip of the neck.

Claim.—In combination with the frusto-conical stopple *a*, wire *c*, hinge *d*, and latch *e*, the wire *c*, when bent into two parts, each passing through the stopple, and thus serving both to hold the stopple down and to prevent any sidewise tipping movement.

No 56,030.—R. E. MONAGHAN, West Chester, Penn.—*Sash Fastening.*—July 3, 1866.—
The rollers are attached to spring pieces rooted to the casing, and are pressed against the corrugated edges of the sash to sustain it.

Claim.—The application of alternating inclined planes attached to both sides of window sash, doors, gates, and other perpendicular slides, and the use of wheels, or rollers, and springs operating on both sides of the sash, doors, gates, and other perpendicular slides, and acting as braces to support the sash, doors, gates, and other perpendicular slides, for the purposes above set forth and described.

No. 56,031.—CHARLES MOORE, Sr., Trenton, N. J.—*Woven Bag.*—July 3, 1866.—Explained by the claims and cut.

Claim.—First, the invention and improvement in bags or sacks, increasing the filling or weft of the cloth gradually from the place of tying near the top to the bottom, to make the bag thickest, heaviest and strongest where it is subject to the greatest strain and the most wear, substantially as described.

Second, in combination with a bag woven gradually thicker from the place of tying to the bottom, the thick welt or edge around the mouth or open end of the bag, formed substantially as described for the purpose set forth.

No. 56 082.—HERBERT A. MORSE, Canton, Mass.—*Clothes Sprinkler.*—July 3, 1866.—
The flaring flange around the perforated end of the sprinkler prevents the trickling of the water down the exterior surface.

Claim.—In a clothes sprinkler constructed as described and provided with the opening D, the combination and arrangement of the flange C, substantially as described and for the purposes set forth.

No. 56,033.—JAMES M. MUHLIG, Cambridge, Mass.—*Sifter and Strainer.*—July 3, 1836.—
The concave is made of shiftable sieves of varying fineness; the material is stirred by curved wings attached to disks on the revolving shaft.

Claim.—The combination and arrangement of the curved metallic beaters, and the changeable sliding sieves and strainers, and the devices for securing them in position, as described for the purpose specified.

No. 56,034.—WILLIAM R. NICHOLS, Philadelphia, Penn.—*Car Spring.*—July 3, 1866.—
The cap slides on guide rods and bears upon the bent plates, which are joined at their ends and enclosed in the case.

Claim.—First, a car spring constructed of a series of elliptical plates of steel, each of which is complete in itself and of such size that, being arranged concentrically, they shall fit snugly upon one another, substantially in the manner set forth.

Second, in combination with the above described form of springs, the case D, cap E, and rods C, when constructed and arranged for use, substantially in the manner and for the purpose set forth.

No. 56,085.—JAMES A. NIMAN, Mansfield, Ohio.—*Car Coupling.*—July 3, 1866.—Each draw-head has its own link permanently attached by a bolt and capable of being held horizontal to enter the other draw-head and raise the swinging coupling pin, or of being laid back in the chamber of the draw-bar.

Claim.—The combination of a pair of draw-bars, each constructed with a swinging pin A, an oscillating bar B, a link bed Q Q. a link holder D D, and bolt C, substantially as described, and each provided with a coupling link, as and for the purpose described.

No. 56,086.—JOSEPH W. NORCROSS, Middletown, Conn.—*Row Lock.*—July 3, 1866.—
The row lock is detachable and is fastened to its bed plate on the gunwale by lugs and catches ; the inside of the hollow metallic thole pin is cut away and exposes the wooden core to the contact of the oar in rowing.

Claim.—First, the bed plate D, with cylindrical sockets or studs E, in combination with the plate C and bed plate A, constructed and operating substantially as and for the purpose described.

Second, the wooden thole pins F, in combination with the segmental metal sockets E, constructed and operating substantially as and for the purpose described.

Third, the latch A and lugs c d, in combination with the plates C D and studs or sockets E, constructed and operating substantially as and for the purpose described.

No. 56,087.—JOSEPH W. NORCROSS.—Middletown, Conn.—*Tackle-block Sheave.*—July 3, 1866.—The metallic rim is connected by spokes to the central hub and the annular wooden checks are sustained in position by flanges on the inner edge of the rim which overlap the periphery of the cheeks.

Claim.—A sheave composed of a metal pulley B and wooden cheeks, as a new article of manufacture when composed of the shoulders d or the rim a, to support the cheeks C and the projections or spurs e on the rim, to retain the cheeks in position, substantially as described.

No. 56,088.—OSCAR D. PADRICK, Shelbyville, Ind.—*Harrow*—July 3, 1866.—The rear portion of the frame is hinged to the forward part, and is extensible laterally to adjust the breadth of tilth.

Claim.—The application of hinged harrow carrying-arms C C to a rigid triangular harrow frame A, in combination with the extensible braces D D, which are provided with means for fastening them together, substantially as described.

No. 56,089.—REES PALMER, West Chester, Penn.—*Sheet Metal Water Wheel.*—July 3, 1866.—Each bucket is shaped of a single piece of metal, and they are fastened consecutively on the rim, which is continuous and forms the inner side of each bucket.   ●

Claim.—In combination with the sheet-metal band A, supported by the arms D, as described, the sheet-metal buckets B when the said buckets are constructed and applied to the said band, in the manner and for the purpose described.

No. 56,090.—JAMES PATTERSON, New York, N. Y.—*Tuyere.*—July 3, 1866.—The hollow cylinder receives the blast from the bellows, and the head has a flaring flange to hold the coal, and a tuyere which is perforated for the passage of the air, and may be dropped to clear it of cinders.

Claim.—The combination of the rod G, with its catch c, the rods E F, and perforated conical valve D, arranged relatively with the hollow base A, and the fire bed B, with the conical flange a, constructed and operating in the manner and for the purpose herein specified.

No. 56,091.—PHILANDER PERRY, Charlestown, Mass.—*Dough Kneader, Meat Pounder, &c.*—July 3, 1866 —The tools are attached as required to the dovetail groove in the under side of the lever, and operate upon the material placed on the bed plate.

Claim.—First, the spring K, in combination with the jointed lever pillar and platform, when constructed and used substantially as described.

Second, the dovetailed groove for the purpose of holding the tools in connection with the movable lever and platform, all constructed and used substantially as described.

Third, the whole machine, being the combination of the above tools, viz: the beef-tenderer, potato masher, meat chopper, pastry roller, and dough kneader, or either of them, with the double-jointed lever, pillar, and platform, when constructed and used substantially as described.

No. 56,092.—C N. POND, Oberlin, Ohio.—*Lamp Chimney, Bottle, or Can Cleaner.*—July 3, 1866.—The elastic cleaner is pivoted in the forked end of the handle, and adjusts itself to the surface against which it is rubbed.

Claim.—The adjustable rubber A, and handle B, when combined substantially as and for the purpose described.

No. 56,093.—CALVIN REED, Springfield, Ohio.—*Mowing Machine.*—July 3, 1866.—The arrangement of the cam wheel and rock shaft for operating the cutters, and of the levers for adjusting the cutting apparatus to pass obstructions, are explained by the claim and cut.

Claim.—First, the cam wheel F, in combination with the rock shaft A, for operating the sickle bar D, when arranged as shown and described.

Second, connecting the sliding lever p to the shoe k, by means of the bolt i, and spiral rod t, when said parts are arranged to operate as herein described.

Third, the combination of the levers z y and p with the pivoted or hinged shoe K, arranged and operating as and for the purpose set forth.

No. 56,094.—CHARLES B. REES and J. B. TEVIS, Philadelphia, Penn.—*Refrigerator.*—July 3, 1866.—The ice drips upon an inclined floor, is received in a pan, and collected in vertical and horizontal pipes in the chamber. Ice-carrying blocks are adjustable in the vertical pipes.

Claim.—First, retaining the drippings from the ice in vertical and horizontal pipes, for the purpose specified and described.

Second, an adjusting carrying-block for adjusting the height of the ice in the vertical pipes, substantially as shown.

Third, the vertical pipes F. and the horizontal pipe E, combined and arranged and situated either on one or both sides of the refrigerator, as specified and described.

Fourth, the vertical pipes F, and the horizontal pipe E, and the drip pan D, combined and arranged and placed either on one or both sides of the refrigerator, substantially as shown.

No. 56,095.—DAVID REESE, Newburg, Ohio.—*Machinery for Tapping Nuts.*—July 3, 1866.—The frame is suspended upon a vertical shaft whose driving wheel actuates the planetary pinions on the vertical spindles around it. The frame is revolved intermittingly. As each mandrel comes to the operator he puts in a tap and nut, the former dropping through the latter into a trough when it has performed its work.

Claim.—First, the standards b, sheeves d, and weights k, in combination with the mandrel-

J, pinion r, gear H, and braces f, arranged and operating in the manner and for the purpose set forth.

Second, the arrangement of the trough C, revolving arms E F, and shaft D, in combination with the gearings D' and e', and weighted stop L, operating in the manner and for the purpose set forth.

Third, the stop L in combination with the revolving frame K, as and for the purpose set forth.

No. 56.096.—G. A. RIEDEL, Philadelphia, Penn.—*Automatic Boiler Feeder.*—July 3, 1866.—The vibrating reservoirs connect by tubular arms and oscillating valves with the pipes leading to the boiler, the traversing weight giving a determinate motion when the centre of oscillation is passed. The reservoirs are alternately supplied from a tank, and discharge their contents into the boiler.

Claim.—First, the two reservoirs F and F', in combination with a chest A, valves B and B', and ports, pipes, and passages, substantially as described, the whole being applied to a steam boiler, and operating substantially as set forth.

Second, in combination with the two vibrating reservoirs, the weight H, so arranged and operating as to prevent the tendency of one reservoir to balance the other.

Third, the arrangement substantially as described of the tank Q and the injector, for the purpose specified.

No. 56,097.—J. F. RIGGS, St. Joseph, Mo.—*Evaporator.*—July 3, 1866.—The ledges extending from opposite sides make a continuous sinuous track for the liquid. The length of the pan may be adjusted by adding or removing sections attached by side rods

Claim.—First, a cast-iron transverse current evaporator, when made in sections, so as to be able to increase the size of the pan by the insertion of one or more sections, substantially as and for the purpose described.

Second, suspending a skimming apparatus over the pan by an elastic contrivance, for the purposes described.

Third, the sections A A A A, the connecting rods B B, the lugs C C C C, substantially as and for the purposes described.

Fourth, the lugs C C C C, the connecting bolts b b, the side bolts h h, substantially as and for the purposes described.

Fifth, a scum trough inclining from the ends towards the centre with an opening covered with any suitable strainer, substantially as and for the purposes described.

Sixth, the sockets O O, the elastic springs T T, the chains t t, the handles s s, the grooves U U, the sliding gates V V, and the hooks X X, substantially as and for the purposes described.

No. 56,098.—GEORGE M. ROBINSON, New Wilmington, Penn.—*Water Box for Tuyeres.*—July 3, 1866.—One end of the open water box forms the fire back, and the other end projects beyond the back of the chimney. The tuyere passes through the water bath.

Claim.—The water box h, with tuyere pipe g passing through it, when constructed and arranged as described, the box being open on top and projecting outside of the back wall or chimney, for the purpose hereinbefore set forth,

No. 56,099.—WILLIAM W. ROBINSON, Ripon, Wis.—*Cutlery.*—July 3, 1866 —A short tang projects into the handle from the heel of the blade, and tangs pass from each bolster down grooves in the sides of the handle and are secured by rivets.

Claim.—Constructing and attaching the blades and handles of knives and forks in such manner that the handle, composed of any suitable material, shall be secured between the tongues C, and to the flange E, substantially in the manner set forth.

No. 56,100.—PHILLIP ROCHE, Binghamton, N. Y.—*Saw Gummer.*—July 3, 1866.—The adjustable lever to which the saw is attached is pivoted to a sliding block working in a slot in the frame and moved by an eccentric lever which feeds the saw toward and from the grindstone. The rake of the teeth is regulated by the adjustment of the saw mandrel in the slot of the lever.

Claim —The combination and arrangement of the adjustable lever B, sliding block D,, eccentric lever C, and indicator H, substantially as described and for the purposes set forth.

No. 56,101.—J. M. ROEBUCK and W. R. REECE, Donaldson, Penn.—*Pump.*—July 3, 1866.—Explained by the claim and cut.

Claim.—In a pump the removable case C, divided by partitions substantially as described, and furnished with valves and openings, as set forth, for the purpose of being readily repaired or renewed without disturbing the rest of the machinery, as set forth.

No. 56,102.—JOHN S. and IRA ROWELL, Beaver Dam, Wis.—*Cultivator.*—July 3, 1866.—A brace bar behind the standard supports it; the brace-bar pin may be made to break by contact of the share with a rigid obstruction.

Claim.—The combination of the slotted beam A, shank B, brace bar C, and bolt D, when the parts are constructed and arranged to operate as and for the purposes herein specified.

No. 56,103.—JACOB L. RUST, Oquaka Junction, Ill.—*Churn.*—July 3, 1866.—The dashers are arranged on line shafts and rotate in different directions.
Claim.—Constructing, arranging, and operating the dashers L M, in the manner and for the purposes herein specified.

No. 56,104.—GEORGE SCHMIDT, New York, N. Y.—*Combined Inkstand, Wafer or Sand Box, Calendar, Letter and Envelope Holder and Pen Rack.*—July 3, 1866.—Explained by the claim and cut.
Claim.—The combination of inkstand, wafer or sand box, calendar, letter and envelope holder, and pen rack, substantially as herein shown and described.

No. 56,105.—CONRAD SCHULLIAN, New York N. Y.—*Faucet.*—July 3, 1866.—The rotation of the handle brings the pressure of a roller upon the stem of a spring valve and admits air to the interior of the cask.
Claim.—Operating the air valve or vent of a faucet by the action of the handle which serves to open and close the plug, substantially as and for the purpose set forth.

No. 56,106.—DANIEL SCHUYLER, Buffalo, N. Y.—*Musical Instrument.*—July 3, 1866.—The performer by a pipe alternately blows into and exhausts air from the chamber of the instrument, and the induced current in each case passes through the reeds in the same direction.
Claim.—A combination of valves and air-chambers so constructed and arranged that by blowing the wind through the air-chambers or drawing the wind back and out of the air-chambers the wind will in both cases pass through the reeds in one and the same direction, substantially for the purpose set forth.

No. 56,107.—GEORGE E. SHAW, Pittsburg, Penn.—*Carbon Oil Fire-tester.*—July 3, 1866.—The oil vessel is placed in a water bath which receives its heat below and has a thick non-conducting jacket. Thermometers are placed in the water and oil, and the testing brand is operated by a spring rod and pivoted lever.
Claim.—First, the water bath D, with its double casing F, and top E, as described, and the pipe C, for the purpose of obtaining heat from the bottom only, as specified.
Second, the combination of the two thermometers M and N with the water bath D, and oil cup H, for the purpose specified.
Third, the shield P, with its notch Q.
Fourth, the combination of the pivoted lever fire-brand holder R with the ring and rod S, and the spring T, arranged as described and acting as specified.

No. 56,108.—THOMAS SHAW, Philadelphia, Penn.—*Quartz Crusher.*—July 3, 1866.—The quartz is beaten upon an anvil by the balls attached by chains to the revolving wheel.
Claim—The employment of metal balls secured by chains to a revolving wheel, the whole constructed and operating for the purpose described.

No. 56,109.—J. HERBERT SHEDD and BENJAMIN WORCESTER, Waltham, Mass.—*Chimney Cap.*—July 3, 1866.—A flanged annulus is suspended above the top of the chimney so as to be rocked by the wind to close it against the cap above and permit the escape of the smoke to leeward
Claim.—First, the combination of a fixed cover above a flue, with a movable shield so adjusted as to be made by the force of the wind to close the opening between the flue and the cover on the windward side while the leeward side is left open.
Second, the combination and arrangement of the base B, with its spherical surface K, and of the cap C, with its curved under surface and projection M, substantially as described.
Third, the ring E, with the tunnel-shaped jets N N, in combination with the base B, and cap C, substantially as described.

No. 56,110.—W. H. SHOEMAKER, Greenwood, Md.—*Washing and Wringing Machine.*—July 3, 1866.—The lower rollers have an endless rubber apron; when the shelf is turned down the clothes pass from the wringer to a basket, and when it is turned up they pass into the tub again.
Claim.—First, the hinged shelf in combination with the wash-tub and the rollers, substantially as and for the purpose described.
Second, the hand-lever J, in combination with the hinged shelf and rollers and wash-tub, substantially as and for the purpose described.
Third, the combination of the two squeezing rollers B B', stretching roller D, rubber belt G, and hinged shelf G', substantially as and for the purpose described.

No. 56,111.—LUTHER SMITH, Rochester, Minn.—*Lifting Jack.*—July 3, 1866.—The rack bar is lifted by a lever pivoted in a strap, and the rack is sustained by a sliding spring catch.

Claim.—First, the catch G, actuated by the spiral spring G', in combination with the jack-bar B.

Second, the arrangement of a lifting jack having the stock A, bar B, lever D, stirrup E, catch G, and spiral spring G', the parts being constructed and combined substantially as and for the purpose set forth.

No. 56,112.—T. D. SMITH, Independence, Ohio.—*Butter-worker and Mold.*—July 3, 1866.—The levers adapted to the working board and bowl are pivoted above each other to the segment frame.

Claim.—The arm F, pivoted or hinged at K, and lever H, pivoted or hinged at N, in combination with the frame A, mould P, and stamp L, arranged as and for the purpose set forth.

No. 56,113.—WILLISON G. SMITH, Carlisle, Penn.—*Car Spittoon.*—July 3, 1866.—The spittoon is inserted in the floor and so arranged that the lid or box closes the opening.

Claim.—In combination with the sides of an inserted box of any shape, even with the floor of the car, the adjustable lid and bottom connected by the bolt and sliding in the guides, as described.

No. 56,114.—RACHAEL SPEAR, Passaic, N. J.—*Nails and Tacks.*—July 3, 1866.—Explained by the claim and cut

Claim.—Making nails and tacks with bodies formed of double cones, or having conical outlines on opposite sides thereof, substantially as described.

No. 56,115.—OTIS W. STANFORD, Lebanon, Ohio.—*Churn.*—July 3, 1866.—The standard which carries the operative devices is slipped into staples on the side of the churn, and retained by a catch.

Claim.—The combination of the standard C, catch D c, pulleys E G, and driver q, constituting a detachable propelling device and arranged to operate in connection with dasher K L, substantially as and for the purposes set forth.

No. 56,116.—LEVI STEVENS, Fitchburg, Mass.—*Treating Gas for Illumination and other Purposes.*—July 3, 1866.—The carburetting vessel contains hydrocarbon liquid and two meter wheels; one is driven by the force of gas from the main and drives the other wheel in the reverse direction, drawing in air, which is carburetted, and mixed with the gas from the first mentioned wheel.

Claim.—First, the combination of two or more gas-meter wheels, propelled by gas or other power, for the purpose of carbonizing atmospheric air and super-carbonizing ordinary illuminating coal-gas, so that when combined and mixed the whole will become an illuminating gas, substantially as and for the purposes described.

Second, the combination of two or more gas-meter wheels, propelled by gas or other power, with a chain elevator, and all with a tank or reservoir for holding hydro-carbon and for receiving and mixing super-carbonized gas and carbonized atmospheric air for the production of an illuminating gas, in the manner and for the purposes described.

Third, the combination of two or more gas-meter wheels, propelled by gas or other power, with a chain elevator a, and the receiver C, and all with the reservoir A, in such manner that the gasoline or other hydro-carbon may be kept at the desired height in the cases of the gas-meter wheels, in the manner and for the purposes described.

No. 56,117.—ISAAC H. STODDARD, Amenia, N. Y.—*Ship's Table.*—July 3, 1866; antedated June 30, 1866.—The table is made in sections attached by rods and actuated by a weight so as to maintain their horizontal position individually.

Claim.—The arrangement of the tables C with the rods F, G, and H, and with the arm g, and weight E, substantially as and for the purpose herein set forth.

No. 56,118.—JOSEPH M. STORY, Cincinnati, Ohio.—*Driving Paddle Wheels.*—July 3, 1866.—Friction wheels driven by the motor are brought into contact with the periphery of a wheel on the shaft of the paddle wheel to impel the latter.

Claim.—Applying the power to the paddle-wheels of boats near the periphery of the same by means of friction wheels and levers, in the manner and for the purpose substantially as herein set forth.

No. 56,119.—WILLIAM D. STROUD, Oshkosh, Wis.—*Broom-heads.*—July 3, 1866.—The encircling straps are connected by a band, and the binders are clamped by bolts which pass through the broom.

Claim.—A broom-head made of flexible bands connected by a rigid strip, with an open space between the bands and an open space around the handle, as described, in combination with the rigid binders D D, and handle C, the whole being constructed and operated substantially in the manner and for the purpose set forth.

No 56,120.—HOMER H. STUART, Jamaica, N. Y.—*Hydraulic Engine.*—July 3, 1866.—The oscillating valve is moved by an eccentric and has ports which connect the respective ends of the double-acting cylinder with the supply and waste pipe alternately.

Claim.—The valve with its passages as shown, when operated by the eccentrically grooved plate in connection with the guide pin and connecting bar, substantially as described.

No. 56,121.—E. G. SUTHERLAND, San Francisco, Cal.—*Process for Refining Oil.*—July 3, 1866.—The blubber is melted in a tank by steam forced directly into the mass, and the melted fat decanted into a vessel of cold water to remove impurities.

Claim.—The within-described process for extracting and refining oil from whales and other marine animals and fishes, all substantially as described and for the purposes set forth.

No. 56,122.—C. W. TALIAFERRO, Keithsburg, Ill.—*Corn Cultivator.*—July 3, 1866.—The plough beams are attached to axles supported beneath the draught frame, which is elevated in the centre to span the corn in the row between the ploughs and wheels.

Claim.—First, the particular manner of constructing the frame of the machine, to wit, of two parallel bars A, connected by semi-circular metallic bars B, with the axles E fitted between or having their bearings in said bars, substantially as shown and for the purpose set forth.

Second, the connecting of the plough beams F F to the axles E, by having the latter formed with slots *a*, to receive clevices *a*ª at the front ends of the beams, with pins *b* passing through the axles and clevices, substantially as and for the purpose specified.

No. 56,123.—J. WARREN THYNG, Salem, Mass.—*Paper Gauge for Printing Presses.*—July 3, 1866 —The gauge is applied to the back or lower edge of the platen and consists of a long bar with clamps holding the gauge rods, which define the position for two sides of the sheet.

Claim.—The combination and arrangement of the graduated grooved bar B, attached to the edge of the platen, the adjusting clamps C C C, and gauge rods F F G, substantially as and for the purpose herein specified.

No. 56,124.—LEVI TILL, Sandusky, Ohio.—*Nut.*—July 3, 1866.—By the dovetail connection the washer is rotatable on the nut, but not removable.

Claim.—So securing a washer to a screw or other nut that, while it can freely turn on the face of the nut, it cannot become disengaged therefrom, substantially in the manner described.

No. 56,125.—ROBERT B. TOLLES, Canastota, N. Y.—*Binocular Eye-pieces for Optical Instruments.*—July 3, 1866.—The eye-piece is so constructed and applied to the object glass as to divide the optical pencil transmitted to the latter, and forms, as to each part of the divided pencil, a real or virtual image of the object beyond the place of division.

Claim.—The construction of an eye-piece in such a manner that the division of the optical pencil necessary to give binocular vision of an object is effected in the eye-piece itself; restricting the claim, however, to that form of a binocular eye-piece in which two real or virtual images of the object are formed in the eye-piece after such division of the pencil has taken place.

No. 56,126.—HOWELL TOPPING, Marion, N. Y., and MERRITT GALLY, Auburn, N. Y.—*Fixture for Ratchet Wheels for Lamps.*—July 3, 1866.—The spring presses the wick-raising wheels against the wick. The pressure is withdrawn while the wick is being inserted.

Claim.—The application of a spring to the shaft of the ratchet wheels of a lamp top, pressing the wheels against the wick, in combination with slots in the sides of the lamp top, allowing the movement of said shaft for different degrees of compressure, substantially as herein set forth.

No. 56,127.—JAMES WASH, Mount Sterling, Ill.—*Beehive.*—July 3, 1866.—The bee entrance is at the lower end of one of the supporting legs. Other openings into the hive have wire-gauze coverings, and grease boxes beneath to catch insects which attempt to enter.

Claim.—The combination of the vats C, trunk D, door *e*, tube E, and leg *g*, table A, and hive B, constructed and arranged in the manner and for the purpose herein specified.

No. 56,128.—CHAUNCEY L. S. WALKER, Newark, N. J.—*Mosquito Bar.*—July 3, 1866.—The sections of the frame are provided with wire-cloth, and are relatively adjustable for varying widths of windows, by slots and screws.

Claim.—The frame adjustable by the slots *a*, and screws *e*, in combination with the wire-cloth upon both parts of the frame, as herein specified.

No. 56,129.—D. M. WESTON, Boston, Mass.—*Reciprocating Pumps.*—July 3, 1865.—Explained by the claim and cut.

Claim.—The cushions or springs C C, made of rubber or other elastic material, or their equivalents, so placed as to form part of the connection between the piston of a pump and the driving gear, substantially as above described, for the purpose of relieving the pump and machinery from jar or concussion.

No. 56,130.—NORMAN W. WHEELER, Brooklyn, N. Y.—*Condenser and Refrigerator.*—July 3, 1866.—The fluids involved, both the refrigerator and the refrigerated, are caused to hug the outer surfaces of their chambers by introducing them in a given direction; the former is ejected against the inner surface of the cylinder, and the latter against the outer surface of the containing shell, and secondarily against the intervening plate, whose other side is traversed by the cold current; the accumulation of the latter is prevented by the injection of air.

Claim.—First, in an apparatus for surface condensation or refrigeration, producing the desired result by directing the refrigerating fluids into the refrigerating vessels in such a way that they shall form their sheets or strata on the inner surfaces thereof, which cling to such surfaces under the influence of centrifugal forces. when such vessels are prevented from filling with liquid, substantially as above described.

Second, in an apparatus for surface condensation or refrigeration, facilitating the object sought by directing the substances to be condensed or cooled into enclosed spaces, surrounding the refrigerating vessels in such a way that they will not impinge primarily against the principal refrigerating surfaces, but against other enclosing surfaces under the influence of centrifugal force, substantially as and for the purpose above set forth.

Third, in an apparatus for surface condensation or refrigeration, the mode herein described of preventing the refrigerating liquid from accumulating in the interior of the refrigerating vessels, to wit, by means of injecting air into such vessels by mechanical means, substantially in the manner above described.

No. 56,131.—LUTHER C. WHITE, Waterbury, Conn.—*Kerosene Burner.*—July 3, 1866.—The deflector is formed of a circular piece of metal with an oval opening and inclined sides, and so placed above the wick tube to deflect the flame.

Claim.—Combining the deflector H, constructed substantially as described, with the wick tube, cap, and shell, when the same shall be combined substantially as shown for the purposes specified.

No. 56,132.—NELSON B. WHITE, South Dedham, Mass.—*Bed Spring.*—July 3, 1866.—The base of the conoidal spring is held to the slat by springing its lower coil beneath metallic overlapping flanged plates secured to the slat.

Claim.—The arrangement and application of the overlapping clasps C C, made as described with the slat B, the base coil of the spring and the projection *a* to enter such slat, the whole being substantially as and for the purpose set forth.

No. 56,133.—ROBERT WILDE, Philadelphia, Penn.—*Spinning Mule.*—July 3, 1866.—The device simplifies the arrangement for putting the roll and scroll wheels in and out of gear during the movements of the carriage.

Claim.—The sliding rod E, levers F G and H, and the weight D, or its equivalent, the same being constructed, arranged, and applied so as to operate substantially as and for the purpose described.

No. 56, 134.—ALBERT S. WILKINSON, Pawtucket, R. I.—*Horseshoe.*—July 3, 1866.—The calk and toe clip are riveted together, and are fastened to the shoe by a hooked plate and wedges.

Claim.—The clip B, in combination with the bar A, calk C, rivets *f f*, and keys *e d*, the whole being constructed and operated substantially in the manner and for the purpose set forth.

No. 56,135.—CHARLES A. WILSON, Cincinnati, Ohio.—*Steam Gauge.*—July 3, 1866.—The construction of the diaphragm with intersecting radial and concentric corrugations permits the use of a more durable material, and gives a more uniform range of action.

Claim.—The diaphragm A for steam-pressure gauges, when constructed with the intersecting corrugations, as and for the purposes set forth.

No. 56,136.—GEORGE D. WOODWORTH, Chicago, Ill.—*Grain Cooler.*—July 3, 1866.—By the described devices air is forced between the stones while grinding and the meal cooled thereby.

Claim.—First, covering the eye of the stone to prevent the escape of the air, in the manner described and substantially as set forth.

Second, the use and employment of the ring O O, in the manner and substantially as set forth.

Third, the combination of the pipes *k* and P, forming one and the same pipe, in the manner and for the purpose set forth.

Fourth, covering the discharge spout, conveyor or elevator with bunting or a similar material, in such a way as to form a surface with one or more re-entering angles, in the manner and for the purpose set forth.

Fifth, the combination of the spouts H I *k* and P, in the manner and for the purpose described.

Sixth, the combination of the tubes H I K P and ring O, in the manner and for the purpose described.

Seventh, the combination of the spouts K P and ring O with plate N N, in the manner and for the purpose described.

No 56 137.—W G WRIGHT, Hornellsville, N. Y.—*Tubular Well.*—July 30, 1866.—A body of fibrous material is placed between the outer perforated tube and the inner cone of wus cloth
Claim.—The use of hemp or any suitable fibrous filtering material as described.

No 56,138.—T. L. YATES, Utica, N. Y.—*Sawing Machine.*—July 3, 1866.—The lumber is fed under a feed roller; the saw cuts from the under side and feeds the lumber along without spec.al device for that purpose.
Claim —The combination of the roller E, arm F, slotted bent arm G, applied and operating with the gauging and feeding saw, in the manner and for the purpose herein specified.

No. 56,139 —CHARLES H. BAGLEY, Elgin, Ill., assignor to C. E. MASON and F, S. BARTLETT, same place.—*Safety Pocket* —July 3, 1866.—A plate is attached by pin and hook to the pocket, and its spring is engaged by the perforated shank of a button attached to the pocket-book.
Claim —First, providing the plate A with the spring B, substantially as and for the purpose specified.
Second, the comb'nation of the button or stud E with the plate A, and spring B, substantia.ly as specified.

No 56 140.—W. BLESSING, Jeffersonville, Ohio, assignor to himself and HORATIO B MAYNARD, Washington, Ohio.—*Preparing Cotton Seed for Planting.*—July 3, 1866.—As the seeds come from the gin they are wetted with glue-water and then rolled in sand, &c., to prevent future adhesion.
Claim --The mode of preparing cotton seed for planting by the application of mucilage, sand, and flour, or their equivalents, substantially as and for the purpose set forth.

No 56,141.—THEODORE COOPER, Warwick, R. I., assignor to himself and THOMAS PHILLIPS, Providence, R. I.—*Tool for Turning or Planing Iron.*—July 3, 1866.—The tool is clamped in the stock by the intervening key and wedge-shaped bolt.
Claim —The holder A, cutter B, clamp C, and key D, arranged substantially as described for the purposes specified.

No. 56,142.—J. E. EMERSON, Trenton, N. J., assignor to the AMERICAN SAW COMPANY —*Saw.*—July 30, 1866.—The indented shank of the tooth is attached by groove, nib, and rivet to projections on the edge of the saw plate.
Claim.—Constructing the teeth of a saw with recesses in such a manner that they may be attached to projections from the saw-plate, substantially in the manner herein shown and described.

No. 56 143.—ALVA J. GRIFFIN, Lowell, Mass., assignor to himself and WILLIAM T. VOSE, Newtonville, Mass.—*Apparatus for Producing and Burning the Gases from Petroleum and Water.*—July 3, 1866.—The oil and water are heated in the respective divisions of the chamber over the flame, the gas passing to the burner and the steam to a foraminous coil, whence it issues to the burner.
Claim.—First, constructing the chamber A with a longitudinal partition K, and lateral ribs H, elevated and depressed at alternate ends, substantially in the manner and for the purpose set forth.
Second, in combination with the chamber A, divided into two compartments by a central partition, the p pes D and E, when constructed and arranged in relation to the chamber A, and to one another, as and for the purpose set forth.

No. 56,144.—BARZILLA HARRINGTON, Boston Mass., assignor to himself and J. L. NEWTON, same place.—*Knife for Unhairing Hides.*—July 3, 1863; antedated June 22, 1833 — This doub e-edged knife has one convex side and a dovetail rib on its upper side. A shield with a groove fits upon the dovetail and extends beyond the edge of the knife.
Claim.—A short hair knife with its form of blade b, its rib c, and its guard d, combined and arranged substantially as and for the purpose above set forth.

No. 56,145 —C. LEFFINGWELL, Clarksburg, Ohio, assignor to himself, H. BLANDY, and F. J. L. BLANDY, Zanesville, Ohio.—*Head-block for Saw Mills.*—July 3, 1866.—The pawls slide freely in inclined head stocks so as to move forward on their sliding sash by the motion of the lever and toggle, the knee of the head-block retaining its position by the engagement of the opposite pawls in a fixed rack when the lever is retracted for another effective stroke.
Claim.—First, the pawl blocks C G F H I, when constructed and arranged substantially as herein described and for the purpose set forth.
Second, the combination of the lever K, rod M, rods N and O, rods R and U, levers P and T, rods S and V, and movable racks E, by means of which the knees of the head blocks are worked with each other and with the movable pawls F, substantially as described and for the purpose set forth.

No. 56,146.—LUTHER W. MCFARLAND, Brooklyn, N. Y., assignor to FREDERICK S. OTIS, same place.—*Clasp for Bottom Hoop of Skirts.*—July 3, 1866.—Explained by the claim and cut.
Claim.—The bottom clasp formed with teeth, as specified, for attaching the tape to the hoop and protecting said tape, as set forth.

No. 56,147.—A. M. OLDS, New York, N. Y., assignor to himself and ALBERT MANVEL, Elizabethport, N. J.—*Wrench.*—July 3, 1866.—The movable jaw is secured by a spring brace which abuts against a projection on the jaw and the bottom of a groove in the shank.
Claim.—First, so combining and arranging the spring-tooth D, jaw C, and shank B, that the tooth will gripe or bind against the bottom of the mortise of the shank and retain the said tooth in position without serrations in the shank, substantially as described.
Second, the mortise b, in combination with the shank B, spring-tooth D, and jaw C, constructed and operating substantially as and for the purpose described.

No. 56,148.—J. W. PAIGE, Corning, N Y., assignor to himself and J. L. PAIGE, Rochester, N. Y.—*Coal Stove.*—July 3, 1866.—A flange extends inward from the case and forms a deflecting cover with a central opening above the fire pot, and the volatile matters are again deflected outward by a disk above.
Claim.—First, the arrangement of the concentrating plate E, in combination with the fire pot P, and the outer case of the stove, substantially as shown and described and for the purposes set forth.
Second, the arrangement of the radiating plate F, in relation to the concentrating plate E, substantially for the purposes specified.

No. 56,149.—EDWARD PARKER, Middletown, Conn., assignor to himself and JULIUS HOTCHKISS, same place.—*Gear Cutter.*—July 3, 1866.—The gear cutter can be applied to the slide rest in place of the tool holder, and adjusted for cutting square or bevel gear, or grooves in taps, &c. The cutter is attached to a mandrel which runs between the centres of the lathe, and the arbor carrying the blank is adjustable upon the rest as to height, proximity to the cutter, and angular presentation.
Claim.—The pointed bracket G, in combination with the bed plate F, slide E, with head D, or D*, and arbor A, constructed substantially as and for the purpose described.

No 56,150.—S. B. PIKE, San Francisco, Cal, assignor to himself and ROBERT H. VANCE —*Quartz Crusher.*—July 3, 1866.—Stationary guide plates distribute the quartz under the rollers which rotate above the revolving bed die; the roller mandrels are adjustable vertically and the adjustable gutter around the edge of the revolving plate is traversed by the discharging scrapers.
Claim.—First, the use of the guides L L, when arranged as described, to insure the passage of all the ore under the rollers, substantially as herein specified and for the purpose set forth.
Second, the employment of the horizontal bar H, hub I, and slotted boxes K K, or their equivalents, for the purpose of allowing the rollers F F to accommodate themselves to the varying amount of ore on the revolving plate B, and die C, substantially as described and for the purposes as set forth.
Third, the gutter D, operated by the binders Y Y, and the keys X X, in combination with the revolving plate B B, rollers F F, guides L L, annular die C, and scrapers E, when constructed to operate in the manner specified and for the purposes set forth.

No. 56,151.—LEWIS A. SMITH, Cincinnati, Ohio, assignor to himself and GEORGE BURROWS, same place.—*Shellac Varnish.*—July 3, 1866.—Bicarbonate of soda is used as a solvent.
Claim.—As a new article of manufacture, a varnish compounded of shellac and bicarbonate of soda, in any suitable proportions, substantially as set forth.

No. 56,152.—P. L. WEIMER, Lebanon, Penn., assignor to AURORA IRON COMPANY, same place.—*Machine for making Metal Tubes.*—July 3, 1866.—The machine has four pairs of rolls turning on horizontal axes and three pairs turning on vertical axes. The metal strip by the successive rolls is fed, its edges scarfed, bent into a trough shape, bent tubular, and its edges forced together.
Claim.—First, the arrangement of the bending rollers d d, e e, and f f', in a machine which is constructed and operates substantially as and for the purpose described.
Second, while not claiming broadly the scarfing rollers b b, and the bending rollers c c, I do claim these rollers in combination with the bending rollers d, and the rollers e e, or f f, all arranged and operating substantially in the manner described and for the purpose set forth.

No. 56,153.—GEORGE L. WITSIL, Philadelphia, Penn., assignor to himself, JOHN L. OLCOTT, and WARREN S. SMITH, New York, N. Y.—*Churn.*—July 3, 1866.—The reciprocating, spirally-grooved upper section is rotated by its contact with a tooth on the bridge-piece, and the perforated hollow lower section admits a downward current of air into the cream.

Claim.—The combination in the dash-rod of a churn of a lower hollow section *b*, and an upper spirally grooved solid section *c a*, the former serving to admit a supply of fresh air, while the latter receives a tooth *e*, by which the dash-rod is rotated as it is moved up and down by the swivel-handle G, or other device, substantially as described.

No. 56,154.—JOHN WORRALL, New Haven, Conn., assignor to himself and JESSE CUDWORTH, same place.—*Refrigerator.*—July 3, 1866.—The air passes through holes in the ends of the ice box to the provision chamber, the drip flowing by a passage provided with traps to the exit below the safe; warm air from the chamber passes upward by a pipe and out at the back of the refrigerator, the water trap preventing its return.

Claim.—The closely-fitting ice chamber B, with apertures *a* in its ends, in combination with the provision chamber *c*, tube E, and water pipe *e*, constructed and operating substantially as and for the purpose described.

Also, the water trap *f*, in the interior of the tube E, in combination with the ice chamber B and provision chamber *c*, constructed and operating substantially as and for the purpose set forth.

No. 56,155.—MORITZ HERZOG, Vienna, Austria, and D. L. COHN, London, England.—*Hydrocarbon Burner.*—July 3 1866.—The oil reservoir, clock-work, air-forcing apparatus, pipes and burner are combined in a single case as a hand lamp.

Claim.—The combination in a lamp of the following devices, viz : the hydrocarbon reservoir, the gas-burner, the air-forcing mechanism, connecting pipes, and case, the whole being combined and operating substantially as hereinbefore set forth.

No. 56,156.—WILLIAM MURPHY, Cork, Ireland.—*Liquid and Spirit Meter.*—July 3, 1866.—Weighs and measures alcoholic spirits as it runs from the still. A brake is applied by a float to a disk on the bucket wheel shaft, pressing more or less as the density of the liquor varies. The measurer consists of two chambers, filled and emptied by a faucet operated by a float in the liquid and by balls which traverse in frames and determine the action of the bucket after passing the centre.

Claim.—First, the bevelled plates H, in the inclined plane R, in combination with the bucket wheel I, and sample receiver F, constructed and operating substantially as and for the purpose described.

Second, the method herein described of checking the motion of the bucket wheel I, according to the specific gravity of the spirits, consisting of the hydrometer N, balance weight I, lever L, and pulleys K J, or other equivalent means which will produce the same effect.

Third, the balance funnel B', and tilting frame D', in combination with the balance frame E', and hydrometer G' G', constructed and operating substantially as and for the purpose sent forth.

No. 56,157.—ANDREW P. ANDERSON and BYRON EDWARDS, Princeton, Ill.—*Cultivator.*—July 10, 1866.—The front shares are adjustable laterally and the four shares simultaneously vertically by the driver, whose seat is upon the rear extension of the tongue.

Claim.—The beams D D, in combination with the plough-standards H H, and the plough-standards E E, substantially as and for the purpose set forth.

No. 56,158.—BENJAMIN ARNOLD, Arba, Ind.—*Feeding Trough for Stock.*—July 10, 1866.—Access is had by steps which are folded up at other times. The stock feeder stands on the platform and pours the feed through the central opening on the ridge between the two troughs.

Claim.—The arrangement and combination of the various parts constituting the trough, platform, food receptacle, and steps, all in the manner and for the purposes set forth.

No. 56,159.—JONATHAN AYERS, Canterbury, N. H.—*Harrow.*—July 10, 1866.—Each tooth has a forward cutting edge; the front has divergent cutting wings and the succeeding teeth wings projecting inward.

Claim.—The front tooth B, as made with the inclined knife edge *f*, and the two double curved wings *g g*, as set forth.

Also each lateral tooth as made with the knife-edge *b*, and the curved furrow opening lip *c*, and the curved furrow opening lip or part *d*, arranged as specified.

No. 56,160.—GEORGE E. BAKER, Waukegan, Ill.—*Hanging Bells.* July 10, 1866.—The glass bell is hung upon India-rubber cushions, which isolate it and preserve its tone unimpaired by metallic connection.

Claim—Attaching a bell to, or suspending it upon, its supports by means of elastic cushions, substantially as and for the purposes herein shown and described.

No. 56,161.—HALSEY H. BAKER, New Market, N. J.—*Horseshoe.*—July 10, 1866.—The calk consists of a wooden block with spikes, enclosed in a collar attached by tongues to lugs in a socket of the shoe.

Claim.—First, the attachment of the collar D to the shoe A, by means of the tongues *f*, on the said collar, and the pins or projections *d*, in the socket *a*, substantially as herein set forth for the purpose specified.

Second, the combination of the wood-block E with the collar D, and shoe A, substantially as herein set forth for the purpose specified.

Third, the combination of the hardened pins, spikes, or wedges, with the wood-block E, secured in the collar D, substantially as herein set forth for the purpose specified.

Fourth, the lips or ribs r, formed on the collar D, and fitted upon the bevelled sides of the flange b of the shoe, substantially as herein set forth for the purpose specified.

No. 56,162.—FREDERICK G. BAKES, Vevay, Ind.—Broom Head.—July 10, 1866.—The butts of the corn are clamped around the barbed collar on the handle by a collar and screw band. The broom is clasped at a lower point by a binding wire and transverse screw-threaded hook.

Claim.—First, a broom head or clamp composed of the parts C C', D D', and E E', all combined, arranged, and operating substantially as set forth.

Second, in combination with the binding wires H H, the screw-threaded hook I, cylindrical nut J, and washer K, substantially as described and for the purpose set forth.

Third, in combination with the elements of the two foregoing claims, the barbed sheet-metal cylinder B, for the purpose described and explained.

No. 56,163.—S. V. BECKWITH, Hamden, Conn.—Attaching Tubes to Oilers.—July 10, 1866.—A spiral groove in the tube is engaged by a pin in the neck and tightened by rotation.

Claim.—The combination of the conically formed tube A, with its correspondingly formed seat in the neck B, and secured therein so as to form a self-packing joint, by means of the spiral groove on the one and corresponding pins on the other, substantially as specified.

No. 56,164.—WM. N. BERKELEY, Cedar Rapids, Iowa.—Tool for Adjusting Lathes to Turn Tapering Shafts.—July 10, 1866.—The longitudinally slotted strip has pairs of arms pivoted to each end and adjustable by set screws; the arms rest against adjustable stops.

Claim.—First, the combination and arrangement of the plate A provided with a slot a, and the arms C D. operating as and for the purposes herein set forth and described.

Second, in combination with the above the employment of the gauges or guides E, arranged as and for the purposes specified.

No. 56,165.—H. B. BIGELOW, New Haven, Conn., and GEORGE MURRAY, Cambridgeport, Mass.—Blower.—July 10, 1866.—The fans are sleeved loosely upon a central shaft and are driven by cranks and separate link connection to wrists equidistantly arranged upon a wheel which is eccentrically journalled and has a uniform velocity; each wing, during a part of its motion, moves slowly and forms an abutment for the succeeding and more rapidly revolving wing.

Claim.—The construction and arrangement of the fans, substantially as described, to operate so that one moves with a diminished velocity through a portion of its revolution, acting as a stop for the fan succeeding it more rapidly revolving, substantially as herein described.

No. 56,166.—BYRON BOARDMAN, Norwich, Conn.—Tool.—July 10, 1866.—This is an aggregation of four tools; a monkey-wrench, pipe-tongs, claw-hammer, and screw-driver.

Claim.—First, the combination of a cam a with the movable or fixed jaw-head of a monkey-wrench, so applied as to form thereof a pipe-wrench, substantially as described.

Second, the manner herein described of securing the pipe-wrench cam within a recess so that this cam will be firmly sustained by the solid metal surrounding it during the operation of turning a cylindrical object and allowed to play loosely when released, substantially as described.

Third, securing the nut g within an oblong slot i in the handle A by means of the fixed pin j, substantially as described.

Fourth, the combination of the permanently attached nut g and the screw-driver bit h formed on or applied to the shank C' of a monkey-wrench, substantially as described.

Fifth, securing a steel bit h to the end of the softer metal shank C' of the monkey-wrench, substantially as described.

No. 56,167.—GEORGE BOLDT, Chicago, Ill.—Fulminating Composition.—July 10, 1866.—Composed of fulminate of silver, sulphur, tin, gunpowder, charcoal, and gum water.

Claim.—A fulminating substance to ignite cartridges by filling with it small cavities in the back part of conical projectiles and using an igniting needle, said fulminating substance consisting of the ingredients above enumerated and prepared, and mixed as above described and specified.

No. 56,168.—PETER BORN, New York, N. Y.—Carpet-bag Frames.—July 10, 1866; antedated June 29, 1866.—Explained by the claim and cut.

Claim.—Making the jaws and covers of frames of carpet bags, valises, &c., of a series of veneers or layers of thin wood fastened together, in the manner herein specified.

No. 56,169.—ROBERT D. BRADLY, Preston, Md.—Shovel and Tongs.—July 10, 1866.—A shovel is attached to one leg of the tongs.

Claim.—The combination of shovel and tongs, constructed and operating as described.

No. 56,170.—A. C. BRIGGS, North Easton, N. Y.—*Horse Hay Fork.*—July 10, 1866.—The prong is pivoted to the lower end of the bar which slides upon the shank; it slips through a slot in the latter and its extended position is sustained by a catch which is disengaged by a trigger rope to discharge the load.

Claim.—First, the combination of the tines C with the bars A and D, and catch E, arranged and operating substant'ally as herein set forth.

Second, constructing the tine C with a shoulder, as described, when used in combination with the bar D, and slide *a*, substantially as and for the purpose specified.

No. 56,171.—JOHN BRIZEE, Alvarado, Cal.—*Transmitting Motion.*—July 10, 1866; antedated July 3, 1866.—The devices are intended to transmit the motion of a spring to machinery such as sewing and washing machines, &c. The oscillating frame acts as a regulator and is adjusted by the motions of the plate which supports the springs of the pendants.

Claim.—First, the combination of the oscillating frame W, pendants X X, and spring A', with the crank M', all arranged to operate as and for the purposes specified.

Second, in combination with the parts above specified, the adjustable or rising and falling plate Y, for the purpose explained.

No. 56,172.—A. P. BROWN, Worcester, Mass., assignor to L. W. POND.—*Feed Stop for Lathes.*—July 10, 1866.—The spline is set at a given point to stop the feed. The revolution of the shaft turns the collar and the feed gear till the latter passes over the feather and by intermediate devices unclutches the feed gear.

Claim.—First, the combination with the feed rod of collar J, and gear I, and adjustable spline K, substantially as set forth.

Second, the combination with feed rod C of ears G G', spiral spring *d*, clutch collar J, gear I, and adjustable spline K, substantially as set forth.

Third, the combination with the feed rod of an engine lathe of a feed-stop device, constructed and arranged as described, whereby the feeding or forward motion of the cutter or tool frame can be automatically stopped at any desired point without stopping the feed rod, substantially as set forth.

No. 56,173.—ROBERT BULLOCK, South Mills, N. C.—*Cultivator.*—July 10, 1866.—The oblique plates which scrape the soil and weeds from the corn are suspended below the frame and adjusted by stay rods.

Claim.—The combination of the stock with the weeding shears, the iron rod braces, the eye-holes, and iron bolts, when these several parts are arranged, constructed, and adjusted substantially as described.

No. 56,174.—J. P. CADMAN, Freeport, Ill.—*Gate.*—July 10, 1866.—The gates are suspended on levers and traverse thereon laterally when the levers are vibrated vertically. This motion is obtained by cords, by a lever and rod, or by a treadle-rod actuated by the wheels of a wagon.

Claim.—First, the ways D D, provided with cogs *o o*, when pivoted to the uprights C C, and beam F, as and for the purposes specified.

Second, the posts H K and H, provided with pulleys J and *t*, cords *m* and *m´*, and rod *f*, working the ways D D, arranged in the manner substantially as and for the purposes set forth.

Third, the treadle rods G, provided with semi-hoops *g*, and weight *k*, working the rod *f*, as and for the purposes set forth.

Fourth, the gates A and A, attached to the ways D D by the strips and pulleys *e*, secured by the posts P P, in combination with the posts H K and H, with pulleys J and *t*, cords *m* and *m´*, lever E, and rod E', in the manner arranged, substantially as and for the purposes herein specified.

No. 56,175.—WILLIAM HENRY CAMPBELL, Brooklyn, N. Y.—*Animal Trap.*—July 10, 1866.—Explained by the claim and cut.

Claim.—The combination with the pivoted platform F, and receptacle A, of the guards G and plates H, which prevent the rats from making their escape by jumping back after the platform has tilted, as described.

No. 56,176.—J. A. CARLISLE and GEORGE A. BOWERS, Elgin, Ill.—*Cheese Vat.*—July 10, 1866.—The cheese vat is placed in a water bath heated by a furnace arranged to cause a circulation of the water and have a direct flue communication when the required heat is attained.

Claim.—First, cavity or opening T, in combination with valve A, packing U'', levers 1 1'', and rod H, the whole constructed, arranged, and operated substantially in the manner and for the purpose set forth.

Second, pipes O and *a a''*, when constructed, arranged, and operated with regard to its cavities or openings, substantially in the manner and for the purpose set forth.

Third, pipe *h*, packing U, and nut *l*, the whole constructed, arranged, and operated substantially in the manner and for the purpose described.

No. 56,177.—JOHN H. CHASE and J. M. TIFFANY, Montgomery, Ill.—*Grain Separator.*—July 10, 1866.—The revolving screen has longitudinal slats whose openings have inclined or hopper-shaped throats to induce the coincidence of the major axis of the grain with the line of the aperture and facilitate its passage there through.

Claim.—In a grain-separating machine one or more hollow cylinders, having narrow openings or passages in their convex surface, with V-shaped grooves upon the interior corresponding with the direction of said passages so as to keep the kernels of grain in the cylinder lengthwise over the said passages, substantially as and for the purposes herein shown and described.

No. 56,178.—OTIS N. CHASE, Boston. Mass.—*Microscope.*—July 10, 1866.—The slip containing the mounted specimen is held by a rubber strip to one end of the glass tube while the eye-glass is mounted in the other and retained by a ring of rubber.

Claim.—First, a hollow transparent chamber A, open at one or both ends, in combination with the lens F, substantially as described.

Second, the projections C C, or their equivalents, in combination with the elastic band D, for the purpose set forth, substantially as described.

No. 56,179.—ROBERT A. CHESEBROUGH, New York, N. Y.—*Purifying Petroleum, &c.*—July 10, 1866.—The bone-black is heated to 220° Fah., to expel the moisture, and is used while warm from the process.

Claim.—The heating of bone-black by dry steam or otherwise, previous to using the same for filtering hydro-carbon oils.

No. 56,180.—EDSON P. CLARK, Northampton, Mass.—*Indelible Pencil*—July 10, 1866.—The filling is composed of nitrate of silver, black lead, calcined gypsum, and lampblack or asphaltum ; it is caused to adhere to the groove in the wood by shellac dusted in and gently heated.

Claim.—The employment of the ingredients, in combination with the nitrate of silver, substantially as and for the purpose set forth.

No. 56,181.—JOHN N. CLARKE, Cincinnati, Ohio.—*Horseshoe.*—July 10, 1866.—The ridges and indentations extend across the face of the shoe.

Claim.—Forming the sole of a horseshoe, substantially as described and shown ; that is to say, with transverse and diverging ridges A, alternating with hollows B, which are pierced at or near their crowns with the nail holes C.

No. 56,182.—DANIEL CLARKE, Ipswich, Mass.—*Coffin.*—July 10, 1866.—Projections on the inside of the coffin are, by a sliding movement of the lid, engaged by slotted plates dependent from that portion of the lid below the hinges.

Claim.—The combination as well as the arrangement of the series of studs b b b b, and the series of notched catches C C C C, with the cover and case of the coffin.

No. 56,183.—WILLIAM COGSWELL, Ottawa, Ill.—*Corn Planter.*—July 10, 1866.—The tongue and seat are connected to the middle portions, and the sides are attached thereto by sockets and set screws. The cut-off rubbers, against which the pockets of the seed cylinders rotate, and the bearings of the latter are supported by screw caps beneath the hopper bottoms.

Claim.—First, the provisions for adjusting the width of planting and for dividing the machine, as described, by means of the divided axles and the divided connections at the fore part, as described and represented.

Second.—The piece G, constructed substantially as shown, affording a bearing for the seat supports, sockets for the divided axles, and pivoted attachment for the tongue.

Third, the piece H, constructed as described. affording adjustable connections to the forward seats and beams, the loop for the rising and falling tongue, and attachment for the rear seat supports.

Fourth, the combination of the hopper m, with its lugs n n, and the cap o, the latter holding in position the cut-off rubbers p and the journal boxes s s of the oscillating cylinders.

No. 56,184.—J. E. D. COMSTOCK, New York, N. Y.—*Holding Books and Manuscripts.*—July 10, 1866.—Explained by the claims and cut.

Claim.—First, the combination of the lazy tongs B, with the clip A, or its equivalent, and strips C D, substantially as and for the purpose specified.

Second, the extension strips C' and D', in combination with the strips C D, and the lazy tongs B, substantially as specified.

Third, the combination of the clip A and strip D, substantially as described.

Fourth, the combination of the clip A, lazy tongs B, and strip C, substantially as specified.

Fifth, the clip A, when constructed in the manner shown and described, being in itself a clip of a new construction.

No. 56,185.—ROBERT CONARROE, Camden, Ohio.—*Sorghum Mill.*—July 10, 1866.—The apparatus is arranged in front of the rollers and the throats are self-adjusted to the size of the passing stalks to strip the leaves therefrom.

Claim.—The series of pivoted stripper jaws, arranged in reference to the row of openings and provided with springs, substantially as and for the purpose described.

No. 55,186.—JOHN CRAWLEY, Perryville, Ind.—*Governor Stop Valve.*—July 10, 1866.—The oscillations of the governor rod rotate the valve stem to move the piston valve relatively to the space between the ends of the divided pipe and regulate the steam aperture; detached from the governor the lever may be made to close the valve.

Claim.—First, the divided pipe A *a*, globe C, and pipe B, tubular valve I *j j*, and stem P, arranged and operating substantially as herein shown and described.

Second, the valve stem *p* with cams *h h k k*, the fixed cams *f f*, stop-lever L, and detachable governor rod *n*, arranged and operating as and for the purpose set forth.

No. 56,187.—GEORGE P. DARROW, Cincinnati, Ohio.—*Composition for Moulders' Match Plates.*—July 10, 1866.—Plaster of Paris, 40 parts; iron-filings, 20 parts; sal-ammoniac, 1 part, dissolved in water.

Claim.—The manufacture of moulders' match-plates and follow-boards by the use of plaster of Paris, iron-dust and sal-ammoniac in solution, substantially as and for the purpose set forth.

No. 56,188.—JUSTUS DAY, Murray, N. Y.—*Machine for Holding Sheep for Shearing.*—July 10, 1866.—The sheep is seated upon and its legs buckled to the revolving table, its body and neck fastened by the clamps proceeding from the standard. The adjustments permit a varied presentation and the pins on the table receive the fleece.

Claim.—First, the forked arms C and D, operating for the purpose and in the manner set forth.

Second, the revolving table in combination with the buckles and straps I I, and the pins J J J, &c., operating in the manner and for the purpose described.

No. 56,189.—GEORGE DEAL, Wayne township, Ohio.—*Horse Rake*—July 10, 1866.—To the axle are hinged two arms, which extend forward therefrom and form the bearings of the rake head. These arms are free to rise and fall according to the undulations of the ground, their motion being limited at will by set-screws placed underneath their forward ends.

Claim.—The combination and arrangement of the forwardly projecting hinged arms I I, adjusting screws or their equivalent H H, tooth roller D, and teeth L L, substantially as and for the purpose herein specified.

No. 56,190.—E. F. DIETERICHS, Philadelphia, Penn.—*Button-hole.*—July 10, 1866.—The plate is laid over the button-hole and its lips bent over the edges of the fabric, forming a metallic covering thereto.

Claim.—A metal binding for button-holes, said binding consisting of a number of plates arranged adjacent to each other and compressed to the fabric, all substantially as set forth.

No. 56,191.—CHRISTOPHER G. DODGE, Jr., Providence, R. I.—*Nail Hammer.*—July 10, 1866.—Explained by the claim and cut.

Claim.—A hammer, Fig. 1, constructed with a sheath D in the handle thereof for a screw-driver, so that when the hammer and the screw-driver are united, as shown in Fig. 1, the handle of the latter will form a continuation of the handle of the former, and when the two tools are not so connected each can be used separately, the whole article being substantially as specified.

No. 56,192.—WILLIAM M. DOTY, Janesville, Wis.—*Wagon Jack.*—July 10, 1866.—The clamping shackle binds the movable bar to the standard and the lever operates beneath the axle of the wagon.

Claim.—The arrangement of the clamp E and standard A, operating in combination with the sliding bar C and lever arm F, without serrations, on the sliding bar C, substantially in the manner and for the purpose represented and described.

No. 56,193.—NICHOLAS DOWNES, Syracuse, N. Y.—*Filter.*—July 10, 1866.—Improvement on his patent No. 46,646, of February 27, 1865. The cylindrical filter is placed in a water chamber and the water drawn from the space above the perforated disk which maintains the filtering material in position.

Claim.—The combination of the cover F, having the collar F', the reservoir B, and cleansing tube C, and with or without the perforated disk D, when severally constructed and arranged for use, substantially in the manner and for the purpose set forth.

No. 56,194.—J. F. DUNHAM, Fayette, Iowa.—*Grain Separator.*—July 10, 1866.—The larger offal collecting at the outer edge of the screen is thrown out by a revolving winged shaft, the perforations in the plate allowing the passage of the smaller matter.

Claim.—First, the perforated plates E G, fitted in a shoe A of a threshing machine, at the outer end of its screen B, in connection with the spout D, and inclined parts *b c* of the bottom of the shoe, substantially as and for the purpose set forth.

Second, the rotating frame or discharger I, in combination with the perforated plates E G. arranged and applied to the shoe A, and in relation with the screen B, substantially as and for the purpose specified.

No 56,195.—SETH C. ELLIS, Jersey City, N. J.—*Wood Boring Machine.*—July 10, 1866.— The bit holder is placed in a jointed frame with a movable arm, and is susceptible of presentation in various positions without affecting the operating devices.

Claim.—The jointed frame F G, bit-stock *m*, pulleys *g k l r*, belts *u s*, and handle H. the whole combined and arranged in relation with each other, and the driving shaft E and table B, substantially as and for the purpose herein specified.

No. 56,196.—J. ALBERT ESHLEMAN, Philadelphia, Penn.—*Neck-tie Holder.*—July 10, 1866.—The ends of the neck band are secured in front by a spring catch and socket, the former having previously been connected by its stud with the button-holes of the collar. A transverse band is tied in front.

Claim.—First, a plate secured to a neck band and adapted for the reception and retention of a detachable tie, substantially as described.

Second, the combination of the socket A, or its equivalent, and a stud *a*, as and for the pur pose specified.

Third, the socket A, with its projections *i i*, and the spring catch B, with its recesses *e e*, when constructed and adapted for attachment to each other and to a neck band, substantially as set forth.

No. 56,197.—G. W. FERRIS, Quincy, Ill.,—*Refining Spirituous Liquors.*—July 10, 1866.—The tanks have false bottoms, tight covers, and are charged with charcoal covered with muslin. The liquor flows from an elevated reservoir and passes through the tanks consecutively.

Claim.—The within-described process of refining spirituous liquors, by passing the same under pressure through a series of filtering tanks, substantially as set forth.

Also, the arrangement of a series of filtering tanks placed gradually lower and wer, and provided with tightly-fitting covers and with connecting pipes, substantially as and for the purpose described.

No. 56,198.—G. W. FERRIS, Quincy, Ill.,—*Preparing Charcoal for Filtering.*—July 10, 1866.—The wood is cut into rectangular blocks one and a half by one inch square, heaped, covered with earth, burnt in the usual manner, raked out and sprinkled with corn meal while incandescent.

Claim.—The within-described process of preparing charccal by treating the same with corn meal, substantially as and for the purposes set forth.

No. 56,199.—EDWARD FIELD, Cincinnati, Ohio.—*Bellows.*—July 10, 1866.—The hinged bottom of the double bellows is operated by lever, crank, treadle or belt pulley.

Claim.—The arrangement of shaft R, rod S, lever T, rod U, crank V, pulley W, lever X, treadle Y, and handle X, combined and operating substantially as and for the purpose specified.

No. 56,200.—LUTHER W. FILLEBROWN, Jr., Wayne, Me.—*Spring Seat.*—July 10, 1866.— The elasticity of the seat is regulated by the vertical adjustment of the slats to which the ends of the bow springs are attached.

Claim.—The improved elastic seat made as described, viz: of the adjustable frame C, the series of bow or arch springs D D D, the frame A, and the flexile covering E, arranged and combined together, substantially as set forth.

No. 56,201.—L. F. FRAZEE, South Amboy, N. J.—*Life Raft.*—July 10, 1866.—Pontoons encased with wood are connected by framing and furnished with hinged grating, side rails, &c.

Claim—The combinations of the drums E and casings C as described, cross-bars D, and hinged rails G, all constructed and arranged in the manner and for the purpose herein specified.

No. 56,202.—WILLIAM FREDERICK, Ashland, Penn.—*Quilting Frame.*—July 10, 1866.— The movable part may be so adjusted as to permit access to either side.

Claim.—First, an improved quilting frame formed by combining and arranging a set of rollers G H and nuts I J K L with each other, and with the legs or supports A B C D of the frame, substantially as described and for the purpose set forth.

Second, the combination with the main part of the frame of the removable and adjustable part M N O, constructed and arranged substantially as herein described and for the purpose set forth.

No. 56,203.—CHARLES GEISSE, Taycheedah, Wis.—*Illuminating Apparatus.*—July 10, 1866.—A permanent elevated reservoir of inflammable liquid supplies the various generators where the gas is evolved and whence it passes to the respective burners : an auxiliary burner gives the preliminary heating to the generator.

Claim —First, the combination and arrangement of the central tube D with the burner C, substantially in the manner described and for the purpose set forth.

Second, combining and arranging the movable branch pipe E with the burner C, for starting the generation of gas in the latter, substantially as described.

Third, the sliding sockets *i j*, arranged and operating substantially as and for the purposes set forth.

Fourth, the combination and arrangement of the globular radiator F with the central tube D, substantially in the manner described and for the purpose set forth.

Fifth, the combination of the pipe G with the burner C, substantially as and for the purpose set forth.

No. 56,204.—CHARLES GEISSE, Taycheedah. Wis.—*Gas Heater.*—July 10, 1866 —The liquid passes from an elevated reservoir to the chamber, where it is heated by a central tube, with a burner beneath ; the gas escapes by an annular orifice, and the flame is deflected by the ball.

Claim —First, the combination and arrangement of the central tube D with the heater C, substantially in the manner described and for the purposes set forth.

Second, the combination of the movable branch pipe E with the heater C for starting the generation of gas in the latter, substantially as described.

Third, the combination and arrangement of the globular radiator F with the central tube D of the heater C, substantially in the manner described.

Fourth, the combination of the conical mouth-piece *a* with the branch pipe B, when constructed and arranged to operate in relation to the heater C, substantially as described.

Fifth, constructing the central tube D with cog teeth, as represented, when used in connection with a pinion for elevating and depressing said tube, substantially as described and for the purpose specified.

No. 56,205.—PHILO M. GILBERT, Kewanee, Ill.—*Press for the Manufacture of Sugar.*—July 10, 1866 ; antedated May 3, 1866.—The sugar is placed in a cylinder, which has a grooved and perforated bottom. The screw follower has rollers at its lower end, and presses out the molasses, which drips into a pan beneath.

Claim.—The arrangement of the head G. with its rollers H H, the screw D, the press box F, the disks I *a* and *c*, block J, slide K and frame A, for forming a press, substantially as and for the purpose set forth.

No. 56,206.—SAMUEL D. and SAMUEL GILSON, Oswego Falls, N. Y.—*Peat Machine.*—July 10, 1866.—The crude peat is thrown into a hopper, and ground by revolving knives, which pass it through the grating to the chamber below ; thence it is driven by the revolving-sliding piston into the moulds of the wheel, where it is formed into triangular prisms, and discharged on to the endless apron.

Claim.—First, the combination of the curved arms *d* of the manipulator C with the rectangular grates *g'* for the purpose specified.

Second, the construction of the grates *g'* with one part vertical and the other part horizontal, for the purpose specified.

Third, the case D, constructed of elongated shape to allow the lever or arm F to revolve unobstructed, substantially as described.

Fourth, the revolving press E, with the sliding lever F working through it when made, arranged, and operated substantially as specified.

Fifth, the mode of relieving the journals of the press E, by extending the shaft of the press at *l'* beyond the slot through which the lever F works, substantially as described.

Sixth, the safety chamber G, arranged below the discharge *g'*, and so constructed that the lever F may slide over it on its side walls, and allow a part of the peat while under the greatest pressure to pass back below the end of the lever, in order to avoid accidents, substantially as described.

Seventh, the revolving mould I, when so made, arranged, and operated as to form logs of uniform length and triangular in their cross sections, for the purpose substantially as specified.

Eighth, the machine for mixing, pressing, and moulding peat and other plastic substances, consisting of the mixing apparatus, the revolving press E, with its sliding arm F and the revolving mould, constructed, arranged, and operated substantially as described.

No. 56,207.—N. W. GORDON, Waupun, Wis.—*Coupling for Vehicles.*—July 10, 1866.—The dovetail tenon on the bolster plate is received by a corresponding mortise in the axle plate ; the entrance being effected in a certain position, the bolster is rotated and the parts locked.

Claim.—First, the combination of the plate C with the bolster B.

Second, the combination of the plate D with the axle A.

Third, the combination of plates C and D, constructed and operating substantially as described and for the purpose set forth.

No. 56,208.—JOHN GREEN, Lowell, Mass.—*Machine for Printing Textile Fabrics.*—July 10, 1866.—Each cam rotates in its own reservoir, and presents a surface of color to the surface of the printing cylinder or to the passing fabric.

Claim.—First, the use and application of two or more furnishing cams D and E, or their equivalent, to the working or printing surface of a printing cylinder, said cams being constructed with one, two, or more furnishing surfaces v, substantially as and for the purpose specified.

Second, printing two or more colors with one cylinder, each color being deposited upon the printing surface of the cylinder by a separate furnishing cam or its equivalent, substantially in the manner set forth.

Third, the arrangement of the cams D and E and cylinder C, whereby two or more colors may be printed on the same piece of cloth or other material, substantially as herein set forth.

Fourth, the arrangement of the cams D and E and the cylinder C, when used separately or in conjunction with the common printing rolls 10 11 12, as set forth.

Fifth, in combination with the furnishing cams D and E for furnishing more than one color to the printing cylinder C, the detachable sections of such cylinder, arranged and made to operate substantially as and for the purpose specified.

Sixth, printing textile or other fabrics, or yarns to be knit or woven into fabrics, in two or more colors by passing them between the cylinder C and the cams D E, substantially as set forth.

No. 56,209.—F. C. GRIDLEY, Hudson, Wis.—*Weather Strip for Doors.*—July 10, 1866.—The strip is made of one piece, cut out and bent in such form as to be operated by the contact of the closing door.

Claim.—The elevator shown at 2, when constructed as described, and secured by means of the plate H, substantially as set forth.

No. 56,210.—CATHERINE A. GRISWOLD, Willimantic, Conn.—*Skirt-supporting Corset.*—July 10, 1866.—Explained by the claim and cut.

Claim.—A boddice made upon a metallic form, opening before and behind, adjustable in the rear, and having shoulder-straps and fastenings for the support of the skirts, all substantially as shown and described.

No. 56,211.—WILLIAM A. GUYER, New York, N. Y.—*Invalid Bedstead.*—July 10, 1866; antedated May 1, 1866.—The sectional bed bottom is so hinged as to be capable of various positions, and the lower portions of adjustment as to length. The foot piece is supported by a catch piece in the rack on the side rails, and may be dropped to assimilate the bed to the form of a chair.

Claim.—The combination of the three parts B C D, adjustable shaft E, and slotted strips k k, constructed and arranged to operate substantially as and for the purposes set forth.

No. 56,212.—JOEL HAINES, Middleburg, Ohio.—*Cheese Cutter.*—July 10, 1866.—The graduated semicircles around the rotating platform indicate the size of the sector to obtain a given weight of cheese in varying sizes of the latter.

Claim.—The combination of the rotating platform, provided with the adjustable indicator-point, and the series of graduated arcs around the margin of the platform, substantially as and for the purpose described.

No. 56,213.—JOSEPH C. HAINES, Lewistown, Penn.—*Bridle.*—July 10, 1866.—The rein is attached to a check strap, which is pulled through the bit ring, and draws the bit up into the angle of the mouth.

Claim.—Providing the driving rein with a shifting bearing H, arranged and operating substantially as and for the purposes herein set forth.

No. 56,214.—FREDERICK HANDSCHUH, Allentown, Penn.—*Combined Chair and Fan.*—July 10, 1866.—When the chair is rocked the levers are alternately operated, and motion is communicated to a fan shaft supported from the upper part of the chair back.

Claim.—First, the levers B and C, as constructed, the cords D and E, and the rod F, with fan attached, the several parts being combined and used as and for the purpose specified.

Second, the bar K and the shaft c, when used with the rod F, substantially in the manner and for the purpose herein specified.

No. 56,215.—P. H. HARDY, Terre Haute, Ind.—*Washing Machine.*—July 10, 1866.—Explained by the claim and cut.

Claim.—The combination of the corrugated false bottom G, the movable side pieces E E, set in angular grooves, and the beaters D, movable from a point in the cover B, substantially as described.

No. 56,216.—B. S. HARRINGTON, Boston, Mass.—*Spring Bed.*—July 10, 1866.—Explained by the claim and cut.

Claim.—A spring bed, the springs of which are fastened to the sides of the bars *b b*, and are operated upon by bent levers in such a way that the bed bottom presses upon one arm of the levers, causing the other arm of said levers to press upon the springs, substantially in the manner and for the purpose set forth.

No. 56,217.—JAMES J. HARRISON, St. Michael, Md.—*Composition for Preserving Meat.*—July 10, 1866.—Composed of salt, 8; alum, 4; saltpetre, 4 parts.
Claim.—The composition, consisting of the ingredients in about the proportions and for the purpose described.

No. 56,218.—ROGER HARTLEY, Pittsburg, Penn.—*Pump.*—July 10, 1866.—The plunger is packed in the middle of the cylinder.
Claim.—In double-acting plug plunger pumps the combination of the cylinder A, having a stuffing box B, with the cylinder N and the gland T, constructed and arranged as described and for the purpose specified.

No. 56,219.—THOMAS HAWKINS, Auburn, N. Y.—*Beehive.*—July 10, 1866.—The sectional arrangement permits the portions to be divided into two perfect hives, or combined as required.
Claim.—The arrangement of the chambers C, box E, chambers C', the base H, and the tie M, the whole constructed in the manner and for the purpose herein specified.

No. 56 220.—F. HAZARD, Mauch Chunk, Penn.—*Machinery for Weaving Wire Rope.*—July 10, 1866.—The rim is pivoted eccentrically to the reel and revolves with it; the frames which support the spools in the reel are so actuated by the cranks as to preserve in a horizontal plane the axes of the spools.
Claim.—First, the flanges B B' B'', in combination with the axle A spool frames, C C'. cranks *c c'*, and ring E, constructed and operating substantially as and for the purpose described.
Second, the joints *b b'*, in combination with the flanges B B', frames C C', cranks *c c'* and ring E, constructed and operating substantially as and for the purpose set forth.
Third, the arm *j*, and set-screw *h*, in combination with the eccentric disk *d*, cranks *c c'*, frames C C', and shaft A, constructed and operating substantially as and for the purpose described.

No. 56,221.—C. W. S. HEATON, Belleville, Ill.—*Pantaloon Protector.*—July 10, 1866.—Metallic strips attached to a band are placed within the lower edge of the pantaloons so as to receive the wear of the boot and other abrading objects.
Claim.—First, the use of separate metallic strips, or their equivalents, in connection with a pliable band, or when attached directly to the garment, when such metallic strips are placed perpendicular with the leg of the wearer, substantially as described and for the purpose set forth.
Second, the use of separate metallic strips, or their equivalents, when attached as described, or in any other way substantially the same, either by passing the pliable band through the strips of metal, the metal strips being made like buckles, or in any other way, substantially in the manner and for the purpose set forth.

No. 56,222.—GEORGE L. HEIDLER, York, Penn.—*Horse Rake.*—July 10, 1866.—The teeth are attached to ferrules upon a rod in front of the axle, and in the rear of the latter are placed two bars connected by rods arranged in pairs and bracing the teeth laterally. The upper bar acts as a pressure bar.
Claim.—The arrangement of the independent teeth attached to the head in front of the axle, and the pressure board to guide and press the teeth to the ground, constructed and operating as herein described.

No. 56,223.—ROBERT HENEAGE, Buffalo, N. Y.—*Hose Coupling.*—July 10, 1866.—Studs on one section of the coupling prevent the disconnection of the sections by turning one upon the other after being locked by their flanges and the screw sleeves.
Claim.—In combination with the several parts of the coupling, constructed and arranged substantially as described, the stops P P, as and for the purposes set forth.

No. 56,224.—JAMES and HENRY A. HOUSE, Bridgeport, Conn.—*Rotating Hook of Wheeler & Wilson's Sewing Machine.*—July 10, 1866.—By the chamfering of the hook the loop is enabled to pass the loop check more readily, and when passed to escape lightly from the hook.
Claim.—Chamfering the Wheeler & Wilson sewing machine hook, substantially in the manner and for the purpose set forth.

No. 56,225.—ZEBULON HUNT, Hudson. N. Y.—*Culinary Sink.*—July 10, 1866.—A portion of the inclined bottom is corrugated; the discharge end has a removable cover, a strainer, and a water trap.

Claim.—First, the combination of the corrugated or fluted drainer A and removable slotted piece B with and forming part of the bottom of the sink, substantially as and for the purpose set forth.

Second, the flaring or funnel-shaped strainer C, in combination with the drop partition or curtain *a a*, substantially in the manner set forth.

No. 56,226.—JOSEE JOHNSON, New York, N. Y.—*Fruit Jar.*—July 10, 1866.—The bent yoke catches beneath the bead on the neck of the jar and rests on the inclined ridges of the cover which enters the neck.

Claim.—First, the semi-ring C and cross-strap C', arranged to operate relatively to the body and cover of a self-sealing can or jar, substantially as herein specified.

Second, the partial spirals *b¹ b²*, arranged opposite to each other on the cover, in combination with a securing piece C C', substantially as and for the purpose herein set forth.

No. 56,227.—E. O. JONES, Oakwood. Mich.—*Horse Rake.*—July 10, 1866.—The forward forked ends of the handle embrace the narrow portions of the grooved uprights and the notched pendants receive the ends of the teeth to hold the rake in working position.

Claim.—First, the standards C C, grooved as described to receive the slotted ends of the levers F F, when used with the said levers and the rake shaft, as and for the purpose specified.

Second, the levers F F, provided with the slotted arms G G, when used to catch the ends of the rake teeth, as and for the purpose specified.

No. 56,228.—SAMUEL F. JONES, St. Paul, Ind.—*Water Elevator.*—July 10, 1866.—As the bucket ascends a stem rising from the valve is engaged and tipped by a pendant from the frame and the valve opened, discharging the contents of the bucket into the spout.

Claim.—First, operating the valve S, through the medium of the standard *o* and pendant *g*, when constructed and arranged substantially as shown and described for the purpose set forth.

Second, in combination with the foregoing the standard *o* and pendant *g*, the spouts *z* and D and the screen T, when constructed and arranged as shown and described for the purpose set forth.

No. 56,229.—DANIEL JUDD, Hinsdale, N. Y.—*Excavator.*—July 10, 1866.—The scoops are pivoted and gudgeons to pedestals beneath the wagon frame, and are rotated and discharged by chains and winches operated as required by hand cranks.

Claim.—First, the combination of the excavator E with side supports *d* and frame A, substantially as described.

Second, constructing the ends of the excavators with gudgeons *c c* and stops *e e*, substantially as described.

Third, the forked guides *h h* applied to the excavator, in combination with the flanged pulleys *j j* applied to the winding shafts G, substantially as described.

Fourth, the combination of the pawl and pinion applied to a vibrating arm with the excavator E, spur wheel *k*, and shaft G, substantially in the manner and for the purpose described.

No. 56,230.—JULIUS J. JUSTIN, Milwaukee, Wis.—*Beehive.*—July 10, 1866.—The adjacent boxes are united by communicating openings. Slats compose the bottom of the hive and an inclined metallic flange arrests the entrance of the bee moth.

Claim.—The combination of the boxes A B, slotted as described, slats G, and projecting guards H, arranged and operating in the manner and for the purpose herein specified.

No. 56,231.—J. E. KARELSEN, New York, N. Y.—*Diamond Holder*—July 10, 1866.—The diamond holder and the notched glass breaker slip out of the respective ends of the case.

Claim.—A diamond holder composed of a tube A, provided with two sleeves *b f* moving in notched slots *d g* and containing the head B and the glass breaker C, as a new article of manufacture.

No. 56,232.—A. C. KASSON, Milwaukee, Wis.—*Cooking Apparatus.*—July 10, 1866.—The box has a hinged cover, a bottom whose flanges fit the boiler holes, a cook stove, and a perforated shelf revolved by clockwork. The fire shines directly into the box, and the arbor of the revolving plate gives motion to the axis of a coffee roaster.

Claim.—First, a cooking apparatus composed of a close box, capable of being thrown open as above shown, having openings in its bottom which are provided with flanges that fit in the boiler holes of a stove or range, substantially as described.

Second, the box A, made substantially as described with flanged openings on its bottom and with a hinged cover at its top, in combination with the perforated plate H, substantially as described.

Third, the combination of the box A, the perforated plate H, and the clockwork which gives rotary motion to said plate, substantially as above described.

Fourth, the adjustable bearing bar L, in combination with the roughened hub J of the revolving plate H, substantially as described, for the purpose of supporting and revolving a coffee roaster, or other cylindrical article.

No. 56,233.—JOHN KING, Ansonia, Conn.—*Oiler.*—July 10, 1866.—For reaching shafting, &c., in an elevated position the oiler is placed on the end of a handle and the oil ejected at the goose-neck by a plunger actuated by a wire and a trigger lever on the handle

Claim.—The combination of the oil vessel A, tube E, having plunger or piston H and discharge spout F, when said plunger is connected to a lever M, or its equivalent, and the whole together is constructed and arranged so as to operate substantially in the manner described and for the purpose specified.

No. 56,234.—OBADIAH B. LATHAM, Seneca Falls, N. Y.—*Oil Well Tube.*—July 10, 1866.—A flexible packing exterior to the tube may be expanded by forcing a conical collar behind a flexible annulus. The tube is held from revolving by a chisel-formed lower termination.

Claim.—First, the cylinder C and band i, varying from a true circle and arranged in relation to the cylinder H and an external packing device, substantially as and for the purpose described.

Second, the sac f, when used in combination with the parts H F G E D d, as and for the purpose set forth.

Third, the chisel G, when used in combination with the described apparatus, for the purpose set forth.

Fourth, the whole apparatus, arranged as described.

No. 56,235.—JACOB LEBEAU, Cincinnati, Ohio.—*Paring Knife.*—July 10, 1866.—The blade is fastened to the handle by set-screws and the throat adjusted by the gauge plate; a bent portion of the blade forms a scoop.

Claim.—First, the combination of the handle A B and blade E, when the latter is adjustable on the former in its own plane, substantially as and for the purposes set forth.

Second, the combination of the gauge piece D with the handle A B and blade E, substantially as and for the purposes set forth.

Third, the scoop G, in combination with the blade E and handle A B, as and for the purposes specified.

No. 56,236.—C. M. LIGHTNER, Harrisburg, Penn.—*Corn Harvester.*—July 10, 1866.—Two rows are cut at one time by stationary cutters in the sides of the openings, and a rotating cutter above, which reaches across both openings. The corn is received on a platform of rods in the rear, which discharges it by turning the bar to which the rods are attached.

Claim.—First, the platform provided with openings as shown, and having the knives c attached, in combination with the centrally located revolving knives f, arranged and operating as shown and described.

Second, the combination and arrangement of the pivoted bar D, having the bent bars o attached, as shown with the hook m, all arranged to operate as set forth.

No. 56,237.—ANDREW J. and HENRY LINEBARGER, Jackson, Ill.—*Grain Separator.*—July 10, 1866.—The floor of the shaking shoe has vibrating rake heads, which lighten up the passing grain, &c., and prevent choking.

Claim.—The shaker a, in combination with the rake heads g, constructed and operating substantially as described.

No. 56,238.—JAMES LOCHRIDGE, Danville, Ind.—*Washing Machine.*—July 10, 1866.—Explained by the claim and cut.

Claim.—The combination of the rubbing board resting upon springs, the movable rubber 9, with the rollers 1 2 3, springs 8 8, and strips 6 6, for the purpose and in the manner specified.

No. 56,239.—JAMES W. LYON, Brooklyn, N. Y.—*Machine for Drilling and Countersinking Umbrella Tips.*—July 10, 1866.—A portion of the lathe, carrying two drills and a holding device, slips to and from a drill at right angles to the former, and adapted to bore out the base of the tip, while the side drills bore and countersink the transverse hole. When the sliding portion retires from the stationary drill a stationary punch passes through the bottom of the hopper, feeds in another blank, and pushes out the finished tip.

Claim.—First, the stationary pusher, and the hopper with its rear entrance, bottom groove, and short tube in front, in combination with the sliding frame and main frame, or shears, substantially as described.

Second, the combination of the holding mechanism with the feeding mechanism and the sliding bed of a lathe, the mechanism and combination being such, substantially as herein described, as to grip and hold the blank during the forward movement of the sliding bed, to release and discharge the finished blank, and feed a new blank during the back motion of the sliding bed.

Third, in combination with the socket drill on the main frame, or shears, the cross drill for drilling and countersinking the cross holes, when combined with the sliding bed and operated by cams on the main frame, substantially in the manner described.

Fourth, in combination with the socket drill on the main frame, the feeding mechanism and the transverse or cross-drilling mechanism, the whole being arranged and combined, substantially as hereinbefore described, to feed, hold, drill, countersink, and discharge blanks supplied to the machine by the back and forward motions of the sliding bed, substantially as set forth.

No. 56,240.—CHARLES MACRAE, New York, N. Y.—*Automatic Gas Cock.*—July 10, 1866.—A wheel with four peripheral recesses is placed on the axis of the gas cock and held by a catch; the latter is connected by intermediate devices with a clock, and released by pre-arrangement at a given hour, turning the cock 90°.

Claim.—First, operating an ordinary gas cock by means of a spring, or its equivalent, applied directly to the cock, substantially as and for the purpose set forth.

Second, in combination with a gas cock, operated as above described, the stop wheel G and pawl *f*, so arranged as to be released by the movements of the clock-work, substantially as herein described.

Third, in combination with a gas cock operated as described, the hands *b* and *c*, arranged to operate as and for the purpose set forth.

Fourth, the stationary dial, provided with the two graduated circles, in combination with the hands *b c* and *d*, when said parts are arranged to operate in connection with a gas cock, operated as herein described.

No. 56,241.—WILLIAM MANNING, Chelmsford, Mass.—*Corn-cake Machine.*—July 10, 1866.—A former is provided to receive the parched, pulverized, prepared corn, and the frame has any number of knives and clearers to indent and finish the cake in squares as required.

Claim.—First, the stationary knives *c*, and adjustable clearers or hoofs *e*, as herein described and for the purpose set forth.

Second, the former *l*, in combination with the adjustable clearers or hoofs *e*, and knives *c*, for the purpose specified and in the manner set forth.

No. 56,242.—JAMES R. MAXWELL and EZRA COPE, Cincinnati, Ohio.—*Steam Engine.*—July 10, 1866.—The piston performs the functions of valves, having interior passages and dispensing with exterior valve gear. The cylindrical valve in the piston has closed ends and openings in its sides and bottom.

Claim.—First, the piston head of a steam engine longer than its stroke, in the manner and for the purpose herein described.

Second, the piston head of a steam engine longer than its stroke, with channels and ports in its sides, in combination with a cylinder having corresponding added length and ports leading to and from the main steam valve, in the manner and for the purpose substantially as described.

Third, the cylindrical piston valve C, with closed ends, and steam openings through its sides and bottom whereby to operate within the main piston, substantially as described.

Fourth, the independent steam pipe and valve P P, for admitting steam to the main piston B, in cylinder A, in the manner and for the purpose herein described.

No. 56,243.—ABNER MCOMBER, Schenectady, N. Y.—*Artificial Limb.*—July 10, 1866.—Within a cavity in the foot is hung a vibrating bar attached by a cord to the leg to assist in regulating the vertical movement of the foot. The hinge joints have divided bushings of hard wood, tightened by screw and wedge as they become worn.

Claim.—First, giving a yielding bearing to the lower end of cord U, by means of the elastic band *g*, and vibrating bar, substantially as above set forth.

Second, placing the elastic band *g*, and vibrating bar *e*, in a movable frame *d*, and also fitting said frame in a cavity G, in the bottom of the foot, substantially as shown.

Third, bushing the hinges of artificial limbs with divided bushings Y, of lignumvitæ or other hard wood, and taking up the wear of the bushing by means of a wedge Z, operated substantially as above described.

No. 56,244.—J. C. MERRIAM and W. N. WHEEDEN, Boston, Mass.—*Embossing Press.*—July 10, 1866.—The vibrating arm moves vertically within a guide case, a spring making the recoil motion, and set screws on the sides regulating its position. Beneath the guide case is a place to receive the paper when the impression is to be distant from the edge.

Claim.—The combination and arrangement of the channelled guide arm F, and its adjusting screws, with the dies A B, and their stationary and movable arms C D, applied together substantially as specified.

Also, the arrangement of the elevating screw *a* of the movable die arm, viz., in the guide arm, combined and arranged with the stationary and movable arms of the two dies, as set forth.

Also, the arrangement of the sheet-receiving recess or space G, with the guide arm, the dies, and their supporting arms, arranged and combined substantially as specified.

No. 56,245.—RUFUS S. MERRILL, Boston, Mass.—*Lamp Burner.*—July 10, 1866.—Explained by the claims and cut.

Claim.—First, the combination in a fluid burner, and with a collecting chamber surrounding the wick tube, of an extinguisher or cap, under the arrangement and for operation substantially as hereinbefore set forth.

Second, forming the collecting chambers of a conical jacket secured to a concentric sleeve adjustable by friction upon the wick tube by means of an annular partition plate, substantially as and for the purposes herein set forth.

No. 56,246.—JAMES W. MILES, Hubbardston Village, Mich.—*Grinding Mill.*—July 10, 1866; antedated July 3, 1866.—A collar is keyed upon the spindle and connected by rods to the ring which is socketed in the runner; each arm of the collar is connected to a block which has an elastic bearing in its seat.

Claim.—Transmitting motion from the driving power through the mill spindle to the runner, by means of jointed driving bars F, operating tangentially against elastic bearings D, or their equivalent, substantially in the manner and for the purpose as herein described and set forth.

No. 56,247.—JAMES MIXTER, Lowell, Mass.—*Bolt Heading Machine.*—July 10, 1866.—The beading die has notches, and is so connected with the driver that the operator, by means of a treadle, can upset the head of the bolt by one stroke, or gradually.

Claim.—The notches *a n* on the carrier H, operating in conjunction with the driver D, or their equivalents, whereby the same end is accomplished, substantially as above set forth and described.

No. 56,248.—GEORGE A. MITCHELL, Turner, Me.—*Safety Pocket.*—July 10, 1866.—Explained by the claims and cut.

Claim.—A pocket made capable of expansion and contraction, by the provision of folds in its sides and bottom and provided with metallic protectors as defensive armor, constructed and arranged substantially as and for the purpose specified.

Also, in combination therewith the yielding cover, constructed and arranged substantially as described.

No. 56,249.—ALVAN MORLEY, Delaware township, Iowa.—*Stock Yard.*—July 10, 1866.—The panels of the pen are mounted on wheels.

Claim.—The construction of a stock yard or grazing pen, constructed substantially as and for the purposes specified, named a "self-moving stock yard."

No. 56,250.—HERMANN MUND and ERDMANN HOFFMAN, Chicago, Ill.—*Lamp Wick Regulator.*—July 10, 1866.—The wick-regulating lever maintains its place without a pivot.

Claim.—The wick controller *g*, constructed and applied to the burner in the manner and for the purpose herein set forth.

No. 56,251.—L. B. MYERS, Elmore, Ohio.—*Pencil Holder.*—July 10, 1866.—Explained by the claim and cut.

Claim.—An elastic socket capable of receiving and adjusting itself to hold a lead pencil of larger or smaller size, and having a hook or suitable device or means for securing it to the garments of a person, constructed and operating substantially as specified.

No. 56,252.—L. M. NEWBURY, Sparta, Wis.—*Apparatus for Cleaning Boots and Shoes.*—July 10, 1866.—The butts of the broom corn or other material are clamped between the stationary and movable jaw by means of the screw bolt.

Claim.—An apparatus for cleaning boots and shoes, formed by combining the hinged clasp A C D with the brush H, the whole being constructed and arranged substantially as described and for the purpose set forth.

No. 56,253.—JAMES W. NEWMAN, Boston, Mass.—*Neck Tie.*—July 10, 1866.—The bow is folded and fastened by an elastic loop, whose edges pass through a slit in the supporter and are turned over in its rear.

Claim.—First, the neck tie, consisting of the parts A B B′ B″ and B‴, in combination with the supporter C, secured together by the parts D D′, substantially as described.

Second, in combination therewith the elastic loop F, applied substantially as described.

No 56,254.—DANIEL NILES, Fly Creek, N. Y.—*Planing Machine.*—July 10, 1866.—The cutter revolves in fixed bearings; the tapered shingle is fed upon a hinged adjustable frame, whose feed-rollers have yielding bearings.

Claim.—Planing tapering shingles on one side by means of the revolving cutter in fixed bearing, and feeding the shingles to such cutter upon a hinged adjustable frame, with feed-rollers that are in yielding bearings, and arranged to operate substantially as described and for the purpose set forth.

No. 56,255.—JOSHUA NORTON, 3d, Chicago, Ill.—*Grain Dryer.*—July 10, 1866.—Improvement on patent of Dole and Fraser, September 1, 1863. The grain while passing through the heated current of air is subjected to constant agitation; the two cylinders revolve together, and their alternate flanges receive and discharge the grain which tumbles upon them, passing through the space between the concentric cylinders discharging through the stationary head.

Claim.—First, the concentric, non-perforated cylinders C and *c*, having the ribs or flanges F attached thereto, as shown and described.

Second, in combination with the cylinder constructed as described, the stationary head I, constructed and operating as set forth.

No. 56 256 —EDWARD A. F. OLMSTEAD, New York, N. Y —*Railroad Sweeper.*—July 10, 1866.—The brushes are revolved by gearing on their shafts and on the wheel axles, or by belt and pulleys ; are adjustable vertically ; may be preceded by oblique and vertically adjustable scrapers ; set obliquely to clear the horse track, or revolve in line with the rail to clear the wheel track.

Claim.—First, the arrangement of the brush bars in combination with the arms S and shaft O, as described.

Second, the combination of the brush bars, arms U, and shaft P, in the manner and for the purpose herein represented and described.

Third, the arrangement of the rods W, levers X, and screw block B, in combination with the shaft P bearing in the slotted arms M N, in the manner and for the purpose herein described.

Fourth, the combination of the scraper F, rods G H, arm I, lever M, in the manner and for the purpose herein described.

Fifth, the arrangement of the brooms R T, gearing O' P' R' S', brooms V' W', pulleys A² B², in combination with the frame A and wheels E F G H, constructed and operating in the manner and for the purpose herein specified.

No 56,257.—ZIBA PARKHURST, Milford, Mass.—*Burr-box for Burring Machines.*—July 10, 1866.—The curved lip extends somewhat over the upper feed roller to prevent the burrs, &c., from falling back upon it, and also to guide them into the box.

Claim.—The improved burr receiver or box, as provided with one or more guards B C D, whether stationary or adjustable, and arranged substantially as hereinbefore described, the same being to arrest the waste filaments of fibrous material, so that they may be caused to fall back upon the burring machine, while the burrs or foreign matters are in the act of being discharged into the box.

Also, the combination and arrangement of the lip E with the box and its guard or guards, as specified.

No. 56,258.—HOMER PAMELEE, Philadelphia, Penn.—*Machine for Twisting and Winding Fibres.*—July 10, 1866.—This machine is for twisting single strands of flax, hemp, &c. The material lies loosely in a trough from which it enters between the twisting rollers. The receiving spool is rotated by the positive revolution and weight of the toothed guide wheels, the strand being guided between them. The bearings of this spool are automatically shifted that its axis may incline a little to insure the better laying of the coil ; the coil as laid gradually slides the spool upon its axis, first in one and then in the opposite direction. The barrel of the spool is hollow and compressible to allow the putting on and off of the heads and removing the material when wound.

Claim.—First, the two rollers E and E', arranged to revolve on their own axes and around a given centre, in combination with a trough D, or its equivalent, for receiving the fibre, the whole being arranged and operating substantially as and for the purpose herein set forth.

Second, the rollers E and E', each being composed of a block *w*, with a central cavity *x* and a tube *y* of gum elastic or equivalent material, substantially in the manner and for the purpose described.

Third, the combination of the rollers E and E' and the levers 4 and 4', or their equivalents, whereby the simultaneous inward and outward adjustment of the two rollers is attained.

Fourth, the said rollers E and E' and levers 4 and 4', in combination with the tubular shaft F.

Fifth, the combination of the barrel for winding the strand of twisted fibres with the wheels M, the latter being arranged to guide the strand in a direct course, all substantially as and for the purpose described.

Sixth, the said wheels M, and the train of wheels L I I' and I'', or any equivalent train of wheels, in combination with the arm J.

Seventh, a barrel arranged to revolve on and to traverse longitudinally an axis situated obliquely to the course taken by the guided strand, all substantially as and for the purpose described.

Eighth, the shaft R, on which the barrel turns and traverses, and the levers *q* and *q'*, or their equivalents, in combination with the cams R', on the said barrel, and the stops R'' and R''', the whole being arranged and operating substantially as and for the purpose specified.

Ninth, the barrel composed of the tube P, cut longitudinally, and the heads Q Q, on applying which to the tube the latter is expanded, the contraction of the tube taking place on detaching the heads, all substantially as and for the purpose herein set forth.

No. 56,259.—C. C. PARSONS, Boston, Mass.—*Soap.*—July 10, 1866.—Composed of petroleum residuum, alkalies, and animal or vegetable oils or resin.

Claim.—As a new manufacture, soap, in which the described petroleum residuum is one of the ingredients.

No. 56,260.—HOLLIS M. PEAVEY, Swanville, Me.—*Machine for Digging Potatoes.*—July 10, 1866.—The potatoes are raised by a pointed broad share, and passing to the rear are received upon a hinged vibrating grate : a hoe with angular projections is vibrated above them to assist in the separation.

Claim.—The combination as well as the arrangement of the compound hoe G and its operative mechanism with the scoop E and the vibratory separator F applied to a wheel carriage, and provided with mechanism for operating the said separator, as explained.

No. 56,261.—CLARK M. PLATT, Waterbury, Conn.—*Button.*—July 10, 1866.—Explained by the claim and cut.

Claim.—The button formed of a single piece of metal with the edge turned over, and with one central hole, as a new article of manufacture, as specified.

No. 56,262.—H. H POTTER, Carthage, N. Y.—*Burglar Alarm.*—July 10, 1866.—The opening door vibrates the lever and gives a stroke upon the bell for each notch on its serrated arm.

Claim.—The bell C, or its equiva'ent, lever I having toothed arm L, bell-hammer D and springs G and K, when arranged and combined together substantially as described and for the purpose specified.

Also, the slotted upright M of the bed plate A, as and for the purpose described.

No. 56, 263.—WILLIAM POUNTNEY, Brooklyn, N. Y.—*Manufacture of Cruet Bottle.*—July 10, 1866.—The mouth and lip are formed by portions which press upon the exterior and interior surfaces simultaneously, the plunger being retracted by a lever and the jaws by unclasping the bow of the tool.

Claim.—First. combining the plunger E and lip and mouth-former G (made or united in one piece) with the jaws B B, the said part G having a rotary motion with the cruet bottle back and forth, arranged and operating substantially in the manner and for the purposes described; also, the combining and uniting the lip and mouth-former with the neck-former or plunger in one piece.

Second, the construction of the stopper mould with a sliding bottom, plunger, or plug to push up the stopper by pressing with a lever, combined and operating in the manner and for the purposes described.

No. 56,264.—JOEL B. PRATT, Corning, N. Y.—*Knife Sharpener.*—July 10, 1866.—The knife edge is drawn in the angle formed by the bevel edge of the circular cutter and the standard.

Claim.—The arrangement of the wheel with the standard, the bevel wheel resting by its axle on a wooden frame and lapping two upright metallic posts, all arranged and combined as set forth.

No. 56,265.—J. QUINN and C. SUMMERS, Columbus, Ohio.—*Snap Hook.*—July 10, 1866.—It is cast in two pieces; the tongue-piece is held in place by a spring which keeps its slot against the pin on the shank.

Claim.—As a new article of manufacture the within-described snap hook, constructed, arranged, and operating substantially as set forth.

No. 56,266.—ROWLAND J. RATHBURN, Poplar Ridge, N. Y.—*Attachment to Cooking Stoves.*—July 10, 1866.—A pivoted shelf, supported on standards near the rear of the stove ; can be laid up against the pipe or placed horizontally to hold dishes.

Claim.—The combination with the stove of a shelf C, constructed and arranged so as to accomplish the purposes specified.

No. 56,267.—W. F. REDDING, Saratoga Springs, N. Y.—*Window Blind.*—July 10, 1866.—A spring on the spindle of the slat maintains the angular position at which the slat is set.

Claim.—The combination of the spindle G, having collar H and spiral spring F, when arranged in and connected to the slat and frame of a blind, substantially as and for the purpose described.

No. 56,268.—JOHN C. RHODES, East Bridgewater, Mass —*Machine for Capping Tacks.*—July 10, 1866 —The frame of three rotary wheels has a central stationary shaft. The lower plate supports dies, and the upper one has corresponding plungers. As the frame revolves each plunger is depressed by a cam ; each die has a discharger operated by a cam and spring to expel the capped tack.

Claim.—The combination of the rotary frame A and its dies C, plungers D, their springs E E and clearers g, and their operative mechanism, with the stationary cam G, the whole being arranged substantially in manner and so as to operate as and for the purpose hereinbefore described.

Also, the combination of the rotary frame A and its dies C, plungers D, and clearers g, with the stationary cam G and the spring A and stud i, to operate as specified.

No. 56,269. GEORGE T. RIDINGS, Shelbyville, Mo.—*Tire Machine.*—July 10, 1866.—The tire is fastened at one point in a vise, and at another is attached the clamp block shown in the cut; blows of a hammer on the block upset the heated portion of the tire intervening between the block and the vise.

Claim.—The combination of the block A B with the eccentric clamp C, when employed as and for the purpose set forth.

No. 56,270.—GEORGE T. RIDINGS, Shelbyville, Mo.—*Rotary Tire Heater.*—July 10, 1866.—A rotary, metallic, cylindrical drum encloses the tire while heating. The tire is held by spokes and segmental plates which are adjusted by a rotary shaft and screw which engages cogs on the spokes.

Claim.—First, the arrangement of the above described tire-holder and drum in a vertical position, the tire-holder turning on a horizontal axis, and the whole operating in combination with a common forge-fire, substantially as set forth.

Second, the combination and arrangement of the adjustable tire-holder with the drum by means of the pillow blocks D^2, and pawls t, and the adjustable plates g, as and for the purpose set forth.

Third, the combination and arrangement of the spokes a^1 and a^2, and the segmental plate b, with the endless screw c^1, and wheel c^2, as and for the purpose set forth.

No. 56,271.—ADDISON ROBBINS, Orange, Mass.—*Railway Frog.*—July 10, 1866.—The pivoted tongue and the portions of the two rails next to its point are made movable by the wheels so as to be adjusted by a train approaching in either direction.

Claim.—The combination as well as the arrangement of the movable angle piece G, and auxiliary rails C D, and their stopping devices, applied together, substantially as described, with the crossing rails of the main and turnout tracks, as specified.

No. 56,272.—G. M. ROSS and WILLIAM H. WEST, New York, N. Y.—*Tobacco Pipe and Cigar Holder.*—July 10, 1866.—The tube in the stem can be brought into connection with the smoke passage in the shank of the bowl, or by rotation, with a smaller tube or cigar holder.

Claim.—First, the combination of the pipe A and cigar socket or holder E, substantially in the manner and for the purpose specified.

Second, the enlargement b of the bore or smoke passage c c of the stem B, as and for the purpose herein specified.

No. 56,273.—C. RUNDELL, Chicago, Ill.—*Hay Stacker.*—July 10, 1866.—A telescopic tube with a flange in one section to extend it when revolved.

Claim.—First, the herein described tubular stack-pole, consisting of the sections A A' D, the same being constructed and operating as and for the purpose set forth.

Second, the plate B in combination with the sections A and D, as and for the purpose specified.

No. 56,274.—JOHN L. RUSSELL, Pella, Iowa.—*Cotton Seed Planter.*—July 10, 1866.—Explained by the claims and cut.

Claim.—First, the wedge-shaped furrow opener F, suspended in an adjustable manner by steadying studs g g, and adjusting screws H H, and having a rotating trash cutter f, substantially in the manner and for the purpose set forth.

Second, the teeth or small shovels m, in connection with and occupying a position between the furrow-opener F, and the closing roller N, substantially as and for the purpose described.

Third, the toothed rollers C and D, with or without the flaring ends c' c' d' d', arranged one above the other in the seed box, and operating as shown and explained, substantially in the manner and for the purpose specified.

Fourth, the sliding roller and frame N n, scraper p, elevating lever q', and stop-chain l, arranged and operating as shown and explained.

No. 56,275.—JONAS RYMOND, Erwinna, Penn.—*Water Elevator.*—July 10, 1866.—A notch on the brake lever engages the notched hub on the end of the windlass shaft when rotated in one direction, and allows the shaft to rotate under frictional contact when the tooth is vibrated out of engagement therewith.

Claim—The combination of the crank D with the ratchet wheel B, the hub E, and the spiral brake spring c, constructed and arranged as and for the purpose herein described.

No. 56,276.—WILLIAM H. SANGSTER and THEO. C. SPENCER, Buffalo, N. Y.—*Distilling Petroleum.*—July 10, 1866.—The vapor from the still is passed into a body of water and is thereby condensed.

Claim.—First, the combination of the tank, Fig. 2, or its equivalent, the tube C, and faucet E, with a still for distilling petroleum, when constructed as and for the purposes herein substantially set forth.

Second, the method herein described of condensing the vapor of petroleum by passing it directly through the water.

No. 56,277.—RUDOLPH SCHMIDT, New York, N. Y.—*Harness Bell.*—July 10, 1866; antedated June 26, 1866.—The clapper has radiating arms and is attached to the boss by flat metallic springs.

Claim.—First, forming the spring for the clapper of flat pieces of metal, in the manner and for the purposes specified.

Second, constructing the clapper with three or more radiating arms, in combination with the spring carrying such clapper, as specified.

No. 56,278.—HEBER G. SEEKINS, Elyra, Ohio.—*Bed Bottom.*—July 10, 1866..—The longitudinal cords are attached to rollers which are tightened by rotation, and the middle cords have a higher attachment in anticipation of the greater tendency to sag at that point.

Claim.—The rollers B, provided with mortises *a*, arranged in an ascending and descending series, as and for the purpose described in combination with cords C, and pawl and ratchet D, all substantially as described.

No. 56,279.—J. W. SHIVELEY, New York, N. Y.—*Railroad Rail.*—July 10, 1866.—The tread neck and flanges of the rail are scarfed together and sustained laterally by a double fish-bar and chair.

Claim.—First, providing the rail ends with angular tongues *a a a a*, and angular recesses *b b b b*, when fitted and operating together as herein shown and described.

Second, the combination with the above-mentioned tongues and recesses of the double-cheek and lock bar C, substantially as herein shown and described.

Third, the combination of the said tongues, recesses, and cheek-bar with the chair A, substantially as herein shown and described.

Fourth, the recesses A A¹ A², in combination with the rails, the cheek-pieces, and the chair, substantially as and for the purpose herein shown and described.

No. 56,280.—JACOB SHINELLER and JOHN BRISLIN, Temperanceville, Penn.—*Horseshoe Machine.*—July 10, 1866.—The revolving disk has a die and a clamp, also a lever, provided with a swaging and punching die and a cutter. The devices are used in combination with guides, cams, springs, and gearing.

Claim.—First, the revolving disk *w*, provided with die S and clamp *y*, when used in combination with the adjustable guide J, or its equivalent, as herein described and for the purpose set forth.

Second, the lever *g*, provided with the swedging and punching die 8 and cutter 6, when used in combination with the revolving disk *w*, die S, clamp *y*, and guide J, as herein described and for the purpose set forth.

Third, the cam *i*, provided with the lug 11, when used in connection with the lever *g* and the press tool 4, as herein described and for the purpose set forth.

Fourth, the disk *o* and spring *p*, when used in combination with the wheels or gear *m* and *n*, as herein described and for the purpose set forth.

Fifth, the feeding guide *l*, provided with friction roller *z*, when used in combination with the cam *k*, spring *u*, and disk *w*, provided with die S and clamp *y*, as herein described and for the purpose set forth.

No. 56,281.—GEO. M. SIMONDS, Lynnfield, Mass.—*Securing Axes on their Handles.*—July 10, 1866.—Explained by the claim and cut.

Claim.—The combination of the expander C, the screw D, and the step E, applied to the tapering chamber B of the axe helve, the whole being arranged and so as to operate substantially as specified.

No. 56,282.—A. P. SMITH, Stirling, Ill.—*Window Blind.*—July 10, 1866.—The connecting rod of the slats is held in position by the catch and notched bar.

Claim.—Connecting the slat rod D to the blind frame through a catch and rack or notched bar, when all are arranged together, and so as to operate substantially in the manner described and for the purpose specified.

No. 56,283.—HECTOR CRAIG SMITH, Dublin, Ind.—*Harness for Horses.*—July 10, 1866.—The draught and hold-back straps are brought together and attached to the shafts at a single point on each side of the animal.

Claim.—First, the hitching device, consisting of a link composed of the following members, to wit, the horizontal bar E, vertical end rods F G, and hitching pin H, together with their accessories or devices, substantially equivalent, all arranged to operate in the manner and for the purpose herein described.

Second, in combination with the elements of the preceding clause, the cross-head K and springs *i j*, or their mechanical equivalents, for the object explained and set forth.

No. 56,284.—ISAAC SMITH and W. D. TEWKSBURY, New York, N. Y.—*Revolving Hose Nozzle.*—July 10, 1866.—Explained by the claim and cut.

Claim.—The double oblique swivel joint, so constructed that the tip is susceptible of arrangement at an acute or other angle, or in a right line with the body, or that it may be adjusted tangentially to the circumference of the body, in combination with the revolving collar F on the body of the nozzle, substantially as and for the purposes specified.

No. 56,285.—MATHIAS SMITH, Lake, Ill.—*Street Scraper.*—July 10, 1866.—The scraper is adjusted vertically in guides by the lever, and has side boards which prevent the escape of the earth at the ends.

Claim.—First, the combination of the adjustable scraper H, the guides F, and lever J, arranged and operating substantially as and for the purposes set forth.

Second, in combination with the said adjustable scraper H, the employment of the adjustable side boards L, arranged as and for the purposes described.

No. 56,286.—E. M. SORLEY, Neenah, Wis.—*Cultivator.*—July 10, 1866.—The bars of the frame are capable of adjustment to assume various shapes. The sides are connected by a pivoted brace and connecting rods, and the teeth present a convex edge to the soil.

Claim.—First, the jointed adjustable cross bar B, with arms b b, in combination with the shifting braces c c c e and the side frame A, constructed and arranged substantially as and for the purposes herein described.

Second, the construction and arrangement of the shifting screw-headed and sabre-shaped harrow teeth or cultivators with hollow backs, in combination with the frame A and the adjustable centre cross-bar B, with its draught arms b b, applied and operated as herein stated.

No. 56,287.—HENRY D. SPRAGUE, Portland, Me.—*Coffin.*—July 10, 1866.—Explained by the claims and cut.

Claim.—First, the attachment of a mirror to the lower side of the described kind of coffin lids, as and for the purposes set forth.

Second, the combination of the jointed lid, brace, and mirror, in the manner and for the purposes set forth.

No. 56,288.—ESEN STARR, Royal Oak, Mich.—*Cultivator.*—July 10, 1866.—The rear standards are laterally adjustable on the curved, transverse brace-bar, and the front standard is adjustable as to its raking angle

Claim.—First, the curved or segment bar B at the rear end of the beam A, in combination with the curved standards J J, provided with upper bent ends a* to abut or fit snugly against the rear side of said bar B, and to which they are secured by bolts, substantially as shown and described.

Second, the standard F, bent or curved, as shown, and secured in position by a brace-rod or bar H from the beam A, substantially as and for the purpose specified.

No. 56,289.—D. C. STOVER, Lanark, Ill.—*Cultivator.*—July 10, 1866.—The plough beams are pivoted forward to a transverse bar of the carriage frame, and are supported from a bridge piece and standards on the axle. The middle ploughs are separately adjustable laterally and vertically.

Claim.—The arrangement of the uprights c c and bar d with the pendant bars J J and plough beams D D L L.

No. 56,290.—P. L. SUINE, Shirleysburg, Penn.—*Package Case for Plants.*.—July 10, 1866.—A perforated metallic wrapper for plants, with tuck and loop to connect the edges.

Claim.—A package case made substantially as herein described, for the purposes specified.

No. 56,291.—WILLIAM SYKES, Glenham, N. Y.—*Removing Vegetable Matter from Wool.*—July 10, 1866.—The wool is treated with diluted sulphuric acid, and the acid neutralized by lime water.

Claim.—The described process, consisting in plunging the wool into an acidulous solution, followed by treatment with lime water and subsequent drying.

No. 56,292.—WILLIAM A. TORREY, Mont Clare, N. J.—*Manufacture of Belting, Hose, &c.*—July 10, 1866.—Canvas is first coated with a stratum of vulcanizable rubber compound, and secondly with a similar compound of gutta-percha ; the coatings are then vulcanized.

Claim.—The combining and applying of the India-rubber and gutta-percha compounds, substantially in the manner and for the purposes above set forth.

No. 56,293.—JULIUS VERCH, Albany, N. Y.—*Candle Holder.*—July 10, 1866.—Intended especially for railroad cars. The candle holder is suspended by chains and springs, which admit of its being drawn away from the chimney and permit the removal of the candle.

Claim.—First, the suspension of the candle holder E, with its apparatus, by the chains a a and the barrels c c, with their springs c e, as described.

Second, the candle tube H, entirely closed on the sides and bottom, with its basin h, serving to prevent the access of grease dripping to the spring S, as described.

Third, the method of securing the cap M to the candle tube H by the screw collars f and k, in the manner and for the purpose described.

Fourth, the combination of the outer tube E, the candle tube H, and the cap M, with the globe T and its attachments to the candle holder, in the manner and for the purpose described.

No. 56,294.—ROSWELL WAKEMAN and JOSEPH L. BALLANCE, Port Deposit, Md.—*Baling Hay.*—July 10, 1866.—The cut feed is compressed and, together with its end slats, is bound by hoops.

Claim.—Putting up short cut hay or straw into compact bales for feed, substantially in the manner herein described and for the purposes set forth.

No. 56,295.—G. W. WARREN, Macomb, Ill.—*Cultivator.*—July 10, 1866.—The inner frame, to which the rear plough standards are attached, is pivoted to the rear of the carriage frame, and adjustably supported by a screw upon the forward portion of the latter. The pendent bar which supports the middle plough standards is adjusted vertically by a lever, and the standards are moved laterally by handles on their curved rear extensions.

Claim.—First, the inner frame, pivoted at its rear end to the main frame, having the two rear ploughs attached permanently thereto, and having the pivoted bars T secured to it in the manner shown.

Second, the long standards I, having the front shovels attached thereto secured to the pivoted bars T underneath the axle and frame, and arranged as set forth.

Third, the screws c, arranged to adjust the depth of the ploughs, in the manner shown and described.

No. 56,296.—OSCAR F. WARREN, Pembroke, N. Y.—*Plough Cleaner.*—July 10, 1866.— The vibrating clearer is pivoted to the beam, and operated by a handle to remove weeds from the front of the sheth.

Claim.—First, the adjustable levers O G F, so connected and arranged that the clearing lever may be operated either in front of a coulter or in front of the plough standard, for the purpose and substantially as set forth.

Second, connecting the operating lever O to the plough handle, and within easy and convenient grasp of the hand of the ploughman, so that the said operating lever will also serve as a handle to guide and hold the plough, substantially as set forth.

No 56,297.—JOSEPH C. WATE, Wilton Junction, Iowa.—*Wind Wheel.*—July 10, 1866.— The doors open by the impulse of the wind, which disengages the spring catches ; the brake is applied to check undue speed.

Claim.—The wind doors A, in combination with the springs D and brake K, substantially as herein specified.

No. 56,298.—GEORGE WATT, Richmond, Va.—*Plough.*—July 10, 1866.—The sheth has flanges for the attachment of the mould board, share, and landside ; the handles are socketed in a block on the bent beam, to which lateral adjustment of the landside is made to adjust the plough to the land.

Claim.—First, the frame a b i, cast in one piece, and constructed as described, in combination with the mould board e, landside g, and point k, the whole being constructed and operating in the manner and for the purposes set forth.

Second, the curved beam c, which is constructed to impinge upon the frame behind the central point of resistance of the mould board, substantially as described.

Third, the beam c, in combination with the brace m and adjustable clamps n, for affording lateral adjustment to throw the plough in and out of land, substantially as described.

Fourth, in combination with the curved beam c, the slotted block b', key c'', and wheel journalled to and between the plough handles, substantially as described.

Fifth, the wheel d' journalled to and between the plough handles, substantially as described.

No. 56,299.—J. S. WEAVER, Dayton, Ohio.—*Ditching Machine.*—July 10, 1866.—A pointed, curved, and broad-winged share, with side-bars, raises the soil, which is carried off by an endless elevator and dumped on the sides of the ditch ; the depth is regulated by a wheel and ratchet, and indicated by a dial attachment. The wheels span the ditch.

Claim.—First, the combination of the arch or frame E with the sills F G, the forward axle C, the king-bolt I, and friction wheel H, the parts being constructed and arranged substantially as described and for the purpose set forth.

Second, the combination of the guide wheel L, standard K, and lever M with each other, with the frame of the machine, and with the forward axle C, substantially as described and for the purpose set forth.

Third, the combination of the cog wheel X with the rear axle D, and with the elevator belt W, constructed and arranged substantially as described and for the purpose set forth.

Fourth, the combination of the crank wheel C' and pinion wheels B' with the toothed side bars S and T of the elevator, substantially as described and for the purpose set forth.

Fifth, the combination of the guide and stiffening bars P and R with the side-bars S and T of the elevator, and with the inclined grooved side-bars N and O of the frame of the machine substantially as described and for the purpose set forth.

Sixth, the combination of the index G, dial H', and cog wheel D' with the pinion wheel B', substantially as described and for the purpose set forth.

Seventh, the combination of the spring F' and catch E' with the cog wheel D', substantially as described and for the purpose set forth.

Eighth, the shovel or cutter I', constructed as described, in combination with the side-bars S and T of the elevator, substantially as described and for the purpose set forth.

No. 56,300.—ALLEN A. WEBSTER, Tremont, Ind.—*Sheep Rack.*—July 10, 1866.—The shelter tables rest on adjustable pins, and the box has feed drawers.

Claim.—As an invention and improvement in a sheep rack and cattle feeding rack, the rack pins denoted in the accompanying drawing, and referred to in the foregoing specifications by the letters E E E E E E E E E; and these rack pins meeting and constructing shelter tables B B, appearing in draught and specification, and also feed drawers denoted in the accompanying drawing, and referred to in the foregoing specification by the letters A A.

No. 56,301.—R. T. M. WELLS, Roxbury, Mass.—*Railroad Car.*—July 10, 1866.—Improvement on his patent of January 30, 1865. A ratchet is keyed on to the front axle and encompassed by a loose segment; the tongue is hitched to a sliding bar, which, by intermediate devices, acts upon the ratchet wheel in starting, and is returned by springs as the speed is checked, ready for another effective stroke.

Claim.—The ratchet D keyed on the front axle B of the truck, and encompassed by a segment F placed loosely on said axle, in combination with the pin or pawl b, slide bar H, springs J J', one or more, and the chains I I', all arranged to operate in the manner substantially as and for the purpose set forth.

No. 56,302.—EDWARD WHITEHEAD, Cincinnati, Ohio.—*Car Brake.*—July 10, 1866.—The rotation of the vertical screw-shaft operates the fourfold sets of toggles, brings the jaws against the edges of the rail and the shoes to bear upon its surface.

Claim.—A car brake adapted to embrace the rails of the track, constructed substantially as described, of clamps E F G H, hinged as shown, and operated simultaneously by means of nut M and screw N, as set forth.

No. 56,303.—EDWARD WHITEHEAD, Cincinnati, Ohio.—*Pavement.*—July 10, 1866.—The congeries of blocks is arranged in a frame curb, and the individual blocks rest upon each other's shoulders to expand the area of pressure.

Claim.—First, the combination of the wooden paving blocks, constructed as herein described, with upper members A, lower members A', shoulders C, and ledges B, all arranged and adapted to operate as and for the purposes specified.

Second, the arrangement of congeries of wooden blocks A A' C B, with the metallic frame curb or casing D D' d, as set forth.

No. 56,304.—WILLIAM N. WHITELEY, Jr., Springfield, Ohio.—*Pitman Head and Crank Wrist Box for Harvesting Machines.*—July 10, 1866.—Explained by the claims and cut.

Claim.—First, the chilled box pitman head K, constructed with the internal screw k, in combination with the jam nut L, and pitman rod M, substantially as described.

Second, the pitman head K, when cast in one piece over a taper chilled pin and provided with the internal screw k, in combination with the jam nut L, pitman rod M, and chilled hollow spindle A, substantially as described.

No. 56,305.—E. P. WHITNEY, Stamford, Conn.—*Meat and Vegetable Cutter.*—July 10, 1866.—The knives are set adjustably on the disk and remove a slice, which is slit into concentric rings or spirals by points protruding through the disk.

Claim.—The arrangement of the knives D with slots d, screws e, lances E, in combination with the disk A, constructed and operating in the manner and for the purpose herein specified.

No. 56,306.—J. L. WIGGIN and E. FOLSOM, South New Market, N. H.—*Lubricating Apparatus.*—July 10, 1866.—Oil displaced from the bulb passes by minute openings through the tube to the journal.

Claim.—The shaft lubricating apparatus, made in manner and for application to a shaft bearing or box, substantially as hereinbefore described.

No. 56,307.—ALBERT S. WILKINSON, Pawtucket, R. I.—*Horseshoe.*—July 10, 1866.—The pintles by which the wings of the shoe are hinged to the middle portion consist of taper rivets with washers, and the parts have clips to embrace the hoof.

Claim.—First, hinging the bar C to the bar A by means of the tapering rivet g and washer f, let into the upper surface of the bar C, all substantially as described and indicated in sheet 1 of drawings.

Second, the combination of the bar clips d d with a toe clip a, heel clips e e, and one or more hinge joints g g, substantially in the manner and for the purpose set forth.

No. 56,308.—ALBERT S. WILKINSON, Pawtucket, R. I.—*Horseshoe.*—July 10, 1866.— Adapted for contracted hoofs. The expanding screw is protected from wear and blows by its elevation, and operates between the raised heel clips; it is fastened by jam nuts against the bar clips.

Claim —First, the method herein described for protecting the expanding screw *e* from injury, viz., the attaching of the expanding screw to the clips above the shoe, and also the locating of said screw in between the heel clips, substantially in the manner and for the purpose specified.

Second, the combination of the shoe A A, heel clips C C, bar clips D, expanding screw *e*, jam nuts *f f*, in the manner and for the purpose represented and described.

Third, the combination of the curved heel clips C C and bar clips D D, substantially in the manner and for the purpose set forth.

No. 56,309.—ALBERT S. WILKINSON, Pawtucket, R. I.—*Horseshoe.*—July 10, 1866.— The bar of the light steel shoe is slotted radially to allow a certain degree of expansion, and a continuous clip conforms to the wall of the hoof.

Claim.—A continuous clip B B′ *b*, constructed as described, in combination with the notched bar A, the whole being constructed and operated substantially in the manner and for the purpose set forth.

No. 56,310.—ALBERT S. WILKINSON, Pawtucket, R. I.—*Horseshoe.*—July 10, 1866.— The toe calk is placed under the inner edge of the bar instead of the outer; a short calk remains in front, and both consist of projections on a plate fastened to the shoe.

Claim.—First, the combination of the toe calks *b* and *b³* with the plate B, when connected with and attached to the shoe, in the manner described.

Second, locating the toe calk at the inner edge of the shoe, when constructed and applied in the manner described.

No. 56,311.—ALBERT S. WILKINSON, Pawtucket, R. I.—*Anvil for Making Horseshoes.*— July 10, 1866.—The anvil corresponds in shape and size to a hoof, and has shanks which permit its adjustment in the natural or reverse position.

Claim.—The shaping anvil A *a*, in combination with the studs or projections B *b*, the whole being constructed and operated substantially in the manner and for the purpose set forth.

No. 56,312.—AARON W. C. WILLIAMS, Hartford, Conn.—*Machine for Labelling Spools.*— July 10, 1866.—The spools are fed consecutively from a spout, and fall endwise into the peripheral pockets of a vertical revolving cylinder; the exposed disk of the spool receives gum from a roller, then passes under a vertically reciprocating tube, containing a pile of labels; the bottom one of the pile adheres to the spool, which is passed under a roller to secure the adherence of the paper, and is presently discharged.

Claim.—First, the recesses *a* in the carrier B, arranged relatively to the feeder C and the roller D, substantially as and for the purpose herein set forth.

Second, the combination of the carrier B, carrying the spools in succession, as specified, with the labelling device G, adapted to present the labels thereto, substantially as and for the purpose herein set forth.

Third, the label-holder G, in combination with the mechanism represented, or its equivalent, for rendering its action automatic, substantially as herein specified.

Fourth, the pressing roller I, in combination with the spool carrier B, labeling device G, sizer D, and spool feeder C, substantially as and for the purpose herein specified.

Fifth, the inclined surface J′ on the vibrating arm J, in combination with the label carrier G, spool carrier B and recesses *b*, arranged as represented, so as hold the spool carrier firmly in position while the label is applied, and liberate it by the retreat of the label carrier to allow the succeeding spool to be presented, substantially as herein specified.

Sixth, the catch K, in combination with the spool carrier B, lever J, and label carrier G, adapted to operate substantially as and for the purpose herein specified.

No. 56,313.—SAMUEL R. WILMOT, Bridgeport, Conn.—*Dies for forming Bead on the Edge of Sheet Metal Vessels.*—July 10, 1866.—The lamp bottom is placed in the lower die, and its rim is bent over into an annular bead by the circular groove in the descending die.

Claim.—First, the dies A B, constructed, arranged, and operating substantially as specified.

Second, the groove *e* in the die B, formed as described, and acting in connection with the recess *a* of the die A for turning the edge of the blank, substantially as set forth.

Third, the combination of the elastic presser C with the dies B and A, substantially as herein set forth for the purpose specified.

No. 56,314.—DAVID S. WOOD, Delavan, Wis.—*Pump Piston.*—July 10, 1866.—Expanding packing rings are placed loosely in a concentric channel around the piston head, and their vertical play on the downward stroke allows grit to pass underneath them with the water.

Claim.—The two part loose packing D, in combination with the channel L J M and the piston head B, substantially as and for the purpose specified.

No. 56,315.—G. B. WOODWARD, Bolivar, N. Y., and M. L. SMITH, Scio, N. Y.—*Wheel.*—July 10, 1866.—The hub has a central box and a boss, with a flaring shoulder; on this the spoke-bearing rim is clamped by a shouldered annulus and a nut, which screws upon the box.

Claim.—First, the combination of the box C, cast with the end piece *b*, central piece *a*, end piece *b'* and nut B, both end pieces being provided with a flange *d*, all constructed and operating in the manner and for the purpose herein specified.

Second, the spokes C, secured to the felloes and part *a* of the hub, as shown, when used in combination with the flanges *d* at the inner ends of the parts *b b'*, substantially as and for the purpose specified.

No. 56,316.—JAMES WOOLEVER, Peoria, Ill.—*Grain Separator.*—July 10, 1866.—Explained by the claims and cut.

Claim.—First, opening the side of the shaking shoe like a door hung on hinges, and closing it and holding it firmly by means of the two rods at the top and bottom with nuts and screws as shown at *m m*, Fig. 1; the rods and nuts are shown at *k k* and *l l*, Fig. 1, the section of which is shown at *k l m*, Fig. 7, thus holding the screens when inserted firmly in any desired position, without slides or wedges or grooves.

Second, the metallic strip connecting the screens together as shown at *a*, Fig. 9, allowing the screens to change their relative position, as shown by the dotted lines *a b* and *a c*, Fig. 9, and allowing one or more of the screens to be removed by passing the head of the screw through the slot, as shown at *a* Fig. 9.

Third, the arrangement of the two elevators C and D, with the shoe N and box E, as constructed, for the purpose of more effectually separating and conveying away from the machine the different grades of seed, as is herein fully set forth.

Fourth, the flange roller shown at Fig. 10 to support and carry the shaking shoe A, Fig. 1, as shown at *c d*, Fig 1.

Fifth, the position of the screens in the lower shaking shoe N, as indicated by the dotted lines, the lower screen L being pushed down as far as the dotted line *g*, and delivering its grain into the box E, and the upper screen K being pushed down as far as shown by the dotted line *f*, and delivering the superior grade of grain at the front of the mill, the screens being held in their position by a rod and nut, as shown at *n*, Fig. 1.

No. 56,317.—K. F. WORCESTER, Nashua, N. H. and J. B. PERKINS, Hollis, N. H.—*Flour Sifter.*—July 10, 1866.—The horizontally rotating shaft has arms with stirring fingers and brushes, and has an intermitting vertical motion by the contact of a radial pin, with a cam disk alternating with a spring.

Claim.—The combination with the fingers *c* and brushes *d*, secured to the shaft of the machine as herein set forth, of the cams and pins, and spring or equivalent mechanism for giving a constant up and down motion to the same, the whole being constructed and arranged for operation substantially as and for the purposes herein shown and described.

No. 56,318.—ROBERT M. YORKS, Schoolcraft, Mich.—*Washing Machine.*—July 10, 1866.—The rubbing board moves in a groove, formed between the inclined bed-frame and yielding upper bars; the rubber is reciprocated upon the bed of rollers, and is moved alternately by a treadle and a recoil spring.

Claim.—First, the combination of the wash-board G, the treadle J, and the springs K, connected to each other by ropes I I', substantially as described.

Second, the cross-bar M working in grooves, formed by the yielding bars L L and the upper edges of the frame B, substantially as described.

No. 56,319.—CARL FRIEDRICH ZIMMERMAN, Philadelphia, Penn.—*Accordeon.*—July 10, 1866.—Certain distinguishing keys which are placed between the consecutive octaves give the same tone in either drawing or compressing the wind chest.

Claim.—The distinguishing keys K', giving the same tone both in drawing and compressing, and arranged in combination with the keys K, substantially as and for the purposes set forth.

No. 56,320.—FRANCIS C. COPPAGE, Terre Haute, Ind., assignor to himself and WILLIAM COPPAGE, same place.—*Harvester.*—July 10, 1866.—The zig-zag flanges on the disk attached to the main axle oscillate the suspended frame, and give a reciprocating motion to the pitman of the cutter bar.

Claim.—The zig-zag wheel F, situated on the axle of the wheels B B, as described, in combination with the frame H, pallets J J, levers I I, link K, pitman L, and cutter-bar N, the whole being constructed, arranged, and operated in the manner and for the purpose set forth.

No. 56,321.—P. W. GATES, Chicago, Ill., assignor to himself and D. R. FRAZER, same place.—*Shoe for Stamping Machinery.*—July 10, 1866.—The stamp has a chilled bottom and soft metal stem ; the hard metal is cast in a chill and the soft metal upon it in sand, which constitutes the upper portion of the mould.

Claim.—As a new article of manufacture a solid shoe for stamping machinery, produced by casting hard and soft metal together while both are in a molten state, the soft metal forming the stem of the shoe, while the hard metal forms the body of the shoe, substantially as described.

No. 56,322.—J. LEWIS GEROLDSEK, Livingston, N. Y., assignor to HENRY W. LIVINGSTON.—*Paint.*—July 10, 1866.—Composed of water, blue vitriol, potash, glue, alum, litharge, linseed oil and aquafortis, colored to suit.

Claim.—The composition of materials and the process of compounding the same, substantially as set forth in the foregoing specification.

No. 56,323.—JAMES B. GOODING, Waltham, Mass., assignor to STILLMAN WHITE, same place.—*Making Calipers or Dividers.*—July 10, 1866.—The blank is bored at the spring end and then sawed from point to bow ; the legs are then spread and the tool shaped, fitted, and tempered.

Claim.—The hereinbefore described new or improved process of making spring dividers, calipers, or other like articles, the same being accomplished by swaging, boring, sawing, and bending open the blank preparatory to further finishing and hardening it, as specified.

Also, the improved article of manufacture as so made.

No. 56,324.—JOHN B. MITCHELL, Portland, Maine, assignor to himself and C. M. PLUMMER, same place.—*Hose Coupling.*—July 10, 1866.—The swivel fits against a collar on the nipple, and its flanges are engaged by the sleeve, which also screws upon the back screw on the other nipple, pressing the ends of the sections against the intervening annular packing.

Claim.—The combination in a hose-coupling of the swivel 2, back screw *f*, nipple 1, having shoulders *a*, the centre clutch 8 with spaces *d* and projections *c*, channels *i* and steady pins *e*, all substantially as and for the purpose set forth.

No. 56,325.—ROBERT PARKS, Philadelphia, Penn., assignor to E. J. SPANGLER, W. E. LOCKWOOD and E. D. LOCKWOOD, same place.—*Envelope Machine.*—July 10, 1866.—Improvement on patent of S. E. Pettee, March 2, 1859. The paper is fed in a continuous strip of a given width for a given size envelope. First, are made transverse incisions which answer for a portion of the division between the adjacent envelopes ; the rectangular crease is made determining the size of the envelope, slits made from the corners of the latter to the edges of the paper ; the included flaps are folded over and paste applied ; the superfluous edge strip is cut off and the angular division is made between the adjacent envelopes ; the envelope is bent on the folding line and passed between rollers, to be afterwards dried and have its flap gummed.

Claim.—First, the cross-head F, carrying knives and creasers, arranged substantially as described, and having a uniform reciprocating motion imparted to it, substantially as set forth for the purpose specified. *

Second, the combination of the cross-head F, guide-rods D or their equivalents, yoke C, and crank-shaft B.

Third, the plate G, made detachable from the cross-head, and having cutters and creasing plates, arranged substantially as set forth.

Fourth, the plate G', with its knives, the whole being made detachable from the cross head, for the purpose specified.

Fifth, the plate H, with its grooves and openings, the whole being made detachable from the stationary plate E of the machine, for the purpose described.

Sixth, the combination of the detachable plate I and its cutting edges *q q*, with the detachable plate J and its knife *r*.

Seventh, the bars P and P', arranged and operating substantially as described.

Eighth, the stationary folders R, constructed, arranged, and operating substantially as set forth.

Ninth, the adjustable connecting rod M', rendered adjustable on the crank, substantially as and for the purpose described.

No. 56,326.—DAVID PATERSON, Jersey City, N. J., assignor to AUSTIN R. PARDEE, same place.—*Flange.*—July 10, 1866.—The radial slots receive the bolts, and the bolt head or nut rests upon the ribs, level with the central hub.

Claim.—A flange provided with radiating slots *b*, and ribs *c*, rising from the edges of said slots to a level with a central hub *a*, substantially as and for the purpose described.

No. 56,327.—SAMUEL PERRY, Troy, N. Y., assignor to CHARLES H. FORT, West Troy, N. Y.—*Furrowing Plough.*—July 10, 1866.—Explained by the claim and cut.

Claim.—The adjustable standard C, having the flat share K attached to its lower end, in combination with the mould boards D D attached to said standards by links or joints, and provided with arms E E, which pass through the beam and are secured by a set screw F, the whole being constructed and operated in the manner and for the purpose set forth.

No. 56,328.—MARTIN C. REMINGTON, Auburn, N. Y., assignor to himself and A. O. REMINGTON, Weedsport, N. Y.—*Grain Fork.*—July 10, 1866.—By the arrangement described the bow and brace rods may be detached when not required in use, and to pack the tool for transportation.

Claim.—The construction and combination of the bow C and brace rod D applied to the fork, substantially as and for the purpose specified.

No. 56,329.—ISAAC SMITH, New York, N. Y., assignor to himself and W. H. HAIGHT, same place.—*Chuck.*—July 10, 1866.—The stock of the chuck terminates in a conical, threaded head which opens or closes the jaws which are threaded, and slide in grooves in the conical shell.

Claim.—The conical screw and sliding threaded gripping jaws D, arranged with reference to each other and with the conical body B and case C, substantially as herein set forth for the purpose specified.

No. 56,330.—ISAAC SMITH, New York, N. Y., assignor to himself and W. H. HAIGHT, same place.—*Chuck.*—July 10, 1866.—A screw collar on the body of the chuck slips a sliding conical sleeve under the levers connected to the jaws. The end of the drill shank rests in a conical socket.

Claim.—First, the combination of the screw-collar C, sliding collar D, inclined planes E, and levers I, substantially as herein set forth for the purpose specified.

Second, the arrangement of the rotating collar C,. sliding collar D, and levers I, with reference to each other and to the sliding gripping dog B and body A, substantially as herein set forth for the purpose specified.

Third, the internally conical centring device F applied within the body A of the chuck, in the manner substantially as herein specified.

No. 56,331.—JOHN P. SMITH, Hudson, N. Y., assignor to himself and JOHN B. LONGLEY, Hudson, N. Y., and W. H. SHUTTS, Claverack, N. Y.—*Machine for Raking and Loading Hay.*—July 10, 1866.—Guards are placed at the ends of the rake to prevent the escape of hay, and elastic strips between the teeth to arrest fine hay.

Claim.—First, the guards N, in combination with the rake and endless elevating apron, substantially as herein set forth for the purpose specified.

Second, the elastic plates n, arranged between the teeth and with regard to the endless elevating apron, substantially as herein set forth for the purpose specified.

No. 56,332.—AMAZIAH S. WARNER, Springfield, Mass., assignor to himself and HENRY REYNOLDS, New Haven, Conn.—*Machine for Drawing Cartridge Shells.*—July 10, 1866.—The shells partially formed are placed upon the horizontally rotating table which carries them into the curved guiding groove, from whence they are consecutively pushed over the die by the feeding device.

Claim.—The combination of the means substantially such as herein described, for moving and feeding the forms with the die and punch, substantially as and for the purpose specified.

Also, in combination the disk or wheel for moving the forms, the guides for directing the forms, the carrier, or equivalent therefor, and the punch and die, substantially as and for the purpose specified.

No. 56,333.—GEORGE A. WHIPPLE, West Pittsburg, Penn., assignor to himself and JACOB PAINTER, same place.—*Puddling Furnace.*—July 10, 1866.—The bed of the furnace has a double bottom, the lower plate being of corrugated metal; water is introduced through the boshes, and is converted into steam in the chamber thus formed.

Claim.—Constructing puddling, boiling, and heating furnaces with a chamber or space under the bottom plates or hearth, so constructed and arranged substantially as hereinbefore described, as that steam may be generated or introduced therein, and come in contact with the under side of the bottom plates and interior surface of the boshes or chills, so as to withdraw from them a portion of their heat, and thus aid in protecting them from the destructive action of the heat of the furnace.

No. 56,334.—MARY E. FRANCISCO, Lake Mills, Wis., administratrix of the estate of HENRY FRANCISCO.—*Sleigh Knee.*—July 10, 1866.—Explained by the claim and cut.

Claim —A hollow cast metal sleigh knee, which is constructed with bracket bearings on its ends, having closed sides adapted to receive the frame-work of a sleigh, substantially as herein described.

No. 56,335.—JOHN BOWDEN, Mitcham, England.—*Bellows.*—July 10, 1866.—The boards are attached by webs of leather to the inside of the wind chest; the lower board is lifted by the lever, and the upper one by wind received through the valved partition.

Claim.—The combination and arrangement of the chambers, diaphragms, valves, rock shaft, and connections, substantially as herein described and shown in the drawings an nexed.

No. 56,336.—PROSPER CARLEVARIS, Turin, Italy.—*Producing Light.*—July 10, 1866.— Spongy oxide of magnesium is rendered iridescent by the flame of the oxy-hydrogen blow-pi e.

Claim.—The employment of chloride of magnesium and of the carbonate and other salts in general of magnesia reduced to an indecomposable state, as and for the purposes herein described, by being introduced in a flame composed of oxygen in combination with hydrogen, or of such other gaseous compounds as are hereinabove set forth.

No. 56,337.—DANIEL JOSEPH FLEETWOOD, Birmingham, England.—*Manufacture of Spoons, Forks, &c.*—July 10, 1866.—The lower die is vertically adjusted upon wedges which are operated by set-screws. Each die contains certain parts of the cameo and intaglio configuration and functions.

Claim.—First, the employment or use in the manufacture of spoons, forks, and other articles of tools B, each of which is partly die and partly matrix, substantially as and for the purposes described.

Second, the die-holder C, with set screws *s t*, and wedges *a*, in combination with tools B, constructed and operating substantially as and for the purpose set forth.

No. 56,338.—LOFTUS PERKINS, London, England.—*Apparatus for Heating and Cooling Air, &c.*—July 10, 1866.—The ends of the sealed tubes project into the furnace, and the steam generated is condensed at the other portions and returns to the heated ends.

Claim.—The use of tubes sealed at both ends and containing water or other volatilizable liquid in heating and cooling atmospheric air and other aëriform bodies, and in heating ovens and in heating and ventilating buildings, as herein described.

No. 56,339.—JOSEPH TANGYE, Birmingham, England.—*Lathe for Cutting Screws.*—July 10, 1866 —Upon one lathe bed are three head and tail centres : gearing operates them simultaneously, and the cutters upon a common slide cut three screws at once.

Claim.—First, the combination in a screw-cutting lathe of a single slide rest, with two or more tools or cutters, which operate simultaneously upon two or more different screw blanks, all constructed and arranged substantially as described.

Second, the combination in a screw-cutting lathe of a head stock and poppet head, each provided with two or more centres, when constructed and arranged substantially as herein described.

Third, in a screw-cutting lathe constructed and arranged as herein described, the special gearing actuated by the driving pulleys for the purpose of rotating two or more screw blanks simultaneously in the same direction as hereinbefore described.

No. 56,340.—JOHN COOPER, Dublin, Ind.—*Churn.*—July 10, 1866.—The dashers are formed of bent rods set obliquely upon the axis.

Claim.—The bowed or bent beaters F, when constructed and arranged substantially as and for the purpose set forth.

No. 56,341.—THOMAS V. PHELPS, Worcester, Mass.—*Ladies' Dress Skirt Elevator.*— July 10, 1866.—A tape is tacked at both ends to the hem of the dress, and when drawn around and between the two upper rings is clamped by their pressure against each other.

Claim.—First, the combination of rings C D and B with the tape A, substantially as set forth.

Second, the combination of the looping device above described with the bottom of the skirt of a dress, substantially as set forth.

No. 56,342.—JOHN G. BAKER, Philadelphia, Penn.—*Measuring Faucet.*—July 10, 1866.— The eccentric in the chamber is revolved by a crank, and the sliding piston ejects at each registered revolution the known contents of the chamber.

Claim.—First, the revolving cylinder D, when arranged eccentrically to the cylinder A, and enclosing the feed port I, and in combination with the rotary piston G, when constructed and arranged as a measuring faucet, substantially as described.

Second, the cylinder D, when enclosing feed port I, and provided with a wide slot *a* for the escape of the fluid from the centre outward, substantially as described.

Third, the graduated dial plate Q, in combination with the reversible index P, worm wheel O, and worm N, as arranged in relation to a measuring faucet, all substantially as and for the purposes set forth.

No, 56,343.—JAMES ADAIR and H. W. C. TWEDDLE, Pittsburg, Penn.—*Evaporating and Distilling Liquids.*—July 17, 1866.—Hot carbonic acid is generated in the furnace and conducted by a pipe to the still, where it is passed through the liquid; the vapor from the still is conducted by a pipe to the worm of the condenser.

Claim.—First, the mode of distilling or evaporating petroleum or other liquids by passing through or over the liquid to be distilled or evaporated heated carbonic oxide or carbonic acid, substantially as and for the purposes described.

Second, the combination of the air-tight furnace through the fire in which air and steam or either of them are forced, with the still or boiler for holding the liquid to be distilled or evaporated, and the pipes connecting the furnace and still or boiler, constructed and operating substantially as and for the purpose hereinbefore described.

Third, the air-tight furnace A, constructed substantially as described for the production of carbonic oxide or carbonic acid, to be used in the making of artificial combinations or mixtures of carbon with other fluids or solid bodies.

No. 56,344.—ISAAC AVERY, Ottawa, Ill.—*Cultivator*—July 17, 1866.—The elevated tongue and frame rest upon the axle, and the plough beams are attached by universal joints to pendants upon which they are laterally and vertically adjustable. The stay chains pass through pedestals below the forward cross-bar of the frame.

Claim.—First, the attaching of the plough beams A* to pendants a*, of the cross-bar C, by means of universal joints D D', substantially as and for the purpose specified.

Second, the combination of the plough beams A*, universal joints D D', doubletree or evener C, trace chains F, and pulleys e*, all arranged to operate in the manner substantially as and for the purpose herein set forth.

No. 56,345.—SAMUEL H. BARNS, New York, N. Y.—*Extension Corset Spring.*—July 17, 1866.—The leaves slip on each other and are set as required by screws.

Claim.—A corset spring, consisting of the parts B, provided with pins b, and slotted springs B², riveted as shown and having suitable clasps C, and headed rivets D, and of form corresponding to the body of the wearer, all constructed and operating in the manner and for the purpose herein represented and described.

No. 56,346.—J. D. BARTON, F. S. ROGERS, and D. FISHER, Kalamazoo, Mich.—*Anvil and Vice Combined.*—July 17, 1866.—The pivoted jaw is operated by a treadle lever and the combined jaws form an anvil.

Claim.—The upright shaft B, and levers C and E, in combination with the several anvil appliances, constructed and arranged substantially as described.

No. 56,347.—BURROUGHS BEACH, West Meriden, Conn.—*Sash Fastening.*—July 17, 1866.—The legs of the levers are pivoted in a plate attached to the casing, and the rough soles of the feet are pressed by springs against the edge of the sash to sustain it and are pressed asunder by a thumb-piece to relieve it.

Claim.—A sash supporter consisting of the arms A, in combination with the lever plate K, and springs E, when arranged together so that the said plate will act upon the said arms, substantially as described and for the purpose specified.

No. 56,348.—HENRY H. BEACH, Rome, N. Y.—*Grain Dryer.*—July 17, 1866.—A series of inclined, curvilinear, perforated plates traverse the vertical flue and compel the falling grain into a sinuous track in contact with the ascending current of heated air.

Claim.—First, the within described grain dryer, composed of the inclined perforated plates B B', &c., and flues G and A, the whole being arranged substantially as and for the purpose herein set forth.

Second, in combination with the above, the vanes z z', &c., arranged substantially as specified.

No. 56,349.—JOSIAH BEARD and MOSES FAIRBANKS, Boston, Mass.—*Bottle Stopper.*—July 17, 1866.—Explained by the claim and cut.

Claim.—A protecting cap in combination with the stopper and fastening wire, passing through both the said cap and stopper, as described.

No. 56,350.—CHARLES BEIDLER, Allentown, Penn.—*Plough.*—July 17, 1866.—The vertical and horizontal vibration of the beam relative to the standard is adjusted, and the depth of furrow and width of land thus regulated by the set bolt, jam nuts, and segment bar at the rear of the beam.

Claim.—The segmental guide bracket h, in combination with the screw rod g, set nuts j, handles C C, and beam A, and operating in the manner and for the purpose substantially as herein shown and described.

No. 56,351.—A. BLOMQUIST, New York, N. Y., and C. CROOK, Yonkers, N. Y.—*Marine Car.*—July 17, 1866.—The platform is supported upon flotative revolving drums and propelled by the paddle.

Claim.—The arrangement of the drums B B C, and paddle D, in combination with the platform A, constructed and operating in the manner and for the purpose herein specified.

No. 56,352.—SILAS R. BOARDMAN, New York, N. Y.—*Water Drawer.*—July 17, 1866.—The bottom of the bucket has several valved openings and spouts beneath. The trigger of the spout next to the discharge trough is depressed by the contact of its rod with a cleat in the curb.

Claim.—First, a well bucket having three or more valves in the bottom thereof and arranged at equal distances from each other, each valve being provided with a stem so arranged and operated that the ascent of the bucket will open those and those only that are upon that side of the bucket presented to the curb spout, as and for the purpose specified.

Second, in combination with a series of valves arranged around the bottom of the bucket as described, a corresponding number of spouts attached to the bottom of the bucket, as and for the purpose set forth.

No. 56,353.—M. C. BOGIA and H. B. TAYLOR, Philadelphia, Penn.—*Plaster.*—July 17, 1866.—Explained by the claim and cut.

Claim.—A plaster consisting of mustard or other material or composition permanently confined between layers of textile or other fabric, substantially as and for the purpose described.

No. 56,354.—WILLIAM BRANT, Paris, Ill.—*Mechanical Movement.*—July 17, 1866.—A reciprocating rotary movement is imparted to the shaft by a band from the oscillating segment.

Claim.—The mode of imparting a reciprocating and alternate rotary movement to the shaft G, by means of pulley D, and thong E, or devices substantially equivalent, all arranged to operate in the manner and for the purpose set forth.

No. 56,355.—S BREWER and W. W. WINTER, Cortlandville, N. Y.—*Well Pipe or Tube.*—July 17, 1866.—The openings in the lower section of pipe are temporarily closed while driving by an expansive shield which is removable when the depth is reached.

Claim.—The device consisting of the springs B B B, the shield A, and the rod D, all in combination, as and for the purposes herein shown and described.

No. 56,356.—JOHN BRIGGS, Roxbury, Mass.—*Apparatus for Preparing Starch, Size, &c.*—July 17, 1866.—The materials are mixed in a tank with a concentric inner foraminous cylinder; beaters in the latter and stirrers in the former are revolved by the shaft to which they are attached. The paste is removed by a pump operating in a steam jacketed pipe.

Claim.—In combination with stirrers, the tank d, and foraminous cylinder e, all operating together for the purpose set forth.

Also, the steam jacketed pipe s, when provided with the screw o, and arranged to operate substantially as described.

No. 56,357.—CHARLES BROWN and C. McGHIE, Chicago, Ill.—*Beer Faucet.*—July 17, 1866.—The beer is foamed by the rapid discharge at the lower central opening of the spigot of what remains in the barrel of the faucet after the communication with the cask is shut off.

Claim.—The plunger B, provided with the hollow stem C, having the holes c' c' and e therein, as shown, in combination with the stem D, having the spiral grooves d cut therein, when said parts are arranged to operate in connection with the body of the faucet, as and for the purpose set forth.

No. 56,358.—JOHN H. BROWN, New York, N. Y.—*Toy Sled.*—July 17, 1866.—The rudder is turned by its tiller connected by rods to the treadle, and the brake prong at the end of the rudder is operated by a wire connected to a button on the withers.

Claim.—The combination of the button G, rods e f, and rudder E, arranged with the horse D and sled A, and operating in the manner and for the purpose herein specified.

No. 56,359.—J. S. BROWN, Washington, D. C.—*Horse Hay Fork.*—July 17, 1866.—Each section of the shouldered lance point has a shank, and the points are united by links which form a toggle vibratable by a trigger rod. When separated the shoulders of the points catch the hay, and when collapsed the shoulders are covered and the hay slips off.

Claim.—First, the employment of movable bars D D to cover and uncover fixed barbs or shoulders C C, substantially as and for the purposes herein specified.

Second, a divided shaft A to be opened in dovetail or inverted wedge form, and closed in connection with the uncovering and covering of the barbs, by movable bars D D, substantially as and for the purposes herein set forth.

No. 56,360.—O. C. BROWN, Iberia, Ohio.—*Clothes Dryer.*—July 17, 1866.—The blocks of each section are fastened to straps and are mutually sustaining like the stones of an arch; they may be wound on an axis to bring the standards closer together.

Claim.—First, broadly a clothes drying rack consisting of a series of suspending rods, bars or equivalents, attached to flexible supports to adapt the rack, as a whole, to be wound upon an axis or windlass in any manner substantially as described.

Second, a flexible clothes rack consisting of the straps C C', blocks D D', and supporting bars F, all combined and operating substantially as described.

Third, in combination with the above, the frames A A', and windlass a', arranged and operating substantially as described.

No. 56,361.—THOMAS W. BROWN, New York, N. Y.—*Lamp Bracket.*—July 17, 1866.—
The lamp and shade or reflector are separately supported from projecting portions of the
bracket.
Claim.—The improved socket plate made with the recess and its openings and the semicircular bearing arranged with the projection of such plate, substantially as specified.
Also, the application of the reflector supporter *d* to the socket plate B, instead of applying it to the ring arm in the usual manner, the same presenting advantages in the casting of
the ring and its arm.

No. 56,362.—H. L. BUCKWALTER and J. A. BUCKWALTER, Kimberton, Penn.—*Horse
Power.*—July 17, 1866.—The main wheel has inner and outer cogged rims to engage the
pinions of the two counter shafts, and also has the sprockets which are engaged by the
transverse rods of the endless chain.
Claim.—First, in the construction of horse power the combination in one wheel of the
sprockets which engage the shafts of the chain and the cogs which communicate motion to
the counter shaft, substantially as described.
Second, placing two counter shafts in gear with the cog wheels of the machine, one within
and one without their rims, in combination with the belt wheel, the same being so made and
arranged that the belt wheel may be changed from the one to the other at the pleasure of the
operator, substantially as described.

No. 56,363.—FRANKLIN M. BUELL, Truxton, N. Y.—*Roofing Cement.*—July 17, 1866.—
Composed of raw coal tar and sand.
Claim.—As a new article of manufacture and sale the paint or composition herein described.

No. 56,364.—JOHN BURNS, Providence, R. I., assignor to himself and JOSEPH W. BAKER,
same place.—*Coffin.*—July 17, 1866.—Explained by the claim.
Claim.—Combining with a wooden coffin of the usual construction a lid of marble, or other
equivalent material, substantially as described for the purpose specified.

No. 56,365.—SAMUEL G. CABELL, Quincy, Ill.—*Crimping Machine.*—July 17, 1866.—
Two fluted cylinders, the lower in fixed bearings, the upper vertically adjustable; one or
both hollow for the reception of a heated iron.
Claim.—First, the combination and arrangement of an iron F in one or each of the hollow
fluted cylinders A, substantially in the manner and for the purpose as herein set forth.
Second, the sliding pivoted cap plates G, as arranged in combination with the fluted cylinders and irons, substantially in the manner and for the purpose as herein set forth.
Third, the slotted curved spring B, screw rod C, projecting arm D, and grooved collar as
arranged in their connection with the upper fluted cylinders and vertical tongued bars *b*, and
operating substantially in the manner and for the purpose as herein set forth.

No. 56,366.—WILLIAM F. CALDWELL, Oxford, Maine.—*Potato Digger.*—July 17, 1866.—
The potatoes in the hill are scooped up and elevated to the sifter, which is partially supported
by springs, and has a horizontal reciprocating motion imparted by an eccentric moved by
gearing from the grooved wheels. By the vibration of the lever the digger is raised from the
ground, and the frame tilted back to discharge its contents on to the apron for conveyance
to the sifter.
Claim.—First, the combination and arrangement of the geared wheels *d* and *c*, shaft *e*,
eccentric *f*, and connecting rod *n*, as and for the purposes herein described, the said wheel
c, shaft *e*, eccentric *f*, and the sifter *s* being attached, as set forth, to the tilting frame F and
the shaft *e*, being also employed to give motion to the endless apron *k*.
Second, the combination and arrangement of the arms *g h* and helical spring *i* to hold the
sifter, as described.
Third, the arrangement of the tilting frame F upon the shaft E, for the purpose herein set
forth and described.

No. 56,367.—ROBERT CARTER, San Francisco, Cal.—*Teapot.*—July 17, 1866.—The
bottom of the interior chamber of the teapot is made convex to prevent violent ebullition in
the water jacket.
Claim.—First, the bottom N N, figure 2, of the inner case H, figure 2, being formed convex
toward E, figures 1 and 2, the bottom of the outer case D, figures 1 and 2, for preventing the
violent ebullition of the water contained in K K K, figure 2, when boiling, as would ensue
if the bottom of H, figure 2, was flat.
Second, without confining to any particular shape, size or material, the general combination of the two cases, with their surroundings and appurtenances, as in this specification
shown, for the purposes described and in the manner substantially herein set forth.

No. 56,368.—SETH P. CHAPIN, Atlantic, N. J.—*Implement for Opening Sheet Metal Cans.*—
July 17, 1866.—Explained by the claim and cut.
Claim.—The cutter B, curved in its cross-section and provided with sloping cutting edges
a' or *a³*, as described, when secured upon a handle or stock provided with a shoulder *d* to
operate substantially as herein set forth for the purpose specified.

No. 56,369.—E. G. CHORMANN, Philadelphia, Penn.—*Skates.*—July 17, 1866.—The two portions of the divided runner are relatively adjustable to adapt them to varying sizes of boots for ice or parlor use.

Claim.—First, the combination of the plate A and its runner C, the plate A' and its runner C', and the screw B and sliding block c, or equivalent device, whereby the runners may be adjusted at any required distance from each other, the whole being constructed and arranged substantially as described.

Second, the combination, substantially as illustrated in figure 4, of the adjustable plates A A' with the rollers, tor the purpose described.

No. 56,370.—GEORGE CLARK, JR., Boston, Mass.—*Machine for Shelling Peas.*—July 17, 1866.—The peas are fed through the aperture in the face plate, the pods are pinched between the rollers, the peas drop out, and the scraper removes adhering particles.

Claim.—The combination of rotating rollers, face plate and screw clamp, whether with or without the scraper, for the purpose of expressing peas and other seeds from their containing vessels, when the same are constructed and used substantially as described.

No. 56,371.—CHARLES D. CLINTON, Peoria, Ill.—*Car Coupling.*—July 17, 1866.—Explained by the claim and cut.

Claim.—The combination of the oblique-faced hook B, spring C, and eye E, the latter serving as a stop for the end of the spring, and constructed and arranged to operate together in the manner and for the purposes herein specified.

No. 56,372.—CHARLES COBB, Plymouth, Mass.—*Cordage Machine.*—July 17, 1866.—The cord as made is delivered to the spool from the end of the arm whose swivel, lying between the adjacent coils and against the side of the coil last formed, guides the cord and also forces to one side or the other the spool and the drum to which it is attached, the spool traversing and not the guide.

Claim.—The combination and arrangement of the self-adjusting guide with the layer arm and the notch thereof, such guide being to operate with the laying drum substantially as set forth.

No. 56,373.—ALEXANDER COLE, Lockport, N. Y.—*Spring Bed Bottom.*—July 17, 1866.—The hangers on the ends of the slats rest upon springs coiled around spindles on the bed rails.

Claim.—The combination of the slats C C, hangers E, guide rods c c, coiled springs s s, and stops or cross-pieces H H', the whole arranged and operating substantially in the manner and for the purpose set forth.

No. 56,374.—E. G. CONNELLY, Jasper, Ind.—*Churn Dasher.*—July 17, 1866.—The semicircular dashers reciprocate vertically and alternately, and valves on their upper surfaces open as they descend.

Claim.—The construction of the dasher C and C', with the valves g and g', with either a double or single dasher, operating in the manner and for the purpose substantially as set forth in the above specifications.

No. 56,375.—A. J. COOLEY, Chardon, Ohio.—*Horse Hay Fork.*—July 17, 1866.—The half blades, which united form a spear point, are capable of outward vibration by means of the links and the longitudinally reciprocating rod on the stock on whose end the blades are pivoted. A spring catch on the stock locks the blades in either position.

Claim.—First, the arrangement of the arms C, shanks A A', and links a, with the catch F, spring d, and notch c, as and for the purpose substantially as set forth.

Second, the hooks B' B", with the connecting ropes or chains in combination with the loops G G', shanks A A', and bands D, substantially as and for the purpose set forth.

No. 56,376.—HENRY CORDES, Belleville, N. J.—*Invalid Bedstead.*—July 17, 1866.—A bowl in the depression of the mattress connects by a valved tube with the chamber pot. The tube passes through the lid of the vessel, and the valve yields to the fecal discharge.

Claim.—An invalid bed formed by combining the pipes B and G, the plates E C D and I the sheet F, valve G, and spring L, with each other, and with the bed or mattress, substantially as described and for the purpose set forth.

No. 56,377.—FRANCIS T. CORDIS, Long Meadow, Mass.—*Tool Holder.*—July 17, 1866.—The perforated horizontal ledge of the bracket has a slit diaphragm of India-rubber, which sustains by adhesion the handle passed through it.

Claim.—As a new article of manufacture, the holder, constructed substantially in the manner herein set forth.

No. 56,378.—J. C. COULT and J. ROACH, San Francisco, Cal.—*Apparatus for Treating Ores.*—July 17, 1866.—The consecutive devices for exposing the metalliferous vapor to the contact of water are explained in the claims and cut.

Claim.—First, the pipe C, connecting with a furnace, and having a wide opening entering the condenser E, thereby imparting a greater distribution of the fumes as they enter said condenser, or water tank, and equally spreading the fumes over the water, substantially as described and for the purposes set forth.

Second, the tank E, with an inclined bottom, and the partitions *b b b* in the inverted tank or cover of the same, and the adjusting screws F F attached thereto, substantially as described and for the purposes set forth.

Third, the perforated diaphragm G, having sufficient openings to equal the opening of pipe C, where it enters the condenser E, as before stated; likewise the water bottom G' and G', over which the fumes collect and are drawn into a fan or pump; also giving a water bottom H to the fan or pump, thereby bringing the fumes again in contact with the water for a long distance, and extracting all that it may be desirable to collect before allowing an escape into the chimney, substantially as described and for the purposes set forth.

No. 56,379.—B. F. COWAN, New York, N. Y.—*Stove-pipe Damper.*—July 17, 1866.—This zonular shell is revolvable in a corresponding enlargement of the pipe to close or open the passage.

Claim.—First, the rotating spheroidal valve damper above shown, constructed and operating substantially as described.

Second, the rotating damper above shown in combination with openings in both sides of that part of the pipe within which the damper revolves, substantially as described.

No. 56,380.—BENJAMIN CRAWFORD, Allegheny, Penn.—*Pump for Deep Wells.*—July 17, 1866.—The central rod is operated quickly by a lever, cam and pendant, to close and open the lower valve; a check valve relieves the upper working valve of pressure; the gas is separately discharged, a trap preventing its outflow with the oil or water.

Claim.—First, the detached rod *t*, in combination with the lower valve *q*, for the purpose of keeping the lower valve closed on the down-stroke of the piston.

Second, the combination and arrangement of the lever *y*, and valve rod *t*, with the cam *a'*, and pendant *d'*, for raising and lowering the valve rod *t*, to relieve the lower valve *q* of its pressure when the up-stroke begins, and hold it down on the commencement of the down-stroke, substantially as described.

Third, the combination of the check valve *h*, and gas pipe *j e*, with the working valve of a pump, constructed and arranged substantially as and for the purposes hereinbefore described.

Fourth, in its arrangement with the devices described in the third claim, the trap *c* in the flow pipe to prevent the passage of gas in that direction, substantially as described.

No. 56,381.—GEORGE CROMPTON, Worcester, Mass.—*Woven Fabric.*—July 17, 1866.—Explained by the claim and cut.

Claim.—A textile fabric, woven with braided threads, substantially as described.

No. 56,382.—JOSHUA DAVIS, Schenectady, N. Y.—*Egg Beater.*—July 17, 1866.—The axis of the rotating egg beater is eccentric with that of the revolving pan in which the eggs are placed.

Claim.—First, an eccentric beater in combination with a revolving pan or vessel substantially as and for the purpose set forth.

· Second, the three bevel wheels B C E, of differing diameters, one of which is adapted for carrying a pan or vessel, in combination with a revolving eccentrically arranged stirrer or beater, substantially as described.

No. 56,383.—CATHARINE DITTENHAFER, Canton, Ohio—*System of Cutting Dresses.*—July 17, 1866.—The patterns are shown in the cut; the system cannot be briefly described.

Claim.—The within described patterns and system of cutting ladies' and children's dresses, sacques, and basques, when used in the manner substantially as herein specified.

No. 56,384.—JOHN B. DOUGHERTY, Rochester, N. Y.—*Slide Valve.*—July 17, 1866.—The pressure of the valve is received on rolling instead of rubbing surfaces; the rollers traverse on ways and a steam chest is dispensed with.

Claim.—First, the arrangement of the exhaust port *e*, inlet ports *a a* and *m*, in combination with the rollers *r r*, and the steam pipe *p*, which combination and arrangement avoids the necessity of a relieving or balance plate.

Second, the combination of the rollers *r r* in slide valves, with the bars *f*, when the same are used without a steam chest, as and for the purposes shown and described.

No. 56,385.—JOHN B. DOUGHERTY, Rochester, N. Y.—*Slide Valve.*—July 17, 1866.—A bulb on the back of the relieving plate receives the steam pipe; ports in the plate are alternately connected to receiving ports in the cylinder by the reciprocation of the perforated slide valve.

Claim.—The arrangement of the ports *c* and *e*, in combination with one or more ports through the relieving plate P, and the exhaust port *a*, substantially as and for the purposes set forth, when the valve is used without a steam chest.

No. 56,386.—HENRY DORER and JAMES STORMS, Buffalo, N. Y.—*Elevator Bucket*—
July 17, 1866.—The edge of the sheet-iron cup is bent over to give its margin double thickness; the margin is then riveted to a rim frame of malleable iron cast with a transverse stay
piece to give it rigidity.
Claim.—An elevator bucket constructed as herein described.

No. 56,387.—SAMUEL S. DURBON, Lebanon, Ind.—*Pump.*—July 17, 1866.—An internally
cogged ellipse is moved by a pinion on the hand-crank shaft and is the means of moving
both plungers, which have separate induction chambers and a common eduction opening.
Claim.—The tubular valve seats 6 6, the spindle gum valves 7 7, the self-adjusting leverage 13, with valves 15 15, the self-adjusting gum piston, composed of 1 2 and 3, and the
elliptic L with the eccentric L', all arranged and operating substantially as and for the purpose set forth.

No. 56,388.—ZOHETH SHERMAN DURFEE, Pittsburg, Penn.—*Flask for Casting Steel
Ingots.*—July 17, 1866.—Explained by the claim and cut.
Claim.—The mode of casting ingots of steel or other metal by pouring or tapping such
metal upon a piston, in a mould so arranged and constructed that, as the metal is continuously
introduced, the piston may be caused or permitted as continuously to descend and be followed
by the metal, while at the same time the metal already poured, or the greater part thereof,
remains at the same or nearly the same height in the mould, that portion successively being
introduced flowing through that already poured, and folding outward against the surface of
the mould, at or near the surface of the piston, as the piston gradually descends in the mould.

No. 56,389.—RUFUS DUTTON, New York, N. Y.—*Harvesting Machine.*—July 17, 1866.—
By the described arrangement the track board is adjusted and held at the required elevation
and allowed to rise over obstructions.
Claim.—The construction and arrangement of the track-board cap D, in combination with
the grass shoe and its projecting spur *a*, and the track-board and its spur *c*, the whole
arranged and operating substantially as and for the purposes set forth.

No. 56,390.—B. F. ELLS, Dayton, Ohio.—*Fruit Can.*—July 17, 1866.—The under side
of the cap has a covering of cement which is softened by the heat of the lip and adheres
thereto.
Claim.—The flanged top A provided with sealing wax, as set forth and used with the can
B, in the manner and for the purpose described, whereby a can is formed, which, when filled
with fruit, will seal itself, substantially as specified.

No. 56,391.—MARTIN R. ETHRIDGE, Lock's Mills, Maine.—*Boot and Shoe.*—July 17,
1866.—The insole is combined with two welts, a cushion sole, a wooden sole, a layer of
shellac or gutta-percha, and an iron sole to form a water-proof boot.
Claim.—The combination as well as the arrangement of the two welts *a b* with the insole
B, the upper and the outer sole D.
Also, the combination and arrangement of the metallic cap sole E, with the wooden outer sole
D, the two welts *a b*, the insole B, and the upper A, arranged and applied together, sub-
stantially as set forth.
Also, the arrangement and combination of the cushion C, with the insole B, outer sole D,
and the upper A, disposed together, substantially as set forth.
Also, the combination of the perforated cap sole E and the gutta-percha sole E', or its
equivalent, applied to the wooden outer sole D, as set forth.
Also, the combination and arrangement of the layer *d* of shellac, or its equivalent, with
the wooden sole, the two welts, the insole and upper, arranged and applied together substantially as explained.

No. 56,392.—SIMEON F. EMERSON, Seville, Ohio.—*Gate.*—July 17, 1866.—When half
opened the gate is balanced on a roller in a pivoted post upon which it is then horizontally
rotated 90°.
Claim.—First, the horizontal arm E, of the pivoted hinge C, operating with the top board
D of the gate, substantially as described and for the purposes set forth.
Second, the combination of the roller H and the arms G, having projecting ends or lugs,
with the top rail I of the fence, and with the top board D of the gate, substantially as
described and for the purpose set forth.
Third, the combination of the guide bar K and arms J with the post A and with the gate,
substantially as described and for the purpose set forth.

No. 56,393.—CHARLES A. ENSIGN, Naugatuck, Conn.—*Machine for making Corded Binding for India-rubber and other Fabrics*—July 17, 1866.—By this machine a strip or ribbon
which is to enclose a cord is guided into the machine, the cord laid in its centre, the strip
folded lengthwise upon it and the parts of the strip beyond the cord pressed closely together;
the cord during the process is received in the grooves of the rollers.
Claim.—An organized, automatically operating machine, substantially such as described,
for making binding for India-rubber or other fabrics.

No. 56,394.—PHILO S. FELTER, Cincinnatus, N. Y.—*Lock.*—July 17, 1866.—The key-hole is covered by a hard metal guard which fills the space between the plates and is fastened by a tumbler; the curved end of the latter enters a notch in a disk to allow the guard to be moved.

Claim.—First, the combination of the wheels E and F, tumbler D, and key-hole guard or cover C, arranged and operating together, substantially as described and specified.

Second, the combination of the wheels E and F, tumbler D, key-hole cover C with the arbor H and dial G, arranged and operating substantially as described and specified.

Third, and in combination with the subject-matter of the above, the detachable plate K, arranged as described for operating the lock without recourse to the numbers of the set by which it is locked, substantially as described and specified.

No. 56,395.—JOHN H. FIELD, Saugerties, N. Y.—*Operating Ordnance.*—July 17, 1866.—The chassis has an endless screw moved by levers, pawls and racks, and engaging with a traverse rack to give horizontal adjustment to the gun; the eccentric bearing of the screw permits it to be turned out of engagement with the rack, if desired.

Claim.—The combination of the circular rack D and endless screw G, mounted on an eccentric shaft *g*, and operated by levers and the double acting pawls A A, substantially as and for the purpose herein specified.

No. 56,396.—MATTHEW FLETCHER, Louisville, Ky.—*Steam Generator.*—July 17, 1866.—The dish-shaped pan beneath the tube sheet is connected by a pipe with the water jacket, against whose inner surface the fire is deflected by the bottom of the pan. The hollow frustum in the chimney is for the protection of the steam drum above the water line.

Claim.—First, the arrangement of the vertical steam boiler with the round pan I and water leg *a*, as herein described and for the purposes set forth.

Second, the cone M in the chimney *d*, substantially as described and for the purpose set forth.

No. 56,397.—ELIAS T. FORD, Stillwater, N. Y.—*Harvester.*—July 17, 1866.—The claims explains the construction and arrangement of the parts of the main and supplemental frames and gearing, the means for adjusting the height of the cut, and for raising the outer end of the hinged cutting bar over obstructions.

Claim.—First, the frame C, hinged to the front extremities of arms D D, in combination with the rod E, adjusting bar F, and pole section Q, embracing the tube B, substantially as described.

Second, the left arm D, forming the pillow block or frame, constructed as described and provided with the bearings *e f*, hanger R, and universal box S, and arranged in relation to the tube B and frame C, substantially as described.

Third, the lever K, constructed as described, and pivoted at *v v* to standards on the shoe L in combination with the flange tops $v^3 v^3$ formed on said standard, in the manner and for the purpose specified.

Fourth, the arrangement of the adjustable rod F, hanger R, box S, bar E, lugs U U, shoe L³, and lever *a*, in combination with the tube B, arms D D, and frame C, in the manner and for the purpose herein specified.

No. 56,398.—JAMES B. FORSYTH, Roxbury, Mass.—*Rubber Rollers for Wringing Machines.*—July 17, 1866.—The rubber is cured on a hollow metallic core which gives a more uniform access of heat to each part.

Claim.—Curing rollers of India-rubber or other vulcanizable gum on a hollow core, substantially as and for the purpose described.

No. 56,399.—GEORGE P. and GEORGE F. FOSTER, Mohawk, N. Y.—*Breech-loading Fire-arm.*—July 17, 1866.—The cartridge is introduced from the front end of the swinging breech block, and as the latter is closed the pintle is forced back; as the block swings open the pintle is thrust forward, loosening the cartridge shell for subsequent ejection.

Claim.—The pintle K, constructed and operated substantially as described, that is to say, being forced to the rear by the back pressure of the cartridge in loading, driven forward by the impingement of its rear end upon a projection on the abutment or its equivalent, and sustained by the spring L in the annular groove, in position to hold the cartridge case free for subsequent retraction or rejection.

No. 56,400.—F. H. FURNISS, Crestline, Ohio.—*Spittoon for Railroad Cars.*—July 17, 1866.—The contents are discharged by depressing the valve with a rod introduced above.

Claim.—First, constructing a spittoon with a valve-seat C and valve C', as set forth.

Second, the hollow stem B' and spring D in combination with the valve C' and body B, as and for the purpose herein set forth and described.

No. 56,401.—E. C. Gero, Galesburg, Mich.—*Horseshoe.*—July 17, 1866.—This shoe for contracted feet has an outward, bearing spring at the toe, and two springs which tend to spread the heels.
Claim.—The shoe d d, with spring a, springs b b and pads c c, constructed and used substantially as and for the purposes herein set forth.

No. 56,402.—Cyrus F. Gillett, Sparta, Wis.—*Bag Holder.*—July 17, 1866.—Explained by the claim and cut.
Claim.—The ring C, applied within the funnel A, for the purpose of holding the upper end of a bag, substantially in the manner described and shown.

No. 56,403.—Russell S. Gladwin, Meriden, Conn.—*Machine for Grinding Cutlery,* &c.—July 17, 1866.—The horizontal disk has radial channels for the knife blades or blanks, and revolves beneath a grindstone; it has a tipping motion at that point to taper the blades in the line of their length and from the back toward the edge.
Claim.—In combination with a revolving grindstone and roller, or its equivalent, placed opposite its grinding point, an interposed table, with suitable recesses for holding the knife or other blank to be ground, and series of cams under said table, and operating in connection with the stone and the roller, substantially in the manner and for the purpose set forth.

No. 56,404.—H. Goodrich and G. R. Edwards, Shawneetown, Ill.—*Steam Engine.*—July 17, 1866.—To avoid the dead point the crank is revolved by two pitmen connected by rods, &c., to three double-acting pistons in the cylinder; the motion of the pistons in contrary directions is harmonized by the intervention of a rock-shaft connected to one pitman.
Claim.—The combination and arrangement of the movable piston heads 8 8 8, with the piston rods 12 12 12, ports A A and B B, with the tubes c c, pitmen 4 5 6, and rock shaft arm 1, substantially in the manner and upon the principle as herein set forth.

No. 56,405.—Francis Granger, Lockport, Ill.—*Harrow.*—July 17, 1866.—The pentangular barrow frame is drawn by a tongue attached to a flanged roller which traverses a central circle; the pressure of the weight forces the teeth on that side further into the ground, and by resistance causes the barrow to revolve.
Claim.—The form of the harrow; also, a draught pulley running inside a circle in such a manner as to leave the harrow in position to revolve freely.

No. 56,406.—James Guckian, Camden, Ohio.—*Implement for Stripping and Cutting Sorghum.*—July 17, 1866.—The movable jaw is opened and the stalk gripped; a motion downward strips off the leaves, and a draw motion cuts the stalk at the ground.
Claim.—An implement for stripping and cutting sorghum, and other analogous uses, constructed with a fixed blade B, a movable blade or jaw C, and a lever D, or its equivalent, said parts being respectively constructed and the whole combined for use, substantially as set forth.

No. 56,407.—G. Gunther, New York, N. Y.—*Musical Attachment to Bird Cages.*—July 17, 1866.—The weight of the bird starts the music, the box being previously wound and its stop connected to the lever on which the perch rests.
Claim.—The application to a cage A of a musical device, such for instance as an ordinary music box, in combination with a suitable lever b and bar c, substantially as and for the purpose described.

No. 56,408.—Edwin R. Hall, Buffalo, N. Y.—*Coal Hod.*—July 17, 1866.—A covered coal hod with a spout; a rod on the bail holds the cover down while discharging.
Claim.—The spout B, constructed substantially as described, in combination with the cover C and bail D, provided with cross-wire e, or its equivalent, arranged and operating as set forth.

No. 56,409.—Edward Hamilton, Chicago, Ill.—*Steam Safety Valve.*—July 17, 1866.—The spring safety valve is locked within a spherical case, but may be lifted at pleasure by the lever which rests below the step.
Claim.—First, the combination and arrangement of the valve e, provided with the stem F, spiral spring a, and set screw E, with the case D, all located within the case A, as shown and described.
Second, in combination with the valve e, arranged as set forth, the lever G, arranged to operate as set forth.

No. 56,410.—B. J. Harrison and J. Condie, New York, N. Y.—*Folding Chair.*—July 17, 1866.—The bar at the lower ends of the posts braces the back by resting against the legs when in use, and otherwise forms a handle.
Claim.—The transverse bar G, so arranged in relation with the pivoted back seat rail C, the back E F, and the legs B, as to serve as a brace to hold the back in position when the chair is opened, and as a handle by which the chair may be carried when closed, substantially as herein set forth.

Also, a folding chair of the crossed legs B A, flexible seat D, back E F, pivoted back seat rail C, and transverse bar G, the whole constructed, combined and arranged substantially as herein set forth.

No. 56,411.—CHARLES T. HARVEY, Tarrytown, N. Y.—*Railroad.*—July 17, 1866 —A cable traverses continuously between the rails, and is clutched at will by the conductor on the car which is propelled thereby. The cable has spring ferrules to be engaged by the clutch. The details are explained by the claims and cuts.

Claim.—First, a coupling clutch for connecting a car, or other vehicle or body, to a moving cable, which is jointed so as to be capable of opening and releasing the cable, and has its divisions which clasp the cable or the heads thereof so shaped as to become of less diameter toward the forward end, substantially as described.

Second, jointing the divisions of that part of a clutch which engage the cable so that they can be raised separately clear of the cable guide, substantially as described.

Third, a coupling clutch whose divisions swing on the rod on which the clutch slides in combination with springs I I', or their equivalents, whereby the clutch and the vehicle are relieved from sudden shocks when connected to a moving cable, substantially as set forth.

Fourth, the pendulous buffers for bringing a clutch into engagement with the cable, when the clutch is made in two or more parts, substantially as described.

Fifth, the cam shafts and their cams G in combination with the buffers, substantially as shown.

Sixth, hinging the divisions of a divided clutch upon a rod or shaft parallel with the length of the car or other vehicle to which it is applied, substantially as described.

Seventh, the use of a hollow coupling clutch which connects a car or other vehicle to a moving cable by embracing or straddling the cable and its ferrules, in combination with a shaft on which it slides, substantially as described.

Eighth, placing an elastic cushion or cushions, or their equivalents, in the interior of the heads or ferrules of a moving cable, when such ferrules are jointed, substantially as described.

Ninth, giving a conical form to that part of the clutch which enters into the cable guide, so that when it receives one of the ferrules of the cable it lifts it out of frictional contact with the guide, substantially as described.

Tenth, making the ends of the ferrules of the cable of a conical form, substantially as described.

No. 56,412.—C. P. HAWLEY and E. B MURDOCK, East Galway, N. Y.—*Gate.*—July 17, 1866.—The gate is rotated by levers and rod connections to the swivel bar on the gate post.

Claim.—The levers J K H L M, and connecting rods N O R S and T U V W, constructed and arranged as herein described, in combination with each other, with the supporting posts A B C, and with the gate G, substantially as herein described and for the purpose set forth.

No. 56,413.—CHARLES HESS, Cincinnati, Ohio.—*Combined Piano, Couch, and Bureau.*— July 17, 1866.—Explained by the claim and cut.

Claim.—A combination of piano, couch, and bureau, arranged and operating substantially as represented and set forth.

No. 56,414.—I. W. HOAGLAND, New Brunswick, N. J.—*Pistons for Deep Well Pumps.*— July 17, 1866.—After a limited depression the valve stem is by rotation made to connect with the frame of the foot valve to withdraw it.

Claim.—The combination of the valve G, rod C, shoulder B, neck D, guards I, and walls E, arranged with a pump cylinder, and operating in the manner and for the purpose herein specified.

No. 56,415.—A. H. HOOK and H. B. ADAMS, New York, N. Y.—*Eraser.*—July 17, 1866.— The blank is swaged out of sheet metal, and the blade has a barrel or semi-cylindrical shank to attach it to a holder.

Claim.—Forming erasers substantially as and for the purposes herein described.

No. 56,416.—W. L. HORNE, Batavia, Ill.—*Steam Water Power Device.*—July 17, 1866.— By the described devices the chamber is alternately filled with steam and with water drawn from the reservoir below, when the steam is condensed.

Claim.—The arrangement and combination of the float d, chamber S, condenser n, perforated pans y and q, slats p, and connected by pipes u and r, as herein described and for the purpose set forth.

No. 56,417.—W. L. HORNE, Batavia, Ill.—*Fire Escape.*—July 17, 1866.—The platform is suspended by ropes on the outside of the window, and is gradually lowered by the rotation of the windlass upon which the ropes are wound.

Claim.—The arrangement and construction of the windlass E, with its ropes H and J, square frame C, with its brake P and its rollers G, when arranged and combined to operate as herein described.

No. 56,418.—ROBERT B. HUGUNIN, New York, N. Y.—*Rollers for Clothes Wringers, Washers, &c.*—July 17, 1866.—The cloth is pressed by the rods into the longitudinal grooves of the cylindrical core, and the rubber placed thereon and vulcanized.

Claim.—The elastic rollers herein described, made by vulcanizing rubber or equivalent gum, upon raw rubber, prepared cloth, or wire cloth, or both combined, the cloth being first wrapped around the central core, and the rods or their equivalents secured within the said cloth and grooves of the core, substantially in the manner and for the purposes specified.

No. 56,419.—LIVERAS HULL, Charlestown, Mass.—*Lathe for Turning Whip Stocks.*—July 17, 1866.—The described devices are for holding and turning the stock, and for giving the direction to the cutters to insure the taper form.

Claim.—For the purpose set forth, the combination of the two adjustable pattern bars I K, the furcated levers G H, the carriage F, its ways or guides E E, the mandrel A and chuck C, or the equivalent of the latter, the cutter q, and the self-adjusting Y-piece l; and also the combination of the same and the slide w.

Also, the combination and arrangement of the adjustable throat lever or piece s with the cutter q, when applied to the upper furcated lever so as to be adjustable thereon, as specified.

No. 56,420.—LAFAYETTE HUNTOON, Milford, Mass.—*Piston.*—July 17, 1866.—The springs are connected by a rod which passes through the hub of the piston, and their ends bear against the sections of the divided ring piston.

Claim.—The combination of the separate springs C C with a connection G, or its equivalent, substantially as described, whereby half of the excess of pressure of one spring may be transferred to the other, so as to equalize the pressure of both on the rings, as specified.

No. 56,421.—MARY E. HURLEY, Baltimore, Md.—*Needle for Caning Chairs.*—July 17, 1866.—Explained by the claim and cut.

Claim.—A needle A for caning chairs, having an eye b through the front end a, constructed and operating substantially as shown and described for the purpose set forth.

No. 56,422.—SAMUEL L. LATTA, Ligonier, Ind.—*Adjustable Store Shelves.*—July 17, 1866.—Explained by the claims and cut.

Claim.—First, the adjustable cleats C, and thumb-screw rods E, operating on bevelled guides D, for the adjustment of the shelves A A, substantially in the manner and for the purpose as herein specified.

Second, the screw lever rod F, and screw nuts G, as arranged in connection with the shelves, and operating in the manner and for the purpose substantially as herein specified.

No. 56,423.—THEODORE C. LAW, Green Island, N. Y.—*Combined Stove Hook, Hammer, &c.*—July 17, 1866.—A combined wrench, grate shaker, nail-claw, screw-driver, hammer, and stove-hook.

Claim.—The household implement, combining the appliances substantially as described.

No. 56,424.—CHARLES LIVINGSTON, Redwood City, Cal.—*Cutter for Wood Planing Machines.*—July 17, 1866.—The tonguing and grooving cutters are arranged obliquely, their edges against a wedge-shaped guide block, through which is adjusted the narrow receding bit.

Claim.—The arrangement of the cutters C and G upon a suitable cutter head, having a wedge-shaped centre piece B, substantially as and for the purposes described.

No. 56,425.—THOMAS E. LOCKWOOD, Cincinnati, Ohio.—*Churn.*—July 17, 1866.—The crank, gearing, and intermediate adjustable connections give a variability to the stroke of the dasher.

Claim.—The arrangement of spur wheel F, pinion E, crank-shaft D, and pitman G, in combination with the adjustable lever I, when provided with the series of apertures K L and M, all arranged to operate substantially as and for the purpose herein described and set forth.

No. 56,426.—DONALD R. MACLENNAN, Cincinnati, Ohio.—*Sawing Machine.*—July 17, 1866.—The guide rod works in an oscillating socket on the driving shaft, and the saw-raising devices are made removable.

Claim.—First, the rocking socket G, mounted directly on the driving shaft A, in combination with the guide rod F, for the purposes specified.

Second, the arrangement of the lever L, removable bracket N, socket n, rod M, and roller box K, relatively to each other and to the sawing apparatus A B C D E, as and for the purposes set forth.

No. 56,427.—J. F. MAGUIRE, East Boston, Mass.—*Artificial Hand.*—July 17, 1866.—The thumb and fingers are respectively actuated by the motion of a slide which is moved and fastened in position by the other hand.

Claim.—Connecting the fingers D of the hand to and with the slide R, having a thumb nut U, through angular lever arm N, connecting rod L, and cross-head F I, substantially as herein described, and for the purpose specified.

Also, in combination with the above, connecting the thumb V with the slide R, through a spring arm W, substantially as and for the purpose described.

No. 56,428.—JOSEPH MARCHANT, Cambridge City, Ind.—*Straw Cutter.*—July 17, 1866.—To the fly wheel is attached an adjustable eccentric which is connected to the rock-shaft and feed pawls, and serves to vary the length of the chaff.

Claim.—The arrangement and combination of the balance wheel A, adjustable-plate wheel B, thumb screw C, rod E, pawls F and G, ratchet wheels H H, and rocker arm or shaft I, constructed and operating substantially as and for the purpose set forth.

No. 56,429.—OSCAR F. MAYHEW, Indianapolis. Ind., assignor to WILLIAM H. WEEKS and G. M. LEVETTE, same place.—*Furnace for Puddling, Heating, &c.*—July 17, 1866.—The throat of the furnace is near the top of the incandescent fuel, and air is admitted between the fire back and the bridge, and also above the latter.

Claim.—First, the construction and arrangement of the throat or opening C, and air passages F and H, when placed in such relation to the incandescent fuel as to operate in the manner and for the purpose substantially as set forth.

Second, the damper G, in combination with the air passage F and throat C, when arranged as and for the purpose substantially as set forth.

Third, the zigzag divisions of the air passage F, in combination with the throat C, when arranged as and for the purpose substantially as set forth.

Fourth, the upper air-passage H, in combination with the throat C, when arranged as and or the purpose substantially as set forth.

No. 56,430.—I. W. McGAFFEY, Chicago, Ill.—*Machine for Planting Cotton Seed.*—July 17, 1866.—Explained by the claims and cut.

Claim.—First, the rotating flanges in the seed box, for moving and agitating the seed, constructed and arranged in the manner and for the purposes specified

Second, the rotating fingers, in combination with the flanges or agitators in the seed box, arranged and operated as shown.

Third, the construction, arrangement and combination of the fingers and adjustable slide for regulating the quantity of seed discharged, substantially as specified.

No. 56,431.—HENRY MEYERS, Hyde Park, and A. WEBB, Scranton, Penn.—*Extracting Bungs from Barrels.*—July 17, 1866.—Explained by the claim and cut.

Claim.—A bung A, provided with a staple b, and depression d, as a new article of manufacture.

Also, the hook d, in combination with a screw or lever, and with the staple b, in the bung A, substantially as and for the purpose set forth.

No. 56,432.—THOMAS S. MITCHELL, Pittsburg, Penn.—*Railroad Switch.*—July 17, 1866.—The flanges of the wheels rise upon and depress plates which by rock-shaft and lever connections operate the switch rail.

Claim.—First, the automatic switch-moving apparatus, composed of the bar D, links E F H, bar I, levers R R, shafts r r, arms S S, frames T T, and weights U U, or their equivalents, when they are arranged and operating as specified.

Second, the pieces of steel V V, in combination with the rail a' and frame T.

Third, operating a switch automatically by the action of the weight of the train itself on the frames T T, in the manner described, and for the purpose of preventing such train from running off the track.

No. 56,433.—ALBERT MOORE, San Francisco, Cal.—*Quartz Mill.*—July 17, 1866.—The adjacent sections are scarfed together to avoid a straight line of junction. The grooves on the face become gradually shallower as they recede from the eye, and the outer portion of the face is "land."

Claim—First, in combination with the radial feeding furrows B B B', the plain surface beyond the ends of the furrows, substantially as described, for the purposes set forth.

Second, the manner of breaking the joints in constructing and laying the shoes and dies, so that no continuous straight lines shall be employed from the feed centre of the muller to its circumference, substantially as described, and for the purpose set forth.

No. 56,434.—W. B. MOORE. Philadelphia, Penn.—*Broom Head.*—July 17, 1866.—The butts of the corn are clamped by an eccentric roller in the bow, and the shank of the latter is screwed into a handle drawing the butt of the broom into the cap.

Claim.—The combination of the cam or eccentric roller with the bow, and with a lever for turning said roller to clamp the broom corn or other material between the bow and cam roller; and this, whether the lever for turning the roller be the screw for holding the broom, cap and handle together, or whether it be a separate or removable lever, substantially as described.

No. 56,435.—W. E. MORRISON and W. L. BETTS, Funkville, Penn.—*Pump for Deep Wells.*—July 17, 1866.—An inverted cage attached to the piston rod to catch falling rivets, &c.

Claim.—Attaching to the piston or sucker rod of a pump, and above the upper valve, secured to it, a perforated receiver, substantially as herein described and for the purpose specified.

No. 56,436.—S. E. and G. L. MORSE, Harrison, N. J.—*Sounding Apparatus.*—July 17, 1866.—Explained by the claim and cut.

Claim.—First, arranging fluids of different specific gravities in a vessel or vessels, so that when sunk in water or submitted to pressure otherwise a mark of the amount of compression of one or more of these fluids at the greatest depth, or at the point of greatest compression, is retained for inspection on the return of the instrument to the operator, substantially as described.

Second, the arrangement of two liquids having unequal specific gravities, with a meter tube in a vessel closed except at one end of the meter tube, in such a way that external pressure, caused by the descent of the instrument in water, or otherwise, will force a portion of the lighter liquid through the heavier liquid into the body of the vessel to supply the vacancy there made by the compression of its contents, and that when, under a relaxation of the external pressure, caused by the ascent of the instrument in water or otherwise, the expansion or reaction of the liquids in the body of the vessel will force the heavier liquid into the meter tube, to the amount of the compression, thus forming a meter of the compression, and, by inference, of the greatest depth to which it has descended, substantially as described.

Third, the introduction of a minute quantity of air or other elastic fluid into the vessel containing the liquids, as described in the clause next preceding, to make the instrument sensitive as a meter of depth in comparatively shallow water.

Fourth, the application to the bathometer of a meter tube, so constructed that the liquids can easily pass each other in the bore of the said meter tube, thereby enabling the operator to restore them to their original position for a new operation merely by turning the instrument, substantially as described.

Fifth, attaching a bag of India-rubber, or other suitable flexible material, to the outer end of the meter tube, for the purpose of preserving the exact quantities of the fluids in the vessel, as at first adjusted, and of enabling the operator, by pressure upon the bag, to discharge the contents of the meter tube into the vessel, and therefore to use a meter tube of small bore, substantially as described.

Sixth, attaching a buoy and weight to a bathometer in such a way that when the instrument or its appendage touches the bottom the weight shall be detached, and allow the buoy to carry the instrument to the surface, substantially as described, thereby dispensing with a line.

Seventh, the method of releasing a submerged buoy, by causing a small weight attached to the long arm of a lever to support on the short arm the larger weight, which sinks the buoy, till the smaller weight, touching the bottom, is supported thereon, thus causing the short arm, no longer counterpoised, to fall and discharge the greater weight, substantially as described.

Eighth, attaching to a bathometer a rod or pole in such a way that on its return to the surface of the water it will attract attention at a distance so as to facilitate the recovery of the apparatus of which it forms a part, substantially as described.

No. 56,437.—ELI P. NEWBANKS and H. M. POWEL, Lincoln, Ill.—*Scaffold.*—July 17, 1866.—The rollers of the brackets rest against the building and the vertical adjustment is by cords and windlass.

Claim.—The brackets as constructed in combination with the scaffold board A and braces H and G, the same being used substantially in the manner and for the purpose herein specified.

No. 56,438.—JEREMIAH L. NEWTON, Boston, Mass.—*Stays, Springs, and Extensors in Wearing Apparel.*—July 17, 1866.—Explained by the claim.

Claim.—As a stay, extensor, or spring in wearing apparel, raw hide cut in strips or otherwise adapted for giving stiffness to and supporting corsets, stays, waists, and skirts of dresses and other articles of wearing apparel, as and for the purpose above set forth.

No. 56,439.—FREDERICK NISHWITZ, Williamsburg, N. Y.—*Horse Hay Fork.*—July 17, 1866.—The point is pivoted to the staff and vibratable by the sliding bar, whose catch retains it in an extended or laterally projecting position or releases it to discharge the load.

Claim.—First, the combination with the shank, the tine, and the traversing bar of the sliding collar G, all arranged and operating substantially as described.

Second, the combination with the shank and traversing bar of the locking lever, when constructed and arranged as and for the purpose described.

No. 56,440.—F. NISHWITZ, Williamsburg, N. Y., and B. S. HYERS, Pekin, Ill.—*Horse Hay Fork.*—July 17, 1866.—Explained by the claims and cut.

Claim.—First, the combination in a horse hay fork of two S-shaped prongs or tines pivoted near their centres to move in parallel planes, so arranged that when entering the hay, the lower arms of the tines unite to form a spear to penetrate more easily, and when expanded the hay is grasped in two separate bundles between the lower arm of one prong and the upper arm of the other respectively, substantially as described.

Second, the arrangement of the tines, pivoted on opposite sides of the rigid shank or draw-bar, as described, for the purpose of avoiding clogging.

Third, the combination of the tines, pivoted to the shank, with the sliding collar, toggles, and stops, substantially as described, for the purpose of locking the tines when hoisting.

Fourth, the combination with the shank, the tines, and the sliding collar of the toggle links, when arranged to operate as a stop to limit the backward movement of the tines in entering the hay, substantially as described.

No. 56,441.—A W. OLDS, Green Oak, Mich.—*Fence.*—July 17. 1866.—The stakes are crossed above the upper rail and support.the rider, which is secured thereto by wire.

Claim.—The braces E E when secured to the uprights B B, as described, in combination with the upper rail and binding wire H, as and for the purpose set forth.

No. 56,442.—NORMAN OLIN and E. L. HOPKINS, Homer, Mich.—*Washing Machine.*—July 17, 1866.—The rubber is connected by springs to the shaft which travels in the grooved sides of the tub. The bed is made of an endless band resting on rollers.

Claim.—The combination with each other of the endless belt or apron D, passing over the rollers B, the rubber E, shaft F, carrying the rollers G G and the springs I I, arranged and operating substantially as described.

No. 56,443.—JOHN PORTER, Ruggles, Ohio.—*Machine for Folding Fleeces of Wool.*—July 17, 1866.—The twines and slotted strap are laid on the table, the fleece placed thereon, doubled in by the hinged sides, the strap turned over and drawn tight by the roller; the twines protruding through the slots in the strap are then tied and the fleece released.

Claim.—The sectional table B B' C C' and leaf L, in combination with the brace G, strap I, and roller E, when arranged in the manner and for the purpose set forth.

No. 56,444.—WM. L. POTTER, Clifton Park, N. Y.—*Roofing Cement.*—July 17, 1866.—Explained by the claim.

Claim.—An improved composition for roofing and similar uses, formed by mixing raw coal tar and powdered clay with each other, substantially in the manner described and for the purposes set forth.

No. 56,445.—WILLIAM PRUETT, Kokomo, Ind.—*Tenon Machine.*—July 17, 1866.—The two reciprocating cutters form the sides and shoulders of the tenon at the same time and are operated by the lever which also feeds the tail-block and the stuff.

Claim.—First, the hereinabove described device for feeding the tail-block toward the cutters with the upward motion of the cross head by means of the lever G, cam lever O, rod P, teeth Q, and pawl P' attached to the tail-block N, the said several parts being constructed and the whole arranged for use substantially as set forth.

Second, in combination with the knives L and K, so arranged as to cut the shoulders and sides of the tenon at the same time, a device for giving a forward feed to the tail-block, actuated by the same lever that communicates motion to the knives, substantially in the manner set forth.

No. 56,446.—M. QUINBY and J. C. STURDEVANT, Skinner's Eddy, Penn.—*Broom Head.*—July 17, 1866.—The butts of the broom corn are enclosed in a leather bag, which is then clamped between the jaws on the end of the handle and secured thereto by a screw-bolt.

Claim.—The movable jaw H, furnished with a hinge A and a shank C, in combination with the stationary jaw G, binders D D, and screw B, as described and for the purposes set forth.

No. 56,447.—CHAS. L. RAHMER, Brooklyn, N. Y.—*Hat.*—July 17, 1866.—The sweat band is attached to a hoop suspended by bent straps inside the hat but not touching it.

Claim.—The combination of the band *a* and bent arm *b* with the sweat lining B. applied to the hat A, forming the space *f*, all in the manner and for the purpose herein specified.

No. 56,448.—THOMAS L. REED, Providence, R. I.—*Socket Coupling for Gas Fixtures.*—July 17, 1866.—The rubber packing of tubular form has flanges at each end, one composed partly of non-elastic substance and having the inner portion of the same bevelled outward.

Claim.—Forming the packing of the coupling with two flanges and an intervening space externally, and a swelling ridge internally, substantially as described for the purpose specified.

Also, making that flange of the packing by which it is confined in the shell of some comparatively inelastic material, substantially as and for the purpose specified.

No. 56,449.—ORRIN REEVES, Greenport, N. Y.—*Clothes Wringer.*—July 17, 1866.—The upper friction roller has a spring pressure upon the lower one.

Claim.—The steel springs 8 8, and the adjustable journal box *g*, the rollers G G and friction rollers B, the several parts being constructed, combined, and arranged, as and for the purpose herein described and represented.

No. 56,450.—CYRUS W. SALADEE, Newark, Ohio.—*Padlock.*—July 17, 1866.—The tumbler affords a hook for grasping the nose of the shackle : a projecting arm to which the spring is attached ; a stud upon one side for holding it in position ; a stud upon the opposite side for guiding the key, and an annular, eccentric raised wall surrounding the key stud, against which the key bit acts to force the tumbler around against the tension of the spring to release the hook from the shackle.

Claim.—First, as constructed, the tumbler A, with the guard ring C, attached as described, and operating as set forth, in combination with the spring E, for the purposes set forth and described.

Second, the key stud X and short stud 8 on tumbler A, constructed as described and for the purposes set forth.

No. 56,451.—CYRUS W. SALADEE and WILLIAM ARMSTRONG, Newark, Ohio.—*Padlock.*—July 17, 1866.—The wheel hasp is pivoted in a central hub, the lower portion extending down to a spring, which forms a stop.

Claim.—First, the wheel hasp A or its equivalent, constructed and operating in the manner and for the purpose substantially as shown and described.

Second, the centre pin or pivot C, in combination with the wheel hasp A, in the manner and for the purpose substantially as shown and described.

Third, the shoulder H or its equivalent in combination with the hasp A and spring B, or its equivalent, in the manner and for the purpose substantially as shown and described.

Fourth, locking the wheel hasp A by taking hold of the notch O, or its equivalent, in the manner and for the purpose substantially as shown and described.

No. 56,452.—RUFUS C. SANBORN, Ripon, Wis.—*Safe.*—July 17, 1866.—A square, outer metallic shell has a series of concentric metallic cylinders, with intervening spaces for air and water.

Claim.—First, the combination of the case A with the cylinders B C D, constructed and arranged substantially as and for the purpose herein specified.

Second, the use of the vessels for holding water when used in connection with the cylinders as herein fully set forth.

Third, the arrangement of the box E with the cylinders and outer case A, substantially as and for the purpose herein set forth.

No. 56,453.—JOHN SCHNEIDER, Williamsburg, N. Y.—*Manufacture of Lager Beer.*—July 17, 1866.—Boiling, unfermented wort is poured through the hops contained in an air-tight vessel, then cooled in a coil and added to fermented beer.

Claim.—First, the above-described process and production of an improved lager beer, substantially as described and set forth.

Second, the peculiar manner of extracting the essence or flavor of hops by means of the boiling wort or unfermented beer, and mixing the same with the fermented beer for the purpose substantially as set forth and described.

No. 56,454.—SILAS C. SCHOFIELD, Freeport, Ill.—*Combined Seeder and Cultivator.*—July 17, 1866.—The pivoted cam rod actuates the seed stirrer ; the hand lever operates the feed slide ; the extended axles afford points of attachment for the outer plough beams.

Claim.—First, the bifurcated double cam rod H *h h*, suspended by a swinging link *k*, and operated by an odd number of pins *i i*, substantially in the manner and for the purpose set forth.

Second, the combination of the agitating rock shaft J, with an actuating cam rod H, substantially in the manner and for the purpose specified.

Third, the compound lever M *m* for operating the seed slide *r*, as herein shown and explained.

Fourth, the stay braces or re-enforcing rods *s e*, in combination with the extended axle ends *f f*, when employed as draught wrists for attaching the outside plough beams E E, substantially in the manner and for the purpose set forth.

No. 56,455.—JOSEPH SCHOTT, Chicago, Ill.—*Pen and Eraser Combined.*—July 17, 1866.—Explained by the claim and cut.

Claim.—The combination of the folding drawing pen, with sliding eraser, the whole arranged as above described and for the purpose herein specified.

No. 56,456.—LEOPOLD SEEBERGER and N. LEVY, Cincinnati, Ohio.—*Baling Press.*—July 17, 1866.—The devices for varying the speed of the descent of the follower, and for displaying and fastening the sides, are explained in the claims and cut.

Claim.—First, the provision in a bailing press of the sliding shaft J, so arranged as to allow a fast or slow motion of the follower by coupling or uncoupling a train of spur wheels D E F G and pinions *c d e f g*, in the manner described and set forth.

Second, a bailing trunk, all of whose sides T are hinged to the bottom or floor R of said trunk, in the manner specified.

Third, in combination with the elements of the clause immediately preceding, the staples U, catches V, and stops W, all arranged and operating as and for the purpose described.

No. 56,457.—S. SHEPHERD and A. M. GEORGE, Nashua, N. H.—*Machine for Polishing Enamelled Paper.*—July 17, 1866.—The elastic apron is made of rubber, velvet, or other material. The burnishing roller may be revolved in a direction opposite that of the apron. The spring-pressing plate serves to spread the paper evenly upon the apron, and the lateral reciprocation of the roller prevents the formation of ridges or creases in the paper.

Claim.—First, the combination of the metallic burnishing roller G, endless apron F and table B, when the burnishing roller revolves at a higher velocity than that of the endless apron, substantially as herein set forth for the purpose specified.

Second, providing an elastic bearing for the paper under the burnishing roller by making either the apron or the table elastic, substantially as herein set forth.

Third, giving the burnishing roller G a reciprocating movement transversely to the endless apron simultaneously with its rotary motion, substantially as herein set forth for the purpose specified.

Fourth, the pressing plate T, applied in relation with the burnishing roller G, endless apron F and table B, substantially as herein set forth for the purpose specified.

No. 56,458.—HENRY SMITH and HIRAM F. SNOW, Dover, N. H.—*Beverage.*—July 17, 1866.—Made by fermenting the following mixture: American sarsaparilla, life of man, prince's pine, water, sugar, oil of spruce, oil of checkerberry, oil of sassafras, and molasses.

Claim.—A beverage prepared from the ingredients and substantially in the proportions and manner herein specified.

No. 56,459.—J. H. SMITH, Pineville, Penn.—*Hollow Auger.*—July 17, 1866.—The cutter is attached to one of the jaws and the latter are adjusted simultaneously by a right and left hand screw to vary the size of the tenon.

Claim.—The frame or stock A and the two adjustable jaws D D', operated by the right and left screw F, and the cutter G, all constructed and arranged to operate in the manner substantially as and for the purpose herein set forth.

No. 56,460.—ATKINS STOVER, New York, N. Y.—*Screw Wrench.*—July 17, 1866.—The worm operates the movable jaw by traversing a slotted spindle and engaging the rack on the bar of the wrench.

Claim.—First, the travelling worm F fitted upon the rod E and working in a screw thread made upon the back of the bar of the wrench, in combination with the movable jaw C and bar A, substantially as specified.

Second, the combination of the rod E, worm F, slot *e*, pin *f*, movable jaw C, bar A, and stationary jaw B, substantially as shown and described.

No. 56,461.—W. PAINE and R. E. CAVINESS, Fairfield, Iowa.—*Broom Head.*—July 17, 1866.—Jaw plates hinged to the handle are clamped against the brush and retained by side bars and screw bolt.

Claim.—The plates A, having the flanges *a* and the teeth *t* hinged to the handle by means of the staples *c*, in combination with the clamps *w* and bolt D, all arranged as shown and described.

No. 56,462.—WASHBURN PEABODY, Dixmont Centre, Maine.—*Harness.*—July 17, 1866.—The hooks on the hip straps serve to hold up the driving reins.

Claim.—The arrangement substantially as described of the two rump hooks A A, with the back strap of a harness, the same being for the purpose specified.

No. 56,463.—O. C. PHELPS, New York, N. Y.—*Adjustable Hand-cuff.*—July 17, 1866.—The pivoted short arm moves upon the segment rack and the spring bolt in the former engages the teeth in the latter.

Claim.—The spring *f*, spiral spring *g*, and sliding bolt *e*, arranged so that said bolt shall catch into notches on the inner or concave side of the bow or long section *a*, substantially as described.

No. 56,464.—ELMORE W. TAYLOR, Franklin, Ind.—*Evaporator.*—July 17, 1866.—The pan is arranged above the furnace and the drum in the rear connects by boiler flues with the chimney; dampers in the rear chimney and at the rear of the boiler flue command the course of circulation.

Claim.—The reversing of the heat from the furnace, which heat plays on the bottom of the pan and passes through the pan C by means of small flues.

Also, the regulating of the heat by means of the shut-offs H and N.

Also, the drum E and the movable connection flues L L, &c., and the stationary ones M M, &c.

No. 56,465.—ALEXANDER L. THORP, Vandalia, Mich.—*Portable Picket Fence.*—July 17, 1866.—Square pickets are inserted in the round slots; the latter are made oval where the junction of the sections is on an uneven surface.

Claim.—The slots *a* in the rails A as constructed, and the picket *b* as arranged therein, in combination with the cross-pieces D as constructed, substantially in the manner and for the purpose as herein set forth.

No. 56,466.—WILLIAM TIBBALS, South Coventry, Conn.—*Revolving Fire-arm.*—July 17, 1866.—The front face of the breech is recessed and an annular lip formed to hold the cartridges from moving forward in the cylinder. The cylinder stop is prolonged to act as an anvil in exploding the cartridge.

Claim.—First, recessing the front face of the breech B to receive the smaller rear end of the cylinder D, when said recess is provided with the annular flange *c*, substantially as shown and described.

Second, the removable anvil *a*, or its equivalent, when constructed and arranged to operate as and for the purpose set forth.

Third, the annular flange *c*, or its equivalent, whether used with or without the anvil *a*, for the purpose of holding the cartridges in the cylinder, as described.

No. 56,467.—A. W. TODD, Chicago, Ill.—*Railroad Station Pump.*—July 17, 1866.—The cylinder is submerged in a cistern; the water therein is forced out and discharged into the tender by steam from the locomotive admitted above the floating piston, which is guided by the stay rod.

Claim.—The arrangement of the cylinder B with the stay rod I, cork *n* and *o*, cock E, spigot J F, handle H, being secured to the cylinder B at K, pipe C, substantially upon the principles and in the manner herein set forth.

No. 56,468.—F. W. TULLY and T. REECE, Philadelphia, Penn.—*Pump.*—July 17, 1866.—The reciprocation of the plunger is produced by connection with a trammel.

Claim.—First, the combination of the disk I with its slots L L', blocks *l l'*, and vibrating link N, with a single or double-acting lift and force pump, constructed substantially in the manner set forth.

Second, the crab or saddle D with its fixtures *d* and *e*, in combination with the foregoing and with the pipe C for attaching and giving support or steadiness to the pump, substantially as described.

No. 56,469.—PHILIP VAN BUSSUM, Henderson, Ky.—*Washing Machine.*—July 17, 1866.—The rubber against which the rollers press is in two semicircular hinged portions. The weighted arm forces over one of the rollers.

Claim.—The slatted rotating or semi-rotating cylinder B, in combination with the concave E, formed of the parts *e e*, connected by hinges *f*, and attached by hinges *g* to arms A projecting from shafts F F and having a weight H applied, all arranged substantially in the manner as and for the purpose set forth.

No. 56,470.—W. POWELL WARE, New York, N. Y.—*Calendar.*—July 17, 1866; antedated June 29, 1866.—Explained by the claim and cut.

Claim.—The dial *b* containing the days of the month in seven radiating columns, the dial *a* denoting the days of the week, and the dial *c* indicating the months, and visible through an opening in the dial *b*, when constructed and arranged in the manner and for the purposes herein set forth.

No 56,471.—R. M. WEBB, New York, N. Y.—*Burglar Alarm.*—July 17, 1866.—The insertion of a key from the outside disengages the piece attached to the lock and connected to the alarm; the latter is sprung by the disconnection.

Claim.—The combination of the tube E, rod F having a swivelled piece J, and spiral or other suitable spring H, with the key-hole of a lock or door, when arranged together and with regard to such key-hole, and connected to a bell or other alarm, so as to operate substantially in the manner and for the purpose described.

No. 56,472.—CALVIN J. WELD, West Wardsboro', Vt.—*Machine for Fluting Wash-boards.*—July 17, 1866.—The claim describes the devices that feed the board to the cutters, clamp it in position, and automatically release and reclamp it as it is fed to be fluted; also, the devices for operating the carriage and returning it for another effective motion.

Claim.—First, the feeding arm V attached to the feeding shaft P, in combination with the slot Z in which it moves, for feeding the blanks for a new cut during the return movement of the carriage, substantially as described.

Second, the springs R R for lifting the carriage out of gear at the end of its forward movement, in combination with the lugs *b b'* and slots or recesses C in the top rail of the boxes S S, substantially as described.

Third, the combination of the springs R R for lifting the carriage with the spring U for effecting its return movement, substantially as described.

Fourth, the stop lever W' with its stops W, made and operated as shown, in combination with the adjacent holder M, substantially as described.

No. 56,473.—C. C. WEBBER, Calmar, Iowa.—*Stove Pipe Drum.*—July 17, 1866.—The inner telescopic pipe is capable of longitudinal adjustment toward the damper and in combination with it commands the course of circulation.

Claim—An adjustable pipe F operated by the rod G, or its equivalent, and employed in conjunction with the flues A B C and damper D to make a direct or indirect communication through the drum, as and for the objects specified.

No. 56,474.—JOSEPH WELCH, Philadelphia, Penn.—*Loom.*—July 17, 1866.—To effect the change in the operations of the loom from plain weaving to the weaving of colored fabrics the lower sides of two of the heddle frames are connected by means of a cord passing under a fixed pulley beneath them, and the two upper sides of the same frames are connected by means of two cords with hooks attached; these cords pass over and are fixed to the usual top roller in addition to the usual connecting cords. These hooks may be changed from one to the other of the inner heddle frames as desired.

Claim.—Giving the described different motions to the heddles of the loom for the purposes specified, by means of the hooked cords or straps A B on the roller C, or their equivalents, operating in combination with the pulley E, or its equivalent, substantially as and for the purposes described.

No. 56,475.—ALBERT S. WILKINSON, Pawtucket, R. I.—*Horseshoe.*—July 17, 1866.—A thin metallic plate is riveted between the upper and lower plates of the shoe and makes an inward protruding flange to protect the sole of the hoof.

Claim.—The combination of the shoe A and web B, having its inner edges curved, in the manner and for the purpose set forth.

No. 56,476.—ALBERT S. WILKINSON, Pawtucket, R. I.—*Horseshoe.*—July 17, 1866.— The grooved rubber sole is clamped by rivets between a narrow plate which occupies the groove and an upper plate to be attached to the hoof. The pointed ends of the rivets come flush with the tread.

Claim.—First, the metal plates A *a* in combination with the rubber or other elastic sole D and rivets *c c'*, as illustrated by figures 1 and 2 of sheet 1, substantially as described.

Second, the hidden calkins *c c c*, operating substantially in the manner and for the purpose set forth.

No. 56,477.—ALBERT S. WILKINSON, Pawtucket, R. I.—*Horseshoe.*—July 17, 1866.— The shoe has a high, wide toe clip and curved heel clips united at the crease of the heel.

Claim.—The bar A, in combination with the toe clip *a*, and heel clips *a¹ a²*, as indicated in figures 1 and 2, the whole being constructed and operated substantially in the manner and for the purpose set forth.

No. 56,478.—F. H. WILLIAMS, Syracuse, N. Y.—*Stench Trap.*—July 17, 1866.—The ball valve closes the aperture against pressure which might drive back the liquid from the bend of the siphon.

Claim.—The siphon B, provided with a floating valve E, in combination with the sink or sinks in a house or building and with the pipe or pipes leading to the sewer, substantially as and for the purpose described.

No. 56,479.—F. H. WILLIAMS, Syracuse, N. Y.—*Stench Trap.*—July 17, 1866.—Explained by the claim and cut.

Claim.—The inclined apron C, tray D, and valve E, in combination with the sink A, constructed and operating substantially as and for the purpose described.

No. 56,480.—GEORGE WILLIAMS, Sterling, Colorado.—*Ore and Timber Car for Mines.*— July 17, 1866.—The hinged end doors may be folded over the top; when closed are supported against the ends of the sides and fastened by a catch, which is tripped by contact with a post at the proper spot for discharge.

Claim.—First, the construction of the doors with a wider portion *b*, to adapt them to be supported by the sides of the car, substantially as described.

Second, a car constructed with end doors adapted to be folded over the top for the purpose of converting it into a timber car.

Third, in combination with the above a trigger C, provided with an inward projection, adapted to be tripped by the post D.

No. 56,481.—GEORGE WILLIAMS, Sterling, Colorado.—*Elevator.*—July 17, 1866.—The bucket is supported underneath by a bail, which tips it when the bucket rollers are deflected by a guide piece; the contents are discharged into a chute, or a water trough, or it may be filled with timber, &c., for the mine.

Claim.—First, the elevating bucket E, with the discharging levers F F, applied to the bottom of the bucket, substantially as described.

Second, and in combination with the above the deflecting rollers D, and curved guide ways K K*, arranged and operating substantially as described.

Third, the adjustable sections J*, employed to enable the bucket to be discharged at different heights, substantially as described.

Fourth, the hinged chute O*, in combination with the levers O² and p, operating substantially in the manner and for the purpose described.

Fifth, the bucket E, in combination with the hook W, or its equivalent, the roller U, substantially as described.

No. 56,482.—L. H. WOLFF, Detroit, Mich.—*Trunk.*—July 17, 1866.—The cleat sustains the bottom and sides, affords bearings for the rollers and protection for the corners.

Claim.—A new article of manufacture intended for a cleat for a trunk, made of metal and constructed substantially in the manner above described.

No. 56,483.—JAMES O. WOODRUFF, Albany, N. Y.—*Apparatus for Applying Liquids to Casks.*—July 17, 1866.—The cask is suspended between adjustable disks in the pivoted frame. A nozzle in the bung hole is connected by a flexible tube with the reservoir containing the composition for coating the interior of the cask.

Claim.—First, the process for applying liquids to the interior of casks so as to penetrate into the pores of their bodies, by the employment of condensed air cold or at the temperature of the atmosphere, as described.

Second, the apparatus described in the within specification to effect the process of forcing liquids into the pores of cask bodies, that is, the frame B, suspended on its axis E, the disks C, with their screw rods R, the flexible tube H, with its nozzle J, and tube k, substantially as described and for the purposes set forth.

No. 56,484.—JOSHUA BARNES, New York, N. Y., assignor to ISAAC A. SINGER, same place.—*Spring Bed Bottom.*—July 17, 1866.—Double coils upon the bed rail and the end of the slat, respectively, are linked together, and both tend to keep the slat elevated.

Claim.—First, in combination with a bed slat a wire spring, having two parallel coils at the base and two parallel coils at the top, the coils at the base C turning adversely to those at the top B, substantially as above described.

Second, in combination with the two adverse springs, the hook or hinge, substantially as above described and for the purposes set forth.

Third, the combination of the cross bar I, rod E, pin D, and slat A, with the wire springs, as above described.

No. 56,485.—JAMES A. BAZIN, Canton, Mass., assignor to himself and A. B. HALL, West Roxbury, Mass., C. SCOTT and W. J. TOWNE, Newton, Mass.—*Machine for making Cordage, Webbing, &c.*—July 17, 1866.—Improvement on Bazin's patent of June 29, 1858. Its object is to avoid the wear incident to the continual striking of the ends of the curved arms of the irregular shaped gears against the shafts of the spool frame.

Claim.—First, in a machine for making cordage, webbing, and other similar fabrics, so actuating the spools by mechanism, consisting essentially of the revolving platform K, furnished with a series of gears L M N, sliding plates P, and recesses O, in combination with the toothed ring B, and a series of carriers V, with their spool frames, that each stand will be carried around two stationary ones, and thereby form an interlocking twist, as set forth.

Second, the above-described mechanism in combination with the rack W, for the purpose described.

Third, the sliding plates P, operated by a cam wheel Q, in combination with the platform K, and a series of carriers V, with their spool frames and spools, operating substantially as set forth.

Fourth, the combination of the gear L with its shaft h, gears S R, and cam wheel Q, for operating the sliding plates P, as described.

Fifth, adjusting the cam wheel Q, by means of the eccentric pin s, on the gear R, as set forth.

No. 56,486.—BURROUGHS BEACH, West Meriden, Conn., assignor to himself and E. A. THORP, North Haven, Conn.—*Castor Bottle.*—July 17, 1866.—The salt or sugar which has become caked in the bottle may be comminuted by revolving the spiked shaft or bent wire therein.

Claim.—The combination, with a castor bottle, of a shaft or spindle extending through the same in the direction of its length, and arranged to be turned therein in the manner and for the purpose described.

No. 56,487.—SMITH W. BULLOCK, Elizabeth, N. J., assignor to the BULLOCK ORE DRESSING MACHINE COMPANY, New York, N. Y.—*Quartz Mill.*—July 17, 1866; antedated July 3, 1866.—The annular revolving trough is suspended on springs and driven by gearing from the shaft on which are two crushing wheels.

Claim.—First, the combination of the rotating trough D with the crushing wheels G G, and gear wheels E and F, so as to govern the rotary motion of the trough while its vertical action is independent of, and disconnected from, the gear wheels.

Second, the application of springs to the adjustable bed, so arranged as to form a binding link or tie between the supports of the crushing wheels G G and the supports of the trough D, each of the several features being arranged and operating substantially as and for the purposes herein set forth.

No. 56,488.—ESEK BUSSEY, Troy, N. Y., assignor to himself and CHARLES A. McLEOD, same place.—*Boiler for Cooking Stoves.*—July 17, 1866.—Explained by the claims and cut.

Claim—First, a water reservoir or tank, constructed of cast iron, and entirely covered or coated upon the inner and outer surfaces thereof by zinc, or an alloy of zinc, substantially as aforesaid, in combination with a cooking stove, in the manner substantially as herein described and set forth.

Second, the water reservoir or tank for cooking stoves, constructed entirely of cast iron, and then covered or coated upon the inner and outer surfaces with zinc, or an alloy of zinc and tin, in the manner, and by the means, and for the purposes, substantially as herein described and set forth.

No. 56,489.—WILLIAM C. DODGE and R. D. O. SMITH, Washington, D. C., assignor to W. C. DODGE and W. S. KING.—*Machine for Filling Cartridges.*—July 17, 1866.—The cartridges are placed in a sliding drawer and are brought under the supply tube, which, together with the measuring slide, are operated by the alternating action of a hand lever. The measuring tubes of this slide are capable of adjustment, their capacity being graduated by an index; above them a supplementary sliding charger is employed to relieve the powder from the varying pressure of the contents of the feeding hopper. The machine is placed within a covered box for security against accident.

Claim.—First, a machine for filling cartridges, in which the powder is entirely enclosed during the operation.

Second, a series of measuring tubes, so arranged that they can be adjusted to contain a greater or less quantity, at will, substantially as described.

Third, the combination and arrangement of the lever P for operating the slides with springs of different tension, for causing the slides to operate alternately, with a single movement of the lever, substantially as and for the purpose set forth.

Fourth, the auxiliary charger C, arranged in relation to the hopper bottom B and the tubes Y, and operating in connection therewith, substantially as described.

Fifth, the bars *a*, or their equivalents, arranged over the openings *b* of the hopper bottom, substantially as and for the purpose set forth.

Sixth, providing the slide C with the raised rim *n*, with or without the spring *r*, as shown and described.

Seventh, the combination of adjusting devices for regulating the charges, with the index lever *s* and graduated plate T', as and for the purpose set forth.

Eighth, the slide H, provided with cells for receiving the cartridge cases, substantially as described.

Ninth, the drawer G', in combination with the cover I', provided with the tubes *t*, arranged to operate as and for the purpose set forth.

Tenth, the frame or guide M for inserting the cases in the cells of the drawer G', substantially as described.

No. 56,490.—EDWARD DUFFEE, Haverhill, Mass., assignor to himself and GEORGE A. KIMBALL, same place.—*Screen for Gas Purifiers.*—July 17, 1866.—Explained by the claim and cut.

Claim.—Constructing screens for dry coal-gas purifiers of strips of thin wood, crossing each other, and either notched together or interlaced in the form of basket work, supported by a framework of metal, or its equivalent.

No. 56,491.—LUTHER B. EDGECOMB, Troy, N. Y., assignor to himself and VAN RENSSELAER POWELL, same place.—*Drawbridge Signal.*—July 17, 1866.—The adjustable signal bearer is rotated by rack and pinion actuated by the motion of the bridge, and the motion is transmitted by oscillating arms and double rods.

Claim.—First, the combination of a rack *a*, a pinion *b*, and a swinging or adjustable signal bearer *g*, or their equivalent operating devices, with a drawbridge, in the manner substantially as shown, and for the purpose of operating the signals of drawbridges in the manner as described.

Second, the combination of a spring *d* and a stop *f* with a pinion *b*, with its stud *e* and a signal bearer *g*, in the manner substantially as shown and for the purpose set forth.

Third, the combination of the swinging or adjustable signal bearers *g* and *i*, with the yokes or double transmitting arms *k* and *k'*, connected by means of ropes or wires *l l*, in the manner substantially and for the purpose as shown.

Fourth, the combination of a signal and its swinging or adjustable bearer *i*, and the double transmitting arms or yokes k^1 and k^2, with each other, and with the double arms or yoke k^3, and its adjustable signal bearer m, arranged relatively to, and in the manner substantially as described and for the purpose set forth.

No. 56,492.—JAMES H. FOWLER and A. J. FRENCH, Waterbury, Conn., assignors to the AMERICAN FLASK AND CAP COMPANY.—*Machine for Trimming Percussion Caps.*—July 17, 1866.—A feed plate is intermittingly moved over the revolving carrier and supplies caps to the pockets near the circumference of the latter. As the caps in the carrier are brought in succession over a receptacle in the plate beneath. they are seized by a revolving spindle while a cutting blade moves forward and trims their edges.

Claim.—First, the feed plate C, (one or more,) provided with a series of feed holes h, which are arranged in curved lines to correspond to the position of the holes b, in the conveyer B, and operating in combination with said conveyer, substantially as and for the purpose described.

Second, the revolving conveyer B, in combination with the plunger b', the trimming dies E, and knife F, constructed and operating substantially as and for the purpose set forth.

Third, the supporting plate D, and spring f', in combination with the conveyer B, constructed and operating substantially as and for the purpose described.

Fourth, the stationary bracket h', and pin g', in combination with the spring f', supporting plate D, and conveyer B, constructed and operating substantially as and for the purpose set forth.

Fifth, the reciprocating slide c, and dog c^*, in combination with the supporting plate D and conveyer B, constructed and operating substantially as and for the purpose described.

Sixth, the revolving trimming dies E, and clamping spring s^*, in combination with the reciprocating knife F, constructed and operating substantially as and for the purpose set forth.

Seventh, giving to the cutting edge of the trimming knife an upwardly inclined position, substantially as and for the purpose described.

No. 56,493.—MARTIN FREE, Philadelphia, Penn., assignor to ALFRED LOUDERBACK, same place.—*Loom for Weaving Slat Blinds.*—July 17, 1866.—The introduction of the wooden slats into the shed is effected by the rotation of a wheel which winds and unwinds a leather strap having spring nippers at its free end to receive the slat; the strap rides across the shed resting on the race. The rotary or shear cutters for trimming the ends of the slats are operated by the movements of the lay.

Claim.—First, the application to a slat-blind loom of the slat-feeding device D d' d^4, the same being constructed, arranged, and operated substantially as and for the purpose described.

Second, the employment or use in a slat-blind loom of either or both of the automatically acting cutters E E', the same being applied so as to operate substantially as and for the purpose described.

No. 56,494.—W. J. GORDON, Philadelphia, Penn., assignor to J. S. MASON & Co., same place.—*Machine for Securing the Ends of Strips of Sheet Metal together.*—July 17, 1866.—The disks on the parallel axes have projections whose coincidence and operation is such as to cut two slits longitudinally in the strips and bend the cut portions; these are subsequently spread to a width greater than the space from which they are cut and thus retained.

Claim.—First. the roller H with its flange c, in combination with the disk F, and its lugs a a', the whole being arranged for joint action. substantially as and for the purpose described.

Second, the detachable disk G with its flange b, in combination with the disk F.

No. 56,495.—LEWIS F. GRANT, Thomaston, Conn., assignor to himself, JOSEPH R. BROWN, and LUCIEN SHARPE, Providence, R. I.—*Making Gear Cutters.*—July 17, 1866.—The mandrel is set in the lathe out of the true centre and directly underneath it. The mandrel is then oscillated by a handle the length of one tooth against the tool which shapes it eccentrically to the centre of the cutter and gives it a properly inclined clearance.

Claim.—Forming the blades of cutters for gear cutting with the tools and in the manner substantially as described.

No. 56,496.—JOHN GREACEN, Jr., and AMBROSE FOSTER, New York, N.Y., and JOHN H. COOPER, Philadelphia, Penn., assignors to the AMERICAN BUILDING BLOCK COMPANY, New York, N. Y.—*Machine for Moulding Materials Admitting of Cohesion.*—July 17, 1866.—The hammer being elevated and the mould depressed, the feed-box is brought forward and the three parts elevated, exposing the opening in the mould from which the anvil has withdrawn; the hammer now falls and packs the material in the mould, the feed-box is withdrawn shaving the upper surface, the hammer falls again, the mould is depressed, leaving the block on the face of the anvil.

Claim.—First, the use in combination with a machine constructed substantially as described, of two or more cylinders and pistons connected to separate operating parts of the machine and operating in unison, substantially as and for the purpose specified.

Second, the use, in combination with the above described machine, of cylinders having openings *c*, arranged in respect to and communicating with the exhaust ports, substantially as and for the purpose set forth.

Third, the combination of the reciprocating feed box J, and its opening *n*, with the sharpened edge *z*, for the purpose described.

Fourth, the combination of the mould box and anvil so adapted to each other that the former can descend while the anvil and formed block remain stationary, for the purpose explained.

Fifth, the combination of the mould box I, connecting rod R, arm Q, and shaft M', the whole being arranged and operating substantially as and for the purpose herein set forth.

Sixth, the spring T or its equivalent, interposed between the mould box and any stationary part of the machine, and operating as and for the purpose herein set forth.

No. 56,497.—BERTRAND J. HOFFACKER, New York, N. Y., assignor to himself and AUGUST W. STEINHAUS, same place.—*Lamp.*—July 17, 1866.—Explained by the claim and cuts.

Claim.—The improved lamp, consisting of the oil well or base A, movable neck C, chimney stay D E, and glass tube G, the mantle J, in combination with the usual parts of a lamp, substantially as herein described.

No. 56,498.—FRANCIS HOVEY, New York, N. Y., assignor to himself and CHARLES H. CLAYTON, same place.— *Stamp Cancelling Press.*—July 17, 1866.—The press may be operated by percussion of the plunger or rotation of the screw.

Claim.—The hollow screw C in combination with the central stem *e*, plunger and frame of a cancelling stamp or press, substantially as herein set forth for the purpose specified.

No. 56,499.—LUTHER JACKSON, Newark, N. J., assignor to EDGAR FARMER and WILLIAM H. CLEVELAND, New York, N. Y.—*Trunk.*—July 17, 1866.—The spring stops on the end of the expanding tray retain it in elevated position by resting on the end pieces of the trunk.

Claim.—First, the spring stops or braces *b* in combination with the inside cover C of a trunk, constructed and operating substantially as and for the purpose described.

Second, the bellows-shaped or expanding tray or cover C in combination with the trunk A, constructed and operating substantially as and for the purpose set forth.

No. 56,500.—CHARLES W. KIMBALL and MARY E. NORCOTT, Springfield, Mass., administratrix of the estate of W. J. NORCOTT, deceased.—*Valve for Relieving Steam Cylinders of Water.*—July 17, 1866.—The stem of the valve is raised by the wedge beneath, which moves in guides: the motion is occasional or produced by mechanical connection with some moving part of the engine.

Claim.—The combination of the valves P P, and the rod A, formed as described with the cylinder of a steam engine, and suitable mechanism for operating the same, substantially in the manner and for the purpose herein set forth.

No. 56,501.—GEORGE B. LEGO, New Haven, Conn, assignor to DAVID K. BROWN, same place.—*Umbrella.*—July 17, 1866.—The handle is made in two portions, uniting in a sleeve and fastened by a spring catch ; the handle end of the staff is removable.

Claim.—An umbrella having the handle in two sections connected by a sleeve and attachment, substantially as described.

No. 56,502.—D. B. TIFFANY, Xenia, Ohio, assignor to himself and ROBERT E. RICHARDSON, same place.—*Boring Wells and Laying Pipes.*—July 17, 1866.—The perforated, tubular shank has a solid or screw point and an upper hollow screw which traverses in the threaded socket in the upper bar of the frame.

Claim.—The boring tube D in combination with the tubular screw C and nut B, arranged to operate in the manner and for the purpose set forth.

Also, in combination with the tubular screw C and rod D, the adjustable ring *e*, as shown and described.

No. 56,503.—WILLIAM M. WRIGHT, Baltimore, Md., assignor to himself and JAMES E. PILKINGTON, same place.—*Apparatus for Carburetting Gas.*—July 17, 1866.—Vertical porous partitions are placed in the box and the current of air passed in a sinuous course through the openings, which are alternately at the top and just above the liquid level.

Claim.—The use of the division plates covered with cloth or its equivalent material, with their openings, notches, and adjustment in an enclosed box, so to form a continuous air-tight chamber when the lower part of the box is charged with a fluid, as and for the purpose described.

Also, the use of the flat tube or its equivalent arrangement for preventing the heavier portions of the fluid hydro-carbon from remaining at the bottom of the box, thereby securing a uniform volatility from all the fluid till it is consumed, as and for the purpose described.

Also, a combination of division plates, openings, notches, cloth coverings, and flat tube in an enclosed box, substantially as and for the purpose described.

No. 56,504.—WILLIAM BOULTON and JOSEPH WORTHINGTON, Burslem, England, assignors to MALKIN & CO., same place.—*Manufacturing Tiles, &c.*—July 17, 1866.— Moulds of different designs are placed in succession upon the ram, the space or remaining space in the mould is each time filled with material and pressed by a counter-mould, the tile is thus pressed and burnt.

Claim.—The improved mode and arrangement of apparatus for forming encaustic tiles, slabs, and other articles, substantially as above described and represented in the accompanying drawings.

No. 56,505.—JULES EDMOND FOURNIER, Courville, France.—*Manufacture of Acetate of Lead.*—July 17, 1866.—Explained by the claims.

Claim.—First, the direct manufacture of acetate of lead or salt of Saturn by the process herein described, that is to say, by means of pyroligneous acid distilled or saturated with lime and decomposed by sulphuric or muriatic acid, then saturated with lead or its derivatives, evaporated, crystallized, pressed, or submitted to the action of a turbine, and finally redissolved, evaporated, decolorized, clarified, and recrystallized.

Second, in the process of manufacturing acetate of lead, as described, specially the employment of the turbine or press, as and for the purposes set forth.

No. 56,506.—CHARLES WILLIAM JONES, Cheltenham, England.—*Adjustable Stock of Firearms.*—July 17, 1866.—The cheek plate of the stock is adjustable to preserve its natural position of ease in firing at long or short range; to the forward end of the stock proper is a lever arm to receive a telescopic back-sight.

Claim.—First, dividing the butt of the arm in such manner and so combining and connecting the parts together by suitable mechanism that a double and simultaneous motion of the parts of the butt is obtained at will; the one action causing an extension of the heel, straightening the stock, and thus allowing the butt to rest against the shoulder, the other motion extending a cheek piece to form an efficient rest for the cheek in the altered form of the butt, in the manner more fully described herein and illustrated in figures 1 and 2,

Second, the lever arm *p* designed to receive an orthoptic or telescopic back-sight.

No. 56,507.—RUDOLPH KECK, Chicago, Ill., assignor to F. W. WOOD, SOLON D. STANBRO, JOSIAH WARREN, and JULIUS H. ROYCE, same place.—*Silvering Mirrors.*—July 17, 1866.— The glass plate is laid on the inclined edges of the trough so that only its lower face will be touched by the solution and the reducing means.

Claim.—The method of precipitating upon glass plates nitrate of silver, or other suitable substance or substances, by means substantially such as herein described, or any other equivalent means.

No. 56,508.—P. C. P. L. PREFONTAINE, Paris, France.—*Storing Hydrocarbon Liquids and other Materials.*—July 17, 1866.—Close metallic vessels bound together like a raft are submerged in a cistern; the raft is decked and tubes for filling and emptying are attached.

Claim.—The system of warehousing, as herein described, by immersing in water any number of vessels or receptacles of any form or dimensions hermetically sealed and joined together, arranged in combined or separate series, strengthened and held together as herein described, and covered either by a roof or by planking on the top and around the sides as described, the said system being adapted and fitted for the warehousing of all liquids and other substances.

No. 56,509.—WILLIAM ALTICK, Dayton, Ohio.—*Machine for Pulling up Old Cotton and Corn Stalks.*—July 24, 1866.—The shield over the rollers is for the protection of the gearing; the arch straddles the stalks, the plates form a throat to direct them to the rollers, which tear them up and cast them on one side of the machine.

Claim.—First, the combination of the two rollers M M, when one is made rigid and the other flexible or yielding in its bearings, substantially as and for the purpose specified.

Second, the arch or bow D when used with the frame pieces or bars A A and the rollers M M, as and for the purpose herein specified.

Third, the arrangement of the shield V with the arch D and rollers M M, substantially as and for the purpose set forth.

Fourth, the plate S constructed as set forth and arranged under the rollers, as and for the purpose described.

No. 56,510.—MAURICE ANDRIOT, Mount Washington, Ohio.—*Carriage Jack.*—July 24, 1866.—The lever is pivoted in the standard and while it is being depressed the chock is lifted by the trigger; the release of the latter allows the chock to bind the lever in the position it has attained.

Claim.—The arrangement of the standard A C, fulcrum pin D, lever E F, self-locking chock G, and trigger K.

No. 56,511.—JOHN N. ARVIN, Valparaiso, Ind.—*Cultivator.*—July 24, 1866.—The inside shovel beams are pivoted by a universal joint to the frame to give them lateral adjustability in addition to their capacity, in common with the other shovels, of being raised vertically by the hand lever and intermediate devices.

Claim.—The arrangement of the joints *b*, universal joints H, links O, arms P, and chains J in combination with the curved plough beams G M and shaft E, operating in the manner and for the purpose herein specified.

No. 56,512.—JAMES E. ATWOOD, New York, N. Y.—*Hoop Skirt.*—July 24, 1866.—The lower portion of the hoop consists of metallic strips suspended from the horizontal hoops. The absence of the lower hoops removes the danger of stepping upon the crinoline or catching it against obstacles.
Claim.—Constructing a hoop skirt of horizontal hoops and the pendants B B B, combined and arranged substantially as described and set forth.

No. 56,513.—GEORGE BACKETT, New York, N. Y.—*Salve.*—July 24, 1866.—This salve for boils, cuts, sores, &c., is composed of lead plaster one and a half pound, Burgundy pitch one and a quarter pound, yellow rosin one half pound, black pitch two ounces, gum galbanum one ounce.
Claim.—The salve made of the several ingredients and mixed together in or about the proportions stated, for the purposes specified.

No. 56,514.—F. A. BALCH, Hingham, Wis.—*Stove Pipe Drum or Heat Radiator.*—July 24, 1866.—The upper and lower drums are divided by horizontal partitions, which have intercommunication by the concentric pipes; the distinct chambers thus formed are occupied respectively by the heated air, &c., from the stove and the vital air to be warmed.
Claim.—First, the pure air chambers G G' connected with each other by the pipes H H which pass through the smoke pipe C C, substantially as shown.
Second, the pure hot air chamber G' provided with the register holes J J, the valved flue K, and the valve L, substantially as described.
Third, a radiator with the pure air chambers G and G', the smoke chambers F and F', the smoke pipes C C and the pure air pipes H H, constructed and arranged substantially as described and shown.

No. 56,515.—SILAS D. BALDWIN, Chicago Ill.—*Sash Supporter.*—July 24, 1866.—The elastic ball becomes jammed between the inclined face of the recess and the window casing, and thereby sustains the sash until it is retracted by the hand rod.
Claim.—The combination of the elastic ball *a*, a spring *b*, and retracting rod *c*, with the case C, provided with the inclined planes, arranged and operating substantially as set forth and specified.

No. 56,516.—ARTHUR BARBARIN, New Orleans, La.—*Bottle Stopper.*—July 24, 1866.—The ring is stretched and placed under the lip, the bow, being of thicker material, pressing over the stopper to hold it in place.
Claim.—The combination with stoppers for bottles, jars, and other receptacles, of the elastic fastening device, the whole being constructed and arranged for operation substantially as herein described.

No. 56,517.—JOHN N. BAXTER, Greensburg, Ind.—*Horse Rake.*—July 24, 1866.—Two sets of teeth extend from opposite sides of the frame; the thills are attached to the latter by journals moving in slots in the end pieces thereof.
Claim.—A frame provided with two sets of rake teeth D D, projecting from opposite sides, in combination with thills E, attached to the end pieces C C of the frame by pins or journals *d*, secured to the inner ends of the thills and passing through oblong slots *c* in the end pieces, substantially as and for the purpose set forth.

No. 56,518.—JAMES BRAIDWOOD, Wilmington, Ill.—*Dumping Car.*—July 24, 1866.—The car runs on and is secured to the framework, which is tilted to dump the contents of the car.
Claim.—The frame *c*, constructed substantially us herein recited, in connection with a rail track, and for the dumping of the cars, all constructed and operated as described.

No. 56,519.—JESSE BRIGGS, Stuyvesant, N. Y.—*Clamp for Holding Saws.*—July 24, 1866.—The pivoted jaw is clamped against the stationary one by a cam lever.
Claim.—The construction and arrangement of the frames A A', jaws B B, hook projections *c'*, pins *c*, cam C, and lever D, in the manner herein described and represented.

No. 56,520.—GEORGE N. BRIGHAM, Montpelier, Vt.—*Churn.*—July 24, 1866.—The arrangement of the breakers and the form of the dashers produce opposite currents.
Claim.—First, the double or forked beaters, they having ribs or raised beads on both edges, as herein described, the same being so constructed as to operate in combination with reverse angular breaks on both sides and ends of the receptacle for containing the cream, so as to produce currents and counter currents toward the centre of the revolving shaft, for the purposes herein set forth.

C P 19—VOL. II

Second, the construction and arrangement of the beaters B B, with their bevel side openings *b b*, and ribbed edges *a a*, the breaks *c c*, and breaks *d d*, top breaks *e e*, for the purpose of churning and working butter, substantially as and for the purposes herein specified.

No. 56,521.—R. D. BROWN, Covington, Ind.—*Harvester Rake.*—July 24, 1866.—The rake head is driven by endless belts beneath the grain platform; at the outer or grain end of the groove traversed by the guide arm of the rake head is a forked pin to assist in turning the teeth of the rake to a vertical position for removing the grain. In connection with the reel is a ratchet and pulley frame for adjusting the tension of the reel belt.

Claim —First, the arrangement of the forked pin Q, located at the end of the slot R, as herein described and for the purposes set forth.'

Second, the arrangement and combination of the ratchet P, and pulley N, with the reel M, for the purpose of tightening the belt as herein described.

No. 56,522.—ROBERT BRYSON, Schenectady, N. Y.—*Reaping Machine.*—July 24, 1866.— The automatic rake reciprocates across the platform at right angles to the line of draught and is supported upon a platform hinged to the draught frame; the devices are explained by the claims and cut.

Claim.—First, the arrangement of the gimbal or universal joint *d*, with the two parts *c c'* of an extensible shaft, when one part of the said shaft drives an endless belt, which operates a reciprocating rake arranged to move through a slatted platform, and the other part is attached to the draught frame and transmits the motion of the driving wheels to the gearing which drives the rake, in combination with a hinged joint harvester, all in such manner that one part of the extensible shaft maintains an unchanging position with respect to the grain platform, and the other part thereof maintains an unchanging position with respect to the draught frame, as set forth.

Second, the construction and arrangement of the parts *c c' d*, for the purpose of forming an extensible joint shaft for a hinged joint harvester with a rake attachment, substantially as herein described.

Third, the construction, arrangement, and combination of the rake-head carrier G, rails *g' g'* and *g²*, rake head *i*, spring slide *h²*, spring catches *j j'*, pin *g*, and endless belt *f f*, substantially as and for the purpose set forth.

Fourth, the combination of the slot *e³*, adjustable pin *e²*, endless rake moving belt *f f*, and reciprocating rake *h h'*, substantially as and for the purpose described.

Fifth, the combination of the rake-head carrier G, rake *h h' i*, spring catches *j j*, spring slide *h²*, and endless belt, substantially as and for the purpose set forth.

Sixth, the rectilinear moving rake spur-gears *d d'*, and section *c'*, of an extensible shaft, arranged on a hinged joint platform as described, in combination with the gimbal or universal joint *d* section *e* of extensible shaft, bevel gears, and draught frame, all arranged and operating in the manner herein described.

No. 56,523.—HIRAM BURKE, Mineral Point, Ohio.—*Clothes Washing Rubber.*—July 24, 1866.—This block is a substitute for the hands and has projecting rubber strips; it is hinged to an arm which is pivoted to a piece secured to the wash-board, and has a cup to carry up suds which trickles through holes in the block on to the clothes.

Claim.—An improved clothes-washing rubber, formed by combining with a rubber board A, having rubber flanges *a'* of a plate D, or its equivalent, a handle *c*, the hinged and pivoted arms F and G, substantially as described and for the purpose set forth.

No. 56,524.—EDMUND F. BURROWS, Mystic River, Conn.—*Garter.*—July 24, 1866.—Explained by the claim and cut.

Claim.—A device for holding ladies' garters in their places upon the stockings, consisting of a flexible strip or band *a*, covered with any suitable material or fabric, and provided with raised edges *e e*, the said device to be clasped around a lady's leg, underneath the stocking, and the stocking held up by a common garter encircling the stocking directly over the band, substantially as shown and described.

No. 56,525.—ESEK BUSSEY, Troy, N. Y.—*Cooking Stove.*—July 24, 1866.—Explained by the claim and cut.

Claim.—A three-flued cooking stove, having the central flue extended so as to enclose on the sides and bottom the culinary boiler or hot-water reservoir B, the latter being so arranged as to rest upon or against the edges of the sides of said central flue, so as to constitute the interior side or wall of the same, substantially as set forth.

No. 56,526.—JOHN M. BUTCHER, North Lewisburg, Ohio.—*Varnish.*—July 24, 1866.— Composed of alcohol, one quart; shellac, eight ounces; camphor, one ounce; Prussian blue, one-half ounce; drop black, one-quarter ounce; and lampblack, one-eighth ounce.

Claim.—The compounding of the several ingredients hereinafter named, in the proportions named, in the manner pointed out, and for securing the advantages enumerated.

No. 56,527.—J. T. CAPEWELL, Woodbury, Conn.—*Sewing Machine Guide.*—July 24, 1866.—The leather or other material has its edges lapped evenly, preparatory to being stitched together, as for reins, &c.

Claim.—A guide made of conical or tapering shape from end to end, and provided with suitable ways or guides for the edges of the strap or material passing through it, so that when the strap issues from the smaller end of the said guide, its edges will be lapped or folded over each other, either more or less, substantially as herein described and for the purpose specified.

No. 56,528.—K. S. CHAFFEE, Cambridge, Mass.—*Apparatus for Making Charcoal.*—July 24, 1866.—A pipe surrounds the kiln and connects by tubes with the interior; pyroligneous acid escapes through the tubes and is condensed in the pipe.

Claim.—The application of the condenser to the kiln by extending such condenser as a pipe around the kiln, and supporting it by means of series of branch pipes leading from it into the kiln, and combining with such condenser a discharge pipe b, to extend from it, as set forth.

Also, the above-described arrangement of the condenser with respect to the kiln, viz., so as to encompass it and connect with it, substantially as described.

No. 56,529.—ROBERT A. CHESEBROUGH, New York, N. Y.—*Filling for Safes.*—July 24, 1866.—The boneblack is ground and heat-dried before using.

Claim.—The use of the boneblack for filling in between the inner and outer walls of the safe or vault to render the same fire-proof.

No. 56,530.—SAMUEL CHILD, Jr., Baltimore, Md.—*Vapor Stove.*—July 24, 1866.—The mode of applying the valve rod is explained by the claims and cut.

Claim.—First, in apparatus for generating heat in vapor stoves as above described, regulating the supply of fluid to the retort or heating chamber, in the manner and by the means hereinbefore specified; that is to say, by locating the valve which regulates the flow of the oil or other fluid at or near the point where the fluid enters the said retort, substantially as and for the purposes herein set forth.

Second, in combination with the retort or heating chamber of a vapor stove and valve seat located at or near the point of junction of said retort, with the pipe which connects it with the fluid reservoir, as specified, the valve constructed and arranged so as to operate on the axis of the said pipe, substantially as and for the purposes herein shown and described.

No. 56,531.—JOHN K. COOK, Richmond, Ind.—*Expanding Frame for Soldering Fruit Cans.*—July 24, 1866.—The expansible framework is an arrangement of arms and wings operated by centre shaft and enables the top and bottom of the can to be fitted and fastened in place, and being afterwards collapsed, is withdrawn at the upper opening.

Claim.—The arrangement and combination herein described of an expanding frame for soldering fruit cans, capable of being withdrawn through the hole in the top of the can, when finished, as and for the purposes substantially as set forth and described.

No. 56,532.—GILBRETH DAWSON, Rockville, Conn.—*Spinning Jack.*—July 24, 1866.—The lever brake rests on a pulley on the drum shaft and is operated by the carriage; when the latter is pushed in, the brake is raised from the pulley, and drops when the roping gear is thrown out.

Claim.—First, stopping the roping drums in spinning jacks from slipping round or continuing their rotation after the roping gear has been thrown out, by means of a brake acting automatically on a pulley placed on the drum shaft, substantially as described.

Second, the brake lever H, in combination with the elbow lever C, substantially as described.

Third, the combination of the brake, the lever C, and the shoe D, substantially as described.

Fourth, the combination of the brake, the lever C, and the slide J, constructed and operated substantially as described.

No. 56,533.—P. M. DEVOS, New York, N. Y.—*Medical Compound.*—July 24, 1866.—The compound is placed in a pad or belt to be worn on the body.

Claim.—A medical compound or composition when formed of such materials as will impart to it the characteristics herein described, and when used substantially in the manner and for the purposes specified.

Also, a medical compound made by mixing camphor, nux moschata, or nutmegs and capsicum, or red pepper, in combination with any suitable disinfectant, whether one or more in number, and when mixed together in or about the proportions named, and used substantially as and for the purposes specified.

No. 56,534.—OWEN DORSEY, Newark, Ohio.—*Reaping Machine.*—July 24, 1865.—Instead of a discharging rake the platform itself reciprocates parallel with the finger bar and tilts to discharge the grain on the ground in rear of the main frame. The grain as it is cut is meanwhile received on a supplemental slatted platform which drops into recesses of the main platform when the latter is restored to its place.

Claim.—First, combining a rectilinear reciprocating platform with a vibrating fender, in such manner that the grain, after it falls upon the fender, shall be deposited upon the platform, conveyed, and by the latter delivered upon the ground at one side of the machine, substantially as described.

Second, the combination of a rectilinear reciprocating platform, which is composed of slatted bars, with a vibrating slatted fender, substantially as described.

Third, automatically delivering the cut grain from one side of the machine by means of a platform which has a rectilinear and vibrating movement, substantially as described.

No. 56,535.—J. H. DOUGHTY, New York, N. Y.—*Combined Blacking Case and Night Chair.*—July 24, 1866.—Explained by the claims and cut.

Claim.—First. the box B, containing the blacking case, the dressing case, and the night chair, in combination with the seat A, constructed and operating substantially as and for the purposes described.

Second, the box holder *c'*, in combination with an ottoman, chair, stool, or other similar article, arranged as a blacking case, substantially in the manner set forth.

Third, the sponge cup *c*, in combination with an ottoman, chair, stool, or other similar article, arranged as a blacking case, substantially in the manner described.

Fourth, the adjustable boot-jack J, or J', in combination with an ottoman, chair, stool, or other similar article, arranged as a blacking case, substantially as and for the purpose set forth.

No. 56,536.—JASON DOW. Biddeford, Me.—*Signal Tower.*—July 24, 1866.—The signal tower is elevated upon its sectional post by means of ropes and pulleys and is stayed by guy ropes.

Claim.—A signal tower, constructed and operated in the manner substantially as shown and described, and for the purpose set forth.

No. 56,537.—JOHN DUBREE, Drumore township, Penn.—*Hitching Strap.*—July 24, 1866.—The hitch strap is passed through the ring on the head-stall, and thence through the bit ring and upward: pulling back on the strap draws the bit into the angle of the mouth.

Claim.—The simple strap G, for the attachment of the hitching strap F, when said strap G is connected with the bridle and bit, in the manner and for the purpose specified.

No. 56,538.—ALBERT DUNN, Plainfield, N. J.—*Wagon Jack.*—July 24, 1866.—The depression of the lever straightens the bent leg and raises the other upon which the axle is supported by a pin.

Claim.—The combination of the bars or frames A and B, or their equivalents, and handle lever D, when constructed, arranged, and connected together, so as to operate substantially in the manner described, and for the purpose specified.

No. 56,539.—WILLIAM C. DUNN.—La Porte, Ind.—*Hardening Iron.*—July 24, 1866.—The plough-shares are hardened by being plunged at a red heat into a bath composed of salt, 12 pounds; saltpetre, 4½ ounces; borax, 3½ ounces: nitric acid, 2 ounces; muriatic acid, 1½ ounce; soft water, 4 gallons; supercarbonate of soda, 3½ ounces; sal ammoniac, 2½ ounces; sulphuric acid, 1½ ounce.

Claim.—The process herein described of treating or hardening the cast-iron parts of ploughs, cultivator shares, and similar articles.

Also, as a new article of manufacture. plough mould-boards, land-sides, or shares, when made of cast iron, treated in the manner herein described.

No. 56,540.—WILLIAM A. EHLMAN, Milwaukee, Wis.—*Combined Chair and Desk.*—July 24, 1866.—The back of the chair is turned over forward to form a desk and is supported upon what were formerly the arms; bent metallic rods support the desk in front.

Claim.—The combination of the chair seat A. back E, having eyes K, side arms D, having eyes J, uprights or supports *c*, and hook arms H, or their equivalents, when all connected and arranged, so as to allow the back to be swung down into a horizontal position, or nearly so, and there supported, substantially as and for the purpose described.

No. 56,541.—B. F. ELLIOTT, Cedar Rapids, Iowa.—*Grape Trellis.*—July 24, 1866.—The different parts of the trellis are hinged to each other and may be turned down and folded together with their vines attached, for winter protection.

Claim.—The side frame C, and upper frames D, in combination with the cross-bars E, or any other suitable fastening device for holding the said upper frames D in a horizontal position, or nearly so, when attached or connected together, and to any suitable bed frame or supports of the ground, substantially as and for the purpose described.

No. 56,542.—A. B. ELY, Boston, Mass.—*Insulator for Telegraphs.*—July 24, 1866.—The described device prevents the accumulation of a film of a moisture between the hook and the racket.

Claim.—First, a flanged disk on the insulating book, when constructed and arranged in reference to the hole in the bracket, substantially in the manner and for the purpose set forth.

Second, the combination of the bracket and hole with the pin, hook and disk, and arranged with or without flanges, substantially in the manner and for the purpose set forth.

No. 56,543.—CHARLES ENGELSKIRCHER, Buffalo, N. Y.—*Hand Lantern.*—July 24, 1866.—The skeleton frame is fastened to the cap and is sprung and clasped around the base.

Claim.—Connecting the chimney cap C, the glass or globe part B, and the metallic base A, together, by means of the skeleton frame D E F, the said skeleton frame being so constructed and connected with the said parts that the vertical wires D shall be permanently attached to the chimney cap, and the metallic base A shall be fastened to the glass or globe part by means of the spring band or clasp F', and released therefrom when the said spring band is unhooked, and the glass or globe part may be retained within the skeleton frame when the metallic base A is removed, substantially as described.

No. 56,544.—LEVI W. FIFIELD, Holderness, N. H.—*Knitting Machine Needle.*—July 24, 1866; antedated July 13, 1866.—The looping book is at the end of the pivoted latch instead of at the end of the needle shank. The latch lies partly in a slot in the shank. The work when forced back on the shank depresses the tail of the latch and lifts the book out of the slot so that it may receive the yarn from the next loop; when the loop is to be discharged its advancing movement depresses the hooked end of the lever so that it may ride over it and off from the needle.

Claim.—The needle as made of the slotted shank A and the hooked lever B, constructed, arranged, and applied together, substantially in the manner so as to operate as described.

No. 56,545.—HENRY FISHER and MILTON BALL, Canton, Ohio.—*Reaping Machine.*—July 24, 1866.—The grain is dropped upon the ground by vibration of the platform on which it falls when cut. The devices are explained by the claims and cut.

Claim.—First, the combination of the slotted arm F and chain F', attached to the hinged wind board C, for the purpose of adjusting the rod D vertically and horizontally, substantially as and for the purpose set forth.

Second, in combination with an overhung reel and cutter bar B, the hinged board C and rod D, attached at the main frame end only to an oscillating arm F, substantially in the manner and for the purpose set forth.

No. 56,546.—JUNIUS FOSTER, Long Branch, N. J.—*Instrument for Measuring Tires of Wheels.*—July 24, 1866; antedated July 15, 1866.—A guide which traverses against the side of the rim assists in maintaining the measuring wheel in place.

Claim.—The guide *h*, fitted as specified, in combination with the measuring wheel *b*, for the purposes and as set forth.

No. 56,547.—TALBOT T. FOWLER, Washington, D. C.—*Scales for Weighing Ice.*—July 24, 1866.—The arrangement of the beams and connecting links is shown in the cut.

Claim.—The links I *m* and *n*, when connecting the bar F, the scale beam B, and weight-beam E, arranged substantially as and for the purposes specified.

No. 56,548.—JOEL GARFIELD, Groton, Mass.—*Planing Machine.*—July 24, 1866.—The upright shafts of the cutter-heads have revolving guides which operate as feed rollers, and also determine the size of the stuff that passes between the cutter-heads.

Claim.—The combination of the feed and guide rolls *c c* with the gears *d* and *d'*, constructed and operating substantially as specified for the purposes set forth.

Second, the combination of the frame J K with the slides *m m*,'the shafts L and F, constructed substantially as described for the purposes set forth.

Third, the combination of levers O and N N with springs S S and the shaft Q, operating substantially as described for the purpose set forth.

No. 56,549.—EUGENE GAUSSOIN, Baltimore, Md.—*Apparatus for Treating Ores with Chlorine.*—July 24, 1866; antedated July 11, 1866.—The revolving barrels are arranged within a casing which has charging and discharging openings. Chlorine gas is introduced and withdrawn through the walls of the casing and the barrels connect by their hollow journals with the passages.

Claim.—First, the enclosing walls and floor, forming chambers in which the barrels revolve and from which the fluid contents are removed by drains from the sides, and the solid by an aperture at corner of the arch.

Second, the combination of the hollow axles and perforated walls forming a series of connections from the generator, from barrel to barrel, and ultimately the discharge apertures at the summit.

Third, the arrangement of the barrels with their operating gearing, so that their respective openings are in revolution presented alternately to the openings of the ones next in series above and next below, to afford the means of discharging as described.

Fourth, the combination of the revolving barrels and the openings J and wall openings H, as and for the purpose described.

Fifth, the combination of the valves K with drains Y Y W, as and for the purpose described.

No. 56,550.—FRANCIS GAY, Bedford, Ohio.—*Farm Gate.*—July 24, 1866.—The gate is launched lengthwise, rising upon its pivoted brace piece, upon which when erect it may be partially rotated.

Claim.—The standard D, the pedestal E, and the pin or stem F, as arranged and in combination with the gate A, in the manner and for the purpose herein set forth.

No. 56,551.—GABRIEL GRAFFHAM, Lawrenceville, Ill.—*Beehive.*—July 24, 1866.—The pit is arranged to catch the bee moth and its larvæ and prevent their re-ascent into the hive; blocks close the entrance when the floor over the slatted hive bottom is withdrawn.

Claim.—First, a bee-hive which combines in its construction the following elements, viz., a pit N, and sloping shelves N′, a case A, separated from the pit by a grated bottom M, and having a porch A′, closed by a sliding door D, and a cover E, with caps F F′, and a drawer L, located above the porch, the several parts being constructed and the whole arranged for use, substantially as set forth.

Second, the wedge-formed stoppers K, when used for closing the slots in the front of the case after the removal of the slide I, substantially as set forth.

No. 56,552.—JOHN R. GROUT, Detroit, Mich.—*Reverberating and other Draught Furnaces.*—July 24, 1866.—Explained by the claims and cut.

Claim.—First, in a reverberating or other draught furnace so arranging the atmospheric passage ways *a a′ a″*, and *b b′ b″*, in the bridge, wall, and arch of the furnace, that the air passing in thin currents shall be heated by contact with the walls, and introduced from above and below into the compartment D, in converging currents of the full width of the throat C ; when mingling with the unconsumed carbonized gases evolved from the fuel in the fire room B, their complete combustion and perfect diffusion will be effected, substantially in the manner set forth.

Second, the combination of the plate *f*, valve *g*, rod *h*, and lever *i*, in the lower air passage, and the similar combination in the upper air-passage or their equivalents for the regulation of the passage of air through the atmospheric passage ways *a* and *b*, substantially as and for the purposes set forth.

Third, constructing the bridge C across the lower atmospheric passage way for the protection of the valve, substantially as set forth.

No. 56,553.—CHRISTOPHER GULLMANN, Poughkeepsie, N. Y.—*Water Elevator for Wells.*—July 24, 1866.—The windlass drives the blower which discharges air into the downcast shaft and ventilates the well ; a pendulum regulates the downward motion of the bucket.

Claim.—First, the mouths i′, on the hollow shaft i, arranged relatively to the bucket and to the rope, or its equivalent, and to the loose sleeve G, connected by a clutch to the shaft D, so as to perform the double function of retarding the descent of the bucket and ventilating the well, substantially in the manner herein specified.

Second, the oscillating part K, so mounted and arranged relatively to the bucket, and its connection liberated for descent, as described, that it shall retard the descent of the latter, substantially in the manner herein specified.

Third, the well bucket arranged to descend automatically, the revolving mouths i′, and the oscillating part K, and the several connecting members of the mechanism, combined and arranged to effect the retardation of the descent of the bucket and the ventilation of the well, substantially as herein specified.

No. 56,554.—JOHN HABERMEHL, Wheeling, West Va.—*Chair.*—July 24, 1866.—The chair bottom turns on metallic straps which enclose the back round at the top of the legs, and is held in its different positions by springs connected to the front round.

Claim.—The combination of the seat, metallic loops, and cross piece of the rear legs, constructed as described.

Second, combination of loops C and rod A, in a chair constructed to turn as described.

Third, combination of spring E and rod A, in a chair constructed as described.

No. 56,555.—DANIEL HARRIS, Canaan, Me.—*Horse Hoe.*—July 24, 1866.—The share has a middle portion inclining forward and sides having an angular inclination; it is followed y wings which have a lateral adjustment by means of segments supported by the rear of the beam.

Claim.—The share C, constructed or formed with sides *a a*, inclined both transversely and longitudinally, and also formed with a central longitudinally inclined surface *b*, having a horizontal position in its transverse section, in combination with the adjustable mould boards E E, pivoted to the rear of the share C, and retained in position by the clamp F, and bars *e e*, all arranged substantially in the manner and for the purposes set forth.

No. 56,556.—E. B. HARRIS, Wilmington, Ill.—*Force Pump.*—July 24, 1866.—Two reciprocating cylinders with valves are operated by a brake; posts erect upon the bottom of the cistern support stationary valve-seats.

Claim.—The arrangement of the well A, cylinders C C, valves D D, vertical rods E E, disks F F, valves G G, partitions H H, valves I I, and trough J, operating in the manner nd for the purpose herein specified.

No. 56,557.—JAMES HAYDEN, Exeter, Wis.—*Ration Feed Box.*—July 24, 1866.—The slides in the conductor regulate the amount of feed passed to the trough.

Claim.—First, the ration box when constructed, arranged, and used in connection with the feed box A, substantially as herein set forth and described.

Second, the gate or slide E, when constructed and used substantially as and for the purpose set forth.

Third, the measure box S, and slides, when constructed, arranged, and used in connection with the ration box and reservoir box, substantially in the manner and for the purpose described.

Fourth, the reservoir box X, when used in connection with the measure box, substantially in the manner and for the purpose set forth.

Fifth, the rod and nut used in connection with the gate or slide E, when the whole are constructed, arranged, and used substantially as and for the purposes set forth.

Sixth, the opening N, connecting the feed box A with the ration box B, when combined, arranged, and used in connection with the gate or slide E, substantially as and for the purpose set forth. Said ration feed box may be made double for two or more horses or other animals, as shown, or single for one horse or other animal; the several parts of the single or double ration feed box, as a whole, being substantially the same.

No. 56,558.—JOHN W. HENDLEY, Washington, D. C.—*Sand Bellows.*—July 24, 1866.—Explained by the claims and cut.

Claim.—First, the arranging of the sand box above the bellows so that it may be operated by the movement of the upper board or plate of the bellows, substantially as herein recited.

Second, the connecting of the box to the pipes, and the nozzle to the sand and air pipes, by the elastic pipes, constructed and operated substantially as set forth.

Third, in combination with the nozzle and the conducting pipe, the lever J, constructed and arranged so that the parts may be operated as described.

No. 56,559.—P. M. HENDRICK and JOHN J. CHATTAWAY, Springfield, Mass.—*Gun Swab.*—July 24, 1866.—The disk of rubber is expanded laterally by vertical compression and maintained in that condition by the springs which engage the shoulders of the recesses.

Claim.—First, the use of a swab of rubber, or other similar elastic material, when the same is expanded laterally by vertical compression within the barrel for the purpose of cleaning the same, substantially as herein set forth.

Second, the combination of the springs b b with the other parts of the device for the purpose of holding the swab in place when compressed and expanded, substantially as herein described.

No. 56,560.—R. S. HOLETON, Niles, Ohio.—*Water Wheel,*—July 24, 1866.—The wheels discharge outward, receiving water from the penstock between them.

Claim.—The arrangement of the penstock G within the flume C and the wheel within the said penstock, in combination with the cap A, side openings e e, gate G' below the wheels, lever a and rod b, in the manner and for the purpose set forth.

No. 56,561.—HORACE HOTCHKISS, Plainfield, N. J.—*Machine for Making Metal Tubes.*—July 24, 1866.—The mandrel of peculiar shape is combined with a series of pairs of grooved rolls, one pair of peculiar construction; the effect is to bend a plate of metal to a cylindrical shape, its edges abutting against each other.

Claim.—First, in machines for bending plates of metal into convex or tubular forms, the combination of the guide spindle M constructed as described with a system of guides of suitable form for the different stages of the work, and a system of rolls, or their equivalent, between which the work is formed into the required shape, substantially as described.

Second, the guide spindle M, constructed and applied substantially as and for the purpose described.

Third, the rolls N O, constructed and operating in the combination shown, substantially as described

No. 56,562.—T. L. HOUGH, Philadelphia, Penn.—*Truss.*—July 24, 1866.—The pad is connected by a pivoted arm to the metallic band; a flat spring is fastened to the pad and its free end operates a plate attached to the band.

Claim.—The arm C pivoted upon the journal c having the spring a attached thereto with its free end operating against the plate x, substantially as shown and described.

No. 56,563.—D. HUESTIS, Cold Spring, N. Y.—*Pavement.*—July 24, 1866.—Iron boxes with depressions at their junctions are filled with concrete, which is partially retained in place by flanges above and below.

Claim.—The grooved street pavement herein described, the same consisting of the boxes A with the bottom flanges c and dovetail spaces d with suitable filling, the upper edges being bevelled and forming grooves when the boxes are combined and give hold to the feet of the animals, and adapted for a railway track, as specified and shown.

No. 56,564.—HUBBIL B. HUTCHINS and WASHINGTON HORTER, Philadelphia, Penn.—*Knife Scourer.*—July 24, 1866.—The stand has grit-catching grooves and the lever has a perforated grit box; the knife is reciprocated between the rubbers.

Claim.—As an improved article of manufacture the knife and fork cleaner or scourer, described and set forth.

No. 56,565 —SETH T. HUTCHINGS, North Anson, Me.—*Last.*—July 24, 1866.—The block is fastened to the last by spring plates whose countersunk holes shut over projections on the surface of the last.

Claim.—The self-operating spring clasp d and projections e, combined and operating together to hold and to release the last block, substantially as described.

No. 56,566.—JOSEPH A. JACOBS, Pittsfield, N. H.—*Stove Pipe Damper.*—July 24, 1866.—The grates are made to register or the reverse by sliding the rod, and their relative distance is adjusted by rotating the rod.

Claim.—A heat regulator composed of two grates A B, which are connected by sliding and revolving crank shaft C, substantially as and for the purpose described.

No. 56,567.—EUGENE N. JENKINS, Chicago, Ill.—*Lantern.*—July 24, 1866.—The intermediate band facilitates the connection between the base and the globe.

Claim.—First, the band D provided with a plate or disk E for supporting a lantern globe, substantially as set forth.

Second, the combination of the band D, disk E, and springs a or ledges e with the base C, substantially as and for the purposes specified.

No. 56,568.—M. W. JENKS, Richmond, Ind.—*Water Drawer.*—July 24, 1866.—The enclosing curb has a detachable spout and a swinging tripper which arrests and empties the bucket.

Claim.—The arrangement of the several parts in combinaion, as hereinbefore specified and set forth.

No. 56,569.—CHARLES H. KEENER, Baltimore, Md.—*Hat Rack.*—July 24, 1866.—The hat is passed through the loop and its brim rests against the wall.

Claim.—The hat rack consisting of the ring A with loop a hanging in eye b, substantially as described for the purpose specified.

No. 56,570.—C. A. KELLOGG, Elyria, Ohio.—*Bed Bottom.*—July 24, 1866 —The elastic strap is looped around a link attached by a pin to the rail and its ends gripped into grooves near the end of the slat.

Claim.—The staple D, pin E, and belt or strap F in combination with the gripe G and slat B, as and for the purpose substantially as set forth.

No. 56,571.—JAMES B. KELLY, Kendallville, Ind.—*Turn Table.*—July 24, 1866.—The conical flanged supporting wheels traverse the circular tracks, and their axles are connected to a ring which revolves around the centre post.

Claim.—First, the yoking ring G in combination with conical rollers and a concentric rail or rails, substantially as and for the purpose herein described.

Second, the conical flanged wheels and bevelled rails in combination with the central ring G, fixed centre post E' and a turning table E, substantially as described.

No. 56,572 —T. J. KINDLEBERGER, Eaton, Ohio.—*Water Wheel.*—July 24, 1866.—The principal and auxiliary wheels are arranged on one shaft, at some distance apart and with an intervening partition. The gates are connected and admit the water to the scroll chutes of the respective wheels; water may be shut off from the auxiliary wheel.

Claim —First, the circular bar d', connecting links e and f, gates a' and b' and guide boxes d'', combined and arranged as above described and for the purpose set forth.

Second, the worm w, rack x, arms u, valve v, with disk t, combined and operating as above shown and for the purpose set forth.

Third, the main driving wheel C, auxiliary wheel D, both upon main driving shaft B, in combination with chutes d and b and gates a' and b', for the purpose above specified.

No. 56,573.—JULIUS KLAMKE, New York, N. Y.—*Robe.*—July 24, 1866.—Explained by the claim and cut.

Claim.—A travelling or other robe, of fur or other material, having pockets or receptacles for the hands and feet, or either, as herein described and represented, so that it may be used as a garment without interfering with any or all of its uses as a robe, as set forth.

No. 56,574.—RUDOLPH H. KLAUDER, Philadelphia, Penn.—*Renovating Faded Fabrics.*—July 24, 1866.—Dye the fabric and then print upon it designs in opaque colors.

Claim.—The herein set forth combination of the processes of dyeing and opaque printing as a new and improved method of renovating worn or faded woven fabrics, whereby the described improved effects are produced, as and for the purpose specified.

No. 56,575.—JOSEPH G. KONVALINKA, Astoria, N. Y.—*Watch and Locket Case.*—July 24, 1866.—The spring catch for retaining the hinged cover is upon the stem, and is released by pressing upon the plunger while holding the watch by the ring.

Claim.—First, the movable head C, fitted or mounted on a fixed pin A, substantially as and for the purpose above specified.

Second, the spring D in combination with the movable head C, and fixed pin A, and operating substantially as and for the purpose above specified.

Third, the catch G, when it is movable, i. e., sliding up and down, and operating substantially as and for the purpose above specified.

Fourth, the spring M, bent externally over the cap H, and operating substantially as and for the purpose above specified.

Fifth, the bridge O, fastened externally upon the cap H, substantially as and for the purpose above specified.

No. 56,576.—NOAH W. KUMLER, Dayton, Ohio.—*Preventing Sealing Wax from Adhering to Moulds.*—July 24, 1866.—The interior surface of the mould is coated with mercury; may be mixed with chalk and applied by friction. The mould may be first washed with a solution of cyanide of silver or muriate of tin.

Claim.—The application of quicksilver in the manner and for the purposes herein respectively set forth.

No. 56,577.—PERLEY H. LAWRENCE, Springfield, Mass.—*Rock Drilling Machine.*—July 24, 1866.—The lower portion has a sleeve sinker, and is attached to the other by a section with two screw flanges. An intervening spring collar moderates the jar.

Claim.—First, attaching to the lower end of a drill pipe a weight or sinker, when the same is arranged in the manner and operated as and for the purpose herein described.

Second, placing the spring A, of rubber or its equivalent, between the pipe X and sinker B, when the same is arranged substantially in the manner and for the purpose herein set forth.

Third, connecting the parts of the sinker B by means of the joint C D, substantially as herein described.

Fourth, attaching the piston G to the frame of the machine L by means of the rod H, pin K, and collar J, and using it in combination with the pipe X, in the manner and for the purpose set forth.

No. 56,578.—MARTIN LEIPPE, Lancaster, Penn.—*Cigar Press.*—July 24, 1866; antedated February 23, 1866.—The form boards have grooves cut in their faces, or the recesses are formed by plates inserted edgewise into the board. The damp cigars are pressed into shape between the boards under pressure.

Claim.—The form boards 1, 11, 111, 1111, constructed and employed substantially in the manner shown and for the purpose specified.

No. 56,579.—C. L. LOCHMAN, Carlisle, Penn.—*Porcelain Printing Frame.*—July 24, 1866.—Designed to enable the operator to examine the picture while printing upon the rigid plate.

Claim.—First, the combination of the slotted lid B with the movable bars G G, mounted with leather, gum, or other elastic cushion to grasp the two opposite edges of the porcelain plate, in the manner shown and described and for the purpose set forth.

Second, the combination of the frame B, negative holder F F, spring a, and hinged lid B, with its movable bars G G, and springs f f, arranged, constructed, and operating in the manner substantially as shown and described.

Third, the movable bars G G, with their accompanying screws and burs moving in slots as represented, or their equivalents.

Fourth, a movable negative holder F F, with spring a, and fastening screw E.

No. 56,580.—SYLVANUS D. LOCKE, Janesville, Wis.—*Grain Binder.*—July 24, 1866.—This binding attachment for reapers cannot be briefly described other than substantially in the words of the claim.

Claim —First, a binding machine operated by hand or by power taken from a harvester, provided with a cam cylinder B, and the cam slides K, operating sets of arms as M N R. alternately, and a sheaf-discharging arm F', combined with a disengaging coupling C, and a self-acting disengaging arm E, foot lever I, binding arms T U, and the friction reel A', when arranged and used in the manner and for the purposes herein set forth and described.

Second, disengaging couplings of grain-binding machines by means of the disengaging arm E, when constructed with or without the shaft spring E', as set forth herein and described.

Third, the cam cylinder B, when constructed substantially as described, and used to operate the working parts of a grain-binding machine, substantially in the manner as herein set forth and described.

Fourth. the cam slides K, when constructed substantially as described, with or without the friction roll, and used to communicate motion to the working parts of a grain-binding machine, substantially as herein described.

Fifth, the foot lever I, or equivalent device, used to raise or remove the disengaging arm of a grain-binding machine, so as to allow the couplings to be engaged, as herein described and set forth.

Sixth, the combination of a back-acting disengaging coupling, with a shaft spring and the disengaging arm E, the combination operating so as to allow a backward motion to the harvester without affecting the process of binding or operating the parts of a binding machine, substantially as set forth.

Seventh, the combination of the back-acting disengaging coupling and shaft spring, with a disengaging arm and a disengaging lever, substantially as set forth.

Eighth, the combination of a revolving cam cylinder and its moving mechanism with the vibrating arm R, for operating a twisting or tying device, the parts being constructed and operated substantially as herein set forth.

Ninth, the combination of a revolving cam cylinder with vibrating binding arms, and a vibrating arm operating a twisting or tying device, arranged and operating as described, whereby the binding arms and the twisting or tying arm are worked alternately, substantially as set forth.

No. 56,581.—JOHN MABBS, Isle Royal Mines, Mich.—*Quartz Crusher.*—July 24, 1866.—The feed table revolves between the mullers at a slower rate, and the ore is swept therefrom by a jet of water and a traversing plough.

Claim.—First, the feed table J, mounted in the tubular shaft H, in combination with the mullers F, and main shaft B, constructed and operating substantially as and for the purposes described.

Second, the plough L in combination with the feed table J, tubular shaft H, and horizontal shaft E, constructed and operating substantially as and for the purposes set forth.

Third, the tank O in combination with the plough L, feed table J, and mullers F, constructed and operating substantially as and for the purposes described.

No. 56,582.—JAMES F. MAGUIRE, East Boston, Mass.—*Tuyere.*—July 24, 1866.—The blast opening in the upper part of the tuyere has a surrounding water chamber, and the hopper below is closed by a cone and weighted lever.

Claim.—A tuyere, constructed substantially as described and for the purpose set forth.

No. 56,583.—M. H. MANSFIELD, Ashland, Ohio.—*Machine for Threshing and Hulling Clover.*—July 24, 1866.—Fans are arranged at the ends of the threshing cylinder to draw the dust into the machine to be conducted to its legitimate discharge outlet. The shoe has two overlapping inclined boards, the interval between which affords another duct for the escape of dust and light tailings; the latter is collected separately.

Claim.—First, in a threshing or clover hulling and threshing machine, which employs a fan G, for blasting or blowing away chaff, dust, and other foreign substances, the construction and arrangement of the dust chambers b', situated within the frame of the machine, apertures b, fan b², and discharge passage at a', all substantially as and for the purpose described.

Second, the construction of the shoe E, with the imperforated boards e e' g, said boards being arranged as described and shown for the purpose set forth.

No. 56,584.—HUBBARD A. MARTIN, Jeffersonville, Ind.—*Plough.*—July 24, 1866.—The beam and standard are of angle iron, the clevis adjustable in and out of land by springing its tail into the notches of the plate.

Claim.—First, the wrought iron angle beam A, connected to the mould board a by the angle bar D, and the rod E, all constructed and arranged substantially as and for the purpose set forth.

Second, the clevis H, provided with an upper elastic plate e, in combination with the notched plate g, attached to the beam, substantially as and for the purpose specified.

Third, the wrought iron handles B B' in combination with the angle beam A, substantially as and for the purpose set forth.

No. 56,585.—SILAS B. MAULSBY, Muncie, Ind.—*Evaporator.*—July 24, 1866.—A series of gradually descending compartments with intervening strainers. The finishing pans are supported on a crane, and may be removed as required.

Claim.—First, the graduating self-straining step pan I, constructed substantially as herein described, in combination with the furnace, for the purposes set forth.

Second, the revolving finishing pans N, supported by and revolving upon cranes, substantially as herein described, in combination with the step pan I, and with the furnace, for the purposes set forth.

No. 56,586.—ROBERT W. McCLELLAND, Springfield, Ill.—*Wagon Hub.*—July 24, 1866.—The wooden hub has a pipe boxing, and is clamped by bolts between metallic disks.

Claim.—First, constructing the hubs of vehicles of wood for receiving the tenons of the spokes, and encasing the same by metallic disks, substantially in the manner and for the purpose set forth.

Second, in combination with the disks C and C' the pipe boxing D, arranged substantially as and for the purpose set forth.

Third, in combination with the spindle E, flange C², and pipe boxing D, the cap I, substantially as set forth.

Fourth, in combination with the wooden hub A, and metallic disks C and C', the bolts H, or their equivalent, substantially as and for the purposes set forth.

No. 56,587.—GEORGE W. McGILL, Washington, D. C.—*Metallic Paper Fastener.*—July 24, 1866; antedated January 24, 1866.—Explained by the claim and cut.

Claim.—The within described paper fastener, formed of a single piece or strip of metal bent in a T-shape, the ends of the strip being in close contact, and pointed so as to make only a single hole in the papers, which it is designed to connect, the two ends opening from each other after passing through the paper, and confining said papers between said ends and the arms of the T, substantially as set forth.

No. 56,588.—JAMES A. McGILLIRRAE, Dyer, Ind.—*Press.*—July 24, 1866.—Improvement on his patent of January 10, 1865. The wheel has a metallic rim, and is connected to the windlass shaft by metallic spokes. Other details are cited in the claims.

Claim.—First, the connecting rim i of the wheel P, with the shaft Q of the windlass, by means of the metal spider R, in combination with the loose drum U, on shaft Q, and the slide V, or an equivalent fastening, to engage with the arms j of the spider, substantially as set forth.

Second, the attaching of a metal rim h to the flange g of the wheel P in combination with the slide O, substantially as and for the purpose specified.

Third, the bar S attached to the rim i of wheel P by a joint l, in combination with the cleats m m on the said rim, the slide O, and the inclined curved bar T attached to the framing of the windlass, substantially as and for the purpose set forth.

No. 56,589.—JOSEPH McKNIGHT, Pomeroy, Ohio.—*Harvester Cutter Sharpener.*—July 24, 1866.—The angular blades are adjustable on each other to accommodate the bevel of the knife blade or cutter.

Claim.—The right-angular arms C D in combination with the rod B, handle A, nut b, head a, constructed and arranged in the manner and for the purpose herein specified.

No. 56,590.—JOHN McLELLEN, Chambersburg, Penn.—*Buckle.*—July 24, 1866.—A wedge-shaped catch slides on the frame and binds the end of the strap against the guard.

Claim.—The plate A, with its guards B and C, in combination with the sliding catch F, the whole being constructed and arranged for the reception and retention of straps x y, substantially as described.

No. 56,591.—WILLIAM T. McMILLEN, Cincinnati, Ohio, and EDWARD P. CONRICK, Delavan, Wis.—*Clothes Wringer.*—July 24, 1866.—The rollers are driven by the pinions on each end of the counter-shaft.

Claim.—The counter-shaft F, having pinions E E' at both ends in the described combination with the pair of doubly geared wringer rolls A C C' and B D D' for the purpose explained.

No. 56,592.—WILLIAM M. MERRIEL, Jefferson, Ind.—*Sash Fastening.*—July 24, 1866.—A rack on the sash is operated by a pinion as usual, but may be locked by a sliding key-bolt and a spring which engages the teeth.

Claim.—The application of a spring F, of suitable construction, in combination with the screw-bolt G, to the cog wheel D, whereby the window is prevented from moving unless force other than its own gravity is exerted upon it, substantially as specified.

No. 56,593.—ISAAC L. MILES, Charlestown, Mass.—*Apparatus for Printing on Bottles.*—July 24, 1866.—Elastic type are arranged in a chase upon the frame, and the bottle rolled thereon, being guided by stops and bars to secure its proper presentation.

Claim.—The within-described apparatus, consisting of the adjustable bed C, with its elastic type-block D, ways M, and gauge I, operating substantially as and for the purpose set forth.

No. 56,594.—EZRA MILLER, Brooklyn, N. Y.—*Railroad Car.*—July 24, 1866.—Improvement on his patents Nos. 38,057 and 46,126, explained by the claims and cut.

Claim.—First, constructing the platform of railroad cars in a horizontal plane with the car beds, and sustaining such platforms by means of trussed rods, substantially in the manner described.

Second, the cross timbers a a' a^2 applied to the two intermediate longitudinal platform beams C′ C′, substantially as and for the purpose described.

Third, the construction of spring buffers and couplings, substantially as herein described, to produce compression between cars which are coupled together, so that the spring buffers and couplings shall constantly act together to prevent shocks and jerks in starting, stopping, or running trains, said buffers and couplings being arranged substantially as set forth.

Fourth, constructing the hooks D partly of cast and partly of wrought metal, substantially as described.

Fifth, chilling the abutting faces of the coupling hooks D, substantially for the purpose described.

Sixth, facing the abutting surfaces of the buffer heads with a metal which is harder than that of which the heads are formed, substantially as described.

No. 56,595.—CHARLES E. MORRIS and JOHN ENON, Bridgeport, Penn.—*Granulating Furnace Slag.*—July 24, 1866.—The slag is run into a receiver containing water. The agitator keeps the water in circulation and discharges the granulated slag.

Claim.—Granulating furnace slag, by running it in its hot fluid condition from the furnace directly into any suitable receiver containing cold water, substantially in the manner described.

No. 56,596.—JOSEPH M. NAGLEE, Philadelphia, Penn.—*Extracting Specimens of Liquor.*—July 24, 1866.—This "velinche" has an elastic bulb with valves which operate to exhaust air from the tube and expedite the flow of the liquor thereinto.

Claim.—The combination of the reservoir tube A, and its valve t, with the elastic air vessel C, and its valves c c', the whole being constructed and operating substantially as and for the purpose described.

No. 56,597.—JOSEPH M. NAGLEE, Philadelphia, Penn.—*Siphon.*—July 24, 1866.—The elastic bulb has valves and operates to exhaust the air from the siphon. The adjustable leg supports the end above the sediment in the cask.

Claim.—First, the elastic air vessels c, and valve c c', combined with a siphon, substantially as and for the purpose described.

Second, the combination of the adjustable rod g with the short arm of a siphon, substantially as set forth for the purpose specified.

No. 56,598.—C. NICKERSON, Chenoa, Ill.—*Wind Wheel.*—July 14, 1866.—The wind acts first upon the more extended wings and then upon those of the inner series.

Claim.—A wind wheel composed of two series or sets of fixed wings or sails C C′, secured between heads B B, one set or series projecting out from the heads further than the other set or series, and placed alternately in position, substantially as shown and described.

No. 56,599.—CHARLES NORTON, New Haven, Conn.—*Dies for Making Eye-bolts for Vessels.*—July 24, 1866.—The iron has an enlargement forged upon it, and this is exposed between the dies to be shaped. One pair of dies is adapted to finish an eye-bolt, made by bending and welding round iron.

Claim.—The combination of the lower die A B and C with the upper die A′ B′ and C′, when constructed, arranged and fitted for making eye-bolts, substantially as herein described.

No. 56,600.—THOMAS G. ODELL and BOYD GLOVER, Camp Point, Ill.—*Hand Spinning Machine.*—July 24, 1866.—The spindle is driven by the multiplying gearing and band from a hand crank. It is vertically adjusted to regulate the tension of the band, and may be attached to the edge of a table.

Claim.—First, the arrangement of the frame D, made as described, the cog wheel C, pinion B, band pulley A, and spindle F, the whole forming a portable spinning machine for domestic use, substantially as above set forth.

Second, in combination with the above, the adjustable plate G for holding the spindle, made and applied to the frame D, as described.

No. 56,601.—EDMOND C. OTIS, Voluntown, Conn.—*Anti-friction Carriage Axle.*—July 24, 1866.—The anti-friction rolls beneath the spindle rotate in contact with the inner surface of the hub.

Claim.—The combination of the two rolls B B, fixed to the axle A, so as to operate within the hub D, substantially as and for the purpose specified.

No. 56,602.—HENRY W. PELL, Rome, N. Y.—*Dies for Forming Heads of Wrenches.*—July 24, 1866.—The bar is not upset, but is swaged into shape by treatment in successive sets of dies.

Claim.—The improvement in the manufacture of wrenches herein described, the same consisting in subjecting the bar of iron from which the wrench is to be made to the action of the consecutive set of dies, substantially as described, and in the manner and for the purpose set forth.

No. 56,603.—GEORGE G. PERCIVAL, Brooklyn, N. Y.—*Inkstand and Calender Combined.*—July 24, 1866; antedated July 19, 1866.—A perpetual calendar surrounds the body of the instand; the days of the week are placed on a movable annular slip; when the latter is adjusted the days of each month upon which those days of the week fall will be found underneath.

Claim.—The calendar constructed and arranged as herein specified in the described combination with the inkstand A.

No. 56,604.—GEORGE PEUGEOT, Buffalo, N. Y.—*Hand Lantern.*—July 24, 1866.—Two knobs of glass projecting from the globle are embraced by the loops of the wire frame; the said wires may be attached to a ring beneath a shoulder of the glass.

Claim.—The manner of attaching the glass globe to the wire frame by means of the vertical wires B C, hooking on to the knobs G G, or into or under the bottom of the glass as shown at L M, substantially as described.

No. 56,605.—OSCAR PLACE, Brooklyn, N. Y.—*Flour Packer.*—July 24, 1866.—The described devices are for packing farina, &c., in small packages; the tubs are arranged with sliding plates at top and bottom to control the filling and stopping off.

Claim.—The arrangement of the perforated sliding plates R S, provided with lever T, having adjustable bearing U, perforated plates M N H I, in combination with the sliding tubes K, applied with the movable frame G and sliding frame L, operating in the manner substantially as described and for the purpose set forth.

No. 56,606.—FREDERICK R. POLLARD, Canaan, N. H.—*Shackle for Carriage Tongues.*—July 24, 1866.—A pivoted drop catch in a mortise at the front end of the tongue prevents the ring of the neck yoke from slipping off.

Claim.—A pivoted catch combined with the end of a carriage tongue, substantially in the manner and for the purpose herein set forth.

No. 56,607.—V. T. PRIEST, Decatur, Ill.—*Collar for Drill Rods.*—July 24, 1866.—The sections being screwed together are locked by a mitre key, key seat and adjustable band.

Claim.—First, the combination of the secion A' and grooved bed section A, connected by a screw joint with the inclined or mitred key I and key seat C, the adjustable band F and groove H, substantially as described.

Second, the grooves D and recesses E in the upper section or rod A, in combination with the band F, having internal pins G, substantially as described.

No. 56,608.—THOMAS C. PROSSER, Bay City, Mich.—*Brick Making.*—July 24, 1866.—The bricks are composed of hydraulic lime one part, common slaked lime one part; and three to seven parts sand or sand and gravel, moistened, mixed, moulded and dried.

Claim.—The forming of the materials, (in which hydraulic lime is one of them,) proportioned or varied as above into separate and individual bricks, as described, to be used for and applied to the purposes hereinbefore set forth.

No. 56,609.—TREAT T. PROSSER, Chicago, Ill.—*Culinary Boiler.*—July 24, 1866.—The water as heated in the outer shell rises to the annular chamber and affords hot water at the faucet before the whole contents of the vessel are equally heated.

Claim.—First, the shallow chamber C below and connected with the inner and main chamber E, by the opening a, and with the upper and exterior reservoir B, substantially as and for the purpose set forth.

Second, the combination and arrangement of the chambers and movable cover, for the purposes hereinbefore set forth.

No. 56,610.—JAMES PROUD, New York, N. Y.—*Shirt Collar Attachment.*—July 24, 1866.—A plate with a pear-shaped opening is attached by hinged pin to the back of the collar; the larger portion of the opening is slipped over the collar button and the collar moved slightly to embrace the shank of the slot of the plate.

Claim.—The plate C, having openings D E and hooks F, made as described for the purpose specified.

No. 56,611.—FREDERICK REYNOLDS, Newark, N. J.—*Manufacturing Harness Nails.*—July 24, 1866.—The heads are made bright by acid, are dipped in solder, attached to a strip of metal, detached and swaged to the desired form.

Claim.—An improved mode of forming and plating the soft metal heads of harness nails, substantially as herein described.

No. 56,612.—JOHN W. ROBERTS, New Monmouth, N. J.—*Securing Buttons to Garments.*—July 24, 1866.—The staple has two barbed points which are sprung within the openings in the back of the button or in the washer, and retain the same by expansion.

Claim.—The spring staple B, with barbed ends *b b*, in combination with the slotted button back or the washer E, operating substantially as described.

Also, in combination with the barbed staple and elastic washer D, applied substantially in the manner and for the purpose set forth.

No. 56,613.—WILLIAM T. ROGERS, Quincy, Ill.—*Gang Plough.*—July 24, 1866.—The described arrangement permits the ready adjustment of the ploughs as to depth and direction of their throw, the substitution of cultivators, likewise adjustable, and the adaptation of the seat to the changed position of the supports.

Claim.—First, the manner, as hereinbefore set forth, of securing gang ploughs or cultivator beams upon a carriage that can be used to support either or both by means of the hangers F, guides F', and braces *f f*, or their equivalents, in combination with the rods E E', arranged and operating substantially as and for the purpose described.

Second, the seat bars B, with their clevises L, or an equivalent, in combination with the manner as hereinbefore set forth, of regulating the seat to suit the inclination of the bars by means of the rocker M and adjusting standard *m*, with supporting and locking pins, or their equivalents.

No. 56,614.—P. H. ROOTS and F. M. ROOTS, Connersville, Ind.—*Cross-head for Blowers.*—July 24, 1866.—The slats are fastened to faces on the cross-heads and afterwards dressed to shape.

Claim.—A piston, constructed of cross-heads A, fastened to a shaft B, in combination with wooden lags or strips C, which are secured to the cross-heads, substantially as and for the purpose set forth.

No. 56,615.—JAMES J. RUSS, Worcester, Mass.—*Knife and Scissors Sharpener.*—July 24, 1866.—The middle plate is adjustable between the oblique side slots and forms one side of each cutting jaw.

Claim.—The combination of the stand or holder A, having inclined slots E and the sharpener plates B, when arranged and connected together substantially as and for the purpose described.

Also, the stand or holder A, slotted in an angular direction and notched at F, in combination with the sharpener plate B, the whole together forming a combined knife and scissors sharpener, substantially as and for the purpose described.

No. 56,616.—CYRUS W. SALADEE, Newark, Ohio.—*Padlock.*—July 24, 1866.—The vibrating locking tumbler is protected by a shield-plate through whose circular slot a portion of the key bit passes; the latter interlocks with the plate and reaches a stud on the tumbler. Pins on the shield plate necessitate corresponding wards in the key bit.

Claim.—First, the shield-plate A, with key slot D, in combination with the lock plate B and spring J, constructed and operating as described and for the purposes set forth.

Second, the stud F and spring J, as arranged in combination with the lock plate B, and slot *d'*, and hasp C, and covers S of the key hole, operating as described and for the purpose set forth.

Third, arranging the wards 9 and 10 on the shield-plate A, for the purposes set forth and operating as specified.

Fourth, the key constructed with hook I, as described, in combination with key slot D, and lock plate B, and spring J, constructed and operating as set forth.

Fifth, the stud F, or its equivalent, in combination with the lock plate B and spring J, in the manner and for the purpose substantially as shown and described.

No. 56,617.—CYRUS W. SALADEE, Newark, Ohio.—*Padlock.*—July 24, 1866.—The hooked tumbler plate which locks the hasp has a protecting shield-plate with a vertical annular flange to exclude burglars' tools. The key bit engages a tumbler pin which projects through a slot in the shield-plate.

Claim.—The shield plate N provided with the guard ring C, and otherwise constructed in the manner and for the purpose substantially as shown and described.

No. 56,618.—L. SAUTER, Jersey City, N. J.—*Finger, Scarf, and Napkin Rings.*—July 24, 1866.—By the rotation of the outer slotted hoop the ornaments of the inner band may be displayed or hid.

Claim.—The annular sliding band *c*, furnished with opening *f* and applied in combination with the hollow body *a b*, furnished at its outer circumference with openings *c'*, substantially as herein set forth for the purpose specified.

No. 56,619.—LEWIS G. SAYRE, Cincinnati, Ohio.—*Over-check Driving Rein.*—July 24, 1866.—The check reins are attached to a separate bit which is pulled upward against the palate; the check reins pass between the eyes over the forehead and are connected to the gag reins.

Claim.—The provision in connection with a bit A and bridle of the ordinary form of the independent upward-bearing bit F, suspended from the over-check G in the described combination with the check or safety rein I, substantially as set forth.

No. 56,620.—T. P. SHAFFNER, Louisville, Ky.—*Method of Packing Nitro-glycerine.*—July 24, 1866.—The interior vessel contains nitro-glycerine and water; it is isolated from the outer case by an intervening jacket of plaster of Paris, and interposed springs lessen the jar upon the central vessel.

Claim.—First, the placing between a bottle containing nitroleum, nitro-glycerine, or other liquid combustible compound, and an outer casing or box, India-rubber or caoutchouc or other material to serve as springs for the purpose of lessening concussion upon the said liquid substance by an exterior force resulting from a fall or otherwise, substantially as hereinbefore described.

Second, the application of plaster of Paris powder or of other equivalent non-conductor of heat and non-explosive or combustible substance when saturated with the liquids hereinbefore mentioned, in combination with the arrangements or parts, substantially as hereinbefore described.

Third, the use of metallic bottles for the purpose of confining the nitroleum, nitro-glycerine, or other explosive liquid, in combination with the arrangements and parts, substantially as hereinbefore described.

No. 56,621.—M. R. SHALTERS, Alliance, Ohio.—*Hame Fastener.*—July 24, 1866.—Each piece has loops for attachment to the hames, and the pieces respectively have lipped slots and a hook; the pivot of the hook rests in the lips, the point projects through a slot, and the vibration brings the line of strain through the pivot.

Claim.—First, the loop B provided with lips and slots, the hook A pivoted to the arms *a a*, arranged and operating conjointly as and for the purpose substantially as set forth.

Second, the hook A, finger *c*, and thumb-piece F, in combination with the pin D, arms *a a*, and loop B, arranged as and for the purpose set forth.

No. 56,622.—W. ANTHONY SHAW, New York, N. Y.—*Die for Making Tin-lined Lead Pipe.*—July 24, 1866.—Excess of lead is allowed to escape until the tin and lead are of uniform density and respond in proper proportion to the pressure; a jet of water is applied below the die to cool it.

Claim.—First, insuring a lining of tin of uniform thickness by providing an escape for the lead, either through the cylinder, die, or ram.

Second, the die A, in combination with the pipe H, when the two are constructed and arranged in relation to each other substantially as described.

No. 56,623.—CALVIN SHEPARD, Kattelville, N. Y.—*Driving Well Tubes.*—July 24, 1866.—The tube is shod with an angular enlarged point, and is sunk by rotation, while a weight bears upon the suspended platform.

Claim.—First, the combination of the tube A, constructed as described, with the flange F, the collar E, and platform D suspended therefrom, all arranged and operating in the manner and for the purpose herein specified and shown.

Second, the platform D supported on the flange F, as shown in combination with the well tube, substantially as described.

No. 56,624.—ALBERT R. SHERMAN, Natick, R. I.—*Lubricating Journal Boxes.*—July 24, 1866.—The caps enclose the ends of the journal box and revolve with the shaft. The oil forced out at the ends of the box is returned by scrapers to the journals.

Claim.—The caps D and scrapers *c*, in combination with the shaft C and box A, constructed and operating substantially as and for the purpose described.

No. 56,625.—JOHN SNARE, New York, N. Y.—*Water-proof Fabric.*—July 24, 1866.—Thin laminæ of mica are cemented to the fabric or leather by starch, glue, varnish, or solution of India-rubber.

Claim.—The water-proof or compound fabric adapted to the purposes specified, and formed of laminæ of mica cemented to flexible material, as specified.

No. 56,626.—M. B. STAFFORD, New York, N. Y.—*Peat Machine.*—July 24, 1866.—Each hinged section of the endless chain of moulds has the half of adjacent moulds; each pocket receives its charge from the hopper, passes under a rammer, and opens to discharge the block.

Claim.—First, the constructing of the moulds *c* of two longitudinal parts or halves connected by hinges and so arranged as to form an endless chain of moulds to work over rollers and receive the peat or other substance to be compressed and to discharge the same after being compressed by passing over the roller at the discharge end of the framing, substantially as set forth.

Second, the plunger F, operated as shown in combination with the endless chain of moulds, substantially as and for the purpose specified.

Third, the hopper M, provided with one or more partitions *l* provided with teeth *m* at their lower edges. in combination with the endless chain of moulds C, substantially as and for the purpose set forth.

No. 56,627.—JOSEPH A. STANSBURY, Baldwinsville, N. Y.—*Pump.*—July 24, 1866.— The pistons are attached to the stem in a rectangular relative position, and alternately assume the vertical, effective presentation by the guiding action of a rod. The stem passes through the rotary hub; the pistons revolve in an annular chamber.

Claim.—The combination of the right-angled wings *g g*, attached to and turning with the same shaft, and the guide G, when said parts are used in connection with a pump, substantially as herein specified.

Also, the butment made up of the parts *m* and *n*, in combination with the wings *g g*, operating substantially as herein set forth.

Also, the spring *l*, in combination with the wings *g g* and guide G, operating substantially as specified.

Also, in combination with the wings *g g* and the shaft *f*, the hub D provided with the flange *d*, substantially as described.

Also, the arrangement as a whole consisting of wings *g g*, guide G, butment *m n*, hub D, and spring *l*.

No. 56,628.—ROBERT S. STENTON, Brooklyn, N. Y.—*Wrench.*—July 24, 1866.—The screw is turned by a milled head and has a bearing on a projection of the shank.

Claim.—Arranging the jaws upon a straight shank, whether the former be perpendicular or inclined to the latter, and operating the movable jaw by a screw supported at its lower end in a step formed in the solid metal of the shank, and with a rosette, or its equivalent, for turning the same located contiguous to said step, all constructed substantially a sset forth.

No. 56,629.—SIMON STEVENS, New York, N. Y.—*Process of Burning Gas.*—July 24, 1866.—Steam is mixed with the gas either before or during combustion.

Claim.—The mixture of steam with coal gas or other gases produced by distillation of hydrocarbon substances, or their equivalents, so as to render it more useful for the production of heat and light, as herein described.

No. 56,630.—J. SEVERANCE STEWART and SAMUEL B. PIERCE, Homer, N. Y.—*Curtain Fixture.*—July 24, 1866; antedated July 15, 1866 —The hinged block operates as a break in adjusting the curtain.

Claim.—First, the break block G, constructed substantially as and for the purposes herein set forth.

Second, the combination of the roller C, the break E, the cord L, with weight H attached, the several parts being arranged substantially as and for the purposes specified.

No. 56,631.—NORMAN C. STILES, Meriden, Conn.—*Adjustable Pitman for Presses, Punches, &c.*—July 24, 1866.—The contacting surfaces of the pitman and disk have semi-circular openings at unequal distances admitting of the insertion of the key in such openings in the opposite faces as may be brought to correspond.

Claim.—The two unequally spaced series of grooves *m n*, arranged to operate together by the aid of one or more keys G fitted into any desired pairs of grooves, so as to compel the same to coincide and to hold the parts very firmly with great nicety of adjustment, substantially in the manner and for the purpose herein set forth.

No. 56,632.—THOMAS B. STOUT, Keyport, N. J.—*Table.*—July 24, 1866.—The central box forms a place of deposit for table linen; the leaves are so folded and supported as to permit the table to assume various shapes and sizes.

Claim.—The supports *f f h h* applied to the end leaves and to the framework, substantially as and for the purposes herein specified.

Also, the combination of the supports *f f h h*, governors *m m*, and slide bearings *g g i i*, substantially as and for the purpose herein set forth.

Also, the combination and arrangement of the battens C C, coupling pins *c c*, and leaves A A and D D, substantially as and for the purpose set forth.

No. 56,633.—MICHAEL O. SULLIVAN, Thompson's Station, Ill.—*Method of Destroying Lice on Trees.*—July 24, 1866.—A hole is bored into the tree just above the roots, is filled with blue vitriol and charcoal and the hole plugged.

Claim.—The ingredients herein described, when compounded substantially as and for the purpose set forth.

No. 56,634.—L. TAYLOR, Jordan, Wis.—*Water Elevator.*—July 24, 1866.—The bucket is journalled in a carriage, is tipped to be emptied, and travels on a track which leads from the well to the house.

Claim.—First, the arrangement and combination of the carriage N, carrying a water bucket, strands K, receptacles E H, pipe G, rope J, and windlass I, for elevating water to the upper apartments of a house, substantially as shown.

Second, the carriage N, shown in figures 1 2 and 4, having wheels O, pulleys O², an opening *e*, to receive the head R′ of the float, and locking pins to lock the head when the carriage is drawn upward along the strands K, substantially as described.

Third, the float R placed over the bucket, substantially as described, and having a head R′, with a pulley to allow it to be suspended by rope J, as shown.

No. 56,635.—THOMAS TAYLOR, Washington, D. C.—*Reducing Oxide of Lead.*—July 24, 1866.—The oxide of lead is placed in an iron kettle, covered with a sheet of waste iron and heated to 800° F.

Claim.—First, the protoxide of lead as a flux in the reduction of lead dross, substantially for the purpose and in the manner herein set forth.

Second, the use of iron as a deoxidizer of the protoxide of lead, substantially for the purpose and in the manner herein set forth.

No. 56,636.—CHARLES F. TESTMAN, Portland, Oregon.—*Gold Separator.*— July 24, 1866.—An air-tight chamber contains a series of upright boxes and intervening spaces for the circulation of hot air from the furnace. The chambers communicate with inclined passages which terminate in pipes directly over the sieve. Fine particles of ore pass through the sieve into the chamber and thence between rollers in a trough containing mercury.

Claim.—Three things: first, the process of drying dirt in the boxes *a a* and *b b*, by means of the fire *e*; second, the application of the springs *o o* to the roller *m*; and third, the method of constructing the amalgamating pans *r r*, and procuring thereof the gold dust in the quicksilver pockets by means of the continued revolution of the stirrer *q q*.

No. 56,637.—DANIEL M. THOMAS, Dowagiac, Michigan.—*Pump.*—July 24, 1866.—The pendulum is connected by gears, segments and elastic straps with the plunger rods; the plungers reciprocate in cylinders between its induction chambers.

Claim.—First, the arrangement of the plunger chamber B, so as to communicate with the induction chamber E and its upper end and the side passage D, which leads to the receiving chamber G, substantially as described.

Second, in combination with a force pump, which is constructed with upper and lower inflow chambers, leading to the main piston chambers, the application of a safety valve *k* to the vertical discharge pipe H, substantially as described.

Third, the vibrating frame C, connected to the working beam L by means of flexible connections *l l*, in combination with the segments J J′ and pendulum J², for operating the pump pistons, substantially as described.

No. 56,638.—W. McK. THORNTON, Clinton, Wis.—*Trace Buckle.*—July 24, 1866.—The pin attached to one of the horizontal bars of the frame passes through a hole in the strap and is capped by a spring tongue-plate.

Claim.—The frame A, constructed of two longitudinal bars *a a*, and two transverse bars *b b′*, in combination with the pin *g* and spring tongue C, all arranged in the manner substantially as described.

No. 56,639.—C. C. TORRENCE, Ripley, Ohio.—*Lock for Securing Throttle Valves.*—July 24, 1866.—The dog engages in notches of the handle of the lever and is held by a spring. By raising the dog to the upper notch it is free from the slotted guide and the valve can be moved, and conversely.

Claim.—The combination of the lock D, slotted guard C, and the valve lever A, substantially as described.

No. 56,640.—HARVEY TRUMBULL, Central College, Ohio.—*Broom Head.*—July 24, 1866.—The butts of the broom corn are fastened in a clamp and the shank screws into the end of the handle, drawing the brush within the cap.

Claim.—The jaws A and C, the screw *b*, the nut *d*, and ferrule E, the whole arranged and constructed in the manner and for the purpose substantially as herein described.

No. 56,641.—JOSEPH C. TUCKER, San Francisco, Cal.—*Sewing Machine.*—July 24, 1866.—Applied to machines of the Grover & Baker class, to make them capable of stitching two or more parallel rows for quilting, &c. On the ordinary rocking needle arm is affixed another needle holder, adjustable by rack and pinion to or from the first needle; supplemental pressers are similarly attached to and made adjustable on the usual fixed arm, and supplemental revolving loopers are adjustable upon hangers underneath the table; the parts coincide for joint operation.

Claim.—First, the combination of a rocking, perforating needle-carrying arm, with one or more adjustable perforating needle-carrying arms, substantially as described.

Second, the combination of the lower thread-carrying looper, working in fixed bearings, with one or more loopers the bearings of which are capable of being adjusted substantially as and for the purpose set forth.

Third, the combination with the rocking arm of a sewing machine provided with a perforating needle of one or more adjustable-needle carrying arms above the table, and a looper working in fixed bearings below the table, and one or more loopers in adjustable bearings for making parallel lines of stitching, substantially as described.

Fourth, in combination with the rocking arm of a sewing machine carrying a perforating needle, and provided with one or more adjustable-needle carrying arms, a stationary arm provided with one presser hold in fixed bearings, and one or more in adjustable bearings, substantially as described.

No. 56,642.—F. B. VAN VLECK and G. NICHOLS, Plainfield, N. J., assignors to themselves and G. D. MERRILL, same place.—*Sash Fastener.*—July 24, 1866.—Pressure on the thumb piece vibrates the bent lever and withdraws the bolt; the latter is returned by a spring when released.

Claim.—The thumb piece *e* passing through a mortise in the plate *c*, and connected to the bent lever *f*, in combination with the lifter *d*, bolt *g*, and spring *k*, the parts being arranged and acting as and for the purposes set forth.

No. 56,643.—FLORENCE L. VEERKAMP and CHARLES F. LEOPOLD, Philadelphia, Penn.—*Braiding Machine.*—July 24, 1866.—Instead of the usual serpentine course, the spools travel in concentric tracks in opposite directions; the threads being made to cross each other by means of cam-shaped wires, which force them outward and out of radial slots in the plate; when not so forced outward the line of thread from its spool to the central braiding point is within the path of the spools.

Claim.—First, in a braiding machine, two sets of spools M and I, caused to traverse in contrary directions in concentric annular paths when the threads of the two sets of spools are made to cross each other and be plaited by the devices herein described or any equivalent to the same, for the purpose specified.

Second, the plate F with its radial recesses *y*, in combination with the cam plate P and its wires or projections *u*, the whole being constructed, arranged, and operating substantially as and for the purpose herein set forth.

Third, the combination substantially as described of the shuttle or spool-carrier A and its guard rod K, for the purpose specified.

No. 56,644.—F. W. VOSMER, Cincinnati, Ohio.—*Washing Machine.*—July 24, 1866.—The pivoted beater has rollers on its face; the under side of the lid is corrugated so as to form a wash-board in its reversed position.

Claim.—First, a batter consisting of the parts J N O P P', in combination with the external lever L and connecting arm M, all arranged and operating in the manner herein described and set forth.

Second, the corrugated lid D, hinged to a permanent support E G, and otherwise arranged substantially as herein set forth to adapt it for use as a wash-board.

No. 56,645.—SAMUEL WARRINGTON, Philadelphia, Penn.—*Writing Pen.*—July 24, 1866.—Explained by the claim and cut.

Claim.—A pen A, having curves *c* and *c*, and flanges *z z*, when the said curves and flanges are formed and arranged in respect to the nib and shank of the pen, as and for the purpose described.

No. 56,646.—ALBIN WARTH, Stapleton, N. Y.—*Sewing Machine.*—July 24, 1866.—Adapted to make a chain stitch on the Wheeler & Wilson class of machines, and to feed by a lateral motion of the needle. The vibrating, pivoted needle-arm is operated by a projection thereon acting against inclined pads. A spring steadies this arm against a spontaneous movement when free of the inclines, and a set-screw may hold rigidly the arm and spring when the ordinary feeding device is retained. The lip braces the needle when it is required to feed; the guard prevents the loops, in passing off the revolving hook, from jumping up and again being caught by the hook; the brush prevents the bobbin from flying round when the machine stops suddenly; to make the chain-stitch a sliding hook replaces the ordinary feed dog; the protector covers as a cap the point of the revolving hook, and prevents its interfering with the needle loops.

Claim.—First, the arrangement of a friction spring in combination with the vibrating needle arm, constructed and operating substantially as and for the purpose set forth.

Second, the arrangement with said spring of a set screw, or other equivalent fastening, in combination with the vibrating needle arm, constructed and operating substantially as and for the purpose described.

Third, the arrangement of a lip *s*, extending from the needle holder on the back of the needle, substantially as and for the purpose set forth.

Fourth, the guard *g*, applied to the top edge of the bobbin holder M, substantially as and for the purpose set forth.

Fifth. the friction brush e', or its equivalent, in combination with the bobbin K, and bobbin holder M, constructed and operating substantially as and for the purpose described.

Sixth, in combination with a Wheeler & Wilson sewing machine, when such machine is so constructed that the needle is made to feed the material, the devices herein shown, or their equivalents, for producing a chain stitch.

Seventh, the protector n'. in combination with the revolving hook I, and chain stitch mechanism, constructed and operating substantially as and for the purpose set forth.

Eighth, the side-surface cam N', in combination with the chain-stitch slide h', spring j, and stop lever q', constructed and operating substantially as and for the purpose described.

No. 56,647.—JAMES H. WEBBER, Charlestown, Mass.—Lamp Shade.—July 24, 1866.—
The fingers clasp the chimney, and are attached below to a ring on which rest the inner edges of the plaited shade; the ring at the upper part of the shade rests in the notches of the finger flanges.

Claim.—In combination with the ring a, and its fingers b, the wings d, provided with recesses e, for holding the ring f, substantially as set forth.

No. 56,648.—HARRY WHITE, Oneida Castle, N. Y.—Shingle Machine.—July 24, 1866.—
The blank is held by clamps until planed on two sides and to a taper form by two adjustable knives held in a reciprocating frame; the blank is then seized by a gripper and discharged.

Claim.—The combination and arrangement of the automatic feeding plate V, the forked guide rod c, with the adjusting rods R R, the whole being arranged for joint operation, substantially as described.

Also, adjusting the knives to shave the shingles in the form described, by the means substantially as described.

No. 56,649.—THOMAS WHITE, Quincy, Ill.—Heating Stove.—July 24, 1866.—On each side are lateral, reverting flues, connected by a back flue; the exit opening is at the side, and has a damper which is closed to direct the heat into the circuitous course.

Claim.—The arrangement in a heating stove of the straight flues E F G H and I, in combination with the exit aperture and pipe C, substantially as and for the purpose above described.

No. 56,650.—JAMES M. WILLCOX, Glen's Mills, Penn.—Safety Paper.—July 24, 1866.—
Explained by the claim.

Claim.—Paper, having intermingled or united with the fibres of the sheet during the stage of the transformation from pulp to paper, or at any other time when such a thing can be done, of detached fibres or shreds different from the ordinary fibres in such a way as to group or locate the introduced matter on any part or parts of the sheet, while the remainder is left free or comparatively free from it, thereby forming one or more streaks or drops or clouds, or giving a general direction to said introduced fibres, or thereby producing any other distinctive mark or marks in the sheet or note.

No. 56,651.—FRANK A., JOHN H. and DANIEL G. WILLIAMS, Cincinnati, Ohio.—
Cabinet Maker's Scraper.—July 24, 1866.—The scraper bit is clamped in the stock by a plate and screw, and set toward the plate by an adjusting screw.

Claim.—A scraper, consisting of the blade A, stock B b, mouth piece C, set-screw D, and clamping screws E E, all constructed and arranged substantially as and for the purpose herein specified.

No. 56,652.—JOSHUA H. WILLIAMS, East Craftsbury, Vt.—Potato Washer.—July 24, 1866.—A shaft and revolving arm agitate the potatoes in the bucket, and the grating restrains them while the water is poured off.

Claim.—The combination of the grate D with the pail A, revolving shaft C, and sweep E, constructed and arranged in the manner and for the purpose herein specified.

No. 56,653.—HENRY WILSON, Paterson, N. J., and JAMES WILSON, New York, N. Y.—
Fastening for Bottles.—July 24, 1866.—The ends of the wire loop are socketed in the lip of the bottle, and the bow rests upon the cork.

Claim.—The sockets a, in the bottle A, in combination with the strap B, substantially as and for the purpose described.

No. 56,654.—JOHN N. WOLFE, Lancaster Ohio.—Water Wheel.—July 24, 1866.—The wheel is arranged outside an inner cylinder, into which the water is inducted from beneath; gated openings on each side of the cylinder admit the water to the wheel.

Claim.—First, the buckets B, constructed as herein set forth, in combination with the openings a a, substantially as specified.

Second, the combination of the gates b b, constructed and operated as described with the chamber C and buckets B, substantially as herein set forth.

No. 56 055.—TWENTYMAN WOOD, Westport, Conn.—*Coal Oil Burner.*—July 24, 1866.—The socket which receives the flange of the chimney is connected by parallel levers to its base, so as to be removable therefrom, retaining the vertical position of the chimney.

Claim.—First, giving to the upper section of the shell a combined vertical and lateral movement, substantially as shown for the purpose indicated.

Second, combining with the upper and lower sections of the burner the parallel levers attached as shown, when the same shall be combined substantially as herein described and for the purposes specified.

No 56,656.—JAMES A. WOODBURY, Boston, Mass.—*Planing Machine.*—July 24, 1866.—The yielding feed rollers are operated as to height y gear, and held in place by weights: they are capable of vertical motion without affecting the vertical position of the weight, and allow the rolls to yield at either end.

Claim.—First, so combining the yielding feed roll in a planing machine, with the weighted levers which control it and when said roll is weighted and geared, so as to raise both ends of it at once, as that when the board runs out the weight of the levers shall be removed from said feed roll, and leave it simply suspended to or by the screws, so that it can be raised or lowered without raising or lowering the weight of the levers, and when constructed and operating substantially as described.

Second, so combining and arranging the yielding feed roll of a planing machine with the gear for raising and lowering it, and when weighted as above claimed, as that while both ends of said roll will raise together by the gearing, yet either end thereof can yield or rock in the line of its length, to conform to the varied thickness of the edges of the boards passed through under it, substantially as described and represented.

No. 56,657.—WILLIAM E. WORTHEN, New York, N. Y.—*Rotary Valve.*—July 24, 1866.—The rotating valve has a cavity extending from its periphery to its face, and has a perforated sleeve around it adjusted by a screw to vary the cut-off.

Claim.—A rotating steam valve, provided with a cavity extending from the periphery to the face of the valve, as described, in combination with a seat, substantially such as described, and proper appliances as specified for holding the valve on its seat.

Also, a rotating valve provided with two cavities substantially and as described in combination with proper appliances for holding the valve on its seat, a valve seat, and a chest provided with a steam passage, all substantially such as described, and all operating in combination as set forth.

Also, in combination with a rotating valve and a steam passage, an adjustable cut-off ring, the combination being substantially such as specified and acting substantially as set forth.

No. 56,658.—FRANCIS WRIGHT, Galesburg, Ill.—*Piston Rod Packing.*—July 24, 1866.—The sleeve is inserted in the cylinder head, held by a gland, and forms a packing for the piston rod; it has a steam recess and a bevelled lead gasket supported by bushing at one end, and at the other end, and communicating, a chamber with packing rings having double inclined faces.

Claim.—First, the gasket c in combination with the bushing b; sleeve d, and steam chamber e, constructed and operating substantially as and for the purpose described.

Second, the steam chamber e and channels i in combination with the packing rings f, sleeve d, and follower g, constructed and operating substantially as and for the purpose described.

Third, the double inclined packing rings f, as and for the purpose described.

No. 56,659.—CHARLES D. YOUNG and JAMES McLEAN, Waterloo, N. Y.—*Grinding Mill.*—July 24, 1866.—The blast is directed tangentially along the furrows of the bed stone and withdrawn from the upper part of the casing.

Claim.—The blast tubes E E, having their ends b b opening outward in opposite directions in the extremities of the furrows of the bed stone to distribute the blast properly employed in combination with an exhaust tube H connected with the same fan case G for the extraction of moisture, as herein set forth.

No. 56,660.—GEORGE H. YOUNG, Charlestown, Mass.—*Marine Car.*—July 24, 1866.—A flotative belt of hinged pontoon sections is driven by a motor and traverses over drums; it answers the purposes of a float and a propeller.

Claim.—The articulated pontoons or floats arranged in the form of one or more endless aprons, and travelling over suitable drums, in combination with the car A, constructed and operating substantially as and for the purpose described.

No. 56,661.—JOHN YOUNG, Adrian, Ohio.—*Churn.*—July 24, 1866.—The dasher is reciprocated vertically; its upper and lower surfaces have concentric grooves with communicating apertures.

Claim.—The dasher D, formed with the concentric channels d d' and with perforations or apertures G G, communicating with said channels in the manner and for the purposes explained.

No. 56,662.—PETER YOUNG, El Paso, Ill.—*Sulky Plough*—July 24, 1866.—The plough is suspended by cords beneath the frame. and tongue of the two-wheeled carriage and is adjustable vertically and laterally.

Claim.—First, the cords *j* and *i*, sliding rod *k*, lever F, and yoke *m*, all arranged and operating as and for the purpose set forth.

Second, in combination with the above the steadying lever *n*, arranged and operating substantially as herein shown and described.

No. 56,663.—NICHOLAS ZILLIER, Newcastle, Del.—*Screw Plate.*—July 24, 1866.—The dies are supported by a circular holder within the opening in the centre of the stock and their outer ends bear against cams. The oscillation of the die holder moves the dies radially by traversing them upon the cams, and an indicator shows the size to which they are set.

Claim.—An improved screw plate formed by combining with the two handled plate A, the die holder B, the dies D, the spring C, and the cap E, the parts being constructed and arranged substantially as herein described and for the purpose set forth.

No 56,664.—GEORGE ZORGER, Greensburg, Ind.—*Wheat Drill.*—July 24, 1866.—Motion from the driving wheel is communicated to the screw feeders which operate in the respective hoppers and are adjustable simultaneously on each side of the median line. Slotted plates secure the shares to the frame and sliding plates in the hoppers adjust the discharge opening

Claim.—First, the means employed for adjusting the arms F F, to wit, the rods G G attached at their outer ends to the rear ends of the arms E and connected at their inner ends to opposite sides of a wheel H on a vertical shaft I, which has an elastic handle or lever J attached to it, engaging with a notched semi-circular bar K, substantially as shown and described.

Second, the two wheels B B supporting the front end of the bar A, in combination with the gearing *b c d E l k k*, all arranged as shown and described for rotating the screws N N O as set forth.

Third, the slotted plates S in the hoppers P, provided with the slides T, for the purpose of regulating the flow or discharge of the seed, as described.

Fourth, the securing in proper position of the seed conveying spouts Q Q R to the arms F F and bar A by means of the slotted plates U, substantially as shown and described.

No. 56,665.—C. F. BINDER, Philadelphia, Penn., assignor to himself and J. BINDER.—*Liquid Glue.*—July 24, 1866.—Glue is extracted from bones by hot water, clarified by lime, alum, and molasses, and concentrated to the proper consistency.

. *Claim.*—A liquid glue produced in the manner and. by the process substantially as herein described.

No. 56,666.—HENRY BRADBURY, Berlin, Conn., assignor to NEAL, WILLCOX & Co., Southington, Conn.—*Snap Hook.*—July 24, 1866.—The tongue and the shank have a barrel case around the pintle upon which the spring of the hinge is coiled.

Claim.—A snap hook formed with a transverse cylinder or opening containing the spring, in combination with the snap or latch *c* and its end plates or fork *e*, enclosing the said transverse cylinder, and composing the spring joint of the snap, substantially as set forth.

No. 56,667.—J. F. CHAMPLIN, Aurora, N. Y., assignor to himself, S. B. THOMSON, and D. C. CORBIN.—*Corn Planter.*—July 24, 1866.—The reciprocating seed slide is moved by the lugs on the wheel and returned by the end spring ; a lever moves it out of the range of the lugs to stop the feed ; the cultivator frame is suspended beneath the carriage by hinged pendants and a bail which serves to raise it when required.

Claim.—First, the combination of the cam spring H, lugs J, and spring I for the purpose of operating the slide G, substantially as described.

Second, the arrangement of the lever N, in connection with the cam spring H, for the purpose of moving the cam spring beyond the touch of the lugs J when desired, substantially as described.

Third, in a machine for planting corn in hills, in which the plough frame is made separate from the main supporting frame, suspending the plough frame under the main frame by means of a pendent hinged connection to the forward end of the main frame, in combination with a rear upward projection bail or handle, (in near proximity to the driver's seat,) so that the driver can conveniently lift and suspend the ploughs from the ground when turning round at the end of the rows (or otherwise) and again drop the ploughs to the ground as required, substantially as described.

No. 56,668.—P. C. CLAPP, Dorchester, Mass., assignor to himself and COTTON C. BRADBURY, Milton, Mass.—*Shears.*—July 24, 1866.—The short, powerful blades near the axis are for cutting wire, whalebone, &c.

Claim.—The scissors as made with the auxiliary blades *e f*, arranged and combined with the blades *a b* and their handles *c d*, substantially as specified.

No. 56,669.—JACOB A. CONOVER, New York, N. Y., assignor to the EMPIRE BREECH-LOADING FIRE-ARM COMPANY.—*Breech-loading Fire-arm.*—July 24, 1866.—The hammer has a curved back and throat so as to be operated without exposing to the access of dirt the mortise in which it works.

Claim.—The hammer F formed with a curved back and throat, said curves being concentric with the axis *f* of the hammer, in combination with the curved upper side of the projection *i* and rear upper edge of the mortise G, substantially as and for the purpose specified.

No. 56,670.—ALEXANDER CUTLER, Malden, Mass., assignor to CHARLES H. HAYWARD, same place.—*Vulcanizing India-rubber in connection with Leather.*—July 24, 1866.—Explained by the claim

Claim.—The improved process of treating leather and rubber during the vulcanizing of the latter, such consisting in the employment of air in the vulcanizing chamber or furnace in sufficient quantity to prevent the heat thereof from injuring the leather without materially impairing its vulcanizing effect on the composition of rubber and sulphur.

No. 56,671.—JOSEPH W. DOUGLAS, Middletown, Conn., assignor to W. and B. DOUGLAS, same place.—*Pump.*—July 24, 1866.—The double acting pump has a side pipe and a hollow piston rod ; the water on the downward stroke passes through the piston, and on the upward stroke is raised by the piston.

Claim.—The combination of the diaphragm D, hollow piston rod B, having a perforated enlargement B', and piston G, as described, valve H, with its spindle *d* and guide fingers *c*, cylinder L and side pipe I, provided with valves J and T', all arranged and operating substantially as described for the purpose specified.

No. 56,672.—ROBERT HENEAGE, Buffalo, N. Y., assignor to himself and J. D. SHEPARD, same place.—*Smut Mill.*—July 24, 1866.—A series of hopper-shaped revolving disks, with ribs on their upper sides, throw the grain centrifugally upon the concentric rings, which drop it upon the next disk, until it reaches the outermost, where it is collected by a hopper, and passed to a similar arrangement on a lower level.

Claim.—The rings *f f* of the rotating disks E, when provided with radial or tangentially inclined ribs *e e*, or their equivalent, in combination with the stationary rings *h* and ribs *i*, arranged and operating substantially as and for the purpose set forth.

Also, in combination with the above described device, the vertical ribs *p p* on the interior of the case, together with the hopper-shaped diaphragms B B, arranged and operating substantially as described.

Also, the guard ring *g*, in combination and concentric with the ribbed ring *h* for the purpose of deflecting the rebounding grain beyond the inclined ring beneath, arranged substantially as specified.

No. 56,673.—CHARLES MCLEAN, Boston, Mass., assignor to himself, T. C. HARGRAVES and CHARLES MITCHELL.—*Maple flavored Sugar and Sirup.*—July 24, 1866.—An extract obtained by boiling maple wood is added to other sugars, &c., to confer the peculiar flavor.

Claim.—The within described new manufacture.

No. 56,674.—JAMES M. MERRITT, Buffalo, N. Y, assignor to himself and JOHN W. A. MEYER, same place.—*Stencil Numbering Apparatus.*—July 24, 1866.—The frame has a series of apertures, and the numbers on the slides are brought into consecution thereat as required.

Claim.—The improved stencil numbering apparatus herein described, consisting of the plate or frame A, with apertures *b c d* and the figure slides 1 2 3, and the guide *f*, or its equivalent, constructed and arranged substantially as described.

No. 56,675.—STEPHEN R. PARKHURST, Bloomfield, N. J., assignor to EMILY R. PARKHURST, same place.—*Machine for Picking and Cleaning Cotton and Wool.*—July 24, 1866.—The burrs and other foreign matter are removed, and the fibre condensed into a sliver ready for the carding machine. The claims and cut explain the devices.

Claim.—First, constructing the toothed rollers *b* and *c*, with separate teeth set into grooves, and secured as described.

Second, the picker cylinder, formed of a series of longitudinally grooved bars, containing separate teeth and intermediate filling pieces, substantially as specified.

Third, the cylinders *d f* and *g*, in combination with the strippers *h* and *i*, substantially as and for the purposes specified.

Fourth, the brush blower *l* and condensing cylinders *m'*, in combination with the picker cylinder *d* and cylinder *f* or *g*, substantially as set forth.

Fifth, the rollers *r r*, in combination with the condensing cylinders *m' m'* and oilers *t t*, substantially as set forth.

Sixth, in a picker for wool and other fibre, arranging the stripper and toothed cylinder over the picker cylinder, so that dust and foreign substances shall fall into the space in which the picker cylinder revolves, and be thrown out by the centrifugal action of said cylinder, aided by a current of air, substantially as set forth.

No. 56,676.—J. N. PEASE and G. LEWIS, Panama, N. Y., assignors to the METROPOLITAN WASHING MACHINE COMPANY.—*Clothes Wringer.*—July 24, 1866.—Explained by the claims and cut.

Claim.—First, the method of gearing wringer rolls, as herein shown and described; that is to say, by the employment in connecting with the pinions or cog-wheels of the upper and lower rolls of a third or auxiliary gear wheel; the whole being so arranged that while the relative positions of the said pinions to each other may constantly vary, they shall bear permanent or fixed relations to the auxiliary gear.

Second, supporting one of the wringer rolls in upright disks, the said roll having its bearings placed eccentrically to the said disk, in combination with the auxiliary gear, when arranged to revolve upon the axis of said disks, the whole being arranged for operation as herein shown and set forth.

Third, in combination with the herein described arrangement of gearing rolls, the crossbar, or the mechanical equivalent thereof for connecting the disks which support the movable roll, so that the said disks may be moved upon their axes in unison, and maintain the parallelism of the rolls in the movement of the one to and from the other, substantially as herein shown and described.

Fourth, the combination and arrangement of the spring and disks supporting the movable roll, substantially as herein shown and set forth, so that the rolls are kept together with a yielding pressure, which may be regulated as described.

Fifth, the herein described device for holding the wringer to the side of the tub, the same consisting of bell-crank levers pivoted on the machine, in combination with an adjusting rod, the whole being arranged for operation substantially as herein shown and set forth.

No. 56,677.—HENRY PENNIE and E. A. LELAND, New York, N. Y., assignors to HENRY PENNIE.—*Gas Stove.*—July 24, 1866.—The gas is admitted into the lower chamber, bounded by the inner cylinder and the surrounding frustum; it is ignited above the wire gauze in the air chamber, dives and ascends into the central, inverted frustal shell.

Claim.—First, the burner or burners located within the stove, and burning air and gas, in combination with the openings *h* for the admission of air to support the flame and produce the draught, and with a chamber above the burner, constructed and arranged substantially as described, by which combination and arrangement the flame is carried downward and toward the opening of egress, as set forth.

Second, in combination with the burner, the openings *h* for draught and the chamber above, substantially as specified, the employment of the very small aperture *j*, arranged essentially as set forth, for the purpose of admitting a comparatively small amount of air to mix with the volatile products of the flame, and assist the consumption of such products as they are carried downward and over the flame by which the burner is made use of to consume its own products of combustion, substantially as described.

No. 56,678.—MILTON ROBERTS, St. Paul, Minn., assignor to himself and JOHN A. LLOYD, same place.—*Spring Bed Bottom.*—July 24, 1866.—The thin wooden slats are strained in the direction of their length till the required tension is attained.

Claim.—The straining screws *b b*, or their equivalents, in combination with a thin slat bed bottom, substantially as described, for the purpose of increasing or diminishing at will the tension of said slats.

No. 56,679.—J. F. and G. W. TAPLEY, Springfield, Mass., assignor to themselves and G. D. TAPLEY, same place.—*Paper Cutting Machine.*—July 24, 1866; antedated February 5, 1866.—The cutter traverses the edge of the arc, and is associated with an embossing or printing roller with inking attachment.

Claim.—First, the method herein described of cutting paper and similar substances in the form of an arc of a circle, for collars and other purposes, by means of a revolving or circular knife, made to travel in an arc of a circle, or similar curve on which the paper is to be cut, substantially in the manner herein set forth.

Second, arranging the knife *b* in the handle *d*, so as to be adjustable by means of the set screws *e e*, substantially in the manner and for the purpose herein described.

Third, arranging the indenting or printing wheel D in connection with the cutting knife *b*, substantially as herein set forth.

Fourth, in combination with the wheel D, the spreading roll or rolls *g* and inking plate E, when arranged substantially in the manner and for the purpose herein set forth.

No. 56,680.—WILLIAM and WILLIAM H. TERWILLIGER and JOHN S. LOCKWOOD, New York, N. Y.—*Uniting Iron and Steel.*—July 24, 1866.—The surfaces of the metals to be joined are coated with a fluxing mixture of borax and saltpetre combined with a hydro-carbon.

Claim.—First, the welded combination of iron and steel plates to make the shell of a safe, for safety against burglarious attacks.

Second, the process of welding iron and steel plates by the use of the composition of borax and saltpetre in paint form laid on the surfaces to be united, heated not above 1,500° F., and rolled with great pressure, to make the best weld possible in the materials for burglar-proof safes.

Third, interposing a steel plate between two iron plates, with the use of the welding composition and process above described, to make economical materials for burglar-proof safes.

Fourth, interposing a plate of iron between two plates of steel, with the use of the welding composition and process above described, to make the strongest and best materials for burglar-proof safes.

Fifth, constructing and preparing the materials for a burglar-proof safe by rolling and punching while hot, so that the parts of it can be put together after transportation in the manner described.

Sixth, making a burglar-proof safe in mutually fitting parts and numbered, so that, from a stock of the materials on hand, a safe of the desired size and strength could be put together in a few minutes in the manner described.

No. 56,681.—JACOB WOODBURN, St. Louis, Mo., assignor to himself and THOMAS SCOTT, same place.—*Carriage Wheel.*—July 24, 1866.—The major diameter of the spoke tenon is coincident with the plane of the rim and exerts no splitting pressure therein.

Claim.—An oval or elliptical-shaped tenon for wheel spokes, in combination with a round-shaped mortise hole in the wheel rim therefor, substantially as herein described and for the purposes specified.

No. 56,682.—WILBUR F. WRIGHT, Nashua, N. H., assignor to himself and EDWIN BLOOD, Newburyport, Mass.—*Machine for Polishing Enamelled Paper.*—July 24, 1866.—Two sets of rollers are employed, the burnisher roller of each set revolving at a higher velocity than its supporting roller, and one burnisher faster than the other.

Claim.—First, the combination of the two sets of rollers H I and F G when the roller I revolves at a higher velocity than the roller G, substantially as herein set forth, for the purpose specified.

Second, the combination and arrangement of the pressing and smoothing rollers C D, the burnishing rollers G I, and supporting rollers F H, substantially as herein set forth, for the purpose specified.

No. 56,683.—CHARLES LEHMANN, Bienne, Switzerland.—*Watch.*—July 24, 1866.—The spring is wound and the hands set through the stem of the pendant; a single inflexible rod operates each mechanism. For setting, the stem is partially retracted and the watch is wound by rotating the stem at its normal projection.

Claim.—The arrangement of the clutch c c' in combination with the rod or stem t, constructed as described, and capable of being connected mediately or immediately with the wheel which controls the mainspring, and with the minute wheel of the watch, substantially as herein set forth.

No. 56,684.—THOMAS C. CRAVEN, Albany, N. Y.—*Saw for Cotton Gins.*—July 24, 1866.—The teeth are swaged and finished to a rounded shape to remove edges which tend to break the fibre.

Claim.—A saw for cotton gins formed with rounded teeth of the character specified, as and for the purposes set forth.

No. 56,685.—THOMAS M. and AMBROSE G. FELL, New York, N. Y., assignors to themselves and WILLIAM BELL.—*Manufacture of White Lead.*—July 24, 1866.—The ores of lead are calcined in a reverberatory furnace, the resulting oxide dissolved in nitric acid; sulphuric acid added, precipitating sulphate of lead, treated with an alkali which deprives it of a portion of acid and converts it into a sub-sulphate, which may be used as a substitute for the ordinary white lead.

Claim.—First, the treatment of sulphate of lead with alkaline substances, or their salts, in the manner and for the purposes substantially as above described.

Second, the treatment of the sulphate of lead with the carbonates of either potash, soda, or lime, followed by the alkaline substances or their salts, in the manner and for the purposes substantially as above described.

Third, the treatment of sub-sulphate of lead with the carbonate of soda or potash, in the manner and for the purposes substantially as described.

Fourth, the manufacture of white lead from ores of lead, or metallic lead, by the use of nitric and sulphuric acids, in combination with alkaline substances, or their salts, either with or without the prior treatment of carbonates of potash, soda, or lime, in the manner and for the purposes substantially as above set forth.

No. 56,686.—E. BUSSEY, Troy, N. Y.—*Cooking Stove.*—July 24, 1866.—Explained by the claims and cut.

Claim.—First, the annular surrounding and downward projecting flange D, or any equivalent thereof, in combination with the boiler or reservoir A, in the manner substantially as and for the purposes herein described and set forth.

Second, the apertures d in the boiler or reservoir A, in combination with the exit flue or flues in the rear end of a cooking stove, in the manner and for the purposes substantially as herein described and set forth.

Third, the arrangement and combination of the lid or cover E with the reservoir A, so that the water or moisture on the under side thereof, by reason of condensation of steam, may and shall pass or drip into said boiler A, in the manner substantially as herein described and set forth.

Fourth, the arrangement and employment of the intermediate and vertical plate F in combination with the said reservoir A, in the manner and for the purposes substantially as herein described and set forth.

No. 56,687.—A. ADLER, Paris, France.—*Apparatus for Working Hides.*—July 31, 1866 —
The table moves in a frame and is covered successively with cork or vulcanized rubber, felt and cow hide; it passes under a roller armed with two sets of spiral blades which commence at the middle of the roller and extend to its ends.

Claim.—The machine for working and preparing skins; constructed and arranged for operation substantially as herein set forth and described.

No. 56,688.—AMBROSE ALEXANDER, Middleville, Mich.—*Washing Machine.*—July 31. 1866.—The beater is vibrated by a compound lever arrangement and its lower end supported by anti-friction rollers.

Claim.—As new the employment of dash board I, with supporting friction rollers J, in combination with a compound leverage, A D E F and G, for operating the same, as substantially described.

No. 56,689.—R. J. ALGEO, Kalamazoo, Mich.—*Trace Buckle.*—July 31, 1866.—The axial bolt of the tug cap which binds the trace against the bar of the frame has motion in the slots of the frame to allow the trace to be bound or loosened.

Claim.—The slotted sides of the frame C, in combination with the bolt B and tug cap D, constructed as described and operating in the manner and for the purpose specified.

No. 56,690.—H. A. AMELUNG, New York, N. Y.—*Drying House.*—July 31, 1866.—The beams upon whose hooks the meat is suspended are severally movable by tackle and supported by cleats; the more solid particles rising from the fire are arrested by perforated diaphragms in the furnace.

Claim.—First, the application of one or more soot catchers i, in combination with the grate h and diaphragm g, in the interior of the furnace F, in the manner and for the purpose substantially as herein set forth.

Second, the movable beams E, in combination with the hoisting tackle e, f, or its equivalent, and with the drying or smoking house A, constructed and operating substantially as and for the purpose herein shown and described.

No. 56,691.—GUSTAVUS ANTON, Philadelphia, Penn.—*Feather Covered Parasol.*—July 31, 1866.—Explained by the claims and cut.

Claim.—As a new article of manufacture, a parasol having a top or covering composed of feathers, secured to a central piece of wood or other suitable material, substantially in the manner set forth.

Also, in combination with the improved covering, the described tilting motion of the same upon the stem when the latter is made adjustable in length, substantially as and for the purpose described.

No. 56,692.—FRANCIS ARNOLD.—Haddam Neck, Conn.—*Meal and Flour Sifter.*—July 31, 1866.—The scrapers are attached to a head and vibrate in contact with the meshes of the semi-cylindrical sieve.

Claim.—The metallic plates E E' connecting the paddles C and D, in combination with the adjustable sieve B and handle F when arranged and used as and for the purposes set forth.

No. 56,693.—VANTUYL BABCOCK, Marshall, Mich —*Gate.*—July 31, 1866.—The upper rail of the sliding gate is supported upon a roller, and its heel end has a traversing hanger whose upper and lower rollers traverse on the edges of the guide rail.

Claim.—The arrangement and combination of the rail G, saddle pieces I, rollers R¹, R², and supplemental posts E and J, with an ordinary gate and fence, substantially in manner and for the use herein specified.

No. 56,694.—RICHARD C. BLAKE, Cincinnati, Ohio.—*Steam Gauge.*—July 31, 1866.—The spiral corrugations give a greater range of elasticity.

Claim.—Spiral corrugations in the diaphragm spring of a steam gauge, substantially in the manner and for the purposes set forth.

No. 56,695.—JOHN BRENEMAN, Mount Joy, Penn.—*Portable Fence.*—July 31, 1866.—Explained by the claim and cut.

Claim.—The combination and construction of the two panels of a fence so that the upper and lower rails R¹ R' of the one will pass between those of the other, the ends of the rails of every alternate panel provided with a short piece S R, forming an open space for a key board K, passing through the overlapping ends and firmly uniting them, in the manner and for the purpose shown and specified.

No. 56,696 —J. O BROWN, A. INGHAM, and F. T. LOMOST.—Massillon, Ohio —
Harvester Rake.—July 31, 1866.—A reciprocating rake is employed to rake the grain together upon one end of the tilting platform, when it is dumped in a compact gavel upon the ground. An apron connected with the finger beam and with a roller on the front edge of the platform serves as a cut-off when the platform is tilted, said apron being wound upon the roller by the action of the rake lever as the platform falls to receive the grain.

Claim.—First, the rake R, pin r, weight t, and slide K, in combination with the tipping platform A, guide O, and notches u u', arranged as and for the purpose substantially as set forth.

Second, the roller H, apron H', and cords e' I', or their equivalents, in combination with the pulleys L J and lever J', substantially as and for the purpose described.

Third, the shaft C, arms E F, and slide K, in combination with the arm f, lever J', and platform, arranged and operating substantially as and for the purpose specified.

Fourth, the platform A, roller H, and apron H', in combination with the cords e' I', lever J', and arm f, substantially as and for the purpose described.

Fifth, hanging or pivoting the platform A to the rear end of the shoes by means of the arms E and F, when said platform is provided with the rake R, grooves i, and slats j, substantially as and for the purpose specified.

Sixth, attaching the arm P to the swath board D', in combination with the cords I', or equivalent lever J, and platform, as and for the purpose set forth.

No. 56,697.—THOMAS BROWN, Albany, N. Y.— *Redyeing the Cushions of Car Seats.*—July 31, 1866 —The cushions are exposed to the dye in a vat and then steamed. The steam box has a perforated shelf and a high cover. The edges may be protected by a clamp while the sides are dyed of a different color.

Claim.—First, exposing the cushions after the color has been applied to them to the action of steam, substantially as and for the purpose set forth.

Second, the boiler with a perforated shelf a and close fitting cover b, in combination with a furnace B and with the cushions to be steamed, substantially as and for the purpose described.

Third, the frame C with adjustable sides c d, constructed and operating substantially as and for the purpose set forth.

No. 56,698.—ROBERT BRYSON, Schenectady, N. Y.—*Harvester.*—July 31, 1866.—Improvement on his patent of March 31, 1863. Explained by the claims and cut.

Claim.—First, pivoting the forked ends of the harvester pitman rod R to an adjustable strap p of a two-part pitman box s, substantially as and for the purpose described.

Second, constructing the frame D substantially as described, in combination with supporting this frame upon the axle B of two drive wheels outside of a hinged frame C, substantially as set forth.

Third, the application of guards G to the inside gear A' of the driving wheels, substantially as described.

Fourth, the arrangement of the lever J so that it forms an intermediate connection between the hand lever E and the finger beam, and its inner long arm slides upon the lower surface of the platform plate g; the said lever J and the hand lever E being applied to a harvester having two hinged frames C D and a hinged cutting apparatus, all substantially as described.

Fifth, the arrangement of the double tree K, staple h, pin k', hook i, chain m, and hook j, in the manner and for the purpose herein described.

No. 56,699.—A. S. CAMERON, New York, N. Y.—*Piston Packing.*—July 31, 1866.—A spiral wire with an expansive tendency occupies a groove of corresponding character and fits against the inside of the cylinder.

Claim.—The spiral packing wire b in combination with the piston A, substantially as and for the purpose described.

No. 56,700.—ANDREW CAMPBELL, Brooklyn, N. Y.—*Bed Recoil Springs for Printing Presses.*—July 31, 1866.—The springs ease the shock of the reciprocating type bed when the direction of its motion is changed at each end of its run.

Claim.—First, so applying the bed recoil springs of a printing press that they are always in contact or connection with the bed of the press through levers operating the springs, substantially as described.

Second, so applying the bed recoil springs of a printing press that the bed when running faster in one direction than the other, may have the required degree of recoil given to it in either direction by one set of springs, substantially as described.

No. 56,701.—ANDREW CAMPBELL, Brooklyn, N. Y.—*Printing Press.*—July 31, 1866.—The wheel has outer and inner spur-gears which actuate the pinion alternately; the studs on the latter have rollers which pass groove cams; when the pinion has reached the end of the segment gear one roller passes through the groove while the other swings over and enters the second groove, the first passing over the cheek, and the pinion is reversed. This is repeated at the end of the opposite segment. The shape of the bearing of the front guides prevents the rolling incident to bad fitting and assists in obtaining a better register.

Claim.—The mode of converting the rotary to a reciprocating or rectilinear motion as above described, or its mechanical equivalent, for the purposes se, forth.

Also, the V-shaped bearing T as applied to the front guide of printing presses, substantially as described and for the purpose set forth.

No. 56,702.—PETER F. CAMPBELL, Jersey City, N. J.—*Dry Dock Indicator.*—July 31, 1866.—The hands of the dial are moved by an endless band and a float.

Claim.—The combination with the section or compartment of a dry dock of the floats and an indicating apparatus, substantially as and for the purposes set forth.

No. 56,703.—E. P. H. CAPRON, Springfield, Ohio.—*Brick Machine*—July 31, 1866.—The clay after being pressed in the mould is re-pressed between an upper hinged flap and pressure roller and a lower follower, the latter eventually discharging it when the flap is lifted by its actuating rods.

Claim.—First, the combination of the pressure roller P with flap O.

Second, the combination of the hinged flap O with its sliding rods

Third, with the follower N, the combination of the rod *j* when the latter is provided with an articulated lever K at its upper end to raise the lid, the whole being constructed and arranged as described.

No. 56,704.—GEORGE W. CARLETON, Brunswick, Me.—*Composition for Settling Coffee.*—July 31, 1866.—Compounded of coarse-grained sugar, white of egg, and chloride of sodium.

Claim.—A compound for clarifying coffee, made substantially as herein specified.

No. 56,705.—SAMUEL CARY, Centreville, La.—*Boring and Grinding Apparatus.*—July 31, 1866.—This drilling, boring, and grinding apparatus in combination with screw-feed mechanism, is for the purpose of producing salt from the mine in a pulverized condition.

Claim.—First, the application and use of tempered steel notched or toothed plates, secured so as to be adjustable to the arms of the metal flanch to form a boring and grinding mill, for the purposes herein set forth.

Second, the drilling, boring, and grinding apparatus in combination with the screw-feed mechanism and driving machinery, as and for the purposes specified.

No. 56,706.—BRANTLY CHALFANT, Williamsport, Penn.—*Automatic Boiler Feeder.*—July 31, 1866.—A barrel divided by longitudinal partitions is pivoted in an oblique position and rotated automatically by the gravity of its contained water. Each chamber on reaching its uppermost position receives water from a reservoir, and on reaching its lowermost position its respective ends communicate with the boiler through separate pipes, so that as soon as the water in the boiler is low enough to admit steam to the upper pipe the water in the feed chamber will flow into the boiler through the lower pipe.

Claim.—The many-chambered barrel A placed in an oblique position between suitable bearings and provided with pipes E F G H, substantially as and for the purposes set forth.

Also, the self-tightening key E', in combination with the standard D, cap B, and barrel A, constructed substantially as and for the purposes described.

No. 56,707.—WILLIAM CHAPPELL, Buffalo, N. Y.—*Stove Pipe Top.*—July 31, 1866.—The main pipe has a removable cap, branching off near the top, and at right angles are two T-shaped pipes, easily turned to present their mouths to the wind, and held in position by springs. The main pipe may also be rotated upon a joint near the floor or roof.

Claim.—The two T-pipes C C, provided with holes *f f*, or their equivalent, in combination with the spring *e*, and adjustable pipe A, provided with elbows *c c*, the whole arranged and operating substantially in the manner set forth.

No. 56,708.—GEORGE CLARK, Buffalo, N. Y.—*Grain Dryer.*—July 31, 1866.—The hot air is introduced by separate concentric cylinders to annular chambers one above the other, whence it is impelled through the perforated outer wall into the space where the grain falls; this is divided into corresponding chambers whose bottoms are filled with apertures having screw threads, so that the grain is turned over in passing down; the outer wall of this drying space is also perforated, and between it and the external wall of the dryer is a space; suitable dampers at the bottom regulate the flow of hot air into the several hot-air chambers. Dampers are arranged in the furnace to preserve the proper draught when the outer door is opened.

Claim.—First, the construction and arrangement of grain-drying perforated cylinders and two or more hot-air chambers in such relation to each other that the hot-air chambers shall be heated centrally within the cylinders (the body of grain to be dried being outside of the chambers,) and the hot air supplied centrally to each chamber by means of hot-air conducting pipes so as to issue from all parts of the chambers and pass directly through and at right angles (or nearly so) to the direction of the body of grain passing between the cylinders, substantially as described.

Second, placing and using screws, or equivalents, in the grain space between the cylinders so that the grain must pass through these screws and thereby be turned over or changed in the position of its kernels in reference to the inner and outer cylinders, and thereby insure all parts of the grain to be acted upon equally by the hot air, substantially as described.

Third, dividing the inner perforated cylinder into two, three, or more hot-air chambers, each chamber being separate and independent of the other, and each having distinct hot-air flues, so that the hot air in each chamber may be regulated and controlled independently of the other, for the purposes and substantially as set forth.

Fourth, placing and arranging the said perforated cylinders and hot-air chambers within an outer stack, so that an evaporation space shall be formed between the larger cylinder and the outer stack, and evaporation from each chamber be discharged directly therein, substantially as set forth.

Fifth, the construction, application, and use of an inner furnace door or valve Z, opening inwardly, for the purpose and substantially as described.

Sixth, a weighted, conical valve P placed at the top of the drying cylinders to insure an equal distribution of the grain into all parts of the grain-drying space, substantially as described.

Seventh, in a grain dryer, constructed substantially as herein described, the arrangement therewith of the valves T and movable disk T', for the purposes set forth.

No. 56,709.—WILLIAM P. CLARK, Boston, Mass.—*Draught Cock for Soda Water Apparatus.*—July 31, 1866.—The two passages are operated by a single screw-stem which actuates consecutively the respective valves ejecting a central jet and then an annular jet.

Claim.—A soda cock constructed with an induction pipe H, and two sets of eduction pipes J and L, and two valves F and G, actuated successively by a common stem E, and resting upon different valve seats, said several parts being respectively constructed and the whole combined and arranged for operation, substantially as set forth.

No. 56,710.—JOHN W. CLARKE, Kingston, Wis.—*Measuring Funnel.*—July 31, 1866.—The sides of the funnel are graduated, its contents are discharged by a screw and a spigot closes its exit tube.

Claim.—A funnel provided with the screw D arranged to operate substantially as and for the purpose set forth.

No. 56,711.—J. W. CLARKE, Kingston, Wis.—*Sulphur Duster.*—July 31, 1866.—The sulphur is fed from a hopper carried forward by a roller, and driven by a fan blower through the foraminous nozzle upon the plants whose vermin are to be exterminated.

Claim.—First, the fan B mounted in a suitable case and arranged to operate in combination with the spout F, hopper E, and feed wheel a, or their equivalents, substantially as shown and described.

Second, in combination with the nozzle H, with its perforated cover and the valve s, arranged and operating as set forth.

Third, the auxiliary tube m, arranged to operate in connection with the spout F, as set forth.

No. 56,712.—JOHN WEBSTER COCHRAN, New York, N. Y.—*Packing Projectiles for Rifled Ordnance.*—July 31, 1866.—The rear end of an elongated projectile being formed with grooves or recesses, is surrounded with a thin copper band fitted or compressed into a neck or annular groove around the projectile, which annular groove is afterwards filled up with fibrous packing and covered with a coil of wire.

Claim.—First, the band b, saturated fibrous material f, and coiled wire d', in combination with each other and with the circumferential and longitudinal grooves in the projectile, substantially as and for the purpose herein specified.

Second, the grooves c c, arranged with reference to the grooves a a and m, for the reception of depressions of the expanding band, as and for the purpose herein set forth.

No. 56,713.—ELISHA T. COLBURN, Boston, Mass.—*Paddle Wheel.*—July 34, 1866.—Grooved wheels support and secure the eccentric guide whose arms are connected to cranks on the buckets to feather them.

Claim.—The improved arrangement of the guide wheel c, and the bearing wheels d d, the latter under such arrangement having the wheel e between them, as set forth.

No. 56,714.—L. T. CONANT, New Lisbon, Ohio.—*Cloth Guide and Binder Gauge for Sewing Machines.*—July 31, 1866; antedated July 25, 1866.—Explained by the claims and cuts.

Claim.—First, the base plate A, with its gauging lips C C, slot D, binder slot F, upright post

F, and screw G in combination with the arms M M, binders J, and spring N, as and for the purposes specified.

Second, the adjustable inclined arms M M, in their combination with the base plate A, upright post F, regulating spring N, and binders J, as and for the purposes specified.

Third, the seamless clamping binders J, or an equivalent, with its regulating nut I, and separating block K, in combination with the base plate A, and inclined arms M M, as and for the purposes specified.

Fourth, the regulating spring N, or its equivalent, in combination with the inclined arm M M and binder J, all operating as and for the purposes specified.

No. 56,715.—THOMAS P. CONARD, West Grove, Penn.—*Portable Shaving Case.*—July 31, 1866.—The lamp and cup are hinged to the box which contains the toilet articles.

Claim.—A case or box constructed to receive the various implements or appurtenances necessary or desirable in shaving, together with a heating apparatus, substantially as described.

No. 56,716.—WILLIAM F. CONVERSE, Harrison, Ohio.—*Car Spring.*—July 31, 1866.—Explained by the claim and cut.

Claim.—The combination of the concave heads F F', double-faced collet A, annular elastic disks B B', and connecting bolt C, all constructed and arranged to operate as and for the purposes specified.

No. 56,717.—E. P. COOLEY, New York, N. Y.—*Broom.*—July 31, 1866.—The butt ends of the corn are scarfed, tied with cords, bent and fastened in a metallic head around a wedge-shaped handle.

Claim.—The combination of the cords A, having knotted ends B, the conical cap C, and pointed handle F, with the stalks arranged and operating substantially in the manner and for the purpose herein represented and described.

No. 56,718.—FERNANDO E. COOMES, Berlin, Wis.—*Combined Cradle and Chair.*—July 31, 1866.—The upper portion of the back is shifted to the front, the arms with attached portions of the seat are moved out to form ends for the cradle, and a piece is placed in the hiatus of the seat.

Claim.—The extension bottom A A, as used in combination with the part B, as arranged in connection with C and the holes F F, substantially as described and for the purposes specified.

No. 56,719.—GEORGE E. CORBIN and JOHN W. PUGH, Grand Rapids, Mich.—*Water Wheel.*—July 31, 1866.—The water is introduced in a whirl above, passes between inclined chutes to buckets inclined in an opposite direction and is discharged below.

Claim.—The combination of the buckets b, and wheel D, cylinder E, with chutes e, and tube F, all constructed as described, and winged spout G, arranged and operating in the manner and for the purpose herein specified.

No. 56,720.—JOHN J. COWELL, Chicago, Ill.—*Trunk Brace and Hinge.*—July 31, 1866.—The leaves of the hinge are fitted to the trunk and lid respectively, and the pivoted brace holds up the lid when open, or lies within a slot when closed.

Claim.—The hinge composed of the parts A and B, constructed substantially as specified, when used in combination with the bar C, the parts operating as and for the purpose set forth.

No. 56,721.—J. B. CROSBY, Boston, Mass.—*Raisin Seeder.*—July 31, 1866.—The raisins are impaled upon closely-set wires on the cylinder by revolution against a rubber cylinder: a scraper removes the seeds from the surface and the pulp is subsequently drawn off the wires and collected.

Claim.—The employment of closely-set wires in combination with a bed or presser, for the purpose of forcing out of raisins or similar dried fruit the seeds or stones thereof, by the impalement of the pulp of the fruit on the wires as specified.

Also, in combination with the above, of seed remover or a pulp remover, or both, arranged to operate substantially as set forth.

No 56,722.—JOHN G. CROSS, Brattleboro', Vt.—*Milk and Cheese Rack.*—July 31, 1866.—The sections on the rack revolve on a central post, are enclosed by a screen, and have ventilators above and below.

Claim.—The revolving rack in sections, and the manner of enclosing rack in screen, with ventilators at top and bottom.

No. 56,723.—P. A. DAILEY, New York, N. Y.—*Mirror.*—July 31, 1866.—The circular glass and its wooden back are secured by a flexible metallic rim with threaded ends which are screwed into the handle.

Claim.—The combination of the handle A with the metal frame C, glass and back, when constructed as and for the purposes and substantially as described.

No. 56,724.—JAMES N. DAVIS, Cincinnati, Ohio.—*Water Closet.*—July 31, 1866.—The hinged seat has the usual opening, and in raising brings down the foot board which becomes its support; side and curved front pieces enclose the space and the odor.

Claim.—The vibratory seat A, having a curved front piece C, and side piece c c, cover or screen D, having arms E E hung to the arms F F, with friction slides f f, and the swinging platform H, having levers J J, or their equivalents, when arranged together so as to operate substantially in the manner and for the purpose described.

No. 56,725—JOHN DAVIES, Baltimore, Md.—*Furnace for Smelting Copper.*—July 31, 1866.—A metallic shield is interposed between the descending flue and the part of the furnace nearest thereto.

Claim.—In furnaces for smelting copper, interposing between the hearth or interior of the furnace and the descending flue that leads into the tunnel, a metallic or other equivalent stopper or plug, to prevent the molten metal, should it break through at that point, from running into and choking up the tunnel, as described.

No. 56,726.—NICHOLAS L. DAVIS and ROBERT O. HEWITT, Rutland, Vt.—*Railway Chair.*—July 31, 1866.—Besides the flanged chairs upon the tie, long break-joint clip-pieces embrace the lower flanges of the rails and are united by bolts beneath the rails.

Claim.—The method, herein described, of joining rails and holding the same on to the ties by the employment, in combination with chairs C, or spikes, or their equivalent, of either of side plates P, bolted or clamped together, substantially as herein shown and described.

No. 56,727.—H. G. DAYTON, Maysville, Ky.—*Hot Air Furnace.*—July 31, 1866.—A series of concentric annular air flues are placed above the combustion chamber and communicate with the external atmosphere through pipes extending into each flue near its bottom. The external jacket of the furnace is filled with coal ashes. A water pan above the fire-pot is fed from an external communicating vessel.

Claim.—First, the concentric series of hot-air flues a b c, in combination with the combustion chamber e and jacket K, all constructed and operating substantially as and for the purpose described.

Second, the air-supply pipes f g h, in combination with the hot-air flues a b c, and combustion chamber e, constructed and operating substantially as and for the purpose set forth.

Third, the water vessel M, in combination with the concentric flues a b c, combustion chamber e, jacket K, and tank N, all constructed and operating substantially as and for the purpose described.

No. 56,728.—GEORGE DECKMAN, Malvern, Ohio.—*Churn Dasher.*—July 31, 1866.—Additional motion is given to the cream by combining double concave and concavo-convex perforated dashers.

Claim.—An improved churn dasher, formed by the combination of the double concave disk B and the concavo-convex disk C with each other, and with the handle A, the whole being constructed and arranged substantially as herein described and for the purpose set forth.

No. 56,729.—AUGUSTUS DESTOUY, New York, N. Y.—*Sewing Machine for Sewing Boots and Shoes.*—July 31, 1866.—The table which supports the last holder is upon a weighted lever and is arranged to be always kept horizontal, though rising and falling as the work demands. A rubber-covered roller feeds the last-holder, which rests upon it and the table by gravity only, being free to turn in any direction; the awl, also, when in the material has a lateral movement to assist in feeding. Both the standards supporting the last are adjustable so as to admit of raising or lowering or turning the last to any desired position to present it to the stitching devices. The auxiliary needle slides on the side face of the hook-needle, opening and closing its barb, and entering the perforation with it.

Claim.—First, the self-adjusting table or platform for the support of the material to be sewed, the same being arranged to exert a yielding pressure against a sewing gauge, substantially in the manner and for the purposes herein set forth.

Second, the combination of a wheel feed in the adjustable platform or table, with an awl feed, the two operating conjointly, in the manner and for the purposes set forth.

Third, in combination with a double feed, as described, a dog to guide the work, in the manner and for the purposes set forth.

Fourth, the employment in a sewing machine, such as described, of adjustable standards to support the last, in the manner and for the purposes set forth.

Fifth, in combination with the herein described machine for sewing boots and shoes, a reservoir to contain wax, or other suitable substance, together with a heater, substantially as and for the purposes set forth.

Sixth, the thread carrier, revolving intermittingly in one direction only, in combination with the hook, operating substantially as herein described and for the purposes set forth.

Seventh, in combination with the hook and awl, the auxiliary needle, when constructed and arranged for operation as herein shown and described.

No. 56,730.—ALBERT L. DEWEY, Westfield, Mass.—*Feed Motion for Sewing Machines.*—July 31, 1866.—The coiled spring is free at one end, but its other end is secured by a screw to a hub or sleeve which is formed in one piece with the feed wheel. The shaft fits snugly the bore of the spring, and when turned in one direction its frictional contact causes the spring to bite and tighten upon it; its motion in the opposite direction frees it from this biting action, and allows it to turn without carrying the hub and feed wheel with it. A friction spring presses on the feed wheel to prevent it from turning during the return movement of the rocker arm.

Claim.—The spring E and hub D, applied to shaft A, substantially as shown, and all arranged to operate in the manner and for the purpose set forth.

No. 56,731.—J. F. DODGE, Newark, N. J.—*Wrench.*—July 31, 1866.—The backs of the jaws are shaped to grasp hexagonal nuts.

Claim.—The double-jawed wrench, with one pair of the jaws cut away to fit hexagonal nuts, and otherwise constructed substantially as described.

No. 56,732.—LEVI DODGE, Waterford, N. Y.—*Apparatus for Bleaching Paper Pulp and Drying Paper.*—July 31, 1866.—The rotary boiler in which the paper pulp is bleached has a smooth exterior surface in order that it may be used at the same time for drying paper which is passed over it.

Claim.—First, the method, substantially as herein described, of bleaching the straw or other paper stuff in a revolving steam cylinder, and of drying the made paper, whereby these two operations are effected simultaneously in one and the same apparatus —the steam used to dry the paper on the cylinder serving at the same time to bleach the material in the cylinder, as set forth.

Second, the process herein described of drying paper in sheets, on a drying cylinder, in one revolution thereof; that is, by so regulating the velocity of the revolutions of the drying cylinder, with respect to its diameter, and the thickness of the paper operated on, that the paper being carried around the cylinder once may be dry and ready to be removed.

Third, the revolving bleaching boiler, when the same is constructed with a smooth cylindrical surface, and one or more manholes in the sides or caps of the boiler for the introduction into and removal from the boiler, of straw or other paper material, as set forth.

Fourth, in combination with the said cylinder, or boiler, the use of an endless apron, or band, and doffers for operation as a drying cylinder, substantially as set forth.

No. 56,733.—AUGUSTUS L. DRAKE, Richmond, Me.—*Washing Machine.*—July 31, 1866.—The rubber is reciprocated upon the slotted bed by means of the revolving gearing and connecting parts, and its pressure is graduated by the suspensory devices.

Claim.—First, the operating of the reciprocating rubber by means of the gearing, pitman, lever, and arm, the latter being connected to or applied to the rubber by a pivot and upright guide, arranged as shown, so that the rubber may work in a plane, or with a rocking motion, as set forth.

Second, the drum N and cord L in combination with the reciprocating rubber I, spring M, and cord L', substantially as and for the purpose specified.

Third, the crank j, in combination with the toothed plate O for retaining the drum N in position, as described.

No. 56,734.—A. A. DUNK, Manchester, N. H.—*Repairing Files.*—July 31, 1866.—The file teeth are covered by means of a roller with a water and acid proof varnish, composed of asphalt, black pitch, burgundy pitch, rosin, and beeswax, dissolved in spirits of turpentine.

Claim.—The process of sharpening and renewing files, substantially as above described, by covering the tops of the teeth only with a protecting coating, and then immersing the file in acid until the intervals are sufficiently deepened.

No. 56,735.—A. ELLIS and O. ALBERTSON, Salem, Ind.—*Animal Trap.*—July 31, 1866.—By the arrangement of connecting devices, doors at either end of the trap are closed when the animal steps upon the platform, and are reopened (or set) as the animal passes through a dividing trap door into an inner chamber.

Claim.—The arrangement of the connecting rods C C, platform G with its spring H, shouldered spring-arm I, trap door O, and arm P, with the boxes A and M, operating in combination with the swinging doors B, all constructed substantially in the manner as and for the purpose herein specified.

No. 56,736.—W. A. ELLIS, Ashtabula, Ohio.—*Machine for Forming Plough Handles.*—July 31, 1866.—The grooves in which the wrists of the carriage traverse are modified by shifting the curved piece which curves the track in one position, and by filling the curve makes it straight when placed in the other position.

Claim.—The groove a' made adjustable by the removable piece I, in combination with the sliding frame B and revolving cutter, arranged and operating substantially as described.

No. 56,737.—ANDREW A. EVANS, Boston, Mass., assignor to JAMES A. WOODBURY, same place.—*Paper, Cuff or Wristband.*—July 31, 1866 —Explained by the claims and cut.
Claim.—As a new article of manufacture, a wristband or cuff made of long fibre paper, substantially such as is above described.
Also, making said wristband or cuff reversible, substantially as and for the purpose described.

No. 56,738 —JOHN K. FERGUSON, Portland, Ky.—*Steam Condenser.*—July 31, 1866.—The exhaust steam is discharged into the lower part of a horizontal cylinder, and condensed by a shower of cold water from perforated pipes in the upper part. The water of condensation flows into a chamfered box from which it is taken by force pumps ; check valves prevent the return of the water toward the condenser.
Claim —First, the cylinder B, provided with a series of perforated pipes C C, and valves D D', and used with the exhaust pipe A, and cold water pipe at a, substantially as and for the purpose specified.
Second, the box E', provided with a series of chambers and valves as described, when used with the cylinder B and force pumps M and L, substantially in the manner and for the purpose set forth.

No. 56,739.—HENRY FISHER, Canton, Ohio.—*Horse Hay Fork.*—July 31, 1866.—The lower end of the shorter bar is slotted and the end of the other passes through it ; the upper end of the shorter bar is looped around the longer bar and catches in notches therein ; a lever attached to the longer bar throws the loop out from the notch, when the shorter bar drops and the load is discharged.
Claim.—The bars A and B, with crooked points, being provided with the lever C, when arranged and used substantially as and for the purpose herein set forth.

No. 56,740.—JOHN FLINN, Philadelphia, Penn.—*Bed Bottom.*—July 31, 1866.—The bottom portion of the wire spring is bent and coiled so as to embrace the slat upon which it is imposed.
Claim.—In combination with a bed bottom, spira. spring B C, extending and bending the wire of the bottom coil b' of the same, so as to produce the spring clamps b^1 b^2 c' c' c^2, substantially in the manner described and set forth for the purpose specified.

No. 56,741.—A. FRAZEE and L. W. SMITH, Canandaigua, N. Y.—*Spring Bed Bottom.*—July 31, 1866.—Hooks upon the end slats of the bed bottom and the rail respectively are united by elastic bands.
Claim.—The combination and arrangement of the cross-bars C C, cleats E E, elastic bands G G, and loop hooks D D and H H, substantially as and for the purpose herein specified.

No. 56,742.—ISAAC H. GARRETSON, Richland, Iowa.—*Making Brick.*—July 31, 1866.—Dry clay is fed into the moulds in small quantities and beaten and tamped compactly until it becomes one homogeneous solid mass.
Claim.—Making brick, tile, and similar articles, by the tamping process—that is to say, by feeding in the clay or other material in small quantities, and tamping or beating each small quantity thus fed in before any more material is added, as herein set forth.
Also, the mechanism, constructed and operating substantially as herein described, for the purpose of making brick and similar articles.

No. 56,743.—ALEXANDER GORDON, Rochester, N. Y.—*Straw Cutter.*—July 31, 1866.—The downward pressure of a wooden spring exerts an upward pressure on the feed roller through the interposition of a pivoted yoke.
Claim.—The relative arrangement of the spring s with the yoke y, the latter being pivoted to the frame of the machine by the pivots g, which are located at a point intermediately between the bar c and the pivots f, to which the supporting bars r are hinged, for the purpose set forth, the parts acting conjointly, in combination with the upward cut.

No. 56,744.—WILLIAM GOWEN, Warsaw, Wis.—*Washing Machine.*—July 31, 1866.—The vertical shaft of the rubber is supported in a cross-bar whose spring bolts lock into keepers on the inside of the tub.
Claim.—The washing machine, constructed as herein described, with cross-bar D, sliding rods e e, sockets d, springs f, shaft C, and rubbers B b, all combined and arranged to operate substantially as and for the purpose set forth.

No. 56,745.—JAMES GRIBBEN, Allegheny, Penn.—*Dies for making Square Bolt Heads.*—July 31, 1866.—Two opposite corners of the die are enlarged to allow the metal to swell out under the action of the heading punch, which is of the same shape. After the first operation the blank is turned a quarter of a revolution, when the side dies operate upon the swelled portion, giving more clearly defined corners to the head.

Claim.—The use of dies for making square-head bolts, the heading cavity of which is enlarged, at two opposite corners, beyond the dimensions of the bolt-head to be formed, while the two remaining corners are of the required shape and dimensions, or distance apart, in combination with a heading tool or plunger, so shaped as to fit closely into or against those last-named corners, while the remaining corners of the heading tool or plunger are enlarged substantially as and for the purposes hereinbefore set forth.

No. 56,746.—JOHN R. GROUT, Detroit, Mich.—*Cupola and Blast Furnace.*—July 31, 1866.—Instead of fire-brick lining are cast iron boshes with flanges for connecting and fastening in place, and made hollow to permit a flow of water through them to preserve their integrity.

Claim.—First, constructing the boshes B of a cupola or blast furnace with metallic chambers *g*, so arranged that a current of cold water may flow through them, and without an internal lining of fire-brick, or other refractory substance, substantially in the manner and for the purposes set forth.

Second, so arranging two or more chambers *g*, in combination with the flanges *b* and *d*, and plate *c*, that the chambers surrounding the boshes may be removed without disturbing the superior brick work C, substantially as set forth.

No. 56,747.—C. L. HART, Mattoon, Ill.—*Sorghum Stripper and Cutter.*—July 31, 1866.—The orifices in the plate are hexagonal outwardly and circular at the side next the spring-stripping jaws which embrace the stalk and remove the leaves.

Claim.—First, the use of the plate A in a sorghum stripper when perforated, substantially in the manner herein described and for the purpose set forth.

Second, the frame or covering B and spring bars C in combination with each other and with the perforated plate A, substantially as described and for the purpose set forth.

Third, the stripping tubes D, constructed and arranged as herein described, in combination with the spring bars C and with the perforated plate A, substantially as described and for the purposes set forth.

No. 56,748.—DAVID GREENE HASKINS, Cambridge, Mass.—*Steam Generator.*—July 31, 1866.—By means of the internal or external series of perforated pipes, jets of gas or inflammable material are applied to the surface of the steam generator.

Claim.—First, the combination of a series of perforated pipes with the exterior or heating surfaces of steam generators for the purpose of utilizing gases in the generation of steam, substantially as herein described.

Second, the combination of the boiler A with the tubes B B and series of pipes *a* and *b*, substantially as and for the purpose specified.

Third, the combination of the boiler A, the casing E, and interposed series of pipes *b*, substantially as and for the purpose specified.

No. 56,749.—MILTON H. HILBURN, Wilmington, Ill.—*Sickle Head for Harvesters*—July 31, 1866.—The sickle-bar head has conical points which enter corresponding sockets in the forked arms of the pitman.

Claim.—A sickle head to be used in mowers and reapers, constructed substantially as described, with the conical journals *d d* upon the lug *c*, in the manner and for the purpose specified.

No. 56,750.—EMIL HISS, Delaware, Ohio.—*Paint Brush.*—July 31, 1866.—The bristles are clamped around the butt of the handle by a draw band and tightening screw.

Claim.—A paint brush provided with an adjustable draw band C, substantially as and for the purpose specified.

No. 56,751.—JOSEPH F. HODGSON, Washington, D. C.—*Roofing.*—July 31, 1866.—The edges of the sheets are bent over angular blocks attached to the sheathing and are secured in dovetail grooves by fusible metal which forms inverted wedges and prevents the retraction of the edges.

Claim.—First, in the construction of metallic roofing, securing the edges of the sheets of metal in dovetail grooves by means of a fusible metal, substantially as described.

Second, the use of blocks B, having bevelled edges applied to the sheathing of the roof for the purpose of supporting the sheets of metal and forming dovetail grooves for receiving the edges of said sheets, and also the fusible metal, substantially as described.

No. 56,752.—EDWIN HOYT, Stamford, Conn.—*Tobacco Pipe.*—July 31, 1866.—The smoke passes through apertures in the diaphragm to a lower chamber, whence it enters openings in a sliding tube which leads to the stem; saliva passes down the said tube to a central pipe and removable collecting chamber.

Claim.—The sliding perforated tube E, in combination with the perforated diaphragm C, tube D, and bore *a*, substantially as and for the purpose specified.

No. 56.753.—FRANCIS M. HUBBARD, Ripon, Wis.—*Spring Seat for Carriages*—July 13, 1866.—The outer ends of the lever rest upon the edge of the bed and its inner ends are connected by elastic bands to the seat, which also rests by rollers upon the levers.

Claim.—A device for giving elasticity to the seats of vehicles by means of the levers D D, fulcrums C, and elastic bands E, combined and arranged substantially as and for the purpose set forth.

No. 56,754.—JOHN M. HUDSON, New York, N. Y.—*Construction and Rigging of Trestle-trees for Vessels.*—July 31, 1866.—An auxiliary trestle-tree supports the extended heel of the topmast, which is keyed thereto by a fid. The lower yard is slung from a bridle band, which passes over and around the mast head and is trussed to a band around the topmast.

Claim.—The placing below the upper trestle-trees A a new pair of trestle-trees B, on the lower mast X, with the projections D on the ends, and securing the trestle-trees B with the iron band U, and extending the topmast Y, so that the heel C with fid P, going through the heel C, will rest on trestle-tree B, and take against the projections D, instead of resting on the trestle-trees A, which now opens with the iron gate F, to facilitate sending the topmast Y up and down, substantially in the manner as herein described.

Also, the bridle band E, over the mast head, as herein described.

Also, the combination of the foregoing with the clew lines and spilling lines, for the purposes and objects herein described.

No. 56,755.—WILLIAM HUNT, New York, N. Y.—*Fruit Jar.*—July 31, 1866.—The glass cover of the earthen vessel allows the contents to be seen and is fastened upon the intervening packing material by flexible links attached to ears on the jar and by a flat transverse key over the cap.

Claim.—First, the within described preserve can composed of a body of pottery ware and a cover of vitreous material, fitted to each as shown, and adapted to withstand the temperature of filling and to exhibit the contents without opening the can, substantially as herein set forth.

Second, the ears *a a*, arranged on the neck or contracted portion of a preserve can, substantially in the manner and so as to form attachments for the links D, as herein set forth.

Third, the flexible links D, adapted to operate in connection with the turning key E, or its equivalent, as described, when said links are permanently attached so as not to be lost on unsealing the can, and are hinged so as to be turned down when out of use, substantially in the manner and for the purpose herein set forth.

Fourth, the flat turning key E, having portions cut away at *e*, in combination with a preserve can A, and cover B, and arranged to induce two different pressures upon the cover by turning upward one edge or the other of the key, as and for the purposes herein set forth.

No. 56,756.—EDWARD H. JACKSON, Boston, Mass.—*Dumping Car.*—July 31, 1866.—The weight of the car is transferred from the wheels to laterally projecting hubs which run upon a frame and allow the car to upset, discharge, and return.

Claim.—First, attaching to the sides of a dumping car a hub or projection D, as and for the purpose substantially as specified.

Second, the combination of a dumping car provided with a hub or projection D on either side, with a frame B, substantially as and for the purpose specified.

No. 56,757.—A. L. JEWELL, Waltham, Mass.—*Sprinkling Syringe for Gardens.*—At the delivery end of the syringe is a perforated disk which is loosely fitted within a guard frame and serves the purpose of a valve and rose.

Claim.—The combination and arrangement of the foraminous valve, its opening and seat with a syringe, the same being to operate substantially as described.

No. 56,758.—CHARLES JONES, Boston, Mass.—*Wool Oiling Machinery for Carding Machines, &c.*—July 31, 1866.—Explained by the claims and cut.

Claim.—First, in combination with carding or other kindred wool-preparing machinery, and arranged over the feed-apron of the same, a dripping oil tank having a traverse motion with respect to the line of feed of the wool, substantially as set forth.

Second, in combination with carding or other wool-preparing machinery, and arranged over the feed apron of the same, a dripping oil tank having both a traverse motion with respect to the line of feed of the wool and a rotary movement, substantially as set forth.

No. 56,759.—JONATHAN M. JONES, East Taunton, Mass., BARNERD SPAULDING, Port Richmond, N. Y., and SYLVESTER PARKINS, Providence, R. I.—*Manufacture of Sheet and Bar Iron.*—July 31, 1866.—The wrought iron is melted in a covered crucible with a composition consisting of nitrate of lead, muriate of antimony, bone dust, and plumbago. When the metal is perfectly fused it is stirred and the flux removed, after which the metal is run into moulds and made into sheets in the ordinary manner.

Claim.—The improved process for the manufacture of iron, substantially as herein described and for the purposes set forth.

No. 56,760.—ROBERT V. JONES, Canton, Ohio.—*Railroad Rail.*—July 31, 1866.—Explained by the claim and cut.

Claim.—The top rail provided with a tongue E, upon each side of which are V-shaped grooves, said tongue being made concave on its side below its centre, and decreasing in size or width from z z to a a, when used with the flanges D D, with bevelled edges and straight sides, substantially as and for the purpose herein specified.

No. 56,761.—WILLIAM and A. G. KELSEY, Delavan, Wis.—*Washing Machine.*—July 31, 1866.—The block which contains the concave bed of rollers is hinged and may be erected against the end of the tub.

Claim.—The combination of the hinged roller block a a and the swinging rubber c with the wash tub A A, for the purpose of converting it when desired into a rinsing tub, arranged and operated as herein specified.

No. 56,762.—SAMUEL M. KING, Lancaster, Penn.—*Bridle Bit.*—July 31, 1866.—Adapted to receive a round rein to draw the bit upward against the palate and prevent its being held by the teeth.

Claim.—The extended ends D constructed with grooved rollers E and round aperture J, as herein described and for the purposes set forth.

No. 56,763.—WILLIAM P. KIRKLAND, San Francisco, Cal.—*Bilge Water Gauge.*—July 31, 1866.—The float rises upon the bilge water and indicates by graduations above the height of the water. The float is confined in a box which has lower openings for water and apertures above for air.

Claim.—First, the aprons C applied to the perforated starboard and larboard sides of the box A, substantially as and for the purpose described.

Second, the disk f of glass or other suitable material in combination with the float B and index rod g, constructed and operating substantially as and for the purpose specified.

No. 56,764.—CHRISTOPHER F. KNAUER and WILLIAM WARWICK, Pittsburg, Penn.—*Sad Iron.*—July 31, 1866.—The sad-iron has a chilled face and an inclined handle which is more convenient to the grasp.

Claim.—The method of constructing sad-irons, substantially as herein specified and set forth.

No. 56,765.—EMILE LAMM, New Orleans, La.—*Preparing Gold for Dentists.*—July 31, 1866.—Eighteen carat gold is dissolved in nitro-muriatic acid and precipitated by boiling with sugar; it is then washed in ammonia and purified by heat.

Claim.—The use of saccharine substances to precipitate gold from its solution in the manner and by the process above described or by any substantially equivalent process, thereby forming a mass of crystal shreds, extremely useful and convenient for dental and other purposes.

No. 56, 766.—JOHN W. LARMORE, Harrison, Ohio.—*Field Fence.*—July 31, 1866.—The metallic post is socketed in a foundation stone and receives a running rider; the running wires are held by clips on the post.

Claim.—First, the metallic post A having the vertical series of flexible ears or clips D, to receive and secure the wires in the manner described.

Second, a field fence composed of the following elements, to wit: a metallic post A having a bottom tenon B, to enter a stone foot or base F, and a top tenon C, to enter a wooden rider G, and having a series of ears or clips D, for the reception of suitable wires E.

No. 56,767.—WORLEY LEAS, Kokomo, Ind.—*Belt Coupling.*—July 31, 1866.—The ends of the belt are connected by two metallic plates and a link; the latter has a central rib and a leather strip to form a continuous bearing surface.

Claim.—First, a belt coupling consisting of a link D and bent metallic plates E E connected together so as to form a joint at each side of the link, substantially in the manner specified.

Second, the rib or back D' formed on the link D, and employed to form a continuous bearing for the leather as described.

Third, in combination with a belt coupling constructed as herein described, the leather covering F F' and G to prevent slipping when the coupling is upon the pulley.

No. 56,768.—A. LINDSAY, Malone, N. Y.—*Quartz Crusher.*—July 31, 1866.—The revolving crusher wheels travel on a circular bed and the ground quartz is brushed on to a series of vibrating sieves which sort it; the coarser particles are elevated and returned to the crusher.

Claim.—First, the combination of the rollers E with the axles e, pins f, and upright shaft B, substantially as set forth and in the manner described.

Second, the combination of brushes or scrapers G with bars g and g', springs g*, friction rollers g'', rollers j' and cams j, substantially as shown and described.

Third, returning the coarse quartz to the crushers by means of sieves or separators *x*, trough *n'*, box M, endless apron N, hopper J, and pipe *t*, substantially as shown and described.

Fourth, the devices for raising and lowering the brushes or scrapers 3, consisting of cam *j*, rod *i''*, shell *i'*, and spiral spring *i*°, substantially as herein shown and described.

No. 56,769.—HENRY LOTH, Philadelphia, Penn.—*Table and Stool.*—July 31, 1866.—The three legs are mutually hinged and support the two semicircular leaves. When the tripod is folded together the leaves are doubled up and sustained by hinged connection to that member of the tripod nearest thereto.

Claim.—The described folding table or stool having its three legs C D E and top A B combined as shown, and relatively arranged to fold up in the order and for the purpose set forth.

No. 56,770.—C. E. LYON, Worcester Mass.—*Extinguisher and Regulator for Lamps.*—July 31, 1866.—By a cam motion the sleeve is raised and near the end of its vertical range the extinguisher is vibrated, capping the wick tube.

Claim.—First, a lamp burner provided with a combined regulator and extinguisher, substantially such as herein described as a new article of manufacture.

Second, the sleeve C adapted to act separately as a regulator and in combination with the cap E as an extinguisher.

Third, the cam shaft D and hook *c* with its shoulder *d* in combination with the sleeve C, wick tube B, and cap E, constructed and operating substantially as and for the purpose described.

No. 56,771.—ALVIN C. MASON, Springfield, Vt.—*Gearing for Churns.*—July 31, 1866.—The master-wheel has two circles of cogs, and its bearings are capable of being shifted to bring either set in contact with the pinion on the dasher shaft according to the speed required.

Claim—First, the gear-wheel B and H in combination with the pinion I, so that the motion of the beaters may be reduced or accelerated for the purpose and substantially as described.

Second, the plate D, or its equivalent, in combination with the wheels B and H and pinion I, substantially as herein set forth.

No. 56,772.—WILLIAM B. MASON, Boston, Mass.—*Operating Hand Punches, Shears, &c.*—July 31, 1866.—One lever forms the base on which the stationary die is fixed and on which the slide works as the rocking lever and the toggle link are vibrated.

Claim.—The above described machine for operating punches, dies, shears, &c, the combination and arrangement of the levers A and B with the link C, substantially as described.

No. 56,773.—ARIADNA B. MERCIER, Providence, R. I.—*Clothes Washer.*—July 31, 1866.—The water boils up through the central tube, flows through the clothes and the perforated plate, which is supported on feet above the bottom of the boiler.

Claim.—The combination of a perforated plate with a stopper in the manner set forth and for the purpose specified.

No. 56,774.—BENJAMIN F. MILLER, New York, N. Y.—*Stove Pipe Drum.*—July 31, 1866.—An air chamber of conical or double conical shape is suspended within a correspondingly shaped enlargement of the flue, by means of pipes which form air induction and eduction openings.

Claim.—The radiating drum *c d* and interior chamber *e* with the pipes *g h*, constructed substantially as and for the purposes specified.

No. 56,775.—WARREN P. MILLER, San Francisco, Cal.—*Attaching Burners to Lamps.*—July 31, 1866.—The shank of the burner slips within a socket in the base and is retained by a spring clamp which clasps grooves in the shank.

Claim.—The application of the grooved shank *c* as shown at *d*, the socket *b*, and spring *s*, or their equivalent, when made to operate substantially in the manner described.

No. 56,776.—J. A. MINOR, Middletown, Conn.—*Pocket Tablet.*—July 31, 1866; antedated July 19, 1866.—The cover is made of a piece of pliable material cut and folded so as to contain the leaves which rotate on a pin and the pencil which slips within its sheath.

Claim.—A case for a pocket calendar constructed with two elastic or yielding sides *a a*, one of which is provided with a pin *d*, to pass through and secure a series of cards B to the case, so that said cards may be turned within and out from the case and readily adjusted to and detached from the same, substantially as described.

Also, the rounded back *b*, when used with the yielding or elastic sides *a a* to serve as a socket for the pencil C.

Also, the blank A, as represented in Fig. 4, for the purpose specified.

No. 56,777.—DAVID A. MITCHELL, Chicago, Ill.—*Bridge.*—July 31, 1866.—The bridge has piers, cable and suspended track from which the draw section is suspended at a lower

level and the draw section is capable of being run within the framework of the adjacen
sections to open the draw for the passage of vessels.

Claim.—First, suspending the draw for bridges from trucks which run on a railway, supported on framework by suspension cables on towers, placed either above or below the framework, at a height sufficient to allow steamboats and other river craft to pass freely under the
structure that supports the draw as herein described.

Second, the collar braces F F, constructed and arranged as and for the purposes specified.

Third, so placing the stirrups, suspension rods, and angular braces, that the strain upon
the draw and the other portion of the bridge structure is equalized upon the suspension cables,
in the manner and for the purpose herein set forth.

No. 56,778.—THOMAS MITCHELL, Albany, N. Y.—*Steam Generator.*—July 31, 1866.—As
the volume of steam decreases in the dome the piston beneath the weighted beam descends
and turns the cock of the water-supply pipe, injecting water into the generator.

Claim.—The arrangement of the stem B′, and beam C C of Fig. 3, and cock B of Fig. 2,
substantially as and for the purpose set forth.

No. 56,779.—RICHARD MONTGOMERY, New York, N. Y.—*Corrugated Metallic Plate.*—
July 31, 1866.—One series of corrugations crosses the other at right angles thereto.

Claim.—The plate or plates of rolled wrought metal doubly corrugated, substantially as
described.

No. 56,780.—J. OWEN MOORE, Washingtonville, N. Y.—*Apparatus for Cooling Milk.*—
July 31, 1866.—The milk flows down a spiral track between the two metallic vessels, is cooled
in its passage by the contents of the outer and inner vessels, and is discharged.

Claim.—First, forming a spiral channel for the purpose set forth by inserting a coiled wire
a between the walls of the vessels B and C, substantially as shown and described.

Second, an apparatus for cooling milk or other liquids, formed by combining with each
other the vessels A B and C, pipes *e f* and *h*, trough *i*, strainer *k*, coiled wire *a* and pipe *g*,
substantially in the manner and for the purpose herein shown and described.

Third, constructing a cooling apparatus in such a manner that the cooling liquid may
overflow from the inner vessel C to the outer vessel A without coming in contact with the
milk contained in the intermediate vessel B, substantially as and for the purpose shown and
described.

Fourth, the combination of the annular trough *i* with the strainer K and vessels B and C,
substantially as described.

No. 56,781.—WILLIAM A. MORSE, Philadelphia, Penn.—*Polishing Box.*—July 31, 1866.—
The box contains a polishing powder which is dusted out through the lid and used beneath
the face of the pad.

Claim.—A polishing pad substantially as described in combination with a box having a
perforated lid for the purpose specified.

No. 56,782.—M. D. MULFORD, Jr., New Providence, Iowa.—*Beehive.*—July 31, 1866.—
The entrance opening flares inward and may be contracted by an additional mouth-piece with
coincident flaring sides. The general shape of the hive is explained in the claims and cut.

Claim.—First, the hive A, having its top and bottom inclined as shown, with its lower
walls made double and provided with movable frames C, having their top bar inclined as
set forth.

Second, the additional mouth-piece *c*, provided with the opening *n*, in combination with the
piece *a*, having the opening *m*, when said pieces are arranged in relation to each other and
to the hive as shown and described.

No. 56,783.—JOHN MUNN, Columbus, N. J.—*Portable Apparatus for Heating and Melting
Roofing Material.*—July 31, 1866.—The fire box consists of a cylinder which heats the air
in the chamber below the pan and discharges at the end flue.

Claim.—First, the pan B and fire box or cylinder C, so arranged relatively to each other as
to form an intervening air chamber *c* between them, whereby the air is heated and applied to
the pan instead of a direct flame as heretofore, substantially in the manner and for the purpose
as herein set forth.

Second, the sliding valves F F, in combination with the box A and air chamber *c*, for
regulating the degree of heat in its application to the pan by the admission of cold air,
substantially in the manner as described.

Third, the arrangement of the pipe D, the fire box C, and pan B, substantially in the
manner and for the purpose as described.

No. 56,784.—F. MUGATROYD, Cleveland, Ohio.—*Condensing Steam.*—July 31, 1866.—
The exhaust steam or the bilge water from the hold is drawn into the chamber and condensed
or removed by an induced current of water in the tube placed outside of the vessel in the
direction of the length of the latter and below the water line.

Claim.—First, the chamber D, funnel shaped pipe B'', and valve *e*, combined with the device for discharging bilge water, arranged in the manner and for the purpose set forth.

Second, the arrangement of the connecting rod or link *d*, cranks A and *b*, and valves *e a*, for the purpose of automatically exhausting into the chamber D, below the water line or outboard, according to the direction of the vessel, in manner and for the purpose described.

Third, the chamber D, funnel-shaped pipe B'', and valves *e a*, combined with a device for discharging bilge water, as and for the purpose set forth, below the water line.

No. 56,785.—NICHOLAS MURPHY, Washington, D. C.—*Buckle.*—July 31, 1866.—The tongue of the buckle is attached to the pivoted portion which shuts between the upturned edges of the frame piece; springs retain the tongue piece and strap in position.

Claim.—The combination and arrangement of the two pieces A and B, when the tongue *a* and pivots *b* are arranged as specified, substantially as and for the purposes set forth.

No. 56,786.—H. NAYLER, Pekin, Ill.—*Sash Fastening.*—July 31, 1866.—Spring books are attached to the window casing and catch against or into notches in the sashes to hold them closed or at the desired height.

Claim.—The combination of the catch C, with its head *b*, and spring *d*, arranged with the piece *a*, applied and operating substantially as specified.

No. 56,787.—CÆSAR NEUMANN, New York, N. Y.—*Hoop Skirt.*—July 31, 1866.—The hoops are associated in groups with wider intervals between the groups.

Claim.—The hoop skirt having its wires arranged in sections or clusters, each section comprising two or more wires placed near together, in separate pockets, substantially as described, as a new article of manufacture.

No. 56,788.—JOSEPH NEWMAN, Falmouth, Maine.— *Wood Bending Machine.*—July 31, 1866.—By the action of the reciprocating rack upon the strap attached thereto, the wood is bent upon the segment former.

Claim.—Operating the mould *c*, or form for bending, by means of the metallic bending scrap *f*, which is attached to the mould at one end and to the reciprocating rack *g*, at the other, all constructed to operate substantially as described.

No. 56,789.—WILLIAM H. NUTTING, Orange, Mass.—*Stove Pipe Damper.*—July 31, 1866.—This damper may be revolved in the pipe as usual, or when occupying a transverse position an opening may be made by sliding the valve to open certain slots in its face.

Claim.—The combination and arrangement of the series of starts *d d d* and the series of notches *g g g g* with the damper and register slide, and the swell D, and the journal C, applied to the damper, as set forth.

No. 56,790.—ALFRED A. OAT, Philadelphia, Penn.—*Safe Lock.*—July 31, 1866.—Cannot be briefly described other than substantially in the words of the claims.

Claim.—First, the interlocking spring slides 2, and 5 7, and 11 13, and 15, 16 and 18, constructed and arranged in relation to each other, and to the sliding blocks K K K K, which are respectively connected to the sliding spring stops F F F F, substantially in the manner described for the purpose of operating the said stops, and thus fixing and releasing the said main bolts E E E E of the lock as described.

Second, securing the flush plug 2', in the plate B, of the lock, by means of the interlocking spring slides, 1 3 4 6 8 9 10 12 14 17, when the same are arranged in relation to each other, and to the interlocking spring slides 2 and 5 7 and 11 13 and 15 and 16 and 18, substantially as described and set forth.

Third, retaining or fastening the plug 1', in the plate C, by means of the rack bolt J, operated by means of the pinion *j*⁴, spring *j*⁵, and spring slide 19, substantially as described and set forth.

Fourth, securing, releasing, and supporting in its retracted position, while holding back the four main bolts E, as described, the retracting plug O, by means of the two spring bolts J' J', cam *j*³, and spring slide 20, the same being constructed and arranged to operate together substantially as described.

No. 56,791.—ABRAHAM W. OVERBAUGH, New York, N. Y.—*Initial Studs.*—July 31, 1866.—The separate initial plate is fastened into the rim by a plate attached to the shank of the button.

Claim.—The application of the changeable initial plate in the manufacture of buttons, pins, ear-rings, and other jewelry or ornaments generally, as herein above described.

No. 56,792.—GEORGE T. PALMER, Brooklyn, N. Y.—*Water Cooler.*—July 31, 1866 : antedated July 23, 1866.—A water reservoir in the upper portion, an ice box beneath, a cooling chamber between; pipes lead from the reservoir through the ice chamber to the discharge faucet.

Claim.—The reservoir *c*, cooler *d*, and pipes *g g*, combined and arranged substantially as and for the purpose shown and described.

No. 56,793. HENRY PEARCE, San Francisco, Cal.—*Quartz Mill.*—July 31, 1866.—The revolving frustum is journalled eccentrically with the axis of the concave shell and has horizontal flanges extending below the latter.

Claim.—The construction of a conically-shaped crushing mill with an eccentric motion as herein described, for the purposes and in the manner substantially as set forth.

No. 56,794.—SAMUEL M. PERRY, Plainfield, N. Y.—*Corset.*—July 31, 1866; antedated July 20, 1866.—Explained by the claim and cut.

Claim.—A corset having one or more jointed clasp plates, so constructed essentially as herein specified, that the top ends of said clasp plates may swing outward and downward when the tops of said clasp plates are unloosed and afford a ready access to the parts of the wearer's person thereby exposed, while the bottom ends of said clasp plates serve to clasp the corset sufficiently when the top is unloosed.

No. 56,795.—J. W. PETTY, New Orleans, La.—*Tobacco Pipe.*—July 31, 1866.—The pipe is divided longitudinally and the sections are clasped together in a light, metallic, removable framework.

Claim.—The combination of the sections of the bowl and stem, A d, and of the mouthpiece e, with the framework a b c f i, when the several parts are constructed and united as described, for the purpose set forth.

No. 56,796.—WM. PHELAN, Peoria, Ill.—*Steam Generator.*—July 31, 1866.—The firechamber flues, rear chamber and reverting flues are attached together and are removable from the shell which forms the exterior wall of the boiler.

Claim.—In combination with an outer shell or jacket, the removable arrangement of fire-box, double set of flues, and flue chamber O, so attached to the outer shell as to permit the space between them to be used as a water and steam chamber I.

Also, in combination with the said shell surrounding the said removable arrangement, steam chambers A B, on the sides of the boiler, substantially as described.

No. 56,797.—E. F. PRENTISS, Philadelphia, Penn., and C. C. PARSONS, Boston. Mass.—*Apparatus for Purifying and Deodorizing Whiskey, &c.*—July 31, 1866.—A chamber above the still is charged with a neutral porous material upon which an alkaline or other purifying solution is distributed by a rose at the end of the trap pipe ; the vapor passes up through the saturated material, and the spent solution is discharged by a pipe.

Claim.—First, the distributor H, constructed and arranged in the manner and for the purpose substantially as shown and described.

Second, the shield F, constructed and arranged in the manner and for the purpose substantially as shown and described.

Third, the pipe b, arranged and operating in the manner and for the purpose substantially as shown and described.

Fourth, the trap tube a, provided with a distributor H, and the case C, containing neutral materials, in combination with the shield F, and the pipe b, or their equivalents respectively, substantially as described, the whole to be used in connection with a still.

No. 56,798.—BENJAMIN PRICE, Leesville, Ohio.—*Plough.*—July 31, 1866.—The handles and beam are shiftable and either the right or the left-hand plough may be brought into operation in hillside ploughing.

Claim.—The jointed beam A A' attached to a front and rear mould-board, and points or hill-side plough, constructed and operating substantially in the manner and for the purposes set forth.

No. 56,799.—CARL RECHT, New York, N. Y.—*Ventilator and Shade for Lamps.*—July 31, 1866.—The ventilator tube above the shade conducts off the heated air, &c., to an exterior exit; a reverse current is arrested by a valve.

Claim.—First, the valve as arranged and described substantially.

Second, the combination of the valve and shade with the ventilating tube as constructed.

No. 56,800.—HENRY C. RICHARDS, Cincinnati, Ohio.—*Protector for the Corners of Stairs and Rooms.*—July 31, 1866.—The corner piece prevents the accumulation of dust and facilitates cleansing.

Claim.—As an article of manufacture, the corner protector, constructed of metal, wood, or other suitable material, as and for the purposes herein described.

No. 56,801.—WILLIAM ROEMER, Newark, N. J.—*Travelling Bag.*—July 31, 1866.—The staple is secured to the covering plate of one jaw and its ends embrace the shut jaws to relieve the strain on the lock.

Claim.—A frame for travelling bags, having staples J, and strap E, adjusted on the top thereof, relieving the lock from strain as described, constructed, combined, and arranged as herein specified.

No. 56,802.—EDWARD ROWSE, Augusta, Me.—*Steering Apparatus.*—July 31, 1866.—
The described apparatus keeps the ropes taut at all parts of the stroke of the tiller.
Claim.—The arrangement of the rib *c*, ropes *f f*, sheaves *i i*, tiller *b*, pulleys *k*, and rudder head B, with its pivoted stud *d*, operating in combination with the windlass E, in the manner and for the purposes herein specified.

No. 56,803.—JULIUS SCHLEISINGER, New York, N. Y.—*Hoop Skirt.*—July 31, 1866—The hoops turn up in front and are attached to vertical strips, to which are likewise attached the hoops of a small inner skirt.
Claim.—First, the combination of the adapter E with the hoops B B' and strips *a*, constructed and operating substantially as and for the purpose described.
Second, turning the ends of the hoops B' up, and securing them to the strips *a*, substantially as and for the purpose set forth.

No. 56,804.—P. F. SCHNEIDER, Hartford, Conn.—*Revolving Cartridge Box.*—July 31, 1866.—The pouch has a series of cartridge-containing tubes divided into sections transversely revolving upon a common axis. The cartridges are consecutively discharged at a lower aperture by an intermittent action, the succeeding cartridge in the same tube being retained by a fixed bridge.
Claim.—First, the stationary bridge W, in combination with the casing D and sections F E, arranged relatively with the discharge opening V, applied and operating substantially as and for the purpose represented and described.
Second, in combination with the shaft I, and sectional tube cylinder C. the ratchet wheel M, spring pawls N O, and lever Q, constructed and operating substantially as described for the purpose specified.

No. 56,805.—M. SCHWALBACH, Milwaukee, Wis.—*Sewing Machine.*—July 31, 1866 —Relates to the take-up of the needle thread, the feed motion and its adjustment, and the arrangement and action of the shuttle carrier and its connection with the feed device. The plate on the shuttle carrier serves to lift and carry back the feed dog, and also to hold the carrier and shuttle close up to the face of the fixed plate or curtain.
Claim.—First, the combination of the take-up rod *d* with the needle bar, when it is rigidly fastened to a rotating pin or pivot placed in the top of said bar and loosely fitted to slide in a rotating pin or pivot placed on a standard *r*, substantially as and for the purpose above described.
Second, the elbow feeding lever N, carrying an adjustable feed-propelling screw at its upper end and having curved branches O P on its lower end, between which the crank pin of the shuttle carrier vibrates, substantially as set forth.
Third, the combination of the shuttle carrier, the feeding plate E, and the elbow lever N, the whole operating in conjunction substantially as described.
Fourth, the plate V, constructed substantially as above described, for holding up to the curtain U the shuttle carrier for holding the feed plate R in proper position, and for moving the feed plate backward when it is in its lowest position, substantially as described.

No. 56,806 —ELIPHALET S. SCRIPTURE, Williamsburg, N. Y.—*Shoe String Fastener.*—July 31, 1866.—The surplus end of the string is wound around a spring plate riveted to the shoe.
Claim.—A shoe-string fastener, composed of a tilting spring button A, corrugated spring washer B, and rivet C, substantially as and for the purpose set forth.

No. 56,807.—JOHN F. SHEARMAN, Brooklyn, N. Y.—*Compressing, Condensing and Extending Metals.*—July 31, 1866.—The metal is immersed in water in a tight vessel and percussion applied to the column of liquid in the tube.
Claim.—First, the percusso-condensation or extension, or both together, of solid or hollow bodies, commonly called hammering, when affected by the intervention of practically non-elastic fluids or liquids between the hammer and the body to be operated on, substantially as set forth.
Second, the operation of the hydraulic hammer, whether it be applied to change or not to change the shape of the article to be treated.

No. 56,808.—GEORGE A. SHERLOCK, Boston, Mass.—*Lock.*—July 31, 1866.—An auxiliary lock is applied to one or both sides of the door to lock the spindle when the latch is projected.
Claim.—The application of each or either of the locks G H to the knob spindle of the spring bolt B, so as when the bolt *g* of the lock is thrown forward it shall lock the spindle, or prevent it from being revolved by force applied to either of the knobs.
Also, the combination of the spring bolt and its spindle with two isolated locks arranged on opposite sides of the door, and constructed in the manner and so as to operate with the spindle, substantially as specified.

No. 56,809.—ABRAHAM and DAVID SHORT, West Liberty, Ohio.—*Lounge.*—July 31, 1866.—The back is lowered and sustained in a horizontal position, the rocker being supported by props.

Claim.—The combination of the leg C, rod D, and pawl rod E with the back A' for adjusting the position of the latter, substantially as shown and described.

No. 56,810.—THOMAS R. SINCLAIRE, New York, N. Y.—*Car Starter and Brake.*—July 31, 1866.—Improvement on his patent of December 19, 1865, and March 27, 1866.· The clamps and clutches on the longitudinal shaft limit the motion of the power-accumulating spring, and prevent breakage.

Claim.—First, the employment or use of the collars Q Q, one or more placed on the shaft F, in combination with the nut P and screw *g* on the shaft F, substantially as and for the purpose set forth.

Second, the clamps R R, with or without the teeth *i j j'*, one or more in combination with the collars Q Q and nut P, all placed on shaft F, and arranged substantially as and for the purpose specified.

Third, the arched bars *b b*, in combination with the frame E, shoe or brake levers S, and arms *l*, all arranged in the manner as and for the purpose specified.

Fourth, the pivot *e*, in combination with the levers J and arched bars *b*, substantially as and for the purposes stated.

Fifth, the shoe levers S, applied to the frame E, as shown and provided with springs Z and stops or pins *n*, substantially as and for the purpose set forth.

Sixth, the eccentrics A*, applied to the springs Z for the purpose of graduating their pressure, as described.

Seventh, the operating of the shoe levers S from the shafts W, by means of the chains V, rods U X, chains Y, and pulleys *m m*, all arranged to operate substantially in the manner as and for the purpose specified.

Eighth, an elastic lever J', in combination with the lever J, substantially as and for the purpose set forth.

Ninth, in combination with the levers J J', the draught hooks O, constructed in the form of a fork, or branched to admit of the levers J', passing through them as described.

No. 56,811.—EARLE A. SMITH and DWIGHT L. SMITH, Waterbury, Conn.—*Buckle.*—July 31, 1866.—The buckle is made in two parts ; the pinching bar of the lever is under the frame.

Claim.—The combination of the bar *e* of the lever with the bar B of the frame, whether the eye parts of the hinges are on the lever part, (as in Figs. 1, 2 and 4,) or on the frame part (as in Figs. 5, 6 and 7,) when the bar *e* is made to pinch the running part of the strap between itself and the edge, or reverse, or under corner of the central bar B of the frame, and the buckle is constructed and fitted for use substantially as herein described and set forth.

No. 56,812.—GEORGE SMITH, Brooklyn, N. Y.—*Coal Scuttle.*—July 31, 1866.—The bottom, hopper, and base are united by screws through their flanges.

Claim.—A coal scuttle composed of three removable parts, when constructed and arranged substantially as and for the purposes herein described.

No. 56,813.—JOHN A. SMITH, Waupacca, Wis.—*Weather Strips for Doors.*—July 31, 1866.—The leaf is attached by an elastic strip to the cleat on the door. The hinging portion admits of the depression of the leaf and by its resiliency acts as a spring to raise it when the door is open.

Claim.—The threshold weather strip made of the two parts *a* and *b*, with edges convex and concave, as described, when united by hinges operating also as springs, substantially as specified.

No. 56,814.—REES B. SMITH, Mount Pleasant, Ohio.—*Paint.*—July 31, 1866.—A ferruginous earth, known locally (Jefferson county, Ohio) as Smith's mineral, is roasted, pulverized, and with the addition of pigments, if required, is made into a paint with oil, &c.

Claim.—The compound described as a new and useful composition for paint.

No. 56,815.—REES B. SMITH, Mount Pleasant, Ohio.—*Composition for Welding and Brazing.*—July 31, 1866.—A ferruginous earth is mixed with a small percentage of manganese.

Claim.—The fluxing or welding composition substantially as described.

No. 56,816.—SIDNEY SMITH, Greenfield, Mass.—*Heating Apparatus.*—July 31, 1866.—Air is supplied through the perforated walls and an intervening space after the fire is started and the communication through the bottom cut off.

Claim.—First, a fire chamber constructed in accordance with the principles and substantially in the manner herein set forth.

Second, the combination of the perforated walls A and G, constructed as described, to form a fire chamber.

Third, the combination of the perforated walls A and G with the close bottom D, substantially as and for the purpose set forth.

Fourth, the combination of the perforated walls A and G with the partition E, and damper M, substantially as and for the purpose set forth.

No. 56,817.—REES B. SMITH, Mount Pleasant, Ohio.—*Composition for Facing Moulds.*—July 31, 1866.—Composed of a powdered ferruginous earth, eighty parts, and coke and charcoal dust, of each from five to fifteen parts.

Claim.—The composition above described, as a "facing" powder, for use in the process of casting.

No. 56,818.—REES B. SMITH, Mount Pleasant, Ohio.—*Composition for Roofing.*—July 31, 1866.—Composed of coal tar (2) and a ferruginous earth (5) found in Jefferson county, Ohio.

Claim.—The composition for roofing, consisting of the ingredients in about the proportions described.

No. 56,819.—WILLIAM H. SMITH, Sparta, Wis.—*Bag Holder.*—July 31, 1866.—The mouth of the bag is folded over the edge of the hoop and there clamped by the larger hoop on the bottom of the hinged hopper.

Claim.—The hoops E, and hinged hopper C, arranged to operate substantially as set forth.

No. 56,820.—JEROME B. STARK, Fisherville, N. H.—*Mortising Machine.*—July 31, 1866.—The machine is attached to the timber and moved upon its base piece by a rod attached to the operating lever, and operating by pawls and racks on the respective portions to advance it between cuts.

Claim.—The feeding mechanism or combination as described, the same consisting of the two racks e f, the pawls p p, the pawl levers E F, the lifter G, and the cords r r, the whole being arranged and applied as explained to the supporter B, the frame C, the chisel shaft, and its operative lever, connected with the said shaft by means of the recessed block K, and the spring catch, as specified.

No. 56,821.—P. A. STECHER, New York, N. Y.—*Safety Pocket.*—July 31, 1866.—The straight side of the semicircular pocket is vertical; the lower end of the stiffening strip arrests the withdrawal of the watch, which lies beneath the partition seam.

Claim.—First, the partition seam a, and rounded or inclined seam b, in combination with the pocket A, and its mouth B, substantially as and for the purpose described.

Second, the recess d, in combination with the stop D, and pocket A, constructed and operating substantially as and for the purpose set forth.

No. 56,822.—CHARLES E. STELLER, Chicago, Ill.—*Bog Cutter.*—July 31, 1866.—The cutters are attached to the longitudinal bars, are all inclined backward, and alternately to the right and left. The machine is dragged over the ground to cut and level it.

Claim.—First, the frame A A' B B', constructed and arranged substantially as and for the purposes set forth.

Second, the arrangement of the cutters D, in four or more transverse rows, two of the rows inclining to the right and rear in alternation with two inclining to the left and rear, the rear rows cutting through the spaces left by the front rows, substantially as set forth and shown.

Third, the combination of a frame A A' B B', constructed substantially as specified, with transverse rows of obliquely placed cutters or cutter teeth, the rows inclining alternately to the right and left, substantially as set forth and shown.

Fourth, the combination and arrangement of the brace bars C, the beams or bars A A' B B', and the oblique cutter teeth D, substantially as shown, set forth, and specified.

No. 56,823.—ANDREW STEVENS, West Milton, Ohio.—*Drying Apparatus.*—July 31, 1866.—The sides of the trays fit together to form a part of the drying passages; the air is admitted behind the flue, which has a horizontal floor between the upper and lower drying chambers.

Claim.—First, arranging the cleats I and trays J of a drying apparatus, so that they may form one side of a hot air flue L, as described.

Second, the vertical cold air passage H, when placed between the smoke pipe and housing of a drying apparatus, substantially as described for the purpose set forth.

Third, in the described combination the vertical cold air passage H, smoke pipe E E' F, and the concave deflecting plates D and G, arranged and operating as explained.

Fourth, the horizontal branch F, when arranged to pass between two sets of drying trays, as and for the purpose set forth.

No. 56,824.—MICHAEL STEVENS, Smithville, Ohio.—*Press for Cider Mills*—July 31, 1866.—The board has two sets of grooves on each side at right angles to each other and conducts the cider from the pomace.

Claim.—The conducting board H, when constructed and used in a press box, substantially as described and for the purposes set forth.

No. 56,825.—A. C. STONE, Steeleville, Penn.—*Machine for Rolling Metal.*—July 31, 1866.—The metal passes through dies before it reaches the taper groove in the rolls; one-half the die is stationary and the other portion is moved up by a cam.

Claim.—The combination of the dies D D with the rolls C C, constructed and operating as described and for the purposes already set forth.

No. 56,826.—JOHN STONE and SAMUEL BLOCKER, Sr., Plattsburg, Mo.—*Fence*—July 31, 1866.—The posts and braces are framed into the sills and the parts secured by keys and wedges, without nails.

Claim.—First, the construction, combination, and arrangement of the sills A, uprights B, braces G, and wedges D and H with each other, substantially as described and for the purpose set forth.

Second, the combination of the horizontal bars C and blocks E with each other and with the upright bars B, substantially as described and for the purpose set forth.

Third, the combination of the keys F or equivalent, with the upright bars B and with the upper horizontal bar C, substantially as described and for the purpose set forth. •

No. 56,827.—SAMUEL S. STONE, Troy, N. Y.—*Paper Collar.*—July 31, 1866.—The band has vertical slits, open at the lower edge and terminating below the fold at enlarged openings.

Claim.—First, a paper or combined paper and cloth turn-over shirt collar, having its neck band B slitted in the manner substantially as and for the purpose set forth.

Second, a turn-over shirt collar of paper, or paper and cloth combined, having only the exterior surface of the turn-over part A, colored or ornamented, as specified.

The slitted neck band B, provided with button holes C, of the form shown, substantially as and for the purpose specified.

No. 56,828 —TURNER STROBRIDGE, Pittsburg, Penn.—*Adjustable Lock Keeper.*—July 31, 1866.—The face plate is made movable to adjust it to the door, whose lock bolt shuts into it.

Claim.—First, the oblong or slotted screw holes b b, when used for the double purpose of adjusting a keeper and fastening the same to the casing of the door.

Second, the combination of the movable face plate and stationary body of a keeper with slotted and regular screw holes, and ratchets or stops forming a keeper, substantially as shown and described.

No. 56,829.—JAMES SUTHERLAND, East Hampton, Mass.—*Self Acting Mule.*—July 31, 1866.—To prevent the yarn becoming slack, and from "kinking" when the carriage "strikes in," an air pump is secured to the mule-head, from which air is permitted slowly to escape before a piston which is driven forward by the force of the same spring or other power which carries the faller wire upwards. An arm on the faller rod operates the piston, and every action of the piston causes the turning of a screw which gradually elevates the part against which this arm strikes to conform it to the height attained by the cop.

Claim.—First, controlling the ascent of the faller wire in spinning machines by the resistance of a body of confined air, substantially as above described.

Second, in combination the cylinder E, having its lower end open and a valve applied to its upper end, the lever I and the arm B projecting from the shaft of the faller wire C, substantially as described.

Third, the combination of the tripping rod H with the piston T, table G, and the arm B of the faller-wire shaft, substantially as described.

Fourth, the screw rod R and nut G', in combination with the tripping rod H and lever I, substantially as described.

Fifth, the screw rod R and nut G', in combination with the piston, whose stroke is shortened by the rising of the nut, substantially as described.

Sixth, the combination of the arm B of the faller-wire shaft with the tripping rod H and lever I, substantially as described.

No. 56,830.—MARY A. TAYLOR, Cincinnati, Ohio.—*Wash Boiler.*—July 31, 1866.—Several vessels with perforated bottoms combine to occupy the wash boiler and may be piled on each other, their flanges fitting for that purpose.

Claim.—A plurality of receptacles D E, each constructed with a perforated bottom and adapted to fit within a wash boiler and operate in connection with each other in the manner and for the purposes herein described.

No. 56,831.—CHARLES WESLEY THOMPSON, Batavia, Ill —*Truss.*—July 31, 1866.—The twin pads are attached to a plate whose pressure is adjusted by the curved screw, set-nut, and plate projecting from the hoop.

Claim.—First, making a truss pad of two separate pieces secured side by side to a plate on which they are allowed to oscillate, substantially as set forth.

Second, the screw-threaded curved arm D, passing through a guide and stop G, and having a nut F thereon, substantially as described.

No. 56,832.—JOEL TIFFANY, Albany, N. Y.—*Manufacture of Paper Stock*—July 31, 1866.—The pressure is attained in the digester without a corresponding increment of heat by forcing in a body of hot or cold air by means of a pump.

Claim.—First, the employment of pressure obtained by forcing into the vessel containing the stock to be treated, air cold or hot, so as to obtain any degree of pressure necessary to force the caustic liquor into contact with every part of the stock, in combination with the caustic liquor so used, substantially in the manner and for the purpose above set forth and described.

Second, the use of condensed cold air forced into the vessel containing stock, producing the necessary internal pressure upon the stock by heating and expanding the air within the vessel, in combination with the caustic liquor so used, substantially in the manner and for the purpose above set forth and described.

No. 56,833.—JOEL TIFFANY, Albany, N. Y., and HARRISON B. MERCH, Fort Edward, N. Y.—*Bleaching Paper Stock.*—July 31, 1866.—Chlorine gas is forced by an air pump into the vessel containing the pulp.

Claim.—The process of bleaching paper stock under pressure produced by forcing a weak chlorine gas, with or without atmospheric air, into a close vessel containing the stock, in combination with the use of a solution of chlorine with which the stock to be bleached is saturated, substantially in the manner and for the purposes above described.

No. 56,834.—C. A. TODD, New York, N. Y.—*Paddle Wheel.*—July 31, 1866.—Obliquely set curved floats are attached by radial arms to the hubs of the wheel.

Claim.—The arrangement of the curved floats B C, in combination with the radial arms D, constructed and operating substantially in the manner and for the purpose specified.

No 56,835.—JESSE TUCKER, Adrian, Mich.—*Water Wheel.*—July 31, 1866.—The water enters at the side of the annular curb above; the wheel receives it within and delivers it outward and downward.

Claim.—A horizontal water wheel provided with a bottom having a series of inclined issues D, and also provided with an upright rim or flange E, having a series of curved taper issues F, in connection with the conical hub G on shaft B and the cylinder A* over the wheel, all arranged substantially as shown and described.

No. 56,836.—L. W. TURRELL, SAMUEL STANTON, and L. C. WARD, Newburgh, N. Y.—*Safety Attachment for Gas Pipe.*—July 31, 1866; antedated May 7, 1866.—Gauze diaphragms cross the aperture and prevent the passage of flame.

Claim.—The combination of the gauze disks C C C, rings D D, and coupling A B, arranged and applied in the manner and for the purposes specified.

No. 56,837.—W. VAN VALKINBURGH, Smithville, N. Y.—*Car Coupling.*—July 31, 1866.—The springs of the draw head reduce the jar due to concussion and sudden tension, and the side springs avoid the lateral jar due to the swinging motion of the cars. The latch which retains the coupling shackle is attached to a sliding gate moved by a lever on the upper part of the car.

Claim.—First, the arrangement of the springs I J, rod C, springs E, pivoted draw head B, and spring L, in combination with the frame of the car, substantially in the manner and for the purpose herein described.

Second, the pivoted draw head B and spring L, operating in combination with the curved catch Q and sliding frame M, constructed and arranged in the manner and for the purpose specified.

Third, the combination and arrangement of the pivoted draw head B, spring L, springs I J, spring E, sliding rod C, sliding frame M, link N, and lever O P, constructed and operating substantially as and for the purpose herein represented and described.

No. 56,838.—W. G. WARD, Savona, N. Y.—*Nail Hammer.*—July 31, 1866.—The nail is gently held by the spring under the poll of the hammer till it is stuck into place ready for driving by a one-armed carpenter.

Claim.—Holding the nail by means of the groove in the hammer, in combination with the head block and spring, when constructed to operate substantially as described and for the purpose set forth.

No. 56,839.—DAVID R. WARFIELD, Muscatine, Iowa.—*Corn Planter.*—July 31, 1866.—The levers which actuate the seed slides are vibrated by the cams attached to the wheel, which is driven by the contact of its spokes with the ground. The spokes, also in connection with spiral springs, mark the ground as they revolve.

Claim.—First, constructing the driving wheel G, with spurs I, when used in combination with the cams H H, and levers R R, for actuating the sliding seed valves of a corn-planter, substantially as set forth.

Second, in combination with the spurs I, the plates K, arranged substantially as and for the purposes set forth.

Third, the combination of the wheel G, and spurs I, and frame A, with the frame B, and seat L, substantially as and for the purposes set forth.

Fourth, in combination with the wheel G, and spurs I, the track clearer Q, substantially as and for the purposes set forth.

Fifth, the levers R R, in combination with the supports N, and lever P, substantially as and for the purposes set forth.

No. 56,840.—JAMES WOLFENDEN, Jersey City, N. J.—*Tool and Rest Holder for Lathes.*—July 31, 1866.—The slotted tool-holding plate is adjustably attached to the slide; the shaft to be turned passes through the centre and towards it the tools and rests converge, being adjusted by set screws and operated by a scroll cam which engages notches on the tool slides.

Claim.—The combination of the slotted plate B, tool-holders C, with segmental threads gearing in scroll D, and plate A, provided with the adjustable guide slides E, when arranged and operating in the manner and for the purpose herein described.

Also, the segmental slots *a*, in the brackets B, for the purpose set forth.

No. 56,841.—A. H. WOODRUFF, Lansing, Iowa.—*Slide Valve.*—July 31, 1866.—The angular valves are fitted to corresponding seats containing induction and exhaust apertures. Steam is admitted to recesses in the valve to partially counterbalance the pressure on their outer surfaces.

Claim.—First, the angular valve, either double-acting or single-acting, in combination with a correspondingly angular seat, substantially as and for the purpose described.

Second, the recesses *f f'* in the exhaust side of the valve seat, in combination with the angular valve, substantially as and for the purpose set forth.

Third, the recesses *e e'* in the steam side of the valve seat, in combination with the angular valve, substantially as and for the purpose set forth.

No. 56,842.—THOMAS D. WORRALL, Central City, Colorado.—*Furnace for Desulphurizing Ores.*—July 31, 1866.—The ore is passed from the hopper to a chamber where a revolving stirrer delivers it to a zigzag furnace flue and a hearth, where it is subjected to the action of the fire and an air blast. It passes thence into a current of water, which carries it to the separator.

Claim.—First, in a desulphurizing furnace, used in combination with a steam engine, to operate a blower or quartz pulverizer, or both, so combining and arranging the steam generating furnace with the desulphurizing furnace, that the flame and other products of combustion escaping from the steam generating furnace shall pass into and through the desulphurizing furnace and supply the flame and heat necessary for effecting desulphurization therein, substantially as described.

Second, condensing flame by means of a blow-pipe or pipes upon a hearth over which pulverized quartz is passed for the purpose of desulphurizing the same, substantially as described.

Third, in combination with a hearth upon which flame is condensed by means of a blow pipe or pipes, for the purpose herein described, a fan-blower and pipe, or other suitable air generator, for the purpose of forcing air through said blow pipe or pipes to condense the flame upon the hearth, substantially as described.

Fourth, an inclined or zigzag flue, with top of soapstone, metal, or other suitable substance, for the purpose of securing a heated surface over which pulverized quartz is passed, for the purpose set forth.

Fifth, the horizontal flue I, with top plate of soapstone, metal, or other suitable substance, in combination with the stirrer or scraper L, for the purpose described.

Sixth, the hopper B, in combination with the worm screw L', and stirrer or scraper L, substantially as and for the purpose described.

Seventh, the hopper B, in combination with the flue J J under the same for the purpose of drying and heating the pulverized quartz before leaving the hopper.

Eighth, in combination with the stirrer or scraper L, and the horizontal flue I, the apertures O (one or more) through both the top and bottom plates of said flue, for the purpose of delivering the pulverized quartz down upon the heated plate covering the inclined flue H, substantially as and for the purpose set forth.

Ninth, in combination with the hearth G, the sluice or water-course W W, for the purpose of conveying the pulverized quartz from said hearth to a buddle, arastra, or shaking table, as described.

No. 56,843.—WILLIAM C. WREN and WM. BARKER, Brooklyn, N. Y.—*Apparatus for the Manufacture of Gas from Petroleum.*—July 31, 1866.—The consecutive retorts are connected by pipes; the petroleum is vaporized in the first, and decomposed in the successive retorts, which are charged with ashes or other absorbent; the gas is collected in suitable vessels.

Claim.—The process herein described of producing gas, to wit: by a combination of one or more heaters and super-heaters (not less than one of each) continuously connected with each other by pipes, such heaters and superheaters fitted and filled as described in the foregoing specification, and by the peculiar combination, arrangement, and graduation of two or more fires, not less than two, as shown in specification, with an addition of more heaters, superheaters and fires, as the quantity of gas to be produced may require.

No. 56,844.—J. E. WILSEY, Chicago, Ill., and D. FORBES, Scotland.—*Spring Bed Bottom.*—July 31, 1866.—The rails rest upon X-shaped end pieces bolted at the intersection, and the side rails have pins which move in slots as the springs expand and contract.

Claim.—The end rails B B, in combination with upper and lower frames, and the notched crossed slats F F, in combination with spiral springs G G, the whole arranged substantially as above described and specified.

No. 56,845.—LEONARD B. ALDEN, Cincinnati, Ohio, assignor to JOHN WALKER, same place.—*Egg Beater.*—July 31, 1866.—The motion of the loaded handle imparts a rapid intermittent rotation to the shaft and its agitators.

Claim.—The combination of the plate D, lever L, rack K, pinion I, shaft G and lug F, all constructed and arranged to operate substantially as and for the purposes set forth.

No. 56,846.—A. M. BACON, Washington, D. C., assignor to himself and GEORGE E. H. DAY, same place.—*Breech-loading Fire-arm.*—July 31, 1866.—The oscillating chambered breech block is actuated by the motion of the hammer through the intervention of a diagonally grooved cylinder.

Claim.—The oscillating chambered breech-block plunger and magazine combined, as constructed and arranged with the hammer operating against the rear end of the chamber, and the mode of oscillating the same by the diagonally grooved wheel, as described.

No. 56,847.—E. P. H. CAPRON and J. F. WINCHELL, Springfield, Ohio, assignors to themselves and T. W. and H. J. MILLER, same place.—*Brick Machine.*—July 31, 1866.—A strip of clay is forced through a tube under the pug mill by means of a plunger, is received on an endless rubber belt, cut into bricks by a series of vertically moving moulds and removed on a sliding frame. Devices for lubricating the presser, securing cloth to the face of the latter, removing the bricks, operating the moulds, adjusting the pressure, and regulating the thickness of the strip of clay, are explained in the claims and cut.

Claim.—First, the combination of the tubes H and plungers B', operated by the oscillating arm *m*, substantially as shown and described.

Second, the division box D, with the adjustable mouthpiece *e*, as set forth.

Third, the combination and arrangement of the endless belts L, with the tubes H, for receiving and conveying the strip of clay, as set forth.

Fourth, the combination of the mould box E and follower F, when connected by means of the bolts *b*, and springs *d*, and otherwise arranged to operate as shown and described.

Fifth, the use of the set screws *o*, or their equivalents, when arranged substantially as described, for the purpose of adjusting or regulating the pressure on the brick, as set forth.

Sixth, providing the moulds with the groove z, and oil cup *e'*, as set forth.

Seventh, the sliding frame *h*, arranged to operate as described, for the purpose of delivering the brick from the machine, as set forth.

Eighth, the combination of the cam wheel K, lever I, and the plunger stems *a a'* and *d'*, for the purpose of giving to the moulds E, and the followers F, the movements herein described.

Ninth, a brick presser or follower having cloth secured to its face by means of the band *g*, fitting in a recess formed therein, as shown and described.

No. 56,848.—J. D. CHASE, DENISON CHASE, and JEFFERSON CHASE, Orange, Mass., assignors to themselves and DANIEL POMEROY.—*Water Wheel.*—July 31, 1866.—The chute has a conical contraction at its lower end, and the buckets have an upward extension above the rim. The wheel is adjusted to the scroll by the vertical and lateral motions of the supporting bridge tree.

Claim.—First, the construction of the buckets H H, as arranged in relation to the central hub and enclosing cylinder, substantially as and for the purpose herein specified.

Second, the extension of the upper edges of the buckets H H upward within the conical part *a* of the curb A, in combination therewith, substantially as and for the purpose set forth.

Third, the combined arrangement of the bridge tree I, adjustable as described, in the chairs M M, and the enclosing curb cylinder G, and the wheel therein, substantially as and for the purpose herein specified.

No. 56,849.—DAVID DALTRY, Philadelphia, Penn., assignor to himself and JOHN PARKER, same place.—*Rod Coupling.*—July 31, 1866; antedated July 13, 1866.—Explained by the claim and cut.

Claim.—The combination of the tapering enlargements *a a'*, of the two tubes, the clamp pieces C and C', and the sleeve D, the whole being constructed substantially as and for the purpose specified.

No. 56,850.—ROLLIN DEFREES, Washington, D. C., assignor to himself and JOHN D. DE-FREES, same place.—*Rotary Pump.*—July 31, 1866.—A central slotted cylinder has a piston which is alternately retracted by contact with an abutment and projected by a spring so as to occupy the annular space in the outer cylinder. The water enters at the base when the piston is out, and ascends above a middle septum in the slit, the latter being the channel of ingress and egress.

Claim.—First, the induction and eduction ports through the rotating cylinder which carries the sliding vane by means of which the water enters and departs from the pump chamber in a direct vertical line through said rotating cylinder, substantially as described.

Second, forming the top and bottom of the pump chamber by means of the plates or disks *o o* and *m m*, recessed for packing as described.

Third, opening and closing the ports by means of the sliding vane, substantially as described.

Fourth, the water passages *l l*, through the vane, whereby the pressure of the column of water in the well tube is admitted behind the vane to equalize the pressure of the water on the inner and outer sides of said vane, and obviate resistance to the action of the spring.

No. 56,851.—JAMES B. DRAKE, Picture Rocks, Penn., assignor to DRAKE, SILL & HUTSON.—*Horse Hay Fork.*—July 31, 1866.—A cutting plate is connected by pivots to the lower ends of two parallel bars, so that by sliding one bar on the other the plate will be retracted into its penetrating position or projected laterally to raise the hay. A cross-bar pivoted to both bars and operated by a cord throws them into either position, and a trigger, also operated by a cord, releases them to discharge the load.

Claim.—First, the combination with the adjustable parallel bars A A', of the stationary cutter B, and pivoted cutter B', all arranged to operate substantially as described.

Second, the adjustable cutter B', formed with a circular cutting edge *b'*, and lifting points 2, substantially as described.

Third, in combination with the above the bar C, pivoted to the bars A A', and provided with studs to enable the said bar C to be operated by a cord or otherwise,

Fourth, in a hay fork and knife, constructed as herein described, the latch F and stop G, or their equivalent, for the purpose set forth.

No. 56,852.—M. P. EWING, Rochester, N. Y., assignor by mesne assignments to H. B. EVEREST and GEORGE P. EWING, same place.—*Apparatus for Distilling Petroleum.*—July 31, 1866.—The oil is introduced continuously and evaporated by a steam coil and jacket at the bottom of the retort. A duplicate system of condensers communicating with a common water tank above and a common receiving tank below connect alternately with the retort to maintain a constant vacuum and continuous distillation.

Claim.—First, the combination of a continuous feed with a vacuum still for petroleum, operating substantially as and for the purpose herein set forth.

Second, the combination of a jet condenser with a vacuum still for petroleum, operating substantially in the manner and for the purpose herein specified.

Third, the combination of the two or more condensers H H with each other and a vacuum still A, in such a manner that the action may be alternated from one to another to preserve the vacuum and to allow the constant running of the still, as set forth.

Fourth, the arrangement as a whole, consisting of the retort A, condensers H H, connected by the tubes G G, and pipes I L, operating substantially as and for the purpose specified.

No 56,853.—BEN FIELD, Albion, N. Y., assignor to himself and GEORGE M. PULLMAN, Chicago, Ill.—*Sleeping Car.*—July 31, 1866.—A hinged board on the back of the seat is raised to form a support for the couch which is lowered on to it from above; the hiatuses in the head and foot boards are filled by triangular pieces.

Claim.—First, in combination with the couches of a convertible sleeping car the hinged board B when attached to the back of a seat and capable of being adjusted, substantially in the manner and for the purpose set forth.

Second, in combination with the upper couch C, one or more intermediate pieces G connecting the head and foot pieces F F' and the partitions E, substantially in the manner and for the purposes set forth.

No. 56,854.—S. HODGINS, St. Louis, Mo., assignor to himself and STEPHEN BLACKIE, same place.—*Fastening for Paper Collars.*—July 31, 1866.—An elastic band unites the ends of the paper collar to the shirt button.

Claim.—An elastic strip provided with one or more button holes and one or more hooks when arranged in relation to a paper collar, substantially as described.

No. 56,855.—WILLIAM HUGHES, Waupun, Wis., assignor to himself and JOHN FIELD-STAD.—*Life Boat.*—July 31, 1866.—The divided chambers on each side occupy a portion of the length of the boat amidships. According to their contents they are intended to act as buoys or ballast.

Claim.—A boat having the two chambers A and B on each side, the former being a water chamber open at bottom *a* and at top *a'*, as also the latter being an air chamber open at bottom *a''*, the two having a partition *c* between them, arranged, constructed, and co-operating in the manner described and for the purpose set forth.

No. 56,856.—E. C. LITTLE, St. Louis, Mo., assignor to EVELINE LITTLE, same place.—*Covering for Steam Pipes and Boilers.*—July 31, 1866.—Explained by the claim and cut.

Claim.—Covering steam pipes and boilers with a coating of plaster of Paris cement with or without a wrapping of canvas, for the purpose of retaining the heat and preventing its loss by radiation, in the manner herein described.

No. 56,857.—JOHN R. MARTIN, Boothbay, Me., assignor to SAMUEL K. HILTON, Portland, Me.—*Fishing Line Sinker.*—July 31, 1866.—The sliding shank of the hook is held in position by a side plate and screw, so as to hold fast the swivel of the line or to allow it to open to detach the sinker.

Claim.—The removable staple bolt or slide inserted in the top of the sinker beneath the plate, with a mouth or space for the admission of a swivel or line which is opened or closed at pleasure as described above, by means of which the sinker is readily detached from the line.

Also, the connection of a plate of hard metal with a body of soft metal, each of which is made separate, as well as the movable staple, and either of which may be supplied anew at pleasure.

No. 56,858.—THOMAS CATO MCKEEN, Irvington, N. J., assignor to the NEW YORK SUBMARINE COMPANY.—*Apparatus for Buoying Vessels.*—July 31, 1866.—Explained by the claim and cut.

Claim.—The use and application of the air reservoir or receiver A in combination with the bags or buoys F, whether connected together directly by the use of pipes or by the use of the intermediate main C, and the method of constructing the air receiver, the buoys and netting, and of inflating the buoys by means of compressed air ; the application to and use with the buoys of the common self-acting safety valve, made to yield or discharge at a certain pressure ; and the application and use of the hole and its head to and with the buoys, the whole arranged and operating substantially in the manner and for the purposes above set forth.

No. 56,859.—WILLIAM A. MIDDLETON, Harrisburg, Penn., assignor to himself, DAVID J. BROUGHER and GEORGE H. HAMMER.—*Broom Head.*—July 31, 1866.—A skeleton cap, side clips and screw bolt enclose the butts of the corn around the handle and the interstices are filled with cement.

Claim.—First, the skeleton B in combination with the clips C and handle A, when said parts are arranged as set forth.

Second, in combination with the metallic frame, the application of cement, substantially as set forth, for the purpose of securing the brush in place.

No. 56,860.—FRANKLIN PERRIN, Cambridge, Mass., assignor to himself and D. C. PERRIN, Boston, Mass.—*Bleaching Palm Leaf, Straw, &c.*—July 31, 1866.—The material is placed on racks and lowered into a chamber charged with pure sulphurous acid gas, which is admitted at the bottom and passes through the chambers consecutively ; the tops of the chambers are secured by water joints.

Claim.—The improvement in bleaching palm leaf, cane, straw, and similar fibrous bodies, substantially as described.

No. 56,861.—PASCAL PLANT, Washington, D. C., assignor to himself and PETER HANNAY, same place.—*Coffee Roaster.*—July 31, 1866.—Two cylinders with wire gauze diaphragms are hinged together and held closed by their handles. The coffee is contained between the foraminous diaphragms.

Claim.—The cylinder A and B, either singly or combined, when used in combination with wire gauze or perforated plates E E and openings *a a*, arranged in the manner substantially as and for the purpose set forth.

No. 56,862.—MILON PRATT, Meriden, Conn., assignor to himself and CLEMENS DARNSTAEDT, same place.—*Machine for Smoothing Ivory Key Boards.*—July 31, 1866.—The bed carries the ivory plates and slides back and forth under a cutter head mounted on a vertical spindle.

Claim.—The combination of the reciprocating bed B adapted for the securing upon it of the ivory key pieces, the rotary head C provided with cutters *a a* arranged obliquely in pairs, and the guide ways A, all constructed, arranged, and employed as specified for the smoothing of the ivory surface of key boards.

No. 56,863.—E. F. PRENTISS and R. A. ROBERTSON, Philadelphia, Penn., assignors to said PRENTISS, W. D. PHILBRICK, and W. J. PARSONS.—*Process for Purifying and Deodorizing Whiskey.*—July 31, 1866.—The vapor from the still is passed through a porous material charged with an alkaline solution or other purifying agent.

Claim.—The process of purifying and deodorizing alcoholic liquids by passing them while in a vaporous state through the interstices of a porous, perforated, cellular, granulated or otherwise finely divided neutral material, kept wet with a solution of alkali or of alkaline salts, or of the substances having an equivalent purifying action, in the manner and for the purpose substantially set forth.

No. 56,864.—E. F. PRENTISS and R. A. ROBERTSON, Philadelphia, Penn., assignors to said PRENTISS, W. D. PHILBRICK, and W. J. PARSONS.—*Apparatus for Purifying and Deodorizing Whiskey.*—July 31, 1866.—The vapor from the still is passed through drawers whose absorbent contents are saturated with a purifying solution supplied by means of traps.

Claim.—The trap tubes E, or their equivalents, in combination with the neutral material K, contained in one or more drawers or cases, in the manner and for the purpose substantially as shown and described, the whole being used in connection with a still, for the purposes herein set forth.

No. 56,865.—HIRAM ROBBINS, Cincinnati, Ohio, assignor to himself and THOMAS H. FOULDS, same place.—*Clothes Wringer.*—July 31, 1866.—Explained by the claims and cut.

Claim.—First, the reversible scolloped and countersunk spring O P P' Q Q', in the described combination with the perforated and slotted posts A B, beam D, relaxing screw E, upper roll J, and rods M M'.

Second, in combination with the above, the spring N, interposed between the two pressure rods M and M'.

Third, the combination of the spring washer or cushion S, the beam D, and the head of the screw E.

No. 56,866.—W. E. TICKLER, E. T. MARSHALL, and DANIEL M. MARSHALL, Pierceton, Ind.—*Car Coupling*—July 31, 1866.—The coupling pin is dropped automatically as the link of the other car enters the draw head and is uncoupled by the lever which raises the gate to which the pin is attached.

Claim.—The rising and falling pin D, connected with a sliding frame composed of the rods B B and bar C, in combination with bar D* and F, the former being provided with a weight G, and both connected with a shaft E, the sliding bar J, connected with a pendent frame I, on shaft E and the plate K, all arranged to operate substantially in the manner as and for the purpose set forth.

No. 56,867.—WM. F. WENISCH, New York, N. Y., assignor to himself and JOHN WENISCH, Newton, N. Y.—*Fire Kindling.*—July 31, 1866.—Composed of rosin, saw dust, and sand, formed into blocks.

Claim.—A composition for kindling wood or coal fires, formed by combining rosin, dry saw dust, and dry sand with each other in the proportions and in the manner substantially as herein described and for the purpose set forth.

No. 56,868.—AUGUST TONNAR, Eupen, Prussia, assignor to SIGISMUND DREY and MORITZ ROSENHEIM, New York, N.Y.—*Machine for Drying and Cleaning Grain, &c.*—July 31, 1866.—The grain is alternately distributed centrifugally by the disks and gathered by the agitated conical sieves for a similar consecutive action in the descending series. The sieves have covers, are agitated by eccentrics, and the process is conducted in an ascending column of heated air.

Claim.—First, the conical perforated agitators J, constructed substantially as specified in combination with the centrifugal disks K and hot-air pipe L, for the purposes and as specified.

Second, the arrangement of the chute 8 and valve h in combination with the chute D, and conical agitators, for the purposes and as set forth.

No. 56,869.—RUSSELL JENNINGS, Deep River, Conn.—*Machinery for Swaging the Heads of Screw Augers.*—July 31, 1866; antedated January 31, 1866.—The jaws of the gripping dies are arranged vertically, one stationary, the other movable; their faces grasp the twisted auger and present the upper end to a hammer, which swages the points and lips at one blow.

Claim.—First, the operating of the heading die by the rotating shaft B, through the medium of the loose driving wheel E, provided with the pins g, the sliding wheel F placed on a square shaft B, and provided with a projection i, bevelled at one end, in connection with the sliding rod H and the fixed inclined lip a*, all arranged substantially as set forth.

Second, the forming die, consisting of the portions l l' D, constructed and operating substantially as described.

Third, the combination of the heading die D, mould or female die L, toggle M, and driving or operating mechanism, so arranged that the driving shaft B may at the will of the operator be connected with and disconnected from the continually rotating driving-wheel E, substantially as described.

No. 56,870.—H. J. ALVORD, Detroit, Mich.—*Disinfecting Commode.*—July 31, 1866.—
The covered pail contains a pan for the feces and one for the disinfectant, which deodorizes
the mephitic vapors.
Claim.—The arrangement and combination of the pans D and E, perforated annular flange
C, and cover F, with the bucket or casing A, or their equivalents, substantially as and for the
purposes described.

No. 56,871.—M. J. ALTHOUSE, Waupun, Wis.—*Pump.*—August 7, 1866.—The piston
works in a bushing of glass, stone, or metal secured within a wooden pump barrel by elastic
rings.
Claim.—The inserting of a glass, stone, or metallic tube, or lining, into the barrel of a
wooden pump, and firmly holding it there by means of rubber or other elastic rings, in the
manner and for the purpose set forth.

No. 56,872.—STEPHEN D. ARNOLD, New Britain, Conn., assignor to himself and W. F.
ARNOLD, same place.—*Bridle Bit.*—August 7, 1866.—Metallic straps are passed around the
bridle rings, and screwed or otherwise fastened into the ends of a tubular bit.
Claim.—The combination of the tube bit *a* with the clasp *d*, ring *c*, constructed and arranged
substantially as and for the purpose described.

No. 56,873.—GEORGE H. BABCOCK, Providence, R. I.—*Anchor Stopper.*—August 7, 1866.—
A tumbler sustaining the cable in the plane of its axis is easily rotated and held in either its
open or closed position by a sliding lever.
Claim.—First, in anchor stoppers, the employment of a rotating tumbler B, adapted to
receive the link C, or its equivalent, on a point or points lying in, or nearly in, the axis of
rotation, substantially as and for the purpose herein set forth.
Second, in combination with the rotating tumbler B, the sliding rod D, substantially as and
for either or both the purposes above specified.
Third, in combination with the rotating tumbler B the stationary mousing piece *a'*, substantially as and for the purpose herein set forth.
Fourth, an automatically locking anchor stopper, consisting of the rotating tumbler B, the
sliding rod D, or equivalent device, and mousing piece *a'*, substantially as herein described.

No. 56,874.—C. L. W. BAKER, Hartford, Conn.—*Coal Scuttle and Sifter.*—August 7,
1866.—The sifter fits a round aperture in the scuttle lid, is supported by a flange and vibrated
by a handle. A hinged cover over the spout prevents the escape of dust, but permits the
pouring out of coal or ashes.
Claim.—As a new improved article of manufacture, the scuttle *a*, sifter *d*, when constructed
and arranged substantially as and for the purpose as described.

No. 56,875.—CLARA A. BARTLETT, Oakland, Cal.—*Side Saddle.*—August 7, 1866.—The
near side horn is hinged so as to be turned down out of the way in mounting or dismounting.
Claim.—A side saddle having one of its horns arranged thereon and attached thereto, so
as to be operated substantially in the manner described and for the purpose specified.

No. 56,876.—D. W. BASHORE, Erie, Penn.—*Washstand and Desk.*—August 7, 1866.—The
hinged lid of the washstand is inclined to adapt it for use as a desk, and the space in which
the pipe or receptacle for waste water may be placed serves also to accommodate the feet of
the writer.
Claim.—The arrangement of the water-heating tank B with the other two tanks C and D
in a washstand, and the construction of the waste-water space E, to adapt the stand to use
as a writing desk, as specified.

No. 56,877.—CALEB BATES, Kingston, Mass.—*Crushing, Rolling, and Kneading Machine.*—
August 7, 1866.—A frame pivoted to arms swinging concentrically above the trough, and
secured at any height with either a rigid or yielding pressure, carries the kneading and rolling devices, and is reversible to bring either into action.
Claim.—First, the swinging bars *c c*, provided with the reversible bars *e e* containing the
rollers G H, and arranged as shown to admit of either roller G H being used as the nature of
the work may require, substantially as and for the purpose set forth.
Second, the slides I I in combination with the spring F and screws E, as and for the purpose set forth.
Third, the perforated receptacle J applied to the bars *c c*, in combination with the reversible
bars *e e* and rollers G H, substantially as and for the purpose specified.
Fourth, the combination of the receptacle A, provided with a curved bottom, in combination with the bars *e e*, rollers G H, and swinging bars *c c*, all arranged to operate substantially
in the manner and for the purpose set forth.

No. 56,878.—WILLIAM BATTELL, Quiney, Ill.—*Gang Plough.*—August 7, 1866.—The
front and rear ends of the plough beams are raised by separate systems of levers, both in

convenient reach of the driver, and are adjusted laterally by sliding their clevises upon a transverse rod.

Claim.—First, the attaching of the axles of the wheels B B of the machine to the rear parts of the bars C C, the front ends of which are attached by hinges *a* to the front part of the frame A, in connection with the segment bars D attached to the rear parts of the bars C and the levers E E attached to the bars D, all arranged substantially as and for the purpose specified.

Second, the arrangement of the curved bars L attached to the plough beams by links M, guides N with rollers *j* fitted in them, and the levers O, all arranged to operate substantially in the manner as and for the purpose herein set forth.

Third, the construction of the clevises H, as shown and described, to admit of the adjustment of the plough beams, as set forth.

Fourth, the thimbles *g*, provided with the set screws *h*, in combination with the clevises H, rod I, and adjustable stays Q, substantially as and for the purpose set forth.

No.56,879.—JOHN BAYLISS, New York, N. Y.—*Twyers.*—August 7, 1866.—The air is heated by passing through a chest exposed to the direct action of the fire, and is discharged through a nozzle protected by a water jacket.

Claim.—The tuyere A. consisting of the water chamber B, connecting pipes D E, water reservoir C, elbow pipe H, air chamber I, and pipe J, and having an opening G, combined and operating substantially. as and for the purpose represented and described.

No. 56,880.—C. F. BAYLOR, Clinton, N. J.—*Grain Cleaner.*—August 7, 1866.—A rapid reciprocating motion is communicated to the screen by a lever vibrated by a zigzag groove in the driving wheel.

Claim.—The arrangement of the wheel H with its groove *h*, lever F, screen frame D with its screens *b b'*, as described, pressure roller H. and rollers C C, constructed and operating in the manner and for the purpose herein specified.

No. 56,881.—WILLIAM BEACH, Philadelphia, Penn.—*Privy Seat Cover.*—August 7, 1866.—The lid is turned down by stepping on a treadle connected to a crank at one end of its axis, and reclosed by a weight.

Claim.—The cover or lid B hinged to the under side of a privy seat, and operated by means of a treadle, substantially as and for the purpose described.

No. 56,882.—W. L. BEARDSLEY, Binghamton, N. Y.—*Last.*—August 7, 1866.—The instep block is secured by a bolt projecting upward from the last, and detached by introducing the last-hook through a vertical hole in the block.

Claim.—Placing the bolt and spring in the body of the last in combination with the position of the vertical opening D, through the heel of the instep block, and the mode of unlocking and detaching said block, as described.

No. 56,883.—JOHN BELL, Lancaster, N. Y.—*Stave Cutting Machine.*—August 7, 1866.—The knife is mounted in a stationary frame, and the bolt drawn across it by a sliding hopper in which it descends by gravity at each stroke. The upper side of the knife edge is bevelled to prevent splitting the wood or cutting the edges of the stave of unequal thickness.

Claim.—Forming the knife with a bevel on the upper side, and combining the knife, when so constructed. with the frame A and reciprocating bolt hopper, substantially as and for the, purposes set forth.

No. 56,884.—ALBERT C. BETTS, Troy, N. Y.—*Tool for Holding and Driving Staples for Wire Fences.*—August 7, 1866.—Staples are placed in a case around a guide rod. A spring follower presses them against the side of a driver, each retraction of which admits a staple in front of it in position for driving.

Claim.—A device for holding staples for the convenience of driving the same, composed of a case in which the staples are placed, a slide and spring, and a sliding bar which is actuated by a hammer for driving the staples, all being arranged substantially as shown, so that when one staple is driven by striking the bar, and the latter is moved back, a succeeding staple will be adjusted or thrown in line with the bar for the purpose of being driven, as set forth.

Also, the placing of the sliding bar G in a hinged cap F, arranged with the case A, so that when said cap F is opened the bar G will be out of the way and the end of the case left open for the ready insertion of the staples.

No. 56,885.—CHARLES E. BILLINGS, Windsor, Vt.—*Die for Swaging Pistol Frames.*—August 7, 1866.—The dies are adapted to remove the displaced metal with a clean cut, and with little force, instead of compressing it into the body of the frame.

Claim.—The cutting dies herein described for forming pistol and rifle frames, formed with cavities *c c*, and otherwise constructed as specified.

No. 56,886.—JOHN BLACKIE, New York, N. Y.—*Electric Telegraph.*—August 7, 1866.—By the application of the switch, two batteries or two cells of the same battery may be made to neutralise each other, and thus render them temporarily inoperative and still permit an impulse to be communicated.

Claim.—The construction and application of a switch to a line connecting two batteries, in such a manner that the electric current between the batteries may be reversed or transferred from one to the other of the poles of said batteries at will, whereby the batteries shall be made to neutralize each other, and thus remain dormant for the time being, substantially as set forth.

No. 56,887.—J. C. BLYTHE, Perry, N. Y.—*Flour Bolt.*—August 7, 1866.—The bolt has external strengthening hoops and internal radial partitions terminating in longitudinal ribs near the periphery, but leaving sufficient space for the finer particles of meal to pass constantly in contact with the cloth.

Claim.—The combination of the partitions E and hoops D, either or both, with the arms B, ribs C, and cloth of a flour bolt, when the said parts are constructed and arranged substantially as herein described and for the purposes set forth.

No. 56,888.—GUSTAVE BONNET, New York, N. Y.—*Horseshoe.*—August 7, 1866.—A toe of crescent form is connected to two arms extending obliquely backward to the heel on either side of the frog. Three clips, hinged respectively to the toe and to the extremities of the oblique arms, are hooked over a rubber band encircling the upper part of the hoof.

Claim.—First, the peculiar shape of my shoe, as shown in Fig. III.

Second, the rubber band F in the combination and for the purpose specified.

Third, the combination of the shoe with the clamp D, the hooks E E, and the band F, as and for the purpose specified substantially.

No. 56,889.—WILLIAM BOYNTON, Auburn, N. Y.—*Tapping Barrels.*—August 7, 1866.—In screwing the faucet into a socket in the barrel it drives the metal plug before it, until apertures in the faucet shank register with corresponding apertures in the socket to permit the escape of liquor.

Claim.—First, the solid plug F for shutting off the contents of the cask, as above set forth.

Second, closing the end of the faucet G by means of a solid plug and projecting therefrom the tenon J, for the purpose specified.

Third, the apertures in the thimble A, marked 1 2 3, and the corresponding apertures in the screw portion of the faucet H, marked 4 5, &c., when used as and for the purpose specified.

No. 56,890.—ISAAC BRADLEY, Hartford, Conn.—*Breech-loading Fire-arm.*—August 7, 1866.—The spring slide or locking bolt in the stock operates with a swinging breech-piece to hold it in place when open and lock it when closed.

Claim.—The arrangement of the spring slide I in the stock A, operating with the breech-piece G, provided with the lug M, in the manner and for the purpose herein specified.

No. 56,891.—R. MOSS BRECKENRIDGE, West Meriden, Conn.—*Stove Pipe Damper.*—August 7, 1866.—The axis wire is bent into a recess in the cast iron damper to secure its rotation, and at its end forms a handle and a spring catch to hold it in either position.

Claim.—First, the rod A, combined and arranged with the damper plate C, substantially as and for the purposes herein set forth.

Second, the spring handle B at the upper part of the rod A, combined with the rod A and damper plate C, substantially in the manner and for the purpose herein shown and described.

No. 56,892.—S. O. BRIGHAM, San Francisco, Cal.—*Body Conformator.*—August 7, 1866.—A "body" of gum elastic adapted to be drawn tightly about the person and conform automatically thereto, is provided with rows of spurs which puncture the pattern paper that is laid thereon, and thus indicate the lines for seams.

Claim.—An apparatus or implement for the cutting and fitting of ladies' dresses and other garments, which, when applied to the person, will adjust itself thereto, and is provided with any suitable means for indicating the line or lines of the seams for the garment to be cut, substantially as herein described.

No. 56,893.—ROBERT BROWN, Dayton, Ohio.—*Reaping and Mowing Machine.*—August 7, 1866.—The arms of the combined rake and reel revolve about a common centre, and have also independent centres of motion or pivots upon which they may vibrate vertically in moving about their common centre to discharge the grain from a quadrant-shaped platform. The rake arms are rotated by means of their attachment to a crown wheel which is dished to receive within its circumference a peculiarly formed cam, which acts upon the projecting heel end of the rake and reel arms to give them the necessary rising and falling motions to pass over the main frame, and to remove the grain. The finger bar is hinged to the coupling arm so as to be allowed to conform to the surface of the ground, but may be made rigid in its connection by means of a metal strap bolted over the joint.

Claim.—First, the combination of a crown wheel, which is adapted for receiving and forming independent bearings for vibrating rake and reel arms, with a cam F, which is fixed to the post E, around which said wheels turn, and which is so constructed as to act upon the inner projecting arms of the rake and reels, substantially as described.

Second, dishing or opening the crown wheel D^2 for the purpose of receiving the cam F and allowing of depression of the inner projecting ends of the reel and rake arms, substantially as described.

Third, the construction and combination of the cam F and guard F', for the purposes substantially as described.

Fourth, in a combined reaping and mowing machine, having a rake attachment substantially as described, providing means, substantially as described, for making flexible or rigid the joints at e', substantially as specified.

Fifth, the arrangement of the bending and gathering board L', directly on the jointed drag bar L, the said board extending along the whole length (or nearly so) of the drag bar, and serving to assist the rake and reel in getting the grain on and off the platform without interfering with the motion of the platform, substantially as described.

No. 56,894.—WILLIAM H. BROWN, Worcester, Mass.—*Bag Fastener.*—August 7, 1866; antedated August 2, 1866.—A jointed metallic clamp is provided at one end with a series of notches, and at the other with a lever, near the fulcrum of which is a link to be hooked into either of the notches. The lever is then turned back upon the clamp and sprung beneath it to prevent accidental release.

Claim.—First, the combination and arrangement of the segments A B and C, the lever D, and link E, substantially as described.

Second, the method described of securing the clasp against accidental release.

No. 56,895.—WILLIAM H. BROWN, Stockwell, Ind.—*Fence.*—August 7, 1866.—Stakes and double riders are applied at the centres of the zigzag panels. The ends of the panels rest on blocks and are connected by clasps, beneath which are inclined braces resting on the ground.

Claim.—First, the combination of the blocks B and C, and inclined corner stakes D with the panels A, when the blocks, stakes, and panels are constructed and arranged substantially as herein described and for the purpose set forth.

Second, the combination of the long poles or rails E F and stakes G with the panels A of the fence, substantially as described and for the purpose set forth.

Third, combining the lower top rails E with the tops of the uprights a^2, substantially as described and for the purpose set forth.

No. 56,896.—N. A. BUHLE, New York, N. Y.—*Composition for Grinding and Polishing.*—August 7, 1866.—Composed of blood from which the serous particles have been removed, 5; linseed oil, 3; emery or other polishing powder, 10 parts; mixed and moulded.

Claim.—A compound for grinding and polishing, made as herein set forth.

No. 56,897.—EDWIN B. BUTLER, New Britain, Conn.—*Machine for Making Eyelets.*—August 7, 1866.—The intaglio dies are formed in rows in a plate of steel, which is, by a reciprocating feed band, advanced successively to a set of punches corresponding in number and position with the dies in each rank, and then to a corresponding set of ejectors.

Claim.—First, the employment of the movable die plates c, having two or more rows of dies d, substantially as and for the purpose described.

Second, the male and female dies, constructed and operating as described.

Third, the employment of the clearers q, in combination with the die plates c, arranged and operating substantially as and for the purpose described.

Fourth, the employment of the slide plate n and pawl n', in combination with the steps or pins e and die plates c, substantially as and for the purpose described.

Fifth, in combination with the male and female dies the clearer A', substantially as described.

No. 56,898.—SAMUEL G. CABELL, Quincy, Ill.—*Operating Steam Engines.*—August 7, 1866.—As the piston reaches the end of its stroke the valves in the piston head are opened by contact with the cylinder head, and admit steam from behind the piston to that space about to receive the incoming steam.

Claim.—First, providing for the use of a portion of the spent steam in the cylinder to fill the vacuum on the opposite side of the piston previously to the admission of fresh steam for its return motion, by means substantially as herein specified.

Second, the arrangement of a valve or valves in the piston of a steam engine, substantially as and for the purpose set forth.

No. 56,899.—S. G. CABELL, Quincy, Ill.—*Piston for Steam Engine.*—The pressure of steam upon the flexible, corrugated disks on either side of the piston head forces the margin steam-tight against the internal surface of the cylinder.

Claim.—The combination of the flexible disks D with a steam or other piston, operating substantially as herein specified.

No. 56,900.—JOSEPH CANTNER and MICHAEL ULRICH, Millheim, Penn.—*Horse Netting.*—August 7, 1866; antedated July 5, 1866.—The bights of the lashes are passed through their respective holes in the strap, and are retained by the running cord. Thimbles units the adjacent lashes and display the reticulation.

Claim.—Connecting the lashes B B with the straps A A by means of looping through single holes *a*, and hol ing them by locking lashes or cords C, substantially as and for the purpose herein specified

Also, the metallic clasps D D, in combination with the lashes B B, substantially as and for the purpose herein set forth.

No. 56,901.—JOSEPH CARLIN, Cincinnati, Ohio.—*Wheel.*—August 7, 1866.—The hub dishes toward the centre where a projecting flange divides the spokes, which are arranged alternately on each side of it.

Claim.—The arrangement of hub A, having the described concave or dished periphery C C and central collar B, which supports on their inner sides two sets of straddling spokes F F, substantially as and for the purpose set forth.

No. 56,902.—WILLIAM H. CATELY, New York, N. Y.—*Needle-feed Sewing Machine.*—August 7, 1866.—The claims describe certain arrangements of the vibrating and reciprocating needle, and the shifting of its bearing point for varying the length of feed, in connection with a rocking looper pivoted on a sliding bar, the looper being rocked one way by a fixed arm plate, and returned by a spring; the bar is reciprocated positively by a crank-pin working in a slot in an arm on the bar.

Claim.—First, the combination of the mechanism for operating the needle and varying the extent of the feed with the mechanism for operating the looper, the same being constructed, arranged and operating substantially as described.

Second, the arrangement of the looper slide J, so as to be operated directly by the crank pin of the shaft G, in combination with the arrangement of the needle, so as to be operated directly by the crank pin of the shaft C, substantially as described.

Third, the construction and arrangement of the plate K, looper J', and spring *s*, in combination with a needle, which is hung upon a crank pin, substantially as described.

No. 56,903.—HENRY A. CLARK, Boston, Mass., and HENRY J. GRISWOLD, Norwich, Conn.—*Writing and Drawing Card.*—August 7, 1866; antedated August 1, 1866.—Explained by the claim.

Claim.—A card or tablet, first printed upon and the printing or surface covered with a transparent water-proof composition, so that the surface will receive pencil or ink marks which can be rubbed or washed off without defacing the printed matter, lesson, or design, substantially as described.

No. 56,904.—J. S. CLARK, Auburn, Mass.—*Butter Tongs.*—August 7, 1866.—Tongs with broad indented plates to grasp the pats of butter.

Claim.—First, the combination of the parts marked A B C and D, constructed and arranged in relation to each other substantially as and for the purposes set forth.

Second, in combination with the plates or pads D of the butter tongs as described, the indentations or ridges *c*, formed in or upon said plates, as and for the purposes set forth.

No. 56,905.—CHARLES W. COPELAND, New York, N. Y.—*Tube Thimble.*—August 7, 1866.—The semi-cylindrical thimble occupies the upper portion of the tube and does not form an obstruction for ashes to collect against.

Claim.—The construction of the calorimeter thimble, substantially as herein shown and described, so as to govern the calorimeter and reduce the draught without obstructing the lower part of the tube, as set forth.

No. 56,906.—JOHN DABLE, Chicago, Ill.—*Machinery for Unloading Railroad Cars.*—August 7, 1866.—Explained by the claims and cut.

Claim.—First, the horizontally vibrating arm *d* constructed with a loop *d'*, crosshead *d'*, and provided with a pulley *c²* and anti-friction rollers, in combination with the bridge bearing *d³*, all operating substantially as described for laying the rope evenly upon the drum *c'*.

Second, the construction and manner of application of the housing E E, in connection with the machine herein described for unloading railroad cars and other receptacles of their contents, substantially as set forth.

Third, the combination and arrangement of the swivel coupling *e c'*, drum C', rope *b*, and hinged frame G, substantially in the manner and for the purpose described.

Fourth, the combination and arrangement of the spring friction device J, drum C', rope *b*, and hinged frame G, substantially in the manner and for the purpose described.

Fifth, the combination and arrangement of the V form spring stop *b²*, clutch lever D, connecting rod D², lever D¹, drum C', and ropes *b b¹*, substantially as and for the purpose described.

Sixth, connecting the frame or shears G to the sill A¹, by means of a rod *g*, and providing at the same time for adjusting said frame laterally and establishing it at any desired point, substantially as described.

Seventh, the combination of the convex ends *v w* of the pulleys *h h* with the convex surfaces of the plates or jaws *j j* of the hinged frame G, substantially as and for the purpose described.

Eighth, the combination with the drum C', friction brake J, clutch lever D, spring stop *b²*, connecting rod D², lever D¹, rope *b b¹*, and hinged frame G, substantially as and for the purpose described.

No. 56,907.—WILLIAM EDWARD DAVIS, Jersey City, N. J.—*Propeller Screw.*—August 7, 1866.—Each blade is made of a blank of a general circular shape; a slit being made which enables the perforated arms to be so bent as to connect to the shaft at distant points, the blade assuming a spiral form.

Claim.—First, the propeller screw blades A A cut out of flat metal plates and bent up to shape, substantially as herein described.

Second, the mode of connecting and fastening the separate screw blades upon the shaft or a propeller in the manner and for the purposes substantially as herein described.

No. 56,908.—JOHN DEGNON, Cleveland, Ohio.—*Governor.*—August 7, 1866.—The longitudinal position of the propeller in the cylinder, due to its rate of speed, determines by the connecting rack and segment the position of the damper in the draught pipe.

Claim.—The combination of the propeller B revolving in water, in the cylinder A, the spindle C, and rack D, applied with the toothed segment E, attached to the spindle *d*, operated through the medium of the gearing *a b*, substantially in the manner and for the purpose represented and described.

No 56,909.—ALBERT L. DEWEY, Westfield, Mass.—*Stop Device for Sewing Machines.*—August 7, 1866.—The band pulley moves in but one direction; a reverse motion of the shaft loosens it in the pulley and it fails to communicate motion thereto.

Claim.—The spring D applied to shaft B, either with or without the bubbed pulley A E, substantially as shown and used, all arranged to operate in the manner substantially as and for the purpose set forth.

No. 56,910.—WILLIAM M. DOTY, E. P. DOTY, and ELLIS DOTY, Janesville, Wis.—*Washing Machine.*—August 7, 1866.—Improvement on their patent of January 30, 1866. Springs on each of the fulcrum pins of the oscillating washboard extend in opposite directions so as to be wound by the depression of the handle. The projecting cheeks attached to the handle keep the water from splashing out at the ends of the tub.

Claim.—First, the combination of the spiral springs *c d*, fulcrum pins *b*, and oscillating washboard B, constructed as described and operating substantially as and for the purpose specified.

Second, in combination with the above the projecting cheek pieces C with chamfered edges, constructed and operating substantially as and for the purpose described.

Third, the cleats F fastened to the ends of the tub and provided with sockets to receive the upper ends of the legs E, which are fastened to the lower edges of the ends of the tub with screws or bolts, substantially as and for the purpose set forth.

No. 56,911.—AMOS DURANT and HENRY GRISSIRN, Stockton, Cal.—*Washing Machine.*—August 7, 1866.—The pendulous dasher is reciprocated in the suds box by means of parallel levers.

Claim.—The application and combination of the levers C C and D D, working from different centres and connected with the plunger E, and worked by reciprocating motion, as herein set forth.

No. 56,912.—CHARLES K. EHLE, Greenbush, Wis.—*Fanning Mill.*—August 7, 1866.—The blast impinges upon the stream descending from the hopper, removing the chaff by an upward current; the sorting and screening are performed by a succession of riddles.

Claim.—First, constructing the drum with its front side extended upward, and its top extended forward and upward, so as to form a throat through which the wind is discharged immediately beneath the hopper and above the screens, substantially as described and for the purpose set forth.

Second, the combination of the cross-bar I and perforated screen board J with each other, and with the shoe G of the mill, substantially as described and for the purpose set forth.

Third, the combination and arrangement of the sieve P, trough R and discharging spout, with each other, with the shoe G, and with the side O of the mill, substantially as described and for the purpose set forth.

Fourth, the combination of the sieves K L, trough M, and spout N with each other, with the shoe G, and with the side O of the mill, substantially as described and for the purpose set forth.

Fifth, the combination of the inclined board S, trough T, and spout U with each other, with the shoe G, and with the side O of the mill, substantially as described and for the purpose set forth.

No. 56,913 —HOSEA ELLIOTT, Globe Village, Mass.—*Picker Motion for Power Looms.*—August 7, 1866.—To secure an uniform pick motion whatever may be the speed of the cam shaft, the shuttle is thrown independently of the shaft by means of a strong spring and loose lifting crank which falls suddenly through a portion of its revolution and allows the spring to act to move the staff inward.

Claim.—First, in combination the loose crank N,' sweep N, the lever R, and the spring P, for giving motion to the picker staff, substantially as described.

Second, operating the picker staff of a loom by means of the appliances that act independently of the driving shaft, substantially as described.

Third, in combination the cam U, the arm G, and the cord F, for drawing the picker staff inward, substantially as described.

No. 56,914.—CHARLES ELVEENA, San Francisco, Cal.—*Tinting Photographs, &c.*—August 7, 1866.—The picture being protected, the background is tinted by the exposure of the paper in a close box to the fumes of tobacco.

Claim.—The mode herein specified of tinting surfaces for use in the arts by the action of smoke or fumes within a closed chamber, as specified.

No. 56,915.—JOHN EVANS, Virginia City, Nevada.—*Safety Cage for Mines.*—August 7, 1866.—The upward vibration of the lever permits the spring dog to engage the opening in the side of the cage and arrests its fall.

Claim.—The employment or use of the lever A, spring bar B, with the friction roller e and guide pin d, or their equivalents, when arranged substantially as described and for the purpose set forth.

No. 56,916.—WM. M. EVERETT, Malden, N. Y.—*Apparatus for Tempering Chisels.*—August 7, 1866.—The supporting shelf is fixed within the tub and a bar raised for the tools to lean against.

Claim.—The combination of the table B and rack D with the tub or water bath A, for the purposes hereinbefore set forth.

No. 56,917.—P. H. FERL and W. LARKINS, Detroit, Mich.—*Fishing Apparatus.*—August 7, 1866.—The net is spread upon the bolts at the extremities of the radial telescopic spokes and "cast" by the simultaneous withdrawal of all the bolts.

Claim.—First, a net of the form and construction herein substantially described.

Second, the construction and combination of the tube A, the arms F, the bolts M, with the springs, ropes, and pulleys, substantially as herein described.

Third, the combination of the tube A, the arms F, the bolts M, and the springs, ropes, and pulleys, substantially as herein described, with the net for the purpose of spreading, casting, and sinking the same, substantially as described and for the purpose hereinbefore set forth.

No. 56,918.—WESLEY S. FERRIER, Indiana, Penn.—*Churn.*—August 7, 1866.—The driving apparatus is attached to the lid; the central and sleeve shafts of the respective gears are detachably socketed into the respective dasher shafts which rotate in different directions.

Claim.—The combination and arrangement of the dashers or beaters B B, shafts c c', with the shafts F F', and gear wheels G G, gear wheel H, all for the purposes and substantially as herein described.

No. 56,919.—JOSEPH FIELDHOUSE, Taunton, Mass.—*Pipe Welding Furnace.*—August 7, 1866.—This compound furnace has one main and one auxiliary chamber of combustion, each chamber being provided with a separate gate. It differs from ordinary furnaces of this class mainly in the arrangement of the fuel-feeding throats and in having a floor composed of bars instead of a solid floor at the bottom of what is usually termed the oven.

Claim.—The arrangement of the fuel-supply throats c c c c of the welding beat chamber in the partition G, which separates it from the oven, and with respect to the fuel-supply throats D D D D of the fire-place of the oven, substantially in manner as specified.

Also, the combination and arrangement of the series of bridges C C C with the oven and its fire-piece, the welding heat chamber and the two series of fuel-supply throats c c c c, D D D D, arranged in the side wall of the oven fire-place and in the partition wall G, substantially as specified.

No. 56,920.—J. T. FOSTER, Jersey City, N. J.—*Pen and Pen Holder.*—August 7, 1866.—The pivoted pen has two effective ends and its shank for the time being is held to the barrel by the case, when the sleeve is retracted even with the offset in the slot.

Claim.—A single or double-pointed pen connected with the holder by a swivel joint, constructed as and for the purposes set forth.

Also, in combination with the above, the offsets in the case, substantially as and for the purposes specified.

No. 56,921.—SAMUEL T. FOWLER, Brooklyn, N. Y.—*Boot and Shoe*—August 7, 1866; antedated July 27, 1866.—Guard plates and screws attach the India-rubber facing to the shoe sole.

Claim.—Securing to the soles of boots and shoes India-rubber soles by means of screws, the washers upon said screws being expanded into guards, as shown in the drawing and described herein.

No. 56,922.—JOHN GOULDING, Worcester, Mass.—*Spinning Mule and Jack*—August 7, 1866.—In this class of machines the stretching and winding of the yarns are effected alternately and intermittingly. This is an improvement on his patent of May 2, 1865. which is an upright machine. The roving spools are stationary, while the delivering jaws travel in a vertical path to and from the stationary spindles. By means of a set of stationary jaws and the travelling jaws the weight of the roving extending from the spools is not permitted to affect the unwinding of the rovings from the spool. The travelling jaws are operated by ropes during the movement of the carriage. To regulate the tightness with which the yarn is wound in cops upon the spindles, a grooved pulley, a tightening pulley, and an endless rope (which may slip more or less upon its pulleys when desired) are used. The fifth clause relates to means for compensating for variations in the length of the endless rope used for transmitting motion to the spindles during winding. The sixth clause embraces the means for regulating the speed of the spindles during the winding. The seventh clause is for giving a rapid motion to the spindles during the stretching and hard twisting, and also for stopping the same by means of a tightening pulley. The eighth clause covers devices for preventing a loss of time after the hard twisting is completed and before backing-off is commenced consequent upon the momentum of the parts still continuing to turn the spindles forward. The ninth clause embraces devices for permitting the variable mechanism that regulates the backing-off or winding of the yarns to be re-set rapidly after one set of cops is spun and before a new set is commenced.

Claim.—First, the combination in a spinning machine of the following instrumentalities, viz: the carriage and jaws for the rings and the stationary turning-spool support, substantially as set forth.

Second, the combination in a spinning machine of the following instrumentalities, viz: the stationary jaws, stationary turning-spool support, travelling jaws and carriage, substantially as set forth.

Third, the combination in a spinning machine of the following instrumentalities, viz: the travelling carriage, the shaft thereof, and ropes for turning the said shaft, substantially as set forth.

Fourth, the combination in a spinning machine of the following instrumentalities, viz: the drum for imparting motion to the spindles, pulleys, endless rope, tightening pulley, and adjustable weight, substantially as set forth.

Fifth, the combination in a spinning machine of the following instrumentalities, viz: the drum for imparting motion to the spindles, pulleys, endless rope, tightening pulley, adjustable weight, and second tightening pulley, substantially as set forth.

Sixth, the combination in a spinning machine of the following instrumentalities, viz: the regulating wire, friction brake, and drum for imparting motion to the spindles, substantially as set forth.

Seventh, the combination in a spinning machine of the following instrumentalities, viz: the drum for imparting motion to the spindles, pulleys, endless belt, tightening pulley, and cams, substantially as set forth.

Eighth, the combination in a spinning machine of the following instrumentalities, viz: the drum for imparting motion to the spindles, friction brake, and cam, substantially as set forth.

Ninth, the combination in a spinning machine of the following instrumentalities, viz: the arm of the backing-off mechanism, shifting screw and section nut, substantially as set forth.

No. 56,923.—F. N. FROST, New Britain, Conn.—*Bridle Bit.*—August 7, 1866.—The bit rings are connected by a chain covered with a closely coiled wire.

Claim.—A bit for horses or other animals made of a closely coiled or spirally wound metallic wire, substantially as described.

Also, in combination with the above, the use of a chain, or its equivalent, as and for the purpose specified.

No. 56,924.—AUSTIN FULLER, Plymouth, Ind.—*Hay Loader.*—August 7, 1866.—Explained by the claim and cut.

Claim.—The hereinbefore described arrangement of a hay loader, consisting of a derrick C mounted upon the centre of one side of the rack, and sustained by the transverse timbers B and B' so as to turn freely in all directions and by a guy I attached to a post D upon the opposite end of the timbers and restrained from turning by another guy H, substantially as described.

No 56,925.—A P GARRETSON, Ripley, Ohio, and J M. HOFFMAN, Miami, Ind.—*Loom.*—August 7, 1866.—The endless belt to which the hooks that operate the picker staff are attached. passes over two fixed rollers and insures a proper action of the hooks. To the lathe is secured an endless belt, which passes over a roller on the breast beam and operates the pattern shaft The square block on said shaft is met by the square end of the lever latch, which is pivoted to the breast, rides over a pin on the sword of the lay, and at the proper period prevents the shaft from turning too far forward.

Claim.—First. the endless belt C and picker A with the hooks H and side binders G fastened in the breast beam B. and the pin K, in the lathe J, all arranged as described for the purpose of vibrating the shuttle.

Second, the shedders M with their projections N and elevators P passing through them, with the arms Q and reacting pulley T on the cam shaft R, when arranged and combined as herein described and for the purpose set forth.

Third, the square block W on the shaft R with the falling latch X, when operated by the lathe as herein described.

No. 56,926.—STEPHEN L. GEORGE, Decatur, Mich.—*Clothes Dryer.*—August 7, 1866.—Explained by the claim and cut.

Claim.—The arrangement and combination of the crank pulley C, tightening pulley K. grooved pulleys D D, with the posts B B, constructed and operated as herein described and for the purposes set forth.

No. 56,927.—NAPOLEON B. GOUSHA, Baltimore, Md.—*Hydrant.*—August 7, 1866.—The valve is raised by rotating the stem; and after closing, the contents of the discharge pipe are drawn into the lower chamber by raising the wrench rod and the inverted cup attached thereto.

Claim.—First, the perforated hollow screw valve D, in connection with the flange c, leather washer f, as constructed and arranged in combination with the chamber B, neck a, and supply pipe C, substantially in the manner and for the purpose herein set forth.

Second, the combination of the perforated hollow screw valve D with the wrench rod F and cup of valve e, substantially in the manner and for the purpose as set forth.

No. 56,928.—W. W. GRIER and R. H BOYD, Hulton, Penn.—*Scissors.*—August 7, 1866.—A sharp ripping hook on the back of one blade.

Claim.—Providing shears or scissors with the ripping hook a, substantially as and for the purpose set forth.

No. 56 929.—J. A. GRIGGS, Charleston, Ill.—*Head Block for Saw Mills.*—August 7. 1866.—The log is set to the saw by a hand rod attached to a toggle-joint lever, working a rock shaft to which are attached pawls that engage in a rack to feed the head-block forward.

Claim.—The setting of logs to circular saws by means of a bar or handle passing over the log and saw, to within convenient reach of the sawyer, and connected by a toggle and rock shaft to pawls which engage with racks on the head-blocks, substantially as herein shown and described.

No. 56,930—NICHOLAS GROEL, Newark, N. J.—*Travelling Bag Frame.*—August 7, 1866—Strengthening plates are bent around the jaws. where they are hinged together.

Claim.—The angular, encasing, metallic band D, made as described, for strengthening the hinged portions of a travelling bag frame, substantially as and for the purpose specified.

No. 56,931.—JOEL HAINES, West Middleburg. Ohio.—*Show Case.*—August 7, 1866.—The rack and shelves are fitted conveniently for placing articles upon them, and are revolved to display their contents.

Claim—The revolving shelves and rack or frame constructed and arranged substantially as described for the purpose set forth.

No 56,932.—C. H. HALE, Fayetteville, N. Y.—*Washing Machine.*—August 7, 1866.—Covered rollers are secured by side springs in a swinging frame, and work over square ribs in a perforated concave board.

Claim.—The arrangement of the perforated and ribbed washboard and covered rollers E, in combination with the adjustable springs K, as herein described, and for the purposes set forth.

No 56,933.—MARSHALL P HALL, Manchester, N. H.—*Key Fastener.*—August 7. 1866.—The key is secured against ejection, or turning from the outside, by a bar inserted in the lower part of the keyhole and clamped to the bow of the key.

Claim.—The devices herein described, that is to say, the notched piece a, or its equivalent, entering the keyhole at one point, and secured to the key at another point by the clamp c, the washer w, and the nut n, or their equivalents.

No. 56,934.—B. I. HARRIS, Harrisburg, Penn.—*Fruit Jar.*—August 7, 1866.—The can is sprung out of circular form in forcing the. ends of a cross-bar into a groove within the mouth. The cover being forced on to its seat by a screw passing through it and into the cross-bar braces the can to prevent the escape of the bar.

Claim.—The open-mouthed can A, provided with corrugations and groove arranged and used with the cross-bar B, and top C, substantially in the manner and for the purpose herein specified.

No. 56,935.—HORACE HARRIS, Newark, N. J.—*Perpetual Calendar.*—August 7, 1866; antedated July 25, 1866.—The supporting frame is marked with the days of the week and the slide with the days of the month, arranged in rows so as to be adjusted once for each month.

Claim.—The table B, arranged to slide by the days C, substantially in the manner and for the purpose set forth.

No. 56,936.—SAMUEL HARRIS and D. HARRIS, Shippensburg. Penn.—*Horse Hay Fork.*—August 7, 1866.—A pair of forks is mounted on the end of a lazy-tongs frame pivoted between two pairs of longitudinal bars whose upper ends are connected by a cross-bar. The upper end of the lazy-tongs is hooked to a lifting rod which passes down through the cross-bar and held by a trigger which is withdrawn by a cord to discharge.

Claim.—The two pairs of bars *a a*, pivoted at their upper ends to a cross-bar *c*, in combination with the levers B B', and forks C C, connected or applied to the bars *a a*, as shown, and provided at their upper ends with the hook *f*, and slotted bar E, or other equivalent fastening to connect the rod A with the levers B B', all arranged substantially as and for the purpose set forth.

No. 56,937.—A. B. HARTILL, New York, N. Y.—*Fishing Line Reel.*—August 7, 1866.—The reel is mounted on steel centres, one of which is adjustable by a milled head and secured by a stop screw.

Claim.—The set-screw *a*, in combination with the fixed stud *b* of the reel, substantially as and for the purpose specified.

No. 56,938.—CHARLES H. HELMS, Poughkeepsie, N. Y.—*Leather Chamfering Machine.*—August 7, 1866.—The cutter is mounted in the bevelled face of the projecting rim of the wheel and the leather is supported by a guide having vertical and horizontal adjustment.

Claim.—First, the cutter wheel, having a projecting rim with a bevelled face in combination with a cutter arranged in the said rim, substantially as hereinbefore set forth, for the purpose of chamfering or scarfing pieces of leather.

Second, the combination of a standard composed of two parts, substantially as hereinbefore set forth, with a guide plate, for the purposes as hereinbefore described.

Third, the method of clamping or holding the piece of leather at its back side, and back of the edge of the cutter, substantially as hereinbefore described, in combination with the cutter wheel and guide plate, for the purposes hereinbefore set forth.

No. 56,939.—HIRAM W. HAYDEN, Waterbury, Conn.—*Magazine Fire-arm.*—August 7, 1866; antedated August 3, 1866.—A sliding breech-pin is secured by a locking brace, and both are connected to a groove in the hammer so as to be retracted in the act of cocking, but on raising the hammer to full cock they escape from their groove and close the breech chamber, carrying into it a cartridge which was automatically elevated from the magazine.

Claim.—First, the blocks *e*, formed as a T-head to the shank *e'*, and both occupying grooves in the breech pin *d*, in combination with the housing *c c'*, having notches for the reception of the ends of block *e* at each side of the breech pin, as and for the purposes specified.

Second, the cross block upon a shank passing into a groove in the breech pin, in combination with a spring to throw said block down into place as the breech is closed, as set forth.

Third, the slide exploding punch fitted as specified, and actuated by forcing the said cross block down to its place as set forth.

Fourth, constructing the hammer with the latch and groove, taking the pin on the said cross block, as and for the purposes specified.

Fifth, the combination of the hammer, the cross block, the breech pin and the spring *n*, as specified, whereby the said hammer lifts the said cross block, and draws back the breech pin in the act of cocking the piece, and the breech closes itself, as set forth.

Sixth, the combination of the breech pin and sliding cartridge conveyer with the lever *t*, actuated substantially as and for the purposes specified.

No. 56,940.—CHARLES E. HOFFMAN, Jersey City, N. J.—*Scale.*—August 7, 1866.—This is a two-panned weighing scale having an arrangement of levers suspended upon adjustable knife-edge bearings, and cannot be briefly described other than in substantially the words of the claims.

Claim.—First, making the knife edges H' J' adjustable in slots, or their equivalents, in the levers C' and G, and arranged to operate relatively to said levers and their connections, substantially as and for the purpose specified.

Second, making the lever G in a single bar, with openings to receive the links I' I', and to support them centrally therein, by which arrangement to construct the scale at less expense and with less liability of derangement.

Third, receiving the post M' within openings in the levers C' and G, so that these parts shall mutually steady and support each other, substantially as herein specified.

Fourth, the rigid posts e' e³, and straight knife edges g' g⁴, in contradistinction to flexible links, and hollowed or notched knife edges, and arranged to operate substantially in the manner and with the advantages herein specified.

No. 56,941.—MARCUS L. M. HUSSEY, New York, N. Y.—*Machine for Holding and Filing Documents.*—August 7, 1866.—The papers are pressed against the front flanges of the case by a detachable plate supported by springs.

Claim.—An improved letter file having a removable or sliding door E, which may be used as such door and also as a paper folder or cutter.

No. 56,942.—SAMUEL HUTCHINSON, Griggsville, Ill.—*Gang Plough.*—August 7, 1866.—The front of the main frame is raised or lowered by a lever to regulate the angle and penetration of the ploughs, and a second lever is employed to hold the ploughs down to their work or raise them above the ground when required.

Claim.—First, the elevating or adjusting of the frame A of the machine in a vertical direction, in order to regulate the depth of the penetration of the ploughs, by means of levers H H, provided with lower segment ends a, and secured to the sides of frame A, in combination with the pendent pins E E, attached to frame A, and passing loosely through the axle D, substantially as shown and described.

Second, the plough frame I, fitted within the frame A, and connected by a chain O with a plate N, attached to a shaft L, over frame I, whereby the plough frame and ploughs may be raised when desired, and the ploughs when at work retained in the ground, substantially as set forth.

No. 56,943.—GEORGE N. JACKSON, Chicago, Ill.—*Sliding Slate for Computation Cards.*—August 7, 1866.—The slate extends across the width of the card and slides vertically thereon, so as to reveal any desirable portion and conceal the part below.

Claim.—The combination with the tubular card of a computing slate arranged to slide thereupon, substantially as set forth.

No. 56,944.—WILLIAM H. JACOBY, Xenia, Ohio.—*Convertible Stereoscope.*—August 7, 1866.—The eye pieces are adapted for connected or separate use, and are readily secured or detached by clamps engaging with their margins.

Claim.—First, the eye piece A A' B D D', formed for separate use, or in combination with a stereoscopic case, substantially as set forth.

Second, the arrangement of eye piece A A' B D D', lips E F, chamber G, and spring H, for the purpose explained.

Third, in the described combination with the elements of claim second, the receptacle I I' J J', substantially as set forth.

No. 56,945.—CHARLES H. JAMES, Cincinnati, Ohio.—*Dentists' Vulcanizer.*—August 7, 1866.—The cover of the flask is secured by a screw in the centre of a yoke, whose hooked ends pass through notches in a flange near the top of the flask, and engage under the flange when the yoke is turned.

Claim.—The yoke F F' f, central set screw G, and notched flange B b, arranged and operating substantially as and for the purpose specified.

No. 56,946.—ANDREW JAMISON; Taylorstown, Penn.—*Steam Engine Valve.*—August 7, 1866.—The valve is of saddle-shape above and below, and works between two seats with openings of equal area to balance the pressure of steam.

Claim.—The saddle-shaped valve C and its seats B and D, made of corresponding shape thereto, when arranged together, substantially in the manner and so as to operate as and for the purposes specified.

No. 56,947.—BARTON H. JENKS, Bridesburg, Penn.—*Hoisting Apparatus.*—August 7, 1866.—The car is drawn up by a rope, the pinion on the car engaging the rack alongside the inclined track. The rope actuates a shaft having a worm thread engaging the rotating gear, and the car remains stationary if the rope break.

Claim.—First, providing for raising or lowering a carriage which is mounted upon an inclined track or perpendicular shaft by means of ropes or chains, and at the same time to so construct the devices which act upon and move the wheels of said carriage, that should the hoisting rope or chain break, the carriage will remain in a steady and safe position upon the track, substantially as described.

Second, the means substantially as herein described for holding the upper end of the carriage down upon the inclined track, and preventing lateral displacement.

No. 56,948.—John Johnson, Saco, Maine.—*Dredging Machine.*—August 7, 1866.—An exhausted receiver in the barge is connected by an adjustable pipe and flexible connections with a spout, which is adapted to suck in the mud, &c., upon which it rests, and discharge it into the receiver for removal and subsequent discharge at the lower valve.

Claim.—First, the system or mode of dredging, consisting of an air-tight, flexible supply tube, an air-tight floating barge or vessel, and suitable pumps for removing air or water from the vessel, substantially as herein described.

Second, in combination with an air-tight floating barge, the discharge valve V, and the bulkheads that allow the water to flow over them into the pump well, as it is displaced by the entrance of sand or other dredged material.

Third, the cross-tube K L, in combination with the main supply tube, as specified.

Fourth, the combination of the cross-tube K L, with its bearings I I', for the purpose of raising and lowering the tube without disturbing the position of the orifice through which the mud or sand passes. .

Fifth, the combination of the cross-tube with the cap or stopper C, so fitted that it can be taken off for the purpose of removing obstructions, substantially as herein set forth.

No. 56,949.—William Jones and M. H. Collins, Chelsea, Mass.—*Aero-gas Burner.*—August 7, 1866.—The burner is in two parts with air inlets at the junction, regulated in capacity, or entirely closed by screwing the upper section down on the lower. The tip is a removable cap with small radial apertures for escape of gas for consumption.

Claim.—First, the improved aero-gas burner; that is, one made with or having a means of closing, and more or less opening its air inlet or inlets, as specified.

Second, the combination of the removable deflecting dome or cap C, made with eduction orifices in it and near its base or lower part, with the aero-gas burner as specified.

Third, the peculiar mode as described, in which to construct the aero-gas burner, viz., by uniting its two parts A B by screws *a b*, and arranging the air-inlet holes of upper part A with respect to such screws in manner as specified.

No. 56,950.—Albert Joyner, Elton, Wis.—*Washing Machine.*—August 7, 1866.—The pendent arms of the rubber are retained in their bearings by short hangers connected by a cross-bar.

Claim.—The manner substantially as shown of keeping the journals on which the rubbing concave is suspended in their bearings.

No. 56,951.—Oliver S. Judd, New Britain, Conn.—*Snap Hook.*—August 7, 1866.—A spring coiled around the pivot draws upon the heel of the latch to close and hold it.

Claim—Securing one end of the spring *g* in the heel of the shank of the hook or latch, while the other end bears against the opposite side of the chamber, and thereby produces a pulling motion, substantially as and for the purpose described.

No. 56,952.—Anson Judson, Brooklyn, N. Y.—*Lamp Chimney*—August 7, 1866.—The portion of the chimney nearest the flame forms a segment of a sphere, and its upper part a conic frustum.

Claim.—Constructing a lamp chimney in the form hereinbefore set forth to prevent the fracture of the chimney by the unequal exposure to heat to which other forms are subject.

No. 56,953.—J. W. Keene and W. E. Snediker, Utica, N. Y.—*Broom Head.*—August 7, 1866.—The ferrule is slotted so as to be compressed upon the handle by screwing down its cap. A nut within the ferrule draws up an axial rod connected by oblique arms with the articulated lower hoop, so as to expand it edgewise, and compress it transversely on the broom.

Claim.—First, the extension rod F, in combination with the links L L and the straps M M M M, or their equivalents, constructed and operating substantially as described.

Second, the head A, in combination with the arms G G, or their equivalents, constructed and operating substantially as described.

Third, the slotted ferrule C, the nut E, and the cap D, in combination substantially as described and for the uses and purposes mentioned.

No. 56,954.—Horace B. Kinney, Leonardsville, N. Y.—*Machinery for Making Forks.*—August 7, 1866.—A sliding head block carries a pair of pivoted die blocks, which have an opening between them, and dies upon their ends. The blank is held in suitable framing upon an adjustable yoke in front of the dies, which open as they move forward, taking the middle tine of a three-tined fork between them, bending the other two outward, the shape of the dies being such that the first operation of the dies spreads the tines partly open, and a second spreads them wide open, so that they may be operated upon by a forge hammer. Other suitable dies and yokes are substituted for bending four-tined forks in a similar manner. A pair of shears is also arranged in connection with the machine for splitting the blanks preparatory to bending.

Claim.—First, spreading and swaging the split blanks of two or many-tined forks by means of a die block and supports, constructed and operating substantially as described.

Second, the adjustable yoke support K, or its equivalent, in combination with a wedge-shaped reciprocating spreading die, substantially as described.

Third, sustaining the tang and the tines of a fork blank during the act of forming the shoulders against and between supports, constructed and arranged substantially as described.

Fourth, the movable shoulder plates *g g*, in combination with back supports N, and a reciprocating die or dies H, substantially as described.

Fifth, the yoke K³, with shoulders on it, in combination with a reciprocating shaping die H², or its equivalent, for shaping the forks after their tines have been drawn out under the hammer, substantially as described.

Sixth, the combination of splitting shears J with a fork-bending and swaging machine, constructed substantially as described, said shears being actuated by the driving shaft of the swaging dies, substantially as described.

Seventh, the gauges S S' to support the fork blanks in proper position during the operation of the shears in splitting these blanks, constructed and arranged substantially as described.

Eighth, the dies constructed and arranged substantially as described for making rounded or square shoulders at the junction of the tines of the fork blank with the tang or head, as set forth.

Ninth, the guide blocks P P, in combination with adjustable shoulder plates *g g*, or their equivalents, supported substantially as described, and reciprocating dies H H, the guides P P closing the dies H H in their forward movement toward plates *g g*, substantially as described.

No. 56,955.—ADOLPH F. KUHLMANN, Glenhaven, Wis.—*Washing Machine.*—August 7, 1866.—The clothes pass first under a soap roller rotated by the oscillator of the rubber frame, and then between feed rollers, which deliver them uniformly to the rubbers.

Claim.—First, the soap roller *k* in combination with the feed board *h* and oscillating shaft C, constructed and operating substantially as and for the purposes described.

Second, the feed rollers *g g* in combination with the oscillating rubbers E, wash-board *s*, soap roller *k*, and feed board *h*, constructed and operating substantially as and for the purpose set forth.

No. 56,956.—T. S. LA FRANCE, Elmira, N. Y.—*Steam Engine Governor.*—August 7, 1866.—Explained by the claims and cut.

Claim.—First, a governor for a steam engine, having balls G supported upon the vertical arms of nut levers E, which with the divergence of the balls, when in action, raise the weight D and actuate the valve rod, substantially in the manner set forth.

Second, in the mechanism of such a governor, the arms C attached to the spindle B, and forming the fulcrum for the nut levers E, substantially as set forth.

Third, arranging the balls G upon the vertical arms of the nut levers E, so as to regulate their action by altering their distances from the fulcrum, substantially as set forth.

Fourth, in combination with the flange D' and lever E, the friction roller F, the said parts being arranged substantially as and for the purpose set forth.

Fifth, in combination with the ball D and balls G, the spiral spring I, substantially as and for the purpose set forth.

No. 56,957.—JOHN S. LASH, Philadelphia, Penn.—*Step Ladder.*—August 7, 1866 —By the construction of the hinge-joint and the arrangement of the connecting bars, the legs are extended laterally while being opened to support the ladder.

Claim.—First, the combination of the eye plates *c*, socket plates *d*, extension braces F F, and hinged centre bar G, the whole constructed and operating substantially as and for the purpose set forth.

Second, providing the top step C of a step ladder with a hinged leaf C', substantially as described.

No. 56,958.—JOHN W. LATCHER, Albany, N. Y.—*Centre Chill for Car Wheels.*—August 7, 1866; antedated July 24, 1866.—The centre chill is constructed in two parts, one having a hole through it longitudinally to receive a tenon ; on the other an annular groove is formed around the chill in the centre for supporting the sand core which forms the oil recesses in the wheels.

Claim.—The employment of the metallic centre chill or core C C', constructed and held substantially as shown and described, in combination with the periphery chill, as set forth.

No. 56,959.—W. LEONARD and J. J. JOHNSTON, Allegheny City, Penn.—*Washing Compound.*—August 7, 1866.—Composed of sal soda, 32; lime, 16; borax, 3; spirits camphor, 1 part.

Claim.—The compound herein described, compounded of the ingredients named, and in the quantities specified, said ingredients being manipulated and treated in the manner and for the purpose herein described and set forth.

No. 56,960.—P. LUGENBELL and T. BARNES, Greensburg, Ind.—*Horse Rake.*—August 7, 1866.—Explained by the claims and cut.

Claim.—First, a horse rake for gathering hay, &c., when constructed with wheels C C and a revolving axle D, and two or more rows of teeth E E attached thereto, and having, also, reciprocating bars I I corresponding in number with the rows of teeth for checking the revolution of the axle and teeth, as they are successively brought into action, substantially as set forth.

Second, in combination with the bars I attached to the axle, the wings M attached to the wheel and guides K, with the flange K' for alternately checking and permitting the revolution of the axle and teeth, substantially in the manner set forth.

Third, in combination with the bars I, the lever F and bell crank H, arranged substantially in the manner and for the purpose set forth.

No. 56,961.—J. H. LUTHER, Petroleum Centre, Penn.—*Grab Tool for Oil Wells.*—August 7, 1866.—The serrated jaws of the hollow cylinder are driven down upon the tool to be extracted, and by their inclined rear faces and locking collar, grasp the same when the grab is lifted. To detach the grab the collar is lifted by the hook.

Claim.—The grab herein described, the same consisting of the hollow cylinder or tube A, provided with grab jaws B connected to a common loose collar D and spiral spring, or its equivalent, when all constructed and connected together, substantially as described and for the purpose specified.

No. 56,962.—JEREMIAH MAEBY, Portage, Wis.—*Flood Gate.*—August 7, 1866.—When the water rises a sufficient height the pivoted gate is tilted and the water allowed to flow beneath it.

Claim.—The tilting gate C pivoted so that the pressure of the water thereon shall cause it to open and close in the manner substantially as herein shown and described.

No. 56,963.—CHARLES S. MARTIN, Milwaukee, Wis.—*Wagon.*—August 7, 1866.—India-rubber springs are interposed between the axle and the bolster. Pistons beneath the bed pass through the bolster and rest upon the springs so as to give an elasticity to the bed up and down between the standards.

Claim.—First, in combination with the bars G, having shoulders resting upon the plates S, and the plates S, the tapering India-rubber springs E, substantially as and for the purpose set forth.

Second, constructing the hind bolsters of a wagon with a recess E', for the purpose of receiving an India-rubber spring, and with or without the strengthening plates and bands M, substantially as set forth.

Third, the double cups I and the caps K in combination with the India-rubber springs, in the form of frusta of cones or pyramids, the several parts being constructed and arranged for use, substantially in the manner and for the purpose set forth.

Fourth, in combination with projections upon the bolster plate P, a corresponding depression upon the top of the plate covering the caps K, substantially as and for the purpose set forth.

No. 56,964.—W. B. MASSER, Sunbury, Penn.—*Coffee Roaster.*—August 7, 1866.—The wire-cloth cylinder has an extended foot flange by which it is supported over the stove hole; a bent wire is revolved to stir the contents.

Claim.—A coffee roaster, composed of a wire-cloth receptacle, with a flange fitted upon it and provided with a stirrer and handle, substantially as shown and described.

No. 56,965.—JAMES D. MATTHEWS, Bowling Green, Ohio.—*Sawing Machine.*—August 7, 1866.—The frame is dogged to the standing tree and the saws, with teeth facing each other, operated upon opposite sides of the trunk, being pressed towards each other by springs as they are reciprocated.

Claim.—The saw blades hung to a common holder in such a manner as to open from and close upon each other, in combination with a frame A, or its equivalent, to receive the said saw-blade holder, when arranged together and operating substantially in the manner and for the purpose described.

No. 56,966.—JAMES MATTIX, Kokomo, Ind.—*Belt Coupling.*—August 7, 1866.—Explained by the claim and cut.

Claim.—The hereinabove described device for adjusting the length of belt couplings without removing the clasps B B, by the insertion of a three-sided link d', the ends of which are riveted by a yoke E attached by the screws F, substantially in the manner set forth.

No. 56,967.—EDWARD MAYNARD, Tarrytown, N.Y.—*Eye Glass.*—August 7, 1866.—The pivoted nose piece rests upon the nose and keeps the glasses from tipping.

Claim.—A stay or nose piece in combination with the frame of eye-glasses, substantially in the manner and for the purpose herein set forth.

No. 56,968 — WILLIAM McCORD, Sing Sing, N. Y.—*Horse Rake.*—August 7, 1866.—The rake is adapted to the varying heights of draught animals by the adjusting standards on the frame and thills. As the rake teeth are raised to discharge the load the forward teeth compress the hay in the windrow.

Claim —First, setting up on the four supports D and E an adjustable frame A, upon which the whiffletree is placed, in the manner and for the purpose above described.

Second, the gathering fingers C, to aid in depositing the hay in a compact row, in the manner described.

No. 56,969.—O. C. McCUNE, Darby Creek, Ohio.—*Corn Planter.*—August 7, 1866.—The seed slides and valves on each side are operated simultaneously by means of the lever and the connected gearing and rods.

Claim.—The combination of the lever Q, toothed segment P, pinion O, wheel N, slides K, and valves e, all arranged as and for the purposes set forth.

No. 56,970.—PRESTON McQUAID, Wenona, Ill.—*Machine for Marking Ground for Planting.*—August 7, 1866.—Three sharp-rimmed wheels are placed in a frame and revolve at corn-planting distance apart, impressing the ground; the middle wheel is in a vertically adjustable frame to enable it to conform to the surface of the ground and the weight of the driver is imposed thereon.

Claim —The frame A with the draught pole B attached, and having two marking wheels C C fitted within it on axles working in fixed bearings, in combination with the adjustable or rising and falling wheels D fitted within the frame A, and connected with the driver's seat F, substantially in the manner as and for the purpose herein set forth.

No. 56,971.—HARRISON B. MEACH, Fort Edward, N. Y.—*Pulping Fibre.*—August 7, 1866.—The process dispenses with the use of strong alkaline solutions and excessive pressure.

Claim.—First, reducing fibrous substances to a pulp by means of chlorine gas under pressure, in combination with an alkaline solution.

Second, the use of chlorine gas under pressure for the purpose of dissolving the gluinous substances during the first process in the preparation of paper stock from wood or other fibrous substances.

Third, using chlorine gas under pressure in dissolving silica in fibrous substances and then converting said substances into pulp and retaining the silica in the pulp.

Fourth, reducing fibrous substances to a pulp for the manufacture of manilla paper by means of chlorine gas under pressure in combination with the solution herein described.

No. 56,972.—S. MELLINGER, Jr., Mount Pleasant, Penn.—*Fruit Gatherer.*—August 7, 1866.—A pair of hinged toothed jaws, one movable by a cord and the other having an attached bag to catch the fruit. The head has a pair of swivel joints at its junction with the handle which enable it to be presented in any direction.

Claim.—The combination of the toothed jaws or frames A A², hinged together with a bag or receiver secured to one when the whole is swivelled or pivoted to a suitable handle so as to swing in right angular planes thereon, substantially as and for the purpose described.

Also, the trigger and cord connecting it with the jaw or frame A², as and for the purpose specified.

No. 56,973.—JAMES MERKEL, Mount Pleasant, Iowa.—*Paddle Wheel.*—August 7, 1866.—The paddles revolve with a slotted cylindrical casing and are consecutively radially protruded through the slots by means of an inner stationary cam groove, which is traversed by anti-friction rollers on the frames of the respective paddles.

Claim.—The combination of the stationary cam plate M, anti-friction rollers R R, and radially reciprocating paddles with a drum H G, having the spaces between the paddles filled and all constructed and arranged to operate substantially as described.

No. 56,974.—RUFUS S. MERRILL, Boston, Mass.—*Lamp.*—August 7, 1866.—The tube is slit and its flanged end sprung within the rim of the collar and there retained.

Claim.—First, the combination with a collar, fitting over the top of a lamp of a central tube secured to said collar and extending therefrom toward the bottom of the lamp, substantially as and for the purposes herein shown and described.

Second, the combination with a lamp collar provided with a groove, or the equivalent thereof as herein described, the cylinder or tube having a flange around one end and longitudinal slits or grooves in its sides, the whole being arranged for operation as set forth.

No. 56,975 —BENJAMIN F. MILLER, New York, N. Y.—*Stove Pipe Damper.*—August 7, 1866 —A double collar with an outer flange forms a connecting inner sleeve for the adjacent sections of pipe and contains the pivoted damper.

Claim —The divided case f, fitted for the reception of the damper and for setting between the lengths of stove pipe, said case being provided with flanges attached together, as and for the purposes set forth.

No. 56,976.—CHARLES MILLER, St. Louis. Mo.—*Harness Motion for Looms.*—August 7, 1866.—Each heddle frame is lifted vertically by two levers, whose outer ends bear upon the under side of the ends of said frame ; the inner ends of this pair of levers are connected so that both shall be simultaneously and equally acted upon by the strap or device which pulls them downward. The heddles descend by gravity.

Claim—The arrangement and application of a set of levers D^1 D^2 to produce the loom harness motion, substantially as set forth.

No. 56,977.—M. V. MILLER, Manchester, Penn., and GEORGE HENRY, Steubenville, Ohio.—*Axle Box.*—August 7, 1866.—The lower portion of the box is maintained by the springs in contact with the wrist of the axle and the escape of the oil is prevented.

Claim.—First, the plate H, slots G, in combination with the springs F, journal box B B', and case A, when arranged as in the manner and for the purpose set forth.

Second, the combination of case A, journal box B B', axle or journal X springs F F, stops G, and plate H, constructed and arranged substantially as shown and described and for the purpose set forth.

No. 56,978.—T. H. MILLER, Allentown, Penn.—*Apparatus for Cooling Liquors.*—August 7, 1866.—The box has a close fitting lid and nests surrounded by ice, and adapted to receive bottles.

Claim.—The metallic-lined box A provided with the lid D, with rubber strips *g*, and tubes C to allow the bottles to be placed, forming a receptacle between each tube for the ice F, the whole being arranged as and for the purposes herein set forth.

No. 56,979.—G. MOODY and W. P. HALL, Piqua, Ohio.—*Heating Stove.*—August 7, 1866.—This stove has an adjustable grate, outer doors, and internal communicating chambers to which external air is admitted.

Claim.—First, the lower hot-air chamber E, in combination with the upper hot-air chamber or oven H, with a flue and air passages for conducting heated air from the lower to the upper chamber, substantially as shown and described.

Second, in combination with the hot-air chamber N', and the perforations *n*, the hinged adjustable grate with the bail or pawl *l*, and the ratchet *l'*, substantially as shown and described.

Third, in combination with the lower chamber E, and the upper chamber H, with flue and air passages communicating between them as described, the movable plate G', as and for the purpose described.

Fourth, in combination with the flue *w'* and throat *t*, the damper I to perform the double function of a damper and scraper, substantially as described.

No. 56,980.—CHARLES N. MORGAN, Granby, Mass.—*Wrench for Nuts of Carriage Axles.*—August 7, 1866.—The circular plate has two opposite clamping jaws operated by a right and left hand screw, and a square socket formed in a boss at the centre of the plate. It is used by applying the socket to the nut upon the end of the axle and by means of the screw and jaws clamping the plate to the rim of the hub, then by turning the wheel the nut is screwed on or off as may be desired.

Claim.—The device for attaching the nut to the wheel, consisting of jaws *c c*, screws D, and plate E, provided with the socket G, the whole combined and arranged in the manner and for the purpose herein described.

No. 56,981.—RUFUS W. MORSE, East Berlin, Conn.—*Chuck.*—August 7, 1866.—The nut has grooves across the end for the jaws ; the stock has a funnel-shaped opening against which the inclined back portion of the jaws bear, whereby they are forced inward to grip the shank of the tool as the nut is screwed into the stock.

Claim.—The combination of the case *g* and sliding jaws *e* with the stocks *a a'*, substantially as and for the purpose described.

No. 56,982.—G. R. NEBINGER, Lewisberry, Pa.—*Fruit Dryer.*—August 7, 1866.—The shelves are so arranged as to give a sinuous course to the heated air. A shield is placed over the heater, into which the exit flue extends.

Claim.—First, a dry-house for fruit, having the side walls made double, with an interior opening at the top and an adjustable exterior opening at the bottom, as shown in Fig. 1, in combination with the hollow rear wall D, having the air passages and valve *m* arranged as shown in Fig. 2.

Second, arranging the racks *e*, so as to form the passages for the hot air at the opposite ends alternately, for the purpose of causing it to pass both over and under all the racks, as set forth.

Third, in combination with the racks *e*, arranged as shown and described, the furnace B, with the inclined shield C, when constructed and arranged to operate substantially as set forth.

Fourth, the extension of the smoke pipe F within the furnace B, as and for the purpose set forth.

No. 56,983.—ROBERT H. NICHOLAS, Chicago, Ill.—*Artificial Leg.*—August 7, 1866.—
The swivel hinge connection and springs give security, elasticity, and resilient contacting
surfaces to the prosthetic ankle articulation.

Claim.—First, the combination of the jointed connection C D, the cushion G, and spring
E, arranged and operating in the manner and for the purposes specified and shown.

Second, the combination of the cushion G, spring E, jointed rod C D, and the springs H I,
arranged and operating as and for the purposes set forth.

Third, in combination with a jointed connection C D, allowing a free movement of the
foot, the arrangement of the two cords F, and springs S, operating substantially as and for
the purposes described.

Fourth, the combination and arrangement of the connection C D, cushion G, springs E,
cords F, and springs S, operating substantially as set forth for the purposes specified.

Fifth, the recess L, in the lower part of the leg, when extending around upon the sides
thereof, as described, in combination with the heel L, constructed with a corresponding
projection extending around upon the sides of the foot, arranged and operating as specified
and for the purposes set forth.

No. 56,984.—W. H. NOBLES, St. Paul, Minn.—*Dredging Machine.*—August 7, 1867.—
Adjustable stanchions or supports are attached to the side of a steam dredge, and rest upon
the bed of the stream to steady the vessel in its operations. A grappling hook is combined
with the plough or dredging apparatus for the purpose of removing rocks, logs, &c., from
the bed of the river or harbor.

Claim.—First, the construction of the adjustable stanchions E, attached to a boat, as
herein described and for the purposes set forth.

Second, the arrangement and combination of the chain G, pulleys and grappling irons K,
when arranged and combined with the revolving ploughs B, as herein described and for the
purposes set forth.

No. 56,985.—ELI ODELL, Winterset, Iowa.—*Fence.*—August 7, 1866.—The projecting
ends of the rails of adjacent panels are scarfed together, and are laterally supported by braces.
At the corners the rails are notched into each other. The connection between the slats and
rails permits the former to remain vertical when the latter conform to an inclined surface.

Claim.—First, pivoting the rails T or bars U to the cleats Q or pickets P, in such a manner as to adapt the panels to the irregularities of the ground, as described, in combination
with the notches M N, for locking the panels together, as specified.

Second, the hooks M N for coupling the panels, in combination with the self-sustaining
brace posts and pivoted rails, made and operating as described.

Third, the transverse brace and supporter, consisting of the long braces E E, the collar
beam H, the short braces F F, and the spur posts G G, or their equivalents, as described.

Fourth, the manner of locking and securing the corners, as shown at R R R and S S S,
in connection with prolonging the boards to supply the want of a collar beam, as shown at J.

No. 56,986.—SOLOMON OPPENHEIMER, Peru, Ind.—*Joint for Frames and Legs of Tables.*—August 7, 1866; antedated August 2, 1866.—A round socket receives the leg, and the
clawed flanges grasp the wings.

Claim.—A metal clamp for joining the frame and legs of a table, having a socket A for
receiving the legs and wings B, with claws *a* for receiving the rails C, and also lips *b*, or
equivalent devices, for attaching the top of the table, substantially as set forth.

No. 56,987.—JOHN G. PAGE, Rockford, Ill.—*Automatic Gate.*—August 7, 1866.—The
approaching carriage, by passing over vertical arms attached to weighted levers, slides the
gate down below the surface of the road, where it is retained by spring catches. As the carriage recede its wheels depress a second lever, which reverses the weighted levers and closes
the gate.

Claim.—First, in combination with the vertically reciprocating gate A, the employment of
the levers D and arms E, arranged and operating substantially as and for the purposes set
forth and described.

Second, in combination with the above, the levers G and arms E, arranged substantially
as and for the purposes specified.

Third, in combination with the levers D, weighted upon their short arms, as described, the
employment of the springs *a a*, or their equivalent, arranged substantially as set forth.

No. 56,988.—AUSTIN FORD PARK, Troy, N. Y.—*Electro-chemical Telegraph.*—August 7,
1866.—The paper strip, just before passing under the recording needle, is moistened by contact with a wheel revolving in a reservoir of suitable liquid.

Claim.—The recording of telegraphic signs by electro-chemical action, in a chemically
prepared wet or moist line or path made in a strip or fillet of paper, as the latter is moved
along to receive the telegraphic signs, substantially as herein set forth.

No. 56,989.—L. PHLEGER, Philadelphia, Penn., and GEORGE G. LOBDELL, Wilmington, Del.—*Obtaining Oil from Wells.*—August 7, 1866; antedated August 2, 1866.—A continuous water pipe extends down into the well, and is passed through a furnace above ground to effect the circulation of hot water.

Claim.—Inducing the flow of oil from a well by melting the paraffine by the application of heat from hot water, conveyed in a circuit through a pipe, substantially as shown and described.

No. 56,990.—D. B. PIPER, Winchendon, Mass.—*Sewing Machine.*—August 7, 1866.—The yielding connection compensates for any want of parallelism between the upper and lower shafts, and prevents the loop being drawn too tightly upon the looper.

Claim.—The combination of the elastic connection, substantially as described, (consisting of the yoke D, the rod E, and the spring F, or their equivalents,) with the crank or crank wheel G of the looper shaft, and with the cam H applied to the cam shaft, as described.

No. 56,991.—MARTIN ROBBINS, Cincinnati, Ohio.—*Molasses Pitcher.*—August 7, 1866.— The hollow handle communicates above with the interior, and is surrounded by a hollow bulb of rubber, by compressing which air is forced into the pitcher to eject liquid through the spout.

Claim.—First, the pitcher, provided with a perforated handle, to which is attached a rubber bulb, for the purposes and substantially as described.

Second, the spout B, when extending nearly to the bottom of the pitcher, in combination with the handle C and bulb F, substantially as described.

No. 56,992.—C. ROBINSON and J. C. MARSHALL, Springfield, Mass.—*Boot Heel.*—August 7, 1866.—This removable boot heel consists of a metallic shell and bottom plate, with an interposed strip of rubber to give elasticity.

Claim.—The combination of the India-rubber strip C with the shell A and movable bottom B, arranged so that the edges of the said bottom press against and are supported by the strip, and is covered with or composed in part of leather, substantially as and for the purpose herein specified.

No. 56,993.—JOHN B. ROOT, New York, N. Y.—*Trunk Engine.*—August 7, 1866.—The piston is attached to an elongated trunk extending through both heads of the cylinder. The pitman passing through the trunk is attached to the end thereof most remote from the crank shaft, and this end works in guides secured to the cylinder head.

Claim.—The arrangement of the extended trunk G, connecting rod K, cross-head or connection J, and guide L, in relation with each other, and with the cylinder, piston, and crank shaft, substantially as herein described for the purpose set forth.

No. 56,994.—E. A. G. ROULSTONE, Roxbury, Mass.—*Musquito Bar.*—August 7, 1866.— The net frames are secured by springs on their vertical edges, fitting over guide rods or within grooves in the window frame. The springs are retracted by means of rods from within when the bar is to be removed.

Claim.—A mosquito bar for windows when constructed with a spring o or n, applied to one or both sides thereof to hold the bar in position with relation to the sash above or below it, in combination with projecting edges or flanges to fit against the window beads c or m, all substantially as set forth.

Also, in combination with a bar made with these projecting edges and springs, as set forth, so applying one of the springs n or o that it may be operated from within the bar, to insert or remove the bar, substantially as set forth.

Also, in combination with a bar so constructed with springs o o', and with a window sash, the employment of the vertical rods f for the better securement of the bar, and to enable it to be raised from, and lowered into, position, substantially as described.

No. 56,995.—ANSELL P. ROUTT, Orange Court House, Va.—*Dough Kneader.*—August 7, 1866.—The dough is kneaded by passing between toothed rollers, and the rollers are cleared by stationary knives beneath.

Claim.—The combination of the two rollers, provided respectively with the round-headed projections and the straight-sided pins, as and for the purpose described.

Also, in combination with the above, the cutters L L, substantially as and for the purpose described.

No. 56,996.—A. NEWTON and FRANK FAVRO, Worcester, Mass.—*Wrench.*—August 7, 1866.—The screw-threaded portion of the bar projected from the outer jaw is enclosed within the handle of the other jaw. A hollow nut screwed upon the end of the bar enters into the end of the handle, a shoulder upon it abutting thereon; it is secured against being withdrawn by a pin which passes through the shell of the handle and into a concentric groove in the surface of the enclosed portion of the nut.

Claim.—In combination with the upper movable or hammer jaw and the stationary lower jaw and handle of a wrench, constructed substantially as described, the tip nut in the end of said wrench, when arranged and operating as herein shown and for the purposes set forth.

No. 56,997.—RUFUS NORWOOD, Baltimore, Md.—*Smoking Pipe.*—August 7, 1866.—The bulb has a perforated and hinged cap; into its interior fits a hollow sliding stem, provided at its lower end with a sponge box; at the upper end of the bulb is screwed a detachable cap through which the stem slides.

Claim.—First, a cap B, constructed and applied to the stem and bowl of a pipe in such a manner that the stem is connected by it to the bowl, and at will may be disconnected or adjusted, substantially as and for the purpose described.

Second, the combination of a cap a, cap B, and a sliding stem C, having a plunger applied to it, substantially as described.

Third, the combination of a hollow plunger stem C' with a pipe having a detachable cap B, substantially as described.

No. 56,998.—WILLIAM SAUSSER, Hannibal, Mo.—*Knife.*—August 7, 1866.—The shank of the blade is so slotted that it can be readily slipped on or off the customary pivot in the handle. A lever device is used to retract the spring when the blade is to be inserted or remove .

Claim.—First, a knife blade provided with a slot, substantially as and for the purpose described.

Second, the spring holder, for the purpose described.

Third, the back spring B, constructed and operating substantially as described.

No. 56,999.—EJLERT O. SCHARTAN, Philadelphia, Penn.—*Lamp Chimney.*—August 7, 1866.—The lower part of the chimney has the form of an urn. The converging top fits within it, and has an external collar resting on the flanged upper end of the lower section.

Claim.—The construction and combination of the glass chimney, jointed and constructed as herein described and for the purposes set forth.

No. 57,000.—J. B. SLAWSON, New Orleans, La.—*Fare Box.*—August 7, 1866.—Between the receiving aperture at top and the money drawer at bottom are a series of inclined plates with sharpened edges presented upward and downward, so as to arrest any instrument inserted to abstract money, but offer no resistance to the descent of the fares.

Claim.—First, the inclined planes b c d, arranged in zigzag form, and provided with knife edges g, in combination with the box A, constructed and operating substantially as and for the purpose described.

Second, the recess h at the lower end of the inclined planes c d, substantially as and for the purpose set forth.

No. 57,001.—A. and G. SMITH, Flint, Ind.—*Smut Machine.*—August 7, 1866; antedated August 2, 1866.—The grain is subjected to the blast of the fan in passing off a revolving disk between the hopper and smutter, again within the smutter, and again in passing off a revolving disk in the discharge spout.

Claim.—First, the spouts C, and p, having the perforations or openings a and revolving disk i, arranged as shown, in combination with the fan G, for the purpose of extracting the dust and other light refuse from the grain, previous to the latter's entering the smutter, as set forth.

Second, the shaft g, having the disk k attached thereto, and located so as to receive the grain as it falls from the smutter D, in combination with the duct or passage l and fan G, all arranged and operating in the manner and for the purpose herein set forth.

No. 57,002 —J. B. SMITH, Dunmore, Penn.—*Coal Screen.*—August 7, 1866.—The concentric screens are provided with imperforate concentric conveyors, from which the matter separated by each is deposited in a separate receptacle.

Claim.—Combining with a series of graduated meshed screens a series of smooth conveyors, so that as soon as the coal is separated it shall be carried out of the machine without any further unnecessary agitation, which only produces waste, substantially as herein described.

No. 57,003.—J. G. SMITH, Battle Creek, Mich.—*Bed Bottom.*—August 7, 1866.—A double frame has between its upper and lower parts spiral and semi-elliptic springs secured by metal straps.

Claim.—The loops or clasps with the elliptic and spiral spring in combination with the bed bottom, substantially as and for the purpose set forth.

No. 57,004.—RUEL SMITH, Bangor, Me.—*Eraser.*—August 7, 1866.—The back of the eraser is made convex both longitudinally and transversely, so as to serve as a burnisher.

Claim.—The new article of manufacture of a bookkeeper's eraser and burnisher, made and operating as hereinbefore set forth.

No. 57,005.—WILLIS W. SOWLES, Manlius, N. Y.—*Harvester.*—August 7, 1866.—Explained by the claim and cut.

Claim.—The arrangement of devices for operating the cradle or receptacle R so as to discharge the grain delivered thereto from the cutters of the machine, the same consisting in hanging their sides *a* (one of which is weighted) so as to swing, said sides being connected to a common beam or lever *e* by rods *d¹ d¹*, upon which beam rests a horizontal lever bar A, having a pendent arm *o* with its stud *q*, in combination with the snail *t* and friction pulley *o* of the apron shaft J, or its equivalent, of the machine, the whole being arranged together and operating substantially in the manner described.

No. 57 006.—J. N. STANLEY, Brooklyn, N. Y.—*Gas Main.*—August 7, 1866.—The pipes to convey gas into the hydraulic mains are cast on the outside of the latter, communicating with them below the water line, and leaving the main pipe unobstructed for the flow of gas.
Claim.—As an improvement in hydraulic mains for gas works, the combination with the elliptical or cylindrical main pipe C of the external supply tubes D D D, cast in one piece, with the pipe *c*, and communicating with the lower part of the latter, as herein specified and for the purposes set forth.

No. 57,007.—WILLIAM W. STILLMAN, Mount Hawley, Ill.—*Cultivator Plough.*—August 7, 1866.—The shank of the rotary coulter is pressed up through a socket clamp, and provided with a collar and set-screw to secure it in any position.
Claim.—The shank B, the collar C, and the clamp E, arranged and used substantially in the manner and for the purpose set forth.

No. 57,008.—WARREN H. STONE, St. John's, Mich.—*Gathering Fruit.*—August 7, 1866.—A tubular conductor suspended from the tree delivers the unbruised fruit into a suspended receiver.
Claim.—The combination of the flexible tube E, apron D, with the frame A, or its equivalent, all for the purposes and substantially as herein described.

No. 57,009.—WILLIAM MONT STORM, New York, N. Y.—*Ripening Liquors.*—August 7, 1866; antedated July 24, 1866.—The revolving closed vessel has inner projecting flanges to agitate the liquid in the oxygenated atmosphere.
Claim.—Subjecting liquors to violent agitation in a closed vessel while subjected to a pressure of oxygen gas, the liquors at the same time being made to constitute part of a galvanic circuit, substantially in the manner and for the reasons set forth.

No. 57,010.—A. J. TEWKSBURY, Haverhill, Mass.—*Presser-foot of Sewing Machines.*—August 7, 1866.—This device allows the proper pressing and feeding action for two overlapping pieces of cloth which are to be stitched together without the necessity of having both of them extend under the whole bottom surface of the presser. The supplemental spring presser bears only upon one piece of cloth, and the main presser bears upon the overlapped portion of both pieces, thus allowing the feeding device to move them equally.
Claim.—The devices hereinbefore described, consisting of an arm A, screw S, and spring P, and affixing the same to the ordinary solid pressure pad for the purpose of changing it into a flexible pressure pad, and applying the whole in connection with the feed wheel of a sewing machine, as and for the purpose set forth.
Also, in combination with the above, the holes 1 2 3, whereby the degree of pressure of the spring P upon arm A may be adjusted for different thicknesses of material.

No. 57,011.—R. A. THOMAS, Damascus, Cal.—*Drilling Machine.*—August 7, 1866.—By means of the stanchion post with its adjustable section and the vertically adjustable horizontal bearing beam the drill is adjusted for operation in the drift. The sleeve on the drill stock forms a support at the disconnecting point.
Claim.—First, the combination in a machine for drilling rock in drifts and tunnels of the stanchion post B, the bar C, that supports the feed screw and drill, and the adjustable box H, constructed and arranged substantially as described.
Second, the square box L fitted to the stocks of the drill and feed screw, substantially as described, so that they can be connected at pleasure.

No. 57,012.—GEORGE H. TIER, Fremont, Ohio.—*Life Boat.*—August 7, 1866.—The carrying portion of the boat consists of a well with an intervening air chamber between it and the sides.
Claim.—First, the air chambers E E, with the compartments *a a a*, as constructed and arranged with the sides B B, for the purpose and in the manner set forth.
Second, the curving deck F, the bottom C, as arranged in combination with the curving sides B B, the sides D D, chamber E, in the manner and for the purpose substantially as described.

No. 57,013.—CHARLES H. TOMPKINS, U. S. A.—*Stretcher.*—August 7, 1866.—The stretcher is folded for transportation, is mounted on wheels and springs or folding legs, and its bed portions are adaptable to the necessities of varying cases.
Claim.—First, the frame A of a hand carriage stretcher, constructed with a transverse

joint which is provided with bolts *a a* for stiffening it, and is also constructed with hinged legs B B and with jointed leg-rests J J, which are adjustable, and with a head rest D, which is also adjustable, all substantially as herein described and for the purposes set forth.

Second, the open hand carriage stretcher frame A, in combination with the adjustable divided leg-rests J J and sliding bottom G, substantially as and for the purpose set forth.

Third, attaching the canvas bottom G at one end to the hinged frame D, and at the other end to a sliding bar H having locking braces *d d*, or their equivalents, applied to it, substantially as described.

Fourth, the combination of the jointed supports D, adjustable head rest D, and the jointed stretcher frame A, substantially in the manner and for the purposes described.

Fifth, the application of a flexible arm rest K to the stretcher, substantially as described.

Sixth, the hand carriage stretcher frame A, constructed with a transverse joint near its middle, with an adjustable bottom G, and with adjustable and sectional rests J J for the legs of a person, substantially in the manner and for the purpose described.

Seventh, connecting the stretcher to transporting wheels by means of the sliding fastening, or its equivalent, which is so constructed that the wheels can be attached or detached at pleasure, as set forth.

Eighth, constructing the axletree of the carriage wheels of the stretcher of folding sections jointed together, and provided with means for stiffening the joints, substantially as described.

Ninth, attaching the carriage wheels to the loose screw collars *p p* upon the axletree, so that these wheels can be removed or applied at pleasure without the use of detachable nuts or other similar devices, substantially as described.

No. 57,014.—HENRY H. TRENOR, New York, N. Y.—*Car Coupling.*—August 7, 1866.— A coupling bar secured beneath the platform extends its whole length and receives the strain, relieving the car itself.

Claim.—The combination with the trucks of a coupling bar extending throughout the whole length of the car, so that the cars shall be relieved from the strain due to the traction of the whole train, substantially as herein shown and set forth.

No. 57,015.—HENRY H. TRENOR, New York, N. Y.—*Car Brake.*—August 7, 1866.— Improvement on his patent of October 17, 1865. The mechanism for applying the brakes simultaneously to each side of the wheel is vertically actuated by the windlass and intermediate devices.

Claim.—The arrangement and alternating connection of the brakes, as herein described, by means of rods actuated by cranks mounted on horizontal or vertical axes, and vibrated through the intermediary of chain and windlass, or the mechanical equivalent thereof, by hand, steam, or other power.

No. 57,016.—H. W. C. TWEDDLE, Pittsburg, Penn.—*Composition for Lining Barrels.*— August 7, 1866.—The addition is intended to prevent the decomposition of the glue used in lining barrels.

Claim.—The use of common glue or gelatine, mixed with a small quantity of bichloride of mercury, as a coating for the interior of barrels for holding petroleum, turpentine, and other articles.

No. 57,017.—GABRIEL UTLEY, Chapel Hill, N. C.—*Lamp and Candle Stand and Holder.*— August 7, 1866.—The half sockets on the ends of the arms are adapted to clasp a candle or the stem of a lamp.

Claim.—The lamp or candle stand, constructed substantially as described, with a base and spring arms with socketed termination.

No. 57,018.—OLIVER VANORMAN, Ripon, Wisconsin.—*Machine for Tenoning Spokes.*— August 7, 1866.—The slicing knife is placed upon a lever pivoted at one end and with adjustable rests and holders.

Claim.—First, the combination of the knives G and I, the levers C, the spring F, the adjustable gauge K, the adjustable face plate H, the crane L, and adjustable presser M, with each other, and with the bed piece A, substantially as herein described and for the purpose set forth.

Second, the combination of the knives G and I, the lever C, the spring F, the adjustable gauge K, the adjustable face plate H, and the adjustable side presser or holder P, substantially as described and for the purpose set forth.

No. 57,019.—W. F. WARBURTON, Philadelphia, Penn.—*Hat.*—August 7, 1866.—The sweat band is secured by eyelets to the hat at certain points, leaving a space between itself and the hat for the ascent of air. When the hat is soft a metal band keeps it distended on the level of the sweat leather.

Claim.—First, a sweat band secured in the interior of a hat or cap, at intervals, and arranged to be yielding or elastic at a number of points, in the manner and for the purpose described.   ♦

Second, the above in combination with the band or ring C.

No. 57,020.—WM. WARNER and E. S. REDSTREAKE, Philadelphia, Penn.—*Gas Apparatus.*—August 7, 1866.—The gas is passed through hydrocarbon oil in a vessel heated by a lamp; within the vessel are two annular spaces formed by perforated diaphragms and filled with porous material.

Claim.—First, the combination of spiral or other springs with the disks D D and D D, arranged and operating substantially in the manner and for the purpose above set forth.

Second, the combination of the induction pipes B a, exit pipe C, and disks D with the vessel A, the whole being constructed and arranged in relation to each other for joint operation, substantially as described and for the purposes specified.

No. 57,021.— HERVEY WATERS, Boston, Mass.—*Rolling Apparatus.*—August 7, 1866.—The several pairs of grooves are so shaped as to successively draw out the bar from a square sectional to a rhomboidal form and thence back again. The tongs for holding the blank slide toward and from the rolls upon a rod, adjustable stops preventing the blank from entering too far between the rolls. The tongs, by an arrangement of mechanism, are permitted to have one-quarter of a revolution, and thus to that extent turn the blank. The rod which supports the tongs is itself placed upon a carriage which has a movement parallel with the axes of the rolls, and by a system of levers and pawls this movement is arrested before each pair of grooves as the blank is brought successively in front of them.

Claim.—The combination with a suit of roller dies of a lateral or cross carriage for holding the work and guiding it to the successive dies.

Also, the arrangement or system of mechanism for moving the carriage laterally, and retaining it in proper position for the action of each pair of dies upon the blank.

Also, the system of die grooves equiangular and rhomboidal in respective joint cross-section, substantially as described.

Also, in combination, the rolls b b, having a suit of die grooves, with the index m, the holder g, the index i, and the stops l, for determining the position of the blank with relation to the dies to which it is presented.

No. 57,022.—JAMES WEED, Muscatine, Iowa.—*Ventilated Vault for Wine, Potatoes, &c.*—August 7, 1866.—A vault with impervious walls, and having floors, stairs, bins, and a non-conducting covering around the entrance.

Claim.—The construction of preserving chambers, having water-proof walls beneath the surface of the ground, said chambers having communication with each other and means of access to them, and surmounted by a covering of some non-conductor of heat, substantially as herein described.

No. 57,023.—JOHN M. WEHKLY, Somerville, N. J.—*Harvester Cutter.*—August 7, 1866.—The cutters are severally attached to the bar by slipping laterally after their book ends have passed through mortises.

Claim.—The mortised clamping plate E, in combination with the notched or hook-shank knives C C and slotted knife frame A, substantially as and for the purpose herein specified.

No. 57,024.—CORYDON WHEAT and CHARLES BUNGE, Geneva, N. Y.—*Fruit Basket.*—August 7, 1866.—The blank is formed of one piece, and when folded into box shape is bound by a hoop and clips.

Claim.—The band C' and clasps D and D, when made and used for the purpose herein set forth, in combination with the body of the basket, when composed of one piece of stuff.

No. 57,025.—NATHANIEL T. WHITING, Lawrence, Mass.—*Pitcher.*—August 7, 1866; antedated August 3, 1866.—The lower lip catches the drip and returns it by a pipe to the pitcher.

Claim.—The dripping tube formed in two pieces, one fixed and the other so attached that it can be taken off when the pitcher is to be washed, in combination with the external dripping cup or lip.

No. 57,026.—HARRY WHITTINGHAM, New York, N. Y.—*Hot Air Furnace.*—August 7, 1866.—The combustion chamber is surrounded by an air chamber to which air is admitted from below and through which extends a series of vertical and horizontal flues; the former conduct the cold air to the latter.

Claim.—First, the arrangement of the combustion chamber B, with a fire pot and with a series of vertical and horizontal air flues, in combination with the air chamber C, smoke chamber D, and hot air chamber E, all constructed and operated substantially as and for the purposes described.

Second, the openings e, in the bottom of the air chamber C, in combination with the horizontal flues c, and vertical tubes b, in the combustion chamber B, constructed and operated substantially as and for the purposes described.

Third, the air tubes K, passing from the chamber C, through the smoke chamber D, into the hot air chamber E, constructed and operated substantially as and for the purpose set forth.

No. 57,027.—HARRY WHITTINGHAM, New York, N. Y.—*Hot Air Furnace.*—August 7. 1866.—The flame is deflected from the interior of the cone and strikes the circular series of vertical pipes which receive air at their lower ends and discharge it into the surrounding drum. An annular evaporator surrounds the drum, is supplied from an exterior reservoir, and emits a jet of steam into the ascending column of heated air.

Claim.—First, the cone F, in combination with the fire pot A, drum E, and air pipes H, constructed and operating substantially as and for the purpose set forth.

Second, in combination with the above, extending the air pipes H through the bottom of the heating drum, substantially as and for the purpose described.

Third, the annular evaporator I', in combination with the jacket D and heating drum E, constructed and operating substantially as and for the purpose set forth.

No. 57,028.—HARRY WHITTINGHAM, New York, N. Y.—*Hot Air Furnace.*—August 7, 1866.—The central chamber above the fire pot discharges air upwardly, and connects by upper and lower pipes with the surrounding chamber, which receives air below. The flue space takes a sinuous, reverting course in contact with the sides of the consecutive air chambers.

Claim.—First, the central air chamber C, with pipes *a b*, in combination with the fire pot A and annular air chamber E, constructed and operating substantially as and for the purpose set forth.

Second, the arrangement of a series of annular smoke chambers D F H, with a series of air chambers C E G, fire pot A, and escape pipe *b'*, all constructed and operating substantially as and for the purpose set forth.

Third, the annular bead or chamber *h*, in combination with the escape pipe *b'*, and with the smoke chamber H, constructed and operating substantially as and for the purpose described.

No. 57,029.—ALBERT S. WILKINSON, Pawtucket, R. I.—*Horseshoe.*—August 7, 1866.—The bottom shoe, which is exposed to wear, is fastened to an upper shoe attached to the hoof by the clamps and bands described in the claims and cut.

Claim.—First, the combination of the double shoes A *a*, toe clip B, with its loop *c*, curved heel clips *b b*, with buttons *g g*, clamping bands *f f*, clamping screw *h*, all constructed as described and operating in the manner and for the purpose herein represented and described.

Second, the jam nut *i*, (Fig. 2,) in combination with the clamping or retaining band *f*, and clamping screw or bolt *h*, substantially in the manner and for the purpose described.

Third, the rubber padding *e e*, in combination with the heel clips *b b*, and toe clip B, substantially in the manner and for the purpose described.

No. 57,030.—ALBERT S. WILKINSON, Pawtucket, R. I.—*Horseshoe.*—August 7, 1866.—The middle projections on the tread of the shoe are the most salient; the toe and heel calks are brought into action by the rolling of the shoe.

Claim.—A round bottom or rolling shoe A, having a toe calk *a'*, or having toe and heel calks *a' a a*, substantially in the manner and for the purpose set forth.

No. 57,031.—J. E. WILSEY and D. FORBES, Chicago, Ill.—*Bed Bottom.*—August 7, 1866.—Explained by the claim and cut.

Claim.—The combination and arrangement of the slats B and A, the latter being provided at their ends with the rabbeted cross-bar D, for the reception of the slats F, with the spiral springs G and H interposed, as shown and described.

No. 57,032.—ELISHA WILSON, New Haven, Conn.—*Telegraph Sounder.*—August 7, 1866.—The sounding tube may be open at both ends, and the size is so diminished that the closing of the aperture to regulate the sound requires the minimum power of the electro-magnet.

Claim.—First, the employment of the open valve stop *v*, or any equivalent therefor, to control the sounding at the end, at the mouth, or at the side of air, gas, or vapor sounding instruments for telegraphic communication.

Second, modifying the mouth and throat, and generally reducing the dimensions of air, gas, or vapor sounding instruments in order thereby to diminish the amount and force of the current until it ceases to essentially interfere with the free action of the armature lever *l* and valve directly or indirectly opposed to it, and also to economize the sounding medium used, and to improve the tone for rapid utterance for telegraphic communication.

Third, the combination of the valve *s s*, and cord *d d*, to both raise and fill the reservoir in one act, for the same purpose.

Fourth, the combined use of two or more reservoirs in connection, or with each instrument of a small expansive reservoir or gas bag, that the supply may be continuous while the reservoir is being raised and filled to supply air for telegraphic sounders.

No. 57,033.—DANIEL WINER, Lockport, N. Y.—*Propelling Apparatus for Vessels.*—August 7, 1866.—The spiral wings are attached to hollow cores which rotate on horizontal axes and act as continuous propellers in chambers the length of and above the floor of the vessel.

Claim.—The inflated cylinder A, provided with helical blades the whole length, in combination with the enclosing trunk B, with open ends, arranged and operating substantially as set forth.

Also, in combination with the above-described device, uniting two or more trunks B B together, and enclosing the angular spaces *g g'* above their points of junction, and a horizontal line touching their peripheries, to form a series of air chambers, substantially as shown and described.

No. 57,034.—CHARLES BAER, New York, N. Y., assignor to himself and J. WILLIAM KRIETZ, same place.—*Type-setting Machine.*—August 7, 1866.—The type are arranged in a circular series of radial cases which part with their contents to the revolving type channel when the latter is brought into correspondence therewith in the required sequence. The line of type in each case is impelled centrally by a spring pusher, and the type in the appropriate case is grasped and held by spring hooks on the revolving type channel, which receives it from the chamber at the end of the case into which it was pushed as the key was depressed.

Claim.—First, the revolving type-receiving channel G, in combination with a series of radiating type cases C, constructed and operating substantially as and for the purpose set forth.

Second, the lips *e*, at the mouth of the type cases C, in combination with the prongs *g'*, on the revolving type channel G, constructed and operating substantially as and for the purpose described.

Third, the cams *g*° on the forked mouth of the revolving type channel G, in combination with the type cases and the line of type contained therein, substantially as and for the purpose set forth.

Fourth, the cams *r*, between the mouths of the type cases C, to act in combination with the line of type in the revolving type channel G, substantially as and for the purposes set forth.

Fifth, the spring hooks *h*, and pushers *e*, in combination with the keys F, and type cases C, and lips *e*, constructed and operating substantially as and for the purpose described.

Sixth, the spring hooks *s* and spring bearer *u*, in combination with the revolving type channel G, connected and operating substantially as and for the purpose set forth.

Seventh, the adjustable galley K, with the sliding rake M, in combination with the revolving type channel G, constructed and operating substantially as and for the purpose described.

No. 57,035.—CHARLES BARNES, Cincinnati, Ohio, assignor to WARDEN, RENSFORD & Co., same place.—*Steam Jet Pump.*—August 7, 1866.—Jets of steam are delivered from a nozzle fixed concentrically in the water passage, so as to act within a cylinder of water.

Claim.—A steam jet pump constructed substantially as described, with jet perforations *d*, and a direct water passage A B C, as set forth, and for the purpose specified.

No. 57,036.—THOMAS BEACH, Freeport, Penn., assignor to himself and STANLEY R. MOORHEAD.—*Caster.*—August 7, 1866.—The socket of the caster is surrounded by an annular cup surmounted by a hood secured by shoulders on the attaching arms and preventing the passage of insects.

Claim.—First, the cap *d*, secured to the socket barrel by the shoulders *n n*, or other equivalent device, and the annular cup *b*, the two being constructed, arranged, and combined substantially in the manner and for the purposes above set forth.

Second, a furniture caster having a socket *a*, an annular cup *b*, a cap *d*, and springs *m m*, arranged and combined substantially in the manner and for the purposes above set forth.

No. 57,037.—JOHN F. BOGARDUS, Brooklyn, N. Y., assignor to himself, JOSEPH ANDERSON and THOMAS K. SCHERMERHORN, same place.—*Packing Ring for Pistons of Steam Engines.*—August 7, 1866.—The transverse joints are kept tight by an intermediate ring with inclined edges so fitted between two packing rings as to wedge them against the heads of the piston, and pressed outward only by the contact of the packing rings, the latter being expanded by customary springs.

Claim.—The arrangement of the packing rings *d d* and *e*, with the edges that come in contact inclined, and the ring *e*, narrower at the back than on the face, and not so thick as the rings *d d*, so that the rings *b* and springs *c c* shall only act to expand the ring *e* by the contact of the rings *d* with its inclined edges, for the purposes set forth.

No. 57,038.—JEREMY BRADLEY, Cedar Falls, Iowa, assignor to himself, WALTER PASHLEY and S. B. HEWETT, Jr.—*Excavator.*—August 7, 1866.—The scoop is drawn forward in the earth; the contents are passed toward the rear by a reciprocating scraper raised by an endless traversing belt, dropped upon a transverse travelling apron and dumped alongside of the ditch.

Claim.—First, combining with the scoop of a ditching or excavating and grading machine a reciprocating shovel, which is so arranged that it will automatically keep the forward part of the scoop clear of earth, substantially as described.

Second, the combination of the scoop D', and elevator *a'*, with a shovel R, operating substantially as described.

Third, the arrangement of the ploughs E E, in front of the scoop D', in combination with a shovel R, and a contrivance for elevating earth, substantially as described.

Fourth, sustaining the forward part of the machine upon transporting wheels J, by means of the beams F F', chain g, and windlass h, the latter being supported upon a post K, which passes through a slot in said beam F', substantially as described.

Fifth, connecting the transverse beam F to the bolster G, by means of a jointed rod G', substantially as described.

Sixth, transmitting motion to the endless discharging apron from the upper roller a of the elevating apron a', by means of a shaft C⁴, a gimbal joint b³, and spur wheels c c', substantially as described.

No. 57,039.—ALBERT D. CHASE, Reading, Penn., assignor to himself and AMOS T. HUBBARD, Philadelphia, Penn.—*Bed Bottom.*—August 7, 1866.—A hook shackle on the slats is connected by a cross pin and elastic band to a rod on the bedstead rail.

Claim.—The hereinbefore described mode of suspending the slats of beds by means of the double hooks B, rod C, loop D, cross-piece E, and plate F, the said several parts being respectively constructed and combined for use, substantially as set forth.

No. 57,040.—WILLIAM B. DODDS, Cincinnati, Ohio, assignor to himself and NEIL MAC-NEALE, same place.—*Safe Lock.*—August 7, 1866; antedated February 7, 1866.—The arm of the bolt is projected through a slot in the edge of the door to engage the keeper in the frame.

Claim.—The L-shaped bolt pivoted in brackets secured to the frame of the door, and moving on a vertical axis under the impulse of the lock and connecting bar, substantially as described.

No. 57,041.—JAMES M. FERRELL, Philadelphia, Penn., assignor to himself, WILLIAM H. SINER and CHARLES H. DEDHICK.—*Brick Mould.*—August 7, 1866.—The followers in the series of moulds are simultaneously advanced to discharge the bricks by means of the cam rods which are geared together and operated by the handles. The follower returns to its seat on the sharp edges of the stops.

Claim.—First, the combination of the rods C D, gears F F', cams E, ways i i', handles G, bottoms L, and frame A, in the manner and for the purpose substantially as shown and described.

Second, the V-shaped stops h h', arranged and operating in the manner and for the purpose substantially as shown and described.

No. 57,042.—CHARLES P. GEISSENHAINER, Pittsburg, Penn., assignor to himself and JOSEPH GRAFF, same place.—*Machine for Making Nuts and Washers.*—August 7, 1866.—The two tubular stationary dies are arranged in line; one of them has an internal solid punch and an external sleeve punch, which act against the opposed tubular die to punch the hole and cut off the nut respectively.

Claim.—The combination of the stationary dies d and f with the hollow die or sleeve g and punch h, which move together, arranged and operating for the manufacture of nuts, substantially in the manner hereinbefore described.

No. 57,043 —EDWARD L. KEELER, Alleghany, Penn., assignor to himself and JOSEPH GRAFF, same place.—*Machine for Bending Chain Links.*—August 7, 1866.—The blank laid across converging dies is carried into oblique grooves between them by a mandrel projecting from the edge of a revolving wheel. The dies then compress the link around the mandrel in proper shape for welding, and the mandrel is retracted within the wheel to discharge the link.

Claim.—The combination of the mandrel or mandrels a, projecting from the periphery of a revolving disk C, the converging dies D D' and knife set so as to bevel the ends of the link, constructed and operating substantially as and for the purposes hereinbefore described.

Also, the converging dies D D', having grooves o, slightly inclined in opposite directions in combination with the mandrel or mandrels for the purpose of bending over the ends of the link without closing the link, substantially as and for the purposes hereinbefore set forth.

No. 57,044.—JOHN LYLE, Newark, N. J., assignor to himself and COTTON H. ALLEY, same place.—*Saw Set.*—August 7, 1866 —The gauge is arranged to suit the size of the teeth, the guard in accordance with their thickness, and the set-screw to limit the motion of the punch according to the set required.

Claim.—A saw set constructed and arranged substantially as herein described and for the purpose set forth.

No 57,045 —RANSOM LYON, West Troy, N. Y., assignor to himself, A. SHILAND and E. JOSEPH GERDORN, same place.—*Riveting Machine.*—August 7, 1866.—Upon the bed are two jaws—one stationary, the other movable; the hinge to be riveted is set in a vertical position between them and the movable jaw brought up by a projecting arm upon the sliding frame acting against an incline upon the back of said movable jaw. The hammer is mounted upon the movable frame and operated in the usual manner.

Claim.—First, the combination and arrangement as described of stationary frame and sliding frame with hammer E, which is elevated by the arms S S on the driving shaft and forced down by the spiral spring J, and rotated as described by ratchet wheel K, in connection with the levers M and N, which are acted upon by the crank on the driving shaft; these parts, or their equivalents, arranged and operating as and for the purpose set forth.

Second, the arrangement of the adjustable bed-plate T, jaws P P, opened by spiral spring inserted between and closed by the action of the arm W in connection with the sliding frame, the whole arranged and operating as set forth.

Third, the combination of the adjuster R, spiral spring J, rubber packing F, with hammer E, substantially as described.

No. 57,046.—SALMON W. PUTNAM, Fitchburg, Mass., assignor to THE PUTNAM MACHINE COMPANY.—*Sawing Machine.*—August 7, 1866.—A number of circular saws are mounted in a revolving frame by which either one of them may be brought into working connection with the belt, and presented at any desired height above the tables.

Claim.—So constructing and operating the frame I, which carries one or more saws, that it can be revolved entirely around and bring each saw to its desired position in respect to the table, when arranged substantially as described and for the purpose set forth.

No. 57,047.—T. K. REED, East Bridgewater, Mass., assignor to ELMER TOWNSEND, Boston, Mass.—*Sewing Machine.*—August 7, 1866.—A hook is passed with a reciprocating rotary movement about the crochet needle so as to seize the double of the loop and carry it to the unbarbed back of the needle.

Claim.—The method of preventing the hook from catching into the bow of the loop through which the needle is passing by swinging the loop around on the needle, substantially as described.

Also, combining with the hook needle and other mechanism the reciprocating rod and its finger, the rod receiving its motion and the finger acting on the thread, substantially as set forth.

No. 57,048.—THEODOR ROSENTHAL, New York, N. Y., assignor to JULIUS SOLMSON, JULIUS MEYER and JACOB SCHWAB, same place.—*Neck-tie Supporter.*—August 7, 1866.—The tie is attached to an elastic plate which passes beneath the collar and has a pair of elastic jaws projecting downward from its lower edge to embrace the button.

Claim.—The plate or spring A provided with projecting spring jaws *b*, which form a clasp to embrace the shirt button, substantially as and for the purpose specified.

No. 57,049.—CHARLES H. SAWYER, Hollis, Me., assignor to himself, T. J. LITTLE and SAMUEL C. RUNDLETT, same place.—*Cattle Hitch.*—August 7, 1866.—The bow has hooked ends passing through slots in a circular block and secured by a smaller block passed between the ends of the bow so as to engage its hooks. The tying cord is passed around both blocks.

Claim.—The combination of the block B, clapper C, bow A, and cord H, when arranged in the hitch as set forth, all constructed and operating as herein specified.

No. 57,050.—JUDD STEVENS, Chicago, Ill., assignor to CHICAGO SPADING AND DITCHING MACHINE COMPANY.—*Ditching Machine.*—August 7, 1866.—A cylinder armed with radial spades is revolved around a stationary cam, which is arranged to force the spades vertically into the ground, then throw them into horizontal position, and when they reach the requisite height tilt them to discharge their loads of earth.

Claim.—First, so attaching and arranging a series of shovels in the periphery of a wheel M, that said wheel may have a forward slip upon the shovels, substantially as and for the purposes described.

Second, in combination with the shovels and wheel aforesaid, the arrangement of the stationary curved bearing O, operating substantially as and for the purposes specified.

Third, in combination with the series of shovels and wheel M, the arrangement of the pivoted arms *m* and cam-bearing T, for the purpose of locking said shovels to the wheel, as and for the purposes set forth.

Fourth, the arrangement of the cam Q, with the wheel and shovels aforesaid, for the purposes and in the manner described.

Fifth, in combination with the wheel M and shovels N, the arrangement of the cam W, operating substantially as and for the purposes specified.

Sixth, the arrangement of the cam Y with the shovels N, for the purposes set forth.

Seventh, attaching the shovels to the wheel by means of the jointed arm *a b*, substantially as and for the purposes described.

No. 57,051.—ALEX. C. WADE, Paris, Ill., assignor to himself and GEORGE THILLMAN.—*Churn.*—August 7, 1866.—The beaters severally work in their own ends of the churn against end boards and a common, central, perforated breaker.

Claim.—The double-winged beater I J K, journalled to the lid D, as described, in combination with the removable frame L and its accessories N N' O O' O'', all arranged and operating as explained and set forth.

No. 57,052.—ALEX. C. WADE, Paris, Ill., assignor to himself and GEORGE TRILLMAN.—
Washing Machine.—August 7, 1866.—The swinging double beaters are operated by a hand
lever, and pivoted in the hinged box lid. The concave series of rollers is placed in a frame
removable from the box.

Claim.—First, the provision, in a washing machine, of the double batlet H, consisting of
the two series of rods I and J, and pivoted to permit a swinging motion, substantially as
described and set forth.

Second, the double batlet H I J, when journalled to the lid C, in combination with the
removable roller frame L, the whole being arranged and operating in the manner herein
explained and set forth.

No. 57,053.—GEORGE M. WHITE, New Haven, Conn., assignor to himself and T. B. CAR-
PENTER.—*Curtain Fixture.*—August 7, 1866.—The vertical cords on each side take a turn
round the pulleys on the ends of the rollers, and each end of the shade is adjustable to the
place desired.

Claim.—The combination of the two cords ff with one or more rolls provided with pulleys
corresponding to the said cords f, and arranged with a compensating spring, and so as to be
operated substantially in the manner and for the purpose set forth.

No. 57,054.—J. C. WYBELL, West Meriden, Conn., assignor to himself and F. N. BIXBY,
same place.—*Turning Lathe.*—August 7, 1866.—The tool holder is adjustable to cut different
sizes, and pivoted to turn 180°.

Claim.—The lathe attachment, constructed, arranged, and operated substantially as and
for the purposes set forth.

No. 57,055.—ABEL ATHERTON, Lowell, Mass., administrator of estate of PATRICK FLINN,
deceased.—*Sad Iron Heater.*—August 7, 1866.—The reticulated tube containing the perforated gas pipe is placed between angular abutments in the chamber, and the flame is
directed against the sides of the pillars by the plate above.

Claim.—First, the angular pillars jjj, in solid connection with the top plate f of the fire
box to the bottom plate g, as herein described. •

Second, the top plate f of the fire box, so constructed as to throw the flame of gas as
herein described.

Third, the flues o and pp, located at the back end of the fire box, for the purpose described
and in the manner set forth.

Fourth, the flattened, perforated gas pipe l, and spiral springs $n n$, for the purpose herein
described.

No. 57,056.—GEORGE W. BACON, London, England.—*Constructing Blocks or Plates for
Printing Maps.*—August 7, 1866.—A comparatively thin tablet of box wood is mounted
upon a block and engraved. In the positions for the lettering the tablet is cut away and
type of the same thickness attached to the backing.

Claim.—The combination of the block or sheet A, having an engraved surface, with the
electrotype letters mounted upon a block B, in the manner substantially as described.

No. 57,057.—JAMES GRESHAM, Manchester, England.—*Injector for Boilers.*—August 7,
1866.—For the purpose of discharging the air from the water supply pipe, previous to starting the Giffard injector, a supplementary steam jet is used to produce a partial vacuum, to be
supplied with water through the injector when it is started.

Claim.—The arrangement of mechanical devices, as herein set forth, whereby a supplemental steam jet may be produced for raising water, in combination with the ordinary steam
jet of the Giffard injector for forcing water, in the manner and for the purpose as herein set
forth. •

No. 57,058.—SAMUEL CARY, Centreville, La.—*Boring Apparatus.*—August 7, 1866.—By
the hollow stem and spring valve air is admitted below the mass of loosened earth when
withdrawing.

Claim.—The combination and application of the centre point spring valve with the double
spur lip, cutter blades, semicircular scoop, and tubular shank for coupling to for any desired
depth, substantially as herein described, for the purposes specified.

No. 57,059.—WILLIAM BATTY, Troy, N. Y.—*Uniting Iron and Steel for the Manufacture of
Railroad Rails.*—August 7, 1866.—The welding of the iron and steel is assisted by a flux
composed of borax, sal-ammoniac, common salt, and charcoal.

Claim.—The employment as a flux of the compound or mixture herein mentioned and
described, substantially in the manner and for the purposes hereinbefore set forth.

No. 57,060.—E. BUSSEY, Troy, N. Y.—*Water Reservoir for Cooking Stoves.*—August 7,
1866.—The cast metal boiler with enamel lining is provided with a permanent place in the
construction of the stove.

Claim.—The reservoir, boiler, or tank A, constructed entirely of cast iron or other cast metal, and covered or coated upon the inner surfaces thereof with or by an enamel *b*, when applied to and combined with cooking stoves, in the manner substantially as and for the purposes herein described and set forth.

No. 57,061.—A. S. ACKER, Albion, N. Y.—*Pitman Box.*—August 14, 1866.—The driving wrist is formed with a ball and flange, on which is fitted a cap having an oil reservoir to lubricate the joint and apertures to receive the forked end of the pitman.

Claim.—The combination of the cap C. as described, part B, with the reservoir E opening into the inner face thereof, the flanged ball *d'* and the pitman A, arranged and operating in the manner and for the purpose specified.

No. 57,062.—R. B. ANGUS, Tremont, Penn.—*Mould for Casting Throttle Valves.*—August 14, 1866.—The two blocks when brought to the proper position within the cylinder by the set screws form a mould for casting the valve in such shape as to require no fitting.

Claim.—The blocks B C and screw clamp D, in combination with the cylinder A, all constructed and operating substantially as and for the purpose described.

No. 57,063.—J. S. and T. B. ATTERBURY, Pittsburg, Penn.—*Making Smoke Bells.*—August 14, 1866.—The glass is blown into approximate shape within a divided mould, and finished by swaging over a former.

Claim.—The method substantially as herein described of producing glass smoke bells and other articles, consisting in blowing the glass in divided moulds, and then shaping the bells over conical fluted formers, substantially as set forth.

No. 57,064.—CHARLES H. BAGLEY, Elgin, Ill.—*Lamp.*—August 14, 1866.—The char is removed from the wick by a metallic scraper, pivoted concentrically with the curved top of the wick tube, and deposited in a receptacle therefor.

Claim.—First, so arranging or hanging within the top of the lamp in which a wick is used, a metallic or other suitable scraper or trimmer that it can be moved forward and backward over the said wick at the upper end of the wick tube, substantially as and for the purpose described.

Second, in combination with the above, so placing a receptacle upon the wick tube of a lamp and below its upper end that it can be readily removed or replaced at pleasure, substantially as and for the purpose specified.

Third, forming the upper end of the wick tube of a circular or curved shape, in combination with a slot in the lamp cap piece, and the receptacle J, or its equivalent, when arranged together, substantially as described and for the purpose specified.

No. 57,065.—B. A. BAILEY, Lewiston, Me.—*Spinning Flyer.*—August 14, 1866.—The objects are to render the flyers more durable in wear, convenient in adjustment, and neat in appearance.

Claim.—Constructing the abutting check of the spring *h* from its own wire at the upper end thereof, substantially as shown and described.

Also, in combination with the shaft *c* and spring *h*, the clutch *g e*, when constructed, arranged, and operating substantially as described.

Also, the combination with the part *e* of the clutch of one or both of the stops *f*, as described.

No. 57,066.—MILTON G BAKER, New Burlington, Ohio.—*Garden Rake.*—August 14, 1866.—The rake handle is pivoted to the head, is adjustable in the plane of the latter, and has a spring catch to secure it at any angle. Knife edges are formed on the ends of the rake head, the outer faces of the end teeth and the fronts of the intermediate teeth.

Claim.—Pivoting or jointing the handle A to the head B of a garden rake, substantially as and for the purposes herein specified.

Also, the flattened and sharpened ends of the head B, for the purpose set forth.

Also, the peculiar construction and arrangement of the two outer teeth G G, for the purpose set forth.

Also, the construction and arrangement of the middle teeth H H, for the purpose specified.

No. 57,067.—CORTLAND BALL and J. W. HOUGHTELIN, Detroit, Mich.—*Grubbing Machine.*—August 14, 1866.—The lever frame is fulcrumed to an axle mounted on wheels, and carrying ratchet wheels, provided with a clamp for attachment to the grubs, and rotated by pawls on the lever.

Claim.—The arrangement of the drum C, ratchets D D, lever frame E, pawls G G, clamp or chain H H, applied in combination with the axle A of the wheels B, all constructed and operating in the manner and for the purpose described.

No. 57,068.—D. C. BAUGHMAN, Tiffin, Ohio.—*Hitching Horses to Seeding Machines.*—August 14, 1866.—The double-tree is pivoted on a screw rod, confined to the tongue by nuts, so that it may be set up or down.

Claim.—First, pivoting the double-tree A to an inclined rod *a*, which is supported at its upper and lower ends, and made adjustable in a direction with its length, substantially as described.

Second, the adjustable screw rod *a* having the double-tree pivoted to it, and provided with adjusting set screws *c c'*, and a lower brace C, substantially as described.

No. 57,069.—WILLIAM BEACH, Albany, N. Y.—*Malt Shovel.*—August 14, 1866.—The shovel is made to permit the replacing of either the blade or handle, without abandoning the other.

Claim.—The construction of a wooden shovel by making the handle with a cross-bar separate from the blade, and fastening them together with screw bolts, or rivets and straps, so that they may be separated from each other, in the manner and for the purpose described.

No. 57,070.—DAVID BEARLY, Newcastle, Ind.—*Vegetable Cutter.*—August 14. 1866.—The knives are set in a horizontal disk revolving beneath a segmental hopper and driven by a pinion shaft, bevel gear, and crank.

Claim.—The arrangement of a vegetable cutter having the gearing C D, frame H, curb C, and cover C', with a revolving disk G bearing adjustable knives, the several parts being attached, connected, and arranged substantially as set forth.

No. 57,071.—H. C. BECKER, New York, N. Y.—*Portable Drug Crusher.*—August 14, 1866.—At one side and near the end of the plate or tablet is hinged a lever, the portion next the hinge being widened for crushing purposes, while the lower side of the portion between the crusher and the handle is sharpened into a cutting edge.

Claim.—Connecting in a drug crusher and cutter, the plate, cutter, and crusher, under the arrangement substantially as set forth.

No. 57,072.—JOHN A. BELL, Lacon, Ill.—*Safety Pocket.*—August 14, 1866.—A bar secured within the pocket holds an elastic thong adapted to grasp and retain any article passed through it into the pocket.

Claim.—The clasp or pin D in the described combination with the bar C and elastic thong A, o admit of the latter being readily applied to the garment or detached therefrom at the will of the user.

No. 57,073.—JOHN S. BETTIS, Hanover, N. Y.—*Farm Fence.*—August 14, 1866.—The rails are supported in wire hooks whose loops fit over the post and are secured in any desired position thereon by wedges.

Claim.—First, the hook C as a means of holding and connecting the rails and post, made either with a single or double loop, substantially as herein described.

Second, the hook or staple C with a single or double loop, made adjustable on the post by means of the key or wedge *d*, substantially as set forth.

No. 57,074.—G. W. BIDWELL, New Haven, Conn.—*Hat Protector.*—August 14, 1866.—Projecting pins on either the brim or crown of the hat support it without letting it touch the place on which it is laid.

Claim.—First, the combination of three or more pins *a* on the top of the hat, when constructed and arranged to operate in the manner and for the purpose herein set forth.

Second, the combination and arrangement of three or more pins *p* upon the brim of the hat, substantially as and for the purpose specified.

No. 57,075.—B. BISBEE, East Pharsalia, N. Y.—*Washing Machine.*—August 14, 1866.—The clothes are placed within a revolving cylinder with internal rubbing ribs and external slats adapted to inject the water.

Claim.—First, the slats B B having their edges bevelled off on opposite sides, as shown, to form the openings for admitting to the interior of the cylinder the suds or water, substantially as shown and described.

Second, the cleats C C secured upon the inside of the slats B B, at an angle therewith or obliquely, substantially as and for the purpose herein specified.

Third, in a cylinder washing machine, the employment of the catch D *d* for automatically locking the lid or door of the machine and holding it when closed, as well as providing a quick means of liberating the door when desired, substantially as herein shown and described.

No. 57,076.—EDWARD S. BLAKE, Pittsburg, Penn.—*Hat Rack and Seat.*—August 14, 1866.—A wire rack is applied to the bottom or end of a seat to support the hat by its brim in a position where it will not receive injury or cause inconvenience.

Claim.—The hat rack *f* in combination with the seat *c*, constructed, arranged, combined, and operating substantially as herein described and for the purpose set forth.

No. 57,077.—ERASTUS BLAKESLEE, Plymouth, Conn.—*Hoop Skirt Supporter.*—August 14, 1866.—The hook for supporting the belt of the skirt is provided with two hooked wires taking into holes on opposite sides of the corset.

Claim.—The combination of the two hooks A and B united together by and combined with a hook or its equivalent C, and arranged in relation to a corset and skirt, substantially in the manner and for the purpose specified.

No. 57,078.—ASA S. BLINN and D. W. HEWITT, Dubuque, Iowa.—*Flooring Clamp.*—August 14, 1866.—The clamping screw works in a head block provided with adjustable gauges, and is operated by a lever and pawls working in a ratchet wheel on the screw.

Claim.—The arrangement and combination of the head block A, arms H, parts C, with the screw B, ratchet wheel T, pawl F, and lever D, when constructed to operate substantially as described.

No. 57,079.—GEORGE B. BRAYTON, Boston, Mass.—*Oil Injector for Steam Engines.*—August 14, 1866; antedated July 30, 1866.—A pump barrel is fitted to slide vertically within the oil reservoir. In its upper position its orifice communicates with the interior of the oil reservoir to receive a supply of oil, and in its lower position with the interior of the cylinder to discharge.

Claim.—The construction by which the oil orifice of the injecting barrel is worked through a packing, inserted and compressed between the reservoir and the steam tube, substantially as set forth.

No. 57,080.—ALBERT BRISBANE, New York, N. Y.—*Roofing.*—August 14, 1866; antedated August 3, 1866.—The boards are pressed into trough shape and set on scolloped rafters. The joints are covered with battens fitted thereto.

Claim.—The construction of roofing or weather boards formed in a curved of trough shape by bending, and laid together and combined with the concave rafters or bearers as herein described.

No, 57,081.—A. H. BURLINGAME, Sparta, Ill.—*Gang Plough.*—August 14, 1866.—The forward ends of the beams are connected to a vertically adjustable carriage by compound joints affording free vertical play to the rear ends and permitting the turning of the implement without raising the ploughs. A rearward projection on the carriage enables the attendant to raise the ploughs at will.

Claim.—First, the combination of the tongue G, pivot *f*, rocking bar G', plough beams D D, and laterally adjustable support J, substantially as and for the purpose described.

Second, the rear under support J, applied and operated substantially as herein described for the purposes set forth.

Third, the perforated rocking bar G', tongue G, plough beams D D, laterally adjustable rear support J, and the device *h j k*, all combined and arranged substantially as described.

Fourth, the combination of the vertical joint *f* and the horizontal joint G' with plough beams D D and a carriage A B B', which is susceptible of being depressed or elevated at one or both ends, substantially as described and for the purpose set forth.

No. 57,082.—CALEB CADWELL, Waukegan, Ill.—*Harvester.*—August 14, 1866.—The teeth of a continuously circulating cutter are connected by recessed links engaged by rotary driving blocks or sprockets, whose journals are lubricated through axial openings; the cutting device is sufficiently enclosed and protected by a cap and base plate. The diameter of the reel may be increased or diminished.

Claim.—First, the circulating cutting device, consisting of the teeth F' and connecting links F', when constructed as herein described and employed in combination with the sprockets G G', in the manner and for the purpose set forth.

Second, the arrangement of the finger beam A, cap plate B, stationary cutter bar E E', plate C, guide D D', plate J, and cutting device F F', all constructed and operating as herein shown.

Third, the rotating blocks or sprockets G G', and hollow journals G² G², when combined and arranged as and for the purpose herein shown and described.

Fourth, the reel consisting of the parallel arms K, chains K', wheels L M, uprights O, and adjustable arms N, all constructed and arranged in the manner and for the objects specified.

No. 57,083.—J. W. CAMPBELL, New York, N. Y.—*Photographic Printing Frame.*—August 14, 1866.—A partial vacuum is produced at one side of the plate to adapt it to be firmly held to the back of the printing frame by the pressure of the atmosphere upon its opposite side; the section of the frame which confines the plate in position is movable upon hinges and furnished with a lock to enable the plate to be readily inserted and secured.

Claim.—First, holding the plate to the back of the photographic printing frame by atmospheric pressure, substantially as and for the purpose described.

Second, the double hinged section C, in combination with the back B, and the frame A, substantially as and for the purpose described.

Third, the lock D, in combination with the double hinged section B, and back B, constructed and operating substantially as and for the purpose described.

No. 57,084.—JAMES M. CARPENTER, Providence, R. I.—*Let-off Mechanism for Braiding Machines.*—August 14, 1866.—This arrangement preserves due tension and allows the spool to rise on its spindle or carrier tube for the purpose of setting free the thread when caught upon or by any portion of the mass wound upon the spool.

Claim.—The combination of the tube T, the stud *f*, and groove *g*, or their equivalent, with the spool, its spindle and the let-off mechanism, or its equivalent.

Also, the improved let-off mechanism, consisting of the gear with one range of teeth and of the dent and wedge lip applied to the spindle and the tension weight, substantially as specified.

No. 57,085.—JOHN W. CARTER, Greenville, N. J.—*Corset Clasp.*—August 14, 1866.—A single piece of metal is bent to form a loop whereby the clasp is fitted upon the corset spring; the leaves are united and furnished with an eye.

Claim.—The armed loop clasps D E F, made in one piece and secured, substantially as described for the purpose specified.

No. 57,086.—AARON CARVER, Little Falls, N. Y.—*Pump for Deep Wells.*—August 14, 1866.—The gas from the well ascends in a separate passage, whence it escapes into the pump cylinder to be discharged with the liquid raised by the piston; the latter has a screw on its lower extremity adapted to engage a threaded socket in the valve box, to release it by rotation from the bayonet-like connection which prevents its displacement by upward pressure during the operation.

Claim.—First, the pump cylinder, in combination with an outer cylinder J, within which it is eccentrically placed so as to form a chamber L between them, substantially as described.

Second, locking the valve box of a pump within the cylinder so that it cannot be displaced by pressure from below, substantially as described.

Third, making the diameter of the neck of the valve box less than that of its body and perforating it as described, to allow any liquid in the valve box to escape and make way for the end of the piston when the latter is screwed into the neck of the valve box, substantially as above set forth.

No. 57,087.—WILLIAM CLARK, Boston, Mass.—*Regulator for Gas Burners.*—August 14, 1866—An adjustable valve, in form analogous to a screw, is fitted inside the burner and enables the delivery of gas to be regulated.

Claim.—In combination with the tip of a gas burner, a regulating device constructed of the frusto-conical valve and valve seat, with the cross-bar *c* and the screw spindle *d* operating together in the manner described.

No. 57,088.—WILLIAM CLEMSON, Middletown, N. Y.—*Saw.*—August 14, 1866.—A slot is formed in the removable tooth or in the blade in such proximity to the fastening clamp that the intermediate portion of metal constitutes a spring to prevent permanent distortion. The segmental clamp is fitted to turn in a recess in the saw plate and its grooved periphery engages a corresponding tongue in the recessed tooth.

Claim.—Slots *f*, one or more, made in the shanks or tangs of the teeth or in the saw plate, or both, and having such a relative position with the rivets, clamps, or fastenings as to form springs or elastic strips E to bear or bind against the rivets, clamps, or fastenings with an elastic or yielding pressure, substantially as described.

Also, the plate or clamp G fitted in the recess F in the saw plate, in connection with the recess D in the shank or tang *a*, substantially as and for the purpose set forth.

No. 57,089.—WALTER R. CLOSE, Bangor, Me.—*Water Wheel.*—August 14, 1866.—The curved buckets are attached at their lower edges to outer and inner annular plates and at their inner edges to a cylindrical hub.

Claim.—The improved water wheel, made substantially as described, viz., with the arched annulus, the series of curved and bent wings, and the shaft tube, or the same and its flange, arranged as set forth.

No. 57,090.—GEORGE E. CLOW, Port Byron, N. Y.—*Thill Coupling.*—August 14, 1866.—In coupling, one end of the cylindrical pin rests on a shoulder alongside the hook; the thill is uprighted, pushed laterally, the flattened portion passing through the slot and then lowered, the pin occupying the socket behind the hook.

Claim.—The socket C, catch *d*, and shoulders *e e*, employed in connection with the thill iron A A' *a*, substantially as and for the purpose specified.

No. 57,091.—WINFIELD R. COE, West Meriden, Conn.—*Spreading Mastic Roofing.*—August 14, 1866.—The mastic is delivered from a travelling box or hopper, spread to any desired thickness by an adjustable board and compressed by a roller, which serves also as a guide.

Claim.—The combination of the guiding and pressure roller B, the box or hopper A, and the adjustable spreading board F with each other, substantially as described and for the purpose set forth.

No. 57,092.—EBENEZER COLEMAN, Woburn. Mass.—*Wheel Tire.*—August 14, 1866.—
The chamfered and overlapping extremities of the tire are formed with teeth and notches,
respectively, which operate in conjunction with the confining bolts to tighten the tire upon
the felloe.
Claim.—The arrangement and combination of the series of notches *c c c*, and teeth *d d*, or
their equivalents, and the slot *e* with the tire laps *a b*, the same being to operate together,
with a felloe and confining bolts, substantially as specified.

No. 57,093.—DOMINIQUE E. COUTARET, New York, N. Y.—*Disinfecting Bone-boiling
Establishments.*—August 14, 1866.—The bones and other animal matter are soaked in a solution of hyposulpbite of soda, and afterward boiled in water containing sulphuric acid.
Claim.—The disinfection of bone-boiling and fat-melting establishments by means of
chemical process and reagents, or any other substantially the same, and which will produce
the intended effect as described.

No. 57,094.—FRANCIS CRICK, Beamesville, Ohio.—*Running Gear of Carriages.*—August
14, 1866.—The leaves of the springs are of horseshoe shape, are united at the middle of the
bow, and at their ends are connected respectively to the axle and carriage bed. The forward
end of the coupling bar is pivoted on the king-bolt, the rear is attached to the hind axle, and
it has a double slider for the lever which limits the locking movement in turning.
Claim.—First, the semicircular or horseshoe springs D and F, constructed as described,
in combination with the axles D and J, and with the body A of the carriage, substantially
as described and for the purpose set forth.
Second, the gearing consisting of the reach M, the lever N, the circular arm K, the horizontal arm L, and king-bolt I, constructed and arranged as herein described, in combination
with the axles E and J, substantially as described and for the purpose set forth.

No. 57,095.—JAMES P. CROSS, Watertown, N. Y.—*Burning Fluid.*—August 14, 1866.—
Composed of gasolin, 40 gallons; gum olibanum, 1 pound; cascarilla bark, ¼ pound; lichen,
¼ pound.
Claim.—The combination of the within ingredients in the manner and about the proportion
described, for the purpose specified.

No. 57,096.—W. R. P. CROSS, Portland, Me.—*Curtain Fixture.*—August 14, 1866.—The
curtain cord is attached by a knot to a wire loop on the sheave, and by a wire and swivel bar
to the barrel of the tassel.
Claim.—First, the combination of the bent wire having the ring and hook with the curtain
roller sheave, as and for the described purposes.
Second, the combination of the wire *h*, having the friction hook *i* and swivel *j*, with a curtain cord, tassel, and roller, in the manner and for the purposes set forth.

No. 57,097.—JAMES DAILEY, Albany, N. Y.—*Liquid Measure.*—August 14, 1866.—The
liquid introduced into the measuring vessel compresses the spring in the outer casing; a
pointer indicates the weight and determines the quantity.
Claim.—First, the combination of the vessel A and B with a spiral or other spring C and
index plates, one or more, D, and indices *b*, substantially as shown and described, for the
purpose specified.
Second, the spiral spring C, in combination with the vessels A and B, substantially as and
for the purpose specified.

No. 57,098.—JOHN WESLEY DENTON, Paris, Ill.—*Trace Buckle.*—August 14, 1866.—The
buckle proper is attached permanently to the hame tug through the medium of a check piece,
which prevents the shifting of the adjustable trace from its fixed position. The clamp of the
buckle, which is hinged to the check piece, is furnished with a broad tongue carrying the
prong, which by the action of a spring is detained in the appropriate hole of the trace.
Claim.—First, the cheek A, clamp E, and tongue F, arranged and adapted to operate as
set forth.
Second, in combination with the elements of the clause immediately preceding the spring
N, attached to the heel of the tongue and operating as set forth.

No. 57,099.—T. DERINGTON, Carbondale, Ill.—*Machine for Turning Wagon Spokes.*—
August 14, 1866.—A spirally grooved feeding and shaping pattern of irregular gain controls
the action of the cutters, adapting the speed of the cutter carriage to the varying irregularity
of the spoke. Auxiliary cutters are arranged on the right and left sides of the main cutters,
and respectively precede the latter according to the direction of its motion, roughing out the
work in advance.
Claim.—First, the feeding and governing pattern, constructed with irregular grooves or
threads, substantially as and for the purpose set forth.
Second, the arrangement of the rollers *b' b'*, straps *b b*, and horizontally adjustable beam

C*, with rail on its top, in a machine which is constructed and operated substantially as described, all for the purpose set forth.

Third, the arrangement of the spring *e'*, with its roller *f'*, upon the cutter carriage and the rail *f* for holding the tooth *c'* up to the patterns, substantially as herein described.

Fourth, the arrangement with the cutter-head of a pulley on both ends of the shaft, and applying belts which move with the cutter carriage on both of said pulleys of the head, all for the purpose of driving the cutter with a more regular and steady motion, as herein set forth.

Fifth, the combination of a cutter-head constructed with right and left auxiliary sets of cutters S S, and a central main or finishing set of cutters, with a reciprocating carriage and the feeding pattern, constructed as described, so that the work of roughing and smoothing is performed at one time, and also during the back as well as the forward movement of the cutter carriage, substantially as herein described.

No. 57,100.—LATHROP DORMAN, Worcester, Mass.—*Azle and Journal for Carriages.*—August 14, 1866.—The hollow journal contains the lubricant, which is conveyed to the external surface by a wick which passes through an opening in the side of the journal.

Claim.—First, the combination with the axle A and projection B of the wrist G, hollow journal D, nut E, and shoulder C, substantially as set forth.

Second, the combination with journal D of nut E and projection B, substantially as set forth.

Third, making the journal D hollow its entire length, and providing it with a wick *f*, which passes through hole *h*, whereby the outer surface of journal D is always kept lubricated, and the end B, if accidentally broken off, can be removed, substantially as set forth.

No. 57,101.—R. N. EAGLE and C. D. SMITH, Washington, D. C.—*Button.*—August 14, 1866.—The head of the button is made of elastic material to adapt it to pass through metallic or rigid button-holes.

Claim.—First, a button or button head made of soft rubber or analogous elastic material, substantially as described.

Second, a button or stud having one or more elastic heads and a shank of different or harder material.

Third, a rubber head attached to a metallic tubular shank by means of a stem *a*, which is held within said tubular shank, substantially as described.

Fourth, the combination of an elastic head A, metal shank B, and disk C, substantially as and for the purpose herein set forth.

No. 57,102.—B. FRANK EARLY, Palmyra, Penn.—*Shoe Cleaner.*—August 14, 1866.—A shoe mat composed of corn husks clamped between the rods of a frame whose parts are movable to permit the husks to be replaced when worn out

Claim.—A frame composed of adjustable side pieces, substantially as shown, in combination with rods having corn husks or other suitable substance doubled or adjusted around them, and clamped or fitted within the frame to form a new and improved mat, for the purpose specified.

No. 57,103.—J. J. ECKEL and I. S. SCHUYLER, New York, N. Y.—*Apparatus for Rendering Tallow, Lard, &c.*—August 14, 1866.—A double-bottomed kettle surmounted by a dome filled with cells, through which water is circulated, and having separate pipes to convey the condensing water and condensed vapors to a common receiver.

Claim.—First, the two kettles B C, the latter, C, being fitted within the former, B, and sufficiently less in diameter to admit of a space *e* between them, in combination with the holes *d* in the kettle C, substantially as and for the purpose set forth.

Second, the rotary stirrer K, placed within the kettle C, when used in connection with the case D or other cover for the kettles, for the purpose specified.

Third, the condensing apparatus composed of the chambers E, or other suitable or equivalent water passages in the tube D, for the purpose of admitting of the kettles being closely covered during the cooking or rendering process.

Fourth, the receiver H, arranged with the pipes G J, substantially as shown and described, to serve as an auxiliary to the condensing apparatus, and also to insure the drawing off of the condensed steam or vapor from D.

No. 57,104.—JOHN J. ECKEL and I. S. SCHUYLER, New York, N. Y.—*Apparatus for Rendering Tallow.*—August 14 1866.—Explained by the claims.

Claim.—First, in rendering tallow, the drawing off of the vapor or vapors from the contents of covered, or confined, or partially confined kettles, by means of a suction or vacuum produced in pipes by any suitable mechanical means.

Second, the injecting of the vapor or vapors drawn off from the kettles into the fire, by which the contents of the kettles are cooked either by the direct application of the fire, or through the agency of steam generated by the same, substantially as set forth.

No. 57,105.—RICHARD H. EMERSON, Fond du Lac, Wis.—*Machine for Cutting Turf.*—August 14, 1866.—The turf is divided into suitable widths by upright knives at the forward end of the machine, and is then detached from the earth by revolving cutters carried by wheels upon the extremities of a rotary shaft, the intention being to obtain the turf in condition suitable for making fences.

Claim.—A machine for cutting turf, having a shaft A, wheels B, stops D, knives C and F, constructed and arranged substantially as herein specified.

No. 57,106.—ELIJAH EVANS, Sparta, Ohio.—*Fruit Picker.*—August 14, 1866.—The segments of spherical shells are pivoted to the holder, have return springs and cutting edges, and are simultaneously vibrated to cut the stalk of the fruit by means of the cords wound upon their axis and drawn by the trigger below.

Claim.—First, a fruit picker in which two quarter spherical jaws C are caused to close by mechanism, operating substantially as set forth.

Second, attaching the jaws C by a single rivet on one side, and by separate rivets on the o her, so as to produce a shearing motion by the junction of their edges, substantially as set forth.

Third, in combination with the jaws C, the lugs C' and pins D², actuated substantially as set forth.

Fourth, the device for actuating the jaws of a fruit picker, consisting of the pulleys D and D', the spring E, cords G C and A, or their equivalents, and bell crank I, attached to the handle A, said several parts being respectively constructed, and the whole combined for use substantially as set forth.

No. 57,107.—CYRUS EVERSOL, Commerce, Mo.—*Screw-driver.*—August 14, 1866.—This screw-driver has four bits set at right angles. The bit stock is swivelled in the handle and connected by a double pawl and ratchet wheel, so that the reciprocation of the handle may turn the bit in either direction desired.

Claim.—First, forming the face or edge of the screw-driver B in the form of a cross with equal arms, substantially as described and for the purpose set forth.

Second, the combination with the handle C of the screw-driver of a ratchet wheel D and pawl E, constructed and operated substantially as described and for the purpose set forth.

No. 57,108.—J. H. FAIRBANKS and FREDERICK ROADS, McKeesport, Penn.—*Valve for the Hulls of Vessels.*—August 14, 1866.—A valve opening outward is placed in the bottom of a vessel and forcibly opened to flood the hold when the cargo is on fire.

Claim.—A valve A applied to and arranged in the hull of a vessel, so as to operate and to be operated substantially in the manner described and for the purpose specified.

No. 57,109.—LEVI FERGUSON, Lowell, Mass.—*Picker for Fibrous Materials.*—August 14, 1866.—The comb plates act in succession upon the material, and each is cleaned by the withdrawal of its teeth through openings in a segmental apron partially surrounding the oblique toothed picking cylinder.

Claim.—The arrangement of the comb plates E, with their arms moving in slots b of the frame, the cam grooves c, and the segmental apron D, operating with the picker cylinder A, all constructed in the manner and for the purpose herein specified.

No. 57,110.—DANIEL FIGGE, Jenner's Cross Roads, Penn.—*Horse Hay Fork.*—August 14, 1866.—The pivoted bars are opened and closed by means of a lever of the first order, which is pivoted to one bar and formed with a loop adapted to play vertically upon the curved upper-extremity of the other bar. The spear-pointed end of the longer bar is formed with a shoulder forming a shield for the hook-shaped extremity of the shorter bar.

Claim.—The bars A and B, constructed as shown and described, and arranged to operate in combination with the lever C, as and for the purpose set forth.

No. 57,111.—CHARLES FORSCHNER, New York, N. Y.—*Machine for Cutting Fat, Lard, &c.*—August 14, 1866.—A close trough has at one end a double set of knives set at right angles, to cut into bars the fat, which is forced between them by a piston. Said bars are then cut into short pieces by knives reciprocated transversely to the trough.

Claim.—First, the arrangement of a series of knives F and J, crossing each other in the end of a trough or longitudinal box C, in combination with a piston D capable of passing into the openings formed by the knives F and J, in the manner and for the purpose substantially as described.

Second, in combination with the above knives N, operating at right angles to the trough C or to the motion of the piston D, in the manner and for the purpose substantially as set forth and specified.

No. 57,112.—CLINTON FOSTER, Galesburg, Ill.—*Transmitting Power by means of Railroad Cars operating upon Rails.*—August 14, 1866.—Rocking levers are so applied to railroad rails as to be oscillated by the wheels of passing trains, and thus communicate motion to machinery

Claim.—First, the oscillating lever B, arranged to operate in combination with the rail A, substantially as and for the purpose set forth.

Second, the use of the rubber springs *h* in combination with the lever E and pitman *n*, as shown and described.

No. 57,113.—F. F. FOWLER, Upper Sandusky, Ohio.—*Corn Harvester.*—August 14, 1866.—A cord employed in conjunction with a reel, and attached to a pivoted lever extending over the platform, holds the stalks in an upright position until a sufficient quantity has been cut to form a shock; the corn being then tightly grasped by the cord, is transferred to the side of the machine and deposited in a standing position upon the ground ready to be bound.

Claim.—In combination with a machine for cutting corn a pivoted holding and transferring lever, with its cord, pulley, hook and reel, for compressing and holding the shock of corn while being transferred from the machine and set up upon the ground, where it may be bound, substantially as herein described and represented.

No. 57,114.—F. F. FOWLER, Upper Sandusky, Ohio.—*Horse Hay Fork.*—August 14, 1866.—The guide rope releases the brace from its engagement with the handle and thus tilts the fork to discharge the load.

Claim.—Locking and releasing the brace G by means of the cord or rope *e*, which is also the guiding and directing rope of the fork, substantially as described.

No. 57,115.—F. F. FOWLER, Upper Sandusky, Ohio.—*Stack Bottom and Feed Rack Combined.*—August 14, 1866.—The base is mounted upon runners, and contains an expanded, open-work frame to contain feed or form a cradle for a hay stack.

Claim.—A portable stack bottom, feed rack, and stock shelter, constructed, arranged, and operating as herein described and for the purpose set forth.

No. 57,116.—ANTON GALLETH, New York, N. Y.—*Feed-wheel of Sewing Machine.*—August 14, 1866.—The claim particularizes the features of this invention, which is an improvement on the inventor's patent No. 55,847.

Claim.—The nose *e* on the jointed feed lever C, in combination with the friction block B and feed-wheel A, constructed and operating substantially as and for the purpose described.

No. 57,117.—H. B. GALLUP and CHARLES WOOD, Watertown, Wis.—*Drying House.*—August 14, 1866.—Explained by the claim.

Claim.—A drying house, having its walls constructed of brick, stone, iron, or other fireproof material, provided with a flooring of metal bars, and arched roof, provided with metal vent pipes, and the interior surfaces of its walls provided with gutters, all arranged substantially in the manner as and for the purpose set forth.

No. 57,118.—GEORGE M. GITHENS, New York, N. Y.—*Water Gauge for Steam Generators.*—August 14, 1866.—A secondary water pipe is employed to cause a constant circulation through the gauge connection to prevent its obstruction.

Claim.—First, the arrangement of the secondary water pipe F with the connection D, in combination with the gauge B, for the purpose of obtaining a constant flow through said pipe and connections, substantially as herein set forth.

Second, the three-way cock E, in combination with the connecting pipe F and D, arranged for the purpose substantially as herein specified.

No. 57,119.—JAMES GLASSON, Brooklyn, N. Y.—*Hoe.*—August 14, 1866.—This hoe is struck up in one piece with the tang, of sheet metal, and has a corrugation extending partly across it and centrally along the tang; the latter, properly curved for the purpose, passes transversely through the handle near its end, and is fastened to the same by a wedge.

Claim.—The hoe constructed and attached to the handle, in substantially the manner specified.

No. 57,120.—ANDREW GOODYEAR, Springport, Mich.—*Machine for Cutting Barrel Hoops.*—August 14, 1866.—A reciprocating knife cuts hoops from the edge of a plank resting on a table which is alternately raised and lowered by a cam to impart an edgewise taper or bevel to the hoops.

Claim.—First, the use of the cam *i* in combination with the gear wheels S¹ S², fly wheel W, and vibrating rod R, arranged and operated relatively with each other, and with the platen table and frame, substantially as and for the uses specified.

Second, the mode of securing the cutter knife C' by clamping it between the half beam A, in combination with backing bolts *c* and nuts *n*, substantially as herein described.

Third, the employment of a spring *j*, in combination with recessed guides L, for safely discharging a hoop *h*, or other cut article, as set forth.

No. 57,121.—HEINRICH GOTTFRIED, New York, N. Y.—*Straw Cutter.*—August 14, 1866.—The straw is cut by an eccentric knife on a shaft mounted longitudinally above the centre of the trough, and communicating by bevel pinions, cam, and lever, with a plate which compresses the straw for cutting and releases it for feeding. The feed is effected by a toothed wheel driven intermittingly by a crank pitman or ratchet lever.

Claim.—First, the circular knife E, placed eccentric on shaft D, the latter being central or nearly so with the feeding trough B, substantially as and for the purpose herein set forth.

Second, the compressing plate L, or its equivalent, in combination with the lever K' and cam K, shaft I, bevel gears H' H, shaft D, and knife E, substantially as and for the purpose herein set forth.

No. 57,122.—WILLIAM A. GOVERN, Holyoke, Mass.—*Cleaning Wool and Woollen Goods.*—August 14, 1866.—Guano in combination with soap is used for scouring wool.

Claim.—The use of the articles above named, or either of them, for the cleansing of wool and woollen goods, whether used alone or in combination with other ingredients, substantially as set forth.

No. 57,123.—H. D. GREEN, Portland, Oregon.—*Manufacture of Illuminating Gas.*—August 14, 1866.—A mixture of two parts of bituminous coal and one part sawdust is subjected to destructive distillation, and the gaseous product passed through a vessel containing heated coal tar, and then purified.

Claim.—The manufacture of illuminating gas from coal and sawdust combined, and subjected to destructive distillation, substantially in the manner and for the purpose set forth.

No. 57,124.—FRANKLIN S. GREGG, Cincinnati, Ohio.—*Taps and Dies.*—August 14, 1866.—Taps and dies are forged out of wrought iron, turned and chased, and are then case-hardened by heating to redness in a box filled with equal parts of leather scrap, sand, and salt.

Claim.—The process of manufacturing case-hardened wrought-iron taps and dies, substantially as described.

No. 57,125.—GEORGE W. GREGORY, Binghamton, N. Y.--*Pulley Attachment for Raising Weights.*—August 14, 1866; antedated February 14, 1866.—The grapple is adapted to hold automatically to a beam above, and has at the lower end of its arms eyes to receive the supporting rope of a pulley and sockets to receive handles to unclasp and adjust it.

Claim.—The combination of the grapple and pulley, and the arrangement of the several parts whereby to facilitate the action and move the apparatus from place to place, substantially as and for the purpose herein set forth.

No. 57,126.—CHARLES GSCHWIND and JOHN GRETHER, Union Hill, N. J.—*Cigar Holder and Hat Hook.*—August 14, 1866.—This device is suspended by means of a hook, and has pivoted jaws to hold a hat by the rim, and small springs to hold a cigar.

Claim.—A device composed of two jaws a b, sliding ring f, springs e, and sharp pointed hook g, all connected and operating substantially as and for the purpose set forth.

No. 57,127.—CHARLES HALL, New York, N. Y.—*Machine for Adjusting Chain Cables.*—August 14, 1866.—The frame has a clamp link-holder and a hook, between which the chain is stretched by screw power.

Claim.—The combination of a stretching frame for sustaining the strain incident to stretching the chain with a clamp link-holder, and with means of applying force to the chain held by the link-holder, substantially as set forth.

Also the combination of a stretching frame for sustaining the strain incident to stretching the chain with a hook link-holder, and with means of applying force to the chain held by the hook link-holder, substantially as set forth.

Also, the clamp link-holder, with recessed jaws, constructed substantially as set forth.

Also, the hook link-holder, recessed at its inner side, substantially as set forth.

No. 57,128.—ENOCH HALLETT, Hillsdale, Mich.—*Brick Machine.*—August 14, 1866.—The followers for pressing the clay into the moulds are operated by levers. The mould is moved in and out by the reciprocation of the carriage by chains and segment, which also raise the follower when the moulds are about to be withdrawn.

Claim.—First, the combination of the levers R T applied to the follower F, substantially in the manner as and for the purpose set forth.

Second, the combination and arrangement of the segments C C, chains Q Q', and the segments D D, and chains J J, for the purpose of operating the mould carriage L and follower F, as set forth.

No. 57,129.—WILLIAM H. HART, Jr., Philadelphia, Penn.—*Neck-tie Supporter.*—August 14, 1866.—The plate has flanges which pass under the lappel of the collar, and a crotch which rests upon the shank of the button. An elastic loop in front holds the neck-tie.

Claim.—In combination with a neck-tie supporter A the loop C and catch D, as arranged, substantially as described.

No. 57,130.—M. P. HATHAWAY, Mankato, Minn.—*Cutting Apparatus for Harvesters.*—August 14, 1866.—The cutters are attached to an endless belt or apron placed at the front of the platform, said belt passing over and under a series of pulleys and driving wheel so arranged as to cause the cutters as they move in opposite directions to pass each other with a close-drawing cut. The pulleys at the other end adjust the belt to the proper working tension.

Claim.—First, the endless cutting belt B, provided with cutters H, having diagonal or oblique cutting edges, and arranged relatively with the platform or bar A, to operate in the manner substantially as and for the purpose set forth.

Second, the combination of the belt B, wheel C, wheel D, rollers E, sliding bar F, guides *a a*, screw G, arranged and operating in the manner and for the purpose herein specified.

No. 57,131.—JOHN H. HEYSER, Hagerstown, Md.—*Wind Mill.*—August 14, 1866.—A spiral flanged disk wheel is set in a wall and the opening graduated by shutters so connected as to operate simultaneously from opposite sides.

Claim.—The sail *a b c d*, slides *e e*, cords *p p*, pulleys *g g*, and fulcrums H H, in combination with the wind wheel, all constructed and arranged as specified, and for the purposes substantially as described and set forth.

No. 57,132.—GIBBONS G. HICKMAN, Coatesville, Penn.—*Washer for Bolts.*—August 14, 1866.—A lip on the washer occupies a slot in the bolt, and another lip is turned up against the nut to keep it from rotating.

Claim.—Providing the washer with a lip or lips, to be bent into a groove in the bolt, either with or without the lip *c⁵*, to be bent against the side of the nut, according to the circumstances of the case, substantially as and for the purpose described.

No. 57,133.—AARON HIGLEY, South Bend, Ind.—*Railroad Car Brake.*—August 14, 1866.—Windlasses are provided for the respective ends; their effect is to force a clutch wheel into action which winds up a chain against the power of a spring whose effect, when released, is to rotate the wheels in the direction they were travelling when arrested.

Claim.—First, the combination of clutch D, pulley F, and clutch lever G, the dog I, ratchet wheel *b*, and pulleys E E', as and for the purpose set forth.

Second, the spring H', swivel I', and chains *c e*, in combination with the chains *d'* and spring *d*, as and for the purpose set forth.

Third, the clutch pulleys E E', friction clutch D, in combination with the ratchet wheel *b* and pulley F, substantially as and for the purposes set forth.

No. 57,134.—JACOB HILLS, Haydenville, Mass.—*Basin Faucet.*—August 14, 1866.—The valve rests with its elastic disk on the upper end of the base tube, its shank being stepped in a socket of the screw shaft above.

Claim.—The above described improved faucet as made with the pivot *f* and step *g*, the valve C and the key D, applied to the parts A and E, as specified, and with the globe B of the discharging tube B' applied to the base tube A, in manner and so as to be capable of being revolved thereon, substantially as specified.

No. 57,135.—L. HOFFSTADT, Philadelphia, Penn.—*Vulcanizing Flask for Dentists.*—August 14, 1866.—Attached to the top of the vulcanizing vessel is a segmental ring which expands when the temperature of the vessel is too high, stops the flow of gas and taps a bell to attract the attention of the operator.

Claim.—First, providing the boiler A with a flange *a*, in combination with the cover E and hook-shaped screws *b*, constructed and operated substantially as and for the purpose described.

Second, the regulator F, in combination with the boiler A and dial G, constructed and operating substantially as and for the purpose set forth.

Third, the lever catch H and stop-cock D, in combination with the regulator F, adjustable screw *h*, and boiler A, constructed and operating substantially as and for the purpose described.

Fourth, the regulator F, boiler A, lever catch H, stop-cock D, hammer *m*, and bell *n*, all constructed and operating substantially as and for the purpose set forth.

No. 57,136.—CYRUS B. HOLDEN, Worcester, Mass.—*Punch.*—August 14, 1866.—The punch is pivoted to one handle and operated by a lever of the second class upon the other handle.

Claim.—The slotted lever A, pivoted arm C, and lever B, in combination with each other and with the punch, constructed and arranged substantially as and for the purpose herein set forth.

No 57,137.—JAMES HOLDEN, Philadelphia, Penn.—*Coal Scuttle.*—August 14, 1866.—Explained by the claim and cut.

Claim.—As a new article of manufacture, a coal scuttle made of flat wooden sides, shaped as described, in combination with the sheet iron casing B B', substantially as described.

No. 57,138.—SOLOMON T. HOLLY, Rockford, Ill.—*Grain Binder.*—August 14, 1866.—The gavel is compressed by a flexible strap which is passed round by a ring carrier; the latter, by a pair of fingers carries the encircling wire which is cut by shears, and the ends twisted together; the fingers are then relaxed, and the strap unclosed, releasing the sheaf. The devices cannot be briefly described other than substantially in the words of the claims.

Claim.—First, the combination, in a binder, of the drum and arbor of the compressing strap with a ratchet wheel and pawl, substantially as set forth.

Second, the combination, in a binder, of the ring carrier with shear blades for cutting the binding material.

Third, the combination, in a binder, of the travelling fingers or forceps for gripping the end of the binding material with shear blades for cutting it, substantially as set forth.

Fourth, the combination, in a binder, of the travelling fingers or forceps, and the instrument for carrying them around the gavel, with a stationary projection in the track of the binding material for retaining it in its proper place, substantially as set forth.

Fifth, the combination, in a binder, of the travelling forceps with a stationary inclined block for operating them, substantially as set forth.

Sixth, the combination, in a binder, of the travelling forceps with a heel extending across the opening between the jaws, substantially as set forth.

Seventh, the combination, in a binder, of the jaws of the travelling forceps with a guard, substantially as set forth.

Eighth, the twister, with radial jaws, projecting from concentric shafts, constructed and operating substantially as set forth.

Ninth, the twister, with radial and hooked jaws projecting from concentric shafts, constructed and operating substantially as set forth.

Tenth, the combination of the twister, with its driving shaft, by means of pinions and cog-wheels, and the spring connection, substantially as set forth.

Eleventh, the combination, in a binder, of the travelling forceps with a movable driver for the band, substantially as set forth.

Twelfth, the combination, in a binder, of the movable driver for the band, with a ring carrier for carrying it round the gavel, substantially as set forth.

Thirteenth, the combination, in a binder, of the travelling forceps and movable driver with a directing instrument to guide the two extremities of the band together, substantially as set forth.

Fourteenth, the combination, in a binder, of the spool or reel of the binding material with the ring carrier by frictional contact, substantially as set forth.

Fifteenth, the combination, in a binder, of the swinging arm of the spool of the binding material with a movable bearing, and with a curved chair for said bearing, substantially as set forth.

Sixteenth, the combination, in a binder, of the ring carrier and spool of the binding material with mechanism for moving said spool slightly from the wheel that is in friction contact with it, substantially as set forth.

Seventeenth, the arrangement, in a binder, of the spool of the binding material and the mechanism for relieving its pressure upon the instrument that resists its turning by friction, in such manner that the turning of the spool, when the pressure is relieved, is sufficiently resisted by friction to neutralize the movement of the spool, substantially as set forth.

Eighteenth, the arrangement of the guide for the binding material, spool, and friction tension mechanism in such manner that the drawing of the material through the guide tends to move the spool from the friction tension mechanism, substantially as set forth.

No. 57,139.—THOMAS HUCKAUS, New Baltimore, N. Y.—*Box for Tobacco and Matches.*—August 14, 1866.—A portion of the box is divided off by a partition, to form a match safe, which has an end opening covered by a flange of the lid.

Claim.—A tobacco box A, divided into two compartments a b, one for holding smoking and the other for chewing tobacco, and also provided with a match receptacle D, substantially as herein described.

No. 57,140.—BUTLER J. HUNTER, Ledyard, N. Y.—*Harvester Pitman.*—August 14, 1866.—Improvement on C. Wheeler's patent of February 9, 1864. The head of the pitman is connected by swivel pieces to the parts holding the boxes of the wrist to secure freedom of motion.

Claim.—The swivel piece C, constructed with a chamber or recess for the reception of the pitman head B, substantially as and for the purposes set forth.

Also, the combination of the swivel piece, having a recess, with the boxes F and G and bolts E, substantially as described.

No. 57,141.—GIDEON HUNTINGTON, Almont, Mich.—*Tire-shrinking Machine.*—August 14, 1866.—The tire is clamped in two places by the wedges on the upper side of the movable blocks; the latter are drawn together, upsetting the tire, by the rotation of a circular bed plate which engages with the under surfaces of the blocks.

Claim.—The combination of the platform A A, movable heads B B, wheel L, with eccentric threads G G, segments H H, cogs I I, and block E, with the loops or bevelled mortises and self-acting keys, when made and used as above described and for the purpose herein set forth.

No. 57,142.—EDWIN F. HURD, Johnsonville, N. Y.—*Machine for Making Axes.*—August 14, 1866.—A series of dies are arranged in the bed beneath, and the reciprocating block above. They cut off the blank for the bit, shape it and weld it to the poll, which is held between the dies by means of a mandrel in the hands of the attendant. At the side of the machine is a punch for trimming the eye, and a trip hammer with suitable dies for trimming the head.

Claim.—First, the adjustable compound guide and gauge *v*, constructed and operated substantially as described.

Second, the movable gauge *v'*, used in connection with the dies *w'*, constructed and operated substantially as herein recited.

Third, the arrangement, on the same bed or platform, of the tilt hammer with dies and the machine herein set forth.

No. 57,143.—ROBERT HUYCK, Sheboygan Falls, Wis.—*Harness.*—August 14, 1866.—The effect of the springs on the tongue and the straps which connect them to the collars is to avoid the jar of the lateral swaying of the tongue.

Claim.—In combination with the wagon tongue A, the springs E E, and straps F F, for the purpose set forth.

No. 57,144.—LEWIS T. ILGEN, Cedarville, Ill.—*Clothes Dryer.*—August 14, 1866.—The frames have rounds and are pivoted together on the lazy-tongs principle; the end sections are prolonged to act as legs.

Claim.—The arrangement and combination of the slats *a b* with the rods *c c' c''* and elbow stops *d*, substantially as and for the purpose set forth and described.

No. 57,145.—HIRAM INMAN, Hagaman's Mills, N. Y.,—*Lathe for Turning Wagon Hubs.*—August 14, 1866.—Attached to a lathe for turning wagon hubs are rotary and fixed cutters in a carriage having the proper stops to limit the cut.

Claim.—First, the rotary cutters P P and fixed cutters *s s*, attached to a carriage L on the frame A of the machine, and arranged in connection with the fixed heads B B' and arbors C C', between which the block Q is centred, substantially as and for the purpose set forth.

Second, the carriage T, with cutter W and stops *j j* attached, in connection with the arbors S S' in the fixed heads R R', substantially as and for the purpose specified.

No. 57,146.—PETER H. JACKSON, New York, N. Y.—*Ship's Windlass.*—August 14, 1866.—The chain wheel is rotated with greater or less power by changing the relative speed of the capstan and chain wheel, and the passage of the chain in running out from the wheel is regulated by friction.

Claim.—First, the sleeve *h*, connected by the gearing *p q r* and *s*, with the shaft *c*, in combination with the capstan barrel *f* and keys *i i* for connecting the said capstan barrel and sleeve to give to the said shaft *c* and windlass a slower movement than that of the capstan, or allow of the separate movement of the capstan, as set forth.

Second, the head *l* on the upper end of the shaft *c*, in combination with the keys *o o* and capstan barrel *f*, and gearing *p q r* and *s*, as and for the purposes set forth.

Third, the friction band *u*, applied to the inner recess of the chain wheel *t*, as and for the purposes set forth.

Fourth, the combination of the friction strap *u*, blocks *v* and 2, and cam 3, applied substantially as and for the purposes set forth.

No. 57,147.—DAVID JACOBY, Mendota, Ill.—*Plough.*—August 14, 1866.—An ordinary plough is attached beneath the wheeled carriage; its depth of furrow is regulated by a screw in front; its lateral motion by a lever connected to the handles and within reach of the driver, and the jumping of the plough prevented by a hinged pedestal depending from the carriage and resting upon the beam.

Claim.—For the arrangement, and combination of the screw P and its lever by which the end of the plough is elevated or depressed, and thus a deep or shallow ploughing effected, the lever L for directing the lateral motion, and the hinged paddle G for regulating the vertical motion of the plough, with the diagonally set wheels, as herein described.

No. 57,148.—B. A. JOHNSON, Jeffersonville, Ind.—*Street Lantern.*—August 14, 1866.—The body of sheet metal is fastened to the base of cast iron, which is specially adapted for said junction; the latter is connected by curved branches to a ferrule which fits the top of the lamp post.

Claim.—A street lantern, having its base A made of cast iron or other suitable metal, and its body B of sheet tin or other suitable sheet metal, substantially as described and for the purpose specified.

Also, the ferrule or sleeve F, secured to the bottom or base A of a street lantern by means of arms E, substantially as described.

No. 57,149.—W. J. JOHNSON, Newton, Mass.—*Lamp Chimney Cleaner.*—August 14, 1866.—Explained by the claims and cut.

Claim.—First, constructing the two arms of a lamp-chimney cleaner of a single piece of wire, or its equivalent, bent into the requisite shape, substantially as set forth and for the objects specified.

Second, uniting the two arms A A by means of the spring B, or its equivalent, substantially as and for the purpose described.

Third, the loop or guide *d*, substantially as and for the purpose described.

Fourth, confining the material of which the brush is formed, by means of the serrated plate G and clasp E, substantially as and for the purpose described.

No. 57,150.—HENRY M. JONES, West Meriden, Conn.—*Latch Fastening.*—August 14, 1866.—The bolt of the latch is made of a semicircular shape and falls by its own weight, interlocking itself within the notch in the side of the window frame.

Claim.—The reversible latch, consisting of the semicircular bolt *g*, with its square end *q* and bevelled end *r*, handle *l*, slotted plate *b*, and end piece *c*, arranged and operating in the manner and as herein described.

No. 57,151.—DANIEL KELLY, Slatersville, R. I.—*Umbrella.*—August 14, 1866.—The ribs are expansible by a sliding joint, and when extended are so held by a spring catch. The handle may be shortened by folding.

Claim.—The flanges *d* and *d'* and spring catches *e*, in combination with the two parts *c c'* of the ribs of an umbrella, substantially as and for the purpose set forth.

No. 57,152.—FRANK KETCHAM, Elizabeth Township, Penn.—*Gate Latch.*—August 14, 1866.—A wheel is attached to the gate that runs up the inclined part of the catch when the gate is shut, to support it and prevent sagging.

Claim.—The combination of the bracket C, cap D, wheel F, post B, and gate A, substantially as shown and described.

No. 57,153.—JOHN KLINE and ANTON SCHMACKER, Cincinnati, Ohio.—*Reclining Chair.*—August 14, 1866.—The back is hinged near the seat, and a given inclination is maintained by the engagement of notches beneath the arms, with spring teeth on the front posts. A slide in front forms a foot rest. Spring cords erect the back.

Claim.—First, the arrangement of seat A, hinged back B C, hinged and sliding arms D E, stationary posts F, studs G, notched channel H J, spring latch I and retracting thongs or springs K, for the purpose set forth.

Second, the sliding rest L adapted to support the seat when the chair is used in a sitting posture, and to support the feet when in a reclining posture.

No. 57,154.—CHARLES LANG, New York, N. Y.—*Ladies' Imitation Collar.*—August 14, 1866.—A lady's collar made by the machine and process described in patent No. 53,991, to Charles Lang.

Claim.—An improved article of manufacture, being an imitation of ladies' lace collars, made of paper linen, cotton-lined paper, paper mixed with cotton threads or linen threads, cotton cloth, linen cloth, or similar material, made in the manner substantially as described, by one continuous operation, by embossing the design on said paper or similar material, and by removing the elevated parts thereof while lying on the die, for the purpose substantially as described.

No. 57,155.—JOSEPH LEVY, Chicago, Ill.—*Medicine.*—August 14, 1866.—Medicine for stomach and bowel complaints, to be taken warm, and with addition of red wine if desired.

Claim.—The medicine prepared from decoctions and juices or powders of young oak tree rind, camomile flowers, parsley, black radish rind, peppermint, and cinnamon, in the manner and proportions herein described and specified.

No. 57,156.—B. B. LEWIS, New York, N. Y.—*Snap Hook.*—August 14, 1866.—The snap hook is detached by pulling on the rein in its rear, but not by force applied from the bit end.

Claim.—A snap hook attached to a hitching strap, and constructed and applied substantially as shown and described, to admit of the detachment of the hook from the ring by the simple pulling of the rein or line to which the hitching strap is connected, as set forth.

No. 57,157.—AUSTIN LEYDEN, Atlanta, Ga.—*Sewing Machine.*—August 14, 1866.—The object of this invention is to adapt the machine for making either the lock stitch or the single-thread chain stitch, or the chain stitch having another interlaced thread running through it. It is shown applied to the Wheeler and Wilson class of machines, a common spool within a spool case being used instead of the disk bobbin. A rod extending through the axis of the shaft carries at its end a crank which operates a spring loop detainer, when said rod is by the hand moved endwise, and by detaining the loop of needle thread, permits it to be taken by the needle at its next descent, to form the chain stitch. When the locking thread is used, at the same time the interlaced chain stitch is formed, and when the detainer is drawn out of action, the ordinary lock stitch is formed.

Claim.—The combination with the detainer F, of the crank H, and bar or rod G, substantially as described.

No. 57,158.—THOMAS LIPPIATT, New York, N. Y.—*Rose Engine Lathe.*—August 14, 1866.—A dictator and cutting tool arranged upon a common support that slides in either direction as desired, giving to the cutting tool a motion the same as that imparted to the dictator by the pattern, which is secured to and revolves upon the same axis with the article under treatment.

Claim.—The combination of the movable dictator A and cutting tool r, projecting arm m, sliding rest x and sliding block P, arranged relatively to each other, and operating in the manner as and for the purpose herein specified.

No. 57,159.—ALONZO LIVERMORE, Ashland, Penn.—*Fishway.*—August 14, 1866.—Explained by the claim and cut.

Claim.—Constructing a fishway or other conduit, with division walls built transversely to its length, and openings through said division walls and tubes or lips projecting therefrom in an oblique direction against the current, by which construction the velocity of the current can be decreased as may be desired.

No. 57,160.—PETER LUGENBELL, Greensburg, Ind.—*Ditching Machine.*—August 14, 1866.—The periphery of the cylinder is depressed into the soil, and consists of two rims, radial blades in the plane of the axis, and side plates which are moved in line with the axis by contact with a stationary cam, and exclude the soil from the pockets laterally.

Claim.—First, the rotary excavator, composed of the cylinder K, with blades j and rims L L attached and provided with movable plates N, operated through the medium of the stationary cam J, substantially as and for the purpose set forth.

Second, the adjustable curved strips P P in combination with the excavator, when the same is constructed substantially as and for the purpose specified.

Third, the coulters Q Q in combination with the rotary excavator, substantially as and for the purpose set forth.

Fourth, the pendent plate O, in combination with the rotary excavator, as and for the purpose specified.

Fifth, the combination of the rotary excavator with an adjustable frame F, arranged as shown, and adjusted through the medium of a screw, substantially as described.

No. 57,161.—GEORGE F. LYNCH, Milwaukee, Wis.—*Car Brake.*—August 14, 1866.—The levers and chains connected to brakes are operated automatically by the plungers when the cars come in contact.

Claim.—First, the ratchet or pawl bar m in connection with the bearer p, or the inclined face of the cross-bar o, by which the brakes are locked as applied at the one end or tripped by the bumper at the other end, as they are arranged and operated substantially as herein recited.

Second, the extra bar v, in connection with the bumper or draw bar, constructed and operated as and for the purposes substantially as set forth.

Third, the combination of the cross-heads, guide bars and slides, by which the motion of the bumper or extra bar is communicated to the levers, and through them or their equivalents to the brakes, as described.

Fourth, the use of the hand or ratchet wheel and stem, in combination with the other means herein recited for taking up the slack and wear, substantially as set forth.

No. 57,162.—HERMANN MADENHEIM, Brooklyn, N. Y.—*Fountain Pen.*—August 14, 1866.—Explained by the claims and cut.

Claim.—First, the combination of the hollow piston rod b, piston a, ink retainer c, movable barrel e, and pen d, all constructed and operating substantially as and for the purpose described.

Second, the longitudinal partition i in the ink retainer c, substantially as and for the purpose described.

Third, the stops j on the inner end of the body of the penholder, or on the end of the barrel e, for the purpose set forth.

Fourth, the air channels k in the body of the barrel, as and for the purpose described.

No. 57,163.—RICHARD MARTIN, Brooklyn, N. Y.—*Piston Rod Packing.*—August 14, 1866.—By the described device a surface is obtained partly vegetable fibre and partly metallic.

Claim.—A steam packing made by combining with the body of said packing a covering consisting of wire and fibrous material interwoven, intertwisted, or interbraided together to form said covering, substantially as set forth.

No. 57,164.—HUGH L. MCAVOY, Baltimore, Md.—*Apparatus for Carburetting Air.*—August 14, 1866.—The air-forcing drum revolves in a vessel within the revolving carburetting cylinder, which is surrounded by the outer casing. The central shaft has a displacing chamber around it. A regulator above equalizes the flow of gas. The interior of the air-drum has fibrous material.

COMMISSIONER OF PATENTS. 1067

Claim.—First, the air-forcing chamber C, revolving in the tank B, which is surrounded by the reservoir A, in which the carburetting cylinder, or its equivalent, revolves.

Second, the inverted chamber F, vibrating upon the axis and actuated by the varying pressure of the gas contained therein to operate as a regulator, substantially as described.

Third, the displacing chamber *c*, in the chamber C', operating as described.

Fourth, the carburetting cylinder D, provided with cups or receptacles on its inner side, and with reticulated or permeable surface on its periphery.

Fifth, the chamber B, containing the fluid in which the air-forcing cylinder revolves, and separating it from the fluid in the reservoir.

Sixth, the hanging strands or tapes *h'*, in the chamber C, as and for the purpose described.

No. 57,165.—THOMAS McDONALD, Roxbury, Mass.—*Stuffing and Currying Leather.*—August 14, 1866.—Stuffing to be used in currying leather; composed of India-rubber, 3 ounces; beeswax, ½ ounce; fish or neat s-foot oil, 10 pounds; tallow, 11 pounds.

Claim.—The employment of caoutchouc or a solution thereof, substantially as described, with the oil and tallow or fatty matters used in currying leather.

Also, the employment of beeswax and resin, or either, and a solution of caoutchouc with the oil and tallow or fatty matters and in stuffing leather, they being combined as set forth.

Also, buffing the leather with a slicker while such leather is damp, and subsequently coloring such leather and applying the stuffing to it as specified, not meaning to claim the application of the stuffing to the leather, and subsequently carrying on the operations of stuffing and coloring as set forth.

No. 57,166.—WILLIAM McFISHBACK, Union, Ohio, assignor to himself and ERASMUS TUCKER.—*Harrow.*—August 14, 1866.—Each rock shaft carries a row of teeth which are cleaned when vibrated between the cleaner-rods in the rear. The shafts are connected for simultaneous clearance, and the whole are mounted on a wheeled carriage.

Claim.—First, the arrangement in the plane of a barrow frame of two or more rock shafts, each being provided with teeth and cleaners, as described, in combination with coupling bars and levers for vibrating said shaft simultaneously, the whole being arranged for operation as shown and described, so that the teeth of the one shaft may be cleared by the cleaners of the next succeeding shaft.

Second, in combination with the rock shafts provided with teeth and cleaners, as described, the stationary cleaners attached to the cross-beam in rear of the rocker shafts for the purpose of clearing the teeth of the rear rock shaft, as herein shown and set forth.

Third, the cleaners constructed as herein described—that is to say, each cleaner being formed of one continuous piece of metal, which is bent to form the jaws between which the harrow tooth passes, the rear end of the cleaner forming the base by which the same is secured in its position.

Fourth, in combination with a harrow frame, having a vibratory or rocker frame provided with teeth and cleaners as above set forth, the wheels mounted in said frame on a fixed axle, in the manner and for the purposes herein shown and set forth.

Fifth, in combination with a barrow constructed as set forth, the hereinbefore described device for marking furrows—that is to say, the rock shaft provided with shovel teeth and guide, arranged and operating as and for the purposes herein shown and specified.

No. 57,167.—CHARLES F. MOELLER, Newark, N. J.—*Lantern.*—August 14, 1866.—By the described hooks and springs, the guards are attached to the rim, and the ring to the base.

Claim.—A lantern having a hook J, flanges E, spring hook S, slide F, spring catch W, and book V, adjusted, combined and arranged as herein specified.

No. 57,168.—C. F. J. MOLLER and C. LATHAM SHOLES, Milwaukee, Wis.—*Shoe Brush.*—August 14, 1866.—A metallic flange, with an edge and a point, is attached to the edge of the brush.

Claim.—The construction and arrangement of the scraper, in combination with the brush as herein described, consisting of the end *f*, for cleaning the edge of the sole; the point *e*, for cleaning between the sole and upper; the curvilinear edge from *c* to *d*, and attached to the side of the brush, for the purposes set forth.

No. 57,169.—E. L. MORRIS, Boston, Mass.—*Hoop Skirt.*—August 14, 1866.—The lower hoops are inserted in a cloth band, which is detachable, and from which the hoops may be removed when it is desirable to wash the band.

Claim.—In combination with a hoop skirt, and so as to form part of the same, a cloth band so applied to the hoops *c*, as to be detachable therefrom, the band being provided with pockets and with hoops or wires sliding loosely therein, and the ends of the band being made to lace or fasten together, all substantially as described.

No. 57,170.—S. A. MORT, Dayton, Ohio.—*Ironing Board.*—August 14, 1866.—The pressing board is pivoted to and supported by legs, and may be collapsed to occupy but little space.

Claim.—The combination of the legs G, the pressing block H, neck B, rest *d*, shoulder *c*, pin *i*, fly *f*, and rod J, all arranged as and for the purpose herein set forth.

No. 57,171.—JOHN MOTT, Danville, Cal.—*Double Revolving Plough*—August 14, 1866.—
The right or left hand ploughs are locked and unlocked by a lever, and brought into use as
required. The standards, &c., are constructed to avoid clogging, and washers are placed
under the arms of the brace to increase the width of furrow.

Claim.—First, the peculiar depressions in the front and back standards *a a b b*, and extension of the arms *d d*, of the front standard conforming to the mould boards, and the forked
brace *c*, for strengthening the ploughs, as described.

Second, the forked washer *f*, and adjustable washers *m m*, and lever *e e*, and the peculiar
shape of the outer ends of the set screws *l l*, substantially as described and for the purposes
set forth.

No. 57,172.—JAMES H. MULHALL, Albany, N. Y.—*Preventing the Freezing up of Gas
Pipes.*—August 14, 1866.—Explained by the claim and cut.

Claim.—The employment at the junction of the lamp-post gas pipe and of the service pipe
at the bottom of the post, of a reservoir to hold alcohol or other fluid uncongealable at the
lowest temperature of the climate, said reservoir to be formed, fitted and arranged substantially as described.

No. 57,173.—WILLIAM H. MYERS, Norwich, Conn.—*Railroad Car Roof.*—August 14,
1866; antedated July 9, 1866.—The roofs are strengthened, by longitudinal metallic strips
and metallic nosings on the ends of the roof.

Claim.—In roofs for railroad cars and other structures combining the strips *c*, which form
the bodies of such roofs, with metallic bands and nosings, substantially as above set forth.

No. 57,174.—HENRY NAPIER, Elizabeth, N. J.—*Disinfecting Compound.*—August 14,
1866.—Composed of alumina, phenic acid, perchloride of iron, sulphite of lime and magnesia
or soda.

Claim.—First, the combination of phenic acid with alumina, for the purposes set forth.
Second, the combination of a metallic perchloride with alumina, for the purposes set forth.
Third, the combination of a solid sulphite with alumina, for the purposes set forth.
Fourth, a disinfecting compound, made as herein described.

No. 57,175.—ALFRED NOBEL, New York, N. Y.—*Explosive Compound.*—August 14.
1866.—The nitrine becomes solid below 55° Fahr., and is not so readily decomposed as the
non-solidifying nitro-glycerine of Sobrero.

Claim.—Nitrine, or crystallizing nitro-glycerine, produced by the mixture of glycerine, sulphuric acid, and nitric acid, free or nearly free from hyponitric acid, for the purpose specified.

No. 57,176.—LEWIS F. NOE, New York, N. Y.—*Carpenters' Bench Hook.*—August 14,
1866; antedated August 2, 1866.—The hooks at the respective ends of the plate are adapted
to hold thick or thin boards, the plate being held against the thrust of the plane by the under
hook in the bench mortise.

Claim.—A reversible bench hook, constructed and operating substantially as herein set
forth.

No. 57,177.—MORRIS OPPER, New York, N. Y.—*Loom for Weaving Cloth with Swells or
Gores.*—August 14, 1866.—The ordinary "sectional take-up" is dispensed with, and in its
stead a single roller is used, extending across the whole width of the cloth and operated as
usual by a ratchet movement at every beat of the lay; a series of pressure rollers borne down
on the take-up roller by springs, and linked to levers connected with the jacquard, are lifted
and depressed, so that the pressure of some will cause a take-up of a corresponding width
of cloth, and where their pressure is relieved, the take-up will be suspended.

Claim.—The pressure rollers operated by the lay of the loom, substantially as described,
and controlled by the jacquard, or equivalent therefor, in combination with the take-up roller,
substantially as and for the purpose specified.

No. 57,178.—JOHN B. PAGE. Chicago, Ill.—*Machine for Converting Reciprocating into Rotary Motion.*—August 14, 1866.—A segment rack on the axis of the rotating shaft engages
alternately with racks above and below the axis, and then reciprocates the frame to which
the said racks are attached.

Claim.—The arrangement and combination of the racks D, supports F, with the lever K,
and pinion H', substantially as described and set forth.

No. 57,179.—HENRY W. PAINTER, New Haven, Conn.—*Carriage Pole.*—August 14.
1866.—A metallic bar is inserted in a longitudinal groove beneath the pole to prevent sagging.

Claim.—The combination of the bar B with the pole, inserted in the manner and for the
purpose specified.

No. 57,180.—JOHN W. PARSONS, Owenton, Ky.—*Head Block for Saw Mills.*—August 14,
1866.—The double rack has two sets of pawls to set the knee forward by operating a lever

and a cam that is so arranged as to release at will both or either head block from the action of the pawls in the racks, so that they can be run back.

Claim.—First, the setting mechanism, consisting of the double rack D and pawls F *f* F' *f*, capable of simultaneous or separate action, substantially as set forth.

Second, the cam K, lever L, rod *l*, and tongue M, adapted for the optional release of both or either head block, in the manner explained.

No. 57,181.—E. M. PAYNE, Waverly, N. Y.—*Bed Bottom.*—August 14, 1866.—The canvas is stretched by rollers, ratchet, and pawl, and is arranged in a frame which rests upon semi-elliptic springs which rise from the foundation.

Claim.—The combination of the half elliptic springs D, with the upper and lower frames B and A of the bed bottom, with rollers C, pawls F, ratchet wheel E, and canvas G, substantially as described and for the purpose set forth.

No. 57,182.—ABNER PEELER, Webster City, Iowa, assignor to himself, W A. CROSBY and N. P. CHIPMAN.—*Machine for Writing and Printing.*—August 14, 1866.—Reading matter is printed in characters upon paper by one movement without movable type or press. The type plate has an arrangement of raised letters, and the paper is moved over them, being depressed upon the desired letter by a pin in consonance with the indications of the thimble on the table of letters corresponding in arrangement with those on the type plate. Spacing and intervals follow the depression of the thimble.

Claim.—First, printing reading matter by means of a self-adjusting type plate having a compound movement, and a lever press, substantially as described.

Second, the lever C, and pin E, combined with the hinged block F, when used for the purposes specified.

Third, the finger plates R and S, with their several adjuncts as described, or their equivalents, for the purpose of moving and adjusting the type plate.

Fourth, the hinged blocks F and I, constructed and operating substantially as and for the purposes set forth.

Fifth, the sliding beam O, ratchet and pawl N, and paper holder P, constructed, combined, arranged, and operating substantially as and for the purposes specified.

Sixth, the entire machine constructed, combined, and arranged substantially as described.

No. 57,183.—SHERMAN PETRIE, Buffalo, N. Y.—*Derrick for Raising Sunken Vessels.*—August 14, 1866.—Chains from each mast pass over each side of the vessels and are slung beneath the wreck. Power being exerted simultaneously on each chain, the strain is equally distributed and the careening of the vessels avoided.

Claim.—The double set of lifting tackle C C, and chains G, G', arranged, combined, and operating substantially as described, by which the sunken vessel may be raised, and at the same time the derrick vessels maintained in their upright positions, as set forth.

No. 57,184.—FREDERICK P. PFEIFFER, Philadelphia, Penn.—*Alloy.*—August 14, 1866.—Composed of lead, 98 parts; copper, 1; tin, $\frac{1}{4}$; antimony, $\frac{1}{4}$; bismuth, $\frac{1}{4}$.

Claim.—The aforesaid new and improved alloy metal, plated or covered, and particularly the addition of bismuth, antimony, and tin, to lead and copper, in the order, proportions, and for reasons as already stated, using for that purpose the aforesaid metals, or any other substantially the same, and which will produce the intended effect.

No. 57,185.—THOMAS J. PRICE, Auburn Ky.—*Combined Seeder and Cultivator.*—August 14, 1866.—The adjustable teeth of the cultivator portion are succeeded by a revolving spiked roller whose teeth are cleaned by a comb which projects into their intervals.

Claim.—First, the combined cultivator and seed coverer, constructed and arranged substantially as herein shown and described, that is to say, having adjustable, interchangeable teeth, capable of being shifted and secured in different parts of the frame, as and for the purposes set forth.

Second, in combination with the adjustable with adjustable and interchangeable teeth, as described, the spiked roller when constructed and arranged for operation as and for the purposes herein set forth.

Third, the combination in a cultivator or seed coverer, constructed and arranged as herein described, of the roller with a stationary spiked cross-bar to clean the roller and prevent its becoming clogged, substantially as herein set forth.

No. 57,186.—RICHARD RAVEN, New York, N. Y.—*Piano.*—August 14, 1866.—Beneath and supporting the principal are supplementary sounding boards secured to the case of the instrument.

Claim.—Introducing beneath the principal sounding board of a piano a drum or sounding chest, composed of one or more sounding boards, on which the principal sounding board is supported, with their edges secured permanently to the case of the instrument, substantially as shown.

No. 57,187.—C. R. RAND, Dubuque, Iowa.—*Heating Furnace.*—August 14, 1866.—In connection with an open fireplace provided with tubular grate bars, and air passage for entrance of external air, are numerous passages nearly level with the bottom plate of the fireplace. The circulation through the passages is governed by dampers attached to spindles whose handles are on the level of the floor.

Claim.—First, the open furnace, provided with the air passage G, in combination with the lateral air flue D, located in the upper part of the fireplace, as shown and described.

Second, in combination with the air passages G and D, the hollow or tubular barred grate U, arranged and operating as set forth.

Third, in combination with the open furnace, the air chamber R, and the vertical flues P and P', arranged as shown and described.

Fourth, the arrangement of the smoke flues I I'' and H H', as set forth.

Fifth, the adjustable water vessel K, located within the hot-air chamber, as shown and described.

Sixth, the dividing chamber or passage F, when provided with the valves *i*, and deflecting plates *k*, arranged as herein set forth.

No. 57,188.—LOUIS C. RODIER, Springfield, Mass.—*Steam Pump.*—August 14, 1866.— The valves are actuated by steam and interior devices. A valve stem at each end of the cylinder is actuated by the piston head forcing the valve into a reservoir, admitting a cushion of steam to the piston and steam to the valve chest to operate the valve. The water valves are hinged and hung upon the plugs, singly or in pairs. The exit plug when both valves are attached has a longitudinal and lateral partition.

Claim.—First, the arrangement in each end of a steam cylinder of the valve *b*, with a steam reservoir L, and port *f*, so as to be operated by the piston to admit steam to the main valve chest, substantially in the manner and for the purpose set forth.

Second, the construction of the stem of the valve *b*, at each end of the cylinder, in such a manner that it shall be moved, reversing the main valve before the piston reaches the end of the cylinder, so as to cushion the piston, as set forth.

Third, the arrangement of the ports *h* and *s*, respectively, in the main valve and its seat, so as to communicate with the exhaust port and the valve chest, substantially as set forth.

Fourth, the arrangement of the water valves hung on trunnions in the plugs, substantially as set forth.

Fifth, the construction of the upper plug divided into three parts by means of partitions communicating with either end of the cylinder and with the delivery pipe and air chamber, as shown and described.

Sixth, the arrangement of the yoke *u*, operating substantially as set forth.

No. 57,189.—JOHN B. ROOT, New York, N. Y.—*Piston Packing Ring.*—August 14, 1866; antedated August 8, 1866.—The single packing ring is cut at one point, bevelled on both sides of its outer edge, and fitted in a rebate groove of the piston head. The groove permits it some longtitudinal motion, and it packs alternately against the respective sides of the containing groove as well as at its periphery.

Claim.—A packing ring bevelled at its outer edges, as shown at *e e*, and herein described, and fitted to the piston and cylinder, to operate as herein set forth.

No. 57,190.—JOHN B. ROOT, New York, N. Y.—*Trunk Engine.*—August 14, 1866.—The metallic packing in the grooves around the trunks forms a joint against the inside of the cylindrical guides which extend longitudinally from the cylinder.

Claim.—The arrangement of the packing rings within the trunk in combination with the cylindrical guides F F, attached to the cylinder, substantially as and for the purpose herein specified.

No. 57,191.—JAMES J. RUSS, Worcester, Mass.—*Planing Machine.*—August 14, 1866.— The presser foot is adaptable to the horizontal or inclined surfaces of the stuff to be planed, being hinged to the head block and adjusted by set screws.

Claim.—The pressure foot for planing machines, arranged and constructed to operate substantially as described.

No. 57,192.—WILLIAM A. RUSSELL, Lawrence, Mass.—*Manufacture of Paper.*—August 14, 1866.—The coarser, stronger material is faced on each side with a finer layer, the union being effected by rollers while the material is in an adhesive condition.

Claim.—The improved fabric or manufacture of paper as composed of a coarse, stronger and more opaque material (such as Manilla hemp or Manilla rope, Sisal grass, jute or gunny bagging) and two superficial or finishing layers of a finer and whiter material (such as linen or cotton or a mixture of both,) applied, arranged, and combined together, under circumstances and in manner as hereinbefore explained.

No. 57,193.—ANDREW V. RYDER, Germano, Ohio.—*Horse Hay Fork.*—August 14, 1866.— The head is tubular and to it is hinged a handle having a holding tine pivoted to its upper extremity. Within the handle is fixed a tumbling catch which engages with the upper end of an arm, rigidly attached to the fork head; when the same is thrown up to bring the tines in working position, the tumbler is held by a spring catch controlled by the attendant.

Claim.—First, the above-described arrangement of the tines A and D, with the tubular head B, arms C, and rope E, substantially in the manner and for the purposes set forth.

Second, the arrangement of the locking lever G, with the tumbler H, and dog I, substantially as described.

No. 57,194.—VAN BUREN RYERSON, New York, N. Y.—*Desulphurizing Ores.*—August 14, 1866.—The ores are heated in a muffle in the presence of a current of air; behind each muffle is a passage in which binoxide of nitrogen is generated, which mixes with the air and sulphurous acid passing from the muffles; the mixture is driven by fans into receivers in company with a steam jet. The receivers are charged with ore previously desulphurized in the muffles. The sulphurous acid is converted into sulphuric acid and combines with the base metals in the receiver; the sulphates are dissolved out by water leaving the gold free; the silver may by the usual method be afterwards precipitated from the solution of mixed sulphates.

Claim.—The process, substantially as described, of subjecting ores which have been desulphurized and their base metals oxidized, to the action of the gases evolved from the process of desulphurizing another charge of such ores in the presence of atmospheric air, and mingled with binoxide of nitrogen and jets of steam, for the purpose specified.

No. 57,195.—HERMAN SCHMIDT, New York, N. Y.—*Hand Screw Clamp.*—August 14, 1866.—Movable nuts are so applied to the screw of the clamp that the screw can be released from the nut and allowed to slide freely on the jaw in either direction.

Claim.—The movable nuts *c c*, in combination with the screw B B', and jaws A A', of a carpenter's clamp, constructed and operating substantially as and for the purpose described.

No. 57,196.—CHARLES SCHOTT, Nashville, Tenn.—*Lubricator.*—August 14, 1866.—A ratchet wheel revolves within the oil chamber, and its dippers drop the oil into a tube which conducts it to the journal. The ratchet is operated by a pawl and lever, the latter being vibrated by a cam upon the shaft.

Claim.—The arrangement of the ratchet wheel *c*, carrying the dippers, the pawl *e*, lever B, and cam I, in relation to the journal box, in the manner and for the purpose herein set forth.

No. 57,197.—JEAN SEGONDY, St. Louis, Mo.—*Meat Cutter for Sausages.*—August 14, 1866; antedated July 31, 1866.—The slide regulates the area of the discharge aperture, and by this means the delivery of the meat.

Claim.—The sliding head blocks G, and the slide I, when constructed and employed as and for the purpose set forth.

No. 57,198.—THOMAS SHARP, Carlisle, Penn.—*Car Seat and Couch.*—August 14, 1866.—When the backs of the seats are raised and arranged horizontally to form a couch, they are supported from beneath by hinged rods and from above by rods suspended from the roof. Extra cushions make the original seats continuous to form a lower couch.

Claim.—First, providing means substantially as described, whereby hinged backs of car seats are elevated after they are swung over to a horizontal position, and after being thus elevated are retained in position, either permanently or with only a slight freedom to move up and down, all for the purpose set forth.

Second, the arrangement of the suspension rods *g g* and springs *h*, in combination with the hinged vertically-adjustable backs of car seats, substantially as and for the purposes described.

Third, the jointed or folding cushions D in combination with supports *a*, formed on the seat frames beneath the bottoms of the seats D', whereby the cushions D when not in use are sustained out of the way of the foot rests C, substantially as described.

Fourth, the combination of the cushions D or D', springs *s s*, and slats E E, substantially as and for the purposes set forth.

No. 57,199.—E. SHORKLEY, Lewisburg, Penn.—*Horse Hay Fork.*—August 14, 1866.—Explained by the claims and cut.

Claim.—The case or sheath A provided with an arm or projection B at its upper end, in connection with the hooks F F connected with the rod C and case or sheath A, and operated through the medium of the lever D, or its equivalent, substantially as and for the purpose set forth.

Also, the particular arrangement of the hooks F F, links *g g*, and arms *h h*, to operate in the manner substantially as and for the purpose specified.

No. 57,200.—J. B. SKINNER, Rockford, Ill.—*Plough.*—August 14, 1866.—Instead of being rigid the coulter swivels in its socket and is vertically adjustable by washers beneath the beam.

Claim.—First, swivelling the coulter or cutter for ploughs or cultivators in sockets or brackets, so attached to the beam as will permit the coulter or cutter a lateral and vertical adjustment, substantially as and for the purpose set forth.

Second, giving the swivelled coulter or cutter both a vertical and horizontal adjustment, substantially in the manner and for the purpose set forth.

No. 57,201.—THOMAS J. SLOAN, New York, N. Y.—*Braiding Machine.*—August 14, 1866.—In order to reduce the range of motion of the tension weight and thereby to lessen the liability of breaking the thread by jerks, the bar of the carrier has three thread eyes and the weight has two; the thread thus taking four courses up and down and so causing the weight to traverse only one-fourth of an inch for every inch of thread delivered.

Claim.—The second guide on the carrier bar and the second guide on the sliding weight, in combination with the first guide on the carrier bar and weight heretofore used, substantially as and for the purpose described.

No. 57,202.—CHARLES D. SMITH, Washington, D. C.—*Button-hole.*—August 14, 1866.—A metalli edged button-hole.

Claim.—A button-hole having a continuous metallic binding, substantially as and for the purpose specified.

No. 57,203.—MORRIS H. SMITH, Brooklyn, N. Y.—*Curing Pork.*—August 14, 1866; antedated August 8, 1866.—The hams to be cured are supported upon bars arranged in rows one above the other in a room with a tight floor. The hams thus arranged are sprinkled with salt and the brine trickles down from one series to another, is collected, pumped up and reflowed upon the hams until they are cured.

Claim.—The suspension of hams and shoulders during the process of curing with the butt upward and the shank down in such a way that the brine dripping from one ham will fall on that beneath as herein described, using for that purpose parallel bars or any other means substantially the same, and which produce the intended effect.

No. 57,204.—THEODORE F. SNOVER, Menasha, Wis.—*Clothes Dryer.*—August 14, 1866.—A semicircular series of bars are supported in the holder in either an erect or expanded radial position.

Claim.—The combination of the plate B with its shouldered collar D, the slotted plate F, and bars H with studs a, constructed as described and operating in the manner and for the purpose specified.

No. 57,205.—ADAM SOWDEN, Cincinnati, Ohio.—*Saw-toothing Machine.*—August 14, 1866.—Improvement on the patent of Ward Eaton, May 16, 1854. Adjustable gauges are secured to the shearing and toothing ends of the gate; the said gauges travel down along with the cut plate or saw and on the return stroke the work is freely released and sticking prevented.

Claim.—The arrangement of shearing jaws J J' and toothing jaws K K', in the described combination, with the adjustable travelling gauges N, as and for the purposes set forth.

No. 57,206.—Cancelled.

No. 57,207.—SILAS L. SPENCER, Hopewell Cross-roads, Md.—*Potato and Drill Machine.*—August 14, 1866.—An endless traversing belt has a series of pockets which pass under a striker to remove superfluous sets and open to drop their sets into the planting share, which precedes the covering shares and the roller, which acts as a driving wheel.

Claim.—The arrangement of the double hopper J K and strike k, in combination with the delivering boxes e e e, in the opening joints i i i, endless belt c c, rollers C D, driving mechanism E F, and frame A, constructed and operating in the manner herein described for the purposes specified.

No. 57,208.—G. D. SPOONER and J. F. F. HALE, Rutland. Vt.—*Car Coupling.*—August 14, 1866.—The draw-head has two entrances and the pin is supported by a stop in proper relation to either.

Claim.—The double stop c in combination with the horizontal partition a, and coupling pin b of a car coupling, constructed and operating substantially as and for the purpose set forth.

No. 57,209.—E. L. STAPLES, Nashville, Tenn.—*Table.*—August 14, 1866.—The top being slipped away exposes the inclined kneading trough, which occupies the position of a drawer when the table is closed.

Claim.—First, the inclined moulding board or kneading trough D, in the top of the table, substantially as described.

Second, the combination of the sliding top E with its attached leg and the inclined moulding board on the top of the table, substantially as and for the purpose herein specified and set forth.

No. 57,210.—ELI T. STARR, Philadelphia, Penn.—*Vulcanizing Flask for Dentists.*—August 14, 1866.—The rings of the flask are held between the upper and lower plates by screw-bolts and nuts; the plaster of the mould is attached to the plate by occupying the countersunk holes therein.

Claim.—First, the combination of the two rings A B and interchangeable plates *s e*, substantially as herein set forth, for the purpose specified.

Second, the holes *f* bevelled or countersunk at their outer ends, in combination with the covering plate A and a ring A or B, substantially as herein set forth, for the purpose specified.

Third, the slotted lugs *g*, recessed as described, of the plate *s*, in combination with the T-bolts *j*, nuts *r*. and perforated lug *k* of the plate *e*, substantially as herein set forth for the purpose specified.

No. 57,211.—CHARLES L. STEVENS, Galesburg, Ill.—*Elevating Water into Railroad Tanks.*—August 14, 1866.—The train passing over a section of rail on a platform supported on a system of compound levers, by the vibration of the latter operates piston rods which pump water into an elevated tank.

Claim—First, the rail platform A operating on a series of compound levers C C C, and so connected with the rods *a a a* of the platform, and the shafts E E E as to impart, by means of the passage of a train of cars, sufficient motion thereto to operate one, two, or three pumps, as may be desired, substantially in the manner and for the purpose as herein set forth.

Second, the rods *a a a* of the platform constructed with the friction rollers *e e e* on their ends and operating in slots D in the bed sills B, under and against the rails, so as to give the proper direction and position to the platform in its elevation and depression, substantially in the manner and for the purpose as herein set forth.

Third, depressing the rail platform and retaining it in that position by means of the lever *h*, on the shaft being locked into the upright spring plate *i* by the the brakesman or man in attendance, substantially in the manner and for the purpose as specified.

Fourth, the fulcrum lever A″, vertical rod and lever connecting therewith, in combination with the shaft E, substantially in the manner and for the purpose as herein specified.

Fifth, the weights F F and levers *f f*, on shafts, as arranged to elevate the rail platform on every succeeding depression of it by the cars of the train when passing thereon, substantially in the manner and for the purpose specified.

No. 57,212.—CHARLES F. STILZ and JOHANN C. KNOEPKE, Philadelphia, Penn.—*Preserve Can.*—August 14, 1866.—The tin preserve can has a conical neck closed by a cast iron stopper with a gum ring around it and has a bar above secured to ears on the can.

Claim.—First, the combination of the tin can A, the conical flanged neck *d*, the cast iron stopper E, gum ring G, ears K K′, and spring bar M, when arranged and operating substantially as described.

Second, the combination of the tin can A, the conical flanged neck *d*, the cast iron stopper E, gum ring G, loops *o o*, cross-bar N, and thumb-screw *p*, when arranged and operating substantially as described.

No. 57,213.—ENOS STIMSON, Montpelier, Vt.—*Grater for Carrots and other Vegetables.*—August 14, 1866; antedated July 30, 1866.—The root is held in a concave and is rasped by the cylindrical grater which rotates in contact with it.

Claim.—A rotary grater with an open end attached to the end of an arbor operated by gears and crank in connection with the frame and rests, as described and for the purpose above specified.

No. 57,214.—H. STOCKTON, Newport, R. I., and W. S. SMOOT, Washington, D. C.—*Saddle.*—August 14, 1866.—The saddle-tree, consisting of pommel, side bars, and cantle, has the usual staple for the suspension of the stirrup straps, and is secured to the saddle cloth by loops and surcingle, dispensing with flaps, girths, and padding. The end of the stirrup loop is hinged and opens to release the strap.

Claim.—First, a saddle-tree highly arched underneath, flat on top, opened between the bars, and covered and bound, substantially as herein described.

Second, in combination with a saddle such as described, and without padding, a saddle cloth having a leather re-enforce *c*, substantially as and for the purpose set forth.

Third, the combination with the stirrup loop of a hinged locking and unlocking lever *z z*, operated without a spring, substantially as set forth.

No. 57,215.—TIMOTHY TERREL, Spring Hill, Ohio.—*Lard Lamp.*—August 14, 1866.—The teeth of the wheel engage with the wick, and by raising the lid of the lamp and placing a finger on the wheel to move it, the wick can be regulated.

Claim.—A lamp for burning lard and other similar substances, having a cylinder body or fountain A, with a toothed wheel E, fitted concentrically within it, on a horizontal shaft D, to act upon the wick F, substantially as herein shown and described.

No. 57,216.—TIMOTHY TERREL, Spring Hill, Ohio.—*Plough Cleaning Attachment.*—August 14, 1866.—The bent end of the sliding rod pushes the trash away from the coulter.

Claim.—The sliding rod D, applied to a plough, substantially in the manner as and for the purpose herein set forth.

No. 57,217.—N. Spencer Thomas, Painted Post, N. Y.—*Process for making Extracts.*—August 14, 1866.—Jets of air are passed through the bottom of the evaporating pan in which the concentration is effected.

Claim.—The within-described process of concentrating extracts or other substances by the action of jets of air injected at or through the bottom of the vacuum pan in which the evaporation of said substance is to be effected, substantially as set forth.

No. 57,218.—N. Spencer Thomas, Painted Post, N. Y.—*Apparatus for Extracting Tanning Matter from Bark*—August 14, 1866.—The material rests upon a false bottom and is treated by jets of steam; the hinged bottom of the vat opens to discharge the spent material.

Claim.—First, the apparatus herein described, consisting of a tank A, movable top B, hinges *a'*, catch *h*, tie rods *a*, pipes *b f*, rose *c*, bottom C, perforated false bottom *d*, rim *e*, and radiating channels *g*, when constructed and arranged in the manner and for the purposes specified.

Second, the process herein described of extracting bark or other materials by the continuous application of steam and intermittent application of water in an apparatus of the construction specified.

No. 57,219.—N. Spencer Thomas, Painted Post, N. Y.—*Composition for Removing Incrustations from Boilers.*—August 14, 1866.—Composed of hemlock bark 1, corn meal 2 parts.

Claim.—A compound for removing and preventing incrustations in steam boilers, made of the ingredients herein set forth and mixed together or applied in about the proportions specified.

No. 57,220.—Thomas J. Thorn, Skaneateles, N. Y., and Jonathan Dennis, Jr., Washington, D. C.—*Faucet.*—August 14, 1866.—The plug has two holes, one for the discharge and the other for a shaft which operates the valve at the inner end of the discharge opening, and also a cleaner which removes obstructions from the face of the strainer.

Claim.—A faucet stock or tube fitted for insertion in a cask, tank, boiler, or other vessel, with two holes through it, and a valve at its inner end, worked by a rod or screw passing through one of the holes, and independent of the hole through which the water or liquor is drawn or flows.

Also, in combination with the above claimed faucet a strainer arranged beyond the valve of the faucet.

Also, in combination with the strainer arranged beyond the valve, an arm arranged to sweep or clear the strainer, substantially as described.

Also, the extension of the valve rod through the strainer to work the arm that sweeps the strainer.

No. 57,221.—Matthew Tschirgi and Louis Kammüller, Dubuque, Iowa.—*Beer Cooler.*—August 14, 1866.—Plates between the tubes separate the films of liquid which flow down the opposite sides.

Claim.—Cooling beer on its way from the "cool beds" to the fermenting tun, by causing the beer to flow in an unbroken sheet over both sides of a series of cold water pipes, arranged one above another, and provided with thin blades *g*, substantially as described.

No. 57,222.—J. W. Tyson, Lower Providence, Penn.—*Cultivator.*—August 14, 1866—The standards of the cultivators or ploughs work up through the slots in the bars whose lateral adjustment regulates the distance from the row of plants or the width between the furrows.

Claim—The mounted frame, provided with the laterally adjustable and longitudinally slotted bars E, as and for the purpose set forth.

No. 57,223.—Charles Van Dyeck, Nashville, Tenn.—*Spring Mattress.*—August 14, 1866.—The waists of the springs are attached by a continuous diaphragm to the extension frame, and their enlarged end coils bear upon the inner surfaces of the padded cover.

Claim.—A spring mattress constructed with a frame A, diaphragm *b*, and spiral springs *a a*, enclosed in a casing *d*, the several parts being constructed, secured, and arranged for use substantially as set forth.

No. 57,224.—P. C. Van Slyke, Bloomfield, Ind.—*Sorghum Evaporator.*—August 14, 1866.—The pan has compartments which communicate at their ends by pipes so as to make the flow continuous.

Claim.—First, in evaporators connecting their divisions by pipes running from one to the other, below their bottoms, substantially as described.

Second, perforating the connecting pipes at their lowest points, substantially as described.

No. 57,225.—Maurice Vergues, New York, N. Y.—*Apparatus for Augmenting Sound in Piano-fortes.*—August 14, 1866.—The sounding boards are secured to the frame by bolts which admit of tightening; the lower is the thicker and is perforated. The objects are volume of sound, economy of space, durability.

Claim.—First, clamping two sounding boards upon the flange of the frame at a distance within two inches of each other, to render them solid with the case and to admit of tightening them when the case shrinks.

Second, the use of the strips D I and G, to prevent a jar of the board on the frame.

Third, using in the sounding board the proportion as to the thickness of 2 of the upper to 3 of the lower one, which is perforated to equalize them in strength.

No. 57,226.—W. B. WALKER and N. D. HARTLEY, Salem, Iowa.—*Hand Loom.*—August 14, 1866.—This is an improvement on Henderson's patent of March 14, 1865. The object is to simplify the means for driving the picker staff; to provide means for adjusting the speed of the treadle shaft with respect to the beats of the lay, so as to require more or fewer beats to one revolution of the shaft, to improve upon the construction of the treadle shaft and the pulley thereon, the latter being made in sections and removable to substitute another.

Claim.—First, the driver strap b, described and shown in Fig. 5, made substantially as described, so as to be stiff except in the part that is pivoted to, and that embraces the head of the staff, for the purpose above set forth.

Second, in combination with the picker staff B, the said driving strap b q' made as shown, and the spring h for holding the staff stationary while the strap is limbered up, substantially as described.

Third, the depressors c c, having rounded shoulders b', their bottoms being inclined or bevelled in two directions, and having planes formed thereon corresponding in number with the number of treadles in the loom, substantially as described.

Fourth, the reciprocating rod g attached to the lay, whose hooked end engages with holes in the treadle shaft to hold the shaft stationary when the lay is still, substantially as shown.

Fifth, making the pulley I in sections, so that it can be removed from the shaft in order to change from one sized pulley to another, substantially as shown.

Sixth, the stop r, connected with the rake-up rod E, which gives it a vibrating movement for the purpose of acting as a stop to the pins R, and so aiding in making the shaft move around or rotate a regulated distance at each reciprocation of the lay, substantially as set forth.

Seventh, the combination of the treadle shaft, having movable pins R, ratchet K, and a loose pulley and pawl constructed as shown, with the adjustable bar S which operates it from the lay, whereby a loom can be operated with three leaves of heddles, and with other odd as well as even numbers, by changing the position of the pins, adjusting the height of the end of the bar S on the lay, changing the number of ratchet teeth, and changing the size of the pulley, substantially as shown.

No. 57,227.—EDGAR A. WARD, Gallipolis, Ohio.—*Hay Press.*—August 14, 1866.—The pressure is alternately to the respective ends of the chest. The shaft revolves continuously and is made to engage alternately with the sections which have the spiral grooves on which are wound the ropes which work the follower.

Claim.—The shaft D with cube a attached, and the spiral sectional drum E, in combination with the lever L, ratchet R, and ropes or chains m and n, the whole operating as described and for the purposes set forth.

No. 57,228.—GEORGE I. WASHBURN, Worcester, Mass.—*Steam Engine.*—August 14, 1866.—The two cylinders are placed side by side; the effective spaces in each receive steam through ports in the partition from spaces in the other cylinder which serve as steam chests, the passages from which are controlled by valves secured to and operated by the piston rods of the respective engines.

Claim.—The cylinder A, constructed as described, with a diaphragm or partition at its central portion, traversed by the piston rod and combined with the pistons as specified.

Also, the combination of the cylinders A B, constructed and operating as described.

No. 57,229.—D. P. WEBSTER, Upper Gilmanton, N. H.—*Spring Bed Bottom.*—August 14, 1866.—The upper coil of the spring upon which the slat rests is fastened to the slat by a clamp with hooked ends.

Claim.—The single wire clamp e, with its hooks f, catching under the top coil of the spring and holding it up against the slat, substantially as described, for the purpose specified.

No. 57,230.—ALFRED WEED, Boston, Mass.—*Machine for Cutting Rasps.*—August 14, 1866.—Improvement on Weed's patent of March 17, 1866, in the several devices by which it is adapted to the cutting of rasps instead of files. To this end, the chisel stock is pivoted to the rock shaft so that the chisel may be vibrated not only in a vertical plane as usual, but also in a horizontal plane, thus cutting the teeth in curved lines across the blank. The forward feed motion is intermittent, taking place only after the completion of one row of teeth and before the commencement of another.

Claim.—Producing the teeth of rasps in arcs of parallel circles, by giving to the tool and the rasp-blank movements, substantially as described.

Also, the means of giving lateral vibration to the cutter, consisting of the lever f, when pivoted to its rocker shaft j, the lever q, and the cam n, and spring s, combined to operate substantially as described.

Also, the compound compensating jaw in lever g, formed by the two cylinders t and u, slotted and arranged substantially as specified.

Also, the means for intermittingly feeding the blanks, consisting of cam a', rocker b', the spurs on ratchet p', and latch k', arranged to operate the ratchet f', and thereby to move the blank carriage k.

Also, the combination with the presser holder of the rod p', weight n', or its equivalent spring and treadle o', arranged and operating together as described.

No. 57,231.—GEORGE E. WEST, Indiananapolis, Ind.—*Driving Fence Posts.*—August 14, 1866.—The framework is mounted upon wheels and provided in front with vertical guides for the hammer and with a cross-beam at the rear serving as a fulcrum for the lever to which the hammer is attached and which is raised by a cam.

Claim.—Raising and dropping the weight C by means of the cam G, pins $S S$, and wheels H and I, when mounted on wheels and made portable, and arranged and operated in the manner and for the purpose substantially as set forth.

No. 57,232.—SIDNEY S. WHEELER and DANIEL B. MANLEY, Danbury, Conn.—*Machine for Pouncing Hats.*—August 14, 1866.—By a system of gearing and belts motion is given to converging mandrels, having conical extremities with emery surfaces, and to a horizontal hat block revolving at less speed than does the emery wheel opposed to it; the emery wheel is journalled in a swinging frame for different presentations, and the hat-block mandrel is journalled in a frame arranged to advance and recede automatically.

Claim.—First, the pouncing cylinder L, or its equivalent, adapted to move at a comparatively high speed, in combination with the hat block a, adapted to move at a comparatively low speed, operating substantially as described, whether the motion be rotary or otherwise.

Second, carrying a rotating cutting cylinder or wheel for pouncing bats in a vibrating frame, so that such cutting cylinder or wheel can act both on the crown and sides of hat bodies, which are stationary or are moving forward in a straight line, substantially as above shown.

Third, giving an advancing motion, and at the same time a rotary motion, to a hat block, upon which hat bodies are placed to undergo the operation of pouncing, substantially as and for the purpose above set forth.

Fourth, in combination in machines for pouncing hat bodies, the carriage b. the hat block a, so mounted as to be capable of rotary motion, the worm O, cog wheel P, and rack V, for moving the carriage forward, and a suitable spring for driving the carriage back when it is released from its driving wheel P, substantially as above shown.

Fifth, in combination, a conical emery rotating surface h, for pouncing the brims of hat bodies, and a conical supporting surface 3, to hold the brim up to the cutting surface, substantially as shown.

Sixth, in combination, a conical rotating cutting surface h, for pouncing the brims of hat bodies, and conical feeding rolls with elastic surface g g, for feeding the brim to the cutting or pouncing surface, substantially as shown.

No. 57,233.—THOMAS R. WHITE and WM. G. BEDFORD, Philadelphia, Penn.—*Fabric for Tubing, &c.*—August 14, 1866.—The fabric has a warp of parallel wires, a weft of threads, and is covered with rubber or other water-proof material.

Claim.—A fabric composed of threads and wires, interwoven and coated with a water-proof material, substantially as described.

No. 57,234.—A. A. WILCOX, Fair Haven, Conn.—*Shoe Nail.*—August 14, 1866.—The collar on the nail limits its penetration and the finishing "lift" is driven upon the projecting portion.

Claim.—A shoe nail, provided with a collar a, or its equivalent, substantially in the manner herein described, as a new article of manufacture.

No. 57,235.—ALBERT S. WILKINSON, Pawtucket, R. I.—*Horseshoe.*—August 14, 1866.—The clip is continuous around the upper bar of the shoe, and the calk continuous around the lower bar; the bars are riveted together.

Claim.—The continuous clip c, in combination with the bar A, and continuous calk d, whether the latter is situated on the inner or outer edge of the shoe, all constructed as illustrated and described in Figs. 1 and 2.

No. 57,236.—ALBERT S. WILKINSON, Pawtucket, R. I.—*Calk for Horseshoes.*—August 14, 1866.—The arms of the calks reach around the sides and rest upon the upper edge of the shoe; the calk is held in position by a wedge.

Claim.—An adjustable calk in B' B b, constructed and fastened substantially in the manner and for the purpose specified.

No. 57,237.—E. WILKINSON, Jr., Mansfield, Ohio.—*Soldering Eave Troughs.*—August 14, 1866.—A bed is made of the size and shape of the trough to be soldered, having clamps to correspond ; into this the bent-up sheets are laid and clamped ; the bed is allowed to rock in its bearings so as to cause the solder to flow over the seam.

Claim.—The frame F and flanges *v*, in combination with the concave bed B, having clamps C C, all constructed and arranged together, substantially as and for the purpose specified.

No. 57,238.—ALEXANDER C. WILLS and HUGH SHARP, Marlton, N. J.—*Manufacture of Vinegar.*—August 14, 1866.—The pomace stands in a heap till it ferments, when it is thrown into a cistern containing the residuum of the distillation of cider; after standing a few days the vinegar is expressed.

Claim.—Utilizing the refuse of apple presses and cider stills, by treating these substantially as described, to form vinegar.

No· 57,239.—JACOB D WINSLOW, Wilmington, Del.—*Shutter Fastening.*—August 14, 1866.—A notched lever pivoted to a bar which is firmly secured to the wall.

Claim.—The notched lever B, hung upon the bar A, in combination with the staple or equivalent device, projecting downward from the lower edge of the shutter E, all being arranged and operating as set forth. •

No. 57,240.—ALVAH WISWALL, New York, N. Y.—*Door Spring.*—August 14, 1866.—A U-shaped spring attached to the door bears upon an anti-friction roller attached by an arm to the jamb leaf of the hinge and tends to hold the door closed or wide open according to the position of the latter.

Claim.—The U-shaped spring D, attached to the door or gate, in combination with the roller *d*, attached to a plate *c*, projecting from the door-frame, or from the part *a* of the hinge attached thereto, and all having such relative position with each other as to operate in the manner substantially as and for the purpose set forth.

No. 57,241.—GEORGE M. WOOD, Decatur, Ill.—*Door Bolt.*—August 14, 1866.—The part sliding upon the plate and the plate itself have corresponding longitudinal slots through which the fastening screws are passed.

Claim.—The semicircular hollow bolt A, provided with a longitudinal slot *a*, in its inner side, and a hole for a knob on its outer side, in combination with the plate B, having a longitudinal slot *b* in it for the screws D to pass through and secure the plate and bolt to the door or other article to which the bolt is to be applied, substantially as shown and described.

No. 57,242.—SYLVESTER WOODBRIDGE, Benicia, Cal.—*Machine for Tilling the Soil.*—August 14, 1866.—The spade handles pass through guide slots in an upper bar, and they receive their motion by attachment to cranks revolved by connection with the drum. The depth is regulated by the vertical adjustment of the tilting frame, which carries the crank shaft.

Claim.—First, the tiller frame B', tiller shaft G, and guide plate or shaft K, in combination with the crank on the tiller shaft G, and crank shaft F, and connecting rods H, substantially as and for the purpose set forth.

Second, the truck frame A, and driving wheel or cylinder C, spur gear D, (whether internal or external,) and pinion E, with or without intermediate gearing, in combination with the means hereinbefore described and set forth of operating agricultural implements by cranks, rods, guide plates, or shafts, substantially as set forth.

No. 57,243.—WILLIAM P. WOODRUFF, New York, N. Y.—*Packing for Piston Rods.*—August 14, 1866 ; antedated August 9, 1866.—Around a braided gasket is a canvas covering and a strip of tinfoil held in position by thin brass plates attached to the canvas; the latter keeps the fibres of the gasket from becoming insinuated between the rods and its gland, and the foil confers an anti-friction quality.

Claim.—A steam packing made of a gasket *a*, braided or otherwise produced, by hemp or other suitable material, and protected by a flexible covering B, in combination with a strip of tinfoil fastened to its inner circumference by one or more strips of brass, substantially as and for the purpose set forth.

No. 57,244.—FRED. W. BACON, Washington, D. C., assignor to himself and D. A. STRONG, same place.—*Culinary Steamer for Boilers.*—August 14, 1866.—The steamer fits within the water kettle and has a double bottom with perforations and conductors arranged to permit the free ascent of steam and descent of water of condensation.

Claim.—The construction of steamers for culinary purposes, provided with the perforated diaphragm D, or its equivalent, the concave centre *c* or its equivalent, and the conical bottom B, with the tubes *t t t* or their equivalents, made substantially in the manner and for the purposes set forth.

No. 57,245.—WILLIAM BESCHKE, Philadelphia, Penn., assignor to himself, P. H. VANDER WEYDE, and L. STRAUS, same place.—*Using Explosive Liquids for the Production of Light and Heat.*—August 14, 1866.—Absorbent or capillary materials are used to prevent the accidental escape of the liquid, or the communication of explosion.

Claim.—Filling up the whole or a part of the interior of the lamps, bottles, cans, barrels, tanks, reservoirs, &c., (the former to burn, and the latter to retain and to transport gasoline or other explosive fluid,) with saw-dust, cotton, asbestos, beads, gravel, shot, &c., and with wire gauze or perforated thin plate, for the purposes specified.

No. 57,246.—HENRY BAILEY, Brooklyn, N.Y., and EDWIN L. BAILEY, Philadelphia, Penn., assignors to HENRY BAILEY, Philadelphia, Penn.—*Drop-handle Urn Cock.*—August 14, 1866.—The stop-pin to limit the rotation of the plug projects from the socket into a transverse in the plug. A longitudinal groove in the plug permits its insertion over the pin.
Claim.—The plug B, having a right angular groove D D', operating with the fixed pin C, or its equivalent, on the face of the tapering hole of the body A, provided with nut E, and washer F, substantially as described for the purpose described.

No. 57, 247.—GEORGE BUTTERFIELD, Boston, Mass., assignor to DAVID ROBINSON, Jr., Dorchester, Mass.—*Shoe Lacing Fastening.*—August 14, 1866; antedated August 1, 1866.—The plate has an inclined slot so formed in it that when applied to an article to be laced, the strain draws it toward the angle of the slot and fastens it in position.
Claim.—The lacing plate A, provided with an eye b, having an angular slot a extending from it, substantially as and for the purpose described.
Also, the combination of such plate A with a series of lacing or eyelet holes c, the plate being arranged with respect to the holes c, and to secure the lacing d, substantially as shown and described.

No. 57,248.—PHILIP B. CURTIS, Amesbury, Mass., assignor to himself and ALBERT P. SAWYER, same place.—*Whiffletree.*—August 14, 1866.—The trace hooks are attached to straps whose inner extremities are held at the centre of the whiffletree by a bolt, which is kept in place by a spring, but which may be withdrawn by the driver to detach the horses.
Claim.—The arrangement and combination of the trace attachment straps B C, and the bolt f, and its holders e c, with the whiffletree A, and its end loops b b, the whole being substantially as specified.
Also, the combination of the safety spring with the whiffletree, the bolts f, its holders, and the attachment straps B C, applied to such whiffletree, as specified.

No. 57,249.—JOHN DECKER, Sparta, N. J., assignor to himself and CHARLES W. WARD-WELL, Brooklyn, N. Y.—*Slide Bolt.*—August 14, 1866.—A series of tumbler bars are applied to the slide bolt, and are brought into correspondence by a key to enable the stud under the bolt to retract them.
Claim.—The bolt B, fitted in a case A, and provided with a projection d at its rear side, and with a shoulder or stop e at its outer side, in combination with the tumblers D, notched as shown, to receive the projection d, and the oblong slot b in the case A, all arranged to operate substantially in the manner as and for the purpose set forth.

No. 57,250.—ORLANDO V. FLORA, Madison, Ind., assignor to himself and JOHN G. MOORE, same place.—*Carpenter's Vice.*—The nut of the clamp screw is fixed to the rear end of a rack bar, whose teeth engage with a catch in the fixed jaw, so that, by raising and slipping the bar, the jaws may be adjusted to any width.
Claim.—The combination of the fixed catch B and the rack bar G, movable up and down, to clear and engage with the catch, substantially as and for the purpose herein specified.

No. 57,251.—JACOB FOX, Minersville, Penn., assignor to CARR & CO., Philadelphia, Penn.—*Machine for Flattening and Punching Umbrella Ribs.*—August 1, 1866; antedated August 1, 1866.—The ribs placed in rotating notched disks are fed by intermittent motion, between anvils and slides, by which they are first clamped and flattened and punched in three places.
Claim.—First, the slides D, with their flattening, punching, and indenting dies, arranged in line and parallel with each other, in combination with similarly arranged counter dies, the whole operating as set forth, for the purpose specified.
Second, the combination of the above with shaft G, and notched disks H and H', the whole being arranged and operating substantially as and for the purpose herein set forth.
Third, the combination of the slides D, and their punching and flattening dies i, with the holding rod d, and its spring c.

No. 57,252.—CHAS. GOTY and AUG. GUILLEMIN, Cincinnati, Ohio., assignors to themselves, CHARLES FIX and JOSEPH OPPINHEIMER.—*Utilizing Tinners' Waste.*—August 14, 1866.—To remove the tin from the iron the scraps are heated in contact with rosin and lycopodium supplied from a hopper by means of a fan.
Claim—First, the process of removing tin from iron, substantially as set forth.
Second, the furnace, substantially as described and for the purpose set forth.

No. 57,253.—SAMUEL A. GRANT, Burlington, N. J., assignor to the BURLINGTON MANU-FACTURING COMPANY, same place.—*Ladies' Fan.*—August 14, 1866.—Explained by the claim.

Claim.—As a new article of manufacture, a fan having its body and handle made in one piece, and of one and the same veneer or sheet of wood, substantially as herein described, and for the purpose specified.

No. 57,254.—P. A. LA MENT, New York, N. Y., assignor to herself and ELIZA LA MENT, same place —*Apparatus for Fitting and Measuring Dresses.*—August 14, 1866.—A skeleton frame or shape adjustable on the body, so as to be set thereon and then removed to a former which may be expanded or contracted to the proper size, that a garment may be fitted thereon.

Claim.—A skeleton frame or "shape," consisting of a series of bands B and C, when so arranged and connected together that they can be separately and severally adjusted, sub-stantially in the manner described, and for the purpose specified.

No. 57,255.—ERNST MARX, New York, N. Y., assignor to himself, HERMANN DITTRICH and JULIUS KEER, same place.—*Swinging Chair.*—August 14, 1866.—The seat is attached to a pivoted arm, which vibrates horizontally to place the seat out for use, or beneath the piano when out of use.

Claim.—The leg E, and jointed and curved arm D, supporting the seat, and operating substantially as described.

No. 57,256.—WHITNEY W. MEGLONE, Nashville, Tenn., assignor to himself and LEAN-DER STEDMAN, same place.—*Instrument for Removing Effervescing Fluids from Bottles.*—August 14, 1866.—A tube is introduced through the cork; the liquid enters holes near its lower end and is discharged at the goose neck.

Claim.—First, the fluid-extracting instrument a C b, constructed and operating in the man-ner as shown and described.

Second, the combination of the tube or instrument a C b, with the cork D, or its equiva-lent, and bottle A, operating in the manner set forth.

Third, the combination of the several parts just named with the auxiliary tube t, operating substantially as set forth.

No. 57,257.—WILLIAM NORDHOFF, Baltimore, Md., assignor to WM. KNABE & Co., same place.—*Piano.*—August 14, 1866.—The agraffes are secured to an L-shaped brass plate which is fastened to the wrest plank and the iron frame.

Claim.—The plate B, secured to the wrest plank C, in combination with the frame A, when said parts are constructed and arranged to operate substantially as and for the purpose herein set forth.

No. 57,258.—TIMOTHY J. POWERS, New York, N. Y., assignor to J. P. FITCH and J. R. VAN VECHTEN, same place.—*Machine for Priming Metallic Cartridges.*—August 14, 1866; antedated July 31, 1866.—The machine deposits the fulminate in the flanges of metallic cartridges by centrifugal action and contains devices for carrying the cartridge to a proper position, giving it a rotary motion, and then by means of an automatic feeder dropping the requisite quantity of the fulminate paste into the cartridge, where it is immediately thrown into the flange or recess provided for it.

Claim.—First, in a machine for depositing the fulminate priming in cartridge shells, a feeder which descends into a box, vessel, or reservoir containing the priming, in a fluid or semi-fluid state, picks up the requisite quantity of priming therefrom, and again descends into or over the cartridge shell, to deposit the priming therein, substantially as herein described.

Second, suddenly arresting the feeder in its descent over or into the cartridge shell, sub-stantially as herein specified, for the purpose of insuring the complete discharge of the prim-ing therefrom.

Third, the combination of a feeder, a priming reservoir, and a cartridge-shell carrier, opera-ting substantially as herein specified.

Fourth, commencing the rotary or spinning motion of the shell about its axis before the deposit of the priming therein, substantially as and for the purpose herein described.

No. 57,259.—HENRY REED and WILLIAM P PENNEWELL, Middletown, Mo., assignors to themselves and WALTER CALDWELL, same place.—*Corn Planter.*—August 14, 1866.—The wire teeth on the side pieces scatter the falling seed.

Claim.—The short-iron teeth or wires z, when attached to the pieces A A', for the purpose of scattering the seed.

No. 57,260.—E. A. ROBINSON, Waterbury, Conn., assignor to the UNITED STATES BUT-TON COMPANY, same place.—*Button.*—August 14, 1866.—The jagged edges of the hole in the blank fasten the ends of the loop against the mould and prevent their withdrawal.

Claim.—A button having the loop secured to and in the mould by means of a metallic plate or blank D, having a ragged edge or lip E, substantially in the manner described.

No. 57,261.—M. S. SAGER, Washington, Ohio, assignor to S. B. YEOMAN, same place.—
Machine for Making Sheet Metal Pans.—August 14, 1866.—For bending up the sides of
a square metal pan and folding over the excess of metal at the corners. The sheet is laid
between a base plate and a former of corresponding shape, and bent up against the sides of
the "former" by levers hinged to the edge of the plate. Two other levers, hinged respectively
to the contiguous ends of the first two levers, serve to turn over the surplus metal at the corners.
Claim.—The combination of the sliding mould block M, and stationary bed plate B having
lever arms or benders H hinged to the sides, and secondary benders or lever arms K hinged
to the benders H, for the purpose of bending or laying over the projecting corner pieces of
the metal sheet, arranged and operated substantially as described.

No. 57,262.—CORNELIUS and ZACHARIAH WALSH, Newark, N. J., assignors to CORNE-
LIUS WALSH.—*Carpet Bag Frame.*—August 14, 1866.—The interior frame has a longitudinal portion which is tenoned into the jaws and connected by side pieces to the portions
through which the hinge rod passes; the jaws as well as the frames are pivoted upon the
hinge rod.
Claim.—The combination of the jaws A with the interior frames, consisting of the wooden
pieces C C, metallic hinge pieces I I, and plates D, the latter secured by tenons F F to the
jaws, and both jaws and inner frames pivoted upon the bolt B, the whole constructed as
described and represented.

No. 57,263.—WILLIAM WINTER, Philadelphia, Penn., assignor to himself, ISAAC TOWN-
SEND, and THEODORE H. BEACHER, same place.—*Clamp for Clothes Lines.*—August 14,
1866.—An elastic split ferrule.
Claim.—A clothes line spring clamp, consisting of a hollow cylinder, open at each end,
divided along one of its sides, and having its inner edges bevelled or rounded off, all as set
forth and described.

No. 57,264.—JOHN DALE, Manchester, England.—*Manufacture of Pigments.*—August
14, 1866.—Explained by the claim.
Claim.—First, the production of a pigment by decomposing the known pigment, satin
white, by means of chloride of barium, or strontium, so as to replace, or partially replace,
sulphate of lime, by sulphate of barium or strontium.
Second, the production of a pigment by using caustic baryta or strontia instead of, or
partly instead of, the lime ordinarily used in making satin white.

No. 57,265.—L. P. R. DE MASSY, Paris, France.—*Filter Press.*—August 14, 1866.—The
material to be filtered is placed between the walls of the filtering chamber and an enclosed
rubber tube or cylinder, the latter is expanded by the introduction of water, and the expressed
liquid is forced through the perforated and gauze-guarded discharge tubes.
Claim.—A filter press having a corrugated surface covered with wire cloth or gauze
against which the material from which the liquor is to be extracted is pressed, all substantially as and for the purpose described.

No. 57,266.—F. A. LANGE, Glashutte, Saxony.—*Repeating Watch.*—August 14, 1866.—
Explained by the claims and cuts.
Claim.—First, the pusher G, in combination with the levers H I, when constructed and
operating substantially as described.
Second, the lever I, in combination with the pusher G, lever H, pinion E, and hammer F,
when constructed and operating in the manner substantially as set forth.
Third, the lever H, when constructed with the projections M and N, and the slot S, substantially in the manner and for the purposes described.
Fourth, the lever H, in combination with the lever I, snail wheel Q, quarter wheel P,
a ring W, and spring R, when constructed and operating in the manner substantially as set
forth.
Fifth, the lever I, when constructed with the rack K, the rack L, the screw T, and the
spring R, substantially in the manner and for the purpose described.
Sixth, providing the star or hour wheel O with twenty-four teeth, in the manner and for
the purpose set forth.
Seventh, the hour wheel O, in combination with the snail wheel Q, quarter wheel P, and
spring Y, substantially as set forth.
Eighth, the hammer F, when provided with the pawl a, on the sliding stem b, and con-
structed substantially as and for the purposes set forth.
Ninth, the hammer F, in combination with the spring X, as described.
Tenth, the wheels D and E, in combination with the levers G H and I, and the hammer
F, when constructed substantially as and for the purposes set forth.
Eleventh, the combination and arrangement of the pusher G, levers H and I, pinion E,
and wheels D and C, the lever Z, springs R and W, hour wheel O, snail wheel Q, quarter
wheel P, spring Y, with the hammer F, and spring X, substantially as and for the purposes
set forth.

No. 57,267.—JOHN MEDHURST, Bermondsey, England.—*Apparatus for Reefing and Unfurling Sails*—August 14, 1866; English patent March 4, 1866.—The foot of the sail is attached to, and in furling is wound upon, a roller whose bearings are in standards on the yard below : the upper part of the sail is bent to the upper yard, which partially descends as the halliards are lowered and the roller rotated by chains passing below to the deck, causing the roller to rise to meet the yard above.

Claim.—The combined arrangement of the parts for reefing and furling a sail on a roller *c c*, substantially as herein described.

No. 57,268.—JEAN THEODORE SCHOLTE. Paris, France.—*Gas Meter.*—August 14, 1866.—The rotating drum is enclosed within a casing and rotated by the pressure of the gas against the spiral partitions ; the revolutions are indicated in the ordinary manner; the drum is partially filled with a constant quantity of water ; the remaining space is a definite measure of capacity and is emptied at each revolution of the drum ; a valve having floats attached cuts off the supply of gas when the water falls in the drum.

Claim.—First, the apparatus for measuring gas, as herein shown and described, the same consisting of the following elements combined : 1st, a drum with interior spiral partitions actuated by the gas, as herein shown and set forth ; 2d, a valve, together with floats so arranged that it may open and close automatically ; 3d, a regulating or discharge pipe for carrying off all water in excess of the proper level.

Second, in apparatus for measuring gas or other fluids, as specified, the drum, provided with interior spiral partitions and actuated as herein shown and set forth.

No. 57,269.—JASPER H. SELWYN, Grasmere, England.—*Breech-loading Fire-arm.*—August 14, 1866 ; English patent October 12, 1865.—Explained by the claim and cut.

Claim.—The improvements in breech-loading fire-arms described and represented—that is to say, making the charge chamber of a conical or spheroido-conical figure, and making an annular groove or depression at the breech end of the barrel, into which groove or depression the end of the cartridge case is expanded on discharge, and retained so as to be started or drawn on the separation of the charge chamber from the barrel.

No 57,270.—CARL FREDERICK UHLIG, Chemnitz, Saxony, assignor to GEORGE FREDERICK WILLIAM PABST, same place—*Harmonium.*—August 14, 1866.—The accordeon is combined with the piano-forte; the mechanism of the former is vertically adjustable, and when raised the reed valves are opened by the depression of the keys of the stringed instrument, sounding the melodeon in concert by a single impulse.

Claim.—First, the combination of the flexible pipe *l*, air movable chamber B, and bellows D, substantially in the manner represented and described.

Second, the combination of the air chamber B, spring catch *m*, arranged with lip *p*, operating with the button *q*, substantially as aud for the purpose herein represented and described.

Third, the combination of the lever F, rod *n*, air chamber B, provided with the spring catch *m*, operating with lip *p*, substantially in the manner and for the purpose represented and described.

Fourth, the combination and arrangement of the keys *a*, with pins *a'*, lever *j*, with pins *a²*, air chamber B, spring catch *m*, lip *p*, button *q*, rod *n*, lever F, flexible pipe *l*, bellows D, constructed and operating substantially as described, for the purpose specified.

No. 57,271.—ISAAC ADAMS, Jr., Boston, Mass.—*Coating Metals with Metal.*—August 21, 1866.—Nickel is applied by electro-plating or otherwise.

Claim.—Rendering gas tips and other similar articles anti-corrosive to heat or moisture by surfacing them with nickel, substantially as set forth.

No. 57,272.—LEVI W. ALBEE and SIMEON W. ALBEE, South Charlestown, N. H.—*Bee-hive.*—August 21, 1866.—Explained by the claims and cut.

Claim.—The external case or box A, provided with doors B at its sides and ends, in combination with a box D, provided with comb frames F, and with doors E at its sides, and arranged so as to slide within A, substantially in the manner and for the purpose set forth.

Also, the hanging of the comb frames F on hinges *g*, in combination with the pins or stops *h* and screws *i* in the side doors E of the box D, substantially as and for the purpose specified.

Also, the exit passage *k*, in the top plate of the box D, in combination with the bar H and pivoted slats *l*, provided with holes *l*, and the holes *m*, in the inner sides of the boxes G, substantially as and for the purpose specified.

No. 57,273.—S. R. ANDRES, Troy, N. Y.—*Eraser and Burnisher.*—August 21, 1866 ; antedated June 30, 1866.—The rubber which erases and the brush which removes the scourings are conveniently attached to the same holder.

Claim.—The combination of a piece of rubber or other erasive material B with a brush C, both set in a suitable frame A, and arranged in a manner to accomplish the purpose of the invention, substantially as herein specified.

No. 57,274.—FREDERICK ASHLEY, New York, N. Y., assignor to JAMES EDGAR and J. E. CAVAN.—*Letter File.*—August 21, 1866.—The upper portion has a spring, and when raised exposes the point of the lower portion on which the letters are impaled.

Claim.—Securing the upper portion G of the hook to a plate H, arranged in a suitable groove or guide-way of the frame A, or its equivalent, in combination with a spiral or other suitable spring, arranged and operating together, substantially as and for the purpose described.

No. 57,275.—GUILLAUME AYMARD, New York, N. Y.—*Manufacture of Tobacco Bags.*—August 21, 1866.—The green bladders are brined, washed, tanned in a solution of alum and salt, inflated, dried, softened by rubbing between corrugated boards, and exposed to the fumes of sulphur.

Claim.—The mode herein specified of preparing bladders for tobacco bags, &c., by which they are rendered soft and pliable, as set forth.

No. 57,276.—W. W. BATCHELDER, New York, N. Y.—*Vapor Burner.*—August 21, 1866.—Movable plates beneath the perforated floor of the vaporizing chamber adjust the size of the slot above the wick tube so as to proportion it to the height of the wick and prevent smoking.

Claim.—First, regulating and steadying the flame of a coal-oil or other similar lamp in the manner and by the means herein specified—that is to say, by interposing between the vaporizing chamber and wick tube or holder of a lamp burner, constructed as above described, a movable and adjustable diaphragm, as and for the purposes hereinbefore shown and set forth.

Second, the movable and adjustable diaphragm, as herein described, the same consisting of a perforated stationary plate, in combination with movable plates of segmental or other suitable shape, constructed and arranged for operation substantially as and for the purposes herein set forth.

Third, in combination with the vapor chamber and surrounding dome and the wick tube of a coal-oil or other burner as described, the movable and adjustable diaphragm, constructed as herein specified, the whole being arranged for operation substantially as set forth.

No. 57,277.—CHARLES BEACH, Penn Yan, N. Y.—*Cider and Wine Press.*—August 21, 1866.—A conical, perforated hopper, divisible in the plane of its axis, is occupied by a conical plunger of corresponding proportions.

Claim.—The platen B with the receptacle C, when made and used as specified and for the purpose set forth.

No. 57,278.—WILLIAM W. BEACH, New York, N. Y.—*Skeleton Tumbler.*—August 21, 1866.—The skeleton tumbler has a hole in the bottom, a stopper therefor, graduated indications of capacity on its sides, and by certain accessories becomes a filter, drinking cup, butter print, &c.

Claim.—The construction of the skeleton tumbler as the basis of a series of apparatus, described and represented in the several figures of the drawings.

No. 57,279.—JOSEPH W. BIER and JOHN B. WAMPLER, Selbyville, Ill.—*Device for Burying Weeds and Stubble while Ploughing.*—August 21, 1866.—A frame attached to the beam collects the trash and turns it over so as to be covered by the furrow slice.

Claim.—The herein described devices denominated "a weed burier," the same being attached to a plough beam, in the manner and for the purpose herein set forth.

No. 57,280.—JOHN BLACKIE, New York, N. Y.—*Machine for Pressing Tobacco.*—August 21, 1866.—The channeled belt receives the tobacco from the cells of a frame and conveys it to grooved pressing rollers which reduce it to a strip of the required proportions, which is cut transversely by the spiral knives of a diagonally set roller.

Claim.—First, the apron B, having a series of channels of varying or uniform width, for receiving and conveying the tobacco to the pressing rollers, substantially as set forth.

Second, in combination with the apron B, constructed as described, the belts m, when said parts are arranged to operate as and for the purpose set forth.

Third, the combination of the grooved cylinder C, adjustable roller e' and pressing roller D, substantially as shown and described.

Fourth, a cutting apparatus, constructed and operating substantially as set forth.

Fifth, the frame H, provided with the cells o and a sliding bottom I, arranged and operating substantially as and for the purpose set forth.

No. 57,281.—GEORGE D. BLOCHER, Indianapolis, Ind.—*Cooler.*—August 21, 1866.—The central ice chamber is surrounded by an annular chamber for milk, each having separate discharge pipes and lids; a butter tray fits in the upper part of the milk chamber.

Claim.—The central ice chamber D and covers E and F, arranged as shown, and in combination therewith the milk chamber B, and butter vessel C, substantially as set forth.

No. 57,282.—THOMAS BOWERS, Zanesville, Ohio.—*Wooden Screw.*—August 21, 1866.—
The thread of the wooden screw has its bearing surface at right angles to the axis of the
screw and a cylindrical surface on its periphery.

Claim.—As an improved article of manufacture the wooden screw, which works in a nut
or matrix, constructed with a spiral thread which has a bearing surface d perpendicular to the
axis of the screw, a cylindrical surface c, which is parallel to said axis and inclined surface
b, which forms an obtuse angle with the cylindrical surfaces c and e, all as described and
represented, and for the purpose set forth.

No. 57,283.—THOMAS J. BROWN, Clñ. Ohio.—*Corn Husker.*—August 21, 1866.—The
talon is placed over the ball of the thumb and attached to a palm plate on the flexible band.

Claim.—A corn husker constructed with a flexible band A, to which is fastened the plate
A, covering the entire palm of the hand, and having attached thereto the curved point C, projecting toward the fingers, substantially as set forth.

No. 57,284.—WALTER BRYENT, Boston, Mass.—*Apparatus for Suppressing Effluvia from
Drains.*—August 21, 1866.—A stink trap whose flanges descend beneath the surface of the
water in the basin and the gutters around it.

Claim.—The combination of the top or strainer $g\,g$, formed with the flanges or lips $f\,f$ and
$h\,h$, with the groove or gutter $c\,c$ and basin $e\,e$, as described and for the purpose specified.

No. 57,285.—D. H. BURKET, Half moon, and J. C. GRAY, Putneyville, Penn.—*Apparatus
for Treating Petroleum.*—August 21, 1866.—The steam pipe is connected by a stuffing box
with the vessel in which is a vertical, tubular, revolving shaft, with hollow arms and radial
stirring blades.

Claim.—First, distributing pipes N N and wings W W, when constructed and operating
with pipe M and tank A, substantially in the manner and for the purposes set forth.

Second, the packing box E, when constructed and operating with box F and connection pipe
D, substantially in the manner and for the purposes set forth.

No. 57,286.—SAMUEL CASEBEER, Roseburg, Oregon.—*Plough Coulter.*—August 21,
1866.—The foot of the coulter is stepped into the landside, and its point lies upon the nose of
the share.

Claim.—The application to ploughs of the aforesaid coulter, in the way and manner herein
described.

No. 57,287.—WILLIAM CHICKEN, Boston, Mass.—*Feeding Device for Sewing Machines.*—
August 21, 1866.—The inner wheel is set eccentrically on the same axis with the outer wheel.
The motion of the lever in the direction of the arrow acting on the inner lever pivoted to the
eccentric, tightens the wedge in the space between the wheels and gives the feed motion.

Claim.—The said friction feed apparatus as composed of the levers C E, the wedge D, and
the two wheels A B, arranged and applied together, substantially in the manner and so as to
operate as specified.

No. 57,288.—SAMUEL CHILD, Jr., Baltimore, Md.—*Vapor Stove.*—August 21, 1866.—The
tubular valve rod forms the conduit for the admission of the hydrocarbon liquid, and by its
longitudinal motion graduates or stops the supply thereof to the chamber where it is ignited.

Claim.—The combination with the retort pipe and heating chamber of a vapor stove of the
tubular valve rod operated upon the axis of the said retort pipe, as described, under such an
arrangement that the said rod, while regulating the supply of fluid to the heating chamber,
shall also constitute the medium through which the fluid is conducted into the said chamber,
substantially as herein shown and set forth.

No. 57,289.—JOHN J. CHRISTIAN, Yonkers, N Y.—*Manufacture of Boots*—August 21,
1866.—Ventilating holes in the upper are cased in by a cap which has a tubular extension to
the top of the boot.

Claim.—The arrangement of the ornamental cap A, vamp or upper leather B, in combination with the perforations $b\,b\,b$, strip C C, forming an air passage, substantially as and for
the purpose herein set forth.

No. 57,290.—CHARLES S. CLARK, Huntsburgh, Ohio.—*Roof.*—August 21, 1866.—Each
side has a gable, and the crown of each gable is connected by a ridge to the common central
peak.

Claim.—An improved roof having as many ridges meeting in a peak at the centre, and as
many gables as there are sides to the building, the lowest points of the rhomboidal parts or
surfaces of the roof being at the corners of the building, substantially as described and for
the purposes set forth.

No. 57,291.—CHARLES A. CODDING, Battle Creek, Mich.—*Washing Machine.*—August
21, 1866.—The elastic arms on the rear end of the reciprocating rubber pass beneath a roller
and impose a pressure upon the clothes.

Claim.—The horizontal spring arms *h i*, in combination with the roller D, substantially as and for the purpose described.

No. 57,292.—E. H. CRAIGE, Brooklyn, N. Y.—*Hanging Pitmans, &c.*—August 21, 1866.—Especially adapted for hanging the pitman rod to the treadle and crank shaft of a sewing machine. The lower end of the rod has a V-shaped bearing to receive a correspondingly shaped wrist which permits a certain amount of lateral play and is secured by an adjustable block.

Claim.—First, forming the bearings or journals of connecting or other rods in machinery of a V or other equivalent shape thereto, substantially as herein described, for the purposes specified.

Second, the adjustable slides or bars I, constituting a portion of the bearings or journals of connecting and other rods in machinery, when arranged upon the said rods so as to be susceptible of adjustment, substantially as and for the purposes described.

No. 57,293.—FRANCIS W. CROSBY and WOODHULL HELM, New York, N. Y.—*Apparatus for Desulphurizing Ores.*—August 21, 1866.—Shelves attached to a reciprocating bar are adjusted lengthwise of an inclined flue and drop their contents through the throat of the furnace and past a perforated steam pipe into a chamber.

Claim.—The combination of a series of sliding tables or shelves with the interior of an inclined hot-air or gas-conducting flue, substantially in the manner and for the purpose herein set forth.

Also, in our improved apparatus, so combining the lower extremity of its inclined flue with the eduction flue of the furnace as that the ore or other material falling from the inclined plane shall drop in a thin sheet over the mouth of the said eduction aperture, substantially in the manner and for the purposes herein set forth.

No. 57,294.—JESSE W. DANN, Columbus, Ohio.—*Bending Carriage Thills.*—August 21, 1866.—The thill is clamped to a bending strap and its end wedged tight in a former, to which it is bent and fastened by another wedge. The devices hold it rigidly to the strap and eventually to that and the "former."

Claim.—First, in the process of bending shafts for vehicles, the wedge *g*, for the purposes specified.

Second, the wedge *k*, for the purpose specified.

Third, the back or outer projection on the step *c*, for the purpose specified.

Fourth, the ears *b b*, and the bevelled wedge *e*, together with the block *d*, for the purpose specified.

Fifth, broadly, the means as herein set forth for tightening and holding the strap upon the shaft during the operation of bending, substantially as specified.

No. 57,295.—EDWARD E. and ALBERT B. DICKERSON, Oshkosh, Wis.—*Railroad Switch Lantern.*—August 21, 1866.—The lantern has numerous glasses of different colors set in tubes so that only the one turned directly in front is visible; it is operated by the motion of the switch bar so as to indicate which of several divergent tracks is open.

Claim.—First, providing a lantern A, having glasses of different colors, with a series of tubes or blinders E, arranged substantially as and for the purposes herein specified.

Second, imparting the requisite rotary movement to said lantern by the reciprocating motion of the switch bar *c*, substantially as shown and described.

No. 57,296.—HIRAM DILLAWAY, Sandwich, Mass.—*Glass Press.*—August 21, 1866.—The plunger is hung on a sliding head moving upon guides and operated by connection with a rock shaft. The cover of the mould is held down by followers and the mould slipped in and out at the requisite periods.

Claim.—First, the plunger or follower L, hung on a sliding head or frame moving upon guides E E, and connected through rods U with crank arms of a shaft W, arranged in slotted bearings, to the opposite ends of which crank arms rods *z* are connected at one end, and at their other hung upon fixed pivots X, substantially as and for the purpose specified.

Second, the mould B⁴, arranged and connected with the shaft for operating the plunger in such a manner as to be moved under and away from the line of movement of the plunger, substantially as and for the purpose described.

Third, the head plates M and R, rods O and Q, cross-bar P, spring R³, when all arranged and connected together and to the plunger so as to operate substantially as and for the purpose set forth.

No. 57,297.—HENRY F. DOUGHERTY, Monmouth, Ill.—*Hand Jack for Congress Gaiters.*—August 21, 1866.—The shoeing horn forms a guide for the foot while the back of the boot is grasped and raised by the catch.

Claim.—First, the bar A and catch B, operating substantially as described and for the purpose set forth.

Second, the combination of the bar A and catch B, with the shoeing horn M, substantially as described and for the purpose set forth.

No. 57,298.—JACOB DREISÖRNER, New York. N. Y.—*Hydraulic Engine.*—August 21, 1866; antedated August 2, 1866.—Two circular disks on the same shaft have a rotary reciprocation in their cylinders; one piston has a peripheral cavity divided by a partition, against the respective sides of which water pressure is exerted alternately. The second piston is moved by the first, and has a spiral groove on its periphery, closed at each end, and occupied by a sliding block which receives a longitudinal motion therefrom. The spiral channel is the passage for the water and has suitable valve induction and eduction arrangements; the sliding block becomes a piston.

Claim.—First, the arrangement of a revolving piston having a concentric groove in its circumference, divided by a partition and placed in a suitable cylinder provided with a slide valve, and a stationary wedge (between the two parts) made to work perfectly tight in said concentric groove, in combination with a second revolving piston, placed in a suitable cylinder and provided with a spiral groove in its circumference, in which a piston block, held by said cylinder, is made to move perfectly tight, said spiral groove communicating at each end through passages closed by self-acting valves with the ends or heads of the cylinders, as well as with a central passage provided in shaft, and communicating with a pressure box connected with the slide-valve case of the first-mentioned cylinder, the whole being combined together and operating in such a manner that the piston with the spiral groove shall force water or other fluid into said pressure box so as to produce a pressure therein, which said pressure shall act upon either the one or the other side of the partition in the concentric groove in the first-mentioned piston, in such a manner as to rotate said piston backward and forward around its axis, and which said motion shall be communicated to the second piston operating the same in the manner specified.

Second, the construction of the revolving piston B, arranged with a concentric groove F in its circumference, having a partition J, and placed in a cylindrical case C, provided with a suitable slide valve and case, in combination with a stationary block G, fitting tight in said groove F, and held fast in the cylindrical case C, and situated between the ports *a* and *b*, the whole being arranged and combined in the manner and for the purpose described.

Third, the construction of the revolving piston H, provided with a spiral groove L in its circumference, working in a cylinder E, and arranged with suitable passages closed by self-acting valves N N' and P P', forming communications between the ends of said spiral groove L and the ends or heads K K' of the cylinder and the reservoir D, connected with said cylinder as well as with the passage M made in the shaft A, and, through the same and its connecting chamber R and valve T, with a strong box or chamber Z, and operating in combination with the piston block Q, made to fit tight in said spiral groove L, and held fast in a groove *h* in the cylinder E, in the manner and for the purpose substantially as set forth and described.

Fourth, the arrangement of the channel way M in the shaft A, communicating with the ends of the spiral groove L in the piston H, and through the chamber R and valve T with the pressure box Z, when constructed in the manner as specified.

Fifth, the arrangement and construction of the piston block Q, provided with suitable packing and fitting tight in the spiral groove L, and held fast by means of slides or friction rollers working in the groove *h* made in the cylinder E, capable of a motion sideways or longitudinal with the cylinder and operating in combination with the spiral groove L, in the manner and for the purpose substantially as described.

Sixth, the construction of the packing rings or bands *p p'*, on the sides of the concentric and spiral grooves acting diagonally toward the grooves and to the surface of the piston, and operated by suitable springs, and arranged in the manner and for the purpose set forth.

Seventh, the manner of packing the surfaces at the ends of the spiral groove L, as well as the surface in the partition J in the concentric groove F, by means of plates z z and *n*, operated by suitable springs situated below said plates, as well as by springs *v* acting against the ends of the packing plates z z, in the manner and for the purpose described.

Eighth, the arrangement of a piston B, provided with a concentric groove F and a piston H, provided with a spiral groove L situated upon the same shaft, or its equivalent, and working in suitable cylinders, in the manner and for the purpose substantially as set forth and described.

No. 57,299.—JOHN P. DRIVER, Marengo, Iowa.—*Fountain Lamp.*—August 21, 1866.—Means are provided to force oil into the lamp from a reservoir in the base by atmospheric pressure; the arrangement of the conduits through which the oil passes are set forth in the claim.

Claim.—First, a groove or trench *g*, Fig. 6, cast in the outside of a glass fount and leading from the top to the bottom of the fount, to be of sufficient depth and width to neatly take in and imbed and hold in place an external supply pipe.

Second, a fluid or oil duct *n o*, Fig. 4, cast in the side of a glass fount, to be and operate substantially as the upper portion of a supply pipe, to be so enlarged and arranged at the lower end that a metallic pipe may be securely fastened to it.

Third, a fluid or oil duct *n o*, Fig. 4, made by inlaying or imbedding a glass tube in the side of a glass fount, with the upper end to open and enter the fount in or about the neck, the lower end terminating in the stem *y y*, said duct or inlaid pipe to operate and be to all intents and purposes as a part of the supply pipe.

Fourth, the upward extension *n o* of the supply pipe *m n o*, so that it shall discharge the oil or fluids forced through it into the upper side of the fount F, whether said elongated pipe be within the fount or curve up around the outside of it, connecting with the inside through or about the collar.

Fifth, the enlarged aperture through the stem or bottom of the fount, or as the entrance of the duct *n o.*

Sixth, the thimble-shaped mouth *b*, the valve *i*, in the air pipe N, Fig. 3, including the said air pipe N, or its equivalent, substantially as specified.

Seventh, the combination in a fountain lamp of a reservoir R in the base, either an external or internal supply pipe *m n o*, which shall discharge the oil into the top or upper side of the fount F, whether it be a separate pipe or a duct fixed in or to the side of the fount, the cork packing *c c*, the cement bed *t t*, the bellows D, the air pump B, the blow pipe P, the air pipe N, the groove *g*, and oil duct *o*, or their equivalents, substantially as and for the purpose specified.

No. 57,300.—CHRISTOPHER DUCKWORTH, Mount Carmel, Conn.—*Shuttle Binders for Looms.*—August 21, 1866.—The tongue is made separate to facilitate its removal and the substitution of a new one, in case of breakage.

Claim.—First, the shuttle binder, composed of the parts B C, when these parts are constructed substantially as described.

Second, the construction of the removable piece C, with a cupped receptacle for the spring which produces pressure upon the tongue B, substantially as described.

No. 57,301.—AUGUSTUS S. EDDY, Smithville, N. Y.—*Evaporator.*—August 21, 1866.—The compartments of the pan communicate by siphons; the supply from the reservoir is regulated by a float.

Claim.—The conducting spout C, cut-off gate *b*, lever *e*, and strainer *f*, operated by the float D, connecting rods *g* and *h*, balance beam E, substantially as herein described, in combination with the receiving tub or tank A, the evaporating pans B B B, and siphons *k k k*, for the purpose herein set forth.

No. 57,302.—JACOB EDSON, Boston, Mass.—*Capstan.*—August 21, 1866.—The capstan sheave is revolved in a central spindle by the pinions on the stationary disk; these are rotated by the central wheel, whose extended upper ratchet disk is engaged by pawls in the capstan head.

Claim.—First, the combination of the pawls *r r* and *s s*, so arranged as to operate in opposite directions to each other, with the ratchet or toothed plate *o o* and ratchets *q q*, as herein above described and for the purpose specified.

Second, in a capstan, forming the head in which the running gears are placed in two disks united by sustaining bridges, the two disks being united so as to form substantially one piece, as specified.

Third, in a capstan, constructing the ratchet or toothed plate *o o* and gear *n n* in one piece as described.

Fourth, in a capstan, casting the base *a a* and shaft or spindle *b b* hollow and in one piece to *e*, as specified.

No. 57,303.—A. H. EMERY, New York, N. Y.—*Hydraulic Press for Peat, Brick, &c—* August 21, 1866.—This rotating disk is fitted with a series of mould disks which are filled with peat by means of a charger which has cells corresponding in number and position with the moulds in the disks. These cells receive the peat from the hopper, and the charger is moved so as to bring the cells over the moulds in one of the disks. A series of dies are then forced into the cells by means of a hydraulic press, driving the peat into the moulds in the disk below. The disk is then rotated so as to bring the moulds into the proper position, and other dies are forced into the moulds from above and below, pressing the peat in two directions. These dies are operated by hydraulic presses.

Claim.—First, the combination and use of the rotating disk A, in combination with the dies 2 2 2, &c., and presses O' and C, as and for the purposes herein specified and set forth.

Second, the combination and arrangement of the disk A, dies 2 2 2, &c., with two or more presses O and O', and for the purposes herein specified and set forth.

Third, the arrangement of the rotating disk A, dies 2 2, &c., in combination with the three presses O O' and O², or their equivalents, as and for the purposes herein described and set forth.

Fourth, the combination and use of the presses O' and 4, or their equivalents, as and for the purposes herein described.

Fifth, the construction and use of the compound press O', essentially as and for the purposes herein described and set forth.

Sixth, the construction and use of the beam C' in combination with the cylinder 4', as and for the purposes herein described and set forth.

Seventh, the construction and use of the plungers *b b'*, &c., with the concave ends, essentially as and for the purposes herein described and set forth.

Eighth, the combination and arrangement of the plungers *b²*, sponges 8, and oil cup 9, as and for the purposes herein described and set forth.

Ninth, oiling the dies 2 2 2, &c., as and for the purpose herein described and set forth.

Tenth, the combination and arrangement of the presses H and O', as and for the purposes herein described and set forth.

Eleventh, the combination and arrangement of the hopper Q, charger R, and dies 2 2, &c., as and for the purposes herein specified and set forth.

Twelfth, the combination and arrangement of the disk A, wheels *u r y z* and their axes, the pawl 15, rack W, and press V, essentially as and for the purposes herein described and set forth.

Thirteenth, the combination of the press 10, pin 11, and disk A, essentially as and for the purposes herein described and set forth.

No. 57,304.—HIRAM B. ESTY, Houlton, Maine.—*Purifying Spruce Gum.*—August 21, 1866.—The gum is melted and strained through sieves in a close chamber heated by a steam coil and hot air.

Claim.—The above-described mode of purifying resinous gum, the same being by means of a close chamber, and one or more sieves therein, and by heat introduced within such chamber, by means substantially as described.

Also, the apparatus for effecting a purification of a resinous gum, the same consisting of the close chamber or vessel, one or more sieves placed therein, and a means of introducing heat into such chamber, the whole being substantially as specified.

Also, especially, for the purification of resinous gum, the employment of steam in the close chamber, with one or more sieves or strainers arranged therein for straining the gum when melted and subjected to the action of the steam.

No. 57,305.—J. J. FAIRBANKS, New York, N. Y.—*Machine for Forming and Cutting Skirt Springs.*—August 21, 1866.—The longitudinal profile of the reel corresponds with that of the skirt for which the springs are intended; a set of cutters sever the wire after it is wound upon the reel, and a traversing guide directs the wire so that it is wound spirally upon the reel; means are provided for adjusting the diameter of the reel to form springs of different sizes.

Claim.—First, a machine for measuring and cutting off wire into various lengths from a continuous piece, consisting of a rotating reel with attached cutters and a traversing guide combined, substantially as herein set forth, for the purpose specified.

Second, the adjustable pins or screws *i*, in combination with the reel B D, substantially as herein set forth, for the purpose specified.

Third, the bar J, furnished with spurs *u* and fixed upon a handle *s* and used in connection with the reel B D, substantially as herein set forth, for the purpose specified.

No. 57,306.—DANIEL FLAGG, Concord, N. H.—*Water Elevating Device.*—August 21, 1866.—The arrangement forms a foot valve for the pump above; a vessel is inverted over the upwardly extended induction tube in the lower chamber.

Claim.—The arrangement of the vessel A, vessel C, and tube D, in combination with the suction pipe B, constructed and operating as herein described.

No. 57,307.—JOHN FLETCHER, Newark, N. J.—*Fabric to be used as a Substitute for Japanned Leather.*—August 21, 1866.—The fabric is coated with boiled oil and lampblack, the paper applied to the adhesive surface and afterwards varnished.

Claim.—A fabric produced of muslin, silk, leather, or other suitable material, united with paper by means of the compound herein specified, and coated with leather japanner's varnish as specified.

No. 57,308.—JOHN C. Ford, Cambridge, Mass., assignor to D. L. & W. M. RICE.—*Paper Collar Machine*—The concave and convex rollers are adapted to shape a turn-down collar, and the guide to induct the collar has a central blade upon which the collar folds.

Claim.—The combination of the convex and concave rolls *c* and *d* and the guides or guiding surfaces *e f g* and *h*, when arranged to operate substantially as described.

No. 57,309.—LAVINIA H. FOY, Worcester, Mass.—*Hoop Skirt.*—August 21, 1866.—The glazed cloth gives sufficient rigidity to prevent the lower hoops turning over and does not stain.

Claim.—First, the glazed cloth supports for the bottom hoops in combination with hoops painted or covered with some insoluble coating, as and for the purpose stated.

Second, the combination with the bottom hoops of a hoop skirt of glazed cloth supports or coverings and strips of stiff paper or other suitable material, substantially as and for the purpose set forth.

Third, the combination with the bottom hoops of a hoop skirt of stiffened supports, substantially as and for the purpose set forth.

No. 57,310.—SMITH D. FRENCH, Wabash, Ind.—*Escapement in Watches.*—August 21, 1866.—The face of each pallet forms an arc of a circle of which the fulcrum of the lever is the centre, and the angular portions of the teeth strike the pallets on their concave sides.

Claim.—First, so constructing the pallet arms J J that they shall extend over and above the escapement wheel, and that the pallets shall be projected downward from their ends across the path of the teeth of the escapement, thereby increasing the distance between the pallets and their staff, substantially as and for the purpose above set forth.

Second, the adjustable cross-bar E in combination with the pallet arms J, operating in the manner and for the purpose herein specified.

No. 57,311.—JAMES B. GRANT, New York, N. Y.—*Apparatus for Distilling Oil.*—August 21, 1866.—The still has a series of flues communicating by pipes with the fire chamber and chimney respectively. It is set in masonry, has a flue space around it, and is connected by a pipe with the condenser.

Claim.—First, the employment in apparatus or machinery for distilling and refining petroleum and other oils of a series of heating pipes set in the form of two cones, the bases of which meet and through which the products of combustion from the furnace pass, all substantially as herein described.

Second, the employment in the same apparatus for condensing the vapors of petroleum and other oils of a cylinder enclosed in another cylinder, with a space between them for the circulation of cool water and suitable pipes for receiving and discharging the vapors and oil in and from the internal cylinder, and the water in and from the external cylinder, all constructed substantially as herein described.

Third, the general arrangement, combination, and method of operation of the apparatus or machinery, substantially as and for the purposes herein described.

No. 57,312.—A. B. GREENWALT, Baltimore, Md.—*Spring for Carriages, &c.*—August 21, 1866.—Explained by the claim and cut.

Claim.—A spring having the general conformation represented in Fig. 2, and formed with the curved part c', substantially as and for the purpose set forth.

No. 57,313.—JOHN GRIBBEN, Alleghany, Penn.—*Dies for Bolt-heading Machines.*—August 21, 1866.—The upper and lower edges which form the socket are cut away a little more than the thickness of the head of the bolt in the direction of the length of the same, so that the head, when being staved up, may have liberty to spread upon two sides while it is formed on the other two sides, after which it is turned a quarter of a revolution and subjected to another action of the dies.

Claim.—The dies for making square-headed bolts constructed substantially as hereinbefore described—that is to say, so that when brought together they will enclose a cavity in which to form the head, of which cavity two opposite sides are removed for a space in the direction of the length of the bolt equal to the thickness of the head, but otherwise enclosing all sides both of the blank and of the upsetting punch.

No. 57,314.—THOMAS GRIFFIN, Roxbury, Mass.—*Self-sealing Button-hole Patch for Paper Collars.*—August 21, 1866.—A patch of paper and cotton cloth, having a button-hole formed in it, gummed on one side for attachment to a collar, the button-hole of which has been torn out.

Claim.—First, the prepared detached self-sealing button-hole patch for mending paper collars when they become broken at their button-holes, as a new article of manufacture and sale, substantially as herein described.

Second, the construction of the prepared self-sealing button-hole mender of paper and cotton, substantially in the manner described, so that it shall have sufficient stiffness for the purpose intended.

Third, mending broken-out button-holes of paper collars by means of a self-sealing patch, substantially as described.

No. 57,315.—EMILE GROUX, Rome, N. Y.—*Clock Escapement.*—August 21, 1866.—The escape wheel of a clock is combined with two pendulous pallet arms or levers, suspended at each side of the escape wheel. The pallets act by gravity against the rod of the pendulum from alternate directions; they swing inward far enough to reach the impulse teeth of the escapement, and the pallet arms are raised alternately, withdrawing their pins from the locking teeth.

Claim.—The combination of the pendulous pallet arms or levers with the escapement, substantially as herein shown and described.

No. 57,316.—D. C. GUTTRIDGE and W. F. ROGERS, Canton, Ohio.—*Block and Tackle Check.*—August 21, 1866.—A pivoted eccentric or segment, with a weighted arm, is so arranged on the block as to jam the rope against the sheave and arrest its motion.

Claim.—The lever E, constructed substantially as specified, and used in connection with the block A, as and for the purpose set forth.

No. 57,317.—Robert Hagen, St. Louis, Mo.—*Ice and Coal Box.*—August 21, 1866; antedated August 3, 1866.—An ice tray is fitted in the coal box.

Claim.—The box A, in combination with the ice tray D and the doors *a a' b b' c* and *d*, when constructed and operated as and for the purpose set forth.

No. 57,318.—A. A. Hardy, Pittsburg, Penn.—*Umbrella.*—August 21, 1866.—A disk or enlargement upon the handle to rest upon the hand and support the umbrella.

Claim.—First, an umbrella or parasol having a supporter A, constructed and arranged substantially as shown and described.

Second, making said supporter adjustable, as and for the purpose set forth.

No. 57,319.—Thomas Harper, West Manchester, Penn.—*Machine for Boring Wagon Hubs.*—August 21, 1866.—Improvement on the patent of Isaac Munden, December 11, 1848. A bearing is provided for the master wheel used in the machine, so that the gear may be prevented from sinking below the pitch line of the cogs.

Claim.—The use of the bearing A' for the master wheel B', said bearing and wheel being constructed, arranged, and operating with relation to the various parts as herein described and for the purpose set forth.

No. 57,320.—Alexander Harroun, Jr., Onondaga, N. Y.—*Broom Head.*—August 21, 1866.—The two parts of the skeleton cover are joined at their lower edges by jointed metallic arms, and when brought together are attached to the handle by a screw thread and ferrule.

Claim.—The combination of the two metallic parts joined together by the arms *b b b b*, with the guides, the metallic pins, and the shank furnished with screw thread and ferrule, all constructed substantially as and for the purpose set forth.

No. 57,321.—Edwin Heald, Washington, D. C.—*Float Valve for Cisterns.*—August 21, 1866.—The valve of the induction pipe is opened or closed by the vertical motions of the float.

Claim.—The construction and combination of the float A, piston B, guides D D, valve C, and double screw cylinder H, forming the valve seat C, all as herein described and for the purposes set forth.

No. 57,322.—Garet G. Heermance, Claverack, N. Y.—*Construction of Wells.*—August 21, 1866.—The surface water is cut off from the interior of the tube by a partition. The space outside the tube is filled with filtering material. Collars on the lower section protect the strainers.

Claim.—First, the manner of constructing the well or hole, as set forth.

Second, the construction of the strainer section with horizontal shoulder collars *d d''* and vertical ribs or strips *i i*, the said strainer section having the wire gauze or other finely perforated material placed outside of the pipe and above the sand chamber thereof, substantially as described.

Third, the collars *d d''* or *d d' d''*, on the strainer section, substantially as and for the purpose described.

Fourth, trapping the water by the partition *e*, filtering it in its descent by the material *g g*, and discharging it through the pipe after it has entered through a strainer section, which is near the lower end of said pipe, substantially as described.

Fifth, the collar *e*, in combination with an apparatus, such as herein described, for the purpose set forth.

No. 57,323.—George Herrick, Nashville, Tenn.—*Wrecking Car.*—August 21, 1866.—The crabs fasten the car in its position on the track. The boom is stepped into the mast, and moved by tackle which connects their respective upper ends. The mast is stepped in the car platform, and braces by framing and stay rods.

Claim.—First, the crabs F, in combination with the windlass H and crank A, applied with the track G, and operating in the manner and for the purpose herein represented and described.

Second, the arrangement of the framing I, mast J, with its ends, as described, and boom K joined thereto, tackle L, tackle N, and the hoisting rope *k*, in combination with the car A, constructed and operating in the manner and for the purpose herein specified.

No. 57,324.—Benjamin Hinkley, Troy, N. Y.—*Roofing Composition.*—August 21, 1866.—The roof, after being covered with a composition made by adding lime or hydraulic cement to hot coal tar, is dusted with sand and whitewashed.

Claim.—A roofing composition composed, applied, and coated with whitewash, as herein described.

C P 26—VOL. II

No. 57,325.—FRANK A. HOWARD, Belfast, Me.—*Mitring Machine*—August 21, 1866 — The reciprocating head has on either side knives adjustable to any angle to trim the ends of the wood presented thereto upon the adjustable bed plates

Claim.—The reciprocating head C, with its adjustable knives or cutters g g', in combination with adjustable rests D D', constructed and operating substantially as and for the purpose described.

No. 57,326.—H B. HOWE and W. J. MACKRELL, New York, N. Y — *Stop Motion for Braiding Machine.*—August 21, 1866 — In machines which braid a covering over skirt wire, if the wire fail to be properly fed while being covered, imperfect work is made, and the braid extends down lower and lower, coming closer to the standards which support the wires. By placing a finger in such position that any of the threads so lowered shall come in contact with it, it will cause the tappet lever to tilt, and thus throw the latch against the stop lever and stop the machine. If the stop lever fail to act, the latch comes in contact with a stop secured to the frame, and the machine is stopped.

Claim.—First, the tappet lever E and hinged latch g, in combination with the stop lever F of a braiding machine, constructed and operating substantially as and for the purpose set forth.

Second, the adjustable finger e, in combination with the tappet lever E and with the threads of a braiding machine, constructed and operating substantially as and for the purpose described.

No. 57,327.—WILLIAM H. ISAACS, Terre Haute, Ind.—*Evaporator.*—August 21, 1866 — The pan has holders against which the skimmer rests when inverted. The skimmer has handles which rest in optional grooves to maintain its adjusted position to catch the scam in the pan.

Claim.—The arrangement with the fan of the holders F F, substantially as and for the purpose specified.

Also, the handles D D, in combination with the skimmer and the notches, or their equivalent, in the pan, substantially as and for the purpose described.

No 57,328.—HANNAH S. C. IWERSON, New York, N. Y.—*Head Dress for Ladies.*—August 21, 1866.—The lady's hair is opened out beneath the crown and a disk with hair attached is fitted against the head; the natural hair is then drawn backward over the disk and is combed and dressed together with the attached hair.

Claim—The foundation piece b, carrying the length or braid of hair, and fitted in the manner specified, so as to be introduced within the natural hair, for the purposes and as set forth.

No. 57,329.—NATHANIEL JENKINS, Boston. Mass.—*Water Gate.*—August 21, 1866; antedated August 15, 1866.—The core of the plug is attached to the screw stem, and is shod with an elastic membrane which adapts itself to the shape of the ridge seat in the water way.

Claim.—For a water gate, the arrangement of a valve, consisting of a non elastic core O, preferably hollow, clothed with elastic material p, and its combination with the ledges d of the valve seat, operating substantially as described.

No. 57,330.—BARTON H. JENKS, Bridesburg, Penn.—*Cop Winding Machine.*—August 21, 1866.—The construction and arrangement of the parts in this improvement admit of winding the shuttle bobbins directly from the hank without the need of spooling, while the bobbin automatically stops when filled and may be conveniently removed and another substituted.

Claim.—First, the hollow spindle C, in combination with the bobbin spindle C' when the same are constructed, arranged, and operate in the manner and for the purpose herein described.

Second, the combination and arrangement of the hollow spindle C, bobbin spindle C', and hank holders B B', substantially in the manner and for the purpose herein described.

No. 57,331.—ISAAC H. JOHNSON, Long Reach, West Va—*Hay Press.*—August 21, 1866.—The central master wheel rotates the surrounding wheels, whose screw rods give vertical motion to the follower in the box. The speed or power of the follower is obtained by the disposition of the connecting gearing. A grating allows the escape of juice when fruit is the subject.

Claim.—First, a double-acting press, consisting of a follower working in a box in either direction on three horizontal or perpendicular fixed screws, the force and velocity of which machine are capable of being varied within certain limits at pleasure, in the manner and for the purposes described.

Second, providing a press constructed as described with a grating g, arranged as and for the purpose set forth.

No. 57,332.—JOHN JOHNSON, Saco, Maine.—*Automatic Calculator for Scales.*—August 21, 1866.—The value is seen upon a fixed circular disk, through holes in a revolving disk, upon which is marked the rate per pound, at a series of spirally arranged openings.

Claim.—The fixed plate F, the revolving disk G, and the index I, in combination with the spiral spring balance or other weighing apparatus, for the purpose of indicating the total value of any article that is weighed upon the scale.

No. 57,333.—ROBERT V. JONES, Canton, Ohio.—*Lining for Oil Barrels*—August 21, 1866.—The barrels are coated internally with a composition of slippery elm bark, white glue, gum copal, and turpentine.
Claim.—The within-mentioned ingredients when mixed together and used as and for the purposes herein specified.

No. 57,334.—GEORGE KEATING, Thomaston, Maine.—*Mitre Box*—August 21, 1865.—The arms of the mitre box are separately adjustable upon the bed to saw the mitres of any required angle.
Claim.—An adjustable mitre box for sawing bevelled work or mitres for rhomboidal figures of any desired angles, constructed and arranged substantially as herein described.

No. 57,335.—JACOB KINZER, Pittsburg, Penn.—*Skate.*—August 21, 1866.—The skate is formed of cast iron in one piece.
Claim.—As an article of manufacture a cast-iron skate, substantially as shown and described.

No. 57,336.—H. KOELLER, Camp Point, Ill.—*Hand Spinning Machine.*—August 21, 1866.—The arms being so secured upon their shaft or centre that they may be turned thereon and tightened at will, admit of tightening or releasing the driving belt with facility and enable the driving wheel and its crank to be adjusted to different heights to suit the stature of different operators. The whole may be screwed to a table.
Claim.—The swivel standard B forming the bearings for the shafts *d e*, and the bifurcated arm C, forming the bearings for the spindle E, in combination with the screw clamp A, constructed and operating substantially as and for the purpose described.
Also, the driving wheel D, and India-rubber disk *f*, standard B, arm C, spindle E, and screw clamp A, all constructed and operating substantially as and for the purpose set forth.

No. 57,337.—ANDREW J. LAIRD, Middletown, Penn.—*Horse Hay Fork.*—August 21, 1866.—The upper horizontal part of the bent tripping lever, when the fork is locked in position for hoisting its load, is passed through the eye at the upper end of the fork ; the tripping rope passes through the eye by the side of said horizontal part of the lever and is always in position to act thereon to straighten the prongs and discharge the load.
Claim.—First, in combination with the sliding rod C, the lever F with its upper horizontal arm adapted to project through the ring H, substantially as described and for the purpose specified.
Second, the combination of the bars A A, sliding rod C, tines D D, link or connecting rod E, and lever F, all arranged and operating substantially as described.

No. 57,338.—EZRA B. LAKE, Bridgeport, N. J.—*Bed Bug Trap.*—August 21, 1866.—The flanged cup to arrest and catch the bugs forms a shoe for the foot of a bedstead post.
Claim.—An improved bed-bug trap, the same consisting of the box with circular flange A, partitions E F, forming compartments B C D, grooved block G, cap H, and trough J, constructed and arranged substantially as herein described and for the purpose set forth.

No. 57,339.—F. J. LATHAM, New York, N. Y.—*Anchor.*—August 21, 1866.—The flukes of an anchor are pivoted in the stock and vibrate on either side of the latter to an extent determined by the contact of the crown piece with the stock.
Claim.—In combination with the shank constructed substantially as shown, the use or employment of the stock, flukes, and crown piece, as and for the purposes fully indicated.

No. 57,340.—M. M. LATTA, Goshen, Ind.—*Fracture Bed.*—August 21, 1866.—A section of the bed below the posteriors slides laterally to bring the bed pan beneath the patient.
Claim.—The bed having a central portion C, consisting of the slide moving in guides F, or rollers G, and the sliding chamber holder H provided with the sections B D and supports E, operating substantially in the manner and for the purpose represented and described.

No. 57,341.—Z. W. LEE, Blakely, Ga.—*Plough.*—August 21, 1866.—The flanges of the brackets are attached by shackles and wedges to the central beam; the obliquity of the standards is determined by braces clamped to the brackets in which they are pivoted.
Claim.—The combination of the shank E, bracket F, pivoted arm J, shackle K, and wedge L, all arranged and operating substantially as and for the purposes herein explained.

No. 57,342.—WILLIAM LIEBER, New York, N.Y.—*Button-hole Cutter.*—August 21, 1866.—The lower portion terminates in a U-shaped duplication. Upon the lower part of this the anvil is adjusted. Upon the extremity of the upper part is pivoted the upper handle, the cutting blade being suspended by a slot upon a pin therein; the cutting blade is thus held between the upper handle or blade and the upper part of the U-shaped duplication of the lower portion.

Claim.—The slotted knife *d*, operating in combination with the grooved bracket *a*, provided with stud *f*, arranged with the adjustable anvil *c* on bracket *b*, substantially as represented and described.

No. 57,343.—EDWARD F. LIGHT, Worcester, Mass.—*Journal Box.*—August 21, 1866.—The lower section of the box has a longitudinal oil chamber with end flanges to retain the oil; an annular central chamber is charged with cotton, and inclined flanges lead superfluous oil toward the mid-length of the spindle.

Claim.—Making the lower half A of a journal box with a groove or channel *b*, and chamber or recess C, in combination with the upper edges, with inclined planes *a*, and the ends with flanges *c* and *e*, substantially as set forth.

No. 57,344.—N. P. LINDERGREEN, Boston, Mass.—*Children's Sled.*—August 21, 1866.—Explained by the claims and cut.

Claim.—The brake levers C, hung to and upon the sides of a children's sled, substantially as and for the purpose described.

Also, connecting the rope of the sled to the brake levers C, through short pieces I, substantially as described and for the purpose specified.

No. 57,345.—ROBERT B. LINTHICUM, Lexington, Ill.—*Corn Harvester.*—August 21, 1866.—To a cross-bar underneath one end of the shock receiver is attached a curved supporting arm, upon whose other end are friction anti-rollers, which travel upon a curved track, when the receiver is turned round to discharge the shock. With the receiver is used a compressing cord, one end of which is hooked to one side of the receiver, and the other crowing the first to a lever hinged to the other side of the same; by the movement of these the shock is compressed for binding. The machine is supported at each side upon caster wheels, pivoted to vibrating arms.

Claim.—First, the combination of the curved arm J, cross-bar L, wheels *m*, shock receiver H, and curved track K, arranged and operating substantially as shown and described.

Second, in connection with a shock receiver H, the employment of lever Q, with the hooks *l* and *q*, and the compressing cord *x*, as and for the purposes specified.

Third, the arrangement of the pivoted wheels W and vibrating bar T with the frame B of a harvesting machine, as and for the purposes set forth.

No. 57,346.—WILLIAM LITZENBERG, Macomb, Ill.—*Horseshoe.*—August 21, 1866.—The double flanged plates are reversible, and their projections are used optionally as the means for attachment to the shoe, or as the calk proper.

Claim.—Combining with the main part A of the shoe, the removable and reversible calks B and C, constructed and applied substantially as herein described and for the purpose set forth.

No. 57,347.—C. L. LOCHMAN, Carlisle, Penn.—*Funnel.*—August 21, 1866.—The funnel has a plug valve at its nozzle, a tapering plug at its neck, and an air discharge tube on its side, with a projecting shield on its upper end.

Claim.—First, a funnel provided at the spout with a spigot or valve, worked by the upright rod H for opening and closing the same, substantially as set forth.

Second, a funnel having a valve at its nozzle, and a screw washer or fastener at its neck, substantially as described.

Third, in combination with a funnel having a metallic or an elastic washer E E, and a valve or a stopple at the spout, an air tube B B and shield D, for the purpose specified.

No. 57,348.—JAMES R. MADISON, Oneida, Ill.—*Washing Machine.*—August 21, 1866.—A reciprocal rotary motion is communicated to hanging beaters by a rack, pinion, and pitman.

Claim.—The rubbers 1 2 3, operated by the crank C and the pitman D, the cogged rack F, in combination with the wheel S and fly-wheel A, when the same are constructed in the aforesaid combination and for the purposes set forth.

No. 57,349.—CEPHUS MANNING, Chillicothe, Ohio.—*Tempering Steel.*—August 21, 1866; antedated August 8, 1866.—The steel is tempered in a bath of linseed oil.

Claim.—The use of raw linseed oil, in the manner and for the purpose specified.

No. 57,350.—ADAM MARCHINGTON, Upland, Penn.—*Stop Motion of Looms.*—August 21, 1866.—The devices named are designed to stop the loom when the weft thread breaks or gives out; in this case the bars of the grid on the lay are allowed to pass the weft-fork, and a lever is thus allowed to drop so low as to be struck by the wiper, which pushes it and the bar forward until the shoulder strikes a lever, and by the intervention of another lever unlocks the shipper.

Claim.—First, the vibrating lever F, in combination with the wiper K on the shaft B, all arranged and operating substantially as described.

Second, placing the vibrating lever F on a sliding bar J, substantially as described.

Third, in combination with the wiper K, the lever F, the sliding bar J, and the levers M and N, substantially as described.

No. 57,351.—MARK M. and FRANK MARTIN, Aurora, Ind.—*Railroad Car Seat.*—August 21, 1866.—The seat is hinged to the frame, and secured in the erect or more recumbent position by means of catches on the latter, which engage pockets on the former. The frame may be raised for reversing the presentation of the seat by detaching the hook and disconnecting the foot sockets of the legs from the pins in the floor.

Claim.—First, a car seat composed of two distinct parts, to wit, the body A and supporting frame D, when hinged or otherwise connected together, substantially as herein described and for the purpose set forth.

Second, in combination with the body A and supporting frame D, the adjustable supports G G' and locking device J J', all arranged as and for the purpose explained.

Third, in combination with the elements of the first claim, the sockets *f f*, anchoring pins K K', hook L and staple N to enable the reversal of the seat, and securing it in either position in the manner described.

No. 57,352.—L. G. MASON, Worcester, Mass., assignor to himself and S. S. BARBER, Fitchburg, Mass.—*Friction Pulley.*—August 21, 1866.—The hinged arcs are projected radially to engage by frictional contact with the interior of the rim, and conversely by means of hinged slides, moved by a cam fork attached to the shipper piece.

Claim.—First, the combination of a hinged friction lever E with the slide *d*, frame C and shipper piece G, with a cam fork *f*, substantially as and for the purpose set forth.

Second, making flange F' with slots to receive the slides *d*, in combination with providing arms D with slots *b*, for the purposes stated.

No. 57,353.—G. B. MASSEY, Mobile, Ala.—*Turn Table for Railroads.*—August 21, 1866.—The driving wheels of the locomotive when on the turn-table rest upon the upper edges of and rotate the endless belts which drive the drums; the pinions upon the lower ends of the drums revolve the central wheel and the axis of the turn-table.

Claim.—A railroad turn-table, provided with endless metallic belts G, fitted on drums E E, the shafts of one or more of which have pinions F on their lower ends, in combination with the fixed wheel B at the bottom of the pit in which the turn-table is fitted, and into which wheel the pinions gear, substantially as and for the purpose set forth.

No. 57,354.—HIRAM S. MAXIM, Boston, Mass.—*Iron for Curling Hair.*—August 21, 1866.—The hollow curling rod is heated by air or steam from the generator over the gas-burner, which is supplied with gas through the tubular handle.

Claim.—In combination with a hair-curling iron or rod, and so as to form a part thereof, a flame burner and chamber, and a fuel or gas chamber for heating the iron or rod, substantially as described.

Also, in combination with the flame chamber and curling rod, the steam-generating chamber, arranged to operate substantially as set forth.

No. 57,355.—JAMES McCRACKEN, Bloomfield, N. J.—*Engine for Reducing Rags, &c., to Fibre.*—August 21, 1866; antedated August 8, 1866.—Owing to the conical form of the wire gauze washer, the velocity of every part of its periphery is proportionate to its distance from the centre of the machine, and the water thus allowed to be drawn off equally all across the trough. The parallelism of the stationary bed of knives with each other, in conjunction with the divergence of those on the conical rollers, effects a shearing cut from the centre of the machine until a given rotating knife has passed over the middle bed knife, and after passing that point effects a cut by it toward the centre.

Claim.—First, the combination in an engine of similar character to what is known as a rag engine for the reduction of fibrous stock of a circular or annular stationary trough, one or more conical rollers and a conical washer, substantially as and for the purpose herein specified.

Second, the combination of a circular or annular stationary trough, one or more conical rollers, and one or more series of stationary straight knives *h h*, arranged in planes parallel with the shaft or shafts of the roller or rollers, substantially as and for the purpose herein specified.

No. 57,356.—FREDERICK MILLER, Newark, N J.—*Stopper for Bottles.*—August 21, 1866.—
A cap with a hollow hemisphere of elastic material is linked to a band around the neck; the
joint is formed by the indentation of the packing by the edge of the lip. •
Claim.—A bottle stopper, composed of a metal cap or socket D. containing a packing E of
India-rubber or other suitable elastic and water-proof material, and connected by metal rods
or straps c e to a metal band C around the neck of the bottle the rods or straps being attached
to the band by a joint connection to admit of the packing being pressed or shoved over and
off from the mouth or nozzle of the bottle, substantially as shown and described.

No. 57,357.—GEORGE MONTGOMERY, Canton, Ill.—*Medicine.*—August 21, 1866.—
Balsam for pulmonary disorders, composed of tincture of lobelia, tincture blood root, landanum,
tincture of anise, tincture of sassafras, hoarhound, pleurisy root, molasses, and the No. 6 of the
botanic dispensatory.
Claim—A pulmonary balsam, as herein described, when compounded of the ingredients
specified, substantially as described.

No. 57,358.—RICHARD MONTGOMERY, New York, N. Y.—*Grate Bar.*—August 21, 1866 —
The grate bar has a central solid portion and fimbriated plates on its sides; these are held
in relative position by tubes and have projections to support adjoining bars.
Claim.—First, the fimbriated corrugated plates or sheets C C, projecting from the sides of
the bar A. constructed and arranged substantially as described.
Second, the hollow tubes d d. in combination with the corrugated fimbriated plates C C,
substantially in the manner and for the purpose set forth.
Third, the notches e e, in combination with the plates or sheets C C, substantially as described.

No. 57,359 —RICHARD MONTGOMERY, New York, N. Y.—*Construction of Ships.*—August
21, 1866.—The framing is of corrugated wrought iron covered with iron plates and planking.
The parts of the frame are secured together by lapping, bolting, and by angle pieces.
Claim—First, the combination of the keel D, ribs E, and keelson B, when arranged and
secured together, substantially as set forth.
Second, the combination and arrangement of the keel D and keelson B in the manner set
forth, for forming the bow and stern post of vessels.
Third, covering the frame of a vessel thus constructed, first with a sheathing of iron sheets
or plates, over which is placed the planking, the three being united together by bolting or
otherwise.
Fourth, connecting the ends of two beams when they are required to be united in any other
than a right line with each other to complete the required structure, substantially as described
and set forth in figures 3 and 4.

No. 57,360—CHARLES G. MOREMEN, Brandenburg, Ky.—*Corn Harvester.*—August 21,
1866.—The cradle or bed which receives the stalks as they are cut consists of two rock-shafts
armed with fingers and placed centrally in relation to the frame, so as to receive the stalks
from both cutters and to discharge the same when sufficient has accumulated to form a shock,
by a partial rotation of the shafts which depresses the fingers.
Claim.—First, the provision in a corn harvester of a cradle consisting of the rock-shafts P
P' and tines p p', arranged in relation to each other and to the cutters F and G, and main
frame A, and operating substantially as described and set forth.
Second, in combination with the cradle P P' p p', the retaining and liberating mechanism
R R' S S' and T t, as and for the purpose explained.

No. 57,361.—F. A. MORLEY, New York, N. Y.—*Device for Cutting Corn from the Cob.*—
August 21, 1866.—In this improvement the four cutters whose edges overlap are placed at
the end of elastic arms which spread apart as the ear of corn is forced in by the head-block
that is pushed forward in a groove.
Claim.—First, the four knives C C C C, forming an elastic cutting ring, their cutting
edges c r being overlapped, each knife having its side edge c on the outside of the lap and the
opposite edge r on the inside of the lap, and having oblique cutting edges in connection with
the bed piece G, having a trough or groove k, and sliding head-block H, as herein shown and
described.
Second, gauges or guides D, in connection with the knives C and set screws e, for regulating the depth of the cut and allowing all sizes of ears to be run through, as herein shown
and explained.

No. 57,362.—O. A. MOSES, Charleston, S. C.—*Blowpipe.*—August 21, 1866.—The blast tube
is connected with an air chamber supplied with air by means of a tube leading from a blowing apparatus. The air chamber is adjustably attached to a lever by whose vertical motions
the jet tube may be fixed at any angle to the wick of the lamp. The jet tube may be made

to approach or recede from the wick by another lever which slides the air chamber in the slot in the former lever.

Claim —First, the slot B and slider a, and the movements imparted to the same, operating in the manner and for the purpose hereinbefore stated.

Second, the forked lever E and set screws G and H, or their mechanical equivalents, and the standard F, and the movements imparted to the same, in the manner and for the purpose specified.

Third, the clamps R S and connecting rods Q Q', and the application of the screw arrangements O P Q c N M F d and K L, or their mechanical equivalents, operating substantially in the manner and for the purpose hereinbefore specified.

Fourth, the axle of the lamp V, the levered screw X and curved washer Y, and the slot in which they play, or their mechanical equivalents, operating in the manner and for the purpose herein described, or any other substantially the same.

No. 57,363.—J. H. MUMMA, Harrisburg, Penn.—*Straw Cutter.*—August 21, 1866.—The disks have ribs on their inner faces and recesses which form sockets for the adjustable plates to which the knives are attached.

Claim.—The disks D D' with their ribs e e and recesses f, adapted for the reception of the adjustable plates E, the whole being constructed and arranged substantially as described.

No. 57,364.—RICHARD MURDOCH, Baltimore, Md.—*Beam of Letter Scales.*—August 21, 1866.—Explained by the claim and cut.

Claim.—The letter scales provided with a graduated beam having the figures upon its flat upper side, substantially as described.

No. 57,365.—JOHN MURPHY, New York. N Y.— *Spring for Railroad Cars.*—August 21, 1866.—Around the rubber spring is a wrapping of fibrous material whose fibres lie obliquely; the whole surface is then covered with rubber and the spring thus formed is vulcanized.

Claim —First, the yielding fibrous envelope B, so arranged around a mass of rubber A as to support the exterior while the rubber in its interior is free from fibre, all substantially as and for the purposes herein set forth.

Second, a rubber and fibrous spring composed of the interior mass A, and an exterior layer C, and ends D¹ D² of rubber, in combination with a yielding fibrous envelope or support completely imbedded in and covered by the rubber, substantially as and for the purposes herein set forth.

No. 57,366.—WILLIAM NEELY, Sandy, Ohio.—*Fence.*—August 21, 1866.—The slots in the two faces of the angular iron posts receive the ends of the bars of the respective panels

Claim.—An improved fence formed by the combination of the iron posts A, constructed as herein described with the rails B, substantially as and for the purpose set forth.

No. 57,367.—HERMAN NITZSCHE, Philadelphia. Penn.—*Cane Handle.*—August 21,1866.— A bent wire passes through the metallic disk at the mitre joint of the two sections of the handle and its ends are secured to a cap piece and the cane respectively.

Claim.—Holding together the parts composing an angular shaped handle for canes, &c., and directing the lengths of the handle forming the angle by means of the bent screw rod a, the nuts b¹ b² and the angle piece b, substantially as herein specified and described.

No. 57,368.—H. OGBORN and A. T. CHAPIN, Richmond, Ind.—*Store-pipe Damper.*—August 21, 1866 --The damper has three disks and is attached to an axis by which it may be rotated to a position transversely across the flue or parallel thereto; the middle disk is the largest and has a central opening controlled by a sliding plate.

Claim.—First, the bar or rod B, slots I I and sliding valve C, when used for the purposes and in the manner set forth.

Second, the plates A and D D, in combination with the bar or rod B, slot I and sliding valve C, when arranged in the manner and for the purposes set forth.

Third, the journals M M, box E E, bearings K K, boxes N N, and rod B, in combination with slots I I, rivets H H, and groove F, when the same are arranged, combined, and operated in the manner and for the purposes set forth.

No. 57,369.—M. W. OWENS, Waterford, Penn.—*Broom-head and Clamp* —August 21, 1866.—The side plates are attached to the handle and are clamped against the broom-corn by slotted bars and screws and bounding edge pieces, until the permanent screw plates are attached.

Claim.—First, an improved broom-head, formed by combining the wooden head B, and metallic clasps.C, when constructed as herein described, with each other and with the bars E bolts F, and corn G, substantially as described and for the purpose set forth.

Second, the clamp or packer H, constructed as herein described, when used in combination with the broom-head B C and corn G, substantially as described and for the purpose set forth.

No. 57,370.—WILLIAM P. PATTON, Harrisburg, Penn.—*Eraser Holder.*—August 21, 1866.—Clamping plates to hold an eraser blade by a set screw are attached by a socket and ferrule to a lead pencil.

Claim.—First, the construction of the clamping plates as shown in figures 2 and 5, and their combination with an eraser *b*, as shown in figure 4, substantially for the purpose herein set forth.

Second, the subject of the first claim in combination with a rigid or non-elastic ferrule *g*, substantially as and for the purpose specified.

Third, the combination of the plates (figures 2 and 5,) eraser *b*, the screw *c*, or its equivalent, and the ferrule *g*, with a lead pencil, substantially in the manner set forth and described.

No. 57,371.—SAMUEL PERRY, New York, N. Y.—*Ventilating Boot.*—August 21, 1866.—The fetid air passes through the perforated inner sole to a space between the soles and thence by tubes between the double upper to outlets above.

Claim.—The arrangement of the perforated inner sole B, and outer sole C, forming the canal E, in combination with the grooved plate H, communicating with the said canal, and the outer air, all constructed and operating in the manner and for the purpose herein specified.

No. 57,372.—LUKE A. PLUMB, Biddeford, Me.—*Tea and Coffee Pot.*—August 21, 1866.—The pot has a central flue and a perforated cover ; the contents are heated by a lamp beneath, whose chimney occupies the said flue.

Claim.—The flue B within the pot A in combination with the tube E, provided with openings covered with mica F, and the lamp D, all arranged substantially as and for the purpose specified.

Also, the radiator G in combination with the flue B and pot A, substantially as and for the purpose set forth.

No. 57,373.—CHARLES H. PORTER, Providence, R. I.—*Stove-cover Lifter.*—August 21, 1866.—The finger above the hook is opened or closed by a trigger and spring.

Claim.—A lifter for stove covers, having a covering plate, or its equivalent, for its lifting hook or end, arranged so as to be operated substantially in the manner described and for the purpose specified.

No. 57,374.—WILLIAM PREISS, New York, N. Y.—*Folding and Plaiting device for Sewing Machines.*—August 21, 1866.—The apparatus is attached to the sewing machine by screws ; the blade adjusted to bring the line of sewing the right distance from the edge ; the small plates adjusted for the width of plait ; a fold of cloth is inserted between the edge of the blades, and the hinged arm turned down and held by friction.

Claim —First, the curved bar A in combination with the blade C and D, constructed and operating substantially as described and for the purposes set forth.

Second, the combination and arrangement of the curved bar A, folding blades C and D, and the hinged holder E, the whole constructed and operating substantially as described and specified.

Third, the combination of the hinged holder E with the blades *c* and *d*, and guides 3 and 4, substantially as described and specified.

No. 57,375.—E. T. PRINDLE, Aurora, Ill.—*Piston Packing.*—August 21, 1866.—Steam, admitted through openings in the sides of the piston, presses against the bases of wedges which expand against the surface of the cylinder, the rings in the peripheral groove of the piston.

Claim.—First, the combination of the sectional bevelled packing rings *a a* and wedge rings *b b*, placed within annular recesses formed in the circumference of a piston. so that steam, acting through perforations *c c*, through the followers of the piston, shall effect the uniform expansion of the packing, substantially as described.

Second, the combination of the two sets of packing rings *a a*, the wedge rings *b b*, and springs *g g g*, with the skeleton rim A of the piston, and the perforated followers B B', substantially as described.

No. 57,376.—WILLIAM L. RAHT, Baltimore, Md.—*Treating Metalliferous Ores.*—August 21, 1866.—Air is forced through the mass of fused metal to remove sulphur, arsenic, antimony, &c. Apparatus similar to the "Bessemer" may be used.

Claim.—The within-described process of expelling from metalliferous ores sulphur, arsenic, or antimony, by treating the mat or regulus run from such ores in the manner set forth.

No. 57,377.—SILAS G. RANDALL, Providence, R. I.—*Manufacture of Elastic Springs.*—August 21, 1866 ; antedated August 8, 1866.—The ends of a flexible tube are closed, and being bent into a coil, it is placed between the surfaces where its elasticity and that of the contained air is required. ·

Claim.—The use of flexible tubing brought into the form and acting in the manner herein described, and for the purposes set forth.

No. 57,378.—GEORGE T. REED, Philadelphia, Penn.—*Broom and Brush Head.*—August 21, 1866.—Two metallic plates are united with a wooden head-block, and are held together by a metallic grooved slide, which is screwed upon a wooden block, in which the handle is socketed.

Claim.—The combination of the handle F, head block H, clamp I, in two parts, containing one or more ribs, dovetail slides A A, and connection by two screws B B.

No. 57,379.—UEL REYNOLDS, New York, N. Y.—*Pole-iron Socket for Carriages.*—August 21, 1866.—The neckyoke is attached by a hinged clasp to a socket in the pole.

Claim.—The clasp *d* fitted as specified, in combination with the socket, for the pole of carriages, &c., as and for the purposes set forth.

No. 57,380.—THOMAS C. ROBBINS, Philadelphia, Penn., assignor by mesne assignments to R EDGAR HASTINGS, same place.—*Gold-beating Apparatus.*—August 21, 1866.—The vertical movement of the beater is rendered partially independent of the movements of the sliding cross-head, by allowing play to the collar on the stem in the stirrup of the cross-head.

Claim.—The vertically-guided hammer *d*, in combination with the reciprocating cross-head G, or its equivalent, the whole being constructed and operating substantially as and for the purpose described.

No. 57,381.—NORMAN C. ROBERTS, Burlington, N. Y., and EZRA W. BADGER, Otsego, N. Y.—*Hop-vine Support.*—August 21, 1866.—Wires are stretched from pole to pole, and support intermediate rods and their hinged bifurcated extensions, which latter are adjustable as to position.

Claim.—The arrangement of the rods A and B, the same being connected together substantially as described, whereby the weight of the upper sections, as well as the vine, is made to devolve upon the lower sections, and thus give a firm or solid character in order to prevent swaying by the wind or sagging from the weight of the vine.

No. 57,382.—GEORGE W. ROGERS, New York, N. Y.—*Bottle Stopper.*—August 21, 1866.—Attached to the cork is a tube with a threaded interior occupied by a screw-plug whose lower valve closes the opening in the cork, and whose upper flange presses a packing ring on the top of the tube.

Claim.—The rectangular recessed valve seat *c c* in the bottom of the screw-threaded cavity, in combination with flange plate *g″*, also used as a valve seat, substantially as described.

No. 57,383.—ROBERT and THOMAS ROSS, Middlebury, Vt.—*Chuck*—August 21, 1866.—The jaws are moved simultaneously through their individual connection by nut, screw, and pinion, with a common central gear; the nuts which engage with the lugs of the jaws are prevented from turning by long, rigid, parallel pinions; but when an eccentric adjustment of the jaws is required by the release of the pinions, the jaws are movable individually.

Claim.—The long pinions *g* or their equivalents, in combination with the nuts *b*, screws *c* and jaws B of a chuck, constructed and operating substantially as and for the purpose described.

No. 57,384.—TOBIAS ROYER, Lancaster, Penn.—*Fly Flap.*—August 21, 1866.—The horizontal cross-strands are plaited into and with the longitudinal strands instead of using slits or holes punched through said strands.

Claim.—The manufacture of fly flaps, or nets, when made by plaiting or braiding the horizontal and longitudinal cords or braids with or into each other, substantially in the manner specified and shown.

No. 57,385.—JESSE B. RUMSEY, Washington, D. C.—*Car Coupling.*—August 21, 1866.—The entering link opens the jaws, dropping the coupling pin into place; the link is held horizontal by the pressure of a spring plate against its end.

Claim.—The eccentric spring B, spindle C, coil spring F, plate H, bracket G, and pin D together with a metal case A, when constructed and arranged in the manner herein set forth.

No. 57,386.—WILLIAM H. SANGSTER and JUSTIN C. WARE, Titusville, Penn.—*Valve for Steam Engines.*—August 21, 1866.—A bent tube forms the valve, its open ends moving on the seat in connection with the ports; a lip on each end of the valve passes through the port into the cylinder and is struck by the piston, which thus actuates the valve and reverses the steam.

Claim.—The steam-chest, the steam cylinder, and the sliding pipe G, constructed and arranged to operate in the manner and for the purpose herein specified.

No. 57,387.—JAMES F. SAYER, Macomb, N. Y.—*Stave Jointer.*—August 21, 1866.—A stave is clamped in shape beneath the pivoted double edge knife; by swinging the knife on the stave it cuts from the centre of the stave to the end, and as the knife is swung back cuts the other half of the stave.

Claim.—The double-acting knife B, in combination with the table having the curved side *a*, groove *r*, plane C, and clamp *f*, attached, when constructed to operate substantially as described and for the purposes set forth.

No. 57,388.—JOSEPH S. SEAMAN, Pittsburg, Penn.—*Rolling Iron or Steel.*—August 21, 1866.—The rod is passed diagonally through the rolls, they revolving in the same direction, giving a rolling motion to the rod and thereby twisting the grain; the groove of the rolls is so constructed and arranged relatively to the inclination of the rod that a bearing of greater length is obtained upon the rod than when it runs through in the ordinary way.

Claim.—First, giving the grain of iron or steel a twist by rolling it on its axis, under compression, in the manner and by means substantially as and for the purposes hereinbefore described.

Second, subjecting metallic bars to rolling compression between parallel bearing surfaces, of greater length than the tangential bearing point given by rolls when the metal is passed between them at right angles to their axes, such bearing surfaces being obtained by causing the metallic bar to pass between grooved cylindrical rolls at an angle to their axis other than a right angle; said rolls being constructed substantially as and for the purposes hereinbefore described.

No. 57,389.—FRANK B. SEELY, Johnson's Creek, N. Y.—*Egg Tester.*—August 21, 1866.—The light passes through the eggs which are arranged in the openings, and their comparative translucency and consequent freshness is indicated by the image in the mirror below; the observation is made through a hole in the lid, and a slide isolates certain rows if desired.

Claim.—The combination of the side C with the holes *a a* of the cover B and the mirror *c*, the whole arranged and operating substantially in the manner and for the purpose specified.

No. 57,390.—E. D. SEELY, Brookline, Mass.—*Burning Fluid.*—August 21, 1866.—Composed of white oak bark, two pounds; alkanet root, two pounds; common salt, two pounds; alcohol, one pint; cyanide of potassium, one ounce, to be added to three gallons naphtha to render it non-explosive.

Claim.—First, the within described compound, which I term "red kerosene," as a new article of manufacture.

Second, the within described process of rendering naphtha non-explosive by treating the same substantially in the manner herein set forth.

No. 57,391.—HENRY SELICK, Lewiston, Penn.—*Apple Parer and Corer.*—August 21, 1866.—The apple is placed on the fork and rotated by the crank; the left hand guides the paring knife; the lever is vibrated, moving the slides together, impaling the apple on the corer, when the fork stops and the block continuing to advance pushes the apple on to the quartering knives. The slides are then returned by the lever.

Claim.—First, the combination of the slides C and F, the fork E, the block D, and lever G, with each other, and with the sides of the box A, the parts being constructed and arranged substantially as described and for the purpose set forth.

Second, the circular rest M, in combination with the arm M, arranged in the manner and for the purpose specified.

Third, the combination of the coring knife or tube H, and quartering knives I J K and L, constructed as described, with the box A, fork E, block D, and slide C, substantially as and for the purpose set forth.

No. 57,392.—ELIJAH SHAW, Milwaukee, Wis.—*Machine for Clamping and Stretching Leather.*—August 21, 1866.—The leather or fabric is stretched and clamped over a former while the seams and the adjoining edges are burnished flat. The devices are explained in the claims and cut.

Claim.—The clamping and stretching of fabrics and other articles connected by sewed or stitched seams, for the purpose of burnishing and trimming the seams, by means of a bell and jaws *g g*, arranged in such a manner that by the action of a treadle, or its equivalent, the jaws *g g* will first clamp and hold the article firmly, and the bed then rise in order to stretch the article on the same, substantially as set forth.

Also, the arrangement of the pivoted arms K K with jaws *g* attached, cross-bar or slide M, treadle H, and springs L L N, substantially as and for the purpose specified.

Also, the slide D, provided with the inclined surfaces *a a* and connected to the treadle H by the strap G, in combination with the inclined surfaces e e at the under side of the bed I, substantially as and for the purpose set forth.

Also, the combination of the bed I, provided with the inclined surfaces e e, slide D, provided with the inclined surfaces a a, and connected to the treadle H by the strap G, the pivoted arms K K provided with the jaws g; and the cross bar or slide M connected with the treadle and arranged to operate in connection with the arms K, all substantially as and for the purpose herein shown and described.

No. 57,393.—E. F. SHERMAN, Chicopee Falls, Mass.—*Corn Sheller*—August 21, 1866.—Ear corn is fed by the belt, passes between the toothed cylinder and concave, and the shelled corn is collected in a trough.

Claim.—The combination of the cylinder A, bonnet C, teeth a, belt F F, pulleys G G, and trough M, arranged and operating substantially as described.

No. 57,394.—GEORGE B. SIMPSON, Washington, D. C.—*Apparatus for Reducing Ores.*—August 21, 1866.—A crucible is supported over a furnace by brick-work, and connected at each end with a gas pipe. The ore is placed in the crucible with a flux, and the open end of the crucible closed by fire brick. The crucible is then subjected to heat, and a current of gas from the pipe is allowed to flow through it.

Claim.—First, the apparatus consisting of the crucible, tube, or vessel, in combination with the fire-brick soap-stone, or other material resistant of heat as a covering for top and bottom, and the frame of similar materials to hold the parts together as a whole.

Second, the gas pipes with the gas burner, in combination with the crucible, frame, and gasometer.

Third, the process of resolving metallic ores in an air-tight crucible, tube, or vessel, heated externally, in combination with common coal gas, petroleum gas, spirit gas, or any other known inflammable gas.

Fourth, the use of salt, borax, saltpetre, soda, potash, or any other known salt or alkali, either dry or in solution as a flux, in combination with the crucible, metallic tube, or vessel, the frame, gas pipes, and the gases, for the purposes and uses specified.

Fifth, the process of resolving the metallic ores in the absence of the oxygen of the atmosphere, and in the presence of a superabundance of carbon, by means of heat externally applied, the gases, salts, and alkalies internally applied, in combination with the galvanic or electric current, if necessary, and the apparatus, substantially as hereinbefore described.

No. 57,395.—EDMUND SMITH, Jr., and ALONZO CHASE, Worcester, Mass.—*Bed Bottom.*—August 21, 1866.—The ends of the wire are secured to the rail, and after being severally coiled the loop is attached to the slat.

Claim.—The triangular bracing form of the coil springs, connected at one point to the slats S S, and the base to the bar or frame D, substantially as above set forth and described.

No. 57,396.—GEORGE L. SMITH, Brooklyn, N. Y.—*Grate.*—August 21, 1866.—The triangular plates of each series are arranged with intervening washers upon a bearer, and are vibratable by force applied to an extension bar beneath.

Claim.—First, a grate which is composed of a number of sections, each one of which is made up of reversible plates so constructed that when one surface is burnt out another surface can be presented, substantially as described.

Second, a grate which is composed of a series of vertical plates applied to an oscillating bearer, substantially as described.

Third, the construction of triangular grate plates a a, with guards H H, or the equivalent thereof, for protecting the bearer K, substantially as described.

Fourth, arranging reversible right-angled grate plates upon movable bearers K, with spaces between them for allowing of a free circulation of air over the bearers, substantially as described.

Fifth, the construction of the end plates G, with arms, in combination with the oscillating bearers and connecting link or rod d, substantially as described.

No. 57,397.—D. B. SNYDER, Millville, N. J.—*Car Coupling.*—August 21, 1866.—A socket receives the end of the draw-bar; the latter is retained in the draw-head by a hook, which is raised or depressed by cams above, operated by a wheel on the platform.

Claim.—First, the buffer B, with its socket X and lever E, constructed and adapted for the reception and retention of the bar D, substantially as described.

Second, the shaft G, with its cam c, combined with the lever E, and with the within described operating devices or their equivalents, substantially as and for the purpose described.

No. 57,398.—GEORGE R. SOLOMON, Jr., and JOSEPH SOLOMON, New York, N. Y.—*Ticket Register.*—August 21, 1866.—In this case the tickets are printed upon a long ribbon or band, which is rolled upon the central shaft of an enclosing drum. One head of the drum has a hinged cover, and a given number of tickets having been wound on the shaft the drum is closed and locked and the conductor held responsible for the contents.

Claim.—The ticket box A B, with a slit *e* and elastic delivering rollers G G, spring pawl *j*, and roll holding shaft *f*, all arranged and operating substantially in the manner and for the purpose described.

No. 57,399.—ALBERT SONNEKALB and JOHN W. LIEB, Newark, N. J.—*Carpet Bag Frame.*—August 21, 1866.—Each jaw is made of a single piece of sheet metal, the inward flange being notched at the mitre joint and bent over.
Claim.—In a travelling bag frame, made with box jaws of equal size and closing into each other, the mitre jointed elbows *b b*, in combination with the hinges, when constructed and arranged as described.

No. 57,400.—CHAUNCEY SPEAR, Hopewell, N. Y., assignor to himself, W. MARKS, G. L. ARCHER, and W. L. PARKHURST.—*Composition for Roofing, &c.*—August 21, 1866.—Composed of peat, 2 parts; white clay, 2; gypsum, 1; petroleum residuum or gas tar, 1.
Claim.—The composition for roofing, paving, or other purposes, consisting of peat, gypsum, coal ashes, or vegetable mould, combined with clay and tar, substantially in the proportions set forth.

No. 57,401.—THOMAS H. SPENCER, Providence, R. I.—*Blacking Box.*—August 21, 1866.— The box is attached to an extension of the handle, and the lid to a pivoted button on the short post.
Claim.—The combination of the box with the handle A, joined at B, and furnished with a short arm C for holding the cover, all as described and for the purpose set forth.

No. 57,402.—A. F. STAYMAN, Baltimore, Md.—*Tobacco Pipe.*—August 21, 1866.—Explained by the claim.
Claim.—A tobacco pipe bowl or stem composed of cork, substantially as described.

No. 57,403.—CHARLES A. STERLING, New York, N. Y.—*Fluting Machine.*—August 21, 1866.—A fluting machine composed of a flat corrugated bed and a segmental corrugated presser.
Claim.—A fluting machine composed of a corrugated bed A, and a correspondingly corrugated segmental presser B, substantially as set forth.

No. 57,404.—EZRA STILES, Springfield, Mass., assignor to himself, J. R. MAURICE and W. F. GOODWIN.—*Safety Car Truck.*—August 21, 1866.—The safety car truck is provided with inside wheels on the same axles with the ordinary wheels, together with a central wheel and axle, which is caused to revolve when the car is thrown from the track, thereby applying the brakes and ringing a bell.
Claim.—In combination with an ordinary car truck, the inside wheels C C C C, and the central axle D, and wheels E E, when arranged and operating substantially in the manner and for the purpose herein set forth.

No. 57,405.—OREN STODDARD, Busti, N. Y.—*Shingle Machine.*—August 21, 1866.—By the arrangement of devices the shingles are alternately cut thick and thin at opposite ends, or may be changed to cut of equal thickness from end to end, as required
Claim.—First, the sliding frame *f t*, lever E, arms I and I', in combination with eccentric *d'* and knife D, arranged as and for the purpose set forth.
Second, the slides *m m'*, spring *l*, arms *t t*, and cam V, in combination with the dogs *i i'*, ratchet wheels *n' n'*, and feed rollers *g A*, substantially as and for the purpose set forth.
Third, the lever L', head K, provided with vertical and diagonal grooves 1 2 3 4, and shifter X, in combination with the arms *t t* and cam V, substantially as and for the purpose described.
Fourth, the set screw *o''*, arm *k''*, guides *g g*, lever L', and arms *t t*, arranged and operating in the manner and for the purpose specified.

No. 57,406.—RICHARD HENRY TAYLOR, Goose Creek, Va.—*Hominy Machine.*—August 21, 1866.—The bent wire beaters are arranged spirally on the shaft, and leave no portion of the interior unreached; the screen permits the escape of meal.
Claim.—First, so arranging the beaters upon a rotating shaft as to sweep the entire surface, or nearly the entire surface, of the shell or hollow cylinder, substantially in the manner and for the purposes set forth.
Second, the combination of the beaters B and screen C for discharging the meal from the machine, substantially as described.

No. 57,407.—W. H. TAYLOR, Pittsburg, Penn.—*Combined Screen and Weighing Device.*— August 21, 1866.—In this device the coals are first screened upon an inclined shelf covered with wire gauze, and thence fall upon an inclined trough suspended under a weighing scale. When the door of the trough is dropped the coals fall into a receptacle.

COMMISSIONER OF PATENTS. .

Claim.—The inclined screen D, in combination with the trough E, suspended from the scales platform C, and provided with the do r F, all arranged with a suitable framing A, to operate in the manner substantially as and for the purpose herein set forth.

No. 57,408.—DANIEL TERRY, Wakeman, Ohio.—*Fence.*—August 21, 1866.—The rectangular frame has supporting feet and horizontally stretched wires from post to post.

Claim.—A fence, constructed with a frame A B C, sustaining the wires F, having the posts A inserted in the ground, or supported upon foot pieces D, and attached to one another by pins E, the said several parts being respectively constructed and the whole arranged for use substantially as set forth.

No. 57,409.—WILLIAM H. TOWERS, New York, N. Y.—*Preparing Raw Hide for the Manufacture of Various Articles.*—August 21, 1866.—The raw hide is dipped in melted sulphur and moulded while in a heated condition.

Claim.—The treatment of raw hide with sulphur, or any combination of, or equivalent to, sulphur, for the purpose of producing the material and the effects before described.

No. 57,410.—THOMAS TULLY, Litchfield, Ill.—*Tire Shrinking Machine.*—August 21, 1866.—The tire is clamped in two places by eccentric jaws mounted respectively in a stationary and a sliding head; the latter is drawn toward the former by an eccentric, upsetting the tire.

Claim.—The combination of the curved grooves G G, eccentric dogs H H, eccentric lever D E, and fixed and sliding blocks F and B, when constructed and arranged to operate as and for the purposes herein specified.

No. 57,411.—LORENZO B. TUPPER, New York, N. Y.—*Grate Bar.*—August 21, 1866.—Explained by the claim and cut.

Claim.—The grate bar, formed with a straight, or nearly straight, surface for the fuel, and with a supporting rib having compound corrugations, as and for the purposes set forth.

No. 57,412.—P. H. VANDER WEYDE, Philadelphia, Penn.—*Supplying Air to Air Chambers.*—August 21, 1866.—A second air chamber attached to the induction pipe has a small opening by which air is admitted from without. By the operation of the pumps air is gradually withdrawn from this chambe and conducted to the principal air chamber attached to the delivery pipe.

Claim.—The combination of air chambers H and F, the valves P and N, and stop-cocks R and G, all arranged in the manner described so as to supply the constant loss of air taking place in the air chambers of force pumps.

No. 57,413.—N. H. VOSBURGH, Coxsackie, N. Y.—*Neck Yoke.*—August 21, 1866.—A metallic strap and ring are hinged together and attached to one side of the leather loop, which passes around the neck yoke and slips over the end of the tongue.

Claim.—The metal bar or strap and eye, connected by a joint or hinge, and applied or secured to the leather loop of a neck yoke, substantially as and for the purpose herein set forth.

No. 57,414.—A. J. WALKER, Lowell, Mass.—*Spice Holder.*—August 21, 1866.—The cylinder has chambers for a variety of spices which are reached by an opening in the cover as the wheel is rotated beneath. The drawers are for culinary powders, &c.

Claim.—The combination of the chambered wheel A and holder G provided with lid I, and having drawer J and box L arranged therewith, substantially as described for the purpose specified.

Also, the wheel A, having partition plates C, slotted centre tube D, and spindle E, with stud b, in combination with the stand or holder G, substantially as and for the purpose specified.

No. 57,415.—ROBERT L. WALKER, Globe Village, Mass.—*Brick Machine.*—August 21, 1866.—The reciprocating piston is fitted within a box or tube placed underneath the cylinder or case of the pug mill and has its ends made in a tapering form. Rollers are applied to the delivery ends of the box.

Claim.—The reciprocating piston or plunger H, fitted within a box or tube G placed underneath the cylinder or case A of the mud or pug mill, and communicating therewith, and having its ends beyond the part in which the piston or plunger works of taper form, so as to cause the clay to be compressed as it is forced out through and from the box or tube, substantially as shown and described.

Also, the rollers f, applied to the ends of the box or tube G, and arranged to operate substantially as and for the purpose specified.

No. 57,416.—R. W. WARE, Chicago, Ill.—*Crutch.*—August 21, 1866.—Near the lower end of the staff is a spiral spring, and below this a block with a projecting spur. This latter is covered or exposed by sliding up and down a metallic sleeve or encircling cylinder, which is retained in either position as desired by means of a bayonet joint.

Claim—The combination of the sliding tube C, provided with the spur E, the spiral spring F, and the adjustable tube A, when said parts are arranged to operate as herein shown and described.

No. 57,417.—JAMES WEATHERS, Greensburg, Ind.—*Wagon Brake.*—August 21, 1866.— The brake lever is pivoted to a bar on the hounds, operated by a cord which winds on a crank shaft, and locked by a trigger which engages a disk on said shaft.

Claim.—The arrangement of the bar F, pivoted or bolted to the hind bound, with the rod G, block H H, cord J, shaft K, provided with wheel *a* and lever L, constructed and operating as and for the purpose herein specified.

No. 57,418.—NELSON J. WHITE, Lawrence, Mass., assignor to SAMUEL C. WOODWARD, same place.—*Steam Engine Oil Cup.*—August 21, 1866—The inner cup is charged with oil while the steam connection is severed; when the cup is raised the steam comes in above, the oil trickles out below and passes to the steam chest.

Claim.—First, the arrangement of two cups, one within the other, substantially as described.

Second, the arrangement of the cup or casing A, cup B, apertures D E and *c*, whereby to inject oil or other substance into a steam chest, or other part occupied by steam, through the same aperture by which the steam is taken to the top of the oil or other substance.

No. 57,419 — WM. N. WHITELEY, Jr., Springfield, Ohio.—*Harvester.*—August 21, 1866.— Explained by the claims and cut.

Claim –First, the reversible, adjustable driver's seat with an adjustable, reversible standard, located on the main frame between the driving wheels, substantially as described.

Second, the plates C^3 on the pinion shaft, provided with two or more hardies or arms, and also provided with an eccentric slot or edge, in combination with the pawl, for the purpose of releasing and holding the pawl from the ratchet and releasing the pinion when desired.

Third, the arrangement, in combination with a harvesting machine having a hinged cutting apparatus and a removable self-raking attachment, of the raker's stand, substantially as described, whereby the attendant is enabled to remove the gavels by hand when the self-raking attachment is removed, as specified.

Fourth, a harvester frame, mounted on two driving wheels, in combination with the divider K', platform and finger bar hinged to the frame and combined with a reel which acts independent of the rake, which rake is arranged substantially as described, so as to move the grain heads foremost at intervals sideways and backward over the platform.

Fifth, the combination of a rake, and reel independent of the rake, on a harvesting machine with a hinged finger bar, substantially as described, the rake moving the grain heads foremost, sideways, and backward, independent of the reeling mechanism.

Sixth, in combination with a reel arranged to traverse on its shaft, the sleeve R⁴, and connecting rod S, which adjusts and holds the reel properly over the cutters throughout all the vibrations of the finger bar.

No. 57,420.—ALBERT S. WILKINSON, Pawtucket, R I.—*Horseshoe.*—August 21, 1866.— Instead of bevelling off the inner upper edge of the shoe, a narrow upper plate is attached to a wider lower plate.

Claim.—Forming a horseshoe of a narrow upper plate, and a broad lower plate attached one to the other by tapering rivets, substantially as shown and described.

No. 57,421.—HOSEA WILLARD, Vergennes, Vt.—*Machine for Raking and Loading Hay.*— August 21, 1866.—The hay is elevated between the teeth attached respectively to the upper and lower endless chains and further raised by the teeth of the upper in connection with the inclined board. The upper endless elevator is adjustable relatively to the surface of the ground and adhering hay is picked off at the summit.

Claim.—First, the hay elevator, composed of the endless chains F F P P, having rods K Q attached, provided with teeth *f m*, in combination with the guide frame R, all arranged in connection with or applied to the frames A D mounted on wheels, substantially as and for the purpose set forth.

Second, the toothed shaft S, in combination with and arranged relatively to the endless elevators and guide frame, substantially as and for the purpose specified.

Third, the adjustable plates L L in which the shaft H is fitted, arranged to vibrate upon the axle C, and applied as shown for adjusting the teeth *f* of the chains F, higher or lower, as may be required.

No. 57,422.—JOHN WILLARD, Norwich, Conn.—*Instrument for Opening Tin Cans.*— August 21, 1866.—The knife cuts the metal and has a shoulder projecting from each side to limit its penetration.

Claim.—The can opener, constructed of one piece of metal with a blade B, transverse shoulders *b b*, and a handle, as herein specified and shown.

No. 57,423.—GEO. M. WOOD, Decatur, Ill.—*Lock.*—August 21, 1866.—The sliding latch is operated by an oblique projection on the spindle when moved at right angles to the lock; it acts against a roller and withdraws the sliding latch from the keeper. The latch is locked by a wedge key which passes through a mortise in the spindle.

Claim.—First, the combination of the slide latch B with friction rollers D E E, in order to admit of the free movement of the slide or to obviate friction, constructed substantially as shown and described.

Second, the combination of the roller C, roller D E E, roller C', and slide latch B, operating with the arbor G, with oblique projector I, substantially as described for the purpose specified.

Third the locking device composed of the sliding bar K, working in a mortise *a* in the spindle or arbor, and operated in the manner shown, or in any equivalent way, substantially as and for the purpose herein set forth.

No. 57,424 —JAMES B. WOOD, Lansingburg, N Y.—*Steam Valve.*—August 21, 1866.—Designed to allow the valve to be ground into its seat. The screw stem of the valve works in a sleeve which is socketed in the outer casing and removable therefrom when the cap is taken off.

Claim.—The outer casing H, inner and independent nut G, having projections interlocking, with the notches in the said outer casing and valve stem F, when combined together, substantially as and for the purpose described.

No. 57,425.—JOSEPH S. WOOD, Philadelphia, Penn.—*Gas Regulator.*—August 21, 1866.—The case is filled with water to an outlet, and within is a tight receiver open at the bottom. Two bent tubes enter the casing and terminate under the receiver above the level of the water; the tube through which the gas enters the receiver is covered with a ring, which forms a valve seat for the conical valve which is suspended from the receiver.

Claim.—First, passing the gas into the floating receiver C by the bent pipe E, and valve *h a*, and out by the bent pipe F, arranged and operating substantially as described.

Second, using the pipes E and F as guides for the rising and falling of the receiver C, substantially as described.

Third, the combination of the vessel A, pipes E and F, receiver C, and valve *h*, arranged and operating substantially as described.

No. 57,426 —A. WOODARD, Bangor, Maine.—*Suspender.*—August 21, 1866 —The tag is attached to the upper end of a spring, which expands and contracts in a sheath at the end of the suspender strap.

Claim.—The arrangement of spring *a a*, stirrup or loop C, and sheath B B', when constructed and arranged to operate in manner substantially as and for the purposes specified.

No. 57,427.—EDWARD H. WOODWARD, New York, N. Y.—*Extracting Oil from Fish.*—August 21, 1866.—The fish is digested in water under a pressure of several atmospheres and the oil thereby extracted.

Claim.—Submitting fish to the action of a digester, as herein set forth, and extracting the oil therefrom without pressure, substantially as herein described.

No. 57,428.—J. E. WOOTTEN, Philadelphia, Penn.—*Winding Apparatus for Inclined Plane.*—August 21, 1866.—The drums and pulleys are arranged to afford additional friction surface for the winding rope by which the car is raised or lowered.

Claim —The drums or pulleys F and F', and pulleys G and H, the whole being arranged for the reception and guidance of the rope in the course, substantially as described for the purpose specified.

No. 57,429.—A. J. WORKS, Fair Haven, Conn.—*Apparatus for Burning Liquid Hydrocarbons.*—August 21, 1866.—The retort within the fire-box communicates with a steam boiler and water tank by means of a pipe, and with the tank containing the hydro-carbon fluid by another pipe. The flow of the hydro-carbon liquid is regulated by a thermal bar attached to the cock.

Claim.—First, the apparatus B, when composed of three departments, Q S and T, with its branch pipes, one above and the other below the perforated plate *t t'*.

Second, the reservoir H, pipe T, cock I, and pipe J.

Third, the pyrodynamic apparatus Z, in connection with the cock X, for the purpose specified.

No. 57,430.—THOMAS D. WORRALL, Central City, Colorado.—*Apparatus for Desulphurizing Quartz.*—August 21, 1866.—The furnace communicates by X-shaped flues with those surrounding the muffle, which discharge into a spiral flue within the casing, through which the quartz passes from the hopper. The products of combustion pass from the spiral flue into a side flue, and from thence into the chimney. The quartz passes from the casing into the muffle, and on to the distributor, through the opening, into the X-shaped flues, where it is subjected to intense heat created by the blast from fans. The gases may be passed to a condenser or the chimney.

Claim.—First, operating a blow-pipe in a confined space or flue, up which flame is passing, for the purpose of intensifying the heat through which metal-bearing substances, in a pulverized or partly pulverized condition, are passing, for the purposes set forth.

Second, operating blow-pipes up flues that form a junction, so that when the flames meet they may be condensed upon each other, and thus intensified, for the purpose of desulphurizing and oxidizing metalliferous ores passing through said flames, as set forth.

Third, an X-shaped flue, so constructed that the fire, starting from the extreme points at the base, must meet in the centre of the flue, and this whether used with or without blow-pipes, for the purpose set forth.

Fourth, the furnace A, with open sides communicating with flues, in connection with the blow-pipes H H H H, and the X-shaped flues, substantially as set forth.

Fifth, the V or diamond-shaped receiver, with perforated base, for the purpose of heating quartz or other metal bearing substances when passing over its inner surface while the fire is passing over its outer surface, and of delivering the same either into flues below on to heated plates through simple flame, through flame condensed upon itself by means of two or more blow-pipes playing from opposite directions, or upon a hearth upon which flame has been condensed by blow-pipes.

Sixth, the revolving fan distributors in the V or diamond-shaped receiver in flues in a muffle furnace or in an open chimney stack, for the purpose of suspending pulverized quartz and other metal-bearing substances in their downward descent, and of distributing the same in or upon the heated surfaces, or through flame, for the purposes set forth.

Seventh, in combination with the V-shaped receiver D D, and the spiral furnace I, a continuous muffle furnace of any shape or dimensions, horizontal, semi-horizontal, or perpendicular, through which ores containing sulphur or other volatile agents may pass, for the purpose of simple desulphurization, or for the purpose of driving off sulphur, arsenic, or any other chemical agent which it may be desirable to save for scientific or commercial purposes.

Eighth, one or more inverted V-shaped plates, either firmly built in the flue or suspended by hinges, at the distributing end of a muffle or other furnace or ordinary spout, for the purpose of distributing pulverized quartz falling upon it, in the manner and for the purposes set forth.

Ninth, the V or diamond-shaped receiver D D, in combination with the inverted V-shaped distributor E, for the purpose set forth, or any other similar purpose.

Tenth, the spiral furnace, with either a double or single flue, for the purpose of securing a slow and gradual descent of pulverized quartz or pyrites while fire is ascending in or under said flues.

Eleventh, so constructing said spiral furnace and the conducting flues connected therewith, that while heat and flame are ascending one flue and the quartz, sulphurets or other metal-bearing substances are descending the other, said substances shall not only be freed from their sulphur for the purpose of metallurgical success, but the sulphurous gases and other volatile agents may be collected and converted into any chemical or commercial agent of which they may be made to form parts.

Twelfth, desulphurizing ores, and driving from them arsenic and other chemical agents, for the purpose of securing successful amalgamation and chlorination or smelting, and simultaneously with this converting the sulphurous gases, arsenic, or other agents into useful articles for chemical or commercial purposes.

Thirteenth, conducting the gases arising from the combustion of carbonaceous substances which have been used to supply heat for the desulphurizing furnace, into a receiver, to be united with the sulphurous gases, for the purpose set forth.

Fourteenth, the furnace N, connected with the conducting pipes M and K', for the purpose of supplying any deficiency of carbonaceous gases that may be lacking from furnace A, for the purpose set forth.

Fifteenth, the use of carbon oil for the purpose of supplying the equivalents of carbon necessary to the manufacture of the chemical compounds, as set forth.

No. 57,431.—SILAS D. YERKS, Downington, Penn.—*Combined Tongs and Poker.*—August 21, 1866.—Explained by the claims and cut.

Claim.—First, a combined fire or cinder tongs, poker and stove, or range cover lifter, substantially as herein described.

Second, a combined cinder or fire tongs and poker, substantially as described.

Third, a combined cinder or fire tongs and stove or range cover lifter, substantially as described.

No. 57,432.- -WILBUR I. ARMSTRONG, Rockford, Ill., assignor to himself and SOLOMON DWIGHT, same place.—*Gate.*—August 21, 1866.—The gate has balance weights, and is supported on levers pivoted to it and to the sill. When the cord is pulled the gate rises and opens, and passing the centre descends to its full open position, and conversely.

Claim.—The combination of a gate, opened and closed on parallel levers, with balance weights to assist it in opening, and prevent its receiving injurious jars in closing, substantially as set forth.

No. 57,433.—JOHN AUSTIN, Rockford, Ill., assignor to ALEXANDER AUSTIN.—*Horse-shoe.*—August 21, 1866.—The band clip which fastens the shoe to the hoof is tightened by screws and nuts; auxiliary hook clips penetrate the wall of the hoof, and are secured by their screw shanks and nuts to the shoe.

Claim.—First, the band B, constructed and attached to the shoe, in the manner substantially as shown and described.

Second the clip C, having its upper end provided with a curved hook for taking hold upon the hoof, with its lower end screw-threaded, as set forth, in combination with the nut, as shown and described.

Third, the enlarge hole or recess r, in the upper side of the shoe, in combination with the band B, or clip C to permit the flat portion of the band or clip to be drawn down therein, in tightening up the shoe, as described.

Fourth, the guards d, for protecting the nuts which secure the rear ends of the band, as shown and described.

No. 57,434.—JOHN JOSEPH BENDER, New York, N. Y., assignor to himself, HENRY J. BANG and GEORGE MULLER.—*Double-action Piano.*—August 21, 1866.—Two sets of strings and appendages are arranged respectively above and below the single set of keys by which they are operated. Each key is pivoted near its mid-length and by its front downward and rear upward motion actuates the hammers, &c., of the respective actions.

Claim.—The above construction and arrangement of an instrument combining two piano-fortes, which are played together at one and the same time, and by one set of keys, substantially as described and set forth.

No. 57,435.—WILLIAM DARKER, West Philadelphia, Penn., assignor to himself and JOSIAH B. THOMPSON, Philadelphia, Penn.—*Braiding Machine.*—August 21, 1866.—Two sets of bobbins revolve in circular paths in opposite directions about a common centre. Vibrating thread guides cause each thread of one set to cross the paths of those of the other set. The claims cite devices for effecting this movement; for driving the upper bobbin carriers by means of two alternately reciprocating bolts, and the means for lubricating said bolts and the rocking arms which actuate the vibrating thread guides.

Claim.—First, the vibrating thread carriers e, for carrying the threads of the lower set of bobbins, made and arranged substantially as described.

Second, the rock shafts d, their side arms f g, and the cam j, for rocking said shafts, substantially as and for the purpose above described.

Third, the application to a braiding machine of the reciprocating bolts P, and cams R, for driving the upper set of bobbins, substantially as described.

Fourth, lubricating the bolts P, by means of the circular groove b, one or more in the disk L, substantially as described.

Fifth, the lubricating tube i, suspended from the disk L, for applying a saturated wick to the arms g of the rock shafts, substantially as described.

No. 57,436.—GEORGE P. DARROW, Cincinnati, Ohio, assignor to himself and JOSEPH HARGRAVE.—*Plough Clevis.*—August 21, 1866.—The socket is slotted, permitting the clevis pin to drop into place when it is tightened by a partial revolution engaging its sections of threads with the threads in the socket.

Claim.—Forming the clevis and bolt by casting the same with interrupted threads, in the manner and for the purpose set forth.

No. 57,437.—MAHLON S. DRAKE, Newark, N. J., assignor to himself and DAVID THOMP-SON, same place.—*Hat-ironing Machine.*—August 21, 1866.—From the shaft near the base of the frame ascends an oscillating frame, whose upper rotating shaft carries at its respective ends a hat block and a cam to regulate the approach of the blocked hat to stationary irons. There are other automatic arrangements for ironing the brim.

Claim.—First, in a machine for ironing hats an oscillating frame B, to which is attached the hat block and hat in such manner that the side of the hat shall as it revolves be kept in contact with a stationary iron, substantially in the manner set forth.

Second, in combination with the face plate X, the spring catches t t for securing the hat block to the face plate, substantially as set forth.

Third, the oval holder g, when used as a cam and in combination with the wheel b and oscillating shaft h, substantially as and for the purpose set forth.

C P 27—VOL. II

Fourth, in combination with the reciprocating iron F, the plate d for supporting the brim of the hat against the pressure of the iron, when arranged substantially as set forth.

Fifth, in combination with the oval hat-block holder g, the spring f and set screw e, for securing the hat, substantially as set forth.

Sixth, so arranging the mechanism for actuating the reciprocating iron F and face plate H supporting the hat block that the hat block shall be turned as the iron is ascending, and remain stationary while the iron is traversing the brim, substantially in the manner set forth.

Seventh, the iron F when attached to an oscillating frame E, and pitman J moved by a crank K, and actuated by mechanism so arranged that the iron F shall traverse one part of its course away from and return in contact with the hat, substantially as set forth.

Eighth, the guide block S' of the same oval as the hat, attached to a shaft on the oscillating frame B and guided by a stationary adjustable arm u, in such manner as to retain a stationary iron attached to the adjustable arm m, in constant contact with the hat, substantially as set forth.

Ninth, so arranging the irons that the iron for the side of the hat shall be attached rigidly to the main frame when in operation, and the irons for the upper side of the brim and crown, the latter of which, at least, is attached to the oscillating frame, shall be adjustably controlled by cords and weights, substantially in the manner set forth.

No. 57,438.—Louis Frühinsfeld, Newark, N. J., assignor to himself and William O. Headley, same place.—*Frame for Travelling Bags.*—August 21, 1866.—The metallic jaws are of curved form and the ends tapered toward the hinge.

Claim.—The combination of the jaws A A', curved as shown, with taper ends b, in the manner and for the purpose herein specified.

No. 57,439.—Joshua F. Hammond, Providence, R. I., assignor to Henry Staples, Barrington, R. I.—*Lifting Jack.*—August 21, 1866.—The bent pivoted arm forms with the stirrup link a toggle joint; this straightens as the bar is raised and maintains the position until the toggle is flexed.

Claim.—The combination of the two upright posts A and B, the one fixed and the other movable vertically, lever D, stirrup E, and the pin F, the whole arranged, combined, and operating as above described.

No. 57,440.—Charles M. Hodges, Mansfield, Mass., assignor to himself, Willard O. Capron, and Nathaniel Whitmore, same place.—*Scythe.*—August 21, 1866.—Explained by the claim and cut.

Claim.—A scythe in which the blade or cutting portion A and the back or holder B are made in separate pieces, and secured together by screws which pass through holes in the two back pieces and open slots in the blade, as set forth.

No. 57,441.—Benjamin Mannon, Newport, Ky., assignor to himself and Isidor Kann.—*Hair Crimper.*—August 21, 1866.—The hair being wound upon the two legs of the hair pin, is pressed by the bar and the latter clasped to the legs of the pin.

Claim.—In the described combination with a hair pin A A', the bearing bar B b, and clasp C c c' c'', for the purpose set forth.

No. 57,442.—Donald McDonald, Albany, N. Y., assignor to himself, Noel E. Sison, and Henry Q. Hawley.—*Apparatus for Carburetting Air.*—August 21, 1866.—The air inlet and outlet apertures at the respective ends of the air-forcing drum are covered with wire gauze which is carried below; the level of the hydro-carbon liquid is thus kept wet and parts with its moisture to the passing air.

Claim.—The combination of a cylinder or drum constructed like the drum of a wet gas metre, with fine wire gauze, or its equivalent, attached to said drum and covering the outlet openings of its measuring chambers, substantially as above described and for the purpose above set forth.

No. 57,443.—George Miller and George Reichert, New York, N. Y., assignors to John Reichert and Dominicos Rottkamp.—*Billiard Game Keeper.*—August 21, 1866.—The register has duplicate mechanisms. The count is made by drawing down the cord, registering one notch at each pull and striking a bell simultaneously. Another lever registers five at a time and gives a different alarm. The devices of ratchets, &c., are detailed in the claim.

Claim.—First, the combination of the ratchet wheels f m, pawls i g thereto attached, arms k k', and operating levers A h' on either side of the division plate, arranged and operating substantially as and for the purpose herein described.

Second, the combination of the ratchet wheels, pawls thereto attached, operating levers, and the arms, hammers, and bells, or their equivalents, on either side of the division plate, arranged and operating substantially as and for the purpose herein described.

Third, the levers A h', with their toes arranged and operating substantially as and for the purpose herein described.

Fourth, the combination of the ratchet wheel *f*, pawl *g*, and lever *h*, either with or without the bell attachment, arranged and operating substantially as and for the purpose herein described.

Fifth, the combination of the ratchet wheel *f*, pawl *g*, arm *s*, lever *h h'*, either with or without the bell attachment, arranged and operating substantially as and for the purpose herein described.

No. 57,444.—LEVI SCOFIELD, Farmington. Wis., assignor to himself and JUSTON B. WAITE, same place.—*Loom.*—August 21, 1866.—The devices cited are: First, for changing the movements of the heddle frames to weave different kinds of fabric. Second, affixing the shuttle operating treadle by a universal joint, so that it may be shifted alternately to actuate both picking levers. Third, operating the harness by a vertical rack on the rising and falling bar actuated by the lay. Fourth, employing the said bar to give motion to a revolving arm for operating a horizontal slide to give the treadle its lateral motion. Fifth, the means by which the bar is guided vertically, and also forced upward without jar. Sixth, the structure of the slide bar and its connections through which it receives motion; and, seventh and eighth, certain general combinations of some of these parts.

Claim.—First, a cam shaft provided with longitudinal slots or mortises to be filled with cams and blanks, or their equivalents, in combination with cams and blanks for the purpose of adjusting the cams in the shaft for weaving different kinds of cloth, substantially as and for the purpose described.

Second, a treadle which has one end pivoted to the framework of the loom, while the other end has a combined lateral and vertical movement when used to throw a shuttle from a right to a left and from a left to a right direction alternately, substantially as described.

Third, stepping bar M, or its equivalent, when used to impart motion to a cam shaft and cams, and to the leaves of a harness, when the whole are constructed and operated substantially as and for the purpose described.

Fourth, stepping bar M, or its equivalent, in combination with a sliding bar S and treadle R, when constructed together and operated substantially in the manner and for the purposes described.

Fifth stepping bar M, or its equivalents, in combination with guides *g*, or their equivalents, when the whole are constructed, connected together, and operated substantially as and for the purposes described.

Sixth, sliding bar S, or its equivalent, in combination with treadle R, elbow lever K, and lever *l*, or their equivalent mechanism, when constructed, connected together, and operated substantially as and for the purposes described.

Seventh, a combination and arrangement of the batten stepping bar, rack and pinion, pawl and ratchet, cam shaft and cams, and bases for the leaves of the harness, when the whole are constructed and arranged substantially as and for the purposes described.

Eighth, a combination and arrangement of the cam shaft, having motion imparted to it substantially as described. A treadle having a vertical and horizontal motion imparted to it, substantially as and by the mechanism described, or its equivalent, when the whole are connected together and operated as and for the purpose described.

No. 57,445.—G. SEITZINGER, Ottowa, Ill., assignor to himself and JOHN ARMSTRONG, same place.—*Grain Gate.*—August 21, 1866.—The rod which forms the pintle is bent and passes across and beyond the middle of the door, the projection forming a catch which locks into a recess in the frame. The pintle moves endways to lock and disengage the catch.

Claim.—The arrangement of the catch *b*, of the rod C and gate B, with the recesses *d e* of the frame A, and operating in the manner and for the purpose herein described.

No. 57,446.—GEORGE SHIPMAN, West Bridgewater, Mass., assignor to AZEL HOWARD, same place.—*Eyeleting Machine.*—August 21, 1866.—The eyelet magazine is a cylinder containing a reciprocating brush and having an adjustable gate through which eyelets of any size may be urged into the chute, through which they are to be delivered singly to the upsetting device. The details cannot be briefly described other than substantially in the words of the claims.

Claim.—The application and arrangement of the two levers E F together, and to the standard A, the eyelet punch, the magazine and chute, substantially as and for actuating such magazine and chute, substantially as described.

Also, the application and arrangement of the spring *a* with the standard A and the two levers E F, applied together and to such standard, and the punch, magazine and chute, substantially as specified.

Also, the combination as well as the arrangement of the three adjustable stops *k l n* with the letters E F, when arranged and applied together, and to the frame of the machine, the punch, the magazine and its chute, substantially in the manner as specified.

Also, the combination and arrangement of the elastic buffer X with the standard A, lever E, and the magazine and chute applied to such lever.

Also, the combination and arrangement of the adjustable guard K with the eyelet magazine and its ports, the chute and rotary brush, arranged together as specified.

No. 57,447.—ALEXANDER C. STOCKMAR, New York, N. Y., assignor to self and WILLIAM S. SEE.—*Hand Vice or Clamp.*—August 21, 1866.—The jaws are opened or closed by the traversing nut and connecting links; the cheeks of the jaws are presented outward or inward.

Claim.—In combination with the limbs, to which are attached the jaws, the use or employment of the nut, screw and toggles, when the same shall be constructed and operated substantially as and for the purpose set forth.

No. 57,448.—JOHN H. VICKERS, Norwich, Conn., assignor to BACON ARMS COMPANY, same place.—*Revolving Fire-arm.*—August 21, 1866.—The revolving cylinder of barrels is so supported by a band jointed to the stock that while the barrels revolve freely within it they may also be tilted up at the rear for loading.

Claim.—The pivoted bracket C, arranged and operating in combination with the band H, encircling the barrel loosely, and with the axial pin E, in the manner and for the purpose herein specified.

No. 57,449.—JAMES M. WARD, New York, N. Y., assignor to himself and JOHN D. GILBERT, same place.—*Hydrant.*—August 21, 1866.—The valve is moved by rotating the nut upon its stem, which passes through the diaphragm at the junction of the upper and lower sections of the penstock.

Claim.—The sliding valve F E, elevating nut D d, diaphragm $l l j k$, and ports $i k g$, arranged and operating substantially in the manner and for the purpose set forth.

No. 57,450.—MAURICE ABORD, Paris, France.—*Brick for Ceilings*—August 21, 1866.—The tubular bricks are provided on their sides with recesses and lips and on the lower part with indentations to facilitate the hold of the plaster thereon. The bricks are strung on beams or girders provided with projections to fit the lateral recesses in the brick.

Claim.—First, constructing ceilings of tubular or hollow bricks supported or suspended on beams or girders, and strung thereon to form a continuity of surface, essentially as herein set forth.

Second, a tubular or hollow brick having recesses d whereby to suspend it on the beams or girders, the upper and lower lips of said recesses being so proportioned that the underlapping ones will meet those of the adjacent row and conceal the girder, substantially as shown and described.

Third, grooving or indenting the lower surfaces of the bricks to facilitate the hold of the plaster thereto, substantially in the manner as specified.

No. 57,451.—JAMES MOORE CLEMENTS, Birmingham, England.—*Sewing Machine for Stitch ng Button-holes.*—August 21, 1866.—Two threads are used, one carried by an eye-pointed needle above, and the other by a double-pointed shuttle below the table; and in connection with these, to form the stitch, the following devices are employed, viz: a book below the table, a barbed book which works through the button-hole and thread-drawing books above the table The machine cannot be briefly described other than substantially in the words of the claim and its references to the cut.

Claim.—First, the vibrating hooks i^5 i^6, arranged and operating substantially as described, and employed for the purpose of drawing the needle silk into a horizontal position in order to supply the barbed book, as and for the object set forth.

Second, the combination of the eye-pointed needle, the circular grooved book d^3, and the barbed hook c^4, constructed and operating as and for the purpose set forth.

Third, in combination with the above the detaching book g^3, arranged and operating substantially as described.

Fourth, the pendulous bar k^3, employed in conjunction with the hooks i^5 i^6, to feed the needle silk to the barbed book, as set forth.

Fifth, the arrangement of the circularly feeding device O^2, spring O^{14}, arm 4, pin 3, shouldered lever O', and hand lever O, as and for the purpose specified.

Sixth, the plate r which carries the needle slide, pivoted to the frame, and adjustable by means of a set screw r^2, in combination with a barbed hook, as and for the purpose described.

Seventh, the combination and arrangement of the several mechanical parts herein described and represented, and mentioned in the preceding claims, or the mere equivalents thereof, forming improved machinery to be employed for sewing, stitching, or embroidering, substantially as herein set forth and specified.

No 57,452.—H. BAKER, G. F. HOLMES, and R. D. KING, Cortland, N. Y.—*Machine for Straining Cream.*—August 21, 1866.—The strainer is suspended within the churn and the cream forced through the meshes by the revolving rubbers.

Claim.—First, the flange for sustaining the strainer on any given size churn.

Second, the cone-shaped strainer and centre pivot, in combination with the scroll rubbers, substantially as set forth.

Third, the scroll rubbers in combination with the shaft, crank, cross bar and fastenings, as herein described and for the purpose set forth.

No. 57,453.—LORENZO HEMPSTEAD and LESTER S. HILLS, Hartford, Conn.—*Safety Bridge for Railroad Cars.*—August 21, 1866.—The floor and sides of the bridge which unite the platforms of adjacent cars are made on the expanding lattice or lazy tongs principle; they are contractible, to be taken on board, or are swung in on pivots when the cars are detached.

Claim.—First, the combination of the lattice-work floor *e* and railing *f*, to constitute an expansion bridge for railroad cars, when the same is constructed and arranged substantially as described.

Second, the combination of the swing table *b* with the floor *c* substantially as described.

Third, the combination of the hinge plates *e* with the expanding railing, as and for the purpose described.

No. 57,454.—ALBERT ADAMS, Springfield, Mass.— *Preventing the Coating of Pipes used in Mash Tuns.*—August 28, 1866.—The formation of crust on the steam coil used in mash tuns is prevented by passing cold water through the coil while the mash is being withdrawn from the tun.

Claim.—The process herein described of passing cold fluid through the pipes ordinarily used for heating a semi-fluid substance, when the same is used substantially in the manner and for the purpose herein set forth.

No. 57,455.—WALTER AIKEN, Franklin, N. H.—*Machine for Ironing Hosiery*—August 28, 1866.—The drums are steam-heated and driven by gearing; the steam is passed through the upper one and by a flexible connecting pipe to the lower, and is thence educted; springs support the bearings of the lower steam drum.

Claim.—The hosiery ironing machine made as described, viz: of the two hollow drums or cylinders, the flexible connection pipe, the driving shaft and gears, the stationary and movable boxes, tubular journals and stuffing boxes, arranged with and applied to a frame A, substantially as and for the purpose and to operate as specified.

No. 57,456.—CHARLES H. ALSOP, Middletown, Conn.—*Snap Hook.*—August 28, 1866.—The two hooks are riveted together and swivelled to a strap loop; each forms a mousing for the other and the ring is slipped between them as in the case of a split ring.

Claim.—The hooks B B secured together at one end, one upon the other, and swivelled at such end to the eye frame E, substantially as and for the purpose described.

No. 57,457.—CLARK ALVORD, Westford, Wis.—*Horse Collar.*—August 28, 1866.—The upper ends of the collar are fastened together by an elastic coupling strap above the neck pad.

Claim.—The mode of fastening the tops of horse collars by an elastic coupling, and for the purposes mentioned as above described and shown.

No. 57,458.—HAYDN M BAKER, Rochester, N. Y.—*Manufacture of Bicarbonate of Soda and Potash and Hydrochloric Acid.*—August 28, 1866.—Nitrate of soda and chloride of lead are obtained by the double decomposition of chloride of sodium and nitrate of lead, the chloride of lead being afterward decomposed by means of oxide of magnesium, forming oxide of lead and chloride of magnesium. The nitrate of soda obtained in the first operation is heated with silicic acid, in order to form a soluble silicate, and nitric acid escaping during the process is led into the vessel containing the oxide of lead formed in the second process, to form nitrate of lead. The soluble silicate formed is charged with carbonic acid, which unites with the alkali of the silicate and precipitates the silicic acid, leaving the carbonate in solution. The chloride of magnesium formed in the second operation is heated to drive off the hydrochloric acid which is condensed, leaving behind the oxide of magnesium which may be used again to decompose the chloride of lead.

Claim.—The application of the principle of double decomposition of chloride of sodium or potassium and nitrate of lead, for the formation and manufacture of nitrate of soda or potash, and the subsequent application to the manufacture of the decomposition of the said nitrate of soda or potash with silicic acid, for the purpose of forming soluble silicate of soda or potash, together with the utilization of the nitric acid liberated by said decomposition in the formation of nitrate of lead with the recovered oxide of lead.

Also, the application of the processes herein described and set forth, for the recovery of the oxide of lead from the chloride of lead with oxide of magnesium, and subsequent recovery of magnesia by distillation from chloride of magnesium.

Also, the application of carbonic acid under any degree of pressure for the decomposition of silicate of soda or potash, in the manner herein described for the purpose of forming carbonate or bicarbonate of potash or soda.

No. 57,459.—A. BALDING, Madison, Ind.—*Evaporator.*—August 28, 1866.—The evaporator has a skimming pan which empties into a scum box, under which is a cooling chamber.

Claim.—The pan, arranged as described, with the hinged skimmer C, scum pan G, cooling opening J, substantially as described and represented.

No. 57,460.—WILLIAM B. BARNARD, Waterbury, Conn.—*Lamp Chimney Cleaner.*—August 28, 1866.—Explained by the claims and cut.

Claim.—As a new article of manufacture a lamp-chimney cleaner having one or more arms B B' hinged to a handle A, and so combined with a spring D as herein described, so placed above the hinge or hinges, as to force said arms outwardly, all substantially in the manner and for the purpose herein set forth.

Also, corrugating the pads E E of a lamp-chimney cleaner, substantially in the manner and for the purpose herein set forth.

No. 57,461.—DAVID L. BARTLETT and GEORGE H. JOHNSON, Baltimore, Md.—*Brick.*—A .gust 28, 1866.—Bricks are moulded with bosses and corresponding recesses on the respective sides.

Claim.—A brick having bosses projecting from one face thereof and counterpart recesses formed in its opposite face, each at a point midway between the sides, centre, and end of the brick, substantially in the manner and for the purpose herein set forth.

No. 57,462.—D. L BARTLETT and GEORGE H. JOHNSON, Baltimore, Md.—*Construction of Double Cylindrical Structures.*—August 28, 1866.—Iron tie-plates are laid between the courses of brick at regular intervals, and are secured to each other by rods passing through them within the air chamber which intervenes between the concentric walls.

Claim.—Granaries, reservoirs, towers, &c., constructed of bricks formed substantially as herein described and laid in double concentric walls in combination with metallic tie-plates, substantially in the manner herein set forth.

No. 57,463.—GEORGE A. BEARD, Cavetown, Md.—*Mould Board for Ploughs.*—August 28, 1866.—The extended surface of the mould board prevents choking in briers or tall grass.

Claim.—The elevation and enlarged extension of the mould board of the plough, as above described, and nothing else or more.

No. 57,464.—JACOB H. BEIDLER, Lincoln, Ill.—*Post Office Delivery Box.*—August 28, 1866.—Explained by the claims and cut.

Claim.—First, the combination of the lever lock bolt C with its hooked end *h*, the corrugated plate D with the notch *b*, and the sliding plate E with its projecting flange *f* and staple *d*, all constructed, arranged, and operating substantially as and for the purposes described.

Second, the lever lock bolt C in combination with the corrugated plate D, or its equivalent, when so arranged that said lock bolt may be operated to unlock the door from the outside by means of a key, and from the inside by a movement of the lever, substantially as shown and described.

Third, the combination of the sliding plate E with its staple *d*, the lever lock bolt C and the bell F, so arranged that the turning of the key to unlock the door will cause the lever C to strike and ring the bell, substantially as described.

Fourth, the inclined plane *j* of the corrugated plate D in combination with the lever lock bolt C, so arranged that when the front end of said lock bolt is depressed by turning the key, it will strike the inclined plane *j* and start the door open, substantially as described.

No. 57,465.—J. C. BELL, Pawnee City, Nebraska.—*Treating Sugar Cane.*—August 29, 1866.—Explained by the claim.

Claim.—The within described process of treating sugar cane previous to grinding, by exposing the same to the action of boiling water or steam, or both combined, substantially as and for the purpose described.

No. 57,466.—GEORGE W. BISHOP, Stamford, Conn.—*Ratchet Drill.*—August 28, 1866.—A friction clutch on the top of the feed screw is so arranged that when the pressure upon the drill will not cause it to cut, the friction clutch will rotate the feed screw until the friction of the cutting becomes greater than the clutch, when it will slip and the screw will stop turning.

Claim.—The screw G, fitted in an internal screw thread in the arbor A, in connection with the friction device composed of the slotted tube J, and band K, or their equivalents, for connecting the screw with the head I, substantially as and for the purpose specified.

No. 57,467.—J. L. BOOTH, Rochester, N. Y.—*Railroad Rail.*—August 28, 1866.—Explained by the claim.

Claim.—A rail for railroads composed of the iron body A, and steel cap B, when the latter is rolled and shrunk on the body, in such a manner as to unite the parts closely, as a unit or whole, but still allow them to be easily separated and replaced, substantially as herein set forth

No. 57,468.—WILLIAM F. BOND, Philadelphia, Penn.—*Artifical Rubber.*—August 28, 1835.—Composed of borax, shellac, glue, flour, linseed oil, and molasses.

Claim.—A compound made of shellac, glue, and borax, substantially as and for the purpose described.

Also, a compound made of glue, shellac, borax, and flour, substantially as and for the purpose set forth.

Also, a compound made of glue, shellac, borax, flour, and linseed oil, substantially as and for the purpose described.

Also, a compound made of glue, shellac, borax, and molasses, substantially as and for the purpose set forth.

Also, a compound made of glue, shellac, borax, molasses, flour, and linseed oil, substantially as and for the purpose described.

Also, a compound made of glue, shellac, borax, molasses, flour, linseed oil, and emery, substantially as and for the purpose set forth.

No. 57,469.—HEZEKIAH BRADFORD, New York, N. Y.—*Apparatus for Drying Peat.*—August 28, 1866.—The long chamber is heated by a furnace near one end, and is traversed by carriages upon whose shelves the peat is spread in thin layers. The carriages move in a direction opposite to the current of air, which when heated is carried back to the furnace.

Claim.—First, a series of cars, moved gradually through a heated chamber, and provided with ranges of platforms, holding the peat to be dried, and then cooled, or partially cooled, by the action of the air, substantially as set forth.

Second, the cars for drying peat formed of a series of sectional platforms, in the manner specified, to facilitate the reception and discharge of peat, substantially as set forth.

No. 57,470.—J. A. CALDWELL, Springfield, Mass.—*Car Ventilator.*—August 28, 1866.—The door of the ventilating aperture in the car is closed by a catch, and the pivoted shutters on the outside are actuated by the resistance of the air and glance off the cinders, &c.

Claim.—The arrangement of the wings D D, on the outside of the case C, in combination with the rod b, packing g, spring d, and catch c, substantially as described.

No. 57,471.—G. J. CAPEWELL, West Cheshire, Conn.—*Machine for Grinding and Polishing Buttons.*—August 28, 1866.—Holders for the buttons are arranged within a common rotary head plate which brings the succession of buttons first to the grinding wheel, and then to the polishing wheel, completing the process in one continued operation by one machine.

Claim.—First, grinding and polishing buttons by means of a machine, having suitable holders for the buttons, and so arranged and operated as to subject the buttons to the action of the grinding and polishing surfaces, substantially in the manner described.

Second, imparting a rotary movement to the button holders while subjected to the action of the grinding and polishing surfaces, as and for the purpose specified.

Third, so constructing and arranging the button holders with regard to the machine, that, as the machine is operated, the buttons shall be automatically delivered therefrom after having been both ground and polished, or either ground or polished, substantially as described.

Fourth, so constructing and arranging the button holders, with regard to the machine, that, as the machine is operated and the button holders in turn pass to the grinding or polishing surfaces, as the case may be, they shall be automatically so operated upon as to sufficiently lift to clear the edges thereof, and thus prevent their impingement against the same, and then lowered thereto, or brought to bear thereon, with sufficient pressure to produce the desired grinding and polishing of their surfaces, substantially as described.

Fifth, the combination with a common head plate, having a series of one or more holders suitable for the reception of the buttons to be ground and polished, and to which head plates a rotary or other proper movement is imparted, of the grinding and polishing wheels or surfaces, arranged and operating with regard to the said head plate and each other, substantially in the manner described and for the purpose specified.

Sixth, holding the buttons, in their holders, upon their sides or edges, by means of a spring or other device, suitably arranged therefor, and substantially as and for the purpose specified.

Seventh, the peculiar construction of the button holders herein described, the same consisting of the hollow shaft l, with its collar o, in which the buttons are placed and held, surrounding casing s, having a coiled or other suitable spring upon its inside, cap or head t, and centre spindle u, passing entirely through the hollow tube l with a spring w, the whole being arranged together as described and operating as the head plate is revolved, and the upper ends of the button holders pass over and under the fixed arms f⁴ and g⁴, substantially in the manner and for the purpose specified.

Eighth, the combination of the head plate, having arranged upon it a toothed disk Q, and stationary plate T, secured to the said toothed disk Q, by means of a set screw V², or other suitable device, with the pinion wheels 8 of the button holders, arranged and operating together substantially in the manner and for the purposes described.

No. 57,472.—PETER CARBACH, Cleveland, Ohio.—*Melodeon.*—August 28, 1866.—A combination of devices to produce a swell action by the motion of the pedal.

Claim.—First, the bellows G, valve *a*, exhaust pipes *i i'*, and valve opening *b b' b"*, in combination with the valve *c*, jointed arm I, link J, and lever J', and pedal D, as and for the purpose set forth.

Second, the bellows G, arm M, and link *f*, in combination with the two-armed lever N', valve N, and spring *g*, as and for the purpose set forth.

Third, the valve *a*, spring *a'*, bellows G, in combination with the induction pipe *b"*, reserve bellows F, exhaust bellows C C', as and for the purpose set forth.

No. 57,473.—I. T. CARPENTER, Thompsontown, Penn.—*Fruit Gatherer.*—August 28, 1866.—The prongs are covered with rubber and book off the fruit, which falls into the mouth of the conveyor.

Claim.—First, the shank C, provided with prongs D D, which said prongs are covered with India-rubber, or its equivalent, and constructed substantially as herein represented.

Second, the cords F F, arranged with the prongs and the ring or loop E, for the purpose of conducting the fruit to the conductor G, substantially in the manner herein set forth.

No. 57,474.—MERRILL E. CARTER, Syracuse, N. Y.—*Machine for Grooving Lumber.*—August 28, 1866.—Two inclined grooving burrs cut the dovetail sides and a horizontal cutter on an upright shaft makes the head of the T groove.

Claim.—The combination of the angular cutters D D, and the horizontal cutter G, operating substantially in the manner and for the purpose herein set forth.

No. 57,475.—H. W. CASWELL, Yarmouth, Me.—*Device for Shrinking Tires.*—August 28, 1866.—The tire is fastened in two places in the clamps of two parallel bars; the latter being drawn together by screw bolts, the hot intervening portion of the tire is upset.

Claim.—The two parallel bars A A, connected by the screw rods B B, and provided with the clamps C C, all constructed and arranged to operate in the manner substantially as and for the purpose herein set forth.

No. 57,476.—JOHN H. CHAPMAN, Utica, N. Y.—*Grappling Iron.*—August 28, 1866.—For suspending the pulley blocks of horse hay forks, &c., from rafters, the weight of the blocks, &c., maintaining the grasp. The grapple is elevated by a fork which holds it by the handles and detached by a spreader which loosens its grasp.

Claim.—First, the grappling irons with the hooked ends add extension points, or their equivalent, substantially as described for the use and purpose mentioned.

Second, the elevating implement with or without the detaching part, constructed and operating substantially as described for the uses and purposes mentioned.

Third, the grappling irons and the elevating implement with or without the detaching part in combination, substantially as described and for the uses and purposes mentioned.

No. 57,477.—ROBERT CHESNUT, Richmond, Ind.—*Feeding Trough.*—August 28, 1866.—Explained by the claim and cut.

Claim.—A feeding trough divided into a series of compartments communicating with one another, and so arranged as to receive the feed simultaneously from a tank located above the trough, the supply being regulated by a valve G, actuated by a lever H, the several parts being respectively constructed and arranged for use substantially as set forth.

No. 57,478.—ANGELUS M. CLARA, Whitney's Point, N. Y.—*Hay Loading Wagon.*—August 28, 1866.—A longitudinally sliding bar under the tongue is connected by a rope to a roller beneath the bed; from this roller a rope passes over a crane to the loading fork. The draught on the rope applies the rubbers to the wheels.

Claim.—The shaft G, provided with the pulley F, with the ropes L D attached respectively thereto, in combination with the slide bar M, applied to the draught pole N, the upright pole B, with arm C, projecting from it, and the levers H, in which the shaft G is fitted, and having the brake or shoe bar J attached, all arranged and applied to a wagon to operate substantially as and for the purpose specified.

No. 57,479.—MARIUS C. C. CHURCH, Parkersburg, West Va.—*Tank for Petro'eum.*—August 28, 1866.—On the top of a cylindrical tank is a chamber provided with a safety valve by means of which the tank proper may be kept constantly full of the petroleum.

Claim.—The combination of the tank A, of the chamber D, provided with the safety valve C, and connected by orifices as described and for the purpose of the storage and transportation of petroleum or other liquid.

The tank A, segment shaped in transverse section, the flat upper surface forming the floor of a secondary chamber.

No. 57,480.—DANIEL COLE, Ornell, Penn.—*Animal Trap*—August 28, 1866.—The weight of the animal, as the treadle trips the trigger, drops the platform and raises the entrance door. The animal passes the hanging door into the other chamber.

Claim.—The treadle *c*, arranged within the box A, with the rod *d*, springs *s* and *n*, door *e*, windlass *y*, and cords *z z*, by which means the animal is lowered into the under chamber of box A, closing the door through which it entered and imprisoning itself after passing through the gates *g* into the supplementary chambers, when arranged in the manner herein specified.

No. 57,431.—JESSIE CONVER and JOHN BORTHWICK, Philadelphia, Penn.—*Heat Radiator.*—August 28, 1866.— Larger disks with central apertures are arranged alternately with smaller disks with marginal flue space between them and the drum. The heated air follows a circuitous route.

Claim.—First, the heat radiator composed of metallic disks permanently attached together and combined with the body or fireplace of any stove or any device or devices, substantially the same, for the purpose and in the manner above described.

Second, the combination of the heat radiator with cylinder P', for the purpose and in the manner aforesaid described.

No. 57,432.—A. J. CURTIS and D. J. ROBERTS, Swanville, and W. CURTIS, Monroe, Me.—*Horse Rake.*—August 28, 1866.—To each end of the rake head a bar is attached provided with a rack upon its under side and having its forward end slotted. The crank arms of a shaft are fitted in these slots, and by rotating the shaft by means of a hand lever attached thereto, the racks are forced to engage with pinions upon the wheel hubs, the bars are forced forward, and the rake tilted.

Claim.—The arrangement and combination of the gears on the wheels with the rack bars and their operative mechanism as described, applied to the rake head and the axle, the whole being substantially as specified.

No. 57,433.—H. W. CURTIS, Lockport, Ill.—*Fanning Mill.*—August 28, 1866—The sieves of the mill can be changed from end shake to a body shake; the former motion for chaffing grain, the latter for screening it.

Claim.—The combination of the bar J, shafts I, pins *m n*, and suspenders L, with the shoe V V, shaft H, and slot K, substantially as described.

No. 57,434.—CHARLES DANIEL, Lamonte, Mo.—*Cultivator.*—August 28, 1866.—The plough beams are suspended from the carriage at front and rear and are adjustable vertically and laterally. Certain adjustments are recited in the claims.

Claim.—First, the jointed handled and vibrating cultivator *h h* in combination with the plough beams *c c*, the axle B, and bounds *a a*, constructed and operated substantially as and for the purposes herein described.

Second, the suspended plough beams *c c*, in combination with the bounds *a a*, and the stirrups *d d*, constructed and operated substantially as and for the purposes set forth.

Third, the arrangement of the oblique hanging bounds *a a*, in connection with the centre beam A, the plough beams *c c*, the side braces *e e*, and the axle B, constructed and applied substantially as and for the purposes herein specified.

Fourth, the arrangement of the swinging seat C on the centre beam A, in combination with the vibrating cultivators *h h*, and their jointed handles *m m*, applied in connection therewith, substantially as and for the purposes herein described.

No. 57,435.—J. C. DICKEY, Saratoga Springs, N. Y.—*Clothes Wringer.*—August 28, 1866.—Explained by the claim and cut.

Claim.—Two cores with alternate depressions and elevations meshing with each other through their entire length, in combination with the rubber covering that conforms both externally and internally to the inequalities of the core.

No. 57,486.—ANDREW DOUGHERTY, Brooklyn, N. Y.—*Printing Press.*—August 28, 1866.—Improvement on his patent of August 9, 1859. Explained by the claims and cut.

Claim.—The combination in a printing press of the following instrumentalities, viz., the printing cylinder, reciprocating carriage for flat printing surfaces, inking apparatus therefor, impression cylinder for curved printing surfaces, inking apparatus therefor, and carriage for the inking apparatus, all operating in the combination substantially as set forth.

Also, the combination in a printing press of the following instrumentalities, viz: the printing cylinder, reciprocating carriage for flat printing surfaces, two impression cylinders for curved printing surfaces, arranged at opposite sides of the main printing cylinder, two inking apparatuses therefor, and two carriages for the inking apparatuses, all operating in the combination substantially as set forth.

No. 57,487.—SAMUEL R. DUMMER, New York, N. Y.—*Hand Truck.*—August 28, 1866; antedated August 17, 1866 —Flexible studs on the peripheries of the wheels and elastic washers on their journals prevent noise in travelling.

Claim.—First, the combination of the elastic studs *b*, and wheels D, constructed and arranged in the manner and for the purpose herein specified.

Second, the combination of the flexible washers *h*, and hubs *g*, arranged in the manner and for the purpose herein specified.

No. 57,488.—D. M. DUNHAM, JEREMIAH WEBB, and ALBION WEBB, Bangor, Me.— *Horse Rake.*—August 28, 1866.—A ratchet wheel is attached to one of the supporting wheels, upon which is loosely fitted a sleeve provided with an arm connected by a rod to the rake head. A pawl is pivoted to this arm, which, by means of spring levers and a bent rod, is forced to engage with the ratchet at pleasure, thereby carrying forward the arm and tilting the rake. A stop upon the ax'e disengages the pawl at the proper time, and a bent lever is so connected to the rake head that when used to raise the rake, it may be locked to hold the rake in an elevated position for transportation.

Claim.—First, the combination of the ratchet wheel H, pawl N, or equivalent, and sleeve J, with each other, and with the wheel F, to which they are attached and by which they are operated, and with the shaft C, which they operate, substantially as described and for the purpose set forth.

Second, the combination of the springs P and O, and bent rod R, with the pawl N, for the purpose of causing the pawl to engage with the ratchet wheel, substantially as described.

Third, the combination of the stop S with the pawl N, for the purpose of disengaging the pawl from the ratchet wheel, substantially as described.

Fourth, the combination of the bent lever U with the frame of the rake and with the shaft C, substantially as described and for the purposes set forth.

No. 57,489.—LOUIS ELSBERG, New York, N. Y.—*Apparatus for Preparing Peat.*—August 28, 1866.—A cylinder has within it a shaft provided with knives and communicates with a steam generator by means of tubes. At the lower part of this cylinder are two openings which allow the peat to pass into the sectional mould, which is fitted with a piston.

Claim.—The combination in a machine of the following implements, viz: the agitator, agitator chamber and steam pipe, all operating in the combination substantially as set forth.

Also, the combination in a machine of the following implements, viz: the chamber for the material, steam delivery pipe, and reciprocating piston press, all operating in the combination substantially as set forth.

Also, the combination in a machine of the following implements, viz; the agitator, agitator-chamber, steam delivery pipe, and reciprocating piston press, all operating in the combination, substantially as set forth.

Also, the combination of the piston and piston chamber of the press with a discharge passage composed of sections all operating in the combination substantially as set forth.

Also, the combination of the piston and piston chamber of the press with a discharge passage whose interior is tapering, all operating in the combination substantially as set forth.

No 57,490.—RUDOLPH EICKEMEYER, Yonkers, N. Y.—*Steam-engine Governor.*—August 28, 1866.—The escapement and balance make isochronal vibrations and are driven by the motor whose speed is to be governed. The upper stem being revolved causes the thread to turn the worm wheel, which is connected to the escapement and has a regulated speed; variation in the speed of the stem causes the threads thereon to bear up or down on the cogs of the wheel and depress or lift the lower stem connected to the valve.

Claim.—First, a governor consisting of an escapement and a balance making isochronal vibrations connected with the valve, by which the speed of the motor is controlled, so as to regulate the motive power, substantially as herein described.

Second, the combination of the escapement D, balance H I, and friction spring *d*, or their equivalents, with the valve, in such manner that the latter may cease moving after the supply of motive power has been entirely shut off, and before the normal speed of the motor is restored, substantially as herein set forth.

Third, the arrangement of the lever L, with graduating weight *s*, connected with the escapement and balance by means of a loose strap *w*, substantially as and for the purpose herein specified.

Fourth, the spring *d*, applied to produce friction between the escapement wheel and its shaft, and to insure the falling of the pallets of the verge or anchor on the teeth of the said wheel, substantially as herein specified.

Fifth, the counterbalance lever N, in combination with the valve and escapement, substantially as and for the purpose herein specified.

Sixth, the governor consisting of the escapement and balance, the friction spring *d*, weights *s*, counterbalance *h*, stops *c' d*, and their connections, the whole constructed, combined, and applied substantially as herein specified.

No. 57,491.—J. J. ENSLEY, New York, N. Y.—*Apparatus for Generating Gas.*—August 28, 1866.—The gas generator is arranged within a furnace and fitted with a perforated cylinder which contains the animal substance; the generator is connected with a series of condensers and these with a gas holder. A chamber, with a pump for removing condensed liquid, is placed between the condensers and the gas holder.

Claim.—The construction of the perforated charge cylinder or cylinders *d*, open at the inner end, and the horizontal close retort cylinder or cylinders *c*, substantially as and for the purpose herein specified.

Also, the condensing chambers *k k*, connected by the pipes *l l*, and provided with discharge cocks *m m*, when used in combination with the enclosing water tank B, substantially as described.

Also, the combination and arrangement of the tight barrel *q* and pipes *p r* with the pipe *n*, for the purpose of pumping off the condensed water that gathers without admitting air or allowing escape of gas, substantially as set forth.

No. 57,492.—WARREN R. EVANS, Thomaston, Me.—*Coal and Ash Sifter.*—August 28, 1866.—The sifter acts in the top of the box, and the pins on its lower plate agitate the ashes as the plate is vibrated.

Claim.—First, the combination of the ash receiver B, with its perforated plate C, with the vessel A, substantially as specified.

Second, in an ash sifter, constructed substantially as set forth, the pins *c c*, in combination with the plate C, substantially as and for the purpose specified.

No. 57,493.—WILLIAM FREDERICK, Ashland, Penn.—*Boot and Shoe Edge Parer.*—August 28, 1866.—The stock is grasped by both ends and has an adjustable cutter and gauge-block.

Claim.—An improved boot and shoe edge parer, formed by combining the holder B, cutter A, and block E with each other, the said parts being so constructed and arranged substantially as herein described and for the purpose set forth.

No. 57,494.—CHAUNCEY W. FULLER, Earlville, Ill.—*Funnel Measure.*—August 28, 1866.—The valve at the lower end of the cone is retracted by a thumb lever and reclosed by a spiral spring Transverse rods indicate quantity.

Claim.—A measuring funnel constructed with a valve C, raised by the bent rods D, with a spiral spring E, and lever F, and having also transverse rods B, for measuring quantities, said several parts being respectively constructed and arranged for use in relation to the funnel and to one another, substantially as set forth.

No. 57,495.—E. A. GILES, New York, N. Y.—*Stem-winding Watch.*—August 28, 1866.—Either side of the case of the stem-winding watch is opened by pressure upon the head of the winding arbor, the spring-retaining catch of either lid being retracted by a stud on the adjustable sleeve.

Claim.—The sleeve *e* and pin *f*, combined with each other and with the winding arbor *c*, pendant ring C, and spring catches *n*, substantially as herein set forth for the purpose specified.

No. 57,496.—GEORGE P. GOODWIN, Lowell, Mass.—*Twine Spool and Stand.*—August 28, 1866.—The disks on the axial shaft are bushed and afford bearings for the centres on which the spool rotates; one of the centres is movable to regulate the friction.

Claim.—As a new article of manufacture a twine spool and stand composed of the hanger, bushings, friction centres, and thumb screws, all arranged to operate substantially as and for the purpose set forth.

No. 57,497.—D. A. GORHAM, Lawrence, Mass.—*Whiffletree.*—August 28, 1866.—The traces are disconnected by the partial rotation of the rocking shaft, to whose hooked ends they are hitched.

Claim.—First, the revolving trace hooks D and rod or bar C, constructed as described, in combination with the whiffletree A, substantially as described and for the purpose set forth.

Second, connecting the whiffletree A to the forward part of the carriage, in such a way that the horse can be released from the carriage in an instant whenever necessary, substantially as described and for the purpose set forth.

Third, the spring pin H, constructed as described, in combination with the strap F, slotted bar G, and lever E, substantially as and for the purpose set forth.

No. 57,498.—E. B. GRAFF, Baltimore, Md.—*Fastening for Railroad Rails.*—August 28, 1866.—The saddle piece has flanges which are bolted to the tie and laps upon the cut-away ends of the adjacent rails, the top of the saddle piece being flush with the wheel track.

Claim.—A railroad rail A, having its ends constructed as set forth, in combination with the saddle D, plate P, and bolt fastenings, as described, the whole being constructed, arranged, and operated substantially in the manner and for the purpose described.

No. 57,499.—JOHN P. GRUBER, New York, N. Y.—*Pump*—August 28, 1866.—The hinged valve closes the spiral duct and prevents the reflux of the water when the shaft ceases to rotate.

Claim.—The application of a valve *b* to a rotary spiral flange for the purpose of elevating water, substantially as described.

No. 57,500.—WILLIAM H. HALSEY, Hoboken, N. J., and MAURICE FITZGIBBONS, New York, N. Y.—*Sewing Machine.*—August 28 1866.—This is an improvement on Hudson's patent of November 1, 1859. Its object is to regulate the feed and also to support and carry the spool of thread with the case or handle. The feed is effected by motions of the looper, which is prevented from feeding except at the proper time, by means of a cam projection on the inner face of its arm coming in contact with a stationary pin.

Claim.—First, the combination and arrangement of the spring E and arm D with its cam end *a*, constructed and operating substantially as and for the purposes set forth.

Second, forming the handle of the machine so that it will contain and support the spool, substantially as described.

No. 57,501.—FREEMAN HANSON, Hollis, Me.—*Turn Table.*—August 28, 1866—This turn table rests upon an inclined base, so that when the car comes upon its more elevated portion the weight causes it to turn upon its axis; brakes and catches stop and retain it in the required position.

Claim. "A turn table operated in an enclosed pit by the weight of the car or engine to be turned, as and for the purposes described.

No. 57,502.—MARK E. HANSON, Newport, Me.—*Apparatus for Generating and Burning Gas from Petroleum, Naphtha, &c.*—August 28, 1866.—The liquid hydro-carbon is detained by the ridges in the generator so as to consume its crudities and is passed thence to the burner, whose orifice is closed by a plug when not in use.

Claim.—The hereinbefore described arrangement of a gas generator and burner, consisting of the generator *a a* with the ribs *b*, the eduction pipe *d*, perforated cap *f*, together with the hinged spring *h*, and pin *i*, the said several parts being constructed and the whole combined for use, substantially as and for the purpose set forth.

No. 57,503.—WILLIAM W. HARDER, New York, N. Y.—*Lock.*—August 28, 1866.—The flat key enters the slot in the rotary post and its bits arrange the double hooked tumblers so as to free the basp from engagement therewith.

Claim.—The combination of the double hooked tumblers D E F G, key and rotating slotted post L, when constructed, arranged, and operating as and for the purposes set forth.

No. 57,504.—ANDREW HARTMAN, Canton, Ohio.—*Car Coupling.*—August 28, 1866,—The throat is formed by upper and lower spring-jaws, which yield to the link and admit it to be engaged by the gravitating hook, which is raised for disengagement by a crank rod and pallets.

Claim.—First, the arrangement of the book D, the crank rod E, rod F with its stops, and the plate G, constructed and used as and for the purpose specified.

Second, the arrangement with the bumper A of the springs B and C, with their ends forming portions of bottom and top of the receding mouth, as and for the purpose specified.

Third, the arrangement of the crank shaft I provided with prongs J J, with the spring C, and hook D, the several parts being constructed substantially as and for the purpose specified.

No. 57,505.—JOSEPH S. HAVENS, T. M. JOHNSON and C. W. HOWE, Buffalo, N. Y.—*Uterine and Abdominal Supporter.*—August 28, 1866.—The copper and silver rings of the uterine supporter are mounted upon a vertically adjustable rod, attachable rigidly or loosely to its joint on the supporting rod; the latter depends from the abdominal pad and is fixed at any inclination by a cogged segment and set screw.

Claim.—First, in providing the joint of the uterine supporter with a rack, a set screw and a segment of a pinion, as described, by means of which it may be either locked in position or swung loosely upon the rod U', as desired.

Second, in making the tube P with the supporting ring L, adjustable vertically by means of the set screw Q and tube P', as described.

Third, the combination of the ring M with the ring L of said supporter, when constructed as and for the purposes set forth.

No. 57,506.—HENRY HERBERT, Cincinnati, Ohio—*Pipe Tongs.*—August 28, 1866.—Different notches in the side of the shank are optionally engaged by the teeth on the head of the pivotal bolt, conferring a variability in the capacity of the jaws. An adjustable stop limits the closing of the handles.

Claim.—First, the adjustable pipe tongs constructed with a toothed pin C G, substantially as and for the purposes set forth.

Second, the adjustable gauge consisting of the sliding collar O, and set screw N, when constructed and arranged to operate as and for the purposes specified.

No. 57,507.—HENRY HISE, Ottowa, Ill.—*Tug Buckle*—August 28, 1866; antedated February 28, 1866.—The upper portion of the frame is pivoted to the lower, has a spur which engages the holes in the trace strap, and is depressed by a spring.

Claim.—As an article of manufacture a buckle composed of the two parts A and C, when the former has a projecting pin *a*, and is secured to the latter by means of a spring G attached to C, said frame C having at its end L a loop to which the tug strap is attached, so that the other part of the frame C remains uncovered, all said parts being constructed in the manner and operating substantially as described and for the purpose set forth.

No. 57,508.—E. R. HOPKINS, New York, N. Y.—*Lock*—August 28, 1866.—The bolt is operated by pushing or retracting the knob, whose shaft is connected to the inner one of a series of vertical concentric rings which are each connected to the one next in series and adjusted by rotation of the knob shaft: the prolongation of the latter actuates a series of annular disks to bring their projecting pins into such correspondence with holes in the rings as to allow the longitudinal motion of the shaft and the freedom of the bolt.

Claim.—The combination with the handle shaft and the bolt of a lock of two separate series or sets of concentric rings or disks, one of which series, by its holder or frame, is so connected with the handle shaft as to always turn with it, while the other series is susceptible of being brought in connection with or disconnected from the said first series, when the two series of rings or disks are so constructed and arranged with regard to each other and with the handle shaft and lock casing as to allow the bolt of the lock to be drawn out or thrown in only by properly moving the handle therefor, after said disks have been brought to certain position with regard to each other, substantially as herein described and for the purposes specified.

Also, adjusting the operating parts of the lock herein described for throwing its bolt in accordance with a graduated disk attached to the handle shaft, and turning in connection with it and the annular fixed graduated ring of the lock casing, when such graduations of the said disk and ring are so arranged with regard to each other that only one of the said graduations of both the ring and disk can coincide with each other at one and the same time, substantially as herein described and for the purpose set forth.

No. 57,509.—GEORGE W. and ELISHA HOPKINS, Brooklyn, N. Y.—*Automatic Steam Valve.*—August 28, 1866.—The cylinder piston controls the valve by opening ports in its rear, admitting steam to act upon the valves, the said valve-operating ports being closed at other times by sleeves which slip upon contracted portions of the valves.

Claim.—The combination with the piston valve, constructed substantially as described, of the sliding rings S S', arranged and operating in connection with the valve heads and their necks to produce the operation of the valves by steam from the cylinder, essentially as herein set forth.

No. 57,510.—HENRY B. HORTON, Ithaca, N. Y.—*Calendar Clock.*—August 28, 1866.—Cannot be briefly described other than substantially in the words of the claims.

Claim.—First, the cam or sliding surfaces on the month cam I, for the purpose of sliding or revolving the thirty-one-days wheel the distance of one day, for the object described: and the said surface for one day's advance of the said wheel for any other purpose connected with the supernumerary days of a calendar clock.

Second, making a four, eight, twelve, or more year wheel, containing virtually the uses and position of the wheel P, moved by the pin or part P *c*, or equivalent, and controlled by the bent rod *a*, or equivalent, and its surfaces P *a* and P *b*, &c., as described; and the wheel P when made of eight slots, six plain and two irregular surfaces as described, or other correlative number of parts, periphery, slots, and surfaces when made substantially as described. And the February cam *c c'* constructed on the year cam Z, for leap year, as described, both when used in combination with the wheel P, and when used with any other device or mechanism in place of the wheel P.

Third, the immediate contact, or other suitable connection, of the pawl *n a* with the pawl K, for the purpose of rendering its action more sure, as described; and the so shaping the upper surface of the pawl *n a* as to act as a cam surface with the pawl K, and thus be in mutual relation at all times to each other, as described.

Fourth, attaching the lever S to the weight lever *n*, in such a manner as to receive the action of the eccentric A by the rod B, and by its cam surface S *c* carry the pawl K over the teeth of the thirty-one-days wheel as described; and the lever S, or its equivalent, for the purpose of raising a pawl over one tooth of the thirty-one-days wheel, and also of the year wheel and of the day of the week wheel, while in the act of making their changes. And the raising or tilting of the stop pawl K or correlative pawls of the week and year wheels over the teeth acted on by the same, when produced by the fall in changing of their respective rods or levers.

Fifth, making the pawl T with a slot in it for receiving the lever of the month cam, thus simplifying and making more sure in action the devices connected therewith.

No. 57,511.—HENRY B. HORTON, Ithaca, N. Y.—*Clock Case.*—August 28, 1866.—Two cast frames form the front and back respectively of the case and hold in their marginal grooves the continuous metallic side plate; the holes in the castings are bushed with wooden plugs for the clamping screw bolts.

Claim.—First, the combination of the two cast frames, one for the back and the other for the front of the case, with the interposed piece or part between them, as described.

Second, holding the cast frames and middle piece together by the bolts or rods G, or other equivalent device, as described.

Third, the use of the wooden plugs in connection with the connecting rods or bolts for the described purposes and uses, and the use of the said wooden plugs in the holes of the frames for adjusting the fronts and backs when the rods are not used.

Fourth, the combination of frames and interposed middle piece, rods, and wooden plugs in the said holes, the same making a whole, as described.

No. 57,512.—HORACE HUBBELL, New Haven, Conn.—*Hydraulic Engine.*—August 28, 1866.—Water is injected into the cylinders alternately, the sliding induction connection being actuated by the reciprocating segment. The water is discharged from each cylinder when the change in the induction is made.

Claim.—The combination of the cylinders and their appendages with the sector and its appendages, when the whole is so constructed, arranged, and fitted, that the jet of water will alternately be forced into the lower end of one cylinder to act on and elevate its piston, while the lower end of the other cylinder will be entirely open for the discharge of the water and the descending of the piston, substantially as herein described and set forth.

No 57,513.—ROLAND C. HUSSEY, Milford, Mass.—*Cutting Board.*—August 28, 1866.—The end of the grain forms the face, and the sections are clamped in a metallic frame.

Claim.—The manner of clamping together two or more seasoned sections A B, that is, by using two bars C, and rods D, carrying nuts c, by which the section may be held closely together, substantially as specified.

No. 57,514.—CORNELIA F. INGRAHAM, Indianapolis, Ind.—*Sewing Machine Bobbin.*—August 28, 1866.—The opposite disks of the Wheeler & Wilson disk bobbin are separable from each other and reunitable by means of a central screw upon one, which enters a corresponding socket on the other; the usual central hole, to allow of winding, is undisturbed. The operator is thus enabled to reach and examine a broken thread.

Claim.—Constructing the bobbin of a sewing machine substantially in the manner and for the purpose above set forth.

No. 57,515.—GEORGE H. JOHNSON, Baltimore, Md.—*Vault Light.*—August 28, 1866.—The inverted metallic girders have intervening elongated glasses, the lines of junction being covered with strips to exclude moisture and secure the glasses in place.

Claim.—The combination of inverted or centrally enlarged girders e c, with elongated glasses A A, in the construction of a vault cover, substantially in the manner and for the purpose herein described.

Also, the arrangement and combination of retaining strips over and upon the joints of the glasses and frame of an illuminated vault cover, to secure said glasses and protect the joints from moisture, substantially in the manner herein set forth.

No. 57,516.—NIELS JOHNSON, Ripon, Wis.—*Pump.*—August 28, 1866.—A combined air and water pump. Each stroke of the piston, while it raises water through the hollow piston rod, also, by means of a central air chamber provided with suitable valves, introduces into the ascending column of water a small portion of air, to aid the discharge of water.

Claim.—The arrangement of chambers K and M, with the sucker G, and valves E H and I, substantially as and for the purpose herein specified.

No. 57,517.—JACOB O. JOYCE, Dayton, Ohio.—*Centrifugal Machine.*—August 28, 1866.—The rotary feeder has hollow perforated arms and revolves in a direction opposite to that of the foraminous cylinder.

Claim.—In combination with a centrifugal sugar separator, a revolving feeder, operating substantially in the manner and for the purpose described.

No. 57,518.—WM. A. L. KIRK, Hamilton, Ohio.—*Steam-engine Valve.*—August 28, 1866.—
The main valve has a continuous rotary motion in its seat, its ports acting in conjunction with those in the seat and those in the cut-off valve, which has its seat in the main valve and is stationary yet adjustable therein.

Claim.—First, the combination of the valve chamber A, and continuously revolving valve H, provided with ports, substantially as described and for the purpose specified.

Second, the combination of the cut-off valve L with the chamber A, valve H, and its ports, constructed and operating substantially as and for the purpose specified.

No. 57,519.—G. KAMMERL and D. L. BOLLERMANN, New York, N. Y.—*Dress Elevator.*—August 28, 1866.—By the means of the supplementary cords the lifting cords are pushed downward to depress the skirt.

Claim.—The combination of the supplemental operating cords *k*, with the lifting cords *n*, and the operating cords *g*, substantially as herein set forth for the purpose specified.

No. 57,520.—CHAS. KUGLER, Cadiz, Ohio.—*Horse Rake.*—August 28, 1866.—When the toggle is straight the rake is held in working position, when it is depressed and bent the rake is lifted; foot triggers actuate it in each case.

Claim.—The combination of the jointed parts *d d'*, front foot-piece *e*, rear foot-piece *e'*, cross-bar G, and arm F', of the rake head D, arranged and operating substantially as described for the purpose specified.

No. 57,521.—TAYLOR D. LAKIN, Hancock, N. H.—*Handle for Cutlery.*—August 28, 1866.—Explained by the claim and cut.

Claim.—The central cast metal plate A, cast on a core to form a longitudinal taper opening *a*, to receive the tang B of the blade, and also cast with projections *b c d* at each side, in combination with the side-piece D D, secured to A by rivets, and the tang also secured in the opening *a* by rivets, substantially as and for the purpose herein set forth.

No. 57,522.—JOSEPH D. LEACH, Penobscot, Maine.—*Machine for Holding the Heads of Casks.*—August 28, 1866.—Disks are swivelled in levers which are actuated by a treadle motion, and the cask-head is clamped between the disks while its edge is trimmed.

Claim.—First, the disks O and P, in combination with the pin W, constructed and operating substantially as and for the purposes specified.

Second, a heading holder, having levers B and C, spring E, ratchet I, treadle H, disks O and P, and pin W, constructed, combined, and arranged substantially as and for the purposes set forth.

No. 57,523.—J. LITTLE and S. W. LITTLE, Patoka, Ind.—*Evaporator.*—August 28, 1866.—The pan has communicating compartments furnished with inclined skimming shelves to collect the scum and with strainers. The pan and furnace are supported upon a rocker frame whose inclination is adjusted by an extensible leg.

Claim.—First, the central supporting framework E, in combination with the arch B, pan D, and adjustable leg F, constructed and operating substantially as and for the purpose described.

Second, the flaring sides of the arch, with recesses or grooves *a*, in combination with the pan D, substantially as and for the purpose set forth.

Third, the skimming shelves *g*, in the pan G, as described.

Fourth, the V-shaped shelves *i*, in the compartment H, as set forth.

Fifth, the straining shelf *k* and box *l*, in the compartment I, as and for the purpose described.

No. 27,524.—HENRY O. LOTHROP, Milford, Mass.—*Machine for Applying Screw Mouth pieces to Cans, &c.*—August 28, 1866.—The mouth-piece is held by the clamp while the can top is being secured to it; the flange on the lower edge of the mouth-piece is passed through the hole in the can and spread over on the inside of the cap plate.

Claim.—The mouth-piece supporting machine, as composed of the jawed plates and the shelf, made, arranged, and combined substantially as specified.

Also, the arrangement and combination of the lever K, with the handle I, and the spring catch H, applied to the two jawed plates as specified.

No. 57,525.—WILLIAM LOUDEN, Fairfield, Iowa.—*Hay-stacking Device.*—August 28, 1866.—The mast is braced by a stay rod with lateral supports; the rope is attached to a pivoted sweep and the speed is multiplied by a pulley arrangement which also swings the crane arm toward the stack in unloading and back again as the fork descends.

Claim.—First, the bracing of the upright A, of the crane by means of the bars or braces B C C, arranged substantially as described.

Second, the sweep M, provided with the pendent pin *e* in connection with the arm J, and fork tackle, all arranged to operate substantially as and for the purpose set forth.

Third, the arranging of the crane and tackle relatively with the arm J, and with the stack and load, in such a manner that the fork in ascending and descending will, under the pull of the tackle, swing from the load over the stack, and vice versa, substantially as described.

No. 57,526.—ARCHIBALD and JAMES LONDON, Boston, Mass.—*Water Closet Valve.*—August 28, 1866.—A screw plug regulates the size of the water passage connecting the chambers above and below the valve, which has also a water cushion formed by a concentric groove in its seat, communicating by a small channel with the passage between the two chambers.

Claim.—The combination of the passage *a*, through the spindle or post G, with the regulating screw M, provided with a recess *e*, as described.

Also, the combination of the cushioning channel *e*, and its opening *b*, with the valve B, and its seat I, when combined with the valve D and the chamber A, so as to operate therewith, as specified.

No. 57,527.—JOHN K. LOWE, Cleveland, Ohio.—*Bronzing Machine.*—August 28, 1866.—The bronze is taken from the box by a canton-flannel roller and passed to the fur roller, which deposits it upon the passing sheet of prepared paper; superfluous bronze is removed by another roller.

Claim.—First, the bronze box F, feed roller F', and handle G in combination with the bronzing fur roller B, as and for the purpose set forth.

Second, the revolving brush E, and cleaning roller D, in combination with the sheet roller C and bronzing roller B, arranged in the manner and for the purpose substantially as set forth.

Third, the belts H H' and pulleys *d e e' f f a' b b'* in combination with the sheet roller C, arranged and operating in the manner and for the purpose set forth.

No. 57,528.—JOHN and WILLIAM H. LUCAS, Philadelphia, Penn.—*Composition for Putty.*—August 18, 1866.—Explained by the claim.

Claim.—The composition of ground marble, whiting, and linseed oil to form a superior putty, substantially in the manner hereinbefore described.

No. 57,529.—SEBEUS C. MAINE, Boston, Mass.—*Screening and Sifting Apparatus.*—August 28, 1866—The horizontal shaft upon which the sieve rests is agitated by a cam at the end outside the case of the sieve.

Claim.—As an improvement in apparatus for screening and sifting for family use, the shaft B, constructed with its guide plates E, or their equivalents, for receiving and operating a sieve or screen of any size or form, substantially as described.

Also, for family use, a sieve or screen with its removable bottom in combination with a receptacle provided with a shaft operated by a cam C, or its equivalent, substantially as set forth.

No. 57,530.—W. M. MARTIN, New York, N. Y.—*Railroad Wedge Rail.*—August 28, 1866.—This wedge-shaped rail is driven in between the main rail and the upwardly projecting flange of the chair.

Claim.—The wedge-shaped auxiliary rail with a curved face and three downward-bearing surfaces, one bearing on the clamp D, one the bottom of the chair, and one, which is curved lengthwise, upon the lower flange of the rails, substantially as and for the purposes herein set forth.

No. 57,531.—SYLVESTER G. MASON, Elbridge, N. Y.—*Pump Piston.*—August 28, 1866.—The side segmental valves with interior chambers fit and reciprocate within the piston, alternately opening and closing the induction and eduction apertures above and below.

Claim.—The combination of the segmental side valves D D with the central plane-sided chamber G, said parts being arranged in connection with the enclosing rim A and cover C, as described, and the whole operating as set forth.

Also, the construction of the segmental valves with plane sides and a circular rim or edge, as described, for the facilities of manufacture, as set forth.

No. 57,532.—WILLIAM McARTHUR, Philadelphia, Penn.—*Carpet Cleaning Machine.*—August 28, 1866.—The breadth of carpet is passed over an inclined bed formed of a steam coil, and there subjected to vibrating beaters; it thence passes to a revolving brush cylinder.

Claim.—First, the combination, substantially as described, of a steam pipe G, or other equivalent heating apparatus, with a carpet-beating machine, for the purpose described.

Second, the combination of the beaters H, their spiral springs and the roller F, with its pins *n*, the whole being arranged and operating substantially as and for the purpose specified.

Third, the inclined plates *i i*, beaters H, brush J, and rollers B B' *c c' d* and *e*, the whole being constructed and arranged for joint operation, substantially as and for the purpose specified.

No. 57,533.—JAMES McBRIDE, Alleghany City, Penn —*Machinery for Cutting Oval Holes in Boiler Heads.*—August 28, 1866; antedated August 17, 1866.—The table which supports the plate to be bored has a reciprocating motion in the line of the major diameter of the oval or eliptical hole required.

Claim.—The combination and arrangement of the wheels *k* and *j*, shafts *h* and *i*, disk *g*, table *c*, pitman *e*, wrists *o*, slots *f*, and cutter X, combined, arranged, and operating substantially in the manner herein described and for the purpose set forth.

No. 57,534.—JAMES McBRIDE, Alleghany City, Penn.—*Machinery for Cutting Oval Holes in Boiler Heads.*—August 28, 1866; antedated August 17, 1866.—The shaft which carries the tool has a reciprocating movement in the line of the major diameter of the hole required

Claim.—The combination and arrangement of the wheels *l m n* and *o*, shafts *h i* and *k*, disk *p*, slot Q, wrist *q*, pitman *e*, head piece *f*, cutter *g*, constructed, arranged, combined, and operating substantially as herein described and for the purpose set forth.

No. 57,535.—JAMES McCALVEY, Philadelphia, Penn.—*Hoisting Machine.*—August 28, 1866.—Grip levers on the frame of the cage are so arranged that the breaking of the hoisting rope throws them into engagement with the ratchets on the vertical timbers of the shaft.

Claim.—First, the combination of the grip levers I, springs K, keys L, and springs M, arranged and operating in relation to each other and to the permanent ratchets E, substantially in the manner hereinbefore described and for the purpose set forth.

Second, the combination of the trip levers H with the yoke F, beam C, and grip levers I, substantially as described and for the purpose specified.

Third, the combination of the spring O with the beam C and yoke F, for giving an instantaneous drop to the latter when the rope or chain breaks, and thus throwing the grip lever I instantly into connection with the permanent ratchets E, to securely lock the cage A, substantially as described.

No. 57,536.—JAMES McCRUM, Locust Grove, Ohio.—*Drilling Attachment for Turning Lathes.*—August 28, 1866.—The thickness of the plate to be drilled being ascertained by the calipers, the slide is adjusted so that a head in its outer end forms a stop for the bar through which the sliding rod passes and limits the motion of the latter.

Claim.—The sliding rod I with the spring O and weighted cords L, either or both, applied to it in connection with an adjustable stop mechanism composed of the bar K attached to the rod I and a head *i* on the slide Q, all being arranged and applied to the puppet head of the lathe to operate in the manner substantially as and for the purpose set forth.

Also, the calipers P, applied to the tube E and slide Q in combination with the drilling attachment, substantially as and for the purpose specified.

Also, the particular arrangement of the slide Q, tube E, vice J, spring O, and weighted cords L with the puppet head C of a lathe or drilling mandrel, substantially as and for the purpose set forth.

No. 57,537.—DAVID McCURDY, Ottowa, Ohio.—*Farm Gate.*—August 28, 1866.—Explained by the claim and cut.

Claim.—As an improvement in gates the arrangement of the post A, roller *d*, and pin *e*, in combination with the extension bar *f*, latch *n*, slotted swivelled post *l*, roller *m*, bars *a*, and post A², constructed and operating in the manner herein specified and described.

No. 57,538.—JOHN S. McGLUMPHY, Wind Ridge, Penn.—*Wagon Brake.*—August 28, 1866.—The raising of the tongue in descending a hill or stopping draws upon the brake levers and applies the rubbers to the hind wheels.

Claim.—In combination with the rubbers F and levers D, the bifurcated bar C, and adjusting nuts E, the same being respectively attached to the lower end of the king bolt B and to the brake levers D, substantially as and for the purpose set forth.

No. 57,539.—WARD McLEAN, New York, N. Y.—*Seal Lock.*—August 28, 1866.—In this lock the paper seal is held down upon the plate through which the key is to pass to reach the tumbler, by a shutter, hinged at one end and when closed fitting within a raised frame upon the lock plate; the shutter has a hook projecting from its free end which is grasped by a projecting portion of the tumbler at the same instant that another portion of the same tumbler engages with the nose of the hasp.

C P 28—VOL. II

Claim.—A lock whose bolt has two catches which simultaneously engage with or disengage from the hasp and from the seal cover so that a seal cannot be removed from the lock without unlocking the same, substantially as described.

No. 57,540.—GEORGE MEIN, Williamsburg, N. Y.—*Banjo.*—August 28, 1866.—The inner rim is adjustable by screws and is forced against the head to tighten it, and conversely. The stem is attached to the head by a sliding dovetail joint.

Claim.—First, the interior rim C, in combination with the exterior rim A, and with the head B of the banjo, substantially as described and for the purpose set forth.

Second, the manner of attaching the stem I to the rim A, with a dovetailed groove and strip, substantially as herein described and for the purpose set forth.

No. 57,541.—JAMES H. MELICK, Albany, N. Y.—*Threshing Machine.*—August 28, 1866.— The apron and the independent spring slats which form the rubber yield to the passing grain and then assume their normal position at their adjusted points of distance from the cylinder.

Claim.—First, a series of bars, each fitted to yield separately and radially and forming the concave or rubber in combination with the revolving threshing cylinder, as set forth.

Second, a yielding incline d, between the feeding table and the concave or rubber, substantially as and for the purposes set forth.

Third, the arrangement of the slides k, and springs i, and adjusting screws l, to the yielding bars f of the concave or rubber, as specified.

No. 57,542.—E. MICHAELS, Palermo, Me.—*Tool for Splitting Bark upon Trees.*—August 28, 1866.—The tool has a bar and adjustable handle upon it; a hooked three-edged cutting or slitting point splits the bark into strips, which are then in condition to peel off.

Claim.—A tool for peeling off the bark of trees, consisting of the shank A, the three-edged knife B, handle C, two hand rollers a and E, substantially as shown and described.

No. 57,543.—PATRICK MIHAN, Boston, Mass.—*Apparatus for Carburetting Air.*—August 28, 1866.—The tubes are covered with fibrous material and contain spiral brushes. Their ends are supported in apertures cut in the rotary end disks, which are enclosed in a vessel provided with an inlet for atmospheric air and an outlet for the carburetted air.

Claim.—First, in apparatus for carburetting air, conducting the air to be treated through an inlet tube M, placed about the shaft of the apparatus and terminating near the remote end of the apparatus in a cone-shaped mouth M', so as to increase the distance to be traversed by the air after it is delivered within the generator, substantially as shown.

Second, in an apparatus for carburetting air, the use of open tubes supplied with bristles, or their equivalent, revolving through or in a bath of hydro-carbon and exposed, on emerging therefrom, to currents of air, substantially as shown.

Third, the cover Q, with its outlet pipe Q', its cones P and O, and tubes U, constructed and arranged substantially as shown and described.

Fourth, forming a space V, for receiving the carburetted air, in combination with the tubes C and cone O, and tubes U, substantially as shown.

Fifth, covering the tubes C with an absorbing material for the purpose of exposing to the fresh air a surface or surfaces wet with hydro-carbon, while such air is on its way to the end of said tubes, substantially as shown.

No. 57,544.—RICHARD S. MILLER, Battle Creek, Mich.—*Stove Pipe Damper.*—August 28, 1866.—A sectional damper, conical in form, is suspended in a stove pipe, a rod extending diametrically across the pipe; the sections are so hinged as to be easily opened or closed.

Claim.—First, the construction and arrangement of the two inverted half cones C C', pivots b b, crank shaft B g, and link h, in combination with the stove pipe A, as herein described and shown for the purpose set forth.

Second, the combination of the pivots b b, notches e e, and rock shaft B g, in the construction of a damper formed of two inverted half cones and arranged in a stove pipe, all as herein described and for the purpose set forth.

No. 57,545.—WILLIAM K. MILLER, Canton, Ohio.—*Burial Case*—August 28, 1866.— Wooden coffins are carbonized in a chamber placed over a furnace and charged with a hydro-carbon.

Claim.—Making and carbonizing burial cases of wood, as and for the purpose substantially as herein described.

No. 57,546.—JACOB E. MOELLER, Terre Haute, Ind.—*Water Elevator.*—August 28, 1866.—In raising the bucket the spring catch on the handle engages the outer ratchets when the catch and the pawl are retracted the windlass revolves without rotating the handle, the brake regulating the speed of descent of the bucket.

Claim.—The combination of the bucket E, strap D, pulley C, ratchet wheel R and *m*, crank K, and spring catch *n*, the whole being constructed, arranged, and operated substantially in the manner and for the purpose set forth.

No. 57,547.—ISAAC MORRIS and C. M. MORRIS, Fair Haven, Conn.—*Wrench.*—August 28, 1866.—An adjustable rack in the movable jaw engages the stationary rack on the shank; the latter is enlarged at its junction with the fixed jaw.

Claim.—The combination of the movable jaw and vibrating rack with the main rack, and enlargement of the bar, near the stationary jaw, when the whole is constructed and fitted for use, substantially as herein described and set forth.

No. 57,548.—J. J. MORRIS, New Bedford, N. J.—*Carriage*—August 28, 1866.—Each hub is constructed in one piece with its half of the axle; the sections of the latter are embraced by a sleeve; the weight of the bed rests by intervening anti-friction rollers upon the axle. The traces are attached to spring hooks on the shafts.

Claim.—First, the independent axle C, having the hub and wheel attached permanently thereto, and revolving with the wheel, substantially as set forth.

Second, the independent axles constructed as set forth, in combination with the sleeve H, when arranged to operate as shown and described.

Third, in combination with the axles C, as described, the friction rollers *a*, arranged and operating as set forth.

Fourth, the combination of the trace *o*, hook *h*; and spring *m*, when said parts are arranged in connection with the vehicle, as shown and described.

No. 57,549.—ROBERT MORRIS, Salem, Ind—*Reaping Machine.*—August 28, 1866—An endless apron delivers the grain continuously to a gavelling reel having an intermittent rotary motion. The shaft of said reel is provided with a head or end having four sides which are successively acted upon by a flat spring, which resists the rotation of the reel, and the tension or pressure of which is regulated by a set screw, thereby regulating the quantity or weight of grain in the buckets of the reel, necessary to the overcoming the resistance of the spring. An index is attached to the spring adjustment to indicate the weight of the gavel thus discharged.

Claim.—First, the square *h*, and spring H, operating with the discharger F, substantially as and for the purpose specified.

Second, the index K, and set screw *i*, in combination with the spring H, arranged with the bar G, and square *h*, substantially as described for the purpose specified.

No. 57,550.—JOHN W. MUNGER and W. O. JONES, Portland, Me.—*Dry Dock.*—August 28, 1866.—A central basin has outer communications with tide water and connects with docks around it which are of less depth and have railways on their floors. Between the docks are cisterns containing water artificially raised or conducted thereinto and used to float the vessel in the basin to a higher level than that of the flood tide and such as to allow it to pass into one of the high level docks.

Claim.—First, the combination and arrangement of the docks D 1 2 3 4 5 6 with the channel B and central basin A, as and for the purposes se: forth.

Second, the arrangement and construction with the docks, as described, of the cisterns 7 8 9 10 11 12, in the manner and for the purposes set forth.

Third, in combination with the docks so arranged and constructed, the ways and carriages, outlets and gutters, as and for the purposes set forth.

No. 57,551.—HENRY B. MYER, Philadelphia, Penn.—*Apparatus for Generating Illuminating Gas.*—August 28, 1866.—The vaporizing vessel is divided into four chambers In the lower one are the coil and the disk which distributes the air through the carburetting liquid. The third chamber is a reservoir to contain the gasoline which is supplied to the lower chamber as required, by a valve operated by a float. The air is forced by series of air pumps which are operated by weights. The rising of the gas-holder stops the wheels of the air pumps, so as to limit the production of gas to the amount needed. The second chamber is filled with lime water to wash the carburetted air received from the lower chamber.

Claim.—First, the arrangement of the chambers A B C D, in their relation to each other, for the purpose specified.

Second, in combination with the chambers A B, the pipe *b* and valve *b'*, constructed and operating substantially as and for the purpose set forth.

Third, the herein described construction and arrangement of the pumps E F, the same being adapted to the use of water as a packing, as specified.

Fourth, such a structure of the pump, as shown in Fig. 18, whereby it is adapted to the use of quicksilver as a packing, as specified.

Fifth, in combination the perforated coil C', double disk *d*, and perforated disk I, arranged to operate as specified.

Sixth, in combination with the chambers A B C, the glass tube indicators a' b''' k, as and for the object set forth.

Seventh, operating the valve b' by means of the float B', as and for the purpose herein specified.

Eighth, the arrangement of the pumps E F, train of wheels G, ratchet wheel R, lever Q, and gas holder O, as and for the purpose herein specified.

No. 57,552.—WALES NEEDHAM, Rockford, Ill.—*Harvester.*—August 28, 1866.—This arrangement of devices is for holding and regulating the tension of the reins, for the purpose of dispensing with a driver and enabling the raker on his seat or stand on the machine to control the team, by means of cranks, while raking the grain from the platform.

Claim.—First, the head H', spool K, face plate J'', in combination with the arbor m, crank P, spring i, reins a' a'', arranged as and for the purpose set forth.

Second, the arm B, rest E, lever I', tooth e', spring d, in combination with the spool and reins, arranged as and for the purpose specified.

Third, the standard A and arm B, in combination with the serrated coupling C, and rest E, as and for the purpose set forth.

No. 57,553.—JOHN NEFF, Jr., Pultney, N. Y.—*Fruit Picker.*—August 28, 1866.—A gripping jaw is attached to one of the blades and by pressure against a flat portion of the other blade holds the fruit by the stem after it is plucked.

Claim.—The blades A and B, when made substantially as specified, also the gripper C and spring D, when made and applied as specified, and used for the purpose set forth.

No. 57,554.—WILLIAM S. NELSON, St. Louis, Mo.—*Hat.*—August 28, 1866.—A light band encircles the head, upright side frames are attached to this and converge at the crown to form the hat frame. The brim is sustained by springing obliquely down from this frame.

Claim.—The arrangement of the band A, the connected side frames and rim frames with the hoops E and G, substantially as described and for the purpose herein set forth.

No. 57,555.—NELSON C. NEWELL, Springfield, Mass.—*Covered Button.*—August 28, 1866.—The cloth is clamped between the periphery of the disk and the inturned flange of the backing into which it is pressed.

Claim.—The disk C, formed in such a manner that when inserted in the button it may be expanded laterally, substantially in the manner and for the purpose herein described.

No. 57 556.—ISAAC H. NEWTON, Oakfield, Mich.—*Saw Mill.*—August 28, 1866.—A frame is operated by a crank shaft and arms that elevate two levers, by which the log is turned upon the head block of the carriage.

Claim.—The bars G, arms H, and frame I J K, combined with each other and arranged in relation to the head and tail blocks B B', substantially as herein set forth for the purpose specified.

No. 57,557.—DAVID H. NICHOLS, New Richmond, Ohio.—*Sorghum Stripper.*—August 28, 1866.—The stalk is pinched and the leaves removed by passing against the edges of the stationary and movable slides; the latter are arranged on each side of a central plate and actuated individually by springs.

Claim.—The combination of the sliding plates F, stationary plate B, provided with apertures C, and the springs D, the whole being constructed, arranged, and operating substantially as herein set forth for the purpose specified.

No. 57,558.—JOHN W. OTTO, St. Louis, Mo.—*Piano.*—August 28, 1866.—The frame, wrest-plank, plane of stringing, &c, is inclined, the rear part resting on casters on the floor; the key-board is in the usual position and the wrest-plank elevated above it. The hammers strike above the top edge of the sounding-board and the damper movement is transmitted through it.

Claim.—As an improvement in the piano-forte in general, and also in the grand piano-forte in particular—First, the particular manner in which the case of the piano (including wrest-plank, iron frame, plane of stringing, and sound-board) is inclined, contracted, elevated, and the area of the sound-board enlarged, for the purpose of attaining convenient dimensions, a more effective sound-board, and other improvements resulting from this transformation, and claimed herein separately, substantially as described.

Second, the removable end casings F F, as described and for the purpose specified.

Third, the removable back casing g, as described and for the purpose set forth.

Fourth, the key lid K, its joints n n, and their connecting rod m, arranged as described and for the purposes specified.

Fifth, aperture r, in the sound-board, serving as passages through which the damper movement is transmitted, as set forth.

Sixth, the soft-pedal arrangement by means of jointed rail J, and India-rubber slips K, or its equivalent, substantially in the manner and for the purposes set forth.

Seventh, set-offs M M, on each side of the key-board, in the manner set forth and for the purpose specified.

No. 57,559.—J. D. C. OUTWATER, New York, N. Y.—*Disinfector.*—August 28, 1866.— The glass vessel has a screw cap, with hollow extended arms bent in opposite directions and having openings in the ends. The vessel contains the volatile disinfectant, and is supported in a frame in such a manner that it will revolve freely when heat is applied beneath and the vaporized contents are ejected.

Claim.—The combination of the glass vessel C, with a depression on its bottom, and a metal cap furnished with hollow arms, the whole being adapted to revolve n adjustable bearings, in a frame, substantially as described and represented.

No. 57,560.—JOSHUA D. PATTON, Davenport, Iowa.—*Bed Bottom.*—August 28, 1866.— The cords are attached to one end of the frame, and passing through the other are secured by heads.

Claim.—A bed bottom composed of a series of cords or other lines c, secured at one end to the frame of the bed bottom and passing loosely through the other, having heads E upon their outer ends, substantially as and for the purpose described.

No. 57,561.—N. PETRE, New York, N. Y.—*Medicine.*—August 28, 1866.—Composed of laudanum 1 part, paregoric 3, and tincture of myrrh 4.

Claim.—The within described medical composition, made of the ingredients in the manner and proportion as set forth.

No. 57,562.—MICHAEL PORTER, C. E. JENKINS, and G. F. JENKINS, Terre Haute, Ill.— *Cultivator.*—August 28, 1866.—The forward end of the beam is adjustable in the swivelled post which depends from and is adjustable in the axle.

Claim.—The combination of the adjustable swivel brackets a, plough beams E E, and axle A, arranged and operating in the manner as and for the purpose herein specified.

No. 57,563.—H. S. POTTER, Fairfield, Iowa.—*Corn Cultivator.*—August 28, 1866.—Ex-plained by the claims and cut.

Claim.—First, the frame A, with the driver's seat B, placed on its rear end, and connected at its front end to the hounds D of the draught pole in front of the axle E, substantially as and for the purpose set forth.

Second, the frame J, in combination with the plough beams L L, connected by swivel joints to the rod K, above the axle E, arranged in connection with the frame A, to operate substantially as and for the purpose specified.

Third, the lever Q, attached to the front part of the frame A, and applied to or arranged in connection with the frame J, and the loop l, with roller m, substantially as and for the purpose set forth.

Fourth, the connecting of the plough standards G M to the frame A and beams L, by means of the pivoted bars b g, provided with wheels c h, substantially as and for the purpose specified.

No. 57,564.—T. T. PROSSER and JAMES LAWSON, Chicago, Ill.—*Knife and Scissors Sharpener.*—August 28, 1866.—The edge of the blade is placed downward in the notch and drawn across the edge of the teeth of the steel cutter.

Claim.—The toothed disk B, in combination with the frame A, having the notch c formed therein, when said parts are arranged to operate as and for the purpose set forth.

No. 57,565.—PAUL PRYIBIL, New York, N. Y.—*Saw Mill.*—August 28, 1866.—The band saw is strained by raising the bearing of the upper drum; it is directed and steadied by an adjustable guide whose cheeks bear with an elastic pressure against the sides of the saw.

Claim.—First, the changeable guide F, for the purpose set forth.

Second, the construction of the guide F, with shoulder e, screw f, nut h, spring j, and jaws c d, substantially as and for the purposes described.

Third, the yielding spring arms b, or their equivalent, in combination with the adjustable journal box of the shaft B, substantially as and for the purpose set forth.

No. 57,566.—SAMUEL S. QUEST, Wellsburg, West Va.—*Car Coupling.*—August 28, 1866.—The coupling hooks are raised and disconnected by the combination of levers.

Claim.—The combination of the levers G G' H J K and M, together with the inclined plane N, when properly constructed and adapted to each other, so as to secure the object proposed, substantially as hereinbefore explained.

No. 57,567.—W. F. QUIMBY, Wilmington, Del.—*Blacking.*—August 28, 1866.—Composed of four parts of finely pulverized coal dust, and one part of molasses, with enough water to form the mass into paste.

Claim.—A blacking composed of the within ingredients in the manner and about the proportion substantially as described.

No. 57,568.—GEORGE RAY, Kinderhook, N. Y.—*Potato Digger.*—August 28, 1866.—The vines are cut and removed, the potatoes are dug, screened, sorted, and deposited in baskets on the machine by rotary cutters, a lifting shovel, and endless open apron, a shaking riddle, and an elevating apron with buckets which discharge them into a second shaking riddle, which sorts them and delivers the qualities into separate baskets.

Claim.—First, the horizontal rotating cutters C, fixed cutters C', and guard C*, arranged in combination with each other and with the shovel plough D, substantially as herein set forth for the purpose specified.

Second, the suspended plate D*, combined and in relation with the vibrating screen D', and with the buckets of the endless elevating apron x x', substantially as herein set forth for the purpose specified.

Third, the endless elevating apron x x', furnished with buckets as described and arranged in relation with the vibrating screen G and hopper F', substantially as herein set forth for the purpose specified.

Fourth, the arrangement of the platform H, at the rearmost end of the machine and in relation with the hopper F' and vibrating screen G, substantially as herein set forth for the purpose specified.

Fifth, the suspension of the shovel plough D from the vertically pivoted frame B' to enable the said plough to be turned laterally when desired, substantially as herein set forth.

Sixth, the lever k, short levers or arms j, links i*, and chains i', arranged with reference to each other and with the bars i, of the plough D, and the rearwardly projecting end of the draught pole B, substantially as herein set forth for the purpose specified.

Seventh, the supplemental tilting frame F sustaining the platform H, and so arranged upon the rearmost end of the machine and combined with suitable operating mechanism, that its rearmost end will be raised simultaneously with the elevation of the forward end of the machine, substantially as herein set forth for the purpose specified.

No. 57,569.—EBENEZER RAYNALE, Birmingham, Mich.—*Propeller for Canal Boats*—August 28, 1866.—The valves of the reciprocating propeller open toward the rear, bearing against the water in their rearward effective stroke and opening as the propeller is drawn forward.

Claim.—A vessel constructed with a submerged recess A at the stern, divided through part of its length by a partition B, and having in one of its compartments A' a reciprocating propeller D, constructed with valves E, the said parts being respectively constructed and arranged for use, substantially as set forth.

No. 57,570.—JOSHUA REGESTER, Baltimore, Md.—*Valve for Water Closet.*—August 28, 1866.—The cup valve is supported in the valve box upon the half round end of a rocking bar to which a lever is applied and is closed by gravity and the pressure of the water.

Claim —First, constructing a valve G of the parts g h and i, and arranging it within a box through which water flows, substantially as described.

Second, the combination of the enlarged cupped portion g, having flattened spaces l, and the reduced cylindrical guide h, with the packing i, said parts being arranged within a chamber A, and above the outlet chamber, substantially as described.

Third, the stem D, constructed with the part d on it as described, in combination with the device h j, in the manner and for the purpose described.

No 57,571.—N. B. REYNOLDS, Auburn, N. Y.—*Machine for Grinding Tools, &c.*—August 28, 1866 —Combined with the grindstone is an instrument that receives a vibrating or tapping motion against the face of the stone to prevent the glazing of the latter. The blank under treatment is shown as the tapping instrument, its clamp being vibrated by a cam beneath, operated by a crank rod.

Claim — In combination with a revolving grindstone, a tapping instrument, which, on being vibrated, hacks the surface of the stone, and thus prevents it from glazing or polishing, constructed and operating substantially as herein described.

Also, in combination with a pair of clamping tongs that are pivoted to a plate or rest traversed by guiding ways which extend past the face of the stone, a tappet or cam m, for vibrating said tongs upon its pivots, substantially as herein described.

No. 57,572.—FRANCIS E. RUTH and JOS. DE LONG, Upper Sandusky, Ohio.—*Stovepipe Drum.*—August 28, 1866.—Inside the drum is an inverted truncated cone around the exit pipe; the air takes a circuitous course except when the damper is opened to allow a straight exit.

Claim.—First, the combination of the interior tapering or thimble-shaped tube F with the exterior drum A and interior central pipe I, substantially as herein described and for the purpose set forth.

Second, the combination of the disk-shaped damper J, rod K, and lever L with each other and with the outer cylinder A, the interior tapering tube F, and the interior pipe I, substantially as described and for the purpose set forth.

No. 57,573.—GEO. H. SANBORN, Boston, Mass.—*Churn.*—August 28, 1866.—The two sets of perforated floats are revolved in different directions above the radially corrugated plate at the bottom of the central shaft.

Claim.—The arrangement of the two sets of floats 1 and 2 and 3 and 4, in the manner and for the purposes set forth when combined with the truck C, the shaft A, and the cream box B, as shown and set forth, as also the use of the arms *k k k,* and the projections *l l l,* when arranged and used as shown.

No. 57,574.—JAMES SARGENT, Rochester, N. Y.—*Lock.*—August 28, 1866.—The revolving tumbler is furnished with a socket or opening of sufficient size, when turned to the proper position, to allow the stem of the bolt to fall therein and become unlocked.

Claim.—The rotating tumbler I, when separated and isolated in action from the permutation wheels, and so arranged that any inward pressure upon the bolt will be exerted on the bearing of said tumbler, and have no action nor effect upon the said permutation wheels, substantially as and for the purpose herein specified.

Also, in combination with the turning tumbler I, the cog bar H and the lever G, arranged and operating as herein set forth.

Also, the combination and arrangement of the combination wheels C, cam disk E, pivoted lever G, cog bar H, and turning tumbler I, the whole operating as herein specified.

No. 57,575.—HERRMAN S. SARONI, Marietta, Ohio.—*Steam Generator.*—August 28, 1866.— The steam generator has vertical tubes and a coil of pipe through which the feed water passes. The burners are arranged beneath the lower tube sheet upon a coil of pipe and the oil therein is heated by conductors which extend up from the jet to the tube sheet.

Claim.—First, the arrangement of the feed water pipe in a coil beneath the boiler in and around the burners, as and for the purpose described.

Second, the combination, substantially as described, of the boiler, the branch pipe, the feed water pipe, and the burners, for the purpose of securing a circulation in the boiler, as set forth.

Third, the arrangement of the burner with its conductors in contact with the boiler, so that the boiler forms the heater cap for the burner.

No 57,576.—HERRMAN S. SARONI, Marietta, Ohio.—*Feed Water Heater for Steam Generators*—August 28, 1866.—The burner is placed beneath the drum and within the coil of pipe; the rods which connect it with the drum conduct heat to vaporize the oil in the chamber beneath the burner. The water is forced into the lower portion of the coil and passes into and through the drum to the steam generator.

Claim—First, the generator drum A, arranged between the feed pump and boiler, as described.

Second, the arrangement of the drum and burner so that the drum serves as the heater cap for the burner, as described.

Third, the arrangement of the drum, the burner, and the feed pipe, substantially as and for the purpose described.

Fourth, the combination, substantially as described, of a series of generators or drums, each having its own burner and entrance and exit pipe, but all having the same main feed pipe and escape pipe.

No 57,577.—HERRMAN S. SARONI. Marietta, Ohio.—*Apparatus for Steaming Vegetables.*— August 28, 1866.—The water of condensation and the contents of the retainer do not percolate or fall into the boiler.

Claim.—As a new article of manufacture, a steaming apparatus, having one or more apertures, each or all of them covered with a hood, as above described, connecting with a boiler for the admission of steam, and a separate pipe for the escape of the water of condensation.

No. 57,578.—HERRMAN S. SARONI, Marietta, Ohio.—*Vapor Burner.*—August 28, 1866.— Explained by the claims and cut.

Claim.—First, a vapor burner having a series of conductors surrounding the jet, but without a heater cap, substantially as and for the purpose described.

Second, the combination of the burner and conductors with removable ring and heater cap, substantially as and for the purpose described.

Third, as a new article of manufacture, a vapor burner or heater having the upper or heat generating portion made of metal of high conducting power, while the lower or fluid conveying portion is made of metal of lower conducting power, as described, for the purpose of concentrating the heat at the generating or vaporizing point.

No. 57,579.—THOMAS SAULT, New Haven, Conn.—*Steam Engine Slide Valve.*—August 28, 1866.—Instead of bearing with full force on its seat the valve is suspended from the axles of rollers, the bearing plates of the saddle pieces being adjustable to regulate the pressure.

Claim.—First, a valve suspended from the axles of the rollers, substantially in the manner and for the purpose herein set forth.

Second, the adjustable saddle plates m m', adjusting screws n n', and bridge pieces l l', in combination with the slide valve, rollers, and axle, substantially as described for the purpose specified.

No. 57,580.—WILLIAM SCHNEBLY, Hackensack, N. J.—*Grain Meter.*—August 28, 1866.—The grain in passing from the hopper to the receptacle below, flows over a winged wheel whose revolutions are indicated by a registering device. Fixed to the apparatus is a time-keeper whose start and stop attachment is operated by a slight rise and fall of the hopper as the grain commences or ceases to flow.

Claim.—A grain meter so constructed and combined with a velocity registering apparatus that the velocity which a body of grain moving through the meter imparts to its grain wheel will be correctly registered, whereby the weight of the grain which has thus passed through the machine may be correctly ascertained.

Second, a grain hopper and trough supported upon a shaft or journals, upon which they may oscillate in combination with a grain wheel, so that a quantity of grain moving in said hopper and trough and upon said wheel will have the effect to start, operate, and stop a velocity registering apparatus, by which means the weight of different qualities of grain may be correctly determined.

No. 57,581.—CONRAD SEABAUGH, San Antonio, Texas.—*Fence*—August 28, 1866.—The ends of the rails are scarfed together and are secured by the loops of the continuous wire which is rove through and through the post.

Claim.—First, an improved fence formed by securing the rails B to the posts A, with wires C, said wires being arranged and applied substantially as described and for the purpose set forth.

Second, securing the ends of the rails B to the posts A, by a series of receivers formed of a single wire C, extending from the bottom to the top of the post, said wire being attached to the post either by passing it through the post or securing it with nails, substantially as described and for the purpose set forth.

Third, forming each receiver of a separate piece of wire, the ends of which are passed through the post and secured thereto, substantially as described and for the purpose set forth.

No. 57,582.—ORAN W. SEELY, Buffalo, N. Y.—*Churn.*—August 28, 1866.—Radial, perforated. removable dashers with intervening washers are secured upon the shaft by a wedge.

Claim.—The removable and adjustable dashers when used in combination with one or more washers, as herein substantially set forth

Also, the wedge and openings, in combination with the removable and adjustable dashers and washers, for the purpose and substantially as herein described.

No. 57,583.—JOHN F. SEIBERLING, Akron, Ohio.—*Harvester.*—August 28, 1866.—The cut-off rod is pivoted in the dividing boards, is used in connection with a dropping platform, and is so operated that its first motion is downward and away from the leaning grain.

Claim.—The arrangement of the cut-off rod D with the crank arm G and arm I, substantially in the manner described, so that the first movement of the rod in falling will be downward and away from the leaning grain, substantially as and for the purposes set forth.

No. 57,584.—G. H. SEYMOUR and W. B. BARNARD, Waterbury, Conn.—*Button-hole Cutter.*—August 28, 1866.—A set screw on one part of the shears bears against a washer on the other portion to regulate the relative lateral adjustment of the two parts.

Claim.—One or more set screws so combined with one-half or division of a pair of shears as to work through said division against the bearing surface of the other division, to determine the lateral adjustment of the two, and of the cutting blades thereof, with reference to each other, substantially in the manner as herein set forth.

Also, an adjustable bearing strip or washer upon one division of a pair of shears combined with two set screws in said division, bearing against said strip or washer, substantially in the manner and for the purpose herein described.

No. 57,585.—JOHN SHELLENBERGER, Hampshire, Ill.—*Shuttle Carrier for Sewing Machines.*—August 28, 1866.—This apparatus is to be secured on the end of the lower rock shaft, in place of a looper arm removed, in order to make the lock stitch, on chain stitch machines. The hinged locking cap serves to allow the introduction and removal of its shuttle, to aid in keeping the bobbin from flying out, also to hold both bobbin and shuttle in place.

Claim.—The shuttle carrier A, made substantially as described, with a socket near its rim for the shuttle, and a hinged gate D, which confines the shuttle and covers the bobbin, said gate being provided with suitable means for locking and unlocking the same, as above set forth.

No. 57,586.—SIMEON SHERMAN, Weston, Missouri.—*Machine for Sawing Stone.*—August 28, 1866.—By the rotary adjustment of the bed different sides are presentable to the screw, and, by tilting, it is presented obliquely. Other adjustments are controlled by side patterns, springs, and sleeves.

Claim.—First, the bed K, hung or arranged substantially as shown and described, to admit of being turned and also of being adjusted out of a horizontal plane, for the purpose specified.

Second, the spring N, and patterns M M, in combination with the frame F, and the frame G, provided with the saws H, substantially as and for the purpose set forth.

Third, the sleeve Q, applied to the shaft B, in combination with the cords or chains I, counterpoise E, and suspended frame F, containing the saw frame G, substantially as and for the purpose specified.

No. 57,587.—A. G. SMITH, Jersey City, N. J.—*Lamp.*—August 28, 1866.—The socket has a double character; one end containing the wick tubes, and the other end having a candle socket.

Claim.—The reversible burner B, constructed substantially as described, to be used in lamps or lanterns, as set forth.

No. 57,588.—PHILO SOPER, Buffalo, N. Y.—*Calipers.*—August 28, 1866.—One of the feet of the calipers is made yielding, so that when applied to an object it will be moved outwardly if the measure to which the instrument is set is exceeded by the dimension to be taken. The motion of the foot is transmitted to an index which indicates the excess on a graduated scale.

Claim.—First, placing a yielding foot on one of the legs of a calipers, substantially as above set forth.

Second, the supplementary index D, for indicating the movements of the yielding foot, substantially as described.

Third, the combination of the yielding foot A, its arm H, and the thumb set screw J, substantially as set forth.

Fourth, the combination of the graduated table C, the index D, and the yielding foot A, substantially as shown.

No. 57,589.—THOMAS E. SPARKS, Norwich, Conn.—*Lamp.*—August 28, 1866.—The cigar lighter is so constructed that while resting on a table the flame will be low, but on lifting the lamp from the table the flame is raised by means of an elevating wire.

Claim.—The combination, as well as the arrangement, of the elevator or wire i, with the lamp A, and the tube b, applied to the wick tube of such lamp, in manner and so as to operate substantially as described.

Also, the combination of one or more standards g g, or the equivalent thereof, with the lamp A, the tube B, and a mechanism applied to such tube for suspending or supporting it, whether such mechanism be composed in part of the flame guard B, or the same and the disk e, or be otherwise properly constructed so as to hold up the tube and allow the lamp and its wick to descend within the said tube for reduction of the flame, as specified.

Also, the combination of the flame guard B with the tube b, and the lamp.

Also, the combination of the disk e with the flame guard B, the tube b, and the lamp.

Also, the combination of both the elevator and the bail (or their equivalents) with the lamp and the slide tube b, provided with means of arrest in the upward motion of such tube on and applied to the wick tube so as to operate with respect to the wick thereof, substantially as specified.

Also, the combination of the flame guard B, or the same and the disk e, with the elevator and the bail, (or their equivalents,) the lamp and the slide tube b, provided with means of arresting the upward movement of such tube and applied to the wick tube, and so as to operate with respect to the wick, substantially as specified.

No. 57,590.—WILSON and HIRAM SPERRY, Tivoli, Ill.—*Hedge Trimmer.*—August 28, 1866.—The elevated horizontal cutter is reciprocated and the vertical, circular cutter is rotated, the one trimming the top and the other the side of the hedge, by power derived from the driving wheel and intermediate gearing.

Claim.—The hinged beam *d*, having the knife bar attached thereto, as arranged, whereby it may be elevated or depressed by the lever *g* and vertical notched bar *h*, substantially in the manner and for the purpose as herein set forth.

Second, the arrangement of the circular saw M, and circular bar N, in connection with the finger bars *n*, arranged substantially in the manner and for the purpose as herein set forth.

Third, the application of a circular plate N, provided with saw teeth and suitable fingers *s*, when arranged and operating substantially in the manner described.

Fourth, the combination of the circular plate M with the knife bar *c*, the one operating simultaneously with the other, substantially in the manner as described.

Fifth, the arrangement of the foot lever C, on the shaft of the driving wheel B, substantially in the manner and for the purpose as herein described.

Sixth, the arrangement of the driving wheel B, large cog wheels D, two smaller ones G G, large bevelled cog wheel H, and smaller bevelled one I, on crank shaft J, substantially in the manner and for the purpose as herein set forth.

No. 57,591.—CHARLES E. STAPLES, Worcester, Mass.—*Combined Sewing Work Holder and Scissors Sharpener.*—August 28, 1866.—The work is held by a cam and in connection with the holder is a steel guarded slot in which the scissors may be sharpened.

Claim. A combined cam-formed cloth holder for sewing purposes and a scissors sharpener all constructed substantially as described.

No. 57,592.—EDWARD J. STEPHENS and HIRAM E. GREEN, Pawtucket, R. I.—*Machine for Printing Yarn.*—August 28, 1866.—Improvement on their patent of March 14, 1865. The yarn is only exposed to pressure between the ridges of the fluted roller and the opposite portions of the smooth roller and takes up the color only at those points from the surface of the printing rollers.

Claim.—The use in machines for printing yarn of the plain roller B, and fluted roller B' in combination with each other, and with suitable color rollers, substantially as and for the purpose set forth.

No. 57,593.—W. W. ST. JOHN, St Louis, Mo.—*Grubbing Machine.*—August 28, 1866.— The pole is attached by the hounds to the axis of the rollers when being transported, but presses directly on the periphery of the roller when in action.

Claim.—The combination and arrangement of the roller A with the hounds or braces B B, the springs *b b*, and the pole C, substantially as and for the purpose set forth.

No. 57,594.—IGNATIUS STOFFEL, Washington, D. C.—*Artificial Arm*—August 28, 1866.— Improvement on his patent of January 10, 1865. It is designed for use in connection with a natural elbow. The springs which represent the tendons are attached to the end phalanges near the joints and dispense with guide rings; the thumb lever is pivoted at the second joint, the trigger maintains the fingers in position; the loops and band are for the attachment of the arm to the stump and upper arm; the fingers and thumbs are actuated by a lever at the prosthetic elbow articulation, whose vibration is effected by the movement of the stump of the fore-arm.

Claim.—First, the construction of the last or outer articulations or joints, or means of operating the outer or last phalanges, by the steel bands without the guide, substantially as shown and described.

Second, pivoting the thumb and lever connecting therewith at the second joint, in the manner and for the purpose substantially as described.

Third, the construction, arrangement, and operation of trigger *n*, in combination with levers *z z*, connecting rods *h* and *k*, substantially as described.

Fourth, the adjustable band *o*, when constructed and arranged as described

Fifth, the bent or bell-crank lever *l*, constructed and arranged to operate as and for the purpose described.

Sixth, the loops or rings *s* and *t*, when arranged as and for the purpose described.

No. 57,595.—RANSOM E. STRAIT, West Oneonta, N. Y.—*Pump.*—August 28, 1866.—A perforated thimble is applied to the lower perforated end of the pump tube; as the openings do not correspond, the water takes a sinuous course in entering and deposits its impurities.

Claim.—In combination with the suction tube A B, having apertures P P, the thimble F, with its openings O O, constructed and operating in the manner and for the purpose set forth.

No. 57,596.—E. C. STRANGE, Taunton, Mass.—*Egg Beater and Liquor Mixer.*—August 28, 1866.—Explained by the claim and cut.

Claim.—The combination of the piston C, tube or cylinder A, and one or more perforated or wire gauze plates B, substantially as herein described and for the purpose set forth.

No. 57,597.—WM. A. SUTTON, New York, N. Y.—*Skate.*—August 28, 1866.—By turning the thumb screw the spurs are drawn against the front of the heel, clamping the latter against the central spur in its rear.

Claim.—The clamping bar E, furnished with an inwardly projecting spur *m*, when pivoted at or near the centre of the heel plate C, in combination with the adjustable bars D, furnished with spurs *e*, substantially as described.

No. 57,598.—BILLY TODD, Reading, Penn.—*Lady's Fan.*—August 28, 1866.—Explained by the claim and cut.

Claim.—A fan consisting of the handle 1, body 5, and spring joint 2 3 and 4, constructed and adapted for operation substantially as described.

No. 57,599.—ROBERT VAN HORN, Northfield, Ohio.—*Expanding Cheese Hoop.*—August 28, 1866.—By turning the crank and roller the belt is wound upon or unwound from the latter and the diameter of the cheese contracted or expanded.

Claim.—The herein described devices for contracting or expanding the hoop, consisting of the strap D and roller E, or their substantial equivalents, the several parts being constructed, arranged, and operating as and for the purpose set forth.

No. 57,600.—RICHARD VOSE, New York, N. Y.—*India-rubber Car Spring.*—August 28, 1866 —Explained by the claim and cut.

Claim.—Springs for railway cars and other purposes, formed by the combination of disks, rings, or transverse strips or layers of elastic or non-elastic fibrous material, with India-rubber, or its equivalent, by cementing or vulcanizing the same in contact with such material, all substantially in the manner and for the purpose herein set forth.

No. 57,601.—DANIEL T. WARD, Cardington, Ohio.— *Washing Machine.*—August 28, 1866.—The washboard is composed of bars resting on elastic surfaces in a frame traversing beneath a roller in the suds box; a second roller is pivoted in arms to press upon the clothing on the bed.

Claim —First, the washboard F, constructed and arranged with reference to the vessel A, substantially as described, and for the purposes specified.

Second, the combination of the washboard F and the roller B, substantially as and for the purpose described.

Third, the arms S and T, roller z, spring s, when arranged in connection with the washboard F and roller B, substantially as and for the purpose specified.

No. 57,602.—OSCAR WARDEN, Cincinnati, Ohio.—*Fluid Lens.*—August 28, 1866.—The lantern is designed for producing "stage effects;" the lamp and its reflector are adjustable behind the bulb containing the colored fluid.

Claim.—The lamp E, base F, and reflector T, connected and moving together, when used in combination with the guide ways *b*, screw G, segment bar K, and plano convex fluid lens A *a*, all constructed and operating as and for the purposes specified.

No. 57,603.—THEOS. WEAVER, Harrisburg, Penn.— *Whip Socket.*—August 28, 1866; antedated August 13, 1866 —-Disks close so much of the aperture above the bosses of the handle as to prevent the withdrawal of the latter until the detents are withdrawn by a key.

Claim.—First, the interposition of a rigid disk or disks between the inner walls of a whip socket tube, so as to lock in it the butt of a whip stock, that it cannot be removed without the use of a key, or other equivalent instrument, substantially as herein described.

Second, the construction of the locking thimble, as shown in Figs. 2 and 3, with the crescent B, springs R R R, and keyhole C; also the construction of the thimble chamber, as shown in Fig 6, with the crescent H, the notches X X X X' X', and key guard O, substantially as herein set forth.

Third, the combination and arrangement of the subjects of the second claim, when the thimble is so operated by the key shown in Fig. 7 as partially to open or close the gibbous aperture of the tube, substantially in the manner and for the purpose as herein shown and explained.

No. 57,604.—M. D., E C., & A. WELLS, Morgantown, West Va —*Seed Distributor.*— August 28, 1866.—The plunger has an adjustable pocket, and carries the seed past the elastic cut-off, dropping it upon the ground within the foot piece.

Claim.—The box A, constructed as described, the shaft B, provided with the slot *a*, or its equivalent, block *b*, strip *d*, and flanges S S, the whole constructed, arranged, and operating substantially as herein set forth.

No. 57,605.—J. A. WELSH, Xenia, Ohio.—*Grain Hulling Machine.*—August 28, 1866.— The cylinder is placed within a casing, so constructed that the feed of the grain is controlled and regulated; the transfer cylinders are so arranged as to shift the grain in its passage; the cylinders are lined with a composition of rough material of sand and cement.

Claim.—The combination of the feeding tube B and the conical end or part *s* of the drum *d*, or its equivalent, the amount fed in being regulated by a gate *b*, substantially as and for the purpose herein specified.

Also, the transferring cylinders K L M, constructed and operating substantially as and for the purpose herein set forth.

Also, the lining of cemented sand, or its equivalent, secured to the grates F F, substantially as and for the purpose herein specified.

No. 57,606.—ISAIAH M. WEST, Wilmington, Ohio.—*Churn.*—August 28, 1866.—The lower perforated segments of the dasher are hinged to the staff, and assume a horizontal position on the downward stroke.

Claim.—The combination of the dasher ring A and apertures *d d*, whether conical or not, in the dasher boards B B, substantially as and for the purpose herein specified.

No. 57,607.—ROLLIN WHITE, Bridgeport, Conn.—*Ordnance and Fire-arms.*—August 28, 1866.—The cylindrical bed piece or seat of the projectile has a longitudinal movement in the chamber in the rear of the bore, and has an accelerating charge within itself, which is fired after the bed piece and projectile are moved forward by the main charge. Vents are applied in front of the accelerating chambers in the barrel in advance of their muzzles.

Claim.—First, the ring F, of larger circumference than the projectile, applied in combination with the enlarged chamber *a a*, to form a seat for the projectile, and to traverse in the enlarged chamber and carry the ball to the rear of the bore of the barrel, substantially as and for the purpose herein specified.

Second, providing the accelerating chamber constructed in the barrel of a piece of ordnance or fire-arm with vents *m* opening into the barrel in front of the muzzle of the said chamber, substantially as and for the purpose herein specified.

No. 57,608.—FRANK WICKS, Upper Sandusky, Ohio.—*Horse-power Apparatus for Elevating Hay, &c.*—August 28, 1866.—The horse is hitched to the sweep, which traverses continuously, but a shipper and brake are so arranged as to disconnect at will the power of the team from the hoisting wheel.

Claim.—In combination with the sweep and hoisting or rope wheel, a clutch, and a brake lever operating therewith, substantially in the manner and for the purpose described.

No. 57,609.—FRANK WICKS, Upper Sandusky, Ohio.—*Horse Hay Fork.*—August 28, 1866.—The fork is self-setting when thrust into the hay; the devices are explained in the claims and cut.

Claim.—The hinged self-acting brace F, for setting the fork and holding it in proper carrying position under its load, substantially as described.

Also, the combination of the hinged bail D, its shank E, and the brace F, pivoted to said shank, and carrying a roller *d*, as and for the purpose substantially as herein described.

Also, in combination with the roller *d* in the brace, and the recess *c* in the shank C, the pivoting of the trigger at a point behind and below said recess, so that when unrestrained it will swing out of the way and allow the brace and its roller to find their proper setting positions, substantially as described.

Also, in combination with a trigger that is operated to throw the brace out of its seat, by means of a cord passing over a pulley in rear of said seat, and arranged in the shank in rear thereof, the controlling devices *i i*, for restricting the motion of the trigger on each side of its centre of motion or pivot, substantially as described.

Also, in combination with the self-acting hinged brace F, a permanent notch, stop, recess, or its equivalent, for the purpose of receiving and retaining the brace, thereby holding the fork in proper position while elevating its load, substantially as described.

No. 57,610.—FRANK WICKS and F. F. FOWLER, Upper Sandusky, Ohio.—*Tackle Block for Hay Elevators.*—August 28, 1866.—Explained by the claim.

Claim.—A hoisting block, the sheave of which is covered, where the rope runs in contact with it, with lead, or a composition of lead with other soft metals, for the purpose of coating or glazing an oiled rope when used therewith, and to prevent it from cutting, chafing, and wearing, substantially as described.

No 57,611.—ALBERT S. WILKINSON, Pawtucket, R. I.—*Stoppings for Horses' Feet.*—August 28, 1866.—The "stopping" may be of cork, an air-cushion, or of rubber; in the latter case it has a heel flap.

Claim.—First, a sponge stopping A, having a V-shaped recess *a* cut in its rear edge for receiving the frog of the foot, substantially in the manner and for the purpose specified.

Second, an air-cushion stopping B, substantially in the manner and for the purpose set forth.

Third, an elastic pad stopping C of rubber, constructed either solid or hollow, and secured in place by a web D D, substantially in the manner and for the purpose set forth.

Fourth, a heel flap *d*, attached to a web covering the sole of the foot, constructed and operating substantially in the manner and for the purpose described.

No. 57,612.—GEORGE L. WILLCOX, Hebron, Conn.—*Sawing Machine.*—August 28, 1866.—The frame is set over the log, and dogged thereto. The saw is reciprocated by the crank, and passed toward the log by the spring, which is adjusted by the dogs and ratchet.

Claim.—The arrangement of the double dog piece M, ratchet N, handle O, and springs L and P, with the guide piece J, when arranged for joint operation, substantially as described, and for the purposes specified.

No. 57,613.—JOSEPH WILSON, Manchester, N. H.—*Steam Trap.*—August 28, 1866.—Thermal action on the rod actuates the valve, and the latter by traversing in its seat, avoids straining of the parts.

Claim.—The arrangement herein set forth, whereby the valve may enter and traverse in its seat, sleeve, or cylinder, and thereby prevent the straining, bending, or otherwise deranging some part of the apparatus.

Also, in combination with a valve arranged to enter and traverse in its seat or sleeve, the rod which operates the valve, substantially as described.

No. 57,614.—CALEB WINEGAR, Union Springs, N. Y.—*Excavating Machine.*—August 28, 1866.—The shovel is hinged to the handle, and is operated and held in either a straight or inclined position, by means of a notched bar on the back of the handle.

Claim.—The hinged shovel A, operated by means of the handle C, notched bar B, and spring D, not confined to the precise arrangement, but by equivalent means, substantially the same, to accomplish the same object.

No. 57,615.—SAM'L YATES, Clarence, Iowa.—*Machine for Manufacturing Eave Troughs.*—August 28, 1866.—A bar of wood, half oval in its cross section, is suspended by its ends on pivots; bands are hinged to the flat side of the bar, by which the metal is held in position, the bands being locked in position by spiral springs.

Claim.—The entire combination of the half oval bar A, the bands B, the spiral springs C, and each and every part of the machine, which combination greatly facilitates the process of making tin spouting for houses, lessens the labor, and secures more perfect spouting, when finished.

No. 57,616.—GEORGE T. ALLAMBY, Bangor, Maine.—*Door Bolt.*—August 18, 1866.—The spring bears against the short socketed arm on the bolt, the shank of the knob engages the socket; the bolt is locked by turning the knob shank into a curve of the slot in the casing.

Claim.—The combination of the socketed arm *f*, or its equivalent, with the bolt A, applied to the case as explained, the separate knob C, its shank *d*, and the curved slot *b*, made in the bolt case, the whole being arranged substantially in the manner and so as to operate as described.

Also, the combination of the covering plate *e*, the slot *b*, the shank *d*, the knob C, and the socketed arm *f*, as applied to the bolt and its case, the whole being arranged substantially in the manner and to operate as explained.

No. 57,617.—EDWIN ALLEN, Norwich, Conn.—*Machine for Making Envelopes.*—August 28, 1866.—Two sets of folders are arranged upon a horizontal table, by whose rotation they are brought consecutively opposite to a "blank" feeder and caused to fold the blanks as the rotation proceeds. The mechanism for folding and delivering cannot be briefly described other than substantially in the words of the claim.

Claim.—First, arranging the folding apparatus of an envelope machine to revolve about an upright axis, substantially as herein specified.

Second, the combination in an envelope machine of one or more sets of revolving folding apparatus and stationary feed board or feed table, substantially as and for the purpose herein described.

Third, the combination in an envelope machine of one or more revolving plungers, a fixed feed table or feed board, and a die plate which moves a certain distance with the said plunger or plungers, and afterwards returns to a fixed position relatively to the said feed board or feed table, substantially as and for the purpose herein described.

Fourth, the sliding pins *n* and *n'*, operating conjointly to deliver the envelopes from the folding apparatus, substantially as herein set forth.

Fifth, the nipping jaws *u' u'*, so arranged as to swing to and from the table A to remove the envelopes from the folding apparatus, substantially as herein set forth.

Sixth, the curved and inclined stationary cams *j', s u v* arranged in relation to the wings *c' d' e' f* of the folding apparatus, substantially as herein set forth for the purpose specified.

Seventh, the brake *f²* applied and operating in combination with the laterally moving die plate H, substantially as herein set forth for the purpose specified.

Eighth, the arrangement of the pivoted bars b^3 and their pins c^3 with reference to the downwardly projecting pin a^3 and to the stationary cam e^3, to produce the lateral movement of the die plate H, substantially as herein set forth.

Ninth, the nipping jaws u' w' swinging on a vertical axis, and arranged with reference to the rotating cam L, bar e'', friction roller f, and arm N, substantially as herein set forth for the purpose specified.

Tenth, the spring y'', curved arm z', and spring catch b'', arranged with reference to each other and to the nipping jaws u' w', projection d', and stationary cam K, substantially as herein set forth for the purpose specified.

No. 57,618 —C. W. BALDWIN, Boston, Mass., assignor to himself, W. D. RICHARDS and W. E. RUSSELL, same place.—*Car Coupling.*—August 28, 1866 —The pivoted drawhead has a counterpoise weight: its coupling block is rotated by the contact of the link and is retained by a gravitating catch bar.

Claim.—The employment of the balanced lever B, when made and applied substantially in the manner and to operate as before described.

Also, in combination with the said lever B, the revolving block C, and locking bar D, substantially as set forth and explained.

Also, the peculiar construction of the block C, as made with the recess d, shoulders d^1 d^2, and notch g', all as described.

No. 57,619 —JACOB BERGEN, Plain Township, Ohio, assignor to himself and PETER KAUFMAN.—*Cultivator.*—August 28, 1866.—By the rotation of the rock shaft the arms of the bent axles are vibrated and the frame adjusted vertically on the wheels. The V-standards of the rear shovel are attached to the rear cross-bars.

Claim.—First, the peculiar arrangement of the bent axle H G F G H, in connection with the frame A and the wheels G, substantially in the manner and for the purpose specified.

Second, the peculiar arrangement of the back braces E and frame A, in the manner and for the purpose specified.

No. 57,620.—JOHN F. BOGARDUS, Brooklyn, N. Y., assignor to himself, JOSEPH ANDERSON, and T. K. SCHERMERHORN, same place —*Apparatus for Turning Shafts* —August 28, 1866 —The stationary ring is supported on pedestals and fits closely to the outer surface of a ring within it, each being formed in two parts held together by bolts passing through projecting flanges. The inner ring has flanges projecting inward from its rim upon which the cutters are arranged, and a continuous row of short cogs on the outer surface of the ring gear, with a worm wheel working in a mortise made through the outer ring.

Claim —First, the rings c and d, divided and otherwise constructed and fitted together as described, in combination with worm wheel f, or other equivalent means for imparting a movement of rotation to said ring d.

Second, the arrangement of the rests k and l and tools i and o with a rotating ring, substantially as described.

No. 57,621.—AMOS CALL, Springfield, Mass., assignor to THE BEMIS & CALL HARDWARE AND TOOL COMPANY, same place.—*Pipe Wrench.*—August 28, 1866.—The spring keeps the front side of the nut against the screw shank and holds the movable jaw to its work.

Claim.—The combination of the spring F with the parts H and E of the wrench, when arranged and operating substantially in the manner and for the purpose herein described.

No. 57,622.—CHARLES A. CONVERSE and SAMUEL S. HOPKINS, Norwich, Conn., assignors to THE BACON MANUFACTURING COMPANY, same place.—*Revolving Fire-arm.*—August 28, 1866.—The chambered cylinder revolves on an axial pin and is held in position longitudinally by a ring which bears against a shoulder on the cylinder and is fastened by a set screw in a bracket below.

Claim.—First, the ring D encircling the pepper box A in front of its shoulder g, in combination therewith and with screw E and bracket G, arranged and operating in the manner and for the purpose herein represented and described.

Second, in combination with the base pin F about which the cylinder revolves, the ring D, the many chambered cylinder A, and the bracket C, substantially as above shown.

No. 57,623.—J. W. CONWAY, Madison, Ind., assignor to himself and WILLIAM CONWAY, same place.—*Propelling Street Cars.*—August 28, 1866.—Explained by the claim and cut.

Claim —The endless screw I G, in combination with the driving or crank shaft N and the axle D of one pair of the wheels of a carriage or car, substantially as and for the purpose specified.

No. 57,624.—HENRY H. COVERT, Detroit, Mich., assignor to himself and SMITH R. WOOLLEY, same place.—*Starting Cars.*—August 28, 1866 —A pinch bar or lever is applied beneath a railroad car so as to act immediately upon the wheel, when the traction power is applied to the draught bar.

Claim.—The pinch bar or lever D, one or more, applied to a railroad car, arranged in connection with a toggle E and bar I, or their equivalents, for the purpose of acting upon the wheel or wheels for the purpose of starting a car under the action of the team or other motor, substantially as described.

Also, the draw bar K, having a spring M upon it and provided with the projecting rod N passing through a slot in the box or case L, in connection with the notch e in bar I, substantially as and for the purpose set forth.

No. 57.625.—GEORGE W. CROWE, Cincinnati, Ohio, assignor to himself and BENJAMIN MARTIN, same place.—Wagon Brake.—August 28, 1866.—The brake lever acts simultaneously and equally upon the two rubber bars beneath the bed.

Claim.—An improved wagon brake, formed by combining the lever brake bars D and F, the connecting lever I, connecting rod or bar J, and the operating lever K with each other, the said parts being constructed and arranged substantially as herein described and for the purpose set forth.

No. 57,626.—JOSEPH DARBY, Cortlandville, N. Y., assignor to himself and STEPHEN BREWER, same place.—Roofing Cement.—August 28, 1866.—Composed of ground plaster or gypsum, 1 part; sand, 2; water lime, 2; added to hot coal tar, 2.

Claim.—An elastic cement or composition of matter composed of the aforesaid herein named ingredients, as and for the purposes herein set forth and described.

No. 57,627.—J. E. EMERSON, Trenton, N. J , assignor to AMERICAN SAW COMPANY, New York, N. Y.—Saw.—August 28, 1866.—Explained by the claim and cut.

Claim.—The teeth B, fitted at the back by tongue and groove joints f g to the shoulders a of the saw plate, in combination with rivets, screws, or keys C applied to the heels c of the teeth, when the parts are constructed and arranged substantially as herein represented, so that in inserting the teeth the shoulders a must be first placed in position and the heels afterward introduced laterally to the saw plate and there secured by the rivets C, as explained.

No. 57.628 —HENRY J. FERGUSON, New York, N. Y., assignor to SELAH REEVE, Brooklyn. N. Y.—Brick Machine.—August 28, 1866.—The clay is driven by curved arms from the pug mill to the press box, in which is a vertical plunger that forces it through a grating into moulds beneath, which are fed to their place by a pusher. The plunger and pusher are severally connected to a disk on the working shaft by which they are actuated The claims detail specific adjustments.

Claim.—First, the combination of the rotating disk J. having a centre cam h and outer eccentric or cam-shaped ring i, with the pitman I to the plunger H, and slide K to the pitman L, for operating the plunger and pusher, substantially as described.

Second, the combination of the disk J with its radial slide f, pitman I, and plunger H, adjustably connected as described through a slot in the standard c by keys d fitting in key ways or grooves e, as herein set forth.

Third, the slotted levers R S, pivoted together by joint pins p, fast and loose collars q r, bosses s, and lock nuts t, substantially as specified.

Fourth, the combination of the pitman L with the arm P on the rock shaft Q. geared together by a joint pin u arranged to fit in a notch at the back of the pitman and held in gear therewith by a catch w and spring z, essentially as and for the purpose or purposes herein set forth.

No. 57,629.—HENRY FOULKES, Utica, N. Y., assignor to JOHN H. CHAPMAN.—Breast Strap Hook —August 28, 1866.—The throat is partially closed by a screw which is swivelled on the shank.

Claim —The securing or closing the opening of the hook by the use of a screw in the manner substantially as described.

No. 57,630.—JAMES E. GILLESPIE, Boston, Mass., assignor to GILLESPIE GOVERNOR COMPANY, same place.—Hydraulic Governor.—August 28, 1866.—A centrifugal pump, operated by the machinery to be governed, forces a liquid into a vertical cylinder in which a plunger is loosely fitted and sustained against gravity by a certain velocity of the fluid forced upward. The fluctuations in the velocity of the liquid are indicated by the plunger, whose motions are transmitted to the gate or valve supplying the power.

Claim.— The combination of a governing device with the cylinder d, its piston f, and the spring n, or the equivalents of these, when arranged to operate together, substantially as specified.

Also, in combination with the foregoing, a passage capable of variation, when so arranged as that by such variation the rate of speed of the movement of the piston, under influence of its spring, may be changed.

Also, combining a governor and the object governed thereby in such a manner that, while the connection between the governor and the governed object can be broken to prevent movement of the latter in one direction, the connection causing movement of the governed object in the reverse direction remains operative, substantially as shown and described.

No. 57,631.—WM. J. GOULDING, Providence, R. I., assignor to the GOODENOUGH HORSE-SHOE COMPANY, New York, N. Y.—*Machine for Bending Horseshoes.*—August 28, 1866.—The blanks are creased, cut to a length, and placed seriatim in front of the forming block on the table, whose motion draws a blank between the rollers and bends it around the former. The shoe is raised and removed by a stationary hook.

Claim.—First, the revolving flat bed or table B, provided with suitable forming blocks G and guides b b, in combination with suitable stationary bending instruments F F', arranged substantially as described, for the purposes specified.

Second, the combination of the stationary finger H with the forming block G and the said revolving bed B, arranged substantially as described, for the purposes specified.

No 57,632.—HENRY C. GRIGGS, Waterbury, Conn , assignor to SMITH & GRIGGS, same place.—*Buckle.*—August 28, 1866.—One of the inwardly-projecting curved portions of the frame is reflected outward to confine the loop at the side of the duplex tongue, by which the waist of the frame is embraced.

Claim.—The combination of the frame part with the tongue part when they are constructed, connected, and fitted to operate substantially as herein described and set forth.

No. 57,633.—HENRY HENSEL, Carver, Minn., assignor to himself and LOUIS SÜLTER, same place.—*Life-boat.*—August 28, 1866.—The cylindrical self-poising boat is pivoted in a surrounding flotative frame which has a propeller worked by a crank in the cylindrical chamber.

Claim.—The air chambers G and H, surrounding the self-adjusting cylinder boat A, operating independently on the centre shaft J, arranged and combined with end propellers K, for the purpose of steering and turning the boat quickly, at the same time preventing the boat from sinking, giving the greatest safety to the passengers as a buoyant life-boat.

No. 57,634.—GEORGE J. HILL, Buffalo, N. Y., and STEPHEN GREENE, Philadelphia, Penn., assignors to STEPHEN GREENE and H.G. LEISENNING, same place.—*Printing Press.*—August 28, 1866 —Adapted for card printing in colors. The cards are pushed by a feed slide from one box and received into another which descends as the cards are fed from the former; the cards are printed in transitu by the descent of the cross-head which carries the forms.

Claim.—First, the vibrating cam levers, constructed and arranged for the operating of a guided cross-head D, substantially as herein set forth.

Second, the guides M M', composed of plates constructed and arranged for vertical self-adjustability, to suit cards differing in thickness and for the retention and guidance of the cards, substantially as set forth.

Third, the combination of the said guides M M' with the ticket or card box N.

Fourth, the movable receiving box N', where the vertical position of the same is controlled by the quantity of cards in the box N, through the medium of the devices herein described or any equivalent to the same for the purpose specified.

No. 57,635.—EDWARD B. JUCKETT, Roxbury, Mass., assignor to himself and HUNNE-MAN & CO.—*Pump.*—August 28, 1866.—An improvement upon the inventor's patent of January 3, 1865. The cylinder is open at both ends and placed in a segmental case with three vertical partitions and two radial ones; it has two valve chambers, two induction and one eduction chamber.

Claim.—Improved double-acting force pump, or peculiar arrangement of the valve chambers D D G, the induction chambers or passages E F, and the pump barrel A, as described, such valve chambers to be provided with valves and valve openings, and such barrel to have a piston to operate as specified.

No 57,636.—NELSON KING, Bridgeport, Conn., assignor to O. F. WINCHESTER.—*Magazine Fire-arm.*—August 28, 1866.—Improvement on Smith & Wesson's patent February 14, 1854; improved by Henry, October 16, 1860. The metallic cartridges are carried from the magazine tube to the barrel; the shell is retracted by the movement of the trigger-guard, which is connected to the breech-pin, whose sleeve-hook positively engages the flange of the cartridge.

Claim.—The arrangement of the retracting hook a^z with its sleeve d^z upon the breech-pin L, so as to retract the cartridge or shell, substantially in the manner herein described.

No. 57,637.—CHARLES KINKEL, New York, N. Y., assignor to CHARLES WEHLE, Hoboken, N. J.—*Steam Generator.*—August 28, 1866.—Steam from the boiler is admitted to a vertical cylinder above, and actuates a piston whose rod passes into the boiler. A disk on the lower end of the piston, by its vertical motion, agitates the water.

Claim.—The new method herein described for preventing explosions of steam boilers by a combination of a steam boiler with the mechanism, substantially as described, which keeps the water in the boiler in constant and regular motion.

No. 57,638.—MONROE MORSE, Franklin, Mass., assignor to himself and AARON H. MORSE, same place.—*Hat Pressing Machine.*—August 28, 1866.—The hat is pressed into a concavity in a steam cylinder by a solid elastic presser actuated by toggles, &c., readily understood by reference to the claims and cut.

Claim.—The combination as well as the arrangement of a solid elastic presser with the mould, its head, and mechanism for expanding such presser by pressure against it, substantially as specified.

Also, the application of the head C to the mould by means of centres, as described, so that such head, while being withdrawn from the mould, shall be brought into an inclined position in order to facilitate the application of a hat to the presser.

Also, the combination of the bar G, the slides I I, and their pins with the toggles or progressive levers *o p q q r* and the foot lever *s*.

Also, the combination and arrangement of bolts *w w*, catches *v v*, and levers *y y*, applied to the mould and the vibratory head, as described.

Also, the combination of the adjusting screw *g* and its nuts *i i* with the bar G, the vibratory head C, and disk E, and presser D.

No. 57,639.—JAMES F. ROWLEY, WILLIAM M. SLOANE, and JAMES E. WOODRUFF, Buffalo, N. Y., assignors to themselves and JOHN D. CROSS.—*Apparatus for Carburetting Air.*—August 28, 1866.—The vessel is divided into several compartments and has a cover which fits in an annular water chamber, forming a tight joint. Into the lower chamber passes a pipe from the blowing apparatus. The second chamber is charged with slaked lime and contains an air distributor communicating with the chamber below. The chambers above are filled with fibrous material. The carburetted air escapes through a pipe above. Heaters are placed at any point in the air-supply and gas-distributing pipes.

Claim.—An apparatus for carburetting air or gas having a reservoir B for hydro-carbon, a lime chamber D, and chambers E and F containing porous substance, all arranged and constructed in the manner and for the purposes substantially as herein set forth.

No. 57,640.—ELNATHAN SAMPSON, Lansingburgh, N. Y., assignor to ALFRED CLARK HITCHCOCK, Troy, N. Y.—*Platform Scale.*—August 28, 1866.—Explained by the claims and cut.

Claim.—First, the location, under the platform of a scale, of a series of bell-crank levers E with yokes F and suspension links *c c*, which connect said yokes with co-operating and connecting parts for supporting the platform of the scale, substantially as herein described and for the purpose set forth.

Second, the arrangement of a series of bell-crank levers with their long arms in upright position below, and placed transversely to the length of the platform or to the trackway over the platform of the scale, substantially in manner as herein described and for the purpose set forth.

Third, when arranged in combination with the lever E, with the yoke F, fulcrum standard D, and platform timbers B B, the supporting plate or frame J, or its equivalent device, constructed and arranged substantially in the manner and for the purpose as herein set forth.

Fourth, the arrangement relatively to and with each other, and with the platform timbers, of the respective bell-crank levers with yokes, supporting plates or frames, and fulcrum standards, as herein described, when the same are so connected with each other by connecting parts as to operate together for the purpose of a platform scale, substantially in manner as herein described.

No. 57,641.—ELNATHAN SAMPSON, Lansingburg, N. Y., assignor to ALFRED CLARK HITCHCOCK, Troy, N. Y.—*Weigh-lock Scale.*—August 28, 1866.—The combination of direct sustaining levers and bell-crank levers supports a cradle for receiving the vessels to be weighed, and transmits the weight through other levers to the scale beam. The sustaining levers are arranged in horizontal parallel positions to the side walls of the lock chamber.

Claim.—First, the combination of the bell-crank lever F and the direct sustaining levers C and C', when said sustaining levers are arranged in horizontal parallel positions to the side walls of the lock chamber, substantially as described, and for the purposes as set forth.

Second, arranged in connection with the bell-crank levers F F in manner as described, the combination of the bell-crank levers *c c'* and transmitting lever *i*, substantially as and for the purpose herein set forth.

Third, connecting the devices arranged on opposite sides of the lock chamber, and constituting a weighing apparatus for weigh-lock scales, by arranging the mechanical means employed for that purpose under the floor of the lock chamber, in manner substantially as herein described, and for the objects as set forth.

Fourth, arranged and connected with the bell-crank levers F F in manner as described, the combination of the bell-crank levers *t t'* with the bell-crank connecting lever *s*, for the purpose of transmitting the weight of the vessel resting upon the cradle to the scale beam, in the manner substantially as herein described.

Fifth, in combination with the bell-crank levers *c c'*, and transmitting lever *i*, the manner herein described of connecting the weighing apparatus of the opposite sides of the lock chamber and the bell-crank levers *t t'* and *s*, all arranged with reference to each other, sub-

·tantially as herein set forth, the arrangement of the said direct sustaining levers for the purpose of a weigh-lock scale, in manner substantially as herein described.

No. 57,642.—JOHN SLOAN, Philadelphia, Penn., assignor to himself and JOHN W. JONES, same place.—*Ventilated Larder.*—August 28, 1866.—A water-tight box is sunk in the ground floor of a cellar, is suspended from pulleys and counterpoise weights, has inlet of air through a charcoal chamber, and exit by ventilator tube.

Claim.—The combination of box A A, air pipes K K, and discharge pipe K' K', charcoal box A A', with larder F F, in the manner as herein described, or any other, and substantially the same, so as to obtain the desired and intended effect.

No. 57,643.—SCOTT A. SMITH, Philadelphia, Penn., assignor to himself, G. V. CRESSON, and GEORGE W. HUBBARD, same place.—*Steam Generating Apparatus.*—August 28, 1866.—Annular water chambers are so attached as to form a cylindrical combination, whose interior constitutes the furnace, ash pit, and flues; the sections are attached at the bottom to a water supply pipe, and discharge at the top by individual connections into a common steam pipe.

Claim.—The method, substantially as described, of constructing a steam generating apparatus, by combining a series of annular generating vessels with a bottom connecting reservoir.

No. 57,644.—JOHN BLAKE TARR, Chicago, Ill., assignor to himself and P. F. MERRIHEW, same place.—*Machine for Making Cast-steel Castings.*—August 28, 1866.—The metal moulds are so arranged that the top part serves as the follower of a press, and is operated upon by screws. The top fits closely into the matrix, and is provided with ingates for the metal, which are closed by slides when the mould is full. The pressure is applied to the metal while in a melted state, by means of the screws, with sufficient force to expel the air and gas from and solidify the metal.

Claim.—First, subjecting the cast-steel, while in a liquid state and within the mould which is to give it the desired form, to such a degree of pressure as will cause the expulsion of air and gas from it and render it more solid, by means substantially as described.

Second, constructing the bottom of the mould, at those points which are directly under the ingates, of adjustable blocks of plumbago, or other analogous refractory substance, substantially in the manner and for the purpose described.

Third, the manner, substantially as herein described, of constructing the ingates with cutters on them, when said ingates are applied and operated as described, for the purpose set forth.

Fourth, the use of adjusting screws, in conjunction with mould sections A B C and the hooks *g h*, or their equivalents, substantially as described.

No. 57,645.—T. W. TERRY, Baltimore, Md., assignor to himself and GEORGE W. HOFF, same place.—*Safety Watch Pocket.*—August 28, 1866.—The pendant of the watch is behind the staple, which prevents the withdrawal by pulling on the chain; but when the staple is turned the watch itself may be lifted from the pocket.

Claim.—The combination of the plates A B, hinged staple *c*, and button *e*, constructed and operating substantially as and for the purpose described.

No. 57,646.—GEORGE L. WHITE, Cumberland, R. I., assignor to W. W. DUTCHER & Co., Milford, Mass.—*Temple for Looms.*—August 28, 1866.—The claim and drawing explain the invention, which is shown applied to the toothed-roller temples known as Dutcher's. The object is to insure the correct position of the temple at the termination of each of its forward movements caused by the pull of the cloth, and so render it less liable to become worn and to work unsteadily.

Claim.—The combination of the tooth *g* and the notch *f*, or their mechanical equivalent or equivalents, with the temple, its arm D, spring F, and frame E, the said tooth and notch being for the purpose hereinbefore explained.

No. 57,647.—J. W. WILDER, New York, N. Y, assignor to himself and EBENEZER BUTTERICK, same place.—*Operating India-rubber Springs.*—August 28, 1866.—The spring is confined in a protecting case to preserve it from doubling up when pressure is applied.

Claim.—The protecting case B, and plunger C, in combination with a piece of India-rubber A, substantially as and for the purpose described.

No. 57,648.—GEORGE BOWER, Ashton-under-Lyne, Eng., and JOHN QUALTER, Barnesley, Eng.—*Piston for Steam Engine.*—August 28, 1866.—The packing rings are expanded against the inner surface of the cylinder by leaf springs attached to wedge pieces, which are driven outwardly by wedges passed through one disk of the piston head.

Claim.—In combination with springs composed of a series of plates, the wedges *e*, and wedge pieces *d'*, combined and arranged to operate in the manner and for the purpose substantially as herein described.

No. 57,649.—C. M. E. DU MOTAY and C. R. MARÉCHAL, Metz, France.—*Bleaching Animal and Vegetable Fibres.*—August 28, 1866.—The fibre is treated with water holding in solution permanganic acid, or a soluble permanganate, and then placed in a bath composed of water, binoxide of hydrogen, and hydrochloric acid, where it remains a sufficient length of time, and is then removed and washed.

Claim.—First, the method of bleaching vegetable and animal fibres and tissues by the employment of manganates, permanganates, and permanganic acid, substantially as herein described.

Second, the method of producing the said manganates and permanganates, substantially as herein shown and described.

No. 57,650.—C. F. HALL, Toronto, Canada West.—*Lumber Wagon.*—August 28, 1866.—The bind and forward bolsters have rollers, and that of the former is rotated by a spoke lever or locked to a plate on the bolster by a rod which passes through holes in each.

Claim.—The rollers A, with perforated projecting ends, the perforated plate E, in combination with the boxes B of the running gear of wagons, provided at its forward end with rollers G and bearings H, all constructed and operating substantially as described for the purpose specified.

No. 57,651.—R. GOTTGETREU, Munich, Bavaria, assignor to —— LENNIG and —— CLEMM, Philadelphia, Penn.—*Producing Etching Grounds.*—August 28, 1866.—A delicate and intricate ground-work for the face of a note is obtained by coating the plate with a peculiar etching ground. To the plate is applied a solution of a salt which covers it with a net-work of fine crystals; these are washed with absolute alcohol to remove water and are then by a delicate process covered with a resin preserving the distinctiveness of the crystals. The plate is then dipped in an acidulous solution, is dried and warmed, and is ready for etching.

Claim.—The entire process, as above, preparatory to etching the design or designs on plates or surfaces of any description, said designs being procured by crystallization of solutions.

Also, the process of using crystallizing solutions for the production of images on surfaces that can be etched, in the usual method of engravers.

No. 57,652.—JOHN FOWLER, Jr., Cornhill, England, WILLIAM WORBY, Ipswich, England, and DAVID GREIG, New Cross, England, assignors to WILLIAM P. TATHAM, Philadelphia, Penn.—*Steam Plough.*—August 28, 1866.—Explained by the claim and cut.

Claim.—In machinery for actuating agricultural implements by steam power, combining the two drums, which alternately wind up and let off the rope by which the agricultural implement is drawn, with the driving shaft of the steam engine, or equivalent motor, by means of the cogged or toothed wheels on the drums, the two sets of pinions on the driving shaft, and the clutches and friction straps, or the equivalents thereof, substantially as and for the purpose specified.

No. 57,653.—ROBERT FOWLER, London, England, and ROBERT WILLIAM EDDISON, Leeds, England, (executors of JOHN FOWLER, Jr.,) DAVID GREIG and RICHARD NODDINGS Leeds, England, assignors to WILLIAM P. TATHAM, Philadelphia, Penn.—*Steam Plough.*—August 28, 1866.—Relates to machinery for hauling agricultural implements across fields by means of a rope upon a drum driven by steam power. A vibratory motion is given to the end of the guiding lever, by differential wheels operated by the rotation of the drum, so as to lay the rope regularly on the latter.

Claim.—In guiding the laying of the rope on to the periphery of the drum, in machinery for drawing agricultural implements by steam power, combining the guiding lever for guiding the rope with the flanched drum for drawing and winding the rope by means of the cam and differential wheels, substantially as described and for the purpose specified.

Also, connecting the guiding lever with the winding drum, so that in addition to having an up and down motion to lay the rope properly on the face of the drum, its guiding end shall be free to revolve around the drum, and thus adapt itself to the angle at which the rope may be hauling, substantially as described.

No. 57,654.—FREDERICK OAKLEY, London, England, assignor to JOHN COLLINS, New York, N. Y.—*Burglar Alarm and Door Fastener.*—August 28, 1866.—The holder is fastened to the floor and projects beneath the door, whose opening touches the trigger, releases the hammer and explodes the cap.

Claim.—First, the revolving eccentric block k, with pins l attached to it, in the manner and for the purpose herein shown and described.

Second, the combination of the barrel or powder chamber f, with the set screw g, nipple e, and bars A and B, substantially as herein shown and described.

No. 57,655.—ASA R. REYNOLDS, Auburn, N. Y.—*Tempering Knife Sections for Harvesters, &c.*—August 28, 1866.—The blade is placed upon an anvil and a weight or hammer dropped upon it under the following conditions: the anvil is slightly yielding so as to produce-a reaction of the hammer; the hammer has an ascertained proper weight and fall and expends the force of its blow most largely at the edge of the blade and gradually diminishing toward the poll.

Claim.—Hard tempering the edges of steel blades or sections by means of a blow from a hammer or drop press, delivered under conditions and with a reaction substantially such as herein described.

Also, supporting the blade, section, or blank, to be hard tempered by a drop die or hammer, upon or by its edges, upon an under die, having inclined or sloping faces, substantially as herein described.

Also, means for forcing or directing a blast of air upon that portion of the blade which is not to be highly tempered, when used in connection with a hammer or drop press conditioned for giving a blow and for reacting, substantially as herein described.

No. 57,656.—ALBERT ALDEN, New York, N. Y.—*Handle for Brushes.*—September 4, 1866.—The upper section is hinged and is maintained by the sleeve in a straight or oblique adjustment.

Claim.—The sleeve B, which slides on the upper part of a handle, in combination with one or more notches *c* in the top edge of the lower hinged part A of said handle, constructed and operating substantially as and for the purpose described.

No. 57,657.—SAMUEL F. ALLEN, Chicago, Ill.—*Axle Box.*—September 4, 1866.—The oil cup beneath the axle is supported by hinged plates which enter grooves in the axle box and rest at their contacting sides upon the set nut.

Claim.—First, securing oil cellars in place in their boxes by means of removable bottoms, which are constructed and applied substantially as described.

Second, the combination of the fixed pin *h* and the jam nut *g* with the hinged plates C C, and a suitable cushion, which is interposed between said plates and the bottom of the oil cellar, substantially as described.

No. 57,658.—JOHN BAILOR, Cannon City, Minn.—*Car Coupling.*—September 4, 1866.— The draw head is so attached beneath the draw bar as to be uncoupled by a lateral strain when the car to which it is coupled leaves the track. The lateral pressure revolves the draw bar, pries out the roller from its seat, and detaches the pin which connects the draw head to the draw bar.

Claim.—The application to railway car couplings of wheels, springs, swivel links, or shackles, in combination and as herein described and for the purposes specified.

No. 57,659.—SETH W. BAKER, Providence, R. I.—*Woven Fabric.*—September 4, 1866.— This fabric is designed to be used for counterpanes, blankets, carriage robes, &c., and is made of many plies interwoven as per Gujer's patent of May 18, 1858, the whole being of cotton or linen, excepting only the weft of the out or surface ply, which is of wool.

Claim.—The fabric above described, produced by combining, by means of the mode of weaving described, the body of the fabric formed wholly of cotton or linen, with a face or surface, the weft of which is woollen and which is interwoven with said body or central portion on one or both sides of the same, in the manner and for the purpose set forth.

No. 57,660.—J. H. BARLEY, Longwood, Mo.—*Cultivator.*—September 4, 1866.—The middle ploughs are pivoted to the front of the frame, and are adjusted laterally by the handles, which are pivoted in a cross-bar resting upon standards on the outer plough beams. The inner plough standards depend from a cross-bar supported by plates on the outer beams.

Claim.—The iron bars D D attached to the plough beams A A, to support the cross-piece ÆE a suitable distance above the plough beams, in connection with the plough standards F F pivoted to the cross-piece, substantially as and for the purpose specified.

Also, the curved handles K K pivoted to the adjustable bar J, and connected to the bars F, which are pivoted in the bar H and secured to the plough standards F F, substantially as and for the purpose set forth.

No. 57,661.—CHARLES J. BAYER, Poughkeepsie, N. Y.—*Railway Switch.*—September 4, 1866.—If the switch be wrongly placed by the approaching car, the flange of the wheel presses against the connecting bar and restores the switch; the edges of the connecting bars are bevelled to fit against the main rails.

Claim.—The connecting of the switch rails C C to the main and branch rails A' B' by means of the bars D D, substantially as shown, so that the switch rails will be adjusted or moved by the action of the car wheels on the bars D D, as described.

Also, in a railway switch, adapted to operate substantially as herein described, having the adjoining sides of the rails A' B' and the bars D D bevelled or inclined to form a lock for the bars D D, as set forth.

No. 57,662.—FREDERICK C. BEACH, Stratford, Conn., and ALEXANDER A. C. KLAUCKE, Washington, D. C.—*Knife and Fork.*—September 4, 1866.—The handle has a receptacle for pepper or other condiment, and a perforated end with a valve operated by a trigger.

Claim.—The combination with a knife or fork of a receptacle for pepper, salt, or other condiment, in any manner substantially as herein described.

No. 57,663.—S. I. BEELER, Wales, Ill.—*Soap.*—September 4, 1866.—Composed of quick lime, 1 pound; sal soda, 2 pounds; saltpetre, 2 ounces; borax, 8 ounces; alum, 2 ounces; animal grease, 4 pounds; water, 3 gallons.

Claim.—The manufacture of soap by the combination of the ingredients in the proportion and manner substantially as herein specified.

No. 57,664.—L. HARROD BELL, Carmichael's, Penn.—*Coupling Joint for Well-boring Shafts.*—September 4, 1866.—A bolt occupies the coinciding longitudinal slots in the contacting ends of the rods, which are coupled by a screw joint. The bolt is secured by a spring, and prevents unscrewing until withdrawn.

Claim.—The device for preventing the two sections from separating by becoming unscrewed, namely, the mortises k and i, bolt D, and spring E, all arranged and operated substantially as set forth.

No. 57,665.—JOSEPH C. BENZINGER, Catonsville, Md.—*Catamenial Sack.*—September 4, 1866.—Explained by the claims and cut.

Claim.—First, the extended flap M, extending backward from the trough of the sack so as to cover the nates and to carry the straps W W beyond that part of the person, substantially as set forth.

Second, the sack, made substantially as described, with a flap M and a trough N, in combination with an elastic girdle, substantially as described.

No. 57,666.—DOUGLAS BLY, New York, N. Y.—*Artificial Leg.*—September 4, 1866.—The ankle articulation has interlocking straining stirrups attached to the respective parts, and elastic contacting pads in the socket. The toe piece is hinged to the prosthetic foot, and a plate on its lower posterior edge is connected by a bolt to a reaction piston spring.

Claim.—First, a set of springs G G, in combination with the leg A and foot B, forming a universal joint, substantially as specified.

Second, arranging the series of springs G G near the periphery of the socket C, surrounding the centre of motion, in such a manner that the central space will be left open, when said springs are employed between the foot and leg to produce reaction by compression, substantially as described.

Third, the arrangement of the plate H, provided with the bearing l and the bolt m, connecting with the spring I when used in combination with the toe joint k in such a manner as to work over it and produce the necessary leverage, substantially as set forth.

No. 57,667.—CORNELIUS BOLLINGER, Harrisburg, Penn.—*Grinding Mill.*—September 4, 1866.—The fan blower has spiral wings attached to a sleeve on the spindle: is separately revolvable in its case, and delivers the air upward and through the driver to the grinding space between the stones, whence it passes up and out above the stones through a central aperture in the case.

Claim.—First, mounting the fan K loosely upon the spindle C, to adapt it to be rotated independently of the latter, substantially as and for the purposes set forth.

Second, the driver I, formed with an air passage or passages i i, to enable the air from the fan to be discharged between the stones through the driver, substantially as described.

Third, the combination with the hoop L of the circular plate L', to form the ventilation passage L², as and for the purposes specified.

No. 57,668.—J. W. BOOSINGER, Marine, Ill.—*Cultivator.*—September 4, 1866.—The fronts of the beams are attached by universal joints to the forward transverse bar of the frame, and at the rear are operated laterally by a foot lever and connecting straps.

Claim.—First, the clevis D' and strap D², when constructed and employed substantially as described and set forth.

Second, the combination of the plough beams D with the clevis D'; also, the combination of the said beams with the devices E a a' $a²$, for the purposes and in the manner substantially as described.

No. 57,669.—JOHN K. BOSWELL, Richmond, Ind.—*Dry House.*—September 4, 1866.—The lower part is occupied by the heater, and the pipes surmounted by a grated platform, upon which vessels may be deposited containing substances to be heated or dried; in the upper part are slats on which to hang clothes, &c.

Claim.—First, the rectangular heater G, when the same is provided with the cylindrical valves H H', as and for the purposes set forth.

Second, the combination of the rectangular heater O, the valves H H', the openings K K', and the connecting pipe *b'*, all arranged and operating substantially as and for the purposes set forth and described.

Third, the arrangement of the movable lattice platform F and rectangular heater G, and valves H H', substantially as set forth.

No. 57,670.—JAMES A. BOYER, Greensburg, Ind.—*Refrigerator Building for Preserving Fruits, &c.*—September 4, 1866.—By the described devices the air is cooled, deprived of moisture, its thermal and hygrometric condition indicated, and itself conducted through the structure, which has double walls, non-conducting sides, and containing chambers.

Claim.—First, the combination and arrangement of the pipes B B, situated one in the lower the other in the upper part of the chamber C, the bellows D, or its equivalent, situated as described, and the series of V-shaped air passages under the ice chambers, so that the air will be drawn from the lower part of the chamber cooled by the condenser, and forced into the upper part of said chamber, substantially as described.

Second, in combination with the above parts, the absorbing chamber G, constructed and operating as described.

Third the construction and arrangement of the cooling or condensing apparatus, as shown, to wit, having a series of ducts or air passages *a a*, inclined in V-form, and placed underneath the ice chamber, having a V-shaped bottom, the whole being constructed and operating substantially in the manner and for the purpose set forth.

Fourth, the arrangement of the absorbing chamber as shown, to wit, having said chamber provided with a partition plate *e* and pipe H, constructed and operating substantially as described.

Fifth, in combination with the pipes B B, or chambers *b* *c*, connected therewith for ingress or egress of the air, a thermometer and barometer, for the purposes described.

No. 57,671.—A. M. BRUSH, Clayton, N. Y.—*Organ Reed.*—September 4, 1866.—Explained by the claim.

Claim.—An organ reed made of silver, either alone or mixed or alloyed with one or more metals.

No. 57,672.—OEORGE S. CARLISLE, Columbus City, Iowa.—*Portable Fence.*—September 4, 1866.—One brace is pivoted to the other, and the latter to the post.

Claim.—The arrangement of the braces F F and end posts A, relating to each other and operating in the manner and for the purpose herein described.

No. 57,673.—ALONZO CHACE, Syracuse, N. Y.—*Electric Telegraph.*—September 4, 1866.—By the described devices the electric circuit is broken in the block wherein the wires are secured, and is switched off to an instrument temporarily attached. The pole has one metallic connection with the wire on which it is hooked, and by its prong rotates the spring in the block, makes a disconnection therein, and connects itself with the other end from that allied to the hook.

Claim.—First, in combination, the devices above shown for breaking the electric connection upon a telegraph line—that is to say, the block N and its appurtenances, and the connecting pole X and its appurtenances, made and applied substantially as above set forth.

Second, the block N and its spring T, in combination with the wires that compose a line of telegraph, substantially as above set forth.

No. 57,674.—JOHN CHANDLER, Cold Water, Mich.—*Well Tube.*—September 4, 1866—A gauze filtering section at the lower end of the tube, above the cone point, is supported by a spiral coil and perpendicular rods.

Claim.—The perpendicular loop rods *a a*, as described, in combination with the wire gauze tube *d d*, inside supporting spiral coil *e e*, cone point C, and tube A, substantially as and for the purposes set forth, thereby adding great strength and security to the lower tube.

No. 57,675.—GEORGE L. CHAPIN, Chicago, Ill.—*Fastening the Slats of Venetian Shutters.*—September 4. 1866.—A bolt beneath the bar of the blind slats locks it in position.

Claim.—The arrangement and combination of the plate F, bolt J, screw C *a*, and the notched band K, substantially as set forth and described.

No. 57,676.—BARNABAS CLARK, Mackinaw, Ill.—*Corn Planter.*—September 4, 1866.—The machine plants in check rows, is thrown in operative condition by a clutch on the axle which has cells to carry the corn to the hopper, which deposits it behind the share, to be covered, by the roller, in the rear.

Claim.—First, the clutch G, in combination with the axle F, cells *c*, and wheel B B', arranged and operating in the manner and for the purpose herein specified.

Second, the marker P, when applied to or used in combination with the loose axle F, substantially as and for the purpose specified.

Third, the attaching of the openers J and colters L to the bars K K', applied to the frame A, and connected with a foot bar O, substantially as and for the purpose set forth.

Fourth, the ratchet Q, on the loose axle F, in connection with the pawl R, attached to the lever S, when said parts are used in combination with the marker P, and the seed distributing device, all arranged substantially as and for the purpose specified.

No. 57,677.—WILLIAM CLAYTON, Bristol, Conn.—*Attaching Knives to their Handles.*—September 4, 1866.—The blade has a recessed bolster, and a tang secured by a thread and nut. The ferrule and guard cover the joint.

Claim.—The recessed bolster a, of the blade A, in combination with the tang D and nut d, arranged with the ferrule c, guard C, and handle B, in the manner and for the purpose herein specified.

No 57,678.—G. W. COLE, Canton, Ill.—*Machinery for Cutting Stalks in the Field, preparatory to Ploughing.*—September 4, 1866.—The hooks, attached to hanging posts, are in the advance, and are maintained in position by certain devices. Their duty is to straighten out the cornstalks parallel with the line of motion of the machine. The rotating cutter wheel has its bearings in a vertically adjustable frame.

Claim.—The combination of a cylinder of cutters O, and its supporting frame I, with the main frame A, when said frame I is hinged at its rear end to the frame A, and has a vertical adjustment at its front end, operating substantially as and for the purpose set forth.

Second, the hook m, constructed as described, in combination with the hanging posts j, arranged substantially as and for the purposes set forth.

Third, the notched or toothed open standard Q, lever s, spring catch e, rod g, and staple h, in combination with the frames A and I, all substantially arranged as and for the purposes set forth.

Fourth, the levers r and rods p, in combination with hooks m, arranged substantially as and for the purposes set forth.

Fifth, the spring fulcrum s, in combination with the frame A and lever r, arranged substantially as and for the purposes set forth.

Sixth, the curved arms y and lever w, as described, in combination with the levers r and catch q, all arranged substantially as and for the purposes set forth.

No. 57,679.—THOMAS COLE, Marshalltown, Iowa.—*Washing Machine.*—September 4, 1866.—The upper roller has its bearings in a frame whose downward pressure is caused by a yielding lever beneath the bed.

Claim.—First, the yielding lever A A, as applied to the washing machine, substantially as herein described.

Second, the fulcrum wedge K, in connection with the yielding lever A A, as applied to the washing machine, substantially as herein set forth.

No. 57,680.—HENRY B. COMER, Pittsburg, Penn.—*File-Cutting Machine.*—September 4, 1866.—The height to which the hammer is lifted, and consequently the force of the blow, is regulated by giving a greater or less projection to the stub upon the hammer bar, against which the tappet strikes. By an arrangement of mechanism the point of the chisel immediately after each blow, and before it is removed from the block, is thrown a little forward, and by this movement advances the blank to the right position for the reception of another blow.

Claim.—First, moving the file blank and its bed, through the medium of the cutting tool, by means substantially as herein described and for the purpose set forth.

Second, the adjustable lifting arm m, when used in combination with the lifter 3, hammer f, mandrel d, and cutter 9, as herein described and for the purpose set forth.

Third, the tool holder B, provided with spring o, set screw 5, and piece e for holding the mandrel d, said holder being used in combination with the lever 10, spring g, lifter n, cam l, eccentric lifter 2, constructed, arranged, combined, and operating substantially as herein described and for the purpose set forth.

No. 57,681.—J. C. CONNOR, Dover, N. H —*Clothes Dryer.*—September 4, 1866.—The bars are jointed together on the lazy-tongs principle, are displayed for use and collapsed for economy in stowage.

Claim.—The bars V W X, constructed as described, in combination with each other, with the horizontal bars C I J and Y, and with the end pieces A B F G of the clothes dryer, substantially as described and for the purpose set forth.

No. 57,682.—MATTHIAS P. COONS, Brooklyn, N. Y.—*Apparatus for Generating Gas.*—September 4, 1866.—Division walls of wire gauze divide the interior of the retort, the outer spaces being occupied by asbestos; the contained hydrocarbon is vaporized by a lamp beneath; the retort and lamp are supplied with liquid from a tank. The former has a safety-valve and a gas discharge pipe leading to a cooling chamber.

Claim.—The internal arrangement of the retort B, with the cylinder D, and in the lamp, Fig. 4, the wick tube J, and also the connecting tubes as arranged, marked L L.

Also, the tube S, in the manner and for the purpose set forth and described.

Also, the application of the safety-valve I, for the purpose and in the manner described, and in combination therewith the application of the stop-cock, in the manner and for the purpose described.

Also, generating gas from combustible fluids by introducing the same into a generating retort by capillary attraction, for the purpose and in the manner herein set forth and described.

No. 57,683.—GILBERT E. CORBIN, St. John's, Mich.—*Bag Holder.*—September 4, 1866.— The holder jaws are pivoted to an inclined table and are expanded by a treadle and connecting cord.

Claim.—The arrangement of the board *a* and expanding holders *m s*, operating substantially as described and represented.

No. 57,684.—F. A. DEUTENBERG, Pittsburg, Penn.—*Tuyere.*—September 4, 1866.—The vertical hollow cylinder has an orifice in its dome-shaped upper end for the escape of the blast, and a sliding plate at the bottom to remove the ashes. The bellows pipe enters the side of the cylinder in an upwardly curved direction. A water jacket surrounds the dome.

Claim.—First, the vertical centre-blast blacksmith's fire tewel or tuyere, having the chamber H, with its dome E, in which is the hole J, and an annular channel K, at its top part, substantially as and for the purpose specified.

Second, the bend upward B C of the pipe A, in combination with the receiver H, dome E, and door G, as described and for the purpose specified.

No. 57,685.—OLONZO R. DINSMOOR, Auburn, N. H.—*Fruit Gatherer.*—September 4, 1866.—The padded jaws grasp and detach the fruit, which falls into the flexible conveyer.

Claim.—The improved fruit gatherer made substantially as described—that is, as consisting not only of the hole, the padded annular and disk jaws, and mechanism or means for opening and closing the disk jaw, but of the cloth or flexible conductor and the tubular cushion, arranged as specified.

No. 57,686.—SILAS R. DIVINE, New York, N. Y.—*Apparatus for Carburetting Air.*— September 4, 1866.—The chamber has porous partitions which descend nearly to the bottom; the porous material absorbs the hydrocarbon and permits the passage of air, which is carburetted by the contact.

Claim.—The use of the chambers B B B B, when placed one within another and composed of porous or perforated walls, substantially as described.

No. 57,687.—JOSEPH DIXON, Jersey City, N. J.—*Galvanic Battery.*—September 4, 1866.— The functions of the porous diaphragm, through which the acid percolates, and the negative plate, are combined in a graphite cell within the zinc cylinder; the latter is placed in the dilute acid bath and isolated from the graphite by rings of vulcanite.

Claim.—First, the combination in a galvanic battery of the porous diaphragm and negative metal or element in one and the same cell, substantially in the manner and for the purposes hereinbefore described.

Second, the graphite cell composed of pure plumbago and clay, or other material of which plumbago is the conducting ingredient, when combined in the proportions, substantially as hereinbefore stated.

No. 57,688.—JAMES W. DONALDSON, DANIEL SHEETS, and ALLEN C. MILLER, Suisun, Cal.—*Gang Plough.*—September 4, 1866.—The ploughs are attached in the usual receding order to a rhomboidal frame with diagonal bars and braces parallel to the beams. The plough standards are bolted to and lapped over the tops of the beams. The frame is adjustably supported on three wheels, two in front and one in the rear, on the "land."

Claim.—First, the peculiar construction of the frame in order to obtain sufficient strength for a number of ploughs by placing the diagonal bars A² A³ between the parallel bars A A' and the cross-braces *a* a¹, substantially as described.

Second, the manner of attachment of the ploughs to their respective portions of the frame by means of the bent braces C C C C passing over the top of said frame-work, as herein shown in combination with the ploughs, substantially as described.

Third, the adjustable wheel E and scraper *f*, with the wheels G G, scrapers N H; also, the turn plates H H, and connecting rods *h h'* with upright bars J J, jointed at the turn plate H, in combination with the sweep L, substantially as described and for the purpose set forth.

No. 57,689.—RICHARD A. DOUGLAS, Chicago, Ill.—*Brick Machine.*—September 4, 1866.— The clay is ground between rollers revolving at different velocities, is pressed by plungers in stationary moulds, the pressed bricks raised therefrom by followers and automatically removed while a fresh charge is substituted.

Claim.—First, the combination of the series of grinding rollers *a* and *a'*, being so geared that one of each pair shall revolve faster than the other, with the hopper *d* and the charger G, when arranged to operate as shown and described.

Second, the charger G provided with the chamber *g*, and the lubricating reservoir *h*, arranged to operate in connection with the hopper *d*, substantially as set forth.

Third, the plungers 1 2 3, &c., arranged to operate in combination with the moulds J and the followers 1' 2' 3', &c., when said plungers and followers are operated by the cams *l m* and *n*, as shown and described.

Fourth, the arrangement of the cam wheels N and the accompanying mechanism for operating the charger G, as set forth.

Fifth, the means substantially as shown for adjusting the height of the plate L and its followers.

Sixth, the means of adjusting the movements of the charger G by means of the slotted arm *r* and pin *v*, substantially as shewn and described.

No. 57,690.—ARTHUR DOYLE, New York, N. Y.—*Propeller for Steamships.*—September 4, 1866.—The frames of the horizontally moving propeller floats are attached by parallel motion arms and links to the ends of vibrating levers, and are reciprocated by a connecting rod from the motor.

Claim.—First, the combination of the vertical buckets *a a* with the side beams *b b*, the upright arms *c c*, the radial arms *h j*, and the crank *s*, with the shafts *f*, as applied either to the side or stern of a ship, in connection with the system of balance beams and oscillating bars, constructed, arranged, and operated substantially as and for the purposes herein described.

Second, the combination of the buckets *a' a'* with the side beams *b' b'*, the main central upright arm *c'*, and the half arms *c² c²*, the radial arms *h' j'* and *j²*, the slide *z*, and the crank, *s'*, with the shaft *f'*, as applied to the side of a ship, in connection with the system of balance beams and oscillating bars, constructed, arranged, and operated substantially as and for the purposes herein described.

No. 57,691.—JOHN S. DRAKE, New York, N. Y.—*Artificial Leg.*—September 4, 1866.—The frame is made of malleable cast-iron, with projections to correspond with the natural processes. Straps unite the anterior and posterior portions of the knee frame, and keep the covering in place. The toes are of spring metal, and attached separately to the foot; a spring at the ankle articulation depresses the heel.

Claim.—First, the malleable cast-metal frame for artificial limbs, formed in the manner and for the purpose specified.

Second, the straps *s i* of the cast-metal frame A, applied in the manner and for the purposes set forth.

Third, the spring metallic frames for the toes, each attached by a separate rivet or screw, so as to be movable, as set forth.

Fourth, the curved metal spring *n*, introduced at the ankle joint, with its ends attached to the heel D and frame C, and acting to keep the toes of the foot from dropping, as set forth.

No. 57,692.—JAMES B. EADS, St. Louis, Mo.—*Operating Ordnance.*—September 4, 1866.—Explained by the claim and cut.

Claim.—The combination of a pair or more of gun carriages, so connected as to preserve the parallelism of the axes of the guns with a corresponding number of curved tracks, so arranged as to cause the said axes to vibrate upon a given point in the embrasure, substantially as described.

No. 57,693.—THOMAS C ENTWISTLE, New York, N. Y.—*Transmitting Motion.*—September 4, 1866.—The driving pulley is placed outside of the frame in a sleeve, which revolves on a stationary shaft, to which, inside the bearing, is fastened a bevel gear. Attached to said sleeve is a yoke, which is sleeved on the shaft, and has a mitre wheel, revolved by its engagement with the stationary gear, and revolving the third gear in the series two revolutions for one of the driving pulley.

Claim.—The combination and arrangement of three bevel gears C D E and a revolving yoke F, or its equivalent, to operate substantially as herein specified.

No. 57,694.—J. A. ESHLEMAN, Philadelphia, Penn.—*Neck-tie Supporter.*—September 4, 1866.—Improvement on his patent of November 15, 1864. The scarf is folded around the frame of the holder, whose prongs are tucked beneath the lapel of the collar, the spring piece connecting with the button fastening.

Claim.—The holder, composed of the plate A, and arms *b* and *b'*, the whole being constructed and arranged for the reception of a scarf, substantially as and for the purpose described.

No. 57,695.—S. H. EVERETT, Milton, Ohio.—*Lubricating Bush.*—September 4, 1866.—The perforated lubricating bush forms a sleeve upon the axle inside the box.

Claim.—The lubricating bush D, having orifices *a a*, or their equivalents. when used in carriage boxes, or in bearings for machinery, in the manner substantially as described, and for the purposes specified.

No. 57,696.—MATTHEW FLETCHER, Louisville, Ky.—*Steam Carriage.*—September 4, 1866.—One car carries the boiler; steam connection is made throughout the train, and each propelling wheel has its rotary engine.

Claim.—First, the application of a rotary steam engine to each propelling wheel, for stability of carriage, avoiding dead centres, and enabling the driver to have at his command with ease, and by the power of steam to back, turn, or advance.

Second, the arrangement of the engine, piston, and wheel, operating together (or independently) with the piston and wheel on the opposite side of the carriage, for the purpose set forth.

Third, suspending the whole weight of carriage and engine to the axle.

No. 57,697.—SAMUEL GARDINER, Jr., New York, N.Y.—*Turning On and Shutting Off Gas by Electricity.*—September 4, 1866.—A wheel is connected with one pole of a battery, and provided with alternate conducting and non-conducting points, which are rotated in contact with the other pole of the battery, so as to give intermittent pulsations to the armature of an electro-magnetic machine placed in the circuit and operating a ratchet wheel on the axis of the stop-cock which controls the flow of gas in the main. In connection with the wheel is a crank, by which it is revolved, and a graduated circle which indicates the position of the wheel, and bears a certain correspondence with that of the stop-cock.

Claim.—First, the wheel D, provided with the non-conducting and conducting surfaces, substantially as described, and rotated by a crank whose revolutions are registered by or upon a dial, substantially in the manner and for the purpose described.

Second, the dial, with its light and dark segments, or portions to indicate, in connection with a certain opening or place on the dial plate or other fixed object, the position of the stop-cock or other object, for the movement of which the apparatus is designed, substantially as described.

Third, the revolving arm and spring tooth M *m*, operating in connection with the stud *c'*, or its equivalent, to rotate the wheel D, by engagement with the cogs thereof, as described.

Fourth, the combination of the shaft B , wheel D, spring tooth *m*, arm O, and spring detent N, as and for the purpose described.

Fifth, the arrangement of the key G, and spring keys *g g' g''*, with the portions *d d' d''*, constituting the wheel D.

Sixth, the arrangement of the shaft B, studs *r r*, wheel S, and detent *s*, the pinion T, spur wheel U, and dial V, operating substantiantially as and for the purpose described.

No. 57,698.—FRANCIS GAY, Bedford, Ohio.—*Farm Gate.*—September 4, 1866.—The gate is attached by the toggle arms to the pivoted post, and rests upon a roller on the long arm, so as to rise in opening, and be capable of rotation when erect.

Claim.—First, the shaft D, arms H and J, and gate A, when the several parts are combined and operate as and for the purpose set forth.

Second, the standard D, face plate E, and wheel G, in combination with gate A, as and for the purpose specified.

No. 57,699.—WILLIAM GEAHR, New Holland, Penn.—*Cultivator.*—September 4, 1866.—The shovel beams are pivoted horizontally by their erect posts in the frame above, so as to be adjustable relatively to the line of draught, to present the V-form with the apex towards the front or rear.

Claim.—The independent, adjustable, and reversible beams D D, in combination with the upright E F, when connected with a suitable framework, substantially in the manner and for the purpose specified.

No. 57,700.—GEORGE T. GIFFORD, Monmouth, Ill.—*Cultivator.*—September 4, 1866.—The depth of tilth is regulated by the position of the seat and the lateral adjustment of the middle ploughs, by the shifting lever.

Claim.—First, the arrangement of the frames B and A, and movable pivot L L, for balancing, substantially as described.

Second, the combination of the lever F, cross-bars E and C, and ploughs G G, with the inside frame, for the purposes set forth, and substantially as described.

Third, the arrangement of frames B B and A A, by which the weight of driver supports or tends to lift ploughs, substantially as described.

Fourth, the slide S, operating in the axles as described, and for the purpose set forth.

No. 57,701.—JOHN L. GILL, Jr.—Columbus, Ohio.—*Smelting Furnace.*—September 4, 1866.—A boiler is attached to the upper part of the furnace; the steam generated is passed into the smoke-stack to increase the draught.

Claim.—Constructing a cupola or smelting furnace in such a manner as to allow of a part of the upper portion (of such cupola or smelting furnace) being made from a hollow steam boiler for generating steam to be used in the production of a blast, or for any other purpose, as described above.

No. 57,702.—SIMON GOLDSTONE, Philadelphia, Penn.—*Cap.*—September 4, 1866.—Explained by the claim and cut.

Claim.—A cap, having a series of eyelet holes through the back, opening from the exterior into the space contained between the body and the band, and series of eyelet holes through the band, opening from the interior of the cap into the same space, and having an oil silk perspiration shield, the whole arranged and operating with respect to each other substantially as is herein specified and described.

No. 57,703.—WILLIAM T. GRANT, Jacksonville, Ill.—*Mop and Scrubber.*—September 4, 1866.—The brush has an elastic strip inserted into its side, and a mop cloth attached, which is looped to the handle when not required.

Claim.—The combination of the brush head A, the rubber strip B, mop F, handle C, bar E E, cylinder D, catch or pin O, and arm *g*, as and for the purposes set forth.

No. 57,704.—JAMES T. GUTHRIE, Leesburg, Ohio.—*Lock.*—September 4, 1866; antedated August 17, 1866.—The lock has two bolts—the one parallel with the other—which are moved simultaneously by the key, but in opposite directions, except by a certain mode of procedure known to the operator, by which the ends of the key are alternately thrust into the lock, when both bolts may be retracted or projected.

Claim.—First, the two bolts B and B′ having spring checks C and C′, pin *d*, and springs *b* and *b′*, arranged and operating as above described and set forth.

Second, guard bolt D, sliding incline D′, lever F, spring G, and catch spring *i*, constructed as above described, and for the purpose set forth.

No. 57,705.—JONES GUTHRIE, Wilmington, Del.—*Shears for Cutting Bolts.*—September 4, 1866.—The short arm is pivoted eccentrically to the long one, and connected to the sliding shear by a link; its movement in the direction of closing the jaws is regulated by a set screw.

Claim.—The combination of the parallel levers A A, the joint B, with the connecting plate E, and movable knife G, operating against an opposite knife J, and the set screw L, all constructed and arranged as herein described, for the purposes set forth.

No. 57,706.—E. R. HALL, Ilion, N. Y.—*Horse Rake.*—September 4, 1866.—The rake is hung in swinging bars, and to the head is attached a tubular arm, in which is placed a lever whose lower end is held by a spring over a pin upon the rake head, to stop the revolution of the rake. The rake may be raised bodily by a lever and chain, and sustained in its elevated position by a pawl.

Claim.—First, the suspending or fitting of the rake G between the swinging bars F F, connected to the front end of the frame A, and having an arm H connected with the rake head, substantially as and for the purpose set forth.

Second, the lever I fitted within the arm H in connection with the pins *f f* on the rake head *e*, when said parts are applied to a suspended rake G, substantially as and for the purpose specified.

Third, the lever L on the axis J, provided with the arm *g*, connected with the arm H by a chain K, in combination with the pawl M attached to lever L, and the rack N secured to a board on the front part of the seat supports *a a*, substantially as and for the purpose specified.

No. 57,707.—GEORGE W. HALL, Augusta, Mich.—*Combined Sower and Drill.*—September 4, 1866.—The machine is adapted for sowing grain, grass seed, and fertilizers, all, or either. The furrowing wheel is followed by the drill tubes and covering shares, which have devices for vertical adjustment; the sowing devices have valves operated by connection with the driving wheels.

Claim.—First, the pendent frames M carrying the furrowing wheels S, and drills N, for the purposes substantially as described.

Second, a machine for sowing different kinds of grain, when provided with pendent frames carrying the furrowing wheels and drill tubes, substantially as and for the purposes herein set forth.

Third, the combination and arrangement of the levers J K U and Y connected to the slides in the seed box L, and agitator X in the box O with the elbow lever D and cam *e*, all for the purposes and substantially as herein set forth.

Fourth, the ploughs when constructed and operated as herein shown.

No. 57,708.—WILLIAM HALSTEAD, Trenton, N. J.—*Manufacture of Artificial Fuel.*—September 4, 1866; antedated August 10, 1866.—Composed of peat and coal-dust formed into blocks and dried by artificial heat; other combustible substances such as petroleum, asphaltum, rosin, &c., may be added.

Claim.—The combination, mixture, and treatment of the ingredients above mentioned, substantially as above described and intended to produce the same effect.

No. 57,709.—E. HAMBUJER, Detroit, Mich.—*Folding Chair.*—September 4, 1866.—The hinged bed is unfolded from the seat of the chair and supported upon legs, while the cushion is supported in an inclined position for a pillow.

Claim.—First, the head part D, having legs E, in combination with the parts C B, hinged to each other, and with frame A, provided with pivoted legs F and braces *f'*, arranged and operating substantially as represented and described.

Second, the screw socket C, in combination with the legs F and frame of the part B, substantially as described and for the purpose set forth.

No. 57,710.—HENRY HARRIER, Indianapolis, Ind.—*Grain Binder.*—September 4, 1866.—A grain gatherer (in shape like the letter S) revolves on a central shaft within a concave grain receptacle, and seizes the gavel at the centre; while leaving the beads interlocked, the other portion of the gavel is divided into two equal parts, which are twisted together to form it into a sheaf. The twisting is resisted by a spring device, which operates to catch the ends of the straw to interlock or tuck them.

Claim.—The cylinder V, with its square shaft W, the grain gatherer T, and the concave U. Also, the combination of the spring *s* with the arrangement L and the lever R.

No. 57,711.—JAMES HARRISON, New York, N. Y.—*Illuminated Sign.*—September 4, 1866.—The letter is studded with glass cups, whose protrusion renders them visible when viewed obliquely.

Claim.—The combination of the close glass cups C with the block or raised work B of the letters or devices to be shown, and with the back ground A, substantially as herein described and for the purpose set forth.

No. 57,712.—PHILO W. HART, Stamford, N. Y.—*Power Loom.*—September 4, 1866.—Improved means for automatically ejecting the shuttle from its box when its yarn breaks or gives out, and supplying a full bobbin in its place while the loom is in operation; cannot be briefly described.

Claim.—First, the sliding plate *f* applied to work through the lay of the loom, and in combination with the movable trap at the bottom of the shuttle box, substantially as and for the purpose herein described.

Second, the fingers *w w* attached to the lay of a loom and operating in combination with the movable trap at the bottom of the shuttle box, substantially as and for the purpose herein set forth.

Third, the spare shuttle box I attached to the breast beam or framing of the loom, having a movable trap at its bottom and operating in combination with a shuttle box having a movable trap at its bottom on one side of the lay, substantially as and for the purpose herein described.

Fourth, the combination of the spare shuttle box working on fixed brackets J J, or their equivalent, attached to the breast beam or other fixed portion of the loom, the rock shaft K, carrying the fingers *w w*, and furnished with a pin Z and the stationary arm L, the whole operating substantially as herein set forth.

Fifth, the sliding plates N M, in combination with each other, with the lay and one of the shuttle boxes thereon, and with the spare shuttle box, substantially as and for the purpose herein described.

No. 57,713.—B. & HEALY, Cohocton, N. Y.—*Gate.*—September 4, 1866.—The notch in the upper bar of the gate sets upon the curved metallic plate, which unites the two posts and forms a hinging point; the lower bar rests against a socket.

Claim.—First, the curved metallic strip K, in combination with the notched top rail of the fence, with the gate post B and with the perpendicular support L, substantially as described and for the purpose set forth.

Second, the combination of the supporting piece O with the gate post B, the perpendicular support L, and with the shortened bottom rail G of the gate, substantially as described and for the purpose set forth.

No. 57,714.—MARSHALL D. HIGLEY and DANA L. COLUMBIA, Morrison, Ill.—*Raking and Binding Attachment to Reapers.*—September 4, 1866.—A brief description other than that contained in the claims is impracticable.

Claim.—First, in an automatic rake for a harvester the combination of the eccentrics D and E, connecting rods G and D², bell crank G¹, and oscillating rake arm H, said parts being respectively constructed substantially as subscribed.

Second, the oscillating rake arm H and parallel rod H^1 adjustably attached to the levers H^2, when used in combination with the eccentric D and connecting rod D^2, substantially as set forth.

Third, the wheel I, with the track I^2, and depressions I^3 and I^4, when used for actuating the levers N^1 and M^2 respectively, substantially as and for the purpose set forth.

Fourth, the twist head L, when constructed in two parts, L^1 and L^2, the part L^2 being arranged to turn on a pivot; the opposed faces of the parts being perpendicular to the axis of rotation, and the said parts being constructed and arranged substantially as set forth.

Fifth, in combination with the twist head L, the two clutches N N attached to levers N^1 N^1, which have their fulcrums near the centre, and are opened and closed by a spring O and cam I, substantially as set forth.

. Sixth, in combination with the twist head L, the nippers M M^1, one being fixed and the other movable, when they are respectively constructed and the movable one actuated substantially as set forth.

Seventh, in combination with a device for binding the sheaf the revolving arm R^1, for throwing the sheaf from the platform, substantially as set forth.

Eighth, the cam C having a dead point a b, when used in combination with, and for the purpose of giving motion to, the binding arm G^1, substantially as set forth.

Ninth, the shield u and u^2, when used in combination with an automatic binding mechanism, substantially as and for the purpose set forth.

No. 57,715.—ASA HILL, North Providence, R. I.—*Guard for Railway Crossings.*—September 4, 1866.—The bar is pivoted to a sill and is elevated so as to form a barrier across the track by the vibration of the rock-shaft arm.

Claim.—An improved safeguard or barrier for railroad crossings, composed of a bar applied to uprights at each side of the roadway in such a manner that the bar may be raised or lowered by means of cranks or pivoted arms, in the manner substantially as herein shown and described.

No. 57,716.—SPENCER HINTON, Jackson, Mich., assignor to WITHINGTON, COOLEY & Co., same place.—*Lathe for Turning Scythe Snaths.*—September 4, 1866.—The revolving wheel has a hollow shaft in which the snath is fed to be turned, and comes in contact with knives in sliding frames which yield to the inequalities in the size of the blank.

Claim.—First, the two plates G H, with hollow bugle-mouthed journals J J' respectively on each, and a pulley P, on one, G, fastened together in such a position, with the pieces L L between them, and diametrically opposite to each other, so that the journals J J' will be in a line with each other, and retain certain gauge and knife holders between them, substantially as and for the purpose specified.

Second, the two gauge and knife holders B B moving toward and from the centre of rotation between the plates G H, substantially in the manner as and for the purpose set forth.

Third, so fitting the ends of gauges where the parts of the circle pass each other (when the gauges are closed up, as in Fig. 1) into each other, that as they open they will tend nearer to form a complete circle, and when fully open the circle will be complete, as shown in Figs. 3 and 4.

Fourth, attaching the knife and feed knife to gauge and knife holders, as herein described.

No. 57,717.—ISAAC N. HODSON, Mount Pleasant, Iowa.—*Loom.*—September 4, 1866.—The movements of the batten effect the movements of the harness and the shuttles. The heddle frames are directly lifted and lowered by double cranks in the ends of the cam shafts or rollers, which are intermittingly oscillated by a projection on the batten. The picking levers are alternately operated by their free passage in one direction, but not in the opposite one, over the heads of swinging dogs which are free to swing one way, but are restrained by a pin when falling back to position.

Claim.—First, the cranks a a', two or more, and cam rollers F, in combination with the batten G and heddle frames E, constructed and operating substantially as and for the purpose set forth.

Second, the hinged swords h h', and dogs k k', in combination with the batten G and buttle blocks j j', constructed and operating substantially as and for the purpose described.

No. 57,718.—DAVID HOIT, Fort Wayne, Ind.— *Wood-boring Machine.* — September 4, 1866.—By the described devices the auger is moved up or down, and horizontally, while in motion, its depth being gauged by a sliding collar.

Claim.—First, the auger shaft C, shafts c and D, and guide posts B and B', arranged and operating as described.

Second, the shaft L, screws J J', nuts m m', and yoke M, as and for the purposes set forth.

Third, the arrangement and combination of the parts herein described for giving the auger of a boring machine a perpendicular and horizontal motion independently or simultaneously, in the manner and for the purposes herein set forth.

No. 57,719.—JACOB HOLLINGER, Millersburg, Ohio.—*Cultivator.*—September 4, 1866.—The shares are attached to an expansible triangular frame, and the handles secured to the central longitudinal beam.

Claim.—The herein described construction of cultivators, consisting of the beam A, curved bars B B, braces D, shears C B' B', and handles E, several parts being constructed, arranged, and operating as and for the purpose set forth.

No. 57,720.—E. F. HOLLOWAY and J. W. HUDELSON, Knightstown, Ind.—*Bed Bottom*—September 4, 1866.—The loop of the wire sustains the slat and the prolongations of its side coils are inserted behind a bar secured to the bedstead rail.

Claim.—In combination with the rails A, the strips C, springs B, and removable slats D, the said several parts being respectively constructed and the whole arranged for use substantially as set forth.

No. 57,721.—NOAH P. HOLMES, Indianapolis, Ind.—*Apparatus for Preserving Milk.*—September 4, 1866.—The cylindrical vessel has double walls with an intervening body of charcoal. The central ice-chamber and annular milk chamber have separate lids with ventilators in each.

Claim.—The can 1, with its double lining 7 7, for charcoal, its cylinder with ice 5, separate lids 2 4, and ventilators 3 3, for the purpose described, and all arranged substantially as set forth.

No. 57,722.—T. L. HOUGH, Philadelphia, Penn.—*Truss.*—September 4, 1866.—The pad arm has its axis on the end of the band, and the pad is pressed upward and inward by the flat metallic spring attached to the hub and the pad arm.

Claim.—The combination of the hub D, spring m, arm B, and journal a, arranged to operate as and for the purpose herein set forth.

No. 57,723.—W. H. HUGHES and H. L. LENT, Peekskill, N. Y.—*Stand and Mirror.*—September 4, 1866.—The mirror is attached to a post which slides vertically in an aperture in the table and post, and has a counterpoise weight to maintain its vertical adjustment.

Claim.—A combined toilet stand, or its equivalent, and mirror, when the latter, by its staff or rod H, is hung or suspended to the said stand by means of a cord J, pulley or pulleys K, and weight M, substantially as described and for the purposes specified.

No. 57,724.—HERBERT A. HUMMER, Franklin Township, N. J.—*Plough.*—September 4, 1866.—The landside has a dovetail projection on its inner side which fits in a corresponding groove in the mould board, and the connection is maintained by a key.

Claim.—Uniting the mould board and landside of the plough by a concealed joint, contructed and arranged substantially as and for the purpose described.

No. 57,725.—GEORGE H. HURD, St. Louis, Mo.—*Plate for Artificial Teeth.*—September 4, 1866.—The bottom part of the plate, instead of being rounded off, is widened out to a flange which reaches to the lip; it is held by the muscles of the lip or by pressure of the air if a partial exhaust has been obtained beneath it.

Claim.—The plate B, when constructed with the flanges b, either with or without the suction cavities b', so that artificial teeth may be fitted into mouths of bad formation, and secured there, either by suction or by muscular power, or by both.

No. 57,726.—GEORGE H. HURD, St. Louis, Mo.—*Dental Mould.*—September 4, 1866.—Explained by the claim and cut.

Claim.—Constructing dental moulds or impression frames A, so that their edges b and b' will be wide enough apart to take an impression of the lip-muscles and tongue-shelf at the same time, substantially as herein described and set forth.

No. 57,727.—JOHN JANN, New Windsor, Md.—*Burning Fluid.*—September 4, 1866.—Composed of benzine, thirty-three gallons; sweet oil, one-half pint; and oil of vitriol, two quarts.

Claim.—The combination of benzine, sweet oil and oil of vitriol, in about the proportions and for the purpose described.

No. 57,728.—BENJAMIN F. JEWETT, Malone, N. Y.—*Manufacture of Pot and Pearl Ash.*—September 4, 1866.—The potash lye is boiled with black muck until it assumes the appearance of black salts. The mass is then converted into pearlash in the usual manner.

Claim.—The process of manufacturing potashes and house ashes into pearlashes by the use of black muck, substantially as herein specified.

No. 57,729.—ALGERNON K. JOHNSTON, New York, N. Y.—*Apparatus for Carburetting Gas*—September 4, 1866.—The box has a series of wires extending across the top upon which shavings are hung to absorb hydrocarbon fluid, which is evaporated by the gas passing through the box.

Claim.—The use of the materials above described, for the purposes set forth.

No. 57,730.—J. D. KELLOGG, Jr., Northampton, Mass.—*Churn.*—September 4, 1866.—By the inclined surfaces the cream is driven forcibly through the perforations of the dasher.

Claim.—The dasher, provided with the opening *b* and the inclined surfaces *a c*, sloping in different directions, and operating as described.

No. 57,731.—ZEBULON S. KELSEY, Huntington, Ohio.—*Fruit Gatherer.*—September 4, 1866.—The pivoted jaws on the end of the pole are opened and closed by the motions of a sleeve, to which they are connected by rods; the fruit drops into a net beneath.

Claim.—The construction and arrangement of the fruit gatherer as herein set forth.

No. 57,732.—JOHN KINGSBURY, Ravenna, Ohio.—*Car Coupling.*—September 4, 1866.—The arrow-head bolt of one draw-head is grasped by a pair of hooked bars on the other; the hooks are opened by a cord which passes to a shaft on the platform; resilient springs are applied to each draw-head, and the coupling is disconnected by lateral deflection when a car leaves the track.

Claim.—First, the arrangement of the jaws C, when pivoted together, and to the adjustable stay B, in combination with the standard E, springs J D, chain F, and windlass, as specified.

Second, the hook D', and adjustable stays K, when arranged and pivoted as set forth, in combination with the springs N *d*, and jaws C, as and for the purpose set forth.

No. 57,733.—D. P. LACEY and J. A. BARTLETT, Oxfordville, Wis.—*Sash Lock.*—September 4, 1866.—The tumbler and bolt act independently in supporting the sash, but are withdrawn by one turn of the knob, thus freeing the sash; the latter is held in position by the contact of the tumbler and locked shut by the bolt.

Claim.—The combination and arrangement of the tumbler *a b f h*, lock bolt *d j k*, and spring *m* and *l*, substantially as and for the purpose set forth.

No. 57,734.—ALEXANDER LADD, St. Lawrence, N. Y.—*Corn Planter.*—September 4, 1866.—The box has a perforated bottom, a false bottom with a bevelled cut-off, and a perforated feed slide operated by a cord and spring. It is designed to be attached to a hoe handle.

Claim.—First, the slide B, provided with a hole *e*, in combination with the box A and the hole *f*, in the bottom *c* thereof, when said parts are arranged as shown and described, to admit of the dispensing with the ordinary strike or cut-off for depriving the hole *e* of superfluous corn or seed, as set forth.

Second, in combination with the box A and slide B, arranged as shown, the false bottom E, having its lower end bevelled or chamfered at its under side, substantially as, and for the purpose specified.

No. 57,735.—CHARLES LEE, Winchester, Ohio.—*Fence.*—September 4, 1866.—The fence boards are attached by cast metal loops to the vertical metallic rods.

Claim.—First, the posts A, when constructed substantially as herein described and for the purpose set forth.

Second, the combination of the loops C, when constructed as herein described, with the posts A and boards B, substantially as and for the purpose set forth.

Third, the combination of the key or wedge E, with the posts A, loop C, and boards B, substantially as described and for the purpose set forth.

No. 57,736.—JAMES LEFEBER, Cambridge City, Ind.—*Machine for Tenoning Spokes.*—September 4, 1866.—The arbor to which the mortising or tenoning tool is attached revolves in a vertically adjustable frame, and the spokes are presented consecutively thereto by the rotation of the hub, which is chucked upon the carriage.

Claim.—First, supporting the gear frames P or *j* upon the movable frame S, and providing for their vertical adjustment thereon, substantially as described.

Second, in combination, the movable frame S, the gear frame P, and the carriage D, substantially as described.

No. 57,737.—JOSEPH M. LIPPINCOTT, Pittsburg, Penn.—*Lubricating Oil.*—September 4, 1866.—The object is to form a lubricating oil from petroleum without the use of expensive animal oils.

Claim.—First, the reduction of the gravity of hydrocarbons, or petroleum oil, by the admixture of pine tar, substantially as above set forth.

Second, the use of pine tar in the manufacture of lubricating oils of any desired gravity, in combination with hydrocarbons or petroleum.

Third, the use of pine tar in the manufacture of lubricating oils, in combination with hydrocarbons or petroleum, animal oils, tallow, or fatty matter of any description.

No. 57,738 —JOHN S. LIPPS, Brooklyn, N. Y.—*Barrel for Petroleum, &c.*—September 4, 1866.—Inside the barrel is an air induction tube attached to the upper head, and with a rose on its lower end. The head has also an eduction nozzle for the air charged with vapor. Connection being made with a blower, the barrel serves as a machine for carburetting air.

Claim.—A barrel for hydrocarbon liquids, provided with an air pipe *c* and escape orifice *a*, substantially as and for the purposes described.

No. 57,739.—SYLVANUS D. LOCKE, Janesville, Wis.—*Grain Binder.*—September 4, 1866.— The mechanism for discharging the bound bundle from the platform of the binder is identified by the claims and cut.

Claim.—First, the combination and arrangement of the part C, pitman O, constructed substantially as described, crank *t*, shaft A, standard J, and head Y, when the whole are constructed, arranged, and used substantially as and for the purpose set forth.

Second, the combination and arrangement of the part C, pitman O, constructed substantially as described, shaft A, standard J, and head Y, and shaft spring F, when the whole are constructed, arranged, and used substantially as and for the purpose set forth.

Third, the combination and arrangement of the part C, pitman O, projection *m*, cylinder B, pin D', shaft A, crank *t*, standard J, and head Y, when the whole are constructed, arranged, and used substantially as and for the purposes set forth.

No. 57,740.—M. J. LOURRENTZ, Leavenworth, Kansas.—*Washing Machine.*—September 4, 1866.—In one compartment of the tub is a reciprocating corrugated rubber acting in connection with a gang of rollers with a spring pressure in their rear; in the other division is a reciprocating pounder, actuated in common with the rubber by hand lever and a counterpoise reaction lever.

Claim.—First, the reciprocating rubber I, operated from a rock shaft J, as shown, in combination with the pressure rollers C, arranged with springs E, connected with adjustable bars or slides F, substantially as and for the purpose herein set forth.

Second, the pounder M, connected with the rock shaft J through the medium of the tubular rod N, sliding rod O, and spring P, substantially as and for the purpose specified.

Third, the operating of the rock shaft J through the medium of the toothed segments *m n*, counterpoised lever K, and band lever L, all arranged substantially as described.

No. 57,741.—C. M. LUNT and W. F. LUNT, Biddeford, Maine.—*Fruit Picker.*—September 4, 1866.—Jaws on the end of the pole are met by sliding jaws above which are operated by cord and spring; when the jaw rises a distended cord delivers the fruit into a basket suspended beneath.

Claim.—An instrument for picking fruit, constructed and operating substantially as shown and described; that is to say, the combination of the handle A, rod B, tines *d*, spring *b*, cords *c*, apron C, and basket-supporting hook *f*, substantially as shown and described.

No. 57,742.—JOHN MARQUIS, San Francisco, Cal.—*Slop Hopper.*—September 4, 1866.— The inner hopper has a marginal trap in a trough of the casing and a lower one in the double cup, whose recess beneath forms a trap over the exit pipe.

Claim.—First, the construction and arrangement of the stationary hopper E E and movable hopper D D, substantially as described, and for the purpose set forth.

Second, the bowl or pan G A', or its equivalent, placed upon standards in the bottom of the lower ledge in the outer hopper, or attached to the inner hopper D, and which forms, together with the lower portion of the movable hopper, and the upper portion of the connection pipe I, the trap A'' A'', substantially as described, and for the purpose set forth.

Third, arranging the inner hopper in the stationary or outer hopper so as to form the upper trap A, as herein specified and for the purpose set forth.

No. 57,743.—MARTIN MARTINS, New York, N. Y.—*Metal Frame for Pianos.*—September 4, 1866.—The metallic frame is strengthened on the under side with tension rods which counteract the tendency of the strings on the upper side to bulge the frame.

Claim.—First, the tension screw rods *A* and springs *j*, in combination with the frame A, constructed and operating substantially as and for the purpose described.

Second, the L-shaped plank *t*, in combination with the lips *e f* of the frame A, and with the tension screw rods *k*, constructed and operating substantially as and for the purpose described.

No. 57,744.—H. MAXELL, E. FESSLER, and H. FESSLER, Canton, Ohio.—*Railroad Switch.*— September 4, 1866.—The lock-up switch box has a window and signal lantern, the latter moved simultaneously with the gear segment attached to the switch rod. The segment is rotated by a crank and pinion when the door is open.

Claim.—First, the switch box A with inner box N and windows C D, with signal T, arranged in the manner substantially as and for the purposes set forth.

Second, the semicircular wheel E, wheel R, wheel H, crank K, lever J, and spring b, arranged within the switch box A, as and for the purposes herein specified.

Third, the shaft M, attached to the wheel E by means of the arm P and metallic plate e, working the rod F, when arranged and used as and for the purposes set forth.

No. 57,745.—J. R. McALISTER, Richville, N. Y.—*Carriage Gearing.*—September 4, 1866.—The coupling pole is dispensed with. The wagon bed is tied on the hind axle and fore carriage by brace rods. The front bolster is pivoted to the axle so as to allow freedom of vertical motion to a stiff tongue without affecting the bolster and the parts imposed upon it.

Claim.—First, the brace rods G G² secured to the wagon body at one end and at their others respectively to the hind axletree and the head block E of the front spring of said body, substantially as and for the purpose described.

Second, the swinging frame circle O, of the front head block E, in combination with the plate or circle P fixed to the front axletree, the two being connected together, substantially as described and for the purpose specified.

No. 57,746.—ROBERT W. McCLELLAND, Springfield, Ill.—*Refrigerator for Liquids.*—September 4, 1866.—A box with doors contains a sliding frame to support the barrel which is connected by a tube with a cooling chamber surrounded with ice. The ale is drawn from the cooling chamber through a faucet.

Claim.—A refrigerator for cooling ale, beer, and other liquids, arranged so that the casks may be supported upon slides E resting upon the ways D in the upper part of the chest A, and the liquids be conducted by a flexible pipe G into a receiver I, enclosed in the cooling tub H, and then drawn for use through a faucet M passing through the small doors O, said several parts being constructed and arranged substantially as set forth.

No. 57,747.—THOS. B. McCONAUGHEY, Newark, Del.—*Guard Attachment for Cultivators.*—September 4, 1866; antedated August 28, 1866.—A guard plate on an arm pivoted to the side of the cultivator intervenes between the young corn and the outer share on that side to keep clods from the plant.

Claim.—The application of a guard or guards to a cultivator, substantially in the manner as and for the purpose herein set forth.

Also, the pivoting the bar F to which the plate or guard G is attached between plates E E: secured to the cultivator near its front end and provided with a rest b, substantially as described.

No. 57,748.—JOHN McKINLEY, Bethesda, Ohio.—*Plough.*—September 4, 1866.—The double-ended reversible point is capable of being used in eight different positions, and has slots in each pyramidal end for the insertion of the foot of the coulter and the forward end of the share.

Claim.—First, the point e, constructed substantially as described.

Second, the combination of the point e with the share e, coulter b, and mould board a, substantially as herein set forth.

No. 57,749.—G. H. MELLEN and J. C. HAZLETON, Washington, D. C.—*Burning Fluid.*—September 4, 1866.—Composed of naphtha, 40 gallons; carbonate of soda, 3 pounds; oil of sassafras, 4 ounces; alum, 2 pounds; gum camphor, ¼ pound; slippery elm, 2 pounds; hydrate of lime, 2 pounds; and essence of tar, 1 ounce.

Claim—An illuminating oil composed of the several ingredients named and of the proportions substantially as set forth.

No. 57,750.—JACOB H. MENDENHALL, Cerro Cordo, Ind.—*Churn.*—September 4, 1866.—The horizontally revolving dashers of the lower set are serrated and the upper radial arms are vertically adjustable on the dasher shaft by a spring stop.

Claim.—First, the combination of the dasher H, shaft G, adjustable gear wheels E and F, crank shaft B, and crank or cranks D with each other and with the box A and cover C, when said parts are constructed and arranged substantially as herein described and for the purposes set forth.

Second, the combination of the gathering board J and spring catch K with the dasher shaft G, substantially as herein described and for the purposes set forth.

No. 57,751.—NATHAN L. MILBURN, St. Louis, Mo.—*Revolving Ordnance.*—September 4, 1866.—The radiating barrels are attached to a central hub, which has trunnions on which it is vertically vibrated for elevation and depression by the screws near the loading platform.

Claim.—First, the arrangement of the radiating series of barrels to revolve upon a central pin b furnished with trunnions c c, and applied to operate substantially as herein specified.

Second, the combination of the curved bar L, the screws M M, the rock shaft K, and the clamps p q, the whole applied in combination with the barrels and carriage, to operate substantially as herein set forth.

No. 57,752.—L. B. MOORE, Janesville, Wis.—*Corn Cultivator.*—September 4, 1866.—The shovels have a lateral reciprocation produced by gearing from the driving wheel.

Claim.—The construction of a corn cultivator by the combination and arrangement of the various parts, substantially as they are described in the foregoing specification, or their mechanical equivalents, when used to produce the said automatic reciprocating motion of the said levers *j j* and shovels X X, as specified.

No. 57,753.—J. G. MORGAN, Colton, N. Y.—*Machine for Shaving Hoops for Casks.*—September 4, 1866.—The knives are attached to the stationary and pivoted bar respectively and the hoop is drawn between them and up the inclined slotted table by nippers attached to a strap that winds on a rotating axis.

Claim.—First, the combination of the stationary bar C, the pivoted bar E, the knives H, levers J and L, and gauge K, substantially as herein described and for the purpose set forth.

Second, the combination of the inclined slotted table M, the nippers O, the sliding block P, strap S, clutch *t* V, lever W, and drum T on shaft U, with the stationary and pivoted bars C and E, substantially as herein described and for the purpose set forth.

No. 57,754.—ELLIS NORDYKE and ADDISON H. NORDYKE, Richmond, Ind.—*Grinding Mill.*—September 4, 1866.—The metallic eye has lugs to secure it in place in the runner and a bail to rest upon the top of the spindle.

Claim.—The herein-described metallic eye for millstones, when constructed and operating as described.

No. 57,755.—EDWIN OSBORNE, Philadelphia, Penn.—*Tool for Making Wooden Legs.*—September 4, 1866; antedated August 23, 1866.—For making the countersinks in the ankle joints of wooden limbs. A plug fits into the hole that is bored through the ankle piece and has a guide slot in its upper end which conducts the shaft of the revolving burr cutters and gauges the depth of their cut.

Claim.—The plug D, in combination with the burr cutters, constructed, arranged, and operating substantially in the manner and for the purpose specified.

No. 57,756.—EDWIN OSBORNE, Philadelphia, Penn.—*Boring Tool for Making Wooden Legs.*—September 4, 1866; antedated August 23, 1866.—The boring bit has a reamer attached which cuts out the enlarged portion and has a disk to limit the penetration and insure uniformity.

Claim.—The combination of the O G or curved blades H with the burr or other cylindrical cutter for forming the orifice in the ankle portion of the artificial legs.

Also, the combination of the head piece I with the curved blades H and burr E for insuring perfect uniformity in depth and gauge of cavity, substantially as described.

No. 57,757.—NELSON PECK, Jay, N. Y.—*Earth Scraper.*—September 4, 1866—The axle of the carriage is bent and passes behind the scraper which is pivoted to the tongue and by a lever is thrown into scraping or discharging adjustment in connection with the motion of the team.

Claim.—First, an improved scraper formed by combining the lever G, bars F, bars I, levers J, and bars K with each other and with the frame L, tongue B *b'*, scraper A, and draught bar D, substantially as described and for the purpose set forth.

Second, the combination of the wheels N and axle O with the frame L of the scraper A, when the axle O is made and attached substantially as herein described and for the purpose set forth.

No. 57,758.—JOHN PECKHAM, New Haven, Conn.—*Buckle.*—September 4, 1866.—The wire is passed through the loops of the frame, the ends bent towards each other, and turned at right angles.

Claim.—The combination of the frame A and tongue B, formed and hinged together, in the manner herein set forth.

No. 57,759.—JOHN PFEIFER, Philadelphia, Penn.—*Handle for Coal Shovels.*—September 4, 1866.—The shank of the shovel passes through the handle, and is prevented from turning therein by radial flanges.

Claim.—The construction of the handle with the metallic neck B, shoulder *d*, and vanes *f f*, in combination with the wooden handle D, substantially as and for the purpose herein specified.

No. 57,760.—BURRILL and EDWIN PICKERING, West Milton, Ohio, and BARTON PICKERING, Montgomery county, Ohio.—*Pump Piston.*—September 4, 1866.—The lower cup-shaped portion has an inner perforated disk, which forms the valve-seat and confines the edge of the flaring leather packing.

Claim.—First, the vertical part of the packing piece A, having an inclined surface, as represented, for the purpose of holding the flaring packing F, when combined with the rod C and valve seat B, substantially as described and represented.

Second, the arrangement of the pieces A B, packing F, valve E, and pump rod C, substantially as described.

No. 57,761.—ROBERT G. PIKE, New York, N. Y.—*Lighting Gas by Electricity.*—September 4, 1866.—Explained by the claims and cut.

Claim.—First, a plate for deflecting and spreading the gas as it comes from the burner before striking it with a spark, so as the more readily to mingle the air with it before striking, and also for the purpose of directing the gas to the place of striking, substantially as described.

Second, the combination of the metallic gauze *a*, or perforated plate, with the tube or cap or curved plate, and also with the deflector or spreader, substantially in the manner and for the purpose described.

Third, the metallic button, or its equivalent, upon the deflecting plate, operating substantially as described.

Fourth, the combination of the gauze *e*, deflector D, and boss *n*, substantially as described.

No. 57,762.—JAMES K. P. PINE, Troy, N. Y.—*Neck-tie.*—September 4, 1866.—A card has a painted imitation of a neck-tie, and is secured in the usual manner.

Claim.—The imitation neck-tie, herein described, adapted for use with a turn-down collar, and consisting of paper of any desired quality or thickness, the surface being ornamented by printing, embossing, painting, staining, or otherwise.

No. 57,763.—ANSON H. PLATT, Ann Arbor, Mich.—*Floor Covering.*—September 4, 1866.—Composed of varnished wall paper united by paste to strong roofing paper, and rendered water-proof by means of a composition of linseed oil, rosin, and hydrocarbon oil.

Claim.—First, the application of paper printed in water colors to heavy base paper, previously made water-proof by the use of the "water-proof compound," as a substitute for oil cloths and carpets, as herein described, under the head of "hand-made variety of paper floor covering."

Second, the application of ornamental figures printed in water colors directly upon heavy, strong paper, previously made water-proof by the use of the "water-proof compound," (or by any other similar compound,) combined with other articles to form a paint, as herein described, under the head of "factory-made variety of paper floor covering."

Third, the "water-proof compound" and "enamel coating," as herein described, for the uses and purposes herein specified.

No. 57,764.—JOSEPH H. PULTE, Cincinnati, Ohio.—*Apparatus for Spreading Cement.*—September 4, 1866.—A hopper filled with liquid cement is placed above a rotary drum, around the inner surface of which passes the cloth to be coated ; the adjustable side gauges the quantity, and a roller presses the cloth against the drum.

Claim.—First, a cement-spreading machine, whose hopper B is provided with an adjustable gate C, arranged and operating substantially as herein described and set forth.

Second, in combination with the adjustable gate C the trowel D, as and for the purpose explained.

Third, in combination with the elements of the two preceding clauses, the set screw E, or its mechanical equivalent, operating as herein explained and described.

Fourth, in combination with the drum G the pressure roller I, for the purpose set forth.

No. 57,765.—STEPHEN R. RAMSDELL, Providence, R. I.—*Fender for Carriage Wheels.*—September 4, 1866.—Explained by the claim and cut.

Claim.—First, a rotating fender, provided with a projection at one end and a recess at the other, for the reception of an adjustable centre pin, in order that said roller may be placed in or removed from its bearings, or adjusted therein with facility, in the manner described.

Second, a bracket having arms provided with bearings for said rotating fender, and set at such an angle with the side of the carriage on which it is placed that the wheel when in contact with said fender shall present to it as large a portion of the surface of its rim as possible, or, in other words, shall be nearly or quite at right angles therewith, substantially as set forth

No. 57,766.—F. J. RAUSSCHERT, Buffalo, N. Y.—*Fruit Gatherer.*—September 4 1866.—By the described devices the canvas is adjusted to the body of the tree, and spread to receive the fruit and conduct it to the basket.

Claim.—The combination of the strap A, pole D, trough F, having spout H, and canopy B, when all constructed and arranged together, substantially as and for the purpose described.

No. 57,767.—JOHN H. REED, New Haven, Conn.—*Car Coupling.*—September 4, 1866.—The swinging coupling pin is attached by a link to a rock shaft, by which it is lifted out of engagement with the coupling link; the pin is held in this position by a spring block, and when lowered is vibrated by the contact of the coupling link, and falls automatically into engagement therewith.

Claim.—The combination of the toggle or coupling pin *c*, rock shaft and crank *k* and *m*, and the spring E, with its appendages *d*, when the whole is constructed, arranged, and combined substantially as herein described and set forth.

No. 57,768.—WILLIAM A. RICHARDSON and HENRY D. WARD, Worcester, Mass.—*Lamp Extinguisher.*—September 4, 1866.—The pivoted cap is vibrated by an operating rod so as to cover the end of the wick tube.

Claim.—The combination of loop B, cap or cover C, and operating rod D, with the tube A and top of the lamp, substantially as and for the purposes set forth.

No. 57,769.—CHARLES RICHMOND, Worcester, Mass.—*Tool.*—September 4, 1866.—The movable jaw of the wrench is held by a spring pawl acting in conjunction with a series of notches on the edge of the shank. In the end of the handle is a socket and spring pawl which receive and hold either bit of a double-bitted tool.

Claim.—The improved compound tool, consisting of the wrench socket and double bit, all constructed and arranged substantially as herein described.

No. 57,770.—E. R. RISON, Kinmundy, Ill.—*Bed Bottom.*—September 4, 1866.—The wires are attached at their ends to vertical rods which rest against resisting springs and confer elasticity upon the wires.

Claim.—The combination of the wires or cords C, the upright supporting pieces B, the gum elastic springs F, the plates E, and screws D, with each other and the frame A of the bed bottom, substantially as herein shown and described and for the purpose set forth.

No. 57,771.—BENJAMIN ROACH, Melrose, Mass.—*Shaft Coupling.*—September 4, 1866.—Disks on the ends of the shafts have recesses occupied by the flanges on the sides of the intervening plate. The disks and plates are clamped together by bolts.

Claim.—The arrangement and combination of the disk D and its ribs *b b*, with the two coupling heads C C, provided with grooves *a a* arranged in them to receive the ribs, as set forth.

No. 57,772.—WILLIAM H. ROBBINS, Richmond, Ind.—*Hog Trough.*—September 4, 1866.—The swill is placed in an upper trough and discharged into the lower one at all points simultaneously. The feed openings on the sides are alternate.

Claim.—First, the construction of trough A with the key or wedge bar C and lever B, all arranged and operating as described.

Second, the equal distribution of the feed to each hog through the length of the trough at the same time and in equal portions.

Third, this device of alternating the opening in the side of the trough, that more hogs can be accommodated in the same space than if they were all allowed on one side of the trough at the same time.

Fourth, the manner of constructing the trough so that the hogs cannot get into the feed, and each one be entirely alone in his mess, all operating in the manner and for the purpose substantially as set forth.

No. 57,773.—C. D. ROBERTS, Jacksonville, Ill.—*Ground Roller.*—September 4, 1866.—The rollers are connected by links, and their adjustable bearings allow them to conform to the irregularities of the ground.

Claim.—First, supporting the outer ends of the axles of rolls C C, in the hinged or vibrating boxes *a a*, and the inner ends of the same in free ends of bars *b b*, permitting the rolls thereby to adjust themselves to the inequalities of the ground, substantially as described.

Second, the connecting link or bar *c'*, in combination with rolls C C, hinged boxes *a a*, and bar *b*, when the rolls are arranged one in the rear of the other, as and for the purpose specified.

No. 57,774.—C. A. ROSE, Columbus, Ga.—*Material for Kindling Fire.*—September 4, 1866.—The acerose leaves of the pine are gathered and bundled for kindling.

Claim.—As an improved article of manufacture a fire kindler made of compressed pine leaves, as herein described.

No. 57,775.—FRIEDERICH SHALLER, Hudson, N. Y.—*Watchmaker's Lathe.*—September 4, 1866.—Explained by the claim and cut.

Claim.—The standard B, provided with the slot F and formed in one piece with the base C and arms D. in combination with the pins E, bar G, and rest I, the whole being constructed and arranged substantially as herein set forth for the purpose specified.

No. 57,776.—E. SAFFORD, Boston, Mass.—*Card Rack.*—September 4, 1866.—The slats have T-headed flanges which slip into corresponding grooves in the board.

Claim.—First, the peculiar method of shaping and holding in position the slats *s s' s'' s'''*, &c., substantially as described and for the purpose set forth.

Second, the combination of the slats, made and secured as described, with the board B and frame A, substantially as described and for the purpose set forth.

No. 57,777.—GEORGE S. SALSBURY, Clarendon, N. Y.—*Training Grape Vines.*—September 4, 1866.—Laterals from the main stem are bent over and become rooted in the ground, forming braces to sustain the vine.

Claim.—The peculiar manner of training and trimming the grape vine so as to make it self-sustaining, forming its own trellis, substantially as set forth, claiming the described method in its broadest sense.

No. 57,778.—HANS HENRIK SENNIKSEN, Richmond, Ind.—*Stove Pipe Drum.*—September 4, 1866.—The drum connects directly with the fire flue, unless the damper is closed when the current is deflected into side pipes, which lead into the drum into which it dives, and is then reverted upward into the exit flue.

Claim.—The combination of the pipes B and B', D and D', d and d', and the damper C, when arranged and operated as set forth and described.

No. 57,779.—JOHN L. SHAW, Fort Wayne, Ind.—*Railway Crossing.*—September 4, 1866.—The upper sills are imbedded in the lower; the ends of the short pieces forming the square are underlaid beneath the corners, which consist of bent rails.

Claim.—The railway crossing, consisting of the bed plates A, lapped and united as described, and used in combination with the rails B, substantially as described.

No. 57,780.—G. V. SHEFFIELD and J. F. COBURN, Hopkinton, Mass.—*Machine for Piercing Leather.*—September 4, 1866.—The leather is fed first to the piercing tool, and then to the die which forces the wire into the openings, the two operations being performed simultaneously in consecutive openings. The wire is fed by a cam acting on feed wheels, is cut by a chisel, and placed in proper position for the driving die.

Claim.—The combined arrangement of the feed wheel R, piercing tool X, hammer Y, feed rollers F², sliding chisel R², or its equivalent, substantially as herein described, and as and for the purpose specified.

No. 57,781.—JOHN W. SIBBET, Cincinnati, Ohio.—*Dies for Bolt Heading Machines.*—September 4, 1866.—The complete die is formed of four blocks. The two blocks on each side, constituting one-half longitudinally of the pair of dies, are bolted together, permitting the one to be turned upon the other, so that any one of four dies in the neck block may be placed in line with any one of four similar dies in the shank block, the other side being correspondingly arranged.

Claim.—First, improved dies, formed in sections and upon the several faces thereof, constructed, arranged, and combined with each other substantially as herein described and for the purposes set forth.

Second, the combination with the above of the headers K, constructed substantially as described and for the purpose set forth.

No. 57,782.—GEORGE W. SIGERFOOS, JOSEPH J. SANDS, and GEORGE FRY, Potsdam, Ohio.—*Gate.*—September 4, 1866.—The gate is supported on rollers, and is guided and sustained above by a hook fastened in the post.

Claim.—The combination of the posts B C D F, rollers m, stay piece n, and gate A, substantially as described and for the purposes set forth.

No. 57,783.—CHARLES G. SMITH, Chelsea, Mass.—*Machine for Making Metal Tubes.*—September 4, 1866.—A series of pairs of rolls, used in connection with a stationary mandrel of peculiar shape, by which the sheet of metal is first bent to the form of a trough, with one edge higher than the other; then into a tube, with its marginal surfaces bent up and in contact. The projecting edge is next bent over, and, lastly, the lap is bent down upon the tube and flattened.

Claim.—In combination with a stationary triblet or mandrel mechanism for feeding the plate, mechanism for bending part of the plate into a tubular form over the surface of the triblet or mandrel, and mechanism for forming the opposite edges of the plate into a lap joint, the whole operating together to form the plate into a tube, substantially as described.

No. 57,784.—WILLIAM S. SMITH, Noyesville, Ill.—*Machine for Sinking Hollow Piles*—September 4, 1866.—The contents of the hollow pile are ejected by a current of air, the discharge pipe having a flexible covering.

Claim.—The method of excavating solid materials from the interior of hollow piles by means of a current of air, using for this purpose the flexible discharge pipe, as herein described.

No. 57,785.—YOUNGS W. SMITH, Bristol, N. Y.—*Fruit Gatherer.*—September 4, 1866.— The canvas is lapped around the tree and buttoned; the corners are adjustably supported on stakes to trim the canvas, which delivers the fruit through an opening to a basket beneath.

Claim.—The combination of the endless adjusting cords *f f* and loops *k k*, with the canvas A and bracing stakes B B, operating substantially in the manner and for the purpose herein specified.

No. 57,786.—JOHN SOLTER, Baltimore, Md.—*Tumbler Washer.*—September 4, 1866.—To place the inverted tumbler in position the lever is moved, which turns on the jet to cleanse the inside of the tumbler.

Claim.—The employment of a lever K, or its equivalent, operating the valve stem *s'* and valve *s*, when in combination with the rim *n*, for holding the tumbler, arranged substantially as and for the purposes set forth.

No. 57,787.—DANIEL E. SOMES, Washington, D. C.—*Generating Gas for Motive Power.*— September 4, 1866.—A mixture of various substances, such as gunpowder, hydrocarbon oils, nitro-glycerine, &c., is converted into gas for use as a motor by plunging it into a heater cylinder.

Claim.—First, combining nytro-glycerine with alkali, and converting the same into gas to be used as a motive power.

Second, combining any kind of oil or fatty matter with alkali, and converting the same into gas to be used as a motive power.

Third, compressing gas, air, water, steam, or any other liquid or volatile substance, substantially as and for the purpose herein described.

Fourth, the apparatus herein described, or its equivalent devices, for compressing gas, air, water, steam, or any liquid or volatile substance, and using the same as a motive power.

No. 57,788.—JAMES F. SPENCE, Williamsburg, N. Y.—*Apparatus for Carburetting Air.*— September 4, 1866.—Within the carburetting vessel are two drums, which draw in the air through the hollow axles and discharge it at their peripheries. The entering air is heated. The oil is supplied from a tank surrounded by a steam jacket, and the quantity is regulated by a float and stop valve.

Claim.—First, the case A, provided with two or more air-or steam wheels B B, working in the liquid, in conjunction with each other, substantially as and for the purpose set forth.

Second, heating the oil before it enters the machine by the jacket E, surrounding the supply tank D, in combination with a hot-air or steam pipe *b*, or any other suitable means, substantially as and for the purpose described.

Third, the hot-air chamber F, in combination with the burner *a*, case A, and jacket E, constructed and operating substantially as and for the purpose set forth.

Fourth, the float *b*, provided with a valve *f*, in combination with the liquid supply pipe *g* and case A, constructed and operating substantially as and for the purpose described.

No. 57,789.—WILLIAM S. SPRATT, West Manchester, Penn.—*Horse Hay-fork.*—September 4, 1866; antedated August 17, 1866.—The tines are attached to the plunger, and worked by the lever and link by means of a rope, which passes over a sheave to the end of the lever. The plunger has a cross arm guided by the side bars of the frame.

Claim.—The combination and arrangement of the rod *b*, provided with guide *i* and prongs *g* and *h*, link or rod *c*, lever *d*, pulley *o*, when used in connection with the frame *a a*, constructed, arranged, and operating in the manner herein described and for the purpose set forth.

No. 57,790.—ALVIN G. SQUIRE, Pelham, Mass.—*Manufacture of Scythe Stones.*—September 4, 1866.—A steel wire is fitted into a groove around the edge of a scythe stone or rifle, by means of which the edge of the scythe may be burnished after sharpening.

Claim.—The cast steel band and the mode of attaching it to the stone or wood, prepared as above stated, and to be used in connection with such stone or wood for the purpose of sharpening scythes, edge tools, and other implements requiring a sharp edge.

No. 57,791.—M. B. STAFFORD, New York, N. Y.—*Construction of Jointed Moulds.*—September 4, 1866.—The mould is jointed in the middle, but the floor is attached to one section only, so that the brick is not defaced at the junction or hinging line.

Claim.—A jointed mould composed of two parts *b b*, connected together and constructed substantially as herein shown and described, so that when said parts are closed a smooth interior is obtained, and the article or substance compressed and moulded without leaving any crease, impression, or ridge, as set forth.

No. 57,792.—ADDISON F. STILLWELL, Fayette, Iowa.—*Cultivator.*—September 4, 1866.— Forward of the rear plough are two lateral shares which are attached by a brace frame to the beam, and are adjustable as to presentation and relative distance by the vibration of the jointed standards and the pivoted share.

Claim.—The bar E, beam A, and cross-bars G, in combination with the bars I, projections *b*, spurs *f*, shares J and brace rods K, all arranged to operate as and for the purposes set forth.

No. 57,793.—S. W. STOCKTON, Philadelphia, Penn.—*Attaching Artificial Teeth to Bases.*—September 4, 1866.—The tenon on the porcelain block is embraced by the plastic material of the base, and when the latter is hardened connects the two securely.

Claim.—Securing artificial teeth and gums to plastic bases by means of the tenons *d*, arranged along on the rear part of that portion of the porcelain blocks which projects inward just above the teeth, substantially as shown in the drawings and herein described.

No. 57,794.- HENRY C. STOLL, Mokena, Ill.—*Churn.*—September 4, 1866.—The vertical vibration of the lever acting upon the twisted blades of the dasher shafts rotates them in alternate and diverse directions. The blades may be disconnected and a vertical motion imparted to the dashers by link connection with the lever.

Claim.—First, the arrangement and combination of the twisted parts B with the lever C and standard A as set forth.

Second, the combination of the support D, slide L, and dasher rod F, substantially as described.

No. 57,795.—J. N. STURTEVANT and HARVEY E. JONES, McGregor, Iowa.—*Tanning.*—September 4, 1866.—The hides are steeped in a bath composed of water, salt, and sulphuric acid, after which they are coated with a composition of lye, tanner's oil, and neat's-foot oil, and are placed in an oven through which a current of hot air is passed in order to saturate the hides with said composition.

Claim.—The within described process for tanning leather, when used substantially as herein specified.

No. 57.796.—GEORGE W. THOMPSON, Ripley, Ohio.—*Plough.*—September 4, 1866.—The double ended and reversible mould board is attached by a universal joint to the standard, and is braced in the rear by pivoted supporting bars.

Claim.—First, the attaching of the mould boards F* F* to the standard E by means of the universal joint composed of the swivel bolt *a* and hinge or joint *b*, substantially in the manner as and for the purpose set forth.

Second, the brace F' applied to the beam A and land side F', substantially as and for the purpose specified.

Third, the combination of the land side F, standard E, and the mould boards F* F* attached to the standard by the universal joint, substantially as and for the purpose set forth.

Fourth, the fastenings composed of the pivoted bars G G attached to the beam A, substantially as and for the purpose specified.

No. 57,797.—GEORGE FREEMAN UNDERHILL, New York, N. Y.—*Rock Drilling Machine.*—September 4, 1866.—The drill is raised, dropped and intermittingly rotated by the revolution of the crank, the upper portion of the framework being adjustable to give the required angle of presentation to the drill.

Claim.—First, the divided framework A hinged together, having the drill rod H arranged in its upper section B, in combination with the clamping devices for securing the sections of the framework together, substantially as and for the purpose described.

Second, the arrangement of the sliding or lifting frame M, tappet shaft Q, and the polygonal-shaped drill rod H, substantially as and for the purpose described.

Third, the eccentric V of shaft Q, yoke or collar W, lever Y, arm A², frame B², ratchet wheel D² and pawl E². when all arranged and combined together, so as to operate upon the drill rod, substantially in the manner and for the purpose specified.

Fourth, the use of rubber cushions upon the under side of the base portion of the framework, for the purpose specified.

No. 57,798.—MAURICE VERGNES, New York, N. Y.—*Propelling Apparatus for Boats.*—September 4, 1866; antedated August 3, 1866.—The paddle shaft is pivoted at a certain point to a revolving crank, and at its upper end is jointed to an arm whose pivot of vibration slides longitudinally of the vessel. The effect is to give a vertical dip and to retract without lifting water.

Claim.—First, the erection of the supporting arms movable at base and apex, spread at the base for solidity and strength, in combination with the oar arms to guide, steady, and support them, in the manner and for the purpose described.

Second, hinging the supporting arms upon a carriage set on a rail, which can be moved to and fro to vary the dip of the oar in the water, and for the purpose of removing the oar from the water, in the manner described.

No. 57,799 —GEORGE I. WASHBURN, Worcester, Mass.—*Steam Engine.*—September 4, 1866.—The cylinder has double-acting pistons on each side of the diaphragm which is traversed by the piston rod; a spool valve reciprocates in a chamber alongside, admits steam to each piston, and is operated by the connection of its rod to the piston rod.

Claim.—The arrangement in the diaphragm cylinder of the two pistons on the same rod, operated as described.

Also, in its arrangement with the double cylinder and pistons, the single valve controlling the steam openings, substantially as described.

Also, the arrangement of the valve chest F, double cylinder A A', and side chests or pipes K L, the latter communicating, each by a single port, with the chest F, and simultaneously, by duplicate ports, with the spaces on corresponding sides of the two pistons.

No. 57,800.—WILLIAM H. WATSON, Yonkers, N. Y.—*Machine for Pressing Tobacco.*—September 4, 1866; antedated August 21, 1866.—The endless chain has pockets to receive the wads of tobacco, which are passed beneath the pressure blocks on the upper endless belt, the operation being effected in a chamber, traversed by currents of heated air.

Claim.—First, the chain 5, constructed and operating substantially as described, for the purposes specified.

Second, the pressing blocks 36, constructed and operating substantially as described for the purposes specified.

Third, in combination with the chain 5, constructed and operated substantially as shown, the pressing blocks 36, constructed and operating substantially as shown, for the purposes shown.

Fourth, the heating chambers, constructed and operating substantially as described for the purpose specified.

Fifth, in subjecting the tobacco or other substance to be pressed to the influence of heat while under pressure, as shown for the purposes designated.

No. 57,801.—GEORGE WATT, Richmond, Va.—*Whiffletree.*—September 4, 1866.—When force is applied the curved portions are somewhat straightened, and confer an elasticity.

Claim.—First, the construction of a double, single, or treble tree, so that it by means of one or more of its bent sides shall form an elastic connection between the draught animal and the object (wagon, plough, &c.,) as described.

Second, the attachment of the double tree by its longest side to the plough beam, as and for the purpose described.

No. 57,802.—W. J. WELLS, Sidney, Ohio.—*Weeding Hoe.*—September 4, 1866.—The curved shank has a cross-bar, with a bent and a straight pointed prong.

Claim.—The cross-arm D having its bottom F and slides G bevelled and pointed at its ends, substantially as and for the purpose described.

Second, in combination with the above the side arm E, for the purpose specified.

No. 57,803.—MILTON WHIPPLE, Medina, N. Y.—*Driving-Rein Holder.*—September 4, 1866.—A spring hook behind the dashboard to hold the reins.

Claim.—The device herein described, consisting of the parts A and B, constructed so as to operate substantially as described, and designed for holding the driving reins of horses while temporarily leaving a carriage.

No. 57,804.—THOMAS H. WHITE, Orange, Mass.—*Knife Sharpener.*—September 4, 1866.—Two cutting disks are pivoted to the stock, which has likewise a finishing strap attached.

Claim.—The arrangement and combination of the circular disks B B', with the stock A, and the strap or bone C, substantially as described and for the purpose set forth.

No. 57,805.—WILLIAM J. WILCOX, New York, N. Y.—*Cooling Lard.*—September 4, 1866.—An ice-cooled blast is driven through the lard.

Claim.—The within described method of cooling lard by passing over or through the same one or more impelled currents of cold air, substantially as and for the purpose described.

No. 57,806.—HENRY F. WILLSON, Elyria, Ohio.—*Evener for Poles for Wgons, &c.*—September 4, 1866; antedated August 15, 1866.—When the evener is at right angles to the pole the fulcrum pin is in the middle of the length of the evener, but if one end of the latter be advanced it becomes proportionately shortened.

Claim.—The radial c and stationary pin or bolt b, in combination with curved slot a and stationary pin d, the whole being constructed in the manner and for the purpose set forth and described.

No. 57,807.—GEORGE W. WILSON, Abingdon, Ill.—*Car Coupling.*—September 4, 1866.—Each draw head has a pivoted hook and a socket to engage and be engaged by corresponding devices on the draw head of the other car; to uncouple, the hooks are raised by cams.

Claim.—Constructing a car coupling of two double slotted blocks A A, with hooks *b b* on one of the prongs *a a* of each block, fitting into corresponding grooves in the prongs *a' a'* of the other block, combined with the shackle *c c* and the cams *f f*, constructed, arranged, and operated as and for the purposes herein described.

No. 57,808.—O. F. WINCHESTER, New Haven, Conn.—*Magazine Fire-arm.*—September 4, 1866.—Improvement of the Smith and Wesson patent, February 14, 1854. Improved by Henry, October 16, 1860. The metallic cartridges are carried by the movement of the trigger guard from the magazine tube beneath the barrel; the cartridges are placed in an inner removable magazine tube. The forward motion of the trigger guard is arrested at a certain point by a pivoted stop.

Claim.—First, constructing the tube or magazine, substantially in the manner described, so that the inner tube may be removed, in combination with the carrier E, breech pin L, and barrel A, as and for the purpose specified.

Second, the combination of the stop S', lever H and carrier block E, when arranged to operate substantially as and for the purpose specified.

No. 57,809.—BENJAMIN L. WOOD, Taunton, Mass.—*Carriage Thill.*—September 4, 1866.—The book of the thill catches on the pin of the clip, and is preserved from disconnection therefrom by a strap which passes through eyes in the thill hook, and the lower bar of the clip, respectively. A strap between the thill hook and clip pin prevents rattling.

Claim.—The improved shaft or pole connection as made with a hook *c*, and an aperture *d* therein, arranged with the start-bolt *a*, and to receive a strap, or its equivalent, as specified.

Also, the arrangement of the safety strap G, to pass through the aperture of the hook, as described.

Also, the combination of a strap, or its equivalent, to go through the eye of the book, with such hook, and the shaft, or its equivalent.

Also, the arrangement and application of the anti-friction strap I with the shaft, or its equivalent, and the hook *c*, provided with an aperture *d*, as and for the purpose described.

No. 57,810.—L. E. WOODARD, Cohocton, N. Y.—*Wagon Brake.*—September 4, 1866.—The rubber bars are pivoted to the coupling pole, and are operated by an eccentric, rotated by the longitudinal motion of the tongue in backing.

Claim—The combination of the eccentric L, rod M, pole H, and friction roller with the brake *c*, bars I I, when constructed for the purposes and substantially as herein described.

No. 57,811.—ALFRED WOODWORTH, North White Creek, N. Y.—*Water Elevator.*—September 4, 1866.—The crank is pivoted to a shell upon the roller, and when revolved, binds the heel of its shoe against the roller, to rotate it; a fulcrum catch, which is brought into play when required, removes the heel, and brings a frictional pressure of a toe upon the roller when the bucket is descending.

Claim.—The pulley E, keyed on one end of the windlass shaft *a*, and having a cylindrical shell F placed over it, with a handle or crank G pivoted in the shell, and provided with a shoe *c* at its inner end within the shell, in combination with the pivoted bar or stop H on the curb, all arranged to operate substantially in the manner as and for the purpose set forth.

No. 57,812.—THOMAS D. WORRALL, New York, N. Y.—*Apparatus for Carburetting Gas.*—September 4, 1866.—The carburetting chamber is formed in the gas pipe, or an enlargement of the gas pipe, as shown. The chamber may be filled with fibrous material if desired.

Claim—First, introducing into a gas pipe carbon spirit, for the purpose of enriching, purifying, or increasing the quantity of common gas, water gas, or common air.

Second, introducing into a gas pipe fibrous material of any desired or suitable kind, for the purpose of drawing up or letting down carbon spirit, so as to vaporize said spirit, for the purposes set forth.

Third, the use of a large gas pipe, into which smaller ones conduct, or out of which they convey any kind of gas, so as to form a reservoir in which said gas can be detained for a long time, while being enriched by the vapors of carbon spirit, or other carbonizing fluid.

Fourth, inner casings of gas pipe of any desirable device, made to hold carbon spirit or other carbonizing fluids, and also to contain fibrous material for holding in suspension and vaporizing the same, while ordinary gas, water gas, or common air is passing through, over, or under them.

Fifth, the gas pipe Fig. 1, with chamber in the bottom for holding any carbonizing material, for the purposes described.

Sixth, the gas pipe No. 2, with chamber and fibrous material stretched horizontally along it, and from which the ends of other fibrous material drop into the carbonizing fluids, and convey them by capillary attraction to those stretched along it, for the purposes set forth.

Seventh, the gas pipe No. 3, with holes drilled in or through the top, for the purpose of suspending wicking or other fibrous material that shall hang in carbon spirit, and drawing up said spirit, for the purposes set forth.

Eighth, the gas pipe No. 4, in which strips of wood or wire run along the top of the pipe, on the inside, either across or in a longitudinal direction, for the purpose of holding wicking or other fibrous material, while the lower ends of the same are immersed in the spirit or fluid, for the purpose set forth.

Ninth, the gas pipe No. 5, in which is inserted an inner casing or tube perforated with numerous holes, and through which cotton wicking or other suitable material is drawn, so as to form a perfect retina or network. in which the carbon vapors are thrown off, and through which any kind of gas may at the same time pass, for the purpose of being enriched or multiplied thereby.

Tenth, the gas pipe No. 6, in which is contained a smaller perforated pipe or tube, around which and through which cotton or other fibrous material is passed, and over the whole of which a series of broad bands of wicking or a continuous apron is passed, for the purpose set forth.

Eleventh, the perforated tin or wire gauze inserted in a gas pipe, for the purpose of distributing the gas to any or all parts of the pipe, as set forth.

Twelfth, gas pipe Fig. 8, with longitudinal partitions, with or without coverings of fibrous material, to insure that the gas to be enriched may run back and forth, for the purposes set forth.

Thirteenth, gas pipe Fig. 9, in which a series of partitions or chambers, each separate or all connected by apertures, is formed, and over which fibrous material is laid, in the manner and for the purpose described.

Fourteenth, gas pipe Fig. 10, with loopholes suspended from the top, through which cotton or other fibrous material may pass and be suspended in carbonaceous fluids, for the purposes set forth.

Fifteenth, gas pipe Fig. 11, in which is a spiral or screw-shaped pipe, cased or surrounded with fibrous material, around which gases and the vapor of carbonizing fluids may pass, for the purposes set forth.

Sixteenth. gas pipe Fig. 12, inside of which is a wire tube, around, along, and across which fibrous material may be stretched, and from which it may hang suspended, for the purposes set forth.

Seventeenth, the compound gas pipe Fig. 13, two or more in any way connected together, for the purpose set forth.

Eighteenth, gas pipe Fig. 14, in which several chambers are cast or otherwise constructed, so as to contain carbon spirit or other carbonaceous fluid, and in which said fluids may be transmitted from chamber to chamber, by means of fibrous material, or in which in any other way the fluids may be vaporized, for the purpose described.

Nineteenth, the arrangement, in combination with any of the devices, of a gas burner that can be turned and lighted under the reservoir, for the purpose set forth.

Twentieth, the use of each of the devices seen within the gas pipes Fig. 5, Fig. 6, Fig. 9, Fig. 10, Fig. 11, and Fig. 14, for use in any other box or chamber, as well as in gas pipes.

No. 57,813.—WILLIAM YAPP, Cleveland, Ohio.—*Eave Trough.*—September 4, 1866.—The trough has an upper brace, and an inner clamp, which are united by a bolt to the arm, which is secured to the building.

Claim.—The brace B, with one end forming a loop, and the other a lap, in which is formed a concave c, in combination with the clamp C, pivot a, trough A, and arm D, in the manner and for the purpose substantially as set forth.

No. 57,814.—A. ZINK, Lancaster, Ohio.—*Lifting Jack.*—September 4, 1866.—The supporting link rests in one of the vertical series of notches in the back of the post, and the lever when depressed, is locked by the pawl above.

Claim.—The shape and construction of the lever jack, when arranged with the shifting link and adjustable catch, as secured thereto, as herein described, and for the purposes set forth.

No. 57,815.—JACOB BALLARD. New Antioch, Ohio, and THOMAS J. MAGER, Cincinnati, Ohio, assignors to themselves and PAUL HULTS, New Antioch, Ohio.—*Ditching Machine.*—September 4, 1866.—The forward carriage straddles the ditch, and the rear supporting-wheel runs in the ditch behind the cutting and elevating mechanism The share is supported by coulters, which cut the sides of the ditch, and delivers the furrow slice to the guides upon which it rises and the mouldboards, which deliver it on the side of the ditch. Adjustments for varying depths are recited in the claims.

Claim.—First, the arrangement of sliding mouldboard J, lower and upper guides D and I, elevating mechanism K L, and adjustable brace P, or their mechanical equivalents, substantially as set forth.

Second, in the described combination, the beam A, sloping sheath B, share C, and the coulters G and H, as and for the purpose set forth

Third, the clinometer attachment W X, in combination with a supporting truck T, and regulating screw U, for the purpose explained.

Fourth, the shiftable handle N, and ditch wheel O, secured and operated as set forth.

No. 57,816.—JOHN W. DRAPER, Wilmington, Del., assignor to himself and ARTHUR C. STOWELL.—*Railway Chair.*—September 4, 1866.—The cheeks embrace the rail, to which they are bolted, and with the latter set within the enclosing flanges of the chair.

Claim.—The combination of the plate A, its bevelled lugs c c', cheeks C C', and bolts D D', the whole being constructed, arranged, and adapted to the rail, substantially as and for the purpose specified.

No. 57,817.—JOSEPH EVANS, Newark, N. J., and ROBERT H. SEYMOUR, Bloomfield, N. J., assignors to HENRY SEYMOUR, Elizabeth, N. J.—*Pruning Instrument.*—September 4, 1866.—The diamond-shaped blade is pivoted to a push-rod, and acts in conjunction with a stationary book on the end of the pole; the knife is rocked on its pivot by a guide-book, which traverses a groove near one edge of the knife, and gives it a "draw" cut.

Claim.—First, constructing the cutting blade C with a groove a along its upper back edge, and so arranging in combination therewith a cap or clamp D, carrying a pin b, that the blade will be guided so that it will operate upon the twig with a drawing cut, in a direction from the bend in the hook, substantially as described.

Second, the spring lever E E', in combination with the cutting blade of a pruning knife, constructed and applied substantially as described, whereby the knife may be operated in a quick and easy manner.

Third, the combination with each other of the diamond-shaped knife C, hook B, rod F, and spring levers E E', arranged and operating substantially as herein shown and described.

No. 57,818.—JOHN FAHRNEY, Boonsboro, Md., assignor to himself and SAMUEL FAHRNEY, same place.—*Faucet.*—September 4, 1866; antedated August 23, 1866.—The molasses follows the draught of the piston, and fills the chamber; during the return of the plunger the molasses flows through the piston, and is ejected at the second draught.

Claim.—First, the cylinder A, with valve H, in combination with piston I, with its openings a a, its rod G, sliding through and turning in its centre, as shown at c d, and the circular plate J, attached to end of the piston rod, as and for the purposes described

Second, in combination therewith, the cut-off piston F, neck E, with its lower wall removed as shown at M, as and for the purposes described.

Third, the nut H, with its projection e, plate L, and gauge screws f f, as and for the purposes described.

No. 57,819.—LUCIUS H. GOFF, St. Albans, Vt., assignor to THOMAS C. WINSLOW, same place.—*Barrel Lifter.*—September 4, 1866.—Explained by the claim and cut.

Claim.—The barrel lifter herein described, the same consisting of a notched bar or lever B, having hooks C C hung upon it, and moving under bands D, or their equivalents, substantially as herein described, and for the purpose specified.

No. 57,820.—W. A. HORRALL, Washington, Ind., assignor to himself and McCRELLIS GRAY, same place.—*Whiffletree.*—September 4, 1866.—The curved whiffletree is attached by a bow to a catch on the thill; when the catch is withdrawn the tree rotates and allows the traces to slip off its ends.

Claim.—The combination of the lever H, spring I, arm G, spring K, bar J, band D, and whiffletree B, with each other, and with the cross-bar a' of the thills, when said parts are constructed and arranged substantially as herein described and for the purpose set forth.

No. 57,821.—CHARLES M. HYATT, Albany, N. Y., assignor to LANSINGH & OSBORNE, same place.—*Clasp for Holding Necktie and Shirt Collar together.*—September 4, 1866.—The clip embraces the band of the collar, and presents spurs, which hold the necktie in position.

Claim.—The within-described attachment for securing the necktie to the collar, consisting of the clip A B, and of spurs p, combined and arranged substantially as set forth.

No. 57,822.—ARMON KING, Utica, N. Y., assignor to himself and JOHN H. CHAPMAN, same place.—*Moulding.*—September 4. 1866.—When the matches are brought together they and the pattern become a mould in which the core may be formed in place; by removing the matches the other part of the pattern may be moulded in the usual way.

Claim.—Match plates constructed substantially as described and used in connection with patterns for forming cores for sand moulds, in the manner set forth.

No. 57,823.—MATTHEW B. MASON, Aurora, Ind., assignor to himself and GEORGE W. HARRIS, same place.—*Piston for Steam Cylinders.*—September 4, 1866.—The steam is admitted by the motion in alternate directions of the double-headed valves into the hollow piston, and passes through the perforated ring against the inner surface of the expansible sectional ring and the break-joint middle ring imbedded therein.

Claim.—First, the arrangement of the L-shaped packing rings I I', and middle ring J, as described and for the purpose specified.

Second, the grooved and perforated ring G, constructed as and for the purpose specified.
Third, the double headed valves H, constructed and arranged substantially as described and for the purpose specified.

No. 57,824.—GEORGE MUNGER, New York, N. Y., assignor to himself and J. W. SCHERMERHORN, same place.—*School Desk and Seat.*—September 4, 1866.—The wooden portions have dovetail mortises and the metallic frame co-acting tenons by which the different pieces are united.

Claim.—As an article of manufacture, a desk consisting of the standards A, brackets *a*, both having their top edges dove-tailed, grooved seat B, grooved top C, grooved back strips *c*, grooved shelf E, with stops *f*, constructed and combined as and for the purpose specified.

No. 57,825.—NEWTON A. PATTERSON, Kingston, Tenn., assignor to himself and R. K. BYRD, same place, and M. L. PATTERSON, Knoxville, Tenn.—*Receiving and Discharging Freight.*—September 4, 1866.—The frame bridge-way is made in sections and has tracks above and below, which are travelled by cars propelled by an endless rope.

Claim.—First, a frame bridge-way made in sections and provided with upper and lower tracks D E, substantially as described and for the purpose set forth.

Second, the combination of the cars G, constructed as described with the sectional bridge-way, and with the revolving cylinders F, substantially as and for the purpose set forth.

No. 57,826.—JOHANN N. PETELER, Sheppach, Bavaria, assignor to ALOIS PETELER, New Brighton, N. Y.—*Portable Railroad.*—September 4, 1866.—The arrangement cannot be briefly described.

Claim.—First, a portable railroad composed of sections A, turnouts B, supporting frames C, bridges D, crossings E, and one or more turn tables F, all constructed, combined, and operating substantially as and for the purpose set forth.

Second, the combination of the perforated studs *a* and turn table F, provided with rails *b*, constructed and operating substantially as described, for the purpose specified.

No. 57,827.—COLIN MATHER, Manchester, England, assignor to CHARLES P. BUTTON, New York, N. Y.—*Machine for Boring Wells.*—September 4, 1866.—The drill is suspended by a flat band and rotated between strokes by a ratchet arrangement above the drill bar. The drill band passes over a pulley on the end of a piston rod whose reciprocation, in connection with the vertical steam cylinder, gives the saltatory motion to the drill. The band is wound on a roller rotated in either direction by a steam engine. The stroke of the piston is regulated by adjustable tappets. The force of the blow is regulated by an adjustable clamp. An adjustable table receives the sand pump and expedites the discharge of its contents.

Claim.—First, the adjustable clamp I, in combination with the drum C, and rising and falling pulley J, constructed and operating substantially as and for the purpose described.

Second, the steam cylinder M and pulley J, in combination with the clamp I and drum C, constructed and operating substantially as and for the purpose set forth.

Third, the adjustable table P, in the reservoir N', in combination with the sand pump O, constructed and operating substantially as and for the purpose described.

No. 57,828.—COLIN MATHER, Manchester, England, assignor to CHARLES P. BUTTON, New York, N. Y.—*Drill for Boring Wells.*—September 4, 1866.—The cutting portion has removable flaring chisels, the boring bar has a guide cylinder and a reamer, and is turned by the ratchet teeth on a sleeve attached to the stirrup by which it is suspended from the drill rod.

Claim.—First, the reamer E, in combination with the drill rod A and cutters C, constructed and operating substantially as and for the purpose set forth.

Second, the sleeve F, with ratchet teeth i i', in combination with the circular racks j j' and drill rod A, constructed and operating substantially as and for the purpose described.

No. 57,829.—JULIAN A. FOGG, Stockport, England.—*Coffin.*—September 4, 1866.—The sides, ends, &c., of the coffin constitute panels united in the mortised metallic angle pieces.

Claim.—First, in the construction of coffins and burial cases the employment of the metallic corner pieces, joints, or connections C, as described, whereby great durability and strength are given to the coffins or burial caskets, and a provision for an elaborate ornamentation of the same at little expense, substantially as specified.

Second, as an improvement in coffins and burial cases, the arrangement of the plate upon an edge of one of the sides or ends of the coffin, or on a bracket or shelf secured thereto, in such a manner that the plate will be visible whether the lid be open or closed, substantially as specified.

No. 57,830.—COLIN MATHER, Manchester, England, assignor to CHARLES P. BUTTON, New York, N. Y.—*Sand Pump.*—September 4, 1866.—The lower clack-valve seat is attached by a central bolt to the top of the cylinder and is removable to discharge the contents of the latter. The bucket has a valve, is sleeved upon the central rod and attached to the suspension rod.

Claim.—The movable seat *b*, clack *a*, and rod *c*, in combination with the barrel A and bucket B, constructed and operating substantially as and for the purpose described.

No. 57,831.—PIERRE FRANÇOIS MILLOT, Paris, France.—*Water Wheel.*—September 4, 1866.—The chute delivers the water on the interior of the wheel at both sides of the radial arms, which, by attachment to the buckets, secure the wheel to its shaft. These arms spring from both ends of the shaft and converge near the buckets. Each bucket is much higher at the outer than at the inner periphery, and thus the water is retained until its weight has ceased to be effective.

Claim.—First, the combination of a series of buckets, open internally and externally to receive the water internally in the manner described, upon each side of the arms, and discharge it externally, and a series of arms attached at or near the middle of said buckets to allow the water to be introduced on both sides of them, and connecting said buckets to the shaft, as set forth.

Second, the combination of a series of buckets open as aforesaid to receive water internally and discharge it externally, with the shaft of the wheel and the arms B B, the said arms B B being set at an angle to each other converging from points separated and distant from each other at the inner end to the middle of the bucket, as set forth.

Third, the arrangement in combination with a wheel adapted thereto of two separate spouts in such a manner as to discharge water into the interior openings between the buckets on each side of the arms of the wheel, substantially as described.

Fourth, the combination with the internally and externally open buckets to receive the water internally and discharge it externally as described, of the projecting flange *e*, to retain the water on its entrance into the buckets, as set forth.

No. 57,832.—EDMUND MOREWOOD, London, England.—*Coating Sheet Iron with Tin and other Metals.*—September 4, 1866.—Within the bath of molten metal is a carriage whose compartments receive successively sheets of iron from a slide above and carry them to the delivery mechanism, which consists of a pair of books to seize the sheets consecutively and deliver them to the rollers by which they are removed from the bath.

Claim.—First, the slide B, to receive the sheet or piece of metal to be coated, in combination with a receptacle C, within the bath of coating metal, to convey said sheet or piece of metal to the point of delivery, substantially as specified.

Second, the delivery rollers F, in combination with the receptacle C, and an elevating apparatus to raise the sheets or pieces of metal to the delivery rollers F, substantially as set forth.

Third, wipers or rubbers G, in combination with delivery rollers F, to act upon the coating metal previous to the delivery of the sheet or piece of coated metal, for the purposes and as specified.

Fourth, a slide or receptacle in a bath of melted coating metal, to receive the sheets or pieces of metal at one place and convey them to a different place in said bath, where said sheets or pieces are delivered upward automatically from said bath, as set forth.

Fifth, in combination with an apparatus for coating sheets or pieces of metal, substantially as described, a pair of delivery rollers, one of which is set in yielding bearings so as to provide for varying thicknesses of the sheets, or of the coating, as set forth.

No. 57,833.—HUGH SELLS, Vienna, Canada West.—*Cider Mill.*—September 4, 1866 —Projections on each side of the case at the ends of the enclosed crushing rollers facilitate the discharge of the ground pulp and juice. The ribs of the pomace cage are secured in the dovetail notches of the hoops.

Claim.—First, the projections *k k*, in combination with the case B, forming the passages *i*, substantially as and for the purpose specified and described.

Second, the dovetail notched rings G, substantially as described for the purpose specified.

No. 57,834.—WILLIAM LOUIS WINANS and THOMAS WINANS, London, England.—*Apparatus for Trimming Ships.*—September 4, 1866; English patent December 21, 1863.—The trimming ballast of the vessel is a heavy object moved by steam or other motor.

Claim.—The employment, for the purpose above described, of a movable weight, operated either by steam power or hydraulic power, or by gearing connected with the propelling engines, as herein set forth.

No. 57,835.—WILLIAM LOUIS WINANS and THOMAS WINANS, London, England.—*Coupling for Propeller Shafts for Ships.*—September 4, 1866; English patent June 20, 1863.—Explained by the claim and cut.

Claim.—Coupling shafts by means of a block or plate provided with grooves, in which are inserted cross-pieces or T-pieces, attached to the ends of the shafts to be coupled, as herein set forth.

No. 57,836.—WILLIAM LOUIS WINANS and THOMAS WINANS, London, England.—*Propeller for Spindle-shaped Ships.*—September 4, 1866; English patent June 22, 1863.—Two propellers are used with a spindle-shaped hull; one at either end of the central or larger compartment of the vessel, and adjacent to the end portions.

Claim.—As aforesaid, the application of two large screw propellers to spindle-shaped ships or vessels, in the manner and for the purposes herein set forth.

No. 57,837.—H. M. CARPENTER, Grand Rapids, Mich.—*Cutting Dresses.*—September 4, 1866.—The marked and graduated patterns have numbered openings, but their application cannot be briefly described.
Claim.—The use of the patterns, constructed and applied as shown and described, for cutting dresses for women and children.

No. 57,838.—WILLIAM G. ADAMS, Boston, Mass.—*Hulling Rice, &c.*—September 11, 1866.—The rice is strained in a tight cylinder, and ejected by steam pressure.
Claim.—The process of treating grain, seeds, &c., with reference to the removal of the husks or coverings therefrom, substantially as set forth.

No. 57,839.—WILLIAM H. ALLARD and ROBERT W. THOMAS, Portage, Wis.—*Self-operating Floodgate.*—September 11, 1866.—The L-shaped gates are pivoted eccentrically in vertical bearings. Until the water rises above the breasts the force of water keeps the gates closed, but above this level the water tends to open them.
Claim.—The gate B, pivoted eccentrically and arranged to operate in combination with the guards or breasts C, substantially as and for the purpose set forth.

No. 57,840.—R. N. ALLEN, Cleveland, Ohio.—*Piston for Deep Well Pumps.*—September 11, 1866.—The upper and lower chambers of the piston are partially divided by a diaphragm which checks the upward motion of the ball valve; the oil ascends between the edges of the diaphragm and the inner surface of the packing around the piston.
Claim.—The piston of an oil pump when constructed with two chambers separated by the valve check *b'*, and a loose flexible packing L, attached only at the upper and lower ends, with exterior spaces B' for the passage of the oil around the valve check between the metallic cylinder and the packing, substantially in the manner and for the purpose set forth.

No. 57,841.—WILLIAM J. ANDREWS, Columbia, Tenn.—*Cultivator.*—September 11, 1866.—A barrow and scraper on one side of the row of plants, and on the other side a knife, making intervals in the row by its reciprocating motion. The gauge determines the penetration of the cultivator devices.
Claim.—First, the harrows *g* and scrapers N, attached to adjustable frames L, applied to the front part of the framing A, in connection with the reciprocating thinning hoe I, all arranged and applied to a mounted frame in the manner substantially as and for the purpose set forth.
Second, the gauge K, applied to the draught pole J, in combination with the lever O, attached to the rear end of the framing and connected with the rear end of the draught pole, substantially as and for the purpose specified.

No. 57,842.—A. ARMSTRONG, Gillespie, Ill.—*Wheat Drill.*—September 11, 1866.—By the hand lever the front end of the machine is raised from the ground. The feed slides of the corn hoppers and that of the wheat hopper are actuated by cams on the roller, having respectively an intermittent and a constant reciprocating motion.
Claim.—First, the lever P and R, together with the post or standard Q, in combination with the pole S, for the purpose of raising the runners from the ground, for the purpose and substantially as described.
Second, the cams *c* and *c'*, in combination with the levers D and D', slides F and H, the cams C, lever D, having a continuous alternate motion, and the cams C and lever D', having an intermittent motion, all for the purpose and substantially as set forth.

No. 57,843.—BENJAMIN BACON, Morrison, Ill.—*Striking Works for Clocks.*—September 11, 1866.—The pins of the vertically reciprocating slide rest in notches in the cam and count wheels of a clock, when the slide is in its lowest position, and prevent striking until they are raised at the proper intervals to allow the striking, after which the detent again assumes its position.
Claim.—In combination with the count wheel and cam of a clock, a slide S, provided with pins *a b*, or their equivalent, arranged and operating substantially as and for the purposes herein specified.

No. 57,844.—F. A. BALCH, Hingham, Wis.—*Whiffletree.*—September 11, 1866.—When the draught is relieved the ends of the whiffletree are retracted; this prevents the traces becoming detached and gives an elastic resistance in starting.
Claim.—The combination of the bar A with the bars C and D and springs E and F, substantially as described and for the purpose set forth.

No. 57,845.—ROBERT BARTLEY, Norwalk, Ohio.—*Weather Strip.*—September 11, 1866.—A roller with adjustable bearings is inserted in the lower edge of the door or French window.
Claim.—The roller E, in combination with the slide *m*, shaft *l*, spring O, and post *p*, for

the purpose of rendering the space between the bottoms of doors or windows and thresholds or window sills air-tight, and at the same time of rendering the roller capable of passing over carpets or floors without hindrance or wear, substantially as described.

No. 57,846.—A. L. BAUSMAN, Minneapolis, Minn.—*Expanding Scraper for Cleaning Gun Barrels.*—September 11, 1866.—The expanding spring scraper is attached to the end of a rod and operated by a plunger having pins which move in a slotted and notched sleeve.

Claim.—The rod A, or its equivalent, having a slotted and notched sleeve C, hung upon pins or studs of its lower end, and a spindle I, capped at its end in combination with the spring or scraper arms D, when the several parts are arranged and connected together substantially in the manner and for the purpose described.

No. 57,847.—FORDYCE BEALS, New Haven, Conn.—*Snap Hook.*—September 11, 1866.—The spring hook is pivoted to the shank, and when closed its end is received in a depression of the latter.

Claim.—The hook A pivoted to the shank of the eye B, and arranged so as to operate substantially in the manner and for the purpose herein set forth.

No. 57,848.—WILLIAM M. BEEMAN, Hartford, Conn.—*Register for Regulating Timekeepers.*—September 11, 1866.—One register is used to record the deviation of the watch for a given time, and the other to record the distance which the regulator was previously moved ; the two afford data in connection with a subsequent observation for correctly regulating the time keeper.

Claim.—The registers B b and C c, in connection with the regulator A a, substantially as and for the purpose herein specified.

No. 57,849 —LEANDER K. BINGHAM, New York,N.Y.—*Composition for Printers' Rollers.*—September 11, 1866: antedated September 5, 1866.—Composed of glue, molasses, acetic or nitric acid and alum or sugar of lead.

Claim.—First, a compound of glue, molasses, or other saccharine matter and alum or other equivalent material, substantially as and for the purpose described.

Second, a compound of glue, molasses, or other saccharine matter and nitric acid or other equivalent material, substantially as and for the purpose set forth.

Third, a compound of glue, molasses, or other saccharine matter, alum or other equivalent material, and nitric acid, and other equivalent material, substantially as and for the purpose specified.

No. 57,850.—HENRY BLACK,Carrollton, Ill.—*Hoisting Apparatus.*—September 11,1866.—When the load has reached its elevation a pawl is raised which allows the carriage to traverse on the track and transport the load horizontally and return when the load is discharged. The brake moderates the rate of the hoisting rope in running over the pulleys

Claim.—First, the combination of the carriage D, having the wheels C C arranged to travel upon the rod or track. B, and the automatic pawl I, and automatic brake G, or their equivalents, when used as and for the purpose set forth.

Second, the construction of the brake G, in an adjustable manner, so it may be set to any required distance from the sheave E.

No. 57,851.—EVERAL BRADLEY, Clyde, N. Y.—*Harrow.*—September 12, 1866.—The hinged wings of the triangular barrow converge forward towards the line of motion so as to pass easily over inequalities, and may be turned over upon the central barrow.

Claim.—The combination with the triangular central barrow A, of parallel wings B B, converging forward in the direction of motion, the connection of said parts being made by the hinges g h, in such a manner that the wings can be detached at any time, or turned up over the centre to weight the same, the whole arranged and operating as and for the purpose herein specified.

No. 57,852.—BENJAMIN BRITTEN, Galena, Ill.—*Dam and Levee.*—September 11, 1866.—The dam is constructed in alternate layers of wood, clay, cement, &c., the down-stream elevated ends of the planks being supported by intervening blocks.

Claim.—First, the improved mode of constructing dams and levees, substantially as herein described and set forth.

Second, the combination and arrangement of the timbers A, planks C, and clay B, or equivalent, in the construction of dams and levees, substantially as herein described and for the purposes set forth.

No. 57,853.—J. B. BROWN, Medina, Wis.—*Sheep and Cattle Rack.*—September 12, 1866.—The sides, division boards, and top are shiftable to adapt it as a rack for feeding cattle, or a sheep rack with a cover for their protection.

Claim.—The combination of the pivoted or hinged boards D D, hinged division strips C C, troughs B B, and frame A, all constructed, arranged, and operating substantially as herein shown and described.

No. 57,854.—JAMES J. BROWN, Madison, Wis.—*Thill Coupling.*—September 11, 1866.—
The forked end of the thill has two cheeks united by a bolt which is placed in position in its
seat between the book of the clip bar and one leg of the bow; the button is vibrated, passing through a notch in the cheek, and is locked by bringing the thill to the working position.

Claim.—First, the thill iron D. having its rear end divided and provided with the cheeks
E, one or both of which are provided with a notch m for the button to move in, as and for
the purpose set forth.

Second, forming the cavity for the reception of bolt o and its packing e by means of the
bar B, having its front end bent as shown, and the front leg of the clip C, as herein shown
and described.

Third. the button A pivoted to the clip C, in combination with the cheek F, provided with
the notch n, arranged to operate as and for the purpose set forth.

Fourth, securing the button A' by means of the stem b and nut c, when used in connection with the packing e surrounding the bolt o, as shown, for the purpose of tightening up
the packing as shown in Fig. 4.

No. 57,855.—HENRY B. BUCH, Litiz, Penn.—*Metallic Hub.*—September 11, 1866.—The
disks which embrace the spokes are clamped between the flanges of the tubular box and the
collar nut.

Claim.—A metallic hub formed by the union of the pipe-box C with its flange d and prolonged screw-pipe D together with the flanged nut E e, cap F f, all held on the pipe-box in
combination with the annular disks G G', headed bolts H, and their nuts h, all combined and
arranged in the manner and for the purpose specified.

No. 57,856.—M. H. BUCKNALL, Darien, Wis.—*Cultivator.*—September 11, 1866.—The
cultivator frame is suspended from a lever which rests on the axle, and is regulated vertically by the adjustment of the lever on the segment bar. The shares are attached by sleeves
to pendent posts, and are braced in their rear.

Claim.—First, the lever frames B B, attached at their front ends by joints a a to the front
end of the frame A, in combination with the segment guides C C and the stop pins d, or their
equivalents, substantially as and for the purpose herein set forth.

Second, the combination of the teeth J, standards I, rivets o, shoulder k, screw nut l,
braces m, and frame A, substantially as described for the purpose specified.

No. 57,857.—R. B. BURCHELL, Brooklyn, N. Y.—*Lock.*—September 11, 1866.—The tumbler stud and key stud are upon the line passing through the centre of the bolt; the free
ends of the tumblers act as detents against the inner end of the bolt head when projected,
until they are properly placed by the action of the key bits, and enter the hollow bolt head
as it is withdrawn.

Claim.—The reversible tumblers set and turning upon a stud that is upon the central
line or plane of the bolt and key stud, the ends of said tumblers projecting beyond and
entering notches at the inner side of the bolt head when the bolt is retracted, substantially
as set forth.

No. 57,858.—JOHN BURNHAM and WILLIAM C. LATHROP, La Salle, Ill.—*Cultivator.*—
September 11, 1866.—The plough frames are attached by an arched yoke, which permits independent motion. Their clevises embrace posts shackled to the carriage.

Claim.—The attaching of the plough frames G to an upright mounted frame A by
means of joints d and bars F, arranged as shown to form a universal joint connection to admit of the vertical and lateral movement of the ploughs, substantially as described.

Also, the connecting of the two plough frames G G by means of a yoke H', having its
ends pivoted to bars I' I' which are secured horizontally on the frames G G by pivots t, to
admit of the frames G rising and falling independently of each other, as set forth.

Also, the duplex double tree arrangement, composed of the two double trees D E, attached to the draught pole C and connected by the rods a, substantially as described.

No. 57,859.—THOMAS BYRNE, New York, N. Y.—*Dredging Machine.*—September 11,
1866.—Wings extend laterally from the vessel and utilize the power of the current in forcing
the dredging vessel over the bar or shallow.

Claim.—The obtaining of power for the purpose of dredging or excavating channels or
sand bars by means of water gates or wings attached to steam or other vessels, as herein
described.

No. 57,860.—THOMAS S. CARD, Cortland, N. Y.—*Attaching Roofing to Buildings.*—September 11, 1866.—The covering consists of felt prepared in coal tar; a narrow strip is laid
along the eave, secured by tacks on an under fold; a wider strip similarly secured overlaps
the former; a third strip of a full width follows, and the remaining courses are laid, overlapping like shingles.

Claim.—The hereinbefore-described process of constructing and covering the roof of buildings, in the manner and with ingredients compounded substantially as set forth.

No. 57,861.—R. W. CARPENTER, Brattleboro', Vt.—*Musical Instrument.*—September 11, 1866; antedated September 2, 1866 —The pedal has its ordinary vibration to work the bellows, and by a little preponderance to one side causes the jointed part to bear sideways upon a lever connected with the swell.

Claim.—The double-jointed pedal susceptible of a double or duplex movement in combination with the lever *e*, substantially as described.

No. 57,862.—CYRUS C. CARTER, Exeter, Ill.—*Machine for Sowing Wheat and other Grain.*—September 11, 1866.—The claims describe the arrangement of furrow openers, seed-conveying tubes, seed box, slides, shaft, lever, and springs for manipulating the machine.

Claim.—First, the shaft Q with the lever R attached, in combination with the springs P and runners or furrow openers O, all arranged to operate substantially as and for the purpose set forth.

Second, the sliding or adjustable inclined board K with the seed-conveying tubes M attached, and arranged in relation with the seed box E substantially as and for the purpose specified.

Third, the pendants or agitators J, attached to the rock shaft I, operated from one of the wheels C, as shown, in combination with the perforated bottom *a* of the seed box E and the perforated slides *c* G, substantially as and for the purpose set forth.

No. 57,863.—CHARLES H. CARVER, Taunton, Mass.—*Barrel or Cask.*—September 11, 1866.—Inside the bung hole of a barrel a piece of wood is secured by a screw whose head rests upon a metallic and a rubber washer on the outside of the metal lining.

Claim.—The combination, as well as the arrangement, of the screw or faucet-receiving block C with the metallic vessel B, or the same and the barrel encompassing such vessel.

Also, the combination, as well as the arrangement, of the rubber washers, or their equivalents, and the metallic washer with the block C and the vessel B, or the same and the barrel A, the said washers being to form with the head of the vessel B and with a faucet, as described, a tight joint at the ajutage of the vessel B.

Also, the improved manufacture of barrels as made of the wooden barrel A and the lining vessel B, made of tinned or plate iron, as specified, arranged together as set forth.

No. 57,864.—ALBERT CHRIST, California, Ohio.—*Revolving Fire-arm.*—September 11, 1866.—Two distinct circles of cartridges are operated by one trigger. A barrel is adapted to each circle, and each has its feed finger: they are fired consecutively by a single hammer whose punch operates on a circle, which brings it in collision with the flanges of the cartridges in each circle. The feed finger operates on the heads of the cartridges, of which there are two in the outer row to one in the inner.

Claim.—First, the combination of the two circles of cartridge chamber I, two feed fingers N N', and single hammer M, constructed, arranged, and operating as and for the purposes set forth.

Second, propelling a revolving breech by means of one or more fingers N, adapted to operate on the heads of the metallic cartridges in the manner explained.

No. 57,865.—WILLIAM CHRISTIAN and J. H. MORROW, New York, N. Y.—*Knife Cleaner.*—September 11, 1866.—The blade of the knife is worked back and forth between the yielding soft wood surfaces; a hopper supplies cleaning material, and a drawer receives the detritus.

Claim.—First, the employment, in connection with the box, of the two yielding boards D and E, the upper board D having formed in it the supply hopper H, and the whole arranged to operate substantially as set forth.

Second, the employment, in combination with the cleaning mechanism, of a waste drawer B and fork platform or rest *f*, the whole arranged to operate substantially as set forth.

Third, the employment, in connection with the case, of holding devices I K, as specified, for the purpose set forth.

No. 57,866.—M. C. C. CHURCH, Parkersburg, West Va., and EDWARD H. KNIGHT, Washington, D. C.—*Tank for Petroleum, &c.*—September 11, 1866.—The tank has an air chamber in any convenient part provided with a safety valve. A bent tube extends from the petroleum tank into the air chamber, the ends terminating a short distance above the bottom of the tank and air chamber The tank is filled with petroleum, and enough is placed in the air chamber to cover the end of the bent pipe under all circumstances of expansion and contraction, so as to prevent the accession of air to the main chamber.

Claim.—A tank for the transportation or storage of petroleum or other liquids, provided with an auxiliary chamber and connecting pipe, operating substantially as described.

No. 57,867.—FRANK CLEMENS, Lafayette, Ind.—*Whiffletree.*—September 11, 1866.—The traces are attached to clevis pins in the ends of the whiffletree, and are disengaged by the rotation of a crank rod which withdraws bolts and allows the springs to raise the clevis pins.

Claim.--The crank shaft C, springs G G, and pins F F, in combination with straps I I and whiffletree B, for the purposes and substantially as described.

No. 57,868.—RIENZI L. CLEVELAND, Durand, Ill.—*Hand Corn Planter.*—September 11, 1866.—The cover is formed with wings which constitute it a hopper when open; a flexible brush over the stationary bottom and a feed slide gauge the seed.
Claim.—First, the combination of the lid L with the box A and brush head N, substantially as set forth.
Second, the slide *o*, in combination with the chamber D, and stationary bottom C, and brush G, as and for the purpose set forth.

No 57,869.—T. COLDWELL, Matteawan, N. Y.—*Apparatus for Stripping Files.*—September 11, 1866.—The file is held by adjustable jaws upon a slightly rotating bed supported by springs. The cross-bar, to which the stripping tool is attached, is connected at each end by a connecting rod with a crank upon the end of the driving shaft, and is reciprocated in guides upon a frame hinged at one end to the driving shaft, and which may be elevated to raise the stripper off the file through the instrumentality of a rock shaft and a system of levers at the other end.
Claim.—First, the reciprocating, sliding, stripping tool I, arranged and operating with holders N², substantially as described for the purpose specified.
Second, in combination with the above, the frame or guideway H H, for the stripping tool, hung so as to allow the said tool to be lifted from the file blank or blanks upon which it is operated without arresting its movement, substantially as and for the purpose specified.
Third, the holders N² for the file blanks, having sliding, adjustable clutches K at each end for securing the blanks in place for the same, substantially as and for the purposes specified.
Fourth, swivelling the blank holders N² to the frame M, supported upon spiral springs N, or their equivalents, at each end, as and for the purpose described.

No. 57,870.—JOHN E. CRANE and JESSE FOX, Lowell, Mass.—*Compound Projectile.*—September 11, 1866; antedated September 5, 1866.—Successive layers of iron are cast upon a cast-iron shot or shell.
Claim.—Constructing shells and shot for ordnance, substantially in the manner and upon the principle herein set forth, for the purpose of increasing the fragments from shells and avoiding the shrink at the centre of shot, as specified.

No. 57,871.—JOHN M. CRAWFORD, Newcastle, Ky.—*Garden Plough.*—September 11, 1866.—This machine is suspended on wheels, and is to be propelled by manual power; it has interchangeable tools, hoe, rake, plough, scraper, roller, &c., which are secured by wedge or set screw in the slotted beam.
Claim.—The garden plough or cultivator, consisting of the slotted beam A B H I, fore and hind wheels J and K, fastening G or P, scrapers L M, guiding and propelling handle N, and interchangeable shares or cultivating instruments, substantially as set forth.

No. 57,872.—SAMUEL CROCKER, Oakland, Iowa.—*Fence.*—September 11, 1866.—The inclined lateral braces are halved together at their intersection in the slotted post, and are tied by a cross-rod making an A. The bars of the adjacent panels interlock at the angles of the fence.
Claim.—First, the combination and arrangement of the stakes or braces C and binding bar D with the overlapped ends of the adjacent panels A and B, substantially as herein described and for the purpose set forth.
Second, securing the ends of the adjacent panels B and E at the corner or angle of the fence to each other, by combining the projecting ends of the boards *b'* of the panel B, the end uprights *e'* of the panel E, and the keys or wedges F, with each other, substantially as herein described and for the purpose set forth.

*No. 57,873.—C. O. CROSBY, New Haven, Conn.—*Machine for Pointing Wire.*—September 11, 1866.—The mandrel in this instance terminates in a jaw, to which another jaw is pivoted; the latter being held closed upon the former by a spring. Thus, the mandrel revolving, the wire held by the jaws will revolve with it except when clamped by the dies, at which time the jaw spring will yield and permit the jaws to continue their revolution without the wire, and without exerting upon the latter a tortive strain sufficient to twist it.
Claim.—A device for holding and rotating the wire, rod, or similar article, in combination with compressing dies or hammers, when the holding device is constructed to operate so as to avoid injury to the said wire, rod, or article being drawn by twisting, substantially as herein set forth.

No. 57,874.—CARY K. and S. F. DANIEL, Covington, Ky.—*Composition for Roofing.*—September 11, 1866.—For roofing, paving, and building blocks; composed of hydraulic cement, bitter water from salt works, pitch or petroleum residuum, Spanish whiting, coal, sand and coarse gravel.

Claim.—The composition herein described, composed and compounded as described for the purposes stated.

No. 57,875.—JOHN and SAMUEL DANNER, Canton, Ohio.—*Spring Bed Bottom.*—September 11, 1866.—Explained by the claim and cut.

Claim.—Supporting the longitudinal slats upon spring bearers or cross pieces, and holding them thereto so that they may move endwise thereon by means of the staples, long slots, and cord or twine, or their substantial equivalents, as herein described and represented.

No. 57,876.— SAMUEL DAVIS and WINDSOR LELAND, Chicago, Ill.—*Clutch-hook for Slaughtering Purposes.*—September 11, 1866.—The clutch chain is noosed around the gambrel, and the hog is raised by a tackle hook which engages the eye on the noose hook; the upper hook is suspended from the inclined rod, the hog is stuck and slides toward the scalding tub, where he is disengaged.

Claim.—First the combination of the hooks A and C, provided with the eyes or rings *a c*, respectively, and the chain B, operating substantially as specified and for the purposes set forth.

Second, the combination of the above-described device with an inclined rod or track F, so arranged upon its supports as to allow the hook A to slide upon the same, substantially as and for the purposes described.

Third, the combination and arrangement of the catching device A B C, the inclined rod F, and hook and chain H I, operating substantially as herein described and for the purposes specified.

No. 57,877.—WILLIAM A. DEVON, Port Richmond, N. Y.—*Tackle for Raising and Lowering Boats.*—September 11, 1866.—In addition to the ordinary davits to which a tackle is applied is a central or intermediate davit; the ropes of the end tackles are rove through a block suspended from the intermediate davit, and are hauled in or slacked together.

Claim.—The method herein described of raising and lowering ship's boats, by reeving the ropes of the end tackle through a block connected with an intermediate or centre davit, and afterward giving the loose ends or portions of said ropes bite on or attaching them to a rotating barrel, substantially as specified.

No. 57,878.—O. P. DILLS, Falmouth, Ky.—*Wheel Plough.*—September 11, 1866.—A seat supported by a wheel is attached by an adjustable outrigger and draught rod to the plough beam.

Claim.—The slotted adjustable bars C F, with wheels D E attached, and the brace rod K, provided with the loop or eye *k*, all arranged and applied to a plough, substantially in the manner as and for the purpose set forth.

No. 57,879.—EDSON DOE, Newbury, Vt.—*Throttle-valve Lever.*—September 11, 1866.—. The handle of the throttle-valve lever has a screw thread, which clamps the two lugs upon a segment which is fast to the boiler and holds the valve in any desired position.

Claim.—The handle D, formed with a threaded extremity and employed in connection with the throttle lever, substantially as and for the purposes herein specified.

No. 57,880.—H. W. DOPP, Buffalo, N. Y.—*Wick Burner for Heating and Cooking.*—September 11, 1866.—Vaporized water under pressure is applied to the centre of the flame of a circular wick burner. The devices are explained in the claims and cut.

Claim.—First, the application of a current of steam or water vapor in combination with a wick burner for burning any inflammable oil, for cooking and heating purposes, substantially in the manner herein described.

Second, the mode of raising and lowering the wick, as described.

Third, the employment of cylinder M in combination with a stove for cooking or heating purposes, and wick burner or burners, so constructed that the same shall serve to create a draught sufficient to supply the flame with atmosphere, and at the same time conduct the heat directly to the cooking utensils used, as set forth.

Fourth, constructing the cylinder M so as to prevent the loss of heat by radiation, as shown in Fig. 2 and described.

Fifth, the window N in combination with cylinder M.

No. 57,881.—OLIVER DROUILLARD, Detroit, Mich.—*Raising Sunken Vessels.*—September 11, 1866.—The central camel over the vessel acts alternately with the side camels to raise the vessel by degrees. The chains whereby the vessel is slung to the central camel are attached by bridles; the side camels are connected by slinging chains, which pass transversely under the vessel after it has been raised clear of the bottom by the central camel.

Claim.—First, the central camel A, Fig. 1, in combination with the lateral camels A' A'', Figs. 2 and 3, all constructed, arranged, and to operate in the mode and for the purposes above described and set forth.

Second, the air boxes L L', Figs. 1 2 and 3, in combination with the central and lateral camels, all constructed, arranged, and to operate in the mode and for the purposes above described and set forth.

Third, the bridle H, Figs. 1, 4, and 5, having its attached chain stops G'' G''', constructed, arranged, and to operate in the mode and for the purposes above described and set forth.

Fourth, the bridle H, having its attached chain stops G'' G''', in combination with the central camel, all constructed, arranged, and to operate in the mode and for the purposes above described and set forth.

Fifth, the bridle H, having its attached chain stops G'' G''', in combination with the air boxes L L', all constructed, arranged, and to operate in the mode and for the purposes above described and set forth.

No. 57,882.—C. F. Du VALL, Milwaukee, Wis.—*Side Gear for Threshing Machines, &c.*— September 11, 1866.—The miter wheel on the end of the tumbling shaft has its bearings in a plate adjustable in a horizontal plane on an axis coincident with the centre of the shaft carrying the mitre pinion.

Claim.—The bed plate B, traversing plate C, shafts E and F, pillow block D, and set screws H H, or their equivalents, combined and arranged substantially as and for the purpose described.

No. 57,883.—ALVAH EATON, Madison, Wis.—*Sugar Cane Mill.*—September 11, 1866.— The frame which carries the rollers for crushing is suspended by binging at one end, and having levers that work in notches in supports at the other, so that the weight of the mill assists in giving pressure to the rollers.

Claim.—First, the employment or use, in a cane-crushing mill, of a jointed frame or a frame hinged at one end, in connection with a roller having its bearings on arms one end of which rests on the frame, and all arranged in such a manner that the gravity of the mill is rendered subservient in assisting in giving the necessary pressure to the rollers, substantially as set forth.

Second, the rock shaft J, provided with the arms L L, or their equivalents, and the projections m m in connection with the links n n, attached to the frame, substantially as and for the purpose specified.

Third, the adjustable boxes G in which the rollers h of the arms E are fitted, for the purpose of virtually lengthening and shortening said arms, as specified.

Fourth, the lips k at the outer surfaces of the side pieces A A, in connection with the lips l at the upper ends of the uprights I I, as and for the purpose set forth.

No. 57,884.—JOEL B. EDWARDS, Knightstown, Ind.—*Medical Compound.*—September 11, 1866.—An antiseptic for chronic catarrh, ozena, and ulcerous sores; composed in the weakest form of permanganate of potash, $\frac{1}{8}$ grain, and chloride of soda, 1 ounce.

Claim.—The combination of the within-named ingredients, to form an antiseptic compound for the purpose described, substantially as set forth.

No. 57,885.—WILLIAM D. FISHER and WILLIAM HOLLY, Freeport, Ill.—*Fruit Dryer.*— September 11, 1866.—The fruit is exposed on a lath and canvas tray in an oven over a furnace; the latter and the oven have parallel discharge flues for the volatile matters from the fire and the moist air from the oven respectively.

Claim.—First, the double flue H K in combination with the furnace and with the drying chamber of a fruit dryer, substantially as described and for the purpose set forth.

Second, the fruit tray, formed of lath B and canvas C, in combination with the furnace and drying chambers of the fruit dryer, substantially as described and for the purpose set forth.

No. 57,886.—PETER FREUTZ, New Albany, Ind.—*Device for Carrying Hose.*—September 11, 1866.—The handles have hose gripping jaws, and are closed like tongs by a few links of chain attached to a swivel handle for carrying.

Claim.—The pivoted handles A A, with bands B B, chains D D, plate E, and eccentric hook H, arranged in the manner substantially as and for the purposes herein set forth.

No. 57,887.—E. G. GALE, Holland, Mich.—*Bow-pin for Ox-yokes.*—September 11, 1866.— The bow is secured in the yoke by a rack bolt passing through the end of the bow upon the top of the yoke, and operated by a lever and segment.

Claim.—Securing the bows of an ox-yoke in and to the yoke by means of toothed rack or ratchet bolts, in combination with toothed sectors engaged with such bolts, when arranged together and upon the yoke, substantially as described.

No. 57,888.—JAMES H. GANO, Tremont, Ohio.—*Corn Husker.*—September 11, 1866.—The rings fit upon the fingers, and the projecting tusk comes in opposition to the thumb to grasp the husk.

Claim.—A corn husker, having upon it the rings *a b*, curved or swelled portion *c*, and projection *d*, and worn upon the fingers, as and for the purpose herein set forth.

No. 57,889.—ROLAND R. GASKILL, Mendota, Ill.—*Sulky Plough*—September 11, 1866.—
The seat is supported on the axle, stayed by a rod from the diagonal brace, and is horizontally adjustable; the plough is suspended at the rear by levers and a link, and at the front by a chain and a lever pivoted to the beam, by which devices its vertical position is regulated.
Claim.—First, a sulky plough, having the driver's seat K so attached as to permit it to oscillate freely horizontally, substantially as and for the purposes set forth.
Second, the hereinbefore described mechanism for adjustably suspending the plough A by a system of levers C E and F, and a chain F″, said several parts being respectively constructed and the whole combined substantially as set forth.

No. 57,890.—SAMUEL D. GILSON, Oswego Falls, N. Y.—*Manufacture of Artificial Fuel.*—
September 11, 1866.—Peat is moulded in perforated blocks of triangular prismatic form.
Claim.—The new article of manufacture herein described, constituting fuel made from peat or peat and other substances combined, in hollow or serrated triangular form, essentially as specified.

No. 57,891.—CYRENIUS GLEASON, Buffalo, N. Y.—*Yeast Cake Dryer.*—September 11, 1866.—The cakes are exposed on trays made of slats and separated by intervening strips.
Claim.—First, the lattice-work frame, when constructed as and for the purpose herein substantially set forth.
Second, in combination therewith the removable pieces marked B, as and for the purposes described.

No.57,892.—FREEMAN GODFREY,Grand Rapids,Mich.—*Furnace for Calcining Gypsum.*—
September 11, 1866.—The furnace contains two rotary boilers, placed one over the other and provided with rotary conveyers. The gypsum is fed through the boilers consecutively, and is thence emptied upon a cooling surface and packed in barrels.
Claim.—First, in an apparatus for calcining gypsum, one, two, or more rotary boilers, substantially as and for the purpose specified.
Second, a distributor B², arranged and operating substantially as described.
Third, a spiral conveyor *b′* located within the heater, and operating substantially as described.
Fourth, a cooler G, constructed, arranged and operating substantially as described.
Fifth, the combination and arrangement of the fire arch A, heaters B B′, rings or collars *a*, friction rollers *g*, spiders *e*, and driving shafts D D, as and for the purpose specified.

No. 57,893.—J. M. GOFF, Ionia, Ill.—*Machine for Cutting Standing Corn Stalks.*—September 11, 1866.—The bearings of the cutting cylinder are in posts supported on a pivoted frame with ground wheels. The adjustment of the frame gauges the depth of penetration of the knives.
Claim.—The knife or cutter cylinder B connected to the draught pole A, substantially as shown and described, in combination with the pivoted frame C, provided with the wheels D D, and with or without the hooks G G, all being arranged to operate substantially as and for the purpose set forth.

No. 57,894.—THEODORE GRUNDMANN, St. Anthony, Minn.—*Apparatus for Making Vinegar.*—September 11, 1866.—The vessel has a series of troughs arranged within it around the sides, each trough having a tortuous channel through which the wash is passed. The troughs communicate with each other at the ends and the last trough empties into a vessel.
Claim.—The arrangement of a series of troughs or channels C C′ C″, &c., provided with covers, through which the liquid to be acidified, and also a sufficient quantity of air to produce acidification, is admitted, substantially as and for the purpose described.

No. 57,895.—CHRISTOPHER GULLMANN, Poughkeepsie, N. Y.—*Carpet Fastener.*—September 11, 1866.—The carpet is clamped to the floor by the serrated edge of a bar which is fastened down by a screw.
Claim.—Holding the toothed rod A against the carpet and floor by means of the screw having a curved neck *b*, substantially as described for the purpose specified.

No. 57,896.—HENRY HAMMOND, Hartford, Conn.—*Manufacture of Spectacle Frames.*—
September 11, 1866.—Two straight parallel bars are cut from a sheet of metal, turned inward at their extremities and joined together at or about the middle of their length by a curved bridge piece. The two bars, before being bent around to form the bow or eye-piece, are twisted so as to turn their bent extremities to a right angle from their first position, and these extremities are then subjected to the action of dies, which form upon the flattened surface of each a protuberance upon which the temple is to be hinged, and through which the screw is to be passed to connect the two halves of the bow and the temple together.

Claim.—First, the method herein described and shown of forming spectacle fronts in one piece out of sheet metal.

Second, the studs *d* on the ends A, serving as a pivot for the temple, and through which the screw passes to fasten the ends A one to the other.   •

No. 57,897.—A. S. HANSON, Milan, Mich.—*Pump.*—September 11, 1866.—The handle has a plunger rod attached on each side of its pivot and these connect with two pumps at different levels, the lower discharging into the upper.

Claim —The combination with the main cylinder or stalk A of the pump, divided into two chambers C and D of the cylinders E and L, respectively, having valves F and G and M and N, and pistons H and O, connected to a common handle K, when all arranged together and operating substantially in the manner described and for the purpose specified.

No. 57,898.—GEORGE HASECOSTER, Richmond, Ind.—*Loom for Weaving Slatted Window Shades.*—September 11, 1866.—The devices claimed are employed to deliver each separate slat in its proper position within the shed of the warp. Cannot be briefly described.

Claim.—First, the slat trough 5 and slat rest 6, in combination with the slat carrie 1, all arranged and operating as and for the purposes set forth and described.

Second, the cord 7, spring 16, slide 11, and top 12, when so arranged and operating as to carry and deliver the slats within the shed, in combination with the slat carrier 1, substantially as set forth.

Third, the combination of the pulley 8, cord 7, and sliding clutch L', when operated as and for the purposes set forth.

Fourth, the shaft *i*, pinion E, drum U, and arms Z and Z', arranged as and for the purposes set forth.

Fifth, the combination of the star wheel T, levers W W, connections V V', and treadles *t t*, as set forth and described.

Sixth, the shaft *n'* and its curved bearings *d d d*, constructed and operating as herein set forth and described.

Seventh, the combination of the batten *a a*, reed *c*, and shaft *d'*, with its curved bearings *d d d*, when arranged and operating as and for the purposes set forth.

Eighth, the friction straps *l l* and slides *m m*, in combination with the warp cylinder R, as set forth and described.

Ninth, the combination of the arm *z*, lever *y*, levers P and P', and ratchet wheel *o*, all arranged and operating as and for the purposes set forth.

No. 57,899 —LAWRENCE W. HEELAN, Petersburg, Ill.—*Hame Tug and Buckle.*—September 11, 1866.—The tug is hooked to the hame ring and the trace is attached to a loop on the tug plate by a spring stop.

Claim.—First, the tug A, when constructed with the necking *a'*, hook *a*, and mortises *a²* *a³* and *a⁴*, substantially as described.

Second, the buckle C and its spring *c³*, when constructed and employed substantially as herein described and set forth.

Third, the combination of the tug A and the buckle C for the purpose of attaching the trace of a harness to the hames thereof, substantially as herein described and set forth.

No. 57,900.—JOHN HERGET, St. Louis, Mo.—*Cooler for Beer, &c.*—September 11, 1866.— The outer and inner vessel are filled with ice and communicate by short tubes which cross the intervening space; over these tubes the beer descends in a shower from a pan above.

Claim.—First, the combination and arrangement of the chambers A B C and the sub-vault A², substantially as herein described and set forth.

Second, the basin D, when constructed with a perforated bottom, and otherwise so arranged as to discharge its contents either in a shower of drops or a number of very small streams down into the cooling chamber B.

Third, the cooling tubes B, when employed as herein described and set forth

Fourth, the construction and arrangement of the pipes E and E', as herein described and set forth.

No. 57,901.—ADOLPH HEUSTON, San Francisco, Cal.—*Paddle Wheel* —September 11, 1866.—The paddles are pivoted and have hooks which are engaged by arrow-headed, pivoted, counterpoised catches, which act in combination with the stops on the wheel arms, the latter limiting the vibration of the paddles.

Claim.—The combination of the paddle A, the pivots B B, and hooks C C, with the stop E and lever F, the regulator I, all substantially as and for the purpose specified.

No. 57,902.—CHARLES HIBBARD, Chicago, Ill.—*Spring Dirk Knife.*—September 11, 1866.—The blade is projected longitudinally from the handle by a spiral spring; catches on the handle maintain it in either its projected or retracted position.

Claim.—First, the catch bar D D, having catches *c c'* on its ends, in combination with the blade B and spiral spring C, substantially as described.

Second. the pin *g* in combination with the blade B and spring C, substantially as and for the purpose specified.

No. 57,903.—JAMES G. HOLLIDAY, Wheeling, West Va.—*Moulder's Flask.*—Septembe 11, 1866.—Locking hooks on the ends pass through mortises in the sides and are fastened by wedges. The relative position of the upper and lower frames is secured by the engaging tongue and socket.

Claim.—First, the combination with the frames A A' of the metallic hooks E *e*, when formed with the studs or projections *e'* and employed in connection with the keys F and plates G to connect together the several parts of said frames, substantially as described.

Second, coupling or connecting together the two parts of the flask by means of the plates H H and I and tongue J, when said plates H H and tongue J are provided with the studs or projections *h' j'*, and arranged to operate as described.

No. 57,904.—SOLOMON T. HOLLY, Rockford. Ill.—*Grain Binder.*—September 11, 1866.—This belongs to that class of binders in which a " ring carrier " is used ; the various improved devices for securing the several necessary operations cannot be briefly described other than substantially in the words of the claims.

Claim.—The combination of the ring carrier with a frame composed of two metallic ring frames, constructed and operating substantially as set forth.

Also, the combination of a detachable holder for the compressing strap, with a revolving carrier, and with an instrument for operating the strap holder, so that the strap holder is operated to release the strap while it is carried by the revolving carrier, substantially as set forth.

Also, the combination of one of the jaws of the forceps (for holding the binding material) with a slide, operating substantially as set forth.

Also, the combination of the spool for the binding material with the driving shaft of the revolving carrier by means of a spring connection, operating substantially as set forth.

Also, the combination of the spool of the binding material with an eye guide having the form of a coil.

Also, the combination of the pulley guide of the binding material with a movable arm, and with a fastening therefor, operating substantially as set forth.

Also, the combination of the pulley guide of the binding material with a curved tongue casing, operating substantially as set forth.

Also, the combination of the movable driver for the binding material with a detachable fastening to hold the driver in position, for supporting the binding material, substantially as set forth.

Also, the combination in a binder of the following instrumentalities, viz: the revolving carrier and a band-securing instrument secured to it, so that the latter is caused to travel around the gavel, substantially as set forth

Also, the combination in a binder of the following instrumentalities, viz: the travelling band-securing instrument and a movable shield plate, substantially as set forth.

Also, the combination in a binder of the following instrumentalities, viz: the travelling band-securing instrument and a yielding pressure holder operating upon the extremities of the band, substantially as set forth.

Also, the combination of the stock of the movable shield plate and guard (or of either of them) with the travelling band securing instrument by means of a locking mechanism, so that the stock and band-securing instrument are connected and moved together, substantially as set forth.

Also, the combination of the travelling band-securing instrument with movable nippers, operating substantially as set forth, to hold the extremities of the band.

Also, the construction of the tying bill with lips flaring at its hinder side, substantially as set forth.

Also, the combination of the tying bill with two sets of actuating mechanism, the first to turn it forward and the second to turn it backward, substantially as set forth.

No. 57,905.—IRA HOLMES, South New Berlin, N. Y.—*Pole and Post Puller.*—September 11, 1866.—A lever pivoted to a post has a shackle on its end which grips and lifts the hop pole.

Claim.—An improved self-adjusting pole and post puller, formed by combining the hook C, constructed as described, with the lever B and standards A, substantially as and for the purpose set forth.

No. 57,906.—HENRY HOOVER, Hemlo, Ill.—*Cultivator.*—September 11, 1866.—The wheeled cultivator has five shovels; the inner plough, those on its flanks and the outer ones being adjustable by special devices cited in the claim.

Claim.—First, the arrangement and combination of the central beam O, shovel standard J', shovel Q', with the central roller F and shaft 5, when constructed substantially as and for the purpose specified.

Second, the segment guides J, having the grooves S, in combination with the inner beams Grollers and shovel standards D'', shovels Q'', cams L and shaft 5, substantially as described al set forth.

No. 57,907.—STEPHEN D. HORTON. Peekskill, N. Y.—*Founder's Cleansing Mill.*—September 11, 1866.—The barrel is formed of perforated metallic staves secured between two flanged heads, and overlapping each other at the joints.

Claim.—An improved cleaning mill, the body of the staves of which are slotted with rectangular hopper-shaped slots or holes, substantially as described, and for the purpose set forth.

No. 57,908.—WILLIAM HOTINE, Brooklyn, N. Y.—*Machine for Moulding and Shaping Dough into Loaves or Crackers.*—September 11, 1866.—A sheet of dough is fed by an intermittent motion, and the strips cut off by a revolving knife pass to the grooved moulding cylinder which rolls the strips upon the stationary concave, where they are gradually separated into pellets, leaving a skin over the whole surface of each. The pellets are ejected from the moulder by cups, flattened between plates, discharged on to a table below, and carried forward to the docker, which stamps and finishes them.

Claim.—First, in combination with a grooved cylinder and shell, the use of nipples or moulding points of any suitable form, set in the grooves, for the purpose as described.

Second, the combination of the roller D with the shell E, made and arranged in the manner and for the purpose as described.

Third, in combination with a grooved cylinder and shell, the use of the revolving cups F F and H H, or their equivalents, for the purpose of placing and discharging the pellets, as described.

Fourth, the use of flatting plates G G, in combination with a moulding apparatus, such as herein set forth, whereby the pellets are received direct from the moulder, for the purpose as herein described.

Fifth, in combination with an apparatus for moulding and shaping crackers, &c., the forked or serrated sliding bar I, by which the cakes are brought up in an even line on the table, for the operation of the docker.

No. 57,909.—CHESTER B. HUNTING, Clinton, Ill.—*Plough.*—September 11, 1866.—The cutting disk is attached to the landside and is revolved by the passage over it of the sod.

Claim.—A cutter in the form of a disk, and attached to a plough so as to cut from bottom to top, for the purposes and substantially as herein described.

No. 57,910.—GEORGE R. HUNTLEY, Taunton, Mass.—*Faucet.*—September 11, 1866.—The nozzle is attached to the faucet by an exterior screw clamp, and is adjustable to discharge in any direction.

Claim.—First, the screw clamp E and trunnions F, in combination with the nozzle D, for the purpose and substantially as described.

Second, the faucet A, in combination with the adjustable nozzle D and screw clamp E, all for the purposes and substantially as described.

No. 57,911.—S. HURT and J. DOUGLASS. Prentice, Ill.—*Spike Drawer.*—September 11, 1866.—The lever bar is hinged and can be used bent or crooked so as to lift the spike engaged by its claw, by a depressing or a lifting action.

Claim.—Constructing a claw bar or spike drawer of two parts connected together by a joint, in such manner that the implement may be used either in an extended position or with one part at an angle with the other, substantially as and for the purpose specified.

No. 57,912.—A. DE KUHN, New York, N. Y.—*Sounding Board for Pianos.*—September 11, 1866.—The sounding board has a box form and is attached by gluing the lower edges of the sides and end to the bottom of the piano case.

Claim.—A sounding board made in a box form, detached from the sides of the case, substantially as described and for the purposes set forth.

No. 57,913.—B. ILLINGWORTH, Freeport, Ill.—*Churn.*—September 11, 1866.—Semi-cylindrical buckets are attached to the radial arms and strike the cream in an inverted position, ejecting it through their openings.

Claim.—The arms B B, provided with the buckets D D, constructed as described, upon the shaft C, when arranged within the churn box A, and operating in the manner substantially as and for the purposes specified.

No. 57,914.—JOHN H. IRWIN, Chicago, Ill.—*Lantern.*—September 11, 1866.—The bottom of the lantern is attached to the frame by a hinge and catch, and when opened the glass globe may be removed.

Claim.—Hinging the guard of a lantern to the bottom casing thereof, and securing th parts together opposite the hinge by means of a catch, or its equivalent, substantially e herein specified and shown.

No. 57,915.—JOHN H. IRWIN, Chicago, Ill.—*Retort for Generating Gas.*—September 11, 1866.—The retort within the outer case is heated by burners beneath, supplied through a branch gas pipe. Air and oil are admitted by their respective pipes to the retort, and the air conducted thence, charged with gaseous vapor, by pipes leading to the burners.

Claim.—First, providing a retort for generating gas and refining petroleum, arranged as described, with an air vent or tube, substantially as and for the purposes specified.

Second, in combination with a retort of the described arrangement and construction, the employment of a reservoir A and connecting pipe *a*, arranged and operating as and for the purpose described.

Third, the combination and arrangement of the reservoir A, connecting pipe *a*, retort B, provided with an air vent as described, pipe *c* and receiver E, substantially as and for the purposes set forth.

Fourth, in combination with a retort B, the employment of the auxiliary vapor pipe and burners *d*, arranged and operating substantially as shown and described.

No. 57,916.—A. W. JACKSON, Centralia, Ill.—*Piston Packing.*—September 11, 1866.—Inside the hollow piston is a valve ring operated by steam admitted through one disk and closing the apertures in the opposite disk by an alternate action. The admitted steam expands the peripheral packing against the inside of the cylinder.

Claim.—The combination and arrangement of the perforated plate C, perforated plate B, having projections *b'*, valve ring E, grooved ring F, ring G, and projecting plate H, constructed and operating substantially as and for the purpose set forth.

No. 57,917.—LEWIS R. JENKINS, Philadelphia, Penn.—*Belt Guide.*—September 11, 1866.—The belt from a driving shaft is directed to a shaft or machine placed at an angle to said driving shaft by the adjustment of the hangers on their bolts and the pulley frames on their centres.

Claim.—The adjustable hangers A and A', with their adjustable frames B and B', and pulleys C C', the whole being constructed and arranged substantially as described for the purpose specified.

No. 57,918.—JOHN S. JENNINGS, Buffalo; N. Y.—*Combined Knife and Fork.*—September 11, 1866.—The cutting blade is pivoted to a sleeve which slides upon the shank of the fork and has a handle whereby a reciprocating motion is given to it by one finger.

Claim.—First, a combined one-handed knife and fork, consisting of the blade H, pivoted to the sleeve F, with s; ring *o*, or its equivalent, and finger lever I *k*, in combination with a fork, operating substantially as set forth.

Second, in combination with the above-described parts, the thumb bearing or socket L, operating in the manner and for the purposes described.

No. 57,919.—JOHN JONES, Baltimore, Md.—*Brick Machine.*—September 11, 1866.—The mould is gauged to hold a charge of clay sufficient to form a brick when compressed. A vertical slide is acted on by a cam to give a continuous upward motion to the bottom of the mould; an apparatus removes the brick from the turn table; the moulds are oiled and the bricks discharged.

Claim.—First, the combination of the shoulder P, and large teeth Q, to overcome the inertia, and the small teeth R, and the entire arrangement by which the bed-plate J receives its alternate motion.

Second, the vertical slides U and *f*, in connection with the movable bottom of the moulds, whether the slides are acted on by cam, eccentric, or lever, for the purpose of giving a perpendicular pressure to the brick, or raising the bricks when pressed perfectly perpendicular from the mould.

Third, the entire arrangement as described in the specifications for clearing bricks to the off-bearer, and also for the oiling apparatus.

Fourth, the arrangement by which the off-bearer receives its alternate motion and communicates the same to the bricks, as described.

Fifth, the application of this machine for pressing clay, peat, or any other substance capable of being moulded under pressure.

No. 57,920 —WILLIAM KEARNEY, Belleville, N. J.—*Valve for Water Pipes.*—September 11, 1866.—The valve is vertically operated by a worm wheel and screw shaft, and when opened slides into a box beneath the main; the water in the box is discharged by a faucet.

Claim.—The combination and arrangement of the valve C, box A, flanges B B, cock *b*, screw D, worm wheel H, and screw I, all arranged to operate substantially in the manner as and for the purpose specified.

No. 57,921.—JOEL F. KEELER, Pittsburg, Penn.—*Scale Beam.*—September 11, 1866.—The beam has two poises on its upper and lower edges respectively, each adjustable relatively to its own line of figures, and capable of passing each other.   The counter weight holder at the end of the beam has divisional plates to hold, respectively, the balancing weights of the scale itself, the weights that balance the larger denominations in the weight of the car, and the net weight.   The poises balance the smaller denominations of the tare and net.

Claim.—First, constructing scale beams and their appendages so as to show both the larger and the smaller denominations of the net and of the tare weighed, substantially as described.

Second, constructing scale beams so that two or more poises may be made to pass each other along the same beam, usually its entire length, substantially in the manner described.

Third, dividing the counter weight holder *d'* into sections, when two or more of said sections are marked or figured, substantially as described and for the purposes set forth.

No. 57,922.—G. H. KIDNEY, Cleveland, Ohio.—*Washing Machine.*—September 11, 1866.—The reciprocating beaters are moved by cranks on a common shaft, and act against racks, which are adjusted according to the amount of clothes in the tub.

Claim.—The adjustable racks H I, button *b*, in combination with the pressure boards F, springs K, and beaters, as and for the purpose specified.

No. 57,923.—DAVID KINSER, Sardinia, Ohio.—*Evaporator.*—September 11, 1866.—The pan has a marginal ledge to collect the scum which boils over.

Claim.—First, the marginal ledge C D, formed upon or adapted to be attached to the common hemispherical boiling kettle.

Second, the kettle A B C D or its equivalent, as a new article of manufacture.

No. 57,924.—E. N. KINGSLEY, Minneapolis, Minn.—*Beehive.*—September 11, 1866.—The case is placed over an excavation, and has doors, bee entrances, moth traps, ventilating openings, and spaces for the use of the aggregated colonies contained therein.

Claim.—First, the case A, constructed with doors or slides E E, notched at their lower edges, and having notched slides F in front of them, for the purpose of closing the bee entrances or wholly or partially opening them, as may be desired.

Second, the passage *a* in the rear of the case between its wall and the frames, opening into the excavation D, for running off the moisture from the boxes, and supplying warm air to the case, substantially as described for the purpose specified.

Third, the securing and arranging of the top and bottoms *h'* in the boxes or sections H, so that a space will be allowed all around them for ventilation and the evaporation of moisture, substantially as described.

No. 57,925.—T. F. KUMS and W. W. BURSON, Rockford, Ill.—*Starting Street Cars.*—September 11, 1866.—The lever is moved by the motive power from a vertical to a horizontal position, and through a crank and arm operates a pawl and ratchet to rotate the wheel.

Claim.—First, the combination and arrangement of lever D, crank E, thrust bar F, pawl I, ratchet H, and tie piece J, or their equivalents, operating substantially as described and for the purpose set forth.

Second, the combination and arrangement of the lever D, with spring K and stud X, or their equivalents, substantially as described.

No. 57,926.—PETER A. LA FRANCE and J. D. DINSMORE, Elmira, N. Y.—*Head Rest for Railway Cars.*—September 11, 1866.—The curved head rest is supported by adjustable jaws, having acute and obtuse angles ; the head rest has slots and a stud by which it is fastened to the adjustable jaws, and has also a pad attached by elastic straps.

Claim.—First, a portable head rest, constructed with adjustable jaws A A', supporting a curved head rest C, said parts being respectively constructed and combined for use, substantially as set forth.

Second, the semicircular head rest C, when constructed with slots for receiving a stud B, by which it is adjustably attached to the jaws A A', substantially as set forth.

Third, the pad E, suspended upon elastic straps E', when used in combination with a head rest, substantially in the manner and for the purpose set forth.

Fourth, a portable head rest for attachment to the backs or arms of the seats of railway cars, when so constructed that the jaws by which it is so attached shall form, the front piece an obtuse, and the hind one an acute, angle with the top face to which the head rest is attached, substantially as set forth.

No. 57,927.—GEORGE LAUTER and JACOB KAUTZ, Vincennes, Ind.—*Extension Table.*—September 11, 1866.—The two parts of the extension table are connected by folding and sliding rails having sectional legs, which afford intermediate support for the table, whether extended or not.   The rails admit of lateral or longitudinal enlargement.

Claim.—First, the application of the half legs B B to a longitudinally and laterally extensible frame of an extensible table, substantially as described.

Second, connecting the outer ends of the sliding rail sections *d d* together by means of jointed cross pieces *e e*, substantially as described.

No. 57,928.—R. WALCOT LAWRENCE, New York, N. Y.—*Dumb Waiter.*—September 11, 1866.—In addition to the counterpoise-weight tackle is a continuous rope passing over a sprocket wheel, and having a lower weight suspended by pulleys to regulate the speed of the dumb waiter.

Claim.—The brake cord S applied in combination with the pulleys T on the adjustable cross-bar U, and the pulley R on the shaft Q, substantially as described for the purpose specified.

No. 57,929.—L. L. LEE, Milwaukee, Wis.—*Water Gauge Float.*—September 11, 1866.—The float is a spherical sheet of sheet copper, made of two pieces brazed together, and afterwards lightly hammered all over its outer surface to condense, stiffen, and strengthen the metal.

Claim.—A float, hammer-hardened in the manner and for the purpose substantially as described.

No. 57,930.—ROBERT LEE, Cincinnati, Ohio.—*Knob Latch.*—September 11, 1866; antedated September 2, 1866.—The latch bolt is fastened to the yoke by a screw pin, the head of which is exposed so as to be reached on the outside of the case by a screw driver. By turning the pin partly around the latch bolt may be withdrawn and replaced in a reversed position. An auxiliary spring is introduced for increasing the force by which the latch is thrown out, which is utilized by a particular disposition of the screw pin in its relation to the shank of the bolt.

Claim.—First, the reversible latch C D d d', detachable slide E F f f, shiftable pin G g g' g'', and spring H, or their equivalents, arranged and operating as set forth.

Second, combining with the above elements the auxiliary spring H' and shallow depressions g''' d'' d''', for the purpose stated.

Third, in the described combination the indexed and reversible pin G a g g' g'' and the cap B, having the hole b and numerals 1 2 3, or devices, substantially equivalent.

No. 57,931.—DAVID LIPPY, Mansfield, Ohio.—*Car Coupling.*—September 11, 1866.—Explained by the claim and cut.

Claim.—A car coupling composed of a swinging or pendent draw pin B or B', fitted in a draw head A or A', in combination with fixed rods G G and a shaft extending down in front of the draw head and connected either with the draw pin or with a sliding ledge I in the draw head, and all arranged in such a manner that the link or shackle C may, when a car is thrown from the track, be liberated by the movement of the shaft acted upon by the rods G G, substantially as shown and described.

No. 57,932.—J. FRED. LLEWELLYN, Louisville, Ky.—*Condenser.*—September 11, 1866.—The vessel has a cap for ice and fits over the top of the ordinary evaporating vessel. The inner surface has a spiral trough to collect the condensed liquid and lead it to the discharge pipe.

Claim.—A condensing apparatus for use by druggists and others in separating alcohol and other volatile liquids, constructed and operated substantially as above set forth.

No. 57,933.—HAMILTON D. LOCKWOOD, Charlestown, Mass.—*Elastic Bulb Syringe.*—September 11, 1866.—The neck has a button which binds the elastic neck of the syringe bulb against the flange and makes a tight joint.

Claim.—The combination of the neck c, button d, and flange e, for securing the elastic bulb, or the tube, to the metal connector, substantially as set forth.

No. 57,934.—EDWARD LYMAN, New Haven, Conn.—*Gauge for Screw-cutting Tools.*—September 11, 1866.—The gauge has angles by which is determined the inclination of the point of the cutting tool, and also the inclination of the tool when arranged in the post for cutting the thread.

Claim.—A gauge combining within itself a guide for forming the cutter with guides for cutting the cutter, substantially in the manner and for the purpose herein set forth.

No. 57,935.—FRANCIS A. MACK, Niles, Mich.—*Well Tube.*—September 11, 1866.—Explained by the claim and cut.

Claim.—Forming the induction openings of iron well tubes of oblique incisions e e, (formed by cutting from the inside of the tube,) having protruding lips or edges, with alternating depressions to prevent the admission of sand or other solid particles within, substantially as set forth.

No. 57,936.—PETER N. MAINE, Olmsted Falls, Ohio.—*Receiving and Delivering Mail Bags on Railway Cars.*—September 11, 1866.—The bag is held by a hook on a platform post and received by a barbed hook on a swinging platform of the car. The bag for delivery is detached from its hook and dropped on the platform by the engagement of an arm with the platform post.

Claim.—First, the adjustable hook C pivoted to the shank and spring *e'*, in combination with spring hook L, the shaft B, weighted arm D, hook F, and staple I', as and for the purpose set forth.

Second, the shaft B, in combination with the arm O, frame A, lever G, and staple I, arranged as and for the purpose set forth.

No. 57,937.—D. P. MATHEWS, Winthrop, Mass.—*Tree Protector.*—September 11, 1866.— The inner trough is charged with liquid through the opening in the roof.

Claim.—An improved tree protector, made as described, viz: of the tubes A D and the two conic frusta B C, arranged and applied, with respect to each other, as specified.

No. 57,938.—THOMAS MAYOR. Pawtucket, R. I.—*Flyer for Roving Frames.*—September 11, 1866.—Explained by the claim and cut.

Claim.—Forming upon the shoulder of the presser to a flyer and inclined plane *a*, and combining the same with a stud pin *b*, or other fixed stop in the side of the flyer, substantially as described for the purposes specified.

No. 57,939.—B. J. McAFEE and J. H. WIRT, Delphi, Ind.—*Ground Drag.*—September 11, 1866.—Notches are cut on the sides of the drag and cross-bars are attached at an angle of about fifty degrees with spaces between them for the earth to pass through.

Claim.—The sides A, in combination with the cross-bars B, when constructed and arranged, in relation to each other, in the manner and for the purpose set forth.

No. 57,940.—H. L. McAVOY, Baltimore, Md.—*Apparatus for Carburetting Air.*—September 11, 1866.—Within the casing is a chambered wheel attached loosely to the central shaft and driven by means of the air from the air-forcing wheel. A wheel provided with a series of buckets revolves in an oil supply chamber and empties the oil upon a spout, from whence it flows into the carburetting chamber. The air is forced from the latter into an inverted vessel in the regulator and from thence escapes to the burners.

Claim.—First, the air-forcing wheel B and the attached cylindrical casing of perforated metal or frame covered with fibrous or porous material, arranged and operating substantially as described.

Second, one or more meter wheels C, covered by permeable casing C', revolving independently by the pressure of the gas from the forcing wheel, being separated from the chamber B'' by the partition D, and connected therewith by pipe E, substantially as described.

Third, the feeding apparatus H, operating as described and connected to any part of the machine, in combination with the partitions G G and shelf or gutter I, substantially as described.

Fourth, the regulator chamber J, and its inverted vessel L, as described, in combination with the connecting pipes K K F F' and E' E''.

Fifth, the casing or box M, arranged on the outside of the machine for containing the gearing, substantially as described.

Sixth, the perforated metal or wire-gauze diaphragms in the air inlet, for the purpose described.

No. 57,941.—THOMAS B. McCONAUGHY, Newark, Del.—*Governor for Horse Power.*— September 11, 1866.—As the balls fly out their arms press against the swinging arms attached to the loose collar and force the latter against the fixed collar, causing the former to rotate and wind up a cord attached to a lever, the shoe on whose opposite end is thereby brought in contact with a friction wheel.

Claim.—The governor herein described, the same consisting of the ball arms F hung up on a common centre pin *a* of the governor shaft, provided with arms I connected together by a spring J, in combination with swinging arms L of the fixed collar K of the governor shaft and loose collar M, when combined and arranged together and connected with the governor shaft so as to operate substantially in the manner described and for the purpose specified.

No. 57,942.—WM. R. McCUTCHEON, Washington, Iowa.—*Churn.*—September 11, 1866.— The dashers are respectively attached to the sleeve and central axis, which are rotated in different directions. Each dasher has oblique openings, and the upper one is set obliquely on its shaft.

Claim.—First, the arrangement of the shafts and dashers herein set forth, whereby the one or the other dasher may be operated as required.

Second, the arrangement of the dasher *p* in the oblique position in relation to the dasher *o* as herein recited.

Third, the arrangement of the spaces or holes *v* and *w* in the dashers as and for the purposes described.

No. 57,943.—B. McDEVITT, Belvidere, Ill.—*Railroad Chair.*—September 11, 1866.— The chair has three flanges. The base of the rail is embraced between one side flange and that on the sliding portion of the chair, which is keyed up by a wedge bearing against it and the outer flange of the chair.

Claim.—The combination of the chair A, made in two parts or sections B and C, and the wedge G, when constructed, combined, and arranged together, substantially as and for the purpose described.

No. 57,944.—JOHN McINNES, Waverly, Ill.—*Hame Tug.*—September 11, 1866.—The length of the trace is adjusted by the engagement with the hooks on the edges of the hame of the cross heads pivoted on the end of the trace.

Claim.—The combination of the metallic plates C and D; also the link E with the tug A and trace B, substantially as described.

No. 57,945.—THOMAS McINTIRE, Portsmouth, Ohio.—*Cotton-tie Fastening.*—September 11, 1866.—Explained by the claim and cut.

Claim.—A new article of manufacture, viz: iron hoops for cotton bales cut to the required length and having a clasp or buckle, constructed substantially as described, permanently attached by a rivet at one end so as to form a hinged joint, the whole being covered by a protecting covering of paint or varnish substantially as described.

No. 57,946.—HORACE D. MEAD, Wayne, N. Y.—*Farm Gate.*—September 11, 1866.—The half gates are moved in vertical planes to the right and left, being attached to oscillating bars and counterpoise I by weights.

Claim.—First, the weights G and G', when applied as and for the purpose specified.

Second, the arrangement of the supports F, when made and applied substantially as set forth.

No. 57,947.—HARRISON B. MECH, Fort Edward, N. Y.—*Apparatus for Packing Straw into Boilers.*—September 11, 1866.—The straw is driven by the descending plunger into the boiler through a slot in the diaphragm which closes after the plunger is withdrawn.

Claim.—First, the plunger a and shaft G, in combination with the cylinder E and boiler A, substantially as described.

Second, the cylinder E and diaphragm d d, substantially as and for the purpose specified.

Third, the combination of the frame o o and F, reciprocating bar H H, shaft G, cylinder E, diaphragm d d, with the boiler A, substantially as and for the purpose specified.

No. 57,948.—WARREN P. MILLER, San Francisco, Cal.—*Bottle Valve.*—September 11, 1866.—Explained by the claim and cut.

Claim.—The valve c, when made to operate in the case b by means of the weight g, substantially as shown and described.

No. 57,949.—RICHARD MONTGOMERY, New York, N. Y.—*Aerial Railroad.*—September 11, 1866.—The aerial railroads are constructed of corrugated iron upon double arches, so that an upper and lower track can be used at the same time. A track is sustained upon a single support, a curved continuation of the sleeper of the roadway. The horizontal cross-ties of the double track have a double curve.

Claim.—First, the construction of an aerial railroad with two tracks, one above the other, substantially as described.

Second, the construction of the horizontal portion or cross-ties of an aerial railroad in the form of a double curve, substantially as shown in figure 2, plate II, and in the manner herein described.

Third, the construction of an aerial railroad upon a single column, substantially as shown in figure 3, plate II, and in the manner herein described.

No. 57,950.—S. A. MOODY, New York, N. Y.—*Ladies' Dress Elevator.*—September 11, 1866.—Short cords are attached at intervals within the dress skirt. A cross-bar at the end of each cord is passed through one or more rings attached to the dress in circles below.

Claim.—The arrangement of the key or toggle b affixed to the dress by a short cord, substantially as described, and a series of rings c for looping up the skirts of ladies' dresses, as and for the purposes herein described.

No. 57,951.—H. L. NARAMORE, Cummington, Mass.—*Sled Brake.*—September 11, 1866.—Pushing back on the tongue vibrates the lever and projects below the bottom of the runners the spurs pivoted thereto.

Claim.—The levers L L, rods J J, and elbow levers I I, in combination with the brake bars H H and sled A, substantially as and for the purposes described.

No. 57,952.—W. D. NICHOLS, Chicago, Ill.—*Cultivator Plough.*—September 11, 1866.—The side beams of the plough are hinged to the middle beam, and are maintained at their lateral adjustment by slotted braces and a set screw, the forward ends of the scrapers traversing in a slot of the centre beam.

Claim.—First, connecting the mould boards or ploughs A A with a hinge joint, substantially as described.

Second, the slot *s*, or its equivalent, in combination with the mould boards or ploughs, substantially as described.

No. 57,953.—WILLIAM NICHOLS, Elmira, N. Y.—*Steam Engine Governor.*—September 11, 1866.—The cams on the ball arms act directly upon the weight and spring, as the balls rise and fall with variations in the speed of the spindle.

Claim.—The arrangement, substantially as described, of the balls D, arms D', cams D'', and weight E, for the purpose set forth.

No. 57,954.—ROBERT C. NOURSE, Corydon, Ind.—*Evaporator.*—September 11, 1866.— The pan is divided into compartments, and stirrers, which incline in two directions, are operated in gangs by means of crank arms. The pan has also skimmers attached to cranks, so arranged as to lift the skimmers out of the sirup as they move forward, and dip them into the sirup as they return.

Claim.—First, the stirrers B, arranged in two rows and applied to a common rod C, connected by an arm to a crank shaft, substantially as and for the purpose set forth.

Second, the skimmers F, operated from the crank shaft E, in connection with the inclined planes H, all arranged and applied to an evaporating pan, substantially as and for the purpose specified.

Third, the combination of the stirrers and skimmers, applied to an evaporating pan, to operate in the manner substantially as and for the purpose set forth.

No. 57,955.—SAMUEL B. B. NOWLAN, New York, N. Y.—*Machine for Varnishing Cloth.*— September 11, 1866.—The cloth is passed from the let-off roller beneath a roller in the steam-heated size vat; thence to a steam-heated table, where the varnish is spread by a revolving brush, and thence over the rollers of the drying frame, where it is exposed to jets of air from a perforated pipe, and from which it is wound on a take-up roller.

Claim.—First, the steam table C, arranged underneath the cloth and in relation with the brush D, substantially as herein set forth for the purpose specified.

Second, the adjustable sliding frame G, applied in combination with the vat B, substantially as herein set forth for the purpose specified.

Third, the vat B, pressure rollers J s, and frame G, arranged in relation with each other and with the steam table C. and brush D, substantially as herein set forth.

Fourth, the arrangement of the blower E, and perforated serpentine pipe F, with reference to the cloth as it passes from the coating or saturating devices, substantially as herein set forth for the purpose specified.

No. 57,956.—SAMUEL B. B. NOWLAN, New York, N. Y.—*Waterproof Composition.*—September 11, 1866.—The fabric is saturated with an aqueous solution of alum and salt, and is then treated with a composition of paraffine, borax water, and gum tragacanth.

Claim.—A waterproof composition, made substantially in the manner herein described.

No. 57,957.—CHARLES PARKER, Meriden, Conn.—*Vise.*—September 11, 1866.—The slide is made of wrought iron and secured to the movable jaw by casting the latter around its end.

Claim.—The wrought metal bar D, in combination with the cast metal jaw C, when constructed and united substantially in the manner and for the purpose specified.

No. 57,958.—JOHN E. PATTISON, Houma, La.—*Treating Cane Juice.*—September 11, 1866.—The gutters on the inner periphery of the revolving cylinder carry up the juice and drop it in a continuous shower, exposing it to the action of the sulphurous acid gas which traverses the cylinder.

Claim.—The revolving cylinder A, inclined on its axis, furnished with lifting buckets *d*, and adapted to the passage of a body of gas, substantially as described.

Also, in combination therewith, the draught wheel, operating as described.

No. 57,959.—FRANCIS PEABODY, Vevay, Ind.—*Sled.*—September 11, 1866.—The raves and cross-bars which form the bed of the sled are connected at their intersections by eye bolts, and the bed is similarly connected to the runners. The sled is thus rendered flexible in shape, assuming a rhomboidal form when one runner meets an obstacle.

Claim.—First, the eye bolts J, or their mechanical equivalents, in combination with the raves E F, transverse beams G H, and cross-bars M N, all arranged and operating substantially as herein described and specified.

Second, in combination with the eye bolts J and self-adjusting frame E F G H M N, the flexible diagonal braces T T', for the purpose specified.

Third, in combination with the elements of the two preceding clauses, the double-ended runners A *a a* B *b b'*, as described and set forth.

No. 57,960.—JAMES PERRY, Leeds, Mass.—*Composition for Filling Wood.*—September 11, 1866.—The ingredients are rubbed into the pores of the wood in order to give it a hard surface capable of receiving a high polish.

Claim.—First, the use of silicious marl, or infusorial earth, for the purpose of filling and polishing wood, substantially as herein set forth.

Second, the combination of silicious marl with any or all of the substances herein named: sulphate of zinc, muriate of ammonia, gum arabic, gum tragacanth, oil, substantially in the manner and for the purpose herein set forth.

No. 57,961.—ALMERON PIERCE, Olmsted, Ohio.—*Combined Harrow and Marker.*—September 11, 1866.—The smooth roller in the forward frame rotates the toothed barrow in the rear frame, which is hinged and secured by a segment bar in the required position. Bars for marking out the ground for corn planting are attached to the frame.

Claim.—First, the toothed roller D and smooth roller C, in combination with the frames A A', as and for the purpose substantially set forth.

Second, the wheels E E' and F, the arms G, and curved arm H, as arranged, and in combination with the rollers D and C, and the frame A A', for the purpose and in the manner described.

No. 57,962.—A. PIERCE, Olmsted Falls, Ohio.—*Car Coupling.*—September 11, 1866.—The coupling link is engaged by a transverse bolt passed horizontally through the beam of the car; the coupling bolt is withdrawn by a lever, and held open for coupling by a spring, which, by means of a knob, is released when the cars come in contact.

Claim.—The arrangement of the lever C, sliding bolt G, and spring b, in combination with the sliding catch e, springs f, and bumper E, arranged and operating as and for the purpose set forth.

No. 57,963.—ALFRED C. PLATT, Oberlin, Ohio.—*Apparatus for Drying Photographs.*—September 11, 1866.—Curved grooves in the shelves bend the prints in one direction, and the proximity of the shelves bends them in the other.

Claim.—A device for holding a print or picture in a curved or convex shape while drying, substantially as described.

No. 57,964.—BENJAMIN F. PORTER, Manchester, N. H.—*Table Stand for Food.*—September 11, 1866.—The stand containing shelves for dishes has a gauze covering and revolves on its axis to present its sliding door in any direction.

Claim.—First, the revolving table stand, consisting of the end plates B B, rods C C, gauze netting d, shelves F, pin G, base H, cross bar I, posts J, and catch L, constructed and arranged in the manner and for the purpose herein specified.

Second, in combination with the above, the sliding door E, arranged and operating as herein described.

No. 57,965.—D'ARCY PORTER, Cleveland, Ohio.—*Sheep Rack and Shed Combined.*—September 11, 1866.—The two feed racks are placed in a shed whose hinged sectional roof is supported on angular end-framing, and is convertible into an open shed.

Claim.—A sheep rack and fold, constructed substantially as and for the purpose herein set forth.

No. 57,966.—T. T. PROSSER, Chicago, Ill., assignor to himself, D. A. KIMBARK, D. W. WELLS, J. A. EASTMAN, and G. W. GILLETT.—*Screw.*—September 11, 1866.—Explained by the claim and cut.

Claim.—A screw having its threads of a convex form, with a concave form of the space between them, when combined with a tapering form from the commencement of the threads to the point.

Also, the combination of a pyramidal front with the threads of a two or three-threaded screw, when the screws and intermediate spaces are constructed substantially as set forth.

No. 57,967.—STEPHEN PUFFER, Oxford, N. Y.—*Gate.*—September 11, 1866.—The gate slides open, traversing on a roller and is vibrated 90°, being received upon the inclined stationary side supports. The friction roller on the extended bar traverses between two boards of the fence and assists in operating the gate.

Claim.—First, two or more stationary guide supports for guiding the gate from the friction roller, and sustaining its weight when opened, substantially in the manner and for the purpose herein set forth.

Second, the large friction roller C, having its axis on either of the rear posts of a gate, and smaller friction roller c, when arranged and combined with stationary guide supports D D', substantially in the manner and for the purpose herein set forth.

No 57,968.—TILLMAN RAMS, Keokuk, Iowa.—*Plough.*—September 11, 1866.—The land edge of the share has a socket to receive a prolongation of the landside.

Claim.—The share A, provided with the socket C, in combination with the bar secured to the landside S, when constructed as and for the purposes and substantially as described.

No. 57,969.—JACOB REESE, Pittsburg, Penn.—*Reducing Metallic Oxides and Refining the Metal therefrom.*—September 11, 1866.—Improvement on inventor's patent of June 19, 1866; the ore or metallic oxide is fused in a cupola, with or without carbonaceous fuel, and by the aid of a blast of air; the melted ore is then run into a reducing chamber placed under the cupola and below the influence of the atmospheric blast, and is there deoxidized by means of a hydrocarbon.

Claim.—First, the manufacture of iron or other metals from their ores in a furnace, is which the ore, having been fused with the aid of a blast of atmospheric air, is then reduced to its base with a hydrocarbon.

Second, the process of fusing iron ores and other metallic oxides by the aid of a blast of atmospheric air in a furnace, and the waste heat from the reducing chamber, without the use of other fuel, or with so little fuel as not to deoxidize the ore in melting, and running the melted ore immediately into a receptacle below the influence of the atmospheric blast, in which it is deoxidized by the injection of any hydrocarbon into the molten mass, in the manner substantially as hereinbefore described.

Third, the use of an atmospheric blast for fusing metallic oxides, in combination with the use of hydrogen or carbon, or a hydrocarbon, for the reduction of the melted oxide to its metallic base, substantially as hereinbefore described.

Fourth, the use of the compound of water, or steam, and oil in a liquid or vaporous condition, in the proportions hereinbefore described, as a new material for the refining of melted metal.

Fifth, combining with a cupola or melting furnace one or more reducing chambers placed for the reception of the melted ore below the point at which the atmospheric blast is introduced into the furnace, for the purpose of running the melted ore directly from the furnace into a reducing chamber in which it may be deoxidized, carbonized and refined, away from the influence of the atmospheric blast, substantially as hereinbefore described.

Sixth, connecting the reducing chamber or chambers to the bottom of a cupola or melting furnace, by a hinge or pivot, so that they may be swung away from the furnace without impairing their connection with the vessel containing the reducing agent.

Seventh, the use of the hollow shaft c, in combination with the reducing chamber B and connecting passage i, for the purpose of hanging the reducer to the cupola, and allowing of the introduction into the reducing chamber of the liquid or gaseous agents for the reduction of metallic oxides.

Eighth, the combination of the water tank E with one or more tanks for holding the liquid, deoxidizing or carburetting agents, for the purpose of applying the pressure of live steam, or of a head of water, for forcing the deoxidizing agents into the reducing chamber, substantially as hereinbefore described.

Ninth, constructing the tuyere forming the communication between the interior of the reducing chamber and the cell below it, of fire clay or other refractory material, in combination with a tubular metallic screw stem projecting therefrom, substantially as and for the purpose hereinbefore set forth.

Tenth, the use of lime or oxide of calcium as a lining for reducers or furnaces, wherein ores or metals are decomposed or refined.

No. 57,970.—JOHN REID, Knoxville, Md.—*Mariner's Compass Scale.*—September 11, 1866.—The compass indicates by the needle the position of the ruling edge; is fixed by centre pin to use by horizontal rotation. By the side index allowance is made for local variation.

Claim.—The combination of the compass and the scale, substantially as described.

Also, in combination with the above, the arrangement for setting the compass to a given variation, for the purpose described.

Also, in its combination with the compass and scale, the centre pin, operating substantially as described.

No. 57,971.—JOHN M. RILEY, Newark, N. J.—*Caster.*—September 11, 1866; antedated September 2, 1866.—The parts are made separately, are united by pressure in dies, and a hole is drilled through the hub for the axis pin.

Claim.—Making caster wheels by preparing the rim, the plate and the hub in separate pieces, and uniting them to each other under pressure, substantially as above set forth.

No. 57,972.—THOMAS J. ROCKEY, McElhattan, Penn.—*Self-acting Wagon Brake.*—September 11, 1866.—The rubbers are attached to compound levers on the cross-bar of the bounds, and are operated by a strap passing around a sheave in the tongue to the neck-yoke of the horses. The strap may be locked to prevent this action when backing the wagon and the springs in the rear restore the rubber levers to their normal position of inaction.

Claim.—The arrangement of the strap *h*, chain *g*, rod *t*, springs *b b*, and locking devices *d e f*, constructed and operating substantially as described and represented.

No. 57,973.—G. A. ROLAND, Wasioja, Minn.—*Wind Wheel.*—September 11, 1866.—The wind wheel, its shaft and the ordinary steering vane are supported on a horizontally rotating table: the obliquity of the sails is regulated by a centrifugal governor acting through connecting devices upon the axes of the sails.

Claim.—First, the arrangement of devices for turning the sails, consisting of the governor M, levers C and D, shaft K, hollow shaft E, loose head F, pin *a*, oblique slot *b*, ring *z* and sails B, all substantially as and for the purpose set forth.

Second, in combination with the foregoing, the windlass *j* and rope *p*, substantially as and for the purpose set forth.

Third, in combination with the wind wheel as herein described, the turn table L upon rollers R, and having a flange *f* and dogs *d*, substantially as and for the purpose set forth.

No. 57,974.—TIMOTHY ROSE, Cortlandville, N. Y.—*Separating Sand from Water in Wells.*—September 11, 1866.—The sand arrested by the wire cloth of the inner tube, falls and passes out by way of the grooves in the stem of the lower pointed section.

Claim.—The combination of the inside tube with the strainer of wire cloth or its equivalent, and the flange to cut off the inferior chamber, together with the movable point with the grooved stem, by which is secured open passage downward through the end of the main tube, substantially as herein described and for the purposes set forth.

No. 57,975.—DANIEL SAGER, Albany, N. Y.—*Wagon Brake.*—September 11, 1866.—The oscillating rubbers are operated by lever connection with the sliding tongue when the latter is held back.

Claim.—First, the arrangement and combination of the pole A, with the lever F and connections *g h*, directly with the brake-bar in rear of the wheels, essentially as herein set forth.

Second, the construction substantially as shown and described, of the pendent shoes H on the ends of the brake-bar, by reducing the top portions, and so that when the shoes are freely suspended the upper extremities *s* of the lower or rubbing portions thereof will be on a level or thereabout with the centres of attachment of the shoes, essentially as and for the purpose or purposes herein specified.

No. 57,976.—BENJAMIN F. SANBORN, Boston, Mass.—*Machine for Cutting Belt Lacing.*—September 11, 1866.—The knives stand in a gang upon the bed with their edges inclined towards the leather, which is smoothed by the pressure bar before the cutters divide it into strips.

Claim.—The arrangement of the bed plate A, pressure bar F and standard B, in combination with the vertical knives with inclined edges K, substantially as described.

No. 57,977.—RUFUS S. SANBORN, Ripon, Wis.—*Clothes Pin.*—September 11, 1866.—The jaws are closed by an elastic band and the lateral displacement of the joint is prevented by a plate inserted at right angles to the axle.

Claim.—The clothes pin constructed with an elastic band passing through apertures in the jaws and two or more times around them, and with a plate or plates inserted in the joint to prevent lateral motion, substantially as described.

No. 57,978.—F. SCHENCK, San Antonio, Texas.—*Gun Lock.*—September 11, 1866.—This hair trigger arrangement is adapted to be attached to a Colt pistol without remodelling the latter, and is explained in the claims and cut.

Claim.—The spring dog *j* and slide *h*, in combination with the hair trigger B, and hammer H, constructed and operating substantially as and for the purpose described.

Also, the yielding tooth *a* in the main trigger to form a point d'appui for the hair trigger mechanism, substantially as herein set forth.

No. 57,979.—DANIEL W. SHARES, Hamden, Conn.—*Sorghum Stripper.*—September 11, 1866.—The blades are opened by pressing their curved ends against the stalk, allowing the latter to reach the jaws which are drawn together by the spring and remove the leaves by motion along the stalk.

Claim.—The combination of jaws B and C, constructed and arranged so as to operate substantially in the manner and for the purpose herein set forth.

No. 57,980.—JACOB SHAW, Hinckley, Ohio.—*Machine for Forming Lock Joints in Sheet Metal.*—September 11, 1866; antedated September 2, 1866.—This machine is used for tightening or consolidating the seams after the margins of the two sheets of metal have been interlocked; the detail of its arrangement is explained in the claims and cut.

Claim.—The vibrating beam C, die plate F, die E, and wedge W, constructed and operating in relation to each other, in the manner and for purposes substantially as herein specified.

Also, the arms B and D, in combination with the stirrup L, adjusting gauge P, and loop T, substantially as and for the purpose set forth.

No. 57,981.—G. W. Shawk and S. C. Morley, Cleveland, and S. D. Cushman, New Lisbon, Ohio.—*Telegraphic Fire Alarm.*—September 11, 1866.—The magnets are arranged on opposite sides of the lever and of its axis, and act in conjunction; the cushion prevents the rebound of the ball, and the arm is extended to vary its sweep according to the strength of the current.

Claim.—The adjustable bell extension arm H', adjustable post K, cushion K', in combination with lever H, magnets and armatures, as and for the purpose set forth.

No. 57,982.—W. C. Sherwood, Buffalo, N. Y.—*Removing Obstructions from Oil Wells.*—September 11, 1866.—The acid is lowered to the spot in a glass bottle, which is broken upon the tool, or in a chamber of smaller diameter than the bore, and provided with a valve.

Claim.—The method of loosening and removing or wholly destroying such obstructing tools, tubing, or other obstructions which may be lodged in artesian and other wells, by the application and use of sulphuric acid and other equivalent acids, substantially as herein set forth.

No. 57,983.—T. P. Sibley, Oberlin, Ohio.—*Combined Sheep Shed and Rack.*—September 11, 1866.—The hay and grain are separated, so that each sheep receives its food in separate apartments ; grain below, hay above, and a roof over all.

Claim.—The combination of the shed G I J, the hay crib D, and grain trough B with each other, the said shed, crib, and trough being constructed and arranged substantially as herein described, and for the purposes set forth.

No. 57,984.—William Simpson, Z. Oman, and S. Foreman, Nottowa, Mich.—*Turning Lathe.*—September 11, 1866.—The rod has adjustable tappets, by contact with which the carriage is made to operate a clutch drum on the gig-back shaft and gig back the tool stock of the latter.

Claim.—The arrangement of the tappet rod M, with tappets O O', by contact with which the carriage is made to operate the clutch drum on the gig-back shaft F, substantially as and for the purpose described.

No. 57,985.—Err and Dwight Skeels, Springwater, N. Y.—*Sheep Rack.*—September 11, 1866.—The slats of the open side of the rack are removable, and their spade-shaped lower portions abut; a grain trough on vibrating arms is lowered into feeding position.

Claim.—The construction of racks for feeding sheep with the sides made up of removable slats 'g g, fitting in grooves f f of the timbers c d, and retained in place by abutting and filling the space at the bottom, the whole operating in the manner and for the purpose specified.

Also, the combination and arrangement of the grain troughs o, and hinged arms n, with the boards k k, forming a central alley, in the manner and for the purpose herein set forth.

Also, the special construction and arrangement of the portable sheep rack, as herein specified.

No. 57,986.—Thomas Slaight, Newark, N. J.—*Lock for Doors of Baggage Cars.*—September 11, 1866.—The vibrating spring-latch is locked by a spring arm beneath ; in unlocking the detent arm is retracted by a key, which arranges the tumblers, and draws a slide to which the detent is pivoted.

Claim.—The combination of the tumblers j and slide E, provided with pin k and the spring arm F, operating with the shouldered arm d, having the thumb piece c, substantially as described for the purpose specified.

No. 57,987.—Alfred E. Smith, Bronxville, N. Y.—*Axle-box for Wagons.*—September 11, 1866.—A serrated ridge around the box secures its rigidity of position in the hub.

Claim.—The new manufacture of axle-boxes for wagons and other vehicles by forming on their butt-ends a raised shoulder, having a series of teeth cut therein to engage in the back end of the hub, in contradistinction to the use of lugs to secure the boxes in the hub.

Also, the making of the axle-box perfectly cylindrical throughout the length of it, so that by boring the hub with a true centre, the box will fit it without the necessity of cutting any grooves or channels in the hub, for the purposes hereinbefore set forth.

No. 57,988.—William P. Smith, Louisville, Ky.—*Lantern.*—September 11, 1866.—By its peculiar arrangement and construction the lantern is capable of being folded into a compact form for transportation.

Claim.—First, two convex top pieces B B', constructed, arranged, and operating as and for the purposes set forth.

Second, in combination with the top pieces B B', the handle consisting of the parts J J *j j*, when constructed and operating as and for the purpose herein explained.

Third, in combination with the elements of the first claim, the clasp I for preserving the shape of the lantern either in its opened or closed condition, as described.

Fourth, a folding lantern having one side made of two distinct pieces E E', connected together by the hinge *e*, substantially as specified.

Fifth, the shiftable candle holder L, in combination with the rod K K, when constructed, arranged, and operating substantially as specified.

No. 57,989.—S. E. SOUTHLAND, Jamestown, N. Y.—*Covered Tub for Butter and Lard.*—September 11, 1866.—The rabbeted cover is pressed upon the elastic ring on the rim of the tub by a wedge key beneath the binding bar, which is secured by ears to the tub.

Claim.—First, the combination of a transverse sliding key F, with the cover B and binding bar D of a butter tub or other vessel, substantially as herein described and for the purposes set forth.

Second, the formation of a key seat C in the cover B and binding bar D for the reception of the key, substantially as described.

Third, the ears E, constructed as herein described, in combination with the tub A, binding bar D and cover B, substantially as and for the purpose set forth.

No. 57,990.—JOHN STEPHEN and WILLIAM ZELLER, Womelsdorf, Penn.—*Pie Rimmer.*—September 11, 1866.—Explained by the claim and cut.

Claim.—The within described rimmer as an article of manufacture, consisting of a handle provided at one end with a rotary cutter, secured to an irregular or ornamental roller and the corrugated wheel E, and at the other end with a butter cutter or print, substantially as represented.

No. 57,991.—A. W. STOKER, Petersburg, Ill.—*Plough.*—September 11, 1866.—The threaded upper end of the standard passes through the beam, and the latter is adjusted vertically by nuts above and below. The rear end of the beam is adjusted laterally by nuts on the screw rod which connects the handles. One handle is bolted to an angle piece which connects it to the standard and mould board.

Claim.—First, a plough having its beam A pivoted and adjusted upon the standard *a*, as shown, and also having its rear end secured and adjusted upon the rod *b*, in the manner herein set forth.

Second, in combination with the standard *a* set in from the landside, as shown, securing the handle C to the elbow iron or brace *n*, when arranged as shown and described.

No. 57,992.—STEPHEN STOUT, Tremont, Ill.—*Sulky Plough.*—September 11, 1866.—The carriage has two wheels running in the furrow and on the land, respectively. The plough is pivoted to the carriage, and is laterally braced thereto ; the tongue is attached to the carriage at the pivoted point of the plough, and to the latter by a universal clevis joint. The button keeps the plough out of the ground when desired.

Claim.—The attaching of the draught pole K to the plough beam F by means of the universal joint L and flexible strap M, in combination with the mounted frame in which the plough beam is fitted, substantially as and for the purpose specified.

Also, the button I applied to the bar C, and in relation with the handle K of the plough, substantially as and for the purpose set forth.

Also, the brace rod N applied to the plough beam F and to the mounted frame, substantially as and for the purpose specified.

No. 57,993.—STEPHEN STUCKY, New Albany, Ind.—*Steam-generator Safety Valve.*—September 11, 1866.—The double-headed puppet valve covers openings which lead from the inner chamber to the box in which the weighted lever works and which communicates with the escape pipe.

Claim.—The valve-box A, provided with an inner box F, valve C, lever D, with weight E, inlet B, and throat G, the whole being constructed and arranged in the manner substantially as and for the purpose herein set forth.

No. 57,994.—SAMUEL M. SWARTZ, Millheim, Penn.—*Corn Planter and Plough Combined.*—September 11, 1866 —The machine may, after planting is over, be used as a plough. The operative parts are thrown out of action by the depression of the forward cross bar connected to the levers, and unshipping the gears. The devices for regulating the passage of the seed are explained in the claims.

Claim.—The arrangement of the levers M and their rung N with the plough, and with the clasp *f* and shipper *h*, so that when the ploughs are raised up the driving mechanism will be thrown out of gear, and when let down will put itself into gear, substantially as described.

Also the ribs *n* and openings *o* on and through the seed slides, so that the grains can stand edgewise between the ribs and drop edgewise through the holes as described.

Also, in connection with the ribbed seed slides, the spring fingers *m* for arranging the grains between the ribs.

No. 57,995.—J. B. SWEETLAND, Pontiac, Mich.—*Horse Power.*—September 11, 1866.— The cogs of the master wheel engage the worm threads on the shaft beneath, which connect by mitre gearing with shafts running in different directions but geared together and uniting their power upon the clutch-joint, upon which the tumbling shaft is to be attached.

Claim.—First, the two shafts E E situated upon opposite sides of the wheel D, and both provided with right threads, or both with left threads, as and for the purpose specified.

Second, the shafts F F lying under the wheel D, both in the same direction and turning in opposite directions, as and for the purpose set forth.

Third, providing the rim of the wheel D with holes as shown, as and for the purpose specified.

Fourth, the arrangement of the shafts E E, gear wheels G H, and shafts F F, with the wheel D, as constructed, as and for the purpose set forth.

No. 57,996.—CHESTER W. SYKES, Suffield, Conn.—*Marine Balloon.*—September 11, 1866.—The balloon, formed of a series of inflated tubes, has a suspended passenger car and a marine guide boat with a rudder at each end worked by a tiller above. Sails connecting the marine guide with the balloon are adjustable at the latter and trimmed as required.

Claim.—A broad, flat balloon of tubular or sectional formation to be held before the wind similar to a kite, in combination with sails and guide, substantially as before described.

No. 57,997.—GEORGE TAYLOR, Camden, N. J.—*Malt-kiln Floor.*—September 11, 1866.— The floor of the malt-kiln is formed of sheet-iron trays laid in break-joint arrangement, and of such length as to cover two or more of the spaces between the joists. The marginal flanges of the sheets are secured to bars resting upon the joists and by wires to the joists.

Claim.—First, constructing the plates A of such length as to cover two or more spaces between the joists B, and so arranging them as to make the plates of each row break joints with the plates of the adjoining rows to give additional strength and stiffness to the floor, substantially as described.

Second, the combination of the clips E with the plates A and joists B, by means of the wires F, or their equivalent, substantially as and for the purpose set forth.

Third, confining the flanges *a* of the plate A together by means of bolts, rivets, or pins, to give permanency to the plates and stiffness to the floor, substantially as specified.

No. 57,998.—J. TAYLOR and R. M. LAFFERTY, Three Rivers, Mich.—*Broom Head.*— September 11, 1866.—The handle passes through the wooden cap of the flexible sheath and rests on a slot in a cross-bar below. Tightening screws pass in from each side of the metallic sheath and engage in the central bar to clamp the side plates against the sheath.

Claim.—First, the combination of the clamping springs C C and adjusting screws *c c*, with the centre bar D attached to the sheath A so that the ends of the screws are within the sheath, substantially as described.

Second, a centre bar D arranged with a central socket for holding the tapering end of the broom handle, substantially as described.

Third, the combination of the block B, handle C passing through it, flexible sheath A and centre bar D, in which the handle rests, as described.

No. 57,999.—J. P. THOMPSON, Kirkville, Iowa.—*Composition for Roofing.*—September 11, 1866.—Composed of coal tar, 1 gallon; pulverized coal, 25 pounds; potters' clay, 12¼ pounds; slaked lime, 12¼ pounds.

Claim.—A composition of matter compounded and prepared substantially as and for the purpose set forth.

No. 58,000.—G. H. TIFT, Morrisville, Vt.—*Meat Mangler.*—September 11, 1866.—The axis of the operative corrugated roller is maintained in its bearing by a block which is detachable by removing a key to withdraw the roller.

Claim.—The construction of the movable block K with key M and staples L, when arranged and combined with the rollers C, spring D, and set screw G, to operate as herein described and for the purposes set forth.

No. 58,001.—ANDREW R. TULLY, Harlem, N. Y.—*Wagon Top.*—September 11, 1866.— The side straps to which the back curtain is fastened have vertical spring plates attached inside to maintain their shape and afford an additional attachment for the curtain knobs.

Claim.—First, the springs F, and their combination with the hinge *f*, wagon frame A, rear bow E, and knobs *a*, substantially as herein shown and described and for the purpose set forth.

Second, the manner of attaching the knobs *a* by riveting them to a metal strap concealed beneath the leather strap H, substantially as shown and for the purpose specified.

No. 58,002.—WILLIAM C. TURNBULL, New York, N. Y.—*Machine for Purifying Gas.*—September 11, 1866.—The metallic cylinder has perforated partitions, each of the latter having a rotary stirrer just above it The lime and water are let in from a tank and pass down through the perforated plates in the cylinder in a fine shower. The gas passes upward through the cylinder and escapes through a pipe into the driers which are made like the ordinary dry-lime purifiers, and are filled with quicklime.

Claim.—First, the combination of the tank D, pipe C, and agitators *d c'*, with the cylinder A, perforated plates *a*, and stirrers *a'*, for the purpose of purifying the gas, constructed and operating substantially in the manner described.

Second, the combination of the cylinder A with pipe H, boxes I, and perforated shelves *i'*, for the purpose of drying the gas after its purification, substantially in the manner herein shown and described.

Third, the application of unslaked lime for the purpose of depriving the purified gas of all its hygrometrical moisture, thereby rendering the gas chemically dry, and increasing its illuminating power, substantially as herein described.

No. 58,003.—WILLIAM TUTTLE, Jr., Dowagiac, Mich.—*Fence.*—September 11, 1866.—The notched extended ends of the rails in one panel catch upon the notched rails of the next panel inside of the post.

Claim.—Providing the adjoining ends of a panel fence with a locking device, substantially as and for the purpose described.

No. 58,004.—HENRY UNDERWOOD, New York, N. Y.—*Lap Joint for Belting.*—September 11, 1866.—The lap joint is formed by combining a separate slip of leather with the thinner parts and with the overlapped ends of the jointed parts of the belting. The overlapping ends of the strip are cut in thongs and laced to the body of the overlapped part of the belting.

Claim.—First, the combination of the strip C with the thinner pieces A B of the belting, the ends of said pieces being overlapped and secured to each other, substantially as described, for the purpose specified.

Second, the manner of lacing the overlapping ends to the overlapped part of the belting, substantially as herein described and for the purposes set forth.

No. 58,005.—P. H. VANDER WEYDE, Philadelphia, Penn.—*Distilling Petroleum and other Liquids.*—September 11, 1866; antedated August 8, 1866.—The still is connected with a tubular condenser having a device for drawing the vapor from the still through said condenser; the latter is provided with a series of exits for the condensed products, distributed at different distances throughout its length, in order to separate the oils of different specific gravity. This process is applicable to the distillation of petroleum, alcoholic spirits, &c.

Claim.—First, the production of a partial vacuum by suction produced in the still by a pump, either between it and the condenser or at the end of the tube intended for the escape of the non-condensable products, which vacuum may be filled by those non-condensable products, vapor, air, or steam, led to and admitted from the other end of the apparatus, provided with a safety-valve.

Second, the peculiar arrangement of the fractional condenser and its collecting and separating boxes, producing at one single operation liquids of different degrees of volatility, as above described.

Third, a series of stop-cocks at different places in the still and condenser, for the purpose of admitting a current of air or of conducting the escaping gases and vapors or steam over or through the liquid during the process of exhaustion.

No. 58,006.—HENRY S. VROOMAN, Hoboken, N. J.—*Mechanical Toy.*—September 11, 1866; antedated September 2, 1866.—The body is suspended in part upon driving wheels on which it traverses; the legs are actuated by springs and connecting gearing, and imitate the natural motions, while a rising and falling motion of the body is given by the eccentricity of the wheels.

Claim.—First, an automatic animal figure having four movable legs operated by a coil spring, in combination with a wheel or wheels C, substantially as shown, when such wheel or wheels are used, upon which to suspend and propel forward the body of such figure, so as to leave the legs free to be operated to imitate the running movements of a quadruped, as shown.

Second, the use or employment of a wheel or wheels upon which animal automatic figures are suspended and propelled forward, of an irregular or eccentrical form, so as to give a rising and falling movement to the head and fore shoulders of such animal moving figure at each stride or movement of the legs.

No. 58,007.—W. G. WARD, Savona, N. Y.—*Clothes-pin.*—September 11, 1866. —The hinged jaws, when closed, are fastened shut by a button.

Claim.—A clothes-pin consisting of the parts A B and C, all constructed, combined, and operating substantially as shown and described.

No. 58,008.—JOHN S. WARREN, Port Chester, N. Y.—*Water Wheel.*—September 11, 1866.—The wheel is surrounded by a scroll and revolves on a vertical shaft. The vertical buckets are not covered and are set radially on the shaft; their upper and lower edges taper outward and are cut away near the shaft to allow of free discharge, which may take place both above and below the wheel.

Claim.—The upper and lower inclined or conical plates *a a* of the wheel with the buckets *b* scolloped at their inner edges, in combination with the scroll A, all arranged substantially as and for the purpose specified.

No. 58,009.—J. C. WHARTON, Nashville, Tenn.—*Condenser.*—September 11, 1866.—Hollow disks inclined in alternate directions are connected at their edges and form a zigzag conductor for the liquid which passes through them consecutively.

Claim.—An improved condenser, formed of a series of hollow disks D, constructed and connected with each other, substantially as herein described and for the purpose set forth.

No. 58,010.—JAMES D. WHELPLEY and JACOB J. STORER, Boston, Mass.—*Amalgamator.*—September 11, 1866.—The outer cylinder is supported on a shaft attached by a hub to an internal plate. The interior of the cylinder is coated with mercury; the pulp being introduced during rapid rotation, is spread over the interior surface by the centrifugal force, and the metallic particles are seized and amalgamated by the mercury.

Claim.—First, an amalgamator or separator, consisting of a hollow revolving cylinder C, partially closed at the ends by centrally pierced disks B, and revolved at a rate sufficiently high to more than overcome the influence of terrestrial gravity on matters placed within it, substantially as described.

Second, the arrangement of the supply pipe P and discharge pipe P', in connection with a revolving cylinder C carrying fluid, and a tank T to secure the supply and discharge of fluid or pulp, substantially as described and for the purpose stated.

No. 58,011.—JAMES D. WHELPLEY and JACOB J. STORER, Boston, Mass.—*Machine for Cleaning and Brightening Particles of Precious Metals.*—September 11, 1866.—The cylindrical vessel is connected with a hopper at one end, an exhaust pipe at the other end, and has a series of rotary agitating arms attached to a shaft passing through the said cylinder. The hopper has a grating and a feed brush. Air may be admitted to the cylinder through a grating.

Claim.—First, brightening metallic particles in finely pulverized and desulphurized ores, when such brightening is effected on the principle of mutual attrition in a cylinder A, alternately closed during the brightening process, and opened to set free the charge by means of valve *d* in the exhaust pipe C, intending to claim for this end the principle of alternately closing and opening the cylinder, so as to do the work in a close cylinder, as well as the combination of cylinder A, valve *d*, and exhaust pipe C, for the purpose and substantially as described.

Second, the combination with feed hopper B of a fine grating to prevent the passage of any but very fine dust into the cylinder, and the further combination with the said feed hopper, and rendered necessary by the exclusion of any but very fine dust from the charge of the machine and the revolving brush *i* and its discharging pin K, operating substantially as and for the purpose described.

No. 58,012.—JAMES D. WHELPLEY and JACOB J. STORER, Boston, Mass.—*Roasting Desulphurizing and Smelting Ores.*—September 11, 1866.—The flux is blown in a finely divided state upon the heated ores by means of a blast of air.

Claim.—The introduction of finely powdered chemical re-agents, floated on an air or steam blast into an atmosphere of heat containing coal in aerial or air-borne combustion, substantially as and for the purpose described.

No. 58,013.—J. WHITESIDE, Coesse, Ind.—*Fence.*—September 11, 1866.—The panels are laid zigzag and are locked together at the corners by pieces which form the longer sides of obtuse-angled triangles with the lines of the panels. The pieces rest against the end posts of each panel, abut upon the second uprights of the said panels, and are locked by a vertical clamp piece, staple and pin.

Claim.—An improved portable field fence, constructed and arranged substantially as herein described and for the purposes set forth.

No. 58,014.—THOMAS T. WIER, Gallatin, Mo.—*Wagon Brake.*—September 11, 1866.—The rubber bar is operated by a rod connected to the neck yoke so as to brake the wagon wheels when the horses back. The brake is operatable by the hand lever which engages a rack on the tongue to lock the rubbers against the wheels.

Claim.—First, the combination with the neck yoke K, tongue J, forward bolster F, and brake bar P of a rod M, jointed in two places to accommodate it to the movements of the tongue and forward part of the running gear of the wagon, substantially as described and for the purpose set forth.

Second, the combination of the iron guide brace T with the brake bar P, and with the rear axle C and bolster E, substantially as described and for the purpose set forth.

Third, the combination of the lever U and rack W, with the brake rod M, and with the tongue J, substantially as described and for the purpose set forth.

No. 58,015.—DICKSON WILSON and HENRY FAIRBANKS, Adams county, Ill.—*Hand Spinning Machine.*—September 11, 1866.—The framework is held together by wooden screws and may be readily adjusted and taken apart for repairs, &c.; the wheel revolves on the ends of screws or pins, thus affording facilities for oiling, removal, or adjustments, to compensate for wear.

Claim.—The pedestal D with the uprights or standards e e, the driving wheel A, the cords H H, the spindle C, the wheel B, the screws f f, and the centre screws or pivots G, combined and arranged as above described, or substantially so, for the purposes therein set forth.

No. 58,016.—LUTHER J. WOODRUFF, Mohawk, N. Y.—*Carriage Seat.*—September 11, 1866.—Explained by the claim and cut.

Claim.—A carriage seat having metallic corner pieces, substantially as and for the purpose described.

No. 58,017.—C. S. BELL, Hillsboro', Ohio, assignor to himself and JOSEPH K. MARLAY, same place.—*Steam Generator.*—September 11, 1866.—The inclined tubes are arranged in layers over the fuel in the furnace and are connected horizontally and vertically.

Claim.—The tubes B B' B'' arranged in rows one above the other and over the fire chamber A, as shown, in combination with the sections or pipes C C' C'' and E E' E'', constructed, arranged, and applied substantially in the manner as and for the purpose set forth.

No. 58,018.—T. B. BUNTING, New York, N. Y., assignor to himself and R. L. DELISSER, same place.—*Bale Fastening.*—September 11, 1866.—The flanges, which are punched up to form a central slot, are rolled over to form rounded edges for the bends of the band loops.

Claim.—The rolled edges c of the slotted frame, substantially as described for the purpose specified.

No. 58,019.—GEORGE B. CURTIS, Walcott, N. Y., assignor to P. K. BRONSON.—*Water Elevator.*—September 11, 1866.—Two self-dumping buckets are employed and their respective rollers are placed in engagement with the raising mechanism by a clutch operated by a shifting lever, maintained in position by a spring bolt.

Claim.—First, the arrangement in double-bucket water elevators, substantially as shown, of the single tilting lever D and its spring bolt I in combination with the sliding clutch C, for the purposes set forth.

Second, the spiral spring s and its locking bolt I in combination with the tilting bar D, substantially as and for the purpose herein shown and described.

No. 58,020.—M. P. EWING, Rochester, N. Y., assignor to H. B. EVEREST and GEORGE P. EWING.—*Material for Lubricating.*—September 11, 1866.—The petroleum is distilled in a retort by means of a steam coil and the residuum is used for a lubricator.

Claim.—As a new manufacture an oil product as above described, when produced from crude petroleum by the evaporation therefrom of the lighter hydrocarbons in vacuo by the use of steam, or its equivalent, to prevent burning, substantially as herein set forth.

No. 58,021.—M. P. EWING and H. B. EVEREST, Rochester, N. Y., assignors to H. B. EVEREST and GEORGE P. EWING.—*Apparatus for Distilling Petroleum, &c.*—September 11, 1866.—The retort is supplied with oil by a pipe, has a steam coil in the bottom and is connected by a neck with a condenser surrounded by a water jacket. A water-jet pipe passes through the neck and terminates in the upper part of the condenser. The air is exhausted from the apparatus and the oil forced into a receiving tank by an air pump.

Claim.—First, the pipes h h', overlaying each other as specified, and forming a compact body, when employed in combination with a vacuum still having a continuous feed, substantially as herein set forth.

Second, the perforated pipes i i for injecting steam into the body of oil when combined with a vacuum still having a continuous feed, substantially as specified.

Third, the chamber k, in combination with the retort A and neck or tube C, for catching the overflow, as herein set forth.

Fourth, the pipe l and receptacle H, provided with the cocks m n o, or equivalent, when combined with the chamber k for discharging the overflow without impairing the vacuum, as specified.

Fifth, the combination of the enclosing water tank I with the jet condenser B, substantially as and for the purpose herein set forth.

No. 58,022.—SAMUEL C. GOODSELL, New Haven, Conn., assignor to himself and BENNET HOTCHKISS, same place.—*Flute, &c.*—September 11, 1866.—The mechanical arrangement of the keys enables $G^\sharp$, $B\flat$, and $C\sharp$ to be produced without using the thumb or fourth finger of the left hand.

Claim.—The within described mechanical arrangement of the keys, substantially as and for the purpose specified.

No. 58,023.—B. A. GRANT, Mount Pleasant, Iowa, assignor to himself and J. B. COATT, same place.—*Hand Spinning Machine.*—September 11, 1866.—This invention relates to that class of spinners in which the carriage bearing the rolls runs in and out and the frame of the spindles is stationary. The devices cited are for the purposes of intermittingly delivering by friction the sliver to be spun, operating the delivering jaws, varying the fineness of the thread and operating the carriage.

Claim.—First, the combination of the arm T upon the car with the U lever *n m* and the spring catch S', or its equivalent, arranged and operating as and for the purposes described.

Second, in combination with the U lever *n m*, the employment of the lever *p* and arm V, arranged and operating as and for the purposes set forth.

Third, the combination of the jaw G, spring *b*, arm *i*, and catch *c d*, or its equivalent, arranged and operating as and for the purposes shown and described.

Fourth, in combination with said jaw G, spring *b*, and catch *c d*, the arrangement of the pin *r*, or its equivalent, as and for the purposes specified.

Fifth, the arrangement of the spring *s* and *t* and arm W with the arm T of the car B, operating as and for the purposes set forth.

No. 58,024.—JOHN B. KERSHAW, Indianapolis, Ind., assignor to himself and JAMES C. DICKSON, same place.—*Loom.*—September 11, 1866.—The treadle cams forming the pattern cylinder are adjustable on the axis by means of holes and pins so as to vary the pattern. Springs serve to raise the harness promptly when the cams permit it.

Claim.—The combination and arrangement of the variable pattern cylinder X, constructed as described, with the treadles 6 6, jacks 9 9, and springs 10 10, the whole operating as and for the purpose set forth.

No. 58,025.—WILLIAM T. NICHOLSON, Providence, R. I., assignor to THE NICHOLSON FILE COMPANY, same place.—*File.*—September 11, 1866.—Explained by the claim.

Claim.—As a new article of manufacture, a round or curved surfaced file, the teeth of which are severally distinct, as when cut by hand, but the rows of which are arranged in wave lines, substantially as described.

No. 58,026.—GEORGE S. REYNOLDS, Tunbridge, Vt., assignor to himself and FRANCIS A. CUSHMAN, Lebanon, N. H.—*Reaping and Mowing Machine.*—September 11, 1866.—The improvements in the finger bar, the connection of the pitman and sickle bar, the driving gear, the finger bar braces, and the mode of attachment of the track clearer to the finger bar are explained in the claims and cut.

Claim.—First, constructing the finger bar A of taper form in its transverse section, and with a lip B at its outer end, in combination with the track board or clearer C attached to the lip B by the pivot bolt *a*, and provided with a pin or stop *b*, substantially as and for the purpose set forth.

Second, the securing of the pitman or connecting rod F to the sickle E by means of the eye *g* at the outer end of F, and a slide *j* attached, and a pin *h* projecting eccentrically from a rounded portion of a rib *i* on the sickle, substantially as herein shown and described.

Third, connecting the finger bar A to the main frame H by means of the bars I I and rod J, constructed, arranged, and applied substantially in the manner set forth.

Fourth, the taper driving shaft K, in combination with the taper or conical bearing tube L, substantially as and for the purpose specified.

No. 58,027.—HIRAM ROBBINS, Cincinnati, Ohio, assignor to himself and THOMAS H. FOULDS, same place.—*Roller for Wringers.*—September 11, 1866.—Explained by the claim and cut.

Claim.—The elastic roll B of rubber, vulcanized upon the longitudinally flanged or ribbed and perforated shaft A *a d*, substantially as set forth.

No. 58,028.—D. W. RUST and GEORGE BUZZEE, East Hampton, Mass., assignors to THE EAST HAMPTON RUBBER THREAD COMPANY, same place.—*Machine for Cutting Rubber into Threads.*—September 11 1866.—The cylinder around which the rubber to be cut is wound is covered with "mixed" or partially cured rubber, which serves as a foundation for the knife to cut against and prevents the surface of the roller from absorbing water and becoming irregular.

Claim.—Covering a cylinder with "mixed" rubber or other equivalent material, as a foundation upon which to cut rubber threads, substantially in the manner and for the purposes herein specified.
p

ELBRIDGE J. STEELE, New Britain, Conn., assignor to himself and E. A. TAYLOR, same place.—*Window Shade Fastening.*—September 11, 1866.—The cord passes freely through a slot in the bed plate which is attached to the window casing, and is held in any position by the pressure of a lever clasp thereon.

Claim—First, the constructing of a window shade fastening with a bed plate *a*, in the under side of which is a recess for receiving the cord *g* between it and the window casing, a hole *s* for the cord to pass through over the shoulder *f*, with a crease or corrugation in it to prevent slipping, combined with the spring lever clamp *b*, substantially as and for the purposes herein described.

Second, the lever clamp *b* with a recess in it for receiving the cord under the lower limb, in combination with the spiral spring *d* and the bed plate *a*, constructed and operated substantially as and for the purposes herein specified.

No. 58,030.—RICHARD WALKER, Batavia, N. Y., assignor to himself and PETER BROAD BROOKS, same place.—*Carriage Wheel.*—September 11, 1866.—The spokes are lengthened to tighten them against the rim of the wheel and the felloes spread to maintain them tight against the tire by screwing out the ferrule on the spoke, thereby expanding the felloes and driving a wedge between their ends.

Claim.—First, the combination of the ferrule E, sleeve G, nut H, screw F, and band I with each other, and with the spokes B and felloes C, substantially as described and for the purpose set forth.

Second, the combination of the wedge or wedge-shaped nut J with the felloes C, cap or band I, and screw F, substantially as described and for the purpose set forth.

No. 58,031.—B. WHYSALL and J. PORRETT, Port Jervis, N. Y., assignors to themselves and M. M. LIVINGSTON, Brooklyn, N. Y.—*Tempering Springs*—September 11, 1866.—The bath is composed of linseed oil, resin, lampblack, and vitriol.

Claim.—First, the process of tempering springs, wire, &c., substantially as herein described.

Second, for the purpose of hardening, cooling, and tempering springs, wire, &c., a bath composed of the chemical agents herein specified.

Third, in carrying out the process, the employment of a vat A, having a water chamber B all around it, and pipes *a b* letting into said chamber, whereby a constant current of cold water may be kept up for cooling the liquid in the vat A, as herein shown and described.

No. 58,032.—GEORGE WOODS, Cambridge, Mass., assignor to HENRY MASON, EMMONS HAMLIN, LOWELL MASON, Jr., and D. J. MASON.—*Musical Instrument.*—September 11, 1866.—Each stop has a pneumatic bellows, and they are operated through the main bellows by channels governed by a perforated slide valve which is common to all. The valve is connected to an index that moves therewith over a register plate and the valve itself is moved by any convenient means. The passages through the valve and in the valve seats are so arranged that combinations of stops may be made.

Claim.—First, the slide valve E with openings so arranged as to allow communication between the exhaust chamber or pump or bellows of a musical instrument and one or more of the bellows or pneumatic stop levers G, when the valve is moved to a proper position, substantially as described.

Second, the combination of the pipe F, perforated on its bottom, with the air passage J leading to the exhaust chamber C and the passages K of the pneumatic stop levers G, substantially as shown.

No. 58,033.—E. B. WRIGHT, Cincinnati, Ohio, assignor to E. D. FINCH and WILLIAM H. RUSSELL, Kokomo, Ind.—*Soda Fountain.*—September 11, 1866.—An ice-cooled reservoir containing the soda water is placed above the fountain and connected by an ice-cooled coil with the fountain.

Claim.—The provision of the elevated reservoir A, pipe B, coil B', refrigerating chamber C, and fountain D, combined and adapted to operate as set forth.

No. 58,034.—LOUIS FOUILLOUX, Clermont Ferrand, France.—*Curing Rheumatism.*—September 11, 1866.—Oxygen gas is applied externally, in conjunction with heated atmospheric air, to the parts affected.

Claim.—The within described process of treating the part or parts of the body affected with oxygen gas and heated air, substantially as and for the purpose set forth.

No. 58,035.—GUSTAV KÖTTGEN, Barmen, Prussia.—*Seamless Pocket.*—September 11, 1866.—These pockets are woven by the process and machinery used in the production of seamless bags; but the division between the several pockets is single for a part of the way, so that when cut apart the band hole is found on the upper side of each pocket.

Claim.—A seamless pocket so woven with others that when separated its mouth will be made on the upper side and adapted to be used for the purpose specified, substantially as described.

No. 58,036.—EMILE SABATHÉ and LOUIS JOURDAN, Paris, France.—*Impregnating Substances with Preservative Material.*—September 11, 1866.—Fabric, leather, &c., are placed in a cylinder charged under pressure with material which preserves them when exposed to the air or to fresh or salt water. The material consists of a metallic soap and is introduced through an opening at one end fitted with a bag filter. The cylinder communicates with a condensing worm and receiver.

Claim.—Preserving vegetable or animal substances, whether in a raw or manufactured state, by means of a soap, as herein described, having a metallic base, insoluble in water but soluble in volatile liquids, or by heat or distillation, and applied in such soluble condition to the article or thing to be preserved by it, substantially as herein set forth and explained.

No. 58,037.—JACOB B. THOMPSON, Middlesex county, England.—*Coating Iron and Steel with Gold, Silver, &c.*—September 11, 1866; English patent, October 7, 1865.—Iron or steel first receives a coating of pure iron from a solution, and then a coating of gold, silver, platinum, or copper, deposited by means of electricity.

Claim.—The process hereinbefore described of coating an article of iron or steel with a precious metal, by first coating the article with pure iron and then with the precious metal, substantially as set forth.

No. 58,038.—WILLIAM LOUIS WINANS and THOMAS WINANS, London, England.—*Steam Generator.*—September 11, 1866; English patent, June 23, 1866.—By the described arrangement the passage of the heated vapors, &c., from the fire may be regulated by the engineer so as to pass the whole or any portion through the superheater, or direct to the chimney, the dampers being simultaneously moved by the rotation of a ring to which they are connected.

Claim.—The combination and arrangement of the valves g, valves g', sector pieces h, ring i, tangent screw j, block k, bevel wheel l, the circular tubes f, dome d, space a, tubes b, whereby the heated vapors may circulate among the tubes f, or pass through the space e surrounding the dome, as may be desired, substantially as described for the purpose specified.

No. 58,039.—WILLIAM S. COOPER, Philadelphia, Penn.—*Steam Engine Valve.*—September 11, 1866.—By means of the guiding rim in the neck the stem is held erect while the valve is ground into its seat.

Claim.—The guiding ring m applied to the chest and spindle of a stop valve, substantially in the manner described.

No. 58,040.—DAVID E. ADAMS, Alleghany City, Penn.—*Making Bolts.*—September 18, 1866.—The end of the blank is staved into an oval form for the head and square shank before the heading and squaring operations are performed.

Claim.—Staving up or enlarging and forming into an oval or elliptic form (when viewed in cross sections) that part of the rod of iron intended for the head and square of the neck of bolts, said staving or enlarging into said oval or elliptic form being done previous to the formation of the squaring of the neck and forming of the head of the bolt, which squaring and heading is performed at a subsequent operation.

No. 58,041.—HOMER ADKINS, Plymouth, Ill.—*Car Coupling.*—September 18, 1866.—The frame holds the link horizontal until it is coupled by the spring pin in the other draw-head. The cross-bar catches on a rack on the draw-head and the coupling link rests on the cross-rod.

Claim.—The adjustable frame K with cross bar h, and cross rod g attached. In combination with the rack bar I, spring J, and link G, all arranged and applied to the draw-head A', to operate substantially as and for the purpose set forth.

No. 58,042.—JAMES O. ALTICK, Dayton, Ohio.—*Globe Steam Valve.*—September 18, 1866.—The packing around the stem bears upon the shoulder of the valve to hold the latte to its seat, and the screw plug in the valve nut rests upon the packing.

Claim.—The arrangement of the shouldered valve stem of the valve B with reference to the packing nut D, nut C, and valve seat F, for the double purpose of packing the stem and holding the valve against its seat, in the manner substantially as described.

No. 58,043.—SETH A. ANDREWS, Farmers' Valley, Wis.—*Alarm Lock.*—September 18, 1866.—The shackle is fastened by screws whose heads are exposed; one of them is connected with a trigger which is released when the screw is tampered with, and this in turn throws back a loaded barrel so as to explode a cap, project a ball, and explode the magazine simultaneously.

Claim.—First, the combination of the sliding barrel X, the trigger or dog I, the lug M of the barrel, and the spring P, substantially as described.

Second, the bolts Z Z', constructed substantially as described, one being a short bolt, in combination with the chains J of the trigger arm H, substantially as described.

Third, forming a magazine S in the lock at one side of the barrel chamber, substantially as described, with a priming orifice R in the path of the sliding barrel, substantially as described.

Fourth, covering the magazine with a cover J weakened by grooves or their equivalents, substantially as described.

No. 58,044.—ERNST ARLETH, Cincinnati, Ohio.—*Apparatus for the Manufacture of Vinegar.*—September 18, 1866.—The tub has a false bottom from which tubes extend to near the top, being covered at their upper ends with caps. The upper chamber is filled with shavings. From the head extends a pipe which terminates in a perforated tube in the lower chamber. The air is supplied to the tub through an adjustable side pipe.

Claim.—First, the tub A having a chamber C in combination with tubes F and conveying pipe J, all constructed as above described and for the purpose set forth.

Second, the adjustable cold-air pipe E, constructed as above described.

Third, the cold-air pipe E in combination with the tub A, as above set forth.

No. 58,045.—AARON ARMSTRONG, Gillespie, Ill.—*Combined Seed Drill and Cultivator.*—September 18, 1866.—The three devices are so arranged that either the planter and roller, or the seed drill and roller, may be used together to the exclusion of the rest of the machine, or the corn planter and seed drill may be both removed and knives fastened to the roller to be used in cutting old cornstalks to prepare the ground for tillage.

Claim.—First, the combination of the roller D with the sliding plate k and lever E, as and for the purpose set forth.

Second, the roller B in combination with the corn planter C D k F, the stalk cutter M, and the seed drill G I H, when the same are all constructed upon one common frame A, from which any of the parts except the roller B may be removed at pleasure.

No. 58,046.—JOHN E. ATWOOD, Mansfield, Conn.—*Spindle for Spinning Frames.*—September 18, 1866.—The sleeve connected with the "whirr" and spindle prevents the spindle from being drawn out of line; the stud on the suspended rod serves to hold the spindle in place; by lifting the rod the spindle may be removed.

Claim.—First, the inclined tube D provided at its upper end with a reservoir g in combination with the socket B, and spindle a b, when said socket is surrounded by a sleeve carrying the "whirr," substantially as herein set forth for the purpose specified.

Second, the bar or rod E suspended within the tube D, and in relation with the annular shoulder r formed around the lower end of the spindle a b, substantially as herein set forth for the purpose specified.

No. 58,047.—JOHN A. BALZART, Piqua, Ohio.—*Churn.*—September 18, 1866.—The dasher is suspended by means of a locking device, which encloses the neck of the dasher shaft, and is locked by a slide.

Claim.—The combination with the suspended dasher of the pivoted arms F F, forming a collar in the groove of the shaft D and the slide G, for retaining the said collar in position, substantially as described.

No. 58,048.—BENJAMIN BARNEY, Time, Ill.—*Machine for Cleaning Grain.*—September 18, 1866.—The grain is delivered from a hopper to a longitudinally vibrating riddle, and thence passes the whole length of the agitated chest and cockle screen; it is delivered into a chute, where an upward blast removes light offal.

Claim.—The rotary fan C, hopper D, shoe E, provided with the riddle K, chute L, and cockle screen M, and having a longitudinal shake motion communicated to it by the cam or eccentric F, and spring H, in combination with the suction blast spout N, all arranged substantially as and for the purpose set forth.

Also, the valve O, applied to the spout N, and provided with the adjustable or sliding weight g, substantially as and for the purpose specified.

Also, the extension support J, formed of the two parts d d', connected by the screw e, for the purpose of varying the inclination of the shoe E, as set forth.

No. 58,049.—WILLIAM N. BARR, Richmond, Ind.—*Cane Stripper.*—September 18, 1866.—The guides are parallel to each side of the rectangular frame, and in them are crescent-shaped knives actuated by springs and embracing the stalk.

Claim.—First, the combination of several sets of strippers in one frame, when constructed, arranged, and operating in such manner, substantially as herein described, that one spring acts upon two stripping edges or cutters.

Second, the arrangement of the parallel guide pieces f' h h', in combination with the strippers B C D E and springs F G H, all constructed and operating substantially in the manner set forth.

No. 58,050.—C. F. BAYLOR, Trenton, N. J.—*Grain Screen.*—September 18, 1866.—The offal which passes through the meshes of the upper screen is delivered separately from the grain, which latter passes over the second screen, losing a further amount of offal, which is deposited with the other screenings.

Claim.—The combination with the sieves A B and chute or conducting board C of the conductors E E, all arranged to operate in the manner and for the purpose herein set forth.

No. 58,051.—EDWARD B. BINGHAM, Newark, N. J.—*Paper Making Machinery.*—September 18, 1866.—Explained by the claim and cut.

Claim.—The arrangement of two felts and four pressure rolls, as herein described, whereby the web is pressed three times between the two felts, in combination with the arrangement of the doctors P Q within the felts, to remove the water from the rolls D E', as herein set forth.

No. 58,052.—JOHN S. BODGE, La Porte, Ind.—*Grain Separator.*—September 18, 1866.—The covers rest upon the grain and exercise some control upon the larger grains to prevent their entering the perforations in the sieve, while the smaller grains and offal are allowed to pass when the shoe is agitated.

Claim.—First, the construction of covers E E' E'', of suitable thickness, covering the screens or sieves B B' B'', with chutes H between, corresponding in number with the smooth or imperforated surfaces C, and opening obliquely or sloping downward nearly or quite their entire length, said chutes being so located that when grains, seeds, or other substances shall pass downward through them, they will fall or be precipitated directly upon the smooth or imperforated surfaces C.

Second, constructing and locating the covers E E' E'' in separate sections, with barriers F at their upper angles, and rectangular slips G G' attached to the ends of said covers to elevate them the requisite height above the screens or sieves B B' B'', and to keep them the proper distance apart to allow the passage of grains, seeds, or other substances in such manner that the oblique or sloping chutes H, through which the grain finds its way upon the smooth or imperforated surfaces C, above each screen or sieve, will retain their proper proportions and positions relative to the surfaces C, and above each screen or sieve.

Third, the flaps or aprons L, composed of flexible, textile, or other fabric, such as rubber, rubber cloth, oil cloth, painted cloth, or any preparation or fabric, so attached to the covers E E' E'', or to separate pieces to occupy their places, as to adapt themselves to the inequalities of the screens or sieves B B' B'', and to grains, seeds, or other substances which may be passed between said flaps or aprons and the screens or sieves, in such a manner as to prevent the longer or broader grains, seeds, or other substances from assuming an oblique or perpendicular position while passing over the screens or sieves, thus compelling said longer or broader grains, seeds, or other substances to pass over the screens or sieves flatwise, while the shorter, narrower, and smaller grains, seeds, and other substances are permitted to pass through them when the screens or sieves are in motion.

Fourth, the building or constructing of covers E E' E'', substantially as herein shown and described, so that they will expand or contract as equally in all parts as possible, and not be liable to warp or twist, thereby preventing the irregularities in their form which would otherwise occur.

Fifth, securing the flaps or aprons L to the covers E E' E'' by means of the clip plates A', as shown and described.

No. 58,053.—P. J. BORIS, Boston, Mass.—*Skirt Elevator.*—September 18, 1866.—A tape is lift the dress is fastened to the lower hem, thence passes through a sheath to the waist, where a catch acts as a ratchet to hold it at any elevation.

Claim.—The combination and arrangement of the clasp B, the tape C, and the sheath D, with a lady's skirt, substantially as described.

No. 58,054.—JOHN F. BOYNTON, Syracuse, N. Y.—*Liquids for Carburetting and Enriching Gases, and for Illuminating and Heating.*—September 18, 1866.—The lighter oils from coal tar, which have an excess of carbon, are combined with the light oils of petroleum, which have an excess of hydrogen.

Claim.—First, the improved liquids for carburetting or enriching gas or air produced by combining one or more hydrocarbons, derived from the petroleum, or equivalent, with one or more hydrocarbons of the coal-tar series.

Second, carburetting or enriching coal-gas, water-gas, or air, by combining with them the vapor of a liquid, made by combining hydrocarbons of the coal-tar series with those of the petroleum series, substantially as herein described.

Third, the method or process herein described of manufacturing liquids for carburetting or enriching gas or air, by combining one or more hydrocarbons derived from petroleum, or its equivalent, with one or more hydrocarbons of the coal-tar series.

No. 58,055.—JOHN F. BOYNTON, Syracuse, N. Y.—*Safety Apparatus for Gas Machines and Carburetters.*—September 18, 1866.—Two concave disks are united at their edges, forming a chamber within which a series of perforated disks are secured, dividing the chambers into compartments which may be filled with shot or other good conductor of heat.

Claim;—A safety device, composed of a body of cooling material, as metal or other good conductor of heat, interposed between the burner and carbureiter, and so arranged as to present an extensive surface or series of surfaces to the carburetted air on its passage to the burner, thereby keeping it at a temperature below the point of ignition.

No. 58,056.—AUSTIN BRONSON, New York, N. Y.—*T Square.*—September 18, 1866.—Explained by the claims and cut.

Claim.—First, the bevel stock D, so constructed and applied to the T square that its sides will project beyond each side of the said square, and thus enable the instrument to be employed for making right or left hand angular lines without interfering with the head B of the square, substantially as herein set forth.

Second, the tangent screw *e*, sliding frame *g*, and adjusting screw *f*, arranged with reference to each other and with the square A B, bevel stock D, and sector C, substantially as herein set forth for the purpose specified.

No. 58,057.—BENJAMIN BROWNELL, Chicago, Ill.—*Hot Air Furnace.*—September 18, 1866.—The flame. &c., from the fire pot is deflected by the dome plate above against the vertical air pipes, which are of greater diameter at their upper end, and thence passed up into the smoke drum and exit flue.

Claim.—The gradually enlarged flues E, arranged and combined with the dome-shaped casting G, in a hot-air furnace, substantially as and for the purposes herein set forth.

No. 58,058.—J. E. BUERK, Boston, Mass.—*Alarm Clock.*—September 18, 1866.—This alarm is operated by the revolution of a watch key on the central arbor of a watch, or it may be actuated by an opening door; in the former case a tooth on the watch key engages with a numbered notch on the serrated segment, which is moved by the revolution of the key until the flat part of the segmental disk comes opposite to the stop lever, and the arm is released.

Claim.—First, the arm *l* and segmental disk *i*, in combination with the stop lever *e* of the alarm movement, constructed and operating substantially as and for the purpose set forth.

Second, the serrated segment *m* in combination with the tooth of the watch key *o*, and with the alarm movement, constructed and operating substantially as and for the purpose described.

No. 58,059.—WILLIAM BURSON and D. C. BURSON, Salineville, Ohio.—*Churn.*—September 18, 1866 —The dasher has oblique, radial perforated blades ; a gathering board with oblique perforations is dropped into transverse position in the box when required.

Claim.—First, the dasher C, when constructed as herein described, in combination with the shaft B, gear wheels H and G, crank J, frame F, cover D, and box A, the said parts being constructed and arranged substantially as and for the purposes set forth.

Second, the combination of the perforated gathering board M with the box A and dasher C, substantially as described and for the purpose set forth.

No. 58,060.—WILLIAM F. CALDWELL, Oxford, Maine.—*Planter.*—September 18, 1866.—The plough and coverer are pivoted to the carriage. The feed slide is operated by contact of a pawl connected with the feed slide, with a ratchet on the carriage wheel.

Claim.—First, the combination and arrangement of the hopper *e*, spring *f*, pawl *d*, ratchet *c*, lever and link *m*, all constructed, applied, and operating as herein set forth and described

Second, connecting the plough and coverer to the carriage body in the manner described.

No. 58,061.—JAMES CHAMBERS, Boonsboro', Md.—*Cooking Stove.*—September 18, 1866.—A supplementary portion is added, when required, to the rear of the stove to afford enlarged cooking space.

Claim.—The combination of the parts D and E, when the latter is constructed and arranged in relation to the former, as shown and described and for the purpose set forth.

No. 58,062.—JAMES T. CHAMBERS, Utica, N. Y.—*Safety Pocket.*—September 18, 1866.—The pocket is attached by plates to the garment ; a spring above the watch has a needle which pointedly punctures the person of the wearer when the watch is surreptitiously pulled

Claim.—First, the pocket A, provided with a spring arm G upon its inside, having a sharppointed or needle arm H hung to its upper end, when combined together and arranged so as to operate substantially in the manner and for the purpose described.

Second, in combination therewith, the springs O upon the inside of the pocket A, for the purpose specified.

Third, in combination therewith, the perforated spring or flexible bands E, secured to the pocket A, for the purpose described.

No. 58,063.—DANIEL P. CHESBROUGH, Lansingburg, N. Y.—*Receptacle of Waste Water for Well Tubes.*—September 18, 1866.—The waste water runs into a reservoir provided with a float valve and pipe, which returns it back into the well tube.

Claim.—The receptacle A, having float I and valve G, in connection with the waste-water pipe of a well tube, when arranged with regard thereto and connected with the well tube, substantially as and for the purpose described.

No. 58,064.—WILLIAM J. CHRISTY, Philadelphia, Penn.—*Many-barrelled Fire-arm.*—September 18, 1866.—The barrels swing to expose their rears for loading, and by this motion the hammers are cocked. A safety-guard is locked over the trigger when releasing the barrels to charge them, and remains in place until removed for firing. Two or more hammers can be operated by the trigger, and it may be adjusted to operate upon them consecutively.

Claim.—First, the application to the trigger E of a vibrating arm d and lever d', constructed in such a manner as to effect the release of the hammers in succession, substantially as described.

Second, the application of a guard G to the trigger E, in combination with a contrivance which will lock this guard in front of the trigger when the breech of the barrel is exposed, substantially as described.

Third, in combination with a barrel which has an end play, and which is pivoted to the frame of the piece, the spring latch b, slides k and s, and a trigger guard G, arranged so as to operate substantially as described.

No. 58,065.—M. C. CLARK, Appleton, Wis.—*Sheep Rack.*—September 18, 1866.—The feeding device is a rack and a trough beneath; the sides which prevent access of the sheep when in one position vibrate toward each other in opening, and form a roof.

Claim.—The combination of the trough B, feed receptacle D, and covers E and F, with each other and with the frame A of the rack, when said parts are constructed and arranged substantially as herein described and shown, and for the purposes set forth.

No. 58,066.—HENRY CLARKSON, Peekskill, N. Y.—*Handle for Sad Irons.*—September 18, 1866; antedated September 5, 1866.—A lever pivoted at the bottom of the dowel attached to the handle is furnished with a spur, which engages in the slot in the body of the flatiron and attaches the handle to the latter.

Claim.—The lever D, with its toe d and nose e', in combination with the stirrup e, handle B, flange b, socket a, and iron A, all combined and operating substantially as and for the purpose described.

No. 58,067.—FREDERICK CLOSS, New Haven, Conn.—*Boot Heel.*—September 18, 1866.—The heel revolves to prevent unequal wear at any one point.

Claim.—A revolving heel for boots and shoes when the same is constructed with a metallic piece B, and ring D, said parts being respectively formed, combined with the other portions of the heel and arranged for use as and for the purpose set forth.

No. 58,068.—GEORGE W. CODDINGTON, Middletown, Ohio.—*Saw Mill.*—September 18, 1866.—The trestle supports the log until the tail-block strikes a spring, releasing the support which is turned over, and the tail-block passes over it; when the carriage in its backward movement passes off the support it again assumes an upright position.

Claim.—First, the trestle D, constructed and arranged as herein described in combination with the carriage C, and operating substantially as and for the purpose set forth.

Second, the combination of the spring F, catch G, chain H, and trestle D, with the carriage C, substantially as described and for the purpose set forth.

No. 58,069.—Z. E. COFFIN, Newton, Mass.—*Capstan Windlass.*—September 18, 1866.—The capstan, by a pinion on its axis, rotates the windlass below; the barrels, with their gears, are sleeved upon the shaft, and have an auxiliary support in anti-friction rollers on the side toward the bow of the vessel.

Claim.—First, the employment of the capstan a, and its system of gears j k l, in combination with gears r s, and the windlass barrels t u, substantially in the manner and for the purpose set forth.

Second, the arrangement of the barrels t u z z, and the gear s, to turn upon the shaft V, this shaft being a fixed one and not allowed to rotate, substantially as described.

Third, the two or more rollers y, in combination with and to support the windlass barrels and shaft t u and V, substantially as described.

No. 58,070.—ELISHA T. COLBURN, Boston, Mass.—*Sad Iron.*—September 18, 1866.—The wings are hinged and the device is clasped around the legs of the handle.

Claim.—The heat-reflecting guard C, made substantially as described, so as to be applied to the sad iron in manner and for the purpose specified, such guard being composed of the plates a a, the standards b b, and the legs c c, and having the notches d d formed in the parts a a, as explained.

No. 58,071.—LUCIUS M. B. COLEMAN, Danby, N. Y.—*Wagon Shackle.*—September 18, 1866.—Wear is transferred from the shackle bolt to the outer surface of the conical washers and the sockets of the ears in which they are fitted. The bolt is square, the washers fit upon it, the nut and head of the bolt fit against the washers and tighten them to prevent rattling.

Claim.—First, the use of washers or metallic cylinders in the ears of a wagon shackle, and transfer as far as possible the wear in the shackle joint to the surfaces made by the periphery of the same and the said ears, when virtually made as described.

Second, the described bevelling of the washers and of the corresponding holes in the ears of the shackle for the purpose of tightening the shackle from time to time as described.

Third, the combination of the described device or devices for preventing motion about the iron bolt, but allowing the same about the washers or cylinders, and of the washers or cylinders in the eyes of the ears of the shackle with the wearing surfaces in the said ears, the same making a whole as described.

No. 58,072.—MALCOLM M. COPPUCK, Philadelphia, Penn.—*Tobacco Pipe.*—September 18, 1866.—Explained by the claim and cut.

Claim.—A tobacco pipe consisting of the open-bottomed fire and tobacco chamber A, neck piece d, exhaust chamber B, and stem C, the same being arranged and combined together as and for the purposes described and set forth.

No. 58,073.—NICHOLAS C. DECKER, St. Louis, Mo.—*Horse Rake.*—September 18, 1866.—The rock shaft, lever, and springs are arranged in connection with a revolving rake supported on a wheeled frame to control the rake and discharge the load. The rake head is supported from the carriage in vertical pendants provided with curved slots, which allow the rake head to rise when the load is to be discharged.

Claim.—First, the construction and arrangement of the rock shaft C and its springs c^3 and c^4, its levers c and c', and the bar c^2, as described and set forth.

Second, the arrangement of the segmental grooves b in the posts B, for the purpose of rendering easy the revolution of the rake.

No. 58,074.—WM. DENNISON, Washington, D. C.—*Valve Gear for Steam Engines.*—September 18, 1866.—The openings in the valve seat are on a line with the axis of the cylinder; the valve swings on its axis above, its motions being generated by the reciprocation of a pin on the cross-head, which traverses in a cam on a segment of a cylinder pendent from a rock shaft which is an extension of the valve rod.

Claim.—First, the cam H arranged to operate in such a way as to cause the valves of steam cylinders to move continuously, while the piston of the steam cylinder be in motion, by a stud or pin on the piston rod or other attachment producing the same result, substantially as set forth.

Second, a steam cylinder having its openings and valves arranged on a line with the axis of the cylinder, substantially in the manner set forth.

Third, the cam H, the rock shaft F, and the steam cylinder with its steam valves E E′, combined and operating as and for the purpose set forth.

No. 57,075.—JOSEPH DILLIER, Greensburg, Ind.—*Horse Rake.*—September 18, 1866.—Explained by the claim and cut.

Claim.—A horse rake having the teeth E attached to a hinged beam F, retained when down by a book G, and disengaged and raised by a lever I, the said several parts being respectively constructed and combined for use by intermediate mechanism, arranged substantially as set forth.

No. 58,076.—JOHN K. DIRNER, Honesdale, Penn.—*Machinery for Cutting Key Seats.*—September 18, 1866.—The support of the arm which carries the cutter and its operating mechanism is pivoted to a vertical slide, and is adjustable to any inclination to the plane of the hole through the hub or wheel in which the key seat is to be cut.

Claim.—The spindle X in combination with the plate P, pivoted to the flanges O and adjusted by set screws R R, constructed and arranged, substantially as described, for the purpose specified.

No. 58,077.—LEVI A. DOLE, Salem, Ohio.—*Tuyere.*—September 18, 1866.—The blast and water pipes are cast in one piece.

Claim.—A tuyere which is composed of blast pipe A, and a chambered portion B, cast in one piece, substantially as described.

No. 58,078.—CHARLES W. DUNLAP, Brooklyn, N. Y.—*Ice Breaker.*—September 18, 1866; antedated September 5, 1866.—The metallic band around the handle makes it efficient as a hammer.

Claim.—The ice breaker formed with a metal band around the larger portion of the wooden handle of the ice pick, as and for the purposes specified.

No. 58,079.—RHESA H. DUNNING, North San Juan, California.—*Rifle Box.*—September 18, 1866.—The gravel and water pass over riffle bars in the flume, whose rear extends nearly to the side of the sluice box. The part of the flume directly over the riffle box is cut away between the riffle bars, forming a communication with the box below, which slides in the sluice and is provided with a dam near one end.

Claim—The riffle box C, dam-wall D, or their equivalents, to be employed for saving gold at any point where there is a divide of waters, substantially as described.

No. 58,080.—J. EIBERWEISER and R. WEBER, Cincinnati, Ohio.—*Straw Cutter.*—September 18, 1866.—The fodder is fed automatically by an endless apron intermittingly revolved, passes beneath a pressure roller and a block, which gives an intermittent pressure coincident with the stroke of the knife, which is raised by a crank rod and swerved into a "draw" cut by the pivoted arm above.

Claim.—The arrangement of a knife H, suspended by two arms F F', pressure-board I, bent lever T T', feeding mechanism C D E V W X, and cams 8 and U, for the purposes explained.

No. 58,081.—CHARLES ELVEENA, New York, N. Y.—*Protecting Animals from the Heat of the Sun.*—September 18, 1866.—The occiput is guarded by a shield containing wet bran.

Claim.—The method herein specified of protecting animals from the heat of the sun by a shield constructed substantially in the manner specified.

No. 58,082.—SAMUEL P. ETTER, Scotland, Penn.—*Cultivator.*—September 18, 1866.—The ploughs are moved laterally and vertically by the compound levers, and the connections described.

Claim.—First, the compound levers R and Q, connected to the plough beams by the braces or rods U U, when constructed and operated for the purposes and substantially as described.

Second, the compound levers R and Q, rods U U, and plough beams H H, in combination with the pendants G G, and pendent guides J J, substantially as and for the purposes shown and described.

No. 58,083.—JOHN FEHRENBATCH, Indianapolis, Ind., assignor to himself and ALEX. INGLIS, same place.—*Hand Propeller for Small Boats.*—September 18, 1866.—The band lever and shaft may be effectively turned in either direction. Each mitre-wheel has ratchet and cogs, and when the lever is reciprocated, they alternate in actuating the mitre pinion on the propeller shaft.

Claim.—The combination of the bevel wheels 3 and 8, the ratchets 2 and 11, the dogs 4 and 10, and springs 5 and 9, with the upright shaft 1, and lever 17, and with the pinion 7 on the shaft 13, which propels the screw, all applied for the purpose of propelling small boats.

No. 58,084.—M. FLEISHER, Philadelphia, Penn.—*Skate.*—September 18, 1866.—The forward clamps have straps and bear against the soles of the boot. The heel clamps are worked by a lever, which is held in place by notches in the foot rest.

Claim.—First, the notched foot rest C, for holding the end of the lever K, combined with the clamps J G, of the heel rest B, substantially in the manner and for the purpose represented and described.

Second, the combination and arrangement of the clamps O, bar Q, straps R, notched rest C, lever K, clamps G, heel rest D, constructed and operating in the manner and for the purpose herein represented and described.

No. 58,085.—A. F. FLETCHER, Athol, Mass.—*Pump.*—September 18, 1866.—The lower part has posts with slots upon which the upper part is locked by means of a corresponding projection thereon, confining the packing between the two parts.

Claim.—A pump box constructed in two or more parts so arranged and combined as to secure the parts to each other and the packing of leather or other suitable material to the pump box, in the manner substantially as and for the purposes described.

No. 58,086.—MATTHEW FLETCHER, Louisville, Ky.—*Rotary Steam Engine.*—September 18, 1866.—The inner cylinder revolves within the casing; the segments are attached to the pistons, and reach from the interior cylinder to the ring, which is concentric to the cylinder, but eccentric to the drum, in which it is fitted. The metal packing rings on the bush press against the end of the drum.

Claim.—First, the method of reducing the side strain on the rotative shaft through the connection of cylinder with eccentric ring flyers, flyer ends, and half rounds c c c c c c in drum M, substantially as described.

Second, the arrangement and combination of the bushes P P, and small metal packer rings f f, by which means soft or elastic packing is avoided from pressing on the revolving shaft.

No. 58,087.—HENRY R. FOOTE, Oil City, Penn.—*Apparatus for Generating and Burning Vapor of Hydrocarbon Oils.*—September 18, 1866.—Passing through the bottom of the metallic oil vessel are pipes terminating in burners, surrounded by a hood which confines the heat to the bottom of the vessel. A coiled steam pipe extends around and supplies steam to the oil vessel. Burners are placed under said coil to heat it; the coil may contain iron filings.

Claim.—First, the combined retort and gas holder, constructed substantially as described.

Second, the coil of metallic tubing charged with iron filings, or their equivalents, and the heaters connected with the gas holder for the purpose of generating hydrogen gas by the decomposition of steam, substantially as and for the purpose set forth.

Third, the tubes at the bottom of the retort with supply pipes elongated so as to extend into the gas holder, as described.

Fourth, the arrangement of burners connected with the gas holder, substantially as described.

Fifth, the hood attached to the bottom of the retort for the purpose of protecting the lower burners, as described.

No. 58,088.—M. FOREMAN and J. R. MATHEWSON, Philadelphia, Penn.—*Amalgamating Gold with Mercury.*—September 18, 1866.—A jet of steam drives the pulp through a pipe which passes it beneath a column of mercury in a lower vessel.

Claim.—First, the amalgamation of gold with mercury by circulating pulverized auriferous ore combined with water upward through a body of mercury, substantially in the manner described.

Second, heating the mass of auriferous ore and water by a jet of steam, which induces the above mentioned circulation, substantially as specified.

No. 58,089.—JOSEPH FOWLER, Rahway, N. J.—*Ore or Quartz Crusher.*—September 18, 1866.—The reciprocating crushing jaws are attached to arms hinged to a yielding eccentric shaft having weighted levers, so that the jaws may release their bite on a substance that will not crush.

Claim.—First, the yielding eccentric bearing shaft f or f' and weighted lever g or g', applied in substantially the manner specified, to keep the jaws at their lower ends towards each other, but allow them to open or yield, as and for the purposes set forth.

Second, the combination of the jaws d and k, connecting rods $l\,l$, and cranks 1 and 2 2, when the jaw d is connected directly to the crank 1, and receives the movement specified, for the purposes set forth.

Third, the links e, in combination with the eccentric yielding bearing f and moving jaw d, as and for the purposes set forth.

No. 58,090.—ROBERT G. FOWLER, Olney, Ill.—*Car Coupling.*—September 18, 1866.—The arrow-headed pivoted draw-bars have springs in their rear, couple by sliding on each other, and are uncoupled by a lifting rod or automatically by extreme deflection when a car leaves the track.

Claim.—The jointed coupling bars G G, having springs I and wedge-shape heads $g\,g$, which slip upon and catch behind each other, permitting the automatic uncoupling by lateral or vertical deflection of one car relatively to the other, under the circumstances described.

No. 58,091.—LEANDER FOX, New York, N. Y.—*Clasp for Mail Bags.*—September 18, 1866.—A metallic clasp with raised direction letters on each side and having a swivel hinge which permits it to be reversed and exposed in another direction when the mail bag is to be returned.

Claim.—The adjustable metal clasp B with permanent raised letters, operating on a joint C and swivel D, as herein described and for the purposes set forth.

. No. 58,092.—JOSEPH FRENCH, Pittsburg, Penn.—*Apparatus for Inserting Fusible Plugs in Steam Generators.*—September 18, 1866.—A ring is screwed into the flue or fire-box of the boiler with a conical seat in it to receive the fusible plug; a tubular rod extends up through the shell of the boiler, through which the plug is inserted; a screw closes the upper end of the tubular rod.

Claim.—The combination of the ring c for holding the fusible plug in the flue, the tubular rod e, and screw f, for closing the bore of the rod e, constructed and arranged substantially as and for the purposes hereinbefore described.

No. 58,093.—JAMES FRERET, New Orleans, La.—*Water Elevator.*—September 18, 1866.—This machine operates on the principle of Hero's fountain, which may be briefly stated: Pressure exerted by a column of water in an air-tight cistern, being transmitted by the air it compresses to another air-tight cistern containing water, will cause the water to rise in a pipe communicating with the latter to a height above its level nearly equal to the height of the first mentioned column. This invention is an improvement on the Hungarian machine at Chemnitz, and has valves in connection with floats to make the operation automatically continuous.

Claim.—The combination of the double or compound valves C' D' and C'' D'', with the cisterns G and F, and simple valve C, when these several parts are constructed and conjointly operate, substantially as described for the purpose set forth.

No. 58,094.—JOHN H. GILBERT, Roxbury, Mass.—*Bridge.*—September 18, 1866.—The arched iron bridge is formed by combining the flanged arch plates, angle sill plates, perforated segmental side plates, and the transverse floor beams.

Claim.—The combination of the angle plates D, pierced side plates C, floor beams E, flanged arch plates B, constructed and operating substantially as described for the purpose specified.

No. 58,095.—HENRY GORTNER, Deavertown, Ohio.—*Apparatus for Separating Gum from Cane Juice.*—September 18, 1866.—The pan is divided into compartments by partitions having semicircular slots covered by disks ; the latter are rotated by levers and have pipes which slide in the slots and form a communication between the compartments..

Claim.—An improved apparatus for separating the gummy substance from cane juice, formed by the combination of the disks B and pipes b' with the partitions a' of the apparatus, the various parts being constructed and arranged substantially as herein described and for the purposes set forth.

No. 58,096.—CHARLES HALL and EMIL HUBNER, New York, N. Y.—*Bolt-heading Machine.*—September 18, 1866.—A rotary table carries the socket dies for holding the blanks ; upon the stationary frame above the table are arranged a set of preparing dies and a set of finishing dies. The socket dies alternately present the blanks to the preparing dies, when the head is partially formed by staving up the blank, and to the finishing dies where it is still further staved up and hammered on the sides by a pair of sliding hammers ; the blank is intermittingly rotated by mechanism underneath the table.

Claim.—First, the combination in a bolt-blank machine of the following instrumentalities, viz : the tubular die, carriage, preparing dies, and set of finishing dies, consisting of an upsetting die and side dies, substantially as set forth.

Second, the combination in a bolt-blank machine of the following instrumentalities, viz : the upsetting die, side heading dies, turning tubular die, and cam tappets which operate the side heading dies twice for each operation of the upsetting die, substantially as set forth.

Third, the combination in a bolt-blank machine of the following instrumentalities, viz : the heading die or dies, turning tubular die, and carriage, substantially as set forth.

Fourth, the combination in a bolt-blank machine of the following instrumentalities, viz : the tubular die, upsetting die, gauge, and cam, substantially as set forth.

Fifth, the combination in a bolt-blank machine of the following instrumentalities, viz : the tubular die, side dies, gauge, and cam, substantially as set forth.

Sixth, the combination in a bolt-blank machine of the following instrumentalities, viz : the tubular die, upsetting die, side dies, gauge and cam, substantially as set forth.

Seventh, the combination in a bolt-blank machine of the following instrumentalities, viz : the heading die or dies, tubular die, gauge and cam, and carriage, substantially as set forth.

Eighth, the combination in a bolt-blank machine of the following instrumentalities, viz : the heading die or dies, turning tubular die, gauge, and cam, substantially as set forth.

No. 58,097.—THOMAS HALL, Bergen, N. J.—*Lamp Burner.*—September 18, 1866.—The cap is mounted on the burner on two semicircular braces, which gives the chimney a parallel motion in raising.

Claim.—The connections A and B and stop C, substantially as described for the purposes set forth.

No. 58,098.—JESSE HANFORD, Lexington, Mass.—*Granulating and Drying Sugar.*—September 18, 1866.—The outer annular steam drum rests on trunnions, and at its ends has arms which support the longitudinal shaft connected by radial arms and oblique wings with an inner cylinder. A granulating roller has bearings in a hanger loosely suspended from the shaft.

Claim.—The combination of the rotary hollow drum or cylinder I and the tubular heater, arranged and connected substantially as described.

Also, the combination of the rotary or hollow cylinder I, the crushing roller K, and the tubular heater, arranged and applied together substantially in manner and so as to operate as and for the purpose described.

Also, the combination of the journals e e and their supporting standards B B of the tubular heater with such heater and the rotary cylinder, or the same and the crushing roller, applied and arranged together, substantially as specified.

Also, the combination and arrangement of the series of helical wings i i with the cylinder and the heater, applied in manner and so as to operate as specified.

No. 58,099.—GEORGE HASECOSTER and J. STEPHENS, Richmond, Ind.—*Feed-water Heater.*—September 18, 1866.—Within the cylinder is a vertical rod sustaining the receiving pan, and the series of perforated pans from which the water successively drips to the lower, and being heated in transit by the steam in the cylinder, the lime in the water being collected in the removable pans.

Claim.—The combination of the rod B, ————; the receiving C, flanged perforated plates D E F, constructed and arranged as described; the receiving C, flanged perforated plates D E F, A, substantially in the manner and for the purpose herein specified, washers I with the cylinder

No. 58,100.—JOHN A. HITCHINGS, Denver City, Colorado.—*Desulphurizing Ore.*—September 18, 1866.—The crucible is supported within the furnace which has a dome-shaped top communicating with a chimney. The crucible has openings in this bottom for the removal of the metal and the slag, and within the dome is a tank for water to be used in detaching the slag from the sides of the crucible.

Claim.—First, the arrangement of the crucible with its dome-covering sectional lid and discharge openings M P, substantially as and for the purpose described.

Second, the combination with the crucible of the water-supply tank K, as and for the purpose described.

No. 58,101.—CHARLES B. HOARD, Watertown, N. Y.—*Spinning Machine.*—September 18, 1866.—The twisting devices are placed in proximity to the bite of the drawing rollers to enable them to seize hold of and draw the twisted roving evenly from the point of delivery of the feed rolls to the draw rolls and to deliver the same properly drawn to the mechanism which gives the final twist.

Claim.—First, a removable tube or spindle for twisting roving while being drawn, when the twist is imparted from one side of the tube or spindle and the roving is conveyed to the bite of the draw rollers from the centre of the tube or spindle, substantially in the manner described.

Second, the combination of a removable twisting tube or spindle, constructed substantially as described, with a revolving tube driven by a whirl, substantially in the manner set forth.

Third, the combination of a removable tube or spindle with the whirl tube and drawing rollers, when constructed, arranged, and operating substantially in the manner and for the purpose set forth.

No. 58,102.—HENRY W. HOLLY, Norwich, Conn.—*Pencil Holder.*—September 18, 1866.—The collar collapses the split tube and makes it grasp the crayon or pencil.

Claim.—The holding stem or tube A, provided with a single slit c, formed longitudinally in one side thereof, in combination with the sliding ring or collar C, substantially as herein set forth for the purpose specified.

No. 58,103.—GEORGE L. HOWLAND, Topsham, Me.—*Hoisting Apparatus.*—September 18, 1866.—The bars slide longitudinally upon each other and are supported by a connecting chain which passes over a pulley. A lever on one bar acts by raising and returning stirrups upon the rack teeth of the other to raise and depress them relatively, and the rack bar actually. The lever is locked by a stop which restrains its motions.

Claim.—An improved hoisting apparatus, formed by combining with each other the bars A B, the levers D H, the pawls G I, the spring K, and the stop L, the parts being constructed and arranged substantially as described and for the purpose set forth.

No. 58,104.—W. W. HUBBARD, Philadelphia, Penn.—*Machine for Making Nuts.*—September 18, 1866.—A rotating table carrying a series of holding dies operates horizontally beneath a vertical slide which carries corresponding dies for cutting, punching, swaging, and a discharging die. At one side of this table are arranged, at right angles to each other, a pair of sliding dies which operate upon two sides of the nut against a right-angled anvil to finish the said sides. The table rotates intermittingly, and all the operations upon the nut are performed while it is at rest. It discharges a finished nut at each movement.

Claim.—First, the revolving disk J, with its recesses z, in combination with a series of punches and dies arranged in a circle, the whole being constructed and operating substantially as and for the purpose described.

Second, the cross-head E and its arm L, in combination with the shaft H, and its rib w, the whole being constructed and operating substantially as and for the purpose specified.

Third, the combination of the boxes or casings c, punches d, and gibs and keys, the whole being arranged substantially as and for the purpose set forth.

Fourth, the carriers M and M', with their adjustable hammers, the whole being arranged in conjunction with the rotating disk J, to operate on the blank, substantially as and for the purpose described.

Fifth, the combination of the carriers M M', lever T, and arms and levers p q, the whole being arranged and operating substantially as and for the purpose set forth.

Sixth, the slide l and spring 5, or its equivalent, in combination with the cutters d k.

Seventh, the slide m, operating in combination with the punch d², hammers S S', and disk J, substantially as and for the purpose described.

No. 58,105.—JEROME KIDDER, New York, N. Y.—*Magneto Electric Apparatus.*—September 18, 1866.—The battery current is first thrown through one helix or system, then interrupted and thrown through the other. The effect of an electro-magnetic apparatus, whereby the battery current is first thrown through the invention.

Claim.—First, the construction helices or systems of helices, to thrown successively in opposite the primary current so as to develop induced currents successively in opposite directions, substantially as herein described.

Second, the arrangement of the parts of an electro-magnetic apparatus, whereby the currents from two helices or systems of helices are made to flow in succession, one immediately after the other, in a uniform direction, substantially as herein set forth, such currents being either of the same character, made thus more rapid in succession, or of different degrees of tension or concentrative or diffusive influence.

Third, for adjusting the screw opposed to the vibrating spring of an electro-magnetic apparatus, the use of the arrangement as represented by *v y*, operating substantially in the manner described.

No. 58,106.—N. KIEFFER, New Orleans, La.—*Medicine.*—September 18, 1866.—A sweetened and colored aromatic and bitter tonic tincture of twenty roots, spices, herbs, berries, flowers, and barks.

Claim.—The medical compound, composed of the ingredients herein named and mixed together, in or about the proportions named.

No. 58,107.—PARLEY LAFLIN, Warren, Miss.—*Oiler.*—September 18, 1866.—The nozzle of the oiler has a forked projection to guide it when oiling spindles.

Claim.—The bifurcated guide E, in combination with the tube of an oiler, substantially as and for the purpose set forth.

No. 58,108.—ABRAHAM LANDIS, Colesburg, Iowa.—*Rack for Hauling Wood.*—September 18, 1866.—The sleepers, sills, and stakes are bound together by the bent stirrups and are loosed by raising the stakes.

Claim.—The wood-rack herein described, the same consisting of the sills A, sleepers B, and stakes C, all secured together by the loops or stirrups D, in the manner and for the purpose explained.

No. 58,109.—HENRY M. LE DUC, Washington, D. C.—*Anti-friction Journal Box.*—September 18, 1866.—The anti-friction rollers around the shaft are prevented from shifting endways by the circular flanges on the axle and in the box which project within the peripheral grooves in the rollers.

Claim.—The rims or projections G F within the hub or box and around the axle, respectively, extending into annular grooves in the concentric series of larger rollers to maintain them in position longitudinally, substantially as described.

No. 58,110.—THOMAS LLOYD, Muncy, Penn.—*Horse Hay-fork.*—September 18, 1866.—The sliding bar has two prongs pivoted to its lower end, and is linked to the main bar below the prongs. It is operated by a lever attached to its upper end, is pivoted to the main bar, and is guided by a pin at its lower end moving in a slot in the main bar.

Claim.—The employment, in combination with the main bar A, of pivoted fingers D D', and arms E E', operated by means of a vertical rod C, and lever B, the whole arranged substantially as set forth.

No. 58,111.—LEVI W. LOOMIS, Homer, N. Y.—*Apparatus for Upsetting Tires.*—September 18, 1866.—The tire is clamped in two places between stationary and movable clamps, respectively. The latter has attached a shouldered plate which is slipped on the base-plate by the contact of a lug on the lever.

Claim.—The movable plate B, provided with the notched strip C, and operated by a shoulder of the lever D, and arms Z Z, to force said plate against the body A, to upset the tire when held between cams and ratchets, substantially as herein set forth.

No. 58,112.—WILLIAM H. LUCE, Hampton, Illinois.—*Shovel Plough.*—September 18, 1866.—Each fork of the beam supports a wing of the double mouldboard, and has a handle attached.

Claim.—The general construction and form of the beams A, and handles C C, in combination with a double concave mouldboard B, substantially as described.

No. 58,113.—ORAZIO LUGO, New York, N. Y.—*Apparatus for Distilling Petroleum, &c.*—September 18, 1866.—The still has a goose-neck communicating with two consecutive condensers, the latter one by a pipe communicating with the rotary pump, the exit pipe of which terminates in a receiver. Air is supplied to the interior of the apparatus by means of a branch pipe from a main air pipe. The vapors escaping from the receiver are conducted to a condensing coil.

Claim.—First, the admission of air or gas into the goose-neck or exit pipe of the still, substantially as and for the purpose herein specified.

Second, varying the point of admission of the air or gas B into the still and goose-neck or exit pipe as the process of distillation progresses, substantially as and for the purpose herein set forth.

No. 58,114.—WILLIAM MAGROWITZ, New York, N. Y., assignor to himself, WILLIAM FREEDMAN and JAMES DAVIS.—*Spirit Meter.*—September 18, 1866. —The meter wheel revolves within a tight vessel, and is divided into chambers of definite capacity. A pipe from the still empties into the upper chamber for the time being of the wheel. A roller attached to levers is lifted by a float at the proper time to allow the meter wheel to revolve. The journal of the wheel is connected with a registering apparatus. A side chamber contains a hydrometer with an indicator needle attached, which marks the specific gravity of the liquid on a traversing paper.

Claim.—First, the arrangement and construction of the wheel A, provided with chambers or cavities, in combination with the roller E, and float G, operating in the manner substantially as described.

Second, the arrangement of the levers D, with the roller E, in combination with the lever F, and float G, and operated by set screws *b* and *d*, in the manner and for the purpose set forth.

Third, the arrangement of the paper rollers P P' and the manner of operating the roller P, by means of the friction roller O, for the purpose substantially as set forth.

Fourth, the manner of operating the friction roller O, from the shaft of the wheel A, by means of the gearing 2 and 3, flanch-wheel 4, pin 8, and pawl lever R, when arranged and combined in the manner and for the purpose specified.

Fifth, the box K, in combination with the pipe *g*, and cylinder L, the hydrometer M, with the needle N, at its upper end, for the purpose substantially as set forth.

Sixth, the frame S, in combination with the needle N, and the manner of operating said frame S, and needle N, substantially as and for the purpose described.

Seventh, in combination with the hydrometer M, and its needle N, the endless paper wound on rollers P' and P, and lined in the manner described, and operating together substantially for the purpose specified.

Eighth, the friction pulley *h*, with its lever *k*, in combination with the measuring wheel A, for the purpose set forth.

Ninth, the combination of the measuring wheel A, with its registering device, the hydrometer M, provided with a marking needle N, the paper rollers P and P', with paper lined in the manner described; the friction rollers O and O', and the frame S, when arranged and operating together so as to mark the quantity and the weight or specific gravity of alcohol, or other liquid which is made to pass through the box V, or through the machine, substantially in the manner as set forth and specified.

No. 58,115.—SAMUEL McCAMBRIDGE and EDWARD G. MARTIN, Philadelphia, Penn.— *Car Brake.*—September 18, 1866.—The continuous brake chain which passes beneath the train of cars has branch chains which pass around drums, whose interior coiled springs draw back the chain when the brakes are released.

Claim.—The combination of a series of coil springs with a train of cars and the continuous chain, which operates the brake levers by means of the cylinders B, shafts C, and chains F, the several parts being arranged and operating substantially in the manner described and for the purpose specified.

No. 58,116.—E. P. McCARTHY, San Francisco, Cal.—*Ore or Quartz Crusher.*—September 18, 1866.—The tappets upon the stamp rod are made of India-rubber, shod with steel plates to receive the wear of the eccentrics.

Claim.—The use of a rubber tappet A, steel shod, the steel shoe B, plate E, and bolts F F, combined in the manner and for the purposes set forth.

No. 58,117.—WILLIAM M. McCOY, Bloomingdale, Ind.—*Holding rolled work while being stitched.*—The work lies in a grooved bed and is held down at each end of the bed by a curved band operated by a set screw.

Claim.—A device for holding rolled work while being stitched, constructed and operating substantially as specified: that is to say, the bed piece A having a groove a along its upper face, and provided with a tightening band B operated by a set screw C, substantially as shown and described.

No. 58,118.—JOSIAH McFARLAND, Clinton, Ill.—*Blowpipe.*—September 18, 1866.—A reservoir for compressed air is interposed between the force pump and the flexible tube, which terminates in a nozzle.

Claim.—In blowpipes a detachable air chamber A in combination with the flexible tube having a suitable mouth-piece for directing the current of air or gas, and a force pump, all constructed and operated substantially as described.

No. 58,119.—F. W. McMEEKIN, Morrison's Mill, Fla.—*Plough.*—September 18, 1866.—The standard and brace are constructed of a single bar of wrought iron doubled and bent.

Claim.—The standard C, constructed of a single metal bar doubled and bent so as to have two diverging arms *a a'*, and an inclined loop *b*, in combination with the landside and mouldboard, all arranged to form a new and improved plough, as set forth.

No. 58,120.—JOHN METZENDORF, New York, N. Y.—*Safety Attachment for Pockets.*—September 18, 1866.—One jaw is attached by eyelets to the pocket and the other clasps over the pendant of the watch.

Claim.—A watch-pocket protector composed of two jaws A and B, flexibly connected together at one end, and furnished with a catch at the other end, and provided with eyelets or other means of attaching it to the pocket, all substantially as herein described.

No. 58,121.—JACQUES MEYER, Williamsburg, N. Y.—*Saddle.*—September 18, 1866.—The loop of the stirrup strap is attached to the tree by a spring pin which works in a socket and renders it easily removed.

Claim.—The device formed of the case or thimble *c* enclosing the spiral spring around the pin *d*, for the purpose of attaching and detaching stirrup straps to a saddle, constructed and applied in the manner herein described.

No. 58,122.—F. MEYROSE, St. Louis, Mo.—*Lantern.*—September 18, 1866.—A dome-shaped cap with a register fits over the wick tube extending to the outer part of the lamp. The wick raising device is so enclosed that any leakage from the wick will be conducted to the outside of the lantern.

Claim.—First, enclosing the wick elevating rod or shaft a^3 with a tubing or casing a^2, for the purpose of conducting any moisture or fluid that may escape on the said rod outside of the guard of the lamp or lantern.

Second, the cap F either with or without the ventilators, substantially as described.

No. 58,123.—S. MILLER, Urbana, Ohio.—*Edge Plane.*—September 18, 1866.—The bit is fixed ; a set screw and spring operating upon the tang of the face plate regulate the aperture of the throat and determine the thickness of the shaving.

Claim.—The exposing of the cutting edge of bit B always evenly and securely, at one operation, as described, or any other way substantially the same.

No. 58,124.—ROBERT B. NEVENS, Lowell, Mass.—*Snow Plough.*—September 18, 1866.—The triangular plough has on its longest side a metallic face which projects downward below the framework to hold the plough in position while the mouldboard pushes the snow to one side of the road.

Claim.—First, the combination and arrangement of the fence A, and facing *d*, sides B B, levelling beam C, and mouldboards E E, substantially as and for the purpose set forth.

Second, in combination with a snow plough constructed and arranged as above stated, the employment of the auxiliary brace chains F and H, connected with the centre draught chain G, substantially in the manner and for the purpose specified.

No. 58,125.—D. J. NOBLE, New Boston, Ill.—*Corn Cultivator.*—September 18, 1866.—The inner ploughs are adjusted laterally by the foot of the operator, their standards being loosely hung from the frame. The four ploughs are simultaneously raised by the vibration of a pivoted supporting bar.

Claim.—The adjustable foot pieces L L connected to the plough standards I I, and arranged with brace K, in combination with the mode of attaching said standards to the frames C and F, substantially as and for the purpose herein specified.

No. 58,126.—EMILE NOUGARET, Newark, N. J.—*Machine for Pouncing Hats.*—September 18, 1866 — The cones are covered with sand-paper, and are mounted on shafts which have their bearings in a revolving disk in connection with a swivel plate and an adjustable revolving block on which the hats are presented to the pouncers. By the combined action of the feed and pouncing rollers the brim is subsequently revolved automatically and operated upon by the pouncing cones.

Claim.—First, the swivel disk B carrying the block D, in combination with the adjustable disk E carrying the pouncing rollers *k k*, constructed and operating substantially as and for the purpose described.

Second, the gears *l m* and revolving shaft F in combination with the pouncing rollers *k k* and block D, constructed and operating substantially as and for the purpose set forth.

Third, the brake *p* in combination with the disk E, spring *n*, pouncing rollers *k k*, and block D, all constructed and operating substantially as and for the purpose described.

Fourth, the rollers G G and supporting brackets *t*, constructed and operating substantially as and for the purpose set forth.

No. 58,127.—JOHN C. PARKER, Chicago, Ill.—*Locomotive.*—September 18, 1866.—When steam is shut off the spring valves close upon the ends of the exhaust pipes. The valves are connected to a sliding rod by means of toothed sectors, and the sliding rod is held in position by a pin from which it is liberated by the movement of the reversing lever of the engine, when the spring closes the valves.

Claim.—First, providing the nozzles of the exhaust pipes of a locomotive engine with valves which shall be allowed to close by a movement of the reversing lever, substantially as described.

Second, the combination of the draw rod D and valves b b with the spring f, and a contrivance which will effect the tripping of said rod when the engine is reversed, substantially as described.

No. 58,128.—G. C. PATTISON, Baltimore, Md.—*Anchor.*—September 18, 1866.—Guards upon the frame of the anchor with pivoted flukes prevent the fouling of the cable therewith.

Claim.—Providing the flukes of an anchor having pivoted arms with stationary guards, substantially as and for the purpose herein described.

No. 58,129.—S. PETTIBONE, Corunna, Mich.—*Straw Cutter.*—September 18, 1866.—The pitman of the fly wheel and the shank of the gauge plate are attached to opposite sides of the axis of the operating knife lever. The gauge plate is adjustable to vary the length of chaff, and the pitman in the crank wheel to determine the length of stroke.

Claim.—First, the application of the fly-wheel B in combination with crank wheel C, pitman D, and lever E, substantially as and for the purposes described.

Second, the adjustable box or bearing F in combination with the lever E, gauge G, and throat I, for the purposes and substantially as herein shown and described.

Third, the mode of securing the gauge plate G to the lever E, as and for the purposes and substantially as herein set forth.

Fourth, the mode of securing the lever E to the pivot by means of mortise and key, substantially as herein shown and described.

No. 58,130.—POMPEIUS PHILLIPI, Beardstown, Ill.—*Fastening for Gate.*—September 18, 1866.—By means of the lever the gate is drawn towards its upper hinging eye on the heel post, and is thereby tipped vertically, raising the extended latch bar from its catch on the shutting post.

Claim.—The arrangement of the hooked rod J and lever K in combination with the bar B, posts E F, and pin D, constructed and operating in the manner and for the purpose herein specified.

No. 58,131.—C. F. PIDGIN, Boston, Mass.—*Shirt Bosom.*—September 18, 1866.—Explained by the claim and cut.

Claim.—A shirt bosom cut or hollowed out at its sides B and ends C D in combination with the strips F, substantially as and for the purpose described.

No. 58,132.—CHARLES L. PIERCE, Buffalo, N. Y.—*Ash Sifter.*—September 18, 1866.—The hopper has a hinged bottom, drops its contents on to an inclined sieve, and is operated by gearing.

Claim.—The enclosing case A and removable charger D, the said case being fitted to receive and support the charger, and the charger having a hinged bottom, so that its contents may be discharged into the sieve, the inclined open-end sieve B, the operating gearing, and the ashes box E, combined as herein described.

No. 58,133.—CHANDLER POOR, Dubuque, Iowa.—*Dental Mallet.*—September 18, 1866.—The shank of the plugger is set in a piston, which passes through the tubular mallet; the latter is struck against a shoulder on the piston, and returned by the recoil spring.

Claim.—The sliding mallet i, piston a, spring f, and socket b, by means of which to apply force to a common plugger or point in filling teeth. The claim has no reference to said plugger or point, except in the manner of applying force to the same.

No. 58,134.—TREAT T. PROSSER, Chicago, Ill., assignor to himself, G. W. GILLET, J. A. EASTMAN, D. KIMBARK, jr., and D. H. WELLS, same place.—*Machine for Making Screws.*—September 18, 1866.—A pair of parallel shafts having heads upon their ends are geared together, so as to rotate in opposite directions at equal speed. Upon these heads at each side are mounted roller dies, which have grooves parallel to their axes of the size of the threads required; the blank is fed along between these heads over an anvil which supports it, and the dies strike it at the same time on each side and form the thread.

Claim.—First, the method of forming the threads on screws by means of revolving swages or dies, constructed and operated substantially as and for the purpose set forth.

Second, the grooved rollers a, when arranged and operating as described, for the purpose of forming threads on screws.

Third, the combination of the revolving swages or dies *a a* with the anvil or rest *u*, substantially as set forth.

Fourth, the chuck E, constructed and operating as herein described.

Fifth, the combination of the hollow shaft F, chuck E and rock shaft *d*, provided with the arms *f* and *c*, when arranged to operate as and for the purposes set forth.

Sixth, in combination with the shaft F and hopper H, the follower I, arranged to operate as described, for the purpose of feeding the blanks into the shaft F, as set forth.

No. 58,135.—A. PUTNAM, Owego, N. Y.—*Corn Planter.*—September 18, 1866.—The feed bar is reciprocated by cams on the driving wheel and a recoil spring. The covering roller in the rear of the share is vertically adjustable.

Claim.—First, the drill tooth G, wheel F, and adjustable slide bar E, arranged and operating as described.

Second, in combination with the above, the arrangement of the cams D, bar A, spring K and valve J, as and for the purpose specified.

No. 58,136.—J. WYATT REID, New York, N. Y.—*Car Brake.*—September 18, 1866.—The brake shaft on the platform is connected by chains and rods with rubbers adapted to the forward and rear wheels of the truck, so as to operate them simultaneously, the chain passing over a sheave in one brake bar. Clearers depend from the rubbers to remove obstructions from the rails.

Claim.—First, the combination of chains, rods, and pulleys for operating the brakes, the whole substantially as described.

Second, the combination with the car brakes of the guards *r r*, constructed and applied in the manner and for the purpose set forth.

No. 58,137.—SAMUEL G. REYNOLDS, Bristol, R. I.—*Machinery for Making Nails.*—September 18, 1866.—That portion of the blank which is to constitute the head is supported while it is being staved up, so that it will not bend; the mechanism consists of a pivoted arm, the end of which bears against the nail blank just in advance of the header, and which has a socket forming a portion of the die for the head when it has arrived at the termination of its movement.

Claim.—First, giving lateral support to that portion of the nail blank which is to be upset to form the head during the operation of heading the nail by means of the radial supporting bar A, arranged to co-operate with the header H, substantially as herein described, for the purpose specified.

Second, combining and arranging the gripping die in a nail-making machine with the movable cutting shear, as described, so that the former shall change its position relatively to the latter, for the purpose of lowering the blank during the swaging and heading operation below the cutting edge of the shear, in the manner and by the means substantially as herein set forth.

No. 58,138.—E. B. ROBINSON, Portland, Maine.—*Flatiron.*—September 18, 1866.—Explained by the claim and cut.

Claim.—The combination and described arrangement on a flatiron of the glass or porcelain handle H and the hinged guard and reflector D, as and for the purposes herein set forth.

No. 58,139.—D. P. ROOD, Warsaw, N. Y.—*Surcingle.*—September 18, 1866.—An elastic section is interposed between the lengths of webbing in the surcingle.

Claim.—An elastic B, made of rubber or other suitable material, in combination with the webbing A, so as to cause the surcingle to adjust itself to the varying size of the horse, for the purposes and substantially as herein described.

No. 58,140.—D. C. ROSIER, Clarkson, N. Y.—*Machine for Harvesting Beans.*—September 18, 1866.—The bean stalks are cut below the surface of the ground by horizontal L-shaped cutters, and are raised from and divested of earth by a trailing device, consisting of diverging prongs.

Claim.—The arrangement of the L-shaped cutters C, in combination with the elevating or adjusting devices, in the manner and for the purposes set forth.

No. 58,141.—ALFRED V. RYDER, New York, N. Y.—*Trunk.*—September 18, 1866.—A portion of the upper part of the trunk folds back on the top, and has drawers; metallic stays brace the angle of division.

Claim.—A trunk provided with a hinged portion A, formed by the vertical and horizontal cuts *a b* in the main portion, and hinged to the part B, so that it may, when turned up or opened, rest upon B, the parts A B being provided with drawers, and the other parts D provided with a lid or cover, substantially as shown and described.

Also, in combination with the hinged portion A the metal stays or plate E, applied to the ends of the trunk, and having the handles F attached substantially as set forth.

No. 58,142.—WILLIAM R. SATTERLY, Port Jefferson, N. Y.—*Hook for Davit-fall Blocks.*—September 18, 1866.—When the hook is lifted by the cord and link, the ring of the eye-bolt is positively disengaged therefrom.

Claim.—The combination with the davit-fall block of the hook D, link H, and cord I, operating substantially as described.

No. 58,143.—SAMUEL SECRIST, West Liberty, Ohio.—*Corn Harvester.*—September 18, 1866.—The stalks are gathered to the cutters by two reels, with interlocking arms, and after being cut are forced through a narrow passage to a revolving circular platform, surrounded by hoops, so arranged that one-half may be opened outward for the discharge of the shock; the stalks are held upright in this receptacle by a semicircular spring upon the top of the hoops. To a post upon the main frame is pivoted a lever, which operates a clasping device by which the shock is lifted for discharge. Two small reels at the front of the frame, revolving in perpendicular planes, pick up broken stalks.

Claim.—First. the reels *j j* with their arms engaging with each other, in combination with the box G, substantially as described for the purpose specified.

Second, the reels *k k* in front of the sides G G for gathering up the leaning and broken stalks, and passing them on to the cutters, constructed substantially as herein described.

Third, the arrangement of the revolving platform H, hoop *o*, discharging gate, half-hoop *r*, spiral spring *t*, constructed and operating substantially as and for the purpose specified.

Fourth, the lever *p* and the clamp *q*, suspended by the standard I over the platform H, to gather the shock together at the top and set it off, standing upright on the ground, constructed substantially as herein described.

No. 58,144.—HENRY SIDLE, Minneapolis, Minn.—*Washing Machine.*—September 18, 1866.—The clothes are caught by the oblique arms on the revolving axis, dashed against the vertical rollers on the inside of the tub, and drawn rapidly through the agitated water.

Claim.—The box A with its vertical rollers B B, arranged with the top C and shaft E, with its oblique arms F F, substantially in the manner and for the purpose herein specified.

No. 58,145.—THOMAS SINNOTT, Brooklyn, N. Y., and JAMES MCINTYRE, New York, N. Y.—*Tuyere.*—September 18, 1866.—A series of radial rings in the blast chamber cause the blast to travel back and forth in passing around and into the central pipe leading to the fire and thus become highly heated.

Claim.—First, a series of wings or divisions around the blast pipe, with openings at alternate opposite ends to cause the air or blast to travel back and forth within the tuyere, for the purposes and as set forth.

Second, the valve *l* attached to the block *m*, in combination with the blast pipe *a*, for the purposes set forth.

Third, the movable nozzle *n* in combination with the tuyere, as and for the purposes specified.

No. 58,146.—FRANCIS H. SMITH, Baltimore, Md.—*Brick Kiln.*—September 18, 1866.—The middle flue beneath the floor extending outside the walls has a furnace at each end and lateral communications with ash pits, whence the heat ascends through the gratings into the kiln; middle doors at the feed mouths regulate the draught.

Claim.—First, the middle flue A beneath the floor, extending outside the walls, to connect with furnace F, and having lateral communications with ash pits.

Second, the furnace placed at the mouths of said flue, heating all the air that passes into the kiln, creating the hot blast.

Third, the middle door *e*, in combination with the flues A and C and the furnace F.

No. 58,147.—ANDREW STARK, Topeka, Kansas.—*Cultivator.*—September 18, 1866.—Explained by the claim and cut.

Claim.—The pivoted bars L L having the driver's seat N attached to them, in combination with the plough beams H H, connected at their front ends to the front part of the frame of the machine, and the plough beams and bars connected by chains *h*, or their equivalents, substantially as and for the purpose specified.

No. 58,148.—JESSE STOW, Geneva, Ohio, and JAMES WHITE, Cleveland, Ohio.—*Roofing Cement.*—September 18, 1866.—Composed of coal tar, 2; clay, 2; water lime, 1; slaked lime, 1; sand, 1; pulverized coal or coke, 1; and mineral naphtha to give consistency.

Claim.—A plastic roofing cement composed of the ingredients herein named and compounded, as specified.

No. 58,149.—STEPHEN G. STURGES, Newark, N. J.—*Amalgamator.*—September 18, 1866.—The barrel amalgamator has a pocket to retain the amalgam and distribute it to the ore as the barrel revolves.

Claim.—The flute or pocket *t*, when attached to a reciprocating or revolving cylinder, in the manner and for the purpose substantially as shown.

Also, the bolts W extending across as supports to the cylinder, when used in combination with the pocket attached to the cylinder.

No. 58,150.—J. W. SUMMERS, Tarr Farm, Penn.—*Deep Well Pump.*—September 18, 1866.—The piston is suspended from the rod by a ball joint. The upper end of the piston flares toward the periphery to conduct rivet heads, &c., into the piston.

Claim.—First, suspending the piston of a pump from the pump rod by means of a ball and socket joint, substantially as described.

Second, the cylindrical stop G having its upper edge bevelled as shown, for the purpose of catching rivets and other objects and directing them into the piston, substantially as described.

No. 58,151.—ALVAH SWEETLAND, Syracuse, N. Y.—*Saw Filing Machine.*—September 18, 1866.—The saw is clamped to a rack which is moved by the pinion beneath. The swing table can be adjusted to or from the file to regulate the depth of the teeth. The file is socketed in holders which are reciprocated vertically in guides by connection with a crank beneath.

Claim.—The swing bar, the adjustable table, the rag, the dog, the lever, the rag-wheel, the straight file, the cylinder, the clasp, and the connecting rod, the whole being arranged and combined substantially as and for the purpose set forth.

No. 58,152.—ROYAL H. THORN, Syracuse, N. Y.—*Chuck.*—September 18, 1866.—The jaws are operated by a screw having a right and left hand thread, and are susceptible of adjustment laterally so as to hold the article to be turned eccentrically when desired.

Claim.—Clamping the jaws C C' by the screw D and allowing the same to play freely through the body of the chuck, for the purposes set forth.

No. 58,153.—SAMUEL H. TIMMONS, Lafayette, Ind.—*Bottle Stopper.*—September 18, 1866.—The base piece is attached to the cork by a hollow screw-shank and has a screw flange on its upper edge for the attachment of the graduated cup.

Claim.—First, a cup, graduated or otherwise, fitting a base piece to be attached to the cork or neck of a bottle of any size.

Second, the base piece, provided with an entering shank for insertion into a cork, and with a flange for the reception of the cup, substantially as described.

Third, a graduated cup attached to the stopper or neck of a bottle for the purpose described.

No. 58,154.—JOEL E. TODD, Middletown, Conn.—*Artist's Stretcher.*—September 18, 1866.—The mitre joints have inserted springs which render them self-expanding.

Claim.—Making the angle of artists' stretchers self-adjusting by means of springs F, or equivalents therefor, substantially as herein described.

No. 58,155.—WILLIAM VAN ANDEN, Poughkeepsie, N. Y.—*Machinery for Making Railroad Chairs.*—September 18, 1866.—Arranged in one machine are an anvil, former, and bending and swaging jaws whereby a flat plate of metal may be bent and swaged into a railroad chair having a projecting base and flanges, the fibres of which are in a line at right angles to the rail when arranged together.

Claim.—First, the dies G G, arranged to take hold of the metal plate at its edges, and constructed and operating so as to form the lip flanges and a projecting base on said plate, all substantially in the manner and on the principle herein set forth.

Second, the devices *c' c'*, in combination with a plunger, for curving or slightly bending the plate of metal, the same being arranged and operating substantially as described.

Third, forming on a metal plate, which has been previously curved or bent, and from which a rail chair is to be made, a projecting base, and flanges standing at right angles or nearly so with the base of the plate, ready for the action of the finishing dies, by means of swaging dies G G, in combination with an anvil *a*, and a former *e*, substantially as described.

Fourth, the combination with the anvil *a* and dies G G, of the elevated pieces *a¹ a¹*, so as to form a box die conforming to the shape of the base of a chair blank, substantially as and for the purpose described.

Fifth, the combination of the plunger or former *e*, swaging dies G G, and finishing dies J J, working in a manner and for the purpose of forming a metal chair having a continuous lip, substantially as described.

Sixth, making from a plate of metal, by one machine, and by a succession of operations in the said machine, a chair, substantially as represented in Fig. 8, the means for doing this being constructed and operating substantially in the manner and on the principle herein set forth.

Seventh, discharging a chair by means of the discharger D, passing from the rear to the front of the chair, and then drawing the chair from the former on its backward motion, the said discharger being arranged and operated substantially as described.

Eighth, as a new product, a swaged rail chair, such as represented in Fig. 8, the fibres of said chair being transverse to the length of the rail, which is to be supported by it, substantially as described and for the purpose set forth.

No. 58,156.—RICHARD VOSE and WILLIAM TOSHACH, New York, N. Y.—*Machinery for Coiling Springs.*—September 18, 1866 —The stationary inclined guide nearly encircles the mandrel upon which the spring is wound, to give the required pitch; two or more forming wheels bend the spring around the mandrel.

Claim.—The inclined edged stationary guide plate S, or its equivalent, in combination with the cylindrical mandrel D', and one or more swaging and forming wheels and rollers,. revolving in unison, all substantially in the manner and for the purpose herein set forth.

No. 58,157.—JAMES E. WEAVER, Temperanceville, Penn.—*Oil Can and Oiler.*—September 18, 1866.—A partition runs vertically from near the top to the bottom of the can; a pipe attached to the exterior communicates with the division nearest to the handle.

Claim —The arrangement of the division d and tube c, when used in connection with the body a and conductor b b', as herein described and set forth.

No. 58,158.—D. N. WEBSTER, Geneva, Ohio.—*Hay Rack.*—September 18, 1866.—The rack is placed upon wagon gears, and after being filled and drawn to the bay or stack, is lifted bodily and the load discharged by opening the hinged bottom; the latter is subsequently replaced and the rack loaded on to the wagon gears.

Claim.—First, the sliding bar H. spring P, cross sills G G', the levers f and i, in combination with the sections F F', as arranged in the manner and for the purpose set forth.

Second, the sides A, as hinged and arranged in combination with the sections F F' and ends D, for the purpose and in the manner herein described.

Third, the ends of the rack D D, the catch b, spring e, and the loops a, as arranged and in combination with the sides A and sections F F', in the manner and for the purpose as substantially set forth.

No. 58,159.—A. A. WILDER, Detroit, Mich.—*Ventilating Pipe for Stoves and Heaters.*—September 18, 1866.—A pipe open at the end near the floor enters the exit pipe of the stove at a little distance from the heater, for the purpose of ventilation; a damper is located near its lower end.

Claim.—The arrangement of the supplementary pipe B, with its portion a a, as described, with the elbow joint of a stove-pipe, combined and operating in the manner and for the purpose specified.

No. 58,160.—JOB T. WILLIAMS, Philadelphia, Penn.—*Toy.*—September 18, 1866.—The figure is stamped and a portion of the plate bent down to make a lateral support.

Claim.—Constructing rocking toys out of single pieces of tin or other metallic plates, substantially in the manner above described.

No. 58,161.—JAMES F. WINCHELL, Springfield, Ohio.—*Gate.*—September 18, 1866.—The brace for the support of the gate extends from the heel in the rear to the upper front corner, and has a slot, bolt and jamb screw by which the out end of the gate is held in elevated position while opening.

Claim.—First, the combination with the pivoted bars a, the slotted brace C, secured by the bolt a and nuts l, when arranged to operate as and for the purpose set forth.

Second, the rubber washer i, when used in combination with the brace C and gate, as and for the purpose set forth.

No. 58,162.—F. M. WOODS, York, Ill.—*Machine for Wiring Sheet Metal Pans.*—September 18, 1866; antedated September 2, 1866.—A series of movable or adjustable sections are placed in a frame and arranged with an adjustable pressure wheel in such a manner as to operate upon pans of varying sizes.

Claim.—The sliding section C, in combination with the adjustable grooved wheel D, arranged in suitable framing, to operate in the manner substantially as and for the purpose, herein set forth.

No. 58,163.—LUCY BROAD, St. Louis, Mo., assignor to CHARLES A. BROAD.—*Disinfectant.*—September 18, 1866.—A fumigating disinfectant composed of gum camphor four ounces, rosin three pounds, gum myrrh four ounces, pine sawdust three pints, coal tar six ounces, and chloride of lime eight ounces.

Claim.—The combination of the materials herein described in the proportions specified, or their chemical equivalent, for the purpose of producing a fumigating disinfectant.

No. 58,164.—E. P. CHASE, Rockland, Me., assignor to himself and JOHN EATON, same place.—*Steam Generator.*—September 18, 1866.—A feed-water heater is arranged above the annular generator and the superheating pipe within it. The feed-water supply is governed by a regulator in accordance with the pressure of steam. A plate in the rear of the grate slides above the fire to direct the heated gases toward another exit flue and away from the generator.

Claim.—First, the arrangement of an annular steam generator C, substantially as and for the purpose set forth.

Second, the combination of the annular generator C with the heater B, substantially as and for the purpose described.

Third, the superheating pipe or pipes p, in combination with the annular generator C, and situated in the flue formed by the inner shell of said generator, substantially as and for the purpose set forth.

Fourth, the cap g, weighted lever F, and diaphragm l, in combination with the generator C and pump E, constructed and operating substantially as and for the purpose described.

Fifth, the slide M, channels t, chamber u, and additional smoke-pipe L', in combination with the grate D and generator C, constructed and operating substantially as and for the purpose set forth.

No. 58,165.—CHARLES M. DRENNAN, Boston, Mass., assignor to WILLIAM P. and ISAAC GANNETT, Roxbury, Mass.—*Egg Beater.*—September 18, 1866.—A diaphragm of wire gauze rests midway of the cylindrical beater, supported by two wires extending from the cover, by lifting which it is removed.

Claim.—The gauze diaphragm c, when supported and inserted by the wires d d, attached to the cover, as and for the purpose specified.

No. 58,166.—ANDREW J. EDGETT, Hornellsville, N. Y., assignor to himself, JOHN W. FERRY and A. GRAVES, same place.—*Combined Corn Planter and Broadcast Seeder.*—September 18, 1866.—The corn box and grain box are ranged alongside and may be connected when containing a supply of the same seed; the feed slides of each are operated by cams on the respective wheels, those of the planter at longer intervals. Either may be thrown out of operation. Wires below the feed slide of the broadcast sower scatter the grain.

Claim.—First, the construction and combination of a corn planter with a broadcast seed sower, so that either machine can be used, substantially as described.

Second, the distributing wires I, in combination and arrangement with the grain box and slide of a broadcast seed sower, for the purpose and substantially as set forth.

No. 58,167.—H. B. GREGG, Camden, Ohio, assignor to himself and JAMES GABEL, same place.—*Clamp for Planking Floors.*—September 18, 1866.—The ratchet bar is pivoted to a clutch on the joists and supports the lever, by which the sliding press-block is pressed against the edge of the flooring plank.

Claim.—The clamp A, the bar B, the guide A, the press block C, the lever M, the bar N, the pawl f, and the plate P, the whole constructed, arranged, and operating as and for the purpose herein described.

No. 58,168.—IRA W. HAMLET, Nashua, N. H., assignor to himself and HENRY J. CHAMPMAN, same place.—*Neck-tie.*—September 18, 1866.—Hooks attached to the neck-tie catch beneath the edge of the collar and maintain the tie in place.

Claim.—The new manufacture or neck-tie as made with hooks B B, applied to and arranged with the part or cravat A, substantially as set forth.

Also, the combination of the hook B, the cravat A, and the stiffener or plate C.

Also, the combination of the cravat hook and stiffener, the same being for the purpose set forth.

No. 58,169.—JAMES G. HOLLIDAY, Wheeling, West Virginia, assignor to himself, WILLIAM HASTINGS, J. HARLAN, and R. A. MCCABE, same place.—*Composition Roofing.*—September 18, 1866.—Composed of coal tar, forty-two gallons; petroleum residuum, two gallons; acid tar, a half pint; finely ground brick clay, one and a half bushels; and refuse lime from gas-works, one and a half bushels.

Claim.—An improved composition roofing, formed by combining coal tar, acid tar, still bottom of petroleum, finely ground brick clay, and refuse lime from gas-house, in the proportions and in the manner substantially as herein described and for the purposes set forth.

No. 58,170.—THOMAS J. JONES, Summit, N. J., assignor to CHARLES J. EAMES and WESLEY WELTY, New York, N. Y.—*Attachment for Pumps.*—September 18, 1866.—The edge of the inverted bowl is raised above the bottom of the cistern by a central pedestal. The bowl has double walls and an opening in the outer one, by which air, &c., may be introduced to clear the induction of obstructions.

Claim.—A suction attachment by which a pump may be constantly supplied with water, and the introduction of solid substances to choke its action may be prevented, all substantially in the manner above described.

No. 58,171.—CLEMENT KEEN, Philadelphia, Penn., assignor to KEEN & Co.—*Covering for Floors.*—September 18, 1866; antedated September 2, 1866.—Sheets of manilla or other tough paper are cemented in a semi-pulpy state to both sides of burlaps or other similar textile fabric.

Claim.—A floor covering consisting of burlap or other equivalent textile fabric and paper or paper pulp combined, substantially as set forth.

No. 58,172.—DANIEL B. LACEY, Mott Haven, N. Y., assignor to himself and ISAAC A. and THOMAS T. LACEY, Jersey City, N. J.—*Weight.*—September 18, 1866.—Explained by the claim and cut.

Claim.—The construction of weights with an outer case of sheet metal and a filling of "slag," and a malleable or wrought metal ring or shank, around which the slag is cast, all substantially as herein set forth, for the purpose specified.

No. 58,173.—E. C. LITTLE, St. Louis, Mo., assignor to EVELINE LITTLE, same place.—*Moulders' Flask.*—September 18, 1866.—The hinges and dowel-pins cast upon the protecting frames of the drag and cope permit the latter to separate without lateral distortion when a match-plate is placed between them.

Claim.—The combination of the hinges B B and the pins C C with the protecting frames D D, when cast together in one piece, and so arranged and applied in connection with the cope and drag of moulders' flasks as to permit a match-plate pattern, a, to be placed between them, without derangement by side movement, substantially as herein described.

No. 58,174.—JOHN MULCHAHEY, Springfield Mass., assignor to himself and CHARLES MULCHAHEY, same place.—*Bottle Stopper.*—September 18, 1866.—The India-rubber stopper and metallic cap are held by hinge and cap to a collar on the neck of the bottle. Above an orifice in the stopper is a spring valve.

Claim.—First, the arrangement of a safety-valve in the stopper of a bottle, substantially in the manner and for the purpose set forth.

Second, the use of a rubber or similarly elastic stopper, when the same is fitted in a case, A, which is hinged at one side, and fastened by a catch or latch at the other, substantially as set forth.

Third, holding the latch or catch b, by means of the pivoted bar H, which is also held in place by means of the spring K, or other suitable means, substantially as described.

No. 58,175.—ROBERT NUTTY, New York, N. Y., assignor to himself and JOHN SCOTT, same place.—*Drilling Machine.*—September 18, 1866.—The carriage moves on a track and the drills operate in a vertical, horizontal, or intermediate position. The drilling apparatus is adjustable upon and supported by a beam from the carriage, and consists of a steam cylinder whose hollow piston rod carries the drill rod. The regulating devices are explained in the claims and cut.

Claim.—First, the piston cylinder of a steam, atmospheric, or other suitable engine, with the piston of which a drill rod is suitably connected, so hung that it can be adjusted to enable the drill to be brought to bear against the surface of the rock or other surface to be drilled, in any desired direction, whether in a vertical or horizontal plane, or in any intermediate plane, substatantially as described.

Second, a piston cylinder, through which a drill rod is operated, hung upon the boom or supporting beam therefor, in such a manner that it can be moved thereon and set at any desired position, according to the point of the rock or other surface against which the drill is to act, substantially as and the purpose described.

Third, the pronged lever L[5] hung to the outer end of the drill boom, and having an extension arm P[5] swivelled or pivoted to it, substantially as described and for the purpose specified.

Fourth, the piston cylinder hung by trunnion pins, to and in a frame W, that by trunnion pins is suspended in the eyes of screw rods S, having screw nuts T[2], said screw rods passing loosely through the boom K, or its equivalent, substantially as and for the purposes set forth.

Fifth, the extension frame C[3], secured to the bottom of the piston cylinder, in which frame slides a cross-head E[3], carrying the drill rod N, and connected with the piston head of the cylinder, in connection with which it moves, substantially as and for the purpose described.

Sixth, the arrangement of the bevel pinion H[3] on the drill rod ratchet pinion I[3], hung in stationary bearings of the frame C[3], and spring pawl L[3] secured to sliding cross-head E[3], when arranged and connected together so as to operate upon the drill rod as the cross-head moves forward and backward, substantially as described and for the purpose specified.

Seventh, the tappet wheel G[4], walking beam J[4], having drill rod B[4] suspended in its outer end connecting link or piece M, double crank shaft O[4], rod Q[4], connected with crank arm S[4], carrying a pawl which engages with the ratchet wheel U[4] of the said drill rod B[4], when the several parts are combined and arranged together so as to operate upon the drill rod, substantially in the manner and for the purpose described.

Eighth, the arrangement and construction of the framework of the machine, the same consisting of the parallel horizontal slotted platforms D and H, connected to a common centre post G, and supported at suitable points by uprights J, for holding the drill booms, the whole being supported upon suitable wheels or friction rollers, and arranged and connected together, substantially in the manner described and for the purpose specified.

No. 58,176.—WILLIAM A. and THOMAS F. POWERS, Brooklyn. N. Y., assignor to WILLIAM A. POWERS.—*Piston Packing.*—September 18, 1866.—The packing rings have annular rabbets on their outer sides, and are held in place on the central ring by perforated flanges, through which steam is admitted to expand them against the cylinder. A rabbet formed in the central ring receives a flange formed on the packing rings to prevent the passage of the steam around the T-shaped break-joint pieces under the ends of the packing rings.

Claim.—First, the packing rings e g, constructed with annular angular grooves or rabbets in their outer sides, and combined with the "bull ring" A and perforated flanges B, substantially as herein set forth for the purpose specified.

Second, the annular groove c', formed at the innermost edge of the rabbet c, and arranged in combination with the rabbeted packing ring e g and the perforated flange B, substantially as herein set forth and for the purpose specified.

Third, the packing pieces m, constructed as described in combination with the tongues t, on the rings e, substantially as and for the purpose herein specified.

No. 58,177.—JOHN A. QUICK, South Danville, N. Y., assignor to himself and CHARLES R. HOLLIDAY, same place.—*Plough.*—September 18, 1866.—The point of the plough is rotated by a shaft and miter-gear connection with the supporting wheel in the rear.

Claim.—The combination with the plough having mouldboard and landside of the conical rotating point H, shaft F, gearing I M, and supporting wheel J, operating substantially as described.

No. 58,178.—TIMOTHY K. REED, East Bridgewater, Mass.—*Shoe Shanking.*—September 18, 1866.—The knife or knives are so arranged in reference to the bed and gauge pieces that by turning the stock after each cut or by carrying it from one knife to another, the successive cuts are made in alternately angular directions across the material. Two bed pieces are so arranged with reference to a knife working in connection with each that the chamfering or scarfing cuts are alternately made by the two knives at opposite angles to the edge of the stock.

Claim.—So combining and arranging a knife or knives and bed and guide places that in scarfing one edge of each piece of the shanking, the opposite scarf of the next piece is formed thereby, substantially as described.

Also, combining with a bed which holds the stock in position to be cut square or at one angle to its supported edge, a bed which supports the stock in position to be cut at an angle to the opposite side, substantially as described.

Also, conjointly and specifically the provision for cutting stock of various thicknesses and into various widths for scarfing the material to a greater or less degree, and for inclining one of the scarfed edges more or less, all substantially as specified.

No. 58,179.—THOMAS SAVILL, Philadelphia, Penn., assignor to C. JONES and CADBURY, same place.—*Low-water Detector for Steam Generator.*—September 18, 1866.—To afford means for removing deposits of mud, &c., from the case with facility the hollow spindle has valves which close the entrance into the boiler or that into the eduction passage; in the former case opening a way, in connection with openings in the stem and the valve chest, for the escape of mud collected in the chamber below the gauge tube.

Claim.—The hollow spindle C, with its valves and openings in combination with the valve chest A, its valve seats and opening b, the whole being constructed and operating substantially as and for the purpose specified.

No. 58,180.—JOHN B. SCOTT, Hyattsville, Md., assignor to himself and GEORGE HALL, Prince George county, and S. MOSS, Baltimore, Md.—*Burning Fluid.*—September 18, 1866.—Composed of naphtha, 40 gallons; potatoes, 50 pounds; lime, 4 pounds; sal soda, 4 pounds; and curcuma, 3 pounds.

Claim.—The use of the above described ingredients, compounded as and for the purpose herein specified.

No. 58,181.—SIDNEY M. TYLER, Brooklyn, N. Y., assignor to the EMPIRE SEWING MACHINE COMPANY, New York, N. Y.—*Sewing Machine.*—September 18, 1866.—The object is to give the shuttle a slow movement and gradual stop, at the time the needle is out of the cloth and the stitch is being tightened, and a rapid motion at the other end of its movement.

Claim.—The rocking shaft g, crank k, link l, and shuttle driver m, arranged and acting on the shuttle in the manner specified, in combination with the needle bar n and cam q, for giving the specified motions to the needle, as and for the purposes set forth.

No. 58,182.—SIDNEY M. TYLER, Brooklyn, N. Y., assignor to the EMPIRE SEWING MACHINE COMPANY, New York, N. Y.—*Sewing Machine Shuttle.*—September 18, 1866.—The projection tapers upward and also towards the needle side of the shuttle, and affords a rest for the prolonged end of the spring thread-detainer, and yet allows a free delivery of the loop from the projection; the action of the spring is to carry back the thread before releasing it in the direction of the feed, sufficiently to take it out of the way of the descending needle.

Claim.—The tapering rearward projecting point 3 of the shuttle, in combination with the spring thread detainer *c* extending to the rear of the heel of the shuttle, as and for the purposes set forth.

No. 58,183.—DANIEL WHITLOCK, Newark, N. J., assignor to himself and JAMES M. SEYMOUR, same place.—*Machine for Cutting the Corners of Paper in the Manufacture of Boxes.*—September 18, 1866.—The guides maintain the knives in proper position when resisted by the paper. By the adjustment of the bed and the use of the gauges the corner openings of the boxes may be cut as required.

Claim.—First, the providing of the knife or cutter I, with pendent bars *g g*, to serve as guides for the same, substantially in the manner as and for the purpose set forth.

Second, the adjustable bed-piece F, in combination with the knife or cutter I, substantially as and for the purpose set forth.

Third, the graduating of the edges of the opening *d* in the bed piece F, in combination with the adjustable gauges G G, substantially as and for the purpose set forth.

No. 58,184.—THOS. WHITTEMORE, Cambridgeport, Mass., assignor to EDMUND G. LUCAS, same place.—*Railway Chair.*—September 18, 1866.—Explained by the claim and cut.

Claim.—The combination and arrangement of the two wedged jaws C C, and the wedged socketed base plate F, constructed, arranged, and applied together, substantially as and so as to operate as specified.

No. 58,185.—JAMES F. WINCHELL, Springfield, Ohio, assignor to himself and JOSEPH LEFLER, same place.—*Fruit Jar.*—September, 18, 1866.—The jar has a cover with a recess sunk in the top, and a bead beneath to rest upon the packing ring. To the yoke is attached a lever with projections to fit in the recess and a cam lever to depress it.

Claim.—First, the circular bead *b*, on the under side of the cap B, for the pupose herein described.

Second, the pressure lever D, having projections *c*, and cam-lever E, in combination with the bridge C, and cap B, when arranged to operate substantially as described.

No. 58,186.—EUGENE BOURQUARD, Paris, France, assignor to himself and PIERRE BOISSET, same place.—*Instrument for Cutting Teeth.*—September 18, 1866.—Explained by the claim and cut.

Claim.—A perforated tube in connection with a handle or any other contrivance for closing the open end of the tube, together with a sponge or other porous substance to be used as above described. The sponge may be dispensed with and the perforated tube filled with sugar or any other sweet substance of such consistency as will not require a sponge to hold it.

No. 58,187.—FREDERICK T. ACKLAND, HENRY G. MITCHELL, and MUSTAPHA MUSTAPHA, Zagazig, Egypt.—*Cotton Gin.*—September 18, 1866.—The feeding bar, by its reciprocating motion, forces the cotton toward the ginning roller and doctor, or fixed knife.

Claim.—The application and use to and in the Macarthy cotton gin, or other cotton gins of a like character, of a feeding bar or surface having a rectilinear reciprocating motion along the grid or grating, for the purpose of pushing the cotton up to the ginning roller and "doctor," substantially as hereinbefore described and illustrated by the annexed sheet of drawings.

No. 58,188.—ALEXANDER ANDERSON, London, Canada West.—*Cultivator.*—September 18 1866.—The axle is stayed by rods from the tongue, and the frame is vertically adjustable beneath by chains and levers. The wheels are attached to half axles, which are extensible upon a full-length, slotted splice piece.

Claim.—First, the mode of suspending the cultivator frame beneath the axle by means of the chains G G, rods R R, and levers J K K, so arranged, as described, to give it the necessary lateral and vertical play.

Second, the slotted extension axles F′ F′, counterpart central portion F, and bolts S S, constructed and operating as described and represented.

No. 58,189.—THOMAS WALKER, Birmingham, England.—*Apparatus for Indicating the Speed of Vessels.*—September 18, 1866.—The resistance of the water on the vanes causes the spindle to rotate, and motion is communicated, by means of a worm screw and pinion, to a train of wheels and indicator hands within the cylinder, whereby the number of revolutions of the cylinder is registered, and the speed of the vessel approximately determined.

Claim.—The adaptation or combination of means forming apparatus for indicating the speed of vessels, whereby the wheel-work and index are placed in front of the vanes of the rotator, and they are definitely acted upon to have motion given to them in the rotation of such vanes, and in which the rotating vanes and the chamber containing the wheel work are immediately connected, substantially as explained.

No. 58,190.—A. H. EMERY, New York, N. Y.—*Machine for Slicing and Drying Peat.*— September 18, 1866.—Attached to a wheeled vehicle is a transverse slicing knife, which, by the force of the draught, cuts and raises a broad belt of peat; this is delivered from the serrated rear edge of the apron behind the knife to the communicating toothed breaker which is revolved by gearing from the drive wheels.

Claim.—First, the combination for cutting peat to facilitate its drying of a knife or slicer attached to or carried by a wheeled vehicle and breaker set in motion by the draught, for operation together, substantially as specified.

Second, the combination with the knife or slicer 9 and its pronged stock or cover of a revolving toothed breaker, arranged and operating together essentially as and for the purpose or purposes herein set forth.

Third, connecting the knife stock with the frame of the vehicle by means of slotted side uprights, made capable of vertical adjustment through the frame, as specified.

Fourth, the combination with the sliding uprights of the knife stock of the levers 2 2, screw nut or box 2 3 and its adjusting screw, for action together, as shown and described.

Fifth, in combination with a revolving toothed breaker the hangers 5, supported by or on the axle of the running wheels, also carrying the revolving shaft of the breaker with its gear and made adjustable relatively to the frame of the machine, substantially as specified.

No. 58,191.—SILAS GRENELL, Mokena, Ill.—*Rotary Harrow.*—September 18, 1866.—The tongue is attached to the axis of the barrow and has a wheel traversing around a hub on the same. A weight box on a bar attached to the tongue traverses a circular track on the harrow, and by depressing one side causes the barrow to revolve.

Claim.—A revolving barrow, when constructed with a frame A and circular plate or ring C, upon which the draught is applied by the wheel I attached to the tongue, the harrow turning upon a wrist pin E, and having an arm G carrying a weight box K resting upon the track B on the friction wheel K, said several parts being respectively constructed and arranged for use substantially as set forth.

No. 58,192.—JOHN R. MICKEY, Chicago, Ill.—*Churn.*—September 18, 1866.—The wheels move on vertical axes and their cogs engage.

Claim.—The combination of the cog dashers B and B', when constructed substantially as and for the purpose set forth.

No. 58,193.—BENNING ROWELL, Elmira, N. Y.—*Fence.*—September 18, 1866.—A-shaped braces support the panels, whose second bars are lapped and notched into the cross-piece of the brace; the upper bars lap across the apex. The middle of each panel is hooked to a short post.

Claim.—A portable fence so constructed that the boards or rails A¹ and A² supporting the panels shall be interlocked and sustained upon posts B, and be held in place by intermediately placed hooks C fastened to short posts D, and arranged to operate substantially as set forth.

No. 58,194.—ESTES ABBOTT, Chagrin Falls, Ohio.—*Milk Can.*—September 25, 1866.— The bottom and top are respectively made of disks struck into form and are applied to the cylindrical portion; the bottom is spun on, enclosing a hoop and the top slips inside.

Claim.—First, forming the cylinder A, bottom D, and cover, Fig. 5, each of whole or entire pieces, as herein set forth, when used in the manner and for the purposes specified.

Second, the bottom D, constructed substantially as described, in combination with the body of the cylinder A and hoop G, as and for the purpose specified.

Third, the cover, Fig. 5, constructed as shown, and provided with the tube J, as and for the purpose set forth.

Fourth, the hoop, Fig. 6, constructed as described, to wit, straight on its inner edge and curved on its outer edge, in combination with the body of the cylinder A, as and for the purpose set forth.

No. 58,195.—JACOB ALBERT, Baltimore, Md.—*Combined Button and Loop for Traces.*— September 25, 1866.—The trace passes through the loop over the whiffletree and is secured by the button on the shank which passes through a hole in the trace.

Claim.—The herein described device whereby the ordinary harness trace can be used with whiffletrees of different construction, as described, the same consisting of a bevelled loop, curved shank and button, formed in one piece, under the arrangement and for operation substantially as herein set forth.

No. 58,196.—E. G. ALLEN, Boston, Mass.—*Railway Frog.*—September 25, 1866.—The railroad frog is formed by the combination of upper and lower plates with interposed supports.

Claim.—The combination and arrangement of the plates A and B with their supports E and F, whether made elastic or not, secured together substantially in the manner and for the purpose specified.

No. 58,197.—SAMUEL ANDREWS, Cleveland, Ohio.—*Distilling Oil.*—September 25, 1866.—The cylindrical retort of boiler iron is set in brick-work over a cylindrical chamber connected with the fire-box. Inclined flues at the sides of the chamber near the front connect with the chimney.

Claim.—First, the fire chamber C and reverberatory chamber F, in combination with the throat C' and the openings I, in the manner and for the purpose substantially as set forth.

Second, the reverberatory chamber F, in combination with the retort A, as and for the purpose specified.

Third, the flues J, when separated from the walls of the retort by the wall J', as and for the purpose set forth.

No. 58,198.—MAURICE ANDRIOT, Mount Washington, Ohio.—*Carriage Jack.*—September 25, 1866.—The lever being pivoted at the required hole and then depressed is maintained in effective position by the wedge in the rear of the standard.

Claim.—The key F, formed with a straight back *f* and two oblique faces F' F'' converging together at *e*, as and for the purposes specified.

No. 58,199.—GEORGE H. BABCOCK, Providence, R. I.—*Darning Last.*—September 25, 1866; antedated September 15, 1866.—The case is used to put in a stocking while darning, and is made of two hinged portions.

Claim.—As a new article of manufacture a light smooth last, or partial last, in the form of a box, substantially as described and for the purpose herein set forth.

No. 58,200.—COLLINS B. BAKER, Troy, N. Y.—*Brick Machine.*—September 25, 1866.—The scraper is operated by a rock shaft provided with laterally adjustable wheels, in combination with ways, bevelled at their extremities; its office is to "strike" the filled moulds by scraping off and levelling their upper surfaces after they have been shoved from underneath the press box.

Claim.—The striker or scraper K, operated from the rock shaft J, as shown, and provided with the laterally adjustable wheels L L, in combination with the ways *g g*, bevelled at their inner ends, as shown at *h h*, and having oblique bars *l l* at their front ends, and all arranged to operate in the manner substantially as and for the purpose set forth.

Also, the strip M, in combination with the scraper K, substantially as and for the purpose specified.

No. 58,201.—ROBERT J. BARR, Philadelphia, Penn.—*Trap for Removing Water from Steam Heating and Evaporating Apparatus.*—September 25, 1866.—The water of condensation is received in a vessel suspended on one end of a counter-balanced lever; as it descends, a valve in the trunnion is opened, and the pressure of the steam above, ejects it until its rising closes the eduction.

Claim.—First, the hollow journals E E', vessel H, beam L, and weight G, as and for the purpose specified.

Second, the valve V, pipes C and E, and vessel H, as and for the purpose described.

Third, the pipes E E' F F' vessel H, and pipe N, arranged and operating as set forth.

Fourth, the chambers *s s'*, stop *a*, valve V, and pipe C, in combination, for the purpose set forth.

Fifth, broadly, the herein-described steam trap, when constructed and operating substantially as and for the purpose specified.

Sixth, operating the valve V, that is, opening and closing the same by the accumulation and discharge of water, as specified.

No. 58,202.—T. B. BELFIELD, Philadelphia, Penn.—*Steam Radiator.*—September 25, 1866.—The radiator is composed of a number of steam-heated hollow sections, having a zigzag form; the spaces intervening between the sections are air passages' made tortuous by inclined ribs projecting from the sections.

Claim.—First, a radiator, composed of a number of sections of the zigzag form described, for the purpose specified.

Second, the combination of the above with the inclined ribs, for the purpose specified.

No. 58,203.—CHRISTOPHER S. BENJAMIN, Kalamazoo, Mich.—*Composition for Filling the Pores of Wood.*—September 25, 1866.—Composed of gum copal, 1 pound; turpentine, 1 quart; alcohol, 1 gallon; shellac, 2 pounds; resin, 2 pounds; beeswax, 4 ounces; japan drier, 1 gill; thicken with starch.

Claim.—The composition for filling the pores of wood, preparatory to varnishing, substantially as herein described.

No. 58,204.—Charles P. Benoit, Detroit, Mich.—*Piston Rod Packing.*—September 25, 1866.—Corrugated packing rings are placed within the box and next the piston rod, and are enclosed by an elastic coil depressed by a cup-lipped gland bearing within it a chamber for containing a lubricator.

Claim.—First, the corrugated or grooved packing rings D D, in combination with the elastic coil of packing or stuffing E, the cup-lipped flange G, the annular space around the piston rod C, or the lubricating chamber H, with its connected oil cup I, and the screw bolts K K, with the springs b b, under the nuts c c, all constructed, arranged, and related to each other, substantially in the manner and for the purposes herein described.

Second, the lubricating chamber H, combined with the piston rod C, the flange G, and the packing rings D D, substantially as and for the purposes herein described.

Third, the cup-lipped flange G, combined with the elastic coil packing E, and the corrugated rings D D, constructed and operated substantially as and for the purposes herein described.

No. 58,205.—Edmond Bigelow, Springfield, Mass.—*Sirup Stand and Soda Fountain.*—September 25, 1866.—The case has a hinged cover, kept in place by projections, and contains earthen sirup jars, and a cooling chamber for the soda water, each surrounded by ice. The sirup faucets are lined with glass, and secured to the jars by a screw and rubber packing.

Claim.—The combination of sirup jars with faucets, substantially as and for the purposes herein specified.

Also, the employment of plastic gum, as herein set forth, for connecting the faucet with a sirup jar and case, by means of the metallic bushing, gum, and screws, substantially as described.

Also,.the corner stops a, for holding the divided cover in place, so that either section can be raised without the other part being displaced, substantially as specified.

Also, the setting of the sirup jars, substantially as herein specified, and fixing the same in place by means of the projection at c, and the gum packing.

No. 58,206.—Aaron C. Baldwin, Boston, Mass.—*Rotary Steam Engine.*—September 25, 1866.—The disk is placed centrally upon the shaft; is reduced in thickness toward its periphery, and has two pistons attached, one on each side, and rotating in annular spaces in the cylinder. An abutment to each piston is moved outward and inward by an eccentric, to permit the passage of the piston. Valves are arranged to leave the exhaust open during the entire movement of the piston. Packing prevents leakage of steam around the pistons and abutments. That portion of the cylinder between the receiving ports, through which the piston moves without being under pressure of steam, is recessed, so as to prevent frictional contact.

Claim.—First, the central disk a, diminished in thickness toward the periphery, with its pistons a¹ a², arranged thereon relatively to each other, and to the correspondingly tapered abutments C, substantially in the manner and for the purposes herein set forth.

Second, the depression or groove z in the part B³ for supporting the end of the abutment, and preventing the escape of steam past that point, substantially as and for the purposes herein set forth.

Third, the within described pistons, arranged as specified, the same being of a tapering or wedge form, with the base of the wedge presented inward, and sliding on the fixed inner surface B³, substantially as and for the purpose herein set forth.

Fourth, the within described combination and arrangement of the reciprocating valve R, and the four ports m n P and Q, whereby the exhaust port is kept open during the whole revolution of the engine, while the steam port is opened and closed as required, for the intermittent admission of the steam at each revolution, and adapted to allow of reversing, or working in the opposite direction, substantially as herein set forth.

Fifth, the metal ring V, and yielding ring v, in combination with the disk a'' and the inner cylinder B³, for the purpose of preventing the escape of steam inward toward the shaft of the engine, substantially as herein set forth.

Sixth, the adjusting screws w, or their equivalents, in combination with the rings V v, disk a, and inner cylinder B³, and adapted to adjust the compression of the elastic packing v, substantially as herein set forth.

Seventh, the recesses in the rubbing surfaces between the ports m and n, so as to reduce the friction of the piston through that part of its revolution, substantially as herein set forth.

No. 58,207.—James Bing, Philadelphia, Penn.—*Shoe for Car Brake.*—September 25, 1866.—The dovetail lug on the back of the sole is inserted in a corresponding socket in the shoe, and maintained in place by gravity and the friction of the wheel.

Claim.—First, the shoe B and sole A, united together without the help of any key or bolt, and so that the sole, by this off-centre tendency, resulting from the peculiar shape of lug C, can remain constantly in close contact with its shoe.

Second, the combination of packing Z Z with lug C and shoe B, for the purpose before described.

Third, the combination of shoe B, sole A, close clevis R, metallic lining L, and packing Z Z, the whole combined, constructed, and arranged substantially as above specified.

No. 58,208.—W. H. BLACKMER and E. R. CARPENTER, Clermont, Iowa.—*Water Wheel.*—September 25, 1866.—The contracting water passages are formed by the walls of the blocks, and deliver the water to the radial wings of the wheel, which has a lower oblique discharge.

Claim.—The combination of the radial buckets O×C of the wheel B, blocks C' C' C', water passage D, all constructed and arranged in the manner and for the purpose herein specified.

No. 58,209.—JOHN F. BOYNTON, Syracuse, N. Y.—*Apparatus for Carburetting Gas.*—September 25, 1866.—A series of wooden pegs, slotted at the ends, and wound with fibrous material, are supported in the carburetting vessel. A frame divides the carburetter into chambers, through which the gas passes consecutively on its passage to the burner. The automatic filler consists of a vessel closed at top, and discharging by a tube near the bottom of the carburetter.

Claim.—First, an apparatus for carbonizing gas and air, being a gas-light multiplier, as herein described.

Second, a detachable reservoir, to operate substantially as described, for filling the carbonizer with liquid.

Third, the filling reservoir D, in combination with the cocks J and K.

Fourth, the combination of ligneous material with fibrous material.

Fifth, the compound capillary action of ligneous material with fibrous material.

Sixth, the perforated base-board, or its equivalent.

Seventh, the use of pegs supported from a base-board.

Eighth, the arranging of the fibrous material parallel with the ligneous material, in the manner described.

Ninth, the slot in the top of the pegs for securing the fibrous material.

Tenth, the securing of the fibrous material at the bottom of the peg, by its being driven with the peg into a perforation of the base-board.

Eleventh, the combination of ligneous and fibrous material, producing a compound capillary action, attached to a base-board, forming a cage, and so arranged that it can be placed in and taken from the reservoir or carbonizing chamber.

·Twelfth, a compound perforated partition, so arranged as to divide the carbonizing chamber into an internal and external apartment.

Thirteenth, the arrangement of the compound perforated partition, that it may be removed and replaced in the carbonizing chamber without disturbing the cage.

Fourteenth, a double partition, so arranged that the compound capillary action of the ligneous and fibrous material may take place between its walls.

Fifteenth, the making of these partitions of any material capable of producing capillary action.

Sixteenth, the use of capillary action of any material of which these partitions may be constructed.

Seventeenth, the using of ligneous material for a partition.

No. 58,210.—THOMAS W. BROWNING and P. C. HARD, Wadsworth, Ohio.—*Hemming Guide for Sewing Machines*—September 25, 1866.—The same screw secures the hemmer, both to the table and to the adjustable slotted under-plate of the folder. The swinging top-plate, when turned aside, allows the easy insertion of the cloth under and over the plate, and it is then turned back to hold the cloth in place.

Claim.—The combination and arrangement of the hemmer B with the under-slotted plate, the middle plate, with its tongue T, and the upper pivoted plate D', as and for the purpose set forth and described.

No. 58,211.—LEONARD A. BURNHAM, Gloucester, Mass.—*Fishing Line Sinker.*—September 25, 1866.—The supporter extends in opposite directions at right angles to the spindle of the hawse; the spindle is inserted in a tube, which is enclosed in a socket hinged to the connecting rod, which goes up through the sinker.

Claim.—The improved swivel hawse, made of the parts C D G, constructed, arranged, and combined with the spindle B, and the sinker S', substantially as above set forth.

No. 58,212.—WILLIAM A. BUTLER, New York, N. Y.—*Reflector for Lanterns, &c.*—September 25, 1866.—The corrugations of the side portion run in the direction of its length, and those of the rear plate are concentric.

Claim.—The combination of the conical side reflector A, the concave back reflector B, and the lens D, all constructed, arranged, and operating as and for the purpose specified.

No. 58,213.—GEORGE J. CAPEWELL, West Cheshire, Conn.—*Lamp Burner.*—September 25, 1866.—The wick is raised for lighting without removing the chimney, by a spur wheel I engaging the notched inner wick tube, and is then lowered to its normal condition. Air is admitted to the flame through indented openings around the chamber.

Claim.—The toothed holder tube G for the wick of a kerosene lamp, in combination with the bent spring K, or its equivalent, and pinion I, when arranged together, and with regard to the ordinary wick tube of such lamps, substantially as herein described, and for the purpose specified.

Also, forming the air-openings C to the air chamber of a kerosene or other lamp, as herein described and shown in the drawings, for the purposes set forth.

No. 58,214.—J. M. CARR, Omaha City, Nebraska.—*Seeding Machine, &c.*—September 25, 1866.—The seed is dropped from the pockets of the revolving roller, and is covered by the cultivating toothed roller, which revolves in and is attached to a vertically adjustable frame, hinged to the axle of the carriage.

Claim.—The swinging or adjustable frame I, attached to the axle *d*, as shown, and provided with the cylinder K, having knives or cutters *l* attached, in combination with a seed-dropping mechanism, all arranged and applied to a mounted frame, substantially as and for the purpose specified.

Also, in combination with the above, the adjustable wheels J J attached to the lower end of the frame I, substantially as and for the purpose set forth.

No. 58,215.—EDWIN CHESTERMAN, Roxbury, Mass.—*Wringer and Washing Machine.*—September 25, 1866.—The rolls are made upon a metallic and wooden core, and are revolved by a worm wheel and pinions. By the substitution of a zigzag cam for the worm, a reciprocating motion of the rollers may be obtained.

Claim.—First, the combination and arrangement of the slotted frame A, dovetail rail F, spring bar G, springs A, fingers K K′, sockets J, capped wringers B, toothed wheel D, and worm E, substantially as described, for the purpose specified.

Second, making the shafts of the rollers of a skeleton or spider, the angles between whose radial arms are filled in with blocks of a larger size than the spaces, substantially as above described.

Third, the cams E′, made substantially as described, for giving a reciprocating rotary motion to the elastic rollers, so as to produce friction on the clothes placed between them, substantially as set forth.

No. 58,216.—SAMUEL B. CHILDS, Syracuse, N. Y.—*Running Gear of Railway Cars.*—September 25, 1866.—One wheel revolves on the axle, and has an adjustable collar and metallic wedges which form bearings.

Claim.—The adjustable collar C and set screw D, and the metallic wedges or box F, in the wheel, when arranged and combined with the car axle, as herein described and for the purposes set forth.

No. 58,217.—JAMES J. CLARK, East Chester, N. Y.; and HENRY SPLITDORF, New York, N. Y.—*Insulating Wires of Helices.*—September 25, 1866.—The wire is insulated by passing it through varnish and then through dry powder.

Claim.—Making helices for electro-magnetic and magneto-electric machines from wire insulated by passing it through any powdered material, such as powdered glass, stone sand or paper pulp, after the wire has been passed through any sticky fluid, and before the sticky fluid becomes dry, as hereinbefore described.

No. 58,218.—BAXTER CLOUGH, Amherst, Ohio.—*Gate.*—September 25, 1866.—The gate is counterbalanced in rear of the pivotal point in the braced frame.

Claim.—The gate B, arranged and hung as specified, in combination with the guys C C′, plates B′, posts A′ and counter-braces D D′, as and for the purpose set forth.

No. 58,219.—W. S. COFFMANN, Coldwater, Mich.—*Broadcast Seeding Machine.*—September 25, 1866.—The apparatus is to be carried and worked by the operator; a gate regulates the feed, and the issuing grain is scattered by beaters rotated by gearing and a hand crank.

Claim.—First, the operating of the revolving scatterer by means of gearing, arranged in the manner substantially as shown and described.

Second, the inclined bottom *a*, in combination with the valve D, adjusted through the medium of the lever and gauge, substantially as and for the purpose specified.

Third, the plate *d* to retain or hold the valve D in place, substantially as and for the purpose set forth.

Fourth, the combination of the box A with the frame B, provided with legs *a*, substantially as and for the purpose specified.

No. 58,220.—THOMAS COLE, Cedar Hill, Ohio.—*Sorghum Evaporator.*—September 25, 1866.—The pan is divided by a partition, and the compartments are severally connected by pipes with a box partially divided by a plate, and having ribs on the bottom to arrest sediment.

Claim.—First, the process herein described whereby the raw juice, after being exposed to a high temperature immediately upon entering the evaporator, then conveyed away rapidly to a precipitating chamber not exposed to heat, and finally introduced at the cool end and discharged at the hot end of the chamber, where the final boiling and separation take place, substantially in the manner and with the effect described.

Second, in combination with an evaporating pan, the precipitating pan D, substantially as and for the purpose described.

Third, the evaporating pan A, divided by the partition a, substantially as and for the purpose described.

Fourth, the evaporating pan A, constructed with two apartments, in combination with the precipitating-pan D, substantially as and for the purpose described.

No. 58,221.—JOSEPH H. CONNELLY, Wheeling, West Va.—*Manufacture of Gas.*—September 25, 1866.—Petroleum and lime water contained in tanks above are conducted by their respective pipes and unitedly fall upon the spout, which discharges them upon the mass of coal from which the gas is generated.

Claim.—First, the forcing of the petroleum or its products, viz., residuum and crude benzole, in about equal quantities, in conjunction with lime water, into the retort with a jet of steam, when the retort is charged with coal, substantially as described.

Second, introducing the petroleum or its products, in conjunction with lime water, into the retort, when charged with coal by means of the pipes F G, provided with seals or traps, substantially as described.

Third, the iron spout C, attached to the manhead and employed to conduct the liquids back to the centre of the retort, substantially as herein set forth and for the purpose specified.

Fourth, the combination of oil and lime water for desulphurizing and purifying in the manufacture of gas from coal, substantially as described.

Fifth, introducing the petroleum or its products, in conjunction with lime water, into the retort, when charged with coal by means of the common receiving pipe E, inserted into the retort at or near the back end, either with the seals, traps, or steam jet, substantially as described.

No. 58,222.—C. C. CONVERSE, New York, N. Y.—*Baling Press.*—September 25, 1866.—The follower is depressed by the pressure of the roller at the end of the extension bar, which traverses to and fro, as rocked by the levers. The downward position of the follower is sustained by screws.

Claim.—The rock shaft E, actuated by the levers F, or other equivalent means, and provided with the extension pendent bar G, having a roller I at its lower end, in connection with the follower or platen C, screws J J, and press box A, all arranged to operate substantially in the manner as and for the purpose set forth.

No. 58,223.—L. M. CRANE, Ballston Spa, N. Y.—*Paper Collar.*—September 25, 1866.—Explained by the claim.

Claim.—As an improved article of manufacture a paper shirt collar having a thin sheet or layer of gutta-percha interposed between the layers of paper of which the collar is constructed or made, substantially as described.

No. 58,224.—ANSON D. CROCKER, Boston, Mass.—*Bail for Kettles and other Vessels.*—September 25, 1866.—The ears straddle the rim of the vessel, and are held in place by adjustable straps connecting with a hoop below the swell of the kettle, and adapted thereto.

Claim.—First, in combination with a bail or handle, adjustable clamps constructed and arranged to fit upon the rim of and hold the receptacle to which they are attached, as described.

Second, in combination with the above, the vertical bands g g and horizontal band k k, as described and for the purpose specified.

Third, making the bands g g and k k adjustable by the means herein above described and for the purpose specified.

No. 58,225.—DAVID CUMMING, Jr., New York, N. Y.—*Slate Frame.*—September 25, 1866; antedated September 17, 1866.—Explained by the claim and cut.

Claim.—A slate frame when constructed with a mitre joint, having a tongue and groove, or similar device, and held together by means of a continuous screw, substantially as set forth.

No. 58,226.—LEWIS CUTTING, San Francisco, Cal.—*Stop Motion for Spinning Machines.*—September 25, 1866.—The cam, by its continuous revolution, vibrates the forked lever, which, by means of a link, moves to and from the notched slide. This notch, fitting a corresponding-projection on the lower slide, causes it to move coincidently with the other slide unless obstructed. When the sliver breaks the trumpet falls back, its lower end rises, and the pin in the lower slide coming against it, arrests the movement, which compels the upper slide to ride over a projection and be lifted, and so raise one end of the lever, whose other end is thus released from the notch in the rod, whose spring then acts to force it back and ships the belt.

Claim.—The bar J and slide N, combined and operating substantially as and for the purposes described.

No. 58,227.—WILLIAM DAMEREL, Brooklyn, N. Y.—*Umbrella.*—September 25, 1866.—The umbrella is closed by an elastic band embracing all the stretchers, and sustained in position thereon by rests attached thereto.

Claim.—First, an umbrella held closed or shut by means of an elastic band placed around the stretchers of their frames, substantially as shown and described.

Second, rests F, as arranged for holding the band or spring E in its proper place on or over the stretchers of an umbrella, in all conditions, whether closed or extended, substantially as described.

No. 58,228.—FRANCIS DANZENBAKER, Bridgeton, N. J.—*Churn.*—September 25, 1866.—By a turn of the stopper the valve can be brought into or out of coincidence with the opening in the tube to open or close the air duct.

Claim.—The combination of the tubular handle A, adjustable stopper B, or its equivalent, and valve D, when constructed and arranged to operate substantially as set forth.

No. 58,229.—MAHLON D. DICKENSON, Pilesgrove, N. J.—*Machine for Assorting Potatoes, Coal, &c.*—September 25, 1866.—The inclined sieve-frames are suspended by rods in a frame, and vertically agitated by resting on revolving crank shafts.

Claim.—The crank shafts A A, sieves B B, and suspension rods E E, when combined and arranged as and for the purpose set forth.

No. 58,230.—BERNARD DOUGLAS, New York, N. Y.—*Boot Jack and Blacking Case.*—September 25, 1866.—Explained by the claim and cut.

Claim.—A combined boot-jack and blacking case constructed by combining the two unequal parts of the case A with the hinges a, blocks f, and cover d, substantially in the manner and for the purpose herein shown and described.

No. 58,231.—J. B. DUANE, Schenectady, N. Y.—*Reclining Chair.*—September 25, 1866.—The back, seat, and front are pivoted together so as to take a normal sitting position or recline to any extent.

Claim.—Pivoting the seat bars to the front and back bars of the chair, by means of the transverse rods A A, for the purpose described, in combination with the arms g g, arranged directly over said bars e e, and pivoted to the inner sides of the front and back bars, substantially as described.

No. 58,232.—KILIAN EGGER, South Cortland, N. Y.—*Extracting Cream from Whey.*—September 25, 1866.—The whey is exposed in zinc pans; salt is added, and the cream skimmed therefrom.

Claim.—The process above described for extracting the cream from the whey, substantially as above described.

No. 58,233.—NORMAN J. ELDRED, Chicago, Ill.—*Stove-pipe Damper.*—September 25, 1866.—The exterior disk fits the pipe and has a square opening occupied by a hoop, in which is a plate slipping up and down on a rod as the axis of the disk is rotated.

Claim.—The disk A with lugs x z cast upon it, and a square hole or opening cast in it, in which opening is placed a hoop B, its axis being transverse to that of the disk together with the damper plate D operating loosely between one side of the disk and the hook, the several parts being arranged and constructed as and for the purpose herein specified.

No. 58,234.—OLIVER ELLSWORTH, Boston, Mass.—*Bit Stock.*—September 25, 1866.—The jaws are hinged at their lower ends to the bit-shank, are expanded by a spring when not closed upon the bit by the traversing sleeve nut.

Claim.—The arrangement of the jaws C C pivoted on the pins D D, with spring L, stationary screw P, and traversing nut G, the whole constructed as described for the purposes set forth.

No. 58,235.—L. C. ENGLISH, Corning, N. Y.—*Stump Extractor.*—September 25, 1866.—The hinged leg is adjustable in its divergence by means of a tie brace and retaining catch, the socket joint of the ankle permitting the foot to conform to the surface of the ground.

Claim.—First, the derrick frame A A B jointed at its upper end, and provided with an adjustable brace, as and for the purpose described.

Second, in combination with the derrick frame A A B, the pivoted foot at the end of the leg B of the frame, as and for the purpose described.

Third, the block E and catch e so arranged as to render the brace adjustable to the angle formed by the supports A A and B, and avoid a breaking strain on the supports, as described.

No. 58,236.—FRANCIS FARQUHAR and R. E. DOAN, Wilmington, Ohio.—*Evaporator.*—September 25, 1866.—The fire box and flues are placed within the evaporating pan to expose all the fire-heated surface to the liquid under treatment.

Claim.—A sugar evaporator having its fire box and flues arranged so as to be surrounded with the juice to be evaporated, substantially as and for the purpose herein specified.

Also, the arrangement of the direct and return flues C D D and E in relation to the introduction and flow of the juice to be evaporated, so as to separate the impurities to be skimmed off, substantially as herein set forth.

Also, the evaporating space 8''' between the flues D D E and the fire box, as described.

Also, the flue connections L L between the fire box B and flues D D, for the purpose set forth.

No. 58,237.—DANIEL FASIG, Rowsburg, Ohio.—*Straw and Hay Knife.*—September 25, 1866.—The tang of the knife slips in the stock and is operated by a lever linked to the latter.

Claim.—A cutting device for cutting hay or straw from the stack A, having the tang e of the knife or cutter C fitted in it and operated by a lever D, all constructed and arranged substantially as herein shown and described.

No. 58,238.—BENJAMIN FISH, Mechanicsburg, Penn.—*Tuyers.*—September 25, 1866.—The grate and cap are movable relatively to change the openings of the blast; a guide plate below attached to the grate conducts the blast.

Claim.—First, the tongued grate B, constructed and arranged as herein described, in combination with the air chamber A of the tuyere, substantially as and for the purpose set forth.

Second, the cap E, constructed as herein described, in combination with the grate B, substantially as and for the purpose set forth.

No. 58,239.—CLARK FISHER, Trenton, N. J.—*Promoting Combustion in Furnaces.*—September 25, 1866.—Explained by the claim and cut.

Claim.—The employment of fans or other suitable air engines having pipes leading from them into the chimneys, and directed upward for the purpose of promoting the combustion of fuel in furnaces by forcing cold air directly into the chimneys at points which are above the heating surfaces, but near the base of the chimneys, at a greater velocity than that which would result from natural draught, substantially as described.

No. 58,240.—FREDERICK FITZGERALD, Cincinnati, Ohio.—*Vault Cover.*—September 25, 1866.—Between the lights are congeries of pyramidal elevations, whose intervening depressions drain into gutters in the belly of each corrugation.

Claim.—The vault cover or sidewalk plate, having the gutters b, the diagonally scored and convex eminences c, and corrugations d, as and for the purpose set forth.

No. 58,241.—JONES FRANKLE, Amesbury, Mass.—*Instrument for Measuring Distances.*—September 25, 1866; antedated September 13, 1866.—Two parallel telescopes are connected by a rod of a determinate length at right angles to their axes. One of the telescopes is fixed, the other is pivoted to allow its extremity to describe a small arc, measured by a wedge which slides between the edge of the support and the free end of the telescope, and graduated to indicate the distance from the base line to the intersecting point of the lines of collimation of the two telescopes.

Claim.—The combination of the inclined plane o, graduated with distances, and the telescopes d and e, arranged substantially as herein set forth.

No. 58,242.—JAMES D. FRARY, New Britain, Conn.—*Manufacture of Knives and Forks.*—September 25, 1866.—A solid bolster is cast upon the tang, the metal flowing through a hole in the latter to connect the two halves of the bolster. Metal is poured into the countersunk rivet holes, which in cooling contracts longitudinally, and draws the scales tight upon the tang.

Claim.—Forming a bolster a'', upon a blade a, having perforations in the centre thereof through which the metal is poured to form a solid bolster, substantially as described.

Also, securing the handles c, made in two parts, having holes therein to correspond with those formed in the blade, and countersunk so that the metal poured therein will firmly gripe the parts together, substantially as described.

No 58,243.—WILLIAM J. FREEMAN, Spring Fork, Mo.—*Combined Seeder and Cultivator.*—September 25, 1866.—The seed slides are actuated through intermediate levers by cams on the driving wheels, and drop the seed behind the forward cultivators and in front of the wheels. The middle set of ploughs are laterally adjustable.

Claim.—First, the combination of the hopper C and beam C' with the slide C², and the levers E and E', and lugs x, on the wheel B, when constructed as and for the purpose set forth.

Second, the frame A and the wheels B B', combined and arranged with cultivator hereinbefore described, substantially in the manner set forth.

Third, the frame A, and the wheels B B', combined and arranged with the seed-planter hereinbefore described, substantially in the manner set forth.

Fourth, the posts D', and the rock shaft D¹, of the ploughs D, in combination with the frame A, substantially as described.

No. 58,244.—ALONZO FRENCH, Philadelphia, Penn.—*Lantern.*—September 25, 1866.—On the outer face of the globe base is a salient spiral projection which screws into the shell of the lantern bottom.

Claim —The globe base C, and the shell or case B, combined and operating substantially as herein specified and described.

No. 58,245.—HENRY WM. FULLER, Brooklyn, N. Y., assignor to BRUEN MANUFACTURING COMPANY.—*Sewing Machine.*—September 25, 1866.—Improvement on Sibley's patents of March 19, 1864, and June 13, 1865. Designed to adapt the Wheeler and Wilson machine to the making of the three-threaded stitch, (composed of a lock-stitch and double looped-stitch,) patented September 5, 1865, to Sibley, without the necessity of dispensing with the ordinary ring-slide of the Wheeler and Wilson machine, and of substituting instead of it another ring forming part of the looping attachment.

Claim.—The combination of the attachment described with the cloth-plate, needle, rotating-hook, bobbin ring-slide, and other operative parts of a Wheelerand Wilson sewing machine, constructed and operating together substantially as and for the purposes set forth.

No. 58,246.—WILLIAM GILBERT, Detroit, Mich.—*Roofing.*—September 25, 1866.—Composed of a layer of fibrous material cemented between two layers of ordinary roofing felt.

Claim.—First, the employment of a layer or network of swamp grass, hemp, straw, or any other tough, fibrous material placed between the layers of the material employed for the roofing, substantially as described.

Second, for holding together the materials of which prepared roofing is made, a cement composed of distilled coal tar, and common slaked lime, as herein specified.

No. 58,247.—CHAUNCEY H. GUARD, New York, N. Y.—*Attaching Hubs to Axles.*—September 25, 1866.—Antedated September 17, 1866.—On the inner side of the hub are circular bearing-plates, united near their margins, and enclosing a journal disk secured on the end of the axle. The spindle is attached to the outer of the bearing-plates.

Claim.—The attachment of a hub to an axle by means of a convexed journal disk, formed substantially as herein described, and combined with bearing-plates secured to the inner end of the hub, and embracing said disk, substantially in the manner and for the purpose herein set forth.

No. 58,248.—JOEL HAINES, West Middleburg, Ohio.—*Fuel Dumper.*—September 25, 1866 —The suspended fuel-box is divisional, has discharge boards closed by contact with the sides of the truck, and fastened by bolt rods.

Claim.—The combination with the sloping sided truck, and the derrick of the suspended fuel-box, having discharge boards closed by contact with the sides of the truck, and fastened and opened substantially as described.

No. 58,249.—JAMES M. F. HALL, Davenport, Iowa.—*Medical Compound.*—September 25, 1866.—A decoction prepared by boiling together two pounds of red root (Ceanothus Americanus) and one gallon of cider vinegar.

Claim.—The composition or compound as above prepared and described, and to be used as a medicine as herein set forth.

No. 58,250.—NATHAN HARPER, Newark, N. J.—*Fan.*—September 25, 1866.—The interlaced wooden slats are bound on the edge, and faced with paper, &c.

Claim.—A fan constructed of wood as described, and re-enforced by paper or cloth, substantially as and for the purpose specified.

No. 58,251.—B. I. HARRIS, Harrisburg, Penn.—*Stovepipe Damper.*—September 25, 1866.—Explained by the claim and cut.

Claim.—The swelled joint composed of the sections A' A', when the lower section is provided with perforations or holes, and the upper section with an opening or door, substantially in the manner and for the purpose herein specified.

No. 58,252.—SAMUEL HARRISON, Philadelphia, Penn.—*Grate Bar.*—September 25, 1866.—A group of ribs are cast together; some are deep and others shallow; the objects are facility in casting, freedom from choking in use, and economy of material.

Claim.—A grate bar consisting of a deep and shallow rib or ribs, combined in one casting, substantially as and for the purpose set forth.

No. 58,253.—THOMAS S. HATHAWAY, Decatur, Ill.—*Match Safe.*—September 25, 1866.—
The shoulder of the slide receives a match which, when raised, comes in contact with a friction spring, and is thereby erected and grasped by the head, so as to be ignited when it is jerked out of the friction clamp.
Claim.—The vertical slide B, with its shoulder or recess *a*, in combination with the receiving chamber A and friction spring *b*, constructed and operating substantially as and for the purpose described.

No. 58,254.—ROBERT HENEAGE, Buffalo, N. Y.—*Grain Dryer.*—September 25, 1866.—
The grain enters upon a heated cone and is segregated by a revolving flanged disk. The distributing disks have annular flanges acting in connection with guards which intercept the grain and hoppers which return it toward the axis. The operation takes place in an upward hot blast, aided by air eduction exhaust pipes.
Claim.—First, the heating cone I in combination with the induction chamber H and spout or hopper G, said cone being supplied with hot air, or its equivalent, substantially in the manner and for the purpose set forth.
Second, the pulverizing and distributing disk L, constructed and operating substantially as described.
Third, the desiccating disks F, provided with the inclined annular flanges *d d* and air passages and shields *i j*, in combination with the curtains *e c*, or their equivalent, substantially as shown and described.
Fourth, the combination and arrangement of the desiccating disks F F, constructed as above described, with the diaphragms C C, the whole operating substantially in the manner and for the purpose specified.
Fifth, the induction hot blast tube K and exhaust tube O in combination with the series of sections or divisions of the cylinder, each provided with a desiccating device, and said tubes having distinct connection with each division, substantially as set forth.

No. 58,255.—THEODORE A. HOFFMAN, Beardstown, Ill.—*Respirator.*—September 25, 1866.—The air is filtered through the gauze covering the mouth and nostrils.
Claim.—The gauze B, enclosed cotton C, covering F, and elastic strand E, combined and provided with the elastic strap D, and operating substantially as described for the purpose specified.

No. 58,256.—GOODRICH HOLLAND, Willimantic, Conn.—*Machine for Stretching Silk, &c.*—September 25, 1866.—This arrangement, besides constituting a double machine and providing for the stretching of several threads at the same time, admits of winding from the bobbins to the first rollers, the winding and stretching from the first to the second rollers, and the subsequently winding upon the bobbins. During the last operation, new threads are being wound upon the drums preparatory to being stretched.
Claim.—The arrangement of the two stretching drums B C, the bobbins E F, or their equivalents for delivering the unstretched silk to the first stretching drum, and the take-up apparatus, for taking up the stretched silk, substantially as herein set forth.

No. 58,257.—CHAUNCEY M. HOOKER, Hartford Conn.—*Dentists' Tooth Plugger.*—September 25, 1866 ; antedated May 1, 1866.—The point of the plugger continues to press upon the metal in the cavity of the tooth, being actuated by the tension of the spring while the tube is reciprocated and acts by concussion on the end of the stem.
Claim.—The arrangement of the spiral spring G, or its equivalent, to give the reciprocating movement as described, in combination with the tube A, head B, rubber stop *b*, stem D, collars F and H, in the manner substantially as described.

No. 58,258.—HENRY HOWARD, Springfield, Mass.—*Steam Radiator,*—September 25, 1866.—Each section has a circuitous passage, and at one end is associated with a series of similar sections, their coincident openings making a continuous pipe with inclined plates to direct the flow into the chambers consecutively. The projections on the outside of the heater force the air into a tortuous course and increase the radiating surface.
Claim.—A steam radiator, consisting of one or more steam chests or sections, each having the form of an extended endless pipe or flue, so arranged as that the steam shall enter and leave the same at one extremity thereof, through apertures pierced thereon opposite to each other all substantially in the manner and for the purpose herein set forth.
Also, breaking the exterior surface of said extended endless steam chests or flues into parallel ribs or flanges obliquely disposed at opposite inclinations, substantially in the manner and for the purpose herein set forth.
Also, the use of diaphragms placed centrally between the induction and eduction apertures pierced at opposite points in a radiating steam chest or flue as hereinbefore described, substantially in the manner and for the purpose herein set forth.

No. 58,259.—J. L. HUBBELL and E. SHERMAN, Fairfield, Conn.—*Hay Elevator.*—September 25, 1866.—The truck is maintained in position above the wagon till the fork is raised by the engagement of its latch with a catch on the beam; the latch is then tripped by a plate on the rope, and draught on the rope causes the carriage to traverse; a pawl binds the rope upon the sheave and holds the load suspended till the trigger of the fork is pulled.

Claim.—The combination of the pawl I, the latch G, and the catch H, with the truck B, constructed and arranged to operate substantially in the manner and for the purpose specified.

No. 59,260.—JOHN S. HULL, Cincinnati, Ohio.—*Petroleum Cooking and Heating Apparatus.*—September 25, 1866.—Around the burner is a tube in form of a conic frustum open at both ends, its inferior end downward: above this is another in a reversed position. The series may be continued up as far as desired.

Claim.—The open inverted conical tube D, around the burner C, in combination therewith and with the tubes E E, arranged and operating substantially as and for the purpose herein specified.

Also, the succession of open tubes E E, formed, arranged, and operating substantially as and for the purpose herein set forth.

Also, the combination of a burner provided with open tubes D E F, as described, with a cooking or heating apparatus or utensil.

No. 58,261.—JOHN S. HULL, Cincinnati, Ohio.—*Burner for Vapor Lamps.*—September 25, 1866.—The expanded cap is exposed to the flame, and has a channel inside running around near its circumference with connecting grooves.

Claim.—The forming of the cap C, internally with an annular recess and with channels and grooves, in the manner substantially as herein shown and described, in order to spread the burning material or cause a large volume of the same to be exposed to the flame for the purpose specified.

Also, the fluted or corrugated plug e, in combination with annular, serpentine, or circuitous passages in the cap C, substantially as and for the purpose set forth.

No. 58,262.—JAMES W. HYDE, Lewiston, Ill.—*Measuring Scale for Coffins.*—September 25, 1866.—Explained by the claim and cut.

Claim.—A scale or rule for the purposes specified, the same consisting of the form A, having legs B F, of different lengths tapered and graduated substantially as described, and the cross-bar E, with graduated arms G G, as and for the purpose specified.

No. 58,263.—JONATHAN JOHNSON, Kent, Ind.—*Manger.*—September 25, 1866.—The matters which pass through the grating of the manger are sorted by screens beneath and deposited in drawers.

Claim.—First, the combination with the manger A of the coarse screen B, substantially as described and for the purpose set forth.

Second, the combination of the screen bottom drawer E with the drawer frame D and manger A, substantially as described and for the purpose set forth.

Third, the combination of the screen drawer F, constructed as described, with the drawer frame D and manger A, substantially as and for the purpose set forth.

No. 58,264.—J. H. JOHNSON, Paducah, Ky.—*Steam Blower.*—September 25, 1866; antedated September 10, 1866.—The exhaust steam is conducted to a chamber and from thence by tubes directed into the flues of the generator.

Claim.—The exhaust pipe C, the receiving chamber B, and the conical tubes c, when combined and operated as herein described and set forth.

No. 58,265.—LEVI KEILER, Catawissa, Penn.—*Lamp Chimney Cleaner.*—September 25, 1866.—The India-rubber bulb is inflated through an opening in the handle to fill the chamber of the article to be cleaned.

Claim.—The adjustable cleaner, of any material, constructed and combined with a hollow stem or handle, to be inflated and operated as herein described and for the purposes set forth.

No. 58,266.—ZENAS KING, Cleveland, Ohio.—*Bridge.*—September 25, 1866.—The chord near its end is connected with a parallel metallic rod, and at the point of intersection with the rod the chord diverges, forming an angle with the end-plate, to which both are attached.

Claim.—The chord B, with the rod B', so that at the point of connection, d', of said chord and rod the chord shall enter the plate D' at an angle in combination with the counter and main braces, thereby rendering the structure less liable to fracture, the whole being constructed as and for the purpose as herein described.

No. 58,267.—ISAAC LAMPEUGH, Springfield, Ill.—*Shearing Apparatus.*—September 25, 1866.—The stationary blade is fitted in the rabbeted face of the lower section; the working blade is hinged to the rabbeted face of the upper section; the faces of the sections give support to the blades.

Claim.—The improved portable shearing machine, constructed as herein described and shown, as a new article of manufacture.

No 58,268.—RICHMOND A. LEEDS, Stamford, Conn.—*Gate Fastening.*—September 25, 1866.—Each sector-shaped tumbler has a trigger, and when down engages a keeper attached to the post.

Claim.—A latch for gates, &c., consisting of two sector-shaped plates D, hung eccentrically within a common casing or its equivalent, in combination with the fixed catch H, substantially as herein described.

No. 58,269.—JOHN B. H. LEONARD, Meriden, Conn., assignor to himself, LESLIE A. BELDING, same place, and JAMES D. FRARY, New Britain, Conn.—*Manufacture of Knives and Forks.*—September 25, 1866.—The tang is enclosed in a mould which is filled with a fluid metal; when the latter becomes set upon its outer surface, the still fluid interior portion is poured off through the gate at the extremity of the handle.

Claim.—The mode or process described of forming around and upon the tang of a knife or fork a hollow soft metal handle, substantially as described.

No. 58,270.—A. A. MARKS, New York, N. Y.—*Card Holder.*—September 25, 1866.—The edges of the cards are grasped between the flanges of the spring disks and the face of the board.

Claim.—The combination of the holding disk C, with spring F, to actute the same, as and for the purpose set forth.

No. 58,271.—DANIEL MASTEN, Bingham, N. Y.—*Level.*—September 25, 1866.—The level is to be attached to a stock and operates by gravity. The annular flange has a graduated circle of degrees. A wheel runs freely in the annular groove. From the centre of a wheel projects an axis, from which is suspended a semicircular bob with pointers to mark the horizontal and perpendicular.

Claim.—The combination of the ring A with its internal flanges A', piece B, and wheel C, substantially in the manner and for the purpose set forth.

No. 58,272.—JOHN W. MASURY, Brooklyn, N. Y.—*Packing Vessel for Petroleum.*—September 25, 1866.—The wooden case contains a metallic vessel; the divided lid of the former fits around the bail of the latter. Screws unite the case to its lid.

Claim.—First, so constructing a casing that when the can is enclosed therein the handle of the can will serve for both.

Second, the top I, with its openings z z, to receive the lugs E E of the can, constructed and arranged in the manner and for the purpose herein specified.

No. 58,273.—C. E. McBETH, Hamilton, Ohio.—*Boring Attachment to Lathes.*—September 25, 1866.—At the end of the hollow spindle is a bearing into which the end of a mandrel carrying a cutter is fed by the tail stock of the lathe. The wheel to be bored is carried by the face plate in the usual manner.

Claim.—An improved boring attachment to turning lathes formed by combining a thimble E, bush G, and cap or ring F, these parts being constructed and arranged as herein described with each other, with the hollow spindle B, and with the mandrel H, substantially as described and for the purpose set forth.

No. 58,274.—JOHN McDONALD, Hardin, Ill.—*Grain Drill.*—September 25, 1866.—The depth of the independently hinged openers is governed by weights. The seed-slide is operated by cams on the wheel hub.

Claim.—The independently hinged cutters A, levers G, weights H and tubes J, in combination with the slide P, lever M, and pins N, when constructed and arranged substantially as and for the purpose set forth.

No. 58,275.—JOHN McMURTRY, Lexington, Ky.—*Steam Generator.*—September 25, 1866.—The feed water pipe conducts the water from the lower part of the boiler to a series of evaporating pans in the upper portion of the boiler, from whence it overflows and trickles down through the steam, being thereby heated. Buckets are arranged in the generator to carry up the water and deliver it into the pans.

Claim.—First, the combination of the supply pipe C with the evaporating pans D D and E E, the latter having corrugated bottoms, the whole constructed and operating as herein set forth.

Second, the combination of the hinged cups J J, the arms n, the axle K, the stuffing box P, and the pulley L, the whole constructed and operating substantially as herein specified.

No. 58,276.—GEORGE W. MERCHANT, La Porte, Ind.—*Beehive.*—September 25, 1866.—The spare honey-box is placed over the entrance to the main hive and has a bottom entrance, so that the bees enter the box without passing through the main hive, and the honey-box may be removed without disturbance.

Claim.—First, the combination of the hive A and frame B with the honey-box C, in such a manner as to admit of the removal of spare honey in box C without disturbing the bees in the breeding hive, and also of the examination of the brood combs without disturbing the bees in the honey-box C, substantially as represented and described.

Second, the spare honey-box C, arranged over the bee entrance or passage E, substantially as and for the purpose set forth.

No. 58,277.—S. P. METZ and MARTIN ROHRER, McDonaldsville, Ohio.—*Corn Planter.*—September 25, 1866.—The ploughs throw soil upon the seed and the roller flattens it above them. The depth of covering furrow is adjusted by the standard over the roller frame.

Claim.—The upright B, frame C and support E, with roller D in combination with the plough beam A, the whole being constructed and arranged as set forth.

No. 58,278.—E. M. and J. E. MIX, Westfield, N. Y.—*Padlock.*—September 25, 1866: antedated September 15, 1866.—The seal plate swings on slides on a plane with the face of the lock and is fastened shut by a pin in one of the tumblers, which is only released by actuating the tumbler by a key whose introduction breaks the seal; the construction of the hinge of the escutcheon and the spring by which it is kept closed is explained in the claims.

Claim.—First, securing the seal plate G in position over the seal recess by means of the pin i on the tumbler C, so that when locked it cannot be released without moving the tumbler, substantially as set forth.

Second, in combination with the hinged escutcheon K, the spring l, pin m, and bevelled ear pieces o o, whereby the spring is made to constitute a self-closing hinge for the escutcheon, arranged and operating substantially as shown and described.

Third, the arrangement of the seal plate G in relation to the face A of the lock, and the escutcheon K, whereby the plate moves on a plane between the two, in the manner and for the purposes set forth.

Fourth, the combination and arrangement of the vertically rising escutcheon K with the recess f and swinging seal-plate G, whereby the former is made to conceal and protect the latter, substantially as shown.

No. 58,279.—J. U. MUELLER, Detroit, Mich.—*Handle for Stoves.*—September 25, 1866.—Explained by the claim and cut.

Claim.—A handle made of wood or other bad conductor of heat, and secured to one end of a radiating coil, the other end being connected to the door of a stove, all substantially as herein shown and described.

No. 58,280.—ANDREW NARAMOR, Utica, Mich.—*Hay Rack for Wagons.*—September 25, 1866.—The ladder is adjustable, and in connection with the stakes, the pulleys, and the pole, holds the load on the bed.

Claim.—The adjustable stakes B C and ladder e, in combination with the rack a, box A, and the pulley or windlass at each end, arranged and operating as and for the purpose substantially as set forth.

No. 58,281.—M. V. OLRY, Philadelphia, Penn.—*Cap for Bottles.*—September 25, 1866.—The neck of the bottle is coated with cement and the cap is placed over it to receive a stopper or lid.

Claim.—The within-described neck cap for bottles, the same consisting of the two tubes A and b, connected together at the top and constructed for attachment to the neck of a bottle, substantially in the manner described.

No. 58,282.—H. S. PALMER, Norvell, Mich.—*Machine for Raking and Loading Hay.*—September 25, 1866.—The endless elevator has a series of laterally sliding rakes for the purpose of carrying the hay inward as it is elevated, the same being used in combination with a rake for gathering the hay from the ground.

Claim.—The endless, flexible apron or elevator provided with reciprocating teeth, as above described, in combination with a rake for gathering and depositing hay upon the load, for the purposes and substantially as set forth.

No. 58,283.—ISAAC E. PALMER, Hackensack, N. J.—*Frame for Mosquito Canopies.*—September 25, 1866.—The pivoted extension arms are maintained in their displayed position by the screw-clamping disk above the hub on which the arms are hinged.

Claim.—The combination of the screw cap B, skeleton centre piece A, and radial arms C, substantially as herein described, whereby the said arms may be folded either in an upward or downward direction.

No. 58,284.—EZRA PERIN, Connersville, Ind.—*Roller and Seed Planter Combined.*—September 25, 1866.—The seed slide is operated and the seed dropped by a wheel which falls into the check furrows previously made at right angles; ploughs in the rear of the roller cover the seed.

Claim.—First, the combination on one frame of a seed planter, rollers B, and furrowing ploughs C, when arranged substantially as set forth.

Second, the combination of the wheel I, rod H, bell crank G, and slide F, with the seed box E, the parts being constructed and arranged substantially as set forth.

No. 58,285.—J. W. PETTENGILL, Rockford, Ill.—*Moving Cars on the Track.*—September 25, 1866.—The clamp is attached to the car; when the lever is raised the grapple slips forward on the rails and when it is depressed the grapple bites the rail and the car is drawn forward.

Claim.—The lever D with arm or handle E attached, and a rod F with clamp F secured to it, in connection with a grapple A, constructed and arranged as shown and described, or in any equivalent way, so that it will grasp the rail and slip or slide thereon in one direction under one movement of the arm or handle E and remain fixed or stationary on the rail under the opposite movement of the arm or handle, substantially as and for the purpose set forth.

No. 58,286.—DAVID N. PHELPS, San Leandro, Cal.—*Animal Trap.*—September 25, 1866.—A thick wire forms the jaws and mainspring of the trap; at one end of the wire is an indentation, against which the other end is held by means of a smaller wire which sets and operates the trap.

Claim.—A trap constructed of wire or iron A, with mainspring B, semicircle C, parallel straight arm D, with bent head F E', said arm crossing the semicircle and forming the hold' when sprung; when set, held in place by the bent arm H, and sprung by bait H', substantially as described and for the purposes set forth.

No. 58,287.—MORRIS POLLAK, New York, N. Y.—*Watch-chain Hook.*—September 25, 1866.—The hook is divided, the points stand in opposite directions, and the parts are united by a rivet so as to move apart sidewise; the pendent ring of the watch is hooked one way into one hook and the other way into the other book, after which it is closed.

Claim.—The book for watch-chains and other articles, formed double, with the points standing in opposite directions and attached to each other by a pin or rivet that allows said hooks to separate sidewise, in the manner and for the purposes set forth.

No. 58,288.—ISAAC T. PRICE, Leesville, Ohio.—*Churn.*—September 25, 1866.—The base board is double, the slotted upper portion admitting the vibration of the flexible supports of the churn box above.

Claim.—The springs G, in combination with a cream vessel, substantially as and for the purposes set forth.

Also, the base board E E', constructed and operating substantially as and for the purposes set forth.

No. 58,289.—W. F. QUINBY and GEORGE G. LOBDELL, Wilmington, Del.—*Rotary Digger.*—September 25, 1866; antedated September 10, 1866.—The curved tooth is flat in front and ribbed at the back; it fits a recess cut in the cross-bar adapted to the back of the tooth, against the front of which latter a plate is secured by bolts.

Claim.—First, the use in rotary diggers of teeth bent to the curve described, for the purpose specified.

Second, the curved tooth a, having the sectional form described, for the purpose set forth.

Third, the tooth a, adapted to a groove in the cross-bar B and secured thereto by a plate. D, as and for the purpose herein specified.

No. 58,290.—D. R. REED, Orangeville, N. Y.—*Sheep Holder.*—September 25, 1866.—A fixed bed and movable head rest between vertical wheels provided with clamps to hold the feet.

Claim.—The removable or detachable head rest F in combination with the fixed bed-piece H and the wheels D D', provided with clamps E and fitted on the screw shafts C C', all arranged substantially as and for the purpose set forth.

No. 58,291.—FREDERICK REHORN, New York, N. Y.—*Sad-iron Handle.*—September 25, 1866.—The foot of the handle slides beneath dovetail lugs on the iron and is retained by a pivoted lever and catch lug. A wooden handle embraces and is pinned to the metal.

Claim.—First, the handle B, consisting of the top cross-bar d, standard b, and bottom cross-bar e, cast in one piece, and applied in the manner and for the purpose specified.

Second the grooved wooden handle C, in combination with the top cross-bar of the metal handle B and with pins or other suitable fastenings, substantially as and for the purpose set forth.

No. 58,292.—UEL REYNOLDS, New York, N. Y.—*Body Top for Carriages.*—September 25, 1866.—A clip at the end of the body loop clasps the spring bar.

Claim.—The clip body loop, constructed as and for the purposes specified.

No. 58,293.—WILLIAM RHODES and GREENVILLE HAZLEWOOD, Bloomfield, Iowa.— *Ladies' Saddle-tree.*—September 25, 1866.—The cantle and pommel are united by side-bars and the seat suspended therefrom.

Claim.—A ladies' saddle-tree which is so constructed that while its bellying portions B B accurately conform to the horse's back without being stuffed or padded, the upper surfaces of said portions are elevated above the side-bars A A, all along the length of said bars, so as to form the requisite raised seat without being padded or stuffed on said surfaces; but this is only claimed when the elevated bellying portion and the bars of each side of the tree are of one piece and the two side-bars are united by straps, all substantially as set forth.

No. 58,294.—GEORGE W. RICHARDSON, Troy, N. Y.—*Steam Safety-valve.*—September 25, 1866.—As soon as the valve is lifted an increased area is exposed to the pressure of the steam to overcome the increased resistance of the spring under compression.

Claim.—A safety-valve with the circular or annular flange or lip *c c*, constructed in the manner or substantially in the manner shown, so as to operate as and for the purpose herein described.

No. 58,295.—JOSIAH S. RICKEL, Geneseo, Ill.—*Corn Planter.*—September 25, 1866.—As the plough is raised the seed-slide is depressed and the corn dropped; the jumper-share following covers it with earth.

Claim.—The lever frame H, with seed-slide G attached, in connection with the corn box or hopper F, arranged and applied to the corn coverer or jumper, substantially in the manner as and for the purpose herein set forth.

No. 58,296.—ALFRED RIX, San Francisco, Cal.—*Candlestick.*—September 25, 1866.—The candle-holder has two jaws, regulated to any size of candle by means of a trapezoid wire, which slips in sleeves on the outside of the jaws.

Claim.—First, the use of the jaws B B, one or both movable, or their equivalents, arranged with a space Z on each side of the candle, by which the candle can be adjusted vertically by the thumb and finger.

Second, the trapezoid *a b c d*, by which to operate the jaws, all constructed substantially in the manner and for the purposes set forth.

No. 58,297.—E. J. ROBINSON, Canton, Ohio.—*Saw.*—September 25, 1866.—Explained by the claim and cut.

Claim.—A saw constructed with elongated teeth in the manner described, for the purpose set forth.

No. 58,298.—E. S. ROBINSON, New York, N. Y.—*Car Wheel.*—September 25, 1866.—The side-plates are cast in one piece with the hub and the cross-pieces which connect their peripheries; the encircling tire is then secured by rivets parallel to the axis.

Claim.—The plate *a a* and cross-piece *c* and hub *b* cast in one piece, forming the open space as shown, in combination with the tire B, secured by webs D, substantially as described, for the purpose specified.

No. 58,299.—JOHN B. ROOT, New York, N. Y.—*Boring and Pumping Apparatus for Oil Wells.*—September 25, 1866; antedated September 10, 1866.—The engine cylinder is shifted laterally on the frame to remove it from its position above the well when tool, tubing, or pump is to be withdrawn.

Claim.—The combination of the horizontally-movable direct-action steam engine, the horizontal platform and guides on the derrick, and the boring-bar or drill-rod and pump, substantially as and for the purpose herein specified.

No. 58,300.—SHERMAN H. ROSE, Wheeler, N. Y.—*Fence.*—September 25, 1866.—The panels are supported by pins passing through holes in the posts and slots in the panels.

Claim.—The pins *a b* passing transversely through the posts A, in combination with the slots *f*, arranged in the ends of the removable panels B, substantially as herein set forth for the purpose specified.

No. 58,301.—P. A. ROYCE, Buffalo, N. Y.—*Mechanical Medicator.*—September 25, 1866.— Topical application of vapor, or liquid, or of powder mechanically suspended in air, are applied externally or internally by means of a double-acting pump connected to reservoirs of the medicament and delivering the same in a jet upon the part affected.

Claim.—First, the apparatus consisting of the pumping cylinder B, chambers F K, provided with induction and eduction valves and orifices, with the supplementary or generating chamber T connected with chamber F by the passage Y, the whole combined, arranged, and operating substantially in the manner and for the purpose herein set forth.

Second, in combination with the above-described apparatus, the perforated air-pipe G and sliding-cap W, operating in the manner and for the purpose specified.

Third, in combination with the said apparatus, the ozone generator, Fig. 14, constructed and operating substantially as described.

No. 58.302.—JOHN T. RYAN, Brooklyn, N. Y.—*Manufacture of Soap.*—September 25, 1866.—Vegetable gums and mucilage are combined with saponaceous compounds, to impart mildness, emolliency, and transparency, and prevent shrinking.

Claim.—The combination of solutions or mucilages obtained from Algaceæ Linaceæ Acacia, or such other vegetables or vegetable substances as do not turn blue when tested with iodine, with soaps or saponaceous compounds, substantially in the manner and for the purpose herein set forth.

· No. 58,303.—JOHN TAYLOR RYAN, Brooklyn, N. Y.—*Silicated Liquids for the Manufacture of Soaps, &c.*—September 25, 1866.—Vegetable gums, mucilage, &c., are combined with the silicates of potassa, soda, ammonia, alumina, or magnesia, for the purpose of reducing the specific gravity of said silicates and imparting viscidity.

Claim.—First, the combination of any vegetable, gummy, or mucilaginous substances or solution, with any one or more of the silicates of potassa, soda (neutral or alkaline,) ammonia, alumina, or magnesia, substantially in the manner and for the purpose herein set forth.

Second, the combination of animal gluten with vegetable mucilages or gums and with the silicates of potassa, soda, ammonia, alumina, or magnesia, substantially in the manner and for the purpose herein set forth.

No. 58,304.—CYRUS W. SALADEE and WILLIAM VEACH, Newark, Ohio.—*Mechanical Movement.*—September 25, 1866.—The wheel has an uneven number of projections and reciprocates the pitman arranged diametrically thereto. Anti-friction rollers on the pitman receive the impact of the projections, and the loops of the pitman slip on the axis of the wheel.

Claim.—First, broadly supporting and operating the pitman C K, (Figs. 1 and 2, Plate 1,) across and upon a line with the centre E of the corrugated driving wheel A, substantially as and for the purpose set forth.

Second, the arrangement of the corrugated wheel A in combination with the pitman C K and friction rollers B B¹, or their equivalent, in the manner and for the purpose substantially as shown and described.

No. 58,305.—GEORGE T. SAVARY, Groveland, Mass.—*Pea Sheller.*—September 25, 1866.—The contact of the rubber ring rotates the upper roller and preserves its distance.

Claim.—Combining a roller of a machine, as described, by the employment of a ring of vulcanized rubber, or other elastic or suitable material, under such an arrangement that while motion is communicated from one roll to the other by friction of contact there shall also be a sufficient space between the two rolls to effect the shelling of the peas, as herein described and set forth.

No. 58,306.—ALONZO SEDGWICK, Poughkeepsie, N. Y.—*Adjustable Wrench.*—Septembe 25, 1866.—The studs on the jaws are engaged by the grooves in the slotted shank; the jaws are expanded by a spiral spring and contracted by being drawn into the sleeve.

Claim.—The slotted shank having a groove T, in combination with the jaws B B, constructed and arranged as described.

No. 58,307.—DANIEL SEXTON, San Gabriel, Cal.—*Rotary Valve.*—September 25, 1866.—The steam is admitted to the cylinder on one side of a central partition in the thimble valve and exhausted on the other. The pressure thus exerted on the receiving side over that on the exhaust side is received upon a temper screw, which bears upon the end of the valve shaft.

Claim.—The arrangement of the ports c c′, port d, valve B, having partition E′, stem C, lever D, temper screw G, ports e e′, ports f f′, and cylinder A, constructed and operating substantially as and for the purpose represented and described.

No. 58,308.—J. W. SHANKLAND, Summerfield, Ohio.—*Fence.*—September 25, 1866.—The posts are supported on metallic blocks and foundation stones, and are braced laterally by horizontal arms and stakes.

Claim.—The arrangement of the post A, iron blocks E E, and foundation stone D, as described.

Also, the mode of securing the arms and post in position by the combination of blocks H, the keys I, arms G, and blocks F on the post A, substantially as described.

No. 58,309.—SIMEON SHERMAN, Weston, Mo.—*Water Wheel.*—September 25, 1866.—A lip on the inner edge of the scroll floor directs the water upward. The buckets are alternately long and short, and are attached to a shell adjustable on the hub of the wheel. The register gate below regulates the outflow.

Claim.—First, the upright ledge a, at the inner edge of the bottom scroll B, in combination with a wheel A, fitted within the scroll and exceeding the former in depth, substantially as and for the purpose set forth.

Second, the buckets C C′ placed alternately on the exterior of the wheel, and constructed and arranged in the manner substantially as shown and described.

Third, the attaching to or casting the buckets C C′, with a shell D, fitted on the body of the wheel and arranged with set screws E, by which the buckets may be adjusted higher or lower on the wheel, substantially as and for the purpose specified.

Fourth, the register G, constructed and applied as shown and described, to regulate the flow of water through the draught tube F, and prevent the upward reflux of water therein.

No. 58,310.—ALBERT M. SMITH, Brooklyn, N. Y.—*Button.*—September 25, 1866; ante-dated September 17, 1866.—The central pin pierces the garment, and the spiral coil is screwed into it to prevent direct retraction.

Claim.—A button having a compound shank composed of a central pin and a spiral fasten-ing revolving around said pin, all substantially as described.

No. 58,311.—W. MORRIS SMITH, Washington, D. C.—*Measuring Stopple for Canisters.*—September 25, 1866.—The cylinder of definite capacity at the top of the canister is filled, ro-tated and withdrawn without pouring from the mouth, and the introduction of more air than equivalent to the matter withdrawn.

Claim.—First, so constructing an oscillating or rotating measuring device for canisters that their contents shall not be exposed to the external air during the act of measuring and dis-charging the measured quantity, substantially as herein set forth.

Second, an oscillating measuring chamber C, or its equivalent, adapted to serve as a cut-off or stopple for a canister, and also as a means for measuring the contents of the same, sub-stantially as described.

No. 58,312.—LOUIS SOEHLMANN, Jersey City, N. J.—*Pepper Caster.*—September 25, 1866.—The finger lever on the outside of the perforated top of the caster agitates the arms on the inside to assist the discharge of the pepper.

Claim.—The finger bar D, the fan E, and the series of wipers G, when in combination with a pepper caster A, substantially as and for the purposes described.

No. 58,313.—WILLIAM B. SOUMEILLIAN, Philadelphia, Penn.—*Attachment for the Legs of Billiard Tables.*—September 25, 1866.—The foot has a sole of rubber and its upper section is rotated on a screw to raise or depress the leg of the table. The nut is turned by a spanner.

Claim.—The combination of the vulcanized India-rubber or other equivalent substance for a step, with a metallic base, hollow screw pillar, plate nut, and friction disk, substantially as and for the purposes herein set forth.

Also, in combination with the device above described, the double armed nib wrench or hand lever for operating the same, in the manner and for the purposes specified.

No. 58,314.—CHARLES L. SPENCER, Providence, R. I.—*Needle Threader.*—September 25, 1866.—The head of the needle is passed into a tube which presents the flat side of the head to the threading eye. The cap is raised and the threaded needle withdrawn without running it through the tube.

Claim.—A needle threader, the threading channel of which is divisible, all substantially as described.

No. 58,315.—CALEB S. STEARNS, Marlboro', Mass.—*Machine for Cutting Leather.*—Sep-tember 25, 1866.—The die stock is sleeved in a hinged frame, can be adjusted over any por-tion of the table beneath, and is depressed by a descending block.

Claim.—Hanging the cutting-out die to a movable frame, so that it can be brought over any portion of the table L, in combination with the presser block M, or equivalent device for pressing it down, substantially as set forth.

Also, the hollow shaft H, with its swivelling collar I, in combination with the die K and the rod g, substantially as described.

No. 58,316.—E. I. STEARNS, Honesdale, Penn.—*Siphon.*—September 25, 1866.—Explained by the claim and cut.

Claim.—A siphon, having its discharge end or nozzle enlarged, and provided with a sponge or other suitable filtering medium, substantially as described and for the purpose specified.

No. 58,317.—JOHN W. STILES, New York, N. Y.—*Means for Raising Water by Steam.*—September 25, 1866.—The water and steam pipes are connected by a brace bar, and the for-mer is closed by a valve while being cleansed of sediment by the steam jet.

Claim.—First, the arrangement of the nozzles or ends of the steam or water pipes, substan-tially as and for the purpose specified.

Second, arranging within the water or eduction pipe a valve, or its equivalent H, for the purpose of securing a back pressure in such pipe A, and the space which surrounds the lower end or nozzle B, thereof, by which any obstructions may be removed, substantially as described.

Third, the arm or brace E, or its equivalent, substantially as and for the object specified.

No. 58,318.—R. B. SUMMERS and S. DEMENT, San Jose, Ill.—*Plough.*—September 25, 1866.—The plough is attached in front to a crank adjusted by a lever and segment ratchet-bar, and regulated laterally in the rear by a spring rod passing through guide-plates on the beam.

Claim.—First, the manner of attaching the beam B to lever and crank G, and raising or lowering the plough D by the lever and crank G and ratchet F, as set forth.

Second, the rod g and spring h, as described and for the purposes set forth.

No. 58,319.—ANDREW THOMPSON, Ottumwa, Iowa.—*Rotary Cultivator.*—September 25, 1866.—The shaft of the rotary digger is suspended by screw rods from the bridge of the frame and is rotated by gearing from the ground wheel.

Claim.—First, the pendent rods J J, provided with bearings for the cylinder G, said bearings being guided in slots in the frame E, for the purposes and substantially as herein shown and described.

Second, the revolving cylinder in combination with the pendent rods J J and crowned braces H H, substantially as and for the purpose herein shown.

Third, the frame E, provided with the slots which guide the bearings of the pendent rods J J and braces, substantially as herein shown.

No. 58,320.—DANIEL J. TITTLE, Albany, N. Y.—*Corn Harvester.*—September 25, 1866.—In the upright side frames of the carriage are two pairs of rotary cutters, vertically adjustable by screw threads on their shafts. In rear of the cutters is an adjustable platform to receive the butts of the stalks. The reel is supported by slotted bars which are vertically adjustable, and the draught frames are connected to the axle outside of the supporting wheels.

Claim.—First, the revolving cutters herein described, arranged in respect to and operated independent of each other, as set forth, in combination with a machine for harvesting cane or corn stalks.

Second, the screw shafts F, in combination with the cutting apparatus of a corn or cane harvester, for the purpose of adjusting the height of the said cutting apparatus from the ground.

Third, in combination with the revolving cutters G, the adjustable platform P, substantially as and for the purpose set forth.

Fourth, the slotted bars M, in combination with the reel N, and cutters G and H, arranged as described, of a cane or corn harvesting machine, for the purpose of adjusting the height of said reel.

Fifth, in combination with the side bars A A, the platform B, provided with a vertical flange as shown, for the purpose of stiffening said platform and frame.

Sixth, connecting the draught frames a of a corn and cane harvesting machine, as herein described, to the outer ends of the main axles, so that each of said axles shall be supported at a point outside of the bearing wheel, substantially as and for the purpose specified.

Seventh, in combination with the main frame and cutting apparatus of a harvester for cane and corn stalks, the curved brace plate Y, to strengthen and support the lower end of the frame which contains the cutting apparatus.

Eighth, in combination with the main frame A of a harvester for cane or corn stalks, as herein described, the vertical frame D, substantially as and for the purpose set forth.

No. 58,321.—N. TREADWELL, New York, N. Y.—*Apparatus for Supplying Gas on Steamboats and other Vessels.*—September 25, 1866.—Gas is forced into the diaphragm holder by a pump which may be used for forcing the gas to burners by changing the valves.

Claim.—The stop-cocks or valves a b c d, in combination with the gas-supply pipe D, pump J, discharge pipe G, pipe C, and gas holder A, all constructed and operating substantially as and for the purpose described.

No. 58,322.—WILLIAM S. VAN HOESEN, Saugerties, N. Y.—*Sash Supporter.*—September 25, 1866.—An elastic ball lies in an inclined groove in the sash stile, and a metallic pin passes through the stile and into the jamb. When binding in the contracted portion of the groove the ball upholds the sash.

Claim.—The loose bolt or pin C, arranged relatively with the inclined recess B, in the manner described, and employed in combination with a ball or roller A, for the purpose of locking and sustaining a window sash as herein explained.

No. 58,323.—WILLIAM VEBER, Jr., Shingle Creek, N.Y.—*Plough.*—September 25, 1866.—The clearer is bent around the standard and coulter, is reciprocated by attachment to a wrist on the plough wheel, and acts to push obstructions from the coulter.

Claim.—The rod F, attached to a plough, when constructed and operated as herein shown, substantially and for the purpose as described.

No. 58,324.—WILLIAM P. WAGE, Bane Centre, N. Y.—*Sad Iron Heater.*—September 25, 1866.—The removable shield has a pipe to fit the flue flange of the lamp stove and slots for the handle of the sad irons when they are slipped in and out.

Claim.—The stove A, constructed as described, with an air chamber provided with a metallic case B, lamp C, and pipe E, substantially as and for the purpose herein set forth.

1234 ANNUAL REPORT OF THE

No. 58,325.—GUSTAVUS A. WARNER, Portland, Oregon.—*Flour Packer.*—September 25, 1866.—The revolving shaft has a spiral blade, and revolves in a cylinder placed in the barrel, which is supported upon a platform counterpoised by weights; as the barrel is filled the platform descends, raising the weight until it strikes a stop on the frame which cuts off the flour by actuating a spring gate.

Claim.—The arrangement of the stop a^2, spring sliding gate S, spring frame v, pulley d, and weight D, in combination with the carriage C, constructed and operating in the manner and for the purpose herein described.

No. 58,326.—G. F. WATERS, Waterville, Maine.—*Pruning Instrument.*—September 25, 1866.—The curved cutting blade is eccentrically pivoted to the stock of the sharp-edged book, and is oscillated by a lever handle and connecting rod.

Claim.—First, the combination of the eccentric cutter D with the hook B, when the parts are constructed and arranged to operate in the manner and for the purposes herein specified.

Second, the lever E in combination with the cutter D, substantially as and for the purpose herein specified.

No. 58,327.—THOMAS V. WEYMOTH, New York, N. Y.—*Envelope Machine.*—September 25, 1866.—The seal flaps of the blanks are gummed simultaneously with the end flaps, to save time in drying. The table is supported by hinged arms to facilitate the introduction of the blanks under the gummers. A curved guide fitting the edges of the blanks prevents their distortion. The seal flap, when folded, bears upon a wing which intervenes between it and the body of the envelope, and prevents adhesion. The blanks are protected by lips from the joint oil of the folding wings. The gummed envelopes are kept separate by the radiating blades of the traversing endless apron, to prevent adhesion, and are retained by a curved rail till they reach the receptacle, which has a movable follower to compress the envelope. A forked lever, acting in connection with a die, imprints the blank between gumming and folding. Dies on the creasing plunger, and the table which supports the flaps while folding, imprint the blank, which may also be printed by type while in the folder.

Claim.—First, gumming the seal flaps of the blanks for envelopes simultaneously, or nearly so, with the lower or end flaps, or during the time while the blank passes from the gumming to the folding mechanism, and by mechanism substantially such as herein described, or any other suitable mechanism which will produce the same effect.

Second, the arrangement of a curved guide T, in combination with the table A*, constructed and operating substantially as and for the purpose described.

Third, causing the seal flap, when folded, to bear on one or more of the folding wings, or on parts or projections of said wings, substantially as and for the purpose described.

Fourth, the protecting lips f^*, in combination with the joints of the folding wings, constructed and operating substantially as and for the purpose specified.

Fifth, the raised surface at or near the edge of the wing which folds the lower flap, substantially as and for the purpose described.

Sixth, the endless apron Q, with radiating plates or arms j', in combination with a suitable gumming and folding mechanism, constructed and operating substantially as and for the purpose set forth.

Seventh, passing the endless apron Q, at its receiving end, over a square or polygonal shaft l', substantially as and for the purpose described.

Eighth, the curved rail o', or its equivalent, in combination with the apron Q, constructed and operating substantially as and for the purpose set forth.

Ninth, the receiving box R and follower S, in combination with the discharging end of the endless apron Q, constructed and operating substantially as and for the purpose described.

Tenth, the lever arm O, in combination with the carrying platform N, and with a suitable die inserted or secured to said lever arm or to the platform, or to both, substantially as and for the purpose set forth.

Eleventh, the arrangement of dies s on the creasing plunger and on the folding table, or on either, substantially as and for the purpose described.

Twelfth, the types u, arranged in an arm r, and operating in combination with folding table P and plunger I, substantially as and for the purpose set forth.

No. 58,328.—WILLIAM H. WHITE, Kent Island, Md.—*Hat.*—September 25, 1866.—The edge of the hat brim is turned over, confined by a draw-string, and distended by an enclosed hoop.

Claim.—First, in combination with an open draw casing, or other open casing, placed at the outer edge of the brim of the hat, as described, the hoop or form of metal, or other suitable material, for forming and holding in shape said brim, constructed and arranged for operation substantially as herein shown and set forth.

Second, as a new article of manufacture, the hat or cap constructed as herein described and set forth.

No. 58,329.—WILLIAM H. WHITE, Kent Island, Md.—*Hat.*—September 25, 1866.—Adjustable crown and brim rings, are kept apart by a spring to distend and give form to the hat crown.

Claim.—First, as a frame for shaping hats and caps made of soft or limber material, the combination of the crown and rim hoops with a distending spring, which, at the same time, shall admit of being compressed, so that the two hoops may be brought into juxtaposition, as and for the purposes set forth.

Second, the combination of the distending spring, as above set forth, with the adjustable crown and rim hoops, so arranged that the frame may be adjusted to hats of different dimensions, as herein shown and described.

Third, in combination with the hat or cap lining, provided at its upper end with draw casing and strings, as described, the hoop or form for shaping the top of the lining, as herein shown and set forth.

No. 58,330.—J. L. WINSLOW, Portland, Maine.—*Shaft Coupling.*—September 25. 1866.—
The bevelled flanges of the collars upon the ends of the sections of shafting are embraced by semi-annular pieces, held together by a bolt, the head and nut of which are sunk within a circumferential groove and covered by an annular plate.

Claim.—The combination of the flanges bevelled and slotted as described ; the collars fitting thereupon ; the bolt and metallic hoop, all constructed, secured, and operating to form a coupling for shafts, as herein set forth.

No. 58,331.—J. E. YOUNGMAN, Rockford, Ill.—*Evaporator.*—September 25, 1866.—The pans are arranged upon a furnace with double walls. The grate has shifting bars raised alternately by cogs on a shaft, to dislodge the ashes. The juice is boiled in the upper pan, which has double walls forming a flue space. It then passes by a conduit into the straining chambers, and from these into the second pan, and thence into the finishing pan, which has a strainer. Under the last pan is a slotted register partition, which separates it from the fire chamber when required.

Claim.—First, the pans S U Z, when arranged in relation to each other as described, in combination with the perforated strainers and skimmers connected with said pans, substantially as and for the purpose set forth.

Second, the skimmer V' and strainers W X, in combination with the pans U, when arranged and operating substantially as and for the purpose set forth.

Third, the dampers, arranged as described, in combination with the fire-box A, and grate bars Fig. 5, as and for the purposes set forth.

Fourth, the chamber 1'' in combination with the pans S, double flues, arranged in the manner and for the purpose set forth.

Fifth, the flue L, with double sides and bottom in combination with the dampers D' M and pans, as and for the purpose set forth.

Sixth, the vents d' and pans S, in combination with the conductor e, strainer V, and pan U, as and for the purpose set forth.

No. 58,332.—ARTHUR W. BROWNE, Brooklyn, N. Y., assignor to himself and JOSEPH L. MOSS, New York, N. Y., and EDWARD W. MOSS, Brooklyn, N. Y.—*Button.*—September 25, 1866.—The loop of the shank rests behind the cloth, and the barbed ends pass through a hole in the button, and, expanding, are there retained. By pressure on a stud on the central face of the button, the barbed ends are pressed together, and may be disengaged.

Claim.—First, the combination of the button A. having shouldered opening a, flanged stud b, having conical depression in its bottom, and slotted plate c, substantially as described for the purpose specified.

Second, the stud or pin b, for the purpose of detaching the button from the dress, and its combination with the wire spring fastening B, constructed substantially as herein shown and described.

No. 58,333.—SAMUEL BURR, New York, N.Y., assignor to himself and DAVID CONLAN, same place.—*Fan Attachment for Sewing Machines.*—September 25, 1866.—Explained by the claim and cut.

Claim.—The employment of the vibratory fan L, when arranged with the supporting standards I I, and driving arms k h, and the treadles of the machine, all as hereinbefore specified.

No. 58,334.—W. H. BURRIDGE, Cleveland, Ohio, assignor to HENRY CARTER. Aylmer, Canada.—*Potato Digger.*—September 25 1866 —The flanged edge of the wheel runs alongside the row ; the hill is dug by the fork and thrown against the rake, the dirt falling through the intervals ; the potatoes are carried up between the rake and the belt on the wheel, over to the front part, and discharged from the table over the beam into a bag or in a row on the ground.

Claim.—First, the rake I, wheel B, provided with flanges and cutters, belt K, fork b, in combination with adjustable table G and guide G', for the purpose and in the manner set forth.

Second, the share F, saw-toothed cutter e, scraper M, section N, and adjustable guide G', when arranged to operate in the manner and for the purpose described.

Third, the wheel B, spuds e, scraper M, share F, adjustable table G and guide G', when arranged in the manner and for the purposes set forth.

No. 58,335 —S. FREDERICK CHARLES, Dahlonega, Ga., assignor to himself and S I. RUS-
SELL.—*Amalgamator.*—September 25, 1866.—The inclined "panners" are suspended by
rods, are oscillated by suitable machinery, and discharge into a trough which conducts the
ore and water to a grinder, where the ore is ground, and is thence carried to a series of
amalgamating boxes, each of which consists of a case containing a series of copper pans
placed one above the other.

Claim.—First, the so combining a panning machine, a regrinding machine, and an amalgamator, that the gold shall be thoroughly extracted by the continuous and connected action
of all of them, substantially as set forth.

Second, the adjustable partition B, constructed and operated substantially as and for the
purposes set forth.

Third, the deep radiating channels *j* in the lower surface of the grinder K.

Fourth, the combination of an upper revolving grinder with an oscillating lower one, constructed and operating substantially as specified.

Fifth, the case or shell M of the grinders, in combination with the rollers I and the reciprocating arm or lever O.

Sixth, the amalgamator R, constructed so as to form one large and one narrow compartment, by the insertion of the removable and adjustable partition *o*, substantially as specified.

Seventh, the revolving frame provided with amalgamating pans suspended below the surface of the auriferous mass, constructed and operating substantially as and for the purposes
specified.

Eighth, the location of the flue below the series of amalgamators, substantially as and for
the purposes specified.

No. 58,336.—DWIGHT M. CHURCH, Derby, Conn., assignor to himself and GEORGE R.
BAILEY, same place.—*Ladies' Skirt.*—September 25, 1866.—Explained by the claim and cut.

Claim.—An improved lady's skirt, formed by combining a detachable lower part *c* with
the body or main part A of the skirt, by fastenings concealed by a tuck *a'*, substantially as described, and for the purpose set forth.

No. 58,337.—HORACE CLIFT, Mystic river, Conn., assignor to E. BURROWS BROWN,
Groton, Conn.—*Churn.*—September 25, 1866.—The blades are inserted in the ends of the
radial arms; the shaft is secured in place by a pin at one end and a clutch at the other. A
current of air passes through the churn.

Claim.—The construction of the dasher of radial arms *a a*, having independent blades *c*
arranged at right angles to them, substantially as described.

Also, the combination of the button-head pin A and turn button i with the tenon collar *d*,
shaft E, and key pin *e*, in the manner and for the purpose described.

Also, the arrangement of the rotary dasher, constructed as herein described, in combination
with the air ducts J J, and exit air pipe N, arranged as described, all for the purpose set forth.

No 58,338.—A. B. CRAWFORD, Piqua, Ohio, assignor to himself, JOHN O'FERRALL, and
THOMAS L. DANIELS, same place.—*Threshing Machine.*—September 25, 1866.—An angular
dividing board is placed in the interval between the grain belt and straw carrier. Arresting
pins on a movable frame agitate the passing straw. The peculiar construction of the links of
the carrying chain is described in the third claim. The forked rod is attached to the shaking
shoe to secure steadiness.

Claim.—First, the angular dividing board L, applied and operating substantially as described and represented.

Second, the arresting pins K, on the movable frame M, as described.

Third, the link, Fig. 2, with its corresponding pintle and socket, the socket for the reception
of the journal of the round and lug limiting its rotation by the impingement of its projection *e'*,
as described.

Fourth, the forked rod for shaking the shoe, as described and represented.

No. 58,339.—JOHN C. EGGLESTON, Waterbury, Conn., assignor to himself, E. M. HITCH-
COCK, and G. W. BEACH.—*Lubricating Apparatus.*—September 25, 1866.—A toothed wheel,
actuated by a pin in the end of the axle. carries oil to the journal by means of a brush.

Claim.—First, the brush D, and toothed wheel A, combined and operating substantially
as described for the purpose specified.

Second, the pintle *a*, in combination with the wheel A, brush D, and oil chamber B, all
substantially as and for the purposes set forth.

No. 58,340.—DENNIS FRISBIE, New Haven, Conn., assignor to himself and SAMUEL C.
GOODSELL.—*Hoisting Apparatus.*—September 25, 1866.—The hoisting chain is suspended
over pulleys on the truck, and the hoisting hook connected by a block thereto, so as to allow
the hoisting chain to traverse in the sheaves as the carriage is moved out and in on the arm
of the crane, by a secondary chain.

Claim.—The traversing frame C, when supported by the truck wheels D, running on the
horizontal arm B of a crane, and moved by the pulley E, through the spur wheels G and H,

and pinion K, and rack L, and bearing suspended the load, attached to the movable pulley P, from pulleys M and M', over which the chain Q passes, said several parts being arranged substantially as set forth.

No. 58,341.—SAMUEL C. GOODSELL and DENNIS FRISBIE, New Haven, Conn., assignors to themselves and D. P. CALHOUN.—*Cotton and Hay Press.*—September 25, 1866.—The ropes which draw upon the chains and raise the follower are wound upon a snail drum and have a differential speed, decreasing as the follower rises. The ratchet and attendant devices, as well as the eccentrically hinged doors, are described in the claims and cut.

Claim.—First, in cotton, hay, and other like presses, the combination with a movable platen operated by levers as described, of a differential drum or shaft for imparting motion to the levers at a speed inversely proportionate to the resistance, as herein shown and set forth.

Second, in a cotton, hay, or other like press, the combination of the following elements: 1st, a movable platen; 2d, levers to operate the platen; 3d, a differential drum to operate the levers.

Third, in combination with the pawls for actuating the ratchet wheel, the oscillating or rocking trips for disengaging the pawl from the wheel, under the arrangement shown and described, so as to be operated by the eccentrics upon which the said pawls are mounted.

Fourth, in combination with a stationary or inwardly yielding end plate, as described, a door, or doors, mounted eccentrically upon hinges in such manner that they may be moved laterally to become engaged with or disengaged from the said end plate, substantially as herein shown and described.

No. 58,342.—RICHARD H. GRAY, Greenville, Ala., assignor to himself and S. ABRAMS, same place.—*Lifting and Pressing Screw.*—September 25, 1866.—The screw traverses in a nut enclosed by a two-part sleeve in the divided beam; the nut bears upward against the sleeve, anti-friction balls being interposed between their surfaces.

Claim.—The combination and arrangement of the cast iron screw B b and cast hollow nut D, provided with cup G, with the two-part cast collar H, provided with flanges f and i, the whole being constructed and operating with friction balls I, in the manner and for the purpose set forth.

No. 58,343.—DANIEL B. HALL, Bucksport, Me., assignor to himself, M. G. WILEY, and C. J. COBB, same place.—*Recumbent Chair.*—September 25, 1866.—The back recedes, the segment guides in the frame directing its motion, which stretches the sacking and raises the foot-rest by the combination of levers beneath.

Claim.—The above-described improved recumbent chair, consisting of the supporting frame A, the movable back frame B, the sacking C, the leg rest, and the levers, arms, and rollers, as explained, or their mechanical equivalents, constructed, arranged, and applied substantially in manner and so as to operate together as hereinbefore specified.

No. 58,344.—WILLIAM H. HALSEY, Hoboken, N. J., assignor to DAVID N. ROPES.—*India-rubber Neck-tie.*—September 25, 1866.—Explained by the claim and cut.

Claim.—The arrangement of the band of metal, or its equivalent, in a neck-tie, formed of hard India-rubber or similar material, under the external strip or band passing around the bow and neck-tie, for confining together and holding the several parts of the neck-tie and preserving the external band from fracture, substantially as above described.

No. 58,345.—MOSES HAWKINS, Derby, Conn., assignor to R. M. BASSETT, T. S. BASSETT, and MOSES HAWKINS.—*Shaft Coupling.*—September 25, 1866.—T-headed keys occupy longitudinal slots in the divided axle and prevent torsion when retained by the sleeve, which also forms the bearing.

Claim.—The use, in combination with the surrounding sleeve and the slotted shaft's ends, of a key E, so shaped and the whole so arranged as to lock the shaft's ends longitudinally together and key their both in the sleeve, substantially as set forth.

Also, forming the journals of the shafting of the coupling sleeves C, substantially as hereinbefore described.

No. 58,346.—BENNET HOTCHKISS, New Haven, Conn., and HENRY SHATTUCK, Hamden, Conn., assignors to BENNET HOTCHKISS.—*Forge Furnace.*—September 25, 1866.—A forge furnace is combined with a reverberatory furnace. A current of air is passed through the hollow furnace doors and conducted by tubes to the ash pit. A blast of air from a pipe is forced into the furnace when the doors are opened.

Claim.—First, combining a forge A with a reverberatory furnace C, in the manner substantially as described for the purpose specified.

Second, applying a current of air, whether set in motion by an artificial blast or by the chimney draught, to cool the furnace doors and to furnish a draught to the fire, in the manner substantially as described.

Third, the use of an air blast to prevent the flames and gases from belching from the furnace when the door is opened, arranged and applied substantially as described.

Fourth, the combination of a cistern K', arranged as described, with a furnace A, for the purpose of converting into vapor a body of water by the waste heat of the furnace, and applying such vapor to stimulate combustion, substantially as set forth.

Fifth, combining with the escape flue J' a space L for the circulation of air, and connecting the same with the furnace for aiding the draught of the fire, substantially as described.

No. 54,347.—SAMUEL J. KELSO, Detroit, Mich., assignor to himself and JAMES EDGAR, New York, N. Y.—*Ciphering Machine.*—September 25, 1866.—Adapted for adding, subtracting, and multiplying. For the two former, wheels are used which revolve on studs beneath a perforated plate. The face plate has semicircular slots numbered on each side, for addition and subtraction respectively. The actuating mechanism is a series of compound pawls, and the train of wheels are connected decimally. The multiplying device is in a carriage fitted to the case. The action cannot be intelligibly described in brief.

Claim.—First, the face plate A, with segmental slots a a^1 a^2, and orifices e e^1 e^2, in combination with the wheels b b^1 b^2 and pointer c, all constructed and operating in the manner herein shown and described.

Second, the jointed rods h i, with noses g and teeth k, in combination with the toothed wheels b b^1 b^2 and pins f, constructed and operating substantially as and for the purpose set forth.

Third, the disengaging slide q, in combination with the carrying mechanism, constructed and operating substantially as and for the purpose described.

Fourth, the multiplying slide D, with strips u u, in combination with the adding mechanism, constructed and operating substantially as and for the purpose set forth.

No. 58,348.—GEORGE H. MANLOVE and J. P. GREEN, Chicago, Ill., assignors to GEORGE H MANLOVE — *Corn Harvester.*—September 25, 1866.—The machine cuts the ears from the stalks, husks, and delivers them into a receiver. Upon a frame vibrating upon the forward axle are yielding rollers, between which the stalks pass and are carried back to a reciprocating cutter, which removes the ear. The latter is then carried by an endless belt to the husking cylinder; here it is held by wire guards over the cylinder until the husks are stripped off, when it falls into a box beneath.

Claim —First, the forked frame K, in combination with the roller R, yielding rollers P Q, and cutters S, substantially as described.

Second, and in combination with the above, the endless belt N, substantially as described

Third, the husking cylinder armed with teeth, as described, and bent rods i k h, in combination with the frame K, cutter S, and endless apron f, all constructed and operating substantially in the manner and for the purpose set forth.

Fourth, and in combination with the above, the partitioned box V, substantially as described.

Fifth, the frame A, lever L, standards M N, forked frame K, with its rollers P Q R, cutters S, and endless belt f, in combination with the busking cylinder p, rods h h i i, and partitioned box V, the whole being constructed and operated substantially in the manner and for the purpose set forth.

No 58,349.—GEORGE H. MANLOVE and J. P. GREEN, Chicago, Ill., assignors to GEORGE H. MANLOVE.—*Corn Harvester for Cutting and Shocking Corn.*—September 25, 1866.—The cut stalks fall upon a guard, from which they are taken by an endless elevator, carried upward, and discharged into a receiver upon a turn-table, where the shock is bound, the table turned, and the shock discharged therefrom in a standing position. A brief description of the several devices by which this is accomplished is impracticable.

Claim —First, the guard H, in combination with the rods M, pins o o for receiving the stalks, curved rods I, and cutters a, substantially as described.

Second, and in combination with the above, the endless apron, cutters, and reel, situated on the adjustable frame L, the whole being constructed and operated substantially in the manner and for the purpose set forth.

Third, the table A', connected to the frame of the machine by a universal joint, and having rods e e attached to the said table, and rods e' e' attached to the hinged piece g, and cords or bands ff, the whole being constructed and operated substantially in the manner and for the purpose set forth.

Fourth, the guard H, cutters a, rods M and I, and pins o o, in combination with the endless conveyer N and table A', constructed as described, the whole being operated substantially in the manner and for the purpose set forth.

No. 58,350.—ALONZO McMANUS, New Britain, Conn., assignor to NORTH AND JUDD MANUFACTURING COMPANY, same place.—*Snap Hook.*—September 25, 1866.—The spring is placed beneath the bridge, its end extending into the recess at the rear of the book. The bridge being flattened down binds the spring in place.

Claim —Constructing the recess a, and combining therewith the solid bridge c, arranged transversely over the spring D, in connection with the slot b, substantially in the manner and for the purpose as herein described.

No. 58,351.—JOSHUA MONROE, New York, N. Y., assignor to himself and JETUR GARDINER, same place.—*Artificial Limb.*—September 25, 1866.—A trifurcated strap connects the thigh with the lower leg. The principal tendon connecting the thigh with the foot is of rigid metal. The "tendo achillis" is fastened at the exterior of the heel, and its upper end is connected to the posterior portion of the ankle by a button which slides freely up and down in a slot made perpendicularly in the ankle. The toe joint is a leaf hinge, explained in the claim and cut.

Claim.—First, the elastic side straps *a a*, in combination with straps *d b* and with the limb A, constructed and operating substantially as and for the purpose set forth.

Second, the arrangement of the tendon C made of rigid material, in combination with a knee joint of an artificial leg, constructed and operating as and for the purpose described.

Third, the button *o* and mortise *p*, in combination with the ankle tendon D, constructed and operating substantially as and for the purpose described.

Fourth, the toe plate *q*, applied in combination with the toe joint, substantially as and for the purpose set forth.

No. 58,352.—GEORGE ROOS and MICHAEL WHITE, Buffalo, N. Y., assignors to themselves and CHARLES W. DANIELS.—*Pad Hook.*—September 25, 1866.—The jointed spring hook prevents the accidental disengagement of the bearing rein.

Claim.—Forming a pad hook for harness in two parts or jointed sections *a b*, the one rigidly attached to the tree and the other movable on its pivot axis, in combination with a spring *d* and bearing *e*, or its equivalent, constructed and arranged substantially as herein set forth.

No. 58,353.—H. L. TAYLOR, Fredonia, N. Y., assignor to himself and CHARLES W. DANIELS.—*Shifting Cutter Thill*—September 25, 1866.—The draught bar on the runners has three eye-bolts, and the thill bar an equal number. The thill is shiftable to allow the horse to travel in the centre or in one of the double tracks, and 1 is retained in place by a spring.

Claim.—The combination of the rod E, double-acting spring I, or its equivalent, and eye-bolts *a b e* and *d e f*, arranged and operating substantially in the manner and for the purpose herein set forth.

No. 58,354.—MAURICE VERGNES, New York, N. Y., assignor to himself and ALPHONSE PERRIN.—*Lubricating Cup for Steam Engines.*—September 25, 1866.—The oil cup discharges into a chamber communicating with the air. A one-fourth revolution of the chamber around the stem closes the upper communications to the chamber, and opens another through the stem to the valve chest.

Claim.—First, the arrangement of the movable oil vessel or reservoir F, in combination with the cup B and oil passages *a* and *b*, and their connections with such reservoir E, the whole arranged and operating substantially as and for the purposes set forth.

Second, the arrangement of the openings *g g* in the fixed plate *c*, in combination with the openings 3 3 in the oil vessel E, substantially as and for the purposes set forth.

No. 58,355.—CASSIUS M. WERNER, Rockford, Ill., assignor to himself and EDWIN A. BIGELOW, same place.—*Horseshoe.*—September 25, 1866.—A band encircles the front part of the hoof having at each end projections to enter the hoof and T-catches to enter the shoe. A clip holds the band in front.

Claim.—First, the combination of the band F with the shoe A, constructed, arranged, and operating substantially in the manner described for the purpose set forth.

Second, the combination of the ribs J with the band F, arranged and operating substantially in the manner and for the purpose set forth.

No. 58,356.—EUGENE CANDLER, London, England.—*Faucet.*—September 25, 1866.—A removable valve and valve seat are placed at its inner end to prevent introduction of liquids from without. The air inlet is opened and closed by the rotation of the spigot.

Claim.—The tap, as above described, having the body *p*, plug *f*, air inlet *e*, valve *i*, and valve seat *g*, all arranged substantially as and for the purpose set forth.

No. 58,357.—F. A. LAMONTAGUE, Montreal, Canada.—*Card Case.*—September 25, 1866.—The cards are elevated by a spring, and pushed consecutively from the slit in the end of the case by a traversing spring catch.

Claim.—First, the spring slide *o*, provided with bevelled teeth so hung in the race *m* and combined with the spring *c* as to press and hold the card on its forward motion and slide back over the cards without contact, substantially as described for the purpose specified.

Second, the spring *e* and pin *f*, in the forward part of the box A, guiding the card in its discharge, and preventing more than one card from being forced out at the same time, substantially as described for the purpose specified.

Third, the elliptic springs *a a*, slotted plate *b*, having lugs *c*, spring *e* and pin *f*, slotted plate *i* and toothed spring slide *o*, arranged in combination with the slotted lid B and case A, provided with guide and holding laps *s* and grooves *d*, operating substantially as described for the purpose specified.

No. 58,358.—GUSTAVE MICHELET, Brussels, Belgium.—*Scouring Wool.*—September 25, 1866.—The "suint" (natural grease) obtained from soaking the wool is made the agent in scouring it.

Claim.—The combined process for extracting from the wool itself all the elements necessary for cleaning scouring, or removing the suint from the same, and of obtaining by the same operation liquids free from foreign matters, which would decrease the value of the products derived from them, and sufficiently concentrated for forming these sub-products with advantage, substantially as above described.

No. 58,359.—JOHN PETRIE, Jr., Rochdale, England, and JAMES TEAL, Towerly, England.—*Lifting Cylinder of Wool Washing Machines.*—September 25, 1866.—An improvement on Petrie & Taylor's patent of February 7, 1865. The levers which cause the teeth to protrude from and recede within the cylinder are caused to operate by having their inner ends within a circular groove on a disk mounted on a crank at the centre of a fixed shaft. The cylinder is concentric with and revolves about this shaft.

Claim.—The grooved disk g, right angular arms i, and levers k, in combination with the crank f, shaft c, prongs m, and cylinder a, constructed and operating substantially as and for the purpose described.

No. 58,360.—LEWIS DRESCHER, Matanzas, Cuba, assignor to GUSTAVUS MEYER, same place.—*Pump.*—September 25, 1866.—The double-headed, concave-faced piston reciprocates in a horizontal cylinder, whose upper edge is slotted to admit the operative arm of the rock shaft. The packing disks are sprung into place on the piston heads. The induction is at the ends, eduction above; each opening valved. .

Claim.—The cup-shaped pistons B, with packing disks f, in combination with spring disks g, applied and operating substantially as and for the purpose set forth.

No. 58,361.—GUSTAV STOBWASSER, Berlin, Prussia, assignor to E. DOUGLAS, SON & COMPANY, New York, N. Y.—*Lamp.*—September 25, 1866.—To prevent spilling the oil, the lower part of the fountain and the socket in which it is inserted are so arranged that the fountain cannot be inserted in or removed from the socket without first closing the discharge port.

Claim.—The pin i, attached to the lower cylindrical part e of the fountain, and having a cylinder f fitted loosely on e, and provided with a vertical rib a", having a notch b" made in it, and a horizontal slot c", in combination with the vertical groove or recess b, and the horizontal groove c in the cylinder G, which is secured in the socket F, the cylindrical part e of the fountain, and the cylinder f having holes g h made in them, and all arranged to operate substantially in the manner as and for the purpose set forth.

No. 58,362.—JOHN ABSTERDAM, New York, N. Y.—*Condenser for Steam Engines.*—October 2, 1866.—A partial vacuum is obtained in the condensing vessel by means of steam jets direct from the generator, and so arranged as to exhaust the air and vapor therefrom: a column of water is aided by a jet of steam, which acts upon the mouth of the water discharge pipe so as to increase the momentum of the outflowing water and create a partial vacuum.

Claim.—First, a condenser for condensing the exhaust steam of steam engines and other steam vessels, wherein the vacuum is produced and maintained by the direct action of one or more jets of steam from a steam boiler or generator, thereby condensing the exhaust steam in vacuo and creating and maintaining a vacuum in said condenser by the direct action of steam instead of the ordinary air pump, substantially as above described.

Second, a condenser for condensing the exhaust steam of steam engines and other steam vessels, wherein the vacuum is produced and maintained by the direct action of steam aided by the columns of water of injection and discharge, dispensing with the use of the air pump, substantially as herein specified and set forth.

No. 58,363.—JAMES ADAIR, Pittsburg, Penn.—*Spring Pen Rack.*—October 2, 1866.—A spiral wire is attached by its ends to a bed-plate and receives pen holders, &c., between its coils.

Claim.—The spring pen rack having its springs arranged close together, made self-connecting or in continuity, and formed of wire properly curved and having its bed and fastening rod as made, used, and applied, all substantially as and for the purpose set forth.

No. 58,364.—ROBERT ADAMS, Cincinnati, Ohio.—*Slide Lacing and Shoe Fastener.*—October 2, 1866.—The strings pass through a jointed slide fastener, which being moved forward unlaces the fastening, and being moved backward tightens the lacing.

Claim.—First, the jointed slide fastener C, as constructed and operating for the purpose set forth.

Second, so arranging the lacing D, that it acts as a guide for tongue B and fastener C.

Third, slide fastener C, tongue B, as arranged in combination with lacing D, as constructed and operating for the purposes set forth.

No 58,365.—CHARLES L. ALEXANDER, Washington, D. C.—*Paper Fastener.*—October 2, 1866.—The plates, or the ends of the plate, are passed to their mid-length through an incision in the paper, and the ends on each side bent in different directions and flattened against the paper.

Claim.—A clasp constructed substantially as described, whether of two plates or of one plate, so that the attachment is made by the passage of the plate or plates through an incision, and the turning down of the plate or plates upon the surface of the paper or other material.

No. 58,366.—JACOB F. ANDREWS, New Providence, Penn.—*Sewing Machine.*—October 2, 1866.—These improvements refer to means and arrangements for driving the machine, feeding and driving the shuttle; the latter are explained in the claim. The fly wheel revolves in a horizontal plane, being operated from a double-armed treadle and two connecting rods which converge and play upon a pin on the wheel. The shaft which imparts motion to the various parts of the machine is vertical and carries this fly wheel.

Claim.—First, the arrangement of the verge F, moving on its pivot 6, in combination with the connecting rod 4, and cam E, in the manner and for the purpose specified.

Second, the feed bar C, when operated by the heel 5 of the needle bar B, in combination with the verge F' and its connecting rod 4, constructed and arranged in the manner set forth and shown.

Third, the shuttle bar D, with its pivot d and elliptical terminus U', forming an elliptical opening which surrounds the vertical shaft G, said shaft having an arm 9, with or without a friction pulley 3 at its end, in contact with the inner edge of said ellipsis U', the whole operating in the manner and for the purpose specified.

Fourth, the treadle H, with its ears A and two connecting rods M M united around the crank pin i upon the horizontal driving wheel, all connected in such a manner as to play freely with the motions of the wheel or arm I, in the manner shown and specified.

Fifth, in a sewing machine the horizontal fly-wheel or its equivalent, when in connection with a vertical shaft provided with a cam E and arm 9, arranged and operating in the manner set forth.

No. 58,367.—F. W. ARVINE, New Haven, Conn.—*Skate.*—October 2, 1866.—The runner is pivoted at the toe and acts as a lever to tighten the straps upon the boot and clamp the skate thereon when hooked to the heel piece. The link of the lever connects to a sliding plate beneath the sole, to which plate the instep straps are looped.

Claim.—First, adapting the runner of a skate to serve as a lever, in conjunction with clamps or other fastenings for securing a skate to the foot, substantially as described.

Second, pivoting the runner A to the foot stand at one end, and providing a latch fastening for said runner at its other end, substantially as described.

Third, applying the instep and toe straps to a sliding plate G G', operating substantially as described.

Fourth, the heel spurs a, in combination with a sliding plate G G', having toe and instep straps applied to it, substantially as described.

Fifth, the construction of the bottom slide of two portions G G', which are adapted for adjusting the toe and instep straps of a skate.

No. 58,368.—EMMET R. AUSTIN, Norwalk, Conn.—*Water Elevator.*—October 2, 1866.—The buckets are strung upon the endless chain and are actuated by a wheel whose periphery has a depression and flanges to engage the buckets.

Claim.—First, the wheel having the concave rim composed of two parts with an open space between them and the arms C, constructed as set forth.

Second, in combination with the wheel constructed as described the buckets B, as shown and described.

No. 58,369.—J. S. BENEDICT, Bedford, Ohio.—*Fence Gate.*—October 2, 1866.—The lower bar of the gate rests upon the grooved end of a short post when closed, and slips therein when traversing to balance on its middle hinging point.

Claim.—The rest A, in combination with the gate roller C and hinge B, when arranged and operating conjointly as and for the purpose specified.

No. 58,370.—A. M. BOUTON, Newark, N. J.—*Composition for Removing Ink from Type.*—October 2, 1866.—The ammoniacal liquor distilled from bone in the manufacture of boneblack is used in combination with pearlash to remove the ink from type.

Claim.—The use of these substances in any combination for cleansing purposes, substantially as set forth in the foregoing.

No. 58,371.—M. BOYES, Pocahontas, Ill.—*Medical Compound for Hog Cholera.*—October 2, 1866.—Composed of sugar of lead, 6 grains; chlorate of potash, 20 grains; Prussian blue, 5 grains, and starch, ¼ oz.

Claim.—The within described medical compound made of the ingredients and in the proportions substantially as herein described.

No. 58,372.—FREDERICK H. BROWN, Auburn, N. Y.—*Cotton Seed Planter.*—October 2, 1866.—One of the droppers is adjustable laterally to regulate the distance apart of the rows planted. The depth is regulated by a bar running on the surface of the ground. The seed is intermittingly agitated by stirrers and is fed by au endless toothed chain.

Claim.—First, the adjustable nature or character of the machine as to regulate the space between rows, in combination with the cotton seed planter, as above specified.

Second, in cotton seed planters, regulating the depth that the seed are deposited.

Third, the mode described of constructing the endless chain and its combination with the cotton seed planter, as above set forth.

Fourth, the peculiar movement given to the shaft X, above described, when used for the purpose above set forth.

No. 58,373.—JOHN T. BUDD, New York, N. Y.—*Hot-air Furnace.*—October 2, 1866.—A regulated supply of heated air is furnished to the fire above the fuel by means of openings from a surrounding casing and a register door. The gases, &c., pass by lateral sinuous flues to the exit, or more directly from the crown of the drum, according to the position of the damper.

Claim.—First, the dome d, pipe e, and damper, in combination with the two ranges of return pipes f at the sides of the furnace, connected to the pipe e by the lateral connections g, as and for the purposes set forth.

Second, the air jacket h, surrounding the fire pot, in combination with the shield k, for allowing a regulated quantity of heated atmosphere to pass into into the gases above the fire for their consumption, as set forth.

No. 58,374.—JOSEPH CARLIN, Cincinnati, Ohio.—*Preserving Flour, Grain, &c.*—October 2, 1866.—A porous body is suspended in the bin or vessel to absorb the moisture of the grain, &c.

Claim.—The mode of preserving flour and grain from souring by suspending masses of unglazed pottery (or its equivalents) within the barrel, bin, or other receptacle, substantially as set forth.

No. 58,375.—JAMES W. CAHILL, Madison, Ind.—*Soda Fountain.*—October 2, 1866.—The pump cylinder is guarded by a perforated partition from the ice chamber. The water is forced by a piston to the main pipe, the delivery ends of which are respectively a nozzle for plain ice water, and a nozzle in which an effervescent powder lies upon a perforated diaphragm in the discharge cup.

Claim.—First, the cup U V, made of two portions screwing together, and with perforated bottom W, constructed and operating as and for the purpose described.

Second, the perforated shield K, extending nearly to the top of the water chamber to keep the ice away from the pump, pump rod, and crank.

Third, the branching head O Y X, provided at the respective ends with the plain nozzle T, and the soda cup nozzle T', substantially as described.

No. 58,376.—DEWITT C. CAREY, Baltimore, Md.—*Ruffling Attachment for Sewing Machines.*—October 2, 1866.—The needle bar as it rises lifts the ruffler by means of a strap. The encased spring as the needle descends forces the ruffler to push and gather the cloth under the path of the needle. The outer spring serves to hold the cloth as gathered until stitched, and prevents its being drawn back as the ruffler retreats. A regular feed is used. The screw and slot allow of an adjustment of the fold or gather by varying the length of the connecting strap

Claim.—First, the curved or segmental guide A, in the described combination with the bar B, carrying a ruffling plate L.

Second, the spring R, when employed in connection with the ruffling plate L, and bar B, substantially as and for the purposes set forth.

Third, the connecting strap T, connected to the head of the needle bar or other suitable needle-operating mechanism, to actuate the ruffling device, substantially as described.

Fourth, in combination with the aforesaid ruffling mechanism A B L, the slotted bar S and clamp screw C, for varying the fineness of the ruffling.

Fifth, the spring detent g, constructed and employed as described to hold the work from being drawn back by the return motion to the ruffler.

No. 58,377.—EDWIN M. CHAFFEE, Providence, R. I.—*Manufacture of Water-proof Hose.*—October 2, 1866.—The inside of the fabric is first covered with rubber cement, and a tube of India-rubber, varnished on the inside to render it non-adhesive, is drawn through it. The parts are then united by rolling.

Claim.—First, the method substantially as herein described of lining hose of woven or other fabric with India-rubber, gutta-percha, or other flexible and vulcanizable gum on the inside, by means substantially such as set forth, or by any other equivalent means.

Second, the use of the tin tube or its equivalent for the purpose herein described.

Third, strengthening the lining by the longitudinal threads or their equivalent, substantially as described.

No. 58,378.—H. W. CHAMBERLAIN, Jersey City, N. J.—*Cord and Line Reel.*—October 2, 1866.—The knotted end of the cord is held by a button. The cord is wound on the reel by the binary revolution of the axes of the two handles, which are secured by slide plates.

Claim.—First, the reel A, and two handles *n m*, arranged and operating substantially in the manner and for the purpose set forth.

Second, the adjusting or securing slides I and P, in combination with the reel A, for securing the handles in the reel, as and for the purpose set forth.

Third, the combination of the cord clamp or holder J *j*, with a reel adapted to operate substantially as specified.

No. 58,379.—ROBERT A. CHESEBROUGH, New York, N. Y.—*Burning Oil for Fuel.*—October 2, 1866.—The oil is passed into a pan filled with bone black and placed beneath the boiler.

Claim.—First, the use of bone black as an absorbent bed for oil when the same is used as fuel.

Second, the combination of bone black and hydrocarbon or other oils for use as fuel.

No. 58,380.—EDWIN CHILDREN, Liberty, Wis.—*Cultivator.*—October 2, 1866.—The tongue has a handle in the rear and is pivoted to the front of the carriage frame, and also by a link to the frame to which the ploughs are clamped. The vertical adjustment is by treadles, ropes and pulleys.

Claim.—The pivoted draught pole D, in combination with the pivoted bars H H, lever G, and the plough beams L L, connected to the bar I, which is pivoted to the bars H H, all arranged substantially as and for the purpose set forth.

No. 58,381.—JOHN G. CLARK, Middletown, Ohio.—*Machine for Planting Cotton Seed.*—October 2, 1866.—To insure separation and regularity of discharge of the seeds the seeds are agitated by a stirrer. The stationary teeth in the hopper stand tangentially to the cylinder and act as a rake to keep back the mass of seeds, while a portion are acted upon by the toothed cylinder which carries them through the reciprocating slide.

Claim.—First, the toothed cylinder B, in combination with the toothed reciprocating slide or slides, constructed, arranged and operating in the manner substantially as described.

Second, the combination of the agitator with the cylinder B and ribs *e e*, operating substantially as specified.

Third, the hopper teeth arranged tangentially in relation to the cylinder, in combination with the positively operating devices for separating and discharging the seed, substantially as and for the purpose described.

No. 58,382.—HENRY B. COMER and JOHN DENTON, Pittsburg, Penn.—*Lightning Rod.*—October 2, 1866.—The sectional shape is a square with wings projecting diametrically from diagonally opposite corners.

Claim.—A new article of manufacture, to wit: a lightning rod consisting of a central core from which extend two thin wings, said core and wings being made in one piece and twisted, the whole being made by means and constructed in the form substantially as herein described and for the purpose set forth.

No. 58,383.—JAMES COOK, Collinsville, Ohio.—*Cultivator.*—October 2, 1866.—The ploughs are vibrated laterally within certain limits by the handles connected to an adjustable bar pivoted in front to the beam.

Claim.—First, the intermediate frames A and *a a'*, constructed in the manner described, in combination with the standards *d* and *d'*, beam B, and drag bars *f*, arranged, connected and operating in the manner and for the purpose specified.

Second, the upper frame *a a'*, and spring connection *b*, in combination with the plough handles and notches or gains, to limit the oscillating motion of the ploughs, in the manner and for the purpose substantially as described.

No. 58,384.—THOMAS J. CORNELL, Decatur, Ill.—*Gang Plough.*—October 2, 1866.—The ploughs vibrate backward when subjected to a breaking strain by an obstruction. A catch maintains their normal position under ordinary pressure. A spring restores them to normal position when the obstruction is passed. An eccentric segment is the lifting agent in raising the ploughs when in motion.

Claim.—First, the plough standard I, journalled on a horizontal axis K, to vibrate rearward under the circumstances described.

Second, the catch Q in combination with the standard I, operating as described.

Third, the spring R, in combination with the standard I, operating as described.

Fourth, the eccentric segment W, in combination with the beam or beams of the plough, and operating substantially as described.

No. 58,385.—JOHN CROCO, Holmesville, Ohio.—*Potato Drill.*—October 2, 1866. -The cams on the wheel make an audible click as a signal to drop the potatoes, which are discharged through a gated opening in the hopper. A share in the rear covers the tubes.

Claim.—The hopper E, the spout M, the bar N, the spring *d*, and knobs *g*, the whole constructed and arranged as and for the purpose herein specified.

No. 58,386.—H. N. DALTON, Pacheco, Cal.—*Gang Plough.*—October 2, 1866.—The plough frame is pivoted behind the carriage, the front end being regulated by the band lever, and the rear by the upper beam, which rests on the plough frame and is controlled in front by pins in a post on the carriage frame.   When the lever is thrown forward the points of the ploughs and the front end of the beam rise until the latter strikes the pin, when the rear of the plough beam commences to rise, clearing the soil entirely.

Claim.—First, the plough frame F, parallel arms E. beam H, links *b c*, vertical bar I, and pins *d e*, combined and operating substantially as described, for the purpose specified.

Second, the lever J, arms E E, beam H, and frame F, combined and operating substantially as described, for the purpose specified.

No. 58,387.—ROYAL E. DEANE, New York, N. Y.—*Carving Table.*—October 2, 1866.—The dishes are set in a hollow trough whose longitudinal division causes the current of hot water to pass along one side and return on the other to the discharge pipe.

Claim.—A carving table having a top A, constructed in the form of a shallow vessel divided by a longitudinal central partition plate *b*, one or more, with a space *c* allowed at one end to form a communication between the two compartments *d d'*, and the latter made to communicate by means of pipes *e e'*, with the water back or heater of a range or furnace, substantially as herein shown and described.

No. 58,388.—CHARLES M. DEDRICK, Temperanceville, Penn.—*Apparatus for Removing the Wire from Soda Water Bottles.*—October 2, 1866.—The strip attached to the table has a catch to engage the wire on the cork while a fulcrum arm holds the neck of the bottle.

Claim.—The apparatus consisting of the base or lever A, fulcrum C, and catching point B, when said apparatus is used for the purpose of removing wire off the corks of soda or mineral water bottles, the whole being constructed, arranged, and operating substantially as herein described and set forth.

No. 58,389.—GEORGE W. DOOLITTLE, Lincoln, Ill.—*Cultivator.*—October 2, 1866.—The frame of the carriage rests by a brace piece on the axle.   The operator guides laterally by foot pressure on the treadles.   The shares have a semicircular piece removed between their upper and lower rear corners to ease the draught.

Claim.—First, the form of the plough mould as herein described, for the purposes specified.

Second, the arrangement and combination of the guiding mechanism herein described.

Third, the iron axle S, and the supporting brace N, constructed and operating as herein described.

Fourth, the combination of all the parts, operating substantially in the manner herein described.

No. 58,390.—HARRISON DOOLITTLE, East Cleveland, Ohio.—*Fork for Digging Potatoes.*—October 2, 1866.—The tines are attached by lugs to countersunk slots in the head. The handle is reversible to constitute the tool a fork or a rake, and in the former case is braced by a bar, and has a fulcrum foot in the rear of the head.

Claim.—First, constructing the cross-head A, adapted to receive the tines in the manner described.

Second, the tine B, constructed as herein described, in combination with the cross-head.

Third, the adjustable support or fulcrum D, in either of the forms represented, in combination with the handle and shank C.

Fourth, the employment of the guard rod E, for the purpose herein specified.

Fifth, converting the fork into a rake in the manner herein described, and the combination of the fork and rake, for the purposes herein set forth.

No. 58,391.—WILLIAM E. DOUBLEDAY, Brooklyn, N. Y.—*Making Hats.*—October 2, 1866.—The edge of the fabric is firmly connected to the brim ring while the crown is stretched by a crown block, the fabric being softened by steam.   The hat is retained in shape until cool or dry.

Claim.—First, the ring brim block, to which the edge of the felt or other material is to be connected by pins or otherwise, in combination with the crown block applied to stretch the fabric to shape, as set forth.

Second, in combination with the brim ring and crown block, the brim block or its equivalent to connect the brim ring and crown block, so that they may be removed and retain the hat or similar article while cooling or drying, as set forth.

Third, the method herein specified of forming or shaping hats and bonnets by heating and softening the felt or other material by steam, and then stretching and shaping said fabric at one operation by blocks, before it cools or dries, substantially by means set forth.

No. 58,392.—A. M. DUBURN, Chicago, Ill.—*Lantern.*—October 2, 1866.—The metal base-band of the globe rests on springs, and is withdrawn vertically when the hinged cap is raised. The wick-raising shaft is dislocated when the lamp is retracted downward.

Claim.—First, the arrangement of the globe G as shown, so that it may rest on spring supports, and be capable of being drawn through the band C when desired.

Second, the wick adjusting mechanism, composed of the two rods *k m*, fitted respectively in the lamp burner and band C, and provided with the square *l*, and three-sided socket *n*, all constructed, combined, and arranged as shown, to admit of the wick being raised and lowered without removing the lamp from the lantern, and at the same admitting of the removal of the lamp whenever desired.

No. 58,393.—EDWARD DUFFEE, Haverhill, Mass.—*Screen for Dry Gas Purifiers.*—October 2, 1866.—Explained by the claim and cut.

Claim.—In a screen for dry coal-gas purifiers, made up of crossed or interlaced thin strips of wood, supported by a framework, grooving the outer edges of each frame, so that the strips shall be imbedded in or sunk below the surface of each outer edge to prevent abrasion, and to allow the frames to abut closely together, substantially as set forth.

No. 58,394.—F. B. DUFFY, Sparta, Wis.—*Bed Bottom*—October 2, 1866.—The end slats are suspended by elastic bands and stirrups from hooks on the rails, and the longitudinal slats rest in elastic loops beneath the end slats.

Claim.—A spring bed bottom, consisting of the slats *c* resting at their ends on the elastic strap E, secured to the bar *a* by the staples *e*, and the whole secured to the bedstead by means of the hooks *m* and staples *o*, as set forth.

No. 58,395.—J. B. EDGELL, J. W. and E. A. ALEXANDER, Independence, Iowa.—*Plough.*—October 2, 1866.—The plough is attached to a post which depends from a beam jointed to the tongue, so that when the hinging point is deflected upward the plough is raised from the soil, and conversely, the required position being maintained by a segment bar. The beam rests on a carriage axle, which also supports the seat post.

Claim.—First, the construction of a plough carriage for supporting ploughs, of the axle-tree A, wheels B, beam C, seat C', beam D, pivoted pole G, hand lever H, and connecting rod *g*, substantially as described.

Second, the pendant E applied to the beam D of an adjustable plough carriage, which is constructed substantially as described.

Third, pivoting a plough to an adjustable beam of a carriage, so as to operate substantially as set forth.

No. 58,396.—WILLIAM EDSON, Boston, Mass.—*Coffee Pot.*—October 2, 1866.—The water in the lower chamber is forced by the evolution of steam into the chamber above, and returns through the coffee and strainer, when the steam condenses after the pot is removed from the fire.

Claim.—As an improved article of manufacture, a coffee pot provided with a fixed diaphragm E, pipe P, strainer F, and otherwise made as shown and described, and for the purpose set forth.

No. 58,397.—PETER A. ENSIGN, Adrian, Mich.—*Air Engine and Motor.*—October 2, 1866.—The smoke flue of the furnace is conducted into a vertical open cylinder, in which is a wind wheel, rotated by the upward current and connected to exterior gearing and a governor, which actuates a damper in the chimney.

Claim.—First, the combination and arrangement of the cylinder A, suitably elevated, wind wheel D, vertical shaft E, shaft K, perforated stationary tube J, governor H, valve F, and heater B, substantially as described, for the purposes specified.

Second, the heater B and stove C, arranged alongside of and connected to the cylinder as above described, in combination with the said cylinder and the wind wheel, substantially as shown.

No. 58,398.—LUTHER ERVING, Brooklyn, N. Y.—*Oven for Gas Cook Stoves.*—October 2, 1866; antedated September 23, 1866.—The heated air from the gas-burner is deflected by the inverted cone, and circulates around the shelves and between the walls of the oven.

Claim.—First, the inverted conical deflector E, when used in connection with an oven F, and a gas-burner, substantially as and for the purpose set forth.

Second, constructing the oven with double walls to form flues or draught passages, when said oven is provided with a deflector E arranged relatively with the draught passages, so as to guide or deflect the products of combustion into the flues, and at the same time admit of the heat from the flame rising directly into the oven, as set forth.

Third, the shelves G in the oven F, arranged in such a manner that they have open spaces on every side to allow the heat to pass, in the manner and for the purposes described.

No. 58,399.—FREDERICK ETZOLD, Union Hill, N. J.—*Sewing Machine Shuttle.*—October 2, 1866.—The object is to graduate the thread tension as desired, and to retain the tension spring securely in the position to which it is adjusted by locking the adjustable notched eccentric with the curved free end of the spring.

Claim.—The toothed eccentric cam D applied in combination with the spring C, and in relation with the bobbin B and shell A of the shuttle, substantially as herein set forth for the purpose specified.

No. 58,400.—D. S. Forney, Wytheville, Va.—*Privy.*—October 2, 1866.—The receptacles in the close privy chamber are portable and provided with covers, which are always closed when the privy is not in use, and which are adapted to receive disinfecting agents to prevent the escape of effluvia in the intervals of using and during removal.

Claim.—First, the portable receptacle F for excrementitious deposits, in combination with the box W connected by a pipe or tube to the receptacle F, and provided with the cover L and lids d d, the whole being constructed, arranged, and operated as described.

Second, a swinging cover M, or its equivalent, so arranged upon the excrement receptacle that it can be either swung away from or over its mouth, by means substantially as and for the purpose specified.

Third, the box cover or lid M, having its bottom perforated or made of wire netting, for the purpose specified.

Fourth, the swinging cover or lid S for the seat opening of the privy, operated as described, whether arranged to swing in conjunction with the cover of the excrement receptacle or not.

No. 58,401.—Andrew I. Frisbie, St. Mary's, Ohio.—*Hay and Grain Protector.*—October 2, 1866.—The hay and grain cover is built of boards in the form of a pyramidal frustum, with a projecting cover.

Claim.—The longitudinal hay and grain cap, as herein described.

No. 58,402.—Martin Fryer, Greenbush, N. Y.—*Oar Swivel.*—October 2, 1866.—In addition to the vertical and horizontal vibrations, this improvement permits the oar to be feathered by turning on its longitudinal axis.

Claim.—The combination of the swivel A with the ring B and its slot b, and the ring C with its pin e, operating together in the manner and for the purpose described.

No. 58,403.—Richard J. P. Goodwin, Manchester, N. H.—*Surgical Splint.*—October 2, 1866; antedated September 24, 1866.—Bands are arranged above and below the articulation of the limb and are joined by straps whose hinge is fastened by a set screw and slot. The bands open to receive the limb, have removable portions, and are covered with water proof material.

Claim.—First, the adjustable band D, one or more, between the stationary bands B or B', in combination with the guide-rods a or a', constructed and operating substantially as and for the purpose described.

Second, the segmental slots c and set screws d, in combination with the hinge joints C which connect the two parts of the splint A, substantially as and for the purpose set forth.

Third, the arrangement of hinge joints h in the middle of the bands B B' D or nearly so, substantially as and for the purpose described.

Fourth, the application to the bands of waterproof pads, substantially as and for the purposes set forth.

No. 58,404.—W. C. Goodwin, Hamden, Conn.—*Fish Hook.*—October 2, 1866.—The application of a spiral spring to the shank of the hook tends to press the bait toward the barb and prevent the exposure of the point.

Claim.—The combination of the fish hook with the spiral or helical spring when they are constructed, put together, and made fit for use, substantially as herein described.

No. 58,405.—John C. Gould, Boonton, N. J.—*Door and Gate Spring.*—October 2, 1866.—The coil spring is attached to the door and presses upon a lever engaging at one end an eye bolt on the door and at the other looped on an arm which is hooked to an eye bolt in the door and slips in a guide on the post; a notch on the arm engages the guide and maintains the open position.

Claim.—The plate D, spring F, or its equivalent, lever G, arm H, with the guides I and M, and the notch J, substantially as set forth and for the purposes named.

No. 58,406.—Christian Grün, New York, N. Y.—*Elevator for Ladies' Skirts.*—October 2, 1866.—The spring jaws are pivoted to posts on the slotted plate, through whose openings small folds have been pulled to be clamped by the pressure of the jaws.

Claim.—The mortises a a' b' in the V-shaped bed plate A, in combination with the eyes e on the spring jaws and with the spring e, constructed and operating substantially as and for the purpose set forth.

No. 58,407.—Charles Guidet, New York, N. Y.—*Belgian Pavement.*—October 2, 1866.—The lines of junction of the blocks are not at right angles; their faces are rhomboidal; no joint is parallel to the line of travel.

Claim.—The employment of rhomboidal blocks in which the angles are unequal and the planes of their sides are not at right angles, arranged substantially as and for the purpose set forth.

No. 58,408.—D. Hagerty, Baltimore, Md.—*Melting and Moulding Tinners' Solder.*—October 2, 1866.—The mould plates slide in grooves beneath the reservoir from which they are filled. The furnace is inside the reservoir and the molten metal passes therefrom, when the gate is raised and fills the hollow in the plate, which is moved beneath by a crank, pinion and rack.

Claim.—The combination of reservoir E with its interior furnace *e* and grate G, mould plates B B, toothed wheel C, and table A, constructed and arranged in the manner substantially as shown and described and for the purpose set forth.

No 58,409.—WILLIAM HAMILTON, Chicopee, Mass.—*Lubricating Apparatus.*—October 2, 1866.—The oil cup is suspended below the journal and its contents are carried up by a ribbon which revolves with the shaft.

Claim.—An oil cup divided into separate compartments as described, when said cup is combined with a chain or ring B, by means of which the oil is carried up to lubricate the axle or shaft, in the manner and for the purpose herein described.

No. 58,410.—THOMAS R. HARTELL, Philadelphia, Penn.—*Ornamenting and Lettering Glass.*—October 2, 1866.—Explained by the claim and cut.

Claim.—Ornamenting and lettering glass objects by making depressions in the same while the glass is in a plastic state, and filling or lining these depressions with white plaster while in a plastic state, as and for the purpose set forth.

No. 58,411.—S. R. HATHORN, Castleton, Vt.—*Snow and Ice Guard for Roofs of Buildings.*—October 2, 1866.—The frame is attached to the sheathing of a roof beneath the shingles and carries horizontal longitudinal bars to arrest the sliding of ice or snow.

Claim.—First, a snow and ice fender composed of the supporting pieces D and fender rods *e*, and combined together substantially as and for the purposes set forth.

Second, the snow and ice fender constructed and arranged substantially as herein described, so that it may be secured to the roof of a building before the shingles or slates are put on, as set forth.

No. 58,412.—JERRAH HAYWARD, Greene, N. Y.—*Horse Hay Rake.*—October 2, 1866.—The triangular frame of the rake is hinged to the axle. The operating lever has a cam at its lower end, which vibrates the tilting frame and raises the rake, which is also connected by a bar with the lever.

Claim.—The combination of the lever F with the lever E, main braces C C, pitman D, and rake B, when made and operated substantially as and for the purposes set forth.

No. 58,413.—EDWARD HEATH, Fowlerville, N. Y.—*Ditching Machine.*—October 2, 1866.—The longitudinally reciprocating excavator of dirt and the laterally oscillating scraper which removes it from the side of the ditch are operated by pitman connection to a cross-head, reciprocated by power from the main vertical shaft, which also winds in the rope by which the progression is accomplished. A second share and scraper follow in the wake of the former.

Claim—First, a ditching machine so constructed as that the cross-head will impart a longitudinal motion to excavators and a laterial motion to the scraper at the same time, substantially as and for the purposes herein described.

Second, the cross-head J, Pitman I, and crank H, in combination with the bar E², lever H², and ratchet wheel C², all for the purposes and substantially as herein set forth

Third, the excavator S, pitman L L in combination with the spring catch T and rock shaft P, substantially as and for the purposes described.

Fourth, the scraper R in combination with the excavator S, for the purposes and substantially as described.

No. 58,414.—H. M. HEINEMAN, Williamsburg, N. Y.—*Fastening for Neckties.*—October 2, 1866.—A split pin is hinged at its lower and smaller end. A ring slides on the jaws to clasp or release the collar; the cravat is attached to one of the jaws.

Claim.—The fastening A, consisting of the two jaws B and C, and clasp or slide E, arranged and connected together, and constructed so as to operate substantially in the manner and for the purpose specified.

No. 58,415.—JOSHUA HENRY, Steubenville, Ohio.—*Churn.*—October 2, 1866.—Air is admitted through one trunnion by a tube which has a bell-shaped flange around it near the end on the inside of the revolving churn box.

Claim.—The arrangement of the metal tube *d*, passing through the journal *c'*, and into and beyond the bottom of the funnel *a*, when used in combination with the revolving churn box A, substantially as and for the purposes specified.

No. 58,416.—AUGUST HERMANN, New Haven, Conn.—*Apparatus for Discharging Bilgewater from Vessels' Holds.*—October 2, 1866.—A rotary pump in the bottom of the vessel discharges outwardly, its shaft being revolved by gearing on deck; valves in the induction and eduction openings of the wheel and a slide which admits bilgewater to the wheel chamber prevent the reflux.

Claim.—The apparatus, substantially as described, consisting of a vertical axle 1 K, to which is attached a hollow cylinder R R, with projecting chambers T T T, containing valves U U U, which open on the said axle and hollow cylinder being rotated, said apparatus being operated in the manner substantially as described for the purpose of freeing vessels from water.

No. 58,417.—THOMAS N. HICKOX, Brooklyn, N. Y.—*Cap for Mucilage Bottle.*—October 2, 1866.—The punched-in flanges around the opening clasp the brush handle.
Claim.—The sheet metal cap for mucilage bottles, &c., formed with a spring flanged hole for the brush handle, as and for the purposes set forth.

No. 58,418.—CHARLES HOLTZ, Chicago, Ill.—*Car Coupling.*—October 2, 1866.—The coupling pin is raised by pressure upon the bumpers, upon collision, and falls when the pressure is removed.
Claim.—First, so arranging the coupling pin, in combination with the adjustable or movable bumpers D. or their equivalent, that pressure upon said bumpers will raise up the coupling pin, and allow the coupling link to enter the draw-head when the cars are run together.
Second, so arranging the springs S, or their equivalent, in combination with the coupling pin. that said pin is caused or permitted to drop through the coupling link to couple the cars when pressure upon the aforesaid bumpers is removed.
Third, the combination of the bumpers D, the movable bar E, the pin d, sliding bar c, bent lever b, and coupling pin a, arranged and operating substantially as specified and shown.
Fourth, the combination and arrangement of the bumpers D, cross-bar E, springs S, pin d, slide c, lever b, and pin a, operating substantially as specified and set forth.
Fifth, in combination with the pin a, bent lever b, and slide c, the employment of a projection e, and hook g, for uncoupling the cars, substantially as specified.
Sixth, in combination with said slide c, the employment of a spring h, as and for the purposes described and shown.
Seventh, the employment of the levers or arms H, and cord I, in combination with the shaft f, and book g, arranged and operating as and for the purposes set forth.

No. 58,419.—CHARLES W. HOWARD, Philadelphia, Penn.—*Clothes-line Clamp.*—October 2, 1866.—The line is pinched between a jaw on the pivoted arm and the lug around which it is wound; strain on the line clenches the grip on the cord.
Claim.—A device for supporting a clothes-line, consisting of the pieces A and B, or their equivalents, constructed and combined to operate together so as to pinch and hold the line by the tensional strain of the latter, substantially as described and set forth.

No. 58 420.—H. A. HOYT, Mott Haven, N. Y.—*Stamp Moistener.*—October 2, 1866.—The stamp is laid with its gummy back on the wet sponge in the cup and the spring lid pressed upon it.
Claim.—An improved stamp moistener formed by combining the spring B, and cover C, constructed and arranged with the sponge D, and sponge cup A, substantially as described and for the purpose set forth.

No. 58,421.—WILLIAM C. HURD, New York, N. Y.—*Composition for paint.*—October 2, 866.—Explained by the claim.
Claim.—A compositon for painting compounted by the addition of powdered quartz to oil, lead, zinc, and other materials ordinarily employed in the manufacture of paints, substantially as set forth.

No. 58,422.—ELIAS S. HUTCHINSON, Baltimore, Md.—*Automatic Feed for Carburetters.*—October 2, 1866.—The valve is kept closed by the weighted arm except when the float slides down upon the rod; when the liquid in the vessel falls below the proper level the float comes in contact with the stud on the revolving shaft and opens the valve.
Claim.—A valve to regulate the supply of liquid from the reservoir to the chamber, which is operated from the revolving shaft, or a projection thereon, which strikes against an object brought within reach by the sinking of the fluid in the chamber.

No. 58,423.—GEORGE M. JACQUES, Boston, Mass.—*Composition for Destroying Insects.*—October 2, 1866.—Explained by the claim.
Claim.—The combination of extracts, solutions, and distillations of tobacco, including the nicotine oils, and ammonia, and soaps, in the manner and for the purpose above described.

No. 58,424.—EUGENE N. JENKINS, Chicago, Ill.—*Sad-Iron Heater.*—October 2, 1866.—The cover has a suspended bent wire which forms a foot when the iron is removed; when the iron is replaced its front part holds up the wire and the cover shuts down.
Claim.—The wire with the feet f, bend g, and end h, constructed substantially as and susceptible of being operated in the manner and for the purposes herein recited.

No. 58,425.—BARTON H. JENKS, Bridesburg, Penn.—*Seasoning Wood.*—October 2, 1866.—Green wood is deprived of its moisture by placing it in an air-exhausted, strong cylinder. Heat may be applied during the operation.

Claim.—The process, substantially as herein described, of seasoning or drying woods.

No. 58,426.—JOB JOHNSON, Brooklyn, N. Y.—*Oyster Rake.*—October 2, 1866.—The double rake, evenly balanced on a rope, is let down on the bed, when a catch being released, the jaws are closed by a spring and the action of raising; any obstruction grasped by the rakes may be released by drawing on the releasing cord. The teeth are clamped in by an angle-iron strip over their back and top: an opposing clamp bar has a recess for each tooth. Diagonal fenders pass from the teeth to the head frame.

Claim.—First, the lever in combination with a pair of rakes, hinged together substantially as specified, for keeping the rakes apart as they are lowered, and allowing the rakes to close after they touch at the bottom of the water, as set forth.

Second, a line or cord passing from the end of one handle to the rake on the other handle, for opening the rake when it becomes necessary to disconnect the same from any article in the water, as specified.

Third, the mode of constructing the metallic rake head with the angle iron, receiving the teeth clamped to the same, substantially as set forth.

Fourth, the metallic fenders formed of the bars *o* and rods *n*, the parts being united to the rake head *c*, in the manner specified.

No. 58,427.—THOMAS JOHNSON, Elmira, N. Y.—*Watch.*—October 2, 1866.—The improvements relate to the adjustment of the cock to the plate of the watch by a pillar and pin; controlling the hair-spring stud by a segment screw; the arrangement of the regulator of the click work on the lever and of the arbor of the main-spring barrel; a detent stop to prevent the strain of the key, at the termination of the winding operation, from going on to the train: an auxiliary stop work and hook for the spring; the toothed wheel is placed around the mid-length of the barrel: the cap jewels are suspended upon the upper and lower pivots of the balance wheel instead of being affixed rigidly.

Claim.—First, the combination and arrangement of the cock with the plate and its pillar and pin, when said pin is inserted in a direction running lengthwise with the cock, substantially as herein shown and described.

Second, the combination and arrangement of the hair spring stud with the segment screw *e*, when constructed and operating substantially as herein shown and described.

Third, the combination of the arms or levers *f g* with the hair-spring *d*, constructed and operating substantially as herein shown and described.

Fourth, the combination and arrangement of the clicks and springs with their ratchet, constructed and operating substantially as herein shown and described.

Fifth, the arrangement of the clicks and click springs combined, and ratchet, with the bridge or cap enclosing them, substantially as herein shown and described.

Sixth, the combination of the face ratchet *u* with the stop work, constructed and operating substantially as herein shown and described.

Seventh, the male stop *p*, when constructed with a lip *s*, stud *q* and bevelled slots *o' o'*, as herein shown and described.

Eighth, the hook and auxiliary stop work *x*, constructed and operating substantially as herein shown and described.

Ninth, the arrangement of the teeth upon the centre of the periphery of the barrel containing the main spring, substantially as herein shown and described.

Tenth, the combination of a suspended cap jewel, by means of an elastic bearing, with a pivot, substantially as herein shown and described.

Eleventh, the construction and arrangement of the jewel in the spring *a''*, substantially as herein shown and described.

Twelfth, the method of forming in a watch movement suspended elastic end bearings to the pivot, substantially as herein shown and described.

No. 58,428.—JAMES M. KEEP, New York, N. Y.—*Card Rack.*—October 2, 1866; antedated September 18, 1866.—The springs are attached to removable bars in a frame in an alphabetical series. Each rests upon the swell of the spring next above.

Claim.—First, the springs for holding cards, when constructed and arranged relatively to each other, and to the tablet or other surface to which they are secured, as herein described, so that each spring shall bear upon the swell or crown of the succeeding spring immediately above, as and for the purposes set forth.

Second, the combination in a card rack of the springs, constructed and arranged as above described, with the frame A, bars B, and alphabetical reference or tablet, substantially as hereinbefore shown and set forth.

Third, constructing a card rack of separable and detachable parts, as described, and so that said rack may be easily taken apart and readily readjusted, substantially as and for the purposes herein set forth.

No. 58,429.—O W. KELLOGG and H. R. HILL, Ripon, Wis.—*Billiard Register.*—October 2, 1866.—For counting games. The motion of the pendulum actuates the disks by means of the spring, pawl. and ratchet; a coinciding colored spot on each disk indicates at numbered openings on each side the number of games played.

Claim.—First, the dial plates A A, with disks D D, bar M, ratchet E. and pendulum F, arranged in the manner substantially as and for the purposes herein specified.

Second, the pendulum F, attached to the count wire when used as and for the purposes set forth.

No. 58,430.—GEORGE R. KELSEY, West Haven, Conn.—*Belt Buckle.*—October 2, 1866.— The smaller frame is hinged upon the larger one, and this upon the socket book. In use the hook engages one end of a belt, the other end is doubled around the smaller frame; tension brings the jaws together.

Claim.—The combination of the bow, Fig. 3, lever, Fig. 4, and hook, Fig. 5, when they are constructed, put together, and fitted, for use substantially as herein described and set forth.

No. 58,431.—ISAAC KENNEDY, Ithaca, N. Y.—*Plough.*—October 2, 1866.—The mould-board has at its rear end a conical roller which can be adjusted to turn the furrow more or less. The land-rest has an eye in which both handles are stocked.

Claim.—First, making the wheel at the rear end of the mouldboard adjustable by means of a frame or other devices at the top and bottom of the said wheel, one or both, by means of which to evert. set on edge, throw completely over, or otherwise regulate the furrow by the use of the said wheel and frames described.

Second, the combination of the wheel, or equivalent device, and frames with the mouldboard and the V-shaped handles meeting in one eye on the land-rest, the same making a whole as described.

Third, the so combining together the wheel and immovable part of the mouldboard, and shaping each to the other, that they shall maintain a constant relation to each other in whatever position the wheel may be placed, as described.

No. 58,432.—GEORGE KICHERER, Brooklyn, N. Y.—*Fence.*—October 2, 1866.—Staples are cast into sections of fence through which a rod is passed for their support; braces on the rods give firmness to the fence.

Claim—As a new article of manufacture, the panels, fitted with staples or eyes for the reception of tie rods and braces, substantially in the manner described and for the purpose specified.

No. 58,433.—JOHN F KIRKWOOD, Thistle, Md.—*Let-off and Take-up for Looms.*—October 2, 1866.—The object of these devices and their arrangement is to cause the let-off to be coincident with the take-up, notwithstanding the gradual diminution of yarn on the beam. The same cam operates both sets of devices. The strain of the yarn on the whip roll, by means of its connections with the other parts, determines the amount of let-off.

Claim.—First, the combination of the lever V and spring U with the whip roll T, arm S, and rod Q, for the purpose of increasing the speed of the let-off devices conformably with the decrease of thread upon the yarn beam, substantially as described.

Second, the arrangement of the lever F, cam E', and arm P, for imparting motion to both the let-off and take-up mechanism, as described.

No. 58,434.—TOBIAS KOUN, Hartford, Conn.—*Machine for Twisting, Stretching, Cleaning, and Reeling Silk and other Threads.*—October 2, 1866.—The thread is twisted from the spools by fliers, then macerated in a trough, coiled for tension around rods, coiled again around each of the rollers on the reciprocating frame which cleans it, and finally is wound on the reel.

Claim.—The combination of the mechanism for twisting with the means for stretching and the mechanism for cleaning, substantially as described.

Also, the combination of the mechanism for twisting with the means for stretching and the mechanism for reeling, substantially as described.

Also, the combination of the mechanism for twisting with the apparatus for macerating, the means for stretching, the mechanism for cleaning, and the mechanism for reeling, substantially as described.

No. 58,435.—EBENEZER G. LAMSON, Shelburne Falls, Mass.—*Drilling and Quarrying Stone, &c.*—October 2, 1866.—The tools are operated by a compound positive and spring motion moderating the positiveness of the pitman crank by the elasticity of the bow spring. The machine is moved in either direction by worm gear actuated by a reversible pawl.

Claim.—Connecting the chisels, drills, or other cutting instruments for working in or on stone, to the crank wheel, or its equivalent raising and lowering mechanism, through the intervention of a bow spring and strap, substantially as and for the purpose described.

Also, in combination with a stone-cutting, channelling, tunnelling, or quarrying machine that is moved along upon a track or ways while operating upon the rock or stone, a reversible pawl, the worm gear, and clutch, so that it may be moved along in either direction upon the track, or stopped thereon at will while the cutter or tools continue to operate, substantially as describe .

No. 58,436.—JAMES LEFEBER, Cambridge City, Ind.—*Hollow Auger.*—October 2, 1866; antedated August 23, 1866.—Adjustable cutters are attached to the radial wings cast on the exterior of the auger head.
Claim.—A hollow auger, substantially as above described, with radial wings on the exterior of its body, and with bits or cutters fastened adjustably to such wings, as above shown.

No. 58,437.—JOHN LEMMAN, Cincinnati, Ohio.—*Bedstead Fastening.*—October 2, 1866.—The socket part of the fastening is attached to the post by sliding down into a suitable groove.
Claim.—Inserting in a bedstead post a segment of circle B, in a mortise with bevelled sides and lower end, for the purpose of securing it to the post A, without any other fastening, substantially as described.

No. 58,438.—H. L. LOWMAN, New York, N. Y.—*Pick and Axe Combined.*—October 2, 1866.—A pick and chopping axe are arranged in one tool, the axe is straight and is secured to the handle without the use of a wedge.
Claim.—As a new article of manufacture, a compound tool, consisting of an axe and pick united, the eye being of elliptical form, and of even size throughout its length, as described and represented.

No. 58,439.—J. B. LYON and ARNOLD DOLL, Cleveland, Ohio.—*Induction Coil for Electromagnets.*—October 2, 1866.—Explained by the claims and cut.
Claim.—First, making the coil for the main or direct circuit of two wires of dissimilar metals, as copper and iron, laid side by side in alternation, as specified.
Second, laying the coil for the induction current in two sections of about equal lengths of wire, one section being upon the core of the coil, and the other upon the outside of the main coil, the two sections being of one continuous wire, as herein set forth, in combination with a primary coil composed of dissimilar metals, as specified.

No. 58,440.—ANDREW MABON, Philadelphia, Penn.—*Skate.*—October 2, 1866.—Projections at the heel and toe of the boot fit in recesses in the skate runner, which is kept in position at the heel by a hinged plate.
Claim.—A skate having a projection d, hinged plate f, and recess g, adapted for the reception of a pin on the boot to which the skate is to be attached, the whole being constructed substantially as described.

No. 58,441.—FRANCIS A. MAKEPEACE, Worcester, Mass.—*Fastening for Window Blinds.*—October 2, 1866.—Two catches are supported by the same pivot, and have notches which engage the pin. A spiral spring keeps them closed, and a locking plate upon the blind prevents their being opened from the outside.
Claim.—First, the construction and arrangement of the double spring jaws, when constructed and operating substantially as set forth.
Second, the locking plate in connection with the double jaws, when constructed and operating in the manner and for the purposes substantially as set forth and described.

No. 58,442.—SIDNEY MALTBY and J. R. BROWN, Dayton, Ohio—*Lifter for Stove Covers, &c.*—October 2, 1866.—The tongs has two pairs of jaws standing at different angles, but opening and shutting simultaneously; the lower jaw in each case has ribs upon its face; the handles end in books.
Claim.—First, a stove lifter composed of the parts B B' and jaws D D and E E, constructed and arranged as and for the purpose shown.
Second, providing one jaw in each set with ribs.
Third, the books on ends of reins in combination with the reins and jaws, all substantially as and for the purposes set forth.

No. 58,443.—JAMES E. MCBETH. New Orleans, La.—*Safety Gunlock.*—October 2, 1866.—To prevent accidental discharge, the hammer is contained within the stock of the piece. The tail of the hammer is curved, and is engaged by the book on the upper end of the curved cocking trigger, which is locked by the rear trigger when the latter is placed in horizontal position and the hammer down.
Claim.—First, the cocking trigger B, formed with the curve b, in combination with the hammer A, formed with the curve a, substantially in the manner and for the purposes herein described.

Second, the notch *h* and spring *k*, (on B,) with the projecting point *o* (on D) and the projecting point *p*, (on C,) substantially in the manner and for the purposes herein described.

Third, the combination of the notch *h*, spring *k*, projecting points *o* and *p*, sear C, sear spring F, trigger D, and trigger B, substantially in the manner and for the purposes herein described.

Fourth, the combination of B and C, substantially in the manner and for the purposes herein described.

Fifth, the combination of the several parts A, B, C, and D, substantially in the manner and for the purposes herein described.

No. 58.444.—REUBEN MCCHESNEY, Ilion, N. Y.—*Breech-loading Fire-arm.*—October 2, 1866.—The breech pin falls by its own gravity as the hammer is moving to half-cock, and is again raised into position by means of a projection on the hammer as the latter is raised to full-cock. Other devices are explained in the claims and cut.

Claim.—First, lifting a swinging breech piece from an open to a close position entirely by the immediate action of the hammer, or of some projection thereof, without any intermediate device, substantially as herein specified.

Second, the relative arrangement and combination of the breech piece and hammer in such a manner that the breech piece may descend by its own weight and open the breech before the hammer reaches half-cock, and be raised again to close the breech while the hammer is passing from half-cock to full-cock, substantially as and for the purpose herein set forth.

Third, the combined arrangement of the breech piece, locking bolt, or its equivalent, and the hammer, in such manner that the breech piece shall be held locked independently of the hammer during the entire descent of the latter, then unlocked by the action of the hammer before the same reaches half-cock in ascending, and again locked by the action thereof at or before reaching full-cock.

Fourth, the sliding firing pin *q y*, arranged to slide in the side of the breech piece, substantially as above described.

Fifth, the dog *g* of the hammer piece of the bolt F, substantially as and for the purpose above described.

Sixth, the dog 6, in combination with the pendent plate *f* of the sliding bolt F, substantially as and for the purpose above described.

Seventh, the hook *p* of the hammer, in combination with the firing pin, constructed and operating substantially as and for the purpose above described.

Eighth, the mode above described of securing the end of the main spring, to wit, by fitting its end in a groove on the screw pin *n*, in combination with the projection 8 and recess 7, substantially as above described.

No. 58,445.—JOHN MCMURTRY, Lexington, Ky.—*Tie for Cotton Bales.*—October 2, 1866.—The ends are looped around each other, and lie between the band and the expansive cotton.

Claim.—The hoop-iron cotton bale tie A, constructed and operating in the manner and for the purposes herein set forth.

No. 58,446.—DUSTIN F. MELLEN, New York, N. Y.—*Griping Screw Blanks.*—October 2, 1866.—The device for holding screw blanks while the same are being threaded consists of a hollow spindle divided longitudinally and containing two pairs of forceps, the one which grasps the blank turning upon a stationary axis, while the other pair, which operates the first, turns upon an axis which passes through an elongated hole or slot in the spindle, and may be adjusted to suit the holding forceps to blanks of different sizes by an outer ring and jam nuts. Other devices are explained in the claims.

Claim.—A hollow spindle in which forceps or some other analogous device is placed, made in two parts by a longitudinal division, substantially as and for the purposes set forth.

Also, the adjustable fulcrum 8, for the adjustment of the forceps, regulated by the screw nut around the spindle, by which it is shifted and held, substantially as and for the purposes set forth.

Also, inserting folding springs 6 within the spindle, to open the jaws of the forceps, constructed, arranged, and combined with the forceps and spindle, as and for the purposes set forth.

Also, the hardened steel bearing 10, combined with the spindle and box 12, substantially as herein specified.

Also, the hardened bearing and box 15 and 16, in connection with a forward bearing that will not allow end chase, constructed and arranged substantially as and for the purposes specified.

Also, the sliding ring 18 and hardened ring 19, for closing the forceps, substantially as herein set forth.

No. 58,447.—EDWARD MELLON, Scranton, Penn.—*Attaching Tires to Wheels of Locomotives.*—October 2, 1866.—Explained by the claim and cut.

Claim.—The wheel with the curved flange upon the inner edge, in combination with a tire with a rounded corner to fit said curved flange, as set forth.

No. 58,448.—S. MERRICK, New Brighton, Penn.—*Construction of Iron Railroad Cars.*—October 2, 1866.—The edges of the corrugated metallic panel are clasped by compressing bolts, between the salient and receding faces of the vertical plates attached to the stanchions of the car.

Claim.—An iron railroad car, having its panels F secured in position by metal stanchions C D, provided respectively with a groove *a*, and rib *b*, and connected together, and to wooden stanchions B, by bolts E, substantially as herein shown and described.

No. 58,449.—JOSHUA MERRILL, Boston, Mass.—*Cask, Barrel, &c.*—October 2, 1866.—Explained by the claims and cut.

Claim.—In combination with the joints of a cask suitable for holding and transporting liquids, having the common straight or plain joints, a coating or stuffing of glue or similar gelatinous cement applied to the joints before putting the cask together, substantially as described.

No. 58,450.—JOSHUA MERRILL, Boston, Mass.—*Cask, Barrel, &c.*—October 2, 1866.—Explained by the claims and cut.

Claim.—The improved cask, substantially as described, having lapped joints formed by a lip and rebate, substantially in the way and for the purposes described.

In combination with the lapped joints of lapped-jointed casks, substantially as described, a coating or stuffing of glue, or similar gelatinous cement, applied to the lapped joints, substantially as described, whereby said joints are firmly glued or cemented together.

In combination with the lapped joints of lapped-jointed casks, substantially as described, a coating or stuffing of shellac, rosin, or other similar resinous cement, applied to the lapped joints, substantially as described, whereby the lapped joints are more securely protected against leakage.

No. 58,451.—JOSHUA MERRILL, Boston, Mass.—*Cask, Barrel, and Keg.*—October 2, 1866.—Explained by the claim and cut.

Claim.—As an improvement in casks, barrels, and kegs used for holding and transporting liquids, the improved cask, substantially as described, having its joints made with two or more tongues, and two or more corresponding grooves at each joint, substantially in the way and for the purposes hereinbefore described.

In combination with the joints, a cask having its joints made with two or more tongues, and two or more corresponding grooves to each joint, and a coating or stuffing of glue, or similar gelatinous cement, applied to the said joints, substantially as and for the purposes described.

In combination with the joints of a cask having its joints made with two or more tongues, and two or more corresponding grooves to each joint, a coating or stuffing of shellac, rosin, or similar resinous cement, applied to the said joints, substantially as and for the purposes described.

No. 58,452.—JOSHUA MERRILL, Boston, Mass.—*Barrel and Keg.*—October 2, 1866.—Resinous cement is applied to the contiguous faces of the barrel staves and to the edge of the head before setting up.

Claim.—In combination with the joints of a cask, suitable to hold and transport liquids, having plain or straight joints, a coating or stuffing of shellac, rosin, or other similar resinous cement, applied to the joints, and substantially in the way and for the purposes described.

No. 58,453.—CHARLES S. MERWIN and CHARLES A. METCALF, Dubuque, Iowa —*Combined Foot-Warmer and Reflecting Lamp.*—October 2, 1866.—Explained by the claims and cut.

Claim.—First, the employment, in a combined foot stove and lantern, of a reflector C, substantially as and for the purpose specified.

Second, the combination of the glass plate *d*, the lamp *c*, and reflector C, substantially as specified.

No. 58,454.—SAMUEL MERWIN, Springfield, Mass.—*Prepared Paste.*—October 2, 1866.—Composed of flour, 2 lbs ; common salt, 1 oz.; alum, ¼ oz.; corrosive sublimate, 6 grs.; and acetic acid, ½ oz.

Claim.—As a new article of manufacture, the substance herein described.

No. 58,455.—WALLACE A. MILES, Meriden, Conn.—*Spice Box.*—October 2, 1866.—By turning the slotted lid it may expose the perforated plate beneath, or otherwise form a tight cover.

Claim.—The covers A, having perforations in a radial line, registering with the slotted cover B, of a box, operating substantially as described for the purpose specified.

No. 58,456.—DAVID R. MILLER, Harrisburg, Penn.—*Dust and Shaving Conveyer for Planing and Sawing Machines.*—October 2, 1866.—The case is so placed as to receive the shavings from the rotary cutter, which are discharged by forcibly striking the inclined sides of the tube, assisted by the current of air generated.

Claim.—The conveyer A made of metal or wood, provided with the hooks F F, flanges C and D, with opening B, when constructed substantially as described, and used for the purposes herein specified.

No. 58,457.—BERNARD MORAHAN, Brooklyn, N. Y.—*Sand Pump.*—October 2, 1866—A loop formed on the piston rod encloses the bail on the top of the cylinder; each portion has valves opening upward, and the piston rod is made the means of operating and withdrawing the cylinder.

Claim.—The split linked piston rod E, in combination with the piston C, operating with the bail B firmly secured to the cylinder A, as and for the purpose specified.

No. 58,458.—G. W. MORSE, New York, N. Y.—*Paint for Ships' Bottoms.*—October 2, 1866.— Composed of antimony, 80 parts; lead, 15; cement-copper, 5; naphtha, 1; benzine, 1; tar, 2.

Claim.—A paint composition made of the ingredients above specified.

No. 58,459.—CHARLES MULCHAHEY, Springfield, Mass.—*Lamp.*—October 2, 1866; ante dated September 15, 1866.—The material in the trough around the wick tube absorbs oil which trickles down from the wick.

Claim.—The application to an oil lamp of an absorbent or retaining material around outside of the tube *c*, substantially in the manner and for the purpose described.

No. 58,460.—JONAS MULL and ASA D. REED, Troy, N. Y.—*Spider.*—October 2, 1866; antedated September 23, 1866.—An annular frame flaring downwardly has gains to receive side projections on the upper rim of a skillet placed within it. Gains in the skillet admit projections beneath the lid. The skillet may be raised at the handle, the side projections supporting it.

Claim—The employment of a spider for culinary purposes, constructed with an outward projecting flange I, with apertures D upon the upper edge thereof, and with the projections G', so as to suspend the said spider back of the centre thereof, in combination with the outer rim A, and surrounding or annular chamber A', each being arranged in the manner substantially as herein described and set forth.

No. 58,461.—HEZEKIAH MONROE, Fall River, Mass.—*Fastening for Window Blinds.*—October 2, 1866.—When the blind is swung the upper plate strikes the pin, vibrating it and striking a stud in the lower plate, causing it to engage with the pin on the sill.

Claim.—The combination of the two plates H and I, when constructed and connected together so as to operate substantially in the manner described and for the purpose specified.

No. 58,462.—G. W. NELLIS, Richmondville, N. Y.—*Folding Table.*—October 2, 1866.—The hooks being detached, the hinged legs are doubled within the frame, which is then folded upon itself.

Claim.—The table having a folding top A, slides K, sockets L, hinged legs H, brace rod *g*, and button I, arranged and operating substantially as described for the purpose specified.

No. 58,463.—E. B. NOURSE, Eaton, Ohio.—*Alphabet Frame.*—October 2, 1866.—Each letter of the alphabet, on the side toward the deaf person, is covered by a shutter: these are successively opened by a trigger in the rear, so as to spell words and sentences to the eyes of the spectators.

Claim—The frame constructed as described, having spring pads connected by link, or equivalent, to shutters covering letters, and operating substantially as described and represented.

No. 58,464.—BENJAMIN OLDFIELD, Newark, N. J.—*Loom for Weaving Narrow Wares.*—October 2, 1866.—The provision for throwing the racks out of gear with the shuttles, permits them to be separately taken out and the remainder of them kept in action.

Claim.—First, making the racks *f* adjustable by means of movable strips, or any other equivalent means which will produce the same effect, for the purpose of throwing them out of gear with the pinions *c*, as set forth.

Second, supporting the sleys or reeds C by springs, which will allow of depressing said reeds, for the purpose of removing them as desired, substantially as described.

No. 58,465.—JOHN H. J. O'NEILL, New Haven, Conn.—*Carriage Shackle.*—October 2, 1866.—The coupling pin lies within the semicircular recesses in the jaws, the upper one of which is pivoted and retained by a cam beneath its rear projection and held in position by a split traverse pin.

Claim.—The combination of the base A and the lever C, hinged thereto with a cam F, pivoted to the base A, constructed and arranged to operate substantially in the manner and for the purpose specified.

No. 58,466.—WILLIAM ORR, Jr., and GEORGE F. WRIGHT, Clinton, Mass.—*Apparatus for Holding Paper Ribbon.*—October 2, 1866.—The paper is furnished in long rolls, is pasted by rollers in uncoiling, and severed by a cutter.

Claim.—The use of the shears *f*, or any device which will sever paper, in connection with the paper ribbon *a* and paste rolls *d* above mentioned, as and for the purpose specified and shown.

No. 58,467.—WILLIAM S. PADDOCK, Albany, N. Y.—*Fastening for Trunks, Boxes, &c.*—October 2, 1866.—The edge plate of the lid has pins whose heads pass through holes in the edge plate of the box and through a slotted plate beneath, which, when shot by the key, engages the heads, to prevent the lifting of the lid, locking it all along its length.

Claim.—The combination of the plates A B C with the lock D, all constructed and arranged substantially as herein described, and constituting a fastening for trunks, bags, valises, and boxes.

No. 58,468.—J. J. PARKER, Marietta, Ohio.—*Packing for Deep Well Tubes.*—October 2, 1866.—Two rings of leather intervene between the pump and the well tubes; in the annular space between the rings is a leather bag inflated with water from a pipe above.

Claim.—First, the sleeve F, pin *g*, and slot *m*, in combination with the safety band H, in the manner described.

Second, the outer and inner tubes A B, the bands C, and the inner metallic sheet D, in combination with the tube E, in the manner and for the purpose described.

'No. 58,469 —J. J. PARKER, Marietta, Ohio.—*Well Packing.*—October 2, 1866.—A coil of hose in the space between the pump and well tubes is distended with air or water introduced from above.

Claim.—The tube A, in combination with the expansible metal case B and lining C, all substantially in the manner and for the purpose described.

No. 58,470 —J. J. PARKER and E. D. PARKER, Marietta, Ohio.—*Machine for Making Dies for Braid and Embroidery.*—October 2, 1866—The sliding cutter is attached to a moving bar having on its free end a stationary cutter. By depressing the bar the sliding cutter clamps the pin against the stationary cutter, and holds it while being forced into the block of wood which forms the face of the stamp, and then cuts it smoothly off.

Claim.—The sliding cutter D and stationary cutter E, in combination with the moving arm B or bar, or their equivalents, by which the pins or wire are set and cut, as set forth.

Also, the turn-table L, set in the frame of the stamping machine, and used to regulate the width of the blocks and the length of the wires constituting the stamps, in the manner described.

No. 58,471.—ANDREW PATTERSON, Birmingham, Penn.—*Apparatus for Carburetting Air, &c.*—October 2, 1866.—The pan is inclined and has partitions with communicating openings at alternate ends and at their lower edges. Liquid is supplied above, and the air forced in below passes through the chambers consecutively and is carburetted by contact with the liquid.

Claim.—The combination and arrangement of the divisions *c c c* within the adjustable inclined pan *a*, with their alternate submerged passages *d d d*, the whole being arranged and applied substantially in the manner and for the purposes set forth.

No. 58,472.—P. PERDEW and H. A. PERDEW, Seal, Ohio —*Filter for Sirup.*—October 2, 1866.—The box has a series of compartments communicating with each other alternately at the top and bottom, and containing straw or other filtering material.

Claim.—First, the use of two or more filter boxes for the purpose of cleansing sorgho sirup.

Second, the downward and upward flow of the sirup in passing through, as set forth.

Third, the combination of the filter boxes, below the apron, with the other filter boxes, for the purpose set forth.

No. 58,473.—HARVEY and ALVAH PHELPS, Albany, N. Y.—*Apparatus for Cutting and Stamping Soap.*—October 2, 1866.—The soap is fed intermittingly along the table by a rack and pawl actuated by a lever; a frame carrying a vertical wire moves transversely across the table and cuts off the bars, which are stamped by the die-block, which is depressed by the treadle; knives trim the ends of the bar to a length.

Claim.—First, the mode herein described for cutting and stamping soap, substantially as set forth.

Second, the arrangement of the feed guide E, cutting frame F, sliding transversely to said feed guide, in combination with the stamping bar H and edging knives ff, the whole constructed and arranged in order to operate as set forth.

Third, the employment of a cutting wire a tensioned in a transverse sliding frame F when used in combination with an intermitting feed, as shown and described and for the purpose specified.

No. 58,474.—CHARLES W. PIERCE, Albany, N. Y.—*Grate Bar.*—October 2, 1866.—The upper portion of the bar is detachable from the lower part, allowing of its removal when burnt out.

Claim.—A compound grate made by the combination of an upper grate D, formed as shown, resting by its outer bars upon the entire upper surface of the outer bars of a lower grate A formed as shown, the grates being held to each other by dovetailed joints formed along the outer edges of their outer bars at f, in the manner and for the purposes set forth in this specification.

No. 58,475.—EMMETT QUINN, Washington, D. C.—*Storing Petroleum, &c.*—October 2, 1866.—The framework is to be attached to a pier or between two boats and is filled with the barrels containing oil; its purpose is to protect the barrels by keeping them immersed.

Claim.—A dock or crib with its appurtenances in which to store petroleum or other oils in barrels or casks, constructed and used in the manner substantially as described.

No. 58,476.—JACOB K. REINER, Line Lexington, Penn.—*Cultivator.*—A lateral adjustment of the beams is made by a movement of converging slide bars.

Claim.—The bars G G', having the handles H H attached and connected by the screw J, provided with the nut K, in combination with the bars F F, attached to the inner sides of the beams A A, and passing through eyes a, at the end of a bar L, secured to the under side of the front bar G, with the clamp bar M, attached to the under side of the rear bar G, all arranged substantially as and for the purpose set forth.

No. 58,477.—WILLIAM RESOR, Cincinnati, Ohio.—*Heating Stove.*—October 2, 1866.—The air enters the side, traverses the fire space, passes down between a fire plate and the stove side, then beneath the fire chamber and out through the smoke flue.

Claim.—The arrangement of chamber A B C, interior descending flue E, bottom flue F, neck G, external ascending flue H, collar J, and registered draught inlet K L, substantially as and for the purpose set forth.

No. 58,478.—GEORGE H. REYNOLDS, New York, N. Y.—*Steam Engine.*—October 2, 1866.—The stay rods which hold the guides for the cross head bind the cylinder and the shaft bearings together to resist the rending strain The link block is fitted by "shimming," and is adjustable in the slotted link by means which allow it to be moved to and fastened in the position required.

Claim.—First, the rods D, arranged as represented relatively to the cylinder E, cross-head F, bed piece C, and binders C, so as to receive the crushing and rending strain between the cylinder and the binders in the line of their axes, as herein set forth.

Second, arranging the link H to hold the link block rigidly in the desired position thereon, substantially in the manner and for the purposes herein set forth.

Third, the pinion O, carried on the link block, in combination with the rack h on the link, and with means for confining the link firmly to the block in any desired position, so as to prevent loose clay between the parts while the engine is working, substantially as and for the purpose herein set forth.

Fourth, the compound link block M L K, adapted to allow of adjustment by shimming, substantially as herein set forth.

No. 58,479.—SAMUEL H. RHOADES, Clyde, Ohio.—*Driving Pump.*—October 2, 1866.—The sections of the pump are secured by collars; the joints sustain the valves in position; the drill head corresponds in diameter with the screw collars.

Claim.—The point I, with an enlargement or projection e', corresponding in size to the sides of the tubular sections and collars, in combination with the said sections and collar valves E D, and rod G, arranged and operating in the manner and for the purpose set forth.

No. 58,480.—EDWARD RICHMOND, Brooklyn, N. Y.—*Whip Rack.*—October 2, 1866.—The edges of the slots in the elastic material grasp the whip by its stock or lash, and suspend it.

Claim.—First, in a rack or frame for holding whips, composed of plates of wood or other suitable material, with an interposed sheet of rubber or other elastic substance, as specified, the V-shaped openings or slots formed in the edge of said rack for holding the whip by its tip, the same being constructed and arranged substantially as herein described.

Second, as a new article of manufacture, a rack or frame composed of the materials herein described, in which the circular holes *a* and *c* are combined with the V-shaped openings formed in the edge of said rack, the whole being constructed and arranged for operation substantially as shown and set forth.

No. 58,481.—KARL RIEDEL, Guttenburg, N. J.—*Lamp.*—October 2, 1866.—The absorbent filling is saturated with burning fluid, the wick being supplied by capillary attraction. The hollow handles condense and return accumulating vapors. The non-absorbent material above the absorbent prevents the transmission of heat from the burners to the oil.

Claim.—The reservoir A, having hollow handles H, with suitable absorbent material D, and non-absorbent material E, in combination with the gauze cylinder F, constructed, arranged, and operating substantially as described for the purpose specified.

No. 58,482.—MICHAEL RIEHL, Philadelphia, Penn.—*Bookbinder's Paper Cutter.*—October 2, 1866.—The knife stock is adjusted by a gauge above the bed. The foot is pressed upon the treadle, clutching the shaft and revolving the main wheel, one of whose pintles at every semi-revolution engages the notch in the pendent bar, connected by an elbow lever to the cutter; the latter is again raised by a counterpoise weight. The bed is raised coincidently by the cams, &c., beneath.

Claim.—First, the wheel L, provided with pintles, in combination with the pendent lever N, and elbow lever O, and counter-balance *k*, substantially as and for the purposes described.

Second, the cams a^2 and a^3 in combination with the wheels o^2 and o^3, and tables C, for the purposes and substantially as described.

Third, a paper-cutting machine by which the paper is pressed by the action of the table raising against the cutter, and so constructed that when the cutter has passed through the paper the table and cutter instantly recede from each other, for the purposes and substantially as herein described.

No. 58,483.—A. J. ROGERS and W. D. RINEHART, Pittsburg, Penn.—*Machine for Making Spikes.*—October 2, 1866.—The spikes are rolled out from point to head, by a wheel carrying the adjustable dies, acting in connection with a reciprocating bed die having projections to correspond with those of the rotating dies to form the points and heads. The spikes are released by the depression of one side of the lower die, which is made in sections for that purpose.

Claim.—First, forming the head of the spike by rolling it from point to head between a flange roll and reciprocating die, constructed and operating substantially as hereinbefore described, for the purpose of enlarging the head without bending or upsetting.

Second, the use, in combination with the die *f*, forming the bottom and one side of the cavity for forming the spikes, of the poppets or side dies *m*, constructed and operating substantially as hereinbefore described.

Third, the use, in combination with the poppets or side dies *m*, of the drum F, and table H, for elevating, and the roller *q* for depressing the poppets, constructed and operating substantially as hereinbefore described.

Fourth, the use of the wedges *k* in combination with the frames *h*, and die *f*, for the purpose of adjusting the die *f* to the flanged roll B, substantially as hereinbefore described.

Fifth, the use of the rollers *l l*, in the housing frame pressing against one side of the die table E, and the rollers *s s*, in the adjustable frame I, pressing against the poppets or side dies on the other side of the table E, in combination with the die table E, die *f*, and poppets *m*, for the purpose of keeping the table in proper position, and preventing the spreading of the poppets, substantially as hereinbefore described.

No. 58,484.—JOHN ROUSE, Port Gibson, N. Y.—*Roofing Material.*—October 2, 1866.—The tile is formed of lime and sand mortar, and is saturated with asphaltum, petroleum, or coal tar.

Claim.—First, the within described process of producing a tile by moulding it from sand and lime mortar in moulds lined with paper or cloth, and afterward saturating it with coal tar, asphaltum, or petroleum, substantially as and for the purpose described.

Second, a tile made of sand and lime mortar, with bevelled edges, and saturated with coal tar, asphaltum, or petroleum, as a new article of manufacture.

No. 58,485.—JOSEPH ROY, Boston, Mass.—*Machine for Forging Horseshoe Nails.*—October 2, 1866.—The nail rod is confined in intermittingly quadra-rotating shafts in a frame subject to longitudinal movement. The nail is formed and cut off between two reciprocating hammers and an anvil having vertical adjustments for those purposes.

Claim.—The combination as well as the arrangement of the stationary anvil C, the movable anvil D, the two hammers F G, and the mechanism for supporting and revolving the nail rod and moving it lengthwise, as set forth, such movable anvil and hammers being provided with mechanism for operating them, substantially in the manner as specified.

Also, the combination as well as the arrangement of the cutter E, with the anvil C, the movable anvil D, the two hammers F G, and the mechanism for supporting and revolving the nail rod and moving it lengthwise, as set forth, such movable anvil, hammers, and cutter being provided with mechanism for operating them, substantially as hereinbefore specified.

Also, the mechanism or combination for supporting and revolving the nail rod and moving it lengthwise, as set forth, the same consisting mainly of the shaft S, the box T, the rail U, the lantern wheel V, the cam X, the projection Y, the spring z, the head a^2, the slotted post w', the rack bar H, with its toothed racks, the pawls b' c', the slider u, the spring y, the cam a, the shaft k, the crank u', the rod t', arm s', shaft r', arm q', pawls o' p', connecting rod r', lever l', and link m', the whole being substantially as specified.

Also, the combination of the cams p', the levers q q', and springs r r, or their mechanical equivalents, with the hammers F G, as arranged and combined with the stationary and movable anvils C D, and mechanism for supporting and operating the nail rod, substantially as described.

No. 58,486.—NELSON SAFFORD, Pleasant Valley, Vt.—*Hand Planter.*—October 2, 1866.— The seed receptacle on the hoe has slides worked by a cord from the hand, and by recoil springs; the seed cavities are varied in size by the introduction of elastic annular bands. The seed is spread by falling upon convex plates beneath the drop holes.

Claim.—First, the elastic bands b, one or more, fitted in the tube or hole E of a seed dropping slide for the purpose of varying the capacity of the same, substantially as and for the purpose set forth.

Second, the scatterers, constructed as shown, or in any equivalent way, when applied to a hoe seed planter, substantially as and for the purpose specified.

No. 58,487.—NATHANIEL C. SAWYER, Boston, Mass.—*Separating Crude Emery from Foreign Substances.*—October 2, 1866.—Explained by the claims and cut.

Claim.—First, the process of cleaning emery and separating it from extraneous softer minerals and other substances by motion and friction in revolving hollow cylinders or receptacles, in connection with the introduction of currents of water or air into such cylinder, or receptacles, to facilitate the separation and escape of such extraneous materials, substantially as set forth.

Second, the process of polishing and rounding emery, by subjecting it, when ground or crushed, to friction in hollow revolving cylinders or receptacles without currents of air or water, substantially as set forth.

No. 58,488.—J. P. SCHMUCKER, Ashland, Ohio.—*Wood-bending Machine.*—October 2, 1866.—The blank is steamed, and its end being clamped between a segment former and an elastic outer band, is placed beneath a roller on the bed, and caused by a screw to travel beneath the roller, assuming the shape of the "former," to which it is eventually clamped at the latter end, and so remains till dry.

Claim.—First, the table G, clamps K, plate M, and former P, when constructed and operating as and for the purpose specified.

Second, in combination with the table G, plate M, and former P, the guide blocks S, constructed as specified.

No 58,489.—REUBEN SHALER, Madison, Conn.—*Turn-table for Baker's Ovens.*—October 2, 1866.—The circular table is revolvable on its base, which sets on any part of the oven floor.

Claim.—The herein described turn-table, as a new article of manufacture.

No. 58,490.—A. G. and P. M. SHULTS, Avoca, N. Y.—*Money Drawer.*—October 2, 1866.—The treadle knob raising the weighted lever beneath the floor releases the catch, when a spring throws the drawer open. The gravitation of the lever again shuts the drawer.

Claim.—First, the combination of the loaded lever E, and locking mechanism, or their equivalents, with the drawer C, operating substantially as and for the purpose specified.

Second, the spring D, in combination with the drawer C, and locking mechanism, or their equivalents, substantially as and for the purpose set forth.

Third, the combination of the lever E E, rod G, spring catch I, bracket J, cord L, and spring D, with the drawer C, substantially as and for the purpose set forth.

No. 58,491.—THOMAS SILVER, New York, N. Y.—*Governor for Steam Engines.*—October 2, 1866.—The governor controls by the combined action of a momentum wheel with speed-limiting vanes and a spring. The usual sliding sleeve is dispensed with, and an oscillating sleeve is arranged to transmit power from the motor to the wheel, and variations of speed of the latter from it to the valve. The angular presentation of the vanes is adjustable while in motion by their connection with a sleeve on the shaft.

Claim.—First, the oscillating sleeve fitted to the governor spindle, and so geared therewith that the driving power is transmitted through it to the momentum wheel and its attached vanes, and so connected with the spring and with the regulating valve that the controlling influences of the wheel vanes and spring are transmitted through said sleeve to the valve, substantially as herein described.

Second, connecting the spring with the oscillating sleeve by means of an eccentric segment or sector, so arranged as to make the action of the spring always uniform or equivalent to that of a weight, substantially as herein specified.

Third, the connection of the speed-limiting vanes with a sleeve or piece which is adjustable longitudinally in relation to the governor spindle, substantially as and for the purpose herein set forth.

No. 58,492.—ALBA F. SMITH, Norwich, Conn.—*Signal for Railroad Drawbridges.*—October 2, 1866.—Upon releasing the draw or unlocking it by means of levers, the signals are raised to indicate that the draw is open. The devices are explained in the claims and cut.

Claim.—First, in drawbridges mechanically connecting the locking devices D¹ D²; with the danger signal or signals, so that the draw cannot be liberated without the danger signal being properly set in advance of the commencement of the movement of the draw from the safe position, and the safety signal cannot be shown until the bridge is locked in the safe position, all substantially as and for the purpose herein set forth.

Second, in combination with the above, so connecting the night and day signals that both shall be operated at the same time.

Third, in combination with the above, the locking of the signals by means F⁷ R r¹ r⁸. or their equivalents, so as to necessarily continue to indicate danger during the whole period that the draw is out of the safe position, substantially as and for the purpose herein specified.

No. 58,493.—SAMUEL SMITH, Philadelphia, Penn.—*Cooking Stove.*—October 2, 1866.— The boiler-hole plate is movable upon a circular series of rollers which traverse a groove in the top plate of the stove; holes in the groove allow dust to fall through.

Claim.—First, the movable plate B, with its boiler holes and cross pieces, the whole being applied to the top of a cooking stove or range, substantially as and for the purpose herein set forth.

Second, the annular cavity x in the plate D, the said cavity containing any desired number of rollers, in combination with the plate B and its annular rib m.

Third, the perforations t in the plate D, beneath the rollers q, for the purpose described

No. 58,494.—STERRY SMITH, Salem, Mass.—*Grate Bar.*—October 2, 1866.—The grate bars are cast in sets, each bar alternately connected to the bars on either side; every third bar is projected downward in an inverted arch form.

Claim—A grate bar composed of a series of parallel, longitudinal bars, connected together by lugs placed alternately on opposite sides of the next adjoining bar, in combination with the curved bars H H', alternating with the bars G G', as shown and described.

No. 58,495.—EDWARD SNEIDER, Baltimore, Md.—*Horseshoe.*—October 2, 1866.—The supplemental roughing shoe is attached to the upper shoe by toe clips and a sliding screw-block which nips its rear edge. Pins in the rear prevent displacement and a plate protects the operative parts.

Claim.—First, a supplemental shoe B, provided with the inclined toe pieces B' B', sliding clamp E, and screw bolt F, whereby said supplemental shoe is to be clamped to the fast shoe, substantially as described.

Second, the split spring-washer G, applied to the bolt F, to prevent the same from working loose, substantially as described.

Third, the studs a a, formed on the rear extremities of the supplemental shoe B, and employed to prevent the lateral displacement of the same, substantially as described.

Fourth, the combination with the supplemental shoe B, of the clamp E, guides C C, and projecting plate B², all constructed and arranged substantially as and for the purpose herein specified.

No. 58,496.—T. S. SPEAKMAN, Philadelphia, Penn.—*Safety Valve in Water Backs for Ranges.*—October 2, 1866.—The upper portion of the water back has a spring safety valve, and an escape pipe to permit steam to escape under extra pressure.

Claim.—The combination, substantially as described, of the valve B, with the water back A, for the purpose specified.

No. 58,497.—JAS. H. SPENCER, Philadelphia, Penn.—*Manufacture of Floor Coverings.*— October 2, 1866.—The fibres of burlap are laid in one direction by brushing; the prominent fibres removed by shearing; it is then wetted and passed between calendering rollers, and finally printed or painted in colors.

Claim.—Burlap, treated and printed, or otherwise colored, substantially in the manner and for the purpose herein described.

No. 58,498.—ELBRIDGE G. STANLEY and JAMES GOODRICH, Fitchburg, Mass.—*Angular Saw Gauge.*—October 2, 1866.—The saw guide face-plate is connected to the slide by means of links so that the guide moving in the centres of the links is adjustable to any angle.

Claim.—The links F F, when used for and applied to the purpose specified, to wit: to connect the plates B and C, constructed substantially as described and set forth, or in any similar manner.

No. 58,499.—GEORGE STEAD, Brooklyn, N. Y.—*Roofing Cement.*—October 2. 1866.— Paint skins are softened by boiling in a solution of caustic potash, linseed oil being added during the boiling; mineral paint is then added and the mixture is strained and put up in vessels for use in pointing and painting roofs and other out-door surfaces.

Claim.—A roofing cement, compounded and prepared substantially as herein described, and for the purposes set forth.

No. 58,500.—CLINTON STEEN, Athens. Ohio.—*Plough.*—October 2, 1866.—The shovel is adjustable up or down by means of a slot in the knee and separate holes for the entrance of a stud under the shovel. The handles are bolted in sockets on the side of the sheath.

Claim.—The novel construction of the plough knee and the mode of attaching the handles, as set forth in the above specification.

No. 58,501.—FRANCIS STEFANI, New York, N. Y.—*Setting Stones in Jewelry.*—October 2, 1866.—The bezel consists of two fixed and two removable claws, the latter one held by screws.

Claim.—The mortised jewel frame A, adjustable clamping fingers e e, and retaining screws d d, constructed and arranged substantially as and for the purpose herein described.

No 58,502.—A. STEWART, Cincinnati, Ohio.—*Top for Fruit Cans.*—October 2. 1866.— The groove for the cement is formed of an annular piece of metal spun into proper shape and attached to the brim of the can.

Claim.—A rim, or trough A, for a metal fruit can, constructed and applied in the manner herein shown and described.

No. 58,503.—C. STIERLE and JOHN C. BAER, Cincinnati, Ohio.—*Steam Engine Globe Valve.*—October 2, 1866.—The valve seat is separable from the valve chamber. The valve is raised or depressed by means of corresponding screw threads upon the stem and a collar; the latter is held fast except when the valve is ground, in which case it is loosened and acts as a guide to the stem.

Claim.—The independent valve seat E, for the valve H, substantially as described.

Also. the combination of the valve stem I, independent screw nut J, and hub or plug L, connected together, substantially as and for the purpose specified.

No. 58,504 —W. H. STRAHAN, Philadelphia, Penn.—*Tool Rest for Grindstones* —October 2, 1866.—The base of the rest is adjustable laterally and longitudinally on the frame of the grindstone and supports a holder in which the tool is clamped and which is adjustable as to the angle of presentation, vertically and horizontally.

Claim.—First, the combination of the adjustable foundation plate A, plate B, the adjustable plate D, and bar E.

Second, the combination of the bar E, its adjustable guard I, arm G. movable block H, and adjustable collar H', the whole being arranged substantially as set forth for the purpose specified.

Third, the guard I, constructed for adjustability and retention on and removal from the bar E, substantially as described.

Fourth. the clamping pieces K, constructed and adapted to the plate B, with serrated edge, and to the foundation plate A, substantially as and for the purpose herein set forth.

No. 58,505.—JOHN TAGGART, Philadelphia, Penn.—*Skirt Supporter.*—October 2, 1866.— The skirt is buttoned to the belt and the latter suspended by strings and stays from a yoke upon the shoulders.

Claim.—Yoke y, double stays c^1 $c^{1'}$, c^2 $c^{2'}$, c^3 $c^{3'}$. single stay c'' c''', belt with buttons M, endless strings 8 P, when combined and contrived together for the purpose and in the manner above described.

No. 58,506.—A. H. TAIT and JOSEPH W. AVIS, New York, N. Y.—*Steam Generator.*— October 2, 1866 —The sections are riveted to the circumferential flanges of the T-rings, their edges resting against the radial flanges thereof. Bars unite the radial flanges of the inner and outer portions, and longitudinal stay bolts bind adjacent T-rings.

Claim.—First, the T-rings C, in combination with the cylinders A B, made in sections and connected to said rings, substantially as and for the purpose described.

Second, the longitudinal stay bolts D, in combination with the rings C and cylinder A B constructed and operating substantially as and for the purpose set forth.

No. 58,507.—JOHN A. TAPLIN, Carthage Landing, N. Y.—*Machine for Cutting Wood.*— October 2, 1866.—The knife sets in slots in the heads upon the bent shaft, and is fastened by bands and keys. The bent shaft gives an enlarged throat and prevents choking.

Claim.—In a rotary wood-cutting machine, the circular head, constructed as herein described, for fastening the cutter and the crank shaft, each constructed and arranged substantially as and for the purposes herein described.

No. 58,508.—JOHN S. P. TAYLOR, Oxford, Ohio.—*Lap Skiver.*—October 2, 1866.—As the strap is drawn across the edge of the knife the eccentric gauge block swings on its bearing and gradually descends, tapering the strap toward the end. Each part is adjustable to vary the operation within given limits.

Claim.—First,the combination with the knife C, or its equivalent, of an eccentric swinging or vibrating gauge, constructed substantially as and for the purpose set forth.

Second, the combination with the knife C, and gauge D, of the springs B, and screws F, or equivalent device, for adjusting the said gauge in relation to the said knife, substantially as described.

Third, the combination with the gauge D, and knife C, of the spring plate K, constructed and arranged substantially as herein set forth.

No. 58,509.—JOHN THEAT, Detroit, Mich.—*Tuyere.*—October 2, 1866.—The cap and stationary plate below it have square holes in their centres, and three others arranged at equal distances in a circle around them. The cap is made to rotate so as to admit the air through all the holes, or shut it off from the outer holes. It also has a flange projecting over the walls of the air chamber, and is convex on the top so as to shed the slag.

Claim.—The combination of the convex or conical cover A, provided with a flange projection A', overlapping the walls of the mouth of the tuyere with the stationary disk B and chamber L, when constructed and arranged substantially as and for the purpose set forth.

No. 58,510.—J. PATTON THOMPSON, Philadelphia, Penn.—*Baby Jumper.*—October 2, 1866.—The seat is suspended by an elastic strap and a hook from a pulley above, which traverses horizontally on a bar suspended by hangers from the ceiling.

Claim.—First, the hangers a a', rod A. traversing pulley C, straps D and E, and the elastic cord G, when combined and arranged substantially as herein specified and described.

Second, the hook l, and ring b, straps D and E, elastic cord G, and the chair H, when combined as specified and described.

No. 58,511.—JOHN TILTON, New York, N. Y.—*Preserving and Curing Meat.*—October 2, 1866.—Explained by the claim and cut.

Claim.—Curing meats by first cutting them from the bone then expelling the water by pressure, then mixing the curing composition, and smoking loose on shelves or in the canvas bags, substantially as set forth.

No. 58,512.—PETER H. VANDER WEYDE, Philadelphia, Penn.—*Double Still for Petroleum.*—October 2, 1866; antedated September 21, 1866.—The still is placed directly over a fire box, and has a d.me communicating with a condenser. The condensing coil terminates in a pipe directly over a funnel leading to a second still situated over the flue of the fire chamber. The dome of the second still communicates with a condensing coil which terminates in the pipes over a funnel communicating with the first still. Liquid is withdrawn from the second still, or either condenser, according to the quality required.

Claim.—A double still, in which all the defects and objection against other double stills are corrected, in the manner described.

No. 58,513.—WILLIAM WALSH, WILLIAM WALSH, Jr., and M. J. WALSH, Brooklyn, N. Y.—*Hat-finishing Press.*—October 2, 1866.—The jointed dies open and close with the vertical motions of the leaves of the table in which they are seated. The blocks of unequal sections are divided on an inclined line, and the superincumbent platen descends upon the brim. A single shaft actuates all the operative parts.

Claim.—First, the press table F, with dies H and leaves K. hung on adjustable hinges so that the hinges can be raised and lowered to suit dies of different heights, substantially as described.

Second, the combination of the table F, having sectional dies H, with a platen E, the dies and platen being operated from the same shaft, substantially as described.

Third, the hinged dies H, constructed in two unequal sections, for finishing and pressing hats, substantially as and for the purpose specified.

Fourth, the overlapped hinges I, in combination with the leaves K and tables F, which contain the dies H, substantially as described, for the purpose specified.

Fifth, the sectional dies H, in combination with the platen E, substantially as described.

Sixth, the sectional hat-blocks P, divided obliquely from near the centre of the tip to opposite side, substantially as described, for the purpose specified.

No. 58,514.—L. A. WARNER, Freeport, Ill.—*Cider Mill and Fruit Press.*—October 2, 1866.—This mill is so arranged that the pomace is periodically pressed and discharged while the grinding is continued. A plunger is reciprocated in the press box which receives the pomace from the grinder, the cider passing through the strainer, and the cheese being subsequently withdrawn through an opening.

Claim.—First, the press K, piston L, strainer C, slides *p* and *e'*. core driver M, rods *i* and *f*, curved lever *f* and *o*, and jointed lever L', arranged in the manner substantially as shown and described and for the purpose set forth.

Second, the arrangement of the attachment of the jointed lever L' to the rear end of the press by the spring *n*, and adjustable stays *n'*, and connecting rod S, for the purpose and in the manner set forth.

Third, the wheel H, friction roller J, as arranged in combination with the levers *f o* and S, for the purpose and operating in the manner described.

No. 58,515.—J L. WEAVER, Davis, Ill.—*Washing Machine.*—October 2, 1866.—The upper and lower rollers are geared together, and turned by a hand crank. The clothes are passed on an endless apron between the rollers. The operating parts are attached to an inner frame, which is lifted from the tub by a treadle lever and connecting side rods. The lower roller is pressed upward against the driving roller by transverse levers connected to its respective ends.

Claim.—First, the combination of the frame G H I with the side and end boards of the washing machine and with the roller frame J K L, substantially as described and for the purpose set forth.

Second, the combination of the roller frame J K L with the side boards B and C, and with the pressure rollers M and N, substantially as described and for the purpose set forth.

Third, the levers U, constructed as described, in combination with the links or bars T, the cross-piece K of the roller frame, and the elastic ropes or bands V, substantially as described and for the purpose set forth.

Fourth, the combination and arrangement of the rods F', levers G' H', with the top frame G H I, substantially as described and for the purpose set forth.

No. 58,516.—VICTORIA QUARRE WEDIKIND, Philadelphia, Penn.—*Engraving Copper, &c.*—October 2, 1866.—The plate is covered with a coat of German white, the design is drawn upon it, removing the white in portions. All is now covered with engraver's "ground," and after a wax wall is built, dilute nitric acid is poured on, which corrodes the metal where the ground is underlaid by the German white, and leaves the design in relief.

Claim.—The production by chemical process of engraved plates forming designs in relief by the following process:

Covering the copper plates with German white, tracing of designs therein, covering the white layer with engraver's varnish and treating it with nitric acid, substantially as above described.

No. 58,517.—DANIEL K. WERTMAN and WILLIAM H. REINBOLD, Centralia, Penn.—*Machine for Gathering Clover Seed.*—October 2, 1866.—The frame carrying the teeth and reel is raised or depressed by means of a lever irrespective of the gathering box.

Claim.—First, the peculiar setting of teeth *a a a* on axle *m*, in the manner and for the purpose above set forth.

Second, the combination of lever P, with uppers *u*, reel R', and axle *m*, when constructed and operated in the manner and for the purpose above set forth and. described.

Third, the independent connection of teeth and uppers and lever from body or box of machine, allowing the teeth to be worked irrespective of the box, as aforesaid set forth and described and for the purpose mentioned.

No. 58,518.—WILLIAM WESTLAKE, Chicago, Ill.—*Coffee Pot.*—October 2, 1866.—The upper portion of the pot is made cylindrical to fit the periphery of an obtusely conical metal strainer, which is operated by a handle.

Claim.—First, in combination with the cylindrical portion of a coffee pot, the scraper and strainer, constructed and operated as herein set forth.

Second, the combination of the bar *d*, having the recesses *e*, with the wires *f*, and projection *g*, for guiding and locking the scraper and strainer, as set forth.

No. 58,519.—CHARLES H. WESTON, Lowell, Mass.—*Bench Hook.*—October 2, 1866.—The bench hook is elevated by the rotation of the thumb screw in its rear, actuating a lever beneath, upon which the bench hook is supported.

Claim.—A bench hook, when its several parts are constructed and arranged to operate substantially as described.

No. 58,520.—AMOS WHITTEMORE, Cambridgeport, Mass., assignor to CAMBRIDGE HORSE NAIL COMPANY.—*Machine for Making Horseshoe Nails.*—October 2, 1866.—The hammers reciprocate in right lines (instead of vibrating, as usual) and strike alternately upon the ends of dies that are applied to the oscillating working beam, which actuates the hammers; the latter are so connected as to counterpoise each other, and a certain elasticity prevents injurious concussion. The nippers operate automatically; one pair feeds the nail rod to be acted on by the hammers and delivers it to the other, which gripes, sustains, and then retracts it; the shears cut off the nail while the head is supported.

Claim.—First, the oscillating dies, operating in conjunction with hammers which move back and forth in a right line, and which are arranged in inclined positions, and operate substantially as described.

Second, the combination of two hammers B B', with the oscillating die-head or beam A³, and connecting rods b³ b², arranged to operate substantially as described.

Third, the combination of the die beam A³, inclined hammers B B', eccentric a, and yoked pitman rods a', the latter being guided by the driving shaft A², substantially as described.

Fourth, the application of spring or yielding guides b², to the reciprocating hammers, substantially as described.

Fifth, the rocking feed nippers D, constructed and operated so as to gripe the nail rod and then deliver this rod between the nippers E, in a position to be acted upon by the hammers and dies, substantially as described.

Sixth, the application of a support and guide f², to the nippers D, for guiding the end of the nail rod up to the gauge F, substantially as described.

Seventh, the adjustable gauge F, in combination with the rocking feed nippers D, substantially as described.

Eighth, the pair of shears d⁵ d⁶, constructed and operated substantially as described, and arranged in front of the point where the nails are forged, and in such position as to allow of the feeding of the nail rod to a proper position by the nippers D, substantially as described.

Ninth, communicating motion to the shears, and also to the feeding nippers, from the driving shaft A², through a cam shaft C', by means substantially as described.

Tenth, the employment of the reciprocating retracting nippers E, so arranged and operated as to take the nail rod from the feeding nippers, hold it until the nail is finished, and then retract the nail from the hammers and dies, substantially as described.

Eleventh, the combination of the hooked rod or lever j with the nippers E, substantially as and for the purpose stated.

Twelfth, effecting the opening and closing of the nipper jaws g g' and hook k, substantially as described.

Thirteenth, connecting the nippers E to a rectilinear reciprocating rod g³, which passes transversely through the frame plate A A', and is operated by means of a cam on the cam shaft C', and a spring, substantially as described.

Fourteenth, the adjusting screw g⁴, or its equivalent, applied to the rod g³, for regulating the length of its strokes, substantially as described.

Fifteenth, the combination of the hook or stop d⁷, on the lower shear blade d⁵, with the retracting nippers E, substantially as and for the purposes described.

Sixteenth, the combination of rocking feed nippers D, the retracting nippers E, and the shears, with the devices for forging horseshoe nails, said parts being arranged substantially as described.

No. 58,521.—AMOS WHITTEMORE, Cambridge, Mass.—*Machinery for Making Split Spikes.*—October 2, 1866 —The spikes are produced with longitudinal splits, their faces bevelled to spread in driving. The spike rod is split by shears while being laterally supported; the bevel is made by cutting off the corners. The series of actuating devices cannot be briefly described.

Claim.—First, so constructing the splitting devices, that in the act of splitting the metal both split portions shall be supported laterally, and be separated in the plane of the cut, substantially as described.

Second, the combination of a device for severing the spike blanks from the rods, with a device which will split the spikes in a direction with their length, constructed and arranged substantially as described.

Third, the combination of the movable blocks E E', with the splitting and cutting-off mechanism, and with the finishing mechanism, substantially in the manner and for the purpose described.

Fourth, the bevelling cutters G' H, or their equivalents, which will bevel the ends of a split spike blank, before its split ends are closed, substantially as described.

Fifth, the combination of splitting and bevelling cutters, in a machine for producing split spikes, constructed and arranged substantially as described.

Sixth, the combination of the heading tool J, with machinery constructed and arranged substantially as herein described for splitting and forming the spike ready for its action, as set forth.

Seventh, the employment of the compressing head L', to close the forked ends of the split rods or spikes, substantially as described.

Eighth, making split spikes of the construction herein described, by means substantially as described, and which operate on the principle set forth.

No. 58,522.—ABEL WHITTOCK, Danbury, Conn.—*Oven.*—October 2, 1866.—The air reverberates through a double flue above and partly around the oven, so as to expose it longer to the absorption of its heat by the baking chamber.

Claim.—The jacket or reverberator E, in combination with the baking chamber of an oven, substantially in the manner and for the purpose set forth.

No. 58,523.—JOHN A. WIEDERSHEIM, Philadelphia, Penn.—*Caster for Furniture.*—October 2, 1866.—The cup of the caster forms a socket for the flanged foot of the leg; an axial stud on the former occupying a depression in the foot of the latter.

Claim.—The combination and arrangement of the armed revolving socket A, and flanged cap D, provided respectively with depressions d, and projections c, and operating substantially as described and for the purpose specified.

No. 58,524.—SAMUEL W. WILCOX. Mendon, Mass.—*Lamp Burner.*—October 2, 1866.—By depressing the lever, plates are raised above the sides of the wick tube, and extinguish the light by intercepting the flow of air to the wick.

Claim.—The arrangement and application of the two separate gates or extinguishers, with respect to the flat wick tube, and a lever or its equivalent applied to such tube and the body of the burner.

Also, the arrangement as well as the combination of the two gates or extinguishers with the wick tube and its auxiliary deflector, the whole being substantially as described.

No. 58,525.—DAVID WILLIAMSON, New York, N. Y.—*Breech-loading Fire-arm.*—October 2, 1866.—A pin in the stock traverses a longitudinal groove in the barrel when the latter is slipped forward for charging. It is held in firing position by a spring catch, which only admits of retraction when the hammer is at half-cock. The spring catch is retracted by a thumb-knob beneath the stock.

Claim.—First, the detached spring stop a, used with the lever G, and the under side of the barrel B, being so constructed and arranged as to act as a stop for the barrel and a direct retractor, substantially as specified.

Second, the pin z, upon trigger E, when used with stop a, for the purpose of locking the barrel and preventing it from being thrown up to or away from the breech except at half-cock, substantially as set forth.

No. 58,526.—JOHN I. WILLIAMS, Meridian, Miss.—*Baling Press.*—October 2, 1866.—The follower is depressed upon the hay by the vibration of the lever to which it is pivoted, the force being applied by a rope from a capstan. The rope passes over the end of a lever whose supporting roller traverses on a curved track.

Claim.—The combination of the capstan M, the horizontal draught lever I, the roller f, on the track L L, the driving or compressing lever H, the follower G, and the tumbling bars F, constructed, arranged, and operated substantially as and for the purposes herein described.

No. 58,527.—SAMUEL R. WILMOT, Bridgeport, Conn.—*Clasp for Hoop Skirts.*—October 2, 1866.—The plate has a scalloped edge except near the ends, where it has straight portions which meet in folding, the scallops, interlocking.

Claim.—First, a hoop skirt clasp, the longitudinal edges of which are straight near the ends, and for the rest of their length so formed into tongues and recesses that when folded upon the hoop, the tongue of one edge shall fit into the recess of the other, and vice versa, substantially as described.

Second, the punctured spurs or projections e in said clasp, substantially as and for the purposes set forth.

No. 58,528.—HERMAN WINTER, Williamsburg, N. Y., and FREDERICK W. NEWTON, South Orange, N. J.—*Machine for Preparing Peat for Fuel.*—October 2, 1866.—The mixing chamber supports a series of dies and contains a steam-heated grinder. The peat passes from the mixer to a second chamber, from whence it is forced by the plunger through the dies. The grinder is operated by gear wheels and the plunger by eccentrics on the same shaft.

Claim.—The combination in a peat machine of the preparing chamber and press in such manner that the material of the former constitutes a part of the frame of the press, substantially as set forth.

Also, the combination in a peat machine of the following instrumentalities, viz: the preparing chamber, grinding instruments, and press, all operating in combination substantially as set forth.

Also, the combination in a peat machine of the following instrumentalities, viz: the preparing chamber, internal heater, and press, all operating in combination substantially as set forth.

Also, the combination in a peat machine of the following instrumentalities, viz: the preparing chamber, grinding or kneading instruments, internal heater and passage thereto, all operating in combination substantially as set forth.

Also, the combination in a peat machine of the following instrumentalities, viz: the preparing chamber, press, grinding instruments, and pipe for admitting steam into the preparing chamber, all operating in combination substantially as set forth.

Also, the combination in a peat machine of the following instrumentalities, viz: the preparing chamber, press, internal heater, and pipe for admitting steam into the preparing chamber, all operating in combination substantially as set forth.

Also, the combination in a peat machine of the following instrumentalities, viz: the preparing chamber, press, grinding instruments, internal heater, and pipe for the admission of steam into the preparing chamber, all operating in the combination substantially as set forth.

No. 58,529.—JOHN D. WOODBURY, Wilson, N. Y.—*Sheep Rack.*—October 2, 1866.—
The pivoted boards in one position form a hay manger and in the other a roof over the grain troughs.

Claim.—The slots *p p* in combination with the pins *o o* and bolts *r r* for the purpose of adjusting the side boards C, so as to form a roof, arranged and operating as described.

Also, the dovetailed notches *s s* in combination with the side boards C C, when provided with slots *p p* for retaining the said boards in place, substantially in the manner set forth.

No. 58,530.—H. T. WOODMAN, Dubuque, Iowa.—*Filter and Cooler Combined.*—October 2, 1866.—The water passes up through a filter chamber and through an apperture into the cooling chamber, from whence it is drawn.

Claim.—The arrangement of the water chamber B, filter C, with top and bottom, perforated plates *c d*, pipe *a*, receiving and supplementary chamber D, and pipe E, with walls A, constructed and operating in the manner and for the purpose specified.

No, 58,531.—CHARLES E. WRIGHT, Auburn, N. Y.—*Cherry Stoner.*—October 2, 1866.—
A corrugated roller revolves in a trough formed by a curved perforated plate; the roller forces the pits through the perforations, and discharges the "sarcocarp" separately.

Claim.—The combination of the corrugated cylinder B with the curved perforated bed plate D and springs F, when used as and for the purpose specified.

No. 58,532.—WILLIAM K. WYCKOFF, Ripon, Wis.—*Composition of Matter.*—October 2, 1866.—For blacking leather and protecting iron; composed of the refined product of a heated mixture of petroleum residuum and chloride of sodium, with hard soap, rosin, carburet of iron, "drop" or ivory black, and sometimes tallow.

Claim.—First, the use of residuum, as above described and for the purposes specified.

Second, the treating and refining process, substantially as herein described.

Third, the within described composition made of the ingredients set forth and mixed together, substantially in the manner and for the purposes specified.

No. 58,533.—GEORGE W. ZEIGLER, Tiffin, Ohio.—*Attaching Cultivator Teeth to the Frames.*—October 2, 1866.—The tooth has a zonal joint in the socket beneath the beam and is clamped by bolt and washer. It has a capacity for universal adjustment within given limits.

Claim.—First, securing a shovel or cultivator tooth to its frame by means of a universal joint, in such manner that the shovel can be inclined either laterally or longitudinally with respect to its frame, substantially as described.

Second, the concavo-convex shank *a* formed on or secured to a shovel or tooth, substantially as and for the purpose described.

Third, the combination of the shank *a*, concave plate D, convex bolt head *g*, nut *c*, and a shovel C, or its equivalent, substantially as described.

No. 58,534.—CHRISTIAN ZIMMERMAN, Collinsville, Ohio.—*Plough.*—October 2, 1866.—
The cultivator has means for lateral adjustment of the ploughs and for adjusting the tongue vertically. The share, mouldboard and ground bar are made in one piece. A plate on the sheath protects young plants from rolling clods.

Claim.—First, the manner of attaching the tongue to the plough by means of the perforated plates *a a* and pins *b b*, arranged and operating in the manner and for the purpose described.

Second, the combination of the protecting plate *c*, enclosing the landside of the plough, with the mouldboard and landside made of one piece of metal, substantially as described for the purpose specified.

Third, the hinged draw bars *d d* with their adjusting nuts and screws, in combination with the adjusting frame composed of screw rods *e e* and cross bar *g* with their nuts and adjustable handles, arranged and operating in the manner and for the purpose set forth.

No. 58,535.—W. W. ARMINGTON, New Haven, Conn., assignor to E. A. KELSEY, Meriden, Conn.—*Clothes Pole.*—October 2, 1866.—Explained by the claim and cut.

Claim.—Making a clothes line supporter *a* in two or more parts with the clasps *b*, catch *c*, substantially as and for the purpose described.

No. 58,536.—JAMES E. AULD, Buffalo, N. Y., assignor to himself and JOHN M. LAYTOS, same place.—*Match Safe.*—October 2, 1866; antedated September 15, 1866.—One match at a time falls into the delivery chamber. A slide catches by a nib against the end of the match and pushes it through a hole in the end of the box, where the prepared end is struck by an igniter and the splint held in position for use.

Claim.—The igniter D so constructed with a pocket match safe that it shall act against the prepared end of the match as it is thrust from the safe and ignite it, and afterward hold it until extinguished, substantially as set forth.

No. 58,537.—WILLIAM BELCHER, New Haven, Conn., assignor to himself and F. G. HICKERSON, same place.—*Cigar Light.*—October 2, 1866.—Strips of paper are folded over the end of the cigar and secured by mucilage; one portion is prepared with material which is ignited by friction when the projecting strip is drawn out.

Claim.—The cigar lighter herein described, constructed and attached to the cigar, substantially as and for the purpose specified.

No. 58,538.—JEREMIAH CAMPBELL, Lancaster, Penn., assignor to himself and JOHN CAMPBELL, same place.—*Cigar Presser.*—October 2, 1866.—The heads of adjoining slats are connected by slotted links and the outside ones to the followers; a filling-in frame determines the position of the cigars in the frame. Wedge blocks act in conjunction with the screws of the press.

Claim.—The construction and application of the flat links E, with their bole e, and slot f, when employed in the manner and for the purpose specified.

Also, the followers I' I'', and headed slats D, when combined, linked, and operated substantially in the manner and for the purpose specified.

Also, the open filling-in frame G g, with its notches h, constructed and operating in the manner and for the purpose set forth.

Also, the key blocks W, in combination with a double side pressure and vertical pressure, given in the manner and for the purpose specified.

No. 58,539.—I. S. CROLL, North San Juan, Cal., assignor to himself and QUARTUS RICE, Nevada, Cal.—*Quartz Crusher.*—October 2, 1866.—The mullers have a planetary motion around the central axis, traversing in an annular trough.

Claim.—Giving the mullers a positive motion on their own axis by means of the gears D, and G G, or their equivalents, while said mullers are driven around in the pan by the rotation of the frame which carries the shafts of the mullers.

No. 58,540.—THOMAS B. DE FORREST, Birmingham, Conn., assignor to the AMERICAN PRESS AND CLASP COMPANY, Bridgeport, Conn.—*Clasp for Hoop Skirt.*—October 2, 1866.—The indentations make the parts mutually retentive.

Claim.—The two clasps d d, in combination with the indentation c, so as to secure the two ends of the hoop, substantially in the manner herein set forth.

No. 58,541.—PHINEAS C. ELLSWORTH, Venice, N. Y., assignor to himself and M. SALIZBIERY, same place.—*Hay Elevator*—October 2, 1866.—Partial revolution of the handle trips the connection between the lever and the reel, so as to allow the latter to revolve without backing the lever in reeling off the rope.

Claim.—The arrangement of the reel, the lever J, the cord E, and the trip handle K, constructed substantially as specified, for the purpose set forth.

No. 58,542.—JOHN FELLOWS, Chicago, Ill., assignor to himself and ALBERT CARD, same place.—*Shifting Rail for Carriage Seats.*—October 2, 1866.—Mortises in the brace bars attached to the seat receive the tenons on the skeleton base frame of the shifting top; pivoted keys on the frame lock them in position.

Claim.—First, the braces or supports D, when provided at their upper ends with heads having slots or mortises on the plane of the seat, and fitted to the seat so as not to project beyond, substantially as and for the purposes specified.

Second, the arrangement and combination of the braces D, with the spurs or projections e e and d d, with the rail C, and carriage seat, substantially as and for the purposes specified.

No. 58,543.—E. J. FRASER, Erie, Penn., assignor to himself and ORANGE NOBLE, same place.—*Rotary Spading Machine.*—October 2, 1866.—The cylinder has projecting tines and is rotated by their penetration of the soil while the carriage advances. By a clutch on the main shaft a pinion is engaged with a rack by which the digger is lifted. Gearing from the cogged rim of the digger rotates a shaft with pulverizing spokes. Scrapers remove adhesive soil from the smooth-faced cylinder.

Claim.—First, the smooth-faced cylinder C, set with rows of teeth or spades m, and hung on the free shaft a a, in combination with the shifting clutch d, and the vertical rack c, and pinion *b*, for raising and lowering the cylinder, constructed and operated substantially as and for the purposes herein described.

Second, the rotating pulverizing arms *p*, in combination with the spading cylinder C, and connected therewith by the gear wheels *A k l*, operated by the epicycloidal wheel F, on the chine of the cylinder, constructed and operated substantially as and for the purposes herein specified.

Third, the spring scrapers *r*, and the friction roller or bearer E, in combination with the spading cylinder C, constructed and operated substantially as and for the purposes herein described.

No. 58,544.—ISAAC T. GREEN, Milford, Conn., assignor to himself and C. H. BERRY, Brooklyn, New York.—*Hat Pressing Machine.*—October 2, 1866.—The hat rests against the exterior form. A perforated disk is placed within the crown, and two perforated arcs, thicker near the crown than near the brim, rest against the sides. The central block presses outward and downward, and has a perforated plate which rests in the brim. Steam is diffused upon the hat through the perforations; or, a wet cloth being placed in the centre, hot air is forced upon it.

Claim.—First, the application of steam to the hat while the same is in place between the presser and the die, as and for the purpose described.

Second, the employment of a wet sheet, in combination with the die and presser, while the operation of pressing a hat progresses, substantially as and for the purpose set forth.

Third, the perforated shield C, in combination with the presser and die, substantially as and for the purpose described.

Fourth, the perforated shield C, and the wet sheet D, in combination with the presser and die, as set forth.

Fifth, making the perforated shield tapering, or thicker at the bottom, as and for the purpose described.

No. 58,545.—ROBERT L. HALL, Lowell, Mass., assignor to himself and JOSEPH G. RUSSELL, Boston, Mass.—*Bed Bottom.*—October 2, 1866.—The slats of the berth bottom have holes near each end for the reception of the rubber springs which rest on the rubber washers upon the transverse rails.

Claim.—The combination of the rubber springs *i*, metallic washers *g*, and the circular holes *e*, as herein described, and for the purpose set forth.

No. 58,546.—WILLIAM T. HOWARD, Baltimore, Md., assignor to himself and ISAAC McKIM CHASE, same place.—*Boiler Gauge Cock.*—October 2, 1866.—The siphon tube has a try-cock in the bend, an outer exit, and is vertically adjustable in its collar to bring its inner end to the height of water in the chamber.

Claim.—First, the adjustable siphon or tubes E E', employed in combination with the chamber A, and a try-cock or valve G, to indicate the water level in the boiler, substantially in the manner specified.

Second, the combination of the scale J with the adjustable gauge tubes, substantially as set forth.

No. 58,547.—JOHN HULL and WM. P. ANDERSON, Hackettstown, N. J., assignors to themselves and HENRY J. HULL, same place.—*Horse Power Brake.*—October 2, 1866.— The brake is supported by a device hinged to the frame and resting on the bolting, so that if the latter runs off the brake comes into action upon the wheel.

Claim.—First, the hooked lever shaft *d* and flanged roller G thereon, operating relatively to the belting D for the release of the brake E in its connection with the lever by the rod *g*, substantially in the manner and for the purpose set forth.

Second, the brake E and weight H combined, operating automatically by the dropping of the flanged roller when the belting slips off from the wheels, substantially in the manner and for the ur ose as herein set forth.

Third, the construction of the box *a*, block *c*, hooked lever shaft *d*, and flanged roller G, as arranged and applied to the standard F of the side of the platform, substantially in the manner and for the purpose specified.

No. 58,548.—J. G. KAST, New York, N. Y., assignor to L. L. ARNOLD, same place.— *Lock for Trunks.*—October 2, 1866.—The rod has hooks which engage the hasps of the lid; is projected by a spring, withdrawn by a key which screws on to its threaded end, and is retained in retracted position ready for the closing of the lid by a spring stop.

Claim.—First, the sliding spring bolt B, provided with two or more hooks *e d* and with a suitable fastening for the key C at its end, in combination with the staples *e f* and with said key, constructed and operating substantially as and for the purpose set forth.

Second, the spring A and notched collar *i*, in combination with the sliding bolt B, key C, and bit *j*, constructed and operating substantially as and for the purpose described.

Third, the spring stop *k*, in combination with the staple *e* and spring bolt B, constructed and operating substantially as and for the purpose set forth.

No. 58,549.—EDWIN R. KERR, Kewanee, Ill., assignor to himself and J. L. PLATT, same place.—*Coal-dumping Apparatus.*—October 2, 1866.—The several chutes have doors, which when opened become delivering spouts in continuation of the chutes, and are counterpoised by weights; the inner doors swing open, and are held in position by notch bars.

Claim.—The chutes B, placed in a shed or building and provided with two doors C E, the outer ones E being so arranged as to serve when lowered or opened as a continuation of the chutes, substantially as shown and described.

Also, counterpoising the outer doors E by means of weights, substantially in the manner as and for the purpose specified.

Also, the hanging of the inner doors C at their upper ends, in combination with the spring bolts D at their lower ends, when said doors are applied to chutes B, for the purposes herein set forth.

No. 58,550.—CHARLES E. LANGMAID, Woburn, Mass., assignor to himself and GEO. A. LANGMAID, Stoneham, Mass —"*Cast Off*" *of Waxed Thread Sewing Machines.*— October 2, 1866.—Instead of the usual rigid "cast off" for covering and opening the book of the hook needle and casting off the loop, a hinged foot is placed on the end of the "cast off," which can adjust itself to the side of the needle and to its hook, even though the "cast off" be sprung out of proper position, thus avoiding uneven wear and the cutting of the thread.

Claim.—A hinged cast-off foot for sewing machines, which adjusts itself to the side of the hook and wears evenly, to prevent cutting the thread, constructed substantially as herein described.

No. 58,551.—JOSIAH LONG, Morristown, Ind., assignor to JACOB G. WOLF, same place.— *Grain Screen.*—October 2, 1866.—The convoluted wheel acts between the rollers to give a reciprocating motion to the shoe.

Claim.—The cam wheel T, in combination with the friction roller 8 and box D, the whole constructed and operating substantially as herein described.

No. 58,552.—HENRY McCLURE, Terre Haute, Ind., assignor to himself and JAMES ELLIS, same place.—*Furnace and Boiler.*—October 2, 1866.—By the mode of setting the heat strikes all the boilers but the first above the line of deposit of sediment, the adjustment being attained by means of movable plates, which form a succession of fire bridges between the boilers.

Claim,—An improved furnace and boilers, formed by combining the slides O, rock shaft P, rock arms R, links or bars 8, and boilers F G I J K L with each other and with the furnace flue curved beneath the boilers, when said parts are constructed and arranged substantially as herein described and for the purposes set forth.

No. 58,553.—ADOLPH MILLOCHAU, New York, N. Y., assignor to the AMERICAN RETORT COMPANY, same place.—*Manufacture of Illuminating Gas.*—October 2, 1866.—The interior casing contains coal, and traverses on wheels in the outer casing, in which the products of distillation are reheated for their more perfect conversion into illuminating gas.

Claim.—First, a separate casing for containing the coal or other material to be distilled, in combination with the main retort within which said casing is introduced, as and for the purposes specified.

Second, the method herein specified of subjecting the products of distillation to a heating operation within the retort in which they are generated, as and for the purposes set forth.

No. 58,554.—J. OSTERHOUDT, New York, N. Y., assignor to himself and WM. B. DUBOIS, same place.—*Opening Tin Cans.*—October 2, 1866.—The disk of tin used to cover the can has a projecting dip, which is left free, as a handle whereby to strip off the cover.

Claim.—First, a device for closing a can or other vessel, consisting of a cover or cap a to be separated from the said vessel by rolling it upon any suitable instrument by the aid of a tag or lip c, formed in one piece with the cover or cap, and left loose in soldering the latter to the can, as herein explained.

Second, in combination with a cover constructed as aforesaid, a recess in the top plate B of the can for the reception of the tag c, as explained.

No. 58,555.—GEORGE M. RICHARDSON, Barre. Mass., assignor to himself and N. L. JOHNSON, Dana, Mass.—*Machine for Pressing Bonnets.*—October 2, 1866.—The blocked bonnet slides horizontally into the die shell, which divides centrally for the reception of the former. The shell has a steam jacket, and the sections are closed by a screw passing through flanges on the upper sides of the dies; a plate covers the junction to prevent pinching the bonnet.

Claim.—First, employing the screw C, in combination with the dies A A and the bonnet block E. constructed and operating substantially as and for the purposes herein described.

Second, the metal strip f within the dies A A for covering the division between them, as herein described.

Third, the spring c, in combination with the block E, as herein described.

No. 58,556.—WM. H. SALISBURY, Providence, R. I., assignor to himself and JESSE A. LOCKE, Boston, Mass.—*Wool-oiling Machinery for Carding Engines, &c.*—October 2, 1866.—In this improvement the oil reservoir which traverses and distributes the oil across the apron of the carding machine has its outlet tube provided with two cocks, one adjusted by the overseer, the other opened and closed by the attendant on starting and stopping the machine. The tube projects a short distance above the bottom of the reservoir to prevent the sediment from passing out, and has a hinged strainer over its top. The oil is kept in a fluid state by a steam pipe, which extends across the apron and lies directly under the path of the reservoir.

Claim.—First, the steam pipe K, in combination with the oil reservoir k, substantially in the manner and for the purpose set forth.

Second, in combination with the oil reservoir k, the tube f, with its regulating cocks g h and strainer i, for the purpose specified.

No. 58,557.—T. T. SHAWCROSS, Allisonville, Ind., assignor to himself, L. D. WYATT, and E. D. McMANAMA, same place.—*Corn Planter.*—October 2, 1866.—The grain box has an agitator and an oscillating seed wheel in its bottom. The dropping is performed by the thumb of the operator, a spring restoring the seed roller to position.

Claim.—First, the box l, the wheel 19, with its cavity 21, and the agitator 20.

Second, the spring 12, levers 13 14 15 16 18 and 17, all arranged and operating substantially as described for that purpose.

No. 58,558.—CHARLES D. SMITH, Washington, D. C., assignor to himself and JOHN A. WIEDERSHEIM, Philadelphia, Penn.—*Combined Eraser and Pencil Sharpener.*—October 2, 1866.—In the butt of the eraser handle is a conical socket with a blade to sharpen pencils.

Claim.—The combination with an eraser of a pencil sharpener, having substantially the characteristics herein specified.

No. 58,559.—LEVI STEVENS, Fitchburg, Mass., assignor to himself and NORMAN C. MUNSON, Shirley, Mass.—*Apparatus for Carburetting Air.*—October 2, 1866.—The pans are in a vertical series, and fit within the cylinder; to the bottom of each is soldered a scroll plate, which rests in the pan beneath; each pan is charged with liquid; the air is introduced below, and passes in contact with liquid through each spiral passage, and then through the pipe in the floor above, winding in and out alternately until it reaches the exit at the crown of the dome.

Claim.—First, the cups or disseminators B B, constructed and operating substantially as described.

Second, the combination of one or more cups or disseminators B B with the cylinder A, for the purposes described.

Third, the arrangement of two or more cups or disseminators with each other, substantially as described.

No. 58,560.—ANDREW WALKER, Claremont, N. H., assignor to himself and J. P. UPHAM, same place, and C. $E_A{}_m M_A N$, Conway, N. H.—*Potato Digger.*—October 2, 1866.—A scraper removes the vines and superfluous earth from the hill. This is followed by two shares that turn the furrow inward. The potatoes are taken up by rotating arms and deposited on a screen.

Claim—The combination of the cutter e, ploughs f, and cylinder of prongs g, with the grate or screen k, and mechanism for actuating said cylinder, substantially as specified.

No. 58,561.—GEORGE V. WOODS, Belchertown, Mass., assignor to AUSTIN WHITE, same place.—*Neck-tie, Collar, and Bosom Combined.*—October 2, 1866.—The paper bosom, collar, and neck-tie, or either two of them, are stamped out of one piece.

Claim—First, making or forming a collar, bosom, and neck-tie from a single piece of paper, cloth, or other suitable material, substantially as set forth.

Second, making the part or neck-tie C, substantially as set forth.

Third, making the part of bosom A, and part or neck-tie C, together or from a single piece of material, substantially as set forth.

Fourth, making the collar or part B and neck-tie C together or from a single piece of material, substantially as set forth.

Fifth, the combination with the end c of the collar B, substantially as shown in the drawings.

Sixth, the piece D, in combination with the bosom and collar, or either, for the purposes set forth.

No. 58,562.—F. J. BOLTON, London, Eng.—*Signal Code for Electric Telegraph.*—October 2, 1866.—Words and letters are codified by numerical arrangement, and the figures are symbolized by dots and dashes.

Claim.—The herein described code of signals for communicating intelligence or transmitting messages by the electric or magnetic telegraph, and the method of arranging and compiling the same, substantially as set forth.

No. 58,563.—GEORGE KEEBLE, Suffolk, Eng.—*Stop Cutter for Cutting Continuous Sheets of Paper into Shorter Ones* —October 2, 1866.—After the cam and segment have passed their point of contact with the arm and gear respectively, the coil spring is at liberty suddenly to act upon and draw downward the lever, and so cause the dog instantly to engage with the toothed wheel and stop the drum while the cutting is being done. One sheet of paper is cut at each revolution of the wheel and its toothed segment and cam.

Claim.—The combination of the lever *k*, connecting strap *i*, rod E, key *o*, collar G, tube B, standard A, cap C, and the spring D, or its equivalent, the whole arranged substantially as and for the purpose set forth.

No. 58,564.—ROBERT W. THOMSON, Edinburgh, Scotland.—*Steam Generator.*—October 2, 1866.—Vertical tubes are arranged in the cylindrical upper portion of the boiler above the bulbous portion which projects downward into the fire chamber, and which forms an annular throat between itself and the furnace wall.

Claim.—The constructing of vertical boilers with fire gas tubes arranged in a circle or circles, and combined with a vessel which projects down into the furnace from within such circle of tubes, and which is enlarged below its attachment to the tube plate, substantially as hereinbefore described.

No. 58,565.—SWAIN WINKLEY, New York, N. Y., assignor to himself and ASA BIGELOW, Jr.—*Construction of Railways.*—October 2, 1866.—The foundation consists of a corrugated base plate and a flat cap supporting the rail, and with a cross-tie uniting the foundation plates of the sister rails.

Claim.—The construction of the corrugated base plates *a*, combined with the cap pieces *b* and tie rods *c* when applied to railways, substantially in the manner herein described.

No. 58,566.—P. M. ACKERMAN, Webster, N. Y.—*Gate.*—October 9, 1866.—The crane is pivoted to the heel post of the gate, and its upper, horizontal bar has rollers which support the gate and in which it slides preparatory to being rotated to an open position.

Claim.—First, suspending the sliding gate upon the swinging crane, substantially in the manner and for the purposes herein shown and described.

Second, the swinging crane or frame constructed, arranged, and operating substantially as herein shown, and for the purposes set forth in combination with the sliding gate.

No. 58,567.—F. S. ALLEN, New York, N. Y.—*Manufacture of Gunpowder.*—October 9, 1866.—Sawdust, &c., is saturated with a boiling solution of binoxide of manganese, chlorate of potash, nitrate of potash, and ferrocyanide of potassium.

Claim.—The within described process of saturating paper or other substance with an explosive compound by boiling the same together, substantially as set forth.

No. 58,568.—ROBERT ALLISON, Port Carbon, Penn.—*Boring Machine.*—October 9, 1866.—The boring mandrel and its operating mechanism are arranged upon a circular bed, constituting a portable machine, which may be secured, by clamping-bar and bolts, to the article to be bored, instead of chucking the latter in a lathe.

Claim.—First, a portable boring machine for boring or reaming shaft holes in cog wheels, spiders, pulleys, &c., constructed and operated substantially as herein described and represented.

Second, the base plate or ring A, in combination with the vertical boring arbor B and guide bar M, substantially as and for the purposes herein specified.

Third, the vertical boring arbor B, in combination with the base plate A, clamping bar M, or its equivalent, and the feeding apparatus, arranged and operated substantially in the manner and for the purposes herein set forth.

Fourth, the guide bar M M, in combination with the base plate A and the boring arbor B, arranged and applied substantially as and for the purposes herein described.

No. 58,569.—EMANUEL ANDREWS, Williamsport, Penn.—*Securing Crosscut Saws to their Handles.*—October 9, 1866.—The saw is embraced between the hooked end of a bolt, which traverses the handle axially, and a groove in a revolving ferrule upon the handle. The rotation of the handle secures it to the saw by means of a screw-thread on the bolt and a nut in the handle.

Caim.—First, the removable stop E and series of holding places B³, arranged for joint operation in fastening and liberating a saw plate A, substantially as and for the purpose herein specified.

Second, a saw handle B, and so arranged as to allow the removal and introduction of the saw blade A by a turning and lateral movement, without the necessity for passing the end of the blade through the bolt, all substantially as and for the purpose herein specified.

No. 58,570.—FRANCIS ARNOLD, Haddam Neck, Conn.—*Vegetable Cutter.*—October, 9, 1866.—The revolving disk wheel has knives and throats, through which latter pass the pieces cut from the roots in the tapering hopper.

Claim.—The wheel D, with the knives cast upon one of its faces and with openings through which the pieces of vegetables and roots pass when said wheel is used with a frame, constructed substantially as herein set forth and for the purposes described.

No. 58,571.—WM. ARROUQUIER, Worcester, Mass.—*Staging.*—October 9, 1866 —The outer frame has an upper, braced stage and a lower suspended stage, and its uprights are bolted to vertical pieces inside the building.

Claim.—The combination and arrangement with pieces C C' of the supporting pieces D D, braces E E, and adjustable bottom pieces F F, with their rods G, for supporting the foot-boards H I, from the window of a building, as shown and described.

No. 58,572.—EDWIN DWIGHT BABBITT, New York,N.Y.—*Pen-holder.*—October 9,1866.— The pen slips into the socket on the end of the pen-holder, and its rear is depressed by a spring to cause it to assume an angle with the holder.

Claim.—A pen-holder constructed substantially as herein shown and described.

No. 58,573.—HALSEY H. BAKER, New Market, N. J.—*Horseshoe.*—October 9, 1866.— The toe and heel calks have hook pieces which catch beneath projections on the shoe proper, when the wooden plugs are driven in; the latter being split and expanded by the operation to secure them in position. A layer of resilient material rests between the shoe and the calk, and ice spurs are attached to the latter when required.

Claim.—First, the attachment of the collar B to the shoe A, by means of lips *i* formed upon the tongues *h* of the said collar, hooked upon the shoulders *b'* by the driving into its place of the filling of the calk, substantially as herein set forth, for the purposes specified.

Second, the wedge-shaped projections *f*, so arranged in combination with the tongues *h* and lips *i* of the collar B, as to insure the rigid locking of the lip *i* upon the shoulders *b'*, by the act of driving the filling of the calk into its place, substantially as herein set forth, for the purpose specified.

Third, the opening or openings *e*, so arranged in relation with the shoulders *b'* and with the tongues *h* and lips *i* of the collar B, as to enable the said lips *i* to be disengaged from the shoulder *b'*, substantially as herein set forth, for the purpose specified.

Fourth, the ice calks D, arranged in relation with the collar B, and with the wood or other suitable filling of the calks, substantially as herein set forth, for the purpose specified.

Fifth, the combination of the layer or thickness *m* of India-rubber or other elastic material with the collar B and shoe A, substantially as herein set forth, for the purpose specified.

No. 58 574.—ARTHUR BARBARIN, New Orleans, La.—*Metallic Band Fastening*—October 9, 1866.—The band is looped around one horn and riveted, and being looped over the other is retained by pressure of the cotton; it is detached by partial rotation.

Claim.—The herein described device for fastening metallic bands, the same consisting of a segmental plate recessed so as to have a C shape, and constructed and arranged for operation as set forth, so that the partial revolution of the plate shall effect the release or loosening of the band, substantially in the manner herein specified.

No. 58,575.—HARVEY BARTON, Black Earth,Wis.—*Dumping Wagon.*—October 9,1866.— The wagon bed is divided into several compartments; each has a hinged bottom which is locked by its own device, and individually operated to discharge its portion of the load by connections leading to its lever by the driver's seat.

Claim.—The combination of the drops A B C, rock shaft D, pawls E F G, plates H I J, levers *d e f*, and springs K L N, substantially as shown and described.

No. 58,576.—HIRAM BARTON, East Hampton, Conn.—*House Bell.*—October 9, 1866.— The hammer is fixed in an oscillating spring disk, to which is also attached a draw-wire; a single pull and release of the draw-wire causes two strokes on the bell.

Claim.—The arrangement and combination of the vibrating cap-plate spring and stop with the bell and hammer, constructed and operating substantially as described.

No. 58,577.—O. L. BASSETT, T. R. BEARSE, and W. B. WILBER, Taunton, Mass.— *Machinery for Making Nails and Tacks.*—October 9, 1866.—The arm or book is arranged to operate with one of the movable cutters, to grasp the blank immediately after it is cut, and transfer it to the gripping dies, which hold it while being headed.

Claim.—First, the combination of a carrier or bearer for the tack blank with any one of the cutters, be they more or less in number, used for cutting the tack blanks, when arranged and so as to operate together, substantially in the manner and for the purpose specified.

Second, the carrier or bearer for conveying the tack blanks to the die to be headed, attached to either one or the other of the arms to which the cutters are attached, in combination with a forked or other suitable lever, arranged with regard to the said carrier and so as to operate upon it in conjunction with the said cutter arm, substantially as and for the purpose described.

No. 58,578.—THOMAS BEALE, New Milford, Ill.—*Cultivator.*—October 9, 1866.—By means of the lever the share and the scraper are reciprocated and the earth stirred and hilled against the plants.

Claim.—The bars A A having the bar B pivoted to them, with the spade or shovel C attached to the front end of the latter, in combination with the bar E connected to the bar F, which is pivoted to A A and connected to the pivoted bar B through the medium of the rod *b* and guide bar D, and the scraper and hilling device G pivoted to the front ends of the bars A A, and operated from one of the pendants *c* by the rod H, substantially as and for the purpose set forth.

No. 58,579.—JAMES BENSON, Bell Air, Ohio.—*Sand Pump.*—October 9, 1866.—A loop is formed in the piston rod to embrace the bail of the cylinder; this bail is hinged to admit of the connection and separation of these parts.

Claim.—The hinged bail I operating in combination with the linked piston rod E of the sand pump herein described, as and for the purpose specified.

No. 58,580.—ISAAC A. BEVIN, Chatham, Conn.—*Gong Bell*—October 9, 1866—The hammer is rotated on its pivot by pulling the draw-wire, and is returned by the spring; a direct and a recoil stroke are made by a single manipulation.

Claim.—The construction of the gong-striking apparatus, substantially as herein shown and described, so that the pull will act between the spring and the striker, all as set forth.

No. 58,581.—JESSE C. BOYD, Milroy, Ind.—*Shovel Plough*—October 9, 1866.—The side beams are pivoted to the fore-bar and adjusted laterally at the rear by slotted plates and set screw.

Claim.—The beams A A in combination with the upright shovels and handles, when connected to the beam C by means of the swivels H H, and attached to the beam B by slotted bars F F and set screw *t*, arranged substantially as specified.

No. 58,582.—NATHANIEL A. BOYNTON, New York, N. Y.—*Base Burning Stove.*—October 9, 1866.—The reservoir has an immediate cover, but is suspended as a frustum, open at both ends, above the fire-pot. The exit flue is below the top of the reservoir.

Claim.—The combination of an outer case or exterior cylinder with an interior cylinder or case, which is a reservoir of fuel above the fire-pot, the space between the outer case and the reservoir forming a combustion chamber with which the reservoir is in free communication at its upper portion.

No. 58,583.—ERNST BREDT, New York, N.Y.—*Galloon Trimming for Under Garments.*—October 9, 1866.—Explained by the claim and cut.

Claim.—A galloon trimming for under garments and similar articles, formed with a corded and scalloped edge to a fluted or corrugated fabric woven with a heading or centre, as specified.

No. 58,584.—ERNST BREDT, New York, N. Y.—*Galloon Trimming for Garments.*—October 9, 1866.—A galloon trimming formed of white cotton or linen has corrugations or flutes of plain surface, connected at their back portions by warps running through the woven fabric, so that the flattening out of the corrugations in washing is prevented.

Claim.—The galloon trimming for linen and cotton garments formed with flutes or corrugations, connected in the manner and for the purposes set forth.

No. 58,585.—L. C. BRISTOL and C. T. ALVERSON, Victor, N. Y.—*Beehive.*—October 9, 1866.—The hive has two boxes, one within the other, and the ventilating openings in the inner box are above those in the outer one, and made sufficiently large for the bees to pass through them. The side pieces in the comb frames stand diagonally to the sides of the hive.

Claim.—The relative arrangement of the ventilators C with the ventilators *a* and double hives A B, as shown and described and for the purposes set forth.

No. 58,586.—NATHANIEL BROCKWAY, Cambria, N. Y.—*Sawing Machine.*—October 9, 1866.—The saw is pivoted at one end to a hanging sectoral frame. The wood is held by a drop bar, and the frame oscillated by hand.

Claim.—The segment *i*, and gibs *j*, and saw *m*, in combination for the purposes described.

No. 58,587.—CHARLES F. BROWN, Ashby, Mass.—*Safety-valve Tender.*—October 9, 1866.—This is a connection between the lever of a safety-valve and the boiler, to operate in place of a weight, and also to assist in the discharge of steam when the pressure is too high in the boiler; the device is reset automatically

Claim.—The valve D with its appendages, the inner cylinder B, the frame F, the outer cylinder I and hanger K, or their equivalents, when in combination with the boiler and safety valve of a steam engine, substantially as and for the purposes set forth and described.

No. 58,588 —C. P. BROWN, Shortsville, N. Y.—*Seed Planter.*—October 9, 1866.—The bottom of the seed box is a series of hoppers. The seed disk admits of adjustment by a screw which springs in a clamp bar in which it is journaled. A wind guard protects the discharging grain from currents of air.

Claim.—First, the arrangement of the wind guard *g*, or its equivalent, with the delivery wheel W, substantially as and for the purpose set forth.

Second, the adjustable clamping bar E in combination with the wheel W and the case C, substantially as shown and described and for the purpose specified.

Third, the arrangement of a series of complete hoppers in the bottom of the grain box of seed drills in combination with the distributors, substantially as and for the purposes herein shown and described.

No. 58,589.—WILLIAM BREITENSTEIN, New York, N. Y.—*Power Loom for Weaving Concave and Convex Surfaces.*—October 9, 1866.—A brief description other than substantially in the words of the claims is impracticable.

Claim.—Operating the sectional take-up roller by levers, which are set by the jacquard and operated to take up the woven cloth by the beat of the lay or the motion of some equivalent part of the loom, substantially as and for the purpose specified.

Also, in combination with the sectional take-up roller, the series of pressure rollers and the frame in which they are hung with its connections, so that while each can yield independently of the others, the whole of them can be lifted from the take-up roller at once, or the tension of their springs increased, substantially as and for the purpose specified.

Also, the thread pullers, in combination with the jacquard and the intermediate mechanism by which the jacquard is made to determine when the said thread pullers shall be thrown in or out of action, substantially as and for the purpose specified.

Also, connecting the finger of the thread puller by means of a spring with the mechanism by which it is operated, so that it shall liberate the thread so soon as it shall have been pulled with sufficient tension, substantially as specified.

Also, the adjustable bar on the lay, in combination with the levers that operate the sectional take-up, and the bar which gauges the extent to which the said levers shall be pulled by the jacquard, substantially as and for the purpose specified.

Also, the shuttle carriers for operating the shuttle, or the equivalents thereof, as distinguished from the fly-shuttle system, in combination with the sectional take-up, or the equivalent thereof, substantially as and for the purpose described.

No. 58,590.—JULIA P. BROWN, Boston, Mass.—*Application of Bedsteads to Apartments.*—October 9, 1866.—A recess is formed in the ceiling for the reception of a bedstead, which is supported by cords, pulleys, and counterbalance weights.

Claim.—An apartment constructed with bedstead-receiving recess, formed in or at its ceiling, or upper part, and having a bedstead applied to such recess by counterbalancing devices, substantially as and for the purpose described.

No. 58,591.—ALFRED C. BRUSH, Norwalk, Conn., and GEORGE C. WHITE, Danbury, Conn.—*Dyeing Hat Bodies, &c*—October 9, 1866.—The hat bodies are dyed after they have been rendered sufficiently firm by felting to sustain the necessary manipulation, and before they have become hardened by the felting process.

Claim.—The process of dyeing hat bodies, substantially as hereinbefore set forth, by applying the dye or coloring matter to them after the sizing has been commenced and before it is completed.

No. 58,592.—H. L. and J. A. BUCKWALTER, Kimberton, Penn.—*Measuring Faucet.*—October 9, 1866; antedated September 30, 1866.—The eccentric piston rotates in the cylinder, the intervening space being of a known capacity. A gate sleeved on the eccentric projects into the spout and alternately opens and closes the eduction An index wheel outside is moved by a stud on the shaft of the eccentric, and registers the revolutions.

Claim.—First, the arrangement of the regulating screw M, grooves N, eccentric B, and cylinder A, constructed and operating in the manner and for the purpose herein described.

Second, the register wheel with two projecting cogs *m m* and a broad cog *p* to stop the revolution of the spindle of the faucet, and an open space *n* to allow the spindle to revolve freely when a measurement is not required, substantially as shown.

Third, the arrangement in relation to the foregoing of the follower G, constructed as described, with the keepers 3 3 of the cylinder A and spindle J, in the manner and for the purpose herein described.

No. 58,593.—ANDREW J. BURKE, Grundy Centre, Iowa.—*Tally Box for Measuring Grain.*—October 9, 1866.—The primary motion is given by means of a thumb piece whose shaft has a pawl actuating a ratchet wheel. A train of gearing rotates a shaft and sleeve, and the hands point to the figures on the concentric graduated circles.

Claim.—The arrangement of the ratchet wheel C, operated by the pawl *f* and provided with the index *r*, the pinion O gearing into the wheel D on shaft *e*, the latter being provided with the index *t* and pinion *n*, operating the index *l* by means of the wheel B secured to the tubular shaft *i*, all operating as and for the purpose set forth.

No. 58,594.—W. H. BURNAP and J. D. BRASSINGTON, New York, N. Y.—*Telegraph Insulator.*—October 9, 1866.—The inverted cup has a shank by which it is secured to the post, and a socket to hold the end of the pin; between the latter and the edge of the cup is an annular non-conducting disk.

Claim.—First, a non-conducting ring or disk, fitted in the base of a hollow shell insulated below the point at which the pin is secured, substantially as and for the purpose herein set forth.

Second, the non-conducting disk M *m*, fitted with springs P¹ P², and adapted to be confined in the base of the shell A, substantially in the manner and for the purpose herein set forth.

Third, the within-described combination and arrangement of the shell A H, confining material J, inverted cup O, and removable non-conducting disk M, as and for the purpose herein set forth.

No. 58,595.—CHARLES CAMP, Buffalo, N. Y.—*Elevator Bucket.*—October 9, 1866.—The body of the bucket is of elastic material and the mouth of metal.

Claim.—First, the elastic or flexible elevator bucket composed of the materials described, or the equivalents thereof.

Second, in combination therewith the metallic rim A, as and for the purpose specified.

No. 58,596.—CHARLES CHAVANNE, New Orleans, La.—*Dough Mixer*—October 9, 1866.—The dough is placed in a bowl which revolves on a vertical axis, while the kneading arms revolve therein on a horizontal axis.

Claim.—The revolving of the bowl E and the arms *a a a a* in the different directions at the same time, and the scraper M, as set forth and fully described.

No 58,597.—JOHN CLARRIDGE, Pancoastburg, Ohio.—*Double Shovel Plough.*—October 9, 1866.—The plough has a lateral adjustment by slide racks and slide bar, and is held to any adjustment by tooth blocks and wedges.

Claim.—The combination and arrangement of the arms L and M, the toothed blocks N and P, and the wedges or keys O and R, with the beams B and A, the plough-head G, and the handle I, substantially as herein described and for the purpose set forth.

No. 58,598.—JOHN CLARRIDGE, Pancoastburg, Ohio.—*Single-row Corn Planter.*—October 9, 1866—The corn in an upper hopper is conveyed by an elevator and pipe to a lower hopper, from whence it is dropped by a reciprocating seed slide, which is worked by hand through a rod and lever. The hopper stem has a sliding joint to admit of adjusting the pitch of the plough.

Claim.—First, the combination and arrangement of the elevator I, shafts H and F, pulleys E and *c*, band D, drive wheel B, and spout J, with each other, with the hopper G and with the frame A of the machine, substantially as herein described and for the purpose set forth.

Second, the combination of valve T, lever X, and slide rod A' with each other, with the hollow plough shank K, and with the handle B', substantially as described and for the purpose set forth.

Third, the combination of the concavo-convex head *k'* of the hollow shank K with the concave bed plate *a³*, substantially as herein described and for the purpose set forth.

No. 58,599.—THOMAS J. CLOSE, Philadelphia, Penn.—*Machine for Making Mouldings*—October 9, 1866.—For the manufacture of composition moulding. The material is driven from the cone-shaped hopper by the spiral compresser, and is received on an endless apron which carries it to the die wheel and delivers it on to an inclined table. The throat of the hopper is regulated by controlling devices cited in the sixth claim. The conveyer is scraped to remove adhering particles, and the die wheel oiled to prevent adherence of the composition.

Claim.—First, the cone-shaped hopper, formed by combining the inner case J and outer case K with each other and with the frame A of the machine, substantially as herein described and for the purpose set forth.

Second, the combination with the hopper J K of the conveyer W, substantially as described and for the purpose set forth.

Third, the combination of the die wheel A', constructed substantially as described, with the conveyer W and with the frame A of the machine, substantially as described and for the purpose set forth.

Fourth, the combination with the conveyer W, of an adjustable slide or scraper B', constructed and arranged substantially as described and for the purpose set forth.

Fifth, the combination with the hopper J K, of the screw H I, constructed and arranged substantially as described and for the purpose set forth.

Sixth, the combination of the valve E', arm F', levers G' I', and cam K' L', with each other, with the hopper J K, and with the frame A, of the machine, substantially as described and for the purpose set forth.

Seventh, the combination of the rubber or cleaner P' and scraper T' with each other, and with the conveyer W, substantially as described and for the purpose set forth.

Eighth, the combination of the oil cup D' with the hopper J K, and with the die wheel A', substantially as described and for the purpose set forth.

No. 58,600 —THOMAS M. COFFIN, Plymouth, Mass.—*Faucet.*—October 9, 1866.—The valve is opened by the depression of the stem, which is furnished with packing to prevent leakage in this position.

Claim.—As a new article of manufacture, the faucet herein described, consisting of the barrel A, provided with a valve seat a', and an eduction nozzle C, the neck D, connected to the barrel by a screw joint a m, to afford access to the valve seat, the stem E, cast in one piece, with the external head G, and threaded at g, to fit an internal screw thread in the barrel A, and the valve E', mounted removably on the end of the stem E, all as herein specified and for the purpose explained.

No. 58,601.—MUXSON COLE, Colebrook, Conn., and DAVID COLE, Kent, Conn.—*Machine for Raking and Loading Hay.*—October 9, 1866.—The gathering rake teeth are attached to a frame pivoted to the carriage and resting upon springs; the toothed endless apron for elevating the hay is thrown into operation by a clutch on the axle actuated by a horizontal lever on the carriage.

Claim.—First, the gathering rake, arranged in combination with the rods r, springs h, and bar b', substantially as described.

Second, in combination with the above, the endless elevator and clutch arrangement, all arranged and operating substantially as and for the purposes set forth.

No. 58,602.—M. H. COLLINS, Chelsea, Mass., assignor to himself and WILLIAM H. HOLLAND, same place.—*Rotary Steam Engine.*—October 9, 1866.—The engine has two rotary pistons on the central hub, and three radial sliding abutment gates. The piston hub has two steam chambers with passages leading out of them through the outer end and the circumference of each. The hub chambers communicate with the valve chest by an axial chamber and a conduit formed in each of the ends of the cylinder. Steam pressure is exerted on each side of the central shaft simultaneously.

Claim.—The combination and arrangement of the chambered drum F, made substantially as described, with the two pistons and their gates and ports d d, arranged in the case or cylinder, as specified.

Also, the arrangement and combination of the springs q r with each of the yokes p, the same being as and for the purpose specified.

Also, the construction of each piston, and that part of the case against which it operates, viz: curved on their peripheries, substantially as represented, in combination with the cylindrical drum, arranged with respect to them, as specified.

No. 58,603.—MICHAEL H. COLLINS and WILLIAM H. HOLLAND, Chelsea, Mass.—*Paddle Wheel.*—October 9, 1866.—The hub has three circular plates; from each of these extend branching arms connected to as many rings joined by radial perforated plates to the marginal rings on the rim of the wheel. Buckets are arranged between the rim plates, presenting their obtuse angles in alternate directions, and secured to each other and to the inner ring, on whose circumference they are placed

Claim.—The arrangement and combination of its obtuse angular plates or paddles r r, with the three series of end and middle connections n p n, and the three wheels composed of the rings i k l, and the three series of spokes applied together and to a hub, as described.

Also, the combination and arrangement of the external rings m m with the paddles r r, the three series of connections n p n, and the three wheels composed of the three rings i k l, and the three series of spokes, applied together and to a hub, as described.

No. 58,604.—WILLIAM CONANT, Geneva, Ill —*Steam Generator.*—October 9, 1866 —Steam is generated in a convoluted pipe in the furnace, and conducted to the superheating receiver, which is arranged within a contracted flue through which the products of combustion pass. The furnace and flue are bounded by a lining of non-conducting material.

Claim.—First, the generating conductor A, and superheating receiver C, the latter arranged within a contracted flue through which the products of combustion pass in close contact with its surface, substantially as and for the purposes described.

Second, in combination with the generating pipe and receiver. arranged as described. the lining D, employed to confine the products of combustion around the receiver, and prevent the radiation of heat.

No. 58,605.—J. J. CONLEY; New York, N Y.—*Lathe for Turning Shafting.*—October 9. 1866.—The shaft to be turned is fed through the hollow mandrel, at one end of which the cutters are arranged, which turn it off as it is fed along. A bush in each end of the mandrel forms a bearing for steadying the shaft.

Claim.—First, the annular block A*, so applied in combination with the tubular spindle B and cutter head C as to support and steady the shaft during the operation of turning the same, substantially as herein set forth.

Second, the attachment of the smoothing cutter *u* to the annular block A*, substantially as herein set forth for the purposes specified.

Third, the annular block B* fitted into the rearmost end of the tubular spindle B, and operating in conjunction with the annular block A*, to steady and support the shaft while being turned, substantially as herein set forth.

Fourth, the combination of the hollow spindle B, cutter-head C, and the two sets of grooved feed-rollers D and E, one set for pushing and the other for drawing the shaft during the operation of turning the same, the whole arranged and operating substantially as herein set forth, for the purpose specified.

Fifth, the swing or self-adjusting shafts C', furnished with tangent screws *f'*, and arranged in relation with the worm wheels *t* and grooved feeding rolls, substantially as herein set forth, for the purpose specified.

No. 58,606.—JOHN B. COOLIDGE, Boston, Mass.—*Gas-burner.*—October 9, 1866.—The gas-burner is made of a slit piece of metal so that by the operation of a screw the orifice for the flame may be enlarged or compressed.

Claim.—The split tip *c* in combination with the exterior tube *b*, operating substantially as described.

No. 58,607.—J. COOPER, Mount Vernon, Ohio.—*Evaporator.*—October 9, 1866.—The doors slip in guides which are attached by ears to the openings in the partitions of the pan.

Claim.—The ears *c*, in combination with the guide strips *b*, gate openings *d*, and partitions D, contracted and operating substantially as and for the purpose described.

No. 58,608.—WILLIAM COUGHLAN, Baltimore Md.—*Soda Fountain.*—October 9, 1866; improvement on his patent of October 25, 1853.—The fountain can be used in an inverted position and the liquid is gauged by a blow-off cock.

Claim.—The valve I and opening J in the stud C at the side of the fountain, at or near the desired level, to act as a gauge, and also as a blow off opening for the discharge of air from the fountain, substantially as described.

Also, the auxiliary tube H, attachable or detachable from the faucet, to permit the fountain to be used in the upright or inverted position, substantially as described.

No. 58,609.—HENRY DALE, Philadelphia, Penn.—*Loom.*—October 9, 1866.—These improvements relate to devices for operating the harness. The pattern cylinder shaft being free at one end admits of changing the cylinder readily for changing the pattern. The graduation in the sizes of the pulleys placed in series on the same shaft allows a true vertical pull upon each of the jacks, which are arranged side by side beneath.

Claim.—First, the within-described arrangement with respect to the frame A and jacks of the shaft I, for the reception and removal of the cylinder J, or its equivalent.

Second, the combination and arrangement on the loom of the cam shaft C, wheels *x* and *s*, and shaft I, as specified.

Third, the combination and arrangement of the graduated pulleys *a a*, the jacks G, and the heddle cords, as set forth.

No. 58,610.—J. DALTON, Williamsburg, N. Y.—*Knitting Machine.*—October 9, 1866.—This machine makes a combined warp and knit stitch, applicable for covering cords and skirt wire, or for lamp wicks, shoe lacings, &c. The lower or warp thread is not caught by the needles. The middle or main thread and the remaining thread are both taken by the outer line of needles, while the inner line of needles takes the main thread only. The levers have provision for adjustment, and may be removed at will. The cylinder is adjustable relatively to the cam and the toes of the needles. The adjustability of the segments provides for any wear in the disk, which they serve to hold in position.

Claim.—First, the method herein described of forming a combined warp and knit stitch by the action of a series of needles which rise and fall in the needle cylinder, in combination with a revolving cam and spool-carrier, substantially as described for the purpose specified.

Second, the detached levers G, with their curvatures *s*, in combination with the movable ring I, substantially as described for the purpose specified.

Third, the reciprocating needles *n*, arranged in a circular series, one portion thereof placed within and concentric to that in which the others are placed, or with those of each series at an inclination from a vertical line, substantially as described, for the purpose specified.

Fourth, the segmental projections *e*, on the upper edges of the levers G, in combination with the top plate or ring I, and fulcrum *a*, constructed and operating substantially as and for the purpose set forth.

Fifth, the adjustable segments *f*, in combination with the cylinder H, levers G, and cam D, constructed and operating substantially as and for the purpose described.

No. 58,611.—ALEXANDER M. DAMON, Lowell, Mass.—*Warp Dressing Frame.*—October 9, 1866.—All the comb guides or raddles are reciprocated by the same bar, whose motions are derived from the gearing and controlled by the oblique end guides.

Claim.—The combination of the slotted bar P, constructed as described, with the raddles C, and the actuating gearing G W, when operating as and for the purpose described.

No. 58,612.—F. S. DAVENPORT, Jerseyville, Ill.—*Gang Plough.*—October 9, 1866.—The plough frame is pivoted to a hinged board which is supported on the axles, one of the latter being changeable, to adapt its wheel to run in the furrow or on a level. The ploughs are adjusted by a compound lever arrangement, secured in position by a ratchet segment. A brake, brought into contact with the off-wheel, assists in raising the ploughs from the ground by erecting the hinged board. The tongue is adjustable, and is secured to the footboard. The rear plough may be lifted separately, and the rear of the frame is lifted from the ground after the front part has been considerably elevated.

Claim —First, the lever P, rod Q, and brake R, arranged and operating as and for the purpose described.

Second, the hinged board G, in connection with the reversible axles, substantially as and for the purpose described.

Third, the lever O, and quadrant N, for regulating the depth of the furrow, substantially as and for the purpose specified.

Fourth, lifting the hind part of the machine by means of the lever or arm I, in connection with the chain J, wheel K, and lever L, these parts operating together substantially as and for the purpose described.

Fifth, binging the footboard M to the plough-frame, as described.

Sixth, securing the tongue or draught-pole to the footboard M, in the manner and for the purpose described.

Seventh, the sliding plough-standard B', guide-block O*, lever A*, and notched seat-standard C, when used together and in connection with the other parts.

Eighth, connecting the lever L with the tongue or draught-pole by fastening it to the foot-board, the whole operating together substantially as and for the purpose set forth.

No. 58,613.— JAMES DAVIS, Loami, Ill —*Combined Roller and Harrow.*—October 9, 1866 —The barrow is adjustable vertically by a lever and rotated by gearing from the axle; it is followed by a roller on the same frame.

Claim.—First, the revolving barrow L, in combination with the pinion K, gear wheels J and H, and traction wheel B, for the purposes and substantially as described.

Second, the levers O O, in combination with the barrow L, gear wheels K J and H, substantially as set forth.

Third, the roller E, in combination with the pendants F F, frames A and L, all for the purposes and substantially as described.

No. 58,614.—JOB A. DAVIS, Great Bend, Penn.—*Needle Feed of Sewing Machines.*—October 9, 1866 —The rise and fall of the needle slide operates an auxiliary feed "helper" or assistant, having a roughened sole, which bears upon the cloth immediately in advance of the needle when the feed takes place, thus preventing the cloth from bunching; the descent of this "helper" causes the rise of the presser foot, and *vice versa.* The foot of the "helper" passes through a slot in the presser foot, and it partially embraces the needle at the period of feeding.

Claim.—First, in sewing machines using a needle-feed, the application and use, in combination with the needle-bar and needle, of the needle assistant or helper bar G, for keeping the cloth smooth and preventing its gathering or bunching as the feed takes place, such bar being placed before the needle and so arranged as to move up and down upon its fulcrum, and operating substantially as and for the purposes set forth.

Second, the combination of such helper bar with the pressure bar, so arranged in respect to each other that as one descends the other rises, and *vice versa*, and operating substantially as and for the purposes set forth.

Third, the arrangement of the slot *m* in the helper bar, and the pin *f* upon the pressure bar, or their equivalent, so that the descent of the needle-bar or its equivalent will force down the helper bar and elevate the pressure bar, substantially as and for the purposes set forth.

Fourth, operating the helper and pressure bars substantially as described, from or by means of the needle-bar, or its equivalent, substantially as and for the purposes set forth.

No. 58,615.—AUSTIN G. DAY, New York, N.Y.--*Artificial Caoutchouc* —October 9, 1866; antedated September 29, 1866.—Animal and vegetable oils are combined with coal tar, asphaltum, &c., and with sulphur.

Claim.—Mixing, heating, and sulphurizing vegetable and mineral oils, in combination with gum, resins, and resinous compounds, to form a composition to be used as a substitute for caoutchouc or India-rubber, substantially in the manner and for the purpose herein set forth.

No. 58,616.—CHARLES M. DAY, Ann Arbor, Mich.—*Hand Woodsaw.*—October 9, 1866.—The saws are connected by a pitman and lever to a spring rock shaft, to assist in the return stroke.

Claim.—The combination of the rock-shaft K, provided with the springs *o*, with the bifurcated lever I, and pitman J, connected to the saws P, when arranged to operate as shown and described.

No. 58,617.—JOSEPH DICK, junior, Canton, Ohio, assignor to himself and F. W. GLENN.—*Harvester.*—October 9, 1866.—The arrangement of mechanism for driving the rake, the manner of securing the reel-pulley to its shaft, and the specific arrangement of lever segment, cord, and pulley for raising the cutting apparatus, is identified by the claims.

Claim.—First, the sliding raker-shaft E, or its equivalent, and the joint-ball *g*, in combination with the driving pulley B, or other suitable case, when the latter has a cylindrical axis within the hanger A, entirely independent of the said joint-ball *g*, as shown and described.

Second, the arrangement, in combination with the sliding raker-shaft, of the ball *g* and the pulley B, or other suitable case, within the hanger A, the latter constituting cylindrical bearings for the axis of the said case or pulley B, as set forth.

Third, the arrangement of the segments G and G' upon the vertical sleeve *f*, and the segmental pinions C and C' upon the horizontal driving-shaft E' of the raker, as shown, so as to constitute collectively an entire circle of gearing, as shown and described.

Fourth, the combination of the detachable pulley T with the sleeve or ferrule S, having one or more locking pins *c*, substantially as and for the purposes set forth.

Fifth, the arrangement of the elevating lever L, ratchet O', head Q, chain U, and pulley V, in combination with each other and the brace of the shoe, as and for the purpose set forth.

No. 58,618.—JOHN B. DOUGHERTY, Rochester, N. Y.—*Shingle Machine.*—October 9, 1866.—The machine has an automatic clamp and a tilting frame; the bolt is fed automatically to the circular saw at a comparatively slow speed, and returned at a faster rate. The saw is formed of circular segments around the blades, which form a chute to carry off the saw-dust. At the end of its back stroke the counterbalance rises, unclamps the block which drops, and is reclamped when the effective stroke commences. The belt frame is oscillated by a crank which is operated by a worm-wheel and belt from the saw shaft.

Claim.—First, the construction and relative arrangement of the saw guard R and U, in the manner shown and described, to facilitate the removal or readjustment of the saw in the machine, and for conducting the saw-dust from the machine, substantially as set forth.

Second, the arrangement of the counterbalance W, shaft *l*, pinions *k*, racks *z*, and clamping bars *g*, in combination with the cams *m* and *n*, substantially in the manner and for the purposes set forth.

Third, the arrangement of the screw gears and crank G with the slotted arm H, attached to the axial shaft C of the bolt or clamping frame B, substantially as and for the purposes shown and described.

Fourth, providing the saw S in the within-described machine with a collar *j*, having a conical bore to fit the mandrel, as and for the purposes set forth.

Fifth, the arrangement of the pivoted or swinging circular track W, as and for the purposes set forth.

No. 58,619.—RALPH C. DUNHAM, New Britain, Conn.—*Buckle and Ring.*—October 9, 1866.—The metal is protected and ornamented by a covering of vulcanite.

Claim.—A buckle or ring composed of a metal core *a* and a covering *b* of India-rubber or other vulcanizable gum, substantially as and for the purpose described.

No. 58,620.—C. R. DURFEE, Rochester, N. Y.—*Horse Collar.*—October 9, 1866.—The draught bar has hame tugs on its end, and is connected by branches to the plate which rests over the "collar-place" of the shoulder, a pad intervening.

Claim.—The horse collar herein described, consisting of the curved draught bow or bar B, plates P, pads A, and connecting strap or pad F, the several parts being constructed, arranged, and operating substantially in the manner herein shown, and for the purpose set forth.

No. 58,621.—ZEBINA EASTMAN, Chicago, Ill.—*Railway.*—October 9, 1866.—The axles have bed plates affixed to them which are coupled together by a bar and king bolts: they have also diagonal reaches connecting them, by which the inclination of one axle in a curve causes a like inclination in the other one. When three axles are used, the fore and hind bed plates have each a segmental rack which works in a similar rack upon the bed plate of the central axle, and tends to thrust it laterally to its proper position on the curved track.

Claim.—First, the axles *e e*, each constructed in one piece, in combination with the tables or bed pieces *d d*, connecting rod or reach *h*, and transverse reach *i*, the whole being constructed, arranged, and operated substantially in the manner and for the purpose set forth.

Second, the axles *e e y* in combination with the reach *b*, toothed frames *r s s*, and transverse reaches *i i*, the whole being constructed and operated substantially in the manner and for the purpose set forth.

No. 58,622.—RUDOLPH EICKEMEYER, Yonkers, N. Y.—*Sewing Sweat Linings into Hats.*—October 9, 1866.—To avoid piercing the sweat band at the part in contact with the head, and to stiffen the angle of the brim and sides, the band is sewn in by a diagonal seam while the sweat band is turned out.

Claim.—The sewing in of the sweat lining of a hat by stitches passing once through the lining and through the hat body diagonally to the brim and sides, without being whipped over the edge of the lining, substantially as herein specified.

No. 58,623.—LEWIS ELLIOTT, Jr., New Haven, Conn.—*India-rubber and Leather Sole.*—October 9. 1866.—The layer of India-rubber is smaller than the surface of leather, leaving a marginal strip of the latter for sewing; the two are united by vulcanizing the rubber, while the two layers are associated under pressure.

Claim.—A water-proof sole for boots and shoes, formed of two thicknesses, one of India-rubber or its compounds, the other of leather, the two being united firmly in the manner specified.

No. 58,624.—ALVA E. ELLIS, Friendsville, Ill.—*Beehive.*—October 9, 1866.—The lower bottom consists of pivoted slats, and over it is a floor which has a gauzed opening therein for ventilation. A central, horizontal, movable division has openings for the passage of bees and for ventilation.

Claim.—The pivoted slats C at the bottom of the case A, the hive in connection with the bottom B, provided with an opening covered with wire cloth *a* and the openings *e'* in the sides of the partition D, substantially as and for the purpose set forth.

No. 58,625.—THEODORE R. FANCHER, Norwalk, Conn.—*Steam Engine Globe Valve.*—October 9, 1866.—An annular piece of India-rubber is fitted into a circular groove in the valve face for contact with the seat; the rubber is retained by an encircling screw ring. The valve-head has free rotation on the stem.

Claim.—The arrangement of the adjustable rubber ring C and screw ring E, substantially as and for the purpose described.

No. 58,626.—R. A. FISH, Worcester, Mass.—*Spike Puller.*—October 9, 1866.—The lever is pivoted in a frame which has at one end a fulcrum and at the other a claw, to act in conjunction with a similar claw upon the end of the lever to seize and withdraw the nail.

Claim.—The combination of lever A, with its bill hook end *a*, with the base piece B, having a bill hook projection *d* and flattened rear part D, constructed and arranged for joint operation, as set forth.

No. 58,627.—JARED B. FLAGG, New Haven, Conn., and GEORGE STORER, New Britain, Conn.—*Stretcher for Canvas.*—October 9, 1866.—The mitre joints have dowel pins, and are expanded by the wedges, the pins in the open centre of the latter preventing their falling out.

Claim.—The perforated wedge *b* in combination with the groove *d*, pins or screws *e*, and frame, substantially as and for the purpose described.

Also, in combination with the above, the metal or wood pieces *c*, substantially as and for the purposes described.

No. 58,628.—JOHN W. FORSYTH, Leesburg, Va., assignor to himself, JOHN W. and NELSON HEAD.—*Churn.*—October 9, 1866.—The operative parts are supported on posts erected on the platform, and the churn is secured to the latter by hooks, while the dasher is vertically reciprocated therein.

Claim.—The combination of the platform F with the posts E E', their connecting cross-piece G, wheel A, pinion B, fly-wheel I, pin N, shaft C and O, the vessel M, the hook *a a*, securing the lid L, the slotted strips D D D, the hooks *t t t*, the staples *h h h*, as is described and for the purpose set forth.

No. 58,629.—CHARLES H. FROST, Peekskill, N. Y.—*Sad-Iron Heater.*—October 9, 1866.—Explained by the claim and cut.

Claim.—Bending up or flaring the front ends of the covers of sad-iron heaters, so as to allow the nose or point of a sad iron to be inserted beneath said covers to raise them, substantially as set forth.

No. 58,630.—FRANCIS FRYE, Time, Ill.—*Winnowing Machine.*—October 9, 1866.—The eccentric on the main shaft works in a pivoted yoke and agitates the shoe longitudinally; the wings of the fan are attached to the shaft, and act upon the light offal beneath the shoe, and also in the fan case through which the grain passes.

Claim.—First, the eccentric E, lever D, in combination with the hopper B and screen C, substantially as and for the purposes set forth.

Second, the lever D, eccentric B, in combination with the fans G, for the purposes and substantially as herein shown and described.

No. 58,631.—WILLIAM H. GATES, Louisville, Ky.—*Dental Drill.*—October 9, 1866.—The pointed bulbous head of the reamer is channelled with grooves bearing effective edges, which are undercut and recede spirally toward the stem to discharge the cuttings.

Claim.—A drill head having two or more of the longitudinally curved external surfaces C, extending from B to H, in conjunction with an equal number of cutting edges, undercut by deep grooves running backward from the end spirally or diagonally to the stem or axis, substantially as described and for the purpose set forth.

No. 58,632.—C. L. GILPATRIC, South Dedham, Mass.—*Nutmeg Grater.*—October 9, 1866.—The grater revolves in a stock, and the nutmeg is pressed against it by a spiral spring which bears against a removable cap, covering the opening by which it and the spice are introduced.

Claim.—The case A as constructed in combination with the spool B secured in said case, and provided with the perforated covering D, in which is cut an opening a, which discharges the ground nutmeg, which passes into the spool, and by means of which the spool may be cleansed internally, the several parts being arranged as and for the purpose herein specified.

No. 58,633.—WILLIS D. GOLD, Philadelphia. Penn.—*Wrench.*—October 9, 1866.—The back of the wrench shank is serrated where the movable jaw plays. The jaw is retained in position by a spring catch, and the use of the wrench serves to tighten the movable jaw upon the shank. A light pressure on the end of the spring plate frees the movable jaw.

Claim.—The arrangement of the spring-toothed lever F in combination with the movable jaw E, operating with the rectangular shank C, in the manner and for the purpose herein described.

No. 58,634.—M. J. GOODWIN, Boston, Mass.—*Railway Track Clearer.*—October 9, 1866.—The clearer is suspended in front of the wheel, and is connected with the brake, so as to operate the latter when colliding with an obstacle.

Claim.—Combining with the car truck and brakes of a railway car the clearers, hung from the brakes in front of the car truck, and at a short distance above the surface of the track rails, substantially as described.

No. 58,635.—ALEXANDER GORDON, Rochester, N. Y.—*Horse Hay Fork.*—October 9, 1866.—To the lower end of the tubular case is pivoted a lifting prong, the inner end of which is a toothed segment. A rack bar within the case engages with the teeth upon the prong; the upper end engages with a similar segment upon an operating hand lever pivoted to the case. A spring catch at the upper end of the case engages with notches in the bar to hold the prong extended, and is tripped to release the load.

Claim.—In combination with the point P the toothed adjusting bar b, the hand lever E, and the locking latch D, they all operating conjointly in the manner and for the purpose specified.

No. 58,636.—SAMUEL J. GOUCHER, Philadelphia, Penn.—*Shovel.*—October 9, 1866.—Strengthening strips of sheet metal are riveted to the top of the blade on each side of the handle.

Claim.—Strips E, secured to the blade and handle straps of a shovel, as and for the purpose set forth.

No. 58,637.—SIMEON L. GOULD, Skowhegan, Me.—*Nail Machine.*—October 9, 1866.—Improvement on Berry's patent, February 14, 1860.—Designed to enable two or more nail plates to be cut by the cutters of the rotary wheel. The slide plate and eccentric are arranged in front of the circular cutter head, which carries two or more series of cutters, consisting of long rectangular plates placed in corresponding sockets and confined by clamp screws.

Claim.—The arrangement of the slide plate k and the eccentric h, with the vibratory cutter, carrying levers, and the cutter wheel and its shaft, or the same and the feed rollers and guide bar, as set forth.

Also, the arrangement of the rotary cutters, their sockets, clamping and adjusting screws, and the wheel or cutter bed, as specified.

No. 58,638.—DANIEL GRAVES, Seneca, Ill.—*Cooking Stove.*—October 9, 1866.—Griddle holes are arranged in the top plate above the fire pot, and several in the broad annular hearth, which is heated by the downward caloric current between the inner and outer frustums of the stove body. Segmental reflectors encase the hearth and an oven space around the waist of the stove.

Claim.—First, the combination and relative arrangement of the frustum plates B C F and H, substantially as herein shown and described, in the formation of cooking stoves, for the purposes set forth.

Second, the broad, annular hearth G, when arranged to be heated by the caloric current passing under it, substantially as and for the purposes shown and described.

Third, the double frustum B C, which constitutes the inner draught flue, and gives a radiating direction to the caloric current as shown.

Fourth, the segmental reflector plates or cases R, in combination with the frustum stove, constituting an annular heating or baking oven, as shown and described.

No. 58,639.—GEBHARD HAGENMEYER, Big River, Cal.—*Miner's Fuze Lock.*—October 9, 1866.—A spur on the spring attaches the lock to the fuze when the hammer is set. The dog is pulled by a long cord from a distant position of safety, releasing the hammer, which explodes the cap and lights the fuze.

Claim.—As an improved fuze lock, the arrangement of the hammer *c*, sere or dog *g*, spring *e*, pin *j*, and barrel *b*, relating to each other and operating in the manner as and for the purpose herein specified.

Also, the pin *j* in combination with the spring *e*, when used for the purpose herein represented and described.

No. 58,640.—F. K. HAIN, Renova, Penn.—*Axle-box Cover.*—October 9, 1866.—One of the bearing lugs is made open on top and the other closed ; their inner faces are oblique. A spring is compressed by one trunnion in the closed eye or bearing, and pushes the gravitating cover laterally so as to lock beneath the other bearing.

Claim.—The cover B, trunnions *a b*, spring *f*, eyes *d e*, with inner inclined planes, the eye *d* being open and provided with notch *g*, and the box A, combined and arranged substantially as described, for the purpose specified.

No. 58,641.—J. HENRY HAYWARD, New York, N. Y.—*Paper Skirt for Ladies.*—October 9, 1866.—The skirt is made of paper, the pieces joined by gum and stayed by cords, &c.

Claim.—First, as a new and useful article of manufacture, the paper skirt as herein described, made by the combination of one or more sections of sheet paper, of any and every kind, quality, and color, water-proof and fire-proof inclusive, arranged adhesively together, and strengthened by means of cords, tapes, or other suitable material and devices, as fully described.

Second, the peculiar manner of arranging and connecting the said skirt at the waist by the means and for the purposes specified.

Third, the joining of two or more skirts made, as above described, on the same band, bodice, or waist hoops, as above set forth.

No. 58,642.—GEORGE M. HEIM, Brownsville, Ind.—*Washing Machine.*—October 9, 1866.—The barrel-shaped washing machine is oscillated upon an axis below its periphery, by link connection of its upper portion to a revolving crank.

Claim.—The combination of the supporting frame A B, rod C, fly-wheel F, crank H, connecting rod I, arm J, and barrel D, all constructed and arranged to operate substantially as and for the purposes described.

No. 58,643.—GEORGE W. HERSEY, Greenbush, Wis.—*Tanning.*—October 9, 1866.—The skins are soaked in a solution of salt, soft soap, and water. They are then limed, bated in water, bran, and sour milk; then tanned in a composition of water, terra japonica, glauber salts, sulphuric acid, and common salt. For tanning with the hair on, water, sulphuric acid, nitric acid, borax, alum, and glauber salts are applied to the flesh side of the skin.

Claim.—First, soaking the hides or skins in salt water mixed with soft soap previous to liming, substantially as and for the purpose described.

Second, the use for tanning leather of a liquor containing glauber salts and common salt in combination with terra japonica, substantially as and for the purpose described.

Third, also applying to the flesh side the sulphuric acid, glauber salts, and alum and borax to all furs, with swab or brush, or paste to the same.

No. 58,644.—DANIEL M. HOLMES, Brooklyn, N. Y.—*Manufacture of Cream Crackers.*—October 9, 1866.—The crackers are composed of wheat flour, white sugar, lard or butter, eggs, cream of tartar, bicarbonate soda, and cream.

Claim.—A cracker composed of the ingredients in the proportions named, and treated in the manner substantially as set forth.

No. 58,645.—CHARLES HUIE, Lockport, N. Y.—*Punch for Horseshoes.*—October 9, 1866.—The tool for punching the nail holes in the shoe has a projection beside the punch, to act as a gauge and as an abutment to restrain the edge from spreading.

Claim.—A punch, the point *e* of which is provided with the flange *f*, serving both as a guard and gauge, when constructed and operating substantially in the manner and for the purpose set forth.

No. 58,646.—E. S. HUNT, Weymouth, Mass.—*Rocket.*—October 9, 1866.—The fuze of the rocket charge tube is ignited by friction on the withdrawal of a prepared strip pressed by a spring against the igniting composition.

Claim.—The arrangement of the spring *m* and its case *m'* with the rocket charge tube A and its priming vent *h*.

No. 58,647.—ALLEN HUSTON, Cincinnati, Ohio.—*Hub.*—October 9, 1866.—The hub is cast in two concentric portions, the inner one having a circumferential grooved ridge, and the outer one a groove to contain the latter. The outer piece has mortises for the reception of spokes, and oil holes connecting with the groove on the inner piece.

Claim.—Casting a hub in two concentric parts, one about or within the other, so that while the parts are inseparable each may revolve freely in relation to the other, substantially as described.

No. 58,648.—THADDEUS C. JOY, Titusville, Penn —*Construction of Oil Tanks.*—October 9, 1866.—The boiler iron tanks are formed of sections having exterior flanges by which they are bolted together.

Claim.—The method, substantially as described, of constructing metal tanks for the purpose of rendering them portable.

No. 58,649.—ABRAHAM B. KING, Camden, Ohio.—*Feed Cutter and Box.*—October 9, 1866.—A removable cutting box is used in combination with a feed box, and connected by pins so as to allow the former to discharge into the latter. A pivoted arm spans the top of the cutting box, and its throat admits of adjustment to the knife by means of a slotted wedge.

Claim.—First, the arrangement of box or manger A *a a'*, shiftable cutting apparatus E, and hinging and supporting devices C C', D D', F F', G, substantially as set forth.

Second, the bar H pivoted to the top of the trough, for the purpose set forth.

Third, the combination of knife I, adjustable wedge J, screw K, and stump L, for the purpose explained.

No. 58,650.—GEO. L. KING, Philadelphia, Penn.—*Driving Pipe for Oil Wells.*—October 9, 1866.—The sections of pipe are attached together by screw-threaded sleeves and between the sleeves, sections of pipe are shrunk on, and the whole turned to an equal diameter. The annular cutter is laid with steel, has the same outside and inside diameter as the finished pipe, and is bevelled from the outer edge.

Claim.—First, the combination of the inner pipe B, strengthening bands D, and thimbles C, the several parts being constructed and arranged in relation to each other substantially in the manner hereinbefore described and for the purpose specified.

Second, the combination of the shoe or cutter A with the lower end of the pipe B, and lower strengthening band D, substantially in the manner described and for the purpose above set forth.

No. 58,651.—JUDAH LEVY, Philadelphia, Penn.—*Hoop for Skirts.*—October 9, 1866.—Each hoop consists of two light wires bound together at intervals and yielding to external pressure by springing apart and allowing the skirt to collapse at that point.

Claim.—A skirt, each hoop of which consists of two light wires, rigidly clasped together at intervals, as and for the purpose set forth.

No. 58,652 —N. C. LINCOLN, Brunswick, Maine.—*Medicine.*—October 9, 1866.—Composed of powdered laurel, 4 ounces; pulverized sugar, 1 ounce; spirits camphor, ¼ ounce, and essence of checkerberry, ¼ ounce.

Claim.—The compound of ingredients for a catarrh medicine, mixed in the manner and proportions above described.

No. 58,653.—CARL A. LINDNER, Cincinnati, Ohio.—*Carpet Stretcher.*—October 9, 1866.—The handles of the serrated jaws are prolonged into curved arms, on which slides a bridle connected to a draw spike.

Claim.—The carpet stretcher composed of the hand-spike A and grapnel B, substantially as set forth.

No. 58,654.—FERDINAND LINDNER, Dayton, Ohio. — *Apparatus for Making Button-holes.*—October 9, 1866.—The spring clamp holds the material during the process of working the button-hole, acts as a needle guide, and is released by pressure upon backward projections.

Claim.—A button-hole regulator, made of steel or any other suitable material, having the jaws A B, bow spring H, levers D D, and handles G G, as herein described and for the purposes set forth.

No 58,655.—J. W. & S. A. LIVINGSTON, Hartford, Conn.—*Sash Supporter and Fastener.*— October 9, 1866.—Within a cavity in the sash a roller and stout spring are so arranged that the spring forces the roller against the frame with sufficient power to hold the sash at any point; when the sash is down the roller drops into a notch in the frame, and a pin is inserted between the roller and sash to lock the latter.

Claim.—The friction roller D and spring *f*, when combined and arranged substantially as and for the purpose specified.

Also, in combination with the foregoing the use of a pin K, or its equivalent, as and for the purpose described.

No. 58,656.—HERMAN S. LUCAS, Chester, Mass.—*Blasting Cartridge.*—October 9, 1866.— The fuze enters the cylinder of compressed gunpowder, and the cartridge is enclosed in a water-proof sack.

Claim.—First, a cartridge for blasting, made of solidly compressed gunpowder, containing either nitrate of potash or nitrate of soda, or granulated, or of any other suitable explosive material capable of being safely compressed into a suitable form for blasting purposes, and provided with any suitable device for igniting the same from the interior.

Second, the combination of a cylindrical cartridge of solidly compressed gunpowder or other explosive materials for blasting, with a central perforation extending partly or wholly through the same, constructed in the manner and for the purpose above described.

Third, the combination of a cartridge of solidly compressed gunpowder or other explosive materials for blasting with a fuze, when said fuze is inserted into its interior, in the manner and for the purpose set forth.

Fourth, the combination of a cartridge of solidly compressed gunpowder for blasting purposes, with an envelope of paper or other textile material made impermeable to water, or with envelopes or casings of sheet metal, earthenware, or wood, the same being attached to the fuze in the manner and for the purpose set forth.

No. 58,657.—H. H. MANSFIELD, South Canton, Mass.—*Metallic Fastening for Buckles and Straps.*—October 9, 1866.—A piece of sheet metal is lapped around the back bar of the buckle and embraces the end of a leather strap, when another strip of metal is wrapped around the strap and metal laterally, and the two riveted together.

Claim.—First, the metallic plate or bushing A, constructed substantially as and for the purpose herein set forth and described.

Second, the manner of attaching the strap C to the buckle B, substantially as herein set forth and described.

No. 58,658.—JACOB McCLURE, Rockland, Maine.—*Hollow Auger.*—October 9, 1866.— The cutter bars are adjusted upon inclined faces by a screw on the shank, and retained by set-screws which traverse slots in the bars. The "centre" is adjustable to suit the cutters.

Claim.—Adjusting the cutters to different sizes of tenons by means of inclined cutter bearing pieces, sliding upon inclined supporting pieces and made adjustable thereon, and used in combination with a receding centre, arranged to operate substantially as described.

No. 58,659.—COLE McCREA, Leavenworth, Kansas.—*Pruning Knife.*—October 9, 1866.— The branch is cut as it is pulled into the throat between the edge of the hook and that of the circular revolving cutter. To cut grafts at a considerable height an extensive handle is attached and the cut made by a thrust.

Claim.—First, the combination with a pruning hook of a revolving wheel A, substantially as and for the purpose described.

Second, the combination with a pruning hook of an extension handle B, substantially as specified.

No. 58,660.—CHARLES McKEE, San Francisco, Cal.—*Composition for removing Incrustation from Steam Boilers.*—October 9, 1866.—Composed of camphor, 3 ounces; potash, 3 ounces; ammonia, 1 ounce; alum, 1 ounce; water to dissolve, and mixed with two gallons of petroleum.

Claim.—The combination of camphor, potash, ammonia, and alum with petroleum or other oleaginous matter, substantially as described and for the uses and purposes hereinbefore set forth.

No. 58,661.—SCOVIL S. MERRIAM, Springfield, Mass.—*Submarine and Torpedo Boat.*— October 9, 1866.—The cigar-shaped boat has water-tanks which are emptied by force pumps to increase its flotative capacity. The torpedo is attached to the end of the bar in an air-tight apartment with a man-hole in the bottom; the torpedo bar is then rotated 180°, bringing the shell in advance of the bow. Ballast attached to cords is lowered to the bottom to anchor the vessel. A screw revolved by hand or engine is the motor. The steersman looks from eye-slits in a dome amidships.

Claim.—The construction of the lower portion or bottom of a submarine vessel, of heavy cast iron bed-plates containing the water tanks, in combination with the ends of the vessel, and arranged substantially as and for the purpose herein set forth.

Second, the arrangement of the rope or cable, guide pulley and windlass with gearing, for the purpose of operating the suspended ballast in a perfectly air-tight box, operating and being operated substantially in the manner and for the purpose described.

Third, in combination with a submarine vessel, the arrangement of a torpedo bar near the bow, at the bottom of the vessel, and the manner of operating said bar from the inside of the vessel, in the manner substantially as described.

Fourth, the arrangement of a chamber X capable of being closed perfectly air-tight, and surrounding one or more of the doors in the bottom of the vessel, for the purpose substantially as specified.

Fifth, the construction of a submarine vessel, consisting of a heavy cast-iron bottom plate with an iron or copper hull, in combination with the water tanks arranged in the bed plate, the air chambers around the side, top, and ends of the working compartment, the suspended ballast weight, the screw propeller, worked either by hand or by a compressed air engine, and the torpedo bar with exploding shell at its end, when the whole is arranged and combined in the manner and for the purpose substantially as set forth and described.

No. 59,662.—EDWARD MIDDLETON, Cleveland, Ohio.—*Fire and Burglar Alarm.*—October 9, 1866.—The alarm is held in silence by means of cords stretched to doors, &c.; the fracture of a cord frees the alarm mechanism by raising a catch from the cogs of a wheel on the spring shaft.

Claim.—The spiral spring Q, yokes R, cross-pieces X, arm Y, levers 8, and cords Y', as arranged, and in combination with the spring E, wheels F K K', hammer o, and bell P, arranged in the manner and for the purpose set forth.

No. 58,663.—JONATHAN S. MILLER, Everton, Ind.—*Attaching Thills or Tongues to Vehicles.*—October 9, 1866.—The thimble socket on the end of the thill is held by a cone and socket in the respective jaws of the clip, which are bolted together.

Claim.—The combination of a clip formed in two parts, C and C', brace G, and bolt E, when said several parts are respectively constructed and arranged for use substantially as set forth.

No. 58,664.—WARREN P. MILLER, San Francisco, Cal.—*Saw.*—October 9, 1866.—The teeth are annular segments, and occupy segmental sockets of corresponding radius, cut on the edges of a screw blade so as to leave a gap of say 50°, at which the teeth project. Ridge and groove on the blade and tooth respectively preserve lateral position.

Claim.—An insertable tooth for saws, when said tooth is constructed upon lines having a true circle and comprising more than one hundred and eighty degrees of the circle, and inserted into a cavity in the saw plate of a shape to fit said tooth, substantially as described.

No. 58,665.—G. I. MIX, Wallingford, Conn.—*Manufacture of Spoons.*—October 9, 1866.—The spoon blank is cast in the form usually attained after several rolling, annealing, and trimming operations, receiving an advanced shape at a single operation, and avoiding the loss consequent upon the usual protracted process.

Claim.—The spoon blank cast and subsequently rolled, as described, as an article of manufacture.

No. 58,666.—HIRAM W. MOORE, Bridgeport, Conn.—*Annealing Furnace.*—October 9, 1866.—The car wheels are piled upon supporting rings at the bottom of the case, so that a passage is formed by the holes through the hubs for cold air, and another passage around the tread of the wheels for the draught for burning the charcoal, which is distributed upon the perforated flanges of the ring interposed between each wheel.

Claim.—The openings a a at the base of the annealing case, in combination with the annular flue F, and perforated plate J for admitting atmospheric air to promote the combustion of the carbon in contact with the wheels, substantially as described.

Also, in combination with the case A, the opening b, the horizontal air flue G and vertical flue space H' for conveying air to and cooling the hubs of the wheels, substantially as described.

Also, in combination with the annealing furnace for containing a pile or series of wheels, the series of perforated and flanged rings to be placed between said wheels for regulating and controlling the combustion of the charcoal therein, as and for the purpose described.

No. 58,667.—GEORGE G. W. MORGAN, Washington, D. C.—*Paper Fastener.*—October 9, 1866.—Explained by the claim and cut.

Claim.—A paper or other fastener formed out of a rhomboidal blank, and bent into a bow or staple form, and capable of piercing and cutting its way into or through the paper or other material, and of being bent down or clinched by the thumb and finger or hand of the user, and overlap each other, as shown at the line b, figures 2 and 5, without the use of any special instrument for inserting and clinching it, all as herein described and represented.

No. 59,668.—WILLIAM HENRY MORRISON, Indianapolis, Ind.—*Window Curtain.*—October 9, 1866.—A double curtain passes over a roller at the top of the window. The two parts are pierced with holes, and when they are adjusted by the roller so that these holes do not register with each other the sun is kept out, while a current of air is allowed to pass.

Claim.—The combination of the roller E, and the double curtain F F, having apertures *x x x z' z' z'* in its opposite parts, substantially as described, when the combination is used for the purposes specified.

No. 58,669.—HIRAM NASH, Cincinnati, Ohio.—*Clothes Wringer.*—October 9, 1866.—The journal slot of the upper roller is concentric with the journal of the motive wheel on the crank shaft, which insures the proper relative position of the spur wheels in all positions of the roller. The roller is depressed by a plate spring. The frame is locked to the tub by a pivoted pin, which forces out the upper ends of the levers, and clasps their lower ends against the tub.

Claim.—The combination of the enclosing case D, made up of the plate *f* and cover *h*, and provided with the concentric bearing *k k*, with the set of gearing 1 2 3 4, arranged as described, the whole used in connection with the rollers B B, substantially in the manner and for the purpose specified.

Also, the tightening buttons or wedges *p p*, in combination with the clamps G G and standards A A, arranged and operating as set forth.

No. 58,670.—S. D. OGBURN, Springfield, Tenn.—*Hemming Guide for Sewing Machines.*—October 9, 1866.—The smaller curved slits in the vertical plate, opening each into the larger one, admit of easily turning hems of different sizes, while the pin forms a guide for the edge of the hem, and regulates its width.

Claim.—The plate B, arranged as shown, with several curved slits as measures of quantity for different widths of hem.

Second, in combination with the above the removable pin E in the plate D, as and for the purpose described.

No. 58,671.—SAMUEL PAGE, McAllisterville, Penn.—*Evaporator.*—October 9, 1866.—The juice from the heating pan near the chimney is conducted to the evaporating pan through a pipe and strainer, and from thence to the finishing pan, which has a cooling chamber beneath it, with a valve to shut off the caloric current.

Claim.—The arrangement of the pans A E G with the connecting pipe C, strainer B and the furnace flues, substantially as described and represented.

No. 58,672.—JOHN T. PARKER, Farmington, Maine.—*Well Borer.*—October 9, 1866.—The pod is divided into two parts, hinged together; the auxiliary section is expanded beyond the necessary diameter of the upper part of the tube by which it entered by pressure on the stem and resistance of the material.

Claim.—The combination of the slotted part B, part A, stop pin *e* and hinge *c*, operating substantially as described for the purpose specified.

No. 58,673.—MORITZ PINNER, New York, N. Y.—*Paper Collar and Bosom.*—October 9, 1866.—The article when made in one piece is cut in a peculiar shape, indented on the lines of fold, and stayed by connecting strips.

Claim.—First, a substitute for a shirt collar and a shirt bosom, such substitute being made in whole or in part of paper cut in one or more pieces, and made to fit or adjust itself around the neck and over the collar bone and chest of the wearer, substantially by the means and in the manner herein set forth and described.

Second, combining the collar part A with the bosom part B of the above invention by means of one or more strips *g*, for the purpose of strengthening, connecting, or holding in place such parts A and B, or either of them, all substantially as herein set forth and described.

Third, bending, creasing or indenting on line *d* the product embraced in the above invention, for the purpose herein set forth and described.

Fourth, making the above described article open and adjustable on any side or part of the neck of the wearer.

Fifth, printing on and embossing the above described article, or the material of which it is made, in whole or in part, all substantially as herein set forth and for the purposes specified.

No. 58,674.—GEORGE A. REYNOLDS, Rochester, N. Y.—*Fruit Jar.*—October 9, 1866 —The jar has a metallic cover, with a vent hole in the top. A clamp closes the vent. The ends of the clamps are turned up, forming hooks with inclined flat surfaces, which tighten the clamp as it is turned.

Claim.—The arrangement of the clamping bar B, having the inclined planes or hooks *c*, formed upon its ends, and provided in the centre with a fixed elastic packing *t* for the vent formed in the apex of the cover in this class of self-sealing cans or jars, as and for the purpose shown and described.

No. 58,675.—W. M. RICE, Boston, Mass.—*Heel Iron.*—October 9, 1866.—Explained by the claim and cut.

Claim.—As a new article of manufacture the above described heel irons, to wit, heel irons made thickest on that part or side which usually wears out fastest, and bevelled on the inside, as described, so as to be held on to the heel by the leather nailed or pegged in the inside of said irons.

No. 58,676.—J. M. RITER and L. J. FARQUHAR, Pittsburg, Penn.—*Machine for Punching Sheets of Metal.*—October 9, 1866.—The machine is for punching rivet holes in boiler plates, and the improvements are for moving the plate under the punch between its successive strokes, at regulated distances, in either straight or curved lines for cylindrical or taper boilers or flues. The mode of application and regulation cannot be briefly described.

Claim.—First, the combination of the adjustable diagonal rack E, with the rack table D, and its ratchet *h*, for the purpose of regulating and varying at pleasure the length of each separate movement of the rack table.

Second, the combination of the rack table D, and the turning frame K, attached thereto by a rivet, for the purpose of producing a curvilinear movement.

Third, the combination of the rack table D, and frame K, pivoted thereto, with the adjustable angle plate L, and rollers *g' g''*, or other similarly arranged bearing surfaces, constructed substantially as and for the purposes hereinbefore set forth.

Fourth, also in combination with the devices specified in the third claim, the slide *q*, with screw J, and spiral springs *u*, constructed and arranged substantially as hereinbefore described, for the purpose of pressing the frame K against the rollers.

No 58,677.—ARCHIBALD H. ROWAND, Allegheny, Penn.—*Car Spring.*—October 9, 1866.—The continuous steel strap is wrapped on a "former" of elliptical cross section, and the elliptic coil is then permanently compressed at the points where it is intersected by the conjugate diameters.

Claim.—An elliptic spring, composed of a single strip of steel, formed in the shape substantially as shown, so that it will retain its form without the use of clamps.

No. 58,678.—ESAU ROWING, Parkersburg, West Virginia.—*Steam Generator.*—October 9, 1866; antedated September 23, 1866.—The bank of water tubes is arranged above the fire and inclined upward toward the transverse steam drum, with which they are severally connected. By their inclination the relative proportion of steam space in the tubes is increased toward their upper ends.

Claim.—So arranging a series of inclined tubes *a*, with reference to the drum *f*, that the water line intersecting the drum at its centre shall leave a gradually increasing steam space in the tubes *a*, from the lower or furnace end of said tubes to the point of their junction with the drum, substantially as and for the purposes set forth.

No. 58,679.—RUFUS S. SANBORN, Ripon, Wis.—*Bed Bottom.*—October 9, 1866.—The ends of the slats are connected by elastic loops to staples on the head and foot rails; cords of India-rubber are interwoven with the slats at right angles, and are attached to the side rails.

Claim.—A bed bottom, formed by interweaving with the ordinary slats, cords or bands of India-rubber or other suitable elastic material, substantially in the manner described and shown.

No. 58,680.—L. B. SAWYER, Charlestown, Mass.—*Observatory.*—October 9, 1866.—Surrounding a tower is an annular car suspended on ropes which pass over sheaves to a partial counterpoise within the tower, and thence to a windlass by which the movement of the car is accomplished. Arms extend from the tower top for attachment of guys.

Claim.—First, the employment, in combination with a tower, of a car of an annular form working upon guides upon the exterior of the tower, and a hoisting apparatus for raising and lowering the car, substantially as described.

Second, the employment, in combination with the tower, of the projecting arms of the frame D, to which the guys are attached, by which the car is permitted to rise to the top of the tower without coming in contact with them, substantially as described.

Third, the annular counterpoise working within the tower in combination with the annular car working upon the exterior of the tower, substantially as described.

No. 58,681.—SAMUEL H. SCHENCK, Zionsville, Ind.—*Machine for Splitting and Skiving Leather.*—October 9, 1866.—A skiving block is pivoted near one side to two checks to which a knife frame is journaled; attached to the skiving block is a roller which is used as a backing in splitting leather. Abutments on the checks check the knife frame at its extremes of position. A set screw adjusts the frame and roller to the knife.

Claim.—First, the checks 7 7, the adjustable skiving block 1, the knife 2, and its attachment to the lever 3.

Second, the roller 5, in connection with the skiving fences and the set screw 6, all arranged and operating substantially as set forth and described.

No. 58,682.—CHARLES C. SCHMITT, New York, N. Y.—*Chair.*—October 9, 1866.—The back and seat cover is of one piece of cloth, and is stretched upon a roller at the front of the seat which has a ratchet wheel and a pawl by which the length of lining cloth is adjusted to vary the inclination of the chair. Brace straps with buckles stretch from the front of the chair to the top of the back. The back and legs are pivoted together as in a camp chair. Ratchet plates on the back in combination with a cross-bar serve to hold the chair to any adjustment.

Claim.—A chair frame susceptible of being adjusted to any inclination desired, from a horizontal to a vertical plane, or nearly so, in combination with a seat and back composed of a continuous strip of cloth so hung to the chair frame that it can be lengthened or shortened, in the manner herein specified and for the purpose set forth.

No. 58,683.—J. G. SCHWEMMER and T. MUELLER, Philadelphia, Penn.—*Fly Fan.*—October 9, 1866.—The flappers are suspended above and in front of the rows of joints of meat, and are connected by bell cranks to a rod which is intermittingly raised by a cam beneath, actuated by gearing from a spring.

Claim.—The flaps A A, when operated by the described cam movement through a sliding rod E, bell cranks B B', and links e c, and when arranged for the purpose of giving access to the articles thereby protected from insects, substantially as set forth.

No. 58,684.—A. G. SHAVER, New Haven, Conn.—*Eraser and Burnisher.*—October 9, 1866.—A burnisher tip is attached to the handle of the eraser by a slot and rivets.

Claim.—In combination with the blade or handle of an eraser or desk knife, a burnisher tip, substantially as described.

No. 58,685.—SIMEON SHERMAN, Weston, Mo.—*Steam Generator.*—October 9, 1866.—The generator has two concentric chambers and an intervening return flue. Feed water enters the central chamber, thence downward through pipes to the annular chamber, where it is raised by a revolving system of perforated troughs, and poured over the exterior surface of the fire flue. The central chamber has mud collecting pipes, and an exterior mud discharge pipe and valve.

Claim.—First, the central chamber F and outer chamber, connected by the vertical pipes I, which operate as described, in combination with the tubes K and mud valves L for the removal of sediment.

Second, the revolving arrangement of perforated troughs for dripping the water upon the heated surface, as described.

Third, the plates R, attached to the heated surface for the retention of the water, as and for the purpose described.

No. 58,686.—SAMUEL SHUCK, Bedford, Penn.—*Manufacture of Cigarettes.*—October 9, 1866.—The prepared paper tubes of the cigarettes are run upon a cylinder filled with tobacco, which is forced from the cylinder into the tube by a piston rod.

Claim.—First, the method herein described of filling a wrapper with fine-cut tobacco by means of a tube and piston, the wrapper being drawn over the tube and receiving the core as the same is pushed out of the tube by the piston, substantially as shown and described.

Second, the method herein described of forming the wrapper preparatory to filling by means of the forming plate, Fig. 2 and the hinged plate, Fig. 3, substantially as described.

Third, the within-described method of filling and packing the wrapper by inserting the fine-cut tobacco through a tube enclosed in the wrapper, and moving the said tube up and down in the wrapper during the process of filling, as described:

No. 58,687.—D. M. SMITH, Springfield, Vt.—*Paper Holder.*—October 9, 1866.—The wire has a coiled spring hinge. Both ends are projected; one is turned into a socketed flange, and the other into a flange with a central pin.

Claim.—The construction and arrangement of the coil spring provided with a point B that pierces the eye or hole C, and all from one piece of wire, substantially as and for the purposes described.

No. 58,688.—AMOS W. SNOW, Norwich, Conn.—*Lock.*—October 9, 1866.—At the lower part of the latch bolt is an arm which enters slots in the spring tumblers, when they are arranged coincidently by the key, whose bits are constructed for the specific duty.

Claim.—The arrangement of the latch bolt C, with its arm f, the spring tumblers b b, as described, and guard plate l, constructed and operating, in combination with a suitable handle and key, in the manner herein specified.

No 58,689.—JACOB A. SPEAR, Braintree, Vt.—*Horse Rake.*—October 9, 1866.—The supporting boards above and below the thill afford a platform for the attendant to stand upon, and braces for the feet while operating the rake. The sections of the rake are attached by hinged reaches to a rock bar, which is operated by a foot frame, and to the rear bar of this frame the sections are also attached by flat bars of iron, which pass through circular slots in the rear bar of the frame to allow the rake to conform to the ground.

Claim.—The arrangement upon the carriage A B C of the boards E J and bars I I¹ I², as and for the purpose herein described and represented.

Also, in combination with the above, the hinged rock shaft F, levers G G, connecting bars G' and P, reaches M, bars Q, and the sectional jointed rake O O' R R S, when constructed and arranged in the manner and for the purpose specified.

No. 58,690.—JOHN SPRINK, Council Bluffs, Iowa.—*Hair Restorative.*—October 9, 1866.— Composed of onions, 2¼ pounds; turnips, 1¼ pound; salt, 2 ounces; burdock root, 2 pounds; oil bergamot, ¼ ounce; Cologne spirits, ½ ounce; and rain water, ¼ ounce.

Claim.—The proportionate quantities of the ingredients, as compounded and made for a vegetable hair tonic, substantially in the manner and for the purpose as herein specified.

No. 58,691.—JOHN M. STANYAN, Milford, N. H.—*Dough Mixer.*—October 9, 1866.—The cover sets within the pan, and is locked by bolts which engage staples on its rim. The axis of the disk cover is then placed in its bearing, and the pan rotated by the handle in a plane at right angles to its axis as a frustum.

Claim.—The improved dough mixer, made substantially as described, viz., of the pan and cover with the handle socket and the journal applied and arranged in manner as specified, such cover and pan being provided with suitable connections, as set forth.

No. 58,692.—G. N. STEARNS, Syracuse, N. Y.—*Vice for Holding Saws.*—October 9, 1866.—One of the two jaws is attached to the work-bench. To the arms of this jaw is attached an eccentric lever, which, in connection with a spring, opens and closes the vice.

Claim.—The herein described vice, as a new article of manufacture, for the purposes set forth.

No. 58,693.—L. O. STEVENS, Pekin, Ill.—*Corn Cultivator.*—October 9, 1866.—The axles are attached to the base of an arched frame which supports the transverse shaft on which the operating levers are pivoted. The plough frame is on a hinged, rearward extension of the tongue, is raised and lowered by the levers, and the middle ploughs are laterally adjusted by the handles, which are within reach of the driver on the counterbalance seat.

Claim.—The frame D, arms N N, and beams A A, combined and operating substantially as described for the purpose specified.

Also, the curved or arched bars M M, in connection with the frames K K for supporting the shaft L, substantially as and for the purpose specified.

No. 58,694.—JAMES STIMPSON, Baldwinsville, Mass.—*Clamp for Wringing Machines.*—October 9, 1866.—Attached to the posts which fit the inside of the tub are clamp bars adjustable to fit against the outside of the tub, and retained in position by set screws.

Claim.—First, the adjustable rods H, attached to or cast with horizontal bars G, which are fitted loosely on the pendent bars *a* of the end pieces A, and provided with screws I to bear against the edges of sector projections J, for the purpose described.

Second, the springs K, in combination with the adjustable rods H, set screws L, and pendent bars *a*, arranged as and for the purpose set forth.

No. 58,695.—JAMES A. STRONG, North Wolcott, Vt.—*Sheep Shears.*—October 9, 1866; antedated September 30, 1866.—At the heel of each blade, on the outer sides and edges thereof, are raised flanges which perfect the hand hold.

Claim.—First, the flanges A' B', in combination with the blades A B and spring C, as and for the purposes herein specified.

Second, the swelled or turned-up front A*, arranged relatively to the flange A', blades A B, and spring C, substantially as and for the purpose specified.

No. 58,696.—SAMUEL TAYLOR, Burlington, Me.—*Beehive.*—October 9, 1866.—The removable sections are arranged vertically, and have a sliding framework of slats at the top and bottom of each for regulating communication between them, or closing any one section for removal.

Claim.—A beehive constructed of a series of sections A A' A'', in combination with sliding frames E, applied or fitted to the sections, and constructed substantially in the manner shown and described for the purpose set forth.

No. 58,697.—HENRY A. TOZIER, Littleton, Maine.—*Beehive.*—October 9, 1866.—An opening from the upper to the lower section of the hive is closed by a board smaller than the opening, to leave apertures for the passage of bees and for ventilation.

Claim.—The removable plate or board D' placed within the body A of the hive, substantially in the manner as and for the purpose herein set forth.

No. 58,698.—CHARLES W. WAILEY, New Orleans, La.—*Cotton Tie.*—October 9, 1866.—The plate is indented in two places, producing projecting flanges, which diverge from each other, and affording slots through which the ends of the hoops pass.

Claim.—The metallic tie or buckle A, when constructed as described for the purpose set forth.

No. 58,699.—JOHN H. WAIT, Portsmouth, Ohio.—*Balance Slide Valve.*—October 9, 1866.—The operations of the slide valve are performed by two slide plates of similar construction and movement, (one on each side of the seat block containing the ports,) which balance each other and are supported vertically on rollers.

Claim.—The valve block A and steam channels C C' of the two balanced slide valves D D, constructed, arranged, and operating substantially as and for the object specified.

No. 58,700.—WILLIAM H. WALRATH, Chittenango, N. Y.—*Machine for Pressing Brick.*—October 9, 1866.—The wheel has moulds on its upper surface and revolves beneath the pug mill, from which the clay is forced into the moulds successively. The bricks are elevated by travelling followers against the pressure of the bracketed plates above, and eventually raised to the top of the table, swept therefrom by the revolving blade, and received on a stationary table.

Claim.—First, the arrangement of a stationary table H, upon the upper surface of the revolving mould table C, in combination with revolving blades g, which are so arranged as to sweep the bricks from the followers F upon said stationary table, substantially as described.

Second, in combination with the pug mill B and revolving mould table C, the upward moving followers F and a plate G, which is sustained by means of the brackets J J, bolted upon the frame A, so that the resistance to the upward pressure of the followers will be sustained by said plate brackets and frame, substantially as described.

Third, providing for adjusting the elevated portion of the track E at the point where the made bricks are discharged from the mould, by means substantially as described.

Fourth, communicating motion to the mould table directly from a horizontal driving shaft L, in combination with the shaft C' and spur wheels b^1 b^2, so that the stirring shaft of the pug mill as well as the mould table shall receive motion from said main shaft, substantially as described.

Fifth, communicating motion to the revolving blades g from the shaft L, which moves the mould table C, substantially as described.

Sixth, in combination with a revolving mould table C, the device which will discharge the bricks from the follower plates upon a stationary table H, or its equivalent, substantially as described.

No. 58,701.—CHARLES P. WALTER, Aston township, Penn.—*Plastering Mould for Cornices.*—October 9, 1866.—The mould is adapted to work stucco cornices in the inner angles of walls. It has slotted plates, by which the adjustable guides are expanded or drawn together to fit the angle of the wall.

Claim.—First, the combination of the "former" B with the guides A A', in the manner and for the purpose substantially as shown and described.

Second, the slotted plate E and screws c, or their equivalents, in combination with the "former" B and the guides A A', whereby the mould is made adjustable to angles of any degree, substantially as described.

No. 58,702.—HORATIO WHITING, New York, N. Y.—*Reaping Machine.*—October 9, 1866.—The grain is delivered in a continuous swath behind the carriage, with the heads towards the near side, by means of an endless toothed belt, which catches the heads and draws them beneath the carriage, where they are disengaged from the belt by the inclined board, through an opening in which the belt passes.

Claim.—The adjustable toothed chain or belt N in combination with the inclined plate M, arranged and operating substantially as described.

No. 58,703.—RUEL W. WHITNEY and ABNER C. STOCKIN, South Berwick, Maine.—*Mop Head.*—October 9, 1866.—The yoke is pivoted in the handle, and forms a jaw to hold the mop against the cross-head when the sleeve is slipped upon the shank.

Claim.—The improved mop head is made as described, viz., of the bearer, the yoke, and the collar, constructed, arranged, and applied together and to the handle, substantially as specified.

No. 58,704.—D. E. WHITON, West Stafford, Conn.—*Chuck.*—October 9, 1866.—In the front end of the body of the chuck is a scroll disk which has a gear wheel upon its inner side, into which a pinion works to rotate it in either direction. The outer end of the pinion shaft is secured in its place by a hollow nut. A hole through the centre of the latter forms a bearing for the shaft.

Claim.—The chuck consisting of the jaws B, scroll disk D, pinion F, and hollow shouldered nut M, combined and operating substantially as described for the purpose specified.

No. 58,705.—WILLIAM WILSON, Boston, Mass.—*Cooking Stove.*—October 9, 1866.—A plate, removable in separate parts, is stretched horizontally across the stove beneath the top, leaving a smoke flue between the said plate and the oven top. An air chamber is attached to the smoke flue above the stove, encasing a damper valve, the raising of which admits air into the stove chimney to ventilate the apartment and moderate the fire.

Claim.—First, in cooking stoves the use of the removable intermediate plate *e e* and slat *f f*, arranged as described and for the purpose specified.

Second, in combination with the above, the arrangement of the draught chamber *k k*, and for the purpose specified.

Third, the use of an elevating damper *r r*, arranged and operating in the air box *o o*, substantially as and for the purpose herein set forth.

No. 58,706.—JAMES F. WINCHELL, Springfield, Ohio.—*Corn Harvester.*—October 2, 1866.—The cutting apparatus consists of two circular cutters revolving on either side of a fixed, pointed, double-edged blade. The discharging device consists of a swinging rack, held in position by a spring upon the main frame and an arm hinged thereto and engaging with a stop upon the frame, from which it is lifted by a lever operated by the foot of the driver, when the weight of the gavel is sufficient to swing the rack for the discharge of the same.

Claim.—First, the cutting apparatus consisting of the revolving disks R P, or their equivalents, in combination with the stationary blade L, when said parts are arranged to operate as set forth.

Second, the combination of the tilting rack D, stop bar *a*, and lever *b*, and spring H, when arranged to operate as herein shown and described.

No. 58,707.—S. M. WIRTS and F. SWIFT, Medina, Mich.—*Grain Separator.*—October 2, 1866.—The riddles and screens receive two separate longitudinal motions from a disk on the jaw shaft through a rock shaft and connecting rods. The hopper bottom has a fine screen for the separation of grass and weed seeds and dust from the grain. A convex-bottomed trough having a lateral motion receives the grass seed, &c., from the hopper and conveys it to the sides of the mill. The upper screen receives a vertical motion from a roller on the seed trough.

Claim.—First, the movable shaking spout or trough L, operated by means of the bar K, and the shoe C, substantially as herein specified.

Second, the screens *a* and *b* as constructed and combined in the supplemental shoe, and with the carrier board *e*, as and for the purpose specified.

Third, the combination of the rod O with the main shoe and supplemental shoe for regulating the inclination of the latter, as and for the purpose set forth.

Fourth, the employment of the rod *i* in combination with the lower screen I, for the purpose of adjusting the inclination of said screen and imparting a bounding motion to it, substantially as set forth.

Fifth, the supplemental shoe pivoted at its inner end to the main shoe so as to allow of adjustment of its outer end, as and for the purpose specified.

No. 58,708.—MARTIN WOLF, New York, N. Y.—*Take-up for Narrow Ware Looms.*—October 9, 1866.—By reason of the slack of the fabric caused by suspending the weight between the breast-beam and the take-up roll any irregularity in the thickness of the woof will be regulated without personal attention, producing greater evenness and regularity in the texture of the goods.

Claim.—The arrangement of a movable roller F, attached to a sliding block C, and operated or acted upon by a weight, in combination with a stationary roller G, when applied to the woven material at any place between the breast-beam roller A and the take-up roller B, in the manner and for the purpose substantially as set forth and described.

No. 58,709.—THOMAS D. WORRALL, Central City, Colorado.—*Speculum.*—October 9, 1866.—The vibrating arms are hinged to the tube and separately or simultaneously operatable. The jointed portion enters beyond the pelvic strait and is covered with an elastic membrane to prevent pinching the coat of the vagina between the plates of the closing valves.

Claim.—First, so constructing a vaginal speculum that the motion of its valves shall be confined exclusively to that portion of the vagina which is inside the pelvic bone.

Second, so constructing a speculum that the whole of its valves may be worked simultaneously, or one or more separately, at the pleasure of the operator.

Third, the use of rubber or other flexible material either securely fastened to or loosely surrounding a valved speculum and operating with it for the purpose set forth.

Fourth, the screw F, in combination with the joint J, and the nut C, for the purposes set forth.

Fifth, the joint K, in combination with the screw F, and the nut C, for the purposes set forth.

Sixth, the ring D, in combination with the tube A, the nuts H, the nuts C, the screws F, the joints J and K, and the valves B, for the purposes set forth.

No. 58,710.—J. K. ANDREWS, Antrim, Ohio, assignor to himself and J. C. TILTON, Pittsburg, Penn.—*Railway Car Window.*—October 9, 1866.—The dusty air which would enter the open window of the car is deflected or turned by curved screens which slip horizontally in the window jambs.

Claim.—The curved screens K K, with their convex side turned inward, in combination with the curved grooves M M, situated between the panels H and E.

No. 58,711.—SAMUEL BAXENDALE, Boston, Mass., assignor to himself, THOMAS H. DUNHAM, Boston, Mass., and SAMUEL B. THAXTER, Abington, Mass.—*Machine for Picking and Opening Fibrous Material.*—October 9, 1866; antedated September 23, 1866.—The cage is of wire; the fibre approaches in separate channels to form distinct bats; a pressure roller above slightly condenses the several bats or slivers as they leave the cage and pass to the delivery rollers.

Claim.—A series of divisions or partitions so placed as to form passages or channels leading to the wire cage of a picking, blowing, or separating machine, and causing the floating fibres to be deposited upon the cage in separate slivers or fillets, substantially as herein described and for the purpose specified.

No. 58,712.—S. V. BECKWITH, Hamden, Conn., assignor to himself and T. B. CARPENTER, New Haven, Conn.—*Adjustable Shelf for Stove-pipes, &c.*—October 9, 1866.—The shank of the shelf is pivoted in a socket attached by a band to the stove pipe.

Claim.—The combination of the adjusting band B, the socket C, and the shelf D, provided with a shank F, constructed and arranged so as to be adjustable, substantially in the manner set forth.

No. 58,713.—W. N. BRAGG, Richmond, Va., assignor to himself and WM. H. TRAINHAM, same place.—*Adjustable Railway Car Seat.*—October 9, 1866.—The chair seat is hinged to its front supports, and the rear legs have lugs which rest in a slotted scroll, so as to permit the chair to be adjusted in a sitting or recumbent position. The foot-rest is hinged to the chair support, and may be turned over and stowed beneath the seat.

Claim.—The arrangement and combination of the slotted scroll D and rest B with the adjustable foot-rest F and foot-lever G, when constructed and operated as herein described and for the purposes set forth.

No. 58,714.—CORNELIUS L. CAMPBELL, Binghamton, N. Y., assignor to WASHINGTON W. WHEATON, same place.—*Machine for Fitting Axle Spindles to Skeins of Wagons.*—October 9, 1866.—The machine is so arranged as to clamp the end of the axletree, and to reduce it to any required conical form by means of an adjustable cutter rotating around it. The cutter frame is journalled on a mandrel whose centre engages the end of the axle.

Claim.—The manner of fitting the arms or spindles of axletrees for wagons to cast-iron skeins or thimbles, by means of the revolving slide cutter J in combination with the adjustable way or guide plate L, Fig. 3, the feed screw E, and the hinged nut K, substantially as and for the purposes described.

No. 58,715.—THOS. F. CHRISTMAN, Wilson, N. C., assignor to himself and WILLIE DANIEL, same place.—*Elevator.*—October 9, 1866.—Buckets for the elevation of building material are fixed upon an endless chain. The chain is worked by a windlass at its upper end.

Claim.—The arrangement and combination of the base frame A B, cylinder C, guide rollers a a on the bent arms b b, with the top windlass frame D E, windlass F G, chain pulleys h h, adjustable endless chain H H, and elevating buckets I I, to operate at various heights, substantially as and for the purposes herein set forth.

No. 58,716.—C. P. CROSSMAN, West Warren, Mass., assignor to himself and PEMBROKE CHURCHILL, same place.—*Pen Rack and Bill Holder.*—October 9, 1866.—Explained by the claim and cut.

Claim.—First, the pen rack B in combination with a bed piece A, substantially as specified. Second, the bill holder C in combination with the pen rack B and bed piece A, substantially as specified.

No. 58,717.—SOLOMON CROWELL, Syracuse, N. Y., assignor to himself and JAMES B. RUE, same place.—*Connecting and Supporting Stove-pipe.*—October 9, 1866.—A rod passes through an ear depending from each coupling, and is tightened by a nut at each end to hold the pipe rigid.

Claim.—The coupling b and the flange on the coupling, c, and the tension rod e, when the same are constructed, combined, and used in the manner as substantially set forth and described.

No. 58,718.—GEORGE W. DAVIS, Fitchburg, Mass., assignor to SILAS PRATT and E. S. RUSSELL.—*Composition for Pavements, &c.*—October 9, 1866.—Composed of gas-tar, chloride of lime, hydraulic cement, ground slate, sand, calcined gypsum, brown cement, charcoal dust, copperas, lime, and coal ashes.

Claim.—A compound for covering walks, and for other similar purposes, composed of the above-named substances, substantially in the proportions specified.

No. 58,719.—CHESTER F. DEAN, St. Johnsbury, Vt., assignor to himself, HORACE PADDOCK, HALSEY R. PADDOCK, and MOSES E. BARRETT.—*Grinding Mill.*—October 9, 1866.—The grinding is done by a corrugated cylinder in a hopper having a foraminous bottom; the corrugations have notches arranged helically around the cylinder.

Claim.—The combination of the grinding cylinder, the stationary grinders, and the foraminous apron, arranged as specified.

No. 58,720.—CHAS. DENTON, Pekin, Ill., assignor to himself, SAMUEL E. BARBER, and SAMUEL F. HAWLEY, Decatur, Ill.—*Harvester.*—October 9, 1866.—The grain-delivery spout or extension of the platform is connected with said platform by means of springs, which admit of its adjustment, to adapt it and the endless apron passing over both spout and platform to the delivery of the grain at any desired height. Cogged racks and pinions are arranged in connection with the cutter and thrust frames for adjusting the height of the cutters from the ground.

Claim.—First, the attaching of the spout P to the front part of the framing by means of springs g g, substantially as and for the purpose described.

Second, the arrangement of the racks a a on the guide I, attached to frame A, in combination with the pinions J J, secured to the pole or tongue G, as shown and described, for adjusting the sickle higher or lower, as described. •

No. 58,721.—JOHN H. DUCK and ELIAS K. WHITCOMB, Elgin, Ill., assignors by mesne assignments to JOHN H. DUCK and JAMES T. WHIPPLE, Chicago, Ill.—*Pipe and Fixture for Wells.*—October 9, 1866.—The tubular screen is suspended by means of flanges which rest on a projection within the main tube, when the latter is lifted above the point to give access to the water.

Claim.—The tube or screen S, suspended within the main tube L, by means of flanges r r", in combination with the several parts of the within described device, for the purpose specified.

No. 58,722.—ERASTUS S. FRENCH, Templeton, Mass., assignor to JOSHUA W. PARTRIDGE, same place.—*Wood Scraping Machine.*—October 9, 1866.—The upper and lower scrapers are adjusted in the head, and the board is forced between them, facing both sides simultaneously. The board rests on a bed, and is pushed by a driver whose reciprocating motions are induced by a draught chain and a recoil spring or cord. The motions of the winding drum are controlled by a shipper.

Claim.—The combination as well as the arrangement of the guide E, the reciprocating driver G, the two scrapers, and the mechanism for advancing and receding the driver, and clutching and unclutching its drum, with respect to the driving shaft, the whole being so as to operate as explained.

No. 58,723.—D. C. GUTTRIDGE, Canton, Ohio, assignor to himself and WILLIAM CLUFF, Stark county, Ohio.—*Car Coupling.*—October 9, 1866.—The rod has handles on its extremities which reach beyond the sides of the car; the coupling pin is suspended from a prong on the rod, and is raised or lowered by turning the handles, or by a rod which reaches to the roof of the car.

Claim.—The rod C, provided with the prong F, and connected to the coupling pin p, when used in combination with the movable coupler B, and rod g, as herein set forth.

No. 58,724.—E. R. HOLFORD, Westford, Wis., assignor to himself, ABIAH KINGSLEY, and CLARK ALVORD, same place.—*Corn Planter.*—October 9, 1866.—The seed dropping is done by a cam on the main axle, through a lever and bar connecting with the seed slides; the ploughs can be raised, and the seeding arrangement thrown out of action by levers.

Claim.—First, the cam D, in combination with the lever E, and slides or valves G, for the purposes and substantially as described.

Second, the cam D, in combination with the bar F, levers I and I', for the purposes and substantially as herein set forth.

Third, the arrangement of the levers I' and R, for the purpose of elevating and lowering the ploughs, substantially as herein described.

No. 58,725.—O. L. HOPSON, Waterbury, Conn., and H. P. BROOKS, Wolcottville, Conn., assignors to TURNER & CLARK MANUFACTURING COMPANY, same place.—*Buckle.*—October 9, 1866.—The frame of this buckle consists of a single endless wire; each side is bent inward, and to the unbent parts is affixed the tongue, which also consists of a single wire ending in two points.

Claim.—First, the buckle frame, formed with double bends in the sides, between the main portion and the loop, one portion of said double bend at each side of the frame forming the axis for the tongue, substantially as specified.

Second, the wire tongue formed with eyes, connected by the central portion of the wire, in combination with the aforesaid buckle frame, having double bends in its sides, substantially as set forth.

No. 58,726.—JEROME HOYT, Stamford, Conn., assignor to himself, EDWIN HOYT, and LAFAYETTE FARRINGTON, same place.—*Sawing Machine.*—October 9, 1866.—The saw frame has a longitudinal reciprocating motion on a gravitating frame. The teeth are in the cross-cut form to cut both ways.

Claim.—In sawing machines, the combination of the grooved uprights *a a*, sliding frame B, saw frame D, guides *i j*, spring pawl H', and rack J, when arranged and operating substantially as described for the purpose specified.

No. 58,727.—ELIAS S. HUTCHINSON, Baltimore, Md., assignor to himself and HUGH L. MCAVOY, same place.—*Apparatus for Generating Steam.*—October 9, 1866.—An air blower, condensed air reservoir, and a carburetting chamber, supply gaseous vapor to the boiler for the generation of steam. The air is forced into the reservoir by a pump actuated by the engine, and the gas may be burned beneath the open ends of the vertical boiler tubes whose spherical enlargements increase the heating surface.

Claim.—First, the combination of. an air-forcing apparatus and carburetter with the engine and the boiler, for the purpose described.

Second, in combination with the engine, air-forcing apparatus, carburetter, and boiler, the chamber for reserve of compressed air, as and for the purpose described.

Third, the boiler tubes, constructed with a spherical or equivalent enlargement, as and for the purpose described.

No. 58,728.—E. M. LANG, Portland, Maine, assignor to himself and ISAIAH GILMAN.— *Solder-casting Machine.*—October 9, 1866.—The reservoir-containing the melted metal is hinged by a bar to a standard on the frame, so as to be elevated when required. The metal passes through an opening in the floor into the pockets in the periphery of the mould-wheel revolving beneath.

Claim.—First, the combination and arrangement of the balance wheel A, geared wheels B and C, and mould D, substantially as set forth.

Second, the combination and arrangement of the receptacle E, and cooler mould D, or their equivalents.

Third, the combination of the receptacle E, bar *k*, joint *f*, uprights *m* and *n*, arranged in the manner and for the purposes described.

No. 58,729.—P. H. LAWLER, Rochester, N. Y., assignor to himself and D. W. ROCHE.— *Barrel Machine.*—October 9, 1866.—The head-clamping frame is attached to an arm having an oblique axis, causing the feed of the head in the proper direction to the trimming saw. The feed movement is made by a treadle, the head being automatically clamped by the said movement, and released on the return movement of the frame, which is caused by gravity. The head stuff is rotated by a band and gear wheels.

Claim.—First, hanging the clamping frame F upon an axis arranged diagonally with relation to that of the clamping collars, and parallel with relation to that of the saw, as and for the purposes set forth.

Second, the arrangement of the adjustable tappet bar or cam *m*, with the pin *n*, substantially in the manner and for the purposes shown and described.

Third, the arrangement with the sliding mandrel I, in the swinging frame F, of the spring S, the pivoted lever N, and cam M, substantially as shown, and for the purposes set forth.

No. 58,730 —ELI J. MANVILLE, Waterbury, Conn., assignor to O. L. HOPSON and H. P. BROOKS, Litchfield, Conn.—*Machine for Pointing Wire.*—October 9, 1866.—Two sliding dies are maintained by a cap in a slot or groove cut across the end of the mandrel, and are simultaneously forced toward one another and upon the interposed wire by a series of pairs of toggles with which the dies in their revolution come in contact. A conical pointed screw passes through the cap, enters a conical hole in the die, and limits the expansion of the dies.

Claim.—First, the shaft *a*, formed with a cross mortise or slot containing the dies *c c*, in combination with two or more pairs of toggle blocks around the said shaft, said pairs of toggles acting alternately upon the dies, in the manner substantially as set forth.

Second, the conical or tapering pointed screws *i i*, in combination with the cap *f*, dies *c*, shaft *a*, and toggle blocks *d d*, all arranged as and for the purposes set forth.

No. 58,731.—WILLIAM MCDONALD, Albany, N. Y., assignor to himself, DONALD MCDONALD, and NOEL E. SISSON, same place.—*Gas Meter Register.*—October 9, 1866.—To prevent surreptitous turning back of the hands the wheels intermediate between the bellows and register will give a forward motion to the latter when turned in either direction.

Claim.—First, one or more intermediate wheels employed to communicate motion from the measuring bellows to the registering wheels, and so connected to the latter that either a forward or backward motion of the intermediate wheel or wheels will communicate a forward motion to the registering wheels.

Second, the arrangement and combination of the wheels F and G, when constructed and operating in conjunction with the wheels A B C D and E, substantially as and for the purposes set forth.

No. 58,732 —GEORGE R. METTEN, St Louis, Mo., assignor to HORACE BALDWIN, Painesville, Ohio.—*Fountain Pen.*—October 9, 1866.—The pressure upon the pen when writing causes an admission of air into the ink receptacle, in proportion to the extent of pressure and consequent amount of ink used.

Claim.—First, in atmospheric fountain pens causing an automatic flow of ink by reason of the act of writing, substantially as described.

Second, in atmospheric fountain pens supplying the place of the ink discharged from the ink reservoir with air during and by reason of the act of writing, substantially as described.

Third, so constructing a fountain pen that the act of writing will open and close a venthole for the admission of air into the ink reservoir, substantially as described.

Fourth, so constructing a fountain pen that the amount of ink discharged from the ink reservoir, and the amount of air admitted therein, shall be in proportion to the stroke of the pen, whether light or heavy, substantially as described.

Fifth, attaching the pen *g* to an oscillating or hinged disk *c*, substantially as and for the purpose described.

Sixth, the rod *k*, in combination with the disk *c*, substantially as and for the purpose described.

Seventh, the spring *n* and cushion *n'*, in combination with the rod *k*, substantially as and for the purpose described.

No. 58,733.—JEREMIAH L. NEWTON, Boston, Mass., assignor to himself and WILLIAM WICKERSHAM.—*Manufacture of Japanned Leather.*—October 9, 1866.—The hide is unhaired, stretched to smoothness, and when dry treated with Japan varnish. It may be bathed in a volatile liquid to remove a part of the grease, or have a preparatory breaking to make it supple.

Claim.—First, the treatment of raw hide with Japan varnish, or other suitable varnish, as and for the purpose above set forth.

Second, the preparatory process by immersing the hide in some volatile substance, in combination with and preparatory for Japan varnishing, substantially as above set forth.

Third, the subjecting of raw hides to a breaking process as a preparation for and in combination with japanning, all substantially as described and for the purposes set forth.

No. 58,734.—JOSEPH G. ROCKWELL, Cortland, N.Y., assignor to CALVIN EATON, Webster, N.Y.—*Ladder Hook.*—October 9, 1866.—The hook acts automatically to clasp the two sections of an extensible ladder together at any point desired, and is self-releasing on the raising of the upper section.

Claim.—The extension ladder hook, constructed and operating substantially as shown and described, and for the purposes herein set forth.

No. 58,735.—GERARD SICKLES, Boston, Mass., assignor to himself, JAMES W. PRESTON, and RUFUS S. LEWIS.—*Water Meter.*—October 9, 1866.—The pistons and their valves are separately actuated by sliding yokes, in connection with a weight and inclined planes, so that as the pistons are reciprocated by the influx and efflux of the water the valves are moved automatically in opposite directions, closing the ports of one piston and opening those of the other at each alternate movement.

Claim.—First, operating the valves by means of two sliding yokes, in connection with a weight, or its equivalent, substantially in the manner herein set forth.

Second, the sliding yoke B, attached to the pistons, and provided with the wedge-shaped projections E, or their equivalent, for the purpose of changing the position of the valves, substantially as and for the purpose set forth.

Third, the yoke C, provided with the horns or projections *b b*, as and for the purpose set forth.

Fourth, the weight H, or its equivalent, in combination with the inclines D, as and for the purpose specified.

Fifth, the combination of the yoke C with the levers F and valves *s*, as and for the purpose set forth.

Sixth, the combination of the yoke B, the yoke C, levers F, valves *s*, and weight H, or its equivalent, when operating substantially as and for the purpose specified.

No. 58,736.—M. G. SMITH and WILLIAM P. STEVENS, Kingston, Penn., assignors to M. G. SMITH, same place.—*Dumping Car for Coal Mines.*—October 9, 1866.—The carriage is raised by the rope until the platform passes above the fan, when the rope is slacked and the carriage settles down and rests at one end upon the arms of the fan, which throws it and the adjustable guides forward into an inclined discharging position. To return, by lifting on the rope the carriage is raised, and the guides erected; the movement also acts upon the toothed rod and intermediate devices, which move the fan clear of the carriage and allow the latter to descend.

Claim.—First, the tilting or adjustable guides G' G, constructed and operating substantially as shown and described.

Second, the use of the fan J, or its equivalent, for the purpose of tilting or inclining self-dumping carriages, so as to dump or unload the car without removing it from the carriage while slacking off, substantially as herein shown and described.

Third, the combination of a fan, or support, with the adjustable guides, substantially as herein shown and described.

Fourth, the combination of the weighted arm L with the fan *s*, or support, substantially as herein shown and described.

Fifth, the combination of the dog N with the weighted lever L, and fan J, substantially as herein shown and described.

Sixth, the combination of the hook *n* with the arm L, and fan J, substantially as herein shown and described.

Seventh, the combination of the latch S with the arm L, substantially in the manner herein shown and described.

Eighth, the combination of the bar O, and arm P, with the fan J, and guide G', substantially as and for the purpose herein shown and described.

No. 58,737.—CHRISTOPHER M. SPENCER, Boston, Mass., assignor to SPENCER REPEATING RIFLE COMPANY, same place.—*Magazine Fire-arm.*—October 9, 1866.—The engagement of the forked guide with the notched stop on the face of the carrier block arrests the downward movement of the latter.

Claim.—The combination of the forked cartridge guide *m* with the stop *s*, in the face of the carrier block, for the purpose of arresting the movement of the latter at the proper point for the introduction of a cartridge, substantially as set forth.

No. 58,738.—CHRISTOPHER M. SPENCER, Boston, Mass., assignor to SPENCER REPEATING RIFLE COMPANY, same place.—*Magazine Fire-arm.*—October 9, 1866.—When temporarily changed from a self-loader and acting as a hand-loader, the carrier block is checked so that it does not uncover the forward end of the magazine tube, and the shell drawer is governed by a swinging stop, which checks its backward movement so that it shall be in advance of the flange of the cartridge as it is pressed into the chamber by hand.

Claim.—Controlling the action of the shell drawer and of the carrier block by means of the swinging stop z, in combination with the cartridge guide m, in the manner set forth, for the purpose of converting the arm from a self-loader into a hand loader, or vice versa, as described.

No. 58,739.—N. BAILLY and C. DURAND, France, and G. H. MESNARD, Wandsworth, England, and Z. POIRIER, Lambeth, England.—*Anti-Friction Device for the Azles of Machinery.*—October 9, 1866.—A cylinder runs freely on the axle, or on a lining fixed thereto, supported on spheres which work upon a shoulder on the axle, the ends of the cylinder having openings to receive the spheres.

Claim.—The combination and arrangement of the several parts for the production of an anti-friction bearing for rotating shafts, substantially as herein set forth.

No. 58,740.—WILLIAM GOODMAN, St. Johns, New Brunswick.—*Ship's Windlass.*—October 9, 1866.—The pawl arms are moved by the undulating cam when the latter is connected to the vertical capstan shaft by the engagement of the clutch collar, which has teeth on its under surface and periphery.

Claim.—First, the cam F', loose on the capstan spindle, in combination with the coupling ring or collar D, whereby the cam is connected with the spindle, substantially as described.

Second, the coupling collar D, made with lateral teeth E, and vertical teeth Q, substantially as described.

Third, the combination of the revolving cam F with the pawl arms I, substantially as described.

Fourth, the combination of the grooved upper collar G of the cam, with the teeth E and Q of the coupling collar, and the keys S, substantially as described.

No. 58,741.—RICHARD C. MANSELL, Ashford, England.—*Car Wheel.*—October 9, 1866.—The retaining rings or tire fastenings not only hold the tires to the bodies of the wheels, but also sustain the parts of the tread which project over the sides.

Claim.—The improvements in the construction of wheels for engines and vehicles used on railways, by the adaptation and application thereto of retaining rings or tire fastenings, formed with outer flanges to support the parts of tires which project over the sides of the bodies of wheels, and also the exclusive use of the adaptation and application to wheels of retaining rings or tire fastenings, made from flat bars of metal, the whole substantially as herein set forth, described and illustrated in and by the annexed sheet of drawings.

No. 58,742.—JOHN WEEMS, Johnstone, England.—*Construction of Ships.*—October 9, 1866; English patent, November 20, 1865.—The hull of the ship is double. Water is admitted at openings near the keel, and air at openings near the gunwale, displacing the water on either side to increase buoyancy. When the jacket is heated by steam the barnacles are detached.

Claim.—Obtaining and distributing buoyancy or weight in ships by regulating a pressure of air or other fluid on the water in which they float, substantially as herein described.

Also, the removal of parasites from ships' bottoms after the manner hereinbefore described.

No. 58,743.—WILLIAM LOUIS WINANS, London, England. and THOMAS WINANS, Baltimore, Md.—*Application of Rudders to Spindle-shaped Hulls.*—October 9, 1866; English patent, December 8, 1863.—The leaves of the rudder are attached to a vertical post, and descend from the hull at each side of the propeller shaft in advance of the bracket which forms the bearing for the hub of the propeller.

Claim.—First, the rudders *d*, placed between the brackets *c*, and the midship of the spindle-shaped hull, its leaves occupying respective sides of the single propeller shaft, substantially as described, for the purpose specified.

Second, the central rudder *d*, placed between the brackets *c*, and the midship of the spindle-shaped hull, when arranged with more than one propeller shaft, substantially as described for the purpose specified.

No. 58,744.—WILLIAM LOUIS WINANS, London. England, and THOMAS WINANS, Baltimore, Md.—*Propeller.*—October 9, 1866; English patent, June 20, 1863.—The spindle shape of the vessel is preserved fore and aft, the fore part of the hub of the propeller forming a continuation of the hull, and its rear extension corresponding to that of the prow. The wings are attached to the flanged hub which partially encloses the bearing in which the propeller shaft is packed.

Claim.—First, the adaptation to and combination with a spindle-shaped vessel, such as was invented by Ross and Thomas Winans, and for which letters patent were granted to them October 26, 1858, No. 21,917, of one screw propeller placed at one end of the vessel, with its shaft coinciding with the centre line or longitudinal axis of the vessel, or nearly so, the outside diameter of the hub of the propeller at one end corresponding with the outside diameter of the end of the vessel, and the other end of the propeller hub being continued to a point, thereby completing the spindle form of the vessel.

Second, the combination of the flanged hub *f*, shaft *d*, webs *f f*, projection *h*, propelling blades *i*, and feathers *l l*, arranged with the grooved portion *b* of the body *a*, of the spindle-shaped hull herein described, substantially as and for the purpose specified.

No. 58,745.—P. H. ROOTS, Connersville, Ind.—*Rotary Blower.*—October 9, 1866.—The surface of the prominences of the pistons is relatively much increased over that of the depressions, causing a longer lap on their abutting surfaces.

Claim.—First, the rotating abutments A B. each constructed with two or more pistons D, and recesses E, constituting arcs of equal radius, but whose chords are of unequal length, the chords of the arcs of the pistons D being made of greater length than those of the recesses E, so as to cause the co-acting pistons to lap at and near the angles, substantially as herein described.

Second, finding the centres and radii for the pistons and recesses of the rotary abutments A B, in the manner described, with reference to figure 3 of the drawings.

No. 58,746.—W. C. ALLISON, Philadelphia, Penn.—*Tank for Containing and Transporting Petroleum.*—October 16, 1866.—The tank in form of a railroad car is made of metallic plates with a wooden casing to which it is attached in a few places. The tank casing has an outer wall and an intervening air space. Vapor from the oil is conducted by pipes to the chamber above the tank and mixed with air before escaping.

Claim.—First, a vat or reservoir, having an outer casing of wood and a thin petroleum-proof lining of metal, so suspended within the casing and detached from the sides and bottom of the same that it can readily, and without danger of rupture, yield and accommodate itself to any twisting or other distortion of the vessel, as set forth.

Second, the combination, as described, of a tank or reservoir, consisting of an outer casing of wood and a petroleum-proof lining with the frame of a car.

Third, the air space between the tank and sides and roof of the car, for the purpose described.

Fourth, the perforated pipes M′, or their equivalents, forming a communication between the interior of the tank and the ventilated space N′, beneath the roof.

Fifth, causing the mixed air and gas to pass through the perforations of the ventilator W′ before it reaches the external air, for the purpose described.

Sixth, the roof G of the tank, with its transverse beams *q*, both being covered with a petroleum-proof lining, substantially as described, and the transverse beams serving to prevent undue agitation of the contents of the tank, as set forth.

No. 58,747.—GEORGE S ANDERSON, Jeffersonville, Ind.—*Machine for Peeling Willow.*—October 16, 1866.—Pressure rollers revolve upon the periphery of a disk, at different angles to the axis of the latter. Scrapers are also so placed that the willow following the curved track is acted on by them, in connection with the rollers and disk.

Claim.—The wheel E, rollers H, brakes 8 and V, when constructed, combined, and arranged to operate together, substantially in the manner and for the purpose specified.

No. 58,748.—E. A. and A. C. APGAR, Philadelphia, Penn.—*Geographical Map.*—October 16, 1866.—The geometrical expression of the continent, &c., is given by a system of triangulation with bisecting and other lines. Special features and facts are indicated by symbols.

Claim.—First, the use for map drawing of such geometrical figures as are constructed by taking in each case some one line as a measuring unit, by means of which the lengths of other lines about the figure are determined.

Second, the trisecting and bisecting of certain lines about our geometrical figures for the purpose of determining the positions of certain prominent points along the coast lines of the continents.

Third, as original, that symbolic language for maps in which dots and lines, arranged substantially as described, are used to represent certain numbers, whether of population of cities, or the height of isolated hills, mountain peaks, or plateaus, in feet or miles, or other units of measurement.

No. 58,749.—T. G. ARNOLD, New York, N. Y.—*Screen for Gas Purifiers.*—October 16, 1866.—The perforated metal shelves which hold the lime in gas purifiers are coated with zinc, to prevent corrosion.

Claim.—The new manufacture of galvanized metal gas sieves, in contradistinction to ungalvanized iron gas sieves, for the purposes hereinbefore set forth.

No. 58,750.—VARNUM G. ARNOLD, Providence, R. I.—*Egg Beater.*—October 16, 1866.—The cylinder has a funnel mouth, and midway of its length is fixed a spiral coiled plate which offers its edges to the eggs reciprocated therein.

Claim.—The combination of the cylindrical can, provided with a funnel-shaped mouth and a broad base with a series of cutters spirally arranged and fixed to the inside of the can, or to a rim fitting inside the can.

No. 58,751.—FREDERICK ASHLEY, New York, N. Y.—*Carpet Stretcher and Tack Holder.*—October 16, 1866.—While the prongs hold the edge of the carpet in place, the tack is held by the pivoted spring jaw against a tooth of the holder, and when driven the latter is withdrawn, the jaw opening to allow it to disengage from the tack.

Claim.—The device, for the purpose specified, consisting of the toothed bar B, and the spring arm F, with its notched end G, bearing against the inner side of the notched jaw H of the bar B, and operating in the manner and for the purpose described.

No. 58,752.—NATHAN E. BADGLEY, New York, N. Y.—*Machine for Planting Cotton Seed.*—October 16, 1866.—The shoe in front levels the surface, and the draught piece is attached to it. The opener forms the furrow. The hopper is made of two flanged frustums, united at their bases. A part of the perimeter of the inner stationary hoop is perforated, and the seed passes through it and a revolving hoop, driven by contact of the frustum flanges with the ground. The seed is covered by spring bar in the rear.

Claim.—First, the construction of the base V, and its connection with the handles.

Second, the manner of constructing the draught piece D, with its fastenings.

Third, the construction of the hopper with its several hoops and its attachment to the cross piece T, as herein described.

Fourth, the attachment of the cover M, to the base and the rod R, with its coiled spring N.

Fifth, the combination of the several parts as herein described and substantially set forth.

No. 58,753.—ALEXANDER BADLAM, Sr., San Francisco, Cal.—*Washing Machine.*—October 16, 1866.—The slatted bed has depressions above, concentric with the axis of oscillation of the pendulum dasher, which has a weighted top.

Claim.—The combination and arrangement of the water box with curved slats *a*, metal dogs *e e* (serving as weights,) handle *e'*, and dash-boards *c c*, the whole being constructed and arranged for joint operation, substantially as described.

No. 58,754.—HORACE BAKER, Cortland, N. Y.—*Barrel Machinery.*—October 16, 1866.—The staves are clutched in their proper position in a frame and jointed by an annular cutter wheel. The churn is suspended and revolved by its ends in such manner as to admit of hand planing on the outside and automatic planing by means of a feeding bar carrying a knife within.

Claim.—First, the annular wheel O, and knives O', in combination with the swinging frame P, when respectively constructed and arranged for use, substantially as set forth.

Second, the combination and arrangement of the annular guide F, and pulley G, for suspending and revolving the barrel of a churn, with the guide bar H, carrying the plane I, and operated by an automatic feed, substantially as set forth.

No. 58,755.—WM. C. BAKER, New York, N. Y.—*Use of Hydro-carbon Liquids for Transmitting Heat.*—October 16, 1866. — Hydro-carbon liquids are employed for a circulating medium in heating apparatus, instead of water, on account of the liability of the latter to become frozen at low temperatures.

Claim.—The employment of hydro-carbon liquids to circulate in heating surfaces, as and for the purposes set forth.

No. 58,756.—HIRAM BARKER, Aurora, Ind.—*Washing Machine.*—October 16, 1866.—The revolving box has six sides, and has a corrugated surface for rubbing the clothes, which are caught and dropped by the independently acting finger of the interior shaft, and subjected to the dashing action of water raised by a scoop-flange inside.

Claim.—The construction of the watering tub A, the movable shaft D, with its pins, the flange E, and ribs F, the whole being arranged and operating in the manner herein set forth.

No. 58,757.—ELIAS BASCOM, New York, N. Y.—*Globe.*—October 16, 1866 —This globe is made of flexible and may be of transparent material; when not in use it is folded into a small compass. When used it is expanded and its spherical form retained by means of parallel wires fixed within at proper intervals between the poles. A lamp or hollow axis, with illuminating jets, lights it from within.

Claim.—The construction of a transparent or opaque globe, when arranged with adjustable wires and end plates, in combination with an illuminated axis, as herein described and for the purposes set forth.

No. 58,758.—AUGUST BASSE, Quincy. Ill.—*Wood Turning Lathe.*—October 16, 1866.— A frame carrying the shaft and gearing of a rotary cutter upon a slide table, is arranged so as to give position varying between horizontal and perpendicular and oblique directions to the cutter shaft.

Claim.—The arrangement of the carriage end J², stand J³, and stand K, for the purpose of supporting the cutter shaft K′, and permit it to be set and fastened at such a position or angle as may be desired, when constructed and operating substantially as described.

No. 58,759.—ALONZO BELL, Washington, D. C.—*Whiffletree.*—October 16, 1866.—The traces of the span of horses are attached to the ends of the doubletree and to a bar pivoted to the centre of the doubletree.

Claim.—As the distinctive feature of this improvement, the application of a combination swingletree and clevis to the centre of double whiffletrees, whereby a direct and equalized strain is brought to bear on the centre of the carriage, so that by this application or combination of movement the traces shall have free play, and equal and steady draught imparted to the centre of the carriage, and the present continual leverage of one horse against the other obviated.

No 58,760.—G. N. BEARD, St. Louis, Mo.—*Hoop Lock.*—October 16, 1866.—A cast iron slotted jaw is riveted to one end of the hoop, and its oblique conformation allows the latter to enter edgewise and then settle into the seat at right angles to the length of the hoop.

Claim.—A hoop lock formed with a rectangular slot, a¹, connecting with and forming a part of curved slots a² and a³, substantially in the manner and for the purpose herein set forth

No. 58,761.—JAMES M. BENT, Wayland, Mass.—*Creasing or Ornamenting Leather.*— October 16, 1866.—The leather is passed between a creaser having circumferential ridges and an adjustable pressure roll, and is adjusted and guided by a lever and guide plate.

Claim.—The revolving creaser I, in combination with the self-adjusting pressure roll K, operating substantially as described for the purpose set forth

Also, in combination with the above the lever Q, or its equivalent, substantially as and for the purpose described.

Also, the gauge *l* in combination with the creaser I, and pressure roll K, substantially as set forth.

No. 58,762.—JAMES M. BENT, Wayland, Mass.—*Punching Leather.*—October 16, 1866.— The punch has an annular sleeve die which makes a circular imprint around the opening, a pin beneath forcing the pellet into the throat of the die. The punch and die stock are reciprocated by a wrist and cam wheel, a certain elasticity of the latter on its shaft providing for the requirements of varying thicknesses of leather.

Claim.—The revolving punch D, with its die E, substantially as and for the purpose set forth.

Also, in combination with the above, the pin *g*, for clearing the punch D, substantially as described.

Also, the spring 6 or its equivalent, for the purpose of causing the die E to adapt itself to leather of varying thickness, substantially as set forth.

No. 58,763.—JACOB BENZ, Philadelphia, Penn.—*Sound Board for Pianos.*—October 16, 1866.—Two sounding boards are connected together by strips of wood, said strips being perforated or cut away at intervals, to allow of the circulation of air and sound. The fibres of the wood in the two boards run obliquely to their length and in different directions relatively to each other.

Claim.—The construction and combination of two different sound boards with traverse-running wood fibres, and provided with supporting ribs and air passages, substantially and for the purpose as described and set forth.

No. 58,764.—HERMANN BERG and ANDREW BLESSING, Springfield, Mass.--*Gas Burner.*—October 16, 1866.—The gas is admitted to the side of the annular chamber, avoiding a heavy shadow.

Claim.—As a new article of manufacture, the Argand burner, constructed in the manner herein set forth.

No. 58,765.—GEORGE W. BISHOP, Stamford, Conn.—*Friction Clutch Pulley.*—October 16, 1866.—The arm and dog revolve with the shaft, and the dog is brought in contact with the inner face of the pulley. to impart rotation thereto, by means of a shipper whose conical face presses against a set screw and presses out the dog.

Claim.—The arm C, and dog D pivoted thereto, parallel with the pulley A, operating in combination with the sliding sleeve E, substantially as described and for the purpose specified.

No. 58,766.—JAMES BROUGHTON, Lambertsville, N. J.—*Piston Packing.*—October 16, 1866.—The packing rings are arranged on each side of the circumferential division plate, and are expanded by spiral springs. The joints of the rings have T-shaped break-joint keys, with grooved beads to allow the passage of steam around the inside of the rings.

Claim.—First, the arrangement of the body A, hub B, division plate D, follower E, rings *b b'*, grooved T-shaped keys *c* springs *f*, in the recesses *g*, combined and operating in the manner and for the purpose herein specified.

Second, the grooves *d* in the keys *c*, which close the joints of the packing rings, for the purpose set forth.

No. 58,767.—JOHN BROUGHTON, New York, N. Y.—*Lubricator for Steam Engines.*—October 16, 1866.—The oil is poured in at the side opening while the valve is closed, the eduction opening closed and the valve lifted. When equality of pressure is established inside, the oil leaks down by the longitudinal slits in the screw thread on the stem of the valve.

Claim.—The combination and arrangement of the reservoir A, and valve stem E, having vertical openings *p p*, and made to screw into the shank B, with the nipple F, tubular cap G, and air chamber K, the whole being constructed and operated substantially in the manner and for the purpose set forth.

No. 58,768.—THOMAS W. BROWN, New York, N. Y —*Grindstone Journal Box.*—October 16, 1866.—The roller-box cover has projecting plates hollowed beneath to protect the roller journals from dirt or grit. The cover is held down by a duplex spring catch, whose ends enter holes in the lower frame above lips on the cover.

Claim.—The improved grindstone journal box as made with the wheel-journal caps *c c c c*, arranged and combined with the wheel cover C, so as to extend over and about the wheel journals, substantially as and for the purpose specified.

Also, the arrangement and application of the duplex spring catch E, with the projections *d d* from the cover C, and with the sides of the box A, as specified.

No. 58,769.—JOHN H. BRUEN, Elmira, N. Y.—*Sinking and Tubing Wells.*—October 16, 1866.—A spiral lever is attached to the base of the perforated inner tube. The borer has a square socket into which the boring rod sets, said rod being carried above the tubing for rotation. When sunk low enough the outer tube is withdrawn, exposing the perforated inner tube.

Claim.—A tube and boring bit for sinking and tubing wells, consisting of a tube A, and internal perforated tube B, to the base of which is attached a spiral bit C, and having a socket in the bottom of the tube B for receiving the point of the rod D, said several parts being respectively constructed and combined for use substantially as set forth.

No. 58,770.—CHARLES H. BUTTERFIELD, Sturbridge, Mass.—*Egg Beater.*—October 16, 1866.—The cylindrical glass vessel is narrowed in the middle, and contains a series of plates rigidly attached to the stopper, and presenting their edges to the eggs when shaken longitudinally.

Claim.—As an improved manufacture, the glass egg beater jar as made, with the contraction as arranged at or near its middle, the same being as and for the purpose or objects as hereinbefore set forth.

Also, an egg beater as composed of the case contracted at its middle, as represented, and a liquid rotator arranged within the contraction, and connected to the stopple of the case by means substantially as set forth.

No. 58,771.—FRANCIS H. CARNEY, Boston, Mass.—*Car Seat Indicator.*—October 16, 1866.—A pocket to be attached to the back of the seat to hold the railroad ticket, and with a slip "Engaged" to be drawn out when temporarily absent.

Claim.—The car-seat indicator, constructed substantially in manner and for the purpose hereinbefore described.

No. 58,772.—NATHAN CHAPMAN, Hopedale, Mass.—*Coffee Mill.*—October 16, 1866.— The bottom of the hopper surrounds the top of the mill case, and holds it together, one or both being eccentric to adjust the relative proximity of the runner and the stationary grinding disk.

Claim.—First, locking or fastening the top of the mill case together, by making the bottom of the hopper to surround the top of the case, when constructed and operating substantially as described.

Second, making the bottom of the hopper eccentric, or the top of the case eccentric, or both, for the purpose of adjusting the top of the case to make the mill grind fine or coarse, substantially as described.

No. 58,773.—JAMES E. CHEASEBRO, Buffalo, N. Y.—*Sulky Plough and Harrow.*—October 16, 1866.—A slide bar projecting downward from the sulky pole has an adjustable sliding eye which is connected to the plough clevis. The rear of the plough is attached to a foot-board which is hung to the rear end of the tongue hounds. The seat is attached to the connecting bar of the foot-board. A harrow may be attached to the pole and hounds in a somewhat similar manner.

Claim.—First, the combination and attachment of a plough to a sulky in such manner that the plough beam shall pass under the axle of the sulky and project forward, and the plough handles project in rear of the axle, and in convenient grasp of the ploughman as he sits upon his seat, substantially as set forth and described.

Second, the combination of the guide stirrup B with the slide G, for the purpose of forming a connection of the forward end of the plough beam with the sulky, substantially as set forth.

Third, connecting the rear end of the plough to a brace or foot-board D, projecting from and in rear of the axle, for the purpose and substantially as described.

Fourth, the driver's seat A⁴ and foot-board D, projected and supported in rear of the axle, for the purpose and substantially as set forth.

Fifth, the combination of a harrow M with a sulky, for the purpose and substantially as described.

No. 58,774.—WILLIAM CHESLEY, Cincinnati, Ohio.—*Globe Valve.*—October 16, 1866.— The stem above the valve reciprocates in a boss forming the floor of the chamber above, which is closed by a screw plug. The object is to prevent accumulation of obstructions at the stem.

Claim.—The construction and arrangement of the boss *d*, cap *b*, and plug *a* with reference to the valve stem C, for the purpose and as herein set forth.

No. 58,775.—WILLIAM CHURCHILL, St. Louis, Mo.—*Governor Valve for Steam Engines.*— October 16, 1866 —The throttle valve stem passes through the sleeve of the governor valve. Horizontal radial arms from the sleeve have on their ends valve faces with perforations coincident with the steam supply passages. The valve stem of the governor is attached to one of the radial arms, which it moves by the revolution of its own screw stem in the gland above.

Claim.—First, the arrangement of the throttle and governor valves in the manner substantially as set forth.

Second, the combination of the nut G, stem F, and spring H, whereby to secure the action and regulation of the governor in accordance with the demands of power and speed.

No. 58,776.—HENRY W. CLARKE, Newport, R. I.—*Setting Fence Posts.*—October 16, 1866 —The fence post is surrounded beneath the soil by a metallic frustum filled with pebbles, and the top closed by cement to prevent access of moisture to the inside.

Claim.—The arrangement and application of the hollow frustum B, its cement or head D, and the mass of gravel E, or its equivalent, with a post A, the whole being substantially as and for the purpose set forth.

No. 58,777.—CUMMINGS P. COLBY, Lancha Plana, Cal.—*Mill for Crushing Quartz.*— October 16, 1866.—The stamps are raised consecutively by the eccentrics on the shaft, operating through the levers and stirrups in which rest the springs attached to the stamp rods. The stamps return by the force of the spring and their own momentum.

Claim.—The combination of the eccentrics B', with collars and spindles and springs *a*, arranged to operate the stampers, substantially as described.

No. 58,778.—ROBERT CONARROE, Camden, Ohio.—*Straw Cutter.*—October 16, 1866.— The knife plate has pins at its ends which enter grooves in the eccentric wheels whose revolution reciprocates the knife. Anti-friction rollers on the pins reduce the friction.

Claim.—First, the combination of the eccentric *e*, pin *d*, and roller *d'*.

Second, the combination of the grooved eccentric cams F, on the shaft G, with the guides and frame C D, and knife E, of a straw cutter, substantially in the manner and for the purpose set forth.

No. 58,779.—SOLON COOLEY, Oakwood, Mich.—*Wool Press.*—October 16, 1866.—The side cheeks and end boards are hinged to the table surrounding the fleece, which is doubled up between them. The table portion has grooves for the string previously placed, as has also the follower, which is pressed upon the wool, and retained by spring catches and ratchets till the fleece is tied.

Claim.—The arrangement of the bottom board A, the sides B B, and ends C C', as constructed with the follower H, spring arms G G, rack bars F, and hooks D D, substantially as and for the purpose herein specified.

No. 58,780.—JOHN CRAM, Chicago, Ill.—*Shaft for Rubber Rollers for Wringing and Washing Machines.*—October 16, 1866.—The shaft for the rubber has gains on both sides and alternately at right angles to those preceding and following; the flattened portions left are traversed by pins which project as far as the general diameter of the shaft.

Claim.—Constructing a shaft A, with a series of recesses and corresponding pins or projections arranged and operating substantially in the manner and for the purposes herein specified and described.

No. 58,781.—CHARLES CROLEY, Dayton, Ohio.—*Ladder.*—October 16, 1866.—The ladder is supported by feet and braces, and has an extension section hinged to blocks which slide in guides in the side pieces, and are fastened by the engagement of a round with books on the upper end of the lower section.

Claim.—First, the sliding pieces *h h*, connected to the ladders A and B, substantially as and for the purposes specified.

Second, the combination of the projections *g g*, the hooks *i i*, and step *s*, substantially as and for the purpose described.

Third, the base pieces C, the braces D, and jointed bar E, when constructed and arranged with reference to the ladder B, in the manner substantially as described and for the purpose specified.

No. 58,782.—E. N. CUMMINGS, Colebrook, N. H.—*Heating Stove.*—October 16, 1866: antedated October 4, 1866.—The fire box and upper chamber are connected by serpentine flues whose upper ends are regulated by independent dampers, one of which controls the openings under a pot hole in the top plate.

Claim.—A stove for heating purposes made substantially as above described, its upper and lower parts A E being connected by serpentine flues whose openings in the upper part E are controlled by two independent dampers, substantially as shown.

No. 58,783.—J. M. CURRIE, Washington, Iowa.—*Projectile for Ordnance.*—October 16, 1866.—This conical projectile has a circumferential depression gradually diminishing toward the point; in this is cast a packing ring of soft metal.

Claim.—The projectile A, with the conical point and tapering rear, having the packing ring B applied as shown and described.

No. 58,784.—JAMES WARREN DAVIS, Washington, D. C.—*Device for Hanging Wall Paper.*—October 16, 1866.—The end of the paper has a turn on the roll at the upper end of the frame, and is held by the spring jaw until released by pulling on the trigger.

Claim.—The roller D. having a yielding surface, the clamping bar E, the frame A B C, the pivoted arms F, the bell crank G, and the spring rods J *m*, the whole arranged and operated substantially in the manner and for the purpose herein described and represented.

No. 58,785.—D. A. DICKENSON, Baltimore, Md.—*Machine for Harvesting, Husking and Shelling Corn.*—October 16, 1866.—The machine cuts two rows simultaneously by reciprocating kuives on each side; the stalks are taken by the attendant and presented to the cutter, which removes the ear; the ears drop on to the busking roller and from thence pass to the sheller where the cobs are dropped and the corn sacked. The motions are all derived from the ground wheels.

Claim.—First, a machine for cutting the stalks from the hill or row, separating the stalk from the ear, husking the ear and shelling it, when the different pieces or parts thereof are constructed, arranged, and operated substantially as herein recited.

Second, combining with cutting and busking or shelling machines the arrangement of the means or parts constituting the apparatus for cutting the stalk from the hill or rows, when constructed and operated substantially as set forth.

No 58,786 —WALLACE DICKINSON, Brooklyn, N. Y.—*Car Wheel.*—October 16, 1866.—The elongated hub has a central bushing secured by a disk and bolts, and enclosing an oil chamber, to which access is had by tubes closed (after filling) by screw plugs.

Claim.—The elongated hub H, having flanges *s s'*, provided with a bush *d*, having cavity *t*, and openings *m m' n*, and washer *w*, all constructed and arranged substantially as described and for the purpose set forth and shown in the accompanying drawings.

No. 58,787.—W. W. DOANE and W. P. BURR, Brewer, Me —*Skid for Supporting Barrels* —October 16, 1866.—The cheeks are framed together, and on their upper edges have bearings for the rollers which support the barrel.

Claim.—The improved hogshead supporter, or combination and arrangement of rollers and skids, made and applied substantially as specified.

No. 58,788.—R. L. DODGE and E. M. WALKER, Gallatin, Mo.—*Gang and Subsoil Plough.*—October 16, 1866.—The tongue is pivoted on a front cross bar of the frame, to which the plough beams are attached. A curved plate projecting upward from the axle traverses a slot in the tongue and has pin holes for regulating the depth of furrow through the raising or depression of the fore end of the tongue. Set nuts allow a vertical adjustment in either or both of the ploughs, so that, while one is turning a first furrow, the other may be "subsoiling."

Claim.—The construction and arrangement of the pole H in connection with the standard I and axle B, so that it may be elevated and lowered, substantially as described.

Second, the pole H, when hinged to the cross-bar of the frame so as to form a lever to raise the ploughs, in combination with the plough beams E E and plough C C', when constructed for the purposes and substantially as described.

No. 58,789.—W. C. DODGE, Washington, D C.—*Composition for Walks, Pavements, &c.*—October 16, 1866.—A basis, two inches deep, of stone, shells, and raw coal tar is covered with a composition of fine gravel, twenty parts, sand ten, ashes ten, sulphur two, with coal jar to make a paste; cover with sand and roil.

Claim.—The composition and process herein described, when applied as and for the purposes set forth.

No. 58,790.—WM. C. DODGE, Washington, D.C.—*Magazine Fire-arm.*—October 16, 1866.—The magazine tube slides beneath the barrel in ferrules, and has a spiral spring to eject the cartridges into the carrier in the rear. A spring catch holds the cartridges in the tube until it is placed, when the catch is raised, allowing free exit. A spring catch, operated by a thumb-piece, projects through the side of the frame and holds the tube in place. The two cheeks attached to the stock form two sides of the cartridge-receiving chamber.

Claim.—The sliding tube B, with the spring *g* attached and sliding in the groove *k*, in combination with the barrel A and breech-frame C, when said parts are arranged to operate as and for the purposes set forth.

Second, in combination with the sliding tube B, the spring catch *c*, located inside of the breech-frame C, and arranged to be operated from the outside, as shown and described.

Third, forming the chamber for the reception of the cartridges, at the rear end of the tube B, by means of the pieces *m*, or their equivalents, substantially as described.

No. 58,791.—F. P. DUPRAZ, S. M. DUMONT, and JOHN DICKASON, Vevay, Ind.—*Steering Apparatus.*—October 16, 1866.—The heliscal pulleys arranged on each side of the drum on the tiller wheel shaft give an increasing power to the latter as the tiller diverges from its fore and aft position.

Claim.—First, the intermediate sheave, or double spiral and drum D E, constructed substantially as set forth and for the purpose specified.

Second, the arrangement of the wheel A, drum B, rope or chain C, drums E E', pulleys D D' F, and tiller G, forming a progressive-power steering apparatus, as described.

No. 58,792.—ZONETH S. DURFEE, Philadelphia, Penn.—*Hat Box and Valise.*—October 16, 1866.—In the bottom of the valise is a hat box with a cover opening downward and having a box attached which occupies the inner portion of the hat.

Claim.—Combining a hat box with a modification of the common travelling bag or valise, substantially as and in the manner described and shown in the accompanying drawings.

No. 58,793.—JOHN ECKERT, Madison, Ind.—*Water Cooler.*—October 16, 1866.—The cast metal chamber has a screw thimble for the attachment of the faucet, a sheet metal cylinder above, and the whole enclosed in an outer jacket with an intervening non-conductor.

Claim.—The sheet-metal chamber A. cast bottom B b, and thimble C c, the whole being combined and adapted to operate as set forth.

No. 58,794.—W. C. S. ELLERBE, Camden, S. C.—*Instrument for Transplanting Plants.*—October 16, 1866.—The cylinder attached to the handle isolates the plant from the surrounding soil. The plant is then raised, and, being placed in position, is removed from the cylinder by the annular pusher piston, operated by the central rod.

Claim.—An improved plant transplanter formed of a cup A, handle C, and of a pusher D and rod G, constructed and combined with each other, substantially as herein described and for the purposes set forth.

No. 58,795.—SAMUEL F. ESTELL, Richmond, Ind.—*Animal Trap.*—October 16, 1866.—The weight of the rat on the hinged section of the floor releases the detent of the vertically sliding door. As the rat passes under and raises the inclined wire gate, the door is again raised and set by the devices in the chamber above.

Claim.—A rat trap in which the self-setting devices as set forth and described are placed in an apartment of said trap immediately above and separate from the body of the trap, substantially as and for the purposes herein mentioned.

No. 58,796.—RICHARD P. ESTEP, Cincinnati, Ohio.—*Valve for Steam Engines.*—October 16, 1866.—The three-faced valve oscillates in and coincides in form with the frustal chamber, whose ports connect with the steam cylinder. The pressure on the valve is nearly balanced, and it is kept to its seat by a coil spring acting longitudinally on its stem.

Claim.—The balanced yielding and adjustable three-winged valves K L, (one or both,) arranged and operating in the manner substantially as described.

No. 58,797.—H. C. FAIRCHILD, Brooklyn, Penn.—*Wagon Brake.*—October 16, 1866.—The inner ends of the break bars are attached to a reciprocating slide beneath the wagon bed, and their projecting ends come in contact with the wheel, when they are locked by a catch.

Claim.—An improved wagon brake formed by combining the spring catch F, slide C, straps G J and H, lever I, hinged brake bars E, rests K, and stops L with each other and with the wagon body A, the parts being constructed and arranged substantially as herein described and for the purpose set forth.

No. 58,798.—JAMES M. FATE, Boonsboro, Iowa.—*Corn and Cane Planter.*—October 16, 1866.—A pivoted walking beam is worked by a treadle and raises the beam carrying the seed droppers; the forward drag bar of the beam is pivoted below the tongue.

Claim.—First, attaching the seed dropping devices to a frame or beam F in combination with a transporting frame, substantially as described.

Second, the vertically swinging beam F in combination with a lever G and a treadle H, substantially as described.

Third, the combination of the lever K with a suspended beam or frame F, substantially as described.

Fourth, pivoting the forward end of the drag bar F', of the beam F, to the draught pole C, substantially as described.

Fifth, suspending the beam F, or its equivalent, from the main frame centrally, substantially as described.

No. 58,799.—HENRY FESSLER and HENRY MAXELL, Canton, Ohio.—*Device for Opening Furnace Doors.*—October 16, 1866.—The furnace door is opened by a treadle on an arm passing beneath the floor. A spring beneath the arm serves to close the door on the release of the treadle.

Claim.—First, the door F, pin a, bar E, and shaft C, arranged and used substantially as and for the purpose herein specified.

Second, the arrangement with the shaft C, of the connecting link e, lever G, foot piece I, and spring J, substantially in the manner and for the purpose set forth.

No. 58,800.—GEORGE A. FITCH, Kalamazoo, Mich.—*Priming Cartridges.*—October 16, 1866.—A stem is attached to the back plate of the cartridge projecting forward and capped at its front end so that the stock of the hammer will explode the cap against the ball. Slow burning powder may be placed in front to start the bullet and a "quick" powder behind. By the addition of fulminate at the root of the stem the charge is exploded at both ends, or the stem may be parted and capped at its centre for lighting the charge at that point

Claim.—First, igniting the cartridge at the front by means of the stem a, in combination with the accelerating charge, as shown in Fig. 1, and as herein described.

Second, igniting the charge at both front and rear, as shown in Fig. 2, and as herein set forth.

Third, providing a cartridge with the stems *a* and *a'*, when arranged to act in combination, as shown in Figs. 3 and 4, for the purpose of igniting the charge at the centre, substantially as set forth.

No. 8,801.—A. . FLETCHER, Athol, Mass.—*Pump.*—October 16, 1866.—Circumferential ridges inclined on their upper sides catch beneath the heads of studs on the partial revolution of the pump barrel. The connection between the handle and bucket is made by twin side rods kept together by a vertical staple dropped into eyes on their insides; guides restrain the valve from lateral divergence.

Claim.—First, the combination of the loose collar C with the pump barrel A, and both with the posts E, and all with the bottom plate F, constructed and operating substantially as described and for the purposes specified.

Second, the combination of the double rod M M, when the parts are connected together by the staple S, and with the pump box G, and with the pump handle P, constructed and operating substantially as described.

Third, the guides H for the valve applied to the lower plate of the valve box, substantially as described.

No. 58,802.—M. FOREMAN, Philadelphia, Penn.—*Steam Generator.*—October 16, 1866.—The series of double spheres of cast-iron, having attaching necks, are held together by tubes and nuts, through which air passes; the air operates to keep these tubes cool, thereby lessening their expansion and preventing the opening of the joints of the boiler section.

Claim.—First, the tubular bolts C, combined with and adapted to the system of spheres A, substantially as and for the purpose herein set forth.

Second, the manner described of arranging bolts in respect to the spheres so as to prevent the sinking of the same.

No. 58,803.—CHARLES FORSTER, Pittsburg, Penn.—*Oil Well Drill.*—October 16, 1866.—A shaft within an outer pipe has curved cutters, having a horizontally reciprocating and a rotary motion by means of a vertically reciprocating shaft and a pawl working upon a ratchet on the cutter shaft.

Claim.—First, the combination in a drilling tool of curved cutters *d*, having slots *e*, and oblique edges *f*, substantially in the manner and for the purposes above set forth.

Second, the combination in a drilling tool of the head *a*, the springs *s*, the plunger *b'*, the ratchet *r*, the pawls *p*, for the purpose of producing a rotary motion and communicating the same to the cutters *d*, and barrel *z*, the whole being constructed and arranged substantially as and for the purposes above described.

Third, the combination in a drilling tool of the abutment *g*, screw box *m*, screw *o*, and adjustable jam nut *o'*, used and operated substantially as and for the purposes above set forth.

Fourth, the use of the ring or collar *c''* in combination with the shaft *b* of a drilling tool furnished with expanding cutters, as a gauge to indicate the degree of spread of the cutters within the chamber of a well.

No. 58,804.—J. W. FOUST, Evansburg, Penn.—*Hay Raker and Loader.*—October 16, 1866.—The drum which carries the endless elevator is put in or out of operation by clutches upon the ends, which are moved into or out of engagement with ratchets on the wheel hubs, by operative levers connected to a rocking lever, and actuated by hand.

Claim.—The arrangement of the clutches H with the wheels D D and shaft G of the reel or drum E, in conjunction with the levers M, attached to the lever N, and all arranged to operate in manner substantially as and for the purpose set forth.

No. 58,805.—JOHN FRIDY, West Donegal township, Penn.—*Cultivator.*—October 16, 1866.—The pivot frame at the front of the beams has journal bearing for a wheel; lateral adjustment of the outside beams is accomplished by means of hooked bolts by which they are attached to a bar at their rear.

Claim.—The construction of the adjusting bar D, fixed in its centre to the central beam 2, and provided with a series of holes G, for the hook bolts E, supporting and embracing the side beams 1 and 3 in combination with the pivots A, when supported between the plates P and *p*, in the manner and for the purpose shown and specified.

No. 58,806.—WILLIAM B. FRUE, Houghton, Mich.—*Shaking Table.*—October 16, 1866.—The rocking table has lateral concavities furnished with agitating pins and vertical discharge openings. The table has longitudinal and vertical motion from cranks on two shafts having different speeds and connected by a belt over pulleys on said shafts.

Claim.—First, the undulating top G forming a series of concave troughs, substantially as and for the purpose described.

Second, the agitators A and discharge openings i in the several troughs of the top G, substantially as and for the purpose set forth.

Third, the crank shafts B and F and links a c in combination with the table G, constructed and operating substantially as and for the purpose described.

No. 58,807.—CHARLES GEISSE, Taycheedah, Wis.—*Boiler for Gas Heater.*—October 16, 1866.—This is a detached heater for use with a vessel to which the direct heat cannot be applied, and has central and annular fluid chambers communicating together by pipes and warmed by the application of a burner or caloric current to the intervening flue space.

Claim.—The combination of the heaters A B arranged and operating in relation to each other and to the vessel or vessels containing the fluid or other matter to be heated and to any known appliance for the combustion of oil, coal oil, or other inflammable fluid, gas or vapor, substantially upon the principle and in the manner hereinbefore set forth.

No. 58,808.—JAMES E. GILLESPIE, Boston, Mass.—*Steam Trap.*—October 16, 1866.— Below a steam pipe is connected a pipe coiled into a helical spring and communicating with a suspended water chamber having at bottom an inwardly opening valve whose stem projecting below is raised by contact with an abutment, when the weight of water stretches the coil.

Claim.—A steam trap, constructed and applied substantially as herein set forth.

No. 58,809.—JOHN GILLMORE and AARON WICKS GILLMORE, Utica, Penn.—*Sinking Well Tubes.*—October 16, 1866.—The vertical ribs or flanges which surround the drill bar near its lower end enter the notches of a collar attached to the inside of the well tube and hold the lower section from rotating while disconnecting the rod.

Claim—The drill B with the flanges 4 5 6 7 in combination with the collar E and the tube D D, when the same are constructed as described in the aforesaid combination, for the purposes set forth.

No. 58,810.—HENRY GOULDING, Dedham, Mass.—*Device for Lowering Boats.*—Octobe 16, 1866.—At each end of the boat a short chain is fastened to the boat and to a stud, the bight of each chain engaging the hook of the davit fall-block above. The ends of the chains are held on the studs by pivoted levers, which are simultaneously withdrawn by mutual connection to a lever amidships.

Claim.—The device for the purpose specified, consisting of the stud b, lever d, chain c, chain g, and pivoted lever h, arranged and operating as described.

No. 58,811.—GEORGE D. GREENLEAF, Depauville, N. Y.—*Radiating Stove and Drum.*— October 16, 1866.—The heated gases, &c., are discharged into a bell-shaped receptacle within, and are deflected nearly to the bottom of the drum; a rod projecting outside the stove opens or closes the bell top to operate as a damper. Vertical plates within an upper drum force the current in a circuitous direction on the closing of two dampers connected to them.

Claim.—The pipe C leading from the stove A and the drum B placed on the stove and enclosing pipe C in combination with the partition plates D D and dampers E E, all arranged substantially as and for the purpose specified.

Also, the partition plates H H' in the pipe F forming the draught passages f g h and arranged with the dampers I I', as shown in combination with the drum B and the parts contained therein as specified, and all arranged to operate in the manner substantially as and for the purpose set forth.

No. 58,812.—JOSEPH W. GRISWOLD and JOHN SIGWALT, JR., Chicago, Ill.—*Machine for Making Paper Collars.*—October 16, 1866.—The collar is cut out, the button-holes made and the imitation stitching embossed in one descent of a platen operated by a treadle. The machine is adjustable in all its parts to form collars of different sizes.

Claim.—First, the combination in one machine of the adjustable stitching plate M, punches or their equivalents a, and adjustable end clips l a, arranged and operating substantially as and for the purposes specified.

Second, in combination with the above the employment of the longitudinal shears T U, arranged and operating as and for the purposes set forth.

Third, constructing the stitching plate M in three parts, one M' removable and one or more adjustable, substantially as herein specified and for the purposes set forth.

Fourth, the combination of the adjustable stitching plate, button-hole punches, and end clips arranged and operating as and for the purposes specified.

Fifth, the arrangement of the adjustable width gauges S S with the shears T U, as and for the purposes set forth.

Sixth, in combination with the adjustable stitching plate, punches, and end clips, the employment of the adjustable feeding guide R, as and for the purposes described.

Seventh, providing the bed plate B with the longitudinal opening V, arranged as and for the purposes specified and shown.

No. 58,813.—C. H. HALL and JOHN ELLIS, New York, N. Y.—*Distilling Apparatus.*—
October 16, 1866.—The two retorts placed one over the other are supplied and connected by
tubes and are heated by means of flues. The vapors pass from the retorts by separate pipes
to a condenser through which passes a coil; this connects with a pipe leading to a second
condenser which has a series of cooling tubes. The residuum in the retort is drawn off by
means of a pipe at a low level.

Claim.—First, the arrangement of two or more retorts A B through which the liquid to be
distilled passes in a thin stratum, substantially as and for the purpose described.

Second, the flues C E in combination with the retorts A B and fireplace D, constructed
and operating substantially as and for the purpose set forth.

Third, the pipes G G′ leading from the retorts A B to one and the same condensing chamber
H, substantially as and for the purpose described.

Fourth, the inclined condensing chamber H′ in combination with the inclined condensing
chamber H and retorts A B, constructed and operating substantially as and for the purpose
set forth.

Fifth, the residuum tank L with pipes W u s in combination with one or more retorts,
constructed and operating substantially as and for the purpose described.

Sixth, passing the vapors through a closed vessel containing a pipe or pipes through which
cold water passes, said vessel being provided with one or more discharge pipes to draw out
the condensed liquid of any desired gravity, substantially as set forth.

No. 58,814.—JOEL A. HALL, Columbus, Ohio.—*Garden Cultivator.*—October 16, 1866.—The
handles are so pivoted that a lateral movement can at any time be given to two hoes, which
work horizontally in the ground at each side of a shovel plough. The axle is attached to a
plate and curved downward to receive the axle pins.

Claim.—First, the cross handles or levers E attached to the side pieces D D in combination
with the hoes K K, axle B, and wheels A A′, substantially as described.

Second, the plate T in combination with axle B, for the purpose and substantially as described.

No. 58 815.—JOHN R. HALSEY, Newark, N. J.—*Boat for Travelling on Ice.*—October 16,
1866.—The boat is set upon sleds at the bow and stern, is adapted to run upon the ice of a
river, and is propelled by wheels with pointed projections on their peripheries. Details are
explained in the claims and cut.

Claim.—First, the arrangement of the geared wheels c and d for giving motion from the
crank shaft to the main shaft in combination with the lifting lever e, or its equivalent, for
adjusting the position of the main shaft and its driving wheels, the whole operating substantially as and for the purposes set forth.

Second, the application and use of the guard springs 8 in front of the runners, substantially as and for the purposes set forth.

Third, connecting the movable or swinging sleds and the stern rudder with the steering
mechanism substantially in the manner described, so that the said sleds and stern rudder may
be simultaneously actuated by such mechanism, as and for the purposes herein shown and set
forth.

Fourth, the application and use, in combination with a rudder hinged as described, of a
spring to act upon such rudder and keep the same in contact with the ice, for the purposes
set forth.

Fifth, a steam ice boat, its several parts constructed, arranged, and operating substantially
as and for the purposes set forth.

No. 58,816.—D. JONES HAPPERSETT, Coatesville, Penn.—*Car Truck.*—October 16,
1866.—The inner pedestal rests upon the inner hub of the wheel and supports the truck upon
the wheel in case of breakage of the axle.

Claim.—The inner pedestal D, attached to the bottom of the truck, and provided with a
recess, partially enclosing the upper portion of the hub on the inside of the wheel, so as to
form a socket therefor, in case of the breaking of the axle.

No. 58,817.—TREEMAN M. HARDISON and JOHN A. HOOPER, South Berwick, Me.—*Mop
Head.*—October 16, 1866.—The toothed bar of the stirrup is drawn within the slot of the mop
head, and secured by a collar which draws the shanks to the handle.

Claim.—The combination of the movable wire frame C with the sliding ring D, and
grooved handle A, when constructed and operating as and for the purpose specified.

No. 58,818.—ANDREW HARTMAN, Canton, Ohio.—*Railroad Switch.*—October 16, 1866.—
The switch rails are moved by a lever and rod, or automatically by a nose on the cow catcher,
which by contact with compound levers restores the continuity of the track if the switch be
misplaced.

Claim —First, the turning lever J, when used with the bar D, rails C C, and a catch under said rails, as and for the purpose specified.

Second, the spring G, when used with the rails C C, the lever J, and catch under the rails C, as and for the purpose herein specified.

Third, the arrangement of the levers H and I, the rails C C, and bar D, with the turning bar J, spring G, and lever E, for operating the switch automatically, as well as by hand, substantially as specified.

No. 58,819.—NATHAN HAWKES, Appleton, Me.—*Combined Cultivator and Ditcher.*—October 16, 1866.—This implement is for hoeing, subsoiling or shallower ploughing, or the planting of potatoes or grain. A two-faced mould board is followed by wings or shares attached or following in the rear. Bars trailing in the rear mellow the ground. A toothed disk adjustable as to depth has side pins to work the slides at the bottom of the seed hoppers. A draw bar is pivoted in the beam, which for deep ploughing is carried up a rack at the fore end of the beam, and held to any one of the adjusting notches by a wedge

Claim.—All the various parts, constructions, combinations, and arrangements hereinbefore described, for planting, hoeing, digging and ditching, except so far as the mould boards A, the beam B B, and the handles C C, figure 1; are like those of the common double mouldboard plough.

No. 58,820.—J. S. HAZARD, Newport, R. I.—*Instrument for Removing Wires from Bottles*—October 16, 1866.—Two levers are pivoted together, one carrying a fork which fits around the neck of the bottle, and the other a hook which seizes the wire and pulls it from over the cork.

Claim.—The implement herein described, the same consisting of the fork-shaped bar E, and hook lever G, when combined together, substantially in the manner and for the purpose described.

No. 58,821.—MICHAEL HICKEY, Boston, Mass.—*Cask and Barrel.*—October 16, 1866.—Grooves in the outer edges of the staves form dovetail recesses to hold corresponding strips when the barrel is put together.

Claim.—Rabbeting out the edges of the staves, and heading and filling the rabbets with a spline or strip (glued, cemented or otherwise,) covering the joints between the staves or pieces of heading, and in combination with the splines, the pins glued, cemented or otherwise, as aids in fastening the splines.

Also, the pins between the staves in the chine of the cask, glued, cemented or otherwise

Also, cutting the rabbets under, and making the splines or strips dovetailing, substantially as described.

No. 58,822.—S. P HILDRETH, Mount Vernon, Ohio.—*Lamp Stove for Dentists.*—October 16, 1866.—The lamp is placed beneath the body of the stove and its flame enters beneath the cylinder which holds the vulcanizer. At other times the jacket is removed and a lid substituted.

Claim.—The jackets F, in combination with the body B, of the stove, having collar D and pipe G, substantially as described, for the purpose specified.

No. 58,823.—ARNOLD HOEPPNER, St. Louis, Mo.—*Apparatus for Making Vinegar.*—October 16, 1866.—The vessel is divided into a series of compartments containing shelves whose perforations direct the wash from one to the other throughout the series, while exposed to an artificial current of air.

Claim—First, the combination of a series of shallow vessels in which, by surface oxygenation, the acidification of the wash is effected.

Second, the combination of the vessels $A^1 A^2 A^3$, &c., their overflow openings c^2, arranged at diagonally opposite ends, substantially as set forth.

Third, the separation of each vessel $A^1 A^2 A^3$, &c., into compartments C $C^1 C^2$, &c., the same connecting by apertures d, substantially as set forth.

Fourth, the combination of the vessels $A^1 A^2 A^3$, &c., with shutters f, as and for the purpose set forth.

No. 58,824.—ADOLF H. HIRSH, New York, N. Y.—*Manufacture of Sugar from Corn.*—October 16, 1866.—The corn is treated with dilute sulphuric acid at an elevated temperature until the grain is softened, after which it is ground and the mass placed in a sieve which retains the husks and passes the milky portion to a series of inclined gutters, in which the starch is deposited. After washing with a weak solution of ammonia and then with pure water, the starch is mixed with water until it assumes the consistency of cream, and is treated with a heated mixture of water, sulphuric acid, sulphate of alumina and coke or vegatable charcoal. After the starch has been completely converted into sugar, lime is added to the tank to neutralize the acid, after which the mass is passed through a filter and the sirup evaporated in the ordinary manner.

Claim. First, the application of diluted acid at an elevated temperature in the process of making starch from maize and other cereals, for the purpose of making sirup and sugar therefrom, substantially in the manner set forth and specified in the first manipulation.

Second, treating saccharine liquid with alumina and charcoal, coke or boneblack combined, substantially as and for the purpose set forth.

Third, the within-described process of manufacturing sirup and sugar from corn or other grain, consisting of three subsequent manipulations, substantially each as set forth.

No. 58,825.—AMELIA B. HOFFMAN, Roxbury, Mass.—*Dust Pan.*—October 16, 1866.—
The dust pan has a lid, a handle for the base and an inclined plane with offset at front edge.

Claim.—The combination of the box E, the lid A, the inclined edge or apron B, and the handle D, all as and for the purpose described.

No. 58,826.—JOHN W. HOLLINGSWORTH, Seymour, Ind.—*Animal Trap.*—October 16, 1866.—The weight of the rat on the platform by the bait springs the trap, dropping the door behind him and darkening the chamber; as the rat passes the wire gate into the next chamber the trap is reset.

Claim.—An improved animal trap, formed by the combination of the rocking lid C, the stationary cap B, the lifting gate L, the treadle K, cam wheel J, spring I, cranks G and E, and connecting rod F with each other, and with the box A, substantially as herein described and for the purposes set forth.

No. 58,827.—GEORGE W. HOLLY, Low Moor, Iowa.—*Hanging Doors.*—October 16, 1866.—The door slides open and shut by a horizontal movement, the upper ends of the pivoted bars attached thereto slide in vertical guides by the rotation of the arms connected by rods to the bars.

Claim.—The arms *f f'* and rods *e*, arranged and operating relatively with the bars C F and door A, substantially as described, for the purpose specified.

No. 58,828.—N. HOMES, Leona, N. Y.—*Attaching Handles to Saws.*—October 16, 1866.—
The monkey wrench may be used as a saw-set and is attachable as a handle to the saw by means of pins on its jaws which enter notches in the blade, being tightened in position by a threaded shank.

Claim.—The saw handle B, constructed and arranged substantially as described and for the purpose set forth.

No. 58,829.—ELI H. HOWARD and ALBERT J. MANCHESTER, Providence, R. I.—*Sponge Cup.*—October 16, 1866.—The sponge cup is surrounded by an annular water chamber and connects therewith at an opening on a low level. Water is admitted to the sponge by allowing access of air to the water chamber.

Claim.—A sponge dish, constructed substantially as herein described.

No. 58,830.—LIVERAS HULL, Charlestown, Mass.—*Braiding Machine*—October 16, 1866.—This machine is intended for making two or more distinct braids connected at their edges; thus admitting of different colored stripes. Each racer for this purpose goes only through its own course of race circles, one of which circles is common to the two carriers.

Claim.—An improved compound braiding machine, constructed in manner and so as to operate as described, viz: as composed of the carriers A B C D E and their gears, the race circles I K L M N, the spring cams *i k*, and the recessed and cammed plates R R, the whole being arranged as set forth.

No. 58,831.—HENRY HUNGERFORD, New York, N. Y.—*Carpet Stretcher.*—October 16, 1866.—This stretcher admits of folding into small space. Its construction is explained by the claims and cut.

Claim.—First, the arrangement of the legs D and E, in combination with the pieces A and B, combined and operating substantially as and for the purposes set forth.

Second, the construction of the pressure foot C, substantially as described, so that the same can be reversed in position and be adapted to take hold of or act upon carpets or cloths of different kinds, as and for the purposes set forth.

Third, a carpet stretcher, constructed and operating substantially as and for the purposes set forth.

No 58,832.—CHARLES S. HUNTINGTON, Black River, N. Y.—*Horse Rake.*— October 16, 1866.—The rake is attached by links to the reaches by which it is drawn, and the supporting platform is placed upon independent bars, which are hinged to both the axle and the rake frame, so as to maintain a horizontal position however the rake itself may be affected by the irregularities of the ground. A bar with bent arms at each end is laid across the thills to support the rake for transportation, the ends of its draught frame resting upon the arms.

Claim.—First, as a carriage or riding attachment for revolving rakes, the detachable carriage herein described, the same being composed of—1. The wheels and the axle upon which they revolve. 2. The reachers or draught bars by which the carriage is drawn. 3. The platform bars arranged so as to maintain a horizontal position during the rising and falling of the reaches and rake frame.

Second, attaching the reachers and the platform bars of a rake carriage, as described, to the frame of a revolving rake, by hinge and link connections, or the equivalents thereof, in such manner that while the rake frame is free to vibrate, and the rake head to revolve, the platform bars shall maintain their level during such vibration and revolution, substantially as described.

Third, in combination with the thills and side bars K of the rake frame, the cross-bar R, constructed as herein described for holding the rake and frame from the ground.

No. 58,833.—RUSSEL N. ISAACS, New York, N. Y.—*Fishing Rod.*—October 16, 1866.— Enamel is applied to the guides through which the line passes on its way from the reel to the tip.

Claim.—The application to the metallic guides and tip of fishing rods an enamel, or covering of glass, porcelain, or any similar vitreous substance, to protect the line from friction and wear, substantially as described.

No. 58,834.—GOODMAN JENSEN, Brooklyn, N. Y.—*Ditching Machine.*—October 16, 1866; antedated October 7, 1866.—The rotary excavator wheel has spiral flanges on a conoidal hub, and lifts the soil, which is carried off by transverse conveyers and delivered upon the sides of the trench to form an embankment; the whole is mounted upon a floating scow propelled by a stern wheel.

Claim.—First, the rotary conoid excavator, formed with the series of scraping blades 1 J, and with the flange 2, at its largest end, against which the soil is scraped, in combination with a stationary trunk l, and scrapers to convey the soil from the conoid excavator, and deliver the same in the manner set forth.

Second, the arrangement of the gearing and pulleys for actuating the scrapers and rotary excavator, combined with said rotary excavator, substantially as specified.

Third, in combination with the rotary excavator and transverse conveyers, the stern wheel or propeller, and the boat or scow, the parts being fitted substantially as and for the purposes specified.

No. 58,835.—MELVIN JINCKS and F. ALTMEYER, Dansville, N. Y.—*Hearse.*—October 16, 1866.—The car is held by a hook until it receives its load with which it then travels forward, the foot resting on the spur of the rear platform and the cam bar thrown forward. In discharging, the latter is rotated backward, lifting the foot and throwing it out so as to be readily withdrawn.

Claim.—The combination of roller F, the bar H, board G, ear B, and the rails c, the whole constructed and operating in the manner and for the purpose herein specified.

No. 58,836.—ROBERT KERSHAW, Philadelphia, Penn.—*Producing Variegated Threads.*— October 16, 1866.—This thread is used in making fancy cassimeres, &c. As the central thread or cord is spun or twisted and wound, the lapping thread (of any other color) is guided to it by slots in a tube through which the cord is conducted : a sectional pinion on the driving shaft causes the delivery of the cord to be intermitted as desired, or the cord may be drawn back a short distance from time to time, or the thread guide may be reciprocated.

Claim.—Producing a variegated thread, by imparting to a central thread, or to a guide which conveys a lapping thread to the central thread, such a varying, irregular, intermittent or reversing traversing motion as will cause the lapping thread to be wound on the central thread in different quantities at different points.

No 58,837.—HENRY KEWLEY, Madison, Ohio.—*Stopwater for Oil Well Tubing.*—October 16, 1866.—A leather sack, enclosing a metal coil spring and filled with sand, is expanded between flanges on the well tube: a ring with friction rollers is interposed between the turning flange and the sack top. Dogs laterally projected by springs prevent the rotation of the sack.

Claim.—First, the spring H, the sack E, and the tube F, and the flange G, in combination with the washers I and J, in the manner and for the purpose set forth.

Second, the dogs K K, the springs M M, in combination with the sleeve F, in the manner set forth.

Third, the sleeve F, the sack E, and the washers J and I, in combination with the tube A, for the purpose and in the manner substantially described.

No. 58,838.—HENRY KRAUT, St. Louis, Mo.—*Adjustable Measure.*—October 16, 1866.— Either the receiving cylinder or the hollow measuring piston is movable, and adjustable as to the range of its longitudinal motion which discharges the contents. Its capacity at any given prolongation is indicated on a graduated bar which has an adjustable stop.

Claim.—The combination of the movable and adjustable receiving cylinder C, and hollow measuring piston A.

No. 58,839.—STEPHEN R. KROM, New York, N. Y.—*Apparatus for Separating Metals from Ores.*—October 16, 1866.—The framework supports a hopper, below which is an inclined trough with a perforated bottom through which a current of air is driven by the bellows beneath. The adjustable gates retard the upper stratum. The tailings are discharged from spouts on each side of the trough and the metal by the spout at the end.

Claim.—First, in combination with the intermittent blowing means and with two or more passages for the escape of the separated materials, the employment of one or more gates G, so arranged as to act on the material to retard the upper strata without retarding the lower strata as it passes down the incline, substantially in the manner and for the purpose herein set forth.

Second, in connection with the above, the perforated beds E and E¹, arranged at different levels, as herein specified, so that the material shall be separated or partially separated into distinct layers on the upper bed, and be accumulated in thicker strata and separated and led away in independent streams by the gate or gates on the lower bed, as herein set forth.

No. 58,840.—THOMAS C. LAMB, Chicago. Ill.—*Paint.*—October 16, 1866.—Textile fabrics are dyed without mordants and put into small pieces suitable for use in the application of the color as a cosmetic.

Claim.—The application of the textile fabrics or woven cloth, dyed with any dyeing stuffs not fixed, to the purposes of painting the human flesh.

No. 58,841.—RALPH G. LAMSON, Brownsville, Vt.—*Hay Rake.*—October 16, 1866.—The ends of the strap on the end of the pivoted lever have notches which engage with pins upon the rake head to hold it in working position; by a lateral motion of the lever the pins are disengaged and the rake rotates.

Claim.—The notched straps J, in combination with the spring lever H, operating with the pins K, in the rake head A, constructed and arranged in the manner and for the purpose herein specified.

No. 58,842.—GEO. W. LANE and J. A. BOLLES, Baltimore, Md.—*Means for Raising Sunken Vessels.*—October 16, 1866.—Spars or cradles are shackled to sling chains beneath the hull, and casks filled with air attached to said spars, until sufficient flotative power is attained. The water is expelled from the casks by air pumps above, after being attached to the spars.

Claim.—The combination of the spars B. chains A a, and pontoons or casks C, with the tubes and cocks, all operating substantially as and for the purpose specified.

No 58,843.—ALFRED B. LAWTHER and GEORGE F. LETZ, Chicago, Ill.—*Water Meter and Motor.*—October 16, 1866.—Each of the three radiating cylinders contains a movable piston attached by connecting rod to a common central crank; on the crank shaft is a rotating slide valve which acts in conjunction with the ports to admit water to the cylinders successively and allow the eduction of the water therefrom.

Claim.—Three or more cylinders radiating from a central chamber enclosing pistons connected with a central crank secured to a rotary valve or its equivalent, operating to regulate the flow of water to and from said cylinders, substantially in the manner and for the purpose herein set forth.

Also, three or more single-acting pistons when so combined with a central crank as to operate a rotary valve, and thus regulate and control a flow of water under pressure against said pistons, substantially in the manner and for the purposes herein set forth.

No. 58,844 —Z. W. LEE, Blakely, Ga.—*Cotton Bale Tie.*—October 16, 1866 —The band is made of a single length of hoop iron without notches or added pieces. One end is looped for the convenience of pulling tight by a hook, and lies beneath the other portion of the band, which in turn is bent and folded beneath its fellow.

Claim.—The metallic band B, having the bend b at one end and the loop b' at the other end, and applied substantially in the manner and for the purpose described.

No. 58,845.—B. E. LEHMAN, Bethlehem, Penn.—*Stop Cock.*—October 16, 1866.—The spigot has a transverse hole for the passage of the liquid when in coincidence with the passage way, and has also a lower valve for water, and an upper one for air when the condensed contents of the hollow plug are discharged to prevent injury by the freezing of the liquid.

Claim.—First, the plug made open at its lower end, with a transverse water passage, as and for the purpose set forth.

Second, the stuffing box D, in combination with the case A, and plug B, open at its lower end, and provided with the bevelled edge E. substantially as described.

Third, the waste valves a b, in combination with the plug B, constructed and operating substantially as and for the purpose set forth.

No. 58,846.—ADOLPHUS LIND, San Francisco, Cal.—*Water Wheel.*—October 16, 1866.—
The chutes above direct the water upon the curved buckets of the wheel below, and are closed,
to limit their capacity, by plates lowered by apparatus above.
Claim.—In combination with the stoppers H H¹ H², the stationary wheel C, and revolving
wheel A, substantially as described.

No. 58,847.—EDWARD A. LOCKE, Boston, Mass.—*Identifying Mark for Casks and
Boxes.*—October 16, 1866.—The metal label plate has a flange for expansion into the side of
a containing cavity, or is retained by a supplemental ring over its periphery, similarly attached
by its edge.
Claim.—A circular or other suitably shaped plate provided with identifying marks and
having a lip turned down from its edge, substantially as described.
Also, the employment of a circular or other proper shaped plate provided with such lip
driven into the wood at the sides of the cavity or depression in which the plate is inserted,
substantially as described.
Also, combining with such attaching plate one or more rotary identifying rings or plates,
operated in connection with the attaching plate, substantially as set forth.
Also, confining the attaching and identifying plates or rings together, so that while held
together they may be respectively rotated, substantially as described.

No. 58,848.—LEWIS E. LOCKLING, Perrysburgh, N. Y.—*Fence.*—October 16, 1866.—The
twisted metallic rod has loops for the reception of the rails, and is suspended from an arched
rod spanning the fence and planted in the soil or a foundation stone; boxes clasp the joining
ends of adjacent rails.
Claim.—The combination of the standard A with the suspended twisted rod C with loops c c,
supporting the boards or rails. &c., of the panel, substantially as described.
Also, in combination with the above, the box E, for holding the adjacent ends of the rails,
&c., substantially as described.

No. 58,849.—SAMUEL M. LONGLEY, Hudson, N. Y.—*Joint for Railroad Bars.*—October
16, 1866.—A short section of rail within the chair has side grooves for keys which lap past
the rail ends on each side. The upper edges of the ends of the rails are notched and overlapped by T-head projections on the short section.
Claim.—The rail section or connection C, having key ways b b, and overlapping the rails
at their top, in combination with the chair B and rails A A, reduced at their ends, the whole being
united by keys c, and constructed and arranged to establish the joint substantially as shown
and described.

No. 58,850.—SILAS H. LORING, Lawrence, Mass.—*Hose Coupling.*—October 16, 1866.—
The projections on the outer cylinder interlock with a groove in a collar on the inner cylinder.
Expansible packing is enclosed between collars and is forced against the outer cylinder by
compression. The hose is brought over the convex circumferentially grooved ends of the
coupling and confined by a screw-threaded ring.
Claim.—The expansion packing, Figs. 8 and 9, in combination with the dogs H and slots
C, in band D of Fig. 1, and the d gs I and slots J of Fig. 2, with the oval ring L, and
collar M, for the purposes herein set forth and described.

No. 58,851.—GEORGE D. and HORACE A. GOODRICH, Joliet, Ill.—*Clay Pipe Dies.*—October 16, 1866.—The core mould is revolvable upon a fixed shaft and protects from fracture the
passing cylinder of clay.
Claim.—The improvement in the machines of any construction used for the manufacture
of pipes by which a rotary or spiral motion is given to the clay or other suitable material in
process of manufacture, consisting in the die, with its revolving core detached from the feed-
in shaft and placed upon a stationary shaft, and its funnel-shaped face plate, constructed and
operating substantially as herein described and specified.

No 58,852.—JAMES W. MALOY, Boston, Mass.—*Marble Polishing Machine.*—October 16,
1866.—A longitudinally reciprocating and intermittingly rotating cylinder formed of one
piece or of a series of collars placed upon a shaft is used in connection with an inclined
table.
Claim.—First, the reciprocating cylinder B, when constructed and operating as and for
the purpose described.
Second, the grooved sleeve D, provided with a pawl p, or its equivalent, as described.
Third, the combination of an inclined table or platform H with the cylinder B, substantially
as described.
Fourth, the combination of cylinder B, grooved sleeve D, and the means of connecting the
same for joint operation, substantially in the manner and for the purpose specified.

No. 58,853.—JAMES W. MALOY, Boston, Mass.—*Machine for Cutting Granite.*—October
16, 1866.—The slide table consists in part of a rotatable disk for the support of the stone in
position to be worked by the rotary cutter.

Claim.—First, the combination of the revolving circular disk C with the sliding portion B of the table, as and for the purpose specified.

Second, the combination of the sliding table B, revolving disk C, and tool D, when constructed and operating substantially as set forth.

No. 58,854.—T. J. MARINUS, JAMES WHAIT, and WILLIAM WHAIT, Independence, Iowa.—*Carriage Harrow.*—October 16, 1866.—The harrows are attached to the tongue in front and have also attachment to rock shafts by whose rotary movement they may be raised.

Claim.—The barrows G G', in combination with the wheels A A and frame C D D', rock shaft F F', supporting link H, and chains L M N, when the several parts are constructed and operated in the manner and for the purpose set forth.

No. 58,855.—JAMES S. MARSH, Lewisburg, Penn.—*Plough.*—October 16, 1866.—The upper extension of the mould-board has a concave depression formed in its front to prevent choking by the accumulation of trash at that point.

Claim.—Constructing the mould-board of a turn plough with an upper extension *b*, having a concave depression *a* formed in it above the highest point of entrance into the ground, substantially as described.

No. 58,856.—JAMES S. MARSH, Lewisburg, Penn.—*Reaping Machine.*—October 16, 1866.—Relates to the arrangement of the seat beam relative to the frame and drive wheel; to the manner of combining the rake and reel with the obliquely placed cutting apparatus; to the particular construction of the main shoe-frame, and to the arrangement of gearing in relation thereto; explained by the claim and cut.

Claim.—First, arranging a seat F' upon a beam which is supported upon the outer end of the axle of the driving wheel, to allow a person to ride and control the machine when a continuously revolving combined rake and reel is mounted upon it, substantially as described.

Second, the mode of attaching the seat beam F to the draught pole, in conjunction with a device which is applied to said beam for adjusting the height of cut, substantially as described.

Third, the combination of a circularly sweeping rake or reel with a finger beam arranged substantially as described, so that the grain at the inner divider corner can be reached by the reel arms.

Fourth, the construction of the metal frame C, with a shoe C', substantially as described.

Fifth, adapting the metal frame C to support and serve as bearings for the spur wheels c^2, c^3, and also for the rake and reel arms, substantially as described.

Sixth, arranging the gearing which gives motion to the sickle upon a frame which is located on the grain side of the driving wheel and in advance of the raking and reeling attachment, substantially as described.

No. 58,857.—JOHN McCLELLAND, Washington, D. C.—*Street Washer.*—October 16, 1866.—A flanged box below contains the stop-cock, the discharge pipe connection and the pipe through which the key reaches the stop-cock; its flanges prevent its being raised by frost.

Claim.—First, the short case B, as constructed in combination with the stop cock A, pipes C and D, top plate or cap F, and key rod G, as described, for the purposes herein set forth.

Second, the flanges E E, on the case B, which covers the stop-cock A, to prevent it from being moved or lifted up by the action of the frost from the top, as described.

No. 58,858.—ROBERT McCONNEL, Lawrenceville, Penn.—*Carpenter's Bench.*—October 16, 1866.—The jaw slides are adjustable ratchet bars retained by toothed pawls in the post. The final pinch is given by a circular disk cam on the swivelled hub, acting against a contrate toothed disk on the outside of the jaw.

Claim.—First, the bench vice consisting of the toothed slides D E, jaws F, double inclined parts G and H, spring J, armed catches K L, bar P, arranged and operating substantially as described for the purpose specified.

Second, holding the toothed slides by means of the toothed catches K L, provided with arms *k' l'*, operating substantially as described for the purpose specified.

No. 58,859.—M J. McCORMICK, New York, N. Y.—*Vault Light.*—October 16, 1866.—The sections of the circular vault cover are alternately open and glazed with lenses. The cover is adjustable above a lower circle, partly glazed and partly open so as to bring the light sections into coincidence and leave ventilating openings or close the latter.

Claim.—The adjustable plate C, provided with openings *i i'*, the former being provided with lenses *j'*, in combination with the openings *a a'* in the plate A, the openings, *a* being covered with glass plates B, and the plate C fitted over the openings in the plate A, and all arranged substantially as and for the purpose set forth.

The ledges or ribs *d d'*, provided with grooved or concave upper surfaces, when used in connection with the plates A and C, substantially as and for the purpose specified.

No. 58,860.—O. C. McCUNE, Darby Creek, Ohio.—*Wool Press.*—October 16, 1866.—The fleece is laid upon the table, the slotted belt brought over it and attached to the treadle lever, whose depression draws the belt and brings up the twine-carrying fingers through the slots in the belt and over the fleece.

Claim.—First, the roller L, provided with the slotted teeth *b*, for the purpose of tying wool, substantially as herein described.

Second, the lever G, in combination with the roller L, lever O, and slotted curved teeth *b*, substantially as and for the purposes described.

Third, the strap D, in combination with the roller L, and teeth *b*, substantially as herein set forth.

No. 58,861.—JAMES McGEARY, Salem. Mass.—*Apparatus for Carburetting Air.*—October 16, 1866.—The air is admitted by a pipe, is deflected by a dome, and passes through openings in the floor of the same into the inner chamber, which has columns of wire cloth, covered with saturated fibre, and enclosed by a roof and sides of the same; the air passes out at openings near the floor of the inner chamber, through the fibrous clothing, and collects beneath the dome-shaped cover to which the exit pipe is attached.

Claim.—The casing A, for liquid hydrocarbons, in combination with the pipe for conducting the gas or air into such casing to act upon the hydrocarbons, when such a pipe terminates in a perforated dome D, substantially as and for the purpose described.

Also, the vertical tubes G, in combination with the above, substantially as and for the purpose specified.

Also, the surrounding cylinders or casings I and M with the outer one M, covered with a web O, in combination with the gas pipe B, terminating in a dome D, either with or without the vertical tubes G, substantially as described and for the purpose specified.

Also, feeding the liquid hydrocarbons to the web covering of the perforated cylinder at a point near the bottom of the cylinder I, substantially as and for the purpose described.

No. 58,862.—E. McKINNEY, Clarksville, Tenn.—*Coal Oil Stove.*—October 16, 1866.—The oil floating upon the water is pressed upward toward the burner by the elevated column of water from the reservoir.

Claim.—The feeding or supplying of petroleum to stoves or furnaces by means of water or other fluid of a greater specific gravity than the petroleum, placed in an elevated tank or reservoir above the petroleum chamber, and communicating with the latter in such a manner as to feed the petroleum to the fire pan or fire chamber by static pressure, substantially as described.

No. 58,863.—JAMES H. MEARNS, Philadelphia, Penn.—*Furnace.*—October 16, 1866.—When the upper grate is agitated by the rock shaft the ashes fall on to the lower one, which acts as a sifter, allowing the dust to pass and delivering the cinders into a pan; the cloud of finer particles is drawn into the chimney.

Claim.—First, the grates P and J, in combination with the shaft H, its arms *e e'*, and link *d*, or their equivalent, the whole being arranged within the ash pit of a heater, and operating substantially as and for the purpose described.

Second, the flues or openings *m m*, communicating with the ash pit and with the casing above the fireplace, for the purpose specified.

Third, the combination of the crank shaft I, the sliding bars *i i*, and the shaft H, its arms *e e'*, and link *d*, as and for the purpose set forth.

Fourth, the adjustable bar K, in combination with the lower grate or sifter J, substantially as and for the purpose specified.

No. 58,864.—W. R. MEINS, Boston, Mass.—*Apparatus for Supplying Liquor to Centrifugal Machines.*—October 16, 1866.—The vessel has an exterior gauge tube extending from near the bottom to the top, and a flexible eduction tube which terminates in a rose.

Claim.—For use with a centrifugal sugar-bleaching machine, a portable liquoring bucket having a construction substantially as described.

No. 58,865.—DANIEL MENDENHALL, Fairfield, Iowa —*Hand Loom.*—October 16, 1866.—The bracket is loosely connected to the batten and slides on a guide pin; in its motion it lifts the heddles in order by forcing into its upper slot the pin, which with its heddle has been previously partially lifted by the cylinder. The sub-treadles are so arranged that the tappets act alternately on them and on the upper treadles, the jack serving to turn the cylinder.

Claim.—First, the sliding bracket L, with its openings and wedge M, and the pins *v v v v*, all constructed and operating in the manner and for the purposes set forth.

Second, the combination and arrangement of the treadles 1, 2, 3, 4, and the sub-treadles *z z z z*, with the tappet cylinder *o*, and jack *i*, as and for the purposes set forth.

Third, the hinged levers *m m'*, and tappets *x n*, in combination with the batten and with the rocker arm I, head H, cords *g g*, and picker staff F F, said several parts being respectively constructed and the whole arranged for use as described.

No. 58,866.—DANIEL MENDENHALL, Fairfield, Iowa.—*Tree Protector.*—October 16, 1866.—A leather trough is fastened by belt and buckle, conforms to the irregular shape of the tree, and holds water to prevent the passage of insects.

Claim.—The arrangement and combination of the parts herein described, constituting a fruit-tree protector, substantially as set forth and described.

No. 58,867.—S. F. MERRITT, New York, N. Y.—*Eye-glass Suspender.*—October 16, 1866.—A spring hook is attached by the usual brooch pin to the garment, and is used to suspend the eye-glass.

Claim.—As a new article of manufacture, an eye-glass-holder, consisting of the shank A, spring hook B, bars C, eye a, and pin D, substantially as described for the purpose specified.

No. 58,868.—CHARLES F. MIETZSCH, Philadelphia, Penn.—*Filter.*—October 16, 1866.—The liquid descends through the filtering material, and passing down through a mass of charcoal and a lower filter ascends to the outlet, which is below the level of the upper filter.

Claim.—A filter having two or more chambers or compartments containing filtering material, so arranged in respect to each other, to an outlet pipe, and to a lower chamber that the fluid to be filtered must pass downward through the material in one compartment to the lower chamber, and then upward through the material in the other chamber to the outlet pipe, without rising above the surface of the filtering material in the second chamber, as and for the purpose described.

No. 58,869.—DAVID T. MILLER, Dayton, Ohio.—*Sorghum Evaporator.*—October 16, 1866.—In one end of the pan is placed a coil of steam pipe to keep the sirup at that end of the pan in a state of ebullition, and cause a circulation of the juice from one end of the pan to the other.

Claim.—The evaporator, constructed, arranged, and operating substantially as and for the purpose set forth.

No. 58,870.—J. R. MOFFITT, Chelsea, Mass., and F. D. HAYWARD, Malden, Mass.—*Protecting Rubber Articles.*—October 16, 1866.—The surface of the rubber is protected from the air by coating with paper applied before vulcanization.

Claim.—Protecting the surfaces of articles made of caoutchouc or gum elastic compounds by surfacing them, substantially as set forth.

No. 58,871.—HIRAM E. MOON and JOSEPH DOAN, Wilmington, Ohio.—*Ditching Machine.*—October 16, 1866.—The progression of the machine is performed by a rope on the driving shaft of the excavating wheel; the power is applied by a sweep on the main shaft; the revolving digger is flanged around its outside; revolving arms carry brushes to clear the tooth cavities. The frame is carried on rollers, which have flanges to cut out the sides of the ditch; the front end of the frame has marking revolving cutters; a steering block slides in the ditch at the rear of the frame. The outside force of the sweep counterbalances the side draught.

Claim.—First, the ditching wheel F, constructed with the cutters *f f*, open throats inside thereof, and projecting rim *s*, bounding the whole outer periphery of the wheel, arranged and operating substantially as and for the purpose herein specified.

Second, the cleaner G, in combination with the ditching wheel, arranged and operating substantially as herein set forth.

Third, the marking knives S S, in combination with the ditching wheel, substantially as described.

Fourth, the combination and arrangement of the cutting or penetrating guide wheel P, feeding rope N, anchor H, and spool E, substantially as and for the purpose herein set forth.

Fifth, the rudder T, provided with the bridge *t*, for the purposes specified.

Sixth, such a combined arrangement of the sweep gearing and ditching wheel that the twisting force of the power applied and the side draught of the machine shall counteract and nearly counterbalance each other, substantially as set forth.

No. 58,872.—WILLIAM MOREHOUSE, Buffalo, N. Y.—*Mop Holder.*—October 16, 1866.—One of the jaws is cast solid with the ferrule, the other is pivoted and closed against the stationary one by a spring collar.

Claim.—First, the combination of a loose and a fixed clamping jaw, the former being connected to the ferrule B of the latter by means of a hooked shank *b*, substantially as described.

Second, the construction of a ferrule B upon the shank of the jaw A for receiving the handle C, and also the shank of the movable jaw A', substantially as described.

Third, the combination of a sliding collar D, which is acted upon by a spring *c*, with the fixed and movable jaws A A', substantially as described.

No. 58,873.—CHARLES MOYER, Jr., Coopersburg, Penn.—*Wood-bending Machine.*—October 16, 1866.—The flexible metallic band binds the strip upon the "former," and has an eye on its end through which the strip is passed to be grasped by the clasp on the roller, which forms the front curve of the runner.

Claim.—The link *e* in the flexible strap C, to operate in combination with the former A and roller *a*, or their equivalents, substantially as and for the purpose described.

No. 58,874.—JOHN MURPHY, New York, N. Y.—*Roller for Clothes Wringers.*—October 16, 1866.—The wringer is composed of layers of fabric and vulcanite. The fabric is woven with an oblique weft, is cut bias, and the threads of the wrapper when wound have an equal obliquity to the axis of the roller.

Claim.—The fabric C mounted within the mass of gum so as to form a compound elastic roll, substantially of the character and for the purpose herein set forth.

No. 58,875.—JOHN MURRAY, New York, N. Y.—*Whistle for Steam Engines.*—October 16, 1866.—To vary the intonations, and thus communicate audible signals to brakemen or helmsmen, the lever is adjusted vertically, bringing the sliding sleeve of the bell nearer to or farther from the issuing orifice of the steam jet.

Claim.—The sliding sleeve H, the gimbal or levers I K, the post O, and spring P, when in combination with the whistle of a steam or other engine, substantially as and for the purposes described.

No. 58,876.—CARLISLE C. MYERS, Sterling, Ill.—*Corn Planter.*—October 16, 1866.—The seed slides are connected to levers having lips to close the lower end of the seed tubes while scattering grains are falling.

Claim.—The levers F, provided with lips *g*, mounted upon the outside of the tubes *f*, and operating in connection therewith, and with the slides E, as and for the purpose set forth.

No. 58,877.—DAVID MYERS, Chicago, Ill.—*Car Brake.*—October 16, 1866.—When the bell cord is pulled from the rear a catch is thrown in such position as to prevent tension on a rope leading to a brake-operating device; but when the cord is pulled from the forward end by the engineer, the brakes are put on, as the catch will not be put in operation.

Claim.—First, controlling the movements of the sliding block B, by means of oscillating levers F, substantially as and for the purposes shown and described.

Second, operating the levers F, by means of the arms H, friction wheels G, and springs I, for holding the arms upon the said wheels, substantially in the manner and for the purposes specified.

Third, the combination of the above mentioned parts with the sliding block B, and the cord *b*, connected with the apparatus beneath the car, arranged and operating substantially as specified and for the purposes set forth.

Fourth, in combination with the above, the arrangement of the ball, and operating substantially as and for the purposes shown and described.

No. 58,878.—WALTER NAGEL, Philadelphia, Penn.—*Mortising Machine.*—October 16, 1866.—The cutter head has an oscillating motion from a cog wheel and segmental rack, operated by a bell crank on the main shaft. The guides of the bed are carried by pivoted parallel bars, allowing an even adjustment by T-arms, secured by set screws in segmental slots.

Claim.—First, operating the rotary reciprocating cutter head M, through the medium of the bars L L, connected to the cutter head and cross head K, as shown and actuated by the segment rack H and the pinion I, on the shaft J, or their equivalents, substantially as and for the purpose set forth.

Second, the attaching of the guides or ways *g g*, of the bed P, to adjustable parallel bars S S', arranged substantially as shown and described, for the purpose of adjusting the stuff R to be mortised in a proper relative position with the cutter head.

No. 58,879.—HEZEKIAH NAYLOR, Pekin, Ill.—*Caster for Furniture.*—October 16, 1866.—The caster ball rests against the faces of three spheres, whose axes are arranged in a horizontal plane and form an equilateral triangle above and around the caster ball.

Claim.—A caster in which the caster ball B turns on three friction balls C, which are placed in the angles of an equilateral triangle, turning on journals which have bearings on both sides of said balls, in recesses formed for the purpose in the top of the caster, substantially as described.

No. 58,880.—JOSEPH P. NOYES, Binghamton, N. Y.—*Drop Press.*—October 16, 1866.—The hammer is suspended from a crank and raised by the revolution of the shaft, when a drum on the latter is engaged by a lever clutch within it; when the hammer reaches its full height the clutch is tripped loose and the hammer falls; the clutch then again becomes operative. A cord releases the clutch and hammer at any time.

Claim.—First, the arrangement of the lever catch *d*, shoe *e*, and arm *h*, in combination with the drum G, crank shaft C, and hammer I, constructed and operating substantially as and for the purpose described.

Second, the spring top K, in combination with the lever catch *d*, drum G, and hammer I, constructed and operating substantially as and for the purpose set forth.

No. 58,881.—W. H. PIERSON, West Jersey, Ill.—*Beehive.*—October 16, 1866.—Explained by the claim and cut.

Claim.—The combination and arrangement of the case A, grooved and perforated partitions a, breeding boxes B, with perforation b, spare boxes D, with perforations c, entrance f, box E, alighting board a*, tube F, and tube G, substantially as described, as and for the purpose specified.

No. 58,882.—CHARLES PINDER, Lowell, Mass.—*Egg Beater.*—October 16, 1866.—Loops of sheet metal are suspended from the cover, inside the case, and cut the contents, when the cylinder is agitated longitudinally.

Claim.—The combination with the case A and cover B of the hoops or strips C C (two or more,) the latter so arranged as to accomplish the purposes herein specified.

No. 58,883.—THOMAS D. POWERS, Rochester, Wis.—*Car Coupling.*—October 16, 1866.—The entering link presses back a sliding block, which detaches the hook of the coupling-pin lever; the lever vibrating, the pin couples with the link. Guide springs hold the link in horizontal position.

Claim.—The sliding block G, having the guide springs A B attached, the pivoted lever D, with the bolt C attached, and the spring catch F, when arranged to operate as shown and described.

No. 58,884.—H. A. RAINS, Nashville, Tenn., and A. P. ADAMS, Newark, N. J.—*Cart Saddle.*—October 16, 1866.—A metallic bridge with moulded edges sets upon the tree of the saddle, and the girth hooks are covered by metallic hinged flaps.

Claim.—First, a cap and flap of metal constructed and arranged substantially as shown.

Second, in combination with metallic clasps and flaps, a metallic covering for the bridge piece, with the moulding rolled or wrought upon it, and made to imitate patent leather in whole or in part, substantially as herein described.

No. 58,885.—GEORGE W. RAITT, Cincinnati, Ohio.—*Dress Guard for Carriages.*—October 16, 1866.—A cylindrical case has a flap, pierced for attachment to the dash-board, and is slotted for the passage of an apron that is wound upon a central spring shaft. The end of the apron may be drawn out for attachment to the seat by hooks.

Claim.—An extensible guard or screen for attachment to carriages, substantially as set forth.

No. 58,886.—F. RAMSEY and JAMES MILLER, New York, N. Y.—*Tool for Cutting Boiler Tubes.*— October 16, 1866.—The hollow mandrel has an enlarged hollow end carrying sliding cutters, which are driven out by a conical-headed central rod that is projected or withdrawn by a screw thread upon it and the mandrel. An adjustment of the collar allows the operations of the cutter at any point desired. The cutter is driven by a belt or cord.

Claim.—First, the combination of the shank D, screw shaft M, with tapering end F, cap J, cutters L L, and spring N N, constructed and operating substantially as described, for the purpose specified.

Second, the flanged adjustable collar O in combination with the shank D, substantially as and for the purpose specified.

No. 58,887.—JOHN O. REILLEY, Baltimore, Md.—*Machine for Making Spikes.*—October 16, 1866.—The levers which actuate the vertically moving dies are thrown out of action by bending upward their jointed ends, so as to place them beyond the range of the cams. The gauge intercepts the end of the bar and determines the length of the blank, and is retracted before the header advances. The cutter cuts off the blank after it is gripped by the side die.

Claim.—First, in combination with the vertically moving dies U and f and the levers W b, the pivoted arms Y c, operating substantially as described and represented.

Second, the gauge K, arranged in relation to the moving die D and the header, as described.

Third, in combination with the moving die C, adjustable for various lengths of spikes, the cutter O, arranged in ways alongside the stationary die, and operating against the plain face of the moving die after the iron is gripped between the dies, substantially as described

No. 58,888.—A. O. REMINGTON and V. R. STEWART, Weedsport, N. Y.—*Water Elevator.*—October 16, 1866.—The flanged belt pulley is clutched by a sliding feather-keyed collar, which is kept in lock by a spiral spring. The depression of a lever pivoted in a cross-bar unlocks the clutch and brakes the revolution of the belt pulley as the bucket descends.

Claim.—The stationary pulley F, fitted loosely on the shouldered axle B, but without lateral motion, operating in combination with the sliding collar I and lever L, whereby the bucket will be elevated or lowered in a line vertical to the pulley F, in the manner substantially as and for the purpose specified.

No. 58,889.—C. ROSENBERRY, Chicago, Ill.—*Cork Extractor.*—October 16, 1866.—By the sliding wire the spring claws are clasped upon the cork, which is then extracted from the bottle.

Claim.—The sliding wire I, with its ring B, for the purpose of closing or loosening the claw M, in combination with the same, the whole constructed and operating in the manner herein described and specified.

No. 58,890.—WM. M. RUNYON, R H. HALLER, and D. B. MORRIS, Oskaloosa, Iowa.—*Medicine for Hog Cholera.*—October 16, 1866.—Composed of madder, sulphur, rosin, saltpetre, black antimony, copperas, assafœtida, black pepper, and arsenic.

Claim.—A composition of matter, compounded of the above enumerated ingredients, or their chemical equivalents, and prepared for use, substantially in the manner and for the purpose set forth.

No. 58,891.—CHARLES F. RUSET, Communipaw, N. J.—*Revolving Cylinder Engine.*—October 16, 1866 —The drum rotates in bearings, the steam passing in and out at the respective trunnions. The cylinders oscillate in bearings in the heads of the drum, and revolve with it. The reciprocating pistons are connected by cranks to planetary gears, which mutually and severally engage a fixed sun-wheel on the central axis. Power is transmitted by a belt on the drum.

Claim.—The combination of the wheel or drum C, arranged to rotate on suitable bearings, engines carried by said drum, with their revolving shafts G hung therein and arranged relatively to the driving axis as described, planet wheels H, and stationary circular rack or sun-wheel I, substantially as shown and described.

No. 58,892.—CYRUS W. SALADEE and T. R. EDDY, Newark, Ohio.—*Chimney.*—October 16, 1866.—The sections of the chimney are connected by flanges with rims for the joint cement. The section above each floor has a foot flange and receives the section beneath, which is insulated from it and the floor by cement. The chimney top has a similar connection to the sections, but somewhat enlarged.

Claim:—First, constructing chimneys for houses of hollow sections, made of fire-proof clay or other similar material, and joining or cementing the same together in the manner and for the purpose substantially as shown and described.

Second, the manner, shown and described, of making each separate floor of the building sustain its proportion of the weight of the chimney, substantially as and for the purpose specified.

Third, th base B¹ on the top section E of the chimney (Figs. 1 and 2, plate 1) for the su port of the chimney top F, in the manner and for the purpose substantially as shown and described.

No. 58,893.—JOHN F. SANFORD, Keokuk, Iowa.—*Lamp Burner.*—October 16, 1866.—Through openings in the shell of the burner a file can be operated to sharpen the teeth of the spur wheel used for raising the wick. The wheel is journalled in a yoke, which is adjusted toward the wick as the wheel is reduced by frequent sharpenings.

Claim.—First, such a construction and arrangement of parts, that while they admit of the sharpening of the spurs of the wheels which move the wicks, will also permit the lighting of the lamps through apertures in the cages of the burners without the necessity of contrivances for depressing the wicks for that purpose, substantially as above described.

Second, in combination with the apertures through the cage of a lamp burner, so constructed and arranged as to admit the sharpening of the spurs upon the wheels which move the wick without moving or displacing any of the parts, a contrivance for causing these spur wheels to engage with the wick, substantially as and for the purpose above described.

No. 58,894.—LORENZO SAUTER, Jersey City, N. J.—*Breastpin.*—October 16, 1866.—The body may contain hair or a miniature, and has openings which may be exposed or covered by the rotation of the perforated and adjustable disk above.

Claim.—The centrally pivoted shield B, furnished with openings as described, and combined with the body A, provided with suitable ornaments, substantially as herein set forth, for the purpose specified.

No. 58,895.—JAMES J. SAWYER, Woodstock, Conn.—*Dish Washer.*—October 16, 1866.—The vessel has a circular concentric lip rising from its bottom to support the dishes against the side. A revolving dasher throws the water against the dishes. For special uses a false annular bottom is inserted.

Claim.—First, the resting lips B on the bottom plate C of the receptacle A, arranged to operate with the floats G therein, as described, for the purpose specified.

Second, the removable false bottom K, in combination with the receptacle A, as described, for the purpose specfied.

No. 58,896.—PETER SCHULER, Philadelphia, Penn.—*Piano-forte.*—October 16, 1866 —Instead of gluing the sounding-board rigidly to the frame, the large end is slipped into its seat between elastic strips in the rabbet in the blocks of the frame, and the smaller end screwed down to the block at the other end of the case, permitting the changes due to hygrometric conditions without "buckling."

Claim.—Securing the sound board of a piano between elastic or compressible bearings c^1, c^3, so that one end of the same may slide between its bearings, substantially in the manner described and set forth for the purpose specified.

No. 58,897.—AMOS R. SCOTT, Bethel, Ohio.—*Wagon Bow Fastening.*—October 16, 1866.—The upper ring staple on the bow catches over the upper rail; the lower staple, also fast to the bow, has a cam catch holding below the lower rail; the cam lever is secured by a slip ring on the bow.

Claim.—First, the upper fastening E, and the lower fastening F G K H I, when said fastenings are constructed and arranged substantially as described, in combination with the bow of a wagon, for the purpose set forth.

Second, tightening the bow upon the wagon body by means of a lever and cam or eccentric, substantially as described and for the purpose set forth.

No. 58,898.—WILLIAM SELLERS, New York, N. Y.—*Lock.*—October 16, 1866.—The rib or flanged edge of the oscillating bolt traverses the enlarged ends of the slots in the lid and lock plates, and binds upon the flanges at the narrower portion of the lid plate.

Claim.—The bolt B, with the rib f, constructed as shown in figures 4 5 and 6 of the drawings, and arranged to operate with the oblong slotted plate g, for the purpose herein specified.

No. 58,899.—JOHN T. SEVERNS, Burlington, N. J.—*Fruit Box.*—October 16, 1866.—Explained by the claim and cut.

Claim.—As a new article of manufacture, fruit boxes, the sides of which are formed by a single piece of thin steamed wood having rounded corners formed by two or more internal kerfs sawed partly through the board and bottom, constructed and inserted substantially as set forth.

No. 58,900.—SIMEON SHERMAN, Weston, Mo.—*Rotary Engine.*—October 16, 1866.—The piston and shaft rotate in the cylinder, which has a rotary abutment valve actuated by a cam on the shaft to admit and exhaust steam. The condenser beneath the cylinder receives the steam from the exhaust ports, and the cold water is received from a chambered valve on the main shaft, which acts as a pump. To the abutment valve is attached a spring arm, which is strained up or down and locked on the position to adjust the valve for the forward or reverse motions of the piston.

Claim.—The spring W, arranged as described and shown.

Also, the arrangement of the condenser in direct connection with the two exhaust ports, guarded by a reciprocating rotary abutment valve.

Also, the described arrangement of the rotary engine, condenser and chambered valve T, operating as described.

No. 58,901.—EDWIN L. SIMPSON, Bridgeport, Conn.—*Rubber for Dental Purposes.*—October 16, 1866.—The vulcanizing composition consists of linseed oil, sulphur, and gum benzoin.

Claim.—Combining the within described vulcanizing compound with India-rubber, in the proportions herein named, and substantially in the manner and for the purpose specified.

No. 58,902.—EDWIN L. SIMPSON, Bridgeport, Conn.—*Manufacture of India-rubber, Gutta-percha, &c.*—October 16, 1866.—The composition is made by heating together linseed oil, gum benzoin, and sulphur.

Claim.—First, the herein described compound of vegetable oil, sulphur, and benzoin gum, prepared substantially as and for the purpose specified.

Second, combining the herein described compound with India-rubber, gutta-percha, or other similar gum or gums, substantially as and for the purpose specified.

No. 58,903.—CHARLES THOMAS SMITH, Utica, N. Y.—*Bobbin for Spinning Machines.*—October 16, 1866.—The claim defines the invention. The object is durability and compactness.

Claim.—A filling bobbin, the tube or barrel of which is formed or manufactured of sheet zinc with a wooden head or flange, as above described, the bore of the barrel at its tip being made tapering, in the manner and for the purpose as herein set forth.

No. 58,904.—JACOB SNYDER, Wheeling, West Va.—*Puddling Furnace.*—October 16, 1866.—A wrought iron plate is substituted for cast iron, dispensing with the use of "scrap."

Claim.—A puddling or boiling furnace with the bottom of its boiling chamber constructed of wrought iron in a single plate or otherwise, substantially as described.

No. 58,905.—GEORGE W. SPANGLE. Clifton Springs, N. Y.—*Burning Fluid.*—October 16, 1866.—Composed of gasoline, 40 gallons; sal soda, 1 pound; and cream of tartar, 1 pound. The offensive odor may be disguised by a fragrant essential oil.

Claim.—First, the method above described for rendering any of the products obtained from petroleum inexplosive and safe as a burning fluid, by the use of sal soda and cream of tartar, substantially as above described.

Second, the removal of the unpleasant odor of any of the above-mentioned products, by the use of the oil of wintergreen, substantially as described.

No. 58,906.—FREDERICK SPECK, Waynesborough, Penn.—*Water Wheel.*—October 16, 1866.—The wheels are attached respectively to the central shaft and to the sleeve. The water enters on the outside of the outer wheel and is delivered from the latter to the inner wheel, which has a lower discharge.

Claim.—The combination of the outer and inner wheels as constructed and arranged together, whereby the water first passing through the buckets of the outer wheel acts directly upon the buckets of the inner wheel, substantially as and for the purpose set forth.

No. 58,907.—JAMES F. SPENCE, Brooklyn, N. Y.—*Shade for Protecting the Eyes.*—October 16, 1866.—The lunate shade is attached by clips to the flexible fillet.

Claim.—First, the elastic clasp or band A in combination with the shade B, substantially as herein set forth for the purpose specified.

Second, the construction of the shade whereby a space is left for the passage of air between the shade and the forehead, substantially as herein set forth for the purpose specified.

No 58,908.—DAVID I. STAGG, New York, N. Y.—*Seat and Desk.*—October 16, 1866.—The back of the seat is pivoted to allow of swinging the seat upward into a reversed position, when its under side forms a desk. This movement discloses an under seat of proper height for a smaller pupil.

Claim.—The reversible and adjustable back C provided at one end with the desk and seat D and secured between the side pieces A A in combination with the adjustable seat B between the upright side pieces A A, substantially as and for the purpose set forth.

No. 58,909.—W. F. STARNES, Macomb, Ill.—*Machine for Cutting Stalks in the Field.*—October 16, 1866.—The stalks are broken down by a roller, straightened by book bars and cut by a reciprocating knife.

Claim.—First, the roller F and the hooks H arranged as shown, to break down and straighten the stalks, as set forth.

Second, the reciprocating knife R operated by the crank shaft D and the gear wheels E and C, for the purpose of cutting the stalks, as set forth.

No. 58,910.—THOMAS STEAD, Cleveland Ohio.—*Railroad Signal.*—October 16, 1866.—One of the semaphore posts is movable and is capable of presenting its arms and variable signal lights in any direction to suit oblique crossings, &c. The other post is similarly furnished and is stationary. The manipulations are made by wheel and chain, disk, and capstan.

Claim.—First, the herein described arrangement of the posts A B in relation to each other and the track, in combination with the lanterns E and K, when constructed and operated as and for the purpose set forth.

Second, the wheels G and H and chain G' in combination with the post B and capstan H', lever I, and disk I', when arranged and operated as and for the purpose set forth.

No. 58,911.—CARLISLE ST. JOHN, Keosauqua, Iowa.—*Plough.*—October 16, 1866.—The landside is reversible, one end being formed as a coulter, and the other having the same curve as the fore end of the mouldboard. The beam at its rear end has a corrugated slotted plate by which its inclination to the landside may be adjusted laterally.

Claim.—First, a landside that may be changed end for end, on one end of which is a cutter so constructed that the cutter may be used or not as desired, for the purposes and substantially as described.

Second, the corrugated plate G¹ and G², the plate G¹ being provided with a strap and socket in combination with the beam S and brace rod E, for the purposes and substantially as described.

No. 58,912.—SOLOMON STUKEY, Sugar Grove, Ohio.—*Machine for Grinding Cob and Corn.*—October 16, 1866.—The frustum, having horizontal convex knives, revolves in a cylinder with horizontal knives attached to its inside. The grinding faces are studded between the knives.

Claim.—The construction of the conical cylinder and concave with their curved knives G and H and projections K, so arranged as to cut and grind the cob and corn, as herein described.

No. 58,913.—G. TAGLIABUE, New York, N. Y.—*Hydrometer.*—October 16, 1866.—Explained by the claim and cut.

Claim.—A hydrometer having a lump of metal, or other suitable material, firmly secured to the inner surface of the bulb, substantially as and for the purpose described.

No. 58,914.—S. H. TIFT, Morrisville, Vt.—*Spring Bedstead.*—October 16, 1866.—The bed bottom has straps to limit its upward movement and is supported from the box bottom by springs which encircle pins depending from the lower board.

Claim.—The oblong box A as constructed with the top C with its permanent pins D, straps J, springs E, and solid bottom G with its apertures H, when arranged and combined substantially as described and for the purpose set forth.

No. 58,915.—FREDERICK JOHN TINKER, Cincinnati, Ohio.—*Lamp Chimney Attachment.*—October 16, 1866.—The flanged base is held in its socket by catches which lap upon it, and whose shanks are shiftable in eccentric slots to adjust the catches out or in; when adjusted they are locked by nuts beneath the collar.

Claim.—First, the eccentrically slotted collar B C, shifting catch D E, and nut F, combined and operating as set forth.

Second, the provision of the concavity *e* on the under side and notches *e'* on the inner margin of the catch E, for the purpose stated.

No. 58,916.—JOHN TINKEY, New Haven, Conn.—*Door Guard.*—October 16, 1866.—The guard has an elastic cushion and is attached to the base board to receive the impact of the door and hold it open by engaging the keeper placed in the edge of the door. The device may be placed in the jamb to engage the keeper and hold the door closed.

Claim.—The combination of the bolt C, constructed with flanges *d* and *d'*, so as to form shoulders on the said bolt keeper E, constructed and arranged to operate substantially in the manner and for the purpose set forth.

No. 58,917.—JACOB UNGERER, Brooklyn, N. Y.—*Chair.*—October 16, 1866.—A metal frame is sunk within the wooden frame and has a seat of interlaced metallic strips with auxiliary springs beneath.

Claim.—First, the combination of the metal frame B with the metallic seat A, substantially as and for the purpose described.

Second, the springs and spring pads *a* in combination with the seat A and chair frame C, and operating substantially as and for the purpose set forth.

No. 58,918.—ISAAC VAN OLINDA, Brooklyn, N. Y.—*Communicating Reciprocating Motion to Pumps, &c.*—October 16, 1866; antedated October 5, 1866.—The band lever is pivoted at the foot and connected by slotted arms and a collar to the hollow pump shaft, which reciprocates vertically; the collar has slide guides and friction washers.

Claim.—The forked and slotted arm *f* of the lever E, and the slotted vertical guiding standards I, applied in combination with each other and with the collar *r*, pins *d*, and friction rollers *i*, substantially as herein set forth.

No. 58,919.—L. H. VAN SPANCKEREN, Muscatine, Iowa.—*Manufacture of Soap.*—October 16, 1866.—Composed of lye (of 30°,) 11 pounds; grease, 6 pounds; potatoes, 3 pounds; flour, 12 ounces; gum tragacanth, 4 ounces; and the yolks of 8 eggs.

Claim.—A soap compounded and prepared from the ingredients and in the manner substantially as set forth.

No. 58,920.—H. N. WALTER, Norwich, N. Y.—*Skirt Elevator.*—October 16, 1866.—Cords from a band near the lower edge of the skirt pass through a ring near the waistband; being drawn through the rings the cords are held by a catch, whose release allows the dress to assume its normal position.

Claim.—The combination of the plate A, tube *b*, valve D, having a concave part *e*, lever *f*, spring *g*, hood *n*, shield *i*, all substantially operating as above described.

No. 58,321.—CHAS. R. WARNER and MOSES BALES, London, Ohio.—*Sawing Machine.*—October 16, 1866.—The saw shaft admits of attachment to its working slide-block, so that the saw shall cut horizontally or vertically. The main frame may stand in a vertical or horizontal position, in respect to the saw guides, for cutting up or felling timber.

Claim.—The arrangement of the guide R, rocking block P, reversible saw shaft N, and reversible frame D E F, when constructed as and for the purposes set forth.

No. 58,922.—ROBERT WIER, Cohoes, N. Y.—*Toy Walking Figure.*—October 16, 1866.—The feet of the jointed toy figure are pivoted to a revolving disk, and the motion communicated from the legs to the arms by diagonal connection.

Claim.—Combining the body and jointed limbs of a figure with a revolving axle *f*, by means of crank pins *h h* and wire W, or their equivalents, in such a manner as to produce an alternate bending of the knee joints and other movements of the limbs, substantially as herein described and set forth.

No. 58,923.—WM. H. WARWICK, Dunlevy, Ohio.—*Machine for Furrowing Corn Ground.*—October 16, 1866.—V-shod runners are attached to a frame, having shafts and handles. Adjustable plates regulate the depth of furrows; the runners are adjustable at the back end by hinged plates and jam nuts.

Claim.—First, in combination with the vertical plates or pieces, or runners, the slide guards *f*, for limiting the depths of the furrows as recited.

Second, in combination with said pieces or runners, the fender plates *i*, constructed and operating substantially as described.

No. 58,924.—THEODORE L. WEBSTER, Brooklyn, N. Y.—*Tube-sheet Cutter.*—October 16, 1866.—The centre point is projected by a s ring, beyond the face of the annular cutter; the spring and the shank of the centre are socketed in the stock of the cutter.

Claim.—As an article of manufacture, a tool for drilling metals, composed of a circular cutter and a yielding centre, constructed and arranged in the manner described.

No. 58,925.—ALBIN WARTH, Stapleton, N. Y.—*Sewing Machine.*—October 16, 1866.—The feed is by the needle; the chain or lock-stitch, Grover and Baker's stitch, or a coiled lock-stitch may be made at option. A double pointed shuttle describes a circular path in either direction, and may describe two revolutions to one ordinary descent of the needle, or make the same coiled stitch with one revolution by the aid of other devices. The increased velocity is given to the shuttle by sliding multiplying gears into gear with the main shaft. The devices are too numerous to describe.

Claim.—First, providing the oscillating arm, which takes up the slack of the needle thread, with an oblong slot, substantially as described, so as to keep the loop of the needle thread open to let the shuttle pass twice.

Second, the needle *n*, provided with a slotted shank, substantially as shown in Fig. 20.

Third, the rough-surfaced clamp *j*, moving from below in opposition to the spring pressure foot above the goods, and operating substantially as described, to hold the material while the stitch is being finished.

Fourth, the double-pointed shuttle, as shown in Fig. 25, and arranged to work continuously in either direction.

Fifth, the ridge *a'*, on that side of the shuttle which faces the needle, substantially as and for the purpose described.

Sixth, the elastic centre *c'*, operating as described, in combination with the revolving shuttle, constructed and operating substantially as and for the purpose described.

Seventh, the circular ridge near the outer edge of the shuttle race, substantially as described, to allow the shuttle to clear its own thread and to leave the loop of the needle thread free to pass over the shuttle.

Eighth, the button *k*, provided with a series of notches in combination with a suitable stop or latch, and with the shafts F F', and shuttle driver H, constructed and operating substantially as and for the purposes set forth.

Ninth, the back gear M, in combination with the shafts F F', shuttle driver H, and needle *n*, constructed and operating substantially as and for the purpose described.

Tenth, the method herein described of producing a stitch by the combined action of the thread guide I, revolving shuttle S, and needle *n*, operating together substantially as described and shown in Fig. 4 to 7 inclusive.

Eleventh, the method herein described of producing a stitch by the combined action of the thread guide I', shuttle S, and needle *n*, operating together substantially as described and shown in Figs. 8 to 11 inclusive.

Twelfth, the method herein described of producing a stitch by the combined action of the reciprocating thread guide I'', constructed as described, shuttle S, and needle *n*, the shuttle being passed twice through the same loop of the needle thread, as described and shown in Figs. 12 to 15 inclusive.

No. 58,926.—GUSTAVUS WEISSENBORN, New York, N. Y.—*Apparatus for Drying Peat.*—October 16, 1866.—The cars have perforated false bottoms and stirring machinery, and move upon a railway between brick walls, the tops of the cars being connected with said walls by means of winged flaps, forming a flue under and around the sides of the cars, which is traversed by a caloric current. Super-heated steam is let into the space between the bottoms and false bottoms of the cars, and passes up through the peat.

Claim.—First, forming a continuous drying table, by means of the cars D D, which can be matched together or used separately, and arranged relatively to the side walls B, and beds A, or their equivalents, so as to convey the gaseous products of combustion from the furnace, and to utilize the heat therefrom for the purpose of drying peat in lumps, or pulverized, substantially as and for the purpose herein set forth.

Second, in combination with the above, the flaps D', arranged to operate substantially as and for the purpose herein specified.

Third, the hollow stirrer K L, adapted to transmit the heated fluid, and to impart the heat thereof to the peat, substantially as and for the purpose herein set forth.

Fourth, the false bottom I, arranged relatively to the cars D, and to the several other parts, substantially as represented, so as to convey a heated fluid between them and the car bottoms, and to allow a portion to rise through the peat in the several cars, for the purpose herein set forth.

Fifth, to connect a series of cars with an iron pipe, or any other analogous means of transmission, so that the hot fluid may pass from car to car, and through the wet pulverized peat, either upward or downward, or through the sides, substantially the same as specified.

Sixth, the process, substantially as herein described, of forcing superheated or waste steam, or heated air, or waste heat from a furnace, through wet pulverized peat, for the purpose of drying it, as herein specified.

Seventh, superheating the exhaust or waste steam from an engine, and heating air between the false bottom and the bottom of the cars, or on the sides, by the hot products, whether the waste heat from a steam boiler or of a furnace built for that purpose, as herein specified.

Eighth, the means for forcing the cars together, the same consisting of the screws O, or their equivalents, adapted to act on the whole series at a single operation, substantially as and for the purpose herein set forth.

No. 58,927.—ISAAC P. WENDELL, Philadelphia, Penn.—*Railroad Car Box.*—October 16, 1866.—The journal rests upon the curved slotted cover of the oil cup, which has a wedge and an elastic web interposed between itself and the bottom of the box.

Claim.—First, the combination of the oil box E with the box A, and journal B, arranged and operating substantially as described.

Second, the combination and arrangement of the curved plate F with the oil box E, and journal B, substantially as described, so as to answer the triple purpose of an under-bearing for the journal, conducting the waste oil back into the oil box, and serving as a cover to said box to keep the oil in place, as specified.

Third, the combination of the wedge G with the box A, and oil box E, substantially in the manner described and for the purpose set forth.

Fourth, the elastic support H, combined and arranged with oil box E and box A, substantially in the manner described and for the purpose set forth.

No. 58,928.—D. J. WHITTEMORE, Milwaukee, Wis.—*Unloading Grain Cars.*—October 16, 1866.—A section of the track is pivoted at its centre. The tilting of this section by a segmental rack at one end unloads the grain.

Claim.—Unloading cars by the arrangement of means constructed and operated substantially as herein recited.

No. 58,929.—G. M. WOOD, Decatur, Ill.—*Keeper for Bolts.*—October 16, 1866—The keeper of the bolt has slots for the attaching screws, to accommodate it to the sagging of the door.

Claim.—The providing of the keepers of bolts with oblong screw slots, in the manner substantially as and for the purpose set forth.

No. 58,930.—L. E. WOODARD, Cohocton, N. Y.—*Gate Hinge.*—October 16, 1866.—The pintle hole in the lower hinge has a slotted form to allow of raising the forward end of the gate to free the latch bar from the catch.

Claim.—Elongating the eye or pintle hole in the lug D, so that the said lug will be allowed a longitudinal play upon the pintle, substantially as and for the purpose specified.

No. 58,931.—M. A. WOODSIDE, Georgetown, Cal.—*Machine for Washing Ores.*—October 16, 1866.—An endless inclined belt passes over rollers, and receives the ore which passes over the perforated bottom of the box. A stream of water falls from a pipe upon the belt, washes the light particles of ore down toward the lower end of the belt, while the heavy particles are carried up and over the upper roller, from which they are brushed into a receptacle.

Claim.—First, the endless blanket H, and revolving brush K, when arranged substantially as described and for the purpose set forth.

Second, the perforated feed box I, and water pipe J, substantially as specified and for the purpose set forth.

No. 58,932.—JOHN E. WOOTTEN, Cressona, Penn.—*Quarrying Slate.*—October 16, 1866.—Successive vertical strokes are made by the rotary saw, whose frame is lowered by turning screws, and again elevated to make a deeper kerf. The frame being raised, is then laterally adjusted to make a parallel kerf, after which the blocks are detached by wedges. The machine is portable, and the engine driven by steam or compressed air conducted in a flexible tube.

Claim.—First, the quarrying of slate and other like rock by the use of a circular saw or cutter caused to revolve on a portable and adjustable frame, and arranged for operating on the rock, substantially in the manner described.

Second, the combination of the frame A, its adjusting screw rods with casters a, its driving engines and circular saw G, the whole being arranged and operating substantially as and for the purpose herein set forth.

Third, the spring d, arranged on each screw rod between a collar d, on the same, and the caster a, substantially as and for the purpose described.

No. 58,933.—ISAAC ALLARD, Belfast, Me., assignor to himself and R. G. TURNER, same place.—*Wood Sawing Machine.*—October 16, 1866.—The saw is reciprocated by rods from a wrist pin on the fly wheel, and is raised for the insertion of "stuff" to be sawed by a pivoted lever which is retained by a catch. A toothed dog arm is brought down on the wood by a connecting rod and lever, retained by a ratchet bar.

Claim.—First, the pivoted angular lever J, frame F, spring catch O, and connecting rod E, arranged and operating substantially as described, for the purpose specified.

Second, the pivoted bent dog O², rod Q, rod R, and toothed bar T, constructed and operating substantially as and for the purpose specified.

No. 58,934.—HORATIO ANDERSON, Chicago, Ill., assignor to himself and GEORGE W. CUSHING, same place.—*Steam Generator Safety Valve.*—October 16, 1866.—The tension of the spring is regulated by the jam nuts. The valve seat and stem guide are attached to a disk to be bolted on the steam generator.

Claim.—First, the capsular spring c. combined with the valve B, the valve stem d, and the jam nuts h h', constructed and arranged as and for the purposes herein described.

Second, the flange A, and stand b, combined with the valve B, and the capsular spring c, constructed and arranged as and for the purposes herein specified.

No. 58,935.—HAYDN M. BAKER, Rochester, N. Y., assignor to A. M. HASTINGS and ALEXANDER McVEAN.—*Bleaching Fibrous Materials.*—October 16, 1866.—The fibrous material is subjected to the action of chloride of lime in a tight rotary vessel under a pressure of carbonic acid gas; this unites with the lime, setting the chlorine free. The application of other gases is associated.

Claim.—The application to the bleaching of fibrous or other substances of chlorine, hydrogen, oxygen, and sulphurous acid gases in a close vessel under their own pressure while in a nascent or free state, in the manner herein described and set forth, or any other processes substantially the same, and which produce the same intended effects and results herein described.

Also, the use of carbonic (or any other) acid under pressure for the purpose of decomposing chloride of lime in a close bleaching apparatus, in the manner herein described, or any other substantially the same, and which produces the same intended effects.

Also, the application of oxygen, hydrogen, and sulphurous acid under pressure in bleaching.

No. 58,936.—JAMES C. BARLOW, Brimfield, Mass., assignor to himself and J. B. HAMILTON, same place.—*Lid Supporter.*—October 16, 1866.—The device has four plates hinged consecutively in three places; the outer pieces are attached to the edges of the box and lid, the inner ones form a toggle, which, when straightened, holds up the lid.

Claim.—First, the jointed lid supporter attached by its ends respectively to lid and box on the edges of their sides, constructed and arranged substantially in the manner and for the purpose set forth.

Second, the manner of arranging the piece and joints so that they fold up out of the way when the lid is shut down.

No. 58,937.—GEORGE W. BRIGGS, New Haven, Conn., assignor to OLIVER F. WINCHESTER, same place.—*Magazine Fire-arm.*—October 16, 1866.—Improvement on the patent of Smith and Wesson, February 14, 1854. Improved by Henry, October 16, 1860. The magazine tube is attached below the barrel by bands which allow it to be slipped forward for filling with cartridges, and the rear to be connected with the chamber in which the cartridge carrier is worked; the motions are regulated and limited by guides and stops.

Claim.—Constructing and arranging the tube or magazine in combination with the barrel of the arm and the carrier block, so as to be operated substantially as and for the purpose specified.

No. 58,938.—A. M. CONNETT, Madison, Ind., assignor to J. C. MOORE and SARAH A. CONNETT, same place.—*Shingle Machine.*—October 16, 1866.—The block lies in the hopper and a portion projecting below is split therefrom between the stationary and reciprocating knives, which are in the same plane. The blank is then forced to the feed rollers, which carry it between the upper and lower shaving knives and the edging knives; the upper knife is gradually depressed to give the taper to the shingle passing endwise beneath it.

Claim.—Splitting, shaving, tapering, and jointing a shingle by one operation by means of a splitter knife L, lower stationary knife blade X, upper movable knife blade Y, and side knife blades or chisels E² and F² in combination with suitable feed rollers, when they are all arranged together so as to operate and be operated substantially in the manner described.

Also, the stationary knife blade P in combination with the splitter blade I, substantially as and for the purpose specified.

Also, the adjustable piece T, secured to the under side of the splitter blade, substantially as described for the purpose specified.

No. 58,939.—W. H. EARLE, Vineland, N. J., assignor to himself and G. M. BUTTRICK, Barre, Mass.—*Fruit Box.*—October 16, 1866.—The box is formed of two pieces, each of which has two sides and a bottom, V-grooved to enable them to bend, the bottom then being two thicknesses. The upper corners are united by angle pieces which lap over the inner and outer sides and the top.

Claim.—The combination with the upper corners of a box, the sides and bottom of which are made as described, of the metal corner fastening pieces F, substantially as shown and described.

No. 58,940.—THOS. GOODRUM, Providence, R. I., assignor to ALBERT T. MANCHESTER, same place.—*School-boy's Book-binder.*—October 16, 1866.—The package of school books is clamped together by an upper and under strip of wood, having perforations for a clamping cord, whose ends wind on the shaft of a ratcheted disk, having free revolution in the upper strip; a pawl prevents the back rotation of the shaft.

Claim.—A portable book package binder, constructed and operating as described, the article being substantially as herein specified.

No. 58,941.—E. HANEISEN, A. WAGNER, and A. NULSEN, Cincinnati, Ohio, assignors to NULSEN & CO., same place.—*Brick Kiln.*—October 16, 1866.—The connecting opening of each one of the circuit of compartments with a common flue has a damper, and each chamber is charged with tiers of bricks and fuel in the usual manner when first starting the furnace; subsequent charges are burnt with the escaping heat of the burning bricks in other compartments, with the help of supplies of comminuted fuel.

Claim.—First, the method, substantially as described, of burning bricks, &c., by the contact of falling coal dust or other comminuted fuel, with a draught of air which has become heated by traversing the already burnt brick.

Second, the arrangement of the continuous gallery A B, A' B', shifting partitions or partitions D D', and dampers G^1 G^2, &c., or devices substantially equivalent; whereby the operations of pre-heating, burning, and cooling are simultaneously and continuously performed, in the manner substantially as explained.

No. 58,942.—JOHN O. HARRIS, Reading, Penn., assignor to himself and ISRAEL S. RITTER, same place.—*Lantern.*—October 16, 1866.—The flanged foot is perforated for the current of air, which then passes between the lamp and the frustum case into the globe; the latter is surmounted by a double cap whose walls are perforated.

Claim.—The conical base B, attached to the lower part of the glass globe A of the lantern, and having a flange t? at its lower end perforated with holes a b, in combination with the cap E, and jacket F at the top of the glass globe A, all arranged substantially as and for the purpose set forth.

No. 58,943.—JOHN A. HEYL, Boston, Mass., assignor to himself and GEORGE H. BAILEY Hudson, Mass.—*Electric Gas Stop Cock.*—October 16, 1866.—The 'rotary perforated sleeve acts between the perforated stationary tubular axis and the opening in the pipe leading to the burner. The ratchet on the shaft of the sleeve is rotated by a pawl whose intermittent motions are obtained by an armature and electro-magnetic coil.

Claim.—The above explained improved cut-off, consisting of the stationary cylinder G and the rotary tube E, provided with passages a b, arranged in them as described, in combination with the ratchet F and the gas burner conduit, the whole being substantially as and for the purpose and to operate as hereinbefore explained.

No. 58,944.—GEO. W. HURLBUT, Fairhaven, Vt., assignor to himself and ABRAM C. WICKER, same place.—*Manufacture of Paper.*—October 16, 1866.—Explained by the claim.

Claim.—The use of pulverized clay, slate or other suitable stone as a material in the manufacture of paper, to give it body, evenness and finish.

No. 58,945.—DANIEL HUSSEY, Nashua, N. H., assignor to RICHARD KITSON, Lowell, Mass.—*Brake for Cotton Lappers.*—October 16, 1866.—The object is to dispense with leather frictional straps or disks, and by avoiding the consequent variable friction due to atmospheric changes to secure more uniform tension and pressure on the lap when being wound. The friction weight is of metal and adjustable. The spring coming in contact with the teeth of the ratchet wheel serves in connection with the weight to regulate the motion and velocity of said wheel, letting off one tooth at a time, as the diameter of the roll increases.

Claim.—First, the employment of the arm E and pinion gear F, in combination with the bevel gears C and C² and ratchet wheel A, all arranged to operate substantially in the manner and for the purpose set forth.

Second, the gear D on the shaft B, in combination with the bevel gears C and C², pinion F, and arm E, when the said gear engages with the pinion J, to operate said pinion and its connections, substantially in the manner and for the purpose set forth.

Third, the ratchet wheel H, or its equivalent, in combination with the friction pulley I, and friction weight L, arranged and made to operate substantially in the manner, by the means, and for the purpose set forth.

Fourth, the lever G, on the shaft B, when the said lever is formed, arranged, and combined with the pulley I, ratchet wheel H, and friction weight L, substantially as and for the purpose specified.

Fifth, the connecting rod m, or the equivalent thereof, in combination with the arm l, cross lever M, and a weight or spring P, all arranged to operate substantially as and for the purpose specified.

Sixth, the spring R, in combination with the ratchet wheel H, pulley I, and friction weight L, and arranged to operate substantially in the manner and for the purpose explained.

Seventh, the combination of the ratchet wheel A, bevel gear C, arm E, pinion F, bevel gear C², spur gear D, lever G, with arms k and l, ratchet wheel H, or equivalent, the pinion J, friction pulley I, friction weight L, connecting rod m, lever M, weight or spring P, with the shaft B, the whole arranged to operate substantially as and for the purpose set forth.

No. 58,946.—J. G. LANE, Washington, N.Y., assignor to himself and W. J. LANE, same place.—Coffee Mill.—October 16, 1866.—After passing the interlocking toothed surfaces the material passes between plane surfaces to insure the comminution of the product.

Claim.—Having the outermost ridge k' of the grinding surface of case A solid, or without being notched, and extending around the outermost ridge f of the corresponding grinding surface on plate B, so as to serve as a barrier to the too free discharge from the mill of the substance being ground, substantially as herein set forth.

No. 58,947.—JAMES E. F. LELAND, New York, N.Y., assignor to H. A. LELAND, same place.—Wood-turning Lathe.—October 16, 1866.—The stuff upon its centres passes through a spindle in a sliding head; the spindle has a longitudinal motion derived from the pattern cam wheel, whose cogs move the head. The conical end of the spindle gives outward radial motion to the cutter slides; an elastic band causes their contra movement.

Claim.—The slide B carrying the material to be turned, cam or eccentric wheel H, sliding tube I, concentric tube J, having cutters N, when all arranged together, substantially in the manner and for the purpose described.

No. 58,948.—E. C. LITTLE, St. Louis, Mo., assignor to EVALINE LITTLE, same place.—Hinge for Moulders' Flasks.—October 16, 1866.—The hinge plate is attached to the edge of the side and end pieces of the drag and cope, running back from the corner in each direction. The pintle has a flange and the other portion a recess for its reception.

Claim.—The projecting wing b and pintle c, in combination with the male half of the hinge plates, and the female plate with notched corner and its eye d operating together, the plates adapted to lie on the corner edges of the cope and drag, substantially as described, for the purpose specified.

No. 58,949.—HARVEY S. LOPER, New Haven, Conn., assignor to himself, COLLINS, PECK & CO., same place.—Adjustable Frame for Forming Hoop Skirts.—October 16, 1866.—The band block is laterally adjustable on its support, which is vertically adjustable on the base. The ribs are pivoted to adjustable slides in the band and base blocks.

Claim.—The combination of the band block B, adjustable upon its support C, with the bars D, adjustable in the band block and upon the base, substantially in the manner and for the purpose herein set forth.

No. 58,950.—WM. H. MASON, Boston, Mass., assignor to himself and H. K. W. PALMER, Chelsea, Mass.—Piano-forte.—October 16, 1866.—The keys are connected to the next octave key by a lever which causes their hammers to act in unison. By lowering the fulcrum bar of the levers they are made inoperative when the combined effect is not desired.

Claim.—The combination of the lever H and its flexile connections a a, with the two octave keys of a piano-forte.

Also, the combination and arrangement of the tongue I with the lever H, its flexile connections a a, and the two octave keys of a piano-forte.

No. 58,951.—SILAS A. MOODY, San Francisco, Cal., assignor to PHILIP E. DIVINE.—Sod Cutter.—October 16, 1866.—The revolving shaft, armed with disk cutters, has a cover and seat, which assist in forcing the disks into the ground.

Claim.—A series of circular blades or knives upon a shaft or axle arranged to rotate, as described, in combination with the cover C and seat upon the cover, substantially as described.

No. 58,952.—GEO. MUNGER, New York, N. Y., assignor to himself and J. W. SCHERMER-HORN, same place.—*Ink-well Cover.*—October 16, 1866.—The ink-well fits a recess in the desk; its cover has gudgeous, which are capped by sockets on a plate secured to the desk.

Claim.—The semicircular sockets *b* or *b** in the bracket or disk, in combination with the gudgeons *a* cast solid with the cover A, substantially as and for the purpose described.

No. 58,953.—R. F. OSGOOD, Rochester, N. Y., assignor to C. W. KINNE, Cortland, N. Y.—*Sinking Well-tubing.*—October 16, 1866.—The lower, perforated section of the well tube has a perforated sleeve with spiral flanges on its exterior which in driving, by contact with the soil, keep the sleeve rotated against a pin, and the holes not in correspondence. When the tube is rotated in the other direction to open the holes, the flanges prevent the rotation of the sleeve.

Claim.—The combination of the spiral wing or wings *k* with the shank B and tubing A, operating substantially as and for the purpose herein set forth.

No. 58,954.—SAMUEL J. SHAW, Marlboro', Mass., assignor to himself and THOMAS COREY, same place.—*Eyeletted Brace.*—October 16, 1866.—A single piece of metal is stamped so as to form two connected eyelets.

Claim.—As a new article of manufacture, for purposes as set forth, the eyelets and brace, struck or stamped together or in connection from one piece of metal.

No. 58,955.—SAMUEL J. SHAW, Marlboro', Mass., assignor to himself and THOMAS COREY, same place.—*Brace and Lacing Device.*—October 16, 1866.—A plate is riveted to the vamp of the shoe, and is pierced for the two lower lace holes.

Claim.—The combination of the metallic stay as made with the lacing holes, and the arrangement of the lacing so as to go through such holes or eyelets serving to fasten the stay to the upper, the whole being substantially as described, whereby such lacing is made to protect the stay from being torn or separated from the shoe while in use.

No. 58,956.—C. W. SINGER, Anderson Store, Va., assignor to himself and ABEL LAND, Rochester, Ohio.—*Car Brake.*—October 16, 1866.—The rubbers are attached to hinged arms above each wheel and are depressed by rollers which are wedged between the rubbers and the car, by cord connection to the rotating brake shaft on the platform. The rollers return by gravity and the arms are lifted by springs.

Claim.—First, the rollers D D', in combination with the adjustable rubbers A, and springs K, arranged in the manner and for the purpose set forth.

Second, binging the rubbers to the truck plates or frame so as to form inclined planes, thereby allowing the rollers to act as a wedge between the rubbers and plate A', to compress the said rubbers upon the wheels and so that said rollers will move back independently on releasing the brake from the wheels, as and for the purpose set forth.

No. 58,957.—JOSIAH STUBBS, Decatur, Ill., assignor to himself and H. E. FOSTER, Macon county, Ill.—*Washing Machine.*—October 16, 1866.—The washing box is mounted on rockers has a laterally corrugated bottom and a beater block sliding on longitudinal strips. It is operated by a lever.

Claim.—The combination of the corrugated floor K, gravitating beater F, and closed rocking box A B E, all constructed and arranged to operate in the manner and for the purposes set forth.

No. 58,958.—MATILDA C ROOT, HARRIS COLT, and ELISHA COLT, Hartford, Conn. executors of E. K. ROOT, deceased.—*Machine for Dressing Willow for Baskets.*—October 16, 1866.—The revolving drum has slots in which the end of the split switches are held by a clamp operated by a cam. The switches are drawn under cutters and then released.

Claim.—The employment, in combination with the shaving mechanism, of a rotary carriage or bed, to which the foremost end of the switch is fastened and by which the switch is pulled or drawn by the cutters during the shaving operation, as hereinbefore described.

No. 58,959.—HUBURT C. BAUDET, Paris, France.—*Keyed Musical Instrument.*—October 16, 1866.—A "lock" is set perpendicular to each string or group of strings of the same note; it consists of fibrous material, and when the key is touched, is grasped between a cylinder and a roller and being drawn past the string, agitates the same as the hair of the bow upon the strings of the violin.

Claim.—First, the locks *a'*, attached to and in combination with the strings of a musical instrument, substantially as and for the purpose herein specified.

Second, the friction rollers *d*, in combination with the locks *a'*, substantially as and for the purpose herein specified.

Third, the combination with the rollers *d*, or their equivalents, strings *a*, and locks *a'*, of one or more driving cylinders *b c*, and a system of keys *f*, the whole operating substantially as herein specified.

No. 58,960.—AUGUSTE P. BERLIOZ, Paris, France.—*Electro-magnetic Engine.*—October 16, 1866.—The devices are for the production and collection of electric currents; the changes being effected at every sixteenth revolution of the disks, passing a constant succession of positive currents to one wire, and a like succession of negative currents to the other wire. The uncorrected currents pass first from one carbon electrode to the other, and then back, so as to consume them with equal rapidity. Instead of collecting the current by springs bearing u on insulated rings, it is collected from the shafts, which have their bearings in insulated boxes.

Claim.—First, the shaft F, divided into two insulated parts, turning in insulated boxes D D', and connected to the bobbin wire or wires t, all substantially as and for the purpose described.

Second, the combination of the above and the disks G, their bobbins I, and the rings u u', when the wires on said bobbins are connected to the shaft, to each other, and to the said rings, substantially as shown in Figs. 2 and 5, for the purpose specified.

Third, the spring o, its projection n, and the roller m, in combination with a ring H.

Fourth, the spring o', its projection m', in combination with a ring H'.

Fifth, the combination of two or more machines constructed as described, when the said machines are so arranged that when their axles are coupled, by the within described devices or their equivalents, alike currents will be simultaneously generated in all the machines.

No. 58,961.—SERVAAS DE JONG, Paris, France.—*Purifying and Softening Water.*—October 16, 1866.—A composition of anhydrous carbonate of soda, 2 parts; bicarbonate of soda, 1, and liquid silicate of soda, 2.

Claim.—Purifying and softening water by silicate of soda and carbonate of soda, or its equivalent, as set forth.

No. 58,962.—WM. NAYLOR, Lorn Terrace, Mildmay Park, England.—*Steam Safety Valve.*—October 16, 1866.—One end of the bent lever rests, by the interposition of an edged bar, upon the valve: the other end is suspended by a stirrup and spring coil, or conversely raised by a spring bolt beneath: In each case bringing the power of the spring, by the intervention of a lever, to bear down on the valve.

Claim.—The arrangement, substantially as hereinbefore shown and described, in safety valves, of bent levers of the first order acting in combination with a spring or springs, the whole operating in the manner and for the purpose set forth.

No. 58,963.—JAMES ALFRED SHIPTON and ROBERT MITCHELL, Wolverhampton, England.—*Machinery for Forging Pipe Joints and other Similar Articles.*—October 16, 1866.—Converging hammers with shaping faces are used in combination with a fixed die and swage to work the plates of metal into shape and form the weld.

Claim.—The construction and arrangement of machinery or apparatus for shaping and forging metallic articles, substantially as hereinbefore described and illustrated by Figs. 1, 2 and 3 of the drawings.

No. 58,964.—JOHN ABSTERDAM, New York, N. Y.—*Screw.*—October 23, 1866.—The plain unthreaded portion next the point permits the screw to be driven without breaking away the wood in advance of the thread.

Claim.—The above-described wood screw with the plain cylindrical portion between the point and the threaded portion, substantially as and for the purposes set forth.

No. 58,965.—HENRY ADAMS, Seattle, Washington Ter.—*Gate.*—October 23, 1866.—The upper hinge slips on a centre pin on the post and is held to the gate by a bolt through a vertical slot. The lower hinge spans and rests against the gate post.

Claim.—A gate hung to its post by means of a hinge E, which passes through a vertical slot I, and is held to the gate by a nut J, substantially as herein described and for the purpose specified.

No. 58,966.—SHERMAN W. ADAMS, Wethersfield, Conn.—*Hoe.*—October 23, 1866.—The boss on the hoe and that on the handle are fastened together by a locking joint and bolt. The blade is capable of being set as a scuffle hoe, or an ordinary hoe with any desired obliquity to the line of the handle.

Claim.—The combination of the blade a and handle b, when constructed and operating substantially as herein shown and described.

Second, the hoe as above described and set forth as a new article of manufacture.

No. 58,967.—WILLIAM F. ALTFATHER, Johnstown, Penn.—*Feed Cutter.*--October 23, 1866.—The knife frame stands diagonally on the box and is worked by an eccentric. Adjustable jaws on the side of the eccentric work an oscillating frame carrying arms ending in corrugated plates, between which the hay is held when feeding forward.

Claim.—First, the combination of the inclined or diamond-shaped knife sash, connecting rod or bar I and eccentric F with each other and with the driving shaft C, cutter frame B, and box A, substantially as herein shown and described and for the purpose set forth.

Second, the combination of the jaws P and S, bent levers O and R, and pivoted cam lever N with each other and with the cutter box A, support M, and eccentric F, substantially as herein shown and described and for the purpose set forth.

No. 58,968.—WILLIAM R. ANDREWS, Mystic River, Conn.—*Mechanism for Operating the Harness of Looms.*—October 23, 1866.—A crank on the main shaft, by means of its connections, lifts and lowers the jacks, which thus, as their positions are varied by springs which are operated on by the pattern chain or by an incline, actuate the racks; the rise of one causing the fall of the other. The cams on the pinion shafts thus lift or lower the arm of the lever and move the heddles.

Claim.—The above specified new and useful harness-operating mechanism or combination, consisting of the tri-armed lever D, the two cams E F, the gears c c, and racks G H, the spring I, and the rack-elevating mechanism, the whole being arranged together and with a pattern chain and its actuating mechanism, substantially in manner and so as to operate as explained.

No. 58,969.—J. T. ASHLEY, Brooklyn, N. Y.—*Instrument for Extracting Corks from Bottles.*—October 23, 1866.—The jaws, while collapsed by the slide, are passed through the neck of the bottle, and being opened, are then clasped around the cork by the motion of the slide, and the cork with its retractor is drawn from the bottle.

Claim.—The slide F in combination with the tongs A, when arranged thereon so as to operate substantially in the manner and for the purpose described.

No. 58,970.—EGBERT H. AVERY, Belvidere, Ill.—*Apparatus for Moving Buildings.*—October 23, 1866.—To turn the building in any direction, one of the trucks is deflected from the ine of travel by means of wedges driven between the sills and pins rising from the fore end of the truck.

Claim.—The guide keys D in combination with the trucks C' C' B, substantially as set forth.

No. 58,971.—CHARLES F. BARAGER, Candor, N. Y.—*Butter Worker and Packer.*—October 23, 1866.—The ladle on the lever is used in connection with the bowl to work the butter. When the lever is rotated 180° on its universal joint, the packing rammer is attached for use in connection with the tub.

Claim.—The arrangement of the bowl B, vessel I, and slotted lever D with the universal joint E F and stop pins d' e, said lever D being adapted to admit of the attachment of the ladle C and packer J, and the whole operating substantially as described.

No. 58,972.—A. B. BARLOW, Ripon, Wis.—*Pump.*—October 23, 1866.—The cylinder and base enclose between them a plate which contains the valve seats, and has a packing on each side abutting against the flanges of the adjacent parts.

Claim.—The method, substantially as above described, of packing the lower joints of the cylinder and said chamber, by means of a bottom piece I and annular flange or cap N, and the packing material a a, secured by them by the aid of a surrounding flange N', substantially as described.

No. 58,973.—PETER BARNHART, Chillicothe, Ohio.—*Corn Plough.*—October 23, 1866.—The standards are pivoted at their upper ends, and have arms running from thence horizontally, adjusted by means of pins in the beam. An adjustable fender is pivoted to the beam to keep clods from the young corn.

Claim.—The adjustable fender F and beam A in combination with the standards B B, for purposes and substantially as described.

No. 58,974.—JOHN W. BARTLETT, Harmar, Ohio.—*Straw Cutter.*—October 23, 1866; ante-dated October 12, 1866.—A crank on the fly wheel shaft reciprocates the knife, which is hung by the other end to an arm pivoted above; this gives the knife a "draw cut." The toothed feed roller is actuated by a pawl, which is connected to a crank on the fly wheel shaft. The press board has intermittent vertical movement to suit the alternate forwarding and cutting of the feed.

Claim.—First, the arrangement of the fly wheel, fly wheel shaft with two cranks, knife C, oscillating arm D, and standard and guide F, substantially as set forth.

Second, the combination of the crank g, attached to the end of the fly wheel shaft, the lever P, and the bent pawl lever H, with the ratchet wheel and feed rollers, substantially as set forth.

Third, the pawl holder and guide I, constructed and connected substantially as set forth.

Fourth, the hinged board O', with its shaft P', in combination with the bent spring S', substantially as and for the purpose set forth.

No. 58,975.—F. BEARSE and G. E. HOPKINS, Barnstable, Mass.—*Composition for Roofing.*—October 23, 1866.—Composed of sand, 60 parts; coal tar, 20; oxalic acid, 1.

Claim.—The composition as made of the acid and other ingredients, substantially as hereinbefore set forth.

No. 58,976.—M. BRATT, Maysville, Ky.—*Churn.*—October 23, 1866.—As the dasher rises the valve on the upper end of its stem falls and admits air, the valve on the hollow guide rod closing. As the dasher descends the action of the valve is reversed, and the air issues into the milk at the openings in the lower end of the hollow stem.

Claim.—First, the combination of the hollow tube E, having the valve e' at its upper end, and with the hollow dasher handle D, having a valve d' at its upper end, and with the bottom a' of the churn A, substantially as herein described and for the purpose set forth.

Second, the combination of the guide rod or plunger F with the hollow dasher handle D, having a valve d' at its upper end, and with the bottom a' of the churn A, substantially as herein described and for the purpose set forth.

No. 58,977.—CHARLES P. BENOIT, Detroit, Mich.—*Grinding Mill.*—October 23, 1866.—The respective rollers are grooved longitudinally and circumferentially.

Claim.—The machine for crushing grain consisting of the longitudinally grooved roller B, and the transversely grooved cylinder C, arranged to operate substantially as described for the purpose specified.

No. 58,978.—C. C. BELLOWS, New Ipswich, N. H.—*Creasing, Slicking, and Skiving Leather.*—October 23, 1866.—Of two slicking and creasing rollers, the under one is adjustable by a spring lever which supports it. Adjustable skiving knives operate in conjunction with the upper roller, which carries an adjustable creasing wheel, working with a flanged roller on the lower shaft.

Claim.—First, the combination of the slotted standards B, slotted triple-armed lever E, springs I, and rod D, arranged to operate with the roller D, when constructed and applied in the manner and for the purpose specified.

Second, the plate J, having skiving knives d, attached to or formed on it, and applied to the upper roller C, by means of the bars or clamp frame, substantially as and for the purpose described.

Third, the laterally-adjustable creasing wheel F, on the upper roller shaft, operating with the flanged roller G, substantially as described, for the purpose specified.

No. 58,979.—JOSHUA BRIGGS, Peterboro, N. H.—*Piano Stool.*—October 23, 1866.—The pedestal is fastened to its base by an axial bolt. The socket nut which enters the upper end of the tubular pedestal has an exterior thread for the latter and an interior thread for the elevating screw.

Claim—Combining with the pillar c the spindle nut f, when made with a wood screw cut upon its outer surface for securing it permanently to the pillar, substantially as described.

Also, the combination of the pillar c, base a, and bolt i, when the pillar is constructed to receive the bolt through the tube in which the screw spindle plays, and with a seat for the head of the bolt at the bottom of said tube, substantially as set forth.

No. 58,980.—RICHARD C. BRISTOL, St. Clair, Mich.—*Steam Engine Slide Valve.*—October 23, 1866.—The valve traverses on rollers arranged in ranks beneath each longitudinal edge, and resting on tracks. The roller journals are in cheek bars which preserve the relative distances of the rollers.

Claim—In connection with a slide valve, the within-described arrangement of rollers C, mounted concentrically upon the cross-bars C', and between the longitudinal bars C² C⁴, and arranged to operate relatively to the valve, and to the cylinder face, and to the steam-chest, substantially as and for the purposes herein specified.

No. 58,981.—EDMUND BROWN, Chicago, Ill.—*Amalgamator.*—October, 23, 1866.—The vessel has a hollow rotary shaft with apertures at its lower end, and enclosing a stationary tube leading from the hopper above. Teeth are attached to the outer shaft between corresponding teeth on the inside of the vessel.

Claim.—First, the revolving and stationary shaft, with apertures and flange for crowding the quartz out into the lead.

Second, the series of combs attached to the revolving shaft and sides of the kettle, the whole combined and arranged for the purpose specified.

No. 58,982.—EDWARD BUCKLIN, Jr., and SEDGWICK A. SUTTON, North Providence, R. I.—*Window Screen.*—October 23, 1866.—The upper and lower bars are made in sections and are longitudinally extended or contracted by their screw connection.

Claim.—Attaching the screen directly to two supporting rails, D and E', in such manner that the width of the screen may be increased or diminished in the same proportion as the lengths of the rails, as and for the purpose described.

No. 58,983.—JOHN A. CHEATHAM, Nashville, Tenn.—*Farm Gate.*—October 23, 1866.—
By pulling on the lever, the spindle and latch are raised, and the lever acts as a push pole
while passing through, rotating the while on its axis; when opened, the gate is locked by
the catch, and when the lever is released, it assumes the horizontal, raises the catch, rotates
back to its normal position, and the gate is closed by its spring.

Claim.—First, the combination of the lever or levers A A, with the vertical spindle E
controlling the gate and its latch, with the cam-shaped piece O, or its equivalent, substantially as and for the purposes set forth.

Second, the combination of the lever or levers A A, and the spindle E, with the upper disk
L, the trigger K, and latch J, substantially as and for the purpose described.

No. 58,984.—G. CKERTIZZA, New York, N. Y.—*Ladder.*—October 23, 1866.—The upper
end of each section fits the lower end of the one above, and has a bar which enters slots in
the feet of the latter, and slots on its upper ends, which receive the lower round of the section
above.

Claim.—The combination of sides so sloping that the narrow end of one sectional ladder
fits within the wider end of any other, with the slots d d' and c c', and the bars b b and b' b',
substantially as described and for the purpose set forth.

No. 5',985.—PATRICK CLARK, Rahway, N. J.—*Fan Blower.*—October 23, 1866.—The
air is driven by the oblique fans in the cases through which it passes consecutively, and is
ultimately discharged at the axis or periphery. The cases are placed in immediate connection,
the diaphragms leading the air from the periphery of one to the middle chamber of the next.

Claim.—First, the diaphragms C C, when used in combination with a compound fan blower.

Second, the fan wheel F F, when constructed with fans or vanes of the form and arrangement with respect to each other as described.

Third, attaching each fan or vane at its ends to two adjacent arms, as described.

Fourth, the leather packing D D, when combined with the diaphragms C C, as described

No. 58,986.—WILLIAM COLWELL, Chillicothe, Ill.—*Corn Sheller.*—October 23, 1866.—
A pin armed cone shells the corn, a ribbed apron throws out the cobs, which are carried off
by an apron and the grain is winnowed and then elevated.

Claim.—In combination with the cone B, shaft C, and hopper F, the fan N, and elevator
U, for the purposes and substantially as herein set forth.

No. 58,987.—D. G. COPPIN and G. H. CLEMENS, Cincinnati, Ohio.—*Safety Valve.*—October 23, 1866.—The annular weights rest upon levers which are pivoted to the interior of
the case and rest upon the lip of the cup above the valve. The tubular stem of the latter
may be raised by means of its cap and flanged sleeve, to determine the operative condition
of the parts, the steam escaping through the hollow stem.

Claim.—First, the valve C, and tube m, constructed as above described and for the purpose set forth.

Second, the valve C, levers l, and weights D and D', arranged as above described and for the
purpose set forth.

Third, the valve C, levers l, weights D and D', tube m, in combination with annular ring f,
casing B, sleeve n, and cap o, for the purpose above described and set forth.

No. 58,988.—W. H. COX, Virden, Ill.—*Corn Planter.*—October 23, 1866.—The seed is
dropped by a perforated revolving disk. A thimble within the hub and axle carries the
grooved pulley for driving the seed apparatus. The dropping frame may be raised by a lever.
The machine admits of use as a hand dropper, by a slide bar having pawls working the seed
disks by means of ratchet wheels.

Claim.—First, the perforated, horizontal, revolving plates m m, in the hoppers D D, &c
feeding and dropping the grains of corn evenly in combination therewith, and with the bevel
gear wheels h i, and the pulleys b d, connected with and deriving their motion from one of
the driving wheels C, constructed and arranged substantially as and for the purposes herein
described.

Second, the thimble a within the hub of the driving wheel C, for carrying the pulley b, in
combination therewith, and with the stationary axle B, constructed and operating substantially as and for the purposes herein specified.

Third, the arrangement of the side pieces f f, hung upon the axle B, for supporting the
hoppers D D, and raising and lowering at pleasure with the lever F, substantially as herein
described.

Fourth, the slide piece G, with the push and pull pawls O O' for working the seed-dropping
apparatus by hand when adjusted for planting corn in hills, in combination with the revolving perforated plates m m, to which they impart an intermittent motion, arranged and operating substantially as herein described.

No. 58,989.—S. L. CROCKETT and BENJAMIN T. MILLS, Lowell, Mass.—*Machine for Stripping Top Flats of Carding Engines*—October 23, 1866.—The object of this invention is to prevent the lifting pins on the slide from getting out of place, and to avoid the loss of time in waiting for the apparatus to work over to one side of the engine ; the improvement allowing the stripping apparatus to be moved backward to the point desired by reversing the cam.

Claim.—The employment of the lifting and replacing cam formed substantially as herein set forth and shown, and arranged to operate in the manner and for the purpose specified.

Also, in combination with the lifting and replacing cam, formed and made to operate as herein set forth, the two pins 1 and 2, in the slide *h*, acted upon by the cam, in the manner and for the purpose specified.

No. 58,990.—GEORGE G. CROWELL, Lime Rock. Conn.—*Hardening Springs.*—October 23, 1866 —The article to be hardened is coated before heating with a composition of glue and prussiate of potash.

Claim.—The employment of glue or equivalent glutinous animal matter, either alone or in combination with other material, as a hardening compound, when employed substantially in the manner and for the purpose herein set forth.

No. 58,991.—A. M. CULVER, Bedford, Ohio.—*Table and Holder for Shearing Sheep.*—October 23, 1866.—The table is adjustable in height and has at each end a foot shackle which permits the foot of the animal to turn.   One of the shackles has longitudinal movement by a ratcheted slide held by a pawl.

Claim.—The table B, arms C C', pawl and ratchet *a b*, and shackle D, constructed and arranged as and for the purpose specified.

No. 58,992.—PORTER E. CUMMINGS, Sanford, Me.—*Knife Carrier.*—October 23, 1866.—The knife carrier is made with a socket provided with a set screw and a rabbet.

Claim.—The improved knife carrier, made substantially as described, viz : with the knife-shank socket, and the rebate arranged in it as set forth, the said carrier being provided with a set screw, or equivalent means of fixing the knife shank in the socket.

No. 58,993.—OBED DANN, Janesville, Wis.—*Hand Seed Sower.*—October 23, 1866.—The seed receptacle has at the bottom a hole covered by a slide with various sized holes to suit different seeds.   A handle screws into the top of the seed vessel.   The seed is emitted by a vertical shake.

Claim.—First, the combination of the box A and slide E, when constructed, arranged, and used substantially as and for the purpose set forth.

Second, the combination of the box A, cap B, and handle C, when constructed, arranged, and used substantially as and for the purpose set forth.

Third, the combination of the box A, slide E, cap B, and handle C, when constructed, arranged, and used substantially as and for the purpose set forth.

No 58,994.—GARET B. DAVIS and CHARLES B. DAVIS, Freeport, Ill.—*Bed Bottom.*—October 23, 1866.—The camber of the arched bed bottom is assisted by the girders which form chords beneath it. .

Claim.—The strengthening rods or girders E E, in combination with the bow-shaped cross pieces C C, and elastic bands D, substantially as specified.

No. 58,995.—ERNEST DINTER, Boston Mass.—*Table.*—October 23, 1866.—Two portions, corresponding, with the exception of the position of their longitudinal grooves, are slipped upon each other at right angles, interlocking by their grooves the two flat portions, and forming a crucial bearing for the table top, and four legs for its support.

Claim.—The improved table stand as having two parts *a b*, constructed with receiving slots, arranged in them so as to enable them to be applied together, substantially as set forth.

No. 58,996.—E. C. EDMONDS, Buffalo, N. Y.—*Steam Engine Governor.*—October 23, 1866 —The adjustable clutches on the slip shaft are so arranged that a rising or falling of the governor balls brings one or the other into connection with a similar clutch upon one of the free bevel wheels on the said shaft, and operates through the connecting bevel wheel and belt to turn the screw shaft, and advance or retract the governor valve.

Claim.—The combination of the adjustable clutches *t t'*, with the slip shaft G, loose pinions *h h'*, bevel wheel *k*, and spindle C, for producing an intermittent motion to the valve-operating mechanism, substantially in the manner set forth.

Also, in combination with the above, the screw valve shaft *p*, pulley *o*, band *n*, and pulley *m*, for operating the valve *r*, arranged and operating substantially in the manner specified.

No. 58,997.—ALFRED A. ENQUIST, San Francisco, Cal.—*Washing Fluid.*—October 23, 1866 ; antedated October 3, 1866.—Composed of soap, one-half pound ; water, 10 gallons ; ammonia, one ounce ; turpentine, one ounce.

Claim.—The compounding of the ingredients in about the proportion as herein described, in combination with the process, substantially as set forth.

No. 52,998.—HENRY S. FISHER, Oakville, Penn.—*Seal for Jars and Cans.*—October 23, 1866.—Interposed between the can and its lid is a flat ring of India-rubber, coated on each side with a similar ring of fibrous material saturated with wax and rosin.

Claim.—A seal for preserve jars, cans, and other vessels which is composed of India-rubber, or other equivalent substance, lined or covered with a substance which is saturated with an adhesive cement, substantially as described.

No. 58,999.—FREDERIC G. FORD, New York, N. Y.—*Caster for Furniture.*—October 23, 1866.—The screw passes through the central boss and holds it to the leg of the table; the flaring head of the screw holds the collar of the horn to its seat on the boss.

Claim.—First, combining and arranging the caster by leaving a space between the central boss and horn, so that the horn shall be out of contact with the central boss, thereby relieving the caster of much of its friction, substantially as shown and described.

Second, the combination, in a caster, of a central screw and horn, with a plate constructed with the boss B, substantially as above shown and described.

Third, constructing and arranging the central boss of the plate with respect to the screw and horn, so that the said boss shall support the screw and not be in contact with the horn, in the manner and for the purposes substantially as herein set forth.

No. 59,000.—THEODORE F. FRANK, Buffalo, N. Y., assignor to himself and DAVID P. BENSON.—*Apparatus for Inhaling Gases.*—October 23, 1866.—The receiver is connected by a tube with the retort, and has a rubber gas-holder secured to the top. An inhaling tube extends from the top of the vessel, and a valve tube admits air when required.

Claim.—The above described apparatus for generating and inhaling vital oxygen, consisting of the generator B, receiver E. provided with the expanding bag K, induction gas-and air tubes *f* G, with their valves *i i'*, stop cock *h*, and inhaling tube L, combined, arranged, and operating substantially as set forth.

No. 59,001.—C. R. FRINK, Norwich, N. Y.—*Hay Spreader.*—October 23, 1866.—An eccentric wheel is used with each fork, in place of a crank, and is surrounded by a sleeve having a heavy projection upon its outer face, through which the arm of the spreading fork is slipped, being held in place by a grooved plate bolted to the sleeve. The fork arms have spiral springs at their connection with the tines, and their upper ends are connected by hinged rods to the frame.

Claim.—First, the grooved or double-flanged eccentric wheel or disk A, combined with a movable plate *a*, and cap plate *b*, for operating the fork rods B, in their connection with the spiral springs C, substantially in the manner and for the purpose as herein set forth.

Second, the movable jointed arms E, in combination with the brackets F, for relieving the forks of any straining or undue pressure, substantially in the manner and for the purpose as herein described.

No. 59,002.—ROBERT A. GAWLER, Concord, N. H.—*Carpet Fastener.*—October 23, 1866.—The driving shank has a double hook pivoted in its top and catching the edge of the carpet.

Claim.—A carpet fastener consisting of a shank to be driven into the floor of a room, and a hook, horse-shoe shaped, as shown, working in said shank, in the manner described.

No. 59,003.—WILLIAM O. GIBSON, Charleston, S. C.—*Cultivator Plough.*—October 23, 1866.—An angular share has its landside to the corn, and a plough connected to the same frame follows, throwing the earth toward the corn.

Claim.—The two parallel beams A A, connected by two cross bars *a a*, and provided with the guage wheel C, in combination with the bar *d*, provided with the coulter projection *e*, and horizontal blade or knife D, and connected to the beam A, by the standards *e c*, and the plough E, attached to the beam A', all being arranged substantially as and for the purpose set forth.

No. 59,004.—HENRY D. GREEN, Portland, Oregon.—*Manufacture of Illuminating Gas.*—October 23, 1866.—Equal parts of substances containing a large proportion of carbon, such as petroleum, and substances containing an excess of hydrogen, such as wood, are submitted to destructive distillation, either combined or separately, and their results combined.

Claim.—The combination of sawdust with naphtha, petroleum, or mineral oils, in about the proportions named, for the manufacture of illuminating gas, substantially as set forth.

No. 59,005.—JONATHAN J. GREEN, Grand Rapids, Mich.—*Washing Machine.*—October 23, 1866.—The ribs of the concave are connected by a flexible band, and are sprung over the toothed roller, which rotates within, carrying the clothes between the corrugated surfaces.

Claim.—The flexible rubbing concave C, in combination with the cylinder D, constructed substantially as shown and described.

No. 59,006.—J. B. GRIDLEY, Albany, N. Y.—*Brick Machine.*—October 23, 1866.—A horizontal cam wheel on the main shaft operates the vertical plunger, which forces the clay into the moulds, which are intermittingly fed beneath by a pusher actuated by cam projections on the cam wheel. The pusher travels on rollers and is retracted by a weight.

Claim.—First, operating the clay-compressing plunger of a brick machine by means of the grooved wheel I, or its equivalent, so that said plunger shall be depressed gradually but forcibly, and elevated, and arrested or retarded, in its operation at intervals, substantially as and for the purposes specified.

Second, the flanges or projections R, or their equivalents, for giving motion to one or more levers, which feed the moulds to the mud box.

Third, in a brick machine, mounting a wheeled follower or mould-feeding device upon tracks, substantially as and for the purpose specified.

No. 59,007.—JAMES H. GRIDLEY, Washington, D. C.—*Cotton Bale Tie.*—October 23, 1866.—The dovetail end catching between flanges of corresponding shape on a plate, or the other end of the hoop, is retained thereby against retracting tension.

Claim.—First, a cotton bale or other tie, so constructed that the fastening is made by the edges of said tie, in connection with corresponding flanges on the opposite end of the tie, or on a separate plate, having said flanges on it, substantially as described.

Second, the plate B, having flanges *b*, cast or otherwise, formed on its edges, in combination with the ends of the tie, when said ends are cut in dovetail form, substantially as described.

Third, the plate B, having flanges *b* on its edges, and lips *a*, as described, in combination with the tie, having a dovetail end with notches cut in the edges of said dovetail, substantially as described.

No. 59,008.—JOSHUA F. HAMMOND, Providence, R. I.—*Carriage Jack.*—October 23, 1866.—The vertical shaft is furnished with graduated pin holes, and is raised by the hand lever, which has a pivoted connection to the segmental cam bar, acting upon a roller-armed pin in the said shaft.

Claim.—A carriage jack with a fixed standard B, movable standard C, levers E and G, and link F, constructed and combined substantially as set forth.

No. 59,009.—BENJAMIN HANDFORTH, Chicago, Ill.—*Curtain Fixture.*—October 23, 1866.—The sheave on the jamb has a circumferential groove which allows the cord to traverse between it and the plate; a lateral strain on the cord leads the latter into an oblique shallower groove, where it is bound against the plate.

Claim.—In combination with the curtain A, roller R, and cord D, the arrangement of the sheave B, provided with the groove *a*, and one or more oblique grooves *b*, operating substantially as and for the purposes specified.

No. 59,010.—JAMES HARRIS, Kansas, Ill.—*Plough.*—October 23, 1866.—The rear plough is attatched to a laterally extending arm and furnished with a diagonal brace to the beam.

Claim.—The attaching of the rear plough H to the beam A, by means of a bent bar F, projecting laterally from the beam, and having a dovetail groove *d* made in it, to receive the plough standard, in combination with the front plough E, attached to the standard D, which is secured to the beam, substantially as and for the purpose set forth.

No. 59,011.—JOHN HARRIS, Marquette, Wis.—*Broom-head.*—October 23, 1866.—The toothed jaws embrace the brush and their prolonged pintles form straps which are fastened by screws into the edge of the cap.

Claim.—The combination of the arms A, pins B, and serrated bars C, when said arms, pins, and bars are constructed and combined substantially in the manner herein described and for the purpose set forth.

No. 59,012.—J. B. HERMAN, Mount Vernon, Iowa.—*Cultivator.*—October 23, 1866.—The transverse bar has vertically descending arms, to the lower ends of which the plough beams are pivoted. Chains connect the beams to the side of a rolling bar operated by a lever by which the ploughs may be raised. The lower ends of pivoted arms rest upon the beams to prevent their rising, and the action of raising the ploughs removes the impact of these arms. The outer beams are curved outward to give room for lateral play to those more central.

Claim.—First, the arms R, applied to the machine, substantially as shown, in combination with the hinged bar J, chains Q, and plough beams L, all arranged substantially as and for the purpose specified.

Second, the curving of the rear parts of the plough beams L in combination with the inner laterally-adjustable plough beams L, substantially as and for the purpose set forth.

No. 59,013.—JOHN B. HERR, West Lampeter township, Penn.—*Cultivator.*—October 23, 1866.—Attached to the tongue are two transverse beams carrying teeth adjustable laterally in slots.

Claim.—The combination and arrangement of the parallel shovel beams A B, united by the tongue C, handles H, and braces D, when constructed and operating in the manner and for the purpose specified.

No. 59,014.—A. V. HEYDEN, Milwaukee, Wis.—*Shifting Carriage Top.*—October 23, 1866.—The skeleton frame sets within the seat and is fastened to a plate which conforms to and is secured by catches to the seat; two posts pass through the seat and are secured by catches beneath.

Claim.—First, attaching all the supports of carriage top B to a single metal plate C, and fastening the same to carriage seat A, by means of catches D and E E, substantially as and for the purpose described.

Second, a carriage top with all its supports attached to metal plate C, with standards I I passing through seat A, with slots or notches in their lower ends, into which catches F F are locked by means of lever G, and held in position by spring catch H, together with catch D, to hold the middle of plate C firmly to seat A, all in combination, substantially as and for the purpose described.

No. 59,015.—LEWIS F. HILDEBRAND, Chicago, Ill.—*Machine for Cutting Tobacco.*—October 23, 1866.—The inclination of the trough assists the feeding of the tobacco. A roller on the frame presses it beneath the cross bar.

Claim.—The combination of the trough A, face plate C, the roller F, knife D, and the cutting block G, substantially as set forth.

No. 59,016.—WARNER HINDS, Worcester, Mass.—*Horseshoe.*—October 23, 1866.—The jointed shoe has side clips for attachment to the wall of the hoof.

Claim.—The horseshoe, constructed substantially as herein described, as a new article of manufacture.

No. 59,017.—CONSTANTINE HINGHER, New Brunswick, N. J.—*Stem for Tobacco Pipe.*—October 23, 1866.—Improvement on his patent of August 6, 1861. The tube is curved to prevent its discharge of liquid from the cup into the stem when the pipe is laid down. The tube rising from the bottom of the cup has a conical cap and the bottom of the stem has a secondary cap.

Claim.—First, the curved tube c, applied to the upper part of the cup B, and operating in combination with the same, and with the conical cap d, applied the tube b, which rises from the bottom of the cup B, substantially as and for the purpose described.

Second, the secondary cup c, in combination with the lower end of the stem A, and with the cork f, constructed and operating substantially as and for the purpose specified.

No. 59,018.—N. D. HINMAN, Stepney Depot, Conn.—*Hay Elevator and Conveyer.*—October 23, 1866.—The truck runs upon inclined ways. The hay is raised on a chain depending from a pulley in the frame. Power communicated to the hoisting rope raises the package to a certain height, when the truck is released and drawn along the ways to the bay, where the load is automatically released, and a bell rung; the relaxing of the rope allows the truck to return for another load.

Claim.—First, the frame I J, hook N, and cross bar j in combination with the pendent bars H H, rod O, and bar M, all arranged in the carriage B, and used in connection with the pins b, in the ways A, substantially as and for the purpose set forth.

Second, the pawl P, in combination with the thimbles W W', on the chain V, of the hoisting rope G, the rod R, and frame Q, all arranged substantially as and for the purpose specified.

Third, the bent or knee lever T, connected with the frame Q, provided with the spar c', and the signal composed of the bell Y, and hammer Z, substantially as and for the purpose set forth.

Fourth, the adjustable bars D E, placed on the ways A A, and the bar E, provided with the hinged plate X, substantially as and for the purpose specified.

Fifth, the combination of the carriage B, ways A, adjustable bars D E, and pins b, hoisting rope G, provided with the protuberance v*, and the chain V, with thimbles W W applied to it, and the frames I J and Q, rods O R, hook N, cross bar j, and the bent or knee lever T, all arranged to operate in the manner substantially as and for the purpose set forth.

No. 59,019.—HARRISON HODGSON, New York, N. Y.—*Hat Protector.*—October 23, 1866.—Attached to the inside of the crown is a device which projects feet through the crown to support the hat. The feet are projected and withdrawn by the rotation of a plate.

Claim.—First, the combination with a hat of the movable feet C, three or more in number, when so arranged and connected that they are projected beyond the crown and withdrawn simultaneously, substantially as described.

Second, the feet C. in combination with the radial legs D, to whose ends they are attached, substantially as described.

Third, the slotted plate H, with eccentric slots I, in combination with the legs D, which are raised and lowered by giving rotary motion to said plate, substantially as described.

Fourth, the wrench F, in combination with the slotted plate H, to the post L of which it is confined through the agency of the screw G, substantially as above shown.

Fifth, the shell or casing E, which covers and protects the slotted plate, in combination with said plate, and with the legs D, substantially as set forth.

Sixth, the apparatus, substantially as above set forth, in combination with a hat or cap, as above shown and described.

Seventh, the enlargements on the ends of the feet C, substantially as above set forth.

No. 59,020.—PHILIP HOELZEL, New Orleans, La.—*Steam Generator.*—October 23, 1866.— The lower-water chamber is interposed between the furnace and flues, and the wood-work of the vessel has mud valves; the water therefrom passes to the pipes and cylinders, which have a zigzag course beneath the boiler.

Claim.—First, the receptacle *a*, when constructed and arranged substantially as and for the purposes herein recited.

Second, combining with or connecting to a steam boiler the series of cylinders, for beating and purifying the feed water, said cylinders having blow-off pipes and the alternating connecting pipes, arranged substantially as herein set forth.

No. 59,021.—JOHN S. HOWARD, Schenectady, N. Y.—*Car Truck.*—October 23, 1866 — The connecting beam beneath the axle boxes has within it an elliptical spring on which the transverse beam rests. The spring is connected to a bar, to whose ends the axle bearings are pivoted.

Claim —The combination of the beam A, elliptical spring D, pendant F, pivoted bar G, and the axle boxes, and operating substantially as described, for the purpose specified.

No. 59,022.—R. C. HOWARD, Lena, Ill.—*Corn Plough.*—October 23, 1866.—The plough beams are pivoted in the front frame bar, and are connected by arms to a rock shaft which has a lever by which they may be raised. Guide rods from the axle restrict their lateral movement.

Claim.—The guide rods O, in combination with the lever *a''*, and beams F, substantially as described, for the purpose specified.

No. 59,023.—WILLIAM HUNTER, Detroit, Mich.—*Fence Post.*—October 23, 1866.—The pointed fence post has radial flanges beneath the surface of the ground; the inclined lower edges assist their entrance, and when driven they keep the post from turning or tripping.

Claim.—The combination of flanges such as above described with a pointed fence post, in the manner above described, for the purpose of making said post, when driven into the earth, firm and immovable and incapable of oscillation, or any other arrangement of flanges so attached which will secure substantially the same effect.

Also, the application of such flanges and the arrangement thereof to any article of use designed to be fixed in the earth, whether for fence posts, lamp posts, awning posts, or any other article which, by substantially the same means, can be firmly fixed or driven into the earth.

No. 59,024.—DANIEL HUSSEY, Nashua, N. H.—*Let-off for Looms.*—October 23, 1866.— The impelling pawl is driven positively by the crank shaft, and is lowered to operate on the detent by the arm which is connected with the whip roll. The tightening of the warps operates the whip roll in the usual manner.

Claim.—The combination of the cam *o* and the impelling pawl *m*, applied to the crank shaft as described with the arm *r*, and the escapement mechanism applied to the yarn guide and the yarn beam, the whole being to operate in the manner and under circumstances substantially as set forth.

No. 59,025.—ISAAC HUTCHINS, Jr., Wellington, Maine.—*Fumigator.*—October 23, 1866.— Tobacco or other material for furnishing fumes is placed in the cylinder and lighted, and the fumes are expelled by air from the bellows.

Claim.—First, the cylinder or box B, constructed as described, in combination with the bellows A, substantially as described and for the purpose set forth.

Second, the combination of the perforated plate C, with the cylinder or box B, substantially as described and for the purpose set forth.

No. 59,026.—BENJAMIN S. HYERS, Pekin, Ill.—*Sash Fastening.*—October 23, 1866.— The rack being withdrawn, allows the wheel to drop into the wider space below, and the sash is raised; the rack is then released and the spring forces it against the wheel which engages the side of the casing supporting the sash.

Claim.—First, a rack B, secured on the pivot D, when applied to a sash fastener, for the purposes and substantially as described.

Second the spring E, in combination with the rack B, substantially as described.

Third, the toothed wheel A, in combination with the rack B, and spring E, substantially as and for the purposes set forth.

No. 59,027.—GUSTAVUS W. INGALLS, Concord, N. H.—*Reed Plates for Melodeons, &c.*—October 23, 1866.—By the described mode a smoothed and rounded surface is given to the edges of the reed opening, no further dressing being required.

Claim.—The improved reed socket plate, made substantially as described, viz: by punching it through in one direction, and forming the mouth on that side which is opposite to that surface at which the punch is made to enter the metal.

No. 59,028.—HENRY C. INGRAHAM, Tecumseh, Mich.—*Ditching Machine.*—October 23, 1866.—The earth is broken by a double mould-board plough, and thrown by two side ploughs upon a belt which runs around a roller within the first plough, and thence around the elevator wheel in which it confines the soil, and from the top of which it passes to an adjustable pulley above the leading plough. The dirt is discharged to one side by an inclined scraper.

Claim.—First, the double mould-board plough H, single mould-boards J J, and chute F, when attached and pivoted by suitable framework on the axle of the elevated wheel A, substantially as and for the purpose herein set forth.

Second, the wheel A, provided with the flanges *a* and *a'*, in combination with the endless belt Q, tightening pulley X, and pulley I, when arranged to operate substantially as and for the purpose herein set forth.

No. 59,029.—ALFRED IVERS, New York, N. Y.—*Cistern for Water Closets, Urinals, &c.*—October 23, 1866.—The upper cistern has a continual supply of water which escapes through a pipe into the lower chamber. A float, on reaching a certain level, raises a valve and admits a flow of water that is carried violently through a siphon pipe to the pan.

Claim.—First, the receptacles or cisterns *a* and *c*, and siphon pipe *g h*, in combination with a float to cause the delivery of water from the vessel *c*, for the purposes and substantially as specified.

Second, a cistern for water closets, urinals, and sinks, to which water is gradually and continuously supplied, in combination with a float and siphon or equivalent mechanism, to effect a periodical discharge by the action of the water of the contents of said cistern, substantially as set forth.

No. 59,030.—DAVID L JAQUES, Hudson, Mich.—*Pump.*—October 23, 1866.—The pump has a two-valved head, and the valves by settling down when the pump is not operated, allow the water to run from the upper part of the supply pipe into a bowl which keeps the pump primed.

Claim.—First, the plunger *b*, of any suitable length, consisting of heads P and P, and leather packing D, at each end, lifting valve *l*, valve *r*, and opening O, all constructed and operated substantially for the purpose and in the manner set forth.

Second, the combination of plunger *b*, space *y y*, and pipe E, for the purpose and in the manner substantially as heretofore stated.

No. 59,031.—MOSES A JOHNSON and WILLIAM MURKLAND, Lowell, Mass.—*Machine for Moulding, Felting, and Fulling Hat Bodies.*—October 23, 1866.—The bat when formed and only sufficiently hardened to be moved is placed in the metal mould, which is to be as near as practicable of the shape of the finished bat. Steam is admitted through perforations in the mould, and also in the former if desired. The mould and its rim rotate continuously. The "former" has an adjustable reciprocating rotary motion and also an up-and-down vertical one; the shingles serve to crowd down and aid in fulling the bat.

Claim.—Moulding, felting, and fulling, fur, wool, or other fibrous material for hat bodies, in a perforated mould and by a cone or former corresponding thereto, and the direct action of steam upon the material to be so moulded, felted, and fulled by giving a reciprocating, rotary, and a rising and falling motion to the cone or former while acting upon or with the bat, substantially as herein described.

Also, in combination with the cone former having four motions, viz: back and forth, and up and down, given to it, the bowl or mould, having a continuous rotating or intermittent or reciprocating rotary movement, substantially as described.

Also, shingling, creasing, or shouldering the cone or former, substantially as and for the purpose described.

Also, automatically moving the weight out upon the beam lever, so as to increase the force of the blow of the falling cone or former, as the process proceeds, substantially as described.

No. 59,032.—E. B. JUCKET, Roxbury, Mass., assignor to himself and HUNNEMAN & Co, same place.—*Combined Fly Wheel and Crank Shaft.*—October 23, 1866.—The axes of the two shafts are in line, and their fly wheels are connected by screws traversing their webs or arms.

Claim.—The combination and arrangement of the two shafts, their cranks, the fly wheels, and the connecting screw or screws, or the equivalent thereof, the whole being substantially as described.

No. 59,033.—HENRY KAUFFMAN, York, Penn.—*Horse Hay Fork.*—October 23, 1866.—The broad plate with a harpoon point enters the hay, and the hooks expand laterally toward the retaining prongs, and are locked by the toggle links above. When the latter are pulled by the trigger rope, the hooks retreat and the load slips off.

Claim.—First, the dividing iron B, which prevents the hay from clogging and stopping the operation of the hooks in their lateral movement, and is provided with the spear head, which facilitates penetration into the hay.

Second, the combination of the hooks A A, and the removable prongs E E, arranged for joint operation in the manner and for the purpose set forth.

No. 59,034.—WASHINGTON E. KEENE, Lynn, Mass.—*Measuring Funnel.*—October 23, 1866.—The throat of the funnel is governed by a valve attached to an axial hand-rod. A cylinder in the middle has graduations for quantity, and after the funnel has received a charge it is placed above a bottle and the plug withdrawn.

Claim.—For a measuring funnel, the arrangement of the valve and valve rod, consisting of elastic piston *c*, attached to rod *d*, by compressing nuts *e* and *f* within cylinder *b*, constructed as described, with holes *m*, and exterior graduating belts, and combined with and arranged in the axis of funnel A by means of standards *g* and cross-piece *b*, all operating together as and for the purpose described.

No. 59,035.—WILLIAM KEGG, Lassellsville, N. Y.—*Cattle Gag.*—October 23, 1866.—The ring is placed transversely in the mouth and fastened by cords to the horns. Facility is thus given for the removal of an obstruction from the throat or the insertion of a probang to remove wind from the paunch of a hoven animal.

Claim.—A cattle gag consisting of a ring A and arms B B, constructed and applied substantially as and for the purpose specified.

No. 59,036.—GEORGE E. KING, New York. N. Y.—*Fluted Puffing for Shirt Bosoms.*—October 23, 1866.—Explained by the claim and cut.

Claim.—The within-described fluted puffing as a new article of manufacture, and made by fluting, by mechanism, and in a regular manner, a strip of muslin or other material, throughout its length, and compressing and flattening down the extremities of the flutes to form straight and regular borders, on either and opposite sides of the flutes, and afterward machine stitching said borders along and at the union of the borders with the flutes, substantially as specified.

No. 59,037.—A. P. KINNEY, South Carver, Mass.—*Spindle Step.*—October 23, 1866.—The step for the foot of the spindle is placed in a socket, the space between them having lubricating material, and an absorbent material over the openings which lead from the reservoir to the spindle.

Claim—The socket A, in combination with the step B, the latter being provided at its upper end with a cup *d*, which rests on the top of the socket, and communicates with the socket by means of one or more holes *e*, and also provided with a groove to receive an absorbent material *a*, all arranged substantially as and for the purpose set forth.

No. 59,038.—JOHN LAMB, Jeffersonville, N. Y.—*Washing and Wringing Machine.*—October 23, 1866.—Corrugated rubber boards are reciprocated in opposite directions upon each other by turning a crank in one direction. Reversing the direction of rotation moves the wringer also.

Claim—First, the combination of the rubber boards L M, crank wheels P S U, double crank V, and pitmen T with each other, with the crank shaft B, and with the frame A of the machine, substantially as described and for the purpose set forth.

Second, the combination and arrangement of the springs I, slotted stop J, rollers E H, gear wheels C D, and clutch F with each other and with the frame A of the machine, substantially as described and for the purpose set forth.

Third, the combination of the rollers E H and the rubber boards L M with each other and with the frame A of the machine, substantially as herein described and for the purpose set forth.

No. 59,039.—WM. M. LANEHART and JOS. C. KING, Cookstown, Penn.—*Machine for Washing Sand and other Mineral Substances.*—October 23, 1866.—The vessel is divided into compartments with inclined bottoms and elevators. The ore flows into the first compartment with a stream of water, and the heavy particles are carried up the inclined bottom by the elevator, and are deposited in a trough. A current of water flowing into said trough carries the ore into the next compartment, and so on, the heavier matters being finally removed by an elevator, and the lighter passing off at several points in the course.

Claim.—First, the construction and use of the cistern *a*, in combination with an apparatus for washing and raising sand, substantially in the manner and for the purposes above set forth.

Second, the cisterns *a a'*, one or more, in combination with a corresponding number of elevators and spouts, and the trough *f*, the whole being constructed and arranged substantially in the manner and for the purposes above set forth.

No. 59,040.—ALEXANDER LINDSAY and MYRON MOSES, Malone, N. Y.—*Hook Pin for Fastening Wearing Apparel.*—October 23, 1866.—One end of the bent wire forms a shank and pierces the garment; the shorter end forms a spring hook to prevent retraction. A ball moving on the latter assists in retracting it.

Claim.—First, a pin A of spring metal having a part *a* turned back to the body A, and terminating in a hook *b*, arranged and operating as and for the purpose specified.

Second, the pin A *a b*, Fig. 1, and releasing slide or ball D, substantially in the manner and for the purpose described.

No. 59,041.—JOHN S. LASH, Philadelphia, Penn.—*Clothes Wringer.*—October 23, 1866.—The upper roller is connected to its pinion by a crank and pivoted arm, so that a rotary motion is communicated in any position, without any change of position in the pinion.

Claim.—The combination of the jointed compensating link A, and arms *g*, with the non-sliding gear J K and the rollers C C', substantially in the manner and for the purpose described.

No. 59,042.—HENRY and FRITZ MARX, Baltimore, Md.—*Apparatus for Making Paper Pulp.*—October 23, 1866.—Several boxes are arranged around the periphery of the rough-faced revolving grindstone, and in each blocks of wood are forced edgewise against the grinding surface, by followers actuated by gearing and weights, while a stream of clear water is poured upon the surface. The pulpy result is sorted by a series of sieves and discharged into different receptacles, according to quality.

Claim.—The arrangement of the millstone B, the boxes D, followers X, and feed facing, operated by rack and pinion, band, wheel and weight, substantially as described.

Also, the arrangement of the longitudinally-shaking shoe provided with inclined sieves in vertical series with separate points of discharge, substantially as described.

Also, the corrugated or rough-surfaced grindstone in the relation and capacity described.

No. 59,043.—JOHN MASSEY, New York, N. Y.—*Feed Cutter.*—October 23, 1866.—The box has an endless apron at bottom, and is longitudinally adjustable on the frame to the cylinder, which has knives on its periphery.

Claim.—The adjustable frame F in combination with the knife I, arranged with the revolving knife cylinder B, constructed and operating in the manner and for the purpose herein specified.

No. 59,044.—EDWARD MAYNARD, Tarrytown, N. Y.—*Priming Metallic Cartridges.*—October 23, 1866 ; antedated October 3, 1866.—The shell is firmly attached to the solid disk which forms its base, and the latter has a rear chamber for fulminate, communicating by an aperture with the interior of the cartridge, and protected by a thin plate of metal in the rear.

Claim.—A closely-fitting metallic shield or cover in combination with the base of a cartridge when said base is so constructed as to receive solid support from the gun and is primed exteriorly, substantially in the manner and for the purpose herein set forth.

No. 59,045.—OLIVER C. MCCARTY, Haysville, Ohio.—*Gate.*—October 23, 1866.—Chains connect one rail to the central hinge shaft, so that turning the gate open will raise it, and its gravity will reclose it.

Claim.—The application of the wires or chains *c c* to the loop E and to the bar D, so as to rotate the gate and cause it to return to its original position when closed, and also the rod B, about which the gate revolves when opened for egress, and also the post *a*, upon which the gate revolves.

No. 59,046.—J. A. MCKINSTRY, Monson, Mass.—*Mitre Box.*—October 23, 1866.—By sliding the adjustable plates upon each other bevels of any angles can be cut. The guide plate indicates the angle. The saw guide slots are held in the same plane by the guide rod traversing their lower ends.

Claim.—The mitre box, the parts of which are composed of the adjustable plates A A', guide C, plate B, hooked rods D, graduated plate E, guide posts F, uprights G, and expanding rod A*, when arranged and operating substantially as described for the purpose specified.

No. 59,047.—C. H. MERRY, Dunleith, Ill.—*Discharging Grain from Vessels.*—October 23, 1866.—The well may be placed in the hatchway of the vessel, or permanently built in, and reaches from the upper deck to the floor of the hold. Gates to suit each floor are made in the sides of the well, and the grain flows in and is raised by the elevator in the ordinary manner.

Claim.—The combination with a vessel of a removable well A B C D, constructed substantially as herein described and for the purpose set forth.

Second, the combination with the removable well A B C D, of the gates F G, one or more, constructed and operated substantially as herein described and for the purpose set forth.

No. 59,048.—A. O. MILES, Nashua, N. H.—*Lock.*—October 23, 1866; improvement on Eddy & Miles's patent, December 17, 1861.—The adjustable tumblers are fitted in sliding frames which are movable under the action of the key up and down, to vary the relative positions of the tumblers, admitting of the changes being effected by the key alone. The bits of the key have lips to prevent their coming out of their slots, and being lost.

Claim.—First, the combination of the sliding frames C, T-shaped slotted tumblers D, having spring ends a, cross-bars d d', and projection b, of the bolt B, operating with the bits e of the key D, substantially as described, for the purpose specified.

Second, the fitting of the tumblers D in the frames C, by means which will admit of changes being effected in their positions by the action of the key alone when the tumblers are retained or held by the projection b of the bolt B, as set forth.

Third, the forming of the adjustable bits e with lips h at their ends, to prevent the slipping of the bits out from the key, as set forth.

No. 59,049.—ADAM MILLER, Chicago, Ill.—*Mole Plough.*—October 23, 1866.—The cutter bar is pivoted to the carriage, and adjustable to vary the presentation of its foot by means of an upper beam to which it is attached and braced. A mole trails behind the cutter foot and by means of a cord, hooks, and cross-bar, the tiles are drawn in behind the mole.

Claim.—First, the cutter brace E, when attached to a pivoted beam lever and movable coulter, substantially as specified.

Second, the arrangement and combination of the coulter F, cutter brace E, and pivotal lever B, with the standards C, provided with ratchet and pawl and beam A, substantially as specified.

Third, the arrangement and combination of the hooks L L, cords I I, and cross-bar k, with the mole H or G, for inserting two or more lengths or pieces of drain tile, substantially as set forth and specified.

No. 59,050.—W. H. MILLER, Brandenburg, Ky.—*Combined Rake and Spade.*—October 23, 1866.—A section with tines is hinged to the handle and laps upon the other tines when the implement is used as a spade and is locked at an angle to form a rake.

Claim.—The combined rake and spade, made and adjusted substantially as described.

No. 59,051.—THOMAS MITCHELL, Albany, N. Y.—*Steam Generator.*—October 23, 1866.—The feed water passes through a chamber having a safety-valve which allows its escape to the feed tank when under too great pressure : a cock operated by a steam pressure diaphragm regulates the water supply; expansion rods open the doors and admit cold air to the crown sheet.

Claim.—The combination of the means substantially as herein described for generating steam, with the means substantially as herein described for controlling the introduction of the water into the generator by the pressure of steam generated as and for the purpose set forth.

No. 59,052.—LORENZO OLEA MORENO, New York, N. Y.—*Medicine.*—October 23, 1866.—A remedy for dysentery and diarrhœa, composed of calomel, extract of opium, sirup of gum-arabic and cocoa.

Claim.—A medical compound made as and for the purpose described.

No. 59,053.—J. B. MORRISON, Fort Madison, Iowa —*Three-horse Splinter Bar.*—October 23, 1866.—The third whiffletree is attached to a cord which runs around pulleys on the splinter bar and tongue, and connects to the long end of the splinter bar, to equalize the draught on its two ends.

Claim.—The strap, cord, or chain F, with whiffletree G attached and applied to the splinter bar B and draught pole A, as shown, when said strap, cord, or chain is used in connection or combination with the whiffletrees D E, attached to the splinter bar B, and the latter secured to the draught pole one-third the distance of its length out of centre, substantially as shown and described.

No. 59,054.—THOMAS W. H. MOSELEY, Boston, Mass.—*Bridge.*—October 23, 1866.—The segment plate is connected above to an arch with flanges turned upward and downward, and beneath, to chords. The arcs are stepped at their ends upon angular shoes bolted to chords and segment.

Claim.—The improved truss, as composed of the arched plate A, the chord B, the flanges C C, or the same and the end strengthening plates g g.

Also, the combination of the shoes D D, and their adjusting screw -bolts k k, and nuts l l, with the truss made of the arched plate A, the chord B, and the flanges C C, or the same and the strengthening plates g g, the whole being arranged substantially as described.

No. 59,055.—DAVID M. NICHOLS, New York, N. Y.—*Steam Jet for Steam Generator Furnaces.*—October 23, 1866.—The flue has a central divided portion up which jets of steam are directed from a series of nozzles beneath, the annular space around being closed by segmental dampers, which are opened for the passage of air when a natural draught is used.

Claim.—The combination of the steam jet with a divided flue and a valve for one division of the flue, all operating substantially as set forth.

No. 59,056.—R. NICKSON, Akron, Ohio.—*Carriage Top Protector.*—October 23, 1866.—
In place of the ordinary prop block, the elastic bearing is provided for the carriage top.
Claim.—A protector for a carriage top formed of a grooved block A, springs C, and plate
D, substantially as described.

No. 59,057.—GEORGE NIMMO, Jersey City, N. J.—*Grate for Heating Stoves.*—October
23, 1866.—The grate below the base of the coal hopper slides forward for cleaning, and a
plate attached to the grate, and lying within the central chamber, is drawn forward and shuts
the lower opening of the hopper.
Claim.—The sliding grate *b*, and plate *i*, in combination with the closed hopper *c*, fitted
and operating in the manner and for the purposes set forth.

No. 59,058.—JOSEPH B. OAKEY, Indianapolis, Ind.—*Stove Cover, Pot and Pan Lifter.*—
October 23, 1866.—Explained by the claim and cut.
Claim.—A metallic head, constructed substantially as set forth, in combination with a
wood or other handle, in the manner and for the purpose herein set forth.

No 59,059.—F. M. OSBORN, Dover Plains, N. Y.—*Tack Extractor.*—October 23, 1866.—
The claw lever has a pivoted foot which forms a fulcrum in prying out a tack.
Claim.—The hinged support D, in combination with the handle A, having pronged end
B, substantially as described for the purpose specified.

No. 59,060.—MANNING PACKARD, Clarendon, N. Y.—*Gate.*—October 23, 1866.—The gate
is held, and slides back and forth on two rollers, having their bearings in the posts between
the upper two slats, or the next two if it be necessary to raise it in deep snow.
Claim.—Sustaining the weight of a gate and allowing it a free action by means of two
extension bars braced together and running on friction rollers *k k*, situated between them,
the whole arranged and operating as herein set forth.
Also, the employment of a series of extension bars *b c d*, in combination with rollers *k k*,
so arranged that the gate may be adjusted higher or lower at pleasure, substantially as
specified.

No. 59,061.—J. S. PATTERSON, Whitney's Point, N. Y.—*Potato Fork.*—October 23,
1866.—In the rear of the fork is pivoted a stirrup to receive the foot when digging, and act
as a fulcrum rest in raising the soil and tubers.
Claim.—First, pivoting a fulcrum or rest to the lower part of a garden or potato fork,
substantially as herein shown and described.
Second, the fulcrum or rest C, constructed as herein shown and described, in combination
with a garden or potato fork, substantially as and for the purpose set forth.

No. 59,062.—Cancelled.

No. 59,063.—CHARLES PICKERING, St. Louis, Mo.—*Obtaining Lead from Dross.*—October
23, 1866.—The lead is smelted with sulphur, saltpetre, and asafœtida.
Claim.—The within-described process of treating dross and scummings made from lead by
smelting the same together with the ingredients herein specified, for the purpose set forth.

No. 59,064.—LYMAN B. POTTER, Putnam, Conn.—*Intermittent and Expansive Gearing.*—
October 23, 1866.—By removing part of the pinion, and the addition of an oscillating spring
disk carrying a segmental rack, an intermittent and reduced motion is secured.
Claim.—The device for an intermittent and expansive gearing, constructed and operated
substantially as herein shown and described.

No. 59,065.—WILLIAM L. POTTER, Clifton Park, N. Y.—*Roofing Material.*—October 23,
1866.—Two layers of felt or other fabric enclose an interposed layer of the roofing composition patented to same party February 21, 1865, and July 17, 1866.
Claim.—An improved roofing formed by the combination of two pieces or layers A and
C of felt or other suitable fabric, with an interposed layer B of my plastic roofing patented
February 21, 1865, and July 17, 1866, when said roofing is constructed and prepared substantially as herein shown and described and for the purpose set forth.

No. 59,066.—W. J. PRALL, Pomeroy, Ohio.—*Oil Can.*—October 23, 1866.—The tray of
the can has a tube in which the pump stock is sleeved, and a perforated floor for the drip.
The recess in the lid covers the projecting upper works of the pump.
Claim.—The perforated suspended drop C, adapted for the purpose described, perforated
cap D, pump receiving tube E, and cover G, with projecting recess *g'*, in combination with
the body A, substantially as and for the purpose specified.

No. 59,067.—THOMAS PRATT, Valparaiso, Ind.—*Tire-shrinking Machine.*—October 23,
1866.—The tire is clamped at one point by a stationary vise, and at another by a vise which
is movable by means of a screw, so as to upset and shorten the tire between the vises.

Claim.—The combination of the flanged bed piece B, vice E, gripe D, and screw C, when said parts are respectively constructed and the whole arranged substantially in the manner and for the purpose set forth.

No. 59,068.—W. H. PRATT, Davenport, Iowa.—*Heater for Washing Machines.*—October 23, 1866.—A chamber is attached beneath the perforated bottom of the suds box, within which is a tubular space for the reception of heaters to keep the water warm.

Claim.—First, arranging chambers C and D beneath the perforated bottom wash box or tub, substantially as described.

Second, the combination with the heater C D, constructed as described, with a wash tub A B, substantially as set forth.

No. 59,069.—FREDERICK E. RAMM, Philadelphia, Penn.—*Sounding Board for Pianos.*—October 23, 1866.—A spring bar is pivoted on the frame, resting at its inner end upon a rib beneath the sounding board, and at its outer end being depressed by a bolt and nut.

Claim.—Combining with the sounding board of a piano an adjustable bearer or spring by which the sounding board may be raised when from any cause it has sunk or settled below its proper position, substantially as described and for the purpose set forth.

No. 59,070.—WILLIAM RANSOM, Portage, Wis.—*Medicine.*—October 23, 1866.—Composed of quinine, 22 grains; laudanum, 45 drops; sweet spirits of nitre, 45 drops; elixir of vitriol, 45 drops; and water, 3 ounces.

Claim.—The improved medicine, compounded substantially as herein described.

No. 59,071.—GEORGE RAY. Kinderhook, N. Y.—*Planting Machine.*—October 23, 1866.—The seed is taken from a hopper by cups on an endless belt, and dropped into the seed spout, which has a stop-slide, worked by a cam to prevent scattering deposit of seed. Another spout deposits fertilizing matter. The additional platform carries a further supply of seed. The seed ploughs and spout may be raised by a lever near the driver's seat.

Claim.—First, the slide n^2, arranged in relation with the cups n^1 of the carrier belt H, and with the hopper G and tubular standard G', substantially as herein set forth for the purpose specified.

Second, the slide A and elastic rod *f*. arranged in relation with each other and with the cam E, tubular seeding stock C and carrier belt H, substantially as herein set forth for the purpose specified.

Third, the supplementary tubular stock D, furnished with two slides *i j*, and so arranged and operated in relation with the seeding stock C, as to be capable of dropping a fertilizing material into the hill simultaneously with the dropping of the seed, substantially as herein set forth.

Fourth, the platform L, arranged below the rearmost end of the frame A, and in relation with the seat K, and hopper G, substantially as herein set forth for the purpose specified.

Fifth, the bent lever *p*, link *p'*, lever *r*, rod *r'*, and arms *s* and *u*, of the transverse shaft *s'*, so arranged in relation with each other and with the clutch *o*, and the bars B², of the thills or draught pole that the clutch will be thrown out of gear with the seed-conducting mechanism simultaneously with the raising of the forward end of the frame A, substantially as herein set forth for the purpose specified.

No. 59,072.—THOMAS B. RAYMOND, Saginaw, Mich.—*Rafting Pin.*—October 23, 1866.—The wedge is slotted from its edge toward its butt and straddles the raft rope, its width corresponding to the direction of the grain of the log.

Claim.—A rafting pin made in wedge form, and having a notch or channel extending up from its point or edge for receiving and grasping a straight rope, substantially as described and for the purpose set forth.

No. 59,073.—OWEN REDMOND, Rochester, N. Y.—*Steam Plough.*—October 23, 1866.—The main wheel carries excavators operated by a cam and eccentric, and so pivoted that they enter the ground in the direction of their length. The cam is attached to a loaded lever, by which damage from stones is avoided.

Claim.—First, the anchors operating substantially as described, or operating them in any manner by which their protrusion and withdrawal are effected in a somewhat similar way.

Second, the eccentric F.

Third, the movable cam H, lever and weight, or a spring equivalent to the weight.

No. 59,074.—THOMAS REECE, Philadelphia, Penn.—*Portable Propeller and Steerer for Boats.*—October 23, 1866.—The frame contains a spiral propeller and operative gearing and has in its wake a rudder. It is hooked over the stern and clamped to the stern post.

Claim.—The combination of the portable propeller and steering apparatus with the clamps *d* and *e e'*, arranged and operating substantially as and in the manner set forth.

No. 59,075.—JOSEPH REISING, Aurora, Ill.—*Counter Supporter.*—October 23, 1866.—A bent plate is attached above the heel and inside the counter, so as to strengthen the latter and keep it erect.

Claim.—As a new article of manufacture, a steel counter supporter constructed as described and applied to the rear portion of the heel, as and for the purpose specified.

No. 59,076.—FRANCIS N. RICHARDSON, Poultney, Vt.—*Stop Mortise Latch*—October 23, 1866.—A notched bar is pivoted in the lock, and when vibrated so as to clasp the square maudrel of the knob, prevents motion of the bolt. A button on the escutcheon inside of the door works the bar.

Claim.—The employment or arrangement of the stop latch E, catch L, and slot or recess F, in combination with the movable bar or stop H, the whole being arranged and operated in the manner and for the purposes substantially as herein described and set forth.

No. 59,077.—MATTHIAS S. and J. S. RICKEL, Geneseo, Ill.—*Wheat Drill.*—October 23, 1866.—Two sets of hoppers are provided for two varieties of seed. Oscillating agitators pass forward the grain. The ploughs are attached to pivoted bars, which are raised and lowered intermittingly by cams and springs.

Claim.—First, the hopper C, provided with communicating apartments b' b' c c', as shown, plates h h' h'', in combination with the agitator B, suitably operated, and provided with the tubes J and J', arranged substantially as described, for the purpose specified.

Second, the combination of the ploughs K, swinging bars L, springs M', and cam M, constructed and operating substantially as described, for the purpose specified.

No. 59,078.—J. L. ROBARTS, Brunswick, Ga.—*Plough.*—October 23, 1866.—The landside may be attached to the standard by either of its arms.

Claim.—The detachable and reversible V-shaped landside E, secured to the stock C, ubstantially as shown and described.

No. 59,079.—CYRUS W. SALADEE, Newark, Ohio.—*Water Drawer.*—October 23, 1866.—The yoke is pivoted in the well curb and the rope passes through its centre; the bucket becomes locked thereto when raised to the full height, and is tipped by the lever of the yoke to discharge the contents. The detent catch is withdrawn to release the bucket.

Claim.—First, the yoke A A 1, constructed and operating in the manner and for the purpose substantially as shown and described.

Second, the flange or spout C, in combination with the yoke A A 1, substantially as and for the purpose shown and described.

Third, operating the yoke A A 1 and bucket B, by means of the lever D, or its equivalent, as and for the purpose herein set forth.

Fourth, holding the bucket B in position against the underside of the yoke by means of the spring or catch P, or its equivalent, in combination with the groove N, in the manner and for the purpose substantially as shown and described.

Fifth, with four arms K K K K, forming the bale of the bucket, in combination with the cone N and yoke A A 1, in the manner and for the purpose substantially as shown and described.

No. 59,080.—JAMES SANGSTER, Buffalo, N. Y.—*Brick Machine.*—October 23, 1866.—The sides of the mould are perforated and the clay escaping into them is driven back by pins. Air and surplus clay escape through the faces of the pistons which act alternately upon the flat faces of the brick, the second one in order of motion forcing the brick and the first piston against a solid support. The adjustment of the stroke is made with shims. The brick is lifted vertically by an automatic table.

Claim.—First, the openings T T, be the number more or less, when placed sliding within the portion of the mould B, where the brick or material receives its pressure.

Second, the opening J³, in the bottom of the mould B, for the purpose of leaving room for the escape of the surplus clay or material as described.

Third, the openings J¹ and J², in the lower part or sides of the pistons, as and for the purposes described.

Fourth, the pins Z Z, or the equivalents thereof, when used in the mould B, substantially as described.

Fifth, a piston moving and compressing the clay to the point desired, which is then forced by the opposite piston with the brick partly compressed back again to a support, where it remains until the piston which moves it back gives the completing pressure to the brick.

Sixth, the employment of one or more plates R, substantially as described.

Seventh, the lifter U, when constructed with the openings B¹ and B², or the equivalent thereof, when used in combination with the plate C¹, substantially as described.

Eighth, the arrangement of the cam-connecting rod and the stationary guide E, when used to give the irregular reciprocating motions to the mould of a brick machine.

Ninth, the combination of the pistons H and H', as described and set forth.

No. 59,081.—HENRY SCHARCH, New York, N. Y.—*Hanging Carriage Bodies.*—October 23, 1866.—Dispensing with bolsters, the body is connected to the springs by U-shaped body loops, whose bows rest upon the springs, and ends are bolted to the frame of the bed.
Claim.—The attachment of the springs to the body, as and for the purposes set forth.

No. 59,082.—REINHARD SCHEIDLER, Newark, Ohio.—*Door Spring*—October 23, 1866.—The inner ends of the springs are so secured to the tongue as to cause it to take a portion of the strain. A curved projection on the inner end of the tongue forms a bearing upon the cylindrical casing of the coiled spring, whose outer ends are secured to lugs on the plate.
Claim.—First, the tongue C, in combination with the coils c c, secured to the lugs a a, in the manner described.
Second, securing the inner straight ends of the spring coils c c to a stiff tongue C, which is constructed substantially as described.
Third, the combination of the spring coils c c, tongue C, and retaining nuts C′ C′, with the slotted box cover B, substantially as described.
Fourth, sustaining the inner end of the lever C, upon the box B, or its equivalent, by means of a curved flanch f′, formed on said lever, substantially as described.

No. 59,083.—C. W. SCHROEDER, New York, N. Y.—*Apparatus for Drying Dishes.*—October 23, 1866.—The closet has racks to support the plate in dripping position and hot air or water pipes to evaporate the moisture.
Claim.—An apparatus for drying dishes, &c., composed of a box or stand provided with one or more sets of shelves and intermediate heaters, substantially as and for the purpose described.

No. 59,084.—CHARLES SEEFELD, Lomira, Wis.—*Gate.*—October 16, 1866.—The pivoted bar on which the gate traverses is tipped by a foot lever to make it run open, and returning after the foot is removed, forms an inclined plane on which it runs shut.
Claim.—The arrangement of the platform F, with the lever E, with bar D, so operated by weight on the platform F as to give the bar D the requisite inclination, so that the gate opens and closes by its gravity.

No. 59,085.—JAMES P. SINCLAIR, Millport, N. Y.—*Hitching Clamp or Holdfast.*—October 23, 1866.—The iron plate for attaching a hitch strap or rope to a post is bolted to the latter and has a pivoted lever which moves to admit the rope, and then returning clamps it by the pawl and ratchet.
Claim.—The hitching clamps, or holdfast, constructed and applied in the manner described for the purpose specified.

No. 59,086.—EDWARD M. SKINNER, Boston, Mass.—*Medicine Chest.*—October 23, 1866.—The two portions of the box are hinged together and when opened the upper is supported, and the whole interior becomes a series of steps, with holes for bottles, drawers, &c., for pharmaceutical purposes.
Claim.—First, constructing the chest or case in such a manner as to form, when the same is open, a series of shelves, substantially as and for the purpose specified.
Second the combination of a series of perforated shelves C, with the slides D, as and for the purpose specified.
Third, the combination with the chest A B, constructed as described, of the perforated shelves, the slides D, the compartments D′ E, and drawer F, substantially in the manner and for the purpose specified.

No. 59,087.—JOHN E. SMALL, Berlin, Wis.—*Crib and Chair*—October 23, 1866.—The rocking chair is converted into a cradle by lowering the arms which are pivoted to the seat, and adding the side and end pieces which form a barrier around the bed.
Claim.—The joints b b, in combination with the end racks e e, the folding side rack E, and the pieces d d, substantially as and for the purpose set forth.

No. 59,088—EARLE H. SMITH, Bergen, N. J.—*Sewing Machine Shuttle.*—October 23, 1866; antedated October 7, 1866.—This improvement avoids the drilling of a solid piece of metal in making a shuttle of nearly cylindrical form to conform to the shape of the loop of a needle thread, and besides allowing the insertion of the shuttle from the side, also admits of the adaptation of the tension devices within the shuttle. The fixed curved guide is to cause the thread to deliver always at about a right angle to the axis of the bobbin, and the slit in the side is to receive this curved part, and so occupy but little space within the shuttle. This shuttle is intended principally for machines in which the needle is withdrawn during or before the passage of the shuttle through the loop; as for example where the needle is driven by a crank or eccentric.
Claim.—First, a cylindrical shuttle formed of sheet metal, in combination with a bobbin inserted and removed from the side, substantially as described.
Second, the fixed curved guide for the shuttle thread, in combination with a slit in the shuttle, substantially as set forth.

No. 59,089.—R. T. SMITH, Nashua, N. H.—*Transmitting Motive Power.*—October 23, 1866.—The revolving cutter or brush is journalled in a holder with an attached handle. Motion is communicated by cores and pulleys from the revolving wheel on the main post. The tool is suspended by a gimbal joint and has freedom of presentation and motion in any direction.

Claim.—The swivel stirrup J, secured to the swinging rod I, in combination with the cage N, handle O, and shaft g, carrying the brush, cutter, or other article, and connecting with the shaft K, in the stirrup, by bevel gear or other equivalent means, substantially as and for the purpose set forth.

No. 59,090.—CHARLES F. SPENCER, Rochester, N. Y.—*Button.*—October 23, 1866.—On the extremity of the shank is a knob which is passed through two slotted disks in the rear. One of these disks being rotated the slots are no longer in coincidence and the button is fastened.

Claim.—The plates or disks d and e, in combination with the headed stem b, attached to the button, the parts or pieces being constructed and operated substantially as herein recited.

No. 59,091.—N. H. SPENCER, Canandaigua, N. Y.—*Churn Dasher*—October 23, 1866.—The inverted bowl of the dasher has ball valves to the openings on its upper surface, and perforations near its rim.

Claim.—First, forming the dasher A in bowl shape, substantially as herein shown and described.

Second, forming two or more valves in the upper part of the bowl-shaped dasher A, substantially as herein shown and described and for the purpose set forth.

Third, the combination of two or more ball valves C, with the dasher A, substantially as herein shown and described.

No. 59,092.—CHRISTIAN and FREDERICK STATTMANN, Chicago, Ill.—*Cap.*—October 23, 1866.—The band of this cap is extensible by means of elastic connections; the rear half of the covering may be turned over the front part, and thus reveal an entire covering of different material from that at first exposed.

Claim.—First, the employment of the reversible flap A, in combination with a cap whose front and rear parts are composed of different material, arranged and operating substantially as described and shown.

Second, in a cap whose sides are slotted or separated as described, the employment of an elastic insertion or connection D, substantially as and for the purposes specified.

No. 59,093.—MATHIAS STRICKER, Vincennes, Ind.—*Machine for Scalding Hogs.*—October 23, 1866.—The hog lies in a cradle within the tub, whose contents are heated by the furnace below. The hog is removed and tipped out on to a bench, by vibrating the cradle on its hinges.

Claim.—The box or boiler A A, combined with the swinging cradle for submerging a hog in the scalding water within, and turning the scalded hog out on a side bench or table by reversing the cradle, constructed and arranged substantially as herein described.

No. 59,094.—WILLIAM W. TAYLOR, Newark, N. J.—*Stitching Clamp.*—October 23, 1866.—The pivoted jaw is opened or closed by pressure on the respective ends of the toggle treadle, which is pivoted to one leg, and hinged to a plate connecting it with the base piece.

Claim.—First, the combination of the toggle lever D with the lower end of the jaw or leg A of the clamp, and with the foot block H, or equivalent, substantially as herein shown and described.

Second, the combination of the jointed arms C with the jaws A and B of the clamp, substantially as herein shown and described.

Third, an improved stitching clamp formed by the combination of the jaws A and B, jointed arms C, toggle lever D, and foot block H, or equivalent, substantially as herein shown and described.

No. 59,095.—SINEUS E. TOTTEN, Brooklyn, N. Y.—*Wrapper for Needles*—October 23, 1866; antedated October 7, 1866.—A metal magnetized plate is combined with the wrapper, to retain the needles.

Claim.—A wrapper for needles provided with a magnetic attachment, substantially such as herein described for the purpose set forth.

No. 59,096.—JOSEPH TRENT, Millerton, N. Y.—*Balanced Steam Valve.*—October 23, 1866; antedated October 7, 1866.—The tubular valve reciprocates in the steam chest, has a middle recess below to connect the exhaust with the cylinder, and two passages to connect the latter with the interior of the chest. The valve has circumferental packing rings.

Claim.—The above-described construction and arrangement of a tubular slide valve for steam or other engines, substantially as and for the purposes set forth.

No. 59,097.—H. W. C. TWEDDLE, Alleghany, Penn.—*Barrel Hoop.*—October 23, 1866.—
The protuberances on the inner surface of the metallic hoop hold it firmly on the cask.

Claim.—Making metallic hoops for barrels, casks, and similar vessels, with the inner surface provided with longitudinal ridges, so constructed as not to prevent the hoop being driven on the cask, while the projecting edge or edges thus provided cause it to remain in place when driven, substantially as hereinbefore described.

No. 59,098.—EZRA W. VANDUZEN, Cincinnati, Ohio,—*Hanging Bells.*—October 23, 1866.—The bell has an opening through the crown and is clamped between a yoke by which it is hung, and a crown plate to which the clapper is swung. The yoke and plate are bolted together, and the bolts sustain the spring buffers.

Claim.—First, a bell proper, having a crown opening B, in combination with a yoke and crown plate, substantially as and for the purposes set forth.

Second, the arrangement of the flanged or collared yoke C D, perforated crown plate E, and two or more attaching bolts F F'. G G', for combination with an open-crowned bell, substantially as set forth.

Third, in the described combination the crown plate E, attaching bolts F F', and caps H H', for the purposes stated.

Fourth, the crown plate E, having a cast projection, or spade handle, K, for the clapper, as and for the purpose set forth.

Fifth, the perforated crown plate E, in the described combination, with two or more attaching bolts F F', as and for the purpose set forth.

No. 59,099.—EZRA W. VANDUZEN, Cincinnati, Ohio.—*Hanging Bells.*—October 23, 1866.—The bell has an opening in the crown, and is clamped between a plate on the yoke and one on the interior boss, an axial bolt uniting the two. The spring buffers are socketed in the crown plate.

Claim.—First, the arrangement of the flanged and mortised yoke C D H, and bossed and tenoned crown plate E F G, the whole traversed by a single axial bolt J, in the manner set forth.

Second, the combination of sockets L L', beneath the crown plate and buffers I I, and set screws M M', for the purposes explained.

Third, the bossed crown plate E E, either with or without the tenon G, and spade handle I, as and for the purpose set forth.

No. 59,100.—BENJ. WALKER, Birmingham, Conn.—*Bolt-heading Machine*—October 23, 1866.—The heated rod is thrust into the feed opening until it strikes the gauge rod, is clasped between the slides, cut off, and moved laterally opposite the vertically sliding block; by the motion of the toggle the upper heading die is thrust against the end of the blank, and the advance of the side dies shapes the sides of the head, forming a cup around it, in which a second die acts as a swage. Another headed die, and a second advance of the side dies, finishes the head, and the bolt blank is discharged.

Claim.—First, the combination of the clamping heads B', gauge d, and cutter F, arranged substantially as described, whereby the blank is gauged, cut off, and clamped between the said heads B', substantially as herein set forth.

Second,'the dies f, in combination with the heading dies and with the clamping heads B', all arranged substantially as herein set forth, for the purpose specified.

Third, the heading dies z' z w in combination with the clamping heads B', and operating in succession, substantially as herein set forth, for the purpose specified.

Fourth, the construction of the preparatory heading die z', with convex inner sides and acute or recessed corners, substantially as herein set forth, for the purpose specified.

Fifth, the vertically sliding block M, furnished with suitable heading dies, in combination with the horizontally sliding block K, and with the clamping heads B', substantially as herein set forth, for the purpose specified.

Sixth, the arrangement of the sliding bars G H, furnished at their inner ends with dies f, the sliding bars G' I, and the cams G* I*, in relation with each other and with the clamping heads B', substantially as herein set forth, for the purpose specified.

Seventh, the slides C', furnished with the inclined plane J, and arranged in relation with the cutter F, toggle bar d*, and slide C, substantially as herein set forth, for the purpose specified.

No. 59,101.—SAMUEL J. WALLACE. Carthage, Ill.—*Car Coupling.*—October 23, 1866; antedated October 13, 1866.—The coupling pin is suspended from an arm of the horizontal rock shaft, which is moved from the side of the car, or by a rod from the roof. A scoop-shaped guide directs the link into the draw-head for coupling.

Claim.—The arrangement at the end of the car of the horizontal shaft A, arm F, and link G, operating the coupling pin P, substantially as described.

Also, in combination with the above, the rising and falling coupling guide L, operating as described.

No. 59,102.—NICHOLAS WALLASTER, Detroit, Mich.—*Grain-drying Kiln.*—October 23, 1866.—Hoppers receive the grain from the elevator and pass it to both sides, whose trap bottoms pass it to the first chamber. sections of whose perforated floor are vibrated to drop it to the next below, &c. The third floor is flat and imperforate. A spout discharges the grain from the machine, which has an arrangement of hot-air pipes; a furnace is used in addition when the grain is wet.

Claim.—First, the loader D, formed by combining the sections or traps d', the covers d' the bars d⁵, and the screws F with each other and with the frame of the loader, substantially as described and for the purpose set forth.

Second, the combination and arrangement of the sections g' g² g³ of the floors, constructed as described, in combination with each other, with the sides of the drying chamber and with the arms H, and rods I and J, substantially as described and for the purpose set forth.

Third, the combination of the pipe S S' S² S³, constructed and arranged as described, with the canals formed in the sides of the kiln, between its outer and inner walls, substantially as described and for the purposes set forth.

Fourth, the combination of the pipes B' with the openings A' in the outer wall of the kiln, and with the canals through which pass the pipes S' S² S³, substantially as described and for the purpose set forth.

No. 59,103.—A. WASHBURN and T. BRINTNALL, York. Ohio.—*Machine for Shearing Sheep.*—October 23, 1867.—The vibrating shears has a shield to prevent the wool being cut more than once, and is attached by a flexible connection and tumbling rod to a counterbalance arm, which is pivoted upon a standard, in which is placed the driving wheel which gives motion to the knife.

Claim.—First, the swinging arm E, with a counterbalance weight, substantially as described.

Second, the shaft or tumbling rod J, with flexible connections I K operating substantially as described.

Third, the adjustable shield R, covering the knives, and operating substantially as described.

No. 59,104.—ORVILL W. WAY, Troy, N. Y.—*Machine for Making Sheet-metal Pans.*—October 23, 1866.—Four adjustable wings or dies are used in connection with an inner die or platform so constructed as to work or operate the wings. A set of slides and bevel regulators adjust the wings so that rectangular pans of various sizes can be made in the same machine.

Claim.—First, the employment and arrangement of the movable and adjustable dies or wings B in combination with a central and vertically moving platform die C, in the manner substantially as herein described and set forth.

Second, the employment of the downward moving platform or die C, as the means for operating or working the said dies or wings B, in the manner and for the purposes substantially as herein described and set forth.

Third, the said die or wings B, in combination with the sides N, and the bevel regulators P, in the manner and for the purposes substantially as herein described and set forth.

No. 59,105.—MILO WEBB, Chenango Forks, N. Y.—*Machine for Raking and Loading Hay.*—October 23, 1866.—Sleeves are attached to the hubs of the rear wagon wheels, and are surrounded by other sleeves bearing cogs for driving the belt pulleys. The forward end of the supporting frame is clasped to the sleeves on the wagon wheels, thereby attaching the loader to the wagon.

Claim.—The sleeves E' and G, attached to the wheels and arranged to operate in combination with the yoke e, substantially as and for the purpose set forth.

No. 59,106.—THOMAS WELHAM, Philadelphia, Penn.—*Apparatus for Spinning direct from the Doffer of Carding Engines.*—October 23, 1866.—The adjustable spinner bed, by its change of position, is adapted for longer or shorter fibre. The other features are explained in the claims and cut.

Claim.—First, the combination with the doffer B of a carding engine of the adjustable spinner bed R, as and for the purpose described.

Second, the combination with the duffer B of a carding engine of the spinners C, constructed and operating as and for the purpose described.

Third, the arrangement of the spinners C with enclosed rollers S, the adjustable spinner bed R, and spools I, operating as herein described and for the purposes set forth.

No. 59,107.—W. T. WELLS.—Decatur, Ill.—*Hinge.*—October 23, 1866.—The leaves of the hinges attached to the gate are slotted, and admit of the vertical and lateral adjustment of the gate; the thumb nuts being loosened for that purpose.

Claim.—The slotted leaf M of the hinge B, in combination with the thumb screw G, secured as described, and adapted substantially as and for the purpose specified.

No. 59,108.—THOMAS S. WHEELER, Boston, Mass.—*Truss Pad.*—October 23, 1866.—
The pad is attached to the hoop by a screw, which is secured by a head within the elliptical
cup on the back of the pad, so as to have a certain freedom of motion to accommodate the
pad to the surface of the body.

Claim.—The improved joint or spring and pad connection made of the screw *a*, the elongated head *b*, and an elliptical cup or socket *c*, arranged and applied together in manner and
so as to operate substantially as specified; also its combination with the spring and pad of a
truss or abdominal supporter.

No. 59,109.—GEORGE D. WHEELOCK, Freedom, Ohio.—*Tanning.*—October 23, 1866.—
The hides are soaked in a liquor composed of fermented corn meal, common salt and water,
and the hair removed by a liquor composed of soft water, carbonate of soda, lime and lye.
They are then bated in a liquor composed of water, nitric acid, common salt, corn meal and
hen dung, and tanned in a bath composed of water, nitric acid, ammonia, salt, fermented
meal, catechu, and sumac. After tanning, the hides are stuffed with tallow, straits oil, castor oil, beeswax, alcohol, starch, glue and lamp-black.

Claim.—The within described process of tanning hides and skins, by treating the same
successively with the liquors herein set forth.

No. 59,110.—ELI WHITNEY, New Haven, Conn.—*Double Barrelled Fire-arm.*—October 23, 1866.—The breech frame and lock piece are constructed of one piece of metal, adapted
to receive the "patent breech" of the barrels and also the rear end of the hand piece of the
latter, thereby dispensing with several parts. Details are explained in the claims and cut.

Claim.—First, the manner, substantially as herein described, of constructing the lock case
A A' in one piece, and with tang sockets or holes *b b d*, and side apertures for plates *j j*, for
the purpose set forth.

Second, the combination of the pin *e* and the tangs of the breech piece in the construction
of a double-barrel gun, as set forth.

Third, the combination of the removable perforated circular plates *j̶j*, and shafts *g g*, with
the one-piece lock case A A', substantially as described and for the purpose set forth.

Fourth, the construction of the perforated plate *j*, with an off-set K, and passing the tumbler and hammer shaft through it, substantially as described.

Fifth, the arrangement of the partition A', case A, detached triggers *k k'*, scars *s s*, and
single pin *p*, substantially in the manner and for the purpose described.

Sixth, the combination of the socket *d* with the sockets *b b*, when the socket *d* is a continuation of the lock chamber A, and said sockets *b b d* receive tangs *d' b' b'*, of the parts C
C D, substantially as described.

Seventh, the partition A', within the case A, constructed as specified, and serving, in connection with the case A, the several described functions, as set forth.

Eighth, the manner herein described of constructing the lock case A A' in one piece, and
with hard metal bearings, such as described, for the purpose set forth.

No. 59,111.—L. J. WICKS, Bridgeton, N. J.—*Fruit Can.*—October 23, 1866.—The cover
has a threaded shank screwing into a cross bar, which rests beneath lugs in the neck of the
jar. An annular packing beneath the cup rests upon the rim of the jar.

Claim.—The lugs, cover, and cross bar, substantially as arranged and described and for the
purposes set forth.

No. 59,112.—SWAIN WINKLEY, New York, N. Y.—*Railway.*—October 23, 1866—The
rails are tied together by cross bars and rest on base plates with a series of arched corrugations and lugs which embrace the foot flange of the rail on each side. The plates are corrugated in one direction and arched in the other, and have cap pieces on which the rails rest.

Claim.—The construction of the base plates with arched corrugations as described.

No. 59,113.—JOHN WOOD, North Bloomfield, N. Y.—*Horse Rake.*—October 23, 1866.—
This invention relates to devices for locking and releasing the rake. To the handle are pivoted two inclined spring stops, the lower ends of which pass over the two middle forward
teeth; and in rear of these are pivoted to the handle two perpendicular spring stops, the
lower ends of which project over the two middle rear teeth. A bar attached to the front cross
piece extends back between the last stops, passing freely in a slot in the handle, and by its
wedge-shape releases the rake from the rear stops when the handle is raised.

Claim.—The special arrangement of the spring stops *k k* and *o o*, and the bar C, provided
with the double wedge *w*, when the said parts are used in combination with the single handle
B, for controlling the rake head, operating as and for the purpose herein set forth.

No. 59,114.—G. C. WRIGHT, Westfield, Ohio.—*Portable Fence.*—October 23, 1866.—The
boards are pinned between upright slats which are socketed in flanged foundation blocks and
braced by inclined struts.

Claim.—The herein described construction of a fence, consisting of the boards A, battens
B, blocks C. braces D, and pins *a*, substantially as specified.

No. 59,115.—Lewis R. Wright, Cohoes, N. Y.—*Seed Planter.*—October 23, 1866.—The adjustable connections of the seeders on the median line permit the distances between rows to be regulated, a marker indicating the position of the next row. Each seeder has its devices for regulating depth, and throwing in and out of gear. The seed is dropped behind an opener, and covered by a share and roller.

Claim.—First, the arrangement and combination of the levers *e f g k* with the sliding cylinder I and cog wheel D, in the manner and for the purposes substantially as herein described and set forth.

Second, the arrangement of a cog wheel or cylinder D, containing the series of cogs *a b c* upon or near each end of the drive shaft B, and the shafts G G, containing the sliding pinions E E, and the planting cylinders F F, each being arranged and combined in the manner substantially as herein described and set forth.

Third, the mode herein described and set forth for combining and disconnecting two or more seed-planting machines, constructed and arranged substantially as herein described and set forth.

Fourth, the employment of the marking device M, constructed and combined with the arm L, hinged to the frame A, in the manner substantially as herein described and setforth.

No. 59,116.—R. Wright and I. Wright, Franklin township, Penn.—*Machine for Raking and Loading Hay.*—October 23, 1866.—The rake head is fitted in slots in the main frame, and behind it is fitted a spring which keeps the rake to its work, and at the same time allows it to yield to large quantities of hay. An endless elevator raises the hay on to the wagon.

Claim.—The yielding rake head *g* in combination with an endless apron when used for raking hay and conveying it up into the hay wagon, in the manner herein described and set forth.

No. 59,117.—John H. Yager, Trenton, Ohio.—*Manual Power.*—October 23, 1866.—The pivoted brakes are connected by rods to each other and to the cranks of the shaft, and their reciprocation rotates the latter; the condition of the attachments to the crank is such that no dead point is encountered.

Claim.—The double brakes *b b* and *b' b'*, with their rock shafts *c c'*, and connecting rods *f* and *f'*, and *g g' g g'*, in combination with the vertical slots *m m* and *m' m'*, and the crank shaft *a a*, the whole being constructed, arranged and operated substantially as and for the purposes herein described.

No. 59,118—Wilbur F. Arnold, Winthrop, Conn., assignor to himself and P. A. Gladwin, Boston, Mass.—*Currycomb.*—October 23, 1866.—A plate is hinged to one side of the comb back and slotted for the passage of the tooth ribs, which project through it when in use. By swinging the plate outward the teeth are cleaned.

Claim.—As an improvement in currycombs the employment of one or more springs F, for throwing over and retaining the plate D upon the comb, substantially as set forth.

No. 59,119.—Solomon C. Batchelor, Cincinnati, Ohio, assignor to himself and W. C. Davis & Co.—*Stove Door.*—October 23, 1866—The pintles of the hinges are cast to the door and turn in recessed projections from the front of the stove. Inclined steps upon the contiguous faces of the hinge tend to keep the door closed. A lug prevents the door from being thrown off in opening.

Claim.—A stove door hinge composed of open backed sockets C, with the described stepped upper portions, in combination with the correspondingly stepped collar E of the pintle, as and for the purpose explained.

Also, in combination with a stove door hinge constructed as above specified, the lug F, for the purpose set forth.

No. 59,120.—Peter Brooks, New Haven, Conn., assignor to The National Line and Cord Company, same place.—*Machine for Making Fishing Lines and other Small Cords.*—October 23, 1866.—The several threads pass once or more around the conical shaft, which rotates slowly on its axis, and is also revolved about a centre; the form and motion of the feeding bar serving to aid in the twisting. When either of the threads breaks or gets too slack, it fails to hold up the lever, which by its own gravity then falls and causes the belt shipper to stop the machine.

Claim.—First, the feeding bar N, arranged upon the hollow shaft *a*, and combined with the hollow shaft *d*, and the thread guide *f*, constructed and arranged to operate substantially in the manner and for the purpose herein set forth.

Second, the levers *h*, in combination with the respective threads of which the line is formed and the twisting apparatus, constructed and arranged to operate substantially in the manner and for the purpose herein set forth.

No. 59,121.—William Burnett, San Francisco, Cal., assignor to John C. Paige, Stoneham, Mass.—*Safety Valve.*—October 23, 1866.—The weight box is suspended from a bell crank, which is connected by a link to the pivoted lever, whose toe rests upon the valve stem.

The arm of the lever passes through an opening in the valve box, and is vibrated by a slide on the locked cover when it is necessary to test the operative condition of the valve.

Claim.—First, the arrangement of the levers and their attachments, substantially as described and for the purpose specified.

Second, the lever and weight so arranged that the lever passes through and works freely within the body of the weight, substantially as described and for the purpose specified.

Third, the means herein described for relieving the valve of its load to test its operative condition.

Fourth, the arrangement substantially as described for securing the cover of the enclosing case.

Fifth, the cap provided to prevent the valve spindle from being wedged, substantially as herein described.

No. 59,122.—WILLIAM S. COOPER, Philadelphia, Penn., assignor to COOPER, JONES & CADBURY.—*Faucet.*—October 23, 1866.—The valve is operated by a screw spindle around the neck of which is an elastic packing placed below the cap through which the stem passes, and preventing leakage at that point.

Claim.—The hollow projection C, its spring G, cap F, annular packing piece *p*, and collar *n*, of the spindle E, the whole being arranged substantially as and for the purpose herein set forth.

No. 59,123.—H. V. DAVIS and GEORGE W. PEABODY, Amherst, N. H., assignors to GEORGE W. PEABODY, Amherst, and CHARLES B. TUTTLE, Milford, N. H.—*Seed Drill.*—October 23, 1866.—The seed slide is operated by a cam on the axle of the covering roller on which the drill is supported. A marker precedes the drop spout. The machine is driven by hand.

Claim.—The bar A, having the seed box D' secured upon its lower part and the strips B B attached to its lower end, and with the roller C between their rear ends and the furrow opener E, at their front end, in combination with the fixed perforated plate *e*, in box D', and the reciprocating perforated plate *g*, at the rear of plate *e*, operated by the rod E' and cam D from the axis of the roller C, substantially as shown and described.

No. 59,124.—J. N. DENNISSON, Newark, N. J., assignor to himself, F. H. and R. J. GOULD, same place.—*Air-compressing Pump.*—October 23, 1866.—The pistons are attached to cranks set at 180° on the same shaft, and reciprocate in cylinders of varying diameters, the larger having an air induction pipe, and discharging into the smaller, which has an eduction pipe. A water jacket keeps the parts cool.

Claim.—First, a pump composed of two (or more) cylinders of unequal size, provided with pistons connected together by a pipe with suitable valves, as shown, in combination with an air supply pipe leading to the largest, and a discharge pipe leading from the smallest cylinders, substantially as and for the purpose herein set forth.

Second, the jackets *l l'*, communicating with each other by a pipe *m*, and provided with a supply and discharge pipe, in combination with the cylinders A C, and with or without an additional pump *n*, substantially as and for the purpose described.

No. 59,125.—JOHN N. DENNISSON, Newark, N. J , assignor to himself, FRANCIS H. GOULD and ROSCOE J. GOULD, same place.—*Pump.*—October 23, 1866.—Improvement on his patent of February 7, 1865. The pistons are fast on the same rod and reciprocate together in their respective cylinders, provided with induction and eduction ports and valves. Each piston is perforated, and has a disk whose perforations may be made to register with those in the piston and thus render it inoperative, or by closing the holes make it water-tight and operative. The disks are operated by exterior devices: one by a sleeve on the piston rod, the other by a turn-key, with which the disk is made to engage at the end of the stroke.

Claim.—So arranging the pistons of a pump composed of two cylinders having pistons secured to one and the same rod that either or both can be put in operation or thrown out of operation, substantially as and for the purpose described.

No. 59,126.—VALENTINE FOGERTY, Boston, Mass., assignor to himself and PAUL P. TODD.—*Magazine Fire-arm.*—October 23, 1866.—The magazine is filled from a hole in the butt, the upper half of the magazine tube being lifted to allow the passage of the cartridges; a spring closes it when charged and the flanges of the cartridges rest in notches. When the guard lever is moved, and the breech-pin descends, the cartridge lying in the lateral recess rolls on to a plate which lifts it for loading by the forward motion of the breech-pin. Simultaneously with this the upper portion of the magazine tube is slipped forward, its notches bearing against the bases of the cartridges and slipping them collectively one space nearer the loading devices.

Claim.—First, the combination of the magazine tube arranged along the side of the gun stock, as described, with the mechanism for delivering the forward cartridge to the breech-pin or to the end of the barrel, substantially as herein shown and set forth.

Second, making the magazine tube in two jaws laterally hinged to each other so as to permit the easy insertion of the cartridge into the feeding apparatus, and also allow the head of the cartridge to pass through the smallest part of the feeding magazine, as described.

Third, the combination and arrangement of the breech pin D, guard lever G, arm or lever H, and connecting rod E, to give the required movements to the breech pin and magazine, as described.

Fourth, the plate V, for opening and closing the lower end of the magazine, in combination with the lever Y, for pressing open the sliding portion of the magazine, substantially as set forth.

No. 59,127.—THOMAS J. HALLIGAN, New York. N. Y., assignor to himself and SAMUEL SHAFTER, same place.—*Waxed Thread Sewing Machine.*—October 23, 1866; antedated June 14, 1866.—In this machine the lock stitch is made by a curved needle and a shuttle. The needle moves in a straight line to feed. The awl is shorter than the needle, so as not to interfere with the looping of the upper thread or with the shuttle. The presser has a positive grip on the cloth when the needle ascends, and vice versa. ·

Claim.—The vertically oscillating and laterally sliding curved eye-pointed needle arranged above a perforated bed plate of a shuttle sewing machine, and operated substantially as described, for sewing and feeding the material or work, as set forth.

Second, the combination of an awl with an eye-pointed needle, constructed, arranged and operated substantially as specified.

Third, the arrangement of a curved awl with a curved eye pointed needle above a perforated bed plate of a shuttle sewing machine and upon a carrier which receives a right-line movement, for the purpose of feeding the material, and a curved movement in a vertical plane for the purpose of carrying the upper thread through and below the bed plate, substantially as described.

Fourth, the arrangement of the pressure foot and its rod with respect to the curved laterally sliding and oscillating needle, substantially as herein described and shown.

Fifth, the arrangement of the sliding rods I and d, guide H, joint eye d', cam h', spring i, and set screw j, substantially as and for the purpose set forth.

Sixth, the arrangement of the hinged devices L L', lever K, and pressure device D D', substantially in the manner and for the purpose described.

Seventh, the pressure foot controlled by means of a cam N, constructed and arranged as set forth, in combination with the perforating and feeding needle, substantially as described.

No. 59,128.—FREDERICK HODDICK, Buffalo, N. Y., assignor to GEORGE A. PRINCE & COMPANY, same place.—*Melodeon.*—October 23, 1866.—The induction and eduction valves of the bellows are alternately and rapidly opened and closed by the action of the shaft, and a pulsative wind current is conveyed into the chest, giving the tremulous effect.

Claim.—First, a tremolo bellows F, having inlet and outlet valves $f^1 f^2$, constructed and operating substantially as herein described.

Second, a tremolo bellows F, or its equivalent, placed and used in connection and combination with the wind chest and swell valve of reed musical instruments, for the purpose and substantially as described.

No. 59,129.—JOHN KAILY, Canton, Ohio, assignor to himself and W. H. ALEXANDER, same place.—*Carpet Stretcher.*—October 23. 1866 —The toothed head is jointed to the shaft and admits of slight side turning. It is extended by a cord passing beneath a pulley, and a catch retains it in extended form.

Claim.—A carpet stretcher composed of the teeth, hinged and adjustable levers, and cord, the whole being arranged to operate substantially in the manner and for the purpose described.

No. 59,130.—WM. A. L. KIRK, Hamilton, Ohio, assignor to OWENS, LANE, DYER & COMPANY.—*Water Grate.*—October 23, 1866.—The grate bars of the steam-boiler furnace are tubular, and have longitudinal horizontal diaphragms extending nearly throughout their length, forming an extension of the boiler surface in which the water circulates.

Claim.—The tubular grate bars B, each provided with a longitudinal diaphragm D, when the said grate bars are arranged beneath the fire box, and at the lower end of the water leg A of the boiler, in the manner and for the purposes herein specified.

No. 59,131.—J. C. LONGSHORE, Mansfield, Ohio, assignor to himself and JOHN LONGSHORE, same place.—*Stove Cover Lifter.*—October 23, 1866.—Explained by the claim and cut.

Claim.—The combination of a stove, lid lifter D, pincers A B, pot, dish, sadiron, &c. &c., lifters F G, hammer H, and tack puller a, all in one implement, constructed substantially as shown and described.

No. 59,132.—STEPHEN C. MENDENHALL and SIMON SPARKS Richmond, Ind., assignors to STEPHEN C. MENDENHALL.—*Hand Loom.*—October 23. 1866.—An improvement on his patent of August 29, 1865, and relates to the devices for the instantaneous retraction of the treadle-operating cam, and its grooved hub, whereby a single cam of long sweep is enabled to operate any desired number of treadles in succession, and yet return automatically and instantly to the starting position.

Claim.—First, the peculiarly shaped cam E e e' e'' e''', in the described combination with the grooved hub F, constructed as described, and a series of treadles 1 2 3 4, or more, constructed as described, whereby one cam is made to operate all the treadles, and also to keep the treadles depressed during almost the entire revolution of the cam.

Second, the rise K, and stop L, in the described combination with the grooved cam hub F, as and for the purpose set forth.

Third, in combination with the cam E e e' e'' e''' and hub F, grooved as shown and described, the retracting spring G.

Fourth, the finger I, hinged to the cross rail H, and provided with a spring J, for the purpose set forth.

Fifth, in the described combination, the yielding and spring-sustained finger I, and the spring catch M, for the momentary detention of the finger outside of the hub, as set forth. .

Sixth, the releasing cam O, in the described combination with the spring catch M, and finger I.

Seventh, the combination of the grooved cam hub F, and the yielding and spring-sustained finger I.

Eighth, the arrangement of feathered and shouldered shaft A a a', cam E e e' e'' e''', grooved hub F, rise K, stop L, releasing cam O. spring catch M, and retracting spring G.

Ninth, in combination with the elements of the clause immediately preceding, the spurred pulleys C and D, winch C', eyeleted belt or chain B b, pitman P P, and batten Q.

No. 59,133.—JOHN O. MONTIGNANI, Albany, N. Y., assignor to A. TURNER.—*Clothes Rack.*—October 23, 1866.—The end of the shank is made larger than its main body. The hole is bored only the size of the main shank, which is then driven forcibly in. The wood, being elastic, closes over the head of the shank and thus holds it firmly in its place.

Claim. -The construction of clothes pins for portable clothes racks by forming their axes or pivots as described.

No. 59,134.—CHARLES E. MOORE, Roxbury, Mass., assignor by mesne assignments to NEW ENGLAND MODEL COLLAR COMPANY.—*Machine for Folding Paper Collars.*—October 23, 1866.—The collar is stamped into a curved angular groove, by a wedge-shaped die above, and is then grasped and passed endwise between rollers, which fold it upon the line already impressed. A single crank and cam movement actuate the parts.

Claim —First, folding collars by means of the angular groove P, and die N, substantially as described.

Second, the rollers R and Q, arranged and operating in combination with the groove P and die N, substantially as and for the purpose described.

No. 59,135.—H. J. OVERMANN, New York, N. Y., assignor to WM. S. HASCALL, same place.—*Manufacture of White Lead.*—October 23, 1866.—Litharge or finely divided lead is treated with nitric and sulphuric acid. The resulting sulphate of lead is treated with oxalic or acetic acid to destroy the crystalline character of the precipitated sulphate of lead. The division is best effected by heating the lead red-hot and pouring it into water.

Claim.—First, the manufacture of white lead from metal lead, or litharge, or ores of lead, by means of nitric and sulphuric acids or their several equivalents, in combination with pyroligneous or oxalic or similar acids, with or without the use of borax or its equivalent, substantially as described.

Second, also the use of pyroligneous or oxalic or similar acids for destroying the crystalline structure of the precipitated sulphate of lead, with or without the use of borax, or its equivalent, substantially as described.

Third, also, in the manufacture of white lead, dissolving and precipitating lead or its compounds at the same operation in one vessel, substantially as described.

No. 59,136.—ELIAS B. QUICK, Brooklyn, N. Y., assignor to himself and GEO. T. PALMER, same place.—*Trunk.*—October 23, 1866.—Plates with ventilating openings are inserted into the sides of the trunk and are closed by other plates when required.

Claim.—A trunk with a ventilator or ventilators attached, substantially as shown and set forth.

No. 59,137.—HENRY J. and JAMES REEDY, Cincinnati, Ohio, assignors to JAMES REEDY.—*Hoisting Machine.*—October 23, 1866.—The hoisting rope is double and passes under a sheave on a staple around the upper beam of the platform, and the dropping of this staple from breakage of the rope frees the serrated eccentrics for the suspension of the platform. The platform is balanced by a weight on a cord passing over the windlass around a sheave and to the windlass. Two cords communicating to every story serve to lock the drum.

Claim.—First, the mode substantially as described of supporting and elevating a hoisting platform I, by a single rope P, whose ends are secured to opposite extremities of the windlass H, while its bight or middle portion is rove through a sheave O, upon the platform for the purpose described.

Second, counterbalancing a hoisting platform by a weighted cord Q, which being carried horizontally over one end of the windlass, traverses a sheave R, and is carried back and secured to the windlass at or near its mid-length and on the reverse side from the hoisting cable P.

Third, the self-locking and releasing brake U X, arranged and operating substantially as set forth.

Fourth, the provision of the serrated eccentrics Y Y', spring bolts Z Z', and stirrup N, the whole operating as a safety check, in the manner explained.

No. 59,138.—PHINEAS LEESON SLAYTON and CHARLES J. KANE, New York, N. Y., as signors to ALMET REED, same place.—*Circular Loom for Weaving Hats.*—October 23, 1866.—Improvement on Slayton's patents of February 2 and November 2, 1864, for looms for weaving hats. A full description would occupy too much space. Among other features the upper series of partitions, which receive the warp carriers, are lowered and elevated, as occasion requires, within the circumference of the series next below, so as to bring such upper series into proper position to be put into connection with the lower series, by a transferring bridge, over which the warp carriers are conveyed from either series to the other. The frames of these series are also made movable, that the operator may put them out of the way of said bridge. The separating wheels of the lower shuttle carriage are constantly rotated positively by the motion of the carriage. The warp carriers, when out of action, are collected in a circumferential channel below the shuttle carriage, out of which they are successively raised automatically, as needed to form the shed.

Claim.—First, in a circular loom having one or more sets of weaving machinery, elevating and depressing the upper series of partitions a^3 and their attachments, substantially as above described.

Second, elevating and depressing the ring H, carrying the series of partitions a^2, and the ring F, carrying the partitions a', by means of screws S, or their equivalents, substantially as described.

Third, the use of fixed rack U, or its equivalent, by which an independent motion is given to the separating wheels $q\ q'$, of the shuttle carriage, substantially as above described.

Fourth, the primary channel O', beneath the shuttle carriage which receives and retains the stems of the warp carriers n, until removed by the indicator rods r, substantially as described.

Fifth, raising the warp carriers from the primary channel O' into the path of the shuttle carriage by means of an indicating apparatus, made substantially as described.

Sixth, transferring the warp carriers from the lower part of the loom to the upper part, and vice versa, by means of the bridge shown in Figs. 9 and 10, or any equivalent device, substantially as described.

No. 59,139.—SEYMOUR ROGERS, Pittsburg, Penn., assignor to LUMAN ROGERS.—*Horse Hay-fork.*—October 23, 1866.—The hoisting bar slides within a semi-cylindrical case and has a barb pivoted to its lower end, which is thrust outward through a slot in the case by the sliding motion of the bar. The bar is held at any desired point by a cam lever pivoted to the upper end of the case, and fitting notches along the side of the bar.

Claim.—The combination of the elevating rod d, (to which the hoisting rope is attached,) having a barb f pivoted thereto, with the penetrator or sheath a, and cam lever i, constructed and operating substantially as hereinbefore described.

No. 59,140.—DAVID SAWYER, Webster, Ohio, assignor to himself and ROBERT BARKER, Perrysburg, Ohio.—*Ditching Machine.*—October 23, 1866.—The carriage is advanced by the sweeps connected by gearing to the wheels. The rotary excavator is worked by gearing from the same motor, and has plough points and elevator buckets, whose hinged bottoms are fastened by bails, the latter are tripped by projecting fingers which come in contact with a flange at the place of discharge, and are returned by a spring to relock the closed bottom after discharging. The rotary excavator frame is raised and lowered by chain, shaft, and gearing operated by hand crank.

Claim.—First, the wheels B C D and H, and shaft E, as arranged with shaft F e and e, wheels G G' and d d', for the purpose and in the manner set forth.

Second, the wheels I and K and sweep M, as arranged, with the wheels c' d d' G G' and B C and D, for the purpose and in the manner described.

Third, the links N and lever N', rod O, and staple P, as arranged, for the purpose and in the manner specified.

Fourth, the shaft T, chains W, and wheels V', as arranged with the sheave X, wheel V, and frame A, in the manner and for the purpose set forth.

Fifth, the bucket b", cutting edges e", in combination with the head g and spade f, for the purpose as specified.

Sixth, bail d'', arm d''', as constructed and operated by the cam j, slide l, and spring o, for the purpose described.

Seventh, the bottom C'' and arm n, as arranged and operated by the cam m, as, for, and in the manner set forth.

Eighth, the adjustable cam j, slide l, and spring r, as arranged for the purpose and in the manner specified.

No. 59,141.—GEORGE A. SEAVER, New Orleans, La., assignor to ALEXANDER H. SEAVER, Brooklyn, N. Y.—*Cotton Bale Tie.*—October 23, 1866.—One end of the band is folded around the end of the constricted link, and the other is passed over the other end of the link, its edges being pinched in and bent by the latter to prevent retraction.

Claim.—Making the inner sides of the link with a curved or angular form, substantially as described, for the purpose of holding the band by indenting its edges, in addition to th transverse bending when it has been forced into position.

No. 59,142.—EDGAR M. SMITH, New York, N. Y., assignor to MITCHELL, VANCE & CO., same place.—*Feeder for Carburetters.*—October 23, 1866.—The flow of oil is regulated by a valve on a lever actuated by a float in a chamber between the reservoir and carburetter.

Claim.—First, a cased or caged float and valve interposed between a supply barrel or cask, and the gas apparatus or machine, for the purpose of supplying automatically and in uniform quantities the gasoline or other gas-making material from the barrel to the apparatus, substantially as herein described.

Also, the vent tube or pipe d, extending from the cage to the gasometer or service pipe, as described.

No. 59,143.—WILLIAM G. SNOOK and O. C. PATCHELL, Corning, N. Y., assignors to themselves and A. H. GORTON, same place.—*Piston Packing.*—October 23, 1866.—Steam is admitted through the disk to the interior of the piston to expand the packing rings; when a sufficient pressure is attained the interior pistons move a cut-off slide and close the apertures. Excess escapes at a valved opening.

Claim.—First, the combination of the bolt or stem J, perforated slide K, and spring N, with each other and with the grooved part E, ring F and G, and perforated plates B D, substantially as described for the purpose specified.

Second, the combination of the springs L and relief valve M with the port C of the piston head, substantially as described for the purpose specified.

No. 59,144.—CHARLES SWETT, Vicksburg, Miss., assignor to CHARLES G. JOHNSON, New Orleans, La.—*Cotton Bale Tie.*—October 23, 1866.—A block is cast or stamped with parallel slots the width of the hoop, and projections beneath the slots nearly covering the openings. Each slot receives one end of the hoop which is passed into it, and one or both bear against the projection to confer greater tenacity to the bite on the side of the bent hoop as the cotton presses against it.

Claim.—As a new manufacture, the improved fastening block for uniting the ends of metallic bands, when said bands are made to embrace compressed bales of cotton, or other equivalent substance, substantially as herein set forth.

No. 59,145.—N. B. WHITE, South Dedham, Mass., assignor to himself and HENRY B. BAKER, same place.—*Device for Forming the Eyes of Bed Springs.*—October 23, 1866.—The lower coil of the spring is clamped between the two plates, bent around one of the pins, then bent under the lower plate and over the corner at the opposite side, by which means the eye for the slat is formed.

Claim.—The combination of the two plates or jaws C D, the stud E, and the shoulder G, arranged substantially as described.

Also, the combination of the two plates or jaws C D, the stud E, the shoulder G and the shoulder F, arranged substantially as specified.

Also, the combination of the recess h with the shoulders F and G and the stud E, applied to the two plates C D, as specified.

Also, the combination of the plate D with the bending recess e arranged in it, as and for the purpose specified.

No. 59,146.—JAMES WOODFORD and WILLIAM H. BANCROFT, Portland, Wis., assignors to WILLIAM H. BANCROFT and W. L. WARD.—*Steam Generator.*—October 23, 1866.—The wick tube heats the water in the chamber around it. The flame rises into a longitudinal opening in the lower part of the boiler, spreads right and left into chambers, and from thence passes by short transverse flues into a longitudinal central flue, to which the exit flue is connected.

Claim.—First, a boiler D, when constructed with an internal fire chamber F, flues F', and an internal flue F^2, and arranged substantially as set forth.

Second, in combination with a boiler D, a jacket C, surrounding the wick tube B, for heating the water before it is applied to the boiler, substantially in the manner set forth.

No. 59,147.—JAMES HOWARD and EDWARD TENNEY BOUSFIELD, Bedford, England.—
Steam Generator.—October 23, 1866.—The vertical tubes are connected by short, threaded,
wrought-iron sections, or by flanges and bolts, to the upper and lower horizontal cylinders,
which discharge steam and receive water respectively. The vertical tubes are so grouped as
to admit the passage of a person to clean them, and have screens at their upper portions to
direct the caloric current, and protect the portions of·the tubes above the water line.

Claim.—First, a series of tubes *a b c*, coupled and connected substantially as represented
in Fig. 8, for the purpose set forth.

Second, arranging the sections of tubes substantially as herein described, to give access
to the heating chamber.

Third, the screens *f*, for directing the current of heated gases, and for allowing a free circulation of heat around the upper part of the sections of tubes, substantially as and for the
purpose described.

No. 59,148.—RALPH A. JONES and JOSEPH HEDGES, Aylesbury, England.—*Telegraphic
Signal.*—October 23, 1866.—A tablet is formed by inserting in wood pieces of conducting
material according to a specific arrangement; different rows have varying arrangements
as regards relative position and length of the conducting and non-conducting portions to
symbolize letters, words, &c. A stylus with suitable electric connection is drawn over the
requisite series on the tablet, and effects a dash and dot, varied by a long stroke, upon the
traversing paper of the receiving instrument.

Claim.—An alphabet or characters composed of a long stroke used in conjunction with
a dot and dash, forming the several characters of the Morse alphabet, substantially as and
for the purpose described.

No. 59,149.—HENRI LAMOTTE, London, England.—*Apparatus for Rectifying Alcohol.*—
October 23, 1866.—The two boilers are connected together by three pipes, from one of which
extends a tubular purifier opening into the vertical rectifier. The latter is connected by pipes
with a condenser, and it with a refrigerator for the ethereal products. Another pipe from the
boilers communicates with a purifier which' leads to a condensing coil. The rectifier is connected by the remaining pipe with the boilers.

Claim.—The boilers A No. 1 and A No. 2, and rectifying column or analyzer B, in combination with the purifiers F and G, the parts being connected by pipes and regulated by
cocks or taps, in the manner and for the purposes set forth.

No. 59,150.—ERNEST MANGEON, Paris, France.—*Privy Seat.*—October 23, 1866.—The
cover has a downwardly projecting annular lip, which when closed enters a trough containing
liquid to prevent exhalation. The catch valve closes on the opening of the cover by a system
of connections therewith.

Claim.—The combination of the cover D, valve C, and mechanism by which the position
of the cover controls the position of the valve, so that when either is open the other is closed,
all being arranged in connection with the hopper A and conduit pipe B of a water-closet or
privy, substantially as herein set forth.

No. 59,151.—JAMES J. McCOMB, Liverpool, England.—*Applying and Securing Metal
Bands on Cotton Bales.*—October 23, 1866.—The obliquity of the slot in the buckle makes
it bite more decidedly on the hoop, and the diminished length of the slot makes it bind on
the edge of the hoop. The jaws of the grapple draw the band tightly, one resting on the
buckle, and the other over the band with spurs in the bale.

Claim.—First, the peculiar form of the self-acting nipping ties or metal band lock for connecting the ends of metal bands surrounding cotton and other bales, as hereinbefore described
and set forth, and forming the slots or holes through the same in an oblique direction, and
also forming the holes or slots on one end a little narrower than the width of the metal band
by forming the side or sides at an angle, substantially in the manner and for the purposes
hereinbefore described and set forth.

Second, the improved construction of grapple, jointed to and operated by a hand lever
having jointed prongs, substantially in the manner and for the purposes hereinbefore described and set forth.

No. 59,152.—JAMES J. McCOMB, Liverpool, England.—*Cotton Bale Tie.*—October 23,
1866.—The folded ends of the bands are passed through the slots of the buckle, being less
readily withdrawn by tension than the single end of the hoop.

Claim.—The peculiar manner of folding the metal bands as hereinbefore described and
illustrated by figures 1 and 2, substantially in the manner and for the purposes hereinbefore
described as set forth.

No. 59,153.—CARL HENRIK RAMSTEN, Carlskrona, Sweden.—*Device for Lowering and
Detaching Boats from their Davits.*—October 23, 1866.—Each end of the boat is suspended
by a similar device. The hook which engages the eye of the davit fall-block consists of

three pieces, which, while the boat is suspended, maintain their hold, but become disengaged by the change of position when the boat touches the water. The disconnection of one disengages the other.

Claim.—First, the hooks D D, constructed substantially as described of a lever *b*, pivoted arm *c* and hook end *d*, for operation essentially as and for the purpose herein set forth.

Second, the combination with disengaging hooks, as described, at or to opposite ends of the boat, of the connection between the hooks, as formed by the rod E and chains *x n*, for action in the manner described.

Third, the stopper F, in combination with the rod E, chain *y* and lever *b*, for operation together, essentially as specified.

Fourth, in combination with the gripe H, the double-jointed hook *r s*, constructed and arranged to hold the gripe when said hook is lashed by the fall, but relieved therefrom when the fall is released, substantially as herein set forth.

No. 59,154.—DANIEL E. SOMES, Washington, D. C.—*Composition for Forming Useful and Ornamental Articles.*—October 23, 1866.—Composition for moulding statuary, vases, plastering, &c., consisting of soapstone dust, combined with gypsum, sulphur, clay, marble dust, plumbago, litharge, red lead, sand, glutinous substances, caoutchouc, gums, varnishes, &c; various combinations for specific uses.

Claim.—The combination of the substances or their equivalents, as herein described.

'No. 59,155.—CORNELIUS L. and A. B. IRVING, Fort Wayne, Ind.—*Piano-forte Action.*—October 23, 1866.—The two series of keys form octaves in the usual manner. The upper set may be operated independently. The lower set are obliquely connected, so that striking a key on the lower board actuates the key an octave above in the upper key-board, but not conversely, as striking the said upper key fails to connect with the said lower one, owing to the break joint connection of the levers.

Claim.—First, the combination of two key-boards with the octave couplers, constructed and operated in the manner and for the purpose substantially as set forth and described.

Second, the jointed compound levers *n n'*, in combination with rod *m* and connecting rod *o*, for the purpose specified, the whole being constructed substantially as set forth.

Third, curved arm *i*, in combination with key A and key B, the same being constructed in the manner and for the purpose described.

No. 59,156.—CHARLES ALDEN, Newburg, N. Y.—*Extinguishing Fires.*—October 30, 1866.—The building has fire-proof walls and air-tight compartments, divided by air spaces. Pipes for exhausting air from the spaces, and injecting steam thereinto, are provided against the casualty of fire.

Claim.—First, a building composed of a series of air-tight compartments separated from each other by air spaces or air chambers, for the purpose specified.

Second, the arrangement of pipes *d e f g*, in combination with the air spaces C D, and air-tight compartments B, constructed substantially as described.

No. 59,157.—ALEXANDER APPLEBY, Bromfield, Maine.—*Apparatus for Tanning.*—October 30, 1866.—The drum has openings through the heads, and partitions at each end. One part of each partition is inclined, forming openings, across which the slats are placed.

Claim.—The hide-handling drum, as made with inclined partitions, combined with and arranged within it, for the purpose of producing reciprocating movements of the hides, while the drum may be in revolution, as stated.

Also, the arrangement of the grated mouths E in the drum, and with its inclined partitions, substantially as specified.

No. 59,158.—FREDERICK ASHLEY, New York, N. Y.—*Clothes Sprinkler.*—October 30, 1866.—The can has at one end a charging nozzle and cap, near which is a rose for discharging jets of water, when the can is agitated.

Claim.—The arrangement of the spout E, with the perforated plate F and the nozzle C, with the screw cap D, in combination with the reservoir A, combined and operating in the manner and for the purpose herein specified.

No. 59,159.—JAMES S. and THOMAS B. ATTERBURY, Pittsburg, Penn.—*Making Ring Jars.*—October 30, 1866; antedated August 30, 1866.—The rings or annular elevations formed upon the sides of the jar are concave on their inner surfaces to contain color. The jar and its cover are blown separately in moulds, and the covers are afterwards fitted by reheating and reaming out the mouth of the jar.

Claim.—The arrangement of the parts forming a mould for producing jars and other articles of glass, substantially as herein described.

No. 59,160.—W. E. BABCOCK, East Pembroke, N. Y.—*Water Elevator.*—October 30, 1866.—The drum is loose on the shaft, and has a crown ratchet, which connects with one on the crank shaft, upon moving the shaft thereto. This movement is made by a weighted pin in a collar between two flanges on the drum, the pin moving in an oblique slot in an enveloping case; another pin working in a similar slot serves to disconnect the ratchets for lowering the bucket.

Claim.—The shaft B, provided with the ratchet D, in connection with the drum E, placed loosely on the shaft, and provided with the ratchet F and loose collar G, and the case H having the oblique slots *b* made in it through which pins *c c* and *d*, attached to the collar pass, one pin *d* being provided with a weight *e*, and all arranged to operate in the manner substantially as and for the purpose set forth.

No. 59,161.—MOSES F. BAGLEY, Alton, Ill.—*Device for Discharging Bilge Water from the Holds of Vessels.*—October 30, 1866.—The plunger when projected through the side of the vessel exposes an opening towards the stern, through which the bilge water is drawn by the current. The plunger works in a gland tightened by bolts to the bed plate.

Claim.—First, the hollow plunger D, the bed plate B, and the gland C, when constructed and arranged substantially as herein described and set forth.

Second, the plunger D, the lever E, and the fulcrum E', when constructed and arranged as described and set forth.

No. 59,162.—SILAS BARKER, Hartford, Conn.—*Carriage Axle.*—October 30, 1866.—A sleeve is fitted tightly upon the spindle, and is capable of adjustment thereon so as to expose all portions in turn to wear, and avoid the constant wear upon the under side of the spindle proper.

Claim.—The combination of a sleeve C with an axle A, when the said sleeve is constructed and arranged so as to be adjusted thereon, substantially in the manner and for the purpose specified.

No. 59,163.—JAMES T. BARNES, Hudson City, N. J.—*Furniture Caster.*—October 30, 1866.—The spindle has an axle at right angles to its axis, with a wheel at each end.

Claim.—The arrangement of the wheels *a a* in combination with the shoulder shank A. arm B, with its axles *d d*, and the washers *c* and bolt *b*, substantially as and for the purpose herein represented and described.

No. 59,164.—LEWIS BARNES, Waterford, Mich.—*Whiffletree.*—October 30, 1866.—The hook shanks of the double and the ring shank of the whiffletree pass through and are secured to malleable cast-iron plates attached to the front and back; the trace-hook sockets are screwed in the threaded ends of the whiffletrees.

Claim.—The malleable cast-iron plates B, secured to the front and rear sides of the double-tree A by screws *b*, and the shanks *c* of the hooks C, and the malleable cast-iron plates E E secured to the front and rear sides of the whiffletrees D by screws *f*, and the shank *g* of the eye *h*, together with the trace hooks F, provided with rings or bands *j*, having internal screw-threads to screw upon the ends of the whiffletrees, substantially as shown and described.

No. 59,165.—MONROE L. BATTELL, New York, N. Y.—*Quartz Crusher.*—October 30, 1866.—The upper and the lower sets of jaws are operated by pitmans and eccentrics on a shaft placed between them. An intermediate screen receives the comminuted quartz from the first pair of jaws, and passes the coarser portion of it to the second pair.

Claim.—The within described ore-crushing machine constructed with two sets of crushing jaws, and operated by a single intermediate or central shaft, substantially in the manner herein set forth.

No. 59,166.—JOSEPH B. BENNETT, South Brooklyn, N. Y.—*Rotary Steam Engine.*—October 30, 1866.—The eccentric hub revolves in the annular cylinder, and has pistons arranged on yokes, traversing at right angles to each other, and provided at their ends with spring packing-plates, which accommodate themselves to the interior surface of the cylinder. The induction openings are also covered with flexible plates which accommodate themselves to the surface of the hub, and permit the passage of the piston. The engine runs in either direction and exhausts at the bottom.

Claim.—First, the combination, substantially as described, of the spring packing plates F F, with the pistons of a rotary steam engine or rotary pump, for the purpose of greatly reducing the loss of power by friction and other causes, as hereinbefore set forth.

Second, the spring valves G G, arranged and operated substantially as and for the purpose described, in combination with the cylinder B, with its pistons and springpacking plates, as set forth.

No. 59,167.—D. J. BIGELOW, Bone Centre, N. Y.—*Sleigh.*—October 30, 1866.—The bolster has a rocking motion in a vertical plane in the direction of the line of travel, so as to accommodate itself to the wagon-bed bottom when the bob sled rocks forward and backward. The plates forming the joint rest on pieces which are notched upon the benches of the sled.

Claim.—The plates *d* and *e* and bolt *a*, constructed as described and arranged with the bars B B and bolster E, as and for the purpose herein fully set forth.

No. 59,168.—FRIEDRICK BINDER, Baltimore, Md.—*Pruning Shears.*—October 30, 1866.—The cutting edge shuts into a slot in the opposite jaw, which holds the bough steady to the knife.

Claim.—The convex-edged knife, working in a slot in the concave holding jaw, and operating substantially as described.

No. 59,169.—T. F. BINGHAM, Gowanda, N. Y.—*Beehive.*—October 30, 1866.—The upper bars of the comb frames have waxed threads running along their lower sides, and also strips of paper between them to act as guides for the bees in constructing the comb.

Claim.—First, the application of waxed cords *f* to frames B, and to the spare honey boxes, to insure the building of straight combs, as set forth.

Second, the combination of the cap G and case H with the spare honey boxes, walls A A, and comb frames B, all arranged substantially as and for the purpose specified.

No. 59,170.—T. F. BINGHAM, Gowanda, N. Y.—*Beehive.*—October 30, 1866.—The triangular comb frames are clamped together and to their supporting sides by a wire link. The honey boxes rest in an inclined position on the comb frames, which are centrally divided. The apertures between the comb frames may be closed by strips, or left open for the passage of bees to the honey boxes.

Claim.—First, the triangular frame D, divider I, and notched end pieces E, and guides F resting thereon, the triangular end pieces E, slats *a*, links G, bar H, spare boxes J, and slats *g*, when combined and operating substantially as described, for the purpose specified.

Second, the construction of the triangular end pieces E E, with the entrances and vestibules, as set forth.

Third, a triangular divider I, with guides, constructed substantially as set forth.

Fourth, the clamp G, constructed as described, and arranged in connection with a bar H, substantially as and for the purpose specified

Fifth, arranging the comb frames D, end pieces E, divider I, clamp G, cover strips or slats *g*, boxes J, and outer case A, substantially as and for the purpose specified.

No. 59,171.—AMOS A. BISSELL, Lockport, N. Y.—*Deck for Canal Boats.*—October 30, 1866.—The curved sections are arranged so that alternate sections form caps over the spaces between those on each side, and are secured in position, or in a pile, by ropes run through eye-bolts in each.

Claim.—The combination of the series of curved panels, constructed of narrow jointed strips or boards *c c*, segmental ribs B, and cleats *d*, with rope *g* and eyes *f f*, or their equivalents, arranged as described, to form a portable deck covering, substantially as set forth.

No 59,172.—AMOS S. BLAKE, Waterbury, Conn.—*Pulley.*—October 30, 1866.—The axis of the sheave has its bearings in a yoke pivoted in the arms of a frame attached by screws to an object; the universal joint thus formed permits the axis of the pulley to revolve in planes at right angles to each other, and adjust itself to lateral deflection of the cord passing over it.

Claim.—First, journalling a sheave or roller D in a swivel frame C, substantially as and for the purpose specified.

Second, the combination of the swivel sheave frame C with the angular arms B B and bed plate A, substantially as shown and described.

No. 59,173.—EDWARD S. BLAKE, Pittsburg, Penn.—*Stove-pipe Drum.*—October 30, 1866.—The drum, of annular elliptical form, has two openings to reach its interior, and that of the exit pipe for cleansing, and the dust falls into the stove. A damper above the neck permits direct passage to the caloric current, or sends it around the circuit of the drum.

Claim.—First, a radiator so constructed that the soot, dust, and other refuse of combustion will tend by its gravitation to collect to one point when released from the shell of said radiator, substantially as herein described and for the purpose set forth.

Second, the radiator B, provided with the flues R and C, openings O and *f*, and valve J, constructed, arranged, and operating substantially as herein described and for the purpose set forth.

No. 59,174.—JOSEPH BRADDOCK, Indianapolis, Ind.—*Varnish.*—October 30, 1866.—Composed of turpentine, 10 ounces; asphaltum, 3 ounces; carriage varnish, 4 drachms; and 4 ounces lard oil or benzine.

Claim.—A combination varnish compounded from the ingredients named, or their chemical equivalents, substantially in the manner and for the purpose set forth.

No. 59,175.—HUGH BROOKS and JAMES BALL, Zanesville, Ohio.—*Smoke Stack for Locomotives.*—October 30, 1866.—The sparks and ashes are deflected against the wire-cloth, and, falling between the chimney and casing, are conveyed out through the openings below.

Claim—The arrangement and combination of the curvilinear deflectors *e e e e*, and cylinder of wire-cloth *d d d*, with the feathers or guides *f f f*, placed between the inner or smoke tube *a a a*, and the outer case *b b b*, with the perforated base *h h*, when constructed and arranged substantially as herein described and for the purpose specified.

No. 59,176.—JONATHAN BUNDY, West Liberty, Iowa.—*Fence.*—October 30, 1866.—Wires attached to foundation stones pass through the ends of transverse blocks above the ground, and are secured to the fence top.

Claim.—The combination and arrangement of the blocks or cross-pieces H, wires G, and anchoring stones I, with each other and with the fence-posts B, substantially as herein shown and described and for the purpose set forth.

No. 59,177.—R. E. CAMPBELL, New York, N. Y.—*Preparing Burning Fluid.*—October 30, 1866.—Naphtha, dried clay, dried chalk, or dried salt are mixed and agitated; after settling, the naphtha is drawn off from the deposited substances.

Claim.—Treating the first runnings of the distillate of petroleum, petroleum oil, or coal, by passing them through or mixing them with burnt clay, chalk, chloride of sodium, or other equivalent absorbent substances, in the manner and for the purposes substantially as herein set forth.

No. 59,178.—F. B. CARLETON, Jeffersonville, Vt.—*Upper Jaw Bit.*—October 30, 1866.— The auxiliary bit is fastened by a nose-strap to the upper jaw and buckled to the gag-bearing rein.

Claim.—The arrangement with an ordinary bridle and bit of the supplementary bit, which is strapped to the upper jaw of the horse forward of the ordinary bit, substantially as described.

No. 59,179.—A. H. CARPENTER, New York, N. Y.—*Chemical Compound for the Manufacture of Medicated Gas.*—October 30, 1866.—Composed of nitrate of ammonia, 85 parts; nitrate of potassa, 3; iodide of potassium, 1; chlorate of potassa, 5; sulphate of alumina, 3; and biborate of soda, 3.

Claim.—Compounding certain chemical salts in the manner and proportions herein specified, for producing an electro-medical gas to be used and applied in the manner and for the purposes herein described.

No. 59,180.—HIRAM CARPENTER, New York, N. Y.—*Railway Chair.*—October 30, 1866.— The base flange of the rail is supported in the chair upon an India-rubber strip. The chair has one loose cheek, which has a lip entering a groove in the fixed part, and is held to place by a key.

Claim.—The railway chair constructed with an elastic support, combined with a loose jaw that locks into the body of the chair and is tightened by a key, substantially as described.

No. 59,181.—ASA L. CARRIER, Washington, D. C.—*Tool Supporter.*—October 30, 1866.— The flanged disk is slotted to admit the shank of the tool, and, being fastened to a support, sustains the tool by the handle resting upon the disk.

Claim.—The slotted disk B, rim C, and spike or screw A, when constructed, arranged, and used in the mode described so as to constitute a new article of manufacture, for the purpose specified.

No. 59,182.—BENJAMIN H. CHADBOURN, St. Louis, Mo.—*Burning Fluid*—October 30, 1866.—Composed of naphtha, hemlock bark, flaxseed, white oak bark, alkanet root, rock salt, nitre, gum camphor, borax and alcohol.

Claim.—The combination of the ingredients herein described.

No. 59,183.—FRANCIS M. CHALFANT, Morgantown, West Va.—*Manufacture of Sorghum Sugar.*—October 30, 1866.—The juice is filtered through straw and the acid neutralized by alum combined with lime, after which the juice is boiled and skimmed. The temperature is then raised to 220° Fahrenheit, and acetic acid is added to neutralize any alkali that may be present. The heat is then raised to 230° Fahrenheit, and the solution evaporated until ready to crystallize.

Claim.—The process of making sugar from sorghum, or its allies, substantially as above described.

No. 59,184.—GEORGE H. CHINNOCK, New York, N. Y.—*Smoking Pipe.*—October 30, 1866; antedated October 19, 1866.—Each end of the tube which intervenes between the mouthpiece and the bowl has a tubular connection of cork and an interior section of pipe.

Claim.—The pipe-stem consisting of the parts A B C E F and G, combined, constructed and arranged as and for the purposes herein described.

No. 59,185.—FREDERICK C. CLASS, Roanoke, Ind.—*Portable Fence.*—October 30, 1866.— The angular trellis slats of the panels have horizontal confining strips. The posts have extended sill-pieces. The panels are locked together through vertical slots in the posts by blocks to which the panels on each side are keyed.

Claim.—The brace post in combination with its lock and key, also the corner post in combination with the panel, as described.

Also, the lattice panel with supporting bars as giving greater strength as a fence.

Also, the peculiar construction of the lock and key in combination with the brace-post, essentially as described.

No. 59,186.—LEWIS COATES, Collamer, Penn.—*Variable Measure.*—October 30, 1866 —The spring catches on the bottom are operated by projections beneath, and enter notches in the sides of the cylinder at the points adapted to adjust the measure to the required capacity.

Claim.—The notched staves c and catches a, in combination with the movable bottom B and measure A. constructed and operating substantially as and for the purpose described.

No. 59,187.—JOHN F. COLLINS, New York, N. Y.—*Distillation* —October 30, 1866.—The wash is heated in a still, the neck of which terminates beneath a pipe leading to the condenser. The vapor passing from the still into the condenser carries with it a current of air through the annular opening between the neck of the still and pipe leading to the condenser.

Claim.—The process, substantially as above described, of separating and obtaining alcohol or other volatile matters by constantly agitating the "wash," or other contents of the still or retort, by means of a current or currents of steam or gas or air forced into the same, and bringing the vapors in contact with currents of air from without, while passing from the still or retort into the conductor which leads to the worm or condenser, as above set forth.

No. 59,188.—WILLIAM COMPTON, New York, N. Y.—*Vessel for Beer, &c.*—October 30, 1866.—The stationary head, which has the bunghole and faucet, is attached by a flexible tube to a head which moves within the cylinder as the contents of the bag are drawn out.

Claim.—The flexible bag fitted as specified within a vessel, and adapted to the reception of beer and other liquids, and the exclusion of the same from contact with the air, as set forth.

No. 59 189.—JOHN B. CORNELL, New York, N. Y.—*Safe.*—October 30, 1866.—The parallel plates of the walls are held together by metal cast into and filling the spaces between them, and occupying the grooves on the interior surface of each of the outer plates and the holes through the middle one. The width of space separating the plates is regulated by bolts projecting from the outer plates.

Claim.—Uniting parallel and contiguous plates of metal with each other by the act of filling the spaces between said plates with molten iron, or other metal or composition, and the inflow of a portion of the same into channels or recesses of a dovetail shape formed in the inner surface of said plates, but not entirely through the same, substantially in the manner represented in the drawings and herein described.

No. 59,190.—JOHN B. CROWELL, Newport, N. H.—*Stanchion for Cattle.*—October 30, 1866.—The stanchions are drawn to the confining position, and fastened by a spring bolt. The sliding of a bar releases the bolt, when the drawing of the cord opens the stanchions.

Claim.—First, the combination of the bolts L and rope N with each other and with the movable stanchions D and timber B, substantially as herein shown and described and for the purposes set forth.

Second, the combination of the bolts F and springs H with the movable stanchions D and timber B, substantially as described and for the purpose set forth.

Third, the combination of the bar I and lever K with the bolts F and timber B, substantially as shown and described and for the purpose set forth.

Fourth, the combination of the springs E with the stanchions D and timber B, substantially as herein shown and described and for the purpose set forth.

No. 59,191.—JOHN C. DOUGHERTY, Bridgeport, Ky.—*Shovel Plough.*—October 30, 1866.—The vertically adjustable fender is attached to the beam to protect small plants from clods.

Claim.—Each and every part of the fender described as above.

No. 59,192.—J. W. DOTY, Lockport, N. Y.—*Pitman for Harvesters, &c.*—October 30, 1866.—A sleeve embraces the conical wrist pin of the crank, and is attached by a hinge joint to the stirrup on one end of the pitman. The other end of the pitman has a conical wrist, which is adjusted in its socket by a bolt and thumb nut, the latter being retained by a pawl.

Claim.—First, the combination of the forked or pronged pitman head D, secured to the ears or lugs f f by a screw bolt j, taper, conical, or spherical wrist pin A, and box B, the whole arranged substantially in the manner and for the purpose set forth.

Second, the tubular conical projection m at the lower end of the pitman E, in combination with the socket n on the cutter bar G, the bolt H provided with a nut o, having a ratchet attached, with which a pawl p on the pitman engages, substantially as and for the purpose specified.

No. 59,193.—JOHN W. H. DOUBLER, Chicago, Ill.—*Saw filing Machine.*—October 30, 1865 —The saw is clamped in fixed jaws. The file is reciprocated by an eccentric upon a frame, which is connected to a device for moving it any fixed distance from one interval to another between the teeth of the saw. The file frame and file can be set to any angle, in respect to the saw plate.

Claim.—First, the combination and arrangement of the stationary jaws E, a stationary rack F, file guide T, and holder a, wheel V, shaft G, and screw H, as and for the purposes specified.

Second, the combination of the wheels J I, spring catch M, arms L, shaft G, and screw H, with the stationary rack F, substantially as and for the purposes set forth.

Third, in combination with the arm L, the arrangement of the movable slotted arc N and pin or stop o, as and for the purpose described and set forth.

No. 59,194.—JAMES W. DREW, Stockbridge, Mich.—*Wheel Vehicle.*—October 30, 1866.— The wheels are on separate axles, having bearings under anti-friction rollers, and in central boxes. The wagon is turned by brace arms connecting a sliding box on the tongue to the axle bearings. The tongue has a support running back to the bolster.

Claim.—First, the sliding boxes h h, constructed and operating as and for the purpose herein set forth.

Second, the tongue support l, in combination with spring m, constructed and operating substantially as herein specified.

Third, the spring bars g g, boxes h h, tongue support l, spring m, boxes h h, the whole constructed and arranged substantially as herein described.

No. 59,195.—NOAH DREW, Howell, Mich.—*Washing Machine.*—October 30, 1866.—The three corrugated faced plungers are connected to a tri-crank shaft. The suds box may be removed from the frame.

Claim.—The employment of the plungers E, attached as described, in combination with the moulded end board D and a yielding suds box B, arranged and operated substantially in the manner and for the purposes herein specified.

No. 59,196.—ELI DUNCAN, West Milton, Ohio.—*Gate.*—October 30, 1866.—The upper bar has ratchet notches for the reception of a pin connecting the adjustable brace bars at their upper end.

Claim.—A gate composed of a series of horizontal bars, when the upper bar A' is notched near one end, z, to catch a pin d, which connects the upper ends of the braces C C, and arranged with the roller e, in the manner and for the purposes specified.

No. 59,197.—WILLIAM C. DUNN.—Greene, N. Y.—*Training Hops.*—October 30, 1866.—A central post has arms radiating from its top at an upward angle of about 45 degrees.

Claim.—The construction or use of a hop trainer, constructed and used substantially as described.

No. 59,198.—JOHN ELDER, Jr., New York, N. Y.—*Steam Heating Apparatus.*—October 30, 1866.—The heater pipes pass up into the air tube. The steam inlet enters the front of the steam box, passes up the hinder pipes, and returns by the other pipes to the box. Drain and escape pipes emanate from the bottom and side of the box respectively.

Claim.—The coil of pipes f g, extending up in the air pipe a, and connected with the steam box at the lower end of the said coil, substantially in the manner specified.

No. 59,199.—DAVID S. EVANS, Richmond, Ind.—*Fence.*—October 30, 1866.—The posts have bevelled edges against which fit the vertical slats near the ends of the panel. Two or more of the boards on each panel project beyond the posts and hook to the vertical slats of the adjacent panels. The fence has a zigzag form.

Claim.—The arrangement and combination of the wedge post d, the catch m, and bevel s, when used in a portable fence, all arranged and operating as set forth and described.

No. 59,200.—W. W. EWING, Mahoning, Penn.—*Saw-mill.*—October 30, 1866.—The two ends of the saw are clamped at the vertex of two pairs of converging arms, and the tension of the saw is effected by coupling-rods at their other ends. The saw frame is pivoted near the centre and balanced on V-points in more obtuse grooves. The frame receives motion from a pitman below. The saw is hung on V-shaped bearings on the pivot bolts.

Claim.—First, the construction and arrangement of the two pairs of converging vibratory tension beams F F, clamping the ends of the saw between their converging ends, substantially as and for the purpose herein specified.

Second, the combination of the pivot blocks C C, having dovetail shanks, with the adjustable clamp bars B B, having corresponding forms to fit the shanks of the pivot blocks, substantially as herein set forth.

Third, the method of hanging the saws between the tension beams by the V-shaped notches therein, and the peculiarly formed pivot bolts H H, substantially as and for the purposes herein described.

No. 59,201.—LEWIS FAGIN, Cincinnati, Ohio.—*Hanging Mill-stones.*—October 30, 1866.— The cock-head upon which the runner hangs is above the stone, and directly midway of the vertical height of the bearings of the driving lugs, whose surfaces and bearings in the slots of the balance rynd are easy of access to be adjusted and fitted.

Claim.—First, the arrangement of the cock-eye A, cock-head B, openings K in the balance rynd, and driving lugs C C, by which the point of balance is adjusted, as described and for the purpose set forth.

Second, the construction of the balance rynd with openings K K, by means of which the driving lugs and driving surfaces are brought to view and rendered accessible for the purpose of fitting.

No. 59,202.—WICKUM FIELD and ROBERT CARRUTHERS, Bergen, N. Y.—*Gate Hinge.*—October 30, 1866.—The jaw of the hinge is connected at one bearing with the upper and middle plate, and at the other bearing with the lower and middle plate, to avoid disconnection and allow freedom.

Claim.—The combination of the braces *h h'*, or equivalent, with the bearing *b* and jaws *d d'*, operating substantially in the manner and for the purpose specified.

No. 59,203.—WILLIAM H. FIELD, Taunton, Mass.—*Machine for Leathering Tacks.*—October 30, 1866.—The elevators extend through the hopper for depositing the tacks into slides which direct them in proper position to the tacking machine. The heading bar works at right angles to the running of the rails, the course of the latter being governed by a quadrant which allows the header to leather but one tack at each revolution. The shaft consists of half cylinders which work together in bearings, and is operated by a slotted cross-bar which gives it an up and down motion; each downward motion driving two tacks.

Claim.—First, the elevators P, for the purpose of elevating tacks from the hopper J² and depositing them upon the slide K² for the purposes and substantially as herein described.

Second, the peculiar construction of the shaft or rod A² in halves, so that one or more nails or tacks may be leathered at the same time, substantially as herein described.

Third, the straight heading bar, with its end reduced to about the size of the tack and working at right angles with the slide or nail, for the purposes and substantially as described.

Fourth, the quadrant B², in combination with the heading bar, for the purposes and substantially as set forth.

No. 59,204.—ISAAC FISKE, Worcester, Mass.—*Cornets and other Wind Instruments.*—October 30, 1866.—The main pipe is united to the bell by rings fastened on the former and spanning the latter, India-rubber block being interposed. The finger pieces are on the piston rods, arranged at equal distances apart, and operate upon the valve stems by direct cord connection. A bar supports the cylinders of the spring piston rods, and is itself attached to the pipe.

Claim.—First, interposing rubber or some other suitable elastic substance between the attachment or attachments of the main pipe with the bell and the bell of a wind instrument, to give greater freedom to the vibrations of the bell, substantially as set forth.

Second, the combination of ring or rings *a* and rubber *b*, or its equivalent, with the bell A, and main pipe B, substantially as set forth.

Third, the combination and arrangement in a wind instrument of the cylinders in which the piston rods work and the valve stems in such a manner as to obviate the necessity of interposing anything between the valve stems and piston rods in order to operate the stems and valves except a cord, substantially as described.

Fourth, the special arrangement and combination of the valve stems *e f* and *g*, and rods 1' 2' 3', and cylinders 1″ 2″ 3″, whereby the valves, cylinders, and finger pieces are of equal distances from each other, and yet all of the valve stems and valves are operated by cords attached directly to the ends of the rods which move in a line parallel to each other, substantially as set forth.

Fifth, the combination and arrangement with the cylinders 1″ 2″ 3″ of the supporting bar G, as shown and described.

No. 59,205.—THOMPSON FRAME, Barnesville, Ohio.—*Machine for Raking and Loading Hay.*—October 30, 1866.—The rake collects the hay, the toothed roller deposits it on the platform and the reciprocating rakes draw it to the elevator, which carries it up and discharges it into the wagon to which the raking, &c., frame is attached.

Claim.—First, the cam *w* on the shaft *f*, in combination with the reciprocating levers *p q*, the connecting rods *t s*, and the reciprocating cross rakes *n n*, for drawing the hay by their reciprocating motion into the chute C, constructed and arranged as herein described.

Second, the platform E, in combination with the reciprocating rakes *n n* and the chute C, constructed and arranged as and for the purposes herein described.

No. 59,206.—TERAH M. FREEMAN, St. Louis, Mo.—*Manufacture of Vinegar.*—October 30, 1866.—The alcoholic vapors from the still are passed into the wash to raise it to the required heat and thereby utilize the heat of the vapor.

Claim.—The formation of vinegar wash by adding alcoholic vapors to the liquids used and usually containing water, vinegar, or acetic and ferment, substantially as set forth.

No. 59,207.—THOMAS T. FURLONG and DE WITT C. FREEMAN, St. Louis, Mo.—*Safe.*—October 30, 1866.—The safe is enclosed in an air chamber which renders it buoyant in the water; the doorway may be closed water-tight and the safe recovered if it fall overboard, or be launched into the water in case of fire, or sinking of the vessel.

Claim.—The outside case A, the inside safe B, the cap C, and the door E. constructed and hermetically closed, and connected in the manner and for the purposes herein specified.

No. 59,208.—A. C. GALLAHUE, New York, N. Y.—*Clothes Wringer.*—October 30, 1866.—The two rollers have bearings in the two jaws, which are pivoted together in such a way that the rubber spring acts to press the rollers together. Their pressure is limited by the hollow set screws.

Claim.—The tubular springs H, in combination with the hollow screws I, or their equivalents, and rod G, arranged or applied to the upper ends of the parts A B, to operate substantially as and for the purpose specified.

No. 59,209.—E. T. GREEN, Stoneham, Mass.—*Last.*—October 30, 1866; antedated October 20, 1866.—A metallic rim around the sole of the last holds a movable, wooden bottom which can be renewed as it wears out.

Claim.—A metallic pan or basin E, or its equivalent, constructed and arranged for holding a false or movable bottom of a shoe last, in the manner and for the purpose substantially as herein set forth.

No. 59,210.—JOHN GREEN, Joliet, Ill.—*Hand Spinning Wheel.*—October 30, 1866.—The machine has but one cord and one pulley to move the spinning head toward the outer end of the ways. The cord from the main wheel brings it back. The cords thus arranged are not so liable to fly off.

Claim.—The peculiar and particular arrangement of the cords and pulleys described, in combination with the inclined ways b, and for the purposes described.

No. 59,211.—BURTON GREENSIDE, Fort Dodge, Iowa.—*Gate Hinge.*—October 30, 1866.—The lower hinge has segmental racks meshing together. A pin in one, running in curved slots in a plate on the other, prevents dislocking. A weighted lever, connected to an arm at the back of the gate, abuts the latter.

Claim.—First, the cogged hinge E, formed in two parts e^1 and e^2, when constructed and arranged substantially as herein described and for the purpose set forth.

Second, the combination of the bent lever F, connecting rod G, and arm H, with each other and with the gate A, the weight and post D, substantially as described and for the purpose set forth.

No. 59,212.—ISAAC GREGG, Philadelphia, Penn.—*Brick Machine.*—October 30, 1866.—This is an improvement on his patents of 19th May, 1863, and 19th of December, 1865. The tops of the plungers are cleaned off by a rotary brush which precedes a lubricating brush of the same form. The foot of each plunger is equally divided into a projection and recess, which enables each set to be raised by its own raising devices, while it is not affected by those for raising the other set.

Claim.—First, the brush M, so arranged and operated as to clear the upper surfaces of the pistons from superfluous clay in advance of the oiler.

Second, the combination of the revolving brush M and revolving oiler N, in the same box K, substantially as described.

Third, the combination of the heads or flanges g of the rods of the two sets of pistons with the stationary inclined projections d and d', when the said heads and projections are so formed that when the heads of one set of piston rods traverse over one of the said projections, the former will be elevated by the latter, but will remain depressed while traversing over the other projection, all substantially as set forth, for the purpose specified.

Fourth, the combination of the wheel F and F' with the heads of the rods of the two sets of pistons, when the said wheels and rods are recessed or halved, substantially in the manner and for the purpose described.

No. 59,213.—HENRY HAAK, Myerstown, Penn.—*Grain Gauge.*—October 30, 1866.—The measure of certain capacity has a weigh frame, and a graduated scale denoting the amount of flour per bushel. The heavier the contents, the greater the result in flour.

Claim.—First, the measure A', provided with the spring and striker A and B, constructed and operated substantially as described.

Second, the graduated beam F, the bearing beam E, tube D, and stem E', arranged and operated as described.

No. 59,214.—DANIEL and JOSEPH HALL, Wheeling, West Va.—*Puddling Furnace.*—October 30, 1866.—The stack and bed plate of the furnace are supported upon pillars, and the fire brick lining of the stack is made thicker nearer the bottom than at the top. The neck connecting the furnace with the stack is built upon a plate supported upon studs projecting from the pillars which support the stack. Other details are explained in the claims and cut.

Claim.—First, the improved iron-cased smoke stack of unequal diameters at the upper and lower parts, lined with fire brick of unequal thickness, supported on the pillars b b b b, and constructed and arranged substantially as and for the purpose herein specified.

Second, the outer shell or casing d d, constructed and arranged substantially as and for the purposes herein specified.

Third, the wrought-iron fore plate r r, and the recess in the doorway in which it is inserted, in combination with the furnace door p, constructed and arranged substantially as and for the purposes herein described.

Fourth, the wrought-iron side bits *s s*, placed in a recess in the doorway of the furnace, substantially in the manner as herein specified.

Fifth, the horizontal or straight-bottomed neck E, supported on the foundation plate *q*, resting on bearers *i i*, which are sustained by the projections *h h*, on the pillars *b b*, constructed and arranged substantially as and for the purposes herein described.

Sixth, the ribbed binding plates *k k*, in combination with the neck E, constructed and arranged substantially as and for the purposes herein described.

Seventh, the improved inclined fire bridge G, constructed substantially as and for the purposes specified.

No. 59,215.—JOEL A. HALL, Columbus, Ohio.—*Cotton Cultivator.*—October 30, 1866.—The earth thrown by the ploughshares from the plants is returned by the inclined plates following. Hanging levers to a walking beam operate to move the ploughs laterally. The depth of the ploughs is regulated by cords passing over a sheave in an upper bar.

Claim.—First, the combination of the curved blades or scrapers with the ploughs B², substantially in the manner herein shown and described, so as to plough the furrow, cut the weeds, and throw the earth upon the roots of the plants, all as set forth.

Second, the combination of the toggle levers I with the plough beams, substantially as herein shown and described.

Third, the combination of the guide with the toggle levers I, substantially as herein shown and described.

Fourth, the combination of the walking beam and treadles with the toggle levers, substantially as shown and described.

No. 59,216.—C. L. HAMMOND, North Java, N. Y.—*Medicine for Horses.*—October 30, 1866.—A liniment composed of turpentine, 1 pint; linseed oil, 1 pint; origanum, 1 ounce; and oil of vitriol, 2 ounces.

Claim.—The above described ingredients mixed as specified and for the purposes set forth.

No. 59,217.—ANDREW HARTER, Delphi, Ind.—*Grain Tallying Machine.*—October 30, 1866.—Placing the measure on the platform operates on a ratchet wheel, and moves the index finger one mark on the dial. Each revolution of the ratchet wheel moves the number plate one notch, indicating the number of measures.

Claim.—The combination of the platform B B', arm C, and wheels D, with the wheel L, sleeve *h*, and index plate E, constructed, arranged, and operated in the manner substantially as shown and described and for the purpose set forth.

No. 59,218.—NICHOLAS HEADINGTON, Cincinnati, Ohio.—*Railway Chair.*—October 30, 1866.—A bed plate extending from tie to tie supports the rails at their junction. A strap passes through gains in their ends and is keyed to the seat.

Claim.—The railroad chair composed of the seat C, having the pendent ribs or flanges C'' C''', in described combination with the strap F, and gib and key D E, or their equivalents, for the purposes set forth.

No 59,219.—CHRISTIAN HEISTERMAN, Brownsville, Penn.—*Machine for Wrinkling the Insteps of Boots and Shoes.*—October 30, 1866.—The machine has two corrugated surfaces between which the uppers of boots and shoes are pressed so as to permanently and uniformly wrinkle the insteps in any desired ornamental style.

Claim.—Board A, and blocks *a* and A', when constructed and operated by a press, substantially in the manner and for the purposes set forth.

No. 59,220.—JOHN D. and ISAIAH HESS, Union, Ohio.—*Sorghum Stripper.*—October 30, 1866; antedated September 28, 1866.—The top of the stalk is entered between the cutters and grasped by the rollers, which draw it through the stripper, which embraces each side of it and removes the blades.

Claim.—First, the combination of the cutter G with the device herein described for carrying the cane through the cutter or its attachment, to a sorghum mill, substantially as and for the purpose specified.

Second, the arrangement of the frame A, cylinder B, pulley D, rollers C, frame E, spring F, cutter G, and support H, substantially as described and represented. .

No. 59,221.—S. E. HEWES, Albany, N. Y.—*Steam and Air Ejector.*—October 30, 1866.—The apparatus is worked by steam or air to raise or move liquids from a reservoir or well, by the combination of the injector, projector, and ejector, operating together; the adjustable nozzles of the injector govern the induction of air and steam, which descend into the well and pass in an annular current around the upwardly presented nozzle on the end of the tube which contains the liquid to be raised. The ejector at the upper and discharge end causes a partial exhaustion in its rear and assists in the elevation. Its nozzles are also regulatable.

Claim.—First, the combination of the injector W, projector Y, and ejector Z, operating together substantially as described.

Second, the adjustable nozzle *h i*, operating substantially as described,

Third, the expansible or contractible slit nozzle *i'*, constructed as described.

Fourth, the slotted nozzle *h'*, substantially as described.

No. 59,222.—A. J. Hindermeyer, Rohrertown, Penn.—*Welding or Brazing.*—October 30, 1866.—Chalcedony quartz, reduced to powder, is used as a flux in welding or brazing.

Claim.—The use of the herein specified mineral substance as a flux for welding and brazing steel, iron, or other metals.

No. 59,223.—Philip K. Holbrook, Malden, Mass.—*Ink Cup.*—October 30, 1866.—The India-rubber cup fits the wooden socket, and has a metallic rim to which a cover is hinged.

Claim.—The combination of the India-rubber cup B with the stand A, both constructed and adapted to each other, substantially as described so as to form an inkstand which can be cheaply made and easily cleansed.

No. 59,224.—Reuben Hoover, Boonsborough, Iowa.—*Washing Machine.*—October 30, 1866.—The tub has a corrugated bottom and oscillates upon its point of suspension. A strap on the shaft gives a reciprocating turning motion to the latter and to the radially ridged lid. The power is communicated by a rod hooked to the tub bottom.

Claim.—The revolving corrugated jointed cover D, fitting loosely on the shaft E, in combination with the tub, arranged to operate with the rock shaft F and strap H, substantially as described as and for the purpose specified.

No. 59,225.—Thomas S. Hudson, East Cambridge, Mass.—*Cancelling Apparatus.*—October 30, 1866.—A socket in the lower end of the plunger has a type chase, which may have movable type to indicate month and year inside the motto of the chase, which may be the title, &c., of the firm. A central slot in the face of the chase is occupied by one link of an endless chain, whose consecutive links have type corresponding to the days of the month, and moved in the required succession by sprocket wheels in the plunger.

Claim.—The movable type as made with the longitudinal and transverse dovetails and with parallel plane surface sides, as and for the purpose set forth, in combination with the type chase as made with its type sockets or recesses, dovetailed longitudinally and transversely, or so as to hold the type in manner as specified, and with a wheel-receiving space or chamber arranged within it and opening into each of such type sockets.

Also, the ribbon box or case, as constructed, with the ribbon-receiving opening, the removable cap and winding spindle and its holding ring, or the equivalent thereof, the whole being applied substantially as specified.

No. 59,226.—Elias S. Hutchinson, Baltimore, Md.—*Automatic Blast for Carburetters in Railroad Cars, &c.*—October 30, 1866.—A pendulum is suspended from an axis in such a manner that the lateral motion of the car will cause it to vibrate and operate the pump rods, to condense air for the use of a carburetter on board the car.

Claim.—A pendulum suspended in a railroad car, carriage, or vessel, and applied in connection with an air-pump bellows or compressor, to furnish a blast to a carburetter, substantially as described and represented.

No. 59,227.—D. Isaacsohn and Adolph Cohn, New York, N. Y.—*Combined Victorine, Cape, and Cuff.*—October 30, 1866.—To the lappets of the cape are attached cuffs which form a muff when hooked together.

Claim.—A victorine, collar, or cape, made up of a collar or cape A, proper, front lappets a a, having combined at the extremities of the latter, so as to form one with the same, cuffs B B, capable of being made convertible at pleasure into a muff, substantially as specified.

No. 59,228.—James M. Jarrett, Brooklyn, N. Y.—*Water-proof Mail Bag.*—October 30, 1866.—The water-proof bag is buoyed by an air chamber or a body of cork, and after the closing of its water-tight jaws floats bottom upward, exposing a flag staff and a hinged false bottom inside, on which is painted the name or destination.

Claim.—First, a floating compartment or pocket O, in combination with the cover A, arranged with the parts of a mail bag herein described, substantially as and for the purpose specified.

Second, the hinged false bottom L, in combination with the frame B, springs M, and catches N, when arranged with the parts of a floating bag herein described, substantially as and for the purpose specified.

No. 59,229.—Isaac Jones, Camden, N. J.—*Machine for Cutting the Fronts of Books.*—October 30, 1866; antedated October 20, 1866.—The book is clamped between cheeks, operated by side screws so as to occupy a position in the centre of the frame; end and bottom screws determine its longitudinal and vertical position. The gouge or convex cutter is attached to a rack bar which reciprocates horizontally and cuts the front of the book to a curved form.

Claim.—First, the bar P, with its gouge T, in combination with the within-described devices, or their equivalents, for holding the book, the whole being constructed and operating substantially as and for the purpose described.

Second, the combination with the above of an adjustable soft metal or wood strip *d*, substantially as and for the purpose specified.

Third, the adjustable frame C, with its adjustable plates H H', in combination with the traversing bar P and its knife T, substantially as and for the purpose set forth.

No. 59,230.—JOHN HOLBROOK KEATING, Marblehead, Mass.—*Gang Punch.*—October 30, 1866.—A gang of adjustable punches are arranged in a form agreeing to the shape of a shoe upper, which is laid upon them, and the series of holes made by a single impact of the platen above.

Claim.—First, the sliding punch-holding blocks G, in combination with the screws I, within the ways H, operating substantially as and for the purpose specified.

Second, the former P, in combination with the flexible strip O of the punches E, substantially as described, for the purpose specified.

Third, the combination and arrangement of the sliding blocks G, screws I, ways H, rubber O, punches E, former P, and vibrating arm C, substantially as described, for the purpose specified.

No. 59,231.—WILLIAM KIDDOO, Keithsburg, Ill.—*Cultivator.*—October 30, 1866.—Levers are so connected to the plough beams as to move them vertically from end to end. A lever operates on vertical guide rods for lateral setting of the ploughs. The fore ends of the plough beams have draught rollers which have vertical play on a depending draught bar.

Claim.—The plough beam F, in combination with the lever G, or equivalent means, for raising and lowering the plough beam without changing its horizontality.

Also, the combination of the plough beam F, lever G, cord E, and link L, substantially as and for the purpose set forth.

Also, the combination of the plough beam F and draught bar K, substantially as and for the purpose set forth.

Also, the combination of the plough beam F, draught rod K, and guide rod S, for the purpose of retaining the said beam in proper horizontal position.

Also, the adjustable suspended frame N, provided with the lever P and link Q, or their equivalents, for the purpose set forth.

Also, the shield U, suspended by the rods V, so that the ploughs may be raised or lowered without affecting the height of the shields.

Also, the levers G and P, and their attachments, substantially as described, so as to enable the attendant to adjust the ploughs vertically or horizontally without leaving his seat.

No. 59,232.—GEORGE KNIGHT, Boone, Iowa.—*Sulky Plough.*—October 30, 1866.—The fore end of the plough beam has a plate which is adjustable in a vertical slot in a post depending from the fore-bar of the frame. The rear end of the beam can be raised by a lever operating a cord attached to it.

Claim.—The attaching of the front end of the plough beam H to the pendent bar E, through the medium of an adjustable plate F, substantially as and for the purpose set forth.

Also, the suspending of the plough beam H, from the axle C, by means of the cords or chains *c c*, and the retaining or holding of the plough beam so as to prevent it from moving laterally by means of a chain or cord *d*, substantially as set forth.

No. 59,233.—ANGELINA J. KNOX, Boston, Mass.—*Wash Bowl and Water Closet Combined.*—October 30, 1866.—The wash bowl is attached to the door, which swings open and discloses the water closet. The same water cock is used for both. Rubber packing lines the joints to prevent the exit of effluvia.

Claim.—The bowl *d*, shelf *c*, door *b'*, privy bowl *g*, pipe *e*, and packing *c'*, all arranged substantially in the manner and for the purpose set forth.

No. 59,234.—WILLIAM KRÆMER, Cincinnati, Ohio.—*Machine for Making Buttons.*—October 30, 1866.—Cloth buttons on metallic frames are formed by a succession of dies and counter dies coacting in their appointed order with a travelling bed or swage stock.

Claim.—First, making buttons by means of the travelling or shifting die J, moving horizontally and transversely to, and adapted to work in co-operation with, a series of consecutively acting dies moving vertically, substantially as described.

Second, in the described combination with such travelling die, the concentric arrangement of the compound die or punch P P', shouldered counter C C', and compound cam *c c'*, as and for the purpose set forth.

No. 59,235.—DANIEL KUNKEL, Oregon, Mo.—*Washing Machine.*—October 30, 1866.—The floor of the tub has knobs, and the disk on the vertical shaft has pendent and erect arms, the lower acting in conjunction with the knobs, and the upper being acted upon by the teeth in the wheel on the rotary shaft.

Claim.—First, the combination and arrangement of the frame B, toothed wheel E, toothed wheel F, with pendent arms *b*, and tub A with its projection *a*, substantially as and for the purpose specified.

Second, the frame B, forming the bearing for the shaft C and the upper end of the spindle G, arranged with the toothed wheel E and wheel F, provided with the pendants b, in combination with the tub A, having studded bottom, all in the manner and for the purpose specified.

No. 59,236.—HENRY H. LADD, Worcester, Vt.—*Sheep Rack.*—October 30, 1866.—The grain trough is pivoted in a drawer which slips into the frame below the hay rack. When the drawer is pulled out the trough may be upset and cleaned.

Claim.—The combination of the trough C and sliding frame B with each other and with the frame A of the sheep rack, when said trough and sliding frame are constructed and arranged substantially as herein shown and described, for the purpose set forth.

No. 59,237.—CHARLES W. LE COUNT, Norwalk, Conn.—*Lamp Wick.*—October 30, 1866.— Cotton threads running through the felt longitudinally increase its conducting power.

Claim.—A lamp wick composed of felt with longitudinal threads of cotton or other fibrous material running through it, substantially as and for the purpose herein described.

No. 59,238.—GEORGE T. LEWIS, Philadelphia, Penn.—*Manufacture of Acetate of Alumina.*—October 30, 1866.—To the alumina obtained from the decomposition of cryolite, 2 parts, (calculated dry,) add 5 parts acetic acid ; heat, stir, settle, and decant the a etate.

Claim.—The manufacture of acetate of alumina by mixing the alumina extracted from cryolite with acetic acid, substantially as described.

No. 59,239.—GEORGE T. LEWIS, Philadelphia, Penn.—*Manufacture of Sulpho-acetate of Alumina.*—October 30, 1866.—To the alumina obtained from the decomposition of cryolite, 2 parts, (calculated dry,) is added acetic acid, 5 parts, to make an acetate of alumina : to this is added sulphuric acid or sulphate of alumina equal to half the weight of the alumina.

Claim.—The manufacture of sulpho-acetate of alumina by mixing the alumina extracted from cryolite with acetic acid and sulphuric acid, or in the place of the sulphuric acid sulphate of alumina or alum, substantially as described.

No 59,240.—MARSHALL T. LINCOLN, Washington, D. C.—*Sewing Table for Bookbinders.*— October 30, 1866.—The bench is vertically adjustable on the supporting table, so as to be lowered by a screw as the pile of sheets rises ; this maintains the top of the latter at a convenient height for sewing.

Claim.—The adjustable sewing bench A B, constructed and operating substantially as described.

No. 59,241.—CARL LÖFFLER, Hoboken, N. J.—*Cork Pull.*—October 30, 1866.—The spur on the shank is pushed beyond the cork, and the latter is withdrawn thereby.

Claim.—As a new article of manufacture, a cork puller, composed of a thin shank A, handle B, and a tooth C, as described.

No 59,242.—JOHN J. LOVE, New York, N. Y.—*Double-headed Wrench.*—October 30, 1866.— The shank of each outer jaw is connected to the sleeved inner jaw of the other pair. the sleeves slipping on the shanks of the jaws to which they are opposed. The double threads act in conjunction to expand or close each pair simultaneously.

Claim.— An improved wrench formed by the combination of the right and left screw C with the parallel-moving bars A and B, having the jaws a^1 a^2 and b^1 b^2 formed upon their ends, the parts being constructed and arranged substantially as herein described and for the purpose set forth.

No. 59,243.—GEORGE MARTZ, Pottsville, Penn.—*Hoisting and Dumping Coal.*—October 30, 1866.—A projection on the platform collides with the lower edge of the breast board of the chute and tilts the car platform, the curved face of the latter traversing against the breast. When the chock blocks between the wheels arrive at the gaps in the guides the platform tilts.

Claim.--First, the combination of the platform, constructed and operating as described, with the breast of the chute, which tilts the rising platform into an inclined position.

Second, the arrangement of the section blocks L L and the gaps N in the guides, operating as described.

Third, the curved face to the platform acting in combination with the supporting breast H, as and for the purpose described.

No. 59,244.—J. R. McALLISTER, Richville, N. Y.—*Sleigh Brake.*—October 30, 1866.— The tongue is hinged to shackles on the roller and connected by a chain to the toothed segment. As long as draught is applied to the tongue the brake is held clear of the ground, but when the tongue is held back the roller rotates and the brake comes into action.

Claim.—The brake shoes C and chains D, connecting them to the pole G, hung to the roller I, turning in the sled frame, when combined and arranged together, substantially in the manner and for the purpose described.

No. 59,245.—J. R. McALLISTER, Richville, N. Y.—*Machine for Boring Wagon Hubs.*—October 30, 1866.—The hub is held by a chuck revolved by a screw shaft, which causes the motion of the cutter bar. The cutter is carried on an inclined slide bar, so as to give the frustal form to the hub cavity.

Claim.—The boring machine herein described, the same consisting of the chuck B, shaft I, curved-arm M, having slotted end L, and eye N, knife-carrying bar Q, sliding frame S, adjustable clasps U, arranged and operating substantially as described for the purpose specified.

No. 59,246.—JAMES McGEARY, Salem, Mass.—*Decomposing Steam.*—October 30, 1866.—The object is to decompose steam continuously and obtain hydrogen and carbonic oxide for illuminating and heating purposes. The steam is expanded and dried in superheating pipes and thence issues into a clay retort, charged with incandescent fuel; it passes thence into a second superheater and retort, and so on in the required succession, until the conversion is accomplished.

Claim.—First, subjecting steam for decomposition to the action of alternate superheating and decomposing surfaces, in the manner substantially as and for the purpose herein described.

Second, subjecting the resulting gases to the action of bituminous coal, petroleum, or other carbonacious material, when used in the manner and for the purpose set forth.

Third, the apparatus, as shown and described, when used for the purposes set forth.

No. 59,247.—HARRISON B. MEECH, Fort Edward, N. Y.—*Preserving Meat.*—October 30, 1866.—The fresh meat is cut up, placed in clear water in a tight vessel, and subjected to pressure of 60 pounds to the square inch. The water is drawn off and brine substituted which is subjected to a similar pressure.

Claim.—First, the within described process of curing meat by subjecting the same first to a pressure under water and then to a pressure under the antiseptic material used in the process, substantially as and for the purpose set forth.

Second, washing the meat under pressure, substantially as described.

No. 59,248.—DAVID N. MINOR, Bridgewater, Mich.—*Fastening for Barn Doors.*—October 30, 1866.—The first closed door is held by a vertical bar, pivoted at the centre, and entering catches on the door and gains above and below. This bar has projecting pins which hold catches on a vertically moving bar on the outer door.

Claim.—First, the combination of the spring I with the door D and with the standard F, substantially as described and for the purpose set forth.

Second, the combination of the guide O with the sill B of the door frame and with the standard F, substantially as described and for the purpose set forth.

Third, the combination of the three catches J with the standard F, substantially as described and for the purpose set forth.

Fourth, the combination of the sliding bar L, clasps K, and hooks M with each other, with the door E, and with the catches J, substantially as described and for the purpose set forth.

No. 59,249.—JOHN H. MORSE, Peoria, Ill.—*Wind Wheel.*—October 30, 1866.—The regulating fan is kept broadside to the wind by the vane. Changes in the force of the wind oscillate the regulator on its axis, and by rack, sleeve, and connecting rods, rotate the sails on their longitudinal axis, and vary the obliquity of their angle of presentation to the wind.

Claim.—The regulating fan A in connection with its balance weight R, cog wheel M, rack L, steel rod J, clutch K, collar E, lever rods $c c c c c$, attached to fans B B B B B, working in eyes $f f f f f$, in flange of collar E, substantially in the manner and for the purpose specified.

No. 59,250.—L. B. MYERS, Elmore, Ohio.—*Fastening for Knobs for Furniture.*—October 30, 1866.—The knob has two dowel pins which enter the front of the drawer and keep it from turning; it is fastened by a wood screw entering from the inside of the drawer.

Claim.—Fastening furniture knobs to drawers or doors by means of two pins and a central screw, substantially as and for the purposes herein specified.

No. 59,251.—HENRY NAPIER, Elizabeth, N. J.—*Curing Hides and Skins.*—October 30, 1866.—Carbolic acid is substituted for the tannin ordinarily used.

Claim.—The use of carbolic acid, or of creosote, in any form, and either alone or in combination with each other, and with other substances, such as a metallic salt, glycerine, &c., for the purpose herein set forth.

No. 59,252.—WILLIAM NICHOLS, Elmira, N. Y.—*Steam Piston Valve.*—October 30, 1866.—The valve shell has an enlargement in which is cast a cup forming a rest for the valve head, the stem of which is connected to the governor so as to close the steam passages by rising.

Claim.—The valve shell A, constructed as described, being enlarged at a, and provided with cup B, cast as a part of the shell and attached at $b b$, so as to form the enlarged circular port G, in combination with the valve E and rod D, in the manner and for the purposes described.

No. 59,253.—WILLIAM W. S. ORBETON, Haverhill, Mass.—*Corn Popper.*—October 30, 1866.—A frame is attached to the end of the handle, and a cage is pivoted on a vertical axis, so as to be reciprocated by means of a rod connected to a wrist on the cage and worked by hand.

Claim.—In combination with the basket A, and its handle and supporting device or devices, a mechanism or means whereby a reciprocating rotary or partially reciprocating rotary motion may be imparted to the said basket, substantially as and for the purpose set forth.

No. 59,254.—WILLIAM W. S. ORBETON, Haverhill, Mass.—*Bit Stock.*—October 30, 1866.—The end of the shank abuts against a sliding centre piece backed by a spring. The shoulder of the tool is embraced by the jaws on the end of the bit stock ; these are closed or opened by a sleeve which has an inclined slot traversed by a pin, which prescribes the longitudinal motions of the sleeve.

Claim.—The improved bit stock, composed of the body portion A, the furcated head or jaws *a*, the rotary sleeve D, and its operative mechanism or equivalent, the centralizer C, and the spring G, the whole being constructed and combined together in manner and so as to operate as set forth.

Also, the improved centralizer C, constructed in the manner as described, and applied to the bit-receiving socket, and so as to operate with the jaws *a*, as specified, and by means as set forth.

Also, the jaws *a*, constructed of one piece of metal and of the tapering form, and with lips *b*, as described and shown, when combined with the sleeve D, made and applied to the said jaws, in manner and so as to operate therewith, and by means substantially as set forth

No. 59,255.—CHARLES G. OTIS, Troy, N. Y.—*Compound for Feeding Stock.*—October 30, 1866.—The ration is composed of ground corn and oats with the addition of a small amount of flaxseed meal. The whole is compressed into a block.

Claim.—The compound feed of ground grain and oil, or flaxseed meal, compressed into packages for transportation, substantially as described.

No. 59,256.—R. W. PARKER, Woburn, Mass.—*Pulley.*—October 30, 1866.—Improvement on his patent of February 17, 1852. A belt from a motor pulley passes over another pulley and around an idler, thence to the place of beginning. The rim of the intermediate pulley is loosely attached to its web and runs upon an anti-friction wheel below.

Claim.—First, the pulley D, with its movable rim *d*, attached to the arms *e e e e* by the screws 1 2 3 4, through holes in the ends larger than the shanks of the screws, in combination with the friction wheel F, or their equivalents, constructed substantially in the manner and for the purposes described.

Second, the loose self-adjusting roller or detached revolving weight E, in combination with the belt C, and pulley D, constructed and operated substantially as and for the purposes herein set forth.

No. 59,257.—SAMUEL PECK, West Haven, Conn.—*Rubber Attachment to Washboards.*—October 30, 1866.—The rubber block has vertically slotted lugs which slide on guide rods at each side of the washboard.

Claim.—The rails *a*, on the sides of the washboard A, in combination with the slotted ears *b*, on the ends of the rubber B, substantially as and for the purpose set forth.

No. 59,258.—JAMES E. PEIRCE, West Boylston, Mass.—*Trace Connection.*—October 30 1866.—The trace pin has a prong carrying a pivoted arm, with a shield and retaining catch to hold the trace link. It is freed by a thumb knob.

Claim.—The combination, as well as the arrangement, of the shield *b* with the guard D. and its latch, applied to the trace pin, as set forth.

Also, the combination of the finger rest C with the guard D, and its latch, applied and arranged in manner and so as to operate with the trace pin, substantially as specified.

Also, the combination of the notch *a*, in the trace pin, with the shield, the guard, and the spring latch, arranged together and so as to operate with the said trace pin, substantially as hereinbefore set forth.

No. 59,259.—ROBERT PILSON, Laurel, Md.—*Machine for Preparing Cotton for Carding Engines.*—October 30, 1866.—The machine is designed to separate the fibre from dirt, and foreign substances, and is explained by the claims and cut.

Claim.—The combination of two or more sets of drawing rollers with two or more toothed cylinders, when the rollers and cylinders are arranged in the order described, and the teeth of the second and each succeeding cylinder are finer and more thickly set than those of the cylinder immediately preceding it.

Second, the combination of two or more sets of drawing rollers and two or more toothed cylinders, the teeth of the second and each succeeding cylinder being finer and more thickly set than those of the cylinder immediately preceding it, with two or more previous cylinders, through which dust and other impurities may pass, and between the surface of which and a suitable roller the opened fibrous material as delivered from each toothed cylinder, passes, and is partially condensed before it is presented to the action of the next cylinder.

Third, the combination, in a suitable case or apartment, of the previous cylinders $E^1 E^2$, and deflector F, when the said apartment is provided with an exhaust, arranged in respect to the said rollers and deflector, substantially as shown and described.

Fourth, the combination of the previous rollers $E^1 E^2$, and fluted roller I^1, the three operating to condense the loose fibrous material as received from a toothed cylinder, and the roller I^1 so arranged in relation to the roller E^2 as to act as a doffer for that roller.

Fifth, the combination of the fluted drawing rollers B, smooth roller B^1, cleaning knife or bar C, and toothed cylinder D, substantially as and for the purpose described.

Sixth, the tapering lap roller H, as and for the purpose described.

Seventh, the combination and arrangement of the several devices, as a whole, herein described and constructed, and operating to draw, open, clean, condense, and wind into a lap cotton or other fibrous material, ready for the carding machine.

No. 59,260.--WILLIAM H. PINNER, New York, N. Y.—*Apparatus for Preventing the Escape of Gases from Soap Kettles, Rendering Apparatus, &c.*—October 30, 1866.—A large pipe receives the vapors rising from the kettle, condenses the watery vapors, and passes inflammable vapors under the grate bars.

Claim.—The condensing tube *d* and vapor tube *f*, in combination with the kettle and furnace for boiling fats, soap, or other similar substances, for the purposes and as specified.

No. 59,261.—LEWIS POSTAWKA, Boston, Mass.—*Piano Seat.*—October 30, 1866.—The seat is vertically adjusted, without being rotated, by means of the hand wheel, and the sleeve nut upon the screw stem of the seat.

Claim.—The combination of the socket sleeve *a*, the hand wheel *f*, connected therewith by the hub or revolving nut *d*, and the slotted screw *b*, for elevating and depressing piano seats without turning them round, constructed and operating substantially as herein described.

No. 59,262.—JACKSON PRICE, Greenfield, Ind.—*Plough.*—October 30, 1866.—The plough frame admits of raising by a lever. The tongue is released for side swinging, by raising one of two latches at its side. Levers pivoted on the frame and worked by the foot, lock either wheel to form a pivot for turning.

Claim.—First, the arrangement of the plough frame K, and springs Q R, for regulating its motions, substantially as described.

Second, the pivoted tongue D and latches G H, operating substantially as desribed.

Third, the foot levers I J, in combination with the tongue D and latches G H, operating substantially as described.

No. 59,263.—THOMAS D. PRICE, Carrollton, Ill.—*Drill.*—October 30, 1866.—The covering wheel has separated detachable rims with an inner bevel. The revolving seed disk has a brush and spring at the entrance of the seed cavities from the box, to prevent injury to seed or machinery. A seed gate is operated by hand.

Claim.—First, the covering wheel B, when constructed with the adjustable rims b^2 and b^3, substantially as and for the purpose described and set forth.

Second, the covering wheel B, in combination with the disk F, when these two parts are so constructed as to operate conjointly, as herein described and set forth.

Third, the disk F, in combination with the brush H and spring I, for the purpose of preventing the clogging or stoppage of the seeds, as described and set forth.

Fourth, the arrangement of the gate K and its operating devices, substantially as herein described and set forth.

No. 59,264.—JOHN C. RAYMOND, Greenpoint, N. Y.—*Wind Sail.*—October 30, 1866.—The upper portion of the wind sail forms projecting quadrantal hoods presented to different quarters to catch the wind and convey it to the common, central downcast shaft.

Claim.—First, a wind sail provided with four or more wings and centre partitions or gores C, substantially as and for the purpose described.

Second, providing the wind sail with a top which extends beyond the circumference of the barrel, substantially as and for the purpose set forth.

No. 59,265.—E. P. RICHARDSON, Lawrence, Mass.—*Sewing Machine.*—October 30, 1866.—The device is to be attached to a sewing machine to facilitate the stitching by machinery of " 'turned shoes," by holding the sole and upper at such an angle as shall enable the needle to follow the required path. The foot rests in the channel, cut in what is to be (when turned) the inside of the sole, while the guard is caused to press against the sole.

Claim.—The combination of the foot F, and the guard or guide G, arranged to operate substantially as and for the purpose specified.

No. 59,266.—JOHN RIDDELL and BOYD ALLEN, Boston, Mass.—*Nutmeg Grater.*—October 30, 1866; antedated October 18, 1866.—The spherical grater is placed within a casing and operated by a crank; two or more nutmegs are pressed against its periphery and grated at the same time ; the ground spice is discharged at a common spout.

Claim.—First, the spherical grater C, arranged within the casing or chamber E, and operating as and for the purpose specified.

Second, the combination of the spherical grater C with the casing E, and chambers A, as and for the purpose specified.

No. 59,267.—L. P. RIDER, Munson. Ohio.—*Mould Board for Ploughs.*—October 30, 1866.—The mould-board is so constructed that the lower inner corner of the furrow slice shall pass in a straight line along it.

Claim.—The construction and arrangement of the plough mould-board in the manner and for the purpose set forth.

No. 59,268.—J. H. ROSE, Mount Sterling, Ill.—*Chalk-line Winder.*—October 30, 1866.—The line, after being withdrawn for use, is re-wound on the reel by a spring upon the reel shaft.

Claim.—As a new article of manufacture, the line winder herein described, the same consisting of the coil spring *b*, spindle *c*, and reel *d*, in combination with the partitioned box A, and handle B, substantially as and for the purpose specified.

No. 59,269.—AMOS W. ROSS, Northfield, Mass.—*Cultivator and Hoe.*—October 30, 1866.—The wheels and their arms are each adjustable, the latter on the frame, the shares also on their vertical bars. Inclined hoe blades are pivoted to a central share so as to admit of a spreading adjustment, and curved hoe bars connect to the ends of these blades, adjustable at both ends. A knife plate has a draught hook screwed to its shank. Slotted bars pivoted in the side beams are provided for lateral adjustment.

Claim.—First, the combination of the adjustable wheels E, and adjustable supporting arms D, with each other, and with the front and rear ends of the central beam B, substantially as herein shown and described.

Second, the teeth F, and adjustable uprights G, in combination with the cultivator beams A B C, substantially as herein shown and described.

Third, the long hoes H, in combination with the central tooth F', and the rear side teeth of the cultivator, substantially as herein shown and described.

Fourth, the combination of the adjustable curved hoes I with the rear ends of the long hoes H, substantially as herein shown and described.

Fifth, the combination of the rear governors or adjusting rods J with the curved hoes I, and the rear ends of the side beams B C, substantially as herein shown and described.

Sixth, the combination of the central adjustable governor K with the central beam B, substantially as herein shown and described.

Seventh, the combination of the guard knife L and draught hook M with each other, with the forward end of the central beam B, and with the front central tooth F, substantially as herein shown and described.

Eighth, the combination of the slotted adjusting bars N, bolt *n'*, and nut *n²*, with each other, and with the beams A B C, substantially as herein shown and described.

Ninth, a combined horse cultivator and hoe constructed and arranged substantially as herein shown and described.

No. 59,270.—E. A. G. ROULSTONE, Roxbury, Mass.—*Carpet Bag.*—October 30, 1866.—The leather is brought around an angle frame which is attached to the partition, and another piece of metal is bent around the edge and riveted through. A protecting band of metal is placed over the corners at bottom. The lock is attached to the inside of the down-turned edge of the frame.

Claim.—First, the method connecting the open part of each half of the bag leather to its frame, by fastening the edge to the outer surface of the frame, said frame projecting into instead of from the bag, substantially as set forth.

Second, the band *i*, doubled over the edge of each frame, and embracing between its edges the adjacent edges of the frame and bag leather, substantially as set forth.

Third, applying the lock to the inner surface of one of the frames, substantially as described.

Fourth, in combination with a carpet or leather bag having two compartments connected as described, the protecting band *k*, and the fastening of the flap or fall, when made to slide or catch into the frame, as described and set forth.

No. 59,271.—E. A. G. ROULSTONE, Roxbury, Mass.—*Travelling Bag.*—October 30, 1866.—The inner edge of the frame is bent around the edge of the leather, which is enlarged into a bead by an inserted wire. A spring latch on the side is raised by a projecting finger piece.

Claim.—First, a metal bag frame, when constructed and arranged with a groove for receiving and securing the bag leather or body, as described.

Second, a travelling bag, in which the frame is united to the leather or body thereof, as described.

Third, the locking spring device, as described and set forth.

No, 59,272.—E. A. G. ROULSTONE, Roxbury, Mass.—*Trunk.*—October 30, 1866.—Angle pieces inside strengthen the corners ; guards project from the lid over the corners of the body when shut. The corner longitudinal strips are bent at right angles at their ends to lap over the vertical and transverse corner strips. The lower corner guards have projections for attachment and protection of the casters. The hinges are bent to lap around the ends of the trunk. A spring catch holds the lid case into the lid.

Claim.—The employment of the angle frames to support and strengthen the trunk body, when applied to the interior of the body, with each frame bent transversely and longitudinally, as described, and with the side of the body lapped over the end, or vice versa, and riveted to the angle frame, substantially as described.

Also, the guards *e*, when made with extensions *f*, and riveted to the frame *x*, substantially as set forth.

Also, the guards *h* and *i*, when shaped and riveted to the angle frames, substantially as set forth.

Also, making the guards *i* with projections *i¹* and shoulders *l*, to protect the casters *k*, substantially as set forth.

Also, the hinges *m*, when each is bent around and riveted through the back and end of the body to the frame *x*, and is extended below the top line of the lower part *b*, in the manner described.

Also, the spring latches *o*, when made and applied substantially as set forth.

Also, the application of springs *v* to the webbing, substantially as and for the purpose set forth.

No. 59,273.—LORENZO D. RUNDELL, South Westerlo, N. Y.—*Car Brake.*—October 30, 1866.—The pawl is hinged to the end of a pivoted arm in such a way that though it holds tight, it yet may be tripped off without turning the ratchet wheel.

Claim.—The combination of the lever pawl *i* and link *e*, when hinged and pivoted as herein described, and arranged in relation to the ratchet wheel *c*, in the manner and for the purposes herein specified.

No. 59,274.—N. M. SANFORD, Vienna, Ohio.—*Washing Machine.*—October 30, 1866.— From the radially corrugated bottom of the tub arises a central pin, on which a corrugated disk reciprocates by means of a lever, which is slotted for the traverse of pins fixed in the disk.

Claim.—The hinged hub I, arranged with the lever A, and movable brace C, in combination with the post H, substantially in the manner and for the purpose as herein set forth.

No. 59,275.—CHARLES W. SAPPENFIELD, Crawfordsville, Ind.—*Sawing Machine.*—October 30, 1866.—A clutch slides on the belt shaft and communicates the motion to the fly-wheel, whose wrist actuates the reciprocating saw pitman, which is supported by a pendulous bar and guided by rods on the frame.

Claim.—The operating device of a sawing machine herein described, consisting of the clutch I, fly wheel H, shaft B, lever J, crank wheel L, pitmen M and N, swinging pitman O, and guides P and R, arranged and operating substantially as and for the purpose specified.

No. 59,276.—CHARLES H. SAWYER, Hollis, Maine.—*Hitching Device.*—October 30, 1866.— The case is inserted into the manger bar or other object ; its exterior expands upward. The spring being compressed, is slipped in, and is only retracted by a downward pull

Claim.—The combination of the V-shaped spring and case, constructed, arranged, and secured in the modes and for the purposes herein set forth.

No. 59,277.—CHARLES H. SAWYER, Buxton, Maine.—*Halter.*—October 30, 1866.—The clamp piece has three holes : one holds the knotted end of the rope : the bend is passed through the other two holes and is thus prevented from slipping either way.

Claim.—The clamp having the three holes, when applied to a halter, as and for the purposes set forth.

No. 59,278.—HENRY SAYLER, St. Paris, Ohio.—*Whip Socket.*—October 30, 1866.—The spring jaw is retained by a spring catch against its ratchet arm, and released by a key operating on the catch pin.

Claim.—A whip socket provided with the clamping jaws and a lock, when arranged to operate as and for the purpose set forth.

No. 59,279.—E. S. SEGER and J. C. ORMISTON, Erie, Ill.—*Cultivator.*—October 30, 1866.—The standards are hinged to their beams, which have strap joints on the axle, and either one can be raised by a lever connected with a segmental crown ratchet.

Claim.—The application to a corn plough or cultivator of the crotch, beams, and strap, revolving hinge, and pitman rod, iron cranks, ratchet circle, and spring catch lever, attached to the beams by the pitman rod and hinge, to raise and lower the beams and shovels, and the blade hinges to attach the inside shovel standards to the beams ; the brace foot stirrup to guide the inside shovels, and the crutch-bearing seat, as herein described, reference being had to the drawings herewith submitted.

No. 59,280.—J. H. SEYMOUR, Hagerstown, Md.—*Spittoon for Railroad Cars.*—October 30, 1866.—As the cover is opened by rotating horizontally on its pivot, the valve is closed, and conversely.

Claim.—The arrangement, in combination substantially as herein described, of the bowl A, lid or cover B, with the valve C, and rod *f*, when operated automatically by the opening of the lid, essentially as and for the purpose or purposes herein set forth.

No. 59,281.—TAL. P. SHAFFNER, Louisville, Ky.—*Manufacture of Paper.*—October 30, 1866 ; antedated October 17, 1866.—Bright metallic powder is combined with the pulp before it is made into paper, so as to become a part of the body of the paper.

Claim.—First, the depositing distributively in pulp, one or more kinds of metallic powder, immediately before said pulp is woven into paper, the object being to scatter the metallic particles into the body of the paper manufactured from said pulp, substantially as hereinbefore described.

Second, the covering or saturating paper with dissolved caontchouc or India-rubber, for the purpose of holding metallic powder upon the surface of, or for carrying the said powder into the body of the paper covered or saturated, substantially as hereinbefore described.

Third, the manufacturing of paper by placing upon an inner surface thereof a coating of dissolved India-rubber or caoutchouc, either mixed or unmixed with metallic powder, or by spreading the powder over the surface of the India-rubber coating, contemplating the covering of said metallized surface with a film of paper woven thereon from pulp, or by pressing another sheet of paper in such manner as will unite the whole practically as one body of paper, substantially as hereinbefore described.

No. 59,282.—ZACCHEUS B. SHANNON, Port Washington, Ohio.—*Churn.*—October 30, 1866.—The inner box is opened at top and has a circular central opening in the bottom through which runs the shaft of the circular dasher. The dasher has a central opening similar to the box bottom, and also horizontal apertures in its shell, in such direction as to throw the cream against the sides of the box.

Claim.—First, the rotary dasher C, constructed and operating substantially in the manner and for the purposes hereinbefore described.

Second, the rotary dasher C, centre box B, and churn A, constructed and operating substantially in the manner and for the purposes hereinbefore described.

No. 59,283.—JOSEPH D. SMITH, Peoria, Ill.—*Seeding Machine.*—October 30, 1866.—The rear hounds are so connected to a transverse horizontal pivoted bar, as to allow a limited free movement; adjustable sliding pieces limit this movement laterally. Brakes are worked to either wheel by sliding foot blocks. The seed slides are double, and a sliding of one part upon the other changes them from corn to small grain seeders.

Claim.—First, the bar *q*, the pin *e*, or its equivalent, and the bars A' and P, constructed and for forming an adjustment, as herein fully set forth.

Second, the bar *q*, the pin *e*, and the slides *a a*, arranged and constructed as and for the purpose herein specified.

Third, the combination of the slides T and U, constructed and arranged together, as and for the purpose herein specified.

Fourth, the combination of the scraper *g*, the slotted piece *f*, the rod *h*, and the foot piece *i*, constructed and used as and for the purpose set forth.

Fifth, so arranging the hounds H H with the hounds C C, that when the driver changes his position to the rear of the seat, the said hounds H H bear against the under side of the hounds C C, and thus make a rigid machine, as and for the purpose set forth.

No. 59,284.—GEORGE B. SNOW and T. G. LEWIS, Buffalo, N. Y.—*Dental Plugging Instrument.*—October 30, 1866.—The plugging tool presses against the filling in the tooth; pressure on the case makes the tool stock recede, imparting its movement to the lifting bar and hammer, until the bar passes the incline of the wedge, releases its hold on the catch, and releases the hammer which descends under the influence of the spring. The force is adjusted by devices operated by an exterior band.

Claim.—First, causing the tool holder to recede from the hammer immediately after a blow is given, in order to obtain distance between the hammer and the head of the tool holder for a new blow, substantially as described.

Second, placing a spiral spring G, in the top of the case to act upon the hammer in combination with either the adjusting stopper I, or screw cap H, for the purpose of causing the hammer to give heavier or lighter blows, as required.

Third, the combination of the ring R, and stop screw S, and collar *u*, for the purposes and substantially as set forth.

Fourth, constructing the lifting bar D with a bent end, in combination with a receiving hole in the upper end of the tool holder, as shown at *d*¹, and with a notch or shoulder at its upper end, as shown at *d*², to allow it to engage with the stops L L' on the hammer, for the purpose of forming a direct connection between the tool holder and hammer, substantially as set forth.

Fifth, the feather O, in combination with the hammer F, for the purpose of arresting the descent of the hammer, and holding it at that point until again raised, substantially as described.

No. 59,285.—T. G. SPRINGER, Conneautville, Penn.—*Wagon Brake.*—October 30, 1866 — The brake shoe is connected by a pin to the slotted arm of the eccentric, and is operated by the longitudinal movement of the tongue in drawing or holding back. The operating mechanism at the shoe is hooded as security from dirt.

Claim.—First, pivoting eccentrics g g', h, which are constructed substantially as described, to a fixed bar F, and a movable bar E, in combination with brake shoes k k, or their equivalents, substantially as specified.

Second, the hooded brake shoes k, applied to rocking eccentrics, or cams g, substantially as described.

Third, connecting the pivoted eccentrics g g to the sliding brake bar E by means of pins passing through slotted portions h, substantially as described.

No. 59,286.—NATHAN H. SPAFFORD, Baltimore, Md.—*Machine for Assorting Bristles.*— October 30, 1866.—The bristles are placed in the box, butt down, and anchored by the traverse pins. They are assorted by the drawing out of the longest by adjustable nippers on a vertically reciprocating slide, which has automatic grasping and releasing connections.

Claim.—First, the box H, as constructed with the slide v'''' and spring y, as arranged and operated for the purposes set forth.

Second, the knives a', as arranged in combination with the box H, for the purpose set forth.

Third, the combination of the box H, with the feed carriage O, and slide table Y, the whole being constructed, arranged, and operated in the manner substantially as and for the purposes described.

Fourth, the method of regulating the forward feed of the bristles without altering the speed of the main shaft, by means of the cam g, the lever S, and pins i i, the slotted plate k, spring j, and frictional pulley R, the whole being arranged and operated in the manner substantially as set forth.

Fifth, the combination of the box H, or its equivalent, with the jaws G' and G'', or their equivalents, when the former is kept stationary during the descent of the latter to seize the bristles and is afterward fed forward when the jaws near the highest point of their ascent, for the purpose described.

Sixth, the slide G,' and jaws G'', the spring 18, cam 20, shaft 9, and crank K, in combination with the adjustable stops L and 21, the whole being constructed and operated in the manner and for the purposes described.

Seventh, the combination of the jaws G' and G,'' with the arms n, the lever U, and slide bar F, the whole operating in the manner and for the purposes described.

Eighth, the combination of the arms n, of the lever U, and slide bar F, with the friction slide I', and receiving box I, all being arranged and operated in the manner and for the purposes set forth.

Ninth, the box I, as constructed in combination with the slide I', rod 3, and spring clamp c, for the purpose of receiving bristles.

Tenth, the India-rubber 14, or its equivalent, fixed to the jaw G'', when used in connection with the fluted steel 15, or its equivalent, fixed to the jaw G'.

Eleventh, the adjustable rod N, in combination with the slide G, as and for the purpose set forth.

Twelfth, the slide 4, the cam 10, the adjustable slotted standard 5, operating as and for the purpose described.

Thirteenth, the gear P, the rack bar Q, the carriage O, and the thumbscrew w'', in combination with the shaft h, as shown and described.

Fourteenth, the combination of the box H, and its attachment, the platform V, the rack bar S, carriage O, gear P, thumbscrew w'', shaft h, feed wheel R, friction spring W, spiral spring j, lever S, cam D, slotted plate K, jaws G' G'', slide G, connecting rod N, slide 4, lever U, slide F, receiving box and slide I, substantially as and for the purpose set forth.

No. 59,287.—T. K. STERRETT and W. R. FARRELL, Philadelphia, Penn.—*Letter Box File.*— October 30, 1866.—The follower rests upon the pile of papers in the pigeon hole, and is raised or lowered by turning a bar connected to the swinging frame, which slides in staples on the back of the follower.

Claim.—The board C, staples D, the frames F and G, and their extension portions I, slotted bar K, rack N, pinion O, shaft P, and spring R, toothed collar T, notched collar U, arranged with the box having grooves S and L, as described, and operating substantially as and for the purpose specified.

No. 59,288.—OSCAR STODDARD, Jackson Mich.—*Boot and Shoe.*—October 30, 1866.— The lower or wearing portion of the heel is detachable from the upper or permanent portion, and is locked thereto by plates with hooks and a spring catch.

Claim.—Constructing the heels of boots and shoes of two parts, A B, the former part A, being permanently attached or secured to the boot or shoe, and the other part B, made separately or detached, and secured to A by means of a fastening, substantially as shown and described.

No. 59,289.—GEORGE STOVER, Centre Hill, Penn.—*Raising and Lowering Carriage Tops.*—October 30, 1866.—Near the foot of the rigid arm are pivoted the bow plates and a 'sectoral frame. The upper end of the rigid arm has a catch to hold up the said frame, and the frame has a catch to hold the forward arm.

Claim.—Combining with the bows of a buggy or carriage top the hinged arcs and rigid arms, with suitable catches for connecting or disconnecting them, and so arranging them on the inside as that the person occupying the seat may raise or lower the top at pleasure and hold it at half or full up, substantially as herein described and represented.

No. 59,290.—M. L. and O. A. STRAY, Willoughby, Ohio.—*Fruit Basket.*—October 30, 1866.—The bottom of the basket is made of interwoven splints which taper in width from the edge of the bottom towards its centre; the greater width of the splints enables them to overlap each other at the sides and ends.

Claim.—The described basket, when constructed and arranged in the manner specified, being a new article of manufacture.

No. 59,291.—E. DWIGHT STREET, East Haven, Conn.—*Threshing Machine.*—October 30, 1866.—The oscillating beater is depressed by a treadle and elevated by a spring, the two being connected to the ends of a belt which passes over the hub.

Claim.—The combination of the beater D, (one or more,) the table A, treadle E, and spring I, arranged to operate in the manner described.

No. 59,292.—W. STREVELL, Jersey City, N. J.—*Device for Stretching Leather.*—October 30, 1866.—The bar has one fixed and one movable clamp, and between them is a slide bar having a pawl to catch in a ratchet in the main bar, and rubber springs acting on the movable clamp.

Claim.—The combination with the sliding jaws or clamps E, of the cross-bar H, connected therewith by rods I, having rubber or other elastic cushions or springs J, substantially as and for the purpose described.

No. 59,293.—MARCUS A. TARLETON, New Orleans, La.—*Cotton Tie.*—October 30, 1866.—The buckle has one smooth bar over which one end of the hoop iron is looped, and a bar with a spur which engages one of a series of holes in the other end of the band, which then passes inward and lies between the cotton and the other band.

Claim.—First, the tie or buckle A, when constructed and operating as herein described, for the purpose set forth.

Second, the combination of the tie or buckle A with hoop iron, when those parts are united and operate as described, for the purpose set forth.

No. 59,294.—ALBERT H. TINGLEY, Providence, R. I.—*Heat Regulator for Hot-air Furnaces.*—October 30, 1866.—Vessels in the heated air pipes and in those containing the draught or cold air are connected by a tube, and the contained air acts upon two diaphragms, the one to regulate the furnace damper, and the other to indicate the pressure. A plunger worked by a screw regulates the general degree of heat by modifying the size of the air space.

Claim.—First, the combination of the two vessels A C, connected together by the pipe D, substantially as described and for the purpose set forth.

Second, the adjuster E, constructed as described, by means of which the general effect of the expansion and contraction of the air, gas, or expansive fluid in the vessel A, upon the damper R, is controlled, substantially as set forth.

Third, the combination of the indicator L or its equivalent, by which the condition of the fire or the position of the damper is indicated by the expansion or contraction of the air in vessel A, with the vessels A and C, substantially as described and for the purpose set forth.

Fourth, the arrangement of the damper R, and ventilator t, upon the same spindle b, substantially as described and for the purpose set forth.

No. 59,295.—WILLIAM V. WALLACE, New York, N. Y.—*Piano-forte Action.*—October 30, 1866.—Explained by the claim and cut.

Claim.—Making the connections or joints between the key and hammer of a piano action of hard rubber, or its equivalent moisture resisting gum or compound, to prevent swelling and consequent binding of said parts, substantially as described.

No. 59,296.—MAXIMILLIAN WAPPICH, Sacramento, Cal.—*Rudder.*—October 30, 1866.—Explained by the claims and cut.

Claim.—First, providing the rudder blade with slots forming openings through the entire body of the rudder in such a manner as to allow a partial efflux through said openings and thereby prevent the backing of water at the same time that the comparative vacuum on the aft side of the rudder blade in steering is being filled with increased rapidity, for the purpose of more evenly balancing the pressure of the water on the forward and aft sides of the rudder, so as to reduce the strain on the pintles and facilitate the turning and handling of the rudder in steering a vessel.

Second, constructing a rudder of tubes or rounded bars firmly braced and framed, substantially as specified, for the purpose of obtaining with a small rudder blade a great effective steering action, and with a reduction of weight of material an increased strength of rudder.

Third, providing the rudder step and shoe, and the pintles and braces, or their equivalents, with concentric grooves and rings, substantially as and for the purpose set forth.

No. 59,297.—THOS. WATSON and CHAS. PERRY, Brooklyn, N. Y.—*Extension Ladder.*—October 30, 1866.—The ladder frame is supported on wheels, and is attached by its foot to the fore-carriage, in travelling. The ladder sections have clip flanges which make each a guide for its fellow, and are raised by cords which pass to a windlass on the rear of the frame.

Claim.—First, the frame F, constructed as herein described, when used for supporting and operating an extension ladder, substantially as described.

Second, the combination of the windlass N with the rear end of the frame F, for the purpose of raising and lowering an extension ladder, substantially as described.

Third, connecting the rear carriage frame F to the forward part of the truck, in the manner described and for the purpose set forth.

Fourth, constructing the side bars of the ladder in the form herein shown and described, so that the side bars of each part may form slides and guides for the adjacent parts, when raising and lowering the ladder.

No. 59,298.—WILLIS WEAVER, Salem, Ohio.—*Carpet Fastener.*—October 30, 1866.—A double hook catches the edge of the carpet and the loop is hitched over a nail in the floor.

Claim.—The carpet fastener, consisting of a wire with hooked ends b, and bent or turned at the middle to form eyes a, as and for the purpose specified.

No. 59,299.—IRVING E. WESTON, Winchendon, Mass.—*Mop Head.*—October 30, 1866.—The shank of the inner jaw screws into the end of the handle; the collar of the outer jaw revolves on the handle. The motion of the head or handle, relatively the one to the other, closes or loosens the jaw.

Claim.—Uniting the outer jaw of a mop to the socket by means of a collar and projection, and the inner jaw to the socket by means of a screw and thread and so that by turning either the mop head or the handle the inner jaw will travel to or from the outer one; but when the material is clamped between the jaws, then the collar shall be rigid on the socket by means of the screw drawing and holding them tightly together, in the manner and for the purpose set forth.

No. 59,300.—SETH WHEELOCK, Richland, Mich.—*Machine for Sowing Plaster, &c.*—October 30, 1866.—The triangular bar presents an edge upward and reciprocates longitudinally beneath the hopper; its spikes protrude into the latter and stir the fertilizer, which falls through upon each face of the bar below and thence to the ground.

Claim.—The arrangement and combination of the triangular spiked bar D and connected lever L, with the hopper H and bed frame A, substantially in the manner and for the uses herein specified.

No. 59,301.—SAML. H. WHITAKER, Covington, Ky.—*Discarbonizing Furnace.*—October 30, 1866.—The cupola is surrounded at its lower portion by an annular chamber. Two ducts, a portion of a series of tuyeres, connect the cupola and chamber at their bottom. The top of the chamber is traversed by several opposite tuyeres.

Claim.—First, the enclosed auxiliary chamber or chambers B, communicating with the bottom of the furnace, and provided at the top with one or more downward discharging tuyeres, placed out of contact of the molten metal, for the objects stated.

Second, the annular blast chamber B, which surrounds the lower portion of a blast furnace, and is provided with one or more pairs of opposite and downwardly directed tuyeres out of contact with the molten metal, substantially as set forth.

No. 59,302.—WM. N. WHITELY, Jr., Springfield, Ohio.—*Harvester.*—October 30, 1866.—The devices are for strengthening the frame, adjusting its height, bracing the driver's footboard, adjusting the angle of the tongue to the frame, bracing the driving crank shaft, and adjusting the tightening pulley over which the reel cord runs.

Claim.—First, the diagonal back brace E, in combination with the main frame A A B C, and the drag bar D, as and for the purpose set forth.

Second, the sector standard I, constructed in the form shown and described, in combination with the driving wheel and main frame of a harvester.

Third, the sector plate H, provided with the booking flange f, in combination with the curved sector standard I and the main frame of a harvester, for the purpose set forth.

Fourth, the brackets *k*, in combination with the curved sector standards and driver's foot-board K of a harvester, for the purpose of strengthening and supporting said standards.

Fifth, the driver's foot-board K and tool box L, when arranged as shown and described.

Sixth, the adjustable lever O, in combination with the tongue M, for the purpose of controlling and changing the angle of the tongue to the main frame.

Seventh, the levers O and *p*, spring *r*, and pin *m*, in combination with the disk P and tongue M, for the purpose set forth.

Eighth, the spring foot latch Q, in combination with the strap R, tongue M, and main frame of a harvester, for the purpose set forth.

Ninth, the box U, constructed as described, in combination with the crank shaft T, cross-bar C, and diagonal brace E, for the purpose of protecting the shaft T, and strengthening the frame, as set forth.

Tenth, the vertically and laterally adjustable, spring-tightening pulley W', when constructed as described.

No. 59.303.—WILLIAM N. WHITELEY, Jr., Springfield, Ohio.—*Harvester.*—October 30, 1866.—The devices are for giving the sweeping and vertical motion to the oscillating rake head; the spindle, firmly attached to the frame, is the axis of the master wheel, and the socket of the rake crank shaft, which is revolved by exterior connection with the hub of the master wheel; a sector plate journalled on the frame and connected to the axis of the master wheel is the means of adjusting the machine vertically. The rake pulley runs on an axis projecting from a sleeve on the post.

Claim.—First, in combination with the quadrant J and rake N, arranged and operating substantially as set forth, the guides E and F, and arm M, substantially as and for the purpose described.

Second, the stationary hollow spindle T, its outside surface forming the bearing for the master wheel and its inside surface forming the bearing for the rake crank shaft, substantially as shown.

Third, supporting the master wheel upon a stationary hollow spindle secured at one end only, in combination with the rake crank shaft running within the spindle.

Fourth, driving an automatic rake through the centre of the driving wheel, and from the outer side thereof, substantially as and for the purpose described.

Fifth, communicating motion to the rake shaft from the outer side by means of the plate W, or its equivalent, and the clutch pin Y.

Sixth, the combination of the sector plate Q, hollow spindle T, and rake shaft Y, substantially as shown and described.

Seventh, the combination and arrangement of the sleeve C, stationary spindle *f*, projecting from one side of said sleeve, and reel pulley *g*, substantially as and for the purpose set forth.

No. 59,304.—J. M. WILLIAMS, Connersville, Ind.—*Water Wheel.*—October 30, 1866.—The flume has an annular flange through which are chutes of an oblong tubular form, standing at an angle of 45°. The buckets have a vertical curvature, discharging the water downward within the annular ring at the base of the wheel. An annular gateway governs the entrance to the chutes and is moved by a rack and pinion.

Claim.—First, the water wheel constructed as described, combining the disk *a*, and bucket *c*, with an annular flange for a base, in the manner and for the purpose specified.

Second, the combination of the dome chute case with its chutes *d* and gate *g*, arranged as described for the purpose specified.

Third, the combination of the wheel with the chute case and gate, arranged and operating conjointly, as and for the purpose specified.

No. 59,305.—JACOB D. WINSLOW, Wilmington, Del.—*Floor Clamp.*—October 30, 1866.—The block is adjusted and held by dogs whose claws are driven into the joists; wedges are then driven between the block and floor plank to make a tight joint.

Claim.—The combination of the block A, composed of wood or other suitable material, with or without ears or lugs *d d*, with the iron dogs *b b*, and the wedge *l*, as hereinabove described, or any other appliances substantially the same as and for a floor clamp.

No. 59,306.—FREDERICK WOLFF, New York, N. Y.—*Artificial Leech.*—October 30, 1866.—The exhausting tube has a puncturing instrument on the end of a central spring shaft. The puncturer has a tripartite edge making an opening resembling that made by a leech, and is made by the recoil of the spring; after which the piston is retracted in the tube and the blood follows.

Claim.—First, the construction of the mechanical leech with a lancet or puncturing device and with a suction piston in such manner that the lancet can be raised and set independently of the piston, operated to puncture the skin, and then both the lancet and piston raised together, so as to draw the blood within the same air pump tube A, in which the lancet and piston are arranged, all substantially in the manner described.

Second, the elastic cushion *h*, in combination with the stop *b*, substantially in the manner and for the purpose described.

Third, extending the lancet handle through the cap C, and making it capable of being set for use, independently of the piston, substantially in the manner and for the purpose described.

Fourth, the combination of the lancet, hollow piston rod, air pump barrel A, stop b, and spring catch, all arranged substantially as described.

No. 59,307.—HENRY YERTY, Sidney, Ohio.—*Burglar Alarm.*—October 30, 1866.—The swivel gun is directed toward and discharged at any approaching object which may come in contact with a cord; the latter, by its tension, in connection with a rod projecting from the front, swings the gun in the direction of tension and then discharges it.

Claim.—One or more barrels mounted upon a pivot or axis and employed in connection with a rod I, extending forward of the axis E, and having attached to it one or more cords I', whereby the barrels are directed towards an approaching object by the object itself and then discharged, substantially as described.

Also, the combination with the swivelled gun A A, of the shaft G, arm H, rod I, and cord I', the whole being arranged to operate in the manner and for the purposes herein described.

No. 59,308.—JOHN H. AIKEN, Norwalk, Conn., assignor to himself and REUBEN ROWLEY, New York, N. Y.—*Machine for Oiling Wool for Carding Engines.*—October 30, 1866.—A vibrating oil box at given periods passes under the perforated bar or sheet, which descends to get its charge of oil, then rises as the cup recedes; the box being by a cam permitted to drop quickly and to be suddenly arrested, sprinkles its oil upon the wool beneath. A shield or case surrounds the distributing apparatus and protects it from the dust of the carding engine.

Claim.—First, giving to a perforated oil distributor an abrupt or sudden falling or dropping motion, by means of a cam or equivalent device, for the purpose of sprinkling the oil on the wool as hereinbefore set forth.

Second, in combination with an oil distributor having an abrupt or sudden falling or dropping motion, a vibratory oil box, substantially as hereinbefore set forth.

Third, in combination with a vibratory oil box, the crank I, and levers J and K, or their equivalents, for the purposes hereinbefore set forth.

Fourth, in combination with a vibratory oil box, the adjustable shield or case L, for the purposes substantially as hereinbefore set forth.

No. 59,309.—HEMAN A. ASHLEY, Springfield, Ohio, assignor to himself and EDWARD M. DOTY, same place.—*Baling Press.*—October 30, 1866.—The power is communicated to the (dou e) follower by a lever pivoted to the follower at its lower end and to an arm depending from the head of the press at its middle, forming a toggle. A cord attached to the head of the lever passes around sheaves in the head of the press and lever, and under another, to the horse. After the pressing of the bale, the follower and lever are elevated by another cord.

Claim.—The arrangement, herein described, of the base bars A A, press box P S (at or near the level of the ground,) double follower K K, cross-beams a a, vertical toggle levers I J J, cord M, and pulleys L j, all constructed and operating as set forth, to provide for the delivery of the bale at or near the ground.

No. 59,310.—M. L. BALLARD, Canton, Ohio, assignor to BALLARD, FAST & COMPANY, same place.—*Pitman Connection.*—October 30, 1866.—The bar is connected to the pitman by means of a hemispherical socket on the former, and a projection of counterpart form on the latter; a screw bolt whose flanged head has a bearing of corresponding curve is the means of uniting the two parts, and an opening in the crown of the socket permits universal play to the hemisphere in its socket, within certain limits as to extent.

Claim.—A pitman connection formed by the hemispherical head B, fitting into a similarly shaped recess, the crown a, and its opening c, and the flanged screw bolt E, or its equivalent, a pin, combined to operate in the manner and for the purpose described.

No. 59,311.—GEORGE H. CLARKE and H. VAN WAGENEN, New York, N. Y., assignors to GEORGE H. CLARKE, same place.—*Grate Bar.*—October 30, 1866.—On each side of the arched section is one which has an interlocking device, to connect with a similar cluster on the right and left. The cross pins near the ends are divided to permit independent expansion to the sections of the bar.

Claim.—In combination with the arched bar A, the bars B and C, provided with vertical dovetails or similar interlockments, the slitted rods dividing the bars to admit of longitudinal expansion, and the grooves or depressions in the supports a a, the whole constructed substantially as described and for the purpose specified.

No. 59,312.—CHARLES J. EVERETT, New York, N. Y., assignor to LOCKWOOD and EVERETT, same place.—*Rendering Apparatus.*—October 30, 1866.—The noxious gases are passed from the rendering tank, through a superheating coil, and discharged in jets along with a current of air into the furnace above the incandescent fuel. A dome above the blow-pipes concentrates the heat.

Claim.—First, consuming the noxious or offensive gases and vapors from a rendering tank, apparatus or contrivance, by introducing them in, over or under the furnace, along with an artificial current of air induced by the flow of said steam and gas from the rendering apparatus or superheater.

Second, uniting or mixing over the furnace of the consumer or deodorizer, an artificial current of air and the heat of the furnace, with the noxious or offensive gases and vapors, from a rendering apparatus, for the purpose of consuming them without materially increasing the consumption of fuel in the furnace.

Third, the use of jets and blowpipes, applied so as to introduce the noxious gases and vapors, along with a current of air, into, over or under the furnace, substantially as shown, and described.

Fourth, concentrating the heat with the noxious gases and vapors from a rendering tank, along with the current of air over the furnace, and arresting their too rapid ascent by the use of a receiver, to insure their ignition and consumption, as set forth.

Fifth, the use of the pipe P, figure 3, in the tank, substantially as in the manner described, for the purpose specified.

No. 59,313.—GEORGE W. HILL, Deep River, Conn., assignor to himself and C. A. MOORE, West Brook, Conn.—*Spring Bat.*—October 30, 1866.—The striking portion of the bat has longitudinal slits to give it an elasticity; the slits may be filled with India-rubber to increase the resiliency of the striking surface.

Claim.—The within-described improvements in ball bats (or clubs,) substantially as specified and for the object set forth.

No. 59,314.—W. A. HORRALL, Washington, Ind., assignor to himself and RICHARD BRUNER, same place.—*Cotton Seed Planter.*—October 30, 1866.—The bottom of the seed hopper has longitudinal bars, and the seed is carried forward by teeth upon an endless belt, actuated by a roller on the main shaft and an adjustable roller in the rear.

Claim.—The hopper E, with the narrow bottom grating formed by the rods *d*, and the endless belt *e e*, with the teeth *m* combined therewith, constructed, arranged and operating together for planting cotton seeds, substantially as herein described.

Also, the adjustable clevis *h*, in combination with endless belt *e e*, arranged and operating as and for the purposes herein specified.

Also, the combination of the plough *a*, the furrow-opening block *b*, and the furrow coverer *c*, with the hopper E, and the endless belt *e e*, arranged and operating substantially as herein described.

No. 59,315.—B. ILLINGWORTH, Freeport, Ill., assignor to J. B. BYERLY and C. A. SHERTZ, same place.—*Flour Sifter.*—October 30, 1866.—The horizontal shaft has eccentric disks, which rotate in the semicylindrical sieve, and drive the flour through the meshes.

Claim.—The box A, provided with legs having a horizontal bar *a*, with the disks D and rods *b b*, operating upon the semicircular sieve C, when arranged in the manner substantially as herein specified.

No. 59,316.—AMOR D. KENDIG, Safe Harbor, Penn., assignor to himself and JOHN MILLER, same place.—*Harness.*—October 30, 1866.—The driving lines pass around pulleys attached to the bit rein and breeching, respectively, so as to give greater power in restraining the animal.

Claim.—The pulley A, attached to the bit in combination with the pulley B, attached to the breeching, and having the line arranged in connection therewith, as shown and described.

No. 59,317.—ALLEN LAPHAM, Brooklyn, N. Y., assignor to himself and JOB JOHNSON, same place.—*Still for Petroleum.*—October 30, 1866.—The still has several tiers of flues, and by means of dampers the caloric current is cut off from the upper set when the liquid has sunk beneath them as the distillation advances, thus keeping the heated portion of the still below the surface of the oil.

Claim.—The arrangement of the flues *g* and *h*, and dampers applied to the still *c*, substantially as set forth, for preventing the still becoming heated above the liquid therein, for the purposes set forth.

Also, forming the lower portion of the still over the fire narrower than the upper portion, as shown, and combining therewith the flues *g*, as and for the purposes set forth.

No. 59,318.—JAMES N. PHELPS, Brooklyn, N. Y., assignor to himself and JOSEPH BAILEY, same place.—*Telegraph Cable.*—October 30, 1866; antedated October 16, 1866.—Explained by the claim and cut.

Claim.—The employment in a cable of one or more spiral metallic conductors C, wound around a core of India-rubber or elastic insulating material B, which constitutes a loose insulating covering to a central conductor A, substantially as herein described.

No. 59,319.—OWEN REDMOND, Rochester, N. Y., assignor to RUFUS F. OSGOOD, same place.—*Packing for Oil Wells.*—October 30, 1866.—The packing frame has a set screw operated from the surface to fix it at any elevation. The leather packing is expanded by an annular wedge, forced down by a screw pipe, worked by bevel gearing, operated by a chain and rod from the well top.

Claim.—A packing device for artesian wells, packing both the tubing and the sides of the well, when the said device is capable either of being adjusted higher or lower upon the tubing, or vice versa, the tubing adjusted higher or lower within the packing, substantially as specified.

Also, the combination of the hollow wedge G, and screw collars D E, for the purpose of expanding the packing disk *b*, substantially as described.

Also, securing the packing device to the oil tube at any position by means of the screw *r*, provided with a head or rim around which passes a wire, cord, or chain *s*, substantially as specified.

No. 59,320.—JOHN STARKEY, Portland, Me., assignor to BYRON D. VERRILL, same place.—*Curtain Eyelet.*—October 30, 1866.—A rim or turned-over flange on the eyelet plate secures a perforated disk of rubber for holding a knob or button.

Claim.—First, an eyelet of brass, copper, or other metallic substance, holding in an oval or nearly flat rim a disk of rubber or other strong material with an aperture therein for the admission of a knob or button, and attached to a curtain or garment of leather or other material, by means of a washer of brass, copper, or other metallic substance, through which the eyelet passes, and over the inner edge of which the edge of the eyelet is rimmed down, thus holding the curtain or garment firmly and securely between the shoulder of the eyelet and the washer.

Second, the eyelet E, having shoulder S, and terminating in oval rim R, with or without the washer W, and in combination with the disk C, all constructed as described and for the purposes set forth.

No. 59,321.—J. W. THOMPSON, Salem, Ohio, assignor to himself and H. BARNABY, same place.—*Cherry Stoner.*—October 30, 1866.—The cherries pass consecutively from the hopper; the claws penetrate the cherry and grasp the stone ; the pulp is stripped off and ejected separately from the stones.

Claim.—First, the combination of the griping knives *a a a* with the stripper C, arranged to operate substantially as described.

Second, the rotating plate D, upon the shaft D', in combination with the cherry receptacle B, and a stripper C, substantially as described.

Third, the construction of the pivoted knives *a* with notches in them for receiving and holding the cherry stone during the stripping of the pulp from it, substantially as described.

Fourth, the construction of the stripper C of spring segments, adapted for receiving through them the cherry stones, and discharging the same at the opposite end of the cylinder to that from which the pulps are discharged, substantially as described.

No. 59,322.—GEORGE W. TRAPHAGAN, Glenn's Falls, N. Y., assignor to himself and A. M. DECKER, same place.—*Slat Iron for Carriage Tops.*—October 30, 1866.—The ends of the bows are fitted into ferrules, having semicircular extensions upward, and longitudinal screw-threaded holes for the reception of screws in the end pieces.

Claim.—First, attaching the bows to the hinge by screws, substantially in the manner herein shown and described and for the purpose set forth.

Second, the combination of the straps A and finger irons C, with each other, with the bows B, and with the supporting iron or rail D, when the said straps and finger irons are constructed substantially as herein shown and described and for the purpose set forth.

No. 59,323.—RICHARD VAN VELTHOVEN, Philadelphia, Penn., assignor to WILLIAM W. HARDING, same place.—*Photographic Album.*—October 30, 1866.—Improvement on patent of Hazzard and Velthoven, October 17, 1865. Slots are cut in the strips of parchment to which the album leaves are attached, and the flap thus made is turned down over the edges of the album leaf, whose other side is in connection with the uncut edge of the strip.

Claim.—Forming one or more flaps *b* in the strips C, in the manner and for the purpose substantially as shown and described.

No. 59,324.—HENRY WATHEW, Philadelphia, Penn., assignor to THOMAS and GEORGE M. MILLS, same place.—*Machine for Peeling Almonds.*—October 30, 1866.—The thin peel is removed from the scalded almond kernels by passing them between two elastic bands of India-rubber, traversing side by side in the same direction, at different velocities.

Claim.—The two elastic endless aprons A A, with their respective rollers and mountings, when constructed substantially as described and for the purposes set forth.

No. 59,325.—HENRY WATHEW, Philadelphia, Penn., assignor to THOMAS and GEORGE M. MILLS, same place.—*Confection Pan.*—October 30, 1866.—The shaft of the pan is secured to a ring by a universal joint. Its lower end rests in a socket made on the upper face of the wheel, which is rotated by gearing, and carries the shaft around with it. The latter describes two cones connected at their common vertices, which is at the centre of oscillation in the universal joint. A rolling motion is imparted to the pan, which is heated by steam or hot air pipes beneath, communicating by flexible pipes with a furnace or boiler and an escape pipe.

Claim.—The pan A, having the described wabbling motion imparted to it by means or devices equivalent to those herein set forth, and being provided with the described arrangement for heating by steam or hot air, the whole being arranged substantially as and for the purpose specified.

Also, supporting the weight of pan A, at or near its centre of gravity, upon the annular bearing L, substantially as and for the purpose set forth.

No. 59,326.—WM. C. WATSON, Paterson, N. J., assignor to himself and IRA W. GREGORY Brooklyn, N. Y.—*Manufacture of Liquid Glue.*—October 30, 1866.—Composed of glue, acetic acid, whiting, isinglass, and decoction of tobacco.

Claim.—A liquid glue cement composed of the ingredients and about in the proportions specified.

No. 59,327.—H. E. and C. W. WOODFORD, Keesville, N. Y., assignors to themselves and P. S. WHITCOMB.—*Horseshoe Nail Machine.*—October 30,1066.—The intermittingly rotating disk carries upon its periphery a series of dies, fourteen in number, of varying form; each of these in turn answers the purpose of an anvil, upon the face of which a hammer working vertically, forges the two faces of the nail; between each blow of this hammer the nail blank rises slightly above the face of the anvil, when two side hammers operate simultaneously upon the edges.

Claim.—First, the intermittingly rotating anvil provided with dies in connection with the vertical and lateral hammers, and arranged to operate substantially in the manner as and for the purpose herein set forth.

Second, the securing of the dies G in or to the periphery of the wheel F, by means of the dovetail groove in the flange c, to receive one end of the dies and the buttons or clamps t, bearing against the opposite end, substantially as shown and described.

Third, giving the anvil an intermittingly rotating motion by means of the cam C and worm wheel D, constructed and arranged substantially as set forth.

Fourth, the cutters K K', in combination with the intermittingly rotating anvil, substantially as and for the purpose specified.

Fifth, the vibrating bed piece B*, provided with the roller E*, in connection with the plate F*, provided with the roller G*, the toothed wheel C*, having a smooth portion s on its periphery, and the projection s on said wheel, with the pendant M* of plate F*, all arranged to operate substantially as and for the purpose set forth.

Sixth, the screw K* and worm wheel H* in combination with the vibrating bed B* and the plate F*, substantially as and for the purpose specified.

No. 59,328.—L. A. C. ST. PAUL DE SINÇAY, La Vielle Montagne, Belgium.—*Manufacture of Sulphur.*—October 30, 1866.—The sulphurous acid evolved in the working of auriferous sulphurets is passed into a chamber, and from this through tubes into a receiver, from which it passes into the decomposing chambers through a pipe. These chambers are filled with carbon or other reducing substance, and are heated by a furnace. The vapor of sulphur then passes into a condenser.

Claim.—The within-described method of reducing sulphurous acid gas, consisting of a series of retorts e, condensers g, and collecting chambers h, in combination with the main pipe a, and secondary pipes b, substantialy as set forth.

No. 59,329.—R. EATON, Lee, England.—*Locomotive Fire Grate.*—October 30, 1866.—The congeries of grate bars are in steps, and afford more uniform access of air to the fuel. The ash pan has back and front dampers and a perforated guard plate in front to govern the draught.

Claim.—First, a grate composed of a series of grate bars A, either square, round, oblong, or polygonal, and placed one above the other in the form of terraces of gradually decreasing size, substantially as and for the purpose described.

Second, the ash pan D, provided with front guard G, front damper F, and back damper E, in combination with the grate, constructed and operating substantially as and for the purpose set forth.

No. 59,330.—JULIUS ROBERT, Selowitz, Austria.—*Process for Making Extracts.*—October 30, 1866.—The plant is cut into pieces and the extractive matter removed by successive infusions in liquid of gradually increasing strength and decreasing temperature.

Claim.—The within-described process of extracting juice from vegetable substances by subjecting them to "diffusion," substantially in the manner set forth.

No. 59,331.—JAMES STEART,Bermondsey, England.—*Extracting Fibre from China Grass, &c.*—October 30, 1866.—After crushing and bruising the material it is treated, at 210° Fah., with a liquor made by digesting fish in hot water or steam, to remove the saccharine, gelatinous, fleshy, &c., matters from the fibre.

Claim.—The obtaining the fibre from China grass, rhea, or Siam grass, Spanish grass, weed, flax, and other analogous vegetable substances, and the preparing, cleaning, and purifying of goat. camel, and other hair, silk, wool, and other analogous substances, by subjecting the same to the process above described.

No. 59,332 —GEO. FREDERICK WHITE, Hornsey, England, and HARVEY CHAMBERLAIN, London, England.—*Clasp for Belting, &c.*—October 30, 1866.—On the ends of the belting are hinged plates; on the latter are studs, engaged by the heliacal slots in a disk pivoted between them, and whose revolution draws the plates toward each other, and tightens the belt.

Claim.—The apparatus constructed and operated substantially as herein described and represented in the drawings, when applied to the elongation and contraction of articles in the manner and substantially as specified.

No. 59,333.—FRANKLIN N. BULLARD, Worcester, Mass.—*Construction of Glass Bottles.*—October 30, 1866.—The horizontal section of the bottles is trilateral, the shorter sides bearing angles of 90° to each other.

Claim.—The construction or making of glass or other bottles of a triangular form and of such relative greater and lesser angles as that two, four, eight, or multiples of these numbers will pack up in a square form, for economy, facility, and security in packing and transporting them, substantially as described.

No. 59,334.—JOHN F. COLLINS, New York, N. Y.—*Apparatus for Distilling Petroleum, &c.*—October 30, 1866.—The goose-neck is attached to the still so as to leave a space between it and the mouth of the still, by which air may enter.

Claim.—First, so constructing the mouth *a* of the still, and combining it with the goose-neck C, or exit pipe,as to provide for the admission of air around the mouth, substantially as herein set forth, for the purpose specified.

· Second, the construction of the goose-neck C, or exit pipe, with the collecting channel *d*, substantially as herein set forth, for the purpose specified.

Third, the conducting tube *f*, combined in relation with the collecting channel *d* and the exit pipe *e* of the goose-neck, substantially as herein set forth, for the purpose specified.

No. 59,335.—E. A. ADAMS, Boston, Mass.—*Flour Sifter, Mixer, and Kneader.*—November 6, 1866.—The upper end of the vertical revolving shaft is square, and on this end is fitted in turn an agitating plate to force the flour through the sifter, a mixing frame, and a frame carrying kneading rollers.

Claim.—First, the combination of driving mechanism B D E, below the pan G, with the said pan and with sifting, mixing, or kneading mechanism within it, substantially as and for the purposes set forth.

Second, the combination of the central driving shaft F with the mixing pan G and tube g^4, substantially as herein shown and described.

Third, the construction of the agitator with a socket hook, substantially as herein shown and described.

Fourth, the employment of the detachable sifter in combination with the mixing pan and central driving shaft, substantially as herein shown and described.

Fifth, the protecting cone k^3 in combination with the sifter and central driving shaft, substantially as herein shown and described.

Sixth, the combination of the sifter H and the agitator I with each other and with the shaft F, and mixing pan G, substantially as herein shown and described.

Seventh, the employment, in combination with the driving shaft, of a detachable mixing device, substantially as and for the purpose herein shown and described.

Eighth, the employment, in combination with the driving shaft, of a detachable kneading device, substantially as and for the purpose described.

Ninth, the employment, in combination with the central tube or stationary roller, of a kneading roller, which revolves upon its own axis, and also rolls around the said stationary roller, substantially as and for the purpose herein shown and described.

Tenth, The combination of the gear wheels k^2 k^4, or their substantial equivalents, with the driving shaft and the kneading roller, substantially as herein shown and described.

Eleventh, the employment of the adjustable guard plate k^6 with the gear wheels k^2 k^4, substantially as and for the purpose shown and described.

Twelfth, the employment of the frame k^3, in combination with the gear wheels k^2 k^4, substantially as herein shown and described.

Thirteenth, the combination of one or more scrapers with the gear frame k^3, substantially as herein shown and described.

Fourteenth, the lugs upon the gear wheel k^4, in combination with corresponding recesses in the tube g^4, substantially as and for the purpose herein shown and described.

Fifteenth, a mechanism capable of use at will as a sifter or a mixer or a kneader, constructed and operating substantially as herein shown and described.

No. 59,336.—WILLIAM ALLENDERFF, Philadelphia, Penn.—*Apparatus for Cooling Malt Liquors.*—November 6, 1866.—The apparatus has a vertical series of pipes connecting by their ends, and internal closed tubes to reduce the current of water to an annular film. The liquid to be cooled drips over the exterior surface of the water pipes consecutively. The respective positions of the beer and refrigerating liquid may be reversed.

Claim.—The construction of a cooling apparatus by the combination of a series of inside tubes E E with an equal number of outside tubes C C and their end connections D D, all arranged substantially as described in the foregoing specification and for the purpose specified.

No. 59,337.—WILLIAM T. ALTFATHER, Johnstown, Penn.—*Car Truck Standard.*—November 6, 1866.—The hinged standards may be folded beside the frame, and, when upright, form a connection between the frame and the body of the truck. The pins near the lower end of the standards enter slots in the upper sockets for pivots, when folded, and in the lower sockets for support when erect.

Claim.—First, the sockets D and a' provided with grooves f, in combination with the bed timber A and side board B, substantially as herein shown and described, and for the purposes set forth.

Second, the standard E, provided with the pin or bolt w, in combination with the sockets D and a, substantially as herein shown and described.

No. 59,338.—E. H. ASHCROFT, Lynn, Mass.—*Non-conducting Covering for Steam Boilers, Pipes, &c.*—November 6, 1866.—Composed of hair mixed with hydraulic or other cement, or plaster of Paris, or a combination of them.

Claim.—A non-conducting covering for steam boilers, pipes, &c., composed of the materials above named, and applied as described.

No. 59,339.—NATHAN E. BADGLEY, New York, N. Y.—*Cotton Seed Planter.*—November 6, 1866; antedated October 4, 1866.—The seed hopper has holes around its circumference at mid-length, and an embracing boop has similar holes. Four of the holes in each are oblong, and when those coincide the others are closed; in this form the seed is dropped in hills. The opening of the round holes closes all but a similar part of the larger ones; in this form the seed is drilled. A bar from the fixed axle and agitating plates on the side of the hopper manipulate the seed. The draught hook and the handles are attached to the opener frame.

Claim.—First, the construction of the opener frame V with its draught hook D, and the manner of fastening the handles thereto.

Second, the revolving cylindrical, flanged-head hopper around a permanent shaft, with its elevating agitators E and stationary rod R.

Third, the slip hoop S, with holes to regulate the planting either in drill or spots, as herein described.

Fourth, the adjustable coverer L, with its teeth T and conductor N, attached as herein described.

Fifth, the combination of the several parts and devices, as herein described and substantially as set forth.

No. 59,340.—WILLIAM S. G. BAKER, Baltimore, Md.—*Locomotive Engine.*—November 6, 1866.—The steam chest is formed in the bed plate to shorten steam passages, save joints, and lighten the weight of castings.

Claim.—The steam chest D, exhaust openings B I, and steam ports C C' formed within the bed plate G, in connection with cylinders of locomotive engines, in such a manner that the steam supply and exhaust discharge pipes are shortened and number of steam joints reduced, substantially as and for the purpose specified.

No. 59,341.—PARDON BARRETT, Jackson, Penn.—*Revolving Table.*—November 6, 1866.—The table is mounted upon a pedestal, and has an axial screw upon which it rises and falls as it is revolved, a nut in the frame engaging the thread on the screw.

Claim.—A table composed of a stand or support A, having a screw C fitted in its upper end, and an upper part B, provided with drawers and a nut G placed centrally within it for 'the screw C to pass through, and a cap H to fit over the screw, substantially as and for the purpose set forth.

No. 59,342.—LOUIS BAUHOFER, Philadelphia, Penn.—*Treating Cork for Mattresses, &c.*—November 6, 1866.—The cork is cut into shavings and placed in a wire gauge casing permeated by the products of combustion from a furnace. After charring, myrrh, bay leaves, 'or other aromatic substance is thrown into the fire to scent the cork.

Claim.—First, subjecting particles of cork to be used as a stuffing for mattresses, &c., to the action of the products of combustion obtained by burning wood, tan, or other suitable material, substantially as and for the purpose described.

Second, charring or partially burning the particles of cork to be used as a filling material, for the purpose set forth.

Third, subjecting the cork to the action of the fumes or vapors arising from heated aromatic substances, as and for the purpose specified.

No. 59,343.—E. M. BAYNE, Uniontown, Penn.—*Broom-head.*—November 6, 1866.—The corn brush is placed on the corrugated frame which is then drawn within the cap by the longitudinal screw.

Claim.—The combination with the handle C, screw B, and cap A of the ferrule D and nut E, when the said nut is firmly brazed or otherwise securely attached to the said ferrule, so as to be a solid part thereof, substantially as described and for the purpose set forth.

No. 59,344.—J. W. BISHOP, New Haven, Conn.—*Capping Screws.*—November 6, 1866.—The screw head has a central cavity and a correspondingly perforated cap.

Claim.—The combination of a centrally perforated metal cap with a centrally perforated screw head, substantially in the manner herein set forth.

No. 59,345.—D. A. T. BLACK, Ray's Hill, Penn.—*Sleigh.*—November 6, 1866—The ratchet faced wheels are pivoted in levers connected by longitudinal. transverse, and diagonal rods, and are worked by a hand lever to act as either wheels or brakes.

Claim.—The combination and arrangement of the wheel levers F, with shoulders *e'*, and bars I and J, with the sleigh, whereby they are held in their lowered position for wheeling the sleigh by the forward draught of the sleigh, in the manner described, for the purpose specified.

No. 59,346.—WILLIAM BLAKE, Boston, Mass—*Diaphragm Faucet.*—November 6, 1866.—The inlet tube is curved up in an enlargement of the outlet tube to receive the contact of the elastic diaphragm which forms a packing for the screw valve above.

Claim.—The improved diaphragm faucet constructed with the tubular extension F, the chambered body A, the exit pipe or passage G, and the inlet pipe F', arranged together and with the diaphragm B, the valve E, the screw D, and the cap C, substantially in the manner and so as to operate as hereinbefore explained.

No. 59,347.—CHRISTIAN BOEHMER, Jr., Madison, Wis.—*Cock Eye.*—November 6, 1866.—The oval ring has a T-end embraced by a slotted plate, which is lapped and riveted around a socket piece on the tug.

Claim.—First, the socket A, in combination with the trace C, for the purposes and substantially as herein shown and described.

Second, the cock eye provided with a cross-bar, as shown and described, in combination with the socket A, substantially as and for the purposes set forth.

No. 59,348.—JOHN F. BOYNTON, Syracuse, N. Y.—*Composition for Roofing.*—November 6, 1866.—Coal tar is heated to drive off the ammoniacal water, and is mixed with clay, or other materals, to form a roofing cement. The consistency of the tar may be reduced by adding dead oils produced by the distillation of gas tar.

Claim.—First, gas tar, rendered anhydrous, as described, in combination with the dead oils distilled from gas tar, as a material to be used in preparing a roofing cement, by mixing therewith ground clay and other similar substances.

Second, as a roofing cement, a combination of gas tar rendered anhydrous, as described, with pulverized clay and the dead oils distilled from gas tar.

No. 59,349.—JOHN F. BOYNTON, Syracuse, N. Y.—*Vat for Evaporating Salt Water.*—November 6, 1866.—The shallow pan has a partition which affords a bearing for one end of each of the rollers. Spaces for the circulation of the water are left between the ends of said partitions and the sides of the vat. The rollers have recesses on their peripheries which serve as buckets to receive the water, by which they are turned, and also to increase their evaporating surface. The roller shafts have fans operated by the wind.

Claim.—First, the central boards, as specified.

Second, the rollers with multiplied surfaces.

Third, the dark color of the rollers for the absorption of heat, as herein specified.

Fourth, the arms with fans upon the rollers.

Fifth, the arrangement which causes the water, by its gravity, to work its own evaporation, substantially as described.

No. 59,350.—WILLIAM BRANAGAN, Burlington, Iowa.—*Lard Boiler.*—November 6, 1866.—The hemispherical boiler has double walls, with openings for the inlet of steam, the discharge of air and condensed water, and an exit opening, with a strainer for the lard. Stirrers revolve within the pan, and a perforated plate above the steam inlet distributes the steam upon the bottom of the pan.

Claim.—First, the employment of a perforated distributing plate, or its equivalent, between the walls of a double-wall lard boiler, in combination with the steam inlet pipe, substantially as and for the purposes described.

Second, providing a double-wall lard-rendering kettle, A, with stirrer, air cocks, and an outlet pipe having a strainer *h* applied to it, substantially as described.

No. 59,351.—J. P. BROADMEADOW, New York, N. Y.—*Cooking Stove.*—November 6, 1866.—The gates are used alternately to contain fire and kindling material; that with kind.

ling occupies a place under the oven and behind the ash drawer. The fire grate is vertically adjustable by a system of levers and connections.

Claim.—First, forming the stove with a recess H beneath the oven, extending from the fire chamber D to the back part of the stove, and of sufficient capacity to receive and contain one of the fire boxes F or G, substantially as described, and for the purpose set forth.

Second, the combination of the shaft J, levers L, guides N, slides M, supports P, and lever K with each other, with the sides of the stove, and with the fire box, substantially as herein described and for the purposes set forth.

Third, the combination and arrangement of the sliding ash pan S, when constructed as herein described, with the two fire boxes F and G, with the recess H, and with the fire chamber D of the stove, substantially as described, and for the purpose set forth.

No. 59,352.—RHODOM M. BROOKS, Pike county, Ga.—*Portable Revolving Screw Press.*—November 6, 1866.—The press box is pivoted on a step below, and on its screw above. The screw works in a fixed nut, and is attached to the follower, so that the revolution of the press box depresses the follower.

Claim.—The combination of the revolving press box with the outer frame which supports said box for the purpose of making a portable press, the several parts being constructed substantially as and for the purpose specified.

No. 59,353.—DANFORTH H. BROWN, Northfield, Vt.—*Water Wheel.*—November 6, 1866.—The water gate has adjustable plates, which run on friction rollers upon projections of the chute. The step of the wheel shaft is adjustable vertically by a system of levers, operated by a temper screw, beside the wheel frame.

Claim.—The arrangement of the lower friction roller ways f, as projections directly from the chute, as and for the purpose described.

Also, in combination with the gate c the construction of the upper part of the friction roller ways in two parts, f and g, when provided with means for their adjustment relative to each other, as described.

Also, the combination of the slotted cylinder l with the inner cylinder m, its set screw p, the lever r, and moving fulcrum s, operating together, substantially as set forth

No. 59,354.—RILEY BURDETT, Chicago, Ill.—*Reed Musical Instrument.*—November 6, 1866.—The attachment is intended for the upper half of an instrument with two sets of reeds, with a divided air chamber. When the ordinary air passage is closed by the cut-off valve, the air passes through an opening, whose valve receives a rapid reciprocating motion, induced by the pressure of the air and the resistance of a spring, which alternately overcome each other and intermit the flow of air.

Claim.—First, the construction of a tremolo, the valve of which is connected with and acted upon by the arm and spring C, as herein specified and set forth.

Second, the cut-off valve E, when the same is constructed and used in the manner and for the purposes herein described and set forth.

No. 59,355.—GEORGE E. BURT, Harvard, Mass.—*Car Brake.*—November 6, 1866.—The braking is accomplished by connecting the axle with a coiled spring by a clutch operated from the platform. By a contrary motion of the clutch the force of the spring advances the car. The unwinding of the spring is prevented during the transfer of the clutch. A pawl holds the spring disk when winding, but releases from a pressure in a contrary direction.

Claim.—First, the combination of the wheel D, the disk G, the pawl s, and the stud F with the spring s.

Second, the lever T, the spring w, the arm P, and the brake S, in combination with the pivot Y, operating substantially as described, for the purpose set forth.

Third, the spring m, in combination with lever L, substantially as described and for the purpose set forth.

No. 59,356.—GEORGE R. CANNON, Guildford, Ohio.—*Shelving for Wagons.*—November 6, 1866.—Cross benches rest on the upper rail of the wagon bed, and support the side shelving; the benches have additional braces resting on the lower rails.

Claim.—First, the securing the cross-beams B to the top rail of the wagon, substantially as specified.

Second, the manner of securing the planks C to the cross-beams B, substantially as described.

Third, the employment of braces D for supporting the planks C between the cross-beams B, substantially as described.

No. 59,357.—OLIVER F. CHASE.—*Water-proof Sole.*—November 6, 1866.—November 6, 1866.—Explained by the claims and cut.

Claim.—First, a whole sole, having the ball filled with rubber or allied gum, and vulcanized after having been so filled.

Second, a whole sole, having the ball filled with rubber or allied gum, in combination with a leather insole, substantially in the manner described, and the whole secured together by the process of vulcanization as herein set forth.

COMMISSIONER OF PATENTS. 1385

No. 59,358.—LYMAN J. CASWELL, Scott township, Ind.—*Cultivator*—November 6, 1866.—Links and brace rods connect the side beams to the middle beam to allow the longitudinal adjustment of the frame to arrange the share in plough or cultivator fashion. The obliquity of the standards is regulated by angular braces, and a slight rotation effects the shed of the earth from the shares.

Claim.—First, the application of the turning armatures F F F F F to the cultivator, to change the positions of the side shovels, so as to form a shovel plough or a cultivator. The application of the braces G G to sustain the side beams and side shovels in their proper positions.

Second, the application of the braces E E E to the shovel standards, to elevate and depress the shovel points and turn the sod or sward; the application of the curve to the extension mould plates; the mortise in the ends of the shovel standards, and the flattening the points of the shovels.

No. 59,359.—PEREZ C. CLAPP, Dorchester, Mass., assignor to himself and R. W. TURNER, Milton, Mass.—*Last.*—November 6, 1866.—The last-block is attached by a bolt, which slides in a mortise and beneath a plate on the last. The block is detached by the last-hook.

Claim.—Combining with the last-block and last the spring bolt or latch, arranged to operate substantially as described.

No. 59,360.—STILLMAN A. CLEMENS, Chicago, Ill.—*Machine for Threshing Flax.*—November 6, 1866.—A bunch of plants are grasped by their butts and the heads pushed into the throat; the panicles of the plant are stripped from the stalks by the teeth of the cylinder operating above the concave; the balls are crushed and the seed winnowed by the succeeding crushing rollers and vibrating sieve.

Claim.—First, the combination of toothed cylinder b with crushing rollers c and c', or their equivalents, substantially as described and for the purposes set forth.

Second, the combination of cylinder b and crushing rollers c c', with the vibrating sieve shoe h, fan q, and air trunk r, substantially as described and for the purposes set forth.

Third, the crushing device consisting of roller c' and spring plates W, substantially as described and for the purposes set forth.

Fourth, a concave y combined with the described machine, substantially as described and for the purposes set forth.

No. 59,361.—LEVI H. COLBORN, Chicago, Ill.—*Steam Generator.*—November 6, 1866.—The furnace is enclosed in a steam jacket, within which is a revolving generator with upward curved steam pipe and downward curved feed-water pipe leading through the trunnion from the jacket. The piston of the force pump is actuated by the steam through the medium of a ratchet wheel, and its action is dependent upon the pressure and not upon the height of the water in the boiler.

Claim.—First, the arrangement of a cylindrical or spherical inner revolving steam generator, with an encircling exterior stationary boiler or heater, and with communicating pipes or passages between them, so that the inner generator may be supplied with heated water from the outer boiler or heater, substantially as and for the purpose described.

Second, the arrangement of an interior revolving steam generator and an encircling exterior stationary boiler or heater, and the former supplied with water from the latter, a fire grate so located and encased as that the burning products thereon shall impinge directly upon the inside of the outer boiler or heater, and upon the outside of the inner steam generator, substantially as and for the purpose described.

Third, combining with a steam generator a regulator that will preserve a uniform head of steam by setting a pump in motion whenever the steam becomes excessive, or rises above a defined point or pressure, and throw out the action of the supply pump whenever the pressure falls to the defined point, substantially in the manner set forth.

No. 59,362.—LEVI R. COMSTOCK, Macon, Mo.—*Stove Pipe and Damper.*—November 6, 1866.—Interposed in the flue is a cylindrical chamber in which is a damper on a horizontal axis, the flat plates of which when vertical permit direct draught, but when horizontal render the caloric current circuitous.

Claim.—The shape and construction of the semicircular concave cast-iron plates A, with their joints and hinges B, and revolving partitions D on the inside, with chamber E, combined and operated as herein described and for the purposes set forth.

No. 59,363.—LEVI R. COMSTOCK, Macon, Mo.—*Stove Pipe Drum.*—November 6, 1866—The radiator drum, has crescent-shaped partitions parallel and equidistant, and the hollow axial shaft has similar intervening plates, which render the passage circuitous or direct according to adjustment.

Claim.—The spiral radiator, having flanges G on the inside of the cylinder A, and a centre shaft D, with flanges E, set spirally on the shaft with collars J, at the end of the cylinder and pipes B and C, for receiving and discharging the heat, arranged, regulated, combined and operated as herein described and for the purposes set forth.

No. 59,364.—ALBERT CONANT and ISRAEL F. BROWN, New London, Conn.—*Pump.*—November 6, 1866.—The foot valve is independent and has a long stem whose fluting enlarges the water way. It is tilted to discharge the water from the cylinder by lowering the piston upon the prong above the disk.

Claim.—First, the wings J of the single taper stem G, in combination with the loose valve B, arranged and operating in the manner and for the purpose herein specified.

Second, the tilting arm E, in combination with the cap C of the loose valve B, arranged and operating in the manner and for the purpose herein specified.

No. 59,365.—RICHARD COVINGTON, Washington, D. C.—*Shutter Hinge.*—November 6, 1866.—The arc-shaped pieces attached to the leaves are concentric and their contacting faces have semicircular recesses which receive a round locking pin.

Claim.—Constructing the hinge with a locking attachment composed of two notched segments of cylinders lying concentric one with the other, which may be rendered immovable at any given point by the insertion of a pin in any one of the notches, substantially as described.

No. 59,366.—MOSES G. CRANE, Chelsea, Mass.—*Ice Cream Freezer.*—November 6, 1866.—The central spindle carrying the beaters and scrapers is rotated half a revolution in alternate directions by the engagement of the pinions consecutively with the segment gears on the hand crank. The rotation of the cream vessel is restrained by studs which engage a projection and limit its motion to about 170° in alternate directions.

Claim.—Combining with the cylinder b, and the spindle d, rotating therein the scraper or scrapers k, acting alternately as the spindle is revolved in opposite directions to throw the fluid cream against the sides of the vessel and to scrape the frozen cream therefrom, substantially as set forth.

Also, in combination with spindle and the cream cylinder rotating therein, the arms r and projection s operating to arrest the rotation of the cylinder, substantially as set forth.

No. 59,367.—JOHN B. CURTIS, Port Henry, N. Y.—*Extension Table.*—November 6, 1866.—The leaves are hinged to each other and to the sliding frame in such a manner that they fol upon each other, drop down and pass with the frame under the stationary leaves when not in use.

Claim.—Constructing an extension table with extension leaves made in sections and hinged together, and also to sliding frames or drawers for folding together and lying under the top or bed leaves when the table is closed, and unfolding even with the top or bed leaves when the table is opened, arranged and operating substantially as herein described.

No. 59,368.—HENRY A. DANIELS, Thomaston, Conn.—*Sawing Machine.*—November 6, 1866.—The curved saw-blade is strained in a swinging sector, oscillated by gearing and eccentric and depressed by a spring; the log is clamped in the saw-buck by a gate depressed by a treadle.

Claim.—First, the swinging saw frame suspended from shafts K, working in sliding boxes having pressure springs over them in combination with and driven by the eccentric I, all constructed to operate substantially as described.

Second, the horse or buck M, constructed substantially as shown and described in connection with the plate O, and the treadle P, or its equivalent, for the purpose specified.

No. 59,369.—JACOB DOBBINS, Waterloo, Mich.—*Machine for Bending Wooden Hoops.*—November 6, 1866.—Improvement on his patent of July 4, 1865. The grooved feed rollers carry the board from which the hoops are cut by the stationary knives, each of which cuts a hoop tapering in thickness; one hoop is the complement of the other, the first knife being inclined and the second vertical. The hoops are passed by spiral guides to the bending device, consisting of a toothed roller and a concave.

Claim.—The machine for the purpose described, the same consisting of the grooved feed rollers L M N O, yielding platform P, grooved roller V, guide X, inclined knife B', vertical knife A', spiral guide C' D' E', concave guide F', arranged and operating substantially as and for the purpose specified

No. 59,370.—HEZEKIAH DODGE, Albany, N. Y.—*Wrench.*—November 6, 1866.—The set rule and sliding jaw have respectively a right and left hand screw which are engaged by a sleeve nut whose rotation moves the jaw at an increased speed over that obtained by a single thread.

Claim.—The two right and left hand screws E F, combined in relation with each other and with the tubular nut G, sliding jaw C, ferrule D, and shank A, substantially as herein set forth for the purpose specified.

No. 59,371.—JOHN E. EARLE, New Haven, Conn.—*Furniture Pad.*—November 6, 1866.—The metallic ears enter slots in the pads and hold them to their places on the furniture.

Claim.—The combination of the metallic ears C C, and the pad A, when the said two ears are made independent each of the other so as to be made adjustable, substantially as described.

No. 59,372.—B. FRANK EARLY, Palmyra, Penn.—*Broom Head.*—November, 6, 1866.—
Explained by the claim and cut.
Claim.—The combination of the socket C, toothed band D, wires F G H, and wires I, substantially as described for the purpose specified.

No. 59,373.—BARTHEL ERBE, Pittsburg, Penn.—*Door Latch.*—November 6, 1866.—The latch is reversible. The lever which operates the latch is movable on its pivot to allow the latch head to protrude from the case for reversal without being disconnected from other parts of the lock.
Claim.—So connecting the lever c with the case or frame that it may have the supplemental motion and adjustment herein described substantially for the purpose set forth.

No. 59,374.—CHARLES R. EVERSON, Palmyra, N. Y.—*Foot Warmer.*—November 6, 1866.— The footstool has a horizontal pipe heated by a caloric current from the chimney of a lamp beneath.
Claim.—The combination of a kerosene oil lamp having a metallic chimney with a horizontal tube or pipe, substantially as described and for the purposes set forth.

No. 59,375.—ALEXANDER F. EVORY and ALONZO HESTON, La Porte, Ind.—*Boot and Shoe.*—November 6, 1866.—Gores in front permit the expansion of the upper to receive the foot and form inner folds under the external lappets, which contain the lacing eyelets.
Claim.—A shoe when constructed with an expansive gore flap C D, the external fold C of which is attached to and in front of the quarter B, and the internal fold D of which is attached to and in rear of the vamp A, the said several parts and pieces being respectively constructed and the whole arranged for use substantially in the manner and for the purpose set forth.

No. 59,376.—JACOB FEDERHEN, Boston, Mass.—*Croquet Board.*—November 6, 1866.— Explained by the claim and cut.
Claim.—Forming the cushion for a croquet board of one or more catgut strings, substantially as and for the purposes described.

No. 59,377.—JOHN E. FINLEY, Memphis, Tenn.—*Beehive.*—November 6, 1866.—The bottom has a gauze ventilator. Inclined tubes for the passage of bees emanate from the lower corners of the hive and meet near its centre.
Claim.—The combination of the tubes B C with the ventilator A for the purposes herein set forth.

No. 59,378.—JOHN E. FITTS, Candia Village, N. H.—*Gate.*—November 6, 1866—The posts are supported on a bed rail on which one roller runs. The other roller is pivoted in a post on which the gate is swung when it has been slid into balance. Gains in projecting bars hold on the latch post, and a swinging button retains them.
Claim.—The combination as well as the arrangement of the gate, the shaft, the supporting rollers, the bed rail and its posts, the whole being applied together substantially in manner and so as to enable the gate to be operated as specified.

No. 59,379.—AMEDEE FONTAINE, Milwaukee, Wis.—*Fire Alarm.*—November 6, 1866.— A copper wire lengthened by expansion from heat releases a weighted lever that by its weight operates the bell hammer.
Claim.—The detaching lever f, the operating lever g, the lever bar h, the spiral spring i, the expansion wire l, the bell wire m, in combination substantially as described and for the purpose set forth.

No. 59,380.—WILLIAM H. FOWLER, Newburgh, N. Y.—*Door Catch.*—November 6, 1866 — The stop bar is pointed in a recess below the floor, and projects through a slot in the latter. It is connected to a treadle plate by which it is adjusted in the path of the door or otherwise.
Claim.—The combination of the plate D, rod E, bar G, and slotted plate H, operating with the door in the manner and for the purpose specified.

No. 59,381.—JAMES C. FRENCH, Monmouth, Ill.—*Cultivator.*—November 6, 1866.—The plough beams are pivoted to the draught frame, which is carried on wheels. An oscillating disk is pivoted to the tongue, and from its sides connecting rods pass to pivoted plates, to whose lower ends the whiffletrees are attached.
Claim.—First, the combination of the frame piece A A', swivels E E, and drag bars F F, said parts being respectively constructed and the whole arranged for use substantially as set forth.
Second, the combination and arrangement of the tongue B, frame A A', plate C, rods C', and bars C'', substantially as described.

No. 59,382.—P. C. FRITZ, Barrytown, N. Y.—*Cooler for Grinding Mills.*—November 6, 1866.—The fan draws air around and between the stones and over the flour in its passage to bolt.

Claim.—The arrangement of the fan *l*, tube *h*, chamber *a*, and conveyer *d*, in relation to each other and to the stones *a*, and for the purposes specified.

No. 59,383.—EDWARD P. FURLONG, Portland, Me.—*Key Guard.*—November 6, 1866.—A slotted escutcheon on the plate shuts over a flattened portion of the key shank to prevent its being turned by an outsider. A pawl acts as a detent for the escutcheon.
Claim.—First, the hook on the bottom of the escutcheon plate, as and for the specified purposes.
Second, in combination with the flattened shank of the key, the curved slot *b*, hook on the lower end of the escutcheon, and pawl *c*, all constructed and arranged as and for the objects set forth.

No. 59,384.—CARLOS GLIDDEN, Milwaukee, Wis.—*Car Spring.*—November 6, 1866.—The spring for railroad cars is composed of a pile of circular plates corrugated radially and arranged around a stem.
Claim.—Making springs out of disks or circular plates of steel with corrugations in circular lines or concentric with the periphery of the disk or circular plate, substantially as herein set forth.

No. 59,385.—CHRISTOPHER GODDEN, Paterson, N. J.—*Wheels for Vehicles.*—November 6, 1866.—The inner and contiguous parts of the felloes of metal have recesses for the spokes, and alternate projections and recesses on their ends for attachment to each other. The hub is formed with recesses for spokes. The flanged spindle-socket has oil holes proceeding from a chamber within the hub. The oil chamber is supplied by a hole stopped with a screw.
Claim.—First, the metal felloe B furnished with a wooden rim or filling, in combination with the tire C, substantially as herein set forth for the purpose specified.
Second, the shell A furnished with sockets having inclined bottoms in combination with the collar *e*, substantially as herein set forth for the purpose specified.

No. 59,386.—D. H. GOODWILLIE, New York, N. Y.—*Inhaler.*—November 6, 1866.—The spigot has two valves, one of which controls the eduction opening for the time being, whether in communication with the air or the gas reservoir, and the other permitting exhalation into the atmosphere. A 90° turn of the spigot places the mouth cup in communication with the air or the gas holder.
Claim.—First, the stop cock C, with passages *a b* and valves *e f*, in combination with the shank B and cup A, constructed and operating substantially as and for the purpose set forth.
Second, the passages *c d g* in combination with the passages *a b*, valves *e f*, shank B, and cup A, substantially as and for the purpose set forth.

No. 59,387.—JOHN GSCHWIND and CHARLES GSCHWIND, New York, N. Y.—*Car Register.*—November 6, 1866—A "turnstile" is placed on the entering side of the platform, and its revolutions are transmitted by a train of gearing to an indicator. Another revolves only for egress.
Claim.—The combination and arrangement of the shaft C, arms *a*, ratchet *b*, pawl *e*, gearing *g*, shaft *h*, index *i*, dial *i*, with the car A, which is provided with the egress armed shaft F on the opposite side, as and for the purpose specified.

No. 59,388.—GEORGE HADFIELD, Cincinnati, Ohio.—*Medical Vacuum Chamber.*—November 6, 1866.—The instrument is for dry cupping, or producing a partial topical vacuum. The rest is to support the hand to prevent its being drawn against the sides of the chamber, and the hand hole is for the access of the band of the operator.
Claim.—As a new invention the rest E and self-sealing cap F, for the purposes set forth.

No. 59,389.—PETER HAGAN, Elizabeth City, N. J., assignor to WILLIAM N. WALTON, Newark, N. J.—*Lamp Chimney.*—November 6, 1866.—The flat bulb has corrugations to diffuse the light and confer greater capacity for expansion and contraction.
Claim.—A chimney having on its bulb a series of rings or circles, substantially as and for the purpose set forth.

No. 59,390.—ALONZO HALE, Eureka, Ill.—*Sulky Plough.*—November 6, 1866.—The plough beams have chain connections to the axle. The ploughs are vertically adjusted by a lever and rack.
Claim.—First, the jack chains Y Y, connected to the axle B and lever R, in combination with the plough beams C C, when constructed and operated substantially as and for the purposes herein shown and described.
Second, the chains Y Y, in combination with the axle B, lever R, and sector Q, for the purposes herein set forth.

No 59,391.—JOHN HARTLIEB, Reading, Penn.—*Composition for Pavements, &c.*—November 6, 1866.—Composed of coal tar, 6 quarts; asphaltum, 1 pound; sand, 8 quarts; scale iron, 6 quarts; turpentine, 1 pint; and common gravel, 16 quarts.
Claim.—The within described compound for pavement, made as set forth.

No. 59,392.—LEWIS S. HAYES, Greene, N. Y.—*Sawing Machine.*—November 6, 1866.—The depending arm of the L-shaped lever enters a cavity in the slide to which the saw is clamped. The frame is supported on a turn-table for the progression of the saw.

Claim.—The combination of the saw A, slide B, guides C, lever D, and turn-table G, constructed and operating substantially as and for the purpose set forth.

No. 59,393.—ISAAC HELME, Philadelphia, Penn.—*Hydrocarbon Burner for Stills, Engines, &c.*—November 6, 1866.—The coal oil flows from the curved pipe into the annular trough, where it is consumed. The flame is regulated by vertical movement of the circular covering disk. This movement is made by its supporting screw bolt.

Claim.—The cup A, the hoop B, and the disk C, provided with a stem or its equivalent for adjustment, the several parts being constructed and arranged substantially as and for the purpose herein specified.

No. 59,394.—ALBERT HILL, New York, N. Y.—*Regulator for Watches.*—November 6, 1866.—The pointer which carries the pins embracing the hair spring is connected to a bar having a segmental rack, actuated by a small cog-wheel; the shaft of the latter carries a pointer, by moving which a delicate adjustment is obtained.

Claim.—The combination and arrangement of the plate E, having toothed curved ends *b*, and ridge *d*, pointer D, pinion F, pointer G, and arcs *a d'*, relatively to each other, and operating in the manner and for the purpose specified.

No. 59,395.—B. B. HILL, Chicopee, Conn —*Hand Stamp.*—November 6, 1866.—The days and months are formed by salient letters on the peripheries of the disks, which are mounted alongside on the central hub and turned in correspondence with dial plates; the whole is arranged on a reciprocating spring plunger.

Claim.—The arrangement of two or more wheels *a a'*, placed successively upon the hub *e* and the hubs of the wheels *a a'*, when inserted into the case from one side only, and two or more dials *k' k''*, bracket *r*, spindle *s*, stock A, when placed so as to be actuatd in combination with and by a plunger motion, substantially as described.

No. 59,396.—E. G. HOLDEN, Covington, Ky.—*Preserving Fruits and Vegetables.*—November 6, 1866.—The fruit, &c., is placed in a chamber kept cool by a circulating stream of salt water which is refrigerated by the evaporation of ammonia, as in some ice-making machines; such as Carré's.

Claim.—First, preserving fruits, vegetables, &c., by keeping them exposed in suitable chambers or cells to a cool and dry atmosphere, by means of the gas or vapor of ammonia circulated through cooling baths and pipes, generated and applied, substantially as herein described.

Second, the force pump I for maintaining circulation, in connection with suitable apparatus for generating and applying the gas or vapor of ammonia for the purpose herein described.

No. 59,397.—ALLEN T. HUTSON, Richmond, Ind.—*Portable Heater.*—November 6, 1866.—The chimney of the combustion chamber is recurved and discharges into the bottom of the ash pit. A valve is placed in the curvature of the pipe and an air inlet in the floor.

Claim.—First, the pipe C, arranged with the valve E, terminating in the furnace pan G, for the purposes set forth.

Second, the ash pan N, and the furnace pan G, in combination with the pipe C, all arranged and operated as described.

Third, the arrangement and combination of the boiler B, heater frame A, furnace pan G, pipe C, and valve E, all as and for the purpose herein described.

No. 59,398.—JOSEPH HYDE, Troy, N. Y.—*Whiffletree Coupling.*—November 6, 1866.—A ball is introduced through an opening in the thimble which clasps the mid-length of the singletree and then takes its seat in the frustum of a sphere which forms the other portion of the universal joint. The neck is attached to the doubletree.

Claim.—First, the connecting of the whiffletree B to the doubletree A, or any other part or a wagon or other vehicle by means of the ball and socket joint F E, in the manner and for the purposes substantially as herein described and set forth.

Second, the employment of the ball and socket joint F E, in combination with the cylinders, or their equivalents, C D, for the purpose of forming a coupling between the whiffletree and doubletree, or any other part of a vehicle to which it is desired to attach the same, in the manner and for the purposes substantially as herein described and set forth.

No. 59,399.—ROBERT B. JACOBS, Quincy, Ill.—*Gaiter Boot.*—November 6, 1866.—The front and rear of the upper are connected by gores which fold together and are secured by a strap behind the tendon.

Claim.—The combination of the gore D on one side, the buckle F on the other side, and the strap E passing around the back from the gore to the buckle, all as herein shown and described and for the purposes set forth.

No. 59,400.—JOHN JORDAN, Wyandott, Kansas.— *Steam Pump.*—November 6, 1866. Steam admitted above forces down the piston and drives the water from the lower part of the cylinder; the contact of the piston with the tappet on the valve rod closes the steam opening, and water from the discharge pipe rushes in and condenses the steam. The piston rises, opens the steam valve and closes the water valve.

Claim.—The arrangement of the steam pipe A, cylinder B, partition plate D, having opening E, sliding hollow float H, stem F, tappets K K², spring T, valve J, feed pipe L, discharge pipe N and pipe P, in the manner described and for the purpose specified.

No. 59,401.—P. FRANKLIN JONES, Ballston Spa, N. Y.—*Railroad Rail.*—November 6, 1866.—The compound railroad rail has a base piece, having an erect central wedge-shaped core, forming the central portion of the rail; the lower edges of the side sections rest on the horizontal flanges of the base piece and the upper edges are rabbeted on the outside and enclose the top of the central core and side section. A cap with a dovetailed groove slides over the rabbeted edges.

Claim.—A compound rail for railroads, consisting of the base with a central wedge-shaped core, substantially such as described, in combination with the rabbeted side sections, constructed substantially as and for the purposes described.

Also, in combination with the base, having a wedge-shaped core, and the rabbeted side sections, substantially as described; the tread cap, having a dovetail groove, as described and for the purpose set forth.

Also, connecting the sections of the cap rail with the core rail, by means of the cross check pieces, substantially as and for the purpose set forth.

No. 59,402.—HENRY W. JOSLYN, Brooklyn, N. Y.—*Rubber-coated Leather.*—November 6, 1866.—The leather is coated with plastic rubber, which is then exposed to the vulcanizing cooking.

Claim.—The fabric composed of leather, coated with vulcanized India-rubber or guttapercha, in the manner substantially as herein set forth.

No. 59,403.—JAMES E. JOUETT, Brooklyn, N. Y.—*Chair Couch and Stretcher.*—November 6, 1866.—At each end of the apron is a roller by which it is attached to the frame; the latter has two rectangular portions pivoted together and affords a support for the apron, which is stretched from point to point upon it, when the frame is arranged as a chair or as a couch.

Claim.—The employment, in combination with the shifting frame-work and detachable apron, of the securing strap or straps I, the whole arranged to operate as specified for the purpose set forth.

No. 59,404.—ALBERT D. JUDD, New Haven, Conn.—*Picture Nail.*—November 6, 1866.— The screw socket has a stem and washer by which the porcelain knob is secured.

Claim.—The head for picture nails, formed by the screw-nut, having a stem or rivet passing through and securing the porcelain head and ornamental metallic washer, as set forth.

No. 59,405.—SAMUEL S. KAPPEL, Woodhull, Ill.—*Gate.*—November 6, 1866.—The gate is opened and closed by crank bars in the track, connecting by levers to the upper bar on which the frame is hung. The gate is swung up vertically, its vertical and horizontal slats approaching the main bar. A latch is automatically caught and released.

Claim.—First, the combination and arrangement of the crank shaft F, pitman G and pivoted lever H, with the frame A, substantially as herein shown and described and for the purposes set forth.

Second, the loop rods J, in combination with the elbow lever K and rods m and S, and gate bar D, substantially as shown and described.

Third, the latches M and i, arranged so that when the wheels of a vehicle strike the crank shaft, it releases the latch i and throws up the gate and engages the latch M, and when the carriage has passed through and strikes the other crank shaft, releases the latch M, and brings the gate down and engages the latch i, substantially as shown and described.

No. 59,406.—JAMES M. KEEP, New York, N. Y.—*Sprinkler for Powdered Substances.*— November 6, 1866.—Improvement on Keep and Dummer's patent, of July 3, 1866. The valve is at the bottom, and is operated by the spring stem, whose casing is braced from the side of the vessel, and admits the withdrawal of the cup for refilling.

Claim.—First, in an apparatus for sprinkling powdered substances, as described, so arranging the valve-operating devices that the cover of the vessel which contains the pulverulent material may be removed from and applied to said vessel without deranging or interfering with the said valve-operating devices, substantially as and for the purposes shown and set forth.

Second, in apparatus for sprinkling powdered substances, as described, the combination of the valve stem and surrounding spring or equivalent mechanism, with the brace or frame or equivalent means for holding in place the said stem and spring independently of the cover, substantially as shown and set forth.

No. 59,407.—JOHN KEITH, Worcester, Mass.—*Lathe for Turning Pen Handles.*—November 6, 1866.—One end of the stuff is placed in a revolving socket. The first cutter reduces the blank to a fixed size, while the other cutter is guided by a pattern, and turns it to any form.

Claim.—First, the combination with slide piece C, standard D, tool-rest piece F, and slide piece F of the adjustable cutters g and L and collar J, arranged for joint operation substantially as set forth.

Second, the combination and relative arrangement of concaved collar J and cutters g and L, as and for the purpose set forth.

Third, the combination and arrangement of the saw t, with collar J and cutters g and L, as and for the purposes set forth.

Fourth, the combination with tool-rest piece E and slide piece F of spring N, pattern M and friction rolls or pulleys m' and m'', with their flanges m, as and for the purposes set forth.

Fifth, the combination of lever a, shaft b and cam c, with the pieces E and F, as and for the purposes set forth.

Sixth, the combination and arrangement of rods G and K, provided with pins or hooks f and i upon their ends, thumb nuts G' and K', with slotted stands H H', cutters g and L, guides I L' and set screws h and k, as and for the purposes set forth.

No. 59,408.—ALBERT H. and H. P. KENNEDY, Brunswick, Ohio.—*Cattle Poke.*—November 6, 1866.—The coupling pin carries pricks and a spring roller to keep the pricks from the animal, when not called for. The stale is allowed free forward movement, but has a pin limiting its movement backward. The bow is slotted to admit the passage of pricks and spring.

Claim.—First, the improvement by having the pricks d stationary in the roller j, (Fig. 1,) which operate in the manner and for the purpose set forth.

Second, the combination of the roller g with spring e e, as a guard over the pricks d when the poke is not in operation, by which the pricks d can pass, arranged and operating substantially as shown and described.

Third, the slot b through which the pricks d and springs e e can pass, when the poke is removed from the animal, as described.

Fourth, the placing of the stale l on the outside of bow a with a slot b, and on the inside of bow a without a slot, and keys in both ends of roller l, as represented in figure 3, as shown and described.

Fifth, the pin f in bow a (Fig. 1) to keep the stale in place, in the manner shown and described.

No. 59,409.—D. J. KIRKMAN and E. K. GRAY, Winchester, Ill.—*Tightening the Tires of Wheels.*—November 6, 1866.—The ends of the tire are turned in at right angles, and drawn together by a right and left handed screw bolt. The joints of the felloes are cased in iron, and have end nuts for expanding bolts. Slots near the ends of the tire receive the claws of a cant hook to draw up the joints, and are then covered by outer plates.

Claim.—The tire D having slots I near its ends, in combination with the felloes, and operating substantially as described for the purpose specified.

No. 59,410.—HENRY F. KNAPP, New York, N. Y.—*Paper Collar Machine.*—November 6, 1866.—The platen has vertical cutters at its front and end, those of the latter passing through the former, and hence cutting a little later. The platen is separable, and parts to permit the descent of the end cutters, and upon the ascent thereof is pressed together again by a spring.

Claim.—First, the combination with their cutting block or blocks of a longitudinal knife or cutter L, and cross or end cutters K, arranged in their simultaneous or joint descent to intersect each other in their line or lines of cut, by making the one knife or set of cutters to traverse through or across the cutting edge of the other cutter, substantially as and for the purpose or purposes specified.

Second, the combination with the intersecting knives or cutters L and K of a section cutting-block J, the ends of sections n of which are made to yield or slide, and whereby, while an unbroken or closed cutting edge is given to the longitudinal knife, a passage for the end or cross-cut knives into or through it is established, essentially as specified.

Third, the arrangement on an intermittingly rotating feed roller A of creasing and embossing strips a and b, in combination with a cutter or cutters, for afterward shaping and detaching the articles creased and embossed, or either, essentially as herein set forth.

Fourth, the combination of the toothed or notched disk E on the feed roller A, drop catch lever F, lifted at intervals, as described, and constantly-travelling endless belt or strap B for giving an intermittently rotating movement to the said feed roller, substantially as specified.

No. 59,411.—EDWARD H. KNIGHT, Washington, D. C.—*Governor.*—November 6, 1866.—Instead of compelling the balls to rise in a prescribed path, the axes from which the balls swing have automatic adjustment, or free motion, in a horizontal plane, so that the balls may swing with freedom in such directions as are due to the speed and proportions of the parts.

Claim.—A governor having its balls or weights so supported as to afford automatic adjustment or free motion in a horizontal plane to the axis upon which the balls swing.

No. 59,412.—EDWARD H. KNIGHT, Washington, D. C.—*Governor.*—November 6, 1866.—The balls are suspended from spindles which rotate in a horizontal plane as the balls fly outward under increase of speed. The partial rotation of the spindle gives motion to a plate connected with the throttle-valve, &c., which determines the area of steam opening.

Claim.—First, a governor having its ball or balls suspended from vertical spindles, whose partial rotation is caused to actuate a valve and control the area of steam opening.

Second, the combination with said partially rotating spindles of a cam or curved toe which is partially rotated by the vertical motion of the ball.

No. 59,413.—EDWARD H. KNIGHT, Washington, D. C.—*Governor.*—November 6, 1866.—The curved arm, toe, tappet, cam or thread, is so constructed as to give a graduated increasing or decreasing action upon the gland or other object connected to the throttle-valve, as the ball rises under increment of speed. It enables the governor to make a greater or less change in the area of steam opening, with a given change of position of the balls when moving at a low speed, that will be obtained by an equal change when running at a higher speed.

Claim.—A cam attached to the governor arm, or to the spindle from which the governor ball is suspended, so constructed, curved or proportioned as to effect a graduated action upon the device or devices for controlling the area of steam passage, substantially as herein set forth.

No. 59,414.—ROBERT T. KNIGHT, Philadelphia, Penn.—*Spring for Closing Doors.*—November 6, 1866.—The rod spring is recessed in the door, and connects the upper hinge pin with a plate fixed to the door at a lower point. The spring being twisted by the opening of the door, acts by torsion to close it.

Claim.—First, a spring *h*, contained within a recess or chamber formed in the door and attached at one end to a projection on the door, and at the other to a stationary pin on which the door turns, all substantially as set forth.

Second, the detachable grooved strip D' arranged on the edge of the door in respect to the spring, as set forth.

No. 59,415.—A. T. LARGE.—Tomah, Wis.—*Clasp for Safety Pockets.*—November 6, 1866; antedated October 27, 1866.—A metallic self-locking clasp is applied to the mouth of the pocket, and is unfastened by a plunger pin which releases the catch.

Claim.—First, the combination of the rod B, spring *d*, catch *f*, and jaws A A', substantially as specified.

Second, the combination of the spring C with the pin A, catch *a*, rod B, and catch *f*, substantially as specified.

No. 59,416.—A. LARROWE, Cochocton, N. Y.—*Gate.*—November 6, 1866.—The gate rests on a roller, and may be opened by sliding longitudinally, or opened on its centre by swinging on its resting pin, or detached from its posts by removing the fastening pin.

Claim.—First, the friction roller *a*, constructed as shown and described.

Second, the combination and arrangement of the roller *a*, hooks *e*, and pins *c*, with a gate, constructed and operated as herein shown and described.

No. 59,417.—STEPHEN LAVENUE, Alton, Ill.—*Machine for Boring Hubs for Wagons.*—November 6, 1866.—The hub is held at each end by a universal chuck, and through its centre passes a revolving cutter frame, with a guide slide and feed screw. Motion is communicated to the cutter by a crank and bevel wheels on the upper chuck.

Claim.—First, the feed screw working in an eccentric manner, in combination with the cutter guides and revolving cutter, thus securing the smallest possible size of bore, substantially as set forth.

Second, the combination of the screw F and check guide bar F', or its equivalent, to produce a relative motion of the feed screw and cutter, and thereby effect the feed motion, substantially as set forth.

Third, the general combination of the motive shaft C C', feed screw F, and guides D C', and self-adjusting cutter head E, substantially as and for the purpose set forth.

No. 59,418.—ALBERT R. LAWRENCE, Saratoga Springs, N. Y.—*Preventing Sediment in Mineral Waters.*—November 6, 1866.—About one per cent. of a solution of tartaric acid is added to the mineral water to prevent the formation of sediment. A vessel of the solution connects with the pipe by which the bottles are filled, and a sponge in the tube graduates the flow of the acidulated water.

Claim.—In barrelling, bottling, and putting up mineral water, introducing or mixing a very small quantity of tartaric acid with the water to prevent precipitation or sediment of the ingredients held in solution in the water.

Also, connecting a reservoir of tartaric acid, or a solution of such acid, to the pipe which supplies the mineral water. so adjusted as to supply the proper quantity of acid to the mineral water, (to prevent precipitation or sediment,) as the water is drawn through the pipe.

No. 59,419.—JOHNSON LETSON, New Brunswick, N. J.—*Water-proof Shoe.*—November 6, 1866.—Between the vamp and inner lining is a sheet of rubber, leaving an air space between the lining and the rubber.

Claim.—As a new article of manufacture, a rubber boot or shoe, in which the upper is made water-proof and ventilating by interposing between the vamp and lining a sheet of rubber or water-proof water fabric, which, whether attached or not to the vamp, leaves the lining loose, as and for the purposes set forth.

No. 59,420.—JOSEPH LETTEREL, Union, Ind.—*Machine for Tenoning Spokes.*—November 6, 1866.—The spoked hub is put on its clamping pin, and the spoke ends brought consecutively to the hollow auger. The frame is moved forward to the stop by a lever on the shaft of a spur wheel acting on a rack. In boring the felloes a pivoted clamping frame is used and a common auger bit substituted as a tool.

Claim.—A bit and boring apparatus, with shafts P and N, movable frame B', by means of the rack bar M, lever handle and wheel, constructed upon a suitable frame A, when arranged in the manner and used as and for the purposes herein described.

No. 59,421.—CYRUS LEWIS, Nelson, Ohio.—*Fence Jack.*—November 6. 1866.—The lever is pivoted to standards on a base block, and when depressed is locked to the block so as to keep the fence corner raised while placing a stone underneath.

Claim.—The base A, pins d, pieces B, and rod D', as arranged, in combination with the lever C, in the manner and for the purpose as herein described

No. 59,422.—SAMUEL K. LIGHTER, and THOMAS HARDING, Hamilton, Ohio.—*Grain Drill Tube.*—November 6, 1866.—The heliacal wire tube is attached to the frame by a flange having a pipe with a screw thread to suit the inner surface of the tube, or the pipe may be so attached to the wire tube as to allow of it slipping up to pass obstacles.

Claim.—First, the mode of connecting the tube to the frame, as shown in Fig. 2, and herein described.

Second, as a modification, the mode of connecting the tube to the frame, substantially as shown in Fig. 4, to overcome obstructions, in the manner and for the purpose set forth.

No. 59,423.—WILLIAM LOVE and ORION THORNLEY, Indianapolis, Ind.—*Device for Operating Wood-splitting and other Machines*—November 6, 1866.—The axe is attached to a mandrel with a T-head which is vertically reciprocated by the engagement of a slide block pivoted to a wrist on the rotary disk.

Claim.—The arrangement of the T or cross-head D, and sliding block F, in combination with the crank G S and mandrel B, substantially as set forth.

No. 59,424.—THOMAS MAIN, Green Point, N. Y.—*Steam Generator.*—November 6, 1866.—The caloric current passes over a bridge wall which may be of metal and contain water; dives, and then passes up through vertical tubes which extend above the water level. The steam is discharged through a horizontal pipe suspended from the crown plate and perforated on its upper side.

Claim.—First, the horizontal chamber A, combustion chamber D, portion B provided with vertical tubes, water space or bridge wall C, when combined and arranged substantially as herein shown and described.

Second, the arrangement of the perforated discharge pipe P with the tubes of a vertical boiler, substantially as herein shown and described, for the purpose of receiving and discharging the steam in a superheated state, as set forth.

No. 59,425.—G. B. MANLEY, Cogan Station, Penn.—*Hammer for Forging Blooms.*—November 6, 1866.—The anvil has two faces at right angles to each other ; the framing above supports a shaft carrying cams which operate the hammers alternately against the bloom, which is held between the two inclined faces of the anvil. Spring-stops catch into lugs on the hammers, when it is desired to arrest their operation and suspend them above the anvil.

Claim.—First, the combination of the hammers E with the helves D and arms D', arranged with the cam shaft C, whereby the hammers fall alternately on a two-faced anvil A, and operating substantially as described for the purpose specified.

Second, the shouldered hammers E, operating with the spring arms I, lever L, slotted plate M, rod O, lever Q, and forked bracket P, constructed and arranged substantially as described for the purpose specified.

C P 45—VOL. II

No. 59,426.—J. MARCHBANK and W. H. HUMPHREY, Lansingburg, N. Y.—*Faucet.*—November 6, 1866.—The bit is attached to the faucet and is projected or retracted by a rack on its shank within the faucet, actuated by a thumb-screw. A frustal projection on the cap affords means for operating the device by a brace.

Claim.—First, the combination of an adjustable bit C, or other boring device, with the faucet A, in the manner and for the purposes substantially as herein described and set forth.

Second, combining with the faucet A and bit C, suitably constructed shank or projection D, or other device to which a brace may be applied, in the manner and for the purposes substantially as herein described and set forth.

Third, the attaching of the said bit C to the faucet A, in such a manner that the same will be adjustable in the manner and for the purposes substantially as herein described and set forth.

No. 59,427.—CHARLES C. MATHER, Burlington, N. Y.—*Fence.*—November 6, 1866.—The panels are supported by wires which clamp their end slats together, embracing a post between them.

Claim—The arrangement and combination of the posts A, panels B, and wires *d*, when said panels B are capable of being slid out or removed by simply removing the wire fastenings *d*, the whole being constructed and operated substantially in the manner and for the purpose set forth.

No. 59,428.—WILLIAM D. MATTHEWS, Columbia, Tenn.—*Churn.*—November 6, 1866.—Two dashers revolve on a horizontal shaft which passes through a central case containing the driving belt and pulley.

Claim.—A churn provided with a dasher composed of one or more rotary beaters attached to a horizontal shaft or axis which passes through a box fitted vertically within the cream receptacle and enclosing the driving mechanism, substantially as herein shown and described.

No. 59,429.—WILLIAM MAXWELL, Washington, N. J.—*Fishing Net.*—November 6, 1866—The mouths of the funnel net and the bag net are distended by a frame and the latter is anchored and buoyed, presenting the mouth up stream.

Claim.—The double net with a rigid open mouth, when constructed and combined with the buoy and anchor to regulate its position in any depth of water, as herein described and for the purposes set forth.

No. 59,430.—BERNARD MCCOLLUM, New York, N. Y.—*Window Shutter.*—November 6, 1866; antedated October 27, 1866.—The shutter is hinged above and at the middle horizontally; the lower section, or the two sections united by a bolt, may be raised by cords, pulleys, and counterweights to form an awning, or may be fastened shut by a cross bar which engages at each end against the frame and is locked in position.

Claim.—First, the shutter hinged horizontally to the upper side of the window frame and divided transversely into two parts B C, hinged together and arranged with reference to the window frame A, substantially as herein set forth.

Second, the cord *u*, weighted cords *i*, and brackets *f*, arranged in relation with each other and with a horizontally hinged and transversely divided shutter, and with the window frame A, substantially as herein set forth for the purpose specified.

Third, the pivoted cross bar D, cleat *w*, pivoted bar *b'*, screw *a'*, and pin or padlock *d'*, in combination with each other and with the horizontally hinged and transversely divided shutter and with the window frame A, substantially as herein set forth for the purpose specified.

No. 59,431.—F. MCDONOUGH, Chicago, Ill.—*Door Fastener.*—November 6, 1866.—The pointed ends of the levers are thrust into the door and floor respectively as the toggle is straightened, in which condition it is locked by a bolt.

Claim.—The construction and arrangement of the pronged levers A B, in combination with the bolt E, substantially as set forth and described.

No. 59,432.—PETER MCKINLAY, Charleston, S. C.—*Machine for Cleaning Rice.*—November 6, 1866.—Motion is communicated to the rockers from a crank on the main shaft, through a connecting rod and depending arm. The acting faces of the rockers and grain chambers are faced with rubber.

Claim.—A rice cleaning machine when constructed with rockers faced with rubber or analogous substance, and caused to rock upon rubber strips in the bottom of the oblong chambers, and arranged to operate substantially as described.

No. 59,433.—JOHN MCMAHEL, Hamilton, Ohio.—*Straw Cutter.*—November 6, 1866.—One roller has spiral knives and the other spiral abutment corrugations, whose intervals are entered by the knives.

Claim.—The knives arranged spirally upon one of the feed rolls, in combination with the grooved or shouldered upper feed roll, as represented in the drawings at D and E, substantially as described, and operating to give a shearing cut, as set forth, for the purpose specified.

No. 59,434.—HELEM MERRILL, Brooklyn, N. Y.—*Spike.*—November 6, 1866.—The spike has wings upon the two sides of the rounded point, and spiral barbs upon two sides of the shank.

Claim.—As a new article of manufacture a spike pointed and ridged, substantially as described.

No. 59,435.—SAMUEL M. MILLER, Beaver Mill, Penn.—*Grinding Mill.*—November 6, 1866.—The emptying of the hopper allows the wing to rise, which brings the friction pulley in contact with the millstone and revolves its shaft; the latter acts by a pulley and cord to move the belt shifter and stop the mill.

Claim.—The devices of the hinged wing E as attached to the hopper C and weight H, so as to operate the lever M, in combination with the square frame P, friction roller S, pulley T, and lever V, or any equivalent devices, when arranged and operating as herein described and for the purposes set forth.

No. 59,436.—W. D. MONK, Williamsburg, N. Y.—*Curing Hemp, Flax, &c.*—November 6, 1866.—The dried stalks are steeped for twenty-four hours in a solution of water (Javelle water,) three gallons; chloride of lime, two pounds; and chloride of soda, one pound. The stalks are then crushed between fluted rollers, dried and beaten, or shaken.

Claim.—The within described process of curing hemp, flax, and other fibrous plants, by exposing them to the action of Javelle water, substantially as set forth.

No. 59,437.—JOSEPH A. MOORE, Louisville, Ky.—*Plough.*—November 6, 1866.—A slotted flange is attached to the inside of the mould-board, to which the handle is attached by bolts and nuts.

Claim.—The slotted flange, made and described and for the purpose set forth.

No. 59,438.—L. F. MORAWETZ, Baltimore, Md.—*Vertical Solar Camera.*—November 6, 1866.—The rays of the sun are reflected vertically upon the horizontal lenses of the solar microscope or megascope, and the pictures are received upon a horizontal table, either with the amount of light commensurate with the diameter of the lens, or with increased light by means of mirrors, one of which is moved automatically as required by the apparent diurnal motion of the sun.

Claim.—First, an automatic heliostat which is constructed with a reflector that is mounted upon an automatic heliotrope and moved by it, synchronic with the sun, and in a required angle with the plane of the daily course of the sun, for the purpose of reflecting the sun's rays continually every day of the year upon a fixed place of the instrument, in such a manner that the reflected rays shall fall with the same angle of incidence upon the surface of that place, substantially as described.

Second, the combination of a concave mirror M M' with a solar microscope or megalascope and camera, in such manner as to reflect an increased amount of light upon the system of lenses constituting the solar microscope or megalascope, substantially as described.

Third, the combination of two mirrors, both plain, or one plain and the other concave, with a solar microscope or megalascope and camera, arranged so as to operate substantially as and for the purposes described.

Fourth, the combination of the movable mirror m m', the fixed mirror M M', with a solar microscope or megalascope, and a horizontal table C, arranged beneath the same, substantially as described.

No. 59,439.—JAMES F. NEALL and WILLIAM MYERS, Philadelphia, Penn.—*Damper Regulator.*—November 6, 1866.—The hollow piston is connected by a pipe to the boiler and is stationary. The weighted cylinder plays upon the piston by the pressure of steam, and operates the furnace damper.

Claim.—The piston A, and cylinder B, when the same are constructed, arranged, and combined to be operated together by the pressure of steam in a boiler, and the counter pressure of movable weights applied directly upon the said cylinder, substantially as described and set forth for the purpose specified.

No. 59,440.—JOHN NICHOLS, EDWARD C. NICHOLS, and DAVID SHEPARD, Battle Creek, Mich.—*Threshing Machine.*—November 6, 1866.—The deck of the cylinder frame is so curved on its under side as to deflect downward the grain which strikes it. The end of the grain carrier has perpendicular perforations for the escaping grain.

Claim.—So constructing the portion indicated at D, of the deck of a threshing machine cylinder casing, that the threshed grain, and straw when it strikes said deck, after passing under the cylinder, may be deflected downward upon the separating part of the machine, substantially in the manner and for the purpose herein specified.

Also, the combination and connection of the perforated board sieve W with the vibrating grain carrier B, when arranged relatively with the separating device and with the winnowing sieves M, substantially as and for the uses set forth.

No. 59,441.—JOSEP and IGNAZ NEUBURG, New York, N. Y.—*Cooking Apparatus and Refrigerator.*—November 6, 1866.—The vessel has two metal shells, the space between which is packed with pasteboard, and has a similar cover hinged on one side. Channels run around the top edge of the vessel with escape holes therefrom. The packing rests upon ribs below and a drain hole through the bottom carries off moisture.

Claim.—The non-conducting double casing *a b c*, having an annular gutter *m*, openings *i*, opening *g*, and pan *f*, and furnished with the non-conducting cover A, in combination with the central vessel B, the whole constructed and arranged substantially as herein set forth for the purpose specified.

No. 59,442.—GEORGE E. NOYES, Washington, D. C.—*Steam Governor.*—November 6, 1866.—The governor valve is operated by the pressure of steam upon a piston upon its stem. The piston operates in a smaller cylinder, receiving steam from the main cylinder.

Claim.—Operating within a supplementary cylinder a piston E, and combined with the governor valve when so arranged that the steam which operates said piston is taken from the main cylinder A, by passages leading therefrom, substantially as shown and described.

No. 59,443.—WILLIAM ONIONS, St. Louis, Mo.—*Cotton Bale Tie.*—November 6, 1866.—Explained by the claim and cut.

Claim.—The bar C, secured to one end of the bale hoop, and having centre tongue-piece D, and side hooks E and F, in combination with the notches *a a*, and opening *b*, in the other end of the bale hoop, substantially as described and for the purpose specified.

No. 59,444.—WILLIAM ORCUTT, Cambridge, Ill.—*Washing Machine and Churn.*—November 6, 1866.—The two-crank shaft has connecting rods to washing heads on one side, and to the cream plungers on the other. The heads and plungers are pivoted and their movements directed by coiled springs.

Claim.—First, the crank shaft M, pitmen rod *r r'* and *c c'*, guides *d* and *d'*, and springs 21 21, combined and arranged as and for the purpose set forth.

Second, adjusting and holding the washing box C, by means of the pins 16 and 17, and holes 18 and 19, as set forth and for the purpose specified.

No. 59,445.—ISAAC OSGOOD, Utica, N. Y.—*Securing Wheel Hubs to Axles.*—November 5, 1866.—The nut engages the plate on the face of the hub and forms one cheek for the enclosed nut, which is attached to the spindle and maintains the latter in the hub.

Claim.—The arrangement of the nut F, screw socket bushing M, collar G, and washers *a b*, in combination with the shouldered box I of the hub H, and shouldered axle A E, with the screw thread *g*, constructed and operating in the manner and for the purpose herein represented and described.

No. 59,446.—FRANCIS S. PEASE, Buffalo, N. Y.—*Carburetter.*—November 6, 1866.—As a means of dispensing with blowing apparatus on railroad cars, locomotives, &c., a strong reservoir is charged with compressed air, to supply air to the carburetting apparatus for the head light or other illumination or heating.

Claim.—First, the combination with a carburetting chamber of a strong reservoir for containing a supply of compressed air, substantially as and for the purpose described.

Second, the mode of operating the air pump, with a water packing above the valve or valves, substantially as described.

No. 59,447.—JAMES M. PEIRCE, Mokena, Ill.—*Churn and Washing Machine Combined.*—November 6, 1866.—The dasher in one instance and the stirrer in the other are operated by a twisted plate on the shaft, which works in a slotted reciprocating lever.

Claim.—The application of the twisted iron or other metal for giving the reversible rotary motion to churn dashers and the clothes stirrer.

Also, the combination of the reversible rotary and vertical motions given by the twisted iron, as herein described.

No. 59,448.—JOHN H. PITEZEL, Three Rivers, Mich.—*Sewing Clamp.*—November 6, 1866.—The clamp frame is bolted on the seat. The moving jaw runs beneath the seat and is actuated by a strap passing over a sheave to a treadle.

Claim.—First, the bed plate A A', constructed as described.

Second, the combination of the cast-iron bed plate and legs with the wooden clamping jaws, constructed and operating substantially as described.

Third, the combination with the bed plate of the short fixed jaw and the long movable jaw, the strap connection and the treadle, arranged and operating as described.

Fourth, the combination of the fixed and movable jaws with the interposed rubber spring, arranged and operating as described.

No. 59,449.—E. L. PRATT, Boston, Mass.—*Apparatus for Aerating and Mixing Substances.*—November 6, 1866.—The plunger descends with its concavity filled with air from the vessel above the liquid, and disseminates the same in fine jets through the material below, the plunger having orifices for that purpose.

Claim.—The apparatus for aerating fluid or semi-fluid substances, having a construction and mode of operation substantially as set forth.

Also, the process of aerating fluid or semi-fluid substances by carrying air in a body into the body of the liquid and there discharging it in minute or finely divided jets or particles, substantially as set forth.

No. 59,450.—E. L. PRATT, Boston, Mass.—*Sieve.*—November 6, 1866.—The corrugated arms are attached to the central hub and revolve horizontally above the sieve bottom.

Claim.—The combination of the corrugated arms *a a a*, with the centre hub C, all operating within the hoop of the sieve, substantially as specified and described.

No. 59,451.—EDEN REED, Joliet, Ill.—*Postal Wrapper.*—November 6, 1866.—The package is enclosed in a wrapper, buckled or tied, and openings permit the direction to be inscribed upon the enclosed matter.

Claim.—The hole *e e'*, through which the direction may be written on the letter or package enclosed, substantially as described, for the purpose specified.

No. 59,452.—JACOB G. REIFF, Farmersville, Penn.—*Sulky.*—November 6, 1866.—The usual elliptical spring is associated with a spiral spring near each wheel. Braced guide rods from the frame embrace the axle. The whiffletree is pivoted to the vertex of two arms whose hooked ends are attached to the axle, or the hooked ends may be advanced for the traces and the vertex connected to the axle.

Claim.—The arrangement and construction of the elliptic spring K, side sliding guides L, spiral springs M, stays N, and forked reversible singletree E, when combined and operating as herein described and for the purposes set forth.

No. 59,453.—W. A. REX, Newville, Ind.—*Punch.*—November 6, 1866.—The removable cutter tube is fastened by a spring catch in the jaw of the punch.

Claim.—The combination of the spring catch C, and removable cutter tube or tubes B, with each other, and with the punch stock A, when the said parts are constructed and arranged substantially as herein shown and described and for the purpose set forth.

No. 59,454.—C. S. ROBERTS, Lyons, Iowa.—*Cultivator.*—November 6, 1866.—The beams of the two sides are connected together in front by an arched bar, and at the rear by braces. The inner beams are journalled at their ends, allowing lateral oscillation, which is given by the feet. The ploughs are raised by a bifurcate lever, pivoted on a raised part of the axle and retained by a hanging rack. The depth is adjusted by holes in the bent bars, to which the tongue is attached.

Claim.—First, the drag bars G and G', arch E and brace L, when respectively constructed and arranged for use, substantially as set forth.

Second, the combination of the drag bars G, stirrups F, axle B, lever I, and rods I', substantially as and for the purpose set forth.

Third, the curved braces D, when adjustably attached to the framework of a cultivator, substantially in the manner and for the purpose set forth.

Fourth, the combination of the shovel ploughs H, and mouldboard ploughs H', when the inner shovel ploughs are made adjustable and arranged substantially as set forth.

No. 59,455.—E. O. ROOD and S. H. HACKETT, Lodi, Ill.—*Thill Coupling.*—November 6, 1866.—To the end of the thill iron is a shouldered lever and spring, which are passed through the enlarged part of the slot in the axletree iron until the pins rest in the bearings ; the spring then throws the lever forward and its shoulders engage beneath the plate.

Claim.—The combination of the shouldered lever D, and its spring F, attached to the thill, with the slot T of the axletree iron, the whole constructed and operating substantially in the manner herein described and specified, and forming the thill coupling.

No. 59,456.—JAMES ROSS, Somerville, Mass.—*Paper Packing Joint of Steam Engines*—November 6, 1866; antedated October 27, 1866.—Paper, card board, or other similar fabric, saturated with a composition of crude petroleum, benzine, and linseed oil, is used for packing steam joints.

Claim.—The packing for steam and other joints, herein described.

No. 59,457.—THOMAS J. ROSS, Union, N. Y.—*Adjustable Back for Stools.*—November 6, 1866.—The back and arms are attached by brackets to the seat, and the spring pad of the back is adjustable in position.

Claim.—First, the swivel brackets *a*, in combination with the seat A, and with the detachable back B C, constructed and operated substantially as and for the purpose described.

Second, the hinged back C, and slide *i*, in combination with the bed B, screws *k l*, and seat A, all constructed and operating substantially as and for the purpose set forth.

No. 59,458.—E. A. G. ROULSTONE, Roxbury Mass.—*Trunk Moulding.*—November 6, 1866.—The sheet metal which laps over the corners is corrugated and is stiffened on the trunk edge by enclosed wire.

Claim.—As a new article of manufacture, the corner moulding or guard *a* for trunks, made of corrugated metal, formed into shape for application, and strengthened by a wire *b*, substantially as described.

No. 59,459.—D. H. RUCKER, JOHN E. ALLEN, and JACOB S. SMITH, Washington, D. C.—*Ambulance.*—November 6, 1866.—The lower frames are hinged and detachable so as to form seats or beds attached to the sides or resting on the floor. The upper frames make an upper tier of beds or backs for the seats below.

Claim.—First, the within-described ambulance accommodation, consisting of the beds or stretchers *a a' d e g h*, adapted to contain two tiers of recumbent patients, or be converted into one or more backed seats, substantially as and for the purpose described.

Second, the upper hinged beds or stretchers *g h*, employed in connection with the lower beds or stretchers of an ambulance, so as to constitute seat backs when the lower beds are converted into seats, substantially as and for the purpose herein set forth.

No. 59,460.—HENRY RYDER, New Bedford, Mass.—*Tubular Plunger for Candle Moulds.*—November 6, 1866; antedated July 18, 1866.—In the foot of the tip mould is inserted a perforated block of vulcanized rubber which grasps the wick closely, but will expand so as to permit a knot in the wick to pass through.

Claim.—The candle-mould plunger provided with the elastic or expansive mouth piece *c*, so arranged within it or with respect to its tip matrix, substantially as specified.

No. 59,461.—WILLIAM H. H. SAUNDERS, Troy, N. Y.—*Combined Blacking Brush and Box.*—November 6, 1866.—The handle is attached by slot and set screw ; the latter also rests on a lip of the blacking receptacle, which is covered by the spreading brush.

Claim.—First, the combination of the removable and adjustable handle C, having a slot mortise in the front end thereof, so as to receive the thumb-screw D with the brush A, containing the recess K, and with the said thumb-screw D, each being arranged in the manner and for the purposes substantially as herein described and set forth.

Second, the employment of the polishing brush A, containing the concave or other shaped blacking receptacle H', or any equivalent thereof, in combination with the cover and applying brush B, constructed, arranged, and operated in the manner and for the purposes substantially as herein described and set forth.

No. 59,462.—H. K. SCHANCK, Benville, Ind.—*Composition for Roofing.*—November 6, 1866.—Composed of clay 12 parts, and sulphur 1 part, mixed with sufficient unboiled coal tar to form a pasty mass, to be spread upon the naked roof, without fibrous covering thereon.

Claim.—A roofing composition composed of the substances herein specified, in the proportions and compounded in the manner substantially as described.

No. 59,463.—H. H. SCOVILLE, P. W. GATES, and D. R. FRASER, Chicago, Ill.—*Quartz Mill.*—November 6, 1866.—The revolving grinding cylinder is corrugated upon its inside, and contains other cylinders of less diameter, which are corrugated internally and externally, are perforated and roll freely within the outer cylinder. The quartz is fed and discharged through the respective trunnions.

Claim.—First, the corrugated cylinder A, constructed so as to revolve and elevate the quartz, or other substances, in combination with one or more hollow cylinders, such as E and F, which are corrugated and perforated circumferentially so as to admit and conduct the quartz or other substances after they have been elevated into and out of the chambers of such cylinders as E and F, the cylinder A having hollow journals, and the cylinders E and F having an opening in each end, so that the quartz may be fed in and discharged continuously, all substantially as set forth.

Second, a corrugated cylinder E or F, perforated entirely through its shell, substantially as and for the purpose described.

Third, the construction and arrangement of the cylinders A E F, so that the substances to be operated upon are free to pass through the circumferences of the cylinders E and F, and are subjected to a grinding and crushing action between the said cylinders A E and F, substantially as described.

Fourth, constructing the corrugated lining of the cylinder A, which has axial supports, with cups or channels *c*, substantially as and for the purpose described.

No. 59,464.—MOSES SEWARD, New Haven, Conn.—*Machine for Upsetting and Forming Articles from Metallic Rods or Bars.*—November 6, 1866.—On the bench are placed two sets of dies, one set for grasping and holding the bar during the action upon it of the other set of swaging dies ; these are operated by cams on a shaft on one side of the bench.

Claim.—The holding dies *e* and *e'*, with upsetting dies *c* and *c'*, or their equivalents, when arranged and combined so as to upset the metal placed therein in two places at one time, and operating substantially as herein set forth.

No. 59,465.—JAMES B. SHARP and R. M. SEYMOUR, New York, N. Y.—*Suspenders.*—November 6, 1866.—A metallic plate is bent 'around the ends and riveted, or around the respective portions at their intersection, as a fastening.

Claim.—As an article of manufacture a pair of suspenders tipped or fastened by a metallic plate, secured substantially as described.

No. 59,466.—CALVIN W. SHERWOOD, Chicago, Ill.—*School Desk and Seat.*—November 6, 1866.—The desk, shelf, and seat are each jointed to the standard so as to fold compactly and occupy but little horizontal room when so desired. A toggle supports shelf and desk.

Claim.—First, the joint composed of the nave C', and axle B', constructed and operating substantially as and for the purposes specified.

Second, the arrangement and combination of the arms c, naves c', and axle B', with the seat D, and standards A, substantially as specified.

Third, the jointed braces F, when provided with lips a and ledges b, substantially as and for the purposes specified.

Fourth, the combination and arrangement of the ledges b, lips a, and pins d, with the braces F, and hinged shelf K, substantially as specified.

Fifth, the arrangement and combination of the hinged arms H, jointed braces F, and hinged arms G, with the standards A, and desk top J, substantially as and for the purpose specified.

No. 59,467.—THOMAS SIMMONS, Chicago, Ill.—*Filter.*—November 6, 1866.—The filter is portable and adapted to be placed in a cistern or reservoir, when the water is drawn through it by means of a siphon. The strata of filtering material are separated by cloth, and the whole compressed by means of screw and follower.

Claim.—First, the arrangement and combination of the cap K, cloth J, and I I I', cotton W, and follower B, with bar C, screw H, and foot D, substantially as set forth.

Second, in combination with the foregoing, the nozzle o, cap S, stopper T, with the tube R, and pipe M, as described and set forth.

No. 59,468.—M. T. SMITH, Keeler, Mich.—*Plough.*—November 6, 1866.—A concave-faced roller is journalled in an arm which is hinged to the beam. The trash is drawn in and covered by the farrow slice.

Claim.—The roller F and bar D, connected together and applied to the plough beam A, to operate in the manner substantially as and for the purpose herein set forth.

No. 59,469.—JOHN SNELL, Jr., Potterville, Penn.—*Machine for Handling Hides.*—November 6, 1866.—The hides are suspended from rods in a frame, which is supported on rollers and reciprocated in the vat longitudinally by crank and connecting rod.

Claim.—The hanging frame B, adapted for longitudinal reciprocating motion, having bars K and hooks, substantially as described for the purpose specified.

No. 59,470.—DAVID SPORE, Sharon, Wis.—*Bed Bottom.*—November 6, 1866.—The slats are supported at their ends, upon short slanting pieces that are inserted in oblique sockets in the end sills.

Claim.—The combination of the slats D and the slanting pieces C, supported by the oblique sockets in the sill, substantially as described and represented.

No. 59,471—J. H. STARR, Middleburg, N. Y.—*Beehive.*—November 6, 1866.—The hive is set upon a base, slotted at the bottom and sides, and within a case the sides of which are removable.

Claim.—The bases B, constructed and perforated, or slotted, as shown, and placed within a bee house or structure A, provided with detachable or removable sides, substantially as and for the purpose set forth.

No. 59,472.—D. H. STEPHENS, Riverton, Conn.—*Tool for Making Tenons.*—November 6, 1866.—Three saws are placed together upon a revolving mandrel; the two outer saws are of larger diameter and cut the sides and shoulder of the tenon; the centre smaller saw cuts the tenon to the length. The device is applied to making tenons for rule joints.

Claim.—The employment of a third or middle saw beween the two or its equivalent, which cuts the end of the tenon of the length required, in combination with the other two saws.

No. 59,473.—LEVI STEVENS, Fitchburg, Mass.—*Apparatus for Carburetting Air.*—November 6, 1866.—The lower chamber contains gasoline, in which is a float which sustains the perforated cylinder above, which is filled with sponge. The carburetted air escapes by a pipe from the sponge box into the gasometer above.

Claim.—The combination as well as the arrangement of the hydrocarbon-holding vessel A, the float B, the box C, with its absorbent material, the tube D, and the receiver E, provided with the stuffing box e, or its equivalent, the whole constituting an improved hydrocarbon vaporizing apparatus.

Also, the combination as well as the arrangement of the cistern F, or the same and the waterjacket I, with the said hydrocarbon vaporizing apparatus.

Also, the above described arrangement of the airometer G, with the vaporizing apparatus, made substantially as described.

No. 59,474.—LEVI STEVENS, Fitchburg, Mass.—*Apparatus for Carburetting Air.*—November 6, 1866.—The cylindrical vessel is divided into three compartments, viz: a gasoline chamber; a gasometer and a carburetting vessel which is composed of a sponge chamber, air-distributing chamber and au air-collecting space. The gasoline is raised by an elevator and poured from a spout over the top plate so as to saturate the sponge. The gasoline not absorbed returns to the bottom of the elevator by a pipe.

Claim.—The combination as well as the arrangement of the chamber C, the annular space b, the sponge chamber D, the perforated partition H, and the space E'.

Also, the combination of the sponge d, and the extension F, with the chambers E and C, and the sponge t, and the partition H, made and arranged in manner and so as to operate as specified.

No. 59,475.—WILLIAM H. STEVENS, Winona, Minn.—*Cheese Press.*—November 6, 1866.—The hoop is placed in a cylinder whose tight cover is driven down upon the follower in the hoop as the air is exhausted from the cylinder by an air pump.

Claim.—A cylindrical atmospheric cheese press, constructed and operated substantially as described and for the purposes set forth.

No. 59,476.—JOHAN JACOB STUDER, Richmond, Ind.—*Pile Driver.*—November 6, 1866.—The apparatus is supported between and attached to posts. The cylinder is the monkey and slides on guides; inside it is the stationary piston head and hollow piston rod through which steam is admitted above the piston to raise the monkey. The valve is worked by the motions of the cylinder.

Claim.—First, the combination of the valve stem O, yokes i and t', lever f, wrist C', and cylinder C, substantially as and for the purpose set forth.

Second, the arrangement of the valve p, with its chest and the ports u and w, substantially as set forth.

Third, the arrangement of the plates r r', set screws w, and clamps u, substantially in the manner and for the purpose set forth.

Fourth, supporting a pile driver on elastic cushions, substantially as and for the purpose set forth.

No. 59,477.—W. C. TAGGART, Fayetteville, N. Y.—*Washing Machine.*—November 6, 1866.—The oscillating rubber is suspended from an axis which has a vertical adjustment in the slotted standards. The ribs on each of the rubbing surfaces are bevelled, inclining in different directions on each side of the middle of the tub.

Claim.—The grooved bevelled ribs a a, in combination with the bottom of the tub A, and face of the rubbing board C, the bevelled side being placed toward the respective ends of the machine, and operating substantially as described for the purpose specified.

No. 59,478.—JAMES M. TALBOTT, Richmond, Va.—*Tobacco Press.*—November 6, 1866.—The portable hydraulic ram is combined with stationary retaining frames, and a double truck arrangement. One truck carries the hydraulic ram, and is supported on the other truck which runs from one retaining press to the other upon a stationary track, so as to act upon the tobacco in the presses consecutively.

Claim.—First, the retaining frame herein described, the same consisting of the threaded columns C C, head T, bottom B, follower F, and nuts r r, said follower being adapted to be slidden upon the columns so as to compress the tobacco and then held in position by the nuts r, so as to retain the tobacco in its compressed state, substantially as set forth.

Second, the springs e, applied to the truck of the hydraulic ram in the manner described, so as to sustain the same when not in action, but permit it to be depressed to find a firm and solid bearing on the press bottom when the power is applied so as to relieve the axles from pressure, as explained.

Third, a movable hydraulic ram, in combination with a stationary retaining frame, substantially as described.

Fourth, a double truck d d S S, which enables the hydraulic ram to be moved two ways, substantially as described.

Fifth, the combination of the double truck and hydraulic ram with the stationary retaining frames, substantially as described.

No. 59,479.—THOMAS G. THOMPSON, Oswego, N. Y.—*Tool for Making Horseshoes.*—November 6, 1866.—One end of the steel spring is attached to the anvil block; the other end terminates in a steel point for pressing the shoe against the anvil; a chain connects the spring to a foot lever by which it is operated.

Claim.—The combination of the spring-holder chain and foot lever with the anvil block and anvil, all constructed and arranged as and for the purpose described.

No. 59,480.—ESAU D. TAYLOR, Hornellsville, N. Y.—*Steam Engine Slide Valve.*—November 6, 1866.—Each cheek of the valve has arms which interlock alternately, and enclose springs by which the faces of the valve are expanded laterally against the seats; a taper key operated by a set screw regulates the adjustment.

Claim.—First, the valve plates D D', provided with the arms F, in combination with the taper key J, substantially as described.

Second, the valve D D', provided with the arms F and taper key J, in combination with the springs H, to keep the valves in position when steam is not present, and to permit the valves to yield to back pressure in the cylinder.

Third, in combination with the taper key J, the adjusting screw Q, substantially as and for the purpose set forth.

Fourth, the valve D D', in combination with the yoke L and hollow stem M, substantially as and for the purpose set forth.

Fifth, the taper key J, in combination with the steady pins p p P', substantially as described.

No. 59,481.—JAMES TODD and ALBERT G. DOWNER, Fayette, Penn.—*Washing Sand.*—November 6, 1866.—A series of elevators raise the sand into a trough, where water is poured upon it, and it thence runs into another trough, where the muddy water runs off. The elevation and washing are repeated again and again till the sand is clean.

Claim.—A new mode of lifting or raising the sand from the muddy water by means of elevators and elevator boxes, both shaped as herein described, for elevators to work in, with pulleys and drums, and the manner in which each box discharges the muddy water from the sand.

No. 59,482.—W. B. TRUNICK, Pittsburg, Penn.—*Boring Tool for Wells.*—November 6, 1866.—The apparatus is for turning the drill rod a part of a revolution at each stroke of the working beam from which it is suspended. Each motion of the beam actuates a pawl, ratchet and rack bar, the latter gearing into a pinion on the shackle bolt of the drill rod. After a certain number of revolutions a tappet changes the pawl and another ratchet actuates the rack in the reverse direction.

Claim.—First, the self-acting boring tool, turning apparatus, applicable to any ordinary boring rigging, composed of the pinion F, racks H K and L, dogs O and P, lever R, tappets T and U, and catches S S, combined with the lever B, and arranged as described, or their equivalent.

Second, the two racks K and L, oscillating with the beam B, in combination with the stationary shaft M, dogs P and O, lever R, catches S S, and tappets T and U, to obtain a self-acting go and come motion of the rack K and L on the beam B.

Third, the spring V, and set screw Y, in combination with the bar I, for the purpose of regulating the motion of the said bar I on the bar B.

Fourth, turning boring tools automatically by the action of the oscillating beam B itself, by means and with the use of the apparatus herein described, or its equivalent.

No. 59,483.—DANIEL TUTTLE, Plantsville, Conn.—*Trace Lock.*—November 6, 1866.—To attach the trace, depress the button and turn it one fourth round, place the loop of the trace over the button and return it, when it is forced out by the spring, and its ears enter the notches in the ferrule, where it is held fast.

Claim.—The combination of the tongue C, and the ferrule or socket B, constructed with ears d, and notches a, and the spring in the rear of the tongue when the shank E is enlarged within the socket, and so as to operate substantially in the manner and for the purpose specified.

No. 59,484.—ISAAC H. UPTON, New York, N. Y.—*Fire-place.*—November 6, 1866.—On the back and sides of the fire box is a heating chamber, to which air is supplied through openings below, and which delivers its heated air at upper openings into the room; dampers may be opened to connect the air chamber directly with the flue.

Claim.—First, the inclined plate F, in combination with valves or dampers J J, and rods J' J', employed to retain or permit the escape of heat, and arranged as and for the purpose specified.

Second, in combination with the above, the air-heating space or chamber E, formed and arranged substantially as and for the purpose set forth.

Third, in a fireplace for grates, constructed as herein described, the perforated plates L L L', arranged as described, and permitting a free circulation of air to and from the air-heating chamber E, as set forth.

Fourth, the arrangement of the grate chamber B, pipe or flue H, dampers I J J, inclined plate F, and air-heating chamber E, as herein described, and for the purpose specified.

No. 59,485.—CHARLES VAN DYECK, Nashville, Tenn.—*Cigar.*—November 6, 1866.—A loose filling is enclosed in a strong paper tube and covered with tobacco leaf. A mouth-piece of paper is provided with a spongy absorbent for the nicotine.

Claim.—First, a cigar, with a filling of "waste" tobacco, enclosed in a paper wrapper, covered by an exterior wrapping of leaf tobacco, substantially as described.

Second, such a cigar when formed also with a mouth-piece *b*, and absorbent *c*, arranged substantially in the manner and for the purpose set forth.

No. 59,486.—JOSEPH A. VINCENT, Fairburg, Ill.—*Weather Strip.*—November 6, 1866.—The lower part of the upper portion is bent down into a vertical position, and the lower portion is so arranged that when the door is closed it is raised to a vertical position in the rear of and fitting tightly against the upper piece. A rubber packing is fastened to the door jamb.

Claim.—First, the flange C, with its finger D, so applied to a door or window as to receive the upper edge of the weather strip between them, the flange C being bent vertically near its outer edge, and being free to pass over the strip when the latter is down, substantially as shown.

Second, placing the weather strip in front of the saddle of the threshold, and constructing it so that when the strip is not in operation it lies level with the saddle, and when the door is closed the strip F will assume a vertical position, with a dead-air space between the strip and the saddle, substantially as described.

Third, the combination and arrangement of the parts C and F, substantially as described.

Fourth, the side packing E, substantially as and for the purpose above described.

No. 59,487.—J. J. B. WALBACH, Baltimore, Md.—*War Rocket.*—November 6, 1866.—The rocket has wings and a percussion point, and an elbow to secure discharge if the point does not collide. A balancing piece on the threaded tail has spiral projections, which causes it to traverse toward the rear under the impulse of the blazing composition, and preserve the equilibrium as the composition is expended.

Claim.—First, the self-adjusting balancing weight C, in combination with the screw tail B, and main body D D', constructed, arranged, and operating in the manner substantially as shown and described and for the purpose set forth.

Second, the combination of the shell F, tubes *ff*, needle *h*, and arm *e*, constructed, arranged and operating in the manner as shown and described and for the purpose set forth.

No. 59,488.—HERVEY WATERS, Boston, Mass.—*Rolling Die Apparatus.*—November 6, 1866.—Combined with a pair of roller dies are contrivances by which the object is presented to and withdrawn from the rolls, and devices which in case the rolls do not return the object as fast as may be demanded by the positive action of said mechanism, will so adjust themselves as to compensate for the difference in the rates of speed.

Claim.—Combining with the rolls, and a cam and weight in connection therewith, as a tongs or holder of the metal to be rolled, a spring so arranged that it may yield to allow the cam to move the mechanism which actuates the tongs faster than the metal held by the tongs is permitted to move by the action thereupon of the die grooves, substantially as described.

Also, the yielding nippers, in combination with the rolls, and any suitable means for working the tongs.

No. 59,489.—WILLIAM H. WATSON, New York, N. Y.—*Machine for Pressing Tobacco.*—November 6, 1866.—The tobacco is fed between revolving aprons, formed of transverse strips of wood, running over corrugated pulleys and slides. A piercer is automatically forced into the plug end at the time it is cut by a guillotine knife. The rest is then tilted, which raises the edge of the tobacco for removal. The counting is done by revolving numbered disks operated by the main shaft.

Claim.—First, the use or employment of a pressing surface formed by combining a series of bars or plates A, constructed and operating as described for the purpose specified.

Second, the combination with the same of a feeding or pressing surface formed by combining a series of bars or plates B, when combined, constructed and operating substantially as described for the purpose specified.

Third, piercing the tobacco, substantially as shown, for the purpose described.

Fourth, constructing the wheels operating the bars or plates A and B, so that they shall force forward the same, as herein fully described, for the purposes set forth.

Fifth, the cutting apparatus, constructed substantially as described, for cutting the tobacco.

Sixth, the combination of the piercer with a movable table for the purposes shown.

Seventh, combining with a tobacco pressing machine a counting or registering apparatus, for the purposes set forth.

No. 59,490.—M. D. WELLS, Morgantown, West Va.—*Cotton-seed Planter*—November 6, 1866.—The seed slide has upward flanges which stir and divide the seed in the hopper. A longitudinal board rises from the bottom and passes through a slot in the seed slide.

Claim.—The reciprocating bar B, constructed as described, and provided with its flanges D D, when used with the hopper A, with false bottom and dividing board C, in the manner substantially as and for the purposes herein set forth.

No. 59,491.—ALBERT WHEELER, Gloucester, Mass.—*Bush Hammer.*—November 6, 1866.—The frame is made of two parts, with shoulders, and with cavities for bolts, and projections from the base for the support of the cutters which are socketed therein.

Claim.—The shoulder plates A B, with their cavities a^2 and b^2, in combination with the projections $a^2 b^2$, as described, and with suitable cutters, and operating in the manner and for the purpose herein described.

No. 59,492.—B. J. WHEELOCK, Bedford, Ohio.—*Fence Gate.*—November 6, 1866.—The gate is slipped longitudinally till it balances on a roller upon the horizontal bar of the pivoted frame, and is then swung open.

Claim.—The crane swing post G, with the arms L L', in combination with the roller M, and gate A, arranged and constructed substantially as and for the purpose specified.

No. 59,493.—WILLIAM G. WILLCOX, Waterloo, Wis.—*Smut Machine.*—November 6, 1866.—The corrugated segments of cylinders which form the beater wings are arranged to present their concave faces, in scoop form, to the grain as they revolve within the casing.

Claim.—The vertical corrugated beater wings, constructed in the form and manner described, and attached to the radial arms in such manner that the wings will have the relative position in regard to the case, substantially as shown and set forth.

No. 59,494.—EDWARD WILSON, Northbridge, Mass.—*Hold-back for Carriages.*—November 6, 1866.—The hold-back hook is in one piece with the toothed clamps which embrace the shafts and are held by a screw.

Claim.—The clasp A, when adjustable in combination with the hook L, and knob B, when constructed and operating in the manner and for the purposes above set forth and described.

No. 59,495.—WYNANT WITBECK, Troy, N. Y.—*Box Setter for Carriage Wheels.*—November 6, 1866.—The radial arms contain sliding strips worked by a rotating cam, and having lugs and set screws to clamp the wheel by its rim. The axial screw-threaded spindle has a sleeve carrying the cutter of the axle-box chamber.

Claim.—The combination of the cam ring D, and radially sliding bars B B B, having lugs C C C for griping a wheel by its rim, with fixed bearings A A A for a side of the rim of the wheel, and central sockets E F for a rotary endwise movable boring spindle, substantially as herein described.

Also, a boring spindle G, mounted in sockets E F, and having its cutter h' arranged between those sockets and fastened in a slot or mortise i in the spindle, by means of a sleeve l, screw nut j, and screw k, on the spindle, substantially as herein set forth.

No. 59,496.—ALBERT G. WOLF, Mystic River, Conn.—*Strapping Blocks.*—November 6, 1866.—Explained by the claim and cut.

Claim.—An iron strap applied around a sheave block so as to clear the ends of the sheave pin on opposite sides of the centre of said pin, substantially as herein described.

No. 59,497.—GEORGE E. WOODBURY, Cambridge, Mass.—*Hand Stamp.*—November 6, 1866.—The axis of the central wheel extends through the plunger; the outer wheels are sleeved upon it on opposite sides and present flanges by which they are manipulated. Each wheel has detent catches, and an ink ribbon is arranged beneath them.

Claim.—The combination and arrangement of the three type cylinders D E F with the shaft of the centre one extending entirely through the plunger, and the shafts of the other two extending through opposite sides of the plunger, and each shaft provided with a hand wheel to turn and set it, substantially as described; and in combination with the above-named devices, the movable ink ribbon, substantially as described.

No. 59,498.—JOHN WOODVILLE, Cincinnati, Ohio.—*Carpenter's Plane.*—November 6, 1866.—The plane is attached to a frame which moves with it and secures its proper presentation to the board. A hinged screw from the guide plate passes to a complex jam nut which is socketed in the portions attached to the plane. Adjustments permit the edge of the board to be made square, or bevelled to any desired angle.

Claim.—The hinged screw rod H, in combination with the parts D G, and provided with the flanged nut I, wrench J, and lock-nut K, when arranged with the jointing plane herein described, substantially as and for the purpose specified.

No. 59,499.—ADAM YOUNG, Millstadt, Ill.—*Cultivator.*—November 6, 1866.—The arched connecting bars are pivoted to the beams and lap past each other within a socket. To the latter is pivoted spring catches, having pins to traverse the socket, and one of the adjusting holes in the arms. The outer handles are made removable.

Claim.—First, the construction of the beams C C', and their combination with the sockets a, or the handle B', as the case may be, for the purpose of forming the connection between two corn ploughs.

Second, the adjustable clamps D, for the purpose of uniting the two parts of the beams C and C', substantially as herein described and set forth.

Third, the braces E' and the staples e, for the purpose of attaching the handles E to the other portions of the plough.

No. 59,500.—E. B. STODDARD, Worcester, Mass., administrator of C. C. COLEMAN, deceased.—*Breech-loading Fire-arm.*—November 6, 1866.—In closing the breech block the hammer is encountered by a pusher which automatically sets the hammer on the half cock. When the breech block is closed, a swell on the hammer disengages the pusher so as to free the hammer for firing.

Claim.—First, the pusher I, or its equivalent, arranged in combination with the hammer and swinging breech block, substantially as, and for the purpose set forth.

Second, the swell A on the hammer, in combination with the pusher I and swinging breech block C, constructed and operating substantially as and for the purpose specified.

No. 59,501.—STILLMAN A. CLEMENS, Chicago, Ill., assignor to self and JAMES J. WALWORTH.—*Machine for Straightening, Breaking, and Cleaning Flax Straw.*—November 6, 1866.—The described devices are for drawing, straightening, and extending a layer of tangled flax so as to reduce it to a thin even sheet of parallel stalks ready for the breaker, scutcher, beater, and picker, to which it is passed directly and continuously.

Claim.—First, in a machine for drawing and straightening tangled flax straw, used either alone or in combination with machinery for breaking or breaking and cleaning flax, the employment of a toothed cylinder or toothed carrier combined with drawing or drawing and breaking rollers of any kind, and with or without a device or devices for impaling the said material upon said toothed cylinder or carrier, when said cylinder or carrier and drawing rollers are adapted to untangle, draw, and straighten said material, for the purposes set forth.

Second, the employment of a flax-impaling device or devices, substantially such as described, adapted to give impalement of the stalks of tangled flax, substantially as described, upon and between the teeth of a toothed cylinder, or its equivalent, or below the said points, to or near their base, when such toothed cylinder is so combined with drawing rollers or fluted drawing and breaking rollers, as to convey or present the said impaled material to the action of said rollers in the said condition of impalement, for the purposes set forth.

Third, the cylinder d and roller g, combined with the adjacent pair of fluted rollers i i, substantially as described and for the purposes set forth.

Fourth, the cylinder d and roller g, combined with a series of pairs of fluted rollers i i i i, substantially as described and for the purposes set forth.

Fifth, the combination of cylinder d and roller g with drawing rollers, either fluted or plain, in combination with one or more pairs of fluted rollers for breaking flax, substantially as described and for the purposes set forth.

Sixth, the rollers j j, in connection with fluted breaking rollers h h, of one or more pairs, the whole operating substantially as described and for the purposes set forth.

Seventh, the combination of the cylinder d and roller g with drawing and breaking rollers and the rollers j j, substantially as described and for the purposes set forth.

Eighth, the combination of cylinder d and roller g with drawing and breaking rollers and the flax-cleaning cylinder l, substantially as described and for the purposes set forth.

Ninth, the combination of the cylinder d and roller g, or any of the described flax-impaling devices, with drawing and breaking rollers and the rollers j j and the flax-cleaning cylinder l, substantially as described and for the purposes set forth.

Tenth, the toothed cylinder d combined with the adjacent pair of fluted rollers h h, substantially as described and for the purposes set forth.

Eleventh, the cylinder d, connected with drawing rollers, either fluted or plain, in combination with fluted rollers for breaking flax, substantially as described and for the purposes set forth.

Twelfth, the combination of cylinder d with drawing and breaking rollers and the rollers j j substantially as described and for the purposes set forth.

Thirteenth, the combination of cylinder d and drawing and breaking rollers, and the flax-cleaning cylinder l, substantially as described and for the purposes set forth.

Fourteenth, the combination of cylinder d with drawing and breaking rollers and rollers j j, with the cleaning cylinder l, substantially as described and for the purposes set forth.

Fifteenth, the combination of the fluted rollers h h, of one or more pairs of rollers j j, the breast k, and cleaning cylinder l, when the whole operate together, substantially as described and for the purposes set forth.

Sixteenth, the cylinder l, combined with the cylinder y and roller b' b', substantially as described and for the purposes set forth.

Seventeenth, the rollers b' b', in combination with the chute bottom u and cylinder g', substantially as described and for the purposes set forth.

Eighteenth, the combination of gill cylinder y, rollers b' b', and the cleaning cylinder g'.

Nineteenth, the combination of the toothed cylinder *d*, top roller *g'*, drawing and breaking rollers *k k h h*, rollers *j j*, breast *k*, cleaning cylinder *l*, concave *s*, chute bottom *u*, gill cylinder *y*, drawing rollers *b' b'*, breast *c'* cleaning cylinder *g'*, concave *d'*, and deflecting board *l'*, all substantially as described and for the purposes set forth.

No. 59,502.—STILLMAN A. CLEMENS, Chicago, Ill., assignor to himself and JAMES J. WALWORTH —*Machine for Straightening and Threshing Tangled Flax.*—November 6, 1866.— The tangled flax is impaled by the spiked roller against the indented roller above, while the drawing rollers straighten the stalk and feed it to the threshing cylinder, which removes the bolls; the latter then fall between the crushing rollers, which release the seed to be cleaned by the vibrating winnower beneath.

Claim—First, in a machine for threshing or for threshing and winnowing and screening the seeds of flax, or other seed-bearing plants, the combination therewith of a drawing and straightening feed device, consisting of a toothed feed cylinder, adapted for impaling the plants upon its teeth, with or without a device or devices for impaling said material upon said cylinder teeth, and with drawing rollers which draw and straighten said impaled material, and also crush the seed bolls, and deliver the material to said threshing machinery, substantially as described and for the purposes set forth.

Second, the combination of toothed cylinder *d*, top roller *e*, hinged cover *f*, shell *g*, drawing rollers *h h*, table *i*, cylinder *k*, breast *l*, concave *m*, chute bottom *q*, chute cover and deflecting board *t*, chute sides *v r*, sieve shoe *w*, levers *z z*, cams *c' c'*, fan blower *e'*, and trunk *f*, substantially as described and for the purposes set forth.

Third, the combination of the pressure crushing rollers *h' h''* with the described machine, substantially as described and for the purposes set forth.

No. 59,503.—STILLMAN A. CLEMENS, Chicago, Ill., assignor to himself and JAMES J. WALWORTH.—*Machine for Breaking and Cleaning Flax.*—November 6, 1866.—The drawing cylinder has alternately long and short teeth, and the roller above has indentations in the peripheries of its annular rings for impaling the stalks. The scutching and cleaning cylinders have flat, narrow, deep teeth, with inclined front edges, slightly hooked at the ends. The teeth of the picking cylinder are made of pointed round wire, and are inclined backward at their bases, their points curving forward. The concave of the cleaning cylinder is formed of parallel curved grate bars, between which pass the teeth of the cylinder. The fibre is forwarded between each cylinder by chute and toothed shell roller.

Claim.—First, the toothed cylinder D, with long and short teeth, substantially as described and for the purposes set forth.

Second, the top roller E, with indentated annular rings *d d*, substantially as described and for the purposes set forth.

Third, the cylinder teeth *e e* and *k k*, substantially as described and for the purposes set forth.

Fourth, the picker teeth *v v*, substantially as described and for the purposes set forth.

Fifth, the cylinder concave O, substantially as described and for the purposes set forth.

Sixth, the hinged cover *e''* combined with the cylinder D, substantially as described and for the purposes set forth.

Seventh, the combination of the toothed cylinder D, top roller E, hinged cover *e*, fluted rollers F F' F'' F''', plain rollers G G', concave I, chute bottom K, shell roller L, cylinder M, breast N, concave O, chute bottom Q, shell roller R, cylinder S, breast T, concave U, grate chute bottom V, and chute cover and deflecting board W, all substantially as described and for the purposes set forth.

No. 59,504.—W. P. COREY, Amsterdam, N. Y., assignor to himself and D. P. COREY, same place.—*Valve Gear.*—November 6, 1866.—The steam is reversed or cut off at any portion of the stroke, by the change of the position of the block in the slotted link; the link is connected at its lower end to the eccentric on the crank shaft, and the block, connected to the rock shaft of the valve rod, slides in the link to make the desired adjustments—being inoperative when over the pivot of oscillation.

Claim.—The arrangement of the link *a*, supporting pin *c*, rod *i*, and sliding block *k*, arm *d*, rock shaft *e*, and the eccentric *f*, relatively to each other, and with the valve, substantially in the manner and for the purpose herein represented and described.

No. 59,505.—E. L. FERGUSON, Buffalo, N. Y., assignor to himself and CHARLES B. CLARK, same place.—*Sash Holder or Fastening.*—November 6, 1866.—The friction roller is contained in a tapering recess within the stile, and becomes jammed between the sash and casing. The axial pivot of the roller projects through a slot, and is attached to the end of a curved lever, by which the roller may be raised. When the sash is closed, a pin upon it holds the curved lever up and acts as a fastening.

Claim.—The friction roller H and loosely pivoted arm J, in combination with the inclined track *g*, and slot *f*, of the plate D, or its equivalent, and sash A, arranged and operating substantially as set forth.

Also, in combination with the above-described device, the recess *u* and pin *c*, operating substantially in the manner and for the purpose specified.

No. 59,506.—E. B. Forbush and Josiah Letchworth, Buffalo, N. Y., assignors to Pratt & Letchworth, same place.—*Snap Hook.*—November 6, 1866.—The spring plays in a longitudinal mortise in the shank, which is pressed down upon the base of the spring to hold it in position.

Claim.—A snap hook having a longitudinal mortise C, for receiving, holding, and protecting its spring B, when constructed substantially as described.

No. 59,507.—W. J. Gordon, Philadelphia, Penn., assignor to John Haworth, Pittsburg, Penn.—*Roller for Finishing Photographs.*—November 6, 1866.—A chamber or plate piece contiguous to the roller is heated by a lamp beneath.

Claim.—The mode of communicating heat to the cylinders or rollers of photographic or other presses by means of the application of heat to a thin metallic plate or heating box which is placed in close proximity to said rollers or cylinders, as described in the accompanying drawing, or any other substantially the same.

No. 59,508.—Benjamin Hurlburt, Milford, Conn., assignor to L. H. Holt, Hartford.—*Construction of Carriage Seats.*—November 6, 1866.—The back and sides are formed of one plain piece of timber bent in form. The base of the seat is formed in the same manner.

Claim.—A carriage seat in which the back and ends with rounded corners are formed from a single piece and in continuous grain of the wood, substantially in the manner described.

Also, the base C, formed in continuous grain as herein set forth.

No. 59,509.—Stephen R. Krom, New York, N. Y., assignor to Louis F. Therasson, John A. Bryan, James M. Blackwell, and Appolos R. Wetmore, same place.—*Apparatus for Separating Metals from Ores.*—November 6, 1866.—The cylindrical casing is supported upon a frame above the carriage, which travels upon the railway. The casing contains an annular bellows operated by pitmans, and the tube from the hopper passes down into the casing through the passage in the bellows. The carriage supports a tilting ring which is secured by trunnions and contains a grate and a perforated plate. The lower edge of the cylindrical casing has a rubber packing to make a tight joint with the ring.

Claim.—First, operating the feed valve G by means of the double links J J', and its connections, substantially as and for the purpose herein specified.

Second, opening and closing the joints between the fixed ring a and the upper face of the travelling ring C, substantially as and for the purpose herein specified.

Third, the sharp edge of the ring C, when arranged and operated substantially as and for the purpose herein specified.

Fourth, varying the depth of the stratum retained on the perforated bed D, by the employment of the movable rings C², arranged relatively to the bed D and to the shoulder, substantially as and for the purpose herein specified.

Fifth, mounting the bellows G' in close proximity to the bed D, substantially as and for the purpose herein specified.

Sixth, the carriage B and crank M in combination with the bed D and ring C, and with a suitable intermittent suction device, substantially as and for the purpose herein specified.

No. 59,510.—Stephen R. Krom, New York, N. Y., assignor to Louis F. Therasson, John A. Bryan, James M. Blackwell, and Appolos R. Wetmore, same place.—*Apparatus for Separating Metals from Ores.*—November 6, 1866.—The sieve ring is secured by adjustable links to a frame, and rests upon a casing which is contracted beneath, and is attached to a bellows operated by rods and gearing. Around the upper casing is a revolving trough which receives the light particles of ore as they are blown out of the sieve; and the trough revolves the heavier particles are crowded against the scraper and thrown out.

Claim.—First, producing a variable aperture through which the blast produced by the bellows may be discharged so as to reduce the action through the sieve D, as required, substantially in the manner and for the purpose herein set forth.

Second, contracting a portion A³ of the casing between the bed D and the bellows E, substantially as and for the purpose herein specified.

Third, the inclined rods F¹ F², cranks G¹ G², and connecting gear G, in combination with a bellows, and adapted to be used for separating ores and analogous uses, substantially as herein specified.

Fourth, mounting the supporting links B B on centres one side and not under the trunnions c of the ring C, substantially as and for the purpose herein specified.

Fifth, the rotating pan or vessel R, arranged to operate in combination with the scraper s, and the bed D, and ring C, substantially in the manner and for the purpose herein set forth.

No. 59,511.—Josiah Letchworth, Buffalo, N. Y., assignor to Pratt & Letchworth, same place.—*Snap Hook.*—November 6, 1866.—A longitudinal mortise in the middle portion of the shank protects the body of the spring, and holding lips below the mortise are battered down upon the base end of the spring.

Claim.—The combination of the longitudinal mortise A with the lips b, made below the mortise, so as to show a recess or depression g, or with the rivet d, for the purpose and substantially as described.

No. 59,512.—JOHN F. MILLIGAN, St. Louis, Mo., assignor to JOSEPH W. BRANCH and JOSEPH CROOKS, same place.—*Cotton Tie.*—November 6, 1866.—The buckle is riveted to one end of the hoop and has an angular slot through which the other end is passed to be held by its bent shape and the pressure of the cotton.

Claim.—A tie plate B, provided with the pointed retaining projection or stop *b*, when combined with an oblique slot *c*, to receive and secure the free end of the hoop, all sub-stantially as and for the purposes herein described.

No. 59,513.—EDEN T. ORNE, Chicago, Ill., assignor to himself and JOHN P. HART, same place.—*Tool for Opening Cans.*—November 6, 1866.—The adjustable cutter is attached by screw and nut to a sharp pointed stationary cutter, so as to bring a new portion of the edges of each to bear upon the tin as the tool becomes worn.

Claim.—The adjustable cutter C, when constructed to operate against the stationary cutter A, substantially as and for the purpose set forth.

No. 59,514.—FRANK J. PLUMMER, Worcester, Mass, assignor to R. BALL & Co., same place.—*Feed Roller for Planing Machines.*—November 6, 1866.—The boxes of the upper rollers have a hinge connection to the vertical screws by which they are adjusted and are operated simultaneously by gear connection. The roller is drawn down by weighted levers connecting to its boxes, so as to allow a limited vertical action to each end apart from the other.

Claim.—First, adjusting the top feed roll of a planing machine by means of screw rods respectively hinged to the journal boxes of such roll and working in internal screw cylinders so attached to the gear wheels as that, while revolving with the gear wheels, they shall have a sliding or vertical motion independent therefrom, substantially as described and for the purpose set forth.

Second, the combination with the internal screw cylinders attached to the gear wheels, as described, and receiving the screw rods hinged to the feed roll of the lever weights, sub-stantially as and for the purposes set forth.

No. 59,515.—JOHN N. POND, Wakefield, Va., assignor to A. W. HOLT and JOHN L. WHITE, same place.—*Plough.*—November 6, 1866.—The rectangular cutter is presented above the cutting edge of the breast and attached to the beam by levers, which have slots and screws for adjustment.

Claim.—The rectangular cutter A, when arranged, combined, and operated by adjustable levers B and C, to be attached to any ordinary plough, as herein described and for the purposes set forth.

No. 59,516.—JAMES L. PRESCOTT, North Berwick, assignor to himself and S. B GOWELL, Portland, Me.—*Portable Cupboards.*—November 6, 1866.—The frame can be readily taken apart and packed, is fitted with detachable shelves, covered with gauze, and furnished with a bail or handle.

Claim.—The combination of the detachable frame A D, removable shelves C, bail E, handle F, and netting or covering O, all arranged in the manner and for the purpose speci-fied.

No. 59,517.—CHAUNCY SPEAR, Chapinsville, N. Y, assignor to himself, HOLMES C. LU-CAS, Canandaigua, N. Y., and WALTER MARKS, Hopewell, N. Y.—*Car Coupling.*—No-vember 6, 1866.—The extended head of the pin engages a notch in the standard, and is thus held out of contact with the entering link; a jar drops it out of the notch and the pin couples with the link in the draw-head.

Claim.—The combination of the pin D with an extended head or plate F, and the notched frame E, operating as described to temporarily hold the pin and release it by the concussion of the cars.

No. 59,518.—AARON C. VAUGHAN, Philadelphia, Penn., assignor to himself and R. W. PARK, same place.—*Lamp Burner.*—November 6, 1866.—The top of the Argand burner has an adjustable perforated cap, whose upper edge projects inward, a rim extending to the inner edge of the wick. The regulation of the flame is effected by a detachable section of wick tube.

Claim.—First, the combination of the tubular wick, with the perforated casing E and flange *e*, the whole being arranged substantially as and for the purpose described.

Second, the wick tubes B and C, with their tubular wick, in combination with the flange *e*, rendered adjustable to and from the top of the wick tubes, substantially as described.

Third, the detachable continuation *d* of the wick tube B, for the purpose described.

No 59,519.—LEON CARRICABURU, Havana, Cuba.—*Steam Pump.*—November 6, 1866.—The valves are moved by the contact of the piston with projections extending through the ports of the cylinder.

Claim.—The valves F or F′, furnished with toes or projections extending through the ports of the cylinder, so as to cause the valves to be actuated by the piston, substantially as described.

No. 59,520.--CYPRIER MARIE ESSIE DU MOTAY and CHARLES RAPHAEL MARÉCHAL, Metz, France.—*Photographic Process.*--November 6, 1866.—The object is to combine certain chemical compositions more freely acted on by the light, and which, by having a greater affinity for fatty bodies, are better adapted to receive finishing. For this purpose salts of chromic acid more acid than the protochromates and bichromates are used, with the addition of gelatinous or cognate solutions and a covering of soap of silver.

Claim.—The new process for the production of photographic images, capable of being nked with fatty inks, substantially as herein described.

No. 59,521.—JOHN K. FARNWORTH, Alderley Edge, England.—*Sash Fastening.*—November 6, 1866.—A turn button in the window-case has links on its shank connecting with pivoted spring hooks; by turning the button the hooks are disconnected with the notches in the counter-balanced sash, which may then be raised.

Claim.—The lever handle *j j′*, link *s*, and spring catches *q*, in combination with the racks *g* in the edges of the movable sash *d*, as and for the purpose set forth.

No. 59,522.—PIERRE FLAMM, Phlin, France.—*Typography.*—November 6, 1866.—The matrix is formed by types in a revolving disk, which are pressed into it by the movement of a lever, and raised by a spring. The type disk is axially connected to an index disk, by the revolution of which the proper type is brought under the depressor. The matrix frame has longitudinal and vertical adjustment, and the return action of the depresser-lever moves it the space of one letter.

Claim.—First, the combination of the mechanism shown and described for impressing the type in the mould with those which regulate the transverse movement of such mould, so that they may be actuated or operated by the same lever or equivalent means, substantially as shown and set forth.

Second, the combination of the mould and ratchet frame E with the pawl which engages with such ratchet, mounted on a rock shaft as described, and operated substantially in the manner and for the purposes herein shown and set forth.

No. 59,523.—JAMES HIGGINS, Manchester, England.—*Ring for Ring and Traveller Spinning Machines.*—November 6, 1866.—The object of the invention is to obtain a surface of metal of uniform texture, and therefore less liable to a variation in the drag. The ring is cut and fashioned from a seamless tube.

Claim.—As a new article of manufacture the seamless ring herein described for ring and traveller spinning cut from a tube or rod, and finished by swaging or turning, all as specified.

No. 59,524.—DAVID JOY, Middlesbrough, England.—*Valve for Steam Hammers.*—November 6, 1866.—The valve is actuated by steam from the cylinder, and the length of stroke changed by means of a valve, which is adjusted to admit steam from below the piston to the valve chest, so as to move the valve and depress the piston. One of several ports may be thus opened, and thus determine the height to which the piston shall be raised before descending.

Claim.—First, in hammers where steam or other fluid which actuates the hammer is used to move the valve without the use of levers, cams, tappets, or links, regulating the action of the valve so moved by the early or late opening of the port or hole, admitting the pressure upon it by means of the slide O, substantially as described.

Second, regulating the force of the blow of the hammer by means of the holes *m*′ in the cylinder and valve chest, and the channel which connects them, substantially as described.

No. 59,525.—G. H. and E. MORGAN, Edgeware Road, England.—*Carriage.*—November 6, 1866.—The levers, &c., by which the head of the carriage is raised or lowered are contained within the head and lining, or within the frame, so as to be unexposed, and are operated by a lever pivoted near the driver's seat.

Claim.—First, the placing the head joints *b*, or their equivalents, inside of the head of a carriage, and hid by the lining, substantially as herein shown and described.

Second, the employment of mechanism connected to the head joints *b*, or their equivalents, of a carriage, in such manner that the head of a carriage, whether in one or more parts, may be capable of being raised or lowered by a person on the driver's seat or other suitable part of a carriage, acting upon a lever or screw or other equivalent means, in manner substantially as herein shown and described.

No. 59,526.—G. H. and E. MORGAN, Edgeware Road, England.—*Carriage.*—November 6, 1866.—The head is raised and lowered by levers actuated by a screw shaft at the driver's seat. The levers, toggles, &c., are hidden between the cover and lining. Details are cited in the claims.

Claim.—First, the application of a head or cover to a wagonette or other similar vehicle, capable of being raised or lowered as desired. substantially as herein shown and described.

Second, the application to wagonettes or other similar carriages of means or apparatus for raising and lowering the head or cover thereof, which apparatus is capable of being put in motion from the driver's seat or other suitable part of the carriage, substantially as herein shown and described.

Third, the mode of applying the mechanism for raising and lowering the heads or covers of wagonettes and other similar vehicles between the cover and the lining of the carriage, substantially as herein shown and described.

Fourth, the mode of applying side lights *p* to the heads or covers of wagonettes constructed according to this invention in such manner that they shall be capable of rising and falling with the heads or covers thereof, and be guided in their motion in suitable guides, substantially as herein shown and described.

Fifth, the mode of giving motion to the upper parts *a' a'* of the heads or covers of landaus and other similar carriages, substantially as herein shown and described.

Sixth, the mode of constructing the connecting rods *c c* and *e e*, when applied to landaus or other carriages, in two parts connected together so as to afford facility for adjustment, substantially as herein shown and described.

Seventh, the mode of connecting together the connecting rods *g g*, so as to form a rigid frame by means of rods or bars provided with screws at their ends, and fixed to the connecting rods by lock nuts, substantially as and for the purpose herein shown and described.

Eighth, the mode of supporting and working the screw by which motion is given to the apparatus for raising and lowering the heads of carriages, substantially as herein shown and described.

Ninth, the mode of communicating motion from the screw *k* to the connecting rods *g g*, and of limiting the amount of motion in either direction of the nut *i*, substantially as herein shown and described.

No. 59,527.—RICHARD H. OATES, Toronto, Canada West.—*Clothes Dryer.*—November 6, 1866.—The bars of the stretcher frame are socketed in a hub whose base piece forms a cap to shed rain. The inner post slides into the outer one, and is elevated by rack, pinion, and pawl.

Claim.—First, the combination of the casing A, post B, rack E, pinion H, pawl J, and cap C, with the revolving cross-arms M, which carry the clothes line, all arranged and operating as herein set forth.

Second, in combination with the posts A B, the roof casting C, to prevent the rain from beating in between the inner and outer posts, substantially as described.

No. 59,528.—EMILE PEUGEOT and J. B. B. C. LAURENT, Paris, France —*Grinding Mill.*—November 6, 1866.—The cylinder that surrounds the grinding cone is slit entirely asunder, so that it will accommodate itself to the interior cones, and gives way to any substance that cannot be crushed and ground, avoiding fracture.

Claim.—The gap *g* in the concave F, substantially as and for the purpose set forth.

No. 59,529.—BENOIST ROUQUAYROL, Paris, France.—*Regulating the Flow of Gases in Apparatus for Diving* —November 6, 1866.—The vessel for containing compressed air has a chamber with an elastic side which is connected with a valve by a stem. As the amount of air in the chamber is used by the diver, the collapse of the disk opens the valve, and a new supply flows out from the vessel. The density of the air in the chamber and the diver's lungs is always equal, both being subjected to pressure proportioned to the depth. The apparatus is strapped to the back of the diver, and the air is inhaled through a mouth piece.

. *Claim.*—First, the apparatus or regulator, substantially as herein described, the same being composed of a compressed air reservoir, surmounted by an air chamber, the latter being provided with an elastic cover, in the centre of which is placed a regulating rod which acts on the valve, separating the two chambers in such manner as to permit the air from the reservoir to pass in greater or less quantity into the air chamber, according as the elastic cover of such chamber is subjected to more or less pressure.

Second, in the apparatus herein described, the combination with the air reservoir of two regulating chambers for producing a constant and regular flow or circulation, substantially as set forth.

Third, the construction of the mouth closer and valve of expiration, substantially as and for the purposes herein shown and set forth.

No. 59,530.—T. STOREY. Lancaster, England, and W. V. WILSON, London, England.—*Manufacture of Leather Cloth.*—November 6, 1866.—The cloth is covered with solutions of the aniline dyes.

Claim.—The application and use, to and in the manufacture of what are known as American leather cloth goods, of coloring matters, of the nature hereinbefore described.

No. 59,531.—MICHAEL BARRETT, Toronto, Canada West.—*Recovery and Purification of Sulphuric Acid used in Refining Petroleum, &c.*—November 6, 1866.—The waste acid from the agitator is drawn into leaden tanks, and the remaining oil is separated. It is treated while hot with black oxide of manganese, which removes tarry matters; clarified by treatment with clay or fuller's earth. The weak acid obtained is then concentrated in the usual manner.

Claim.—The recovery, purification, and revivification of the sulphuric acid spent and deteriorated in the process of refining petroleum, coal, and shale oils, by means of oxygen gas in the nascent state, by whatever means developed or obtained.

No. 59,532.—GEORGE L. WITSIL, Philadelphia, Penn., assignor to himself and WILLIAM DARMANN, same place.—*Washing Machine.*—November 6, 1866.—The cubical washing vessel revolves upon journals whose axis passes through opposite corners of the cube; the box is provided with diagonal ribs on its inner surface.

Claim.—A washing machine consisting of a revolving cubical box A, with internal ribs B placed on each face of the cube, the middle one diagonal and the others parallel therewith, closed by a door F', when constructed and arranged substantially as set forth.

No. 59,533.—WALTER H. FORBUSH, Buffalo, N. Y., assignor to HENRY G. LEISENRING, Philadelphia, Penn.—*Railway Ticket Printing Press.*—November 6, 1866.—The bed plate is rigidly fixed to the frame. The platen carries the type in two chases to print in different colors, and it receives motion from a connecting rod to a crank on the main shaft. A numbering wheel on the platen frame receives rotation from a pawl operated by a fixed arm. The paper is carried over the bed plate by connecting bars to two endless chains which have intermittent motion from a pawl worked by a cam on the fly wheel. The inking rollers take ink from rollers at each end of the platen and coat the type during the upper parts of the ascent and descent of the p'aten; they are worked by cams from the main shaft. The nippers are raised for the taking and discharge of paper at each end of the platen. The tickets receive the different colors at different impressions, being automatically forwarded during the elevation of the platen.

Claim.—First, the combination of the wedge openers I', or their equivalent, with the nippers G', having an intermittent feed movement, constructed and operated substantially as described.

Second, the draw-out fingers L³, operating in connection with the wedge openers, to remove the sheets from the nippers G', substantially as set forth.

Third, the grooved slides O, attached to the platen parallel to the face of the form, and carrying the inking roller or rollers over the form, in the manner and for the purpose substantially as described.

Fourth, the arrangement of the inking roller cams N⁴, radius arms N⁵, supported upon the vibrating levers N⁶, and carrying the "form" inking rollers N, combined with the grooved slides O, and permanent bearers O³, so that the inking rollers will reach the limit of their forward vibration at the same time the platen reaches the limit of its upward movement, and so that both platen and inking rollers change the direction of their movement at the same time, substantially as described.

Fifth, the fountain ink rollers in combination with segments P, which receive their motion from one of the cranks D², for the purpose and substantially as set forth.

Sixth, attaching the platen to the cross-bead by the combined suspension bolts C', and impression screws C⁶, in the manner and for the purpose set forth.

Seventh, the clamp bars J' and L⁴, (either or both,) arranged and operating substantially as described and for the purpose specified.

No. 59,534.—CHARLES E. ABBOTT, Malden, Mass.—*Lamp Extinguisher.*—November 13, 1866.—An extinguishing plate is hinged to an exterior sleeve which is moved upon the wick tube by a lever. When the sleeve is raised above the wick tube the lid falls upon the top of the wick and extinguishes the flame.

Claim.—The tube c, with its lid or cover d, operated by the lever A, substantially as set forth.

Also, pivoting the lid d at a point above its upper surface so as to insure its falling by its own weight, substantially as set forth.

No. 59,535.—NATHAN ADAMS, Altoona, Penn.—*Machine for Drawing Spikes.*—November 13, 1866.—The gripper jaws are suspended from a lever which has a pivoted fulcrum point upon a post.

Claim.—The combination of the lever B, fulcrum post I, and the guide pin J, with the stock A, guide plate and rod D, and jaws or nippers F, substantially as described and for the purpose set forth.

No. 59,536.—M. J. ALTHOUSE, Waupun, Wis.—*Pump.*—November 13, 1866.—The pump bucket has packing disks expanded by water admitted through passages above. Valves control the said induction.

Claim.—First, providing the apertures *e e*, leading to chamber *b*, containing an expansive packing with valves *g g*, substantially as described.

Second, the combination of means for regulating the inflow of water through apertures *e e*, with the spring *c* and expansible rings *a a*, substantially as and for the purposes described.

No. 59,537.—M. J. ALTHOUSE and P. REIFSNIDER, Waupun, Wis.—*Attaching Thills to Carriages.*—November 13, 1866.—The thill iron has two cheeks, both perforated to receive the cross-head of the thill iron, and one slotted to allow the narrow part of the iron to pass in as the cross-head is moved endways into place while the thills are held erect.

Claim—The thill iron A, provided with the cross-head *c*, in combination with the clip B, provided with the eyes *b*, one of which has the notch and hinge piece *a*, arranged to operate as set forth.

No. 59,538.—ALBERT ANGELL, Newburgh, N. Y.—*Coffee Huller.*—November 13, 1866.—The coffee passes between the roughened surfaces of the cylinder, which revolves within a concave and the roughened surfaces of the spring pads which are attached to the casing.

Claim.—First, the combination with the roughened or serrated hulling cylinder B, of independent spring strippers D, arranged within a hollow segment or trough partly encircling the cylinder, substantially as specified.

Second, the divided spring strippers D, constructed with roughened or serrated fronts, arranged side by side and in a series, one in advance of the other, within a hollow segment or trough C, to which they are secured at their one end for operation in combination with a serrated or roughened hulling cylinder B, essentially as shown and described.

No. 59,539.—T. H. ARNOLD, Troy, Penn.—*Horse Hay Fork*—November 13, 1866.—Guards or retaining arms are rigidly fastened to the main bar which has slotted prongs operated by a lever and arm; when the latter is erect the prongs are in line with the bar and in position for entering the hay, and conversely.

Claim.—The slotted prongs E F, in combination with the bar C, arm D, slotted bar A, and arms G, arranged and operating in the manner and for the purpose specified.

No. 59,540.—JOSEPH N. ARONSON, New York, N. Y.—*Breech-loading Fire-arm.*—November 13, 1866—The sliding block is placed between the nipple and the rear of the barrel and is raised and lowered by a hinge connection with the oscillating trigger guard. When raised the rear of the opening in the block is exposed to receive the flanged cartridge; is fired by a spring pin depressed by the hammer and the shell ejected by a spring piece which pushes against its flange when the breech block is raised.

Claim.—The sliding breech E, in combination with the firing pin or needle O, the cartridge shell discharger H, and the lever I, operating in the manner substantially as and for the purposes described and set forth.

No. 59,541.—CHARLES AUSTIN, Concord, N. H.—*Machine for Stamping Reed Plates.*—November 13, 1866.—A sheet of metal is converted into a series of reed plates formed with sockets for holding the reeds. The plate rests upon dies on the bed and the plungers descend vertically upon it; one is adjustable towards and from the other and cuts the slot, while the other has the shape of the plate without the slot. The bed die of the adjustable plunger die is correspondingly adjustable.

Claim.—The combination and arrangement of the gauge *l*, the two sets of male and female dies and adjustable die carriers applied to a bed and plunger, so as to operate substantially as and for the purpose set forth.

Also, the combination and arrangement of the three adjustable die carriers E G with the bed and plunger, and its fixed or larger die, such die carriers being provided with mechanism for adjusting them, substantially as set forth.

No. 59,542.—D. B. BAKER and P. S. MILLER, Rollersville, Ohio.—*Pulley Suspension Hook.*—November 13, 1866.—Intended to facilitate hanging the pulley of a horse hay fork from a rafter of difficult access. The socket is mounted on the end of a pole, the stop holding the hook in inclined position ready to slip over the rafter; the pole is then withdrawn from the socket.

Claim.—An improved pulley suspension hook formed by the combination of the double hook A, arm B, stop D, and socket C, with each other, the said parts being constructed and combined substantially as herein shown and described and for the purpose set forth.

No. 59,543.—W. R. BALDWIN, Philadelphia, Penn.—*Corn Planter.*—November 13, 1866.—The beams of the plough, planter, and coverer, are attached to a frame in the wheeled carriage and are simultaneously raised or lowered into working position. By gearing from the driving wheel the seed slide is reciprocated, dropping seed through two chutes, and the cover is intermittingly reciprocated to draw earth upon the seed.

Claim.—First, the reciprocating plates *p*, in combination with a plough N, tubes *g*, and with the within described devices or equivalents, for measuring and discharging the grain, the whole being constructed and operating substantially as and for the purpose described.

Second, the boxes K with their openings *i i*, in combination with the slides *m* and their openings *n*, when the latter are of the form described for the purpose specified.

Third, the frame F with its boxes K K, ploughs N N, crank shaft P, and pinion *a*, in combination with the frame C and cog wheel *w*, the whole being constructed and operating substantially as set forth.

No. 59,544.—IRA S. BARBER, New York, N. Y.—*Mouth-piece for Cigars.*—November 13, 1866.—Improvement on the patent of Jonathan Ball, March 25, 1865. The wooden tube has a paper socket and an enlarged opening at its inner end to expose a larger portion of the end of the cigar.

Claim.—A cigar mouth-piece composed of a paper socket and a wooden tube having its longitudinal orifice *c* terminated at its inner end in a recess or chamber *a*, bound by a shoulder *b*, substantially as and for the purpose herein set forth.

No. 59,545.—GEORGE F. BARDEN, Dover, N. H.—*Clothes Pin or Clamp.*—November 13, 1866.—The jaws are drawn together by the bolt and spring; the latter is compressed as the jaws open.

Claim.—The combination of the rubber cushion D and double-headed spindle G with the arms B, arranged in the manner and for the purpose specified.

No. 59,546.—BENJAMIN F. BENNETT, Lockport, N. Y.—*Spring Bed Bottom.*—November 13, 1866.—The ends of the slats are suspended from spiral springs enclosed in cases attached to the head and foot boards.

Claim.—The special arrangement of parts as herein set forth, viz: the cases C screwing to the bedstead and enclosing the springs *g*, the shanks *f* resting therein upon the springs, and the hook *d* and locps *b* connecting with the slats, the whole operating in the manner and for the purposes specified.

No. 59,547.—JOHN BLAIR, St. Louis, Mo.—*Combined Poker and Tongs.*—November 13, 1866.—Explained by the claims and cut.

Claim.—First, a combined poker and tongs made substantially as herein shown and described.

Second, the combination with the rigid bar B of the movable jaw C, rod D, and lever E, substantially as herein shown and described and for the purpose specified.

Third, the combination of the spring catch F with the lever E and handle A, substantially as and for the purpose herein shown and described.

No. 59,548.—JAMES C. BOWE, Urbana, Ohio.—*Awning.*—November 13, 1866.—The awning is attached by its edge to the house and is wound upon a roller which is pivoted to a frame hinged to the jambs of the store front. A sign board is suspended from the frame so as to hang below the awning roller.

Claim.—The combination of the adjustable frame, sign board, canvas, and roller with pulley and cords, constructed and working as herein described.

No. 59,549.—A. D. BOWMAN, New York, N. Y.—*Compound for Making Writing Ink.*—November 13, 1866.—Composed of extract of logwood, 32 parts; bichromate of potash, 6; prussiate of potash, 1; pulverized, mixed with water, and shaken till thoroughly commingled.

Claim.—A compound for making writing ink composed of the ingredients substantially as herein specified.

No. 59,550.—LEVI BROWN, Evans, N. Y.—*Confining Cows while being Milked.*—November 13, 1866.—The two stakes, connected by straps behind the buttocks, are planted crosswise, and occupy positions against the flanks so as to prevent the forward motion of either hind leg or the falling of the cow.

Claim.—The stakes D and D', placed and supported in suitable holes in the stable or stall floor, with or without the strap F and rope G, for the purpose and substantially as described.

No. 59,551.—WILLIAM H. H. BURNHAM, East Homer, N. Y.—*Roofing Cement.*—November 13, 1866.—Composed of coal tar, 1 barrel; lime, 18 pounds; sand, 3 bushels, and wood ashes, 2 pecks. Other proportions for other uses.

Claim.—The within-mentioned ingredients, coal tar, quicklime, quicksand, and ashes, when mixed and used in the manner and for the purpose specified.

No. 59,552.—JOHN BURT, Sturgis, Mich.—*Turning Lathe.*—November 13, 1866.—The hollow revolving mandrel has a cutting knife attached, and runs on hollow arbors; the arbor in which the stick is fed is square and the other has a round hole to receive the stuff after it is turned.

Claim.—First, a hollow arbor so constructed that only the cutter or bit comes in contact with the stick to be rounded.

Second, the hollow bearings B and C, so arranged that one shall receive the square stick and hold it from turning while being rounded, while the other shall receive the stick after being rounded and hold it steady and true, substantially as herein shown and described.

No. 59,553.—SAMUEL G. CABELL, Quincy, Ill.—*Door and Shutter.*—November 13, 1866.—A double course of slats stand angularly to each other. Gauze may be interposed between.

Claim.—First, the arrangement of slats in the panels of a door or shutter either stationary or pivoted, so that they may form a series of Vs slightly overlapping each other, substantially as and for the purpose set forth.

Second, in combination with the V-shaped slats c, the arrangement of woven wire inserted on a plane with the frame of the door or shutter and intersecting the angles of the slats, substantially as herein specified.

No. 59,554.—GEORGE J. CAPEWELL, West Cheshire, Conn.—*Lamp Burner.*—November 13, 1866.—The wick tube is lowered for the purpose of lighting the wick, and when raised is sustained by a spring; the door in the side of the burner is opened for the access of the match by the same motion which lowers the wick tube, and conversely.

Claim.—First, the combination of the slot a, door b, one or more, with the ratchet shaft C, wick D, burner A, and cylinder B, substantially as described, for the purpose specified.

Second, the spring F, for holding the wick tube D, when arranged so that its lower end catches under the lower end of the tube when the tube is raised to its highest point, substantially as and for the purpose specified.

No. 59,555.—GEORGE W. CARPENTER, Medina, Mich.—*Feed Apparatus for Threshing Machines.*—November 13, 1866.—This feeding apparatus is to be attached to the feed chute of a thresher; it has band cutting knives and a spiral flanged roller for spreading the sheaf equally over the apron which carries it to the cylinder.

Claim.—The roller a a, and its band cutters a a, the cylinder B, and its spreaders b b, the roller C, and its teeth c c, the threshing belt D, and the wing gate E, combined, arranged and connected with a threshing machine for feeding the same, substantially as herein described

No. 59,556.—HERMAN H. CHRISTIE, Herkimer, N. Y.—*Pessary.*—November 13, 1866.—Explained by the claim and cut.

Claim.—The curved pear-shaped tube a b c d, perforated at the upper end a, and opened at the lower end within the flange or rim e e, by removing the handle g h, constructed and used in the manner described in this specification.

No. 59,557.—JOHN COFFEY, Monroe, N. Y.—*Plough for Cutting Bogs.*—November 13. 1866.—The pronged sole plate has cutters attached beneath, and is followed by a mouldboard attached to the standard. A coulter may be placed on the left prong of the sole plate.

Claim.—The sole plate D, provided with the prongs a a', with the cutters F F attached, in connection with the beam A, substantially as and for the purpose specified

Also, the mouldboard E, in combination with the sole plate D, cutter F F and either with or without the cutter or coulter G, for the purpose set forth.

No. 59,558.—DE LANCE COLE, Marshall, Ill.—*Sash Fastener.*—November 13, 1866.—The slots run in such directions that the gravitation of the sash, when the plate is held to the stile, will support it. When the window is shut the gravitation of the plate fastens it.

Claim.—The slotted plate H, when hung upon a pin or stud D, of a sash frame, as and for the purpose described.

No. 59,559.—JOHN CONRAD, Centralia, Ill.—*Corn Planter.*—November 13, 1866.—The seed slide operates by a cam wheel on the axle and a return spring. The furrow openers are raised or depressed by a lever.

Claim.—First, operating the perforated seed slide F, from the axle C, through the medium of the lever L, cam K, ratchet wheel J, pawl d', and spring I, arranged substantially in the manner as set forth.

Second, the adjusting or raising and lowering of the shoes or furrow openers M M, through the medium of the rods N N, bar O, and lever P, all arranged substantially as shown and described.

No. 59,560.—WM. M. COOK, Lyons, Iowa.—*Churn.*—November 13, 1866.—The oscillating churn box is supported upon arms and its motions assisted by a spring planted in the base. Stops limit the motion.

Claim.—The churn vibrating upon an axis in combination with a vertical reacting spring planted upon the frame and engaging with the churn substantially as described.

Also, in combination with the above, the deflecting surfaces G and the bolt K.

No. 59,561.—PERLEY D. CUMMINGS, Portland, Me.—*Machine for Rolling File Blanks.*—November 13, 1866.—The blank is clamped between two jaws, one of which is pivoted so as to open and close by the action of a toggle which is actuated by a slide on the frame and a cam on the main shaft. The jaws remain stationary while a roller above is operated by a sliding rack connected by a pitman to a crank on the main shaft, a pinion on the roller engaging teeth on the rack and others on the immovable jaw. The shaft is unclutched after each revolution.

Claim.—First, the combination and arrangement of the wheels A B C, spring *a*, bolt *k*, rod *k*, shaft *i*, crank *l*, connecting rod *m*, and sliding rack *o*, all constructed and operating as and for the purposes hereinbefore set forth.

Second, in combination with the subject of the first claim, the combination of the sliding rack *o*, geared roller *g'*, tracks *k' l'*, with the inclined channel between the same, as and for the purposes set forth.

Third, the combination and arrangement of the wheels A B C, and rod *k*, operating as described, cams 1 and 2, with thrusting beam *r*, toggle *s*, jaw *t*, levers *b' z*, bolt *w*, spring *v*, and spring *r*, all operating as and for the purposes set forth.

Fourth, the combination and arrangement of the screw *e'*, in the projection *y*, with the screw *d'*, on the toggle *s*, as and for the purposes set forth.

Fifth, the combination of the part *h*, sliding rack *o*, geared roller *g'*, tracks *k' l'*, and projection 3, on the roller, for the purpose of submitting the blank to the necessary pressure.

No. 59,562.—EPHRAIM CUTTER, Woburn, Mass.—*Dental Anæsthetic Instrument.*—November 13, 1866.—Each tube is bifurcated so as to reach the inner and outer sides of the jaw simultaneously. The straight tube carries the air blast and thus draws a current of liquid whose rapid evaporation produces cold and local anæsthesia. The lower end of the bent tube is dipped in the liquid and it discharges at its end while the air tube discharges laterally just in advance of it, producing a spray of the liquid.

Claim.—In combination with the tube *a*, having its orifice directly in the end thereof, the tube *b* having its orifice opening laterally directly from the tube, substantially as described.

Also, the bifurcated construction or arrangement of the nebulizing tubes *a b*, substantially as described, when the orifices are arranged in the manner set forth.

No. 59,563.—HENRY G. DAYTON, Maysville, Ky.—*Coal Stove.*—November 13, 1866.—A reverberating fire chamber above the fire pot is supported on the annular base of the air chamber around it, and the latter by lugs on the rim of the fire pot.

Claim.—The arrangement above the fire box K, and within the air-heating chamber C, of the reverberating chamber A, supported upon the plate B, substantially as and for the purpose described.

No. 59,564.—P. S. DEVLAN, Jersey City, N. J.—*Lining for Journal Boxes.*—November 13, 1866—A pasteboard, one-quarter or half inch in thickness and suited in size to the recess of the journal box, is saturated with petroleum or other lubricating oil and subjected to heavy pressure at the time of insertion.

Claim.—Lining journal boxes and other rubbing surfaces with pasteboard saturated with lubricating oil, and then compressed, substantially as and for the purpose described.

No. 59,565.—J. L. DICKINSON, Dubuque, Iowa.—*Valve Device for Steam Engines.*—November 13, 1866.—The throw of the valve is regulated by the length of the arm connecting to the eccentric, and the said length is determined by the governor. The frustal steam valve has a follower at the head and centre screw opposing. The outer end of the valve stem traverses a thimble adjustable in the follower.

Claim.—The follower E, the thimble box F, and the sliding arm K, constructed and arranged substantially as herein set forth, in combination with the governor valve of a steam engine.

No. 59,566—JAMES V. DUNLAP, Hartford, Conn.—*Lamp Shade.*—November 13, 1866.—The spring fingers, which clasp the chimney quadrangularly, consist of two bent wires bowed outward at the top and joined by their ends to the circle in the upper edge of the shade.

Claim.—The shade holder formed with wire springs that are made in pairs united at their upper ends, in the manner and for the purposes set forth.

No. 59,567.—L. H. DWELLEY, Dorchester, Mass.—*Machine for Making Plugs for Barrels.*—November 13, 1866.—The stuff is clamped in a frame moved up by the sliding ratchet which acts upon a pawl. A similar pawl and ratchet holds the "feed" gained. A hollow, reciprocating revolving cutter shapes the plugs. An automatically descending saw severs them. At the rear of the cutter are two saws for tenoning the plugs. A lever raises the pawls on the clutch frame to allow of its retraction.

Claim.—In combination with the reciprocating hollow arbor G, the cutting off saw O, brought up automatically at the required time by the means substantially as described.

Also, the combination of the reciprocating toothed bar J, carriage H, pawls *s v*, and stationary toothed bar I, when constructed and operating substantially as and for the purpose set forth.

Also, the carriage H, provided with the automatic feed, constructed substantially as set forth, in combination with the reciprocating cutting arbor G, and the cutting-off saw O, all operating substantially as described.

Also, the combination of the hollow post M, with its spring catch b', lever z, with its catch a', and pawls x v, all constructed and operating substantially as described for the purpose set forth.

Also, the cutters S, in combination with the reciprocating hollow arbor G, and feeding device, when operating substantially as set forth.

No. 59,568.—CHARLES JAMES EAMES, New York, N. Y.—*Compound for Coating Ships' Bottoms, &c.*—November 13, 1866.—Composed of asphaltum, 7 pounds; naphtha, 4 pounds; white arsenic, ¼ pound; red oxide of mercury, 2 ounces; carbolic acid, ½ pound. Mixed by means of heat and cooled.

Claim.—A compound made of the ingredients herein named, for the purpose described, substantially as specified.

No. 59,569.—WILLIAM T. EISENHART, Doylestown, Penn.—*Bee-hive.*—November 13, 1866.—Two of the sides are fast to the base and the other sides are hinged to them. The comb frames are hinged together, and the series to one of the movable sides so as to swing out and expose the frame like leaves in a book.

Claim.—A bee-hive constructed with two fixed and two hinged sides, with the comb frames connected together by hinges and the outermost frame at one side attached by a hinge to one of the hinged sides of the case, substantially as and for the purpose herein set forth.

No. 59,570.—W. H. ELLIOT, New York, N. Y.—*Hay Fork.*—November 13, 1866.—A support is attached at an acute angle to the fork handle and has a curved foot which forms the fulcrum upon which the load is lifted.

Claim.—First, the employment of support c, in combination with and arranged under the fork and resting upon the ground, substantially as described.

Second, the arrangement of support c, at an acute angle with the fork handle, substantially as and for the purpose specified.

Third, the fulcrum e, when permanently fixed in relation to the fork, by means of support c, and brace d, or their equivalents, substantially as set forth.

No. 59,571.—W. EVANS, Forestville, Conn.—*Wrench.*—November 13, 1866.—The movable jaw is held by a catch which is pivoted to the spring lever and engages the serrated shank of the outer jaw. The catch is raised by a thumb piece.

Claim.—The arrangement of the catch E, lever b, and spring c, when said parts are combined with the movable jaw D, the serrated shank A, and stationary jaw B, substantially as described and for the purpose specified.

No. 59,572.—JAMES EWING, New York, N. Y.—*Clamp for Wash Basins.*—November 13, 1866.—The clamp is struck up from a piece of sheet metal, the dies conferring upon it the requisite corrugations to fit the rim of the pan, the under side of the slab, and the bolt whereby it is fastened.

Claim.—As a new article of manufacture, the clamp c of sheet metal stamped to receive the form, substantially as set forth, for the purpose of securing basins to marble slabs by the nut d, as specified.

No. 59,573.—ALFRED FERRIS, Benville, Ind.—*Composition for Paint.*—November 16, 1866.—Composed of coal tar, 1 quart; sulphate of iron, 1 ounce; rosin or asphaltum, 1 ounce.

Claim.—An improved composition for paint, consisting of the materials in substantially the proportions and compounded in the manner described.

No. 59,574.—THOMAS FIRTH, Cincinnati, Ohio.—*Low Water Detector.*—November 13, 1866.—Explained by the claim and cut.

Claim.—The combination and arrangement of the float and needle, mounted upon opposite ends of a single bent rod, whose outer end is made of small diameter and enclosed in a stuffing box, all as herein specified and represented.

No. 59,575.—MOYER FLEISHER, Philadelphia, Penn.—*Skate Fastening.*—November 13, 1866.—The straps pass through eyes in the clamps and bind them against the edges of the sole when tightened over the foot. The clamps are connected to opposite ends of a pivoted bar to insure equal movement.

Claim.—First, the clamp C, consisting of jaws D, eyes F, and arms G, constructed in one piece and adjusted to the slotted connecting bar B, so as to give a direct sliding motion to the clamps, substantially as and for the purpose specified.

Second, the pivoted connecting bar B, adapted to move the clamps *b* equally, thereby causing the centre of the skate to be at the centre of the foot, substantially as described for the purpose specified.

Third, the arrangement of the strap *c d* in combination with the eyes F of the clamp and guides H, whereby they move in the same line with the clamps, as and for the purpose specified.

No. 59,576.—ANTHONY L. FLEURY, New York, N. Y.—*Apparatus for Diffusing the Vapor of Medical or Aromatic Substances.*—November 13, 1866; antedated May 13, 1866.—The revolving vessel is provided with a screw cap, to which are attached the tubes whose ends are bent in opposite directions and act on the principle of the æolipile as the steam issues therefrom. The cap has also a strainer extending into the vessel, which is heated by a lamp beneath.

Claim.—First, the self-revolving retort A, lid E, pipes I I, and strainer K, when used in combination with the pin C, and pin G, and the lamp I, or the flame of the gas burner, for the purposes specified.

Second, the apparatus B, or its equivalent, when arranged and operating in the manner and for the purposes above specified.

Third, the combination of the lamp-shade supporter or gas-light shade supporter V V, having the pin C, with the apparatus B, the thimble F, pipes E, stopper D, the whole arranged and operating as set forth.

No. 59,577.—FREDERIC FOGELGESANG, Canton, Ohio.—*Plough.*—November 13, 1866.—The beam is attached to the landside handle by two bolts which pass through the handle. The upper one is bent and passes through the beam and an eye in the lower one, where it is secured by a nut.

Claim.—The employment of two rods so bent and joined at the under side of the beam by a screw as to make them a continuous bolt through the beam and handle and firmly fastened by nuts and washers on the outside of said handle, as hereinbefore described.

No. 59,578.—W. S. FOLLENSBEE, Janesville, Wis.—*Grouting Form for Wells.*—November 13, 1866.—The hinged formers when locked together form a cylinder secured by pins at certain joints. On removing the locking pins the sections may be withdrawn inward and removed.

Claim.—The combination and arrangement of the staves *a*, ribs *b*, hoops *f*, and keys *d* and *e*, substantially as and for the purpose set forth.

No. 59,579.—EUGENE FONTAINE and OSCAR A. SIMONS, Fort Wayne, Ind.—*Fire Alarm.*—November 13, 1866.—The expansion of the wire by heat frees the rod, when its spring forces it against the central joint of the toggle arms and allows the spring bar to draw on the arm device.

Claim.—First, the toggle arms C, and spring D, in combination with the spring *c*, rod *a*, and wire *f*, constructed and operating substantially as and for the purpose set forth.

Second, the studs *g*, and tension device *h*, in combination with the wire *f*, bed plate A, supporting the alarm mechanism, substantially as and for the purpose described.

No. 59,580.—JAMES B. FORSYTH, Roxbury, Mass.—*Manufacture of India-rubber Rollers.*—November 13, 1866.—The inner part of the roller is composed of rubber, rubber rags, sulphur, oxide of zinc, magnesia, and lampblack; the outer portion of ordinary soft rubber.

Claim.—A roller for clothes wringers, &c., so made, substantially as herein described, as a new article of manufacture.

No. 59,581.—WILLIAM H. FOULDS, Henderson, Ky.—*Hinge for Window Shutters.*—November 13, 1866.—The leaves have a spiral joint around the pintle and the upper one rises as it rotates, giving the blind a tendency to close by gravity. A catch on the blind crosses the hinge and locks in a recess of the leaf attached to the casing.

Claim.—In combination with the hinge so constructed that the weight of the shutter will close the same automatically, the arrangement of the recess F and catch E, operating substantially as specified and for the purposes set forth.

No. 59,582.—MOSES H. FREEMAN, Summerville, Mass.—*Pipe Tongs.*—November 13, 1866.—A tooth upon the shorter jaw engages with one of a series of notches made in the adjacent edge of the hook jaw lever, which is clasped by a hooked stud on the other lever to hold the parts together. By adjustment of the tooth in different notches, the tongs are adapted for different sizes of pipe.

Claim.—The arrangement of the clasp *e*, the tooth *c*, and the series *d* of notches with the two levers A B, and their jaws *a b*, the whole being substantially as specified.

No. 59,583.—JOSEPH GECMEN, Chicago, Ill.—*Malt Kiln.*—November 13, 1866.—The vessel has a series of perforated shelves supported within it by rods. These shelves are so arranged as to swing upon said rods and dump the grain into the hopper.

Claim.—First, in a malt kiln the arrangement of a series of perforated floors, operating substantially as and for the purposes shown and described.

Second, in combination with the above, the employment of a vertical passage C, a series of openings D, and one or more slides E, arranged and operating substantially as specified and for the purposes set forth.

No. 59,584.—JOHN GIFFORD, Jr., Watertown, N. Y.—*Horse Hoe.*—November 13, 1866.—The adjustable wings in the rear of the share are attached at their forward ends to the standard, and at the rear are adjusted by slotted plates proceeding from the handles and beam respectively.

Claim.—The reversible wings I I, attached to and following the share and adjustably supported from the frame A B, substantially as described and represented.

No. 59,585.—JOHN GIFFORD, Jr., Watertown, N. Y.—*Hay Elevator.*—November 13, 1866.—The horse hay-fork is suspended from the rope which passes through the sleeve; as a means of shortening up the said rope it is run through the sleeve and fastened by a pivoted tooth.

Claim.—The combination with the socket C of the ropes A B, and the pivoted detaining tooth E, operating substantially as described.

No. 59,586.—W. and W. S. GILLETT, Stowe, Vt.—*Mop Wringer.*—November 13, 1866.—The perforated boards are hinged together, the lower one is fixed, and the upper one has a lever and side strips. The mop is pressed between them and drains into a bucket.

Claim.—The arrangement and combination of the hinged perforated boards C and E, when constructed with the side boards B B, stays H, and foot board J, operated by the lever G, as herein described and for the purposes set forth.

No. 59,587.—ELLIOTT P. GLEASON, New York, N. Y.—*Cigar Lighter.*—November 13, 1866.—The cigar lighter receives gas through its trunnions; the jet is decreased as the handle hangs suspended and is increased as it is raised for lighting. The plug is chambered for half its length and the gas pipe is screwed into it. A perforation in the plug connects the interior with a channel on its periphery and in the socket, the channel being regulated by a screw.

Claim.—The self-adjusting gas cock, constructed substantially in the manner described for the purpose specified.

No. 59,588.—ELLIOTT P. GLEASON, New York, N. Y.—*Chimney Holder for Gas Burners.*—November 13, 1866.—The holding springs around the platform for the glass have tongues which also press against the chimney and assist in maintaining its position.

Claim.—The equalizing spring for chimney holders, constructed substantially as described.

No. 59,589.—GEORGE P. GOULDING, DANIEL CLARK, and THOMAS DICKINSON, Buffalo, N. Y.—*Fog Signal.*—November 13, 1866.—An air chamber is charged by a pump on a frame above it, and is connected to a whistle. A system of clock-work and levers is so connected to a stop-cock on the connecting pipe as to give the alarm by a succession of strokes with an occasional intermission.

Claim.—First, the construction of an automatic air whistle in connection with an air pump or pumps and air reservoir, and the application and use thereof on shipboard for the purpose of giving signals to indicate the course of the vessel and the "tack" she is sailing on, substantially as described.

Second, the combination of train of wheels 1 2 3 4, levers k^1, k^2 $c^2 f^2$, and connecting bar g, or equivalents, with an air whistle, for the purpose of opening and closing the valves E^2 and F^2, substantially as set forth.

Third, the combination of the wheels o o', arm p, levers m and n, stop pins j' and r', pawl r, and ratchet bar L (or equivalents,) with a timepiece and air whistle, for the purpose of regulating and controlling the intervals at which the signals shall be given, substantially as set forth.

Fourth, the cam q', in combination with the dial plate q, and bar L, for the purpose of enabling the officer of the deck to set the mechanism so as to give any required signal.

No. 59,590.—E. GRATTEM, Williamstown, Mich.—*Combined Measure and Funnel.*—November 13, 1866.—The funnel has feet and a perforate plunger with a foot valve. The graduations on the stem indicate quantity, and the liquid is discharged by pressure on the cap plate which opens the valve.

Claim.—First, the funnel A, having nozzle a, feet b, and cross-piece c, valve C, lugs d, the graduated perforated hollow stem B, and spring a, arranged and operating substantially as described and for the purpose specified.

Second, the perforated graduated tubular stem B, in combination with the funnel A, herein described, as and for the purpose specified.

No. 59,591.—CHARLES T. GRILLEY, New Haven, Conn.—*Capping Wood Screws.*—November 13, 1866.—The unnicked head is capped and the nick then made in the capped screw by a single operation.

Claim.—In the manufacture of capped screws, the method herein indicated, whereby the nicks in the cap and screw head are formed simultaneously, after the cap has been applied and closed upon the screw, as and for the purpose herein set forth.

No. 59,592.—A. F. GROVE, James Creek, Penn.—*Cultivator.*—November 13, 1866.—The side beams have longitudinal movement by which either plough may be put in the lead.

Claim.—The sliding or adjustable plough or shovel beams C C, applied to the main beam A of the implement, and arranged in connection with suitable levers, or their equivalents, to operate substantially as and for the purpose set forth.

No. 59,593.—MOSES GUTHRIE, Clifton, Iowa.—*Beehive.*—November 13, 1866.—The bottom is perforated ; a second bottom slides into place and is recessed on its upper side for the passage of bees into and inside the hive. The openings leading to the honey box may be closed by a slide.

Claim.—The combination of the rabbeted sliding partition B, perforated bottom D, slats E, perforated bottom H, slide F, and box I, with box A, substantially as described for the purpose specified.

No. 59,594.—ABRAHAM G. HAMAKER, Eberly's Mills, Penn.—*Step for Upright Shafts.*—November 13, 1866.—The foot of the vertical shaft revolves upon and partially between three balls retained by a circumferential fixed ring.

Claim.—The arrangement and combination of a round pointed spindle, revolving upon and with three balls as a revolving step, as herein described and for the purposes set forth.

No. 59,595.—J. F. HARCOURT, Moscow, Ind.—*Wheat Drill.*—November 13, 1866.—The wheel is journaled on a bar pivoted to the beam and adjustable by holes in a standard. The seed is drawn from the hopper by the projections of revolving wheels beneath it. The seed wheels have a yielding plate beneath and the amount of seed is regulated by bars on a rock shaft connection to a finger lever, having a scale on the side of the hopper.

Claim.—First, the concave bottom *l*, in hopper E, provided with the holes or openings *m*, in combination with the toothed wheels *n*, fitted in enclosures *t*, underneath the bottom *l*, the yielding plates *u*, and arms *v*, attached to the shaft *w*, for adjusting the plates *u*, substantially as and for the purpose set forth.

Second, the pivoted standard *c*, in combination with the slotted arm *i*, bearing the shaft C, substantially as described for the purpose specified.

Third, the adjustable yielding plate *u*, in combination with arms *v*, shaft *w*, and index arm *a'*, substantially as described for the purpose specified.

No. 59,596.—S. B. HARTMAN, Millersville, Penn.—*Bridle.*—November 13, 1866.—Two lines extend from the driver's band to the horse's bridle ; the extra reins slip in piping in the ordinary reins and passing through the gag loops, and bit rings are attached to the cheek strap, giving them an extra purchase on the bit.

Claim.—First, the safety check lines or reins I, when such reins are arranged in connection with the bridle, and connected to the bit rings, or their equivalents, so as to operate upon the bit, substantially as and for the purpose described.

Second, the double or looped cheek straps A in combination with the reins I, substantially as described and for the purpose specified.

No. 59,597.—WILLIAM HARVEY, Portland, Me.—*Composition for Printers' Inking Rollers.*—November 13, 1866.—Composed of glue, 17 pounds; molasses, 33 pounds ; nitric acid, 5 ounces ; and sulpuric acid, 5 ounces.

Claim.—The compound of ingredients for printers' rollers, substantially as herein set forth and described.

No. 59,598.—R. HASKET and W. B. COX, West Milton, Ohio.—*Centrifugal Machine for Draining Sugar.*—November 13, 1866.—In the body of the centrifugal machine is placed a concave or conical distributer with wings so as to project the sirup and sugar crystals upon the middle of the wire gauze sides.

Claim.—The distributing device E, when constructed with a plain concave surface as described and represented, or when wings F are attached in the manner as set forth, and arranged with reference to a centrifugal sugar mill, substantially as described.

No. 59,599.—GEORGE H. HAZLETON, Philadelphia, Penn.—*Coated Sheet Metal.*—November 13, 1866.—The sheets of copper are coated with an alloy of tin, lead, and antimony, and are to be used for roofing, gutters, culinary vessels, &c.

Claim.—The use and manufacture of sheet copper, coated, substantially as herein set forth and described.

No. 59,600.—WILLIAM HEATH, Bath, Maine.—*Invalid Bedstead.*—November 13, 1866.—
The bed is divided into four portions, namely, back, seat, and leg supports, the latter jointed
at the knees. The change from the reclining to the sitting posture is made by simultaneously
raising the back and depressing the leg supports, which is done by revolving a shaft whose
gears engage toothed sectors pivoted to the frame and connected to the bed rails.
Claim.—The combination for simultaneously operating or moving the leg and back portions C and E of the bed, the same consisting of the shaft H, its gears G G, the toothed
sectors F F, the arms *ff*, and the spring or band I, the whole being applied to the frame A,
and the said portions C and E, and arranged therewith, substantially in manner and so as
to operate as specified.
Also, the combination and arrangement of the bands K K, with the parts B C D E, and
the mechanism for operating the two parts C E, substantially as described.

No. 59,601.—B. HEIDERICH, Brady's Bend, Penn.—*Car Truck.*—November 13, 1866.—The
ends of the car trucks are supported from the bottom of the car by means of elongated loops
to sustain them in case of the breaking of a wheel or axle.
Claim.—The supporting of the trucks from the bed or bottom of the car by means of the
loops F G, substantially as and for the purpose set forth.

No. 59,602.—PETER C. HEINZ, Funksville, Penn.—*Injector.*—Steam is admitted from the
dome to the central pipe of the injector, and its pressure on a disk below the spring raises the
steam injection pipe, and the puppet valve which controls the annular opening by which gas
escapes. When steam is shut off both the steam and gas openings are closed by the spring.
Claim.—First, a gas injector for furnaces constructed and operating in the manner substantially as herein set forth.
Second, the valve D' in combination with the steam pipe E and gas-supply pipe A, for the
purpose and substantially as described.

No. 59,603.—S. L. HILL, Williamsburg, N. Y.—*Alphabet Block.*—November 13, 1866.—
Each triangular block has a word and a portion of a letter, which form a complete sentence
and a letter when the blocks are properly associated, in which position they may be retained
by a strap.
Claim.—First, the triangular blocks A having portions of a letter on the face near their apex,
and words on their centres adapted to form a square with a complete letter and a complete
sentence, retained together, and operating substantially as described for the purpose specified.
Second, the grooves *b*, in the edges of the blocks, in combination with the spring *a*, constructed and operated substantially as and for the purpose described.

No. 59,604.—J. HINDMAN, Olathe, Kansas.—*Washing Machine*—November 13, 1866.—
The rubber block is suspended by an adjustable arm from a rock shaft, and is reciprocated by
a crank shaft which engages a book piece on the back of the block.
Claim.—The crank shaft C, the vertical shaft D, with its washer J, the arm E, and the rock
shaft F, in combination with the box B, arranged substantially as described for the purposes
specified.

No. 59,605.—HUGO HOCHHOLZER and FRANK DENVER, Virginia City, Nevada.—*Machine
for Tenoning Timber.*—November 13, 1866.—The timber is clamped in revolvable heads carried on a frame having a lateral sliding movement. The cutters have vertical and longitudinal movement, and are actuated by a revolving drum. The same drum gives motion to
the saws which have a longitudinal adjustment. By these devices tenons of various forms
may be cut and the ends sawn off.
Claim.—Clasping or clamping and turning the timber or log by the means and in the manner described, substantially as set forth.
Also, holding and presenting the timber or log to be tenoned to the cutters, by the means
and in the manner substantially as described.

No. 59,606.—FREDERICK W. HOFFMANN, Morrisania, N. Y.—*Machine for Cutting off
Cigars.*—November 13, 1866.—The cigar lies in a trough with a regulating plate to determine
its position longitudinally; a guillotine knife is depressed, cuts off the tip and is raised by a
spring.
Claim.—First, the construction of a plate A. provided with a movable guide B. and guides C
and D, at the forward end of said plate, in combination with a knife G, fast to a rod or plunger F,
moving in an upright tube or pipe E, attached to said plate A, when the whole is arranged
and combined in the manner and for the purpose substantially as described and specified.
Second, the projecting piece *n*, on the guide D, acting in combination with the guide C, on
the flexible or feathering knife blade G, in the manner and for the purpose substantially as
set forth.

No. 59,607.—JOHN N. HOWE, Franklin, N. H.—*Extinguisher for Lamps.*—November 13,
1866.—Through the side of the burners is a tube passing to the upper end of the wick to enable
the flame to be blown out.

Claim.—As an improvement in extinguishers for lamps, the tube E, through which a current of air may be directed to the flame, substantially as set forth.

No. 59,608.—JOHN HUTCHINS, Elmira, N. Y.—*Drill or Well Tube.*—November 13, 1866.—The conical end of the hollow tube has spiral projections to lead it into the soil when revolved, and perforations to permit access of water.

Claim.—The hollow conical drill point A of cast-iron, provided with bevelled holes or slots *a* with the spiral flange *c*, flattened as described, the whole being constructed as described and for the purposes set forth.

No. 59,609.—GEORGE JONES and BEVERLY E. MEAD, Peekskill, N. Y.—*Fastening Door Knobs to Shanks.*—November 13, 1866.—The knob is secured to the shank by a square mortise in the former and a tenon in the latter, and a screw from the outer side of the knob passing into the end of the shank.

Claim.—The fastening of a porcelain mineral or clay door or other knob upon its shank by means of a screw or rivet passing through the knob into the shank, substantially as set forth.

No. 59,610.—WILLIAM ASHLEY JONES, Dubuque, Iowa.—*Vehicle.*—November 13, 1866.—A jointed slide bar has an abutment for the neck yoke and connects to the brake bar. This bar is also connected to a hand lever on the bed. A spring pin is drawn down by a cord through the tongue and slide bar to retain the brake in operation. On relaxation of the cord the pin springs up. The whiffletrees have backward turned trace-hooks, with guards to retain the traces. A cord acts on an upwardly projecting lever to rock the whiffletree and remove the guard, which releases the traces.

Claim.—First, the combination of the jointed rod or bar K with the tongue J, reach H, and brake bar N, when said bar K is constructed and arranged substantially as herein described and for the purpose set forth.

Second, the combination of the bolt or pin V, spring Z, lever W, cord or strap X, and pulley Y, with each other, with the tongue J and with the jointed bar K, substantially as herein described and for the purpose set forth.

Third, the combination of the lever T, rack U, connecting rod S, lever R, and connecting rods P with each other, and with the box I, axle G, and brake bar N, substantially as herein described and for the purpose set forth.

Fourth, the combination of the bent bars D′, hooked rods C′, springs F′ and E′, cords or straps G′, and pulleys H, with each other and with the whiffletrees B′, substantially as herein described and for the purpose set forth.

No. 59,611.—VINCENT E. KEGAN, Roxbury, Mass.—*Manufacture of Saltpetre.*—November 13, 1866.—The potash is spread in thin layers on trays in a room from which the light is excluded, and exposed to the action of the atmosphere, when it absorbs oxygen and nitrogen, and is converted into nitrate of potash.

Claim.—The within-described process of producing nitrate of potassa by treating potassa substantially in the manner set forth.

No. 59,612.—ELI KEITH and DELL BIRD, Lafontaine, Ind.—*Machine for Driving Spokes in Wagon Wheels.*—November 13, 1866.—The standard forms a support for the adjustable portions of the apparatus. The point of the hub rests on a block keyed up by wedges. The butt of the hub rests on a pivoted bar whose carriage is vertically adjustable on the standard by a lever and rod in the rear. The mandrel bolt clamps the pivoted hub rest to a bar on the back of the standard.

Claim.—The arrangement upon the standard B of the adjustable frame K and pivoted rest I, operated substantially as described.

Also, the combination of the adjustable rests I and D, lever G, and mandrel N, constructed and operating substantially as described.

No. 59,613.—L. J. KNOWLES, Warren, Mass.—*Narrow Ware Loom.*—November 13, 1866.—Improvement on inventor's patent, No. 54,742, for weaving narrow goods. The slack in the link allows the lay-rack to remain at rest a short time before its longitudinal movements take place. The mode of hanging the pressure roller allows its ready removal from off the lower one.

Claim.—The arrangement of the heddle-operating cams in circular disks between which the levers extend, when the levers and pins, disks, cams, and cam slots have a relative disposition, substantially as described.

No. 59,614.—EDWARD KRETCHMER, Pleasant Grove, Iowa.—*Beehive.*—November 13, 1866.—The bee trap has swing bars, and is invertible to change the size of the apertures. A groove is cut in the bottom for a moth trap, which may be opened by dropping a hinged lighting board. Sectional frames have varying gains on their different sides, which by connection with the entrance slot may prevent the queen or drones from passing through or impeding the passage of the workers.

Claim.—First, the reversible entrance protector C provided with swinging bars V, supports y y, and front G, all arranged and operating substantially as and for the purpose set forth.

Second, constructing and operating the moth trap, substantially in the manner and for the purpose as set forth.

Third, constructing and operating the sectional adjustable sliding swarming-guard and entrance regulator, substantially in the manner and for the purpose as above set forth.

No. 59,615.—ER. LAWSHE, Atlanta, Ga.—*Lock.*—November 13, 1866.—This lock for freight cars may be operated by different keys, and has a tablet upon which is written the destination of the car, and which is exposed to view by the operation of the key.

Claim.—First, the bolts C and M in combination with the pawl 8, springs P and F, guide-plates *a a*, and lever I, all constructed, arranged, and operating in the manner and for the purpose specified.

Second, the combination with a lock, constructed as described, of a tablet or plate, or its equivalent, when arranged with regard to the locking mechanism of the lock so as to be operated by the key or keys for the lock, substantially in the manner and for the purposes specified.

No. 59,616.—F. LEPPENS, Hartford, Conn.—*Axle Box.*—November 13, 1866.—A ring is shrunk on the axle at the inside of the box, and fits tightly to an upper and under slide-plate. The former is kept to position by gravity, the latter by a spring beneath it. The object is to close the opening toward the shoulder of the axle and exclude dirt from the oil chamber.

Claim.—The combination of the sections G H, extension pieces I, and spring L, with the axle B, with the ring O shrunk thereon, substantially as described and for the purpose specified.

No. 59,617.—WILLIAM C. LESSTER, New York, N. Y.—*Combined Gasolier and Cigar Lighter.*—November 13, 1866.—The gas light stand has within its enlarged ornamental pedestal a reservoir of liquid with projecting tubes for filling and for the sponge torch whereby cigars are lighted.

Claim—First, the combination with a gasolier of a cigar lighter, the fonts of which are supplied with combustible fluid or spirit from a reservoir arranged within the stem of the gasolier, substantially as specified.

Second, the arrangement within the stem of the gasolier of a close combustible fluid or spirit reservoir, having the gas pipe pass through a sleeve in it and communicating with the exterior by a supply pipe and font tubes, essentially as herein set forth.

Third, the combination of the gasolier cigar lighter with its reservoir and jet cup, constructed and arranged substantially as shown and described and for the purpose specified.

No. 59 618.—CHARLES MAHON, Macon, Ga.—*Ticket Holder.*—November 13, 1866.—The coil of stamps is placed in the cylindrical portion, and the edge from which they are to be detached protrudes between the tangential plates, being advanced or retracted by acting on the coil at the opening in the midlength of the cylinder; the opening is afterwards closed by a slide.

Claim.—A pocket case for postage stamps composed of the cylinder *c c'*, feed opening *o*, guide plates *p p'*, and cutters *l l'*, the several parts being combined and arranged as and for the purpose herein described and represented.

No. 59,619.—G. C. MANNER, New York, N. Y.—*Piano-forte.*—November 13, 1866.—A slot in the metallic frame allows the damper lifters to be placed behind the point supporting the string, and the application of "French" damper levers over the bridge is rendered practicable.

Claim.—Placing the damper lifters in a slot of the metal frame behind the point supporting the strings, substantially as and for the purpose described.

No. 59,620.—THOMAS MCCLEARY, Blairsville, Penn.—*Fireplace.*—November 13, 1866.—The basket grate is suspended on two lugs at its ends, which rest in bearings on the jamb; the grate when pulled forward will upset, but when pushed back is locked by the square lugs in the slots of the bearings. The damper above may be swung forward to act as a reflector.

Claim.—First, the construction of a grate B, with open front and ends and with an open elevated back, having air spaces surrounding it, when such grate is supported by journals in such a manner that it can be upset at pleasure, or secured firmly in an elevated position, substantially as described.

Second, the construction of the oblong bearings *c c*, for the flattened journals *b b*, of the grate, substantially as described.

Third, arranging the swinging concave reflector and damper C, above the open grate B, substantially as described.

Fourth, so constructing a grate and arranging it in a fireplace that it can be upset at pleasure and at the same time so that it can be locked in an upright position, by means substantially as described.

No. 59,621.—ROBERT W. MCFARLAND, Monticello, Wis.—*Farm Gate.*—November 13, 1866.—The parts of the gate are pivoted together, and an inclined rail from its front post is connected by a cord, and over sheaves to a lever whose depression opens the gate away from the operator, the slots collapsing as they rise. It is closed by a slight reverse movement of the lever or may be made to gravitate shut.

Claim.—The V-shaped frame lever H, pulleys K, rope T, sliding rail N, when constructed and arranged in combination with the adjustable gate, as herein described and for the purposes set forth.

No. 59,622.—JOSHUA MERRILL, Boston, Mass.—*Cask, Barrel, &c.*—November 13, 1866.—Explained by the claims and cut. The device is intended to apply to old as well as new barrels.

Claim.—The improved cask, substantially as described, having its joints made by matched grooves in the staves, and a separate tongue or key strip of wood driven in to fill the said matched grooves, substantially in the way and for the purposes described.

Also, in combination with the joints of a cask made with matched grooved joints and a separate tongue or key strip, a coating or stuffing of glue, or similar gelatinous cement, between the members of said joints, applied substantially in the way and for the purposes described.

Also, in combination with the joints of a cask made with matched grooved joints and a separate tongue or key strip, a coating or stuffing of shellac, rosin, or other similar resinous cement, between the members of said joint, applied substantially in the way and for the purposes described.

No. 59,623.—BENJAMIN MERRITT, Jr., Newton, Mass., assignor to AMERICAN TREE PROTECTOR COMPANY.—*Tree Protector.*—November 13, 1866.—Improvement on his patent of November 15, 1864, in respect of having several grooves of different sizes on the under side of the segments, which are united by a clamp and suspended around the trunk. The projecting peripheral flange assists in attaching the clamp and suspensory web or cord.

Claim.—The combination of two or more grooves of unequal size, when arranged in the segments of a tree protector, substantially as described.

Also, in combination with the segments *a*, the outwardly projecting flange *f*, arranged as 'and for the purpose specified.

Also, the bead *g*, on the segments *a*, when combined with a corresponding formation of the clamp *c*, as seen at *h*, for the purpose of securely holding the parts of the protector together.

No. 59,624.—ALEXANDER MONROE, Watkins, N. Y.—*Stump Extractor.*—November 13, 1866.—The chain passes from the stump to the clevis on the centre of the lever. The fulcrum of the lever is changed and the lever depressed on alternate sides. The lever has indentations in a plate on its lower side for the shifting fulcrum pins.

Claim.—The arrangement of the clevis D, constructed as described, moving vertically in the grooves *i i*, with the lever C working through it, said parts being used in combination with the double sets of holes *m n*, and pins *p p*, the whole operating substantially as and for the purpose herein specified.

No. 59,625.—JOHN MOORE, Gardiner, Me.—*Machine for Polishing Wood.*—November 13, 1866.—The supporting table for the door has longitudinal and lateral movement on wheels which traverse upon track rails. The rotary polisher runs in a frame separate from the table, and is vertically raised by a treadle to admit the door beneath it.

Claim.—The combination and arrangement of the carriage C, mounted on the wheels *b b*, upon the transverse rails *a a*, with the table D, mounted on the wheels *d d*, upon the longitudinal rails *c c*, or its reversed equivalent arrangement, when used in connection with a revolving rubber G, supported by a sliding arm F, for polishing doors, constructed and operated substantially as herein described.

No 59,626.—D. M. MOURLAND, Little York, Ill.—*Clamp and Gauge for Weather Boarding.*—November 13, 1866.—The frames are secured so as to act as gauges, and support the weather boarding both vertically and horizontally.

Claim.—First, the within-described clamp, consisting of the adjustable gauge and spacing rest C, gauge block B, guide bar A, the clamps *c c*, and the marker D, arranged and operating in the manner and for the purpose herein specified.

Second, the clamps *c' c'*, in combination with the adjustable spacing bar and rest C', for fastening to the studding of the heading of the siding, constructed and operating substantially as and for the purpose herein specified.

No. 59,627.—JOHN M. MULLER, North Becket, Mass.—*Softening Dry Hides.*—November 13, 1866.—The hides are treated with a solution of soft soap and sal soda, in water.

Claim.—Treating hides before tanning in a liquor which is composed of the within-described ingredients mixed together in about the proportions mentioned.

No 59,628.—A. B. MULLETT, Washington, D. C.—*Lock.*—November 13, 1866.—This is an improvement on "Johnson's rotary lock." The flanged head of the key is notched at its edge, and the keyhole has corresponding openings and intervening projections by which latter the door is drawn open after unlocking.

Claim.—The plate, Fig. 4, and its corresponding key, Figs. 2 and 8, made and combined substantially as herein set forth.

- No. 59,629.—ALBERT L. MUNSON, New Haven, Conn., assignor to himself and ARTHUR MOFFATT.—*Revolving Fire-arm.*—November 13, 1866.—The cylinder chambers run clear through and may be loaded from either end, the cartridges ejecting the shells if desired. Either end has pins to be engaged by the ratchet, which remains on the breech and has a pin which enters a hole in the rear, for the time being, of the cylinder. The movable portion of the axial pin, when the cylinder is reversed, slips endwise and projects forwardly into its socket on the barrel piece.

Claim.—First, the reversible cylinder C, in combination with the detached ratchet *e* and plate or collar *o*, or its equivalent, substantially as and for the purpose described.

Second, the centre pin *d*, in connection with the pin *m*, operating substantially as and for the purpose described.

No. 59,630.—WILLIAM B. NICKELSON, Lowville, N. Y.—*Hoop for Curing and Packing Cheese.*—November 13, 1866.—The cheese is cured in a hoop without fibrous covering, and the hoop receives two heads to constitute it a box for transportation.

Claim—The hoop has a covering for the circumference of the cheese, in lieu of bandages in curing, and in connection with covers serving as a box for the cheese during turning, storage, and transportation, substantially as described.

No. 59 631.—MORGAN NOTTINGHAM and WILLIAM DUNCAN, Vinton, Iowa.—*Drill.*—November 13, 1866.—The upper fixed collar is connected to the lower movable collar by double toggle bars. The lower bar of each toggle has a curved cutter on its outer edge.

Claim—The rod A, having fixed collar C, and sliding collar G, connected together through arms E and F, with the latter, F, provided with cutting blades H, substantially as and for the purpose described.

No. 59,632.—CALEB M. OLIVER, Port Carbon, Penn.—*Axle Box.*—November 13, 1866.—The upward pressure of the axle is upon the brass box, and a diaphragm packing is placed between it and the follower on which the spring rests. The box has free vertical movement in the oil chamber. A packing ring inside the bearing is adjusted by set rods passing through the face plate of the oil box.

Claim.—The bearing B and follower E, in combination with the axle box C, the former being arranged in relation to the latter so as to relieve the box of pressure, substantially as described.

No. 59,633.—JOHN K. O'NEIL, Kingston, N. Y.—*Horse Hay-fork.*—November 13, 1866.—The tines are suspended from the lever at one point by a link which unites with them at their point of junction, and are also pivoted by separate links to another point on the lever, so that as the power applied to the hoisting rope is changed from one end to the other of the horizontal lever, the tines will be opened or closed. This change is effected by a tripping hook.

Claim.—Suspending the lifting bar D, and opening bars C C, by the same straight or direct lever B, all operating in combination, substantially as and for the purpose herein specified.

Also, the tripping hook H, provided with the tripper *h*, pivoted thereto at their points *i*, constructed and operating substantially as and for the purpose herein set forth.

No. 59,634.—WILLIAM M. OWEN, Homer, Iowa.—*Wrench.*—November 13, 1866.—The inner jaw is attached to the sleeve. It is held to any adjustment by a pin on a spring lever traversing the sleeve and entering the shank of the outer jaw.

Claim.—The combination of the handle A, the spring lever G, and plug F, with the perforated shank D of the movable jaw, substantially as described.

No. 59,635.—CHARLES PADMORE, Philadelphia, Penn.—*Shirt Stud.*—November 13, 1866; antedated October 27, 1866.—Several studs are connected to a single plate and, in the form represented, answer for the button holes of the shirt and those of the collar.

Claim.—The grouping together of three, four, or more studs or buttons as hereinbefore mentioned, and for the purpose described.

No. 59,636.—EDWIN A. PALMER, Clayville, N. Y.—*Cheese Hoop.*—November 13, 1866.— The follower fits loosely in the hoop, and its lower edge is chamfered off to fit against a ring of triangular section which is fitted within the hoop, and gives a chamfered edge to the cheese.

Claim.—First, the corner taken off the follower, as described in Figs. 3 and 6, at D and c. Second, the little ring D. Fig. 4, or an equivalent, substantially as described and for the purpose therein set forth.

No. 59,637.—J. D. PARROT, Morristown, N. J.—*Churn.*—November 13, 1866.—The churn tub with internal dashers is pivoted on a frame. A pendulum is attached to assist in the oscillation of the tub, which forces the cream through the slots in the bulkheads.

Claim.—The combination of the tub C, stirrup B, pendulum D, bulkheads E, and frame A, when arranged and operating in the manner and for the purpose herein described.

No. 59,638.—CHARLES W. PATTON, Exeter, Ill.—*Wheat Drill.*—November 13, 1866.— The drag bars of the cutters are pivoted at the front end and depressed by springs in a rock bar worked by a foot lever. The agitator is rotated by a belt from the main shaft, and formed so as to present alternately a recess and projection to the seed over and between the seed openings. The latter are varied in size by slides connecting to a lever with a ratchet guide. The cutters are either circular and rotating or segmental and fixed.

Claim.—First, the shaft Y, operated by the foot lever Q, to press the spring P upon the drag bars D, substantially in the manner set forth.

Second, in combination with the shaft Y, lever Q, and springs P, the hinged drag bars D, teeth M, and cutters U or G, substantially as set forth.

Third, the agitator N, when constructed as described.

Fourth, the slide H', operated by springs P', and a lever B, in combination with the bottom H, when constructed substantially as and for the purpose set forth.

No. 59,639.—WORDEN P. PENN, Belleville, Ill.—*Seeding Machine.*—November 13, 1866.— The upper chute is attached to the brace bar of the tooth; the brace connects the drag bar to the upper end of the tooth.

Claim.—First, sustaining the forward end of the chute *f* upon the brace *g'* of the tooth E, when said brace is arranged above the drag bar F, substantially as described.

Second, the combination of the brace *g'* with the drag bar F and tooth E, said brace being located above the drag bar and pivoted to it and the upper end of the tooth, substantially as described.

No. 59,640.—L. C. PENNELL, Portland, Me.—*Skirt Lifter for Ladies' Dresses.*—November 13, 1866.—The cords are attached by tags to the skirt near the hem, and passing outside of the hoops enter through eyelets on tags attached to the tapes; the cords then pass collectively to a tag in front whereby they are operated simultaneously.

Claim.—The attachment to the hoop skirt of tags to the tapes thereof and the cords as described, all constructed, arranged, and operating as and for the purposes indicated.

Also, in combination with the subject of the first claim, the arrangement of the rings i i, on the skirt of the dress, as and for the purposes set forth.

No. 59,641.—STUART PERRY, Newport, N. Y.—*Sawing Machine.*—November 13, 1866.— The two feeding disks are adjustable upon the shaft by means of a series of grooves of varied lengths and a stud or pin fitting in said groove, which together determine the distance between the disks

Claim.—Making one or both of the feeding disks or wheels adjustable upon the shaft by which it is turned, by means of a series of grooves of varied lengths, and a stud or pin, substantially as and for the purpose described.

No. 59,642.—JAMES PHELPS, Red Creek, N. Y., assignor to himself and ISAAC F. MO- SHER.—*King Bolt for Carriages.*—November 13, 1866.—This "king bolt" spans the axle instead of penetrating it; has a shackle bar below, resting shoulders on the axle, and a center bolt above for the attachment of the bed.

Claim.—The projecting shoulders or bearings B B resting on the axle at the fork of the king bolt, for the purpose herein specified.

No. 59,643.—W. W. PHILLER, Port Byron, Ill.—*Cultivator.*—November 13, 1866.—The tongue is pivoted, and has a semicircular cross-bar, furnished with anti-friction rollers, between it and a similar segment beneath. Pins on the cross-bar form rests for the feet in guiding the implement, and a backward continuation of the tongue serves the same purpose when the operator walks.

Claim.—First, the pivoted draught pole C, provided at its rear end with a curved or segment bar D, having friction rollers or wheels *c* inserted within it, and working or resting upon a semicircular way or track E on the frame A. substantially as and for the purpose set forth

Second, the bar or lever Q attached to the rear end of the draught pole C, substantially in the manner as and for the purpose set forth.

No. 59 644.—WILLIAM M. PICKSLAY, Philadelphia, Penn.—*Manufacture of Bars and Articles of Iron and Steel Combined.*—November 13. 1866.—This is an improvement on Charles Sanderson's patent of September 19 1865, and is explained by the claim.

Claim.—The manufacture of bars and other articles of iron and steel combined by applying the steel in a molten state to the iron while the latter is at a welding heat, and subsequently rolling or otherwise working the combined mass.

No. 59,645.—LEONCE PICOT. Hoboken, N. J.—*Comb.*—November 13, 1866.—A straight comb is fortified by inserting longitudinally in its back a metal strip, which is T shaped, and thus covers the edge of the comb and may dip down on each side

Claim.—First, strengthening the back and sides of a comb by forming in the top of said comb a groove, and placing therein and upon said back a metallic brace, substantially as herein shown and described.

Second, as a new article of manufacture, a comb to and in the top of which a T-shaped brace of metal or other suitable material for strengthening the same is fitted and held, substantially as herein described and set forth.

No. 59,646.—CHARLES W. POWELL, Milford, Conn.—*Collar and Neck-tie Attachment.*—November 13, 1866.—The collar and neck-tie are adjusted and secured upon the narrow metallic or rubber band before being placed upon the neck. At suitable points metallic loops and hooks are placed, whereby the collar and neck-tie can be attached. The band may be secured to the shirt by other fastenings.

Claim.—A band having clasps, loops, and spurs, for the attachment of a collar and neck-tie, all arranged substantially as described.

No. 59,647.—ISAAC A. POWELL, Morrison, Ill.—*Horse Collar.*—November 13, 1866.—The pins are attached to the upper end of the collar, and have a long tudinally flaring head, with a catch on one side. The slots in the metal plates upon the cap are formed to allow the upward passage of the pin-head, while a notch operates with the catch to prevent retraction.

Claim.—A horse-collar, when constructed with the lock D and plate C, for securing the same, when said parts are respectively constructed, attached, and combined, substantially as set forth.

No. 59,648.—JAMES POWELL, Cincinnati, Ohio.—*Case for Medallions.*—November 13, 1866.—The metallic medallion is hermetically sealed in its case to prevent tarnishing, between a bowl-shaped back plate and a lunette glass front, luted in a groove of the back plate. The inside of the bowl where it forms the background for the medallion may have a flock covering, and the back has ears for attachment to a frame.

Claim.—First, the hermetically sealed medallion case, constructed substantially as herein described.

Second, in combination with a medallion case, constructed as specified, the flock or cloth dust coating of the face side of the back plate.

Third, the combination and arrangement of concave convex back plate A, groove C, glass front D, and ears E, as and for the purposes specified.

No. 59,649.—DANIEL PUNCHES, Plymouth, Mich.—*Bed Bottom Spring.*—November 13, 1866.—A follower attached to the end of each slat slips in the bars of a stirrup-formed frame, secured to the rails. The bars have coiled springs which are compressed by weight on the slats.

Claim.—The combination of the frames E F with the springs *a a*, when said frames are constructed as herein described, and attached to the ends of the slats and end rails of the bedstead, so as to be free to move in the direction of the length of the slats under extension and contraction, as described and for the purpose specified.

No. 59,650.—JAMES RADLEY, New York, N. Y.—*Locomotive Head Light.*—November 13, 1866.—An air current passes through the passages which surround the tubes, carrying the burning fluid from the tank to the lamp for the purpose of cooling them and preventing the accumulation of heat in the lamp.

Claim.—First, the method of cooling the burning fluid as it passes from the tank to the lamp. by enclosing the connecting oil tube or tubes in a pipe or pipes, through which a current of air is made to pass around such oil pipe or pipes, substantially in the manner herein described.

C P 47——VOL. II

Second, the method of cooling the body of the lamp within the reflector by means of vents adjacent thereto in the air passages, substantially as herein described.

Third, the receiving aperture in front of the head-light case and the air passages enclosing the oil tubes, in combination with a locomotive engine, when so arranged and constructed as to cause the air entering said aperture to pass through said passages when the locomotive is in forward movement, substantially as herein described.

Fourth, the scattering vents through which the air escapes in combination with the air passages and receiving aperture, substantially as herein described.

No. 59,651.—HENRY A. RAINS, Nashville, Tenn.—*Cart Harness Saddle.*—November 13, 1866.—The edge roll of the pad is confined in the groove of a metallic frame by a covering plate, and the latter attached to a bridge-piece, or the pad frame may be attached to wooden pieces and covered with metal. The bridge-piece has loops at each end for the belly band.

Claim.—First, the wooden housing, constructed and arranged as described, in combination with a metallic covering, with the moulding rolled or wrought upon it, as set forth.

Second, grooved framework for attaching the pads to the tree-bars, constructed substantially as described.

Third, a belly-band fastening, constructed and attached as hereinabove set forth.

No. 59,652.—JOHN R. READER, New York, N. Y.—*Railroad Car.*—November 13, 1866.—The guard or fender-frame is supported upon bearings on the prolonged axles of the cars, so that the motion of the car on its springs is not interfered with; plates are hinged to the frame and supplementary plates to the body at each end. Brushes and rollers precede the wheels.

Claim.—First, the guard or fender, constructed with a frame D, supported on the axles of the wheels C, substantially as and for the purpose set forth.

Second, the combination with such guard or tender of the suspended supplemental plates I, hinged to the bottom of the body A, substantially as herein set forth for the purpose specified.

Third, so arranging the opposite parallel sides of the guard or fender that the upper edges thereof will be situated outside of the body A, substantially as herein set forth for the purpose specified.

Fourth, the roller u combined with the brushes S and with the frame D, wheels and body of the car, substantially as herein set forth, for the purpose specified.

No. 59,653.—FRANK REED, Brattleboro', Vt.—*Door Fastening.*—November 13, 1866.—This is intended for folding doors. The vertical latch bolts are retracted simultaneously by turning the handle, which acts upon the yokes in contrary directions. A forked bolt of the lock on the other door secures the device.

Claim.—The combination of the cap piece g and coupling pieces c c with the shaft of the door-knob, and with the arms a a, rods d d, and springs e e, constructed and arranged as and for the purpose herein specified.

No. 59,654.—JOHN J. REED, Polo, Ill.—*Sulky Plough.*—November 13, 1866.—The main frame is pivoted in front to arms attached to the axle. The ploughs are raised by the pressure of the feet on levers pivoted to the axle and connected to the plough frame. A lateral simultaneous movement is given to the ploughs by a pivoted beam acted on by the feet.

Claim.—First, the walking beam G pivoted to the rear end of the tongue or pole F, in combination with the stirrups e e, yoke J, and plough standards K K, substantially as herein shown and described, and for the purposes set forth.

Second, the pivoted pendent bars e² e², and bars R R, in combination with the frame C, substantially as shown and described and for the purposes set forth.

Third, the projecting bars D D in combination with the frame, substantially as herein shown and described.

No. 59,655.—L. RICHARDS and D. LINCOLN, Orangeville, N. Y.—*Fruit Picker.*—November 13, 1866.—The cylinder has a large hole near its upper end, through which the fruit passes, and a tapering slit in which the stem is cut; the fruit drops into the conveyer.

Claim.—First, a hollow cylindrical fruit-picker A, made conical at its upper end, and a large hole or opening and tapering slit near its upper end, with or without the removable bottom B, substantially as described.

Second, the combination of a flexible bag or hose C with said cylindrical picker, for the purposes and substantially as described.

Third, attaching the handle to the cylinder A by means of the staples E, wedge F, and notch f', substantially as set forth.

No. 59,656.—PHILIP RILEY, New Bedford, Mass.—*Securing Shoe-tips.*—November 13, 1866.—The tip is sewn around its margin and then to the sole by a stitch which passes through the loops of the former stitching and not through the leather.

Claim.—Securing tips to the toes of boots and shoes by stitching around and through the margins of the tips a loop or chain stitch, and sewing the upper side of the loop or chain stitch to the sole, as herein set forth and described.

No. 59,657.—DANIEL T. ROBINSON, Boston, Mass.—*Tool for Drawing Nails.*—November 13, 1866.—The lever and fulcrum are pivoted together, and their jaws are clamped together by the act of lifting.

Claim.—Constructing a nail-pulling device in the manner described, so that the jaws will clamp the nail by the action of applying force to the lifting lever in raising the nail from its position, substantially as described.

No. 59,658.—S. W. ROBINSON, Detroit, Mich.—*Escapement for Time-pieces.*—November 13, 1866.—The object is to impart impulses of equal force to the balance at each double vibration. This is effected by a lever acted on by a spring and applied in combination with the escape wheel, the balance, and two detents; the force for unlocking the latter is derived from the hair-springs of the balance and lever, while the power of the hair-spring, acting on the lever, imparts to the balance the desired impulse.

Claim.—The lever B and hair-spring H, in combination with the detents I J, escape wheel A, and balance C, constructed and operating substantially as and for the purpose described.

No. 59,659.—PETER RODIER, Springfield, Mass.—*Loop Check for Sewing Machines.*—November 13, 1866; antedated November 5, 1866.—The spring loop-check passes through an opening in the outer bobbin-holder, and is formed with a curve which bears upon the face of the bobbin at two points to keep the thread in place, the more effectually to prevent its slipping off, and the consequent liability of the hook to take the same loop twice, and so entangle and break the thread.

Claim.—The loop or thread-check C, constructed as described, when combined with and operated by a spring B or B', and used in combination with the parts of a sewing machine, substantially as and for the purpose herein set forth.

No. 59,660.—CHAS. ROGERS, Bergen, N. J.—*Device for Hitching Horses.*—November 13, 1866.—The rings forming the annular strap space are confined between two disks. The hitch strap passes through a slot in the outer ring, and its other end is attached to the inner ring. A spiral spring connecting the inner ring to a central pin tends to draw the hitch-strap within. A staple is secured to the outer ring for attachment to the manger.

Claim.—The combination of the case A, centre piece B, cylinder C, spring D, strap E, rollers G, staple L, and hook H, when these several parts are constructed and arranged substantially as herein shown and described and for the purposes set forth.

No. 59,661.—STEPHEN G. and GEORGE S. ROGERS, Thetford, Vt.—*Paper-making Machinery.*—November 13, 1866.—The object of this device is to prevent the adhesion of the paper to the upper delivery roller, and its consequent liability to wind thereon, as discharged from the endless blanket.

Claim.—The arrangement as well as the combination of the auxiliary roller D with the rollers A B, and the delivery apron or blanket C of a paper-making machine, the purpose of such auxiliary roller being as set forth.

No. 59,662.—LARKIN S. SAFFORD, Hope, Maine.—*Stanchion.*—November 13, 1866.—The stanchion post is pivoted so as to swing horizontally; the movable stanchion bar is pivoted to the lower arm and locked in the upper one.

Claim.—The construction, arrangement, and combination of the parts B F C D and E, so as to allow them to swing in toward or out from the crib on said pins or pivots F, at the pleasure of the animal, when fastened as herein described.

No. 59,663.—CURTIS SATTERLEE, Paris, Ill.—*Horse-rake.*—November 13, 1866.—The locking and releasing devices of the revolving rake are explained by the claim and cut.

Claim.—The combination of the pivoted lever J with its arm K, pivoted lever I, rakehead shaft F', and bar L, strap T, lever S, and post P, constructed as described, and arranged to operate substantially as and for the purpose specified.

No. 59,664.—THEODORE SCHREIBER, Wheeling, W. Va.—*Comb.*—November 13, 1866.—Behind the comb is an adjustable spring pad which presses upon and smooths the hair.

Claim.—The spring pad B, in combination with the comb A, constructed and operating substantially as and for the purpose described.

No. 59,665.—W. B. SEWARD, Bloomington, Ind.—*Sorghum Skimmer.*—November 13, 1866.—The shallow pan with a perforated bottom is deepest in the middle, and each side forms a skimmer edge; a bridge piece and socket afford means for attaching the handle.

Claim.—An improved skimmer A, open at both ends, *a'* and *a'*, so as to operate when moving back and forth, substantially as herein shown and described and for the purpose set forth.

No. 59,666.—ERASTUS SLATER, Girard, Penn.—*Fruit Extension Ladder.*—November 13, 1866.—The ladder is extended by winding the cord on a "round," rotated by a crank. The extension is secured by spring catches, which fall in place on projections under the steps.

Claim.—The arrangement of the sections A B C and clasps D as described, in combination with the windlass F, pulleys G H I, rope J, and catches K and L and springs K' and L', the several parts being constructed and arranged and operated as and for the purpose specified.

No. 59.667.—HENRY SMITH, dec'd, Salem, Mass., by his administrator. GILES K. COATES, Boston, Mass.—*Mechanism for Closing Doors.*—November 13, 1866.—The spring shaft has a spline which slides in a key seat of the spring hub, allowing a longitudinal motion in the shaft, and a wheel on the shaft has teeth entering cavities in the disk to which the door lever is secured. The course of the cavities on the disk insures equal strain on the lever in the varying tension of the spring.

Claim.—The combination of a coiled spring, the power of which is equalized by a movable pinion working in a scroll gear, with the mechanism for closing a door or gate, all constructed and arranged substantially as described.

No. 59,668.—HENRY SMITH, dec'd, Salem, Mass., by his administrator, GILES K. COATES, Boston, Mass.—*Equalizing Spring for Clock Movements.*—November 13, 1866.—Improvement on his patent of January 30, 1866. The object is to combine an equalized coiled spring with a clock movement so as to exert a constant power upon the train. A movable pinion engages with a scroll rack so as to act with a constantly increasing leverage as the spring unwinds.

Claim.—The combination with a watch or clock movement of a coiled spring, the power of which is equalized by a scroll rack and movable pinion, substantially in the manner and for the purpose described.

No. 59,669.—J. NOTTINGHAM SMITH, Jersey City, N. J.—*Foot Press.*—November 13, 1866.—The vertically reciprocating slide is actuated by a system of inclined grooves with anti-friction rollers running therein. The power is communicated from a treadle on a toggle-jointed lever connected to a pivoted arm.

Claim.—The combination of two or more wedge drivers, operating at right angles or transverse to each other, substantially as and for the purpose herein specified.

Also, either simple or compound levers in combination with two or more wedges acting at right angles or transverse to each other, substantially as herein specified.

Also, such a combination of wedge power or of wedge powers and lever powers combined as to produce the final action in either direction, for the purpose set forth.

Also, a wedge or wedges, adjustable in direction when applied substantially as and for the purpose herein specified.

No. 59,670.—JOSEPH NOTTINGHAM SMITH, Jersey City, N. J.—*Hydrant.*—November 13, 1866.—The valve is opened by a pin projecting from the plunger on the lower end of the delivery spout. The spout, at its upper end, passes through a slot in a cap piece embracing the top of the hydrant case. By means of oblique grooves in this cap and corresponding pins in the hydrant case, a rotation of the cap elevates or depresses the spout, and accordingly closes or opens the valve. There is also attached to the principal valve a duplicate of similar construction which operates simultaneously with it.

Claim.—First, the combination and arrangement of the spiral grooves P P, projections X X, and horizontal guide opening *v*, substantially as and for the purpose herein specified.

Second, the guide plate or disk *f*, in combination with the hydrant body and discharge pipe, for the purpose set forth.

Third, the arrangement of the duplicate valves L and M, so as both to close fully against their seats, in combination with the hydrant reservoir and the discharge pipe thereof, substantially as and for the purpose herein specified.

Fourth, the arrangement of the reservoir lining I, in connection with the removable reservoir bottom I, substantially as herein set forth.

Fifth, the combination of the concave valve, the annular soft valve seat, and the ring metallic lining thereof, substantially as and for the purpose herein specified.

No. 59 671.—JOSEPH NOTTINGHAM SMITH, Jersey City, N. J.—*Hydrant.*—November 13, 1866.—The descent of a hollow piston in the discharge pipe brings it in contact with the stem of a spring valve, and pressing it open permits the flow of water up through the piston.

and hollow piston rod at the end of which it is discharged. The piston is moved by a rack and wheel.

Claim.—The combination and arrangement of the spout G, weighted as described, reservoir B and valve I, so that the water when flowing is conducted through said spout without communicating with the reservoir, but when the valve is closed and the water ceases to flow, a communication is opened between the spout and reservoir, substantially as and for the purpose herein specified.

No. 59,672.—J. NOTTINGHAM SMITH, Jersey City, N. J.—*Faucet.*—November 13, 1866.—The inner spout slides in an enveloping cylinder of the outer spout and comes in contact with a packing plate at the bottom of the cylinder. A sliding wedge depresses the outer spout to allow the liquid to flow. On releasing the wedge it springs back and the spigot is closed.

Claim.—The spout B, closing around the end of the barrel A and provided with a packing disk *d*, in combination with the barrel, substantially as and for the purpose herein specified.

Also, the wedge *c*, either with or without the spring *i*, in combination with the spout B, substantially as and for the purpose herein set forth.

No. 59,673.—OTHNIEL J. SMITH, Wauwatosa, Wis.—*Bolt Cutter.*—November 13, 1866.—The stock has a stationary chisel and a movable chisel operated by an eccentric on a slide adjustable by a set screw. The movable chisel is slipped toward the stationary one by the motion of the eccentric and returned by a spring.

Claim.—The combination of the lever, the eccentric, the adjustable slide, moved by a screw, the chisels, one stationary and the other movable, secured by bolts to the respective blocks, one chisel, with sharp shoulder near edge, the spring, and the movable chisel block, all constructed and arranged as described.

No 59,674.—THOMAS F. SMITH, Ohio county. W. Va.—*Grinding Mill.*—November 13, 1866.—A plate over the eye of the upper millstone prevents any air from entering between the stones except what is fed in through a tube with the grain.

Claim.—The employment of an air obstructor, substantially as described, in connection with the tube *n*, hopper *o*, and revolving upper millstone *a*, by means of which a vacuum or partial vacuum is obtained between the upper and lower millstones, in the manner and for the purposes described.

No. 59,675.—SAMUEL W. SOULE and C. LATHAM SHOLES, Milwaukee, Wis.—*Numbering Machine.*—November 13, 1866.—Series of numbers are arranged on parallel bars and are worked by a treadle, all the combinations being produced automatically. The treadle being depressed lowers the platen; a dog operating its ratchet pushes forward the moving column, exposing the number next in series, which is impressed upon the ticket as the treadle is released and the platen rises. The following columns are brought into operation in their decimal order.

Claim.—First, the application of the numerals 1 2 3 4 5 6 7 8 9 0 to a series of plane reverse travelling columns or bars, the ten figures being arranged consecutively on each, as described, for the purpose of producing by their combination any desired number.

Second, the construction and combination of the set dogs *g* and the slides *s s*, by means of which the set dogs and moving dogs are raised from their ratchets, as described.

Third, the combination of a series of moving columns containing the numerals 1 2 3 4 5 6 7 8 9 0 in numerical order with the slides *s s*, moving dog *f*, and set dog *g*, by which they are operated to produce any combination of numbers, as described.

Fourth, the attachment of the lug *o*, whereby the moving dog *f'* is made to carry the other dogs *f² f³*, &c., for the purpose described.

Fifth, the construction of the moving columns *a b c* and ratchet bar *z* with the flange to keep the moving dogs from impinging on their ratchets, as described.

No. 59,676.—CHARLES W. STAFFORD, Saybrook, Conn.—*Ore Crusher.*—November 13, 1866.—The reciprocating crushing jaw is guided in a rectilinear path so that the motion is equal both at the top and base of the jaws, differing in this respect from the rocking motion of a pivoted jaw.

Claim.—The reciprocating jaw H, guided in a rectilinear path by the plate *a* and actuated by eccentric G and lever D D', substantially as and for the purpose herein specified.

No. 59,677.—W. H. STARRY, Middletown, Ohio.—*Bag Holder.*—November 13, 1866.—The hem of the bag is lapped over the stationary curved jaw, and is clamped by the movable jaw, which is depressed by a treadle.

Claim.—First, the bevel clamping jaws E G in combination with ratchet and pawl *b*, co-operating for holding sacks and bags, when constructed and arranged in the manner substantially as described.

Second, the combination of the treadle mechanism for settling the contents of the sack or bag with the clamping jaws and ratchet and pawl, operating conjointly, when constructed and arranged substantially as described for the purpose specified.

No. 59 678.—GEORGE W. STORER, Portland, Conn.—*Tag or Label.*—November 13, 1866.—The label has a loop and a tongue; the former is bent around the object and the tongue is passed through the loop and flattened against it.

Claim.—An improved tag or label, made and applied substantially in the manner described and for the purpose specified.

No. 59,679.—S. C. TALCOTT, Ashtabula, Ohio.—*Elastic Button-hole for Carriage Curtains.*—November 13, 1866.—The perforated disk of rubber is clamped between the annular portion and the turned-over radial flanges of a metallic plate; the ends of these flanges are passed through an opening in the curtain, and are then recurved to clamp the edge of the latter.

Claim.—The tin A, or its equivalent, and the rubber B, as arranged and in combination with the curtain E and lining F, for the purpose and in the manner herein set forth.

No. 59 680.—TEMPLE TEBBETTS, New York, N. Y.—*Apparatus for Creasing Paper Collars.*—November 13, 1866.—The curved creasing knife is elevated and depressed by a crank movement, and the curved gauge is adjustably arranged in its rear, its automatic ascent and descent being dependent upon the motion of the knife, as is also the operation of a trip feeder before the blade.

Claim.—First, a curved gauge in combination with a curved creasing knife, constructed and operating substantially as and for the purpose herein fully described.

Second, a doffer in combination with a creasing knife and gauge, constructed and operating substantially as and for the purpose set forth.

No. 59,681.—HENRY THOMASON, Lafayette, Ind.—*Seeding Machine.*—November 13, 1866.—The seed agitators are actuated by gearing from the ground wheel through a universal jointed bar. The ground wheel may be moved forward out of gear with the seeding mechanism by a lever. The outer seed spouts have lateral movement by a pivot lever and connecting arms.

Claim.—First, the lever K. provided with the taper bar L at its lower end, in connection with the pendent elastic bar F, having the bearing *a* of the shaft E at its lower end. the arm or rod *f*. projecting from the inner side of the bearing *a* and the spring J, all arranged substantially as shown, to admit of the bevel pinion D being thrown in and out of gear with the wheel C.

Second, the seed box P attached to bar I and cross piece r and notched bar I*, so as to be independent of beam A, in the manner specified.

Third, the two bars T T in connection with the beams S S and seed boxes R R and bar G.

Fourth, the seed cups u in connection with seed boxes R R P. arranged in the manner described.

Fifth, the attaching of the seed-conveying tubes W to the beams S S by means of the pins w and clamps w*, passing through blocks a*, substantially as and for the purpose set forth.

No. 59,682.—J. P. THOMPSON, Kirkville, Iowa.—*Composition for Roofing.*—November 13, 1866.—Composed of pulverized limestone 50 pounds, and raw coal tar 1 gallon.

Claim.—A composition for roofing compounded from the ingredients named, and substantially as set forth.

No. 59,683.—DAVID S. TROUT, Arcola, Ill.—*Wheelwright's Machine.*—November 13, 1866.—The auger shaft is revolved by a treadle. The hub of the wheel is clamped on a table having longitudinal movement to or from the auger. The same table has guides for the felloes, and a clamping screw above.

Claim.—The arrangement of the adjustable reciprocating table D of a wheelwright's boring machine, the plate E, plates b b and segments g g, when operated as and for the purpose described.

No. 59,684.—HIRAM TUCKER, Newton, Mass., assignor to THE TUCKER MANUFACTURING COMPANY.—*Venting Cores for Foundry Purposes.*—November 13, 1866.—A spiral and flexible coil of wire is introduced into the core and by the closeness of its coils prevents the material falling through, while its interior forms a vent duct.

Claim.—The described improvement in the art of casting molten metals, by which the cores are better and more easily vented than heretofore.

No. 59,685.—EDGAR B. VAN WINKLE, New York, N. Y.—*Station Indicator for Railways.*—November 13, 1866.—The normal position of the pointer is at zero. The passage of a train gives it one revolution; it commences to retrace its path after the train has passed, and consumes an hour in so doing. The conductor of a succeeding train, if within an hour, can thereby read the time that has elapsed since the preceding train passed.

Claim.—A train of wheels for operating or giving the proper movement to an index or pointer K. which works over a graduated dial plate L. a weight V, or its equivalent, applied to said wheels and to the pulley G, link or rod F, and lever D, for connecting the train of wheels with the lever B, applied to one of the rails A, all arranged to operate substantially in the manner as and for the purpose set forth.

No 59.686.—ALLEXEY W. VON SCHMIDT, San Francisco, Cal.—*Steam Propeller for Boats*—November 13, 1866.—Pipes open fore and aft, and a jet of steam in one or the other direction propels the vessel astern or ahead.

Claim.—In combination with a propeller pipe, arranged either inside or outside the vessel but below the water line, and one or more stationary steam pipes, the ends or nozzes of which are within, and point respectively towards the openings of the propeller pipe, one or more valves or cocks arranged so that a column of steam may be projected at pleasure through either nozzle, thus inducing a current of water through the propeller pipe, as and for the purpose shown and described.

No. 59,687.—H. WALDSTEIN and M. FAUSKI, New York, N. Y.—*Vessel for Cooling Liquids.*—November 13, 1866.—A chamber in the interior of the bottle holds ice or other cooling material, the mouth being covered by a tray.

Claim.—As a new article of manufacture, a bottle or pitcher, provided with a hollow or chamber in its bottom for holding a refrigerating material, the chamber to be closed by a suitable cover or stopper, substantially as specified.

No. 59,688.—W. J. WALKER, Baltimore, Md.—*Apparatus to be attached to Stills to Prevent Frauds on the Revenue.*—November 13, 1866.—The locked box has glass sides, is attached to the end of the still worm, and contains a hydrometer and thermometer. Two tubes lead from the bottom of the box, the one to the high and the other to the low wine tank. A plate serves to direct the stream of spirit against the glass side to clear it of dew or moisture.

Claim.—The connecting of an enclosed vessel, having one or more transparent sides, to the worm of the still, into which the liquor passes, and where its proof is tested by an enclosed hydrometer and thermometer, and from thence passing into the high or low wine tank, as the case may be, so that no one can have access to the liquor from its passage from the worm to its respective tank.

No. 59,689.—HARRISON WEED, New Haven, Conn.—*Suspension Device for Lamps, &c.*—November 13, 1866.—The suspension chain is wound on the fusee by the force of the attached spring, and hangs through an elongated opening in the case on one side of a perpendicular line of suspension, and having teeth on the edge which engage with the links of the chain to hold the article suspended at any desired point.

Claim—Retaining a lamp or burner, which ascends automatically in any position to which it may be adjusted, by the means and in the manner substantially as described.

No. 59,690.—DAVID WERNZ, New York. N. Y.—*Preserving Beer while on Draught.*—November 13, 1866.—The rubber bag is fastened to a bung with vent below. The bag floats upon the surface of the beer and air enters the bag as the beer is drawn.

Claim.—The combination of a flexible bag with a bung, provided with air passages, as described, when said bung and bag are applied at the under side of the cask or barrel, so as to cause the air to enter at the bottom, operating in the manner and for the purpose substantially as described.

No. 59,691.—HENRY P. WESTCOTT, Seneca Falls, N. Y.—*Churn.*—November 13, 1866.—The reciprocating lever is pivoted to a standard on the churn and the main dasher shaft is attached thereto by a pin at the height desired. The other dasher shaft is attached to the main one at the height required, by eccentric clamp.

Claim.—First, affixing the adjustable dasher firmly to its own standard, which is separated from the standard which supports the lower dasher, substantially as and for the purpose described.

Second, fastening the standards of the two dashers together, by a fastening located above the top of the churn, substantially as and for the purpose described.

Third, providing a removable standard for the support of the outer end of the lever, substantially as described.

Fourth, so arranging and attaching the spiral spring as that it will hold the standard which supports the outer end of the lever in place, substantially as described.

Fifth, the combination of the spiral spring with the attachment *h*, when said attachment is so formed and arranged as to allow the operator to vary the upper end of said spring at pleasure, and thereby to increase or diminish the length of the lever by which said spring is drawn out, substantially as and for the purpose described.

No. 59,692.—JESSE S. WHEAT, South Wheeling, W. Va.—*Apparatus for Fleshing and Stoning Hides and Skins.*—November 13, 1866.—The hide is clamped upon the table, which is adjustable horizontally from time to time to bring a new surface to the knife. The latter is attached to a sliding arm and reciprocated by a pitman and crank wheel; the sliding arm is guided upon a beam which is depressed at the commencement of the stroke by an eccentric. The knife is raised at will by a cord.

Claim.—First, the combination of the arm M, pins O, pitman P, strap R, and roller S, with each other, with the sliding arm G, with the treadle T, and with the frame A, of the machine, substantially as described and for the purpose set forth.

Second, the combination of the drop U with the treadle T, substantially as described and for the purpose set forth.

Third, the combination of the strap V, or equivalent, with the drop U, and with the pivoted arm M, substantially as described and for the purpose set forth.

Fourth, the combination of the screw W, or equivalent, with the strap V, and pivoted arm M, substantially as described and for the purpose set forth.

Filth, the combination of the line Z with the drop U, substantially as described and for the purpose set forth.

Sixth, the combination of the cam D, working beam C', and slotted upright bar I. with each other, with the treadle T, crank shaft C, and with the knife beam H, substantially as described and for the purpose set forth.

Seventh, the combination of the set screw rod D', or equivalent, with the working beam G', and with the knife beam H, substantially as described and for the purpose set forth.

Eighth, attaching the knife to a stationary, rigid head shank, substantially as described and for the purpose set forth.

Ninth, the combination of the holding bar J', constructed and operating as herein described, with the grooved edge of the table E', and with the holding cam K, substantially as and for the purpose set forth.

No. 59,693.—J. D. WHELPLEY and J. J. STORER, Boston, Mass.—*Process of Treating Sulphurous Ores of Copper.*—The manipulations cited are substantially as follows: First, minute pulverization; second, aerial combustion in air blast heated flue, and immediate lixiviation in acidulated water; thirdly, burning without special fuel, the supply of heat being derived from the combustion of sulphur by mixing the ores with that view; fourthly, where the third manipulation has produced insoluble oxides by too rapid burning. a reaction is produced, resulting in a soluble sulphate by adding sulphate of iron or sulphuric acid to the finely pulverized mass and baking slowly on the floor of a reverberatory furnace; fifthly, lixiviating with pure or acidulated water, either sulphuric or hydrochloric, according to the nature of the salt produced by the last manipulation, followed by the use of the centrifugal machine; sixthly, the manufacture of revived iron from the oxide or colcothar produced by mixing carbonaceous matter and burning to a red heat without air; seventhly, refining the sponge iron by heating to a welding point without air and passing through squeezers.

Claim.—First, the seven manipulations above set forth, in their order, and with the variations described, as a process for treating sulphurets of copper.

Second, the first, second, third and sixth manipulations and the variations described, as a process for treating sulphurous ores of copper.

Third, the first six manipulations and the variations described, as a process for treating copper sulphurets.

Fourth, the first, second, third, fifth, and sixth manipulations, and the variations thereof caused by omitting the third and sixth, and employing the method described after " seventh," as a process for treating copper sulphurets.

Fifth, the rearrangement of the equivalents of the ore by the heat generated by its own combustion, in presence of oxygen and without other fuel than that contained in itself, substantially as described.

Sixth, the employment for the lixiviation of minerals of the centrifugal drying machine, as described, and the arrangement of the felt lining upon its interior, substantially as described.

Seventh, the revival of iron from iron oxides, by diffusion of gases between carbon and the oxides, at a low degree of heat, and without currents of air, substantially as described.

No. 59,694.—J. D. WHELPLEY and J. J. STORER, Boston, Mass.—*Treating the Mixed Sulphurets of Zinc and Lead.*—November 13, 1866.—First, the pulverized ore is burnt in the inventor's water furnace described in another specification; the results being a soluble sulphate of zinc and insoluble sulphate and binoxide of lead. Second, the sulphate of zinc is drawn off and the sulphate of lead reduced by charcoal in the usual way. The silver, if any, will go with the lead and may be extracted by cupelling. Third, to the sulphate of zinc add chloride of calcium and obtain sulphate of lime and chloride of zinc.

Claim.—First, the first, third, and fourth manipulations, in their order, as a means or method of treating zinc blende.

Second, the first, third, and fourth manipulations, in their order, with the addition of the second, as a method or means of treating associated blende and galena.

No. 59,695.—J. D. WHELPLEY and J. J. STORER, Boston, Mass.—*Apparatus for Feeding Fuel to Furnaces.*—November 13, 1866.—The case is divided into compartments, one of which has a hopper communicating with it by means of a pipe. Within the pipe is a fluted roller for dividing the passing fuel. The chamber contains a rotary disk having cutting arms and breakers attached to it, and the other chamber contains a rotary fan which drives the comminuted fuel through a duct to the furnace.

Claim.—First, the construction of a machine containing a comminuting apparatus for fibrous fuel, substantially as described, in combination with the fan blower of an air blast, as and for the purpose described.

Second, the arrangement of cutting blades O, and air wheel paddles P, upon one or more revolving disks L, in the cutting chamber, substantially as described, and the same in combination with crushing cylinder N, substantially as described and for the purpose stated.

Third, the combination of a register F F' with the air or fuel feed of the fan blower, as and for the purpose described.

No. 59,696.—J. D. WHELPLEY and J. J. STORER, Boston, Mass.—*Process and Machinery for Obtaining Metals and other Products from Ores and Minerals.*—Improvement upon the inventor's patent of June 12, 1864, in which ore is roasted while falling in fine powder through a stack, at the foot of which is water to cool and dissolve any salts contained in the ore. By a system of spray chambers any volatile matters in the gases are condensed.

Claim.—First, the construction of the interior of the tower in the form of a hollow truncated cone, for the purpose of securing perfect combustion and the exposure of all the materials, especially the fuel, to heat and oxygen.

Second, the construction of the head of the furnace with dome and arched flues above the fire boxes forming groins at their springs, substantially as described, for the purpose of forming a focus of combustion near the head of the furnace.

Third, the arrangement of the chimney F and telescopic slide G, with its counterpoise and flanges as drawn, substantially as and for the purpose described.

Fourth, the arrangement of the feed apparatus so as to discharge the ore and coal to be supplied to the air blast on the side of the fan blower A away from the furnace, as and for the purpose described.

Fifth, the division of the horizontal flue into chambers, substantially as described, to secure more perfectly the hot lixiviation of the ores, and the similar division of the conductor L into chambers, as and for the purpose described.

Sixth, the arrangement and combination of the settling tank U with the water bottom and pool by means of water exit and water entrance, and the further arrangement of the propeller or conveyer M, in combination with said water bottom and pool, as and for the purpose described.

Seventh, the employment of a wetting wheel, succeeded by a chemical wheel, to remove dust and gases from air, when said wheels are sufficiently separated to allow the effect of the first to be complete before the air to be purified comes under the action of the second, and the arrangement of a trap or valve in the intermediate conductor to balance the draught and projection of the two wheels.

Eighth, the arrangement of the inclined floor h of the spray chamber, in combination with the overflow a, settling tubs and their overflows b, &c., and with the water chamber of the spray wheel, substantially as described.

Ninth, the employment of oxide of copper, or other reducible protoxides, fed into the head of the furnace, substantially as described and for the purpose stated.

Tenth, the means of brightening gold, herein described, by the employment of heat and instantaneous plunging in water or dilute acid.

Eleventh, the evaporating apparatus, substantially as described, consisting of shallow tanks or vats, forming the bottom of an air flue, through which is drawn or forced an artificial current of air, when employed to evaporate a heated solution of sulphate of copper, which cools as the operation proceeds, in order to effect the crystallization of the salt to its greatest practicable extent.

Twelfth, as a manufacturing process, to effect, from a solution of sulphate or chloride of copper or other soluble metal, the precipitation of pure metal in quantity as distinguished from assay, by the substitution of another metal, such as iron, in the solution; the employment of heat and relative motion between the solution and the precipitating metal, and with or without auxiliary galvanic currents distinct from those of local action, substantially as described.

Thirteenth, the employment of heat and relative motion between the solution and the poles of a battery, to accelerate the action of the galvanic current in electro-precipitation of metals, substantially as described.

No 59,697.—STEPHEN B. WHITING, Pottsville, Penn.—*Rope Driving Machine.*—November 13, 1866.—The gear wheels of the drum maintain the same rate of speed. The continuous rope makes the circuit of the drums, which are inclined and arranged laterally so as to prevent the binding of the rope against the sides of the grooves.

Claim.—The use of two drums, inclined and arranged laterally in respect to each other, substantially as and for the purpose described.

No. 59,698.—SILAS M. WHITNEY, Galesburg, Ill.—*Coupling for Cultivators.*—November 13, 1866.—The nose of the plough beam is attached by eye-bolts to a rectangular frame to which the spindle of the supporting wheel is attached, and which is adjustable laterally in the transverse bridge-bar of the machine.

Claim.—The adjustable rectangular frame C, eye-bolts D and E, and connecting bar B, when said parts are constructed substantially as herein shown and described, in combination with the plough beam A and axletree or frame G, as and for the purpose set forth.

No. 59,699—G. WILLIAMS, West Middleburg, Ohio.—*Fruit Jar.*—November 13, 1866.—
The cover fits over the opening and has two lips, one of which fits in a socket on the top plate
and the other over a screw upon which it is secured by a nut.

Claim.—The combination of the can B, socket *a*, and screw *d* with the cap C, tongues *b*,
and *c*. and nut *e*, for the purpose shown, and operating substantially in the manner herein set
forth.

No. 59,700.—J. H. WILLIAMS, Sandusky, Ohio.—*Cider Mill.*—November 13, 1866.—The
fruit passes between the upper metallic rollers, where it is crushed without bruising the seeds,
and then between a metallic and an elastic roller; the juice is kept back and discharged at
one spout while the pomace is discharged separately.

Claim.—The combination of the two metallic or wooden rollers with the India-rubber or
elastic roller, all arranged to operate substantially as and for the purpose set forth.

No. 59,701.—THEODORE A. WILLIAMSON and CHAS. A. RICHARDSON, Alleghany City,
Penn.—*Rolling Pin, &c.*—November 13, 1866.—The different uses are cited in the claims.
The rolling pin is built up of sections with special adaptations of their surfaces.

Claim.—First, a combined rolling pin, beetle, grater, steak hacker, and butter print, substantially as herein shown and described.

Second, the combination with the rolling pin of a beetle, a grater, and steak hacker, all
constructed and arranged in the manner and for the purpose specified.

No. 59,702.—RICHARD WILSON, Cold Spring. N. Y.—*Gas Pipe Joint.*—November 13,1866:
antedated November 1, 1866—Two ends of gas pipe are united by a short sleeve of rubber distended by a spiral spring. Plates attached to each section are united by a middle
plate and prevent retraction of the pipes while they permit flexibility in one direction.

Claim.—First, the hinge E, spiral spring D, elastic sleeve C, and swinging pipe A, in combination with the pipe B, substantially as and for the purpose set forth.

Second, the rubber sleeve C, spiral spring D, and swinging pipe A, in combination with
the fixed gas pipe B, substantially in the manner and for the purpose specified.

No. 59,703.—ROBERT M. YORKS, Schoolcraft, Mich.—*Corn Planter.*—November 13,
1866.—This is a portable planter for dropping two parallel rows simultaneously at any required distance apart. The rods being over the places for the seed, the boxes are depressed,
rotating the seed rollers by means of the straps, whose springs return the rollers and raise the
boxes when the pressure is withdrawn.

Claim.—First, the rods G G, applied to the seed boxes A A, in combination with the cylinders F, straps *e* H, and springs J, all arranged to operate substantially in the manner as and
for the purpose set forth.

Second, the disks or bottoms *h*, in the holes or seed cells *g*, in the cylinders F, arranged in
connection with the springs J and straps K, substantially as and for the purpose specified.

Third, the straps M, applied to the bars I, when used in combination with the cylinders F
and rods G, substantially as and for the purpose set forth.

No. 59,704.—JOHN K. ANDREWS, Antrim, Ohio, assignor to JOSEPH C. TILTON, Pittsburg, Penn.—*Lamp Burner.*—November 13, 1866.—The outer and inner perforated tubes
have slotted caps, above which the gas is ignited; air enters the perforated sides and slotted
floor to mix with the gas before reaching the point of ignition.

Claim.—A lamp burner composed of two perforated tubes B C, with caps G H, in combination with the perforated chamber D, wick tube F, and vent tube I, constructed and
operating substantially as and for the purpose set forth.

No. 59,705.—LEWIS F. BETTS, New York, N. Y., assignor to himself and L. G. HUSTING-
TON, same place.—*Lantern.*—November 13, 1866.—The band forms the connection between
the globe and the lamp, and being detachable permits the withdrawal of the globe for cleaning or renewal.

Claim.—The loose or detachable band F, arranged in connection with the globe G, the
annular base C, and the lamp D, substantially as shown and described.

No. 59,706.—PIERRE BOURDEREAUX, New York, N. Y., assignor to JOSEPH MERWIN and
EDWARD R. BRAY.—*Breech-loading Fire-arm.*—November 13, 1866.—The breech has provision for several modes of firing, adapted to act upon different descriptions of ammunition
with which the barrel may be loaded; firing pins are adapted for cartridges with fulminate
in their bases and and an ordinary cap nipple for charges of loose powder.

Claim.—The breech B, provided with a fixed centre fire nipple *f*, and a sliding firing pin
or pins *e*, in combination with the adjustable face piece E of the hammer, substantially as
and for the purpose specified.

No. 59,707.—SETH BOYDEN, Newark, N. J., assignor to HENRY H. JAQUES, same place—
Hat Blocking Machine.—November 13, 1866.—The unformed hat body is placed upon a

former consisting of expansible ribs, and by vertical pressure thereon the hat is expanded to the desired form by means that cannot be briefly described.

Claim.—First, a block for the blocking of hats, so constructed that from a form corresponding, or nearly so, to that of the hat previous to being blocked it can be changed or made to assume or brought to the form of an ordinary hat block, by means substantially as herein described and for the purpose specified.

Second, in combination with the above, a presser so constructed that when brought to bear upon the said block on which the hat has been placed to be blocked, it will change the form of such block, substantially as and for the purpose described.

Third, the movable block E, having a double series of ribs M and O hung upon and around the same, one series above the other, in combination with the surrounding stationary or fixed flange or rest L, for the outer ends of the lower ribs M, and the centre fixed cup or head D, substantially as described and for the purpose specified.

Fourth, the presser S, formed of two concentric rings T U, connected together through a series of ribs V, hung to the inner ring and passing through the outer ring, substantially as and for the purpose described.

Fifth, the weight or block H^1, in combination with the presser S, when arranged and combined together, substantially as and for the purpose specified.

Sixth the combination of the double-ribbed hat block E, surrounded by a fixed raised flange L, and pivoted with a fixed centre cap or head D, with the ribbed presser S, having block or weight H^1, when combined and arranged together so as to operate substantially in the manner and for the purpose described.

No. 59,708.—JAMES W. CAHILL, Madison, Ind., assignor to himself and A. S. DAVISON, Cincinnati, Ohio.—*Pump*—November 13, 1866.—The plunger rod has two single acting pistons, with valves therein; moves through one-half the length of the cylinder, to and from the induction ports, which are at its mid-length, and ejects at the ends.

Claim—The arrangement of the plungers L M on the shaft K, provided with valves, and operating as described in the cylinder A, which has its induction openings at its mid-length, and eduction openings at its ends, as described and represented.

No. 59,709.—F. M. CARNES, New York, N. Y., assignor to himself and B. F. S. LOGEN-DYKE, same place.—*Shoe and Stove Brush.*—November 13, 1866.—A revolving brush is mounted on the back of the polishing brush, and its bristles come in contact with a cake of blacking, which is contained in a chamber, and advanced by a piston.

Claim.—First, the rotary brush D in combination with the polishing brush A, arranged or applied substantially as and for the purpose herein set forth.

Second, the movable plate F, adjusted or operated by the screw *b* and nut G, or their equivalents, in combination with the rotary brush D, and cake of blacking E, substantially as and for the purpose specified.

No. 59,710.—WILLIAM B. CHAPMAN, La Salle, Ill., assignor to himself, DAVID L. HOUGH, and WILLIAM F. KEELER, same place.—*Horse Holder.*—November 13, 1866.—The lines are connected to the ring-arm of the outer collar, which slips on the inner collar when the horse backs, but tightens when he moves forward.

Claim.—The arrangement and combination of the toothed collar A, surrounding collar B, having arm *c*, and spring pawl G, and ring E, the whole being constructed and operated in the manner and for the purpose set forth.

No. 59,711.—T. B. DE FORREST, Derby, Conn., assignor to J. N. McINTIRE, Brooklyn, N. Y.—*Hoop Skirt Wire.*- November 13, 1866.—The wire, previous to braiding with cotton yarn, is covered with cotton flock, which is made to adhere by a coating of dissolved rubber.

Claim.—Wire coated with flocks, and then "braided" with yarn, either with or without being afterward glazed, or sized and finished.

No. 59,712.—HENRY N. DEGRAW, Newburg, N. Y., assignor to himself and HENRY WRIGHT, same place.—*Boot-jack.*—November 13, 1866.—The jack has a central support, and the pressure of the foot on the fore end rocks it forward, and draws the jaws together by forcing the pendent projections on the jaws between converging slide arms. A plate of rubber expands the jaws when the foot is removed.

Claim.—The pendent frame D, firmly secured to the foot rest A, to which spring jaws F, provided with pendants H, are pivoted, operating with the slotted sliding frame E, with its inclined bars *d*, in the manner described, for the purpose specified.

No. 59,713.—LOUIS DE L'HOMME and ANGELO LAZZARO, New York, N. Y., assignors to JOSEPH ARATA, same place.—*Cement for Roofing, Pavements, &c.*—November 13, 1866.—Composed of pitch, rosin, chalk, Roman cement, sand, and powdered flint stone.

Claim.—The preparation of the "metallic lava," as above described and its practical use in paving halls, hall-ways, basements, cellars, stables, coach-houses, bath houses, yards, piazzas, sidewalks, garden pathways, terraces, &c.

No. 59,714.—R. OGDEN DOREMUS, New York, N. Y., assignor to the ELASTIC SPONGE MANUFACTURING COMPANY, same place.—*Treating Sponge for Stuffing Mattresses, &c.*— November 13, 1866.—Sponge is cut in small pieces, and is moistened with solution of chloride of magnesium, or other deliquescent salt.

Claim.—The preparation of sponge by moistening it with a solution of chloride of magnesium, or equivalent therefor, substantially as and for the purpose specified.

No. 59,715.—WILLIAM DUCHEMIN, Lynn, Mass., assignor to himself and SULLIVAN E. CLOUGH, same place.—*Sewing Machine for Sewing Together the Soles and Uppers of Shoes.*— November 13, 1866.—This machine makes the chain stitch with a half twist in its loop. The shoe last used has a false bottom, which, on removal of the body, is left in the shoe during the sewing; one last thus suffices for any number of bottoms. The sole is channelled, and sewed to the upper with the assistance of guiding gauges. The loop advancer seizes the loop, and drops it on the loop twister. The tacks are extracted by automatic pincers, which open and close, and rise and descend, at proper periods. The shoe-supporting arm swivels in a post which is upheld by a spring, that it may yield to the demands of the work; but by means of a friction brake it may also be firmly held while the awl punctures or the stitching is being done.

Claim.—In combination with the machinery for sewing, as described, or its equivalent, the shoe supporter or arm B, provided with mechanism for operating it, substantially as described, such arm being to be used in connection with the bottom piece of the last, in manner and for the purpose as hereinbefore explained.

Also, the combination of mechanism for extracting the tacks, substantially as explained.

Also, the combination of the loop advancer M, and the loop twister N, with the hooked needle L, the feeder O, and the thread carrier K, each being constructed and provided with mechanism for operating it, substantially as described.

Also, the combination of the loop advancer M, the loop twister N, the hooked needle L, the awl P, the feeder O, and the thread carrier K, each being provided with mechanism for operating it, substantially as specified.

Also, the combination of the main and auxiliary gauges H and I with the arm B, and mechanism for sewing, substantially as described.

No. 59,716.—JOHN B. FONTAINE, Philadelphia, Penn., assignor to HOFF, FONTAINE & ABBOTT, same place.—*Hollow Pressure Plate.*—November 13, 1866.—The pressure plates are to be heated by steam, for pressing knit or woven fabrics, and have stay rods cast in one piece with the plates. A hole is drilled longitudinally through each stay rod, and a heated rod inserted and riveted down at each end into countersinks in the outer surfaces of the plates. This strengthens them so as to compensate for the fracture of any of the stay rods, which may be occasioned by unequal contraction during the cooling of the casting.

Claim.—A hollow cast-iron die or plate, stayed and riveted in the manner and for the purpose described.

No. 59,717.—JOHN O. HARRIS, Reading, Penn., assignor to himself and ISRAEL S. RITTER, same place.—*Lantern.*—November 13, 1866.—The lamp slips into a frustal chamber with a perforated floor; the cap is double, each part being perforated.

Claim.—A square or quadrilateral lantern having a square base A, with perforated sides and a perforated top with a conical chamber B attached, in combination with a top piece or cap composed of a cylindrical chamber D, perforated with holes or openings *e*, and a jacket E, encompassing D, and perforated at its upper and lower parts, substantially as and for the purpose set forth.

No. 59,718.—CHARLES W. HOLLAND, Fredonia, N. Y., assignor to himself and H. L. TAYLOR, same place.—*Knob Hole for Carriage Curtains.*—November 13, 1866.—The annular ring is inserted around the opening in the cover lining and the strengthening piece, so as to prevent tearing.

Claim.—The application and use of the metallic ring D and strengthening piece C, in combination with knob holes in carriage trimmings, in the manner and for the purpose substantially as herein described.

No. 59,719.—A. J. HOLMES, Saratoga Springs, N. Y., assignor to WELLS L. ROBBINS, same place.—*Picture Frame.*—November 13, 1866.—The central portion and the facets or bevelled sides of the frame are each adapted for the display of pictures or ornamentation.

Claim.—A new article of manufacture, consisting of picture and similar frames, constructed substantially as herein described.

No. 59,720.—H. C. HUNT, Amboy Ill., assignor to himself and C. D. VAUGHAN, same place.—*Cultivator.*—November 13, 1866.—The circular frame is bolted to a nearly rectangular inner frame. The wheel spindles are connected to the frame by a vertical bolt, and have a projecting foot rest by which they are cramped. A connecting bar insures the equal cramping of the two wheels, and a hook renders the bar rigid. The two inner plough beams are of one piece, and curved in front in a U form. The ploughs are raised by a segmental cam, operated b a hand or foot lever.

Claim.—First, the combination of the circular frame A and rectangular frame C, when constructed and arranged substantially as and for the purposes specified.

Second, constructing the two beams of a single piece of wrought iron or wood, bent in the form of the letter U or any equivalent form, provided with the draw loops, as and for the purposes shown.

Third, the employment of the cams P, arranged with respect to the plough beams and frame of a cultivator, substantially as and for the purposes specified.

Fourth, the employment of the rollers *h h*, when arranged with the cams P and beams D, and operating as and for the purposes described.

Fifth, the crank shaft Q, arranged and operating with the cams D, substantially as specified and shown.

Sixth, the employment of a transomed spindle, constructed and operating as herein shown and set forth.

Seventh, in combination with said transomed spindles, the arrangement of the arms J extending parallel with each other forward, so that the connecting bar K will not obstruct the view of the operator, as and for the purposes described.

Eighth, the employment of the hooks *r* or their equivalent, for the purposes specified, in the manner described.

No. 59,721.—E. C. C. KELLOGG, Hartford, Conn., assignor to himself, S. F. BENNETT, and D. H. BURRILL, Little Falls, Conn.—*Gauge for Determining Angles*—November 13, 1866.— The graduated bar has graduated arms, one of them movable and provided with a block, whose edge forms with it an angle of 120° as a gauge for hexagonal prisms.

Claim.—First, the construction of a slide gauge with scales, such as are represented in Fig. 1 and herein described, for slabbing polygonal prisms.

Second, a slide gauge with the several scales on its link A and fixed and movable arms B and C, with points *g g* on the said arms, and with a hexagon gauge on its movable arm, all substantially as herein specified.

No. 59,722.—EDWARD M MANIGLE, Philadelphia, Penn., assignor to GEO. H. HAZLETON, same place.—*Manufacture of Wash Boilers, Kettles, and other Vessels made of Sheet Metals.*—November 13, 1866.—Copper for the bottoms of wash boilers, kettles, &c., is coated with alloys of tin, lead, and antimony, or with tin and lead. The object is to cover the plate with a homogeneous alloy to prevent galvanic action at the point of junction with the tinned iron plate of the sides.

Claim.—The manufacture and use of bottoms for wash boilers and other similar vessels made of sheet copper, sheet brass, or of other equivalent sheet metal, and coated substantially in the manner and for the purpose herein set forth and described.

No. 59,723.—WM. H. MILLER, West Meriden, Conn., assignor to MERIDEN MANUFACTURING COMPANY.—*Breech-loading Fire-arm.*—November 13, 1866.—The improvements refer to means for opening, closing, and locking the breech. The latch is formed on a lever pivoted to the tail piece, and enters above the barrels; lateral projections on the lever lock in the breech pieces. A horizontal bolt enters the rear of the barrel, and is withdrawn or retracted by a vertical bar and a finger piece just in advance of the trigger guard; the same bar elevates the latch above. The bar is retained in this position by a pivoted lever actuated by a spring. The descending barrel releases the detent.

Claim.—First, the lever M, constructed and arranged with the projections *d*, in combination with a corresponding recess *f* and the latch L, so as to operate substantially in the manner herein set forth.

Second, the vertical bar R, in combination with the bolt P and the lever M, constructed and arranged to operate substantially in the manner and for the purpose herein set forth.

Third, the combination and arrangement described of the lever N with the bar R and the barrel of the arm, substantially as and for the purpose described.

No. 59,724.—WM. B. MILNE, Chicago, Ill., assignor to himself and WM. BROPHY, same place.—*Manufacture of Soap.*—November 13, 1866.—To the oils or fats used in the manufacture of soap is added a quantity of farinaceous material; the oils thus prepared are saponified by a mixture of soda and potash.

Claim.—First, the mode of introducing the farinaceous materials by dissolving or saponifying them with the fats or oils without a separate alkaline treatment, substantially as specified.

Second, as a new article of manufacture, a soap made by dissolving or saponifying farinaceous substances and fats or oils with an alkali having a potassium base, and mixed with a soap having its fats or oils saponified with an alkali having a sodium base, substantially as herein set forth and specified.

No. 59,725.—JOSEPH I. PEYTON, Washington, D. C., assignor to J. P. TORBERT, same place.—*Horseshoe.*—November 13, 1866.—The wearing plates of the shoe are secured to the main portion by means of clamps and screws, and with interposed strips of rubber between the plates to secure elasticity.

Claim.—First, securing India-rubber, or its equivalent, and the wearing portion of the shoe, in a groove in the main plate,' substantially as described.

Second, constructing a horseshoe substantially in the manner described, for the purpose of rendering the shoe elastic, and removing or replacing the wearing surface without removing the shoe from the foot of the horse.

Third, the combination of the shoe with the rubber, the yielding blocks or bar, and the screws, substantially in the manner and for the purpose set forth.

No. 59,726.—A. SCHRICK and H. HILDENBRAND, St. Louis, Mo., assignors to themselves, F. C. KRAYER, and C. R. SCHRICK, same place.—*Machine for Filling Horse Collars.*—November 13, 1866.—The open end of the case of the collar is clamped over the stuffing pipe, and its other end held by the toggle nippers. The stuffing rod is then put in operation. The stuffing hopper is automatically reciprocated to present fresh material to the stuffer.

Claim.—First. the combination of the bed plate D with the hopper C, and also with the straining shaft E, as described and set forth.

Second, the nippers G and E', when constructed and employed substantially as described and set forth.

Third, giving to the hopper C a vertical motion, for the purpose of enabling the plunger to reach new material at each successive stroke.

No. 59,727.—A. SCHRICK and H. HILDENBRAND. St. Louis, Mo., assignors to themselves, F. C. KRAYER, and C. R. SCHRICK, same place.—*Machine for Stretching Horse Collars.*—November 13, 1866.—The smaller end of the block has an extending part moved longitudinally by a screw, which is operated by a hand crank. The clamping rope has a twist around a pivoted lever connected by a pawl to a rack on the main block.

Claim.—The combination of the block B, the stretching slide C, the lever E, and its pawl E', and rack e², and the screw D, when constructed and employed substantially as set forth.

No. 59,728.—T. S. SMITH, New Haven, Conn., assignor to himself, S. A. SMITH, and HENRY LINES, same place.—*Carriage Shackle.*—November 13, 1866.—A ball on the end of the thrill iron is held by a screw cap and a socket on the axle clip; elastic packing in its rear gives it the proper frictional rigidity, and prevents rattling.

Claim.—The combination of the ball and socket joint, provided with a packing E, with a strap H, or its equivalent, and constructed with a plate B, so as to be attached to the axle, substantially as and for the purpose herein set forth.

No. 59,729.—DANIEL STRUNK, Janesville, Wis., assignor to himself and FRANKLIN STRUNK, same place.—*Wind Mill.*—November 13, 1866.—The increased force of the wind places the wings at a less effective inclination, and further increment brings a brake lever in contact with the periphery of the fly wheel.

Claim.—First, the ring governor or weight B, in combination with the vertical shaft A and the connecting rods *e* and *g*, the bent levers *f*, and the brackets *h* fastened to the wings *h*, constructed and arranged substantially in the manner and for the purpose herein described.

Second, the combination of the weight B with the lever brake *k*, the friction fly wheel *l*, lever *q*, and check rope *o*, substantially as described for the purpose specified.

No. 59,730.—A. TANNER, Hoboken, N. J., assignor to himself and HENRY J. PHILLIPS, same place.—*Constructing Buildings.*—November 13, 1866.—The walls are made of boards, the edges of which are laid over the different sides alternately; grooves are formed on the upper and lower sides of these projections; vertical coincident holes are formed at intervals to prevent dry rot.

Claim.—First, a building, the walls of which are produced by a series of boards placed one on top of the other, so as to form a zig-zag pile, substantially in the manner and for the purpose set forth.

Second, the gutters *b* in the recesses *a*, substantially as and for the purpose described.

Third, the air channels *c d* in the board B, substantially as and for the purpose set forth.

No. 59,731.—JOH. TURNER, Richmond, Va., assignor to himself and JOHN G. HUNTER, same place.—*Adjustable Nut Box.*—November 13, 1866.—The die of the nut box is made in parts, so arranged that they may be adjusted to different sizes of nuts, the steel slides slipping out or in to acquire the right adjustment.

Claim.—First, the angular slides A, in combination with each other and with the block or frame C, constructed and arranged substantially as herein described and for the purpose set forth.

Second, the combination of the steel slides D with each other, with the angular slides A, and with the block or frame C, constructed and arranged substantially as described and for the purpose set forth.

No. 59,732.—STEPHEN WEST, Trenton, N. J., assignor to WEST & THORN.—*Baking Pan.*—November 13, 1866.—The baking pan is adapted for fancy crackers of cylindrical form, which lie in the semi-cylindrical grooves. The ends of the pan are notched to afford gauges for the rule, which is laid across to determine the length to which the sections are cut.

Claim.—A baking pan, having a grooved or corrugated bottom B and notched end pieces C, as and for the purpose herein described and represented.

No. 59,733.—THOMAS WRIGHT, New York, N. Y., assignor to himself and R. VOSE.—*Broom.*—November 13, 1866.—The bundles of strips are inserted through apertures made in pairs in the base plate, the back plate being fastened over the loops of the bunches, and having sockets for the handles.

Claim.—As an article of manufacture, a wire broom, made substantially in the manner herein set forth.

No. 59,734.—EDWARD DODÉ, Paris, France.—*Metallizing Mirrors.*—November 13, 1866.—Metallic platinum is precipitated in a state of minute division from a solution of the bichloride of platinum by means of oil of lavender; the metal thus obtained is filtered and washed, after which it is mixed with litharge and borate of lead and applied to the glass. The glass is then heated in a suitable oven to fix the metal.

Claim.—First, the peculiar processes or modes of preparing a product to be employed in metallizing glass, with a view to the manufacture or production of looking-glasses, mirrors, and other reflecting surfaces, as hereinbefore described.

Second, the peculiar processes for extracting or removing the salts and acids from the metallizing product, as hereinbefore described.

Third, as a new article of manufacture, a glass or mirror, metallized on its face by the application of platinum, in the manner substantially as hereinbefore described and set forth.

No. 59,735.—WILHELM HUGO, Celle, Hanover.—*Umbrella.*—November 13, 1866.—A T-shaped rib has a longitudinal gutter or cavity to receive the suture of the cover.

Claim.—First, a T-shaped rib for umbrellas or parasols, as a new article of manufacture.

Second, providing the T-shaped rib with a longitudinal gutter or cavity, substantially as and for the purpose described.

No. 59,736.—LOUIS GUSTAVE SOURZAC and LOUIS BOMBAIL, Bordeaux, France.—*Rendering Leather more Durable and Flexible.*—November 13, 1866.—The leather is treated with a compound of siccative linseed oil, 13 ounces; ordinary linseed oil, 13 ounces; orchanet, 1 ounce; mix and reduce it to 17 ounces add dried garlic, 2 ounces, and burnt bread 1 ounce. It is then placed in a bath of red tartar, salt and brazil wood. The process is repeated.

Claim.—The improved process herein described for rendering leather more durable and flexible.

No. 59,737.—G. L. TURNEY, London, England, assignor to SAMUEL A. HARSHAW, New York, N. Y.—*Needle Case.*—November 13, 1866.—Needles are wrapped in fine paper or foil, the eyes exposed. The parcels are then placed in cylindrical boxes the lids of which are removable.

Claim.—The method herein described of putting up needles, by wrapping them up in a paper or tinfoil b, leaving their heads exposed, and enclosing said paper or tinfoil in a box a, substantially as set forth.

No. 59,738.—HENRY WILDE, Manchester, England.—*Magneto-electric Machine.*—November 13, 1866.—The permanent horse shoe magnets are fixed on a cylinder formed of two segmental pieces of cast iron and two intermediate of brass of the same length, bolted together. A hole bored in the cylinder is occupied by the revolving armature of slightly less diameter than its case and of peculiar construction, with a wrapping of insulated wire, and has alternate conducting and non-conducting surfaces exposed to the iron segments. The combination with the electro-magnetic machine and the course of the currents cannot be briefly described.

Claim.—First, the method of constructing magnet cylinders for magneto-electric and electro-magnetic machines, by making them of segmental iron concaves, with intervening strips of wood, brass, or other non-electric material, substantially as set forth.

Second, the combination of a magneto-electric and electro-magnetic machine, constructed and operating substantially as and for the purpose set forth.

No. 59,739.—A. ELY BEACH, Stratford, Conn.—*Receiving and Delivering Letters, Parcels, &c.*—November 13, 1866.—A pneumatic tube is traversed by a car which receives and delivers letters at intermediate stations and at the ends of its route. Devices actuated by the car in passing, or controlled by electro-magnetic apparatus from an office, arrest the car, either for the reception of letters from a receiver, or their deposition at a given point, as the devices may be arranged or actuated by the electro-magnetic connection. Details cannot be explained within reasonable limits.

Claim.—First, the method, substantially as herein described, of automatically collecting letters, parcels, and other freight.

Second, the method, substantially as herein described, of automatically delivering letters, parcels, and other freight.

Third, the employment of the studs and rods, or their substantial equivalents, to operate the delivering and receiving mechanism, substantially as herein shown and described.

Fourth, the employment of an electro-magnetic apparatus in combination with the receiving and delivering mechanism, substantially as and for the purpose herein set forth.

Fifth, the means herein described, or their equivalents, for reducing the speed, substantially as described.

Sixth, in the collection of letters, parcels, and other freight, the employment of a moving box or receiver U, or its equivalent, operating substantially as described.

Seventh, the employment of the adjustable nose, or its equivalent, with the car, substantially as and for the purpose herein shown and described.

Eighth, the combination of a swinging valve with the freight receptacle, substantially as herein shown and described.

Ninth, in the collection and delivery of letters, parcels, and other freight, the employment of the several mechanisms herein shown and described, or their substantial equivalents, operating separately or otherwise, substantially as herein shown and described.

No. 59,740.—PERLEY AINSWORTH, Cape Vincent, N. Y.—*Drill for Wells.*—November 20, 1866.—The separate sections and working parts are attached by short screw-threaded sections of pipe. The grabs are made of two parts attached together at their middle and having catches at their ends to hold on the inner screw-threaded segments after being sprung into place. The rimmer is attached in the same manner as the separate sections of piping. The drill runs inside the rimmer and lower section, and has side openings, above which it is hollow; a valve in the pipe at its upper end retains the matter which has worked through the openings. The grab jars may, with the addition of a clamp-ring, be used as a rope socket.

Claim.—First, the combining the centre bit reamer, auger stem, sand pump, grabs, jar, sinker bar, and rope socket, so that they may be worked and used efficiently, separately or together, as above set forth.

Second, the p au or method of coupling, as above described.

Third, the grab jars, constructed as above described, separately and in combination with the apparatus, as above substantially set forth.

Fourth the tube and reamer C, centre bit B, and valves *a a*, constructed so as to admit of their being used separately or in combination with the other apparatus as a sand pump, pumping the debris or sediment into the main trunk as fast as made, by the action of either the centre bit or reamer, or both, when worked together, as herein above substantially set forth.

Fifth the round reamer constructed as above described, separately and in combination with the apparatus, as above substantially set forth, whether attached to the bottom or the top of the auger stem or the sinker bar, cutting either upward or downward.

Sixth, in combination with the above described apparatus, the centre bit made hollow, as shown at B B, and bevelled at the edges, substantially as and for the purpose set forth.

No. 59,741.—A. G. BURTON and H. W. COVERT, Rochester, N Y.—*Filing Machine.*—November 20, 1866.—The file carriage and the table upon which the object to be filed is placed are connected respectively, by ball and socket joints, to pedestals resting upon the base plate; by this means the file may be adjusted so as to move in a vertical plane or in a plane of any inclination thereto, and the table may be adjusted to bring the surface to be filed to correspond to the direction of movement of the file; in addition to these means of adjustment, the pedestals may be adjusted as to distance the one from the other, by set screws passing through the feet of the same and through elongated slots in the base plate. The forward movement of the file is effected by a cam, and its return by a spring arranged with the file carriage

Claim.—First, the table C, adjustable in the manner described, in combination with the reciprocating file K, substantially as and for the purpose herein specified

Second, the combination of two ball and socket joints E D, adjustable in position, one bearing a file carriage and the other a table or support for the articles filed, for the purpose herein set forth.

Third, the combination of the lateral auxiliary leaves L with the table C, and file carriage H, for the purpose of presenting the articles to be filed to the edge of the file, substantially as specified.

Fourth, the arrangement of the machine as a whole, consisting of the carriage frame B, the file carriage H, the table C, provided with the lateral leaves L, and the universal joints D E, constructed as described, the whole operating substantially as herein set forth.

Fifth, the combination of the cam **m** with the file carriage H, which is provided with a reacting force, so that its returns will economize time and power, as set forth.

No. 59,742.—WILLIAM R. BUTLER, Greenbush, Ill.—*Corn Planter.*—November 20, 1866.—The sets of seed disks are furnished with holes of various sizes for different seeds. The upper seed disk is fixed, and the reciprocal oscillation of the under one brings its cavities alternately in communication with the cavities of the upper disk and the discharge hole.

Claim.—The combination, construction, and arrangement of the boxes F, plates G, regulators G¹, levers H, and bar H¹, substantially as and for the purpose represented.

No. 59,743.—ISAAC J. GRAY, Seville, Ohio.—*Gate.*—November 20, 1866.—The gate is pivoted on a bar which, in turn, is pivoted to the post. The gate rotates 90° on the pivot of the bar till it is horizontal, and then swings upon the ends of the bar as an axis; the descending arm is weighted to counterbalance the gate.

Claim.—The double pivot revolving hinge, and the mode of its application to the hanging of gates, as described.

No. 59,744.—REUBEN HARPER. Philadelphia, Penn.—*Manufacture of Hoes.*—November 20, 1866.—A rectangular plate is cut by a die into two pieces with bifurcated ends. One is adapted to make a hoe with a reverted shank and the other piece has two tongues, which are turned up to form a shank without riveting.

Claim.—The stamping of the plate and the forming of the shank in the manner and for the purpose substantially as set forth.

No. 59,745.—JAMES L. HAVEN and CHARLES HETTRICK, Cincinnati, Ohio.—*Bandelore.*—November 20, 1866.—Explained by the claim and cut.

Claim.—The disks A B, united at their centres by the rivet F and interlocking bosses E, substantially as set forth.

No. 59,746.—EDWARD E. KILBOURN, New Brunswick, N. J., assignor to the NORFOLK & NEW BRUNSWICK HOSIERY COMPANY, same place.—*Machine for Sewing the Seams of Looped Fabrics.*—November 20, 1866.—This machine is for stitching together the selvage edges of knitted fabrics in making shirts, drawers, &c. The loops are suspended on a series of points projecting from a plate. The stitching mechanism is on a travelling carriage. A needle and looper are used, forming a chain stitch, which, in order to give sufficient elasticity to the seam, is made alternately through and over the loops.

Claim.—The combination in a seaming machine of the following devices, viz: the seaming mechanism, straight supporting plate, and feed screw, all operating in the combination substantially as set forth.

Also, the combination in a seaming machine of the devices recited in the preceding claim, with cam collars operating upon the feed screw, substantially as set forth.

Also, the combination in a seaming machine of the needle, the looper, the cam that operates them, the carriage, and the mechanism for causing the carriage and its appurtenances to vibrate, all these devices operating in the combination substantially as set forth.

Also, the combination in a seaming machine of a series of points for holding the loops of fabric to be traversed by the needle, with the reciprocating needle and the looper, and with a cam so formed as to cause the looper, after the passage of the needle through a loop of the fabric to be seamed, to take the loop of needle thread at one side of the needle and withdraw from it at the opposite side thereof, all operating in the combination substantially as set forth.

Also, the combination in a seaming machine of the devices recited in the first claim with a section of a nut that can be disengaged from the feed screw to permit the carriage to be moved quickly back to its starting point, all said devices operating in the combination substantially as set forth.

Also, the combination in a seaming machine of the seaming mechanism with a supporting plate provided with half as many supporting points as the number of stitches made by the seaming mechanism, so that the stitches are made in regular succession through the loops of the fabric and out of them, substantially as set forth.

No. 59,747.—FRANKLIN S. PACKARD, Springfie'd, Mass.—*Stone Sawing Machine.*—November 20, 1866.—The saw frame is hung from swing bars pivoted to plates which have curved guides for vertical movement. The front pair of swing bars are pivoted to the slide plates at their upper ends, and the rear pair at their lower ends, to give a rocking motion to the saw.

Claim.—First, producing the rolling or rocking motion of the saws by the arrangement of the connec ions *d d' g g'* and standards *f f*, or equivalent devices, arranged substantially as herein described.

Second, forming the ways or guides *k k'*, &c., for the slides *e e* and *h h* on a curve, substantially in the manner and for the purpose herein set forth.

No. 59,748.—SAMUEL G. RICE, Albany, N. Y.—*Tobacco Press and Cutter.*—November 20, 1866.—The cover of the box is hinged at the rear, and may be depressed in front to reduce the bulk of the tobacco forwarded. The follower is expansible vertically by a contained spring, and is forwarded by a screw passing through an inside screw-threaded wheel at the back of the box. The tobacco is cut at the box mouth by revolving cutters.

Claim.—First, the cover B applied to a box A by a hinge at one end and a vertical adjusting device at the opposite end for the purpose of pressing the tobacco at and near its point of discharge from said box, substantially as described.

Second, the combination of an expansible follower G with a hinged pressing cover B, press box A, and cutter K, substantially as described.

No. 59,749.—HENRY RYDER, New Bedford, Mass.—*Manufacture of Paraffine Candles.*—November 20, 1866.—Improvement on the patent of H. Leonard and Henry Ryder, February 8, 1859.—The candles are annealed after the surface chilling by a continued refrigeration, a small stream of cold water being allowed to flow through the annealing bath.

Claim.—An improved process, as described, for effecting the annealing or gradual cooling of paraffine in a mould, such consisting in the subjection of the mould containing the melted paraffine to a chilling bath of water of about freezing temperature, and subsequently allowing water of about such temperature to flow into, and warmer water to flow out of, the chilling bath, as explained.

No. 59,750.—BARCLAY SAMSON, Cortlandville, N. Y.—*Medical Compound.*—November 20, 1866.—Medicine for hepatic diseases, composed of gentian, mandrake, colombo, sarsaparilla, and calamus roots, cascarilla and wahoo bark, orange peel, quassia, anise, leptandra, Virginica, soda, molasses, alcohol, and water.

Claim.—A preparation or compound composed of the ingredients above specified, and compounded substantially in the proportions and manner set forth.

No. 59,751.—H. K. TAYLOR and D. M. GRAHAM, Cleveland, Ohio.—*Treating Hydrocarbon Oils.*—November 20, 1866.—The pipe through which the air used in agitating the distillate is passed is connected with an air-tight vessel, in which gas is generated; by this means the gas is carried directly into the oil by the air passing through the pipe. The object of the invention is to increase the specific gravity of the oil.

Claim.—First, the disposing affinity of sulphuric acid, causing a chemical combination of the gases used with the oil, either in connection with our process patented May 22, 1866, or in connection with other processes.

Second, the disposing affinity of sulphuric acid in the treatment of hydrocarbons that its use in connection with other substances, solid, gaseous, or fluid, by means of which the energetic combination of these substances, or parts of them, with the hydrocarbons, is very much increased.

Third, treating oil by means of air and acid gases, substantially as set forth and described for the purposes specified.

No. 59,752.—GEORGE E. WHITMORE, Housatonic, Mass., assignor to himself and E. S. PIXLEY, same place.—*Stop-valve for Steam Pipes.*—November 20, 1866—The valve has a face on its end, which acts in conjunction with a seat at the end of the chamber, a pin entering the opening in the seat. The valve is moved by an eccentric pin on the end of the stem, which has a ground joint and an elastic collar packing at its respective ends.

Claim.—The eccentric pin, in combination with the valve stem for forcing the valve A against and into the valve seat F F.

The choke plug I I, which prevents the valve A and the valve seat F F from being worn by the action of the steam in passing between the valve A and valve seat F F.

The arrangement of the ground joint G G at the inside end of valve stem C C and lead ring H H, or equivalent, at outside end of valve stem C C, which forms a double packing. And the combination of all these parts in the manner and for the purpose specified.

No. 59,753.—GEORGE W. BALDING, Pleasant, Ind.—*Gate and Gate Post.*—November 20, 1866.—The posts have acute edges presented to the gate. The alternate rails project into notches in the posts, so formed as to allow of the gate opening from either end, but in opposite directions; loose pins secure it when shut.

Claim.—First, extending alternately the ends of the boards or horizontal bars, so that they may form sufficient bearings to support the gate at both ends, with or without notches in the boards, as shown in the accompanying drawings.

Second, the form of the posts, having an angle so acute at one corner that they can be notched to receive the alternate projections of the ends of the gate and admit the gate to turn in said notches, so that it can be opened either way, at right angles to the standing line when shut.

Third, the form of the notches in the posts, being an open-hooked notch with square or smooth bottom, so that the gate bearings can turn in them, as necessary, to open the gate as aforesaid, into which it is secured by a loosely fitting pin, reference being had to the said drawings.

No. 59,754.—ARTHUR BARBARIN, New Orleans, La.—*Lighting Gas.*—November 20, 1866.—A stream of pure hydrogen is thrown on the platinum sponge by means of a pipe, and the hydrogen being ignited thereby, ignites the ordinary gas as it issues from the burner.

Claim.—First, in combination with ordinary burners for burning illuminating gas, the application and use of spongy platinum in connection with the means for projecting through or upon it a jet of hydrogen gas, substantially as described, for the purpose of igniting the gas issuing from the burners, as herein shown and set forth.

Second, arranging the hydrogen gas pipe and spongy platinum above and on opposite sides of the gas burner to which they are applied. so that the hydrogen while issuing from its pipe shall traverse the course of the gas discharged from the burner, substantially as and for the purposes shown and set forth.

Third, the method of, and apparatus for, simultaneously igniting two or more ordinary gas burners by the use of spongy platinum, acted on by hydrogen in the manner and by the means herein described, whether the pipes by which the said hydrogen gas is conveyed to the platinum be within or exterior to the pipes by which the illuminating gas is conducted to the burners.

Fourth, the concentric arrangement of the pipes for conveying the illuminating and the hydrogen gases to their respective burners, the pipe by which the hydrogen gas is thus conveyed being within and surrounded by the pipe which conducts the illuminating gas, substantially as herein shown and set forth.

No. 59,755.—HAZEN J. BATCHELDER, Boston, Mass.—*Manufacture of Blanks for Horseshoes.*—November 20, 1866.—Straight rods of metal of various lengths are shaped, creased, and punched so as to make a series of blanks suitable for horseshoes.

Claim.—As an article of manufacture, a bar of metal shaped, creased, and punched as herein described, and as represented in Figs. 1, 2, and 3, of the drawings, the same constituting a series of blanks suitable for horseshoes.

No. 59,756.—DANIEL D. T. BENEDICT, Havana, N. Y.—*Medical Compound.*—November 20, 1866.—Composed of quinine, ipecacuanha, and iodine, to be made into pills for the cure of fever and ague.

Claim.—The combination of the three medical materials herein named, namely, quinine, ipecac, and iodine, in the proportions herein stated, into one medicine.

No. 59,757.—C. W. BIOREN, Philadelphia, Penn.. assignor to LYSANDER FLAGG, Central Falls, R. I.—*Tool Handle.*—November 20, 1866.—The tool socket revolves in the handle and its ratchet face connects by a ratcheted central ring with a similar face on the handle. A sliding splined ring connects either the handle or the socket with the middle ring and insures the turning of the tool in the given direction, a backward turn having no effect.

Claim.—A tool handle, provided with the stationary ratchet d, the double-faced sliding ratchet ring i, and the revolving socket b, having a ratchet on its under face, when said parts are combined and arranged to operate as shown and described.

No. 59,758.—JOB W. BLACKHAM, Brooklyn, N. Y.—*Machine for Felting Hat Bodies.*—November 20, 1866.—The machine has a bed of rollers, receives the bats in a rolled-up condition and fulls or hardens them in their passage, steam and water being applied directly to them. The rollers have convex faces, and are continuously turned in the same direction with equal speed : a corrugated presser is reciprocated longitudinally above the rollers at greater speed, and the hat is felted and fed forward by the conjoint action.

Claim.—First, the combination and arrangement of the reciprocating presser C with the bed of rollers b, so as so treat the bats by the surfaces of the presser and the rollers, substantially as and for the purpose herein set forth.

Second, in connection with the above, so proportioning the extent and velocity of the motion of the presser C to the size and velocity of the rotation of the rollers b, that the hats shall tighten in the roll and tend to stand and roll with only a rotary motion during a great portion of the return movement of the part C. and shall tend to advance to an extent equal to the diameter of one of the rollers during each forward movement, substantially as and for the purpose herein set forth.

Third, in combination with the presser C and rollers b the springs G 1 2, &c., or their equivalents. arranged as herein specified for increasing the force with which the hat is treated as and for the purpose herein set forth.

Fourth, in combination with the presser C and rollers b the shaft I and cords g, or their equivalents, arranged as specified, for graduating the effect of the force in holding down the presser C, substantially as and for the purpose herein set forth.

Fifth, the spring P or its equivalent, arranged to operate in opposition to the gravity of the presser C, so as to reduce the pressure on the bats, substantially as and for the purpose herein specified.

Sixth, the devices substantially as herein shown and described, for putting into and out of action the lateral motion of the bed of rollers b.

No. 59,759.—JOHN C DODINE, Camden, N Y., assignor to himself and J. VICKERS.—*Churn.*—November 20, 1866.—The rotary dasher has radial arms which pass between projecting slats from the churn bottom, and also between the bars of a detachable angular frame.

Claim.—The oblong body A, with its semi-cylindrical bottom, its detachable angular grate composed of the bars z, and its vertical slats t, in combination with the central shaft B, and its paddles g, the whole being arranged and operating as and for the purpose herein set forth.

No. 59,760 —URIAH CUMMINGS, Buffalo, N. Y.—*Lime Kiln.*—November 20, 1866.—The inner ends of the furnaces are enlarged and open into an unpartitioned space within the cupola. The draw flues are arranged below, side by side, one for each fire, and lime may be drawn from either fire as required.

Claim.—In a lime kiln the draw flues G G¹, and enlarged fire chambers D¹, constructed and arranged with an unpartitioned cupola, substantially as described.

No. 59,761.—E. K. DUTTON, Manchester, England.—*Apparatus for Measuring Liquids.*—November 20, 1866.—The two-way cock allows the liquor to flow alternately into and from the receiver and connects with a register. The receiver communicates with an air chamber above it, the compression of the air balancing the weight of liquor and assisting to empty the receiver. A pipe communicates from the air chamber to the supply pipe to keep up the equilibrium by discharging any liquor which had entered the air chamber when the pressure of liquor was at its maximum.

Claim.—First, the chamber g, its openings e and f, and two-way cocks c or their equivalents, in combination with the tubes l m, and receptacle i, the whole being constructed and arranged substantially as described.

Second, the combination of the above with the within-described registering mechanism and dials.

No. 59,762.—SAMUEL F. ESTELL, Richmond, Ind.—*Animal Trap.*—November 20, 1866.—The animal passes through the round side hole and treading on the trap board springs down the doors. The passage into the inner chamber resets the trap.

Claim.—First, the double rack N N, constructed and operating substantially as set forth and for the purposes described.

Second, the shaft H H, the crank m, and arms b b, all made of one single piece of wire, when operating substantially in the manner and for the purposes described.

Third, the rack N N, shaft H H, crank m, arms b b, doors K K, and panels S S, in combination, the box J, the openings A A, and platform B, all arranged substantially as set forth and for the purposes described.

No. 59,763.—M. G. FARMER, Salem, Mass., and G. F. MILLIKEN, Boston, Mass.—*Line Wire for Telegraphs*—November 20, 1866.—The copper wire of superior conductivity is strengthened by a core of iron or steel of greater tensile tenacity.

Claim.—In combination with the instruments making up with the conducting wire a telegraph circuit, a copper wire conductor strengthened with iron or steel, substantially as set forth.

No. 59,764.—CHILION M. FARRAR, Buffalo, N. Y.—*Lock-up Safety Valve.*—November 20, 1866.—Two levers of the second order are used; the weight is non-adjustable; the efficient bearing point of the first lever forms the power on the second, and is adjustable on its bar. A hook allows the weight to be raised to try the working order, and a plate below the escape pipe prevents tampering.

Claim.—First, the arrangement and combination of compound levers D F, movable and adjustable fulcrum J, fixed weight H, and safety valve B, within a suitable lock-up box A, for the purposes and substantially as described.

Second, the hook L, having a long handle projecting upwardly through the box, in combination with the lock-up box, for the purpose and substantially as set forth.

Third, the plate N, placed under the escape pipe and within the box A, for the purpose and substantially as described.

No. 59,765.—HENRY R. FELL and EDWARD PHIFER, Trenton, N. J.—*Cotton-seed Planter.*—November 20, 1866.—The seed is driven from the hopper in specific quantities by the thread on the rotating shaft, the masses of seed being segregated by the revolving stirring rod.

Claim.—The combination with the hopper and spout of the shaft M, provided with the feeding screw P and stirring rods N, operating substantially as described.

No. 59,766.—C. A. HARPER, Rahway, N. J.—*Coal-oil Stove or Gas Heater.*—November 20, 1866.—The caloric currents pass around and down through the inner drum, whence they escape through a horizontal pipe. The air currents pass through vertical pipes from beneath the heater to the inner drum, and from thence to the room through vertical pipes.

Claim.—First, the combination of an exterior and interior drum with draught pipes and air flues, when arranged substantially in the manner and for the purpose set forth.

Second, a shifting heater, movable upon slides, when used in combination with an exterior and interior drum, substantially as set forth.

No. 59,767.—JOSEPH J. and EDWARD HARRISON, Manchester, England.—*Device for Operating the Shuttles of Looms.*—November 20, 1866.—In this invention pickers are dispensed with and the shuttle is driven by rapidly revolving frictional rollers, which intermittingly are brought into contact with its sides.

Claim.—First, the novel and peculiar arrangement of mechanism described, for expelling the shuttle of looms from one box into the other, and vice versa, without the use of the picker as now employed for the same purpose.

Second, the eccentric "swell" and the apparatus in connection therewith for gradually and effectually checking the shuttle upon its being received by the shuttle box, and also for instantly releasing the same upon its expulsion, as hereinbefore described, set forth, and fully illustrated in the drawings attached.

No. 59,768.—A. T. HAY, Burlington, Iowa.—*Preventing Incrustation of Steam Boilers.*—November 20, 1866.—By the connection described an electric current is passed through the boiler to prevent the incrustation arising from the adherance of sediment to the boiler plates.

Claim.—The use of a galvanic battery, or its equivalent, placed outside of the boiler of a steam engine, the two poles of which battery are connected respectively with the opposite ends of the boiler, substantially as and for the purpose above set forth.

No. 59,769.—GEORGE P. HERTHEL Jr., St. Louis, Mo.—*Truss Bridge.*—November 20, 1866.—Explained by the claims and cut.

Claim.—First, a truss bridge or other structure, having flexible joints throughout, when constructed substantially in the manner herein described.

Second, in the construction of truss bridges and other structures, forming the joints between the posts, braces, and cords thereof, or either of them, by interposing a rod or bolt in such joint, whereto said posts, braces and cords shall be so pivoted or otherwise loosely connected as to form a flexible or yielding joint, each part being independent in movement, all substantially in the manner and for the purpose herein set forth.

Third, the arrangement of the braces K, for adjustment of the parts forming the panels, in combination with the keys g^4 for adjustment of the end parts, thus by said combination permitting a change or reproduction of camber, substantially as set forth.

Fourth, the general combination of the parts F, with G H I and K, substantially as and for the purpose set forth.

No. 59,770.—JOHN JANN, New Windsor, Md.—*Tool for Sharpening Sickles of Harvesters, &c.*—November 20, 1866.—Cutters with facets of different angles are attached to the end of the shank adapted to the face of the object from which a shaving is to be cut.

Claim.—The tool holder A B, in combination with substitute cutter plates C C C, of varying angular forms, applied and secured to said holder, as and for the purpose described.

No. 59,771.—ADOLPH S. JOURDAN, Nashville, Tenn.—*Cement.*—November 20, 1866; antedated November 9, 1866.—A composition of equal parts by weight of melted shellac and finely pulverized pumice stone.

Claim.—The herein described cement consisting of shellac and pumice stone, substantially as set forth.

No. 59,772.—ANSON JUDSON, Brooklyn, N. Y.—*Lamp Burner.*—November 20, 1866.—The burner and wick tube are made of a single piece of cast metal of a cheaper character than the sheet brass, which is fashioned by successive manipulations. The parting is at the upper end or rim of the shell, and the wick opening is made by a core.

Claim.—First, making the shell of the burner and the wick tube in a single piece, substantially as and to the effect hereinabove set forth.

Second, making the shell of the burner with vertical slits or openings for the admission of air to supply the flame when said openings are so formed and arranged as to permit the said shell to be cast in a two-part flask, and without cores to form said openings, as hereinabove set forth.

No. 59,773.—STEPHEN R. KROM, New York, N. Y., assignor to APOLLOS R. WETMORE.—*Apparatus for Separating Metals from Ores.*—November 20, 1866.—The ore is supplied to the revolving endless belt from a hopper while a current of air is made to pass through the perforations in the belt. A rapidly vibrating metal plate throws off the tailings, which collect at the lower part of the belt, which carries along the metal and discharges it after passing over the upper roller.

Claim.—First, giving the intermittent action to the blast in a separating machine, substantially as specified by means of the valve M, or its equivalent, operating between the blowing means and the perforated bed, substantially as herein set forth.

Second, the endless traversing perforated bed E, in combination with the means for blowing through the upper half thereof, so as to lift and agitate the material thereon, substantially as herein set forth.

Third, the tube *a*, or its equivalent, of less breadth than the endless belt E, and adapted to carry the fresh material down and deposit it immediately upon the endless belt E, while the lighter material is allowed to traverse past it down the incline, substantially as and for the purpose herein specified.

Fourth, the scraper K, arranged to operate in combination with the endless traversing perforated bed E, and with means for producing intermittent blasts or puffs of air through the same, substantially as and for the purpose herein specified.

No. 59,774.—ALBERT LEACH, Lynn, Mass.—*Cement for Leather.*—November 20, 1866.—Composed of caoutchouc dissolved in benzine, ivory black and sulphuric acid.

Claim.—The composition made of the materials and in the manner substantially as hereinbefore specified.

No. 59,775.—JAMES S. MARSH, Lewisburgh, Penn.—*Coal Stove.*—November 20, 1866.—Flues from an annular air space between the fire pot and the stove jackets pass through the stove or main flue and discharge into the room.

Claim.—The construction of a stove with an upper single heat-radiating wall, and with a lower air-heating space which is formed by the outer wall and fire pot, and which is provided with a pipe leading through the smoke pipe, or some other portion of the stove, where it will be exposed to the escaping products of combustion, as described.

No. 59,776.—ZEPHANIAH MARSHALL, Andersonville, Ind.—*Medical Compound.*—November 20, 1866.—For the cure of fever and ague. Composed of equal quantities of dogwood bark and yellow willow, boiled to a solid extract and made into pills.

Claim.—The within described medical compound for the purpose specified.

No. 59,777.—WM. G. PIKE, Philadelphia, Penn.—*Cut-off Valve Gear.*—November 20, 1866.—This is an improvement on his patent of November 20, 1865. A supplementary steam cylinder is combined with the spindle which is attached to the valve to close the latter after it has been opened by the eccentric. The piston of the dash pot is attached to the piston rod of the said cylinder to prevent concussion. A lifter is attached to the tripping shaft to throw the tripping rods which work the valve out of gear when accident happens to the governor. Adjustable spring rods above the trip rods insure proper action.

Claim.—First, the combination of the double arm G, cylinder *a'*, and dash pot *b'*, in the manner and for the purpose substantially as shown and described.

Second, the spring boxes *l l'*, with their enclosed spring *k k'*, adjusting screws *m m'*, and bearing rods *i i'*, in combination with the rods S S', in the manner and for the purpose substantially as shown and described.

Third, the lifter *q*, arranged and operating substantially as shown and described.

No. 59,778.—J. C. PLUMER, Boston, Mass.—*Dies for Cutting and Pointing Wire.*—November 20, 1866; antedated November 10, 1866.—The set of dies are arranged to cut the metal from two sides of the wire and press the other two sides to a point at the same operation.

Claim.—The device substantially as shown in figures 3 and 4, and for the purpose specified.

No. 59,779.—SILAS S. PUTNAM, Dorchester, Mass.—*Self lubricating Axle-box for Carriages.*—November 20, 1866.—The hub box has an annular chamber running nearly its whole length and holding absorbent material saturated with oil. This chamber connects to the butting ring and the spindle by slots.

Claim.—First, the chamber *a*, with its slots *c*, in combination with the openings *b* at its inner end, constructed and operating substantially as set forth.

Second, the axle-box A, provided with a flange *e*, and furnished with a projection *f*, which forms the outer end of the hub, in combination with the screw nut F, for confining it tightly in place within the hub, substantially as set forth.

No. 59,780.—EDWIN REYNOLDS, Boston, Mass.—*Steam Generator.*—November 20, 1866.—The upper and lower horizontal pipes are connected together by pipes consisting of separate sections placed vertically to each other. Each section consists of a central vertical pipe with bent horizontal pipes connecting therewith.

Claim.—The section *c*, when constructed and arranged so as to operate substantially as described.

Also, in connection therewith, the arrangement of the sediment chambers in the ends of the lower pipes *a*.

No. 59,781.—JAMES T. RITTENHOUSE, Urbana, Ill.—*Churn Dasher.*—November 20, 1866.—Explained by the claim and cut.

Claim.—The combination of the arms A A, made tapering, as above set forth, with the diamond-shaped wedges B B, attached to the ends thereof, for the purpose of agitating the cream in such a manner (forcing it upward and outward) as to admit of a free circulation of air while thus agitated. No claim whatever is made to that part of the dasher designated as the shaft, except in so far as it is necessary to secure, by the use of it, the combination above mentioned.

No. 59,782.—WILLIAM SHOUPE, Saltsburg, Penn.—*Apparatus for Obtaining Oil from Wells.*—November 20, 1866.—Of the two concentric tubes the outer opens into a dome, and the inner delivers the flow of oil. Air or water may be forced into the dome and thence down the well to expel the oil, or gas may be permitted to escape out of it and thence to the furnace. Packings are placed at different depths around the outer tube, which is perforated to admit oil between such points.

Claim.—First, in combination with the well pipes B and D, and a chamber E, the gas pipe F, having a safety-valve applied to it, substantially as described.

Second, sustaining the pump tube D in the well by means of a self-sealing dome E, or its equivalent, which forms a chamber at the top of the well, communicating with the gas space between the two pipes B and D, substantially as described.

Third, the inlet J, applied to the dome E, for admitting of the introduction of water into the well, substantially as described.

Fourth, the means, substantially as herein described, whereby oil is allowed to flow into the outer casing B, at points intermediate between the packing, and at the same time the "surface" water is kept back and not allowed to flow into the well.

Fifth, in combination with a casing B, which is perforated at suitable points, and which encloses the pump tube D, the chamber E, and an outlet pipe which is provided with a safety cock or valve, substantially as described.

No. 59,783.—ROSWELL T. SMITH, Nashua, N. H.—*Apparatus for Binding Circular Paper or other Fans.*—November 20, 1866.—The instrument is to be used for holding the fan while binding the edges, which protrude from between the jaws of the disk and the hinged flap.

Claim.—Constructing a fan binder with the circular disk A, or its equivalent, in combination with the hinged flap B, mounted upon a suitable turn-stand a, substantially in the manner and for the purpose herein described.

No. 59,784.—G. B. SNOW and T. G. LEWIS, Buffalo, N. Y.—*Dental Plugging Instrument.*—November 20, 1866 ; improvement on their patent of October 30, 1866.—The exterior ring has a screw which engages a collar on the tooth-holder to lock the striking mechanism when it is desired to operate by simple hand pressure ; rotation of the ring disengages the holder to be automatically reciprocated to act as a plugger.

Claim.—So constructing a dental plugging instrument that it may be operated automatically to give repeated blows, or its operating parts locked so that it may be used as a hand-pressure instrument, substantially as described.

No. 59,785.—JAMES J. TOBEY, Boston, Mass.—*Scissors Sharpener.*—November 20, 1866.—The rectangular file is attached by a screw to a wooden block to form a tool to act upon the blade.

Claim.—The combination and arrangement of the block A, the steel B, and the screw C, in such a way as to form a scissors sharpener, substantially in the manner and for the purpose set forth.

No. 59,786.—ELBER VAN GIESON, Newark, N. J.—*Type Separator.*—November 20, 1866.—The line of type is placed in a slide box with a spring follower and brought singly to the reciprocating separating joint. The separator is worked by an oblong cam. A spring guide conducts the type in proper position to the catch lever, which carries the type to the rotary distributor.

Claim.—First, the use of the springs or spring levers e and the spring g, constructed, arranged, and operated in the manner and for the purpose specified.

Second, each cam and lever, shown and described, when used in combination with the springs e and g, substantially in the manner and for the purpose herein above set forth.

No. 59,787.—JOHN H. WHITLING, Philadelphia, Penn., assignor to J. BARDSLEY and M. HALL, same place.—*Coating Sheet-iron with other Metals.*—November 20, 1866.—The surface of the sheet-iron is cleaned by immersing it in a bath of dilute hydrochloric acid. It is then dipped in a bath of chloride of zinc in solution in water, and while wet is dipped into the bath of melted tin.

Claim.—Coating sheet-iron plates with tin or other metals, or alloy, by the process substantially as herein described.

No. 59,788.—JOHN and WILLIAM YEWDALL, Philadelphia, Penn.—*Machine for Washing and Drying Wool.*—November 20, 1866.—Instead of passing the wool from the lifting cylinder to an endless conveyer and thence to pressing rollers, a pressing roller is located immediately above the cylinder, squeezing the wool before it leaves the cylinder. The grooved plates being in removable sections, admit of removing any of the toothed bars from the cylinder.

Claim.—First, the hollow cylinder E, with its bars F, and pins *e*, in combination with the weighted pressing roller H, the whole being arranged and operating as and for the purpose described.

Second, the sectional grooved plates G, constructed and adapted to the cylinder E and its bars F, substantially as and for the purpose specified.

No 59,789 —STEPHEN ALBERTSON, New York, N. Y., administrator of the estate of SAMUEL ALBERTSON, deceased.—*Adjustable Damper for Fireplaces.*—November 20, 1866.— The fixed plate of the damper has an extension slide which is adjusted by a rack and pawl to different sized flues.

Claim.—The combination of the extension piece *k*, with the rack and pawl, for the purpose of governing or controlling the sliding plate upon and with the stationary plate, as herein described and set forth.

No. 59,790.—JOHN DAVIS, 2d, Lake Village, N. H.—*Churn.*—November 20, 1866; antedated November 16, 1866.—The conical agitators pass point forward, and, tending to form a vacuum, the air passes through the supply pipe and into the cream.

Claim.—The hollow conical agitators E E, with their air-supplying tubes G G, arranged and operating substantially as and for the purpose herein specified.

No 59,791.—HENRY G. DAYTON, Maysville, Ky., assignor to himself, B. YOUNG, N. C. MORSE, and R. B. WILSON.—*Manufacture of Artificial Fuel.*—November 20, 1866.—Corn cobs are saturated with a composition of rosin two parts, and residuum of the distillation of petroleum one part.

Claim.—As a composition for treating or saturating corn cobs or kindling wood, a compound or rosin and the residuum of the petroleum distillation, in about the proportions described.

Also, as a new article of manufacture, a kindling material composed of corn cobs, treated or saturated with the above composition.

No. 59,792.—DANIEL DRAWBAUGH, Eberly's Mill, Penn.—*Faucet.*—November 20, 1866.— This is intended for the rapid and accurate measuring of viscid liquids, and is explained by the claims and cut.

Claim.—First, the hubs *b* and *c*, with their pistons, constructed and operating substantially as above described—that is to say, so that when one of those hubs has a constant motion that of the other shall be intermittent, substantially in the manner and for the purpose shown.

Second, in combination with the subject-matter of the first claim, the recess *k*, upon one or both the flat surfaces which are swept by the hubs and pistons aforesaid, which said recesses enable the liquid drawn through said faucet to pass outward towards the educt pipe *h*, from between the said pistons, substantially as described.

Third, in combination with the subject-matter of the first claim, the cogged wheel *s*, with its index *z*, and the pin *m*, upon the crank *n*, for the purpose of denoting, automatically, the precise number of revolutions of said crank, substantially as described.

Fourth, in combination with the subject-matter of the first claim, the gate *g*, having its face and the plane of its seat inclined to the axis of revolution of such gate, substantially as and for the purpose described.

No. 59,793.—DANIEL DRAWBAUGH, Eberly's Mill, Penn.—*Faucet.*—November 20, 1866.— The liquid is forced by a screw flange into the measure cylinder. A rod from the piston has a finger to indicate the amount introduced. The cylinder is emptied by a slide gate.

Claim.—First, the combination of the screw flange *b*, spigot A, chamber A, and piston *p*, substantially as and for the purpose set forth.

Second, in combination with the above, an indicator *d*, attached to the piston *p*, and a proper scale on the outer side of the vessel *h*, by which to denote the exact amount of gas within that vessel, substantially as and for the purpose described.

No. 59,794.—WILLIAM H. ELLIOT, New York, N. Y.—*Bedstead Fastening.*—November 20, 1866.—An oblique-faced tenon with an enlarged edge is formed on the end of the rail and occupies a mortise of counterpart form and position in the post. The end of the tenon being entered, the rail by depression is drawn in and its shoulders jammed against the face of the post.

Claim.—First, so constructing the mortise cut in the face of the post, and the tenon cut on the end of the side rail, that the tenon, with its projections, may pass to its place in the mortise by a direct downward and inward movement, substantially as herein set forth.

Second, cutting the bottom of the mortise and the end of the tenon at an angle which corresponds with the direction of the movement of the tenon when it passes into the mortise, as herein shown.

Third, the combination of the several shoulders, herein described, for supporting the side rail, when formed out of the solid material of the bedstead, and operating as set forth.

No. 59,795.—WILLIAM H. ELLIOT, New York, N. Y.—*Bedstead Fastening.*—November 20, 1866.—The tenon on the rail fits within the mortise in the post and is secured by a key driven into a seat cut obliquely in contiguous sides of the tenon and mortise.

Claim.—First, cutting the seat for the key *c*, in the contiguous surfaces of the mortise and tenon, substantially as shown and described.

Second, the key *c*, when placed between the contiguous surfaces of the mortise and tenon and supported as herein shown, in combination with the mortise and tenon, as a self tightening bedstead fastening, substantially as specified.

No. 59,796.—A. W. ELMER, Springfield, Mass., assignor to himself and C. ENSMINGER, same place.—*Skate.*—November 20, 1866.—The foot piece is pivoted to the fore pillar and slides vertically on the hind one. A spring tends to keep the heel raised.

Claim.—An elastic skate, having the foot plate B, and runner A, combined with the spring M, constructed and operating substantially as described.

No. 59,797.—GEO. L. FATTIE, Buffalo, N. Y.—*Burning Fluid.*—November 20, 1866.—Composed of gasoline, 40 gallons; sulphur, 5 pounds; rusty iron, 100 pounds; onions, 1 bushel; and pulverized rosin, 5 pounds.

Claim.—An illuminating burning fluid, compounded substantially as herein described.

No. 59,798.—JAMES BENNETT FORSYTH, Roxbury, Mass.—*Roller for Clothes Wringer.*—November 20, 1866.—Designed to secure the rubber roll firmly to the shaft. Strips are cut bias from a sheet formed by several alternate thicknesses of fabric and rubber: these strips are built upon each other and cemented, forming a sheet which is cut into strips to be wound around the shaft to form a core. The fiber of the fabric is thus exposed at intervals to contact with the shaft.

Claim.—A roll, constructed substantially as described, with a core so formed that the ends of the threads of the cloth will rest upon the shaft for the purpose set forth.

No. 59,799.—RUFUS LAPHAM, New York, N. Y.—*Spring Holder for Bed Bottoms.*—November 20, 1866.—The end of the slat rests upon a spring whose lower coil engages a series of lugs on the edge of the socket whose shank is screwed into the bed rail.

Claim.—First, the device for attaching the spring to its seat, substantially as described.

Second, the combination of the screw seat and spring, substantially as described.

Third, also the combination of the seat, spring, and pin, substantially as described.

No. 59,800.—C. L. LOCHMAN, Carlisle, Penn.—*Photographic Printing Frame.*—November 20, 1866.—Intended for printing porcelain pictures or glass transparencies. The negative is placed in the rabbeted bars after they are properly adjusted; the prepared plate is laid on the lid, the projecting ends of the albumen paper clasped by the bars; the lid is shut and clasped and the plate is ready for exposure to the light.

Claim.—First, on the one side or body of a photographic printing frame, two adjustable rabbeted bars F F, to hold a negative in place, in combination with two or more tightening screws E E E', or their equivalents, constructed and operated in the manner substantially as shown and described, and for the purpose set forth.

Second, the adjustable hinged piece or bar J, with its mortised ends, and supporting screws and burs, constructed and operated substantially in the manner set forth.

No. 59,801.—DANIEL H. MCLEAN, Ilion, N. Y.—*Reel.*—November 20, 1866.—The devices claimed adapt the reel for skeins of varying sizes.

Claim.—In combination with the hinged or pivoted arms of the reel, the washer or nut for spreading and forcibly holding said arms to the skein, and against any tendency to rise on their pivots or hinges, substantially as described.

No. 59,802.—ELI MCMILLAN, Jr., Wilmington, Ohio.—*Churn.*—November 20, 1866.—The reciprocating dashers are operated by distinct hand cranks. A guide pin projecting from the bottom guides the lower dasher, and the rod of the latter forms a guide for the upper one. The churn dashers are intended to alternately approach and recede from each other, forcing the cream around their edges.

Claim.—The combination of the imperforate dashers B C, rods B' C', guiding stem D, flywheels F F', and cranks I I', all constructed, arranged, and operating as and for the purpose specified.

No. 59,803.—DAVID W. STUTSMAN, Upshur, Ohio.—*Medical Compound.*—November 20, 1866.—For the cure of fever and ague. Composed of powdered butternut bark, Cayenne pepper, beef gall and sweet oil.

Claim.—The composition for a fever and ague pill, composed and compounded as set forth.

No. 59,804.—EDWARD WHITEHEAD, Cincinnati, Ohio.—*Iron Gutter.*—November 20, 1866.—The gutter is flush with the walk and has a narrow slot to allow cleaning and prevent bursting from freezing.

Claim.—The cast-iron gutter A A' B C, when provided with the longitudinal slot D, as

No. 59,805.—Francois Ferdinand Auguste Achard, Paris, France.—*Electro Magnetic Car Brake.*—November 20, 1866.—The power is derived from the rotation of the wheels. Upon one axle is placed an eccentric which, at each rotation of the wheel, operates a lever provided with a pawl gearing, with a ratchet wheel on an auxiliary shaft. The latter carries a magnetic cylinder and two drums, which constitute the armatures of the cylindrical magnet and which are connected by chains to the brakes. When the drums are attracted to the cylinder the motion of the eccentric causes them to rotate, and thereby apply the brakes. To render the lever inoperative, it is attached to an arm which forms the armature of a second magnet, and by which the lever is raised. The transfer of the current from the magnet to the magnetic cylinders will bring the brakes into operation.

Claim.—First, the eccentric B, and lever C, in combination with the magnetic cylinder N. flanged drums O, and brakes 5, all constructed and operating substantially as and for the purpose described.

Second, the sliding armatures I, and hand lever G, in combination with the electro-magnet K. lever C, and eccentric B, constructed and operating substantially as and for the purpose set forth.

No. 59,806.—R. N. Allen, Cleveland, Ohio.—*Drying Fruit, &c.*—November 20, 1866.—The fruit is enclosed between the hinged sides of the frames, which are covered with fabric or wire gauze. The frames are arranged in a truck which runs on rollers in and out of the dry house, which has a stove beneath, and a slatted ceiling.

Claim.—First, one or more adjustable frames D, constructed substantially as and for the purposes set forth.

Second, the combination of the truck C with the adjustable frames D, as and for the purpose set forth.

Third, the arrangement of an open or perforated ceiling K, carriage C, and frames D, when used in connection with a drying apparatus, for the purpose set forth.

No. 59,807.—Thomas Appelget, Princeton, N. J.—*Cider Mill.*—November 20, 1866.—The pomace from the grinding rollers falls upon the lower belt, and after travelling a short distance an upper belt comes upon it, and it is carried by the two between the sets of pressure rollers, which are placed above and below the two parts of the belt.

Claim.—The machine constructed to operate substantially as described and in the manner set forth, to press the juice from ground fruit by means of passing the pomace between the endless belts which are geared to work together, and have pressure rollers above and below that part of the belts which have the pomace between them.

No. 59,808.—Frederick Ashley, New York, N. Y.—*Lamp Chimney Cleaner.*—November 20, 1866.—One jaw of the sponge holder is serrated, and is depressed upon its fellow by a spring.

Claim.—So constructing the holder or holders for the sponge, &c., of a lamp chimney or other similar cleaner, that the said sponge, &c., can be attached thereto or detached therefrom at pleasure, substantially as and for the purpose described.

No. 59,809.—John I. Baringer, Germantown, N. Y.—*Smoke Furnace for Curing Meat, &c.*—November 20, 1866.—The fire is made in a portable furnace, with a perforated conical pan, ash pit, hinged cover, and grated opening for the exit of smoke.

Claim.—First, the fire pan J, constructed as described, in combination with the smoke furnace A D, substantially as and for the purpose set forth.

Second, the combination of the covers H G, one or both, with the furnace A D, substantially as described and for the purpose set forth.

No. 59,810.—W. F. Bartlett, Hillsdale, Mich.—*Combined Lantern and Foot Warmer.*—November 20, 1866.—The curved deflector turns the caloric current back, and between two plates beneath the feet, where it is forced to a tortuous course before escaping.

Claim.—The cross flues D', and reflector H, in combination with the lamp G and case A. as and for the purpose set forth.

No. 59,811.—E. H. Beckwith, Westernville, N. Y.—*Churn.*—November 20, 1866.—The dasher is adjustable on its rod by a sliding hub and set screw. From the sliding hub project two curved arms carrying hubs, having inclined wings upon them which acquire a rotary motion. Angular ribs project from the churn bottom.

Claim.—First, the curved hanging arms g g, combined with the metal hubs e e, the wooden bushes h h, the wooden wings f f f, and the sliding hub d on the spindle C, when arranged in a churn, substantially as and for the purposes herein specified.

Second, making the outer ends c' c' of the cross-ribs e e on the bottom of the churn diverging at an angle from them, and higher than they are made, substantially as herein specified.

No. 59,812.—William Belwin, Baltimore, Md.—*Oyster Dredging Apparatus.*—November 20, 1866.—The davit lever by which the dredge is lifted is mounted upon a swivel post upon the gunwale. The dredge has an extra bow to which the davit hook is attached, and the flange and sleeve prevent its becoming jammed between the post and the gunwale roller.

Claim.—First, the davit G pivoted upon the gunwale, and provided with a lever by which the dredge is raised, substantially as described.

Second, the auxiliary bow I of the dredge.

Third, the flange in the gunwale roller.

Fourth, the roller or sleeve e on the post of the davit.

No. 59,813.—D. BEQUERET and E. DUMOULIN, Jamestown, Ill.—*Gang Plough.*—November 20, 1866.—The ploughs are moved vertically by levers. The axle is bent to depress the furrow wheel. The tongue is attached to one side of the centre to suit the position of the horses.

Claim.—A plough composed of a long and a short beam with ploughshares F F', which are operated by levers G G', in combination with the unequal crank axle C, draft pole E, placed in line with the short beam, and with the caster wheel I, all constructed and operating substantially as and for the purpose set forth.

No. 59,814.—FRANK BOWMAN, Waltham, Mass.—*Store Cricket.*—November 20, 1866.—The frame is intended for the support of articles near a stove. On its upper shelf it has rings for pans, and has a floor beneath.

Claim.—The peculiar construction and arrangement of a portable stove cricket, substantially as and for the purpose set forth and described.

No. 59,815.—R. BRADY, New York, N. Y.—*Heater for Soldering Irons.*—November 20, 1866.—The multiplicity of casings with intervening imprisoned air chambers lessens the radiation of heat from the outer walls. The soldering tools rest upon a dishing bed and the lower one is exposed to the heat of the gas jet.

Claim.—First, the cylinder B of the furnace, lined with a series of casings H, having air spaces between them, substantially as and for the purpose described.

Second, the inclined rests or supporters G for the soldering irons, substantially as described.

● No. 59,816.—E. K. BRUCE and J. M. BRUCE, Wilkins, Penn.—*Apparatus for Sealing Fruit Cans.*—November 20, 1866.—A vessel contains the cement, is heated by means of a spirit lamp and has a spout, at the end of which is a valve for controlling the flow of the liquid cement. The fruit can is placed on a revolving table beneath, and as it rotates the liquid cement runs into the groove in which the edge of the cover is placed.

Claim.—First, the cement vessel E, provided with a discharge opening or tube with a cut-off or valve K, in combination with a rotary table or disk O, on which the can to be sealed is placed, and a screw clamp or its equivalent for holding the lid or cover on the can while the latter is being sealed, substantially as shown and described.

Second, the adjustable bar C, fitted on the upright bar A in connection with the adjustable arm D, placed on bar C, all arranged as shown, to admit of the proper adjustment of the cement vessel to the can to be sealed, as set forth.

Third, the central tube F, in the cement vessel E, in combination with the lamp G, fitted in an adjustable support H, when said parts are used in connection with the rotary table or disk O, and clamp, for the purpose set forth.

No. 59,817.—J. W. CAMPBELL, New York, N. Y.—*Machine for Ornamenting Mouldings.*—November 20, 1866.—A strip of cotton is drawn through the plaster hopper and carries a coat of the same forward under moulding rollers. A roller revolving in a box applies the liquid therein to the face of the last moulding roller. A box contains adhesive mixture for the attachment of the ornament.

Claim.—First, constructing a machine for forming long strips of ornaments either independently or to be attached to mouldings, consisting of a combination of the hopper i, wheels g and h, rollers m and s, and wheel p, and operating substantially as described.

Second, the application of the strip of muslin or other material for the purpose of supporting the plastic material, while and after the same is being finished, in the manner specified.

Third, the box r for the purpose of supplying gum to the moulding, so that the ornament may at once be firmly attached to the same, substantially as herein shown and specified.

No. 59,818.—JOSEPH B. CASSEL, Worcester township, Penn.—*Tread-wheel Hoisting Power.*—November 20, 1866.—The tread-wheel has a ratchet rim and sustaining pawls, and by grasping the handles a person is able to throw some muscular force into the work in addition to the mere weight in stepping.

Claim.—The arrangement of the tread-wheel provided with notches and pawl, and the platform with handles, the whole being adapted, as described, to the purpose set forth.

No. 59,819.—WILLIAM CLARK, Valatie, N. Y.—*Harness Rein.*—November 20, 1866.—The driving rein runs over sheaves on the bit rings and the gag loops, and thence passes to the check hook. Stops on the check portion limit the length of the gag rein portion.

Claim.—The combination of the stop D, with the harness reins A, substantially as and for the purpose described.

No. 59,820.—D. L. COLLINS, Antwerp, N. Y. assignor to himself and ELDRIDGE SISS, same place.—*Door Bell and Burglar Alarm.*—November 20, 1866.—The bell may be rung by turning a crank in the door post, which revolves a triangular cam in the works. The mechanism is operated as a burglar alarm through a detachable crank operated by the spring door.

Claim.—The within-described door bell and burglar alarm, consisting of a single hammer D, arm E, piece or tappet F, spring J, cam L, spindle K, curved notched arms L', actuating spring S, spindle P, and projections Q R arranged in connection with the bell A, as and for the purpose specified.

No. 59,821.—J. C. COOK, Buffalo, N. Y.—*Stave Machine.*—November 20, 1866; antedated November 16, 1866.—The blanks are fed between spring feed rollers, and having passed between a fixed and a spring guide, are operated upon by the rotary cutters, and the stave received between adjustable side guides.

Claim.—The combination and arrangement of the self-adjusting feed rollers G G, spring roller guide *a*, fixed guides *b* and *c c*, and the two cutter heads H I, having a different number of cutters, substantially as and for the purpose herein specified.

No. 59,822.—E. COPLESTON, Wrentham, Mass.—*Fabric for Hats and Bonnets.*—November 20, 1866.—The buckram is pressed into the proper form and covered with a cement of boiled linseed oil and white lead, to which the flock is applied.

Claim.—As a new article of manufacture, hats or bonnets made by flocking upon the surface of a foundation or body of buckram or other suitable fabric, substantially as herein described.

No. 59,823.—JOSEPH V. C. CRATE, Waterbury, Conn.—*Rod of Punching Presses.*—November 20, 1866.—A rubber or other yielding disk is inserted between the two joints of the rod. which are connected by means of an axial screw pin in whose annular recess the end of a set-screw works as the disk yields to undue strain.

Claim.—The shaft A, in combination with the screw rod B and spring E, constructed, arranged, and operated substantially as described.

No. 59,824.—WILLIAM A. DEVON, Richmond, N. Y.—*Holding and Adjusting Scaffolds.*—November 20, 1866.—The arrangement of pulleys and windlass for raising or lowering painters' scaffolds is explained by the claim and cut.

Claim.—The combination with the platform C, of the winch barrel G, falls *c c*, of the end tackles E E, intermediate block F, and window hooking frames G G, for operation together, essentially as herein set forth.

No. 59,825.—WILLIAM H. ELLIOT, New York, N. Y.—*Bedstead Fastening.*—November 20, 1866.—A tenon on the end of the rail has an oblique projection on one side which occupies a counterpart groove in the side of the mortise in the post. The tenon is entered perpendicularly to the mortise and then jammed sideways by a flat key.

Claim.—The combination of the mortise *m*, the tenon *e*, projection *i*, depression *i'*, and key *e*, when these devices are so employed as to form a self-tightening bedstead fastening, substantially as herein shown.

No. 59,826.—D. EVERS, Van Wert, Ohio.—*Tenoning Machine.*—November 20, 1866.—The stuff is placed on an adjustable frame and operated upon by the cutter, which consists of two circular saws with cutters between them.

Claim.—The cutter head constructed of the sliding collars *d d*, saws *e e*, cutters *s s*, and mortised bolts *i i*, when arranged and operating substantially as described, for the purpose specified.

No. 59,827.—ASAHEL FAIRCHILD, Independence, Iowa.—*Combined Seeder and Harrow.*—November 20, 1866.—The cam projections on the ends of the roller give reciprocating motion to the barrow bars, seed slide and agitator. The rock shaft is worked by a lever and connected to the frame, and operates to raise the barrow teeth from the ground and move the slide which closes the seed openings.

Claim.—First, the roller A, provided with cams C C, in combination with the harrow bars H and K, for the purposes and substantially as described.

Second, the separate harrow bars H and K, pivoted levers I I, in combination with the slotted lever M, slide P, and agitator O, substantially as described.

Third, the combination of the rock shaft S with the frame G and slide *a*, for the purpose of shutting off the grain when the barrows are raised, substantially as herein set forth.

No. 59,828.—J. C. FAY, New York, N. Y.—*Wind Wheel.*—November 20, 1866.—The wind doors are spread apart or drawn together by means of a governor and a system of levers. The doors operate to lessen the amount of wind on increase of speed. The frame rests on rollers in a circular "way," which allows of its being turned to face in any direction.

Claim.—First, the two vertical wheels I I, placed side by side in the box A, connected with each other by gears r r, and to a power shaft L, by gears J K, substantially as and for the purpose set forth. ,

Second, in combination with the wheels I I, the doors C C, applied to the front end of the box A, as shown, so as to open and close and admit of a greater or less amount of wind to pass through the box and act upon or against the wheels, as set forth.

Third, the governor H, connecting with and receiving its motion from the wheels I I, and connected with the doors C C, as shown, or in any equivalent way, so that the doors will be open and closed and the speed of the wheels rendered uniform under variable degrees of velocity of the wind, as set forth.

Fourth, the power shaft L, connected by a pinion K, and toothed wheel J, with the wind wheels I, in combination with the base M, provided with the annular way t, and the box A, having rollers s attached to its under side to work on the way t, substantially as and for the purpose specified.

No. 59,829.—J. D. FIELD, Keokuk, Iowa.—*Seeding Machine.*—November 20, 1866.—
The revolving coverer is tripped by a rod operated by the hand. The revolution of the coverer works the seeding mechanism. The lower seed cylinder has an inner oscillating cylinder with an upper and lower seed cavity. A fixed scraper of higher elevation follows the rotating coverer.

Claim.—First, the intermittingly rotating scrapers or shovels G G and rods r, in combination with the fixed scraper H, arranged to operate in the manner substantially as and for the purpose set forth.

Second, the arms J J', in combination with the cam I, having the pins d d attached, the arm K, and spring l, for operating the cylinder O, substantially as and for the purpose specified.

Third, the cylinder O, placed within a case L, which communicates with the seed box N, and provided with the holes p p', and an inclined projection t within it, in combination with the bar P, with shoe or cut-off b* attached, all arranged to operate substantially as and for the purpose set forth.

No. 59,830.—CHARLES E. FOWLER, Carmel, N. Y.—*Pocket Match Safe.*—November 20, 1866.—Explained by the claim and cut.
Claim.—An improved match safe A, so constructed that the lid a² may close down over the opening and over the roughened part a³ of the box, being kept closed by its own spring or elasticity, substantially as herein described and for the purpose set forth.

No. 59,831.—E. R. GARD, Chicago, Ill.—*Brick Machine.*—November 20, 1866.—The machine has a rotary mould-block and followers operated by an annular track with spiral projections. When the striking plate meets an obstructing stone in the mould, the wooden pins whereby it is fastened are broken and it becomes detached, as does also the plate following, thereby avoiding fracture.
Claim.—The striking plate M and plate P, when applied or arranged substantially as shown and described, to yield or detach themselves in the event of stones or other foreign substances entering the moulds, as set forth, the whole being constructed and operating in the manner and for the purpose herein described.

No. 59,832.—C. W. GILLIS, San Antonio, Texas.—*Baling Press.*—November 20, 1866.—
The rope is attached to the frame near the windlass, passes over the follower to the other side of the frame and back to the windlass. The ends of the guide bar traverse slots in the side pieces of the frame. A ratchet bar is so pivoted as to be operated on by a pawl to retain the follower in depressed position. The doors are secured by battens, and by hinged arms slotted for a retaining key.
Claim.—First, the upright windlass B, with rope D attached, in combination with two rollers a fitted in sockets F at each end of the follower or plunger E, and the roller e in bar f, all arranged substantially as and for the purpose set forth.

Second, the bar G, attached to the follower or plunger E, with roller c at its ends, working in grooves d in the sides of the press box, substantially as and for the purpose specified.

Third, the adjustable or pivoted racks H, in combination with the pawls I, and the bars J, all arranged and applied to operate in the manner substantially as and for the purpose set forth.

Fourth, the securing of the doors K K', in a closed state, by means of the battens f*, buttons j, in connection with the hinged bars M M, provided with slotted lips m, with a key l driven in them, substantially as shown and described.

No. 59,833.—E. T. GILMORE, Springfield, Mass.—*Preserving Oysters.*—November 20, 1866.—The can has two apartments—the larger for the oysters, the other for the liquor. The parts have separate openings.
Claim.—First, the method herein described of putting up oysters for preservation.
Second, the oyster can, having two apartments B and C, arranged substantially as and for the purpose set forth.

No. 59,834.—ISAAC and I. M. GROSS, New Galena, Penn.—*Wagon Brake.*—November 20, 1866.—The brake is operated by a lever in the rear, by a lever pivoted to the side of the bed, or by a rope and pulleys connected to a hand crank near the driver.

Claim.—The shoe bar B, in combination with the levers J Q, pulley blocks K M rope L, windlass O, lever R, and rod j, all arranged and applied to a wagon, substantially in the manner as and for the purpose set forth.

No. 59,835.—JOHN HAGGERT, New York, N. Y.—*Axle Nut for Wagons, &c.*—November 20, 1866.—The part of the spindle projecting beyond the hub bush has a longitudinal slot with a circumferential continuation by the bush. A ring has projections which slide in the longitudinal part of the slots, and by a partial rotation occupy the circumferential parts forming a bayonet joint. A plate has projections which enter the longitudinal slots and retain the keeper ring in position. A swivel button secures the plate.

Claim.—The combination with the axle arm a, and collar f, of the grooves e q l, projections h, and cap o p, when all are constructed and arranged as herein specified to form a bayonet joint with a shoulder at right angles to the axis, to secure the wheel upon the axle, as explained.

No. 59,836.—STEDMAN W. HANKS, Lowell, Mass.—*Attaching Pictures to Frames.*—November 20, 1866; antedated November 11, 1866.—The clasps embrace the edge of the paper, and the hooks proceed from the clasps to the wires stretched on the back of the frame.

Claim.—The combination of clasps A with hooks or clasps B and wire springs C, supported upon nails or tacks D, operating in the manner and for the purposes set forth.

No. 59,837.—WILLIAM HARSEN, Greenpoint, N. Y.—*Stuffing Box.*—November 20, 1866.—The follower has an outside screw engaging the inside screw of two lugs upon the stuffing box. Circular holes in the top of the follower serve in connection with a key to turn it.

Claim.—The combination of the lugs D D, having threads on their inner faces and forming an open or divided screw box with the screw formation s on the gland C, outside of the socket, for operation together, substantially as specified.

No. 59,838.—F. HENCKEL and WILHELM SECK, Munich, Bavaria.—*Grain-hulling Machine.*—November 20, 1866.—The grain is conducted from the hopper to the first and second annular spaces, from each of which it passes to the next but one below. The revolution forces the grain against the corrugated jacket and blows away the husk. The grain passages are adjusted in size by a slide.

Claim.—First, the method herein described of separating the grain in two or more currents and uniting the same again, consisting of the centrifugal feeder T, channels W, terraces E F, jacket A, and apertures a, substantially as described.

Second, the adjustable slide R, in combination with the apertures a, in the jacket A, and with the several terraces of the revolving drum, when constructed and operating substantially as and for the purpose described.

No. 59,839.—ISAAC L. HOARD, Bristol, R. I.—*Lamp Wick.*—November 20, 1866.—Explained by the claims.

Claim.—First, paper lamp wicks made of paper pulp, substantially in the manner herein described and for the purpose set forth.

Second, the use of paper pulp for the manufacture of lamp wicks, substantially as described and for the purpose set forth.

No. 59,840.—JAMES G. HOLT, Chicago, Ill.—*Making Sand Cores for Axle Skeins and Hub Boxes.*—November 20, 1866.—The core boxes are confined in a frame so as to insure the true position of the cores, by standing in a set upon a horizontal base.

Claim.—First, a frame which is constructed substantially as described and adapted for sustaining core boxes in a fixed position during the operation of making a sand core.

Second, the means substantially as explained for making one or many sand cores upon a sand bed so that the axis of the core shall be perpendicular to said bed, thus insuring the proper centring of the core in the mould for which it is adapted, substantially as set forth.

No. 59,841.—SAMUEL JACKSON, Jr., New York, N. Y.—*Saw-mill.*—November 20, 1866.—The saw slides are held in caps that permit rotation, and the equal rotation of the two ends is secured by racks which mesh into gear wheels traversed by the slides, the racks being connected by a pivot bar. A belt passes over pulleys and connects the upper and lower saw cage together, and the power is communicated to the saw from a cam wheel operating on a slide block attached to this belt.

Claim.—First, the combination of the segmental gears, pinions, movable boxes, caps, and grooved guides for turning the saw while retaining the belt in position when the same shall be constructed and operated substantially as shown.

Second, the combination of the belt, grooved cam wheel, and sliding guide, for the purpose specified.

No. 59,842.—WILLIAM and A. G. KELSEY, Delavan, Wis.—*Boot Tree.*—November 20, 1866.—The independent centre wedge is drawn up between the side pieces by the treadle lever. A rack and catch serve to keep the lever depressed.

Claim.—The independent centre C and pivoted stretchers *a a*, in combination with the follower *e* and rod *d*, all arranged as herein described and adapted to operate upon a boot tree, as set forth.

No. 59,843.—H. KEYES, Terre Haute, Ind.—*Apple Parer.*—November 20, 1866.—The apple is impaled on the revolving fork and the knife made to sweep around automatically as its platform is revolved by gear connection with the hand crank shaft.

Claim.—First, the segmental wheels *d d'*, in combination with the lever brake *h*, and the paring knife arm *f*, constructed and arranged substantially as and for the purposes herein described.

Second, the segmental wheel *i* and the pin *m*, in combination with the eccentric rack frame *k* and rack *j*, constructed and arranged substantially as and for the purposes herein specified.

Third, the combination and arrangement of the bevel wheel *c'* with the segmental wheels *d d'*, the pin *m*, the segmental wheel *i*, the eccentric rack frames *h*, the lever brake *h*, and the knife arm *f*, constructed and operated substantially as and for the purposes herein set forth.

No. 59,844.—JACOB KING, Jr., Fort Wayne, Ind.—*Spring Fish Hook.*—November 20, 1866.—The hooks are pivoted to the side of the case and are sprung by pulling the bait on the end of the plunger. When released the teeth on the spring plunger engage the segment racks on the hooks to rotate them.

Claim.—The tube A, with the hooks C, pivoted to it and provided with the pinions *e* and catches *j j*, in combination with the internal tube D, provided with the spring *e* and rack *d*, in which the pinions *c* gear, and the rod E, provided with the cup disk *h* and spring *i*, all arranged to operate substantially in the manner as and for the purpose set forth.

No. 59,845.—DANIEL KINNEY, Loami, Ill.—*Compound for Curing Foot Rot in Sheep.*—November 20, 1866.—Compound of cantharides, 2 ounces; mercurial ointment, 1½ ounces; butter of antimony, 1 drachm; spirits of turpentine, 3 ounces; tincture of iodine, 2 ounces; gum euphorbium, 5 drachms; sulphate of copper, 2 pounds; lard, 2 pounds.

Claim.—The improved medical compound herein described, for the purpose specified.

No. 59,846.—DENNIS LANE, Montpelier, Vt.—*Saw-mill.*—November 20, 1866.—A lever is attached to a bar having an idler pulley to tighten the working belt of the saw and feed gear, and the foot of this lever has an oblique ridge working in a gain to throw in a friction clutch for the return movement of the carriage. The depression of the lever causes a forward movement of the machinery, and releases the friction clutch.

Claim.—Operating the carriage by means of the lever arm Q, with its lip *a*, and friction roller R, belt K, pulleys L and M, clutch S, hollow loose wheel J, operating shaft J, forked arm V, and slide W, constructed and arranged substantially as described for the purpose specified.

No. 59,847.—JAMES R. LAURENT, Milford, Penn.—*Extinguishing Fires.*—November 20, 1866.—The liquid in the vessel is ejected by the expansive force of gas generated within. The liquid is a solution of common salt and bicarbonate of soda; a chamber above contains alum, which when occasion occurs is dropped into the solution below and generates carbonic acid gas.

Claim.—The solution of bicarbonate of soda and common salt in combination with alum, applied and operating substantially as and for the purpose set forth.

No. 59,848.—JOHN LOCKHEAD, San Francisco, Cal.—*Balanced Slide Valve.*—November 20, 1866—The steam exhausts through the centre of the valve, and a brass ring is fitted between the valve and the outside place of the steam chest. Rubber packing occupies an annular groove under the brass ring.

Claim.—The overhanging edge or flange *e* of the ring D, in combination with opening *b* in the valve B, and with the inner surface of the steam chest cover, constructed and operating substantially as and for the purpose described.

No. 59,849.—RICHARD MAGEE, Philadelphia, Penn.—*Envelope.*—November 20, 1866.—Explained by the claim and cut.

Claim.—An envelope for use, closed at its top and bottom and open at both ends, one of the inner sides of each of the said ends being gummed, and the other provided with a flap gummed in like manner, so that either can be cemented to the enclosures, substantially as described for the purpose specified.

No. 59,850.—JOHN MARCHBANK, Lansingburg, N. Y., assignor to JAMES McQUADE, same place.—*Paint Brush.*—November 20, 1866.—The coil of wire around the butts of the bristles and the end of the handle is continuous and forms two frustums, which when opened respectively admit the insertion of the bunch of bristles and the handle, and are closed and cemented to complete the brush.

Claim.—The coiled wire ferrule C, enlarged at the centre and adapted to admit of the insertion of the bristles, substantially as and for the purpose specified.

No. 59,851.—JOHN MCCRELLISH, Philadelphia, Penn.—*Blacking*—November 20, 1866.—To a decoction of logwood, japonica and barks from black oak, white oak, chestnut oak, and Spanish oak, add copperas, and filter the product, then mix with neat's-foot oil and japan.

Claim.—A blacking or coloring composed of the ingredients herein specified and described.

No. 59,852.—JAMES MCLAREN, Albany, N. Y.—*Wrench.*—November 20, 1866.—The fixed jaw projects backward and has a serrated recess. A pivoted piece having a toothed face acts with this recess as a pipe wrench. The strain upon the pipe serves to increase the grasp. The fixed and movable jaws on the other side of the shank act as an ordinary monkey wrench.

Claim.—The within-described tool as a new article of manufacture.

No. 59,853.—THOMAS E. MCNEILL, Lynchburg, Va., assignor to himself and WM. D. MILLER, same place.—*Cotton Press.*—November 20, 1866.—The two presses are worked by a steam pump, and may be used in conjunction, or separately.

Claim.—First, the cylinder J, pipes I and N, in combination with the cylinder O, for the purposes and substantially as herein shown.

Second, cylinder J, piston L, in combination with the press or follower K, substantially as described.

Third, a hydraulic press provided with or constructed with two cylinders so as to operate or act in conjunction with each other or separately, when constructed and operated for the purposes and substantially as described.

No. 59,854.—CHARLES V. MEAD, Hamilton, N. J.—*Machine for Rolling Rubber.*—November 20, 1866.—The sheet of rubber, with edge turned up, is placed upon the platform and the table caused to travel over it, coiling the rubber into a roll. The table is moved by racks on the top which are engaged by pinions on the rotary shaft, which is depressed by springs.

Claim.—First, rolling rubber by the action of the reciprocating table H and platform A, substantially as herein set forth.

Second, the springs a, in combination with the pinions F, racks G, table H, and platform A, constructed and operating substantially as and for the purpose described.

Third, the weights I, or their equivalents, in combination with the table H and platform A, constructed and operating substantially as and for the purpose set forth.

No. 59,855.—PETER MERKEL, St. Louis, Mo.—*Gang Plough.*—November 20, 1866.—The fore ends of the plough beams are adjustable by sliding blocks in a vertical slot. The ploughs are raised by a windlass connecting to a rack bar having an arm linked to the top of the plough standard. The land wheel is adjustable by a rack and spur wheels.

Claim.—First, the combination of the windlass M. shaft L, with its arm n and p, links q, and plough beams I, substantially as described, for the purpose specified.

Second, the adjusting of the ploughs J higher or lower to suit the depth of furrow required, and also adjusting the wheel C', by means of the shaft G, provided with the pinions f f g, and the rack bars F F, and slide d, provided with a rack, substantially as and for the purpose set forth.

Third, the fitting and securing of the front ends of the plough beams I I in a cross piece H, composed of two parallel bars j j, provided with journals m, fitted in the rack bars F, substantially as and for the purpose specified.

No. 59,856.—ANDREW T. MERRIMAN, Rutland, Vt., and THOMAS ROSS, Middlebury, Vt., assignors to themselves, J. B. REYNOLDS and R. BARRET, Rut'and, Vt.—*Machine for Channelling Stones*—November 20, 1866.—The cutters have direct motion from the piston. The valve is reversed at the blow of the cutters, or in case of no blow being given, it is reversed before the cylinder bottom is touched by the piston. The cutter bar is adjustable on the cylinder bar to suit the depth of groove cut. The whole mechanism is mounted on vertically adjustable rollers and the feed device is operated from the cross-head.

Claim.—First, a stone-channelling machine composed of a gang of cutters in combination with the direct-acting steam cylinders, the automatic valve gear, and provided with a suitable truck frame upon which the boiler and whole apparatus is mounted, constructed and operating substantially as described and for the purpose specified.

Second, the adjustable cross piece h and feed screw j, in combination with the cutter bar K and cross-head I, substantially as described and for the purpose set forth.

Third, operating feed wheel y, from the cross-head I, by means of the rod t', lever r, rod s, and rock shaft u, or their equivalents, operating substantially in the manner described.

Fourth, a swinging frame, operating a gang of cutters in a stone channeling machine, in a position inclined to the direction of the cut, in such manner that both the rear cutters of the gang will cut deeper than the forward cutter.

Fifth, mounting the wheels B and B' on adjustable brackets, substantially as and for the purpose described.

Sixth, the method herein described of changing the valve of the steam cylinder by the concussion of the cutters in striking the rock, consisting of the elbow o' p parallel bar m stop pawl j', rod g', lever r, and valve rod q, or other equivalent means of producing the same effect.

Seventh, the method herein described of changing the valves of the steam cylinder when the cutters do not strike the rock, consisting of the roller or stud $8'$, incl ned projection t', pawl j', rod g', lever r, and valve rod q, or other equivalent means of producing the same effect.

Eighth, the adjustable rod d' and the lever r in combination with the cutter bar K, cross-head I, valve rod q, and feed slide w, constructed and operating substantially as and for the purpose set forth.

No. 59,857.—ERASMUS D MILLER, Dorchester, Mass.—*Cranberry Gatherer.*—November 20, 1866.—The cylindrical vessel has projecting fingers with curved ends.

Claim.—An improved cranberry gatherer, made as hereinbfore described—that is, of a series of wires or teeth and a shallow cylindrical box, or the same and a handle, arranged as specified, the front ends of the teeth, under such arrangement, being back sloped, as set forth.

No. 59,858.—THOMAS S. MINNISS, Meadville, Penn.—*Car Coupling.*—November 20, 1866.—The upper end of the coupling pin is recurved and rests on the inner end of the link to keep it horizontal. A trigger supports the pin when the link is out, and the entrance of the link trips the trigger to drop the pin into place.

Claim.—First, the triggered pin, constructed as and for the purpose specified.

Second, the weighted link holder, operating as set forth.

Third, the combination of the weighted link, triggered pin, and draw-head, constructed as and for the purpose described.

No. 59,859 —HENRY NEUMEYER, Millerstown, Penn.—*Horse Hay-fork.*—November 20, 1866.—The plunger to which the barbs are attached has a rack on one edge, and is moved by a pinion fixed upon a shaft, provided with a crank arm and a spring to force the pinion out of gear whenever desirable. To the upper end of the plunger is pivoted a forked lever which engages with a stop upon the case to hold the fork in position for hoisting the load.

Claim —First, the forked locking lever H with its short arm or prong, adapted to press the catch F in against the notch, and its longer arm releasing it, in combination with the rack bar B and slotted handle D, receiving the pulley E, substantially as described, for the purpose specified.

Second, the combination of the crank M, shaft J, cog wheel K, and spring L with each other and with the tube A and bar B, substantially as described and for the purpose set forth.

No. 59,860.—H. L. OGDEN, Atkinson, Ill.—*Car Coupling.*—November 20, 1866.—The coupling links are raised by pins connecting by chains to their arbors. The links hold to upward projections on the opposite draw-heads. The device forms an automatical attachment.

Claim.—The rings C, attached to the shafts or arbors B, in combination with the rods D, provided with the pins e and the chains or cords d, which pass around the shafts or arbors B and the sliding bars E, all arranged and applied to operate in the manner substantially as and for the purpose herein set forth.

No. 59 861.—WILLIAM ORMSBY, Boston, Mass.—*Hand Screw Clamp.*—November 20, 1866 —The fulcral screw has a free nut, and the clamping screw a jam nut.

Claim.—A hand screw formed of the jaws A and B, the screws C and D, and the nuts E and F, substantially as herein described and for the purpose set forth.

No. 59,862.—GEO. H. REYNOLDS, Peoria, Ill.—*Bran Duster.*—November 20, 1866.—The bran passes between the radially ridged surface of the double frustum and an enveloping jacket of similar form and surface. It then passes between the pin-armed head of the cylinder and the head disk of a revolving frame, which carries inclined serrated plates which work the bran against the wire screen. Cams of the spur wheel above operate to agitate the screen to keep the meshes clear. The bran is finally discharged through one spout and the other products through another.

Claim.—The bran duster, consisting of the ribbed cone E, ribbed encasing cone F, spiked stationary head H, spiked movable head I, winged arms K, bolt frame M, chamber R, arms U, enclosing case T, uprights C', weighted levers D', and levers E , arranged and operating substantially as described, for the purpose specified.

No. 59,863.—J. L. RUMRILL, Hartford, Vt.—*Water Wheel.*—November 20, 1866.—A stationary cylinder is provided with an adjustable collar and placed over the water wheel. The water flows over a scroll plate attached to the cylinder and to radial arms of the wheel, and thence it acts against the series of curved buckets and discharges outward.

C P 49—VOL. II

Claim.—First, the cylinder C, constructed as described, in combination with the radial buckets J and curved buckets N, the whole being constructed, arranged, and operated in the manner and for the purpose set forth.

Second in combination with the above, the collar X and cylinder C, made so as to be adjusted to the inside of the wheel in order to prevent leakage, substantially as described.

No. 59,864 —WILLIAM SHAW, New Gordon, Ohio.—*Cider Mill* —November 20, 1866.—The apples are mashed by a toothed roller and carried by an endless belt to two pressure rollers, the upper one of which is adjustable by pivoted arms. The pomace is carried out by the belt, and the juice drops into a chute and trough, which convey it to a receptacle. Scrapers clear the belt and upper roller, and a revolving brush acts on the belt for the same purpose.

Claim.—The cider mill, consisting of the toothed cylinder B, endless apron D, rollers E E, adjustable pivoted bars F, scrapers H I, brush J, spout N, and chute O, arranged and operating substantially as described, for the purpose specified.

No. 59,865.—A. G. SMITH, Jersey City, N. J.—*Lantern.*—November 20, 1866.—The globe remains fast to the cap when the lamp is detached; the globe has bands at each end with projections, by which, in connection with devices on the cap and base, respectively, it is connected to the latter. The lamp top has a flaring flange to deflect the air, which is again deflected by a flange on the case.

Claim.—First, fastening the glass or globe D into the frame or guards B, substantially as and by the means described, so that when the said guards B and cap A are removed the said globe shall remain fixed within the said guards B, while at the same time it may be readily removed to be cleaned, substantially as and for the purpose set forth.

Second, the combination of the metal rims *b* and *c* with the removable globe D, substantially as and for the purpose set forth.

Third, the burner shaft *s* and pin *g*, in combination with the band *f*, for the purpose set forth.

Fourth, the deflector *l* and square shoulder S in combination, substantially as and for the purpose set forth.

Fifth, the frame or guard B provided with the groove *c* and the band *f*, in combination with the globe D provided with the projection *a*, so that the globe and frame may be securely fastened together and the globe readily removed by passing it down through the band *f*.

No. 59,866.—JAMES A. SMITH, New York, N. Y.—*Tool.*—November 20, 1866.—The jaws on the back of the shank are adapted to hold screw dies, which are capable of reversal to bring different matching faces into contact; dowel pins act as stays, traversing the dies and penetrating the faces of the jaws.

Claim —The combination of the rods *o o* with the dies F F, jaws A A', bar B, and nut C, substantially as and for the purpose specified.

No. 59,867.—ROSWELL T. SMITH, Nashua, N. H.—*Machine for Folding Paper Fans*—November 20, 1866.—Transversely in the centre of a table is arranged a vertically adjustable gauge. Two blades moving horizontally in a circle radiating from the remote end of this gauge, and held by springs to the limit of the desired plait or fold, are drawn apart by a treadle movement, and again pass each other by a spring movement to form a new plait; each successive fold passes beneath the folders.

Claim.—First, the alternately sliding or vibrating folding blades C C', or their equivalents, when used for folding or crimping paper or textile fabrics, substantially as herein set forth.

Second, the alternately sliding or vibrating folding blades C C', in combination with the vertical gauge B, or its equivalent, when used for the purposes substantially as herein set forth.

No. 59,868.—CHARLES J. M. SOHET and H. C. T. MOLVAUT, New York, N. Y.—*Compound for Lighting Cigars,* &c.—November 20, 1866 —Neutral sulphate of iron is treated with diluted nitric acid and precipitated; the resulting oxide is reduced by heat in a tube through which a current of hydrogen is passed. This is combined with a sulphide of aluminium and potassium and a small amount of carbon. A portion of this is placed on the end of a cigar, the breath is inhaled, and by the affinity of certain particles of the composition for oxygen the iron is heated and the cigar lighted.

Claim.—A lighting compound, made as described.

No. 59,869.—WILLIAM P. SQUIRE, Paris, Ill.—*Pump.*—November 20, 1866.—Explained by the claim and cut.

Claim.—The combination of the pump cylinder A with its fixed upper and lower movable valve boxes B B', when the latter is connected through a suitable connecting rod H and other parts with the pump handle J, or its equivalent, and when the flow of water is direct, substantially as and for the purpose described.

No. 59,870.—WILLIAM P. SQUIRE, Paris, Ill.—*Hay Derrick.*—November 20, 1866.—The hoisting rope has a ball attached to it, which engages the link upon the extensor rope at the proper time, when a continued draught serves to further elevate the hay and transport it laterally. The lower pulley carriage has movement on a traverse bar of the frame.

Claim.—First, the car pulley C, constructed as described, in combination with the crane B and ropes D G, substantially as and for the purpose set forth.

Second, the sliding pulley E, constructed as described, in combination with the ropes D G and frame A of the derrick, substantially as and for the purpose set forth and described.

Third, the combination and arrangement of the crane B, car pulley C, ropes D G, fixed pulleys H E, and sliding pulley E with each other and with the frame A of the derrick, substantially as herein described and for the purpose set forth.

No. 59,871.—D. H. STEPHENS, Riverton, Conn.—*Machine for Making Rule Joints.*—November 20, 1866.—The moving arbor is slipped longitudinally by a treadle, and the joint is held between the ends of the arbors. The partial rotation of the cutter frame on the arbors dresses the joint. When the treadle is raised one arbor is withdrawn and the joint released.

Claim.—The cutter head c working on the arbors a, and arranged with the adjustable cutters or knives b, feed screw h, and binding screw n, in connection with the movable arbors a and lever d, or its equivalent, whether operated by hand or other power.

No. 59,872.—LEWIS W. TEETER, Hagerstown, Ind.—*Flour Packer.*—November 20, 1866—The barrel is placed on the stand disk and raised to position. In packing, the barrel descends, and when filled a pin trips the shifter lever and stops the machine.

Claim.—The flour packer, the operating parts of which consist of the platform B, rack slide C D, eccentric G, inclined rack H, slide I, pin F, lever J, spring catch K, spring lever L, slide O, shaft R, gear S T, shaft U, packer X, when constructed and arranged to operate together, substantially as and for the purpose specified.

No. 59,873.—FAYETTE F. TERRY, Port Gibson, N. Y.—*Farm Gate.*—November 20, 1866—The gate has longitudinal and rotary movement on a roller, suspended on an adjustable swivel pin.

Claim.—Suspending the gate by means of the screw rod b and nut c, substantially as and for the purpose specified.

No. 59,874.—JONATHAN S. TIBBETS, Terre Haute, Ind.—*Pulley*—November 20, 1866—The book has a projection which acts as a detent on the cord, and keeps it from running back after the weight is suspended.

Claim.—The cam book A, when constructed substantially as herein shown and described, in combination with the pulley B, substantially as and for the purpose set forth.

No. 59,875—SAMUEL H. TIMMONS, Lafayette, Ind.—*Bottle Stopper and Medicine Gauge.*—November 20, 1866.—The contracted portion fits the neck of the phial. The bulb is graduated to mark its capacity at different points. The flat flange forms a foot.

Claim.—A graduated bottle stopper, or one made with a definite capacity, by which a specific dose may be measured, and the stopper fitted either within or around the neck.

No. 59,876.—A. P. TORRENCE, Columbus, Georgia.—*Treadle.*—November 20, 1866.—The treadle frame is not pivoted, but slides on two vertical guide rods.

Claim.—A treadle, hung and arranged in such a manner as to operate substantially as and for the purpose described.

No. 59,877.—ELSON TOWNS, Morland's Grove, Ill.—*Vibrating Governor.*—November 20, 1866.—The clock is driven by the machine in action, and its motions are regulated by a pendulum and escapement. When the motions of the machine and the clock cease to be isochronous, the differential action reciprocates a rack, whose motion may be communicated in any way to a throttle valve or other object, whereby the access of force is adjusted.

Claim.—First, the combination of the pulley A, pinion wheel K, bevel gear wheels F G H, and frame or box E, with each other, with the driving shaft B, the verge shaft I, the weighted rack L, and the frame D, of the machine, substantially as described, for the purpose of obtaining the difference between the required and actual velocity and the application of said difference, as herein set forth.

Second, the combination of the jointed verge X, constructed as described, with the verge wheel W, with the pendulum B¹ C¹, and vibrating wheel M′, either or both, and with the frame D of the machine, substantially as described and for the purpose set forth.

Third, the combination of the spring W¹, pitman S¹, lever U, and vibrating frame T¹, with each other, with the vibrating wheel M¹, and with the frame D of the machine, substantially as described and for the purpose set forth.

Fourth, the combination of the pawl G¹, ratchet wheel H¹, windlass E¹, and cord I¹, with each other, with the pendulum weight C¹, and with the frame D of the machine, substantially as described and for the purpose set forth.

Fifth, the combination of the spring catch U, and pulley V, with the driving shaft B, with the rack L, and with the frame D of the machine, substantially as described and for the purpose set forth.

Sixth, the combination of the jointed lever O o¹, and stirrup N, with the rack L, and with the fulcrum post P, substantially as described and for the purpose set forth.

No. 59,878.—WILLIAM VALENTINE and THOMAS C. LONGTON, Trenton, N. J., assignors to themselves and CHARLES V. MEAD, Hamilton Township, N. J —*Swing Power.*—November 20, 1866.—The swing is operated by a foot bar connected by pivoted levers to the depending swing arms. The latter are attached to a rock bar, on which one end may give reciprocating motion. The other end of the rock shaft has a drum connected by straight and cross belts to two loose pulleys on a shaft beneath. These pulleys have ratchet connections to two ratchet collars attached to the shaft, so that each in turn operates to rotate the shaft forward, the ratchet connection of the other slipping freely back.

Claim.—First, a swing, constructed substantially as described, in combination with the pulleys B and F, the latter being provided with the arm H¹, by means of which a rocking motion is imparted to the shaft of a washing machine, and a reciprocating motion to the dasher of a churn, substantially as described.

Second, a swing, constructed substantially as described, in combination with the double pulley J, and clutches K K' on the shaft L, by means of which a continuous rotary motion is imparted to the shaft L, substantially in the manner and for the purpose set forth.

No. 59,879.—GEORGE VINCENT, Stockton, Cal.—*Binder for Sewing Machines.*—November 20, 1866.—The binder has a lip for the lower edge only of the binding. The part attached to the presser-foot serves to guide the upper edge. The thickness of the goods, by lifting the presser, thus adjusts this upper guide to the proper elevation.

Claim.—The combination of the plates A and C with the block G. and its spur H, operating substantially as above described and for the purpose herein set forth.

No. 59,880.—WILLIAM S. WATSON, New York, N. Y.—*Hoisting Apparatus.*—November 20, 1866.—The goods are drawn up to a hook on a cord round the windlass. This book is then engaged with the package, which is then disconnected from the first book by the depression of a lever continuation of the hook shank. An enveloping brake assists in the lowering of the package.

Claim.—First, the combination of the rope A, and hoisting tackle B C, with the book a, and its arm g, windlass D, rope E, and hook b, all arranged in the manner described, and employed to permit the weight or thing lifted to be transferred from the rope and tackle to the windlass, substantially as and for the purpose specified.

Second, in combination with the above, the spring brake F, lever H, and nose d, arranged and operating in the manner and for the purpose specified.

No. 59,881.—AMBROSE H. WELLS, Waterbury, Conn.—*Spittoon for Railroad Cars.*—November 20, 1866.—The cover and bottom are attached to the same axis, and when one is opened the other is closed, and conversely.

Claim.—An open-ended spittoon A, having a cover C. and valve I, both arranged to turn upon a pivot E, substantially in the manner described and for the purpose specified.

No. 59,882.—THOMAS J. WELLS, New York, N. Y.—*Peat Machine.*—November 20, 1866.—The peat is placed in the hopper, and falls into the case wherein revolves a horizontal shaft armed with oblique radial cutters which alternate with stationary cutters. Between the two the peat is cut by a shear action, and is pushed forward to the throat, whose spiral ridges act in concert with the spiral flanges on the point of the cutter shaft to deliver the peat into the mould wheel, which is revolved by an intermittent impulse, and discharges the twin blocks from its moulds by automatic followers.

Claim.—First, the combination of the fixed and rotating diagonal arms or knives a b, when placed within a horizontal frame, with spiral propelling blades Q, substantially as described.

Second, making spiral blades or ridges, one or more, in the inside of the discharge neck O, in the direction the converse of the directions of the spiral propelling blades Q, substantially as described.

Third, a stationary die L, one or more, projecting through the movable bottom of the mould, substantially as described.

No. 59,883.—GEORGE WHITAKER, Lewistown, Ill.—*Distributing Table.*—November 20, 1866.—The table for the distribution of mail matter has a series of boxes at its edges, with hooks upon them for the attachment of bags. The sides presented to the operator have racks to hold cards of designation.

Claim.—A table made in a box form. and provided with a series of spouts or boxes E around its sides, either one or more, having hooks a, or other means at their lower ends suitable for suspending a bag or sack or other receptacle thereto, substantially as described and for the purpose specified.

Also, the combination with the table boxes E of the racks or frames I, for receiving and holding the cards indicating the portions of mail or other matter to be placed in the same, substantially as described.

No. 59,884.—D. H. WHITTEMORE, Worcester, Mass.—*Apple Parer.*—November 20, 1866; antedated November 11, 1866.—Cams on the main wheel turn an intermediate wheel by operating on its cogs, and similar cams on the intermediate wheel cause the oscillation of a rack which sweeps the paring knife alternately from the stem to the calyx of one apple, and in a contrary direction on the next. A ridge on the intermediate wheel throws the knife from the apple at the outer position of said knife.

Claim.— First, the cams c e upon the face of the gear C, for the purpose of giving and reversing the motion of the paring knife, substantially as described.

Second, the projection or rib I I upon the gear C, for the purpose of tipping the knife back from the apple to admit of the easy removal of the apple from the fork and replacing of another, as described.

No. 59,885.—N. A. WRIGHT, Prairie du Chien, Wis.—*Milk Strainer.*—November 20, 1866.—The spout with its lip and strainer is attached to the rim of the bucket by spring rods : an elastic strip of rubber packs the joint between the two.

Claim.—The combination of the rubber or other elastic lining a on the inside of a lower section A of a strainer, with the springs b b, arranged substantially as herein shown and described.

No. 59,886.—WILLIAM WRIGHT, New York, N. Y.—*Cut-off Valve Gear.*—November 20, 1866.—The rise or fall of the governor rod causes its partial rotation by a spiral gear at its foot, and this partial rotation depresses or projects the toes by which the induction and cut-off is effected. This effect is produced by means of a longitudinally ribbed part of the stem acting upon racks on the inner sides of the toes. A flexible plate is introduced under the heels of the induction valves with a set screw beneath it to balance the pressure of steam.

Claim.—First, the clogged sliding toe or toes, and the spirally-grooved or threaded longitudinally moving spindle N, in combination with each other and with the cam F, and valves, substantially as and for the purpose herein specified.

Second, the spirally cogged or threaded spindle N, so combined with the governor and the sliding toe or toes of the valve-operating cam as to have a longitudinal movement as the governor rises and falls, and also to be capable of turning independently of the governor, as the latter rises and falls, substantially as and for the purpose herein specified.

Third, the combination of the rolling slide valves I I', with the flexible plates U U', substantially as herein described, for the purpose herein set forth.

Fourth, the set screws u u', in combination with the flexible plates U U', and valves I I', substantially as and for the purpose herein specified.

No. 59,887.—WILLIAM WRIGHT, New York, N. Y.—*Variable Cut-off Valve Gear.*—November 20, 1866.—The lifter has movement on an eccentric cam upon the valve shaft by means of a rack engaging the ribs of a central pin. This pin has a spiral gear connection to the valve shaft, so that the partial insertion or withdrawal of the pin adjusts the lifter

Claim.—The combination of the revolving toothed lifter E, eccentric G, fast to the valve shaft and toothed and spirally grooved or ribbed longitudinally sliding spindle C, arranged for operation together, and with the hollow valve shaft, substantially as shown and described.

No. 59,888.—FREDERICK AUGUSTUS ABEL, Woolwich, England.—*Manufacture of Gun Cotton.*—November 20, 1866.—Ordinary gun cotton is pulped and then solidified by means of hydraulic pressure, after which it is divided into grains by any suitable machinery.

Claim.—First, reducing gun cotton to a pulp, and consolidating such pulp with or without the aid of pressure into the form of sheets, disks, granules, cylinders, or other solid forms, either with or without the admixture of binding materials.

Second, combining with gun cotton reduced to a pulp gun cotton in a fibrous state, and consolidating such mixture into sheets, disks, granules, cylinders, or other solid forms, either with or without the admixture of binding materials.

Third, combining soluble and insoluble gun cotton, either when both are in a state of pulp, or when one is in a state of pulp and the other in a fibrous condition, and consolidating such mixtures into cylinders, sheets, disks, granules, or other solid forms, either with or without the admixture of binding materials.

Fourth, subjecting mixtures of soluble and insoluble gun cotton, either when both are in a fibrous condition or when both are in a state of pulp, or when one only is in a state of pulp, and the other in a fibrous condition, to the action of solvents of the soluble gun cotton, either alone or with the employment of pressure, so as to effect the consolidation of the same.

Fifth, the application to the surface of the consolidated gun cotton of a solution of the soluble forms of gun cotton or of shellac, or other suitable gums or resins.

No. 59,889.—WALTER BRIGGS, Greencastle, Iowa.—*Cane Mill.*—November 20, 1866.— The wooden rollers are fitted with boxes of metal upon axial shafts. Annular projections on the bed pass up into recesses in the lower surfaces of the rollers to keep the juice from the journals. The sliding feeder is vertically adjustable and secured by pins.

Claim.—First, the rollers *a a*, with iron shafts and metal boxes, with concave in the ends of the rollers *a* and *b*, and convex surfaces, operating as described and for the purposes set forth.

Second, the sliding feed H, in combination with the rollers *a a* and *b*, and the concave *a*, operating as described and for the purposes set forth.

No. 59,890.—C. B. BRISTOL. New Haven, Conn.—*Hook and Fastening for Ropes.*—November 20, 1866.—The rope is impaled upon the spurs of the sleeve, and the latter is then clasped around the rope.

Claim.—Casting the clasp A, hook or loop B, and points or spurs *a* and *b* in one piece, when so constructed and shaped that in fitting, the points or spurs *a* and *b* may pass through the rope or cord D, and the edges or lips *c* and *d* may be bent or turned over so as to firmly and entirely clasp the rope or cord, substantially as herein described and set forth.

No. 59,891.—JOHN F. COLLINS, New York, N. Y.—*Manufacture of Alcohol and other Spirits.*—November 20, 1866.—The salt of lime, soda or ammonia is introduced to neutralize the acid of the wash; the formation of acid is checked by the low temperature at which it is distilled. The aqueous vapors are removed by gas mechanically introduced. The object is to obtain alcohol of high proof.

Claim.—First, the treatment of wash or mash, or other substance from which alcohol or other spirits are to be distilled, with phosphate of lime or soda, or carbonate of ammonia.

Second, the distillation of spirituous solutions at a temperature not exceeding 170 degrees Fahrenheit, for the purpose of preventing the formation of acetous and acetic acid, and also the different ethers.

Third, having in connection with the still or other apparatus a pipe, longer or higher than the column of water at the natural pressure of the atmosphere, for the purpose of returning any aqueous vapors to the still.

Fourth, the use of a current of suitable gas to propel the alcoholic vapors, instead of using heat.

Fifth, the use of a circulating current of air or suitable gas between the condenser and the still, for the purpose of preventing oxidation during the process, substantially as and for the purposes described and set forth.

No. 59,892.—WILLIAM COTTON, Loughborough, England.—*Knitting Machine.*—November 20, 1866.—Two or more straight series of barbed needles and two or more series of sinkers are used to make at one and the same time two or more pieces of work, each piece serving, say, for one leg and half the body of a pair of drawers. Every other sinker is moved successively, the alternate ones being moved in a body, thus forming a loop on each needle, but reducing the liability of breaking the thread. Independent grooved instruments serve to take loops from certain of the needles and transfer them to others in widening and narrowing. A detailed account would require too much space.

Claim.—The improvement in operating the sinkers in order for the formation of loops on the main needles, such being by means of the jacks *f*, and a catch bar *h*, or its equivalent, provided with actuating mechanism, substantially as described, whereby certain of the sinkers are operated in succession by the jacks and the intermediate sinkers, or those termed "dividers," as well as the rest or jack sinkers, are subsequently moved in a body, in manner as explained.

Also, the improvement by which the beards of the needles are closed, the same being effected by means of a stationary presser P, and by moving the needles up against it as explained.

Also, the vertical arrangement of the needles *d* relatively to the stationary presser bar P, and the sinkers *e*, arranged horizontally or thereabout, as described, the same being advantageous in rendering the needles capable of being moved up to and away from the presser bar with great ease and little expenditure of power.

Also, the combination as well as the arrangement of the fashioning needles *m*, formed and provided with mechanism for operating them, substantially as described with the comb *i*, and presser P, and the main needles *d*, and the sinkers, provided with mechanism for operating them for the production of a knit fabric, as set forth.

Also, the fashioning needle, constructed as described, viz: with a curve at and near its point, and with the groove in its shank, and extending back from the point, as set forth.

Also, the improvement for operating the thread carriers, viz: a friction brake applied to their supporting rod, and extended from the slur-box sustaining bar, as set forth.

Also, the combination as well as the arrangement of the adjustable stops *j⁴ k⁴*, provided with mechanism for operating them, substantially as described, with the thread carrier or carriers, their supporting rod and the friction brake extended from the slur-box carrier.

Also, the improvement in the construction of the thread carrier, the same consisting in making it next the needle flat, or nearly flat, as represented at *k⁴*, in figures 15, 16, 17, and 18, the same being for the purpose hereinbefore set forth.

Also, the combination of the adjustable stops $j^4 k^4$ provided with mechanism for operating them, substantially as described, with the fashioning needles, and the thread carrier or carriers provided with mechanism or mechanisms for operating them, substantially as explained.

No. 59,893.—C. O. CROSBY, New Haven, Conn., assignor to THE FISH HOOK AND NEEDLE COMPANY, same place.—*Fish Hook.*—November 20, 1866.—The hook is strengthened by being flattened in the line of its strain.

Claim.—Flattening fish hooks in the bend, substantially as and for the purpose described.

No. 59,894.—JOHN B. CROWLEY, Cincinnati, Ohio., assignor to himself and CHAMBERLAIN & Co , same place.—*Hot Water Reservoirs for Stoves.*—November 20, 1866.—The sliding plate forms the rear half of the flue in the back of the reservoir; the said flue forms a communication between the flue hole on the top of the stove and the smoke pipe above.

Claim.—First, the sliding plate D or D', formed with projections d d', and a lug G, in the described combination with the grooves E E', for the purposes set forth.

Second, the combination of the lugs G M, screw bolt or pin H, hooks N, and studs O, for securing the flue plate and cover to each other and to the vessel A, as described.

No. 59,895.—THOMAS B. DE FOREST, Birmingham, Conn.—*Metallic Binding.*—November 20, 1866.—The plate has a notch on each side, is bent into an angular shape, doubled and flattened upon the end of the band, and secured by oblique indentations.

Claim.—Tipping or binding the end of skirt bands with metal, when the said tip is secured to the strap by indentations oblique to the fibre, in the manner substantially as described.

No. 59,896.—THOMAS B. DE FOREST, Birmingham, Conn.—*Buckle.*—November 20, 1866.—The tongue points project through the sleeve. The sleeve envelops the frame ends, and may be made to embrace the end of the strap.

Claim.—First, the combination of the frame A, the tongues B B, and the sleeve D, when the said sleeve D serves to hinge the tongues to the frame, substantially in the manner as herein set forth.

Second, securing the said buckle to its strap by means of the sleeve D around the bar and over the strap, substantially as herein set forth.

No. 59,897.—DANIEL DISHART, Canton, Ohio.—*Washing Machine.*—November 20, 1866.—The two boards reciprocate upon each other, and an adjustable slide piece rests upon the upper one. An uncovered clothes receptacle is at one end.

Claim.—First, the adjustable slide C, with spring attached, in combination with the boards B B, as and for the purposes specified.

Second, the support m and receptacle b in combination with the box A and hinged cover, the whole being arranged and operating substantially as herein specified.

No. 59,898.—H. W. DOPP, Buffalo, N. Y.—*Tubular Wick Burner.*—November 20, 1866.—The interposed water chamber nearly isolates the reservoir from the wick tubes and burner. The inner rings form the throat for the wick, and tend to prevent the passage of heat downward to the tubes. The frustum attached to the middle ring deflects the heat upward. The finely perforated floor of the outer ring debars the downward passage of flame when gas may be escaping.

Claim.—First, the water jacket B and B¹ in combination with a wick burner, for the purposes set forth.

Second, the combination of the two rings I and K with the pipes C and C¹.

Third, the combination of K¹ with a tubular wick burner, when operated as set forth.

Fourth, the combination of the cone I¹ and I, for the purpose specified.

No. 59,899.—H. W. DOPP and CHARLES P. WEISS, Buffalo, N. Y., assignors to themselves and J. FORSYTH, same place.—*Skate.*—November 20, 1866.—A metal rim forms an abutment for the heel behind, and a pivoted clamp pierces and holds it in front. A lever beneath works the heel clamp and the ball clamp simultaneously. Double screwed nuts adjust the length of the connections.

Claim.—First, the combinations of the self-adjusting jaws C C, levers B B, and rod D, as described.

Second, the lever arrangements E and G in combination with rod H.

No. 59,900.—ABRAM FANKBONER, Schoolcraft, Mich., assignor to ALMERON F. CHAPIN, Richmond, Ind.—*Field Fence.*—November 20, 1866.—Improvement on his patent of October 13, 1863. The stakes for the support of the fence are secured by additional battens, and the end pieces by pins.

Claim.—The braces or stakes F, battens C D, pins E, in combination with the end pieces B and fence panels, constructed and arranged as and for the purpose set forth.

No. 59,901.—Thomas M. Fell and Ambrose G. Fell, Brooklyn, N. Y., assignors to themselves and Wm. Bull, New York, N. Y.—*Manufacture of White Lead.*—November 20, 1866; antedated September 25, 1866.—The lead is converted into an oxide by roasting in a reverberatory furnace, and the oxide thus formed is mixed with nitric acid to form it into a pasty mass. Sulphuric acid is then added to convert the nitrate and oxide of lead into the sulphate of lead. The nitric acid may be recovered by pressure, or otherwise, to be again used. The sulphate is boiled in an alkaline solution to deprive it of a portion of its acids and water.

Claim.—First, the production of the sulphate of lead, in the manner and for the purposes substantially as described.

Second, the treatment of the sulphate of lead so produced with an alkali solution, in the manner and for the purpose substantially as described.

Third, the treatment of the sulphate of lead so produced with the carbonate of either potash, soda, or lime, followed by an alkali solution, in the manner and for the purpose substantially as described.

Fourth, the treatment of the sulphate of lead so produced with an alkali compound, in the manner and for the purpose substantially as described.

Fifth, the manufacture of white lead from the ores of lead or the metallic lead, or from the oxide of this metal, by the use of nitric and sulphuric acids, in combination with an alkali solution with or without the prior treatment by an alkali compound, substantially as described.

No. 59,902.—Thomas M. and Ambrose G. Fell, Brooklyn, N. Y., assignors to themselves and Wm. Bell, New York, N. Y.—*Manufacture of White Lead.*—November 20, 1866.—The solution of lead is precipitated by a combination of hydrochloric and sulphuric acids. The precipitate is afterwards treated with an alkali.

Claim.—First, the treatment of a solution of lead with a combination of acids, in the manner and for the purposes substantially as described.

Second, the production of a basic chloro-sulphate, either by separate or complete solution and precipitation, or by treatment in one and the same vessel, in the manner and for the purpose substantially as described.

Third, the treatment of a precipitate so produced with an alkali solution, in the manner and for the purpose substantially as described.

Fourth, the treatment of a precipitate so produced with the carbonate of either soda, potash, or lime, followed by an alkali solution, in the manner and for the purpose substantially as described.

Fifth, the treatment of a precipitate so produced with any alkaline compound, in the manner and for the purpose substantially as described.

Sixth, the manufacture of white lead from the ores of lead, metallic lead, oxide of lead, or litharge, or other substance containing lead, either by separate or complete solution and precipitation, or by treatment in one and the same vessel of a salt of lead with a double precipitant in combination with an alkali solution, either with or without the prior treatment by an alkaline compound, substantially as described.

No. 59,903.—Walter Fitzgerald, Boston, Mass.—*Pegging Machine.*—November 20, 1866.—The last rest has three motions—in a circle, horizontally, and longitudinally on the plate, which is pivoted to move transversely. The peg is cut from the peg-wood strip by the motion of the cutter plate.

Claim.—In combination with the gear plate *b*, upon which the last is mounted, and through which the feed of the shoe is effected, the mechanism which admits of both longitudinal and transverse rocking movement of the shoe, substantially as set forth.

Also, in combination with the peg tube *n*, in the face place *o*, the cutter *p*, placed in the face plate, with its cutting edge disposed as shown, so that the lateral movement of the plate severs the end of the peg-wood in the tube from the strip to form the peg, substantially as described.

No. 59,904.—Ira Flanders, Lafayette, Mich.—*Stump Extractor.*—November 20, 1866.—A slot in the lever permits the adjustment toward or from the fulcrum of the pin from which the link of the chain is suspended. The standards are braced to the sills, which form runners for the transportation of the machine.

Claim.—Providing the lever C with the slot *m*, and the recesses for the pin F, in combination with the said pin F, and link *g*, and lever D, arranged substantially as described and for the purposes set forth.

Second, the combination of the hooks or bolts *j j j j*, the braces *i i i i*, with the bed pieces A A, to operate as set forth.

No. 59,905.—Joseph S. Foster, San Francisco, Cal.—*Pastry Roller.*—November 20, 1866.—The paste-board is revolvable. The roller is moved by a crank and has pinions gearing in side racks, by which its ends are advanced at equal rates. The rack frames are supported by links on the base and are thus vertically adjustable.

Claim.—The combination of the revolving board B, the roller C, the adjustable guide rocks D D, with the rods *e e e e*, operating substantially as and for the purpose specified.

No. 59,906.—H. P. GENGEMBRE, Pittsburg, Penn.—*Skating Floor.*—November 20, 1866 —
The floor is composed of a series of boxes communicating with each other by pipes, and with
a refrigerating apparatus by means of a pipe surrounded by non-conducting material. A rim
is made around the edges of said floor forming a shallow receptacle for water.

Claim —First, producing a sheet of ice for a skating floor by freezing from the under side
a sheet of water, as described.

Second, the floor A, composed of a number of boxes *a a' a''*, &c., in combination with a
refrigerating apparatus R, arranged as described and for the purpose specified.

Third, a skating floor composed of a sheet of ice resting on the floor A, as described or its
equivalent, and kept from melting as specified.

No. 59,907.—THOMAS R. GRANT, Newark, Ohio.—*Piston-rod Packing.*—November 20,
1866.—Around the piston-rod within the gland are sections divided longitudinally and with
abutting shoulders. The ends of the sections are received in cups, and springs keep them
in position.

Claim.—First, a metallic steam packing composed of the sections A, and cups E and F, constructed as described.

Second, in combination with the sections A and the cups E and F, the springs H, as and for
the purpose described.

No 59,908.—J. H. GREENLEAF, New Haven, Conn., assignor to himself and ISAAC N.
DANN, same place.—*Boot Heel.*—November 20, 1866.—The metallic shell of the heel has a
concave plate by which it is attached to the shoe, and a recess in its lower surface to receive a
disk of resilient rubber.

Claim.—The heel constructed as herein described, that is to say, by the combination of the
shell A, the plates B and C, and the elastic face D, substantially as herein set forth.

No. 59,909.—CALEB H. GRIFFIN, Chelsea, Mass., assignor to himself and W. E. P.
SMYTH.—*Boiler Feeder.*—November 20, 1866.—The piston is moved in the cylinder and the
space between the heads is brought alternately in connection with the boiler and the feeder.
Steam from the boiler fills the chamber, which is condensed as soon as the water connection
is reached; when the piston returns, the water flows to the boiler.

Claim.—The arrangement of the pistons *c c'*, chamber *a*, well pipe *e*, steam pipe *g*, port *k*,
reservoir *i*, and feed pipe *h*, with reference to each other and the boilers, whereby to operate,
as and for the purpose set forth

Also, in connection therewith the arrangement of the pipes *n o* whereby to regulate the
height of water in the boiler, as set forth.

No. 59,910.—A. T. HAY, Burlington, Iowa.—*Preventing Incrustation of Sugar or other
Boilers.*—November 20, 1866.—To prevent the incrustation of evaporating pans with the
sediment deposited, the kettles are consecutively encircled by a coil of wire connecting the
poles of a battery.

Claim.—The application of electricity to prevent the formation of scale or incrustation in
evaporating pans or kettles, substantially in the manner herein described.

No. 59,911.—GEORGE HEISS, Lancaster, Penn.—*Cigar Press.*—November 20, 1866.—The
cigars receive a partial pressure in their frames in a vertical series for complete pressure in a
direction perpendicular to the former.

Claim.—The frame A A C C, with its movable ends B, and slats 1 2 3 4, &c., in combination with the rods G, springs and blocks E F, arranged and operating in the manner
and for the purpose specified.

Also, the arrangement of the open frame A C, and its parts, in combination with the top
P, with its slats H, and pins I, and bottom O, substantially made and used for the purpose
specified.

Also, the construction of the filling-in board L, with its handled screws, in combination
with the forked terminus K N, turning press block M, when employed for the purpose and
in the manner specified.

No. 59,912.—WILEY JONES, Norfolk, Va.—*Instrument for Stretching Boots, Shoes, &c,
Lengthwise.*—November 20, 1866.—The fixed abutment is passed to the toe and the movable
one placed inside the heel. Rotation of the rod gives motion and stretches the shoe.

Claim.—The above-described instrument or its equivalent for the purpose of stretching
boots and shoes lengthwise at the toe, and thus securing ease and comfort to the wearer when
the boot or shoe is found to be too short or to cramp the toes of the foot in the manner and
for the purpose specified.

No. 59,913 —SUSAN R. KNOX, New York, N. Y.—*Fluting Machine*—November 20, 1866 —
The fabric is passed between a pair of hollow corrugated rollers, the upper one of which
is elevated and depressed by a lever which has its fulcrum in a bent arm to avoid obstructing
the passage of the fabric. The rollers are removable by detaching the bearing at one end of
the upper roller upon which the spring rests.

Claim.—First, the standard D when constructed substantially as herein described, and located at the ends of the rollers in contradistinction to a point between the ends when said standard is employed in combination with the lever C, as and for the purpose set forth.

Second, the detachable bearing F', in combination with the rollers B B', substantially as and for the purpose specified.

Third, the spring J, arranged as described, and employed in connection with the lever C, bar E, and bearings F F', for the purpose specified.

No. 59,914.—OTTO KROMER and CHARLES OHLEMACHER, Sandusky, Ohio.—*Belt Clasp.*—November 20, 1866.—The tongues are attached to a back bar, penetrate the ends of both straps, and are bent around another bar.

Claim.—The bar or back A, the tongues C, rod or slide B, as arranged in combination with the belt F, for the purpose and in the manner as set forth.

No. 59,915.—PIERRE LALLEMENT, Paris, France, assignor to himself and JAMES CARROLL, New Haven, Conn.—*Veloripede.*—November 20, 1866.—The fore wheel is axled in the jaws of a depending bar, which is pivoted in the frame and turned by a horizontal lever bar. This wheel is revolved by a treadle crank.

Claim.—The combination and arrangement of the two wheels A and B, provided with the treadles F and the guiding arms D, so as to operate substantially as and for the purpose herein set forth.

No. 59,916.—JAMES W. MALOY, Boston, Mass.—*Machine for Polishing Marble.*—November 20, 1866.—The marble slab is placed on the slide and its edge polished by an endless belt, which is tightened by an adjustable idler.

Claim.—First, polishing marble by means of an endless band of felting or other suitable material, arranged and operating as described.

Second, the combination of an endless band, as described, with the sliding platform d, substantially as and for the purpose described.

No. 59,917.—W. H. MASTERS, Princeton, Ill.—*Solar Camera.*—November 20, 1866.—A conical shield close behind the condensing lens in the mouth of the upper chamber, in connection with a diaphragm in front of the smaller lens, prevents the rays of light being concentrated on and setting fire to any portion of the wood work.

Claim.—The shielding cone v and diaphragm w between the lenses t u, substantially as and for the purposes set forth.

No. 59,918.—JAMES P. MCLEAN and JOHN VANDERCAR, Brooklyn, N. Y., assignors to JAMES P. MCLEAN.—*Piston-rod Packing.*—November 20, 1866.—Cork on the outside and strips of lead on the inside, next the piston-rod, are wrapped in a canvas gasket, and form a packing around the piston rod within the gland.

Claim.—The combination of the cork N N and leaden strips 1 2 3, for the purposes substantially as described and shown in the drawings.

No. 59,919.—THOMAS MCQUIRK and ORIN COLE, Millville, Mass.—*Carding Engine.*—November 20, 1866.—The object is to avoid the waste and recarding of the imperfect outside strands of roving as this comes from the doffer, and to deliver the same to a creeper, which delivers them back to be mixed with the new wool.

Claim.—The combination of the doffer of the finishing carding engine with one of the carding cylinders and a conveying mechanism, the whole operating substantially as set forth.

No. 59,920.—CHRISTIAN K. MELLINGER, Manor township, Penn.—*Shifting Seat for Carriages.*—November 20, 1866.—The fore seat is supported on parallel pivoted bars, which admit of a change in its position toward or from the dash-board.

Claim.—The specified combination and arrangement of the top and bottom pieces A B, braces C D, held by pivots P within the boxes or slots a' a'' and b' b'', bevelled or constructed and operating in the manner and for the purpose specified.

No. 59,921.—WILLIAM T. MERSEREAU, Newark, N. J.—*Stair Rod Fastening.*—November 20, 1866.—The fastening ring has a pivoted cap, which closes upon the end of the stair after admitting its passage lengthwise through it.

Claim.—Combining with the ring the tip, for the purposes herein fully indicated.

No. 59,922.—OREN E. MILES, Aurora, Ill., assignor to himself and WM. B. SIGLEY, same place.—*Bolster Plate for Wheeled Vehicles.*—November 20, 1866.—Explained by the claim and cut.

Claim.—The projections M N around the bolt holes near the ends of the plates, and adapted to stand in corresponding holes in the wood work and resist both the lateral and the torsional strains, as and for the purpose herein set forth.

No. 59,923.—GEORGE L. MORRIS, Taunton, Mass.—*Screw.*—November 20, 1866.—Explained by the claim and cut.

Claim.—As a new article of manufacture, a cast wood screw, combining the conical stem under the thread, thread whose convolutions are all of the same diameter, except the last concave faces to the thread, rounded spaces between the convolutions of the thread, and the horizontal cutting edge for the point.

No. 59,924.—JOHN S. MORRIS, Buffalo, N. Y. — *Torpedo.*—November 20, 1866.—A friction rod or wire passes longitudinally through the centre of the torpedo, and is armed with teeth at intervals, which act in combination with fuzes or detonating tubes secured in position by the blocks and intervening stays. The conical openings around the wire at the ends of the case are plugged with wax.

Claim.—A blasting cartridge or torpedo, constructed substantially as described.

Also, the blocks II and stays I, in combination with the detonating tubes *b*, by which the latter are secured in place and maintained in their proper relative position within the case A, substantially in the manner specified.

Also, constructing the plugs C D with the hole for the friction rod enlarged at the outer end, when used in combination with wax *o* or other suitable substance for packing the same, and also constructing the former with the groove *h*, all arranged and operating substantially in the manner and for the purpose described.

No. 59,925.—JOHN H. POCOCK, Chicago Ill.—*Coal Hod.*—November 20, 1866.—The bottom is made of two pieces swaged up and fastened together so as to hold the base and sides. The sides extend upward so as to form by the lips a discharge spout for the coals.

Claim.—First, the bottom made up of the pieces *b* and *c*, the pieces having the flanges for connection to the body of the scuttle and for forming the hoop or bottom bearing surface as a part of them, and constructed substantially as herein set forth.

Second, the lips *a*, when constructed and formed as herein recited, with the side pieces of the scuttle, as described.

No. 59,926.—IGNATIUS RICE, New York, N. Y.—*Comb.*—November 20, 1866.—The curved comb has a metallic spring band secured to the outside of the back or rim.

Claim.—The combination of the elastic spring band with the comb, in the manner and for the purpose set forth.

No. 59,927.—N. L. ROBERTSON, Rockford, Ill.—*Sifter.*—November 20, 1866.—The sieve is reciprocated on the roller bed by the pivoted lever. A removable hopper above the sieve contains the cinders.

Claim.—The sifting apparatus consisting of the box A, provided with the sieve C, resting on the rollers D, and the removable hopper B, all arranged to operate as herein shown and described.

No. 59,928.—SAMUEL J. SEELY, New York, N. Y.—*Roofing and Clap-boarding.*—November 20, 1866.—The sheets of metal are corrugated to resemble weather-boarding, are united by matched or lapped joints, and fastened to the studding, &c., at proper intervals.

Claim.—The sheet metal clap-boarding, bent and united substantially in the manner described, for covering the walls or roofs of buildings, and for other analogous purposes.

No. 59,929.—HENRY SIDLE, Minneapolis, Minn.—*Construction of Houses.*—November 20, 1866.—An air-tight chamber is formed by a strip of board inclined against the building, its lower edge being imbedded in the ground.

Claim.—The casing C, placed in an inclined position and used below the lower windows of the house with its lower edge buried so as to form an air-tight chamber between it and the house, as and for the purpose specified.

No. 59,930.—STEPHEN SPOOR, Phelps, N. Y.—*Gate*—November 20, 1866.—The latch post of the sliding gate has pins for the engagement of the latch bars and a gravitating button for security. By engaging the lower latch over a higher pin, smaller animals only are allowed a passage.

Claim.—Adapting the gate to be elevated for the passage of swine, &c., by means of the two projections *h i*. and pin *d*, and securing the same in place by means of the catch I, hanging free on the outside of the post, the whole arranged and operating substantially as and for the purposes specified.

No. 59,931.—LYMAN S. STEVENS, Waltham, Mass.—*Portable Metallic Smoke-house.*—November 20, 1866.—The cylindrical metal case has a drawer for fire, a mica panel for observation, register air holes and a door above by which the meat is introduced to be suspended from the hooks within.

Claim.—As a new article of manufacture, a portable metallic smoke-house, substantially as described.

No. 59,932.—JOHN STOFER, Cleveland, Ohio.—*Window Brush.*—November 20, 1866.—
The brush handle has a locking hinge joint and an elastic flexor band; it may be locked
at any angle, and is used to enable a person inside a room to clean the outside of the
window.
Claim.—The brush A, jointed handles B C, provided with a locking device and elastic
cord E, arranged substantially as and for the purpose set forth.

No. 59,933.—DAVID WALKER, Newark, N. J.—*Blotting Pad.*—November 20, 1866.—
The block has a grooved edge and the edge of the paper is clamped therein by spring-rods.
Claim—Securing the paper to the holder by the springs, constructed and arranged sub-
stantially as shown.
Also, the projecting edges of the holder, when in combination with the springs.

No. 59,934.—LEVI WILKINSON, New Haven, Conn.—*Screw Clamp for Bending Axletrees.*—November 20, 1866.—This device is to give the spindle, when new, the proper pitch
or to straighten a bent axle. It is explained by the claim and cut.
Claim—The combination of the bar A, and screw c, with the clamp or hook C, when the
whole is constructed, arranged, and fitted to produce the result substantially as herein described and set forth.

No. 59,935.—L. C. WRIGHT, Lockport, N. Y, assignor to WARREN CHRYSLER, same
place —*Clothes Dryer.*—November 20, 1866.—The rack consists of four frames, two of which
are supports and the others are suspended or projected according to their relative position to
the former, which are locked in either position by a key.
Claim.—The combination of the sockets a, and keys b, with the particular arrangement
of the rack, consisting of the bars A and B, of different lengths, pivoted at the top to the bearings D, the whole operating substantially as and for the purpose herein set forth.

No. 59,936.—EDWARD A. L. ROBERTS, New York, N. Y.—*Increasing Capacity of Oil
Wells.*—November 20, 1866; antedated May 20, 1866.—A charge of explosive matter is
ignited in the vicinity of the oil-bearing rock to expand the seams or create new ones, to
enable the oil to flow. The charge of the torpedo is ignited by friction or electricity.
Claim.—The above-described method of increasing the productiveness of oil wells by causing
an explosion of gunpowder, or its equivalent, substantially as above described.

No.59,937.—GEORGE H. ALBRIGHT and WILLIAM R. BURNS, Lancaster, Penn.—*Bridle —*
November 27, 1866.—Attached to each end of the driving line are branch reins, one answering
for a gag rein, passing over the horse's head to the ring of the single bit: the other branch
connects to the same ring of the bit by a spiral spring within a case. The devices correspond on each side.
Claim.—The construction and combination of the gag reins c c', face pieces A, passing,
over the horse's head; one branch thereof connected to the bit by the cased spring F, when
the united branches arise from the single bit, and unitedly connected with an ordinary line
and bridle, in the manner and for the purpose specified.

No. 59,938.—DAVID A. ALDEN, Roxbury, Mass.—*Paper Collars and Cuffs.*—November
27, 1866.—The collar or cuff is made directly from the pulp, being formed between a drum
with flaring elevations and a roller of corresponding shape, the lateral flare and the longitudinal curve being thus fixed in the production of the paper.
Claim.—First, the method herein described of manufacturing flaring paper collars and cuffs.
by forming paper directly from the pulp, so corrugated that it may be cut into collars or
cuffs, with the requisite flare, without subsequent moulding, swedging, or stretching, substantially as described.
Second, as a new article of manufacture, a flaring paper collar or cuff, in which the flare
is produced   the form given to the paper in its manufacture from the pulp, substantially
as described by
Third, as a new article of manufacture, paper so formed and corrugated in its manufacture
from the pulp, that it may be cut into flaring collars or cuffs, substantially as described.

No. 59,939.—CHARLES E. ALLAN, Boston, Mass.—*Faucet.*—November 27, 1866.—The
faucet connects with two pipes or reservoirs and has a single discharge. The spigot has a
bifurcated port, and by adjustment may draw liquid from either, neither, or both of the pipes, or a
combination of the two, in any required proportion respectively; as, for instance, of hot and
cold water for the bath.
Claim.—An improved combination cock formed by the combination of the plug B, with
the body A, of the cock, when the passage ways through said plug are formed and arranged
substantially in the manner herein shown and described and for the purposes set forth.

No. 59,940.—JOHN GAY NEWTON ALLEYNE, Alfreton, England, assignor to ZORETH
SHERMAN DURFEE, Pittsburg, Penn.—*Forging Apparatus.*—November 27, 1864.—The

hammers are adapted to converge upon the same mass of metal successively or simultaneously. The arrangement avoids the time expended in "up-edging," and facilitates the working down or drawing of masses by simultaneous impact on top and sides. The horizontally reciprocating hammers traverse on wheels or oscillate upon points of suspension and the upper hammer may be counterweighted or act against the force of recoil springs.

Claim —First, the combination and arrangement of two or more direct-acting steam hammers, each connected with a separate steam cylinder, so as to be susceptible of simultaneous or alternating, or other relative action, as may be desired, and so situate relatively to each other as that their strokes converge at a central point, at which the iron to be worked is placed, substantially as hereinbefore described.

Second, the combination of two or more horizontal or nearly horizontal direct-action steam hammers, with or without a vertical steam hammer, arranged substantially as hereinbefore described, and connected by links attached to a counter weight for the purpose of producing uniformity of action of the horizontal hammers, and of compensating for the force of gravity of the vertical hammer, when used in connection therewith.

Third, the combination of two or more horizontal or nearly horizontal direct-action steam hammers, with one vertical or nearly vertical steam hammer, so arranged as to strike a mass of iron placed at the converging point of their stroke, when such horizontal steam hammers are both connected by links, or equivalent device, with the vertical hammer, so as to act alternately therewith and produce regularity and uniformity of action.

Fourth, the use of rollers or wheels, in combination with the hammer blocks of steam hammers, for the purpose of sustaining and guiding the hammer blocks in their motion in a horizontal or nearly horizontal direction, substantially as hereinbefore described.

No. 59,941.—ALBERT G. BAGG, Holland Patent, N. Y.—*Cheese Vat.*—November 27, 1866 —The pipes which supply the vat with water are exposed to the heat of the furnace in a chamber within the fire walls. The vat has waste and gauge pipes and the induction pipes have valves. The dog supports the ends of the fire sticks and has openings for the passage of air.

Claim.—First, the combination of the pipes m f J and K with the vat P, said pipes and vat being constructed, arranged, combined, and operating substantially as herein described.

Second, the combination of the above with the furnace A, the heat chamber of said furnace being provided with tiles or bricks 1 and 2, for the purpose of supporting the bottom of the vat, and for distributing the heat evenly through all parts of said heat chamber, as herein described and set forth.

Third, the "fire dog" or grate bar e, provided with openings i, when used in connection with a furnace, combined with a cheese vat, as herein described and for the purpose set forth.

No. 59,942.—ALFRED BARNES, Williamsburg, N. Y.—*Manufacture of Sheet Metal Boxes.*— November 27, 1866.—The bottom and sides are composed of one piece of sheet metal, the blank being bent and lapped into the required shape; the lid is attached by its rolled flange and a wire pintle to the box, and has shut-in flanges at its ends and sides.

Claim.—The struck-up flanged sides of the body A, of the shape herein described for the purpose specified.

No. 59,943.—WALLACE BARNES, Bristol, Conn.—*Tempering Clock Springs*—November 27, 1866.—The spring. previously hardened. is clasped between two corrugated or perforated plates, and then immersed, while so clamped, in a bath of melted lead.

Claim.—The mode or process in the manufacture of springs for clocks, &c., substantially as described.

No. 59,944.—RICHARD T. BARTON, New Haven, Conn.—*Apparatus for Making Paper Boxes.*—November 27, 1866.—A recessed block on the table forms the intaglio die. The cameo die is surrounded by four knives which cut out the rectangular blank, and four corner cutters which remove square pieces from the corners; the central portion of the die pushes the paper into the recess and leaves it there as the cameo rises.

Claim.— First, the formation of the recess E.

Second, the arrangement of the knives A A A A.

Third, the punches C C C C, all in combination and constructed for the purposes herein specified.

No. 59,945.—LOUIS BAUHOEFER, Philadelphia, Penn.—*Cushion or Mattress.*—November 27, 1866; antedated November 3, 1866.—Air bags occupy a position between the upper and lower padded surfaces, and are charged through openings in their ends which extend to the tick.

Claim.—The mattress A, in combination with the air bag B, divided into two or more compartments when the mattress is reduced in thickness in the centre, as and for the purpose described.

No 59,946 —GEORGE H BAUGH, Oskaloosa, Iowa.—*Medicine for Hog Cholera*—November 27, 1866.—Composed of saltpetre, indigo, copperas, cayenne pepper, alspice, sulphur, spanish brown, turpentine, black antimony and oil of sassafras.

Claim.—The compound substantially as above specified, as a medicine for the cure of hog cholera.

No. 59,947.—F. BAUMANN, Chicago, Ill.—*Steam Generator.*—November 27, 1866.—A water jacket within the casing and around the furnace affords a supply to the central chamber whose radiating nozzles distribute jets to the generators, which are charged with loose metallic matter to increase the heating surface. The steam is collected in a central receiver and passes to a superheater in the dome.

Claim.—First, generating steam by the injection of heated water through pipes *c*, into a generator H, in which are placed loose metallic substances, as described.

Second, the combination of a number of heated metallic vessels H, with water-injecting pipes *c*, and outlet pipes *o*, connecting with a common receiver G, arranged to operate substantially as set forth.

Third, the water jacket I, arranged in combination with the generators H and receivers G, as shown and described.

Fourth, the combination and arrangement of a receiving and distributing vessel E, supplied with water from the water jacket I, with a series of generating vessels H, all located within the furnace, substantially as set forth.

Fifth, in combination with the heating and distributing vessel E, generator H, and receiving vessel G, the use of a receiving and equalizing chamber F, arranged to operate substantially as herein described.

Sixth, in combination with a steam generating furnace, constructed substantially as described, the dampers *l* and *l'*, arranged to operate as set forth.

No. 59,948.—ISRAEL BEETISON, New B tain. Conn.—*Expanding Mandrel.*—November 27, 1866 —The tapering mandrel is flattened on three sides, which are fitted with segmental wedges ; these are held in position by a ring and screws and adjusted longitudinally to adapt the mandrel to different sized holes.

Claim.—First, the tapering mandrel A, flattened on three or more sides, in combination with the segmental wedges *b*, fitted on the flattened sides of the mandrel, substantially as and for the purpose described.

Second, the rings *c*, provided with screws *d*, in combination with the segmental wedges *b*, and mandrel A, constructed and operating substantially as and for the purpose described.

No. 59,949.—CHARLES A. BLAKE, Philadelphia, Penn.—*Burglar Alarm.*—November 27, 1866.—The tail of the pivoted lever is placed across the path of the opening door, which vibrates the lever and draws the friction primer connected to the loaded barrel on the plate, which is secured to the door jamb.

Claim.—The block A, primer E, and forked lever H, or lever L, when combined and arranged together substantially as and for the purpose described.

No. 59,950.—JOHN BORTHWICK, Philadelphia, Penn.—*Machine for Sharpening Saws.*—November 27, 1866.—A solid emery ring is clamped between disks attached to a revolving arbor. The face of the wheel is fashioned for th : peculiar shape required. The saw is presented to the wheel lying upon the table, which is adjustable in a vertical plane coincident with that of the revolution of the emery annulus.

Claim.—First, the solid emery ring or rim of any suitable size, for the purpose of sharpening saws, combined with flange K and plate P, and arranged substantially as herein described.

Second, the combination of the hinged rest *g*, constructed as described with screws *y p*. and bed plate L, for the purpose as above described.

Third, the combination of the emery ring with the rest screws and bed plates, as herein described and set forth.

No. 59,951.—ALFRED BOYNTON, WrightTownship. Mich , assignor to EBEN M. BOYNTON. Grand Rapids, Mich.—*Saw.*—November 27, 1866.—The more prominent, angular faced tooth presents an acute edge to the fibre and cuts the kerf; the less prominent tooth with a horizontal cutting edge removes the wood from the kerf.

Claim.—A saw provided with the teeth No. 1, and the cleaner teeth No. 2, arranged and constructed to operate substantially as shown and described.

No. 59,952.—WILLIAM BRANAGAN, Burlington, Iowa.—*Steam Generator.*—November 27, 1866.—The steam boiler has elliptical outer and inner shells; transverse horizontal tubes traverse the fire chamber and connect the water spaces at the sides between the shells; removable side plates opposite the flues give access for cleaning.

Claim.—The construction of a steam boiler of an elliptical outer shell A, an elliptical inner shell B, horizontal transverse communications *g*, and a removable portion or portions G, on the side or sides of the outer shell, all arranged substantially as described and for the purpose set forth.

No. 59,953.—ABEL BREAR, Saugatuck, Conn.—*Trapping Wild Fowl.*—November 27. 1866.—Wires are stretched between posts across the path of the game, which are disabled by contact therewith and fall into the net beneath.

Claim.—First, the combination to form a game trap of supports and stretchers, constructed, arranged, and operating substantially as herein specified.

Second, the combination with such trap of a net D, one or more, substantially as and for the purpose specified.

No. 59,954.—BENJAMIN F. BROWN, Portland, Maine.—*Combined Coal Hod and Sifter.*—November 27, 1866.—A sifter is placed upon the top edges of the parallel sides of the scuttle and is reciprocated thereon.

Claim.—The combination with a hod, of the form described, of the sifter, constructed and set forth as and for the purpose specified.

No. 59,955.—EDMUND BROWN, Chicago, Ill.—*Separating Gold and Silver from Ores.*—November 27, 1866 ; antedated October 20, 1866 —Pulverized ores are forced through melted lead at high temperature to separate the precious metals. The ore in dry powder is placed in the upper vessels and by pressure of air forced in above is driven down the tubes into the melted lead in the tank, which is placed over a furnace.

Claim.—First, the application of compressed air for forcing the quartz down into the lead.

Second, the tubes D D, running down the outside of the kettle, allowing the quartz to become thoroughly heated and desulphurized, the whole combined and arranged for the purpose specified.

No. 59,956.—J. R. BROWN, Winchester, Ind.—*Centrifugal Machine for Draining Sugar.*—November 27, 1866.—The sirup flows from the funnel to the revolving distributor, which feeds it in a film to the foraminous basin with divergent sides. The liquid passes through the openings, while the crystals are detained and are ultimately delivered over the edges into the annular tub.

Claim.—First, the revolving separator H, when constructed in the form of the frustum of a cone inverted, and having its outwardly inclined sides formed of wire gauze or woven wire, and a foraminated plate, substantially as and for the purpose set forth.

Second, the combination of the distributor F E, shaft D, and separator H, when respectively constructed and arranged substantially as and for the purpose set forth.

Third, the combination and arrangement of the tub B, partition C, outwardly flaring separator H, and cover S, arranged to act as a deflector, substantially as set forth.

No. 59,957.—JOHN T. BRUEN and G. M. JACOBS, New York, N. Y., assignors to G. M. JACOBS.—*Machine for Cutting and Re-enforcing Button Holes.*—November 27, 1866.—Paper collars are automatically fed between two die plates. Above the upper plate a fabric to re-enforce the button hole is passed, the under side being sized. A descending die cuts the button hole from each ; a second die in the rear cuts off the re-enforce piece and carries it down upon the collar and presses it to its place, imparting any desired design.

Claim.—First, the suspended die plate D, in combination with the punches *v w*, and main die plate B, constructed and operating substantially as and for the purpose described.

Second, the feeding toes *a* in combination with the die plates D B, and punches *v w*, constructed and operating substantially as and for the purpose set forth.

Third, the hinged dogs i, arm *f*, and pitman rods *g*, in combination with the feeding toes *a*, and die plates B D, constructed and operating substantially as and for the purpose set forth.

Fourth, the pusher bars *b'*, in combination with the die plates B D, constructed and operating substantially as and for the purpose described.

Fifth, the hinged dogs *e'*, and stops *g'*, in combination with the pusher bars *b'*, cross head *b*, and die plates B D, constructed and operating substantially as and for the purpose set forth.

Sixth, the roller *c'*, and apron *d'*, in combination with the die plates B D, constructed and operating substantially as and for the purpose described.

No. 59,958.—CHARLES W. BURDIC, Norwich, Conn.—*Turning and Planing Tool.*—November 27, 1866.—The angular tool is clamped in the oblique slot of the holder, by means of a wedge-shaped block and a set screw.

Claim.—The combination of the holder A, shoe C, cutter B, and set screw D, when the said holder and shoe are constructed and arranged substantially as herein described and for the purposes set forth.

No. 59,959.—J. F. BURGESS, Brooklyn, N. Y.—*Ornamenting Piano-forte Covers, &c.*—November 27, 1866.—An adhesive solution of an elastic gum is laid in a design upon the India-rubber cloth and receives a coating of colored flock, gold, or bronze.

Claim.—First, applying to India-rubber cloth, or such cloth, having a flocked surface, ornamental figures or designs, in the manner and for the purpose substantially as described.

Second, a piano-forte or table cover composed of an India-rubber cloth or such cloth, with flocked surface, having ornamental figures or designs placed thereon, either in flock or bronze or both, substantially as described.

No. 59,960.—CHARLES BURLEIGH, Fitchburg, Mass.—*Rock Drilling Machine.*—November 27, 1866.—An annular projection on the rear end of the horizontally moving piston bar operates the valve of the engine and the rotating device for the drill. The axis of the rotating valve is on a line with the axis of the cylinder. The drill and engine are clamped to the frame upon which they slide.

Claim.—The annular projection M, upon the piston bar E, or an attachment thereto, for the purpose of operating the valve and feeding device or either of them, substantially as set forth.

Second, the stationary rotating device when operated by the movement of the piston bar, substantially as shown and described.

Third, the hollow valve K, with its chest L, arranged to operate in line with the axis of the cylinder, substantially as described.

Fourth, the clamp C, consisting of the box *c*, and pieces A', and B', constructed and operating substantially as described, in combination with the slide or drill frame B.

No. 59,961.—JOSEPH N. BYINGTON, Stockton, Minn.—*Spring Wagon.*—November 27, 1866.—The wagon bottom is supported upon springs beneath connected to the bolsters. To place them out of use and allow the wagon bottom to rest upon the bolster the springs are disconnected below and hooked up beneath the wagon bed.

Claim.—First, hinging the springs F to the bottom D of the wagon, and connecting them to the side of the bolsters A and B, in such a manner that when not in use they may be turned up out of the way, substantially as herein described and for the purpose set forth.

Second, the combination and arrangement of the springs F, cross bars G, links I, bolts J, and catches H, with each other, and with the bottom D, and bolsters A and B, of the wagon, substantially as herein described and for the purpose set forth.

No. 59,962.—ALEXANDER CARMICHEL and WILLIAM BARNEY, Westerly, R. I., assignors to themselves and COTTRELL & BABCOCK, same place.—*Stop Motion for Looms.*—November 27, 1866.—A positive stop arrests the motion of the loom at the same time the belt is shipped, thus avoiding another throw of the shuttle, and preventing the lay from advancing so far as to strike the shuttle, if it be in the shed.

Claim.—First, the combination with a loom of a belt-shifting motion and positive stop arranged for joint operation, substantially as and for the purpose herein specified.

Second, arranging a positive stop in such combination that it shall act against the lay of the loom, substantially as and for the purpose herein specified.

Third, the within-described arrangement of such positive stop in combination as specified, in sight and within convenient reach of the operator, and providing it with a handle K', or its equivalent, so that it may be more easily placed by hand in such a position as to allow the loom to be operated freely by hand, all substantially as and for the purpose specified.

No. 59,963.—G. F. CASE, New York, N. Y.—*Feed Motion for Drills.*—November 27, 1866.—The feed screw has a sleeve connected to it by a spline key so as to revolve with it; the sleeve engages with a worm upon a revolving shaft actuated by cams upon the frame.

Claim.—The sleeve B, which carries the worm *f*, and cam wheel *g*, in combination with the feed screw A, nut D, and worm wheel *h*, all constructed and operating substantially as and for the purpose described.

No. 59,964.—PAUL CEREDO, Montreal, Canada East, assignor to himself and JAMES H. SPRINGLE, same place.—*Ash Sifter.*—November 27, 1866.—The ashes are poured into the hopper and pass to the hinged sieve, which is raised by the tripping cam and depressed by the spring giving it a jerking vertical reciprocation. The cinders descend the inclined screen and emerge at the spout.

Claim.—The arrangement of the screen box B, cam E, spring G, spouts I J, constructed and operating in the manner and for the purpose herein specified.

No. 59,965.—GEORGE T. CHAPMAN, New York, N. Y.—*Rein for Horses.*—November 27, 1866.—The gag rein passes through a loop above the bit ring and is connected to the check piece by the brow band. It is also connected by cross strap to the driving rein, which has a spring section next the bit. The driving and gag reins exert their usual functions until the horse is restive, when severe draught on the driving rein stretches the elastic section and transfers the power to the gag, which draws the bit into the angle of the mouth.

Claim.—The combination of the elastic rein above described with the auxiliary reins and with the curb gag and check reins, as constructed and arranged, for the purpose described and set forth

No. 59,966.—HENRY CLIMER and JOHN D. RILEY, Cincinnati, Ohio.—*Planing Machine.*—November 27, 1866.—The table top has two sections, respectively before and behind the rotary cutter. Each is adjustable longitudinally on its intermediate frame and vertically on its main frame. One frame is adjusted to the rough board, and the other to correspond to the face of the planed board. The tables correspond in height to the faces of the rough and planed portions, respectively.

Claim.—The two beds C C', to receive the material to be planed, operating substantially in the manner and for the purpose set forth.

No. 59,967.—FRANCIS CLYMER, Galion, Ohio.—*Tree Protector.*—November 27, 1866.—The stake has a flat cap, to whose edges are hinged radial depending arms to support a protecting cover for use in severe weather.

Claim.—The cap A, stake E, slats C, and braces D, in combination with the thatch K as, and for the purpose set forth.

No. 59,968.—JOHN H. COATE, West Milton, Ohio.—*Portable Fire Grate and Andiron.*—November 27, 1866.—The three supporting bars are connected by removable grate bars and have side frames and back bars, to form a grate for the burning of small fuel.

Claim.—First, the combination of the back plate L, the side bars A and B, foot E, cross-bars F H G C, and part d⁴ of the bar D, with each other, substantiallly as described and for the purpose set forth.

Second, the combination of the side-grates J and K, cross bars I, and bar D, with the back plate L, side bars A and B, foot E, and cross-bars F H G C, substantially as described and for the purpose set forth.

No. 59,969.—D. N. B. COFFIN, JR., Boston, Mass., assignor to himself and J. D. SPAULDING, same place.—*Capstan and Wind ass.*—November 27, 1866.—The capstan head may be locked to the body by drop pawls to operate it directly, or may be turned by the fulcrum gearing at its foot operating through geared connection to a cog wheel upon the head ; the gearing in this case is connected to the bed plate and raised by pawls. Inclined surfaces operate against projections on the pawls. The cap is attached by lugs entering vertical and horizontal slots in the head. The partial rotation of the cap in one direction serves to attach the head to the body of the capstan, whilst a contrary movement attaches it to the power gearing. The lower end of the spindle admits of attachment by bevel gearing to a windlass which is furnished with a brake, and may be disconnected from the bevel gear.

Claim.—First, the employment of the shaft b, extended from the capstan, in combination with the shafts d or e, with suitable gears, as u f s and t, substantially as described.

Second, locking the fulcrum gear of a capstan to the bed plate by means of bolts movable upward from beneath into contact therewith, substantially as described.

Third, the employment of a series of inclines movable in a circle, and so applied in combination with the fulcrum gear of a capstan as to lift said gear from its position of inaction, to the proper position to be acted on by gears playing into it, substantially as described.

Fourth, fastening the cover of a capstan by means of lug⁷ v, in combination with sockets, or grooves in the hub, substantially as described.

Fifth, the arrangement of the cover fastenings in relation to the locking mechanism of the capstan, so that when the cover is rotated to a stop in one direction the bolts will be adjusted for the simple power and the cover fastened, and when rotated in the opposite direction to a stop the cover will also be fastened and the bolts adjusted for the multiplied power, while in an intermediate position the cover is unfastened and may be removed, substantially as described.

Sixth, suspending the fulcrum gear to the rotating body of the capstan.

Seventh, the arrangement of the gears A Z, pawl wheel F, barrel I, friction L, and chain wheel P, and the points of disconnecting, whereby both barrel I and chain wheel P are brought under control of the friction mechanism and yet used separately for winding, heaving in, &c., substantially as described.

Eighth, the windlass shaft N, in combination with the friction band and barrel of a windlass, substantially as described.

Ninth, the partly-circular heads and sockets in combination with the pawls of a windlass substantially as described.

Tenth, a groove formed under the projecting part of a capstan's base, substantially as and for the purpose set forth.

No. 59,970.—GEORGE COFFIN, Boston, Mass.—*Attaching Shafts to Sleighs and Carriages.*—November 27, 1866.—The portion of the thills to which the harness is attached is socketed in a frame which is coupled to the carriage. By withdrawing keys, the forward portion is detached and the horse freed from the vehicle.

Claim.—The combination of the wedge-shaped keys C, and rods D, with the parts A and B of the shafts, when said parts are constructed with tenons and sockets, substantially in the manner herein described and for the purpose set forth.

No. 59,971.—NATHANIEL E. CORNWALL, New York, N. Y.—*Heat Radiating Attachment for Stove Pipes.*—November 27, 1866.—The caloric current enters the lower heat-chamber at its centre and passes through a concavo-convex central pipe and four outer pipes to the upper heat chamber, and from thence through a central circular exit pipe. This arrangement diffuses the current as its direct course is broken on leaving each chamber. A plate connects the two edges of the lunate pipe.

Claim.—First, the convexo-concave or crescent-shaped pipe *c*, arranged and combined with the chambers *b* and *e*, and the pipes *d d d d*, in the manner and for the purposes set forth.

Second, in connection with the pipe *c*, the close chamber *i*, constructed substantially as described and for the purposes shown.

Third, the chambers *b* and *e*, connected by pipes *d d d d*, and *c*, together with the close chamber, *j*, constructed and arranged to operate in connection with a stove-pipe, in the manner described.

No. 59,972.—GEORGE CROMPTON, Worcester, Mass.—*Harness Motion for Looms.*—November 27, 1866.—The sheaves of the cords by which the harness frames are suspended are so mounted as to be simultaneously raised and lowered. The construction by which a uniform opening of the shed through all the harness is produced, involves a bent or angular form of heddle levers, each of which has one end notched for adjustment of the heddle cords, and also the horizontal position of the acting faces of the lifter and depresser bars and the employment of upper and lower bent and notched heddle levers, and upper and lower heddle-cord guide sheaves.

Claim.—The provision for simultaneous and equal adjustment of all the sheaves of either set, by mounting them in a swinging frame, which is controlled and moved by an adjusting screw or other equivalent device, substantially as set forth.

Also, the construction and arrangement, as shown and described, or the parts by which the heddles are operated to produce the shed.

No. 59,973.—JOSEPH CROWFOOT, Worcester, Mass., assignor to himself and JONATHAN GILL, Fitchburg, THOMAS GILL, Springfield, and WILLIAM GILL, Cambridge, Mass.—*Spinning Machine.*—November 27, 1866.—This machine employs the "Victory" drawing head. The several sets of cone-pulleys admit (for the purpose of operating on wool of different lengths of fibre) of an adjustment of the velocity of the delivery rollers, the twist rollers and the draw rollers.

Claim.—The machine, substantially as described, by which the silver not only can be drawn and twisted but have the twist varied, by means as set forth, or their equivalent.

No. 59,974.—ALPHEUS CUTLER, Pittston, Penn.—*Table Leaf Support.*—November 27, 1866.—The tubular, fixed portion of the flat support has an extensible inner portion retained in extension by a ratchet and pawl.

Claim.—An adjustable extension leg. or table-leaf supporter, constructed as described and operating as set forth.

No. 59,975.—DANIEL W. DAKE, Brooklyn, N. Y.—*Preparing Coloring Matter for Butter.*—November 27, 1866.—The coloring matter of annotto is extracted by heating it, with the oil of butter, to a temperature of 212° Fahrenheit.

Claim.—First, the method or process herein described of preparing a coloring matter for butter, consisting of pure oil of butter, and the coloring matter of annotto, combined as herein described.

Second, as a new composition of matter for coloring butter, pure oil of butter and the coloring matter of annotto, prepared and combined substantially as described.

No. 59,976.—CHARLES F. DAVIS, New Market, N. H., assignor to J. L. NORRIS and G. R. NEAL, South New Market, N. H.—*Oil Cup.*—November 27, 1866.—An insulated chamber in the oil vessel has a pivoted lever and spring stem, by which the valve at the inner end of the nozzle is closed, or opened by pressure on the exterior stud.

Claim.—The combination and arrangement of the said insulator with the oil feeder A, the valve C, and the valve-operating apparatus, substantially as described.

No. 59,977.—JUSTUS DAY, Murray, N. Y.—*Broom Clamp.*—November 27, 1866.—The two levers forming the clamp are fastened by a link at their ends to admit of being reversed, and are furnished with semicircular and semi-oblong openings to hold the corn brush. Slots in the sides of the levers admit wires that are bound around the corn while compressed by the levers.

Claim.—The levers A and B, in combination with the connecting links or straps C C, the slots D D, the chamfered semicircular notches E E and F F, the wide notches K K. the clasp H, and the rods G G, operating for the purpose and in the manner specified.

No. 59,978.—A. DE FIGANIÈRE, Philadelphia, Penn.—*Manufacture of Superphosphate of Lime.*—November 27, 1866.—The fertilizer is fed from a hopper, and falls on the surface of a cylinder whose lower edge revolves beneath the surface of the sulphuric acid in the tank below. The material absorbs the acid, and is scraped off by the upper edge of the discharging spout.

Claim —First, making superphosphate of lime by bringing the powdered guano or other suitable fertilizing material into contact with a surface covered or dampened with sulphuric acid, substantially as herein described.

Second, an apparatus for incorporating sulphuric acid with powdered guano, or other suitable fertilizer, formed by the combination of the acid tank A, the cylinder C, the hopper E, and the scraper F, with each other, substantially as herein shown and described.

No. 59,979.—GEORGE M DENISON, New London, Conn.—*Surface for Washing Machines.*—November 27, 1866 —The rubber block is made in lateral sections, between which the rubber is clamped.

Claim.—The construction of a ribbed or corrugated washing surface of sheet India-rubber folded over the edges of and clamped between strips *f,* substantially in the manner herein set forth.

No· 59,980.—S. F. DIMOCK, Spencer, Ohio.—*Animal Trap.*—November 27, 1866.—The bait is hung beneath the canopy, and beyond the drop door. The weight of the animal on the door depresses it, brings down the hinged flap, preventing retreat, and exposes the way into the trap. When the door is relieved from the weight, the trap resets.

Claim.—The box A, door I, link J, and tin door K, as arranged in combination with the pedal E, in the manner and for the purpose as substantially set forth.

No. 59,981.—THOMAS DOLAN, Albany, N. Y.—*Attaching Locks to Safes, Vaults, &c.*—November 27, 1866.—The lock is contained in the frame, and operates a vertically sliding bar to engage dogs upon the door.

Claim.—A locking mechanism for safes, vaults, &c., consisting of a lock A and notched sliding bar C, applied to the door frame and dogs or catches D, applied to the door, all constructed, arranged, and operating substantially as and for the purpose herein set forth.

No. 59,982.—EDWARD DUFFEE, Haverhill, Mass., assignor to himself and GEORGE APPLETON, same place.—*Pipe Coupling.*—November 27, 1866.—The entering end has two separated annular flanges of a diameter nearly equal to the internal diameter of the enclosing socket. A hole through the socket admits the introduction of cement into the space between the two flanges of the pipe and the enclosing socket.

Claim.—The new or improved pipe joint, consisting of the opening *d,* and the auxiliary flange *a,* the main flange *f,* and the socket head *r,* arranged and applied to the pipes A and B, as set forth.

No. 59,983.—MARY A. DUFFY, New York, N. Y.—*Sewing Machine Gauge for Tucking, Felling, Binding, &c.*—November 27, 1866.—The parts are adaptable by different arrangements for tucking, binding, felling, hemming, and overhand seaming. A second tuck may be creased while the previous tuck is being stitched, and work prepared for a fell without basting. The different adjustments cannot be briefly described.

Claim.—First, the several devices of plate holder B, tucking plate A, folding plate F, and tucking gauge D, or their equivalents, in combination with the presser foot of a sewing machine for the purpose of folding and marking or creasing a tuck, substantially as explained.

Second, the combination of plate holder B, tucking plate A, felling plate F, and felling guide G, or their equivalents, constructed and operating together substantially as and for the various purposes described.

Third, the combination of plate holder B with felling plate F, constructed substantially as and for the purposes described.

No. 59,984.—WARREN W. DUTCHER, Milford, Mass.—*Tool for Setting Temple Teeth.*—November 27, 1866.—The tool has a knob at one end to receive the blow of the hammer; the other end has an inclined surface and tapering hole or socket coincident with the axis of the tool, which leads to and terminates in a hole made transversely through the tool near the end. This end of the tool is magnetic, and the tapering tooth placed point foremost in the socket is held there by magnetic attraction until adjusted to and driven into the hole made for it in the surface of the cylinder, the inclined end of the tool serving to adjust the teeth to a uniform pitch.

Claim.—The rotary temple tooth-setting tool made substantially in manner and so as to operate as and for the purpose hereinbefore specified.

No. 59,985.—REUBEN R. EASTMAN, Granby, Mass.—*Butter Worker.*—November 27, 1866.—The bottom of the trough has angular corrugations, and is traversed by a roller with corresponding teeth. The liquid is removed by a discharge spout.

Claim.—First, in a butter press, the combination of the roller C and plate G, with teeth to mesh, substantially as and for the purpose set forth.

Second, the trough A, in combination with the roller C and plate G, substantially as described.

No. 59.9⁰6.—JOSEPH A. ELIAS, Le Roy, Ohio.—*Composition for Detroying Insects on Trees and Plants.*—November 27, 1866.—The compositions are applied consecutively. The first is composed of strong decoction of tobacco 10 parts, spirituous solution gum aloes 1 part; the second, spirits turpentine 1 part, alcohol 30 parts. Apply with a brush to the trunk.

Claim.—The compositions Nos. 1 and 2, herein described, consisting of the several ingredients in or about the proportions stated, and applied in the manner as herein specified, and for the purposes set forth.

No. 59,987.—HENRY A. ELLIS, Mystic River, Conn., assignor to himself and PEQUOT MACHINE COMPANY, same place.—*Loom.*—November 27, 1866.—The object of these improvements is to allow of the detachment of any one of the heddle-operating levers from its cam without detaching it from its axial bearing; and also to admit of detaching the lever both from its cam and from its bearing without disturbing the other levers. The screw hooks admit of adjusting and levelling the harness without detaching the cords from the books.

Claim.—First, the construction of the lever C with a slot d, pin A, and screw hooks r r, substantially as and for the purposes set forth.

Second, the combination of the slots d, in levers C, with the slots i, in the outer rims of the cams E, substantially as described.

No. 59,988.—FREDERICK ERNST, New York, N. Y.—*Lozenge.*—November 27, 1866.—Composed of permanganate of potash, 1 part; peroxide of barium, 1; sugar, 194; and gum, 4.

Claim.—The deodorizing lozenges as a new article of manufacture, consisting of the ingredients described, for the purpose set forth.

No. 59,989.—JOHN R. EVERTSON. Mount Vernon, Ind.—*Grain Dryer.*—November 27, 1866.—The drying pan is suspended from a frame above the furnace so as to oscillate longitudinally. The grain enters at the hopper at the upper end, and is discharged at a spout after traversing the length of the pan.

Claim.—First, the oscillating slide or pan B, constructed as described in combination with the furnace A and with the frame G, substantially as and for the purpose set forth.

Second, the combination of the cover F and hopper D, constructed as herein described with the slide or pan B, substantially as and for the purpose set forth.

Third, the combination and arrangement of the discharging spout E with the slide or pan B, substantially as described and for the purpose set forth.

No. 59,990.—D. D. FOLEY, Washington, D. C.—*Post Driver.*—November 27, 1866.—The driver is of considerable weight, is raised by the handles and allowed to slide down the post and strike upon the post head.

Claim.—The fence post driver composed of a cap A and sides B, substantially as and for the purpose set forth.

No. 59,991.—THEODORE F. FRANK, Buffalo, N. Y.—*Apparatus for Carburetting Air.*—November 27, 1866.—The two air blowers work alternately and connect by pipes with the carburetting chambers, which contain perforated pipes connected to the former by flexible tubes. The air discharging pipes may be adjusted by rods to any desired height in the carburetting chambers. An elastic bag acts as a regulator, and has a discharge pipe leading to the burners.

Claim.—The combination of the two air-supplying reservoirs, operating alternately and provided with valves b and f, and pipes e, with the series of carburetting vessels D D¹ D², or equivalent, substantially as and for the purpose set forth.

Also, arranging the series of carburetting vessels D D¹ D², on different planes and connecting them by suitable pipes and stop cocks y z, substantially in the manner and for the purpose specified.

Also, in combination with the carburetting vessel D, the perforated air-discharge pipe a, provided with stem r, stuffing w, and flexible pipe t, for adjusting the former, substantially as and for the purpose set forth.

Also, the combination of the air-supplying apparatus and the carburetting apparatus, both constructed as described, with the regulator E, the whole arranged and operating as described.

No. 59,992.—FRANCIS N. FROST, New Britain, Conn.—*Currycomb.*—November 27, 1866; antedated November 15, 1866.—Explained by the claim and cut.

Claim.—In currycombs made of thin plate steel, as described, the shank c, of malleable cast-iron and made in the form shown, so as to bear with equal force laterally across the entire breadth of the comb, laterally as described.

No. 59,993.—WILLIAM GARRARD, Fallston, Penn.—*Cooler for Preserving Butter, Milk and other Articles.*—November 27, 1866; antedated November 23, 1866.—The inner vessel is made of porous unglazed earthenware and contains a glazed vessel. The intervening space is filled with water. The cover has double walls with a water space. An air-tight joint is formed between the cover and the vessel, and the water evaporated from the outer porous surface refrigerates the interior.

Claim.—A cooler made of any kind of unglazed pottery ware of any size or shape, and so constructed that the articles to be kept cool are in a vessel entirely surrounded by water, and therefore air tight and made to combine all the advantages of a more perfect evaporation for the purposes intended, as is herein substantially described and set forth.

No. 59,994.—P. A. GERRY, Dover, Me.—*Sawing Machine.*—November 27, 1866 —The wood is clamped in the buck rollers by a depending arm from a lever connecting by a rock bar to a hanging lever held to place by a pivoted drop arm. The release of the wood automatically raises the saw.

Claim.—First, clamping the wood by means of the lever R, operating with the arm O, roller N, arms T, and swinging arm U, arranged and operating as herein represented and described.

Second, raising and lowering the saw by means of the lever R, arm O, roller N, arm V, spring W, and slide bar X with its guide arm Y, substantially in the manner and for the purpose represented and described.

No. 59,995.—SAMUEL GREGG, Boston, Mass.—*Spectacles.*—November 27, 1866.—The segments are of different foci and their line of junction is curved.

Claim.—Constructing glasses of spectacles where two distinct lenses or segments of lenses are contained in one glass, adapted for seeing near and distant objects, in such manner that the upper edge of the convex lens adapted for seeing near objects shall be concentric with the upper edge of the lens adapted for seeing distant objects, for the purpose of enlarging the field of vision for the latter lens.

No. 59,996.—MICHAEL HABERBUSH and EDWARD KRECKEL, Lancaster, Penn.—*Bridle.*—November 27, 1866.—The combined throat-latch and face-piece prevents slipping the bridle as the draught steam on the hitch strap draws the bit into the angles of the mouth. In driving, a pull on the lines stretches the gum which attaches the driving reins to the rings of the bit and draws upon the face straps and throat latch to pull the bit into the angle of the mouth.

Claim.—First, the construction and combination of a tubulated snap hook C, gum, within a sheath D, formed by the end of a continuous strap, both united and held by a socket head and ring E, jointly passed through the rings of the bit so that the enclosed gum with its snap will book into the said ring of the bit in the manner and for the purpose specified.

Also, the face pieces G G', attached to the rings of the bit and severally passed upward through their respective loops H I, and forming the throat latch K, with its hitching ring L, all combined and arranged in the manner specified for the purpose set forth.

No. 59,997.—THOMAS HALL, Bergen, N. J., assignor to GEORGE B. BUELL, New York, N. Y.—*Sawing Machine.*—November 27, 1866; antedated November 22, 1866.—The groove in the table is designed to receive that portion of the buckle to which the cloth is to be stitched to support and guide it as the cloth is fed in the usual manner.

Claim.—Cutting away a portion of the cloth table of a sewing machine or depressing a furrow therein to admit of sewing buckles or other articles upon garments or fabrics, substantially as described.

No. 59,998.—JAMES W. HANNA, Wabash, Ind., assignor to himself and JAMES H. OSGOOD, Jr.—*Apparatus for Drying Lumber.*—November 27, 1866.—The lumber is first saturated with steam, which enters beneath a plate which deflects it, and is then subjected to dry heat from a steam-heated coil, the heated air being circulated by a blast of air.

Claim.—First, the combined process of steaming and drying lumber, substantially as herein described and for the purpose set forth.

Second, the combination and arrangement with the kiln of the pipe or pipes C. for admitting the exhaust steam, the pipe or pipes E, for admitting the blast of air, and the pipe F, for introducing the live steam, substantially as herein described and for the purposes set forth.

No. 59,999.—E. B. HARDING, Northampton, Mass.—*Pessary.*—November 27, 1866.—The pessary is not supported by pressure upon or distension of the walls of the vagina, but its anterior end rests in a sac in the rear of the os pubis. The spreading ends of the pessary are drawn together by a pair of pivoted hooks or a loop, to enter or withdraw the same.

Claim.—First, a pessary constructed with the loops g h, arranged substantially as and for the purpose set forth.

Second, in combination with a pessary so constructed the levers A B, arranged and operating substantially as set forth.

Third, the instrument shown in Fig. 6, when constructed and operated in the manner and for the purpose substantially as set forth.

No. 60,000.—PHILANDER HARLOW, Hudson, Mass.—*Eyeleting Machine.*—November 27, 1866.—By the revolution of a horizontal shaft beneath the table the fabric to be eyeleted is passed by successive steps beneath the punch and beneath the eyeleting die, and the eyelets are successively placed in position for insertion and striking up, all by an automatic arrangement which cannot be briefly described.

Claim.—First, the combination with the setting die of an eyeleting machine of an eyelet-receiving rod and surrounding sheath or die plate, and mechanism for actuating the same under the arrangement and for operation as herein described, so that the said receiver rod shall constitute the means whereby the eyelet is adjusted to the punched leather and the leather fed forward to the setting die, substantially as set forth.

Second, in an eyeleting machine, the combination with the eyelet-receiver rod, and surrounding sheath or die plate, as described, of the setting die and punch forming the eyelet holes under such an arrangement that by the action of said punch, receiver rod, sheath and setting die, the leather or other material operated on shall be alternately punched, fed forward to the setting die and stamped with eyelets, substantially as herein shown and described.

Third, in an eyeleting machine as described, the combination with the punch for forming eyelet holes, of a reciprocating or sliding plate for sustaining the leather under the action of said punch, arranged and operating substantially in the manner and for the purposes set forth.

Fourth, the adjustable presser foot and its actuating mechanism as herein described for producing an intermittent pressure upon the leather or other material operated on by the punch and eyelet-setting die, substantially as and for the purpose set forth.

Fifth, the mechanism for feeding the eyelets as herein described, the same consisting of a hopper and vibratory arm provided with one or more chutes for conducting and holding the eyelets, and a wheel or disk in the periphery of which recesses are formed for the reception of the said eyelets, the whole being combined and operating substantially in the manner shown and specified.

Sixth, the combination of the above-described eyelet-feeding mechanism with the eyelet-receiver rod and surrounding sheath, under such an arrangement that by the motion of said vibrating arm the eyelets held in the recessed wheel may be fed to the said receiver rod, substantially as herein shown and described.

Seventh, the combination in an eyeleting machine of mechanism for punching, feeding and holding the material operated on, and for feeding and setting the eyelets to the same, under the arrangement and for operation substantially as herein shown and set forth.

No. 60,001.—J. P. HASKINS, Saratoga Springs, N. Y.—*Exhibiting the Gas of Mineral Springs.*—November 27, 1866.—The spring of mineral water is tubed and the water conducted through a glass vessel which exhibits the ascending bubbles of gas as they rise to the surface.

Claim.—The method herein described of exhibiting the gas contained in mineral waters while they are ascending through the same and before reaching the surface.

No. 60,002.—WILLIAM HENDERSON, Glasgow, Scotland.—*Manufacture of Wrought Iron and Steel direct from the Ore.*—November 27, 1866.—The bottom of the furnace is formed of pulverized well burnt coke with coal tar. The pulverized ore, such as pure hematite, is mixed with fine charcoal or mineral coal, lime, and salt. If the ore contains sulphur, manganese is added. The furnace is first heated by gas such as is produced by Siemen's process. The ore is then introduced and heated nearly white hot, but not to fusion. A strong reducing flame is introduced for two hours, and then the metal is gathered into balls with as little access of air as possible and hammered and rolled in the usual manner. To produce steel the heat is continued for a longer time before closing the furnace and applying the reducing flame.

Claim.—First, the several improved processes hereinbefore described for manufacturing wrought iron and steel direct from the ores of iron.

Second, the formation of the bottom of the furnace to be used in the working of the above-described processes with the materials and in the manner hereinbefore described.

No. 60,003.—CHARLES F. HENIS, Cincinnati, Ohio.—*Coal Scuttle.*—November 27, 1866.—The hod is made of sheet metal in two parts with a struck-up corrugated bottom.

Claim.—The coal hod constructed with its body D, fitted upon the shoulder C of the corrugated base B, as and for the purpose set forth.

No. 60,004.—JOSEPH H. HICKS, Brooklyn, N. Y., assignor by mesne assignment to IGNATIUS RICE, New York, N. Y.—*Comb.*—November 27, 1866.—A straight comb is strengthened by a strip of metal inserted longitudinally in a groove in the centre of the back and held by transverse pins.

Claim.—A comb having the strip of metal e, and rivets i, in combination with each other, and with the longitudinal groove in the back of the comb, substantially as herein set forth, for the purpose specified.

No. 60,005.—W. M. HICKS and F. WELKER, St. Louis, Mo.—*Cut-off for Cistern Leaders.*—November 27, 1866.—A chute is so hinged in the rain-water pipe that in one position it will convey the passing water out of the pipe, and in the other position will permit the water to flow downward unobstructed.

Claim.—The arrangements herein described for a cut-off for cistern leaders, consisting of a chamber A and a revolving and adjustable door B set in the side thereof, said door being hinged in the middle, and when turned down itself serving as a spout, for the purpose and in the manner as described.

No. 60,006.—ALEXANDER HILL, Dubuque, Iowa.—*Tanning.*—November 27, 1866.—The hides are placed in a bath composed of soot, 1 pound; sulphuric acid, 1 pound; water, 20 gallons, for 12 hours, and then in a solution of Bengal catechu, 10 pounds; sal ammoniac, 1 pound, with sufficient water; handle daily for 10 days.

Claim.—The within mode of treating hides, substantially as specified.

No. 60,007.—ASA HILL, Norwalk, Conn.—*Ornamenting Marble.*—November 27, 1866; antedated November 14, 1866.—The design is drawn in outline upon the marble, and colored with solutions of the salts of copper, iron, potassa, silver, tannin, &c. The stone is then placed over a vapor bath, which causes the solutions to be absorbed and the coloring fixed.

Claim.—The process of producing figures upon and in marble or other calcareous stones, substantially as herein described.

No. 60,008.—THOMAS M. HILL and S. D. TUTTLE, Eaton, Ohio.—*Machine for Cutting and Raking Corn Stalks.*—November 27, 1866.—The divaricated plough shares are carried on a vertical cutter, and may be raised by a lever. The standards are extensible, and their brace chains are connected to the sliding thill frame. The rake frame is depressed by a spring and raised by a lever.

Claim.—First, the ploughs and cutters M M applied to standards L, which are secured on the axle A of the machine and connected to the sliding thills E by chains and rods *l*, and connected by chains *m* to the frame G, all arranged to operate substantially in the manner as and for the purpose herein set forth.

Second, the rake J attached to the levers H H, and connected to the frame G by springs *f f*, in combination with the ploughs and cutters M M, all arranged substantially as and for the purpose specified.

No. 60,009.—AMELIA B. HOFFMAN, Roxbury, Mass.—*Crumb Remover.*—November 27, 1866.—The scraper is passed over the cloth and gathers the crumbs, which are deposited in the receiver by the rotation of the scraper on its axis.

Claim.—The combination of a pan or receiver A with a blade or scraper B, in the manner and for the purpose described.

No. 60,010.—E. R. HOPKINS, New York, N. Y.—*Machine for Making Hoop Skirts.*—November 27, 1866.—The former revolves horizontally above the reels, which are similarly shaped, and between the two arranged markers and cutters, which are operated automatically.

Claim.—First, a machine in and by which a series of springs, such as are used in the manufacture of hoop skirts, by properly feeding them to the machine, can be respectively, but simultaneously, cut to their proper lengths for encircling the frame or other suitable form used for the building or making the skirt upon, substantially as herein described.

Second, whether combined with the above or not, a marker so arranged and operated as to mark the series of springs, whether more or less in number, used for a hoop skirt, at such points of their respective lengths as correspond to the points of intersection therewith of the various tapes to be used for connecting and binding the series of springs together, substantially as herein described.

Third, the conical or other equivalent shaped feed rollers or drums B B, so hung and connected together as to revolve with their surfaces in contact, or nearly so, with each other and in conjunction, in combination with any suitable knife or cutter blade, when arranged and combined together, substantially in the manner described and for the purpose specified.

Fourth, in combination with the above, either with or without the knife or cutter blade, of a marker of any suitable form, arranged so as to operate substantially in the manner and for the purpose described.

Fifth, the combination with the rollers or drums B of the guides H and I for the springs passing between the said rollers, arranged together substantially as and for the purpose specified.

Sixth, the raising and lowering the marker R used by means of a cam A^2, so constructed or formed as to operate upon the marker, substantially as described for the purpose set forth.

Seventh, so arranging and connecting the marker used with the driving power or shaft of the machine at the proper times that it will be moved against the springs for the purpose of marking them and then draw back from the same, substantially as and for the purpose described.

Eighth, the combination with the feed rollers or drums B B of any suitable form or frame, such as are now used in the manufacture of hoop skirts, or any equivalent for the same, whether a knife or marker, or both together, are used, when said frame is arranged with regard to the rollers, substantially as described and for the purpose specified.

Ninth, a machine in which a series of springs, such as are used for the manufacture of hoop skirts, can be cut to their respective and proper length according to the portion of the skirt they are to encircle, and marked at such points of their length as are to be intersected by the tapes used for binding and connecting the springs together, and also delivered to and wound about and around any suitable form or frame on which the skirt is to be built or made, when the whole is combined together and arranged so as to operate substantially in the manner described.

Tenth, providing a frame or form suitable for the manufacture of hoop skirts upon it, with a series of spring arms V², or their equivalents, having studs projecting from them, when such arms with the studs are so arranged as to prevent the hoop skirt when wound upon the form from slipping thereon, while at the same time they can be moved or swung away so as to offer no obstruction to the withdrawal of the skirt from the said form' after completion or whenever so desired, substantially as described.

No. 60,011.—H. HUFENDICK and E. SPANGENBERG, St. Louis, Mo.—*Bung.*—November 27, 1866.—The outer frame of the bung is screwed into the bunghole, and the inner part contains valves, which allow ingress to air but prevent egress of liquid.

Claim.—First, the application of a valve or valves and capillary passages or any equivalent devices to permit the passage of air into a barrel, but prevent the passages of liquor out of the barrel through the bung, substantially as set forth.

Second, the combination of the top plate B, its projections b¹, and recesses b², and bottom parts C and D, and rubber or leather packing d², respectively, with the upper part of the socket A, its recesses a¹, and projections a², shelving a³, stops a²₁, and the lower part of said socket a, or their equivalents, as and for the purposes set forth.

Third, the application of the screw c and nut b², or their equivalents, when operating on the parts described in the second claim, as and for the purposes set forth.

Fourth, the combination of the ridge b³, on B and the packing c¹ on C, operated in connection with the pin c¹ in the groove b⁴, and the screw c and nut b², to either permit or prevent the passage of air, &c., between the plates A and C, as set forth.

Fifth, the combination of the plate C and holes c⁴, valves c⁵, nut d¹, plate D, and rubber or other packing d², as and for the purpose set forth.

Sixth, the application of the packing d², to produce with the grooves d⁴ and d³ on D, capillary passages, which permit the passage of air, but prevent the passage of fluid, as set forth.

Seventh, the application of a key to the parts of said bung, when operating as and for the purpose set forth.

No. 60,012.—JOHN H. KAVANAGH, Joliet, Ill.—*Car Coupling.*—November 27, 1866.—The coupling is oval in section, and has a guide tube above the bumper. A vertical shaft has an arm whose bent point retains the pin in an elevated position, and a wing on this shaft operates with the link to trip the pin. Another lever has a point to keep the pin up.

Claim.—First, the shaft F, provided with the wing G and trip lever E, connected to the bumper of a car, substantially as herein shown and described and for the purposes set forth.

Second, the lever H, in combination with the coupling pin C, substantially as shown and described and for the purposes set forth.

Third, the tube or guide B, in combination with coupling pin C and shaft F and trip lever E, substantially as shown and described.

No. 60,013.—ROBERT KING, Brooklyn, N. Y., assignor to T. PROSSER & SON, New York, N. Y.—*Boiler Flue Brush.*—November 27, 1866.—Each wire in the heliacal series protrudes at one end to a given distance from the axis, but at its other end is not so far prolonged; that is, they are not held midway of their length, but are more closely associated round the stock than near the periphery where flexibility is required.

Claim.—A heliacal wire brush, in which the wires are arranged to protrude unequally from opposite sides of the core or stock, substantially as shown and described.

No. 60,014.—HENRY KNIGHT, Brooklyn, N. Y.—*Cement Pipe.*—November 27, 1866.—The core is hollow, and has pipes for the introduction and exit of heated fluid. The mould may have chambers for the same use.

Claim.—The combination of a mould, composed of a shell and core, for making a cement article with a pipe, by which a heated fluid may be introduced into the hollow portion of such mould for the purpose of heating it, substantially as set forth.

No. 60,015.—J. KNUPP and H. S. ELLIOTT, New Prospect, Ohio.—*Sheep Shed.*—November 27, 1866.—The shed is made in four similar sections, each having runners for convenience in moving it.   When placed together it forms a tight shed with a central passage and a stable on each side.   By turning up the longitudinal door flap the sheep may reach the hay in the central passage.
Claim.—The construction and arrangement of a shed B, in combination with the trough *d*, doors I, rail N, and runners F, for the purpose and in the manner as set forth.

No. 60,016.—LOUIS KUTSCHER, New York, N. Y.—*Carriage.*—November 27, 1866.—The front box is connected to the lower part of the fifth wheel and to the tongue, and turns with them.
Claim.—Attaching the box F to the lower part *a* of the fifth wheel instead of to its upper part, as usual, substantially as and for the purposes set forth.

No. 60,017 —LEOPOLD LALL, New York, N. Y.—*Clasp Lock for Books, &c.*—November 27, 1866.—A catch on the connecting plate enters the edge of the cover and engages a similar catch on a spring lever within the cover.   It is released by the introduction of a pin key at the end of the cover.
Claim.—A clasp for books, portfolios, albums, &c., having a lock applied to it and inserted in the binding, substantially as shown and described.

No. 60,018.—JONAS LAMPHEAR, Panama, N. Y.—*Washing Machine.*—November 27, 1866.—A small frame contains two fluted rollers journalled in its sides; one of them is backed by springs to press the clothing.   A vibrating arm pivoted above them carries a fluted roller and a plain feed roller.
Claim.—The vibrating frame H, provided with one or more rollers, to operate in combination with the rollers B C, substantially as for the purposes set forth.   .

No. 60,019.—A. T. LARGE, Tomah, Wis.—*Hand Seed Planter.*—November 27, 1866; antedated November 17, 1866.—The seed hopper is attached to the side of the hoe handle and has two seed slides connected to a reciprocating rock lever; the space between the slides furnishing seed for one deposit.
Claim.—The tubes B, provided with the valves C C' arranged with the levers D G and spring H, and applied to the handle of a hoe to operate in the manner substantially as and for the purpose herein set forth.

No. 60,020.—CORNELIUS G. LAZEAR, Norwalk, Ohio.—*Fence.*—November 27, 1866.—The panel ends are halved into the intersection and brace bar of the straddle posts, or hung on the cleats on the corner posts.   Low posts have metal hooks engaging an eye on the lower bar at the centre of the panels.
Claim.—First, the panels A and posts D, when respectively constructed as described, and used in combination with the intermediate posts C, and hooks C', for securing the panels in position, substantially as set forth.
Second, the panels A, in combination with the posts C and hooks C' and with the corner posts B, with cleats B', all of said parts being respectively constructed and the whole arranged for use, substantially as set forth.

No. 60,021.—GEORGE H. LENHER, Richmond, Va.—*Sewing Machine Shuttle.*—November 27, 1866.—The location and motion of the bobbin during the revolution of the shuttle is such that the thread is not twisted or untwisted; but when it is carried to the side of its orbit farthest from the needle hole its thread is delivered by the turning of the bobbin, in a direction contrary to the course of the shuttle; and when the bobbin has passed that point the slack is taken up by being wound around the bobbin, which is then stationary on its axis.
Claim.—First, in rotating shuttles for sewing machines, winding and unwinding the shuttle thread so as to prevent it from being twisted and untwisted during the revolutions of the shuttle, by arranging the bobbin eccentrically to the centre of motion of the shuttle, substantially as shown.
Second, the arrangement of the spring I, collar A, bobbin F, screw G, shuttle C, and plate E, constructed and operating in the manner and for the purpose herein specified.

No. 60,022.—H. L. LOWMAN, New York, N. Y.—*Die for Forming the Eyes of Pickaxes.*—November 27, 1866.—The die consists of a pair of steel plates closed upon all sides, except a space through which the blank (a plain bar) is entered, and another one at right angles to the former, at which the punch is inserted.   The space corresponds to the external form of an elongated eye of a pickaxe.   The punch is wedge-shaped, and when driven in first, bends the bar down until it rests upon the bottom of the cavity, then splits it; the metal, as it is thus displaced, rises up and fills all the angles of the die.

Claim.—The combination of a die A and punch B, constructed substantially as described, for simultaneously punching and giving proper form, both externally and internally, to the elongated eye of a pickaxe, or other tool, with similar elongated eye, the whole operating substantially as set forth.

No. 60,023.—G. H. LUPTON, Cleveland, Ohio.—*Hanging Windows and Doors.*—November 27, 1866.—The side bars have projecting pins which traverse slots in the stiles and hinges, and sustain the pintles. These bars are so hinged to the weather strip as to raise it in a groove in the door when it is closed, and at the same time lower the pintles and door to their depressed position.

Claim.—The link E, and counterbalance G, in combination with the rods D, pins, and window or door, as and for the purpose set forth.

Second, the weather strip H, groove *c*, and sill M, arranged as set forth, in combination with apparatus for raising the window or door, substantially as set forth.

No. 60,024.—JOHN C. MACK, Philadelphia, Penn.—*Boot and Shoe.*—November 27, 1866.—The upper has an opening with a piece of rubber cloth or elastic material fitted therein so as to lie snugly to the foot and expand therewith. This rubber cloth is protected from the weather by an overlapping flap sewed in the boot on one edge and fastened on the other by rubber loops and buttons.

Claim.—First, a boot or shoe having a slit or opening in the centre of the upper, in combination with an elastic fabric, extending across said opening, all substantially as and for the purpose described.

Second, in combination with the foregoing, the elastic bands *c* and buttons *d*, or their equivalents, the whole being constructed and arranged substantially as and for the purpose specified.

No. 60,025.—J. H. MARSHALL, Lockport, N. Y., and O. E. MANN, Somerset, N. Y.—*Ditching Machine.*—November 27, 1866.—The fore end of the excavator frame is vertically adjusted by a triangular frame one angle of which is connected with the main wheels, another with the driving-belt shaft and the frame, and the third is extended to form a lever. A pinion on the lever runs on a rack upon the frame and has a pawl to retain the frame at any height. The excavated dirt is elevated by an endless apron and discharged by a V-slide to each side. The ditch, if deep, is dug at several passages of the machine.

Claim.—First, the triangular lever frame consisting of the parts K *g h*, hinged to the frame A, and connecting with the axis of the wheels B B, caster wheel C, scoop E, in combination with the toothed segment I, pinion *i*, and spring pawl *k*, for supporting, raising, and lowering the frame A, arranged and operating in the manner specified.

Second, the arrangement of the gearing *d e f*, the latter provided with pawl *a*, with the axis *b*, ratchet wheel *m*, and belt G, operated by driving wheels B B, for communicating motion to the endless belt F, substantially in the manner set forth.

No. 60,026.—JAMES B. MARTINDALE, New Castle, Ind.—*Hook and Eye.*—November 27, 1866.—The stirrup-shaped hook has a recurved central portion which catches over the convoluted bar of the eye and rests in two loops of the latter. A central tongue on the eye is advanced beyond the catch loops and is bent downward to catch the hook bar and prevent casual disengagement.

Claim.—A hook, formed as described, and an eye having loops *e e*, and a tongue *d*, when respectively constructed and arranged to operate, when combined substantially as and for the purpose set forth.

No. 60,027.—C. C. MASON, Brookfield, N. Y., assignor to himself and LEVI MASON, Clayville, N. Y.—*Curing Hops.*—November 27, 1866.—Chlorine gas is mixed with the current of heated air and the fumes of the sulphur of the bleaching process. \
Claim.—The improvement in the process of curing hops, in the manner and by the means substantially as described.

No. 60,028.—CURRAN E. MCDONALD, Indianapolis, Ind.—*Apparatus for Obtaining the Measures for Ladies' Dresses.*—November 27, 1866.—The mode of using the graduated strips cannot be briefly described. They have hooks for attachment to the dress and are used in connection with a tape measure.

Claim.—The within-described instrument when the same is constructed as aforesaid, in its said several parts, and operated for the purpose and in the manner substantially as set forth.

No. 60,029.—JOHN MCDONALD, Saratoga Springs, N. Y.—*Brick Kiln.*—November 27, 1866.—The truck floors are formed in steps to furnish horizontal support to the bricks, and the trucks pass down an incline through the furnace, and issue at the lower end. They are received on sub-trucks moving at right angles and are carried to other side tracks. The series of trucks in the inclined furnace is supported by gearing which engages a rack on the under side of the lower one of the series.

Claim.—First, the groove or way G, in the tunnel A, and the cars K K, in combination with their running gear, all arranged to operate substantially as and for the purpose specified.

Second, the combination with the cars K K, and the supplemental cars *m m*, tracks *a*, H I and J, arranged as herein described, and employed to permit the cars to be moved circuitously, in the manner and for the purpose specified.

No. 60,030.—JAMES C. McLELLAND and JAMES GRAHAM, Pittsburg, Penn.—*Broom Head.*—November 27, 1866.—A central ring clamps the inner course of corn. The side edges of the cap are folded back on themselves to give rigidity thereto. Wings depending from the ends of the cap are connected to the lower binding ring. A bolt traverses the cap and handle below the inner ring.

Claim.—First, forming the lower side edges of the cap A double, substantially as herein shown and described, and for the purpose set forth.

Second, the combination of the ring H with the cap A, handle B, and bolt C, substantially as herein shown and described, and for the purpose set forth.

No. 60,031.—A. McMULLEN, Sterling, Ill.—*Hame Tug and Breast Collar.*—November 27, 1866.—The breast strap is stuffed and made in two pieces connected by a snap. A plate upon it holds the breast rings and tug-buckle pieces.

Claim—First, the breast strap A, made in two parts, a^1 and a^2, connected by a strap hook and eye, when said parts are constructed in the form and manner substantially as herein shown and described, and for the purpose set forth.

Second, the breast collar A, in combination with the plates D and J, and the tug clips H, substantially as described and for the purposes set forth.

Third, pivoting the neck-strap tugs K to the parts a^1 and a^2 of the collar A, so that the position of the neck strap on the neck can be changed, when desired, substantially as described and for the purpose set forth.

No. 60,032.—MELCHOR MELLINGER, Dayton, Ohio.—*Foot Stove.*—November 27, 1866.—The flanged disk over the lamp has a central perforation with an inverted funnel through which the flame ascends into the combustion chamber between this disk and the cover. Radial tubes with gauze exits furnish air to the flame in this chamber. The bottom is detachable, and between it and the side is an annular air space.

Claim.—First, the disk H, and annular plate F, provided with the flange *a*, and mouth *b*, in combination with the air tubes *c*, and annular flue E, constructed, applied, and operating conjointly in the manner and for the purpose specified.

Second, the detachable bottom A, with its fastening B, constructed and arranged in combination with the cylindrical body of the stove so as to form an annular air passage for the lamp, substantially as and for the purpose described.

No. 60,033.—GEORGE W. MILLER, Springfield, Mass.—*Breeching Hook.*—November 27, 1866.—The curved spring holds the breeching ring in place within the book on the thill.

Claim.—A breeching hook for the shafts of vehicles, having the spring C constructed substantially in the manner and for the purpose described.

No. 60,034.—J. K. MIZNER, Detroit, Mich.—*Knapsack.*—November 27, 1866.—The knapsack is combined with a haversack. The straps that secure the parts of the sack together when packed and folded are not sewed to the material, but are riveted to each other, and also to the sling straps. The latter pass from the knapsack over the shoulders, beneath the armpits, and unite behind the back.

Claim.—First, the straps D F and H, riveted together and arranged and combined with each other, and with a knapsack, substantially as set forth for the purpose of affording a reliable support for the weight of a knapsack and its contents.

Second, the manner, as described, of slinging a knapsack by means of straps passing over the shoulders, under the arms, and across the back to meet and connect together, substantially as set forth.

No. 60,035.—LAWRENCE MOONEY, Alleghany City, Penn.—*Churn.*—November 27, 1866.—The dashers are so attached to their shaft in inclination and relative position as to drive two streams of cream in a spiral current downward.

Claim.—Arranging the breakers *e* with relation to the shaft *d*, so that they will throw two counter currents of milk or cream, the reaction of which will cause the milk or cream to flow spirally down and around the shaft *d* and thereby gather the butter as it is formed into two balls or rolls, the whole being constructed, arranged, and operating substantially as herein described and for the purpose set forth.

No. 60,036.—ROBERT F. MOORE, Manchester. N. H.—*Nozzle for Hose Pipe.*—November 27, 1866.—A divided hemispherical perforated cap is made to traverse the end of a nozzle by means of double-threaded screw bearings, so that when the caps are brought together a rose jet is produced, and when separated the stream is compact.

Claim.—In combination with a hose pipe or nozzle, a perforated movable cap made in two parts and arranged to traverse on the end of a hose pipe or nozzle, substantially as described to enable the operator to convert a single stream into a shower or sprinkle, and to reconvert it to a single stream without stopping the flow of water from the nozzle.

No. 60,037.—SOLOMON B. MOORE, Lockport, N. Y.—*Roofing Cement.*—November 27, 1866.—This invention consists of coal tar, 5 parts; mineral paint, 1; and limestone dust from marble works, 16.

Claim.—The combination and using the materials as herein specified.

No. 60,038.—THOMAS MOORE, Bloomington, Ill., assignor to JACKSON HUKILL, same place.—*Broom Head.*—November 27, 1866.—The corn is placed in the head with the curved nut at the end of the screw. The upper band is then put on and the clamp nut drawn in by rotating the screw to which the handle is fastened. The lower wire is held by hooks on the cap and key wires from side to side.

Claim.—The combination of the part A with its ears a^2, hooks D, wires E, wire F, crescent nut G, screw H, socket a^1, and handle C, and keys I, in the manner as and for the purpose specified.

No. 60,039.—CALEB H. NEEDLES, Philadelphia, Penn.—*Medicated Troche.*—November 27, 1866.—For the cure of cholera, &c. A composition of camphor, catechu, Jamaica ginger, African capsicum, powdered opium, and the oils of peppermint and cloves, is worked with a mucilage of tragacanth and rolled into sheets.

Claim.—The compound camphor troches, prepared substantially as herein set forth and described.

No. 60,040.—ANDREW P. ODHOLM, Bridgeport, Conn.—*Machine for Dressing the Felloes of Wagon Wheels.*—The wheel is suspended upon an adjustable arbor, which has centring cones to enter each end of the spindle socket. The rim is dressed by revolving cutters. An upright adjustable arbor furnishes bearing for the hub when the periphery of the wheel is dressing. In this case the rim rests upon a plate, and beneath it there is a wheel having projecting arms, which traverse a slot in the plate as guides for the wheel. The spokes as they are revolved depress one arm and raise another.

Claim.—First, the slide D, with adjustible arbor E attached, the rotatory cutter D, and the adjustable plate F, provided with the guide arms 7, arranged so as to rotate as shown, or in any other equivalent way to admit of being removed or adjusted out of the way of the spokes by the action of the spokes themselves in the turning of the wheel, substantially as and for the purpose set forth.

Second, the swinging pendant K, suspended from the adjustable slide L, having an arm I attached in connection with the plate O on slide F and the cutter D, all arranged to operate substantially in the manner as and for the purpose specified.

No. 60,041.—NELSON ORCUTT, Binghamton, N. Y.—*Composition for Making Soap.*—November 27, 1866.—Composed of tallow, 100 pounds; sal soda, 100 pounds, (or soda ash, 35 pounds;) resin, 60 pounds; starch, 4 pounds; wheat or rye flour, 20 pounds; carbonate ammonia, 1 pound; glue, 1 pound; stone lime, 20 pounds; water 32 gallons, with the addition of a small quantity of alum or borax.

Claim.—Making soap from untried or unrendered tallow or grease, and the other ingredients named, the ingredients being in the proportions stated.

No. 60,042.—FREDERICK ORTLIEB, Williamsburg, N. Y.—*Surface Condenser.*—November 27, 1866.—The air and water induction pipes and nozzles are arranged to give downward movement through the condenser. The water is introduced in the form of spray, and is thoroughly mingled with the air in its descent through the pipes, at the bottom of which the water is discharged below while the air is allowed to escape upward through a central pipe. The steam chamber is enveloped by the overflow-water jacket and is traversed vertically by the pipes which contain commingled air and water.

Claim.—First, in hydro-atmospheric condensers the use of air and water combined, when the same is introduced to the body of the condenser in the form of spray or mist, injected directly downward in contradistinction to an upward injection, through the condensing tubes which form said body, substantially as specified.

Second, the combination of the condensing tubes t, air-discharge pipe O, and receiving chamber F, with its delivery pipe X, said pipes O and X being arranged so as to effectually separate the air and water after the same have been injected or passed through the tubes, essentially as herein set forth.

Third, the combination of the upper chamber H, blast nozzles or tuyeres T, perforated pipe or pipes P, and screen S with the condensing tubes *t*, arranged for action together, substantially as specified.

Fourth, the receiving chamber F with its delivery pipe X, arrranged to pass through a stuffing box W, in combination with the tubes *t* and air pipe O suspended from above, essentially as and for the purpose or purposes herein set forth.

Fifth, the combination with devices for producing a spray action of air and water combined at the top of the condenser and downwardly through the tubes thereof of a water base B', hollow jacket A', and circulating pipe N, substantially as specified.

Sixth, in combination with a surface condenser, the arrangement at or near the top thereof of one or more automatic relief or escape valves Q, operating essentially as shown and described.

No. 60,043.—JAMES H. OSGOOD, Jr., Boston, Mass., assignor to himself and JAMES W. HANNA, Wabash, Ind.—*Apparatus for Drying Lumber.*—November 27, 1866.—After being subjected to the action of steam in one chamber the steam is shut off, the partition removed, and the car load of lumber wheeled into another chamber, where it is subjected to dry heat from the furnace.

Claim.—The combination of a steam-heating chamber with a hot air chamber, substantially as and for the purpose set forth.

No. 60,044.—SAMUEL A. OTIS, Boston, Mass.—*Turnout for Railroads.*—November 27, 1866; antedated November 11, 1866.—The main track is not curved, and the return car entitled to the road receives no shock thereat. The switch iron under the near wheels is flat and the wheels run on their flanges, causing them to "cone" and swerve the car to the off side, where the flanges of the off wheels are engaged by the fixed "dummy" and run the car on the turnout.

Claim.—The combination of the device described for turning the car out of the straight line with the fixed points of the dummy switches, as shown and described.

No. 60,045.—ABRAHAM W. OVERBAUGH, New York, N. Y.—*Slate Cleaner.*—November 27, 1866.—The sponge is attached by a rim to a circular plate and is supplied with water from a perforated tube which forms the handle.

·*Claim.*—The plate and tube for moistening the sponge.

No. 60,046.—ISAAC E. OVERPECK, Overpeck's Station, Ohio.—*Corn Sheller.*—November 27, 1866.—The corn is shelled by the revolution of the toothed disk. The grain is conveyed to the grinding hopper, and the cobs are thrown out by a spring worked automatically.

Claim.—The shelling disk A and bisected conical tube *c'*, in combination with discharging spring *m*, arranged and operating in the manner and for the purpose specified.

No. 60,047.—LOUIS E. PAGE, Pontiac, Mich.—*Adjustable Spring Back for Vehicles.*—November 17, 1866.—The lazy-back has hinge attachments to the spring support bars which allow an easy rolling movement.

Claim.—The springs E E at the sides, the springs F F at the back, when constructed as described and connected to the buggy seat by means of the bar B, or its equivalent, the plates G and lazy-back H, in the manner and for the purposes specified.

No. 60,048.—GEORGE PARR, Buffalo, N. Y.—*Manufacture of Cutlery.*—November 27, 1866.—Upon one section of the bolster is a hollow protuberance which passes through a hole in the tang; the other section has a stem which enters the hollow of the protuberance and is split at its extremity and wedged to hold the two sections together.

Claim.—A cast metal bolster for cutlery, made in two parts A and B, and having a prong *a'*, socket *b'*, and wedge *c* or *d*, for the purpose and substantially as herein described.

No. 60,049.—GEORGE W. PARSONS, Harrisburg, Penn.—*Lubricating Axle Journals, &c.*—November 27, 1866.—The screw cap and plunger ·are withdrawn to admit oil. The oil is forced to the axle by the plunger, and retained by the valve at its foot.

Claim.—The screw plug or tube D, in combination with the plunger *e* with its stem and the cap F, constructed as and for the purpose herein specified.

No. 60,050.—JOHN W. PEASE, Belmont, N. Y.—*Brick Machine.*—November 27, 1866.—The endless belt carries mould boxes which have projecting lips and recesses upon their upper surfaces to break the joints between them when passing beneath the rollers. A smaller roller is used for levelling, and a larger for pressure. The pressure roller may have grooves to admit the box edges for extra pressure. The bricks are discharged on a belt by an inclined slide acting on the movable bottoms of the mould boxes.

Claim.—First, the employment in conjunction with an endless chain of mould boxes G, of the levelling roller L, and a pressure roller N, arranged and operating substantially as described.

Second, the grooved face pressure roller P, constructed and operating substantially as described for pressing the clay into the mould boxes.

Third, sustaining the endless chain of mould boxes G upon the supporting bars J and J', by means of offsets *g* on the ends of said boxes, during the filling of the boxes and also during the discharging of the bricks therefrom.

Fourth, discharging the endless chain of boxes, by means of an inclined plate K, acting upon the stems *c* of followers *b*, substantially as described.

Fifth, constructing the mould boxes with offsets on their ends which are adapted to form connections for said boxes, and also means for sustaining the boxes upon the bars J and J', substantially as described.

Sixth, preventing clay or other substance from getting between mould boxes, which are connected together in the form of a chain by means of lips *i*, substantially as described.

No. 60,051.—HENRY PEMBERTON, Alleghany City, Penn.—*Box for Putting up Caustic Alkali.*—November 27, 1866.—The object is to form a box of sufficient strength to bear handling and packing, and capable of resisting the caustic action of the enclosed material.

Claim.—Making boxes or cases for enclosing, preserving, and protecting the aluminates and hydrates of the alkalies, by means of paper, muslin or other suitable fabric, wrapped and cemented together in folds or layers, with silicate of soda or other cement or glue, and externally coated with tar, beeswax, rosin, or other substance capable of resisting the caustic action of such alkalies, and of rendering the fabric of which the case or box is made impervious to moisture or air, substantially in the manner and for the purposes hereinbefore described.

Also, the use of a solution of the silicates of soda or potassa, for saturating, coating, or cementing together paper, muslin, pasteboard, or other fabrics for making boxes, cases, or wrappings for enclosing and protecting the aluminates and hydrates of the alkalies and similar articles requiring protection from air or moisture.

No. 60,052.—DEWEY PHILLIPS, Manchester, Vt.—*Regulator for Reaction Water Wheels.*—November 27, 1866.—The valve gate to the water exit is hung by one corner in such a manner that the centrifugal force from the rotation of the wheel tends to close it and regulates the amount of water by the speed. An adjustable weight enclosed in the gate modifies its action.

Claim.—First, a valve or gate hinged at or near the end of the water way, in a reaction water wheel, to regulate the opening for the escape of the water in proportion to the centrifugal force resulting from the rotation of the wheel acting upon such valve or gate to close the same, substantially as specified.

Second, the valve or gate, hinged as specified, in combination with the elastic or yielding packing *n* and *s*, substantially as specified.

No. 60,053.—L. D PHILLIPS, New York, N, Y.—*Machine for Making Buttons.*—November 27, 1866; antedated November 11, 1866.—The eye of the button is formed of wire, and the ends of the loop inserted in the plastic material simultaneously with the formation and moulding of the button. The machine cannot be briefly described other than substantially in the words of the claims.

Claim.—First, the general construction and arrangement of the entire machine for forming buttons from plastic materials, the several parts constructed and operating, severally and in combination, substantially as described.

Second, the combination and arrangement of the operating lever B, with the two cross-heads or levers D and E, substantially as described, by which the outer ends of such cross-heads are made to approach or recede from each other simultaneously, substantially as and for the purposes set forth.

Third, the construction and arrangement of the movable jaws, in combination with the former or its equivalent for making and shaping the eye of the button, the several parts constructed and operating substantially as described.

Fourth, the construction and arrangement, substantially as described, of the mould to form the button, the same being in two parts, one fixed and the other movable, and the latter shutting over or upon the former, substantially as and for the purposes set forth.

Fifth, the construction and arrangement of the oscillating plate between the mechanism that forms the eye and the movable part of the mould, such plate supporting the former while the eye is being formed, and constituting also a portion of the mould for shaping the button, and by its oscillating motion discharging the buttons as they are formed.

Sixth, the combination of the several parts forming the eye of the button with the parts which mould or shape the button, and their relative motions towards each other, and operating substantially as described, whereby the eye of the button is formed and inserted in the button at the same time the latter is being moulded and shaped.

Seventh, the arrangement of the mechanism for feeding the wire to form the eye, arranged and operating substantially as described.

Eighth, the construction of the arm B, in combination with the pin *s* or their equivalent, for oscillating the plate I, substantially as and for the purposes set forth.

No. 60,054.—SAMUEL G. PIPER, Lynn, Mass.—*Cooking Stove.*—November 27, 1896: antedated November 22, 1866.—A hollow base with a close bottom rests upon and fits into a boiler hole in the top of a cook stove or range; from this a pipe passes to an upper room for heating purposes. In the side of the pipe is a hole with a register to admit a current of air when desired.

Claim.— The combination and arrangement of the vessel B, and the air conduit C, provided with an air-receiving passage b, and a register arranged with respect to it, as described, the whole being for application, as explained, to a stove and to the floor of an apartment over that in which the stove may be situated.

No. 60,055.—LEMAN B. PITCHER, Salina, N. Y.—*Mortar Mill.*—November 27, 1866.— The lime is slaked in a revolving barrel and discharged into a settling vat. From this the paste passes to the curing vat, and from thence to the three revolving mixing cylinders. The "cylinders" (so called) are slightly tapering, and have teeth on their inner surfaces intended to mix the mortar and gradually to forward it to the discharge end.

Claim.—The cylinders A B C and D, with or without any of the attachments which form a part of either of them, made and operated substantially as and for the purposes described.

Also, the mechanical process of making mortar therewith, substantially in the manner described.

No. 60,056.—ARLON M. POLSEY, Boston, Mass., assignor to T. H. FULLER —*Machine for Making Nails.*—November 27, 1866.—Each pair in the series of rollers is arranged at right angles to the preceding, and has dies with sockets of the form of the finished heads of the nails. The headed blanks are dropped into the dies and have the shanks rolled down to the finished form. The nail is operated upon by the successive pairs of rollers, the head in each being seized within the next succeeding pair of rolls before the point is released from the last preceding pair; the successive pairs operate upon alternate faces.

Claim.—A series of rolling dies provided with die grooves having enlargements therein for reception of headed blanks, and otherwise formed substantially as described, when arranged with reference to each other, and so as to operate substantially as specified.

No. 60,057.—ORRIN PRATT, Athol, Mass.—*Sinking Tubular Wells.*—November 27, 1866.— The boring instrument works through and under and carries down with it an internal case. The borer is then withdrawn through said external tube, which remains for the proper placing of the tube and strainer that forms the tubular well. The external case is then also withdrawn, leaving the tube and strainer to occupy the tube so bored.

Claim.—So combining with an external tube or case a boring instrument such as described, so that said instrument will bore a suitable hole for, and carry down with itself, the external tube, which can remain while the borer is withdrawn, and until the tubular well and strainer are introduced, and the external tube then withdrawn from the hole, the operation being such substantially as described.

No. 60,058.—JOEL K. REINER, Line Lexington, Penn.—*Churn.*—November 27, 1866.— The dasher arms are curved; are set obliquely on their shafts, and respectively force the cream to opposite ends of the box. When the butter is formed the communication is opened to the butter chamber and the churn inverted. The buttermilk then escapes through the strainer.

Claim.—First, the two rotating shafts C C', provided with curved arms a a', having an oblique position in their transverse section, and the arms of one shaft having a reverse oblique position to those of the other, substantially as and for the purpose specified.

Second, the butter chamber D, applied to the cream receptacle A, and provided with the slide E, and the strainer or screen F, substantially as and for the purpose set forth.

No. 60,059.—HENRY RENSCH, Quincy, Ill.—*Apparatus for Cooling Beer.*—November 27, 1866.—The annular vessel is supported in a shallow pan and has pipes at the bottom and top to conduct a current of cold water. At the top of the vessel are troughs which distribute the beer in a thin sheet over the inner and outer surfaces of the vessel. A current of air circulates through the vessel.

Claim.—As an improvement, a perpendicular cylinder with double walls, supplied with cold or ice water at the bottom and escaping at the top, in combination with the receiving and distributing troughs at the top, and opening in the bottom of the cylinder for the escape of the beer running down on the inside of the cylinder.

Also, in combination with the cylinder above claimed, the pipe Q, or its equivalent, for supplying air to the interior of the cylinder near the bottom, to carry off the steam or evaporation from the liquid being cooled.

Also, the construction and arrangement of the guiding rims M and N, or flanges, for the purpose set forth.

No. 60,060.—DANIEL T. ROBINSON, Boston, Mass.—*Fruit Press.*—November 27, 1866.— The cloth sack containing fruit is clamped by its ends and twisted by power applied to the clamp at each end so as to wring out the juice. Dogs maintain the clamps in position and the spring arms distend the sack while being charged.

Claim.—The improved implement or machine constructed and operating substantially as above described, consisting of the frame A, spring arms *e e*, dogs *d d*, and clamps D D, or their equivalents.

No. 60,061.—A. ROHM and P. GUETLICH, Lancaster, N. Y.—*Reel or Swift.*—November 27, 1866.—Auxiliary arms form braces to hold the primary arms against lateral strain under rapid motion or sudden stoppage; their projecting ends also form an additional small reel for silk and small skeins, and the central disk of the system of projections forms a convenient ball holder or pocket. The sleeve of the lower disk is vertically adjustable on the shaft, and is retained by spring pawl and ratchet.

Claim.—The combination of the auxiliary arms I, and disk H, with the rimary arms E G, and disks C D, operating substantially as and for the purpose herein set forth.

Also, in combination with the above, the slots *i* and joints *b*, for the purpose of producing an independent adjustment of the lower arms, as set forth.

Also, in combination with the above, the ratchet teeth *g* and collar K, with spring projection *h*, operating in the manner and for the purpose specified.

No. 60,062.—FRANK J. ROTH, Newark, Ohio.—*Piston Packing.*—November 27, 1866.—In an annular recess in the periphery of the piston is a ring whose exterior is divided by two flanges into three grooves which are occupied by packing rings extending beyond the periphery of the piston head, and each expansible by interior springs, and the outer ones by pressure of steam.

Claim.—First, the application of expansible packing rings to the circumference of pistons, in such manner that said rings shall be expanded by the pressure of steam upon such portions of them as are exposed beyond the periphery of the piston, substantially as described.

Second, the arrangement of spring plates *g* to recesses formed in circumference of pistons, substantially as described.

Third, the arrangement of springs *c c*, so as to act upon the ends of the packing rings *b b*, which are made up of segments, substantially as described.

No. 60,063.—E. A. G. ROULSTONE, Roxbury, Mass.—*Shoe Brush.*—November 27, 1866.—The handle of the brush has a recess into which the box of blacking is slipped and retained by a spring.

Claim.—Combining with the handle *d*, and brush block *a*, a spring for holding the box on the brush block, substantially as described.

No. 60,064.—HENRY B. ROWLEY, Rushville, N. Y., assignor to himself and H. S. ROSE.—*Axle Box.*—November 27, 1866.—A collar having one or more chilled or hardened flanges is shrunk on the end of the axle, and the box is made to conform in shape to this collar. A flexible plate is held in the box joint to act as a wiper.

Claim.—First, in combination with the journal box C D, the axle A, constructed with a hardened metal collar B, or collars B a¹, on its ends, the hardened or chilled metal collar or collars being made separate from the axle and shrunk or otherwise fastened thereon, and said collar or collars conforming in diameter to the chamber or chambers of the journal boxes in which they revolve, all substantially as herein described and for the purposes set forth.

Second, the arrangement of the flexible wipers *e e*, in combination with the journal and journal box, substantially in the manner herein described.

No. 60,065.—WILLIAM F. RUNDELL, Genoa, N. Y.—*Hay Fork.*—November 27, 1866.—Explained by the claim and cut.

Claim.—The ferrule C, constructed with a conical part *d*, and a cylindrical part *e*, in combination with the screw tang *b* of the fork B, the key *f*, and handle A, to form a new and improved hay fork, substantially as set forth.

No. 60,066.—WILLIAM F. RUNDELL, Genoa, N. Y.—*Carriage Top.*—November 27, 1866.—The sides of the front bow are bent against the second and attached thereto, the upper part forming a hood.

Claim.—The top frame for carriages, having its front section or bow secured to the section next adjoining, substantially in the manner described and for the purpose specified.

No. 60,067.—D. T. SANFORD, New York, N. Y.—*Machinery for Embossing Brass Tubes.*—November 27, 1866.—The knurls or milling rollers are arranged around a circular head, and their faces may be set at any angle to the axis of the tube; they receive their rotary motion from a system of gearings connected to the main driving wheel on the hollow shaft through which the tube is drawn by the action of the knurls; the pressure of the said knurls upon the tube is effected by spiral springs bearing upon the ends of the radial shafts, which allow them to yield to any inequalities.

Claim.—First, the arrangement of a series of knurls, to which a positive revolving motion is imparted, in combination with a hollow spindle A, constructed and operating substantially as and for the purpose described.

Second, making the knurls yielding in a radial direction, by springs *g*, substantially as and for the purpose set forth.

Third, the thimbles *e*, and boxes E, in combination with the shafts *c*, the ends of which form the bearings for the knurls, substantially as described, so that said knurls can be adjusted at any desired angle toward the axis of the articles to be embossed, without throwing them out of gear with the driving mechanism.

No. 60,068.—WILLIAM H. SANGSTER, Chicago, Ill., assignor to JAMES A. COWLES, same place.—*Lamp*—November 27, 1866.—A spring and train of gears in the base run a fan which drives a current of air to the burner.

Claim.—First, the combination of the sheave *i*, on shaft *t*, the sheave *v* on shaft *h*, driven by the belt *s* that carries upon one end the blower *g*, with the within described mechanism that supplies the power for operating the blower when used in a lamp, lantern, or chandelier, in the manner and for the purpose herein set forth.

Second, the placing of the mechanism *c c* on its side, substantially in the manner described.

Third, attaching the mechanism *c c* with a lamp, lantern, or chandelier, by means of the springs *o o*, substantially as described.

No. 60,069.—W. D. SCHOOLEY, Richmond, Ind.—*Straw Cutter.*—November 27, 1866.—Improvement on his patent of August 22, 1865. The devices are for adjusting the feed. A wrist on the shaft of the revolving cutter actuates the triangular system of rods which actuates the feed wheels. The middle rod of the series is adjustable longitudinally in its box, so as to vary the sweep of its upper end which connects to the radius plate, whose pawl moves the ratchet on the lower feed wheel.

Claim.—First, the vibrating box O, provided with the pawl P and the lever N, fitted in said box, and provided with the rack *j* for the purpose of admitting of said lever being readily adjusted higher or lower to vary the movement of the feed as occasion may require.

Second, the radius plates T, provided with the pawls U. and fitted loosely on the roller shafts *a a'*, in combination with the ratchets M M' on said shafts and the adjustable lever N, substantially as and for the purpose set forth.

Third, the combination of the rod R, operated from the shaft E, by the eccentric Q, the adjustable lever N, fitted in the vibrating box O, provided with the pawl P, the rods S S, radius plates T T, with pawls U attached, and the ratchets M M', all arranged to operate substantially in the manner as and for the purpose specified.

No. 60,070.—FRIEDRICH SHALLER, Hudson, N. Y.—*Blow-pipe.*—November 27, 1866.—The piston forces the air into a reservoir which is connected with the blow-pipe. The working portions are situated below the top of the table so as not to be in the way of the operator. The stroke of the piston is adjustable.

Claim.—First, the arrangement of the blow-pipe J, and the piston blower C B, in relation with each other and with the work table *b*, and air reservoir G, substantially as herein set forth for the purpose specified.

Second, the arrangement of the treadle R, pitman *z*, cranks 5 and *s*, bent connecting rod N, and lever M, in combination with each other and with the spring T, substantially as herein set forth for the purpose specified.

No. 60,071.—WILLIAM C. SHERMAN, Boston, Mass.—*Attaching Carriage Thills.*—November 27, 1866.—The shafts can be attached when the forward ends are depressed, but cannot be detached from the axle clip when horizontal or in working position. The pin of the clip is held in the notch of the thill bar by a forward spring attached to the bar and an elastic pad behind it.

Claim.—The hook E, taking a solid bearing on the lower part of the draw bar, in combination with spring G, shoulder J, and rubber pad C, all constructed, arranged, and operating substantially as described.

No. 60,072.—FRANCIS P. SMILEY, CHARLES K. BECHER, and ISAAC TOWNSEND, Philadelphia, Penn.—*Policeman's Rattle.*—November 27, 1866.—The handle is hinged to a boss on a prolongation of the axis of the ratchet wheel, and may be latched in its rectangular position for use.

Claim.—The application to a policeman's rattle of a jointed handle, constructed and arranged to operate in combination therewith, substantially as and for the purposes described.

No. 60,073.—A. R. SMITH, Delaware, Ohio.—*Flour Bolt.*—November 27, 1866.—Around the reel is a series of cam-shaped blocks which act in connection with a pivoted spring to agitate the bolt as the end of the lever snaps against the blocks consecutively.

Claim.—The reel F, provided with knocker blocks, or a circular series of projections around it from rib to rib, in combination with a pivoted spring lever, operating substantially as described.

No. 60,074.—DEXTER SMITH, Springfield, Mass.—*Cartridge Machine.*—November 27, 1866.—The trimming and beading devices are combined in one machine. The case is placed in an opening in a block and driven forward by a punch into a revolving chuck, where it is trimmed to the required length by a tool advanced by a lever and cam, and is then pushed forward by the introduction of another case into the chuck and placed within a circular disk, by which it is transferred to the other side of the machine, where it is headed.

Claim.—First, the disk P, operated from the shaft B by means of the cams K and O, connected with the ratchet and pawls *g h*, and stop piece *m*, or equivalent mechanism for the purpose of transferring the shells, substantially as herein set forth.

Second, the combination of the cam H, shaft D, and mandrel S, with the cam Y, and tool Z, when arranged and operating substantially as and for the purpose described.

Third, the chuck O, constructed as described and used for the purpose of holding the shells, as set forth.

No. 60,075.—E. C. SMITH, Birmingham, Conn.—*Attaching Draught Poles to Axles.*—November 27, 1866.—The eyes of the cross-bars are made adjustable to suit the clips of different vehicles. The shanks of the eyes slip on a bar attached to the cross-bar of the tongue, and are secured in their adjustment by set screws.

Claim.—The bars G with the eyes F at their rear ends fitted on the iron bar E of the cross-bar B, within mortises or notches D made therein, and provided with the gib or key H and screw I, all arranged substantially as and for the purpose herein set forth.

No. 60,076.—H. L. SMITH, Gambier, Ohio.—*Treating Oils, &c.*—November 27, 1866.—A current of steam is passed into the bottom of the retort, and the crude-filtered petroleum agitated thereby until the condensed water in the retort is equal to the quantity of oil originally placed therein. The volatile portions of the oil are vaporized, and escape by the neck into the cold water tank, where they are condensed.

Claim.—For the purpose of rectifying or refining crude petroleum, or the distillates thereof, the herein described process, which is divided into two parts.

Also, the process of treating crude petroleum or its distillate in a closed retort by the introduction of steam, as and for the purpose specified.

Also, the process of treating cold petroleum or its distillate by agitation with cold air, as and for the purpose specified.

No. 60,077.—T. K. SMITH, Oskaloosa, Iowa.—*Churn.*—November 27, 1866.—The churn stands on a spring-board and receives a saltatory motion from the foot. A funnel inside keeps the cream from dashing out.

Claim.—The spring-board and churn combination, which can be used without a dasher of any kind.

No. 60,078.—JOHN SPEIRS, Boston, Mass.—*Tobacco Pipe Case.*—November 27, 1866.—The case is made of practically fire-proof material, and contains the pipe and its clearing wire; the bowl chamber has a hinged lid to permit the withdrawal of the pipe. The clearer is withdrawn in the reverse direction.

Claim.—The combination of the parts *a b* and B, arranged and applied together substantially as described.

Also, the pipe case, as made of the tubes *e* B, and the bowl parts *a b*, combined and arranged substantially as specified.

Also, the pipe case as composed of the bowl parts *a b*, the stem-holder B, the cup C, and the tube *e*, arranged together as specified.

Also, the combination of the clearer *d* with the cup C, to be applied to the pipe-case stem receiver B, as specified.

No. 60,079.—B. P. STEBBINS, Corry, Penn.—*Still.*—November 27, 1866.—The upper surface of the corrugated cone is exposed to the refrigerating water in the vessel; upon the lower surface the vapors are condensed and are collected in an annular trough, from which they are discharged by a pipe.

Claim.—An improved cone still, formed by combining a corrugated cone-shaped partition B and an inclined circular flange or apron C with each other and with the tank A, substantially as described and for the purpose set forth.

No. 60,080.—J. W. STEED, Minneapolis, Minn.—*Hoisting Jack.*—November 27, 1866.—The cap rotates upon anti-friction rollers on the head of the screw-shaft.

Claim.—The combination of the cap D and the screw B with the flange C and the conical rollers *a*, arranged and operating substantially as and for the purpose set forth.

No. 60,081.—WM. J. STEVENS, New York, N. Y.—*Slide Valve.*—November 27, 1866.—The valve is suspended from the axis of a roller which traverses in a yoke suspended from the flexible diaphragm cover of the valve chest. A nut above graduates the pressure of the valve on its seat; the stop-lugs on the yoke limit its upward movement.

Claim.—First, the arrangement relatively to each other of the slotted yoke C, flexible diaphragm D, roller *b*, stop *f*, and valve A, substantially as shown and described.

Second, the arrangement of the flexible diaphragm D with the yoke C and roller *b*, substantially as specified.

Third, the stops *f* in combination with the yoke C, diaphragm D, and valve A, substantially as and for the purpose set forth.

No. 60,082.—JAMES R. STREET and SIDNEY M. DAVIS, Washington, D. C.—*Faucet.*—November 27, 1866.—The ball valve is kept to its seat by a spring plunger, which is withdrawn to permit the flow of liquid.

Claim.—The combination and arrangement of the barrel 7, cap 5, plunger 3, spiral spring 2, elastic ball-valve 4, raised valve-seat *a*, inlet passage 6, and exit port *c*, all substantially as shown and described.

No. 60,083.—JAMES SWEENEY, New York, N. Y.—*Car Bell.*—November 27, 1866.—The hammer strikes the clamped edge of the bell and is immediately withdrawn by the contact of the studs on its shank with the axial pin of the bell; the shank is somewhat elastic, to permit the hammer to strike after the said contact.

Claim.—The bell B, having the edge chamfered as described, in combination with the frame C and hammer D, provided with the ears *c c*, lug *d*, and projections *e*, the whole being constructed, arranged, and operated in the manner and for the purpose set forth.

No. 60,084.—FRANCIS TAGGART, Brooklyn, N. Y.—*Governor for Steam Engines.*—November 27, 1866.—The steam from the boiler and that from the cylinder operate upon opposite ends of the same piston; variation in their pressure moves the piston longitudinally, and actuates a device which controls the admission of steam to the cylinder, and thus affects the movement of the engine.

Claim.—A regulator for engines, fitted substantially in the manner specified, so that the pressure from the boiler shall act in the opposite direction to the pressure from the engine, and the difference of pressure produce a movement to regulate the supply of steam, substantially as set forth.

No. 60,085.—CHAS. M. TANNER, Mentor, Ohio.—*Scaffold.*—November 27, 1866.—The legs of the platform slip in guides on the lower frame, and the upper portion is raised by ropes, pulleys, and windlasses. The legs of the upper portion are extended by sections provided with sockets for the reception of the feet of the section above.

Claim.—The sliding frames which support the stage or platform H, in combination with the main frame constructed with recesses D, guide plate I, windlass C, hoisting ropes E and E', and additional pairs of framed legs provided with sockets L as herein described, and operating as and for the purpose set forth.

No. 60,086.—R. S. TAY, Medford, Mass., and FRANCIS T. FRACKER, Cambridge, Mass., assignors to Tucker Manufacturing Co.—*Lamp Bracket*—November 27, 1866.—The lamp stand is hinged to its bracket so as to shut against it for convenience in transportation, &c.

Claim.—A side light in which the lamp ring or support is hinged to, and so as to swing up against the hanger plate, substantially as set forth.

No. 60,087.—J. W. TAYLOR, Jr., Dubuque, Iowa.—*Curtain Fixture.*—November 27, 1866.—A metallic disk on the end of the roller is perforated with holes near its edge. Into these holes the pin of a side lever fits, retaining the curtain at any height. The lever is swayed laterally by the cord passing through a ring at the end and temporarily withdraws the detent pin.

Claim.—The perforated wheel *e* attached to the roller of a window shade, in combination with the lever H, when constructed and arranged substantially as herein shown and described and for the purposes set forth.

No. 60,088.—G. L. THOMPSON, New York, N. Y.—*Machine for Pressing Hats.*—November 27, 1866.—By means of a plunger a fluid is pressed against a diaphragm within the hat, which is thus pressed into shape within a mould.

Claim.—The combination of the barrel G and movable piston H with the flexible or elastic core F and die D, constructed and operating substantially as and for the purpose set forth.

No. 60,089.—HENRY ALDERSON THOMPSON, Grant Gipps Land, Victoria.—*Table for Concentrating Ores.*—November 27, 1866.—The material upon the table is kept in a loose state by means of stirrers. A percussive action upon the bed gives it longitudinal agitation, and the lighter, larger particles find their way toward the tail end; water is flowed upon the surface; the work is continuous or in charges, and the vertical adjustments are made by suspension chains and rollers.

Claim.—The combination of the frame A, supports N, and screws R, with the table B, stirrers Q, and stirrer-frame O, substantially as and for the purposes herein shown and described.

No. 60,090.—SAMUEL R. THOMPSON, Portsmouth, N. H.—*Machine for Making Crucibles.*—November 27, 1866.—The piston forces the clay from the cylinder into the mould above. The bottom of the mould corresponds to the base of the crucible, and the core is held in place by a locking plate above. The clay in the mould is cut off from the mass by a wire which runs between the cylinder and mould.

Claim.—First, the combination of the cylinder B, piston head D, screw piston rod E, bevel gear wheels F and G, shaft H, and crank wheel J, with each other, and with the frame A of the machine, substantially as herein described and for the purpose set forth.

Second, the combination of the case K, core L, and cover M, with each other, when said parts are constructed and arranged substantially as herein described and for the purpose set forth.

Third, the combination of the swinging guide bar R, the swinging binding bar T, and the binding and centring screw U with each other, with the cover M, and with the posts O and P, substantially as described and for the purpose set forth.

Fourth, the combination of the bow V, wire v^3, and guide-rod W with each other, and with the cylinders B and K, substantially as herein described and for the purpose set forth.

No. 60,091.—WM. W. TICE. California, Ohio.—*Moulder's Clamp.*—November 27, 1866.—The parts of the flask are clamped by a hook and a sliding plate, which is depressed by an eccentric pivoted to the shank of the hook.

Claim.—The moulder's clamp composed of the hook A B, guard C, and cam-headed lever E E' F.

No. 60,092.—DANIEL J. TITTLE, Albany, N. Y.—*Car Coupling.*—November 27, 1866.—The hook heads of the draught bars are sprung laterally by contact, and their shoulders then become engaged. Slots in the heads and vertical openings permit coupling with an ordinary link and pin. A segment pinion and rack deflect the draught bar to uncouple.

Claim.—The draught bar c, provided with a head i and two hooking shoulders A A, and slots, in combination with the springs l l and e.

Second, the rack bar m and segment n, in combination with the draught bar c, as and for the purposes set forth.

No. 60,093.—JAMES D. VAN HOEVENBERGH, Kingston, N. Y.—*Carriage.*—November 27, 1866.—A bar is clamped on the front axle, carrying at its ends wrist pins for attachment of the front ends of the lower halves of the longitudinal springs. The two parts of the spring are bolted together through a pad of India-rubber.

Claim.—The wrist and sleeve, or equivalent hinge connection between the side springs and the "third axle," substantially as and for the purpose herein specified.

Also, the yoke M and king-bolt I, constructed and arranged substantially as described, to connect the forward axle A and third axle D, for the purpose herein set forth.

Also, the construction of the springs, each composed essentially of the half springs E and G, with the India-rubber spring cushion O between them, substantially as herein specified.

No. 60,094.—LOUIS ALEXIS VELU, EUGENE FRANÇOIS FOSSE, and LOUIS EUGENE ALPHONSE FOSSE, Paris, France.—*Car Brake.*—November 27, 1866.—The brake block is fixed to the main frame and descends nearly to the rail beneath the wheel. The brake is put in operation by a wedge-shaped piece drawn beneath the brake block by toggle levers, actuated by racks.

Claim.—Combining with a lever, or its equivalent, for operating the brake, and with the frame or wheels of a car, or other similar carriage, the brake blocks E and wedges G, constructed and arranged to operate in the manner and for the purpose substantially as described.

No. 60,095.—BRUNO VOLKMANN, New York, N. Y., assignor to FREDERICK VOLKMANN, same place.—*Plough.*—November 27, 1866.—This is a two-wheeled plough, having lateral and vertical adjustment on the axle frame. It is explained by the claims and cut.

Claim.—First, the plough frame or plough cart A B C D E F, constructed substantially as described in combination with the plough beam L, for the purpose set forth.

Second, the small balance beam e, in combination with the screw chain n, the screw n', and plough beam L, substantially as described for the purpose set forth.

Third, the screw tree I in combination with the two horizontal frame pieces d d' and the plough beam L, substantially as described for the purpose set forth.

Fourth, the cart shaft G in combination with the semicircle D, regulating screw S, and with the plough frame A B C D E F, substantially in the manner and for the purpose described.

Fifth, the cart axle D¹ D² in combination with the movable side piece E and regulating screw T, for the purpose of deepening the furrows as required, substantially in the manner described.

No. 60,096.—ELIJAH WAGONER, Westminster, Md.—*Grain Drill.*—November 27, 1866.—The drill share is attached by lugs to a hinged lever above, and to two flexible bars below, with a rubber spring interposed between them. The share will vibrate rearward in contact with fixed obstacles, and afterwards resume its position.

Claim.—First, the combination of the drill A, elastic bars B B, lever F, spring G and link D, substantially as and for the purpose herein specified.

Second, in combination with the above parts, the link D D and lug *a'*, when formed with the shoulders *a² d*, in the manner and for the purpose set forth.

Third, the stem or shank *i'* when formed as a part of the washer I, and employed in connection with the spring G, bars B B and stirrup H, as described.

No. 60,097.—G. B. WALLER, Franklin, Ill.—*Churn.*—November 27, 1866.—The dasher has a chamber at its foot, open below, and from this chamber emanate radial tubes, carrying the cream upward and outward by rotation.

Claim.—First, the combination of two or more inclined tubes K, with the dasher shaft B, substantially as herein shown and described and for the purpose set forth.

Second, the combination of the inverted cup J, and two or more inclined tubes K, with the dasher shaft B, substantially as herein shown and described and for the purpose set forth.

Third, the combination and arrangement of the inclined tubes K, inverted cup J, dasher shaft B, shafts D and H, gear wheels F and G and frame E, with each other, with the cover C, and with the body A of the churn, substantially as described and for the purpose set forth.

No. 60,098.—J. M. WALLIS and E. P. SWEARINGEN, Milton, Iowa.—*Bellows Pump.*—November 27, 1866; antedated November 11, 1866.—The double bellows is arranged on each side of a perforated dividing board, and operated by a rod from the lever above. The flap valve in the dividing board operates alternately with the upper and lower openings in connection with the respective bellows.

Claim.—The pump, consisting of a double bellows with a centre board *a*, having a chamber H, and a valve guarding the openings *j j*, operated and combined with the pump stock, substantially as described and represented.

No. 60,099.—JOHN WALTER, Princeton, Ill.—*Grape Vine Protector.*—November 27, 1866.—The narrow roof is supported on hinged stakes, whose two joints allow of the descent of the roof to the ground for winter protection to the vine. The frame is secured in an elevated position by a pin passing transversely through the roof and stakes.

Claim.—A grape-vine protector, made and operating substantially as herein shown and described.

No. 60,100.—WILLIAM H. WATSON, Yonkers, N. Y.—*Machine for Pressing Tobacco.*—November 27, 1866; antedated October 10, 1866.—The tobacco is passed between endless belts moving in concert. The belts consist of hinged sections covered by continuous sheets of metal, which pass around rollers, and are held firmly to the tobacco by slide frames. Revolving belts retain the tobacco at the sides.

Claim.—First, giving to the tobacco a gradual pressure, substantially as shown, prior to the actual or uniform pressure, for the purpose specified.

Second, retaining the tobacco under a uniform pressure, substantially as shown, for the purpose of giving solidity to the sheet formed.

Third, the use or employment of the side belts, in combination with the main belts, substantially as shown, for the purpose set forth.

Fourth, the use or employment of the chains, in combination with the bolts and springs, for the purpose described.

Fifth, the use or employment of the main belts, in combination with the chains, for the purpose specified.

Sixth, the use or employment of the side belts, in combination with the chains, for the purpose specified.

No. 60,101.—GEORGE L. WHITE, Woonsocket, R. I.—*Loom.*—November 27, 1866.—The purpose of this improvement is to facilitate the "piecing up" or repair of the warp threads by providing a means for holding the lay from falling back, or retaining it in a convenient position for the operator.

Claim.—The combination of the stop cam and the spring bolt, constructed, arranged, and to be applied as specified, and the combination of the same or their equivalents with the lay and its operative mechanism, substantially as described, the purpose and mode of operation of such mechanism being as herein before explained.

No. 60,102.—J. F. WHITE, Brattleboro', Vt.—*Mop Wringer.*—November 27, 1866.—Explained by the claim and cut.

Claim.—The rectangular frame A and yielding roller C, in combination with the hinged or pivoted frame E, roller B and foot rest H, when said frame E is arranged within the said frame A, all as herein described and for the purpose specified.

No. 60,103.—HUGH WHITEHILL, Newburgh, N. Y.—*Machine for Drying Yarn.*—November 27, 1866.—The fan occupies the centre of the cylinder or case, the steam-pipes surround it, and the slatted cylinder revolves outside of and around these; the casing encircles all. The shield prevents the too rapid outflow of the heated air from the interior of the shell or casing.

Claim.—First, the system of heating pipes E, connected at both ends by transverse pipes *f*, forming a continuous passage for the steam through the entire system, arranged and operating with relation to the revolving slatted cylinder G and fan D, substantially as and for the purpose set forth.

Second, the guard or shield I, arranged in relation with the opening B' of the shell or casing B, substantially as herein set forth, for the purpose specified.

Third, the arrangement and combination of the revolving slatted cylinder G with the casing A B, and system of heating pipes E, substantially as and for the purpose specified.

No. 60,104.—SEBA SQUIRE WILES, Santa Clara, Cal.—*Measuring Faucet.*—November 27, 1866.—The liquid enters the measure cylinder, and the indicator attached to the piston rod shows the amount contained. The movement of the cut-off plate shuts off connection from the measure to the barrel, and opens it to the discharge orifice; the latter has an additional cut-off to save drip.

Claim.—First, the faucet composed of the barrel A and the feed and discharge passages C and D, with the valve *d*, operated substantially as described.

Second, in combination with the devices above claimed, the scale *o*, index *b*, and piston B, with its rod *c*, all arranged to operate substantially as and for the purpose set forth.

No. 60,105.—GEORGE L. WITSILL, Philadelphia, Penn.—*Grinding Mill.*—November 27, 1866.—The grinding roller has a single heliscal ridge, and rotates in a spirally ribbed concave. The roller shaft revolves in vertically sliding boxes depressed by set-screws. The device is for grinding quartz, &c.

Claim.—The roller D, when constructed with a single heliscal rib like the thread of a screw, and suspended upon the vertically adjusted shaft B, when used in combination with the hopper A and spirally ribbed concave A', said parts being respectively constructed and the whole arranged to operate substantially in the manner and for the purposes set forth.

No. 60,106.—H. H. WALCOTT, Yonkers, N. Y.—*Breech-loading Fire-arm.*—November 27, 1866; antedated November 22, 1866.—The double-jointed swinging breech-piece forms also the lock frame. Of the two projecting points on the breech-piece proper, one is for the purpose of half cocking the hammer as the breech is opened, and the other to prevent the opening of the breech-piece when the hammer is at full cock. The cartridge shell extractor is operated by a projection on the trigger guard bow. Both ends of the main spring are so secured that it has no other bearing than at its extremities; namely, on the hammer and trigger.

Claim.—First, the lock frames C composed of two parts *a a'* hinged to each other at *b*, and constructed and operating substantially as described.

Second, securing the extremities of the main spring F, in the manner described, within notches U S in the hammer, butt, and sere, respectively, avoiding any intermediate attachment or bearing point, thus adapting the spring to move freely with the pivoted frame, and causing any deflection of the spring to be equally distributed throughout its entire length.

Third, the swell 6 on the back part of division *a'* of the lock frame, for the purpose of bringing the hammer to half cock, in the way substantially as above described.

Fourth, the projection 4 of division *a'* for keeping the lock frame in place and preventing its withdrawal when the hammer is at full cock, substantially as above described.

Fifth, in combination with a hinged lock frame, of the construction herein specified, the shell drawer *e* made with arms *r* and *k*, and operated by the contact of a projection *j* from the guard bow I, as explained.

No. 60,107.—CHRISTIAN WOLF, Rantoul, N. Y.—*Horse-collar.*—November 27, 1866.—The collar consists of a wooden frame carrying the harness attachments and padded behind. It is connected above by a hinged bolt and a spreading block. A leather strap envelops the hinge and block. The lower ends of the collar are connected by a traverse bolt.

Claim.—First, connecting the upper ends of the side pieces A A of a wooden collar together, by means of a hinge *a* and clamping nuts *c c*, in combination with an interposed block for staying said parts A A laterally, substantially as described.

Second, the leather strap D applied to the upper part of the wooden collar, substantially as and for the purpose described.

Third, connecting the lower ends of the side pieces A A together by means of an adjusting screw C, substantially as described.

Fourth, connecting the trace-hooks *k* and rings *s* to the wooden sides A A of the collar, by means substantially as described.

Fifth, in combination with the wooden side pieces A A and padding B, the hinged clamps, movable block *d* or *g*, and the adjusting screw C, all constructed and operating substantially as described.

No. 60,108.—IRA WOOD. Woodstock, Vt.—*Tanning.*—November 27, 1866.—The tanning mixture is a decoction of alum, 7 pounds; glauber salt, 3 pounds; rock salt, 4 pounds; soft water, 10 gallons; sumac, 5 pounds; oak bark, 3 pounds; ground nutgalls, 1 pound; to which is added oil of vitriol, 4 ounces.

Claim.—The tanning mixture, composed of the ingredients mixed together in and about the proportions herein stated, and substantially as and for the purpose described.

Also, the leather, as a new article of manufacture, produced substantially as herein described.

No. 60,109.—D. A. WOODWARD, Baltimore, Md.—*Fluid Lens.*—November 27, 1866.—These lenses are formed of glass plates bent to any degree of curvature and cemented together by rims or bands of metal. The cells thus formed are filled with fluids of different densities.

Claim.—First, the method herein described of bending disks of glass so as to produce a concave or convex lens of spheroidal, parabolic, or hyperbolic profile for the purpose set forth.

Second, the use of two or more pieces of plate, or flattened glass, bent to the required curve or profile, as herein specified, in combination with the cement and rings, or their equivalents, for holding them together, substantially as set forth.

Third, combining with three or more glasses formed into cells, as described, fluids of different index of refraction, that will correct chromatic aberration, substantially as described.

No. 60,110.—WM. WRIGHT, New York, N. Y., assignor to JOHN H. CHEEVER, same place.—*Stamp Quartz Mill.*—November 27, 1866.—The stamps are arranged in quadrangular sets and have central feed. The slide valves are worked by the inclined ends of the piston rods acting on a slide block connecting to the valve lever. The cylinder frame is vertically adjustable to suit the wear of the stamps.

Claim.—First, the arrangement of the stamps in the form of a group, in combination with a central feed opening and surrounding screen and delivery receptacle, substantially as and for the purpose herein specified.

Second, in the direct application of steam-power to stamping mills or hammers, providing for the adjustment of the steam cylinders and their valves, toward and from the battery, substantially as herein described, whereby the wear of the stamps and the battery may be compensated for, and a uniform, or nearly uniform, clearance between the pistons and the tops and bottoms of the steam cylinders may be maintained.

Third, in combination with the above specified provision for the adjustment of the steam cylinders, the attachment of the feeding hopper or hoppers to the said cylinders, substantially as and for the purpose herein set forth.

No. 60,111.—CHARLES O. YALE, New York, N.Y.—*Marking Device for Sewing Machines.*—November 27, 1866.—The object of rotating the pencil is to insure with certainty a distinctly visible mark on the fabric, and without moving the point of the marker lineally, as in making a line or dash. The spring and tightly fitting tube surrounding the pencil-tube admit of adjusting the pressure of the pencil by moving the outer tube up or down.

Claim.—First, giving a rotatory motion to the pencil while resting on the fabric, substantially as and for the purpose herein set forth.

Second, in connection with the above, the supporting and guiding parts E H and the pressure spring I, arranged relatively to each other and to the marking device G for joint operation, substantially in the manner and for the purpose herein set forth.

No. 60,112.—JOSEPH ADAMS, Fairhaven, Vt.—*Railroad Rail.*—December 4, 1866.—Explained by the claim and cut.

Claim.—A railroad rail made in two parts A A', with their top surfaces so shaped as to form a central longitudinal groove c, and having said top surfaces highest at the points where the wheels will bear directly over the centre of the neck of each half rail, the whole being arranged and constructed substantially in the manner and for the purpose set forth.

No. 60,113.—RICHARD ALLEN, Jersey City, N. J.—*Lathe Clutch.*—December 4, 1866.—The bolt to be operated upon is held between two arms, which slide on the head of a T-bolt attached to the face-plate. In securing the clutch to the face-plate the arms are fixed at any distance required.

Claim.—The carrier clutch, constructed and operating substantially as described.

No. 60,114.—M. M. AMMIDOWN, Boston, Mass.—*Drilling Machine.*—December 4, 1866.—The platen has vertical adjustment by an upright screw within the slotted standard; a projection from a collar on the said screw enters a groove in the platen arm. This groove runs all around the socket so as to admit of a complete revolution of the platen on the standard.

Claim.—The combination of the collar E provided with the projection e, and the grooved arm F, substantially as and for the purpose specified.

Second, the combination of the slotted standard B and screw D with the collar E, arm F, and platen C, as and for the purpose specified.

Third, the grooved arm or holder F, as and for the purpose specified.

No. 60,115.—JAMES C. ARMS, Northampton, Mass.—*Embossing Machine.*—December 4, 1866.—The roller has a roughened surface, and is rotated by a hand-crank. Above the roller is a hollow press block having a removable convex-faced plate, with ridges for emboss ing any substance passed between it and the roller. The block is depressed by a pivoted lever having an elastic press-band over the end. The hollow within the block serves to introduce some substance to heat the embossing plate.

Claim.—First, the embossing machine, constructed and arranged to operate as and for the purpose substantially as set forth.

Second, the heater D provided with the removable embossing plate *n*, as shown and de scribed.

No. 60,116.—T. G. ARNOLD, New York, N. Y.—*Chuck.*—December 4, 1866.—The four jaws are pivoted in slots of the head, and have a taper screw on their ends on which a nut turns to clamp the pipe. A centre cup in the head guides the end of the pipe to position.

Claim.—The arrangement of the expanding jaws E pivoted to the solid head piece B, con structed and operating substantially as hereinbefore set forth and for the purposes described.

No. 60,117.—ELIAS ASHCROFT, South Boston. Mass.—*Lubricating Oil Can and Lamp.*— December 4, 1866; antedated November 29, 1866.—From above the perforated diaphragm proceeds the wick spout, and the lubricating spout from below.

Claim.—The combination of the tube D, tube C, plate B, and can A, constructed and arranged in the manner and for the purpose herein specified.

No. 60,118.—FREDERICK M. BAKER, South Reading, Mass.—*Cooking Stove.*—December 4, 1866; antedated November 22, 1866.—This is an improvement on his patent of October 31, 1865. The stove has two ovens on one level and separate flues and dampers so arranged that the caloric currents may be made to pass around both ovens separately or to take a direct course up the chimney.

Claim.—The combination as well as the arrangement of the flues, *m* and H, the dampers D D^2 and D^3, and openings a^2 b^2, with the auxiliary oven G, its flue F, and the main oven, and its flues, B B^1 B^2, and C, the whole being substantially as hereinbefore specified.

No. 60,119.—HAYDN M. BAKER, Rochester, N. Y., assignor to himself and ROBERT J. LESTER.—*Manufacture of Glass.*—December 4, 1866.—The clay retort is charged with 1.2$^{...}$ pounds of sand, previously washed in muriatic acid and then in water; 1,223 pounds of re crystallized nitrate of lead, and 670 pounds recrystallized nitrate of potash. The nitric acid is recovered by a condensing apparatus and used to treat salts of lead and potash to obtain nitrates for another charge. These are dissolved, evaporated and recrystallized.

Claim.—The application to manufacture of the processes herein described, for the produc tion of best flint glass from a mixture of nitrate of potash, nitrate of lead, and silicic acid at elevated temperatures, and the recovery of the nitric acid employed by displacement and distillation in the manner herein described and set forth, or any other process substan tially the same, and which produces the same intended effects or results.

No. 60,120.—NEWELL BARNARD and J. G. SPILLER, Saginaw City, Michigan.—*Appa ratus for the Manufacture of "Salt Block."*—December 4, 1866.—Under the drying racks are troughs which communicate with the lower pan at the rear, and with the upper pan at the front directly over the fire.

Claim.—First, admitting the brine at the forward end, or hottest part of the block D, sub stantially as described and for the purpose set forth.

Second, drawing off the bitter water at the rear end of the block D, substantially as described and for the purpose set forth.

Third, the combination with the lower vat *c*, with the block D, and with the drying rack D, of the trough A, substantially as described and for the purpose set forth.

No. 60,121.—CHARLES J. BARNEY, Edgartown, Mass.—*Bag Holder.*—December 4, 1866.— The bag mouth is brought up through the opening in the platform and confined there by the hinged hopper. The bag bottom rests on the floor or upon a removable rest in the frame.

Claim.—The frame A, provided with a platform B, having an opening C made in it to re ceive the flange or lower part of the hinged hopper D, and used in combination with a remov able platform E, substantially as and for the purpose specified.

No. 60,122.—HIRAM BECKWITH, Grass Lake, Michigan.—*Scaffold Bracket.*—Decem ber 4, 1866.—The horizontal bar is screwed into the building. The tie rod has at its foot a socket to engage the end of the brace bar, and a point to enter the wall and prevent lateral movement.

Claim.—The tie B, the rail C, and the brace D, when constructed and combined substan tially as herein shown and described, for the purposes set forth.

No. 60,123.—JOHN BELLERJEAN, Philadelphia, Penn.—*Lantern.*—December 4, 1866.— This is an improvement on the patent of W. H. Pierce, dated February 18, 1862. The guard wires are connected above and below by sections of rings which engage catch pins on the cap and base and hold the lamp together. The lamp is separated by slipping the zonal ring from the guard wires, and removing them with their sectional connections; the cap and base are then free from the glass.

Claim.—The two-part guard D D, and confining ring *f*, when used in combination with projections *h*, studs *i*, or their equivalent, so as to constitute a means of connecting the cap and base of the lantern, substantially as described.

No. 60,124.—THOMAS E. BELTON, Buffalo, N. Y.—*Steam Generator Heater.*—December 4, 1866.—The heater is interposed between the furnace and the fore end of the flues. The feed is beneath a deflecting plate at its bottom. At each end of this deflecting plate are mud chambers, communicating by small holes with the main part of the heater, and discharge pipes for the mud. The heater is horizontally traversed by flues about mid height, and discharges through flattened, inverted funnel-shaped passages through the crown sheet. A feed pipe from the boiler supplies water to keep up the circuit when the doctor is stopped.

Claim.—The heater F, provided with flues *g g*, and conical discharge ports O O, arranged in combination with a boiler and furnace, so as to operate substantially as set forth.

Also, the mud shields H H, provided with perforations *i i*, in combination with the pipes J, for discharging the mud and other sediment, substantially in the manner specified.

Also, the deflecting and distributing plate *m*, arranged and operated as described.

Also, in combination with the heater F, constructed as described, and leg *c*, of the boiler, when the same extends below the level of said heater, the pipe *q*, arranged and operating substantially in the manner and for the purpose herein set forth.

No. 60,125.—CHARLES C. BEMIS, San Francisco, Cal.—*Grate Bar for Furnace.*—December 4, 1866.—The bars have alternate sections with longitudinal and transverse vertical slots, and transverse plates beneath to deflect the air upward in its passage backward beneath the bars.

Claim.—A furnace grate with bridges or barrier plates, *b c d e f*, placed beneath the said grate at intervals, and increasing in depth to near the flue, the whole arranged and constructed substantially as described and for the purpose set forth.

No. 60,126.—WILLIAM BICKNELL, Hartford, Me.—*Communicating Motion.*—December 4, 1866.—The weighted wheels are pivoted on the main wheel, and the power being applied to the latter causes the central pinion and drum, with their connecting sleeve, to revolve. The speed may be regulated by the relative size of the central and weighted wheels.

Claim.—The new method of transmitting motion, consisting of the arrangement upon fly wheel A of cog or friction wheel *c*, so constructed that a plane passing through the axis of either of them shall preserve the same angle with the horizon throughout the revolution of wheel A, said wheel *c* driving wheel *c'*, substantially as described.

No. 60,127.—BENJAMIN BILLINGS, Lyons, Iowa.—*Fence.*—December 4, 1866.—The rails are traversed by pickets, and for a straight fence have supporting posts driven in the ground. A shoulder on the post has a socket for the picket foot and a tie cap above. For a zigzag fence no posts are used, but a tie block slips over the pickets of two adjoining panels.

Claim.—The arrangement of the post A, picket P, cap D, when constructed, arranged, and operating substantially as and for the purpose set forth.

No. 60,128.—G. E. BINGHAM, Milwaukee, Wis.—*Revolving Flue Cleaner.*—December 4, 1866.—The cylinder has metal ends and wooden sides, and the boiler tubes are admitted through a hinged section of its sides. The incrustation on the tubes is removed by the abrading matter and water with which the cylinder is furnished.

Claim.—The revolving flue cleaner formed by the combination and arrangement of the heads E, wooden bars F and A, binding bands B and C, binding rods G, and shaft I, substantially as herein described and for the purposes set forth.

No. 60,129.—J. B. BLAIR, Philadelphia, Penn.—*Solar Camera.*—December 4, 1866.— The reflector is attached to an adjustable frame secured to an inner cylinder; the latter is made to revolve in the lens cylinder by clock work, so as to reflect the light continually on one spot.

Claim.—The application of an adjustable piece E to the use and purpose substantially as set forth.

No. 60,130.—WILLIAM Y. A. BOARDMAN, New Haven, Conn.—*Wick Inserter for Lamps.*— December 4, 1866.—The plate is divided into fingers at one end, and these are bent slightly to opposite sides alternately, and the ends pointed and bent in. The wick is put between the fingers and the plate driven through the socket, drawing the wick after it.

Claim.—The herein-described instrument for inserting wicks in lamp tubes as a new article of manufacture.

No. 60,131.—JOHN W. BOUGHTON, Chicago, Ill.—*Elastic Strap for Garments.*—December 4, 1866.—Explained by the claim and cut.

Claim.—An elastic strap A, having an attaching plate or its equivalent at each end for application to the garment, substantially as herein shown and described.

No. 60,132.—LEWIS BRIDGE, York, Penn.—*Heating Stove.*—December 4, 1866.—The caloric current passes through one or both of two flues into an annular drum, and these flues are furnished with dampers. 'One of the flues discharges under the escape flue, while the current from the other is deflected around the drum. An air jacket around the furnace discharges beneath the drum, and a flue may carry the air to another room. A sliding vertical grate limits the size of the coal space.

Claim.—The arrangement of a parlor fireplace heating stove, with a vertical cold air space D around the back of the fire chamber A, in combination with the surrounding hood *h h* for utilizing the heat radiated from the rear of the stove, and conducting it into the room, or to an upper chamber when desired, constructed and operating substantially as and for the purpose herein described.

No. 60,133.—S. BROWN and LEON LEVEL, New York, N. Y.—*Apparatus for Detaching Boats.*—December 4, 1866.—The hook of each davit fall-block engages a shackle hook pivoted on a post in the stem and stern respectively. The free end of each shackle hook is held by a loop of a pivoted bar, and the latter, belonging to the respective ends, are tripped simultaneously by a lever and rod connection, allowing the shackle bar to fly up and disengage the boat.

Claim.—The standard C, the bent lever *a*, and the shackle hook *b*, constructed, combined, and operating as a detaching hook, substantially as herein shown and described, for launching boats and for other purposes.

No. 60,134.—WM. BRÜCKNER, San Francisco, Cal.—*Furnace for Desulphurizing Ores.*—December 4, 1866.—The inside of the revolving roasting chamber has spiral ridges running in opposite directions.

Claim.—The internal screw ribs or rifles arranged spirally in opposite directions, so as to convey the ore alternately from end to end of the cylinder, and heat it uniformly.

No. 60,135.—ANGELINE BUTTON, Pontiac, Mich., administratrix of CHARLES A. BUTTON, deceased.—*Buckle.*—December 4, 1866.—The frame has curved fingers forming pivot bearings for the clasp. The clasp has a jaw and cross-bar, which act in opposition to each other to hold the strap.

Claim.—The combination of the clasp C and body or rim A, constructed and connected substantially as and for the purpose herein specified.

No. 60,136.—AARON CASEBEER, Sipesville, Penn.—*Instrument for Destroying Empire Caterpillars.*—December 4, 1866.—The twigs or small limbs are removed by knives attached obliquely to a pole, so as to make a tapering throat between the knife and shank.

Claim.—A knife which is composed of two blades C C, united to a contracted shank *b*, to be used substantially as described.

No. 60,137.—GEORGE CHAMBERS, Ithaca, N. Y.—*Car Replacer.*—December 4, 1866; antedated November 26, 1866.—The inclined plane is placed beside the rail, fitting against and inclined toward it, so as to lift the wheel when traction is applied and bear it off laterally, and bring the tread on to the rail.

Claim.—First, the truss, trunnion, or tool A, made with two surfaces, one for replacing a wheel from the inside, and the other from the outside of the track, and the duplicating the same in one instrument or trunnion, so as to fit any emergency and either direction of motion of the displaced wheel or wheels, as described.

Second, on either a single or doubled inclined plane, with a surface or surfaces suited to replacing a wheel off of the track, the placing or combining therewith an adjustable piece or part, which, while it aids in replacing a wheel off on the inside of the track, is also useful in carrying the flange over the rail when the wheel is off on the outside of the track, as described, and all equivalents thereunto.

No. 60,138.—JOHN C. CHAPMAN, Cambridgeport, Mass., assignor to himself and DAVID W. WESTON, Boston, Mass.—*Chuck.*—December 4, 1866.—This chuck has transverse tapering keys behind loose collars. The article to be turned is screwed on to the chuck up against the collar. It is released so as to be readily unscrewed by loosening the key behind the collar, against which it becomes jammed by turning.

Claim.—The beveled keys A *a*, and the prevention pins *c c c*, in combination with the collars B *b*, operating substantially as above described.

Also, the beveled key A, and the prevention pins *c c*, in combination with the removable bushing E, operating substantially as above described.

No. 60,139.—RICHARD CHESTER, Chicago, Ill.—*Lantern.*—December 4, 1866.—Between the case of the lantern and the lamp is a perforated jacket connecting with an annular apport for the globe. The air which enters the holes in the case is deflected by the diaphragm and passed in a circuitous course to the flame to avoid direct impingement upon and extinguishment of the light.

Claim.—The combination and arrangement with a lantern and its globe of the globe support *d*, perforated jackets *e e'* and diaphragm *e*, arranged and operating as and for the purposes specified.

No. 60,140.—RICHARD CHESTER, Chicago, Ill.—*Lantern.*—December 4, 1866.—Air openings are made in the center of the bottom of the case, and in an up-curved covering plate beneath the lamp holder. The smoke passes through holes in the edge of a diaphragm plate in the cap, and has exit through the end perforations of the horizontal cups and under the bell cover.

Claim.—First, the arrangement of the perforations in the centre of the bottom E, and the perforated inclosure F, arranged beneath the platform C, for the lamp, as and for the purposes described.

Second, the horizontal tubes H, provided with perforated cups and extending within the lantern top as shown, in combination with the diaphragm G, having openings at the corners out of the range of said tubes, as herein specified and set forth.

No. 60,141.—GEORGE H. CHINNOCK, New York, N. Y.—*Toy Building Block.*—December 4, 1866; antedated November 26, 1866.—Explained by the claim and cut.

Claim.—The building blocks of the form of the half of a cube, having five sides, in combination with letters or numerals on their surfaces, which are whole in themselves or bisected diagonally, substantially as shown and described for the purposes specified.

No. 60,142,—WILLIAM W. CHIPMAN, New York, N, Y., assignor to the CHIPMAN MINING Co., same place.—*Manufacture of Paris White and Whiting.*—December 4, 1866; antedated November 22, 1866.—Marl is calcined and exposed to the action of the atmosphere after it has become cool, in order that it may become neutralized by the carbonic acid of the air. It is then washed with water, and the water allowed to flow through a tank, where the whiting is deposited.

Claim.—The manufacture, as herein described, of Paris white and whiting from the earthy material known as marl.

No. 60,143.—GREVILLE E. CLARKE, Racine, Wis.—*Animal Trap.*—December 4, 1866.—Successive vibrations of the pivoted platform actuate the ratchet wheel a certain number of teeth, according to adjustment, before bringing the pin on the wheel in contact with the lever; the vibration of the latter moves the latch from its support, closing the door, which is locked by a notched bar.

Claim.—First, the combination of the pivoted platform E and the arm F, arranged and operating substantially as and for the purposes described.

Second, in combination with said platform E and arm F, the ratchet wheel G, provided with a finger *l*, operating substantially as specified.

Third, in combination with said ratchet wheel G, the arrangement of the spring *s* to hold the wheel from moving back while the arm F recedes, in the manner described.

Fourth, the combination of the spur wheel G, provided with the pin *l* and the lever H, arranged and operating substantially in the manner and for the purposes described.

No. 60,144.—EMMETT COON, Kalamazoo, Mich.—*Vice.*—December 4, 1866.—The jaw dies are secured to the pos s by means of dovetailed keys which jam into correspondingly shaped sockets. The faces of the jaws are beveled or counterpart recesses.

Claim.—First, the adjustable dies F F, with cavities *b b* and keys *c c*, as and for the purpose set forth.

Second, the adjustable dies, Figs. 4 and 5, made with the bevel *n*, horn *d*, and keys *e c*, when used in combination with a vice.

No. 60,145.—E. HALL COVEL, New York, N. Y.—*Centrifugal Pump.*—December 4, 1866.—A wheel, similar to a turbine, at the foot of the revolving stem, and a heliacal flange on the stem itself, raise the water into the angular chamber above, where its vertical motion is abated before discharge.

Claim.—In pumps the combination of a water wheel and screw elevator, when arranged substantially as and for the purpose described.

Also, an angular or irregularly shaped chamber, in combination with a pump cylinder, in which water or other fluid is elevated by a spiral or vertical motion, substantially as described, for the purpose specified.

No. 60,146.—W. G. CRUTCHFIELD, Dayton, Ohio, assignor to himself and JAS. O. ALTICK.—*Force Pump.*—December 4, 1866—A water cup is attached to the fret-cock and allows the escape of steam or gases which may collect in the barrel of the pump without permitting the entrance of air.

Claim.—The arrangement with the stem A of a force pump, of the pipe C, with its cock

No. 60,147.—H. D. DANN, Waupun, Wis.—*Seeding Machine.*—December 4, 1866.—The seed cylinders consist of two disks which interlock and form seed cavities between them. One disk in each pair is fixed, the other movable, and by placing them nearer together the seed cavities are lessened. This is done by means of a bar to which the movable disks are connected. The seed passages from the hopper are limited in size, by slides connected to a rod.

Claim.—First, the seeding cylinders, consisting of the parts B and C, constructed and arranged to operate in combination, as herein described.

Second, attaching the parts B permanently to the axle O, and the parts C to the rod *a*, for the purpose of adjusting the size of the cells, as set forth.

Third, the plates E, provided with the opening *i* and the slides *m*, attached to the bar *b*, arranged to operate in combination therewith, as shown and described.

No. 60,148.—JOHN DAVIES, Manchester, England.—*Treating Threads or Yarns Previous to Weaving.*—December 4, 1866.—The yarn is boiled in the "cops" in a solution of tannin as a mordant, in order that it may readily take the dye when passing from the warpers to the weaver beam.

Claim.—Cops or bobbins of yarn or thread saturated with a solution containing tanning matter, for the purpose specified.

No. 60,149.—FRANKLIN DAVIS, Lawrence, Kan.—*Machine for Scouring Leather.*—December 4, 1866.—The leather is fed over the apron to the revolving cylinder, which is armed with spiral rows of scouring stones. Jets of water are thrown upon the cylinder. A spring raises the apron from the cylinder when not in use.

Claim.—The cylinder B, and the apron C, constructed, arranged, and operating substantially as described, in combination with the frame A, and the water reservoir D, for the purposes set forth.

No. 60,150.—JOSEPH S. DENNIS, Chicago, Ill.—*Sad Iron Heater.*—December 4, 1866.—Gas is introduced by a flexible hose through the top of the iron into the upper chamber and passes through pipes in the diaphragm, at whose exits it is ignited. The sides of the iron have perforations for air supply. The fumes pass off through a chimney above.

Claim.—The chambers D and E, with the inlet and exit tubes, connecting pipes or jets *b* and orifices *a*, constructed and operating substantially as herein described.

No. 60,151.—ALEXANDER T. DE PUY, New York, N. Y., assignor to R. HOE & COMPANY, same place.—*Printer's Galley.*—December 4, 1866.—The metallic strip which lines the inside edge of the galley is made to clasp the latter instead of being secured by screws, &c., to the face of the same.

Claim.—The combination of the metal lining with the frame of a printer's galley, in the manner substantially as herein shown and described.

No. 60,152.—CHARLES DIMMICK, Brockport, N. Y.—*Spade Handle.*—December 4, 1866.—Timber is bent around in the required form and the end fastened to the shank of the handle.

Claim.—Making a spade, or other handle, from a straight piece of wood, formed and bent round in the manner substantially as herein shown and described and for the purposes set forth.

No. 60,153.—ROBERT DIVEN, Brooklyn, N. Y.—*Detachable Flange for Privy Bowls.*—December 4, 1866.—A separate annular detachable locking flange is fastened upon the seat of the privy seat and has projections around the central opening. The cams on the neck of the inserted bowl wedge beneath the flanges and fasten the bowl.

Claim.—The water-closet hopper, constructed with a detached flange B, substantially as herein set forth, for the purpose specified, as a new article of manufacture.

No. 60,154.—M. B. DODGE, New York, N. Y.—*Ore Crusher.*—December 4, 1866.—The soft-iron facing will permit hard objects, accidentally introduced, to imbed to some extent, and thus avoid fracture of the jaws.

Claim.—The application of soft wrought-iron faces to the jaws of a quartz crusher, substantially as and for the purpose described.

No. 60,155.—REUBEN W. DREW, Lowell, Mass.—*Turning Lathe.*—December 4, 1866.—Each of the spindle boxes is formed of one piece of metal. The front one is made forwardly flaring within, and the rear one made backwardly flaring outside, and has longitudinal also to allow contraction. The contraction is accomplished by a nut on its front end drawing it further into its frustal bearing. The spindle is tightened in the front box by nuts on its rear end.

Claim.—Making the front box, or bearing B, tapering, in the manner as and for the purpose set forth.

Also, the spindle D, with its check nuts H I, in combination with the slitted box C.

Also, the slitted box C, with its nut *l*, in combination with the cap J and its set screw *j*, for the purpose described.

No. 60,156.—E. F. DRIGGS, Brooklyn, N. Y.—*Sash Holder and Fastener.*—December 4, 1866.—The stile has a ratchet, and the sash a spring bolt operating therewith to keep the window closed, or to any height desired. The bolt is retracted by a thumb knob.

Claim.—In combination with a window sash A, a ratchet C, provided with teeth as described, and a spring dog or pawl D, provided with a thumb piece E, the whole constructed and operating substantially as described and specified.

No. 60,157.—J. F. DUBBER, Brooklyn, N. Y.—*Manufacture of Springs.*—December 4, 1866.—The former is heated sufficiently to draw the temper of the spring, when the latter is adjusted upon it. While the temper is being drawn it is bent to the shape required.

Claim.—Tempering steel springs and adjusting their shape by means of the former herein described, and in the manner set forth.

No. 60,158.—JOHN C. DUNLEVY, Dayton, Ohio.—*Apparatus for Testing Spirits and Preventing Frauds on the Revenue.*—December 4, 1866.—This apparatus has a test tube placed on the pipe proceeding from the condensing worm. By closing the cocks on the branches of the pipe the spouts from the still will fill the tube, when it is agitated so as to form a "bead" when it settles. This test is to be applied from time to time, and the spouts, by opening either cock, are permitted to flow to the high or low wine tanks respectively.

Claim.—The testing apparatus herein described, composed of the vessel C, with the parts *f* and *g*, or their equivalents, in combination with the pipe A, and the stop-cocks D and E, all substantially as and for the purpose set forth and described.

No. 60,159.—GEORGE EICHENSEER, Waterloo, Ill.—*Horse Power.*—December 4, 1866.—The rim of the wheel has a jointed section to allow of the horse passing within for harnessing to the draught arm. The caster-shafts have extensions above, and are formed below to catch the belt in case of its slipping from the wheel. The pivot shaft of the wheel is moved horizontally on the base block by a screw and band-crank to tighten the belt.

Claim.—First, the arrangement of the shaft B, for horizontal adjustment, thereby tightening the driving belt H, substantially as set forth.

Second, the flexible joint of the parts C and D, as set forth.

Third, the arrangement of the drum E, in sections *e*, and their combination with each other and with the arms D, as set forth.

Fourth, the arrangement of the supporting casters F, and their combination with the segment *e*, as set forth.

Fifth, the caster shaft as a guide for the belt H, after it has slipped, substantially as set forth.

No. 60,160.—CHARLES R. EVERSON, Palmyra, N. Y.—*Stove-pipe Damper.*—December 4, 1866; antedated November 23, 1866.—The damper plate has a circular central opening, which, on a semi-rotation, is covered by a plate connected to it by a slide loop. An arched and perforated plate attached to the damper deflects the caloric current to the sides of the stove-pipe.

Claim.—The combination of the arched plate E*, perforated at *d d*, with the annular plate B, valve C, loop D, and shaft E, all constructed and arranged as and for the purposes specified.

No. 60,161.—FRANCIS M. EVRINGHAM, Lafayette, N. Y.—*Wagon Hay Rack.*—December 4, 1866.—A rope from the rear end of the rack is connected to a rope from a windlass at the front end. The windlass is turned by a hand-crank, and has a lever with pawls operating on ratchets to tighten the rope.

Claim.—The lever A, the drum B, the flanges C, the ratchet D, the catches E E, the winch F, and the rope G, when the same are constructed and operated substantially in the manner and for the purpose specified.

No. 60,162.—WILLARD FARNHAM, Janesville, Wis., assignor to himself and SIMEON H. REYNOLDS.—*Burglar Alarm.*—December 4, 1866.—The wedge block is placed with its point under the door, and any attempt to open it trips the tumbler. The cap is then discharged, and the match lit by the face and crown of the hammer.

Claim.—First, the combination and arrangement of the wedge A, base B B, and spurs *a a*, substantially as and for the purpose set forth.

Second, the combination and arrangement of the wedge A, base B B, tumbler C, spring D, hammer E, dog H, and nipple F, substantially as and for the purpose set forth.

Third, the combination and arrangement of the wedge A, base B B, tumbler C, spring D, hammer E, dog H, and match-holder I, substantially as and for the purpose set forth.

No. 60,163.—LUTHER R. FAUGHT, Philadelphia, Penn.—*Turning Lathe.*—December 4, 1866.—The tail centre spindle of the lathe has a sleeve which is slotted and tapering in front, and screw-threaded at the rear; a nut in the stock operates to draw in the sleeve and tighten it on the spindle. The nut is turned by a lever.

Claim.—An improved binder for lathe spindles, formed by combining the nut E and sleeve D, constructed and arranged as herein described, with each other and with the spindle C

No. 60,164.—JAMES C. FITZGERALD, Willet, N. Y.—*Rotary Cultivator.*—December 4, 1866.—The bent pulverizers are placed spirally on their shaft. The shaft is vertically adjusted by a chain connection to the frame, and receives rotary motion from cog wheels upon the main axle. The gearing can be disconnected by the semi-revolution of the cam bearings of the shaft in its frame.

Claim.—The arrangement of the spirally and inclined armed pulverizer H, resting in the eccentric bearings *k*, when said parts are combined with a vertically adjustable frame G, suspended from the main frame and concentric with the axle, as set forth.

Also, in combination with the frames G and C, the draught chains M, and the gauge arms N, operating substantially as and for the purpose specified.

No. 60,165.—THOMAS M. FOSTER, Union Mills, Penn.—*Oil Well.*—December 4, 1866.— The usual gas-pipe sucker rod is furnished with wooden sleeves at the couplings, confined by ferrules and screw collars.

Claim.—The sleeve B, sections A A, couplings *a a*, connecting piece D, and ferrule E, combined and operating substantially as described, as and for the purpose specified.

No. 60,166.—ANDRÉ FOUBERT, New York, N. Y.—*Distilling and Refining Oils, Wines, and other Liquids.*—December 4, 1866.—The refining column above the still has a progressively lower temperature toward the top. The vapors pass in jets through the successive perforated diaphragms in the column, and the portion condensed in the upper coils of the worm is passed back to the head of the column. A succession of pipes and overflow cups conduct downward the descending liquid returned by the pipe from the worm.

Claim.—First, the column *c*, containing the perforated diaphragms *d d*, in combination with the worm or condenser *g*, and pipe *m*, passing back to the column *c*, as and for the purposes set forth.

Second, the flanges or divisions upon the perforated diaphragms to cause the liquid to circulate from the cup *i* to the pipe *e*, in the manner specified.

No. 60,167.—WILLIAM FUZZARD, Chelsea, Mass.—*Machine for Sizing Fibrous Materials.*— December 4, 1866.—The fabric passes over the roller, whose lower surface revolves in contact with the size in the vat, and is thereby sized. It then passes over the heated cylinder by which it is dried.

Claim.—The combination of the size-distributing roller E and the heated cylinder C, arranged relatively with each other to operate substantially in the manner as and for the purpose set forth.

No. 60,168.—JOHN FYE, Hamilton, Ohio.—*Harness Saddle.*—December 4, 1866.—The terrets and check book are connected by means of detachable bases, with dovetailed rabbets therein.

Claim.—First, the combination of the detachable bisected bases with dovetailed rabbets therein, with the turret feet *m m* and saddletree, when constructed and secured together in the manner and for the purpose specified.

Second, the detachable water book with its saddle *s s*, in combination with the saddletree, when constructed and applied in the manner described.

Third, the turrets with their bases, and the water book with its saddle, in combination with the crupper loop and saddletree, when the several parts are constructed and secured together in the manner and for the purposes set forth.

No. 60,169.—CHARLES O. GARDINER, Springfield, Ohio.—*Centring and Squaring Device.*— December 4, 1866.—One end of the shaft to be operated on is placed in the centring chuck and is held and rotated by three converging blades. The opposite head of the lathe has a drill centre and a squaring tool.

Claim.—First the chuck C, provided with the conical sectional rings or wedges *a*, and nuts for centring and holding the drill *e*, as shown and described.

Second, the sectional rings *n* and cap D, in combination with the chuck C and spindle F, arranged and operating as shown and described.

Third, the grooved arm E, attached to the body C, and provided with the standard G', for holding the chisel H, when arranged in connection with the other parts as set forth.

Fourth, the hollow conical chuck A, in combination with the detachable blades or jaws *k*, all constructed and arranged as and for the purpose set forth.

No. 60,170 —WILLIAM L. GEBBY, New Richland, Ohio.—*Cotton Planter.*—December 4, 1866.—The seed slides are carried in a revolving disk, and the seed is forced into the recesses by a flexible arm, which is actuated by a cam on the axle. These slides are reciprocated by springs and fixed cams, so as to alternately project into the hopper for their charge and withdraw it for deposit. The vertical inclination of the tongue to the frame may be changed to suit the team. The ploughs are adjustable as to the depth of furrow thrown.

Claim.—First, the wheel H, droppers *b b*, springs *d d*, inclined planes I and I', and apron 12, constructed and operating substantially as and for the purposes set forth.

Second, the seed box F, constructed substantially as described.

Third, the plunger K, constructed and operated substantially as and for the purposes set forth.

Fourth, the cam L, in combination with the lever j, arms i i, spring o, and plunger K, substantially as and for the purpose set forth.

Fifth, the combination of lever s, arm t, clutch p, pin r, and spring u, constructed and operating substantially as and for the purpose set forth.

Sixth, the ploughs N N, bars R R, arms S S and T T, and rods z and y, in combination with the lever 2', and arm 2, substantially as described.

Seventh, the shaft U, arm 2, lever 2', rods 4 4, books 6 6, and beams P P, when used in combination, substantially as and for the purpose set forth.

Eighth, the bar 7, in combination with the hounds 8 8, for the purpose set forth.

No. 60,171.—RILEY JAMES GILBERT, Hanover, Wis.—*Gate.*—December 4, 1866.—The rollers at each end of the gate run on trucks upon pivoted bars, which counterbalance each other, and which may be moved by a lever so as to incline either way, and cause the gate to open or shut. The track bar of the latch end of the gate is overhead, and projecting past the post, has a pivoted connection to the other track bar.

Claim.—First, the carrying beams B and C, when constructed, arranged, and used with or without the side cap r, substantially as and for the purpose set forth.

Second, the combination and arrangement of the handles N, levers L, swinging fulcrums M, carrying beams B and C, connecting bar D, gate A, parts F E R, and cross beam I, substantially and for the purpose set forth.

No. 60,172.—CARLOS GLIDDEN, Milwaukee, Wis.—*Plough.*—December 4, 1866.— Explained by the claim.

Claim.—Coating or covering with porcelain or silicious enamelings, substantially as herein set forth, such portions of the metal surfaces of ploughs and other ground-preparing or cultivating and planting implements as come in contact with the earth.

No. 60,173.—GOTTLIEB F. GOETZE, New York, N. Y.—*Machine for Ornamenting Mouldings.*—December 4, 1866.—The strips of moulding are passed between slides under the pattern roller, which is supplied with paint, ink, or gum, by another roller having a face complementary thereto. The strips are moved foward by a serrated wheel.

Claim.—The arrangement of the pattern wheel D, roller F, springs j and l, adjustable plate E, carriage G, slide H, screws n and o, and clamp p, combined and operating in the manner and for the purpose herein specified.

No. 60,174.—E. C. GORDON, Sevastopol, Ind.—*Fence.*—December 4, 1866.—The panel posts are erected upon blocks of burnt clay and supported by wires anchored in the ground. The wire has a turn around a pin which connects the panels together. A wedge driven beneath the pin and wire draws the latter taut.

Claim.—First, the combination of the wire guys D, keys C, wedges F, and anchoring blocks E, with each other and with the panels A of the fence, when said guys, keys, wedges, and blocks, are constructed and arranged substantially as herein described and for the purpose set forth.

Second, the combination of the supporting blocks B, with the panels A, of the fence, when said blocks are constructed substantially as herein described and for the purpose set forth.

No. 60,175.—G. GRAEZZLE, Hamilton, Ohio, assignor to JACOB ROOP and STEPHEN HUGHES.—*Brick Machine.*—December 4, 1866.—The brick earth is fed in at the top of the curb and is tempered by the revolving arms on the shaft and a fixed arm on the curb. The tempered earth is forced through holes in the curb bottom by inclined plates revolving with the shaft, and is received by mould boxes. The knife is carried on a frame that is kept in place by a wooden pin which allows it to give way before a stone.

Claim.—First, the arrangement of the frame L and knife L', in relation to the moulds G, when carried upon an endless apron H; substantially as set forth.

Second, the combination of an endless apron H, having cleats I', with the revolving arms I, when constructed and arranged substantially as and for the purpose set forth.

Third, the combination of the endless aprons H and moulds G, when respectively constructed and arranged substantially as set forth.

Fourth, the openings A', when constructed with adjustable plates K, operated substantially in the manner and for the purpose set forth.

No. 60,176.—J. H. GRAY, Boston, Mass.—*Machine for Picking Millstones.*—December 4, 1866.—The pick helve is pivoted on a slide rod which is journaled in two heads. These have adjustment by two traversing screws having sprocket wheels moved by an endless chain. The pick is set by a groove in a projection from the base plate.

Claim.—A machine for the purpose specified, so arranged and organized that while the pick is guided and controlled as to the direction of its movement, each blow is effected and its force controlled by the operative, substantially as described.

Also, the arrangement and organization of a machine for picking millstones, complete and independent in itself, and fitted to be moved over and to operate anywhere upon the surface of such a stone, to produce lines in any direction without attachment to the millstone centre or spindle, substantially as described.

Also, the flange *j*, when combined with the matter forming the second claiming clause herein, and arranged as and for the purpose specified.

Also, the combination with a frame *a a i q* of the two screws *b b*, and means for simultaneous and equal rotation thereof, when said screws are provided with nuts *f f*, arranged to move a shaft *g*, which bears a pick helve.

Also, the combination with the shaft *g* of the conical sleeve *s* thereon, and the slotted hammer helve.

Also, the combination with a millstone picking machine of a pick set *r* or guide, for the purpose described.

Also, the means described for securing the pick in position and to the pick helve.

No. 60,177.—THOMAS GRAY, Wandsworth, England.—*Treating Hemp, Flax, &c.*—December 4, 1866.—Explained by the claim.

Claim.—The new and useful and improved method of treating flax, hemp, grasses, and other like fibrous substances for manufacturing and useful purposes, in removing the bark or skin and resinous or gummy mucilage and the boon or woody fibres of flax or other like plants while in a wet state, and in neutralizing the alkaline matter left in the fibre previous to bleaching, in the manner hereinbefore set forth, and in bleaching the same with a combination of bleaching liquor and alkaline saponified fat or oil, or with an alkaline solution without the fat or oil, so that the fibres after the process of bleaching is completed are rendered stronger than they were in their natural or original state; and, also, the permeating of the fibre with saponified fat or oil, as herein set forth.

No. 60,178.—ALFRED GWYNNE, New York, N. Y.—*Pipe Coupling.*—December 4, 1866.—The pipe which enters the socket has at its extremity an annular flange, between which and the shoulder at the bottom of the socket is a ring of elastic packing. The two sections are clamped together by screws which pass in an inclined direction through the sleeve or wall of the socket and abut against the annular flange.

Claim.—First, the method of securing or fastening together the ends of water and other pipes by means of the screws 1 1 1, or their equivalents, arranged and operating substantially as and for the purposes set forth.

Second, in combination with such method of fastening the ends of such pipes, the use of an elastic ring or packing, substantially as and for the purposes set forth.

No. 60,179.—FRANÇOIS HAECK, Brussels, Belgium.—*Imparting Age to Wines.*—December 4, 1866.—The wine is placed in a receptacle over a water chamber heated by steam beneath it. The wine chamber has a vertical stirring shaft, and escape pipes for the vapors condensed on the lid, and for the expanded air. The wine is gradually heated to from 113° to 190° Fahrenheit.

Claim.—First, the treatment, substantially as herein described, of wines, spirits, and other distilled liquors, by subjecting them to heat or heat and agitation combined, when the same is effected in a close vessel, or chamber D, gradually heated in the manner described, or in any other equivalent way, and the condensed vapors collected at or near the top of the wine chamber and run off therefrom, essentially as and for the purpose set forth.

Second, gradually heating the vessel or chamber containing the wine or distilled liquor to be treated, by means of steam and water combined, substantially as specified.

Third, the employment within the evaporating chamber D of a stirrer E, in combination with a suitable heating device below said chamber, and condensed vapor collecting channel or receptacle at or near the top thereof, essentially as herein set forth.

No. 60,180.—SAMUEL C. HALL, White Water, Wis.—*Machine for Cutting Sickle Sections.*—December 4, 1866.—The blank is attached by a lever cam to a table, which has sliding movement on a frame, adjustable to any incline desired upon the front of the main frame. The blank is forwarded beneath the cutter by a belt from the main shaft, which connects with a screw rod traversing its supporting table. The cutter is raised by a cam on the main shaft, and its descent may be arrested by freeing a catch lever which connects with a treadle.

Claim.—First, an improved machine, which is adapted for sustaining sickle sections beneath a reciprocating chisel, in such positions that the chisel will form serrations or teeth upon the bevelled edges of said sections, constructed substantially as described.

Second, the combination of an adjustable table I with and adjustable way C, a feeding screw K, a half nut connection S', and a file cutting chisel, or a chisel which is adapted for cutting teeth upon sickle sections, substantially as described.

Third, the construction, substantially as described, of the adjustable way C for receiving and supporting a reciprocating table, and admitting of said table being adjusted at different angles with respect to the cutting edge of a chisel C', substantially as described.

Fourth, the arrangement of the adjustable table I upon the adjustable way C, so as to move at right angles to the chisel arm B beneath the chisel C', substantially as described.

Fifth, sustaining the adjustable table I upon the way C, by means of adjustable bearings, substantially as described.

Sixth, in combination with the adjustable table I and an adjustable way C, the means substantially as described for feeding the said table with a fast or slow movement, according to the size of the teeth required upon the bevelled edges of the sickle sections, substantially as specified.

Seventh, constructing the upper portion of the frame A with a vertical slot through it for receiving and guiding the chisel arm B', and also for receiving a spring G', upon which said arm strikes in its descending strokes, said slot being arranged directly over the axis of the way C, substantially as described.

Eighth, the combination of the spring latch H' and treadle I' with the means herein described for cutting sickle sections or files, substantially as and for the purpose set forth.

Ninth, the sliding adjustable table I with its clamp or clamps N, its screws L and M, nut S', feed screw K, in combination with the adjustable way of a machine for cutting sickle sections or files, substantially as described.

Tenth, the adjustable way C, in combination with the frame A of the machine, both being constructed and arranged substantially as and for the purposes described.

No. 60,181.—THOMAS HANDY, Decatur, Ill.—*Grinding Metal Plates.*—December 4, 1866; antedated November 23, 1866.—On the grindstone frame are longitudinal slide-bars for a carriage, which has transverse ways for the clamping frame. The plate to be ground is adjusted to gauges at one side of the stone and clamped. It is then brought in contact with the stone.

Claim.—First, the combination of the carriage E, carriage G, ways D, screws F, clamp H, and bars e f, arranged and operating in the manner and for the purpose herein specified.

Second, the gauges composed of the bars e f applied to the carriage G, substantially as and for the purpose set forth.

No. 60,182.—THEODORE HARCOURT, Indianapolis, Ind.—*Skate Fastener.*—December 4, 1866.—The bar attached to the recurved rear end of the runner has projections on each side of the heel to hold the rings for the attachment of the instep straps.

Claim.—First, the heel plate C, with the grooved ends C C, section I, made to the rings E E of the instep strap F, as represented in the accompanying drawings.

Second, the combination of the heel plate C when attached to a skate runner and plate, as represented in section I and the drawings.

Third, the combination of the instep strap F, the back strap M, and the rings E E, when constructed and used for the purposes substantially as set forth.

No. 60,183.—GEORGE W. HARRIS, Elizabeth, N. J.—*Gas Retort.*—December 4, 1866.—A superheating chamber beneath the clay retort is connected therewith by a pipe which discharges the steam into the coal within the retort.

Claim.—The combination of the fire-clay retort A and superheater B, when arranged as herein specified, to inject the superheated steam at the bottom of the bed of incandescent coal.

No. 60,184.—SANDY HARRIS, Philadelphia, Penn., assignor to C. THORNTON MURPHY, same place.—*Balance Weight.*—December 4, 1866.—The graduated tri-lateral rectangular frames are pivoted by their ends to the beam, so that by a change of position of the frames on their pivots any weight between the smallest and the aggregate denoted can be determined on the scale.

Claim.—Combining the lever weights with the scale beam in the manner described.

No. 60,185.—G. W. HATCH, Parkman, Ohio.—*Machine for Gathering and Loading Flax, &c.*—December 4, 1866.—The hay is taken up by a rake and carried to the elevator by teeth upon a roller, which is connected to a wheel by a pawl and ratchet. The teeth with their connecting strip are thrown back at the proper time to release the hay.

Claim.—First, the lag g hinged to the roller N, in combination with the elevator I and roller M, for the purpose and in the manner set forth.

Second, the right-angled lever e and cord H, in combination with the lever F, for the purpose and in the manner as substantially described.

No. 60,186.—R. HATHAWAY, Chicopee, Mass.—*Surface and Depth Gauge.*—December 4, 1866.—The gauge pin traverses the head of an arbor having rotary movement in a horizontal socket of the base piece. The arbor and the pin may be fixed in their sockets by set screws.

Claim.—The holder A, or its equivalent, in combination with the gauge rod B hung in one end of a spindle G, and set screw F, and thumb nut J, respectively, for the said gauge rod B and spindle G, when combined and arranged together substantially as and for the purpose described.

No. 60,187.—JOHN W. HEDENBERG, Chicago, Ill.—*Cotton Tie.*—December 4, 1866.—The end of the band is doubled to make a foundation for the attachment of the keys; these have lateral extensions to enter enlargements in the slots of the other end of the band, and guide the key to its place. They can only be dislodged by slackening and torsion of the band.

Claim.—A cotton tie, with one or more keys C, so made and arranged as to require a lateral movement to clasp and to unclasp it, constructed and operated in the manner herein escribed.

No. 60,188.—GRANVILLE HENRY, Nazareth, Penn.—*Fitting Lock-plate to Stock of Fire-arms.*—December 4, 1866.—The rear end of the barrel is attached to a metal frame containing the lock, and let into a notch, so as to be flush with the right side of the stock.

Claim.—The stock D, made in one piece, and cut out at G, in combination with the frame C, having a shoulder E, substantially as and for the purpose described.

No. 60,189.—FRANK A. HILL, Marysville, Cal.—*Seeding Machine.*—December 4, 1866.—The hopper is mounted upon wheels, whose bevel gearing is connected to agitators placed in the bottom of the hopper; the grain falls through the holes upon an angular scatter-board, by which it is spread.

Claim.—The agitators D D, in combination with the bevelled gearing B and C, and connecting rods E E, substantially as described and for the purpose set forth.

No. 60,190.—NOAH P. HOLMES, Indianapolis, Ind.—*Apparatus for Preserving Milk.*—December 4, 1866.—This is an improvement on his patent of September 4, 1866. The case has an annular ice space outside, and water space next. The milk chambers may be divided at mid-height, and an ice chamber interposed. The lids are fitted to contain ice. Any of the chambers may be emptied by cocks connecting to the outside.

Claim.—In combination with the external and internal ice chambers 6 and 8, the use of horizontal partitions for the subdivision of the preserving chamber into compartments for various uses, the entire apparatus being constructed substantially in the manner and for the purpose set forth.

No. 60,191.—EDWIN B. HORN, Boston, Mass.—*Safe.*—December 4, 1866.—An inner cylinder revolves in the outer one. The cylinders have separate locks, and another lock prevents the said revolution. The bottom is perforated to allow free exit to the gases when powder is introduced between the cylinders and exploded.

Claim.—First, as an article of manufacture, a safe made substantially as described.

Second, the holes or perforations in the space between the inner and outer safe, so as to afford an outlet for explosive material, substantially as described and for the purpose set forth.

Third, the wing H, in combination with the lock L″ and the loop M, substantially as described and for the purpose set forth.

No. 60,192.—REUBEN K. HUNTOON, Boston, Mass.—*Steam Engine Governor.*—December 4, 1866.—The governor valve is operated by revolution of the spiral blades on the vertical shaft within a closed cistern, with a stop-cock for occasional discharge. The shaft is stopped in a box which extends upward therein, so far as to allow necessary play.

Claim.—The combination of the bearing C, its passage c and stop-cock d, with the shaft B and the propeller D, and the cistern or vessel A, arranged as and for the purpose set forth.

Also, the combination and arrangement of the deflector e with the cistern A, the shaft B, and the propeller D, arranged as and for the purpose set forth.

Also, the combination and arrangement of the wings a with the cistern A, the shaft B, and propeller D, arranged as set forth.

No. 60,193.—RALPH S. JENNINGS, New York, N. Y., assignor to himself and N. G. KELLOGG.—*Metallic Seal Fastener.*—December 4, 1866.—Explained by the claim and cut.

Claim.—The implement, as a new article of manufacture, for sealing metallic seal envelopes, which is constructed with a spreading portion c, and a closing or riveting portion d, substantially as described.

No. 60,194.—JOHN M. JOHNSON, New York, N. Y.—*Fastening for Buttons.*—December 4, 1866.—The shank has a pointed T-head, one end of which is turned down as a pin. This head enters a slot in the collet, and a partial revolution attaches them together. The button is detached by depression and a reverse movement.

Claim.—A button whose collet is provided with a slot to admit the head of the shank, said collet having upon its inner side and at an angle to the slot a depression on one side and a hole on the other, or a hole upon each, in combination with a spear or T-shaped shank or stud, whose transverse end or ends are provided with a spur at right angles thereto, with or without a spring J, substantially in the manner and for the purpose described.

No. 60,195.—A. C. KASSON, Milwaukee, Wis., assignor to himself and NELSON C. GRIDLEY.—*Hoe.*—December 4, 1866.—The lower edge of the hoe is obtusely recessed from the corners to the centre, and the sides curved toward the handle.

Claim.—A hoe made substantially as herein shown and described, that is to say, constructing the edge proper at an angle, and setting the blade at such angle relatively with the handle that the two opposite sides of the hoe will operate upon the earth, substantially as and for the purpose specified.

No. 60,196.—WENDALL R. KING, Chicago, Ill.—*Baling Press.*—December 4, 1866; antedated August 28, 1866.—The followers of the two divisions of the press are on the opposite ends of the press screw, so that while one is retracted the other is advanced. The crank shaft has sliding adjustment by which different gear wheels are brought in connection, to increase either the speed or the power.

Claim.—The combination of the gearing J F and G I with the screw C and boxes A, when constructed and operating substantially as described.

No. 60,197.—EDWARD H. KNIGHT, Washington, D. C., assignor to IGNATIUS RICE, New York, N. Y.—*Comb.*—December 4, 1866.—The long bowed comb has a strip of metal inserted in a longitudinal groove in its stock fastened by rivets or attached outside and fastened by small bands passed over the edge of the comb; the strip when applied exteriorly is lapped around the ends.

Claim.—First, the combination with a comb of a strip of metal imbedded or inserted in the back of the comb, substantially as herein above set forth.

Second, the combination with a strip of metal A, and the back of a comb, of the bands or hooks D, substantially as and to the effect set forth.

Third, returning the strip of metal A over the ends of the comb, substantially as herein above set forth.

No. 60,198.—JOHN KNOX, Auburn, N. Y.—*Boot and Shoe Iron.*—December 4, 1866.—The space between the back and front guard is adjustable by moving the front guard on the bed piece. The tool has two sides fitted for pegged and sewed work, respectively.

Claim.—First, adjusting the space between the back and front guards, substantially in the manner and for the purpose set forth.

Second, the combination of the short back guard and the long front guard with the long back guard and short front guard, substantially in the manner and for the purpose above specified.

Third, in fore part irons holding the front guards in their place by means of a screw operated through the handle, as above set forth.

No. 60,199.—GEORGE T. LAPE, Summit, N. Y.—*Cast-iron Arch for Bridges, Vaults, &c.*—December 4, 1866.—The voussoirs have an upper curved plate with a central downwardly projecting rib, and an end plate upon one end. The length of end plate is double the breadth of the upper plate, and laps past the contiguous voussoir on one side, forming a connection and bearing. The ends are connected by pins which traverse holes in them and are retained by wedge-formed keys

Claim.—A cast-iron voussoir for the construction of arches and vaults for bridges, subterranean railroads, and similar purposes, formed of a top plate *a*, a rib or stem *b*, and abutting ends *c c*, and fastened with bolts *e*, substantially as herein described.

No. 60,200.—JOHN H. LATIMER, Crystal Lake, Ill.—*Hand Seed Planter.*—December 4, 1866.—The perforation is made in the ground by converging plates, one of which is hinged and backed by a spring to allow of the issuing of the hand staff and passage of seed. An adjustable stop-plate near the point regulates the depth of deposit. The hand staff contains the seed cavities, which are presented alternately to the hoppers and the place of exit as the implement is raised or struck into the soil.

Claim.—First, a seed planter provided with one or two chambers B C and a chamber L, provided with a hinged bottom G, and a spring S, or its equivalent, arranged with a slide D, provided with one or more seed cavities *m c*, operating substantially in the manner and for the purposes specified.

Second, in combination with the above, the arrangement of the gauges *d f*, and the point *e*, as and for the purposes specified.

Third, providing the chamber L with the hinged bottom G and a spring S, substantially as and for the purposes set forth.

No. 60,201.—JOHN E. LAUER, New York, N. Y.—*Substitute for Yeast for Baking Purposes.*—December 4, 1866.—Bone dust is heated with hydrochloric acid, and the resulting solution evaporated to dryness. The mass thus obtained is pulverized and put up in packages for sale. Before mixing with the flour for use, bicarbonate of soda is added.

Claim.—First, the preparation of muriate of phosphate of lime herein described.

Second, the mixture of the above-described preparation of muriate of phosphate of lime with an alkaline carbonate, as a substitute for yeast in raising bread.

No. 60,202.—WILLIAM LEACH and JOSEPH LEACH, Stewartsville, Ind.—*Hand Spinning Machine.*—December 4, 1866.—The machine has three pairs of drawing rollers with interposed twisters, so that the threads are drawn out by a continuous operation. The winding spool has a vertical reciprocating motion to distribute the layers of yarn evenly. It is operated by band crank or treadle.

Claim.—First, as an improvement in a hand spinning machine, the arrangement of two sets of twisters to each thread, each twister having three or more grooves for the purpose of varying the amount of twist to suit different kinds of wool, and the same being placed immediately behind the front and middle rollers so as to retain the twist close up to said rollers, in the manner and for the purpose specified.

Second, also raising and lowering the spools J, by means of the platform K, the connecting rods P and M, the levers O L, and the rock shaft H, substantially as described.

No. 60,203.—HENRY J. LEASURE and JAMES S. GILL, Wheeling, W. Va.—*Cooling Glass Press.*—December 4, 1866.—A stream of water is forced through the plunger to prevent overheating.

Claim.—Cooling the plunger of a glass press with water or other liquid or atmospheric air, substantially as herein shown and described.

No. 60,204.—JAMES H. LEE, Charlestown, Mass.—*Coffee Pot.*—December 4, 1866; antedated November 22, 1866.—The water is put in the boiler and the coffee in the annular space above the wire gauze. The steam formed in the boiler forces the water up the central tube and through the ground coffee into the pot. The pot may then be removed from the boiler.

Claim.—The combination and arrangement of the tubes B and H, the boiler A, the coffee pot D, and the vessel F.

Also, the combination and arrangement of the tubes B H and E, the boiler A, the coffee pot D, and the vessel F.

Also, the combination of the safety bell cover G, and its spring m, the cover C, the coffee holder F, the pipes H E B, or their equivalents, the coffee pot D, and the boiler A.

Also, the combination and arrangement of the seat b, one or more openings d, the flexible cap or washer a, and the screws f g, with the coffee pot D, in boiler A, and the tube B, applied to such boiler and opening into it, as specified.

No. 60,205.—O. G. LEOPOLD, Cincinnati, O.—*Bridge.*—December 4, 1866.—The bridge is constructed of plate and angle iron, with string pieces of flat iron beneath. It is so formed that a cross section of the bridge or of each girder shall present a double T-form. The upper longitudinal pieces may be made hollow, and the upper rib of the girder of hollow wood plated outside. Diagonal brace rods connect the longitudinal parts together at their intersections with the girders.

Claim.—First, the general arrangement and combination of wrought angle iron and plate in a bridge, so as to present in the cross section of the girders and all other essential parts of the bridge the double T-form, substantially as described.

Second, the arrangement and adaptation of the bar D to the central rib a³, in such a manner that the platform or the roadway of the bridge shall be located at or near the line of stability or neutral line of the girder, substantially as described.

Third, and in combination with the above the lateral bracing for the support of the road way, substantially as described.

Fourth, making, in the combination of the double T of the bridge girder, the upper stringer or head of the same of either a flat bar or a hollow tube of any form, substantially as described.

Fifth, making the sills or cross-ties of the roadway of hollow wooden beams instead of solid timber, and covering the same with metal plate, substantially as described.

No. 60,206.—ELISHA W. LEWIS, Philadelphia, Penn.—*Core Box Plane.*—December 4, 1866.—The main frame has a side guide and carries a cutter which may be revolved with its shaft to form a semicircular groove.

Claim.—In combination with the stock A, the rotative tool holder E, carrying a transversely-adjustable cutter F, to which a circular feed motion is given by means of the worm I, and worm wheel H, or in any other equivalent manner.

No. 60,207.—HENRY C. LEWIS, Essex, Conn.—*Countersink Bit.*—December 4, 1866.—The bit shank has a cutter which makes a countersink for the screwhead.

Claim.—A countersink bit constructed in the manner herein described, and so as to operate as and for the purpose specified.

No. 60,208.—SYLVANUS D. LOCKE, Janesville, Wis.—*Grain Binder.*—December 4, 1866.—The binding wire is extended along two arms, which approach each other around the sheaf; the ends of the wire are then carried down a slot in a pulley, which, by its revolution, twists the wire. The ends of the wire are then cut by a knife moved automatically, and the arms again extended.

Claim.—First, the combination of a revolving twisting or tying device, and a reciprocating toothed rack with a vibrating driving arm, substantially as set forth.

Second, the combination of a reciprocating toothed rack and a vibrating driving arm, substantially as set forth.

Third, the combination of a revolving twisting or tying device with a scroll spring in such a manner that the former is returned to its original position by the latter, substantially as set forth.

Fourth, the combination of a reciprocating toothed rack with a scroll spring, in such a manner that each is alternately operated by the other, substantially as set forth.

Fifth, the combination of a revolving twisting or tying device and a scroll spring with a reciprocating cutting device, substantially as set forth.

No. 60,209.—JOHN J. LOOK, Farmington, Maine.—*Wagon Brake.*—December 4, 1866.— The brake bar is attached to the tongue, which has longitudinal play, so that a backward movement of the tongue locks both wheels, but a lateral movement only locks the pivot wheel.

Claim.—The pole C, provided with the tapering enlargement c', as described, in combination with the slotted hounds D, brake bar E, lateral braces F F, and axle A, when the parts are so arranged that by a lateral movement of the pole one of the shoes only is brought to bear on its corresponding wheel, substantially in the manner and for the purpose set forth.

No. 60,210.—J. LUTHER and A. MARSH, Worcester, Mass.—*Window-blind Fastening.*— December 4, 1866.—The catch spring is clamped into its groove by one of the attaching screws of the plate and serves, by catching over the pins, to hold the shutter either open or closed.

Claim.—First, the peculiar formation of spring C, as shown and for the purposes stated.

Second, making a blind fastening of two pieces, a spring and main or bed piece, when the latter is constructed as described, so that it will receive and arrest the blind, when opening or closing the same, without injury to the spring.

Third, the combination with the main or bed piece A, and spring C, of screw a, as shown and described, whereby the screw serves to hold the spring in place, and also answers as a fastening to the blind.

No. 60,211.—EDWARD T. C. LUTTON, Philadelphia, Penn.—*Machine for Shearing Twisted Strands of Wool, &c.*—December 4, 1866.—The yarn is presented lengthwise to the knives, and bears against their approaching edges, being at the same time rapidly twisted, so that its whole surface may be equally sheared.

Claim.—Knives operating substantially as described in combination with the devices herein set forth, or their equivalents, for so guiding and turning twisted strands of fibrous material, that the same may be shorn by the said knives.

No. 60,212.—SEBEUS C. MAINE, Boston, Mass.—*Apparatus for Cooling and Disinfecting.*— December 4, 1866 ; antedated November 22, 1866.—A sheet of porous cloth is passed through a disinfecting solution, and a current of air driven through the saturated cloth to carry the vapor therefrom into the room.

Claim.—The employment of cloth or equivalent porous material for receiving and carrying the disinfecting and cooling liquid through or in contact with a current of air produced by a fan H, or equivalent device, substantially as described.

No. 60,213.—GEORGE MALLORY, Bridgeport, Conn.—*Spring for Hat Brims.*—December 4, 1866.—Explained by the claim and cut.

Claim.—A hat, the brim of which is distended by a covered spring, having enclosed within its covering a cord or an equivalent therefor, substantially as herein set forth.

No. 60,214.—PHILO MALTBY, Kent, Ohio.—*Expanding Cylinder.*—December 4, 1866.— The longitudinal chuck bars are carried on radial arms, which are adjustable in their sockets by means of inclined portions of the supporting centre bars. These centre bars have longitudinal movement in a hollow, concentric cylinder, and are secured to any adjustment by jam nuts.

Claim.—First, the grooved centres H, in combination with the cylinder A and nuts M, substantially as and for the purpose set forth.

Second, the cylinder A and sockets B, in combination with the arms C and bars D, substantially as and for the purpose described.

Third, the grooved centres H and slides F, in combination with the screws E, sockets B, and arms C, substantially as and for the purpose specified.

Fourth, the gib N and centres H, in combination with the set screws O and cylinder A, as and for the purpose set forth.

No. 60,215.—J. J. MARCY, West Meriden, Conn., assignor to EDWARD MILLER, same place.—*Lamp Burner.*—December 4, 1866.—The wick tube has an enveloping outer tube ascending above its top, and perforated with air-holes.

Claim.—First, the combination of the tube A and the case D, when the said case D is enlarged at its mouth, and the said outer case D, constructed with perforations *d*, met its base, substantially as and for the purpose specified.

Second, the combination of the wick tube A and case D, when the said case D is perforated near its mouth and above the wick tube, substantially as and for the purpose specified.

No. 60,216.—JAMES B. MARTINDALE, New Castle, Ind.—*Carpet Stretcher.*—December 4, 1866.—Explained by the claim and cut.

Claim.—A carpet stretcher made with a roller covered with India-rubber with or without a roughened surface, the rolling pressure of which roller is regulated at pleasure, by means of the brake B, substantially as set forth.

No. 60,217.—PAUL FRANÇOIS MAUVAS, New York, N. Y.—*Button.*—December 4, 1866.—The shank of the button head penetrates the outer lap of cloth, and is attached by a perforated disk, which has gains to allow the passage of projections on the shank. A slight turn secures it by removing the projections from the gains. The pin is attached to the inner lap of cloth in the same manner as the head. In buttoning, the pin passes through an opening in the under side of the head, and is secured by a spring plate which has a slot to embrace the pin below its flanged head. A thumb pin releases the attachment.

Claim.—First, the combination with a button shank provided at or near its end, with one or more laterally projecting teeth or studs, of a metal or other plate or equivalent device slotted and perforated as herein described, so that it may be adjusted to or removed from said shank, as and for the purpose set forth.

Second, the combination with a button provided with a tubular shank and locking mechanism as described, of the fastening or buttoning device herein described, the same consisting of a plate provided with a shank secured to the cloth or other material in the manner above indicated, and with a stem grooved or flanged at its upper end so as to engage with the locking mechanism of the button, substantially as herein shown and set forth.

No. 60,218.—THOMAS B. MCCONAUGHEY, Newark, Del.—*Corn Planter.*—December 4, 1866.—This is an improvement on his patent of March 27, 1860. It is explained by the claim and cut.

Claim.—The slide or part D of the corn planter provided with an oblique opening F, with a recess *e* above it for the purpose of stirring or agitating the seed and insuring the filling of F, as set forth, when this is combined with the attachment of the said slide to the part A, by means of the band F, and the motions of said slide are limited by the stops *b* and *c*, as described.

No. 60,219.—JAMES T. MCDOUGALL, San Francisco, Cal.—*Furnace.*—December 4, 1866.—The hearth is divided into two compartments by means of a bridge—the one for smelting, the other for refining. Fuel is fed into the fireplace by a cylinder with a longitudinal opening emptying into a passage on its semi-revolution. The slag from the furnace is removed through doorways. The furnace is set on rockers and is inclined, to aid the discharge of metal through side openings. The metal is introduced by slide hoppers carried on bars. A plate is introduced between the furnace and chimney as an abutment to the former in rocking.

Claim.—First, a smelting hearth of peculiar construction A B, the sloping portion A inclining toward C, its lower portion forming the dam wall or ridge B, running across the hearth of the furnace from side to side, substantially as described and for the purposes set forth.

Second, the half-oval-shaped refining hearth C, conforming in shape to the smelting hearth, where they join, the sole of which has a slight inclination from the flue D toward B, where it has a lower level than the smelting hearth A, for the purposes specified and set forth.

Third, the manner of feeding the flupes and ores to the furnace by the use of the hoppers V V and grooved bars or rods W W, substantially as described.

Fourth, the devices for feeding the fuel to the furnace and depriving it of its moisture by the use of the cylinder N and conducting pipe *c*, or their equivalents, as herein specified and shown.

Fifth, the arrangement of the door hearths H H, for discharging the metal and slag, substantially as described.

Sixth, the manner of binding the said furnace with bands of iron secured to the casing of the furnace and keyed below it, when arranged substantially as described and for the purpose set forth.

Seventh, the concave rockers R R R and convex rails S S S, with chimney shield L and lever sockets *d d*.

Lastly, the within-described improvements, whether employed singly or in combination, in smelting furnaces, substantially as and for the purposes herein specified.

No. 60,220.—WILLIAM T. MCMILLEN, Cincinnati, Ohio.—*Fireplace.*—December 4, 1866.—The back plate of the fireplace has an encompassing jacket, forming an air chamber, and pipes communicating with this chamber for the introduction of cold and exit of hot air.

Claim.—The combination of the deflectors K K with the chamber C, pipes F F', caliducts D E, and flue H, all constructed and arranged as and for the purposes set forth.

No. 60,221.—CHARLES MESSENGER, Chicago, Ill.—*Broom Head.*—December 4, 1866.— The corn is inserted in the perforations of the plate and drawn into the case. Hooks in the plate serve to hold in position the lower rim of the case. A screw socket on the handle has a nut to keep the case in position.

Claim.—The socket A, nut E, and handle A', perforated plate C, arranged in combination with the hooks b and case D, for the purpose and in the manner as specified.

No. 60,222.—J. M. MILLER, New York, N. Y.—*Steam Engine Condenser.*—December 4, 1866.—The exhaust steam chambers are flat in form, and have pipes projecting from their sides. The pipes have closed ends which are rebent inward, like the bottom of a wine bottle, the return extending almost to the communication between the pipe and the chamber. A set of these steam chambers are enclosed in a case. The water of condensation passes from the steam chambers through a coil of pipe in the case to another coil of pipe in a lower chamber furnished with cold water; from this lower coil it is pumped into the main case, and from thence to the boiler. The case has a safety-valve.

Claim.—First, the combination of apparatus for transferring the water of condensation to the boiler highly heated by passing it through the case containing the condenser, said apparatus consisting of the vacuum pump case *e* and pump *n*, that forces the water from the case *e* into the boiler, substantially as and for the purposes set forth.

Also, arranging the coils of pipe *h* within the case *e*, above the base plate *f*, and in combination with the chambers *a*, as specified.

Also, the conjoint arrangement of pipes *h* and pipes *k*, substantially as herein described.

Also, the combination of the chambers *a b*, case *e*, vacuum and feed pumps *l* and *n*, arranged and combined substantially as and for the purposes herein made known.

No. 60,223.—OSCAR F. MORRILL, Chelsea, Mass.—*Faucet.*—December 4, 1866; antedated November 21, 1866.—The diaphragm packing beneath the threaded stem is made of metal. The valve is raised by a spring plate, which acts as a guide.

Claim.—In combination with the metal valve *e*, the metal diaphragm packing *h*, extending over the faucet chamber and held down by the screw cap *g*, substantially as described.

Also, in combination with the diaphragm *h* and valve *e*, the spring lifter *l*, operating to raise the valve from its seat as the follower is unscrewed, said spring being provided with legs or projections which serve to keep the valve in central position with respect to its seat, substantially as described.

Also, the relative arrangement of the valve *e*, diaphragm *h*, spring *l*, shoe *o*, and follower *i*, to effect the raising of the valve from its seat and its closing thereupon, substantially as set forth.

No. 60,224.—OSCAR F. MORRILL, Chelsea, Mass.—*Hydro-carbon Heating Apparatus.*— December 4, 1866.—The plates of metal near the flame are all insulated from the plates communicating with the oil. The annular wick space can be diminished on either side to regulate the length of flame.

Claim.—Giving to the metallic casing *d*, surrounding and protecting the filling *c*, the spheroidal form shown, for the purpose of obtaining large radiating surface from which to dissipate the heat received from the chimney *e*.

Also, connecting the metallic covering *d*, by the filling *c*, with the outer wick tube, when the spaces 1 and 2 intervene, so as to cut off metallic connection, substantially as and for the purpose specified.

Also, connecting the tube *f* with the inner wick tube *b*, by filling in such a manner as to cut off metallic connection by the spaces 3 and 4, substantially as described and for the purpose set forth.

Also, regulating irregularities of the length of the flame from the wick, by contracting the flue opening on one side and increasing it on the other, substantially as and for the purpose specified.

No. 60,225.—L. B. MORRIS, Hopkinsville, Ky.—*Pruning Knife.*—December 4, 1866.— The two sides and the blade are pivoted as usual, the other ends being free. To open the knife the two parts of the handle are swung around in opposite directions until lugs on them engage projections on the blade, in which case they coincide, and are held by a catch. A reverse movement closes the knife, and the catch is again made fast.

Claim.—A knife having its blade provided with the notches or recesses *l* and *d*, and a handle having the projections *i* and *n* arranged to fit therein, as shown and described.

No. 60,226.—CHARLES MULCHAHEY, Springfield, Mass.—*Blacking Brush.*—December 4, 1866; antedated November 25, 1866.—An ordinary blacking box is secured by clamps in a tin case upon the back of the polishing brush. The spreading brush fits over the case and serves as a cover, by means of a tin rim upon its under side.

Claim.—First, attaching to the top of an ordinary polishing brush for boots, shoes, &c., a blacking box fitted in a case B, formed to receive it.

Second, attaching to this box, or to the case which is formed to receive it, the smaller spreading brush *c*, by means of a rim or cover, arranged as shown and described.

No. 60,227.—DAVID MYERS, Chicago, Ill.—*Car Spring.*—December 4, 1866; antedated September 24, 1866.—The bolster is connected to a bar beneath, and this latter bar is connected to the frame by jointed hangers, whose middle joints are drawn together by a spiral spring; this is to give ease of movement and relieve the main springs.

Claim.—The jointed hangers F G, when operated by springs, substantially as and for the purposes herein specified and shown.

No. 60,228.—DAVID MYERS, Chicago, Ill.—*Car Brake*—December 4, 1866.—A pull on the bell cord withdraws the catch bar from the slot in the cam wheel and frees it. The next spring then revolves the cam and draws the brakes into action. The catch bar being withdrawn, the incline spring holds the trip bar in position to trip the jointed pawl on a second pull of the cord to release the brake.

Claim.—First, the combination of the cam wheel C, shaft *m*, chain D and spring or springs F, arranged and operating substantially as and for the purposes specified and shown.

Second, in combination with cam wheel C and shaft *m*, the box M and shives *h*, arranged substantially in the manner and for the purposes set forth.

Third, in combination with cam wheel C, provided with a recess *c*, the bar B and elbow lever A, arranged and operating as described and for the purposes specified.

Fourth, providing the ratchet wheel H with the smooth periphery or flange N, when arranged in combination with the projection *u* upon the point *t* of the pawl P, and operating substantially as described.

Fifth, the arrangement of the arm R, provided with a shoulder, as described, with the elbow lever A and pawl P, operating substantially in the manner and for the purposes specified.

Sixth, in combination with said bar R, the arrangement of the movable elastic support or spring S, operating substantially as and for the purposes herein shown and described.

No. 60,229.—JOHN F. MYERS, Kokomo, Ind.—*Furnace for Steam Boilers.*—December 4, 1866.—A plate behind the grate bars may be raised to stop the flue passage, or lowered beneath the level of the grate bars. The plate is moved by a crank projecting from the furnace side. Doors in the furnace wall admit air behind the plate. A pipe from the fire space to the bridging carries the heat directly up the chimney.

Claim.—First, a furnace for a steam boiler, when constructed with a door or valve B, for closing the throat of the furnace, and with doors through the exterior walls of the furnace for the admission of cold air to the surface of the boiler and flues, substantially in the manner and for the purpose set forth.

Second, in combination with a door or valve B, as described, the pipes G and valves I, arranged to operate substantially as and for the purpose set forth.

No. 60,230.—WALTER P. NEWHALL, New York, N. Y., assignor to himself and HARRIET A. DAVISON, same place.—*Hydrant.*—December 4, 1866.—The drinking water is drawn from the vicinity of the fire plug valve. The overflow from the draught spout descends through a small pipe to the trough below.

Claim.—First, a hydrant in which two or more exterior basins, at different altitudes, are connected by pipes within the casing, substantially as described.

Second, combining a fire hydrant with the said hydrant, substantially as described.

Third, keeping the water in a live condition beneath the fire-plug valve D, or cut-off, by means of the auxiliary eduction pipe F, or its equivalent, substantially as described.

No. 60,231.—A. N. NEWTON, Richmond, Ind.—*Brick Machine.*—December 4, 1866.—The tempered clay is fed through a hopper on to an endless apron, with an intermitting motion by which it is carried under a series of rollers, and by them pressed into a sheet of uniform thickness, and thence under a cutter by which the bricks are cut; they are then deposited upon a board by a lubricated plunger. The cutters are actuated through a spring which yields to hard substances under them. A sieve covers the surface of the sheet with dry sand after it has passed under the reciprocating roller.

Claim.—First, in a machine for making brick, a reciprocating lubricated plunger acting in combination with an exterior reciprocating cutter, substantially in the manner and for the purpose set forth.

Second, in combination with the plunger L and cutter M, the endless apron C, stationary when the cutters are acting on the clay, and moving forward when the cutters are raised, substantially as set forth.

Third, the actuating mechanism of the cutters L, in combination with the same, when so arranged that the pressure shall be applied to the cutters by a spring which will yield to the resistance of solid substances, substantially as set forth.

Fourth, the roller E, when used in combination with the endless apron C, and so geared, that it shall revolve only with the forward motion of the same, substantially as and for the purpose set forth.

Fifth, the reciprocating roller F, when operated substantially in the manner and for the purpose set forth.

Sixth, in combination with the rod I and plunger L, the actuating cam G³, when constructed substantially as and for the purpose set forth.

Seventh, the oscillating screw O, when constructed and operated as and for the purpose set forth.

Eighth, the vertical endless aprons C', running lengthwise with the side of the hopper, when used in combination with the endless apron C and roller E, substantially as and for the purpose set forth.

No. 60,232.—JOHN P. NICHOLS, New Richmond, Ohio.—*Mechanical Movement for Operating Churns, &c.*—December 4, 1866.—The obtusely toothed verge wheel acts alternately in two systems of pivoted levers to reciprocate the dasher.

Claim.—The arrangement and combination of the pallet levers F F', levers H H', and their connections G G' S and T, with the ratchet wheel D, and the dasher of a churn, operating substantially as and for the purpose specified.

No. 60,233.—J. V. HENRY NOTT, Guilderland, N. Y.—*Bag Holder.*—December 4, 1866.—The platform has casters and an upright on which the holder-frame slides. The latter has two horizontal spring arms with outwardly projecting flanges, to hold the bag by its down-turned edge.

Claim.—The frame D, consisting of two bent or curved spring arms F, each having an outward-projecting lip or flange, in combination with the post or standard B, when arranged together so as to operate substantially in the manner described and for the purpose specified.

No. 60,234.—W. D. OSBORN, Boston, Mass.—*Express Wagon.*—December 4, 1866.—Explained by the claims and cut.

Claim.—A wagon made with the rear part of its body offset and depending downward below the bottom of the front part thereof, when combined with a bent binder axle placed directly under the rear part of the body, and when the sides and front end of the body rise above the bottom of that part, all substantially as and for the purpose specified.

Also, the combination with such a wagon body and bent hind axle, located as described, of springs e and bifurcated perch c, as specified.

No. 60,235.—JOHN G. PAGE, Memphis, Tenn.—*Cotton Picker.*—December 4, 1866.—The four picker cylinders vary in diameter, and have equal rotation, giving a variable speed to the surfaces.

Claim.—A series of toothed or armed cylinders B B' B'' B''' placed within a case or box A, provided with a bottom composed of a series of perforated concaves C, the cylinders gradually increasing in size from the feed to the discharge end of the case or box, and their speed of rotation increasing about in proportion to the increase of their dimensions, substantially as and for the purpose set forth.

No. 60,236.—SAMUEL PECK, New Haven, Conn.—*Spring for Hat Brims.*—December 4, 1866.—The drooping or boat shape is given the wire by rolling its edges thin at alternate points—first upon one side, and then the other.

Claim.—Forming springs for hat brims, so as to droop at the front and rear, by curving the wire, substantially as herein set forth.

No. 60,237.—M. M. PETTES, Oxford, Mass.—*Machine for Filing Saw Teeth.*—December 4, 1866.—The saw has longitudinal adjustment in the frame, and the file guide admits of angular or vertical adjustment thereupon. The file is clamped in a handle furnished with side pieces to work in the recessed part of the guide. A file may be made with raised portions to work in the guide.

Claim.—First, so hanging the guide A, or its equivalent, to and upon any frame or holder suitable for being secured upon a saw blade, that said guide can be adjusted to various angles with regard to the length of the blade, substantially as herein described and for the purpose specified.

Second, so hanging the guide A, or its equivalent, to and upon any frame or holder suitable for being secured upon a saw blade, that the said guide can be adjusted in position to vary the pitch of the teeth of the saw blade, substantially as described.

Third, so hanging the guide A, or its equivalent, to and upon any frame or holder suitable for being secured upon a saw blade, that it can be adjusted both to various positions or angles with regard to the length of the blade, and also to accommodate it to various pitches of the teeth of the saw, substantially as and for the purpose described.

Fourth, so hanging the frame E, or its equivalent, to which a guide A, or its equivalent, secured in any proper manner to a yoke or other frame suitable for being secured upon a 'saw blade, that it can be either raised or lowered, or so set as to more or less incline the said guide A, or both, substantially as herein described and for the purposes specified.

Fifth, the guide A, frame E, with circular bar G, and yoke S, when arranged, combined and connected together so as to be susceptible of each and all the several adjustments herein-above described, and substantially as and for the purposes specified.

Sixth, in combination with the above, the spring pawl Y, when arranged upon the yoke frame S, or its equivalent, so as to operate substantially as and for the purposes described.

No. 60,238.—WILLIAM PIMLOTT, Syracuse, N. Y.—Lathe Dog.—December 4, 1866.—One of the jaws is pivoted at the centre of the face plate. The slide bar of the jaws passes through the face plate and has set nuts to regulate the distance of the jaws from said plate. The jaws are moved on their slide by a bolt having a right and left hand screw.

Claim.—The bar B, the screw C, the nut of the jaws D, the jaws E E, the set nuts F F, when the same are constructed and operated substantially in the manner and for the purpose described.

No. 60,239.—ISAAC A. PINNELL, Galva, Ill.—Water Elevator.—December 4, 1866.—The two windlass drums are geared together so that while one bucket is being raised the other will descend. The crank wheel may be meshed with either drum by means of a shifter lever, so as to raise the buckets alternately while turning in one direction.

Claim.—First, the bar P, provided at the ends with slotted arms that engage the shafts of the drums E E', for the purposes and substantially as described.

Second, the bar P, provided with slotted arms Q Q, in combination with the lever N, substantially as and for the purposes herein described.

Third, in combination with the bar P, and lever N, the drums E E', with the gear wheels F F', and the ratchet wheels D D', for the purposes and substantially as described.

Fourth, the wheel B, provided with a ratchet upon its periphery and one upon the inside of the rim, in combination with the drums E E', lever N, and bar P, substantially as herein set forth.

No. 60,240.—A. J. POPE, Strongsville, Ohio.—Churn.—December 4, 1866.—The dasher frame oscillates longitudinally in the box, and its arms pass through slots of their own size in a slide piece of the lid which reciprocates with the frame. A fixed dasher is attached laterally in the centre of the box.

Claim.—The standard C, pendulum D, and handle E, in combination with the arms F, dasher H, and beater G, constructed and operated as and for the purpose set forth.

No. 60,241.—T. K. REED, East Bridgewater, Mass.—Sewing Machine.—December 4, 1866.—The serrated ends of the bolts engage the head of the shuttle tension screw and by depression adjust the tension without removing the shuttle.

Claim.—Combining with the reciprocating shuttle and shuttle race of a sewing machine, a device or mechanism operated by the movement of the shuttle to regulate or change the tension of the shuttle thread, substantially as set forth.

No. 60,242.—LAURENCE REID, New York, N. Y., and DAVID LYMAN, Middlefield, Conn., administrators of the estate of EDWARD H. SWIFT, deceased, assignors to PHINEAS L. ROBINSON and JOSEPH H. PARSONS.—Defecating Cane Juice.—December 4, 1866; antedated November 29, 1866.—To 700 gallons of cane juice is added one pound superphosphate of lime, and then seven pounds of slaked lime dissolved in three gallons of water.

Claim.—The mode herein described of defecating cane juice with superphosphate of lime and slaked lime introducing some of the superphosphate in advance of the lime as herein specified, with or without the final use of the prepared slightly alkaline phosphate of lime described to correct acidity and promote the crystallization of the sugar.

Also, in the defecation of cane juice, the alternate use of superphosphate of lime and slaked lime in small proportions, and in two or more successive increments, as described above.

Also, the combination with the superphosphate of lime in the above described process of one or more of the other defecating agents set forth in patents issued to us of even date herewith.

No. 60,243.—LAURENCE REID, New York, N. Y., and DAVID LYMAN, Middlefield, Conn., administrators of the estate of EDWARD H. SWIFT, deceased, assignors to PHINEAS L. ROBINSON and JOSEPH H. PARSONS.—Defecating Cane Juice.—December 4, 1866; antedated November 29, 1866.—The juice is treated alternately with oxalic acid and lime until completely defecated. The acid is finally neutralized by a slightly alkaline phosphate of lime.

Claim.— The mode herein described of defecating cane juice with acid and slaked lime, introducing some of the acid in advance of the lime, as herein specified.

No. 60,244.—LAURENCE REID, New York, N. Y., and DAVID LYMAN, Middlefield, Conn., administrators of the estate of EDWARD H. SWIFT, deceased, assignors to PHINEAS L. ROBINSON and JOSEPH H. PARSONS.—Defecating Cane Juice.—December 4, 1866; antedated

November 29, 1866.—To the cane juice is added slaked lime mixed with water, after which water highly charged with carbonic acid gas is allowed to pass up through the juice from the bottom of the tank until litmus paper is slightly reddened by the solution.

Claim.—Defecating cane juice with lime and a liquid impregnated with carbonic acid gas, in the manner above specified.

Also, defecating cane juice by slaked lime and the supercarbonate of lime or magnesia with carbonic acid gas, applied as above described.

Also, the combination in the above described process of one or more of the other defecating agents set forth in patents issued to us of even date herewith, with a liquid containing carbonic acid gas, with or without the supercarbonates of lime and magnesia, applied as herein specified.

No. 60,245.—LAURENCE REID, New York, N. Y., and DAVID LYMAN, Middlefield, Conn., administrators of the estate of EDWARD H. SWIFT, deceased, assignors to PHINEAS L. ROBINSON and JOSEPH H. PARSONS —*Defecating Cane Juice* —December 4, 1866; antedated November 29, 1866.—A slightly alkaline phosphate of lime, prepared by mixing 20 pounds slaked lime, 20 gallons of water, and 15 pounds superphosphate of lime. Used to neutralize acid in defecated cane juice.

Claim —The within-described chemical compound adapted for use in the defecation of sugar cane juice, substantially as and for the purpose herein set forth.

No. 60,246.—LAURENCE REID, New York, N. Y., and DAVID LYMAN, Middlefield, Conn., administrators of the estate of EDWARD H. SWIFT, deceased, assignors to PHINEAS L. ROBINSON and JOSEPH H. PARSONS.—*Defecating Cane Juice.*—December 4, 1866; antedated November 29, 1866.—Sulphurous acid gas is agitated with alcohol in a closed vessel until a saturated solution is obtained. From two to six pounds of the above is added to seven hundred gallons of cane juice, and afterwards neutralized with slaked lime.

Claim.—In the defecation of sugar cane juice, the use of the compound of alcohol and sulphurous acid, prepared by impregnating alcohol with sulphurous acid gas, in the manner and for the purpose herein set forth.

Also, the combination with the compound of alcohol and sulphurous acid in the above described process of one or more of the other defecating agents set forth in patents issued to us of even date herewith.

No. 60,247.—CHARLES RICHARDSON, Richmond, Va.—*Game of Battle War Chess.*—December 4. 1866 —The checker squares stand diagonally to the sides of the board and an interrupted barrier is placed across the board near the centre. The figures are named in the claim

Claim.—The board herein described and illustrated, in combination with movable figures representing cavalry, artillery, infantry, a supply train and a citadel, or base of supplies, substantially as shown and described and for the purpose set forth.

No. 60,248.—LOUIS S. ROBBINS, New York, N. Y.—*Preparing Peat and other Substances for Fuel.*—December 4, 1866.—The peat is placed in a vessel connected with a retort containing fuel. The vapors arising from the fuel pass into the vessel and condense in the peat. It may be dried by connecting the vessel in which it is placed with a heated retort containing broken marble through which a current of air is forced.

Claim.—Saturating peat, coal dust, or other substances, either separately or in combination with hot oleaginous vapors, substantially as herein described.

Also, the drying and saturating the peat, coal dust, or other substances, either separately or in combination at one and the same operation, substantially as described.

Also, the method herein described of drying the peat, coal dust, or other substances by the use of heated air, substantially as herein described.

No. 60,249.—C. R. ROBERTS and J. S. HARTZELL, Addison, Penn.—*Fruit Gatherer.*—December 4, 1866.—Explained by the claim and cut.

Claim.—An improved fruit harvester formed by the combination of the movable jaw F, lever G, cords H and K, receiving sack J, stationary jaw E, ring A, shank B, and socket I, when said parts are constructed and arranged substantially as shown and described.

No. 60,250.—DANIEL T. ROBINSON, Boston, Mass.—*Pole for Horse Railroad Cars.*—December 4, 1866.—The brace coupling bar is made adjustable longitudinally by means of a nut on each side of its supporting eye.

Claim.—Combining with the pole *a*, coupling plate or bar *b*, and brace rod or bar *c*, the means or mechanism for adjusting the position of this brace bar relatively to the pole, substantially as set forth.

No. 60,251.—T. R. ROBINSON and R. E. JONES, Providence, R. I.—*Lubricator.*—December , 1866.—The lubricator cup is furnished with an interior screw thread at top, into which is screwed a plug. The centre of the plug projects inward and impinges upon an absorbent gasket to regulate the flow of oil to the journal by adjustment of pressure

Claim.—A lubricating bolster having chamber A, cap C, with or without central perforation E and absorbent B, substantially as described for the purpose set forth.

No. 60,252 —W. B. ROBINSON, Detroit, Mich.—*Slide Valve.*—December 4, 1866.—The steam is exhausted through the body of the valve bridge and passes through an aperture in its upper shell, which is surrounded by a metallic packing ring. The packing ring is forced by springs against the under side of the steam chest cover and the steam exhausted through it.

Claim.—The combination of the counterbalance E, having suitable packing rings, and the bridge D with its passage *b* and valve C C', arranged with the valve chest A and cylinder B, provided with ports *a a'* and operating substantially as described for the purpose specified.

No. 60,253.—A. H. ROCKWELL, Harpersville, N. Y.—*Safety Line for Harness.*—December 4, 1866.—The joint bit has sliding rings to which the overdraw straps are secured, while the check straps are secured to the end rings of the bit, so that when the rein which connects them is drawn, the horse's head will be thrown up, and kicking prevented.

Claim.—The overdraw straps D secured to the rings G, sliding loosely on the joint bit, the check straps C, secured to the end rings of the same bit, operating together in combination with the rein B, in the manner as and for the purpose specified.

No. 60,254.—A. H. ROCKWELL, Harpersville, N. Y.—*Bridle.*—December 4, 1866.—The snaffle bit has two slide rings connected to the opposite ends of a strap passing over the nose, and supported by a strap from the fore piece of the bridle. A rein affixed to one end of the bit passes through the other end, over the neck and then over the part between the two ends of the bit.

Claim.—The combination of the double ring bit H, head stall A, and strap K, when all connected together and applied to a horse or other animal, substantially as and for the purpose described.

No. 60,255 —E. A. G. ROULSTONE, Roxbury, Mass.—*Travelling Bag.*—December 4, 1866.—The edges of the cover are introduced between the plates of the frame, and the rebends of the cover held in a groove of the frame plates. The hinge rod is screwed in the corner pieces of metal. The bolt enters and holds in the rectangular slot of the frame.

Claim.—The combination and arrangement of the inner and outer plates *h i*, making up one-half of the frame, when so constructed and applied to the body *a*, inserted between them, that from one or both of them a lip is turned down which shall protect or cover the edge of the body without either lip passing around or enclosing the edge, the plates and body being secured together substantially as shown and described.

Also, the corner stay piece when constructed and applied substantially as described.

Also, the construction of the bearings *f*, with a screw thread extending only partly through the bearings so as to secure and protect the hinge rod, substantially as described.

Also, so applying the lock to the frame that its bolt works in a slot in the frame, substantially as shown and described.

Also, in a bag made up of two parts *a a*, applied to frame *c d*, forming the bag body from one piece, substantially as set forth.

No. 60,256.—E. A. G. ROULSTONE, Roxbury, Mass.—*Travelling Bag.*—December 4, 1866.—The metal frame has an inward flange, perforated for stitching. The body leather is brought under and completely round the frame, and it and a welt are held by a single row of stitches.

Claim.—The arrangement and manner of connecting together the body, the frame, and the frame cover, substantially as shown and described,

Also, combining with the frame and its covering and the body, the welt *f*, secured to the frame covering and body, substantially as set forth.

No. 60,257.—E. A. G. ROULSTONE, Roxbury, Mass.—*Lock.*—December 4, 1866.—The bolt has a spiral spring which withdraws it when released from the catches.

Claim.—The bolt *e*, constructed to operate in connection with the locking mechanism, substantially as set forth.

No. 60,258.--EDWIN RUSSELL, Naugatuck, Conn.—*Chuck for Holding Buttons.*—December 4, 1866.—The socket is filled with rubber, which is forwarded or retracted by a set screw in the rear.

Claim.—The chuck A, made substantially as above described, with an elastic centre B, as set forth.

No. 60,259.—ROBERT SAFELY, Cohoes, N. Y.—*Machine for Turning Shafting.*—December 4, 1866.—The shaft is retained in a vertical position by means of the sliding holder which feeds it downward through the tubular support; upon the top of the latter are fixed the chuck and cutters, which revolve around the shaft.

Claim.—First, an improved machine which is adapted for turning and finishing shafts when they are supported in an upright position, said machine being constructed and operating substantially as herein specified.

Second, the combination of the horizontal chuck support E, and sliding shaft-holder C, with devices for feeding, turning, and finishing shafts, that are arranged in an upright or vertical position, substantially as described.

No. 60,260.—ROBERT SAFELY, Cohoes, N. Y.—*Machine for Straightening Shafting.*— December 4, 1866.—The plane bed has a heavy plate moving back and forth above it, regulated by guides. The shaft to be straightened is rolled while hot between the reciprocating plate and the table.

Claim.—First, the combination of a reciprocating slide D, which is vertically adjustable with the bed plate A, for the purpose of straightening rods or shafts, substantially as described.

Second, supporting the ends of the slide D, by means of the guides E E, having adjusting screws g g, and steadying screws h h, applied to them, substantially as and for the purposes described.

No. 60,261.—HENRY W. SAFFORD, Philadelphia, Penn.—*Drill for Wells.*—December 4, 1866.—A supplementary boss projects from the shank of the tool by which to withdraw it in case of breakage.

Claim.— Providing the stem or shank a^3, of the drill A B, or other boring or cutting tool used in making deep wells, with a fixed projecting shoulder or supplementary boss b^1, between the upper boss a^2 and the cutting end a^4, substantially as and for the purpose described.

No. 60,262.—CYRUS W. SALADEE and JESSE R. MOORE, Newark, Ohio.—*Sheep Chair.*— December 4, 1866.—The box is hinged to the frame and is open at one end, but has a shallow inclined end piece at the other. The sheep is reared on its hind legs with its back to the bottom of the box, which is in perpendicular position. The head of the sheep is then confined by the traverse pin, and the box vibrated to its horizontal position.

Claim.—The combination of box A and bench S, as described, and pin E, in combination with box A, constructed and operating as specified and for the purposes set forth.

No. 60,263.—JOHN F. SANFORD, Keokuk, Iowa.—*Lamp Burner.*—December 4, 1866.— The shell of the burner is made in an upper and under section, which are attached together by screws. The wick-spur stem is hung in hook journals adjustable by a spring and set screw, and may be raised from the journals when the latter are retracted.

Claim.—First, the application of a wick-spur stem d to adjustable bearings in such a manner that this stem with its spurs can be removed from the burner at pleasure, substantially as described.

Second, constructing the body of the burner of two sections A B, in combination with the wick spurs, which are so applied that they can be detached from their bearings c^1 c^2, at pleasure, substantially as described.

No. 60,264.—A. M. SAWYER, Athol, Mass.—*Composition of Matter for Polishing Metals.*— December 4, 1866.—Explained by the claims.

Claim.—First, the polishing compound of emery and soft vulcanized rubber, made substantially as described.

Second, the forming of the surfaces of polishing or scouring devices by means of a thin layer of the polishing compound before described, united to a backing of soft vulcanized rubber, substantially as described.

No. 60,265.—JOHN SAWYER, Moravia, N. Y.—*Miter Plane.*—December 4, 1866.—The plane works in a groove of the frame and has a cutter at each end. Pivoted adjustable guides hold the stuff at any required angle.

Claim.—The grooved and slotted plate B, and pivoted guide bars C, when used in combination with the plane A, having its iron a^1 a^2 inclined in opposite directions, substantially as described and for the purpose specified.

No. 60,266.—JAMES SERVICE, Greenville, Conn.—*Tool Rest for Lathes.*—December 4, 1866.—The tool stock is elevated or depressed by a screw operated by a pinion and spurwheel; to the latter a wrench is applied through the medium of a shaft projecting upward through the bed of the stock.

Claim.—The combination of the screw D, and pinion and spur wheels B E, with the lathe tool rest, arranged and operating substantially in the manner and for the purpose herein described.

No. 60,267.—WARREN SHAILER, Deep River, Conn.—*Window Fastener.*—December 4, 1866.—The fastener plate has an inclined slot for the traverse of a holding screw by which it is attached to the sash. The slot is so inclined that the descent of the screw causes the plate to bind against the beading and sustain the window. A gain in the plate engages with a pin in the beading to fasten the window when closed.

Claim.—The window fastener, substantially as herein described and represented by Figs. 3, 4, and 5.

No. 60,268.—BENJAMIN SLUSSER, Sidney, Ohio.—*Excavator.*—December 4, 1866.—The machine is supported on four wheels. The rear frame is adjustably attached to the front axle and carries a shovel which precedes an endless apron, whereby the excavated earth is carried to a dumping box suspended below the rear axle. The bottom of the dumping box consists of several hinged doors, operated from the driver's seat by means of ropes connected therewith.

Claim.—First, the combination of the spool L, chain M, wheel N, pinion O, and crank P, when used to regulate the height of the shovel A, adjustably sustained by the front axle, substantially as and for the purpose set forth.

Second, in combination with the driver's seat Q, the cord I''' and rope G' or their equivalents, for opening and closing from the driver's seat, the hinged doors F in the bottom of the box E, substantially as set forth.

Third, the doors F, latches H, and inclined faces H', in combination, when constructed and arranged substantially as set forth.

Fourth, in combination with the doors F, the latch I, lever I', arm I'', and bar K, substantially as set forth.

No. 60,269.—CHARLES B. SMITH, Newark, N J., assignor to WRIGHT & SMITH, same place.—*Friction Pulley.*—December 4, 1866.—The friction arms are carried on pivoted levers and a sliding block on the shaft carries a wedge, which by a movement of the block is inserted between the levers and forces the faces of the friction arms against the pulley rim. A retraction of the wedge allows the springs to act on the arms and free the pulley.

Claim.—The combination of levers E', and E'', block B, sliding block J, wedge H, or its equivalent bolt F, and set screws I' and I'', for the purpose set forth.

No. 60,270.—FRANCIS M. SMITH and EDWIN BRUMFIELD, Albion, N. Y.—*Horse Rake.*—December 4, 1866.—The latches are hinged to the handles, and forwardly extending bars, so that raising the handles frees the teeth and inverts the tooth frame. This movement is further assured by spring arms on the handles, acting upon lips on the tooth bar.

Claim.—The combination of the jointed latches C, straps N N, pendants I, spring braces J, flanged plate K, jointed head B, and teeth L, arranged and operating substantially as described and for the purpose specified.

No. 60,271.—GEORGE Y. SMITH, Plainfield, Ill.—*Hay Rack for Wagons.*—December 4, 1866.—The different pieces are detachable and are attached together upon the wagon gears by hook and head bolts in such manner that they can be handled by one person.

Claim.—First, the combination and arrangement of the hooks 1, and eye or eye bolt 2, with the cross-pieces B and bed-pieces A, substantially as and for the purposes described.

Second, the combination and arrangement of the hooks 3, and eyes 7, with the beams C, and bed pieces A, substantially as and for the purposes set forth.

Third, the combination and arrangement of the hooks 6, the straps 5, bolts 4, with the beams C, and raves D and E, when constructed and operating substantially as and for the purpose described

No. 60,272.—JOHN E. SMITH, Buffalo, N. Y.—*Machine for Cutting Bungs.*—December 4, 1866; antedated November 16, 1866.—A tongue projects circumferentially from the shaft, and the cutters have a groove that will fit on this tongue, and an inclined flange in the rear upon which a sleeve can be screwed to hold the cutters to their shaft.

Claim.—The ring J and nut K, in combination with the groove T, and the flange or ring L, when constructed as and for the purposes described.

No. 60,273.—JOHN P. SMITH, Hudson, N. Y.—*Potato Digger.*—December 4, 1866; antedated November 22, 1866.—The digger share raises the potatoes and earth and passes them to plates which have spaces between them, and perforations for the passage of earth. From these plates the potatoes are passed to a screen which is reciprocated by a crank on one of the wheels.

Claim.—First, the inclined digging screen A, furnished with shares C, and combined with the shaking screen L, substantially as herein set forth, for the purpose specified.

Second, the pitman R, cranked lever P, and sliding bar N, arranged in relation with each other and with the shaking screen L, driving wheel or wheels H, and digging screen A, substantially as herein set forth, for the purpose specified.

Third, the sled c J, arranged in rear of the digging screen A, and underneath the shaking screen L, substantially as herein set forth, for the purpose specified.

Fourth, the arrangement, with reference to the digging screen A, of the arched braces D, beam E, transverse bar F, and wheels H, substantially as herein set forth, for the purpose specified.

No. 60,274.—PHILIP W. SOMERS, Danbury, Conn.—*Machine for Felting Hat Bodies.*—December 4, 1866.—The rolls are fed in between the under side of the corrugated endless belt and the adjustable jointed spring bottom, pass to the spring concave at the other end, then above the belt and between it and the spring top, back to the operator. A pivoted valve may be set to throw out the rolls or to forward them for another round.

Claim.—First, in combination with the endless moving platform of grooved bars or slats, the stationary, yielding, adjustable lower bed concave and upper bed, the whole constructed and arranged to yield and be adjusted substantially as described.

Second, in combination with the lower bed and endless moving bed, the sliding bed with its two inclined concaves, one of them forming, in connection with the endless platform, an adjustable throat, and the other being a receiving table to receive the rolls thrown out of the machine.

Third, the valve located at the entrance of the throat of the machine, in combination with the endless bed, throat, and receiving table, substantially as described and for the purposes set forth.

No. 60,275.—PHILIP W. SOMERS, Danbury, Conn.—*Roll for Felting or Sizing Hat Bodies by Machinery.*—December 4, 1866.—The enveloping cloth is rolled in the reverse direction to the hat body.

Claim.—The new roll adapted for working hat bodies in the roll by machinery, wherein the coils or folds of the body or bodies rolled up within the covering cloth are in a reverse direction from the coils or folds of the covering cloth, substantially as and for the purposes hereinbefore set forth.

No. 60,276.—REUBEN SPARKS, Buffalo, N. Y.—*Steering Apparatus.*—December 4, 1866.—The horizontal hand wheel has a vertical shaft with a system of gearing to act powerfully upon the chain drum to which it is connected.

Claim.—The arrangement of the vertical shaft B, having a pinion D on the lower end thereof, with the pinions E and F, and spur wheel G, connected with the drum H, and chain I I', connecting with the drum H and wheel H', substantially as described.

No. 60,277.—QUINCY STODDARD, Jackson, Mich.—*Machine for Rolling Leather.*—December 4, 1866.—The frames of two pairs of rollers are supported on one standard. One roller of each pair is corrugated, and the other plain. The frames are so pivoted that the two plain or two corrugated rollers can be placed to act in conjunction.

Claim.—The combination of two pairs of plain and corrugated rollers D D', resting in swing frames C C, so arranged that they may be turned to opposite positions to roll plain, round, or half round leather, as herein set forth.

Also, forming the corrugated rollers with the independent spools b b, operating in the manner and for the purpose specified.

No. 60,278.—DANIEL STONER and JOHN SIGWALT, Jr., Chicago, Ill.—*Machine for Boxing Paper Collars.*—December 4, 1866; antedated November 21, 1866.—The catch spring is depressed by the sliding bar, and the ends of ten collars inserted. The shaft is then revolved sufficiently to fold the collars around the mandrel. The box is placed over them and then the catch released.

Claim.—First, a cylindrical form F G, and face plate E, in combination with a rotating shaft or its equivalent, for imparting a rotating motion thereto, substantially as and for the purposes described.

Second, in combination with said face plate and cylindrical form, the adjustable bar H operating as and for the purposes set forth.

No. 60,279.—D. B. TANGER, Bellefontaine, Ohio.—*Steam Generator.*—December 4, 1866.—The pipes are so arranged with respect to the fire box that air or gas can be passed through them and heated. The air is afterwards injected into the boiler, to be mingled with the steam to be used therewith to propel the engine.

Claim.—First, the pipes E F H, in combination with the fire box D, and boiler A, constructed and operating substantially as and for the purpose described.

Second, the pipes I J K, in combination with the fire box D, and boiler A, constructed and operating substantially as and for the purpose set forth.

No. 60,280.—GEORGE A. TAYLOR, LESTER CRANDALL, HORACE L. CRANDALL and JONATHAN LARKIN, Hopkinton, R. I.—*Machine for Making Twine and Small Cordage.*—December 4, 1866.—The "top" bar has grooves upon its upper side as converging strand guides, and its carriage has anti-friction rollers for passage on the way rail. The separate strands are carried on a wire with guide points projecting above it, from a shaft which has freedom to rock to admit the passage of the top bar, but gravitates back by means of a weight.

Claim.—The elevated way or rail A, provided with the suspended carriage B, having the top bar C attached, all arranged substantially as and for the purpose set forth.

No. 60,281.—HENRY R. TAYLOR, Roxbury, Mass.—*Refrigerator.*—December 4, 1866.—
Explained by the claim and cut.
Claim.—A refrigerator provided with a drawer so arranged that when pulled out it will
close the opening in which it slides and exclude the external air from the interior of the refrigerator, substantially as described.

No. 60,282.—FERDINAND TELLMANN, Stamford, Conn., assignor to ELLSWORTH FOX
and W. L. SMITH, same place.—*Tobacco Pipe.*—December 4, 1866.—Explained by the claims
and cut.
Claim.—A tobacco pipe having a nicotine chamber C, interposed between the tobacco
chamber of the pipe and its stem, when such chamber is provided with a valve stem or plug
G, arranged so as to operate substantially as and for the purpose described.
Also, the valve stem G, when so constructed and arranged in combination with the aperture
a, in the bottom of the pipe bowl, as to close and clear the same as it is opened, substantially
as described for the purpose specified.

No. 60,283.—NATHAN THOMPSON, London, England.—*Bottle Stopper.*—December 4,
1866.—The stopper is made of soft wood somewhat compressed and then hollowed out inside.
A metal cap is attached over the top with cementing material interposed between it and the
wood.
Claim.—First, a stopper made of hollow wood capped with metal, substantially such as
described.
Second, a stopper made of hollow wood capped with metal with a layer of material, substantially such as specified, interposed between the wood and the metal, the complete stopper
being substantially such as described.

No. 60,284.—SAMUEL H. TIMMONS, Lafayette, Ind.—*Car Brake.*—December 4, 1866.—
The longitudinal brake rods beneath the cars are connected together by chains so that the
brakes can be applied from the ends, or be brought into operation if any of the car couplings
give way.
Claim.—First, the connecting of the brakes of a train or series of cars by means of rods N
N', or their equivalents, in such a manner that by applying the brakes to the wheels of any
one of the cars, the pull of the locomotive or draught will be transmitted through said connections to the several brakes, and the latter all applied, substantially as shown and described.
Second, the rods N N', or their equivalents, when applied in such a manner as to serve the
double purpose of a brake connection and a car coupling, substantially as set forth.
Third, the screws S, and nuts T, applied to the rods N N', for the purpose of operating or
applying power to the brakes, substantially as set forth.

No. 60,285.—PHIL. TOMPPERT, Jr., Louisville, Ky.—*Book-Mark Holder.*—December 4,
1866.—The strip of metal is bent so as to form a fork to embrace a few leaves, and its upper
end has slots to hold a ribbon marker.
Claim.—The ribbon-receiving slots *a*, in the book marker herein described, whereby a
place of reference may be marked and opened by the ribbon, as and for the purpose specified.

No. 60,286.—SIMEON P. TUTTLE, Decatur, Mich.—*Fence.*—December 4, 1866.—The panels
are supported upon hooks that are screwed into the posts and are held in place by a button
upon the top of the post.
Claim.—The post C, provided with the L-shaped screws or staples D D, and cap F, when
used in combination with the fence sections A or B, constructed as described to form a portable fence, when arranged and used as and for the purposes set forth.

No. 60,287.—GEORGE L. UPTON, Milbridge, Me.—*Drag for Vessels.*—December 4,
1866.—The wings are hinged to the hub so that they can assume a radial or a folded position, and braces connect them to a block which slides on the shaft.
Claim.—The combination of the shaft A, hub B, bevelled wings D, braces F, slide E, constructed substantially as and for the purpose as specified.

No. 60,288.—SAMUEL W. VALENTINE, Bristol, Conn.—*Trunk.*—December 4, 1866.—
A strap from the lid is eyeleted to a strap connected to the body. The said eyelet is so
stamped as to act as a seal.
Claim.—A trunk as made with a sealing strap A, and loop B, to be combined or connected
by an eyelet C, or a rivet, substantially as and for the purpose specified.

No. 60,289.—ANDREW J. VANATTA, Vanatta, Ohio.—*Churn.*—December 4, 1866.—The
shaft carries a rotary dasher whose wings act as slides for the arms of a reciprocating dasher.
An inner sleeve carries the reciprocating dasher and passes through an outer sleeve carrying
the pinion of the rotary arm.

Claim.—First, the combination of the wheel D, and wheel P, with the pinions H and Q, for the purpose of giving the beaters N both a vertical and rotary motion at the same time, so as to break the rotary current of the cream at the sides of the churn, substantially as shown and described.

Second, the dasher N, having a bearing Y on the rod M, so as to have a rotary motion, at the same time a vertical motion, substantially as shown and described.

Third, the beater O, in combination with beater N, substantially as and for the purpose described.

No. 60,290.—HENRY C. VAN TINE, Pittsburg, Penn.—*Refining Petroleum.*—December 4, 1866.—To a barrel of crude petroleum is added a mixture of sulphate of zinc, 5 lbs. ; sugar of lead, 3 lbs. ; bichromate of potash, 4 lbs. ; sulphuric acid from 4 to 25 lbs.

Claim.—The refining of petroleum or carbon oil without the aid of artificial heat, by means of the series of operations hereinbefore described, consisting substantially of the use of sulphuric acid, sulphate of zinc, sugar of lead, and bichromate of potash, or their equivalents, for separating the heavy carbons and impurities, the neutralizing of the acid, and washing with water, combined with the subsequent exposure of the oil thus heated in shallow pans to the action of the atmosphere, substantially in the manner and for the purposes hereinbefore described.

No. 60,291.—H. H. B. VINCENT, Oshkosh, Wis.—*Broom Head.*—December 4, 1866.—The corn is placed between the expanding end and side arms, and the contracting slide ring-forced down to clamp the corn in place. Traverse bolts retain the parts in position.

Claim.—The combination of the projecting sides and edge arms with the sliding band H, and binding rods I, all substantially as and for the purpose set forth.

No. 60,292.—LOUIS VON GUNTEN, Cincinnati, Ohio.—*Lathe Chuck*—December 4, 1866.—The spindle has four spring jaws within the mandrel of the lathe so that it may be drawn in and out by means of a nut in a recess between two scollops in the side of the mandrel. The ends of the spring jaws are tapered, and being drawn against the conical opening in the end of the spindle are closed against the stuff.

Claim.—The grasping jaws *h h h h*, formed upon a stem F, guided by collars J J', and operated by a nut E *e*, confined within the chambered spindle A, which is formed with scollops *a a*, to afford access to the milled head *e* of the nut, all constructed and combined substantially as herein described and for the purposes specified.

No. 60,293.—Z. B. WAKEMAN, Rockford, Ill.—*Window Sash Fastener.*—December 4, 1866.—A roller is eccentrically hung upon the sash so as to pinch against the beading on its descent. A cavity receives the edge of the roller when the sash is closed and a spring lever operates to keep it there to prevent opening.

Claim.—The combination of the roller lever arm D, secured to it, fixed staple or guide G, and stop or rest pin K, when arranged together, as and for the purpose described.

No. 60,294.—W. J. WALKER, Baltimore, Md.—*Apparatus to be Attached to Stills to Prevent Fraud.*—December 4, 1866—The worm of a still is connected with a pipe from which extend two branch pipes. The latter communicate with a tube containing a hydrometer and thermometer. A pipe permits the escape of gas and air.

Claim.—First, the combination of a vessel or tube containing a hydrometer, or a hydrometer and thermometer, with a system of pipes and stop cocks, so as to test spirits, and pass it to the proper tank, substantially as described.

Second, the waste pipe for safety and escape of gas and air in combination with a testing apparatus, as described.

Third, the combination of stop cocks and pipes for testing and distributing the spirits, as described.

No. 60,295.—SAMUEL L. WALKINSHAW, Baden, Penn.—*Steering Apparatus.*—December 4, 1866.—The power shaft has a bevel wheel engaging two pinions, which turn freely on a shaft carrying a cog wheel, which actuates the steering wheel. A sleeve operated by a lever connects either of the free bevel wheels with the shaft, so as to turn the tiller to the right or left respectively.

Claim.—The combination and arrangement of the shaft D and *g*, wheels 1, 2, 3, 4, and C, coupling sleeve 5, jointed lever *m*, friction pulley *f*, and pilot wheel B, constructed, arranged and operating in the manner and for the purpose herein described and set forth.

No. 60,296.—SAMUEL H. WARD, Altona, Ill.—*Attaching Thills to Carriages.*—December 4, 1866.—The coupling bolt slides in place, and a spring attached to the axle carries a plate which stands behind the bolt head to prevent its sliding out.

Claim.—In combination with the thills E, lugs C, and removable bolt F, the employment of the spring G, provided with a lip H, when arranged so as to secure the bolt from slipping out, and also to prevent rattling and wear of the same, as herein set forth.

No. 60,297.—A. E. and J. V. WARNER, Norwalk, Ohio.—*Sawing Machine.*—December 4, 1866.—The log is clamped by levers which rest upon it and are retained by ratchets against which their ends catch. The pitman runs in a guide by which it is raised.

Claim.—First, the special arrangement of the levers *p p*, cross piece T, and lever V, in combination with the racks *j l*, and saw frame, as and for the purpose set forth.

Second, the guide *m*, and spring *n*, in combination with the slide I, spring catch L, and catch I', as and for the purpose set forth.

No. 60,298.—WILLIAM T. WATSON, Nottingham, Md.—*Tobacco Press.*—December 4, 1866.—The keg is clamped between sliding bars, and the follower is operated by a rack which is actuated by spur wheels connected by a shaft to lever wheels worked by hand. A pawl on the ratchet wheel retains the pressure.

Claim.—First, the combination of the sills G, sliding bars I, and rods I', for confining the keg in a tobacco press, substantially as described.

Second, in combination with above parts and the main frame A, the rack B, pinions D D', shafts E', and wheels E, together with the ratchet and wheel F F', arranged to operate substantially as set forth.

No. 60,299.—CLEMENS WEAVER, Easton, Penn.—*Car Coupling.*—December 4, 1866.—The coupling bar ends in a link with an inclined hook to slide up and hold to a similar link on the coupling bar of the other car. The link end of the bar is sustained by a chain connection to a pivoted lever, and the lever has a spring bar connecting with a rack for adjustment.

Claim.—The arrangement of the loop and hook coupling bars *a a*, pivoted to the car frame and connected with the lever *f*, when applied to railroad cars for coupling them together, substantially as herein described.

No. 60,300.—A. WELLS, Morgantown, West Va.—*Cotton-seed Cleaner.*—December 4, 1866.—The jointed slide is covered on the lower side with canvas, and reciprocates in the inclined trough by means of a crank and pitman. The upper end of the slide is armed with ratchet corrugations to draw forward the seed, and is raised by a slide piece and allowed to fall back on the seed at the turn of movement.

Claim.—First, the reciprocating slide C, used in connection with the inclined trough A, substantially as and for the purpose herein specified.

Second, the arrangement of the slide C, and board D, as constructed on their under sides, with the bar G, guide F, and trough A, as and for the purpose herein specified.

Third, providing the under side of the slide with an elastic covering, for the purpose of rolling the seed between it and the bottom of the trough, as and for the purpose set forth.

No. 60,301.—CHARLES WELLS, Cincinnati, Ohio.—*Rack Motion for Hand Press.*—December 4, 1866.—Beyond the horizontal part of the rack are additional down curved sections which engage with the smaller side of an eccentric wheel beside the usual rack wheel. This device is to give more purchase in starting or changing the motion.

Claim.—The rack C, having curved ends E E, in combination with the eccentric pinion F, and ordinary pinion D, of larger diameter, the whole being constructed, arranged, and operated in the manner and for the purpose set forth.

No. 60,302.—WILLIAM WELSH, McHenry, Ill., assignor to himself and ORA C. COLBY, same place.—*Buckle.*—December 4, 1866.—The two frames have connecting pieces and transverse bars. The tongues are pivoted on the bars of one frame and hold against the bars of the other. The outer frame has side loops.

Claim.—The combination and arrangement of the buckle frame A B D, with one or more tongues E, and corresponding cross bars *b*, operating substantially as and for the purpose specified.

No. 60,303.—W. D. WHALEN, Northville, Mich.—*Pendulum for Clocks.*—December 4, 1866.—The pendulum is pivoted centrally and horizontally, to oscillate in a vertical plane, and has two adjustable balls.

Claim.—First, a horizontal pendulum vibrating in a vertical plane, and suspended and operating substantially as and for the purpose specified.

Second, in combination with the above, the adjustable balls G, applied in the manner and for the purpose specified.

No. 60,304.—JOHN WHEELER, Augusta, Me.—*Washboard and Wringer.*—December 4, 1866.—The upper part of the board frame carries a wringer, and to its feet a frame is pivoted to hold it in the more perpendicular position required when the wringer is used.

Claim.—The combination of the wringer and washboard in connection with the folding frame, as and for the purposes herein named.

No. 60,305.—JOHN WHITEHEAD, Oskaloosa, Iowa.—*Hand Loom.*—December 4, 1866.—
The point of suspension of the lathe is made variable, to enable the operator to give a light or hard blow with the same expenditure of power, to accommodate it to the kind of work required to be done.

Claim.—The combination of the adjustable flanged plates *b b*, attached to the lay, and the notched plates *c c* applied to the top of the loom-framing, as herein described and for the purpose set forth.

No. 60,306.—LEVI H. WHITNEY, Vallejo, Cal.—*Training Hops.*—December 4, 1866.—
The wires diverge upwardly, and at a certain elevation pass through eyes in a metal rod and from thence horizontally and equidistantly to their termini.

Claim.—First, the device herein described for training grapes, hops, &c., in such manner as to retain them separate to any desirable width or distance from each other, and to carry them horizontally across the space to the next row opposite, substantially as described.

Second, the shackles or device herein described for securing the strings or cords, when constructed and used in the manner described.

Third, constructing the shackles *b b b*, with longer arms than those of *c c c*, to allow them to drop lower than the latter, to which the upper ends of the cords are attached.

Fourth, the device constructed and arranged as described, for securing the lower ends of the cords over the hills of vines, for the purpose described.

No. 60,307.—SAMUEL A. WHITNEY, Glassborough, N. J.—*Melting Furnace.*—December 4, 1866.—The crucibles rest upon projections on an annular bed within the dome, and are exposed to the caloric current which rises from the central opening connecting with the fire chambers, which are arranged opposite to each other. Each ash pit has a blast opening, and the central flue connects by a double series of holes with the crucible chamber, which discharges into the chimney.

Claim.—First, a furnace composed of the lower portion A, containing the fire places and central flue C, and the superstructure containing the central chamber C', and crucible chamber communicating with each other through contracted passages, all substantially as described.

Second, the combination and arrangement substantially as described of the central distributing chamber C', passages H *h*, and crucible chamber G'.

Third, the projections *z* arranged as a support for the crucibles, substantially as and for the purpose described.

Fourth, the dome-shaped structure F, depressed in the middle and arranged to cover the central chamber C', and crucible chamber G', in the manner and for the purpose specified.

No. 60,308.—WILLIAM M. WHITTAKER, Wallingford, Conn., assignor to himself and B. CHURCH, same place.—*Butter Dish.*—December 4, 1866.—A hemispherical, concentric canopy is pivoted to the side of the dish, so that a semi-rotation will place it within the base part.

Claim—The combination of the pivot *d* and the ears *c* and *a*, when constructed and arranged so as to operate substantially in the manner herein set forth.

No. 60,309.—JAY J. WIGGIN, Cincinnati, Ohio.—*Composition for Furniture and other purposes.*—December 4, 1866.—This composition is made by boiling together coal tar, sand, air-slaked lime and clay.

Claim.—The composition by boiling a mixture of sand, lime, and clay in coal tar or pitch, in the manner and for the purpose substantially as specified.

No. 60,310.—JAY J. WIGGIN, Cincinnati, Ohio.—*Mould for Plastic Material.*—December 4, 1866.—A bed of sheet iron is supported by side and end pieces, which are held together by clamps. The partition piece is made in two parts.

Claim.—The within-described mould, constructed with removable partitions, in the manner and for the purpose shown and described.

No. 60,311.—WILLIAM H. WILEY, Fredonia, N. Y.—*Well and Cistern Filter.*—December 4, 1866.—Two concentric cylinders are clamped between an upper and under disk, by means of an enlarged section of the pump tube. The annular space between the cylinders is filled with filtering material, and the cylinders are perforated on opposite sides, so that the water makes a partial circuit to reach the inner space which connects with the pump tube.

Claim.—A portable well and cistern filter, composed of the sub-plate E, filtering cylinder F G, disk D, met or fastening *j*, and perforated central tube C, combined with and attached to the pump tube B, the whole constructed and arranged substantially as and for the purposes set forth.

Also, the partially perforated sides F and G, when constructed as described, in combination with the porous packing L, base plates D E, and tubes C and B, arranged and operating substantially in the manner and for the purpose described.

No. 60,312.—J. P. WILLMS, Baltimore, Md.—*Pessary.*—December 4, 1866.—The lobes are pivoted on the ends of the arms and hinged together to the end of the central rod. This rod is projected or retracted by the screw upon it, and serves to expand or contract the lobes.
Claim.—The lobes A, opening forwardly relatively to the person, and operated by a swiveled screw and stem between the arms B, substantially as described and represented.

No. 60,313.—S. W. WOOD, Cornwall, N. Y.—*Conveying Grain.*—December 4, 1866.—A pump has an exhaust pipe and conveying pipe so arranged that, when the air is exhausted, the grain will be forced in by atmospheric pressure, and by a return movement of the piston be made to pass through the conveying pipe to the place of delivery.
Claim.—The combination of a pump A B, exhaust pipe C, and conveying pipe D, arranged and operating substantially as and for the purpose herein specified.

No. 60,314.—THOMAS WOODS, Jessamine county, Ky.—*Stone Drill.*—December 4, 1866.—The drill cord passes into the pulley at the groove and under a small pulley within it through the spindle. It then passes over pulleys upon the flyer arm to the spindle near the driving pulleys.
Claim.—First, the eye for the drill rope passing from the ordinary eye of the spindle, in front of the bearing through the side of the spindle, and then through the pulley in the line of the radius to the groove of the circumference, as described.
Second, the arrangement of the rachet wheel and pawl on the spool and spindle, in combination with the band wheel g on the spool, for the purpose of producing the feed motion of the drill, as herein described.
Third, the arrangement of the small pulleys on the arms of the flyers and the pulley at the eye for the rope.
Fourth, the combination of the flyers, the drill, and the other improvements, as herein described.

No. 60,315.—BENJAMIN WORCESTER, Waltham, Mass.—*Scale Pencil.*—December 4, 1866.—The pencil has a fixed collar, with a conical point and a sliding collar, with rounded point elongated laterally. A scale upon the stem indicates the space between the points.
Claim.—First, the combination of a fixed and a sliding point with a pencil, substantially as and for the purposes set forth.
Second, the construction of a marking point with a thin curved edge, and so arranged as to cover and protect the other point when not in use, substantially as and for the purposes set forth.
Third, the combination of pencil, scale, and points as an article of manufacture, substantially as and for the purposes set forth.

No. 60,316.—JAMES M. WORCESTER, Oberlin, Ohio.—*Apparatus for Packing and Tying Wool Fleeces.*—December 4, 1866.—The side flaps are hinged to extend horizontally and receive the fleece, and by turning in to fold the edges. The fleece is compressed by the heads, which are actuated by the hand crank. The opening out of one of the leaves frees the crank shaft, and allows the weights to retract the heads.
Claim.—The sliding heads E, weights J, grooved bars D, and the adjustable leaves C, as arranged, in combination with the lever M and rollers K and H, in the manner and for the purpose set forth.

No. 60,317.—WILLIAM H. WUSSOW, Aurora, Ill.—*Wood Lathe.*—December 4, 1866.—The cutter and the guide roller are fitted in the swinging frame, so that the cutters shall be above the stuff, and the roller below the pattern. The cutters are flanked by circular saws. A cover and conveying chute carry the shavings from the working parts.
Claim.—First, so constructing and arranging the cutting and guiding frame that its forward end, without the aid of extraneous devices, bears with a preponderating force down upon the work to be turned, while the rear end is caused to bear up against the pattern, substantially in the manner herein described.
Second, the construction of the cutter with two circular saws and intermediate finishing cutters, all arranged and operating substantially as described.
Third, the arrangement of the device y y′ with the longitudinally moving frame H and the vertically swinging and longitudinally moving cutter frame K, substantially as herein described.

No. 60,318.—THOMAS YATES, Dubuque, Iowa.—*Corn Harvester.*—December 4, 1866.—The two rails of the double tongue are connected at the front end by an upwardly arched bar, which bends the corn-stalks down. The fingers then seize the ears, and the knives sever them, and they fall into the bed of the wagon. The rear hounds may be disconnected from a transverse bar to allow of the rapid turning of the machine.
Claim.—First, the fixed or stationary fingers f at the front part of the bed of the wagon, and having a reciprocating cutter composed of a series of knives g working underneath it, in

combination with the double draught pole composed of two parts *b b* connected by a bow-shaped bar *c* at their front ends, substantially as described and for the purpose specified.

Second, the pivoted or turning rear axle B', with forked bar D attached. in connection with the catches *i i*, on the bar E, substantially as and for the purpose specified.

No. 60,319.—W. J. ALEXANDER, Manchester, Iowa.—*Hame Fastening.*—December 11, 1866.—The portions are attached to the respective hames; one slips into the other and is fastened therein by the engagement of a spring catch with recesses in the socket. The catch is released by the rotation of a cam on the end of a shaft, and an adjustable pin determines the point to which the catch shall enter the socket.

Claim —First, the hame fastening consisting of the catch piece B and the socket C, with the spring catch D and notches E, respectively, and united to the loops of the hames, substantially as and for the purpose described.

Second. the arrangement of the button F, shaft G, eccentric H, and spring M, operating as described.

Third, the pin L, as and for the purpose described.

No. 60,320.—HENRY W. ANGELL, Waukesha, Wis.—*Manufacture of Brick or Building Blocks.*—December 11, 1866.—Composed of lime and sand, mixed with gravel and stones and moulded into bricks or blocks.

Claim.—A brick composed of lime, sand, small stones, and gravel, prepared and moulded in the manner described.

No. 60 321.—JAMES E. ATWOOD, Trenton, N. J.—*Saw.*—December 11, 1866.—The saw blade has gaps to receive the teeth, which are sprung laterally and driven into their seats, from the rear. V-edges on the seats and corresponding grooves on the edges of the inserted pieces keep the respective parts in position, the rear end being straightened and abutting upon the blade.

Claim.—The teeth H H, when inserted and secured in the manner herein described and for the purposes set forth.

No. 60,322.—HUGH BAINES, Manchester, England.—*Machine for Rolling Metal.*—December 11, 1866.—The two rollers are formed of an assemblage of annuli, capable of being arranged in varying positions. They are severally provided with cameo die patterns and counterpart intaglio recesses or with grooves to act in conjunction; the lower roller acts in conjunction with a reciprocating table. The direction of movement may be reversed by a clutch on the driving shaft which connects it with either of two bevel wheels.

Claim.—First, the combination of the hollow perforated rollers C^x, with the reversible gearing H^x I^x, when constructed, arranged, and connected together so as to operate substantially in the manner described and for the purposes set forth.

Second, in combination with the above, the movable table 7, arranged and operating substantially as and for the purpose specified.

No. 60,323.—WILLIAM C. BARTOL, Huntington, Penn.—*Brick Machine.*—December 11, 1866.—The clay is thrown into the hopper and the motion of the wheels in going to the drying floor brings the lower course of mould boxes beneath the hopper and depresses the follower. A continued motion brings in turn each series of moulds down under the hopper and forwards them to the shelves on the other side. The return to the pit raises the follower and mould frame and so fits it for the next moulding operation.

Claim.—First, operating the machinery for moulding brick from the drive or supporting wheel or wheels of the machine, substantially in the manner herein shown and described.

Second, the combination of the slides H. racks C', cog wheels G, and shaft D, with each other, substantially as herein shown and described, for the purpose of raising and lowering the shelves and moulds, as set forth.

Third, the combination of the catches E'. spring G', and arms H I', with each other and with the racks C', and wheels G, substantially as herein shown and described and for the purpose set forth.

Fourth, the combination of the racks A', and cog wheels I, with each other and with the moulds J, and shaft D, substantially as herein shown and described and for the purpose set forth.

Fifth, the combination of the pins B', connecting rods M', pivoted levers K', and springs L', with each other and with the projections J', of the racks C', substantially as herein shown and described and for the purpose set forth.

No. 60,324.—SAMUEL BAXENDALE, Boston, Mass.—*Batting and Wadding.*—December 11' 1866.—This wadding is designed principally for linings in clothing, quilts, &c., the interior sheet of paper rendering it nearly impervious to air and causing it to hold its place better than if both sides were glazed, as is customary.

Claim.—The batting or wadding composed of a layer of fibrous material attached by any adhesive substance to opposite sides of a sheet of paper, as herein described, the same being a new article of manufacture.

No. 60,325.—J. B. BEALL and B. F. GRIM, Westerville, Ohio.—*Road Scraper.*—December 11, 1866.—The scraper is reversible by the oscillation of a pivoted catch arm whose movement is accomplished by turning a handle which has a spur wheel acting on a rack bar connecting with the catch.

Claim.—First, in a reversible road scraper, constructing the latch *c* to a loosely turning handle piece *b'*, by means which will allow of the movement of said latch by turning the handle piece without removing the hands from either of the handles, substantially as described.

Second, the combination of a vibrating latch *c*, a sliding spring latch *d*, and a movable hand piece *b'*, substantially as described, with a reversible scraper.

No. 60,326.—E. BECKWITH, South Pass, Ill.—*Washing Machine.*—December 11, 1866.— A pendulum is connected with the oscillating rubber as that of the oscillating rubber is operated by a pivoted arm extending longitudinally. The lower rubber board consists of lateral slats with openings between them.

Claim.—First, the combination of the pendulum J, with the bar K, counterpoise L, upright G, and rubber E, substantially as and for the purpose herein shown and described.

Second, the levers C, in combination with the board B, and rubber E, all constructed and operating substantially as herein shown and described.

No. 60,327.—J. W. BEEBE and T. F. LLOYD, Albany, N. Y.—*Apparatus for Distilling Grain.*—December 11, 1866.—The wooden still has several chambers and the vapors therefrom enter the chambered vessels above ; the heavier are condensed and returned to the still and the lighter go on and are condensed in the worm. At the end of the worm is a tube containing a hydrometer and thermometer, the whole covered by a glass dome. This still furnishes strong spirits at one distillation.

Claim.—First, the generater C, constructed with two or more chambers C' C'' C''', for the purpose set forth, substantially as described.

Second, the drop pipe *i*, in combination with the encasing pipe *k*, for the purpose set forth substantially as described.

Third, the indicator E, combining the pipes *n n'* and *q*, together with the permanent hydrometer *z*, and thermometer *t*, the bell glass or dome *o*, and the cock *r*, for the purpose set forth, substantially as described.

Fourth, the combination of the registering metre *s*, or its equivalent, with the indicator, for the purpose set forth, substantially as described.

No. 60,328.—S. I. BEELER, Wales, Ill.—*Soap.*—December 11, 1866.—Composed of lime 1 pound ; sal soda, 2 pounds ; concentrated lye, 1 pound ; borax, ¼ pound ; alum, ¼ pound ; saltpetre, 2 ounces ; animal grease, 10 pounds ; water, 3 gallons ; after having been mixed and boiled it is cooled down and moulded.

Claim.—The use of the ingredients herein named in the proportions and manner substantially as set forth, for the manufacture of soap.

No. 60,329.—JAMES BIRD, New York, N. Y.—*Cast-iron Chain Pulley.*—December 11, 1866 ; antedated December 2, 1866.—The projections on the sides of the groove between which the flat links rest are "chilled" in casting to render them more enduring under the wear of the chain.

Claim.—In cast-metal chain pulleys, making their projections which hold the links of the chain with a "chill," substantially as and for the purpose above described.

No. 60,330.—M. BRAND and C. P. HOFFMANN, Chicago, Ill.—*Mash Machine.*—December 11, 1866.—Attached to the shaft are arms carrying revolving propeller-shaped wings ; adjustable scrapers are attached to the arms by means of bolts.

Claim.—First, the adjustable scraper on bottom of tub.

Second, the vertical propeller-shaped wings, in combination with the machine.

No. 60,331.—W. W. BRATT, Ottawa, Ill.—*Fence Gate.*—December 11, 1866.—The gate is suspended by a slide rail on rollers, which are journalled upon the arms of a pivoted post on which the gate swings.

Claim.—First, the two rollers G G on which the gate slides sidewise.

Second, the guide board K fitting in the circular grooves of the rollers G G.

Third, the guide I at the lower end of the swinging post B.

Fourth, the part L of the gate H back of the rollers G G lying against the fence when closed, substantially as and for the purpose described in the foregoing specification.

No. 60,332.—OTIS BRIDGEMAN, Addison, N. Y.—*Planing Machine.*—December 11, 1866.— The rough lumber is subjected in turn to a circular saw, a rotary planer, and rotary tonguing and grooving cutters.

Claim.—First, the arrangement of the cutter head K, adjustable rollers A², saw N, feed table U, cutter head S and T, frame L², and its flange S², strip P², screw shaft O², arms A², and feed rollers V², substantially as described for the purpose specified.

Second, the arrangement of the frame L² carrying the circular saw blade N and revolving cutter head, suitable either for tonguing or grooving, substantially as described and for the purpose specified.

No, 60,333.—T. E. C. BRINLY, Louisville, Ky.—*Tuyere.*—December 11, 1866.—For the purpose of giving access for the removal of cinders, the cap is secured by books to lugs on the tuyere chamber, and is readily removable therefrom.

Claim.—The cap E, provided with a flange *a* and hooks *e e e*, body A, with the raised centre, forming an annular air chamber C and ears *n n n*, when arranged as herein set forth, and operating as and for the purpose specified.

No. 60,334.—ALBERT BROWN, Troy, N. Y.—*Grate for Stoves.*—December 11, 1866.—Underneath the grate, in the centre, is a hemispherical depression occupied by a corresponding protuberance on a bar attached to the circumferential frame. Upon this centre the grate rotates to sift out the ashes, and is supported when the contents are dumped.

Claim.—First, the permanent bar G, in combination with the frame F and convex centre E, or equivalents, as and for the purposes set forth.

Second, the cavity C, in combination with the cross-pin I and the convex centre E, as and for the purposes set forth.

Third, the grate B working upon the centre E, so arranged as to dump on a line with the shank J at any point within the sphere of its movement, as and for the purpose set forth.

No. 60,335.—JOHN H. BROWN, New York, N. Y.—*Propelling Horse.*—December 11, 1866.—The band levers have connections to two cranks on the rear axle shaft. The fore wheel is guided by the feet.

Claim.—The combination and arrangement of the wheel B, axle C, horse A, wheels D, shaft E, and band levers F F, as herein set forth, operating in the manner and for the purpose specified.

No. 60,336.—JOHN H. BROWN, New York, N. Y.—*Spring Toy.*—December 11, 1866.—The figure is pivoted through its hind quarters, and is oscillated by the weight of the child and the power of the spring acting upon the bifurcated lever. The force of the latter, to adapt it to the weight of the child, is regulated by winding the spring cord on the drum in the forward part of the base frame.

Claim.—First, the lever C, spring D, and regulating drum *b*, in combination with the toy A, constructed and operating substantially as and for the purpose set forth.

Second, graduating the lever C, so that the leverage of the spring can be accommodated to the weight of the child occupying the toy.

Third, supporting the toy on a fulcrum *a* at its back end, substantially as and for the purpose described.

No. 60,337.—JOHN H. BROWN, New York, N. Y.—*Attaching Handles to Boilers and other Vessels.*—December 11, 1866.—The ears have upward slots occupied by the flattened ends of the bail when the latter is in an upright position, and enable the vessel to be tipped thereby.

Claim.—The pear-shaped slots in the ears *a*, in combination with the flattened ends of the bail or handle of the kettle A, constructed and operating substantially as and for the purpose described.

No. 60,338.—GEORGE BRUCE, Corydon, Ind.—*Feeding Device for Carding Engines.*—December 11, 1866.—The invention, described in the claims and cut, dispenses with the constant attendance of the operator in spreading and regulating the feed.

Claim.—First, the pitman G, pawl L, ratchet wheel M, toothed wheels O P, and pinions Q and R, in combination with the aprons B and B', for the purposes and substantially as herein described.

Second, the manner of accelerating or reducing the feed by lengthening or shortening the stroke of the pawl L by means of the lever G and cord J, substantially as herein set forth.

No. 60,339.—SANFORD S. BURR, Dedham, Mass.—*Combined Table and Bedstead.*—December 11, 1866.—The table top is divisible, and when detached and changed in position it forms the head and foot of the bedstead, the leaves forming the head and foot supports. The table-top supports when placed horizontally become extension sections to the bed bottom of the central portion.

Claim.—The combination of the table top I and its connections with the slat frame H, and box A B C D E F and G, for the purpose and operating substantially as above described.

Also, a combination of the sides of the table L M and G, with the top I, and the leaf N, so constructed and arranged by means of the pin sliding in the groove E, that the same can at will be transformed into the bottom end and support of the bed, in the manner and for the purposes substantially as herein described.

Also, the combination of a table with a bedstead in one piece of furniture, adjustable at the will of the operator by means of the pin sliding in the groove E, constructed and operating substantially as herein described.

No. 60,340.—WILLIAM BURT, Marquette, Mich.—*Fastening and Unfastening Drop Doors in Coal Cars.*—December 11, 1866.—The free edge of the door is kept up by cam rollers upon a shaft operated by a lever and fastened by a pin.

Claim.—The combination of the lever B, shaft L, cams c, attached to the said shafts L, pins P, with drop doors a, when the same are constructed and arranged in the manner and for the purpose set forth.

No. 60,341.—JAMES J. BUTLER, Cincinnati, Ohio.—*Rotary Cutting Machine.*—December 11, 1866.—The middle mandrel acts as a holder when the screw is turned; the sleeve is driven by the gearing and has a cutter which is adjustable radially to regulate the diameter of the circular pieces of paper cut thereby.

Claim.—First, the rotary knife or cutter M, and arm L, in combination with the mandrels C and G, for the purposes and substantially as described.

Second, the bevel gearing attached to a crank for driving the mandrels, to which the cutter or knife is attached, substantially as and for the purpose set forth.

Third, the upright screw D, in connection with the mandrel pressing on the disk i, or material, substantially as herein shown and described.

No. 60,342.—SILAS S. CROCKER, Maquoketa, Iowa., assignor to himself and D. R. CROCKER, same place.—*Tap Borer.*—December 11, 1866.—Explained by the claim.

Claim.—The volute-shaped tool with a sharpened, salient spiral edge, with the gimlet point, substantially as described.

No. 60,343.—A. C. CROSBY, Union, Penn.—*Tool for Setting Jewels in Watches.*—December 11, 1866.—This circular steel cutter has a concave end with a circumferential angular edge, which slightly exceeds in circumference the bezel into which the jewel is to be fitted, and by which a circular burr of metal is pushed down upon the jewel.

Claim.—The securing or setting jewels in watch plates by means of a die or tool construed in the manner substantially as shown and described.

No. 60,344.—GEORGE W. DEVOE, New York, N. Y.—*Measuring Liquids.*—December 11, 1866.—The plug levers of the fill cocks of the series of measures are connected together so that one movement opens or closes them simultaneously. The discharge cocks are similarly connected. Whistles connected with the tops of the measures indicate by the passage of air the flow of liquid into the measures. Pivoted funnels gravitate to the mouths of the cans on their presentation.

Claim.—First, the suspension and arrangement of the weighted funnels M, in relation with the series of measuring vessels E, and with the cans C, as they are pushed under the said funnels, substantially as herein set forth, for the purpose specified.

Second, a measuring apparatus consisting of one or more measuring vessels E, arranged in relation with a reservoir B, and furnished with inlet and outlet valves or stop cocks F K, and with a suspended funnel or funnels M, and a whistle or whistles e, substantially as herein set forth.

No. 60,345.—JOHN V. DINSMORE, Milford, Mass., assignor to himself and M. HARRIS, same place.—*Feathering Paddle Wheel.*—December 11, 1866.—The staffs of the paddles occupy radial sockets in the centre wheel and rotate therein to secure the proper presentation of the float. This rotary actuation is attained by the eccentricity of the paddle when the paddle staffs are moved longitudinally, by the contact of rectangularly projecting arms on the paddle staffs with certain plates, with which they come in contact when the wheel revolves.

Claim.—The combination for operating each of the paddles while the wheel is revolving, the same consisting of the cam G, the sliding and rotary shaft F, its bearing b, stud g, notch f, shoulders d e, arm m, cams n and q, and the arm p, and the roller o, or the equivalent thereof, the paddle being applied eccentrically upon the shaft, and the whole being in other respects substantially as hereinbefore described.

Also, the cam G, as made in two parts hinged together and applied to the side of the vessel, so that one of them may be stationary thereon, and the other movable, so as to produce with the wheel, results as above set forth.

No. 60,346.—SAMUEL L. DONNELL, Spring Creek, Tenn.—*Grading Instrument.*—December 11, 1866.—A weight suspended from the bubble block has a vertical aperture through which the shaft passes; lugs are movable radially along the bubble block by means of screws passing through them, and thus the position of the centre of gravity of the balancing weight may be varied until the level is perfectly adjusted.

Claim.—The bubble block D mounted upon a collar G, swivelled upon the stand A, or the equivalent, in combination with the balance ball or weight H, suspended from the said block, and about and around the stand A, substantially as herein described and for the purpose specified.

Also, in combination with the above, arranging either one or both of the lugs J J⁰, in the bubble block D, from which the balance ball H is suspended in such a manner as to be susceptible of being adjusted therein, substantially as and for the purpose described.

Also, the combination with the sights of the bubble block of the thumb or set screws R, and fixed pointer U, arranged substantially as described and for the purpose set forth.

~ Also, the use of a transverse swivelled block V, for the purpose described.

No. 60,347.—M. T. DRAKE, Pleasant Ridge, Ohio.—*Potato Digger.*—December 11, 1866.— The levellers throw the vines and weeds outward. The rotary side cutters cut downward to the share level. The potatoes and earth pass from the chute plough to an endless elevator and fall upon a vibratory screen. The plough frame is adjustable by a screw operated by a hand crank.

Claim.—First, the main frame A, vibrating inner frame A', gauging frame A'', and axle B, constructed substantially as above described and for the purpose specified.

Second, the levellers *m*, constructed and operating as above described and for the purpose set forth.

Third, the levellers *m*, and cutters *k*, in combination with gauging frame A'', substantially as above described and set forth.

Fourth, the plough F, consisting of shovel *q*, share *r*, and mould boards *s*, as above described, when used for the purpose set forth.

Fifth, the shovel *q*, in combination with the vibrating inner frame A', constructed and operating as above described and for the purpose set forth.

Sixth, the endless bucket elevator D, constructed substantially as above described.

Seventh, the pulverizing board 8, in combination with the endless bucket elevator D, for the purpose specified.

Eighth, the irregularly reciprocating screen 1, in combination with the crank shaft 5, constructed and operating as above described and for the purpose set forth.

Ninth, the dumping box 9, in combination with rod 10, and lever 11, constructed and operating substantially as above described and for the purpose set forth.

Tenth, the clutch gearing *f* and *z*, in combination with the levers *y* and *y'*, and eccentric *z*, as above described and for the purposes set forth.

Eleventh, the frames A A' and A'', levellers *m*, circular cutters *k*, plough F, endless bucket elevator D, pulverizing board 8, irregularly reciprocating screen 1, dumping box 9, and clutch gearings *f* and *z*, combined and operating as above described and for the purpose set forth.

No. 60,348.—R. B. DUTTON, Iron Hill, Iowa, assignor to himself and N. C. WHITE, same place.—*Sled Brake.*—December 11, 1866.—The pointed cutter which acts as a brake is operated by a lever attached to a fulcrum post, the depression of the lever being retained by a ratchet bar.

Claim.—The combination and arrangement of the jointed dog G, clasp H, lever F, fulcrum rod E, ratchet bar I, and spring K, with each other, and with the rave D, and runner A, of the sled, the whole being constructed and operated substantially as herein described, and for the purpose set forth.

No. 60,349.—EDGAR ELTINGE, Kingston, N. Y.—*Coal Scuttle.*—December 11, 1866.— The scuttle has a cover hinged to the ears, and the side plates of the cover constitute a spout for directing the discharge of coal.

Claim.—The shield D, when constructed as described and applied as and for the purpose specified.

No. 60,350.—JOHN A. EVARTS, Meriden, Conn.—*Blower.*—December 11, 1866.—The inner cylinder has two encircling straps which connect with a slide plate to alternately open and close the air inlet as required.

Claim.—The combination of the outer cylinder A, the inner cylinder C, and the plate F, when said plate F is connected by a strap or straps G to and so as to be operated by the inner cylinder, and the whole constructed and arranged substantially in the manner and for the purpose specified.

No. 60,351.—WILLIAM FISHER, Ripon, Wis.—*Bed-Clothes Clamp.*—December 11, 1866.— Jaws on the sides and foot of the bedstead are formed of slight bars and are clamped together by ordinary clothes-line clasps, holding the edges of the bed-clothes between the bars.

Claim.—The combination with the spring lever arms F, hinged and jointed together in pairs, of the jaw rails or bars C, connecting the several pairs of said levers, substantially as described for the purpose specified.

No. 60,352.—J. W. FOARD, San Francisco, Cal.—*Vacuum Ventilator.*—December 11, 1866.—The inner end of the funnel tube passes beyond the vertical tube, and the passing air creates a vertical current in the latter.

Claim.—The apparatus above described for raising fluids, gaseous or liquid, by means of natural or artificial currents of air composed of horizontal tubes or pipes *a c*, which are supported by means of the collar *b*, upon the vertical pipe *e*, so as to be free to revolve thereon, and which are guided and maintained in proper position by means of supplementary pipe *d*, that extends from pipe *a* into said vertical pipe *e*, substantially as above shown.

No. 60,353.—M. B. FOOTE, Northampton, Mass.—*Needle Threader for Sewing Machines.*—December 11, 1866.—The frame has an adjustable gauge by which the position of the slotted eye is determined and brought opposite to the eye of the needle. The frame slips upon the needle and is held by a spring.

Claim.—The frame A with its handle B, springs D, and the arms C, to the lower one of which the slotted eye E is secured in combination with the slotted right angular gauge G, all operating substantially as described for the purpose specified.

No 60,354.—LEWIS FOSSEE, Jeffersonville, Ind.—*Circular Sawing Machine.*—December 11, 1866.—The frame carrying the saw and its motive pulleys admits of vertical or longitudinal inclination to saw the edges of planks curved or bevelled.

Claim.—First, the adjustable slotted frame G and set screw *f*, for canting the saw, arranged and operating substantially for the purpose specified.

Second, sluing the plank by means of the segment racks F F', wheels G G', radiating shafts *m m'*, guide rod H, and dogs I I', substantially as described for the purpose specified.

No. 60,355.—VINCENT FOUNTAIN, Jr., Castleton, N. Y.—*Potato Masher.*—December 11, 1866.—The potatoes are placed in the cylinder, and, by the reciprocation of the plunger, are forced through the meshes of the wire work bottom.

Claim.—As a new article of manufacture, a potato masher, constructed as described, to wit: the cylinder or tubular receptacle A, provided with the detachable perforated bottom B and handles, one on each side, and the plunger C, all as shown and set forth.

No. 60,356.—JOHN H. FRENCH, Geddes, N. Y.—*School and Family Slate.*—December 11, 1866.—Explained by the claims and cut.

Claim.—First, the frame A, made with two compartments, one of which contains a slate, while the other is so constructed as to admit of the insertion of any convenient number of cards of pasteboard or other material containing lessons or copies for writing, printing, marking, or drawing, and exercises in arithmetic, either any or all combined, substantially as and for the purpose described.

Second, the arrangement in one or both surfaces of a slate of permanent perpendicular and slope lines, in combination with the ordinary horizontal lines, substantially as and for the purpose set forth.

No. 60,357.—WILLIAM L. GERARD, Fort Wayne, Ind.—*Crib for Children.*—December 11, 1866.—The crib is to be placed on the floor. The panels are hinged together by elastic bands to admit of folding up.

Claim.—The panels A in combination with the elastic hinges B and the mat C, the whole being arranged and constructed in the manner and for the purpose described and set forth.

No. 60,358.—THOMAS GLOVER, Woodbury, N. J.—*Fence.*—December 11, 1866.—The posts and braces are anchored in a horizontal bar beneath the ground. The rails have holes and are impaled upon the posts.

Claim.—First, the post A and plate *a*, in combination with the braces *b b*, and detachable bolt *d*, or its equivalent, the whole being constructed and arranged substantially as and for the purpose described.

Second, the post A and plate *a* with its slots *x x*, in combination with the detachable braces *b b*, enlarged at the lower ends for the purposes set forth.

Third, the shoulder *e* on the post A, for the purposes specified.

No. 60,359.—ARTEMUS W. GODDARD, Clinton, Mass.—*Machine for Pulling Hemp and Cotton Stalks.*—December 11, 1866.—The stalks pass between the endless belts and are supported by them. In their passage they are drawn from the ground by the wheels, whose peripheries are in sections; the sides of the two wheels press against each other on the lower side to grasp the stalks and separate at the rear for their discharge. The side movement of the sections is accomplished by pulleys on cam frames.

Claim.—First, the sectional flexible wheels *d d*, of the form substantially as herein described, or their equivalent, arranged together and operated as described.

Second, the spur belts *e e*, for carrying the tops of the stalks to the rear, or their equivalent, substantially as described.

Third, the rollers *i i i i* and frames *f f*, for keeping the sectional flexible wheels together, arranged as and for the purpose specified.

Fourth, the combination of the whole, consisting of the sectional flexible wheels *d d*, the spur belts *e e*, the rollers *i i i i*, and the frames *f f*, arranged together as and for the purpose specified.

No. 60,360.—H. C. GOODRICH, Chicago, Ill.—*Sewing Machine Guide.*—December 11, 1866.—Explained by the claims and cut.

Claim.—First, the rigid plate B, having its under surface corrugated, as shown when hinged to the guide plate A, as herein set forth, so that it can be folded over back on the plate A when not in use.

Second, the plate B, hinged as shown to the plate A, and having its inner made heavier than its outer end, substantially as and for the purpose set forth.

No. 60,361.—N. W. GREEN, Cortland Village, N. Y.—*Grappling Tool.*—December 11, 1866.—The grapple is intended for withdrawing piles, posts, tubes, pins, &c., and is placed over the object to be withdrawn. The operation of the levers is to close the grapple and to raise it in the slots, carrying the object with it. A release from the levers allows the grapple to slip down upon the object and assume a lower hold on the pile ready for another bitch.

Claim.—First, the self-adjusting grapple B B, the adjustable fulcra C C, in combination with the lever D D.

Second, the frame F F F, including the standard A A, in combination with the grapple B B, the adjustable fulcra C C, and the levers D D, for the purposes described.

No. 60,362.—SAMUEL B. GUERNSEY, Chicago, Ill., assignor to himself and RICHARD S. THAIN, same place.—*Portable Washstand.*—December 11, 1866.—On the top of the washstand are receptacles for hot and cold water and a recess under the hot water division in which a spirit lamp is placed for heating. A cock from each division discharges into the basin, in the bottom of which is a tube entering a secondary tube proceeding from the cover of a slop pail which receives the discharged water.

Claim.—In combination with a portable washstand, a water reservoir permanently attached thereto, having two compartments C D, and a recess E beneath one of them, substantially as and for the purposes specified.

Second, as arranged with the foregoing, the washbasin H, pipes I J, cover K, and can L, as and for the purposes specified.

No. 60,363.—A. W. HALL. New York, N. Y.—*Churn Dasher.*—December 11, 1866.—Explained by the claim and cut.

Claim.—An atmospheric churn dasher constructed with a series of inverted funnels or hollow cones arranged one above the other and attached firmly and air-tight to the stem of the common dasher, in combination with the plate B, at bottom of the stem, substantially as shown and described.

No. 60,364.—AARON S. HADLEY, Boston, Mass.,—*Dish Mop.*—December 11, 1866.—A tussock of cords projects from a socket in the middle of the stock, and others are rove through holes in the stock and unite with the former to constitute a mop.

Claim.—The combination and arrangement of the auxiliary or internal mass of the fibrous material C projecting from the end of the socketed handle A, with such handle and either or both the masses D E of fibrous material extended from the periphery of the head of such handle, as set forth.

No. 60,365.—DEXTER D. HARDY, Cincinnati, Ohio, assignor to THOMAS H. FOULD, same place.—*Rotary Pump and Engine.*—December 11, 1866.—The two pistons revolve in two communicating segments of cylinders, and are geared together to insure similar rotation.

Claim.—The rotary pistons B B', whose peripheries are formed by quarter circles *a a'*, and *b b'*, of different diameters, connected by suitable sides c, in the described combination with the two central inner tongues D D' of the case A, as and for the purpose specified.

No. 60,366.—D. D. HARDY and J. J. MORRIS, Cincinnati, Ohio.—*Rotary Pump and Engine.*—December 11, 1866.—The epicycloidal faces have spring packing plates. Set screws at the ungeared ends of the piston shafts prevent lateral play.

Claim.—First, constructing the rotary pistons B B', of arcs of circles *a a' b b'*, of different diameters, connected by epicycloid faces c, in the described combination with the elastic gibs G, as and for the purpose specified.

Second, as arranged with the above combination, the set screws J, for preventing lateral play of the shafts and pistons, as set forth.

No. 60,367.—BENJAMIN T. HARRIS, Brooklyn, N. Y.—*Tool for Inserting Screw Eyes.*—December 11, 1866.—The machine has a slotted revolving mandrel into which the screw eye is placed; when the mandrel is revolved the screw eye is forced into the wood.

Claim.—The device for inserting screw eyes in picture frames, &c., consisting of the eye holder e, formed substantially as specified, and rotated by the mechanism herein described, substantially as set forth.

No. 60,368.—W. J. HASWELL, Waverly, N. Y.—*Bed Bottom.*—December 11, 1866.—The elastic webbing bottom is secured by elastic straps passing over rollers on the side and end rails and fastened by iron rods beneath the rails.

Claim.—The springs *d*, passing down through mortises in the sides and ends of the bedstead or frame A, in combination with movable rods *e*, straps *c*, and the flexible bottom B, all constructed and operating substantially in the manner and for the purpose set forth.

No. 60,369.—JONATHAN HATCH, South Windham, Conn., assignor to himself, C. SMITH, and H. WINCHESTER, same place.—*Packing Revolving Joints.*—December 11, 1866.—The pipe which conveys the steam into the rotating cylinder has a collar of nearly the inside diameter of the stuffing box. The shaft is packed on both sides of the collar.

Claim.—The combination with the stuffing box *b* and gland C of the pipe B, formed with a collar *e* and provided with packings *f f* on each side or face of said collar, substantially as and for the purpose or purposes herein set forth.

No. 60,370.—THOMAS HAWKS, Rochester, N. Y.—*Condensed Extract of Malt.*—December 11, 1866.—The malt is steeped in warm water at 160° or 170° Fahrenheit. The strength is extracted by repeated infusion. The extract is boiled by means of steam, and a small quantity of salt may be added; the liquor is clarified with gelatine, and then transferred to a vacuum pan and condensed to a thick sirup.

Claim.—Condensing the extract or infusion of malt by evaporation in vacuo, for the manufacture of ale, beer, and other liquors, substantially as set forth.

Also, as a new product, the extract of malt when condensed to a solid or nearly solid substance, substantially in the manner and for the purposes herein set forth.

Also, as a new product, the extract of malt, either with or without the addition of hops, when condensed in vacuo, substantially in the manner and for the purposes set forth.

No. 60,371.—HENRY M. HAYWARD, Boston, Mass.—*Sawhorse or Buck.*—December 11, 1866.—The central bar revolves freely. Face plates at the joint have lugs, which engage and prevent the supports opening too far. The buck may be folded up.

Claim.—The improved sawhorse, as constructed with stop joints, as described.

Also, the application or arrangement of the roller with a sawhorse, so as to operate or turn on its middle connecting rod or bar, in manner and for the purpose as hereinbefore set forth.

No. 60,372.—NATHAN M. HEALY, Flushing, Mich.—*Stump Extractor.*—December 11, 1866.—The upright hoisting bar and levers are arranged on an upright frame, and operated by chains and ropes to lift the stump.

Claim.—The holdfast bars E', slotted blocks C', levers F, and chains G, in combination with the notched bar E, cap piece C, upright D, inclined supports B, and runners A, arranged and operating substantially as herein shown and described.

No. 60,373.—SAMUEL HICKS, Orangeville, Ind.—*Water Wheel.*—December 11, 1866.—The body of the wheel is an inverted cone. The buckets are vertical the greater part of their length, and turned back at bottom. The water is received at the sides and discharged beneath.

Claim.—The wheel having an inverted conical head D, with buckets E attached, curved at the lower ends and secured to a ring or annular bar F, substantially as described for the purposes specified.

No. 60,374.—JAMES A. HOLFORD, Guionsville, Ind.—*Saw Mill.*—December 11, 1866.—One of the belts through which the feed motion is communicated to the carriage runs on conical pulleys whose vertexes are in contrary directions, and a longitudinal shifting of the belt regulates the feed. Stops on the carriage reverse the feed. The head and foot blocks are of metal and have racks beneath, which are engaged by cog wheels. These wheels are are on a single shaft, which has a ratchet wheel moved by the pawl on a lever, actuated by an inclined plate on the carriage.

Claim.—First, the automatic log-setting device, consisting of the head blocks I and K, pinions *o* and *p*, shaft L, ratchet wheel *r*, pawls *r'*, lever *r''*, adjustable rest *t*, and incline S, all constructed substantially as and for the purpose herein shown and described.

Second, the automatic feeding device, consisting of the carriage H, reversing lever *b'*, spring *m*, and stop levers *h²* and *h³*, constructed substantially in the manner and for the purpose herein shown and described.

No. 60,375.—MICHAEL and SIMEON HOUSMAN, Huntington, Ind.—*Corn Sheller.*—December 11, 1866.—This is an improvement on the patent of Michael Housman, No. 29,938, September 4, 1860. The transverse pin has a coiled spring on each end to keep the jaws closed upon the ear.

Claim.—The pin passing through the jaws or blades, and the coiled spring around the same, constructed and arranged as hereinbefore described and substantially as set forth.

No. 60,376.—GUY D. HOWE, Lewisport, Ky.—*Baling Press.*—December 11, 1866.—
The follower arms are actuated by a slide block to which they are pivoted, and the block is operated by a vertically traversing screw. The boxes are on rollers, and each box is attached to the opposite follower, so that both the box bottom and follower give active pressure.
Claim.—The movable boxes C, when used in combination with a suitable press, constructed and operating in the manner and for the purpose herein specified.

No. 60,377.—A. R. and E. A. HUNT, Newark, N. J.—*Street Lantern.*—December 11, 1866.—
The base plate has branches connecting it with a sleeve, which fits over the top of the post. The side frames hold the panes, and are fitted into the base and fastened by flanges to the upper sides of the lantern, which are covered by a cap similarly secured.
Claim.—A street lantern, having its various side frames H and top frames P made of cast iron, and constructed and joined together, substantially in the manner and for the purpose described.

No. 60,378.—G. G. HUNT, Bridgeport, Conn.—*Steam Valve.*—December 11, 1866.—The inner valve cylinder has vertical movement, and its shaft has an arm playing in the slot of an angle plate on the stuffing box to prevent rotary motion in the cylinder.
Claim.—The arm G, applied to the valve rod F and secured thereto by means of a set screw *g*, in connection with the slotted plate *A* attached to the stuffing box *f*, substantially as and for the purpose specified.

No. 60,379.—G. A. JASPER, Charlestown, Mass.—*Steam Pipe.*—December 11, 1866. -In the steam conduit, near the place where the steam does its work, is placed a diaphragm or reducing pipe, which has an orifice through it of such an area as will just supply the required amount of steam. The object is to allow a given volume of steam to pass any given point in the pipe.
Claim.—The register, as a device to be used in combination with a steam apparatus for the purpose of more efficiently controlling and utilizing the steam, substantially as described.

No. 60,380.—JOHN V. JEPSON, Brooklyn, N. Y.—*Tool for Cutting Tubes.*—December 11, 1866.—This device has two cutters, whose edges by a slight movement of the tool may be made to cut around the entire circumference of the pipe. The frame which supports the cutters constitutes at one end a nut through which the threaded extremity of the handle passes, and the holder of the movable cutter is attached thereto, so that by turning the latter around one way or the other the cutters may be adjusted to pipes of different sizes.
Claim.—The combination of the stationary knife D with the adjustable knife F and the jaw A of the wrench, constructed and arranged substantially as described.

No. 60,381.—MOSES A. JOHNSON, Lowell, Mass.—*Fire-proof Felt for Roofing and other purposes.*—December 11, 1866.—Improvement on his patent of March 24, 1863. The felt is covered with silicate of soda on one or both sides, and before it is dry paper is laid on and made a part of it. The silicate serves as a cement between paper and felt.
Claim.—A felted fabric of hair or other material, covered on one or both of its sides with paper and with silicate of soda or soluble glass, and whether painted or otherwise rendered water-proof or not, as herein described.

No. 60,382.—GILBERT D. JONES, New York, N. Y.—*Apparatus for Drying Artists' Materials.*—December 11, 1866.—The material is taken up by the smaller roller, spread on the surface of the steam-heated cylinder, and after being conducted through a heated jacket is scraped off below.
Claim.—The rotating heated drum or cylinder A, having its periphery partially inclosed by a jacket C, forming a hot-air passage *a*, in connection with a spreading knife or trowel G and a scraper or discharging knife H, all arranged to operate in the manner substantially as and for the purpose set forth.
Also, the roller F, when used in connection with the heated drum or cylinder A and trowel G, to operate substantially as and for the purpose herein set forth.

No. 60,383.—HORACE C. JONES, Dowagiac, Mich.—*Basket.*—December 11, 1866.—The basket is strengthened by an inner and an outer hoop riveted together at the point to which the ends of the handles reach. The handles are included in the riveting at this point, and are secured by staples to the upper rim hoop.
Claim.—First, the combination of the inside hoop or band *g* with the outside hoop or band *g'* in the construction of the stave basket, substantially in the manner described.
Second, the manner herein shown and described of securing handles G G to the basket.

No. 60,384.—ELIAS H. KEITH, Peoria, Ill.—*Car Coupling.*—December 11, 1866.—The loops are pivoted to the heads, and their movement is limited by studs. In coupling, the loops pass up an incline on the head and over the hooked end of a bolt. The turning of said bolt, which is done by a lever, frees the loop and uncouples the car.

Claim.—First, the swinging loop B, with projections *g*, and limited by stops *i* and *k*, as described, and for the purpose set forth.

Second, the combination of the hook C, crank D, and rod *d*, and reversing bar *a*, in mortise *m*, or their equivalents, as and for the purpose set forth.

No. 60,385.—BERNARD KEMP, Knoxville, Md.—*Hominy Mill.*—December 11, 1866.— The longitudinally corrugated arms revolve between corrugated arms projecting from the face of the concave. When sufficiently broken, the hominy passes down through a chute to the fan, which blows away the hulls.

Claim.—The arrangement of the frame A with its case D, screens H, door I, board J, fan C, arms E E, and teeth G, in the manner substantially as and for the purposes herein specified.

No. 60,386.—P. C. KIRK and M. PENDERGAST, Lawrence, Mass.—*Wool-oiling Apparatus for Carding Engines.*—December 11, 1866.—This is an improvement on Salisbury's patent. The automatic alternate opening and closing of the cocks, so that one shall be open while the other is shut, is to secure a certainty of delivery of oil when the reservoir is in motion, and a certainty of stoppage when the machine stops, so as not to be dependent upon an attendant. The swinging agitator is brought into action every time it passes one of the pins.

Claim.—Operating the supply cocks of a wool-oiling apparatus by the travelling motion of the reservoir, substantially as set forth.

Also, operating the cocks *j k* by the pinions *l m* and rack I, substantially in the manner and for the purpose described.

Also, the pipe K with its cock *n*, for admitting air into the passage between the cocks *j k*, substantially as set forth.

Also, the agitator or mixer *o*, when operated by contact with the pin *p* projecting from the railing L, for the purpose set forth.

No. 60,387.—RICHARD KITSON, Lowell, Mass.—*Machine for Opening and Cleaning Cotton.*—December 11, 1866.—The cotton passes between the feed rollers; the seed falls between the triangular slats into the seed chamber. The fibre then passes over the curve-topped slats and between the sieve rollers, which receive the dust.

Claim.—First, the fender D arranged beneath the seed rack, substantially in the manner and for the purpose set forth.

Second, forming the communication between the seed chamber C and throat E, so as to allow the air to pass from the seed chamber to the dust trunks K, giving vent to the seed chamber C and preventing the seed being sucked back again with the cotton, substantially as set forth.

No. 60,388.—L. J. KNOWLES, Warren, Mass.—*Narrow Ware Loom.*—December 11, 1866.—This is an improvement on his patent of May 15, 1866.—The lay rack rests for awhile before each longitudinal movement. Each pressure roller is supported by an inclined hanger, hinged to the breast beam, enabling it to be moved off and above its fellow roller. Each set of take-up rollers is formed of a series of annuli and two short cylinders, connected by parallel bars, and the series of rollers are connected by interlocking end devices.

Claim.—The combination of the slot *v*, or its equivalent, with the mechanism, substantially for operating the lay rack, as set forth, such mechanism consisting not only of the link *l*, the straps *e e*, and their guide wheels, as applied to the lay and its rack, substantially as described, but of the crank *n*, the shaft *o*, and the bevel gears *t u*, or their mechanical equivalent or equivalents, operated by the cranked shaft of the lay.

Also, the above-described mode of making each of the series of take-up rollers, viz: of two heads a series of annuli and cross bars or connectors, as set forth.

Also, the application of each of the top or pressure rollers to the loom, viz: by means of an inclined hanger, to operate substantially as described.

Also, the application of one series of take-up rollers to the other series thereof, substantially in the manner as described, that is, by means of the head *c'*, the journal *s'*, socket *d'*, and stud *f*, arranged and applied to the two series, as set forth.

No. 60,389.—JOHN KOPP, Bridgeport, Conn.—*Baby Chair and Table.*—December 11, 1866.—In one position the chair is mounted on the support to make it a high chair for use at an ordinary table, and in the other position it forms a low chair, the support forming a table therefor.

Claim.—The combination of the reversible chair with the table, whereby the chair may be used either as a high or a low chair, substantially as herein set forth.

No. 60,390.—JOEL LEE, Galesburg, Ill.—*Fire Shield.*—December 11, 1866.—The shield is to be run between two houses to prevent the spread of fire. It is mounted on wheels, and when the supports are erected the plates are raised by means of chains and pulleys.

Claim.—The arrangement of the pulleys C C, supports and guides B B, chains G G, and stay braces D D, with the metallic plates A, substantially in the manner and for the purpose specified.

No. 60,391.—GEORGE H. LENHER, Richmond, Va.—*Churn.*—December 11, 1866; antedated November 30, 1866.—Stationary counter dashers are attached to the lid, and a perforated tubular rotating dasher supplies air from without. The rotary motion of the cream is resisted by the stationary dashers. The cover is attached to the churn by means of cams on the lid and ears on the tub.

Claim.—First, the arrangement of the hollow spindle L, perforated rotating tube M, and vertical fixed counter dashers *o o*, constructed and operating in the manner and for the purpose herein described.

Second, fastening on the covers of churns by beams of books on opposite sides of their bodies, and of revolving cams on the edges of the covers, substantially as described.

No. 60,392.—SAMUEL K. LIGHTER, Hamilton, Ohio.—*Harvester Rake.*—December 11, 1866; antedated October 23, 1866.—The rake head is pivoted upon a universal joint and reciprocated by a pitman and crank. During the effective stroke the limber heel is vibrated upward and is non-effective, but dropping at the end of the stroke, runs under the cam on its return and elevates the rake.

Claim.—The arrangement of rake E, universal joint or fulcrum F, limber heel G, fixed cam H, pitman J, and guide K *k*, the whole being constructed, combined and adapted to operate in the manner set forth.

No. 60,393.—JAMES H. LOCKIE, Humphrey, N. Y.—*Brace for Wagon Springs.*—December 11, 1866.—To prevent the surging forward or backward of the bed, its ends are connected by rods to a short lever pivoted on the coupling, while the springs support the bed as usual and are not strained by the surge of the bed.

Claim.—The combination of the short levers C, the connecting rods D and F, and the T braces E and G, with each other and with the reach A and box frame B of the wagon, substantially as herein shown and described and for the purpose set forth.

No. 60,394.—RADCLIFFE B. LOCKWOOD, New York, N. Y., assignor to LOCKWOOD & EVERETT, same place.—*Disinfecting Noxious Vapors from Rendering-Houses, Hospitals, &c.*—December 11, 1866.—The vapors are drawn from the houses in which they are generated and poured through heated tubes and thence through fire or through disinfecting chemicals.

Claim.—Controlling, disinfecting, and deodorizing the noxious gases or vapors generated in rendering-houses, slaughter-houses, hospitals, or sewers, by forcing or drawing said gas or vapors from said buildings, houses, or sewers through a superheating or other furnace, or through a chamber charged with a disinfecting and deodorizing material, by which said gases or vapors are disorganized and rendered innoxious and inodorous, or are consumed.

No. 60,395.—BARKER LOWE, Fall River. Mass.—*Piston Packing.*—December 11, 1866.—The pair of rings has a bevelled spiral interior spring by which they are kept in contact with the interior surface of the cylinder. The rings have a rabbet in their outer edge to receive the flanges on the piston head and the follower, and from the interior of the rabbet a flange projects inwardly and forms a steam tight joint with the flange and follower.

Claim.—The rings *a b*, incasing the head and follower, and provided with the flanges *' b'*, in combination with the bevelled spiral spring *c*, constructed as described, substantially as and for the purpose specified.

No 60,396.—ORAZIO LUGO and T. O. L. SCHRADER, New York, N. Y.—*Distilling Petroleum Oils and Other Substances.*—December 11, 1866.—Air from the force pump is driven through steam-heated jacket around the still. From thence it passes to an oil chamber and thence to the still, where it occupies a submerged chamber and is thence diffused above the liquid. Gauges, &c., and cocks at the various points indicate pressure and heat and permit the necessary connections.

Claim.—First, the admission of air or gas into the still at a temperature equal to or greater than that of the oil or substance undergoing the distilling process, substantially as herein described.

Second, the heating of the air or gas previous to its admission into the still, by the same means or medium employed for the heating of the still itself, and in such manner that the temperatures of the substance undergoing distillation and that of the air or gas admitted to the still will increase or decrease in or nearly in the same ratio, substantially as herein described.

Third, the utilization of the wase heat from the still by its employment to effect a preparatory heating of the air or gas which is to be admitted into the still, substantially as herein set forth.

Fourth, causing the air or gas while in the heated state, but before its admission into the still, to pass through or in contact with an oil of suitable character, substantially as herein specified.

No. 60,397.—JAMES B. LYONS, Cornwall, Conn., assignor to EDWIN MCNIELL, GIDEON H. HOLLISTER, HENRY B. GRAVES, Litchfield, Conn., and H. TUDOR BROWNELL. Hartford, Conn.—*Peat Machine.*—December 11, 1866.—The moulds are formed in sections with racks on each side, into which the pinions work, pushing the moulds forward under the pressing roller.

Claim.—The construction of the moulds C C, they having racks *b b* on both sides for the action of the pinions *a a*, to convey them into and through the receiving box D, as arranged, and operating substantially in the manner and for the purposes herein set forth.

No. 60,398.—JAMES B. LYONS, Cornwall, Conn., assignor to EDWIN MCNIELL, GIDEON H. HOLLISTER, HENRY B. GRAVES, Litchfield, Conn., and H. TUDOR BROWNELL, Hartford, Conn.—*Apparatus for Grinding Peat.*—December 11, 1866.—The iron cylinder has a central shaft with shearing blades arranged spirally upon it. The interior of the cylinder has corresponding blades attached to it. At one end of the cylinder is a hopper, and at the other is a perforated cap through which the ground peat is forced by means of scrapers.

Claim.—The perforated end cylinder H, in combination with the wiping wings I I I, cutting and grinding mechanism F F *h h*, operating to discharge the pulverized mass, substantially as and for the purposes herein set forth.

No. 60,399.—HOMER C. MARKHAM. West Turin, N. Y., assignor to himself and CHARLES G. RIGGS, same place.—*Making Butter from Cheese Whey.*—December 11, 1866.—The acid liquid used is made by boiling whey and then keeping it till it becomes quite sour.

Claim.—The separation of cream or butter from whey by the means of heat and in the use of an acid liquid, substantially as herein described.

No. 60,400.—JAMES W. MCDONOUGH, Chicago, Ill.—*Folding Lounge.*—December 11, 1866.—The seat of the lounge unfolds and makes a bed; the head also of the lounge opens out to form a pillow, supported by resting on the end of the bed rail.

Claim.—The rails A A of the frame of the folding lounge, constructed and operating substantially in the manner herein described and specified, and the folding head H of the lounge, constructed and operating substantially as herein described and specified.

No. 60,401.—B. F. MCKINLEY, Cincinnati, Ohio, assignor to himself and H. R. MATHIAS, same place.—*Head Block for Saw Mills.*—December 11, 1866.—The slide upon the head block has a ratcheted arm on which the spring pawls of a sliding piece operate. The slide piece is moved by a lever connected to it by pivoted arms. Each forward movement of the lever moves the log the thickness of one board, and its return connects the pawls with fresh teeth of the ratchet. The spring pawls are disconnected from the ratchet by a cam, in which case the slide can be moved back upon the head block.

Claim.—The regulating of the movement of the slide B, and knee C, by means of the adjustable head N, or its equivalent, on bar L, substantially as and for the purpose specified.

Also, so adjusting head N, or its equivalent, on bar L, as to bring the pivot centres of link M and the fulcrum of lever K all in line when starting, in combination with a similar arrangement of levers or links on head block A, the centres of which are brought in line in stopping, and the slide I, pawls H H, and rack bar G, all arranged substantially as and for the purpose set forth.

No. 60,402.—F. M. MCMEEKIN, Morrison's Mills, Fla.—*Cotton Gin.*—December 11, 1866.—The cotton is fed over the table to the reciprocating rake; the rollers draw the fibre from the seed. The adjustable plate retains the seed while the rake unravels the fibre from it. The brushes prevent the lint from being drawn under the machine.

Claim.—The combination of the rollers C, roller D, with adjustable bearings, slotted plate E, reciprocating rake G, and teeth F, as described, and brush I, arranged and operating substantially as and for the purpose described.

No. 60,403.—J. H. MEARS and C. W. YALE, Oshkosh, Wis.—*Knuckle Joint.*—December 11, 1866.—The bulb has five curved sides which fit a socket of corresponding shape: by this connection the rotative movement of one is imparted to the other, while a certain play is permitted, though the shafts be not exactly in line.

Claim.—The pentagonal socket, Fig. 1, in combination with the head, Fig. 2 substantially as set forth.

No. 60,404.—JOHN M. MILLER, Hamilton, Ohio.—*Grinding Mill.*—December 11, 1866.—The case around the millstones has a tube connecting with an exhaust fan chamber. The case has an adjustable slide at the hopper spout, where it enters the case, which governs the amount of air that is admitted into the latter.

Claim.—The arrangement of the tube G, blower case H, and receiving box K, in combination with the case E, provided with regulating slide *b*, operating substantially as and for the purpose described.

No. 60,405.—L. H. MILLER, Baltimore, Md.—*Safe.*—December 11, 1866.—The plates are secured together by projecting studs, around which the molten metal is flowed. The studs are secured to each of the plates, but do not extend entirely across the interval, and thus give no indication on the outside of their position.

Claim —First, a burglar-proof wall for safe and other similar purposes, of wrought and cast metal, made and held together by secret studs *a a*, all constructed substantially as described.

Second, the studs *a a*, projecting from the inner surface of one of the wrought metal plates but not extending to the inner surface of a contiguous plate, in combination with a cast metal filling between said plates, substantially as described.

No. 60,406.—CHARLES A. MOORE. West Brook, Conn.—*Tag.*—December 11, 1866.—The twine has a metallic fastening at each end. One end is secured to the tag and the other to the goods.

Claim.—The interposing or placing between metallic fastenings, the thread, tape, or twine, substantially as specified and for the purpose herein set forth.

No. 60,407.—JAMES D. MOORE, Grinnell, Iowa.—*Tatting Shuttle Winder.*—December 11, 1866.—The shuttle is attached to a clasp and rotated above the spool; the spring which serves to carry and revolve the shuttle, also forces its jaws open to receive the thread.

Claim.—The frame A formed of a single piece of metal, and bent into form substantially as described, in combination with the wheels G and H, and spool B, supported in said frame, as set forth.

Also, in combination with said frame, the spring I which holds the shuttle in its place, and opens the jaws of the shuttle to receive the thread freely, substantially as specified.

No. 60,408.—HENRY G. NELSON, LOCKPORT, N. Y.—*Water Wheel.*—December 11, 1866.—The wheel takes water at its periphery and discharges beneath. The buckets have a horizontal upper plate, inner and radial vertical plates, and an inclined lower plate. The water flows between the scroll plates and the gates ; the latter are turned in the direction of the curren when running, but extend from scroll to scroll to shut off the water when required. The gates are operated by a ring plate to which their arms are connected.

Claim.—First, making the arms of the bridge tree grooved for the purpose of conducting water into the step for lubrication, substantially as described.

Second, the adjustable hub E, in combination with the hub D, for the purpose and substantially as described.

Third, a water-wheel bucket having the parts *f¹ f² f³ f³* in combination, substantially as set forth.

Fourth, the stationary scrolls J, and horizontally moving gates L, when arranged with a segment ring N, operating gear and arms Q, and friction rollers S, as a means of opening and closing the gates L, substantially as described.

No. 60,409.—WILLIAM C. NEWKIRK, Piqua, Ohio.—*Oil Can.*—December 11, 1866.—A trough around the bottom of the can catches any overflow of oil ; an extra spout leads drip from the end of the spout to the trough below.

Claim —An oil can with a projecting rim around the outside, either at the bottom or above it, forming a channel, substantially as described, for the purposes herein set forth.

Also, the extra spout B, in combination with the said channel, substantially as described.

No. 60,410.—L. L. NEWMAN, Jackson, Mich., assignor to himself and A. C. ZEARING, Dayton, Ohio.—*Washing Machine.*—December 11, 1866.—The upright oscillating arm has a block connected to it with rollers upon its upper inclined faces. The side plungers are pivoted to the vertical arm and have rollers on their lower sides to act with the rollers in the block to rub the clothing. These side arms have extended vertical faces to press the clothing against the ends of the box.

Claim.—The combination of the parts C C and F, with their several rollers, pieces D, the shaft and box A, when arranged and operated substantially as described.

No. 60,411.—R. J. NUNN, Savannah, Ga.—*Pin for Attaching Wearing Apparel.*—December 11, 1866.—The pin has a head and a lanceolate point.

Claim.—A pin consisting of a shank, a disk, ball, oval, or other head, an enlarged portion and a retaining shoulder, substantially as described.

C P   II

No. 60,412.—SOLOMON OPPENHEIMER, Peru, Ind.—*Damper or Draught Regulator for Stoves.*—December 11, 1866.—This device is to be placed near the fire box, and consists of a cup in the sides of which are apertures controlled by a slide operated by a handle extending to any desired point; the bottom of the vessel has an ash-discharging trap.

Claim.—The cup A, constructed as shown, and provided with the discharge cap E, or its equivalent.

Also, in combination with the above, the arm C, and the handle D, and the connecting rod.

No. 60,413.—J. L. ORDNER, Cleveland, Ohio.—*Dumping Wagon.*—December 11, 1866.— The intermediate frame between the bolsters and box has metal rollers which run in the grooves of metal strips beneath the box, when the bed is launched to the rear for dumping. The box is secured by a hasp and pin.

Claim.—The rollers e, and frame F, when constructed and arranged in relation to each other as set forth, in combination with the described box A, provided with fastenings N, the whole supported upon the bolsters G, when used conjointly in the manner and for the purpose set forth.

No. 60,414.—V. PALMER, Castalia, Ohio.—*Churn.*—December 11, 1866.—The churn dasher has hinged flaps and a perforated cap and is reciprocated in the churn by a lever pivoted to a yoke erected on the base.

Claim.—The perforated dasher I, the adjustable wings I', and checks d, as arranged in combination with the links H and G, and the handle F, in the manner and for the purpose set forth.

No. 60,415.—ENOS H. PECK, Brownhelm, Ohio.—*Farm Gate.*—December 11, 1866.—The pivot rod of the gate is stepped in a socket with anti-friction balls, which gravitate into depressions of the socket to close the gate.

Claim.—The cap d, inclined planes h i, and balls j k, in combination with the centre post B, brace C, and counter balance gate, as and for the purpose set forth.

No. 60,416.—JOHN M. PERKINS and MARK W. HOUSE, Cleveland, Ohio.—*Lamp.*—December 11, 1866.—The wick tube is isolated from the reservoir by an annular space except at the supply pipes which cross the intervening air space, and small tubes which allow gases to pass to the wick. The walls of the air duct are carried up around the lower part of the burner.

Claim.—First, the combination of the annular reservoir A, the annular air chamber D, the burning chamber B, the supply pipes P, and vent tubes V, constructed and arranged substantially as and for the purposes described.

Second, the collar C, in combination with the perforated burner G, when both are so constructed and arranged as to extend the cold air chamber D up around the perforated portion of the burner, substantially as shown and described.

No. 60,417.—BARTON PICKERING, Milton, Ohio.—*Apparatus for Charging Air with Gasoline*—December 11, 1866.—The tank blower forces air through the vaporizer. The vessel containing the gasoline is rotated within the tank, which, together with the holder for the carburetted air, rests upon the vaporizer as a base.

Claim.—First, the tube o², to supply mixed air or gas to the vessel C, from the vessel D, connected substantially as described, for the purpose set forth.

Second, the supply tube o³, connecting the vessels C and D, substantially as described and for the purpose set forth.

Third, arranging the supply vessel C, within the vessel A, substantially as described.

Fourth, the construction of the generating vessel D, with the partitions i i i, having bagging or other suitable material on these surfaces, the orifices k k k, giving an alternating direction to each, the air and gasoline, as described and for the purposes set forth.

Fifth, the vessels A and A', vessels B and B', the vessel C, the vessel D, the tubes e f c h n, the stop cocks o¹ o² o³ o⁴, the whole being constructed and combined substantially for the purposes set forth.

No. 60,418.—FRANCIS A. PRATT, Hartford, Conn.—*Sewing Machine.*—December 11. 1866.—The improvement affords a ready means of adjustment for the wear of the shuttle driver or its race, and to secure steady motions.

Claim.—The employment of the gib or shoe d, and cross-head a, with the way c, substantially as and for the purpose described.

No. 60,419.—MILES PRATT, Boston, Mass.—*Stove Boiler.*—December 11, 1866.—A vertical indentation in the side of the boiler forms one side of the flue, the other portion being formed by an attached plate. The flue thus occupies a part of the boiler space.

Claim.—The improved boiler as made, with a flue arranged against its sides or rear, and its ll as set forth.

No. 60,420.—JOHN P. PRIGE, Philadelphia, Penn., assignor to FRANK K. HIPPLE, same place.—*Welding Iron and Steel.*—December 11, 1866.—Cryolite is pulverized and its impurities removed; it is then used as is customary with borax.
Claim.—The welding of iron or steel by the aid of cryolite, as set forth.

No. 60,421.—EDMUND W. QUINCY, Lacon, Ill.—*Dice Box.*—December 11, 1866.—The table is enclosed by a bell glass and supports the dice. The table is depressed by a thumb piece and elevated by a spring, throwing up the dice which are confined by the glass.
Claim.—The arrangement of the disk D, and its connection with the flat spring E, by means of the piston or rod C, and the open bottom, which enables an easy renewal of the spring when necessary, all as and for the purposes set forth.

No. 60,422.—BARBERY S. RICH, Penfield, N. Y., administratrix of the estate of J. C. RICH, deceased.—*Cultivator.*—December 11, 1866.—The shares have longitudinal adjustment on iron strips, and these strips have lateral adjustment at the rear by an expansible cast metal frame to which the handles are attached.
Claim.—First, the standard frame C, provided with the lugs i' at the top, and the parts *k l l n* at the bottom, when combined with expanding arms E E of a flexible metallic cultivator, the whole operating substantially as and for the purpose specified.
Second, the gauge wheel stirrups, composed of two counterparts *r r*, and provided with the projections *t t*, when combined with the flat side of *a*, of the cultivator frame, as herein set forth.

No. 60,423.—J. P. RILEY, Philadelphia, Penn.—*Hydrant.*—December 11, 1866.—The upper part of the hydrant case is removable and exposes the lower portions to be detached by a wrench and then withdrawn. The plug is hollow and operated by a partial rotation.
Claim.—First, the hollow plug H, and shank *k*, in combination with the box G, substantially as described for the purpose specified.
Second, the combination and arrangement of the stock D, base B, cap C, section A, key K, and its shield plug H, as described, and box G, substantially as described, as and for the purpose specified.

No. 60,424.—DANIEL T. ROBINSON, Boston, Mass., assignor to himself and NATHANIEL JENKINS, same place.—*Bottle Stopper.*—December 11, 1866.—The wooden stopper has a rubber washer secured around its lower edge, and on its upper part are transverse grooves crossing each other at right angles. The binding wire passes through one of these grooves, and in the other is a cam through which the binding wire also passes. By vibrating this cam the stopper is tightly fastened in the bottle.
Claim.—The improved bottle stopper fastening as composed of the slotted stopper or plug B, the cammed lever C or its equivalent, and the wire *e*, the whole being arranged and combined together in manner and to operate as specified.
Also, in combination with the above described arrangements of parts, the rubber washer or covering *k*, essentially in manner and for the purpose as described.

No. 60,425.—FREDERICK O. ROGERS, Niles, Mich.—*Constructing Buildings.*—December 11, 1867.—The walls are formed of a series of horizontal and perpendicular strips, each series laid transversely in respect to the contiguous one on each side. The roof is similar, except that one of the courses has the inclination of the roof.
Claim.—Constructing the walls and roofs of wooden buildings, whole or in sections, by the union of two or more series, or layers, of laths or battens, crossing each other transversely, the said walls and roofs, or sections thereof, being united so as to form a firm structure without the use of additional walls, frame work, or braces, substantially as and for the purpose herein specified.

No. 60,426.—LEVI ROSS, Springfield, Mass., assignor to himself and OLIVER K. PHILLIPS, same place.—*Curtain Fixture.*—December 11, 1866; antedated November 24, 1866.—The ends of the curtain are connected by tapes, which, with it, run over the upper and lower rollers, being moved by cords. The curtain may be adjusted to shade the upper or lower portion of the window or the whole of it.
Claim.—The mode of hanging window curtains herein described by means of the two rolls A B, connected with the curtain by the tapes E E F F, when combined with the two cords G H, and their respective rollers or pulleys *c d*, arranged and operated substantially as herein set forth.

No. 60,427.—NATHANIEL ROWE, Emmettsburg, Md.—*Harvester.*—December 11, 1866; antedated December 4, 1866.—The gearing is brought into a compact form near the centre of the frame, and is all enclosed in a casing made in form like two transversely arranged cylinders, one of which surrounds the main drive axle, and is furnished at each end with a cylindrical flange or collar, to which the forked end or bounds of the tongue plate are attached. A forked slotted lever of peculiar construction and arrangement is employed for adjusting the height of the cutting apparatus.

Claim.—First, arranging the entire gearing at the centre of the machine and surrounding it with a casing E, the gearing and casing being constructed and arranged substantially as set forth.

Second, the detachable piece P, in combination with the shaft N, and casing E, when the latter is constructed with a movable cap that covers the gearing ¦on the axle and the piece P, substantially as and for the purpose set forth.

Third, attaching the braces of the tongue T by a collar to a flange E², projecting from and being a part of the casing E, substantially in the manner set forth.

Fourth, the combination of the driver's seat V, and springs V², with the standards U, when the same are permanently attached to and on the opposite side of the axle from the braces of the tongue.

Fifth, the slotted and bifurcated lever S, when connected with the cutter bar in front and rear, substantially as and for the purpose set forth.

No. 60,428.—C. D. RUTHERFORD, Brooklyn, N. Y.—*Corset.*—December 11, 1866.—The frame is a skeleton, and network covers the bosom.

Claim.—A corset made of a skeleton or open form, in combination with the network F F, the whole being constructed substantially as herein described and for the purpose specified.

No. 60,429.—STEPHEN SARGENT, Lowell, Mass.—*Oil Can.*—December 11, 1866.—The venting chamber of the oil can allows a free discharge of oil and is accessible for cleaning purposes and the removal of obstructions.

Claim.—The combination of the venting chamber A, venting tube C, passage c or g, or their equivalent, with the oil can, the whole arranged substantially as and for the purpose set forth.

No. 60,430.—JOHN GEORG SCHMIDT, Rochester, N. Y.—*Machine for Cutting Bungs for Barrels.*—December 11, 1866.—The cutter is adjustable to cut different sizes of bungs and taper them at the same operation. The central spring-mandrel holds the blank, and the cutter descends at an angle coincident with the taper desired.

Claim.—Adjusting the cutters of a bung cutter by means of the screws g, springs f, and pins k, so as to enable it to cut bungs of different sizes and taper, substantially as described.

No. 60,431.—GEORGE SCOTT and JOHN W. SMITH, Boston, Mass., assignors to selves and JAMES SMITH, same place.—*Skate.*—December 11, 1866.—The middle runner is rounded on the edge and the side runners have acute edges. The heel post has a slot to engage the head groove of the heel screw and the screw is held in position by the foot plate.

Claim.—The arrangement as well as the combination of the rocker B, and the two runners C C, and the foot rest A.

Also, the combination of the foot rest with the inclined toe beak a, arranged on it, and to operate with the beak c, when applied thereto, substantially as set forth.

Also, the mode of applying the heel screw to the skate, viz: by extending it up through the foot rest, and supporting its head in a bearing to extend into the nick of the screw, as described.

No. 60,432.—TAL. P. SHAFFNER, Louisville, Ky.—*Magnet for Telegraphs.*—December 11, 1866; antedated November 25, 1866.—The horseshoe core is wound with copper wire of, say, No. 22 size, and with alternate convolutions of, say, No. 32 size. Each coil is insulated; the larger is connected with a battery and the smaller to the hollow bobbins, through which the armature passes.

Claim.—Making an electro-magnet of the armature by means of a current induced by the main circuit, substantially as herein set forth, in connection with the electro-magnet A A, or its equivalent, for telegraphic purposes.

No. 60,433.—ISAAC MERRITT SINGER, Yonkers, N. Y.—*Sewing Machine.*—December 11, 1866.—The objects of these improvements are as follows : To prevent a slackness of shuttle thread between the shuttle and the work, and to prevent its being split by the needle. To hold the shuttle laterally in place by a spring, and still allow it to move loosely. To maintain the shuttle point, while entering the loop, always in the same position. To facilitate the movements of an oscillating shuttle by a reciprocating instrument. To secure the taking up of the slack of the needle thread. To deliver the thread to the tension device always without strain upon it. To reverse the feed or vary its extent, by the same instrument. To regulate the feed when reversing without examination by the eye.

Claim.—First, the combination in a sewing machine of the following instrumentalities, viz: the reciprocating needle carrier and an oscillating shuttle, having its delivery eye coincident with its centre of oscillation, substantially as set forth.

Second, the shuttle oscillating substantially as described, having a delivery eye arranged at one side of the plane of oscillation of the shuttle point, substantially as set forth.

Third, the combination in a sewing machine of the following instrumentalities, viz: the shuttle, spring, holder, and stop for the spring holder, substantially as set forth.

Fourth, the combination in a sewing machine of the following instrumentalities, 'viz: the block for holding the shuttle in place at one side, and a shuttle guide for holding the shuttle at its other side, so arranged that it guides the shuttle only when its point is in the vicinity of the needle, leaving it free at other times, substantially as set forth.

Fifth, the combination in a sewing machine of the following instrumentalities, viz: the oscillating shuttle drivers, reciprocating connecting rod, and spring acting crosswise to said rod, substantially as set forth.

Sixth, the combination in a sewing machine of the following instrumentalities, viz: the needle holder, oscillating shuttle, take-up lever, cam, and spring, substantially as set forth.

Seventh, the combination and arrangement in a sewing machine of the following instrumentalities, viz: the spool support, thread tension, and thread-slackening mechanism, substantially as set forth.

Eighth, the combination in a sewing machine of the following instrumentalities, viz: the feeding instrument, bar reciprocating crosswise to the movement of said instrument, and turning, slotted, regulating plate, with its slot extended at opposite sides of the centre on which it turns, substantially as set forth.

Ninth, the combination in a sewing machine of the following instrumentalities, viz: the reversing and regulating lever that controls the feed, and two stops, substantially as set forth.

No. 60,434.—FREDERICK H. SMITH, Baltimore, Md.—*Bridge.*—December 11, 1866.— The central post is supported by rods passing to opposite ends of the straining beam. The other posts are supported by the first named bars and separate bars to the ends of the straining beam.

Claim.—Constructing a suspension truss in such a manner that all posts other than the centre post C shall be supported by bars or braces, one of which, as D², extends from the foot of the post C¹, to be supported to an end of the straining beam B, and the other bar or brace D¹, supporting said post C¹, being in a prolongation of the line of the bar or brace D, supporting the centre post C, from the opposite end of the straining beam, substantially as described.

No. 60,435.—STEPHEN SPOOR, Phelps, N. Y.—*Gate.*—December 11, 1866; antedated November 21, 1866.—The gate slides upon a roller carried on a pivot pin on the hinge post.

Claim.—The combination with a vertically revolving roller supporting the upper rail, of a roller support C, having a horizontal joint and composed of guards k k, recessed standards i i, and bearings k k, socket b, and conical bearing pin c, constructed and arranged substantially as shown and described.

No. 60,436.—JOHN STADERMANN, New York, N. Y., and HENRY SAUERBIER, Newark, N. J.—*Breast Protector.*—December 11, 1866.—Explained by the claim.

Claim.—Artificial breasts formed by swaging or striking up, out of one and the same piece of wire gauze or wire cloth, two protuberances of proper dimensions, and having the other portion of the wire gauze or cloth swaged or struck up to conform to the shape of the chest of the wearer, substantially as shown and described.

No. 60,437.—HOMER H. STUART, Jamaica, N. Y.—*Meter for Gas and Liquids.*—December 11, 1866.—The cylinder encloses a shaft with spring valves which revolve within it, being impelled by the passing liquid. As the vanes pass the induction tube they are partially opened by the spring and fully opened by the force of the liquid ; in this position they continue till they reach the eduction, when the fluid is discharged, and by an automatic movement the vanes are folded back upon the hub and pass to the induction tube again by a smaller circle.

Claim.—Rotating the axle of a meter by vanes that are alternately extended and folded or retracted, substantially as described.

Also, by this invention to measure a fluid body by passing it transversely through a cylinder or other circular vessels containing vanes revolving in said cylinder or circular vessel, in such a manner that the said vanes, while moving in the direction of the flowing body, are radii of a circle as large as the inner surface of the cylinder, and while returning through the lower part of the cylinder to the point where the fluid is received they are folded or retracted so as to be the radii of a circle smaller than the periphery of the cylinder.

No. 60,438.—WILLIAM THOMAS and WILLIAM RHOADES, Mukwonago, Wis.—*Cheese Press.*—December 11, 1866.—The cheese is placed between the bed piece and the cross-beam, and its weight causes the pressure upon it. The bed rests upon levers linked to the cross-beam, and the pressure is regulated by engaging the pawls below the bed with certain notches in the supporting levers.

Claim.—The combination of the notched levers I and K with the pawls T T, arms V V, bed piece M, and cross-beam B.

No. 60,439.—ASAHEL TODD, Jr., Pultneyville, N. Y.—*Stamp for Marking Sheep.*—December 11, 1866.—The letters are cast in skeleton form, and have a projecting shank which forms a handle.

Claim.—As an improved article of manufacture, the metallic skeleton marking stamps, constructed as and for the purposes herein shown and described.

No 60,440.—CHARLES TRUESDALE, Cincinnati, Ohio, assignor to himself and WILLIAM RESOR & Co., same place.—*Cupola Furnace.*—December 11, 1866.—The upper series of tuyeres have smaller openings than the lower.

Claim.—The provision in a cupola or blast furnace of the several tiers of tuyeres in an ascending series with diminished issues, substantially as set forth.

No. 60,441.—OTIS TUFTS, Boston, Mass.—*Elevator.*—December 11, 1866.—This is an improvement on his patent of May 28, 1861. The ends of the cords are held in screw sockets, which have adjusting nuts to regulate the length of the cords.

Claim.—Attachment of a suspending rope of an elevator car by means of an intermediate device *f*, adjustable with reference to the car or drum, substantially as described.

Also, in combination with an elevator car, its winding drum and two or more ropes or chains for hoisting and lowering the car, means for relative mechanical manipulatory adjustment of the length and tension of said ropes or chains, arranged to operate substantially as described.

Also, in combination with a winding drum of two rope bed curves, reversed in position with respect to each other, as set forth.

No. 60,442.—OTIS TUFTS, Boston, Mass.—*Elevator.*—December 11, 1866.—This is an improvement on his patent of May 28, 1861. The ropes are severally attached to the pivoted connection of links, which, by means of intervening links, are connected to the car frame. This arrangement allows of the slackness of one or more ropes being distributed among the others.

Claim.—For the purpose of automatically adjusting the strain upon the ropes or chains *b b*, the mechanism herein described, when arranged to operate substantially as specified.

No. 60,443.—OTIS TUFTS, Boston Mass.—*Elevator Guide.*—December 11, 1866.—The car has guide rollers on spring rods, which keep the rollers in contact with the slides. The rollers may be made of elastic material. The slide has a ratchet with vertical re-enforcing flanges.

Claim.—Combining the suspended car of an elevator with the ways or rails which confine it, by means of guides kept by springs constantly in contact with said ways or rails, when said guides are so arranged as to be capable of motion toward and from the rails.

Also, combining with the car and rails of an elevator, guide wheels provided with soft-surfaced peripheries, so as to operate as set forth.

Also, combining with the car and rails of an elevator, guide wheels having elastic peripheries, for the purpose specified.

Also, the rails or ways *a*, provided with ratchet teeth *b*, re-enforced by flanges *c c*.

No. 60,444.—OTIS TUFTS, Boston, Mass.—*Means for Oiling Bodies Rotating around Shafts.*—December 11, 1866.—The hub has packing boxes at the ends, and an annular chamber from which oil holes pass to the shaft. Radially above these oil holes are conveyers to drop the oil thereinto when the hub is at rest.

Claim.—The construction, substantially as shown, of a loose pulley, clutch, or other similar device or the hub thereof, with a removable head or heads, when the removable head or heads are provided with suitable packing, so as to form a closed cavity for containing lubricant directly around the parts to be lubricated.

Also, the oil cups *h*, when arranged within the cavity aforesaid, so as to operate as described.

No. 60,445.—P. H. VANDER WEYDE, Philadelphia, Penn.—*Continuous Percolator and Filtering Machine.*—December 11, 1866.—A series of boxes are arranged on the steps of an endless belt platform. The liquid passes down the side of the boxes, so as to filter upward. By moving the belt platform, the exhausted filters are removed and new ones are placed at the lowest part of the belt.

Claim.—First, the above-described upward-acting percolators, into each of which the liquid enters below the solid material and overflows at the top into the next one.

Second, the placing of them on steps, so that the liquid flows from the more exhausted solid material into that which is later added, coming in its course all the time in contact with fresher material.

Third, the manner described of displacing a series of any number of them on the inclined rail track by turning the wheels A and B, in order to remove only the vessel above containing exhausted solid material and supplying one with fresh material at the lower end, without interrupting the operation, thus effecting a saving in material as well as of time.

No. 60,446.—GEORGE W. WALKER, Boston, Mass.—*Cooking Stove.*—December 11, 1866.—
A hot-air chamber is placed before the fire chamber. The ash pit runs beneath this air
chamber, and by removing the slide division between them, broiling may be done. The oven
extends beneath the ash pit.

Claim.—The construction, substantially as shown and described, by which the space above
the ordinary high hearth or ash pit of the fire box of the common form of cooking stove is
utilized for the purposes specified, said construction consisting of the extension of the top and
sides of the stove over and to the front of said hearth or ash pit, when provided with doors
which enclose the space and retain the heat therein while capable of being opened for the
various purposes connected with the utilization of said space.

Also, in combination with such an enclosed space over the aforesaid hearth or ash pit, of
the enclosure of the space below the same, so as thereby to increase the capacity of the oven
of the stove, substantially as described.

Also, in a cooking stove, the arrangement of the oven with side and front doors and double
walls or plates on either side and at the front of the oven between the front and side doors
thereof, when the spaces enclosed between said double walls open into the lower flue, so as
to become charged with heated air, and thus heating the oven as well as affording the
strength needed in the structure of the stove.

No. 60,447.—J. C. WALKINSHAW, Leavenworth, Kan., assignor to himself and Jos. W.
MCGONIGLE, same place—*Corn Planter.*—December 11, 1866.—The seed slides are
operated vertically by means of a transverse bar from wheel to wheel, and moved by curved
cams on the wheels. The ploughshares are carried on the discharge spouts and are adjustable vertically. Additional spouts furnish other seeds or fertilizers.

Claim.—First, the arrangement of the cams *b b* on the driving wheels A A, in combination
with the dropping bar F, the vertical slides *k k*, the seed compartments *a a*, and the conductors *g g'*, constructed and operating substantially as and for the purposes herein
described.

Second, the lever frame *l l*, connected with the ploughs *h h* for raising and lowering them,
constructed and arranged substantially as shown and described herein.

No. 60,448.—CORNELIUS and ZACHARIAH WALSH, Newark, N. J., assignors to CORNE-
LIUS WALSH.—*Travelling Bag Frame.*—December 11, 1866.—One jaw is bent flatwise, and
has a strengthening plate secured to it by tenons at the end and edges. The other jaw is a
piece bent edgewise and jointed to the former.

Claim.—The combination of jaws A A' with the ends *a*, as described, the bar B with the
tenons *c* secured in the jaw A', all constructed and arranged in the manner and for the purpose
herein specified.

No. 60,449.—WILLIAM T. WATSON, Nottingham, Md. —*Cultivator.*—December 11,
1866.—The teeth are curved forward, and are inserted in metallic plates in the rails.

Claim.—First, the teeth C of a cultivator when constructed as set forth.

Second, the combination of the frame A, teeth C, and metallic sockets D, when said several
parts are respectively constructed and the whole arranged substantially as set forth.

No. 60,450.—PATRICK WELCH, New York, N. Y.—*Type Dressing Machine.*—December
11, 1866.—The type are assembled (letter up) with rules between the rows. The rows are
automatically passed between adjustable knives, and arranged in order after the operation.

Claim.—First, setting up the types in lines standing upon their feet in the manner described
to prepare them to be fed to the machine, substantially as herein set forth.

Second, the combination with a table or bed having a channel for the types formed upon it,
substantially as described, of the knives M M and feed bar D, as and for the purpose set
forth.

Third, the combination with the knife beds L L, forming a portion of a channel for the
types of an adjustable support W, substantially as herein set forth.

Fourth, the combination of the knives N N with the knives M M and feed bar D, substantially as set forth.

Fifth, the combination of the movable guide U and the pusher V with the knife beds L L,
forming a channel for the types and knives M M, substantially as described.

Sixth, the combination of the feed bar D with the frame or pusher *j*, the said pusher being
operated substantially as described.

Seventh, the combination of the stop *u* with the feed bar D and pusher *t*, substantially as
herein set forth.

Eighth, the combination with the movable guide U and pusher V of the yielding guard *y'*
and adjustable guard *u'*, substantially as set forth.

Ninth, the combination with the puncher V and adjustable guard *u'* of the stop or holder *t'*,
substantially as herein described.

No. 60,451.—L. P. WILCOX, Brooklyn, N. Y.—*Lathe Carrier or Dog.*—December 11,
1866.—The saddle is secured in the lathe dog by means of a projection on the saddle, which
enters a right-an led slot in the inside of the do .

Claim.—The saddle or supplementary block B, constructed and fitted to the carrier, substantially as herein set forth for the purpose specified.

No. 60,452.—SAMUEL R. WILMOT, Bridgeport, Conn.—*Lantern.*—December 11, 1866.—Each side of the oil pot in the base of a lantern has cavities large enough to permit the manipulation of the spring catches by which the lamp is attached to the body.

Claim.—The construction of the base of the lantern with an internal oil reservoir, and with cavities *s* on opposite sides of the said reservoir, for the reception and manipulation of the spring catches *n*, by which the base is attached to the body of the lantern, substantially as herein set forth for the purpose specified.

No. 60,453.—JAMES WILSON, Wilmington, Del.—*Stove-pipe Elbow.*—December 11, 1866.—The elbow is made of cast and sheet iron, has a door for the removal of soot, and a slide damper.

Claim.—First, an improved stove-pipe elbow formed by combining the door frame and door C D, the plates A and B, and the part E with each other, substantially as described and for the purpose set forth.

Second, the combination of the sliding damper F with the elbow, substantially as described and for the purpose set forth.

No. 60,454.—JOHN E. WOOTTEN, Cressona, Penn.—*Padlock.*—December 11, 1866.—The catch hasp interlocks with a catch on the bolt cylinder. The end and the inner shoulder of the pin key move the two tumblers to the proper positions to allow their projections entering the wards and permit the unlocking of the catch.

Claim.—First, the combination of the case A, cylinder B, and its arm C with the hasp E, and the catch on the end of the same adapted to a catch on the end of the arm C, the whole being arranged and operating substantially as and for the purpose herein set forth.

Second, the combination of the cylinder B, circular tumblers and their projections, the hollow cylinder *h*, and the cylinder H with its wards, the whole being arranged and adapted to the casing A, substantially as and for the purpose herein described.

No. 60,455.—JAMES YOUMANS and JOHN REED, Davenport, Iowa.—*Hydro-carbon Burner for Heating Purposes.*—December 11, 1866.—The liquid passes from the reservoir to an annular chamber, and from thence up the heated disk stem, where it is vaporized, and passes into a lower chamber, from whence issue the burner tubes. The burners are situated below the disk.

Claim.—A gas generator constructed with a chamber C, divided by a partition D into two compartments E and F, and having also a pipe I, burners L, disk M, and stem N, said several parts being respectively constructed and arranged in relation one to another substantially as set forth.

No. 60,456.—THEODORE ZINCK, New York, N. Y., assignor to himself and F. W. KALBFLEISCH, same place.—*Self-regulating Tension for Sewing Machines, &c.*—December 11, 1866.—An adjustable spring brake bears upon the thread on the spool. The thread passes over an arm projecting at right angles from the brake. Any impediment in delivery causes the thread to pull the arm backward and relieve the pressure of the spring.

Claim.—The arm C, in combination with the friction spring B and bobbin A, constructed and operating substantially as and for the purpose described.

No. 60,457.—JAMES S. ALLUMS, Cusseta, Ga.—*Cotton or Hay Press.*—December 18, 1866.—The box is made of upright timbers braced and tied and is supported upon a frame in the gin-house. The cotton is entered at the top, the door closed, and power applied to the screw beneath the follower, which forms a sliding bottom.

Claim.—The mode herein described of constructing the frame and supporting the same by strong iron rods, in the manner and for the purpose set forth.

Also, the combination of the frame so constructed with the modes of supporting the frame and the screw, in the manner and for the purpose set forth.

No. 60,458.—LEONARD and IRA ANDREWS, Biddeford, Me.—*Blacksmith's Forging Apparatus.*—December 18, 1866.—The trip hammer has a spring in the rear to confer additional force, and the tripping cam is worked by treadle and crank wheel with suitable connection to the cam shaft.

Claim.—The arrangement of the treadle *c*, link *d*, crank shaft *a*, balance wheel *b*, truck *f*, shaft *i*, having the truck *h*, and cam *k*, hammer *m*, and anvil *t*, in order to constitute, when the trucks *f* and *h* are connected by a belt, a machine which can be operated by a single person, in the manner herein set forth and working as described.

No. 60,459.—WILLIAM C. BAKER, New York, N. Y.—*Steam Generator.*—December 18, 1866.—The pipes or tubes through which the water circulates are embraced by a partition which directs the caloric current across such pipes, upward and then downward. Combined with the generator is a supplemental water chamber for aiding the circulation of water, and an upper steam chamber which connects with the upper end of the sinuous system of pipes. The receiving chambers at each end which collect and forward the circulation at each bend, form supports for each other.

Claim.—First, the partition *p*, dividing the front from the rear of the boiler, when combined with a tubular boiler constructed as herein set forth, and directing the heat in its course between the tubes, as and for the purposes described.

Second, the supplemental water chamber *g*, combined with the circulating tubular boiler, as specified.

Third, the additional steam chamber *c'*, connecting with the steam chamber *s*, substantially as and for the purpose set forth above.

Fourth, the connections or bends which bear against and support each other vertically and laterally, by which the tubes are supported, as and for the purposes described.

No. 60,460.—E. P. BARRABÉ, Paris, France.—*Machine for Hulling Wheat.*—December 18, 1866.—Inclined serrated plates are attached to the case of the machine and intervene between similar plates attached to the vertical revolving shaft. The grain is alternately ejected from the latter and returned by the former and is thus subjected to the succession of rasping devices, and is eventually discharged at the bottom of the case and cleaned by a blast of air.

Claim.—First, the case D, and its plates *c*, in combination with the shaft C, and its plates *b*, when each of the latter plates is greater in diameter than the plate next above it, for the purpose specified.

Second, the driving shaft E, and disk *d*, in combination with the shaft C, its adjustable roller *f*, and the lever *k*, or its equivalent, the whole being arranged and operating substantially as set forth

No. 60,461.—IRA W. BARTLETT, Otter Creek, Ill.—*Plough.*—December 18, 1866.—The forward end of the beam is supported by an axle and wheels. By means of a lever within reach of the ploughman, the inclination of the fore carriage is changed to adjust the wheels vertically and thus gauge the depth of the furrow. The coulter bar is maintained at its adjustment by its serrated edge, which is clamped against a toothed bar by a set screw.

Claim.—First, the axle and wheels E G G', combined with the rod H, and lever I, when employed in connection with the beam A, for governing the depth of the plough, as herein set forth.

Second, the combination of the ratchet bar *l*, and set screw *n*, with the toothed coulter bar L, arranged and operating as herein set forth.

No. 60,462.—E. S. BARTHOLOMEW, Westfield, N. Y., assignor to himself and C. H. BALLOU, Cleveland, Ohio.—*Preserving Fruits, Meats, and other Substances.*—December 18, 1866.—The air-tight vessel containing the fruit has exhaust tubes leading from the bottom, and is connected with a purifying vessel into which carbonic acid is passed from the retort. The retort contains ignited charcoal, and is supplied with air by a pump.

Claim.—First, the combination of the air-pump B, and close combustion retort D, receiver and purifier G, provided with the perforated coil *h'*, with the hermetical preserving chamber A, arranged and operating substantially as set forth.

Second, a preserving chamber formed with inclined or converging sides *a a*, in combination with the several eduction pipes *c c*, and main eduction pipe *b*, when used in combination with an exhausting pump B, or equivalent, substantially as and for the purpose described.

Third, the sulphurous acid gas generator E, in combination with the retort D, and induction tube *f*, when used in connection with the air-pump or equivalent, for the purpose described.

Fourth, the employment of sulphurous acid gas in preserving meats, by injecting a small percentage thereof into the preserving chamber, in combination with the nitrogen and carbonic acid gases, substantially as set forth.

No. 60,463.—B. H. BARTOL, Philadelphia, Penn.—*Low Water Detector.*—December 18, 1866.—When the vertical tube becomes emptied of water and charged with steam, the elongation of the thermostatic rod therein, under the increased heat, raises the valve and sounds the whistle.

Claim.—The pipe A, rod *d*, and steam whistle C, constructed and arranged in respect to each other and to a steam boiler substantially as described.

No. 60,464.—C. L. BAUDER, Cleveland, Ohio.—*Invalid Travelling Chair.*—December 18, 1866.—The locomotive chair is propelled by turning the crank which is geared to the forward wheel. Levers actuate the rear wheels, whose axles are pivoted and form the means of guidance. The back and foot piece are connected and are simultaneously moved to a sitting or reclining adjustment.

Claim.—First, the foot piece D, arms E and E', in combination with adjustable back B, and seat B', hinged and hung so as to operate conjointly, as and for the purpose substantially as set forth.

Second, the lips I I', in combination with the arms E E', and hinged foot piece, as and for the purpose set forth.

Third, the bracket C', consisting of two arms n n', in combination with the levers e, connecting rod d, link c, and chair, arranged as and for the purposes set forth.

No. 60,465.—G. and G. T. BENJAMIN, and H. S. WESTON, Millersburg, Ohio.—*Railway Car.*—December 18, 1866.—The motion of the double windows in opening is limited by rods, in combination with a deflector on the window sill, on which the window bottoms slide. The dust from below is kept out of the car by the deflector, and sparks and smoke from the engine are excluded by one of the double windows held partially open by the rod.

Claim.—The double window B B', the rod D, and deflector F, in combination with the car, as arranged in the manner and for the purpose herein set forth.

No. 60,466.—ERASTUS BLAKESLEE, Plymouth, Conn.—*Metal Plated Sole.*—December 18, 1866—The plating being in strips preserves the elasticity of the sole.

Claim.—Plating soles for boots and shoes with strips of metal, when the said strips are formed and arranged so as to completely cover the face of the sole, substantially as and for the purpose specified.

No. 60,467.—C. C. BLODGETT, Watertown, N. Y.—*Horse Hay Fork.*—December 18, 1866.—Within the tubular sheath is a centre rod, to which the claws are pivoted. The rod is locked by a trigger when the claws are extended, and the withdrawal of the trigger allows the claws to recede and the hay to slip from the harpoon fork.

Claim.—First, in combination with the tubular sheath and centre rod of a hay fork, as described, the slotted claws or barbs, elongated at their ends above the pivotal point, so that when the claws are projected from the sheath the said ends shall be brought in contact with the sides of the sheath, substantially as and for the purposes set forth.

Second, the combination with the tubular sheath of a hay fork, as described, of a centre rod or bar provided with flanges, arranged relatively to each other and to the locking mechanism of the fork in such manner that they shall not only guide and centre the rod, but also constitute the means whereby its motion in the sheath may be limited or stopped, substantially as herein shown and set forth.

Third, in combination with the centre rod or bar, arranged as described, the open-topped tubular sheath and the pin or equivalent device for preventing the withdrawal of the rod from and the rotation of the same within the said sheath, substantially as herein shown and set forth.

Fourth, the herein-described device for locking and unlocking the centre rod, the same consisting of a hoop or sleeve F, loosely encircling the sheath and combined with the spring G, and locking pin h, substantially in the manner and for the purposes herein shown and set forth.

Fifth, the guard formed on the centre rod, and constructed and arranged so as to protect the locking and unlocking device, substantially as herein shown and described.

No. 60,468.—JOSEPH BORDEN, Bridgeton, N. J., assignor to T. and J. BODINE, same place.—*Fruit Jar.*—December 18, 1866.—The cover is secured to the jar by a screw and cross bar, the latter fitting in recesses inside of the neck of the jar.

Claim.—The cover B, the set screw D, projecting from the under side of the same, in combination with the cross bar E, serving as a nut for the said screw and adapted to recesses or projections in the neck of the vessel, all as set forth.

No. 60,469.—JOHN H. BOSWORTH, Bath, Maine.—*Carpet Stretcher.*—December 18, 1866.—The tubular extension bar has a toothed head on each end, the teeth on one being arranged in a circle; one head rests on the carpeted floor and the other extends the edge of the carpet to the position for nailing. The jointed bar is rendered stiff by slipping a sleeve upon the hinge.

Claim.—The tubular and serrated head, as constructed and applied to one of the parts of the staff of the carpet stretcher, in the manner and for use as set forth.

Also, the combination of two toothed heads, two bars, and a slide tube, arranged and applied together, substantially in manner as hereinbefore set forth.

No. 60,470.—RICHARD C. BRISTOL, St. Clair, Mich.—*Water and Steam Separator for Steam Generators.*—December 18, 1866.—The steam from the generator is discharged, with what water it contains, into the vessel above, and the steam educted thence by a tube whose opening is above the discharge opening from the generator. The water which is intercepted is returned by a pipe to the generator.

Claim.—First, in combination with the steam generating apparatus A, the vessel C, disconnected portions A¹ B¹, of the steam pipe and the drain pipe C¹, arranged for joint operation in separating the water and steam flowing through the pipe A¹, discharging the water through the pipe C¹, and the steam alone through the pipe B¹, substantially as herein set forth.

Second, the within described arrangement of the vessel C, and its connections relatively to the boiler A, so that the water separated from the steam and descending in the pipe C¹, shall flow directly back to the boiler, without the necessity for intervening mechanism, substantially as and for the purpose herein specified.

No. 60,471.—CHARLES W. CAHOON, Portland, Maine.—*Propelling Vessels.*—December 18, 1866.—The frame rests upon two boats which are pivoted thereto. The undulatory motion of the vessels actuate pumps which imbibe water at the bow and eject it at the stern to impel the vessel.

Claim.—The application of the undulatory motion of the sea to the propulsion of vessels by means of pumps, and substantially as described.

Also, controlling the movement of the connecting rods by which the pumps are actuated, so that the length of stroke of the piston may be governed substantially as described.

No. 60,472.—JOSEPH W. CALEF, Salisbury, N. H., assignor to himself and JOHN R. FOLSOM, Stoneham, Mass.—*Tanning.*—December 18, 1866.—The hides are steeped for 30 or 40 days in a solution of hard hack, sumach, catechu, and glauber salt, with or without borax for hardening, or an arsenical solution for preserving the leather.

Claim.—In the process of tanning, the employment of the ingredients first described, when used as and in the proportions substantially as set forth.

Also, in combination with said tanning ingredients, the employment of the material for hardening sole-leather, substantially as set forth.

Also, the employment of the preservative solution, in connection with the tanning process, substantially as set forth.

No. 60,473.—ALEXANDER CARBNOW, Potsdam, N. Y.—*Churn.*—December 18, 1866.— The dasher has vertical arms, which revolve between other arms projecting from the bottom and from the lid. The cog gearing is attached to supports, which are so hinged as to allow of swinging down to admit of entrance to the churn.

Claim.—First, the arms B B, the standard E E, or their equivalent, as arranged and combined, with a drive and pinion wheel, and connected with a tub or churn, for the purposes herein specified.

Second, the arms k k k k, with the paddles l l l l l l l l, the breakers m m m m, and n n n n, as represented in figs. 2, 3, and 4, or their equivalents, for the purposes herein specified.

Third, the peculiar arrangement of the arms B B, the standard E E, the drive wheel f f, the pinion wheel h h, and the shaft i i, and their peculiar combination to and with each other, for the purposes herein specified.

Fourth, the peculiar arrangement and combination of the paddles l l l, &c., and the breakers m m m m and n n n n, for the purposes herein specified.

Fifth, the adjusting of the drive wheel f f to the pinion wheel h h, for the purposes herein specified.

No. 60,474,—PETER CHICK, Taunton, Mass.—*Air Engine.*—December 18, 1866.—Water is injected, in the form of spray, upon the incandescent fuel in the furnace and the resulting vapors are mingled with the volatile results of combustion. The injection pipe is removable and its jet is broken by a hemispherical plate whose concavity is presented downward.

Claim.—First, the inverted semi-sphere in the upper portion of the fire-box, substantially as shown and described.

Second, the semi-sphere, in combination with the fire-box, substantially as shown and described.

Third, the arrangement of the water injection pipe in such a manner that it may be withdrawn at any time, substantially in the manner shown and described.

No. 60,475.—JOHN H. CHIDESTER, Cleveland, Ohio.—*Computing Machine.*—December 18, 1866.—Nine disks, notched and numbered in their peripheries, are regulated on a shaft within a circumferentially and longitudinally slotted frame, to compute numerically by the turning of the disks in reference to figures on the frame.

Claim.—First, the series of disks D and toothed wheels B, arranged upon the shaft G, in combination with the case A, index openings b b b, and numeral springs a a, arranged and for the purpose set forth.

Second, the notches c and flange F of the disks and wheels B, provided with a series of numerals, in combination with the spring h, recess C, and shaft G, arranged in the manner and for the purpose set forth.

Third, a computing machine, when constructed, arranged, and operating in all its parts substantially as herein set forth.

No. 60,476.—GEORGE W. CHIPMAN, Melrose, Mass.—*Carpet Lining.*—December 18, 1866.—A sheet of wadding is enclosed between two sheets of paper and the edges of the latter cemented.

Claim.—As an improvement in the manufacture of carpet linings, the construction herein described, viz: a lining in which a thin sheet of fibrous material is confined between two sheets of fabric of close texture, by reason of the edges of said sheets being cemented together.

No. 60,477.—JOHN A. COFFEY, London, England.—*Distilling Apparatus.*—December 18. 1866.—In this still for the fractional distillation of petroleum the crude material flows from an elevated tank and traverses a coil in the still. The pipe forming the coil is continued and forms smaller flat coils in the chambers of the column, from each of which are separate draw-off pipes leading to condensers. The still has a pyrometer, and the oil, flowing through the main coil, and then through the series of coils, is discharged at the top of the column, and thence flows back over the coils in the chambers, where the portions of various boiling points are vaporized as the liquid descends to the still, the vapors being drawn off and condensed separately. A scraper keeps the tar from adhering to the still, and as the residuum in the still becomes too thick for further distillation it is drawn off by a pipe in the bottom of the still without stopping the operation.

Claim.—The improved constructions and arrangements set forth in regard to the distillatory portions of such apparatus, claim not being laid to any of the mechanical details thereof *per se* and apart from the purposes of the invention.

No. 60,478.—WILLIAM COGGESHALL, Springfield, Ohio.—*Paints, and Coating Wood, Stone, &c.*—December 18, 1866.—Composed of a mineral found in Ohio, and composed of vegetable bituminous matter, silica, alumina, peroxide of iron, lime, magnesia, sulphate of lime, and chloride of sodium. In applying, the pulverized mineral is spread upon the article, which is previously coated with linseed oil, varnish and lubricating oil.

Claim.—First, the method herein described for coating substances, by the application dry to properly prepared surfaces, of the hereinbefore described crude article, or any equivalent compound, composed substantially as set forth.

Second, the use of the aforesaid crude article, or any equivalent substance, in combination with any coloring matter when applied dry, substantially in the manner set forth.

No. 60,479.—V. and E. COLE, Detroit, Mich.—*Car Coupling.*—December 18, 1866.—The link has a projection at one end with an eye for the coupling pin. The coupling pin has a rubber washer, against which the projection rests to keep the link horizontal. The other coupling pin is supported when the cars are uncoupled by a sliding plate, which at the contact of the heads is driven into position for the passage of the pin.

Claim.—The pin *a*, with its rubber spring *b* and pointed link B, arranged to operate with the headed slide B^x, spring C, and bar E, in the manner and for the purpose herein specified.

No. 60,480.—O. K. COLLINS and WILLIAM B. GROVER, Woodbury, N. J.—*Vapor-burning Stove.*—December 18, 1866.—The burning fluid flows from an upper reservoir to a chamber, receiving heat from the flame by which it is vaporized. Any superabundance of gas passes through a condenser to a lower tank.

Claim.—First, the burner B, consisting of the chamber *y* communicating with an elevated reservoir, the flange *b* and disk *c*, with the intervening annular space *x* communicating with the said chamber *y*, the whole being arranged substantially as and for the purpose described.

Second, the combination of the above with the pipe *f* and valve *g*.

Third the gas-generating burner B, in combination with a pipe F, water-tank G, and valve *i*, and pipes C' and E.

No. 60,481.—JAMES M. CONNEL, Newark, Ohio.—*Door Spring.*—December 18, 1866.—The spring arm is attached to a projection from the hollow slotted mandrel. The latter is surrounded by the coiled wire spring and works on studs in the ends of the case.

Claim.—First, the hollow slotted mandrel C, or its equivalent, provided with a tongue which projects into the arm G, substantially as specified.

Second, the mode of journalling the mandrel C, or its equivalent, on the studs B projecting into the casing K and attached to the base plate A, substantially as described.

· Third. the coils D, in combination with the hollow slotted mandrel C, or its equivalent, the tongue *c* and arm G, substantially as described.

Fourth, the arrangement of the coils as conical frusta upon a core or mandrel of corresponding character, as and for the purpose described.

Fifth, the arrangement of the arm G and the semi-cylindrical portion G', rotating between the guides *k k* on the casing K and occupying in the rear enlargement K, substantially as described.

Sixth, the recesses *e* for securing the tangential prolongation of wire coil, substantially as described.

Seventh, the general combination of parts consisting of the mandrel C, or its equivalent tongue *c*, coiled springs D D, arm G G', casing K, studs B B, and base plate A, substantially as described.

No. 60,482.—DAVID T. CROCKETT, Newark, N. J.—*Hammer.*—December 18, 1866.—
Explained by the claim and cut.

Claim.—A hammer, screw-driver, and tack extractor, combined as shown, formed of one piece of metal, as a new article of manufacture.

No. 60,483.—ISAAC CRUM, Port Union, Ohio.—*Corn Planter.*—December 18, 1866.—The seed receptacle is within the hoe handle; the seed slide has finger grooves for the operating hand. An outer case forms a bottom to the seed cavity in the slide when it is presented to the receptacle for filling.

Claim.—In combination with a hoe, or its equivalent, having tube *a* and aperture *e*, the distributing plate *g*, constructed and operating as above described and set forth.

No. 60,484.—JOHN J. CURRIER and SAMUEL WELLS, Jr., Boston, Mass., assignors to themselves and JAMES H. PLAISTED.—*Apparatus for Finishing and Boxing Paper Collars.*—December 18, 1866.—This apparatus is for pressing and winding up paper collars, finishing the fold previously defined, imparting the curve to fit round the neck, depositing a certain number of collars concentrically in each box, and bringing the empty boxes consecutively into place, all automatically by one movement.

Claim.—First, a machine for finishing and boxing shirt collars made of paper or other material, consisting of the rollers B C and D, cylinder E, piston F, and wheel P, arranged and combined substantially as described.

Second, the combination and arrangement of the cylinder E, piston F, and rollers B C and D, substantially as and for the purpose specified.

Third, the wheel P and the devices for moving the same, consisting of cam 10, rod M, bell crank Q, arm R, ratchet wheels U and X, arm V, lever S, rod L, and spring N, substantially as described.

Fourth, the combination and arrangement of the cylinder E, piston F, and wheel P, said wheel having devices attached for giving it an intermittent motion, substantially as described.

Fifth, the combination and arrangement of the cylinder E, piston, F, shaft G, lever J, ring H, rods L and M, spring N, and cam 10, substantially as and for the purpose specified.

No. 60,485.—JAMES DAVIS, Buffalo, N. Y.—*Saw Mill.*—December 18, 1866.—The arbor of each set of gang saws is fixed on a separate laterally sliding frame. Both frames are moved by a single shaft. A spreading disk follows each saw. The log carriages are moved by a chain passing around a drum, and the latter is so connected with its motor that the feed movement of the carriages is slower than the return movement.

Claim.—The placing of the saw arbors D in sliding frames C C, arranged with gearing and racks, substantially as shown and described, to admit of the saws G on the two arbors D D being adjusted simultaneously toward and from each other by the turning of a single shaft I, as and for the purpose specified.

Second, adjusting the plates M upon the shaft L, by the means substantially as described, in combination with the adjustable slide J, as and for the purpose set forth.

Third, the two carriages O O', provided with the dogs P P, and operated through the medium of the drum R and chains Q Q, all constructed and operated substantially as shown and described.

No. 60,486.—JAMES DAVIS, Buffalo, N. Y.—*Saw Mill.*—December 18, 1866.—The upper bearing roller is connected to levers having adjustable weights, by which the pressure of the roller is regulated. The said roller can be retained in elevated position by racks upon the vertical arms pivoted to the weighted levers.

Claim.—Adjusting the rollers N N', by means of bars *d*, levers O, weights S, notched bars P, guide Q, and weights R, as and for the purpose specified.

No. 60,487.—S. T. DENISE, Branchport, N. J.—*Railroad Switch.*—December 18, 1866.—The short sections of rails are bevelled on top at one end, and have swivelled guide plates at the other to conduct the wheels on to the track.

Claim.—The bars or rails F, provided with swivelled pieces G and H, respectively, substantially as and for the purpose described.

No. 60,488.—A. T. DENISON, Poland, and E. P. FURLONG, Portland, Maine.—*Wearing Apparel Made of Paper.*—December 18, 1866.—Explained by the claim.

Claim.—The new manufacture consisting of articles of wearing apparel, made in whole or in part from paper formed into crinkles or flexures while in a pulpy or semi-pulpy condition, substantially as set forth.

No. 60,489.—DANIEL DENNETT, Buxton, Maine.—*Dumping Wagon.*—December 18, 1866.—The wagon body is pivoted to longitudinal bars which extend from the fore to the rear bolster. The side boards are divided obliquely before the rear bolster to admit of dumping. The tail-board is pivoted to the rear bolster and has oblique sides to agree with the oblique ends of the body sides, forming continuations to the same. A hook-ended arm at each end engages with eyes in the body to hold it in its elevated or horizontal position.

Claim.—First, the tail-board D, with its oblique sides so arranged as to enable the body to tip between the fore and rear axles of the wagon, substantially as described.

Second, the combination of the tail-board D with its oblique side beams A A and rod B, all arranged as described for the purposes specified.

No. 60,490.—EMANUEL DETWILER, Milwaukee, Wis.—*Heat Radiator for Stove Pipes.*—December 18, 1866.—A rectangular coil of flues is arranged in a slotted metallic case, and the stove flue furnished with a damper by which the caloric current may be made to traverse the coil.

Claim.—First, the T-iron ribs or braces A, combined in relation with the flues *d g*, substantially as herein set forth for the purpose specified.

Second, the horizontal flues *d g* and vertical flues E, arranged in relation to each other and with the flues *m n* and the pipe D, substantially as herein set forth for the purpose specified.

No. 60,491.—JACOB J. DETWILLER, Greenville, N. J.—*Igniting Illuminating Signals*—December 18, 1866; antedated December 5, 1866.—Improvemnt on the Coston telegraphic night signal. It is adapted to be fired by a percussion cap in a pistol into whose barrel the smaller end of the signal stock is inserted. The said end has an opening through which the fire of the cap is communicated to the fuse which lights the signal.

Claim.—First, arranging the fuse or quick-match C in connection with the signal and the stock A or B, substantially as shown and described for the purpose specified.

Second, making the stock A and B with a hole *d* in its axis, in combination with the transverse hole or groove *e* to receive the end of the match and the fire from the percussion cap, substantially as set forth.

Third, enveloping the signal and capping the lower end of the stock in a metallic case for the purpose of protecting it from punctures and dampness, substantially as shown in Fig. 12 and as described.

No. 60,492.—EDWARD P. EASTWICK, Baltimore, Md.—*Bone Black Kiln.*—December 18, 1866.—The upper and lower retorts are independently supported in their separate chambers in the kiln, the upper being exposed to less heat and discharging into the one below it.

Claim.—Connecting two or more vertical retorts by means of intervening bed plates and adapters or couplings resting in the floor of the upper chamber of the bone-black kiln, supporting the upper retorts independently and allowing a separate expansion and contraction of the lower retorts, substantially as herein described.

No. 60,493.—SAMUEL S. ELDER, Springfield, Ill.—*Churn, Beer-Cooler, &c.*—December 18, 1866.—The central shaft, carrying radial dashers, and the sleeve carrying the frame dasher, are revolved in different directions by their respective pinions and the single master wheel.

Claim.—First, the combination of the hexangular casing A and dashers D'' and G' revolving in opposite directions, when respectively constructed and arranged substantially as set forth.

Second, the combination of the dashers D'' and G', when carried in opposite directions up a system of shafts F F', and G, and collars D' and D''', arranged and operated substantially in the manner and for the purposes set forth.

No. 60,494.—JAMES EMERSON, Lowell, Mass.—*Comb.*—December 18, 1866.—The comb is curved vertically and horizontally to fit the lip, and has guards to sustain the mustache clear of the mouth while eating or drinking.

Claim.—First, the comb A, when made in proper form to fit the upper lip, with guards in front to hold the mustache, when made substantially as described.

Second, the nippers *i i* in combination with the comb for the purpose of holding the comb in the mustache, as described.

No. 60,495.—BENJAMIN M. ESTERLE, San Francisco, Cal.—*Wagon Wheel.*—December 18, 1866.—On the rims of the wheels are projecting plates which catch upon the edges of the rails of the street-car railroads and enable the wheels to climb thereon.

Claim.—As an improvement in carriage wheels, the use of the plate E, constructed as shown in Fig. 3 of the drawings, so that it may be used on the inside of the front wheels of a wagon and pass or slip over the lock or friction plate fastened to the carriage for the wheels to rub against in turning the wagon.

No. 60,496.—HENRY FEYH, Columbus, Ohio, assignor to himself and GEORGE T. EMORY.—December 18, 1866; antedated September 13, 1866.—*Steam Generator.*—Clusters of tubes of varying sizes are connected at bottom by hollow couplings and pass through the crown plate of the furnace. The inner ones are the longer and smaller in their bore, and being exposed to greater heat, the circulation is upward through them and downward through the larger outer tubes.

Claim.—First, pipes or tubes of different diameters, arranged so as to be exposed to the direct action of the fire, and connected at one end, for producing a forced circulation of water in steam boilers, substantially as described.

Second, the combination of the feeding pipes *c*, leading from the water space and below the water level, with the end couplings G and with pipes leading from said couplings above the water level in the boiler, substantially as described

No. 60,497.—C. D. FOOTE, Fond du Lac, Wis.—*Rock Drill.*—December 18, 1866.—The drill hammer is operated directly from the cylinder by compressed air or steam. The drill is retracted after each stroke. The frame is mounted on trunnions to admit of swinging to any position without moving its supports. The drill can be completely withdrawn from the rock by a longitudinal screw, operated by a hand crank.

Claim.—First, in drilling machines of the character above described, so arranging the cylinder and its attachments that the same shall be fed up to their work, substantially as set forth.

Second, mounting the cylinder and its attachments upon a horizontal frame hinged to the side of the main frame, so that the drilling mechanism may be swung out of line with the hole being drilled without removing the main frame, as set forth.

Third, arranging the cylinder in such a manner as to operate the drill by a blow direct from the end of the piston, substantially as set forth.

Fourth, the rod F, or its equivalent, arranged to operate substantially as set forth.

Fifth, the drill-holding device, with the opening in its side to permit the insertion or removal of the drill, as and for the purposes set forth.

Sixth, the bar *m*, arranged to operate the ratchet wheel T for the purpose of feeding the machine forward, substantially as described.

Seventh, the mechanism so arranged as to raise the drill from the rock at its cutting point and return it again at each blow of the hammer, as herein described.

No. 60,498.—J. FRASER and O. S. GARRETSON, Buffalo, N. Y.—*Stove-pipe Damper.*—December 18, 1866.—The corrugated, concave disks are united by lugs, present their concavities inward, and are rotated by an axial rod so as to present the edge or the face of the damper to the caloric current.

Claim.—First, connecting the two disks of a pipe damper together by means of lugs *f f* and corresponding slots, or their equivalents, in combination with the axial rod C, substantially in the manner and for the purposes set forth.

Second, the two scalloped or equivalently-formed disks B B' combined to form a pipe damper, substantially in the manner and for the purposes set forth.

No. 60,499.—JOHN A. FREY, Washington, D. C.—*Lamp Burner.*—December 18, 1866; antedated December 4, 1866.—The circular burner has square apertures and perforations to reduce the condition of heat to the oil reservoir. The sinuous course for the air prevents sudden puffs from reaching the flame directly. A water reservoir intervenes between the central air tube and the outer air chamber.

Claim.—First, the square apertures L and the perforation M of the circular burner to regulate and keep the burner cool without any metallic tubes passing into or through the oil or fluid.

Second, the outer air chamber B, in combination with the apron N and springs T, as herein described.

Third, the circular guard F on the top of the lamp, in combination with the burner, as herein described, to prevent any sudden transmission of air.

Fourth, the cone-shaped reservoir of water in the centre of the lamp, for the purposes set forth.

No. 60,500.—JEAN FROT, Orleans, France.—*Ammoniacal Gas Engine.*—December 18, 1866.—The generator is supplied with an ammoniacal solution and heated to the proper temperature, setting the gas free from the solution; passing through the condenser it is cooled, and thence into the dissolver it again mingles with the water contained therein to form the solution for resupplying the generator.

Claim.—The herein-described apparatus, by means of which ammoniacal gas may be substituted for steam in motor engines, the same consisting substantially of a combined condenser and dissolver, arranged as described, in which the ammoniacal vapor is condensed and dissolved continuously, as herein described and set forth.

No. 60,501.—ALLEN GILMORE, Fort Atkinson, Wis.—*Potato Digger.*—December 18, 1866.—The vines are caught by bars and the potatoes dragged off by the rakes upon endless chains, and carried over a screen. They pass from thence to another screen, from which they are taken by an elevator and deposited in a box at the front of the machine.

Claim.—First, the combination of shovel screen C, screen D, and revolving toothed shaft *a²*, with carriers *f² f²*, applied to a carriage A, and operating substantially as described.

Second, the arrangement of comb teeth over the shovel C', in combination with the carriers *f² f²*, substantially as described.

Third, the arrangement of a pressure roller g' in front of the carriers f^2 and over the shovel C', substantially as described.

Fourth, sustaining the shovel screw c against backward strain when said screen is suspended at its rear part from a shaft b' by means of segments d' d' and bearings e e, substantially as described.

Fifth, the clearer S, in combination with the carriers $f^2 f^2$, substantially as described.

Sixth, conducting the potatoes from the screen D upward and forward, and delivering them at a point which is near the front part of the machine, by means substantially as described.

Seventh, the use of a rake K for separating the vines from the potatoes, said rake being arranged upon an elevator and caused to discharge the vines over gate J and guard F, substantially as described.

Eighth, the application of a screen k to the inclined bottom of the elevator for separating the smaller from the larger potatoes, substantially as described.

Ninth, the construction of the screen D with a guard F upon its rear end, substantially as and for the purpose described.

No. 60,502.—JOHN SACHEVERELL GISBORNE, Liverpool, England.—*Signalling Apparatus.*—December 18, 1866.—For communicating by visible and audible signals, the latter calling attention to the former. Corresponding visible signals at each end of the line (say the bridge and helm of a steam vessel) are connected by the mechanical apparatus described, so that the motion of one will bring the counterpart sign into view on the other.

Claim.—First, the pulley e, stud c, pointer d, and handle f, arranged substantially in the manner described, as means for giving motion to one or more endless or double-line flexible motion conductors k to give either audible or visible signals, or both.

Second, in combination with the above the spring g, constructed to fall into the notches h, in the manner and for the purpose set forth.

Third, the dials b and i placed close together, substantially as shown on the drawings, so that they are illuminated by one lamp and can be seen at a glance.

Fourth, the employment of one or more flexible endless or double-line conductors k as means for conveying or communicating motion for operating signalling apparatus, and for conveying or communicating the motion of a rudder-stock, substantially as described.

Fifth, the combination of the pulley o, disk q, case n, with openings r, and motion conductors k, either with or without the bell or signalling apparatus, substantially as described.

No. 60,503.—SAMUEL GISSINGER, Lawrenceville, Penn., assignor to himself and DAVID E. HALL, Pittsburg, Penn.—*Clamp for Gluing the Tips on Billiard Cues.*—December 18, 1866.—The metal cylinder has springs which form guide clamps for the end of the cue, which is passed between, and is thus held centrally while the tip is attached by pressure.

Claim.—The springs o, or their equivalents, formed in such a manner as to have a bearing on the cue at two different points of its length, in combination with the conical chamber i, piston e, and spring f, the whole being arranged and operating substantially in the manner herein described and for the purpose set forth.

No. 60,504.—GEORGE P. GORDON, Brooklyn, N. Y.—*Printing Press.*—December 18, 1866.—An additional roller is placed in contact with two others to act as a distributor. A gripper plate operated by a cam removes the sheet from contact with the form. The distributing surface is a rotary disk.

Claim.—First, in combination with the inking rollers M, held and carried in a rocking roller frame, the use or employment of a third or supplemental roller M', whether said third or supplemental roller M' shall vibrate or not, substantially as and for the purposes shown.

Second, the grippers T to relieve the printed sheet or card from the form or types, constructed and operated substantially as shown.

Third, in combination with a revolving ink-distributing table or disk N, the use or employment of the inking rollers M and a third or supplemental roller M', or its equivalent, for the purpose specified.

No. 60,505.—WILLIAM D. GRIMSHAW, Newark, N. J.—*Capstan.*—December 18, 1866.—The pivot bar has a spiral spring upon it to support the barrel. The latter is moved by means of double pawl levers operating to turn bevel wheels engaging gearing on the base of the barrel. The lever pawls may be set to move the barrel in either direction.

Claim.—First, the combination of the shaft b, spring f, nut g, barrel c, and ring d with the base a, in the manner and for the purposes specified.

Second, the pawl o, constructed in the manner specified, in combination with the pointed spring socket r, wheel n, and handspike socket m, as and for the purposes set forth.

No. 60,506.—J. C. GUERRANT and B. J. FIELD, Leakesville, N. C.—*Machine for Engraving.*—December 18, 1866.—The plate to be engraved is attached to a holder, and the tool is secured to an adjustable arm; the vibrations of the latter as the stylus is drawn over the copy are duplicated in the desired proportion upon the plate. The devices are explained in the claims and cut.

Claim.—First, the arrangement, substantially as described, of the adjustable plates E F G with their slots and set screws, in combination with the supporting plate A, for the purpose and substantially in the manner set forth.

Second, the levers O O' and Q Q', with the adjustable arm P and joints *g* and *h*, arranged substantially as set forth, in combination with the adjustable staff L and sleeve M, for the purpose of allowing universal motion and adjustment to said levers O O' and Q Q', as set forth.

Third, the vibrating ark S and pattern frame T, constructed, arranged, and operating substantially as described and for the purpose set forth, in combination with the plate A and its adjuncts, as set forth.

Fourth, the arrangement, substantially as set forth, of the stylus or tracing point *m* and its adjuncts, whereby it will be always kept against the pattern, as described, in combination with the adjustable sleeve U, as set forth.

Fifth, the ring-holder Y, with its set screws *r r* or equivalent device, arranged and made adjustable by means of the staff V and sleeve W, substantially as set forth.

No. 60,507.—CHARLES M. GUSTIN, Saccarappa, Maine.—*Jack for the Manufacture of Boots and Shoes.*—December 18, 1866.—The arms have supports for the toe and heel of the shoe, and admit of rotation and of inclination on the supporting frame.

Claim.—First, the combination of the bolt *c*, and friction washer *o*, to prevent a too easy and ready revolution of the yoke A, upon the bolt in a horizontal plane.

Second, the combination of the crank-stop *d*, sliding in the blocks *e e*, with the holes in the segment C, as and for the purpose specified.

Third, the sliding toe-rest on the top of the arm *f*, when secured at any point in the manner described and for the purpose set forth.

No. 60,508.—WILLIAM HAILES, Albany, N. Y., assignor to himself and S. H. RANSOM.—*Attaching Covers to Kettles, Boilers, Stoves, &c.*—December 18, 1866.—The cover has a projection on which is a pivot pin and a stop. On the top of the kettle is a socket for the pin and a hooked lug. In one position the projection on the cover will pass the hook; the latter, in other positions, prevents the cover becoming detached.

Claim.—First, securing covers to vessels or other objects by a pivot connection, in such manner that the cover will be held down in place by an overhanging hook C, or its equivalent, substantially as described.

Second, the combination of the hook *h*, and the perforated portion *f*, with the projection *a*, and its stud *b*, substantially as and for the purposes described.

Third, the construction of the projection *a*, with a stop *c* and a notch *e* formed on it, substantially as and for the purposes described.

No. 60,509.—JAMES HANLEY, New York, N. Y.—*Spiral Friction Clutch for Machinery.*—December 18, 1866.—A cord from the frame passes spirally several times around the shaft, and is attached to a spring. The latter by extension allows the shaft to rotate in one direction, but the shaft is prevented from rotating in the other direction by the frictional contact of the cord, which acts as a brake.

Claim.—The friction cord, as herein described and applied, substantially to control the movement of machines.

No. 60,510.—HENRY HASSENPFLUG, Huntington, Penn.—*Sawing Machine.*—December 18, 1866.—The two slotted arms between which the saw is strained, have screws and nuts by which the saw may be set at any distance in front of the saw frame, according to the requirements of the stuff or the direction of the cut.

Claim.—The arrangement of the slotted arms with respect to the saw frame, in such manner that the saw may be set at any required distance in advance of the frame, substantially as set forth.

No. 60,511.—FRANK HATCH, La Crosse, Wis.—*Combined Blacking Brush and Box.*—December 18, 1866.—A receptacle for blacking is attached by a spring bar and catch to the back of the blacking brush.

Claim.—The combination of an adjustable and attachable liquid blacking reservoir, as set forth in specifications and drawings accompanying my application.

No. 60,512.—THOMAS HAYES, Cambridge, Mass.—*Filter.*—December 18, 1866.—A body of filtering material is enclosed in a porous cylinder, and inclined between two diaphragms in the chamber. Two sets of valves have their seats in these diaphragms, and are respectively used for allowing the water to flow in reverse directions, passing it through the filter in each case.

Claim.—The passages *c d e f*, and valves *g h i j*, operated by cams F G, all arranged for the purpose of reversing the current of water through the filterer, substantially as set forth.

No. 60,513.—H. HEFLEBOWER, Alexandria, Va., and J. M. REED, London county, Va.—*Grain Separator.*—December 18, 1866.—The grain is passed between a pressure roller and one or more rollers surfaced with a substance to which the cockle alone will adhere. The cockle is brushed from the rollers at a succeeding part of their revolution to re-prepare them for duty.

Claim.—The combination of a pressure roller, with one or more rollers, with an adhesive covering to which the cockle will become attached, and thereby removed from the wheat, substantially as described.

No. 60,514.—WILLIAM HENDERSON, Glasgow, Scotland.—*Treating Ores of Copper and other Metal to obtain Metals and other Products therefrom.*—December 18, 1866.—For treating ores of copper containing oxides, carbonates, &c., the ore is digested with hydrochloric acid or perchloride of iron, from which the copper is precipitated by lime, magnesia, baryta, by any soluble sulphuret, or by iron. The hydrochloric acid is recovered by beating the insoluble residuum of the ore with the hot concentrated liquor containing chloride of calcium, magnesium, or iron. The precipitated copper is smelted in the usual way. Ores containing suboxide of copper are roasted and treated with sulphuric acid containing a small amount of nitric acid. The sulphuric acid for this purpose may be prepared from pyrites, and the acid once used may be recovered by treating the sulphate of copper or zinc produced. Iron powder for precipitating copper and other metals may be produced by beating finely powdered hematite and charcoal, and then grinding the reduced mass of spongy iron.

Claim.—First, the two several improved processes, hereinbefore described, for extracting copper from any ore in which it may be found as a salt of copper, whether iron or other metal be or not be found in such ore.

Second, the improved process and processes, hereinbefore described, for separately obtaining from the sulphuret ores of copper, silver, zinc, or other metal, the copper, silver, zinc, or other metal therein contained, whether the object be to obtain from such sulphurets one only or all of the metals therein contained.

Third, the manufacture of the product hereinbefore denominated iron powder, by the process hereinbefore described, to be used in the processes hereinbefore described as a precipitate.

No. 60,515.—A. HIGLEY, South Bend, Ind.—*Car Brake.*—December 18, 1866.—Improvement on his patent of August 14, 1866; intended to conserve a part of the force of the street car in stopping, to be used in assisting the starting of the car. The action of the brake lever throws a friction clutch into action and locks the wheels, releasing the spring, which throws a pawl against a ratchet and retains the brake in action. The draught of the team releases the pawl and consequently the brake.

Claim.—First, the stirrup L and spring N, in combination with the pawl *k* and ratchet *j*, arranged and operating as and for the purpose set forth. •

Second, the lever G and friction coupling D, in two sections, as set forth, in combination with the pawl *i*, ratchet *b*, and pulleys E E', as and for the purpose described.

Third, the swivel I', bar H, spring I, and chains *c d e*, in combination with the pulleys F E and E', arranged and operating as and for the purpose described.

No. 60,516.—ANTHONY J. HINDERMEYER, Rohrerstown, Penn.—*Flux for Welding, Puddling, and Brazing Iron and Steel.*—December 18, 1866.—Composed of pulverized oyster shells, one part; chalcedony quartz, two parts.

Claim.—The use of the herein-described compound as a flux for welding and brazing, and as a physic for cleansing and improving iron in the operation of puddling.

No. 60,517.—WILLIAM H. HOLLAND, Chelsea, Mass.—*Paddle Wheel.*—December 18, 1866.—The main float has two oblique faces forming an obtuse angle. The auxiliary floats connect them with the circular side frames. The radial spokes which support the floats proceed from the main shaft in three circular series, and the auxiliary radial arms connect the floats with the concentric rings of the frame.

Claim.—An improved paddle wheel or propeller, as constructed with its main and auxiliary floats C D, three series of radial rods and arms, and two series of radial auxiliary radial arms, arranged and combined together and with rings and hubs, substantially as hereinbefore described.

No. 60,518.—WILLIAM H. HOLLAND, Chelsea, Mass.—*Paddle Wheel.*—December 18, 1866.—Explained by the claims and cut.

Claim.—The arrangement of the main and auxiliary floats of the wheel with respect to each other and the side frames, substantially as specified and represented, each main float under such arrangement being extended diagonally or obliquely across the entire wheel, and its auxiliary float being made to extend from the middle of the main float at an acute angle thereto, and joined to one of the side frames, as specified.

Also, the arrangement of each main float and its brace or auxiliary, so as to stand obliquely in the wheel in directions opposite to those of the next adjacent main float and its brace or auxiliary float, the whole being as represented in the drawings and as hereinbefore described.

No. 60,519.—G. H. HULSKAMP, New York, N. Y.—*Piano-forte.*—December 18, 1866.—To enable the hammer to strike nearer the bearing of the string, the agraffe is inserted obliquely into the solid wooden block which is attached to the wrest plank.

Claim.—The combination of the wooden bridge *d* and the wrest plank *h* with the agraffe A, substantially as and for the purpose set forth.

No. 60,520.—JOSEPH HURD. Boston, Mass.—*Repeating Action for Piano fortes*—December 18, 1866.—After the hammer has struck the string it is sustained in a raised position, so that upon very slight elevation of the key the jack-fly will at once assume its position under the hammer-butt, in readiness to give the repeating blow.

Claim.—For the purpose of supporting the hammer of a piano-forte action near its string in position to give a repeating blow, the combination of an elastic support of the hammer-butt with a lever *o*, when this is connected with the key lever actuating the hammer by means of the link *n*.

Also, the employment of the right and left-hand screw in the link *n* for the purpose of adjusting the position of the lever *o*.

No. 60,521.—ISAAC A. ISAACS, Cleveland, Ohio.—*Foot Bath.*—December 18, 1866.—The feet rest upon the imperforate portion of the false bottom, and water to maintain the temperature of the bath is poured in at the side spout, from which it reaches the lower stratum of water in the pail.

Claim—Tube D, funnel *c*, in combination with the perforated bottom B, guard *b*, and pail A, arranged in the manner and for the purpose set forth.

No. 60,522.—HENRY JAHNE, New York, N. Y., assignor to himself, GERRIT SMITH, and ANTHONY J. G. HODENPYL, same place.—*Watch Chain Fastening.*—December 18, 1866.—The watch chain has a padlock, whose hasp fastens to the button hole.

Claim.—A watch chain fastening composed of a case *b*, containing a spring bolt *c*, operated by the guard *f*, and provided with a hinged shackle *a* at one end and a ring at the other end, as a new article of manufacture.

No. 60,523.—JOHN C. JEWETT, Buffalo, N. Y.—*Bar or Slat for Refrigerators.*—December 18, 1866.—The bars which support the ice are of wood, sheathed with zinc.

Claim.—The construction of ice racks for refrigerators of bars of wood sheathed and hermetically enclosed in zinc or other sheet metal, substantially in the manner and for the purposes herein set forth.

No. 60,524.—JAMES J. JOHNSTON, Alleghany City, Penn.—*Tanning.*—December 18, 1866.—The revolving vat has frames for the hides, and communicates through the trunnions with the vats at each end. The hides are stretched on the frames, and the vats are filled with tanning liquid. The air is then exhausted, and the tanning liquid is allowed to flow into and partially fill the hide vat, which is then slowly revolved for a proper length of time. After this the tanning liquor is forced into the hides by means of hydrostatic pressure.

Claim.—First, placing skins of animals in air-tight vats, from which the air has been exhausted and then treated with tanning liquid and agitated, in the manner and for the purpose described

Second, in connection with the above, the application of hydrostatic pressure, in the manner and for the purpose described.

Third, the combination and arrangement of the vats A B and C, furnished with trunnions *o* and *e*, frames *z*, and pressure device, the whole being constructed, arranged, and operating in the manner substantially as herein described and for the purpose set forth.

No. 60,525.—EDWIN LAMPMAN, Catskill, N. Y.—*Centre Table.*—December 18, 1866.—The table top has a segmental leaf and an eccentric supporting disk on which the top turns, so as to support the leaf or allow it to hang.

Claim.—The arrangement of the disk C, as constructed with the leg A, pin or plug *a*, and top D, as and for the purpose herein set forth.

No. 60,526.—SIMON M. LANDIS, Philadelphia, Penn.—*Female Syringe.*—December 18, 1866.—The numerous cavities flaring outward and connected by channels will not be readily choked by resting against the membranes. The apparatus is intended to be attached to the spigot of a hydrant or force pump, by means of an annular membrane in the box adapted to receive the faucet, to which the box is tied by a cord.

Claim.—The funnel-shaped cavities, with or without grooves, of the syringe bulb tube.

Also, the box applicable to spigots of any size, in combination with syringe bulb tube, as herein described.

No. 60,527.—A. LARROWE, Cohocton, N. Y.—*Farm Gate.*—December 18, 1866.—The gate is supported on headed pins and retained by slip pins, so that it may be opened at either end, but in opposite directions.
Claim.—The gate, constructed as shown and used, in connection with the hooks *d* and pins *c*, all arranged to operate as herein shown and described.

No. 60,528.—R. G. LATTING, New Orleans, La.—*Cotton Bale Tie.*—December 18, 1866.— One end of the band is passed into the perfect loop, and the other end is introduced laterally into the open loop, which is strengthened by a hook. The last-mentioned end is passed below a sharp rib, which prevents its withdrawal when the cotton expands against it.
Claim.—A buckle arranged with two loops *d* *g*, constructed as described, in combination with an angular ridge *c*, formed by the depression of the centre of the plate, as and for the purpose described.
Also, the hook *h*, or its equivalent, for strengthening the open loop *g*.

No. 60,529.—WILLIAM W. LEVERING, San Francisco, Cal.—*College Cabinet.*—December 18, 1866.—One of the sides forms a blackboard when in a vertical position, and a desk when let down.
Claim.—A cabinet, constructed as described, and having a door serving for a blackboard and held up by the bars *e′ e′*, substantially as described.

No. 60,530.—EDWARD T. C. LUTTON, Philadelphia, Penn.—*Manufacture of Yarn.*— December 18, 1866.—The superfluous projecting fibres are sheared by giving the yarn a temporary twist, while submitted in passing to the action of shearing blades. The object is to give to yarn of coarser wool the appearance of fine German zephyr.
Claim.—Yarn from the entire surface of which the superfluous projecting fibres have been sheared, for the purpose specified.

No. 60,531.—IRA MANNING, Philadelphia, Penn.—*Machine for Forming Bridle Fronts.*— December 18, 1866.—The guides and pressing board form a corded bead on the upper, and bring it and the sole together ready for sewing.
Claim.—First, the guides B B′, either fixed or graduating, for the purpose herein specified and described.
Second, the guides B B′, in combination with the centres D D′, substantially as specified and described.
Third, the guides B B′, in combination with the centres D D′ and the presser E, substantially as specified and described.
Fourth, the sliding centre board or piece *l*, substantially as specified and described.

No. 60,532.—WILLIAM B. MASON, Boston, Mass., assignor to himself and CHARLES H MOORE, same place.—*Marking Stamp.*—December 18, 1866.—The types have elastic faces, and are placed in a case with an elastic roller in the rear of each one, to enable them to accommodate themselves to uneven surfaces.
Claim.—Making the face of the type elastic, to yield to the small inequalities of the surface printed, in combination with a small elastic base, (less in area than the face of the type,) to yield to the large inequalities of the surface printed. And, in combination with the elastic face and small elastic base, so arranging or holding the solid body of the type in the case that it can rock when required to adapt the surface of the types to the irregularities of the surface printed.

No. 60,533.—B. A. McCONNAUGHEY, New Market, Ohio.—*Detaching Horses from Vehicles.*—December 18, 1866.—The traces are attached to blocks, which are held in sockets of the singletree by spring arms; these connect with a bifurcated lever, so that their inner ends can be depressed and the tugs released.
Claim.—The tugs D D connected to the end of the traces and fitting within the openings therefor, provided and held by the bar *m*, when used in combination with the springs and lever *a* for detaching the horse, substantially as herein set forth.

No. 60,534.—LEONARD H. MILLER, Ottawa, Ohio.—*Type Case.*—December 18, 1866.— The movable boxes are made with a curved side toward the operator, and each box has an edge and side curved over the upper edge of those in proximity to it.
Claim.—First, a type case, when formed with detachable boxes B set into a frame A, substantially in the manner and for the purpose set forth.
Second, in combination with a case A and independent boxes B, a removable partition C, constructed and used substantially as set forth.
Third, the boxes B, when constructed with oval bottoms, substantially as described.
Fourth, the boxes B, when formed with flanges B′ for interlocking the boxes, when placed in the case, substantially in the manner set forth.

No. 60,535.—OLIVER MILLER, Salem, Ohio, assignor to himself and THOMAS D. BALL.—*Pump.*—December 18, 1866.—The working plungers may be driven by a walking beam or by direct action of steam. The reciprocating piston valves in the lower cylinder are operated by the pressure of the water in the action of pumping.

Claim.—First, the chambered chest E, with its valves *e e'*, in combination with the cylinder B B, pistons *b''*, valves *g g'*, and induction pipe A, all arranged in relation to each other and operating conjointly for the purpose set forth.

Second, in combination with the foregoing cylinders C C,' with the respective plungers, chambers D D''', valves I' I'', pipe H, valve *d*, and nozzle G, arranged and operating in the manner and for the purpose set forth.

Third, the chamber J J' of the chest E, valves *e e'*, and openings *m m'*, in combination with the cylinder B, pipes D' D'', and chamber D D''', as and for the purpose set forth.

No. 60,536.—ROBERT R. MILLER and A. W. CARVER, Philadelphia, Penn., assignors to B. HOOPES and C. BORIE, same place.—*Composition for Making Sharpening Stones.*—December 18, 1866; antedated December 5, 1866.—Dissolve four ounces of gum shellac in half a pint of alcohol, and add one pint of a solution of twelve ounces of frosted glue in a quart of water. To this compound add emery or sand to form a paste, which may be applied to strips of wood or other material.

Claim.—The composition for making sharpening stones, consisting of the materials herein described, combined substantially as specified.

No. 60,537.—ADOLPH MILLOCHAU, New York, N. Y., assignor to THE AMERICAN IMPROVED GAS RETORT COMPANY.—*Retort for the Manufacture of Illuminating Gas.*—December 18, 1866.—The inner chamber of the retort is supported by legs on the main floor, and discharges by pipes in the rear downward to the intervening space, thence forward to the front space and the eduction pipe.

Claim.—The platform or false bottom *c*, openings or pipes *k*, and cap plate *e*, in combination with the retort *a*, substantially as and for the purposes set forth.

No. 60,538.—JOHN MONFORT, Jessamine county, Ky., assignor to himself and G. E. BILLINGSLEY.—*Composition for the Cure of Ague.*—December 18, 1866.—Made by steeping ¼ ounce of bloodroot (sanguinaria canadiensis) and ¼ ounce of yellow root (hydrastis canadiensis) in one pint of whiskey.

Claim.—A composition of matter composed, compounded, and prepared substantially as and for the purpose set forth.

No. 60,539.—MARK F. MORSE, Boston, Mass.—*Rotary Sifter.*—December 18, 1866.—The cylindrical sifter has a segmental opening for the introduction of meal. The hand crank has a projection which runs on inclined, abruptly terminating cams upon a circular disk, and which serves, in combination with a spiral spring, to shake the sifter longitudinally.

Claim.—The combination and arrangement of the disk D, its series of cams *g g g*, the stud *i*, and the spring *l*, with the sifting drum, its shaft and supporting frames, the whole being to operate as specified.

No. 60,540.—WILLIAM J. MURPHY, Cork, Ireland.—*Breech-loading Ordnance.*—December 18, 1866.—The plunger rod for forcing the ammunition forward into the barrel has a piston, which reciprocates in an open-ended cylinder by the pressure of water. A valve cock turns the water to either side of the piston as required, and allows a discharge of the same.

Claim.—First, the barrel *a a*, with its opening *d d* and plunger or breech-piece *b*, in combination with the within described devices, or their equivalents, whereby the pressure of the water may be caused to operate the breech-piece, all substantially as set forth.

Second, the combination of a barrel, a movable breech-piece, and a chamber containing a body of water, and so situated that the water will retain the breech-piece in its position during the discharge of the piece.

Third, the barrel *a a*, with its openings *d d*, cylinder *f f*, plug *i i*, and openings *k l m*, or their equivalents, in combination with the plunger *b* and its piston *e*, the whole being constructed and operating substantially as and for the purpose described.

No. 60,541.—JOHN MURRAY, New York, N. Y.—*Whistle for Steam and Other Engines.*—December 18, 1866; antedated October 7, 1866.—The whistle is cast in one piece.

Claim.—The above-described whistle for steam and other engines as a new article of manufacture, substantially as and for the purposes set forth.

No. 60,542.—FRANCIS B. NORTHROP, Newark, N. J.—*Sawing Shingles.*—December 18, 1866.—Two reciprocating sashes, in an inclined position to each other, have each a gang of saws to saw shingles with alternate points and butts.

Claim.—First, sawing a block of wood into shingles or other analogous things, having alternate butts and points, by means of a gang of reciprocating saws, when arranged and operated substantially as described.

Second, the projections *c* on the cross-bars of the two gates, when used in combination with the two reciprocating gates having a gang of saws, operating substantially as described.

No. 60,543.—M. and R. B. PACKARD, North Bridgewater, Mass.—*Shoe Binding*—December 18, 1866.—A strip of even width is folded in the direction of its length, and encloses a cord running through the fold, the joint being cemented.

Claim.—As a new article of manufacture, a shoe binding, having a construction substantially as set forth.

No. 60,544.—E. E. PACKER, Jr., and JOHN DALRY, Philadelphia, Penn., assignors to themselves and EDGAR L. THOMPSON, same place.—*Car Coupling.*—December 18, 1866.—The link-bar hook enters the slot in the draw head, when the latter is rotated by a lever to bring it into the proper presentation, when it is rotated 90° and locked, securing the connection.

Claim.—First, the combination of the coupling rod B with the cylinder C and screw socket *b* of a cylinder A, substantially upon the principle and in the manner hereinbefore described and for the purpose specified.

Second, the combination and arrangement of the lever E, stud *m*, link *n*, and pawl *o*, substantially in the manner described and for the purpose specified.

No. 60,545.—WILLIAM PALMER, New York, N. Y.—*Fastening Blind Slats.*—December 18, 1866.—Applied to the inner edge of the blind stile is a flat and slightly bowed spring, which presses against the ends of one or more of the slats, forces them against the opposite stile, and thus holds them at the required adjustment.

Claim—The adjustable half-elliptic spring D, as arranged and constructed and for the purposes as set forth.

No. 60,546.—ALFRED PARAF, Mulhouse, France.—*Dyeing and Printing Textile Fabrics and Yarns.*—December 18, 1866.—Chromic acid is applied in the dyeing and printing operations by the use of insoluble combinations of chromium capable of developing chromic acid in a moist atmosphere, and is specially useful in the production of aniline or its analogous black or gray colors. The brown binoxide of chromium to be mixed with a salt of aniline for this purpose is prepared by precipitating a solution of chloride of chromium by a solution of chromate of potash. This precipitate, mixed with 5 per cent. of glycerine, is called the "oxidizing paste." A compound of manganese and chromium may be formed by using the mixed chlorides of manganese and chromium instead of the chloride of chromium.

Claim.—The process of developing chromic acid in dyeing and printing operations by the application to the fabric of an insoluble salt of chromium and the subsequent action of a moist atmosphere, substantially as set forth.

No. 60,547.—THOMAS PARKER, Germantown, Penn., assignor to himself and THEODORE RUFE, same place.—*Self-sealing Fruit Can.*—December 18, 1866.—Around the upper part of the tin can are dentations projecting inward. The glass lid has openings and inclined flanges to pass, and then, by the rotation of the cover, to bind against the lugs on the can, while the flange of the lid rests on the elastic ring on the rim of the can.

Claim.—First, the construction of a tin can with a glass lid the full diameter of the can, substantially as and for the purpose set forth.

Second, the laying off the top of the can as a bearing for the rubber ring *r r*, as shown in Fig. 3, substantially as and for the purpose set forth.

Third, the impression of lugs sunk in the body of the can, or their equivalents, substantially as and for the purpose set forth.

No. 60,548.—JULIUS A. PEASE, New York, N. Y.—*Tanning.*—December 18, 1866.—The tanning bath is composed of nutgalls, 4 ounces; sulphate of iron, 2 ounces; and water, 1 gallon.

Claim.—The tanning of hides or skins, substantially as above described.

No. 60,549.—JULIUS A. PEASE, New York, N. Y.—*Tanning.*—December 18, 1866.—Extract of logwood is dissolved in water, and used either with or without other material as a tanning agent.

Claim.—The use of the above-mentioned material for tanning, either alone or in combination with other materials, substantially as described.

No. 60,550.—JOHN G. PERRY, South Kingston, R. I.—*Straw Cutter.*—December 18, 1866.—Explained by the claims and cut.

Claim.—First, the combination of the two-flanged cylinders G and H, so arranged and geared together that the periphery of the flanges of the cylinder H shall move faster than those on the cylinder G, and be so situated in relation to each other that a shear-cut shall be made in the direction of the centre of the shaft G, or radial instead of tangential thereto, as shown and specified.

Second, the feed-roll N, having flanges that work in concert with those on the cylinder G, in combination with one or more cutting cylinders of a hay or feed-cutter, substantially as herein described and for the purposes set forth.

Third, making the flanges against which the knives cut, with projections *t t t t*, substantially as herein set forth and for the purposes specified.

Fourth, the hub *x*, screws *s s*, and stud *c*, in combination with the wheel C and cylinder G, substantially as herein described and for the purposes set forth.

No. 60,551.—E. PETTRYS and T. C. LEGGETT, Chestertown, N. Y.—*Gate.*—December 18, 1866.—The rock-bar has a transverse lever extending over the roadway. By depressing either end of this lever the two pendent arms operate to swing the gate from its balance and release the fastening. It then opens to the side latch post. A reverse movement of the lever releases and closes the gate.

Claim.—First, the arrangement with the gate F, and link hinge E, of the rock-shaft B, and arms D D', substantially as described.

Second, the arrangement of the palette I, plate L, and arm D', as and for the purpose described.

No. 60,552.—CHARLES F. PIKE, Providence, R. I.—*Apparatus for Cooling and Preserving Meats, Fish, Vegetables, and other Substances.*—December 18, 1866.—From the ice chamber descend tubes of downwardly increasing diameter, discharging into horizontal tanks. The chamber, tubes, and tanks, have a free circulation of air which passes around them and out of the refrigerator; the water escapes by a trap.

Claim.—First, the application of the ventilator B, or its equivalent, to the refrigerator, constructed and refrigerated substantially as herein described and for the purposes herein stated.

Second, the making of the pipes or tubes D larger at the lower end than at the upper end, substantially as and for the purpose herein stated.

Third, the construction of the pipes or tubes D, larger at the lower ends than at the upper ends, in combination with the ice-box or receptacle C, and the water tank E, or their equivalents, substantially as herein described

Fourth, the structure of the ice-box or receptacle C, pipes or tubes D, water tank E, and its appendages, provision chamber A, and its fixtures, ventilator B, or their equivalents, substantially as herein described and for the purposes herein stated.

No. 60,553.—S. MONTGOMERY PIKE, Cincinnati, Ohio.—*Filter Cooler.*—December 18, 1866.—The water from the outer vessel passes through the filtering material, which is enclosed between two perforated cylinders, and thence flows to the inside cooler, which has a discharge faucet.

Claim.—First, the outer metallic case B and inner case C, with their series of perforations *b* and *c'*, respectively constructed and arranged as and for the purpose above described and set forth.

Second, the cooler D, having the flutes *d*, in combination with the inner case C, as above described and for the purpose specified.

Third, the outer perforated case B, the inner perforated case C, and cooler D, in combination with the casing A of the filter cooler, all arranged as above shown and for the purpose set forth.

No. 60,554.—BRITTON POULSON, Fort Wayne, Ind.—*Fan, Brush, and Rack.*—December 18, 1866.—Explained by the claim and cut.

Claim.—The hereinbefore described arrangement of parts consisting of the bell-formed base A, rod B, set screw B', head C, and radial arms D, when said base also encloses the actuating clock-work, constructed as described, communicating by means of the crank wheel K, rod L, and arm M, a rotary reciprocating motion to the rod B, and arm D, in the manner and for the purposes set forth.

No. 60,555.—JAMES and JAMES A. PUNDERFORD, New Haven, Conn.—*Manufacture of Leather Hose.*—December 18, 1866.—The gum is spread upon the leather, which is then rolled into form. The gum forms the joint at the contacting parts of the leather, and on its inner and outer surfaces, either or both, and is then subjected to a vulcanizing heat.

Claim.—First, forming the joint in leather hose by the application to the surface of the laps of a vulcanizable gum, and curing the same after such application, substantially as herein set forth.

Second, coating the inner or outer or both surfaces of leather hose with a vulcanizable gum, and curing the same after such coating, substantially as herein set forth.

Third, coating the outer and inner surface and the joint or laps by a single sheet of fabricated rubber, substantially as herein set forth.

No. 60,556.—JOHN RICHARDS, Washington, D. C.—*Combined Poker, Tongs, Wrench, &c.*—December 18, 1866.—The different portions are adapted to the uses cited and the invention is explained by the claim and cut.

Claim.—The construction of the fire tongs when arranged and combined with wrenches C and J, hook H, poker G, and regulating screw K, as herein described and for the purposes set forth.

No. 60,557.—DARIUS S. ROBINSON, Oswego, N. Y.—*Compound Oil for Paint, &c.*—December 18, 1866.—Composed of petroleum, 6 ounces; linseed oil, 2 ounces; India rubber $\frac{1}{16}$ ounce; beeswax, $\frac{1}{16}$ ounce; gutta-percha, $\frac{1}{16}$ ounce; sulphate of zinc, $\frac{1}{16}$ ounce; sugar of lead, $\frac{1}{16}$ ounce. The ingredients are boiled, mixed and cooled gradually.

Claim.—The art of mixing, combining, and compounding the aforesaid articles or ingredients, and making a liquid compound or composition called and designated "elastic oil," possessing the qualities and answering the purposes of pure linseed oil or other pure oils, to be used in painting and all other general purposes for which linseed oil and other pure oils are used.

No. 60,558.—SAMUEL F. ROGERS, Malden, Mass.—*Dryer for Petroleum and Heavy Oils.*—December 18, 1866; antedated December 12, 1866.—The composition is to be added to petroleum and other oils to render them "drying" so as to be suitable to be used for painting. The "dryer" is composed of a resin known as "clay gum," linseed oil, red lead, litharge, black oxide of manganese, sulphate of zinc, and benzole.

Claim.—The within described petroleum dryer, constituted substantially as set forth.

No. 60,559.—WILLIAM B. ROGERS, Chicago, Ill.—*Burning Fluid.*—December 18, 1866.—Composed of naphtha, 40 gallons; caustic soda, 1 pound; alum, 1 pound; salt, 1 pound; manganese, 1 ounce; and water, 4 ounces.

Claim.—A burning and carbonizing fluid which is composed of the several ingredients herein mentioned, mixed together in about the proportions specified.

No. 60,560.—CHARLES ROSS, Jr., and JOHN ROSS, New York, N. Y.—*Steam Generator Flue Brush.*—December 18, 1866; antedated December 6, 1866.—The fused metal is cast around the bases of the wires or cutters with or without a perforated metal blank. Separate sections, carrying cutters or wires as desired, are placed on the handle rod.

Claim.—First, the construction of a brush for cleaning tubular boilers by combining fused metal among wires, as in Figs. 1 and 3, whereby to hold the wires of the brush firm, substantially as set forth.

Second, the arrangement of metal containing wires and cutters so as to form cutters, and a section of brush as shown at E, Figs. 5 and 6, adapted to be put on and off, and of the size to fit the tubes to be cleaned, substantially in the manner and for the purpose as herein set forth.

No. 60,561.—J. F. SAIGER and A. DAVIS, Shelby, Ohio.—*Fruit Picker.*—December 18, 1866.—The wire frame is mounted on a pole, and has an open space to admit the fruit stalk. The bag is drawn over the rim of the frame and retained upon it to catch the fruit.

Claim.—The skeleton spring head B, tension rod C, and adjustable screw nut E, in combination with the hood D, bag F, and handle A, when arranged in the manner and for the purpose described.

No. 60,562.—HERRMAN S. SARONI, Baltimore, Md.—*Steam Generator.*—December 18, 1866.—Above the hydrocarbon burners is a tubular heater, and above this heater a flue space to act as a combustion chamber.

Claim.—First, a shallow water drum or drums or tubular heater or heaters interposed between the burner or burners and the boilers, substantially as and for the purpose described.

Second, the combination of the boiler with the tubular heater, substantially as described, and so arranged that a space shall be left between them, for the purpose set forth.

Third, the combination of the boiler, tubular drum, and burners, substantially as described, so that the drum shall act as a beater cap to the burners and permit the flame to pass to the bottom of the boiler, as set forth.

No. 60,563.—FRANZ M. SCHMITT, Jamaica Plains, Mass.—*Sole-edge Finishing Tool.*—December 18, 1866.—The block is adjustable or reversible in and removable from the head, and has flat faces, projections and gains for polishing and finishing the sole.

Claim.—The block A, provided with opposite flat faces and polishing surfaces and separate projections *c d*, or the same and one or more notches *e*, all substantially as explained.

No. 60,564.—HERMAN F. SCHRODER, Cincinnati, Ohio.—*Pan for Evaporating Sugar.*—December 18, 1866.—In the evaporating pan are one or more hollow revolving disks containing water. The heat of the sirup in the pan vaporizes a part of the water, so as to fill the disks with steam; the disks revolving raise the sirup and expose it to more rapid desiccation and consequent granulation.

Claim.—The provision of one or more hollow disks adapted to receive water and be hermetically closed, the said disks being adapted for attachment to a common evaporating pan or kettle and having a crank or other means of rotation, as and for the purposes set forth.

No. 60,565.—HENRY L. SCOTT, Plessis, N. Y.—*Fruit Gatherer.*—December 18, 1866.—A basket is mounted upon a pole. One of the curved ribs of the basket is prolonged and bent over the mouth of the basket, forming one blade of the shears; the other blade is operated by a cord.

Claim.—The combination with the basket A of the shears E E', when the stationary blade E is formed and applied in conjunction with the flat strip *e'*, to form one of the ribs of the basket, and when the several parts of the instrument are combined and arranged in the manner and for the purpose herein specified.

No. 60,566.—HENRY SEUER, St. Louis, Mo.—*Ventilating Fan for Gas Burners.*—December 18, 1866.—The ventilating wheel is placed above the flame and its rotation agitates the atmosphere.

Claim.—A ventilating.wheel constructed, operated, and applied in the manner as shown and described and for the purpose set forth.

No. 60,567.—TAL. P. SHAFFNER, Louisville, Ky.—*Charging Shells, &c.*—December 18, 1866.—The vials filled with nitro-glycerine are surrounded with gun cotton, which serves as a cushion acting in concert with the India-rubber cushions to prevent the vials from breaking while the shell is passing out of the gun.

Claim.—First, the application of nitro-cotton (known as gun cotton) for the purpose of serving as a cushion and an explosive substance in shells, torpedoes, &c., wherein nitroleum (in chemistry known as nitro-glycerine) or other explosive liquid compounds are used as a charge, substantially as hereinbefore described.

Second, the application of India-rubber as a cushion lining, for the purpose of lessening the concussion upon the nitroleum or other explosive liquid, substantially as hereinbefore described.

Third, the honey-combing of India-rubber with openings between the cells, for the purpose of perfecting the cushion and for the purpose of hastening the spread of the fire throughout the said honey-comb cushion lining, substantially as hereinbefore described.

No. 60,568.—TAL. P. SHAFFNER, Louisville, Ky.—*Graduating Vessels.*—December 18, 1866.—The graduating marks for the major divisions are made in the form of rings, and the minor divisions by shorter marks.

Claim.—The bottle graduated irrespective of size, form, or material, substantially as described and represented.

No. 60,569.—TAL. P. SHAFFNER, Louisville, Ky.—*Electric Fuze.*—December 18, 1866.— The wooden head of the fuze has a protecting recess for the fuze composition and another for the non-conducting cement which surrounds the wires at their point of entrance into the head. The head is enclosed in a flanged cylinder with a cap, and the fuze chamber has a water-proof covering.

Claim.—First, the fuze-head *a* or *d*, with its chambers *b s*, one or both, as and for the purpose or purposes described.

Second, the indented or flanged cylinder *j j'*, with its cap *k*, and head *d*, for the direction of the flame of the fuze, as described.

Third, the mode of attaching the wires to the fuze-head by means of a non-conducting cement inserted into a chamber in said head, or in the cylinder in immediate connection therewith, as described.

Fourth, the protecting water-proof membrane or cover *n*, for closing the mouth of the composition chamber *b*.

Fifth, the water-proof lining to the composition chamber *b*, to prevent access of moisture to the said composition.

No. 60,570.—TAL. P. SHAFFNER, Louisville, Ky.—*Hydraulic Press to Prevent Corrosion.*—December 18, 1866.—The chamber and piston of the hydraulic press are lined with non-corrosive material, such as glass, gutta-percha, &c., so as to permit acids to be contained therein without injury.

Claim.—A non-corrosive lining to the chamber of an hydraulic or other press, and to the piston or that end of it presented to the said chamber ; and this whether the said parts consist wholly of material capable of withstanding the action of acids, or whether only such parts are thus protected and are exposed to the said action.

No. 60,571.—TAL. P. SHAFFNER, Louisville, Ky.—*Manufacture of Gun Cotton.*—December 18, 1866.—Cotton is saturated with the acids more speedily and completely, and is also cleansed more quickly and thoroughly after treatment with the acids, by means of pressure which causes the fluid to permeate the mass of fibre.

Claim.—The process of making nitro-cotton (commonly called gun cotton) or other nitro-fibre under pressure.

No. 60,572.—TAL. P. SHAFFNER, Louisville, Ky.—*Artillery and Mining Blasting.*—December 18, 1866.—Improvement on his patent of December 19, 1865. Blasting charges of nitroleum are placed at such depths and relative distances that when exploded simultaneously by electricity they combine to wrench out the face of the rock.

Claim.—The combination of blasts to be discharged simultaneously by electricity in such manner as will effect a conjunctive force of the respective charges, thereby increasing the disruption of matter beyond what can be obtained by separately discharging the said blasts.

No. 60,573.—TAL. P. SHAFFNER, Louisville Ky.—*Blasting with Nitroleum.*—December 18, 1866.—The charge of nitroleum is divided, one portion being at the bottom of the hole for blasting, and the other for tamping at the top, to resist the exit of the gases evolved by the charge below and increase the disruptive force.

Claim.—First, the combination of nitroleum with sand for the purpose of blasting and distributing the explosive force throughout the drill-hole or space, where the same are employed in the manner and for the purposes described.

Second, for blasting purposes the use and interposition of a column of water between the "tamping" and "blasting" charges, when the same are arranged in the manner and for the purposes described.

Third, as a method of blasting in rock, the adjustment and arrangement of the "tamping" and "blasting" charges in such manner that the former shall be placed at or near the surface or upper part of the drill-hole, while the latter is located at the bottom thereof, or in such a manner as that the gases of the two charges may be united, disrupting the rock in the manner and for the purposes hereinbefore described.

No. 60,574.—M. R. SHALTERS and T. CATERN, Alliance, Ohio, assignors to themselves, SAMUEL RAY and S. THOMAS.—*Hame Fastening.*—December 18, 1866.—The two bars hook into the hames. The pivoted hook on one bar is run through one of the holes in the other, and its long end brought around. The pressure against its shorter end keeps the other end in contact with the bar.

Claim.—First, the bar A, provided with hooks, as seen at each end, and with the hook B secured to and operating with it, as and for the purpose set forth.

Second, the bar C, with links formed in it and used in connection with the bar A, constructed as and for the purpose set forth.

No. 60,575.—JOHN S. SHAPTER, New York, N. Y.—*Rendering and Bleaching Tallow, Lard, &c.*—December 18, 1866.—The process is conducted in an ordinary vacuum pan into which the tallow is run and heated to 200° to 300° Fahrenheit. The vapors are constantly withdrawn by the air pump. When all the vapors are drawn off an alkaline liquor of about 70° B. is added, and the pumps and condensers again operated till all the water is withdrawn.

Claim.—The bleaching of tallow and other fatty matters by subjecting them to the action of alkaline lye while heated in vacuo, substantially as herein specified.

No. 60,576.—W. B. SHELTON, Congruity, Penn.—*Fence.*—December 18, 1866.—The fence post and the upright slat are respectively dovetailed into the horizontal clamp blocks which intervene between the boards. A modification of the clamp secures the ends of the boards at the angles of the fence.

Claim.—First, the combination of the posts A, strip A', boards B, and clamps C, constructed and arranged substantially in the manner and for the purpose set forth.

Second, the mode of forming the corners by means of the post A, strip A', boards B, and clamps D, respectively constructed substantially as set forth.

No. 60,577.—ALLEN SHEPARD, Ashland, Mass.—*Lamp Shade.*—December 18, 1866.—The upper end of the corrugated frustum is embraced between two annular disks which are locked together by the springs, which also embrace the chimney.

Claim.—First, the combination of the support B, and cap C, with the corrugated shade A, when said parts are arranged to operate as set forth.

Second, as a new article of manufacture, the corrugated shade A, provided with the support B and cap C, as shown and described.

No. 60,578.—N. H. SHERBURNE, Elgin, Ill., and J. T. WHIPPLE, Chicago, Ill.—*Well Tubing.*—December 18, 1866.—The tube has a conical point, is perforated transversely at its lower end, and enters a sleeve with similar openings. The sleeve has an inner lip at the upper end, and the lower end of the tube has an outward lip to engage the lip on the sleeve on raising the tube to expose the holes. A screen inside the pipe acts as a strainer. A disk in the bottom of the sleeve has a projecting screw to engage a rod by which it may be vertically moved to clear away sand, &c.

Claim.—The combination of cylinder B, pipe A, screen m, and disk L, the whole constructed, arranged, and operated substantially in the manner and for the purpose described.

No. 60,579.—S. J. SHERMAN, Brooklyn, N. Y.—*Belt Fastening.*—December 18, 1866.—One end of the belt is passed through an adjustable slide which has hooks to engage with the eyelets in the other end of the band.

Claim.—The arrangement of the hooks A, on the adjustable fastening C E D, and adapted for use on belts and waistbands, substantially in the manner and for the purpose

No. 60,580.—ASAHEL M. SHURTLEFF, Boston, Mass., assignor to himself, BENJAMIN S. CODMAN, and F. O. WHITNEY.—*Atomizing Tube.*—December 18, 1866.—The induction end of the liquid tube is split and has a screw plug by which the size of the induction orifice is adjusted. The liquid is raised, and expelled by a current of air across the eduction opening of the liquid tube.

Claim.—Combining with the atomizing tubes,' operating as described, an adjustable plug, or its equivalent, placed in or connected with the tube *b*, substantially as set forth.

No. 60,581.—JOSEPH SIGOURNEY, AZEL T. ROBINSON, and JAMES SHEPARD, Bristol, Conn., assignors to J. SIGOURNEY, AZEL T. ROBINSON, and B. B. LEWIS, same place.— *Hanger for Stove-Hooks.*—December 18, 1866.—This device is for attachment to a stove pipe. It is explained by the claims and cut.

Claim.—First, the band F, with the swaged groove B, when applied to a stove-pipe, in the manner and for the purposes described.

Second, in combination with the foregoing, the cast-metal hook or hanger, the whole constructed and used as set forth.

No. 60,582.—ALONZO N. SMITH, Hallowell, Maine.—*Shoulder Brace.*—December 18, 1866.—Explained by the claim and cut.

Claim.—The combination of the back pieces B B, made with the spaces, the elastic cross-pieces *b* attached as specified, tags A A, and lacings *a*, arranged and operating as and for the purposes set forth.

No. 60,583.—FRANCIS W. SMITH, Philadelphia, Penn.—*Artificial Teeth.*—December 18, 1866.—The platinum strips of double concave form are fashioned by co-acting dies which give the flanged head and cut-off the strip.

Claim.—The use of the flange-headed plates, constructed substantially as described, out of strips drawn from platinum wire, for confining artificial teeth to vulcanized gum or other plates, as above specified and shown.

No. 60,584.—G. TRUMAN SMITH and WILLIAM E. SPARKS, New Haven, Conn.—*Blind Fastener.*—December 18, 1866.—The latch secures the blind on the inside; another latch secures it, when open, to an outside stud. A spring trigger releases the latter latch.

Claim.—The combination of the latch A, provided with the lever B, with the latch C, when constructed and arranged so as to operate in the manner and for the purpose specified.

No. 60,585.—HAMILTON L. SMITH, Gambier, Ohio.—*Refining Hydro-Carbon.*—December 18, 1866.—The air is forced through the still by means of a fan blower, being heated on its passage through the coil over the furnace. The air carries off the volatile portions of the oil into the condenser, and the oil remaining in the still is passed into a filter containing charcoal.

Claim.—The charcoal filterer G, in combination with the receiver D, fan A, heater C, coiled pipe B, arranged and operating as and for the purpose set forth.

No. 60,586.—WILLIAM B. SNOW, Titusville, Penn.—*Pump for Deep Wells.*—December 18, 1866.—Sections of metallic packing form couplings for tubing of larger inside diameter, and sections of piping on the pump rod reciprocate in these couplings. Horizontal grooves are cut in the couplings to form liquid packing to the rod.

Claim.—The grooved metallic packing section or sections B, secured in the tubing of an artesian well, in combination with an elongated piston C, working through the same, the range of which is above and below the said packing section or sections, they being of less interior diameter than the tubing, arranged substantially in the manner and for the purposes herein et forth.

No. 60,587.—HORACE P. STEWART, Oak's Corners, N. Y.—*Boot and Shoe.*—December 18, 1866.—The alternation of welts and taps gradually approximates the heel to the required shape.

Claim.—Forming the heels of boots and shoes by the alternate, or nearly so, arrangement of welts and taps so as to build up the right shape without paring away the material to any considerable extent, substantially as herein specified.

No. 60,588.—HORACE P. STEWART, Oak's Corners, N. Y.—*Manufacture of Boots and Shoes.*—December 18, 1866.—The intaglio die is fixed on the base piece and the cameo die upon a horizontally swinging arm. The upper die is attached to a vertically sliding shaft which is depressed by a blow on the head and raised by an encircling spring.

Claim.—The instrument, as described, for shaping the heel welts, having its supporting arm B of the upper die swing on the base A, which bears the other die, substantially as and for the purpose herein specified.

No. 60,589.—WILLIAM W. ST. JOHN, St. Louis, Mo.—*Combined Gang Plough and Cultivator.*—December 18, 1866; antedated December 2, 1866.—The implement is convertible into either gang plough or cultivator. The draw-beams of the ploughs are pivoted to the frame and admit of vertical adjustment. The frame is expansible laterally.

Claim.—First, the combination of the frame A C C D D¹, with the beams M¹ and F, the draught-rods M, and either the cultivator or ploughs F¹, and their attachments, or the gang-ploughs M⁴, and their attachments, substantially as described.

Second, the combination of the wheel stands B¹, with the frame A C C D D¹, in such a manner as to admit of lateral regulatory movement, substantially as and for the purpose set forth.

Third, the combination and arrangement of the levers M³, with the beams M¹ and chord or chain m⁴, substantially as set forth.

Fourth, the employment of the guiding bars M², when constructed and used as and for the purpose set forth.

Fifth, the attachment of the draught rod M for the plough beams M¹ to the pole P, substantially as described.

No. 60,590.—H. R. STONE and E. N. SCHOULTZ, Greenwich, N. Y.—*Stuffing and Varnishing Wood.*—December 18, 1866.—For "oil finish" it is composed of two quarts double boiled linseed oil, 1 pound beeswax, 2 pounds litharge, ¼ pound of sugar of lead, and 1 pound plaster of Paris. For "varnish finish" the composition is made of 2 quarts benzine or turpentine, 2 pounds plaster of Paris, ¼ pound sugar of lead, ¼ pound beeswax, and 1 pound litharge. Both compositions may be colored with umber, sienna, or other pigment.

Claim.—The compound which we have above described, to be applied either to "oil finish" or "varnish finish," and to be used for the purpose of filling or stuffing the pores or interstices of all kinds of porous woods, thus making a perfect enamelled surface impenetrable to air or water, and which cannot be injured by either.

No. 60,591.—D. W. STOW, Thorntown, Ind.—*Compound for the Cure of Diseases in Hogs.*—December 18, 1866.—Compound of soft soap, 1 quart; saltpetre, 5 tablespoonfuls; sulphur, 5 tablespoonfuls; to be given as one dose to twenty hogs after drawing their cuspid teeth.

Claim.—The compound herein described, in combination with the operation described, as a remedy for the cholera, &c., in hogs, substantially as set forth.

No. 60,592.—THOMAS L. STURTEVANT, Boston, Mass.—*Breech-loading Fire-arm.*—December 18, 1866.—The lever guard raises the breech of the barrel by the operation of a pin in the cam grooves of a plate connected to it. At the raising of the barrel the hammer is half-cocked, and the cartridge shell thrown out by a spring arm acting on its flange. An extensible screw in the hammer may be made to strike a pin which impinges upon the centre of the cartridge.

Claim.—First, in combination with the hammer, trigger, guard-lever, and the barrel applied to the stock, and so as to operate with a stationary breech c, as described, mechanism substantially as hereinbefore specified, whereby, by one movement of the trigger guard-lever, the barrel may be caused to be raised to receive a cartridge, the hammer to be set at half or full cock, and the spent cartridge shell be expelled from the barrel.

Second, the combination and arrangement of the main auxiliary hammers, and a screw whereby the main hammer may be either caused to actuate the auxiliary hammer or be thrown out of action therewith, as occasion may require, and for the purpose hereinbefore explained.

The construction of each of the grooves d, with its rear end open, when such groove is combined with the barrel and the trigger guard-lever, and is to operate therewith, as and for the purpose described.

No. 60,593.—GEORGE W. SYLVESTER, Newark, N. J.—*Filter for Petroleum.*—December 18, 1866.—The cylindrical vessel has a series of slightly inclined diaphragms whose alternating openings are made at opposite sides. At the bottom of the vessel is an inclined perforated screen, supported upon cross pieces. A tube introduces hot air near the bottom of the vessel.

Claim.—First, a petroleum filter, so arranged as to keep the crude oil in prolonged contact with the filtering material, and to admit of the clarified oil being drawn off by a stop-cock, or its equivalent, substantially as herein set forth.

Second, parallel diaphragms or their equivalents, serving to lengthen the pathway of the descending oil, substantially as herein arranged and for the purposes set forth.

Third, the introduction of a current of heated air into the bottom, so as to permeate and agitate the whole mass and assist chemical action.

Fourth, the perforated screen s f b g, with its radiating supports, used as in the manner described and for the purposes set forth.

No. 60,594.—ELISHA H. TOBEY, Watertown, N. Y.—*Railroad Signal Light.*—December 18, 1866.—The signal plate, having glasses of different colors, is revolved by the switchman, and gives an intermittent light. The reflector box is revolvable in a horizontal plane, to present it in different directions. The motion of the signal plate is obtained by gearing with which it engages when raised in its frame.

Claim.—First, the combination with a revolving reflector box, or other receptacle for a signal light, of a signal or dial plate, capable of being rotated in a plane at right angles to the plane in which the said box is revolved, as herein shown and described.

Second, in the apparatus herein described, the combination of the dial or signal plate with the gearing by which it is revolved, under such an arrangement that the raising or lowering of the said plate in the signal frame shall cause it to be thrown in or out of gear, substantially as shown and set forth.

Third, in combination with the main signal frame, a reflector box or other receptacle for the signal light, under such an arrangement that the said box, while sliding vertically in said frame, shall also be capable of being rotated in a horizontal plane, substantially as shown and described.

Fourth, in the herein-described apparatus, the combination with the reflector box and revolving dial plate, whose axes of rotation are at right angles with each other, of the gear mechanism by which the said box and plate are respectively revolved, arranged for operation substantially as shown and set forth.

No. 60,595.—R. S. TORREY, Bangor, Maine.—*Drill for Rocks, Wells, &c.*—December 18, 1866.—The temper rod has a pinion upon it which is engaged by a spur wheel upon the shaft of a ratchet wheel. The ratchet wheel is actuated by a sliding pawl rod operated by the movement of the walking beam. This device turns the drill $90°$ at each stroke.

Claim.—First, the self-operating reversible ratchet J, so constructed that it will reverse its motion with the same stroke and will give the desired motion to the drill at every vibration of the walking beam B, for drilling purposes, in combination with the friction roller and spring L.

Second, the dog or hand K, connecting P, figs. 1, 2, and 7, adjustable slide N, figs. 1 and 2, nut 3, sleeve 2, fig. 6, which is attached to the centre post A, fig. 1, in combination with the circular slanted plate for regulating the stroke of dog or hand K, fig. 1, to each stroke of the walking beam B, in the manner and for the purpose substantially as described.

Third, the reversible shaft H and H', fig 3, in combination with gear H, and pinion i, figs. 1, 2, and 3, the whole operating in the manner and for the purpose set forth.

No. 60,596.—G. G. TOWNSEND, Rochester, N. Y.—*Shoe-peg Float.*—December 18, 1866.— The tooth plate is detachable from the head which is pivoted so as to be adjustable in inclination upon the standard.

Claim.—First, the employment or use of the auxiliary or detachable float plate F, substantially as and for the purposes shown and described.

Second, connecting the said plate F to the head H, by means of the lug *f* and one or more keys *e*, substantially in the manner shown and described.

Third, the arrangement of the spring locking latch or lever C, in combination with the pivoted head H, substantially as shown and described, and for the purposes set forth.

No. 50,597.—JAMES H. WALKER, Bergen, N. J.—*India-rubber Covered Umbrella.*— December 18, 1866.—The cover is made of India-rubber cloth, cut in sections, as in the case of other cloth, and united into form by sewing or cementing.

Claim.—The India-rubber umbrella covering, formed in the manner specified, as a new article of manufacture.

No. 60,598.—WILLIAM S. WALKER, Alexandria, Penn.—*Top for Gas-heating Parlor Stoves.*—December 18, 1866.—Explained by the claim and cut.

Claim.—The detachable top A B for a stove, the said top consisting of the oven b^1, the valve b^2, the two hot-air chambers or flue spaces a^2, a^3, and the recesses a^1 a^1, in the outside plates, for receiving within them the faces of the smoothing irons, as described, the said parts being constructed, arranged, and combined together as and for the purposes described.

No. 60,599.—JOHN R. WATKINS, Baltimore, Md.—*Trace Fastening.*—December 18, 186 .— The trace pin is held by a bayonet fastening. A spiral spring serves to retain the projection in the return notch.

Claim.—The use of a barrel with slot and spiral spring and pin with catch thereon, making a secure and ornamental fastening for a trace to the end of the swingletree.

No. 60,600.—ISRAEL WEINBERG, Philadelphia, Penn.—*Hook and Eye.*—December 18, 1866.—The hook is sewed fast to the garment. The eye has pointed shanks, which are booked into any part of the waistband.

Claim.—The construction of the eye, having curved forked ends E, for the purpose of regulating, tightening, or loosening the waistbands of pantaloons, so as to correspond with the shape of the body and suit the wearer, as herein described.

No. 60,601.—MARSHALL D. WELLMAN, Pittsburg, Penn.—*Open Fireplace.* December 18, 1866.—The sectional dampers limit the supply of air to any desired part of the grate. The reflectors radiate heat into the room.

Claim.—First, the use of a damper or dampers, so constructed and arranged relatively to the grate bars of a fireplace or other furnace as that the area of the opening between the bars for the admission of air into the fire shall be increased or diminished by the operation of the damper or dampers, substantially as hereinbefore described.

Second, the use of a reflector or reflectors placed in front of and within the fire chamber of an open fireplace, substantially as and for the purpose hereinbefore set forth.

No. 60,602.—MARSHALL D. WELLMAN, Pittsburgh, Penn.—*Cooking Stove.*—December 18, 1866.—Explained by the claim and cut.

Claim.—The use in cooking stoves, ranges, and other furnace grates, of dampers, slides, or shutters, in combination with a close fire chamber, so constructed and arranged, substantially as herein before described, as that the air may be admitted below the surface of the fuel to a particular part or portion of the fire chamber, while it is excluded from entering the fire chamber at other points.

No. 60,603.—ROLLIN C. WICKHAM, Pawlet, Vt.—*Milk Can.*—December 18, 1866.—The cylinder extends below the tin bottom, and a wooden disk is inserted below the latter. A hoop strengthens the can at this point. The lower ends of the ears extend downward toward the bottom, and the can is nearly balanced when swung by the bail.

Claim.—The construction and arrangement of milk cans with reference to the supporting bottom and ears, as herein described, to be operated in the manner and for the purposes set forth.

No. 60,604.—ALBERT WINDECK and ANDREW RUNSTETLER, Peoria, Ill.—*Corn Planter.*—December 18, 1866.—The seed passes through curved plates which are actuated by a hand lever and which slide between two concentric plates. Valves in the seed spouts operate to deposit the seed in hills. For planting cotton the oscillating plates are removed and hubs with projecting arms substituted. These hubs receive rotation from eccentrics on the axle. The runners are raised from the ground by a foot lever.

Claim.—First, in a seeding machine the oscillating semi-cylindrical dropper C, in combination with the casing B, when arranged substantially in the manner and for the purpose set forth.

Second, the combination of the dropper C, lever J, and valves K and L, in the runner shank, constructed substantially as and for the purpose set forth.

Third, in combination with the valve crank and lever J, the box I, when said several parts are constructed and arranged as set forth.

Fourth, the combination of the oscillating dropper C, arm N, and handle M, when arranged substantially as set forth.

Fifth, the combination of the treadle G, hinged roller F, and runner P, hinged to the frame at e, substantially as and for the purpose set forth.

Sixth, the runner P, and its shank D, when constructed as set forth.

Seventh, in combination with the dropper C, the adjustable slide c, when constructed and arranged substantially as set forth.

Eighth, in combination with the dropper casing B, the cotton dropper, constructed and operated substantially as described.

No. 60,605.—E. B. WINSHIP, Racine, Wis.—*Pump.*—December 18, 1866.—The packing ring has vertical movement on the radial guide plates. The upper flange has perforations, and the under flange has indentations at its periphery for the passage of water.

Claim.—The combination and arrangement of depressions 1 2 3 4 5 6 7 8, flange W, loose packing B, ribs D, and openings F, substantially as set forth and described.

No. 60,606.—EDMUND F. WOODBURY, Rochester, N. Y., assignor to himself and H. A. STRONG, same place.—*Whip.*—December 18, 1866.—Explained by the claims and cut.

Claim.—First, a whip having the handle or any other portion covered with a knit fabric, substantially as herein described.

Second, covering the handle or any other portion of a whip by drawing on the same a piece of tubular knit fabric and fastening it thereon in any suitable manner, substantially as and for the purpose herein described.

No. 60,607.—THEODORE YATES, Milwaukee, Wis.—*Breech-loading Fire-arm.*—December 18, 1866.—The cartridge block is drawn down into a recess of the stock, to permit the insertion of the cartridge. The block is held in its upper position by a catch upon its actuating pin or upon the trigger guard.

Claim.—First, the construction of the breech C, having the rectangular opening with the overhanging shoulder at the top, and having the wall or shoulder m, for the breech block to rest against, substantially as described.

Second, in combination with the breech C, the block D, having its upper arm pivoted on a line with the upper surface of the bore, and its lower arm abutting against the shoulder *m* in line with the lower surface of the bore, with its upper front corner bevelled, as shown, said block being arranged to operate as herein set forth.

No. 60,608 —C. W. ACKER, Watertown, N. Y.—*Carriage Curtain Eyelet.*—December 18, 1866.—The slit disk of elastic material through which the knob or button passes is clamped between annular plates on each side of the curtain; the plate on one side has circumferential spurs, which pass through the curtain and are riveted down on the opposite plate.

Claim.—The toothed struck-up plate C, and notched struck-up plate D, in combination with the slitted elastic plate E, constructed and applied substantially as described, for the purpose specified.

No. 60,609.—H. M. ALBEE, Webster, Mass.—*Ellipsograph.*—December 18, 1866.—One end of the compass carries the pencil as usual; the other leg has a second foot, which is adjusted at a distance from its fellow equal to the difference between the major and minor axes of the ellipse. In use, the associated feet move along the catheti of a right-angled triangle, and the pencil describes one quarter of the ellipse.

Claim.—The arm D, with its adjustable point *c*, in combination with the compass A, constructed and operating substantially as and for the purpose described.

No. 60,610.—PETER BADORE, Montpelier, Vt.—*Anvil for Swaging Calks for Horseshoes.*—December 18, 1866.—The anvil has a groove across its face in which a bar of metal is swaged to the level of the plane surface of the anvil, acquiring a triangular shape in cross section.

Claim.—The swedge A, for drawing the steel from which the toe calks are to be formed to an edge, when said swedge is constructed substantially in the form herein shown and described.

No. 60,611 — ROBERT BAILEY, Idaho City, Idaho.—*Steam Generator.*—December 18, 1866.—The generator has four principal parts, two central sections, and two external jackets. The connected portions form an entire generator, but are susceptible of separation for the purpose of facilitating transportation and repairs.

Claim.—First, a sectional steam boiler divided into four principal parts, two central sections B' B', and two external jackets B B, constructed, combined, and arranged as herein described.

Second, the combination of the hollow grate bars with the central sections B' B', and the jackets B B, arranged and connected as herein described.

Third, the damper *k*, for giving direct or indirect draught through the boiler, in combination with the fire spaces D and *e e'*, and the exit flue F, arranged and operating as and for the purposes herein described.

No. 60,612.—JAMES BALLARD, Almont, Mich —*Washing Machine.*—December 18, 1866 — The rubber is reciprocated above the corrugated bed by means of the lever whose axis assumes its own vertical position in the bearings, which rise from each side of the tub.

Claim.—The reciprocating rubber B, having the hand lever C connected to it through the medium of the hinged bar F, with elastic cord D, or its equivalent attached, substantially in the manner as and for the purpose herein set forth.

No. 60,613.—ABNER G. BEVIN, Chatham, Conn.—*Sleigh Bell.*—December 18, 1866.— The rivet has two legs with knife edges, which cut into and keep the rivet from turning in the leather.

Claim.—The sharp edged lugs *a*, extending from the sides of the shank A, and resting on the head of the rivet, to the other end of which the bell is secured and adapted to cut into the leather and prevent the turning of the rivet, as and for the purpose specified.

No. 60,614.—JOHN BRIGGS, Louisville, Ky., assignor to himself and E. A. HOLMES, Passaic, N. J.—*Tackle Block.*—December 18, 1866.—The snatch block has a bushing between the sleeve and the pin and also between the eye of the hook and the pin by which it is attached to the block.

Claim.—The combination and arrangement of the sheave pin C, sheave B, collar D, pin G, bushing F, hook E, pins *e e*, and cheeks A A, when all are constructed as herein shown and described.

No. 60,615.—WILLARD P. BROOKS, Fairmount, Minn., assignor to himself and F. B. CRIPPEN, same place.—*Broom Head.*—December 18, 1866.—The plates of the cap are drawn together against the brush and secured by bolts. The handle is forced in and its pointed end is traversed by a bolt which passes below the transverse bolts of the cap.

Claim.—The combination of the metallic point or piece B, the rod C, bolts D, or equivalents, and the wires I, with each other and with the handle A, and socket E F, when said point, rod, bolts, and wires are constructed and arranged substantially as herein described and for the purposes set forth.

No. 60,616.—THOMAS S. BROWN, Poughkeepsie, N. Y., assignor to himself and JOHN P. ADRIANCE, same place.—*Self Oiling and Adjusting Bearing for Machinery*—December 18, 1866.—The immediate bearing of the journal rests on a secondary one in such a manner as to accommodate itself to any inclination of the axle. The axle has a lubricating ring which dips into the oil reservoir beneath.

Claim.—The combination of the bed A, part B, with its bearing surfaces as described, and lips *h*, slotted shell *c*, cast with part B, forming the oil chamber *a*, the grooved part E, slotted caps F, and semi-spherical washers H, arranged with the journal C, provided with the ring D, substantially as and for the purpose specified.

No. 60,617.—WALTER D. BURNET, Newark, N. J.—*Anvil on which to Rivet Trunks.*—December 18, 1866.—The trunk is placed upon an anvil block covered with sheet iron and mounted upon a horizontal pivot upon a vertical frame, and the rivets are clenched against the sheet iron plate. Pins working through the frame enter corresponding holes in the block to hold it at any desired point; the said pins are held in place by springs and drawn out by a treadle.

Claim.—The adjustable rotating block C, secured to a frame A, and arranged with pins *c*, or other equivalent stops, substantially as and for the purpose set forth.

Also, the connecting of the pins with springs and a treadle, to operate in the manner substantially as and for the purpose specified.

No. 60,618.—CHARLES B. CANNON, Keokuk, Iowa.—*Potato Digger.*—December 18, 1866.—The tops are cleared away by a horizontal cutter or by the revolving corrugated and conical rollers. The potatoes are sorted as to size and separately delivered.

Claim.—The improved potato digger, consisting of the fluted wheels *n³ n⁴*, the prongs U, the carrier W, and screens V, the various parts of which are constructed, arranged, and operated substantially as herein described and for the purpose set forth.

No. 60,619.—JOHN CLARRIDGE, Pancoastburg, Ohio.—*Saw Set.*—December 18, 1866.—The saw is passed between an adjustable smooth-faced roller and an adjustable longitudinally ribbed, conical roller. One passage sets the teeth which project to one side, and a return passage sets those turning in a contrary direction. The conical setting roller and the similar guide rollers have a scale by which their adjustment is regulated to suit any size of teeth.

Claim.—An improved saw set formed by the combination of the cylinder B, cones G and L, scales J and O, index pins K and P, set screws E F I N and R, with each other and with the frame A, substantially as herein described and for the purpose set forth.

No. 60,620.—EUSEBIO CORTES, Sagua la Grande, Cuba, assignor to JOSE A. MORA, New York, N. Y.—*Sugar Cane Planter.*—December 18, 1866.—The cane slips are placed in the chute and delivered into the furrow made by the first plough; the two side ploughs cover the cane, and the scraping plates level the ground. The body is vertically adjustable by lever arms carrying the wheels.

Claim.—First, the cane planter consisting of the wire-box C, plough D, ploughs I, hopper J, inclined chute K, handle H, and adjustable wheels A E, substantially as and for the purpose specified.

Second, the lever F and wheel E, in combination with the frame A and plough D, when constructed and applied as herein shown and described.

No. 60,621.—T. H. CUSHING, Dover, N. H.—*Sawing Machine.*—December 18, 1866 —The carriage is curved and runs in a curved course, and the stuff is first operated on by saws and then by rotary cutters. A lever acts on a second lever, and a swinging arm carrying an idler to disconnect the feed gearing and to reverse the feed shaft, for gigging back the carriage.

Claim.—First, the combination of reciprocating jaws J with rotary planers W W, arranged with a carriage or bed K, placed on curved ways or guides *b*, for the purpose of sawing and planing sticks in curved form simultaneously or at one operation, substantially as shown and described.

Second, the friction roller R, attached to the pendent swinging frame S, which is connected to the lever T, as shown, in connection with the two shafts O P, provided with the gears *c d*, the belt *h*, and the adjustable bearing *a*, of the shaft P, arranged with lever T, to operate substantially as and for the purpose specified.

Third, the taper blocks U U, when used in connection with a carriage or bed K, working on curved ways or guides *b*, and reciprocating saws J, substantially as and for the purpose set forth.

No. 60,622.—T. M. DAVY, Jeffersonville, Ind.—*Steam Trap.*—December 18, 1866.—The cover screws on to the cylinder and encloses a valve which is supported by a spring: the latter rests on a movable plug which is raised as required by a cam whose shaft has a crank arm on the outside of the case.

Claim.—First, the stem or upper portion A, and the case B, when combined with the faucet barrel, cam or key H, and the arm or handle J, constructed substantially as described.

Second, the cam H, when used in connection with the adjustable spring seat G, and spring E, substantially as herein shown and described.

No. 60,623.—B. F. ELLIOTT, Cedar Rapids, Iowa.—*Trellis for Grape Vines.*—December 13, 1866.—Improvement on his patent of July 24, 1866. The posts are attached by hinges to their bases, and may be laid down in winter without disturbing the vines attached to the trellis wires.

Claim.—The combination with the hinged uprights A of the wedge-pieces E and staples F, substantially as and for the purpose described.

Also, the rails G hung to the uprights A, so as to turn or swing thereon, substantially as and for the purpose specified.

No. 60,624.—SAMUEL B. FAY, Franklin, Penn.—*Metal Loop for Tags.*—December 18, 1866.—The hooks at each end of the continuous wire by which the tag is attached are hooked into the hole in the tag, so that one loop forms a mousing for the other.

Claim.—The hook for tags consisting of the loop-wire A, with both ends turned or bent into circles, adapted to lie side by side, and applied to the tag, in the manner described, for the purpose specified.

No 60,625.—A. K. and B. H. FOSTER, Hallettsville, Texas.—*Cotton Cultivator.*—December 18, 1866.—The forward share is divided longitudinally, and the sides are adjustable laterally. The central cutter is within the side cutters, and has a lateral reciprocating movement by means of projections on its sustaining bar, which come in contact with cam bolt-heads on the wheel.

Claim.—First, the share E, composed of two parts *d d*, arranged in V-form, with a space *c* between their front ends, and attached to a standard F and to the front ends of the handles B B, in the manner shown and described, or in an equivalent way, to admit of being adjusted at a greater or less distance apart at their front ends, substantially as shown and described.

Second, the reciprocating cutter L, operated from the wheel D through the medium of the screws *i* and the rock-bar I, provided with the arms *p p'*, in combination with the share H, substantially as and for the purpose specified.

Third, the fitting or securing of the screws *i* to the wheel D by means of the concentric annular grooves *c c'* in the side of the rim *b* of said wheel to receive the nuts *a* of the screws *i*, whereby the screws may be readily applied to and detached from the wheel and secured at an equal distance apart, substantially as described.

No 60,626.—WILLIAM GIBSON, New York, N. Y., assignor to HENRY M. JOHNSTON and O. S. FOLLETT.—*Preparation of Paper, &c., for Photographic Use.*—December 18, 1866.—Explained by the claim.

Claim.—The production of an insoluble enamel or surface or size in or upon paper, silk, cloth, fibrous, and textile articles of all kinds, wood, leather, glass, porcelain, earthen-ware, metals, India-rubber, gutta-percha, papier-mache, passe-partout, and compositions by the successive application thereto of an adhesive mixture or body and an astringent mixture or solution, substantially as described.

No. 60,627.—T. L. GOBLE, Orange, N. Y.—*Clothes Pin.*—December 18, 1866.—The waist of the shackle bar passes between the pivoted jaws, and lugs upon its edges engage with channels on the inside faces of the jaws.

Claim.—A clothes pin consisting of two jaws A pivoted or hinged together, in combination with the shackle B, arranged to slide upon the said jaws, substantially as and for the purpose described.

No. 60,628.—EDWIN HARD, Canal Dover, Ohio.—*Sawing Machine.*—December 18, 1866.—When a kerf is completed the driving shaft is temporarily clutched to the log-feeding devices, which advance the carriage until the log strikes the gauge rod and disconnects the feed.

Claim.—Operating the saw carriage by means of the bevel pulleys Y Z, clutch A', lever B', curved lever G', adjusting bar D', and gauge rod E', arranged and operating substantially as described for the purpose specified.

No. 60,629.—A. A. HENDERSON, Naval Hospital, near Norfolk, Va.—*Steam Engine Governor.*—December 18, 1866.—Eccentrics and cams are so arranged upon the revolving shafts which are driven by the engine that any change in the speed of the latter shall operate through them to move the throttle valve.

Claim.—The shaft B, the sleeve C, the eccentric plates D and F, and cam E', the shaft G, with its lug-wheel H and ratchet I, the arm J and the spring S, and the eccentric rod C', with its guide pins *e* and *e'*, when constructed, arranged, and combined substantially as herein described for the purposes specified.

No. 60,630.—JAMES HIGGINS, East Cambridge, Mass.—*Solar Time Indicator.*—December 18, 1866.—The encircling ring has a circumferential slot in one side; an aperture in the middle ring travels in the direction of the said slot. The inner ring is set in regulated position by a screw, and is marked with a time scale, upon which impinges a solar beam.

Claim.—The combination of the ring A, having slots *a* and *c* in it, with the ring B, provided with the scales D D' and set-screw C, the same constituting a solar time indicator, substantially as herein shown and described.

No. 60,631.—LEOPOLD HOPP, New York, N. Y.—*Malt Extract.*—December 18, 1866.—The barley is soaked with a decoction of fennel and malted. To the wort is added the distillate of a decoction of hops, cascarilla, lysimachia purpurea, riphanus niger, radius dictamus albus, and calamus. The resulting compound is then fermented, and the beer treated as usual.

Claim.—First, the within described process of malting by soaking the barley with a decoction of fennel instead of plain water, as set forth.

Second, the beer of health, obtained by mixing the wort obtained by the above-named process with the hygienic ingredients herein set forth.

No. 60,632.—S. M. HUNTER, Terryville, Conn.—*Hydraulic Governor.*—December 18, 1866.—The gate of a water wheel or the cut-off valve of a steam engine is controlled by the action of an engine operated by water or other non-elastic fluid. The latter engine is controlled by a common centrifugal governor, and acts as a governor to the first-mentioned engine.

Claim.—First, regulating the speed of a water wheel or a steam engine by the action of an engine driven by water or other non-elastic fluid, in the manner herein set forth.

Second, the oscillating valve T, constructed substantially as described, in combination with the non-elastic fluid engine.

No 60,633.—BENJAMIN S. HYERS, Pekin, Ill.—*Sash Fastener.*—December 18, 1866.—The friction wheel is made to bear upon the edge of the sash to hold it in any position; it has a small pinion which engages with an inclined rack made fast to the window frame.

Claim.—The catch E, operating within the toothed roller A, in combination with the plate F, inclined toothed back C, and toothed roller A, substantially as herein shown and described and for the purpose specified.

No. 60,634.—JOSEPH IRVING, New York, N. Y.—*Fifth Wheel.*—December 18, 1866.—The safety clip may be drawn up to prevent the fifth wheel from rattling, and while it does not interfere with its motions, the parts are united and strengthened.

Claim.—The safety clip E, consisting of the part *c*, with the semicircular recess and the recessed part *d*, between which the fifth wheel plays in the recesses, having a yielding pressure, and secured and operating in the manner and for the purpose specified.

No. 60,635.—THOMAS IRVING, JOHN McNEIL, GEORGE W. RICH, and CYRUS J. FAIR Elwood, N. J.—*Manufacture of Water and Fire-proof Paper.*—December 18, 1866.—Manila or hemp rope is cut into pieces of two inches length, boiled in a solution of alkali, pulped, saturated in a solution of gum catechu, borax, and rosin sizing. Before the second process rolls the paper passes through a solution of alum or borax, closing the surface of the paper and making it water and fire-proof.

Claim.—First, water-proof paper, prepared as herein described, as a new article of manufacture.

Second, the within-described process of manufacturing water-proof paper, by treating manilla or hemp or other fibrous material with the ingredients and in the manner set forth.

No. 60,636.—HENRY LAMPERT, Nunda, N. Y.—*Apparatus for Scraping Hides.*—December 18, 1866.—Below the scraping knife is a drum, upon which the hide to be scraped is secured. The drum may be vertically adjusted to the knife by means of eccentrics, and is partially rotated after each movement of the knife to bring a fresh portion of the hide into the proper position to be scraped.

Claim.—First, the revolving beam A, mounted on the vertically adjustable shaft B, in combination with the block *g*, carrying the scraper *e* and knife *f*, and with suitable mechanism to impart to said worker a reciprocating motion, substantially as and for the purpose described.

Second, the eccentrics *b*, lever *a*, and serrated arc *d*, in combination with the cleats *c*, shafts B, and beam A, constructed and operating substantially as and for the purpose set forth.

Third, the worker composed of a block *g*, with scraper *e* and knife *f*, in combination with the spring *h*, pitman G, and crosshead F, and eccentric wrist pin *n* on the disk H, constructed and operating substantially as and for the purpose described.

Fourth, the screw rod *j* and nut *k*, in combination with the worker *g* *e f*, spring *h*, and pitman G, constructed and operating substantially as and for the purpose set forth.

No. 60,637.—PETER LOUIS, New York, N. Y.—*Sprinkling Attachment for Brooms.*—December 18, 1866.—A metallic water reservoir rests upon the top of the broom head. It, has a hole for the passage of the handle, and is perforated at the bottom.

Claim.—The within-described sprinkling attachment for brooms, composed of reservoir A. provided with a socket *a*, a number of holes *b* in its bottom, and a vent hole *c* in its top, for the purpose herein set forth and described.

No. 60,638.—JAMES L. MACKEY, Seymour, Ind.—*Rotary Steam Valve.*—December 18, 1866.—The conical chambered plug has an enveloping sleeve fitted into a shell cast with the cylinder. The ports are so arranged as to suit a single or double cylinder engine.

Claim.—First, the valve A, with a V-shaped partition B and apertures *c d a a' b*, in combination with the sleeve C and shell D, constructed and operating substantially as and for the purpose described.

Second, the transverse channel *m* in combination with the hollow plug valve A, V-shaped partition B, and with aperture *j j' k k' l l' c d*, substantially as and for the purpose set forth.

No. 60,639.—F. B. MORSE, Milwaukee, Wis.—*Tightening the Tires of Wheels.*—December 18, 1866.—The ends of the tire have backwardly inclined lips, and the keyheads of bolts operate with these lips to tighten the tire. A central key bolt has separate side pieces that may be removed to give place for the tire ends when tightening.

Claim.—First, the lips *e e* at the ends of the tire C, in combination with the keys D D E, operated through the medium of screws and one or more removable or adjustable keys F, all arranged substantially as and for the purpose set forth.

Second, the arms *f f* on the lips *e*, in combination with the keys E D D, substantially as and for the purpose specified.

Third, the socket B, provided with the partitions *a a*, to form two end partitions to receive the ends of the felloes, and a central compartment to receive the tire-tightening mechanism, substantially as and for the purpose specified.

No. 60,640.—JAMES MURTAUGH, New York, N. Y.—*Life Raft.*—December 18, 1866.—Two concentric tubes of India-rubber may be inflated at separate openings, and are enclosed by a netting which secures them in position and makes a central support for the feet.

Claim.—The expansible rings *a a*, separate and arranged one within the other, with a network covering forming a seating surface and a rest or hold for the feet, in the manner described for the purpose specified.

No. 60,641.—MITCHEL RENZ, Naugatuck, Conn.—*Extension Weeding Hoe.*—December 18, 1866.—The two blades and rakes are attached to the respective portions of the frame. When the blades lap, the tool is at its narrow adjustment, but by slipping them on each other the width is increased and the adjustment maintained by a set screw.

Claim.—First, forming the head of the hoe in two parts, substantially as herein shown and described, so that they may overlap and slide upon each other, as and for the purpose set forth.

Second, forming ways upon the parts A and B of the hoe head to keep them in proper relative position upon each other while sliding back and forth, substantially as herein shown and described.

Third, forming slots in the parts A and B of the hoe head for the reception of the handle,. substantially as herein shown and described and for the purpose set forth.

No. 60,642.—ELEAZER ROOT, Indianapolis, Ind.—*Terrasphere.*—December 18, 1866.—This instrument is intended to exhibit the diurnal and annual movements of the earth, and to illustrate the phenomena dependent theron.

Claim.—First, the stand A and the pillar B, combined with the vertical dial plate C, the horizontal spindle D, carrying the sun G, and the frame H H, carrying the earth K, revolving on its own axis *f f*, in the suspended vernier *g*, and while revolving vertically around the sun G, with the frame H H, keeping its true angle of inclination to the ecliptic throughout her orbit, arranged and operating substantially in the manner and for the purpose herein described and specified.

Second, the zodiacal belt I, in combination with the pillar B, the dial plate C, the frame H H, the sun G, and the earth K, arranged and applied substantially as and for the purposes herein described.

Third, the adjustable moon N, in combination with the earth K and its axis *f f* arranged and applied substantially as and for the purposes herein set forth.

Fourth, the radiating points *c c*, representing the rays of the sun, in combination with the spindle D, the sun G, and the earth K arranged and applied substantially as and for the purposes herein described.

Fifth, the hemispherical night-cap M, in combination with the earth K and its axis *f f*, and the sun G, arranged and applied substantially as and for the purposes herein described and represented.

No. 60,643.—JACOB RUSSELL, Brooklyn, N. Y.—*Corn Stalk Cutter and Corn Husker.*—December 18, 1866.—The stalks, with the ears on them, are passed, butt end foremost between rollers armed with circular and transverse cutters. The ears as they are cut from the stalks fall upon a system of rollers, which strip off the husks, while the ears pass on into a basket.

Claim.—First, the construction of rollers C C', of circular cutters *f f*, and longitudinal cutters *h h*, these latter intersecting the cutters *f*, so as to cut the stalks of corn crosswise at the same time that they are split longitudinally, and at the same time crowd the ears of corn back, substantially as described.

Second, the husking bed, consisting of non-elastic rollers arranged in an inclined plane and rotated substantially as described, in combination with knives on rotary cylinders which allow the stalks to pass between them, but crowd the ears of corn back upon the bed, substantially as set forth.

Third, the inclined bridge G, interposed between the roller C' and the husking bed, substantially as and for the purpose described.

Fourth, the yielding gate J, arranged over the inclined busking bed, so as to operate substantially as described.

Fifth, a machine which is adapted for cutting up corn-stalks, stripping the ears of corn from the stalks, and then separating the husks from the ears, substantially as described.

No. 60,644.—B. D. SHAW, Beverly, Ohio.—*Gate and Door Latch.*—December 18, 1866.—The angular latch pin projects from the post, and is engaged by the catches on the gate. Above these catches is a pivoted lever, so placed that the raising of one of them for the passage of the pin depresses the other and arrests the movement of the gate. The lever gravitates to a horizontal position and forces both catches into a bearing on the latch pin.

Claim.—The combination of the catches D D, and pivoted lever E, constructed so as to operate substantially as herein shown and described.

No. 60,645.—JAMES M. SHEW, Paper Mills. Md.—*Rag Engine for Paper Making.*—December 18, 1866.—The object of the spiral flange is to prevent the collection of rags around the axis and the consequent choking of the cylinder ; its action is to constantly throw off the pulp from the centre to the circumference.

Claim.—Placing a spiral scroll or voluted flange on the ends of the cutting cylinder of a rag-engine of a paper mill, substantially as and for the purposes herein described.

No. 60,646.—PETER A. SNYDER, Jersey City, N. J.—*Adjustable Mitre.*—December 18, 1866.—The arms are set by the corner of the object, and then used to scribe the stuff.

Claim.—The arrangement with the pivoted blades *a b*, leaves *c d*, and flat stock A, of the screw *e*, nut *h*, and groove *f*, when the parts are constructed as to admit of the ready adjustment of the pivoted blades *a b* and leaves *c d*, when said stock is resting on a flat surface, as herein set forth.

No. 60,647.—MATHIAS SPENLÉ, Detroit, Mich.—*Wood Turning Lathe.*—December 18, 1866.—To a lathe having a revolving cutter and forming wheel, an oscillating adjustable frame is attached, which holds a pattern last and the last block to be finished. The pattern is brought in contact with the guide wheel and the unfinished block is presented to the cutter.

Claim.—First, the arrangement of two clamps, which are coupled together, one to contain the pattern last and the other the last to be finished, in combination with the guide wheel F, and cutter-wheel G, constructed and operating substantially as and for the purpose described.

Second, the swinging head B, and adjustable platform *c*, carrying the clamps D D', in combination with the guide wheel and the cutter wheel, substantially as and for the purpose set forth.

No. 60,648.—E. H. STEARNS, Erie, Penn.—*Saw Mill.*—December 18, 1866.—The outer rollers of the carriage have plain faces, and are so inclined from a longitudinal direction as to give the carriage a slight side movement from the saw when gigging back. The grooved pulleys have side play in the boxes for this purpose. The feed and gigging movements are communicated from friction pulleys on a pivoted arm, and this arm is moved by a lever.

Claim.—First, relieving the saw from friction of the log while resting on the carriage B, by means of the flat wheels *a*, grooved wheels *a'*, axles *c*, and boxes *a''*, flat track *b*, and angular track *b'*, constructed and operating substantially as described, for the purpose specified.

Second, constructing a cant support joined to and forming an inseparable part of the saw-guide, in such a manner that when the guide is moved nearer or further from the saw arbor, when changing saws the cant support will necessarily move with it, arranged and operating substantially as and for the purposes herein specified.

Third, giving motion to the log carriage, and reversing the same by means of friction-pulleys *l* and *l'*, when caused to bear upon the pulley *m*, and when rigidly fixed to shafts *f f*, each having a continuous and positive revolving motion from belts applied to pulleys *i i'*, all constructed and operating substantially as described.

No. 60,649.—WASHINGTON H. STEWART, Logansport, Ind.—*Sawing Machine.*—December 18, 1866.—The saw arm has a pawl,which engages a segmental rack to sustain the saw at any height when it is not in operation.
Claim.—The brace E', provided with a ratchet, in combination with the pawl G, saw bar H, pitman J, and crank wheel L, for the purposes and substantially as described.

No. 60,650.—P. L. SUISE, Shirleysburg, Penn.—*Fork and Cake Lifter.*—December 18, 1866.—Either the knife or fork may be placed in position for using; the other is locked in the handle by a spring plate.
Claim.—An improved culinary instrument formed by combining the fork B, and the flat blade C, in one piece pivoted at one end of the handle A, and held in place when reversed for the use of either the fork or the blade by the yoke *c*, under pressure of the spiral spring *g*, arranged and operating as described.

No. 60,651.—BYRON D. TABOR, Wilson, N. Y.—*Sheep Rack.*—December 18, 1866.—The revolving covers in their upper position serve as protectors for the sheep, and in their lower position as guards and guides in filling the grain troughs.
Claim.—An improved sheep rack formed by the combination of the end pieces or frames A B, the binder C, the sides D, of the hay rack, the revolving troughs E, levers F, and revolving levers G H, with each other, the parts being constructed and arranged substantially as herein shown and described and for the purpose set forth.

No. 60,652.—S. J. TALBOT, Milford, N. H.—*Dough Mixer.*—December 18, 1866.—The flaring receiver is hung on trunnions and oscillated by a horizontal lever.
Claim.—An oscillating dough mixer consisting of the can A, double metallic cover B, hoop D, trunnions E, handle F, hooks G H, and frame C, constructed and operating in the manner as and for the purpose specified.

No. 60,653.—SIDNEY VAN AUKEN, Binghamton, N. Y.—*Tool for Cleaning Boiler Tubes.*—December 18, 1866.—The segmental scrapers are carried on spring arms, which keep their edges in contact with the tube. The outer edges of the arms may be made to scrape the tube by turning the rod.
Claim.—The doubly yielding spring arms B, in combination with the segmental scraper C, and cams *a b*, constructed and operating substantially as and for the purpose set forth.

No. 60,654.—J. M. WHITESIDE, San Francisco, Cal.—*Reverberatory Furnace.*—December 18, 1866.—The revolving stirrer is moved by a vertical shaft which is cooled by a stream of water and protected by a cap. All but superheated air, with or without superheated steam, is excluded from the furnace while in operation; the supply of water is effected by an automatic device operated by the rotary driving shaft.
Claim.—First, the arrangement of a revolving stirrer in combination with the oven A, constructed and operating substantially as and for the purpose described.
Second, the protecting cap *b*, in combination with shaft C, which gives motion to the stirrer, substantially as and for the purposes set forth.
Third, the arrangement of the supply tank E, and stop-cock *i*, which is operated automatically from the driving shaft C', in combination with the generator *h*, tube or tubes *g*, and the oven A, constructed and operating substantially as and for the purpose described.

No. 60,655.—GEORGE H. WOODRUFF, Jerseyville, Ill.—*Combined Roller and Harrow.*—December 18, 1866.—The frame may be tilted forward, so that the weight of the barrow will increase the effect of the rollers. The rollers may be removed, the wheels at their ends supporting the frame.
Claim.—First, the employment of the pivoted lever O, for the purpose of elevating the barrow so that only the rollers may be used, substantially as shown and described, and for the purpose set forth.
Second, the wheels B B, in combination with the rollers C C, and frame A, constructed and arranged in such a manner that the rollers may be removed and only the wheels used, or the wheels removed and the rollers moved out to the frame, substantially as herein shown and described and for the purposes set forth.
Third, the beams G G', standing obliquely to the frame A, when the said frame A is provided with wheels B B, and rollers C C, constructed and operated substantially as and for the purposes set forth.

Fourth, the adjustable seat in combination with the frame A, wheels A A, rollers C C, and harrow, for the purpose of regulating the depth of the harrow teeth *e*, substantially as described and for the purposes set forth.

Fifth, attaching the pole or tongue E, by means of bifurcated bar or straps P, to near the centre cross piece K, of the frame A, in combination with the guide *a¹*, substantially as shown and described and for the purposes set forth.

No. 60,656.—JAMES SUGGETT, Cortlandville, N. Y.—*Manufacture of Butter from Whey.*— December 18, 1866.—The warm whey, as it runs from the cheese vat, is put into a cooler and subjected to the action of neutralizing salts, when the cream is suffered to rise. This is worked and washed like butter from cream, and it is then ready for packing.

Claim.—The manufacturing of butter from whey, substantially in the manner herein described.

No. 60,657.—LEVI DECKER, New York, N. Y.—*Cushion for Billiard Tables.*—December 18, 1866.—A cord attached to an elastic strip is placed inside the lining at the upper angle of the cushion, to stiffen it and present a firm, narrow, projecting surface to the ball.

Claim.—The cord E, employed in combination with the elastic strip D and cushion C, in the manner and for the purposes specified.

EXTENSIONS.

WENDELL BOLLMAN, Baltimore, Md.—*Construction of Bridges.*—Patented January 6, 1852, No. 8,624; extended January 5, 1866.

Claim.—The combination of the tension rods *e*, connecting the foot of each strut with the end of the stretcher, substantially as described, by which an independent support is given to the strut carried back directly to the abutment, while at the same time no lateral force or strain is brought upon the abutment, as herein fully set forth.

HENRY JENKINS, Brooklyn, N. Y.—*Ornamental Connection of the Parts of an Iron Fence.*—Patented January 13, 1852, No. 8551; reissued September 6, 1859, No. 807; extended January 9, 1866.

Claim.—Forming the ornament or cast-iron connections for a railing, fence, or other article of iron cast into a divided iron mould, substantially as and for the purposes specified.

ALFRED PLATT, Waterbury, Conn.—*Buckwheat Fan.*—Patented January 13, 1852, No. 8,659; extended January 12, 1866.

Claim.—The method of separating the bulls from the kernels of buckwheat by shaking them on a table or tables made slightly concave and rough, substantially as specified, in combination with a current or currents of air blown over the surface of such table or tables to carry off the hulls, whilst the kernels are retained or held back by the form of the surface of the table or tables, as specified.

MARTHA A. DODGE, Bedford, Mass., administratrix of the estate of GEORGE H. DODGE, deceased.—*Ring Spinner.*—Patented January 27, 1852, No. 8,683; extended January 26, 1866.

Claim.—The combination of the standard or projection B with the ring and traveller, substantially in the manner and for the purpose of removing or loosening waste from the latter, as specified.

BYRON DENSMORE, New York, N. Y.—*Harvester.*—Patented February 10, 1852, No. 8,720; reissued January 28, 1862, No. 1,262; extended January 30, 1866.

Claim.—First, hanging the driving wheel in a supplementary frame, or its equivalent, which is hinged at one end to the main frame while its opposite end may be adjusted and secured at various heights or be left free, as desired, whereby the cutting apparatus may be held at any desired height for reaping, or be left free to accommodate itself to the undulations of the ground, for mowing, substantially as described.

Second, the employment in a harvesting machine of a wheel provided with a crank and lever for the purpose of raising and lowering the outer end of the finger bar to cut high or low, substantially as described.

LOUISA R. KETCHUM, Buffalo, N. Y., executrix of WILLIAM F. KETCHUM, deceased.—*Grass Harvester*—Patented February 10, 1852, No. 8,724; reissued February 28, 1854, No. 259; again reissued June 2, 1857, No. 466; extended January 30, 1866.

Claim.—First, extending the shoe H G from the heel of the rack or finger bar upward and forward, and firmly connecting its continuation with the draught when the finger bar is located as set forth, so that the power by which the machine is drawn shall through the shoe be communicated to and draw forward the heel of the rack or finger bar, thus relieving the great strain which would otherwise come upon the lateral connections of the rack or finger bar with the wheel frame, while the heel is enabled to slide over obstructions, substantially as shown.

Second, when the main wheel and inner end of the finger bar or rack D are located relatively as described, continuing the shoe H G from the heel of the rack or finger bar upward and forward until the upper end of its extension reaches a part of the machine which always runs above the mown grass, and which will keep the said grass down and prevent its rising over the point of the extended shoe, thus aiding the shoe to ride over the mown grass, even when accumulated before it, substantially as shown.

Third, supporting the heel of the rack or finger bar sufficiently near the ground and at a convenient distance laterally from the main wheel by arms extending upwards and forwards, and upwards and backwards therefrom, and connected with the frame or strong bars firmly bolted across the frame in front and rear of the said rack or finger bar, while the said frame and bars are elevated to pass over the cut grass, and the above parts are arranged substantially as shown.

Fourth, supporting the rack or finger bar at the side of and lower than the main frame by means of auxiliary framing in a fixed position at the side thereof, and extending downward and forward so that while the finger bar is held as near the ground as desired and lower than the main frame, the main frame may be nearly horizontal in the line of draught and at any convenient height to avoid clogging or accommodate the diameter of the main wheel as shown; such an auxiliary frame as a whole is shown in the drawings, composed of the bar

C, rods E E I, and rack or finger bar D ; but its details may of course be varied, while the principle of my invention is retained.

Fifth, supporting the rack or finger bar D in its position at the side of and lower than the main frame by extending a strong bar C behind said rack or finger bar firmly supported by said frame, and rigidly connecting said rack or finger bar to said bar C by a straight brace or braces E E: said frame being elevated and said bar being elevated and placed sufficiently in rear of said rack or finger bar, to avoid clogging or lodging of the mown or falling grass against either when said parts are arranged in relation to each other, substantially as shown.

Sixth, supporting the outer end of the rack or finger bar by a rod extending downwards and forwards from the cross-bar C to the finger bar, parallel or nearly so to the face of the main wheel, when the frame and bar C are elevated above the rack or finger bar in the manner and for the purposes contemplated in the last claim to avoid the falling or clogging of the cut grass against such rod, as set forth.

HENRY G. BULKLEY, Kalamazoo, Mich.—*Drying Grain.*—Patented March 2, 1852, No. 8,769; reissued June 27, 1854, No. 267; extended February 21, 1866.

Claim.—The method of seasoning or kiln-drying substances by using steam in a vessel which has an opening communicating with the atmosphere to limit the pressure for the purpose of transmitting caloric to the substances to be seasoned or kiln-dried, or the vessel or vessels containing them, substantially as specified.

MARTHA M. JONES, Staten Island, N. Y., administratrix of SAMUEL T. JONES, deceased.—*Manufacture of Zinc White.*—Patented February 24, 1852, No. 8,756; extended February 23, 1866.

Claim.—The use of a porous or fibrous bag or receiving chamber with porous sides or bottom, or an air-tight chamber with a straining or porous bag adapted to the inside thereof, and used in connection either with a blowing or exhausting apparatus, so that the products of the distillation and oxygenation of zinc, or other volatile metals, may be separated from the accompanying air and gases, which latter will be forced or otherwise drawn through the pores of the cloth bag or chamber, and escape into the atmosphere.

SIMEON SAVAGE, Pomfret, N. Y.—*Machine for Printing Floor Cloths.*—Patented March 2, 1852, No. 8,778; extended February 28, 1866.

Claim.—The arrangement of the printing mechanism, the stamping-down mechanism, and the mechanism for advancing the piece or strip of cloth or of material to be printed and pressed or stamped, such arrangement being as exhibited in the drawings, and as above described.

Also, the combination of the lip bar or plate y, the series of bent levers a' a', &c., the slide bar R' or S, and the bar c', as made and operated, substantially in the manner and for the purpose of seizing the selvage edge of the cloth and moving the piece as described.

Also, the combination of mechanism for operating the coloring carriage, or imparting to it its back and forth movements, and necessary intervals of rest, the said combination consisting of the rotating shaft O, with its circular disks Q R, and their projections i k, the four book bars l l, p p, together with the vibrating bars n o, as applied together and operated, substantially as specified.

CHARLES NEER, Brooklyn, N. Y.—*Canal Lock Gate.*—Patented March 9, 1852, No. 8,789; extended March 8, 1866.

Claim.—First, the opening of the lower gates of a canal or river lock outwards or down stream, in combination with the means described, or their equivalents, for operating them, for the double purpose of saving length in the lock chamber with the same walls, and for allowing the gates to be opened before the chamber is entirely empty, so that the escaping water may carry out with it the boat, raft, or other thing being passed through, with the least possible delay.

Second, the stationary gate at the head of the lock, which forms, with the breast wall of the lock, with the top of which it is level, a recess or chamber, through which the lock chamber may be filled at any desired height above the bottom of the lock, and thus save length of lock wall.

Third, in combination with the stationary gate, the sinking head gate, extending across the lock, and reaching down a little below the top of the stationary gate when the gate is shut, and which sinks or slides into the recess formed in part by said stationary gate, and is on a level therewith when open, for passing boats, &c., for the purpose of saving in the length of the lock chamber an amount nearly equal to the width of the gate.

Fourth, the so placing of an adjustable batten or water strip on the bottom of a lock so that it may be operated upon by the pressure of the water within the lock chamber, and be forced up against the gate when prevented from being closed tigh by an intervening substance, substantially in the manner herein set forth and described. t

SAMUEL T. THOMAS, Laconia, N. H., and ELIZA A. ADAMS, Townsend, Mass., administratrix of EDWARD EVERETT, deceased.—*Pattern Card for Jacquard Looms.*—Patented March 10, 1852, No. 8,810; extended March 9, 1866.

Claim.—The combination of the buttons with the metallic card, as described, the buttons being so riveted or attached to the card as to allow of their being turned, for the purpose of closing or opening the holes to which they are respectively attached.

NICHOLAS TALIAFERRO, Augusta, Ky., and WILLIAM D. CUMMINGS, Maysville, Ky.—*Smoothing Iron.*—Patented March 30, 1852, No. 8,848 ; extended March 12, 1866.

Claim.—The application (substantially described) to a self-heating smoothing iron of a tube or chamber *j* at the bottom of the fire box, provided with a registered mouth or inlet *i*, some distance above the bottom and at its lower portion, with distributing apertures K communicating with the fire, whereby the draught is applied from beneath, and equally at every part, and placed under the control of the operator, without permitting the escape of ashes, or other refuse of combustion.

JOHN McCOLLUM, New York, N. Y.—*Cracker Machine.*—Patented March 23, 1852, No. 8,828 ; reissued May 31, 1859, No. 730; extended March 22, 1866.

Claim.—The combination of adjustable springs with a cracker cutter and its resisting surface or bed, substantially as hereinbefore described and substantially for the purposes hereinbefore set forth.

JOHN M. THATCHER, New York, N. Y.—*Air-heating Stove.*—Patented March 23, 1852, No. 8,832 ; reissued September 11, 1855, No. 327 ; extended March 22, 1866.

Claim.—Making the bottom plates of the flue spaces of air-heating furnaces or stoves for the passage of the products of combustion outward or inward, among or around their passages, inclining inward and downward toward the fire chamber, substantially as described, for the purpose of facilitating the increase of the heating surface without the inconvenience of the accumulation of ashes, soot, and other solid matter on such plates, as set forth.

Also, the combination of the inverted domes or frustums F I M, and plate P, with the short tubes *b b,f f, i i, l l,* connecting them substantially in the manner herein described for the purpose of effecting the connection between the lower ends of the fire or draught flues, and carrying the air through them to the spaces between the cylinders or tubes.

DANIEL SHAW, Elkhart, Ind.—*Smut Mill and Grain Separator.*—Patented April 6, 1852, No. 8,861 ; reissued November 3, 1863, No. 1,564 ; extended April 5, 1866.

Claim.—The combination and joint operation of a smut mill and an exhaust-fan grain separator, substantially as herein specified and set forth.

EBENEZER W. PHELPS, Elizabeth, N. J.—*Moth Trap to Beehives.*—Patented April 6, 1852, No. 8,859 ; extended April 5, 1866.

Claim.—The peculiar construction of the moth trap as herein described, composed of a slide, having the centre groove and two side grooves, and the metallic hinged cover, arranged all as set forth in the specification.

CHARLES T. GRILLEY, New Haven, Conn.—*Capping of Screws.*—Patented April 20, 1852, No. 8,888; extended April 9, 1866.

Claim.—The attachment of a brass, copper, or other suitable metallic cap to, and its combination with, an iron wood-screw, substantially in the manner and by the process described in the foregoing specification, which I conceive to be the only practicable method in which the same can be usefully effected, whereby, and by means of the successive operations of punching or stamping, the nick is first cut through the shell, and then, after being adjusted to the groove or slot in the head of the screw, the sides thereof are driven down into, and made to press closely against the sides of the slot, leaving the bottom of the groove or slot uncovered, so that the cap, when closed round the head of the screw, will preserve its hold without liability to be turned or displaced by the screw driver which works upon the iron surface at the bottom of the slot, and against the covered sides thereof, thereby furnishing to the public, at a comparatively small cost, a wood screw, having all the beauty and finish of a brass, copper, or plated screw, in combination with the greatly superior strength of an iron one. The invention is equally applicable to steel screws, which may be capped in a similar way.

JOEL WHITNEY, Winchester, Mass.—*Feed Apparatus of Planing Machines.*—Patented April 13, 1852, No. 8,881 ; extended April 12, 1866.

Claim.—The arrangement by which the upper feed roll is allowed to yield to any inequalities in the board, and, at the same, draw down upon the surface to which it has yielded in proportion to the resistance to the cutting tools, that is, connecting the fixed shaft with the vertical sliding bearings of the upper feed roll by means of the swinging, inclined, and vertical arms; the gears on the fixed shaft operating the lower feed roll, and also playing into the gears which move the upper feed roll, said latter gears having their bearings in the intersection or joint of the said arms; the arrangement being substantially as herein above set forth.

JOHN M. EARLS, Troy, N. Y.—*Smut Machine.*—Patented April 27, 1852, No. 8,904; extended April 26, 1866.

Claim.—First, the projecting screen chambers, in combination with the arrangements for separating the rubbing chamber from the fan chamber, whereby the grain is prevented from being affected by the blast from the fan chamber while it is passing through the rubbing chamber, and is only brought in contact with the current of air, where it ascends to take away the chaff and other impurities, substantially as herein set forth.

Second, in combination with the scouring surfaces, the beating forks, for the purpose of beating the grain and breaking the hulls while falling from the rubber to the scourers, whereby the berries are more effectually cleaned from adhering impurities, as herein set forth.

WILLIAM SOUTHWELL, West Cambridge, Mass.—*Machinery for Grinding or Polishing Saw Blades, &c.*—Patented May 4, 1852, No. 8,929; extended May 1, 1866.

Claim.—First, the combination of two grindstones, or their equivalents, revolving in the direction herein made known, for the purpose of grinding or polishing two sides of a saw or other article simultaneously, with a reciprocating frame, or its equivalent, for the purpose of holding the article being ground or polished, whereby the tendency of either stone to move the article is counteracted by the action of the other stone, and the same force is thereby required to reciprocate the article in either direction, as described.

Second, the combination of the right and left-hand screws, carriers, and nuts for said screws, movable pedestals or boxes, together with the cross-shaft, worms, worm-wheels, and handles, substantially as set forth, for the purpose of moving two grindstones, or their equivalents, simultaneously against opposite sides of an article being ground or polished, as described.

Third, I do not claim giving an automatic traverse motion to grindstones, but what I do claim is the arrangement of screws, mitres, wheels, handles, eccentrics, eccentric boxes, and movable frame, substantially as herein described, whereby I am enabled at any time to move the grindstones, or their equivalents, entirely across the machine, for the purposes set forth, without interfering with the automatic traversing motion, which is given to the said stones irrespective of their precise position with reference to either saw frame, or either saw, or other articles fixed in said frame.

Fourth, the arrangement in the same machine of two sets of reciprocating frames, either of which can be stopped without affecting the other, and a carriage, whereby the grindstones can be caused to move from one frame to the other, by which arrangement one saw can be ground or polished while another is being adjusted into place.

MOSES G. FARMER, Salem, Mass.—*Electro-Magnetic Alarm Bell.*—Patented May 4, 1852, No. 8,920; extended May 1, 1866.

Claim.—The combination, substantially as herein set forth, of the electro-magnet and armature (or its electro-magnetic equivalent) with the falling ball or spring, and the detents and the lifting cam, or its equivalents, so arranged that when the ball is supported by the armature a slight force only of the electro-magnet is required to trip the ball, which ball in falling acquires sufficient momentum to produce much greater mechanical effects than the magnet alone, the velocity of the ball in falling being still further accelerated by the force of a spring, if desired. The power thus obtained is used in the manner and for the purpose herein described.

THOMAS J. WOOLCOCKS and WILLIAM OSTRANDER, New York, N. Y.—*Speaking Tube.*—Patented May 4, 1852, No. 8,932; extended May 1, 1866.

Claim.—The combination of an alarm valve with a speaking tube or pipe, in the manner and for the purpose substantially as herein set forth.

WILLIAM ALFORD and JOHN D. SPEAR, Philadelphia, Penn.—*Iron Safe.*—Patented May 18, 1852, No. 8,952; extended May 16, 1866.

Claim.—The application of chalk or whiting, which has been subjected to the action of acids, and has been partially deprived of its carbonic acid, the material which is used being in fact the waste or residual matter left from the manufacture of what is called mineral water, after chalk or whiting has been subjected to the action of acids for the purpose of expelling a portion of the carbonic acid; this residual matter, consisting substantially of the substances named in the analysis before referred to, in the construction of double iron chests or safes, in the manner above described, or in any other manner substantially the same.

RENSSELAER REYNOLDS, Stockport, N. Y.—*Power Loom.*—Patented June 1, 1852, No. 8,984; extended May 24, 1866.

Claim.—First, connecting the rocker of each picker staff, made and operated substantially as specified, with the bed on which it rocks by means of an interposed strap of leather, or other flexible substance, attached at the inner end to the bed, and at the outer end to the rocker, substantially as and for the purpose specified.

Second, forcing the shuttle binders inwards against the shuttle while boxing by a gradually increasing force by means of arms on a rocker provided with a spring, which is acted upon by a pin on the connecting rod of the lay, substantially as described.

Third, securing the raw-hide pickers to the inner face of the staffs by means of a leather strap, or the equivalent thereof, embracing and binding the two together, substantially as described, to insure the firm union to resist the rapid blows, and to prevent pieces of raw-hide from breaking and flying, as set forth.

JOHN RIDER, New York, N. Y.—*Manufacturing Gutta-percha.*—Patented June 1, 1852, No. 8,992; extended May 31, 1866.

Claim.—The preparing of gutta-percha for vulcanizing by a preliminary separate heating of it to such a degree as to expel its volatile ingredients, herein specified, which can generally be effected at the high temperatures from 285 to 430 degrees Fahrenheit, substantially as herein set forth.

Also, the process, herein described, of vulcanizing gutta-percha by first heating it to a sufficiently high temperature to expel from it the volatile ingredients, herein specified, which it is believed can be accomplished between 285 and 430 degrees Fahrenheit, and then incorporating with it, substantially as herein specified, a hyposulphite, either alone or in combination with metallic sulphurets, or whiting, or magnesia, or with all of them together, and then subjecting the mixture to a temperature of from 285 to 320 degrees Fahrenheit, all the steps of the said process being performed substantially in the manner herein set forth, at the same time desiring it to be understood that I disclaim the vulcanizing of gutta-percha in all cases, save when it has been prepared for the vulcanizing operation by the aforesaid preliminary heating.

REUBEN DANIELS, Woodstock, Vt.—*Manufacture of Granular Fuel from Brushwood and Twigs.*—Patented June 15, 1852, No. 9,015; extended June 13, 1866.

Claim.—The granular fuel produced from brushwood and twigs by cutting the same into lengths about equal to its average diameter, as herein described, as a new manufacture.

CHRISTOPHER C. BRAND, Norwich, Conn.—*Bomb Lance for Killing Whales.*—Patented June 22, 1852, No. 9,047; reissued August 26, 1856, No. 392; extended June 13, 1866.

Claim.—The mode of sustaining the fuze rope in the fuze tube, and preventing the fire of the charge of the gun from passing by the fuze rope and into the bomb, viz: by metal or metallic plugs, or the equivalents thereof, cast around or made to closely encompass the fuze rope after it has been inserted in the fuze tube, as specified. I do not claim the application of wings or feathers to a shaft or rod to direct its passage through the air, but what I do claim is so making or applying them to the shaft shank, or to the body of the bomb, that not only may they be folded or moved down so as to be capable of entering with the shaft the bore of the gun, but each have an elastic property or spring, such as will cause it to unfold or be thrown outwards immediately after the projectile may be discharged from the gun, such wings being made of vulcanized India-rubber, or any substance or substances which may be deemed an equivalent thereto, inasmuch as such may possess the requisites as above specified.

ALLEN B. WILSON, Waterbury, Conn.—*Sewing Machine.*—Patented June 15, 1852, No. 9,041; extended June 14, 1866.

Claim.—The combination of the bobbin F for carrying one thread with a rotating hook, which is of such form, or forms part of a disk, or is equivalent of such form, as to extend to the loop on the other thread and pass it completely over the said bobbin, whereby the two threads are interlaced together, the parts being arranged and operating in any way substantially as herein set forth.

JAMES SHARP, Brooklyn, N. Y.—*Label Card.*—Patented June 15, 1852, No. 9,039; extended June 14, 1866.

Claim.—The manufacture of label cards, or tickets of cloth and paper stuck and pressed together, substantially as above described.

WILLIAM O. GROVER and WILLIAM E. BAKER, Boston, Mass.—*Sewing Machine.*—Patented June 22, 1852, No. 9,053; reissued July 6, 1858, No. 572; extended June 21, 1866.

Claim.—First, in combination with an upper needle or eye-pointed perforating instrument, a non-perforating eye-pointed instrument so shaped and moved substantially as specified, that it shall spread a loop of the thread it governs while advancing through a loop of the upper needle thread, substantially in the manner and for the purpose specified.

Second, in combination with an eye pointed upper needle, a non-perforating instrument having the function, substantially in the manner specified, of carrying the loop of the upper needle thread out of the location or position in which it was originally seized, for the purpose substantially as set forth.

Third, in combination with an eye-pointed perforating instrument, an eye-pointed non-perforating instrument, substantially such as is described, and performing the two offices of spreading a loop of its own thread while advancing through a loop of the upper needle thread and of changing the locality of the loop of upper thread that it has seized, both offices being performed substantially in the manner and for the purpose hereinbefore described.

Fourth, an eye-pointed upper needle and an eye-pointed needle so arranged and operating as to make a stitch substantially such as is herein represented, in combination with a feed apparatus, one surface of which has motion in four different directions, substantially in the manner and for the purposes described.

GEORGE O. WAY, Claremont, Minn., administrator of LAFAYETTE F. THOMPSON, deceased, and ASAHEL G. BACHELDER, Lowell, Mass.—*Railroad Car Brake.*—Patented July 6, 1852, No. 9,109; extended July 5, 1866.

Claim.—To so combine the brakes of the two trucks with the operative windlasses, or their equivalents, at both ends of the car, by means of the vibrating lever A', or its equivalent or mechanism, essentially as specified, as to enable the brakeman, by operating either of the windlasses, to simultaneously apply the brakes of both trucks, or bring or force them against their respective wheels, and whether he be at the forward or rear end of the car.

ELIAKIM B. FORBUSH, Buffalo, N. Y.—*Harvester.*—Patented July 20. 1852, No. 9,134; reissued July 8, 1856, No. 376; again reissued April 19, 1859, No. 692; again reissued May 23, 1865, No. 1,972; extended July 19, 1866.

Claim.—First, so connecting the cutting apparatus; having a short and separate finger bar to the main frame of the machine that it may be adjusted to different heights for reaping or lowered to the ground for mowing without changing the position of the main frame, substantially as described.

Second, connecting the finger bar to the grain side of the main frame and supporting it by one end only, by means of an adjustable device and the inwardly projecting ends of the cross-pieces of the main frame, substantially as set forth.

Third, the slotted frames K K, and locking bolts i i, applied and used for the purpose and substantially as set forth.

Same; reissue, No. 1,973; extended July 19, 1866.

Claim.—First, making the outer and inner shoes broader in front of the finger bar, as shown at J and m''', for the purpose of bracing the guard fingers laterally.

Second, so constructing skeleton guard fingers and arranging them on the finger bar that they will mutually brace and support each other forward of the finger bar, substantially as set forth.

Third, the bearing piece Z, placed between the outer shoe and guard finger, for the support of the outer end of the cutter bar, substantially as described.

Same; reissue, No. 1,974; extended July 19, 1866.

Claim.—In combination with a short finger bar and a shoe, by which it is connected to the main frame of the machine, and a cutting apparatus located in rear of a line drawn through the front of the driving wheel, a quadrant-shaped platform, so arranged that the cut grain may be delivered therefrom at the side of the platform and in rear of the main frame, substantially as set forth.

Same; reissue, No. 1,975; extended July 19, 1866.

Claim.—First, in combination with a cutting apparatus and a quadrant shaped grain platform, and both located in the rear of a line drawn through the front of the driving wheel, a rake, supported by a pivoted connection on the main frame in rear of the axis of the driving wheel, and so arranged that it will sweep over the platform and deliver the grain in rear of the main frame, substantially as set forth.

Second, a movable fulcrum, upon which the rake is suspended and operated in the manner substantially as described.

JESSE S. and DAVID LAKE, Smith's Landing, N. J.—*Grass Harvester.*—Patented July 20, 1852, No. 9,137; reissued January 1, 1861, No. 9, (1861;) extended July 19, 1866.

Claim.—Attaching or fastening that part of a mowing machine to which the guards or fingers, which support and hold the grass while it is being severed by the cutter or cutters, are attached to the main frame or to an intermediate coupling piece, so that the guards or fingers, or that part to which they are attached and by which they are sustained and supported, will be free to rise or fall bodily: and also to have a lateral rolling or wabbling motion, to enable the cutting apparatus to conform freely to the undulations of the ground over which it is drawn, independent of the up and down motions of the main frame.

Same; reissue, No. 10, (1861;) extended July 19, 1866.

Claim.—First, the combination of a hinged lever seat for the driver, with the main frame and cutting apparatus of a mowing machine.

Second, the combination of a main frame, and an inclined projecting hinged lever seat with a main drive wheel, having no outside support.

Same; reissue, No. 11, (1861;) extended July 19, 1866.

Claim.—First, placing the driving wheel in a mowing machine on the outside of the frame, in combination with a triangular shaped frame on the outside of the wheel.

Second, the combination of the following elements, viz: a triangular main frame, a single driving wheel arranged on the outside of said frame, and a hinged lever seat for the driver, for the purposes set forth.

Third, in combination with the main frame of a mowing machine, a single drive wheel and a leading swivel wheel, so arranged in relation to each other and the cutting apparatus as that the said wheels shall run on lines just outside of the grass to be cut.

Same; reissue, No. 12, (1861.) Extended July 19, 1866.

Claim.—First, the combination of the coupling piece I and finger support G, with an intermediate metallic connection or lever H, for the purposes stated.

Second, the metallic connection or lever H, both as a hinge and a support to the cutting apparatus and finger support.

Third, the combination of the finger support, and the metallic connection or lever H, in such a manner that their lower surfaces shall be flush, for the purposes set forth.

Fourth, in combination with the main frame of a grass harvester, a finger, or guard support, when'ever it may be called, having two independently acting hinges or yielding connections interposed between it and the main frame, in such a manner as that the said finger, or guard support, as it is advanced by means of its hinges or yielding connections, shall be free to rise above or fall below the plane on which the carrying or driving wheel or wheels are passing.

Fifth, so hinging that part of a mowing machine to which the guards or fingers are attached as that it may oscillate or turn on a line at right angles to the line of motion of the machine, for the purpose of raising the points of the guards to adapt the cutting apparatus to the condition of the ground, while it is also free to roll, rock, or wabble on a line parallel to the line of motion of the machine.

Sixth, the combination with the main frame, finger support, and cutting apparatus of a mowing machine, of lever with an adjustable weight thereon, whereby the pressure of the finger support and cutting apparatus upon the ground can be regulated by simply moving said weight upon its lever.

Seventh, the combination of a counterpoise weight, or the equivalent thereof, with that part of a mowing machine, whatever it may be called, to which the guard fingers are attached, to diminish its pressure upon the ground, and thus obviate side draught and friction.

THADDEUS HYATT, New York, N. Y.—*Vault Cover.*—Patented November 12, 1845; reissued April 3, 1855, No. 303; extended seven years; again extended by act of Congress, approved July 26, 1866 (By office August 1, 1866.)

Claim.—Making covers for openings to vaults, in floors, docks, &c., of a metallic grating or perforated metallic plate, with the apertures so small that persons or bodies passing over or falling on them may be entirely sustained by the metal, substantially as described; but this I only claim when the apertures are protected by glass, substantially as and for the purpose specified.

Also, in combination with the grating or perforated cover and glass fitted thereto, the knobs or protuberances on the upper surface of the grating or perforated plates, for preventing the abrasion or scratching of the glass, substantially as specified.

THOMAS CASTOR, Philadelphia, Penn.—*Dumping Wagon.*—Patented August 3, 1852, No. 9,164; extended August 2, 1866.

Claim.—The arrangement of the body on a fixed roller fulcrum, on the frame of the running gear, in such manner that by a slight amount of force the body can be turned to give its under side, which rests on the roller, either a forward or backward inclination, to cause the weight of its load to tend to hold it forward or back, as it is required to carry or to dump the same, substantially as herein set forth.

EDWARD A. PALMER and ADOLPHUS J. SIMMONS, Clayville, N. Y.—*Whiffletree Hook.*—Patented September 7, 1852. No. 9,252; extended Septemper 6, 1866.

Claim.—The head turning upon the shaft to close the hook, the sliding catch to prevent its opening, and the spring within the head acting upon them, the whole combined and operating substantially in the manner specified.

WILLIAM MOORE, Brooklyn, N. Y.—*Door Lock.*—Patented September 14, 1852, No. 9,265; extended September 12, 1866.

Claim.—The tumbler k, enclosed by the dividing plate h, to be operated on solely by the key when entered from the inner key-hole, in combination with the revolving check, or its equivalent, and the bolt, for the purposes and as described and shown.

ROBERT KNIGHT, Cleveland, Ohio.—*Machinery for Bevelling the Edges of Skelps or Metallic Strips, &c.*—Patented September 21, 1852, No. 9,274; extended September 20, 1866.

Claim.—Arranging the rollers in the frame so as to receive a lateral movement, as may be desired ; in other words, giving the rollers end play one over the other, as thereby increasing or diminishing the distance between the bosses, (according to the width of the plate or strip,) and providing suitable means for retaining the same in place.

HENRY C. SMITH, Cleveland, Ohio.—*Lath Machine.*—Patented September 28, 1852, No 9,286 ; extended September 27, 1866.
Claim.—The combination of the method of rotating the log or bolt from which the lath are to be cut, by means of the poppet wheels J J', arranged respectively on the shafts E' E which forms a part of the mandrel at each end of the log, and the gear wheels I I', or their equivalents, moving with equal velocities, so as to prevent any wrenching or twisting of the log on its centres, and to hold it firmly up to the knives while being operated upon by them, and the method of clutching and releasing the log, by means of the dog A, hollow bearing C', for containing the clutch head G. and hollow shaft E', for receiving the rod F, which screws into said clutch, and by which the dog may be driven into the log, or the log released : the whole being arranged and operating substantially in the manner and for the purpose set forth.

LORENZO L. LANGSTROTH, Oxford, Ohio —*Beehive.*—Patented October 5, 1852, No. 9,300 ; reissued May 26, 1863, No. 1,484 ; extended October 4, 1866.
Claim.—First, constructing and arranging the movable comb frames of beehives, in such a manner that when placed in the hive or case they have not only their sides and bottom kept at suitable distances from each other and from the case, substantially in the manner and for the purposes described, but have likewise their tops separated from each other throughout the whole or a portion of their lengths, substantially in the manner and for the purposes set forth.
Second, constructing and arranging movable frames in such a manner that when they are inserted in the hive the distances between them may be regulated at will, substantially in the manner and for the purposes described.
Third, constructing movable frames and arranging them in the hive in such a manner that the bees can pass above them into a shallow chamber or air space, substantially in the manner and for any or all of the purposes set forth.
Fourth, the shallow chamber, in combination with the top bars of the laterally movable frames or their equivalents, and with the perforated honey board upon which to place surplus honey receptacles, substantially as and for the purposes set forth.
Fifth, a movable partition or divider, substantially as described, when used in combination with movable frames, substantially in the manner and for the purposes described.
Sixth, the use of movable blocks for excluding moths and catching worms, so constructed and arranged as to increase or diminish at will the size of the bee entrance, substantially. in the manner aud for the purposes set forth.

OLDIN NICHOLS, West Roxbury, Mass.—*Grinding Mill.*—Patented October 12, 1852. No. 9,330 ; extended October 5, 1866.
Claim.—The pointed projections b on the front edges of the teeth of the cylinder E, when used in combination with the the teeth c c, in the concave formed with concavities in their front edges, substantially in the manner and for the purpose herein set forth.

PETER GEISER, Greencastle, Penn.—*Grain Separator.*—Patented October 19, 1852, No. 9,341 ; extended October 5, 1866.
Claim.—The method herein described of regulating the blast of winnowing machines, by means of a flap on the fan case, arranged and adjusted substantially as herein set forth.
Also, the reciprocating toothed bars G, with the trough A. whose bottom is divided into three portions, the lowermost being tight and acting merely as a conveyer, the middle one acting both as a conveyer and screen to separate the wheat from the straw and allow it to pass into the winnower, and the upper or third portion acting as a conveyer for the straw, and a coarse screen to separate therefrom the heads of unthreshed grain that would not pass through the lower screen ; the teeth of the reciprocating bars moving the straw regularly along the trough and working or shaking the grain and heads so effectually through the screens that none is left to pass off with the straw when it is discharged through the upper end of the trough.

JAMES GREENHALGH, Woonsocket Falls, R. I.—*Counterbalancing Harnesses in Looms.*—Patented November 2, 1852, No. 9377 ; extended October 15, 1866.
Claim.—The construction of the long double heddles or jacks D D, in such a manner and so hanging them on the axle E by a short arm, or its equivalent, that in their vibrations neither end of them shall pass beyond a vertical plane passing through the axle on which they rock or oscillate, so that the weight of the jacks shall be thrown outside of their points of suspension, thus counterbalancing the weight of the harness.

D. D. ALLEN, Adams, Mass.— *Tool for Cutting Pegs out of Boot Soles.*—Patented October 19, 1852, No. 9 340: extended October 11, 1866.

Claim.—The adjustable float or cutter C D E, connected to shank B, by means of the pin or pivoted *b*, which turns loosely in the bearing or standard *a*, so as to permit the float to adjust itself to the proper positions to cut the pegs from the heel to the toe of the boot, in the manner herein set forth.

L. Q. C. WISHART, Philadelphia, Penn.— *Design for Ornamenting Bottles.*—Patented October 25, 1859, No. 1,161; extended October 23, 1866.

Claim.—The ornamental design described, and represented in the drawings, for pine tree tar cordial bottles.

ALBERT GARDNER, for himself, and as administrator of the estate of WILLIAM L. HUNTER; deceased, Cincinnati, Ohio.— *Constructing Ploughs.*—Patented October 26, 1852, No. 9,362, extended October 25, 1866.

Claim.—Bolting the standard mould-board, landside, and share to the block F, or its equivalent, instead of bolting or fastening the parts to each other as has been practiced heretofore; which block F may be connected to the beam by a bolt K, or otherwise, substantially as described and represented.

WANTON ROUSE, Taunton, Mass.— *Self-acting Mule for Spinning.*—Patented November 2, 1852, No. 9,378; reissued March 15; 1853, No. 233: extended October 27, 1866.

Claim.—First, governing the revolution of the spindles in winding the yarn on the cop, and also in backing off during the successive stages of the building by means of a cam B, or any equivalent device of irregular form, circumferentially with the said irregularity varying from end to end; the said cam or equivalent being caused to operate upon the mechanism which drives the spindles in any way that will produce the results herein set forth.

Second, the mechanism for causing the finger *d*, through which the irregular surface of the cam B, or its equivalent, acts upon the mechanism which drives the spindles in backing off and building on, to traverse the said cam, and to be kept close to its surface, consisting of the screws *e* and *k*, the nut *j*, cord or chain *f*, lever G, and stud *h*, operating in combination in the manner substantially as set forth.

STEPHEN C. MENDENHALL, Richmond, Ind.— *Throwing Shuttles in Looms.*—Patented November 9, 1852, No. 9,387; extended November 6, 1866.

Claim.—The combination and arrangement of the spring triggers *f f*, cords *h k*, and treadles 1, 2, 3, &c., so that the depression of any one of these treadles shall release the triggers on the forward movement of the lay and allow the picker staff to actuate the shuttle, substantially as set forth.

STEPHEN C. MENDENHALL, Richmond, Ind., OBED KING and EZRA KING, Salem, Iowa.— *Hand Loom*—Patented November 9, 1852, No. 9,388; extended November 6, 1866.

Claim.—The combination of nerve K, operated by lay inclined plane O, and its guides M M', and adjustable pin W, or their equivalents, combined and operating substantially as described, so that we can operate and vary the number of heddles, substantially as and for the purpose set forth. We are aware that the picker staff has been operated by hooks alternately raised from the shoulders on the picker staff by pins on a vibrating slide operated by grooves in the treadle cam; this we do not claim. But we do claim the combination of the inclined plane Q, on picker staff spring T, and hooks R R, for the purpose of lifting the hooks in the manner and for the purpose specified.

JOSEPH J. COUCH, Brooklyn, N. Y.— *Machine for Drilling Stone.*—Patented November 23, 1852, No. 9,415; extended November 6, 1866.

Claim.—The improvement of making the drill rod to slide through the piston rod, substantially in manner as above set forth.

Also, the combination of the rocker lever K, the wedge M, the bolt P, within the lever, the two cam plates N O, the spring catch Q, the spring and the two projections *c d* as applied to the drill shaft, the carriage or block I, and the slide-ways thereof, and made to operate together and to actuate the drill, substantially in manner as hereinbefore set forth.

DANIEL TAINTER, Worcester, Mass.— *Rotary Knitting Machine.*—Patented November 30, 1852, No. 9,435; extended November 14, 1866.

Claim.—So to combine a draught and take-up roller, and mechanism for revolving it, with a rotary series or set of needles and other mechanism of the above mentioned peculiar kind for knitting, that such draught roller shall rotate simultaneously or with the same velocity with such series of needles, so as to prevent the longitudinal rows of stitches from being produced in helical lines, and the evil consequences resulting to the fabric therefrom.

Also, the arrangement of the draught and take-up mechanism, in connection with the knitting mechanism, supported by two separate frames A T, and also their connection with the mechanism for producing an equal and simultaneous rotation of the frames A T, all substantially as described, whereby there shall not only be no connection between the frames A T, to extend through the fabric, but no projection from the frame A, to come in contact with the presser, stitch wheels, and cam bar, or their respective supports, during the simultaneous and equal rotations of both or either of the said frames A T.

JOHN R. MOFFIT, Chelsea, Mass.—*Grain Separator.*—Patented November 30, 1852, No. 9,432; reissued March 23, 1858, No. 540; and again reissued May 17, 1859, No. 715; extended November 24, 1866.
Claim.—The endless chains *d* composed of metallic links provided with protuberances or depressions, when used in combination with suitable driving chain-gears to impart a positive motion to the straw carrier of a threshing and separating machine, as explained.

Same; reissue, No. 716, B. I; extended November 24, 1866.
Claim.—In combination with a receptacle in which the tailings are deposited by the winnowing apparatus, the arrangement of the screw elevator *o*, in relation to the threshing cylinder for the purpose of returning the tailings to be re-threshed, as set forth.

Same; reissue, No. 717, C. I; extended November 24, 1866.
Claim.—The reversible screen *k²* and delivery spout *l² m¹*, arranged, adapted and constructed substantially in the manner described with and to the discharging spout of the "fanning mill" or "shoe" of a threshing machine so as to be isolated from the winnowing arrangement and made to deliver at either one side or the other of the machine, as set forth.

JOSEPH GUILD, Buffalo, N. Y.—*Mortising Machine.*—Patented November 30, 1852, No. 9,431; reissued December 11, 1855, No. 333; extended November 30, 1866.
Claim.—First, the sliding wrist *o*, connected with the chisel, and also with the driving power in the manner described in combination with the mechanism described, or its equivalent for sliding said wrist, so that the operator can during the motion of the machine vary the depth of cut of the chisel, or cause it to be suspended without disconnecting the driving power.
Second, the combination in a mortising machine, substantially as described, of a treadle and opposing spring or weight connected to a toggle, one end of which being pivoted to the frame, the other is pivoted to a sliding wrist upon a vibrating arm actuated by the power, the said wrist being slid out and in upon the arm with varying power and speed, by the action of said toggle and its attached weight or spring and treadle, as explained, or other equivalents.

ERASMUS A. POND, Rutland, Vt.—*Pill Making Machine.*—Patented December 7, 1852, No. 9,455; extended November 27, 1866.
Claim.—First, moulding or forming pills by means of two cylinders B B, having each a number of recesses *a a* in its periphery, the recesses in one cylinder matching with those in the other, and each matching pair forming a mould of the required form of the pill, the said cylinders revolving in opposite directions, and the pill mass being conducted between them, substantially as herein described.
Second, the bands I I, of India-rubber, or any sufficiently elastic material, passing round or partly round the mould cylinders, for the purpose of expelling the pills from the recesses *a a*, after the moulds are open, substantially as herein set forth.

SARAH DUTCHER, Waukesha, Wis., administratrix of the estate of ELIHU DUTCHER, deceased, and WARREN W. DUTCHER, Milford, Mass.—*Temple for Looms.*—Patented December 28, 1852, No. 9,502; extended December 26, 1866.
Claim.—The arrangement of parts so that the temples have a reciprocating action corresponding with the motion given to the cloth by the beat of the lay, substantially as herein set forth.

DESIGNS.

No. 2,239.—THOMAS R. EVANS, Philadelphia, Penn.—*Quarter of a Balmoral Shoe.*—January 2, 1866.

No. 2,240.—SIDNEY SMITH, Greenfield, Mass —*Plate of a Stove.*—January 2, 1866.

No. 2,241.—LEWIS RATHBONE, Albany, New York.—*Plate of a Stove.*—January 16, 1866.

No. 2,242.—LEWIS RATHBONE, Albany, N. Y.—*Plate of a Parlor Stove.*—January 16, 1866.

No. 2,243.—JOHN B. WARING, New York, N. Y.—*Trade Mark for Pens and Pen Boxes.*—January 16, 1866.

No. 2,244.—JAMES G. ABBOTT, assignor to ABBOTT and NOBLE, Philadelphia, Penn.—*Plates of a Stove.*—January 30, 1866; antedated December 27, 1865.

No. 2,245.—HENRY BERGER, New York, N. Y.—*Composition in Alto-Relievo.*—January 30, 1866.

No. 2,246.—SAMUEL W. GIBBS, Albany, N. Y., assignor to ABBOTT and NOBLE, Philadelphia, Penn.—*Cook's Portable Range.*—January 30, 1866; antedated December 27, 1865.

No. 2,247.—SAMUEL W. GIBBS, Albany, N. Y., ; assignor to ABBOTT and NOBLE, Philadelphia, Penn.—*Cook's Stove.*—January 30, 1866 ; antedated December 27, 1865.

No. 2,248.—JOHN HOGE, and ROBERT D. SCHULTZ, Zanesville, Ohio.—*Trade Mark.*—January 30, 1866.

No. 2,249.—G. I. MIX, Wallingford, Conn.—*Spoon Handle.*—January 30, 1866.

No. 2,250.—DAYTON MORGAN, Chillicothe, Ohio.—*Bust of Abraham Lincoln.*—January 30, 1866.

No. 2,251.—JOHN ROGERS, New York, N. Y.—*Group of Figures.*—January 30, 1866.

Nos. 2,252 and 2,253.—G. SMITH and H. BROWN, assignor to ABBOTT & NOBLE, Philadelphia, Penn.—*Cook Stove, (two patents.)*—January 30, 1866; antedated December 27, 1865.

No. 2,254.—J. H. STONE, Philadelphia, Penn.—*Base of Sheet-metal Vessel.*—January 30, 1866.

No. 2,255.—SARAH F. AMES, Boston, Mass.—*Bust of Abraham Lincoln.*—January 30 1866.

No. 2,256.—WILLIAM S. BELL, Jr., Boston, Mass.—*Paper Collar.*—January 30, 1866.

No. 2,257.—HENRY L. MESERVEY, Boston, Mass.—*Flour Sifter.*—January 30, 1866.

No. 2,258.—JAMES POWELL, Cincinnati, Ohio.—*Medallion of General Grant.*—January 30, 1866.

No. 2,259 to No. 2,262.—THOMAS MERRY, assignor to SAMUEL NEEDHAM, Philadelphia, Penn.—*Fabric Trimming, (four patents.)*—February 6, 1866.

No. 2,263.—JOHN H. BARNES, Brooklyn, N. Y.—*Door Lock.*—February 13, 1866.

No. 2,264.—CHARLES ZEUNER, assignor to M. GREENWOOD & CO., Cincinnati, Ohio.—*Piano Stool.*—February 13, 1866.

No. 2,265.—THOMAS DEVINS, Cambridgeport, Mass.—*Coffin.*—February 20, 1866.

No. 2,266.—F. A. GILES, New York, N. Y.—*Top Plate of a Watch.*—February 27, 1866.

No. 2,267.—WILLIAM H. PAGE, assignor to WILLIAM H. PAGE & CO., Norwich, Conn.—*Alphabet of Letters.*—February 27, 1866.

Nos. 2,268 and 2,269.—JOSEPH SCHEDLER, Hudson City, N. J., assignor to AMERICAN LEAD-PENCIL COMPANY, New York, N. Y.—*Trade Mark, (two patents.)*—February 27, 1866.

Nos. 2,270 and 2,271.—GARRETTSON SMITH and HENRY BROWN, assignors to ABBOTT & NOBLE, Philadelphia, Penn.—*Plates of a Stove, (two patents.)*—February 27, 1866; antedated January 30, 1866.

No. 2,272.—GARRETTSON SMITH and HENRY BROWN, Philadelphia, Penn., assignors to ABBOTT & NOBLE, same place.—*Plate of a Cook Stove.*—February 27, 1866; antedated January 30, 1866.

No. 2,273.—JOHN S. ARMSTRONG, Prairie du Chien, Wis —*Military Cenotaph.*—March 6, 1866.

Nos. 2,274, 2,275, and 2,276.—SAMUEL KELLETT, San Francisco, Cal.—*Architectural Centre Flowers, (three patents.)*—March 6, 1866.

No. 2,277.—ROBERT S. LYON, West Morrisania, N. Y.—*Trade Mark.*—March 6, 1866.

Nos. 2,278 and 2,279.—P. C. CAMBRIDGE, Jr., Enfield, N. H.—*Bedstead, (two patents.)*—March 13, 1866.

No. 2,280.—A. W. FAGIN, St. Louis, Mo., assignor to COLSWELL & CO.—*Trade Mark.*—March 13, 1866.

No. 2,281.—F. A. GILES, New York, N. Y.—*Top Plate of a Watch.*—March 13, 1866.

No. 2,282.—JOHN MARTINO, JACOB BEESLEY, and JOHN CURRIE, assignors to STUART & PETERSON, Philadelphia, Penn.—*Portable Furnace.*—March 13, 1866.

No. 2,283.—SAMUEL M. RICHARDSON, New York, N. Y.—*Hinge Plate.*—March 13, 1866.

No. 2,284.—P. BRADFORD, assignor to SARGENT & CO., New Haven, Conn.—*Coffee Handle.*—March 27, 1866.

No. 2,285.—CHARLES T. MEYER, Bergen, N. J.—*Floor Oil Cloth.*—March 27, 1866.

No. 2,286.—LOUIS SAARBACH, Philadelphia, Penn.—*Tobacco Pipe.*—March 27, 1866; antedated March 6, 1866.

No. 2,287.—GARRETTSON SMITH and HENRY BROWN, assignors to ABBOTT & NOBLE, Philadelphia, Penn.—*Plate of a Cook Stove.*—March 27, 1866; antedated February 27, 1865.

No. 2,288.—WILLIAM S. BELL, Jr., Boston, Mass.—*Paper Cuff and Collar.*—April 3, 1866.

No. 2,289.—W. H. CARTER, Cincinnati, Ohio.—*Alphabet of Letters.*—April 3, 1866.

No. 2,290.—WILLIAM A. GILES, Chicago, Ill.—*Statuette and Clock Case.*—April 3, 1866.

No. 2,291.—WILLIAM A. GILES, Chicago, Ill.—*Clock Case.*—April 3, 1866.

No. 2,292.—WILLIAM A. GILES, Chicago, Ill.—*Bust and Clock Case.*—April 3, 1866.

No. 2,293.—B. S. NICHOLS, Seneca Falls, N. Y., assignor to DOWNS & COS'. MANUFACTURING COMPANY, same place.—*Pump.*—April 3, 1866.

No. 2,294.—SARAH E. COOK, Philadelphia, Penn.—*Ornament for the Head.*—April 24, 1866; antedated March 20, 1866.

No. 2,295.—JOSEPH W. BARTLETT, New York, N. Y.—*Frame of a Sewing and Embroidering Machine.*—May 1, 1866.

No. 2,296.—JOHN H. BELLAMY, Charlestown, Mass., assignor to himself, CYRUS W. STOUT, Boston, and BENJAMIN BROWN, Somerville, Mass.—*Picture Frame.*—May 1, 1866.

No. 2,297.—JOHN H. CLARK and JOHN RHINESMITH, Fort Wayne, Ind.—*Weather-boarding.*—May 1, 1866.

No. 2,298 to 2,300.—FRANKLIN O. DAY and WM. S. STEWART, St. Louis, Mo.—*Trade Mark, (three patents.)*—May 1, 1866.

No. 2,301.—H. H. GROSSKOFF, Philadelphia, Penn.—*Hand Stamp.*—May 1, 1866.

No. 2,302.—JOHN MARTINO, JACOB BEESLEY, and JOHN CURRIE, Philadelphia, Penn., assignors to J. S. CLARK, same place.—*Cook's Range.*—May 1, 1866.

No. 2,303.—JOHN McARTHUR, Philadelphia, Penn.—*Railing.*—May 1, 1866.

No. 2,304.—LOUIS SAARBACH, Philadelphia, Penn.—*Tobacco Pipe.*—May 1, 1866.

No. 2,305.—PIERRE A. BERTHOLD and MAKLOT THOMPSON, St. Louis, Mo.—*Trade Mark.*—May 8, 1866.

No. 2,306.—ISAAC LEVY, New York, N. Y.—*Ace of Spades.*—May 8, 1866.

No. 2,307.—EDWARD A. MARSH, Chicopee, Mass., assignor to GAYLORD MANUFACTURING COMPANY, same place.—*Pen.*—May 8, 1866.

No. 2,308.—C. L. NEIBERG, New Haven, Conn., assignor to SARGENT & CO., same place.—*Coffin Handle.*—May 8, 1866.

No. 2,309.—LOUIS SAARBACH, Philadelphia, Penn.—*Tobacco Pipe.*—May 8, 1866.

No. 2,310.—J. B. SARGENT, New Haven, Conn., assignor to SARGENT & CO., same place.—*Coat or Hat Hook.*—May 8, 1866.

No. 2,311.—JOHN TAYLOR, Philadelphia, Penn.—*Lady's Hood.*—May 8, 1866.

No. 2,312.—ALBERT BRIDGES, New York, N. Y.—*Bracket.*—May 15, 1866.

No. 2,313.—JOSHUA BROOKS, Newton, Mass.—*Card Holder.*—May 15, 1866.

No. 2,314.—SAMUEL W. GIBBS, Albany, N. Y., assignor to SAMUEL H. RANSOM, same place.—*Plate of a Cook's Stove.*—May 15, 1866.

No. 2,315.—SAMUEL W. GIBBS, Albany, N. Y., assignor to SAMUEL H. RANSOM, same place.—*Plate of a Parlor Stove.*—May 15, 1866.

No. 2,316.—ANDREW LITTLE, New York, N. Y.—*Printers' Type.*—May 15, 1866.

No. 2,317.—SOLOMON OPPENHEIMER, Peru, Ind.—*Knife Handle.*—May 15, 1866.

No. 2,318 to 2,320.—SAMUEL H. RANSOM, Albany, N. Y.—*Plate of a Cook's Stove,* (*three patents.*)—May 15, 1866.

No. 2,321 to 2,322.—SAMUEL H. RANSOM, Albany, N. Y.—*Plate of a Parlor Stove,* (*two patents.*)—May 15, 1866.

No. 2,323.—CHARLES BALLINGER, Pittsburg, Penn., assignor to MCKEE & BROTHERS, same place.—*Goblet.*—May 22, 1866.

No 2,324.—AMOS RANK, Salem, Ohio.—*Wheel of a Harvesting Machine.*—May 22, 1866.

No. 2,325—SAMUEL F. STOWE, Providence. R. I., assignor to himself and LEVI H. STONE, same place.—*Trade Mark.*—May 22, 1866.

No. 2,326.—LEMUEL H. FLERSHEIM, Chicago, Ill.—*Trade Mark for Lead Pencils.*—May 29, 1866.

No. 2,327.—ERNEST KAUFMANN, Philadelphia, Penn.—*Pitcher.*—May 29, 1866.

No. 2,328.—G. I. MIX, Wallingford, Conn.—*Spoon or Fork Handle.*—May 29, 1866.

No. 2,329.—GEORGE T. SPICER, Providence, R. I.—*Cook Stove.*—May 29, 1866.

No. 2,330.—JAMES F. TRAVIS, New York, N. Y., assignor to MITCHELL VANCE & CO., same place.—*Chandelier.*—May 29, 1866.

No. 2,331.—JULIUS L. D. SULLIVAN, Somerville, Mass.—*Handle of a Spoon.*—June 5, 1866.

No. 2,332.—SAMUEL E. BARNEY, New Haven, Conn., assignor to THE ELM CITY COMPANY, same place.—*Trade Mark.*—June 12, 1866.

No. 2,333.—JOHN H. BELLAMY, Charlestown, Mass.—*Bracket.*—June 12, 1866.

No. 2,334.—J. M. FLAGG, Providence, R. I.—*Lady's Tucked Paper Collar.*—June 12, 1866.

No. 2,335 and 2,336.—H. FLETCHER, Providence, R. I., assignor to THE FLETCHER MANUFACTURING COMPANY.—*Trade Mark,* (*two patents.*)—June 12, 1866.

No. 2,337.—C. HARRIS and P. W. ZOINER, Cincinnati, Ohio.—*Cook's Stove.*—June 12, 1866.

No. 2,338.—C. HARRIS and P. W. ZOINER, Cincinnati, Ohio.—*Parlor Stove.*—June 12, 1866.

No. 2,339.—C. HARRIS and P. W. ZOINER, Cincinnati, Ohio.—*Coal Store.*—June 12, 1866.

No. 2,340.—SAMUEL RAYNOR, New York, N. Y.—*Envelope.*—June 26, 1866.

No. 2,341.—LEVI SOHL, A. J. SOHL, and DAVID GIBSON, Indianapolis, Ind.—*Trade Mark.*—June 26, 1866.

No. 2,342 to 2,344.—B. G. BRIGGS, Providence, R. I., assignor to himself and T. HILTON, same place.—*Paper Collar, (three patents.)*—July 3, 1866.

No. 2,345.—CHARLES H. DREW, Great Falls, N. H.—*Stove Panel.*—July 3, 1866.

No. 2,346.—CHARLES D. ELLIOT, Cambridge, Mass.—*Paper Collar.*—July 3, 1866.

No. 2,347.—E. C. FOUGERA, Brooklyn, and B. A. VANDERKIEFT, New York.—*Bottle.*—July 3, 1866.

No. 2,348.—EDWARD J. FROST, Springfield, Mass.—*Collar and Cuff.*—July 3, 1866.

No. 2,349.—EDWARD F. KELLY, Chicago, Ill.—*Ornamentation of Tops.*—July 3, 1866.

No. 2,350.—J. MARTINO, J. BEESLEY, and J. CURRIE, Philadelphia, Penn., assignors to A. SHEPPARD, same place.—*Cook's Store.*—July 3, 1866.

No. 2,351.—W. H. REED and PHILIP REED, Philadelphia, Penn.—*Base Ball Bat.*—July 3, 1866.

No. 2,352.—JOHN ROGERS, New York, N. Y.—*Group of Figures.*—July 3, 1866.

No. 2,353.—SAMUEL CONKEY, New York, N. Y.—*Statuette.*—July 10, 1866.

No. 2,354.—NICHOLAS MÜLLER, New York, N. Y.—*Clock Case Front.*—July 10, 1866.

No. 2,355.—CHARLES A. PERRY, Chicago, Ill., assignor to himself and THOMAS DERRY, same place.—*Trade Mark.*—July 10, 1866.

No. 2,356 and No. 2,357.—B. G. BRIGGS, Providence, R. I., assignor to himself, J. A. HANLEY, and T. HILTON, same place.—*Paper Cuff, (two patents.)*—July 17, 1866.

No. 2,358.—B. G. BRIGGS, Providence, R. I., assignor to himself, J. A. HANLEY, and T. HILTON, same place.—*Paper Collar.*—July 17, 1866.
No. 2,359 and No. 2,360.—F. T. FRACKER, Boston, Mass., assignor to THE TUCKER MANU-FACTURING COMPANY.—*Clock Case, (two patents.)*—July 17, 1866.

No. 2,361.—JOHN MARTINO, JACOB BEESLEY, and JOHN CURRIE, Philadelphia, Penn., assignors to LAKE, BEARDER & Co.—*Store.*—July 17, 1866.

No. 2,362.—NOAH POMEROY, Hartford, Conn.—*Door of a Clock Case.*—July 17, 1866.

No. 2,363.—WILLIAM W. ROBERTS, Hartford, Conn.—*Burial Casket.*—July 17, 1866.

No. 2,364.—JOHN H. BELLAMY, Charlestown, Mass., assignor to himself and DAVID A. TITCOMB.—*Bracket.*—July 24, 1866.

No. 2,365 and No. 2,366.—FRANCIS T. FRACKER, Boston, Mass., assignor to THE TUCKER MANUFACTURING COMPANY, same place.—*Chandelier, (two patents.)*—July 24, 1866.

No. 2,367.—FRANCIS T. FRACKER, Boston, Mass., assignor to THE TUCKER MANUFAC-TURING COMPANY, same place.—*Bracket and Lamp.*—July 24, 1866.

No. 2,368.—FRANCIS T. FRACKER, Boston, Mass., assignor to THE TUCKER MANUFAC-TURING COMPANY, same place.—*Clock.*—July 24, 1866.

No. 2,369.—FRANCIS T. FRACKER, Boston, Mass., assignor to THE TUCKER MANUFAC-TURING COMPANY, same place.—*Pendent Light.*—July 24, 1866.

No. 2,370.—EDWARD HOWARD, Boston, Mass.—*Watch Plate.*—July 24, 1866.

No. 2,371.—MOSES A. JOHNSON, Lowell, Mass.—*Felted Goods.*—July 24, 1866.

No. 2,372.—EDWARD LOCHER, Newark, N. J.—*Trade Mark.*—July 24, 1866.

No. 2,373.—R. J. ROBERTS, New York, N. Y.—*Trade Mark.*—July 24, 1866.

No. 2,374.—DAVID SHIRRELL, Buffalo, N. Y.—*Trade Mark.*—July 24, 1866.

No. 2,375.—AMOS HAMLIN, Schoharie, N. Y.—*Metallic Armor of Boots and Shoes.*—August 7, 1866.

No. 2,376.—HENRY HUNERMUND, New York, N. Y.—*Work Basket Stand.*—August 7, 1866.

No. 2,377.—NICHOLAS MÜLLER, New York, N. Y.—*Clock Case.*—August 7, 1866.

No. 2,378.—SILAS S. PUTNAM, Dorchester, Mass., assignor to S. S. PUTNAM & CO.—*Trade Mark.*—August 7, 1866.

No. 2,379.—G. SMITH and H. BROWN, Philadelphia, Penn., assignors to ABBOTT and NOBLE, same place —*Plate of a Stove.*—August 7, 1866 ; antedated July 10, 1866.

No. 2,830.—S. D. ARNOLD, New Britain, Conn., assignor to P. and F. CORBIN, same place, and J. C. SHURLER & CO., Amsterdam, N. Y.—*Coffin Handle.*—August 14, 1866.

No. 2,381 to No. 2,386.—S. D. ARNOLD, New Britain, Conn., assignor to P. and F. CORBIN, same place, and J. C. SHURLER & CO., Amsterdam, N. Y.—*Coffin Trimmings, (six patents.)*—August 14, 1866.

No. 2,387.—ALONZO B. BAILY, Middle Haddam, Conn.—*Coffin Handle.*—August 14, 1866.

No. 2,388.—J. C. GEARHART, Jersey Shore, Penn.—*Fence.*—August 14, 1866.

No. 2,389 to No. 2,426.—HENRY G. THOMPSON, New York, N. Y., assignor to THE HARTFORD CARPET COMPANY, Hartford, Conn.—*Carpet Pattern, (thirty-eight patents.)*—August 14, 1866.

No. 2,427.—SAMUEL W. GIBBS, Albany, N. Y.—*Plate of a Stove.*—August 21, 1866.

No. 2,428 and No. 2,429.—JOHN H. BELLAMY, Charlestown, Mass., assignor to DAVID A. TITCOMB, same place.—*Picture Frame, (two patents.)*— September 25, 1866.

No. 2,430.—HENRY BERGER, New York, N. Y.—*Centre Piece.*—September 25, 1866.

No. 2,431.—L. F. CARTER and W. W. CARTER, Bristol, Conn.—*Clock Case Front.*—September 25, 1866.

No. 2,432.—FRANCIS T. FRACKER, Boston, Mass., assignor to THE TUCKER MANUFACTURING COMPANY, same place.—*Pulley Hall Lamp.*—September 25, 1866.

No. 2,433.—WILLIAM A. GREENE, Troy, N. Y.—*Blacking Brush Handle.*—September 25, 1866.

No. 2,434.—B. B. HILL, Springfield, Mass.—*Cancelling Stamp.*—September 25, 1866.

No. 2,435 and 2,436.—GEORGE HIMROD and CHARLES G. MOULTON, Chicago, Ill., assignors to GEORGE HIMROD.—*Stove Plate, (two patents.)*—September 25, 1866.

No. 2,437.—JAMES R. HYDE, Troy, N. Y., assignor to FREDERICK A. SHELDON and CHAUNCEY O. GREENE, same place.—*Cook's Stove.*—September 25, 1866.

Nos. 2,438 and 2,439.—SAMUEL KELLETT, San Francisco, Cal.—*Centre Piece, (two patents.)*—September 25, 1866.

No. 2,440.—JOHN MATTHEWS, JR., New York, N. Y.—*Drinking Tumbler.*—September 25, 1866.

No. 2,441 to 2,444.—CHARLES T. MEYER, Bergen, N. J., assignor to EDWARD C. SAMPSON, New York, N. Y.—*Floor Oil Cloth Pattern, (four patents.)*—September 25, 1866.

No. 2,445.—LEVI MOSES, Janesville, Wis.—*Burial Case.*—September 25, 1866.

No. 2,446 to 2,455.—ELEMIR J. NEY, Lowell, Mass., assignor to THE LOWELL MANUFACTURING COMPANY, same place.—*Carpet Pattern, (ten patents.)*—September 25, 1866.

No. 2,456.—EDWARD F. PEUGEOT, Chicago, Ill.—*Trade Mark.*—September 25, 1866.

No. 2,457.—LEONCE PICOT, Hoboken, N. J., assignor to THE RUBBER CLOTHING COMPANY, New York, N. Y.—*Child's Long Comb.*—September 25, 1866.

No. 2,458.—JOHN PROTZ, Easton, Penn.—*Snap Hook.*—September 25, 1866.

No. 2,459 to No. 2,466.—LEONHARD SCHULZE, Philadelphia, Penn., assignor to THOMAS MILLS & BRO., same place.—*Candy Figure, (eight patents.)*—September 25, 1866.

No. 2,467.—SAMUEL S. UTTER, New York, N. Y.—*Stove Ornament.*—September 25, 1866.

No. 2,468.—JASPAR VAN WORMER, Albany, N. Y.—*Plate of a Hall Stove.*—September 25, 1866.

No. 2,469.—JOHN T. WEBSTER, New York, N. Y., assignor to EDWARD HARVEY, Brooklyn, N. Y.—*Floor Oil Cloth Pattern.*—September 25, 1866.

No. 2,470.—WILLIAM B. DURGIN, Concord, N. H.—*Fork or Spoon Handle.*—September 25, 1866.

No. 2,471.—A. MILES, Winona, Minn.—*Label.*—September 25, 1866.

No. 2,472.—JOHN H. BELLAMY, Charlestown, Mass., assignor to himself and D. A. TITCOMB, same place.—*Bracket.*—October 2, 1866.

No. 2,473.—JOSEPH BOND, Jr., Newark N. J., assignor to THE LATHROP SEWING MACHINE COMPANY, New York, N. Y.—*Sewing Machine.*—October 2, 1866.

No. 2,474.—JOSEPH BOND, Jr., Newark, N. J., assignor to THE LATHROP SEWING MACHINE COMPANY, New York, N. Y.—*Frame of a Sewing Machine.*—October 2, 1866.

No. 2,475.—AUGUSTUS CONRADT, Philadelphia, Penn.—*Spoon Handle.*—October 2, 1866.

No. 2,476.—HENRY COY, Philadelphia, Penn.—*Dentists' Grinding Lathe.*—October 2, 1866.

No. 2,477.—JOHN H. KNIGHT and JOSEPH P. FARRAND, Detroit, Mich.—*Cigar and Tobacco Safe.*—October 2, 1866.

No. 2,478.—CATHOLINA LAMBERT, New York, N. Y.—*Reel.*—October 2, 1866.

No. 2,479.—THOMAS MOORE and J. S. HALDEMAN, Bloomington, Ill.—*Lamp Flue.*—October 2, 1866.

No. 2,480.—REES MOSS, Philadelphia, Penn.—*Plate of a Cook's Range.*—October 2, 1866.

No. 2,481.—C. L. NEIBERG, New Haven, Conn., assignor to SARGENT & CO., same place.—*Coffin Handle.*—October 2, 1866.

No. 2,482 and 2,483.—LEONCE PICOT, Hoboken, N. J., assignor to THE RUBBER CLOTHING COMPANY, New York, N. Y.—*Child's Long Comb, (two patents.)*—October 2, 1866.

No. 2,484.—AMOS RANK, Salem, Ohio.—*Wheel.*—October 2, 1866.

No. 2,485.—JACOB STEFFE, Philadelphia, Penn., assignor to THOMAS FORSTER.—*Cook's Stove.*—October 2, 1866.

No. 2,486.—J. A. WILSON, Springfield, Mass.—*Burial Case.*—October 2, 1866.

No. 2,487 and 2,488.—JOHN W. CARROLL, Lynchburg, Va.—*Trade Mark, (two patents.)*—October 9, 1866.

No 2,489.—WILLIAM FREUDENAU, St. Louis, Mo., assignor to THE UNION STEAM MILL COMPANY, same place.—*Trade Mark.*—October 9, 1866.

No. 2,490.—CATHOLINA LAMBERT, New York, N. Y.—*Drapery Trimming.*—October 9, 1866.

No. 2,491 and 2,492.—MORRIS M. PEYSER, Boston, Mass.—*Shawl Border, (two patents.)*—October 9, 1866.

No. 2,493.—LEONCE PICOT, Hoboken, N. J., assignor to THE RUBBER CLOTHING COMPANY, New York, N. Y.—*Child's Long Comb.*—October 9, 1866.

No. 2,494.—CHARLES HUSBAND, Taunton, Mass.—*Paper Hangings, &c.*—October 16, 1866.

No. 2,495.—CHARLES T. MEYER, Bergen, N. J., assignor to E. C. SAMPSON.—*Floor Oil-cloth Pattern.*—October 23, 1866.

No. 2,496.—LEONCE PICOT, Hoboken, N. J., assignor to THE RUBBER CLOTHING COMPANY, New York, N. Y.—*Ornamenting Child's Long Comb.*—October 23, 1866.

No. 2,497.—JOHN S. ARMSTRONG, Prairie du Chien, Wis.—*Military Monument.*—October 30, 1866.

No. 2,498.—J. A. CHARNLEY, Providence, R. I.—*Paper Collar.*—October 30, 1866.

No. 2,499.—A. P. DE VOURSNEY, New York, N. Y.—*Coach Lamp.*—October 30, 1866.

No. 2,500.—M. DE VOURSNEY, New York, N. Y.—*Coach Lamp.*—October 30, 1866.

No. 2,501.—WM. FREUDENAU, St. Louis, Mo., assignor to THE UNION STEAM MILL COMPANY.—*Trade Mark.*—October 30, 1866.

No. 2,502.—NICHOLAS MULLER, New York, N. Y.—*Clock Case.*—October 30, 1866.

No. 2,503.—JOHN R. WASLEY, Boston, Mass., assignor to THE WASHINGTON MILLS, same place.—*Table Cover.*—November 6, 1866.

No. 2,504.—CHARLES J. HAUCK, Williamsburgh, N. Y.—*Oil Can.*—November 13, 1866.

No. 2,505.—JOSEPH HARVEY, Philadelphia, Penn., assignor, to HARVEY & FORD, same place.—*Tobacco Pipe.*—November 13, 1866.

No. 2,506.—EDWIN BLAKESLEE, New Haven, Conn., assignor to C. COWLES & Co., same place.—*Coach Lamp.*—November 20, 1866.

No. 2,507.—GEORGE C. BRITNER, Philadelphia, Penn., assignor to HARVEY & FORD, same place.—*Tobacco Pipe.*—November 27, 1866.

No. 2,508.—JAMES H. DOWNS, New Haven, Conn., assignor to C. COWLES & Co., same place.—*Coach Lamp*—November 27, 1866.

No. 2,509.—SAMUEL JACKSON, Jr., Roxbury, Mass.—*Hat Advertising Carriage.*—November 27, 1866.

No. 2,510.—WILLIAM L. LOCKHART and J. C. SEELYE, East Cambridge, Mass.—*Burial Casket.*—November 27, 1866.

No. 2,511.—REUBEN MILLER, Lincoln, Ill.—*Cemetery Monument.*—November 27, 1866.

No. 2,512.—GEORGE SHARP, Philadelphia, Penn.—*Handle of a Fork or Spoon.*—November 27, 1866.

No. 2,513.—GEORGE W. WESTBROOK, New York, N. Y.—*Trade Mark.*—November 27, 1866.

No. 2,514.—CHARLES A. FOSTER, North Providence, R. I.—*Ornamental Picture for Animal Studies.*—December 4, 1866.

No. 2,515.—ROBERF HOSKIN, Brooklyn, N. Y., assignor to EDWARD C. SAMPSON, New York, N. Y.—*Floor Oil Cloth.*—December 4, 1866.

No. 2,516 to 2,518.—WILLIAM RESOR, Cincinnati, Ohio.—*Stove,* (three patents.)—December 4, 1866.

No. 2,519 and 2,520.—WILLIAM RESOR, Cincinnati, Ohio.—*Cook's Stove,* (two patents.)—December 4, 1866.

No. 2,521.—JOHN ROGERS, New York, N. Y.—*Group of Figures.*—December 4, 1866.

No. 2,522.—W. H. WARE, Philadelphia, Penn.—*Bottle.*—December 4, 1866.

No. 2,523.—JOHN H. BELLAMY, Charlestown, Mass.—*Picture Frame.*—December 11, 1866.

No. 2,524.—JOHN H. BELLAMY, Charlestown, Mass.—*Bracket.*—December 11, 1866.

No. 2,525 and 2,526.—WILLIAM A. GILES, Chicago, Ill.—*Clock Case, (two patents.)*—December 11, 1866.

No. 2,527.—DANIEL B. KIMBALL, New York, N. Y.—*Trade Mark.*—December 11, 1866.

No. 2,528.—C. C. KLEIN, Philadelphia, Penn., assignor to THOMAS WOOD, same place.—*Shafting Hanger.*—December 11, 1866.

No. 2,529.—JOHN REYNOLDS, Washington D. C.—*Buggy.*—December 11, 1866.

No. 2,530.—HENRY H. HAYDEN, New York, N. Y., assignor to HOLMES, BOOTH & HAYDEN, Waterbury, Conn.—*Knife Handle.*—December 18, 1866.

No. 2,531.—HENRY HEBBARD, New York, N. Y.—*Spoon Handle.*—December 18, 1866.

No. 2,532.—CHARLES E. HOFFMANN, Philadelphia, Penn.—*Balance.*—December 18, 1866.

REISSUES, 1866.

No. 2,140.—HENRY H. PACKER, Boston, Mass.—*Drill.*—Patented June 29, 1858; reissued January 2, 1866.

Claim.—First, the combination in a ratchet drill of the following instrumentalities, viz : the drill stock, feed nut, feed screw, shell, ratchet wheel, and pawl carrier, substantially as set forth.

Second, the combination in a ratchet drill of the following instrumentalities, viz: the drill stock, feed nut, feed screw, exterior and inner shells, ratchet wheel and pawl carrier, substantially as set forth.

No. 2,141.—NELSON PALMER, Hudson, N. Y., assignee of ISAAC S. SPENCER, Guilford, Conn.—*Threshing Machine.*—Patented September 23, 1856; reissued January 2, 1866.

Claim.—First, the endless feeding belt or apron B, in combination with the threshing cylinder F, as and for the purpose specified.

Second, two or more threshing cylinders with the straight, spiral, oblique, or angular corrugations, working in concert as specified.

No. 2,142.—ABRAHAM STEERS, Medina, N. Y.—*Making Extracts.*—Patented March 11, 1856; reissued January 2, 1866.

Claim.—First, the within described process of separating the soluble and insoluble parts contained in the bark or other substance to be extracted, by first saturating or swelling said substance with the menstruum and exposing the same in its damp state to the action of steam, substantially in the manner set forth.

Second, washing the bark or other substance after the same has been acted upon by the steam, with liquid obtained by the condensation of the steam, substantially in the manner herein specified.

Third, the apparatus composed of a percolator K and receiver M, separated from each other by a perforated diaphragm, or its equivalent, in combination with a metallic cover, supplied with an outwardly opening valve and with a pipe connecting the top of the percolator with the receiver, substantially as described, so that the contents of said percolator can be operated upon, first by the steam generated in the receiver, and then by the percolation of the menstruum.

No. 2,143.—WILLIAM WICKERSHAM, Boston, Mass.—*Machine for Cutting Nails.*—Patented June 26, 1860; reissued January 2, 1866.

Claim.—First, arranging the pairs of cutters, substantially as described, so that the next pair but one from any pair shall be the reverse thereof, to form the opposite side of the same nail, substantially as described.

Second, the placing of each alternate pair of cutters in advance of the others, in the plane of the sheet of metal or further from the axis of the movable cutters than the others, to enable the contiguous nails in the several columns to be cut separately, substantially as described.

Third, the employment of a continuous collective breadth of either movable or stationary cutters, sufficient to extend entirely across the sheet, added to the extent of lateral motion given to the sheet to transfer it from one set of dies to another, substantially as described.

Fourth, forming each series of cutters in separate sections placed side by side and otherwise arranged, substantially as set forth.

Fifth, in combination with the series of cutters, arranged and operating as described, the mechanism for holding the sheet of metal while being cut, and for moving it laterally the distance from one pair of cutters to the pair that co-operates with it, and for feeding the sheet forward a distance equal to the breadth of the nail, substantially as described.

No. 2,144.—CHARLES W. JOHNSON, Waterbury, Conn.—*Power Press.*—Patented November 7, 1865; reissued January 9, 1866.

Claim.—First, the combination described of the gear I and plate P, or their equivalents, constructed and arranged to operate together, substantially as and for the purpose specified.

Second, the combination of the cam S, lever N, and bolt r, in the manner substantially as and for the purpose specified.

Third, the combination of the bolt r and lever N, or their equivalents, substantially in the manner specified, as and for a cut-off or stop motion.

No. 2,145.—ELI W. BLAKE, New Haven Conn.—*Stone Breaker.*—Patented June 15, 1858; reissued January 9, 1866.

Claim.—First, the combination in a stone-breaking machine of the upright convergent jaws, with a revolving shaft and mechanism for imparting a definite reciprocating movement to one of the jaws from the revolving shaft, the whole being and operating substantially as set forth.

Second, the combination in a stone-breaking machine of the upright movable jaw with the revolving shaft and fly wheel, the whole being and operating substantially as set forth.

Third, in combination with the upright converging jaws and the revolving shaft, imparting a definitely limited vibration to the movable jaw, so arranging the jaws that they can be set at different distances from each other at the bottom, so as to produce fragments of any desired size.

No. 2,146.—RUSSEL JENNINGS, Deep River, Conn.—*Auger.*—Patented September 30, 1855; reissued October 3, 1865; again reissued January 16, 1866.

Claim.—The projecting of the floor lips in advance of the cutting spur, substantially as herein described and for the purpose herein set forth.

No. 2,147.—SACKETT, DAVIS & CO., Providence, R. I., assignees of JAMES LANCELOTT.—*Ornamental Chain.*—Patented March 22, 1859; reissued January 16, 1866.

Claim.—First, a sheet-metal chain composed of links, the base of each of which is a polygon of six or more sides, the chain being formed by bending each arm longitudinally at the same angle, or nearly so, with one of the outer angles of the base, so that a cross bar on the extremity of each arm of the next preceding link in the chain shall, when bent down, bear against the angular sides of two of the arms of the next succeeding link, and thereby enable the chain to withstand a strain nearly equal to the cohesive strength of the metal, the article being substantially as specified.

Second, a sheet-metal chain, the arms of whose links are bent longitudinally as described, for the purpose of increasing the strength of the chain and giving to it the appearance of being made from wire instead of from sheet metal.

No. 2,148.—ARIEL B. SPROUT, Hughesville, Penn.—*Horse Rake.*—Patented June 6, 1865; reissued January 16, 1866.

Claim.—First, the use of a foot lever for holding up the rake head of a horse rake, which has the point or centre of vibration of the teeth arranged in rear of said rake head, as herein described and for the purpose set forth.

Second, the combination of the foot lever E with the roller k, said lever being so arranged that it moves back with the depression of the rake head and forward with the elevation of said rake head, and travels upon said roller in its back and forward motion, being rigidly fixed to the machine, as herein described and set forth.

Third, attaching the fulcrum bar F to the cleaners or other rigid parts of the rake, by means of straps g, connecting the two parts of a hinge joint so as to allow a limited amount of vertical play to the bar F, for the purpose herein described and set forth.

Fourth, in combination with the straps g, the movable rings or their equivalents, for the purpose of preventing the vertical play of the bar F, relatively to the cleaners, under the circumstances described.

Fifth, the extension in front of the axle of the cleaners G, which support the rake head, so as by their vertical adjustment to regulate their height of the rake head from the ground at a given elevation to the shafts.

Sixth, the rotating notched pintle bolts H A', with grooves therein corresponding to similar grooves on the lug H', for coiling the spring formed on the end of the tooth until said tooth has acquired the requisite force for holding it in the desired position, said spring being held in its coiled position by the action of the nut on the bolt, as herein described and set forth.

No. 2,149.—MARTIN R. COOK, Jersey City, N. J., assignor by mesne assignments of S. HILL and WM. J. WOOD.—*Gas Holder.*—Patented November 6, 1855; reissued January 23, 1866.

Claim.—In gas holders for locomotives, dividing the vessel into two compartments by an inclosed flexible diaphragm, or the equivalent thereof, when one of the said compartments is provided with a tube or tubes to supply gas to burners, and the other is provided with a suitable aperture for the admission of air or equivalent gaseous fluid, substantially as and for the purpose described.

No. 2,150.—A. COURLANDER CRONDAL, New York, N. Y.—*Cork Hat.*—Patented November 8, 1864; reissued January 23, 1866.

Claim.—Manufacturing coverings for the head of sheets composed of one or more layers of cork, and one or more layers of canvas, muslin, or other textile or flexible material, substantially as herein set forth.

No. 2,151.—PHILO S. FELTER, Cincinnatus, N. Y.—*Lock.*—Patented December 17, 1861; reissued January 23, 1866.

Claim.—First the bar or guard D, provided with the recess a, in connection with the notched disks G, spring F, provided with the projections $b\ d\ d$ and the key H, arranged substantially as and for the purpose herein set forth.

Second, in combination with the subject-matter of the above, the employment of numbered or lettered dials, by means of which the lock may be used as a burglar-proof or common lock, as desired, substantially as set forth.

No. 2,152.—PHILIP KEENAN and EDWARD O'CONNOR, West Manchester, Penn.—*Puddling Furnace.*—Patented November 14, 1865; antedated August 26, 1865; reissued January 23, 1866.

Claim.—The use of iron ore as a "fixing" for puddling or boiling furnaces, when mixed with fire-clay or other refractory material, and used for fixing those portions of the furnace which need protection without previous melting of the "fix."

No. 2,153.—HUGH McDONALD, Pittsburg, Penn.—*Fix for Puddling Furnaces.*—Patented October 17, 1865; reissued January 23, 1866.

Claim.—The use of iron ore as a fixing for a puddling or boiling furnace, when applied as a fix to those parts of the furnace which require protection, and so used without previous melting.

Also, the use of raw or unmelted iron ore as a fixing for puddling or boiling furnaces, when ground or pulverized and mixed into a pasty mass with water or other suitable liquid.

Also, mixing raw iron ore, ground or pulverized, with carbonaceous matter, and made into a pasty or adhesive mass and used as a fixing for puddling or boiling furnaces.

No. 2,154.—NELSON PALMER, Hudson, N. Y.—*Threshing Machine.*—Patented May 16, 1865; reissued January 23, 1866.

Claim.—First, the cylinder *h*, when constructed as described, for feeding the unthreshed straw to the threshing cylinder, as specified.

Second, the guard *g*, in combination with the feeding cylinder *h*, operating as specified.

Third, the corrugated ribbed or granulated thrashing cylinder *b*, in combination with a concave or rubber, ribbed, corrugated, or granulated.

Fourth, the lever *d*, or its equivalent, in combination with the concave *c*, for adjusting the same as set forth.

No. 2,155.—NELSON PALMER, Hudson, N. Y., assignee of P. W. MILLS, Conneaut, Ohio.—*Threshing Machine.*—Patented January 19, 1858; reissued January 23, 1866.

Claim.—First, the threshing cylinder D, one end thereof being of greater diameter than the other and provided with ribs or corrugations, as and for the purpose specified.

Second, the concave E, when so constructed as to fit the cone-shaped threshing cylinder D, the parts and sections thereof being made adjustable in relation to each other, in combination with the adjustable concave F and apron B, as and for the purpose specified.

Third, the arrangement of the screws *k s u b* in their relation to the threshing cylinder D and fan wheel B, and operating as set forth.

No. 2,156.—RANDAL PRATT, Marple Township, Penn.—*Horse Rake.*—Patented January 8, 1856; reissued January 23, 1866.

Claim.—First, the method described of firmly uniting the tooth with the elongated collar, by bending and shrinking the hinging end of the tooth around the collar, substantially as described.

Second, providing the elongated collar with a groove into which the tooth is shrunk, as and for the purpose described.

No. 2,157.—RICHARD S. RHODES and EBENEZER WHYTE, Chicago, Ill.—*Process for Preserving Eggs.*—Patented December 12, 1865; reissued January 23, 1866.

Claim.—The herein described process for preserving eggs from decay, substantially as herein specified.

No. 2,158.—SUMNER SARGENT, Watertown, Mass., assignor through mesne assignments to himself, A. P. KNAPP, and EDWARD MILLER.—*Coal Oil Lantern.*—Patented September 17, 1861; reissued January 23, 1866.

Claim.—The employment of an aperture, or its equivalents, in the lantern case, through which the shaft or its equivalent of the wick regulator extends, so as to be reached outside of the lantern case, said aperture having a slot or lateral passage leading to it, for the introduction of the said shaft or equivalent part of the wick regulator into the aperture, and its withdrawal therefrom, in the act of inserting and taking out the lantern lamp; the whole constituting a convenient arrangement for enabling the wick to be regulated outside of the lantern case, and at the same time keeping it closed so as not to disarrange the draught, substantially as and for the purpose herein specified.

In combination with the above, the plate M, or its equivalent, for covering and uncovering the passage leading to the regulator aperture in lantern case, as set forth.

Also, the arrangement and combination of the perforations *i i* in the base flange of the lamp D, the draught collector *u*, division plates N N, perforated regulating plate P, and guard cylinder R, in the manner and for the purposes herein specified.

No. 2,159.—EDWIN R. STILWELL, Dayton, Ohio.—*Feed-water Heater and Filterer.*—Patented October 4, 1864; reissued January 23, 1866.

Claim.—First, the depositing plates *a a a*, constructed and arranged substantially as described and for the purposes specified.

Second, the arrangement of the steam pipes *m* and *n*, with reference to the plates *a a a*, substantially as described and for the purposes specified.

Third, the combination of the vessel A, the plates *a a a*, the plate *d*, the steam pipes *m s* and *e*, and water pipes *f* and *r*, substantially as described.

No. 2,160.—EDWIN R. STILWELL, Dayton, Ohio.—*Feed-water Heater and Filterer.*—Patented October 4, 1864 ; reissued January 23, 1866.

Claim.—First, the overflow box *c*, the pipe *b*, arranged with reference to the vessel A, substantially as described and for the purposes specified.

Second, the arrangement of the steam pipe *e* to the overflow box *c*, for the purposes set forth.

No. 2,161.—HENRY MOESER, Pittsburg, Penn.—*Printing Names of Subscribers upon Newspapers, &c.*—Patented June 24, 1851 ; extended seven years ; re-issued January 30, 1866.

Claim.—First, the within described mechanical record of names and addresses of subscribers of newspapers, periodicals, &c., or correspondents, to whom it is desirable to send circulars, documents, or other mail matter, said mechanical record being constituted of type, and representing said names, addresses, and so forth, locked up in a standing form or forms, capable of such changes, additions, or alterations in the names or addresses as occasion may require, and also capable of being used in connection with a press for printing said names, addresses, &c., the said form of type being both a record or list of existing subscribers or correspondents which can be referred to from time to time for information, and a means of printing the names and addresses of subscribers, correspondents, &c., substantially as described and specified.

Second, combining with a standing address form or forms, a mechanical record of names, &c., and a press mechanism as described, an automatic feed for moving the forms step by step for the successive address impressions, substantially as described and specified.

Third, combining with the standing address, forms or mechanical record of names, &c., and a printing and feeding mechanism as described, a shield or equivalent device, provided with a slot, to shield and protect the paper from ink except at the point of impression, substantially as described and specified.

No. 2,162.—FRANCIS STABLER, Baltimore, Md.—*Process for Preserving Animal and Vegetable Substances.*—Patented November 14, 1865 ; reissued January 30, 1866.

Claim.—Preserving animal or vegetable substances used for food, when wholly or partially desiccated as above described, by sealing them in air-tight vessels from which the atmospheric air has been expelled by the introduction of carbonic acid gas, or other gas that will not support combustion, substantially as described.

No. 2,163.—DANIEL BARNUM, New York, N. Y.—*Sewing Machine Guide.*—Patented February 12, 1861 : reissued September 13, 1864 ; again reissued January 30, 1866.

Claim.—First, the use of thin sheet or light elastic spring metal for making automatic clamping surfaces extending out from the gauging line of a sewing-machine gauge and in combination therewith, beyond the line of seam to be sewed by a needle, and of thus producing a gentle automatic spring pressure upon the upper surface of flexible material while the same is approaching the needle, and thereby automatically smoothing and holding the said material preparatory to its being sewed outside of the line of seam as well as between it and the line of gauge, substantially as and for the purposes specified.

Second, also the use of thin sheet or light elastic spring metal for making automatic clamping spring surfaces, as specified, when provided with diagonal corrugations, struck up thereon outside of the line of the seam to be sewed by a sewing machine needle, substantially as and for the purposes specified.

Third, also the use of thin sheet or light elastic spring metal for making automatic clamping surfaces, as specified, when provided with the bevelled edges which are turned or struck up as described, to facilitate an easy entrance of varying thicknesses and uneven surfaces of material under the upper clamping surface, substantially as and for the purposes specified.

Fourth, also a sewing-machine gauge having an upper automatic gently clamping surface, in front of the gauging line, which surface, when in use on a sewing machine, will extend over and automatically press upon the upper side of the material which is being sewed beyond the line of the seam as well as between it and the gauge, whether said material be of equal or unequal thickness.

Fifth, and by the use of the means mentioned in the last preceding clause, automatically removing the wrinkles from and smoothing the upper side of any soft or undressed woven fabric of unequal thickness or of uneven surface, and holding and guiding the same without basting while it is approaching the needle and being sewed, both outside of and around the needle as well as between it and the gauge, substantially as and for the purposes specified.

No. 2,164.—ANTOINE CHOPLAIN and PIERRE E. CHOLLET, New York, N. Y.—*Stringing Pianos.*—Patented December 19, 1865 ; reissued January 30, 1866.

Claim.—First, the use of the lever C, substantially as described, in combination with the tension slide B and knife-edge prop T, arranged and operating substantially as hereinbefore set forth, for the purpose of regulating the tension of the strings of piano-fortes.

Second, the use of the lever C, in combination with the indicator E, and dial plate L, or other equivalent devices for registering the tension or tone of the strings of a piano-forte, substantially as hereinbefore set forth.

No. 2,165.—SARAH A. MOODY, New York, N. Y.—*Abdominal Supporter.*—Patented May 3, 1864; reissued January 30, 1866.

Claim.—The corsets or abdominal supporters herein specified, having the front extended down to the line of the pelvis so as to cover the abdomen with elastic plates *b b*, as described, and side lacings as above specified, constructed, arranged and combined, as and for the purposes herein set forth.

Also, in combination with an abdominal supporter, substantially as above specified, the air sacks B, as and for the purposes herein set forth.

No. 2,166.—THE MIDDLETOWN TOOL COMPANY, Middletown, Conn., assignees by mesne assignments of J. R. HENSHAW.—*Self-mousing Hook.*—Patented October 26, 1858; reissued February 6, 1866.

Claim.—Locating the spring of a snap hook, substantially as shown and described, so as to act upon points intermediate between the hinge and hook proper, in combination with forming recesses for holding the spring, as set forth.

No. 2,167.—THE WASHOE TOOL COMPANY, New York, N. Y., assignees of H. L. LOW-MAN.—*Tool.*—Patented June 6, 1865; reissued February 6, 1866.

Claim.—An elliptical socket the opposite sides of which are parallel to each other, and elongated in the line of its axis, in combination with one or more projecting arms or bits merging by curved lines into the socketed head, substantially as described and represented.

No. 2,168.—JONATHAN C. BROWN, Brooklyn, N. Y.—*Machine for Cutting Splints.*—Patented June 21, 1864; reissued September 12, 1865; and again reissued February 13, 1866.

Claim.—Cutting forms from wood in the manner described, when the knives on the revolving cylinder are set at different angles as and for the purposes herein set forth.

Also, combining the knives, rotating and stationary, in one frame, as herein set forth, so as to cause the two cuts that are at right angles to exactly meet, so as to cut the form off perfectly, and at the same time avoid scoring the wood deeper than the part removed, all as above specified.

No. 2,169.—CHARLES GOODYEAR, Jr., New York, N. Y., executor of estate of CHARLES GOODYEAR, deceased.—*Making Hollow Articles of India-rubber.*—Patented April 25, 1848; extended seven years; reissued February 13, 1866.

Claim.—As a new article of manufacture and trade, the hollow, vulcanized India-rubber articles, the external shape of which is produced by internal pressure derived from an elastic fluid.

No. 2,170.—CHARLES GOODYEAR, Jr., executor of estate of CHARLES GOODYEAR, dedeceased.—*Making Hollow Articles of India-rubber.*—Patented April 25, 1848; extended seven years; reissued February 13, 1866.

Claim.—The above described process of making hollow spheres, various hollow toys or other hollow articles of caoutchouc, the same consisting in the employment of a mould, and heat and air, or its equivalent, substantially in the manner and under the circumstances above set forth.

No. 2,171.—HIRAM W. HAYDEN, Waterbury, Conn.—*Machine for Making Kettles.*—Patented December 16, 1851; extended seven years; reissued February 13, 1866.

Claim.—First, the combination of mechanism constructed and arranged substantially as specified, for making kettles and similar articles, in substantially the manner set forth.

Second, the construction of the mandrel f^3, part of which is cylindrical and part fitted with a short screw 13, to take the screw of the hand wheel f^2, so that great pressure may be made at the point desired, while at the same time the mandrel can be easily and quickly moved through a long distance for the purposes and as described and shown.

No. 2,172.—J. WILSON HODGES and P. DE MURGUIONDO, Baltimore, Md., assignees of J. WILSON HODGES.—*Horseshoe.*—Patented July 4, 1865; reissued February 13, 1866.

Claim.—The attachable and removable roughing bar C, provided with calks and countersunk or let into the face of the shoe, substantially as described.

The blank bar E', adapted to occupy the groove B, in the absence of the roughing bar, and secured in a similar manner within the groove.

No. 2,173.—JOHN PHILIP LEBZELTER, Lancaster, Penn.—*Wood Bending Machine.*—Patented February 21, 1865; reissued February 13, 1866.

Claim.—First, the winged or framed side levers E, held by a pivot or hinge affixed by a bolt or plate on each side of the drum, substantially in the manner shown and specified.

Second, the binding straps M, when they are firmly united to a slotted hook, affixed to each end of the same, in the manner and for the purpose specified.

No. 2,174.—JOSEPH F. POND, Cleveland, Ohio.—*Roller for Wringers.*—Patented April 5, 1864; reissued February 13, 1866.

Claim.—The application of canvas, cloth, or other similar material, for the purpose of covering, repairing, and protecting elastic India-rubber or compound rollers, and to prevent the shaft getting loose or turning in the roll, as and for the purpose specified.

No. 2,175.—WM. WESTLAKE, Chicago, Ill.—*Lantern.*—Patented September 26, 1865; reissued February 13, 1866.

Claim.—First, the band *d* in combination with the band *b*, for the purpose set forth.

Second, the band *l* in combination with the band or upright portion of the bottom *e* for keeping the bottom of the globe in place, as herein described.

Third, the means described, or its equivalent, for securing the ends of the upright bars to the horizontal bars of the guards.

No. 2,176.—CHARLES GOODYEAR, Jr., New York N. Y., assignee of JACOB AUTESRIETH, Philadelphia, Penn.—*Eyelet for Lacing Shoes.*—Patented January 6, 1863; reissued February 20, 1866.

Claim.—First, a shoe lacing with its eyelets and cords, constructed and arranged substantially as described.

Second, the metallic lacing, eyelet, or loop, constructed and arranged substantially as herein described, so that the lacing cord shall run through the same without traversing the leather or material of the shoe or other article of wearing apparel to be laced.

Third, the arrangement of the metallic eyelet or loop transversely in relation to the fastening device, as herein described, so that the said eyelets or loops, when fastened on to the leather or material, shall be situate in vertical planes relatively to the surface of the leather or material, as set forth.

No. 2,177.—WILLIAM GEE, New York, N. Y.—*Apparatus for Drawing Soda Water.*—Patented May 19, 1863; reissued February 2, 1864; again reissued February 20, 1866.

Claim.—First, the valve D and its parts *e* G H H' and passage or aperture *g*, in combination with the valve B and its parts *c* E F F', and passage or aperture *h* forming a cock, for the purpose set forth.

Second, the means of drawing soda or mineral water from a small and a large outlet passage or aperture having one connection with a draught tube or soda-water apparatus, substantially as and for the purpose herein specified.

Third, the small passage or aperture *a*, for the purpose of compressing the soda water while being admitted into the large passage or outlet aperture *g*, for the purpose set forth.

Fourth, drawing soda water in a large stream passing first through a smaller passage into a larger passage or space, from which proceeds the larger stream.

Fifth, drawing soda or mineral water in a large and small stream from one nozzle or opening, in connection with a fountain or other apparatus, substantially as herein described.

No. 2,178.—CHARLES S. HAMILTON, Fond du Lac, Wis.—*Distributing Grain to Different Bins.*—Patented June 21, 1864; reissued February 20, 1866.

Claim.—First, the combination with a revolving spout for delivering grain or similar material to different bins of the shaft M or any equivalent device, to enable the attendant to move or adjust said spout, substantially as and for the purpose set forth.

Second, the combination with a revolving spout of an indicator, arranged to show the position of said spout and to enable the attendant to properly adjust the same, substantially as and for the purposes set forth.

No. 2,179.—F. MARQUARD, Rahway, N. J.—*Manufacture of White Rubber.*—Patented December 5, 1865; reissued February 20, 1866.

Claim.—First, the method or process of treating India-rubber, gutta-percha, or other similar gums, with hot water, for the purpose of washing them after they have been previously bleached with chlorine gas, substantially as hereinbefore set forth.

Second, the method or process of treating India-rubber, gutta-percha, or other similar gums, by distillation, after the gum has been bleached with chlorine gas, for the purposes hereinbefore set forth.

Third, the method or process of treating India-rubber, gutta-percha, or other similar gums, that have been previously bleached with chlorine gas and washed and distilled as hereinbefore set forth, by redissolving it in chloroform or other solvent, and mixing with it phosphate of lime and subjecting the compound to pressure in hot moulds to harden and solidify it, for the purposes described.

No. 2,180.—F. MARQUARD, Rahway, N. J.—*Manufacture of White Rubber.*—Patented December 5, 1865; reissued February 20, 1866.

Claim.—First, the method or process of treating India-rubber, or other similar gums, when dissolved in chloroform or other solvent, with caustic ammonia gas or chloride of ammonia, for the purposes substantially as hereinbefore set forth.

Second, the method or process of washing the dissolved and bleached gum as hereinbefore set forth, with hot water, for the purpose described.

Third, the method or process of distilling the dissolved and bleached gum, while in the washing process or by a subsequent process, for the purposes hereinbefore set forth.

Fourth, the method or process of redissolving the rubber or gum obtained by the foregoing operations and combining the same with phosphate of lime or a carbonate of zinc by means of pressure in hot moulds to harden the compound, for the purpose set forth.

No. 2,181.—RALPH J. FALCONER, Washington, D. C.—*Hose Coupling.*—Patented June 7, 1853; reissued February 27, 1866.

Claim.—First, the hose coupling consisting of two parts so constructed and applied that they shall be secured when brought together by a movement transversely to the direction of the water-course, as set forth.

Second, the means herein shown and described by which the parts of the coupling can be linked or held in position to advance and complete the joint, as set forth.

Third, a hose coupling in which one of the parts is set or pressed up against a washer imbedded in or permanently secured to the face of the other part, as herein set forth.

No. 2,182.—AMERICAN SCREW COMPANY, Providence, R. I., assignee of THOMAS J. SLOAN, New York, N. Y.—*Machine for Shaving and Nicking Screws.*—Patented October 21, 1851; extended seven years; reissued March 6, 1866.

Claim.—The combination of the gripping jaws on the rotating mandrel, the shaving tool on the movable tool-post, and the cutter for cutting the nick in the head, substantially as described, to perform the operations of shaving and nicking in succession, and while the blank is gripped in the same jaws.

Also, the two mandrels with their gripping jaws, in combination with the shaving mechanism and the nicking mechanism, arranged as described, so that the operation of shaving can be performed on one blank while the operation of nicking is being performed on another blank, as set forth.

Also, giving to the mandrel or mandrels end play in their boxes, in combination with the permanent rest at the back of the mandrel and with the cutter, substantially as specified, by means of which the same position of the blank relatively to the cutter is obtained for the second shaving operation which it had for the first, as described.

Also, the combination of the gripping jaws on the rotating mandrel, the shaving tool on the movable tool-post, the rest for bearing against the blank to steady it while being acted upon, and the cutter for cutting the nick, substantially as and for the purpose described.

Also, the combination of the gripping jaws on the mandrel for holding the screw blank, the cutter for cutting the nick, and the rest for holding the blank steady while it is being nicked, substantially as described, whereby I am enabled to nick the head while the blank is held in the jaws of a mandrel capable of being rotated.

Also, subjecting the blank, while held in the same jaws, successively to the three operations of shaving, nicking, and reshaving, by the means substantially as herein described.

No. 2,183.—BYRON BOARDMAN, Norwich, Conn.—*Wire Staple.*—Patented March 30, 1858; reissued March 6, 1866.

Claim.—As a new manufacture or commodity, a wire staple adapted for use in making window blinds or screens, and constructed substantially as above described.

No. 2,184.—CHARLES S. BURT, Dunleith, Ill.. assignee by mesne assignment of H. H. LOW.—*Shingle Machine.*—Patented March 16, 1858; reissued March 6, 1866.

Claim.—First, a vertically movable and counterbalanced bolt gate, or frame G, in combination with a circular saw D, which is arranged in a fixed frame and operating substantially as described.

Second, providing a vertically moving counterbalanced bolt frame, or gate G, with a head block K and contrivances for adjusting said block up to and from the saw, when constructed substantially as described.

Third, so constructing a machine for sawing tapering or straight slabs from bolts, that the table or frame upon which the bolts are secured shall be automatically returned by an upward movement, or a downward movement to a position which will admit of the adjustment of the bolt after each cut, by the means substantially as described.

Fourth, the combination of the vertically movable counterbalanced gate and treadle with the head block K, levers L, pawls M, and racks J, arranged and operating substantially as described.

No. 2,185.—THOMAS BRACHFR, New York, N. Y., assignee by mesne assignments of
WARD EATON.—*Machine for Serrating Sheet Metal.*—Patented May 16, 1854; reissued
March 6, 1866.

Claim.—The cutters attached to the reciprocating cutter-stock K—that is to say, the
series of serrated cutters and the straight shear cutter whose cutting surfaces or edges lie in
the same different horizontal planes, in combination with the stationary series of serrated cutters and stationary shear cutter, whose cutting surfaces or edges lie in inclined planes, substantially as set forth.

Also, arranging serrated dies in series so as to operate with a shear cut, substantially as
described.

No. 2,186.—DENNIS LANE, Montpelier, Vt.—*Head Block for Saw-mills.*—Patented September 6, 1864; reissued March 6, 1866.

Claim.—First, the arrangement of the stationary open racks E, wheels F, and shaft G,
constructed and operating in the manner and for the purpose herein specified.

Second, the dog N applied to the segment L and bent or curved as shown, to operate in
connection with the ratchet M and pin *a* of pawl I, substantially as described.

Third, the latch K* applied to the segment L, in connection with the pin *a* of the pane I,
arranged substantially as shown, to regulate the sweep of lever H.

No. 2,187.—JOHN SCHAFFER, St. Louis, Mo.—*Steam Capstan.*—Patented March 31, 1857;
reissued March 6, 1866.

Claim.—First, a capstan with the drum divided in two parts, the shaft R of which rotates
within the drum *c* and *d*, which can be rotated separately or combined and by, or
independently of said shaft, and operated substantially in the manner and by the means
herein described and for the purpose set forth.

Second, the use of the winch heads B, when used in combination with the extension
of the hoisting shaft *y* placed on the gallows frame A, as herein described and for the purpose
set forth.

No. 2,188.—THE ELM CITY COMPANY, New Haven, Conn., assignee of C. O. CROSBY
and HENRY KELLOGG.—*Manufacture of Tape Trimming.*—Patented September 16, 1862;
reissued March 6, 1866.

Claim.—As an improved article of manufacture, the finished tape trimming, constituted
and made substantially as herein described.

No. 2,189.—MERRITT BURT, Cleveland, Ohio.—*Watch.*—Patented September 13, 1864;
reissued March 6, 1866.

Claim.—First, so connecting a pinion of a watch train to the centre shaft or arbor that
said pinion will turn with it, or in the ordinary running of the movements, and at the same
time turn independently of it in case of any sudden recoil or rupture of the main spring or
undue strain upon the levers when being wound up or otherwise, for the purpose specified.

Second, in combination with the centre shaft of a watch train a hollow friction pinion
thereon, so as to turn with or independent of its arbor without clicks or ratchets, substantially as and for the purpose described.

Third, holding or supporting the friction pinion D in place upon its shaft or arbor by means
of a screw nut, substantially as set forth.

Fourth, the combination of a spring washer and nut with a friction pinion and its shaft of
a watch train, substantially as and for the purpose set forth.

No. 2,190.—WILLIS HUMISTON, Troy, N. Y.—*Apparatus for Moulding Candles.*—Patented
April 4, 1854; reissued March 6, 1866.

Claim.—First, the employment of the wick stretcher E, so arranged and combined with
the machine having vertical stationary candle moulds therein that the candle wick within
such moulds shall be uniformly stretched or strained before the material is run or poured into
such moulds, and the friction or strain be removed therefrom before the candles are drawn or
ejected from such moulds in a vertical direction, substantially as herein described and set
forth.

Second, the stretching or straining of the candle wick in each and every of the vertical stationary candle moulds contained in the candle mould machine, at and by one continued or simultaneous operation, when the said wick extends from spools or bobbins below said moulds,
upward, into, and through the centre thereof, and from the lower or tip end of such moulds,
to and into the candles suspended above such moulds, substantially as herein described and
set forth.

Third, the employment of the candle tip bar F, or any substantial equivalent therefor which
shall be so constructed and arranged as to be moved in a lateral direction, up to or against, or
under the tips of the candles drawn or ejected from the stationary candle moulds below, and
thereby come in contact with the tips of the said candles in such manner as to centre the
candle wick in the said moulds, and at the same time hold the said candles thus suspended
during the operation of filling the said moulds with melted material from which to mould can-

dles, and during the cooling thereof and until the wick is cut or severed between the said suspended and moulded candles in said stationary candle moulds, substantially as herein described and set forth.

Fourth, a vertical stationary candle mould, constructed with an inner and annular shoulder *h'*, and with an outer surrounding shoulder *c*, and with a screw and nut at or near the lower end thereof, in the manner and for the purposes substantially as herein described and set forth.

Fifth, the contraction of the lower end of the vertical stationary candle moulds so as to form an inner annular shoulder in the manner and for the purposes substantially as herein described and set forth.

Sixth, the mode substantially as herein described and set forth for attaching to, and combined with the lower end of the vertical stationary candle moulds having an outer surrounding shoulder *c*, and the bottom plate B'', of the surrounding water box, so as to make the same water tight and firm therein, in the manner and for the purposes substantially as herein described and set forth.

Seventh, the employment of the shovel blade cutter J, or any equivalent therefor, and the passing of the same between the rows of the wicks of the vertically suspended candles, so as to cut or sever the two rows of the said wicks, in the manner and for the purposes substantially as herein described and set forth.

No. 2,191.—JAMES BENNETT FORSYTH, Roxbury, Mass., assignee by mesne assignment of himself.—*Machinery for Making India-rubber Hose, Belting, &c.*—Patented September 13, 1864; reissued March 13, 1866.

Claim.—A machine for making hose, round packing cord, wringer rolls, tubing and similar articles of rubber or other similar material, rubber cloth or rubber and cloth, consisting essentially of the rolls D E and M, and operating substantially as described.

No. 2,192.—THOMAS C. CRAVEN, Albany, N. Y., and WILLIAM H. DAVIS, New York, N. Y., assignees of THOMAS C. CRAVEN.—*Combined Hay Spreader and Elevator.*—Patented December 19, 1865; reissued March 13, 1866.

Claim.—The elevating chute, in combination with the raking cylinder deriving motion from the supporting and driving wheels, substantially as and for the purpose specified.

Also, the rotating cylinder deriving motion from the driving and supporting wheels, in combination with the movable teeth projected and withdrawn by an eccentric or equivalent means, substantially as described and for the purpose set forth.

No. 2,193.—CHRISTOPHER DUCKWORTH, Mount Carmel, Conn.—*Power Loom.*—Patented June 28, 1853; reissued July 4, 1865; again reissued March 13, 1866.

Claim.—The combination of a reversible ratchet mechanism with the reversible revolving tappets used in the loom to move the shuttle boxes in a vertical direction, substantially as and for the purpose described.

Also, the combination of the pattern mechanism of the loom with a reversible ratchet mechanism, and the reversible revolving tappets, substantially as and for the purpose described.

No. 2,194.—CHRISTOPHER DUCKWORTH, Mount Carmel, Conn.—*Power Loom.*—Patented June 28, 1853; reissued July 4, 1865; again reissued March 13, 1866.

Claim.—First, giving an alternate movement to the shuttle boxes in a horizontal plane, by means of pawls, reversible tappets, and a contrivance which will automatically control the movements of said pawls, substantially as described.

Second, giving an alternate diagonal movement to the shuttle boxes by means of pawls, reversible tappets, and a contrivance which will automatically control the movements of said pawls, substantially as described.

Third, the combination of reversible tappets with shuttle boxes, which are so applied to the loom that they will admit of being moved, either laterally, vertically, or diagonally, substantially as described.

Fourth, giving an intermittent, oscillating or rotary movement to a shuttle box, actuated by means of pawls and ratchet wheels, which are controlled by a cam surface *t*, or its equivalent, substantially as described.

Fifth, the use of tappets which receive a forward and backward movement or a continuous rotary movement, in combination with many-chambered shuttle boxes at both ends of the lathe, which boxes are connected together by a lever G, and operated simultaneously by means of said tappets, substantially as described.

Sixth, giving a reciprocating movement to many-chambered shuttle boxes of looms, by means of contrivances which are controlled automatically, in such manner that the boxes are moved a greater or less distance by a single vibration of the lathe, so as to throw the shuttles in regular order, or to skip a shuttle, according to the figure which it is desired to weave, substantially as described.

No. 2,195.—LEWIS FRANCIS and CYRUS H. LOUTREL, New York, N. Y., assignees of LEWIS FRANCIS.—*Composition of Matter for Printers' Inking Rollers and for other purposes.*—Patented March 8, 1864; reissued September 27, 1864; reissued February 28, 1865; again reissued March 13, 1866.

Claim.—Combining an alkali or alkalies, or alkaline earths, or any of the compounds of alkalies or alkaline earths, with glue and glycerine, to form a new and useful composition of matter for various purposes.

No. 2,196.—CHARLES OYSTON, Little Falls, N. Y.—*Nozzle.*—Patented August 25, 1863; reissued March 13, 1866.

Claim.—The combination of the dividers or divergers, substantially as described, with a water nozzle, substantially as and for the purpose specified.

Also, making the dividers or spreaders movable on the pipe, substantially as and for the purpose specified.

No. 2,197.—H. O. PEABODY, Boston, Mass.—*Breech-loading Fire-arm.*—Patented July 22, 1862; reissued March 13, 1866.

Claim.—First, the combination of a swinging breech block D, hinged at the rear end with the trigger guard lever E, by means of a pin and slot connection, operating substantially as described, for the purposes specified.

Second, the combination of a swinging breech block D, hinged at the rear end with the retractor F, for the purpose of ejecting the discharged cartridge case, by a continued movement of the guard lever in the same direction which it makes to bring the piece to the position for loading, substantially as described.

Third, holding the breech piece at its respective positions for loading and firing by the use of the notches *j* and *l*, in combination with the spring G and roller *i*, operating in the manner substantially as described.

Fourth, causing the swinging breech block D, when in the manipulation of the arm it has been made to operate upon the retractor F to eject the cartridge shell, to be returned to such position that its top surface *h* shall coincide with the surface of the bore of the chamber, for the purpose of facilitating the introduction of a fresh cartridge, by the means substantially as described.

No. 2,198.—DAVID SHIVE, Philadelphia, Penn.—*Enlarging Photographs.*—Patented March 22, 1859; reissued March 13, 1866.

Claim.—First, a photographic solar camera, swivelled or otherwise jointed, so as to permit the axis of the lenses to be adjusted in horizontal and vertical planes, in conformity with or approximation to the apparent direction of the sun's rays, for the purpose specified.

Second, a photographic solar camera, provided with a condensing lens F, a negative holder, a magnifying lens or combination of lenses, and a paper holder C, the described portions of the apparatus so arranged having an adjustability in reference to each other, and the apparatus itself being adapted for the direct presentation of the condensing lens to the sun's rays, whereby the axes of the lenses are made approximately or truly conformable to the direction of the said rays.

No. 2,199.—HENRY P. SISSON, Providence, R. I.—*Portfolio.*—Patented April 5, 1859; reissued March 13, 1866.

Claim.—A holder A, provided with curved paper file books *b b*, and a hinged spring folder B, or their equivalents, in combination, arranged and operating substantially as described for the purposes specified.

Also, in combination with the folder B, a spring latch *f* and stop *g*, substantially as described.

No. 2,200.—IRA C. STORY and GEORGE W. SKAATS, Cincinnati, Ohio, assignees of IRA C. STORY.—*Running Gear of Street Locomotives.*—Patented November 21, 1865; reissued March 13, 1866.

Claim.—First, the combination of one or more friction rollers or wheels with the driving wheel or wheels of locomotives or land carriages, for the purpose of propelling said carriages.

Second, the adjustable platform, in combination with the friction wheels O and N and the driving wheel C, substantially as described.

Third, reversing the movements of a locomotive by the alternate application to the driven of friction rollers revolving in opposite directions.

No. 2,201.—WASHINGTON VAN GAASBEEK, Mount Vernon, N. Y.—*Sash Stopper and Lock.*—Patented October 4, 1864; reissued March 13, 1866.

Claim.—First, the retaining lever C, reaching beyond the sash, provided at its outer end with a spur *c*, arranged to bite upon the outer face of the sash, in the manner substantially as set forth for the purpose specified.

Second, in combination with the above, the inner spur or biting edge *b*, substantially as set forth for the purpose specified.

No. 2,202.—HORACE WOODMAN, Saco, Me.—*Machinery for Cleaning Top-flats of Carding Engines.*—Patented July 8, 1856; reissued March 13, 1866.

Claim.—First, a cross connecting shaft H', so disposed in relation to the cleausing frame arms *a a'* as to be carried by or to traverse with them, when said shaft is used in combination with mechanism operating in connection with said arms, which produces by means of or through the said shaft, so disposed, conjoint or uniform intermittent reciprocating traversing movements of the two sides of the cleansing frame, substantially in the manner and for the purposes set forth and specified.

Second, a traversing mechanism proper, substantially such as described, and a cleansing mechanism proper, substantially such as described, combined in the manner and for the purposes set forth and described.

Third, a traversing mechanism proper, substantially such as described, a cleansing mechanism proper, substantially such as described, and a locking mechanism proper, substantially such as described, combined in the manner and for the purpose specified.

Fourth, a detent or locking mechanism, constructed substantially in the manner and for the purposes shown.

Fifth, the combination of a traversing mechanism, cleansing mechanism and detent or locking mechanism, with a pulley P', located on a line with the axis of the main cylinder of the carding engine, so that the whole stripping mechanism may be actuated or driven by a single belt acting on the said pulley, substantially as and for the purposes set forth and specified.

Sixth, a brush bar V, and waste pan F, disposed in the upper part of the cleansing frame and carried thereby, in combination with a card-clothed surface or strip of card filleting, so disposed or arranged in reference to the said brush bar as to remove the waste from the same, or cleanse the card of said bar prior or preparatory to the cleansing of each top card, substantially in the manner and for the purposes specified.

Seventh, in combination, the lever O, dogs M M', rod *q*, and sliding clutch N, arranged and operating to reverse the motion of the cleansing frame, substantially as and for the purposes specified.

Eighth, the grooves across the teeth connecting the space or slots between the teeth of the toothed rack, in combination with such teeth, whereby the series of top cards being cleansed is changed, substantially as and for the purposes set forth and described.

No. 2,203.—LYSANDER WRIGHT and CHARLES B. SMITH, assignees of LYSANDER WRIGHT, Newark, N. J.—*Scroll Sawing Machine.*—Patented May 16, 1865; reissued March 13, 1866.

Claim.—First, the vibration weighted lever A or its equivalent, constructed and operating substantially as and for the purposes specified.

Second, the combination of the pulleys C C and strap D with the saw hook E and vibrative-weighted lever A, substantially as and for the purposes specified.

Third, the connecting rod D', in combination with the saw hook E', and vibrative-weighted lever A, operating substantially as and for the purposes set forth.

No. 2,204.—ARCALOUS WYCKOFF, Elmira, N. Y.—*Hollow Auger.*—Patented July 12, 1859; reissued March 13, 1866.

Claim.—First, in an annular auger in combination with a prime cutter *a*, a transverse auxiliary cutter *b*, carried back to the extremity of the stock either longitudinally with the auger or obliquely toward the heel of the next preceding cutter.

Second, in combination with the spiral flange B, bevelling to a thin edge the cylinder at *d*, in front of the base of the prime cutter, for the purpose of giving an outward direction to and carrying away the cuttings, substantially as set forth.

No. 2,205.—AUGUSTUS ADAMS, Sandwich, Ill.—*Corn Sheller.*—Patented August 6, 1861; reissued March 20, 1866.

Claim.—First, in combination with a series of conveying belts B, or their equivalent, provided with cleats or buckets as described, the employment of a series of widening partitions C, arranged and operating substantially as and for the purposes herein specified and shown.

Second, in combination with a series of passages formed upon the conveying belts, or their equivalent, by the described arrangement of the partitions C, the arrangement of a corresponding series of separate feeding throats or passages D, leading down into the hopper to the sheller, whereby the ears of corn are kept separate and delivered endwise into the said sheller, substantially as described and shown.

Third, in combination with a series of separate feeding throats in the hopper of a corn shelling machine, into which the ears are delivered endwise as described, and the shelling wheels or surfaces thereof, the employment of a series of picker or feeding wheels *e*, or their equivalent, arranged and operating substantially as and for the purposes shown and set forth.

No. 2,206.—CHARLES McBURNEY, Roxbury, Mass.—*Utilizing Waste Vulcanized Rubber.*—Patented August 20, 1861; reissued March 20, 1866.

Claim.—The use of the oils mentioned in combination with waste vulcanized rubber and crude gum or rubber, as set forth for the purpose specified.

No. 2,207.—FRANCIS D. HAYWARD, Malden, Mass., and IRA E. SANBORN, Boston, Mass., assignees of JOHN C. BICKFORD.—*Process for Rolling India-rubber Cloth.*—Patented March 19, 1850; extended 7 years; reissued March 19, 1864; again reissued March 20, 1866.

Claim.—The new or improved process, substantially as described, of applying rubber or caoutchouc not previously dissolved by a solvent or solvents, to cloth, the same being accomplished by means of revolving rollers running at different velocities, and being so arranged that the rubber, while the cloth is in the act of passing between two of them, shall be forced and ground into the meshes or pores of the cloth by the faster or fastest roller of the set or series.

Also, the new or improved fabric or article of manufacture as produced by the said new or improved process.

No. 2,208.—SYLVESTER J. SHERMAN, Brooklyn, N. Y.—*Spring for Ladies' Dresses.*—Patented February 25, 1862; reissued March 20, 1866.

Claim.—A spring or busk for clothing having the ends permanently covered and rounded, substantially as and for the purpose herein set forth.

No. 2,209.—WILLIAM B. BERNARD, Waterbury, Conn.—*Lamp Trimmers and Shears.*—Patented December 27, 1864; reissued March 27, 1866.

Claim.—The construction of shears or lamp trimmers, substantially in the manner herein set forth.

No. 2,210.—JOSEPH W. BARTLETT, New York, N. Y.—*Sewing Machine.*—Patented January 31, 1865; reissued March 27, 1866.

Claim.—First, imparting to the looper rod the rocking and sliding motions described, when the parts for giving these motions are arranged and operated substantially as and for the purposes described.

Second, the adjustable sleeve o, constructed and operated substantially as and for the purposes set forth.

Third, in combination the rocking and sliding rod i, sleeve o, cam u, and feed bar s, constructed and operating substantially as described.

Fourth, the combination and arrangement of the rotating cam g, on the driving shaft, with the connecting rod h, rocking shaft i, cam u, and feed bar s, operating as and for the purposes set forth.

Fifth, the presser foot e, cam x, shown in figure 7, and sliding and rocking rod i, when combined and operating substantially as described.

No. 2,211.—JOHN WEBSTER COCHRAN, New York, N. Y.—*Revolving Fire-arm.*—Patented November 10, 1863; reissued March 27, 1866.

Claim.—First, the unloading and cartridge shell expelling piston or plunger d', attached to the frame of the fire-arm so as to work through the recoil shield, substantially as and for the purpose herein specified.

Second, the unloading piston or plunger d', and loading rammer d, connected by a bar or yoke e, to operate substantially as and for the purpose herein specified.

No. 2,212.—ALFRED B. ELY, Newton, Mass.—*Insulating Telegraph Wires.*—Patented January 9, 1866; reissued March 27, 1866.

Claim.—First, insulating telegraph wires and conductors, or their supports, with the material described, substantially as set forth.

Second, the new article of manufacture described, constituting an insulated wire, made substantially as set forth.

No. 2,213.—GEORGE S. HARWOOD and GEORGE H. QUINCY, Boston, Mass., assignees of WM. CLISSOLD.—*Machine for Oiling Wool.*—Patented October 7, 1862; reissued September 13, 1864; again reissued March 27, 1866.

Claim.—First, the oiling or lubricating of wool or other fibrous material, by means of a pressure roller imprinting and diffusing the oil or lubricating mixture which shall have been supplied in requisite quantity to the roller, in the manner substantially as set forth.

Second, a machine or apparatus for oiling wool, consisting of the following elements combined: 1st. One or more reservoirs for containing the oil or oleaginous mixture. 2d. A dipper or dippers and a brush or brushes, the former to convey and the latter to receive the determinate and requisite quantity of oil or oleaginous mixture. 3d. A distributor or distributors receiving oil or oleaginous mixture from the brush or brushes and transferring it to the wool, substantially as set forth.

Third, in automatic wool-oiling machinery, the combination of a tank or reservoir with a dipper or equivalent mechanism for performing the double function of stirring or agitating the oil or lubricating matter in the tank, and of lifting therefrom at each action a quantity of oil or lubricating matter requisite for one oiling operation, and this is claimed only when arranged for operation as described—that is to say, so that the said dipper shall not come in contact with the wool, substantially as set forth.

Fourth, in automatic wool-oiling machinery, combining with an oil tank a dipper, constructed substantially as described, for the more perfect agitation of the liquid, substantially as set forth.

Fifth, in automatic wool-oiling machinery, the combination of an oil tank with a dipper constructed substantially as described, so that the requisite quantity of oil for each operation shall be lifted and conveyed from the tank by adhesion of the oil or lubricating matter to the dipper, substantially as set forth.

Sixth, the combination with an oil reservoir and intermediate dipper and brush, or other equivalent device for conveying oil in requisite quantities to the distributor of a pressure roller arranged immediately in front of the feed roller above the feed apron of carding or other wool preparing machinery, substantially as set forth.

Seventh, in a wool-oiling apparatus, in which the wool is oiled by imprinting the oil as described, in combination with a pressure roller or the equivalent thereof, a brush or brushes charged with oil by means of one or more dipping plates, substantially as set forth.

Eighth, the combination with one or more oil reservoirs and travelling brush or brushes of a plate or plates for charging the said brush or brushes with the requisite amount of oil, and when arranged for action so as to properly agitate the oil or oleaginous mixture, substantially as set forth.

Ninth, in combination with a distributing pressure roller, a brush travelling diagonally over the said roller, that is to say, at an angle with the axis of, but in a plane tangent to the roller, substantially as and for the purposes set forth.

No. 2,214.—LOUIS LUDOVICI and LOUIS LUDOVICI, JR., New York, N. Y., assignees of MORIZ NOWAK.—*Blasting Compound.*—Patented May 24, 1864; antedated March 28, 1863; reissued March 27, 1866.

Claim.—First, the combination of the chlorate of potash, nitrate of potash, and ferrocyanate of potash with each other, and with a substance capable of evolving gases, such as carbon or equivalent materials, substantially as and for the purpose described.

Second, the combination of chlorate of potash, nitrate of potash and ferrocyanate of potash with binoxide of manganese, substantially as and for the purpose set forth.

Third, the use of bicromate of potash, in combination with chlorate of potash and nitrate of potash and ferrocyanate of potash, substantially as and for the purpose described.

Fourth, the application of a solution of chlorate of potash, nitrate of potash and ferrocyanate of potash to paper or other vegetable materials capable of being formed into cartridges, and protected by some water-tight compound, such for instance as that above specified, for the purpose set forth.

No. 2,215.—CHARLES WANZER, New York, N. Y.—*Cement for Slate Roofing.*—Patented June 24, 1862; reissued March 27, 1866.

Claim.—A compound to be used as a cement, composed of grease, pitch or tar, and quick-lime, hydrate of lime, chloride of lime or bleaching powder, or any equivalent thereof, either with or without linseed or other oil, and venetian red or other oclire, substantially as described.

No. 2,216.—THOMAS R. SINCLAIRE, New York, N. Y.—*Starting Street Railway Cars.*—Patented December 19, 1865; reissued March 27, 1866.

Claim.—First, in the application of car starters and brakes to railroad cars, the suspending of the frame C, in which a portion of the operating mechanism is placed directly on the axles D D of the car, the axles passing loosely through said frame, substantially as shown and described.

Second, the arrangement of a spring or a series of springs, working conjointly or in unison, and constituting the motor of a car starter and brake, with the operating mechanism in such a manner that the motor may be applied to either axle of a car, and the spring or springs wound up from either axle, substantially as shown and described.

Third, the sliding clutch F, in connection with the gears E I, when arranged with a spring or springs J, substantially as and for the purpose herein set forth.

Fourth, the levers G G', connected with the clutch F, and applied to the truck, substantially as described, when used in connection with the gearing and spring or springs, as and for the purpose set forth.

No. 2,217.—J. J. ECKEL and ISAAC S. SCHUYLER, New York, N. Y., assignees of JOSEPH SHORT.—*Bleaching Fibrous Substances.*—Patented January 23, 1866; reissued April 3, 1866.

Claim.—First, the washing or cleansing liquid composed of potash or caustic soda and the chloride of sodium, spirits of ammonia, or an equivalent substance, dissolved in water, about in the proportion as set forth.

Second, the cold alkaline solution composed of liquid potassa and the chloride of sodium, spirits of ammonia, or an equivalent substance, and water, about in the proportion as set forth.

Third, the bleaching liquid compound of chloride of lime and sulphuric or other acid, about in the proportion as set forth.

No 2,218.—ANTON MENNEL, New York, N. Y.—*Artificial Limbs.*—Patented December 20, 1864; reissued April 3, 1866.

Claim.—Constructing the shell of an artificial limb of two or more layers of wood, each composed of one or more strips, placed crossways to each other, substantially in the manner and for the purpose herein set forth.

No. 2,219.—W. ANTHONY SHAW, New York, N. Y.—*Lining Pipes with Tin.*—Patented March 10, 1863; reissued April 3, 1866.

Claim.—First, forming a double metallic pipe or tube out of any two of the ductile metals, or their alloys, by pressing them together through, over, on, or in a die in such manner as to make each of said metals form a tube or pipe, one inside of the other.

Second, putting the tin and lead together in a cylinder over a mandrel, and forcing them through a die over said mandrel so as to press a lining of tin in the pipe at the time of its formation.

Third, the manufacture of lead pipe with a lining of tin, by forcing an ingot of tin and an ingot of lead, while over a core, out of a cylinder through a die by hydraulic pressure, as specified.

No. 2,220 —S. L. SIMPSON, New York, N. Y.—*Ruler.*—Patented June 27, 1865; reissued April 3, 1866.

Claim.—First, the springtop *d*, applied in combination with a valve *a*, substantially as and for the purpose set forth.

Second, the index *e* and scale *f* in combination with the link C, connecting the two parts of a parallel ruler, substantially as and for the purpose described.

Third, the adjustable pencil slide F in combination with the radius arm D, compass dial E, and parallel ruler A A', constructed and operating substantially as and for the purpose set forth.

Fourth, the clamping screw *k*, in combination with the arbor *h*, pinion *g*, and disk *j*, compass dial E, and parallel rulers A A', constructed and operating substantially as and for the purpose described.

No. 2,221.—N. F. BURTON, Galesburg, Ill.—*Gang Plough.*—Patented October 29, 1861; reissued April 10, 1866.

Claim.—First, the device for adjusting the beams A A', by means of plate *f*, and clamps *e e*, and bars *g g*, substantially as set forth, whereby the depth of penetration of the ploughs M and I may be changed at pleasure.

Second, the combination of the subsoil plough I, having a long winged mould-board, with the surface plough M, arranged as and for the purpose set forth.

Third, the attaching of the axle D to the beams A A', through the medium of the bar-shaped rod L, in combination with the arm H, attached to the axle D, and having its bearing or fulcrum on the rod L, as herein described, whereby the depth of the penetration of both ploughs may be regulated at pleasure, and they may also be made to run out of the ground when desired.

No. 2,222.—RUDOLPH VOLLSCHWITZ and J. J. SCHLAEPFER, New York, N. Y., assignees by mesne assignments of F. RUDOLPH.—*Lock.*—Patented July 25, 1865; reissued April 10, 1866.

Claim.—First, a lock with a tubular case B, containing a bolt D, and one or more tumblers E, to be operated from either side by a key K, substantially as and for the purpose set forth.

Second, the latch F, in combination with the bolt D and tubular case B, constructed and operating substantially as and for the purpose described.

No. 2,223.—J. J. GREENOUGH, New York, N. Y., assignee of JAMES WARNER.—*Revolving Fire-arms.*—Patented June 24, 1856; reissued April 10, 1866.

Claim.—First, the employment of a pin or other projection, as described, in the shield or rear frame for receiving recoil of the breech, constructed as herein described, at a point just behind and in rear of a chamber that is on a line with the barrel when in position to be discharged, for firing the arm, as and for the purposes above set forth.

Second, making the pin or projection *e* adjustable, substantially as and for the purposes set forth.

Third, the cavities *i i'*, in the battery plate *e*, in such position and of such form as to receive and hold the ball or balls in case of the accidental discharge of any of the chambers not in adjustment with the barrels, as described.

No. 2,224.—REUBEN HOFFHEINS, Dover, Penn.—*Harvester.*—Patented May 20, 1862; reissued April 10, 1866.

Claim.—First, a sweep rake which is mounted upon the heel of the finger beam proper, or upon the inner front corner of the platform of a harvester which has its cutting apparatus and platform hinged to the draught-frame, all in such manner that the rake arm sweeps the platform from front to inner side, and maintains a correct position in relation to the finger beam and platform during the rising or falling movements thereof on the joint or joints by which the finger beam is connected to the draught frame, substantially as set forth.

Second, a rake rotating upon an axis which is perpendicular to the top surface of the platform, or nearly so, and having its arms successively turned up, substantially as and for the purpose described.

Third, the angular rake arms rotated independently of the axis f, and controlled substantially as described, in combination with a guide way which is perpendicular, or nearly so, to the said axis f of the rake head, substantially as and for the purpose described.

Fourth, elevating and depressing revolving rake and reel arms by means substantially as described, whereby I am enabled to dispense with an inclined plane or cam way, as set forth.

Fifth, an inclined standard or support F, or its equivalent, rigidly mounted upon a loosely hinged platform or finger beam and adapted for supporting a sweep rake in an unchanging position in relation to said platform without obstructing the free motion of the platform or finger beam, substantially as described and shown.

Sixth, a standard or support F, which sustains the sweep rake above the draught frame or driving wheel thereof, said standard being mounted directly and wholly upon a hinged finger beam or the platform thereof, substantially as described and for the purpose set forth.

Seventh, a revolving toothed head or crown wheel J, constructed with supports for rake and reel arms, in pairs, said supports being arranged outside and around the axis of said wheel J, substantially as and for the purpose described.

Eighth, in a harvesting machine which has its cutting apparatus hinged or jointed to the main frame in such manner as to allow it to conform at both ends to the undulations of the ground and a rake mounted on the said jointed cutting apparatus, or upon the platform thereof; so constructing and arranging the several parts that the support of the rake can occupy a position outside of the inner drive wheel B, or a position which is between the point of suspension h and the outer divider G, and can also be hung or be suspended below the draught frame, substantially as described.

Ninth, effecting a combination of a rake and reel located substantially as described, and a finger beam and platform, with the main frame, by means of a hinged draw bar b, a hinged brace I, a hinged suspender f, and an extension bracket 2, or their equivalents, substantially as and for the purposes described.

Tenth, the combination of a rake and reel, a yielding draw bar b, inner shoe of the cutting apparatus and hinge joint e, on the draw bar, substantially as and for the purpose described.

Eleventh, preventing a too sudden or abrupt deflection of a rake and reel mounted upon a hinged joint cutting apparatus by carrying the point of suspension beyond the rake support toward the centre of the draught frame by means substantially as described.

Twelfth, a continuously revolving rake which is mounted directly and wholly upon the platform or finger beam so as to rise and fall therewith independently of the draught frame, when said rake is located between the centre of the draught frame and the outer divider and passes in at the front of the machine upon the platform and sweeps around to the inner side of the platform, substantially as described.

Thirteenth, the combination with a double hinged joint combined reaping and mowing machine of a sweep rake which enters at the front of the machine upon the platform, in such manner that the rake and cutting apparatus rise and fall together while reaping; and also in such manner that the rake and platform may be readily removed and the cutting apparatus at its inner and outer ends allowed to float upon the ground, and to accommodate itself at both ends to the undulations of the ground, substantially in the manner described.

Fourteenth, the combination of a suspended hinge joint cutting apparatus of harvesters, and a combined rake and reel which is mounted directly and wholly upon the suspended platform or hinged finger beam, substantially as and for the purpose described.

Fifteenth, controlling the rake and reel arms by an upper and lower guide between which an attachment of the respective rake and reel arms passes, substantially as described.

Sixteenth, the combination of a combined rake and reel mounted upon a hinge joint cutting apparatus, and a yielding belt tightener, substantially as and for the purpose described.

Seventeenth, the employment of a yielding belt or chain tightener, or its equivalent, in connection with harvesters which are constructed with a hinged joint cutting apparatus, substantially as and for the purpose described.

Eighteenth, the pulley support Q, with its pulleys w^2 w^2, in combination with a band or chain N and pulleys w^1 w^1, substantially as and for the purpose set forth.

Nineteenth, providing in a harvester with the rake attached to its hinged finger beam or platform an extensible means for driving the rake which will permit the platform and rake to rise and fall together and accommodate themselves independently of the draught frame to the undulations of the ground, substantially as described and for the purpose set forth.

No. 2,225.—JOSEPH HOLLEN, Blair county, Penn., assignor by mesne assignments to JOHN NESMITH.—*Knitting Machine.*—Patented July 16, 1850; extended seven years; reissued April 10, 1866.

Claim.—Forming the needle in the manner substantially as described, so that each needle shall be capable of being separately projected and withdrawn in the operation of knitting, by the application thereto of mechanism, substantially such as described, for producing these movements.

Also, the above-described means of projecting and withdrawing the separate needles, or any other substantial equivalent of a knitting needle, when used to perform its functions.

Also, the arrangement of the needles on a cylinder in the manner described, in connection with the described means of moving or revolving the cylinder as a needle carrier.

Also, the combination of the jack, the sinkers, and depressers, substantially as described.

Also, the thread bearer v, having an extended sideway motion to and fro, at each stitch, by which it lays the thread across the needle at each stitch, and returns with it to be ready for the next stitch.

Also, the spring vice for regulating the supply of thread to the needle, opened by the rod w, substantially as described.

Also, the particular arrangement and combination of the several parts of the machine by which their various motions are derived from a single crank and screw thread, substantially as described.

No. 2,226.—GAIL BORDEN, South East, New York.—*Condensing Milk*—Patented August 19, 1856; reissued May 13, 1862; again reissued February 10, 1863; again reissued November 14, 1865; and again reissued April 17, 1866.

Claim.—The within described process or method of operation for concentrating and preserving milk by means of the preliminary beating thereof, substantially as described, in combination with the evaporation of the fluid in vacuo.

No. 2,227.—JARVIS CASE, Springfield, Ohio.—*Corn Planter.*—Patented January 16, 1855; reissued November 16, 1858; and again reissued April 17, 1866.

Claim.—First, the seed slide c, lever m, rod o, and slide r, whereby a valve at the seed hopper of a corn planter is so connected with another valve below the seed hopper that by a single impulse or movement a charge of seed shall be dropped from each valve, substantially as described.

Second, the combination of lever I, rock shaft u, weight v, and wire 2, with lever J, whereby the valves of one corn planter may not only be operated to produce the double drop of seed described, but will become convertible at will from a hand planter to an automatic planter, substantially as described.

Third, so arranging and connecting a valve at the seed hopper of a corn planter with another valve below the hopper, in combination with a lever, that the attendant of the machine, as it is moved over the ground by a single throw of the lever, not only discharges the seed from the valve below the hopper at the proper time, but drops a charge from the valve at the seed hopper to the valve below in readiness for the next bill, substantially as set forth.

Fourth, so combining and arranging the mechanism of a corn planter's valves, that the valve in the seed hopper and the valve in the seed tube below the hopper may each be made to drop a charge of seed by a single impulse or movement, substantially as described.

No. 2,228.—THOMAS CRANE, Fort Atkinson, Wis.—*Stump Extractor.*—Patented October 31, 1865; reissued April 17, 1866.

Claim.—First, so constructing a stump-pulling machine that a draught on the chain or cord causes the frame to form an angle which is less obtuse than the angle which it forms at the commencement of the operation, and thus an upward draught on the stump is produced, substantially as described.

Second, the combination of the triangular lifting frame G G with the lifting beam F, substantially as described.

Third, the combination of the tripod lifting frame G G and F, the triangular base A, and a windlass D, operating substantially as described.

Fourth, sustaining the lower end of the lifting beam F upon a rope or chain c of the windlass D, substantially as described.

Fifth, the combination of the pulley E, draught rope a, windlass D, stirrup chain c, and the lifting beam of the tripod, substantially as described.

No. 2,229.—JAMES M. HORTON, Albany, N Y., assignor by mesne assignment to himself.—*Attaching Augers to their Handles.*—Patented July 8, 1862; reissued April 17, 1866.

Claim.—Attaching augers to their handles by means of barrel A, cylinder C, having a central rectangular tapering bore, and provided with grips B and D, that are operated upon by nut N, working on a screw thread upon the cylinder, all arranged to operate substantially as described.

No. 2,230.—JOHN F. SEIBERLING, Doyleston, Ohio, assignee of S. RAY and M. R. SHALTERS.—*Harvesting Machine.*—Patented April 19, 1859; reissued April 17, 1866.

Claim.—First, combining the finger beam with the machine by means of a hinge joint, so arranged as to allow the finger beam or cutting apparatus to be horizontally folded to a position or line at a right angle, or nearly so, to the working position of said beam, for the purpose of transportation.

Second, attaching the finger bar F to the machine by means of the plate G, one end of which is pivoted to the machine as at l, and the other end connected with the finger bar by joints k k, the above parts being in connection with a jointed connecting rod i, to admit of the folding and turning of the finger bar, substantially as described.

Third, placing the driver's seat J on the springs r, fitted in the hollow standards p p q, substantially as and for the purposes set forth.

No. 2,231.—CHARLES C. ALGER, New York, N. Y., assignee by mesne assignments of LEONARD GEIGER.—*Breech-loading Fire-arm.*—Patented January 27, 1863; reissued April 17, 1866.

Claim.—First, these elements in combination, viz: a barrel open at the breech, a solid breech piece swinging in the plane of the barrel, an independent hammer and a shoulder operated thereby, for the purpose of locking or unlocking the breech, substantially as set forth, the combination being substantially as described.

Second, in combination with a barrel open at the breech, a hammer independent of the breech piece, and a solid breech piece swinging in the plane of the barrel, an ear or thumb piece, located substantially as described, by which the breech piece may be operated from above its centre of motion, in the manner set forth, the combination being substantially such as hereinbefore described.

Third, an independent hammer and a swinging shoulder to lock a solid breech piece, swinging in the plane of the barrel, and to receive the strain of the explosion, in combination with a spring which holds a breech piece, closed in the manner and for the purpose specified, the combination being substantially such as described.

Fourth, a solid breech piece, swinging in the plane of the barrel, and a barrel open at the breech, in combination with an independent swinging shoulder (not attached to the breech piece) to receive the strain of the explosion, and a spring to force the shoulder into its locking position, the combination being substantially such as described.

Fifth, in combination with a barrel open at the breech, and a solid breech piece swinging in the plane of the barrel, a swinging locking shoulder, to receive the strain of the explosion, and an ear or thumb piece projecting above the breech frame, by means of which the said shoulder may be withdrawn from the breech piece, to permit the opening thereof, the combination being substantially such as described.

Sixth, in combination with a barrel open at the breech, and a solid breech piece swinging in the plane of the barrel, and a swinging locking shoulder, an ear, or thumb piece, located above the centre of motion of the breech piece, whereby the breech piece can be opened after the shoulder has been withdrawn, the manipulation to open the breech piece being similar to cocking an ordinary gun lock, and the combination being substantially as set forth.

No. 2,232.—WILLIAM WESTLAKE, Chicago, Ill.—*Machine for making Lanterns.*—Patented April 25, 1865; reissued April 17, 1866.

Claim.—First, the plate B, with groove T, and notches P P P, or its equivalent, constructed substantially as described.

Second, the grooved and notched plate B, in combination with a clamping device, substantially as described.

Third, the combination of the grooved and notched plate B with the shaft A, in the manner described.

Fourth, the combination of the shaft A, grooved and notched plate B, with the slides I I, or their equivalents, in the manner described.

Fifth, the combination of the shaft A with the plates B and C, in the manner and substantially as described.

Sixth, the combination of the shaft A, plates B and C, with the springs N N, substantially as described.

Seventh, a former or device constructed substantially as described, upon which to make lantern guards.

No. 2,233.—LEWIS LILLIE, Troy, N. Y.—*Bank and Safe Door Knobs.*—Patented July 5, 1869; reissued April 24, 1866.

Claim.—First, the employment of the switch or bar D and the nut C, or any equivalent thereof, arranged upon and combined with the knob bolt or spindle B, in the manner substantially as and for the purposes herein described and set forth.

Second, the tapering or conical spindle B, in combination with a door of an iron or metallic safe, vault or other structure so as to prevent the lock or lock bolt switch, by which the door thereof is fastened therein, from being driven from such door, from the outside of the same, by any burglar, in the manner substantially as herein described and set forth.

Third, the employment of said tapering or conical spindle B, in combination with the lock case c of the lock F, or any equivalent thereof, in the manner and the purposes substantially as herein described and set forth.

Fourth, the employment of the tapering or conical spindle B, or any equivalent thereof, when used in the manner and for the purposes substantially as herein described and set forth.

No. 2,234.—HENRY REYNOLDS, Springfield, Mass.—*Cartridge Extractor for Fire-arms.*—Patented November 22, 1864; reissued May 1, 1866.

Claim.—First, a cartridge shell ejector, consisting of a lever attached to the frame of the fire-arm, and so constructed and arranged that by a suitable movement a portion of it is made to enter an opening in the chamber or chambers between the breech or rear end thereof and the bottom of the cartridge shell, substantially as and for the purpose herein described.

Second, so bevelling the bottom of the chamber of the fire-arm as to provide for the entrance of the lever between it and the rear end of the cartridge shell, substantially as herein described.

No. 2,235.—THE SPENCER REPEATING RIFLE COMPANY, Boston, Mass., assignee by mesne assignments of C. M. SPENCER.—*Self-loading Fire-arms.*—Patented March 6, 1860; reissued May 1, 1866.

Claim.—First, the combination of the rolling breech E, the lever G, and sliding locking bolt F, the whole fitted and applied substantially as herein set forth.

Second, the slide H, applied to the rolling breech, and operating in combination with the hammer, substantially as described.

Third, the rolling breech E, constructed as described, to operate as a carrier block to receive the cartridge from the magazine and deposit it in the chamber in the end of the barrel, and also to cut off all communication between the chamber and magazine when the piece is loaded.

Fourth, the serrated projection u, constructed, arranged, and operating as described.

Fifth, the tongue J, constructed, arranged, and operating as described.

No. 2,236.—ROLLIN WHITE, Lowell, Mass.—*Breech-loading Fire-arms.*—Patented April 3, 1855; reissued May 1, 1866.

Claim.—First, the movable breech, connected with and operating with the tumbler and hammer, and on the same fulcrum pin, substantially as herein described.

Second, the plate applied, substantially as described, to serve as a guide to conduct the cartridge into the open chamber, and as a guard to prevent the cartridge falling out at the rear of the chamber before the breech is liberated, as herein set forth.

Third, making an aperture into the chamber, and constructing the hammer, or its equivalent, so that it will ignite the charge by striking the cartridge in front of the rear end thereof, substantially as and for the purpose specified.

No. 2,237.—HORATIO ALLEN, New York, N. Y.—*Packing for Tubes of Condensers.*—Patented July 20, 1858; reissued May 8, 1866.

Claim.—Making the joint formed by two metal surfaces (as in the joints of the tubes in the tube sheets of surface condensers and other similar instruments) tight, by inserting between the tube and the tube sheet a tube of seasoned or compressed wood, made either in one or several pieces, relying on the expansion of the wood after being saturated by water to make the joint tight, and on the freedom of the metal tube to move endwise without impairing the tightness of the joint, to avoid injurious results from the expansion and contraction of the metal tube, all substantially in the manner and for the purpose herein set forth.

Also, closing at any desired point the annular space formed where a tube passes through a long cylindrical hole, by surrounding the metal tube at that point by a tube of seasoned or compressed wood (in one or several pieces or staves) of such thickness, that when saturated with water it will fill water tight the annular space at that point, substantially in the manner herein described.

No. 2,238.—SILAS CRISPIN, New York, N. Y.—*Revolving Fire-arm.*—Patented October 3, 1865; reissued May 8, 1866.

Claim.—A fixed barrel and many-chambered rotating and bisected cylinder, with the parts swinging on a hinge of the frame forming the partial cartridge chambers, in each part of the cylinder, with annular recesses to accommodate the fulminate flanch of a cartridge, in combination with the means shown, or the equivalent, for allowing the said fulminate flanch of such cartridge to be exploded, as hereinbefore set forth.

Also, in combination with a frame divided and hinged together as described, the employment of separate fixed centres or studs for the support of the revolving cylinder, one projecting forward from the rear, and the other backward from the forward portion of the said frame, as set forth.

Also, the employment in combination with a hinged frame of the two cylinder sections, the whole so constructed and arranged that the said cylinder sections may be swung or vibrated on the hinge towards and from each other, substantially as and for the purposes set forth.

No. 2,239.—JAMES E. CROSS, JAMES F. DANE, and WILLIAM WESTLAKE, Chicago, Ill., assignees of WILLIAM WESTLAKE.—*Lantern.*—Patented December 12, 1865; reissued May 8, 1866.

Claim.—Connecting the dome or lamp part of a lantern to the guard by means of a ring or rod of the guard and the spring catches, or their equivalents, substantially as and for the purposes herein shown and specified.

No. 2,240.—ALEXANDER HARTHILL, New York, N. Y., assignee of J. A. BAWSEL, Powhatan Court House, Va.—*Tobacco Press.*—Patented May 31, 1859; reissued May 8, 1866.

Claim.—First, the process of pressing tobacco by passing it through the channel or groove C', with the follower B' working therein, substantially as shown and described.

Second, the combination of the rollers B and C with the endless belt or apron I, arranged to operate as set forth.

Third, the combination of the springs S, the elbow levers T T', rods a and x, with the treadles H, arranged to operate as herein described.

Fourth, the roller V, located in a suitable position to lubricate the channel C' or the follower B, and supplied with oil from any suitable reservoir, substantially as set forth.

Fifth, the scraper U, when arranged to operate in connection with the channel C', for the purpose of keeping the latter free from adhering substances, as set forth.

No. 2,241.—F. H. BARTHOLOMEW, New York, N. Y.—*Water Closet Valves.*—Patented January 19, 1864; reissued May 15, 1866.

Claim.—First, in combination with a water closet, a supply valve, valve stem, diaphragm E, and regulating chamber, said parts being so arranged that the valve may be opened by the manual force applied to it, and closed by the pressure of the water and retarded in its closing, substantially in the manner as and for the purpose set forth.

Second, a flexible disk, forming an obstruction in the water passage or way of the cock, to operate in the manner substantially as and for the purpose set forth.

No. 2,242.—R. A. BUNNEL, Rochester, N. Y., assignee of CHARLES F. SPENCER.—*Preserve Jar.*—Patented February 10, 1863; reissued May 15, 1866.

Claim.—A cover or stopper, with a packing around its edge, or periphery, in combination with a jar or can, whose mouth or neck has a cylindrical or slightly flaring inner surface, and an inwardly projecting shoulder or seat below.

Also, a self-retained cover or stopper, with a packing which projects above and below its edge, so as to be impressed between the same and the mouth or neck of the jar or can.

Also, the combination of a jar or can, which has an inwardly projecting shoulder in the mouth or neck thereof, with a cover or stopper, whose packing packs both against the said shoulder and against the inner periphery of the neck.

Also, a packing ring, one edge of which is bent inward and held in a peripheral groove of the stopper or cover.

Also, a cover or stopper, which has an edge or flange below the packing nearly fitting within and in combination with a shoulder in the neck of the jar or can.

Also, the combined arrangement and construction of the double-flanged cover B, packing *b*, and jar-neck seat *a*, one flange *f* of the cover compressing and tightening the packing ring, and the other flange *g* nearly fitting and closing the circle within the seat, substantially as and for the purposes herein specified.

No. 2,243.—MARY JANE MONTGOMERY, New York, N. Y., assignee by mesne assignments of RICHARD MONTGOMERY.—*Iron Ship.*—Patented December 6, 1859; reissued May 15, 1866.

Claim.—First, the use of corrugated metallic beams, whether straight, curved, bent, or sheared, in the construction of ships, steamers, or other vessels.

Second, the use of corrugated metallic plates or sheets in the construction of bulkheads or partitions for forming fire-proof or water-proof compartments in the holds of ships, steamers, and other vessels.

Third, the block or knee *c*, when constructed substantially in the manner and for the purpose set forth.

Fourth, the metallic plate E, when forming the plating of the sides, and, if required, the flooring of the deck, in combination with the corrugated beams to which they are attached, substantially as shown and described.

No. 2,244.—MARY JANE MONTGOMERY, New York, N. Y., assignee by mesne assignments of RICHARD MONTGOMERY.—*Iron Ship.*—Patented December 6, 1859; reissued May 15, 1866.

Claim.—The above-described curved, bent, or sheeted corrugated beam, irrespective of any peculiar curve, bend, or shear, in the construction of ships or other vessels.

No. 2,245.—THE BARTRAM & FANTON MANUFACTURING COMPANY, Danbury, Conn., assignees by mesne assignments of W. B. BARTRAM.—*Sewing Machine for Stitching Button Holes.*—Patented November 7, 1865; reissued May 15, 1866.

Claim.—First, stitching a bar across the end of a button-hole, for the purpose of strengthening the same, by means of devices which produce a lateral reciprocation of the material being stitched while it is also being fed forward, and permit the extent of the said lateral reciprocation to be increased at will by hand while the machine continues in operation.

Second, working and entirely completing a button-hole without the use of a hand needle, by means of devices substantially as herein described.

Third, in combination with the arm H, the wheel I and the eccentric of the driving shaft of a sewing machine, or its equivalents, for the purpose set forth.

Fourth, in combination with the wheels I and I', the arm H and the stop J provided with the projection *n*, substantially as and for the purpose set forth.

Fifth, in combination with the wheel I, the stop J and the plate A, substantially as and for the purpose set forth.

Sixth, in combination with the movable plate A of a sewing machine, the cloth holder O and the circular plate N, as and for the purpose set forth.

Seventh, in combination with the cloth holder O, the stationary guide P and circular piece N, substantially as and for the purpose set forth.

Eighth, in combination with the springs L, the stop J, wheels I I', arm H, and the eccentric of the driving shaft of a sewing machine, substantially as described, for the purpose of producing a zig-zag stitch.

Ninth, in combination with the feed bar b and its eccentric regulating lever, the lever R, for throwing the feed bar entirely out of action, substantially as and for the purpose herein set forth.

No. 2,246.—THOMAS T. TASKER, Philadelphia, Penn.—*Regulating the Temperature of Hot-water Apparatus.*—Patented December 5, 1854; reissued May 15, 1866.

Claim.—First, the combination of a float actuated by the expansion of water, with a draught or smoke damper, as the means of regulating the temperature of a hot-water heating apparatus, substantially as and for the purposes described.

Second, the use of the damper O for admitting cold air over the fire, as a means of cooling the gases of combustion, of reducing the draught, and of cooling the water in the boiler, either alone or in combination with other dampers.

Third, the use of a simultaneous motion for the dampers P and M, and the succeeding simultaneous motion for the dampers P and O, as described.

Fourth, the combination of the three dampers P O M with the float A, together with rods G K and the link R H, for the purpose and object herein described.

No. 2,247.—ARETUS A. WILDER, Detroit, Mich.—*Machine for Sawing and Edging Clapboards.*—Patented October 30, 1855; reissued October 18, 1859; again reissued May 15, 1866.

Claim.—First, the flanged rollers d, with their spring e, or equivalents, in combination with the adjustable back rest b. for the purpose hereinbefore described.

Second, the combination of the circular saw k and the rotary edging cutter n, attached to its arbor with the upper rotary edging cutter g, for the purpose herein described.

No. 2,248.—REUBEN G. ALLERTON, New York, N. Y.—*Water-proof Fabric*—Patented January 23, 1866; reissued May 22, 1866.

Claim.—The water-proof fabric, formed in the manner specified.

No. 2,249.—JOSHUA GRAY, Medford, Mass., S. S. BUCKLIN, Providence, R. I., and WILLIAM G. LANGDON, Malden, Mass., assignees by mesne assignments of JOSHUA GRAY.—*Magazine Fire-arm.*—Patented December 20, 1864; reissued May 22, 1866.

Claim.—First, the rack F arranged below the sector D, for the purpose described.

Second, moving the cartridge carrier from the magazine to the barrel, and vice versa, by passing it through a longitudinal slot in the breech pin, and sliding the latter over it.

Third, so constructing the end of the cartridge lifter as to cover the port of the magazine when the piece is used as a breech loader.

Fourth, the slot L to guide the cover or end K of the cartridge carrier, as described.

Fifth, the cartridge extractor M, in combination with the arm N, provided with the knob e, or its equivalent, as described.

Sixth, the cam R, in combination with the hooked lever S, or their equivalents, for the purpose of withdrawing the cartridge case without uncovering the magazine when it is required to use the piece as a breech-loader.

Seventh, the guide pin H, in combination with the groove I, for stopping and guiding the breech pin, as described.

Eighth, the combination of a breech-loader, a cartridge-shell extractor, and a convertible magazine stop, substantially as described.

No. 2,250.—E. C. HASERICK, Lake Village, N. H.—*Melting and Aggregating Iron Chips, Turnings, &c.*—Patented March 13, 1866; reissued May 22, 1866.

Claim.—The within-described process of protecting iron chips, shavings, turnings, or other fine particles of iron, and preventing them from being burned and blown away by mixing them with water or a solution of clay or earth or other material, in water or with any liquid capable of producing the same result.

No. 2,251.—MOSES A. JOHNSON, Lowell, Mass.—*Felted Fabric.*—Patented January 5, 1864; reissued May 22, 1866.

Claim.—The sheet felted fabric, herein described, having one or both surfaces coated with a film of sizing, as set forth, but the body of the fabric not saturated or impregnated with the sizing material.

No. 2,252.—GEORGE MUNGER, New Haven, Conn.—*Writing Tablet.*—Patented September 27, 1859; reissued May 22, 1866.

Claim.—A new article of manufacture, to wit: an argillaceous-surfaced wood writing slate, which is formed by uniting two or more layers of veneering or thin wood, and then coating the exterior surfaces of the compact mass with a composition of slate, emery or other similar argillaceous material, substantially as and for the purpose set forth.

No. 2,253 —GEORGE ODIORNE, Boston, Mass.—*Apparatus for Carburetting Air.*—Patented November 1, 1864; reissued May 22, 1866.

Claim.—First, so constructing and arranging a float in connection with a suitable air-forcing apparatus as to invariably cause the air to be forced through the liquid hydrocarbon, in the manner herein described—that is to say, by dividing the volume of air into uniform streams which shall flow from or escape through apertures at the periphery of the float, as set forth.

Second, the air-forcing apparatus herein described, the same consisting of a hollow drum having a series of buckets or chambers arranged upon the periphery of the same and communicating with the interior thereof, all as specified.

Third, the combination of the air-forcing apparatus and float, arranged and operating with regard to each other as described.

No. 2,254.—THOMAS PROSSER, New York, N. Y.—*Surface Condenser.*—Patented December 15, 1857; antedated October 31, 1854; reissued May 22, 1866.

Claim.—A condenser consisting of two hollow slabs or other equivalent chambers connected together by concentric tubes, and communicating with each other by means of the annular spaces formed by said tubes, said slabs and tubes being arranged within a cistern, and operating substantially as herein specified.

No. 2,255.—THOMAS PROSSER, New York, N. Y.—*Surface Condenser.*—Patented December 15, 1857; antedated October 31, 1854; reissued May 22, 1866.

Claim.—A surface condenser constructed to obtain distilled water from the condensing water by making the heat of the steam to be condensed, converting a portion of the condensing water into vapor, and afterward condensing such vapor by the cool incoming condensing water by means of a distilling condenser, substantially as herein described.

No. 2,256.—THOMAS THATCHER, Danville, Penn.—*Direct Action Engine.*—Patented December 1, 1863; reissued May 22, 1866.

Claim.—First, the two tappet levers J J', applied and combined with each other, with the valve and with the plungers C C', substantially as herein described.

Second, the rim or casing A A around the exhaust post of the valve seat in combination with the two cavities g and j, in the valve, substantially as and for the purpose herein specified.

Third, the arrangement of the cylinder A, pump cylinders B B', plungers C C' rod E, piston F' and tubes, as constructed and operating in the manner and for the purpose herein described.

No. 2,257.—W E. DOUBLEDAY, Brooklyn, and J. STEWART, New York, N. Y., assignees by mesne assignments of WILLIAM OSBORN.—*Machine for Pressing Bonnets, Bonnet Frames, &c.*—Patented August 19, 1856; reissued February 17, 1857; again reissued March 27, 1860; and again reissued May 29, 1866.

Claim.—First, manufacturing, stretching, or shaping by means of heated dies, the whole of the bonnet frame (or similar article to be worn upon the head,) at one operation, substantially as specified.

Second, manufacturing by stretching, forming or shaping by heated dies, the flaring face piece and side crown of a bonnet, or similar article to be worn on the head, jointly at one operation, substantially as specified.

No. 2,258.—CHARLES PARKER and RUSSELL B. PERKINS, Meriden, Conn., assignees of RUSSELL B. PERKINS.—*Manufacture of Spoons.*—Patented April 1, 1862; reissued May 29, 1866.

Claim.—A spoon, the bowl and handle of which are united, substantially as described.

No. 2,259.—LEWIS C. REESE, Phillipsburgh, N. J.—*Harvester.*—Patented April 10, 1860; reissued May 29, 1866.

Claim.—First, the combination with a series of revolving, rising and falling rake and reel arms, pivoted to a rotating shaft, from which they derive their motion of a guide, which so controls the motion of the rake arms as to cause them to descend into the standing grain in advance of the finger beam, to press the grain back against the cutting apparatus, to sweep across the platform and discharge the gavel at the rear end thereof, and then to rise and move forward in an elevated position out of the way of the falling grain, without projecting beyond the stubble side of the machine, and leave an unobstructed space on the machine upon which the driver may ride.

Second, a combined reel and rake, with its arms so constructed that they shall have a revolving motion around their shaft and a rotating motion around their own axes, for the purpose of turning their teeth downward when raking off and upward when moving forward.

Third, the combination of the rake arms with the guide, by means of the overlapping projections, stays or rods i, substantially in the manner described, for the purpose of holding the rake to the guide during its entire circuit.

Fourth, the combination of a revolving rake, having its arms hinged to its shaft at the inner end with a guide located between the rake head and its pivot, whereby the rake is caused to descend into the uncut grain in advance of the finger beam, to press the stalks back against the cutting apparatus, to sweep across the platform in a horizontal circular path, and discharge the gavel, and then to rise and move forward in an elevated position out of the way of the falling grain, without projecting beyond the main frame or passing over the head of the driver

No. 2,260.—SEYMOUR ROGERS, Pittsburgh, Penn.—*Horse Hay Fork.*—Patented January 24, 1865; reissued May 29, 1866.

Claim.—First, in hay elevators which combine a penetrator and rod with a barb or barbs, attaching such barb or barbs to the rod or lever by which they are operated, substantially as herein described.

Second, so constructing the hay elevator, substantially as herein described, as that the rod or lever to which the movable barb or barbs are attached or are set and discharged shall serve as the handle by which the elevator is suspended and supported.

Third, the notches e and f, in the rod D, in combination with the cap C and the cam H, substantially as and for the purpose hereinbefore set forth.

No. 2,261.—JAMES DENSMORE and AMOS DENSMORE, Meadville, Penn.—*Car for Transporting Petroleum.*—Patented April 10, 1866; reissued May 29, 1866.

Claim.—First, the two tanks B B, or their equivalents, when constructed and operating in combination with an ordinary railway car, substantially as and for the purposes set forth.

Second, the two tanks B B, or their equivalents, when set directly or nearly so over the car trucks, and when constructed and operating in combination with an ordinary railway car, substantially as and for the purposes set forth.

Third, the frame C C C C, the bolts 1 2 3 and 4, and the cleats H H H H, when constructed and operating in combination with tanks B B, and an ordinary railway car, substantially as and for the purposes set forth.

Fourth, the steps F F, the manholes and man-heads D D, the faucets E E, and the run-way G, when constructed and arranged in combination with the tanks B B, and an ordinary railway car, substantially as herein set forth and described.

No. 2,262.—FRANCIS D. HAYWARD, Malden, and IRA E. SANBORN, Boston, Mass., assignees of JOHN C. BICKFORD.—*Process for Rolling India-rubber Cloth.*—Patented March 19, 1850; extended seven years; reissued March 20, 1866; again reissued June 5, 1866.

Claim.—The new or improved process of applying rubber or caoutchouc, when reduced to plastic consistency, on to cloth, substantially as herein described—that is to say, by means of two contiguous rolls revolving in opposite directions, one carrying the cloth to be coated, the other the coat of rubber to be applied; the latter revolving at a higher rate of speed than the former.

No. 2,263.—FRANCIS D. HAYWARD, Malden, and IRA E. SANBORN, Boston, Mass., assignees of JOHN C. BICKFORD.—*Process for Rolling India-rubber.*—Patented March 19, 1850; extended seven years; reissued March 20, 1866; and again reissued June 5, 1866.

Claim.—The cloth coated with rubber, in the manner and by the means hereinbefore described, as a new fabric or article of manufacture.

No. 2,264.—CHARLES D. LEET, DEXTER SMITH, JOSEPH M. HALL, and CHARLES K. FARMER, Springfield, Mass., assignees by mesne assignments of LUTHER C. WHITE.—*Die for Making Lamp Tops, Rivets, &c.*—Patented September 7, 1852; reissued June 5, 1866.

Claim.—The combination of the annular die, supporting punch or mandrel, and presser, substantially as set forth.

No. 2,265.—JOHN C. LOVELAND, Springfield, Vt.—*Dough Kneader.*—Patented January 16, 1866; reissued June 5, 1866.

Claim.—First, the combination of the grooved, fluted, or irregular surfaced rollers B and D, constructed substantially as described with each other and with the frame A, in which they work, as and for the purpose herein set forth.

Second, the combination with the rollers B and D, and with the frame or supports A, of the machine of a pair of inclined aprons or tables K, substantially as described and for the purposes set forth.

Third, the combination of the dish M, with the roller B, aprons K, and supports or frame A, of the machine, substantially as described and for the purpose set forth.

No. 2,266.—WILLIAM K. MILLER, Canton, Ohio.—*Harvester.*—Patented February 8, 1859; reissued June 5, 1866.

Claim.—First, a frame composed essentially of a coupling or drag-bar and a brace, which are hinged to the main frame at two points, and connected in the line of the hinges and furnished with lugs by which they may be united to the shoe at two points, substantially as described.

Also, in connection with a yielding coupling arm and brace, an intermediate lug bolted to said arm, and projecting therefrom at such point as to make it capable of being united to a lug placed at the heel of the shoe, substantially as described.

No. 2,267.—WILLIAM K. MILLER, Canton, Ohio.—*Harvester.*—Patented February 8, 1859; reissued June 5, 1866.

Claim.—The short finger beam hinged to the coupling arm or brace, so that said finger beam may be raised up and folded to the rear in line with the driving wheels, substantially as and for the purpose described.

Also, so constructing and arranging the hinged or pivoted point on which the finger beam is thus folded backward as that the outer or free end of said finger beam in its folded position may conform to the inequalities of the ground over which it is drawn, and independent of the vertical movement of the machine, substantially as described.

No. 2,268.—WILLIAM K. MILLER, Canton, Ohio.—*Harvester.*—Patented February 8, 1859; reissued June 5, 1866.

Claim.—A draw or coupling bar hinged to the main frame, in combination with an adjustable hinge piece and one or more setting or tightening bolts, substantially as described.

Also, a yielding coupling arm, made of two or more pieces or parts, in combination with a pivot and tightening bolts, for adapting one part to the other part thereof, substantially as described.

Also, making the hinge piece adjustable, for the purpose of raising or lowering the points of the fingers or guards, substantially as described.

No. 2,269.—WILLIAM K. MILLER, Canton, Ohio.—*Harvester.*—Patented February 8, 1859; reissued June 5, 1866.

Claim.—The combination of a supporting shoe hinged by its lugs to corresponding lugs on an adjustable hinge piece, which latter is in turn hinged or pivoted to the yielding connection, by which the progressive movement of the finger beam is controlled, substantially as described.

No. 2,270.—WILLIAM K. MILLER, Canton, Ohio.—*Harvester.*—Patented February 8, 1859; reissued June 5, 1866.

Claim.—So constructing and arranging the cutting apparatus of a harvesting machine and its connections with the main frame as that it may be converted from a front to a rear cutting machine by transposing said parts, without taking from or adding to the machine any other parts or pieces than those which constitute the cutting apparatus in either of its positions, substantially as described.

No. 2,271.—CHARLES S. MARTIN, Milwaukee, Wis.—*Attaching Springs to Wagons.*—Patented February 9, 1864; reissued June 5, 1866.

Claim.—A device for suspending and securing India rubber springs by or under the hind axles of wagons, substantially as and for the purpose set forth.

No. 2,272.—CHARLES S. MARTIN, Milwaukee, Wis.—*Wagon Springs.*—Patented February 9, 1864; reissued June 5, 1866.

Claim.—The spring bar A and rods B B, when used for the purpose of sustaining the load on the hind axle of a wagon, in combination with the India-rubber springs F, substantially in the manner set forth.

No. 2,273.—SQUIRE RAYMOND, Venice, N. Y.—*Horse Hay Fork.*—Patented November 11, 1862; reissued June 5, 1866.

Claim.—First, the combination of the fork arms D D', turning or opening at their extreme upper ends, and the levers E E', constructed and operating substantially as described.

Second, the pulleys A G and H, arranged and operating in combination with the rope F, substantially as described, for the uses and purposes mentioned.

Third, the fork arms D D', the levers E E', with the extension g, all used in combination with the rope F and cord J, arranged as and for the purpose set forth.

No. 2,274.—WING H. TABER and THOMAS R. ABBOTT, Lowell, Mass., assignees of W. H. TABER.—*Carpenter's Bench Plane.*—Patented February 28, 1865; reissued June 5, 1866.

Claim.—The mechanism whereby the cutting edge of the bit is adjusted with respect to the face of the plane by springing the iron upon its bed in the plane stock, in the manner and by the means substantially as described.

Also, adjusting the inclination of the cutting bit with respect to the face of the plate by means of the adjustable bed G, when used in combination with the clamping mechanism, substantially as described.

No. 2,275.—JOSEPH W. FOWLE, Boston, Mass.—*Steam Drilling Machine.*—Patented March 11, 1851; extended seven years; reissued June 5, 1866.

Claim.—In a drilling machine in which the drill has an intermittent rotary movement, or a progressive f-ed movement, or both a rotary and feed movement, the attachment of the drill directly to the cross head of the engine or to the piston, or an elongation therefrom, in such manner that the drill is driven by the direct pressure of the motor upon the piston.

No. 2,276.—ALFRED F. JONES, Lexington, Ky.—*Vacuum Apparatus for Treating Diseases.*—Patented September 13, 1864; reissued June 5, 1866.

Claim.—The use of the means above set forth, consisting of a receptacle A in combination with a cape e, or its equivalent, for rendering such receptacle air-tight when to be used for enclosing any portion of the human body, substantially in the manner and for the purpose described.

· No. 2,277.—THE SILVER SKIRT AND WIRE MANUFACTURING COMPANY, New York, N. Y., assignees of T. S. SPERRY.—*Manufacture of Skirt Wire.*—Patented March 7, 1865; reissued June 5, 1866.

Claim.—Skirt wire protected wholly or partially by metal wire, substantially as and for the purpose described.

No. 2,278.—THOMAS ROSS, Middlebury, JOHN B. REYNOLDS, R. BARRET, and A. T. MERRIMAN, Rutland, Vt, assignees by mesne assignments of JOHN TAGGART.—*Machine for Channelling Stone.*—Patented December 4, 1855; reissued June 5, 1866.

Claim.—First, a machine consisting of drills arranged in one or more gangs, and also of mechanism for guiding such gang or gangs of drills, and imparting thereto or causing to be imparted thereto reciprocating movements or the same and longitudinal movements, wherby such gang or gangs when applied to stone may be caused to cut or drill one or more grooves therein, substantially as described.

Also, the combination of one or more standards U U, or the equivalent thereof, with the drills or the gang or gangs thereof, and machinery for guiding and operating the same, such standard or standards being to rest in the groove or grooves in which the drills may be in action, and being for the object or purpose as hereinbefore explained.

No. 2,279.—BENJAMIN F. WRIGHT, Springfield Township, Ohio, assignee by mesne assignments of THOMAS S. STEADMAN.—*Clover and Grass Seed Harvester.*—Patented May 23, 1854; reissued June 19, 1860; and again reissued June 5, 1866.

Claim.—First, in combination with the main frame of a harvester, an axle upon which the cutter's driving wheel revolves, that derives all its connection with the frame through one end, and which end does not cross a vertical plane parallel with and touching the side of this frame nearest to it, a plate from which this axle projects, and a holding mechanism that holds this plate and frame together, and prevents any essential variation in the distance between this axle and the cutter's driving pinion shaft, or in their parallelism while the frame is being raised or lowered in respect to this axle, substantially as and for the purpose set forth.

Second, in combination with the main frame of a harvester, an axle upon which the cutter's driving-wheel revolves, that derives all its connection with the main frame through one end, and which end does not cross a vertical plane, parallel with and touching that side of this frame nearest to it; and a plate from which this axle projects, a holding mechanism that prevents any essential variation in the distance between this axle and the cutter's driving pinion shaft or of their parallelism, while the main frame is being raised or lowered, in respect to this axle; and another holding mechanism by which the attendant is enabled to have this main frame held at different heights in respect to this axle, substantially as and for the purpose set forth.

Third, in combination with the main frame of a harvester, an axle plate which is connected with one end of the axle of the cutter's driving wheel; said plate being wholly between the plane of said wheel, and a plane parallel with and touching that part of said frame nearest to said wheel; a holding mechanism which prevents any movement of this plate other than its movement in the arc of a circle concentric to the axis of the cutter's driving pinion: and a holding mechanism having one portion further forward than the axle of the cutter's driving wheel, and another portion further back than said axle, between each of which and the frame is a portion of said plate, and by which said plate is held to the frame while it is being raised or lowered in respect to said axle, substantially as and for the purpose set forth.

No. 2,280.—L. AUGUSTUS ASPINWALL, Albany, N. Y.—*Potato Digger.*—Patented November 14, 1865; reissued June 12, 1866.

Claim.—First, the screen or screens F F, having a lifting movement and vibrating one rom front to rear, and when two are used vibrating alternately.

Second, the combination of such screen or screens with a plough, having an uninterrupted passage for the earth over its entire upper surface, substantially as set forth in the above specification.

No. 2,281.—LUMAN BISHOP and STEPHEN BREWER, Cortlandville, N. Y., assignees of LUMAN BISHOP.—*Apparatus for Straining Paints and other Materials.*—Patented February 20, 1866; reissued June 12, 1866.

Claim.—First, the strainer G, or its equivalent, as and for the purposes herein shown and described.

Second, the combination of the strainer G, with the tube F and piston B, substantially as and for the purposes described.

Third, the lateral apertures H H H, or their equivalents, in combination with the tube and piston, substantially as and for the purpose herein described.

No. 2,282.—GEORGE COWING, Seneca Falls, N. Y.—*Cylinder Polisher.*—Patented April 14, 1863; reissued June 12, 1866.

Claim.—The expanding arms A A, in combination with the vulcanized scourers B B, substantially as described.

No. 2,283.—ALLEN BEACH, Boston, Mass., assignee of HENRY H. GILMORE.—*Pipe Tongs.*—Patented April 6, 1858; reissued June 12, 1866.

Claim.—First, constructing either member of a pair of pipe tongs with a slot or its equivalent, so that by the relative adjustment of the jaw and pivot the "grip" of the tongs can be accommodated to various sizes of pipes, substantially as herein described.

Second, the combination of an inclined plane or planes, or the equivalent thereof, with the slotted jaw, for the purpose described.

No. 2,284.—CHARLES T. GRILLEY, New Haven, Conn.—*Capping Wood Screws.*—Patented April 20, 1852; extended seven years; reissued June 12, 1866.

Claim.—The application to a nicked screw-head of a cap already nicked, and folding the same upon and around said screw-head by compression, substantially as described.

No. 2,285.—JUDSON SCHULTZ, Ellenville, N Y.—*Tanning.*—Patented April 3, 1866; reissued June 12, 1866.

Claim.—The treating of hides or skins substantially as herein described.

No. 2,286.—BENJAMIN F. STURTEVANT, Boston, Mass.—*Blanks for Shoe Pegs.*—Patented August 16, 1859; reissued June 12, 1866.

Claim.—As of my invention the new article of manufacture herein described, which is a peg blank, having essential characteristics substantially such as herein set forth.

No. 2,287.—WILLIAM B. RHOADS, South Dedham, Mass., assignee by mesne assignments, of NELSON B. WHITE.—*Clothes Wringer.*—Patented March 4, 1862; reissued June 12, 1866.

Claim.—First, the combination with the rolls of a wringing machine of bevel cog wheels for transmitting motion from one roll to the other, substantially as described.

Second, in a machine for wringing clothes, the bevel gears K and N, with i and m, and shaft b, operated by the groove r, arranged substantially as described, and driven by any power in combination with rollers of India-rubber or other suitable material, as set forth.

Third, in a wringing machine a "purchase" cog wheel, that is to say, a cog wheel of any form, mounted upon an independent axis, and employed to give motion to the gearing or other devices which rotate the rolls.

No. 2,288.—W. E. DOUBLEDAY, Brooklyn, and J. STEWART, New York, N. Y., assignees by mesne assignments of WILLIAM OSBORN.—*Machine for Pressing Bonnets, Bonnet Frames, &c.*—Patented August 19, 1856; reissued February 17, 1857; again reissued March 27, 1860; and again reissued June 19, 1866.

Claim.—A pair of dies fitted to press the whole of a bonnet frame, or similar article to be worn upon the head, at one operation, substantially as specified, whether said bonnet frame or similar article is formed of one or of several pieces, and irrespective of the particular size or shape.

No. 2,289.—THE RUSSELL AND ERWIN MANUFACTURING COMPANY, New Britain, Conn., assignee of NATHANIEL WATERMAN.—*Egg Pan and Cake Baker.*—Patented April 5, 1859; reissued June 19, 1866.

Claim.—As a new or improved article of manufacture, the baking pan or arrangement of cups and a handle at each end of the series, all connected together and cast or founded in one united piece of metal, with heat passages between the cups, and longitudinal, transverse or circuitous channels on the under side of the same, substantially as hereinbefore described and set forth.

No. 2,290.—EMANUEL ANDREWS, Williamsport, Penn.—*Machine for Grinding Saws.*—Patented December 16, 1856; reissued June 19, 1866.

Claim.—First, the combination in a grinding machine of the following instrumentalities, viz: the revolving grinder, roller bearing, and spring, substantially as set forth.

Second, the combination in a grinding machine of the following instrumentalities, viz: the revolving grinder, bearing for the article, and turning frame for the grinder, substantially as set forth.

Third, the combination in a grinding machine of the following instrumentalities, viz: the revolving grinder, traversing carriage for the same, bearing for the article, and mechanism to move the grinder along the surface of the article with varying speed, substantially as set forth.

Fourth, the combination in a grinding machine of the following instrumentalities, viz: the revolving grinder, bearing for the article, and pattern holder, of varying thickness at different parts of its length, substantially as set forth.

Fifth, the combination in a grinding machine of the following instrumentalities, viz: the revolving grinder, bearing for the article, and pattern holder, of varying thickness, both lengthwise and crosswise, substantially as set forth.

Sixth, the combination in a grinding machine of the following instrumentalities, viz: the revolving grinder, bearing for the article, and pattern holder, whose thickness at different parts of its length can be varied by adjustment, substantially as set forth.

Seventh, the combination in a grinding machine of the following instrumentalities, viz: the revolving grinder, bearing for the article, and the pattern holder, whose thickness at different parts of its length and breadth can be varied by adjustment, substantially as set forth.

Eighth, the combination in a grinding machine of the following instrumentalities, viz: the revolving grinder, bearing for the article, pattern holder, and spring to maintain a yielding pressure on the article during grinding, substantially as set forth.

Ninth, the combination in a grinding machine of the following instrumentalities, viz: the revolving grinder, bearing for the article, pattern holder, spring to maintain a yielding pressure on the article during grinding, and mechanism to move the said pattern holder past the grinder, substantially as set forth.

No. 2,291.—DANIEL E. PARIS, Troy, N. Y., assignee by mesne assignments of SAMUEL B. SPAULDING.—*Cooking Stove.*—Patented June 22, 1858; additional improvements May 17, 1859; reissued June 19, 1866.

Claim.—First, the extending of one or more of the flues of a square shaped or diving flue cooking stove, so as to conduct the heat and products of combustion under the bottom of a reservoir or water tank and against it before the same passes into the exit pipe, when the said reservoir is placed wholly or partly below the top of the stove, and in rear of the back flue or flues thereof, substantially as herein described and set forth.

Second, a return flue so constructed that the heat will be conducted under the bottom of a reservoir or water tank, and thence back to the exit flue or pipe, and when the whole or a part of the reservoir is placed below the surface of the stove, and in rear of the back flue or flues thereof, substantially as herein described and set forth.

Third, the damper F, so arranged and operated as to throw the heat and products of combustion under the reservoir E, and against it, or by a proper adjustment or change of position thereof to allow the same to pass directly into the exit flue 1, substantially as herein described and set forth.

Fourth, constructing the flues which pass under the oven of a cooking stove in such a manner that the heat and products of combustion will pass under the hearth in front of the stove, substantially in the manner and for the purposes herein described and set forth.

Fifth, the hot air chamber projecting out from and situated immediately in the rear end of the stove and back of the flue or flues connected therewith by means of an aperture or opening in and through the rear end and vertical plate of the cooking stove, so as to permit or allow the heat or heated air or escaping products of combustion to pass through the same into such chamber or its equivalent, and thereafter come in direct contact with the reservoir or water tank, arranged for the purposes substantially as aforesaid.

Sixth, an open seat or outward projecting chamber which shall receive and contain said reservoir or water tank, and formed on the back end of a cooking stove, so that such reservoir shall itself form the covering or a part of the outer casing of the rear flue or flues, and of the said hot air chamber combined just below the same, in the manner substantially as herein described and set forth.

Seventh, the arrangement of an aperture opening, or openings, constructed in and through the rear vertical end plate of the stove, so that the heat, hot air, or escaping products of combustion may by the means thereof come in positive contact with the bottom or other exposed parts or portion of said reservoir or water tank, and thereby heat or warm the water therein, substantially as herein described and set forth.

Eighth, the arrangement of a reservoir or water tank, with the whole or part thereof sustained or supported substantially as herein described and set forth, in combination with an aperture or opening or openings in and through the rear vertical end plate of said stove, for the purposes substantially as herein described and set forth.

Ninth, the arrangement of a hot air or warming or drying chamber or closet directly underneath the bottom of the cooking stove, in combination with the downward projecting bottom x of the horizontal flue or flues of the stove into such chamber or closet, in the manner and for the purpose substantially as herein described and set forth.

Tenth, the suspending and supporting of a cooking stove, without the use of the ordinary legs, upon the surrounding casing or box C, which forms ·the hot air or drying chamber or closet, between the seam and the floor, substantially as herein described and set forth.

No. 2,292.—ADALBERT FISCHER, New York, N. Y.—*Liquid Cooler.*—Patented May 15, 1866; reissued June 26, 1866.

Claim.—The movable heads E, with annular recesses a, in combination with one or more bolts F, annular cylinder A, and vessel B, all constructed and operating substantially as and for the purpose described.

No. 2,293.—WARREN GALE, Chicopee Falls, Mass.—*Straw Cutter.*—Patented March 7, 1854; reissued June 26, 1866.

Claim.—First, the fixed pivot F, on which the moving knife works, provided with a flanch for fastening to the cutter-box, and made adjustable thereon by means of slots and bolts, or their equivalents, subtantially as and for the purpose herein specified.

Second, the arrangement of the adjustable gauge-plate, G, in front of the fixed knife, in such a manner that it shall be raised above the said fixed knife in proportion to the increased distance at which it is adjusted away from the knife, to give a longer cut and *vice versa*, substantially as herein set forth.

No. 2,294.—WILLIAM K. MILLER, Canton, Ohio.—*Harvester.*—Patented July 2, 1861 reissued June 26, 1866.

Claim.—First, the pulleys at or near the lower end of the inner reel support, for the purpose specified.

Second, the single endless belt or chain passing from the driving-pulley g, on the main frame, to and over intermediate pulleys placed to one side of a vertical plane passing through said driving-pulley, and thence to and over the pulley on the inner end of the reel shaft, substantially as and for the purpose described.

No. 2,295.—WILLIAM K. MILLER, Canton, Ohio.—*Harvester.*—Patented July 2, 1861; reissued June 26, 1866.

Claim.—The bent lever s, pivoted at the heel of the shoe, and extending to a point between the driving wheels, and thence connected to a lever near the driver's seat or stand, for the purpose of raising the outer end of the finger beam, substantially as described.

No. 2,296.—WILLIAM K. MILLER, Canton, Ohio.—*Harvester.*—Patented July 2, 1861; reissued June 26, 1866.

Claim.—The brace on the grain side of the machine extending from the front end of the frame to the axle of the driving wheel as a support for the finger beam and the reel post, substantially as described.

No. 2,297.—E. Y. ROBBINS, Cincinnati, Ohio.—*Warming and Ventilating Buildings.*—Patented May 6, 1862; reissued June 26, 1866.

Claim.—First, the arrangement of the hot-air chamber or reservoir of heat for warming the floor and lower part of the walls, in connection with the arrangement for the introduction at the bottom of the room of moderately warmed fresh air, which has not been in contact with the hot metallic surface either of hot-water pipes or steam pipes, or of a stove or furnace, or any other highly heated surface, substantially as above set forth.

Second, the fresh-air chamber F, Fig. 1, under the hot-air chamber C, Fig. 1, and under the reservoir of heat formed by the pipes, mortar, &c., of the radiator W, Fig. 6, all substantially as set forth.

Third, in case of warming the upper rooms by the waste heat of the fire in the lower story, the arrangement of an inner smoke-flue within the brick flue or chimney E, Fig. 3, and the diaphragm F, Fig. 5, for turning the current of hot air rising between this inner smoke-flue and the sides of the chimney inwards under the floor of the upper room for warming it, or any equivalent device, substantially as set forth.

Fourth, in using hot-air pipes, for warming cars or rooms, the making of said pipes in their different parts of different materials and of different shapes, so that their conducting and radiating power shall increase as the distance from the furnace or source of heat in creases and as the temperature of the air within them decreases, so that they shall distribute the heat as nearly uniform as possible throughout their entire length, substantially as above set forth.

Fifth, the horizontal radiator W, Fig. 6, formed of tiles, mortar, cement, or other similar material, (which has acquired, or will acquire, a sufficient degree of hardness,) placed in and forming a part of the floor, in combination with the arrangement for heating it by hot-water pipes or steam pipes, placed under the tiles or other material of which said radiator may be formed.

No. 2,298.—THE GOODYEAR METALLIC RUBBER SHOE COMPANY, Naugatuck, Conn., assignees of THOMAS C. WALES.—*Water-proof Gaiter Shoe and Boot.*—Patented February 2, 1858; reissued June 26, 1866.

Claim.—A new or improved manufacture, or water-proof vulcanized rubber and cloth gaiter shoe, made in manner and with its external layer of cloth and its lining of cloth arranged together and with respect to the remainder or rubber parts or foxing, substantially as specified.

No. 2,299.—DAVID W. CANFIELD, New York, N. Y.—*Combined Shoulder Brace and Suspenders.*—Patented December 16, 1862; reissued July 10, 1866.

Claim.—A combined shoulder-brace and suspenders, in which the ends of the strap A, united in the rear, at which point of junction the back-straps are secured, substantially as described and for the purpose specified.

No. 2,300.—P. TENNEY GATES, Plattsburgh, New York.—*Wristband.*—Patented March 13, 1866; reissued July 10, 1866.

Claim.—First, an elastic wristlet or adjuster, in combination with a pendent cuff of any description connected to the wristlet by elastic cord or otherwise.

Second, a wristband or cuff having both ends finished and button-holed, and thus adapted for reversing, when unattached to any other garment, and arranged substantially as described and for the purpose set forth.

Third, a cuff pendent from an attachment of the arm, substantially as and for the purpose set forth.

No. 2,301.—JOHN C. GOULD, Boonton, N. J.—*Nail Plate Feeding Machine.*—Patented May 12, 1857; reissued July 10, 1866.

Claim.—First, the feed or nipper-rod Z, and the feeding device S S', in combination with the rod T and nose-piece U, as and for the purposes specified.

Second, fixing the bar F² on which the rack F³ is formed upon a shaft Y, journaled within and supported by a vibrating box or bearing L', substantially as and for the purpose set forth.

Third, supporting the nose-piece and feeding device upon a hinged or pivoted shaft M, to adapt them to be turned back to render the cutters accessible as described.

Fourth, connecting the nose-piece with the driving-shaft by means of the two rods V V', adapted to be disconnected as described to admit of the turning back of the nose-piece.

Fifth, the combination of the rock shaft G, vibrating box L', rock shaft Y, bar F³ cogged rim d, and nose-piece U, substantially as and for the purpose specified.

No. 2,302.—GEORGE B. HARTSON and E. J. WOOLSEY, New York, N. Y.—*Centrifugal Machine.*—Patented February 13, 1866; reissued July 10, 1866.

Claim.—Supporting and driving the centrifugal separator from below, substantially as and for the purpose herein shown and described.

Also, the method of constructing the centrifugal separator with a hollow hub in the centre of the lower plate thereof as described, the said hub being provided with a pulley for the driving-belt below the bottom plate, and fitted to run on and combined with the stud of the baseplate, in the manner and for the purpose specified.

Also, the said centrifugal separator in combination with the cap or cover provided with holes or channels covered with funnel-shaped hoods, as described, to force currents of air in and through the said apparatus to aid in effecting the separation as described.

No. 2,303.—WILLIAM KIEFER, New York, N. Y., assignee of E. BURGY and L. GUILLE-MIN.—*Machine for Dressing and Finishing Threads, &c.*—Patented February 7, 1865; reissued July 10, 1866.

Claim.—The arrangement of the gum or size-bath e, wiper g, and flat heater or heaters h i, and a winding frame, constructed and operating substantially in the manner and for the purpose described.

Also, in combination with the gumming device and beater, the winding frame, constructed and operating substantially as and for the purpose described.

No. 2,3C4.—JAMES A. LAWSON, Troy, N. Y.—*Ash Pan Drawer for Stoves.*—Patented June 16, 1863; antedated April 7, 1863; reissued July 10, 1866.

Claim.—First, the ash-pan drawer A, for cooking or other stoves, or for furnaces, having the bail B and the handle C thereto attached, and arranged within an enclosed ash-pit or chamber, and in combination with the fire grate or fire chamber thereof, in the manner and for the purposes substantially as herein described and set forth.

Second, the arrangement of an ash pan drawer, having the bail B combined therewith, within an inclosed ash pit or ash chamber of a stove or furnace, and in combination with the fire grate or fire chamber thereof, in such manner as to receive the ashes and other material falling therefrom and thereafter be removed in the manner substantially as herein described and set forth.

Third, the employment of an ash pan for cooking or other stoves, or for furnaces, constructed in the manner substantially as herein shown and described, in combination with a surrounding or enclosed ash chamber or ash pit, so that no dust or other matter shall escape from or by reason of the falling of ashes or other material from the burning of the fuel upon the fire grate into the room where such stove or furnace is used, in the manner substantially as herein described and set forth.

Fourth, the bail B, in combination with the ash-pan drawer A and with the stops *a a*, and handle *c*, in the manner substantially as and for the purpose herein described and set forth.

No. 2,305.—JAMES SPEAR, Philadelphia, Penn.—*Ash-sifting Pan for Stoves.*—Patented April 15, 1862; reissued July 10, 1866.

Claim.—The application of a sifting pan capable of being vibrated to the hearth of a stove or range, substantially in the manner and for the purpose herein described.

No. 2,306.—THE UNION PAPER COLLAR COMPANY, assignee by mesne assignments of WALTER HUNT.—*Shirt Collar.*—Patented July 25, 1854; reissued to WILLIAM E. LOCKWOOD, April 4, 1865; again reissued to THE UNION PAPER COLLAR COMPANY, July 10, 1866.

Claim.—A shirt collar, or wristband, or bosom, made of fabric composed of paper and muslin, or an equivalent fabric, having a surface covered with enamel, substantially as and for the purpose above specified.

No. 2,307.—THE UNION PAPER COLLAR COMPANY, assignees by mesne assignments of WALTER HUNT.—*Shirt Collar.*—Patented July 25, 1854; reissued to WILLIAM E. LOCKWOOD, April 4, 1865; again reissued to THE UNION PAPER COLLAR COMPANY, July 10, 1866.

Claim.—A shirt collar, bosom, or wristband, made of a fabric composed of paper and muslin, or an equivalent fabric, having a smooth white surface coated with transparent varnish, for the purpose specified.

No. 2,308.—THE AMERICAN METAL COMPANY, West Meriden, Conn., assignees by mesne assignments of ELLIOT SAVAGE.—*Plating, Tempering, and Hardening Iron and Steel.*—Patented December 26, 1865; reissued July 10, 1866.

Claim.—First, heating the metal by immersing it in a bath of melted cyanide of potassium, substantially as described.

Second, cooling the metal by immersing it in a cooling liquid, substantially as described, after it has been heated in a bath of melted cyanide of potassium, substantially as described.

No. 2,309.—JAMES A. WOODBURY, Boston, Mass., assignee of ANDREW A. EVANS.—*Paper Shirt Collar.*—Patented May 26, 1863; reissued July 10, 1866.

Claim.—As a new article of manufacture, a collar made of long fibre paper, substantially such is above described.

No. 2,310.—JAMES A. WOODBURY, Boston, Mass., assignee of ANDREW A. EVANS.—*Paper Shirt Collar.*—Patented May 26, 1863; reissued July 10, 1866.

Claim.—As a new article of manufacture a collar made of long fibre paper, substantially such as above described, and coated or varnished, as and for the purposes above set forth.

No. 2,311.—DAVID A. WOODWARD, Baltimore, Md.—*Solar Camera.*—Patented February 24, 1857; reissued July 10, 1866.

Claim.—First, adapting to the camera obscura a lens, or lenses, and reflector, in rear of the object glass, in such manner that it is made to answer the two-fold purpose of a camera obscura and a camera lucida, substantially as and for the purposes specified.

Second, the arrangement and combination of the condensing lens H, or lenses D' and H, negative slide or holder N, and achromatic lens or lenses E, made adjustable with regard to each other for condensing the sun's rays upon and through the negative, and focusing them upon prepared paper, canvas, or other suitable material for photographic purposes, substantially as described.

No. 2,312 —JOHN C. GARNER, Ashland, Penn.—*Whiffletree Attachment.*—Patented December 26, 1865; reissued July 10, 1866.

Claim.—The plate C, provided with the lips, as shown, and secured to the bar A by the bolt D, in combination with the tube E fitted in the whiffletree, and the plate F, at the front side of the latter, the bolt D passing through the tube E, and all arranged to operate in the manner substantially as and for the purpose herein set forth.

No. 2,313.—RALPH ALLEN, Rock Stream, N. Y., assignee of ALBERT J. ALLEN.—*Steam Gauge.*—Patented October 4, 1859; reissued July 17, 1866.

Claim.—First, the improved pressure gauge, herein described, consisting of the following parts: 1st, the within-described elastic capsule, having faces formed of sheet metal, neither convex nor concave, to be connected with the boiler to receive and be expanded by the pressure of the steam; and 2d, multiplying levers and index, the whole being so arranged as to render available the elasticity of two opposite faces of said capsule at the same time in operating the index through the medium of delicately operating levers, substantially as and for the purpose herein set forth.

Second, mounting such capsule in combination with the other parts, so that the steam is admitted at the centre of one of the said faces or ends, substantially as and for the purpose herein specified.

Third, the employment of the capsule G, of peculiar construction, having the steam admitted at one side and through the centre of that side, and using the flexibility of both sides (such capsule being made of a permanently elastic metal and not injuriously oxidized by steam or water, preferring for that purpose the metal used in making melodeon reeds) in combination with fulcrum block F, lever H, spring P, rod O, rod I, swivel block J, radius bar L, and segment K, having tail piece k, pinion N, index pointer E, dial plate D, and friction pressure spring M, substantially as shown and described.

Fourth, radius bar L, in combination with rod I, swivel block J, segment K having tail piece k, pinion N, index pointer E, and dial plate D, having increasing divisions on its face, substantially as shown and described.

No. 2,314.—T. F. ALLYN, Canandaigua, N. Y.—*Car Spring.*—Patented June 30, 1865; antedated March 28, 1865; reissued July 17, 1866.

Claim.—The construction of a metallic car spring of rectangular, rhombic, or other equivalent-shaped plates, which plates are curved diagonally, and have central and end or corner bearings and fastenings, substantially as and for the purpose set forth.

No. 2,315.—LEVI DEDERICK, Albany, N. Y.—*Hay Press.*—Patented July 14, 1863; reissued July 17, 1866.

Claim.—In the above-described press, the use of two toggle levers placed at opposite sides of the press box, in combination with links having their fixed ends located at or near the bottom of the frame, so the upper ends of the levers will move up on the outside of the press box in pressing a bale.

Also, the use of toggle levers in combination with a suspended follower.

Also, suspending the follower to the upper ends of the levers through the perpendicular slots or openings in the sides of the press box.

Also, the rigid bracket connections, or their equivalents, in combination with the ends of the follower timbers projecting through the slots in the sides of the press, in the manner and for the purpose substantially as described.

No. 2,316.—ELIJAH FREEMAN PRENTISS, Philadelphia, Penn., WILLIAM D. PHILBRICK and WILLIAM J. PARSONS, Boston, Mass, assignees of ROBERT ADAM ROBERTSON.—*Process for Distilling Rock Oil and other Hydrocarbons.*—Patented March 8, 1864; antedated July 31, 1862; reissued July 17, 1866.

Claim.—First, drawing off the residuum during the process of distillation from a steam-heated still, substantially as described.

Second, feeding crude oil into the still through one or more condensers, so that the crude oil serves as a surface condensing bath to the oil vapor coming from the still, and at the same time the crude oil itself undergoes a separate partial distillation before reaching the main still.

Third, simultaneous fractional surface condensation by means of crude oil on its way to the still.

Fourth, the employment of superheated steam for maintaining a series of baths at graduated temperatures in carrying on the process of fractional distillation, substantially as described.

No. 2,317.—ELIJAH FREEMAN PRENTISS, Philadelphia, Penn., and WILLIAM J. PHILBRICK and WILLIAM J. PARSONS, Boston, Mass., assignees of ROBERT ADAM ROBERTSON.—*Apparatus for Distilling Rock Oil and other Hydrocarbons.*—Patented March 8, 1864; antedated July 31, 1862; reissued July 17, 1866.

Claim.—First, the combination of the still 'A, the injecting worm aIV aV aVI $aVII$, and the central tube G G' G''.

Second, roughening the surface of the injecting worm or tube aV aVI $aVII$, to render the ebullition regular and quiet.

Third, the combination of the still A with the series of columns, two or more, each column being set and maintained at the temperature necessary to separate the product condensible at such temperature, whereby, at one continuous operation, the crude oil is separated into the various products due to the condensation at the different temperatures fixed upon.

Fourth, the arrangement of the vapor tubes and oil spaces in columns B B''', or C C''', whereby the crude oil, on its way to supply the still A, is made to act as a condensing bath to the vapors in these columns, coming from the still A.

Fifth, the arrangement of the columns B B''' and C C''', in combination with the still A and the movable exit tube g'', whereby the operation of the still is rendered continuous.

Sixth, the air regulator, or its equivalent, for regulating the temperature of the respective columns, or either of them, in combination with the pipes of supply of the heating and cooling media.

Seventh, the water legs X and the floats Z for regulating the escape of water from the columns.

Eighth, the auxiliary heads V V' for enabling the oil bath in each column to act as a still.

Ninth, the warming of the bottoms of the chambers, which are the bases of the columns, by means of steam chambers, arranged and operating as shown.

Tenth, the warming of the bottom of the column on which the still A is supported, substantially as above described.

Eleventh, the arrangement of the column B B''', so as to act immediately as a still to C C''', and the column C C''', so as to act as a still to DD''', and so on, if crude oil is fed into the main still through more than two columns.

Twelfth, the arrangement of the oil spaces in B B''' and C C''' in combination with the still A, whereby the operation of the still is made continuous, and the fresh oil is introduced at a high temperature into the still A.

No. 2,318.—DAVID WOLF, Lebanon, Penn.—*Reaping Machine.*—Patented January 31, 1865; reissued July 17, 1866.

Claim.—A platform for reapers, constructed in such a manner that it may, either automatically or at the will of the driver or attendant, be made to tilt on a sliding axis arranged parallel with the finger bar, or nearly so, and by said movement throw or discharge the cut grain from it, substantially as set forth.

Also, the construction of the platform of two or more parts, connected by a joint or joints, and arranged to operate in the manner substantially as described.

Also, the arrangement of the crank G and springs c d, substantially as described for operating the platform.

No. 2,319.—ESEK BUSSEY, Troy, N. Y.—*Cooking Stove.*—Patented December 5, 1865; reissued July 24, 1866.

Claim.—First, the outward continuation or extension of the top plate E of a cooking stove over and upon or near to the upper part or top of the boiler or reservoir A, and containing therein an opening or reservoir aperture E' for receiving into said boiler A, in the manner substantially as herein described and set forth.

Second, the supporting of the boiler or reservoir A upon the vertical end plate C', in combination with the top plate E of a cooking stove, projecting or continuing outward with an opening or aperture E' therein, or any equivalent therefor, in the manner and for the purposes substantially as herein described and set forth.

Third, the employment of the boiler or reservoir A, having its upper or open top or part in combination with opening or aperture E' in the top projecting plate E of a cooking stove, in the manner and for the purposes substantially as herein described and set forth.

Fourth, the arrangement in a cooking stove of a culinary boiler and an exit passage for the gases of combustion, both at one end of the stove, and so that said boiler forms a part of the lateral casing on the outer side of a fire flue or fire flues in the end of the stove, below the said exit passage, in the manner substantially as and for the purposes herein described and set forth.

Fifth, the employment and arrangement of the boiler or reservoir A, or any equivalent thereof, within and upon the rear end of a cooking stove, and wholly or partly below the top plate thereof, so that one side of such boiler shall form and complete the lateral casing of the rear end vertical flue or flues below the top plate thereof and the bottom of said boiler, in the manner and for the purpose substantially as herein described and set forth.

Sixth, the construction of the rear end and vertical flue or flues of a cooking stove by means of the boiler or reservoir A and the lower vertical end or boiler supporting plate C', so that the hot air or escaping products of combustion shall come into direct contact with that part or portion of such boiler next adjoining such flue or flues, so as to warm or hea the water or other material in said boiler, in the manner substantially as herein described and set forth.

Seventh, the removing of the lateral or vertical outside casing of a cooking stove, or some part or portion of the same, so that the hot air or heated products of combustion may or shall come into contact with the boiler or reservoir, or some part or portion thereof next adjoining thereto or in combination therewith, in the manner and for the purposes substantially as herein described and set forth.

No. 2,320.—JOHN G. FORD, Philadelphia, Penn., assignee by mesne assignments of JULIUS A. JILLSON and HENRY WHINFIELD.—*Apparatus for Washing and Bleaching Fibrous and Textile Substances.*—Patented October 9, 1855; reissued July 24, 1866.

Claim.—First, the process of washing, cleansing, or extracting gum, dirt, or other similar matter from fibrous and textile substances or materials by inserting them in a closed vessel or receiver and forcing the cleansing or extracting fluids to circulate through the materials by the action of a pump.

Second, the rinsing of the materials by forcing fresh cleansing liquids into and through the fibrous and textile substances and materials, and out of the closed washing or extracting chamber by means of a pump.

Third, the forcing of a bleaching solution to circulate through the mass of fibrous and textile substances and materials contained within a closed receiver or extracting chamber by means of a pump.

Fourth, the combination of a closed receiver or extracting chamber or vessel, with a pump for causing a direct circulation through it, as described.

Fifth, the employment in a closed vessel or receiver of an upper strainer or perforated diaphragm for causing a uniform distribution of the liquid upon the matter treated, in combination with a lower perforated diaphragm or strainer for permitting the circulation of the extracting, cleansing or bleaching liquid.

Sixth, the employment of a closed vessel, for cleansing or extracting, with a fire below, and having a lower perforated diaphragm whereby the material is above the bottom of the boiler and prevented from being acted upon by the fire.

Seventh, forming within a closed receiver or extractor a chamber below the lower perforated diaphragm for cleansing or extracting liquid.

Eighth, the combination of the closed cleansing or extracting vessel, the heater and the pump for forcing the heated liquor to circulate through the mass.

Ninth, the combination of the closed receiver or extracting vessel, the lower perforated diaphragm or strainer, the heater and the pump.

Tenth, the combination of the closed receiver or extracting vessel, the upper and lower perforated diaphragm or strainer, and the pump *f.*

No. 2,321.—JESSE W. HATCH and HENRY CHURCHILL, Rochester, N. Y.—*Machine for Cutting Boot and Shoe Soles.*—Patented January 2, 1855; reissued July 24, 1866.

Claim.—The reciprocating cutter shaft A, having the endless edged knife or die C, attached thereto, when the same is made to perform half a revolution between successive cutting strokes, by means of the segment gear F', or other equivalent means for that purpose, operating substantially as described.

Also, the said reciprocating cutter shaft, when the same is used in connection with the cutting block M, and guide bar J, or their equivalents.

Also, the said cutter shaft A, guide bar J, cutting block M, and discharging plate T, or their equivalents, combined and operating together, substantially as described.

No. 2,322.—JOHN W. LANE, Newton, N. J.—*Heating Stove.*—Patented June 20, 1865; reissued July 24, 1866.

Claim.—First, the fuel chamber C, having its front plate W extending downward, leaving the space O, through which the circuit draught enters the front chambers D' D', substantially as described, for the purpose specified.

Second, the fuel chamber C, having its front plate W extending downward, leaving the space O, and having its back plate G, resting directly upon the bottom plate of the stove and provided with the grate F, substantially as described and for the purpose specified.

No. 2,323.—GEORGE F. CLEMONS, Springfield, Mass.—*Cloth Guide for Sewing Machines.*—Patented June 27, 1865; reissued July 31, 1866.

Claim.—First, in a sewing machine a cloth guide adapted to give adjustably variable pressure upon the material being sewed, for the purposes specified.

Second, in combination with a cloth gauge upon a sewing machine, a spring pressure plate presenting a smooth bearing surface upon the material being sewed, and having means for an adjustment of the pressure for guiding the same toward the gauge face, and for distribution of its pressure upon the material, substantially as and for the purposes set forth.

Third, regulating the pressure of the guide plate by means of the auxiliary plate H, arranged and operating substantially as and for the purposes specified.

Fourth, the combination of the spring plate E and pressure plate H with the cloth gauge C, in the manner and for the purposes shown and described.

Fifth, the combination of the spring plate E and pressure plate H, with the gauge plate C', pivoted to the shank G, substantially as and for the purposes set forth and shown.

Sixth, the combination of the pivoted gauge plate C', and gauge shank G, arranged and operating substantially as and for the purposes specified.

No. 2,324.—GEORGE DUNHAM, Unionville, Conn.—*Machine for Making Nuts.*—Patented June 27, 1865; reissued July 31, 1866.

Claim.—First, constructing and arranging the sizing bar O, so as to act in the threefold capacity of sizing, holding, or gauging, actuated by proper mechanism, substantially as and for the purpose described.

Second, the combination of the conical-shaped recess Q, with the spring or yielding table P, substantially as and for the purpose described.

Third, the combination of the shearing punch L', with the conical recess Q, substantially as and for the purpose described.

Fourth, the employment of the lifting holders S S', substantially in the manner and for the purpose described.

Fifth, the employment of the hammers K¹ and K² or their equivalents, in combination with the holders S S, substantially as and for the purpose described.

Sixth, the clearer bar N, for holding, clearing, and carrying the nut from one point to another, substantially as described.

Seventh, the screw upon the upper end of the punch K, in combination with the threaded socket *i*, substantially as described

Eighth, a machine constructed substantially as described, for cutting the blank, forming the basil, and hammering or finishing the edges of a nut before the punching of the hole, substantially as described.

No. 3,325.—JOHN LEMMAN, Cincinnati, Ohio.—*Hoisting Apparatus.*—Patented July 17, 1860; reissued July 31, 1866.

Claim.—First, the arrangement of the pulleys K U R, belt L, rectangular frame B C D, pulley M, shaft N, and operating in the manner and for the purpose herein specified.

Second, the system of gearing, composed of gear wheels G G, worm wheels H H, worms I I, and racks F F, all arranged and operating substantially as and for the purpose described.

Third, in hoisting machines the use of two worms on one shaft, the one right hand and the other left, each gearing into an appropriate worm wheel, and each furnishing a step for the other, substantially as shown and described.

Fourth, in power hoisting machines the attaching to and carrying the climbing machinery with the platform, the power for operating the same not being located on the platform, in the manner herein described.

No. 2,326.—ELIZABETH BELLINGER, Mohawk, N. Y.—*Composition for Kindling Fires.*—Patented December 13, 1859; reissued August 7, 1866.

Claim.—A self-lighting fire kindler, made substantially as herein described.

No. 2,327.—ALBERT D. CROMBIE, LUCY COLBORN, Baltimore, Md., and JOHN M. D. GREENE, Funkville, Penn., assignees by mesne assignments of WELLS L. COLBORN.— *Cut-off Gear for Steam Engines.*—Patented March, 1, 1864; reissued August 7, 1866.

Claim.—The combination of the sliding block I, the fulcrum *g*, and the rock lever G, or their equivalents.

No. 2,328.—ALBERT D. CROMBIE, LUCY COLBORN, Baltimore, Md., and JOHN M. D. GREENE, FUNKVILLE, Penn., assignees by mesne assignments of WELLS L. COLBORN.— *Cut-off Valve Gear for Steam Engines.*—Patented March 1, 1864; reissued August 7, 1866.

Claim.—The sliding block I, the fulcrum *g*, and the rock lever G, or their equivalent devices, in combination with a steam valve and governor.

No. 2,329.—PURCHES MILES, New York, N. Y., assignee of JOHN J. WEEKS.—*Sausage Stuffer.*—Patented September 19, 1854; reissued August 7, 1866.

Claim.—First, the revolving cylinder and conveyer or conveyers, eccentric to the case or cylinder, in combination with a hopper for containing the meat, and with a tube through which the sausage meat is delivered and stuffed into a gut or intestine, substantially as set forth.

Second, the shaping and protecting tube K, in combination with the delivery tube I, substantially as and for the purposes specified.

No. 2,330.—GEORGE A. PRINCE, CHARLES E. BACON, and CALVIN F. S. THOMAS, Buffalo, N. Y., assignees of JOSIAH A. ROLLINS.—*Melodeon.*—Patented June 3, 1856; reissued August 7, 1866.

Claim.—First, placing and arranging within a melodeon, or other like reed musical instrument, two or more sets of valves, in connection with two, three, or more sets, or parts of sets, of reeds, so that one set of valves will act upon and open another set of valves, by the action of one set of keys and one set of push-down pins, substantially as described.

Second, supporting the front set of valves by a strip of wood *k* or other equivalent, for the purpose and substantially as described.

No. 2,231.—D. WHITTEMORE, North Bridgewater, Mass., assignee of WILLIAM R. LAND-FEAR.—*Pegging Jack.*—Patented November 28, 1865; reissued August 7, 1866.

Claim.—A pegging jack as made with its last supporter sustained by and so as to be capable of being moved or turned laterally on either of two separate centres *d e*, substantially as and for the purpose described.

Also, a pegging jack as made with its last supporter sustained by and so as to be capable of being moved or turned laterally on either of two separate centres *d e*, and also of being tipped or turned longitudinally on a third centre *a*, the whole being substantially as described.

Also, the last supporter A, as made with the two arms *b b*, substantially as represented.

Also, the arrangement of the pins or screws *d e f*, and the slots *g h i*, with the plates B C, arranged and combined with the last supporter A, substantially as specified.

No. 2,332.—TOBIAS J. KINDLEBERGER, Eaton, Ohio.—*Cider Mill.*—Patented May 29, 1855; reissued August 14, 1866.

Claim.—First, a mill for grinding fruit, when constructed with three rollers E G and H, the former of which is placed above the two latter, and so arranged, in relation to the sides of the hopper, that the fruit shall be broken by the longitudinal projections on the roller E, and crushed against the breast piece, and then delivered upon the two crushing or grinding rollers between which the pomace passes, substantially as set forth.

Second, in a cider mill, having the three parallel rollers arranged as shown, in combination with the gearing arranged as described, and each of the grinding rollers revolving, by means of said gearing, at different velocities, as set forth.

Third, an adjustable breast piece I, or its equivalent, by which the space between it and the upper crushing roller can be varied as desired.

Fourth, the scraper *h*, arranged to operate in combination with the roller E, substantially as set forth.

Fifth, constructing the case B, with concave or segmental metallic end pieces B', as and for the purpose set forth.

Sixth, the use of a slatted grate, or its equivalent, to form the bottom of the tub or curb, for the purpose of permitting the juice to drain through it, as set forth.

No. 2,333.—GEORGE HAND SMITH, Rochester, N. Y.—*Process of Making Steel direct from the Ore.*—Patented July 18, 1854; reissued August 14, 1866.

Claim.—The process, substantially as herein described, for converting iron ores directly into steel by subjecting the ore, in the comminuted state, in connection with carbon, and with or without other flux, in a close oven, retort, or equivalent vessel, to a high degree of heat, and when converted, treating it in a reheating furnace to weld and ball the particles, and then hammering, rolling, or squeezing the balls, to express the impurities and complete the welding and compact the mass, as set forth.

Also, in the process of conversion, charging the comminuted ore and charcoal, or other carbonaceous substance, in the cementing oven, or other equivalent vessel, in alternate layers, substantially as and for the purpose specified.

No. 2,334.—GEORGE HAND SMITH, Rochester, N. Y.—*Process of Making Steel direct from the Ore.*—Patented July 18, 1854; reissued August 14, 1866.

Claim.—The combination of the process of deoxidizing iron ore, and carbonizing the metallic particles, substantially such as herein described, with the process of melting in crucibles, substantially as and for the purpose described.

No. 2,335.—ELMER TOWNSEND, Boston, Mass., assignee by mesne assignments of JEREMIAH KEITH.—*Machine for Punching and Eyeleting Shoes, &c.*—Patented December 16, 1862, reissued August 14, 1866.

Claim.—First, the combination, with eyelet setting or clinching tools, of a magazine in which the eyelets are loosely contained, but from which they emerge in upright position for the action of the setting tools.

Second, the combination with the eyelet magazine and eyelet clinching tools of an interposed chute or conductor into which the upright eyelets emerge from the magazine, and by which each in succession is conveyed to position to be seized by the pin of one of the setting tools.

Third, a construction by which the eyelets in the magazine are agitated to cause them to assume positions in which they may emerge from the magazine.

Fourth, so constructing the lower end of the chute that each lowermost eyelet is detained in position until removed by the eyelet pin.

Fifth, combining with an automatic eyelet presenting and setting mechanism an automatic punching mechanism, operating in the same vertical line of operation with the clinching mechanism.

Sixth, combining with the mechanism, which brings each eyelet into position before the clinching mechanism, a yielding pin, projecting from the anvil set, said pin receiving the eyelet and holding it in position for the descent of the upsetting tool, and yielding to the pressure of said upsetting tool, or the pin projecting therefrom.

No. 2,336.—SAMUEL JACOB WALLACE, Carthage, Ill.—*Grain Binder.*—Patented April 12, 1864; reissued August 14, 1866

Claim.—First, a rack *c*, in combination with arm D, or its equivalent, for giving motion to the twister, substantially as described.

Second, the slotted wire holder on binding arm D, formed of bent plates *b b*, substantially as described.

Third, the spring fingers *e e*, and spring *e'*, for carrying the strand to the twister, and releasing the strand, substantially as described.

Fourth, the spring fingers *e e* and spring *e'*, in combination with the slotted twister F, operating substantially as described.

Fifth, the binding arm D, provided with bent plates *b b*, for holding and afterward catching the wire, in combination with the spring fingers *e e*, for introducing the wire into the twister, substantially as described.

Sixth, the binding arm D, provided with bent plates *b b*, for holding and afterward catching the wire, in combination with the twister F, and spring fingers *e e*, for introducing the wire into the twister, substantially as described.

Seventh, the cutter *d*, attached to the binding arm D, and operating in combination with the twister, substantially as described.

Eighth, effecting the several operations of carrying the strand around the sheaf, drawing up the slack of strand, forming the fastening, and severing the sheaf from the machine in the manner described, by the action of the lever D', moved backward and forward, substantially as described.

Ninth, the binding arm D, in combination with the levers D¹ D², the two latter being pivoted to the arm D, and to the draught frame of the harvester, on the grain side of the drive wheel, substantially as described.

Tenth, the levers D¹ D², binding arm D, and bottom plate E, operating as described, and arranged independently of the grain platform, and so that the parts can be operated by the driver, substantially as set forth.

Eleventh, the reel H, in combination with the ratchet *k* and levers J and D², the whole operating substantially in the manner and for the purpose set forth.

Twelfth, the binding arm D, in combination with the levers D¹ D², reel H, plate E, spring fingers *e e*, slotted twister F, and rack *c*, substantially as described.

No. 2,337.—WILLIAM WESTLAKE, JAMES E. CROSS, and JAMES F. DANE, Chicago, Ill., assignees of WILLIAM WESTLAKE.—*Lantern.*—Patented July 18, 1865; reissued August 14, 1866.

Claim.—First, constructing a lamp pot of a lantern with edge or sides extending above the top of the pot and forming a flange *g*, the side and flange forming one piece, so that the globe of the lantern will rest upon this flange, substantially in the manner described.

Second, the oil pot *e*, having its sides *g* extended upward and perforated for air passages, and provided with a flange *d*, so as to form the base of a lantern in one piece, and combined with a detachable globe-guard *a*, as set forth.

Third, the hole *h* with the sliding door *i*, in combination with the recess *j* of the globe, for lighting the lamp, as herein described.

No. 2,338.—RANSOM C. WRIGHT, Meadville, Penn.—*Railway Journal Box.*—Patented November 14, 1865; reissued August 14, 1866.

Claim.—First, the dovetailed projecting flanges D D' and dovetailed slides E F, (the slide being composed of two pieces constructed and united as described,) in combination with the rod *g* for securing the same, the parts being arranged for use, substantially in the manner and for the purpose set forth.

Second, the bearing B with its projection *h*, and chambered recess *b* with or without the bar *m*, substantially as and for the purpose set forth.

No. 2,339.—JAMES F. MONROE, Fitchburg, Mass., assignee by mesne assignments of JOHN D. BROWNE.—*Machine for Paring Apples.*—Patented May 6, 1856; reissued August 21, 1866.

Claim.—The combination with the apple-revolving mechanism, the knife-moving and controlling mechanism, and the frame of a clamp B C for temporarily attaching the apple-paring machine to a table or support, substantially as and for the purpose herein specified.

No. 2,340.—JOHN G. PUGSLEY, New York, N. Y.—*Car Spring.*—Patented August 1, 1863; reissued August 21, 1866.

Claim.—First, the use or employment of a cylindrical spring case, having a central flange that sustains the springs on either side and toward which they yield, as specified.

Second, a series of perforated spring disks or annular metallic plates, in combination with a central and flanged distributor and with concave rings or bearings against which the spring disks rest, substantially as and for the purposes specified.

No. 2,341.—REUBEN SHALER, Madison, Conn.—*Balance.*—Patented November 28, 1865; reissued August 21, 1866.

Claim.—The combination of the supports E and F, the parallel bars G G, arranged substantially in the manner described, with a spring S, or its equivalent, so as to operate substantially in the manner and for the purpose specified.

No. 2,342.—HENRY L. CASE, Jersey City. N. J., administrator of the estate of RICHARD C. ROBBINS, deceased, HENRY L. CASE, and JESSE M. KEEN, same place, and JOHN W. MASON, Brooklyn, N. Y., assignees of RICHARD C. ROBBINS.—*Gas Pipe Joint.*—Patented January 13, 1863; reissued August 28, 1866.

Claim.—A joint for gas and other similar pipes, constructed in the manner herein above described, that is to say, the socket end of a length or section of pipe, having a groove formed in it capable of retaining the packing in three directions, as set forth, a packing ring of soft metal being cast into the said groove, and the plug end or entering end of the adjoining length or section of pipe being made conical and driven into the packing, substantially as hereinbefore described, and said joint being composed of three parts only, as set forth.

No. 2,343.—WILLIAM D. HOOKER, San Francisco, Cal.—*Pump.*—Patented August 15, 1865; reissued August 28, 1866.

Claim.—A vertical and inclined partition h, in combination with the suction valve seats c' c' and discharge valve seats d¹ and d² and the valves a¹ a² b¹ b², the whole arranged as described and for the purpose specified.

No. 2,344.—MARTIN W. POND and HENRY E. MUSSEY, Elyria, Ohio, assignors to MARTIN W. POND.—*Metallic Shield for Breast Straps.*—Patented September 6, 1864; reissued August 28, 1866.

Claim.—First, the seat f and base of horns d d, the same being constructed and operating in the manner and for the purposes substantially as specified.

Second, the curved metallic slide having curved metallic horns, as described, in combination with the projections or base of horns forming the clamping seat f of the shield, the whole being constructed in the manner and for the purpose described.

No. 2,345.—ALBERT J. SESSIONS, Bristol, Conn.—*Bag Frame.*—Patented June 12, 1866; reissued August 28, 1866.

Claim.—First, slitting or punching the strip, substantially as described, so as to form the side and rim of the corner or corners of a frame in one piece of metal.

Second, forming the inside corners square at a, substantially as described.

Third, forming the inside corners of a bag frame square, while the outside corners are flattened, curved, or rounded, by means substantially as described.

No. 2,346.—PETER A. VOGT, Buffalo, N. Y.—*Refrigerator.*—Patented May 22, 1866; reissued August 28, 1866.

Claim.—The door D, constructed as described, combined and arranged with the rack f f of the ice compartment so as to turn down and form a platform in continuation thereof, for introducing and removing ice from the chamber B, substantially in the manner specified.

Also, the arrangement of the induction air passage through the drip pipe l and extension pipe m, in combination with the trap k or its equivalent, whereby the air entering is carried to the top of the ice chamber and cooled in its passage, substantially as set forth.

No. 2,347.—BENJAMIN F. BEE, Harwich, Mass.—*Salinometer.*—Patented January 9, 1866; reissued September 4, 1866.

Claim.—First, the combination in a salinometer of the closed transparent vessel (for containing the liquor to be tested) and a float for indicating the density of the liquid, these two operating substantially as set forth.

Second, the combination in a salinometer of the following instrumentalities, viz: the closed transparent vessel, salinometer float, supply pipe and escape pipe, all operating in the combination substantially as set forth.

Third, the combination in a salinometer of the following instrumentalities, viz: the closed vessel, supply pipe, escape pipe, and air valve at the top of the vessel, all operating in the combination substantially as set forth.

Fourth, the combination in a salinometer of the following instrumentalities, viz: the closed vessel, salinometer float, and guide for the salinometer float, all operating substantially as set forth.

Fifth, the combination in a salinometer of the following instrumentalities, viz: the close d vessel and closed case communicating therewith for the thermometer, both operating substantially as set forth.

Sixth, the combination in a salinometer of the following instrumentalities, viz: the closed vessel, salinometer float, and thermometer, all operating substantially as set forth.·

Seventh, the combination in a salinometer of the closed vessel, with an escape valve, having the valve and screw independent of each other, so that the valve may be turned upon its seat to lighten the joint, substantially as set forth.

No. 2,348.—M. EASTERBROOK, Jr., Geneva, N. Y.—*Harvester.*—Patented May 22, 1866; reissued September 4, 1866.

Claim.—First, the combination of the band lever C, and two loose pinions *p* and *p'*, with the double pinion *b*, and the spur wheel B, arranged and operating substantially as and for the purposes set forth.

Second, the two loose pinions *p* and *p'*, whether they are adjusted with a band lever or other suitable device, in combination with the double pinion *b*, and the spur wheel, substantially in the manner and for the purposes shown and described.

Third, the arrangement of the pinions *p* and *p'*, upon a pivoted hand lever, substantially as shown in Fig. 3, having its axis or centre of motion upon the counter shaft, as and for the purposes set forth.

Fourth, the employment of the pinion *b*, of the counter shaft, made independent of the spur gear B, and connected thereto when desired by one of two intermediate adjustable gear wheels, arranged and operating substantially as and for the purposes set forth.

No. 2,349.—J. C. HILLS, Willoughby, Ohio.—*Construction of Churn Bodies.*—Patented May 24, 1864; reissued September 4, 1866.

Claim.—The above described construction of a churn body, consisting of the sides A A, the sheet A', groove B B', bars C C', and E, when constructed and arranged in the manner and for the purpose specified.

No. 2,350.—ISAAC E. OVERPECK, Overpeck's Station, Ohio —*Manual Power Machine.*— Patented April 25, 1865; reissued September 4, 1866.

Claim.—The two levers *d* and *e*, moving simultaneously toward and from each other, when arranged to operate upon a centre wheel *b*, or its equivalent, substantially in the manner and for the purpose herein specified.

No. 2,351.—WILLIAM EARL, Jr., Nashua, N. H., assignee of THOMAS PYE.—*Spinning Machine.*—Patented February 14, 1865; reissued September 4, 1866.

Claim.—First, the moving of the belt shifter C, in a spinning machine, by the ordinary operation of the machine, at the point or time when the thread is fully twisted as desired, and the jack is about to be returned for the purpose of winding up the thread in such manner that the belt R is partially thrown on to the tight pulley *g*, for the purpose of assisting the spinner in running up the jack and in making a tight bobbin, in the manner and by the means substantially as herein described and set forth.

Second, the moving of the said belt shifter C, by the ordinary operation of the machine, at the moment when the jack is nearly run up to the required or desired point or place for the piecing up of the thread, and just before it is fully wound upon the bobbins, in such manner that the said belt is thrown wholly off from the tight pulley aforesaid, and upon the said loose pulley, in the manner and by the means substantially as herein described and set forth.

Third, the employment of the crooked lever A, the oblique lever B, and the angular lever D, as arranged and combined, and then the whole in combination with the aforesaid belt shifter C, in the manner and for the purposes substantially as herein described and set forth.

Fourth, the employment of the chain E, arranged and combined with the angular lever D, and the slide bolt F, in the manner and for the purposes substantially as herein described and set forth.

Fifth, the combination of the slide bolt F, with the coil or spiral spring H, the bolt stock G, and the shoe *b*, and with a drop or sliding bar *d*, each being arranged and combined in the manner and for the purposes substantially as herein specified, described, and set forth.

Sixth, the drop or sliding bar *d*, and the wire or cord *e*, connected and combined with the faller W, and with the shoe *b b'*, in the manner substantially as and for the purposes herein described and set forth.

Seventh, the employment of the lever A, having an oblique arm *v*, in combination with the lever *u*, and with the carriage of the jack aforesaid, by means of the arm or frame L, containing the friction roller M, each being arranged and operated in the manner substantially as and for the purposes herein specified, described, and set forth.

No. 2,352.—C. W. WARNER, Williston, Vt.—*Horse Rake.*—Patented November 15, 1864: reissued September 4, 1866.

Claim.—First, the joints or hinges *b b*, on arms *a a*, projecting backward from the axle A, in combination with the rake G, all arranged as described to admit of the folding of the rake forward upon the thill bar *a**, substantially as set forth.

Second, the method substantially as described and represented of operating the rake by means of the combination of the foot triggers K K', bar J, and raking pawl *f g*, with the notched bar *h*, on the rake head.

No. 2,353.—SAMUEL L. DENNEY, Christiana, Penn.—*Horse Rake.*—Patented August 4, 1863; antedated April 2, 1863; reissued September 11, 1866.

Claim.—First, the bar *c*, connecting the ends of the thills A, which are suspended from the axle F, in combination with the clearing fingers E, all arranged and operating in the manner described.

Second, the cast hollow casing A', in combination with the arms *a* and *b'*, as and for the purpose set forth.

Third, the guard *g*, constructed and used in the manner described.

Fourth the guard *g*, in combination with the tang *t'*, as and for the purpose set forth.

Fifth, the serrated rim P, or equivalent, secured to the spokes or hub of the rake wheel, in combination with the arm *a*, attached to the rake axle or spindle, the arm *b'*, the rod I, the lever R, the spring X, and the curved standard Y, all arranged in the manner described.

No. 2,354.—JAMES S. MARSH, Lewisburg, Penn.—*Harvester.*—Patented February 10, 1863; reissued September 11, 1866.

Claim.—First, the combined raking and reeling apparatus, which rotates around a vertical shaft, when its arms adjust themselves successively from a horizontal to a vertical position, and when the combined apparatus is so located that its arms swing on hinges which are below the highest point of the drive wheel, and the extent of the sweep of any one of the arms does not interfere with the driver seated on any part of the draught frame, which is outside of the drive wheel, substantially as described.

Second, the construction and adaptation of a combined rake and reel which revolves entirely around a vertical centre, so that it may be applied to the harvester at a point which is on the inside of the drive wheel, and below the highest point of said wheel, substantially as described.

Third, locating the hinges of the respective arms of the combined rake and reel around a centre which is on the inner side of the drive wheel, and below the top of said wheel, substantially as described.

Fourth, attaching each of the respective arms of the combined rake and reel to a hinge or pivot, which is on the inner side of the drive wheel, and below the top of the same, substantially as described.

Fifth, the adaptation of a raking and reeling apparatus combined, which revolves entirely around a vertical centre, for application to the inner side of the draught frame of a harvester, at a point below the top of the drive wheel, substantially as described.

Sixth, the construction of the cam R, of the combined rake and reel, in the manner described and shown.

Seventh, the construction of the crown wheel with boxes for a series of rake and reel arms, in the manner described and shown.

Eighth, the linking devices described, or their equivalents, applied to the arms of the raking apparatus, substantially as described.

Ninth, the use of the inner bearing of the drive wheel as the support of the centre, on which the combined rake and reel revolve, substantially as described.

Tenth, the construction of the shaft or centre P, of the rake and reel bars, and the inner segment of the drive wheel, in one piece, in the manner described.

Eleventh, the combination of the cam R, hinged rake and reel bars, and adjustable links, so as to keep the rake and reel bars firmly in contact with the grain in the field and on the platform, substantially as set forth.

Twelfth, the arrangement of the sliding and turning sping pin *p*, incline *p²*, loose bevel pin Q, and the raking and reeling apparatus, substantially as described.

Thirteenth, the adjustable grain guard K, constructed substantially as described, and applied to the inner front corner of the draught frame.

Fourteenth, the combination of a draft frame, platform, driver's seat, device for adjusting the cutting apparatus, and a continuously revolving rake and reel combined, substantially as and for the purposes set forth.

Fifteenth, the combination of the device T, on the outer divider, and a combined continuously revolving rake and reel, substantially as described.

Sixteenth, so constructing a harvester and a raking and reeling apparatus combined, that a driver can ride on the machine, and from his seat adjust both the rake and reel, and cutting apparatus and platform without stopping the machine, substantially as described.

Seventeenth, so arranging a revolving raking and reeling device, having two or more arms, including the rake as one arm, that the driver can sit on the machine and drive the team, the shaft of the rake and reel being at or nearly at right angles with the grain platform, and the arms of rake and reel not sweeping over the frame on which the driver is located so as to interfere with the driver on his seat, substantially as described.

Eighteenth, the combination of a central shaft, a revolving hub or crown wheel, a cam, and a hinged rake and reel arms, which are bent or curved near their hinging ends, as described, whereby the rake and reel arms, although hinged in rear of the cutting apparatus, are capable of reeling in and raking off grain at the inner front corner of the platform, as well as at the outer front corner thereof, and whereby, also, these arms are caused to incline over toward the grain side of the platform when they rise to their greatest altitude, substantially as described.

No. 2,355.—HIRAM TUCKER, Newton, Mass., assignor to THE TUCKER MANUFACTURING COMPANY, Boston, Mass.— *Process of Bronzing or Coloring Iron.*—Patented December 15, 1863; reissued September 11, 1866.

Claim.—The process of ornamenting iron, in imitation of bronze, by the application of oil and heat, substantially as described.

No. 2,356.—HIRAM TUCKER, Newton, Mass , assignor to THE TUCKER MANUFACTURING COMPANY, Boston, Mass —*Manufacture from Iron in Imitation of Bronze.*—Patented December 15, 1863; reissued September 11, 1866.

Claim.—The new manufacture, hereinabove described, consisting of iron ornamented in imitation of bronze by the application of oil and heat, substantially as described.

No. 2,357.—JOHN A. BASSETT, Salem, Mass., assignor to THOMAS D. WORRALL, New York, N. Y.—*Process and Apparatus for Carburetting Gas for Illumination.*—Patented March 4, 1862; reissued September 18, 1866.

Claim.—First, the combination, substantially as herein described, of the vessel A, in which the gas passes circuitously over the surface of the hydrocarbon liquids to be partly carburetted and cooled by the evaporation of the liquids, and the vessel B, containing a porous substance, and saturated with such liquid, through which the gas subsequently passes, as herein set forth and described.

Second, the gas-regulating valve *j* and float *k*, combined with a gas-naphthalizing or carburetting apparatus, substantially as herein specified, that is to say, with the float floating in the naphtha or other hydrocarbon liquid used for the carburetting processes.

Third, in combination with a carburetting apparatus, a regulator to govern the flow of gas to the burner, substantially as described.

Fourth, the use of carbon spirit or light products of petroleum for the purpose of carburetting air or of enriching and carbonizing any kind of gas.

No. 2,358.—WILLIAM S. CHAPMAN, Baltimore, Md.—*Preventing Rattling in Carriages.*—Patented August 8, 1854; reissued September 18, 1866.

Claim.—First, the employment of blocks of India-rubber, or other equivalent elastic material, so shaped as to be self-sustaining in position when interposed between the ends of the carriage shafts or poles and the "clip" to prevent rattling, substantially as above described.

Second, the use of India-rubber blocks, or other equivalent elastic material, interposed between the ends of carriage shafts and the "clips," in such a way and under such strong compression as to hold the bolts in place independently of the nuts, and also to prevent the rattling of the parts, substantially as above described.

Third, finally, as a new manufacture, a block of India-rubber, or other equivalent elastic material, intended to be used as contemplated, when made substantially in the form described—that is to say, when so shaped that it will remain permanently in place, and perform its functions without the aid of any other special contrivance for that purpose, in the manner above set forth.

No. 2,359.—G. W. HUBBARD, Brooklyn, N. Y., and WILLIAM E. CONANT, Little Falls, N. J.—*Operating Slide Valves in Direct Action Engines.*—Patented January 9, 1855; reissued September 18, 1866.

Claim.—First, so combining a main engine or motor, a supplementary valve-working engine, and their induction and eduction valve or valves, that the movement of the valve or valves of the main engine or motor is commenced and partly effected by the piston of said engine, and completed by the piston of the supplementary or valve-working engine, substantially as herein described.

Second, when two direct-action engines are so combined that the movement of the induction and eduction valve or valves of one is produced by the movement of the piston of the other, the arrangement of the cylinder and piston of one engine within the valve chest of the other, substantially as herein described.

Third, in operating the slide valve in one direct-action engine by the piston of another, so connecting the said slide valve with a tappet-rod operated by an arm on the piston-rod of its own engine, that the said rod and valve may have each a certain amount of motion independently of the other, substantially as and for the purpose herein specified.

Fourth, the arrangement of the valves E and k, the tappet-rod F, and its connections with the said valves, and the cut-off plate j, and stops l l, substantially as described for the purpose set forth.

No. 2,360.—JOHN S. LLOYD, Salem, N. J.—*Hay Hoisting Machine.*—Patented April 24, 1860; reissued September 18, 1866.

Claim.—First, an elevated way or railroad A, in combination with a hoisting or horse hay fork F, arranged to operate in the manner substantially as herein shown and described.

Second, the construction, combination, and arrangement of the fork, cords, levers, pulleys, springs, and railway, the arms E E, to the block D, and the mode of attaching and supporting the railway to the barn or frame, so as to allow the wheels B B, with the attached blocks and fork, to pass freely along the length of the rail.

Third, the post P, as constructed in combination with the pulley, lever, slide, spring, and cord.

No. 2,361.—WALLACE T. MUNGER and J. A. LEGGAT, Branford, Conn., assignees of WALLACE T. MUNGER.—*Knob Latch.*—Patented April 3, 1866; reissued September 18, 1866.

Claim.—First, the follower E, recessed in its rear side, in combination with the bar a of the yoke H, attached to the latch bolt, substantially as and for the purpose specified.

Second, the horseshoe F, link I, and spring M, in combination with the lock bolt D, substantially as and for the purpose set forth.

No. 2,362.—H. D. SMITH, G. F. SMITH and EDWARD W. TWICHELL, Plantsville, Conn., assignees of JAMES P. THORP.—*Wagon-shaft Shackle.*—Patented May 1, 1860; reissued September 18, 1866.

Claim.—The improved manufacture of a carriage-shaft shackle blank, constructed with the projections d d, arranged at or about at the junctions of the arms and body of the blank, substantially as and for the purpose specified.

Also, for making the said blank, the die as constructed with the projection forming recesses, arranged with respect to the portion for swaging the body and arms of the blank, substantially in manner as specified.

No. 2,363.—GEORGE R. BAKER, St. Louis, Mo.—*Machine for Kneading Dough.*—Patented October 10, 1865; reissued September 25, 1866.

Claim.—First, the combination of the dough-kneading chamber B B' and the eccentrically rotating wheel C, whether said wheel rotates on a fixed and rigid or a flexible and yielding axis of rotation, as and for the purposes described.

Second, the combination of the shaft a, with its pivoted arm a', the slot e, and spring f, arranged and operating substantially as and for the purpose described.

No. 2,364.—RICHARD W. CHAPPELL, Chicago, Ill.—*Washing Compound.*—Patented September 19, 1865; reissued September 25, 1866.

Claim.—A washing compound composed of unslaked lime, sal soda, borax, salt of tartar, and ammonia, in the proportions substantially as herein specified and described.

No. 2,365.—R. L. DELISSER, New York, N. Y., assignee of ALOYSE CHEVALIER AUER DE WELSBACH.—*Manufacture of Paper.*—Patented April 21, 1863; antedated November 21, 1861; reissued September 25, 1866.

Claim.—My discovery or invention, as a means of obtaining from the husks, leaves, and stalks of Indian corn fibres in a suitable condition to be spun into thread, is the treatment of such material in a heated alkaline solution, substantially as described, to dissolve and separate the proximates of the plant from the fibres, in combination with the after process of hatchelling, or the equivalent thereof, to get the fibres in a suitable condition for spinning, substantially as described.

No. 2,366—HENRY C. BERLIN and GEORGE H. JONES, New York, N. Y., assignees of THOMAS V. WAYMOTH.—*Machine for Gumming and Printing Envelopes.*—Patented June 12, 1866; reissued September 25, 1866.

Claim.—First, the operation of the hinged table B, in combination with the gummer D, substantially in the manner and for the purpose described.

Second, the operation of the movable separator G in combination with the gummer D, substantially in the manner and for the purpose described.

Third, imparting an intermittent motion to a suitable mechanism combined with the endless apron H and the reciprocating carrier F and gummer D, operating in the manner and for the purpose described.

Fourth, the operation of the finger l' and rollers k' in combination with the apron H and carrier F, substantially in the manner and for the purpose described.

Fifth, putting the gum on the seal flap of an envelope blank by the picker or gummer so that the envelope is raised or held stationary while the balance of the pile is removed, substantially in the manner and for the purpose described.

No. 2,367.—EDMUND H. GRAHAM, Manchester, N. H.—*Picker Staff Motion for Looms.*—Patented October 16, 1860; reissued October 2, 1866.

Claim.—Steadying the rocker of the picker staff on its bed by journals at a right angle to the picker staff, which journals form its centre of motion, substantially as described.

Also, the journal boxes with open ears, in combination with the journals that steady the rocker on the bed.

No. 2,368.—FRANKLIN BENJAMIN HUNT, Richmond, Ind.—*Straw Cutter.*—Patented January 5, 1864; reissued October 2, 1866.

Claim.—First, as my improvement in straw cutters, so attaching the balance wheel to its shaft by a yielding or frictional device that when the knife meets with an obstruction the wheel may continue to revolve for a limited period independent of the knife, until stopped by the frictional device, for the purpose of preventing injury to the knife, substantially as herein described.

Second, the bar c' connected to the shaft n, and carrying the pinions b' and d', in combination with the pinions a' and e' and link f, connected to the shaft p, in substantially the manner specified, whereby the rollers H and G on the shafts n and p are allowed to move apart and the wheels remain in gear, as set forth.

Third, the guide board or plate u, connected to and moving with the frame t of the upper feed roller H, and extending down at the back of the said roller to near a level with its axis, substantially as and for the purpose set forth.

Fourth, the bearing bar z, formed in one piece with the lower halves b of the journal boxes of the cutter shaft C, and extending across from one to the other, in combination with the standing cutter E, attached to said bar z, substantially as and for the purposes specified.

Fifth, mounting the upper feed roller H in a frame formed with slings extending below the lower roller and acted upon by a spring, or its equivalent, in combination with a slotted frame to guide the roller as it moves up or down, substantially as set forth.

Sixth, the hooked slings q, in combination with the yielding feed roller H and spring I, or its equivalent, substantially as set forth, whereby the said feed roller H is limited in its lateral movement, as set forth.

Seventh, the curved slot v, in the frame w, in combination with the feed roller H, slings q, and a hub n', surrounding the axis p of the roller H, and relieving the same from friction against the frame w, as set forth.

Eighth, in combination with the bar z and standing cutter E, made as set forth, the single revolving and diagonal knife D, with its axis placed above the standing cutter E, to act with a slanting and shearing cut, substantially as set forth.

No. 2,369.—P. H. ROOTS, Connersville, Ind.—*Blower.*—Patented September 25, 1860 reissued October 2, 1866.

Claim.—The coacting rotary abutments A B, each consisting of two or more pistons D D and two or more recesses E E, which are arcs of true circles and formed with equal radii, substantially as set forth.

No. 2,370.—BENJAMIN WRIGHT, Hudson, Mich.—*Washing Machine.*—Patented March 28, 1865; reissued October 2, 1866.

Claim.—First, the rocker N and arms G, constructed and used with the box A, substantially as and for the purpose herein specified.

Second, the arrangement of the box A, with the rubber C raised above its bottom when used with the rocker N and arms G, as and for the purpose specified.

No. 2,371.—N. B. WEBSTER, Portsmouth, and ROBERT W. YOUNG, Richmond, Va.—*Prevention of Incrustation in Steam Boilers.*—Patented October 23, 1860; reissued October 2, 1866.

Claim.—The connecting with the interior of a steam boiler a metal electro-negative to the boiler, for the purpose of preventing incrustation in the steam boiler.

No. 2,372.—JAMES SPEAR. Philadelphia, Penn.—*Cooking Stove.*—Patented February 19, 1861; reissued October 9, 1866.

Claim.—The combination of the sifting and ash drawer A and B with the curved or guide plate D, when used in connection with a stove or range, in which the oven extends under the fire-grate.

No. 2,373.—JAMES SPEAR, Philadelphia, Penn.—*Cooking Stove.*—Patented February 19, 1861 ; reissued October 9, 1866.

Claim.—The arrangement of the ash drawer B, having bail G combined therewith in an ash pit or chamber of a stove or range, and in combination with the fire grate or fire chamber thereof, so as to receive the ashes or cinders falling therefrom, and thereafter to be removed, substantially as described.

No. 2,374.—JAMES SPEAR, Philadelphia, Penn.—*Cooking Stove.*—Patented February 19, 1861 ; reissued October 9, 1866.

Claim.—The arrangement of an ash pit under the hearth and in front of a long oven, constructed and operating substantially as described.

No. 2,375.—JOHN F. BOYNTON, Syracuse, N. Y.—*Apparatus for Carburetting Gas.*—Patented September 5, 1865 ; reissued October 16, 1866.

Claim.—First, in an apparatus for carburetting gas, by charging it with the vapors of hydrocarbon liquids, the use of wood as a capillary agent, to draw up the liquid and expose it to evaporation.

Second, in a carburetting apparatus as above described, the use of wood in combination with cotton wicking or other fibrous material, to produce the capillary action necessary to promote rapid evaporation, substantially as described.

Third, so arranging and constructing the cotton wicking or other fibrous material, and its wooden supports, that as the surface of the liquid in the carburetting vessel descends, the number of capillary pores brought into action will be all the while increasing, substantially as described.

Fourth, a combination of wood and wicking, or other fibrous material, so arranged as to form a movable frame or cage, setting into a box, and producing a compound of capillary action of porous and fibrous material, substantially as described.

No. 2,376.—JOHN F. BOYNTON, Syracuse, N. Y.—*Apparatus for Carburetting Gas.*—Patented September 25, 1866 ; reissued October 16, 1866.

Claim.—First, the automatic filling reservoir D, in combination with the tube C, substantially as described.

Second, the base board H, in combination with a series of wooden pegs inserted therein, and supporting fibrous material, to produce a compound capillary action, as and for the purposes described,

Third, the wooden pegs I, wound with cotton wicking J, or other equivalent fibrous material to produce a compound capillary action as described.

Fourth, the base board H, wooden pegs I, and cotton wicking or other fibrous material J, so combined, constructed and put together as to form a movable frame or cage which may be inserted into the carburetting box and removed therefrom, together, as one entire structure.

Fifth, securing the fibrous material at the lower end of the peg, by driving it with the peg into a perforation of the base board, substantially as described.

Sixth, the internal box K, with its partitions K′, constructed and arranged substantially as described.

Seventh, so constructing and arranging said internal box K, that when set in the main carburetting box it will divide the carburetting chamber into an outer and inner apartment, substantially as described.

Eighth, constructing said box K, with its partition walls, of wood, or any other porous substance which will produce capillary action.

No. 2,377.—ADOLPH GEISS, Buffalo, N. Y.—*Gas Burner for Cooking, &c.*—Patented November 28, 1865 ; reissued October 16, 1866.

Claim.—First, the draught and mixing chamber A, in combination and arrangement with the perforated dome D, perforated shell *s′*, and metallic base A², (including gas pipe F,) for the purposes and substantially as described.

Second, in a gas burner for cooking and heating purposes, the thimble C, in combination with the wire gauze dome D, substantially as described.

Third, the combination of the thimble C, wire gauze dome D, and cap B, substantially as set forth.

Fourth, the combination of the outer thimble C, the inner and upper thimble *c*, and dome D, for the purposes and substantially as set forth.

No. 2,378.—BENJAMIN OLDFIELD, Newark, N. J.—*Loom.*—Patented January 23, 1866 ; antedated January 17, 1866 ; reissued October 16, 1866.

Claim.—The application to a batten of two or more shuttles for plain weaving, and one or more figuring shuttles, to operate in conjunction, substantially in the manner and for the purpose herein set forth.

No. 2,379.—BENJAMIN OLDFIELD, Newark, N. J.—*Loom.*—Patented January 23, 1866 ; antedated January 17, 1866 ; reissued October 16, 1866.

Claim.—An upright shuttle, driven by rack and pinion or in any other suitable manner, and which is grooved on each of its sides, and the body part of which is cut away for the quill and provided with a guard *g*, substantially in the manner and for the purpose herein set forth.

No. 2,380.—JAMES DUNDAS, Nemaha county, Nebraska.—*Cultivator.*—Patented February 8, 1859 ; reissued October 16, 1866.

Claim.—First, the combination in a straddle-row cultivator of the following instrumentalities, viz: the two wheels, frame, and a series of ploughs arranged in two gangs with a central space between the gangs, so as to till the soil simultaneously at both sides of a single row of plants which the machine straddles; all of these operating in the combination substantially as set forth.

Second, the combination in a straddle-row cultivator of the following instrumentalities, viz: the two wheels, frame, the series of ploughs arranged in two gangs as aforesaid, and seat for the driver, all of these operating in the combination substantially as set forth.

Third, the combination in a straddle-row cultivator of the following instrumentalities, viz: the two wheels, frame, the series of ploughs arranged in two gangs as aforesaid, and movable stocks, all operating in the combination so that while the wheels limit the penetration of the ploughs, the inner ploughs of the two gangs may be moved laterally to avoid the plants that are out of line in the row, substantially as set forth.

Fourth, the combination in a straddle-row cultivator of the following instrumentalities, viz: the two wheels, frame, the series of ploughs arranged in two gangs as aforesaid, movable stocks, as aforesaid, and driver's seat; all operating in the combination substantially as set forth.

Fifth, the combination in a straddle-row cultivator of the following instrumentalities, viz : the two wheels, frame, the series of ploughs arranged in two gangs as aforesaid, driver's seat and a connection between the movable ploughs, all operating in the combination substantially as set forth.

Sixth, the combination in a straddle-row cultivator of the following instrumentalities, viz: the wheels, frame, series of ploughs arranged in two gangs as aforesaid, and mechanism to permit the ploughs to be raised relatively to the treads of the wheels, all constructed and operating in the combination substantially as set forth.

No. 2,381.—G. H. REYNOLDS and M. A. HINCKLEY, administratrix of D. B. HINCKLEY, Mystic Bridge, Conn., assignees of G. H. REYNOLDS.—*Operating Cut-off Valves.*—Patented February 3, 1857 ; reissued October 16, 1866.

Claim.—First, automatically shutting a cuff-off valve carried on the steam valve, so that so soon as the valve commences to close it will continue its closing motion independent of the motion of the engine, substantially as and for the purposes herein specified.

Second, the inclined dogs H H, arranged to operate in connection with a cut-off valve F, carried on the steam valve B, substantially in the manner and for the purpose herein set forth.

No. 2,382.—ROBERT T. CAMPBELL, Washington, D. C., assignee of T. N. LUPTON.—*Harvester.*—Patented May 8, 1855 ; reissued October 23, 1866.

Claim.—First, the application of a hinged diagonal brace *b*, in front of the finger beam, for the pur ose of sustaining such beam against backward strain and thrust, substantially as set forth.

Second, hinging a diagonal brace *b*, which extends forward of the finger beam, so that its axis of motion shall coincide with the axis of motion of said beam, substantially as described.

Third, sustaining the platform frame and the finger beam by means of a forward diagonal brace *b* and a rear brace *b*^{1}, substantially as described.

Fourth, a hinged diagonal brace which extends forward of the finger beam and is adapted to serve as a guard and also as a means of sustaining the finger beam against backward strain, substantially as described.

Fifth, a diagonal brace which is connected at one end to the finger bar and extended forward of the same and connected by an eye formed on it to the draught frame so as to move concentric with the axis of motion of said finger beam. substantially as described.

Sixth, extending the brace *b*, as claimed in the fifth clause of the claim, under the platform which sustains the fallen grain, substantially as described.

Seventh, an inclined shaft K, supported in bearings *a a*, upon the inner side of the draught frame, and adapted to serve as a driving shaft for the cutters and also as a means for hinging the finger beam to said draught frame and arranged with respect to a horizontal shaft H, or equivalent, substantially as described.

Eighth, in combination with a vibrating finger beam which projects laterally from the grain side of the draught frame and which receives vertical movements independently of the frame from the undulations of the ground over which it is drawn, a reel, or its equivalent, mounted wholly upon the finger beam or its platform for gathering in the standing grain, substantially as described.

Ninth, in combination with the subject-matter of the sixth clause of the claim, means which will carry off the cut grain from behind the cutting apparatus and deliver it from the inner side of the platform, substantially as described.

Tenth, in combination with a finger beam which vibrates independently of the main draught frame and is supported at its outer end upon the ground, providing means which will enable the attendant while riding upon the machine to stop and start the cutting apparatus, and the means employed for conveying the grain away from behind the cutting apparatus, substantially as described.

Eleventh, the employment of the cutters d, placed on the rotating shaft M, in combination with the two sets of fingers P Q, the said parts being constructed and operating substantially as described.

Twelfth, supporting a reel or its equivalent at its inner end by the hinge of the finger beam and at its outer end by a wheel, or other equivalent device, substantially as described.

Thirteenth, supporting both a device which moves the cut grain to one side of and away from behind the cutter, and a device which reels in the uncut grain to the cutting apparatus by means of the axis a, hinged at the inner end of the finger beam and a wheel or its equivalent at the outer end of such beam, substantially as described.

Fourteenth, supporting a reel wholly upon the hinged finger beam or platform which receives the falling grain, in combination with applying the finger beam thus wholly carrying the reel, to one side of the draught frame in such manner that the finger beam and reel together will be supported at one end by a hinge connection and by a wheel, or its equivalent, at the other end, substantially as described.

No. 2,383.—THE HYDROSTATIC PAPER COMPANY, Rochester, N. Y., assignee by mesne assignments of HENRY L. JONES and DUNCAN S. FARQUHARSON.—*Treating Wood, Straw, &c., for the Manufacture of Paper Pulp.*—Patented June 5, 1866; reissued October 23, 1866.

Claim.—First, the subduing of straw, wood, or any fibrous material to be converted into pulp by subjecting the same to the action of alkali liquor of any desirable temperature applied under the hydrostatic pressure of the liquid itself, applied by a force pump or otherwise instead of using steam pressure preparatory to the bleaching of such material in the ordinary method, substantially as described.

Second, the combination with the cylinder A of the pump D and pipe B, substantially as and for the purpose above set forth.

Third, the safety valve K, in combination with the pump D, below the piston or plunger and in direct communication with the pump barrel, substantially as above described.

No. 2,384.—THE HYDROSTATIC PAPER COMPANY, Rochester, N. Y., assignee by mesne assignments of HENRY L. JONES and DUNCAN S. FARQUHARSON.—*Apparatus for Bleaching Paper Pulp.*—Patented March 13, 1866; reissued October 23, 1866.

Claim.—First, bleaching the material to be converted into paper, by subjecting the same to the action of bleaching liquor, applied under pressure, substantially as described.

Second, the combination with the cylinder A of the pump D and pipe B, substantially as and for the purpose set forth.

Third, the combination with the cylinder A of the elevated reservoir E and pipe F, substantially as and for the purposes set forth.

Fourth, in combination with the cylinder A, pump D, and pipe P, the valve O, for relieving the pressure of the liquid, as explained.

No. 2,385.—RADCLIFFE B. LOCKWOOD and CHARLES J. EVERETT, New York, N. Y., assignees of CARROL E. GRAY.—*Apparatus for Rendering Lard, Tallow, &c.*—Patented January 31, 1865; reissued August 8, 1865; again reissued October 30, 1866.

Claim.—First, making a close water jacket in combination with the tank, and a part of it, and arranging said water jacket so made a part of said tank in direct communication with the furnace, so that the water jacket shall intervene between the fire and the tank, and act as a means of conducting and distributing the heat from the fire to and around the substance contained in the tank.

Second, using the steam generated in a close tank from the constitutional water in the fat for the purpose of aiding and controlling the escape of the noxious gases and vapors either to a superheater for consumption in the furnace, or to a deodorizer for the purpose of deodorizing them, in the manner substantially as described for the purpose specified.

Third, controlling and superheating the noxious gases and vapors as they escape from a rendering apparatus by passing them through a pipe or flue leading from said apparatus to a superheater, preparatory to their consumption.

Fourth, controlling the escape of the noxious gases and vapors from a rendering apparatus by passing them through a pipe or flue into a surface condenser, for the purpose of condensing the vapor and absorbing in the water of condensation the noxious gases or as much of them as may be possible, substantially as shown and described.

Fifth, deodorizing the water of condensation holding said noxious gases in solution by passing it through a deodorizer after it leaves the condenser, substantially as described.

No. 2,386.—JOHN MATTHEWS, New York, N. Y., assignee of ALBERT ALBERTSON.—*Bottle Stopper.*—Patented August 26, 1862; reissued October 30, 1866.

Claim.— First, a stopper which is inserted through the mouth of a bottle or other vessel and which when inserted is closed perfectly tight against a seat formed within the bottle itself by pressure in an upward direction.

Second, a prolongation of such stopper by means of a central stem, rod or other extension of the stopper in an outward direction beyond the seat of the valve for the purpose of affording facility for opening the stopper or that of receiving the upward pressure of a spring or other means of drawing the valve to its seat, substantially as herein specified.

Third, the two disks B C, of unequal size, and the interposed flexible disk or diaphragm D, in combination with each other and with a stem or standard A, substantially as herein specified.

No. 2,387.—EDWIN CHAMBERLAIN, Troy, N. Y.—*Whip Socket Fastening.*—Patented August 23, 1864; reissued November 6, 1866.

Claim.—First, a detachable and removable whip-socket fastening attached to the dash or other suitable parts of a land carriage or other vehicle, in the manner and for the purposes substantially as herein described and set forth.

Second, a whip-socket fastening having a clamp or holder B B for a whip socket combined with the jaws A A for receiving and griping a bar or rod in a covered dash or other part of a carriage or other vehicle, substantially as herein described and set forth.

No. 2,388.—ROBERT T. CAMPBELL, Washington, D. C., assignee of THOMAS I. STEALEY.—*Supporting Reels for Harvesters.*—Patented December 15, 1857; reissued November 6, 1866.

Claim.—First, combining with a hinged platform which is free to conform to the undulations of the ground independently of the motions of the draught frame, or of the action of the transporting wheels, a toothed rake which will deliver the cut grain upon the ground in gavels, and a reel or gathering device which will press the standing grain toward the cutters, said rake and reel or gatherer being wholly supported upon the said platform, substantially as described.

Second, combining with a hinged platform a toothed rake, and a reel or gatherer, which are wholly supported upon and move in harmony with said platform, an adjustable hinged connection which will allow of the vertical adjustment of the cutting apparatus to adapt the machine to different heights of cut required, substantially as described.

Third, sustaining a toothed rake and a reel or gatherer wholly upon a platform which is supported at its inner end by a vertically adjustable joint, and at its outer end by a wheel or its equivalent, substantially as described.

Fourth, suspending the hinged platform, which has a toothed rake mounted wholly upon it, from the main draught frame at a point in rear of the cutting apparatus, in such manner that this part of the platform can be adjusted vertically without changing the position of the forward adjustable hinge connection, substantially as described.

Fifth, the combination of a hinged finger beam, a platform and an auxiliary adjustable suspending and sustaining flexible connection in such manner that the finger beam and platform are firmly suspended at their inner ends and are free to conform at their outer ends to the undulations of the ground, independently of the main frame or of the axle of the supporting wheels, substantially as described.

Sixth, in combination with a vertically adjustable hinge joint and hinge movement of the finger beam and cutter bar, and with the crank *d*, for communicating motion to the cutters, the employment of the universal joint *m*, to connect the pitman I with the cutter bar, substantially as described, and the adjustable blocks 1 and 2, for tightening the joint around the crank wrist *d*, substantially as set forth.

Seventh, the combination of a crank shaft O, with adjustable bearings *a a*, the pitman Q, and the oscillating rake S, substantially as described.

Eighth, hanging the reel to the rake frame or platform, and adjusting said reel to different heights by means of braces *w w*, or their equivalents, substantially as described.

No. 2,389.—BENNET HOTCHKISS, New Haven, Conn., assignor through mesne assignments to himself.—*Trip Hammer.*—Patented June 14, 1859; reissued November 6, 1866.

Claim.—First, in combination with a hammer and an actuating mechanism having a definite reciprocation, the elastic spring or springs, whether of air or other material, interposed between the definite reciprocating mechanism and the hammer, substantially in the manner herein shown and described, so that the extent of motion given to the hammer and the force of its blow may be regulated by the speed of the actuating mechanism, substantially as set forth.

Second, the reciprocating pneumatic cylinder, having a hole near its central portion, in combination with a piston rod and hammer, substantially as and for the purposes specified.

Third, adjusting the space between the anvil and the hammer by mechanism, constructed and arranged substantially as specified, so as to accommodate different sized forgings, as set forth.

No. 2,390.—RUFUS S. SANBORN, Ripon, Wis.—*Safe.*—Patented July 17, 1866; reissued November 6, 1866.

Claim.—First, the combination of two or more concentric cylinders or cases B C D, whether in the form herein represented or otherwise, when each cylinder or case is separated from the next one to it within or without, in such a manner that air is allowed to circulate freely all around it, both at its sides and ends, as and for the purpose represented.

Second, the combination of the water vessels F F, or their equivalents, when used with the cylinders or cases B C D, arranged as specified, whereby steam from said vessels may be allowed to circulate freely around the sides and ends of the cases, substantially as and for the purpose herein specified.

Third, the arrangement of the inner box E for containing books and papers, with the cylinders B C D, or their equivalents, in box form, and an outer case A, substantially as and for the purposes herein set forth.

No. 2,391.—J. S. BROWN, Washington, D. C.—*Horse Hay Fork.*—Patented July 17, 1866; reissued November 6, 1866.

Claim.—The employment of a movable bar or bars D D, to cover and uncover fixed barbs or shoulders C C, in combination with a divided shaft A, to be opened in a dovetail or inverted wedge form, and closed in connection with the uncovering and covering of the barbs or shoulders, substantially as and for the purposes herein specified.

No. 2,392.—J. S. BROWN, Washington, D. C.—*Horse Hay Fork.*—Patented July 17, 1866; reissued November 6, 1866.

Claim.—The employment of a movable bar or bars D to cover and uncover fixed barbs or shoulders C C, substantially as and for the purposes herein specified.

No. 2,393.—WARREN GALE, Chicopee Falls, Mass.—*Straw Cutter.*—Patented September 12, 1854; reissued April 3, 1860; again reissued October 25, 1864; again reissued November 13, 1866.

Claim.—The automatic mouth of a feed box, constructed by any means, substantially the same as described, when used in combination with a revolving cutting cylinder armed with one knife, or with several knives, so arranged that one knife shall release its hold upon the material being cut before the following knife shall grasp it sufficiently to hold it, substantially as and for the purposes set forth.

Second, the adjustable bottom mouth piece M, or its equivalent, constructed and operating substantially as and for the purposes set forth.

Third, combining a revolving cutting cylinder, armed with one knife, or with several knives, so arranged that one knife shall release its hold upon the material being cut before the following knife shall grasp it sufficiently to hold it, with a hinged bottom mouth piece of a feed box, substantially as and for the purposes described.

Fourth, an automatically operating mouth to a feed box, in combination with a revolving knife cylinder, armed with one knife or with several knives, so arranged that one knife shall release its hold upon the material being cut before the following knife shall grasp it sufficiently to hold it, when this cylinder is geared to a revolving pressure cylinder, substantially as and for the purposes set forth.

Fifth, making those parts of the pressure cylinder against which the knife or knives are made to cut, by having their edges brought into actual contact therewith, in sections or strips separate from from the body of the cylinder, substantially as and for the purposes set forth.

Sixth, a revolving cutting cylinder, having one or more knives, in combination with a pressure cylinder, having one or more radial flanges, arms, or projections, so arranged that the knife or knives shall, as they revolve, meet the flange, arm, or projection, or either of them, in actual contact, so that the material to be cut is caught between the two, drawn forward, and cut off by the pressure between the knife on one cylinder and the flange on the other, substantially as and for the purposes set forth.

Seventh, the flanged pressure cylinder, arranged and operated substantially as described, when the face of the flange is covered with suitable soft material, to protect the edge of the knife, when used in combination with a revolving cutting cylinder, substantially as and for the purposes set forth.

Eighth, an automatically operating mouth of a feed box, or an adjustable mouth of a feed box, substantially as described, in combination with a revolving cutting cylinder armed with one knife, or with several knives, so arranged that one knife shall release its hold upon the material being cut before the following knife shall grasp it sufficiently to hold it, and with a revolving pressure cylinder armed with one or more radial arms, flanges, or projections, substantially as and for the purposes set forth.

Ninth, a pressure cylinder provided with one or more radial flanges, arms, or projections, and a revolving cutting cylinder armed with one knife, or with several knives, so arranged that one knife shall release its hold upon the material being cut before the following knife shall grasp it sufficiently to hold it, when these cylinders are used in combination with a hinged bottom mouth piece of a feed box, substantially as and for the purposes set forth.

No. 2,394.—LACY, MEEKER & CO., New York, N. Y., assignees of GEORGE H. MEEKER.—*Riding Saddle.*—Patented May 16, 1865; reissued November 13, 1866.

Claim.—The forming of the projections or calf and thigh supports on the skirts of a riding saddle by means of a swaging, striking up or embossing, substantially in the manner as herein shown and described.

No. 2,395.—FRANCIS GRANGER, Lockport, Ill.—*Harrow.*—Patented July 17, 1866; reissued November 20, 1866.

Claim.—A combination of the circle *b*, the flanged friction roller *c*, the arms *d d*, the weight *e*, and friction rollers *i*, arranged and operating in the manner substantially as described.

No. 2,396.—JOHN F. COLLINS, New York, N. Y.—*Distillation.*—Patented October 30, 1866; reissued November 20, 1866.

Claim.—The process, substantially as above described, of separating and obtaining alcohol or other volatile matters by constantly agitating the "wash" or other contents of the still or retort, by means of a current or currents of steam, or gas, or air forced into the same, and bringing the vapors in contact with currents of air from without while passing from the still or retort into the conductor which leads to the worm or condenser, as above set forth.

No. 2,397.—CLARK TOMPKINS, Troy, N. Y., assignee of HENRY BROCKWAY.—*Take-up for Circular Knitting Machines.*—Patented November 8, 1864; reissued November 20, 1866.

Claim.—The combination in the take-up mechanism of a knitting machine of the following instrumentalities, viz: a stationary cam, revolving frame of the take-up mechanism, take-up roll, cam lever, ratchet wheel and vibrating pawl, stop (for preventing the pawl from vibrating,) and variable instrument (for operating the stop,) all operating in the combination substantially as set forth.

Also, the combination in the take-up mechanism of a knitting machine of the following instrumentalities, viz: the revolving frame of the take-up mechanism, take-up roller, cam, ratchet wheel and pawl, endless screw and worm wheel, and variable instrument for controlling the action of the pawl, all operating in the combination substantially as set forth.

No. 2,398.—PHILIP S. JUSTICE, Philadelphia, Penn., assignee of THOMAS SHAW.—*Power Hammer.*—Patented February 27, 1866; reissued November 27, 1866.

Claim.—The combination of a vibratory hammer with the spring and flexible belt, substantially as described.

No. 2,399.—HOBART H. SMITH, Carlisle, Penn., assignee by mesue assignments of C. M. LUFKIN.—*Harvester.*—Patented September 8, 1857; reissued November 27, 1866.

Claim.—First, in a two-wheel side-draught machine of a main frame arranged between the wheels and connected to the main axle with a laterally-projecting hinged cutting apparatus connected to said frame by means of a hinge at its inner or heel end only, in such manner that each end of the cutting apparatus, independently of the other, is free both to fall below and to rise above the plane on which the main carrying wheels are passing, for the purpose specified.

Second, the employment in a two-wheel side-draught machine of a main frame which is connected to the main axle or gear centre, and which extends beyond the periphery of the driving wheel at one end only, in combination with a laterally-projecting hinged cutting apparatus, for the purpose specified.

Third, in a two-wheel side-draught machine a main frame connected at one end to and vibrating about the main axle or gear centre, in combination with a cutting apparatus hinged to the other end of said frame.

Fourth, in a two-wheel side-draught machine a laterally-projecting cutting apparatus hinged to one end of a frame which is connected at its other end to and vibrates about the axle or gear centre of said machine.

Fifth, in a two-wheel side-draught machine a vibrating frame arranged betweeen the wheels and connected to the main axle or gear centre, in combination with a laterally-projecting cutting apparatus which is hinged at one of its ends to said frame.

No. 2,400.—HOBART H. SMITH, Carlisle, Penn., assignee by mesne assignments of C. M. LUFKIN.—*Harvester.*—Patented September 8, 1857; reissued November 27, 1866.

Claim.—First, the combination in a two-wheel side-draught machine of a main frame, a hinged cutting apparatus, and a draft pole or tongue hinged at a point within the periphery of the driving wheel, for the purpose specified.

Second, the combination in a two-wheel side-draught machine of a vibrating frame, a tongue hinged to said frame on a line nearly coincident with the axle, and a laterally-protecting hinged cutting apparatus, for the purpose specified.

No. 2,401.—HOBART H. SMITH, Carlisle, Penn., assignee by mesne assignments of C. M. LUFKIN.—*Harvester.*—Patented September 8, 1857; reissued November 27, 1866.

Claim.—First, in a two-wheel side-draught machine the combination of a main frame arranged between the wheels and a vibrating draw bar, or its equivalent, with a laterally-projecting hinged cutting apparatus, which is free at each end independently of the other end, and of the vertical movement of the main axle to conform to the surface of the ground over which it is drawn.

Second, the employment in a two-wheel side-draught machine of a draw bar, or its equivalent, connected at one end to and vibrating about a shaft or gear centre of said machine, in combination with a main frame arranged between the wheels and a laterally-projecting hinged cutting apparatus.

Third, a draw bar, or its equivalent, connected at one end to and vibrating about the main axle or gear centre, in combination with a laterally-projecting cutting apparatus hinged to the other end of said bar.

No. 2,402.—HOBART H. SMITH, Carlisle, Penn., assignee by mesne assignments of C. M. LUFKIN.—*Harvester.*—Patented September 8, 1857 ; reissued November 27, 1866.

Claim —First, the employment in a side-draught machine of a curved slot and set screw, in combination with a vibrating frame, or its equivalent, which adapts a laterally-projecting hinged cutting apparatus to conform to the surface of the ground over which it is drawn, independently of the vertical movements or vibrations of the main axle of the machine, for the purpose specified.

Second, the employment of a curved adjusting way or standard, located on the vibrating frame or its equivalent, for the purpose specified.

Third, the combination of an adjusting lever with a curved standard or way on which said lever is adjusted or held, for the purpose specified.

Fourth, the combination of a curved way or standard, an adjusting lever, and a means for setting or holding said lever, for the purposes specified.

· No. 2,403.—HOBART H. SMITH, Carlisle, Penn., assignee by mesne assignments of C. M. LUFKIN.—*Harvester.*—Patented September 8, 1857 ; reissued November 27, 1866.

Claim.—First, the employment in a two-wheel side-draught machine of a carrying wheel or roller connected to the inner or heel end of a laterally-projecting hinged cutting apparatus, which is free at said end to conform to the surface of the ground over which it is drawn, independently of the vertical movements or vibrations of the main axle of said machine, for the purpose specified.

Second, the employment in a two-wheel side-draught machine of an adjustable wheel or wheels connected to and in combination with a laterally-projecting hinged cutting apparatus, which is free at each end independently of the other end and of the vertical vibrations of the main axle, to conform to the surface of the ground over which it is drawn, in such a manner that the same may be used both in reaping and mowing, for the purpose specified.

Third, the employment of a rock-shaft lever and an adjusting wheel or wheels, in combination with a hinged cutting apparatus which projects laterally from the main frame by which it is drawn forward over the ground.

Fourth, a laterally-projecting hinged cutting apparatus, which is free at each end independently of the other to conform to inequalities in the surface of the ground over which it is drawn, in combination with a mechanism whereby the attendant is enabled by operating a single lever to raise and lower said cutting apparatus bodily, for the purpose specified.

No. 2,404.—HERBERT W. C. TWEDDLE, Pittsburgh, Penn —*Apparatus for Distilling Coal Oil and other substances.*—Patented February 4, 1862; reissued November 27, 1866.

Claim.—First, distilling hydro-carbon oils, such as petroleum and other substances, under a vacuum or partial vacuum, by the use of steam for producing the vaporization of the article to be distilled.

Second, the use of superheated steam, in combination with the employment of a vacuum or partial vacuum, for the distillation of petroleum and other hydro-carbon oils and similar substances.

Third, the combination of a steam vacuum apparatus, constructed substantially as hereinbefore described, with the oil receiver L and M, for the purpose hereinbefore described.

Fourth, the combination of the vacuum apparatus hereinbefore described, with the steam pipe or pipes in the interior of the still and with the still, for the purpose hereinbefore described.

No. 2,405.—E. G. ALLEN, Boston, Mass.—*Railroad Frog.*—Patented September 25, 1866; reissued December 4, 1866.

Claim.—First, the combination and arrangement of the plates A and B, with their supports H and F, with or without the elastic packing *u*, connected together substantially as and for the purpose specified.

Second, in a truss railroad frog, constructed substantially as herein set forth, the use of the supports H and F, recessed to receive and hold the elastic material *u*, as set forth.

Third, the plate *f*, at the point of the toe piece E, fitted in the hole in plate A, and secured to the support H', by means of bolts, all arranged as shown and described.

No. 2,406.—EDMUND BIGELOW, Springfield, Mass.—*Apparatus for Supplying and Measuring Sirups in Soda Water.*—Patented April 6, 1858; reissued May 4, 1858; again reissued December 4, 1866.

Claim.—The employment of reservoirs in permanent cases or stands, revolving or otherwise, as herein described, with the registering faucets, substantially as and for the purposes herein set forth.

Also, a self-registering apparatus, with an air tube or vent, substantially as herein set forth, combined with a reservoir, as and for the purposes herein described.

No. 2,407.—ADAM S. CAMERON, New York, N. Y.—*Valve Gear for Steam Engines.*—Patented October 3, 1865; reissued December 4, 1866.

Claim.—First, the valves I I', in the heads of the main steam cylinder A, to be operated by the direct action of the main piston B, substantially as and for the purpose set forth.

Second, the construction of the stems of the valves I I', at each end of the cylinder, in such a manner that said valves shall be moved, reversing the main valve before the piston reaches the end of the cylinder, so as to cushion or arrest the motion of the piston, as set forth.

Third, the valve chambers H H', and valves I I', in the heads of the main cylinder A, in combination with supplementary cylinders E E', pistons F F', and main valve C, constructed and operating substantially as and for the purpose described.

No. 2,408.—COLBY BROTHERS & CO., Waterbury, Vt., assignees of GEO. J. COLBY.—*Clothes Wringer.*—Patented December 4, 1860; reissued December 4, 1866.

Claim.—First, the frame A I of a wringing machine with elastic rollers C J, and the springs F, or their equivalents, so as to be self-adjusting in regard to mutual pressure of the rollers, without the use of wedges, cams or screws, substantially as and for the purpose set forth.

Second, the construction of clothes wringers with the tangs E E, pivoted arms I I, and rollers C J, or their equivalents, arranged so that the act of compressing the clothes between the rollers will cause the device to clamp itself firmly to the tub or other article, substantially as herein shown and set forth.

No. 2,409.—CHARLES DION, Montreal, Canada.—*Fire Alarm.*—Patented April 3, 1866; reissued December 4, 1866.

Claim.—The expansion piece A or *a* and bed plate or tube B, or other equivalents, as shown in the different modifications, in combination with the tilting lever D, or its equivalent, and with the falling weight F, or its equivalent, constructed and operating substantially as and for the purpose described.

No. 2,410.—THE GILLESPIE GOVERNOR COMPANY, Boston, Mass., assignee of JAMES E. GILLESPIE.—*Hydraulic Governor.*—Patented January 7, 1862; reissued December 4, 1866.

Claim.—The combination, with a valve or with a water gate, and for the purpose of automatically governing or controlling the position thereof, in order to regulate the flow of fluid past such valve or gate, of a pump, a cylinder, and its piston, and a notched bar, or the equivalent of these, operating together substantially as described.

No. 2,411.—JULIUS GUTTMAN, Great Falls, N. H.—*Artificial Dentures.*—Patented March 6, 1866; reissued December 4, 1866.

Claim.—First, an artificial tooth, or set of teeth, provided with pins set in a zigzag line, substantially as and for the purpose set forth.

Second, loading sections of artificial teeth previous to making them up in sets, substantially as and for the purpose described.

No. 2,412.—EDWARD MILLER, Meriden, Conn., assignee of JOHN J. MARCY.—*Lamp Burner.*—Patented July 21, 1863; reissued December 4, 1866.

Claim.—The combination, with the hinge C, of the rigid curved rod F, fixed to the cone B, projecting downward through the shell A, between the hinge and the wick tube, and provided with a bent end or equivalent stop *b*, which, coming in contact with the under side of the said shell, operates by a tensional strain upon the rod F to limit the turning of the chimney, all as herein described.

No. 2,413.—THE NEW YORK ENGRAVING AND CARVING COMPANY, New York, N. Y., assignee by mesne assignments of JOHN G. PUSEY.—*Machine for Boring and Drilling Gun Stocks.*—Patented February 17, 1863; reissued December 4, 1866.

Claim.—First, arranging a series of tool stocks to radiate from a common centre, in combination with a series of tracers, substantially as specified, whereby all the tools and tracers may be moved together in mortising, boring, or carving, but the tools not in use will, by their divergence, be out of the way, as set forth.

Second, the arrangement of the pulley *m*, in the middle of the circular head *k*, and of the fork *p'*, or its equivalent, for receiving and changing the belt *d'*, in the manner set forth.

Third, the parallel bars *e e'*, each jointed at one end by a universal joint to a fixed support, and at the other end to a movable head, in combination with a cutter and a guide or tracer, substantially as specified, whereby the said cutter and tracer may be freely moved in carving, substantially as set forth.

Fourth, the frame *u*, on centres 10, at right angles, or nearly so, to its length, and receiving the pattern and gun stock, or other article, substantially as specified, whereby the pattern and article to be acted upon can be reversed, to present either side to the tool and tracer, as set forth.

Fifth, a holder fitted on centres, and carrying the pattern and gun stock or other article, and arranged substantially as specified to swing on said centres, while the tool is inletting or cutting the curved parts, in order that said tool may act at right angles to the surface, for the purposes and as specified.

Sixth, rotating the pattern and the article to be carved in parallel planes at right angles to the axis on which they are supported, substantially as and for the purpose set forth.

No. 2,414.—GEORGE F. BLAKE and P. HUBBELL, Boston, Mass., assignees of GEORGE F. BLAKE.—*Machine for Pulverizing Clay.*—Patented November 26, 1863; reissued December 11, 1866.

Claim.—First, in combination with a machine for cleaning and pulverizing clay or other plastic material, the reciprocating wipe or plunger W, constructed and operating substantially as described.

Second, the revolving grate, constructed and operating as described.

Third, the stationary fingers *w*, constructed, arranged and operating as set forth.

Fourth, the arrangement for conjoint operation of the inclined blades or sweeps G, with the horizontal grate G', in the manner described.

Fifth, the arrangement for conjoint operation of reciprocating, oscillating, or rotating wipes, blades or sweeps, with rotating or stationary screens or gratings, substantially in the manner and for the purpose set forth.

No. 2,415.—GEORGE W. CHIPMAN, Boston, Mass., assignee by mesne assignments of JOHN R. HARRINGTON.—*Carpet Lining.*—Patented April 1, 1856; reissued December 11, 1866.

Claim.—As a new article of manufacture, a carpet lining made up of a long or continuous sheet or sheets of stout paper, and a layer or layers of fibrous material applied thereto, substantially as set forth.

No. 2,416.—GEORGE W. CHIPMAN, Boston, Mass., assignee by mesne assignments of JOHN R. HARRINGTON.—*Machine for making Carpet Lining.*—Patented April 1, 1856; reissued December 11, 1866

Claim.—The process of making a carpet lining by progressively bringing into contact a sheet or sheets of stout paper, and a sheet or sheets of batting or soft fibrous material, so that upon contact of the contiguous surfaces they shall be caused to adhere together, substantially as described.

Also, in combining with a roll or rolls upon which the sheet or sheets of paper are wound, feed and presser rollers, operating to compress together and feed the paper and batting, substantially as set forth.

Also, in combination with the paper roll or rolls and feeding and compressing rolls, a paste or cement-applying mechanism, operating substantially as described.

Also, in combination with a wadding-forming apparatus, creasing or fold-forming rolls, operating as and for the purpose substantially as set forth.

Also, in combination with such fold-forming mechanism, the box or platform into which the wadding is delivered in folds, and from which it is removed for baling by letting down the fall, substantially as described.

No. 2,417.—E. VICTOR FASSMAN, New Orleans, La.—*Hoop Lock for Cotton Bales.*—Patented April 18, 1865; reissued December 11, 1866.

Claim.—The plate provided with slots, when the same is constructed with ridges or projections on both sides of the plate, substantially as set forth.

No. 2,418.—SANFORD A. HICKEL, Roane Co., and CALVIN, JAMES, and BENJAMIN F. ARMSTRONG, Jackson Court House, W. Va., assignees of SANFORD A. HICKEL.—*Tanning.*—Patented November 7, 1865; reissued December 11, 1866.

Claim.—The employment or use of manure, in combination with bark or other tanning material, substantially as and for the purpose set forth.

No. 2,419.—JONAS B. AIKEN, Franklin, N. H., assignee by mesne assignments of J. B and W. AIKEN.—*Knitting Machine.*—Patented July 8, 1856; reissued December 18, 1866.

Claim.—First, the hollow circular needle plate having grooves cut on its inner surface, substantially as described for the objects specified.

Second, the horizontal groove *c*, near the bottom of the cone, so arranged in relation to the inclined operating groove that the needles may be retreated thereto, substantially as described, and retained therein when they are not wanted to operate on the fabric knit, in the manner set forth.

Third, the switch *g*, arranged substantially as described, to change the needles from the inclined operating groove to the retreating groove.

Fourth, the use of the two sets of sliding needles in rotary knitting frames called the plain series and the ribbing series, arranged and operated substantially as set forth.

No. 2,420.—RICHARD B. BURCHELL, Brooklyn, N. Y.—*Lock.*—Patented September 11, 1866; reissued December 18, 1866.

Claim.—First, the fixed tumbler stud *g* and key stud *l*, arranged on the line of a plane passing through the centre of the bolt-head, in combination with tumblers of unequal shape and dimension, but so constructed that they can be applied on either side of the key-stud, and be equally operative with a key of different shape or dimension, substantially as set forth.

Second, constructing the head of the bolt as described, and arranging a series of swinging tumblers to operate in conjunction therewith, substantially as set forth.

Third, the tumblers *h i k*, constructed and applied substantially as specified, in combination with the key-stud *l* and bolt-head *c*, constructed and arranged in the manner and for the purposes set forth.

No. 2,421.—C. A. HARPER, Rahway, N. J.—*Portable Lamp Cooking Apparatus.*—Patented May 1, 1866; reissued December 18, 1866.

Claim.—First, the heater E when attached to the bottom of the boiler or its supporting plate by cleats, on which it may be moved without disturbing the other parts of the apparatus, as and for the purpose set forth.

Second, the boiler A of a cooking stove, when constructed with a central vertical pipe B opening through it, and also with transverse pipes F across the same, as and for the purpose set forth.

Third, a cooking apparatus when so constructed that there shall be a direct communication from the heater, through the boiler, into the chamber of the oven without the interposition of bottom plates to the latter.

Fourth, so arranging the boiler and oven that access may be had to the interior of the boiler through the doors of the oven, substantially as set forth.

Fifth, the oven C, when so constructed that the heated air is introduced directly into the body of the oven, and its exit is controlled by the flue-sheet L and damper O, arranged substantially as set forth.

No. 2,422.—THE LESTER OIL MANUFACTURING COMPANY, New York, N. Y., assignees JOHN H. LESTER.—*Composition of Matter for Lubricating Machinery and for other Purposes.*—Patented January 30, 1866; reissued December 18, 1866.

Claim.—The combination of oil with the synovial fluid or other substance obtained from animal matter, substantially as described.

No. 2,423.—JOHN A. MERRIMAN, Chicago, Ill.—*Machine for Cutting Screws.*—Patented August 1, 1865; reissued December 18, 1866.

Claim.—First, in combination with a revolving die-holder, two or more screw-cutting dies provided with inclines and projections, arranged and operated substantially in the manner and for the purposes herein specified.

Second, in combination with a revolving die-holder and two or more screw-cutting dies, constructed substantially as described, the arrangement of longitudinally sliding bearings operating upon said dies, substantially in the manner and for the purposes set forth.

Third, in combination with longitudinally sliding bearings operating directly upon screw-cutting dies, substantially as described, the arrangement of a latch or stop for the purpose of retaining said dies in, or releasing them from, the die-holder at the will of the operator, substantially as herein specified.

Fourth, closing the dies by the use of the bearings K upon the longitudinally sliding cylinder acting directly upon the dies, substantially as and for the purposes herein specified.

Fifth, the sliding bearings K, the levers *i*, and the dies *g*, arranged and operating substantially as and for the purposes herein specified.

No. 2,424.—WILLIAM E. PRALL, Washington, D. C.—*Machine for Picking Cotton.*—Patented February 27, 1866; reissued December 18, 1866.

Claim.—First, a machine for harvesting cotton, constructed and operating substantially as described.

Second, in a machine for harvesting cotton the revolving picking cylinder as a device for gathering the cotton among the branches of the plant, constructed and operating substantially as described.

Third, in a cotton harvesting machine the employment of one or more continuous series of said picking cylinders, so arranged as to be made to pass in close succession through the plant to gather the cotton as described, and also to move progressively along the row of plants so as to operate on all parts thereof, substantially as described.

Fourth, the combination and arrangement of a continuous series of picking cylinders projecting from the face of a wheel set at an angle to the line of progression of the machine or its equivalent, so that in their downward movement the picking cylinders shall pass outside of the plant, then beneath its branches, and up, through, and among the same, substantially as described.

Fifth, the combination of a series of revolving picking cylinders arranged upon a wheel or its equivalent, as described, of a stationary band or other equivalent mechanism that will rotate said cylinders, substantially as described.

Sixth, the combination of the wheel E and its series of picking cylinders, or the series arranged in a manner equivalent thereto, with the carrying wheels C of the machine, so that the picking cylinders shall automatically and simultaneously receive therefrom a rotative movement, as described, and an upward movement through the cotton plant, and a progressive movement along the row of plants, substantially as described.

Seventh, in combination with the series of picking cylinders, arranged as described, the series of toothed arcs and the clearing wheel, or their equivalents, for removing the cotton from the cylinders, substantially as described.

Eighth, the employment in combination of two systems of mechanism for gathering and discharging the cotton, constructed, arranged, and operating substantially as described, and set diagonally face to face so as to operate simultaneously upon opposite sides of the plant, substantially in the manner described.

Ninth, the trough or receptacle for the cotton placed above the cross connection of the machine the within and series of picking cylinders, so as to receive the cotton as it is removed from the cylinders, substantially as described.

Tenth, the manner of combining and arranging the wheels E with their adjuncts, the framing of the machine, the carrying wheels, and the tongue or perch, so as to leave an open space lengthwise through the machine below the cross connections, which enables the machine to pass over the plants without bending them down so as to interfere materially with the gathering of the cotton, substantially as described.

Eleventh, combining with the framing and operative parts of the machine, constructed and arranged as described, the tongue or perch to which the animals may be harnessed, in the manner before described, to draw the machine along and maintain it in an erect position.

No. 2,425.—SILAS T. SAVAGE, Albany, N. Y.—*Heating Stove.*—Patented March 30, 1858; reissued December 18, 1866.
Claim.—In stoves or furnaces a fire-box with a grated back for the admission of the air which circulates through the ash pit and the flue back of the grate, substantially as described.

No. 2,426.—J. STEADMAN, Pecatonica, Ill.—*Punching Machine.*—Patented December 19, 1865; reissued December 18, 1866.
Claim.—First, the socket F sliding through the opening V, operating in combination with the link G pivoted to the right-angled lever at its inner end, whose outer end is pivoted to the connecting-rod J operated by the adjustable right-angular lever P, substantially as described.
Second, the combination and arrangement of the link G which drives the socket and punch F E with the right-angled lever I H, working on the rock-shaft R, and the connecting rod J, arms K N, working on the rock-shaft L and the link O, and hand-lever P, substantially as described.

No. 2,427.—W. STEELE, Wheeling, W. Va.—*Machine for Cutting Staves.*—Patented October 19, 1858; reissued December 18, 1866.
Claim.—First, the swinging frame composed of the pieces C C, bed-plate K and knife E, in combination with cross-tie I, constructed and operated in the manner substantially as shown and described and for the purpose set forth.
Second, the apron M hinged to the bed-plate K, as described, or otherwise attached to the machine in such a manner that it can be held under or back of the knife to support the piece during the process of cutting and then swing down or fall back to allow the piece to drop from the knife.
Third, the combination of the levers L L, and stops B and D, or their equivalents, as described.

No. 2,428.—JAMES GAMAGE TARR and AGUSTUS H. WONSON, Gloucester, Mass.—*Paint for the Bottoms of Ships.*—Patented June 20, 1865; reissued December 18, 1866.
Claim.—First, the composition of paint consisting of oxide of copper, oxide of zinc, and oxide of arsenic, used with a basis of ochre and a suitable medium, substantially as described.

Second, the composition of paint consisting of oxide of copper and oxide of zinc, used with a basis of ochre and a suitable medium, substantially as described.

Third, the composition of paint consisting of the oxide of copper and oxide of arsenic, with a basis of ochre and a suitable medium, substantially as described.

No. 2,429.—EDWARD L. WALKER, Benford's Store, Penn.—*Hay Elevator.*—Patented September 6, 1864; reissued December 18, 1866.

Claim.—First, the combination in a fork for elevating hay or similar material of the following devices: first, a shaft for penetrating such material; second, a barb or barbs for sustaining the material while being elevated; third, a device for relieving the barb or barbs for discharging the material, substantially as set forth.

Second, a penetrator in combination with one or more sustaining prongs or barbs, and a discharging device, substantially as specified, for raising and discharging hay or similar material, substantially as set forth.

Third, the barbs E E, arranged substantially as specified, in combination with the rod F for sustaining said barbs, as set forth.

Lightning Source UK Ltd.
Milton Keynes UK
UKHW020011201118
332599UK00016B/1856/P